ILLUSTRATING THE BASIC GRAPHING PRINCIPLES

Vertical Shift Principle

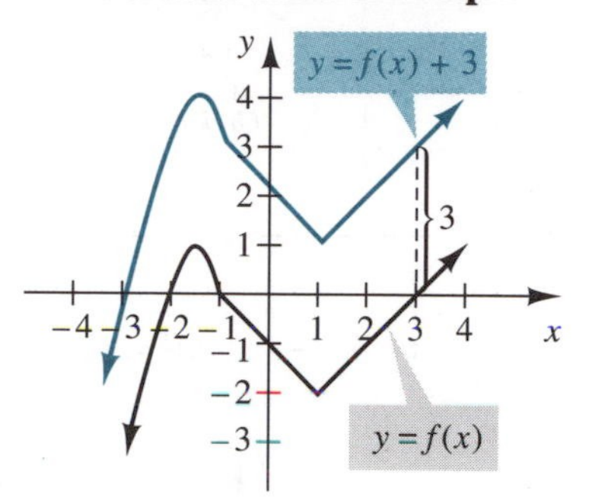

Horizontal Shift Principle

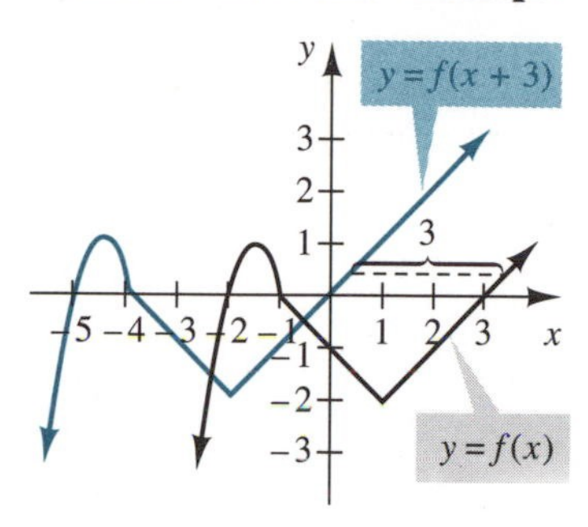

e

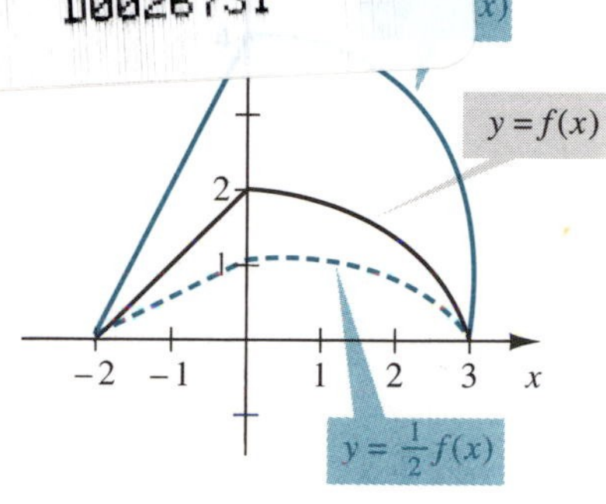

Graphing $-f(x)$

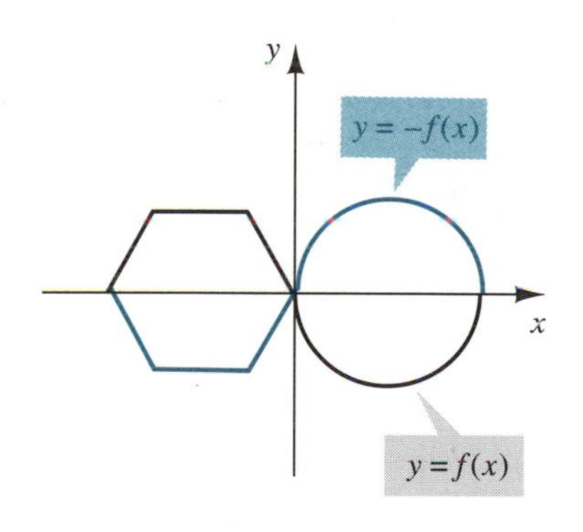

Graphing $f(-x)$

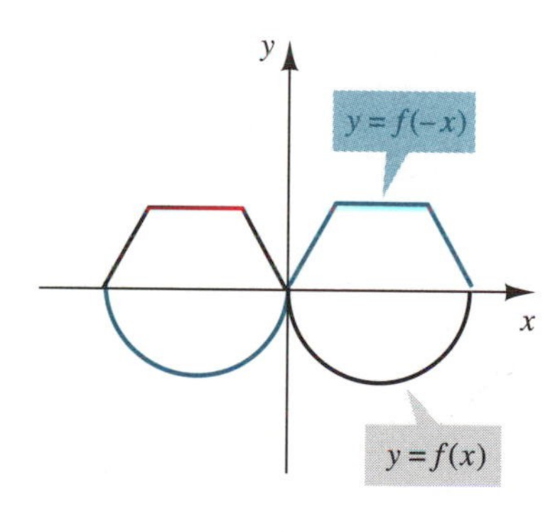

Graphing $|f(x)|$

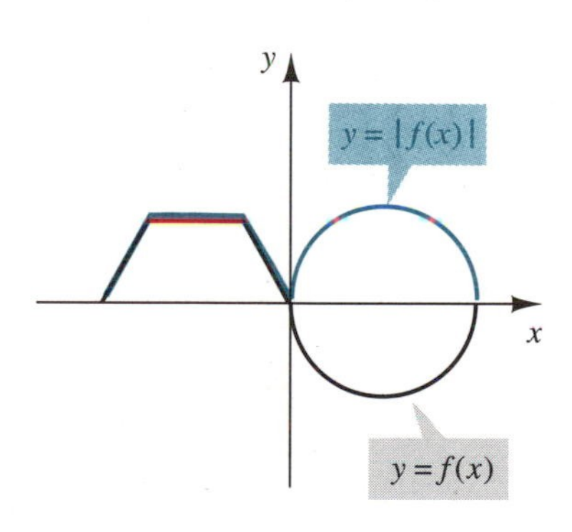

Graphing $f^{-1}(x)$

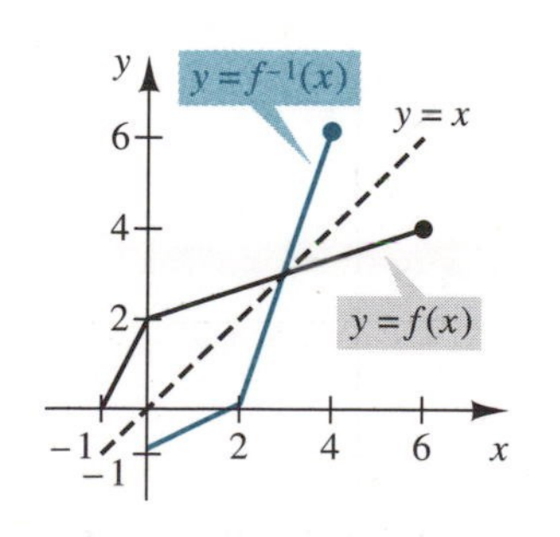

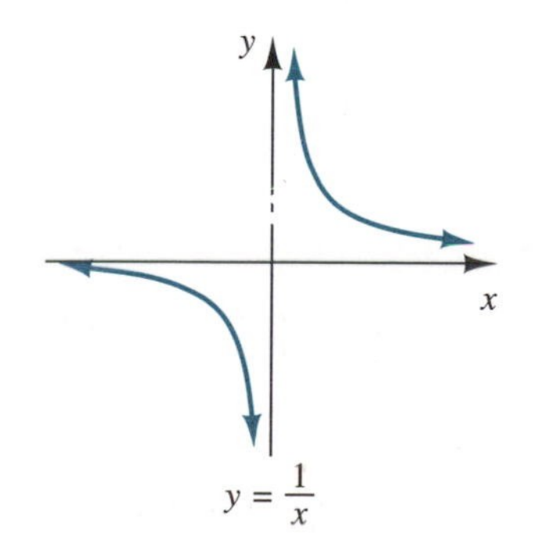

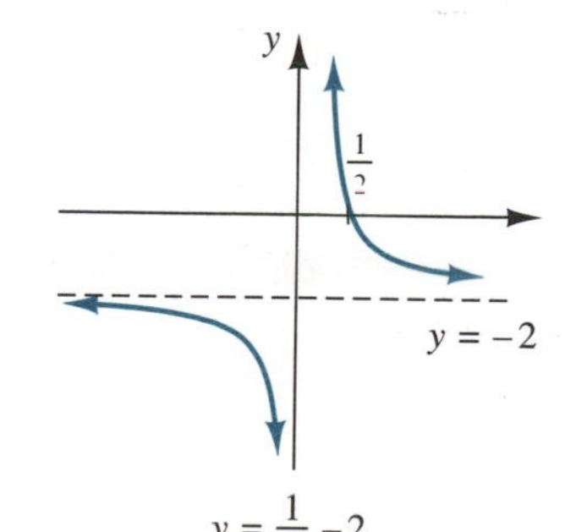

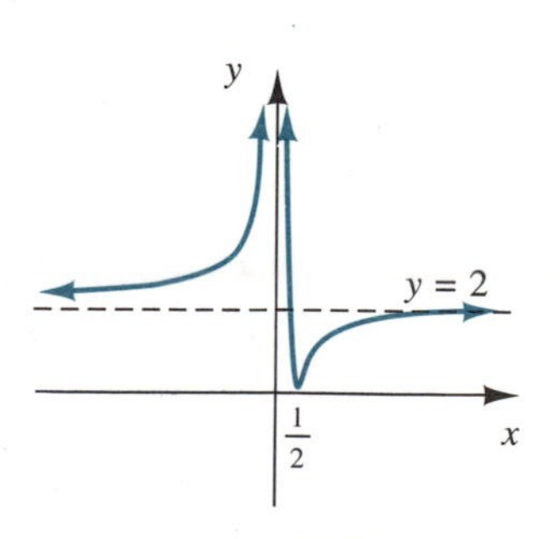

PRECALCULUS

ARTHUR GOODMAN

Queens College of the City University of New York

LEWIS HIRSCH

Rutgers University

PRENTICE HALL *Englewood Cliffs, New Jersey* 07632

Library of Congress Cataloging-in-Publication Data

Goodman, Arthur.
Precalculus / Arthur Goodman, Lewis Hirsch.
p. cm.
Includes index.
ISBN 0-13-716408-4
1. Mathematics. I. Hirsch, Lewis. II. Title.
QA39.2.G66 1994
512'.1—dc20 93-38100
CIP

Editorial production/supervision: Debra Wechsler; Rachel J. Witty, Letter Perfect, Inc.
Development editor: Phyllis Dunsay
Design director: Florence Dara Silverman
Interior design, layout, and cover design: Lisa Jones
Cover art: Marjory Dressler
Buyer: Trudy Pisciotti

Printed in the United States of America
10 9 8 7 6 5 4 3 2 1

ISBN 0-13-716408-4

Prentice-Hall International (UK) Limited, *London*
Prentice-Hall of Australia Pty. Limited, *Sydney*
Prentice-Hall Canada Inc., *Toronto*
Prentice-Hall Hispanoamericana, S.A., *Mexico*
Prentice-Hall of India Private Limited, *New Delhi*
Prentice-Hall of Japan, Inc., *Tokyo*
Simon & Schuster Asia Pte. Ltd., *Singapore*
Editora Prentice-Hall do Brasil, Ltda., *Rio de Janeiro*

CONTENTS

CHAPTER 10
Conic Sections and Nonlinear Systems 636

CHAPTER 10
Sequences, Series, and Related Topics 728

Tables T1–T11

PREFACE

TO THE INSTRUCTOR

The last ten years have been a period of great introspection in the mathematical community, focusing on what and how to teach undergraduates at the introductory calculus level. Motivated by the fact that unacceptably high percentages of students who take calculus do not successfully complete their course sequence, professionals have made various suggestions about how to revamp the calculus curriculum and integrate the new technology into the classroom.

While there are many exciting and innovative ideas being discussed, our experience has been that one of the main factors contributing to students' poor success rate is that they come to calculus without the level of mathematical sophistication necessary to be successful. By this we mean that many students have had insufficient practice in the kinds of thinking required for understanding and applying the concepts of calculus. Even students whose algebraic skills are adequate often do not know how to apply these skills appropriately in calculus. They may have a bag full of tools, and they may even know how to use each tool, but they frequently do not recognize which tools are appropriate to a given situation. This lack of sophistication also manifests itself in the fact that students often think about a problem in only one way—algebraically. Understanding the concepts of calculus requires students to view a problem from two additional perspectives: numerically and geometrically (graphically).

In this text we have made a conscientious effort to afford the student every opportunity to look at questions from a variety of points of view; to take the time to analyze a problem carefully so that he/she clearly understands what is being asked; to reformulate problems in more familiar terms; and to recognize that most mathematical problems require significantly more thinking and less writing. We hope that if students are repeatedly exposed to this philosophy, they will come to understand that they should be investing most of their time and effort into conceptualizing and formulating the problem carefully, rather than focusing on the answer. If they learn to ask themselves the right questions, the answers will take care of themselves.

This text is intended for students who are preparing to take calculus or other courses requiring a similar background. While it is assumed that the students who use this book have taken intermediate algebra, it comes as no surprise to those who have taught precalculus that students arrive with many different levels of algebra skills. Many students never really mastered the material when they learned it, while for others there may be a hiatus of several years between their study of algebra and precalculus. Consequently, the review material in Chapter 1 is rather detailed. Instructors may choose to go over the review material or to assign it to the students to review on their own, and, therefore, begin the course with Chapter 2. However, particular attention should be paid to Section 1.9 on *Quadratic Inequalities* and Section 1.10 on *Substitution*. The material in these sections may be new to a number of your students and is used repeatedly throughout the text.

A perusal of the Contents shows that the topics covered and their order are fairly traditional (as it is in most precalculus books). What we think distinguishes this text is our approach, the ideas we have chosen to emphasize, and some of the special features we have introduced.

This text is replete with figures. In particular, Chapters 2 and 3 alone contain more than 200 figures, so that every idea that benefits from an accompanying figure has one. On many occasions, instead of referring a student to a graph which appeared previously, we repeat the graph to help clarify the discussion.

A great deal of attention has been paid to developing graphing principles and techniques and to encouraging students to familiarize themselves with a catalog of basic graphs and equations. This familiarity will aid students when they get to calculus by allowing students to concentrate on the calculus material and not letting them sit and wonder, "Where did that graph come from?"

Calculators This text has been written under the assumption that students entering a precalculus course have access to a scientific calculator. In the exercise sets we have *purposely* not identified those exercises which require the use of a calculator. We feel that the calculator should be a tool students bring to every exercise (just as they bring their algebraic skills) and through experience, and some guidance in examples, will learn where and when it is necessary or appropriate to use.

Trigonometry As our approach in the text illustrates, we feel that at this level most ideas are presented most effectively by going from the more concrete to the more abstract. Our experience in the classroom tells us that with regard to trigonometry this philosophy, while still basically valid, needs to be modified slightly. Rather than beginning with right-triangle trigonometry, which we have found encourages students to think exclusively in the very restricted "opposite-adjacent-and-hypotenuse" mode, we first introduce radian measure, and then define the trigonometric functions of the general angle in standard position. Once the student has become somewhat familiar with the trigonometic functions in this general setting, we discuss their specific application to right triangles, and reemphasize that the radian measure of an angle is nothing more than a real number. This leads quite naturally to understanding the trigonometric functions as functions of real numbers and developing the more analytic side of trigonometry. Trigonometric applications appear throughout Chapters 6, 7, and 8.

Special Features

1. **Functions and Graphs** In Chapters 2 and 3 we have spent an enormous amount of time in introducing and developing the idea of a function and its graph. Particular attention has been paid to pointing out the connection between the algebraic and geometric interpretations of certain important concepts.

An entire section (2.6) is devoted to the idea of interpreting graphs: developing the ability to look at the graph of an equation (extracting geometric information) and to recognize the key features that describe the relationship given by the equation (an algebraic relationship).

2. **Different Perspectives** boxes appear wherever there is an opportunity to highlight the connection between the algebraic and geometric interpretation of the same idea. See page 150, for example. In this way the student is encouraged to think about mathematical ideas from more than one point of view.

3. **Applications and Mathematical Modeling** We have made a conscientious effort to include word problems and applications wherever possible, and generally speaking, they are integrated throughout the text. However, there are a few places where applications are concentrated.

   Section 3.3 offers a broad introduction to the idea of mathematical modeling, giving the student opportunities to put the function concept to use in a variety of situations. This entire section allows students to practice the skills and gain the experience necessary to use these ideas in calculus.

   Section 3.4, on quadratic functions, includes optimization problems which carry these ideas a step further. Through such problems students can see why we might want to express one quantity as a function of another and presents yet another opportunity to see how the algebraic, graphical, and numerical perspectives of each offer particular insight into a problem.

   Chapter 5 includes a variety of exercises that illustrate the remarkable range of disciplines in which quantities are related by exponential or logarithmic functions.

4. **Graphics Calculators** are clearly a valuable learning tool. Nevertheless, we realize that a majority of students still do not have access to one. Consequently, the text presentation is not dependent upon the use of a graphics calculator. However, we have included problems designed to make use of the graphics calculator (or computer graphics software that may be available in a math lab setting) as a learning tool. We have labeled these in a box with the **GRAFFIX** heading. For the most part, these are problems designed to allow the student to explore and gain insight into upcoming material, or to clarify some of the points of a previous discussion. In addition, many of the exercise sets include exercises specifically designated to be done on a graphics calculator.

5. **Problem Analysis** As mentioned above, we feel very strongly about helping students develop at least the minimum level of mathematical sophistication and maturity necessary for success in calculus. One step we have made in this direction is to take certain problems and present their solutions in a question-and-answer format, so that students can see some of the thought processes involved in approaching and solving new or unfamiliar problems. In this way we hope the student will develop appropriate problem-solving strategies. See Example 7 on page 120 and Example 3 on page 204.

6. **Margin Comments** Throughout the text we have used the margins to add another dimension to the discussion in the text. Recognizing that students are often too

passive as they read mathematics, the questions and comments placed in the margin are specifically designed to more actively involve the student in the development presented in the text.

Ancillaries

Instructors using this text may obtain the following supplements from the publisher:

Instructor's Solutions Manual Prepared by John Garlow, *Tarrant County Junior College,* contains worked-out solutions to all even-numbered problems: (0-13-718693-2).

Test Item File Prepared by Robert Flagg, *University of Southern Maine,* contains hard copy of test questions from PH Test Manager and Make Test for the Macintosh: (0-13-718800-5).

PH Test Manager 2.0 (3.5″ or 5.25″) contains a bank of test questions that can be edited and manipulated to create tests: 3.5″ (0-13-718792-0) 5.25″ (0-13-475427-1) demo (0-13-075763-2).

Make Test for Macintosh is a state-of-the-art Macintosh test generator that makes full use of the superior Mac graphics capabilites as well as the user-friendly interface: (0-13-718776-9) demo (0-13-458118-0).

Graphing Calculator Videos A set of tutorial videos that covers basic principles and operations of the TI-81 and Casio 7700-G graphing calculators for precalculus: (0-13-020017-4).

Students using this text may obtain the following supplements from the publisher:

Student Solutions Manual Prepared by Karen Schwitters, *Seminole Community College,* contains worked-out solutions to all odd-numbered problems: (0-13-718701-7).

X(PLORE) Version 4.0 by David Meredith, is a powerful, fully programmable, symbolic and numeric mathematical processor for IBM, compatibles, and Macintosh computers that evaluates expressions, graphs curves, and solves equations and matrices. This same program may be used by students as they go on to calculus to graph surfaces, integrate and differentiate functions, and solve differential equations.

Book/Disk IBM (0-13-014226-3).
Book only IBM (0-13-014234-4).
Book/Disk MAC (0-13-014200-X).
Book only MAC (0-13-014218-2).

X(PLORE) 4.0 Graphing Tutorial Disk contains templates for performing the activities in the Graffix boxes using X(PLORE) 4.0 by David Meredith: (0-13-718750-5).

Math Master Tutor (3.5″ or 5.25″)

Math Master Tutor Macintosh carefully keyed to the text with page references, Math Master Tutor contains four modes of instruction for each section in the text: Exploration (a text-on-disk feature), Summary (a concise review), Exercises (open-ended, algorithmically generated drills with step-by-step solutions), and Quiz (timed tests that teach students test-taking strategies). Available upon adoption (with cite License) for IBM and Macintosh.

IBM 3.5″ (0-13-718784-X)
IBM 5.25″ (0-13-503798-0)
IBM demo (0-13-018615-5)
MAC (0-13-078775-2)
MAC demo (0-13-075698-9)

Cooperative Learning Booklet is a collection of cooperative learning activities compiled by the authors to be used in the classroom in conjunction with the text: (0-13-718719-X).

Acknowledgments

The authors would like to sincerely thank the following reviewers for their many thoughtful comments and suggestions during the preparation of this text.

Dr. Sabah Al-Hadad, *California Polytechnic State University* (San Luis Obispo, CA)
Dr. Frank Battles, *Massachusetts Maritime Academy* (Buzzards Bay, MA)
Dr. Carole Bauer, *Triton College* (River Grove, IL)
Helen Burrier, *Kirkwood Community College* (Cedar Rapids, IA)
Eunice Everett, *Seminole Community College* (Sanford, FL)
Dr. Paul Fallone, *University of Connecticut* (Hartford, CT)
Judy Kasabian, *El Camino College* (Torrance, CA)
Vince McGarry, *Austin Community College* (Austin, TX)
Lois Miller, *Golden West College* (Huntington Beach, CA)
Richard Nadel, *Florida International University* (Miami, FL)
Jack Porter, *University of Kansas* (Lawrence, KS)
Cheryl V. Roberts, *Northern Virginia Community College* (Annandale, VA)
Kathy V. Rodgers, *Southern Indiana University* (Evansville, IN)
Ken Seydel, *Skyline College* (San Bruno, CA)
Edith Silver, *Mercer County Community College* (Trenton, NJ)
Ara Sullenberger, *Tarrant County Junior College* (Fort Worth, TX)
Bruce Teague, *Sante Fe Community College* (Gainesville, FL)
Faye Thames, *Lamar University* (Beaumont, TX)
Dr. William Tomhave, *Concordia College* (Moorhead, MN)
Dr. Jan Vandever, *South Dakota State University* (Brookings, SD)
Carol Warnes, *University of Georgia* (Athens, GA)
John F. Weglarz, *Kirkwood Community College* (Cedar Rapids, IA)

Gary L. Wood, *Azusa Pacific University* (Azusa, CA)
Mary Yorke, *Eastern Michigan University* (Ypsilanti, MI)

We also acknowledge the assistance of the following colleagues who checked proofs at the galley stage:

Margaret Donlan, *University of Delaware* (Newark, DE)
Anne Landry, *Dutchess Community College* (Poughkeepsie, NY)
David Randall, *Oakland Communtiy College* (Union Lake, MI)

We also are grateful to our students who were kind enough to allow us to class-test this text in its manuscript form. Their criticisms and input were invaluable.

Obviously, the writing and production of a textbook is a collaborative effort. We must thank the staff at Prentice Hall: Rob Koehler, Priscilla McGeehon, and Debra Wechsler, for their efforts during the various stages of this project; Margaret Donlan and Ann Landry for their assistance in working out all the exercises and checking the solutions; and a special thanks to Phyllis Dunsay for her tireless efforts in bringing this text to its present form and to Rachel J. Witty for her extraordinary competence in bringing this project to its completion. Of course, the authors are solely responsible for any errors that remain, and we would be most appreciative if they were brought to our attention.

Finally, we would like to thank our wives, Sora and Cindy, respectively, and our families for understanding the enormous amount of time such a project demands.

Arthur Goodman
Lewis Hirsch

PREFACE

TO THE STUDENT

This textbook is intended to expose you to many ideas and help you to develop a variety of skills you will find useful in your study of calculus. Perhaps the most important of these skills is the ability to approach a problem mathematically: read a question, understand what is being asked, and develop a coherent strategy as to how the problem is to be solved. (Doing these three things will not insure your success, but at least it will give you a fighting chance.) As you read the textbook, you will notice that a great deal of emphasis is given to the idea of *really understanding* what a question is saying and what it is asking.

You should always read a mathematics text with pencil and paper in hand. While we have included most of the steps in the solutions to the illustrative examples, invariably some steps (often algebraic ones) are left out. You should fill these steps in by yourself.

Calculators

The calculator has become an indispensible tool for work, school, and household use. From the basic four-function calculator to the programmable/graphics calculator, most people either own or know how to use a calculator. In this text, although tables do appear in the appendix, we assume that students have access to a scientific calculator—one which has the basic trigonometric functions and the logarithmic function keys—so that it can be used to find these values as well as for basic computations.

As you probably know, not all calculators are the same; for some computations, different calculators require different key-stroke sequences in order to arrive a the intended numerical answer. For example, for many of the standard scientific calculators, to find $\sqrt{27}$ you have to press the sequence [2][7][√‾][=], or simply [2][7][√‾]; for the graphics calculator, however, and for some of the programmable calculators, the sequence must be entered as [√‾][2][7][=]. Another examples: To find sin 23° some calculators require the sequence [2][3][sin][=], while other calculators require [sin][2][3][=].* In this text, when we demonstrate the use of a scientific calculator, we will use the most common type, namely, the one which requires you to enter the number *before* pressing a function key.

In any case, you can see why it is important for you to familiarize yourself with your calculator and to read over the manual that came with it, to determine how to perform certain computations or to use certain keys.

*Most scientfic calculators already have the algebraic order of operations built in. This means that if you want to evaluate the expression $3 + 4 \cdot 5$, you may enter it as [3][+][4][×][5][=], and the calculator "knows" to evaluate $4 \cdot 5$ before adding 3 to get the correct answer, 23.

Exact Answers Knowing how and *when* to use a calculator for a problem can make a difference in the accuracy of your result. In this text, unless otherwise specified, you are expected to give exact answers (answers which are not rounded). If you need to add $\frac{2}{3} + \frac{4}{5}$, then the exact answer is $\frac{2}{3} + \frac{4}{5} = \frac{22}{15}$. If you do this computation on a calculator, you would get an answer of 1.4666667 rounded to seven decimal places. Although this answer is accurate, it is not exact. Additionally, suppose you use a calculator for a problem and you need to enter $\frac{33}{7}$ in your calculator before using it in your computations. If you round $\frac{33}{7}$ to 4.71, you will automatically introduce a rounding error in your calculations. How much this error is magnified and reflected in the solution will depend upon the operations performed with this number, but your answer will not be exact. Even if you enter $\frac{33}{7}$ in your calculator as [3][3][÷][7][=], your calculator will round the number before performing computations, resulting in a more accurate answer, but it still may not be exact.

General Advice on Using the Calculator We list the following general advice for using your calculator while working on problems.

1. When an exact answer such as $\frac{\pi}{4}$, $\sqrt{3}$, or $\frac{3}{7}$ is requested or required, the calculator should *not* be used except, possibly, to *check* the accuracy of your computations.
2. When solving most problems, work out as much of the solution as possible on paper. While the calculator may be necessary for computations performed in the various stages of a solution, you should try to write down as many steps as possible. This helps in your understanding of the problem; and you will be able to check the logic of your solution steps, spot errors, and check that the calculator answer makes sense. In many cases, rounding errors will be minimized if you work the problem out on paper and save the calculator computations for the last step. (Another reason to record as much of your work as possible is that teachers are often more interested in *how* you solve the problem than in your final answer.)
3. You should check your computations and answers with a calculator whenever possible. (If your answer is exact, remember that the calculator will often give a rounded answer.)

Keep in mind that the calculator is a tool that makes the computation of a problem easier. It is *not a substitute for understanding* what you are doing and why.

The Calculator and Accuracy of Measurement Measurements of physical quantities are almost never exact; they are approximations which contain errors. For example, when you use a ruler to measure the length of an object, you know that at some point you have to estimate the measurement to the closest, say tenth, of a unit.

The simplest way to write a number that also indicates the error of measurement involved is by writing fewer or more digits. Recording a measured quantity as 4.2, for example, usually means that we are confident that the estimate is accurate to one decimal place and that its true value lies between 4.15 and 4.25. Recording the same value as 4.20 means we are confident that the value is accurate to two decimal places, and that its true value lies between 4.195 and 4.205. Notice that the precision of measurement is indicated by the number of digits in the recorded value, and that the last digit is always uncertain.

The use of the calculator has made it easy to confuse accuracy of computations with precision of measurement. For example, suppose we want to find the length of the hypotenuse, c, of a right triangle given that the legs measured 16.4 cm and 82.6 cm. By the Pythagorean theorem, we arrive at the value $c = \sqrt{(16.4^2) + (82.6)^2}$. Using a calculator (which can display eight digits), we arrive at $c = \mathbf{84.212351}$. This display gives us the impression of a level of precision which is unwarranted. The legs are only accurate to one decimal place, and these measurements are used to compute the hypotenuse. How can the hypotenuse be more precise than either of the two measurements used to compute it? Consequently, for problems involving computations with measurements, our final answer should never have a greater degree of accuracy than any of the measurements used to compute it. For our triangle problem, this means that the best estimate is $c = 84.2$ cm.

Graphics Calculators and Computer Graphics Packages Throughout this text we stress the importance of being able to understand the relationship between variables or equations by visualizing the relationship through graphs. Graphics calculators and computer graphics packages have made graphing readily accessible. We can graph functions quickly and even estimate points on a graph with a high degree of accuracy.

We realize that a majority of students may not have access to this technology, and, therefore, we have not made this text dependent upon its use. Nevertheless, recognizing that this technology can be a valuable tool for learning, we have included problems designed to make use of this technology.

If you have access to either a graphics calculator, or graphics software, we suggest that you take advantage of the **GRAFFIX** exercises scattered through the text. These exercises are designed to allow you to explore and discover the relationships and concepts discussed in the text for yourselves. You will find that very often the process of exploring and arriving at your own conclusions will help you to better understand the concepts being discussed.

Since there are several models of graphics calculators (and software) available, the **GRAFFIX** exercises are not calculator specific; they contain only general instructions using terminology common to graphics calculators. You should read your calculator manual (or software documentation) carefully to learn how use the graphics utilities you have available. A more detailed, calculator-specific manual was designed to accompany this text and contains additional exercises. For more information, you may contact the publisher.

CHAPTER 1

Algebra: The Fundamentals

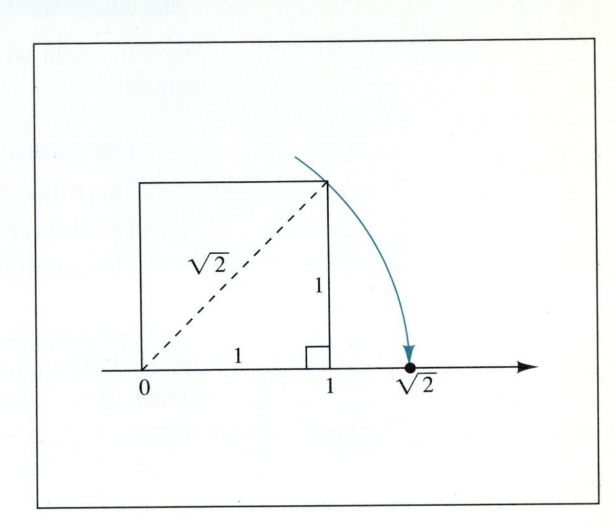

This chapter reviews some of the important and fundamental skills and concepts of algebra necessary for precalculus. It is not designed to serve as a substitute for a solid prerequisite course in algebra but instead it is an overview of the basics of "symbolic arithmetic" as well as a demonstration of how the more rigorous or complex algebra problems are actually approached and solved by using the basic fundamental properties.

1.1 The Real Numbers

We will begin with some of the basic sets of numbers with which you are already familiar:

The **natural numbers** (counting numbers): $N = \{1, 2, 3, 4, 5, \ldots\}$

The **whole numbers:** $W = \{0, 1, 2, 3, 4, \ldots\}$

The **integers**: $Z = \{\ldots, -3, -2, -1, 0, 1, 2, 3, \ldots\}$

The **rational numbers:** $Q = \{\frac{p}{q} \mid p, q \in Z, \text{ and } q \neq 0\}$

We associate the rational numbers with points on the number line (see Figure 1.1). The number associated with a point on the number line is called the **coordinate** of the point.

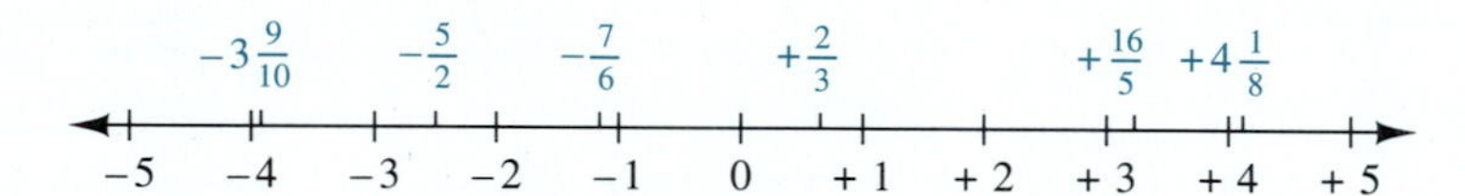

FIGURE 1.1 The number associated with a point on the number line is called the **coordinate** of the point.

After associating each rational number with a point on the number line, we find that there are still points which remain unlabeled. The numbers associated with these unlabeled points are called ***irrational numbers***. To get a better idea of what the irrational numbers look like, we examine rational numbers in decimal form.

You will recall that certain fractions have a finite decimal representation, such as $\frac{1}{4} = 0.25$, whereas others have an infinite decimal representation, such as $\frac{1}{3} = 0.3333\overline{3}$. (The bar means that the digits under the bar repeat without end.) The first case is called a **terminating decimal** (the decimal ends or at some point is followed by zeros); the second is called a **repeating decimal** (the same group of digits in the decimal is repeated indefinitely).

If a number can be written as a quotient of two integers, then it is a rational number. Why does this imply that 2.36 is a rational number?

If a decimal terminates, we can easily recognize it as a rational number. For example, 0.863 is equal to $\dfrac{863}{1000}$. On the other hand, if the decimal repeats, the process of writing it as a quotient of integers is not quite so straightforward. Example 1 illustrates how we can write the repeating decimal $0.189189\overline{189}$ as a fraction.

EXAMPLE 1 Show that $0.189189\overline{189}$ is a rational number.

Solution In order to show that a number is rational, we must somehow represent it as a quotient of integers, p/q. We start as follows:

Let $x = 0.189189\overline{189}$ *We multiply both sides of the equation by 1000. Then*

$1000x = 189.189189\overline{189}$

Now subtract $x = 0.189189\overline{189}$ from $1000x = 189.189189\overline{189}$:

$$\begin{aligned} 1000x &= 189.189189\overline{189} \\ x &= \quad 0.189189\overline{189} \\ \hline 999x &= 189 \\ x &= \frac{189}{999} = \boxed{\frac{7}{37}} \end{aligned}$$

Notice that the repeating decimal portions of the numbers match up exactly. Thus

Note the equation $1000x = 189.189189\overline{189}$ is found by multiplying both sides of the equation $x = 0.189189\overline{189}$ by 1000 (*since* 1000 *is the power of* 10 *needed to move the decimal point over and match up the infinitely repeating decimal portions of the numbers*). ■

Hence, terminating and repeating decimals represent rational numbers. It is a fact that decimals that are both *nonterminating and nonrepeating are* not *rational numbers*. In other words, such a decimal cannot be represented as the quotient of two integers. (Exercise 53 outlines how we might prove that a specific number cannot be represented as a quotient of two integers.)

This set of nonrepeating, nonterminating decimals is called the set of **irrational numbers**. For example, π, $\sqrt{2}$, $\sqrt[5]{3}$ are irrational numbers: We cannot write their exact values as decimals. At best we approximate their decimal values using a table, calculator, or computer and use the symbol $\approx$ to indicate an approximation. For example, $\sqrt{2} \approx 1.414$ ($\sqrt{2}$ is approximately 1.414).

The important thing for us to recognize is that irrational numbers also represent points on the number line (see Exercises 50–52). If we take all the rational numbers together with all the irrational numbers (both positive and negative), we get all the points on the number line. This set is called the set of *real numbers* and is usually designated by the letter $\boldsymbol{R}$.

The real numbers $\boldsymbol{R} = \{x \mid x \text{ corresponds to a point on the number line}\}$.

Unless stated to the contrary, we will assume that we are always working within the framework of the real number system.

Figure 1.2 illustrates the relationship between the sets discussed above.

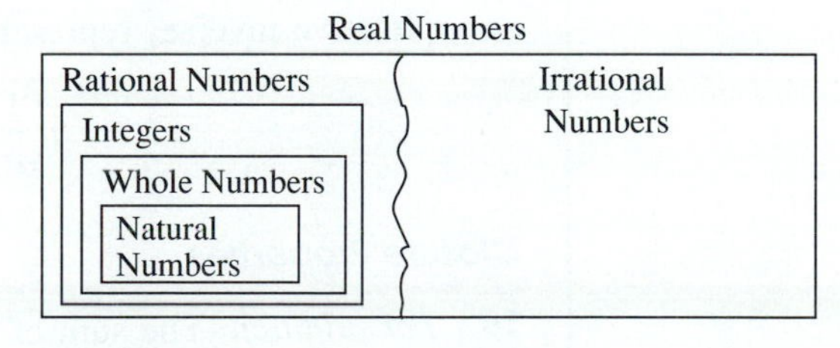

FIGURE 1.2

Properties of Real Numbers

The real numbers, along with the operations of addition (+) and multiplication (·), obey the 11 properties listed in the following box. Most of these properties are straightforward and may seem trivial. Nevertheless, we shall see that these 11 basic properties are quite powerful in that they allow us to do much in simplifying algebraic expressions.

The Commutative Properties

1. *For addition*: $a + b = b + a$
2. *For multiplication*: $ab = ba$

The Associative Properties

3. *For addition*: $a + (b + c) = (a + b) + c$
4. *For multiplication*: $a(bc) = (ab)c$

The Distributive Property

5. $a(b + c) = ab + ac$ or $(b + c)a = ba + ca$

Identities

6. *For addition*: There is a unique real number called the additive identity represented by 0, which has the property that $a + 0 = 0 + a = a$ for all real numbers, a.
7. *For multiplication*: There is a unique real number called the multiplicative identity, represented by 1, which has the property that $a \cdot 1 = 1 \cdot a = a$ for all real numbers, a.

Inverses

8. *For addition*: Each real number a has a unique additive inverse, represented by $-a$, which has the property that

$$a + (-a) = (-a) + a = 0$$

9. *For multiplication*: Each real number a, except 0, has a unique multiplicative inverse, represented by $\frac{1}{a}$, which has the property that

$$a\left(\frac{1}{a}\right) = \left(\frac{1}{a}\right)a = 1$$

Closure Properties

10. *For addition*: The sum of two real numbers is a real number.
11. *For multiplication*: The product of two real numbers is a real number.

In the product ab, a and b are called *factors*; in the sum $a + b$, a and b are called *terms*.

When we state that a set is closed under an operation, we mean that when we perform the operation on two elements in the set, the result will be an element in the set. For example, we could start with the set of whole numbers and develop a system of whole numbers that has the associative, commutative, distributive, and identity property (there are no inverses). This system would be closed under addition and multiplication; that is, sums and products of whole numbers are again whole numbers.

EXAMPLE 2 Show that the set of irrational numbers is not closed under multiplication.

Solution

WHAT DO WE NEED TO DO?	Show that the set of irrational numbers is not closed under multiplication.
HOW DO WE START?	First understand what is being asked. This requires knowing what closure is.
WHAT DOES THE STATEMENT "THE IRRATIONALS ARE CLOSED UNDER MULTIPLICATION" MEAN?	It means that the product of two irrational numbers is an irrational number.
HOW DO WE SHOW THAT IT IS NOT CLOSED?	We are being asked to find an example of two irrational numbers whose product is not irrational.

In mathematics, when we are asked to show, prove, or justify, we usually have to go back to the definitions of the terms that are being used. Often we have to rephrase the problem using a definition.

In this problem we have to ask ourselves what it means for the set of irrational numbers to be closed under multiplication. Reviewing the previous comments on closure, we find it means that the product of two irrational numbers is always an irrational number. Since we want to show this set is ***not*** closed under multiplication, we need only to find one example of a pair of irrational numbers whose product is ***not*** irrational. This is called a *counterexample*. $\boxed{\sqrt{2} \text{ and } \sqrt{8}}$ are two such numbers; they are irrational numbers whose product, $\sqrt{2}\sqrt{8} = \sqrt{16} = 4$, is a rational number. So the set of irrational numbers is not closed under multiplication. ■

Keep in mind that the real number properties apply not only to numbers or single variables, but to more complex expressions as well. For example, the statement

$$(2x^2 + 3x - 1)(x + 2) = (2x^2 + 3x - 1)(x) + (2x^2 + 3x - 1)(2)$$

is an application of the distributive property, where $2x^2 + 3x - 1$ is distributed over $x + 2$.

Order and the Real Number Line

Numerical order is determined by the "less than" symbol, $<$, and can be defined by using the number line:

$a < b$ means that a is to the left of b on the number line.

Hence $3 < \pi$ is the symbolic statement meaning that 3 is to the left of π on the number line as shown in Figure 1.3.

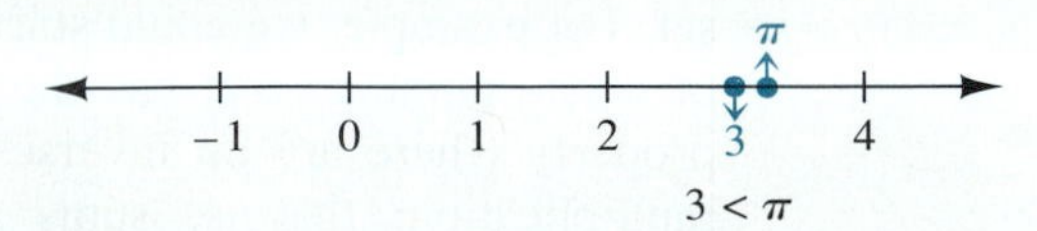

FIGURE 1.3

Algebraically, the symbol $<$ has the following meaning:

> $a < b$ means that $b - a$ is a positive number.

Hence $3 < 8$, *since* $8 - 3 = 5$ is positive; we can also see that $\sqrt{2} < 2$ *since* $2 - \sqrt{2} \approx 2 - 1.414$ is positive.

The "greater than" symbol is similarly defined: $a > b$ *means that* $a - b$ *is positive*.

The inequality symbol $\leq$ means "less than or equal to"; hence, $6 \leq 6$. Similarly defined, $\geq$ means "greater than or equal to." We usually put a slash through equality and inequality symbols when we want to indicate that the statement is not true; for example $5 \neq 4 - 1$ means "5 is not equal to $4 - 1$."

Inequalities using the $<$ and $>$ symbols are called *strict* inequalities, and inequalities using the $\leq$ and $\geq$ symbols are called *weak* inequalities. By the definition of less than and greater than,

$a > 0$ *means that a is positive* and $a < 0$ *means that a is negative*.

The double inequality, $a < x < b$, is used to indicate betweenness. For example, $-3 < x < 6$ means that x is between -3 and 6. The double inequality is actually a combination of two inequalities that must be satisfied simultaneously; that is, $a < x < b$ is a combination of the two inequalities $a < x$ **and** $x < b$, where x satisfies *both* inequalities at the same time. Obviously, for the double inequality $a < x < b$ to make sense, a must be less than b.

Why doesn't it make sense to write $-2 < x < -4$?

Why doesn't it make sense to write $2 < x > 5$?

> **THE TRICHOTOMY PROPERTY**
>
> For $a, b \in \mathbf{R}$, one and only one of the following holds:
>
> $$a < b, \quad a > b, \quad \text{or} \quad a = b.$$

DIFFERENT PERSPECTIVES: *Inequalities*

GEOMETRIC DESCRIPTION

a b

$a < b$ means that a is to the left of b on the number line.

ALGEBRAIC DESCRIPTION

$a < b$ means $b - a$ is a positive number.

Interval Notation

Another way to express sets of numbers described by inequalities is by using interval notation. Interval notation is a convenient and compact way of representing intervals on the number line. We will begin with bounded intervals—i.e., intervals that have two endpoints.

We use parentheses to indicate that an endpoint is *not* included and brackets to indicate that an endpoint is included. Hence, for $a < b$, we have the following.

INTERVAL NOTATION: BOUNDED INTERVALS

Set Notation		*Interval Notation*
$\{x \mid a \le x \le b\}$	is written as	$[a, b]$, called the *closed interval from a to b.*
$\{x \mid a < x < b\}$	is written as	(a, b), called the *open interval from a to b.*
$\{x \mid a \le x < b\}$	is written as	$[a, b)$, called a *half-open interval: closed at a and open at b.*
$\{x \mid a < x \le b\}$	is written as	$(a, b]$, called *a half-open interval: open at a and closed at b.*

Line Graph

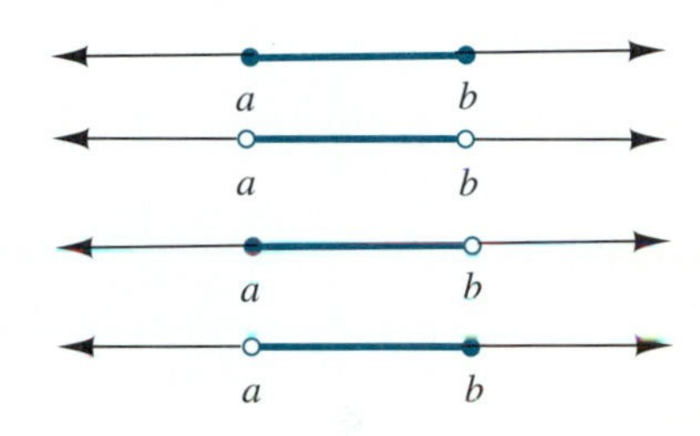

The smaller number is always written left of the larger. Unfortunately, the open interval (a, b) uses the same notation as the ordered pair (a, b). It should always be clear, however, from the context of a problem whether we are talking about an interval or an ordered pair.

When expressing unbounded intervals, lines, or half-lines, we use the infinity symbol, ∞ or $-\infty$, with interval notation, as follows.

INTERVAL NOTATION: UNBOUNDED INTERVALS

Set Notation		*Interval Notation*
$\{x \mid x \ge a\}$	is written as	$[a, \infty)$, called the *unbounded interval from a to (positive) infinity, closed at a.*
$\{x \mid x > a\}$	is written as	(a, ∞), called the *unbounded interval from a to (positive) infinity, open at a.*
$\{x \mid x \le a\}$	is written as	$(-\infty, a]$, called the *unbounded interval from negative infinity to a, closed at a.*
$\{x \mid x < a\}$	is written as	$(-\infty, a)$, called the *unbounded interval from negative infinity to a, open at a.*

Line Graph

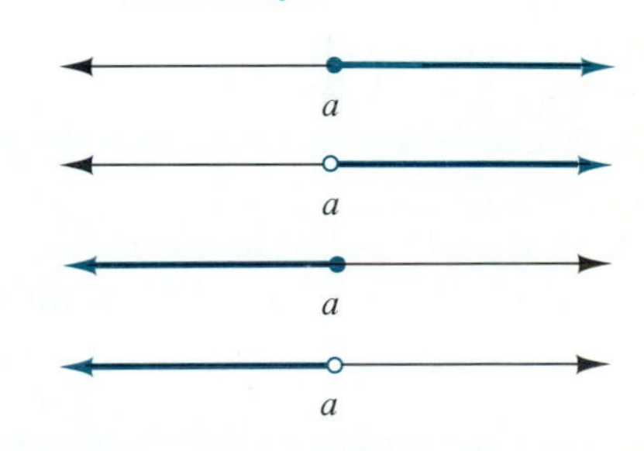

The symbols ∞ and $-\infty$ do not represent numbers; they are simply symbols to remind us that the interval goes on forever, or increases (or decreases) without bound. Therefore, we always write a parenthesis next to the ∞ symbol.

Remember that whenever we use interval notation, we are working within the framework of the real number system. The heavy line indicates that all points on the line are included.

EXAMPLE 3 Graph the following inequalities on the number line, and express the set using interval notation.

(a) $\{x \mid x > -3\}$ **(b)** $\{s \mid s \le 4\}$ **(c)** $\{t \mid -2 < t \le 6\}$

Solution

(a) $\{x \mid x > -3\}$ which is $(-3, \infty)$

(b) $\{s \mid s \le 4\}$ which is $(-\infty, 4]$

(c) $\{t \mid -2 < t \le 6\}$ which is $(-2, 6]$ ■

Absolute Value

Geometrically, the absolute value of a number is its distance from zero on the number line. The absolute value of x is symbolized $|x|$. Hence

$|-4| = 4$ *Since −4 is 4 units away from zero on the number line* (Figure 1.4)

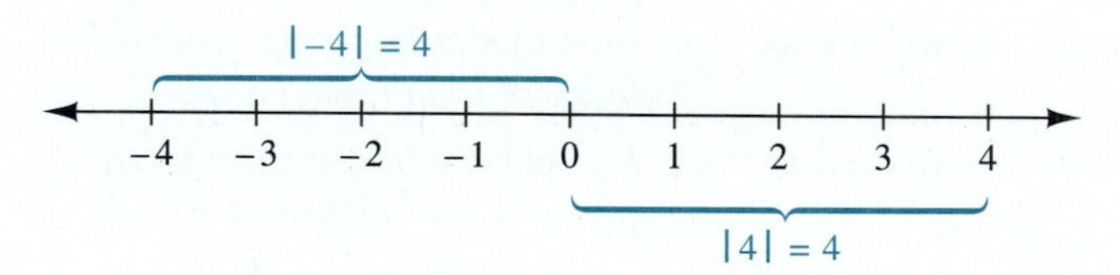

FIGURE 1.4

Also,

$|4| = 4$ *Since 4 is 4 units away from zero on the number line*

Algebraically, we define absolute value as follows:

ABSOLUTE VALUE

$$|x| = \begin{cases} x, & \text{if } x \ge 0 \\ -x, & \text{if } x < 0 \end{cases}$$

A word about this notation. The notation symbolically expresses that the definition of $|x|$ depends upon the value of x: If x is nonnegative ($x \ge 0$), then $|x|$ is simply x; on the other hand, if x is negative ($x < 0$), then $|x|$ is $-x$, which makes $|x|$ positive. For example, $|5| = 5$ *since* $5 \ge 0$, and $|-5| = -(-5) = 5$ *since* $-5 < 0$. Consequently, $|x|$ can never be negative.

EXAMPLE 4 Write the following without absolute value symbols.

(a) $|\pi - 3|$ **(b)** $|3 - \pi|$ **(c)** $|x^6 + 1|$ **(d)** $|y - 3|$

Solution According to the definition of absolute value, in order to evaluate absolute value expressions, we need to know whether the expression within absolute values is positive or negative.

(a) Since $\pi \approx 3.14$, then $\pi - 3$ is positive; hence

$$|\pi - 3| = \boxed{\pi - 3} \quad (\textit{since } \pi - 3 \ge 0)$$

(b) On the other hand, $3 - \pi$ is negative; hence

$$|3 - \pi| = -(3 - \pi) = \boxed{\pi - 3} \qquad (\textit{since } 3 - \pi < 0)$$

Note: $|\pi - 3| = |3 - \pi|$

(c) Although x is a variable, we do know that x^6 must always be nonnegative. Therefore, $x^6 + 1$ must also be nonnegative; hence

Why must x^6 be nonnegative?

$$|x^6 + 1| = \boxed{x^6 + 1} \qquad (\textit{since } x^6 + 1 \geq 0)$$

(d) We cannot determine whether the expression $y - 3$ is positive or negative, so we cannot evaluate this expression. However, we can use the definition of absolute value to rewrite this expression without the absolute value symbols:

$|y - 3| = y - 3$ when $y - 3 \geq 0$ (*that is, when* $y \geq 3$)

$|y - 3| = -(y - 3) = 3 - y$ when $y - 3 < 0$ (*that is, when* $y < 3$)

We can consolidate this as:

$$\boxed{|y - 3| = \begin{cases} y - 3 & \text{when } y \geq 3 \\ 3 - y & \text{when } y < 3 \end{cases}}$$

■

Distance on the Number Line

We can see by Figure 1.5 that on the real number line the distance between two points can be found using the differences of the coordinates; for example, the distance between the points with coordinates 5 and 2 is $5 - 2 = 3$ units.

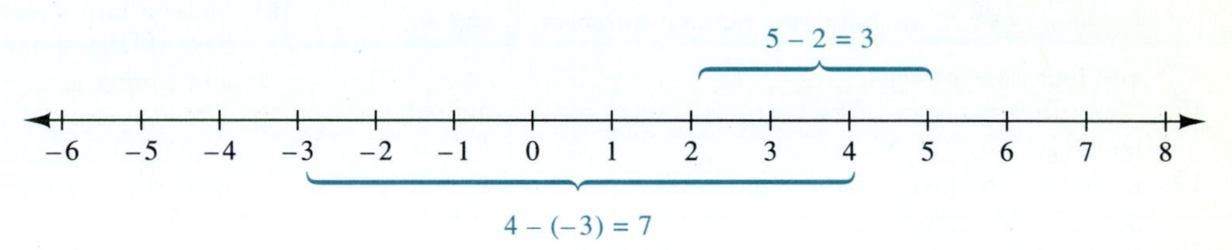

FIGURE 1.5

Since we want distance to be positive, we use absolute value to define distance as follows:

THE DISTANCE BETWEEN TWO POINTS ON THE NUMBER LINE

On the real number line, the distance between points with coordinates a and b is $|a - b|$ or $|b - a|$.

Notice that the distance between 4 and -3 is 7 whether we compute it as $|4 - (-3)|$ or $|-3 - 4|$. The outline of an algebraic proof that $|a - b| = |b - a|$ is discussed in Exercise 54. Note as well that this definition is consistent with the geometric definition of $|x|$ being the distance from 0 on the number line, since $|x| = |x - 0|$.

EXERCISES 1.1

In Exercises 1–6, express the infinitely repeating decimals as a quotient of integers, $\frac{p}{q}$.

1. $0.22\overline{2}$ **2.** $0.3535\overline{35}$
3. $4.55\overline{5}$ **4.** $6.2323\overline{23}$
5. $8.238238\overline{238}$ **6.** $14.354354\overline{354}$

In Exercises 7–16, if a given statement is true, give the property that the statement illustrates. If the statement is not true for all real numbers, write FALSE, and give a counterexample (an example which demonstrates that the statement is not true).

7. $\left(x + \frac{1}{2}\right) + \frac{2}{3} = x + \left(\frac{1}{2} + \frac{2}{3}\right)$
8. $y + (5 + x) = y + (x + 5)$
9. $3 + (xy) = (3 + x)(3 + y)$
10. $5a + 0 = 5a$
11. $[3(xy)z] = [(3x)(yz)]$ **12.** $\left(\frac{3}{4} + x\right)1 = \frac{3}{4} + x$
13. $(x - y + z)(a + b) = (x - y + z)a + (x - y + z)b$
14. $(x + 4) + [-(x + 4)] = 0$
15. $\frac{1}{x^2 + 1} \cdot (x^2 + 1) = 1$ **16.** $a(bc) = (ab)(ac)$
17. Show that the product of two rational numbers is a rational number. HINT: Start with two rational numbers, $\frac{a}{b}$ and $\frac{c}{d}$, and find their product.
18. Show that the sum of two rational numbers is a rational number.
19. Is the sum of two irrational numbers always irrational? If not, give a counterexample.
20. Is the difference of two whole numbers always a whole number? If not, give a counterexample.

In Exercises 21–26, graph each set on a real number line.

21. $\{x \mid x < 4\}$ **22.** $\{x \mid x \geq -5\}$
23. $\{x \mid x > 5\}$ **24.** $\{x \mid -3 < x \leq 2\}$
25. $\{x \mid -8 < x < -2\}$ **26.** $\{x \mid -2 \leq x < 4\}$

In Exercises 27–36, graph the set on the number line and express the set using interval notation.

27. $\{x \mid x > 5\}$ **28.** $\{x \mid x \leq -1\}$
29. $\{x \mid x \geq -5\}$ **30.** $\{x \mid -3 < x\}$
31. $\{x \mid -8 \leq x < -5\}$ **32.** $\{x \mid 0 < x \leq 6\}$
33. $\{x \mid -2 \geq x\}$ **34.** $\{x \mid -3 < x < 4\}$
35. $\{x \mid -9 < x \leq -2\}$ **36.** $\{x \mid 0 \leq x \leq 6\}$

In Exercises 37–44, write each expression without absolute value symbols.

37. $|3 - 5|$ **38.** $|\pi - 3.14|$
39. $|\sqrt{2} - 1|$ **40.** $|1 - \sqrt{2}|$
41. $|x - 5|$ **42.** $|x + 4|$
43. $|x^2 + 1|$ **44.** $|x^4 + 3|$

In Exercises 45–48, find the distance on the number line between each pair of points with the given coordinates.

45. 2 and 3 **46.** 5 and -9
47. -3 and 8 **48.** -8 and -4

QUESTIONS FOR THOUGHT

49. The *transitive property* of inequalities states that if $a < b$ and $b < c$, then $a < c$. Prove this property. HINT: Determine what each statement means by the algebraic definition of an inequality.
50. Locate $\sqrt{2}$ on the number line by the following steps.
(a) Draw a unit square on the number line such that the base of the square is the line segment starting at 0 and ending at 1.
(b) The diagonal of the square divides the square into two right triangles, where the diagonal is the hypotenuse of both right triangles. (See figure below.)

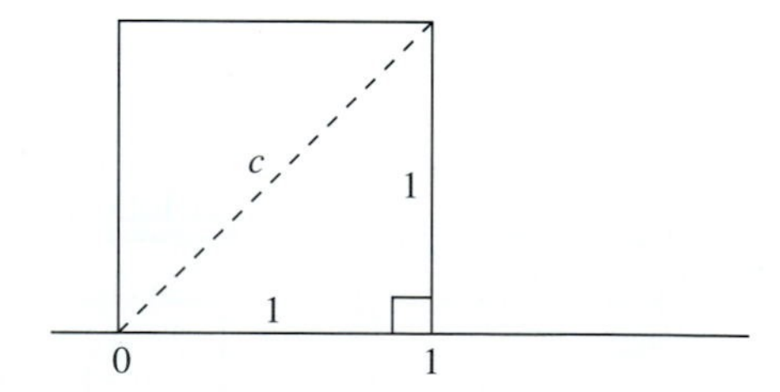

(c) Using the Pythagorean Theorem from geometry, show that the length of the diagonal of the square is $\sqrt{2}$.

(d) Place the point of a compass on 0 and open it to the length of the diagonal.

(e) Rotate the compass down to the number line to locate $\sqrt{2}$ on the number line. (See figure below.)

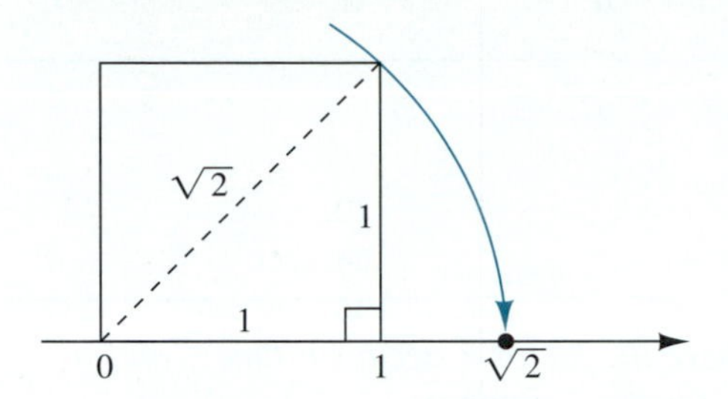

51. Locate $\sqrt{3}$ on the number line using the method described by Exercise 50. HINT: First locate $\sqrt{2}$.

52. Locate $\sqrt{5}$ on the number line using the method described by Exercise 50.

53. The following is an outline of how we might prove that $\sqrt{2}$ is not a rational number:

(a) Show that for all integers, n, n^2 is even if and only if n is even.

(b) Assume $\sqrt{2} = \frac{m}{n}$, where $\frac{m}{n}$ is *reduced to lowest terms*.

(c) Multiply both sides of the preceding equation by n and square each side. Show that this implies that m^2 is even and, therefore, m is even.

(d) If m is even, then we can write $m = 2k$ for some k. Substitute $2k$ for m in the original equation; following a similar approach, show that this implies that n is even.

(e) Why does this contradict the original assumption? What does this mean about the original assumption?

54. Prove that $|a - b| = |b - a|$ for all real numbers a and b. HINT: By the trichotomy property, only one of the following holds: $a = b$, $a < b$, or $a > b$. Check each case by using the definition of $|a - b|$ to determine if for all real a and b, $|a - b| = |b - a|$.

1.2 Operations with Real Numbers

Subtraction and Division

Subtraction and division are defined in the following way:

> ***Subtraction:*** $a - b = a + (-b)$ ***Division:*** $\frac{a}{b} = a\left(\frac{1}{b}\right)$ for $b \neq 0$

Thus subtracting b means *adding the additive inverse of* b; dividing by b means *multiplying by the multiplicative inverse of* b. For example,

$$5 - (+8) = 5 + (-8) = -3, \quad \text{and} \quad \frac{6}{2} = 6\left(\frac{1}{2}\right) = 3.$$

What happens if we try to divide a nonzero number by 0? Let's assume that $\frac{6}{0} = x$. This means that $x \cdot 0 = 6$, and we end up with the contradiction $0 = 6$. For any nonzero number a, $\frac{a}{0}$ will yield the contradiction $a = 0$. Therefore, division of a nonzero number by zero is *undefined*. On the other hand, if $a \neq 0$, $\frac{0}{a} = 0$, since $0 \cdot a = 0$.

If we try to divide 0 by 0, we run into problems for another reason. Let's suppose we let $\frac{0}{0}$ be equal to some number r. Then this means $r \cdot 0 = 0$. But this is true for all numbers r. This means that any number will work; that is, $\frac{0}{0}$ is not unique. Therefore, we say that $\frac{0}{0}$ is *indeterminate* (there is no unique answer). To summarize:

If a is any nonzero number, then

$$\frac{a}{0} \text{ is undefined, } \quad \text{whereas} \quad \frac{0}{a} = 0 \quad \text{and} \quad \frac{0}{0} \text{ is indeterminate}$$

Exponents

We define

> $x^n = x \cdot x \cdot x \cdots x$, *where the factor x occurs n times, and n is a natural number.*

For x^n, x is called the **base** and n is called the **exponent**. A natural number exponent tells how many times x occurs as a factor in the product. x^n is called the ***n*th power of *x***. (Note that $x = x^1$; that is, an unwritten exponent is assumed to be 1.)

To compute or evaluate an expression with a numerical base means to multiply out the expression. For example:

$$3^5 = 3 \cdot 3 \cdot 3 \cdot 3 \cdot 3 = 243$$

$$(-2)^4 = (-2)(-2)(-2)(-2) = +16$$

On the other hand,

$$-2^4 = -(2)^4 = -(2 \cdot 2 \cdot 2 \cdot 2) = -16$$

Keep in mind that the exponent applies only to the quantity to its immediate left. In the expression $3 \cdot 2^4$, we raise 2 to the fourth power and then multiply by 3 to get 48. Similarly, in -2^4, we raise 2 to the fourth power and then use the minus sign to get -16.

Multiple Operations

If there is more than one operation to perform in an expression, there must be agreement as to which operation should be performed first: an order of operations. For example, to evaluate the expression $3 + 2 \cdot 4$, we could arrive at the answer 20 or 11, depending upon whether we choose to multiply or add first. We agree to the following **order of operations**:

1. Start by performing operations in the grouping symbols, innermost first.
2. Then evaluate powers (and roots) in any order.
3. Perform multiplication and division, working left to right.
4. Finally, perform addition and subtraction, working left to right.

For example, to evaluate the numerical expression $4[-5 - 3(-2)^2]$ we would proceed as follows:

Try this problem with your calculator for practice.

$$4[-5 - 3(-2)^2] = 4[-5 - 3(4)] = 4[-5 - 12] = 4(-17) = -68$$

Substitution

Applications of algebra very often require us to substitute numerical values for variables and then to perform the operations with the substituted values. In such a situation we will be asked to evaluate an expression given the values for the variables.

EXAMPLE 1 Given $x_1 = 3$, $y_1 = -7$, $x_2 = -2$, and $y_2 = 5$, evaluate the following.

$$\sqrt{(x_1 - x_2)^2 + (y_1 - y_2)^2}$$

Solution x_1, x_2, y_1, and y_2 are called **subscripted variables**. It is usually a good idea to enclose the values being substituted in parentheses. This helps avoid confusing the original operations. Substitute $x_1 = 3$, $y_1 = -7$, $x_2 = -2$, and $y_2 = 5$ into the expression:

$$\sqrt{(x_1 - x_2)^2 + (y_1 - y_2)^2} = \sqrt{[(3) - (-2)]^2 + [(-7) - (5)]^2}$$

Work inside brackets first.

$$= \sqrt{[5]^2 + [-12]^2}$$

Powers under the radical

$$= \sqrt{25 + 144}$$

$$= \sqrt{169} = \boxed{13}$$

■

Fractions and Their Operations

While a rational number is a quotient of integers, $\frac{p}{q}$ where $q \neq 0$, a *fraction* is a quotient of any two numbers $\frac{a}{b}$ where $b \neq 0$. We know from previous experience with fractions that two fractions may look different but may actually be equivalent. Hence we define what we mean by "=" with fractions as follows:

EQUIVALENCE OF FRACTIONS

$$\frac{a}{b} = \frac{c}{d} \quad \text{if and only if} \quad ad = bc \qquad (b, d \neq 0)$$

Hence $\dfrac{6}{13} = \dfrac{24}{52}$ *because* $6 \cdot 52 = 13 \cdot 24 = 312$.

EXAMPLE 2 Show that $\dfrac{2}{\sqrt{2}} = \sqrt{2}$.

Solution Again, to prove or justify a given statement, we often refer back to a definition. In this example we look at the definition of "=" for fractions, rephrasing the problem using this definition. Expressing $\sqrt{2}$ as the fraction $\frac{\sqrt{2}}{1}$, we find that for $\frac{2}{\sqrt{2}}$ to equal $\frac{\sqrt{2}}{1}$ means that (2)(1) should equal $\sqrt{2}\sqrt{2}$. Since $(2)(1) = 2 = (\sqrt{2})(\sqrt{2})$, we have $\frac{2}{\sqrt{2}} = \sqrt{2}$. ■

A fraction reduced to lowest terms or written in simplest form is a fraction that has no factors (other than ± 1) common to both its numerator and denominator. If we require all our fractions to be reduced to lowest terms, then, rather than resorting to the definition of equivalence, we can observe by inspection that two fractions are equivalent. Hence, the fundamental principle of fractions is usually more useful:

THE FUNDAMENTAL PRINCIPLE OF FRACTIONS

$$\frac{a \cdot k}{b \cdot k} = \frac{a}{b} \qquad b, k \neq 0$$

This principle says that if we divide or multiply the numerator and denominator of a fraction by the same nonzero expression, we obtain an equivalent fraction.

To reduce a fraction to lowest terms often requires us to factor both numerator and denominator and then use the fundamental principle by dividing out factors common to both the numerator and denominator. For example, $\frac{24}{52} = \frac{4 \cdot 6}{4 \cdot 13} = \frac{6}{13}$.

We define operations with fractions as follows:

OPERATIONS WITH FRACTIONS

Multiplication

$$\left(\frac{a}{b}\right)\left(\frac{c}{d}\right) = \frac{ac}{bd}$$

Division

$$\left(\frac{a}{b}\right) \div \left(\frac{c}{d}\right) = \frac{ad}{bc}$$

Addition

$$\frac{a}{b} + \frac{c}{d} = \frac{ad + bc}{bd}$$

Subtraction

$$\frac{a}{b} - \frac{c}{d} = \frac{ad - bc}{bd}$$

Notice that subtraction is still adding the additive inverse, and division is still multiplying by the reciprocal. For example:

$$\left(\frac{2}{7}\right) \div \left(\frac{5}{8}\right) = \left(\frac{2}{7}\right) \cdot \left(\frac{8}{5}\right) = \frac{2 \cdot 8}{7 \cdot 5} = \frac{16}{35}$$

If the denominators of two fractions are identical, we can use the distributive property to find:

$$\frac{a}{c} + \frac{b}{c} = \frac{a + b}{c}$$

Hence, an alternative method for combining fractions with different denominators is to express the fractions as equivalent fractions having their least common denominator (LCD) as denominator and then add them as fractions with the same denominator as above. For example,

$$\frac{5}{18} + \frac{11}{12} \qquad \textit{The LCD is 36, the smallest multiple of both 12 and 18.}$$

$$= \frac{5 \cdot 2}{36} + \frac{11 \cdot 3}{36} = \frac{10}{36} + \frac{33}{36} = \frac{43}{36}$$

Try this problem with your calculator for practice.

EXAMPLE 3 Evaluate the numerical expression.

$$3\left(-\frac{2}{3}\right)^2 + 4\left(-\frac{2}{3}\right) + 5$$

Solution

$$3\left(-\frac{2}{3}\right)^2 + 4\left(-\frac{2}{3}\right) + 5 = 3\left(\frac{4}{9}\right) - \frac{8}{3} + 5 = \frac{4}{3} - \frac{8}{3} + 5 = \frac{4}{3} - \frac{8}{3} + \frac{15}{3}$$

$$= \boxed{\frac{11}{3}}$$

■

Complex Fractions

EXAMPLE 4 Write as a simple fraction reduced to lowest terms. $\dfrac{\frac{3}{5} + \frac{2}{3}}{\frac{3}{4} - \frac{2}{5}}$

Solution We offer two methods of solution. For our first method, we treat the problem as a multiple-operation situation. That is, we treat it as the problem:

$$\left(\frac{3}{5} + \frac{2}{3}\right) \div \left(\frac{3}{4} - \frac{2}{5}\right)$$

$$\frac{\frac{3}{5} + \frac{2}{3}}{\frac{3}{4} - \frac{2}{5}} = \frac{\frac{9 + 10}{15}}{\frac{15 - 8}{20}} = \frac{\frac{19}{15}}{\frac{7}{20}} = \left(\frac{19}{15}\right)\left(\frac{20}{7}\right) = \frac{19 \cdot 20}{15 \cdot 7} = \boxed{\frac{76}{21}} \qquad \textit{(When reduced)}$$

An alternative way to clear denominators of fractions within a complex fraction is to apply the fundamental principle by multiplying the numerator and denominator of the complex fraction by the LCD of *all* simple fractions in the complex fraction.

In this case, the LCD of $\frac{3}{5}$, $\frac{2}{3}$, $\frac{3}{4}$, and $\frac{2}{5}$ is 60:

$$\frac{\left(\frac{3}{5}+\frac{2}{3}\right)\frac{60}{1}}{\left(\frac{3}{4}-\frac{2}{5}\right)\frac{60}{1}} = \frac{\left(\frac{3}{5}\right)\frac{60}{1}+\left(\frac{2}{3}\right)\frac{60}{1}}{\left(\frac{3}{4}\right)\frac{60}{1}-\left(\frac{2}{5}\right)\frac{60}{1}} = \frac{36+40}{45-24} = \boxed{\frac{76}{21}}$$

■

EXAMPLE 5 The *harmonic mean* of n positive numbers, $X_1, X_2, X_3, \ldots, X_n$, is defined as follows:

$$h = \frac{n}{\frac{1}{X_1}+\frac{1}{X_2}+\frac{1}{X_3}+\cdots+\frac{1}{X_n}}$$

(a) Find the exact value of the harmonic mean of 4, 6, and 7.

(b) Find the harmonic mean of 4, 6, and 7, rounded to four places using a calculator.

Solution

(a) Since we have three numbers, $n = 3$, and we substitute $X_1 = 4$, $X_2 = 6$, and $X_3 = 7$ in the given formula.

$$h = \frac{n}{\frac{1}{X_1}+\frac{1}{X_2}+\frac{1}{X_3}+\cdots+\frac{1}{X_n}}.$$

Substitute $X_1 = 4$, $X_2 = 6$, $X_3 = 7$, and $n = 3$ to get

$$= \frac{3}{\frac{1}{4}+\frac{1}{6}+\frac{1}{7}}$$

Multiply the numerator and denominator by the LCD of $\frac{1}{4}$, $\frac{1}{6}$, and $\frac{1}{7}$, which is 84.

$$= \frac{3}{\frac{1}{4}+\frac{1}{6}+\frac{1}{7}} \cdot \frac{\frac{84}{1}}{\frac{84}{1}}$$

$$= \frac{252}{21+14+12} = \boxed{\frac{252}{47}}$$

(b) How you handle this problem using a calculator depends upon what type of calculator you are using. If your calculator has a $\boxed{1/x}$ key (or a $\boxed{x^{-1}}$ key), you can enter the following sequence of keys:

$\boxed{4}\ \boxed{1/x}\ \boxed{=}\ \boxed{+}\ \boxed{6}\ \boxed{1/x}\ \boxed{=}\ \boxed{+}\ \boxed{7}\ \boxed{1/x}\ \boxed{=}\ \boxed{\div}\ \boxed{3}\ \boxed{=}\ \boxed{1/x}\ \boxed{=}$

This sequence of keys computes $\frac{(\frac{1}{4}+\frac{1}{6}+\frac{1}{7})}{3}$, and then computes its reciprocal.

This gives (to 8 places) 5.36170213 which rounds to $\boxed{5.3617}$.

If your calculator does not have a $\boxed{1/x}$, then you have to write down numbers as you perform computations with your calculator. The accuracy of your answer is dependent upon how much you round each computation you write down. For example, you may be able to evaluate the denominator of the complex fraction, $\frac{1}{4} + \frac{1}{6} + \frac{1}{7}$, by entering the following sequence of keys:

[1] [÷] [4] [=] [+] [1] [÷] [6] [=] [+] [1] [÷] [7] [=]

which gives (to 8 places) 0.55952381.

If we write this number down as 0.5595, which is rounded to four places, and use this rounded number in the final computation, we arrive at $3 \div 0.5595 = 5.3619$, which is NOT accurate to four places. ■

EXERCISES 1.2

In Exercises 1–26, evaluate the numerical expressions.

1. $-3 + (-6) - (+4) - (-8)$

2. $-6 + (-2) - (-9) - (-4)$

3. $(-6)(-2)(-3)$

4. $(-5)(-8)(-6)$

5. $-2 - 3.552$

6. $-8 + 5.582$

7. $-4 + 7.29$

8. $-4 - 7.29$

9. $-2[3 - (2 - 5)]$

10. $-6[2 - (5 - 8)]$

11. $2 - (-3)^2$

12. $2(-3)^2$

13. $6 - [4 - (5 - 8)^2]$

14. $9 - \{3 - [6 - 2(9 - 4)^2]\}$

15. $|-6 - 5| - |6 - 5|$

16. $|-4 - 4| - |4 - 4|$

17. $\dfrac{|-9 - 5|}{|-9| - |5|}$

18. $\dfrac{|-6 - 12|}{|-6| - |12|}$

19. $\frac{3}{4} - \frac{2}{3} + \frac{1}{2}$

20. $\frac{3}{5}\left(-\frac{2}{3}\right) - \frac{1}{2}$

21. $\left(-\frac{2}{5}\right)^2 - \frac{3}{4}$

22. $\left(-\frac{3}{4}\right) - \left(\frac{2}{3}\right)^2$

23. $6\left(-\frac{2}{3}\right)^2 + \left(-\frac{2}{3}\right) - 2$

24. $6\left(\frac{1}{2}\right)^2 + \left(\frac{1}{2}\right) - 2$

25. $3\left(-\frac{1}{5}\right)^2 + 2\left(-\frac{1}{5}\right) - 3$

26. $2\left(-\frac{1}{4}\right)^2 - 3\left(-\frac{1}{4}\right) + 8$

In Exercises 27–30 write each as a simple fraction reduced to lowest terms.

27. $\dfrac{3 + \frac{3}{5}}{5 - \frac{1}{8}}$

28. $\dfrac{4 - \frac{2}{3}}{\frac{2}{5} - 6}$

29. $\dfrac{\frac{2}{3} - \frac{1}{2}}{\frac{1}{8} + \frac{2}{5}}$

30. $\dfrac{\frac{3}{5} - \frac{1}{2}}{\frac{7}{10} - 2}$

In Exercises 31–34 evaluate each expression given $x = -1$ and $y = -2$.

31. $2x^2 - 4y^2$

32. $|x - y| - |x| - |y|$

33. $\dfrac{x^2 - 2xy + y^2}{x - y}$

34. $(x - y)^2 - x^2 - y^2$

35. The *geometric mean* of n positive numbers, $X_1, X_2, X_3, \ldots, X_n$ is defined as follows:

$$g = \sqrt[n]{X_1 \cdot X_2 \cdot X_3 \cdots X_n}$$

Find the geometric mean of 5, 8, 7, 9, 7, 8, and 6 rounded to four places.

36. The *harmonic mean* of n positive numbers, $X_1, X_2, X_3, \ldots, X_n$ is defined in Example 6. Equivalently, we can define the harmonic mean as follows:

$$h = \frac{1}{\left[\frac{1}{X_1} + \frac{1}{X_2} + \frac{1}{X_3} + \cdots + \frac{1}{X_n}\right]/n}$$

Find the harmonic mean of 5, 8, 7, 9, 7, 8, and 6 both exactly and rounded to four places.

37. Given $s_e = s_y\sqrt{1 - r_{xy}^2}$, if $s_y = 1.25$ and $r_{xy} = 0.4$, find s_e rounded to two places.

38. Given $s_e = s_y\sqrt{1 - r_{xy}^2}$, if $s_y = 2.24$ and $r_{xy} = 0.73$, find s_e rounded to four places.

39. Given

$$Z = \frac{Z_{r_1} - Z_{r_2}}{\sqrt{\frac{1}{n_1 - 3} + \frac{1}{n_2 - 3}}}$$

compute Z to two places for $Z_{r_1} = 0.50$, $Z_{r_2} = 0.32$, $n_1 = 65$, and $n_2 = 83$.

40. Given

$$t = \frac{\overline{X} - a}{\frac{s_x}{\sqrt{n}}}$$

compute t to two places for $\overline{X} = 90$, $a = 95$, $s_x = 5.2$, and $n = 15$.

41. Given

$$\sigma_r = \sqrt{\frac{1 - \rho^2}{n - 1}}$$

compute σ_r to three places for $\rho = 0.5$ and $n = 100$.

42. Given

$$\sigma_r = \sqrt{\frac{1 - \rho^2}{n - 1}}$$

compute σ_r to three places for $\rho = 0.67$ and $n = 128$.

43. Use the real number properties and the definition of rational addition to prove $\frac{a}{c} + \frac{b}{c} = \frac{a + b}{c}$.

44. Use the *definition of equivalent fractions* to show:

(a) $\frac{1}{\sqrt{2}} = \frac{\sqrt{2}}{2}$ **(b)** $\frac{5}{\sqrt{3}} = \frac{5\sqrt{3}}{3}$

45. Use *the fundamental principle of fractions* to show:

(a) $\frac{1}{\sqrt{2}} = \frac{\sqrt{2}}{2}$ **(b)** $\frac{5}{\sqrt{3}} = \frac{5\sqrt{3}}{3}$

QUESTIONS FOR THOUGHT

46. Consider the expression $\frac{3}{x}$. Discuss the following.

(a) What happens to the value of $\frac{3}{x}$ as x gets larger?

(b) What happens to the value of $\frac{3}{x}$ as x remains positive but gets smaller (closer to 0)?

(c) What happens to the value of $\frac{3}{x}$ as x remains negative but gets closer to 0?

(d) Can $\frac{3}{x}$ ever be zero? Explain your answer.

47. Consider y defined in the following way:

$$y = \begin{cases} x + 1 & \text{if } x < 2 \\ x^2 & \text{if } x \geq 2 \end{cases}$$

This notation indicates that y is defined by two possible rules, dependent upon the value of x: If x is less than 2, use $x + 1$ for y; if x is greater than or equal to 2, then use x^2 for y. Hence if $x = -5$, then, since -5 is less than 2, we use the first rule to find y: $y = x + 1 = -5 + 1 = -4$. If $x = 8$, then, since 8 is greater than (or equal to) 2, we use the second rule to find y: $y = x^2 = (8)^2 = 64$. Find y, defined previously, for $x = 1$ and for $x = 3$.

48. Given

$$y = \begin{cases} x & \text{if } x \geq 0 \\ x - 3 & \text{if } x < 0 \end{cases}$$

find y when **(a)** $x = 4$, **(b)** $x = -3$, **(c)** $x = 0$.
HINT: See Exercise 47.

1.3 Polynomials and Rational Expressions

An **algebraic expression** is an expression obtained by adding, subtracting, multiplying, dividing, and taking roots of constants and/or variables. For example,

$$2x^{-\frac{1}{2}} + 7, \qquad \frac{\sqrt{3x - 4}}{5x^2 - 3}, \qquad 3x^2 + \frac{2}{x} - 1, \quad \text{and} \quad x^3 - 2x + 3$$

are algebraic expressions. In this section we will review two particular types of algebraic expressions: polynomials and rational expressions.

Polynomials

A **polynomial in one variable** is an expression of the form:

$$a_nx^n + a_{n-1}x^{n-1} + a_{n-2}x^{n-2} + \cdots + a_2x^2 + a_1x + a_0,$$

$$a_n \neq 0$$

where the a_i's are real numbers, x is a variable, and n is a nonnegative integer called the **degree of the polynomial**. Each expression a_ix^i is called a **term** of the polynomial.

When a polynomial is written with its terms arranged in descending powers, it is said to be in **standard form**. A polynomial such as $5 + 3x - 4x^2$ written in standard form is $-4x^2 + 3x + 5$. In this form, by the definition, $n = 2$, $a_2 = -4$, $a_1 = 3$, and $a_0 = 5$. Note how the subscripts of a conveniently match the exponents of x.

When written with descending powers, we assume that missing powers of the variable have coefficients of 0. For example, $3x^5 - 2x + 3$ can be rewritten as $3x^5 + 0x^4 + 0x^3 + 0x^2 - 2x + 3$.

A polynomial in more than one variable contains terms such as $ax^my^nz^s$, where a is real, x, y, and z are variables, and m, n, and s are nonnegative integers.

We may classify polynomials by the number of terms making up the polynomial: A monomial is a polynomial consisting of one term, a binomial is a polynomial consisting of two terms, and a trinomial is a polynomial consisting of three terms.

Besides the number of terms, we can also classify a polynomial by its degree. First we define the degree of a monomial: The **degree of a monomial** is the sum of the exponents of its variables. For example, $-3x^2y^3z$ has degree 6, since $2 + 3 + 1 = 6$ (remember $z = z^1$). The **degree of a polynomial** is the highest degree of any monomial in it. For example, $7x^2y^4 - 3x^7y^5z^2$ has degree 14, since the highest degree of any monomial in the polynomial is 14, the degree of the second term; $7x^4 - 4x^6$ has degree 6, since the highest-degree term, $-4x^6$, has degree 6.

The degree of a *nonzero* constant is 0 (we can rewrite 4 as $4x^0$). When the number 0 is considered as a polynomial, we call it the **zero polynomial**; the degree of the zero polynomial is undefined.

A third way to classify polynomials is by the number of variables. The following is an example of a polynomial in two variables written in standard form.

$$x^7y + x^6 + x^3y^2 + x^2y^3 + x$$

The terms are arranged in descending degree order with the powers of x in descending order.

Polynomial Operations

Recall that $x^n = x \cdot x \cdot x \cdot x \cdots x$, where the factor x occurs n times. Hence $3xxxxxyyy = 3x^5y^3$.

Using the definition of exponential notation and the real number properties, we can derive the following rules of natural number exponents:

$$x^n x^m = x^{n+m}, \qquad (x^n)^m = x^{nm}, \quad \text{and} \quad (xy)^n = x^n y^n$$

We use these first few rules of exponents to find products of powers with the same base. Using the associative and commutative properties of multiplication (that is, ignoring order and grouping), we can find the following product:

$$(5x^3y^6)(-6x^4y^2) = (5)(-6)x^3x^4y^6y^2 = -30x^7y^8$$

The distributive property allows us to combine like terms; for example,

$$3x^2y^3 - 8x^2y^3 = (3 - 8)x^2y^3 = -5x^2y^3$$

Addition (and subtraction) of polynomials is simply a matter of removing grouping symbols and combining like terms. For example,

$$(2x^2 - 3) + (x - 4) - (5x^2 - 1)$$

$$= 2x^2 - 3 + x - 4 - 5x^2 + 1 \qquad \textit{Notice } -(5x^2 - 1) = -1(5x^2 - 1)$$

$$= -3x^2 + x - 6 \qquad\qquad = -5x^2 + 1$$

The distributive property, along with the other real number properties and the first rule of exponents, gives us procedures for multiplying polynomials. For example,

$$3x^3y^2(2x^2 - 7y^3) = 3x^3y^2(2x^2) - 3x^3y^2(7y^3) = 6x^5y^2 - 21x^3y^5$$

Keep in mind that variables in equivalent expressions may stand not only for numbers but for other variables, expressions, or polynomials as well. Hence we can apply the distributive property in multiplying $(2x + 5)(x + 3)$ as follows:

$$(2x + 5)(x + 3) = 2x(x + 3) + 5(x + 3)$$

$$(B + C) \cdot A \quad = B \cdot A \quad + C \cdot A$$

Note we let A stand for the binomial $x + 3$ in applying the distributive property. The problem is still unfinished, for we must now apply the distributive property again and then combine terms:

$(2x + 5)(x + 3)$ — *Distribute $x + 3$.*

$= (2x)(x + 3) + (5)(x + 3)$ — *Apply the distributive property again.*

$= (2x)x + (2x)3 + (5)x + (5)3$ — *Simplify and combine like terms.*

$= 2x^2 + 6x + 5x + 15$

$= 2x^2 + 11x + 15$

When multiplying two polynomials, each term of one polynomial is multiplied by each term of the other.

There are several particular products that play a prominent role in mathematics. Knowing these products will shortcut the process of multiplying binomials. We are already familiar with the general forms from basic algebra.

GENERAL FORMS

1. $(x + a)(x + b) = x^2 + (a + b)x + ab$
2. $(ax + b)(cx + d) = acx^2 + (ad + bc)x + bd$

Special products are specific products of binomials that can be derived from the general forms (which, in turn, are derived from the distributive property).

SPECIAL PRODUCTS

1. $(a + b)(a - b) = a^2 - b^2$ Difference of two squares
2. $(a + b)^2 = a^2 + 2ab + b^2$ Perfect square of sum
3. $(a - b)^2 = a^2 - 2ab + b^2$ Perfect square of difference

Special products are important in factoring; in many cases, the quickest way to factor an expression is by recognizing it as a special product. In addition, recognizing and using special products can reduce the time needed for multiplication.

EXAMPLE 1 Perform the following operations.
(a) $(2a - 7b)(2a + 7b)$ **(b)** $(2a - 7b)^2$ **(c)** $(x + y - 3)(x + y + 3)$
(d) $2x(x - 4)^2 - (x + 4)(x - 4)$

Solution The first three problems can be worked by using the general forms or, more slowly, by using the distributive property. The quickest way, however, is to use special products:

(a) $(2a - 7b)(2a + 7b) = (2a)^2 - (7b)^2$ *Special product 1: difference of squares*

$$= \boxed{4a^2 - 49b^2}$$

Keep in mind that when you square a binomial, you should get a middle term in the product.

(b) $(2a - 7b)^2 = (2a)^2 - 2(2a)(7b) + (7b)^2$ *Special product 3: perfect square of a difference*

$$= \boxed{4a^2 - 28ab + 49b^2}$$

Study the differences between parts (a) and (b); also note their similarities.

(c) At first glance, $(x + y - 3)(x + y + 3)$ does not seem to be in special product or general form. However, we can regroup within the brackets to get it into special product or general form. This product is a difference of two squares with $x + y$ as the first term and 3 as the second term. We add parentheses around $x + y$ to help see it more clearly.

$$[(x + y) - 3][(x + y) + 3]$$

Apply the difference of squares.

$$= (x + y)^2 - 3^2$$

Then apply special product 2 to $(x + y)^2$.

$$= \boxed{x^2 + 2xy + y^2 - 9}$$

(d) We follow the same order of operations discussed in Section 1.2 (that is, parentheses, exponents, multiplication and division, and, finally, addition and subtraction).

$$2x(x - 4)^2 - (x + 4)(x - 4)$$

Square the binomial $(x - 4)$ and find the difference of squares $(x + 4)(x - 4)$.

$$= 2x(x^2 - 8x + 16) - (x^2 - 16)$$

Distribute $2x$ and subtract $x^2 - 16$.

$$= 2x^3 - 16x^2 + 32x - x^2 + 16$$

$$= \boxed{2x^3 - 17x^2 + 32x + 16}$$

Note that we multiplied $(x + 4)(x - 4)$ before we subtracted. It is a good habit to retain parentheses to remind us that we are subtracting the entire expression $(x^2 - 16)$. ■

EXAMPLE 2 An open box is to be made from a 1-ft by 3-ft rectangular piece of cardboard by cutting out identical squares of length x from each of the corners of the sheet and then folding up the sides on the dashed lines as illustrated in Figure 1.6. Find the volume of the box in terms of x.

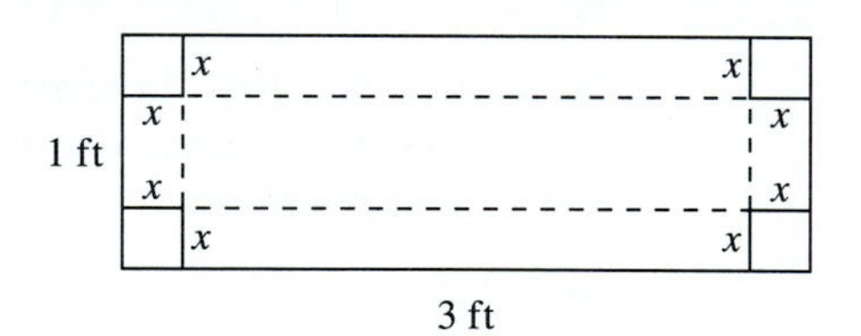

FIGURE 1.6

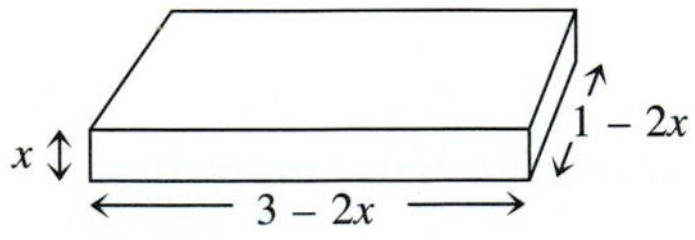

Solution The volume of a box is given by $V = lwh$, where l = length, w = width, and h = height.

By Figure 1.6, if the cardboard is folded up along the dashed lines, then its height is x. The length of the box is $3 - 2x$, since we are eliminating x length from each side of the cardboard. For this same reason, the width of the box is $1 - 2x$ (see Figure 1.7). Hence,

FIGURE 1.7

$$\begin{aligned} V &= lwh \\ &= (3 - 2x)(1 - 2x)x \\ &= (3 - 8x + 4x^2)x \\ &= \boxed{3x - 8x^2 + 4x^3 \text{ cubic feet}} \end{aligned}$$

■

Factoring

To factor a polynomial is to rewrite it as a product of polynomials. The distributive property gives us a method for factoring polynomials as well as for multiplying polynomials.

Multiplying →

$$a(b + c) = ab + ac$$

← *Factoring*

Unless otherwise noted, we will be factoring over the integers; that is, all polynomial factors should have integer coefficients.

The most basic type of factoring is factoring out the greatest common factor. For example, the greatest common monomial factor of $24x^2y^3 - 16xy^3 - 8y^4$ is $8y^3$ because $8y^3$ is the greatest factor common to *all three* terms, $24x^2y^3$, $-16xy^3$, and $-8y^4$. Therefore,

$$24x^2y^3 - 16xy^3 - 8y^4 = (8y^3)(3x^2) + (8y^3)(-2x) + (8y^3)(-y)$$

By the distributive property we get

$$= 8y^3(3x^2 - 2x - y)$$

We can generalize common factoring to more complex expressions, as illustrated next.

EXAMPLE 3 Factor the following completely: **(a)** $3x(y - 4) + 2(y - 4)$
(b) $(x + 2)^2 + (x + 2)$ **(c)** $4x(x - 4)^2 - 2(2x^2 + 1)(x - 4)$

Solution

(a) In order to factor an expression such as $3x(y - 4) + 2(y - 4)$, we note that $y - 4$ is common to both expressions, $3x(y - 4)$ and $2(y - 4)$, and therefore can be factored out, just as we would factor A from $3xA + 2A$.

$$3x \cdot A + 2 \cdot A = A\,(3x + 2)$$

$$3x(y - 4) + 2(y - 4) = \boxed{(y - 4)(3x + 2)}$$

(b) In order to factor $(x + 2)^2 + (x + 2)$, we note that $x + 2$ is the factor common to both $(x + 2)^2$ and $x + 2$. This is like factoring $A^2 + A$ to get $A(A + 1)$.

$$A^2 + A = A\,[A + 1]$$

$$(x + 2)^2 + (x + 2) = (x + 2)[(x + 2) + 1]$$ *Then simplify* $[(x + 2) + 1]$.

$$= \boxed{(x + 2)(x + 3)}$$

(c) Again, to factor $4x(x - 4)^2 - 2(2x^2 + 1)(x - 4)$, first observe that $2(x - 4)$ is common to each term, $4x(x - 4)^2$ and $-2(2x^2 + 1)(x - 4)$. Hence we can factor $2(x - 4)$ out, and are left with $2x(x - 4)$ and $-(2x^2 + 1)$.

$$4x(x - 4)^2 - 2(2x^2 + 1)(x - 4)$$

$$= 2(x - 4)[2x(x - 4) - (2x^2 + 1)]$$ *Next, simplify inside the brackets.*

$$= 2(x - 4)[2x^2 - 8x - 2x^2 - 1]$$

$$= 2(x - 4)[-8x - 1]$$ *We can factor out -1 from $-8x - 1$ to get*

$$= \boxed{-2(x - 4)(8x + 1)}$$

■

The greatest common factor of a polynomial is not always apparent. We often have to take a step or two to put the polynomial in factorable form. First, we may have to group the terms and then factor the groups before it becomes clear what can be factored from the entire expression. This is called **factoring by grouping,** illustrated in Example 4.

EXAMPLE 4 Factor the following completely.

$$3xb - 2b + 15x - 10$$

Solution

$3xb - 2b + 15x - 10$ *There is no factor common to all terms, so group in pairs and factor each pair.*

$$= 3xb - 2b + 15x - 10$$

$$= b(3x - 2) + 5(3x - 2)$$ *Now factor $3x - 2$ from each group.*

$$= \boxed{(3x - 2)(b + 5)}$$

■

Factoring Trinomials

From basic algebra, we should already be familiar with factoring polynomials of the form $Ax^2 + Bx + C$. The simplest cases are those with leading coefficient $A = 1$, that is, when the trinomial is of the form $x^2 + Bx + C$. On the other hand, factoring the general trinomial $Ax^2 + Bx + C$ requires more trial and error.

EXAMPLE 5 Factor the following completely.

(a) $3y^3 - 6y^2 - 105y$

(b) $12a^3 + 2a^2 - 4a$

Solution

(a) $3y^3 - 6y^2 - 105y$ *Don't forget: Always factor the greatest common factor first.*

$= 3y(y^2 - 2y - 35)$ *Next, factor $y^2 - 2y - 35$*

$= \boxed{3y(y - 7)(y + 5)}$

(b) $12a^3 + 2a^2 - 4a$ *Factor the common monomial, 2a, first.*

$= 2a(6a^2 + a - 2)$ *Factor $6a^2 + a - 2$ into $(2a - 1)(3a + 2)$.*

$= \boxed{2a(2a - 1)(3a + 2)}$ ■

Factoring Using Special Products

If we try to factor $9x^2 + 30x + 25$ by trial and error, it may take a while to arrive at the correct factorization. However, if we recognize this polynomial as a form of a special product, we can cut down our labor a bit.

We again list the special products that we have had so far and add two more.

SPECIAL PRODUCTS

1.	$a^2 - b^2 = (a + b)(a - b)$	Difference of two squares
2.	$a^2 + 2ab + b^2 = (a + b)^2$	Perfect square of sum
3.	$a^2 - 2ab + b^2 = (a - b)$	Perfect square of difference
4.	$a^3 - b^3 = (a - b)(a^2 + ab + b^2)$	Difference of two cubes
5.	$a^3 + b^3 = (a + b)(a^2 - ab + b^2)$	Sum of two cubes

EXAMPLE 6 Factor the following completely.
(a) $9x^2 + 30x + 25$ **(b)** $x^4 - y^4$ **(c)** $27x^3 + y^3$ **(d)** $24x^4 - 3x$

Solution

(a) $9x^2 + 30x + 25 = (3x)^2 + 2(3x)(5) + 5^2$ *Perfect square of a sum*

$= \boxed{(3x + 5)^2}$

(b) $x^4 - y^4$ — *Rewrite as a difference of two squares.*

$= (x^2)^2 - (y^2)^2$ — *Factor: difference of squares.*

$= (x^2 - y^2)(x^2 + y^2)$ — *Now factor $x^2 - y^2$.*

$= \boxed{(x - y)(x + y)(x^2 + y^2)}$

(c) $27x^3 + y^3$ — *Rewrite as a sum of cubes.*

$= (3x)^3 + y^3$ — *Factor: sum of two cubes.*

$= (3x + y)[(3x)^2 - (3x)(y) + y^2]$ — *Simplify inside brackets.*

$= \boxed{(3x + y)(9x^2 - 3xy + y^2)}$

(d) $24x^4 - 3x$ — *Factor out the common monomial, 3x, first.*

$= 3x[8x^3 - 1]$ — *Now rewrite $8x^3 - 1$ as a difference of cubes.*

$= 3x[(2x)^3 - 1^3]$ — *Factor as a difference of cubes.*

$= 3x(2x - 1)[(2x)^2 + (2x)(1) + 1^2]$ — *Simplify inside brackets.*

$= \boxed{3x(2x - 1)(4x^2 + 2x + 1)}$ ■

We will now apply what we know about factoring special products to more complex expressions.

EXAMPLE 7 Factor completely:

(a) $x^3 - x^2 - 4x + 4$ **(b)** $x^2 - 4xy + 4y^2 - 9$

Solution

(a) When a negative sign appears between the pairs of binomials we intend to group, we occasionally have to factor out a negative factor so that the binomial factors are identical. For example, in factoring $x^3 - x^2 - 4x + 4$ we would have to factor out -4 from $-4x + 4$:

$x^3 - x^2 - 4x + 4$ — *First separate the pairs*

$= x^3 - x^2 \quad - 4x + 4$ — *Factor out x^2 from the first pair, and -4 from the second pair. [Be careful: check with multiplication.]*

$= x^2(x - 1) - 4(x - 1)$ — *Factor out $x - 1$ to get*

$= (x^2 - 4)(x - 1)$ — *Now factor the expression $x^2 - 4$.*

$= \boxed{(x - 2)(x + 2)(x - 1)}$

(b) If we tried to factor $x^2 - 4xy + 4y^2 - 9$ by grouping in pairs, we would find no common factors. But suppose we group together the first three terms.

$$x^2 - 4xy + 4y^2 - 9 = (x^2 - 4xy + 4y^2) - 9$$

Note that $x^2 - 4xy + 4y^2$ is the perfect square $(x - 2y)^2$, and $9 = 3^2$.

$$= (x - 2y)^2 - (3)^2$$

A difference of two squares.

$$= [(x - 2y) - 3][(x - 2y) + 3]$$

$$= \boxed{(x - 2y - 3)(x - 2y + 3)}$$

■

In general we offer the following advice for factoring polynomials:

Always factor out the greatest common factor first. If the polynomial to be factored is a binomial, then it may be a difference of two squares or a sum or difference of two cubes. If the polynomial to be factored is a trinomial, then (1) if two of the three terms are perfect squares, the polynomial may be a perfect square or (2) otherwise the polynomial may be one of the general forms. If the polynomial to be factored consists of four or more terms, then try factoring by grouping.

Rational Expressions

A **fractional expression** is a quotient of two algebraic expressions, $\frac{a}{b}$ $(b \neq 0)$. We define a **rational expression** as a quotient of two polynomials, $\frac{p}{q}$, provided the denominator is not the zero polynomial. (Remember the zero polynomial is simply 0).

But even if the denominator is not the zero polynomial, we must still be careful about division by zero; a nonzero polynomial can have a value of zero when we substitute certain values for the variable. For example, $\frac{3x - 4}{2x - 1}$ is a rational expression with the nonzero polynomial $2x - 1$ in the denominator. However, since $2x - 1$ is zero when $x = \frac{1}{2}$, the expression $\frac{3x - 4}{2x - 1}$ is not defined for $x = \frac{1}{2}$. In the same way, the expression $\frac{x + y}{x - y}$ is undefined when $x = y$.

Equivalent Fractions

We already defined what we meant by equivalent fractions in Section 1.2. We also mentioned in that section that since we require fractions to be reduced to lowest terms, we can observe their equivalence by inspection. Reducing fractions to lowest terms requires use of the fundamental principle of fractions:

$$\frac{a \cdot k}{b \cdot k} = \frac{a}{b} \quad (b, k \neq 0).$$

Again, a fraction reduced to lowest terms or written in simplest form is a fraction that has no factors (other than ± 1) common to both its numerator and denominator. This requires us to factor both numerator and denominator and then divide out factors common to the numerator and denominator; for example,

$$\frac{x^2 - y^2}{(x - y)^2} = \frac{(x - y)(x + y)}{(x - y)(x - y)} = \frac{x + y}{x - y} \qquad \textit{First factor, then reduce.}$$

Remember, the fundamental principle of fractions allows us to reduce by common factors, *not terms.*

EXAMPLE 8 Express the following in simplest form: $\dfrac{5(x^2 + 2)^2 - 5x(x^2 + 2)(2x)}{(x^2 + 2)^4}$.

Solution Rather than starting this problem by performing operations to simplify the numerator, we begin by factoring the common factor of $5(x^2 + 2)$ from the numerator:

$$\frac{5(x^2 + 2)^2 - 5x(x^2 + 2)(2x)}{(x^2 + 2)^4} \qquad \textit{Factor } 5(x^2+2) \textit{ from the numerator.}$$

$$= \frac{5(x^2 + 2)[(x^2 + 2) - x(2x)]}{(x^2 + 2)^4} \qquad \textit{Now simplify } [(x^2+2) - x(2x)].$$

$$= \frac{5(x^2 + 2)(2 - x^2)}{(x^2 + 2)^4} \qquad \textit{Reduce by a factor of } (x^2+2).$$

$$= \boxed{\frac{5(2 - x^2)}{(x^2 + 2)^3}}$$

Try this problem by simplifying the numerator first.

Approaching the problem in this way cuts down the labor of performing polynomial operations in the numerator. More important, however, if you simplify the numerator as a first step, you may not recognize the polynomial in the numerator as being factorable. ■

Operations with Rational Expressions

We perform arithmetic operations with rational expressions as we would for rational numbers (See page 14):

EXAMPLE 9 Perform the operations and express the answer in simplest form.

(a) $\dfrac{a^2 - ab + b^2}{a^2b - ab^2} \cdot \dfrac{a^3 + a^2b}{a^3 + b^3}$ **(b)** $\dfrac{5x - 10}{x - 4} + \dfrac{3x - 2}{4 - x}$

(c) $-\dfrac{4x(x + 1)}{(x^2 - 2)^3} + \dfrac{1}{(x^2 - 2)^2}$

Solution

(a) $\dfrac{a^2 - ab + b^2}{a^2b - ab^2} \cdot \dfrac{a^3 + a^2b}{a^3 + b^3}$ *Factor.*

$$= \frac{a^2 - ab + b^2}{ab(a - b)} \cdot \frac{a^2(a + b)}{(a + b)(a^2 - ab + b^2)}$$ *Then reduce.*

$$= \boxed{\frac{a}{b(a - b)}}$$

(b) $\dfrac{5x - 10}{x - 4} + \dfrac{3x - 2}{4 - x}$

NOTE: Since $x - 4$ is the negative of $4 - x$, we multiply the numerator and denominator of the second fraction by -1.

$$= \frac{5x - 10}{x - 4} + \frac{(3x - 2)(-1)}{(4 - x)(-1)}$$ *Now the denominators are the same.*

$$= \frac{5x - 10}{x - 4} + \frac{-3x + 2}{x - 4}$$ *Combine numerators.*

$$= \frac{5x - 10 + (-3x + 2)}{x - 4} = \frac{2x - 8}{x - 4}$$ *Factor and reduce.*

$$= \frac{2(x - 4)}{x - 4} = \boxed{2}$$

(c) $-\dfrac{4x(x + 1)}{(x^2 - 2)^3} + \dfrac{1}{(x^2 - 2)^2}$ *The LCD is $(x^2 - 2)^3$.*

Next, rewrite each fraction as an equivalent fraction with the LCD in the denominator.

$$= -\frac{4x(x + 1)}{(x^2 - 2)^3} + \frac{1 \cdot (x^2 - 2)}{(x^2 - 2)^3}$$ *Combine numerators; place over the common denominator.*

$$= \frac{-4x(x + 1) + (x^2 - 2)}{(x^2 - 2)^3} = \boxed{\frac{-3x^2 - 4x - 2}{(x^2 - 2)^3}}$$ ■

EXAMPLE 10 Express as a simple fraction reduced to lowest terms.

$$\frac{\dfrac{3}{x + h} - \dfrac{3}{x}}{h}$$

Solution Using method 2 for simplifying complex fractions as discussed in the previous section, we multiply the numerator and denominator by $x(x + h)$, which is the LCD of $\frac{3}{x + h}$ and $\frac{3}{x}$.

$$\frac{\frac{3}{x+h} - \frac{3}{x}}{h} = \frac{\left(\frac{3}{x+h} - \frac{3}{x}\right)}{h} \frac{x(x+h)}{x(x+h)}$$
Apply the distributive property.

$$= \frac{\left(\frac{3}{x+h}\right)\frac{x(x+h)}{1} - \left(\frac{3}{x}\right)\frac{x(x+h)}{1}}{h[x(x+h)]}$$
Reduce where appropriate.

$$= \frac{3x - 3(x+h)}{hx(x+h)}$$
Now simplify the numerator.

$$= \frac{-3h}{hx(x+h)}$$
Reduce by the common factor of h.

$$= \boxed{\frac{-3}{x(x+h)}}$$
■

EXERCISES 1.3

In Exercises 1–32, perform the operations and express your answers in simplest form.

1. $(3x^2y)(-2xy^2)$
2. $(-7x^2y)(3xy^3)$
3. $(-3xy)^2(5xy)$
4. $(6xy^2)^2(-3x^2y^3)^2$
5. $(2x - 3)(3x + 2)$
6. $(5x - 9)(3x + 2)$
7. $(2x - 7)(3x + 1)$
8. $(2x - 5)(7x + 3)$
9. $(3x - 2)(2x^2 - 3x + 1)$
10. $(2x - 1)(4x^2 + 2x + 1)$
11. $(5x + 3)(25x^2 - 15x + 9)$
12. $(2x + 3)(x^2 - 2x + 1)$
13. $(x - 3)(x - 2)(x + 1)$
14. $(x - 4)(x + 5)(x - 1)$
15. $(x - 3y)^2$
16. $(x - 3y)(x + 3y)$
17. $(x^2 - 7)(x^2 + 7)$
18. $(x^2 + 7)^2$
19. $(2x - 3y)^2$
20. $(2x - 3y)(2x+ 3y)$
21. $(x - y + 7)(x - y - 7)$
22. $(2x + 3 - y)(2x + 3 + y)$
23. $(x - y + 2)^2$
24. $(x + y + 3)^2$
25. $(x - 2)^2 - (x - 2)(x + 2)$
26. $(x - 2)(x + 2) - (x - 2)^2$
27. $5y(y - 5)^2 - (y - 5)(y + 5)$
28. $6x(x - 3)(x + 3) - (x - 3)^2$
29. $2(x - 3)^2 - 2x^2$
30. $5(x - 1)^2 - 5x^2$
31. $2(x + h)^2 + 1 - (2x^2 + 1)$
32. $3(x + h)^2 + 2(x + h) - 2 - (3x^2 + 2x - 2)$
33. In terms of x, find the area of the shaded region given in the accompanying figure.

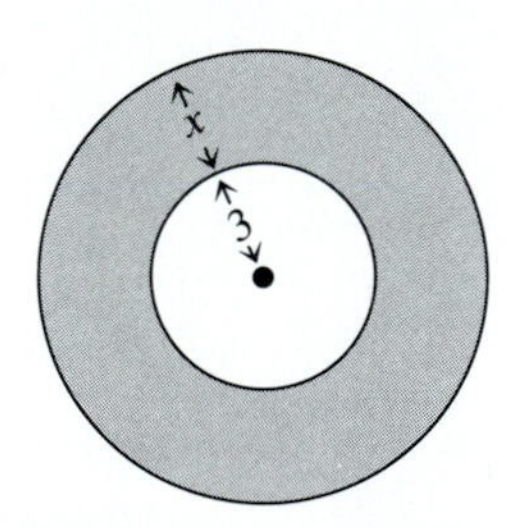

34. An open box is to be made from a 2-ft by 3-ft rectangular piece of cardboard by cutting out identical squares of length x from each of the corners of the sheet and then folding up the sides on the dashed lines, as illustrated in Example 2. Find the volume of the box in terms of x.

35. An open box is made by cutting squares off the corners of a 2-ft by 3-ft rectangular piece of cardboard and then folding up on the dashed line, as illustrated in the figure in Example 2. If the squares to be cut out have length x, find the *surface area* of the box in terms of x.

36. A rectangular garden is surrounded by a path of uniform width 2 feet. If the length of the garden is twice the width, express the total area of the path and garden in terms of the width of the garden.

In Exercises 37–66, factor each expression as completely as possible.

37. $x^2 + x - 20$
38. $6x^2 - 7x - 3$
39. $12x^2 + 10x - 12$
40. $3x^3 - 3x^2 - 6x$
41. $3a(b - 2) - (b - 2)$
42. $(x - 3)^2 - (x - 3)$
43. $5x(x - 1)^2 + 10(x - 1)^3$
44. $3x(x - 2) + 2(x - 2)^2$
45. $3x(x - 2)(2x^2 + 1) - 6(x + 3)(2x^2 + 1)$
46. $4x(x - 2)^2(x - 1) + 8(x - 2)(x - 1)$
47. $ax + bx - 2a - 2b$
48. $x^3 - 5x - 2x^2 + 10$
49. $x^3 - 5x - 3x^2 + 15$
50. $x^3 - 7x - 2x^2 + 14$
51. $x^2 - 9$
52. $x^4 - 16$
53. $x^2 - 8x + 16$
54. $4x^2 + 4xy + y^2$
55. $8x^3 - 1$
56. $27x^3 + 8$
57. $81x^3 - 24$
58. $64x - 27xy^3$
59. $x^3 + 3x^2 - 16x - 48$
60. $x^3 - 3x^2 - 25x + 75$
61. $x^4 - 10x^2 + 24$
62. $2x^3 - 50x + 2x^2 - 50$
63. $x^2 - 2xy + y^2 - 16$
64. $x^2 + 4x + 4 - y^2$
65. $a^5 - a^3 - a^2 + 1$
66. $x^5 - x^2 - 4x^3 + 4$

In Exercises 67–72, simplify the fraction.

67. $\dfrac{2x^3 - 2xy^2}{4x^4 - 8x^3y + 4x^2y^2}$

68. $\dfrac{4x(x + 2)^2 - 2x^2(2)(x + 2)}{(x + 2)^4}$

69. $\dfrac{3(x + h) + 1 - (3x + 1)}{h}$

70. $\dfrac{(x + h)^2 + 2(x + h) - 3 - (x^2 + 2x - 3)}{h}$

71. $\dfrac{(x^2 - 4)^2(-4) - (-4x)(4x)(x^2 - 4)}{(x^2 - 4)^4}$

72. $\dfrac{(x^2 - 3)^2(-6x) - (-3x^2 - 9)(4x)(x^2 - 3)}{(x^2 - 3)^4}$

In Exercises 73–84, perform the operations and express each answer in simplest form.

73. $\dfrac{6x^2 - 7x - 3}{4x^2 - 12x + 9} \cdot \dfrac{2x - 3}{3x^2 - 5x - 2}$

74. $\dfrac{5x - 1}{5x + 2} \div \dfrac{25x^2 - 10x + 4}{125x^3 + 8}$

75. $\dfrac{27x^3 - 8}{9x^2 - 4} \div (9x^2 + 6x + 4)$

76. $\dfrac{6x^2 - 8x}{3x^3 - x^2 - 4x} \cdot \dfrac{4x^4 - 8x^3 - 12x^2}{2x - 6}$

77. $\dfrac{x}{x - y} - \dfrac{x}{y}$

78. $\dfrac{2y + 10}{3 - y} + \dfrac{y + 7}{y - 3}$

79. $\dfrac{2a + 6}{a^2 + 5a + 6} + \dfrac{5a + 1}{a + 2}$

80. $\dfrac{x}{x - 2} + \dfrac{2}{x^2 - 4x + 4}$

81. $\dfrac{3x}{x - 1} + \dfrac{x + 3}{x - 2} + 2x - 3$

82. $\dfrac{3y - 4}{y - 5} + \dfrac{4y}{10 + 3y - y^2}$

83. $-\dfrac{x^2 + 1}{(x - 2)^2} + \dfrac{2x}{x - 2}$

84. $\dfrac{2}{(x^2 - 3)^2} - \dfrac{8x^2}{(x^2 - 3)^3}$

In Exercises 85–90, express each as a simple fraction reduced to lowest terms.

85. $\dfrac{1 - \dfrac{1}{x^2}}{\dfrac{x - 1}{x}}$

86. $\dfrac{x + \dfrac{2}{xy^2}}{\dfrac{1}{x} + 2}$

87. $\dfrac{\dfrac{1}{x + 3} - \dfrac{1}{x + 1}}{2}$

88. $\dfrac{\dfrac{3}{x - 1} - \dfrac{3}{x - 4}}{3}$

89. $\dfrac{\dfrac{5}{x + h} - \dfrac{5}{x}}{h}$

90. $\dfrac{\dfrac{3}{(x + h)^2} - \dfrac{3}{x^2}}{h}$

91. $\dfrac{\dfrac{3x + 1}{x - 4}}{\dfrac{2x + 1}{x + 2} - \dfrac{x}{x - 4}}$

92. $1 - \dfrac{1}{1 - \dfrac{1}{1 - \dfrac{1}{x}}}$

QUESTIONS FOR THOUGHT

93. Describe what happens to the value of the polynomial $x^2 + 2x + 3$ as x gets larger and larger.

94. Describe what happens to the value of $\dfrac{x + 3}{x - 4}$ if:

(a) $x > 4$, and x gets closer and closer to 4.
(b) $x < 4$, and x gets closer and closer to 4.

1.4 Exponents and Radicals

We originally defined x^n as x being taken as a factor n times. Based on this definition, n must be a natural number. It does not make sense for n to be negative or zero. However, we can extend the definition of exponents to include 0 and negative exponents with the following:

Definition of Zero Exponent

$$x^0 = 1 \qquad (x \neq 0)$$

NOTE: 0^0 is undefined

Definition of Negative Exponents

$$x^{-n} = \frac{1}{x^n} \qquad (x \neq 0)$$

As a result of the definition of negative exponents, we have $\dfrac{1}{x^{-n}} = \dfrac{1}{\frac{1}{x^n}} = x^n$.

We find that *changing the sign of the exponent of a power changes the power into its reciprocal*. Thus

$$x^{-3} = \frac{1}{x^3}, \qquad \frac{1}{x^{-4}} = x^4, \qquad \frac{1}{2^{-2}} = 2^2 = 4, \qquad 10^{-5} = \frac{1}{10^5} = \frac{1}{100{,}000}$$

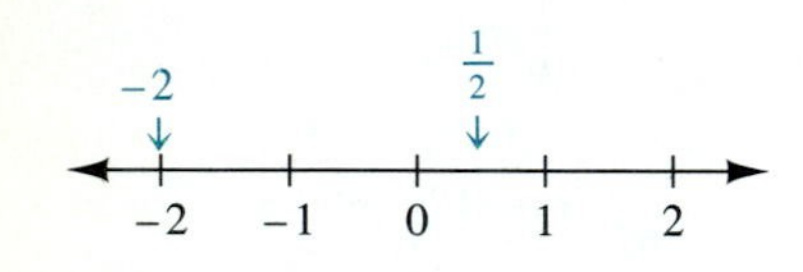

FIGURE 1.8

Do not confuse the sign of an exponent with the sign of the base. For example, -2 is a number 2 units to the left of zero on the number line, whereas $2^{-1} = \frac{1}{2}$ is a positive number, one-half unit to the right of zero, as shown in Figure 1.8.

The rules for exponents are listed in the following box.

RULES FOR INTEGER EXPONENTS

1. $x^n x^m = x^{n+m}$
2. $(x^n)^m = x^{nm}$
3. $(xy)^n = x^n y^n$
4. $\dfrac{x^n}{x^m} = x^{n-m} \qquad (x \neq 0)$
5. $\left(\dfrac{x}{y}\right)^n = \dfrac{x^n}{y^n} \qquad (y \neq 0)$

These rules can be used for shortcutting the process of multiplying and dividing expressions involving exponents.

EXAMPLE 1 Perform the operations and simplify; express each answer as a single simple fraction with positive exponents only.

(a) $\dfrac{(2a^{-2}b)^{-3}(-3ab)^{-2}}{a^{-4}}$ **(b)** $\dfrac{a^{-2} - b^{-2}}{a^{-1} + b^{-1}}$ **(c)** $5(x-6)^{-1} + 5x(x-6)^{-2}$

Solution

(a)

$$\frac{(2a^{-2}b)^{-3}(-3ab)^{-2}}{a^{-4}}$$ *Use rule 3 and rule 2.*

$$= \frac{2^{-3}a^{6}b^{-3}(-3)^{-2}a^{-2}b^{-2}}{a^{-4}}$$

$$= \frac{2^{-3}(-3)^{-2}a^{6}a^{-2}b^{-3}b^{-2}}{a^{-4}}$$ *Use rule 1 and rule 4.*

$$= 2^{-3}(-3)^{-2}a^{6+(-2)-(-4)}b^{-3+(-2)}$$

$$= 2^{-3}(-3)^{-2}a^{8}b^{-5}$$ *Rewrite using positive exponents.*

$$= \frac{a^{8}}{2^{3}(-3)^{2}b^{5}} = \boxed{\frac{a^{8}}{72b^{5}}}$$

(b) Remember the rules of exponents apply to factors, not terms. We must rewrite each *term* in the numerator and denominator using the definition of negative exponents.

$$\frac{a^{-2} - b^{-2}}{a^{-1} + b^{-1}} = \frac{\frac{1}{a^{2}} - \frac{1}{b^{2}}}{\frac{1}{a} + \frac{1}{b}}$$ *Multiply the numerator and denominator of the complex fraction by a^2b^2. (Why?)*

$$= \frac{\left(\frac{1}{a^{2}} - \frac{1}{b^{2}}\right)}{\left(\frac{1}{a} + \frac{1}{b}\right)} \frac{a^{2}b^{2}}{a^{2}b^{2}} = \frac{b^{2} - a^{2}}{ab^{2} + a^{2}b} = \frac{(b - a)(b + a)}{ab(b + a)} = \boxed{\frac{b - a}{ab}}$$

(c) $5(x - 6)^{-1} + 5x(x - 6)^{-2}$ *Rewrite each term using positive exponents.*

$$= \frac{5}{x - 6} + \frac{5x}{(x - 6)^{2}}$$ *The LCD is $(x - 6)^2$.*

$$= \frac{5(x - 6)}{(x - 6)^{2}} + \frac{5x}{(x - 6)^{2}}$$

$$= \frac{5(x - 6) + 5x}{(x - 6)^{2}} = \boxed{\frac{10x - 30}{(x - 6)^{2}}}$$ ■

Scientific Notation

Scientific notation is a way of concisely writing very large or very small numbers. For example, the approximate mass of the earth written in our standard decimal form is 5,980,000,000,000,000,000,000,000 kg. Using scientific notation, we can write this number as 5.98×10^{24}.

DEFINITION A number is said to be in **scientific notation** if it has the form $a \times 10^{n}$ *where* $1 \le a < 10$ *and n is an integer.*

1.645×10^4 is written in scientific notation with $a = 1.645$ and $n = 4$. 6.8×10^{-5} is written in scientific notation with $a = 6.8$ and $n = -5$.

Converting from scientific to standard notation is straightforward: it is simply a matter of moving the decimal point right or left. For example,

$$1.642 \times 10^5 = 1.642 \times 100{,}000 = 164{,}200$$

$$7.3 \times 10^{-4} = 7.3 \times 0.0001 = 0.00073$$

The middle step may be bypassed by observing that the number of places the decimal point is moved is given by the absolute value of the exponent of 10; the direction the decimal point is moved is determined by the sign of the exponent.

Converting from standard to scientific notation takes a bit more thought, but again it is simply a matter of shifting decimal points right or left as demonstrated in Example 2.

EXAMPLE 2 Express the following in scientific notation:

(a) 78,964 **(b)** 0.00751

Solution

(a) 78,964 — *Scientific notation requires the number to have the following form:*

$= 7.8964 \times 10^n$ — *What remains is to find the power of* 10.

Since we moved the decimal point of 78,964 to the left 4 places, we obviously changed the value of the number. To restore the number to its original value, we multiply it by a power of 10 which will move it 4 places right: this is 10^{+4}. Hence,

$$78{,}964 = \boxed{7.8964 \times 10^4}$$

(b) 0.00751 — *In scientific notation, the number must have the following form:*

$= 7.51 \times 10^n$ — *Since we moved the decimal point right* 3 *places, we need to multiply it by the power of* 10 *which will move it back* 3 *places left*: 10^{-3}.

$$0.00751 = \boxed{7.51 \times 10^{-3}}$$

A quicker way to state the above is that the power of 10 restores the decimal point back to its original position. ■

If you tried using a calculator to compute $6{,}500{,}000 \times 950{,}000$, your calculator will probably display 6.175 E 12 or 6.175 12. When an answer is given in this form, it is usually because the answer has more digits than the calculator can display. 6.175 E 12 or 6.175 12 is the way the calculator expresses the number 6.175×10^{12}; likewise a calculator display of 2.3852 E −15 means 2.3852×10^{-15}.

Often computations can be simplified using scientific notation. We first convert all numbers from standard to scientific notation and then perform the computations.

EXAMPLE 3 If the mass of a particle is 4.3×10^{-7} grams, then is the mass of 5 million of these particles closest to 1 gram, 10 grams, 100 grams, or 1000 grams?

Solution To find the mass of 5 million of these particles, we simply multiply the mass of one particle by 5 million = 5,000,000 as follows:

$(4.3 \times 10^{-7})(5{,}000{,}000)$ *Rewrite* 5,000,000 *using scientific notation.*

$= (4.3 \times 10^{-7})(5 \times 10^{6})$ *Separate out powers of* 10.

$= (4.3)(5) \times (10^{-7})(10^{6}) = 21.5 \times 10^{-1}$

$= 2.15$ gram

Of the options given, this is the closest to $\boxed{\text{1 gram}}$

Most scientific calculators have an [EXP] or [EE] which allows you to enter numbers in scientific notation. To perform the above multiplication in scientific notation using a calculator, enter the sequence of keys: [4] [.] [3] [EXP] [7] [+/−] [×] [5] [EXP] [6] [=] to get 2.15 ■

Rational Exponents

Next we extend the definition of exponents even further to include rational number exponents. In order to do this, we assume that we want the rules for integer exponents also to apply to rational exponents and then use the rules to show us how to define a rational exponent. For example, how do we define $a^{1/2}$? Consider $9^{1/2}$:

If we apply rule 2 and square $9^{1/2}$, we get $(9^{1/2})^2 = 9^{2/2} = 9^1 = 9$

So $9^{1/2}$ is a number that, when squared, yields 9.

There are two possible answers: 3 and -3, since squaring either number will yield 9. In order to avoid ambiguity, we define $a^{1/2}$ (called the principal square root of a) as the *nonnegative* quantity that, when squared, yields a. Thus $9^{1/2} = 3$.

We arrive at the definition of $a^{1/3}$ in the same way as we did for $a^{1/2}$: For example, if we cube $8^{1/3}$, we get $(8^{1/3})^3 = 8^{3/3} = 8^1 = 8$. Thus, $8^{1/3}$ is the number that, when cubed, yields 8. Since $2^3 = 8$, we have $8^{1/3} = 2$. Similarly $(-27)^{1/3} = -3$.

Do we have two choices for the value of $8^{1/3}$?

Thus we define $a^{1/3}$ (called the cube root of a) as the quantity that, when cubed, yields a.

We generalize with the following definition:

DEFINITION OF $a^{1/n}$

If n is an odd positive integer, then

$$a^{1/n} = b \quad \text{if and only if} \quad b^n = a$$

If n is an even positive integer and $a \geq 0$, then

$$a^{1/n} = |b| \quad \text{if and only if} \quad b^n = a$$

$a^{1/n}$ is called the **principal nth root of a**. Hence, $a^{1/n}$ is the real number (nonnegative when n is even) that, when raised to the nth power, yields a.

Notice that with even roots we have to be careful about two things: (1) we want only the principal (nonnegative) root of the expression, and (2) since raising any real number to an even power will always yield a nonnegative number, the even root of a negative number cannot be a real number. Hence

$$(16)^{1/2} = 4 \quad \textit{since} \quad 4^2 = 16 \qquad 27^{1/3} = 3 \quad \textit{since} \quad 3^3 = 27$$
$$(-125)^{1/3} = -5 \quad \textit{since} \quad (-5)^3 = -125 \qquad (-16)^{1/4} \text{ is not a real number}$$
$$\left(\frac{1}{81}\right)^{1/4} = \frac{1}{3} \quad \textit{since} \quad \left(\frac{1}{3}\right)^4 = \frac{1}{81}$$

Thus far we have defined $a^{1/n}$, where n is a natural number. With some help from the second rule for exponents, we can define the expression $a^{m/n}$, where m and n are natural numbers and $\frac{m}{n}$ is reduced to lowest terms.

DEFINITION OF $a^{m/n}$

If $a^{1/n}$ is a real number, then $a^{m/n} = (a^{1/n})^m$ (the nth root of a raised to the mth power).

We can also define negative rational exponents:

$$a^{-m/n} = \frac{1}{a^{m/n}} \qquad (a \neq 0)$$

Now that we have defined rational exponents, we assert that *the rules for integer exponents hold for rational exponents as well, provided the root is a real number*—that is, provided we avoid even roots of negative numbers.

By the exponent rules, we find $(a^{1/n})^m = (a^m)^{1/n}$. Hence we can view $a^{m/n}$ two ways:

If $a^{1/n}$ is a real number: $a^{m/n} = (a^{1/n})^m = (a^m)^{1/n}$.

EXAMPLE 4 Evaluate the following:

(a) $27^{2/3}$ **(b)** $36^{-1/2}$ **(c)** $(-32)^{-3/5}$

Solution In general, it is easier to find the root *before* raising to a power.

(a) $27^{2/3} = (27^{1/3})^2 = 3^2 = \boxed{9}$

(b) $36^{-1/2}$ *Begin by using the definition of a negative exponent.*

$= \dfrac{1}{36^{1/2}} = \boxed{\dfrac{1}{6}}$ *Note that the sign of the exponent has no effect on the sign of the base.*

(c) $$(-32)^{-3/5} = \frac{1}{(-32)^{3/5}}$$ *Rewrite using positive exponents.*

$$= \frac{1}{[(-32)^{1/5}]^3}$$ *Find the root first.*

$$= \frac{1}{(-2)^3} = \boxed{-\frac{1}{8}}$$ ■

One point of confusion that frequently arises involves the differences between negative exponents and fractional exponents: Negative exponents involve *reciprocals* of the base, whereas fractional exponents yield *roots* of the base. For example,

$$x^{-4} = \frac{1}{x^4}, \quad \text{whereas} \quad x^{1/4} = \textit{the fourth root of } x$$

$$16^{-4} = \frac{1}{16^4} = \frac{1}{65{,}536}, \quad \text{whereas} \quad 16^{1/4} = 2$$

What are the differences among 4^{-2}, $4^{1/2}$, and $4^{-1/2}$?

As we stated earlier, we defined rational exponents so that the rules for integer exponents were also valid for rational exponents. Thus we simplify expressions involving rational exponents by applying the same rules we used for integer exponents. As with integer exponents, when we are asked to simplify an expression, the bases and exponents should appear as few times as possible.

Unless otherwise noted, we assume all variables represent positive real numbers.

EXAMPLE 5 Perform the operations and simplify. Express your answer using positive exponents only.

(a) $(y^{2/3}y^{-1/2})^2$ **(b)** $\left(\frac{a^{1/2}a^{-2/3}}{a^{1/4}}\right)^8$ **(c)** $2x(x^2 + 4)^{3/5}(x^2 + 4)^{-1/2}$

Solution Make sure you can follow the computations.

(a) $$(y^{2/3}y^{-1/2})^2 = (y^{[2/3]+[-1/2]})^2 = (y^{1/6})^2 = y^{2/6} = \boxed{y^{1/3}}$$

(b) $$\left(\frac{a^{1/2}a^{-2/3}}{a^{1/4}}\right)^8$$ *You may choose to simplify inside parentheses first; however, for this problem, if we apply rules 5 and 3 first, the computations with fractions are simpler.*

$$= \frac{(a^{1/2})^8(a^{-2/3})^8}{(a^{1/4})^8} = \frac{a^4a^{-16/3}}{a^2} = a^{4-(16/3)-2} = a^{-10/3} = \boxed{\frac{1}{a^{10/3}}}$$

(c) $$2x(x^2 + 4)^{3/5}(x^2 + 4)^{-1/2} = 2x(x^2 + 4)^{[\frac{3}{5}-\frac{1}{2}]}$$

$$= \boxed{2x(x^2 + 4)^{1/10}}$$ ■

EXAMPLE 6 Perform the operations and simplify the following: $(5a^{1/2} + 3b^{1/2})^2$.

Solution

$(5a^{1/2} + 3b^{1/2})^2$

For squaring a binomial, we can use the perfect square special product.

$$= (5a^{1/2})^2 + 2(5a^{1/2})(3b^{1/2}) + (3b^{1/2})^2$$

Note squaring $a^{1/2}$ does **not** yield $a^{1/4}$.

$$= 5^2a^{2/2} + 30a^{1/2}b^{1/2} + 3^2b^{2/2}$$

$$= \boxed{25a + 30a^{1/2}b^{1/2} + 9b}$$

■

In basic algebra we are used to factoring polynomials. We can use the same principles of factoring that we learned in algebra to factor expressions with terms containing rational exponents. For example, in factoring $x^3 + x^2$, we know that the greatest common factor is x^2, which is x raised to the smallest exponent appearing in the expression. Hence, $x^3 + x^2 = x^2(x + 1)$.

In the same way, we can factor $x + x^{1/3}$ by noting that the greatest common factor is $x^{1/3}$, which is x raised to the smallest exponent appearing in the expression. Hence,

$$x + x^{1/3} = x^{1/3}(x^{2/3} + 1) \qquad \textit{NOTE}: x = x^1 = x^{\frac{2}{3}+\frac{1}{3}} = x^{2/3}x^{1/3}$$

EXAMPLE 7 Factor the following: $10x(x - 3)^{1/2} + 5(x - 3)^{3/2}$.

Solution In the expression $10x(x - 3)^{1/2} + 5(x - 3)^{3/2}$, each term, $10x(x - 3)^{1/2}$ and $5(x - 3)^{3/2}$, has a common factor of $5(x - 3)^{1/2}$. So we can factor out this common factor and simplify the rest of the expression as follows:

$10x(x - 3)^{1/2} + 5(x - 3)^{3/2}$ — *Rewrite each term with $5(x - 3)^{1/2}$ as a factor.*

$= [5(x - 3)^{1/2}][2x] + [5(x - 3)^{1/2}][x - 3]$ — *Factor out $5(x - 3)^{1/2}$ from each term.*

$= [5(x - 3)^{1/2}][2x + (x - 3)]$ — *Simplify $[2x + (x - 3)]$.*

$= [5(x - 3)^{1/2}][3x - 3]$ — *Factor $[3x - 3]$.*

$= [5(x - 3)^{1/2}][3(x - 1)]$

$= \boxed{15(x - 3)^{1/2}(x - 1)}$

■

EXAMPLE 8 Express as a single simple fraction reduced to lowest terms:

$$2x(x^2 - 2)^{1/2} + x^2(x^2 - 2)^{-1/2}$$

Solution

$2x(x^2 - 2)^{1/2} + x^2(x^2 - 2)^{-1/2}$ — *Rewrite each term using positive exponents.*

$= 2x(x^2 - 2)^{1/2} + \dfrac{x^2}{(x^2 - 2)^{1/2}}$ — *The LCD is $(x^2 - 2)^{1/2}$.*

$$= \frac{2x(x^2-2)^{1/2}(x^2-2)^{1/2}}{(x^2-2)^{1/2}} + \frac{x^2}{(x^2-2)^{1/2}}$$

$$= \frac{2x(x^2-2)+x^2}{(x^2-2)^{1/2}}$$

$$= \boxed{\frac{2x^3+x^2-4x}{(x^2-2)^{1/2}}}$$

A second way to approach the same problem is by first factoring the expression $x(x^2-2)^{-1/2}$ from each term:

$$2x(x^2-2)^{1/2} + x^2(x^2-2)^{-1/2}$$

Rewrite each term with $x(x^2-2)^{-1/2}$ as a factor.

$$= [x(x^2-2)^{-1/2}][2(x^2-2)] + [x(x^2-2)^{-1/2}][x]$$

Use the distributive property.

$$= [x(x^2-2)^{-1/2}][2(x^2-2)+x]$$

Now rewrite using positive exponents.

$$= \frac{x[2(x^2-2)+x]}{(x^2-2)^{1/2}}$$

Simplify the numerator.

$$= \boxed{\frac{2x^3+x^2-4x}{(x^2-2)^{1/2}}}$$

■

Radical Expressions

Radical notation is an alternative way of writing an expression with rational exponents. That is, we define

DEFINITION $\sqrt[n]{a} = a^{1/n}$ where n is a positive integer

$\sqrt[n]{a}$ is also called the **principal *n*th root of *a*.** In $\sqrt[n]{a}$, n is called the **index** of the radical, the symbol $\sqrt{\ }$, is called the **radical** or **radical sign,** and the expression a under the radical is called the **radicand.**

Since $\sqrt[n]{a} = a^{1/n}$, where n is a positive integer, we have

For a a real number and n a positive *odd* integer: $\sqrt[n]{a} = b$ if and only if $b^n = a$.

For a a real nonnegative number and n a positive *even* integer, $\sqrt[n]{a} = |b|$ if and only if $b^n = a$.

Thus $\sqrt[n]{a}$ is that quantity (nonnegative when n is even) that, when raised to the nth power, yields a.

$$\sqrt[3]{64} = 4 \quad \textit{since} \quad 4^3 = 64$$
$$\sqrt[5]{-32} = -2 \quad \textit{since} \quad (-2)^5 = -32$$
$$\sqrt{9} = 3 \quad \textit{since} \quad 3^2 = 9$$

Note that even though $(-3)^2 = 9$, we are only interested in the principal or nonnegative square root.

$\sqrt[6]{-64}$ is not a real number *since no real number when raised to the sixth power will yield a negative number.*

In general we find the following.

For a a real number and n a positive integer,

$$\sqrt[n]{a^n} = \begin{cases} |a|, & \text{if } n \text{ is even} \\ a, & \text{if } n \text{ is odd} \end{cases}$$

For example:

$$\sqrt[3]{5^3} = 5$$
$$\sqrt[4]{(-3)^4} = |-3| = 3 \qquad \text{NOTE: } \sqrt[4]{(-3)^4} = \sqrt[4]{81} = 3.$$

Also note that $(\sqrt[4]{-3})^4$ is not equal to $\sqrt[4]{(-3)^4}$, since $\sqrt[4]{-3}$ is not a real number.

Using radical notation, the following statements are equivalent:

If $a^{1/n}$ is a real number, then $(a^m)^{1/n} = (a^{1/n})^m$

If $\sqrt[n]{a}$ is a real number, then $\sqrt[n]{a^m} = (\sqrt[n]{a})^m$.

EXAMPLE 9 Rewrite as an expression without using radicals:

$$\frac{2x}{\sqrt{x^2 - 5}}$$

Solution

$$\frac{2x}{\sqrt{x^2 - 5}} = \boxed{\frac{2x}{(x^2 - 5)^{1/2}} \quad \text{or} \quad 2x(x^2 - 5)^{-1/2}}$$

■

Simplifying Radical Expressions

Now that we have defined radicals, our next step is to determine how to put them in simplified form. Recall that a radical expression is simplified if exponents of factors of the radicand are less than the index and there are no fractions under the radical. Along with the definition of a radical, the following three radical properties will provide us with much of what we will need to simplify radicals. These properties of radicals are a consequence of their exponential counterparts.

PROPERTIES OF RADICALS

If $\sqrt[n]{a}$ and $\sqrt[n]{b}$ are real numbers:

1. $\sqrt[n]{ab} = \sqrt[n]{a}\sqrt[n]{b}$ **2.** $\sqrt[n]{\dfrac{a}{b}} = \dfrac{\sqrt[n]{a}}{\sqrt[n]{b}}$ $(b \neq 0)$

3. $\sqrt[m]{\sqrt[n]{a}} = \sqrt[mn]{a}$ $(a \geq 0)$

How are the properties of radicals related to the rules of exponents?

Properties 1 and 2 are actually forms of the third and fourth rules of exponents, and property 3 is a result of applying the second rule of exponents: $(a^{1/n})^{1/m} = a^{1/mn}$.

Given the definition of a radical and the three radical properties, we can simplify many radical expressions. Here are a few examples of how we would use the properties to simplify radicals.

EXAMPLE 10 Express the following in simplest radical form.

(a) $\sqrt{25x^4y^2}$ **(b)** $\sqrt[3]{\dfrac{27}{y^{18}}}$ **(c)** $\sqrt[3]{24}$ **(d)** $\sqrt[3]{\sqrt[4]{2x^2}}$ **(e)** $\sqrt[3]{x^3 + y^3}$

Solution

(a) $\sqrt{25x^4y^2}$ *Apply property 1.*

$= \sqrt{25}\sqrt{x^4}\sqrt{y^2}$ *Since $5^2 = 25$, $(x^2)^2 = x^4$, and $(y)^2 = y^2$, we have*

$= \boxed{5x^2|y|}$

Notice that $\sqrt{y^2} = |y|$, yet we can write $\sqrt{x^4} = x^2$ rather than $|x|^2$. Why?

(b) $\sqrt[3]{\dfrac{27}{y^{18}}} = \dfrac{\sqrt[3]{27}}{\sqrt[3]{y^{18}}} = \boxed{\dfrac{3}{y^6}}$

(c) $\sqrt[3]{24} = \sqrt[3]{8 \cdot 3} = \sqrt[3]{8}\sqrt[3]{3} = \boxed{2\sqrt[3]{3}}$

(d) $\sqrt[3]{\sqrt[4]{2x^2}} = \boxed{\sqrt[12]{2x^2}}$

(e) $\sqrt[3]{x^3 + y^3}$ $\boxed{\text{cannot be simplified}}$ *(Why not?)* ■

The first two properties allow us to multiply and divide radical expressions with the same index. If we need to express a product or quotient of radical expressions with different indices under a single radical, we can convert to fractional exponents, as demonstrated next.

EXAMPLE 11 Express $\sqrt[5]{x^2}\,\sqrt[3]{x^4}$ as a single simplified radical.

Solution

$\sqrt[5]{x^2}\,\sqrt[3]{x^4}$ *Rewrite using rational exponents.*

$$= x^{2/5}x^{4/3} = x^{\frac{2}{5}+\frac{4}{3}} = x^{26/15}$$ *Rewrite using radical notation and simplify.*

$$= \sqrt[15]{x^{26}} = \sqrt[15]{x^{15}}\,\sqrt[15]{x^{11}} = \boxed{x\sqrt[15]{x^{11}}}$$ ■

Alternatively, we can rewrite $x^{26/15}$ as $x^{15/15} \cdot x^{11/15} = x \cdot x^{11/15}$ and then rewrite this product using radical notation.

In calculus it is important for you to be comfortable working with expressions involving fractional exponents or their equivalent radical forms.

The distributive property allows us to combine terms with radical factors. For example, we can combine $3\sqrt{2} + 4\sqrt{2}$ as follows:

$$3\sqrt{2} + 4\sqrt{2} = (3 + 4)\sqrt{2} = 7\sqrt{2}.$$

EXAMPLE 12 Simplify $\sqrt{27} - \sqrt{81} + \sqrt{12}$.

Solution Simplify each term first.

$$\sqrt{27} - \sqrt{81} + \sqrt{12} = \sqrt{9}\sqrt{3} - \sqrt{81} + \sqrt{4}\sqrt{3} = 3\sqrt{3} - 9 + 2\sqrt{3}$$

Then combine where possible.

$$= \boxed{5\sqrt{3} - 9}$$

Note we cannot combine $5\sqrt{3} - 9$, just as we cannot combine $5x - 9$. ■

As with multiplying polynomials, we may use the distributive property or special products to multiply expressions with more than one radical term.

EXAMPLE 13 Perform the indicated operations.

(a) $\left(2\sqrt{x} - 3\sqrt{y}\right)\left(2\sqrt{x} + 3\sqrt{y}\right)$ **(b)** $\left(\sqrt{x + y}\right)^2 - \left(\sqrt{x} + \sqrt{y}\right)^2$

Solution

(a) $\left(2\sqrt{x} - 3\sqrt{y}\right)\left(2\sqrt{x} + 3\sqrt{y}\right)$ *This is a difference of squares.*

$$= \left(2\sqrt{x}\right)^2 - \left(3\sqrt{y}\right)^2$$ *Square each term.*

$$= 2^2\left(\sqrt{x}\right)^2 - 3^2\left(\sqrt{y}\right)^2$$

$$= \boxed{4x - 9y}.$$

(b) $\left(\sqrt{x + y}\right)^2 - \left(\sqrt{x} + \sqrt{y}\right)^2$

$$= x + y - \left[\left(\sqrt{x}\right)^2 + 2\sqrt{x}\sqrt{y} + \left(\sqrt{y}\right)^2\right]$$ *Note the differences in the way we handle $\left(\sqrt{x + y}\right)^2$ and $\left(\sqrt{x} + \sqrt{y}\right)^2$.*

$$= x + y - [x + 2\sqrt{xy} + y]$$

$$= x + y - x - 2\sqrt{xy} - y = \boxed{-2\sqrt{xy}}.$$ ■

How would you write $3x - 5y$ as a product using the difference of squares? The difference of cubes?

Example 13(a) illustrates that any expression containing a difference can be factored if we lift our restriction of requiring factors to be polynomials with integer coefficients.

Rationalizing Radical Expressions

In some cases radical expressions are easier to work with if the radical is eliminated from the numerator (or denominator); this is called *rationalizing the numerator (or denominator)*. This can be done using the fundamental principle of fractions.

EXAMPLE 14 Rationalize **(a)** the denominator of $\frac{\sqrt{2}}{\sqrt{5}}$; **(b)** the numerator of $\frac{\sqrt{2}}{\sqrt{5}}$.

Solution

(a) We apply the fundamental principle of fractions and multiply the numerator and denominator by $\sqrt{5}$. We choose $\sqrt{5}$ because multiplying the denominator by $\sqrt{5}$ will give us the square root of a perfect square, which in turn yields a rational expression. Thus

$$\frac{\sqrt{2}}{\sqrt{5}} = \frac{\sqrt{2}}{\sqrt{5}}\frac{\sqrt{5}}{\sqrt{5}} = \frac{\sqrt{10}}{\sqrt{5^2}} = \boxed{\frac{\sqrt{10}}{5}}$$

(b) To eliminate the radical from the numerator, we use the fundamental principle of fractions and multiply the numerator and denominator by $\sqrt{2}$.

$$\frac{\sqrt{2}}{\sqrt{5}} = \frac{\sqrt{2}}{\sqrt{5}}\frac{\sqrt{2}}{\sqrt{2}} = \boxed{\frac{2}{\sqrt{10}}}$$

■

In rationalizing the denominator we try to multiply the numerator and the denominator of the fraction by *the expression that will make the denominator the nth root of a perfect nth power*.

EXAMPLE 15 Rationalize the denominator of $\frac{1}{\sqrt[3]{2x^2}}$.

What principle allows you to "eliminate the radical in the denominator" when you rationalize the denominator?

Solution To change $\sqrt[3]{2x^2}$ into the cube root of a perfect cube, we can multiply the numerator and denominator by $\sqrt[3]{4x}$. Then $\sqrt[3]{2x^2}\,\sqrt[3]{4x} = \sqrt[3]{8x^3} = 2x$.

$$\frac{1}{\sqrt[3]{2x^2}} = \frac{1}{\sqrt[3]{2x^2}} \cdot \frac{\sqrt[3]{4x}}{\sqrt[3]{4x}} = \frac{\sqrt[3]{4x}}{\sqrt[3]{8x^3}} = \boxed{\frac{\sqrt[3]{4x}}{2x}}$$

■

In Example 15 we rationalized the denominator of a fraction with a single radical term in the denominator. In order to rationalize the denominator of a fraction with more than one term in the denominator, we exploit the difference of squares and multiply the numerator and denominator by the conjugate of the denominator; the **conjugate** of $a + b$ is $a - b$.

EXAMPLE 16 Rationalize the denominator of $\frac{2}{3 + \sqrt{5}}$.

Solution

$$\frac{2}{3+\sqrt{5}}$$ *Multiply numerator and denominator by $3-\sqrt{5}$, the conjugate of $3+\sqrt{5}$.*

$$=\frac{2}{3+\sqrt{5}}\cdot\frac{3-\sqrt{5}}{3-\sqrt{5}}$$ *The denominator is a difference of squares.*

$$=\frac{2(3-\sqrt{5})}{(3)^2-(\sqrt{5})^2}$$ *Don't multiply out the numerator yet.*

$$=\frac{2(3-\sqrt{5})}{9-5}=\frac{2(3-\sqrt{5})}{4}=\boxed{\frac{3-\sqrt{5}}{2}}$$

■

EXAMPLE 17 Rationalize the numerator and simplify: $\dfrac{\sqrt{x}+2}{x-4}$.

Solution We can approach this problem in two ways. First we can rationalize by multiplying numerator and denominator by $\sqrt{x}-2$:

$$\frac{\sqrt{x}+2}{x-4}=\frac{\sqrt{x}+2}{x-4}\cdot\frac{\sqrt{x}-2}{\sqrt{x}-2}$$

$$=\frac{(\sqrt{x})^2-2^2}{(x-4)(\sqrt{x}-2)}$$ *Don't multiply out the denominator yet.*

$$=\frac{x-4}{(x-4)(\sqrt{x}-2)}$$ *Now reduce.*

$$=\boxed{\frac{1}{\sqrt{x}-2}}$$

A second way to do this problem is to rewrite the denominator as a difference of squares. Noting that each term of the binomial in the denominator (x and 4) is the square of each term of the binomial in the numerator ($\sqrt{x}$ and 2), we can factor the denominator as demonstrated below.

$$\frac{\sqrt{x}+2}{x-4}=\frac{\sqrt{x}+2}{(\sqrt{x}+2)(\sqrt{x}-2)}=\frac{1}{\sqrt{x}-2}$$

■

EXERCISES 1.4

In Exercises 1–8, perform the operations and simplify. Express answers with positive exponents only.

1. $\dfrac{(x^2y)^2(x^3y)^2}{(xy)^4}$
2. $\dfrac{(3a^2b)^2(2ab)^3}{6a^2b^3}$
3. $\dfrac{(3x^{-1}y^{-2})^{-3}(x^2y^{-1})^3}{(9x^{-2}y)^{-2}}$
4. $\dfrac{4^{-2}16^{-1}}{2^{-4}}$
5. $\dfrac{x^{-1}+2}{x+2}$
6. $\dfrac{x^{-1}+2}{(x+2)^{-1}}$
7. $\dfrac{x^{-1}+y^{-1}}{xy^{-1}}$
8. $\dfrac{x^{-2}+y^{-2}}{(x+y)^{-2}}$
9. Express 7,500,000,000 using scientific notation.
10. Express 0.000000653 using scientific notation.

11. The Sun is approximately 93 million miles from Earth. How long does it take light, which travels at approximately 186,000 miles per second, to reach us from the sun?

12. Scientists estimate that sun converts about 700 million tons of hydrogen to helium every second. If the sun contains about 1.49×10^{27} tons of hydrogen, approximately how many years will it take to use up the sun's supply of hydrogen?

In Exercises 13–16, evaluate the expression.

13. $32^{-3/5}$ **14.** $\left(\frac{1}{8}\right)^{2/3}$

15. $\left(-\frac{8}{27}\right)^{-2/3}$ **16.** $\frac{9^{-1/2}81^{1/4}}{27^{-2/3}3^{-2}}$

In Exercises 17–24, perform the operations and express in simplest form with positive exponents only.

17. $\frac{x^{1/2}y^{2/3}}{x^{-1/3}}$ **18.** $\frac{x^{1/2}x^{-2/3}}{x^{-3/4}}$

19. $(x^2+1)^{-1/2}(x^2+1)^{3/5}$ **20.** $\frac{(x^2+1)^{1/3}}{(x^2+1)^{-2/3}}$

21. $(x^{1/2}+2)^2$

22. $(x^{1/3}+y^{1/3})(x^{2/3}-x^{1/3}y^{1/3}+y^{2/3})$

23. $(x^{1/2}+2)(x^{1/2}-2)$

24. $4x(x^{-1/2}+1)^2$

In Exercises 25–30, express as a single fraction with positive exponents only.

25. $3(x+1)^{-1}-3x(x+1)^{-2}$

26. $2x(x+1)^{-2}-6(x+1)^{-1}$

27. $4x^3(x^2+1)^{-1/2}+4x(x^2+1)^{1/2}$

28. $2x^2(x^2-2)^{-2/3}+(x^2-2)^{1/3}$

29. $\frac{3x^3}{2}(x^3-3)^{-1/2}+(x^3-3)^{1/2}$

30. $\frac{2x^2}{3}(x^2-4)^{-2/3}+(x^2-4)^{1/3}$

In Exercises 31–32, rewrite as an expression without radicals.

31. $\frac{3}{x\sqrt{x^3-3}}$ **32.** $\frac{3x^2}{\sqrt{x^2-1}}$

In Exercises 33–38, express in simplest radical form.

33. $\sqrt{24x^4y^6}$ **34.** $\sqrt[3]{24x^4y^6}$

35. $\sqrt{\frac{16x^8}{9}}$ **36.** $\sqrt{\frac{24x^8}{9x^3}}$

37. $\sqrt{3x^3}\sqrt[3]{81x^5}$ **38.** $\sqrt{\sqrt[3]{5x^2}}$

In Exercises 39–42, rationalize the denominator and express your answer in simplest radical form.

39. $\frac{5}{\sqrt{3x^5}}$ **40.** $\frac{5}{\sqrt[3]{3x^5}}$

41. $\frac{1}{\sqrt{x}-4}$ **42.** $\frac{a-b}{\sqrt{a}-\sqrt{b}}$

In Exercises 43–48, rationalize the numerator and simplify the fraction where possible.

43. $\frac{\sqrt{x^2-3x+2}}{x-1}$ **44.** $\frac{\sqrt{5}}{5x+25}$

45. $\frac{\sqrt{x}-3}{x-9}$ **46.** $\frac{\sqrt{x}+2}{x^2-3x-4}$

47. $\frac{\sqrt{x+h}-\sqrt{x}}{h}$ **48.** $\frac{\sqrt{x+h+3}-\sqrt{x+3}}{h}$

In Exercises 49–64, perform the operations and express your answer in simplest radical form; rationalize denominators where possible.

49. $(3xy^2\sqrt{x^2y})(2x\sqrt{18xy^2})$ **50.** $\frac{xy^2\sqrt{24x^2y}}{x\sqrt{8x^5y}}$

51. $\frac{\sqrt{x^2-2x+1}}{\sqrt{x-1}}$ **52.** $\frac{\sqrt{(x^2-2)^3}}{\sqrt{x^2-2}}$

53. $\sqrt{27}-3\sqrt{18}+6\sqrt{12}$ **54.** $2\sqrt{32}-\sqrt{8}-3\sqrt{2}$

55. $(2\sqrt{3}-5)(2\sqrt{3}+2)$ **56.** $(3\sqrt{x}-2)(2\sqrt{x}+5)$

57. $(\sqrt{a}-5)(\sqrt{a}+5)$ **58.** $(\sqrt{x}-5)^2$

59. $(\sqrt{x}-3)^2-(\sqrt{x-3})^2$ **60.** $(\sqrt{x-6})^2-(\sqrt{x}-6)^2$

61. $\frac{\sqrt{2}}{\sqrt{7}-\sqrt{5}}$ **62.** $\frac{20+\sqrt{60}}{4}$

63. $\frac{12}{\sqrt{6}-2}-\frac{36}{\sqrt{6}}$ **64.** $\frac{20}{\sqrt{7}+3}+\frac{28}{\sqrt{7}}$

In Exercises 65–70, express as a single simple fraction.

65. $2x-\frac{1}{\sqrt{x^2-2}}$ **66.** $5x+\frac{2x}{\sqrt{x+3}}$

67. $3\sqrt{x^2-2}-\frac{2}{\sqrt{x^2-2}}$ **68.** $3x\sqrt{x^2+4}+\frac{5x}{\sqrt{x^2+4}}$

69. $\frac{\sqrt{\frac{2}{x+h}}-\sqrt{\frac{2}{x}}}{h}$

70. $\frac{\frac{1}{\sqrt{x+h+1}}-\frac{1}{\sqrt{x+1}}}{h}$

1.5 The Complex Numbers

When we solve the equation $3x + 5 = 7$ we get $x = \frac{2}{3}$ as a solution. If we were restricted to the set of integers, then this equation would have no solutions. However, if we are allowed to extend to the set of rational numbers, then the solution $x = \frac{2}{3}$ is available. Similarly, in order to solve the equation $x^2 = 3$, we have to "go beyond" the rationals and define the real numbers in order to obtain the solutions $x = \sqrt{3}$ and $x = -\sqrt{3}$.

It may seem as though our system is complete, but there are polynomial equations we are still unable to solve. For example, the equation $x^2 = -5$ has no real number solution, because no real number, when squared, yields a negative number.

In order to obtain the solutions to such equations, we define a system that goes beyond the real numbers. We begin by defining i.

> The imaginary unit, i, is defined by
>
> $$i^2 = -1$$

Hence, $i = \sqrt{-1}$.

Given this definition, we have:

$$i$$
$$i^2 = -1$$
$$i^3 = i^2 i = (-1)i = -i$$
$$i^4 = i^2 i^2 = (-1)(-1) = +1$$

If we continue, $i^5 = i^4 i = (1)i = i$, $i^6 = i^4 i^2 = (1)(-1) = -1$, etc. We find that this cycle repeats itself after i^4, so that any power of i can be written as i, -1, $-i$, or 1.

EXAMPLE 1 Rewrite as ± 1 or $\pm i$: **(a)** i^{39} **(b)** i^{26}

Solution Because $i^4 = 1$, it would be convenient to factor the largest perfect fourth power of i.

(a) $i^{39} = i^{36} i^3$ *36 is the largest multiple of 4 in 39; hence i^{36} is the greatest perfect 4th power factor of i^{39}. We rewrite i^{36} as a perfect 4th power.*

$= (i^4)^9 i^3$ *Since $i^4 = 1$, we get*

$= (1)^9 i^3 = i^3 = \boxed{-i}$

Hence $i^{39} = -i$.

Actually, we find that $i^s = i^r$, where r is the remainder when s is divided by 4, and then rewrite the expression as $\pm i$ or ± 1.

(b) $i^{26} = (i^4)^6 i^2$ *NOTE: dividing 26 by 4 yields a remainder of 2; hence; $i^{26} = i^2$*

$= 1^6 i^2$

$= i^2 = \boxed{-1}$ ■

Using i we can now rewrite square roots with negative radicands as follows:

$$\sqrt{-4} = \sqrt{4(-1)} = \sqrt{4i^2} = \sqrt{4}\sqrt{i^2} = 2i$$

$$\sqrt{-\frac{1}{16}} = \sqrt{\frac{1}{16}(-1)} = \sqrt{\frac{1}{16}i^2} = \sqrt{\frac{1}{16}}\sqrt{i^2} = \frac{1}{4}i$$

We will use i, the imaginary unit, to define a new type of number: a complex number.

DEFINITION A **complex number** is a number that can be written in the form $a + bi$, where a and b are real numbers and i is the imaginary unit. a is called the **real part** of $a + bi$ and b is called the **imaginary part** of $a + bi$.

For example, in the complex number $3 + 4i$, 3 is the real part and 4 is the imaginary part.

If a is a real number, then we can rewrite it as the complex number $a + 0i$. Since we can put any real number into complex form (with the imaginary part equal to zero), we conclude that all real numbers are complex numbers. In other words, the set of real numbers is a subset of the complex numbers.

If a nonzero complex number does not have a real part (real part is zero), then we say that the number is pure imaginary. For example: $3i$, $-4i$, and $i\sqrt{5}$ are pure imaginary numbers.

EXAMPLE 2 Put the following in $a + bi$ form.

(a) $5 + \sqrt{-16}$ **(b)** -5 **(c)** $\dfrac{6 - \sqrt{-3}}{2}$

Solution

(a) $5 + \sqrt{-16} = \boxed{5 + 4i}$ *Since $\sqrt{-16} = \sqrt{16(-1)} = \sqrt{16i^2} = 4i$*

(b) $-5 = \boxed{-5 + 0i}$

(c) $\dfrac{6 - \sqrt{-3}}{2} = \dfrac{6 - i\sqrt{3}}{2}$

$= \dfrac{6}{2} - \dfrac{\sqrt{3}}{2}i = \boxed{3 - \dfrac{\sqrt{3}}{2}i}$ *The real part is 3; the imaginary part is $\dfrac{-\sqrt{3}}{2}$.* ■

Two complex numbers are equal if their real parts are identical *and* their imaginary parts are identical. Algebraically we state this as follows.

EQUIVALENCE OF COMPLEX NUMBERS

$a + bi = c + di$ if and only if $a = c$ and $b = d$

Now that we have defined complex numbers, our next step will be to examine operations on complex numbers.

Addition and Subtraction of Complex Numbers

Addition and subtraction of complex numbers are relatively straightforward.

Addition of complex numbers:

$$(a + bi) + (c + di) = (a + c) + (b + d)i$$

Subtraction of complex numbers:

$$(a + bi) - (c + di) = (a - c) + (b - d)i$$

This says that the real part of the sum (difference) is the sum (difference) of the real parts and the imaginary part of the sum (difference) is the sum (difference) of the imaginary parts.

EXAMPLE 3 Perform the operations.

(a) $(3 + 4i) + (5 - 6i)$ **(b)** $(6 + 9i) - (3 - 2i)$

Solution

(a) We could either use the definition of addition and subtraction:

$$(3 + 4i) + (5 - 6i) = (3 + 5) + (4 - 6)i = \boxed{8 - 2i}$$

Or we could treat i as a variable and combine terms:

$$(3 + 4i) + (5 - 6i) = 3 + 4i + 5 - 6i = \boxed{8 - 2i}$$

(b) $(6 + 9i) - (3 - 2i) = (6 - 3) + [9 - (-2)]i$

$= \boxed{3 + 11i}$ *By using the definition* ■

Products of Complex Numbers

We can treat a complex number as a binomial and multiply two complex numbers using binomial multiplication:

$(a + bi)(c + di) = ac + (ad + bc)i + bdi^2$ *General form 2*

$= ac + (ad + bc)i + bd(-1)$ *Since $i^2 = -1$*

$= ac - bd + (ad + bc)i$ *Rearrange terms so that the real parts are together and the imaginary parts are together.*

Thus we formulate the rule for multiplying complex numbers.

Products of complex numbers:

$$(a + bi)(c + di) = (ac - bd) + (ad + bc)i$$

It is probably not very useful to memorize this rule as a special product at this point. It is better just to multiply two complex numbers as binomials, substitute -1 for i^2, and then combine the real parts and imaginary parts to conform to $a + bi$ form.

EXAMPLE 4 Perform the operations.

(a) $(3 + 2i)(4 - 5i)$ **(b)** $(3 - 7i)^2$ **(c)** $(5 - 2i)(5 + 2i)$
(d) $(3 + i)^2 - 2(3 + i)$

Solution

(a) $(3 + 2i)(4 - 5i)$

$= 12 + (8 - 15)i - 10i^2$ *Since $i^2 = -1$, then $-10i^2 = -10(-1) = +10$.*

$= 12 - 7i + 10 = \boxed{22 - 7i}$

(b) $(3 - 7i)^2 = (3)^2 - 2(3)(7i) + (7i)^2$ *Special product: a perfect square of a difference*

$= 9 - 42i + 49i^2$ *Since $49i^2 = 49(-1) = -49$*

$= 9 - 42i - 49 = \boxed{-40 - 42i}$

Why is the result of Example 4(c) a real number?

(c) $(5 - 2i)(5 + 2i) = 5^2 - (2i)^2$ *Special product: the difference of squares*

$= 25 - 4i^2$ *Since $-4i^2 = -4(-1) = +4$*

$= 25 + 4 = \boxed{29}$ *Note that the result is a real number.*

(d) $(3 + i)^2 - 2(3 + i)$ *Powers first: square $(3 + i)$.*

$= 9 + 6i + i^2 - 2(3 + i)$ *Then multiply.*

$= 9 + 6i + i^2 - 6 - 2i$

$= 9 + 6i - 1 - 6 - 2i = \boxed{2 + 4i}$ ■

Observe that in part (c) of Example 4, the product of two complex numbers $5 - 2i$ and $5 + 2i$ yields the real number 29. The two number $5 - 2i$ and $5 + 2i$ are *complex conjugates* of each other. The **complex conjugate** of $a + bi$ is $a - bi$ (and conversely, the complex conjugate of $a - bi$ is $a + bi$). In general, *the product of two complex numbers that are conjugates of each other yields a real number:*

$(a + bi)(a - bi) = a^2 - (bi)^2$ *Difference of squares*

$= a^2 - b^2i^2$ *Since $i^2 = -1$, $-b^2i^2 = -b^2(-1) = b^2$.*

$= a^2 + b^2$ *Note that $a^2 + b^2$ is a real number.*

We will use this result to help us in finding certain quotients of complex numbers.

Quotients of Complex Numbers

Our goal is to express a quotient of complex numbers in the form $a + bi$. If we have a quotient of a complex number and a real number, where the real number is the divisor, then expressing the quotient in the form $a + bi$ is similar to dividing a polynomial by a monomial. For example,

$$\frac{6 - 5i}{2} = \frac{6}{2} - \frac{5}{2}i = 3 - \frac{5}{2}i$$

Keeping in mind that $i = \sqrt{-1}$ is a radical expression, we will find the rationalizing techniques we used with quotients of radicals useful when working with quotients of complex numbers when the divisor has an imaginary part.

EXAMPLE 5 Express the following in the form $a + bi$.

(a) $\dfrac{4 + 3i}{2i}$ **(b)** $\dfrac{7 - 4i}{2 + i}$

Solution

(a) We will treat this expression in the same manner as we would if we were rationalizing a denominator: Multiply the numerator and denominator by i.

$$\frac{4 + 3i}{2i} = \frac{(4 + 3i)}{2i}\frac{i}{i} = \frac{4i + 3i^2}{2i^2}$$ *And, since* $i^2 = -1$,

$$= \frac{4i + 3(-1)}{2(-1)} = \frac{-3 + 4i}{-2} = \boxed{\frac{3}{2} - 2i}$$

(b) Our goal is to get a real number in the denominator so that we may express our result in the form $a + bi$. We proceed as follows:

$$\frac{7 - 4i}{2 + i}$$ *Multiply the numerator and denominator by the conjugate of the denominator, which is* $2 - i$.

$$= \frac{(7 - 4i)(2 - i)}{(2 + i)(2 - i)}$$

$$= \frac{14 - 7i - 8i + 4i^2}{2^2 - i^2}$$ *Since* $i^2 = -1$, *we get*

$$= \frac{14 - 7i - 8i - 4}{4 + 1}$$ *Note that the denominator is now a real number.*

$$= \frac{10 - 15i}{5}$$ *Then rewrite in complex number form.*

$$= \frac{10}{5} - \frac{15}{5}i = \boxed{2 - 3i}$$ ■

We have defined complex numbers and their operations. Complex numbers, together with the arithmetic operations as we have defined them, obey the same properties as the real numbers (associative, commutative, distributive, inverses, etc.) and form a system called the complex number system, designated by $\boldsymbol{C}$. In Section 1.8 we will see that any second degree equation has its solution in $\boldsymbol{C}$, the complex number system. In Chapter 4 we will see that any polynomial equation has all its solutions in $\boldsymbol{C}$.

EXERCISES 1.5

In Exercises 1–6, express each as ± 1 or $\pm i$.

1. i^{16}
2. i^{35}
3. i^{100}
4. i^{43}
5. i^{4n+3} (n an integer)
6. i^{4n+1} (n an integer)

In Exercises 7–10, write in $a + bi$ form.

7. $3 + \sqrt{-16}$
8. $2 - \sqrt{-12}$
9. $5 - \sqrt{-\dfrac{1}{12}}$
10. $\dfrac{2 + \sqrt{-12}}{2}$

In Exercises 11–32, perform the operations and write the results in $a + bi$ form.

11. $(3 + 2i) + (6 - 8i)$
12. $(7 + 2i) + (3 - 5i)$
13. $(8 - 2i) + (9 + 2i)$
14. $(7 + 6i) - (7 - 6i)$
15. $(2 - 3i)(5 + i)$
16. $(5 + 3i)(2 - i)$
17. $(5 - 6i)(5 + 6i)$
18. $(3 - 8i)(3 + 8i)$
19. $(2 - 3i)^2$
20. $(5 - i)^2$
21. $\dfrac{2 + 8i}{2}$
22. $\dfrac{5 - 15i}{3}$
23. $\dfrac{2 + 3i}{i}$
24. $\dfrac{6 - 2i}{2i}$
25. $\dfrac{3}{2 + i}$
26. $\dfrac{4}{3 - i}$
27. $\dfrac{i}{2 - i}$
28. $\dfrac{i}{2 + i}$
29. $\dfrac{i + 1}{i - 1}$
30. $\dfrac{i - 1}{i + 1}$
31. $\dfrac{6 - 2i}{5 + 2i}$
32. $\dfrac{1 + i}{3 + 2i}$

In Exercises 33–36, perform the operations and write the results in $a + bi$ form.

33. $(2 + i)^2 - 4(2 + i)$
34. $2(1 - i)^2 - 4(1 - i)$
35. $(2 - 2i)^2 + 2(2 + i)$
36. $(1 - 3i)^2 - 2(1 + 3i)$
37. Verify that $3 + 2i$ is a solution to $x^2 - 6x + 13 = 0$.
38. Verify that $2 - i$ is a solution to $3x^2 - 12x = -15$.
39. Express i^{-1} in $a + bi$ form.
40. Express i^{-2} in $a + bi$ form.

QUESTION FOR THOUGHT

41. The number i allows us to represent even roots of negative numbers other than square roots. Show how we can represent $\sqrt[6]{-1}$ using i. HINT: $\sqrt[6]{x} = x^{1/6} = (x^{1/3})^{1/2} = \sqrt{\sqrt[3]{x}}$. How about $\sqrt[10]{-1}$? $\sqrt[14]{-1}$?

1.6 First-Degree Equations and Inequalities in One Variable

First-Degree Equations

An equation is a symbolic statement of equality. That is, rather than writing "Twice a number is four less than the number," we write: $2x = x - 4$. Our goal is to find the solution to a given equation. By **solution** we mean the value or values of the variable that make the algebraic statement true.

In some cases, we may have an equation that is always true. For example $x + 6 = x + 8 - 2$ is *true for all values of the variable for which it is defined.* We call such equations **identities.** On the other hand, there are equations for which there is no solution, such as $x + 2 = x + 1$. Such an equation is called a **contradiction.** Equations that are true for some values of the variable and false for other values are called **conditional.** The conditional equation $3x - 2 = 5x - 1$ is true when x is $-\frac{1}{2}$ and false when x is any other value.

DEFINITION **A first-degree equation in one variable** is an equation that can be put in the form $ax + b = 0$, where a and b are constants and $a \neq 0$.

In other words, a first-degree equation is an equation in which the highest degree of the variable is 1.

Equations that have identical solutions are called **equivalent equations.** The equations $3x - [5 + 2(x - 1)] = 2x + 1$ and $x = -4$ are equivalent since they each have exactly the same solution, -4. Our goal is to take an equation and, with the help of a few properties, gradually change the given equation into an equivalent *obvious equation*, an equation of the form $x = a$ (or $a = x$), where x is the variable for which we are solving.

The **substitution property** states that if $a = b$, then a can be used to replace b and vice versa. The **addition property of equality**, if $a = b$, then $a + c = b + c$, is simply a specific application of the substitution property. This is also true of the **multiplication property**: If $a = b$, then $ac = bc$. (We don't need to specifically state a subtraction or division property, since subtraction and division are defined, respectively, in terms of addition and multiplication.)

The properties imply that: (1) *Adding (subtracting) the same quantity to (from) both sides of an equation will produce an equivalent equation, and* (2) *Multiplying (dividing) both sides of an equation by the same* nonzero *quantity will produce an equivalent equation.*

Now we have two (or four) ways to transform an equation into an equivalent equation. Our goal is to apply the preceding transformations to a given equation until we have reduced it to an obvious equation.

In general, our strategy is to: (1) First simplify each side of an equation as much as possible. If there are fractions in the equation, use the multiplication property to "clear the denominators." (2) Using the addition (and/or subtraction) property, collect all terms containing the variable for which we are solving on one side of the equation. (3) Use the division (or multiplication) property to isolate the variable and get the obvious equation. (4) Finally, check all solutions in the original equation.

EXAMPLE 1 Solve for x.

(a) $820x = 10x + 30(50 - x)$

(b) $\dfrac{3}{x - 1} + 5 = \dfrac{4 - x}{x - 1}$

(c) $\dfrac{8}{x - 5} - \dfrac{5}{x + 3} = \dfrac{3x + 49}{x^2 - 2x - 15}$

Solution

(a)

$820x = 10x + 30(50 - x)$ *Simplify the right-hand side.*

$820x = 10x + 1500 - 30x$

$820x = 1500 - 20x$ *Apply the addition property (add 20x to each side of the equation).*

$840x = 1500$ *Now use the division property (divide each side of the equation by 840).*

$$x = \frac{1500}{840} = \boxed{\frac{25}{14}}$$

Remember to check by substituting $\frac{25}{14}$ for x in the original equation.

(b)

$$\frac{3}{x - 1} + 5 = \frac{4 - x}{x - 1}$$ *Multiply each side of the equation by $x - 1$ to clear the denominators.*

$$\frac{3}{x - 1}(x - 1) + 5(x - 1) = \frac{4 - x}{x - 1}(x - 1)$$ *Multiply.*

$3 + 5(x - 1) = 4 - x$ *Simplify the left-hand side.*

$5x - 2 = 4 - x$ *Apply equality properties.*

$6x = 6$

$x = 1$

It seems as though we have a solution to this equation. However, if we check this solution by substituting the value 1 for x in the original equation, we find undefined terms (zero in the denominator). Thus, the solution we arrived at by applying the properties does not work. Did we make a mistake? No, when we multiply both sides by $x - 1$, we must be sure that $x - 1 \neq 0$ to arrive at an equivalent equation. This means that x cannot equal 1; hence the equation has no solutions.

(c)

$$\frac{8}{x - 5} - \frac{5}{x + 3} = \frac{3x + 49}{x^2 - 2x - 15}$$ *Multiply both sides of the equation by the LCD of all the fractions. Since $x^2 - 2x - 15 = (x - 5)(x + 3)$, the LCD is $(x - 5)(x + 3)$ (note $x \neq 5$, $x \neq -3$).*

Why do we exclude 5 and −3 in Example 1(c)?

$$\frac{8}{x - 5}(x - 5)(x + 3) - \frac{5}{x + 3}(x - 5)(x + 3) = \frac{3x + 49}{(x - 5)(x + 3)}(x - 5)(x + 3)$$

Reduce.

$8(x + 3) - 5(x - 5) = 3x + 49$ *Simplify the left-hand side.*

$3x + 49 = 3x + 49$

At this point we notice that this equation is always true. If we start applying the properties of equality, we would end up with an equation such as $0 = 0$ or $49 = 49$. Since these statements are always true *and since they are equivalent to the original equation,* the first equation must be true for all reals *except* 5 and -3. ■

We employ the same strategies for solving literal equations (equations involving more than one letter).

EXAMPLE 2 Solve the following for the indicated variable.

$$y = \frac{x - 1}{2x + 3} \qquad \text{(for } x\text{)}$$

Solution

$$y = \frac{x - 1}{2x + 3}$$ *Multiply both sides by $2x + 3$.*

$$y(2x + 3) = x - 1$$ *Simplify the left-hand side.*

$$2xy + 3y = x - 1$$ *Now isolate the x-terms (collect all x-terms on one side and all non-x-terms on the other side).*

$$2xy - x = -3y - 1$$ *Now factor x from the left hand side.*

$$(2y - 1)x = -3y - 1$$ *Finally, divide both sides by the multiplier of x: $2y - 1$.*

$$x = \frac{-3y - 1}{2y - 1}, \quad \text{or} \quad x = \frac{3y + 1}{1 - 2y}$$ ■

EXAMPLE 3 Given the formula $P = \frac{kV}{T}$, where k is a fixed constant, what happens to T when P is doubled and V is tripled?

Solution Since we want to examine what happens to T when P is doubled and V is tripled, let's first solve for T:

$$P = \frac{kV}{T}$$ *Solve for T to get*

$$T = \frac{kV}{P}$$ *Doubling P means we use 2P instead of P and tripling V means using 3V instead of V in the formula $T = \frac{kV}{P}$. Hence the new T, which we call T′, is*

$$T' = \frac{k[3V]}{[2P]}$$

$$= \frac{3}{2}\frac{kV}{P}$$ *And since $\frac{kV}{P} = T$,*

$$T' = \frac{3}{2}T$$

Thus, doubling P and tripling V causes T to be multiplied by $\frac{3}{2}$.

EXAMPLE 4 A computer discount store held an end-of-summer sale on two types of computers. They collected \$41,800 on the sale of 58 computers. If one type sold for \$600 and the other type sold for \$850, how many of each type were sold?

Solution If we let x = the number of \$600 computers sold, then $58 - x$ = the number of \$850 computers sold (since 58 computers were sold all together.)

Our equation involves the amount of money collected on the sale of each type of computer. (That is, the *value* of the computers sold.) We compute the amount as follows:

Amount collected from sale of \$600 computers	+	Amount collected from sale of \$850 computers	=	Total amount collected
(No. of \$600 computers sold)(cost of a \$600 computer)	+	(No. of \$850 computers sold)(cost of an \$850 computer)	=	\$41,800
$x \cdot 600$	+	$(58 - x) \cdot 850$	=	41,800

Our equation is $600x + 850(58 - x) = 41{,}800$, which yields $x = 30$. Hence there were

$x = 30$ computers sold at \$600 and $58 - 30 = 28$ computers sold at \$850.

EXAMPLE 5 Art can process 200 forms per hour, and Lew can process 150 forms per hour. How long would it take to process 900 forms working together if Art starts $\frac{1}{2}$ hour after Lew begins?

Solution The underlying relationship is that the quantity of work done is equal to the product of the rate at which the work is done and the time doing the work, or $Q = RT$. If Art can process 200 forms in 1 h, then he can process $200 \times 2 = 400$ forms in 2 h, $200 \times 3 = 600$ forms in 3 h, etc.

We can let t be the number of hours Lew works. Then Lew processes $150t$ forms in t hours. On the other hand, Art starts $\frac{1}{2}$ hour later, so his time working is $t - \frac{1}{2}$ hours; therefore, he processes $200\left(t - \frac{1}{2}\right)$ forms.

Our equation illustrates the contribution made by each one.

$$\underbrace{150t}_{\text{No. forms processed by Lew}} + \underbrace{200(t - 1/2)}_{\text{No. forms processed by Art}} = \underbrace{900}_{\text{Total no. forms to be processed}}$$

This yields

$$350t - 100 = 900$$

Solve for t to get $t = \boxed{2\frac{6}{7} \text{ hours}}$ to process 900 forms.

EXAMPLE 6 If it takes Lew 3 hours to complete a job and Art 2 hours to complete the same job, how long would it take for them to complete the job working together?

Solution This is the same type of rate problem, only the rates are stated slightly differently than in Example 5. To say that it takes 3 hours to complete a job means Lew's rate is $\frac{1}{3}$ of a job per hour. Notice $\frac{1}{3}$ job per hour means in 2 hours, $\frac{2}{3}$ of the job is completed; in 3 hours, $\frac{3}{3}$, or 1 job is completed, etc. Now that we have their rates in familiar form, we can approach it in the same way as in Example 5.

If t is the time it takes for them to complete the job working together, then Lew can do $\frac{1}{3}t$ job in t hours, and Art can do $\frac{1}{2}t$ job in t hours. So altogether, in t hours they do $\frac{1}{3}t + \frac{1}{2}t$ jobs. Since we want to determine how long it takes to complete 1 job, our equation is as follows:

$$\underbrace{\frac{1}{2}t}_{\text{Amount of job completed by Lew}} + \underbrace{\frac{1}{3}t}_{\text{Amount of job completed by Art}} = \underbrace{1}_{\text{One complete job}}$$

Solving for t, we get $t = \boxed{\frac{6}{5} \text{ hours or 1 hour 12 minutes}}$. ■

First-Degree Inequalities

DEFINITION **A first-degree inequality** is an inequality that can be put in the form $ax + b < 0$, where a and b are constants with $a \neq 0$. (The $<$ symbol can be replaced with $>$, $\leq$, or $\geq$.)

In order to solve inequalities, we will need the properties of inequalities listed below. Keep in mind that although conditional first-degree equations have single-value solutions, the solutions to conditional first-degree inequalities are usually infinite sets.

PROPERTIES OF INEQUALITIES

For $a, b, c \in R$, if $a < b$, then

1. $a + c < b + c$ **2.** $ac < bc$ when $c > 0$ **3.** $ac > bc$ when $c < 0$

Thus, to produce an equivalent inequality, we may add (subtract) the same quantity to (from) both sides of an inequality, or multiply (divide) both sides of an inequality by the same *positive* quantity. On the other hand, we must reverse the inequality symbol to produce an equivalent inequality if we multiply (divide) both sides of an inequality by the same *negative* quantity.

EXAMPLE 7 Solve for the variable. Express your answer using interval notation.

(a) $5x + 8(20 - x) \geq 2(x - 5)$ (b) $-2 \leq \frac{5}{3} - 3x < 5$

Solution

(a)
$$\begin{aligned} 5x + 8(20 - x) &\geq 2(x - 5) && \textit{Simplify each side.} \\ 5x + 160 - 8x &\geq 2x - 10 \\ 160 - 3x &\geq 2x - 10 && \textit{Now apply the inequality properties.} \\ -5x &\geq -170 && \textit{Divide both sides by } -5. \\ x &\leq 34 && \textit{Note the inequality symbol is reversed.} \end{aligned}$$

Using interval notation, the answer is $\boxed{(-\infty, 34]}$.

(b) $-2 \leq \frac{5}{3} - 3x < 5$

This double inequality is actually two inequalities that must be satisfied simultaneously: $-2 \leq \frac{5}{3} - 3x$ *and* $\frac{5}{3} - 3x < 5$. We could break the double inequality up into two inequalities, $-2 \leq \frac{5}{3} - 3x$ *and* $\frac{5}{3} - 3x < 5$, solve them, and rewrite the solution as a double inequality. But we can reduce our labor by isolating x in the middle member, as follows:

Solve Example 7(b) using two inequalities.

$$\begin{aligned} -2 \leq \frac{5}{3} - 3x &< 5 && \textit{Multiply each member by 3.} \\ -6 \leq 5 - 9x &< 15 && \textit{Subtract 5 from each member.} \\ -11 \leq -9x &< 10 && \textit{Divide each member by } -9. \\ \frac{11}{9} \geq x &> -\frac{10}{9} && \textit{Note we reversed the inequality symbols.} \end{aligned}$$

Hence: $-\frac{10}{9} < x \leq \frac{11}{9}$ or, using interval notation, $\boxed{\left(-\frac{10}{9}, \frac{11}{9}\right]}$.

EXAMPLE 8 A solution of 10% (by volume) alcohol must be mixed together with a solution of 25% alcohol in order to make 24 liters of a solution that is at least 15% but no more than 20% alcohol. How much of the 10% alcohol can be used to produce a mixture with an alcohol content within the given limits?

Solution This problem is similar in structure to Example 4 in this section. In Example 4 we differentiated quantity (how many) from value (the cost of each), but in this problem we differentiate the quantity (amount) of the solution from the pure alcohol content of each solution. For example, 40 gallons of a 25% solution of alcohol would contain $0.25(40) = 10$ gallons of pure alcohol.

We develop an *inequality* reflecting the fact that the amounts of pure alcohol in each of the solutions to be mixed should sum to the amount of pure alcohol in the final mixture.

If we let x = amount of 10% solution, then $24 - x$ = amount of 25% solution. Our inequality is

$0.15(24)$	$\leq$	$0.10x$	$+$	$0.25(24 - x)$	$\leq$	$0.20(24)$
Total amount of alcohol in 24 liters of a 15% solution		Amount of alcohol in the 10% solution		Amount of alcohol in the 25% solution		Total amount of alcohol in 24 liters of a 20% solution

$$0.15(24) \leq 0.10x + 0.25(24 - x) \leq 0.20(24)$$

Clear decimals: Multiply all three members by 100

$$(100)0.15(24) \leq (100)[0.10x + 0.25(24 - x)] \leq (100)[0.20(24)]$$

$$15(24) \leq 10x + 25(24 - x) \leq 20(24)$$

Simplify each member to get

$$360 \leq 600 - 15x \leq 480$$

Solving for x, we get

$$16 \geq x \geq 8 \quad or \quad 8 \leq x \leq 16$$

Hence **between 8 and 16 liters of 10% alcohol** can be used to produce mixtures within the desired limits. ■

EXERCISES 1.6

In Exercises 1–22, solve for the given variable. For inequalities, express your answer using interval notation.

1. $2 - 3(x - 4) = 2(x - 1)$
2. $3x - [2 + 3(2 - x)] = 5 - (3 - x)$
3. $4x + \frac{2}{3} \leq 2x - (3x + 1)$
4. $5x - 2 > 3x - \left(x - \frac{1}{5}\right)$
5. $5\{y - [2 - (y - 3)]\} > y - 2$
6. $2 - \{a - 6[a - (4 - a)]\} \leq 3(a + 2)$
7. $\frac{2}{3}x - 5 = \frac{3}{2}x + 4(x - 1)$
8. $\frac{3}{4}(2x - 3) = \frac{2}{3}x + 5$
9. $-\frac{7}{y} + 1 = -13$
10. $\frac{2}{x + 3} - 4 = 8$
11. $\frac{3x + 1}{2} - \frac{1}{3} < 1$
12. $\frac{5x - 2}{3} \geq \frac{x + 3}{4}$
13. $-\frac{7}{y} + 3 = 3$
14. $\frac{3}{a - 2} + 5 = \frac{1 + a}{a - 2}$
15. $\frac{2}{x + 3} + 4 = \frac{5 - x}{x + 3}$
16. $\frac{7}{a - 1} + 4 = \frac{a + 6}{a - 1}$
17. $\frac{4a + 1}{a^2 - a - 6} = \frac{2}{a - 3} + \frac{5}{a + 2}$
18. $\frac{5}{y + 3} + \frac{2}{y} = \frac{y - 12}{y^2 + 3y}$
19. $\frac{5}{2x + 1} + \frac{3}{2x - 1} = \frac{22}{4x^2 - 1}$
20. $\frac{1}{3x + 4} + \frac{8}{9x^2 - 16} = \frac{1}{3x - 4}$
21. $\frac{6}{3x + 5} - \frac{2}{x - 4} = \frac{10}{3x^2 - 7x - 20}$
22. $\frac{6}{x^2 - 3x} = \frac{12}{x} + \frac{1}{x - 3}$

In Exercises 23–36, solve for the given variable.

23. $3x + 2y - 4 = 5x - 3y + 2$ for y
24. $\frac{x}{3} + \frac{y}{4} = \frac{x}{2} + 3$ for x
25. $S = 2LH + 2LW + 2WH$ for W
26. $ax + b = cx + d$ for x
27. $A = \frac{1}{2}h(b_1 + b_2)$ for h
28. $A = \frac{1}{2}h(b_1 + b_2)$ for b_1
29. $(3x - 2)(2y - 1) = 0$ for y
30. $(3x - 2)(2y - 1) = 0$ for x
31. $\frac{x - \mu}{s} < 1.96$ for s ($s > 0$)
32. $\frac{x - \mu}{s} < 1.96$ for μ ($s > 0$)
33. $\frac{1}{f} = \frac{1}{f_1} + \frac{1}{f_2}$ for f
34. $\frac{1}{R} = \frac{1}{R_1} + \frac{1}{R_2} + \frac{1}{R_3}$ for R_2

35. $y = \dfrac{3x - 2}{x}$ for x

36. $y = \dfrac{2x + 3}{5x - 1}$ for x

In Exercises 37–44, solve for x. Graph your answer on the number line.

37. $-4 < 3x - 2 < 5$

38. $-5 \le 2x - 5 < 8$

39. $0 < 5 - 2x \le 4$

40. $-1 \le 3 - \dfrac{1}{2}x \le 2$

41. $-\dfrac{3}{2} < \dfrac{1}{3} - x < 2$

42. $0 \le \dfrac{2 - 3x}{5} < \dfrac{1}{2}$

43. $2x - 3 < 4 < 3x - 1$

44. $5x - 2 < 5 \le 2x - 1$

45. If $s = kt^2$, where k is a fixed constant, what happens to s if t is tripled?

46. If $V = \dfrac{kT}{P}$, where k is a fixed constant, what happens to P if V is halved?

47. If $E = \dfrac{k}{d^2}$, where k is a fixed constant, what happens to E when d is tripled?

48. If $F = \dfrac{km_1m_2}{d^2}$, where k is a fixed constant, what happens to F when d is doubled, and m_1 is halved?

49. A truck carries a load of 50 boxes; some are 20-lb boxes, and the rest are 25-lb boxes. If the total weight of all boxes is 1175 lb, how many of each type are there?

50. A merchant wishes to purchase a shipment of clock radios. Simple AM models cost \$25 each, whereas AM/FM models cost \$30 each. In addition, there is a delivery charge of \$70. If he spends \$700 on 24 clock radios, how many of each type did he buy?

51. Orchestra seats to a certain Broadway show are \$48 each and balcony seats are \$28 each. If a theater club spends \$2,328 on the purchase of 56 seats, how many orchestra seats were purchased?

52. A plumber charges \$22 per hour for her time and \$13 per hour for her assistant's time. On a certain job the assistant works alone for 2 hours doing preparatory work; then the plumber and her assistant complete the job together. If the total bill for the job was \$271, how many hours does the plumber work?

53. A manager needs 7,200 copies of a document. A new duplicating machine can make 70 copies per minute, and an older model can make 50 copies per minute. The older machine begins making copies but breaks down before completing the job and is replaced by the new machine, which completes the job. If the total time for the job is 1 hour and 50 minutes, how many copies did the older machine make?

54. An experienced worker can process 60 items per hour, whereas a new worker can process 30 items per hour. How many hours will it take to complete a job of processing 500 items if the new worker begins the job 3 hours before being joined by the experienced worker?

55. How many ounces of a 20% solution of alcohol (by volume) must be mixed with 5 ounces of a 50% solution of alcohol in order to get a mixture of 30% alcohol?

56. Tamara's radiator has a 3-gallon capacity. Her radiator is filled to capacity with a 30% mixture of antifreeze and water. She drains off some of the old antifreeze and refills the radiator to capacity with pure water to get a 20% mixture. How much did she drain off?

57. Cindy wants to invest \$8,000 in two money-market certificates; one yields 9% interest per year and the other, higher-risk certificate yields 14% per year. If she needs to receive an income of at least \$890 a year from her two investments, what is the *minimum* amount she should invest in the higher-risk certificate in order to get her desired income?

58. Lisa's math teacher assigns course grades in the following way: Two exams are given, each worth 20% of the course grade; quizzes and homework combined together are worth 20% of the course grade; and the final exam is worth 40% of the course grade. Lisa received a grade of 85 on the first exam, 65 on the second exam, and had a quiz and homework combined average of 72. What is the minimum score she must get on the final in order for her to receive a grade of at least 80 for the course?

59. Pipe A can fill a pool with water in 3 days, and pipe B can fill the same pool in 2 days. If both pipes were used, how long would it take to fill the pool?

60. Pipe A and B together can fill a pool in 2 days. If pipe A alone can fill the pool in 5 days, how long would it take to fill the pool with pipe B alone?

61. When a bathtub faucet is turned on (and the drain is shut), it can fill a tub in 10 minutes: when the drain is open (and the faucet is turned off), a full tub can be emptied in 15 minutes. How long would it take the tub to fill if the faucet were turned on with the drain left open?

62. How long would it take for the bathtub of Exercise 61 to fill if the tub could empty in 6 minutes when the drain is open?

QUESTIONS FOR THOUGHT

63. If $\dfrac{1}{a} < 0$, what can be said about a?

64. When does $\dfrac{1}{a}$ equal 0? Explain your answer.

65. A student was given the inequality: $\dfrac{3}{x - 2} > 4$. The first step the student took in solving this inequality was to transform it into $3 < 4(x - 2)$. Explain what the student did wrong.

1.7 Absolute Value Equations and Inequalities

Absolute Value Equations

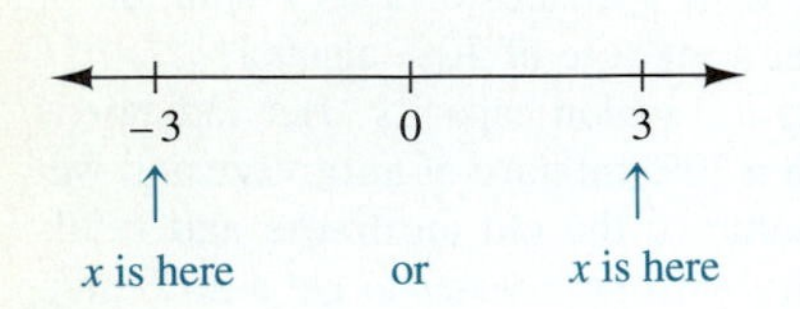

FIGURE 1.9

Consider the equation $|x| = 3$. Recall in Section 1.1 that we visualize $|x|$ as the distance from x to zero on the number line. Hence $|x| = 3$ means that x is 3 units away from 0 on the number line.

Given this definition, we can see by Figure 1.9 that there are two possible answers: $x = -3$ or $x = 3$, since both -3 and 3 are 3 units away from 0.

EXAMPLE 1 Solve the following: $|4x - 7| = 5$.

Solution $4x - 7$ must be 5 units away from 0 on the number line as pictured in Figure 1.10.

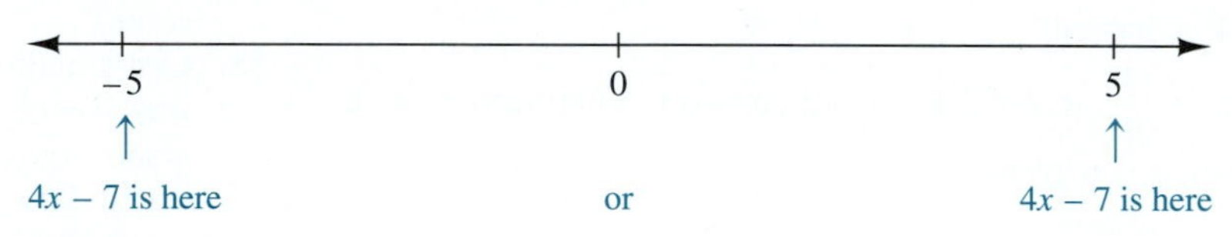

FIGURE 1.10

Therefore,

$$4x - 7 = -5 \quad \text{or} \quad 4x - 7 = 5$$ *We solve each equation to get*

Hence, $\boxed{x = \frac{1}{2} \quad \text{or} \quad x = 3}$. ■

Thus we see algebraically that

> For $a \geq 0$,
>
> $|x| = a$ is equivalent to $x = -a$ or $x = a$

EXAMPLE 2 Solve the following: $|2x - 8| + 4 = 10$.

Solution

$$|2x - 8| + 4 = 10$$ *First we put the equation in the form $|x| = a$ by isolating the absolute value on one side of the equation.*

$$|2x - 8| = 6$$ *This means*

$$2x - 8 = -6 \quad \text{or} \quad 2x - 8 = 6$$ *Solve each equation to get*

$$\boxed{x = 1 \quad \text{or} \quad x = 7}$$ ■

EXAMPLE 3 Solve the following: $|2x - 5| = |3x - 5|$.

Solution In order for the absolute value of two expressions to be equal, they must be the same distance from 0 on the number line. This can happen if either the two expressions are equal or if they are negatives of each other (If $|a| = |b|$, then either $a = b$, or $a = -b$). Hence

$2x - 5 = 3x - 5$ or $2x - 5 = -(3x - 5)$ *Solve each inequality to get*

$$\boxed{x = 0 \quad \text{or} \quad x = 2}$$

■

Absolute Value Inequalities

To solve absolute value inequalities we again refer back to the geometric definition of absolute value. We first note that $|x| < 5$ means that x must lie within 5 units of 0 on the number line. Hence, by Figure 1.11, we see that x must lie between -5 and 5, or $-5 < x < 5$.

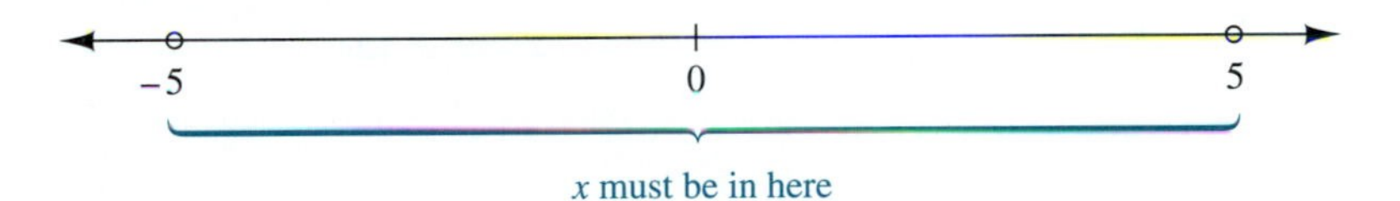

FIGURE 1.11

EXAMPLE 4 Solve for x: $|5x - 3| < 8$.

Solution $5x - 3$ must be less than 8 units away from 0 on the number line, as pictured in Figure 1.12.

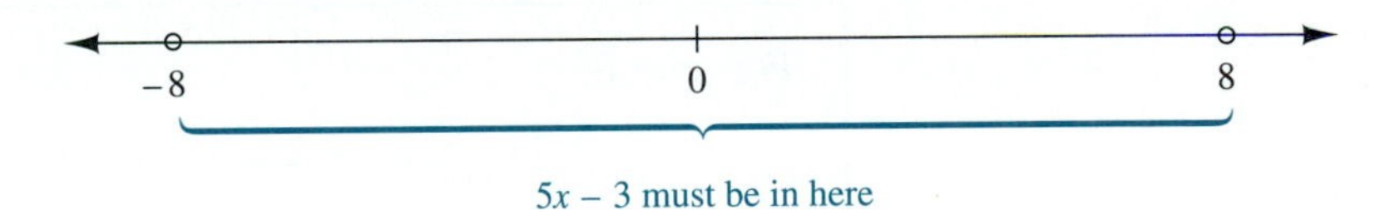

FIGURE 1.12

Therefore, $-8 < 5x - 3 < 8$. *Solving this double inequality we get*

$$-1 < x < \frac{11}{5}.$$

Using interval notation, the solution is $\boxed{(-1, 11/5)}$. ■

The procedure for solving $|x| > a$ is different than the procedure for solving $|x| < a$. This is why it is important for you to be able to visualize the meaning of absolute value.

For example, $|x| > 5$ means that x is *more* than 5 units from 0 on the number line. By Figure 1.13, we see that the solutions lie out at the ends of the number line. We must use two inequalities to describe the solution set: $x < -5$ or $x > 5$. (We cannot use the word "and" between the sets because that implies that x must satisfy both conditions at the same time, which is clearly impossible.)

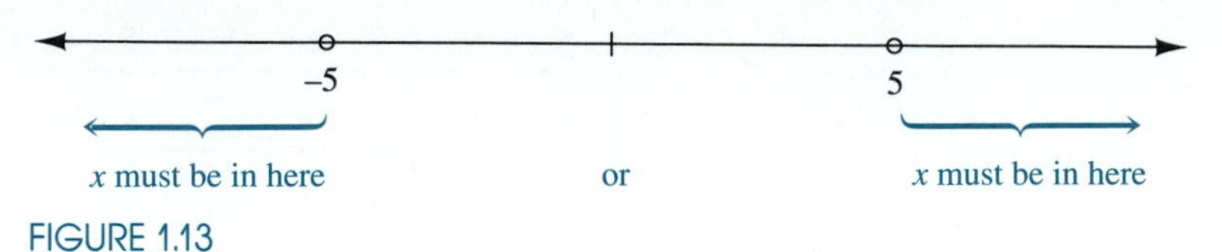

FIGURE 1.13

EXAMPLE 5 Solve for x: $|2x - 9| > 7$.

Solution $2x - 9$ must be more than 7 units away from 0 on the number line, as pictured in Figure 1.14.

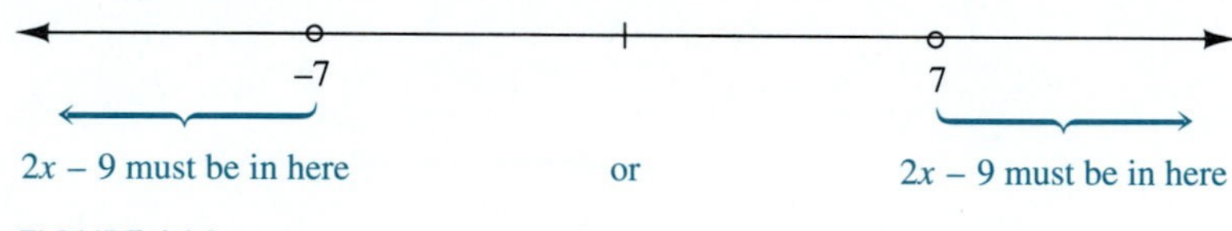

FIGURE 1.14

Therefore,

$$2x - 9 < -7 \quad \text{or} \quad 2x - 9 > 7 \qquad \textit{We solve each inequality to get}$$

$$x < 1 \quad \text{or} \quad x > 8$$

This notation $(-\infty, 1) \cup (8, \infty)$ means the interval $(-\infty, 1)$ **or** the interval $(8, \infty)$.

Using interval notation, we can write the solution as: $\boxed{(-\infty, 1) \cup (8, \infty)}$. ■

Algebraically, we can summarize solving absolute value inequalities as follows.

For $a > 0$,

$|x| < a$ is equivalent to $-a < x < a$.

$|x| > a$ is equivalent to $x < -a$ or $x > a$.

EXAMPLE 6 Solve the following. Express the answers using interval notation.

(a) $\left|\frac{3 - 7x}{2}\right| \le 12$ **(b)** $\left|\frac{2}{3}x + 1\right| > 5$ **(c)** $|9x + 1| + 5 < 1$

Solution

(a) $\left|\frac{3 - 7x}{2}\right| \le 12$ *This translates into*

$$-12 \le \frac{3 - 7x}{2} \le 12$$ *Multiply each member by 2 to get*

$$-24 \le 3 - 7x \le 24$$ *Add −3 to each member.*

$$-27 \le \quad -7x \le 21$$ *Then divide each member by −7.*

$$\frac{27}{7} \ge x \ge -3$$ *Notice the inequality symbols are reversed.*

Using interval notation, we get $\boxed{\left[-3, \frac{27}{7}\right]}$.

(b) $\left|\frac{2}{3}x + 1\right| > 5$ *This translates into*

$$\frac{2}{3}x + 1 < -5 \quad \text{or} \quad \frac{2}{3}x + 1 > 5$$ *Solve each inequality:*

$$2x + 3 < -15 \quad \text{or} \quad 2x + 3 > 15$$

$$x < -9 \quad \text{or} \quad x > 6$$ *which is* $\boxed{(-\infty, -9) \cup (6, \infty)}$

(c) $|9x + 1| + 5 < 1$ *First isolate the absolute value by adding −5 to each side of the inequality.*

$$|9x + 1| < -4$$

At this point we can stop; the inequality states that a quantity in absolute value is less than a negative number. Since absolute values are always nonnegative, this is impossible, and so the answer is: $\boxed{\text{no solution}}$. ■

DIFFERENT PERSPECTIVES:
Absolute Value Equations and Inequalities

GEOMETRIC DESCRIPTION

The absolute value of x, $|x|$ is its distance, from 0 on the number line. For $a > 0$,

$|x| = a$ means x is a units from 0 on the number line:

$|x| < a$ means x is less than a units from 0 on the number line:

$|x| > a$ means x is more than a units from 0 on the number line:

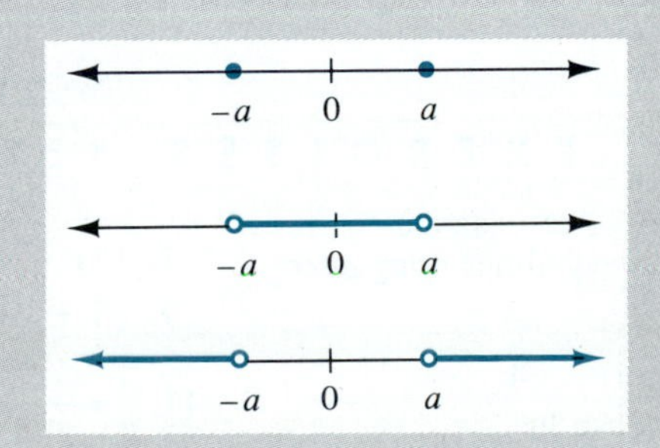

ALGEBRAIC DESCRIPTION

For $a > 0$,

$|x| = a \Leftrightarrow x = a$ or $x = -a$

$|x| < a \Leftrightarrow -a < x < a$

$|x| > a \Leftrightarrow x < -a$ or $x > a$

We complete this section with a few more absolute value properties.

MORE ABSOLUTE VALUE PROPERTIES

1. $|a| \geq 0$
2. $|ab| = |a||b|$
3. $\left|\dfrac{a}{b}\right| = \dfrac{|a|}{|b|}$
4. $|a|^2 = |a^2| = a^2$
5. $|a + b| \leq |a| + |b|$ *(The triangle inequality)*
6. $|a| - |b| \leq |a - b|$

EXAMPLE 7 Using the preceding properties, show that we can write $\left|\dfrac{-2x^2 - 3}{3}\right|$ as a single expression without absolute value symbols

Solution

$$\left|\frac{-2x^2 - 3}{3}\right| = \frac{|-2x^2 - 3|}{|3|} \qquad \textit{By property 3}$$

$$= \frac{|(-1)(2x^2 + 3)|}{|3|} \qquad \textit{Factoring out } -1 \textit{ from } -2x^2 - 3$$

$$= \frac{|-1||2x^2 + 3|}{|3|} \qquad \textit{By property 2}$$

At this point we know that $|-1| = 1$ and $|3| = 3$; since $2x^2 + 3$ is always positive (why?), we conclude:

$$\left|\frac{-2x^2 - 3}{3}\right| = \boxed{\frac{2x^2 + 3}{3}}$$

■

EXERCISES 1.7

In Exercises 1–34, solve the absolute value equations or inequalities. Express the solutions to the inequalities using interval notation.

1. $|x| = 12$
2. $|x| = -4$
3. $|x| < 9$
4. $|x| \geq 9$
5. $|x - 8| \leq 2$
6. $|x + 3| > 5$
7. $|2x - 4| = 7$
8. $|5 - 3x| = 8$
9. $|5 + 2x| < 3$
10. $|7 + 3x| \geq 4$
11. $\left|x - \dfrac{2}{3}\right| > \dfrac{2}{3}$
12. $\left|x - \dfrac{1}{5}\right| \leq \dfrac{3}{5}$
13. $|3 - 2x| < 3$
14. $|4 - 3x| > 5$

15. $\left|\frac{x}{3}+1\right|=\frac{3}{4}$

16. $\left|1-\frac{x}{5}\right|=\frac{4}{3}$

17. $\left|\frac{x}{2}-\frac{2}{3}\right|>\frac{1}{2}$

18. $\left|\frac{x}{5}+\frac{1}{2}\right|<\frac{3}{5}$

19. $\left|\frac{x}{2}-\frac{x}{3}\right|\geq\frac{1}{2}$

20. $\left|\frac{x}{5}+\frac{x}{2}\right|<\frac{1}{2}$

21. $\left|\frac{2x-3}{2}\right|=4$

22. $\left|\frac{1-3x}{5}\right|<6$

23. $\left|\frac{5-x}{5}\right|=\frac{1}{2}$

24. $\left|\frac{3-5x}{2}\right|\leq\frac{1}{2}$

25. $|3x-2|=|-5|$

26. $|2-7x|=|-2|$

27. $|3x-4|=|x|$

28. $|2x-3|=|x-5|$

29. $|3x-2|+3=6$

30. $|1-5x|-4=7$

31. $|5-3x|-4<8$

32. $|2-3x|+5>9$

33. $|3x-2|+3\leq 1$

34. $|1-5x|+4\geq 2$

In Exercises 35–42, rewrite the following as a single expression without absolute value symbols, *if possible*.

35. $|x^2+8|$

36. $|3x^2+10|$

37. $|-4x^2-9|$

38. $|-5-3x^4|$

39. $|3x^3+3|$

40. $|2x^2-5|$

41. $\left|\frac{-5-x^6}{4}\right|$

42. $\left|\frac{7x^4+5}{-3}\right|$

43. Prove that if $\left|\frac{1}{a}\right|<\left|\frac{1}{b}\right|$ and $a, b\neq 0$, then $|b|<|a|$.

QUESTION FOR THOUGHT

44. Prove: $|x|-|y|\leq|x-y|$. HINT: Start with the triangle inequality (property 5). Substitute $x+y$ for a and $-y$ for b.

1.8 Quadratic Equations and Equations in Quadratic Form

Quadratic Equations

A second-degree equation is a polynomial equation in which the highest degree of the variable is 2. In particular, a second degree equation in one unknown is called a **quadratic equation.** We define the **standard form** of a quadratic equation as $Ax^2+Bx+C=0$, where $A\neq 0$.

As with all other equations, the solutions for quadratic equations are values of the variable that make the equation a true statement. The solutions of $Ax^2+Bx+C=0$ are also called the *roots of the polynomial equation* $Ax^2+Bx+C=0$.

THE ZERO-PRODUCT RULE

If $a\cdot b=0$, then $a=0$ or $b=0$.

In solving the equation $Ax^2+Bx+C=0$, if the polynomial Ax^2+Bx+C can be factored, then we can use the zero-product rule to reduce the problem to solving two linear equations. For example, to solve the equation: $6x^2-x-2=0$, we can factor the left-hand side to get $(3x-2)(2x+1)=0$. Hence we can conclude that $3x-2=0$ or $2x+1=0$, which yields $x=\frac{2}{3}$ or $x=-\frac{1}{2}$.

Another method is to apply the square Root Theorem.

THE SQUARE ROOT THEOREM

If $x^2 = d$, then $x = \pm\sqrt{d}$.

This theorem is usually applied to solve quadratic equations that have no first-degree term. For example, in solving the equation $x^2 = 3$, we can apply this theorem to get $x = \pm\sqrt{3}$.

EXAMPLE 1 Solve the following.

(a) $4x^2 + 10x = 6$ **(b)** $\dfrac{y}{y-5} - \dfrac{3}{y+1} = \dfrac{30}{y^2 - 4y - 5}$

(c) $5x^2 - 6 = 8$ **(d)** $(x - 2)^2 = 6$

Solution

(a)

$$4x^2 + 10x = 6$$ *Put into standard form.*

$$4x^2 + 10x - 6 = 0$$ *Factor the left-hand side.*

$$2(2x - 1)(x + 3) = 0$$ *Hence we have*

$$2x - 1 = 0 \quad \text{or} \quad x + 3 = 0$$ *Solving each linear equation, we get*

$$\boxed{x = \frac{1}{2} \quad \text{or} \quad x = -3}$$

We have ignored the factor of 2 when we applied the zero product rule. What allows us to do that?

(b) $$\frac{y}{y-5} - \frac{3}{y+1} = \frac{30}{y^2 - 4y - 5}$$ *Noting $y^2 - 4y - 5 = (y - 5)(y + 1)$, we multiply both sides by the LCD of all the fractions: $(y - 5)(y + 1)$.*

$$\frac{y}{y-5}(y - 5)(y + 1) - \frac{3}{y+1}(y - 5)(y + 1) = \frac{30}{(y - 5)(y + 1)}(y - 5)(y + 1)$$

This yields

$$y(y + 1) - 3(y - 5) = 30$$ *Simplify each side and put into standard form.*

$$y^2 - 2y - 15 = 0$$ *Factor*

$$(y - 5)(y + 3) = 0$$ *This yields*

$$y = 5 \quad \text{or} \quad y = -3$$

In checking the solutions, we find that we must eliminate $y = 5$ as a possible solution as it leads to an undefined expression. Hence the solution is $\boxed{y = -3}$

(c) We note that there is no first degree term, so our approach will be to apply the Square Root Theorem.

$5x^2 - 6 = 8$ — *Isolate x^2 on the left-hand side before applying the Square Root Theorem.*

$5x^2 = 14$

$x^2 = \frac{14}{5}$ — *Applying the Square Root Theorem, we get*

$x = \pm\sqrt{\frac{14}{5}}$ — *or (with the denominator rationalized)* $x = \boxed{\pm\frac{\sqrt{70}}{5}}$

(d) We could multiply out the left-hand side and put it in standard form; however, we would find that the quadratic expression in standard form does not factor with integer coefficients. Since it is in the form of a squared quantity equal to a number, we will apply the Square Root Theorem first.

Try solving $(x - 2)^2 = 6$ by multiplying out the left-hand side first and using the factoring method.

$(x - 2)^2 = 6$ — *Apply the Square Root Theorem.*

$x - 2 = \pm\sqrt{6}$ — *Isolate x (add 2 to both sides).*

$x = \boxed{2 \pm \sqrt{6}}$ ■

Part (d) of Example 1 illustrates that if we can construct a perfect square binomial from a quadratic equation (i.e., get the equation in the form $(x + p)^2 = d$), then we can apply the Square Root Theorem and solve for x to get $x = -p \pm \sqrt{d}$.

The method of constructing a perfect square is called **completing the square.** It is based on the fact that in multiplying out the perfect square $(x + p)^2$, with p a constant, we get

$$(x + p)^2 = x^2 + 2px + p^2$$

Notice the relationship between the constant term, p^2, and the coefficient of the middle term, $2p$: The constant term is the square of half the coefficient of the middle term, or

$$\left[\frac{1}{2}(2p)\right]^2 = p^2$$

Suppose we started out with the equation $x^2 + 6x - 5 = 3$.

$x^2 + 6x - 5 = 3$ — *Let's rewrite this equation as*

$x^2 + 6x \quad = 8$ — *Then we make the left-hand side a perfect square by adding $\left[\frac{1}{2}(6)\right]^2 = 9$ to both sides of the equation to get*

$x^2 + 6x + 9 = 8 + 9$	*Next, we rewrite this equation as*
$(x + 3)^2 = 17$	*We then apply the Square Root Theorem to get*
$x + 3 = \pm\sqrt{17}$	*Finally, we isolate x.*
$x = -3 \pm \sqrt{17}$	

EXAMPLE 2 Solve by completing the square: $2x^2 - 8x + 4 = 6$.

Solution We note that the leading coefficient (in this case, the coefficient of x^2) is *not* 1, and completing the square is based upon squaring a binomial with the resultant leading coefficient being 1. Thus our first step is to make the leading coefficient 1.

$2x^2 - 8x + 4 = 6$	*Divide both sides by 2, the coefficient of* x^2.
$x^2 - 4x + 2 = 3$	*Isolate the constant term on the right-hand side. (This step makes it clearer to see how we complete the square on the left-hand side.)*
$x^2 - 4x \quad = 1$	*Take half the middle term coefficient, square it* $\left(\left[\frac{1}{2}(4)\right]^2 = 4\right)$: *we add 4 to both sides of the equation.*
$x^2 - 4x + 4 = 1 + 4$	*Factor the left-hand side and simplify the right-hand side.*
$(x - 2)^2 = 5$	*Solve for x using the Square Root Theorem.*
$x - 2 = \pm\sqrt{5}$	*Isolate x.*
$x = \boxed{2 \pm \sqrt{5}}$	

■

Solve for x by completing the square for the equation $Ax^2 + Bx + C = 0$.

Unlike the factoring method, all quadratic equations can be solved by completing the square. If we were to complete the square for the general quadratic equation $Ax^2 + Bx + C = 0$, $A \neq 0$, we would arrive at the formula given below.

THE QUADRATIC FORMULA

If $Ax^2 + Bx + C = 0$ and $A \neq 0$, then

$$x = \frac{-B \pm \sqrt{B^2 - 4AC}}{2A}$$

EXAMPLE 3 Solve the following using the quadratic formula: $2x^2 - 2x = -8$.

Solution Before using the quadratic formula, we put the equation in standard form and identify A, B, and C as follows: $2x^2 - 2x + 8 = 0$. Hence $A = 2$, $B = -2$, and $C = 8$. By the quadratic formula, we get

$$x = \frac{-(-2) \pm \sqrt{(-2)^2 - 4(2)(8)}}{2(2)} = \frac{2 \pm \sqrt{-60}}{4}$$

$$= \frac{2 \pm 2i\sqrt{15}}{4} \qquad \textit{Factor and reduce.}$$

$$= \frac{2(1 \pm i\sqrt{15})}{4} = \boxed{\frac{1 \pm i\sqrt{15}}{2}} \qquad \textit{There are no real solutions.}$$

■

EXAMPLE 4 A 20-ft by 55-ft rectangular swimming pool is surrounded by a concrete walkway of uniform width. If the area of the concrete walkway is 400 sq ft, find the width of the walkway.

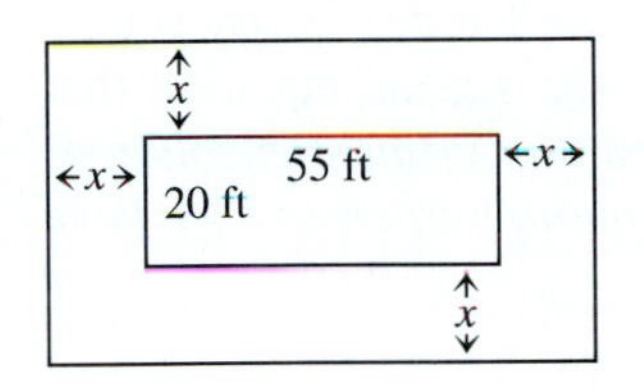

FIGURE 1.15

Solution We can draw a diagram of the pool and walkway, labeling the uniform width of the concrete walkway x as pictured in Figure 1.15.

Looking at Figure 1.15 we can see that the length of the outer rectangle is $55 + x + x = 55 + 2x$ feet, and the width of the outer rectangle is $20 + x + x = 20 + 2x$ feet. Therefore, the area of the outer rectangle (the pool *and* walkway) is given by $(55 + 2x)(20 + 2x)$. Since the area of the pool is $20 \times 55 = 1100$ sq ft, and the area of the walkway is given as 400 sq ft, the diagram shows us that

$$\underbrace{1100}_{\text{Area of pool}} + \underbrace{400}_{\text{Area of walkway}} = \underbrace{(55 + 2x)(20 + 2x)}_{\text{Area of outer rectangle}}$$

Hence our equation is

$$1100 + 400 = (55 + 2x)(20 + 2x) \qquad \textit{Which becomes}$$

$$1500 = 1100 + 150x + 4x^2 \qquad \textit{And is put into standard form as}$$

$$0 = 4x^2 + 150x - 400 \qquad \textit{Solve by factoring:}$$

$$0 = 2(2x - 5)(x + 40) \qquad \textit{Hence}$$

$$x = \frac{5}{2} \quad \text{or} \quad x = -40 \qquad \textit{Eliminate the negative answer.}$$

The width of the walkway is $\frac{5}{2} = \boxed{2\frac{1}{2} \text{ feet}}$.

■

Radical Equations: Equations Reducible to Quadratic Form

The multiplicative property of equality yields the following theorem:

THEOREM 1.1 If $a = b$, then $a^n = b^n$.

If we want to solve an equation containing a radical or fractional exponent such as $x^{1/3} = 5$, or $\sqrt[3]{x} = 5$, then we can apply Theorem 1.1 by raising each side of the equation to the third power:

$x^{1/3} = 5$ *Raise each side to the third power (the power that will yield x^1).*

$(x^{1/3})^3 = (5)^3$ *By the second rule of exponents, $(x^{1/3})^3 = x^{(1/3)\cdot 3} = x^1$.*

$x = 5^3 = 125$

We should note some things about using this theorem before we continue. First, using this theorem (i.e., raising each side of an equation to the same power) will *not* necessarily yield an equivalent equation, as was true with the other equality properties. This theorem guarantees only that solutions to the first equation will show up as solutions to the transformed equation.

For example, if we start with the equation $x = 2$, we observe that there is only one solution to this equation, 2. However, if we square each side of this equation, we will arrive at $x^2 = 4$, which has two solutions: $x = +2$ and $x = -2$. By squaring both sides of the equation, we picked up a solution to the second equation that was not a solution to the first. This extra solution is called an *extraneous solution*. Thus, we must always *check our solutions in the original equation to ensure we have not picked up extraneous solutions*.

EXAMPLE 5 Solve the following.

(a) $\sqrt{3x + 1} + 3 = x$ **(b)** $(x - 2)^{1/5} = 3$

Solution

(a) If we tried squaring both sides now and did it properly, the new equation would then be $3x + 1 + 6\sqrt{3x + 1} + 9 = x^2$. (Note that we still have a radical in the transformed equation.) Our first step should be to isolate the radical:

$\sqrt{3x + 1} + 3 = x$ *Isolate $\sqrt{3x + 1}$.*

$\sqrt{3x + 1} = x - 3$ *Then square both sides of the equation.*

$(\sqrt{3x + 1})^2 = (x - 3)^2$ *This gives*

$3x + 1 = x^2 - 6x + 9$ *Put this quadratic equation in standard form.*

$0 = x^2 - 9x + 8$ *Solve by factoring.*

$0 = (x - 1)(x - 8)$ *Hence*

$x = 1$ or $x = 8$

You must check the answers in the original equation. In this case, we find that $x = 1$ does not work, but 8 does check. The answer is $\boxed{8}$.

(b) $(x - 2)^{1/5} = 3$ *Raise each side to the fifth power.*

$[(x - 2)^{1/5}]^5 = 3^5$ *This gives*

$x - 2 = 243$ *Hence*

$x = \boxed{245}$

Again, if you check you will find that this answer is a solution to the original equation. ■

The equation $4x^{2/3} - 9x^{1/3} + 2 = 0$ is not a quadratic equation; however, we can use the methods discussed in this section to solve such an equation. We use the method of **substitution of variables** as follows:

Let $u = x^{1/3}$. Then $u^2 = (x^{1/3})^2 = x^{2/3}$.

By substituting u for $x^{1/3}$ and u^2 for $x^{2/3}$,

$$4x^{2/3} - 9x^{1/3} + 2 = 0$$ *Becomes*

$$4u^2 - 9u + 2 = 0$$ *This equation can be solved for u by factoring.*

$$(4u - 1)(u - 2) = 0$$ *Hence*

$$u = \frac{1}{4} \quad \text{or} \quad u = 2$$ *But we must solve for x. Substitute back $x^{1/3}$ for u and get*

$$x^{1/3} = \frac{1}{4} \quad \text{or} \quad x^{1/3} = 2$$ *Cubing both sides of each equation yields*

$$(x^{1/3})^3 = \left(\frac{1}{4}\right)^3 \quad \text{or} \quad (x^{1/3})^3 = (2)^3$$

$$x = \frac{1}{64} \quad \text{or} \quad x = 8$$

The equation $4x^{2/3} - 9x^{1/3} + 2 = 0$ is said to be *reducible to quadratic form.* The pattern to watch for is that, ignoring the coefficients, one of the variable expressions should be the square of the other.

The zero-product rule can be generalized to solve higher-degree polynomial equations as demonstrated below.

EXAMPLE 6 Solve the following:

(a) $x^4 + x^2 - 6 = 0$ **(b)** $x^3 + 4x^2 - x - 4 = 0$

Solution

(a)

$$x^4 + x^2 - 6 = 0$$ *Factor the left-hand expression.*

$$(x^2 + 3)(x^2 - 2) = 0$$ *Set each factor equal to 0.*

$$x^2 + 3 = 0 \quad \text{or} \quad x^2 - 2 = 0$$ *Isolate x^2 for each equation.*

$$x^2 = -3 \quad \text{or} \quad x^2 = 2$$ *Take square roots.*

$$x = \pm\sqrt{-3} \quad \text{or} \quad x = \pm\sqrt{2}$$

$$x = \pm i\sqrt{3} \quad \text{or} \quad x = \pm\sqrt{2}$$

$$\boxed{x = \pm i\sqrt{3}, \pm\sqrt{2}}$$

(b)

$$x^3 + 4x^2 - x - 4 = 0 \quad \textit{Factor the left-hand side by grouping.}$$

$$x^2(x + 4) - (x + 4) = 0 \quad \textit{Factor } x + 4 \textit{ from the left-hand side.}$$

$$(x^2 - 1)(x + 4) = 0 \quad \textit{We can still factor } x^2 - 1.$$

$$(x - 1)(x + 1)(x + 4) = 0 \quad \textit{Set each factor equal to 0 and get}$$

$$\boxed{x = 1, -1, \text{ or } -4}$$

■

EXAMPLE 7 Solve the following: $\dfrac{(x - 5)^{1/2}(x + 3)^{1/5}}{(x - 2)^{1/3}} = 0.$

Solution This is a rational expression equal to 0. Provided we keep in mind that $x \neq 2$ (that is, the denominator cannot be 0), we can start to solve this equation by multiplying both sides of the equation by $(x - 2)^{1/3}$. Alternatively, we can simply recognize that the only way a fraction can be equal to 0 is if its numerator is 0 (and its denominator is not 0). Hence if $x \neq 2$,

$$\frac{(x - 5)^{1/2}(x + 3)^{1/5}}{(x - 2)^{1/3}} = 0 \quad \textit{Becomes}$$

$$(x - 5)^{1/2}(x + 3)^{1/5} = 0 \quad \textit{Since the product is equal to 0, we set each factor equal to 0 to get}$$

$$(x - 5)^{1/2} = 0 \quad \text{or} \quad (x + 3)^{1/5} = 0 \quad \textit{Which yield}$$

$$\boxed{x = 5 \qquad \text{or} \qquad x = -3}$$

■

Show that if $q \neq 0$, then the rational expression $\dfrac{p}{q} = 0$ if and only if $p = 0$.

EXERCISES 1.8

In Exercises 1–6, solve for the variable by factoring or using the Square Root Theorem.

1. $0 = y^2 - 65$

2. $x^2 + 7 = 2$

3. $2x^2 - 7x = 15$

4. $2a^2 - 12 = -5a$

5. $x - 3 = \dfrac{1}{x + 3}$

6. $\dfrac{3}{x + 4} + \dfrac{5}{x + 2} = 6$

In Exercises 7–10, solve by completing the square.

7. $x^2 + 2x - 4 = 0$

8. $2a^2 + 4a - 3 = 0$

9. $\dfrac{1}{a - 5} + \dfrac{3}{a + 2} = 4$

10. $3a^2 - 6a + 5 = 0$

In Exercises 11–16, solve by any algebraic method.

11. $(x + 5)(x - 7) = 3$

12. $(t + 3)(t - 4) = t(t + 2)$

13. $\dfrac{3}{x - 2} = x$

14. $\dfrac{1}{x + 2} = x - 4$

15. $\dfrac{x}{x + 2} - \dfrac{3}{x} = \dfrac{x + 1}{x}$

16. $\dfrac{3}{x - 2} + \dfrac{7}{x + 2} = \dfrac{x + 1}{x - 2}$

17. Given $s_e = s_y\sqrt{1 - r_{xy}^2}$, if $s_e = 1.24$ and $r_{xy} = 0.63$, determine s_y to two places.

18. Given $s_e = s_y\sqrt{1 - r_{xy}^2}$, if $s_e = 1.72$ and $s_y = 3.73$, find r_{xy} to two places.

19. Given $Z = \dfrac{Z_{r_1} - Z_{r_2}}{\sqrt{\dfrac{1}{n_1 - 3} + \dfrac{1}{n_2 - 3}}}$

determine the whole number n_1 for $Z = 0.8$, $Z_{r_1} = 0.62$, $Z_{r_2} = 0.52$, and $n_2 = 83$.

20. Given $Z = \dfrac{Z_{r_1} - Z_{r_2}}{\sqrt{\dfrac{1}{n_1 - 3} + \dfrac{1}{n_2 - 3}}}$, find the whole number n_2 for $Z = 1.2$, $Z_{r_1} = 0.85$, $Z_{r_2} = 0.45$, and $n_1 = 14$.

21. Given $t = \dfrac{\overline{X} - a}{\dfrac{s_x}{\sqrt{n}}}$, determine the whole number n for $t = 3.2$, $a = 68$, $\overline{X} = 70$, and $s_x = 3.4$.

22. Given $t = \dfrac{\overline{X} - a}{\dfrac{s_x}{\sqrt{n}}}$, determine $\overline{X}$ to two places for $a = 65$, $t = 1.3$, $s_x = 3.4$, and $n = 50$.

In Exercises 23–28, solve for the given variable.

23. $K = \dfrac{2gm}{s^2}$ for s $(s > 0)$

24. $V = \dfrac{2}{3}\pi r^2$ for r $(r > 0)$

25. $3a^2 + 2b = 5b - 2a^2x - 2$ for a

26. $x^2 - 6xy + 5y^2 = 0$ for y

27. $s_e = s_y\sqrt{1 - r_{xy}^2}$ for s_y $(s_y > 0)$

28. $s_e = s_y\sqrt{1 - r_{xy}^2}$ for r_{xy}

29. The sum of a number and its reciprocal is $\dfrac{13}{6}$. Find the numbers.

30. The product of two numbers is 5. If their sum is $\dfrac{9}{2}$, find the numbers.

31. The length of a rectangle is 5 more than twice its width. If its area is 75 sq ft, find its dimensions.

32. Find the area of a square if its diagonal is 8 inches.

33. A 21-ft by 21-ft square swimming pool is surrounded by a path of uniform width. If the area of the path is 184 sq ft, find the width of the path.

34. A circular garden is surrounded by a path of uniform width. If the path has area 57π sq ft and the radius of the garden is 8 feet, find the width of the path.

In Exercises 35–48, solve for the variable.

35. $\sqrt{x - 2} = x - 4$

36. $\sqrt{3x + 1} + 3 = x$

37. $\sqrt{7x + 1} - 2\sqrt{x} = 2$

38. $\sqrt{3s + 4} - \sqrt{s} = 2$

39. $x^4 - 17x^2 + 16 = 0$

40. $x^3 - 2x^2 - 15x = 0$

41. $x^{1/2} + 8x^{1/4} + 7 = 0$

42. $x^{-2} - 2x^{-1} - 15 = 0$

43. $\sqrt{a} - \sqrt[4]{a} - 6 = 0$

44. $\left(x + \dfrac{12}{x}\right)^2 - 15\left(x + \dfrac{12}{x}\right) + 56 = 0$

45. $x^3 - 9x + 4x^2 - 36 = 0$

46. $x^4 - 81 = 0$

47. $\dfrac{(x - 3)^{1/3}}{(x + 7)^{2/3}} = 0$

48. $\dfrac{(x + 1)^{1/5}(x - 6)^{2/3}}{(x - 1)^{1/2}} = 0$

49. Verify that $2 + i$ is a solution to $2x^2 - 8x + 10 = 0$.

50. Verify that $2 + \sqrt{3}$ is a solution to $x^2 - 4x + 1 = 0$.

QUESTIONS FOR THOUGHT

51. Complete the square for the general quadratic equation $Ax^2 + Bx + C = 0$ $(A \neq 0)$ to verify the quadratic formula.

52. Find an equation whose roots are the following.
(a) 3 and 5 **(b)** 4 and -2 **(c)** $3 - i$ and $3 + i$

1.9 Quadratic and Rational Inequalities

Quadratic Inequalities

A quadratic inequality is in **standard form** if it is in the form $Ax^2 + Bx + C < 0$. (We can replace $<$ with $>$, $\leq$, and $\geq$.)

If we keep in mind that $u > 0$ means u is positive, then solving an inequality such as $2x^2 + 5x - 3 > 0$ means we are interested in finding the values of x that will make $2x^2 + 5x - 3$ positive. Or, since $2x^2 + 5x - 3 = (2x - 1)(x + 3)$, we are looking for values of x that make $(2x - 1)(x + 3)$ positive.

Although we will refer to the real number line, our approach in solving quadratic inequalities will be primarily algebraic. After putting the inequality in standard form, we will determine the sign of each factor of the expression for various values of x. Then we determine the solution by examining the sign of the product. This process is called a *sign analysis*.

Returning to the problem $2x^2 + 5x - 3 > 0$: This translates into finding values of x that make $(2x - 1)(x + 3)$ positive. In order for $(2x - 1)(x + 3)$ to be positive, the factors must be either both positive or both negative. To determine when this happens, we first find the values of x for which $(2x - 1)(x + 3)$ is equal to 0; we call these the cut points of $(2x - 1)(x + 3)$. The cut points are $\frac{1}{2}$ and -3.

Next we draw a number line and examine the sign of each factor as x takes on various values on the number line, especially around the cut points $\frac{1}{2}$ and -3.

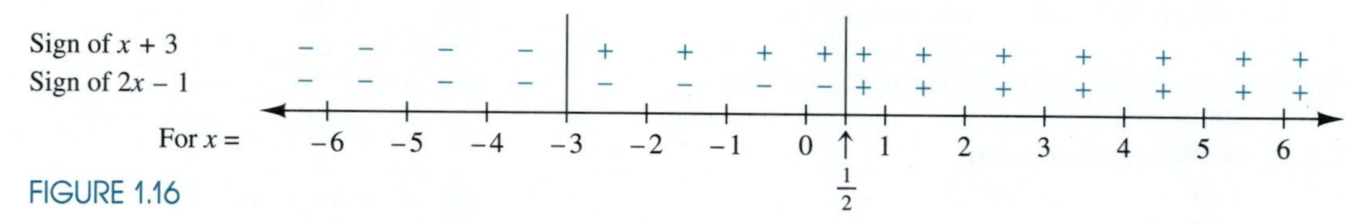

FIGURE 1.16

Figure 1.16 illustrates that the factor $x + 3$ is negative when $x < -3$, and positive when $x > -3$. It also shows that $2x - 1$ is negative for $x < \frac{1}{2}$ and positive for $x > \frac{1}{2}$. Looking at the figure and the signs of the factors, we easily observe that the *product* of the two factors is positive when $x < -3$ or $x > \frac{1}{2}$. See Figure 1.17.

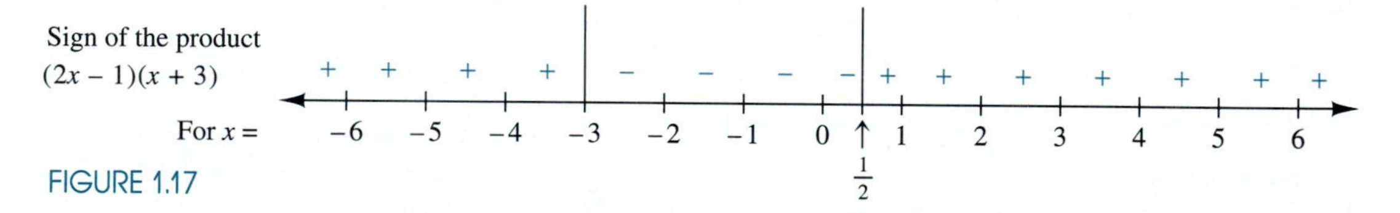

FIGURE 1.17

Hence $2x^2 + 5x - 3 > 0$ (is positive) when x is in either of the two intervals $(-\infty, -3)$ or $(\frac{1}{2}, \infty)$ so the solution is $(-\infty, -3) \cup (\frac{1}{2}, \infty)$.

We can shortcut this analysis by observing that (1) the cut points of the inequalities will break up the number line into intervals, and (2) most importantly, the sign of the product does not change *within* an interval (excluding the cut points); i.e., *if the expression is positive (or negative) for one value within the interval, it is positive (or negative) for all values within the interval.*

Hence, if we want to find the sign of the product within each interval, all we need to do is to test one value within the interval. For example, we found that the cut points for the expression $2x^2 + 5x - 3 = (2x - 1)(x + 3)$ are -3 and $\frac{1}{2}$. This breaks up the number line into three intervals, as shown in Figure 1.18.

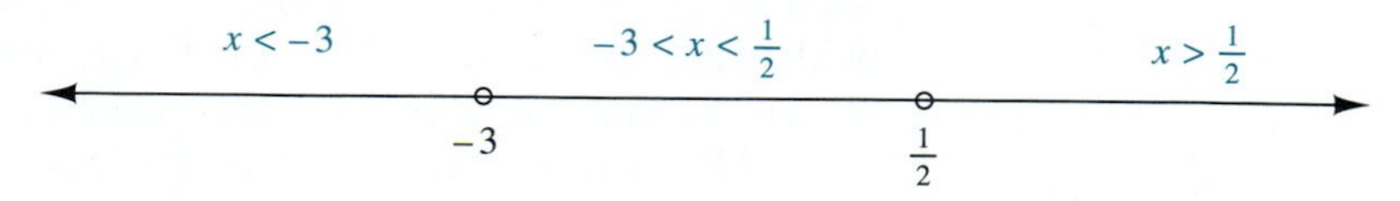

FIGURE 1.18

Review the preceding sign analysis and think about why the product never changes its sign within an interval.

We choose any value in the leftmost interval (i.e., any value less than -3), say, $x = -10$. When $x = -10$, the expression $(2x - 1)(x + 3) = [2(-10) - 1][(-10) + 3]$ is positive and therefore $(2x - 1)(x + 3)$ is positive for *all* values of x in the interval $(-\infty, -3)$.

We choose any value in the middle interval, $\left(-3, \frac{1}{2}\right)$, such as $x = 0$, and note that when $x = 0$ the expression $(2x - 1)(x + 3) = [2(0) - 1][0 + 3]$ is negative; therefore, $(2x - 1)(x + 3)$ is negative for *all* values of x in $\left(-3, \frac{1}{2}\right)$.

Finally, choose any value in the interval $\left(\frac{1}{2}, \infty\right)$, such as $x = 100$, and note that for $x = 100$, the expression $(2x - 1)(x + 3)$ is positive; therefore, $(2x - 1)(x + 3)$ is positive for *all* values of x in $\left(\frac{1}{2}, \infty\right)$. (Note we are not interested in the exact value but only in the *sign* of the evaluated expression.)

Hence the solution is: $(-\infty, -3) \cup \left(\frac{1}{2}, \infty\right)$ since $2x^2 + 5x - 3$ is positive when x lies in these intervals.

On the other hand, the identical analysis also yields the solution to $2x^2 + 5x - 3 < 0$. That is, $2x^2 + 5x - 3$ is negative when x is in the interval $(-3, \frac{1}{2})$.

EXAMPLE 1 Solve: $3x^2 - 7x \leq -2$.

Solution

$3x^2 - 7x \leq -2$	*Put into standard form.*
$3x^2 - 7x + 2 \leq 0$	*Factor.*
$(3x - 1)(x - 2) \leq 0$	*Find the cut points [find x when $(3x - 1)(x - 2) = 0$]. The cut points are $\frac{1}{3}$ and 2.*

Plot the cut points on the number line in Figure 1.19, choose any value within each interval determined by the cut points, and check the sign of the product. We used $x = -10$, $x = 1$, and $x = 10$ as test values for the three intervals, substituted the test values for x into $(3x - 1)(x - 2)$, and found the product to be positive, negative, and positive respectively.

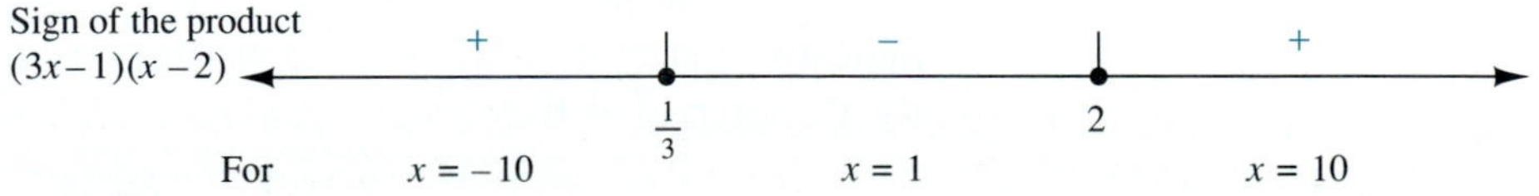

Sign of the product $(3x-1)(x-2)$: $+$ on $(-\infty, \frac{1}{3})$, $-$ on $(\frac{1}{3}, 2)$, $+$ on $(2, \infty)$

For $x = -10$, $x = 1$, $x = 10$

FIGURE 1.19

The product $(3x - 1)(x - 2)$ is negative for x in the interval $\left(\frac{1}{3}, 2\right)$, and since $(3x - 1)(x - 2) = 0$ at $x = \frac{1}{3}$ and $x = 2$, we include the endpoints in the solution of this inequality. Hence all solutions to the inequality $3x^2 - 7x \leq -2$ lie in the interval $\boxed{\left[\frac{1}{3}, 2\right]}$.

Note the endpoints $\frac{1}{3}$ and 2 are included in the interval.

■

EXAMPLE 2 Solve the following: $x^2 - 2x - 2 < 0$.

Solution Since we cannot factor $x^2 - 2x - 2$, we use the quadratic formula to find that its roots are $1 \pm \sqrt{3}$. This gives the cut points for the polynomial $x^2 - 2x - 2$. We use the sign analysis in Figure 1.20 with the test points given. NOTE: $1 + \sqrt{3} \approx 2.7$ and $1 - \sqrt{3} \approx -0.7$.

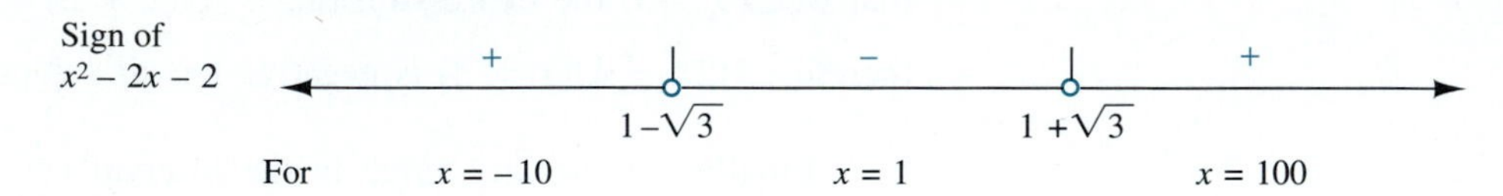

FIGURE 1.20

Substituting the test values -10, 1, and 100 for x in the expression $x^2 - 2x - 2$, we find that $x^2 - 2x - 2$ is negative only when x is in the interval $\boxed{(1 - \sqrt{3}, 1 + \sqrt{3})}$. ■

We apply the same approach to solving higher-degree polynomial inequalities.

EXAMPLE 3 Solve the following: $(x - 2)(x + 2)(x - 5) > 0$.

Solution The cutpoints are 2, -2, and, 5. We analyze the signs in the same way we analyzed quadratic inequalities. See Figure 1.21.

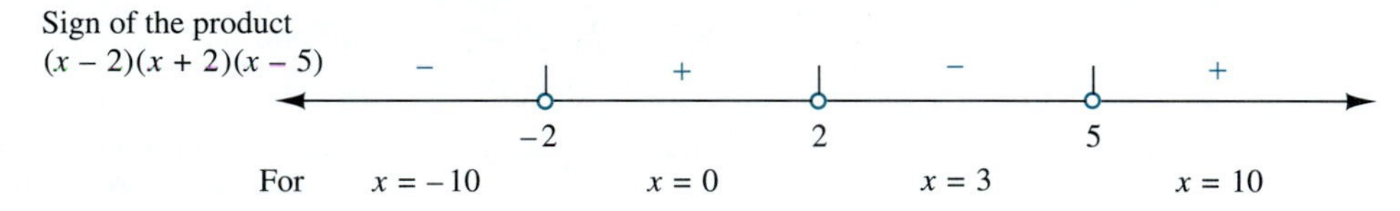

FIGURE 1.21

Hence the solution is $\boxed{(-2, 2) \cup (5, \infty)}$. ■

EXAMPLE 4 For what values of x is $\sqrt{x^2 - x - 2}$ real?

Solution We first note that $\sqrt{x^2 - x - 2}$ is real when the expression under the radical is nonnegative. Hence, the problem reduces to solving the quadratic inequality $x^2 - x - 2 \geq 0$.

$$x^2 - x - 2 \geq 0 \quad \textit{Factor}$$
$$(x - 2)(x + 1) \geq 0 \quad \textit{The cut points are 2 and } -1.$$

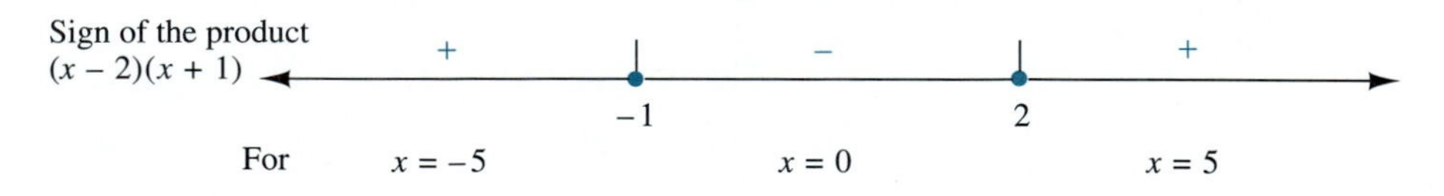

FIGURE 1.22

On the real number line in Figure 1.22 we have the needed information. Hence $x^2 - x - 2 \geq 0$, or $\sqrt{x^2 - x - 2}$ is real, when x is in the interval $\boxed{(-\infty, -1] \cup [2, \infty)}$. ■

Rational Inequalities

We can apply this same method in solving rational inequalities such as

$$\frac{x-4}{x-3} > 0 \quad \text{or} \quad \frac{2x-1}{x-5} \leq 2$$

These inequalities are different than the linear inequalities with constants in the denominator. In attempting to solve $\frac{x-4}{x-3} > 0$, our first inclination might be to clear the denominator by multiplying both sides of the inequality by $x - 3$. With *equations* this approach is appropriate provided we keep in mind $x \neq 3$. With inequalities, however, we need to know whether the multiplier, $x - 3$, is positive or negative in order to determine whether or not the inequality symbol should be reversed. Although we could reason out the answer this way on a case-by-case basis, it is easier to approach this problem in a manner similar to that used for quadratic inequalities, as illustrated in the next example.

EXAMPLE 5 Solve the following. **(a)** $\frac{x-4}{x-3} > 0$ **(b)** $\frac{2x-1}{x-5} \leq 2$

Solution

(a) To solve $\frac{x-4}{x-3} > 0$, we need to find the values of x that make $\frac{x-4}{x-3}$ positive.

In order for this quotient to be positive, the numerator and denominator must have the same sign. ***We now define the cut points of a rational expression to be the value(s) where the numerator is 0 or where the denominator is 0.***

$\frac{x-4}{x-3} > 0$ *The cut points are 4 (where the numerator is 0) and 3 (where the denominator is 0). We plot the cut points on the number line and perform the sign analysis to get Figure 1.23.*

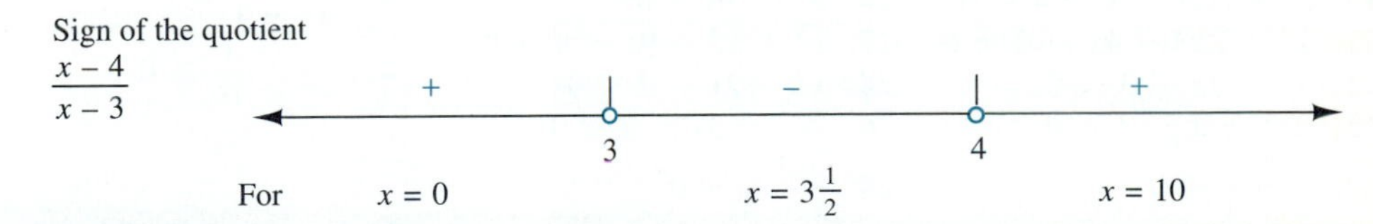

FIGURE 1.23

The solution is $\boxed{(-\infty, 3) \cup (4, \infty)}$.

(b) We must first put this expression in the form $R \le 0$, where R is a rational expression, before we can apply the sign analysis:

$$\frac{2x-1}{x-5} \le 2 \qquad \textit{Get 0 on the right-hand side.}$$

$$\frac{2x-1}{x-5} - 2 \le 0 \qquad \textit{Express the left-hand side as a single fraction.}$$

$$\frac{2x-1}{x-5} - \frac{2(x-5)}{x-5} \le 0$$

$$\frac{2x-1-2(x-5)}{x-5} \le 0 \qquad \textit{Which becomes}$$

$$\frac{9}{x-5} \le 0 \qquad \textit{There is only one cut point: 5. See Figure 1.24.}$$

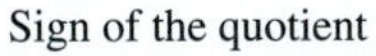

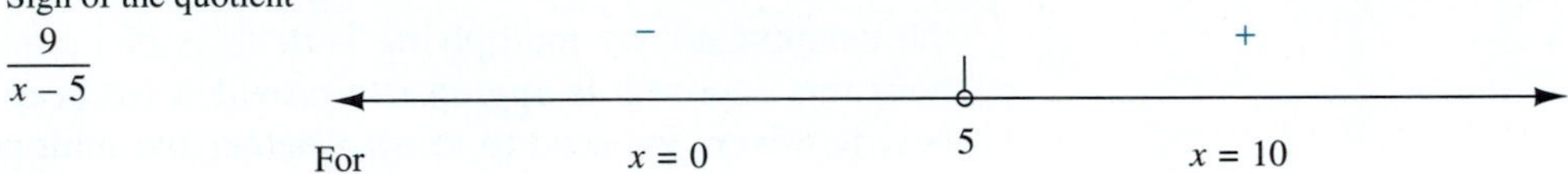

FIGURE 1.24

Notice we exclude $x = 5$. (Why?) Hence, the answer is $\boxed{(-\infty, 5)}$. ■

EXERCISES 1.9

In Exercises 1–50, solve the inequalities. Express your answer using interval notation.

1. $x^2 + 2x - 24 > 0$

2. $x^2 + 3x - 10 < 0$

3. $2x^2 - 3x - 2 < 0$

4. $5x^2 - 14x - 3 > 0$

5. $x^2 - 5x \le 24$

6. $x^2 - 10 \ge -1$

7. $2x^2 - 3x \ge 5$

8. $6x^2 - 5 \le -13x$

9. $15x^2 - 2 \le 7x$

10. $6y^2 + 1 \le 5y$

11. $8x^2 - 6 > -8x$

12. $6y^2 - 12 > -21y$

13. $9x^2 - 6x + 1 \ge 0$

14. $25x^2 \le -10x - 1$

15. $x^2 - 2x > -1$

16. $x^2 + 4x + 4 < 0$

17. $y^2 > 1$

18. $x^2 \le 16$

19. $x^3 + 4x^2 - x - 4 < 0$

20. $x^4 - 16 > 0$

21. $x^3 + 2x^2 - 4x - 8 \ge 0$

22. $x^3 + x^2 - 9x - 9 > 0$

23. $x^2 - 3x - 3 < 0$

24. $x^2 - 2x - 2 < 0$

25. $2x^2 - x - 2 \ge 0$

26. $2x^2 - 3x - 1 \ge 0$

27. $\dfrac{5}{x+1} < 0$

28. $\dfrac{3}{x-4} \le 0$

29. $\dfrac{2x-1}{x+5} \ge 0$

30. $\dfrac{x-7}{x+7} > 0$

31. $\dfrac{x-2}{x+1} < 0$

32. $\dfrac{2x+3}{x-5} \le 0$

33. $\dfrac{5x-1}{x-2} \ge 0$

34. $\dfrac{7x+3}{x-1} > 0$

35. $\dfrac{-2x}{x-1} \le 0$

36. $\dfrac{-4x}{x+1} \ge 0$

37. $\dfrac{3x}{x-1} < 1$

38. $\dfrac{4}{x-1} > 3$

39. $\dfrac{x-1}{x+1} \ge 2$

40. $\dfrac{2x-3}{x-3} \le 1$

41. $\dfrac{x-2}{(x+1)(x-3)} \ge 0$

42. $\dfrac{x+5}{(x-1)(2x+3)} \le 0$

43. $\dfrac{2x}{x^2-1} \le 0$

44. $\dfrac{5x}{x^2-16} \ge 0$

45. $\dfrac{x^2-9}{x+5} < 0$

46. $\dfrac{x^2-4}{x-3} > 0$

47. $\dfrac{-2x}{x^2+1} \le 0$

48. $\dfrac{5x}{x^2+16} \ge 0$

49. $\dfrac{a^2-1}{a^2-16} > 0$

50. $\dfrac{x^2-2x-3}{x^2+x-2} < 0$

In Exercises 51–54, given that $|a| < |b| \Leftrightarrow a^2 < b^2$, use this property to solve the following absolute value inequalities. Express your answer using interval notation.

51. $|2x-1| < |x|$

52. $|3x-1| > |x|$

53. $|x-3| < |x+2|$

54. $|2x+1| > |x+3|$

55. A biologist finds that the size of the population of a certain species of marine life is related to the temperature of the water at a particular location in the following way:

$$N = 1{,}000(-T^2 + 42T - 320)$$

where N is the size of the population and T is the water temperature in degrees Celsius. What must the water temperature range be in order to support a population size of at least 96,000?

56. The concentration of a certain drug in the bloodstream varies with time in the following way:

$$C = \frac{3t}{t^2 + t + 1}$$

where C is the concentration of the drug in the bloodstream in milligram/liter t hours after it is taken orally. It was determined that the drug is effective if the concentration is at least 0.6 milligram/liter. During what time interval is the drug effective?

1.10 Substitution

Up until now we have discussed single equations and inequalities as conditions placed on one variable. On occasion, we may have to deal with more than one equation in which conditions are placed on one or more variables.

EXAMPLE 1 If the perimeter of a square is 32 inches, find its area.

Solution Our first response is to begin with the formula for the area of a square, which is $A = s^2$, where s is the length of a side. This formula requires that we have s, the length of a side, in order to compute the area. We are not given the length of the side, but we are given the perimeter. Hence we find the formula for the perimeter of a square and see if we can somehow use that information to find the length of the side. The formula for perimeter is $P = 4s$.

To find s, given the perimeter, we substitute $P = 32$ in the formula for perimeter and solve for s:

$$P = 4s \Rightarrow 32 = 4s \Rightarrow s = 8 \text{ inches}$$

Now that we have $s = 8$, we can find the area of the square by substituting $s = 8$ in the formula for area, $A = s^2$.

$$A = s^2 \Rightarrow A = 8^2 \Rightarrow A = 64 \text{ sq in.}$$

Hence the area of the square is $\boxed{64 \text{ sq in.}}$ ■

Suppose we are given a task that requires us continuously to find the area of a square given its perimeter. Rather than continuing through the two-step process described in Example 1, it would be more convenient if we had a single formula yielding a direct relationship between the area of a square and its perimeter. We may state this problem as "Find the area of a square given its perimeter" or "Express the area of a square in terms of its perimeter." We will demonstrate that the steps taken in the process of developing this formula are not much different than the approach taken in Example 1.

EXAMPLE 2 If the perimeter of a square is P, find its area, A, in terms of P.

Solution Again, our first response is to begin with the formula for the area of a square, which is $A = s^2$, where s is the length of a side.

This formula requires that we have s, the length of a side, in order to compute the area. We are not given the length of the side, but we are given the perimeter. Hence we find a formula that relates the perimeter to the length of a side. This formula is $P = 4s$.

To find s given P means to solve for s in terms of P. Hence

$$P = 4s \Rightarrow s = \frac{P}{4}$$

Now that we have s (in terms of P), we can find the area of the square by substituting $s = \frac{P}{4}$ in the formula for area, $A = s^2$.

$$A = s^2 \qquad \textit{Substitute } s = \frac{P}{4}.$$

$$= \left(\frac{P}{4}\right)^2 = \frac{P^2}{16}$$

Hence $\boxed{A = \frac{P^2}{16}}$.

Notice that if $P = 32$, then $A = \frac{32^2}{16} = 64$, as we found in Example 1. ■

In finding the area of a square, A, in terms of its perimeter, P, we had to solve explicitly for the side, s, in $P = 4s$, and then substitute $\frac{P}{4}$ for s in the formula $A = 4s^2$. This method of replacing a variable by an expression involving another is called *substitution*.

EXAMPLE 3 Find the area, A, of a square in terms of the length of its diagonal, d.

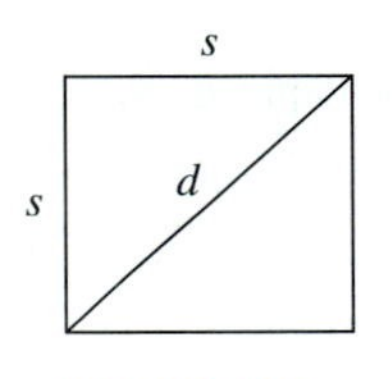

FIGURE 1.25

Solution We approach this as if we were given a number for the length of the diagonal and had to find the area. Again, our first response is to begin with the formula for the area of a square, which is $A = s^2$, where s is the length of a side.

The formula requires that we have s, the length of a side, in order to compute the area. We are not given the length of the side, but we are given the length of its diagonal as d. Hence we need to find a relationship between the side s and the diagonal d. We can draw a picture of a square with a diagonal, labeled d, and side labeled s. See Figure 1.25.

We note that a right triangle is formed with the diagonal as the hypotenuse, and by the Pythagorean theorem, the relationship between its sides and diagonal is

$$s^2 + s^2 = d^2 \qquad \text{which simplifies to} \qquad 2s^2 = d^2$$

We still need s in order to find A, so we solve for s in terms of d:

$$2s^2 = d^2$$

$$s^2 = \frac{d^2}{2} \qquad \textit{We can stop here.}$$

Rather than continuing our attempt to solve for s, we note that we have A in terms of s^2. Since we now have s^2 in terms of d, we can find the area of the square by substituting $s^2 = \frac{d^2}{2}$ in the formula for area, $A = s^2$.

$$A = s^2 \qquad \textit{Substitute } s^2 = \frac{d^2}{2}.$$
$$= \frac{d^2}{2}$$

Hence $\boxed{A = \frac{d^2}{2}}$.

■

EXAMPLE 4 A boat is anchored in the middle of a still lake when its fuel tank begins to leak, causing a circular slick on its surface (see Figure 1.26). If the radius of the slick is growing at a constant rate of 2 ft/min, how much area will the slick cover **(a)** 1 hour from the time it started? **(b)** t minutes from the time it started?

FIGURE 1.26

Solution

(a) In this example we begin with the area of a circle, which is given by $A = \pi r^2$. We need to find the radius. We are given that the radius is growing at 2 ft/min. Using the distance formula, *distance* = *rate* × *time,* we can determine that at the end of 1 hour (60 minutes), the radius is $2 \times 60 = 120$ feet.

Since $r = 120$, the area of the slick after 1 hour is;

$$A = \pi r^2$$
$$= \pi(120)^2$$
$$= \boxed{14{,}400\pi \approx 45{,}238.9 \text{ sq ft}}$$

(b) Again, the area of a circle is given by $A = \pi r^2$. We need to find the radius. We are given that the radius is growing at 2 ft/min. Using the distance formula we can determine that the radius is $r = 2t$. See Figure 1.27.

The area of the slick after t minutes is

$$A = \pi r^2$$
$$= \pi(2t)^2$$
$$= 4\pi t^2$$

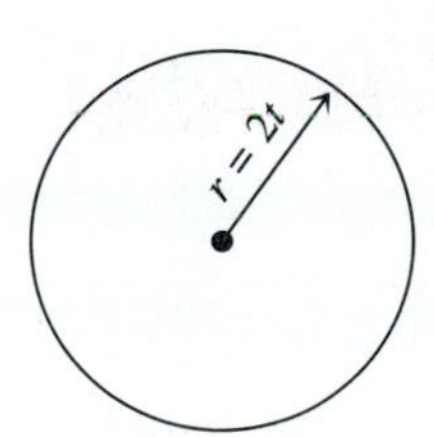

FIGURE 1.27

Hence $\boxed{A = 4\pi t^2 \text{ sq ft}}$.

■

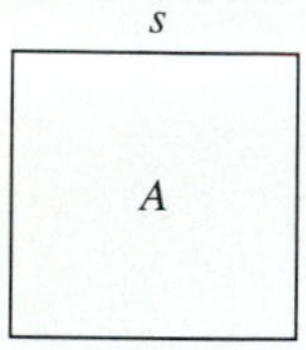

EXAMPLE 5 A farmer wants to enclose a square field with fencing materials: the fencing material costs $6 per foot. If the area of the field is A square feet, in terms of A, how much would it cost her to enclose the field with fencing?

FIGURE 1.28

Solution We draw a diagram (see Figure 1.28) and note the following:

The cost of the fencing is found by multiplying the price per foot of the fencing by the number of feet of fencing. The number of feet of fencing to be used is the perimeter of the field. If we call the perimeter P, since the cost of the fencing is $6 per foot, the total cost of the fence is $C = 6P$ dollars. Since the field is a square, we write out the formula for the perimeter of a square: $P = 4s$, where s is the length of a side.

In order for us to find the perimeter, we need the length of a side. We are not given the length of a side, but we are given the area, and so we look for a formula which relates the area of a square to the length of a side. The formula is: $A = s^2$.

We use the formula for the area to find s in terms of A:

$$A = s^2 \Rightarrow s = \sqrt{A} \qquad \textit{(s must be positive.)}$$

Now we have s in terms of A. We find the perimeter by substituting $s = \sqrt{A}$ in the formula for the perimeter: $P = 4s$ to get

$$\begin{aligned} P &= 4s && \textit{Substitute } s = \sqrt{A} \\ &= 4\sqrt{A} \end{aligned}$$

The cost of the fencing is $6P = 6(4\sqrt{A}) = \boxed{24\sqrt{A} \text{ dollars}}$ ■

EXAMPLE 6 Given that the volume of a sphere is found by $V = \frac{4}{3}\pi r^3$, where r is its radius, find the volume of a sphere in terms of its diameter.

Ask yourself: How would you find the volume given the diameter is 10?

Solution We start out with the volume of the sphere given in terms of its radius as $V = \frac{4}{3}\pi r^3$. Since we want to find the volume in terms of the diameter, we need to find a relationship between the radius and the diameter of a sphere. We already know that if d is the diameter and r is the radius, then

$$d = 2r \Rightarrow r = \frac{d}{2}.$$

Hence, given the diameter, we can find its radius. Now we substitute for the radius in the formula for volume as follows:

$$\begin{aligned} V &= \frac{4}{3}\pi r^3 && \textit{Now substitute } \frac{d}{2} \textit{ for } r \\ &= \frac{4}{3}\pi\left(\frac{d}{2}\right)^3 && \textit{And simplify.} \\ &= \frac{4\pi d^3}{24} \end{aligned}$$

Hence $\boxed{V = \dfrac{\pi d^3}{6}}$. ■

EXERCISES 1.10

1. Find the area of a square if **(a)** its perimeter is 84 inches and **(b)** its perimeter is x inches.
2. Find the perimeter of a square if **(a)** its area is 90 sq in. and **(b)** its area is A square inches.
3. Find the area of a square if its diagonal is 5 feet.
4. Find the area of a square if its diagonal is x feet.
5. Find the diagonal of a square if its area is 84 sq cm.
6. Express the diagonal of a square in terms of its area, A.
7. Find the diagonal of a square in terms of its perimeter, P.
8. Find the perimeter of a square if its diagonal is 12 feet.
9. Express the perimeter of a square in terms of its diagonal, d.
10. Find the perimeter of a square if its diagonal is 4 inches.
11. Find the volume of a sphere given its diameter is 8 feet.
12. Find the surface area of a sphere given its diameter is 8 feet.
13. Express the surface area of a sphere in terms of its diameter, d.
14. Express the surface area of a sphere in terms of its volume, V.

Use the following figure for Exercises 15–18. The circle is inscribed in the square.

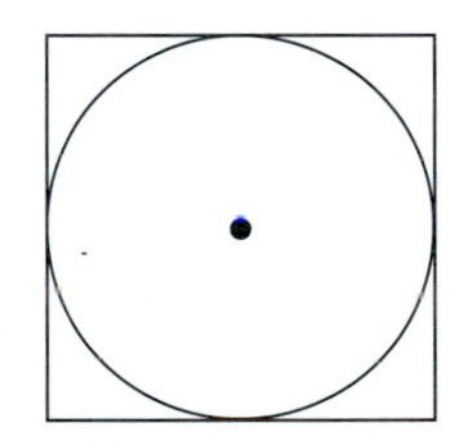

15. If the radius of the circle is 5 inches, what is the area of the square?
16. If the side of the square is 8 inches, find the area of the circle.
17. If the radius of the circle is r inches, express the area of the square in terms of r.
18. If the side of the square is s inches, express the area of the circle in terms of s.
19. The radius of a circle is growing at a constant rate of 3 in./sec. **(a)** What is the area of the circle after 2 seconds (starting at $r = 0$)? **(b)** What is the area of the circle after t seconds (starting at $r = 0$)?
20. The area of a circle is growing at a constant rate of 10 sq in./sec. **(a)** What is the radius of the circle after 2 seconds (starting at $A = 0$)? **(b)** What is the radius of the circle after t seconds (starting at $A = 0$)?
21. The radius of a snowball is growing at a constant rate of 3 in./min. What is the volume of the snowball after 5 minutes (starting at $r = 0$)?
22. The volume of a snowball is growing at a constant rate of 8 cu in./min. What is the radius of the snowball after 5 minutes (starting at $V = 0$)?
23. The radius of a snowball is growing at a constant rate of 3 in./min. What is the volume of the snowball after t minutes (starting at $r = 0$)?
24. The volume of a snowball is growing at a constant rate of 8 cu in./min. What is the radius of the snowball after t minutes (starting at $V = 0$)?
25. A farmer wants to enclose a rectangular field with fencing materials; the fencing material costs \$8 per foot. If the perimeter of the field is 500 feet, how much would it cost to enclose the field with fencing?
26. A farmer wants to enclose a rectangular field with fencing materials; the fencing material costs \$8 per foot. If the perimeter of the field is P feet, how much would it (in terms of P) cost to enclose the field with fencing?
27. A farmer wants to enclose a square field with fencing materials; the fencing material costs \$8 per foot. If the area of the field is 200 sq ft, how much would it cost to enclose the field with fencing?
28. A farmer wants to enclose a square field with fencing materials; the fencing material costs x dollars per foot. If the area of the field is 1,000 sq ft, how much would it cost (in terms of x) to enclose the field with fencing?
29. A gardener wants to enclose a circular garden with fencing materials; the fencing material costs \$2 per foot. If the radius of the garden is 5 feet, how much would it cost to enclose the garden with fencing?
30. A gardener wants to enclose a circular garden with fencing materials; the fencing material costs \$3 per foot. If the radius of the garden is r feet, how much would it cost (in terms of r) to enclose the garden with fencing?
31. A gardener wants to enclose a circular garden with fencing materials; the fencing material costs \$3 per foot. If the area of the garden is 90 sq ft, how much would it cost to enclose the garden with fencing?
32. A gardener wants to enclose a circular garden with fencing materials; the fencing material costs \$3 per foot. If the area of the garden is A square feet, how much would it cost (in terms of A) to enclose the garden with fencing?
33. Two boats leave a harbor at the same time at right angles to each other. If one boat is traveling at 30 mi/h and the other is traveling at 40 mi/h, how far away are the boats from each other after 2 hours?
34. Two boats leave a harbor at the same time at right angles to each other. If one boat is traveling at 30 mi/h and the other is traveling at 40 mi/h, how far away are the boats from each other after t hours?

Chapter 1 REVIEW EXERCISES

In Exercises 1–2, express as a quotient of two integers.

1. $8.252\overline{525}$ **2.** $72.47\overline{247}$

In Exercises 3–4, state the real number property being illustrated.

3. $5 - (x + 4) = 5 - (4 + x)$
4. $(x + 3)(x^2 - 3) = x(x^2 - 3) + 3(x^2 - 3)$

In Exercises 5–8, graph the set on the real number line and express the set using interval notation.

5. $\{x \mid x \geq -4\}$ **6.** $\{x \mid -8 < x \leq -1\}$
7. $\{x \mid x < 3\}$ **8.** $\{x \mid x \geq 0\}$

In Exercises 9–10, write each expression without absolute value symbols.

9. $|1 - \sqrt{5}|$ **10.** $|x - 8|$

In Exercises 11–16, evaluate the numerical expressions.

11. $6 - \{4 - [5 - 2(3 - 9)]\}$
12. $|-5| - |-9| - |-5 - 9|$
13. $2\left(\frac{2}{3}\right)^2 - 4\left(\frac{2}{3}\right) + 6$ **14.** $5\left(-\frac{1}{3}\right)^2 - 2\left(-\frac{1}{3}\right) - 4$
15. $\dfrac{3 - \frac{1}{2}}{\frac{2}{5} - 1}$ **16.** $\dfrac{\frac{1}{5} - \frac{2}{3}}{1 - \frac{3}{4}}$

In Exercises 17–26, perform the operations and express your answer in simplest form.

17. $(-2xy)^2(3xy)^3$ **18.** $3xy^2(2xy^2 - 3y^2)$
19. $(x - 3)(2x + 7)$ **20.** $(3x - 4)(2x - 1)$
21. $(5a^3 - b)(5a^3 + b)$ **22.** $(5a^3 - b)^2$
23. $(3x - 1)(9x^2 + 3x + 1)$
24. $(1 + 2x)(1 - 4x + 16x^2)$
25. $2x(x - 2)^2 - (x + 2)(x - 2)$
26. $(x - 2)^2(x - 3)^2$

In Exercises 27–36, factor as completely as possible.

27. $x^2 - 2x - 15$ **28.** $6x^2 - 11x - 10$
29. $3y(x - 2)^2 + 5(x - 2)$
30. $2x(x + 8)^2 - 2x^2(x + 8)^3$
31. $25x^2 - 30x + 9$ **32.** $a^4 - 81$
33. $a^3 + a^2 - 16a - 16$ **34.** $2x^3 + 6x^2 - 18x - 54$
35. $a^5 - a^3 + a^2 - 1$ **36.** $a^2 + 4a + 4 - x^2$

In Exercises 37–44, perform the operations and express your answer in simplest form.

37. $\dfrac{4x^2 - 12xy + 9y^2}{2x^2 - xy - 3y^2} \cdot \dfrac{3x + y}{6x^2 - 7xy - 3y^2}$
38. $\dfrac{2x^2}{x^2 + 3x + 9} \div \dfrac{2x^3 - 18x}{x^3 - 27}$
39. $\dfrac{3x}{x - 2} + \dfrac{1}{x}$ **40.** $\dfrac{2x - 3}{x + 3} - \dfrac{x - 1}{x - 3}$
41. $\dfrac{3x}{(x - 1)^2} - \dfrac{2}{x - 1}$ **42.** $\dfrac{2x}{x^2 - 1} + \dfrac{2}{x^2 - 2x + 1}$
43. $\dfrac{\frac{1}{x} + \frac{6}{y}}{\frac{1}{x^2} - \frac{36}{y^2}}$ **44.** $\dfrac{\frac{3y - 1}{2y + 1} + \frac{2y}{4y^2 - 1}}{\frac{3}{2y - 1} + 3}$

In Exercises 45–52, perform the operations and express your answer in simplest form with positive exponents only.

45. $(xy^2)^{-2}(xy^{-2})^{-3}$ **46.** $(3x^{-1}y^{-2})^{-1}(2x^{-2}y)^{-3}$
47. $\dfrac{x^{-1}y^{-2}}{xy^{-2}}$ **48.** $\dfrac{x^{-1} + y^{-2}}{xy^{-2}}$
49. $\dfrac{3x^{-2}y^{-2}}{x^{-1}y}$ **50.** $\dfrac{3x^{-1} - y^{-2}}{x^{-1} + y}$
51. $\dfrac{x^{2/3}x^{-1/4}}{x^{-2/5}}$ **52.** $\dfrac{2x(x^2 + 4)^{-3/2}}{(x^2 + 4)^{-1/2}}$

In Exercises 53–54, evaluate the number.

53. $(-64)^{-2/3}$ **54.** $81^{-3/4}$

In Exercises 55–56, express each as a single simplified fraction with positive exponents only.

55. $2x(x^2 + 1)^{1/2} + x^2\left(\frac{1}{2}\right)(x^2 + 1)^{-3/2}$

56. $(x^2 - 1)^{1/3} + \dfrac{2x^2}{3}(x^2 - 1)^{-2/3}$

In Exercises 57–58, rewrite the expression using rational exponents.

57. $3x^2\sqrt{x^2 + 1}$ **58.** $4x^5\sqrt[3]{x^2 - 1}$

In Exercises 59–60, rewrite the expression using radical notation.

59. $2x(x + 1)^{1/2} + x^2(x^2 + 1)^{-3/2}$

60. $(x^2 - 1)^{1/3} + \frac{2x^2}{3}(x^2 - 1)^{-2/3}$

In Exercises 61–62, rewrite in simplest radical form.

61. $\sqrt{48x^{12}y^{15}}$ **62.** $\sqrt{\frac{48x^{12}y^{15}}{16x^4y^8}}$

In Exercises 63–64, express each in simplest radical form with the denominator rationalized.

63. $\frac{5x}{\sqrt{x^5}}$ **64.** $\frac{\sqrt{8}}{\sqrt{6} - \sqrt{2}}$

In Exercises 65–70, perform the operations and express in simplest radical form with the denominator rationalized.

65. $(\sqrt{2x} - 2)^2$ **66.** $(\sqrt{2x} - 2)(\sqrt{2x} + 2)$

67. $(\sqrt{a} - 4)^2 - (\sqrt{a} - 4)^2$ **68.** $\frac{\sqrt{x^2 + 2x - 3}}{\sqrt{x + 3}}$

69. $\frac{\sqrt{8}}{\sqrt{11} - \sqrt{3}}$ **70.** $\frac{x^2 - 7x - 18}{\sqrt{x} + 3}$

In Exercises 71–72, express each as a single fraction.

71. $2 - \frac{2}{\sqrt{x^2 - 4}}$

72. $5\sqrt{3x^2 - 2} - \frac{5}{\sqrt{3x^2 - 2}}$

In Exercises 73–78, perform the operations and express the result in $a + bi$ form.

73. $(3 + 4i) + (2 - 5i)$ **74.** $(6 - 3i) - (2 + 5i)$
75. $(5 + i)^2$ **76.** $(3 - 2i)(3 + 2i)$
77. $\frac{2 + 3i}{2i}$ **78.** $\frac{6 + 3i}{2 - 3i}$

In Exercises 79–86, solve for the variable. For inequalities, express your answer using interval notation.

79. $2 - \{3 - [2(1 - x)]\} = 5x + 3$
80. $3 - 2[2 - (x - 6)] = 2x + 8$
81. $3x - \frac{2}{5}x \geq -2x + 3(1 - x)$
82. $\frac{3x - 2}{5} - 2 < \frac{2x - 3}{2}$
83. $\frac{4}{x^2 - 2x} - \frac{3}{2x} = \frac{17}{6x}$ **84.** $\frac{3x}{x - 1} - 2 = \frac{3}{x - 1}$
85. $2 < 5 - 3x \leq \frac{7}{3}$ **86.** $-2 \leq \frac{5}{2}x + 3 \leq 4$

In Exercises 87–88, solve for the given variable.

87. $C = \frac{5}{9}(F - 32)$ for F **88.** $y = \frac{x - 9}{3x + 1}$ for x

89. Raju has 2 liters of a 60% solution of alcohol. What is the minimum amount of water he should add to the solution in order to have a solution that is less than 40% alcohol?

90. It takes Collin 3 hours to paint a room and it takes Mary $2\frac{1}{2}$ hours to paint the same room. How long would it take for them to paint the room working together?

In Exercises 91–98, solve for the given variable. For inequalities, express your answer using interval notation.

91. $|7x - 2| = 9$ **92.** $|2x - 3| = 4$
93. $|5 - 3x| < 6$ **94.** $|5 - 3x| \geq 6$
95. $\left|\frac{1 - 4x}{3}\right| \geq 5$ **96.** $\left|1 - \frac{4}{3}x\right| < 3$
97. $|x - 1| = |5x + 3|$ **98.** $|x - 2| = |x + 2|$

In Exercises 99–100, write each as a single expression without absolute values if possible.

99. $|-7x^2 - 3|$ **100.** $|-6x^4 - 3x^2 - 4|$

In Exercises 101–102, solve by completing the square.

101. $x^2 - x = 8$ **102.** $2t^2 = 8t + 10$

In Exercises 103–112, solve by any algebraic method.

103. $3x^2 + 8 = 23$ **104.** $6x^2 = 2 - x$
105. $\frac{3}{x - 2} + \frac{2}{x - 1} = \frac{3}{2}$ **106.** $\frac{1}{x} + x = 2$
107. $\frac{6}{x^2 - 2x - 3} + \frac{x}{x - 3} = 3$
108. $\frac{(x - 3)^{1/3}(x + 1)^{1/2}}{(x - 2)^{1/5}} = 0$
109. $\sqrt{3x + 1} + 1 = x$ **110.** $(x - 3)^{-1/3} = 4$
111. $2x^{2/3} - x^{1/3} = 3$ **112.** $\sqrt{5x + 1} - 1 = \sqrt{3x}$

In Exercises 113–114, solve for the given variable.

113. $5x^2 + 7y^2 = 9$ for x
114. $15a^2 + 7ab - 2b^2 = 0$ for a

In Exercises 115–124, solve the inequalities and express your answers using interval notation.

115. $3x^2 - 14x - 5 > 0$ **116.** $2x^2 \geq -5x - 3$
117. $x^2 - 4x + 4 < 0$ **118.** $x^3 + 2x^2 - x - 2 \leq 0$
119. $\frac{3x - 2}{x + 1} > 0$ **120.** $\frac{3 - 2x}{x - 5} \leq 0$

121. $\dfrac{3x-1}{x} \le 2$ **122.** $\dfrac{x}{3x+1} < 5$

123. Two common units of measurements in physics and chemistry are the *angstrom* unit (written Å), which is 10^{-8} cm, and the *micron* unit (written μ), which is 10^{-4} cm. How many angstroms are there in a micron?

124. Carlos throws a ball straight up into the air off a building. The equation $s = -16t^2 + 80t + 44$ gives the distance, s, in feet the ball is above the ground t seconds after he tosses it up.

(a) How high above the ground is the ball at $t = 2$ seconds?

(b) How long does it take for the ball to hit the ground?

125. A circular garden is surrounded by a path of uniform width. If the path has area 44π sq ft and the radius of the garden is 10 feet, find the width of the path.

126. The radius of a circle is growing at a constant rate of 5 in./sec. What is the area of the circle after t minutes (starting at $r = 0$)?

127. A gardener wants to enclose a circular garden with fencing materials; the fencing material costs \$3 per foot. If the radius of the garden is r feet, how much would it cost (in terms of r) to enclose the garden with fencing?

Chapter 1 PRACTICE TEST

1. State the real number property illustrated by $(x^2+3)\left(\dfrac{1}{x^2+3}\right) = 1$.

2. Graph the sets on the real number line and express the sets using interval notation.
(a) $\{x \mid x \le -2\}$ **(b)** $\{x \mid -6 < x \le -2\}$

3. Write without absolute value symbols: $|1 - \sqrt{5}|$.

4. Evaluate the numerical expression: $\dfrac{\frac{2}{5} - 3}{1 - \frac{3}{2}}$.

5. Perform the operations and express your answer in simplest form.
(a) $(3a^3 - 2b)(3a^3 + 2b)$
(b) $3x(x-5)^2 - (x+5)(x-5)$

6. Factor as completely as possible.
(a) $10x^2 + x - 2$ **(b)** $3y(y+3)^2 + 15(y+3)$
(c) $x^3 + 2x^2 - 9x - 18$

7. Perform the operations. Express your answer in simplest form with positive exponents only.

(a) $\dfrac{3}{x+1} - \dfrac{2x}{(x+1)^2}$ **(b)** $\dfrac{\frac{1}{b} - \frac{5}{a}}{\frac{1}{b^2} - \frac{25}{a^2}}$

(c) $\dfrac{x^{-2} + y^{-2}}{(x+y)^{-2}}$ **(d)** $\dfrac{3x^2(x^3-2)^{-1/3}}{(x^3-2)^{2/3}}$

8. Rewrite each in simplest radical form. Rationalize denominators where possible.

(a) $\sqrt{\dfrac{32x^9y^7}{18x^4y^8}}$ **(b)** $\dfrac{5x^2}{\sqrt{2x}}$

(c) $(3x - \sqrt{2})^2$ **(d)** $\dfrac{\sqrt{9}}{\sqrt{10} - \sqrt{7}}$

9. Perform the operations and express in $a + bi$ form.
(a) $(3 - 5i)^2$ **(b)** $\dfrac{2 - 3i}{3 + i}$

In Exercises 10–23, solve for the variable. For inequalities, express your answers using interval notation.

10. $4 - [3(2 - x)] = x + 3$ **11.** $\dfrac{x-1}{3} - 1 < \dfrac{x+3}{2}$

12. $|7 + 2x| = 7$ **13.** $\left|\dfrac{2}{3}x - 2\right| \le 5$

14. $|7 - 2x| > 4$ **15.** $|2 - 3x| = |3x - 2|$

16. $(2x - 1)(x + 2) = 5x + 2$

17. $3x^2 - 2x = 5 - 2x$

18. $\dfrac{2}{x-2} + \dfrac{3}{x+2} = \dfrac{5}{x^2-4}$

19. $\sqrt{x-5} + 4 = x - 1$ **20.** $2x^2 \le x + 3$

21. $(x-1)(x-2) > 2x - 2$

22. $\dfrac{x-2}{2x+1} > 0$ **23.** $\dfrac{x-5}{x} \ge 2$

24. Ken can process 200 forms in 3 hours and Kim can process the same 200 forms in $2\frac{1}{3}$ hours. How long would it take them to process the 200 forms working together?

25. Sandy throws a ball up into the air. The equation $s = -16t^2 + 40t + 96$ gives the distance, s, in feet the ball is above the ground t seconds after she tosses it up. How long does it take for the ball to hit the ground?

26. A farmer wants to enclose a circular field with fencing materials. The fencing material costs \$6 per foot. If the area of the field is A square feet, how much would it cost (in terms of A) to enclose the field with fencing?

CHAPTER 2

Functions and Graphs: Part I

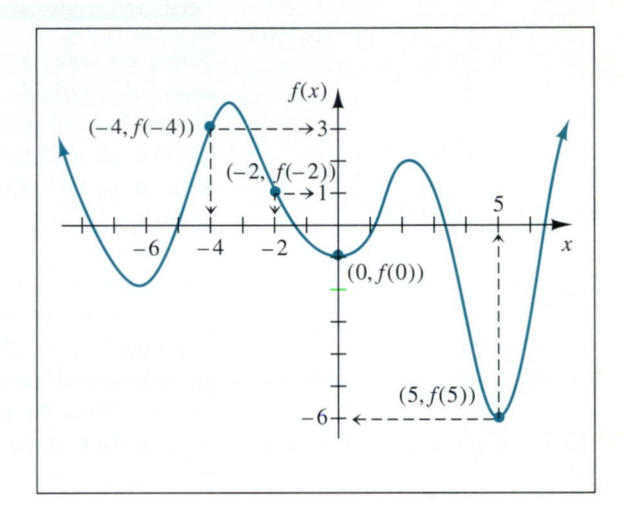

The concept of a function is one of the most important ideas in mathematics. It is an idea that unifies many different branches of mathematics and allows a variety of other disciplines to be "mathematized." We often hear or read statements such as "The inflation rate is a function of the labor force's productivity" or "Life insurance rates are a function of a person's age." We understand such statements to mean that there is a relationship between one of the quantities and the other.

In this chapter we describe different types of mathematical relationships and define precisely what we mean by a function. In addition, we begin a discussion of how to graph various types of equations and functions that will continue throughout the text.

2.1 The Cartesian Coordinate System: Graphing Straight Lines and Circles

When we solve a first-degree equation in one variable, we are trying to find all values of the variable that satisfy a certain condition. Every equation is, in fact, a condition. For example, when we solve an equation such as $3x - 2 = 11$, we are looking for all numbers that satisfy the condition that "two less than three times a number is 11." An equation in one variable imposes a condition on one unknown quantity.

Similarly, a first-degree equation in *two variables* imposes a condition on two unknown quantities. For example, when we solve an equation of the form $y = 3x - 6$, we are looking for all *pairs* of numbers x and y that satisfy the condition that "the y-value is 6 less than three times the x-value." Keep in mind that a single solution to the equation $y = 3x - 6$ consists of *two* numbers—an x-value and a y-value. Thus the pair of numbers $x = 5$, $y = 9$ is one solution to this equation.

In fact, we can generate infinitely many solutions to the equation $y = 3x - 6$ by simply choosing an x-value and then finding the corresponding y-value that makes the pair of numbers satisfy the equation.

For example, if we let $x = -1$, then $y = 3(-1) - 6 = -9$. Thus $x = -1$, $y = -9$ is *one* solution to this equation.

Table 2.1 contains some of the solutions to the equation $y = 3x - 6$. Since we cannot list the infinitely many pairs of numbers x and y that satisfy this equation, how can we exhibit all the solutions? One answer lies in using the Cartesian (rectangular) coordinate system invented by the French philosopher and mathematician René Descartes (1596–1650).

TABLE 2.1

x	$y = 3x - 6$	y
-2	$y = 3(-2) - 6$	-12
-1	$y = 3(-1) - 6$	-9
0	$y = 3(0) - 6$	-6
1	$y = 3(1) - 6$	-3
2	$y = 3(2) - 6$	0

The Cartesian plane is formed by two real number lines perpendicular to each other at their respective 0 points. The horizontal number line is called the x-axis; the vertical number line is called the y-axis. Together they are called the **coordinate axes**, and their common zero point is called the **origin**. Just as we use the real number line to represent the set of real numbers, we can use the Cartesian plane to represent ordered pairs of real numbers. The real number line is often called a one-dimensional coordinate system, since it takes only one number to identify a point, whereas the Cartesian plane is a two-dimensional coordinate system, since it takes two numbers to identify a point.

As shown in Figure 2.1, the coordinate axes divide the plane in four parts, which are called **quadrants**. They are numbered from I to IV in a counterclockwise order, starting in the upper right-hand quadrant. *Note that the points* **on** *the coordinate axes are not considered as being* **in** *any of the quadrants.*

This coordinate system allows us to associate a point in the plane with each *ordered pair* (x, y). The first and second members of the ordered pair are often called the x-coordinate and y-coordinate, respectively. To plot (graph) the point associated with an ordered pair (x, y), we start at the origin and move $|x|$ units to the right if x is positive or to the left if x is negative, and then $|y|$ units up if y is positive or down if y is negative.

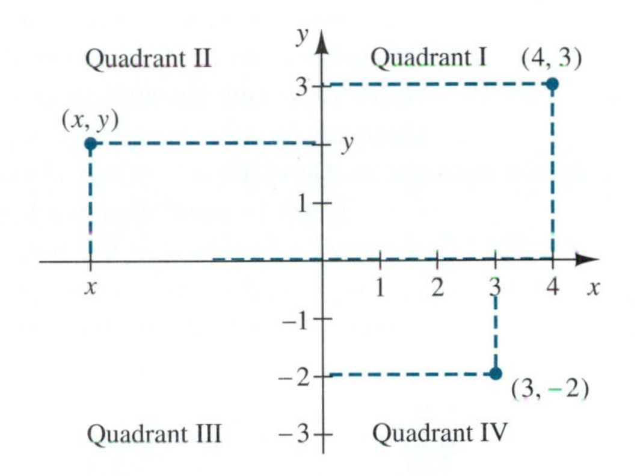

FIGURE 2.1 Finding the ordered pair (x, y) corresponding to a point in a rectangular coordinate system

Just as every ordered pair is associated with a point, so too is every point associated with an ordered pair. Figure 2.1 indicates how we would find the ordered pair corresponding to a particular point. The x-coordinate is found by *projecting* the point vertically to the x-axis and the y-coordinate is found by projecting the point horizontally to the y-axis.

Thus we have a one-to-one correspondence between the set of points in the plane and the set of ordered pairs of real numbers; that is, every point in the plane is assigned a unique pair of real numbers, and every pair of real numbers is assigned a unique point in the plane. For this reason we frequently refer to the point (x, y) rather than the ordered pair (x, y).

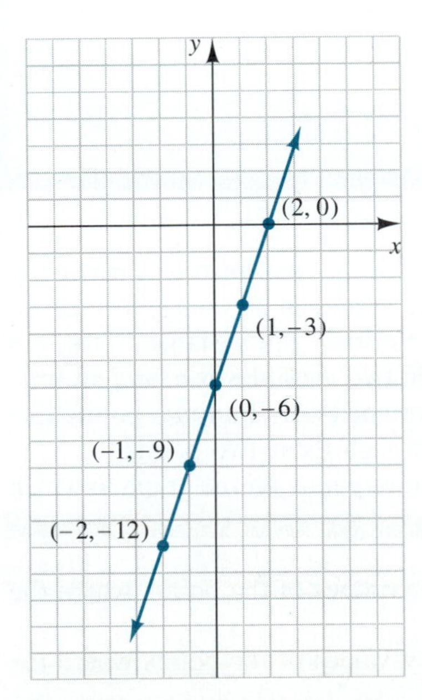

FIGURE 2.2 Ordered pairs that satisfy $y = 3x - 6$

Keeping the meaning of an ordered pair in mind, we recognize that all points on the x-axis are of the form $(x, 0)$. (Why?) Similarly, all points on the y-axis are of the form $(0, y)$. (Why?)

With this coordinate system in hand, we can return to the question, How can we exhibit the solution set to the equation $y = 3x - 6$? If we look back at the pairs of numbers (x, y) appearing in Table 2.1 that satisfy the equation $y = 3x - 6$ and plot these points in a coordinate plane, we see that the points seem to lie on a straight line. See Figure 2.2. By drawing a straight line through the points, we are saying two things: First, every ordered pair that satisfies the equation is a point on the line and second, every point on the line corresponds to an ordered pair that satisfies the equation. Often, we will simply say that *the point satisfies the equation.* We will find the following definition useful.

DEFINITION The **graph** of an equation is the set of all points whose ordered pairs satisfy the equation.

Thus the straight line in Figure 2.2 is the graph of the equation $y = 3x - 6$. In fact, the graph of any first-degree equation in two variables is a straight line. We state this formally in the following theorem.

THEOREM 2.1 The graph of an equation of the form $Ax + By = C$ (where A and B are not both equal to zero) is a straight line.

For this reason a first-degree equation is often called a **linear equation**. The form $Ax + By = C$ is called the *general form* for the equation of a line. As we continue with our discussion of straight lines, we will see that the converse of this theorem is also true; that is, a straight line graph has an equation of the form $Ax + By = C$.

Keep in mind that we have not proven this theorem. (The fact that the points we graphed *seemed* to fall on a straight line does not prove that all the points that satisfy the equation will.) The proofs of this theorem and its converse will be the substance of much of our discussion in the first three sections of this chapter.

GRAFFIX

Use a graphics calculator or computer to graph $y = 2x - 5$. If you have a graphics calculator, use the trace function to determine

(a) y when $x = 2$ **(b)** y when $x = -0.1$

(c) y when $x = 0.4$ **(d)** y when $x = -5$

Describe the graph of the equation $y = 2x - 5$.

As we proceed through the text, we will point out that the form of a particular equation often allows us to recognize what its graph is and, in so doing, to identify particular features of the graph that make it easier to draw. For instance, Theorem 2.1 tells us that the graph of a first-degree equation in two variables is a straight line. Therefore, if we want to graph such an equation, we need only find two points that are on the line and then draw the line that passes through these two points.

Throughout our work in graphing there are certain points on the graph to which we want to pay particular attention.

DEFINITION The ***x*-intercepts** of a graph are the x-values of the points where the graph crosses the x-axis.

The ***y*-intercepts** of a graph are the y-values of the points where the graph crosses the y-axis.

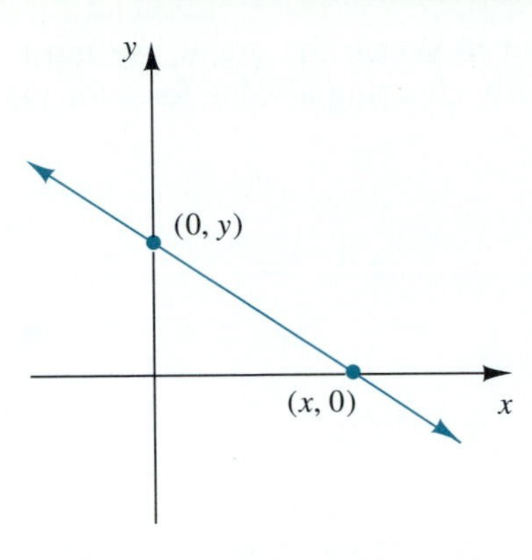

FIGURE 2.3 Finding the x- and y-intercepts of a line

Look back at Figure 2.2. The graph of $y = 3x - 6$ crosses the x-axis at $(2, 0)$ and so has an x-intercept of 2. The graph crosses the y-axis at $(0, -6)$ and so has a y-intercept of -6.

Note that since the x-intercepts correspond to points on the x-axis (and every point on the x-axis has a y-coordinate of 0) x-intercepts can be found by substituting $y = 0$ into the equation and solving for x; similarly, y-intercepts can be found by substituting $x = 0$ into the equation and solving for y. See Figure 2.3. *Whenever possible (and practical), we label the x- and y-intercepts of a graph.*

Let's illustrate how we can use the intercepts to graph a linear equation.

EXAMPLE 1 Sketch the graphs: **(a)** $5y - 2x = 12$ **(b)** $y = -3x$

Solution

(a) We recognize $5y - 2x = 12$ as a first-degree equation in two variables, and so its graph is a straight line. Any two points on the line will be sufficient to allow us to draw the graph. We will find the intercepts, since they are often easy to compute. To find the x-intercept, we set $y = 0$ and solve for x. In other words, we let $y = 0$ in $5y - 2x = 12$ and get $5(0) - 2x = 12 \Rightarrow -2x = 12 \Rightarrow x = -6$. Thus the x-intercept is -6, which means that the line crosses the x-axis at $(-6, 0)$.

Similarly, to find the y-intercept, let $x = 0$ in $5y - 2x = 12$ and get $5y - 2(0) = 12 \Rightarrow 5y = 12 \Rightarrow y = \frac{12}{5} = 2\frac{2}{5}$, which means that the line crosses the y-axis at $\left(0, \frac{12}{5}\right)$.

To sketch the graph of $5y - 2x = 12$, we plot the points corresponding to the intercepts and draw the line passing through them. See Figure 2.4.

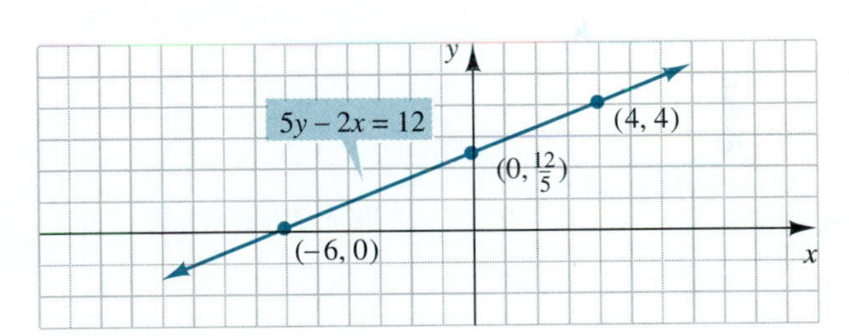

FIGURE 2.4

Because any two points we find determine a line, it is generally a good idea to find a third point as a check. For example we may choose an x-value, say $x = 4$, and find the corresponding y-value.

$$5y - 2(4) = 12, \quad \text{which gives} \quad 5y = 20 \quad \text{and so} \quad y = 4$$

Therefore the point $(4, 4)$ should be on the line, and it is. (See Figure 2.4.)

(b) As in part (a), we recognize that the graph of $y = -3x$ is a straight line. We begin by finding the intercepts. We let $y = 0$ in $y = -3x$ and get $x = 0$. The x-intercept is 0, which means the line goes through $(0, 0)$. Since the line goes through the origin, we automatically know that the y-intercept is also 0.

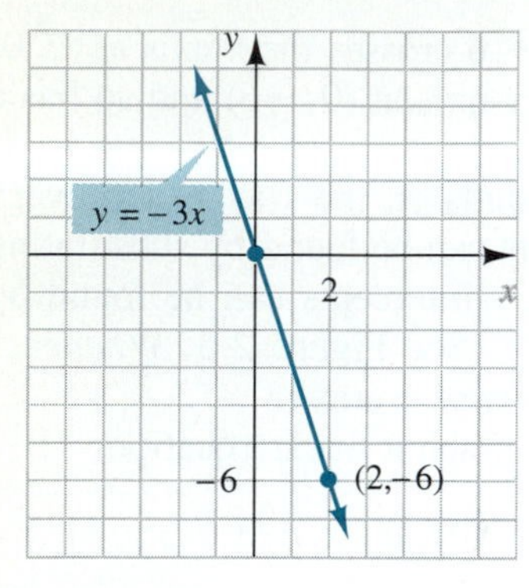

FIGURE 2.5

The intercepts give us only *one* point. In order to sketch the graph, we must find an additional point. We do this by arbitrarily choosing a value for x (or y) and solving for the other variable.

$y = -3x$ *We choose* $x = 2$.

$y = -3(2) = -6$ *Thus the point* $(2, -6)$ *is on the line.*

The graph appears in Figure 2.5. ■

DIFFERENT PERSPECTIVES: *Intercepts*

Consider the geometric and algebraic interpretations of the x- and y-intercepts of the graph of $3x - 2y = 6$.

GEOMETRIC INTERPRETATION

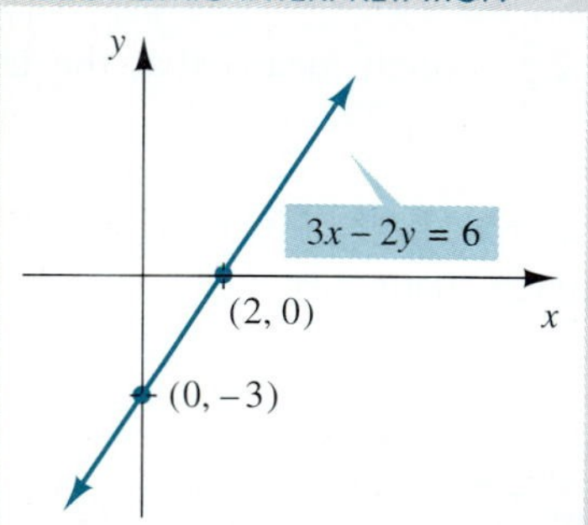

ALGEBRAIC INTERPRETATION

To find the x-intercept, set $y = 0$ and solve for x:

$$
\begin{aligned}
3x - 2y &= 6 \\
3x - 2(0) &= 6 \\
3x &= 6 \\
x &= 2
\end{aligned}
$$

To find the y-intercept, set $x = 0$ and solve for y:

$$
\begin{aligned}
3x - 2y &= 6 \\
3(0) - 2y &= 6 \\
-2y &= 6 \\
y &= -3
\end{aligned}
$$

The line crosses the x-axis at the point (2,0). ↔ The x-intercept is 2.

The line crosses the y-axis at the point $(0, -3)$ ⟷ The y-intercept is -3.

EXAMPLE 2 Sketch the graphs of the following equations in a rectangular coordinate system: **(a)** $x = -4$ **(b)** $y = 2$

Solution

(a) Had we been asked to graph the equation $x = -4$ on a number line, we would have drawn the graph shown in Figure 2.6.

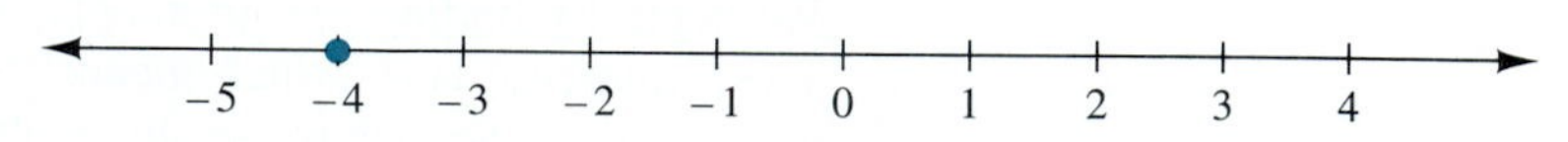

FIGURE 2.6 The graph of $x = -4$ on a number line

However, in this example we are being asked to graph in a rectangular coordinate system. The equation $x = -4$ is in the form $Ax + By = C$, with $A = 1$, $B = 0$, and $C = -4$.

$$Ax + By = C$$

$$1x + 0y = -4$$

As mentioned before, an equation can be viewed as a condition that x and y must satisfy. The equation $x = -4$ imposes the condition that the x-coordinate of any point satisfying the equation must be -4. Since y does not appear in the equation $x = -4$, there is *no* condition on y. In other words, x must be equal to -4, but y can be any real number. This gives a line parallel to and 4 units to the left of the y-axis. The graph appears in Figure 2.7.

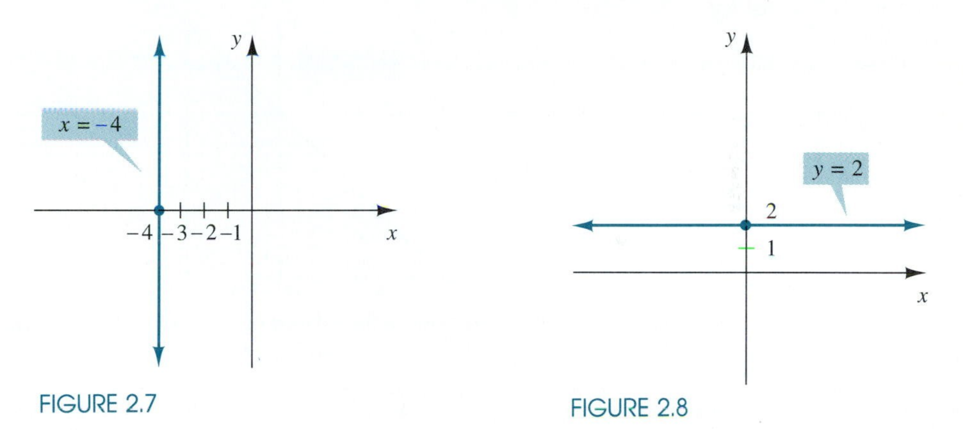

FIGURE 2.7

FIGURE 2.8

(b) Similarly, the equation $y = 2$ fits the general form $Ax + By = C$ with $A = 0$, $B = 1$, and $C = 2$. The equation $y = 2$ imposes the condition that the y-coordinate of any point on the line must be 2, while x can be any real number. This gives a line parallel to and 2 units above the x-axis. The graph appears in Figure 2.8. ■

In general, the graph of an equation of the form $x = h$ is a straight line parallel to the y-axis and passing through the point $(h, 0)$. The graph of an equation of the form $y = k$ is a straight line parallel to the x-axis and passing through the point $(0, k)$.

Graphing Linear Inequalities in Two Variables

On the number line, a one-dimensional coordinate system, the graph of a linear inequality such as $x \geq a$ is a half-*line* (or ray) starting at (or bounded by) the *point* $x = a$ (Figure 2.9).

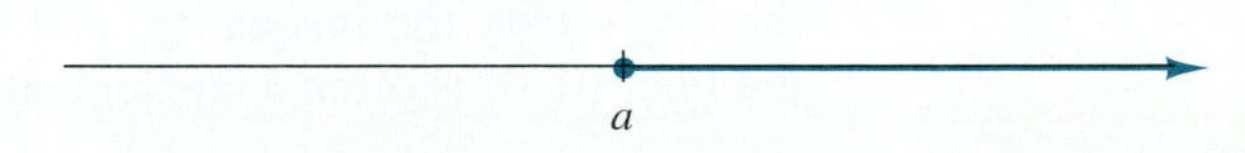

FIGURE 2.9 The graph of $x \geq a$ on a number line

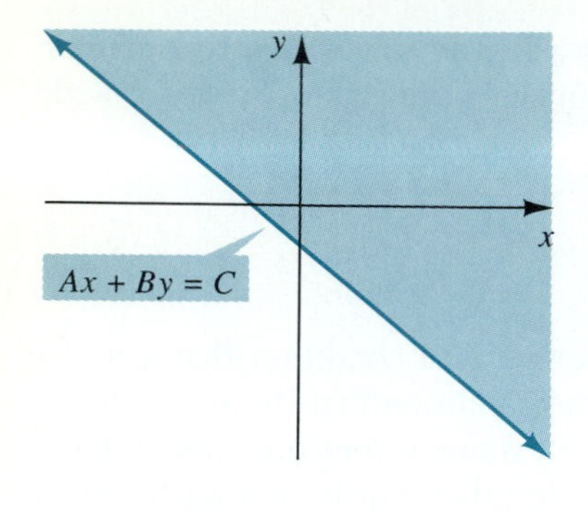

FIGURE 2.10 The graph of a typical solution set for $Ax + By \leq C$

An analogous situation occurs graphing linear inequalities in *two variables* on a two-dimensional rectangular coordinate plane: The graph of $Ax + By \leq C$ is a half-*plane* bounded by the *line* $Ax + By = C$. See Figure 2.10.

Consider the inequality $y - x \leq 5$. What do the solutions look like? If we isolate y and rewrite this equation as $y \leq x + 5$, then we can see that we are looking for all points on the rectangular coordinate system in which the y-coordinate is less than or equal to 5 more than the x-coordinate—all points on or below the line $y = x + 5$. This is pictured in Figure 2.11(a).

On the other hand, the same equation can be rewritten as $y - 5 \leq x$, which can be interpreted as all points on the plane such that the x-coordinate is greater than or equal to 5 less than the y-coordinate—all points on or to the right of the line $y - 5 = x$. This is pictured in Figure 2.11(b).

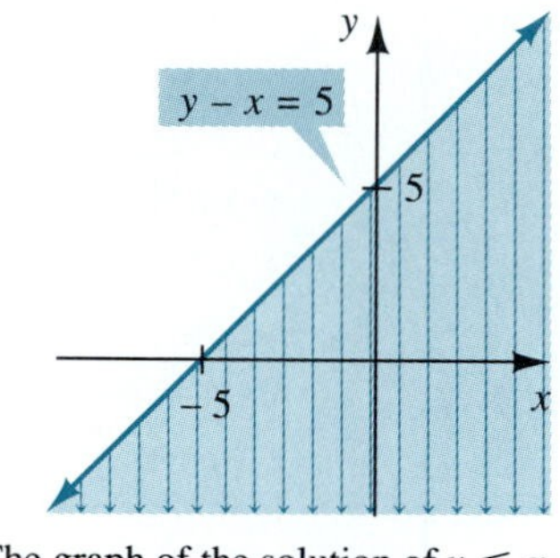

The graph of the solution of $y \leq x + 5$

(a)

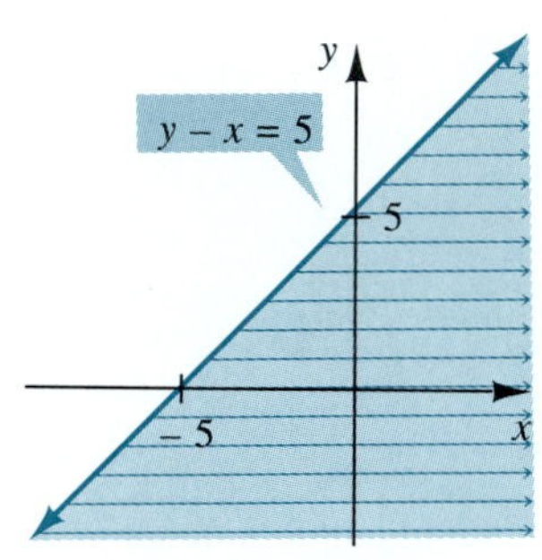

The graph of the solution of $x \geq y - 5$

(b)

FIGURE 2.11

Both views of the problem leave us with the same solution: the half-plane to the right of and bounded by the line $y - x = 5$. The shaded region is the solution. This means that any point lying in the shaded region is a solution to this inequality, and any point not in the region is not a solution. For example, the point $(-1, 2)$ is in the shaded region and $y - x \leq 5$ is true for $x = -1$ and $y = 2$. The point $(-3, 6)$ is *not* in the shaded region and $y - x \leq 5$ is *not* true for $x = -3$ and $y = 6$.

EXAMPLE 3 Graph the solution of $2y - 3x > 12$.

Solution We know that the solution will be a half-plane bounded by the *line* $2y - 3x = 12$. Hence, we first graph the line, as indicated in Figure 2.12. Notice that we draw a dashed line rather than a solid line. The dashed line indicates a strict inequality; that is, the points on the line itself are *not* part of the solution set.

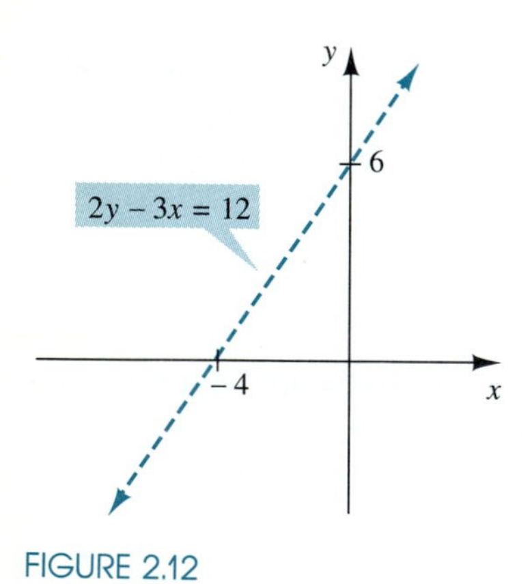

FIGURE 2.12

We know that all solutions lie on either one side of the line or the other side, so we need only test one point *not on the line*. If the coordinates of the test point we choose satisfy the inequality, then all points on that side of the line satisfy the inequality and, therefore, the solution is the half-plane that includes the test point.

If the test point does not satisfy the inequality, then no points on that side of the line satisfy the inequality, and the solution is the half-plane on the other side of the line. $(0, 0)$ is often a convenient test point (so long as it does not fall on the line):

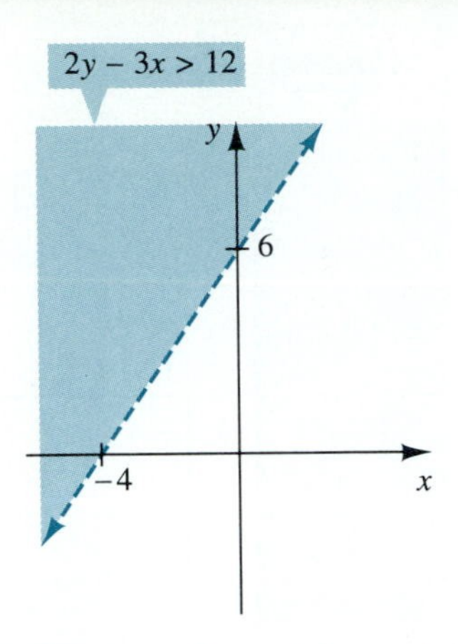

FIGURE 2.13 The solution set to $2y - 3x > 12$

$$2y - 3x > 12 \quad \textit{Substitute } x = 0 \textit{ and } y = 0.$$

$$2(0) - 3(0) \overset{?}{>} 12 \quad \textit{This inequality is false.}$$

Since the inequality is false for the test point, $(0, 0)$, we shade the *other* side of the line, as indicated in Figure 2.13. ■

GRAPHING LINEAR INEQUALITIES

In summary, we graph linear inequalities as follows:

1. Sketch a graph of the equation of the line that is the boundary of the solutions. The line should be dashed if it is a strict inequality ($<$ or $>$) and solid if it is a weak inequality ($\leq$ or $\geq$).
2. Choose a test point: Pick a point not on the line and determine if its coordinates satisfy the inequality.
3. If the coordinates of the test point satisfy the inequality, shade in the half-plane on the side of the line that includes the test point; if the coordinates of the test point do not satisfy the inequality, shade the other side of the line. The solution to the inequality is the shaded side together with the line if it is solid.

EXAMPLE 4 Graph the following inequalities: **(a)** $3x \leq 2y$ **(b)** $x < -1$

Solution

(a) $3x \leq 2y$

1. Sketch the graph of $3x = 2y$ using a solid line.
2. We will use the point $(1, 4)$ as the test point, since it does not lie on the line. We substitute $x = 1$ and $y = 4$ into $3x \leq 2y$. Since $3(1) \leq 2(4)$ is true, we shade in the half-plane on the side of the line that contains the point $(1, 4)$.
3. The graph of the solution set to $3x \leq 2y$ is shown in Figure 2.14.

Why can't we use (0, 0) as the test point?

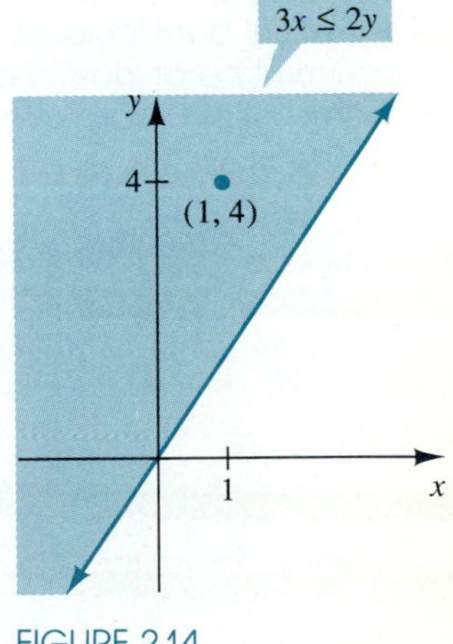

FIGURE 2.14

(b) $x < -1$

1. Sketch the graph of $x = -1$ using a dashed line.
2. We will use the point $(0, 0)$ as the test point, since it does not lie on the line. We substitute $x = 0$ and $y = 0$ into $x < -1$. Since $0 < -1$ is not true, we shade in the half-plane on the side of the line that does not contain $(0, 0)$.
3. The graph of the solution set to $x < -1$ is shown in Figure 2.15.

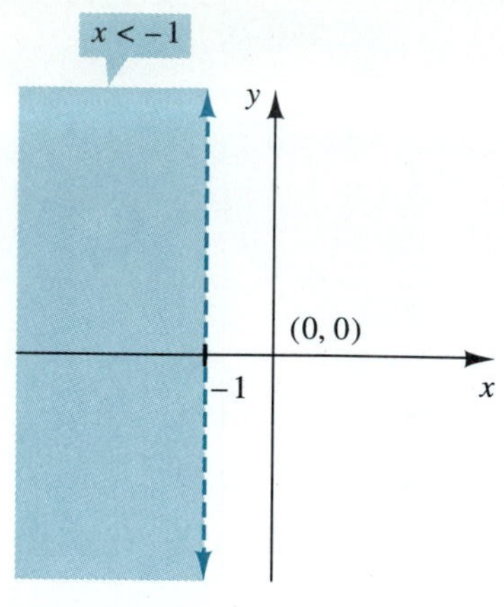

FIGURE 2.15 The graph of the solution set to $x < -1$ ■

How does the graph of $x < -1$ on the number line differ from the graph of $x < -1$ in a rectangular coordinate system?

EXAMPLE 5 A pharmaceutical company develops a drug called Darbane, which substantially reduces the pain of arthritis but has the negative side effect of producing stomach ulcers. The chemists at the company found, however, that the negative side effect can be eliminated if the user takes *at least* 500 mg of antacid with every gram of Darbane. Write an inequality expressing how much of each drug could be taken without the negative side effect; then graph the solution set. (Assume the user can take fractions of milligrams.)

Solution Let $x =$ the amount of Darbane taken by the user, and let $y =$ the amount of antacid taken with Darbane. We convert to grams and note that 500 mg $= \frac{1}{2}$ gram, therefore, in order to have no side effects, the amount of antacid taken should be *at least* $\frac{1}{2}$ the amount of Darbane taken. Hence we have

Do you see why the fact that the amount of antacid being taken must be at least one-half the amount of Darbane taken is translated as $y \geq \frac{1}{2}x$?

$$y \geq \frac{1}{2}x$$

(We check that this equation is correct by noting that if a person takes 1 gram of Darbane, $x = 1$, then he or she must take at least $\frac{1}{2}$ gram of antacid, $y \geq \frac{1}{2}$.) The graph of the solution appears in Figure 2.16. Note that (3, 10) is in the solution set, meaning that 3 grams of Darbane can be taken without the side effect if 10 grams of antacid are taken along with it. On the other hand, (15, 5) is not in the solution set which means that 15 grams of Darbane coupled with 5 grams of antacid will still have side effects. Common sense tells us that negative values for x and y make no sense, so we have sketched only that portion of the solution to $y \geq \frac{1}{2}x$ indicated in Figure 2.16.

In reality, there would be other limitations on drug dosages; although the point (300, 5000) is in the solution set, we would not realistically expect the body to tolerate such large quantities of drugs. In Section 10.6 we will solve simultaneous inequalities and discuss how to handle several restrictions.

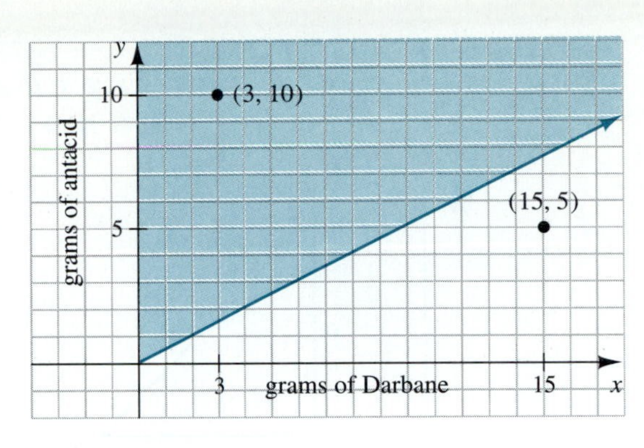

FIGURE 2.16

The Distance and Midpoint Formulas

Although we tend to take the Cartesian coordinate system rather for granted (remember that it is only about 350 years old), it is important to recognize what a revolutionary idea it is. Having a coordinate system allows us to approach geometric problems from an algebraic point of view. The ideas we are now going to discuss are part of a branch of mathematics called *analytic geometry*.

For example, the distance between two points and the midpoint of a line segment are both inherently geometric ideas; nevertheless, with the aid of a coordinate system we can derive an algebraic formula for each.

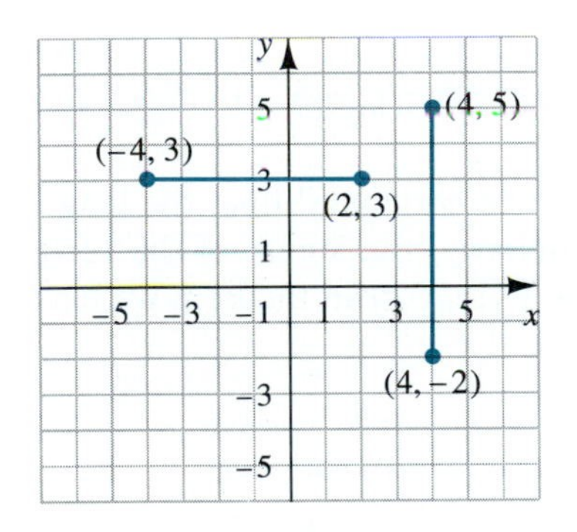

FIGURE 2.17 Finding horizontal and vertical distances

EXAMPLE 6 Find the distance between the given pair of points:

(a) $(-4, 3)$ and $(2, 3)$ **(b)** $(4, -2)$ and $(4, 5)$

Solution We plot the given pairs of points in Figure 2.17.

(a) Since the points $(-4, 3)$ and $(2, 3)$ lie on the same horizontal line, the distance between them is the same as the distance between the points -4 and 2 on the x-axis. As we saw in Chapter 1, the distance between two points a and b on a number line is $|a - b|$. Therefore, the distance between $(-4, 3)$ and $(2, 3)$ is $|2 - (-4)| = \boxed{6}$.

(b) Similarly, the points $(4, -2)$ and $(4, 5)$ lie on the same vertical line, and so the distance between them is the absolute value of the difference of their y-coordinates. Therefore, the distance between $(4, -2)$ and $(4, 5)$ is $|5 - (-2)| = \boxed{7}$.

In the previous example we could simply have found the distances by counting horizontal or vertical units. We chose to take a somewhat more formal approach so that we could recognize that the distance between the points (x_1, y_1) and (x_2, y_1), which lie on the same horizontal line, is $|x_2 - x_1|$, and the distance between the points (x_1, y_1) and (x_1, y_2), which lie on the same vertical line, is $|y_2 - y_1|$.

In order to derive a formula for the distance between *any* two points, we will also need the Pythagorean theorem.

THE PYTHAGOREAN THEOREM AND ITS CONVERSE

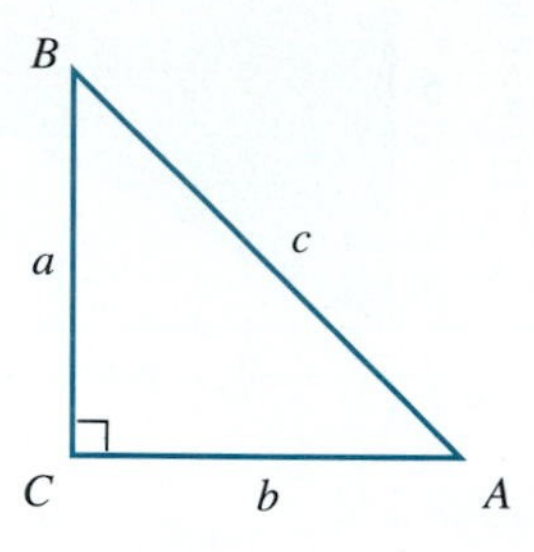

In right triangle ABC labeled as indicated, $a^2 + b^2 = c^2$.

The converse of this is also true. That is, if $a^2 + b^2 = c^2$, then ΔABC is right triangle with c as hypotenuse.

Now suppose we want to find the distance d between any two points (x_1, y_1) and (x_2, y_2). We can drop perpendiculars toward the x- and y-axes, as indicated in Figure 2.18, forming a right triangle whose third vertex is (x_2, y_1).

Why is this third vertex (x_2, y_1)?

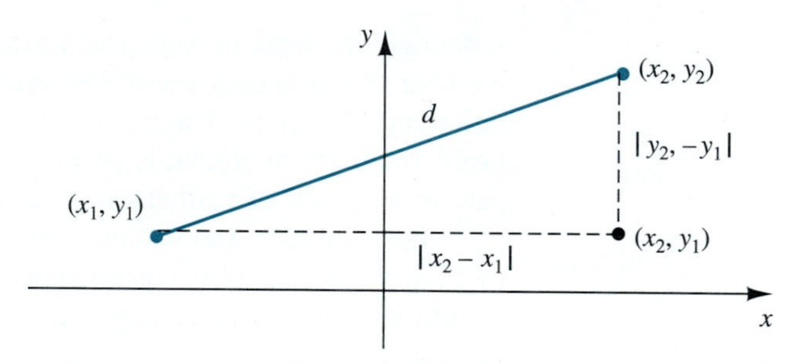

FIGURE 2.18 Finding the distance between two points

Since the points (x_1, y_1) and (x_2, y_1) lie on the same horizontal line, the length of the horizontal side of the triangle is $|x_2 - x_1|$. Similarly, the length of the vertical side of the triangle is $|y_2 - y_1|$. Therefore, by the Pythagorean theorem we have

$$d^2 = |x_2 - x_1|^2 + |y_2 - y_1|^2$$

Since the distance, d, must be positive, we take the positive square root to get

$$d = \sqrt{|x_2 - x_1|^2 + |y_2 - y_1|^2}$$

We leave it as an exercise for the reader to show that $|x_2 - x_1|^2 = (x_2 - x_1)^2$ and similarly for the y's.

$$d = \sqrt{(x_2 - x_1)^2 + (y_2 - y_1)^2}$$

We have thus derived the distance formula.

THE DISTANCE FORMULA

The distance between the points (x_1, y_1) and (x_2, y_2) in the Cartesian plane is

$$d = \sqrt{(x_2 - x_1)^2 + (y_2 - y_1)^2}$$

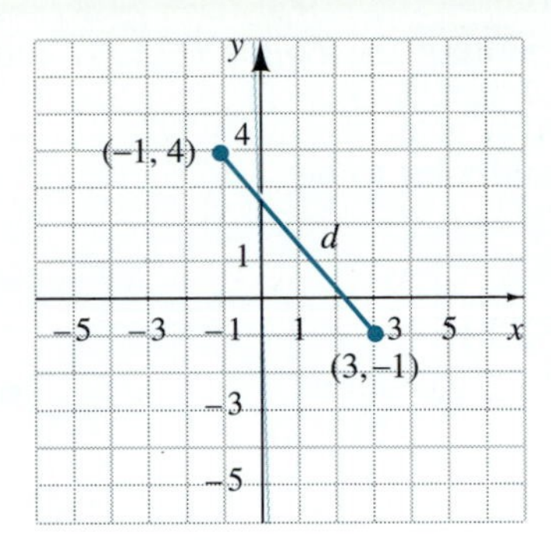

FIGURE 2.19

EXAMPLE 7 Find the distance between the points $(3, -1)$ and $(-1, 4)$.

Solution Even though this example can be done without a diagram, we strongly urge you to draw a diagram to accompany your solution whenever possible. We plot the points in Figure 2.19. We arbitrarily let $(x_1, y_1) = (3, -1)$ and $(x_2, y_2) = (-1, 4)$. Applying the distance formula, we get

$$d = \sqrt{(x_2 - x_1)^2 + (y_2 - y_1)^2}$$

$$d = \sqrt{(-1 - 3)^2 + [4 - (-1)]^2} = \sqrt{(-4)^2 + 5^2} = \sqrt{16 + 25}$$

$$= \boxed{\sqrt{41}} \approx 6.4$$

■

It is worthwhile noting that the distance formula applies equally well to two points that fall on the same horizontal or vertical line. If we apply the distance formula to the points $(4, -2)$ and $(4, 5)$ we get

$$d = \sqrt{(4 - 4)^2 + [5 - (-2)]^2} = \sqrt{0^2 + 7^2} = \sqrt{7^2} = 7$$

which agrees with the result we saw in Example 6.

Sometimes we need to know the midpoint of a line segment. In Figure 2.20 we have drawn a line segment joining the points $P(x_1, y_1)$ and $Q(x_2, y_2)$.

Why do A, B, and C have coordinates $(x_1, 0)$, $(a, 0)$ and $(x_2, 0)$, respectively?

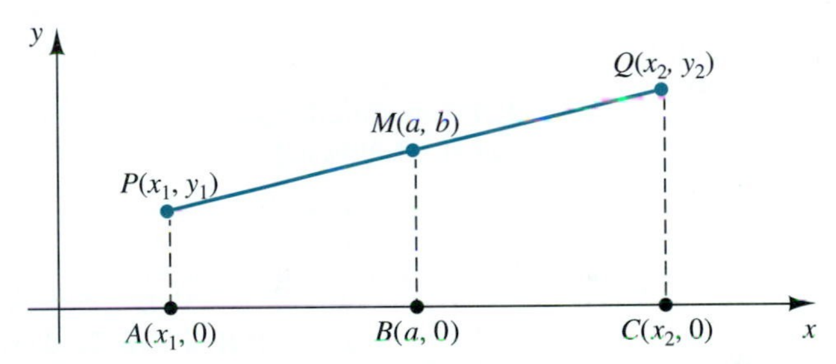

FIGURE 2.20 Finding the midpoint of a line segment

Let $M(a, b)$ be the midpoint of line segment $\overline{PQ}$. We have drawn dashed lines through P, $M(a, b)$, and Q parallel to the y-axis and labeled the points $A(x_1, 0)$, $B(a, 0)$ and $C(x_2, 0)$ on the x-axis.

It is a basic fact from geometry that if parallel lines intercept equal segments on one line, then they intercept equal line segments on any line. In other words, if $\overline{PM} = \overline{MQ}$, then it must follow that $\overline{AB} = \overline{BC}$, which means that B is the midpoint of $\overline{AC}$. But since $\overline{AC}$ is on the x-axis, its midpoint is just the average of the x-coordinates. Therefore, $a = \dfrac{x_1 + x_2}{2}$. Similarly, $b = \dfrac{y_1 + y_2}{2}$. We have thus shown the following.

THE MIDPOINT FORMULA

The midpoint M of the line segment joining the points $P(x_1, y_1)$ and $Q(x_2, y_2)$ is

$$M\left(\frac{x_1 + x_2}{2}, \frac{y_1 + y_2}{2}\right)$$

EXAMPLE 8 Find the midpoint of the line segment joining the points $(-3, 5)$ and $(2, 1)$.

Solution Using the midpoint formula, we find the midpoint to be

$$\left(\frac{-3+2}{2}, \frac{5+1}{2}\right) = \boxed{\left(-\frac{1}{2}, 3\right)}$$

■

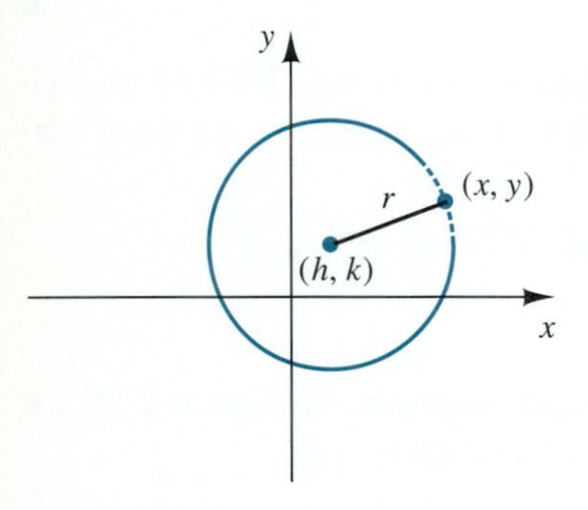

FIGURE 2.21 A circle with center (h, k) and radius r

Circles

A **circle** is defined to be the set of all points in a plane whose distance from a fixed point is constant. The fixed point is called the center, C, and the constant distance from the center to the circle is called the radius, r (where $r > 0$).

Let's place a circle of radius r on the Cartesian plane and center it at the point (h, k). Pick a point in the plane and call it (x, y). See Figure 2.21.

The definition of a circle tells us that in order for (x, y) to be on the circle the distance from the center, (h, k), to (x, y) must be r. By the distance formula, we have

$$\sqrt{(x-h)^2 + (y-k)^2} = r$$

For convenience we eliminate the radical by squaring both sides of this equation to get the following.

> **THE STANDARD FORM OF THE EQUATION OF A CIRCLE**
>
> $$(x-h)^2 + (y-k)^2 = r^2$$
>
> is the equation of a circle with center (h, k) and radius r.

This equation is the **standard form of the equation of a circle**. The center and radius are all that is needed to describe or graph a circle.

EXAMPLE 9 Find the equation of a circle with:

(a) Center $(2, 5)$ and radius 6

(b) Center $\left(\frac{1}{2}, -3\right)$ and radius $\sqrt{2}$

Solution

(a) Looking at the standard form, since the center is $(2, 5)$, we have $h = 2$ and $k = 5$; since the radius is 6, $r = 6$.

$$(x-h)^2 + (y-k)^2 = r^2 \quad \textit{Substitute } h = 2,\ k = 5,\ \textit{and } r = 6.$$

$$(x-2)^2 + (y-5)^2 = 6^2$$

Hence, the equation of the circle is

$$\boxed{(x - 2)^2 + (y - 5)^2 = 36} \quad \textit{which, when multiplied out, is}$$

$$\boxed{x^2 + y^2 - 4x - 10y - 7 = 0}$$

(b) Looking at the standard form, since the center is $\left(\frac{1}{2}, -3\right)$, we have $h = \frac{1}{2}$ and $k = -3$. Since the radius is $\sqrt{2}$, $r = \sqrt{2}$. We have

$$(x - h)^2 + (y - k)^2 = r^2 \quad \textit{Substitute } h = \tfrac{1}{2},\ k = -3, \textit{ and } r = \sqrt{2}.$$

$$\left(x - \frac{1}{2}\right)^2 + (y - (-3))^2 = (\sqrt{2})^2 \quad \textit{Which simplifies to}$$

$$\left(x - \frac{1}{2}\right)^2 + (y + 3)^2 = 2$$

Hence, the equation of the circle is

$$\boxed{\left(x - \frac{1}{2}\right)^2 + (y + 3)^2 = 2} \quad \textit{which, when multiplied out, is}$$

$$\boxed{x^2 + y^2 - x + 6y + \frac{29}{4} = 0}$$

Although both forms of the answer are acceptable, the standard form of the equation of the circle has the significant advantage of making the center and radius easily recognizable. ■

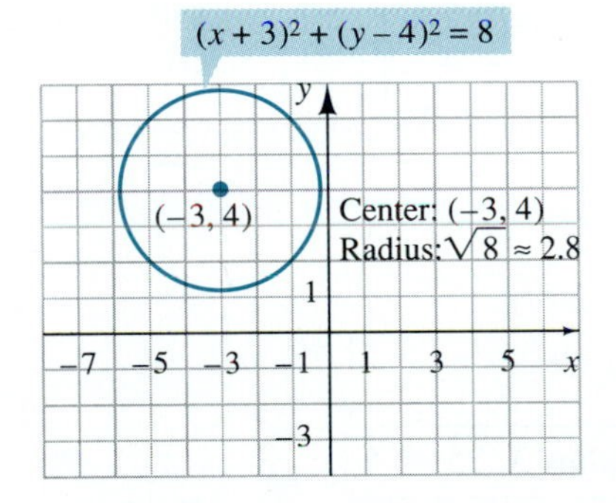

FIGURE 2.22

EXAMPLE 10 Sketch the graph of the following equations:

(a) $(x + 3)^2 + (y - 4)^2 = 8$ **(b)** $x^2 + y^2 = 9$

Solution

(a) We recognize that the given equation is in the standard form for the equation of a circle, $(x - h)^2 + (y - k)^2 = r^2$, and so we can simply read off the values h, k, and r, but we must be careful of the signs.

$$x - h = x + 3 \Rightarrow -h = 3 \Rightarrow h = -3$$

$$y - k = y - 4 \Rightarrow k = 4$$

$$r^2 = 8 \Rightarrow r = \sqrt{8} = 2\sqrt{2}$$

Thus the center of the circle is $(-3, 4)$; the radius is $2\sqrt{2} \approx 2.8$. With this information, we can easily sketch the graph of the circle, which appears in Figure 2.22.

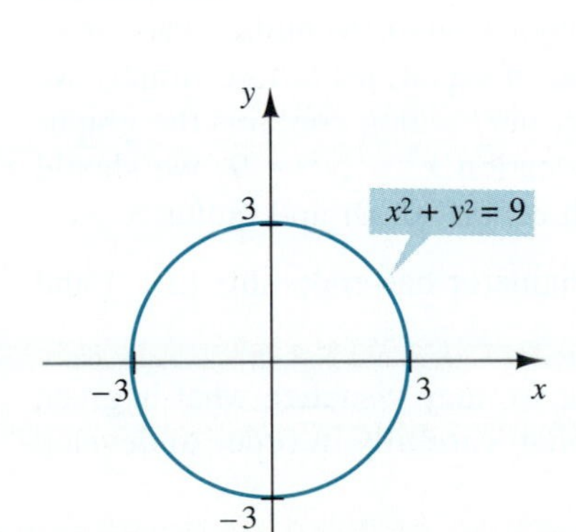

FIGURE 2.23

(b) The equation $x^2 + y^2 = 9$ is also in standard form. (It can be thought of as $(x - 0)^2 + (y - 0)^2 = 3^2$.) Consequently, the center is $(0, 0)$ and the radius is 3. The graph appears in Figure 2.23. ■

Suppose the equation in part (a) of the previous example had not been given to us in standard form, where identifying the center and radius of a circle is straightforward, but rather in its multiplied-out form, $x^2 + y^2 + 6x - 8y + 17 = 0$. Would we recognize that this is the equation of a circle? How do we find the center and radius?

If the equation $(x - h)^2 + (y - k)^2 = r^2$ is multiplied out, we get an equation of the form $x^2 - 2hx + h^2 + y^2 + 2ky + k^2 = r^2$ (remember that h, k, and r are just real numbers). The key feature of this equation is that it is a *second-degree equation in which x^2 and y^2 both appear with coefficient* 1. Thus if we have a second-degree equation in x and y in which the coefficients of x^2 and y^2 are both 1 or can both be made 1, we should recognize that it is the equation of a circle.

So if we are given an equation such as $x^2 + y^2 - 4x + 8y = 5$, we should recognize that it is the equation of a circle, and our goal would be to put it in standard form so that we may read off the center and radius.

The standard form of a circle involves perfect squares; this suggests that the technique of completing the square discussed in Chapter 1 would be useful.

EXAMPLE 11 Find the center and radius of: $x^2 + y^2 - 4x + 8y = 5$

Solution We find the center and radius by completing the square.

$$x^2 + y^2 - 4x + 8y = 5$$ *Group x and y terms together.*

$$(x^2 - 4x \qquad) + (y^2 + 8y \qquad) = 5$$ *Complete the square for each quadratic expression:*

$$\left[\frac{1}{2}(-4)\right]^2 = 4; \quad \left[\frac{1}{2}(8)\right]^2 = 16.$$

Add both numbers to both sides of the equation.

$$(x^2 - 4x + 4) + (y^2 + 8y + 16) = 5 + 4 + 16$$

Rewrite the quadratic expressions in factored form.

$$(x - 2)^2 + (y + 4)^2 = 25$$

Thus, we have a circle with center $(2, -4)$ and radius $\sqrt{25} = 5$. ■

As we proceed through the text, one of our major goals is to build a catalog of basic equations and their graphs—that is, a catalog of equations whose graphs we recognize at a glance. Based upon our work thus far, our catalog contains the graphs of straight lines and circles. If we encounter the equation $x^2 + y^2 = 9$, we should immediately recognize that its graph is a circle with center (0, 0) and radius 3.

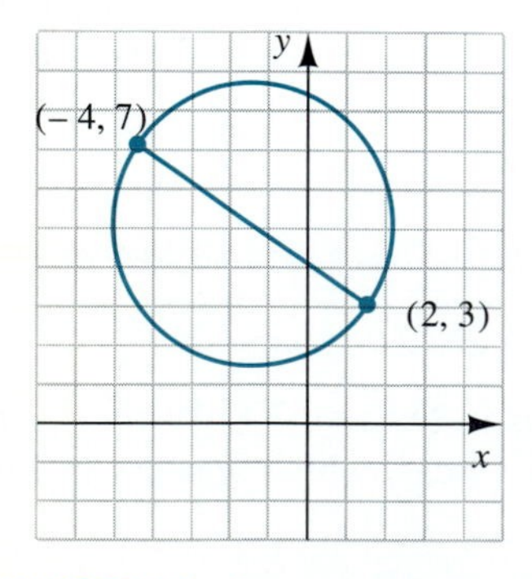

FIGURE 2.24 The circle with diameter having endpoints (2, 3) and (−4, 7)

EXAMPLE 12 Find the equation of a circle whose diameter has endpoints (2, 3) and (−4, 7).

Solution We draw a diagram (Figure 2.24) so that we may visualize what is given and what needs to be found. Let's analyze this problem carefully in order to develop a strategy for the solution.

WHAT DO WE NEED TO FIND?	The equation of the circle with the given diameter.

WHAT IS NEEDED TO FIND AN EQUATION OF A CIRCLE?	The center and the radius.
WHAT INFORMATION IS GIVEN IN THE PROBLEM?	The diameter's endpoints.
HOW CAN I RESTATE THE PROBLEM IN SIMPLER TERMS?	Find the center and radius of the circle given the endpoints of a diameter.
WHAT ADDITIONAL KNOWLEDGE OR INFORMATION DO I NEED IN ORDER TO SOLVE THIS SIMPLER PROBLEM?	That the center of the circle is the midpoint of the diameter (requiring the midpoint formula) and that the radius of a circle is the distance from the center of the circle to any point on the circle, including the given endpoints of a diameter (requiring the distance formula).

To locate the center we find the midpoint between (2, 3) and (−4, 7), which is

$$\left(\frac{2 + (-4)}{2}, \frac{3 + 7}{2}\right) = (-1, 5)$$

Therefore, the center of the circle is (−1, 5).

To find the radius we use the distance formula to find the distance between the center, (−1, 5), and one of the endpoints of the diameter; we use the endpoint (2, 3):

$$r = \sqrt{(-1 - 2)^2 + (5 - 3)^2} = \sqrt{9 + 4} = \sqrt{13}$$

Hence the radius is $\sqrt{13}$.

Can you find the radius using only the endpoints of the diameter?

Having found that the center is (−1, 5), and the radius is $\sqrt{13}$, the equation of the circle is

$$(x + 1)^2 + (y - 5)^2 = (\sqrt{13})^2 \quad \text{or} \quad \boxed{(x + 1)^2 + (y - 5)^2 = 13}$$

■

Two of the major questions on which we will focus our attention throughout the text are:

Question 1: Given an equation, how do we find its graph?

Question 2: Given a graph, how do we find its equation?

In this section we have discussed question 1 as it pertains to straight lines and circles and question 2 as it pertains to circles. That is, given the equations $3x + 5y = 8$ and $x^2 + y^2 = 9$, we recognize that the first is a straight line and the second is a circle, *and* because of this recognition we can sketch their graphs fairly easily. Conversely, if we specify the center and radius of a circle (that is, we know its graph), we can write its equation. In the next two sections we address question 2 as it pertains to straight lines: How do we obtain an equation of a line whose graph we already know?

EXERCISES 2.1

In Exercises 1–14, find the x- and y-intercepts of the given equation.

1. $5y - 4x = 20$
2. $7x - 2y = 14$
3. $x + 3y = 6$
4. $y + 2x = 8$
5. $3x - 8y = 16$
6. $7y + 5x = 10$
7. $2x + 9y + 6 = 0$
8. $6x - 5y + 3 = 0$
9. $x = -3$
10. $y - 5 = 0$
11. $6y = 5x + 8$
12. $4x = 3y - 10$
13. $2x = 3y$
14. $7x - 2y = 0$

In Exercises 15–32, sketch the graph of the given equation in a rectangular coordinate system. Label the intercepts.

15. $7y - 3x = 21$
16. $5x - 4y + 20 = 0$
17. $y = 2x - 8$
18. $y = -3x - 6$
19. $\frac{x}{3} + \frac{y}{2} = 2$
20. $\frac{y}{4} - \frac{x}{5} = 1$
21. $\frac{3y}{2} - x = 6$
22. $\frac{5x}{6} + y = 10$
23. $5x + 2y - 9 = 0$
24. $6y - 3x = 10$
25. $y = \frac{3}{2}$
26. $x = -2$
27. $y = 3x$
28. $y = 3$
29. $5x - 4 = 0$
30. $5x - 4y = 0$
31. $y = \frac{2}{5}x + 4$
32. $y = -\frac{3}{4}x - 6$

In Exercises 33–40, sketch the graph of the given inequality in a rectangular coordinate system.

33. $2y - 3x \leq 12$
34. $5y + 4x + 20 > 0$
35. $4x + 7y > 10$
36. $y - 5x \geq 8$
37. $x + 3 < 0$
38. $y - 4 \leq 2$
39. $y \geq 4x$
40. $3x - 4y < 0$

41. Using d for the vertical axis and t for the horizontal axis, sketch the graph of the equation $d = 5t$.
42. Using D for the vertical axis and p for the horizontal axis, sketch the graph of the equation $D = -20p + 160$.
43. Using s for the vertical axis and t for the horizontal axis, sketch the graph of the equation $s = 0.5t + 15$.
44. Using V for the vertical axis and p for the horizontal axis, sketch the graph of the equation $V = 200 - 8p$.

In Exercises 45–50, find the length and the midpoint of the line segment joining the two points.

45. $A(-3, 4)$ and $B(1, -1)$
46. $P(2, -3)$ and $Q(3, 6)$
47. $R(1, 5)$ and $S(-1, -4)$
48. $E(0, 0)$ and $F(2, 4)$
49. $C(0, 0)$ and $D(a, a)$
50. $T(0, 0)$ and $U(a, 2a)$

51. Draw the triangle ABC with vertices $A(-1, 0)$, $B(4, 0)$, and $C(4, 6)$. What is the area of triangle ABC?
52. Draw the rectangle $ABCD$ with vertices $A(5, 0)$, $B(5, 4)$, $C(-3, 4)$, and $D(-3, 0)$. What is the area of rectangle $ABCD$?
53. Use the converse of the Pythagorean theorem to prove that the three points $A(2, 1)$, $B(7, 2)$, and $C(5, -1)$ are the vertices of a right triangle. What is the area of $\triangle ABC$?
54. Which of the three points $(1, 5)$, $(2, 4)$, and $(3, 3)$ is the furthest from the origin? Would the answer be the same if we interchanged the x- and y-coordinates of each point?
55. Find the value(s) of w so that the points $(0, 3)$ and $(6, w)$ are 10 units apart.
56. Find the value(s) of t so that the points $(3, -2)$ and $(1, t)$ are 4 units apart.
57. Can you find a point $(x, 4)$ that is 2 units from the point $(5, 1)$? Explain.
58. Can you find a point $(-2, y)$ that is 1 unit from the origin? Explain.

In Exercises 59–62, write an equation of the circle with the given center, C, and radius, r.

59. $C = (2, 3)$; $r = 3$
60. $C = (7, -4)$; $r = 5$
61. $C = \left(\frac{1}{2}, 4\right)$; $r = 6$
62. $C = \left(-\frac{3}{4}, -2\right)$; $r = \sqrt{7}$

In Exercises 63–70, identify the center and radius of the given circle.

63. $(x - 3)^2 + (y - 2)^2 = 16$
64. $\left(x - \frac{1}{2}\right)^2 + (y + 3)^2 = 24$
65. $x^2 + y^2 = 16$
66. $x^2 + (y + 2)^2 = 72$
67. $x^2 + y^2 - 6x - 10y = -9$
68. $x^2 + y^2 - 4x + 6y + 4 = 0$
69. $x^2 + y^2 + 6y = 0$
70. $x^2 - 4x + y^2 = 1$

71. Sketch the graph of $(x - 2)^2 + (y + 3)^2 = 4$.
72. Sketch graph of $x^2 + y^2 + 6x - 10y + 33 = 0$.
73. Find an equation of the circle with a diameter having endpoints $(-2, 8)$ and $(4, -5)$.
74. Find an equation of the circle with a diameter having endpoints $(-3, 5)$ and $(-4, 0)$.
75. Find an equation of the circle passing through the point $(2, 6)$, with center $(3, -5)$.

76. Find an equation of the circle passing through the point $(3, -2)$, with center $(2, 5)$.
77. Find the circumference of the circle passing through the point $(-3, 4)$ with center $(5, 2)$.
78. Find the area of the circle passing through the point $(3, -2)$ with center $(5, 2)$.
79. Find the equation of the circle tangent to the x-axis with center $(3, -2)$.
80. Find the equation of the circle tangent to the y-axis with center $(3, -2)$.
81. Find the equation of the circle tangent to the x-axis at $(3, 0)$ and tangent to the y-axis at $(0, -3)$.
82. Find the equations of all circles with radius 6 and tangent to both the x- and y-axes. HINT: Check each quadrant.
83. The circle with center $(0, 0)$ and radius 1 is called the *unit circle*.
  (a) Write an equation of the unit circle.
  (b) Determine which of the following points are on the unit circle.

$$\left(\frac{3}{5}, -\frac{4}{5}\right), \quad \left(-\frac{\sqrt{3}}{2}, \frac{1}{2}\right), \quad \left(\frac{2}{3}, \frac{3}{5}\right)$$

QUESTIONS FOR THOUGHT

84. Does the equation $\frac{x}{4} + \frac{y}{5} = 2$ fit the form of a first-degree equation in two variables, i.e., $Ax + By = C$? If so, what are the values A, B, and C? Are the values A, B, and C unique? Explain.
85. Use the distance formula to prove that the point $M\left(\frac{x_1 + x_2}{2}, \frac{y_1 + y_2}{2}\right)$ is equidistant from the points $P(x_1, y_1)$ and $Q(x_2, y_2)$.
  (a) Does this prove that the point $M\left(\frac{x_1 + x_2}{2}, \frac{y_1 + y_2}{2}\right)$ is the midpoint of the line $P(x_1, y_1)$ and $Q(x_2, y_2)$? Explain.
  (b) Can you use the distance formula to prove that M is the midpoint of the segment $\overline{PQ}$? How?
  (c) Do you think this proof is easier or harder than that offered in the text's presentation of the midpoint formula?
86. Define x- and y-intercepts in two ways:
  (a) In terms of the graph of an equation
  (b) In terms of an equation of a graph
87. How could the distance formula be used to prove that the three points $P(-3, 4)$, $Q(0, 1)$, and $R(3, -2)$ are collinear (lie on the same line)?
88. We have been tacitly assuming that we always use units of the same length along the x- and y-axes; however this is not necessarily the case. Describe how you would choose units along the coordinate axes to sketch the graph of $y = 0.01x + 2$.
89. In our discussion of the derivation of the midpoint formula, we found the x-coordinate of the midpoint can be found by averaging the two x-coordinates. Suppose $x_1 < x_2$; show that we get the same result if instead we add one-half the distance between x_1 and x_2 to x_1.
90. In the derivation of the distance formula we stated that $|x_2 - x_1|^2 = (x_2 - x_1)^2$. Prove this fact. HINT: Recall that $|x| = x$ if $x \geq 0$ or $-x$ if $x < 0$.

2.2 Slope

In the last section, we raised the two general questions: Given an equation, how do we find its graph? Given a graph, how do we find its equation? In the last section, we answered the first question for first-degree equations in two variables. We will repeat these questions for a variety of graphs and equations as we proceed through this book. In this section we begin to answer the second question when the graph is a straight line. The ideas we develop in this section play a pivotal role in calculus as well.

Keep in mind that, as we pointed out before, an equation is a condition that all points on the graph must satisfy. Once we find such a condition, we obtain an equation by translating this condition mathematically.

Suppose we are given any nonvertical line L. We choose any four distinct points on L and label the points $P(x_1, y_1)$, $Q(x_2, y_2)$, $S(x_3, y_3)$, and $T(x_4, y_4)$. We have also formed right triangles PQR and STU. See Figure 2.25.

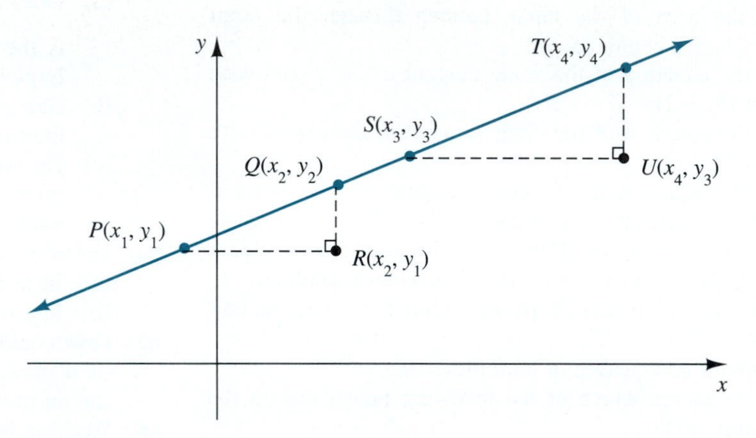

FIGURE 2.25 Triangles *PQR* and *STU* are similar.

Since $\overline{PR}$ and $\overline{SU}$ are parallel, as are $\overline{QR}$ and $\overline{TU}$, we have $\angle QPR \cong \angle TSU$ (they are corresponding angles formed by parallel lines), and $\angle R$ and $\angle U$ are both right angles. Consequently, triangles PQR and STU are similar triangles; therefore, their corresponding sides are in proportion. That is,

$$\frac{|\overline{QR}|}{|\overline{PR}|} = \frac{|\overline{TU}|}{|\overline{SU}|}$$

NOTE: $|\overline{QR}|$ means the length of line segment QR.

$$\frac{y_2 - y_1}{x_2 - x_1} = \frac{y_4 - y_3}{x_4 - x_3}$$

In other words, because triangles PQR and STU are similar, whenever we move from one point on a nonvertical line to another point on the line, the ratio of the change in y-coordinates to the change in x-coordinates remains *constant* for each line. This is exactly what we are looking for—a condition that all points on the line must satisfy.

Will the ratio $\frac{y_2 - y_1}{x_2 - x_1}$ be constant if the points (x_1, y_1) and (x_2, y_2) are on a graph that is not a straight line?

We will actually derive an equation for a line from this condition in the next section. The remainder of this section is devoted to amplifying on the idea that the ratio of the change in y to the change in x is constant. We begin with the following definition.

DEFINITION Let $P_1(x_1, y_1)$ and $P_2(x_2, y_2)$ be any two distinct points on a nonvertical line L. The **slope** of the line L, denoted by m, is given by

$$m = \frac{y_2 - y_1}{x_2 - x_1}$$

Note that based upon the preceding discussion, the slope of a nonvertical line is well defined. That is, every nonvertical line has a unique slope. Regardless of which two points we choose, the ratio of the change in y to the change in x will be constant.

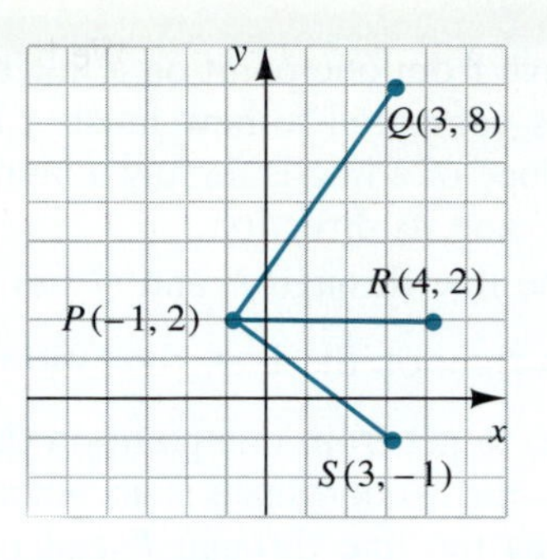

FIGURE 2.26

It does not matter which point is first and which is second, as long as we are consistent for both the x- and y-coordinates.

EXAMPLE 1 Find the slope of the line joining the two points:

(a) $P(-1,2)$ and $Q(3,8)$ **(b)** $P(-1,2)$ and $S(3,-1)$
(c) $P(-1,2)$ and $R(4,2)$

Solution Although it is not always necessary, it is generally a good idea to draw a diagram to help visualize the given information. See Figure 2.26.

(a) Using the points $P(-1,2)$ and $Q(3,8)$ and the formula for the slope of a line, we compute the slope of the line passing through P and Q as

$$m = \frac{8-2}{3-(-1)} = \frac{6}{4} = \frac{3}{2} \quad \text{or} \quad m = \frac{2-8}{-1-3} = \frac{-6}{-4} = \frac{3}{2}$$

Thus the slope is $\boxed{m = \frac{3}{2}}$. Remember to subtract the x-coordinates in the same order as the y-coordinates.

(b) Using the points $P(-1,2)$ and $S(3,-1)$ we compute the slope of the line passing through P and S as

$$m = \frac{-1-2}{3-(-1)} = \frac{-3}{4} = \boxed{-\frac{3}{4}}$$

(c) Using the points $P(-1,2)$ and $R(4,2)$ we compute the slope of the line passing through P and R as

$$m = \frac{2-2}{4-(-1)} = \frac{0}{5} = \boxed{0}$$ ■

What does this number, the slope, tell us about a line? We begin with the following ground rule.

> Whenever we describe a graph, we describe it as we move from *left to right*. In other words, we describe it for increasing values of x.

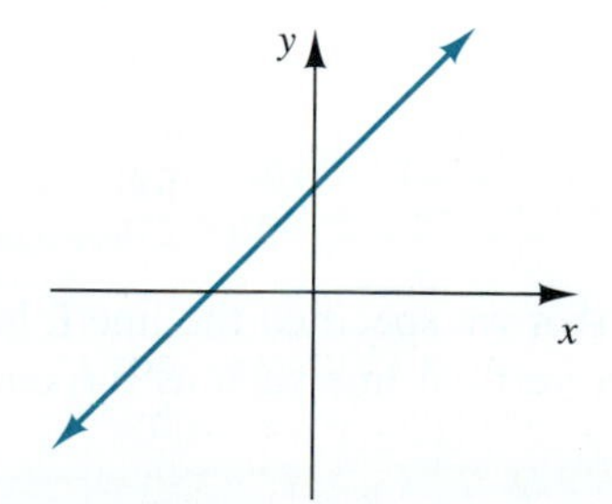

This line is rising.
This means the y-values are increasing as we move from left to right.

(a)

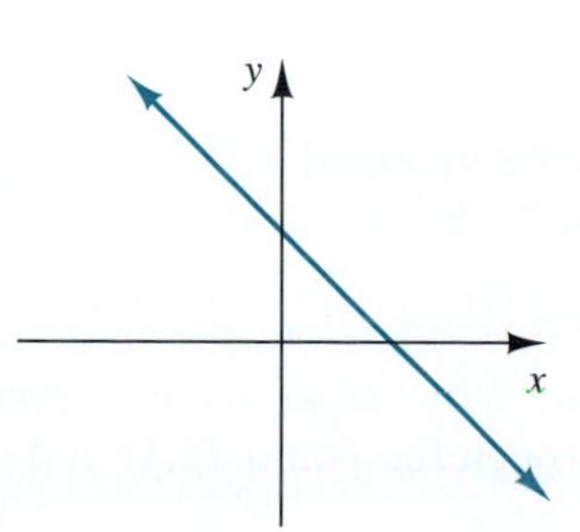

This line is falling.
This means the y-values are decreasing as we move from left to right.

(b)

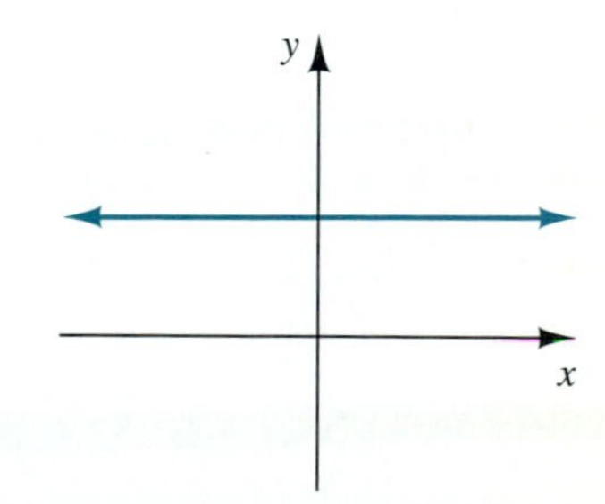

This line is neither rising nor falling.
This means the y-values are constant as we move from left to right.

(c)

FIGURE 2.27

The slope is actually a rate of change. As we move from one point on a line to another, the slope tells us how much y is changing as compared to how much x is changing. What is important to recognize is that the slope of a line is merely a number that gives us information about a line's "steepness" and its direction.

For instance, if we look back at Figure 2.26, the line through P and Q has a slope of $\frac{3}{2}$, which means that as we move from left to right on the line, the ratio of the change in y to the change in x is 3 to 2. Therefore, to get from one point on the line to another point on the line, a change of 2 units in the x-coordinate is accompanied by a change of 3 units in the y-coordinate. Thus the line through P and Q, which has a *positive* slope, rises (goes up) as we move from left to right.

The line through P and S in Figure 2.26 has a slope of $\frac{-3}{4}$, which means that as we move from left to right on the line, a 4-unit change in the x-coordinate is accompanied by a -3-unit change in the y-coordinate (meaning the y-coordinate goes down 3 units). Thus the line through P and S, which has a *negative* slope, falls (goes down) as we move from left to right. In other words, on a line with positive slope, the y-values increase as we move from left to right, whereas on a line with negative slope the y-values decrease as we move from left to right.

What does the slope actually tell us about a line?

Figure 2.28 illustrates lines with various slopes passing through the point (3, 2). Note that the larger the slope in absolute value, the steeper the line.

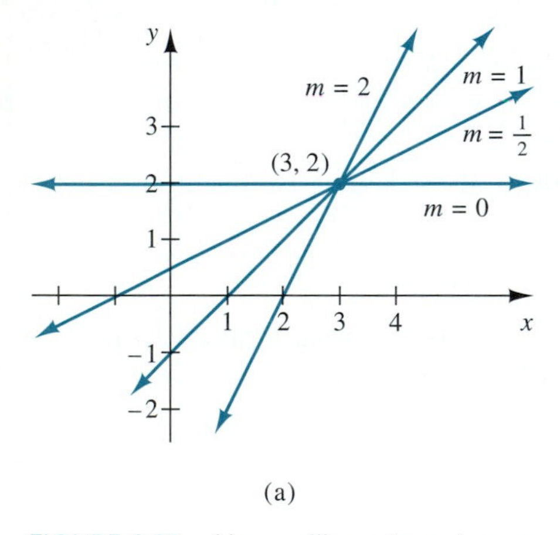

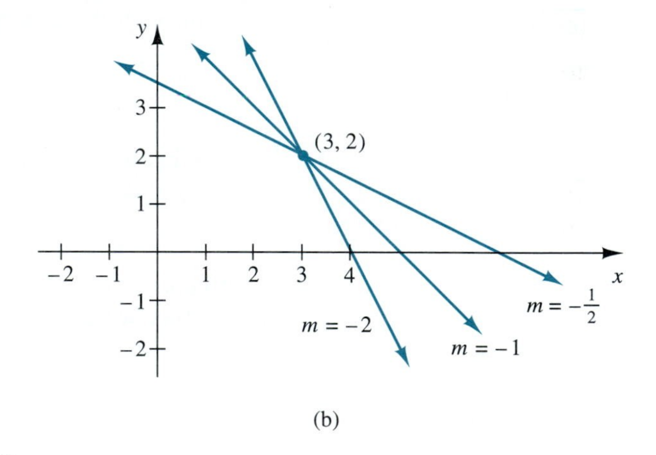

FIGURE 2.28 Lines with various slopes passing through (3, 2)

Looking back at our definition of the slope, note that we specified the line L be nonvertical. Why? If we try to compute the slope of a vertical line such as the one passing through the points (2,1) and (2, 5), we get

$$m = \frac{5-1}{2-2} = \frac{4}{0} \quad \text{which is undefined}$$

Thus *the slope of a vertical line is undefined.*

Be careful not to confuse a line that has slope 0 and is horizontal with a line whose slope is undefined and is vertical. See Figure 2.29.

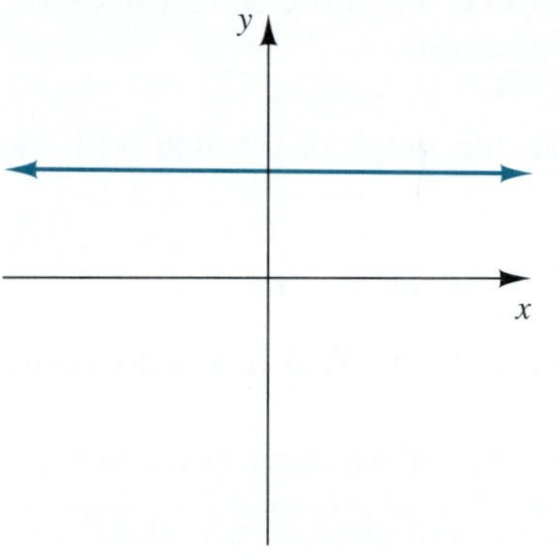

A horizontal line; its slope is 0

(a)

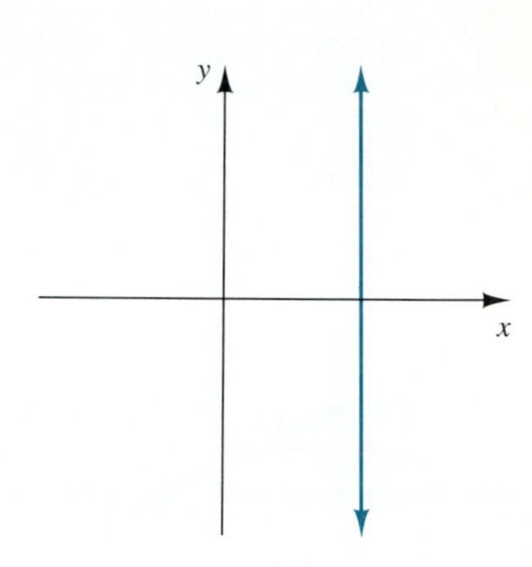

A vertical line; its slope is undefined

(b)

FIGURE 2.29

How does a line of slope 2 compare to a line of slope $\frac{1}{2}$?

How does a line of slope 2 compare to a line of slope -2?

SUMMARY OF SLOPES

1. A line with positive slope rises as we move from left to right.
2. A line with negative slope falls as we move from left to right.
3. A line with zero slope is horizontal.
4. A line with undefined slope is vertical.

EXAMPLE 2 Find the slope of the line $3x - 4y = 18$.

Solution Based upon the definition, in order to find a slope we need two points on the line. We can compute the slope using any two points on the line $3x - 4y = 18$. For convenience, we choose the points containing the intercepts of this line: $\left(0, -\frac{9}{2}\right)$, and $(6, 0)$.

Using the formula for the slope of a line, we get

$$m = \frac{-\frac{9}{2} - 0}{0 - 6} = \frac{-\frac{9}{2}}{-6} = \frac{3}{4}$$

Thus the slope is $\boxed{m = \frac{3}{4}}$.

Calculate the slope of the line using the points $(2, -3)$ and $(10, 3)$.

If we choose any other pair of points that satisfy the equation $3x - 4y = 18$, such as $(2, -3)$ and $(10, 3)$, we will still get the same answer: $m = \frac{3}{4}$. This again illustrates that the slope of a line is independent of the points chosen on the line.

In the next section we will discover another method for determining the slope of a line given its equation. ■

EXAMPLE 3 Sketch the graph of the line with slope $-\frac{1}{2}$ passing through the point $(2, 3)$.

Solution In order to graph the line we need to have 2 points on the line. We are given one point; how do we find a second point? The given slope of $-\frac{1}{2}$ can be thought of as $\frac{-1}{2}$ or $\frac{1}{-2}$. If we think of the slope as $\frac{-1}{2}$, then a 2-unit change in x is accompanied by a -1-unit change in y; that is, to get from one point on the line to another, we move 2 units to the right and 1 unit down. Alternatively, viewing the slope as $\frac{1}{-2}$, we would move 2 units to the left and 1 unit up. Either way we end up with the same line. See Figure 2.30. ■

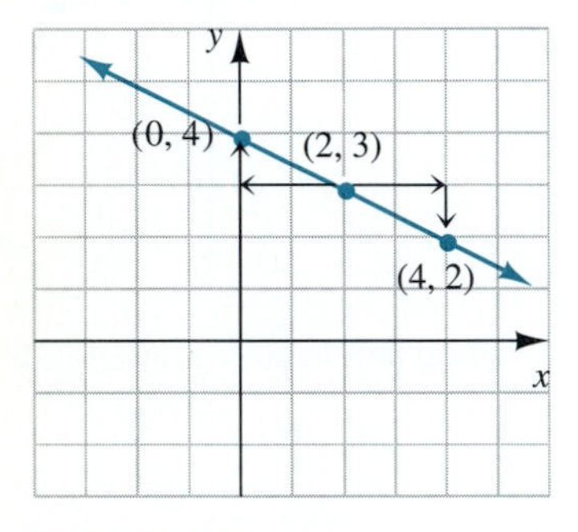

FIGURE 2.30 The graph of the line with slope $-\frac{1}{2}$ passing through $(2, 3)$

What is most important to recognize here is that once we specify one point and a slope, the line is completely determined.

> A point and a slope determine a line.

Throughout your study of mathematics, parallel and perpendicular lines have played a prominent role, and they are special with regard to their slopes as well.

THEOREM 2.2 Let L_1 and L_2 be nonvertical lines with slopes m_1 and m_2 respectively. Then

1. L_1 and L_2 are parallel if and only if their slopes are equal, i.e., $m_1 = m_2$.
2. L_1 and L_2 are perpendicular if and only if their slopes are negative reciprocals of each other, i.e.,

$$m_2 = -\frac{1}{m_1} \qquad \text{or equivalently} \qquad m_1 \cdot m_2 = -1$$

Proof Part 1 of this theorem seems quite plausible. The fact that lines are parallel suggests that they have the same steepness, which in turn means that they have the same slope. A formal proof of this is outlined in Exercise 47 at the end of this section.

Part 2 of this theorem is not nearly as intuitive and we offer a formal proof here. For the sake of simplicity, let's assume that lines L_1 and L_2 intersect at the origin. (A similar proof works in the general case.) Let P and Q be, respectively, the points where the lines L_1 and L_2 intersect the line $x = 1$, as indicated in Figure 2.31.

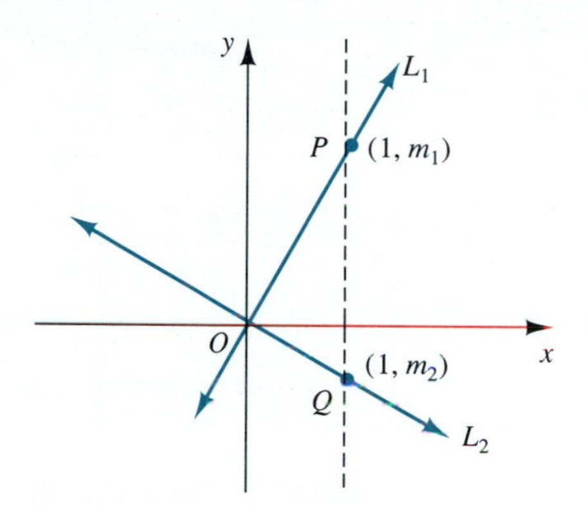

FIGURE 2.31

Since the slope of L_1 is $m_1 = \frac{m_1}{1}$, as x changes by 1 unit, y changes by m_1 units, and so the coordinates of P are $(1, m_1)$. Similarly, the coordinates of Q are $(1, m_2)$.

Now, L_1 is perpendicular to L_2 if and only if $\angle POQ$ is a right angle. Applying the Pythagorean theorem, $\angle POQ$ is a right angle if and only if

$$|\overline{PQ}|^2 = |\overline{OP}|^2 + |\overline{OQ}|^2$$

Applying the distance formula, we get

$$\begin{aligned}
\left(\sqrt{(1-1)^2 + (m_1 - m_2)^2}\right)^2 &= \left(\sqrt{(m_1 - 0)^2 + (1 - 0)^2}\right)^2 + \left(\sqrt{(m_2 - 0)^2 + (1 - 0)^2}\right)^2 \\
\left(\sqrt{(m_1 - m_2)^2}\right)^2 &= \left(\sqrt{(m_1)^2 + 1}\right)^2 + \left(\sqrt{(m_2)^2 + 1}\right)^2 \\
(m_1 - m_2)^2 &= (m_1)^2 + 1 + (m_2)^2 + 1 \\
(m_1)^2 - 2m_1m_2 + (m_2)^2 &= (m_1)^2 + (m_2)^2 + 2 \qquad \textit{Simplifying this equation, we get} \\
-2m_1m_2 &= 2 \\
m_1m_2 &= -1
\end{aligned}$$

as required, which proves part 2 of the theorem in the case where the two lines intersect at the origin. ■

EXAMPLE 4 Find the value of t so that the line passing through the points $A(2, t)$ and $B(5, -2)$ is parallel to the line passing through the points $C(-6, 3)$ and $D(0, -4)$.

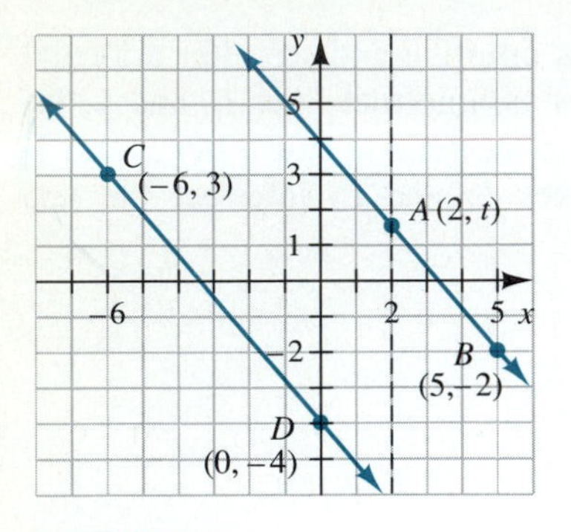

FIGURE 2.32

Solution A diagram will help us visualize exactly what this example is asking. In Figure 2.32 we have plotted the points A, B, C, and D. Note that although we do not know the exact location of the point $A(2, t)$, we do know that it must fall on the line $x = 2$. The example is asking what value of t will make the line through A and B parallel to the line through C and D.

The fact that the lines are to be parallel tells us that the slopes must be equal, which gives us the equation

$$m_{\overline{AB}} = m_{\overline{CD}}$$

$$\frac{t-(-2)}{2-5} = \frac{3-(-4)}{-6-0}$$

$$\frac{t+2}{-3} = \frac{7}{-6} \qquad \textit{Multiply both sides of the equation by } -3.$$

$$t + 2 = \frac{7}{2} \Rightarrow \boxed{t = \frac{3}{2}}$$

■

EXAMPLE 5 Prove that the points $P(-3, 1)$, $Q(-2, -2)$, $R(1, -1)$, and $S(0, 2)$ are the vertices of a rectangle.

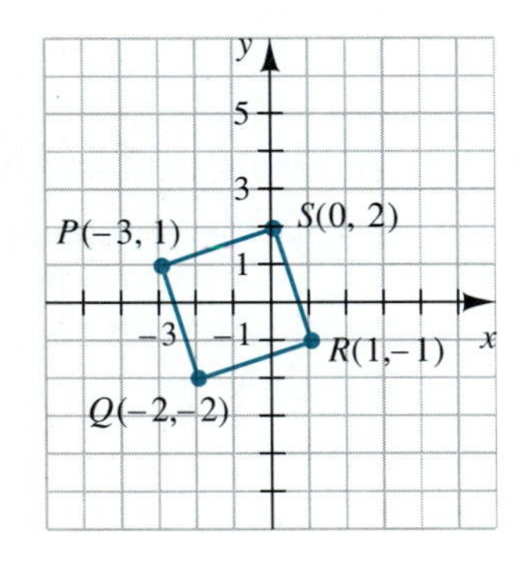

FIGURE 2.33

Solution There are a number of ways we can prove that $PQRS$ is a rectangle. We will use the idea of slope. In Figure 2.33 we draw the quadrilateral formed by the four given points. Let's analyze this problem carefully in order to develop a strategy for the solution.

WHAT DO WE NEED TO DO? (WE USE THE DIAGRAM TO HELP IN THIS ANALYSIS.)	Prove that the four given points form the vertices of a rectangle.
WHAT IS A RECTANGLE?	A rectangle is a parallelogram (opposite sides are parallel); all angles are right angles.
WHAT IS NEEDED TO SHOW THAT A 4-SIDED FIGURE IS A RECTANGLE?	We need to show that the opposite sides of the figure are parallel and the adjacent sides are perpendicular.
HOW DO WE DEMONSTRATE THAT THE SIDES ARE PARALLEL AND/OR PERPENDICULAR?	Find the slopes of the opposite sides. If these slopes are equal, then the opposite sides are parallel. Find the slopes of the adjacent sides. If these slopes are negative reciprocals of each other, then the adjacent sides are perpendicular.
HOW CAN I RESTATE THE PROBLEM IN SIMPLER TERMS?	We can prove that $PQRS$ is a rectangle by showing that sides $\overline{PQ}$ and $\overline{RS}$ have equal slopes; that sides $\overline{PS}$ and $\overline{QR}$ have equal slopes; and that the slopes of $\overline{PQ}$ and $\overline{QR}$ are negative reciprocals of each other.

We find the slopes of the four sides of the quadrilateral $PQRS$.

$$m_{\overline{PQ}} = \frac{1 - (-2)}{-3 - (-2)} = \frac{3}{-1} = -3 \qquad m_{\overline{RS}} = \frac{2 - (-1)}{0 - 1} = \frac{3}{-1} = -3$$

$$m_{\overline{QR}} = \frac{-2 - (-1)}{-2 - 1} = \frac{-1}{-3} = \frac{1}{3} \qquad m_{\overline{PS}} = \frac{2 - 1}{0 - (-3)} = \frac{1}{3}$$

$\overline{PQ}$ and $\overline{RS}$ are parallel since their slopes are equal. Similarly, $\overline{PS}$ is parallel to $\overline{QR}$. Therefore, $PQRS$ is a parallelogram. In addition, since the slopes of the adjacent sides are negative reciprocals of each other, we also know that all the angles are right angles, which makes $PQRS$ a rectangle. ■

In the next section we will complete our discussion of the equations of straight lines.

EXERCISES 2.2

In Exercises 1–6, sketch the line through the given points and compute its slope.

1. $(2, 1)$ and $(5, 6)$
2. $(-3, 2)$ and $(4, -1)$
3. $(0, 3)$ and $(3, 0)$
4. $(-4, 0)$ and $(0, -4)$
5. $(-2, -1)$ and $(1, -3)$
6. $(-3, -4)$ and $(1, 5)$

In Exercises 7–20, find the slope of the line passing through the given points.

7. $(2, 7)$ and $(4, 10)$
8. $(-1, 5)$ and $(2, -3)$
9. $(-4, -3)$ and $(2, 1)$
10. $(-2, 6)$ and $(-4, -8)$
11. $(-4, 2)$ and $(6, 2)$
12. $(-1, 3)$ and $(-1, -5)$
13. $\left(\frac{1}{2}, \frac{3}{5}\right)$ and $\left(\frac{3}{4}, \frac{2}{3}\right)$
14. $\left(\frac{1}{6}, \frac{1}{4}\right)$ and $\left(-\frac{1}{2}, 2\right)$
15. $(2, \sqrt{3})$ and $(4, \sqrt{27})$
16. $(\sqrt{8}, 4)$ and $(\sqrt{18}, -2)$
17. (a, a^2) and (b, b^2) $(a \neq b)$
18. (b, a^2) and (a, b^2) $(a \neq b)$
19. (r, s) and $(r + s, 2s)$ $(s \neq 0)$
20. $(-p, n)$ and $(2p, 5n)$ $(p \neq 0)$

In Exercises 21–26, sketch the graph of the line passing through the given point and having the indicated slope.

21. $(-1, 2)$, $m = 3$
22. $(2, -3)$, $m = -2$
23. $(4, 0)$, $m = -\frac{2}{3}$
24. $(0, -3)$, $m = \frac{1}{4}$
25. $(-5, 1)$, $m = 0$
26. $(-5, 1)$, m is undefined

In Exercises 27–32, determine whether the line passing through the points P_1 and P_2 is parallel to or perpendicular to (or neither) to the line passing through the points P_3 and P_4.

27. $P_1(2, 1)$, $P_2(4, 3)$, $P_3(-2, -1)$, $P_4(-4, -3)$
28. $P_1(-1, 4)$, $P_2(3, 2)$, $P_3(2, -3)$, $P_4(3, -1)$
29. $P_1(-3, -2)$, $P_2(-1, 1)$, $P_3(5, 4)$, $P_4(7, 1)$
30. $P_1(2, 9)$, $P_2(5, 10)$, $P_3(-4, -5)$, $P_4(-1, -6)$
31. $P_1(1, 4)$, $P_2(-3, 4)$, $P_3(-2, 7)$, $P_4(-2, -3)$
32. $P_1(1, 2)$, $P_2(3, 6)$, $P_3(-6, -7)$, $P_4(-3, -1)$
33. Find the value of c so that the line passing through the points $(2, 5)$ and $(-4, c)$ has slope $-\frac{1}{2}$.
34. Find the value of a so that the line passing through the points $(-2, 4)$ and $(a, 1)$ has slope $\frac{2}{3}$.
35. Find the value of t so that the line through the points $(0, t)$ and $(t, -1)$ is parallel to the line through the points $(1, 2)$ and $(2, -3)$.
36. Find the value of c so that the line through the points $(c, 1)$ and $(1, c)$ is perpendicular to the line through the points $(-2, 5)$ and $(3, -4)$.
37. Find the value(s) of h so that the line through the points $(h, 1)$ and $(-1, h)$ is perpendicular to the line through the points $(7, h)$ and $(h, 2)$.
38. Find the value(s) of t so that the line through the points (t, t) and $(3, 4)$ is parallel to the line through the points $(-9, t)$ and $(t, 9)$.
39. Prove that the points $P(-3, 0)$, $Q(1, 2)$, and $R(3, -2)$ are the vertices of a right triangle.
40. Prove that the points $A(0, -4)$, $B(4, -2)$, $C(5, 4)$, and $D(1, 2)$ are the vertices of a parallelogram.
41. Prove that the points $P(-3, -2)$, $Q(1, 4)$, $R(-2, 4)$, and $S(-4, 1)$ are the vertices of a trapezoid.
42. Prove that the points $A(-2, -5)$, $B(2, -4)$, $C(1, 0)$, and $D(-3, -1)$ are the vertices of a square.
43. Let $ABCD$ be the parallelogram with vertices $A(0, 0)$, $B(4, 0)$, $C(5, 2)$, and $D(1, 2)$. Prove that the quadrilateral formed by joining the midpoints of the sides of $ABCD$ is also a parallelogram.
44. Let $ABCD$ be the square with vertices $A(0, 0)$, $B(6, 0)$, $C(6, 6)$, and $D(0, 6)$. Prove that the quadrilateral formed by joining the midpoints of the sides of $ABCD$ is also a square.

45. The accompanying figure illustrates four lines with slopes m_1, m_2, m_3, and m_4. List these slopes in increasing order.

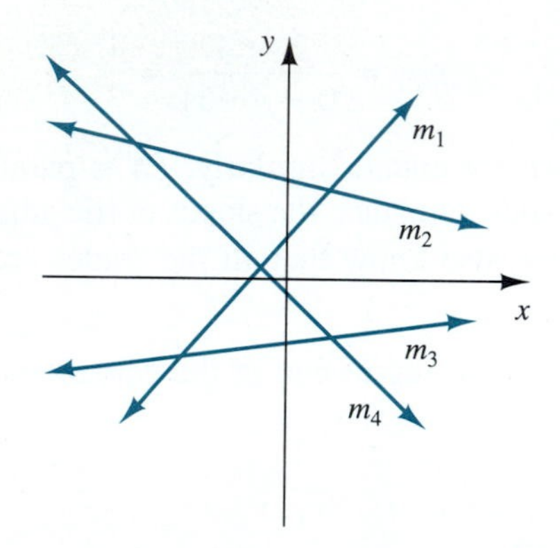

46. The accompanying figure illustrates three lines with equations as indicated.

(a) Show that the values b_1, b_2, and b_3 are the y-intercepts of their respective lines.

(b) List the slopes in decreasing order and the y-intercepts in increasing order.

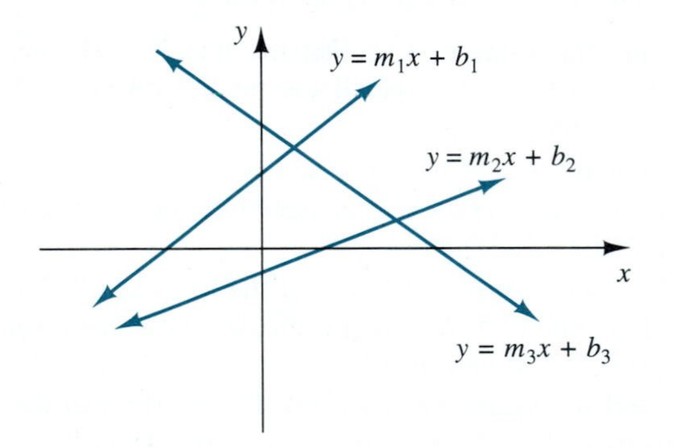

QUESTIONS FOR THOUGHT

47. In this exercise we outline proof of part 1 of Theorem 2.2, which says that two nonvertical lines are parallel if and only if their slopes are equal. Let L_1 and L_2 be two nonvertical lines with slopes m_1 and m_2, respectively, crossed by line M, as in the accompanying figure. We have drawn $\overline{ST}$ and $\overline{PQ}$ parallel to the x-axis, and $\overline{UT}$ and $\overline{RQ}$ parallel to the y-axis. Justify each of the following statements. (See figure, top right.)

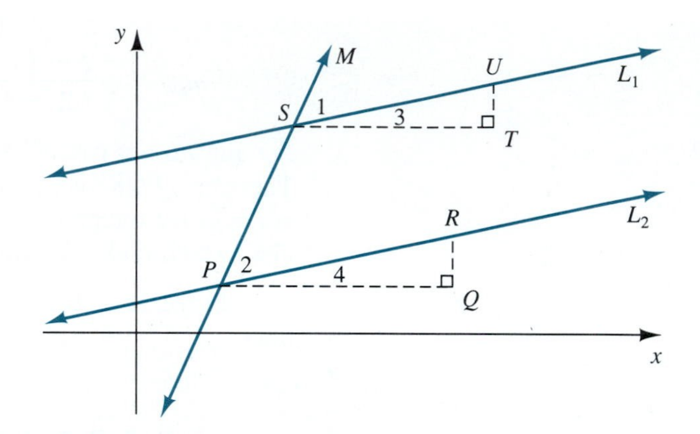

(a) Since $\overline{PQ}$ is parallel to $\overline{ST}$, $\angle MST \cong \angle SPQ$.

(b) L_1 is parallel to L_2 if and only if $\angle 1 \cong \angle 2$.

(c) L_1 is parallel to L_2 if and only if $\angle 3 \cong \angle 4$.

(d) Since $\angle T$ and $\angle Q$ are right angles, L_1 is parallel to L_2 if and only if $\triangle STU$ is similar to $\triangle PQR$.

(e) L_1 is parallel to L_2 if and only if
$$\frac{|\overline{UT}|}{|\overline{ST}|} = \frac{|\overline{RQ}|}{|\overline{PQ}|}.$$

(f) L_1 is parallel to L_2 if and only if $m_1 = m_2$.

48. How would you describe a line whose slope is positive? Negative? Zero? Undefined?

49. How can the idea of slope be used to prove that the three points $(-2, -3)$, $(2, -2)$, and $(6, -1)$ are collinear?

50. What happens to our visualization of a line with slope 3 if we don't insist that the units along both axes be the same? Sketch the graph of a line passing through the point $(2, 3)$ with slope 3 if the units along the x-axis are twice as large as the units along the y-axis. Now draw the graph again if the units along the y-axis are twice as large as the units along the x-axis.

51. Prove part 2 of Theorem 2.2 in the case where the two lines intersect at the point (a, b).

52. As was mentioned before, the slope of a line can be thought of as a rate of change. Suppose that the height h, in meters, of an object above the ground after t seconds is given by the equation $h = 3t + 2$ for $t \geq 0$. Thus when $t = 4$, $h = 14$, meaning that the object is 14 meters above the ground after 4 seconds.

(a) How far has the object traveled from time $t = 5$ to time $t = 8$?

(b) What is the average speed of the object for these 3 seconds? Remember that average speed is computed by dividing the distance traveled by the time elapsed.

(c) Repeat parts (a) and (b) as t changes from time $t = 10$ to time $t = 15$.

(d) Letting t be the horizontal axis and h be the vertical axis, sketch the graph of $h = 3t + 2$ for $t \geq 0$. What is the slope of this line?

(e) How are the average speed of the object and the slope of the line related?

2.3 Equations of a Line

We are now ready to answer the question raised earlier: Given a line, how do we find its equation? Suppose we are given a line L with slope m that passes through the point (x_1, y_1), as indicated in Figure 2.34.

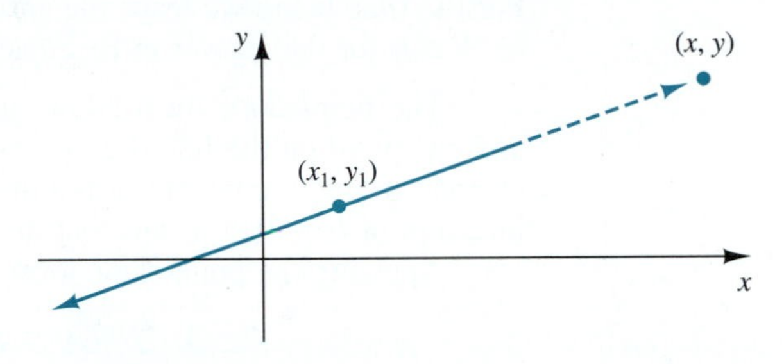

FIGURE 2.34

If (x, y) is any other point, then the condition it must satisfy to be on the line is that the slope of the line as determined by the points (x_1, y_1) and (x, y) must be m. If the slope is m, then the point is on the line, and if the slope is not m, then the point is not on the line.

Algebraically, we translate this condition as

$$\frac{y - y_1}{x - x_1} = m \qquad \text{for } x \neq x_1$$

Multiplying both sides of the equation by $x - x_1$ we get the following.

POINT-SLOPE FORM OF AN EQUATION OF A STRAIGHT LINE

An equation of the line with slope m passing through the point (x_1, y_1) is

$$y - y_1 = m(x - x_1)$$

Keep in mind that in the *point-slope formula,* (x_1, y_1) denotes the *given* point and (x, y) denotes any other point on the line.

EXAMPLE 1 Write an equation of the line with slope $\frac{2}{3}$ that passes through the point $(-2, 5)$.

Solution We are given exactly the information necessary to use the point-slope form for the equation of a line. The given point $(-2, 5)$ corresponds to (x_1, y_1) and the slope $m = \frac{2}{3}$.

$$y - y_1 = m(x - x_1) \quad \textit{Substitute } (-2, 5) \textit{ for } (x_1, y_1) \textit{ and } \frac{2}{3} \textit{ for } m.$$

$$y - 5 = \frac{2}{3}[x - (-2)] \text{ and so the equation of the line is } \boxed{y - 5 = \frac{2}{3}(x + 2)}$$

For the time being we leave the answer in this form. We discuss the different possible forms for the answer in Example 3. ■

The point-slope form allows us to write an equation of a line given its slope and *any* point on the line. Let's see what happens if the given point happens to correspond to the y-intercept of the line; that is, suppose the line has slope m and a y-intercept of b, which means that the line passes through the point $(0, b)$.

Applying the point-slope form we get

$$y - y_1 = m(x - x_1) \quad \textit{Substitute } (0, b) \textit{ for } (x_1, y_1).$$
$$y - b = m(x - 0)$$
$$y - b = mx$$

This equation is usually solved explicitly for y giving the following.

THE SLOPE-INTERCEPT FORM OF AN EQUATION OF A STRAIGHT LINE

An equation of the line with slope m and y-intercept b is

$$y = mx + b$$

EXAMPLE 2 Write an equation of the line with slope -2 passing through the point $(0, -3)$.

Solution Since the line passes through the point $(0, -3)$, we know that the y-intercept is -3. Therefore, from the given information, we can apply the slope-intercept form with $m = -2$ and $b = -3$.

$$\boxed{y = -2x - 3}$$ ■

EXAMPLE 3 Write an equation of the line passing through the points $(1, 2)$ and $(3, -5)$.

Solution This example illustrates that in writing an equation of a line (as long as the line is not vertical), we always have the choice of using either the point-slope or slope-intercept form. In either case we need to compute the slope of the given line. The slope is

$$m = \frac{-5 - 2}{3 - 1} = \frac{-7}{2}$$

If we choose to use the point-slope form, we may use either of the given points.

Using the point $(1, 2)$ we get $\boxed{y - 2 = -\frac{7}{2}(x - 1)}$.

Using the point $(3, -5)$ we get

$$y - (-5) = -\frac{7}{2}(x - 3)$$

$$\boxed{y + 5 = -\frac{7}{2}(x - 3)}.$$

If we choose to use the slope-intercept form, we proceed as follows.

$$y = mx + b$$

We substitute $m = -\frac{7}{2}$. We need to find the value of b.

$$y = -\frac{7}{2}x + b$$

Since we know the line passes through the points $(1, 2)$ and $(3, -5)$, either of these points satisfy the equation. We choose to substitute $(1, 2)$ and solve for b.

$$2 = -\frac{7}{2}(1) + b \Rightarrow b = \frac{11}{2}$$

Therefore, we get

$$\boxed{y = -\frac{7}{2}x + \frac{11}{2}}$$

What are the advantages and disadvantages of using the slope-intercept form?

Although these three answers may look different, they are, in fact, equivalent. If we take the first two answers and put them in slope-intercept form, we get the third answer. In fact, one of the advantages of the slope-intercept form is that everyone's answer comes out looking the same. Even though we have just seen that using the slope-intercept form can involve some extra computation, for most situations the slope-intercept form of the equation is the most useful and is preferred. ■

Show that the first two answers are equivalent to the third by putting them in the form $y = mx + b$.

GRAFFIX

Use a graphics calculator or computer.

1. Graph the following on the same set of coordinate axes: $y = 2x$, $y = 3x$, $y = 5x$. What can you conclude about how m affects the graph of the equation $y = mx$?
2. Graph the following on the same set of coordinate axes: $y = 3x - 1$, $y = 3x$, $y = 3x + 1$. What can you conclude about how b affects the graph of the equation $y = mx + b$?

One of the most useful features of the slope-intercept form is that when an equation of a line is written in the form $y = mx + b$, the slope of the line is readily identified as the coefficient of x.

EXAMPLE 4 Find the slope of the line whose equation is:

(a) $y = -4x + 7$ **(b)** $3x - 4y = 18$

Solution

(a) By comparing the slope-intercept form $y = mx + b$ with $y = -4x + 7$, we can simply read off the slope (as well as the y-intercept).

$$\begin{array}{ccccc} y = & mx & + & b \\ \downarrow & \downarrow & & \downarrow \\ y = & -4x & + & 7 \end{array}$$

Therefore, the slope is $\boxed{-4}$.

(b) In Example 2 of Section 2.2, we found the slope of the line with equation $3x - 4y = 18$ by identifying two points on the line and using the definition of the slope. However, we now have an alternative approach. We may read the slope of a line from its equation, provided that the equation is *exactly* in slope-intercept form.

$$3x - 4y = 18 \quad \textit{We solve this equation explicitly for } y.$$

$$-4y = -3x + 18 \Rightarrow y = \frac{3}{4}x - \frac{9}{2}$$

Therefore, the slope is $\boxed{\frac{3}{4}}$. ■

EXAMPLE 5 Write an equation of the line that passes through the point $(4, 0)$ and is perpendicular to the line whose equation is $5y - 3x = 15$.

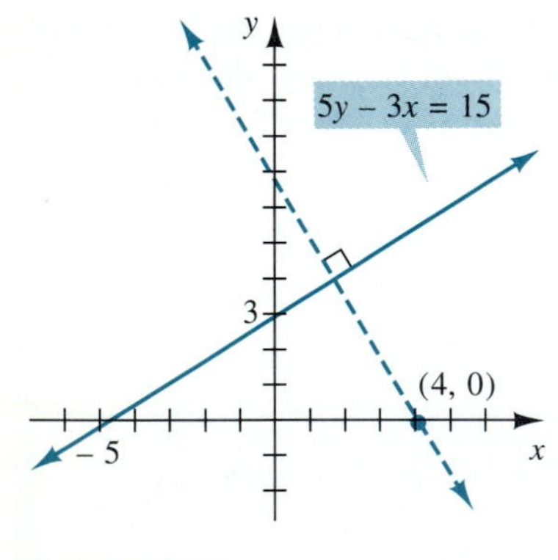

FIGURE 2.35

Solution Again, it is a good idea to draw a diagram to help us visualize what the example is asking. We plot the point $(4, 0)$ and sketch the line whose equation is $5y - 3x = 15$. See Figure 2.35. We have drawn a dashed line perpendicular to $5y - 3x = 15$, which passes through the point $(4, 0)$. It is the equation of this dashed line that we seek.

Let's analyze this problem carefully in order to develop a strategy for the solution.

WHAT DO WE NEED TO FIND?	The equation of a line.
WHAT IS NEEDED TO FIND THE EQUATION OF A LINE?	A point on the line and the slope of the line.
WHAT INFORMATION IS GIVEN IN THE PROBLEM?	The point is given, along with the equation for a line perpendicular to the line whose equation we want to find.

HOW DO WE FIND THE SLOPE OF A LINE GIVEN THE EQUATION OF A LINE PERPENDICULAR TO IT?	The slopes of perpendicular lines are negative reciprocals of each other. If we find the slope of the perpendicular line, we compute its negative reciprocal and we have the slope of the line we want to find.
HOW DO WE FIND THE SLOPE OF THE LINE WHOSE EQUATION IS GIVEN?	Put the given equation into slope-intercept form.

Based upon this analysis we begin by finding the slope of the line whose equation is given.

$$5y - 3x = 15 \Rightarrow 5y = 3x + 15 \Rightarrow y = \frac{3}{5}x + 3$$

So the slope of the given line is $\frac{3}{5}$.

Therefore, the slope of the perpendicular line is $-\frac{5}{3}$.

Now that we have the point $(4, 0)$ and slope $-\frac{5}{3}$, we can use the point-slope form to get

$$y - 0 = -\frac{5}{3}(x - 4)$$

and so our final answer for the equation of the dashed line is

$$\boxed{y = -\frac{5}{3}x + \frac{20}{3}}.$$

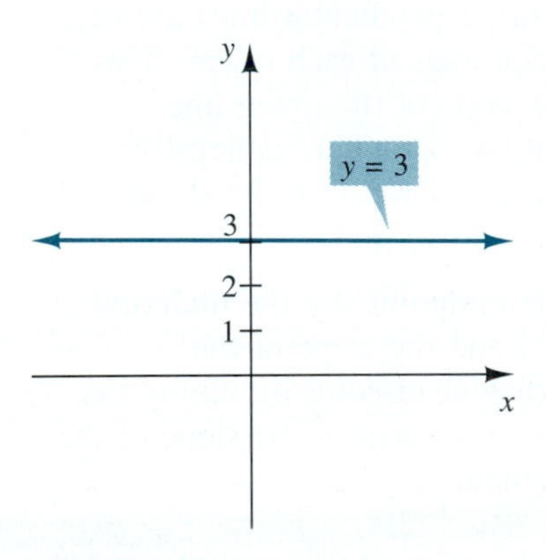

FIGURE 2.36

EXAMPLE 6 Write an equation of the line passing through the given pair of points.

(a) $(4, 3)$ and $(-2, 3)$ **(b)** $(2, -3)$ and $(2, 5)$

Solution

(a) The slope of the line is $m = \frac{3 - 3}{4 - (-2)} = \frac{0}{6} = 0$. Using the point-slope form with the point $(4, 3)$, we get

$$y - 3 = 0(x - 4) \Rightarrow y - 3 = 0 \Rightarrow \boxed{y = 3}$$

Alternatively, we could have recognized from the given points that the line passing through them is horizontal, 3 units above the x-axis; as we saw in Section 2.1, an equation of this line is $y = 3$. (See Figure 2.36.)

(b) If we attempt to compute the slope of the line passing through $(2, -3)$ and $(2, 5)$, we get

$$m = \frac{5 - (-3)}{2 - 2} = \frac{8}{0} \quad \text{which is undefined}$$

Can the equation of a vertical line be put in slope-intercept form?

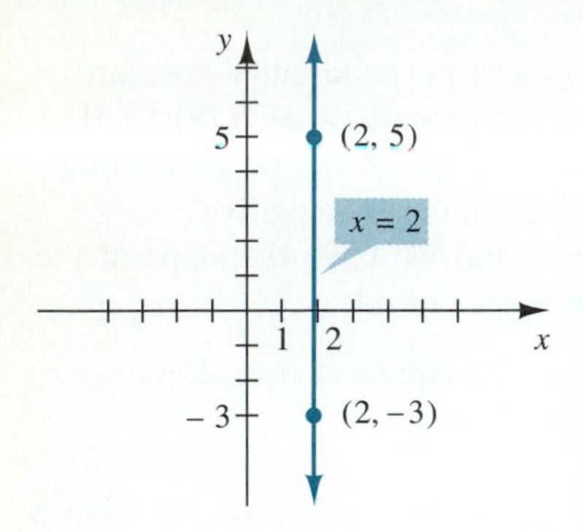

FIGURE 2.37

Consequently we cannot use either the point-slope or the slope-intercept forms, since either form requires the line to have a slope. (This situation explains why the discussion leading up to the derivation of both forms specified a nonvertical line L.) Once we recognize that these points determine a vertical line 2 units to the right of the y-axis (see Figure 2.37), we know the equation of this line is $\boxed{x = 2}$. ■

EXAMPLE 7 Find an equation of the line that is the perpendicular bisector of the line segment joining the points $P(-3, 1)$ and $Q(4, 2)$.

Solution Figure 2.38 shows $\overline{PQ}$ with its perpendicular bisector indicated by the dashed line. Let's analyze this problem carefully in order to develop a strategy for the solution.

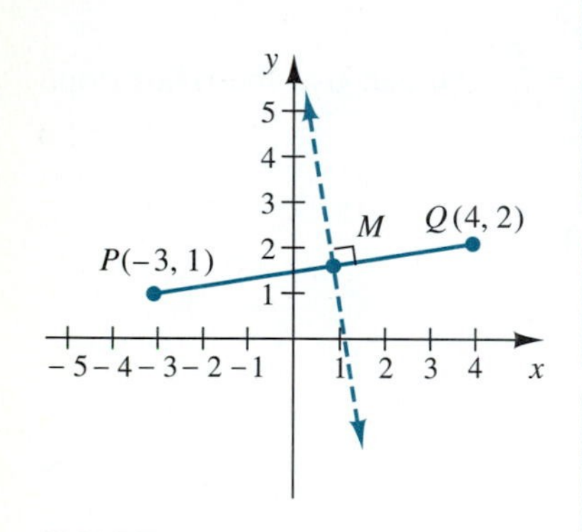

FIGURE 2.38

WHAT DO WE NEED TO FIND? (USE THE DIAGRAM TO HELP IN THIS ANALYSIS.)	The equation of a line.
WHAT IS NEEDED TO FIND THE EQUATION OF A LINE?	A point on the line and the slope of the line.
WHAT INFORMATION IS GIVEN IN THE PROBLEM?	The line we need to find is the perpendicular bisector of a line segment joining two given points.
WHAT IS THE PERPENDICULAR BISECTOR?	It is a line that divides the given line segment into two equal parts and is perpendicular to the line segment.
HOW DOES THIS INFORMATION HELP US TO FIND THE POINT?	Since the perpendicular bisector divides the line into two equal parts, it must pass through the midpoint of the line segment.
HOW DOES THIS INFORMATION HELP US TO FIND THE SLOPE?	Slopes of perpendicular lines are negative reciprocals of each other. If we find the slope of the given line segment, we compute its negative reciprocal and we have the slope of the line we want to find.
HOW CAN WE RESTATE THE PROBLEM IN SIMPLER TERMS?	Find the midpoint (by the midpoint formula) and the slope of the perpendicular bisector by taking the negative reciprocal of the slope of the line segment.

Based upon this analysis, we begin by using the midpoint formula to find the midpoint of $\overline{PQ}$.

$$\text{Midpoint of } \overline{PQ} = \left(\frac{-3+4}{2}, \frac{1+2}{2}\right) = \left(\frac{1}{2}, \frac{3}{2}\right)$$

We find the slope of $\overline{PQ}$ by using the two given points.

$$m_{\overline{PQ}} = \frac{2 - 1}{4 - (-3)} = \frac{1}{7}$$

Therefore, the slope of a perpendicular line will be -7, and we have the following as an equation of the perpendicular bisector of $\overline{PQ}$:

$$y - \frac{3}{2} = -7\left(x - \frac{1}{2}\right) \quad \text{or} \quad \boxed{y = -7x + 5}$$

■

As we continue through the text, we will see that straight lines and their equations play a very significant role.

EXAMPLE 8 Suppose that a shoe store finds that it sells 23 pairs of shoes per day at \$30 per pair and 20 pairs of shoes per day at \$36 per pair. Assuming a linear relationship between P, the price of a pair of shoes, and N, the number of pairs sold, predict the number of pairs of shoes sold per day at \$40 per pair.

Solution We may express the given information as ordered pairs. We are free to express the ordered pairs with P as the first coordinate and N as the second coordinate or vice versa. Reading through the example, it seems reasonable to assume that the number of pairs of shoes sold *depends* upon the price. In sketching the graph of a relationship between two quantities, it is generally accepted that the dependent quantity (in this case the number of pairs of shoes sold) is taken as the vertical axis and the quantity upon which it depends (in this case the price) is taken as the horizontal axis. Consequently, we will write the ordered pairs in the form (P, N).

The given information gives us the ordered pairs $(30, 23)$ and $(36, 20)$, Since we are told that the relationship between price and number of pairs sold is linear, we compute the slope of the line:

$$m = \frac{23 - 20}{30 - 36} = \frac{3}{-6} = -\frac{1}{2}$$

We can use the point-slope form with the point (36, 20) to write an equation describing the relationship between P and N.

$$N - 20 = -\frac{1}{2}(P - 36)$$

Now we may substitute $P = 40$ to determine the number of pairs of shoes sold when the price is \$40.

$$N - 20 = -\frac{1}{2}(P - 36) \quad \textit{Substitute } P = 40.$$

$$N - 20 = -\frac{1}{2}(40 - 36)$$

$$N - 20 = -2$$

$$N = 18$$

Is there an advantage to transforming the equation $N - 20 = -\frac{1}{2}(P - 36)$ into slope-intercept form?

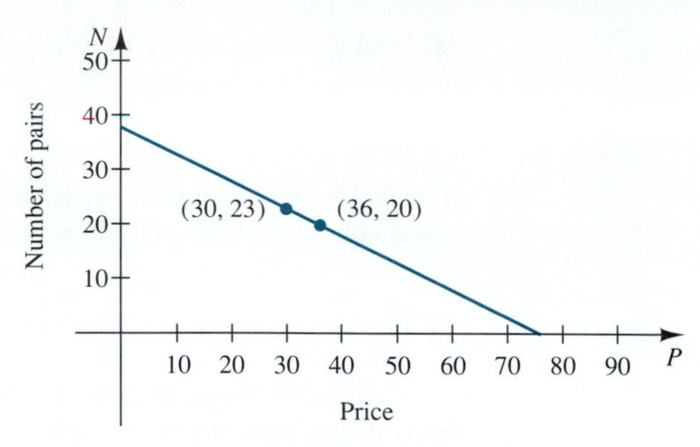

FIGURE 2.39 The linear relationship between the price and the number of shoes sold.

Thus, assuming a linear relationship, the store will sell 18 pairs of shoes per day at \$40 per pair. Figure 2.39 illustrates the linear relationship between the price and number of pairs of shoes sold. Several aspects of this graph are worth noting. First, the graph appears only in the first quadrant, because it makes no sense for either P (the price per pair of shoes) or N (the number of pairs of shoes sold per day) to be negative. Second, N can realistically take on only positive integer values (you can't sell a fraction of a pair of shoes), and so the graph should just be a discreet set of points. Nevertheless, in order to analyze the graph, it is often useful to draw the entire line segment. ■

One final comment: In this section we have proven that the equation of a straight line is a first-degree equation in x and y. We still have not proven Theorem 2.1 (originally stated in Section 2.1), which states that the graph of an equation of the form $Ax + By = C$ (with both A and B not equal to zero) is a straight line. The proof of this theorem is outlined in Exercise 71.

EXERCISES 2.3

In Exercises 1–4, write an equation of the line satisfying the given conditions and sketch its graph.

1. The line has slope 3 and passes through the point $(-1, 2)$.

2. The line has slope $-\frac{1}{4}$ and passes through the point $(0, 0)$.

3. The line has slope $\frac{2}{5}$ and passes through the point $(-2, -3)$.

4. The line has slope -3 and passes through the point $(0, 2)$.

In Exercises 5–22, write an equation of the line satisfying the given conditions.

5. Line L has slope 4 and passes through the point $(-3, 4)$.

6. Line L has slope $\frac{3}{5}$ and passes through the point $(-2, -1)$.

7. Line R passes through the points $(2, 6)$ and $(8, 3)$.

8. Line T passes through the points $(-2, -5)$ and $(-6, -7)$.

9. Line F passes through the origin and has slope 4.

10. Line G passes through the origin and has slope $-\frac{1}{4}$.

11. Line L has slope $-\frac{2}{3}$ and a y-intercept of 4.

12. Line L has slope $\frac{1}{4}$ and an x-intercept of -2.

13. Line R has an x-intercept of 5 and a y-intercept of -1.

14. Line Q has a y-intercept of 3 and an x-intercept of -2.

15. Line V is vertical and passes through the point $(-3, 4)$.

16. Line H is horizontal and passes through the point $(-3, 4)$.

17. Line T has slope $-\frac{1}{3}$ and crosses the y-axis at $y = 5$.
18. Line R has slope $\frac{2}{7}$ and crosses the x-axis at $x = -5$.
19. Line F has no x-intercept and passes through the point $(3, 5)$.
20. Line G has no y-intercept and passes through the point $(3, 5)$.
21. Line L passes through the points $(2, -5)$ and $(-3, -5)$.
22. Line L passes through the points $(-4, 6)$ and $(-4, -1)$.

In Exercises 23–30, find the slope of the line whose equation is given.

23. $y = -2x + 7$
24. $y = \frac{3}{5}x - \frac{2}{3}$
25. $y = \dfrac{2x + 5}{3}$
26. $y = \frac{1}{3} + \frac{7}{4}x$
27. $x = 5y - 1$
28. $6x - 5y = 10$
29. $\dfrac{4x + 3y}{3} = 2$
30. $\dfrac{5x}{4} - \dfrac{4y}{5} = 1$

In Exercises 31–44, write an equation of the line satisfying the given conditions.

31. Line S passes through the point $(1, -5)$ and is parallel to $y = 4x - 7$.
32. Line S passes through the point $(-2, 3)$ and is perpendicular to $y = -4x + 1$.
33. Line M passes through the point $(4, 7)$ and is perpendicular to $y = -\frac{3}{4}x + 9$.
34. Line T passes through the point $(5, -5)$ and is parallel to $y = \frac{4}{5}x + 2$.
35. Line S passes through the point $(5, 0)$ and is parallel to $5y - 4x = 9$.
36. Line R passes through the point$(0, 5)$ and is perpendicular to $7x + 4y = 1$.
37. Line M passes through the point $(7, 4)$ and is perpendicular to $6x - 7y = 6$.
38. Line T passes through the point $(-3, 3)$ and is parallel to $10y - 8x = 7$.
39. Line L passes theough the point $(0, 0)$ and is perpendicular to $y = x$.
40. Line L passes through the point $(0, 0)$ and is parallel to $x + y = 8$.
41. Line L passes through the point $(-4, 5)$ and is parallel to the x-axis.
42. Line L passes through the point $(-4, 5)$ and is parallel to the y-axis.
43. Line T is perpendicular to $2x + 7y = 14$ and has the same y-intercept.
44. Line T is parallel to $6y - 4x = 9$ and has the same x-intercept.
45. Find an equation of the perpendicular bisector of the line segment joining $(-4, 3)$ and $(1, 0)$.
46. Find an equation of the perpendicular bisector of the line segment joining the center of the circle $x^2 + y^2 - 4y = 1$ to the center of the circle $x^2 + 3x + y^2 = 5$.
47. Find an equation of the perpendicular bisector of the line segment joining the points where the line $5y - 3x = 2$ crosses the x-axis and the y-axis.
48. **(a)** Find an equation of the perpendicular bisector of the line segment joining $(1, 2)$ and $(5, 8)$ using the method outlined in Example 7.
    (b) Find an equation of this perpendicular bisector by using the distance formula and verify that you get the same result as in (a). HINT: The perpendicular bisector of a line segment is the line containing all points equidistant from both endpoints.
49. A car rental company charges $29 per day plus $0.30 per mile. Write an equation relating the cost, C, of renting a car for a day to the number of miles, m, driven.
50. A car rental company charges $39 per day plus $0.22 per mile and also requires customers to purchase a collision damage waiver at a cost of $8 per day. Write an equation relating the cost, C, of renting a car for 3 days to the number of miles, m, driven.
51. Each month, a local utility company charges a flat fee of $7 for the first 35 kilowatt-hours or less, and $0.12 for each kilowatt-hour above 35. Write an equation describing the relationship between the dollar amount, A, of an electric bill and the total number of kilowatt-hours, h, used by a customer each month. Sketch the graph of this equation.
52. Each month, a local phone company charges a flat fee of $11.60 for the first 80 message units and 10.6¢ for each message unit above 80. Write an equation describing the relationship between the dollar amount, A, of a phone bill and the total number of message units, m, used by a customer each month. Sketch the graph of this equation.
53. Suppose that a manufacturer determines that there is a linear relationship between P, the profit earned and x, the number of items produced. The profit is $600 on 50 items and $750 on 65 items.
    (a) Write an equation relating x and P.
    (b) What would the expected profit be if 90 items were produced?
54. A quality-control inspector finds that there is a relationship between D, the number of defective items produced, and h, the number of overtime hours put in by the employees. When the employees put in a total of 120 overtime hours, 80 defective items were found; when a total of 70 overtime hours were put in, 60 defective items were found. If the inspector suspects that the relationship between h and D is linear, how many defective items would she expect to find if 90 overtime hours are put in?

55. In physiology, a jogger's heart rate, N, in beats per minute is related linearly to the jogger's speed, s. A jogger's heartbeat is 80 beats per minute at a speed of 15 ft/sec and 85 beats per minute at a speed of 18 ft/sec.
(a) Write an equation expressing N in terms of s.
(b) Using the equation obtained in (a), predict the jogger's heart rate at a speed of 25 ft/sec.

56. An assembly-line worker receives an hourly salary of \$13.50 per hour. In addition the worker receives a piece-work commission of \$0.65 per unit produced.
(a) Write an equation relating the hourly wages, W and the number of items produced per hour, x.
(b) Use the equation obtained in part (a) to determine the hourly wage if the worker produces 5 units per hour.
(c) Use the equation obtained in part (a) to find the number of units per hour a worker must produce if he wants to have an hourly wage of at least \$22 an hour.

57. A business buys a piece of machinery for \$8500. Assuming that the machinery depreciates linearly to a value of zero dollars in 12 years, write an equation expressing the value, V, of the machinery after t years.

58. A small company buys a computer system for \$30,000 and, for tax purposes, uses the linear depreciation method to arrive at a salvage value of \$1800 after 10 years. Write an equation describing the value, V, of the computer system after n years.

59. A car rental agency buys a car for \$12,375 and incurs expenses of \$8.25 per day to operate the vehicle. It then rents the car for \$49.95 per day.
(a) Write an equation for the total cost, C, to buy and operate the car for d days.
(b) Write an equation for the total income, I, earned if this car is rented for d days.
(c) The profit the rental company earns on this car is obtained by computing $I - C$. The *break-even point* for a business is that point at which the income and costs are equal (i.e., the profit is 0). Find the number of days this car must be rented in order for the company to break even.
(d) Sketch the graphs of the equations for C and I on the same coordinate system and give a graphical description of the break-even point.

60. In economics a *demand equation* describes the relationship between the price, p, of an item and the number of items, n, that can be sold at that price. The *supply equation* describes the relationship between the projected selling price, s, of an item, and the number of items, n, that the manufacturer is willing to produce at that price.
(a) Would you expect the demand price, p, to increase or decrease as n increases? Explain.
(b) Would you expect the supply price, s, to increase or decrease as n increases? Explain.
(c) A clothing manufacturer estimates that for a certain swimsuit, the demand equation is given by $p = 100 - 0.05n$ and the supply equation is given by $s = 70 + 0.03n$. Do p and s as given by these equations agree with your answers to parts (a) and (b)?
(d) Sketch the graphs of the demand equation and the supply equation on the same coordinate system. (You will need to label the vertical axis as both p and s.).
(e) Using the graphs obtained in part (d) and given the fact that p and s are prices and that n is the number of items produced, what is the possible range of values that makes sense for p, s, and n?
(f) The price at which the supply and demand are equal is called the *equilibrium price*. Use the equations given in part (c) to find the equilibrium price for these swimsuits.

61. An audio equipment company knows that it can sell 275 portable cassette players each month at a price of \$62.50. Based upon survey information, the company also believes that for each \$2.50 increase in price, 10 fewer units will be sold each month.
(a) Let N denote the number of cassette players sold each month and P be the price per unit. Assuming a linear relationship between the number of players sold and the price per player, write an equation for N in terms of P.
(b) Using the equation obtained in part (a), determine how many players will be sold at a price of \$75.
(c) At what price would the company sell 240 players per month?

62. The cost of producing an item usually consists of two components: **fixed costs,** such as the cost of machinery, insurance, and taxes, and **cost per item,** such as the cost of raw material, labor, or utilities. Suppose the cost, C, of producing x items is approximately given by the equation

$$C = 20x + 120$$

(a) What is the fixed cost? HINT: What is the cost to produce 0 items?
(b) An important quantity in economics, known as the **marginal cost,** is approximately equal to the cost of producing *one* additional item—that is, the difference between the cost of producing x items and $x + 1$ items. Compute the marginal cost for this cost equation.
(c) How much would it cost to produce 250 items?

63. Two popular systems for measuring temperature are the Celsius scale, C, and the Fahrenheit scale, F. Water freezes at 0°C and 32°F; water boils at 100°C and 212°F. Given that the Celsius and Fahrenheit scales are linearly related, find an equation that gives the Celsius temperature in terms of the Fahrenheit temperature. What is the Celsius equivalent to 98.6°F?

64. Use the information given (and obtained) in Exercise 63 to write an equation that gives the Fahrenheit temperature in terms of the Celsius temperature.

65. Anthopologists often extrapolate the appearance of a human being from the parts of a skeleton that they uncover. For example, it has been found that there is a linear relationship between the length, f, of the femur (thigh bone), and the height, h, of the human from whom it came. Suppose that it is known that a femur of length 47.5 cm corresponds to a height of 177.8 cm and that a femur of length 39.1 cm corresponds to a height of 146.3 cm.
(a) Rounding to the nearest tenth, write an equation expressing the height in terms of the length of the femur.
(b) What height would correspond to a femur of length 52 cm?

66. From elementary geometry we know that a tangent line drawn to a circle is perpendicular to a radius drawn to the point of tangency. Use this fact to write an equation of the tangent line to the circle $x^2 + y^2 = 25$ at the point $(-3, 4)$.

67. Sketch the graph of the circle $x^2 + y^2 = 169$ and write an equation of the tangent line to this circle at the point $(5, -12)$. (See Exercise 66.)

68. Sketch the graph of $y = x - 2$, and use the graph to answer the following questions.
(a) For what values of x is the graph above the x-axis?
(b) For what values of x is $x - 2 > 0$?
(c) For what values of x is $y > 0$?

69. Sketch the graph of $y = 2x + 3$, and use the graph to answer the following questions.
(a) For what values of x is the graph below the x-axis?
(b) For what values of x is $2x + 3 < 0$?
(c) For what values of x is $y < 0$?

QUESTIONS FOR THOUGHT

70. Given the line with equation $5y - 7x = 11$, describe two ways to find its slope. Which method is easier?

71. In this exercise we outline a proof of the fact that the graph of an equation of the form $Ax + By = C$, where $B \neq 0$, is a straight line. (Note that if $B = 0$, the equation becomes $Ax = C$, which we know to have a vertical line as its graph.)

Our strategy is to show that *any* two ordered pairs (x_1, y_1) and (x_2, y_2) will give the same slope. It then follows that the graph is a straight line. We can proceed as follows: First show that if (x_1, y_1) is a point satisfying $Ax + By = C$, then $y_1 = \dfrac{C - Ax_1}{B}$, which gives the ordered pair $\left(x_1, \dfrac{C - Ax_1}{B}\right)$ on the graph. Similarly, for (x_2, y_2) we get $\left(x_2, \dfrac{C - Ax_2}{B}\right)$ on the graph. Now show that the slope determined by two such points is independent of x_1 and x_2.

72. There is yet another form for a first-degree equation in two variables: Prove that the equation of a line passing through $(a, 0)$ and $(0, b)$, where $a \neq 0$, $b \neq 0$, can be written in the form

$$\frac{x}{a} + \frac{y}{b} = 1$$

Why do you think this form is called the *intercept form*?

73. Consider the following 10 data points:

$$(-1, -4.8), (0, -2), (0.5, -0.4), (1, 1), (1.5, 2.7),$$
$$(2, 3.8), (2.4, 5.5), (3.1, 7.2), (3.7, 8.4), (4, 10)$$

Draw a straight line that, in your opinion, lies as close as possible to as many of these data points as possible. Estimate the equation of this line.

GRAFFIX

Use your calculator or computer to graph the two given equations on the same set of coordinate axes. Identify the y-intercepts for each and discuss the similarities and differences between the two graphs.

74. $y = x$, $y = 2x$
75. $y = 2x$, $y = 3x$
76. $y = x - 3$, $y = 2x - 3$
77. $y = 2x - 1$, $y = 3x - 1$
78. $y = x$, $y = -x$
79. $y = 2x$, $y = -2x$
80. $y = 5x - 2$, $y = -5x - 2$
81. $y = 3x + 1$, $y = -3x + 1$

Use your calculator or computer to graph the three given equations on the same set of coordinate axes. Identify the y-intercepts for each and discuss the similarities and differences among the three graphs.

82. $y = x$, $y = x - 3$, and $y = x + 3$
83. $y = 2x$, $y = 2x - 3$, and $y = 2x + 3$
84. $y = 5x$, $y = 5x - 1$, and $y = 5x + 2$
85. $y = -2x + 3$, $y = -2x - 2$, and $y = -2x - 4$

2.4 Relations and Functions

Our everyday lives are filled with situations in which we encounter relationships between two sets. For example,

To each taxpayer in the United States, there corresponds a social security number.

To each automobile, there corresponds a license plate number.

To each circle, there corresponds a circumference.

To each item in a supermarket, there corresponds a price.

To each number, there corresponds its square.

To each nonnegative number, there corresponds its two square roots.

In order to apply mathematics to a variety of disciplines, we must make the idea of a "relationship" between two sets mathematically precise. Let's begin with the following definition:

DEFINITION A **relation** is a correspondence between two sets so that to each member of the first set (called the **domain**) there corresponds one or more members of the second set (called the **range**).

Let's consider the relation described by the following table, in which we associate with each person listed, his or her telephone number.

Domain	Range
Sarah	→ 297-4419
John	→ 348-7743
Ginger	→ 348-7743
Lamar	→ 459-8810
Lamar	→ 459-8811

Note that Lamar has two telephone numbers, whereas the number 348-7743 is associated with two people, John and Ginger.

Let's look at another example. Let the domain be the set $D = \{u, v, w\}$ and the range be the set $R = \{4, 7, 11\}$. There are numerous possible relations with domain D and range R, four of which are as follows:

Relation A	Relation B	Relation S	Relation T
$u \to 4$	$u \to 7$	$u \to 4$	$u \to 4$, $u \to 7$
$v \to 7$	$v \to 11$	$v \to 4$	$v \to 11$
$w \to 11$	$w \to 11$	$w \to 7$, $w \to 11$	$w \to 11$

In Relation A, u is assigned 4, v is assigned 7, and w is assigned 11.

In Relation B, u is assigned 7 and v and w are assigned 11.

In Relation S, u is assigned 4, v is assigned 4, and w is assigned 7 and 11.

In Relation T, u is assigned 4 and 7, v is assigned 11, and w is assigned 11.

In fact, any type of correspondence gives us a relation as long as each element in the domain is assigned at least one element in the range.

An alternative way of describing a relation is by using ordered pair notation. For example, we may write the correspondences in relation A as

Relation A	**Ordered Pair Notation**
$u \longrightarrow 4$	$(u, 4)$
$v \longrightarrow 7$	$(v, 7)$
$w \longrightarrow 11$	$(w, 11)$

Thus we may write relation A as the set of ordered pairs A: $\{(u, 4), (v, 7), (w, 11)\}$.

When we use ordered pair notation, we will assume (unless stated otherwise) that the first coordinate is a member of the domain and the second coordinate is the associated member of the range.

Any set of ordered pairs describes a relation. The set of all first (x-) coordinates constitutes the domain, and the set of all second (y-) coordinates constitutes the range. The ordered pairs themselves describe the correspondence.

We can rewrite the relations B, S, and T as follows:

relation B: $\{(u, 7), (v, 11), (w, 11)\}$
relation S: $\{(u, 4), (v, 4), (w, 7), (w, 11)\}$
relation T: $\{(u, 4), (u, 7), (v, 11), (w, 11)\}$

Just as with the arrow notation, each set of ordered pairs defines a relation. Thus we can also define a relation as follows.

DEFINITION A **relation** is a set of ordered pairs (x, y). The set of x-values is called the *domain* and the set of y-values is called the *range*.

EXAMPLE 1 Find the domain and range of the following relations.

(a) K: $\{(-4, 0), (-2, 2), (0, 4), (3, 5), (6, 2)\}$ **(b)** L: $\{(a, b), (b, c), (c, a)\}$

Solution

(a) The domain is the set of first coordinates; therefore, the domain of K, which we will denote as D_K, is

$$\boxed{D_K = \{-4, -2, 0, 3, 6\}}$$

The range is the set of second coordinates; therefore, the range of K, which we will denote as R_K, is

$$\boxed{R_K = \{0, 2, 4, 5\}}$$

(b) Looking at relation L we can see that the set of first coordinates and second coordinates is the same. Therefore, we have

$$\boxed{D_L = R_L = \{a, b, c\}}$$

If the domain and/or range of a relation is infinite, we cannot list each element assignment, so instead we use set-builder notation to describe the relation. The situation we will encounter most frequently is that of a relation defined by an equation or formula. For example,

$$R = \{(x, y) \mid y = 2x - 3\}$$

is a relation for which the range value is 3 less than twice the domain value. It is understood that x represents domain values and y represents range values. Hence $(0, -3)$, $(0.5, -2)$, and $(-2, -7)$ are examples of ordered pairs that are part of the assignment. Often we drop the set-builder notation and simply write that the relation is described by $y = 2x - 3$.

Since a relation can be described as a set of ordered pairs or an equation in two variables, we can use the rectangular coordinate system to specify the relation as well. In Figure 2.40 we have drawn a graph and labeled some of the points on the graph. Since every graph is a set of ordered pairs, it is automatically a relation; *the points on the graph themselves give the relation.*

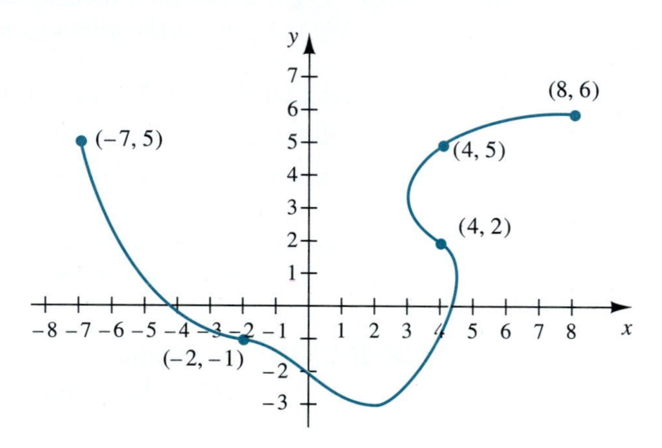

FIGURE 2.40 The graph of a relation

By examining the graph we determine several things. First, each point on the graph tells us which y-value is assigned to each x-value. For example, $(8, 6)$, $(4, 5)$, $(4, 2)$, and $(-2, -1)$ are on the graph, which means that 6 is assigned to 8, 5 and 2 are both assigned to 4, and -1 is assigned to -2.

We can also identify the domain and range of the relation by noting the following: Every x-value between -7 and 8 has a y-value associated with it, whereas x-values to the left of -7 or to the right of 8 have no y-values associated with them (there is no point on the graph to go up or down to). Thus we can see that the domain is the set of numbers between -7 and 8. Alternatively, we can project the points on the graph up or down to the x-axis to see which x-values are in the domain. See Figure 2.41(a).

Given the graph of a relation, how do we find its domain?

As for the range, we can see that the smallest y-coordinate on the graph is -3, the largest y-coordinate on the graph is 6, and every y-value in between is also the y-coordinate of some point(s) on the graph. Therefore, we can see that the range is the set of numbers between -3 and 6. The range can also be seen by projecting the

Given the graph of a relation, how do we find its range?

points on the graph left or right to the y-axis. See Figure 2.41(b). When we read the domain and range of a relation from its graph, we say that we are finding the domain and range by *inspection*.

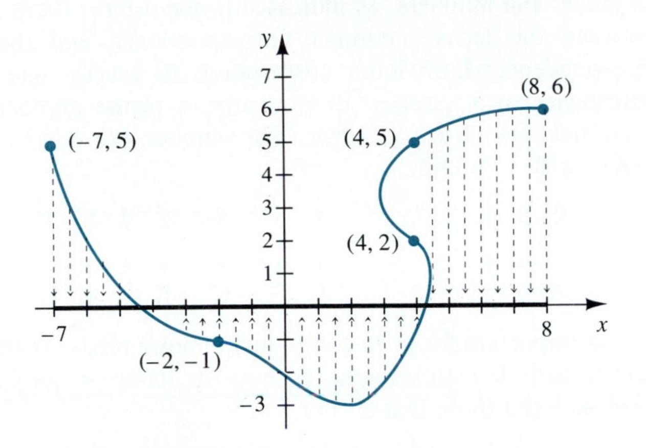

(a) To get the domain we project the points on the graph vertically to the x-axis.
The domain is $\{x \mid -7 \le x \le 8\}$

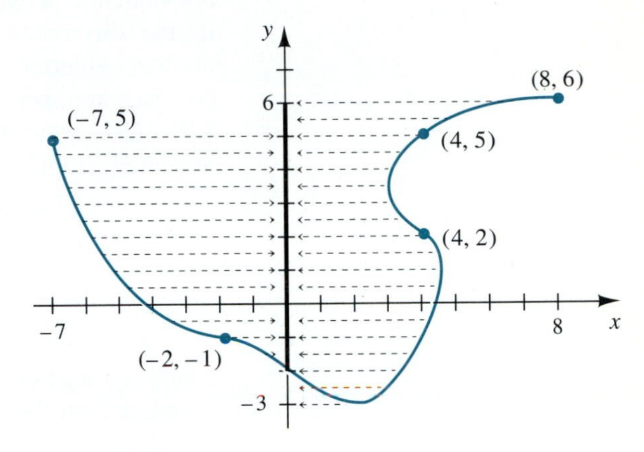

(b) To get the range we project the points on the graph horizontally to the y-axis.
The range is $\{y \mid -3 \le y \le 6\}$

FIGURE 2.41 Finding the domain and range from a graph

To summarize our discussion thus far, we have four ways to express the same relation.

By an Arrow Diagram

$-2 \longrightarrow 4$
$2 \nearrow$ (to 4)
$0 \longrightarrow 0$
$3 \longrightarrow 9$
$5 \longrightarrow 25$

With Ordered Pairs

$\{(-2, 4), (0, 0), (2, 4), (3, 9), (5, 25)\}$

By an Equation

$y = x^2$, for $x = -2, 0, 2, 3, 5$

By a Graph

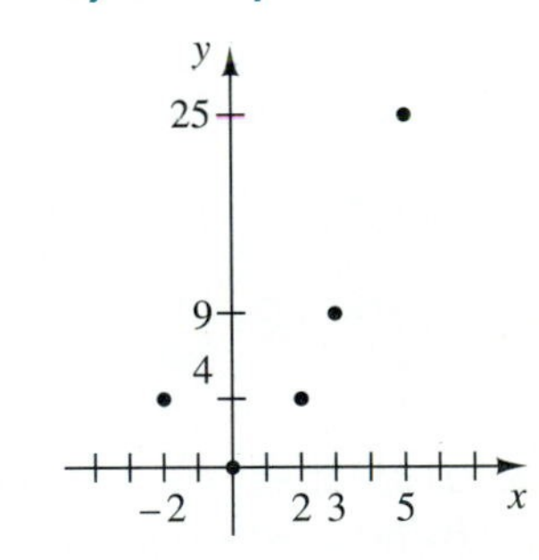

Functions

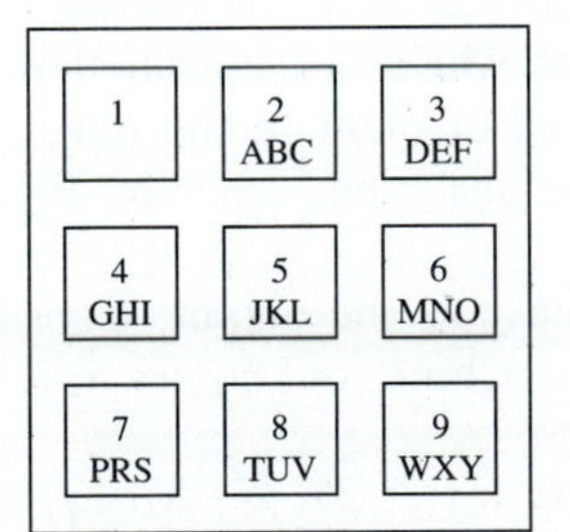

FIGURE 2.42 A typical telephone keypad

Let's consider a real-life relation described by the part of a telephone keypad illustrated in Figure 2.42.

It is very common today to see a business advertise its phone number by saying something like

"To get an over-the-phone price quotation on a new car, dial BUY CARS."

Anyone wanting to call this number can easily decode BUY CARS as the number 289-2277. This is because the telephone keypad sets up a correspondence between the set of letters of the alphabet (except for Q and Z) and the set of numbers $\{2, 3, 4, 5, 6, 7, 8, 9\}$. The correspondence between the two sets is that indicated by the telephone keypad—that is, the letters A,B,C correspond to the number 2; the letters D,E,F correspond to the number 3, etc.

What is important to note is that although the telephone keypad sets up a correspondence between letters and numbers, as indicated in the figure, there is a qualitative difference between the letter → number correspondence and the reverse number → letter correspondence: Each letter corresponds to exactly one number, but each number corresponds to 3 letters. Consequently, a phone number such as 438-2253 cannot be *uniquely* encoded into words. The number 438-2253 can be interpreted as GET CAKE or IF U BAKE.

$$\begin{array}{ccc} 4\ 3\ 8 & 2\ 2\ 5\ 3 \\ \downarrow\ \downarrow\ \downarrow & \downarrow\ \downarrow\ \downarrow\ \downarrow \\ \text{G E T} & \text{C A K E} \end{array} \quad \text{or} \quad \begin{array}{cccc} 4\ 3 & 8 & 2\ 2\ 5\ 3 \\ \downarrow\ \downarrow & \downarrow & \downarrow\ \downarrow\ \downarrow\ \downarrow \\ \text{I F} & \text{U} & \text{B A K E} \end{array}$$

Mathematically it is important for us to distinguish among relations that assign a *unique* range element to each domain element (such as the letter → number correspondence on the telephone) and those that do not.

DEFINITION A **function** is a relation in which each element of the domain corresponds to exactly one element of the range.

Note that every function is a relation, but not every relation is a function.

Looking back at relations A, B, S, and T, which we rewrite here,

A: $\{(u, 4), (v, 7), (w, 11)\}$ B: $\{(u, 7), (v, 11), (w, 11)\}$

S: $\{(u, 4), (v, 4), (w, 7), (w, 11)\}$ T: $\{(u, 4), (u, 7), (v, 11), (w, 11)\}$

A is a function, since each element in the domain, $\{u, v, w\}$, is assigned only one element in the range.

On the other hand, S and T are not functions—in S, w is assigned *two numbers,* 7 and 11; in T, u is assigned 4 and 7.

In B, even though the range element 11 is assigned to two elements of the domain, v and w, it is still a function since each element in the domain, $\{u, v, w\}$, is assigned only *one element of the range*, $\{7, 11\}$: u is assigned only one value, 7; v is assigned only one value, 11; and w is assigned only one value, 11.

If two different x's are assigned to the same y, why is B still a function?

Recalling the telephone relation, we have that the letter → number assignment is a function (a phrase gives you only one number), but the number → letter assignment is not a function (a number may give you more than one possible phrase).

A function may also be viewed as a "machine" into which you put an element of the domain, x, and out of which comes the associated element in the range, y. Figure 2.43(a) illustrates the idea of a general function machine, whereas Figure 2.43(b) illustrates the function $y = x^2 + 4$.

Think about how a calculator acts as a function machine.

Hence we consider x as the input and its associated y as the output. Consider the function $\{(2, 4), (3, 4), (5, 8)\}$. If we put 2 into the machine, out comes 4; if we put in 3, out comes 4; and if we put in 5, out comes 8.

If we try to define a function machine by the relation $\{(2, 5), (3, 8), (3, 9)\}$, then we run into trouble when we put in 3; we cannot be sure if 8 or 9 will come out. Because two outputs can occur from one input, $\{(2, 5), (3, 8), (3, 9)\}$ is not a function.

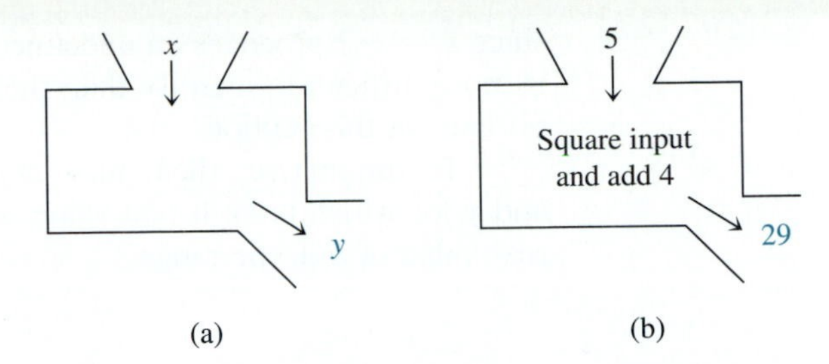

FIGURE 2.43 Viewing a function as a machine

EXAMPLE 2 Determine whether the following relations are functions.

(a) $\{(5, -2), (3, 5), (3, 7)\}$ **(b)** $\{(2, 4), (3, 4), (6, -4)\}$

Solution

(a) Since the domain element 3 is assigned two different values in the range, 5 and 7, it is not a function.

(b) Since each element in the domain, $\{2, 3, 6\}$, is assigned no more than one value in the range; 2 is assigned only 4, 3 is assigned only 4, and 6 is assigned only -4. It is a function. ■

As with relations, we can describe a function with an equation. For example,

$$y = 2x + 1$$

is a function, since each x will produce only one y.

We will find it useful to use the following vocabulary: The **independent variable** refers to the variable representing possible values in the domain, and the **dependent variable** refers to the variable representing possible values in the range. Thus in our usual ordered pair notation (x, y), x is the independent variable and y is the dependent variable.

It is very helpful to view functions within the framework of "pick x (a value in the domain) and determine y (the corresponding value in the range)." For the function $y = 2x + 1$, if we choose $x = 4$, then $y = 2(4) + 1 = 9$, so $(4, 9)$ is an assignment described by this function. In this way we can generate as many ordered pairs as we like. The question that remains is: What is the domain of $y = 2x + 1$? In order to answer this question we need to set up the following ground rule.

> The **natural domain** of a function is the set of real numbers for which the equation or formula is defined *and* that produces real number values in the range. Such a function is called a *real-valued function of a real variable*. In other words, the function must have a real input and a real output. Unless otherwise specified, a function is assumed to have its *natural domain*.

Thus the domain of $y = 2x + 1$ is the set of all real numbers, since there is no restriction on what real number we may choose for x. The range is also the set of all real numbers. On the other hand, the domain of $y = \dfrac{1}{x + 2}$ is all reals *except* -2, since $x = -2$ produces an undefined value for y. In general, the range of a function is more difficult to identify than the domain; we have more to say about the range a bit later in this section.

To summarize, then, for our purposes a function is a relationship between x and y for which to each real value of x in the domain there corresponds exactly one real value of y in the range.

EXAMPLE 3 Determine whether the following equations determine y as a function of x; if so, find the domain.

(a) $y = -3x + 5$ **(b)** $y = \dfrac{2x}{3x - 5}$ **(c)** $y = x^2$

(d) $y^2 = x$ **(e)** $y = \dfrac{4}{x^2 + 1}$

Solution

(a) In order to determine whether or not $y = -3x + 5$ gives y as a function of x, we need to know whether each x-value uniquely determines a y-value. THINK: Pick x, get exactly one y. Looking at the equation $y = -3x + 5$, we can see that once x is chosen, we multiply it by -3 and then add 5. Thus for each x there is a unique (only one) y. Therefore, $y = -3x + 5$ is a function.

As for its domain, we can see that there is no restriction on the values we may choose for x. (Is there any real number you can't multiply by -3 and then add 5 to the result?) Therefore, the domain is the set of all real numbers.

(b) Looking at the equation $y = \dfrac{2x}{3x - 5}$ carefully, we can see that each x-value uniquely determines a y-value. (One x-value cannot produce two different y-values.) Therefore, $y = \dfrac{2x}{3x - 5}$ is a function.

As for its domain, we ask ourselves, Are there any values of x that must be excluded? Since $y = \dfrac{2x}{3x - 5}$ is a fractional expression, we must exclude any value of x that makes the denominator equal to zero. We must have

$$3x - 5 \neq 0 \Rightarrow x \neq \frac{5}{3}$$

Therefore, the domain consists of all real numbers except for $\dfrac{5}{3}$. The domain is $\{x \mid x \neq \frac{5}{3}\}$.

(c) For the equation $y = x^2$, if we choose $x = 3$, we get $y = 9$, and if we choose $x = -3$, we also get $y = 9$. However, this poses no problem as far as $y = x^2$ being a function is concerned. Each x is assigned exactly one y-value, its square. Be careful! The definition of a function requires that each x-value in the domain is assigned exactly one y-value, *not* necessarily that every x-value have a different y-value. Therefore, $y = x^2$ is a function.

Since we can square any number, the domain is the set of all real numbers.

(d) For the equation $y^2 = x$, if we choose $x = 9$ we get $y^2 = 9$, which gives $y = \pm 3$. In other words, there are *two* y-values associated with $x = 9$. Therefore, $y^2 = x$ is not a function.

Look very carefully at parts (c) and (d) to make sure you see why $y = x^2$ is a function, but $y^2 = x$ is not.

(e) Examining the equation we can see that each x-value uniquely determines a y-value and so $y = \dfrac{4}{x^2 + 1}$ is a function.

In trying to determine the domain of $y = \dfrac{4}{x^2 + 1}$, we are again looking at a fractional expression, and so we must again exclude any values that make the denominator equal to zero. In other words, we require

$$x^2 + 1 \neq 0$$

$$x^2 \neq -1 \quad \textit{This statement is true for all real numbers.}$$

In other words, there are no real values of x that make the denominator equal to zero, and so no values of x need to be excluded. Therefore, the domain is the set of all real numbers. ■

EXAMPLE 4 Find the domain of the function $y = \sqrt{3x - x^2}$.

Solution Let's analyze this problem carefully so that we clearly understand what is being asked and can develop a strategy for the solution.

WHAT DO WE NEED TO FIND?	The domain of the function $y = \sqrt{3x - x^2}$.
WHAT IS A DOMAIN?	The real values of x that make y defined and real.
WHEN IS y DEFINED AND REAL?	When the expression under the radical is nonnegative.
HOW CAN I RESTATE THE PROBLEM IN SIMPLER TERMS?	Since nonnegative means greater than or equal to zero, we solve the inequality $3x - x^2 \geq 0$.

Because an even root of a negative number is undefined in the real number system, we need x to satisfy the inequality

$$3x - x^2 \geq 0 \quad \textit{This is a quadratic inequality; we solve it by analyzing signs.}$$

$$x(3 - x) \geq 0 \quad \textit{Sign of } 3x - x^2$$

- - - - - + + + + + + + + - - - - - - -
0 3

Since we want $3x - x^2 = x(3 - x)$ to be nonnegative, the sign analysis shows us that the domain is

$$\{x \mid 0 \leq x \leq 3\} \qquad \text{or, using interval notation,} \qquad [0, 3]$$ ■

A comment is in order here. Students are sometimes heard remarking about a problem like Example 4, "I know how to solve a quadratic inequality, but that is not what the problem asked." Well, in fact, the problem did not directly ask that a quadratic inequality be solved. A problem such as this obliges the reader to reformulate the question into more familiar terms. Our analysis of what the question was asking and what condition x must satisfy led us to recognize that the domain could be found by doing a sign analysis. Analyzing a question and breaking it down into smaller more digestible parts is an idea we will use many times in this book.

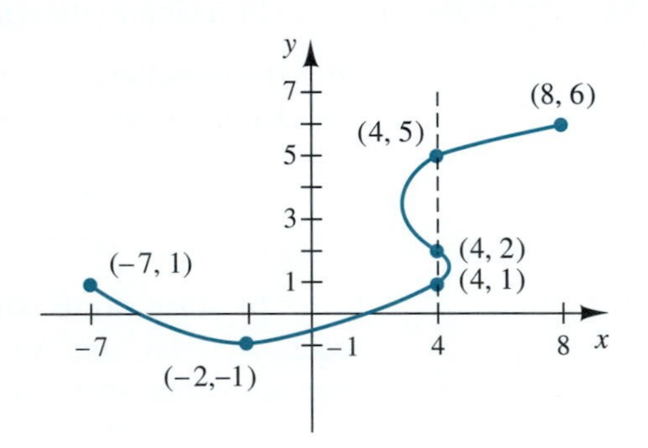

FIGURE 2.44 The graph of a relation

As with relations, we can describe functions by graphs. For example in Figure 2.44, we have drawn a graph and labeled some of the points on the graph. The graph represents a relation—but is it a function? Since the points (4, 1), (4, 2), and (4, 5) are on the graph, there is a value of $x = 4$ that has three y-values associated with it. But a function can only assign one y-value for each x. Therefore, we can see that this relation is *not* a function.

We can generalize this last comment as follows. We can visually determine whether a graph is the graph of a function by using a simple procedure called the **vertical line test.**

THE VERTICAL LINE TEST

If any vertical line intersects a graph in more than one point, then the graph is *not* the graph of a function.

Alternatively, the vertical line test says that any vertical line may intersect the graph of a function in at most one point.

DIFFERENT PERSPECTIVES: *Functions*

Consider the geometric and algebraic descriptions of a function.

GEOMETRIC DESCRIPTION

This *is* a function.

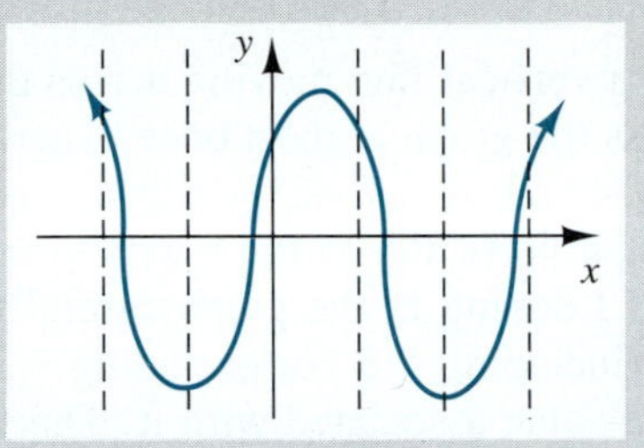

Any vertical line crosses the graph at most once; therefore, each x-value is assigned exactly one y-value.

This is *not* a function

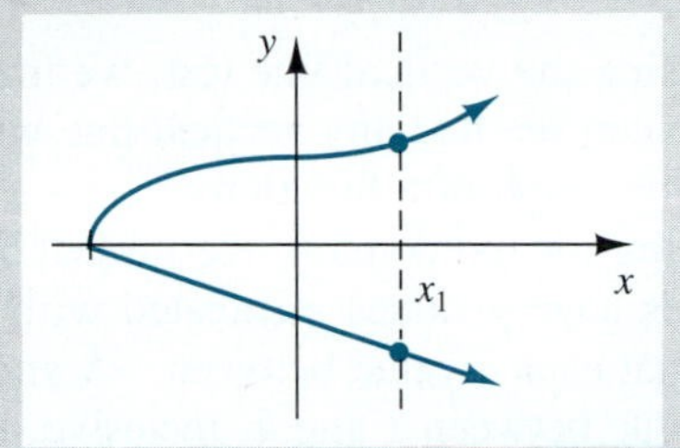

Note that two different y-values are assigned to x_1; this violates the definition of a function.

ALGEBRAIC DESCRIPTION

If a relation assigns more than one y-value to *any* x-value, then it is *not* a function.

EXAMPLE 5 Which of the following is the graph of a function?

(a)

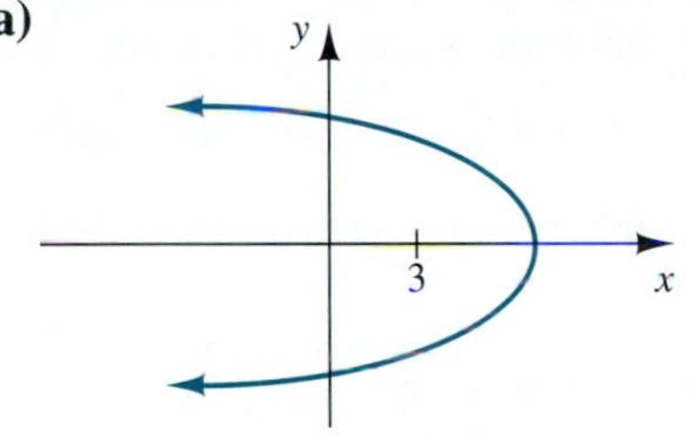

(b)

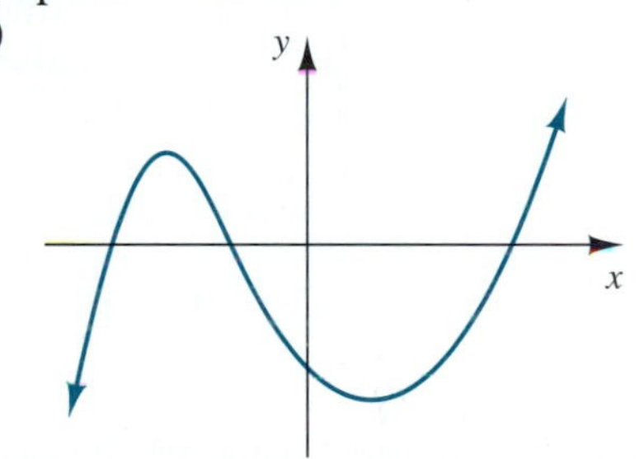

Why is a vertical line used to determine if a graph is a function?

Solution If we imagine a vertical line moving across the graph from left to right we see that the graph in part **(a)** is intersected twice by the vertical line $x = 3$, whereas the graph in part **(b)** is never intersected more than once by any vertical line. See Figure 2.45.

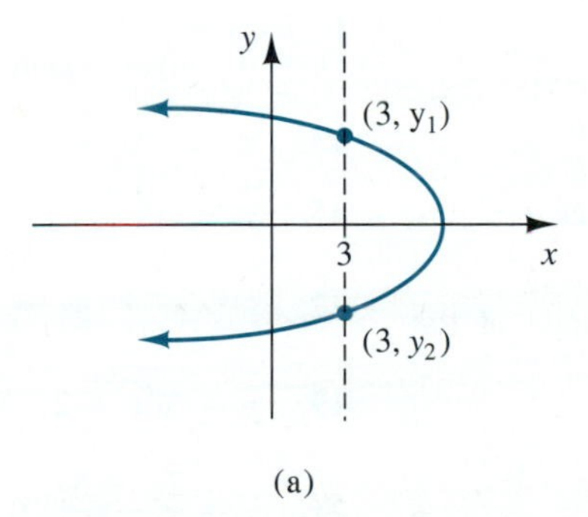

(a)

(b)

FIGURE 2.45 Employing the vertical line test

(a) In Figure 2.45(a), the vertical line shows us that there are two y-values, y_1 and y_2, associated with $x = 3$. Therefore, the graph in part **(a)** is not the graph of a function.

(b) According to the vertical line test, the graph in part **(b)** is the graph of a function. ■

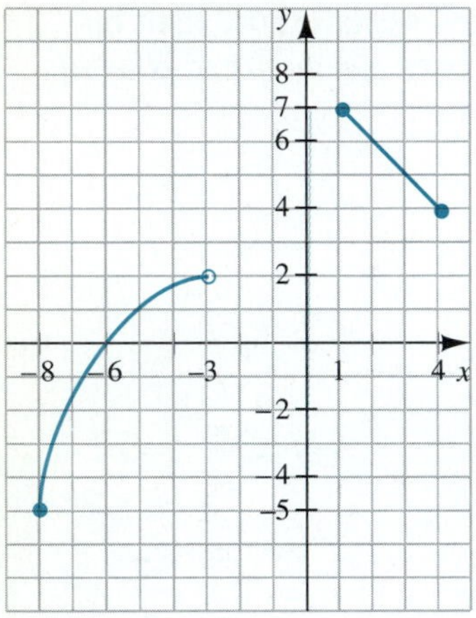

EXAMPLE 6 Given the graph in Figure 2.46, determine if it is the graph of a function and find its domain and range.

FIGURE 2.46

Solution Using the vertical line test, we imagine a vertical line moving across the graph and we can see that any vertical line will cross the graph at most once. Therefore, this is the graph of a function.

In looking for the domain, we project the graph vertically to the x-axis to see which x-values have y-values associated with them. Looking at the graph carefully, we can see that each x-value between -8 and -3 (including -8 but excluding -3) and each x-value between 1 and 4, inclusive, has a y-value associated with it. Therefore, the domain is

$$\boxed{\{x \mid -8 \le x < -3 \text{ or } 1 \le x \le 4\}} \quad \text{or, equivalently,} \quad \boxed{[-8, -3) \cup [1, 4]}$$

In looking for the range, we project the graph horizontally to the y-axis to see which y-values have x-values associated with them. Again looking at the graph carefully, we can see that each y-value between -5 and 2 (including -5, excluding 2) and each y-value between 4 and 7, inclusive, is included. Therefore, the range is

$$\{y \mid -5 \le y < 2 \text{ or } 4 \le y \le 7\} \quad \text{or, equivalently,} \quad [-5, 2) \cup [4, 7]$$ ■

In the next section we will continue our discussion with particular attention paid to functions.

EXERCISES 2.4

In Exercises 1–6, the given arrow diagram defines a relation. Translate the diagram into ordered pair notation. Find the domain and range of the relation. Is the relation a function?

1. 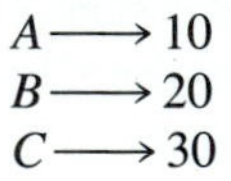 $A \to 10$, $B \to 20$, $C \to 30$

2. 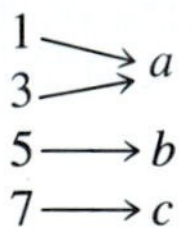 $1 \to a$, $3 \to a$, $5 \to b$, $7 \to c$

3. 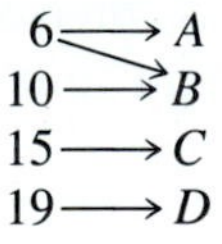 $6 \to A$, $6 \to B$, $10 \to B$, $15 \to C$, $19 \to D$

4. 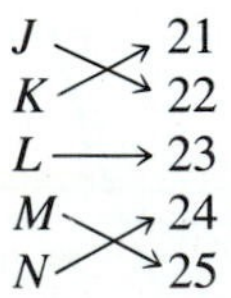 $J \to 22$, $K \to 21$, $L \to 23$, $M \to 25$, $N \to 24$

5. $A \to 7$, $B \to 7$, $C \to 7$, $D \to 7$

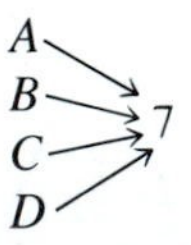

6. $7 \to A$, $7 \to B$, $7 \to C$, $7 \to D$

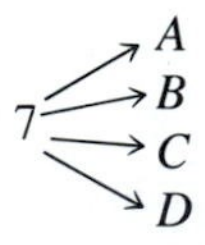

In Exercises 7–12, determine the domain and range of the given relation. Is the relation a function?

7. $\{(-4, -3), (2, -5), (4, 6), (2, 0)\}$

8. $\left\{(8, -2), \left(6, -\frac{3}{2}\right), (-1, 5)\right\}$

9. $\{(-\sqrt{3}, 3), (-1, 1), (0, 0), (1, 1), (\sqrt{3}, 3)\}$

10. $\left\{\left(-\frac{1}{2}, \frac{1}{16}\right), (-1, 1), \left(\frac{1}{3}, \frac{1}{81}\right)\right\}$

11. $\{(0, 5), (1, 5), (2, 5), (3, 5), (4, 5), (5, 5)\}$

12. $\{(5, 0), (5, 1), (5, 2), (5, 3), (5, 4), (5, 5)\}$

In Exercises 13–42, find the domain of the given relation. Does the given equation describe y as a function of x?

13. $y = -3x + 2$

14. $y = x^2 + 3$

15. $y = \dfrac{1}{x - 2}$

16. $y = \dfrac{x + 4}{x + 8}$

17. $y = x^2 - 3x - 4$

18. $y = \dfrac{1}{x^2 - 3x - 4}$

19. $y = \sqrt{x + 2}$

20. $y = \dfrac{1}{\sqrt{x - 4}}$

21. $y = \dfrac{2x - 5}{3x + 4}$

22. $y = \dfrac{6x}{x^2 - x - 1}$

23. $y = \dfrac{|x|}{x}$

24. $y = \dfrac{-4}{x^3 - 9x^2}$

25. $y = \dfrac{5}{\sqrt{x^2 - 9}}$

26. $y = \sqrt{x^2 - 6x + 5}$

27. $y = \sqrt{x^2 + 16}$

28. $y = \sqrt[3]{x - 8}$

29. $y = \sqrt[5]{x^2 - 2x - 3}$

30. $y = \sqrt[4]{x^2 - 2x + 1}$

31. $y = \sqrt{8 - 5x}$

32. $y = 8 - \sqrt{5x}$

33. $y = \sqrt{|x - 4|}$

34. $y = \dfrac{1}{x^2}$

35. $y = \dfrac{\sqrt{x + 5}}{x - 3}$

36. $y = \dfrac{\sqrt{x - 3}}{x + 2}$

37. $y = \dfrac{7}{\sqrt{3x - 2x^2}}$

38. $y = \dfrac{3x}{x^2 + x + 1}$

39. $y = \dfrac{x^2 - 3x - 10}{x - 5}$

40. $y = \dfrac{x^3 - 5x^2}{x}$

41. $y = \dfrac{2x - 8}{x^2 - 16}$

42. $y = \dfrac{3x + 15}{x^2 + 3}$

In Exercises 43–48, determine whether the given graph is the graph of a function.

43.

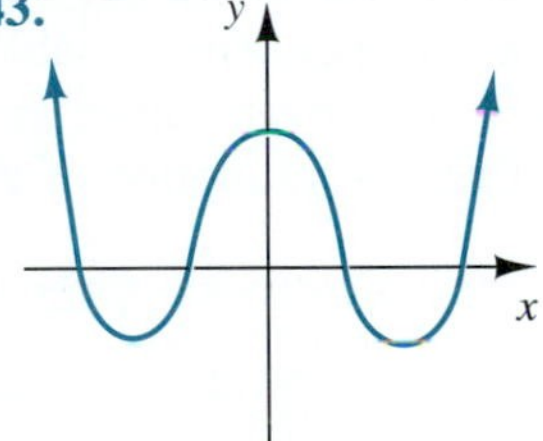

44.

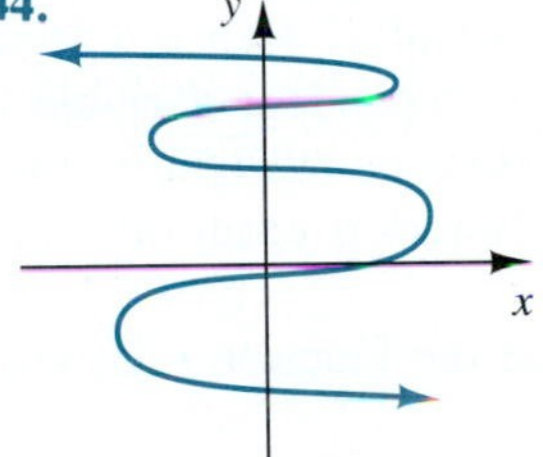

45.

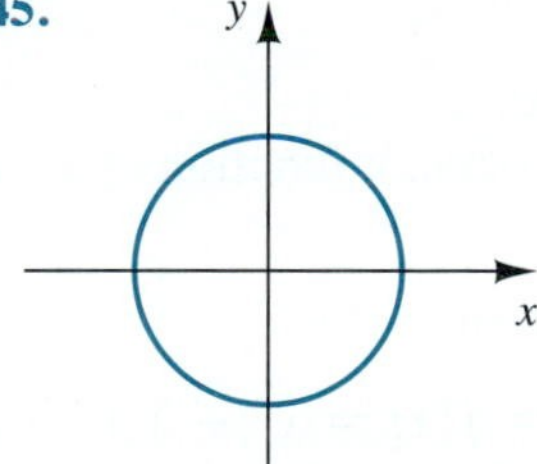

46.

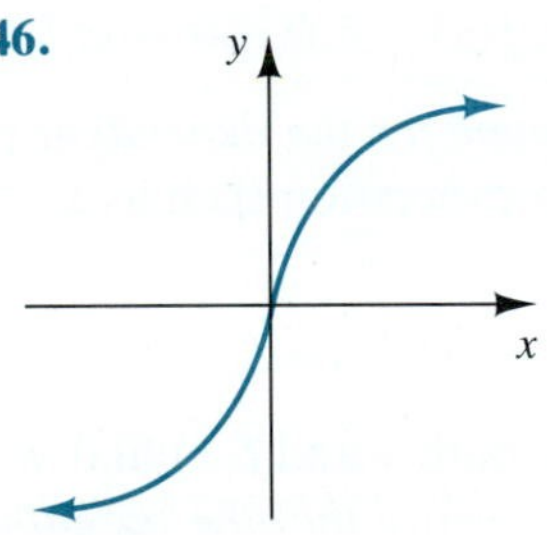

47.

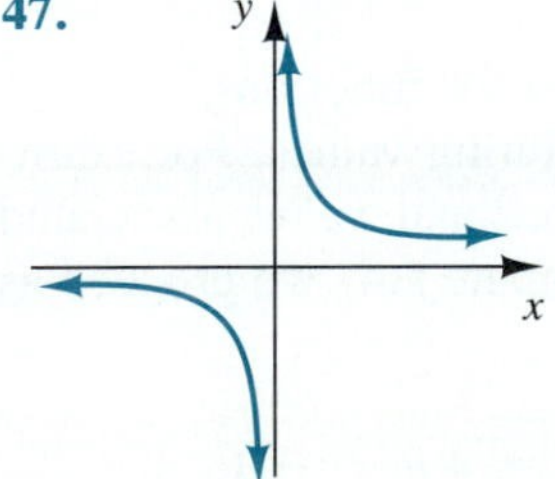

48.

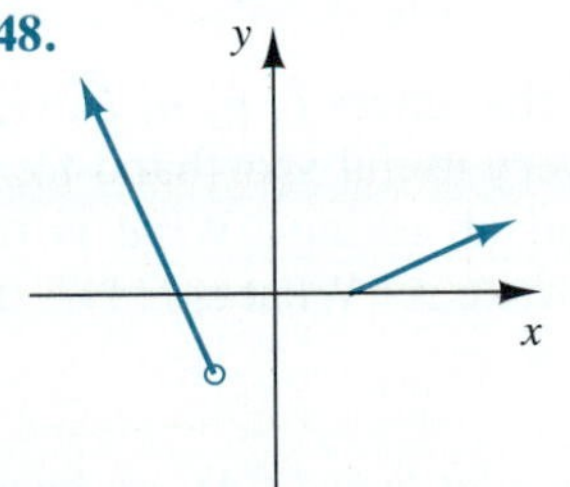

In Exercises 49–54, use the given graph to determine the domain and range of the relation or function.

49.

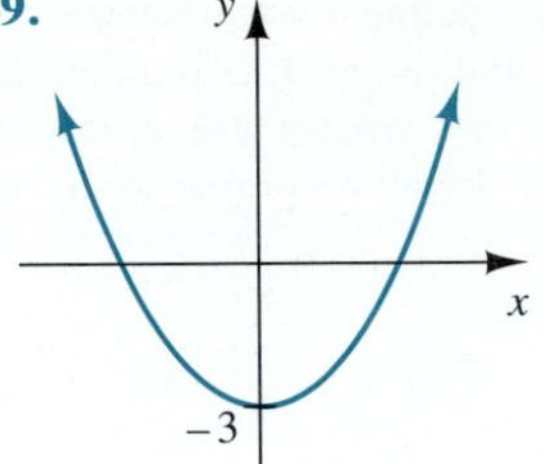

50.

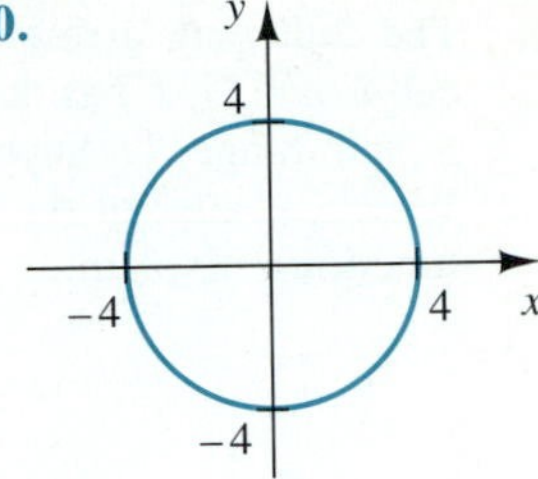

51.

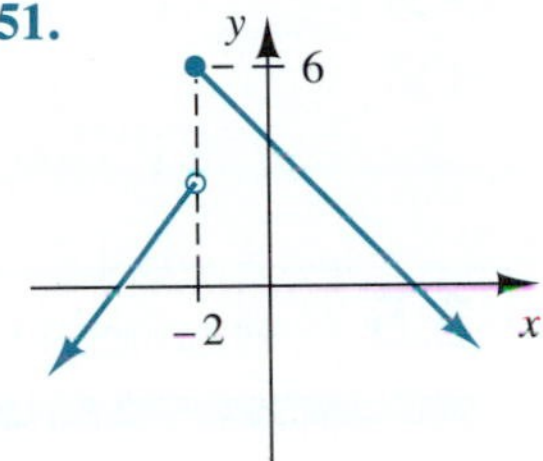

52.

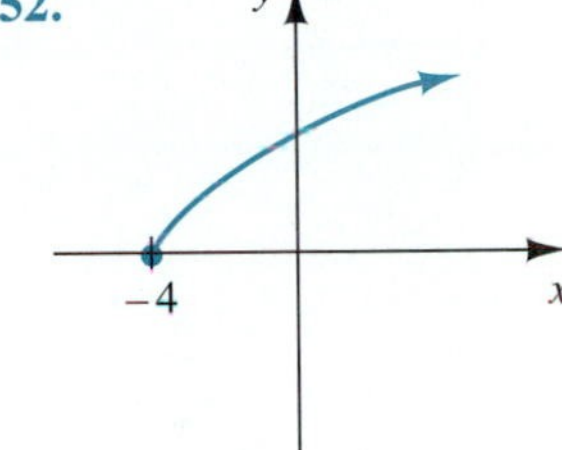

53.

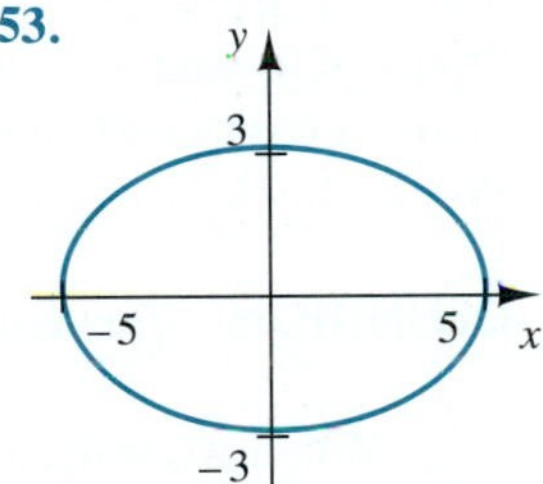

54.

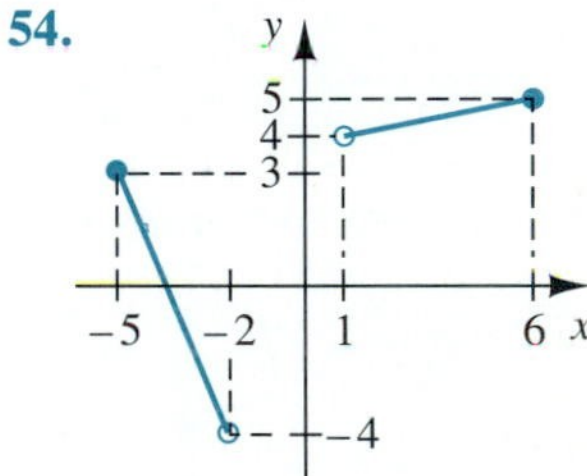

55. Suppose a car rental company charges a rate of \$22 per day plus \$0.11 per mile. Express the cost of a 5-day car rental as a function of the number of miles driven, m.

56. A truck driver drives from point A due east at 50 mi/hr for h hours and then drives due north at 60 mi/hr for another h hours until reaching point B. Express the distance, d, between points A and B as a function of h.

57. A jogger runs in a road race for a total of 6 hours. She runs at an average speed of 8 km/hr for t hours and the rest of the time at an average speed of 10 km/hr. Express the total distance, d, the jogger runs as a function of t.

58. A farmer has 100 feet of fencing, which is to be cut into two pieces. The first piece, of length x feet, will be erected in the shape of a square, and the remainder of the fencing will also be erected in the shape of a square. Express the total area enclosed by the two squares as a function of x.

QUESTIONS FOR THOUGHT

59. State in words what makes a relation a function.

60. The following arrow diagrams define two functions we call F and G: F has domain A and range B; G has domain S and range T. Suppose all the arrows are reversed. Which, if any, of the two new arrow diagrams would be functions? Explain.

F

A B

$a \to 20$
$b \to 10$
$c \to 30$
$d \to 40$

G

S I

$a \to 20$
$b \to 10$
$c \to 30$
$d \to 30$

GRAFFIX

61. Use a graphics calculator or computer to sketch the graphs of each equation.
(a) $y = x^2 - x - 6$
(b) $y = \sqrt{36 - 4x^2}$

62. How would you use a graphics calculator to sketch the graph of the relation $x^2 + y^2 = 4$?

2.5 Function Notation

In the last section we saw that equations such as $y = 3x + 4$ and $y = x^2 - 1$ define y as a function of x. Because these equations are solved explicitly for y, we can say that y is dependent on the value of x. Hence x is called the independent variable ("pick x"), and y is called the dependent variable ("get y"). However, there are many occasions when we wish to state outright or emphasize that y is a function of x. We do this by using a special **function notation**: $f(x)$.

DEFINITION $f(x)$ is the value the function f assigns to x.

When we want to indicate that y is a function of x, we write

$$y = f(x)$$ *$f(x)$ is read "f of x."*

$f(x)$ is merely another name for the dependent variable.

Thus we can describe an expression such as $x^2 - 1$ as being a function of x by writing

$$f(x) = x^2 - 1$$ *The expression $x^2 - 1$ is a function of x.*

(In fact, many times we write both y and $f(x)$ and write $y = f(x) = x^2 - 1$.)

It is very important to recognize that the parentheses in $y = f(x)$ are not *being used to indicate multiplication. Rather, the parentheses are being used to specify the independent variable.*

Most times we will use the letters f, g, h, F, G, and H for functions.

Function notation is a very useful shorthand for substituting values. For example, if $f(x) = x^2 - 1$, instead of asking "What is the functional value associated with $x = 4$?" all we need to write is "What is $f(4)$?" To compute $f(4)$ we proceed as follows:

$f(x) = x^2 - 1$ *In order to find $f(4)$, we substitute $x = 4$ in $x^2 - 1$.*

$f(4) = (4)^2 - 1 = 15$ *$f(4)$ is the value of $f(x)$ when 4 is substituted for x.*

It is useful to think of x as the *input* and $f(x)$ as the *output*. Thus in computing $f(4)$, **4 is the input and f(4) = 15 is the output.**

$$\begin{array}{c} \textit{Input} \\ \downarrow \\ f(4) = 15 \\ \uparrow \\ \textit{Output} \end{array}$$

EXAMPLE 1 Given the function $g(x) = \frac{1}{3}x^3 + 2x - 4$, find each of the following:

(a) $g(-3)$ **(b)** $g\left(\frac{1}{2}\right)$ **(c)** $g(t)$

Solution Keeping in mind that x is the independent variable, $g(x) = \frac{1}{3}x^3 + 2x - 4$ tells us that the rule of assignment for the function g in this example is "take the input, cube it and multiply by $\frac{1}{3}$, and then add 2 times the input and subtract 4." We will follow this rule in all parts of the example, regardless of what the input is.

(a) $g(x) = \frac{1}{3}x^3 + 2x - 4$ *To find $g(-3)$, we replace all occurrences of x with -3.*

$$g(-3) = \frac{1}{3}(-3)^3 + 2(-3) - 4$$

$$= -9 - 6 - 4$$ *Thus*

$$\boxed{g(-3) = -19}$$

Remember that $g(-3) = -19$ means that when $x = -3$, $g(x) = -19$.

(b) $g(x) = \frac{1}{3}x^3 + 2x - 4$ *To find $g\left(\frac{1}{2}\right)$, we replace x with $\frac{1}{2}$.*

$$g\left(\frac{1}{2}\right) = \frac{1}{3}\left(\frac{1}{2}\right)^3 + 2\left(\frac{1}{2}\right) - 4$$

$$= \frac{1}{3}\left(\frac{1}{8}\right) + 1 - 4 = \left(\frac{1}{24}\right) - 3$$

$$\boxed{g\left(\frac{1}{2}\right) = -\frac{71}{24}}$$

(c) $g(x) = \frac{1}{3}x^3 + 2x - 4$ *To find $g(t)$, we replace all occurrences of x with t.*

$$g(t) = \frac{1}{3}(t)^3 + 2(t) - 4$$ *Thus*

$$\boxed{g(t) = \frac{1}{3}t^3 + 2t - 4}$$

■

EXAMPLE 2 Given the function $f(x) = 3x^2 - 4x + 5$, find each of the following and simplify:

(a) $f\left(\frac{1}{t}\right)$ **(b)** $\frac{1}{f(t)}$ **(c)** $f(a + 2)$ **(d)** $f(a) + 2$ **(e)** $f(2x)$ **(f)** $2f(x)$

Solution

(a) $f(x) = 3x^2 - 4x + 5$ *To find $f\left(\frac{1}{t}\right)$, we replace x with $\frac{1}{t}$.*

$$f\left(\frac{1}{t}\right) = 3\left(\frac{1}{t}\right)^2 - 4\left(\frac{1}{t}\right) + 5 = 3\left(\frac{1}{t^2}\right) - \left(\frac{4}{t}\right) + 5 \quad \textit{Thus}$$

$$\boxed{f\left(\frac{1}{t}\right) = \frac{3}{t^2} - \frac{4}{t} + 5 = \frac{3 - 4t + 5t^2}{t^2}}$$

(b) To find $\frac{1}{f(t)}$, we first find $f(t)$ and then take its reciprocal.

Since $f(t) = 3t^2 - 4t + 5$, we have $\boxed{\frac{1}{f(t)} = \frac{1}{3t^2 - 4t + 5}}$

Note the difference between parts **(a)** and **(b)**. In part **(a)** we choose a value t and use its reciprocal as the input, whereas in part **(b)** we choose a value t and then take the reciprocal of its output.

(c) $f(x) = 3x^2 - 4x + 5$ *Keep in mind that x represents the input to the function. It is sometimes helpful to view the function as $f(\ ) = 3(\ )^2 - 4(\ ) + 5$.*

Whatever we substitute into the parentheses on the left (which is being called x), we also substitute on the right. Therefore, we replace all occurrences of x with $a + 2$ and simplify.

$$\begin{aligned} f(a + 2) &= 3(a + 2)^2 - 4(a + 2) + 5 \\ &= 3(a^2 + 4a + 4) - 4a - 8 + 5 \\ &= 3a^2 + 12a + 12 - 4a - 8 + 5 \quad \textit{Thus} \end{aligned}$$

$$\boxed{f(a + 2) = 3a^2 + 8a + 9}$$

(d) $f(a) + 2$ means we are to add 2 to $f(a)$. First find $f(a)$:

$$f(a) = 3a^2 - 4a + 5 \quad \textit{And so}$$

$$f(a) + 2 = \underbrace{3a^2 - 4a + 5}_{f(a)} + 2 \quad \textit{Thus}$$

$$\boxed{f(a) + 2 = 3a^2 - 4a + 7}$$

Note the difference between parts **(c)** and **(d)**. In part **(c)**, we are adding 2 to a and using that as the *input*, whereas in part **(d)**, we are adding 2 to the *output*

$f(a)$. Thus there is no reason to expect $f(a + 2)$ and $f(a) + 2$ to be equal, and as we just saw they are not equal.

(e) $f(x) = 3x^2 - 4x + 5$ *Again don't take x literally; x stands for the input. To compute $f(2x)$, 2x becomes the input; that is, we replace all occurrences of x with 2x.*

$f(2x) = 3(2x)^2 - 4(2x) + 5$ *Thus*

$$\boxed{f(2x) = 12x^2 - 8x + 5}$$

(f) $2f(x)$ means that we multiply $f(x)$ by 2.

$f(x) = 3x^2 - 4x + 5$ *First find $f(x)$; then multiply by 2.*

$2f(x) = 2(3x^2 - 4x + 5)$ *Thus*

$$\boxed{2f(x) = 6x^2 - 8x + 10}$$

How are $f(2x)$ and $2f(x)$ different from each other?

Note the difference between parts **(e)** and **(f)**. In part **(e)**, $f(2x)$, we are doubling x and using that as the input, but in part **(f)**, $2f(x)$, we are doubling the output. Again there is no reason to expect the two results to be equal. ■

EXAMPLE 3 Given the function $H(x) = \dfrac{x + 3}{x - 4}$, find $H\left(\dfrac{1}{x + 1}\right)$.

Solution

$$H(x) = \frac{x + 3}{x - 4}$$

To find $H\left(\dfrac{1}{x + 1}\right)$ we replace each occurrence of x by $\dfrac{1}{x + 1}$.

$$H\left(\frac{1}{x + 1}\right) = \frac{\dfrac{1}{x + 1} + 3}{\dfrac{1}{x + 1} - 4}$$

Simplify the complex fraction.

$$= \frac{\left(\dfrac{1}{x + 1} + 3\right)(x + 1)}{\left(\dfrac{1}{x + 1} - 4\right)(x + 1)} = \frac{1 + 3x + 3}{1 - 4x - 4} = \boxed{\frac{3x + 4}{-4x - 3}}$$ ■

EXAMPLE 4 Given $f(x) = x^3 + 1$, find each of the following:

(a) $f(x) + 4$ **(b)** $f(x + 4)$ **(c)** $f(x) + f(4)$ **(d)** $f(x + 4) - f(x)$

Solution

(a) $f(x) + 4$ requires us to add 4 to $f(x)$.

$$f(x) + 4 = \underbrace{x^3 + 1}_{f(x)} + 4 = \boxed{x^3 + 5}$$

(b) $f(x + 4)$ is asking us to use $x + 4$ as the input for $f(x)$.

$$f(x) = x^3 + 1$$

$$f(x + 4) = (x + 4)^3 + 1 \qquad \textit{We multiply out } (x+4)^3.$$

$$f(x) = x^3 + 12x^2 + 48x + 64 + 1 \qquad \textit{Thus}$$

$$\boxed{f(x + 4) = x^3 + 12x^2 + 48x + 65}$$

(c) $f(x) + f(4)$ is asking us to add $f(4)$ to $f(x)$.

$$f(x) + f(4) = \underbrace{(x^3 + 1)}_{f(x)} + \underbrace{(4^3 + 1)}_{f(4)}$$

$$= x^3 + 1 + 64 + 1 \qquad \textit{Thus}$$

$$\boxed{f(x) + f(4) = x^3 + 66}$$

How are $f(x) + 4$, $f(x + 4)$, and $f(x) + f(4)$ different from each other?

Note the difference between parts **(b)** and **(c)**. We cannot simply add $f(4)$ to $f(x)$ to get $f(x + 4)$. It is important to recognize that in general,

$$f(a + b) \neq f(a) + f(b).$$

(d) In order to find $f(x + 4) - f(x)$, we need $f(x + 4)$, which we found in part **(b)** to be $x^3 + 12x^2 + 48x + 65$. Then we subtract $f(x)$ from this expression and simplify.

$$f(x + 4) - f(x) = \underbrace{(x^3 + 12x^2 + 48x + 65)}_{f(x+4)} - \underbrace{(x^3 + 1)}_{f(x)}$$

$$= x^3 + 12x^2 + 48x + 65 - x^3 - 1 \qquad \textit{Simplify to get}$$

$$\boxed{f(x + 4) - f(x) = 12x^2 + 48x + 64}$$

■

EXAMPLE 5 Given $f(x) = x^2 - x + 1$, find $\dfrac{f(x + 3) - f(x)}{3}$.

Solution An expression of the form $\dfrac{f(x + h) - f(x)}{h}$ is called a *difference quotient* for $f(x)$. The difference quotient plays a pivotal role in calculus. We hint at the significance of the difference quotient in Exercise 74 but restrict our attention here to simplifying difference quotients algebraically, as we do in this example.

We begin by finding $f(x + 3)$ for $f(x) = x^2 - x + 1$.

$$f(x + 3) = (x + 3)^2 - (x + 3) + 1$$

$$= x^2 + 6x + 9 - x - 3 + 1$$

$$= x^2 + 5x + 7$$

Therefore,

$$\frac{f(x+3)-f(x)}{3} = \frac{\overbrace{x^2+5x+7}^{f(x+3)} - \overbrace{(x^2-x+1)}^{f(x)}}{3}$$

Don't forget the parentheses around $f(x)$.

$$= \frac{x^2+5x+7-x^2+x-1}{3} = \frac{6x+6}{3} = \frac{6(x+1)}{3}$$

$$= \boxed{2(x+1)}$$

■

Split Functions

Although all the functions we have examined thus far in this section have had one rule of assignment for all members of their domain, this need not be the case.

For example, consider the following function $f(x)$:

$$f(x) = \begin{cases} x^2 + 3 & \text{if } -3 \le x < 2 \\ 5x - 4 & \text{if } x \ge 2 \end{cases}$$

This means that we are to use the rule $f(x) = x^2 + 3$ for any x in the interval $[-3, 2)$, but we are to use the rule $f(x) = 5x - 4$ for any x in the interval $[2, \infty)$.

If we want to find $f(-1)$, we first note that -1 is in the interval $[-3, 2)$, and so we use the top rule. Thus $f(-1) = (-1)^2 + 3 = 4$.

If we want to find $f(2)$, we note that 2 is greater than or equal to 2, and so we use the bottom rule. Thus $f(2) = 5(2) - 4 = 6$.

If we want to find $f(-4)$, we note that there is no rule for $f(x)$ when x is less than -3. Therefore, -4 is not in the domain and so there is no value associated with it.

Such a function, which has more than one rule of assignment, depending upon the input value, is called a **split function.** The *domain of a split function* is the set of x-values for which there is a rule of assignment. Thus the preceding split function $f(x)$ has as its domain $\{x \mid x \ge -3\}$.

EXAMPLE 6 Given the function $G(x) = \begin{cases} x - 1 & \text{if } -5 \le x < -1 \\ x^3 & \text{if } -1 \le x < 2 \\ 6 & \text{if } x > 2 \end{cases}$

Find: **(a)** $G(0)$ **(b)** $G(5)$ **(c)** $G(2)$ **(d)** $G(-5)$

Solution

(a) Since 0 lies between -1 and 2, we use the middle rule to get $G(0) = 0^3 = 0$. Thus $\boxed{G(0) = 0}$.

(b) Since 5 is greater than 2, we use the third rule to get $\boxed{G(5) = 6}$. Note that the third rule will assign the value 6 to any x value greater than 2.

(c) To find $G(2)$ we note that $x = 2$ does not fall into any of the three categories. $x = 2$ is not in the domain of $G(x)$ and so $\boxed{G(2) \text{ is undefined}}$.

(d) To find $G(-5)$ we use the first rule and get $G(-5) = -5 - 1 = -6$. Thus $\boxed{G(-5) = -6}$. ■

In this section our attention has been focused on understanding and using function notation from an algebraic point of view. In the next section we discuss the graphical interpretation of $f(x)$ notation.

EXERCISES 2.5

In Exercises 1–20, let $f(x) = 5x - 2$, $g(x) = 3x^2 - 4x + 1$, and $h(x) = \sqrt{4x - 3}$. Find:

1. $f(6)$ **2.** $g(-5)$
3. $h(10)$ **4.** $f(-2)$
5. $g(8)$ **6.** $h(0)$
7. $f\left(\frac{1}{3}\right)$ **8.** $g\left(-\frac{1}{2}\right)$
9. $f(x + 2)$ **10.** $f(x) + 2$
11. $g(x - 1)$ **12.** $g(x) - g(1)$
13. $h(x^2)$ **14.** $[h(x)]^2$
15. $g(3x)$ **16.** $3g(x)$
17. $f(4x + 7)$ **18.** $4f(x) + 7$
19. $g(x + h)$ **20.** $f(x - a)$

In Exercises 21–40, given $F(x) = \dfrac{x}{x + 1}$ and $G(t) = \dfrac{1}{\sqrt{t - 1}}$, find:

21. $F(9)$ **22.** $F(-2)$
23. $G(15)$ **24.** $G(6)$
25. $F\left(\frac{1}{2}\right)$ **26.** $G(1)$
27. $G(t^2)$ **28.** $F\left(\frac{1}{x}\right)$
29. $F(-1)$ **30.** $F(x^2)$
31. $F(x + 1)$ **32.** $G(t - 1)$
33. $5G(6)$ **34.** $G(30)$
35. $F(40)$ **36.** $8F(5)$
37. $F\left(\frac{a}{3}\right)$ **38.** $G(t^2 + 1)$
39. $F(x - 1)$ **40.** $G\left(\dfrac{a}{a + 1}\right)$

In Exercises 41–60, let $f(x) = 4x - 3$, $g(x) = \dfrac{1}{x}$, and $h(x) = x^2 - x$. Find (and simplify):

41. $f(5x + 7)$ **42.** $5f(x) + 7$
43. $g(x - a)$ **44.** $f(x) + f(a)$
45. $g(x) - g(a)$ **46.** $f(x) + a$
47. $g(x) - a$ **48.** $f(x + a)$
49. $f(4)h(4)$ **50.** $f(1)g(2)h(-3)$
51. $f[h(4)]$ **52.** $h[f(4)]$
53. $g[h(x)]$ **54.** $h[g(x)]$
55. $h(kx)$ **56.** $kh(x)$
57. $f(g[h(3)])$ **58.** $f(3)g(3)h(3)$
59. $h\left(\frac{1}{x}\right)$ **60.** $\dfrac{1}{h(x)}$

In Exercises 61–66, let $f(x) = 2 - 3x$ and $g(x) = x^2 - 3x + 2$. Find (and simplify):

61. $\dfrac{g(x) - g(5)}{x - 5}$ **62.** $\dfrac{f(x) - f(2)}{x - 2}$
63. $\dfrac{f(x + 3) - f(x)}{3}$ **64.** $\dfrac{g(x + 4) - g(x)}{4}$
65. $\dfrac{g(x + h) - g(x)}{h}$ **66.** $\dfrac{f(x + h) - f(x)}{h}$

67. Given $f(x) = \begin{cases} 3x - 5 & \text{if } x < 1 \\ x^2 - 1 & \text{if } x \ge 1 \end{cases}$
Find: **(a)** $f(-3)$, **(b)** $f(1)$, **(c)** $f(6)$.

68. Given $g(x) = \begin{cases} x^2 - 3x + 2 & \text{if } x \le -5 \\ x + 1 & \text{if } -5 < x < 2 \\ 8 & \text{if } x > 2 \end{cases}$
Find: **(a)** $g(-4)$, **(b)** $g(2)$, **(c)** $g(-6)$, **(d)** $g(4)$.

69. Given $h(s) = \begin{cases} |2s - 1| & \text{if } 0 \le s < 3 \\ \dfrac{s + 1}{s - 1} & \text{if } 3 < s \le 7 \end{cases}$
Find: **(a)** $h(0)$, **(b)** $h(3)$, **(c)** $h(5)$, **(d)** $h(8)$.

70. Given $f(x) = \begin{cases} 3x - 5 & \text{if } -2 \le x < 1 \\ x^2 - 4 & \text{if } 1 < x \end{cases}$
Find the domain of $f(x)$.

71. If an object is thrown upward from ground level with an initial velocity of 50 ft/sec, then its height, h, after t seconds is given by the equation
$$h(t) = 50t - 16t^2$$
(a) Find the height of the object after 1, 2, and 3 seconds.
(b) How many seconds does it take for the object to hit the ground?

72. If an object is dropped from a window 600 feet above ground, then its height, h, after t seconds is given by the equation
$$h(t) = 600 - 16t^2$$

(a) Find the height of the object after 1, 2, and 3 seconds.
(b) Approximately how many seconds does it take for the object to hit the ground?

QUESTIONS FOR THOUGHT

73. Given $f(x) = 5x + 7$, discuss what is *wrong* (if anything) with each of the following:
(a) $f(x + 4) \stackrel{?}{=} 5x + 7 + 4 \stackrel{?}{=} 5x + 11$
(b) $f(x + 4) \stackrel{?}{=} (5x + 7)(x + 4) \stackrel{?}{=} 5x^2 + 27x + 28$
(c) $f(2x) \stackrel{?}{=} 2(5x + 7) \stackrel{?}{=} 10x + 14$
(d) $f(x^2) \stackrel{?}{=} (5x + 7)^2 \stackrel{?}{=} 25x^2 + 70x + 49$

74. Suppose $y = f(x) = 3x - 6$.
(a) Find $f(9)$ and $f(4)$ and explain why this means the points (9, 21) and (4, 6) lie on the graph of $f(x)$.
(b) Compute $\dfrac{f(9) - f(4)}{9 - 4}$. What is the significance of this difference quotient?
(c) Compute the difference quotient $\dfrac{f(x + h) - f(x)}{h}$ for $y = f(x) = mx + b$. What is the significance of the difference quotient for a function whose graph is a straight line?

2.6 Relating Equations to Their Graphs

Suppose we were managing a factory and found that the relationship between the cost of manufacturing items, $C(x)$, and the number of items produced daily by our factory, x, is related in the following way: $C(x) = 10{,}000(-3x^2 + 2x + 10)$. As managers, it would be our goal to understand this relationship or know the "behavior of this function," perhaps in an effort to minimize the cost, $C(x)$. Suppose we are medical researchers who find that the decrease in a person's blood pressure, $D(x)$, is related to the amount, x, of a particular drug taken by the person as follows: $D(x) = \frac{1}{3}x^2(k - x)$ (where k is some positive constant). If we want to know what happens to blood pressure as we change the dosage (perhaps in order to find the appropriate dosage for a particular patient), it is important that we understand this mathematical relationship. In this section we will review how equations are related to graphs. Our goal is to better understand how variables are related in an equation by picturing the relationship with a graph.

Suppose you are given the equation $y = x^2 - x - 6$. How would you describe the relationship between x and y? What do you think happens to y as x changes? Does y always increase as x increases? Can y be positive, or is y negative for all values of x?

We could substitute values for x and see what happens to y such as in Table 2.2. However, a table may not give a clear picture of the relationship between x and y; if we miss some values for x, we may miss important values for y. For example, we may miss the minimum or maximum y-values.

TABLE 2.2

x	-2	-1	0	1	2	3
y	0	-4	-6	-6	-4	0

It would be nice to have a "snapshot" of this relationship—a picture that tells us at a glance how the variables are related. This picture is what a graph is. We learn how to get the graph of $y = x^2 - x - 6$ in Section 3.4.

For now, let's carefully examine the graph in Figure 2.47 and review what it means for a point (x, y) to lie on a graph and how, given the graph of a relationship, we can determine y given x.

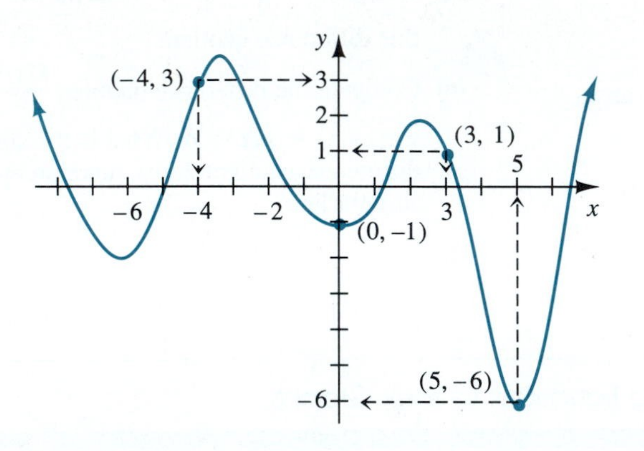

FIGURE 2.47

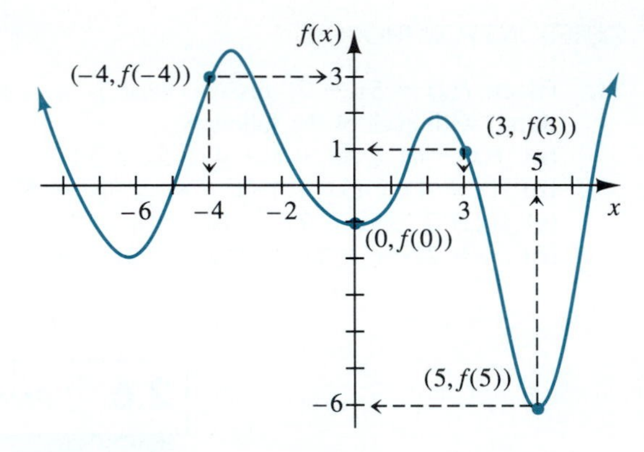

FIGURE 2.48

We can see in Figure 2.47 that the point $(5, -6)$ is on the graph. This means that the graph passes through a point that is 5 horizontal units to the right of the y-axis and 6 vertical units below the x-axis. If we need to find the y-value associated with a particular x-value, we start at the x-value on the x-axis, project vertically upward or downward to the graph, and then project horizontally to the y-axis and read this y-value. Hence if $x = 3$, then we see that for this graph, $y = 1$. Notice that at $x = 3$, the height of the graph (the vertical distance from the x-axis) is 1; likewise, if $x = -4$, $y = 3$.

The x-coordinate gives the horizontal distance from the y-axis, and the y-coordinate gives the vertical distance from the x-axis.

Returning to the graph in Figure 2.47, using functional notation, we can label the graph $y = f(x)$. Then, instead of writing $y = 3$ when $x = -4$, we can write $f(-4) = 3$. Similarly, $y = -1$ when $x = 0$ can be written as $f(0) = -1$. Figure 2.48 repeats Figure 2.47 with the vertical axis labeled $f(x)$ instead of y.

Referring to Figure 2.48, let's summarize the equivalence of these alternatives. See Table 2.3.

TABLE 2.3

(x, y) Notation	$y = f(x)$ Notation	$(x, f(x))$ Notation
$(-4, 3)$	$3 = f(-4)$	$(-4, f(-4))$
$(3, 1)$	$1 = f(3)$	$(3, f(3))$
$(0, -1)$	$-1 = f(0)$	$(0, f(0))$

Again, $f(x)$ is the y-value at x, and gives the "height" of the graph (the vertical distance above or below the x-axis) at x. Hence, $f(a)$ is the height of the graph at $x = a$; the height of the graph at $x = b$ is $f(b)$. See Figure 2.49.

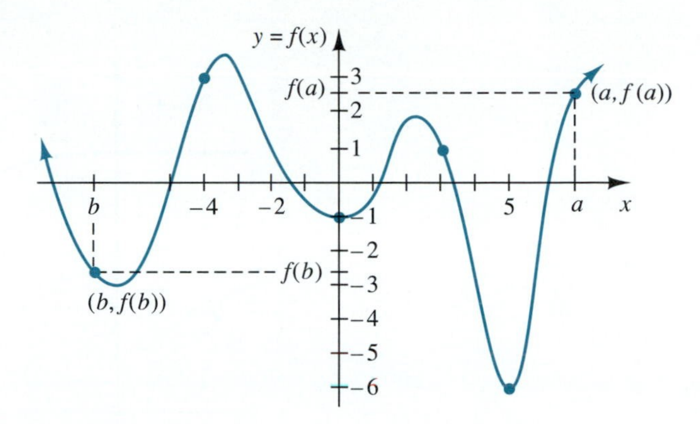

FIGURE 2.49

EXAMPLE 1 Consider the function defined by the graph in Figure 2.50 to find each of the following.

(a) $f(6)$ and $f(-3)$ **(b)** For what value(s) of x is $f(x)$ equal to 2?

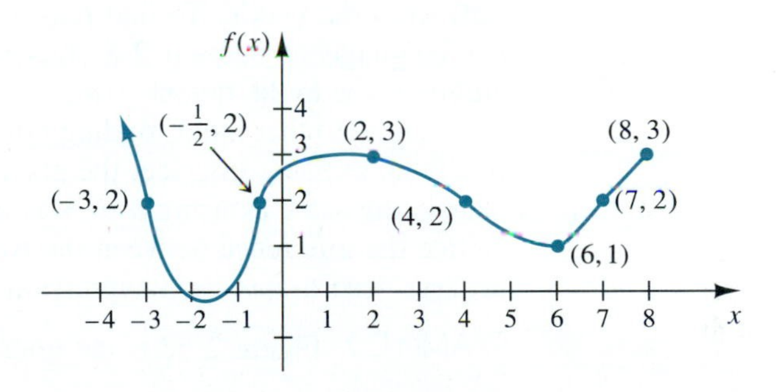

FIGURE 2.50 The graph of $f(x)$

Solution

(a) Note that the vertical axis is labeled as $f(x)$ rather than y. We may think of points on the graph as being labeled (x, y), or, equivalently, $(x, f(x))$. When we write $f(6)$, we mean the value of y when $x = 6$. By looking at the graph we can see that the point $(6, 1)$ is on the graph; therefore, $\boxed{f(6) = 1}$.

We can also see that the point $(-3, 2)$ is on the graph; therefore, $\boxed{f(-3) = 2}$. Remember that based upon the given graph, $(6, 1)$ is the same as $(6, f(6))$ and $(-3, 2)$ is the same as $(-3, f(-3))$.

(b) Asking for what value(s) of x is $f(x)$ equal to 2? is equivalent to asking for what x-values is $y = 2$. By looking at the graph we can see that $\boxed{f(x) = 2 \text{ for } x = -3, -\frac{1}{2}, 4, \text{ and } 7}$. ∎

In the last section we learned that algebraically, $f(a + 3)$ and $f(a) + 3$ are not the same. Keeping in mind that $f(a)$ is the "height" of the graph at a, we can geometrically demonstrate the differences between the two quantities. See Figure 2.51.

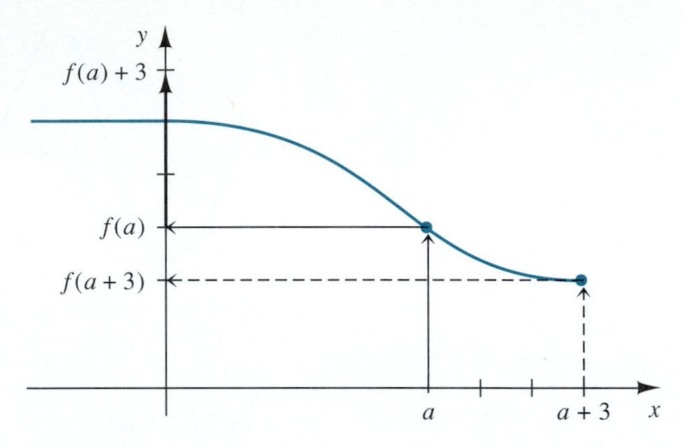

FIGURE 2.51 Comparing $f(a) + 3$ and $f(a + 3)$

Notice that finding $f(a) + 3$ requires us first to find $f(a)$. To find $f(a)$, we first locate a, move vertically up until we intersect the graph, and then project horizontally onto the y-axis. To find $f(a) + 3$, we move vertically up three more units. In the language of Section 2.5, this is equivalent to using a as the input and then adding three to the output, $f(a)$.

On the other hand, finding $f(a + 3)$ requires us first to find $a + 3$, move vertically up until we intersect the graph, and then project horizontally onto the y-axis. This is the same as using $a + 3$ as the input and then finding the output, $f(a + 3)$. Notice the difference between the two expressions in their values as well as the procedures used to find each expression.

EXAMPLE 2 Figure 2.52 is the graph of $y = f(x) = x^2 + 1$.

(a) Find $f(3 + 2)$ and $f(3) + 2$.
(b) Evaluate $f(3 + 2)$ and $f(3) + 2$ to check your answer.

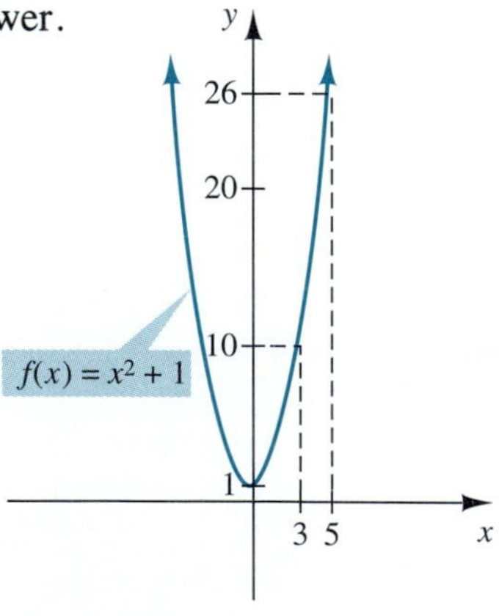

FIGURE 2.52

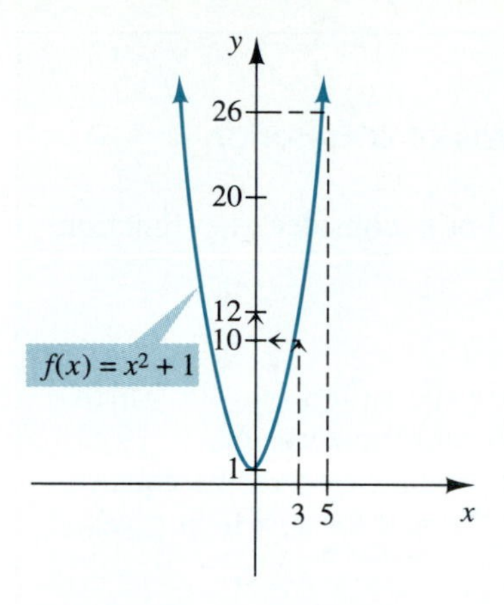

FIGURE 2.53 Finding $f(3 + 2)$ and $f(3) + 2$

Solution

(a) To find $f(3 + 2) = f(5)$, first locate 5 on the x-axis, project vertically upward until you intersect the graph of $f(x)$, and then project horizontally onto the y-axis, which in this case is 26. See Figure 2.53.

To find $f(3) + 2$, start out at 3 on the x-axis, project vertically upward until you intersect the graph, project horizontally onto the y-axis to get 10, and finally move up 2 more units on the y-axis to get 12. See Figure 2.53.

(b) If $f(x) = x^2 + 1$, to evaluate $f(3 + 2) = f(5)$ means to evaluate $f(x)$ at $x = 5$:

$$f(x) = x^2 + 1 \qquad \textit{To find } f(5), \textit{ substitute } x = 5 \textit{ in } f(x).$$

$$f(5) = 5^2 + 1 = 26 \qquad \textit{Which is confirmed by Figure 2.53.}$$

To evaluate $f(3) + 2$ means to evaluate $f(3)$ first and then add 2 to the result:

$$f(3) = 3^2 + 1 = 10 \qquad \textit{Hence,}$$

$$f(3) + 2 = 10 + 2 = 12 \qquad \textit{Which is also confirmed by Figure 2.53.}$$

■

GRAFFIX

Use a graphics calculator.

1. Graph the function $y = f(x) = 3x^2 - 2x + 1$, and use the trace function to
   (a) Approximate the values of $f(2)$ and $f(4)$.
   (b) Approximate the values of x for which $f(x) = 1$ and the values of x for which $f(x) = 4$.
2. Graph the function $y = f(x) = x^3 - 1$, and use the trace function to
   (a) Approximate the values of $f(2)$ and $f(-2)$.
   (b) Approximate the value of x for which $f(x) = 1$ and the value of x for which $f(x) = -3$.

The Zeros of a Function

DEFINITION Given a function $y = f(x)$, a solution to the equation $f(x) = 0$ is called a **zero** of the function.

A zero of a function is defined algebraically—it is a solution to the functional equation $y = f(x) = 0$, i.e., the value of x where $f(x) = 0$. On the other hand, we have defined an x-intercept of a graph as the x-coordinate of a point where the graph crosses the x-axis. Since an x-intercept corresponds to a y-coordinate of 0, each x-intercept of the graph corresponds to a zero of the function. Thus we have a direct relationship between an algebraic concept (a solution to an equation) and a graphical one (a graph crossing the x-axis). See Figure 2.54.

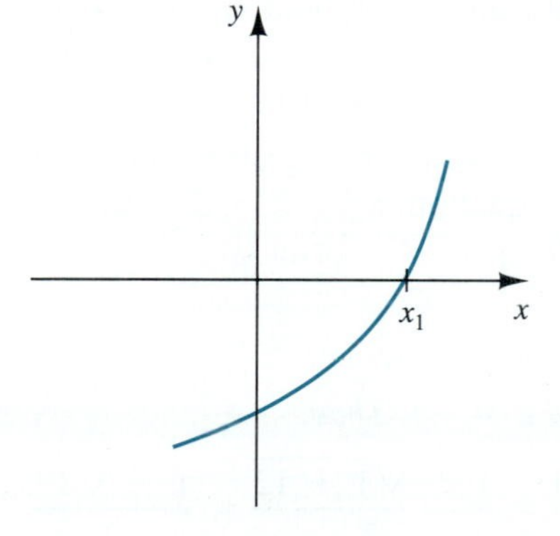

FIGURE 2.54 x_1 is an x-intercept, or, equivalently, $f(x_1) = 0$, and so x_1 is a zero of $f(x)$

DIFFERENT PERSPECTIVES: *The Zeros of a Function*

Consider the geometric and algebraic description of the zeros of a function. Let's consider the function $y = f(x) = (x + 3)(x - 1)(x - 2)$ whose graph is shown here.

GEOMETRIC DESCRIPTION

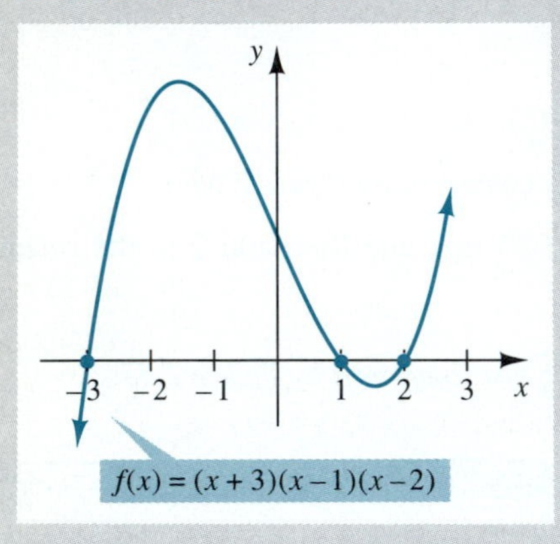

The zeros of a function are the x-intercepts of its graph. In this example, the x-intercepts of $f(x) = (x + 3)(x - 1)(x - 2)$ are -3, 1, and 2.

ALGEBRAIC DESCRIPTION

The zeros of a function are the values of x for which $f(x) = 0$. For example, to find the zeros of $f(x) = (x + 3)(x - 1)(x - 2)$, we solve the equation $(x - 3)(x - 1)(x - 2) = 0$ for x, which gives $x = -3$, 1, and 2.

Using function notation, the x-intercepts of a function are the solutions to the equation $f(x) = 0$; the y-intercept of a function is $f(0)$.

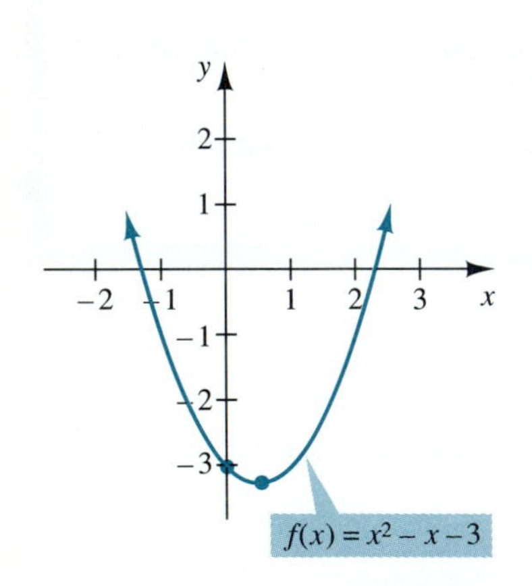

FIGURE 2.55

EXAMPLE 3 Given the graph of $f(x) = x^2 - x - 3$ in Figure 2.55.

(a) Estimate the zeros of $f(x)$ using the graph of $f(x)$.

(b) Find the zeros of $f(x)$ algebraically.

Solution

(a) In estimating the zeros of $f(x)$ by the graph, we are simply looking for the x-intercepts (the x-coordinates where the graph crosses the x-axis). We see two intercepts: One we can estimate as $-1\frac{1}{3}$; the other we can estimate as $2\frac{1}{3}$.

(b) In order to find the zeros of $f(x)$ algebraically, we solve the equation $f(x) = 0$ by algebraic means; that is, we solve the equation $x^2 - x - 3 = 0$. To solve the equation $x^2 - x - 3 = 0$, we use the quadratic formula:

$$x = \frac{-B \pm \sqrt{B^2 - 4AC}}{2A}.$$

$x^2 - x - 3 = 0$ *Hence $A = 1$, $B = -1$, and $C = -3$. Thus*

$$x = \frac{-(-1) \pm \sqrt{(-1)^2 - 4(1)(-3)}}{2(1)} = \frac{1 \pm \sqrt{1 + 12}}{2} = \frac{1 \pm \sqrt{13}}{2}$$

Hence the zeros are $\boxed{\dfrac{1 \pm \sqrt{13}}{2}}$.

How do we find the y-intercept of a function? Can a function have more than one y-intercept?

How do we find the x-intercepts of a function? Can a function have more than one x-intercept?

We may compare our estimates of the zeros in part (a) above with the exact values. Approximating $\sqrt{13} \approx 3.6$, we find the "exact values" for the zeros are

$$x \approx \frac{1 \pm 3.6}{2} = 2.3 \text{ and } -1.3$$

which compare favorably with our estimates from the graph. ■

GRAFFIX

1. Estimate the zeros of $f(x) = 5x^2 - 2x - 1$ by graphing $f(x)$ using your graphics calculator, and use the trace function to locate its x-intercepts.
2. Estimate the zeros of $g(x) = x^3 + 3$ by graphing $g(x)$ using your graphics calculator, and use the trace function to locate its x-intercepts.

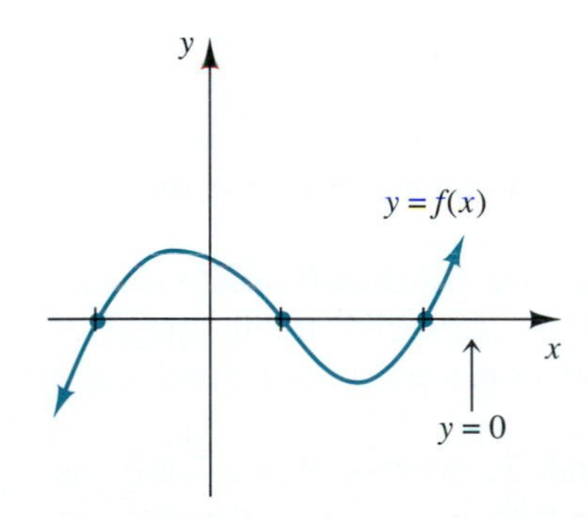

FIGURE 2.56 The solutions to $f(x) = 0$

The solutions to the equation $f(x) = 0$ are found by noting the x-values at which the height of the graph of $f(x)$ is 0. We can restate this by saying that we are looking for the values of x where the graph of $f(x)$ intersects the horizontal line $y = 0$ (the x-axis). See Figure 2.56.

The arrows that appear at the end of the graph indicate that the graph continues in the direction of the arrows, and so Figure 2.56 illustrates all the zeros of $f(x)$.

We can generalize this idea further and state that the solutions to the equation $f(x) = k$ are found by looking for the values of x where the graph of $y = f(x)$ intersects the line $y = k$.

GRAFFIX

Use a graphics calculator.

1. Estimate the solutions to $x^2 - 4x - 5 = -8$ in the following way: Graph $y = x^2 - 4x - 5$ and $y = -8$ on the same set of coordinate axes and use the trace function to estimate the point of intersection of the two graphs.
2. Estimate the solutions to $(x - 2)(x + 1)(x - 7) = 2$ in the following way: Graph $y = (x - 2)(x + 1)(x - 7)$ and $y = 2$ on the same set of coordinate axes, and use the trace function to estimate the point of intersection of the two graphs.

Interpreting the Graph of $f(x)$: General Trends

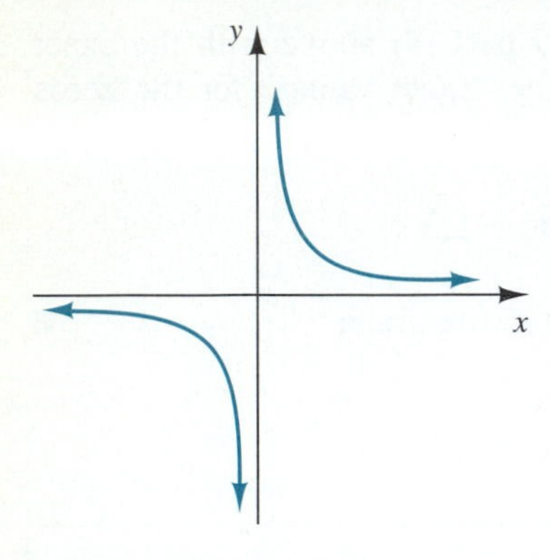

FIGURE 2.57 The graph of a function $y = f(x)$

Consider the graph of a function $f(x)$ pictured in Figure 2.57. The picture tells us a few things about the relationship between x and $y = f(x)$. One thing we can see immediately is that the graph does not cross the y-axis. This tells us that 0 is not in the domain of this function.

Let's see what happens when x is positive (staying to the right of the y-axis). Notice that as x gets extremely large—that is, as we move out further and further to the right, which we denote as $x \rightarrow +\infty$, the values of $f(x)$ (the y-values) remain positive (above the x-axis) but get closer and closer to 0. On the other hand, as x gets closer to 0 *from the right* (x is still positive), the values of $f(x)$ (the y-values) get extremely large, which we denote as $f(x) \rightarrow +\infty$.

To the left of the y-axis (when x is negative), we see that as we move out further and further to the left, which we denote as $x \rightarrow -\infty$, the values of $f(x)$ get closer and closer to 0 but remain negative. On the other hand, as x approaches 0 *from the left* (x remains negative), the values of $f(x)$ approach $-\infty$, which we denote as $f(x) \rightarrow -\infty$.

Examining the graph of $f(x)$ often reveals the general trends of a function for large and small values of x.

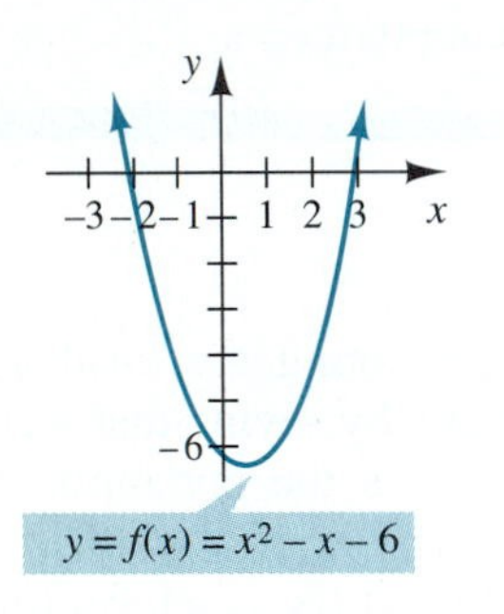

FIGURE 2.58

Interpreting the Graph of $f(x)$: Inequalities

In the preceding case of $f(x)$, we discussed what happens to y when x was positive or negative. A graph can also tell us when—that is, for what values of x—the function is negative or positive.

Let's return to $y = f(x) = x^2 - x - 6$, which we discussed earlier in this section and whose graph appears in Figure 2.58. Look carefully at this graph of $y = f(x) = x^2 - x - 6$. Notice that it tells us the relationship between x and y at a glance.

We can see that $f(x) = 0$ when $x = -2$ and 3, written equivalently as $f(-2) = f(3) = 0$. When the graph dips below the x-axis, $y = f(x)$ is negative, or $y < 0$. This happens when x is between -2 and 3. (Notice that we talk about the behavior of y as x *changes*.) To describe this we would write $f(x) < 0$ when x is in the interval $(-2, 3)$. On the other hand, when the graph rises above the x-axis, $y = f(x)$ is positive, or $y > 0$, and this happens when x is in the intervals $(-\infty, -2) \cup (3, \infty)$. See Figure 2.59.

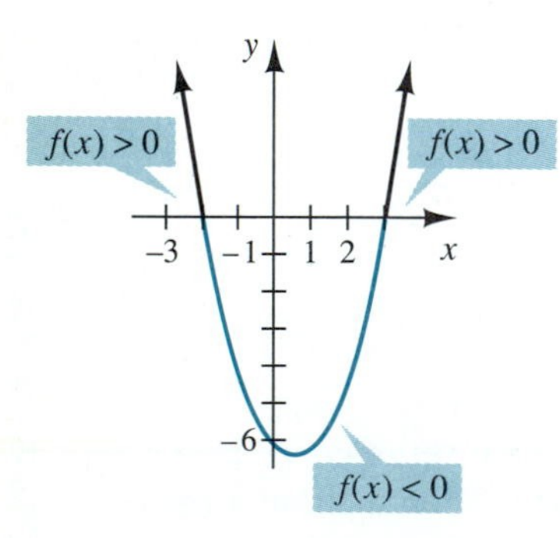

FIGURE 2.59 Interpreting inequalities graphically

Recall that in Chapter 1, we were called upon to solve inequalities such as $x^2 - x - 6 > 0$ or $x^2 - x - 6 < 0$. To solve these inequalities algebraically, we did a sign analysis to determine for what values of x the expression was positive or negative. Looking at the graph of $y = f(x) = x^2 - x - 6$, notice that if we project to the x-axis when y is positive and when y is negative, we arrive at the same result as we do with a sign analysis. See Figure 2.59.

DIFFERENT PERSPECTIVES: *Solving Inequalities*

GEOMETRIC ANALYSIS

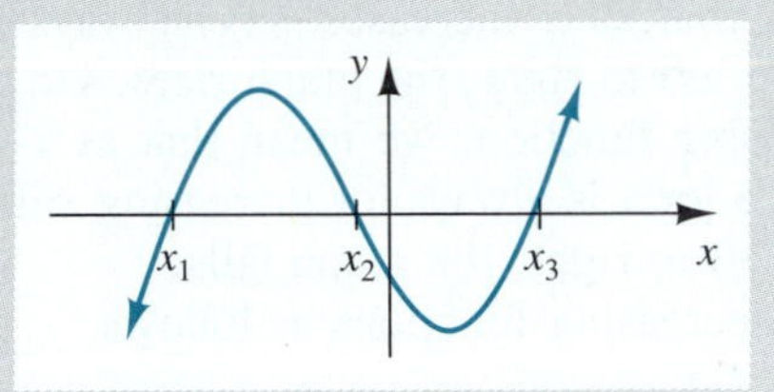

Suppose this is the graph of $y = f(x)$. Solving the inequality $f(x) > 0$ means we are looking for values of x where the graph of $f(x)$ is above the x-axis. Based upon the graph, $f(x) > 0$ for x on $(x_1, x_2) \cup (x_3, \infty)$.

ALGEBRAIC ANALYSIS

Solving the inequality $f(x) > 0$ means we are seeking the values of x for which the expression $f(x)$ is positive and usually utilizes a sign analysis.

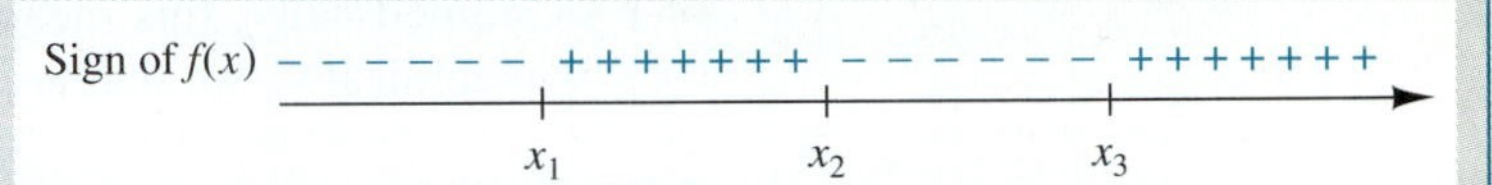

EXAMPLE 4 Use the graph of $f(x) = x^3 - 7x + 6$ in Figure 2.60 to solve the inequality $x^3 - 7x + 6 \geq 0$.

Solution Finding where $x^3 - 7x + 6 \geq 0$ means that we want the x-values of the points where $f(x)$ is positive or zero. We can restate this in terms of intervals where the graph of $f(x)$ is on or above the x-axis. We can see by the graph in Figure 2.60 that $f(x)$ is above the x-axis when x is in the intervals $(-3, 1) \cup (2, \infty)$ and is zero when $x = -3, 1, 2$. Hence the solution to the inequality $x^3 - 7x + 6 \geq 0$ is $[-3, 1] \cup [2, \infty)$. ■

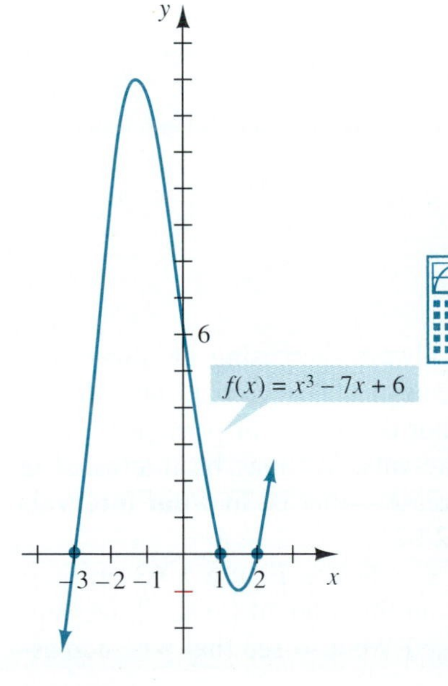

FIGURE 2.60

GRAFFIX

Use a graphics calculator or computer.

1. Estimate the solutions to $2x^3 - 2x \leq 0$ in the following way: Graph $y = 2x^3 - 2x$ and determine the values of x where the graph is on or below the x-axis.
2. Estimate the solutions to $x^4 - 2x^2 + 3x - 2 > 0$ in the following way: Graph $y = x^4 - 2x^2 + 3x - 2$ and determine the values of x where the graph is above the x-axis.

Increasing and Decreasing Functions

When we say that a function $f(x)$ is increasing or decreasing, are we talking about x-values or y-values?

When we talk about a function increasing or decreasing, we are referring to the value of $f(x)$ as x *gets larger*. This is consistent with our previous agreement to describe the graph of a function *as we move from left to right*. Hence, when we say that $f(x)$ is an increasing function, we mean that as x increases, $f(x)$ always increases. Geometrically, this means as we move left to right, the graph rises. On the other hand, when we say that $f(x)$ is a decreasing function, we mean that as x increases, $f(x)$ decreases. (Our frame of reference for x is always for increasing values of x.) Geometrically, this means as we move left to right, the graph falls.

Algebraically, we define increasing and decreasing functions as follows.

DEFINITION A function $y = f(x)$ is said to be an **increasing function** on an interval I if and only if for x_1 and x_2 in the interval, $x_1 < x_2$ implies that $f(x_1) < f(x_2)$. A function $y = f(x)$ is said to be a **decreasing function** on an interval I if and only if for x_1 and x_2 in the interval, $x_1 < x_2$ implies that $f(x_1) > f(x_2)$.

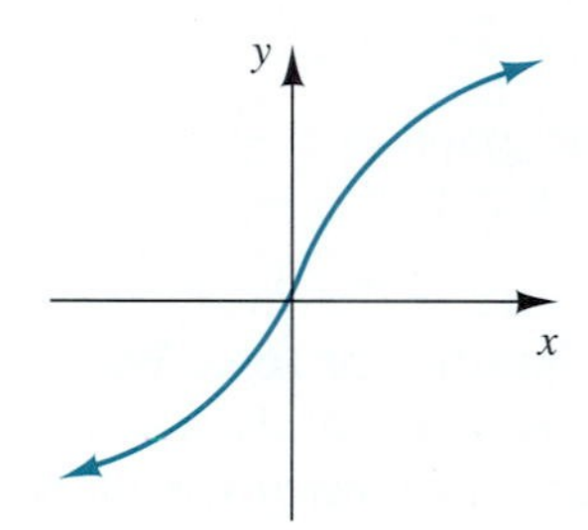

The graph of an increasing function

(a)

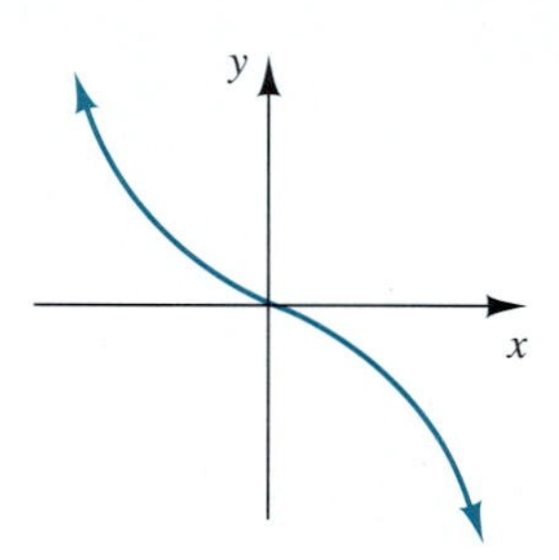

The graph of a decreasing function

(b)

FIGURE 2.61

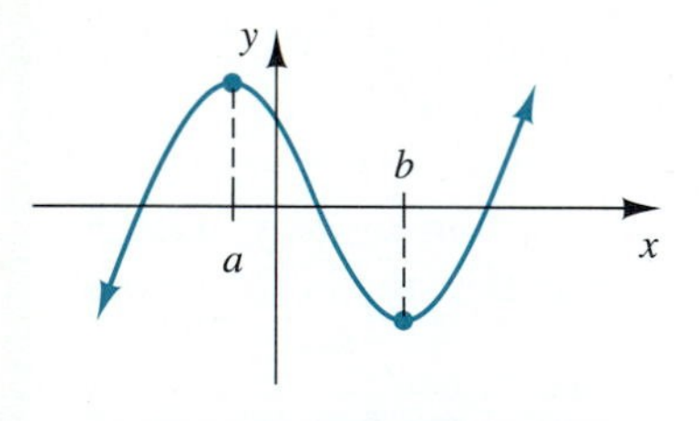

FIGURE 2.62 This function is increasing on the intervals $(-\infty, a]$ and $[b, \infty)$ and decreasing on the interval $[a, b]$.

The functions illustrated in Figure 2.61 are always increasing or always decreasing, but many functions do not fall into this category: They could be constant functions such as $f(x) = 3$ (which neither increase nor decrease), or perhaps they increase on some intervals and decrease on other intervals. We may be interested in where these functions increase and where they decrease—that is, **in what intervals of x is** $f(x)$ increasing and decreasing. See Figure 2.62.

Referring again to the graph of $f(x) = x^2 - x - 6$ (see Figure 2.63 on page 155), we can see that y is decreasing (gets smaller) on the interval $(-\infty, \frac{1}{2}]$. (Again, notice we talk about the behavior of y for values of x.) We also see that y is increasing on the interval $[\frac{1}{2}, \infty)$.

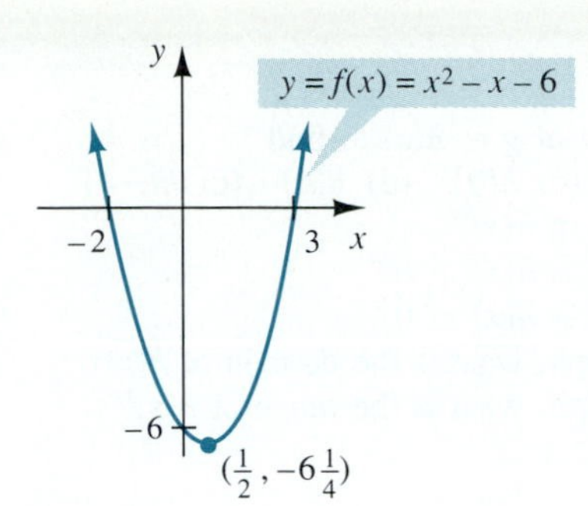

FIGURE 2.63

Notice that at the point $\left(\frac{1}{2}, -6\frac{1}{4}\right)$ on the graph of $y = x^2 - x - 6$ in Figure 2.63, the function changes from decreasing to increasing. A point on the graph of a function where it changes from decreasing to increasing (or vice versa) is called a **turning point** of the graph.

EXAMPLE 5 Use the graph of $f(x)$ in Figure 2.64 to identify where $f(x)$ is increasing and where $f(x)$ is decreasing.

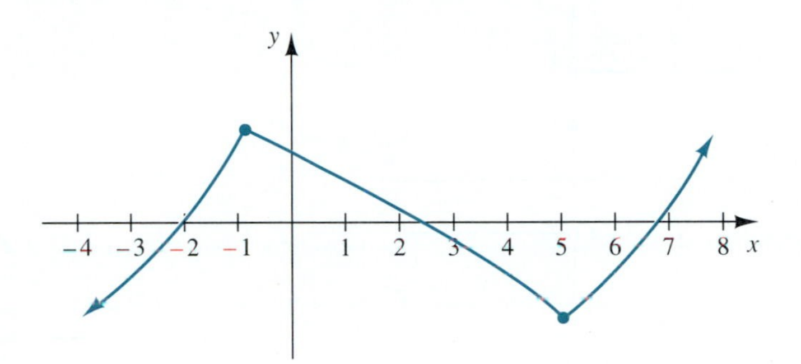

FIGURE 2.64

Solution Finding where $f(x)$ is increasing means that we want the x intervals where the graph of $f(x)$ is rising as we move from left to right. We can see from the graph in Figure 2.64 that $f(x)$ increases when x is in the intervals $(-\infty, -1] \cup [5, \infty)$.

On the other hand, finding where $f(x)$ decreases means locating the interval(s) of x where the graph of $f(x)$ falls *as we move from left to right*. Thus we can see that $f(x)$ is decreasing in the interval $[-1, 5]$. ■

We will continue this discussion of increasing and decreasing functions in a later chapter.

As we have demonstrated in this section, a picture of an equation in the form of its graph yields a great deal of information about the relationship between the two variables in the equation. By graphing an equation we can better understand the nature of the relationship, the general and local trends, and the important values for an equation. In the next section, we continue to examine the relationships expressed by equations by studying their graphs.

EXERCISES 2.6

1. Use the following graph of $y = f(x)$ to find
(a) $f(-4)$ **(b)** $f(-1)$ **(c)** $f(0)$ **(d)** $f(2)$ **(e)** $f(5)$
(f) For what values, if any, is $f(x) = 2$?
(g) For what values, if any, is $f(x) = 1$?
(h) For what values, if any, is $f(x) = -1$?
(i) For what values, if any, is $f(x) = 4$?

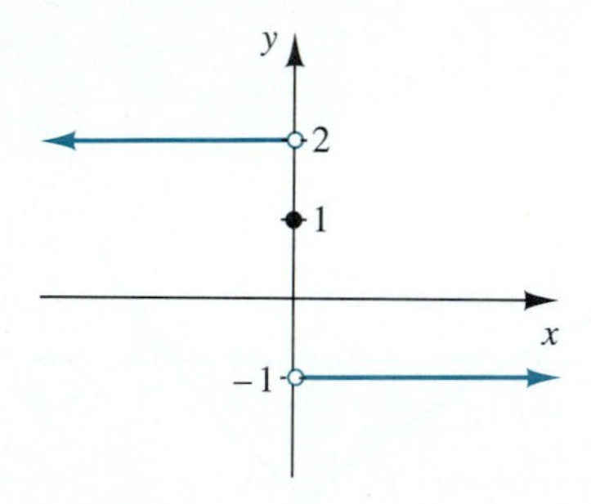

2. Use the following graph of $y = g(x)$ to find
(a) $g(-5)$ **(b)** $g(-2)$ **(c)** $g(0)$ **(d)** $g(1)$ **(e)** $g(3)$
(f) What are the zeros of $g(x)$?
(g) On what intervals is $g(x) > 0$?
(h) On what intervals is $g(x) < 0$?
(i) How many solutions are there to the equation $g(x) = -1$?

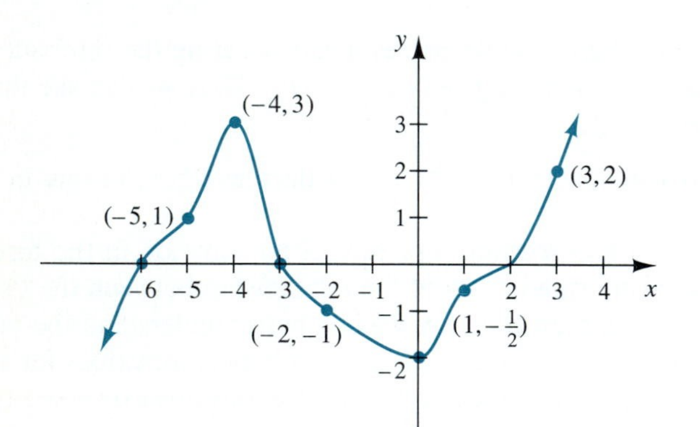

3. Use the following graph of $y = h(x)$ to find
(a) $h(-2)$ **(b)** $h(6)$ **(c)** $h(0)$ **(d)** $h(4)$ **(e)** $h(-4)$
(f) What are the zeros of $h(x)$?
(g) On what interval(s) is $h(x) > 0$?
(h) On what interval(s) is $h(x) < 0$?
(i) Based upon the graph, what is the domain of $h(x)$?
(j) Based upon the graph, what is the range of $h(x)$?

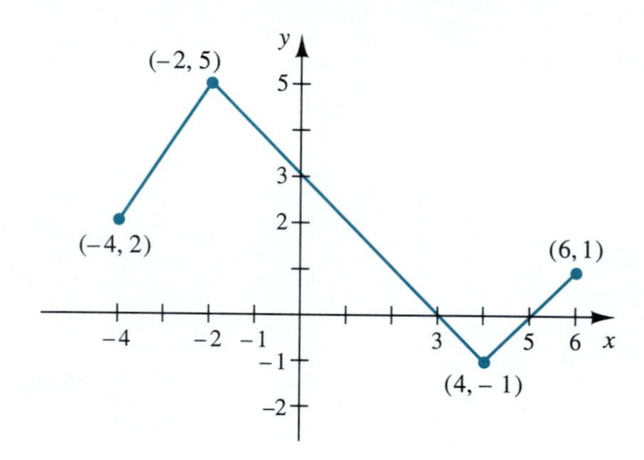

4. Use the following graph of $y = F(x)$ to find
(a) $F(6)$ **(b)** $F(-5)$ **(c)** $F(1)$
(d) How many zeros does $F(x)$ have?
(e) How many solutions are there to the equation $F(x) = 3$?
(f) Based upon the graph, what is the domain of $F(x)$?
(g) Based upon the graph, what is the range of $F(x)$?

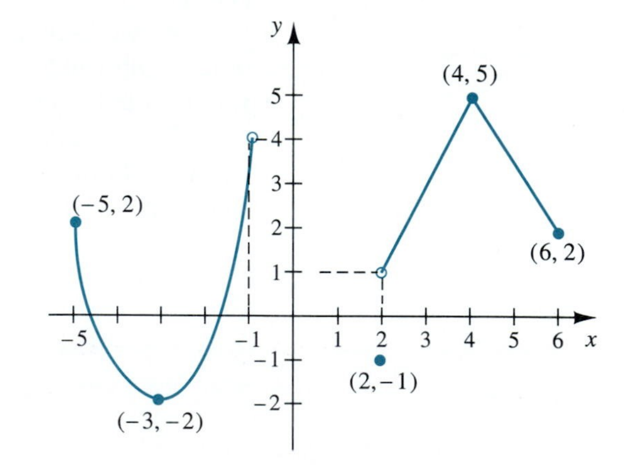

5. Use the following graph of $y = F(x)$ to find
   (a) $F(-7)$ **(b)** $F(-5)$ **(c)** $F(-2)$ **(d)** $F(1) + F(4)$
   (e) $F(1 + 4)$
   (f) Based upon the graph, what is the domain of $F(x)$?
   (g) Based upon the graph, what is the range of $F(x)$?

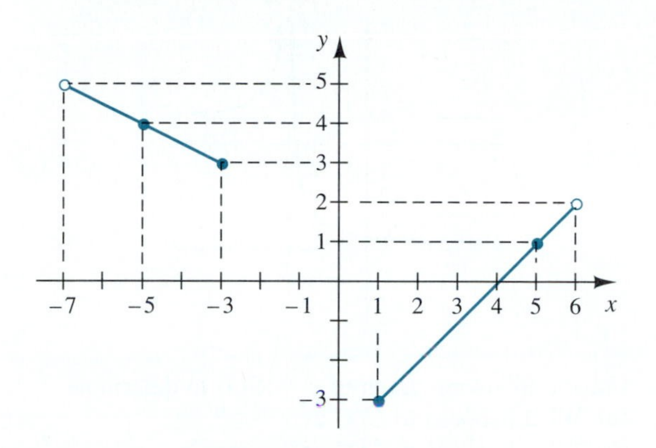

6. Use the following graph of $y = g(x)$ to find
   (a) $g(6)$ **(b)** $g(-3)$ **(c)** $g(3)$ **(d)** $g(-1) + g(1)$
   (e) $g(-1 + 1)$
   (f) On what intervals is $g(x)$ constant?
   (g) Based upon the graph, what is the domain of $g(x)$?
   (h) Based upon the graph, what is the range of $g(x)$?

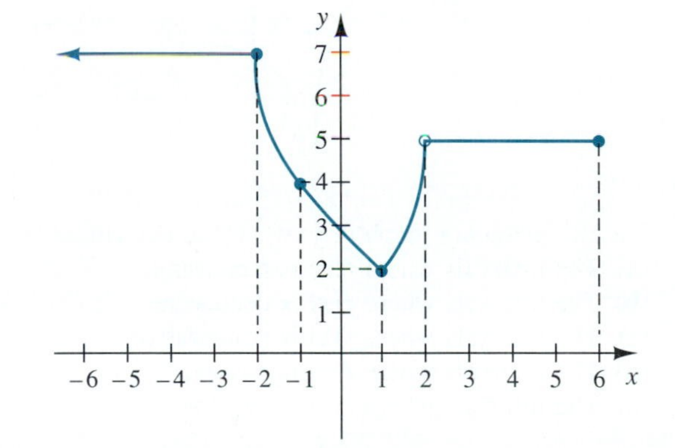

7. **(a)** Use the following graph of $f(x)$ to estimate its zeros.
   (b) Given that this is the graph of $f(x) = x^2 - 2x - 4$, find the zeros of $f(x)$ algebraically and approximate the algebraic answers to the nearest tenth.

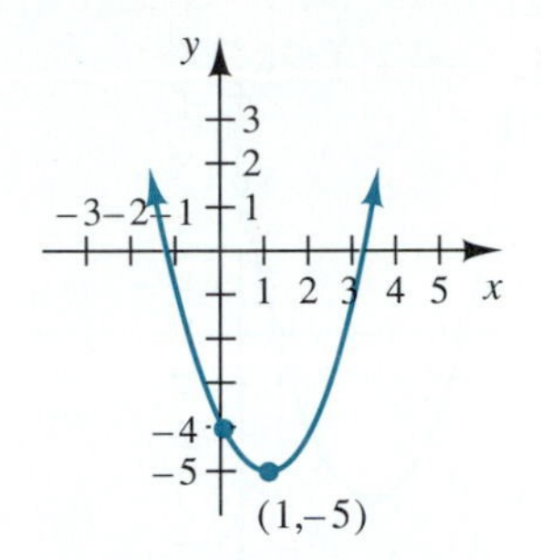

8. **(a)** Use the following graph of $g(x)$ to estimate its zeros.
   (b) Given the graph of $g(x) = -x^2 - 4x + 3$ below, find the zeros of $g(x)$ algebraically and approximate the algebraic answers to the nearest tenth.

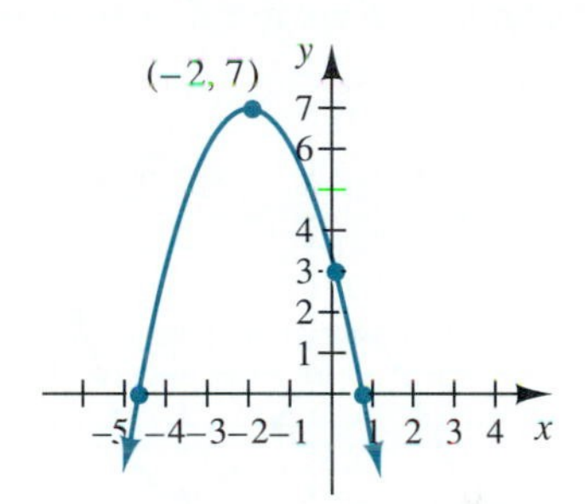

9. **(a)** Use the following graph of $g(x)$ to estimate its zeros.
   (b) Given the graph of $g(x) = -x^2 + 3x + 1$ below, find the zeros of $g(x)$ algebraically and approximate the algebraic answers to the nearest tenth.

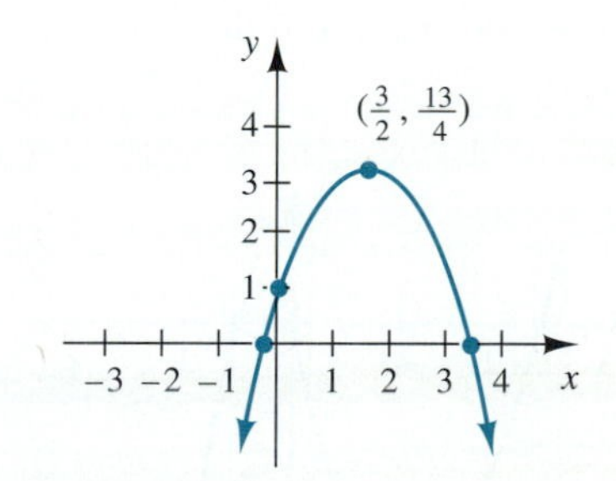

10. **(a)** Use the following graph of $h(x)$ to estimate its zeros.
(b) Given that this is the graph of $h(x) = x^2 + 5x + 3$, find the zeros of $h(x)$ algebraically and approximate the algebraic answers to the nearest tenth.

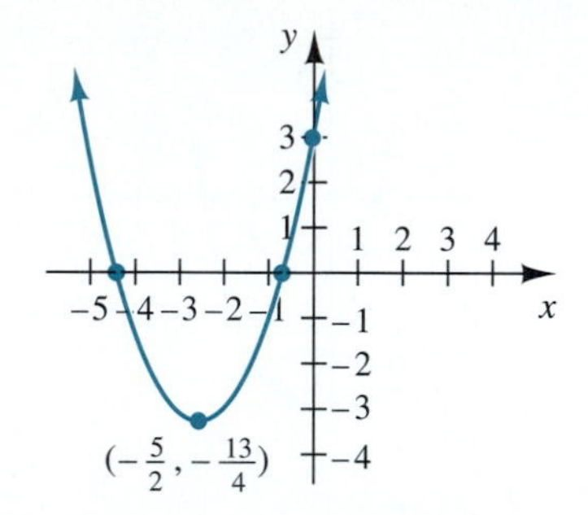

11. **(a)** Use the following graph of $F(x)$ to estimate its zeros.
(b) Given that this is the graph of $F(x) = x^3 - x^2 - x$, find the zeros of $F(x)$ algebraically and approximate the algebraic answers to the nearest tenth.

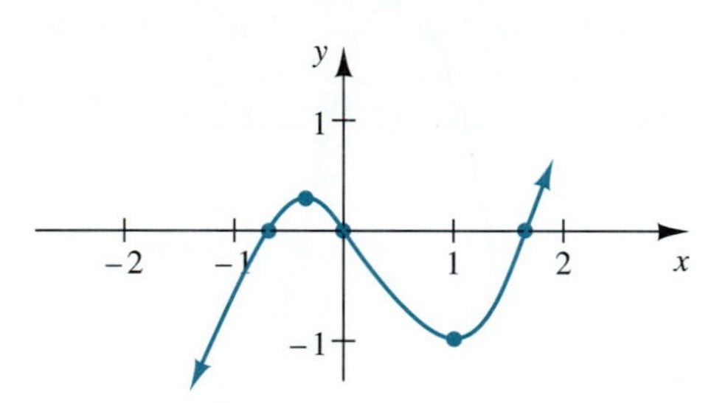

12. **(a)** Use the following graph of $G(x)$ to estimate its zeros.
(b) Given the graph of $G(x) = -x^3 - 2x^2 + x$ below, find the zeros of $G(x)$ algebraically and approximate the algebraic answers to the nearest tenth.

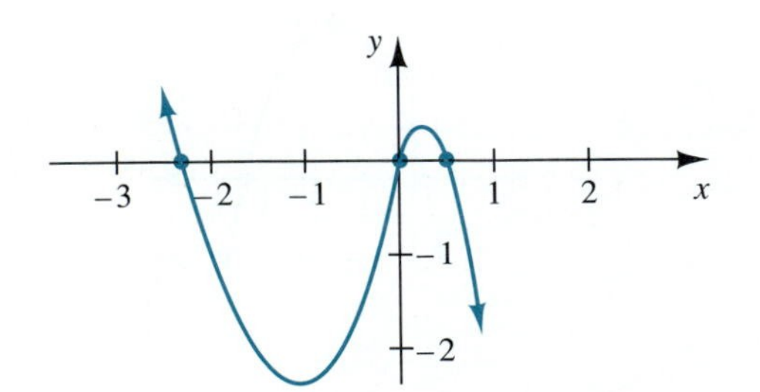

13. Use the following graph of $y = F(x)$ to answer each question.
(a) What happens to $F(x)$ as $x \to \infty$?
(b) What happens to $F(x)$ as $x \to -\infty$?
(c) What happens to $F(x)$ as $x \to 3$?

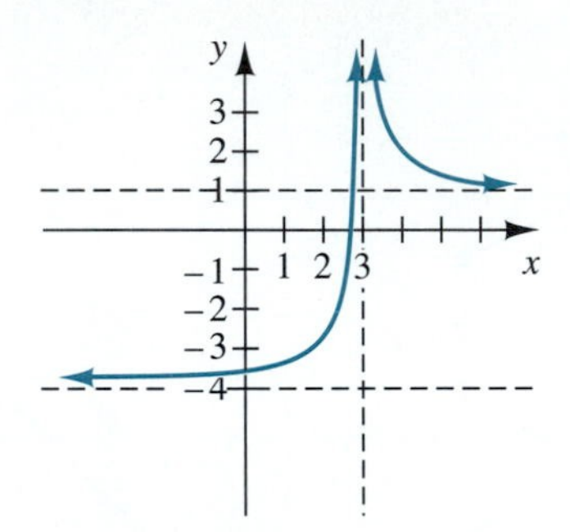

14. Use the following graph of $y = G(x)$ to determine
(a) What happens to $G(x)$ as $x \to \infty$?
(b) What happens to $G(x)$ as $x \to -\infty$?
(c) What happens to $G(x)$ as $x \to -2$?

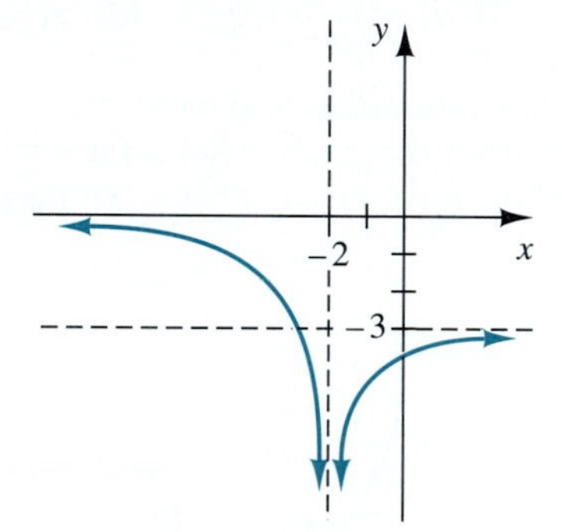

15. Use the following graph of $y = f(x)$ to determine
(a) The intervals where $f(x)$ is increasing.
(b) The intervals where $f(x)$ is decreasing.
(c) The intervals where $f(x)$ is nonnegative.
(d) The intervals where $f(x)$ is positive.
(e) The intervals where $f(x)$ is negative.

16. Use the following graph of $y = g(x)$ to determine
(a) The intervals were $g(x)$ is increasing.
(b) The intervals were $g(x)$ is decreasing.
(c) The intervals were $g(x)$ is constant.
(d) The intervals were $g(x)$ is positive.
(e) The intervals were $g(x)$ is negative.

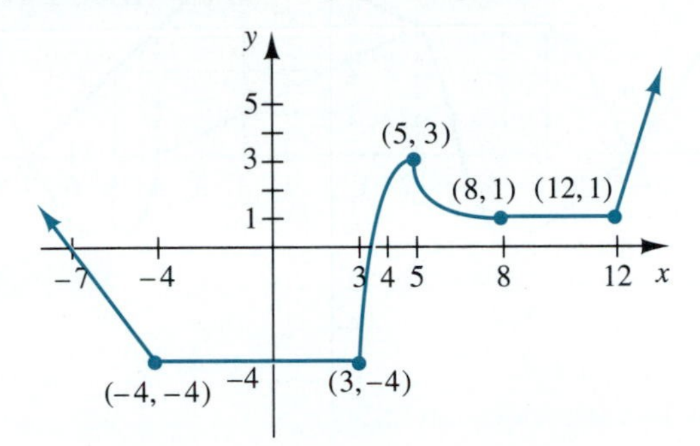

17. Use the following graph of $y = f(x)$ to determine
(a) The intervals where $f(x)$ is increasing.
(b) The intervals where $f(x)$ is decreasing.
(c) The intervals where $f(x)$ is nonnegative.
(d) The intervals where $f(x)$ is positive.
(e) The intervals where $f(x)$ is negative.

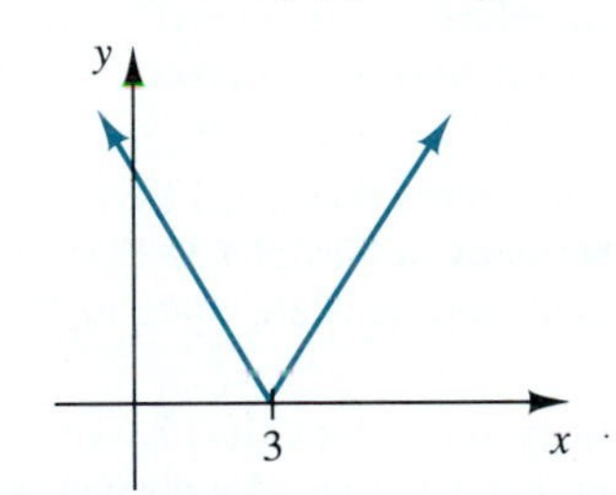

18. Use the following graph of $y = g(x)$ to determine
(a) The intervals where $g(x)$ is increasing.
(b) The intervals where $g(x)$ is decreasing.
(c) The intervals where $g(x)$ is nonnegative.
(d) The intervals where $g(x)$ is positive.
(e) The intervals where $g(x)$ is negative.

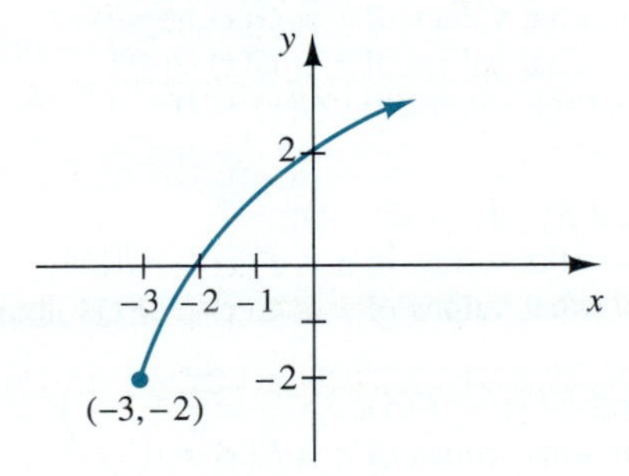

19. Use the following graph of $y = f(x)$ to determine
(a) The intervals where $f(x)$ is increasing.
(b) The intervals where $f(x)$ is decreasing.
(c) The intervals where $f(x)$ is constant.
(d) The intervals where $f(x)$ is positive.
(e) The intervals where $f(x)$ is negative.

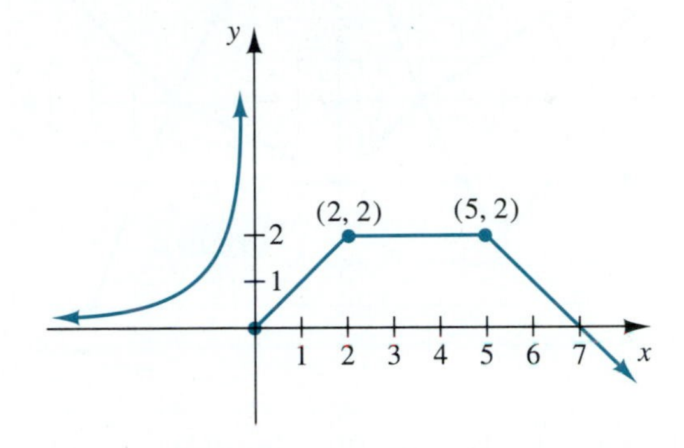

20. Use the following graph of $y = g(x)$ to determine
(a) The intervals where $g(x)$ is increasing.
(b) The intervals where $g(x)$ is decreasing.
(c) The intervals where $g(x)$ is nonnegative.
(d) The intervals where $g(x)$ is positive.
(e) The intervals where $g(x)$ is negative.

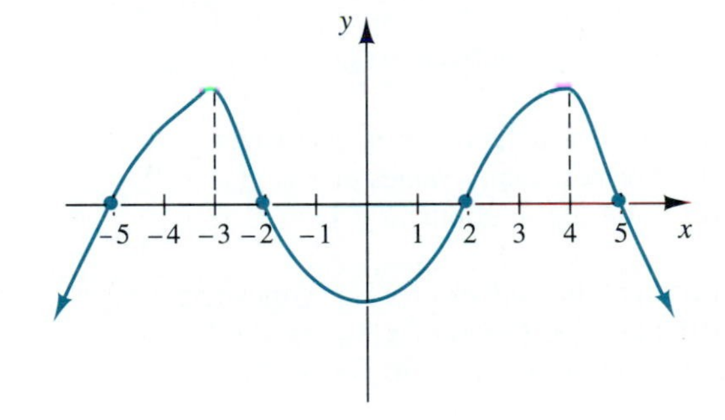

Answer Exercises 21–28 using the graphs in the accompanying figure.

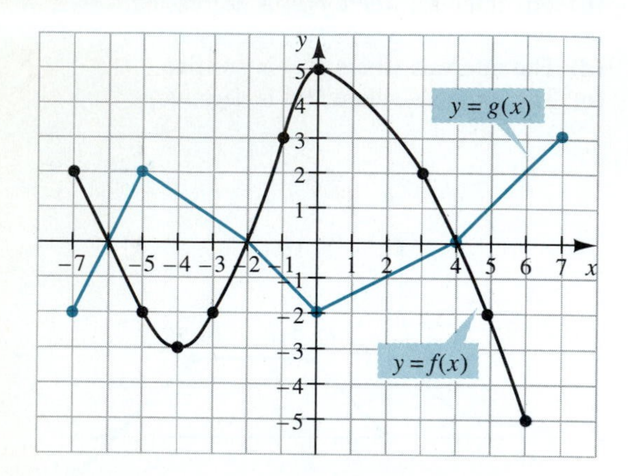

21. Find **(a)** $f(-1)$ **(b)** $f(0)$ **(c)** $f(4)$
(d) For what value(s) of x is $f(x) = -2$?
(e) For how many value(s) of x is $f(x) = 4$?

22. Find **(a)** $g(-1)$ **(b)** $g(0)$ **(c)** $g(4)$
(d) For what value(s) of x is $g(x) = -2$?
(e) For what value(s) of x is $g(x) = 4$?

23. **(a)** How many zeros does $f(x)$ have?
(b) For how many values of x is $f(x) = 2$?
(c) Find the solutions to $f(x) = 6$.

24. **(a)** How many zeros does $g(x)$ have?
(b) For how many values of x is $g(x) = 2$?
(c) Find the solutions to $g(x) = 6$.

25. **(a)** For what values of x is $f(x)$ positive?
(b) For what values of x is $f(x) < 0$?
(c) Find the solutions to $f(x) \geq 0$.

26. **(a)** For what values of x is $g(x)$ negative?
(b) For what values of x is $g(x) \geq 0$?
(c) Find the solutions to $g(x) \leq 0$.

27. **(a)** Find $f(3) - g(-4)$.
(b) For what values of x is $f(x) > g(x)$?
(c) For what values of x is $g(x) - f(x)$ nonnegative?

28. **(a)** Find $g(-6) - f(-1)$.
(b) For what values of x is $f(x) = g(x)$?
(c) For what values of x is $g(x) > f(x)$?

Answer Exercises 29–36 using the graphs in the following figure.

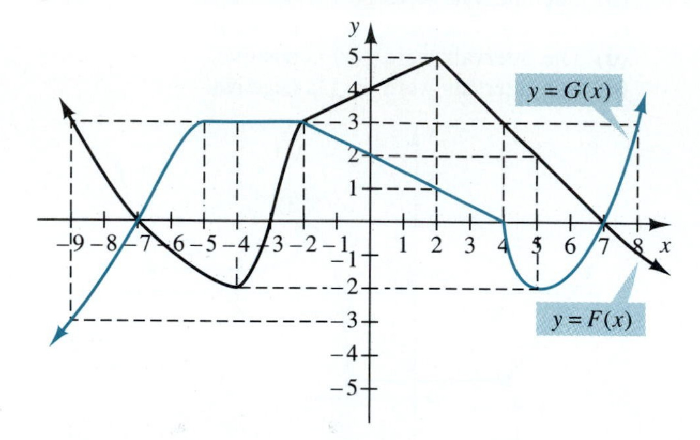

29. Find **(a)** $F(-2)$ **(b)** $F(0)$ **(c)** $F(2)$
(d) For what value(s) of x is $F(x) = 3$?
(e) For what value(s) of x is $F(x) = 0$?

30. Find **(a)** $G(2)$ **(b)** $G(0)$ **(c)** $G(5)$
(d) For what value(s) of x is $G(x) = 3$?
(e) For what value(s) of x is $G(x) = -3$?

31. **(a)** How many zeros does $F(x)$ have?
(b) For how many values of x is $F(x) = 2$?
(c) How many solutions are there to $F(x) = -6$?

32. **(a)** How many zeros does $G(x)$ have?
(b) For how many values of x is $G(x) = 2$?
(c) How many solutions are there to $G(x) = 6$?

33. **(a)** For what values of x is $F(x)$ positive?
(b) For what values of x is $F(x) < 0$?
(c) Find the solutions to $F(x) \geq 0$.

34. **(a)** For what values of x is $G(x)$ negative?
(b) For what values of x is $G(x) \geq 0$?
(c) Find the solutons to $G(x) \leq 0$.

35. **(a)** Find $F(-4) - G(-4)$.
(b) For what values of x is $F(x) > G(x)$?
(c) For what values of x is $G(x) - F(x)$ nonnegative?

36. **(a)** Find $G(7) - F(5)$.
(b) For what values of x is $F(x) = G(x)$?
(c) For what values of x is $G(x) > F(x)$?

Graphs are often used in many disciplines. They allow us to visualize relationships between various quantities. The next several examples illustrate this idea.

37. **Ecology** A Russian biologist, G. F. Gause, formulated the principle of *competitive exclusion,* which states that in any given biological community only one species can occupy any given ecological niche for an extended period of time. (An ecological niche refers to an organism's position in the structure of the ecosystem.) In an attempt to support his hypothesis, Gause conducted a number of laboratory experiments. In one such experiment, Gause used the laboratory cultures of two species of paramecium, *Paramecium aurelia* and *Paramecium caudatum.* When the two species were grown under identical conditions in separate containers, *P. aurelia* grew much faster than *P. caudatum.* When the two species were grown together, *P. aurelia* rapidly outmultiplied the *P. caudatum,* which soon died out. The accompanying figure illustrates the results of this experiment. The horizontal t-axis indicates the number of days that have elapsed since the culture was started. The vertical P-axis indicates the number of paramecium present. Thus a point on one of the graphs indicates how many of that type of paramecium are present after t days. The solid graphs indicate the behavior of the mixed cultures, whereas the dotted graphs indicate the behavior of each type of paramecium alone.

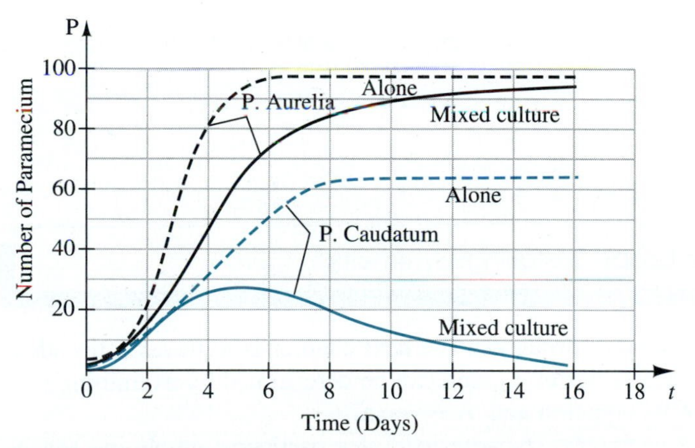

Results of Gause's experiment with two species of Paramecium

(a) Describe what happens over time to each species when grown alone.

(b) Describe what happens over time to each species when they are grown together.

(c) Explain how the graph of the data supports or refutes Gause's principle of competitive exclusion.

38. **Medicine** The accompanying graph illustrates the relationship between the dosage, d, of a particular drug and its effect upon the heart rate, H, in a female monkey. What is the minimum dosage that will maintain a normal heartbeat of 72 beats per minute?

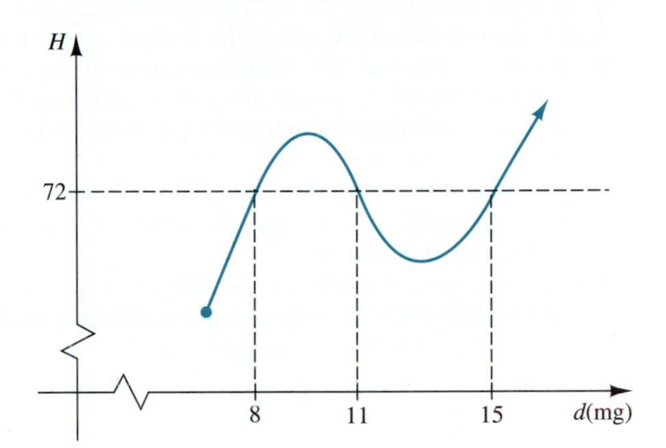

Relationship between drug dosage, d, and heart rate, H

39. **Business** The following graph illustrates the relationship between the profit, P, in thousands of dollars a company earns and the number of line machines in operation, x.

(a) What number of machines maximizes the profit?

(b) What is the maximum profit?

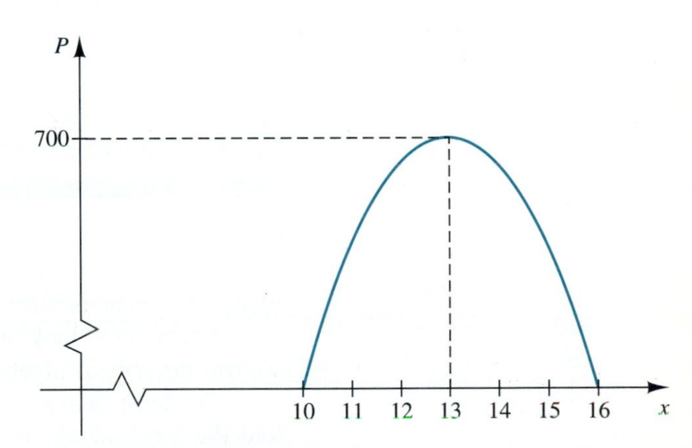

The relationship between profit, P, and number of machines in operation, x

40. Psychology
The accompanying figure contains the graph of data collected in an experiment to measure how deviant verbal behavior can be influenced by using reinforcing techniques and the withholding of social attention. The solid graph indicates how the number of psychotic verbal responses (measured along the vertical axis) changes over a 36-day period (measured along the horizontal axis) as the psychotic responses are reinforced during the first 18 days and the neutral responses are reinforced during the last 18 days. The dotted graph gives the same information for the number of neutral verbal responses.

(a) Describe what happens to the number of psychotic verbal responses as they are reinforced while neutral verbal responses are not.

(b) Describe what happens to the number of psychotic verbal responses as they are not reinforced while neutral verbal responses are reinforced.

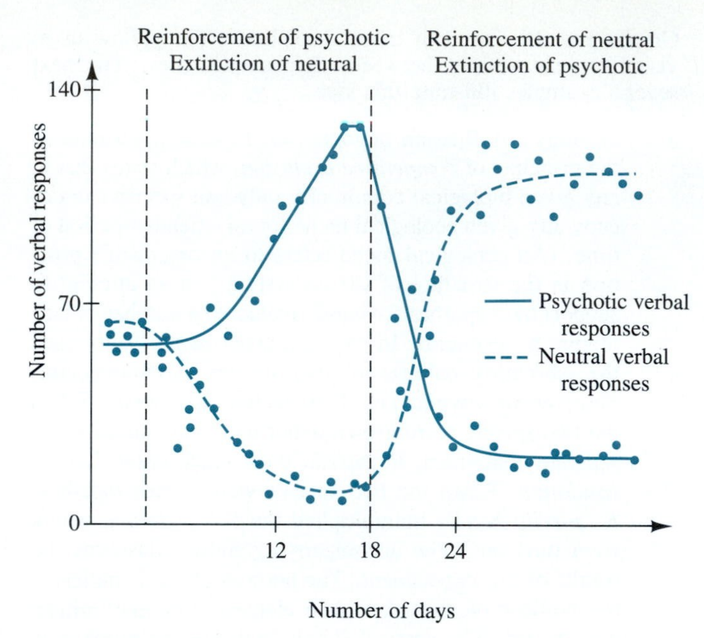

Incidence of psychotic and neutral verbal behavior as affected by social reinforcement techniques

In Exercises 41–46, use a graphics calculator or computer to obtain the graph of the given function and use the graph to determine

(a) the zeros of the function
(b) the intervals where the function is positive
(c) the intervals where the function is negative

Round off to the nearest hundredth where necessary.

41. $f(x) = x^2 - x - 6$ **42.** $f(x) = -x^2 - 2x + 8$

43. $y = \dfrac{x^2 - 1}{x^2 + 1}$ **44.** $y = \dfrac{-1}{x^2 - 2}$

45. $y = x^3 - 3x + 1$ **46.** $y = x^4 - 2x^3 - x + 1$

2.7 Introduction to Graph Sketching: Symmetry

Our main goal in this section and throughout the next chapter is to develop the idea that certain *geometric* characteristics of a graph can be determined by examining the *algebraic* characteristics of its equation and vice versa.

Let us now look at a geometric characteristic of a particular graph and see if we can describe it algebraically. Consider the letter **A** shown in Figure 2.65.

We have drawn a vertical line through the letter and note that if we were to fold the letter along the dotted line, the two halves would coincide. Alternatively, we can say that if a mirror were placed on the vertical line through the **A**, the left half of the letter would be the reflection of the right half and vice versa, as illustrated in Figure 2.66. We say that the letter **A** is **symmetric about** the vertical line through its center, which is called the **axis of symmetry.**

FIGURE 2.65

FIGURE 2.66

Let's consider the graph in Figure 2.67. We observe that if we imagine a mirror on the y-axis, then that portion of the graph to the left of the y-axis is the reflection of that portion of the graph to the right of the y-axis (and vice versa). Such a graph is said to be **symmetric with respect to the y-axis.**

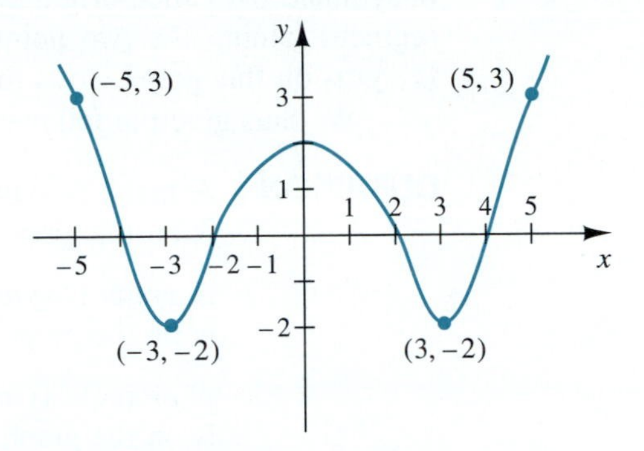

FIGURE 2.67 A graph that exhibits y-axis symmetry

How can we describe this symmetry algebraically? If we look carefully at the graph in Figure 2.67 we can see that the points (5, 3) and (−5, 3) are both on the graph. Similarly, we also have the following pairs of points on the graph: (4, 0) and (−4, 0), (3, −2) and (−3, −2), and (−2, 0) and (2, 0).

We can state this more concisely: In general, if a graph exhibits y-axis symmetry, then whenever a point (x, y) is on the graph, so is $(-x, y)$. This is an algebraic description of y-axis symmetry. In fact, since it is so concise, we generally use this algebraic description to define y-axis symmetry (which we will do in a moment).

Figure 2.68 illustrates two additional types of symmetry. The graph in Figure 2.68(a) is symmetric with respect to the x-axis. Notice that for every point y vertical units on one side of the x-axis, there corresponds another point y vertical units on the other side of the x-axis with the same x-coordinate. Again, we can state this more concisely: Whenever a point (x, y) is on this graph, so is the point $(x, -y)$.

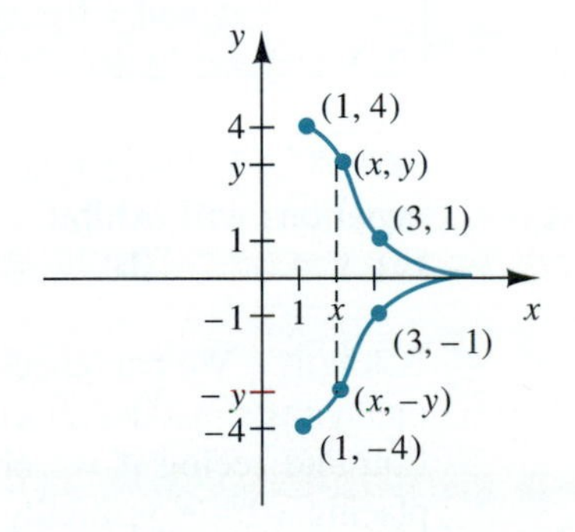

(a) A graph exhibiting x-axis symmetry

FIGURE 2.68

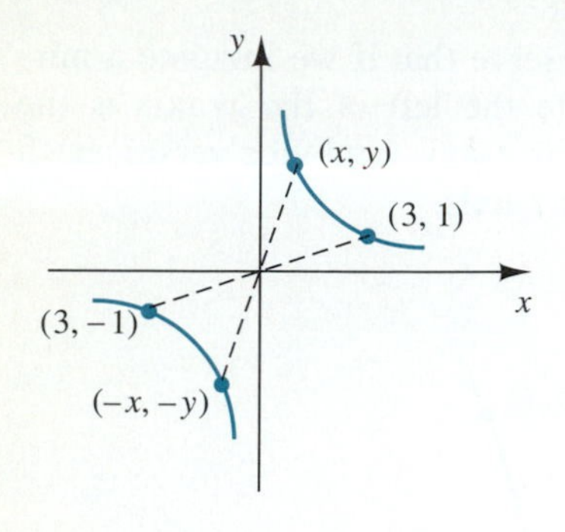

(b) A graph exhibiting origin symmetry

FIGURE 2.68

Although we can easily describe symmetry with respect to a line such as the x- or y-axis, symmetry with respect to a point is a bit more difficult to describe. We cover symmetry with respect to the origin here because it does come up frequently in our discussion of functions.

The graph in Figure 2.68(b) is symmetric with respect to the origin. This type of symmetry is called origin symmetry because the origin is the midpoint of the line segment joining the two points (x, y) and $(-x, -y)$. Note that whenever a point (x, y) is on this graph, so is the point $(-x, -y)$.

We thus give the following definitions.

DEFINITION A graph is **symmetric with respect to the y-axis** if whenever (x, y) is on the graph, $(-x, y)$ is also on the graph.

A graph is **symmetric with respect to the x-axis** if whenever (x, y) is on the graph, $(x, -y)$ is also on the graph.

A graph is **symmetric with respect to the origin** if whenever (x, y) is on the graph, $(-x, -y)$ is also on the graph.

If a graph of an equation is symmetric with respect to the y-axis, then by definition, if (x, y) is on the graph, $(-x, y)$ is also on the graph. Hence, both (x, y), and $(-x, y)$ satisfy the equation of the graph. This means that in the equation of the graph, both x and $-x$ will yield the same y, or in other words, *replacing x by $-x$ yields the same equation*. This is a convenient test for symmetry. In the same way we can use these definitions to establish the other algebraic tests for symmetry.

TESTS FOR SYMMETRY

1. The graph of an equation will exhibit *y-axis symmetry* if replacing x by $-x$ yields an equivalent equation.
2. The graph of an equation will exhibit *x-axis symmetry* if replacing y by $-y$ yields an equivalent equation.
3. The graph of an equation will exhibit *origin symmetry* if replacing x by $-x$ and y by $-y$ yields an equivalent equation.

EXAMPLE 1 Determine what types of symmetry (if any) the graphs of the following equations will exhibit.
(a) $y = x^3$ **(b)** $y = x^2 - 4$ **(c)** $y = x^3 - x^2 - 2x$ **(d)** $x^2 + y^2 = 4$

Solution We can check each of these equations for symmetry by applying the symmetry tests just described. We do this by first replacing x by $-x$ in the original equation and seeing if we obtain an equivalent equation; then we do the same after replacing y by $-y$; finally, we replace x by $-x$ *and* y by $-y$ and see if we obtain an equivalent equation.

Keep in mind that a graph can exhibit more than one type of symmetry, so that we need to apply the various tests individually. However, if a graph exhibits any two of these symmetries, then it must necessarily exhibit the third symmetry as well. For example, if a graph exhibits both y-axis and x-axis symmetry, then it must necessarily exhibit origin symmetry, because reflecting a portion of a graph about the y-axis and then about the x-axis is equivalent to a reflection through the origin. See Figure 2.69.

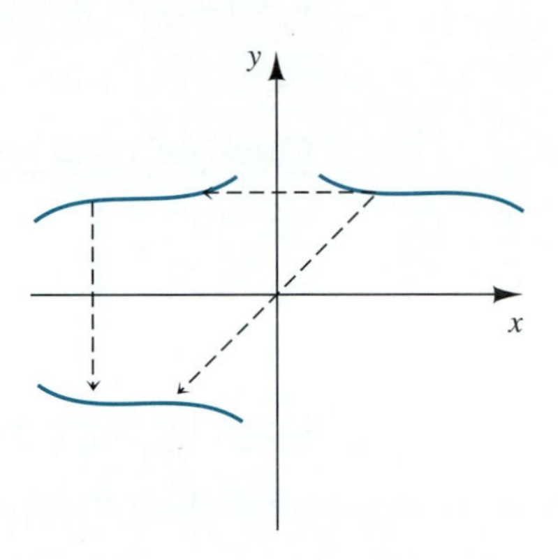

FIGURE 2.69 x-axis and y-axis symmetry implies origin symmetry.

Note that we are answering the geometric questions about symmetry by examining each equation algebraically. Each of the equations in this example is of a type that we are going to discuss in much greater detail later in the text. Even though we have not yet discussed how to obtain the graphs of these equations we include each of their graphs so that we may visually verify the symmetry that we have deduced algebraically.

(a) Check for y-axis symmetry

$y = x^3$ — *To check for y-axis symmetry, we replace x by −x.*

$y = (-x)^3 = -x^3$ — *This is not equivalent to* $y = x^3$. *Therefore, the graph of* $y = x^3$ does not *have y-axis symmetry.*

Check for x-axis symmetry

$y = x^3$ — *To check for x-axis symmetry we replace y by −y.*

$-y = x^3$ — *This is not equivalent to* $y = x^3$. *Therefore, the graph of* $y = x^3$ does not *have x-axis symmetry.*

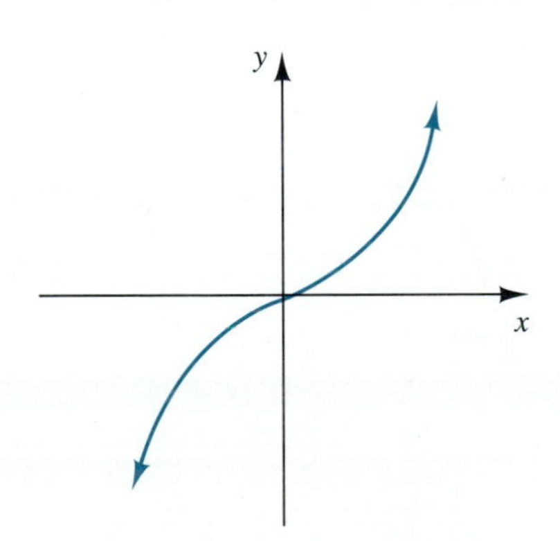

FIGURE 2.70 The graph of $y = x^3$ exhibits origin symmetry.

Check for origin symmetry

$y = x^3$ — *To check for origin symmetry, we replace x by −x and y by −y.*

$-y = (-x)^3$

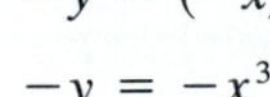

$-y = -x^3$ — *This is equivalent to* $y = x^3$. *Therefore, the graph of* $y = x^3$ does *have origin symmetry.*

The graph of $y = x^3$ appears in Figure 2.70.

(b) Check for y-axis symmetry

$y = x^2 - 4$	*We replace x by $-x$.*
$y = (-x)^2 - 4 = x^2 - 4$	*This* is *the original equation. Therefore, the graph of* $y = x^2 - 4$ does *have y-axis symmetry.*

Check for x-axis symmetry

$y = x^2 - 4$	*We replace y by $-y$.*
$-y = x^2 - 4$	*This is not equivalent to* $y = x^2 - 4$. *Therefore, the graph of* $y = x^2 - 4$ does not *have x-axis symmetry.*

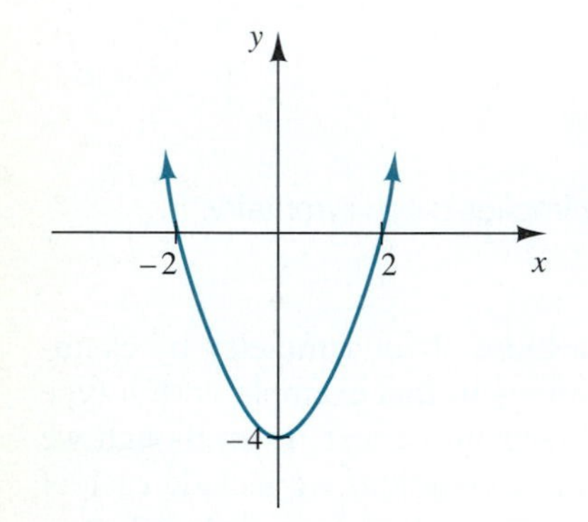

FIGURE 2.71 The graph of $y = x^2 - 4$ exhibits y-axis symmetry.

Check for origin symmetry

$y = x^2 - 4$	*We replace x by $-x$ and y by $-y$ to check for origin symmetry.*
$-y = (-x)^2 - 4$	
$-y = x^2 - 4$	*This* is not *equivalent to* $y = x^2 - 4$. *Therefore, the graph of* $y = x^2 - 4$ does not *have origin symmetry.*

Actually, once we know that the graph of $y = x^2 - 4$ has y-axis symmetry but not x-axis symmetry, it cannot have origin symmetry, and so the test for origin symmetry was unnecessary. Keep in mind that it *is* possible for a graph to have neither x- nor y-axis symmetry but to be symmetric with respect to the origin.

The graph of $y = x^2 - 4$ appears in Figure 2.71.

(c) We may do the three symmetry tests in any order that we like. However, it makes sense to do the x-axis and y-axis symmetry tests first, since if a graph exhibits *only one* of these symmetries, it cannot have origin symmetry, and if a graph exhibits *both* x- and y-axis symmetry, it must have origin symmetry. It is left to the student to verify that the equation $y = x^3 - x^2 - 2x$ fails all three symmetry tests and so the graph exhibits none of these symmetries. The graph of $y = x^3 - x^2 - 2x$ appears in Figure 2.72.

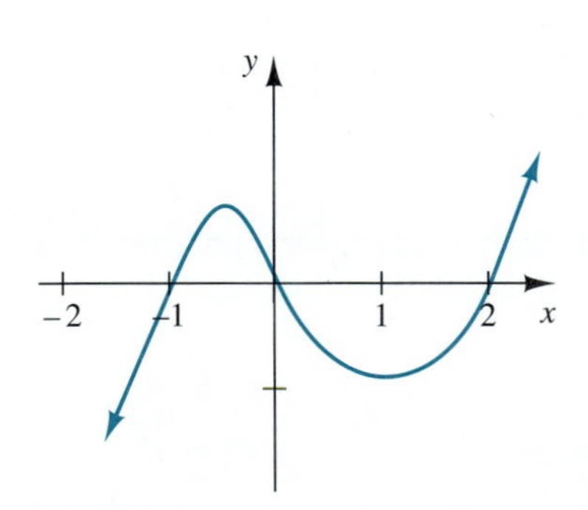

FIGURE 2.72 The graph of $y = x^3 - x^2 - 2x$ exhibits none of the three symmetries.

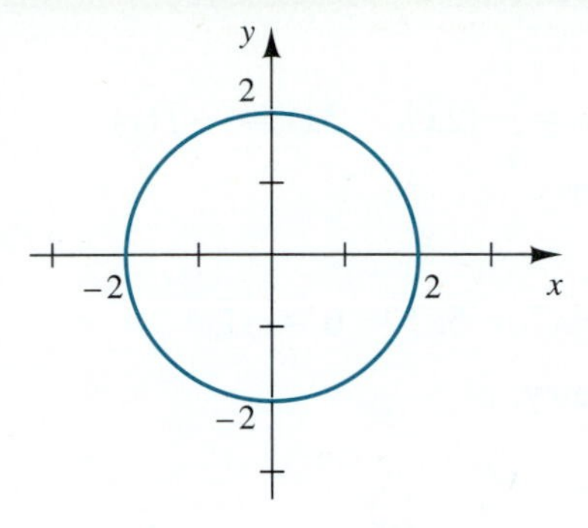

FIGURE 2.73 The graph of $x^2 + y^2 = 4$ exhibits all three symmetries.

(d) It is fairly easy to recognize that if we replace x by $-x$ or y by $-y$ in the equation $x^2 + y^2 = 4$, we will obtain an equivalent equation. Thus the equation $x^2 + y^2 = 4$ exhibits y-axis, x-axis, and, necessarily, origin symmetry. As we saw in Section 2.1 the graph of $x^2 + y^2 = 4$ is a circle with center (0, 0) and radius 2; it appears in Figure 2.73. ■

EXAMPLE 2 Find the point symmetric to the point $(-2, 3)$ with respect to the y-axis, x-axis, and origin.

Solution The point symmetric to the point $(-2, 3)$ with respect to the y-axis is the point with the same "height" as $(-2, 3)$; hence, its y-coordinate is 3. It is the same distance from the x-axis, only on the opposite side of the x-axis; hence, its x-coordinate is 2. The point is (2, 3).

The point symmetric to the point $(-2, 3)$ with respect to the x-axis is the point with the same x-coordinate as $(-2, 3)$, which is -2. It is the same distance from the y-axis, only on the opposite side of the y-axis; hence, its y-coordinate is -3. The point is $(-2, -3)$.

The point symmetric to the point $(-2, 3)$ with respect to the origin is the point with opposite x- and y-coordinates, which is $(2, -3)$. See Figure 2.74. ■

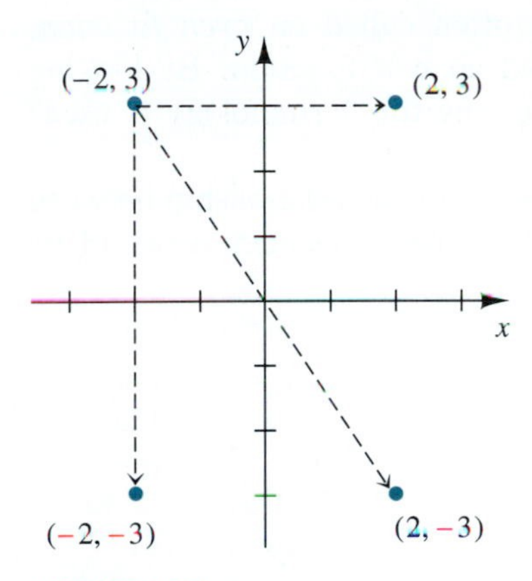

FIGURE 2.74

Up to this point, our discussion of symmetry has been about graphs of equations in general. If we restrict our attention to the graphs of functions we can reformulate some of our results about symmetry.

For example, a graph has y-axis symmetry if the same y-value corresponds to both x and $-x$. Using function notation, we can restate this as: if $f(x) = y$, then $f(-x) = y$, or, more concisely: $f(x) = f(-x)$. A function exhibits origin symmetry if opposite y-values correspond to opposite x-values. Again using function notation: If $f(x) = y$, then $f(-x) = -y$. Noting that $f(-x) = -y = -f(x)$, we have $f(-x) = -f(x)$. Using function notation we restate the symmetry definitions given earlier, as follows.

DEFINITION The graph of a function $y = f(x)$ has y-axis symmetry if $f(-x) = f(x)$.

The graph of a function $y = f(x)$ has origin symmetry if $f(-x) = -f(x)$.

Why don't we talk about x-axis symmetry for the graph of a function?

Thus in order to test a function for y-axis or origin symmetry, we examine $f(-x)$. If $f(-x) = f(x)$, then the graph of $y = f(x)$ has y-axis symmetry; if $f(-x) = -f(x)$, then the graph of $y = f(x)$ has origin symmetry.

EXAMPLE 3 Discuss the symmetry of the following functions:

(a) $y = f(x) = 2x^3 - 3x$ **(b)** $y = g(x) = -x^4 + 5x^2 - 6$

(c) $y = h(x) = 3x^2 - 6x$

Solution In order to determine whether a function has y-axis or origin symmetry, we compare $f(-x)$ with $f(x)$. If they are the same, then the graph of $y = f(x)$ has y-axis symmetry, while if they are opposites, then the graph of $y = f(x)$ has origin symmetry.

(a) For $y = f(x) = 2x^3 - 3x$, we have

$$f(-x) = 2(-x)^3 - 3(-x) = -2x^3 + 3x = -(2x^3 - 3x) = -f(x)$$

Therefore, the graph of $f(x)$ has origin symmetry.

(b) For $y = g(x) = -x^4 + 5x^2 - 6$, we have

$$g(-x) = -(-x)^4 + 5(-x)^2 - 6 = -x^4 + 5x^2 - 6 = g(x)$$

Therefore, the graph of $g(x)$ has y-axis symmetry.

(c) For $y = h(x) = 3x^2 - 6x$, we have

$$h(-x) = 3(-x)^2 - 6(-x) = 3x^2 + 6x$$

which is equal to neither $f(x)$ nor $-f(x)$; therefore, the graph of $h(x)$ has neither y-axis symmetry nor origin symmetry. ■

Why do you think an odd function is called odd?

Why do you think an even function is called even?

A function that exhibits y-axis symmetry is often called an *even function,* whereas one that exhibits origin symmetry is called an *odd* function. By looking back at the results of the last example, can you see why this terminology is used? (See Exercise 36.)

In the next chapter we will continue our discussion of the relationship between the algebraic characteristics of an equation and the geometric characteristics of its graph.

EXERCISES 2.7

In Exercises 1–8, determine whether the given graph has y-axis symmetry, x-axis symmetry, origin symmetry, or none of these symmetries.

1.

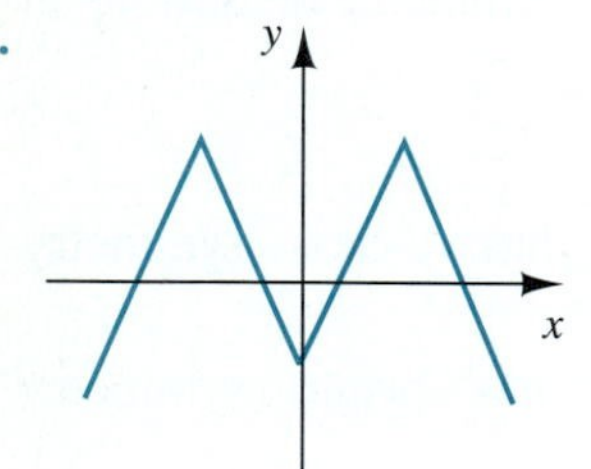

2.

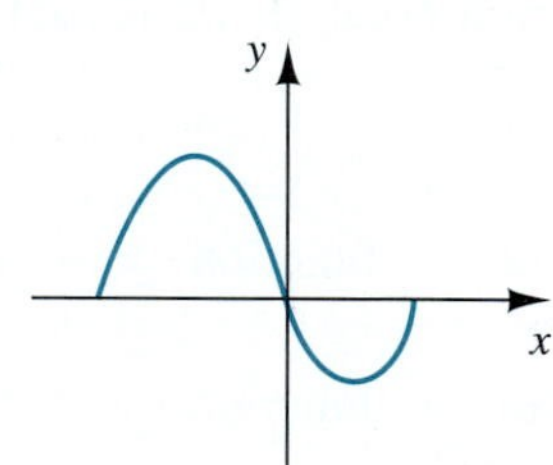

3.

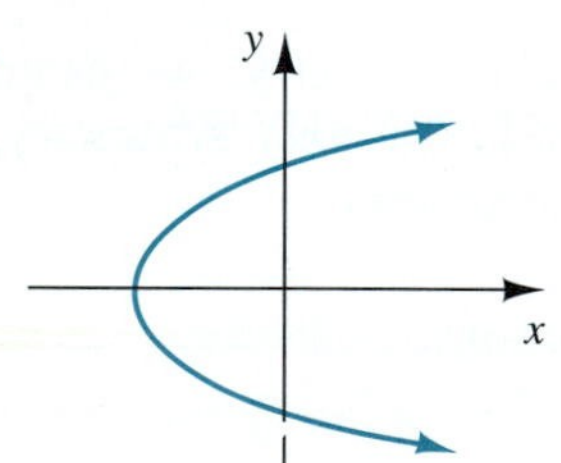

4.

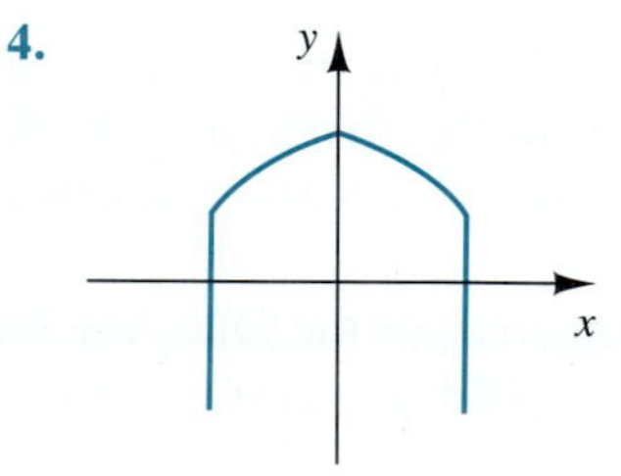

5.

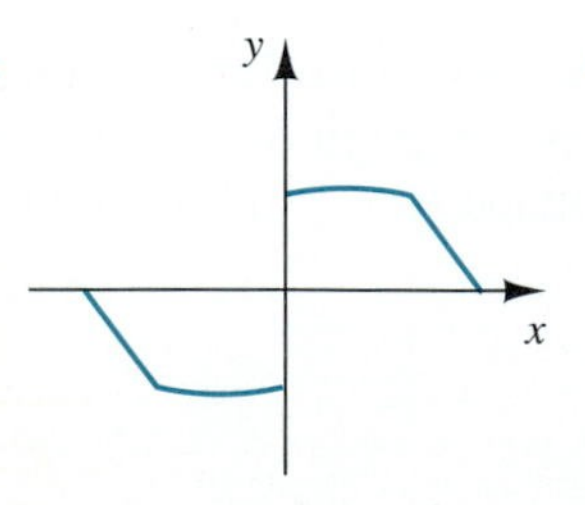

6.

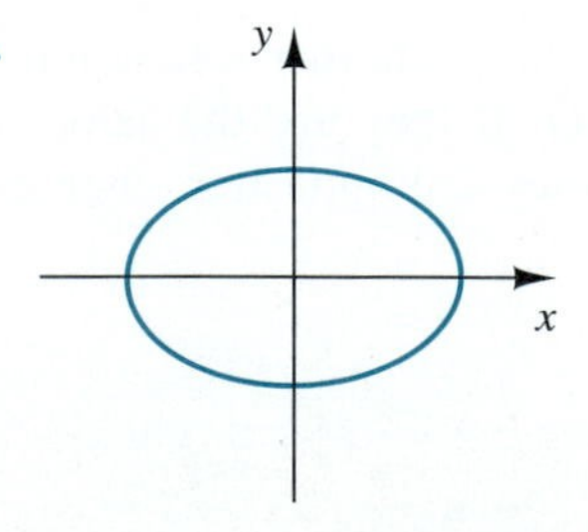

7.

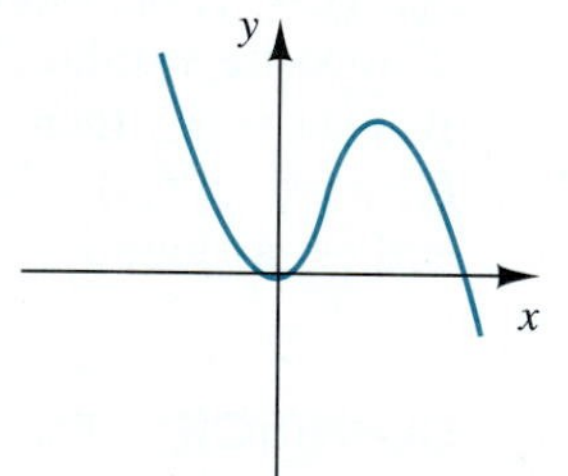

8.

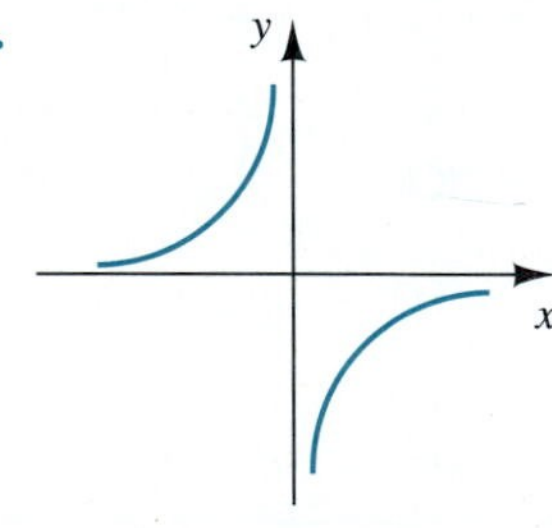

In Exercises 9–12, you are given a portion of a graph. Complete the given graph if it exhibits **(a)** y-axis symmetry and **(b)** origin symmetry.

9.

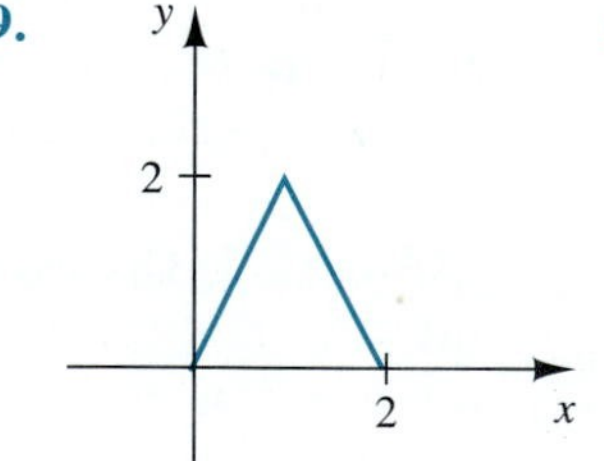

10.

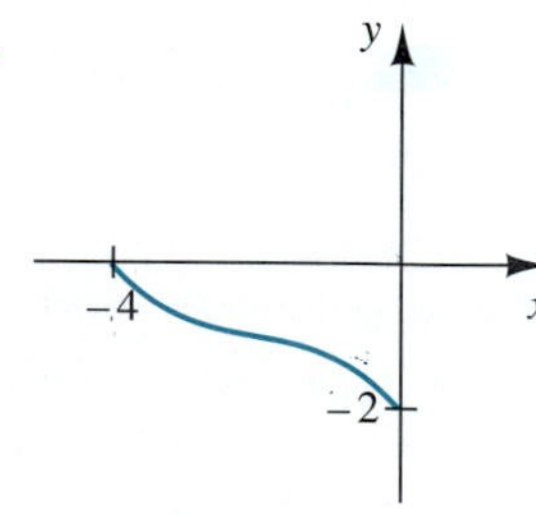

11.

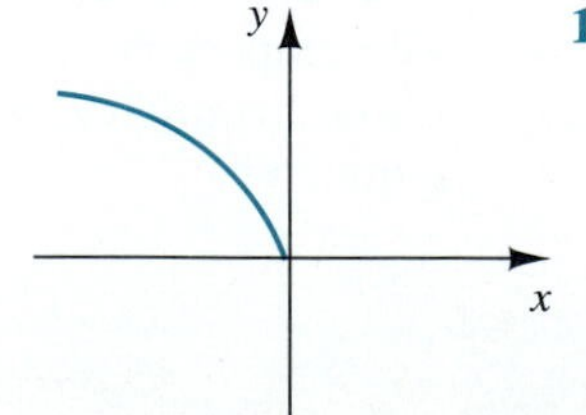

12.

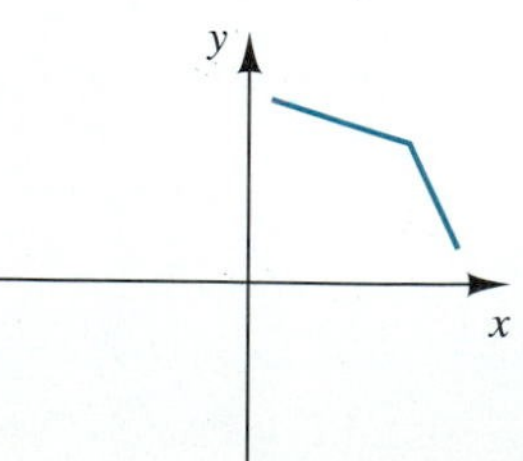

In Exercises 13–16, the endpoints of line segment $\overline{PQ}$ are given. Sketch the reflection of $\overline{PQ}$ about the **(a)** y-axis, **(b)** x-axis, and **(c)** origin.

13. $P(2, 1)$ and $Q(4, 6)$
14. $P(-2, -3)$ and $Q(0, 2)$
15. $P(3, 0)$ and $Q(5, -1)$
16. $P(-4, 0)$ and $Q(0, -4)$

In Exercises 17–34, determine whether the graph of the given function has y-axis symmetry, origin symmetry, or neither.

17. $y = f(x) = 4x^2 - 1$
18. $y = g(x) = -2x^3 + 16$
19. $y = h(x) = x^5 - 2x$
20. $y = f(x) = 6 + x^2 - x^4$
21. $y = f(x) = 2x^2 - 3x + 1$
22. $y = h(x) = 7$
23. $y = F(x) = \dfrac{5}{x}$
24. $y = G(x) = \dfrac{2}{x + 1}$
25. $y = F(x) = -\dfrac{x^4}{x^2 - 1}$
26. $y = G(x) = \dfrac{x^2 + 9}{x}$
27. $y = F(x) = \dfrac{x^3}{x^2 - 1}$
28. $y = G(x) = -\dfrac{x^2 + 9}{x^2}$
29. $y = H(x) = |x|$
30. $y = H(x) = |x - 3|$
31. $y = f(x) = \dfrac{x}{4} - \dfrac{x}{3}$
32. $y = G(x) = \dfrac{5}{x} + \dfrac{2}{x^2}$
33. $y = h(x) = \dfrac{x - 3}{4 - x}$
34. $y = F(x) = \dfrac{x^4 + 1}{2x^2 - 8}$

QUESTIONS FOR THOUGHT

35. We have described the vertical line test for determining whether a graph defines y as a function of x. Suppose we want the graph to also define x as a function of y. What special characteristic must the graph have?

36. As mentioned in this section, a function for which $f(-x) = f(x)$ is called an *even function*, whereas a function for which $f(-x) = -f(x)$ is called an *odd function*. Thus an even function exhibits y-axis symmetry, and an odd function exhibits origin symmetry.
(a) Show that if $f(x)$ is a polynomial made up exclusively of odd powers of x, then it is an odd function.
(b) Show that if $f(x)$ is a polynomial made up exclusively of even powers of x, then it is an even function.
(c) Show that if $f(x)$ is a polynomial made up of both even and odd powers of x, then it is neither an even nor an odd function.
(d) By examining $f(x) = \sqrt[3]{x}$, show that a function can be odd without having odd powers.

37. **(a)** Suppose $f(x)$ and $g(x)$ are even. What can be said about $f(x) + g(x)$?
(b) Suppose $f(x)$ and $g(x)$ are odd. What can be said about $f(x) + g(x)$?
(c) Suppose $f(x)$ is even and $g(x)$ is odd. What can be said about $f(x) + g(x)$?

GRAFFIX

In Exercises 38–43, use a graphics calculator or computer to determine whether the graph of the equations exhibits y-axis or origin symmetry.

38. $y = x^3 + x$
39. $y = \dfrac{4}{x + 2}$
40. $y = x^4 - x^2 - 4$
41. $y = 3x - 6$
42. $y = \dfrac{x - 1}{x^2 + 1}$
43. $y = \dfrac{x}{x^2 - 5}$

Chapter 2 SUMMARY

After completing this chapter you should:

1. Be able to graph straight lines in a rectangular coordinate system. (Section 2.1)
2. Be able to compute the slope of a line. (Section 2.2)
   The slope is the ratio of the change in y to the change in x as we move from one point to another point along the graph. For a line, this ratio is constant.

For example:
In order to find the slope of the line passing through the points $(-3, 2)$ and $(1, -4)$, we use the formula for the slope of a nonvertical line which is $m = \dfrac{y_2 - y_1}{x_2 - x_1}$, and get

$$m = \frac{-4 - 2}{1 - (-3)} = \frac{-6}{4} = \frac{-3}{2}.$$

3. Understand the significance of the slope of a line. (Section 2.2)
Lines with positive slope rise as we move from left to right, and lines with negative slope fall as we move from left to right. A line with zero slope is horizontal, whereas a line whose slope is undefined is vertical.

4. Know that two nonvertical lines are parallel if and only if their slopes are equal, and are perpendicular if and only if their slopes are negative reciprocals. (Section 2.2)

5. Be able to write an equation of a line satisfying certain conditions. (Section 2.3)
A line is determined once we know one point on the line and the slope.
For example:
Write an equation of a line passing through the point $(4, -5)$ that is perpendicular to the line whose equation is $3x + 4y = 7$.
Solution:
The slope of the line whose equation is $3x + 4y = 7$ can most easily be found by getting this equation into slope-intercept form

$$3x + 4y = 7 \Rightarrow y = -\frac{3}{4}x + \frac{7}{4}$$

Thus the slope of the given line is $-\frac{3}{4}$.

The slope of a line perpendicular to the given line will therefore be $\frac{4}{3}$. We can write the equation of the perpendicular line by using the point-slope form for the equation of a line, which gives $y - (-5) = \frac{4}{3}(x - 4)$; our final answer is usually written as $\boxed{y = \frac{4}{3}x - \frac{31}{3}}$.

6. Be able to sketch the graph of the solution set of a linear inequality in two variables. (Section 2.1) The solution of a linear inequality is a half plane bounded by a line. Use a test point to determine on which side of the line all solutions lie.
For example:
In order to sketch the solution set of $4x - 3y > 12$, we first sketch the boundary of the region, which is the graph of the line $4x - 3y = 12$. We sketch a dashed line, since the points on the line are *not* included in the solution set. Next we choose a test point not on the line, say $(0, 0)$, and find that it does not satisfy the given inequality. Therefore, the region containing $(0, 0)$ does not satisfy the inequality and so the solution set is the region not containing $(0, 0)$, The solution set appears in Figure 2.75

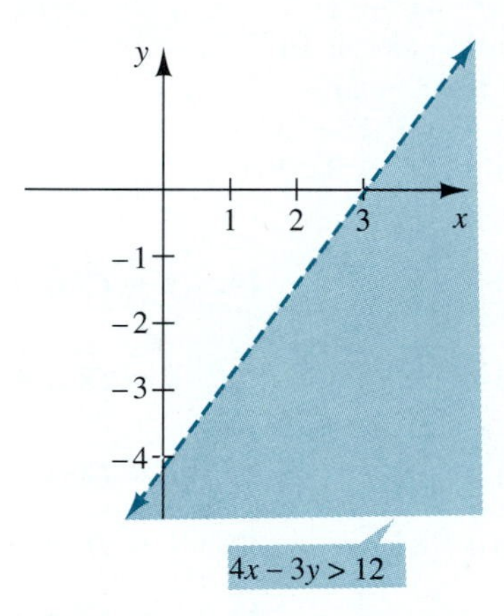

FIGURE 2.75 The solution set of $4x - 3y > 12$

7. Recognize and graph equations of circles. (See Section 2.1) The standard form of the equation of a circle with its center (h, k) and its radius r is $(x - h)^2 + (y - k)^2 = r^2$. Knowing the center and the radius of the circle makes it simple to graph the circle.
Find the center and radius of the circle whose equation is $(x + 2)^2 + (y - 5)^2 = 15$.
Solution:
Comparing the given equation to the standard form $(x - h)^2 + (y - k)^2 = r^2$, we can see that $h = -2$, $k = 5$, and $r = \sqrt{15}$. Therefore,

$$\boxed{\text{the center is } (-2, 5) \text{ and the radius is } \sqrt{15}.}$$

8. Understand the definition of a function and be able to find its domain. (Section 2.4) A function is a relationship between x and y for which to each real value of x in the domain, there corresponds exactly one real value of y in the range.
For example:
Find the domain of the function $y = \dfrac{\sqrt{2x + 5}}{x - 3}$.

Solution:
The domain of this function consists of all real numbers x for which the expression $\dfrac{\sqrt{2x + 5}}{x - 3}$ is defined and a real number. Consequently we require that

$$2x + 5 \geq 0 \Rightarrow x \geq -\frac{5}{2}$$

and

$$x - 3 \neq 0 \Rightarrow x \neq 3$$

Therefore, the domain is $\{x \mid x \geq -\frac{5}{2},\ x \neq 3\}$.

9. Use function notation to compute functional values. (Section 2.5)
When using function notation, it is useful to think of x as the input and $f(x)$ as the output.
For example:
Given $f(x) = -x^2 - 3x + 2$, find **(a)** $f(-6)$ and **(b)** $f(x + 5)$.
Solution:
(a) $f(x) = -x^2 - 3x + 2$ *To find $f(-6)$, we replace x by -6.*

$$f(-6) = -(-6)^2 - 3(-6) + 2$$

$$f(-6) = \boxed{-16}$$

(b) $f(x) = -x^2 - 3x + 2$ *To find $f(x + 5)$ we replace x by $x + 5$.*

$$\begin{aligned} f(x + 5) &= -(x + 5)^2 - 3(x + 5) + 2 \\ &= -(x^2 + 10x + 25) - 3x - 15 + 2 \\ &= -x^2 - 10x - 25 - 3x - 15 + 2 \end{aligned}$$

$$f(x + 5) = \boxed{-x^2 - 13x - 38}$$

10. Be able to read information about a function from its graph. (Section 2.6)

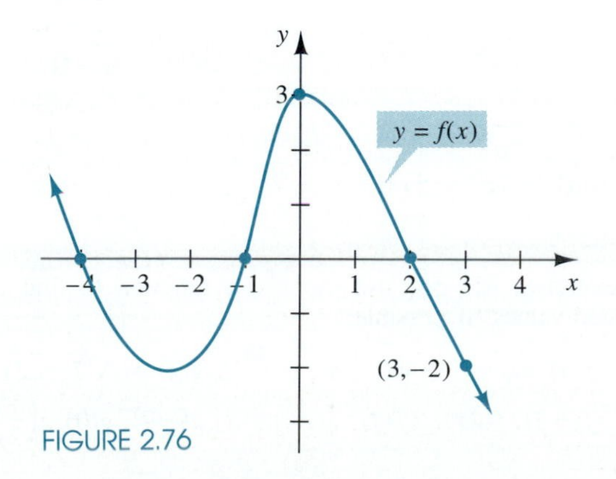

FIGURE 2.76

For example:
Consider the graph of $f(x)$ in Figure 2.76.
From the graph of $f(x)$ in Figure 2.76, we can see, among other things, that $f(3) = -2$; that $f(0) = 3$; that $f(x)$ has zeros at -4, -1, and 2; that $f(x)$ is positive on the interval $(-1, 2)$; and that $f(x)$ is decreasing on the interval $[0, \infty)$.

11. Recognize when a graph exhibits y-axis, x-axis, or origin symmetry. See Figure 2.77. (Section 2.7)

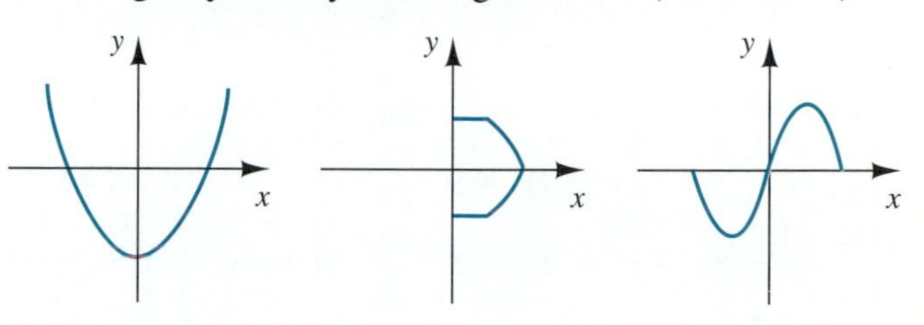

FIGURE 2.77

12. Be able to use the algebraic formula of a function to determine whether its graph will exhibit y-axis or origin symmetry. (Section 2.7)
A function for which $f(-x) = f(x)$ has a graph that exhibits y-axis symmetry.
A function for which $f(-x) = -f(x)$ has a graph that exhibits origin symmetry.
For example:
For $f(x) = x^3 - 4x$, we have

$$\begin{aligned} f(-x) &= (-x)^3 - 4(-x) \\ &= -x^3 + 4x = -(x^3 - 4x) = -f(x) \end{aligned}$$

and so the graph of $f(x) = x^3 - 4x$ exhibits origin symmetry. The graph appears in Figure 2.78.

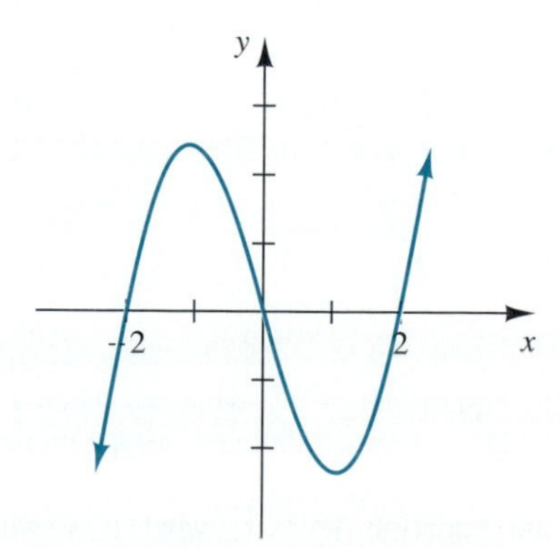

FIGURE 2.78 The graph of $y = f(x) = x^3 - 4x$ exhibits origin symmetry.

Chapter 2 REVIEW EXERCISES

In Exercises 1–18, sketch the graph of the given equation or inequality on a rectangular coordinate system.

1. $4x - 5y = 20$

2. $y = 2x - 8$

3. $3x + 7y + 14 \le 0$

4. $y = -5x$

5. $y = -5$

6. $\dfrac{x}{2} - \dfrac{y}{3} > 4$

7. $3x = 4$

8. $3x = 4y$

9. $x^2 + y^2 = 9$

10. $x + y = 9$

11. $(x - 3)^2 + (y + 4)^2 = 9$

12. $x^2 + y^2 - 6y = 1$

13. $5x + 5y = 20$

14. $5x^2 + 5y^2 = 20$

15. $x^2 - 3x + y^2 + 2y = 2$

16. $3x - 2y < 10$

17. $x < -2$

18. $y \ge \dfrac{3}{2}$

In Exercises 19–22, use the given graph to determine **(a)** whether it is the graph of a function, **(b)** its domain, and **(c)** its range.

19.

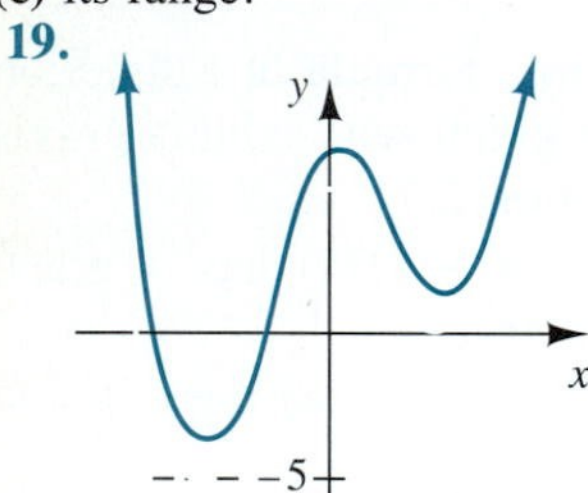

20.

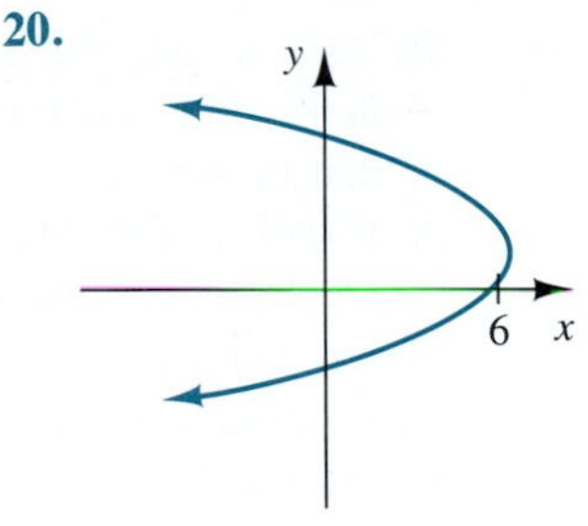

21.

22.

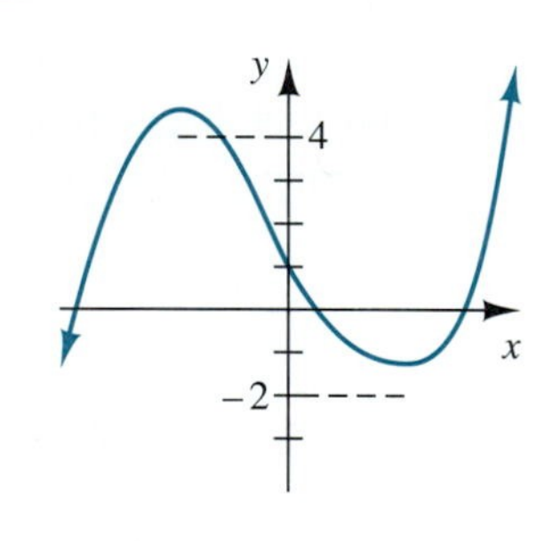

23. Given the equation $y = x^4$, what is y when $x = 2$? When $x = -2$? Does $y = x^4$ define y as a function of x? Explain.

24. Given the equation $y^4 = x$, what is y when $x = 16$? Does $y^4 = x$ define y as a function of x? Explain.

25. Write an equation of the line passing through the points $(-3, 4)$ and $(2, -5)$.

26. Write an equation of the line that passes through the point $(2, 0)$ and is perpendicular to the line whose equation is $4x - 7y = 2$.

27. Find the value(s) of t so that the line passing through the points $(1, -4)$ and $(t, 6)$ is parallel to the line passing through the points $(-3, -1)$ and $(5, t)$.

28. Prove that the points $P(-1, 3)$, $Q(3, -2)$, and $R(13, 6)$ are the vertices of a right triangle.

29. Write an equation of the circle with center $(-4, 1)$ and radius 7.

30. Find the center and radius of the circle with equation $x^2 + (y + 2)^2 = 20$.

31. Find the center and radius of the circle with equation $x^2 + 8x + y^2 - 2y = 8$.

32. Sketch the graph of $x^2 - 4x + y^2 - 6y + 9 = 0$.

33. Write an equation of the circle with center $(-5, 4)$ and radius $\dfrac{1}{2}$.

34. Write an equation of the line tangent to the circle $x^2 + 10x + y^2 = 33$ at the point $(2, -3)$.

35. Write an equation of the circle with a diameter having endpoints $(1, -4)$ and $(0, 5)$.

36. Write an equation of the circle with a radius of 7 and with a center at the point where the line $3y - \dfrac{1}{2}x = 4$ crosses the y-axis.

In Exercises 37–44, find the domain of the given function.

37. $f(x) = \sqrt{5 - 3x}$

38. $g(x) = \sqrt[3]{5 - 3x}$

39. $h(x) = \dfrac{x - 1}{x^2 - 3x - 4}$

40. $F(x) = \dfrac{5}{x^2 + 9}$

41. $G(x) = \sqrt{6x - x^2}$

42. $f(x) = \dfrac{\sqrt{x + 6}}{x - 2}$

43. $F(x) = \dfrac{-2}{\sqrt{x + 4}}$

44. $H(x) = 4x^3 - 5x^2 - x + 7$

In Exercises 45–56, use the given function to find the requested values (if possible).

45. $f(x) = -x^2 + 4x - 1$

$f(-3), f(2x), 2f(x)$

46. $g(x) = \sqrt{7 - 2x}$

$g(-9), g(0), g\left(\dfrac{1}{2}\right)$

47. $h(x) = \sqrt{5}$
$h(2), h(10), h(-3)$

48. $f(x) = 9 - 4x^2$
$f(x + 3), f(x) + 3$

49. $g(t) = \dfrac{2t - 1}{t + 5}$, $g\left(\dfrac{1}{2}\right), g(-5), g(t + 1)$

50. $H(r) = \dfrac{r}{4r^2 + 1}$, $H\left(-\dfrac{1}{2}\right), H\left(\dfrac{1}{r}\right),\quad H(r + 1)$

51. $f(x) = \begin{cases} -2x + 7 & \text{if } x \leq -4 \\ x^2 - 1 & \text{if } -1 \leq x \leq 8 \end{cases}$
$f(0), f(-1), f(-5)$

52. $f(x) = \begin{cases} -x & \text{if } x \leq -2 \\ x^2 & \text{if } -2 < x < 4 \\ \sqrt{x} & \text{if } 4 \leq x \end{cases}$
$f(3), f(4), f(-2)$

53. $f(x) = 5x^2 - 6x + 3;\quad \dfrac{f(x - 4) - f(x)}{4}$

54. $g(u) = 9 - 7u;\quad \dfrac{g(u + h) - g(u)}{h}$

55. $f(t) = \dfrac{-3}{t};\quad \dfrac{f(t + h) - f(t)}{h}$

56. $h(x) = \dfrac{x + 1}{x - 1};\quad \dfrac{h(x - 3) - h(x)}{3}$

57. Use the graph of $y = f(x)$ in the figure to answer the following.

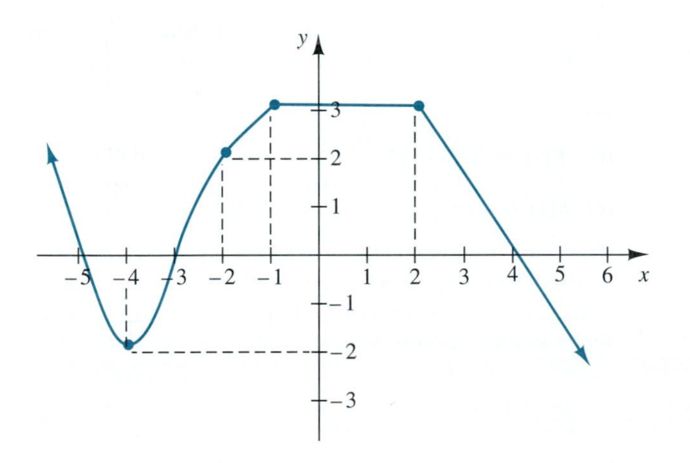

(a) Find $f(-4), f(-2), f(0)$, and $f(1)$.
(b) What are the zeros of $f(x)$?
(c) On what intervals is $f(x) > 0$?
(d) On what intervals is $f(x)$ negative?
(e) On what intervals is $f(x)$ increasing?
(f) On what intervals is $f(x)$ decreasing?
(g) On what intervals is $f(x)$ constant?

In Exercises 58–61, examine the given graph to determine whether it exhibits y-axis symmetry, x-axis symmetry, origin symmetry, or none of these.

58.

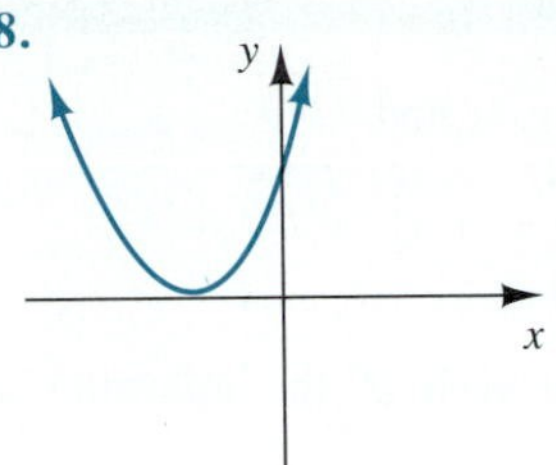

59.

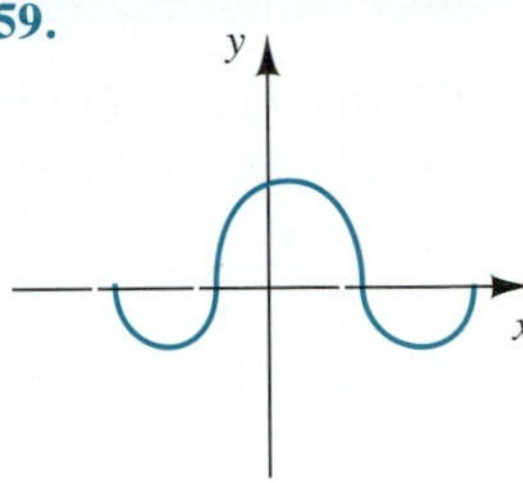

60.

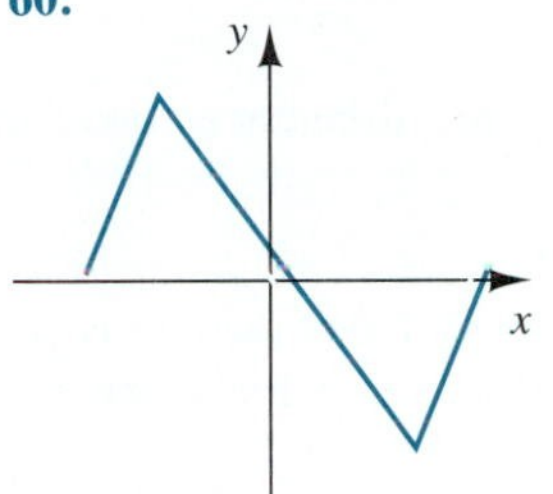

61.

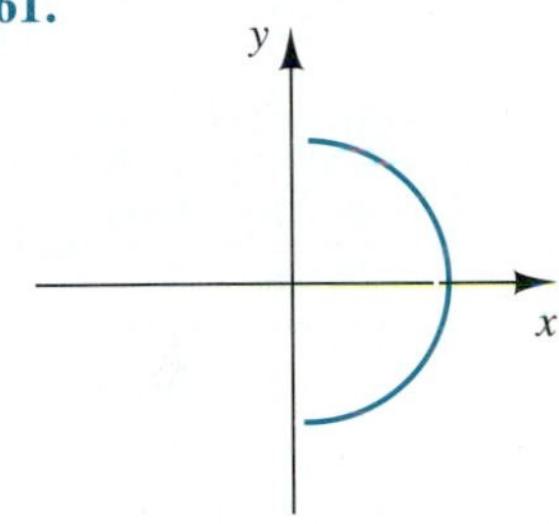

In Exercises 62–67, determine whether the graph of the given function exhibits y-axis symmetry, origin symmetry, or neither.

62. $f(x) = \dfrac{-3x}{x^2 + 1}$

63. $f(x) = 2x^2 - 5x$

64. $g(x) = \dfrac{x + 4}{x - 4}$

65. $h(x) = 7x - x^3$

66. $F(x) = \sqrt{9 - x^2}$

67. $G(x) = \dfrac{x^2 + 4}{x}$

68. An oil company charges \$1.32 per gallon for each of the first 150 gallons of home heating oil, and \$1.21 for each gallon above 150. Express the cost, C, of an oil delivery as a function of g, the number of gallons of oil purchased.

69. A student notes that she received a grade of 76 on her first math test after studying 3 hours for the test, and a grade of 85 on her second math test after studying 5 hours. Assuming a linear relationship between her grades on her math tests and the number of hours she studies, what would her grade be if she studies 6 hours for her third math test?

70. A computer consultant charges \$75 for an initial consultation lasting up to 1 hour, and \$95 per hour for each additional hour devoted to a project. Express the consultant's charge, C, as a function of h, the total number of hours of the consultant's time required.

71. At a certain time, the sides of a square are 6 cm long and increasing at the rate of 2.5 cm/min. Express the area of the square t minutes later as a function of t.

Chapter 2 PRACTICE TEST

1. Given $f(x) = -2x^2 - 5x + 3$, find
(a) $f(-4)$ (b) $f(x + 4)$ (c) $f(4x)$
(d) $f(x^2)$ (e) $\dfrac{f(x + h) - f(x)}{h}$

2. Given $g(t) = \dfrac{2t + 1}{t - 2}$, find each of the following and simplify.
(a) $g\left(\dfrac{2}{3}\right)$ (b) $g\left(\dfrac{1}{t}\right)$

3. Given $F(x) = \dfrac{x}{x - 3}$, find the difference quotient $\dfrac{F(x + 5) - F(x)}{5}$ and simplify.

4. Sketch the graph of $5x - 3y = 20$. Label the intercepts.

5. Write an equation of the line that crosses the x-axis at 4 and the y-axis at -3.

6. Write an equation of the line passing through the point $(-2, -5)$ that is perpendicular to the line whose equation is $3x - 7y + 10 = 0$.

7. Sketch the graph of the equation $x^2 - 2x + y^2 = 0$. Is it the graph of a function? Explain.

8. Write an equation of the circle with diameter whose end-points are $(-3, 4)$ and $(1, -3)$.

9. Use the graph of $y = f(x)$ in the figure to answer the following.

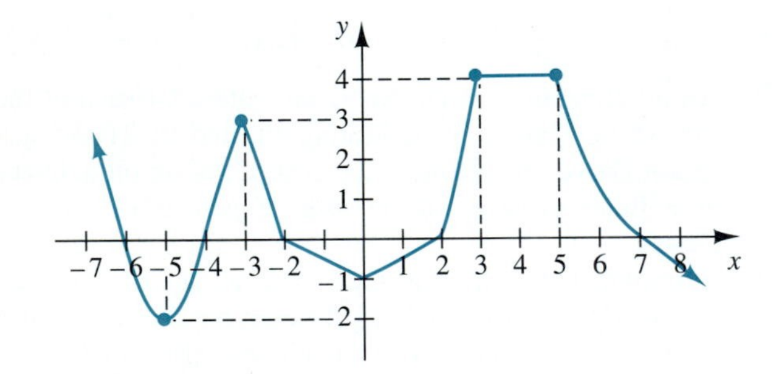

(a) Find $f(-5)$, $f(-3)$, $f(0)$, and $f(4)$.
(b) What are the zeroes of $f(x)$?
(c) On what intervals is $f(x)$ positive?
(d) On what intervals is $f(x) < 0$?
(e) On what intervals is $f(x)$ increasing?
(f) On what intervals is $f(x)$ decreasing?
(g) On what intervals is $f(x)$ constant?

10. The accompanying figure contains a portion of a graph. Complete the remainder of the graph if the graph exhibits

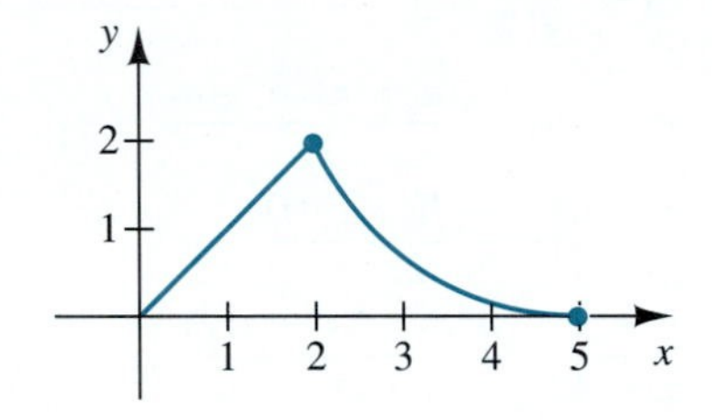

(a) y-axis symmetry
(b) origin symmetry
(c) x-axis symmetry

11. Determine whether the graphs of the following functions exhibit y-axis symmetry, origin symmetry, or neither.
(a) $f(x) = \dfrac{x}{x + 5}$
(b) $F(x) = 3x^2 - x^4$
(c) $h(x) = x - \dfrac{1}{x}$

12. The width of a rectangle is w and the length is 3 less than 4 times the width. A fence costing \$3.50 per foot is to be erected around the rectangle. Express the cost, C, of the fence as a function of w.

CHAPTER 3

Functions and Graphs: Part II

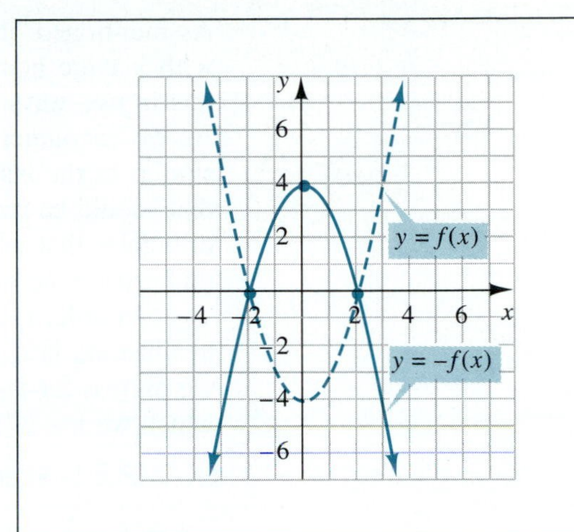

In this chapter we expand upon many of the ideas regarding functions that we introduced in the last chapter. In the first two sections we continue our discussion of graph sketching by describing some basic graphing principles. In Section 3.3 we begin looking at how we can apply the idea of functions to model real-life situations. In Section 3.4 we discuss quadratic functions. Section 3.5 introduces the algebra of functions—that is, how we can produce new and often more complex functions from simpler ones. Section 3.6 concludes the chapter with a discussion of equations that not only define y as a function of x but also define x as a function of y. Such a function is said to have an inverse.

Many of the ideas we introduce and develop in this chapter will be used repeatedly throughout the text as we continue to encounter a wide variety of special functions.

3.1 Basic Graphing Principles

As mentioned previously, one of our main goals is to become thoroughly familiar with a large number of graphs that are frequently encountered in calculus. We do this in two ways. First, we learn to recognize the graphs of certain basic functions. As we encounter these basic graphs, we record them in a catalog. (This catalog is similar to the inside front covers of this textbook.) This catalog will contain graphs that should be recognized immediately. Second, we establish a set of basic graphing principles that will allow us to easily graph variations of our basic graphs. We begin this process here and it continues throughout the text.

In order to demonstrate the various graphing techniques, it is extremely helpful to illustrate how these ideas work with one particular function first. Consequently, we digress for just a moment to determine the graph of the function $y = f(x) = x^2$, which we use as the springboard for our discussion.

EXAMPLE 1 Sketch the graph of $y = f(x) = x^2$.

Solution In general, we try to obtain a graph by analyzing its equation rather than by plotting points. However, since we have to start somewhere, we compute some values to get an idea of what *this* graph looks like. See Table 3.1. Using these points,

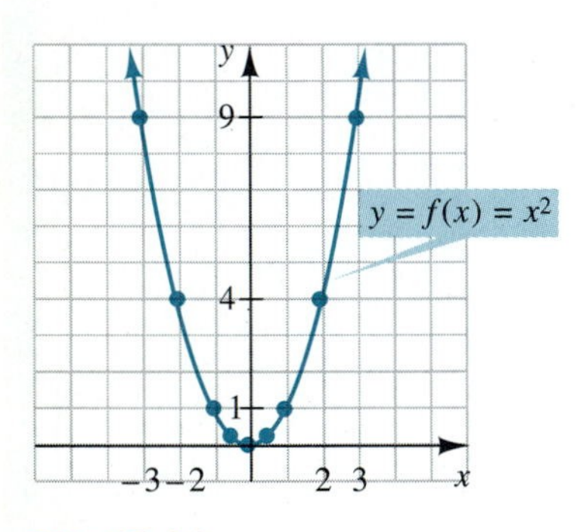

FIGURE 3.1

TABLE 3.1

x	$y = f(x) = x^2$	y
-3	$y = (-3)^2$	9
-2	$y = (-2)^2$	4
-1	$y = (-1)^2$	1
$-\frac{1}{2}$	$y = (-\frac{1}{2})^2$	$\frac{1}{4}$
0	$y = (0)^2$	0

TABLE 3.1 (cont.)

x	$y = f(x) = x^2$	y
$\frac{1}{2}$	$y = (\frac{1}{2})^2$	$\frac{1}{4}$
1	$y = (1)^2$	1
2	$y = (2)^2$	4
3	$y = (3)^2$	9

we draw a smooth curve through them, obtaining the graph appearing in Figure 3.1.* Based upon this graph we can see that the domain is the set of all real numbers and the range is the set of all nonnegative real numbers. ■

We will refer to the graph of $y = f(x) = x^2$ as the **basic parabola,** and it is the next entry in our catalog of basic graphs. The lowest (or highest) point of a parabola is called the **vertex** of the parabola, and the vertical line passing through the vertex is called the **axis of symmetry.** (In Sections 3.4 and 10.2 we engage in a much more detailed discussion of parabolas.)

*The question of the exact shape of the graph—that is, how to connect the points we have found—is a very important one. In order to answer this question precisely, one generally needs to have some of the techniques of calculus available.

GRAFFIX

Use a graphics calculator or computer.

1. Graph the following on the same set of coordinate axes:

$$y = 0.3x^2,\ y = x^2,\ y = 2x^2$$

Describe how changing the coefficient a in $y = ax^2$ affects the graph of $y = x^2$.

2. Graph the following on the same set of coordinate axes:

$$y = 0.5(x^3 - 3x + 1),\ y = x^3 - 3x + 1,\ y = 2(x^3 - 3x + 1)$$

Describe how changing the coefficient a in $y = (x^3 - 3x + 1)$ affects the graph of $y = x^3 - 3x + 1$.

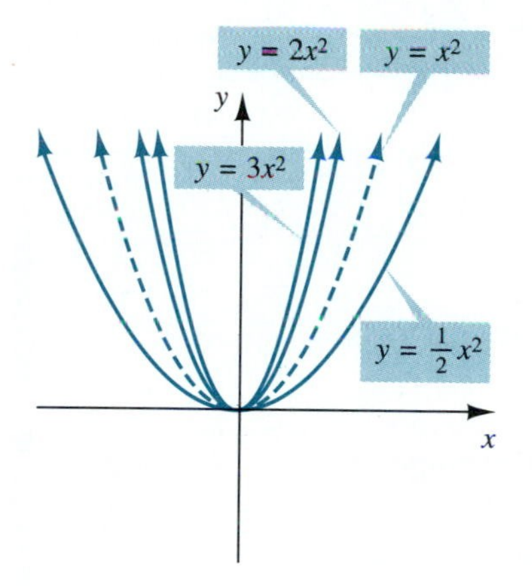

FIGURE 3.2

Suppose we now consider the graph of $y = 2x^2$. Rather than compute a table of values, we recognize that for each x-value, except for $x = 0$, the y-value on the graph of $y = 2x^2$ will be twice as large as the y-value on the graph of $y = x^2$. This gives us a narrower parabola that is somewhat "sharper" at the vertex. The graph appears in Figure 3.2.

Similarly, the graph of $y = 3x^2$ will be even narrower; its graph also appears in Figure 3.2.

For a given value of x, the y-value on the graph of $y = \frac{1}{2}x^2$ is one-half the y-value on the graph of $y = x^2$. Thus for $x \neq 0$, the graph of $y = \frac{1}{2}x^2$ will be wider than the graph of $y = x^2$ and a bit flatter at the vertex. See Figure 3.2. If $a > 0$ (and $a \neq 1$), we say that the graph of $y = ax^2$ is obtained by *stretching* the graph of $y = x^2$.

More generally, the graph of $2f(x)$ is obtained by deflecting the graph of $y = f(x)$ further away from the x-axis, whereas the graph of $\frac{1}{2}f(x)$ is obtained by pulling the graph of $y = f(x)$ closer to the x-axis. In many (but not all) cases, this can be described by saying that the graph of $af(x)$ is wider or narrower than the graph of $f(x)$, depending on the size of a.

The Stretching Principle

(for $a > 0$, $a \neq 1$)

The graph of $y = af(x)$ can be obtained by stretching the graph of $y = f(x)$. In other words, the graph of $y = af(x)$ will have the same basic shape as the graph of $y = f(x)$. If $a > 1$, the graph of $y = af(x)$ is obtained by deflecting the graph of $f(x)$ away from the x-axis. If $0 < a < 1$, the graph of $y = af(x)$ is obtained by pulling the graph of $f(x)$ toward the x-axis.

How does the value of $3f(4)$ compare with the value of $f(4)$?

Figure 3.3 illustrates the stretching principle.

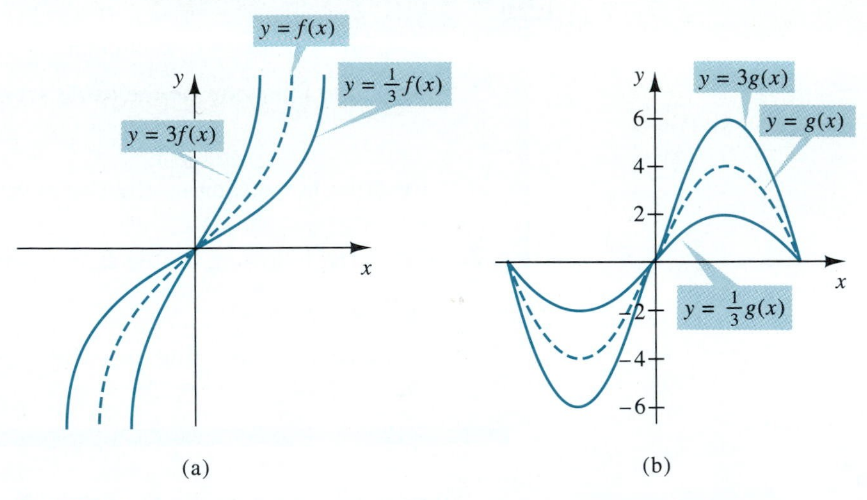

FIGURE 3.3

GRAFFIX

Use a graphics calculator or computer.

1. Graph the following on the same set of coordinate axes:

$$y = x^2, \quad y = x^2 + 1, \quad y = x^2 - 1$$

What can you conclude about the effect on the graph of $y = x^2$ of adding a constant to x^2?

2. Graph the following on the same set of coordinate axes:

$$y = x^5, \quad y = x^5 + 1, \quad y = x^5 - 2$$

What can you conclude about the effect on the graph of $y = x^5$ of adding a constant to x^5?

Let's use our knowledge of the graph of $y = x^2$ to graph other functions.

EXAMPLE 2 Use the graph of $y = f(x) = x^2$ to sketch the following graphs.

(a) $y = g(x) = x^2 + 3$ **(b)** $y = h(x) = x^2 - 4$

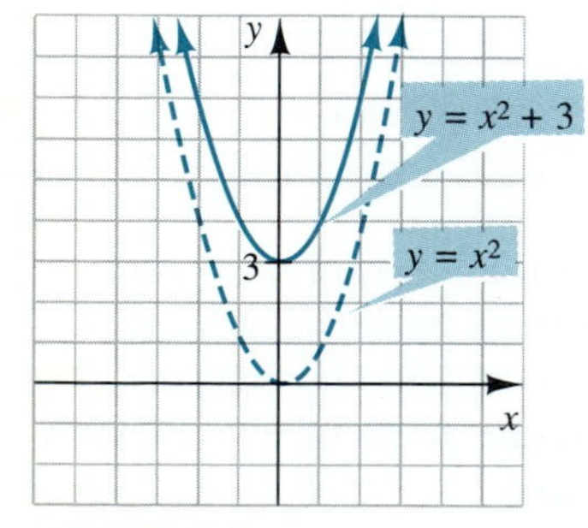

FIGURE 3.4

Solution

(a) Although we could complete a table similar to Table 3.1 for $y = x^2 + 3$, there is a much more efficient way to determine its graph. Let us compare the y-values obtained from the equations $y = x^2$ and $y = x^2 + 3$. For each value of x, the associated y-value obtained from $y = x^2 + 3$ is 3 more than the y-value obtained from $y = x^2$. In other words, for a particular x-value, the point on the graph of $y = x^2 + 3$ will be 3 units above the point on the graph of $y = x^2$. Increasing the y-value on a graph by 3 moves that point up 3 units. To obtain the graph of $y = x^2 + 3$, we shift the graph of $y = x^2$ up 3 units. The graph of $y = g(x) = x^2 + 3$ appears in Figure 3.4. The graph of $y = f(x) = x^2$ is also included and is indicated by the dashed graph.

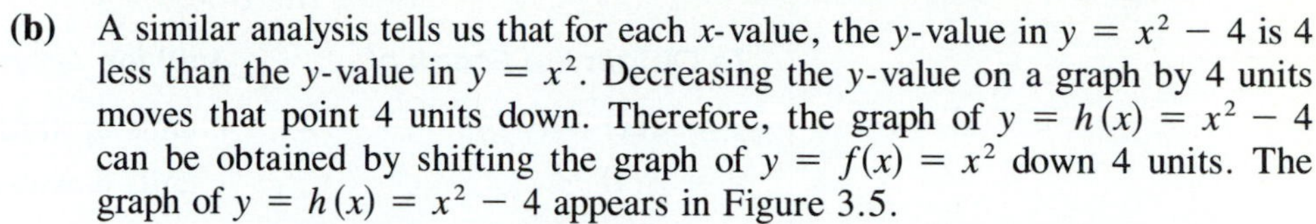

(b) A similar analysis tells us that for each x-value, the y-value in $y = x^2 - 4$ is 4 less than the y-value in $y = x^2$. Decreasing the y-value on a graph by 4 units moves that point 4 units down. Therefore, the graph of $y = h(x) = x^2 - 4$ can be obtained by shifting the graph of $y = f(x) = x^2$ down 4 units. The graph of $y = h(x) = x^2 - 4$ appears in Figure 3.5.

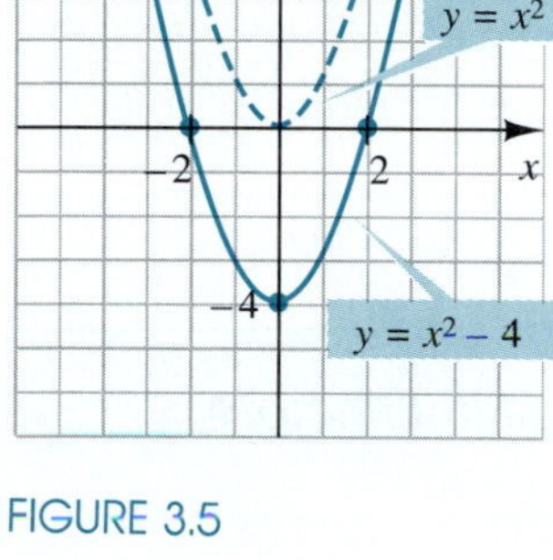

FIGURE 3.5

The graph clearly indicates that there are x-intercepts, which we can find by setting $f(x) = 0$. (We could have said that we are finding the x-intercepts by setting $y = 0$. We are purposely using y and $f(x)$ interchangeably to remind ourselves that the two represent the same quantity.)

$$y = f(x) = x^2 - 4 \qquad \textit{Set } f(x) = 0.$$

$$0 = x^2 - 4 \Rightarrow x^2 = 4 \Rightarrow x = \pm 2$$

These x-intercepts appear in Figure 3.5. The y-intercept is found by shifting the original graph down 4 units. We could have found the y-intercept algebraically by setting $x = 0$ and solving for y, i.e., finding $f(0)$. ■

Let's now apply these same ideas to a slightly more general situation.

EXAMPLE 3 Given the graph of $y = f(x)$ shown in Figure 3.6, sketch the graph of
(a) $y = f(x) - 3$ **(b)** $y = f(x) + 2$

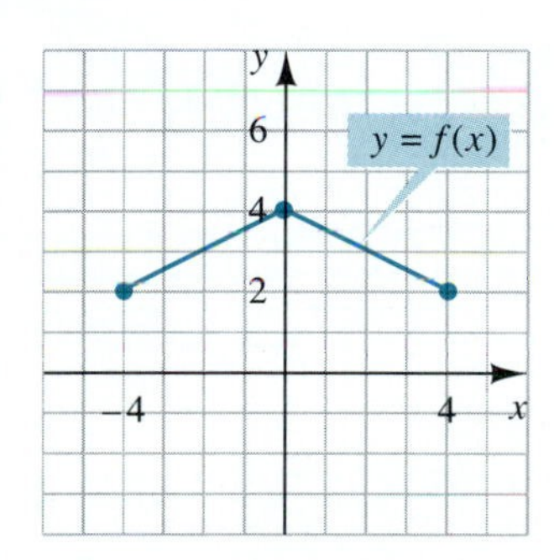

FIGURE 3.6

Solution Using the ideas presented in the last example, we recognize that for a particular x-value, each y-value on the graph $y = f(x) - 3$ will be 3 less than the y-value on the graph of $y = f(x)$, whereas for a particular x-value, each y-value on the graph of $y = f(x) + 2$ will be 2 more than the y-value on the graph of $y = f(x)$. Therefore, to obtain the graph of $y = f(x) - 3$, we shift the original graph of $y = f(x)$ down 3 units, and to obtain the graph of $y = f(x) + 2$, we shift the original graph up 2 units. The graphs appear in Figure 3.7(a) and (b). Again the original function appears as the dashed graph.

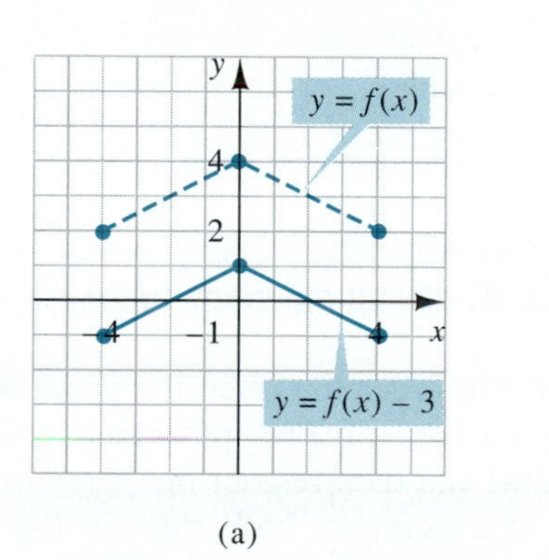

(a)

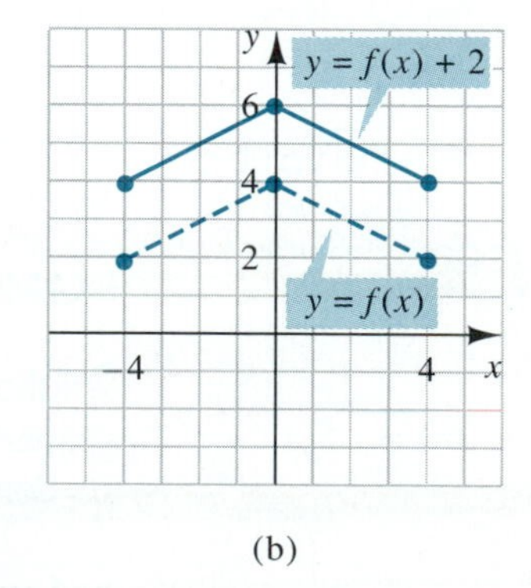

(b)

FIGURE 3.7 ■

We can generalize the results of the last two examples into the following principle of graphing.

In general, how does the value of $f(2) - 5$ compare with the value of $f(2)$?

The Vertical Shift Principle for Graphs

(for $c > 0$)

To Obtain the Graph of:	*Shift the Graph of y = f(x):*
$y = f(x) + c$	c units upward
$y = f(x) - c$	c units downward

DIFFERENT PERSPECTIVES: *The Vertical Shift Principle*

GEOMETRIC DESCRIPTION

To obtain the graph of $y = f(x) + 2$, shift the graph of $y = f(x)$ up 2 units.

To obtain the graph of $y = f(x) - 3$, shift the graph of $y = f(x)$ down 3 units.

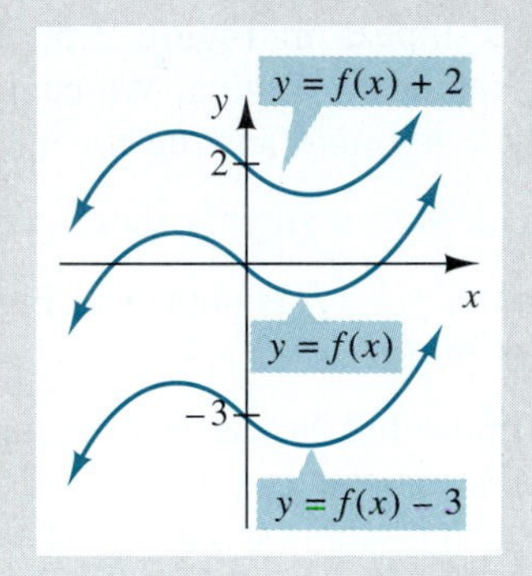

ALGEBRAIC DESCRIPTION

For each value of x, the value of $y = f(x) + 2$ is 2 greater than the value of $f(x)$. If $f(4) = 5$, then $f(4) + 2 = 5 + 2 = 7$.

For each value of x, the value of $y = f(x) - 3$ is 3 less than the value of $f(x)$. If $f(4) = 5$, then $f(4) - 3 = 5 - 3 = 2$.

We can compare the value of $f(3)$ with $f(3) + 2$. Can we, in general, compare the value of $f(3)$ with $f(3 + 2)$?

In Section 2.5 we examined some split functions. Let's look at how we might graph a split function.

EXAMPLE 4 Sketch the graph of $f(x) = \begin{cases} 2x - 1 & \text{if } x \leq 1 \\ -x + 4 & \text{if } x > 1 \end{cases}$. Label the intercepts.

Solution In order to sketch the graph of a split function, we imagine that we are graphing each part of the split function (as if there were no arbitrary restriction on its domain) and then restrict the graph to those x's for which the rule applies.

In this example, both rules in the function $f(x)$ are first-degree expressions, and so the graphs of both parts will be straight lines. In Figure 3.8(a) we have drawn the complete graphs of $y = 2x - 1$ and $y = -x + 4$. For $y = 2x - 1$, we have highlighted that portion of the graph for which $x \leq 1$, and for $y = -x + 4$, we have highlighted that portion of the graph for which $x > 1$. See Figure 3.8(a).

Therefore, the graph of the split function $f(x)$ is obtained by combining the two highlighted pieces. The graph of $f(x)$ appears in Figure 3.8(b).

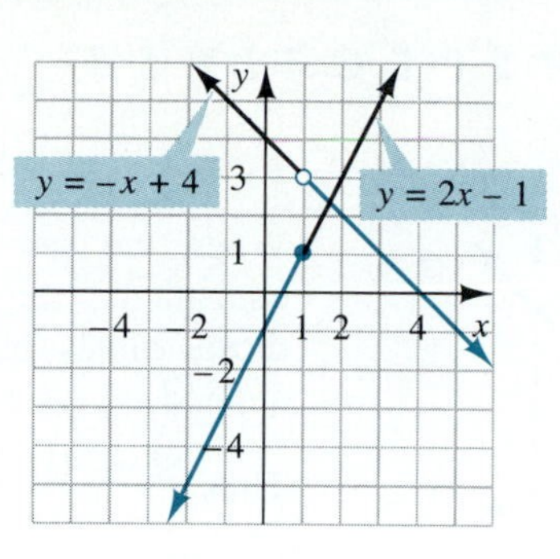

The graphs of $y = 2x - 1$ and $y = -x + 4$

(a)

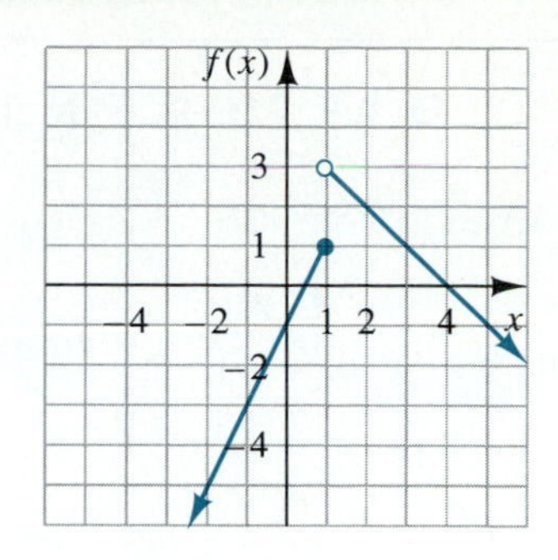

The graph of the split function
$$f(x) = \begin{cases} 2x - 1 & \text{if } x \le 1 \\ -x + 4 & \text{if } x > 1 \end{cases}$$

(b)

FIGURE 3.8 ■

The following example offers a slight variation on the idea of a split function.

EXAMPLE 5 Sketch the graph of $y = f(x) = \dfrac{x^2 - x - 6}{x - 3}$.

Solution Looking at $f(x)$ we can see that its domain consists of all real numbers except $x = 3$. However, if $x \neq 3$, we may simplify $f(x)$.

$$y = f(x) = \frac{x^2 - x - 6}{x - 3} = \frac{(x - 3)(x + 2)}{x - 3} = x + 2 \quad \text{for } x \neq 3$$

Algebraically, we are stating that $\dfrac{x^2 - x - 6}{x - 3}$ and $x + 2$ agree on all values except $x = 3$. Hence, provided that $x \neq 3$, we have $y = f(x) = x + 2$, which means that the graph of $f(x)$ will be a straight line with the single point corresponding to $x = 3$ missing. The graph appears in Figure 3.9.

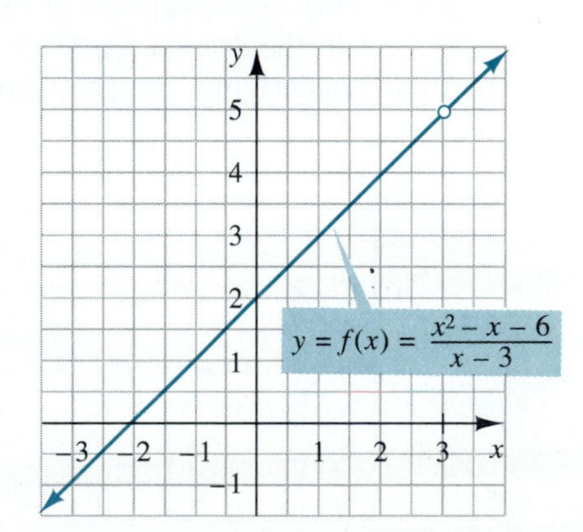

FIGURE 3.9

Keep in mind that $y = \dfrac{x^2 - x - 6}{x - 3}$ is *not* the same function as $y = x + 2$. In order for two functions to be identical, they must have the same rule of assignment *and* the same domain. ■

DIFFERENT PERSPECTIVES: *Equivalent Expressions*

GEOMETRIC DESCRIPTION

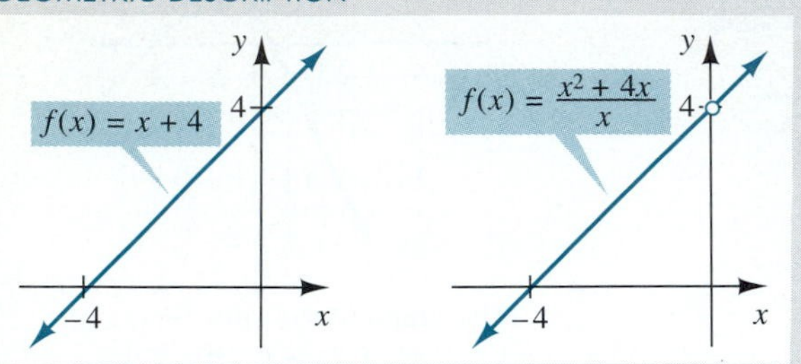

The graph of $f(x) = \dfrac{x^2 + 4x}{x}$ is identical to the graph of $f(x) = x + 4$ except at $x = 0$.

ALGEBRAIC DESCRIPTION

When we say that two algebraic expressions are equivalent, the equivalence can be valid only for those values of x for which both expressions are defined.

Thus $\dfrac{x^2 + 4x}{x} = x + 4$ for all $x \neq 0$.

Let us now add another graph to our catalog and in so doing establish a second graphing principle.

EXAMPLE 6 Sketch the graph of $y = f(x) = |x|$.

Solution We offer two approaches. The first approach begins by recalling the algebraic definition of the absolute value, which is

$$y = f(x) = |x| = \begin{cases} x & \text{if } x \geq 0 \\ -x & \text{if } x < 0 \end{cases}$$

We are, in effect, being asked to graph a split function. The graph of $y = |x|$ will look like the graph of $y = x$ for $x \geq 0$ and will look like the graph of $y = -x$ for $x < 0$. (The graphs of $y = x$ and $y = -x$ are very basic and you should be familiar with them by now.) The graphs appear in Figure 3.10(a) and (b). The graph of $y = f(x) = |x|$ appears in Figure 3.10(c).

In the second approach we obtain the graph of $y = f(x) = |x|$ by looking at the "underlying" graph—that is, the graph of the function without the absolute

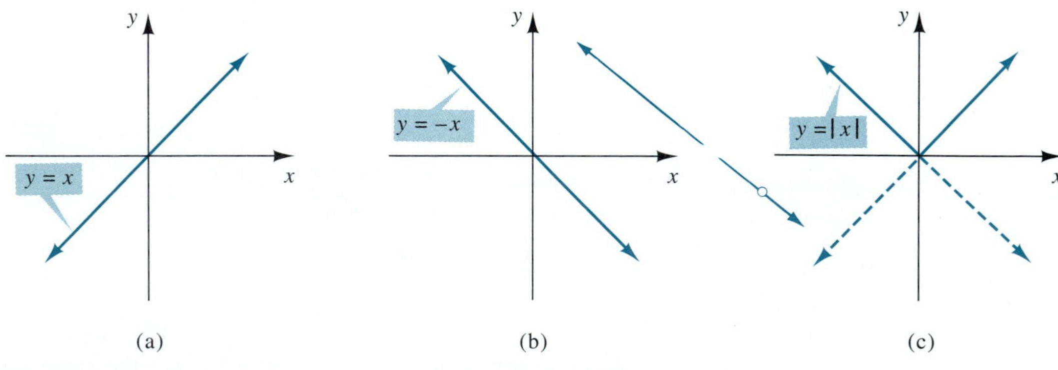

FIGURE 3.10

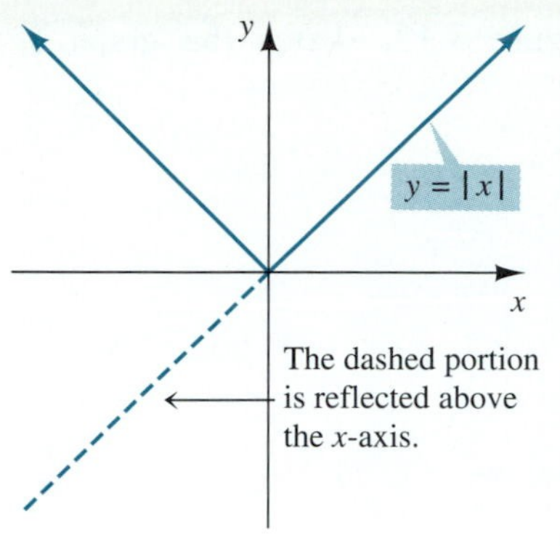

FIGURE 3.11

value—and then analyzing the effect that the absolute value has on the underlying graph. If we drop the absolute value, we get the equation $y = x$, whose graph appears in Figure 3.10(a). How do the y-values on the graph of $y = |x|$ compare with the y-values on the underlying graph $y = x$? For those x-values where x is positive or zero on the underlying graph, the absolute value has no effect. For those x-values where x is negative on the underlying graph, the absolute value makes it positive.

Thus we recognize that the absolute value has the following effect on the underlying graph: that portion of the underlying graph on or above the x-axis is unaffected by the absolute value, whereas that portion of the underlying graph below the x-axis is reflected above the x-axis.

This approach to the graph of $y = f(x) = |x|$ appears in Figure 3.11. Note that the underlying graph of $y = x$ is shown as a dashed line.

The graph of $y = |x|$ should be added to your catalog of basic graphs. ■

GRAFFIX

Use a graphics calculator or computer.

1. Graph the function $y = x^2 - 5$. Then clear the screen and graph $y = |x^2 - 5|$. Observe the differences between the two graphs.
2. Graph the function $y = x^3 - 3x + 1$. Then clear the screen and graph $y = |x^3 - 3x + 1|$. Observe the differences between the two graphs.
3. What can you conclude about the relationship between the graph of $f(x)$ and the graph of $|f(x)|$?

The second approach used in the last example can be used to answer the following more general question. Suppose we know the graph of $y = f(x)$ (which we shall call the underlying graph), how can we obtain the graph of $y = |f(x)|$? Wherever $f(x)$ is nonnegative, the absolute value will leave it unchanged, while wherever $f(x)$ is negative, the absolute value will make it positive. Keeping in mind that $f(x)$ is just another name for y, that y being positive means that the point on the graph is above the x-axis, and y being negative means that the point on the graph is below the x-axis, absolute value has the following effect on the underlying graph.

Principle for Graphing $y = |f(x)|$

To obtain the graph of $y = |f(x)|$, begin with the graph of the underlying function $y = f(x)$.

1. Leave that portion of the underlying graph on or above the x-axis unchanged, and
2. Take that portion of the underlying graph below the x-axis and reflect it above the x-axis.

How does the distance from the x-axis to the graph of $f(x)$ at $x = 5$ compare to the distance from the x-axis to the graph of $|f(x)|$ at $x = 5$?

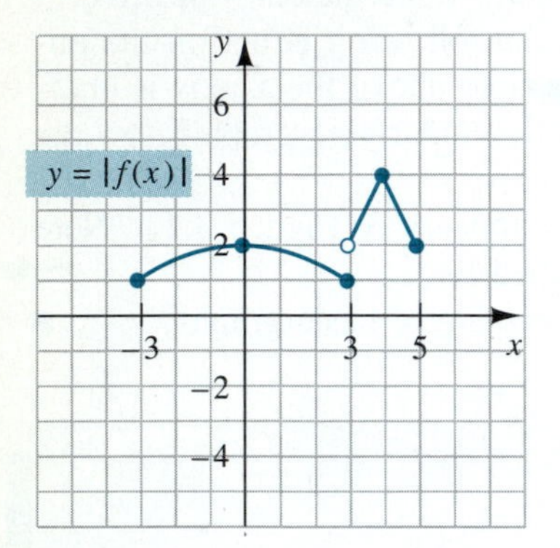

FIGURE 3.13

EXAMPLE 7 Given the graph of $y = f(x)$ in Figure 3.12, sketch the graph of $y = |f(x)|$.

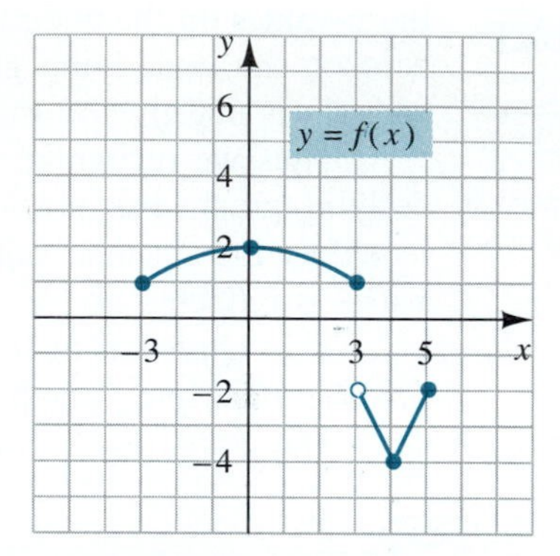

FIGURE 3.12

Solution Applying the principle for graphing $y = |f(x)|$, that portion of the graph on the interval $[-3, 3]$ is left alone because it is on or above the x-axis, while that portion of the graph below the x-axis on the interval $(3,5]$ is reflected above the x-axis. The graph appears in Figure 3.13. ■

EXAMPLE 8 Using the graph of $y = |x|$, sketch the graphs of

(a) $y = f(x) = |x - 2|$
(b) $y = f(x) = |x + 3|$

Solution We could sketch these graphs by using the principle for graphing the absolute value of a function, but instead we will examine the functions from a slightly different point of view, which will lead us yet to another graphing principle.

(a) Looking at the function $y = |x|$, we recognize that y cannot take on negative values. In fact, the smallest y can be is zero, which will occur when the input, x, is zero. This is mirrored in the graph by the fact that the lowest point on the graph of $y = |x|$ is the origin; i.e., $y = 0$ when $x = 0$.

If we now look at the function $y = f(x) = |x - 2|$, we again recognize that y cannot be negative. The smallest y can be is zero, and this will happen when $x - 2 = 0$, or when the input is $x = 2$. Thus the lowest point on the graph of $y = |x - 2|$ will be $(2, 0)$.

Similarly, $|x|$ will be equal to 5 when $x = 5$ or -5, whereas $|x - 2|$ will be equal to 5 when $x = 7$ or $x = -3$. Note that 7 is 2 units to the right of 5 and -3 is 2 units to the right of -5. This suggests that the y-values on the graph of $y = |x - 2|$ will be the same as the y-values on the graph of $y = |x|$ but will occur two units further to the right.

Thus the graph of $y = f(x) = |x - 2|$ can be obtained by shifting the graph of $y = |x|$ two units to the *right*. The graph appears in Figure 3.14. Note that the dashed graph is that of $y = |x|$.

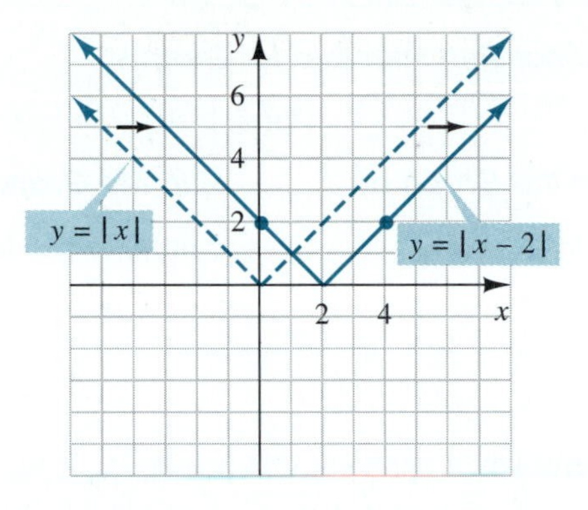

FIGURE 3.14

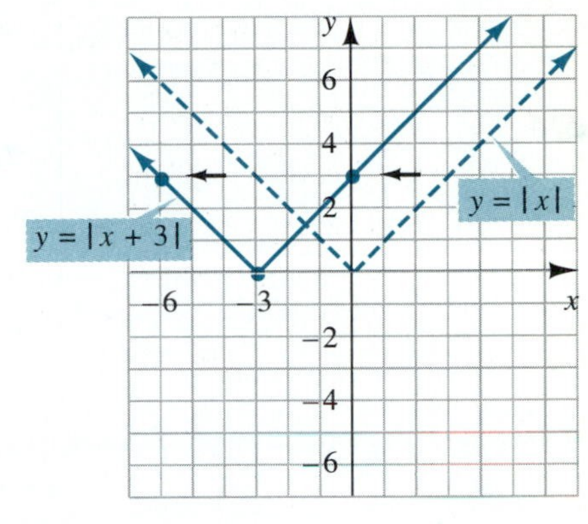

FIGURE 3.15

(b) A similar analysis for $y = f(x) = |x + 3|$ tells us that the lowest point on its graph will occur when $x + 3 = 0$, that is, when the input $x = -3$. This suggests that the graph of $y = f(x) = |x + 3|$ can be obtained by shifting the graph of $y = |x|$ three units to the *left*. The graph appears in Figure 3.15. ■

GRAFFIX

Use a graphics calculator or computer.

1. Graph the following on the same set of coordinate axes:

$$y = x^2, \quad y = (x - 1)^2, \quad y = (x + 3)^2$$

What can you conclude about the effect on the graph of $y = x^2$ of adding a constant to x before squaring?

2. Graph the following on the same set of coordinate axes:

$$y = x^3, \quad y = (x - 1)^3, \quad y = (x + 3)^3$$

What can you conclude about the effect on the graph of $y = x^3$ of adding a constant to x before cubing?

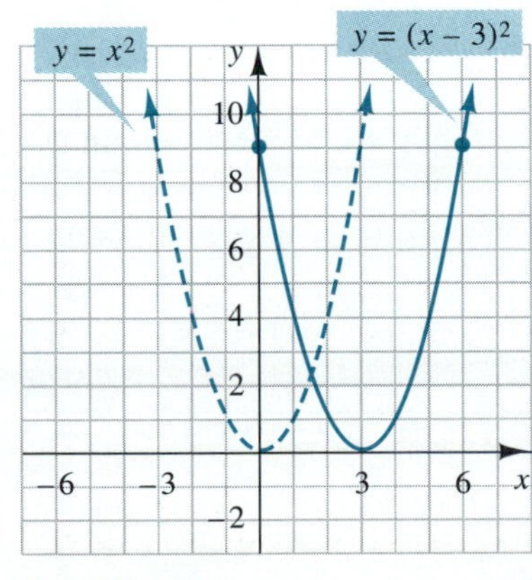

FIGURE 3.16

EXAMPLE 9 Sketch the graph of $y = f(x) = (x - 3)^2$.

Solution Using the same idea as in the previous example, we recognize that since the lowest point of $y = x^2$ occurs when $x = 0$, the lowest point of $y = (x - 3)^2$ occurs when $x - 3 = 0$, or $x = 3$. Thus the graph of $y = (x - 3)^2$ can be obtained by shifting the graph of $y = x^2$ horizontally 3 units to the right. The graph appears in Figure 3.16. ■

We can generalize the result of the last two examples to the following principle of graphing.

The Horizonal Shift Principle for Graphs

(for $c > 0$)

To Obtain the Graph of:	**Shift the Graph of $y = f(x)$:**
$y = f(x + c)$	c units to the left
$y = f(x - c)$	c units to the right

Be careful when applying this principle. If we have the graph of $y = f(x)$ and we want the graph of $y = f(x + 3)$, there is a natural tendency to expect to shift the graph 3 units to the right, which is *incorrect*. As our analysis has just shown us and as the horizontal shift principle describes, we should shift the graph 3 units *to the left*.

In general, is it true that $f(5) > f(3)$?

DIFFERENT PERSPECTIVES: *The Horizontal Shift Principle*

GEOMETRIC DESCRIPTION

To obtain the graph of $f(x + 3)$, shift the graph of $f(x)$ to the left 3 units.

To obtain the graph of $f(x - 2)$, shift the graph of $f(x)$ to the right 2 units.

ALGEBRAIC DESCRIPTION

Consider the value of $f(x)$ at $x = 1$, which is $f(1)$. If we want $f(x + 3)$ to have this same value then $x + 3 = 1 \Rightarrow x = -2$. Thus the value of $f(x)$ at $x = 1$ is equal to the value of $f(x + 3)$ at $x = -2$ (three units *to the left* of 1).

If we want $f(x - 2)$ to have the value $f(1)$ then $x - 2 = 1 \Rightarrow x = 3$. Thus the value of $f(x)$ at $x = 1$ is equal to the value of $f(x - 2)$ at $x = 3$ (two units *to the right* of 1).

Let's apply this graphing principle in the next example.

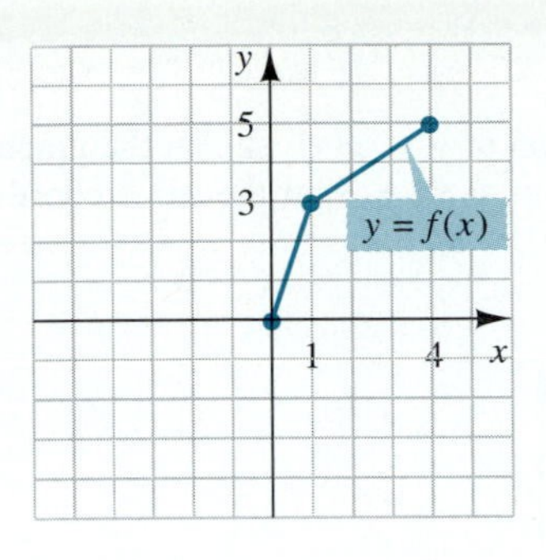

FIGURE 3.17

EXAMPLE 10 Given the graph of $y = f(x)$ in Figure 3.17, sketch each graph.

(a) $y = f(x + 4)$ **(b)** $y = f(x - 2)$

Solution According to the horizontal shift principle, the graph of $f(x + 4)$ is obtained by shifting the given graph of $y = f(x)$ four units to the left, and the graph of $f(x - 2)$ is obtained by shifting the given graph 2 units to the right. The graphs appear in Figure 3.18(a) and (b). Again, the dashed graph is that of the original $f(x)$. ■

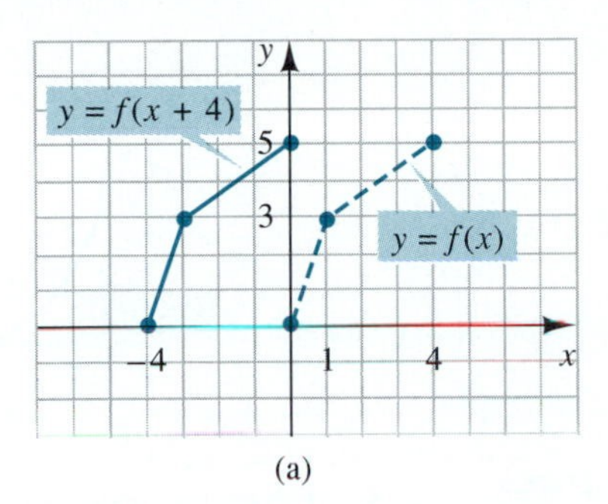

(a)

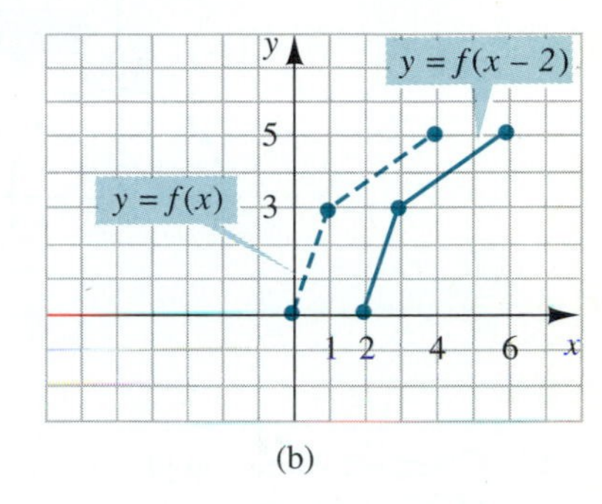

(b)

FIGURE 3.18 Using the horizontal shift principle

EXAMPLE 11 Sketch the graph of

(a) $g(x) = (x - 2)^2 + 4$ **(b)** $h(x) = |x + 2| - 3$

Solution

(a) The key idea here is recognizing that we can obtain this graph by shifting the basic parabola horizontally and vertically. If we let $f(x) = x^2$, then $f(x - 2) + 4 = (x - 2)^2 + 4 = g(x)$. Hence the graph of $g(x)$ is obtained by taking the basic parabola $f(x) = x^2$ and shifting it 2 units to the right and 4 units up. See Figure 3.19.

(b) The key to this problem is to recognize that we can obtain this graph by shift ing the graph of the absolute value function horizontally and vertically. If we let $f(x) = |x|$,then $f(x + 2) - 3 = |x + 2| - 3 = h(x)$. Hence $h(x)$ is obtained by taking the absolute value function and shifting it 2 units to the left and 3 units down. The graph appears in Figure 3.20.

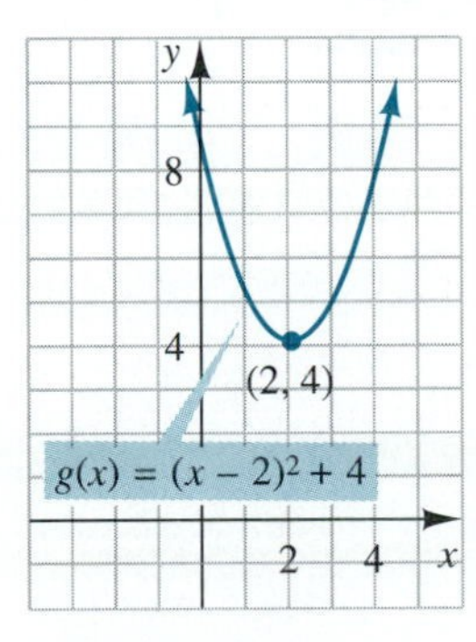

FIGURE 3.19

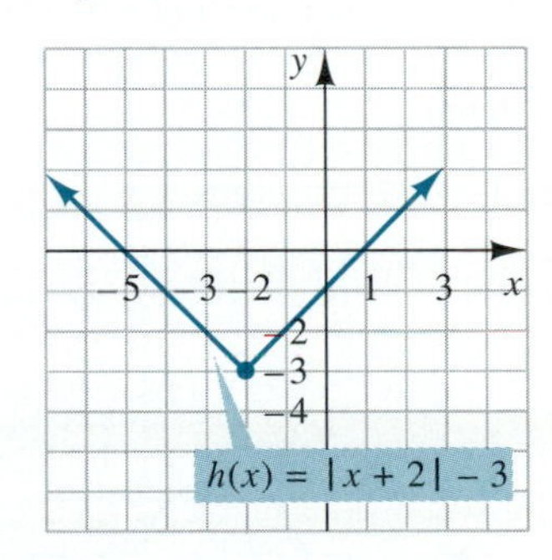

FIGURE 3.20 ■

In the next section, we continue the development of graphing principles.

EXERCISES 3.1

1. Given the following graph of $y = f(x)$, sketch the graphs of $y = f(x) - 3$ and $y = f(x - 3)$ on the same coordinate system.

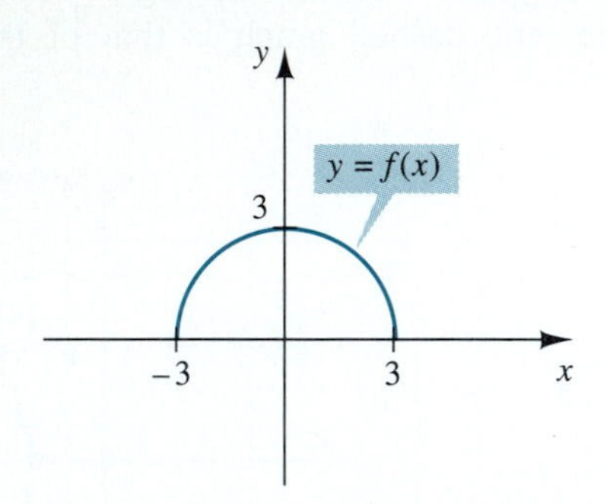

2. Given the following graph of $y = G(x)$, sketch the graphs of $y = G(x) + 2$ and $y = G(x + 2)$ on the same coordinate system.

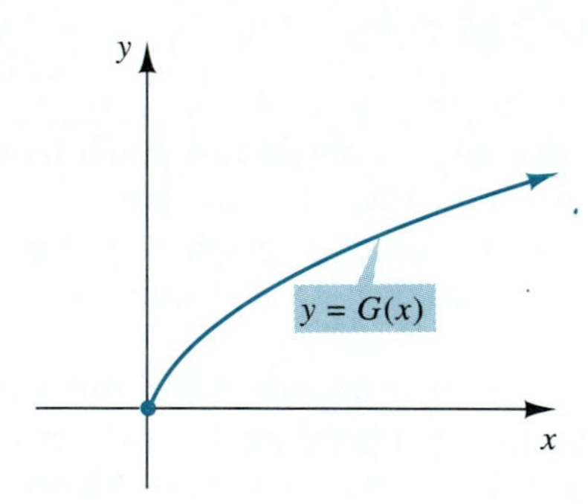

3. Given the following graph of $y = h(x)$, sketch the graphs of $y = h(x) - 1$ and $y = h(x) + 1$ on the same coordinate system.

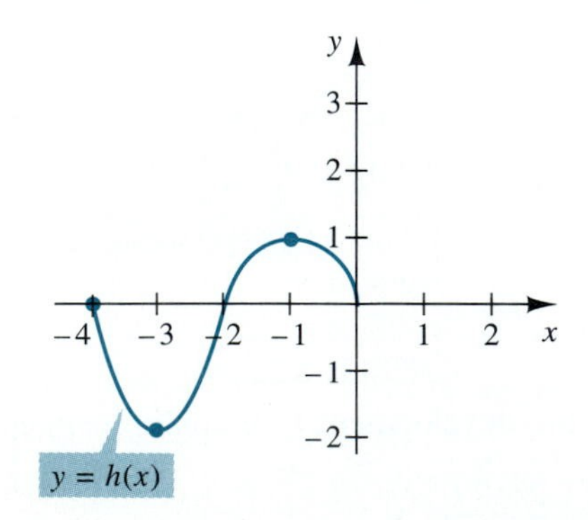

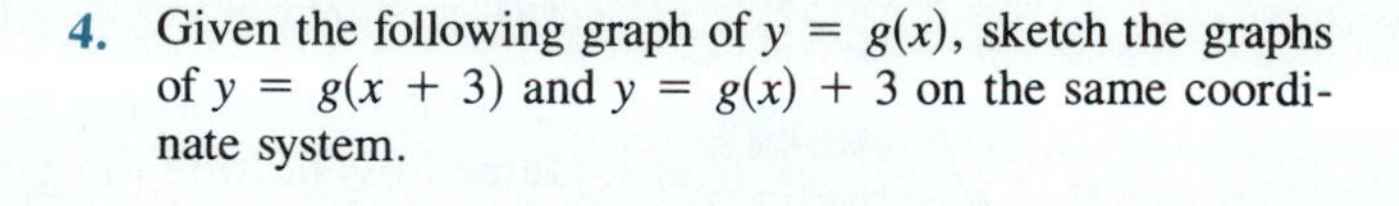

4. Given the following graph of $y = g(x)$, sketch the graphs of $y = g(x + 3)$ and $y = g(x) + 3$ on the same coordinate system.

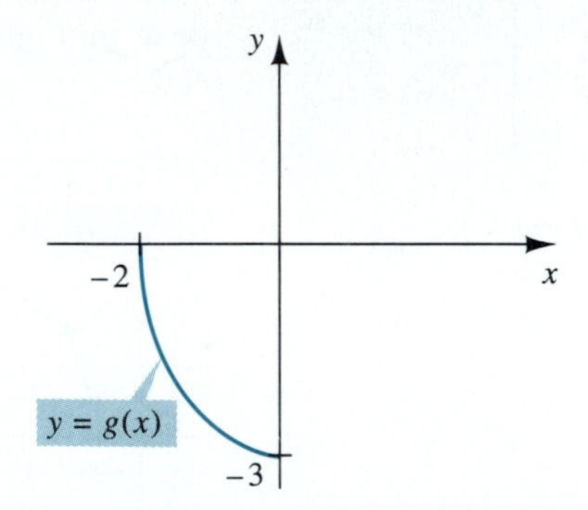

5. Given the following graph of $y = f(x)$, sketch the graph of $y = f(x - 1) + 2$.

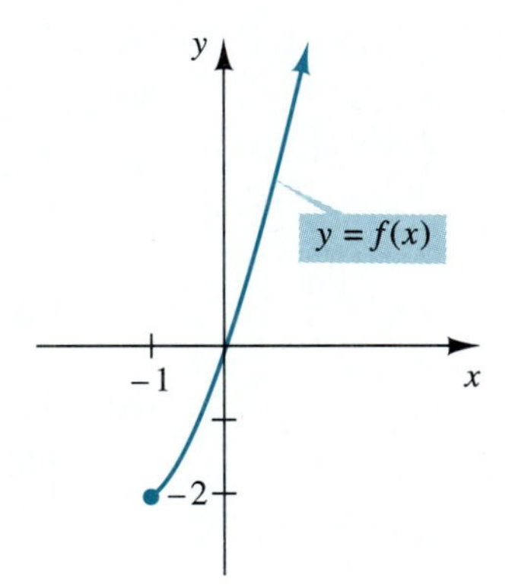

6. Given the following graph of $y = g(x)$, sketch the graph of $y = g(x + 1) - 2$.

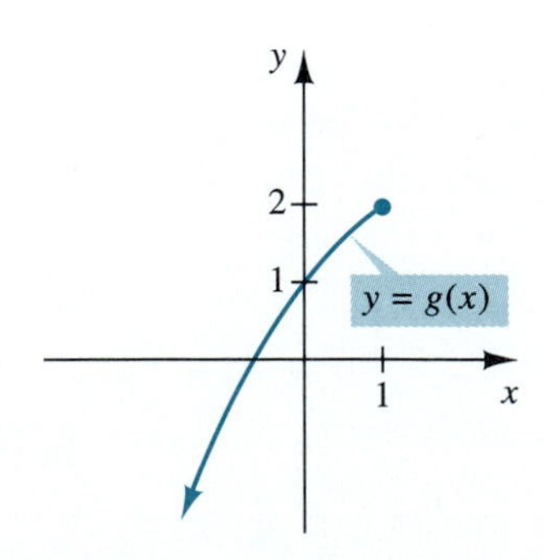

7. Given the following graph of $y = h(x)$, sketch the graph of $y = h(x + 2) + 3$.

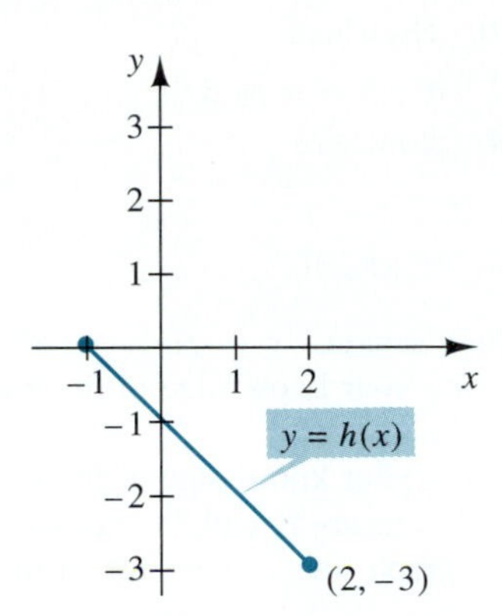

8. Given the following graph of $y = f(x)$, sketch the graph of $y = f(x - 2) - 3$.

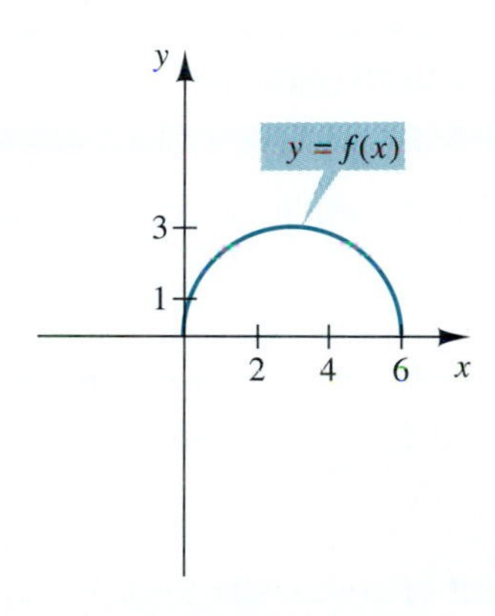

Use the following graph to answer Exercises 9–12.

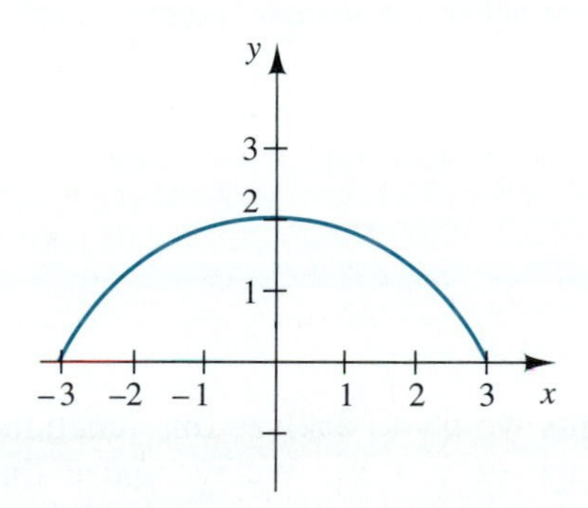

9. If the graph in the figure is the graph of $y = f(x) + 2$, sketch the graph of $y = f(x)$.

10. If the graph in the figure is the graph of $y = f(x + 3)$, sketch the graph of $y = f(x)$.

11. If the graph in the figure is the graph of $y = f(x - 2)$, sketch the graph of $y = f(x)$.

12. If the graph in the figure is the graph of $y = f(x) - 3$, sketch the graph of $y = f(x)$.

In Exercises 13–36, sketch the graph of the given function. Label the intercepts.

13. $y = f(x) = x^2 + 2$

14. $y = f(x) = (x + 2)^2$

15. $y = g(x) = (x - 1)^2$

16. $y = g(x) = x^2 - 1$

17. $y = h(x) = |2x - 5|$

18. $y = h(x) = |2x| - 5$

19. $y = f(x) = |x + 4|$

20. $y = f(x) = |x| + 4$

21. $y = F(x) = (x + 1)^2 - 4$

22. $y = G(x) = (x - 1)^2 + 4$

23. $y = G(x) = (x - 2)^2 - 3$

24. $y = H(x) = (x + 2)^2 + 3$

25. $y = g(x) = |x| - 3$

26. $y = g(x) = |x - 3|$

27. $y = h(x) = |2x|$

28. $y = h(x) = 2|x|$

29. $y = F(x) = |x - 1| + 2$

30. $y = G(x) = |x + 2| - 1$

31. $y = H(x) = |-x + 2|$

32. $y = F(x) = |-x| + 2$

33. $y = f(x) = |x^2 - 4|$

34. $y = g(x) = |x^2 - 6|$

35. $y = F(x) = |(x + 2)^2 - 1|$

36. $y = G(x) = |(x - 2)^2 - 9|$

In Exercises 39–52, sketch the graph of the given function and find its domain and range.

37. $f(x) = \begin{cases} 3x - 6 & \text{if } x \le 1 \\ 1 - x & \text{if } x > 1 \end{cases}$

38. $g(x) = \begin{cases} 2x + 5 & \text{if } x < -3 \\ x & \text{if } x \ge -3 \end{cases}$

39. $f(x) = \begin{cases} x - 4 & \text{if } -2 \le x < 2 \\ 4 & \text{if } 2 \le x < 5 \end{cases}$

40. $g(x) = \begin{cases} -3 & \text{if } 0 \le x < 4 \\ -2x & \text{if } 4 < x \le 6 \end{cases}$

41. $f(x) = \begin{cases} x + 6 & \text{if } x < -6 \\ x & \text{if } -6 \le x < 6 \\ 6 - x & \text{if } x > 6 \end{cases}$

42. $f(x) = \begin{cases} 2 & \text{if } x < -3 \\ |x| & \text{if } -2 \le x < 2 \\ 2 & \text{if } x \ge 2 \end{cases}$

43. $f(x) = \begin{cases} -4 & \text{if } x < -4 \\ |x| & \text{if } -4 \le x \le 4 \\ 4 & \text{if } x > 4 \end{cases}$

44. $f(x) = \begin{cases} 2 & \text{if } x < -2 \\ -x & \text{if } -2 \le x \le 2 \\ -2 & \text{if } x > 2 \end{cases}$

45. $f(x) = \dfrac{x^2 - 4}{x + 2}$

46. $g(x) = \dfrac{x^2 + 3x - 18}{x - 3}$

47. $h(x) = \dfrac{2x^2 + 7x - 15}{x + 5}$

48. $f(x) = \dfrac{3x^2 - 10x + 8}{2 - x}$

49. $F(x) = \begin{cases} 0 & \text{if } x < 0 \\ \dfrac{7}{30} & \text{if } 0 \le x < 1 \\ \dfrac{9}{15} & \text{if } 1 \le x < 2 \\ 1 & \text{if } x \ge 2 \end{cases}$

50. $G(x) = \begin{cases} 0 & \text{if } x < 0 \\ \dfrac{4}{7} & \text{if } 0 \le x < 3 \\ \dfrac{1}{14} & \text{if } 3 \le x < 5 \\ 1 & \text{if } x \ge 5 \end{cases}$

51. $G(x) = \begin{cases} \dfrac{1}{5} & \text{if } 1 < x < 5 \\ 0 & \text{elsewhere} \end{cases}$

52. $F(x) = \begin{cases} 2 & \text{if } -3 \le x \le 3 \\ 4 & \text{elsewhere} \end{cases}$

QUESTIONS FOR THOUGHT

53. Analyze the function $y = f(x) = -x^2$ and obtain its graph by using your knowledge of the graph of $y = x^2$.

54. Analyze the function $y = f(x) = -|x|$ and obtain its graph by using your knowledge of the graph of $y = |x|$.

55. Based upon Exercises 53 and 54, can you describe how to obtain the graph of $y = -f(x)$ from the graph of $y = f(x)$?

56. Describe how you would obtain the graph of $y = (x - 3)^2 + 5$ from the graph of $y = x^2$.

3.2 More Graphing Principles: Types of Functions

By building on the ideas introduced in the last section, we can complete our repertoire of graphing principles.

GRAFFIX

Use a graphics calculator or computer.

1. Graph the following on the same set of coordinate axes:

$$y = x^2 + 2 \qquad y = -(x^2 + 2)$$

2. Graph the following on the same set of coordinate axes:

$$y = x^3 - 1 \qquad y = -(x^3 - 1)$$

3. What would you conjecture is the relationship between the graph of $f(x)$ and the graph of $-f(x)$?

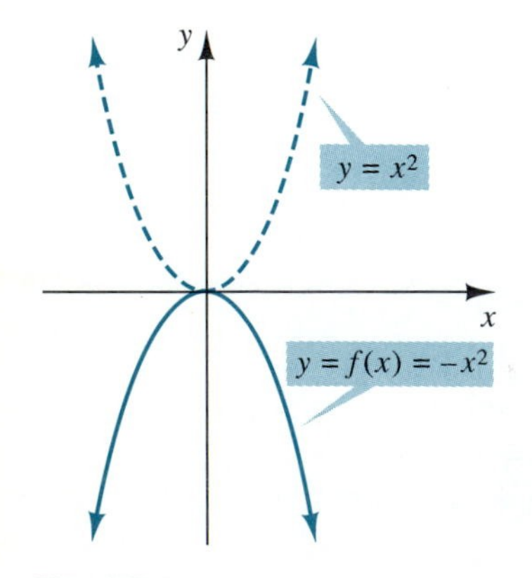

FIGURE 3.21

EXAMPLE 1 Sketch the graph of $y = f(x) = -x^2$.

Solution Again, rather than plot points we try to analyze this function in light of what we already know about the graph of $y = x^2$. We recognize that for each x-value, the y-value obtained from $y = -x^2$ will be the negative of the y-value obtained from $y = x^2$. All the positive y-values on the graph of $y = x^2$ become negative. Therefore, the graph of $y = -x^2$ can be obtained by turning the graph of $y = x^2$ *upside down*. The graph appears in Figure 3.21. ■

EXAMPLE 2 Sketch the graphs of $y = f(x) = 2x + 6$ and $y = g(x) = -(2x + 6)$ on the same coordinate system. How would you describe the relationships between $f(x)$ and $g(x)$ and their graphs?

Solution We recognize that the graphs of $f(x)$ and $g(x)$ are straight lines. (Why?) We find the intercepts for each line; the graphs appear in Figure 3.22.

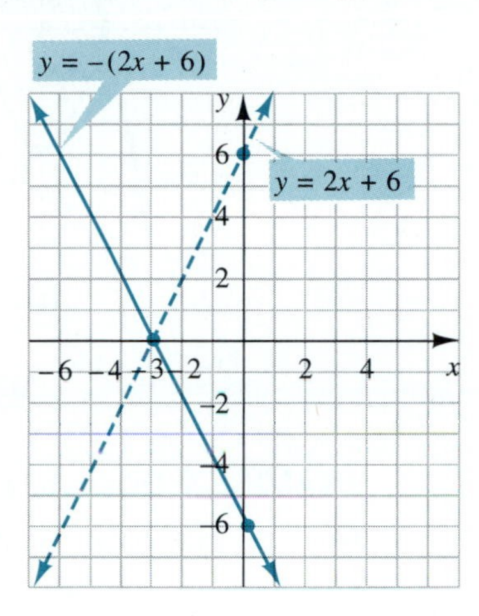

FIGURE 3.22

We note that $y = f(x) = 2x + 6$ and $y = g(x) = -(2x + 6)$ are negatives of each other. This means that the y-values on the graph of $g(x)$ are the opposites of the y-values on the graph of $f(x)$. In other words, wherever the graph of $f(x)$ is above the x-axis, the graph of $g(x)$ is below the x-axis, and vice versa. The graph of $g(x)$ is just the graph of $f(x)$ turned upside down. We say that the graph of $g(x)$ is obtained by *reflecting the graph of $f(x)$ about the x-axis.* ■

We can generalize the results of the last two examples into the following principle of graphing.

> ***Principle for Graphing $y = -f(x)$***
>
> To obtain the graph of $y = -f(x)$ reflect the graph of $y = f(x)$ about the x-axis.

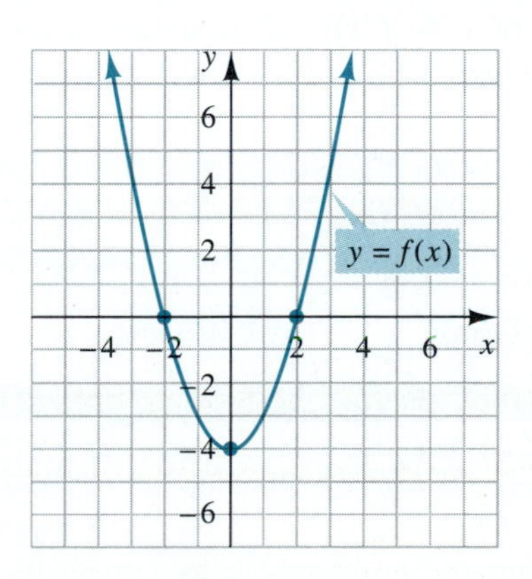

FIGURE 3.23

EXAMPLE 3 Given the graph of $y = f(x)$ in Figure 3.23, sketch each graph.
(a) $y = -f(x)$ **(b)** $y = |f(x)|$

Solution The graphing principles tell us that to get the graph of $y = -f(x)$, we reflect the entire graph of $f(x)$ about the x-axis, whereas to get the graph of $y = |f(x)|$ we take only that portion of the graph that is below the x-axis and reflect it above the x-axis. The graphs appear in Figure 3.24(a) and (b). The graph of the original $y = f(x)$ appears as a dashed curve. Look carefully at the answers to make sure that you see how each graph relates back to the original graph.

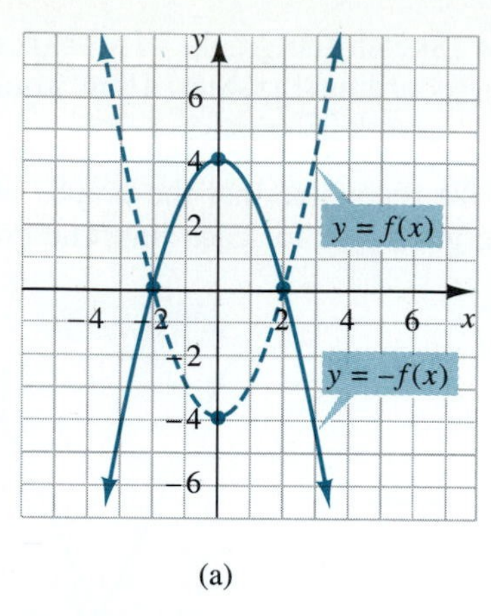

(a)

(b)

FIGURE 3.24

EXAMPLE 4 Given the graph of $y = f(x)$ in Figure 3.25, sketch each graph.
(a) $y = f(x + 2)$ **(b)** $y = f(x) + 2$

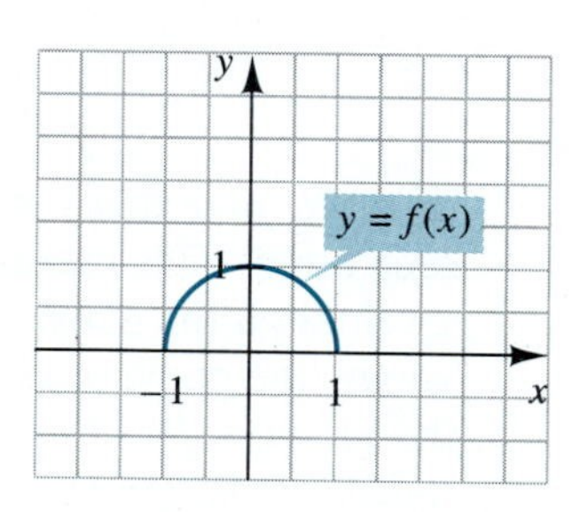

FIGURE 3.25

Solution Let's not forget our earlier graphing principles. Based upon the principles for vertical and horizontal shifts, the graph of $y = f(x + 2)$ is obtained by shifting the graph of $f(x)$ to the *left* 2 units, and the graph of $y = f(x) + 2$ is obtained by shifting the graph of $f(x)$ *up* 2 units. The graphs appear in Figure 3.26(a) and (b).

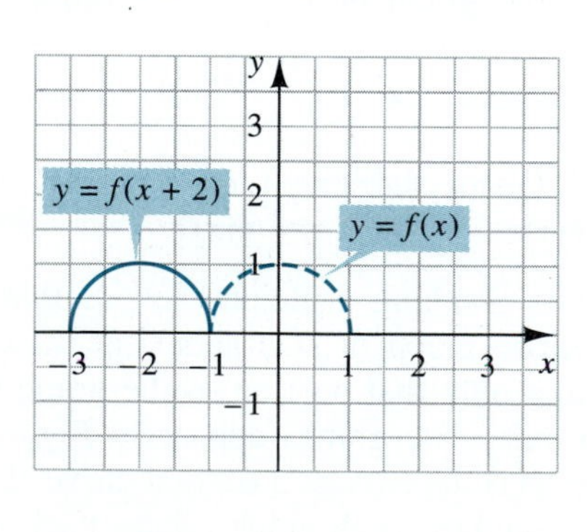

(a)

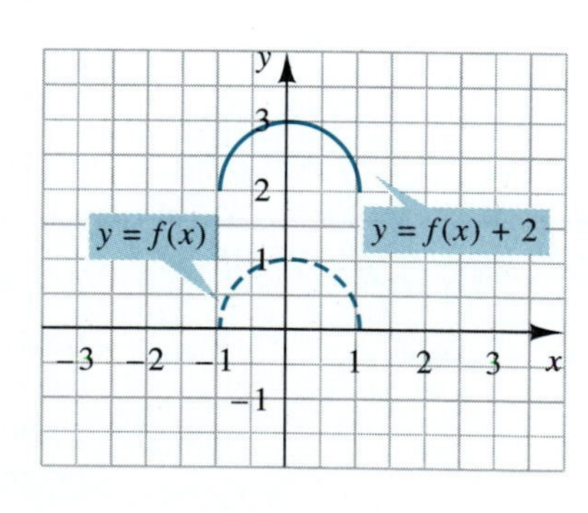

(b)

FIGURE 3.26

EXAMPLE 5 Sketch the graph of $y = 2 - |x|$.

Solution We can rewrite the equation as $y = -|x| + 2$ and employ our graphing principles to the graph of $y = |x|$. We start with the graph of $y = |x|$ (Figure 3.27(a)). The negative sign tells us to reflect the graph of $y = |x|$ about the x-axis (Figure 3.27(b)) and the $+2$ tells us then to shift the graph up two units (Figure 3.27(c)).

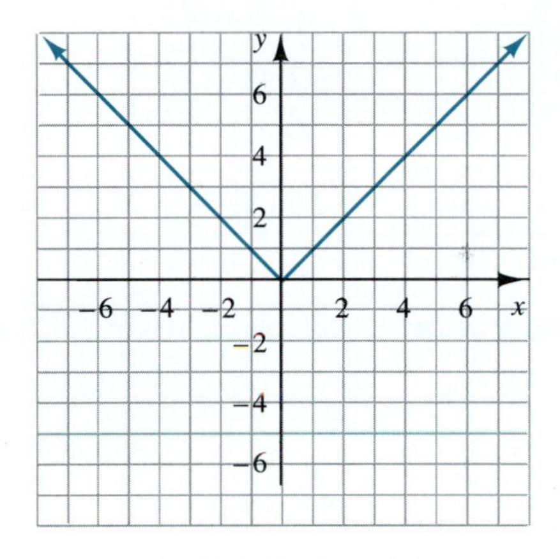

(a) The graph of $y = |x|$

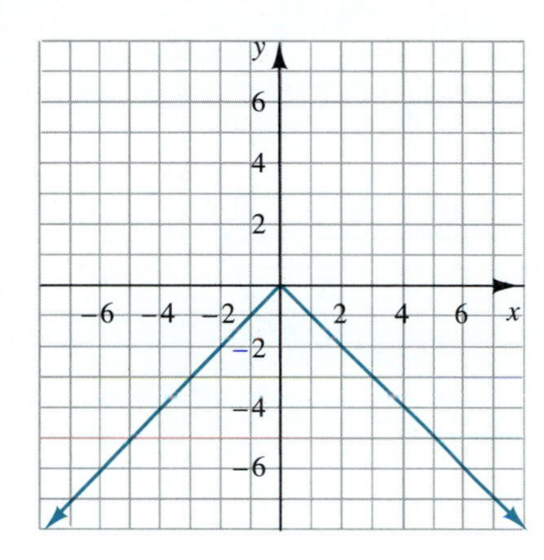

(b) The graph of $y = -|x|$

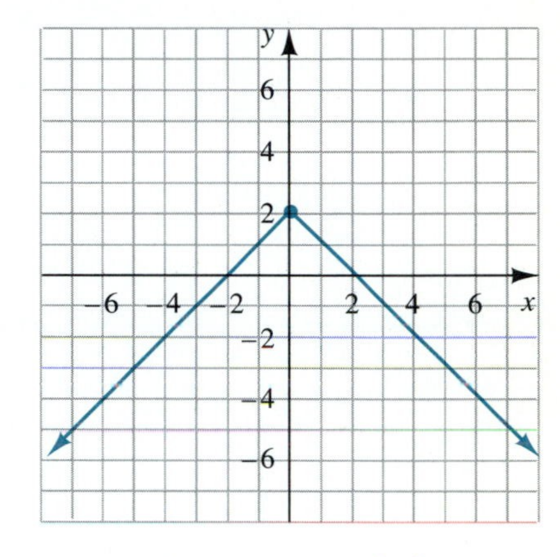

(c) The graph of $y = -|x| + 2$

FIGURE 3.27 The graph of $y = 2 - |x|$

The x-intercepts are found by setting $y = 0$.

$$0 = 2 - |x| \Rightarrow |x| = 2 \Rightarrow x = \pm 2$$

■

In Section 2.7 we saw that a function for which $f(-x) = f(x)$ is symmetric with respect to the y-axis, meaning that the portion of the graph to the left of the y-axis is obtained by reflecting the portion of the graph to the right of the y-axis about the y-axis (or vice versa).

We can generalize this further. Suppose we know what the graph of $y = f(x)$ looks like and we want to sketch the graph of $y = f(-x)$. If the graph has y-axis symmetry, then the graph of $f(-x)$ will be identical to the graph of $f(x)$, but what if the graph does not have y-axis symmetry?

GRAFFIX

Use a graphics calculator or computer.

1. Graph the following on the same set of coordinate axes:

$$f(x) = x^3 + 1 \qquad f(-x) = -x^3 + 1$$

2. Graph the following on the same set of coordinate axes:

$$f(x) = x^5 + 3x^2 \qquad f(-x) = -x^5 + 3x^2$$

3. What can you conclude about the relationship between the graph of $f(x)$ and the graph of $f(-x)$?

Let's suppose $x \neq 0$ so that x and $-x$ are opposites: If x is to the right of the y-axis, then $-x$ is to the left of the y-axis, and vice versa. By replacing x with $-x$ in $f(x)$, we end up on the other side of the y-axis *but at the same height as* $f(x)$. Figure 3.28(a) and (b) illustrate the graph of a function $y = f(x)$ and the graph of $y = f(-x)$.

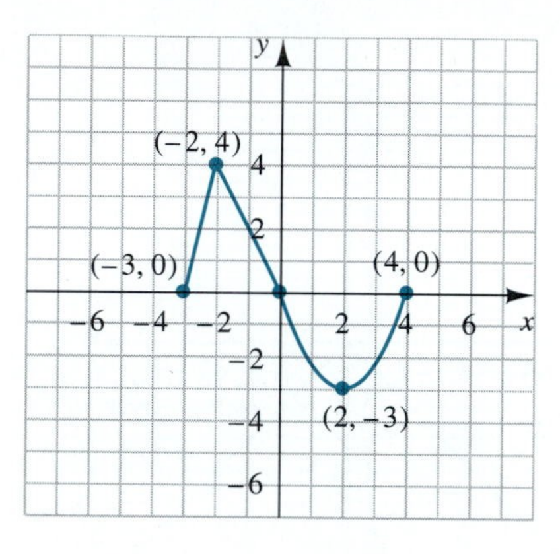

If this is the graph of $y = f(x)$

(a)

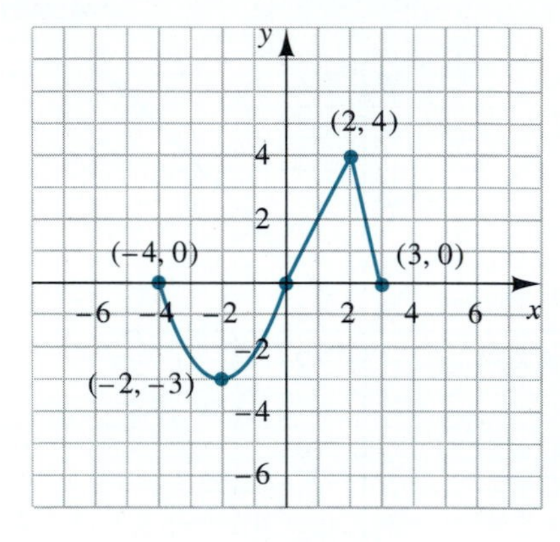

then this is the graph of $y = f(-x)$

(b)

FIGURE 3.28

Pay particular attention to the points that have been labeled. Note that since the point $(2, -3)$ is on the graph of $y = f(x)$, the point $(-2, -3)$ is on the graph of $y = f(-x)$. In other words, if $f(2) = -3$ for $y = f(x)$, then for $y = f(-x)$ and $x = -2$ we have $f(-x) = f(-(-2)) = f(2) = -3$. The same is true for the other points on the graph.

Verbally, we can describe this by saying that what happens on the graph of $y = f(x)$ on the right happens on the graph of $y = f(-x)$ on the left, and vice versa. More precisely, the graph of $y = f(-x)$ is obtained by reflecting the graph of $y = f(x)$ about the y-axis.

Based upon this discussion we can formulate the following graphing principle.

Principle for Graphing $y = f(-x)$

To obtain the graph of $y = f(-x)$, reflect the graph of $y = f(x)$ about the y-axis.

Keep in mind that in order to use this graphing principle, we must know what the graph of $y = f(x)$ looks like.

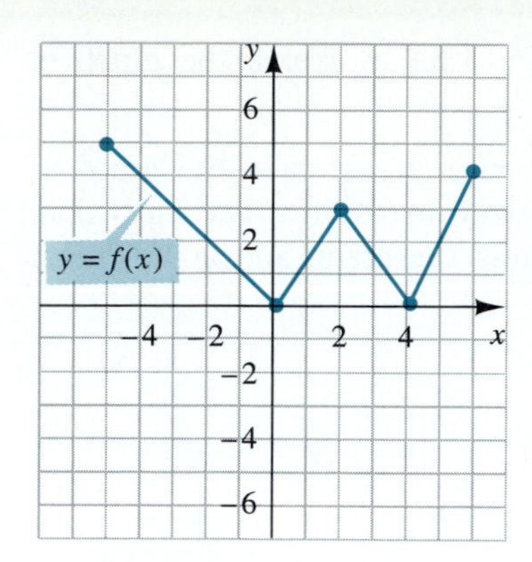

FIGURE 3.29

If the graph of $f(x)$ has y-axis symmetry, what is the relationship between the graphs of $f(x)$ and $f(-x)$?

EXAMPLE 6 Given the graph of $y = f(x)$ in Figure 3.29, sketch the graph of $y = f(-x)$.

Solution Using the graphing principle for $y = f(-x)$, we reflect the given graph about the y-axis to obtain the graph in Figure 3.30.

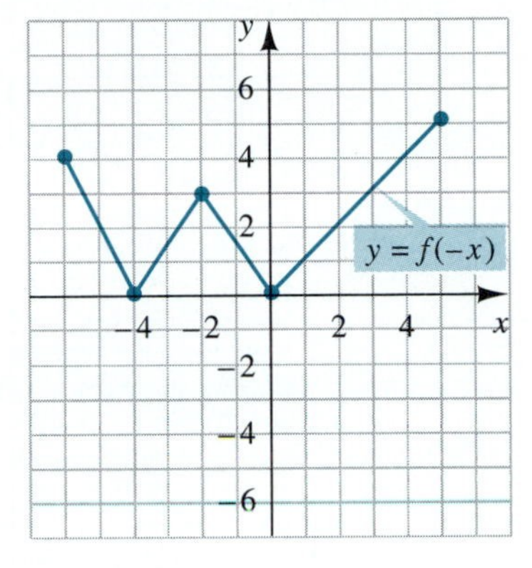

FIGURE 3.30

■

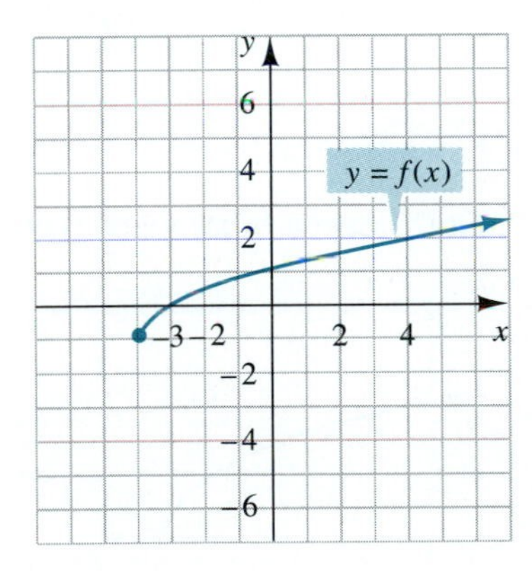

FIGURE 3.31

EXAMPLE 7 Given the graph of $y = f(x)$ in Figure 3.31, sketch each graph.

(a) $y = -f(x)$ **(b)** $y = f(-x)$

Solution It is very important to apply our graphing principles carefully. To obtain the graph of $y = -f(x)$ we reflect the original graph about the x-axis, and to obtain the graph of $y = f(-x)$, we reflect the original graph about the y-axis. The graphs appear in Figure 3.32(a) and (b).

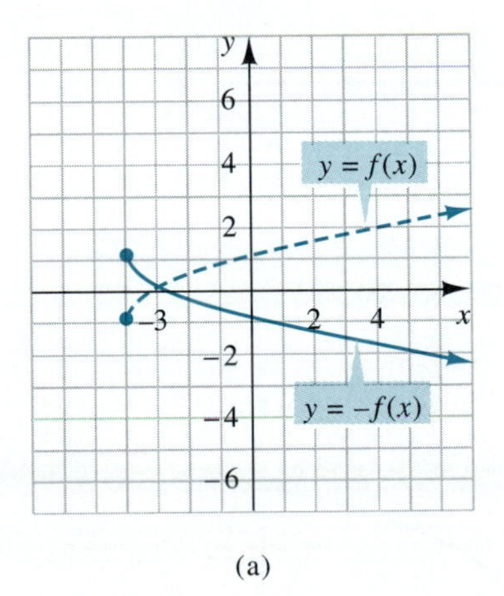

(a)

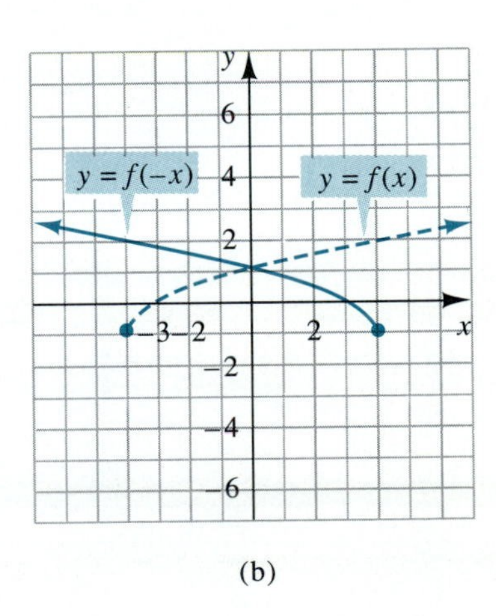

(b)

FIGURE 3.32

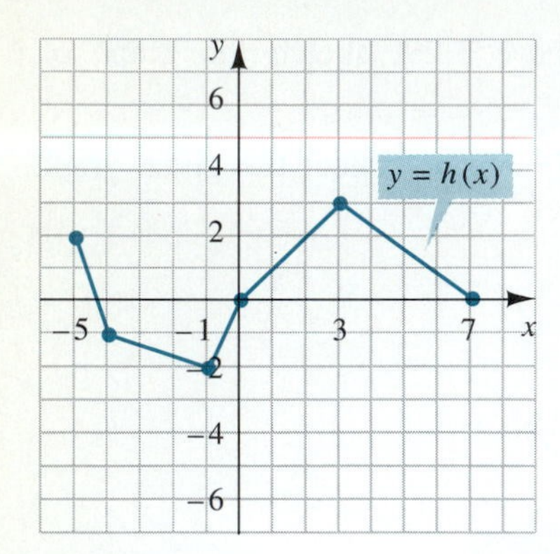

FIGURE 3.33

EXAMPLE 8 Given the graph of $y = h(x)$ in Figure 3.33 to sketch the graph of $y = h(-x) + 2$.

Solution In order to obtain the graph of $y = h(-x) + 2$, we take the graph of $y = h(x)$, reflect it about the y-axis to obtain $h(-x)$, and then shift it vertically up 2 units to obtain $h(-x) + 2$. The two steps in obtaining the graph appear in Figure 3.34.

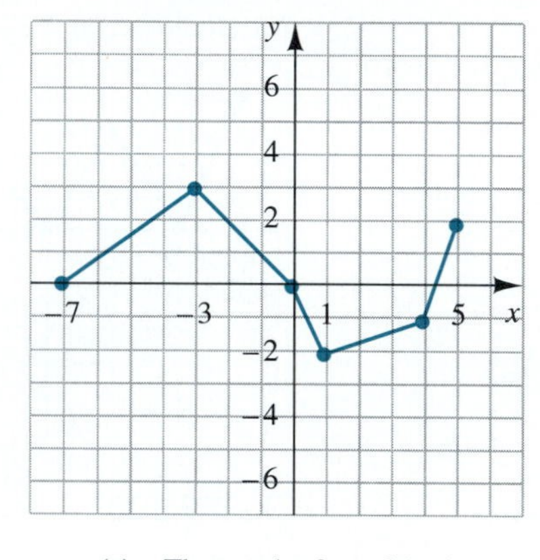

(a) The graph of $y = h(-x)$

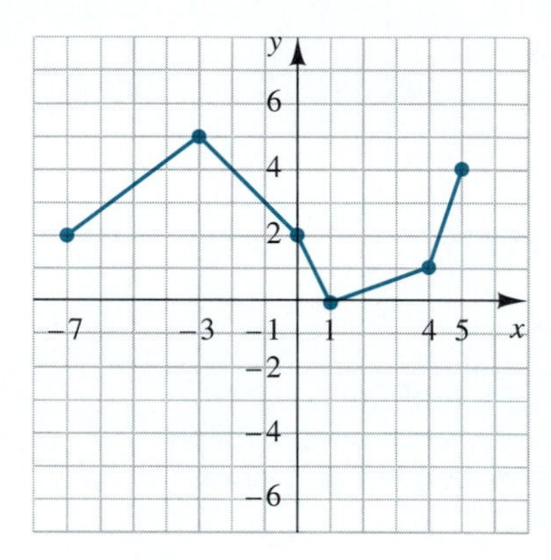

(b) The graph of $y = h(-x) + 2$

FIGURE 3.34

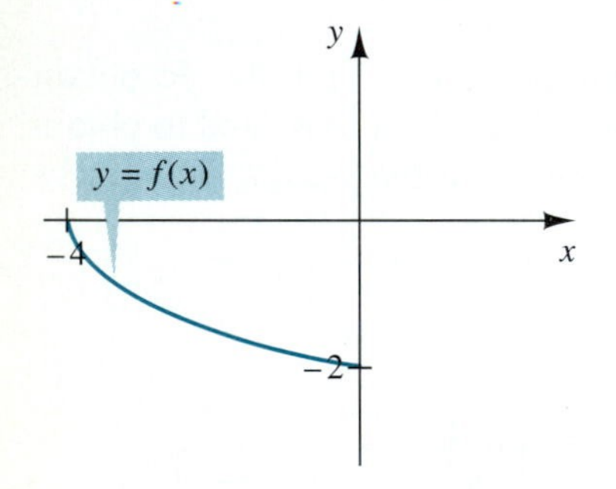

FIGURE 3.35

EXAMPLE 9 Figure 3.35 at the left illustrates the graph of $y = f(x)$.

Describe the graphs in Figure 3.36(a), (b), (c), and (d) below in terms of $f(x)$.

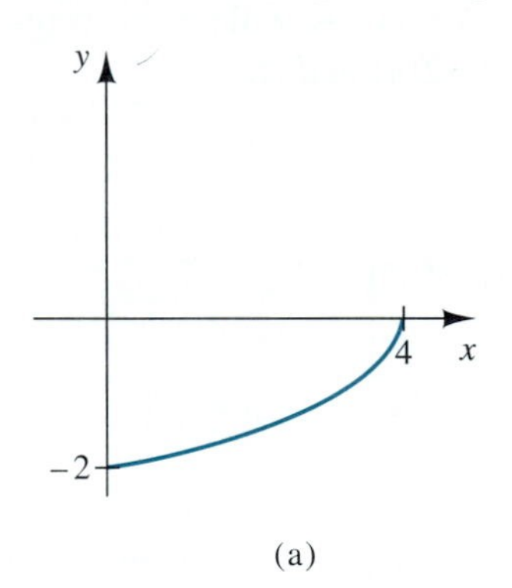

(a)

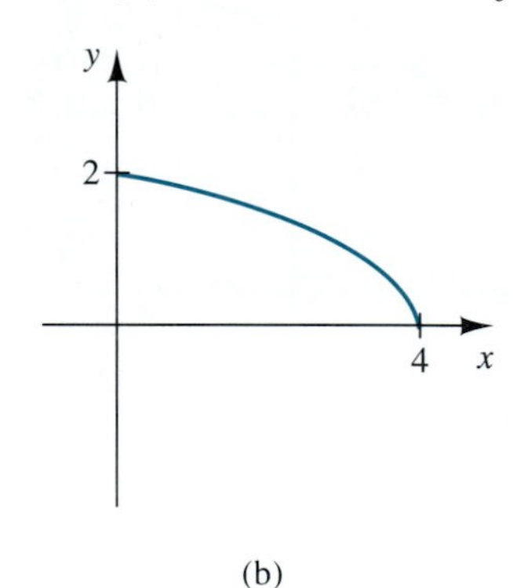

(b)

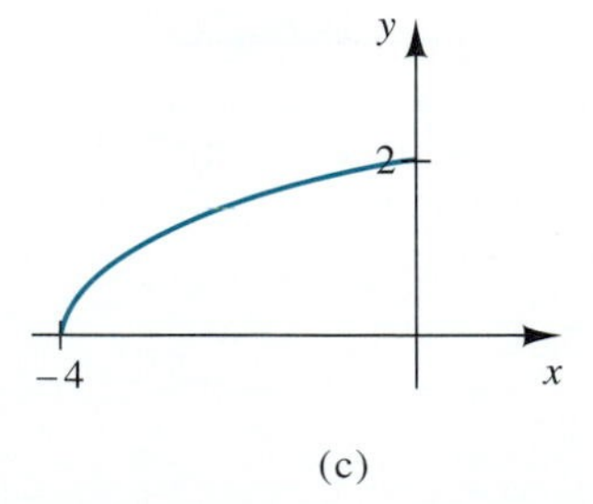

(c)

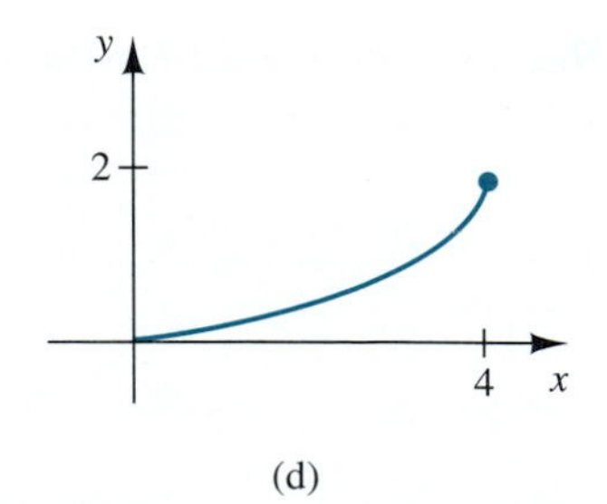

(d)

FIGURE 3.36

Solution

The Graph Given in	*Is Obtained by*	*So its Equation Is*
Figure 3.36(a)	Reflecting the graph of $f(x)$ about the y-axis	$y = f(-x)$
Figure 3.36(b)	Reflecting the graph of $f(-x)$ about the x-axis	$y = -f(-x)$
Figure 3.36(c)	Reflecting the graph of $f(x)$ about the x-axis	$y = -f(x)$
Figure 3.36(d)	Reflecting the graph of $f(x)$ about the y-axis and shifting it vertically up 2 units	$y = f(-x) + 2$

■

For ease of reference we summarize the various graphing principles here, with an illustration of each.

SUMMARY OF GRAPHING PRINCIPLES

The Stretching Principle

(for $a > 0$, $a \neq 1$)

The graph of $y = af(x)$ can be obtained by stretching the graph of $y = f(x)$. In other words, the graph of $y = af(x)$ will have the same basic shape as the graph of $y = f(x)$. If $a > 1$, the graph of $y = af(x)$ is obtained by deflecting the graph of $f(x)$ away from the x-axis. If $0 < a < 1$, the graph of $y = af(x)$ is obtained by pulling the graph of $f(x)$ toward the x-axis.

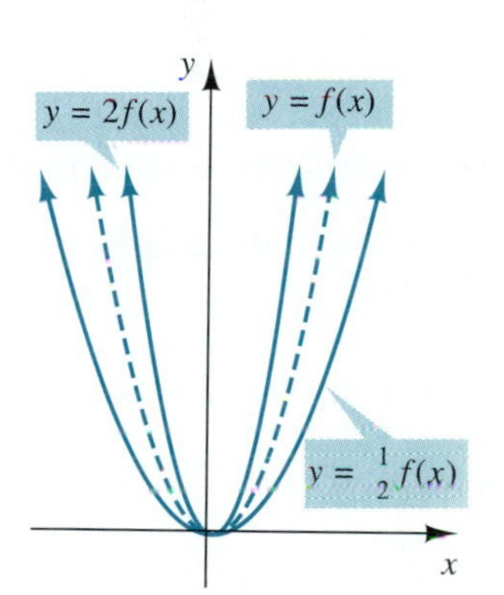

The Vertical Shift Principle for Graphs

(for $c > 0$)

To Obtain the Graph of:	*Shift the Graph of $y = f(x)$:*
$y = f(x) + c$	c units upward
$y = f(x) - c$	c units downward

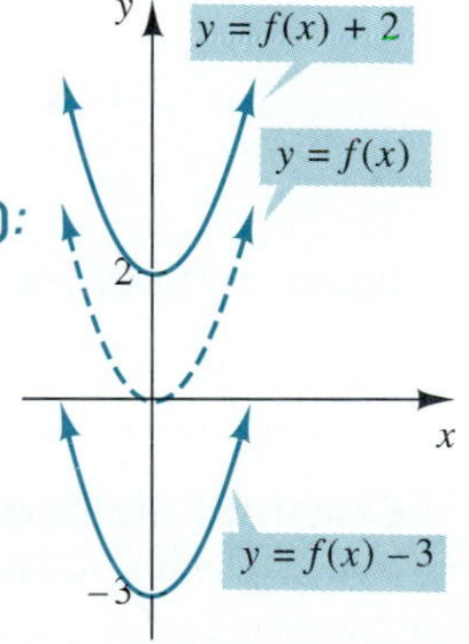

The Horizontal Shift Principle for Graphs

(for $c > 0$)

To Obtain the Graph of:	*Shift the Graph of $y = f(x)$:*
$y = f(x + c)$	c units to the left
$y = f(x - c)$	c units to the right

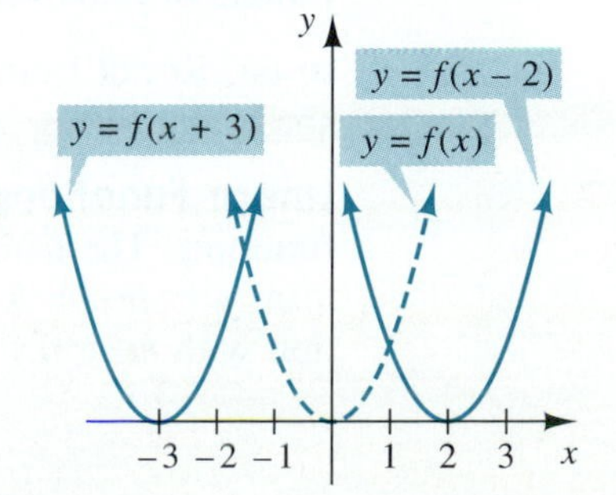

Principle for Graphing $y = |f(x)|$

To obtain the graph of $y = |f(x)|$, begin with the graph of the underlying function $y = f(x)$.

(a) Leave that portion of the underlying graph on or above the x-axis unchanged.

(b) Take that portion of the underlying graph which is below the x-axis and reflect it above the x-axis.

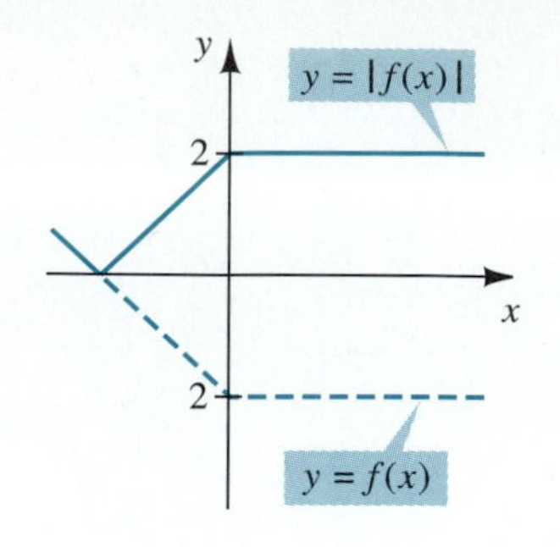

Principle for Graphing $y = -f(x)$

To obtain the graph of $y = -f(x)$, reflect the graph of $f(x)$ about the x-axis.

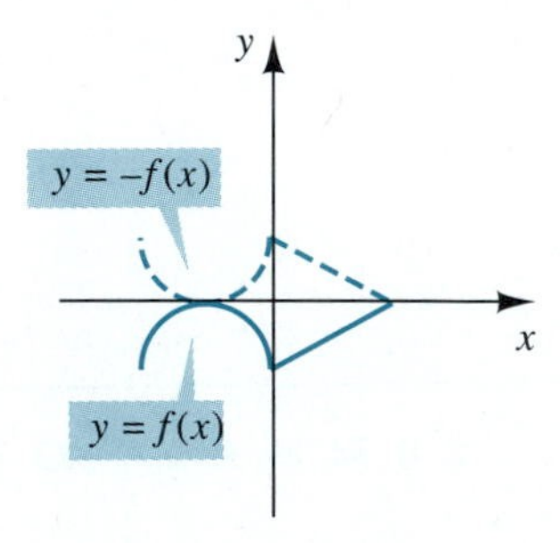

Principle for Graphing $y = f(-x)$

To obtain the graph of $y = f(-x)$, reflect the graph of the $y = f(x)$ about the y-axis.

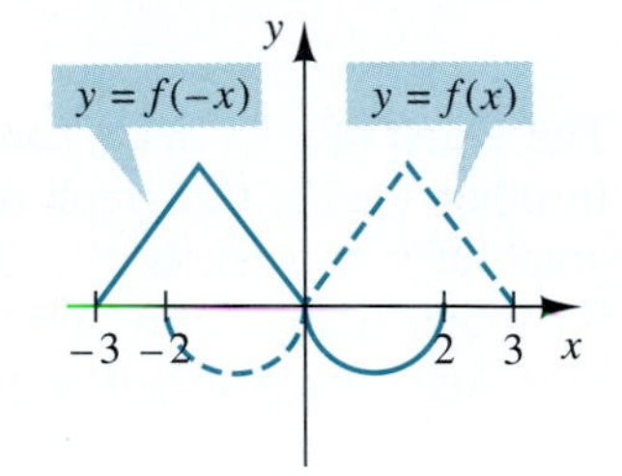

Types of Functions

As we continue to be exposed to a variety of functions and their graphs, it is useful to categorize them for ease of reference.

Constant Functions A function of the form $y = f(x) = K$, where K is a constant, is called a **constant function.** If $y = f(x) = 6$, then this function assigns a y-value of 6 to every input x. Thus $f(2) = 6$, $f(-9) = 6$, $f(0) = 6$, $f\left(\frac{1}{5}\right) = 6$, and so on. Recall from Section 2.2 that the graph of a function of the form $y = K$ is a horizontal line.

Linear Functions A function of the form $y = f(x) = mx + b$ is called a **linear function.** The name is obviously due to the fact that the graph of a function of the form $y = mx + b$ is a straight line. (Note that a constant function is a linear function with $m = 0$.)

Quadratic Functions A function of the form $y = f(x) = ax^2 + bx + c$ with $a \neq 0$ is called a **quadratic function.** (If $a = 0$, then the function becomes $y = f(x) = bx + c$, which is a linear function.) We discuss the graphs of quadratic functions in great detail in Section 3.4.

Polynomial Functions Constant, linear, and quadratic functions are all particular examples of a more general class of functions.

DEFINITION A **polynomial function** is a function of the form

$$y = p(x) = a_n x^n + a_{n-1}x^{n-1} + a_{n-2}x^{n-2} + \cdots + a_2x^2 + a_1x + a_0, \quad a_n \neq 0$$

Each a_i is assumed to be a real number, and n is a nonnegative integer. Such a polynomial function is said to be of degree n.

For example, $f(x) = 6x^3 - 5x + 1$ is an example of a polynomial function of degree 3.

Rational Functions A function that can be expressed as the quotient of two polynomials is called a **rational function**. $f(x) = \dfrac{x - 1}{x^2 + 1}$ and $g(x) = \dfrac{1}{x - 5}$ are examples of rational functions.

Note that every polynomial function, $p(x)$, is also a rational function, since it can be written as $\dfrac{p(x)}{1}$. We discuss polynomial and rational functions in Chapter 4.

Some other functions we have already discussed or will discuss are $f(x) = |x|$, the absolute value function; $f(x) = \sqrt{x} = x^{1/2}$, a radical function; $f(x) = 2^x$, an exponential function; and logarithmic and trigonometric functions.

We conclude this section with a different type of function. It is worthwhile remembering that *any* rule that assigns a single y-value to each x-value is a bona fide function. Let's consider the following function.

DEFINITION The **greatest integer function,** written $f(x) = [x]$, means the greatest integer less than or equal to x.

Thus, for example,

$[6.4] = 6$ Because 6 is the greatest integer less than or equal to 6.4

$[5] = 5$ Because 5 is the greatest integer less than or equal to 5

$[-2.8] = -3$ Because -3 is the greatest integer less than or equal to -2.8

In terms of the number line, the greatest integer function assigns to each x the closest integer at or to the left of x.

EXAMPLE 10 Sketch the graph of $y = f(x) = [x]$.

Solution Since the rule for the greatest integer function is a bit unusual, its graph will be somewhat different from what we have seen thus far.

For instance,

When $0 \le x < 1$, $y = [x]$ will be equal to 0.

When $1 \le x < 2$, $y = [x]$ will be equal to 1.

When $2 \le x < 3$, $y = [x]$ will be equal to 2.

Similarly,

When $-3 \le x < -2$, $y = [x]$ will be equal to -3.

When $-2 \le x < -1$, $y = [x]$ will be equal to -2.

When $-1 \le x < 0$, $y = [x]$ will be equal to -1.

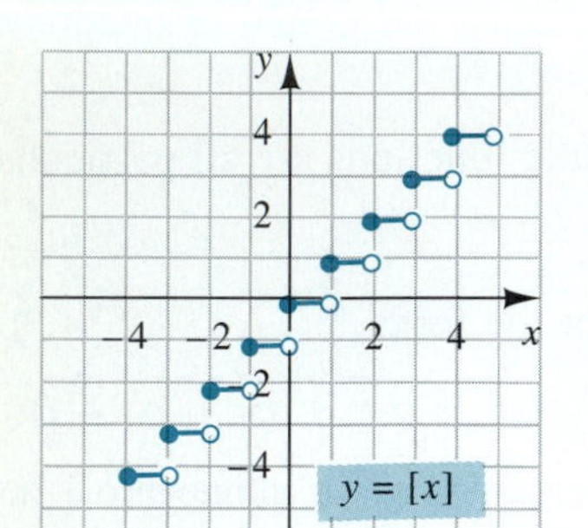

FIGURE 3.37

On each interval of the form $[n, n + 1)$, where n is an integer, $y = [x]$ will be constant and equal to n. Thus the graph consists of a series of horizontal line segments. A representative portion of the graph appears in Figure 3.37. This type of graph, which consists of a series of horizontal line segments, is called a **step function.** ■

In the next section we see how the idea of a function can be applied to describe real-life situations.

EXERCISES 3.2

1. Given the following graph of $y = f(x)$, sketch the graphs of $y = -f(x)$ and $y = f(-x)$.

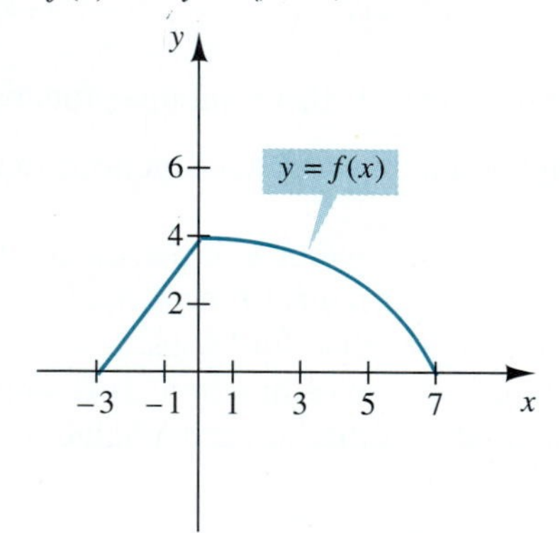

2. Given the following graph of $y = g(x)$, sketch the graphs of $y = |g(x)|$ and $y = -g(x)$.

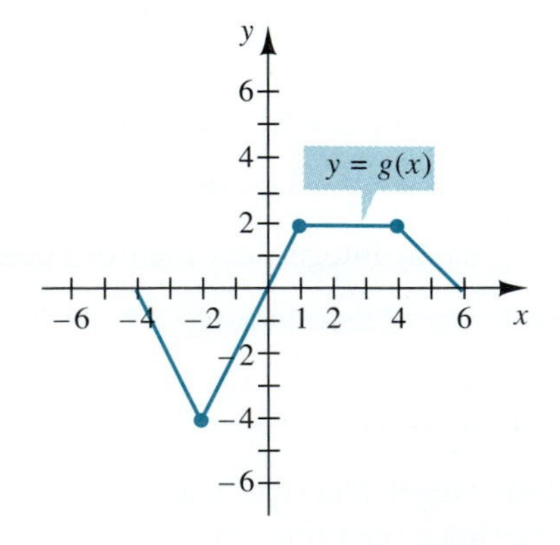

3. Given the following graph of $y = h(x)$, sketch the graphs of $y = h(x) - 2$ and $y = h(x - 2)$.

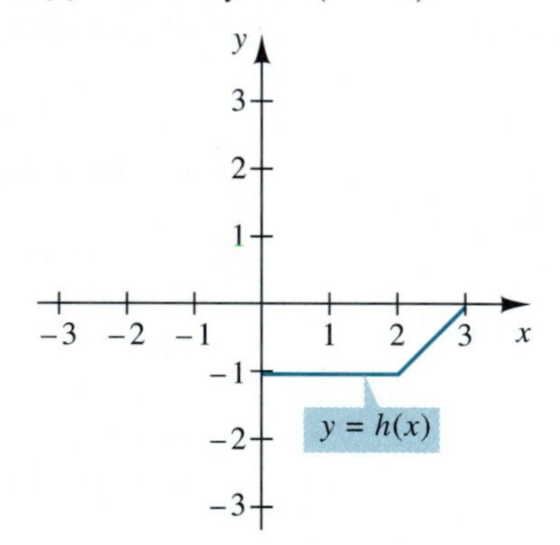

4. Given the following graph of $y = g(x)$, sketch the graphs of $y = g(x + 3)$ and $y = -g(x + 3)$.

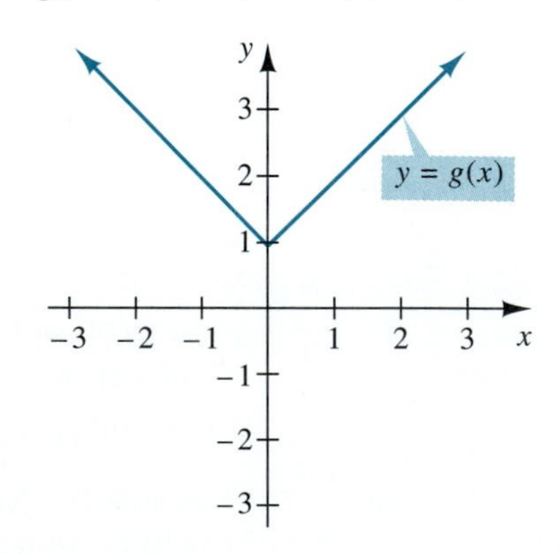

5. Given the following graph of $y = f(x)$, sketch the graphs of $y = f(x - 1)$ and $y = f(1 - x)$. HINT: To get the graph of $f(1 - x)$, reflect the graph of $f(x + 1)$ about the y-axis. Explanation: Replacing x by $-x$ in $f(x + 1)$ gives $f(-x + 1) = f(1 - x)$ and causes a reflection about the y-axis.

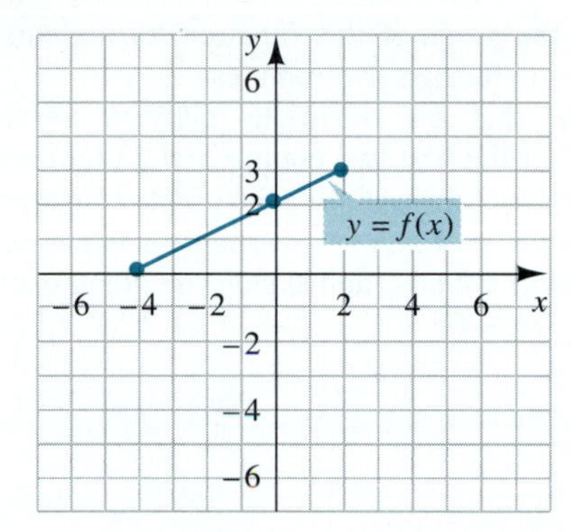

6. Given the following graph of $y = g(x)$, sketch the graphs of $y = g(x - 1)$ and $y = g(1 - x)$. See the hint in Exercise 5.

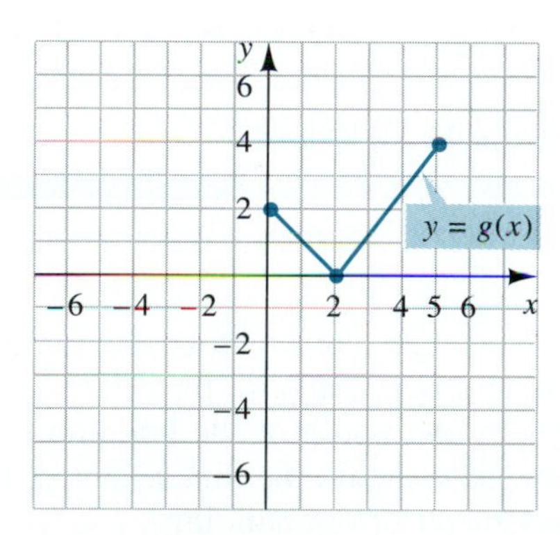

In Exercises 7–12, use the following graph of $y = f(x)$ to obtain the requested graphs.

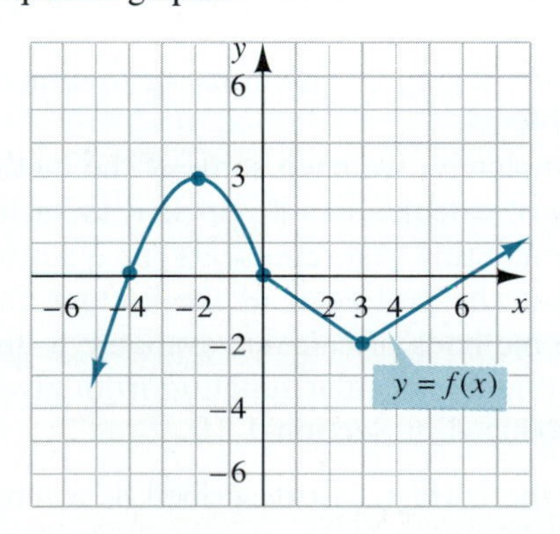

7. $y = f(-x)$ and $y = -f(x)$
8. $y = |f(x)|$ and $y = -f(-x)$
9. $y = |f(x)| + 2$ and $y = |f(x + 2)|$
10. $y = -f(x - 2)$ and $y = f(2 - x)$
11. $y = f(-x) - 3$ and $y = f(-x - 3)$
12. $y = |f(-x)|$ and $y = -f(-x)$

In Exercises 13–28, sketch the graph of the given function. Label the intercepts.

13. $y = f(x) = x^2 + 4$
14. $y = f(x) = (x + 4)^2$
15. $y = f(x) = -x^2 + 4$
16. $y = f(x) = -(x + 4)^2$
17. $y = g(x) = |x - 5|$
18. $y = g(x) = |x| - 5$
19. $y = g(x) = 3 - |x|$
20. $y = g(x) = |3 - x|$
21. $y = f(x) = (x + 1)^2 + 1$
22. $y = f(x) = -(x - 1)^2 - 1$
23. $y = h(x) = |x^2 - 9|$
24. $y = h(x) = ||x| - 3|$
25. $y = F(x) = -(x + 1)^2$
26. $y = G(x) = |16 - x^2|$
27. $f(x) = \begin{cases} |x| - 4 & \text{if } -4 \le x \le 4 \\ 16 - x^2 & \text{elsewhere} \end{cases}$
28. $f(x) = \begin{cases} 2 - |x| & \text{if } -2 \le x \le 2 \\ |x| - 2 & \text{elsewhere} \end{cases}$

Use the following figure in Exercises 29–32.

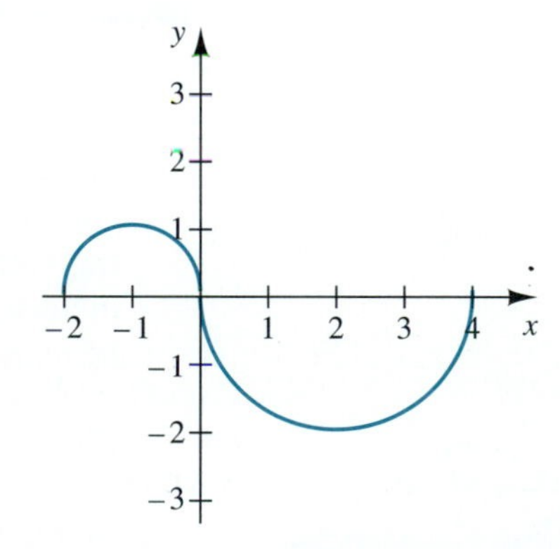

29. If the graph is the graph of $y = f(-x)$, sketch the graph of $y = f(x)$.
30. If the graph is the graph of $y = -f(x)$, sketch the graph of $y = f(x)$.
31. If the graph is the graph of $y = -f(x) + 1$, sketch the graph of $y = f(x)$.
32. If the graph is the graph of $y = f(-x + 1)$, sketch the graph of $y = f(x)$.

In Exercises 33–44, describe in words what you would do to the graph of $y = f(x)$ to obtain the graph of the given function.

33. $f(x) - 4$
34. $f(x - 4)$
35. $f(-x)$
36. $-f(x)$
37. $|f(x)|$
38. $-f(-x)$
39. $-f(x) + 3$
40. $f(-x) - 3$
41. $f(x + 3) + 4$
42. $f(x - 2) - 5$
43. $f(2 - x)$
44. $f(1 - x) + 1$

In Exercises 45–48, use $[x]$, the greatest integer function, to evaluate each expression.

45. $[6]$
46. $[-8]$
47. $[-2.9]$
48. $[6.7]$

In Exercises 49–52, sketch the graph of the given function.

49. $y = f(x) = [x] + 1$
50. $y = f(x) = [x + 1]$
51. $y = f(x) = [2x]$
52. $y = f(x) = 2[x]$

53. Suppose that the cost of a telephone call between Atlanta and Houston is \$0.80 for the first minute and \$0.50 for each additional minute or part thereof. We can use the greatest integer function to express the cost, C, of such a phone call, as a function of the length, m, of the phone call in minutes.

$$C = 0.30 - 0.50[-m] \qquad \text{for } m > 0$$

Sketch the graph of this function for a phone call which lasts no more than five minutes.

54. Suppose that an overnight delivery service charges \$12.50 for the first two pounds and \$3 for each additional pound or part thereof. Use the greatest integer function to express the cost, C, of overnight delivery of a package weighing p pounds, and sketch the graph of this function for $0 < p \leq 6$.

QUESTION FOR THOUGHT

55. Describe how you would graph $y = a(x - h)^2 + k$.

3.3 Extracting Functions from Real-Life Situations

Let us consider the following problem:

> Suppose that a person wants to build a single-story rectangular summer cottage containing 150 square meters of floor space. Suppose further that local building codes require that both the length and width of the building be at least 6 meters. In order to conserve energy, how should the dimensions of the cottage be chosen so as to minimize the perimeter of the building?

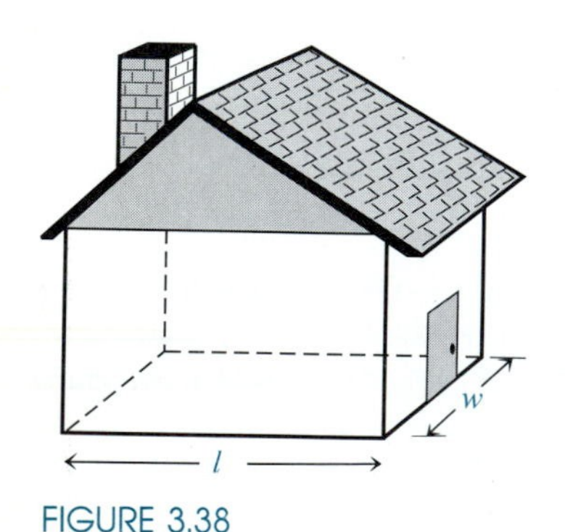

FIGURE 3.38

Figure 3.38 illustrates the cottage. We have labeled the length l and the width w.

Problems of this sort, where we are trying to maximize or minimize a certain quantity, are called **optimization problems.**

In order to solve this problem completely, we need some of the methods of calculus. However, before the techniques of calculus can be applied to such a real-life situation, it is first necessary to write a function that represents the quantity in which we are interested. (In this problem it is the perimeter of the cottage that is to be minimized). Since we do not have the methods of calculus available to us, this section is devoted to the process of extracting a particular function from given information. We expand on the ideas first introduced in Section 1.10.

EXAMPLE 1 Express the perimeter of the cottage just described as a function of its width, w, and find its domain.

Solution The objective is to express the perimeter of the house as a function of w. We have the perimeter, P, given by $P = 2w + 2l$. In order to express the perimeter as a function of one variable, we need to use the fact that the area is to be 150 sq meters.

$$A = lw = 150 \qquad \textit{Solve for } l.$$

$$l = \frac{150}{w}$$

We can now substitute for l in the perimeter equation, which gives us the perimeter as a function of w.

$$P = 2w + 2l \qquad \textit{Substitute } l = \frac{150}{w}.$$

$$P = P(w) = 2w + 2\left(\frac{150}{w}\right) = 2w + \frac{300}{w}$$

All that remains for us to do is to determine the domain for the function $P(w)$. We are told that both the length and width must be at least 6 meters, so that $w \geq 6$. Since $l \geq 6$ and the the area is to be 150 sq meters, we have

$$lw = 150 \Rightarrow w = \frac{150}{l} \leq \frac{150}{6} = 25$$

Why does $l \geq 6$ imply that $\frac{150}{l} \leq \frac{150}{6}$?

Therefore, the perimeter, P, of the cottage as a function of its width, w, is

$$\boxed{P = P(w) = 2w + \frac{300}{w} \qquad \text{for } 6 \leq w \leq 25}$$

The remainder of this section is devoted to examining a variety of examples requiring us to extract a function from a realistic situation. As we proceed, we will pause along the way to outline an approach to this type of problem.

FIGURE 3.39

EXAMPLE 2 An oil rig is leaking oil into the ocean, creating a circular slick of oil on the ocean surface (Figure 3.39). Suppose the diameter is growing at a constant rate of 8 meters/min. Express the area of the slick as a function of t, the time in minutes since the leak began.

Solution

WHAT DO WE NEED TO FIND?	The area A of the circular oil slick in terms of t, the number of minutes since the leak began.
HOW DO WE FIND THE AREA OF THE SLICK?	Since the oil slick is circular, we use the formula for the area of a circle, $A = \pi r^2$, where r is the radius of the circle.

WE HAVE THE AREA IN TERMS OF r, ITS RADIUS. HOW DO WE EXPRESS THE AREA IN TERMS OF t?

We can express the radius in terms of the diameter, and we are given the diameter in terms of the time. In other words, since the radius is half the diameter, we may substitute $r = \frac{d}{2}$ into the formula for A.

$$A = \pi r^2 \qquad \textit{Substitute } r = \frac{d}{2}.$$

$$= \pi\left(\frac{d}{2}\right)^2 = \frac{\pi d^2}{4}$$

We are told that the diameter is growing at a rate of 8 meters/min. If t is the time in minutes since the leak began, we can express the diameter, d, in terms of t by $d = 8t$ (distance = rate × time) and substitute into the formula for A.

$$A = \frac{\pi d^2}{4} \qquad \textit{Substitute } d = 8t$$

$$= \frac{\pi(8t)^2}{4} = 16\pi t^2 \text{ sq meters}$$

Thus the area of the slick as a function of t is $A(t) = 16\pi t^2$. The statement of the problem does not tell us how long this oil leak will last, so we assume that the domain of the function is $t \geq 0$.

$$\boxed{A(t) = 16\pi t^2 \text{ sq/meters} \qquad \text{for } t \geq 0}$$

EXAMPLE 3 A rectangular area of 2500 sq ft is to be fenced off. Two opposite sides of the fencing will cost \$3 per foot; the remaining two sides cost \$2 per foot. If x represents the length of the sides requiring the more expensive fencing, express the total cost of the fencing as a function of x.

Solution

WHAT DO WE NEED TO FIND?

The total cost of the fencing in terms of x, the length of the sides requiring the more expensive fencing.

HOW DO WE FIND THE TOTAL COST?

The cost of fencing is found by multiplying the length of fence by the cost per unit length. Since there are two types of fencing, we need to identify the length of fence at each price. We draw a diagram of the rectangle and label the expensive sides x and the inexpensive sides y.

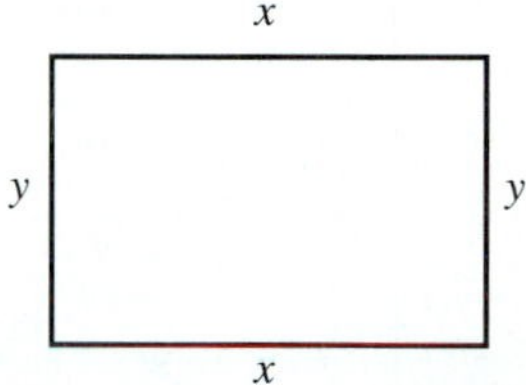

The diagram indicates that we have $2x$ feet of the more expensive fencing; hence the total cost for the \$3 fencing is $3(2x) = 6x$. Since we have $2y$ feet of the \$2 fencing, the total cost of the less expensive fencing is $2(2y) = 4y$. Hence the total cost for all the fencing is $C = 6x + 4y$ dollars.

THE COST IS EXPRESSED IN TERMS OF TWO VARIABLES. HOW DO WE EXPRESS THE COST IN TERMS OF X ONLY?

We need to find a relationship that allows us to express y in terms of x. We know that the area of the rectangle is 2,500 sq ft; hence we can write $A = 2500 = xy$. We solve this equation for y to get $y = \dfrac{2500}{x}$ and substitute into the cost equation.

$$C = 6x + 4y \qquad \textit{Substitute } y = \frac{2500}{x}.$$

$$= 6x + 4\left(\frac{2500}{x}\right) = 6x + \frac{10{,}000}{x}$$

Since x represents the length of a side of the rectangle, the domain for this function is $x > 0$. Thus the answer is

$$\boxed{C(x) = 6x + \frac{10{,}000}{x} \qquad \text{for } x > 0}$$

■

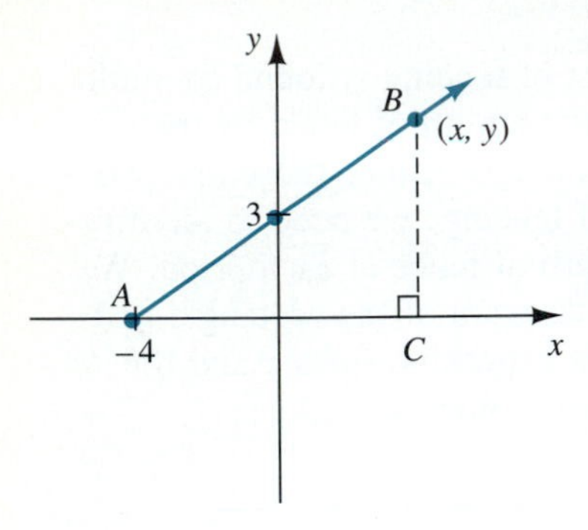

FIGURE 3.40

EXAMPLE 4 Use Figure 3.40 to express the area of ΔABC as a function of x.

Solution First let's recognize that the question makes sense. A point $B(x, y)$ is chosen on the line passing through the points $A(-4, 0)$ and $(0, 3)$, and a perpendicular is dropped from the point (x, y) to the x-axis, forming ΔABC. To each point (x, y), there corresponds one triangle and hence one area. Thus to each x, there corresponds exactly one area of ΔABC, and so the area of the triangle is a function of x.

We begin with the formula for the area, A, of a triangle, $A = \frac{1}{2}bh$. (We often call this the *generic formula*—that is, since we are looking for the area of a triangle we use the formula for the area of *any* triangle.) Next we use the given figure to apply the generic formula to this particular example. We note that the distance from the origin to point A is 4 units and the distance between the origin and point C is x units, so that the length of base AC is $x + 4$. The height BC of the triangle is y. See Figure 3.41. Thus the area formula for this triangle becomes

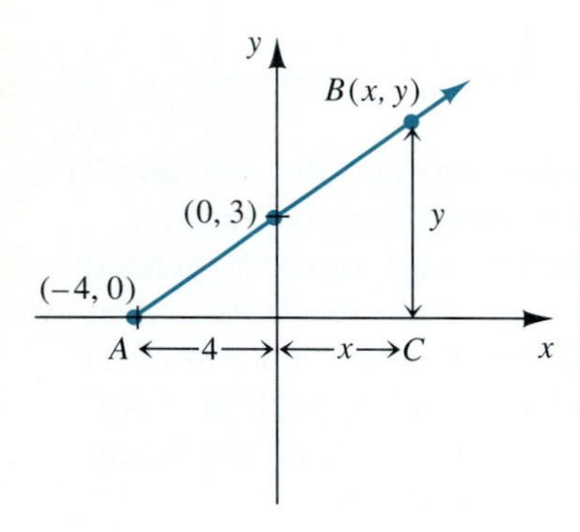

FIGURE 3.41

$$A = \frac{1}{2}(x + 4)y$$

The example asks us to express the area as a function of x. All that remains is to find a relationship that allows us to express y in terms of x. Since the point (x, y) is on the line passing through the points $(-4, 0)$ and $(0, 3)$, the equation of this line gives us the relationship we seek. To get the equation of the line, we first find its slope to be

$$m = \frac{3 - 0}{0 - (-4)} = \frac{3}{4}$$

Since the y-intercept of the line is 3, we have $y = \frac{3}{4}x + 3$ as an equation of the line. Substituting this into the area formula we get

$$A = A(x) = \frac{1}{2}(x + 4)\left(\frac{3}{4}x + 3\right) = \frac{3}{8}x^2 + 3x + 6$$

Based upon the given diagram, the point (x, y) is to be chosen to the right of the point $(-4, 0)$, and so the domain of $A(x)$ is the set of real numbers greater than -4. Thus the answer is

$$\boxed{A(x) = \frac{3}{8}x^2 + 3x + 6, \text{ for } x > -4}$$

■

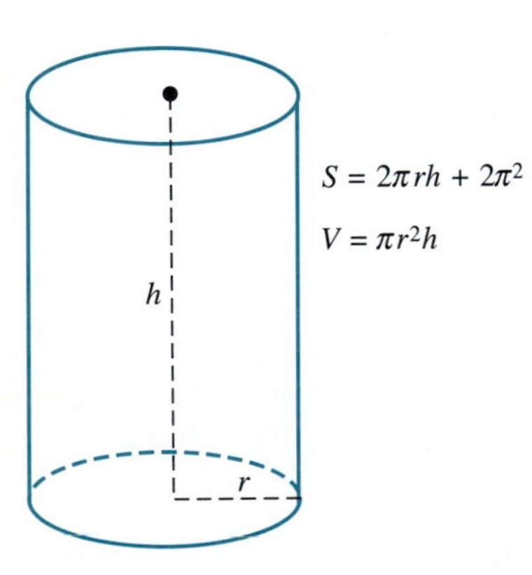

FIGURE 3.42 A closed right circular cylinder with the formulas for its surface area and volume

EXAMPLE 5 Suppose that the surface area of a closed right circular cylinder is 30π sq cm. Express the volume of the cylinder as a function of its radius r.

Solution In order to answer this question we need to know the formulas for the surface area, S, and volume, V, of a closed right circular cylinder. See Figure 3.42. Thus we see that the formula for the volume depends upon both r and h. In order to express the volume as a function of r alone, we need a relationship between r and h.

The fact that we are told that the surface area is 30π sq cm allows us to use the surface area formula to solve for h in terms of r.

$$S = 2\pi rh + 2\pi r^2$$

$$30\pi = 2\pi rh + 2\pi r^2 \qquad \textit{Divide both sides by } 2\pi \textit{ and solve for } h.$$

$$15 = rh + r^2$$

$$\frac{15 - r^2}{r} = h$$

Now we can substitute for h in the volume formula.

$$V = \pi r^2 h \qquad \textit{We substitute for } h.$$

$$V = V(r) = \pi r^2\left(\frac{15 - r^2}{r}\right) \qquad \textit{Thus the volume as a function of } r \textit{ is}$$

$$\boxed{V = V(r) = 15\pi r - \pi r^3}$$

What is the domain of $V(r)$? Clearly, since r represents the radius of the cylinder, r must be positive. However, the height h must also be positive and so from the equation $h = \dfrac{15 - r^2}{r}$, we require that $15 - r^2 > 0$. Solving this quadratic inequality we find that the domain of $V(r)$ is $D_V = \{r \mid 0 < r < \sqrt{15}\}$ ■

EXAMPLE 6 Suppose that a computer manufacturing company determines that it can sell 2000 computers at a price of \$750 and that for each \$25 that the price is increased, 40 fewer computers will be sold. Let n represent the number of \$25 increases in the price. Express the total revenue from computer sales as a function of n.

Solution The revenue is computed by multiplying the number of computers sold by the price per computer.

$$\text{revenue} = (\text{no. of computers sold})(\text{price per computer})$$

This is the *generic formula*. In a situation like this it may be helpful to set up a table with some numerical values for n.

n	Price per computer	No. of computers sold	Revenue
0	750	2000	750(2000)
1	750 + 25	2000 − 40	(750 + 25)(2000 − 40)
2	750 + 2(25)	2000 − 2(40)	(750 + 2(25))(2000 − 2(40))
3	750 + 3(25)	2000 − 3(40)	(750 + 3(25))(2000 − 3(40))
⋮	⋮	⋮	⋮
n	$750 + n(25)$	$2000 - n(40)$	$(750 + n(25))(2000 - n(40))$

Thus if we call the revenue function $R(n)$, we have

$$\boxed{R(n) = (750 + 25n)(2000 - 40n) = -1000n^2 + 20{,}000n + 1{,}500{,}000}$$

Since the number of computers sold is $2000 - 40n$ and this number cannot be negative, we need $2000 - 40n \geq 0 \Rightarrow n \leq 50$. Thus the domain of $R(n)$ is $\{n \mid 0 \leq n \leq 50\}$ ■

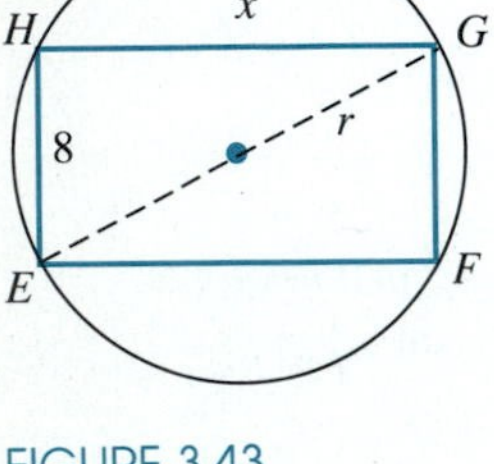

FIGURE 3.43

EXAMPLE 7 Figure 3.43 illustrates a rectangle inscribed in a circle. Express the area of rectangle *EFGH* as a function of the radius, r, of the circle.

Solution Since we are looking for the area, A, of the rectangle, the generic formula for this example is $A = bh$, where b and h represent the rectangle's base and height, respectively. In this example this formula becomes $A = 8x$.

All that remains is for us to find a relationship between x and r. It is a basic fact from geometry that since angle *EHG* is a right angle, the diagonal *EG* of the rectangle is a diameter of the circle. And so we may label the diagonal *EG* as $2r$. (See Figure 3.44.)

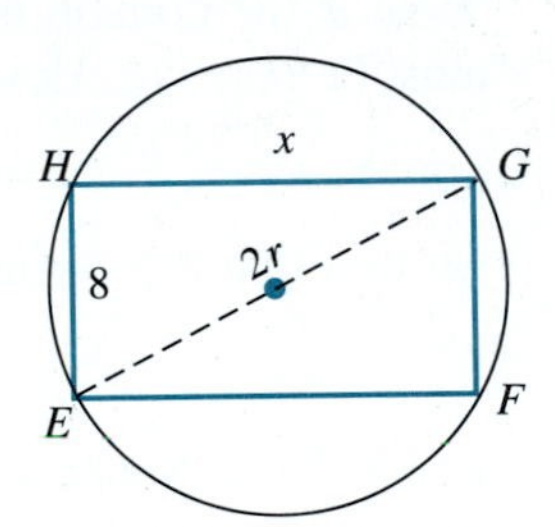

FIGURE 3.44

ΔEGH is a right triangle, and so we may apply the Pythagorean theorem to get

$x^2 + 8^2 = (2r)^2$ *We solve for x.*

$x = \pm\sqrt{4r^2 - 64}$ *Since x is a length, we take the positive square root.*

$x = \sqrt{4r^2 - 64} = \sqrt{4(r^2 - 16)} = 2\sqrt{r^2 - 16}$

Now we substitute for x in the area formula to get $A = 8x = 8(2\sqrt{r^2 - 16})$.

Since the diameter of the circle must be greater than 8, r must be greater than 4 and so the domain of $A(r)$ is the set of real numbers greater than 4.

$$\boxed{A(r) = 16\sqrt{r^2 - 16}, \text{ for } r > 4}$$ ■

This process of extracting a function from specific information describing a real-life situation is often called **mathematical modeling.** Based upon these examples we can suggest the following outline.

OUTLINE FOR OBTAINING A MATHEMATICAL MODEL

1. Read the problem over until you clearly understand what information is given and what is being sought. Often a diagram is extremely valuable in organizing and relating the given information.
2. Write an expression (formula) for the quantity being sought. This is what we are calling the *generic formula*. For example, if a volume is being sought, write a formula for the volume; if a total cost is being sought, write a formula for the cost.
3. If necessary, apply the generic formula to the particular situation under consideration. That is, replace items such as base, height, radius, or price by the letters or values given in the problem or which appear in the diagram.
4. Use the given information to express all variables in terms of the specified variable.
5. Using the results obtained in steps 3 and 4, write the required quantity as a function of the one specified variable.
6. Use the given information or the physical limitations of the problem to determine the domain of the function.

Note that some useful geometric formulas are summarized inside the back cover of the textbook.

EXAMPLE 8 An oil tank is in the shape of a right circular cone with an altitude (height) of 15 feet and a radius of 4 feet. Suppose the tank is filled to a depth of h feet. Let x be the radius of the circle on the surface of the oil. Express the volume of oil in the tank as a function of x.

Solution We follow the given outline.

1. Figure 3.45 illustrates the situation described in the example.

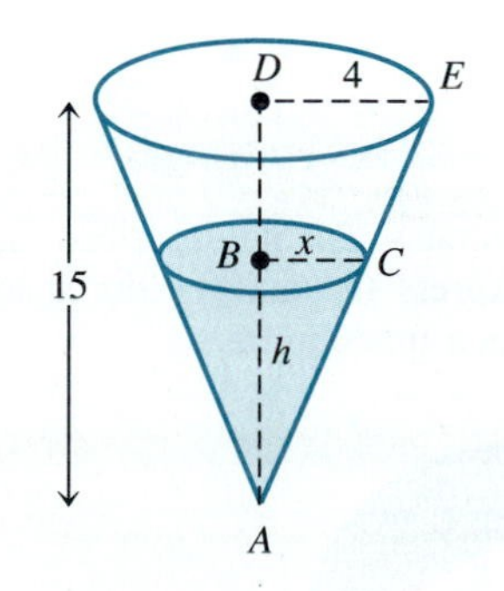

FIGURE 3.45

2. The formula for the volume, V, of a right circular cone is $V = \frac{1}{3}\pi r^2 h$. *(This is the generic formula.)*
3. Thus the volume of oil in the tank is $V = \frac{1}{3}\pi x^2 h$. *(This is the generic formula applied to this particular example.)* In order to express the volume as a function of x, we need to find a relationship between x and h.
4. Looking carefully at Figure 3.45, we can see that $\triangle ABC$ is similar to $\triangle ADE$, and so we may write the following proportion.

$$\frac{h}{x} = \frac{15}{4} \qquad \textit{We solve for } h.$$

$$h = \frac{15}{4}x \qquad \textit{Now we substitute for } h \textit{ in the volume formula.}$$

$$V(x) = \frac{1}{3}\pi x^2\left(\frac{15}{4}x\right) = \frac{5}{4}\pi x^3$$

5. $\boxed{V(x) = \frac{5}{4}\pi x^3.}$

6. Since x is the radius of the circle on the surface of the oil, x can be any value greater than or equal to 0 (if the tank is empty) and less than or equal to the radius of the tank (if the tank is full). Therefore, the domain for $V(x)$ is $\boxed{\{x \mid 0 \le x \le 4\}}$. ■

As we proceed through the text, we will continue to apply these ideas about extracting functions to the wider variety of situations we encounter. In the next section we examine some optimization problems that can be solved without any methods from calculus.

EXERCISES 3.3

In each of the following exercises, extract the required function and determine its domain.

1. Use the accompanying figure.
(a) Express y as a function of x.
(b) Express the area of $\triangle ABC$ as a function of x.

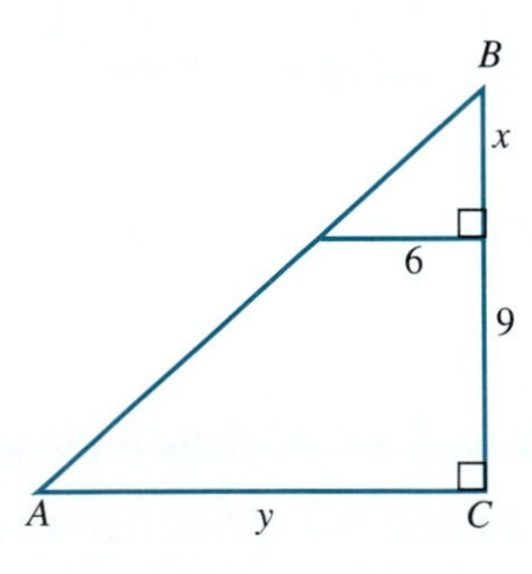

2. In the figure the point (x, y) is on the line passing through the points $(0, 2)$ and $(5, 0)$.
(a) Express y as a function of x.
(b) Express the area of $\triangle ABC$ as a function of x.

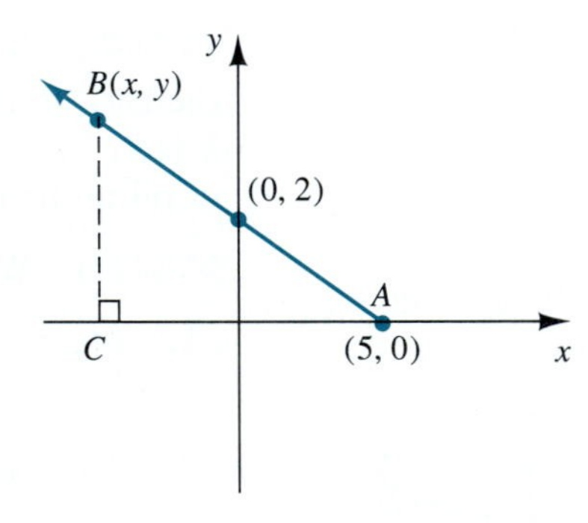

3. The following figure illustrates $\triangle ABC$ inscribed in a semicircle of radius r. Express the area of the shaded portion of the semicircle as a function of r.

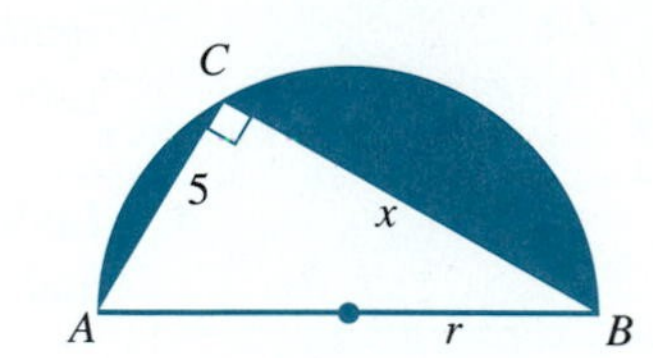

4. The figure illustrates isosceles trapezoid $ABCD$. Express the area of the trapezoid as a function of x.

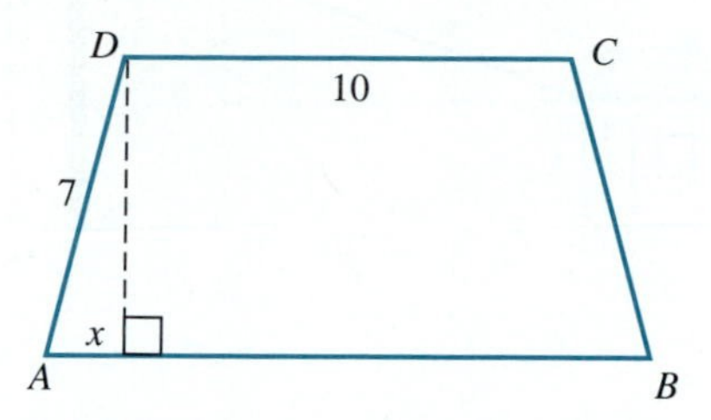

5. A closed box has a square base of side x and a height of h. If the volume of the box is 80 cu cm, express the surface area of the box as a function of x.

6. An open box is to be made from a 30-cm by 40-cm rectangular sheet of metal by cutting out identical squares of side x from each of the corners of the sheet and folding up the sides to form a box, as shown in the figure. Express the volume of the resulting box as a function of x.

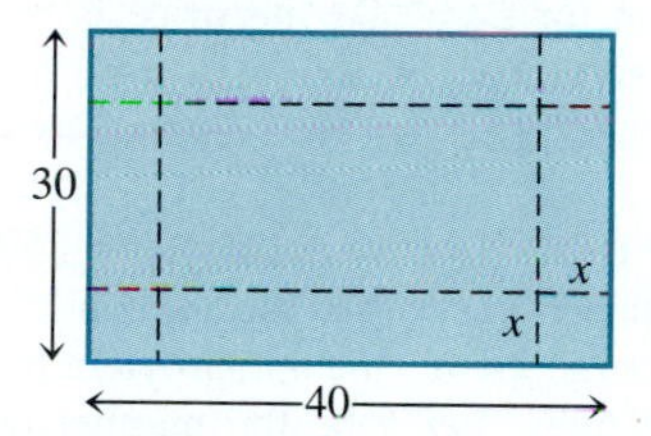

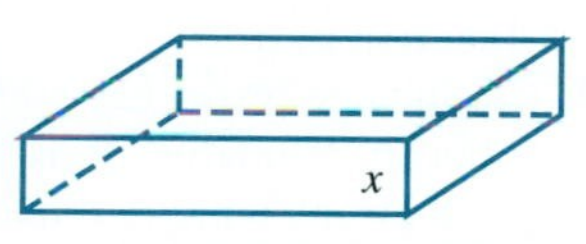

7. Express the area of a circle as a function of its circumference.

8. The volume of a closed right circular cylinder is 20 cu in. Express the surface area of the cylinder as a function of its radius r.

9. A 6-foot man is standing x feet away from a 20-foot pole with a light at the top, as shown in the figure. Express the length of the shadow s cast by the man as a function of x.

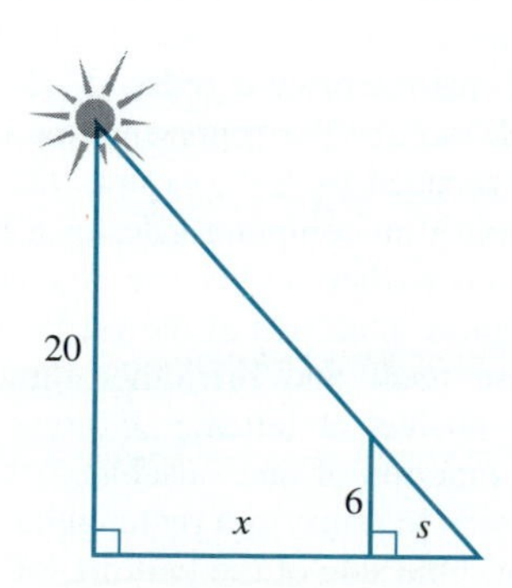

10. A water tank is in the shape of a right circular cone of height 15 feet and radius 5 feet. If the tank is filled so that the radius of the circle at the surface of the water is r, express the volume of the water in the tank as a function of the depth, d. See the following figure.

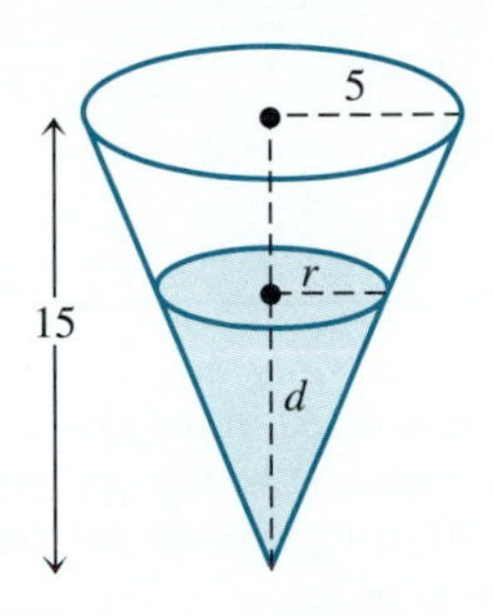

11. The point (4, 1) lies on the line passing through the points $(a, 0)$, and $(0, b)$. Use the figure to express the area of $\triangle ABC$ as a function of a.

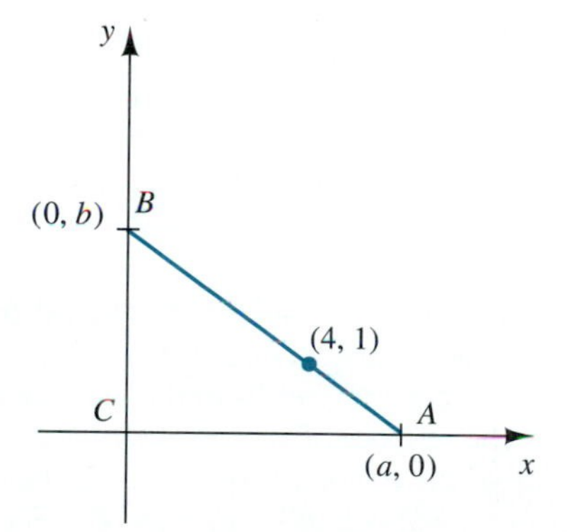

12. The point $D(x, y)$ is chosen on the line joining the origin to the point (4, 7). Use the figure to express the area of rectangle $ABCD$ as a function of x.

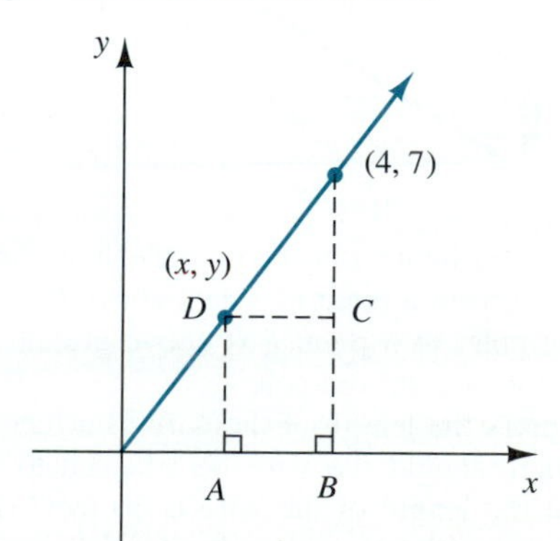

13. A running track is in the shape of a rectangle with a semicircle of radius r at each end, as shown. If the total distance around the track is 200 meters, express the area enclosed by the track as a function of r.

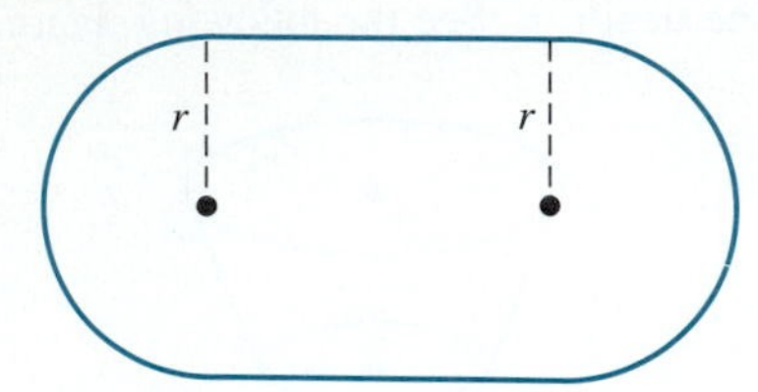

14. A window is constructed in the shape of a rectangle with a semicircle of radius r on top, as shown. If the area of the window is 10 sq ft, express the perimeter of the window as a function of r.

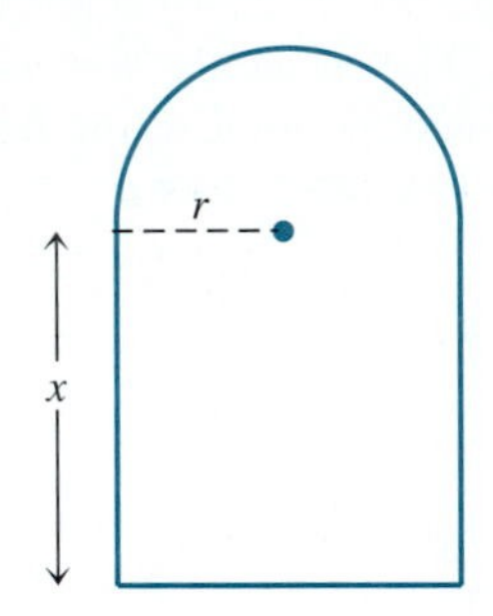

15. At 6:00 A.M. a weather balloon is released from ground level; it rises at a constant rate of 5 ft/sec. An observer is situated at a point 120 feet from the point of release. If t represents the number of seconds that have elapsed after the release of the balloon, express the distance from the observer to the balloon as a function of t. See the figure.

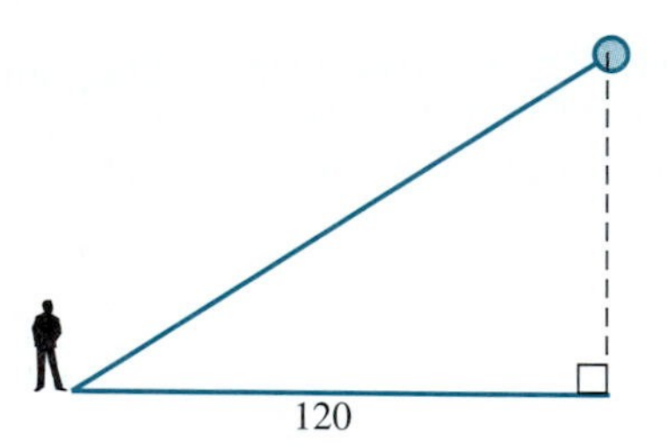

16. The following figure illustrates a telephone line that is to be attached from a point P, x feet above the ground on a telephone pole, to a point 8 ft above ground on a house located 30 feet from the pole.

(a) Express the length of the wire as a function of x.

(b) Where should the wire be attached to the pole so that the length of the wire is 50 feet? (Round your answer to the nearest tenth of a foot.)

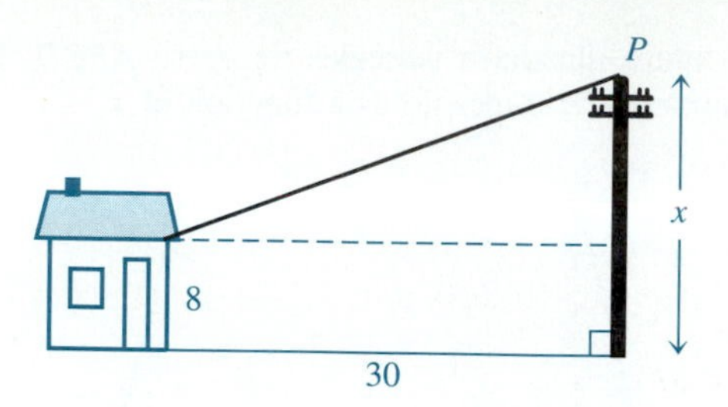

17. A piece of wire 30 cm long is to be cut into two pieces. One piece is to be formed into a square and the other piece formed into a circle.

(a) Express the total area enclosed by the square and circle as a function of the side of the square s.

(b) Express the total area enclosed by the square and circle as a function of the radius of the circle r.

18. Suppose that in Exercise 17 the piece of wire is of length L. Express the total area enclosed by the circle and square as a function of L and s.

19. A shoe manufacturer can sell 3500 pairs of shoes at \$24 each. It is determined that for each 50¢ the price is reduced, 80 more pairs of shoes can be sold. Let n be the number of 50¢ reductions. Express the total revenue as a function of n.

20. An orange grove has 800 trees, each of which yields 750 oranges annually. It is determined that for each additional tree planted in the same grove, the annual yield for each tree drops by 30 oranges. Let n be the number of additional trees planted. Express the total annual yield of oranges as a function of n.

21. A professional basketball team determines that an average of 8,500 people per game will purchase general-admission tickets at a price of \$9 per ticket, and that for each 50¢ added to the price, 225 fewer tickets will be sold. If n represents the number of times the ticket price is increased by 50¢, express the revenue taken in for general admission tickets as a function of n.

22. A small computer manufacturer finds that it is selling 750 computers per month at a price of \$1250 each and that for each \$75 that the price is reduced, 35 more computers are sold each month. If r represents the number of times the price is reduced by \$75, express the revenue generated each month on computer sales as a function of r.

23. A homeowner wishes to enclose a rectangular garden next to the house. One side of the garden will be bounded by the house itself and the other three sides will be bounded by 80 feet of fencing. Express the area of the garden as a function of one variable.

24. A factory needs to fence in a rectangular parking lot next to a building. One side of the parking lot will be bounded by the building itself, so that fencing is needed only for the other three sides. If the parking lot is to have an area of 1000 sq meters, express the length of fence needed as a function of one variable.

25. A farmer wishes to enclose a rectangular plot of land and subdivide, as indicated in the figure. Heavy-duty fencing for the perimeter costs \$5 per foot, and regular fencing for the interior costs \$3 per foot. Express the total area that can be enclosed for \$240 as a function of one variable.

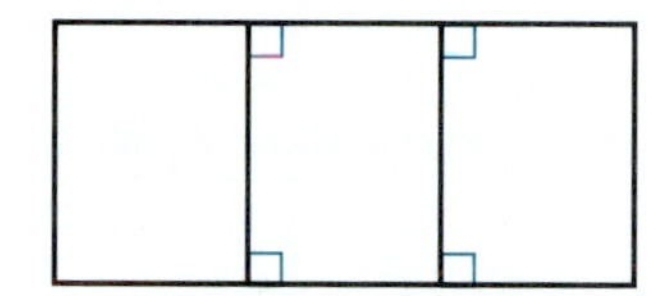

26. If the farmer of Exercise 25 wishes to use the same configuration to enclose an area of 480 sq ft, express the total cost of the fencing as a function of one variable.

27. A partition, as illustrated in the accompanying diagram, is to be set up so that it encloses an office area of 400 sq ft.

(a) Express the length of the partition as a function of x.

(b) If the partition costs \$18 per foot, express the total cost of the partition as a function of x.

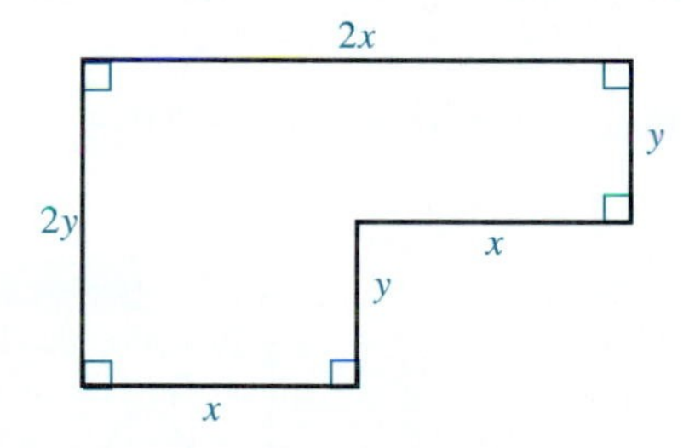

28. A hot air balloon is released from point A on the ground and rises vertically at a rate of 3 meters/min. An observer is situated on the ground 200 meters from the release point A. If t represents the number of seconds after the ballooon is released, express the distance from the observer to the balloon as a function of t.

29. A telephone company needs to lay a telephone cable from point A on one side of a river that is 1 mile wide to point D on the opposite side and 6 miles downriver, as indicated in the following diagram. The company will lay the cable underwater from A to some point C on the opposite shore, x miles from B, and then the cable will be aboveground from C to D. Express the total length of the cable as a function of x.

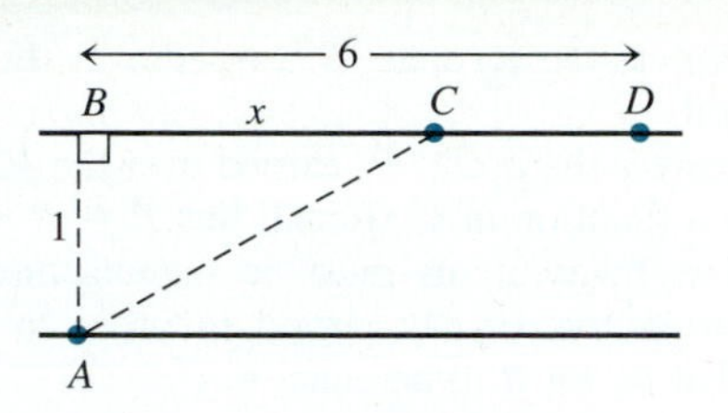

30. A graphics designer is creating a poster whose dimensions are to be 18 inches by 30 inches. The poster will have a uniform border x inches wide all around with an illustration in the center.

(a) Express the area of the border as a function of x.

(b) Express the central illustration area as a function of x.

31. A credit card company charges interest on an account's unpaid balance. The monthly interest rates are 1.65% on the first \$2000 of the unpaid balance and 1.25% on any amount over \$2000. Express the monthly interest, I, as a function of the unpaid balance b.

32. A salesperson is paid a commission of 6% on the first \$20,000 in sales made each month and 10% on any amount over \$20,000 in sales. Express the salesperson's monthly commission, C, as a function of the monthly sales, s.

33. A rectangular sheet of aluminum is 18 feet long by 1 foot wide. It is to be formed into a rain gutter by bending up a strip x feet wide along each edge of the sheet, as illustrated in the accompanying figure. Express the volume of rain the gutter can handle as a function of x.

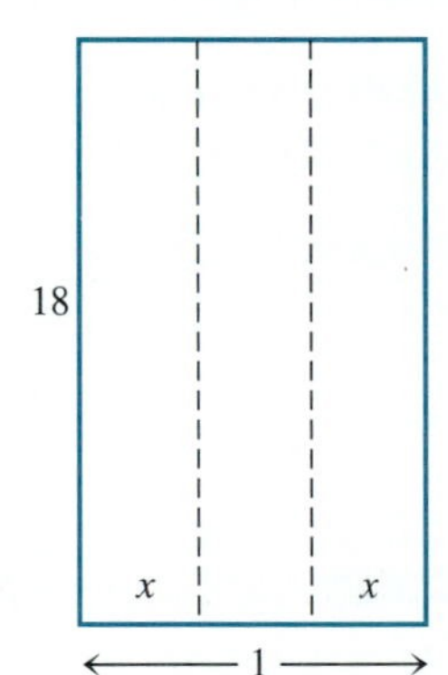

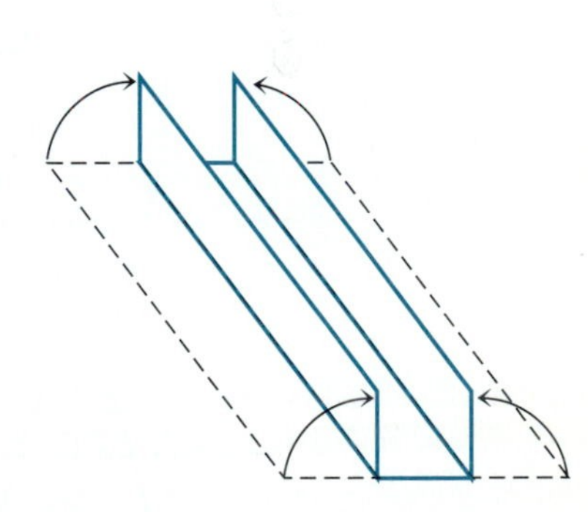

34. A manufacturer produces a product at a cost of \$17.50 per unit. There are also fixed daily costs of \$1900 regardless of how many units are produced. Each unit is sold for \$29.75. Let x represent the number of units produced during a 10-day period.

(a) Express the total cost, C, for producing x items during a 10-day period.

(b) Express the revenue, R, earned over the 10-day period.

(c) Express the profit, P, earned over the 10-day period as a function of x. (Recall that $P = R - C$.)

(d) How many items must be manufactured and sold during the 10-day period in order to break even (that is, for R to be equal to C)?

35. Express the volume of a sphere as a function of its surface area.

36. Express the surface area of a sphere as a function of its volume.

37. Express the surface area of a closed right circular cylinder of height 10 as a function of its volume V.

38. Express the distance, D, between the point $(1, 4)$ and a point (x, y) on the parabola $y = x^2$ as a function of x.

39. Express the distance, D, between the point $(-1, 2)$ and a point (x, y) on the parabola $y = 9 - x^2$ as a function of x.

40. Express the distance, D, between the point $(-2, 4)$ and a point (x, y) on the parabola $y = (x + 2)^2$ as a function of y.

41. Express the distance, D, between the point $(5, 3)$ and a point (x, y) on the parabola $y = -(x - 5)^2 + 3$ as a function of y.

42. Express the distance, D, between the point $(6, 0)$ and a point (x, y) on the curve whose equation is $y = \sqrt{x}$ as a function of x.

43. Express the distance, D, between the point $(0, 2)$ and a point (x, y) on the curve whose equation is $y = 2 - \sqrt{x}$ as a function of x.

3.4 Quadratic Functions

A **quadratic function** is a function of the form $f(x) = Ax^2 + Bx + C$, $A \neq 0$. In Section 3.2 we developed the fact that the graph of $f(x) = a(x - h)^2 + k$ (called a parabola) can be obtained by applying the graphing principles to the graph of the basic parabola $f(x) = x^2$.

EXAMPLE 1 Sketch a graph of the function $y = f(x) = -(x - 3)^2 + 4$.

Solution See Figure 3.46.

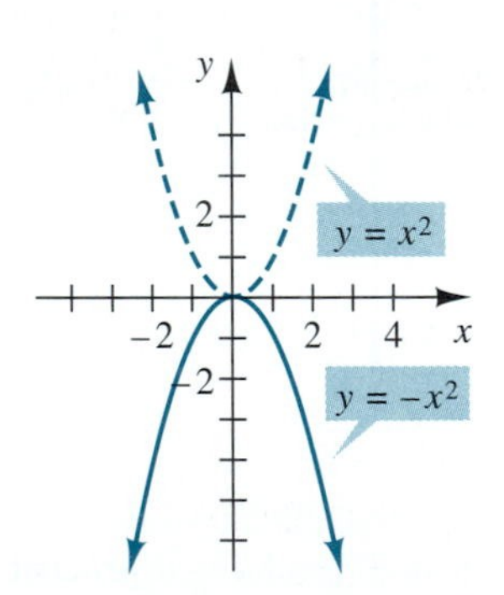

Starting with the basic parabola $f(x) = x^2$, the first negative sign tells us to reflect the graph $f(x) = x^2$ about the x-axis.

(a)

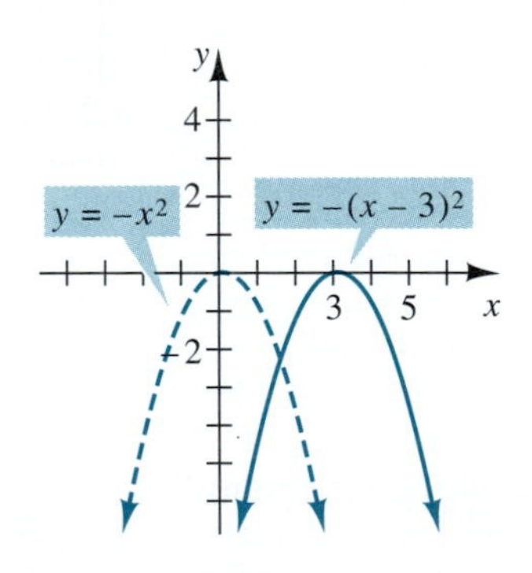

The $(x - 3)^2$ tells us to shift the graph horizontally right 3 units.

(b)

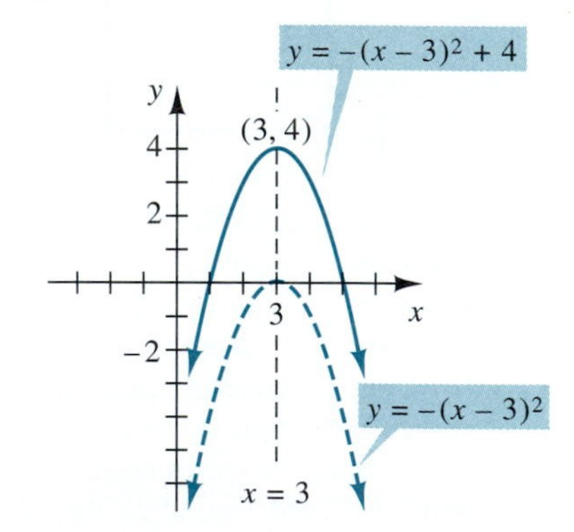

The +4 tells us to shift the graph vertically up 4 units, for the graph of $f(x) = -(x - 3)^2 + 4$.

(c)

FIGURE 3.46

Notice that the vertex is shifted from (0, 0) three units right and four units up to (3, 4), and the axis of symmetry is now the vertical line $x = 3$. As usual, the x-intercepts are found by solving $f(x) = 0$. ■

In general, the graph of $f(x) = a(x - h)^2 + k$, called the **standard form of the quadratic function**, can be found by shifting and stretching (and possibly turning upside down) the graph of $f(x) = x^2$. Observe that in shifting the graph of $f(x) = x^2$ to obtain the graph of $f(x) = a(x - h)^2 + k$, the vertex is shifted from (0, 0) to (h, k), and the axis of symmetry is shifted from $x = 0$ (the y-axis) to the vertical line $x = h$. Hence,

> The graph of $f(x) = a(x - h)^2 + k$ is a parabola with vertex (h, k) and axis of symmetry $x = h$.

It may seem strange that h is preceeded by a minus sign, and k is preceeded by a plus sign: this form is constructed this way in order to easily identify the vertex, (h, k) and axis of symmetry, $x = h$.

As with the circle, a parabola is actually defined geometrically. Using this geometric definition (which we do in Chapter 10), we can derive the fact that the equation of a parabola can be expressed in the form $f(x) = Ax^2 + Bx + C$, where A, B, and C are constants with $A \neq 0$. This is called the *general form of the quadratic function*.

Notice that if we multiply out $f(x) = a(x - h)^2 + k$, we then get $f(x) = ax^2 - 2ahx + ah^2 + k$ which is of the form $f(x) = Ax^2 + Bx + C$, a quadratic function (remember that a, h, and k are constants).

On the other hand, we can take any quadratic function $f(x) = Ax^2 + Bx + C$ and put it into the form $f(x) = a(x - h)^2 + k$ by completing the square as discussed in Chapter 1. For example, to put the equation $f(x) = x^2 + 6x + 5$ into $f(x) = a(x - h)^2 + k$ form, we proceed as follows:

$f(x) = x^2 + 6x + 5$ — *First group the terms involving powers of x together.*

$= (x^2 + 6x \quad\quad) + 5$ — *Next complete the square for the expression within the parentheses. We take one-half the coefficient of x and square it.*

$$\left[\frac{1}{2}(6)\right]^2 = (3)^2 = 9$$

In order to complete the square for $x^2 + 6x$, we want to have $x^2 + 6x + 9$.

Keep in mind that we do not want to add 9 to both sides of the equation because we want $f(x)$ to remain isolated. Thus in order to compensate for the 9 being added inside the parentheses we must also *subtract* 9 inside the parentheses if we are to ensure the transformed function is equivalent to the original function. We have

$f(x) = (x^2 + 6x + 9 - 9) + 5$ — *Regroup and put the parabola in standard form.*

$= (x^2 + 6x + 9) - 9 + 5$

$= (x + 3)^2 - 4$ — *This is the standard form of the parabola with $a = 1$, $h = -3$, and $k = -4$.*

We can now easily identify the vertex of the parabola as $(-3, -4)$.

EXAMPLE 2 Identify the vertex and axis of symmetry of the graph of the function $f(x) = 2x^2 - 6x + 7$

Solution In cases where the leading coefficient A is not 1 we proceed as follows.

$f(x) = 2x^2 - 6x + 7$ *Factor the coefficient 2 from $2x^2 - 6x$.*

$= 2(x^2 - 3x \quad\quad) + 7$ *Complete the square for $x^2 - 3x$: $\left[\frac{1}{2}(-3)\right]^2 = \frac{9}{4}$. We add and subtract $\frac{9}{4}$ inside parentheses.*

$= 2\left(x^2 - 3x + \frac{9}{4} - \frac{9}{4}\right) + 7$ *Now distribute the 2 to the $-\frac{9}{4}$, leaving a perfect square within the parentheses.*

$= 2\left(x^2 - 3x + \frac{9}{4}\right) - 2\left(\frac{9}{4}\right) + 7$ *Put the parabola in standard form.*

$f(x) = 2\left(x - \frac{3}{2}\right)^2 + \frac{5}{2}$ *This is in standard form with $a = 2$, $h = \frac{3}{2}$, and $k = \frac{5}{2}$.*

Hence, the vertex is $\left(\frac{3}{2}, \frac{5}{2}\right)$; the axis of symmetry is $x = \frac{3}{2}$. ■

Since the graph of $f(x) = Ax^2 + Bx + C$, for $A > 0$, is a parabola opening up, the vertex yields the lowest point on the graph of $f(x)$, called the *minimum of* $f(x)$. If $A < 0$, the parabola opens downward and the vertex yields the highest point on the graph of $f(x)$, called the *maximum of* $f(x)$. In Example 2, we found that the vertex of $f(x) = 2x^2 - 6x + 7$ is $\left(\frac{3}{2}, \frac{5}{2}\right)$. This tells us that the minimum value of $f(x) = 2x^2 - 6x + 7$ is $\frac{5}{2}$, and this minimum occurs when $x = \frac{3}{2}$.

Let's approach finding the vertex of the general parabola $y = f(x) = Ax^2 + Bx + C$ from a slightly different point of view. Recall that the axis of symmetry divides the parabola into two identical mirror-image parts. In particular, if there are x-intercepts, *the axis of symmetry must cut through the midpoint between the x-intercepts on the x-axis.*

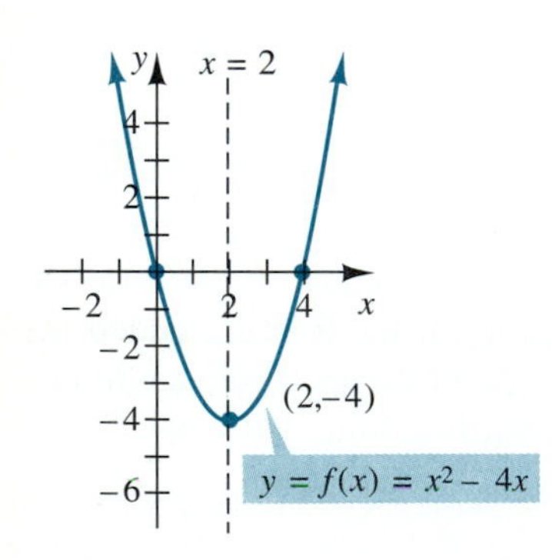

FIGURE 3.47 Note that the axis of symmetry is the vertical line midway between the x-intercepts.

For example, the x-intercepts of $y = f(x) = x^2 - 4x$ can easily be found by factoring. $0 = x^2 - 4x = x(x - 4) \Rightarrow x = 0$ or $x = 4$, and so the axis of symmetry must be the vertical line $x = 2$. See Figure 3.47. Since the vertex falls on the axis of symmetry, we know that the x-coordinate of the vertex is $x = 2$, and we can find the y-coordinate by substituting $x = 2$ into $y = x^2 - 4x$, obtaining $y = -4$. Thus the vertex is $(2, -4)$ as indicated in Figure 3.47.

We can use this same approach to find the x-coordinate of the vertex of any parabola of the form $y = f(x) = Ax^2 + Bx$. We find the x-intercepts by setting $y = 0$ and solving for x.

$$0 = Ax^2 + Bx = x(Ax + B) \Rightarrow x = 0 \text{ or } x = -\frac{B}{A}$$

Since the x-intercepts are $x = 0$ and $x = -\frac{B}{A}$, the axis of symmetry will again be the vertical line that passes through the midpoint between them. Using the midpoint formula, this midpoint has x-coordinate

$$x = \frac{0 + \left(-\frac{B}{A}\right)}{2} = -\frac{B}{2A}$$

Knowing that the axis of symmetry is vertical and passes through the point $\left(-\frac{B}{2A}, 0\right)$, what is its equation?

Thus we have found that the x-coordinate of the vertex of any parabola of the form $y = Ax^2 + Bx$ is $x = -\frac{B}{2A}$.

How are the graphs of $y = Ax^2 + Bx$ and $y = Ax^2 + Bx + C$ related?

The vertical shift principle tells us that the graph of the general parabola $y = f(x) = Ax^2 + Bx + C$ is obtained by shifting the parabola $y = Ax^2 + Bx$ up or down C units. This shifts the y-coordinate of the vertex vertically up or down *but has no effect on the x-coordinate of the vertex*, and so the x-coordinate of the vertex of $y = Ax^2 + Bx + C$ is also given by $x = -\frac{B}{2A}$. Knowing that the x-coordinate of the vertex is $-\frac{B}{2A}$ allows us to find the vertex without the necessity of completing the square. We summarize this discussion as follows:

> An equation of the form $y = f(x) = Ax^2 + Bx + C$, $A \neq 0$, is a parabola with axis of symmetry $x = -\frac{B}{2A}$. The x-coordinate of the vertex is $-\frac{B}{2A}$.

EXAMPLE 3 Graph the function $y = f(x) = -x^2 + 6x - 7$. Label the intercepts, vertex, and the axis of symmetry.

Solution This is the quadratic function $f(x) = Ax^2 + Bx + C$ with $A = -1$, $B = 6$, and $C = -7$. Hence, the x-coordinate of the vertex is

$$x = -\frac{B}{2A} = -\frac{6}{2(-1)} = 3 \text{ and the axis of symmetry is } x = 3$$

Since the x-coordinate of the vertex is 3, the y-coordinate of the vertex is

$$f(3) = -(3)^2 + 6(3) - 7 = 2.$$

Hence, the coordinates of the vertex are (3, 2).
The y-intercept is -7 since $f(0) = -(0)^2 + 6(0) - 7 = -7$.
The x-intercepts are found by using the quadratic formula:

$$x = \frac{-6 \pm \sqrt{6^2 - 4(-1)(-7)}}{2(-1)} = \frac{-6 \pm \sqrt{8}}{-2} = \frac{-6 \pm 2\sqrt{2}}{-2} = 3 \pm \sqrt{2}.$$

Since $\sqrt{2} \approx 1.41$, the x-intercepts are

$$x = 3 + \sqrt{2} \approx 4.41 \quad \text{and} \quad x = 3 - \sqrt{2} \approx 1.59.$$

The graph appears in Figure 3.49.

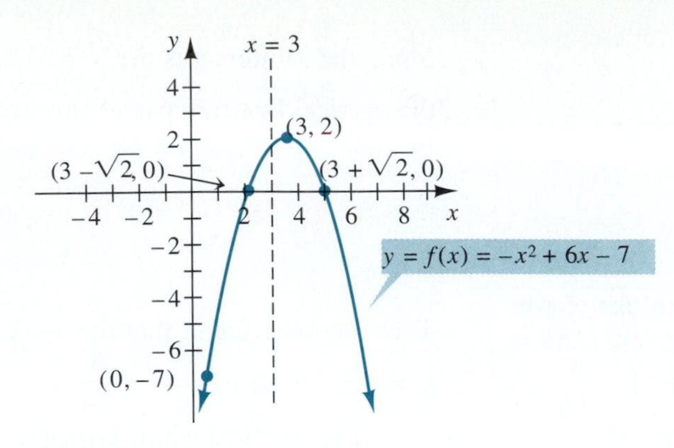

FIGURE 3.48

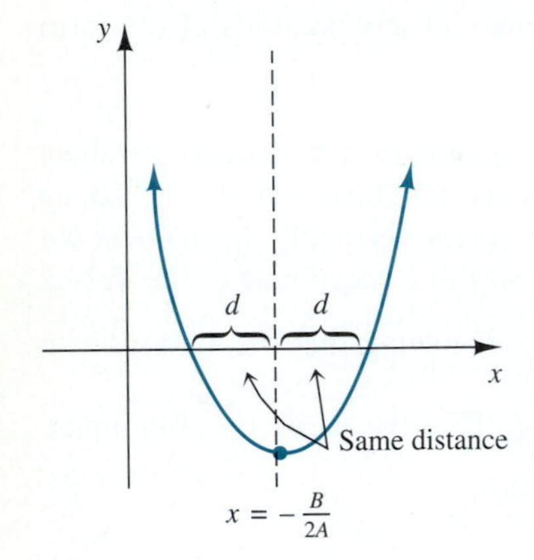

FIGURE 3.49 If a parabola has x-intercepts, they are equidistant from the axis of symmetry.

What does the fact that the roots of $Ax^2 + Bx + C = 0$ are imaginary say about the x-intercepts of the graph of $f(x) = Ax^2 + Bx + C$?

Note that we did not bother to develop a formula to find the y-coordinate of the vertex, since it is much easier to substitute the x-value in the equation to find the y-coordinate.

Let's look at Figure 3.49 and note a few things about the quadratic formula and its relationship to the axis of symmetry, vertex, and x-intercepts of the graph of $f(x) = Ax^2 + Bx + C$. The two solutions to the equation $Ax^2 + Bx + C = 0$ are the x-intercepts of the graph of $f(x) = Ax^2 + Bx + C$. The two solutions are found by applying the quadratic formula, which tells us to add and subtract the same quantity, $\frac{\sqrt{B^2 - 4AC}}{2A}$, to and from $\frac{-B}{2A}$, which is the x-coordinate of the vertex.

When solving a quadratic equation, the quadratic formula tells us that if the expression under the radical, $B^2 - 4AC$, is 0, then there is only one solution to $Ax^2 + Bx + C = 0$. This means the parabola $f(x) = Ax^2 + Bx + C$ has only one x-intercept and therefore its vertex is *on* the x-axis. If $B^2 - 4AC < 0$, then there are two solutions, but the solutions are imaginary; hence, the graph of $f(x) = Ax^2 + Bx + C$ has no x-intercepts. On the other hand, if $B^2 - 4AC > 0$, then there are two real solutions and, therefore, two x-intercepts for the graph of $f(x) = Ax^2 + Bx + C$.

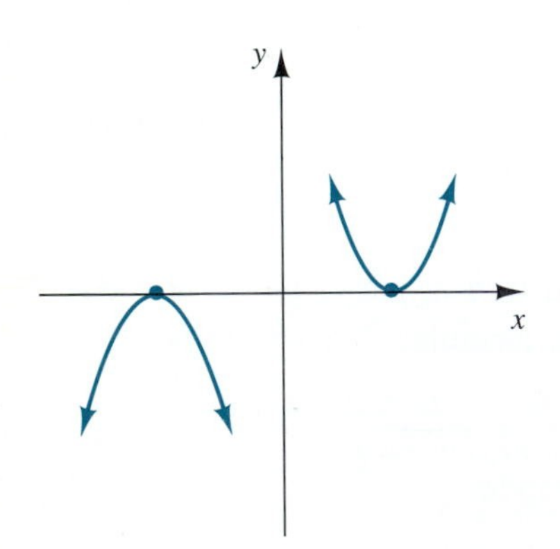

$B^2 - 4AC = 0$ means one x-intercept.

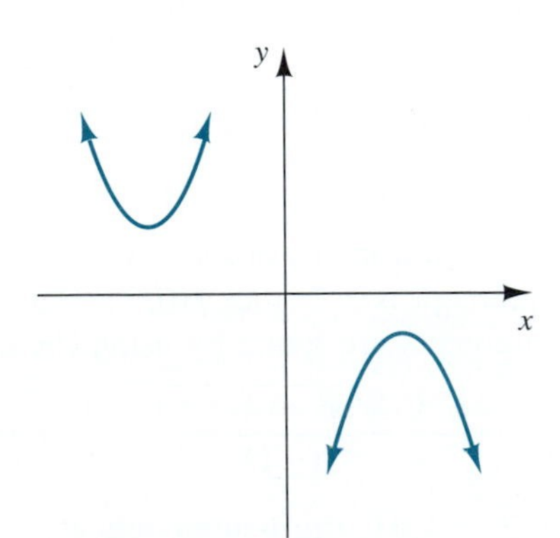

$B^2 - 4AC < 0$ means no x-intercepts.

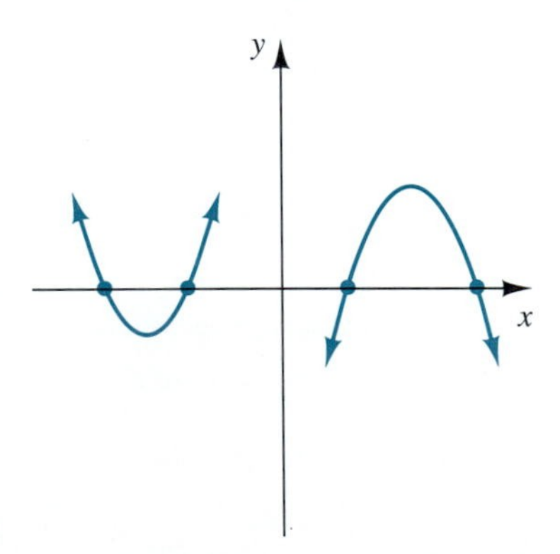

$B^2 - 4AC > 0$ means two x-intercepts.

FIGURE 3.50

The expression $B^2 - 4AC$ is called the **discriminant** of the equation $Ax^2 + Bx + C = 0$. In Figure 3.50 we consider the three possibilities for the discriminant and illustrate some possible graphs of $y = Ax^2 + Bx + C$ for each of these cases. ■

EXAMPLE 4 Sketch the graph of $g(x) = |x^2 - 2x - 3|$.

Solution First we note that we are sketching the absolute value of the quadratic function $f(x) = x^2 - 2x - 3$. Since $f(x)$ is the underlying function, we begin by sketching $f(x)$: its axis of symmetry is $x = 1$; its vertex is $(1, -4)$; the y-intercept is $(0, -3)$; and its x-intercepts are -1 and 3. The graph of $f(x) = x^2 - 2x - 3$ appears in Figure 3.51(a).

To sketch the graph of $g(x) = |x^2 - 2x - 3|$, we leave those portions of the graph of $f(x)$ lying above the x-axis alone, and take those portions of $f(x)$ lying below the x-axis and reflect them above the x-axis, as illustrated in Figure 3.51(b).

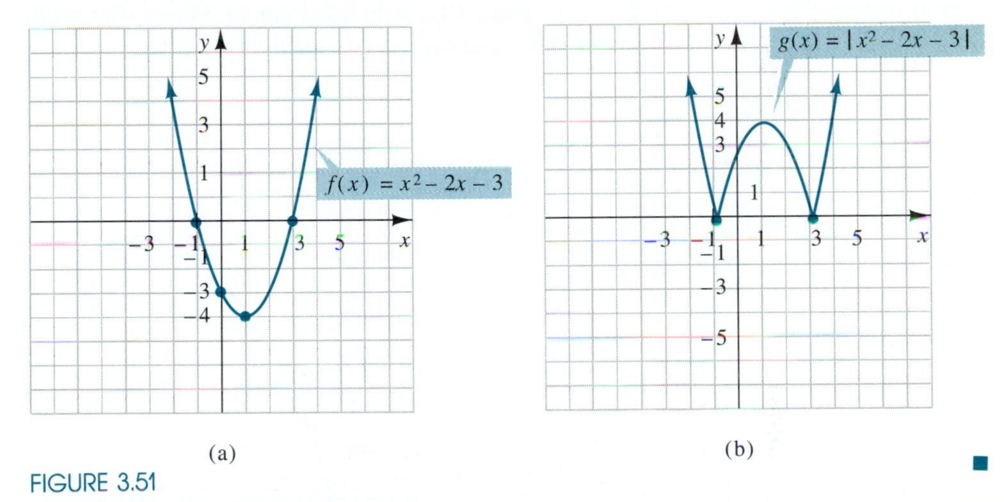

FIGURE 3.51 ■

Sometimes relationships between variables can be described by quadratic functions, and we find that the vertex is an important point to identify. We can apply what we have learned about quadratic functions in precalculus to find this important point.

EXAMPLE 5 The number of items produced weekly by Ajax Inc. is related to the weekly profit in the following way:

$$P(x) = -2x^2 + 96x - 52$$

where $P(x)$ is the weekly profit in hundreds of dollars and x is the number of items produced weekly. How many items must be produced weekly for the maximum weekly profit? What is the maximum weekly profit?

Solution We note that the relationship is quadratic; therefore, if we let the horizontal axis be the number of items produced weekly, x, and let the vertical axis be the weekly profit in hundreds of dollars, $P(x)$, the graph of $P(x)$ is a parabola which opens down. (Why?) The ordered pairs satisfying the equation are of the form $(x, P(x))$. Since the parabola opens down, the vertex is the highest point. This means that the vertex is the point that yields the highest, or maximum, profit, $P(x)$. Thus, finding the number of items that will yield the maximum profit is equivalent to finding the x-coordinate of the vertex of the parabola $P(x) = -2x^2 + 96 - 52$. To find the x-coordinate of the vertex, we use $-\dfrac{B}{2A}$, with $A = -2$ and $B = 96$:

$$x = -\frac{B}{2A} = -\frac{96}{2(-2)} = 24$$

Therefore, Ajax Inc. must produce $\boxed{\text{24 items weekly}}$ to get the maximum profit.

To find what this profit is, we simply find $P(24)$, the value of $P(x)$ when $x = 24$:

$$P(24) = -2(24)^2 + 96(24) - 52 = 1100$$

Since $P(x)$ is in hundreds of dollars, the profit is $\boxed{\$110{,}000 \text{ weekly}}$. The graph of this function appears in Figure 3.52.

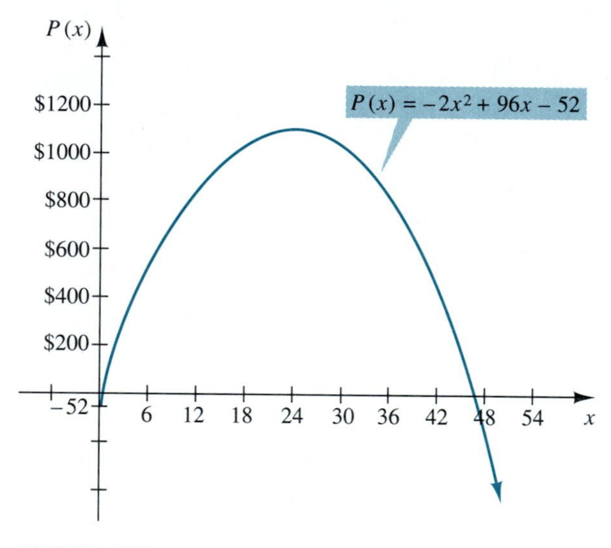

FIGURE 3.52

Since one cannot produce a negative number of items, we have restricted the domain of $P(x)$ to nonnegative values. (In fact, the number of items must be a whole number, but for the sake of simplicity we draw the graph for all real $x \geq 0$.) However, it is possible for the range values, $P(x)$, to include negative numbers (as losses), but note that the maximum value of $P(x)$ is 110,000. Hence the range is all real numbers less than or equal to 110,000. ■

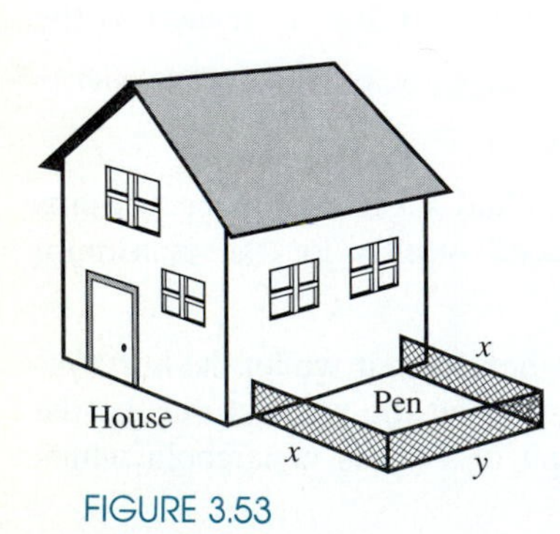

FIGURE 3.53

EXAMPLE 6 Sarah wants to fence in a rectangular dog pen against her house. Since the house is to serve as one of the sides, she needs to fence in only three sides. If she uses 80 linear feet of fencing, what are the dimensions of the pen that would give the maximum area possible? See Figure 3.53.

Solution We adapt the outline given in the last section for obtaining a mathematical model.

1. The diagram is given in the statement of the problem.
2. We write a formula for the area of a rectangle:
$$A = bh$$
3. We apply this formula to the rectangular pen in Figure 3.53:
$$A = xy$$
This is the quantity we are trying to maximize. We would like to express the area as a function of one variable.
4. Since the length of fence to be used is 80 feet, we have
$$2x + y = 80 \Rightarrow y = 80 - 2x$$
5. We now substitute into the area equation.
$$A = xy \qquad \textit{Substitute } y = 80 - 2x.$$
$$A = A(x) = x(80 - 2x) = -2x^2 + 80x$$
which we recognize as a quadratic function.
6. If we graph this function with x as the horizontal axis and $A(x)$ as the vertical axis, we get a parabola opening downward. The vertex of this parabola yields the maximum $A(x)$, or area. We will use the method of completing the square to find the vertex:

$$A(x) = -2x^2 + 80x \qquad \textit{Factor out } -2.$$
$$= -2(x^2 - 40x \qquad) \qquad \textit{Complete the square: } \left[\frac{1}{2}(-40)\right]^2 = 400.$$
Add and subtract 400 *inside the parentheses.*
$$= -2(x^2 - 40x + 400 - 400)$$
$$= -2(x^2 - 40x + 400) + 800 \qquad \textit{Put in } a(x - h)^2 + k \textit{ form.}$$
$$A(x) = -2(x - 20)^2 + 800 \qquad \textit{Hence } h = 20 \textit{ and } k = 800\textit{, and the vertex is } (20,800).$$

This means that the maximum value of $A(x)$, the area of the rectangle, is 800 sq ft, and this occurs when $x = 20$ feet. Since x is the length of one side, the other side labeled y is $80 - 2x = 80 - 2(20) = 40$ feet. Hence the dimensions to produce the maximum area are $\boxed{20 \text{ ft} \times 40 \text{ ft}}$.

Observe that in completing the square for this problem, we simultaneously find both the x and the $A(x)$ values of the vertex. It is instructive to examine the graph of the area function $A(x) = -2x^2 + 80x$, which appears in Figure 3.54. Since the area of the pen cannot be zero or negative, we can see that the domain for x is the interval $(0, 40)$.

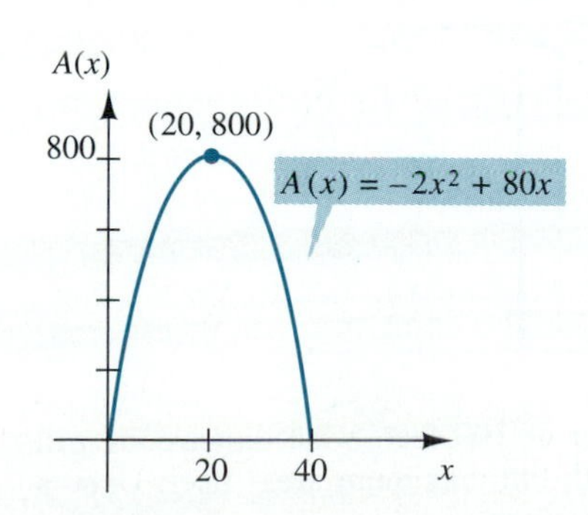

FIGURE 3.54

As we indicated in our discussion of straight lines, when we describe a graph we always describe it as we move from left to right. Thus we can also see that the area increases as x increases on the interval $(0, 20]$ until it reaches a maximum when $x = 20$, and then the area decreases on the interval $[20, 40)$. ■

EXERCISES 3.4

In Exercises 1–34, sketch the graph of the given equation. Label the axis of symmetry, vertex, and intercepts where appropriate.

1. $f(x) = 3x^2$
2. $f(x) = -3x^2$
3. $f(x) = (x - 3)^2$
4. $f(x) = (x + 3)^2$
5. $f(x) = x^2 + 8$
6. $f(x) = x^2 - 9$
7. $f(x) = (x - 3)^2 + 1$
8. $f(x) = -(x + 1)^2 - 1$
9. $f(x) = -3(x - 5)^2 + 4$
10. $f(x) = 3(x - 5)^2 + 4$
11. $f(x) = 2(x - 4)^2 + 3$
12. $f(x) = -2(x - 4)^2 + 3$
13. $y = x^2 - 4$
14. $y = x^2 - 4x$
15. $y = x - 4$
16. $y = x^2 - 4x + 4$
17. $y = 10 - 2x^2$
18. $y = 10 - 2x$
19. $f(x) = 3x^2 - 9x$
20. $y = 4x^2 + 8x$
21. $y = x^2 - 2x - 8$
22. $f(x) = x^2 - 4x - 12$
23. $y = f(x) = x^2 - 10x + 21$
24. $y = f(x) = x^2 - 5x + 5$
25. $y = 2x^2 + 12x + 16$
26. $y = -3x^2 + 6x + 24$
27. $f(x) = -2x^2 + 6x - 4$
28. $f(x) = 3x^2 + 3x + 6$
29. $y = 25x^2 - 1$
30. $f(x) = -3x^2 + 3x - 3$
31. $y = |x^2 - 9|$
32. $y = |1 - x^2|$
33. $y = |6x - x^2|$
34. $y = |x^2 + 4x + 3|$

In Exercises 35–42, find the maximum or minimum value of $f(x)$ if it exists.

35. $f(x) = -x^2 + 4x + 32$
36. $f(x) = -x^2 - 6x - 5$
37. $f(x) = -2x^2 + 6x - 18$
38. $f(x) = -3x^2 + 18x - 24$
39. $f(x) = x^2 - 4x - 32$
40. $f(x) = x^2 - 5x + 6$
41. $f(x) = 2x^2 - 6x + 18$
42. $f(x) = 3x^2 - 18x + 24$

43. A factory finds that its profit, $P(x)$, is related to the number of items it produces, x, in the following way:

$$P(x) = -x^2 + 80x$$

where $P(x)$ is the daily profit in dollars and x is the number of items produced daily.
(a) Sketch the graph of $P(x)$.
(b) How many items must be produced daily in order to maximize the profit?
(c) What is the maximum profit?

44. The profit, $P(x)$, made on a concert is related to the price of a ticket, x, in the following way:

$$P(x) = 10{,}000(-x^2 + 10x - 24)$$

(a) Sketch the graph of $P(x)$.
(b) What ticket price would produce the maximum profit?

45. The daily cost for a table manufacturer is

$$C(x) = 0.02x^2 - 0.48x + 428.8$$

where $C(x)$ is the daily production cost in dollars, and x is the number of tables produced daily. How many tables should the manufacturer produce in order to minimize the daily production cost? What is this minimal production cost?

46. The daily profit earned by the Weldon factory is related to the number of cases of candy canes produced in the following way:

$$P(x) = -x^2 + 160x - 3400$$

where $P(x)$ is the daily profit in dollars, and x is the number of cases of candy canes produced daily. Find the number of cases of candy canes to be made daily in order to maximize the daily profit. What is the maximum profit?

47. Robin fires a rocket upward and the rocket travels according to the equation

$$s(t) = -16t^2 + 864t$$

where $s(t)$ is the height (in feet) of the rocket above the ground t seconds after the gun is fired. How many seconds does it take for the rocket to reach maximum height? What is the maximum height of the rocket?

48. Stacey stands on the roof of a building and throws a ball upward. The ball travels according to the equation

$$s(t) = -16t^2 + 64t + 60$$

where $s(t)$ is the height (in feet) of the ball above the *ground* t seconds after it is thrown. See the accompanying figure.
(a) How high does the ball travel?
(b) How many seconds does it take for the ball to hit the ground?
(c) How high above the ground is Stacey when the ball is thrown?

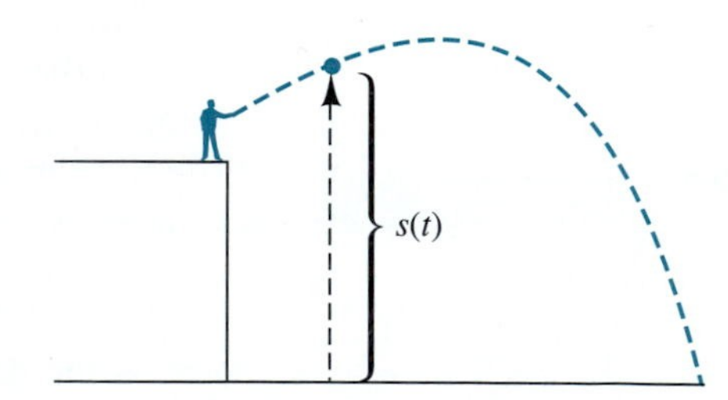

49. For a fixed perimeter of 100 feet, what dimensions will yield a rectangle with the maximum area? HINT: Draw a picture, label one side x, find the other sides in terms of x, and find the equation of the area in terms of x.

50. Repeat Exercise 49 if the fixed perimeter is P.

51. What two numbers whose sum is 104 will yield a maximum product?

52. What two numbers whose difference is 104 will yield a minimum product? Is there a maximum product? Explain.

53. A farmer wants to fence in two rectangular pens as illustrated in the following figure. If the farmer has 800 feet of fencing, what dimensions will yield the maximum area for the pens? HINT: Describe the total area in terms of one variable.

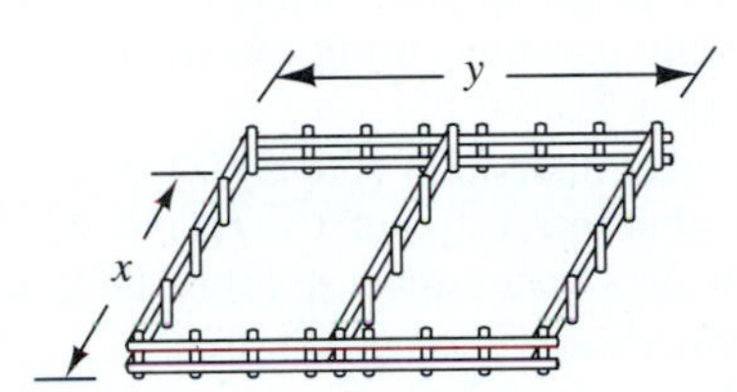

54. Suppose a farmer wants to set up a rectangular fencing scheme as illustrated in the figure. In addition, suppose that due to the strength of prevailing winds, the farmer must use fence that costs \$6 per linear foot for the fence in the north-south direction and \$12 per linear foot for the fence in the east-west direction. What is the maximum area that can be enclosed if the farmer has a total of \$1440 to spend on the fence?

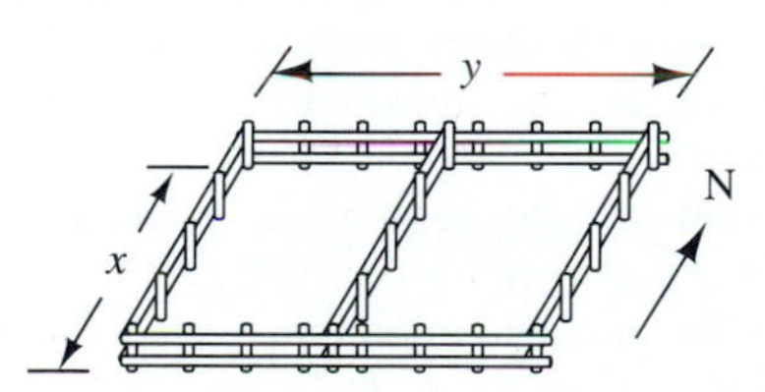

55. In economics, the demand function for a given item indicates how the price per unit p is related to the number of units x that are sold. Suppose a company finds that the demand function for one of the items it produces is

$$p = 10 - \frac{x}{5} \qquad \text{where } p \text{ is in dollars}$$

(a) How many items would be sold if the price were \$7 per unit?

(b) What should the price per unit be if 25 units are to be sold?

(c) Sketch the graph of this demand function.

(d) The revenue function, $R(x)$, is found by multiplying the price per unit p and the number of items sold x ($R = xp$). Find the revenue function corresponding to this demand function.

(e) Sketch the graph of this revenue function.

(f) How many items should be produced in order to maximize the revenue? What is the corresponding unit price?

56. An athletic field is to be constructed in the shape of a rectangle with a semicircle at each end, as indicated in the accompanying figure. The perimeter of the field is to be a $\frac{1}{4}$-mile running track. Find the dimensions of r and x that yield the athletic field with the greatest possible area.

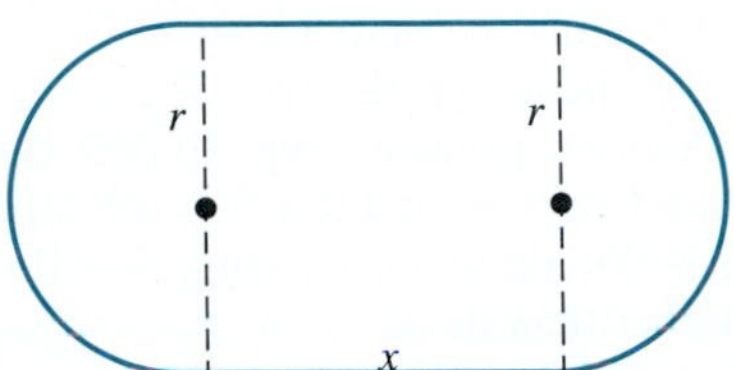

57. A Norman window is in the shape of a rectangle surmounted by a semicircle, as shown in the accompanying figure. If the perimeter of the window is 12 feet, show that the window will have a maximum area when both r and x are equal to $\dfrac{12}{\pi + 4}$, and find this maximum area.

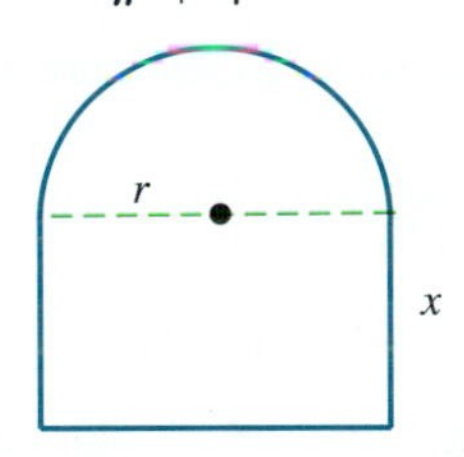

58. A manufacturer finds that the revenue, R, earned on the production and sale of n items is given by the function

$$R = R(n) = 1.56n - 0.0002n^2$$

Keep in mind that a negative number of items cannot be produced; assume that the manufacturer will stop production when the revenue becomes negative.

(a) Sketch the graph of the function $R(n)$.

(b) How many items can the manufacturer produce and not have a negative revenue?

(c) Compute $R(7000)$.

(d) What is the maximum possible revenue and how many items must be manufactured and sold to achieve this revenue?

59. A real estate management firm is handling an apartment complex of 200 units. At a monthly rent of \$800 all the units are occupied. The managers estimate that for each \$30 the monthly rent per unit is increased, five additional units will become vacant. For each occupied unit the management must pay \$320 dollars per month for taxes and maintenance, whereas for each empty unit the management must pay \$170 in taxes. What rent should be charged in order to maximize the monthly profit?

60. Suppose that a motor vehicle consumer ratings service finds that the distance, d, in miles, that a certain car can travel on one tank of gasoline depends on its speed, v, according to the function

$$d = d(v) = 45v - \left(\frac{v}{1.46}\right)^2 \quad \text{for } 10 \le v \le 90$$

What speed maximizes the distance d and consequently minimizes the fuel consumption?

61. In order to encourage the use of mass transit, a local transit authority projects that 36,000 people will use buses if the fare is \$4 and that for each \$0.25 decrease in the fare, 3000 additional passengers will decide to take the bus rather than drive.

(a) What fare should be charged if the transit authority wants to maximize the revenue from the bus fare?
(b) What fare will maximize the revenue if the buses can handle no more than 40,000 passengers?

62. A company finds that if it spends x dollars on advertising, it earns a profit, P, given by the function

$$P = P(x) = 250 + 30x - .02x^2$$

Find the advertising expenditure that generates the greatest profit.

63. For infants, the growth rate g, in pounds per month, can be approximated by the function

$$g = g(w) = kw(21 - w)$$

where w is their present weight and k is a positive constant. At what weight is an infant's rate of growth a maximum?

64. Suppose that a piece of wire 20 cm long is bent into the shape of a rectangle with length x and width y.

(a) Express the width y as a function of x.
(b) Express the area, A, of the rectangle as a function of x.
(c) Use the function obtained in part (b) to show that the area A is a maximum when the rectangle is a square.

65. Given the x-coordinate of the vertex for the parabola $f(x) = Ax^2 + Bx + C$ is $-\dfrac{B}{2A}$, find the y-coordinate of the vertex in terms of A, B, and C.

66. Show that the coordinates of the vertex of the parabola $f(x) = Ax^2 + Bx + C$ are $\left(-\dfrac{B}{2A}, \dfrac{-B^2 + 4AC}{4A}\right)$ by putting $f(x) = Ax^2 + Bx + C$ in standard form, and identifying h and k. HINT: Complete the square for $f(x) = Ax^2 + Bx + C$.

67. Sketch a graph of $f(x) = Ax^2 + Bx + C$ satisfying the following conditions.

(a) $A > 0$, and $B^2 - 4AC < 0$
(b) $A < 0$, and $B^2 - 4AC < 0$
(c) Is it possible for the vertex of the graph described in (a) to be below the x-axis? Why or why not?

68. Sketch a graph of $f(x) = Ax^2 + Bx + C$ satisfying the following conditions.

(a) $A > 0$, and $B^2 - 4AC > 0$
(b) $A < 0$, and $B^2 - 4AC > 0$
(c) Is it possible for the vertex of the graph described in (a) to be above the x-axis? Why or why not?

69. Sketch the graph of $f(x) = 2x^2 - 6x - 8$ and answer the following questions using the graph.

(a) When is $f(x) = 0$?
(b) On what interval(s) of x is $f(x) > 0$?
(c) On what interval(s) of x is $f(x) < 0$?
(d) How does your answer to (b) relate to solving the inequality $2x^2 - 6x - 8 > 0$?

70. Sketch the graph of $f(x) = -2x^2 + 6x + 8$ and answer the following questions using the graph.

(a) When is $f(x) = 0$?
(b) On what interval(s) of x is $f(x) > 0$?
(c) On what interval(s) of x is $f(x) < 0$?
(d) How does your answer to (b) relate to solving the inequality $-2x^2 + 6x + 8 > 0$?

71. **(a)** What is the maximum value of the function $f(x) = -x^2 + 4$?
(b) Use the information in (a) to find the maximum value of the function $F(x) = \sqrt{-x^2 + 4}$.

72. **(a)** What is the maximum value of the function $h(x) = -x^2 + 6x - 8$?
(b) What is the maximum value of the function $H(x) = \sqrt{-x^2 + 6x - 8}$?

73. **(a)** What is the minimum value of the function $f(x) = x^2 - 4x + 7$?
(b) Use the information in (a) to find the maximum value of the function $F(x) = \dfrac{1}{x^2 - 4x + 7}$.

74. **(a)** What is the maximum value of the function $h(x) = -x^2 + 6x - 10$?
(b) What is the minimum value of the function $H(x) = \dfrac{1}{-x^2 + 6x - 10}$?

75. We have noted that if a parabola has x-intercepts, they are equidistant from the axis of symmetry. Suppose the parabola $y = Ax^2 + Bx + C$ has x-intercepts. Verify that the equation of the axis of symmetry is $x = -\dfrac{B}{2A}$ by using the quadratic formula to find the x-intercepts and then using the midpoint formula to find the midpoint between them.

76. What happens to the reasoning in Exercise 75 if there is only one x-intercept?

QUESTION FOR THOUGHT

77. Show that the x-intercepts of the parabola $y = a(x - h)^2 + k$ are $x = h \pm \sqrt{\dfrac{-k}{a}}$. What must the signs of a and k be in order for the parabola to have two x-intercepts? Describe what these signs of a and k mean in terms of the graph of the parabola.

3.5 Operations on Functions

Analyzing the behavior of functions is one of the main topics in calculus. In order to analyze the behavior of a function such as $h(x) = \dfrac{x^2 - 1}{x^3}$, it will be advantageous to view $h(x)$ as the quotient of the two functions $f(x) = x^2 - 1$ and $g(x) = x^3$, that is, to view $h(x)$ as being "built" from simpler functions via the operations of arithmetic.

Just as we can add, subtract, multiply, and divide two real numbers to create another real number, we now define what it means to perform the arithmetic operations on functions. This is often called the *algebra of functions*.

DEFINITION Let $f(x)$ and $g(x)$ be two functions with domains D_f and D_g, respectively. We define the following four functions:

1. $(f + g)(x) = f(x) + g(x)$ — The *sum* of the two functions
2. $(f - g)(x) = f(x) - g(x)$ — The *difference* of the two functions
3. $(f \cdot g)(x) = f(x) \cdot g(x)$ — The *product* of the two functions
4. $\left(\dfrac{f}{g}\right)(x) = \dfrac{f(x)}{g(x)}$ — The *quotient* of the two functions (provided $g(x) \neq 0$)

Since an x-value must be an input into both f and g, the domain of $(f + g)(x)$ is the set of all x common to the domains of f and g. This is usually written as $D_{f+g} = D_f \cap D_g$. Similar statements hold for the domains of the difference and product of two functions. In the case of the quotient, we must impose the additional restriction that all elements in the domain of g for which $g(x) = 0$ are excluded.

EXAMPLE 1 Let $f(x) = 3x^2 + 2$ and $g(x) = 5x - 4$. Find each of the following and its domain. **(a)** $(f + g)(x)$ **(b)** $(f - g)(x)$ **(c)** $(f \cdot g)(x)$ **(d)** $\left(\dfrac{f}{g}\right)(x)$

Solution

(a) $(f + g)(x) = f(x) + g(x) = (3x^2 + 2) + (5x - 4) = \boxed{3x^2 + 5x - 2}$

(b) $(f - g)(x) = f(x) - g(x) = (3x^2 + 2) - (5x - 4)$
$= 3x^2 + 2 - 5x + 4 = \boxed{3x^2 - 5x + 6}$

(c) $(f \cdot g)(x) = f(x) \cdot g(x) = (3x^2 + 2)(5x - 4) = \boxed{15x^3 - 12x^2 + 10x - 8}$
Since D_f and D_g are each the set of all real numbers, so D_{f+g}, D_{f-g}, and $D_{f \cdot g}$ are each the set of all real numbers.

(d) $\left(\dfrac{f}{g}\right)(x) = \dfrac{f(x)}{g(x)} = \boxed{\dfrac{3x^2 + 2}{5x - 4}}$

In order for x to be in the domain of $\dfrac{f}{g}$, we require that $5x - 4 \neq 0$. Thus we have

$$D_{f/g} = \left\{x \mid x \neq \frac{4}{5}\right\}$$

■

EXAMPLE 2 Given $f(x) = \sqrt{x}$ and $g(x) = \dfrac{1}{x^2 - 5x + 6}$. Find $(f \cdot g)(x)$ and its domain.

Solution

$$(f \cdot g)(x) = f(x) \cdot g(x) = \sqrt{x} \cdot \frac{1}{x^2 - 5x + 6} = \boxed{\frac{\sqrt{x}}{x^2 - 5x + 6}}$$

In order for x to be in the domain of $f(x)$, we require that x be nonnegative, whereas for x to be in the domain of $g(x)$, we require that the denominator not be zero, that is

$$x^2 - 5x + 6 = (x - 2)(x - 3) \neq 0$$

We have $D_f = \{x \mid x \geq 0\}$ and $D_g = \{x \mid x \neq 2, 3\}$. Therefore,

$$\boxed{D_{f \cdot g} = \{x \mid x \geq 0, x \neq 2, 3\}}$$

■

Composing Functions

There is yet another way of producing a new function from two given functions.

DEFINITION Given two functions $f(x)$ and $g(x)$, the *composition* of the two functions is denoted by $f \circ g$ and is defined by

$$(f \circ g)(x) = f[g(x)]$$

$(f \circ g)(x)$ is read "f composed with g of x."

The domain of $f \circ g$ consists of those x's in the domain of g whose range values are in the domain of f, i.e., those x's for which $g(x)$ is in the domain of f.

When we compose two functions f and g, we begin with an input value x in the domain of g and get a unique output value $g(x)$ in the range of g. This output value is then used as the input value for $f(x)$, giving the unique output value $f[g(x)]$. Thus $g(x)$ must be in the domain of f.

For example, suppose we have the functions $F = \{(2, z), (3, q)\}$ and $G = \{(a, 2), (b, 3), (c, 5)\}$. The function $(F \circ G)(x) = F[G(x)]$ is found by taking elements in the domain of G and evaluating as follows:

$$(F \circ G)(a) = F[G(a)] = F(2) = z \qquad (F \circ G)(b) = F[G(b)] = F(3) = q$$

If we attempt to find $F(G(c))$ we get $F(5)$, but 5 is not in the domain of $F(x)$ and so we cannot find $(F \circ G)(c)$. Hence $F \circ G = \{(a, z), (b, q)\}$. Figure 3.55 illustrates this situation.

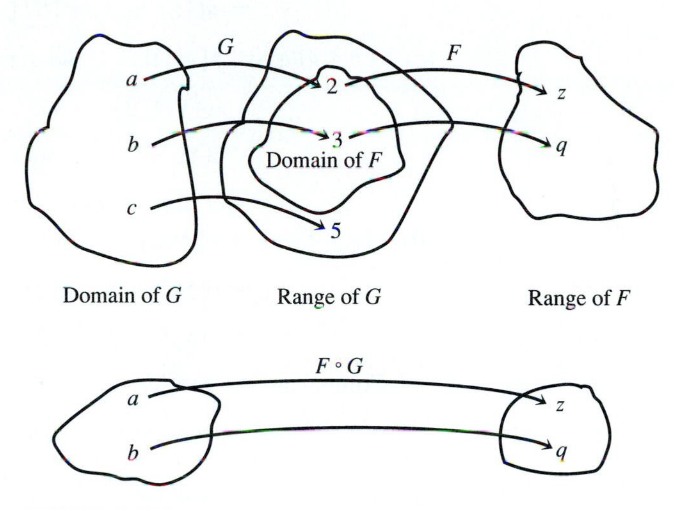

FIGURE 3.55 Diagram illustrating $F \circ G$

Figure 3.56 illustrates the situation for the general function $F \circ G$.

Notice that to evaluate $(F \circ G)(x)$ we first apply $G(x)$.

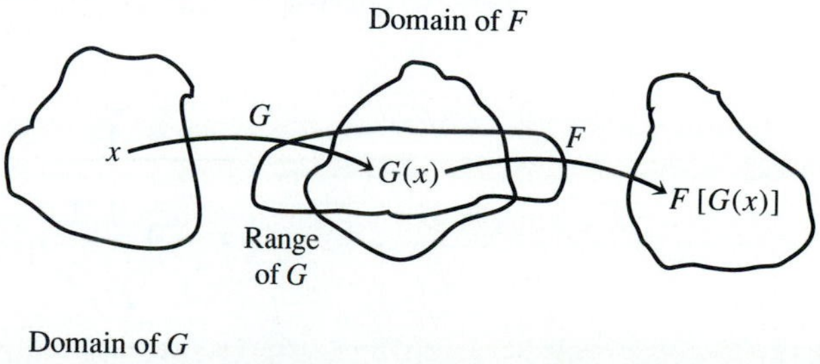

FIGURE 3.56 Diagram illustrating $(F \circ G)(x)$

EXAMPLE 3 Given $f(x) = 5x^2 - 3x + 2$ and $g(x) = 4x + 3$, find

(a) $(f \circ g)(-2)$ **(b)** $(g \circ f)(2)$ **(c)** $(f \circ g)(x)$ **(d)** $(g \circ f)(x)$

Solution

(a) $(f \circ g)(-2) = f[g(-2)]$ *First evaluate* $g(-2) = 4(-2) + 3 = -5$.

$= f(-5)$

$= 5(-5)^2 - 3(-5) + 2 = \boxed{142}$

(b) $(g \circ f)(2) = g[f(2)]$ *First evaluate* $f(2) = 5(2)^2 - 3(2) + 2 = 16$.

$= g(16)$

$= 4(16) + 3 = \boxed{67}$

(c) $(f \circ g)(x) = f[g(x)]$ *But* $g(x) = 4x + 3$.

$= f(4x + 3)$

$= 5(4x + 3)^2 - 3(4x + 3) + 2$

$= 5(16x^2 + 24x + 9) - 12x - 9 + 2 = \boxed{80x^2 + 108x + 38}$

(d) $(g \circ f)(x) = g[f(x)]$ *But* $f(x) = 5x^2 - 3x + 2$.

$= g(5x^2 - 3x + 2)$

$= 4(5x^2 - 3x + 2) + 3 = \boxed{20x^2 - 12x + 11}$

Comparing the results of parts (c) and (d), we see that $(f \circ g)(x)$ is not the same as $(g \circ f)(x)$. ■

EXAMPLE 4 Given $f(x) = \dfrac{x}{x + 1}$ and $g(x) = \dfrac{2}{x - 1}$, find

(a) $(f \circ g)(x)$ and its domain **(b)** $(g \circ f)(x)$ and its domain.

Solution

(a) $(f \circ g)(x) = f[g(x)] = f\left(\dfrac{2}{x - 1}\right)$ *In* $f(x)$, *we substitute* $\dfrac{2}{x - 1}$ *for* x.

$$= \frac{\dfrac{2}{x - 1}}{\dfrac{2}{x - 1} + 1}$$

We simplify the complex fraction by multiplying the numerator and denominator by $x - 1$.

$$= \frac{\dfrac{2}{x - 1} \cdot (x - 1)}{\left(\dfrac{2}{x - 1} + 1\right) \cdot (x - 1)}$$

$$= \frac{2}{2 + x - 1} = \boxed{\frac{2}{x + 1}}$$

Looking at the final answer we can easily see that x cannot be equal to -1. However, it is not sufficient to look only at the final form of $(f \circ g)(x)$. As stated before, x must first be an input into $g(x)$ and so must be in the domain of g. Since 1 is not in the domain of g, (Why?), then 1 is not in the domain of $f \circ g$. Therefore, x cannot be equal to 1. Thus our final answer for the domain of $f \circ g$ is

$$\boxed{D_{f \circ g} = \{x \mid x \neq \pm 1\}}.$$

(b) $(g \circ f)(x) = g[f(x)] = g\left(\dfrac{x}{x+1}\right)$ *In $g(x)$, we substitute $\frac{x}{x+1}$ for x.*

$$= \frac{2}{\dfrac{x}{x+1} - 1}$$ *We simplify the complex fraction by multiplying the numerator and denominator by $x + 1$.*

$$= \frac{2x + 2}{x - (x + 1)}$$

$$= \frac{2x + 2}{-1} = \boxed{-2x - 2}$$

Why is -1 not in the domain of $f(x)$?

Looking at the final form of $f \circ g$, we might mistakenly think that its domain is the set of all real numbers. Again, if we keep in mind that x must first be an input into $f(x)$ and so must be in the domain of f, we see that the domain of $g \circ f$ is $\boxed{D_{g \circ f} = \{x \mid x \neq -1\}}$. ■

Decomposing Functions

Perhaps the most important aspect of the composition of functions is that it allows us to express a given function in terms of simpler functions.

In the composition $(f \circ g)(x) = f[g(x)]$, it is very useful to view g as the "inner function" and f as the "outer function." It is often useful in precalculus and calculus to represent a given function $h(x)$ as the composition of two functions $f(x)$ and $g(x)$. The process of identifying possible functions $f(x)$ and $g(x)$ is called *decomposing* the function $h(x)$. We illustrate the process of decomposing a function in the next example.

EXAMPLE 5 Let $h(x) = \sqrt{x^2 + 1}$. Find two functions $f(x)$ and $g(x)$ so that $h(x) = (f \circ g)(x)$.

Solution Looking at $h(x) = \sqrt{x^2 + 1}$, we can view $x^2 + 1$ as the inside function and the square root function as the outside function. This suggests that we can let $g(x) = x^2 + 1$ and $f(x) = \sqrt{x}$; then

$$(f \circ g)(x) = f[g(x)] = f(x^2 + 1) = \sqrt{x^2 + 1} = h(x) \qquad \text{as required}$$

It is worthwhile noting that this is not the only solution. Instead we may let $g(x) = x^2$ and $f(x) = \sqrt{x + 1}$; then

$$(f \circ g)(x) = f[g(x)] = f(x^2) = \sqrt{x^2 + 1} = h(x) \qquad \text{as required}$$

Although both solutions are correct, the first has the advantage of adhering most closely to our description of g being the inner function and f being the outer function. Keep this in mind when you use the answer key to check the answers to the exercises at the end of this section. ■

If you look back at a number of the examples in Section 3.1 (especially Examples 2 and 11) and Example 1 in Section 3.4, you will see that we graphed the various functions by mentally decomposing them to identify the underlying function. In fact, our vocabulary of the *underlying function* is actually very often synonymous with the *inside function*.

We conclude this section with an application of composite functions.

EXAMPLE 6 Suppose that the radius, r, of a circle is increasing so that its length in centimeters after t seconds is given by the function

$$r = r(t) = 5t + 3 \qquad \text{where } 0 \le t \le 8$$

(a) Use a composite function to express the area of the circle as a function of t.

(b) Find the area of the circle after 6 seconds.

(c) Find how many seconds it will take for the area to be 324π sq cm.

Solution

(a) The area is a function of r; r is, in turn, a function of t, so that if we compose the two functions we will get the area as a function of t. We know that the fo mula for the area of a circle is $A = \pi r^2$. Thus the area of a circle is a function of its radius, and we may write

$$A = A(r) = \pi r^2$$

We are also given that the length of the radius is a function of t, $r(t) = 5t + 3$, and so we may write the area as a function of t as

$$A = A(r) = A[r(t)] = A(5t + 3) = \pi(5t + 3)^2$$

and so the area of the circle as a function of t is $\boxed{A = A(t) = \pi(5t + 3)^2}$

(b) Once we have expressed the area as a function of t we simply substitute $t = 6$ to find the area after 6 seconds.

$$\begin{aligned} A = A(t) &= \pi(5t + 3)^2 && \textit{Substitute } t = 6. \\ &= \pi(5(6) + 3)^2 = \pi(33)^2 \\ &= \boxed{1089\pi \text{ sq cm}} \end{aligned}$$

(c) We want to find the value of t that makes the area 324π. We can use the result of part (a), which expresses the area as a function of t.

$$\begin{aligned} A(t) &= \pi(5t + 3)^2 && \textit{We substitute } 324\pi \textit{ for the area.} \\ 324\pi &= \pi(5t + 3)^2 \\ 324 &= (5t + 3)^2 && \textit{Take square roots.} \\ \pm\sqrt{324} &= 5t + 3 && \textit{We take the positive square root.} \\ 18 &= 5t + 3 \\ t &= \boxed{3 \text{ seconds}} \end{aligned}$$

■

EXERCISES 3.5

In Exercises 1–32, use the following functions f, g, h, r, s, and t.

$f(x) = 2x^2 - x - 3$ $\quad g(x) = 3x - 2$ $\quad h(x) = 5$

$s(x) = \dfrac{1}{x}$ $\quad r(x) = x^3 - 1$ $\quad t(x) = \dfrac{4}{x + 2}$

In Exercises 1–14, find the required value.

1. $(g + r)(2)$
2. $(f - t)(-3)$
3. $(h \cdot s)(-6)$
4. $(r \cdot t)(3)$
5. $\left(\dfrac{g}{f}\right)(4)$
6. $\left(\dfrac{s}{t}\right)(x)$
7. $\left(\dfrac{h}{r}\right)(-1)$
8. $(t - s)(8)$
9. $(f \circ g)(4)$
10. $(g \circ f)(4)$
11. $(h \circ s)(-6)$
12. $(s \circ h)(8)$
13. $(f + g)(x)$
14. $(s \cdot r)(x)$

In Exercises 15–32, use the functions f, g, h, r, s, and t to find the required function and its domain.

15. $(f - g)(x)$
16. $(r + s)(x)$
17. $\left(\dfrac{s}{h}\right)(x)$
18. $\left(\dfrac{h}{s}\right)(x)$
19. $(r - f)(x)$
20. $(s + t)(x)$
21. $(g \cdot t)(x)$
22. $\left(\dfrac{g}{t}\right)(x)$
23. $(f \circ g)(x)$
24. $(g \circ f)(x)$
25. $(r \circ s)(x)$
26. $(s \circ r)(x)$
27. $(h \circ t)(x)$
28. $(t \circ h)(x)$
29. $(r \circ g)(x)$
30. $(g \circ r)(x)$
31. $(s \circ t)(x)$
32. $(t \circ s)(x)$
33. Let $f(x) = \sqrt{x + 1}$ and $g(x) = 2x - 7$.
    (a) Find $(f \circ g)(x)$ and its domain.
    (b) Find $(g \circ f)(x)$ and its domain.
34. Let $F(x) = \dfrac{x - 1}{x + 1}$ and $G(x) = \dfrac{2}{x}$. Find
    (a) $(F \circ G)(x)$ **(b)** $(G \circ F)(x)$
35. Let $f(t) = t^2 + t$ and $g(t) = \dfrac{6}{t - 3}$. Find
    (a) $(f \circ g)(t)$ **(b)** $(g \circ f)(t)$
    (c) $(f \circ f)(t)$ **(d)** $(g \circ g)(t)$
36. Let $f(x) = 3x - 5$ and $g(x) = \frac{1}{3}(x + 5)$. Find
    (a) $(f \circ g)(x)$ **(b)** $(g \circ f)(x)$
37. We may define the composition of three functions as follows:

    $$(f \circ g \circ h)(x) = f\{g[h(x)]\}$$

    Let $f(x) = 2x + 3$, $g(x) = \sqrt{x}$, and $h(x) = \dfrac{1}{x}$. Find
    (a) $(h \circ g \circ f)(x)$ **(b)** $(f \circ h \circ g)(x)$
    (c) $(g \circ f \circ h)(x)$
38. Let $f(x) = x^2$ and $g(x) = \sqrt{x}$.
    (a) Find $(f \circ g)(x)$ and its domain.
    (b) Find $(g \circ f)(x)$ and its domain.
    (c) Are $(f \circ g)(x)$ and $(g \circ f)(x)$ the same function? Explain.

In Exercises 39–44, find two functions $f(x)$ and $g(x)$ so that the given function $h(x) = (f \circ g)(x)$.

39. $h(x) = (x + 3)^3$
40. $h(x) = \sqrt{5x - 3}$
41. $h(x) = \left(\dfrac{x + 4}{x - 1}\right)^2$
42. $h(x) = \sqrt[3]{x^2 - 4x + 5}$
43. $h(x) = \dfrac{1}{x} + 6$
44. $h(x) = \dfrac{1}{x + 6}$
45. Let $f(x) = 5x - 3$. Find $g(x)$ so that $(f \circ g)(x) = 2x + 7$.
46. Let $f(x) = 2x + 1$. Find $g(x)$ so that $(f \circ g)(x) = 3x - 1$.
47. Let $f(x) = 8 - 5x$. Find $g(x)$ so that $(f \circ g)(x) = x$.
48. Let $f(x) = \dfrac{2x + 3}{5}$. Find $g(x)$ so that $(f \circ g)(x) = x$.
49. Suppose a laboratory technician is growing a bacteria culture in which the number of bacteria present, N, depends upon the Celsius temperature, C, of the surrounding air given by the function

    $$N = N(C) = 3C^2 + 250C + 10{,}200 \quad \text{for } 15 \le C \le 40$$

    The Celsius temperature, C, is, in turn, dependent upon the number of hours, h, after the culture begins growing and is given by the function

    $$C(h) = 5h + 15 \qquad \text{for } 0 \le h \le 5$$

    (a) Express the number of bacteria, N, as a function of h.
    (b) How many bacteria are present after 4 hours?
    (c) After how many hours are there 300,000 bacteria?
50. Suppose that the base of a rectangular box is a square of side x and its height is 6 inches. The length of x is dependent upon the number of minutes, t, that have elapsed after $t = 0$ minutes according to the function

    $$x(t) = 25 - t^2 \qquad \text{for } 0 \le t \le 4$$

    (a) Express the volume of the box as a function of x.
    (b) Express the volume of the box as a function of t.
    (c) What is the volume of the box after 3 seconds?

51. A calculator manufacturer sets the price, P, of a calculator at 30% above the cost to manufacture it. The cost for manufacturing n calculators is given by the function

$$c = c(n) = 32n + 370$$

(a) If 1 calculator is produced, what will its price be?
(b) If 10 calculators are produced, what will the price of each be?
(c) If n calculators are produced, express the price per calculator, P, as a function of n.

52. The volume, V, of a sphere of radius r is given by the formula

$$V = \frac{4}{3}\pi r^3$$

The radius is decreasing with time, t, according to the formula

$$r = \frac{1}{\sqrt{t+1}} \qquad \text{for } t \geq 0$$

Express the volume as a function of t.

QUESTION FOR THOUGHT

53. We have seen through a variety of examples and exercises that, in general, $(f \circ g)(x) \neq (g \circ f)(x)$. Can you find a specific example of two functions f and g for which $(f \circ g)(x) = (g \circ f)(x)$? Can you decribe some general examples or categories of functions for which $(f \circ g)(x) = (g \circ f)(x)$?

3.6 Inverse Functions

In our initial discussions we stressed the critical aspect of a function—that to each x there corresponds *exactly one* y. We noted that it is quite possible for a function to assign the same y-value to two different x-values.

In addition, we have been very specific in our designations of the variables we use in defining functions; x is the independent variable and y is the dependent variable, or, as we have been saying, "Pick x, and get y." We have agreed always to list the independent variable (the value we pick) as the first coordinate and the dependent variable (the value we get) as the second coordinate. However, this arrangement is quite arbitrary.

Let's examine the following rather simple function F defined by the following set of ordered pairs:

F: $\{(-2, 2), (0, 4), (1, 5), (3, 7)\}$
with domain $D_F = \{-2, 0, 1, 3\}$ and range $R_F = \{2, 4, 5, 7\}$

Looking carefully at these ordered pairs we note that not only do these ordered pairs define y as a function of x, but they can equally well be viewed as defining x as a function of y. That is, since each y-value has exactly one x-value associated with it, these ordered pairs allow us to "pick y and get x." Let us call this function G; that is, G is the result of using these ordered pairs to define x as a function of y. We would record this function G as

G: $\{(2, -2), (4, 0), (5, 1), (7, 3)\}$
with domain $D_G = \{2, 4, 5, 7\}$ and range $R_G = \{-2, 0, 1, 3\}$

Remember the independent variable is recorded as the first coordinate, so that G is obtained by *interchanging* the x and y coordinates of F.

Let's examine what happens when we compose the functions F and G in either order.

$(F \circ G)(x)$	$(G \circ F)(x)$
$(F \circ G)(2) = F[G(2)] = F(-2) = 2$	$(G \circ F)(-2) = G[F(-2)] = G(2) = -2$
$(F \circ G)(4) = F[G(4)] = F(0) = 4$	$(G \circ F)(0) = G[F(0)] = G(4) = 0$
$(F \circ G)(5) = F[G(5)] = F(1) = 5$	$(G \circ F)(1) = G[F(1)] = G(5) = 1$
$(F \circ G)(7) = F[G(7)] = F(3) = 7$	$(G \circ F)(3) = G[F(3)] = G(7) = 3$

Figure 3.57 illustrates $(F \circ G)(x)$ and $(G \circ F)(x)$ with a diagram. We see that $(F \circ G)(x) = x$ for all x's in the domain of G and $(G \circ F)(x) = x$ for all x's in the domain of F. Functions that have this property are called **inverse functions.**

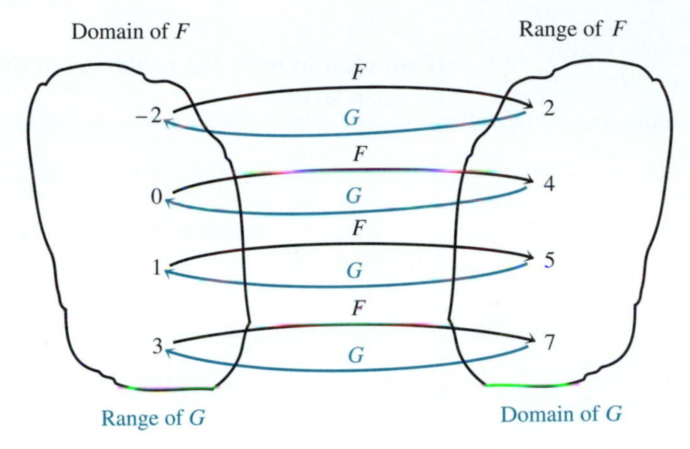

FIGURE 3.57 Diagram illustrating $(F \circ G)(x)$ and $(G \circ F)(x)$

DEFINITION Two functions f and g are said to be **inverse functions** if and only if

$$f[g(x)] = x \qquad \text{for all } x \text{ in the domain of } g$$

and

$$g[f(x)] = x \qquad \text{for all } x \text{ in the domain of } f$$

EXAMPLE 1 Verify that the functions $f(x) = \frac{1}{2}x - 5$ and $g(x) = 2x + 10$ are inverse functions.

Solution According to the definition, in order to verify that f and g are inverse functions, we must check that $f[g(x)] = x$ and $g[f(x)] = x$.

$$\begin{aligned} f[g(x)] &= f(2x + 10) \\ &= \frac{1}{2}(2x + 10) - 5 \\ &= x + 5 - 5 \\ &= x \qquad \textit{As required} \end{aligned}$$

$$\begin{aligned} g[f(x)] &= g\left(\frac{1}{2}x - 5\right) \\ &= 2\left(\frac{1}{2}x - 5\right) + 10 \\ &= x - 10 + 10 \\ &= x \qquad \textit{As required} \end{aligned}$$

Thus we have verified that f and g are inverse functions. ■

Note that based upon this definition, the functions F and G described earlier, which have their x- and y-coordinates interchanged, are inverse functions.

It is customary to denote the inverse function of $f(x)$ as $f^{-1}(x)$, which is read "f inverse of x." Using this notation, we can express the result of Example 1 by writing

$$f(x) = \frac{1}{2}x - 5 \qquad \text{and} \qquad f^{-1}(x) = 2x + 10$$

Important Even though in general we use an exponent of -1 to indicate a reciprocal, inverse function notation is an exception to this rule. Please be aware that $f^{-1}(x)$ is *not* the reciprocal of f.

$$f^{-1}(x) \neq \frac{1}{f(x)}$$

If we want to write the reciprocal of the function $f(x)$ by using a negative exponent, we must write

$$\frac{1}{f(x)} = [f(x)]^{-1}$$

Since f^{-1} denotes the inverse function of f, the definition of inverse functions just given yields

$$f^{-1}[f(x)] = x \qquad \text{for all } x \text{ in the domain of } f$$

and

$$f[f^{-1}(x)] = x \qquad \text{for all } x \text{ in the domain of } f^{-1}$$

As we have noted before, a function may be viewed as a machine that takes an input x and does something to it. Inverse functions are in some sense "opposites" in that they have the property that what one function "does," the other function reverses, or "undoes." Thus when we compose two inverse functions, we get the original input value x back again. See Figure 3.58.

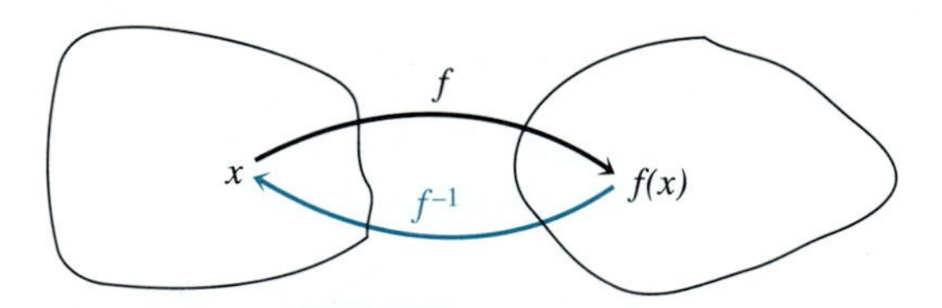

FIGURE 3.58 The interaction of inverse functions

Let us now examine this idea from a more general standpoint. Suppose we have a function, which we know means that to each x there corresponds exactly one y. We may ask if this assignment also defines x as a function of y: If we pick y, does the equation determine a unique value of x?

For example, $F = \{(2, 4), (3, 6), (5, 4)\}$ is a function that has a unique y assigned to each x. Note, however, that it does not define x as a function of y. That is, *each y is not assigned a unique x*. Alternatively, we can say that if we reversed the x- and y-coordinates, the resulting set of ordered pairs, $\{(4, 2), (6, 3), (4, 5)\}$, is not a function.

Why is the set $\{(4, 2), (6, 3), (4, 5)\}$ not a function?

If we consider the function defined by the equation $y = f(x) = x^2$ and we choose $y = 4$, we get $x^2 = 4$, and so $x = \pm 2$. Thus in general we can see that an equation that defines y as a function of x will not necessarily define *x as a function of y*.

Why doesn't $y = x^2$ define x as a function of y?

A natural question to ask is, Can we specify conditions under which a function $y = f(x)$ also specifies x as a function of y? For y to be a function of x, each x is assigned a unique y, and for x to be a function of y, each y is assigned a unique x. Hence there must be a one-to-one correspondence between the elements of the domain (x) and elements of the range (y). This means that we cannot have two x's assigned the same y-value as occurred in the above two examples. In the language of functions we can formulate these ideas as follows:

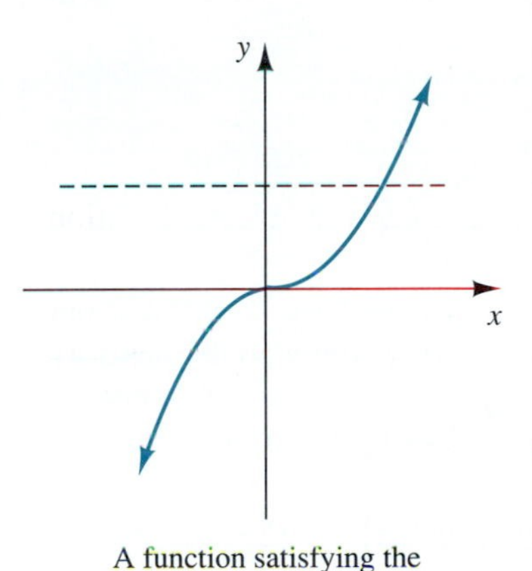

A function satisfying the horizontal line test.

(a)

FIGURE 3.59(a)

DEFINITION A function $f(x)$ is said to be **one-to-one** if and only if for any two distinct domain values, $x_1 \neq x_2$ implies that $f(x_1) \neq f(x_2)$. In words this condition says that different x-values necessarily give different y-values.

Let's examine this definition graphically. Suppose we have a function $y = f(x)$ and we know its graph. If we want the equation $y = f(x)$ also to define x as a function of y, then to each y-value there must correspond exactly one x-value. Recalling that the vertical line test graphically expresses the idea that to each x-value there corresponds exactly one y-value, we recognize that an analogous *horizontal line test* (defined here) will ensure that to each y-value there must correspond exactly one x-value.

THE HORIZONTAL LINE TEST

A function $y = f(x)$ will also define x as a function of y provided that any horizontal line intersects the graph in at most one point.

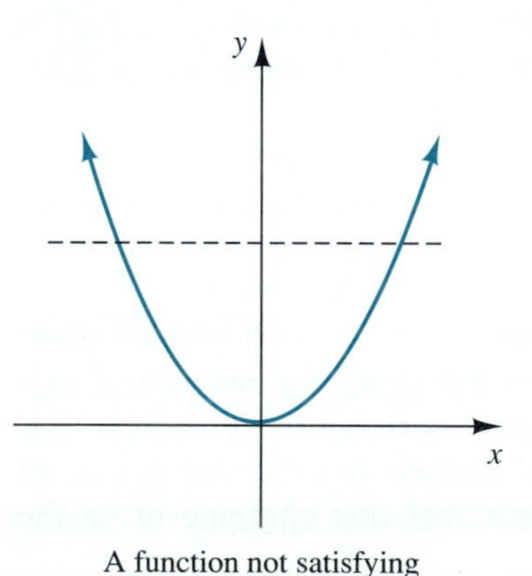

A function not satisfying the horizontal line test.

(b)

FIGURE 3.59(b)

Figures 3.59 (a) and (b) illustrate one function that does satisfy the horizontal line test and one that does not.

Note that although only one of the graphs in Figure 3.59 satisfies the horizontal line test, both are the graphs of functions, since they both satisfy the vertical line test.

Basically, the horizontal line test says that the graph of a function does not repeat any y-values and so the following two statements are equivalent:

1. The graph of $y = f(x)$ satisfies the horizontal line test.
2. The function $y = f(x)$ is one-to-one.

DIFFERENT PERSPECTIVES: *One-to-One Functions*

GEOMETRIC INTERPRETATION

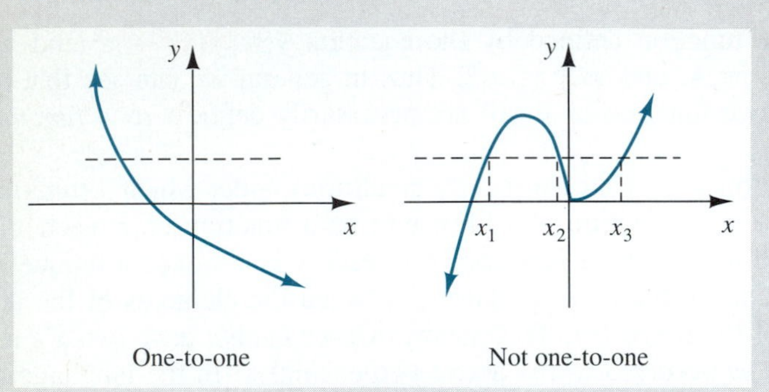

For a function to be one-to-one, any horizontal line can intersect the graph in at most one point.

ALGEBRAIC INTERPRETATION

A one-to-one function necessarily assigns different y-values to different x-values.

In other words, a function *is* one-to-one if and only if $x_1 \neq x_2 \Rightarrow f(x_1) \neq f(x_2)$.

In summary, then, based upon our discussion thus far we have seen that a one-to-one function $y = f(x)$ not only defines y as a function of x, but also defines x as a function of y. Also, the function obtained by viewing x as a function of y gives us the inverse function of $f(x)$. In fact, we can state the following theorem.

THEOREM 3.1 $f(x)$ has an inverse, $f^{-1}(x)$, if and only if $f(x)$ is a one-to-one function.

The theorem we have just stated tells us when a function will have an inverse, but how do we go about actually finding the inverse function of a one-to-one function? Consider the function $y = f(x) = 2x - 1$, whose graph is a straight line of slope 2. This function is certainly one-to-one, and so we can view the equation $y = 2x - 1$ as we usually do, defining y as a function of x or defining x as a function of y.

The ordered pairs $(0, -1)$, $(-1, -3)$, $(2, 3)$, and $(5, 9)$ all satisfy the function $y = 2x - 1$, and based upon our earlier ground rules, we understand this to mean, If x is chosen to be $0, -1, 2, 5$, then the corresponding y-values are, respectively, $-1, -3, 3, 9$.

Keep in mind that the various graphing techniques we have developed are predicated upon the independent variable being represented along the horizontal axis (that is, being the first coordinate) and the dependent variable being represented along the vertical axis (that is, being the second coordinate). In order not to lose all the graphing machinery we have built, we will insist that this continue to be the case.

Therefore, if we now return to the ordered pairs for $y = 2x - 1$ and we want to view them as "pick y and get x," we record them as

$$(-1, 0), (-3, -1), (3, 2), \text{ and } (9, 5)$$

What we are in effect doing is *interchanging the roles of x and y*. As we saw at the beginning of this section, if we interchange the x- and y-coordinates of the points of a one-to-one function, we obtain its inverse function. Since most of the functions we will encounter are defined by an equation rather than a set of ordered pairs, instead of interchanging the coordinates of points, we will interchange the roles of x and y in the equation. However, keeping in mind that we are generally used to having functions solved explicitly for the dependent variable ($y = f(x)$), after we interchange x and y, we want this new inverse function to be solved explicitly for the dependent variable y.

EXAMPLE 2 Given $y = f(x) = \frac{1}{4}x + 3$, find $f^{-1}(x)$.

Solution We first note that the graph of $y = f(x) = \frac{1}{4}x + 3$ is a nonhorizontal line. Thus $f(x)$ satisfies the horizontal line test and therefore has an inverse. Based upon the preceding discussion, we interchange x and y and then solve for y.

$$y = \frac{1}{4}x + 3 \qquad \textit{We interchange } x \textit{ and } y.$$

$$x = \frac{1}{4}y + 3 \qquad \textit{Now we solve for } y.$$

$$4x = y + 12$$

$$4x - 12 = y \qquad \textit{This is the inverse function.}$$

Thus $\boxed{f^{-1}(x) = 4x - 12.}$

It is left to the reader to verify that if we compose f and f^{-1} in either order we will get x. (See Example 1.) ■

Thus we have seen that an equation of a one-to-one function can be viewed as defining y as a function of x or vice versa and that if we interchange the roles of x and y, we obtain a function that is the inverse of the original function.

Keep in mind that since we are interchanging the roles of x and y, we also interchange the domain and range. That is, the domain of the inverse function is the range of the original function and vice versa.

The following box summarizes what we have discussed thus far.

INVERSE FUNCTIONS

In order to find the inverse of a one-to-one function $y = f(x)$,

1. Interchange x and y in the equation $y = f(x)$.
2. Solve the resulting equation for y, obtaining the inverse function.
3. The domain of the inverse function is the range of the original function and the range of the inverse function is the domain of the original function.

EXAMPLE 3 Given $y = f(x) = x^3$, find $f^{-1}(x)$ and its domain.

Solution Again we begin by interchanging x and y, and then we solve for y.

$y = x^3$ *Interchange x and y.*

$x = y^3$ *Take the cube root of both sides.*

$\sqrt[3]{x} = y$ *This is the inverse function. Thus* $\boxed{f^{-1}(x) = \sqrt[3]{x}}$

The domain of the inverse function is the set of all real numbers.

The functions $y = f(x) = x^3$ and $y = f^{-1}(x) = \sqrt[3]{x}$ clearly illustrate the idea of a function and its inverse undoing each other. What the cubing function does, the cube root function undoes. ■

GRAFFIX

1. Use a graphics calculator. Sketch the graphs of $y = \frac{1}{4}x + 3$ (from Example 2), its inverse $y = 4x - 12$, and $y = x$ on the same set of coordinate axes. Can you make a conjecture as to the relationship between the graphs of the first two equations with respect to the graph of $y = x$?
2. Sketch the graphs of $y = x^3$ (from Example 3), its inverse $y = \sqrt[3]{x}$, and $y = x$ on the same set of coordinate axes. You may find it helpful to use **RANGE** to change the viewing window. Can you make a conjecture as to the relationship between the graphs of the first two equations with respect to the graph of $y = x$?

EXAMPLE 4 Let $y = f(x) = \frac{x}{x + 2}$. Find $f^{-1}(x)$ and verify that it satisfies the definition of an inverse function.

Solution

$y = \frac{x}{x + 2}$ *Interchange x and y.*

$x = \frac{y}{y + 2}$ *Now solve for y.*

$x(y + 2) = y$

$xy + 2x = y$ *Which becomes*

$2x = y - xy = y(1 - x)$ *And so we have*

$\frac{2x}{1 - x} = y$ *This is the inverse function.*

$$\boxed{f^{-1}(x) = \frac{2x}{1 - x}}$$

According to the definition, we need to verify that $f^{-1}[f(x)] = x$ and $f[f^{-1}(x)] = x$.

$$f^{-1}[f(x)] = f^{-1}\left(\frac{x}{x+2}\right) = \frac{2\left(\frac{x}{x+2}\right)}{1 - \frac{x}{x+2}} = \frac{2\left(\frac{x}{x+2}\right)\cdot(x+2)}{\left(1 - \frac{x}{x+2}\right)\cdot(x+2)} = \frac{2x}{x+2-x} = \frac{2x}{2} = x$$

and

$$f[f^{-1}(x)] = f\left(\frac{2x}{1-x}\right) = \frac{\frac{2x}{1-x}}{\frac{2x}{1-x}+2} = \frac{\left(\frac{2x}{1-x}\right)\cdot(1-x)}{\left(\frac{2x}{1-x}+2\right)\cdot(1-x)} = \frac{2x}{2x+2(1-x)} = \frac{2x}{2} = x$$

as required. ■

In Example 1 we found that $y = f(x) = \frac{1}{2}x - 5$ and $y = g(x) = 2x + 10$ are inverse functions. Let's graph these two functions on the same coordinate system. The graphs appear in Figure 3.60.

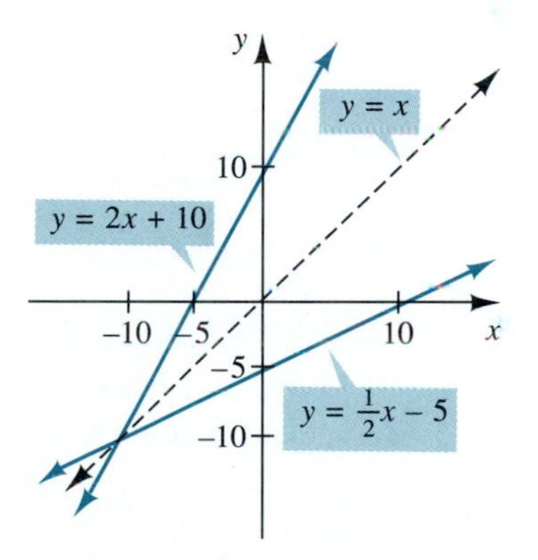

FIGURE 3.60

Note that we have also drawn in the line $y = x$ and that the graphs of the inverse functions $y = f(x) = \frac{1}{2}x - 5$ and $y = g(x) = 2x + 10$ appear to be symmetric with respect to the line $y = x$. In other words, if we placed a mirror along the line $y = x$, the graphs of $f(x)$ and $g(x)$ would be reflections of each other. (Keep in mind that we have already seen symmetry with respect to two particular lines—the x- and y-axes.)

The graphs of inverse functions always exhibit this type of symmetry.

> The graph of $y = f^{-1}(x)$ can be obtained by reflecting the graph of $y = f(x)$ about the line $y = x$ (and vice versa).

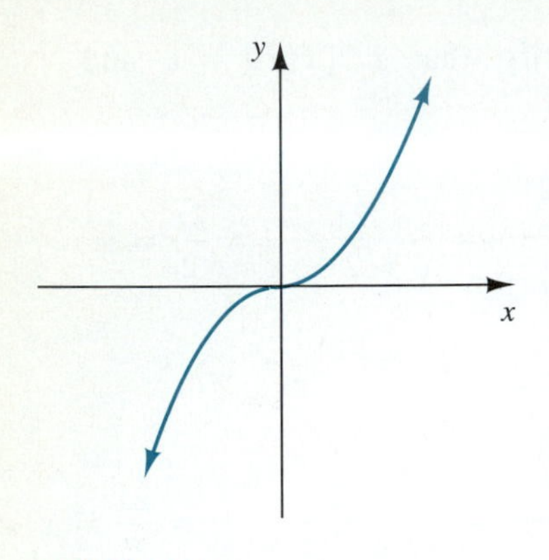

FIGURE 3.61

A more detailed discussion of symmetry with respect to a line appears in Exercise 53 at the end of this section.

EXAMPLE 5 Use the graph appearing in Figure 3.61 to answer the following.

(a) Is this the graph of a function? **(b)** Does this function have an inverse?
(c) Let $f(x)$ be the function whose graph is given. Sketch the graph of $f^{-1}(x)$.

Solution

(a) Since this graph satisfies the vertical line test, it is the graph of the function.

(b) Since this graph satisfies the horizontal line test, this function does have an inverse.

(c) We draw the original graph and the line $y = x$. The graph of $f^{-1}(x)$ is obtained by reflecting the graph of $f(x)$ about the line $y = x$. See Figure 3.62.

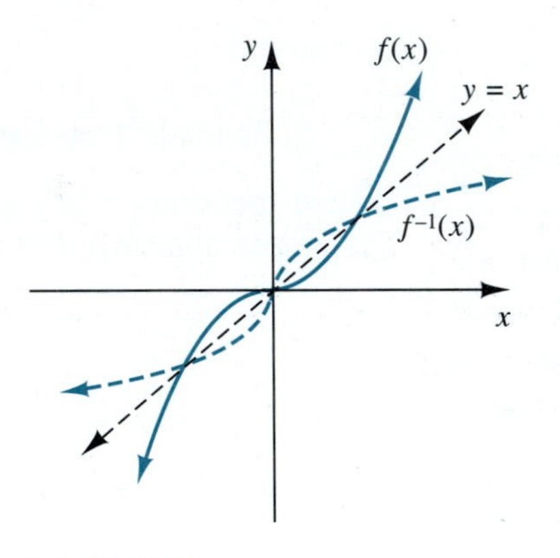

FIGURE 3.62

We will return to this idea of an inverse function in conjunction with a number of special functions we will encounter later in this text.

EXERCISES 3.6

In Exercises 1–6, use the given graph to determine whether the function has an inverse.

1.

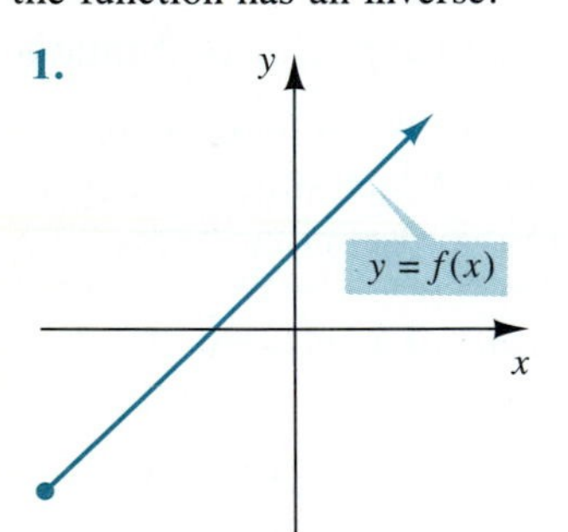

2.

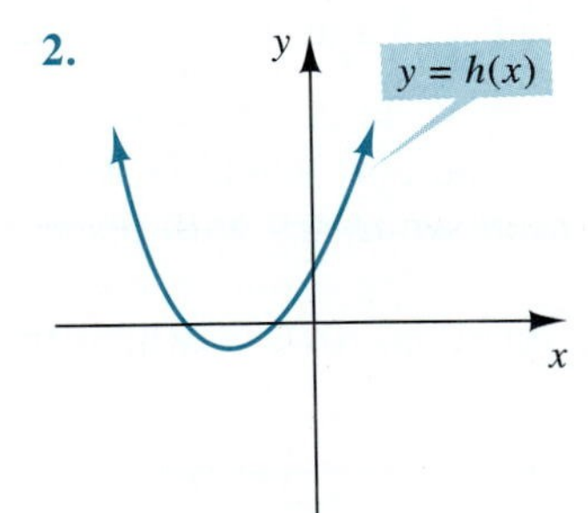

3.

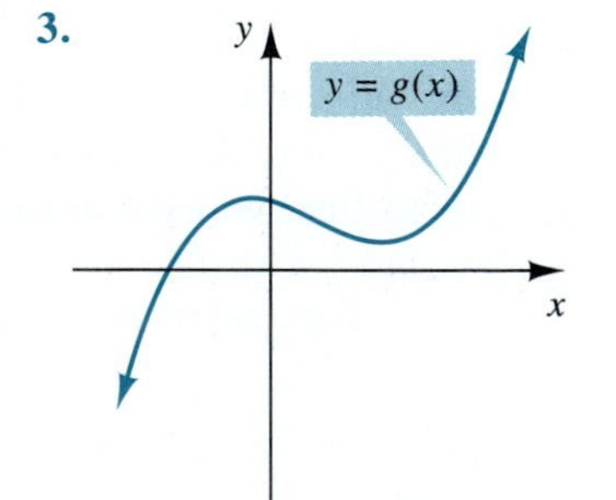

4.

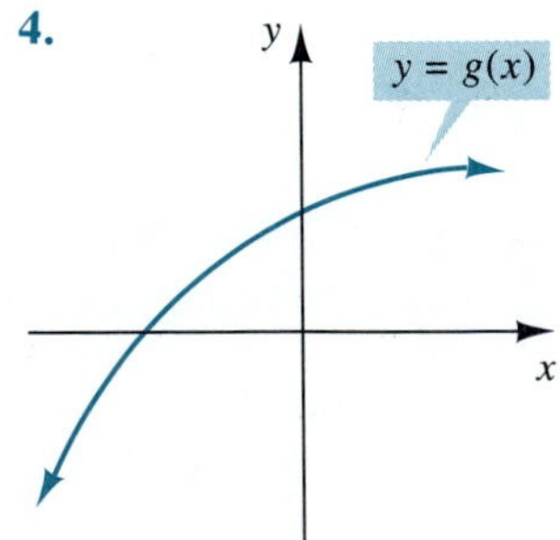

5.

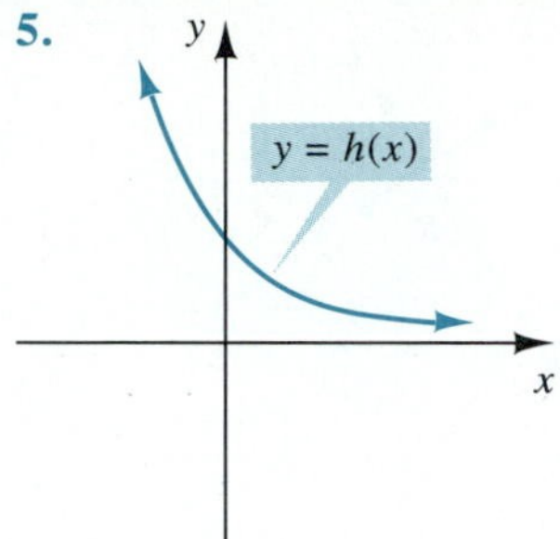

6.

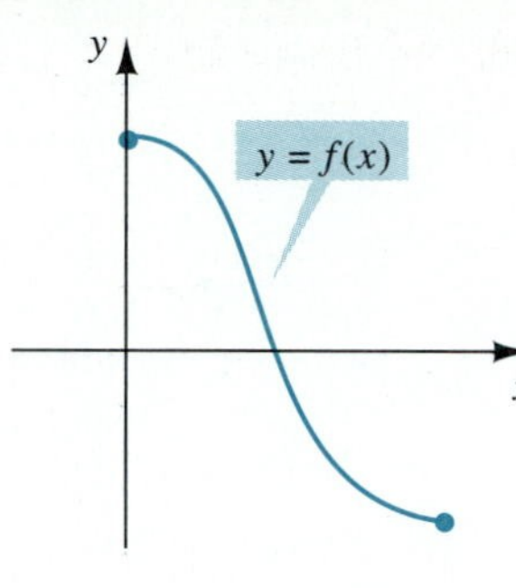

In Exercises 7–14, verify that $f(x)$ and $g(x)$ are inverse functions.

7. $f(x) = x - 1; \quad g(x) = x + 1$

8. $f(x) = 6x; \quad g(x) = \dfrac{x}{6}$

9. $f(x) = x^3; \quad g(x) = \sqrt[3]{x}$

10. $f(x) = 5x + 4; \quad g(x) = \dfrac{x - 4}{5}$

11. $f(x) = \dfrac{1}{3}x + \dfrac{2}{5}; \quad g(x) = 3x - \dfrac{6}{5}$

12. $f(x) = \sqrt[5]{x} + 6; \quad g(x) = (x - 6)^5$

13. $f(x) = \dfrac{4x}{x + 4}; \quad g(x) = \dfrac{4x}{4 - x}$

14. $f(x) = \dfrac{x + 5}{2x + 1}; \quad g(x) = \dfrac{5 - x}{2x - 1}$

15. Show that $f(x) = \dfrac{1}{x}$ is its own inverse function.

16. Verify that $f(x) = \dfrac{x + 1}{x - 1}$ is its own inverse function.

In Exercises 17–32, find the inverse of the given function.

17. $y = f(x) = 5x - 9$

18. $y = g(x) = \dfrac{1}{3}x + 6$

19. $y = F(x) = 2x^3 + 1$

20. $y = G(x) = 8 - x^5$

21. $y = h(x) = \dfrac{1}{x + 4}$

22. $y = H(x) = \dfrac{1}{x} + 4$

23. $y = f(x) = 2\sqrt{x} - 7$

24. $y = F(x) = \sqrt{2x - 7}$

25. $y = h(x) = x^2 - 1$ for $x \geq 0$

26. $y = h(x) = (x - 1)^2$ for $x \geq 1$

27. $y = g(x) = \dfrac{2x + 5}{3x - 2}$

28. $y = G(x) = \dfrac{6x}{x + 3}$

29. $y = f(x) = x^{3/5}$

30. $y = f(x) = -x^{-5/3}$

31. $y = f(x) = x^{-5/7} + 1$

32. $y = g(x) = \sqrt[3]{x^3 - 5}$

In Exercises 33–38, find $f^{-1}(x)$ and verify that $f^{-1}[f(x)] = f[f^{-1}(x)] = x$.

33. $f(x) = 7x - 6$

34. $f(x) = \dfrac{2x - 9}{4}$

35. $f(x) = 1 - \dfrac{3}{x}$

36. $f(x) = \dfrac{4 - x}{3x}$

37. $f(x) = \dfrac{5x + 3}{1 - 2x}$

38. $f(x) = \sqrt[3]{x + 1}$

39. Find the inverse of $y = f(x) = x^2$ for $x \geq 0$. Explain why the restriction $x \geq 0$ is needed. Find the domain and range of both f and f^{-1}.

40. Find the inverse of $y = f(x) = (x + 1)^2$ for $x \geq -1$. Explain why the restriction $x \geq -1$ is needed. Find the domain and range of both f and f^{-1}.

In Exercises 41–48, determine whether the given graph of a function has an inverse. If it does, sketch the graph of the inverse function.

41.

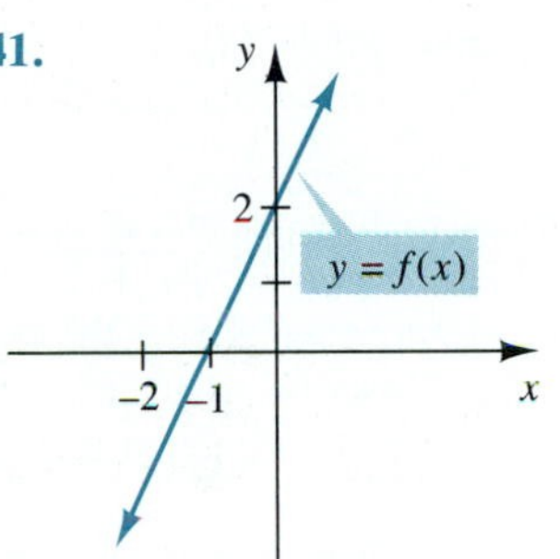

42.

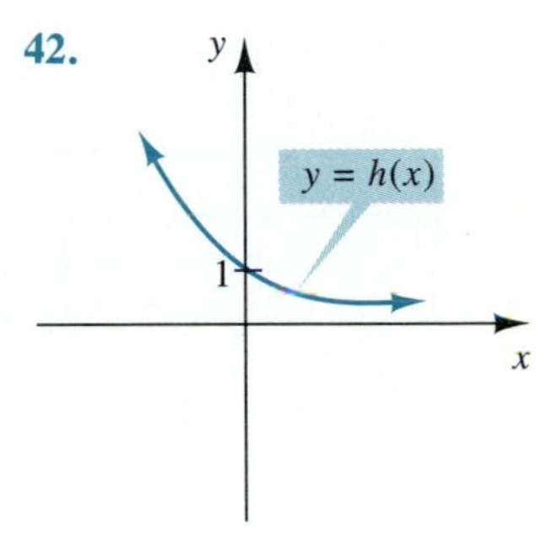

43.

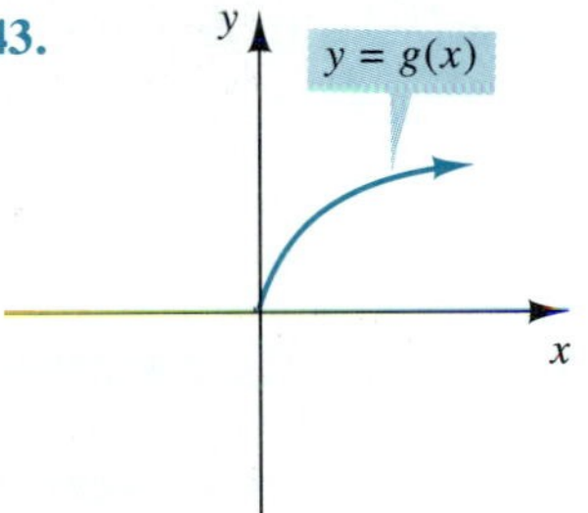

44.

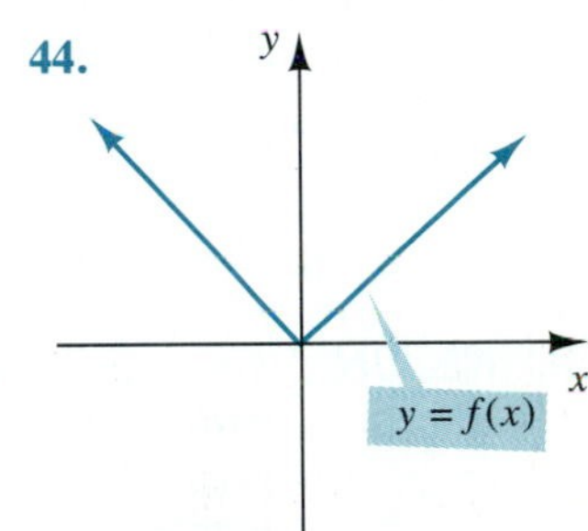

45.

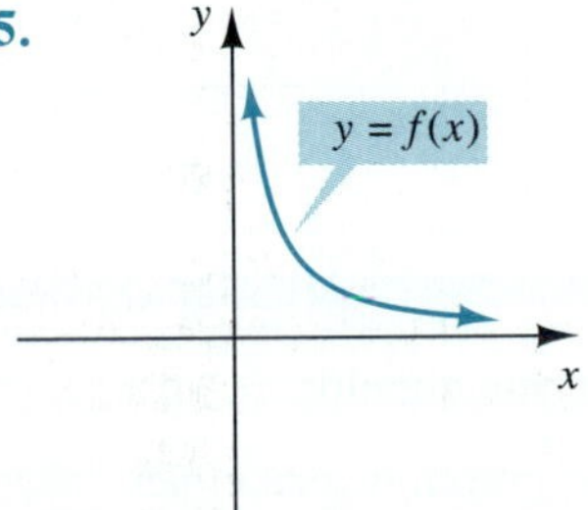

46.

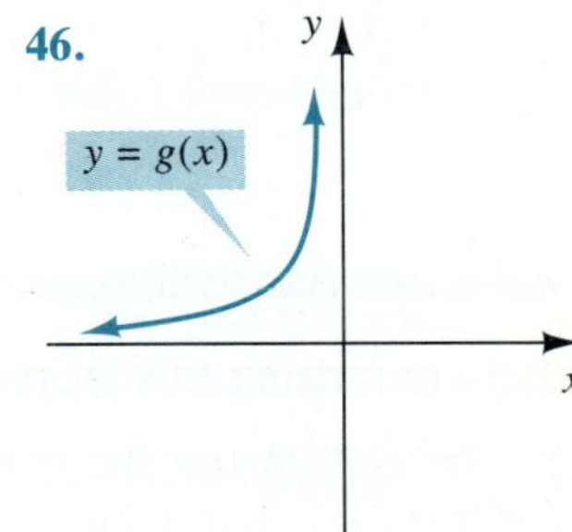

47.

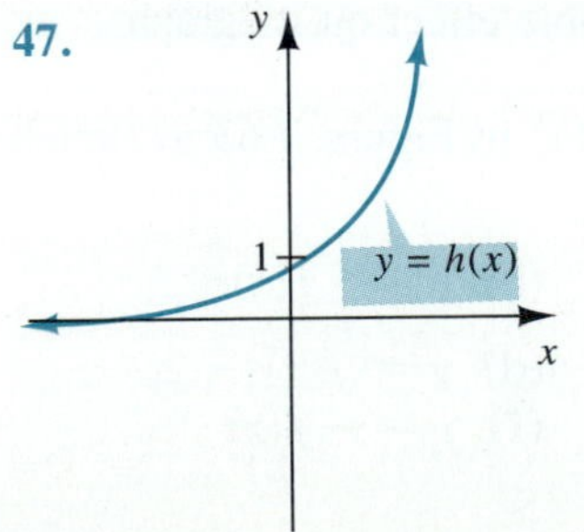

48.

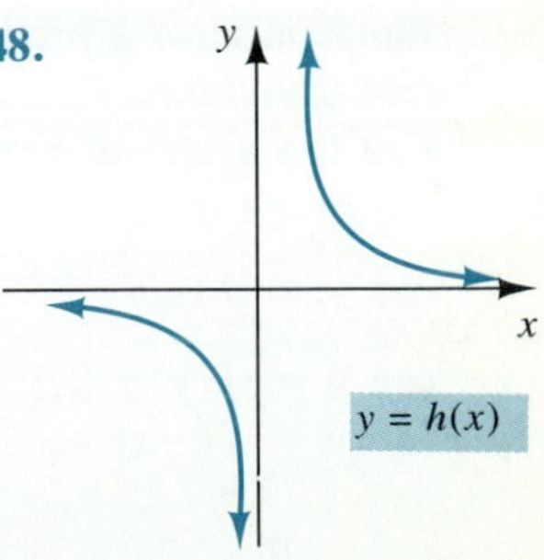

QUESTIONS FOR THOUGHT

49. Suppose $y = f(x)$ has an inverse. Discuss the meaning of the following set of implications.

$$y = f(x) \Rightarrow f^{-1}(y) = f^{-1}[f(x)] \Rightarrow f^{-1}(y) = x$$

50. Once we understand that $f^{-1}(x)$ is the function that has the property that $(f \circ f^{-1})(x) = x$, we may use this property as an alternative approach to finding f^{-1} as follows: Let $f(x) = 5x - 3$; then f^{-1} must satisfy

$$f[f^{-1}(x)] = x \qquad \textit{Use the definition of } f(x).$$

$$5[f^{-1}(x)] - 3 = x \qquad \textit{Solve for } f^{-1}(x).$$

$$f^{-1}(x) = \frac{x+3}{5}$$

Use this approach to find $f^{-1}(x)$ for $f(x) = \sqrt[3]{x+7}$.

51. In this section we outline a procedure for finding the inverse of a one-to-one function. Describe what happens when you try to apply this procedure to a function that is *not* one-to-one.

52. We have stated that the graphs of inverse functions are symmetric with respect to the line $y = x$. This exercise expands upon the idea of symmetry with respect to a line.

DEFINITION Two points P and Q are said to be **symmetric with respect to a line L** if and only if L is the perpendicular bisector of the line segment joining P and Q. The accompanying figure illustrates this definition.

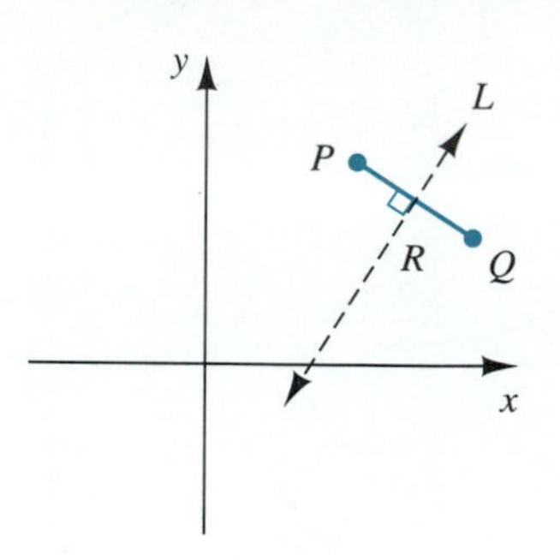

The points P and Q are symmetric with respect to the line L. Note that $\overline{PR} = \overline{QR}$. In Exercise 50 we saw that $y = 5x - 3$ and $y = \frac{x+3}{5}$ are inverse functions. The point $P(2, 7)$ is on the graph of $y = 5x - 3$, and, of course, $Q(7, 2)$ is on the graph of $y = \frac{x+3}{5}$. Verify that these two points are symmetric with respect to the line $y = x$ by using this definition. HINT: First, show that the line joining the points $P(2, 7)$ and $Q(7, 2)$ is perpendicular to the line $y = x$. Second, let $R(a, a)$ be the point where the line segment $\overline{PQ}$ intersects the line $y = x$ and show $|\overline{PR}| = |\overline{QR}|$. Note that it was not necessary to determine the value of a.

GRAFFIX

In Exercises 53–56, find the inverse of the given function, then use a graphics calculator or computer to graph the given function, its inverse, and $y = x$ on the same coordinate system.

53. $y = 3x - 6$

54. $y = x^3 - 1$

55. $y = \frac{x-1}{x+1}$

56. $y = \frac{5}{x} - 1$

Chapter 3 SUMMARY

After completing this chapter you should:

1. Be able to use the basic graphing principles. (Sections 3.1 and 3.2)
Specific algebraic changes made to the equation of a function have a predictable effect on its graph.
For example:
Use the graph of $y = f(x)$ in Figure 3.63 to sketch the graph of

(a) $y = 2f(x)$ **(b)** $y = \frac{1}{2}f(x)$
(c) $y = f(x - 2)$ **(d)** $y = f(x) - 2$
(e) $y = f(-x)$ **(f)** $y = -f(x)$

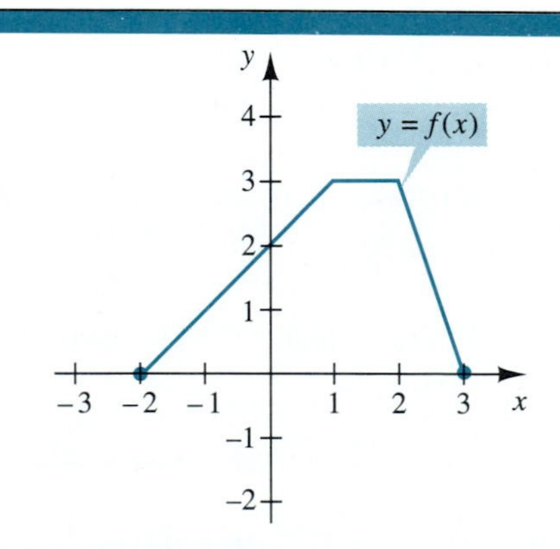

FIGURE 3.63

Solution:

(a) The graph of $2f(x)$ is obtained by deflecting the graph of $f(x)$ away from the x-axis. The graph appears in Figure 3.64(a).

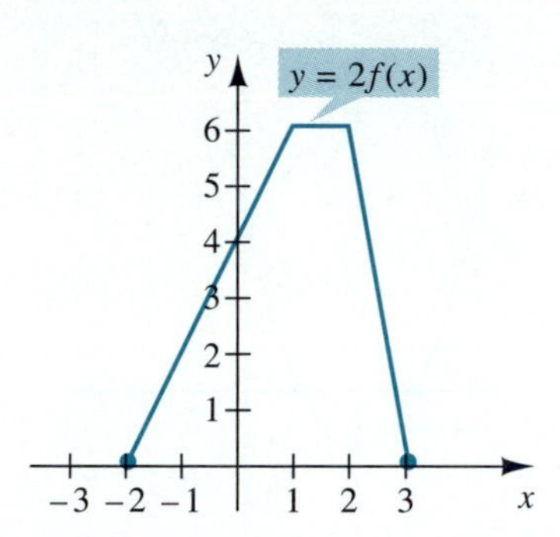

FIGURE 3.64 (a)

(b) The graph of $y = \frac{1}{2}f(x)$ is obtained by pulling the graph of $f(x)$ toward the x-axis. The graph appears in Figure 3.64(b).

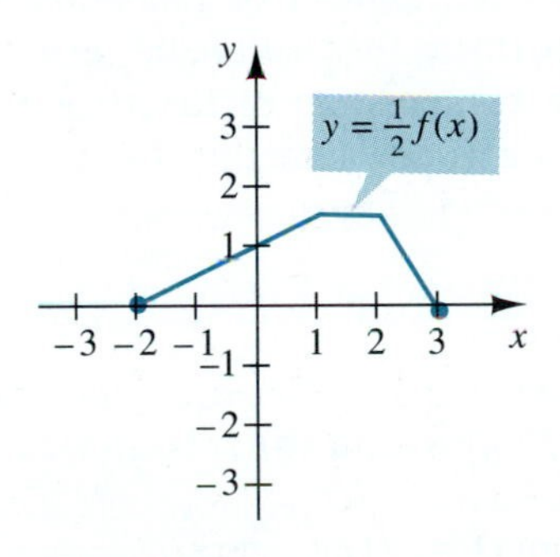

FIGURE 3.64 (b)

(c) The graph of $y = f(x - 2)$ is obtained by shifting the original graph 2 units to the right. The graph appears in Figure 3.64(c).

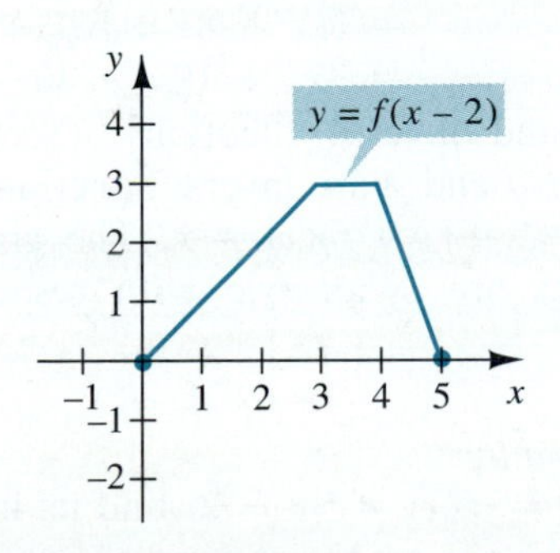

FIGURE 3.64 (c)

(d) The graph of $f(x) - 2$ is obtained by shifting the original graph 2 units down. The graph appears in Figure 3.64(d).

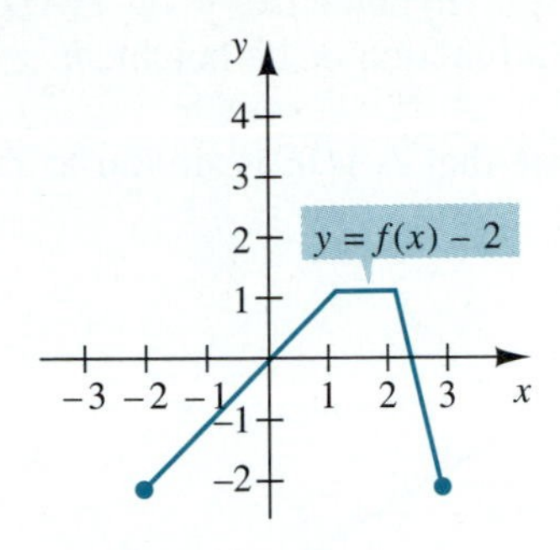

FIGURE 3.64 (d)

(e) The graph of $y = f(-x)$ is obtained by reflecting the graph of $y = f(x)$ about the y-axis. The graph appears in Figure 3.64(e).

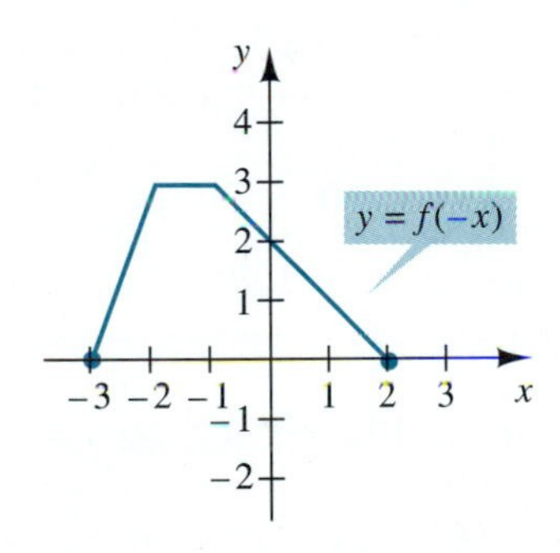

FIGURE 3.64 (e)

(f) The graph of $y = -f(x)$ is obtained by reflecting the graph of $y = f(x)$ about the x-axis. The graph appears in Figure 3.64(f).

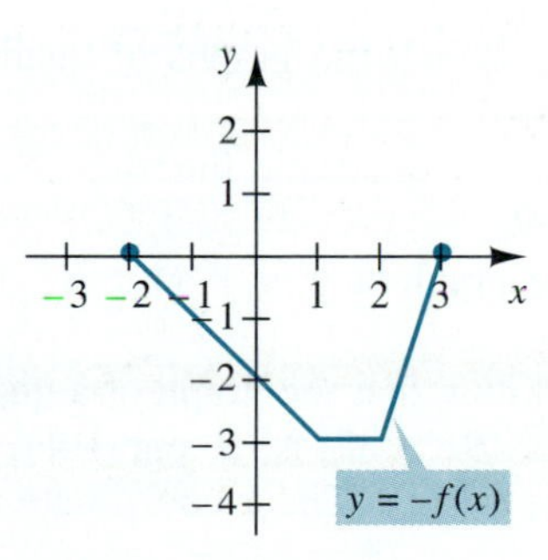

FIGURE 3.64 (f)

2. Be able to extract functions from *real-life* situations. (Section 3.3)
For example:
Express the hypotenuse y of $\triangle ADE$, in Figure 3.65, as a function of its height, h.
Solution:
We can see that $\triangle ADE$ is similar to $\triangle ABC$. Therefore,

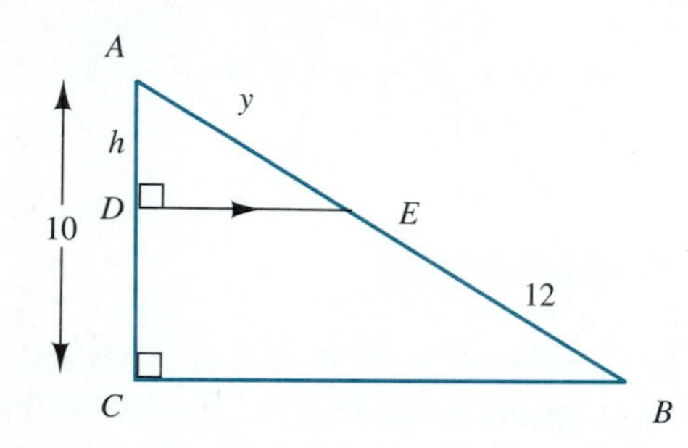

FIGURE 3.65

$$\frac{y}{h} = \frac{y + 12}{10}$$
$$10y = hy + 12h$$
$$10y - hy = 12h$$
$$y(10 - h) = 12h$$

Solving for y we get y as a function of h.

$$y = y(h) = \frac{12h}{10 - h}$$

Since h is the height of $\triangle ADE$, h must be greater than or equal to 0 and less than or equal to the height of $\triangle ABC$ which is 10. Therefore the domain for $y(h)$ is $\{h \mid 0 \le h \le 10\}$.

3. Be able to sketch the graphs of quadratic functions. (Section 3.4)
The graph of a quadratic function is a parabola.
For example:
Sketch the graph of $y = f(x) = x^2 + 4x + 3$.
Solution:
Since we know that the graph of a quadratic function $y = Ax^2 + Bx + C$ is a parabola, we begin by finding the vertex. The x-coordinate of the vertex is $x = -\frac{B}{2A} = -\frac{4}{2(1)} = -2$. The y-coordinate of the vertex is found by computing $f(-2)$. We get $y = (-2)^2 + 4(-2) + 3 = -1$. Thus the vertex is $(-2, -1)$.

We find the y-intercept, $f(0)$, to be 3, and the x-intercepts (found by setting $y = 0$) to be $x = -1$ and $x = -3$. The graph appears in Figure 3.66.

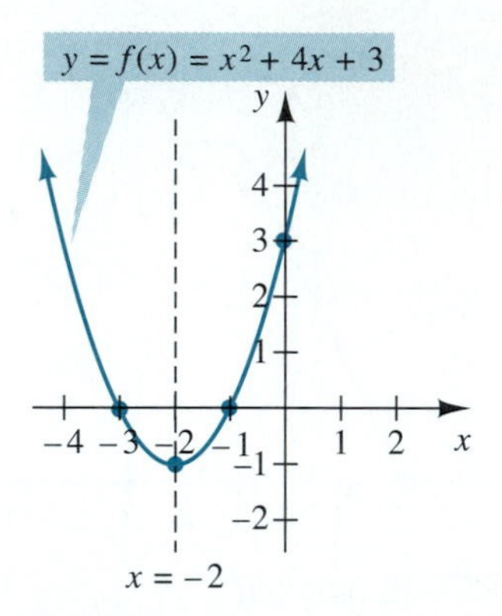

FIGURE 3.66

4. Recognize that the maximum or minimum values of a quadratic function can be found by identifying the vertex of the parabola. (Section 3.4)
5. Understand the algebra of functions. (Section 3.5) You can perform the arithmetic operations on functions. The composition of functions is another way of producing a new function from two given functions.

For example:
Given $f(x) = 3x^2 - 5x + 2$ and $g(x) = x + 4$, find **(a)** $(f \cdot g)(x)$ and **(b)** $(f \circ g)(x)$.
Solution:

(a)
$$\begin{aligned}(f \cdot g)(x) &= f(x) \cdot g(x)\\ &= (3x^2 - 5x + 2)(x + 4)\\ &= 3x^3 + 7x^2 - 18x + 8\end{aligned}$$

(b)
$$\begin{aligned}(f \circ g)(x) &= f[g(x)]\\ &= f(x + 4)\\ &= 3(x + 4)^2 - 5(x + 4) + 2\\ &= 3(x^2 + 8x + 16) - 5x - 20 + 2\\ &= 3x^2 + 19x + 30\end{aligned}$$

6. Understand inverse functions. (Section 3.6) Two functions f and g are inverse functions if and only if $(f \circ g)(x) = (g \circ f)(x) = x$. The graphs of inverse functions are symmetric with respect to the line $y = x$.

For example:
Given $y = f(x) = 5x - 2$, find its inverse function and sketch the graphs of f and f^{-1} on the same coordinate system.

Solution:
In order to find the inverse function of $y = f(x)$, we interchange x and y and then solve for y.

$y = 5x - 2$ *Interchange x and y.*

$x = 5y - 2$ *Solve for y.*

$$y = f^{-1}(x) = \frac{x + 2}{5}$$

Note that the graphs of f and f^{-1} in Figure 3.67 are symmetric with respect to the line $y = x$.

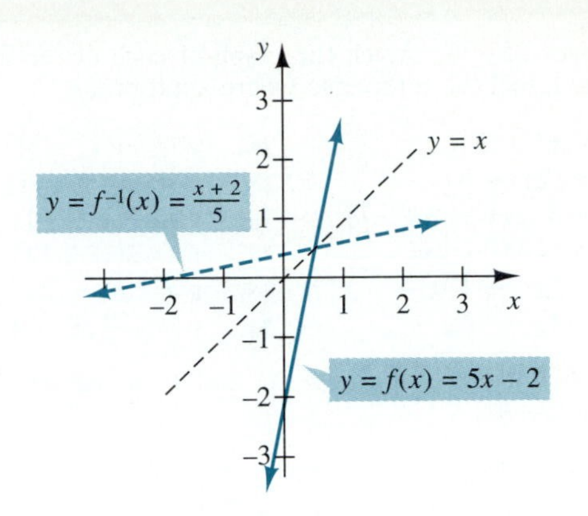

Chapter 3 REVIEW EXERCISES

In Exercises 1–16, use the graphs of $y = f(x)$ and $y = g(x)$ given in the figures to graph each of the following functions.

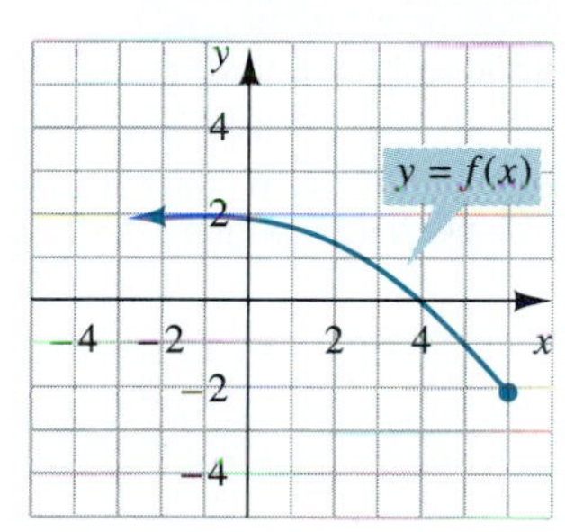

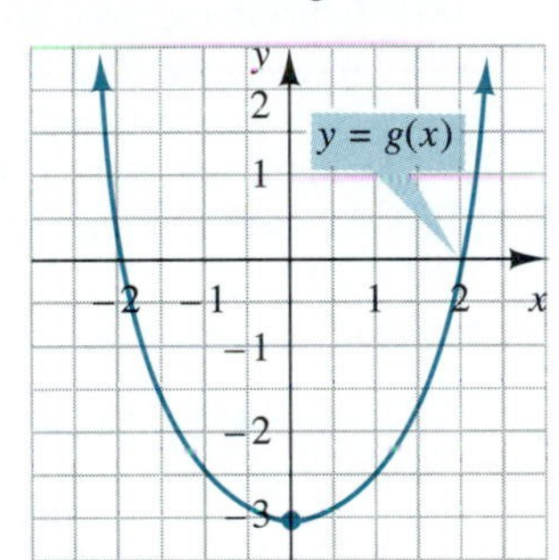

1. $f(-x)$
2. $-f(x)$
3. $-g(x)$
4. $g(-x)$
5. $f(x + 2)$
6. $f(x) + 2$
7. $|f(x)|$
8. $|g(x)|$
9. $g(x) + 3$
10. $-f(x - 2)$
11. $g(3 - x)$
12. $f(x) - 2$
13. $-f(x) - 1$
14. $g(-x) + 3$
15. $-f(x - 4)$
16. $|g(x) - 1|$

In Exercises 17–20, use the graph of $y = f(x)$ given in the figure to determine the equation for each of the following graphs in terms of $f(x)$.

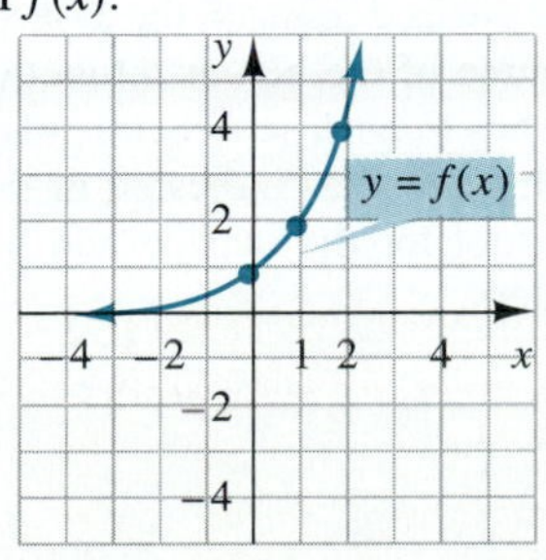

17.

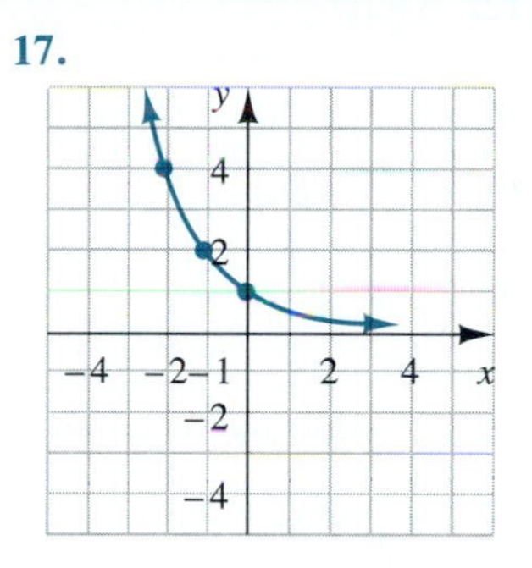

18.

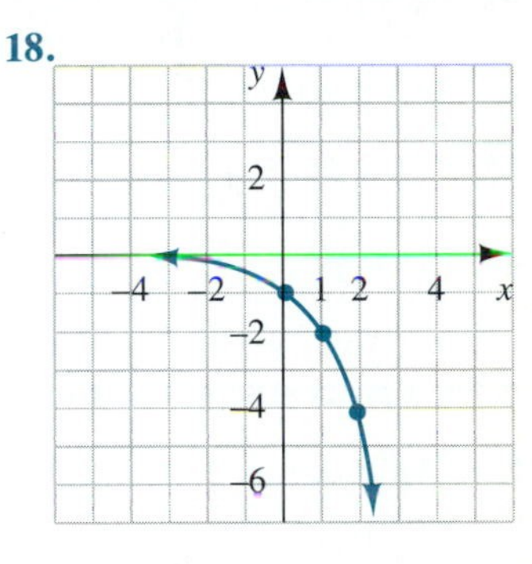

19.

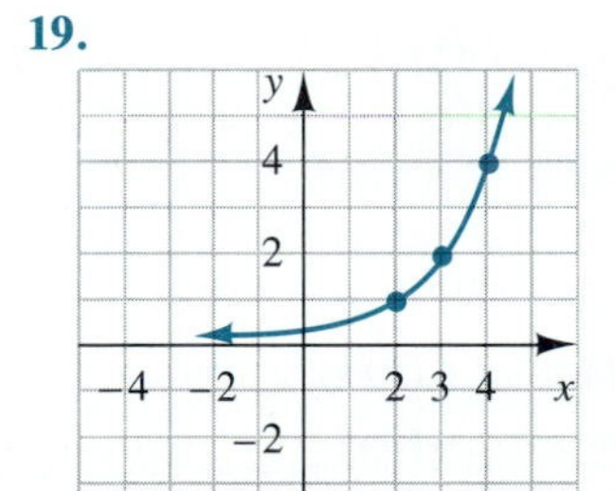

20.

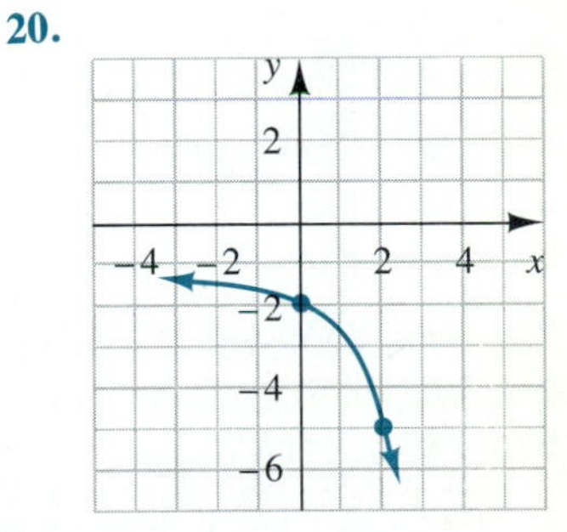

21. Find the vertex and the axis of symmetry of the quadratic function $f(x) = -(x - 4)^2 + 6$.
22. Find the vertex and axis of symmetry of the parabola whose equation is $f(x) = x^2 - 6x + 5$.
23. What is the maximum value of the function $y = -2x^2 - 9x$?
24. What is the minimum value of the function $f(x) = 3x^2 - 4x + 2$?

In Exercises 25–36, sketch the graph of each of the following functions. Label the intercepts where appropriate.

25. $y = x^2 - 5$
26. $y = -(x - 3)^2 - 2$
27. $y = |2x - 6|$
28. $f(x) = |x + 1| - 3$
29. $f(x) = 2x^2 - 3x - 2$
30. $y = -x^2 + 4x - 5$
31. $f(x) = 36 - 9x^2$
32. $y = 2(x - 1)^2 - 8$
33. $y = 2x^2 - 10x$
34. $y = |x^2 - 2x - 3|$
35. $y = 3x^2 - 6x + 2$
36. $y = -3x^2 - 4$

Use the functions f, g, h, r, s, and t as defined here for Exercises 37–70.

$f(x) = 3x^2 - 4x + 1$ $\qquad g(x) = 5 - 2x$ $\qquad h(x) = -3$

$r(x) = x^3 + 8$ $\qquad s(x) = \dfrac{-2}{x}$ $\qquad t(x) = \dfrac{3}{x - 4}$

In Exercises 37–50, find the required value.

37. $(g + h)(2)$
38. $(r - t)(-3)$
39. $(f \cdot s)(-6)$
40. $(t \cdot f)(3)$
41. $\left(\dfrac{g}{f}\right)(4)$
42. $\left(\dfrac{s}{t}\right)(4)$
43. $\left(\dfrac{r}{h}\right)(-1)$
44. $(s - t)(8)$
45. $(f \circ g)(4)$
46. $(g \circ f)(4)$
47. $(h \circ s)(-6)$
48. $(s \circ h)(8)$
49. $(r \circ g)(a)$
50. $(s \circ h)(c)$

In Exercises 51–70, use the functions f, g, h, r, s, and t as defined previously to find the required function and its domain.

51. $(f - g)(x)$
52. $(r \cdot s)(x)$
53. $\left(\dfrac{s}{h}\right)(x)$
54. $\left(\dfrac{h}{s}\right)(x)$
55. $(r - f)(x)$
56. $(s + t)(x)$
57. $(g \cdot t)(x)$
58. $\left(\dfrac{g}{t}\right)(x)$
59. $(g \circ f)(x)$
60. $(f \circ g)(x)$
61. $(s \circ r)(x)$
62. $(r \circ s)(x)$
63. $(t \circ h)(x)$
64. $(h \circ t)(x)$
65. $(g \circ g)(x)$
66. $(t \circ t)(x)$
67. $(t \circ s)(x)$
68. $(s \circ t)(x)$
69. $(s \circ g \circ t)(x)$
70. $(t \circ s \circ g)(x)$

In Exercises 71–74, find functions $f(x)$ and $g(x)$ so that the given function $h(x) = (f \circ g)(x)$.

71. $h(x) = \dfrac{1}{\sqrt{x + 2}}$
72. $h(x) = (3x + 2)^4$
73. $h(x) = (5x - 7)^{-3}$
74. $h(x) = \sqrt[3]{\dfrac{x}{x + 1}}$

75. Suppose a laboratory technician is growing a bacteria culture in which the number of bacteria present, N, depends upon the Celsius temperature, C, of the surrounding air, as given by the function

$$N(C) = C^2 + 125C + 1000 \qquad \text{for } 20 \le C \le 32$$

The Celsius temperature, C, depends in turn upon the number of hours, h, after the culture begins growing, as given by the function

$$C(h) = 4h + 10 \qquad \text{for } 0 \le h \le 6$$

(a) Express the number of bacteria, N, as a function of h.
(b) How many bacteria are present after 4 hours?
(c) After approximately how many hours are there 10,000 bacteria?

76. Suppose that the length of the radius of a circle is dependent upon the number of minutes, t, that have elapsed after $t = 0$ minutes according to the function

$$r(t) = 64 - t^2 \qquad \text{for } 0 \le t \le 8$$

(a) Express the area of the circle as a function of t.
(b) What is the area of the circle after 4 minutes?

77. Use the accompanying diagram to express the area of $\triangle ABC$ as a function of x. The point (x, y) is chosen in the first quadrant on the line through the points $(0, 0)$ and $(3, 4)$.

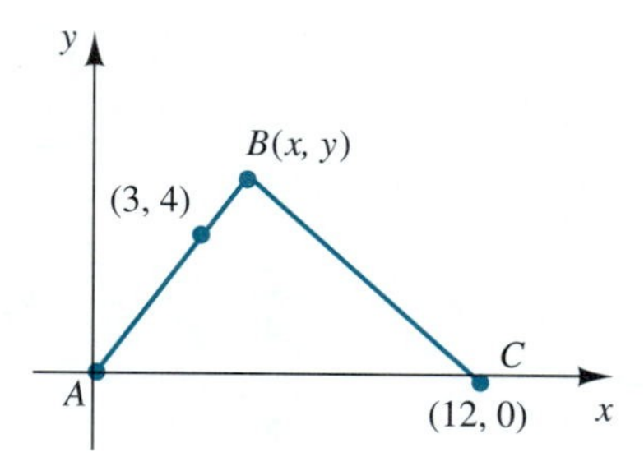

78. A closed right circular cylinder has a surface area of 20π sq cm. Express the volume of the cylinder as a function of r.

79. A closed rectangular box has a base that is x units wide and twice that long. If we let h be the height of the box and the total surface area of the box is 180 sq cm, express the volume of the box as a function of x.

80. A farmer wishes to enclose a rectangular garden and subdivide it with a fence as indicated in the figure. Fencing costs \$6 per foot for the outside and \$4 per foot for the interior dividers.

(a) Express the area that can be enclosed for a total cost of \$240 as a function of one variable.

(b) What is the maximum area that can be enclosed for \$240?

In Exercises 81–82, determine whether the function whose graph is given has an inverse.

81.

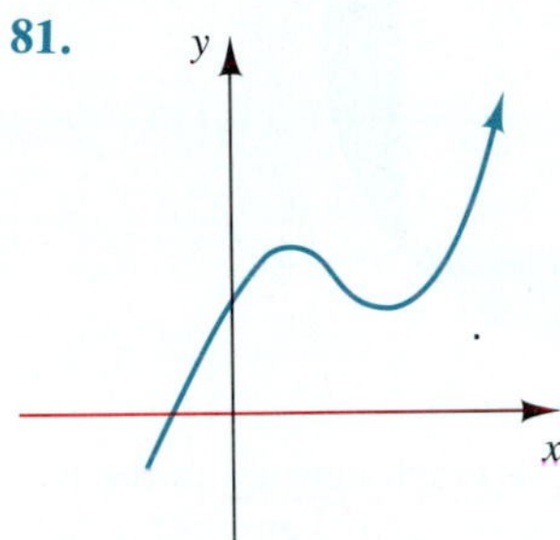

82.

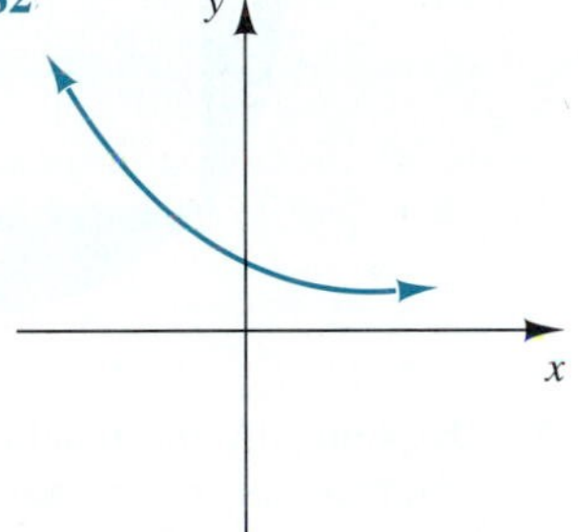

In Exercises 83–86 sketch the graph of the inverse of the given function f. Give the domain and range for both f and f^{-1}.

83.

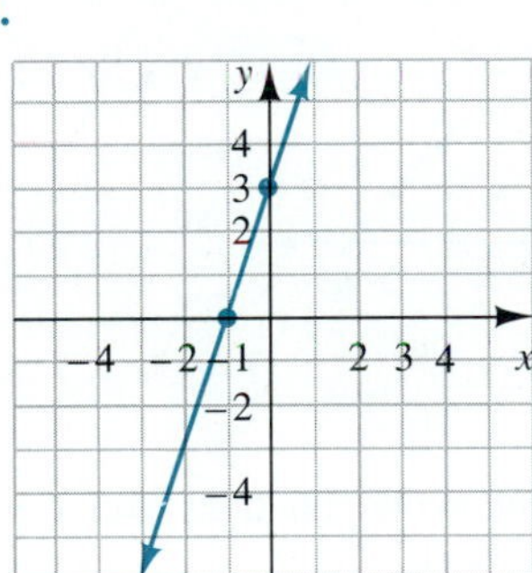

84.

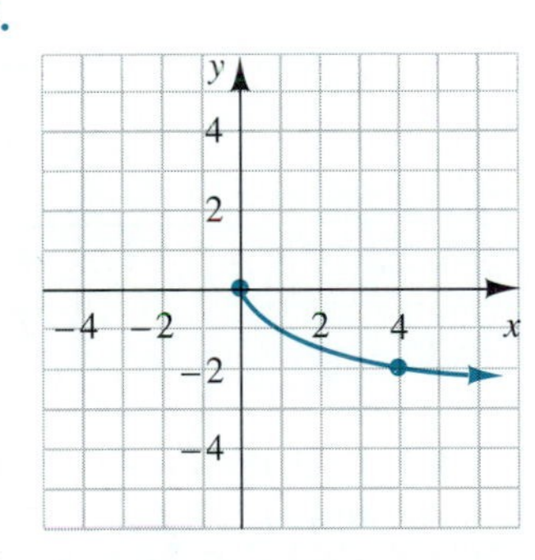

85.

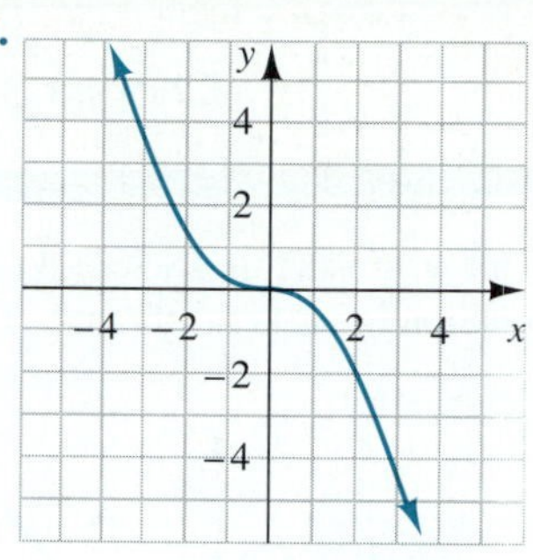

86.

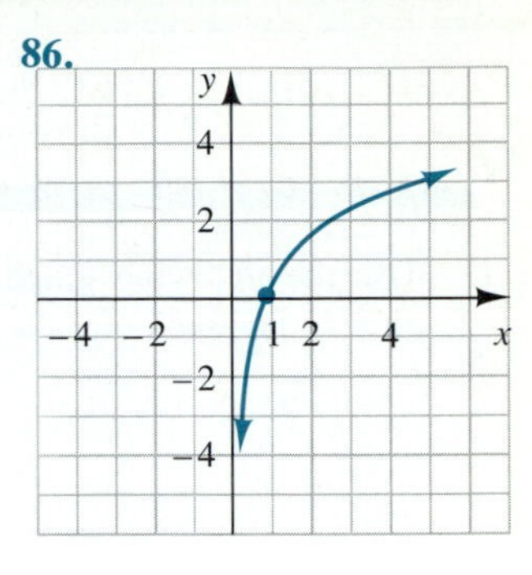

In Exercises 87–92, find the inverse of the given function and verify that

$$(f^{-1} \circ f)(x) = (f \circ f^{-1})(x) = x$$

87. $f(x) = \dfrac{1}{4}x + 5$

88. $f(x) = 8 - 5x$

89. $f(x) = \dfrac{x + 4}{x - 3}$

90. $f(x) = \sqrt{x - 5}$

91. $f(x) = \dfrac{3}{x + 6}$

92. $f(x) = \dfrac{1}{x} - 2$

Chapter 3 PRACTICE TEST

1. Use the following graphs for $y = f(x)$ and $y = g(x)$ to sketch the graph of each function in parts (a) through (f).

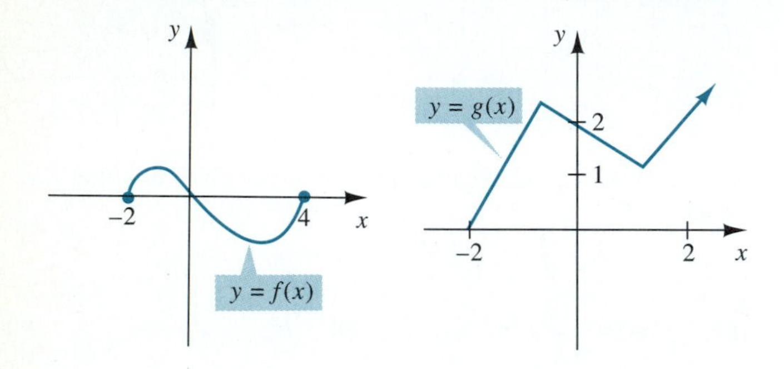

(a) $f(-x)$ **(b)** $-f(x)$ **(c)** $|f(x)|$
(d) $g(x) + 2$ **(e)** $g(x + 2)$ **(f)** $-f(x - 2)$

2. Find the vertex and axis of symmetry of the following quadratic functions.
(a) $y = 3(x + 5)^2 - 3$ **(b)** $y = -2x^2 - 5x + 1$

3. Sketch the graphs of each of the following functions. Label the intercepts where appropriate.
(a) $y = 25 - x^2$ **(b)** $f(x) = |4x + 8|$
(c) $y = -2(x + 3)^2 + 8$ **(d)** $f(x) = x^2 + 4x - 5$
(e) $y = 3x^2 - 48x$ **(f)** $f(x) = -x^2 + 3x - 4$
(g) $y = |2x^2 - 10|$ **(h)** $f(x) = 2x^2 - 5x - 1$

4. Given $f(x) = 7 - x^2$, $g(x) = \dfrac{3}{x + 2}$, and $h(x) = \sqrt{x - 1}$, find each of the following.
(a) $(f \cdot h)(3)$ **(b)** $(g + h)(10)$ **(c)** $\left(\dfrac{f}{g}\right)(-3)$
(d) $(f \circ g)(x)$ **(e)** $(g \circ f)(x)$ **(f)** $(g \circ g)(x)$
(g) $(g \circ f \circ h)(x)$

5. Let $f(x) = \dfrac{2x}{x + 1}$ and $g(x) = \dfrac{x}{x - 1}$.
(a) Find the domain of $(f \circ g)(x)$.
(b) Find the domain of $(g \circ f)(x)$.

6. A rectangle whose length is 12 and width is x is inscribed in a circle of radius r. Use the following figure to express the area of the shaded portion as a function of r.

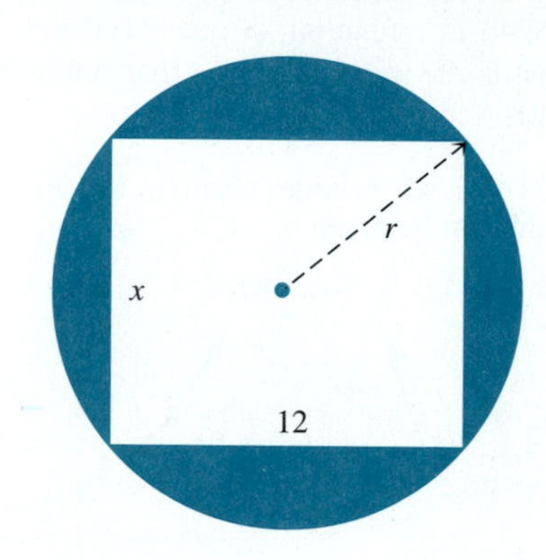

7. Explain why the function whose graph appears in the accompanying figure has an inverse, and sketch the graph of the inverse function.

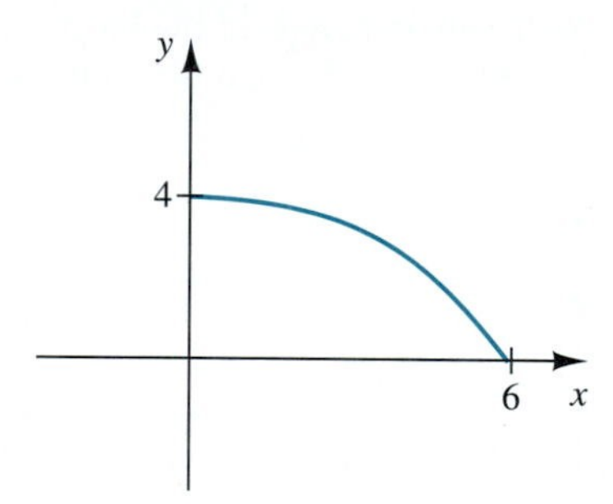

8. Find the inverse of the following functions and verify that $(f^{-1} \circ f)(x) = (f \circ f^{-1})(x) = x$.
(a) $f(x) = \dfrac{5x - 4}{3}$ **(b)** $f(x) = \sqrt[3]{2x + 9}$
(c) $f(x) = \dfrac{2}{x} - 5$

CHAPTER 4

Polynomial, Rational, and Radical Functions

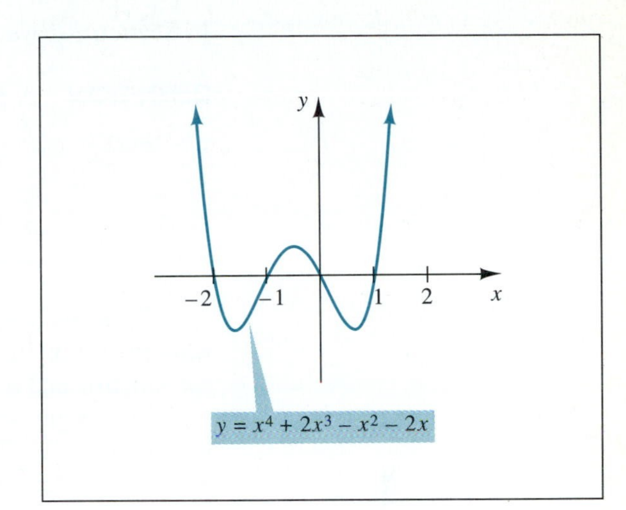

In this chapter we continue our discussion of the different types of functions mentioned in Chapter 3. Many of the ideas introduced in this chapter have wide applications throughout mathematics.

In the first four sections we investigate polynomial functions. We will examine rational functions in Section 4.5, and as we do we compare and contrast their behavior to that of polynomial functions. The chapter concludes with a discussion of radical functions and variation.

4.1 Polynomial Functions

In Section 3.2 we gave the following definition of a polynomial function.

DEFINITION A **polynomial function** is a function of the form

$$y = p(x) = a_n x^n + a_{n-1}x^{n-1} + a_{n-2}x^{n-2} + \cdots + a_2x^2 + a_1x + a_0, \qquad a_n \neq 0$$

Each a_i is assumed to be a real number, and n is a nonnegative integer. a_n is called the **leading coefficient.** Such a polynomial function is said to be of **degree n.**

From its definition we can see that the domain of a polynomial function is always the set of all real numbers. Recall that a polynomial of degree 1 is called a *linear function* and a polynomial of degree 2 is called a *quadratic function*.

We begin by examining the simplest polynomial functions of the form $y = f(x) = ax^n$. These are often called *pure power functions*. If $n = 0$ or 1, then we are dealing with a function of the form $y = a$ or $y = ax$, which we recognize as a function whose graph is a straight line.

GRAFFIX

Use a graphics calculator or computer.

1. Graph the following on the same set of coordinate axes:

$$y = x^2, \qquad y = x^4, \qquad y = x^6$$

Make the viewing rectangle $-3 \leq x \leq 3$ and $0 \leq y \leq 4$.
For even values of n, can you conjecture about how changing n affects the graph of $y = x^n$?

2. Graph the following on the same set of coordinate axes:

$$y = x^3, \qquad y = x^5, \qquad y = x^7$$

Make the viewing rectangle $-3 \leq x \leq 3$ and $-3 \leq y \leq 3$.
For odd values of n, can you conjecture about how changing n affects the graph of $y = x^n$?

We first consider positive even integers n. If $n = 2$, then we are dealing with a quadratic function whose graph we know to be a parabola. In order to get a sense of how these functions behave, we consider the values of $y = x^2$ and $y = x^4$ (Table 4.1) and the graphs that we get from this table (Figure 4.1).

Looking at the graphs we can see that on the interval $(-1, 1)$, the graph of $y = x^4$ is 'flatter' than the graph of $y = x^2$, whereas outside this interval (that is for $x < -1$ or $x > 1$), the graph of $y = x^4$ is 'steeper' than the graph of $y = x^2$. In other words, for $|x| < 1$, the graph of $y = x^4$ is below the graph of $y = x^2$, but for $|x| > 1$, the positions of the graphs are reversed.

TABLE 4.1

x	$y = x^2$	$y = x^4$
-2	$y = (-2)^2 = 4$	$y = (-2)^4 = 16$
-1.5	$y = (-1.5)^2 = 2.25$	$y = (-1.5)^4 = 5.0625$
-1	$y = (-1)^2 = 1$	$y = (-1)^4 = 1$
$-\frac{1}{2}$	$y = (-\frac{1}{2})^2 = \frac{1}{4}$	$y = (-\frac{1}{2})^4 = \frac{1}{16}$
0	$y = (0)^2 = 0$	$y = (0)^4 = 0$
$\frac{1}{2}$	$y = (\frac{1}{2})^2 = \frac{1}{4}$	$y = (\frac{1}{2})^4 = \frac{1}{16}$
1	$y = (1)^2 = 1$	$y = (1)^4 = 1$
1.5	$y = (1.5)^2 = 2.25$	$y = (1.5)^4 = 5.0625$
2	$y = (2)^2 = 4$	$y = (2)^4 = 16$

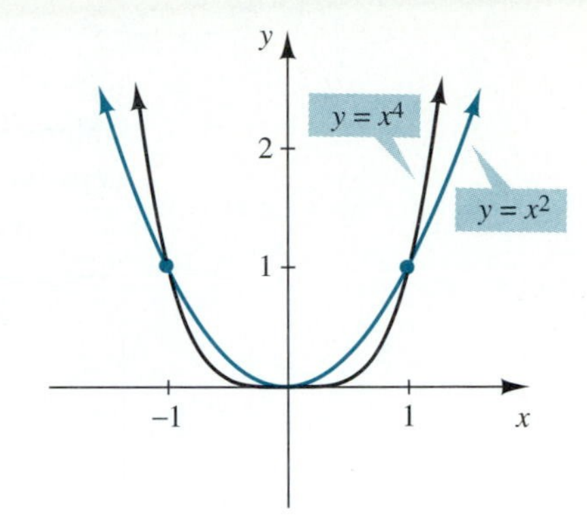

FIGURE 4.1

A moment's reflection should convince us that for all *even* integers $n > 1$, the graph of a function of the form $y = f(x) = x^n$ will have a shape similar to the graph of $y = x^2$. We call this the "x^2 look."*

It is important to recognize that although the graphs of $y = x^n$ for n even are similar, they are not identical. As n increases (but still remains even), the graphs of $y = x^n$ get flatter on the interval $(-1, 1)$ and steeper outside this interval (just as we saw in Figure 4.1).

Now let's examine the graphs of $y = x^n$ for odd integers $n > 1$. We begin with the function $y = f(x) = x^3$. Let's complete a table of values for $y = f(x) = x^3$ and $y = f(x) = x^5$ (Table 4.2) and sketch the suggested graph (Figure 4.2).

TABLE 4.2

x	$y = x^3$	$y = x^5$
-3	$y = (-3)^3 = -27$	$y = (-3)^5 = -243$
-2	$y = (-2)^3 = -8$	$y = (-2)^5 = -32$
-1	$y = (-1)^3 = -1$	$y = (-1)^5 = -1$
$-\frac{1}{2}$	$y = (-\frac{1}{2})^3 = -\frac{1}{8}$	$y = (-\frac{1}{2})^5 = -\frac{1}{32}$
0	$y = (0)^3 = 0$	$y = (0)^5 = 0$
$\frac{1}{2}$	$y = (\frac{1}{2})^3 = \frac{1}{8}$	$y = (\frac{1}{2})^5 = \frac{1}{32}$
1	$y = (1)^3 = 1$	$y = (1)^5 = 1$
2	$y = (2)^3 = 8$	$y = (2)^5 = 32$
3	$y = (3)^3 = 27$	$y = (3)^5 = 243$

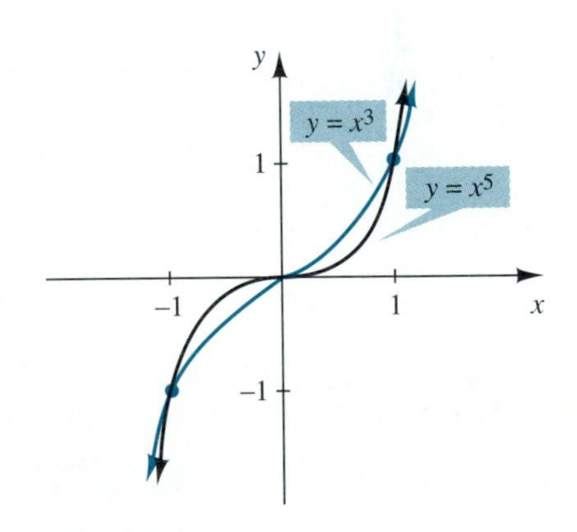

FIGURE 4.2

Again, it is important to recognize that although the graphs of $y = x^3$ and $y = x^5$ are similar, they are not identical. As n increases (but still remains odd), the graphs of $y = x^n$ have a similar shape but get flatter on the interval $(-1, 1)$ and steeper outside this interval (just as we saw in Figure 4.2). We call this the "x^3 look."

*It is very important to realize that although the graphs of $y = x^n$ for n a positive even integer are *similar* to a parabola, they *are not* parabolas. The only function of the form $y = x^n$ that is a parabola is $y = x^2$.

We note the similarities and differences between the graphs of $y = x^2$ and $y = x^3$. For both functions the $|y|$ increases as the $|x|$ increases. (Think about this statement.) For positive x-values, both $y = x^2$ and $y = x^3$ are positive, whereas for negative x-values, $y = x^2$ is positive and $y = x^3$ is negative.

We summarize these results in the following box.

FOR POSITIVE INTEGERS $n > 1$

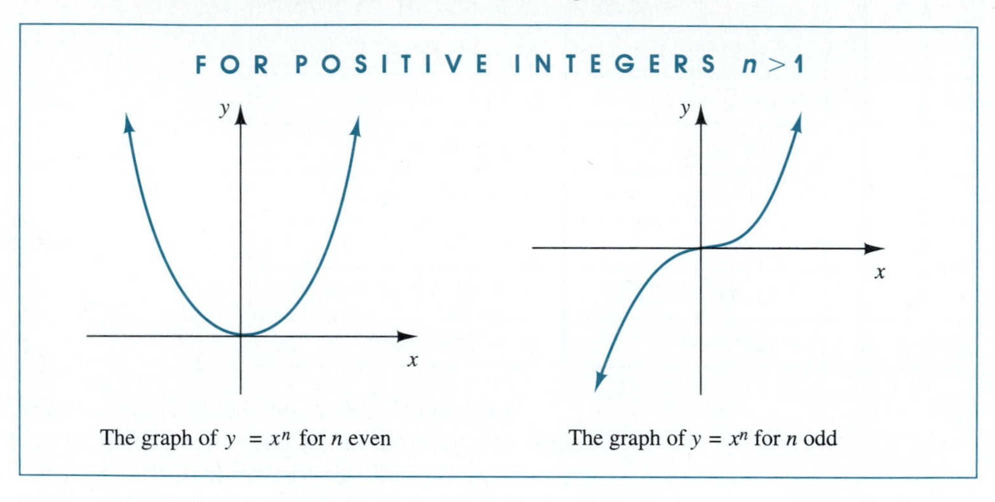

Based upon the stretching principle, we recognize that the description in the box applies equally well to functions of the form $y = ax^n$. The coefficient a merely affects the shape of the graph of $y = x^n$ if a is positive and also reflects the graph about the x-axis if a is negative.

EXAMPLE 1 Sketch the graphs of

(a) $y = f(x) = (x + 2)^4$ **(b)** $y = g(x) = 1 - x^5$ **(c)** $y = h(x) = 5x^3$.

Solution

Compare the graphs of $y = x^4$, $y = x^4 + 2$, and $y = (x + 2)^4$.

(a) The graph of $y = (x + 2)^4$ is similar to the graph of $y = x^4$ (it will have the x^2 look) but is shifted 2 units to the left. By substituting $x = 0$, we find the y-intercept to be $y = 16$. The graph appears in Figure 4.3.

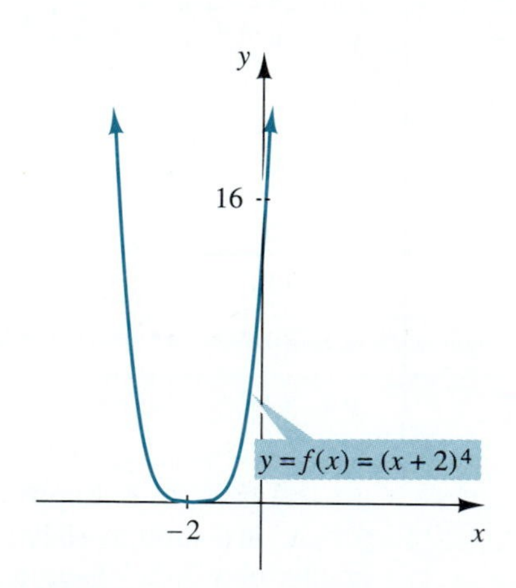

FIGURE 4.3

(b) In order to get the graph of $y = 1 - x^5$, it is helpful to view the equation as $y = -x^5 + 1$. The graph of $y = -x^5 + 1$ can be obtained from the graph of $y = x^5$ (which has the x^3 look) by reflecting it about the x-axis and shifting it 1 unit up. By substituting $x = 0$, we find the y-intercept to be $y = 1$. The x-intercept(s) are found by setting $y = 0$. These steps are shown in Figure 4.4.

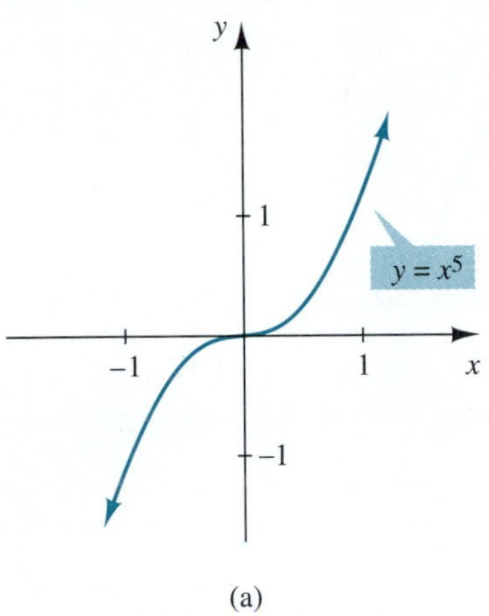

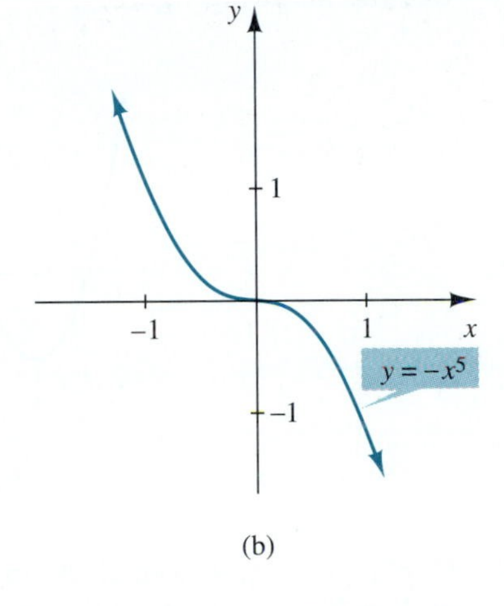

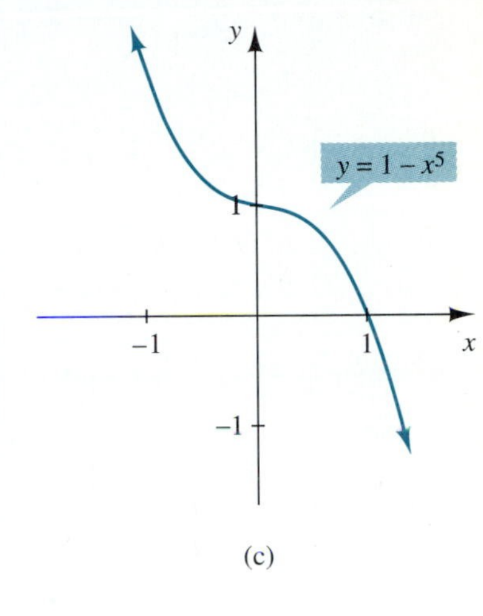

FIGURE 4.4

(c) Based upon the preceding discussion, the graph of $y = 5x^3$ will be similar to the graph of $y = x^3$. The coefficient of 5 stretches the graph of $y = x^3$. The graph appears in Figure 4.5. ■

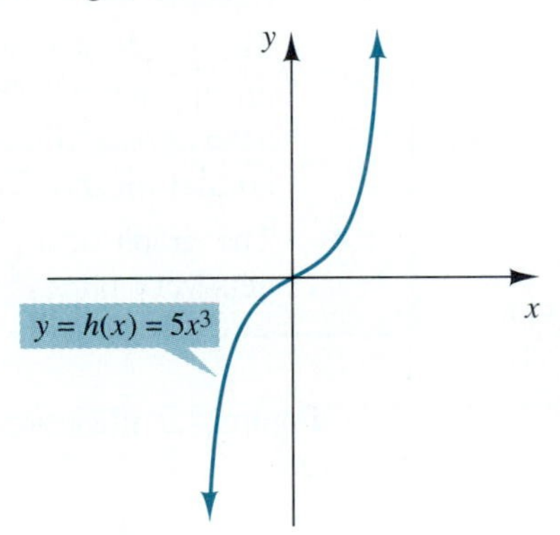

FIGURE 4.5

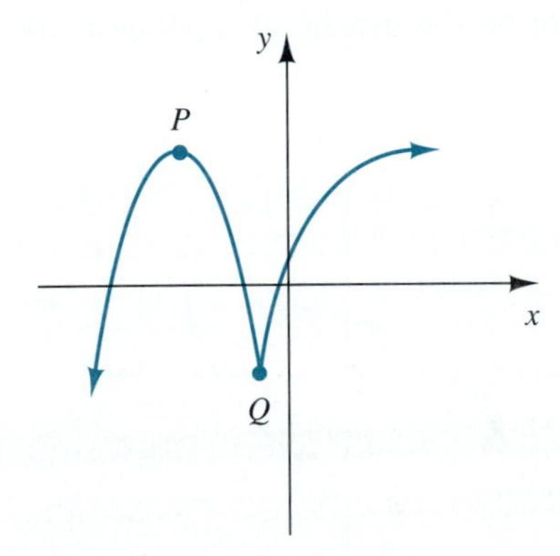

FIGURE 4.6 P and Q are turning points of the graph of $y = f(x)$.

Before looking at more polynomials, we introduce some useful terminology. By a **turning point** of a graph, we mean a point on the graph where the function changes from increasing (graph rising) to decreasing (graph falling) or vice versa. In Figure 4.6, the points P and Q are turning points.

One example of a turning point that we encountered previously is the vertex of a parabola.

In order to sketch the graphs of more general polynomials, we will need to accept some basic properties of the graphs of polynomials. (These properties can be proven using the ideas of calculus.)

PROPERTIES OF POLYNOMIAL FUNCTIONS

1. The graph of a polynomial is a smooth unbroken curve. The word smooth means that the graph does not have any sharp corners as turning points. See the following figure.

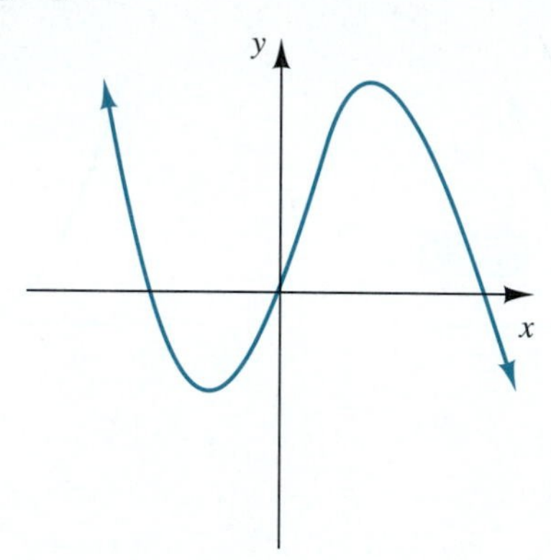

2. If $p(x)$ is a polynomial of degree n, then the polynomial equation $p(x) = 0$ has *at most n* distinct solutions; that is, $p(x)$ has at most n zeros. This is equivalent to saying that the graph of $y = p(x)$ crosses the x-axis at most n times. Thus a polynomial of degree 5 can have *at most* 5 x-intercepts.
3. The graph of a polynomial function of degree n can have *at most* $n - 1$ turning points. For example, the graph of a polynomial of degree 5 can have *at most* 4 turning points. In particular, the graph of a quadratic polynomial (degree 2) always has exactly one turning point—its vertex.
4. The graph of a polynomial always exhibits the characteristic that as $|x|$ gets very large, $|y|$ gets very large.

What does it mean that $|y|$ increases as $|x|$ increases?

Figure 4.7 illustrates two graphs that *could not* be the graphs of a polynomial function.

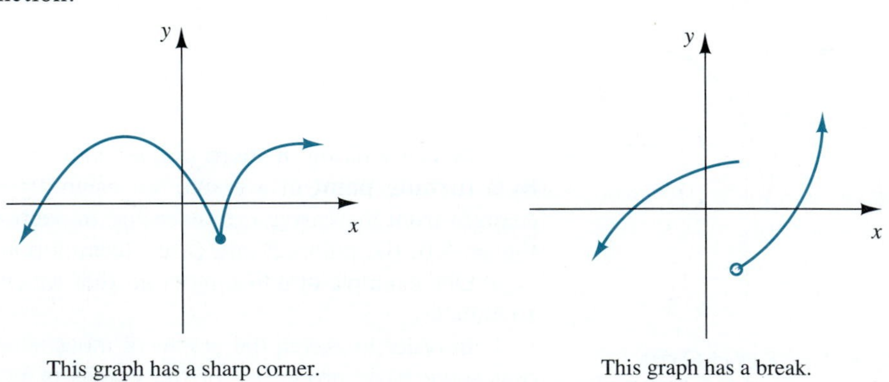

FIGURE 4.7 These graphs cannot be the graphs of polynomial functions.

Our discussion throughout this and the next few sections elaborates on these properties. What is important to recognize is that these properties of polynomials highlight the intimate relationship between the algebraic properties of polynomials (e.g., the number of solutions to the equation $p(x) = 0$) and graphical properties (e.g., the number of x-intercepts of the graph). In fact, we often use information about a graph to find the zeros of a function, and in other cases we use information about the zeros of a function to help us sketch the graph.

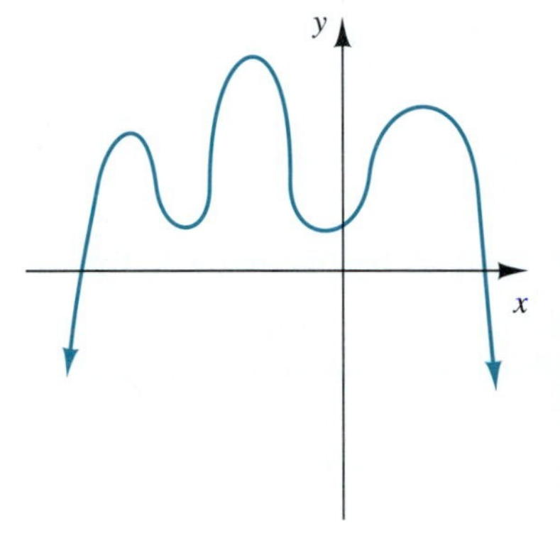

FIGURE 4.8

EXAMPLE 2 Suppose Figure 4.8 shows the graph of a polynomial function. What is the minimum possible degree of the polynomial?

Solution

WHAT DO I NEED TO FIND?	The minimum possible degree of the polynomial whose graph is given.
WHAT INFORMATION DO WE GET FROM THE GRAPH?	The graph is a polynomial function with 2 zeros (x-intercepts) and 5 turning points.
WHAT CONCLUSION CAN BE DRAWN FROM THE NUMBER OF ZEROS?	Property 2 for polynomials tells us that if a polynomial has 2 zeros, then it must have degree at least 2.
WHAT CONCLUSION CAN BE DRAWN FROM THE NUMBER OF TURNING POINTS?	Property 3 for polynomials tells us that if a polynomial has degree n then it can have *at most* $n - 1$ turning points. In other words, the degree of a polynomial must be at least one more than the number of turning points. Since this graph has 5 turning points, the degree of the polynomial must be at least 6.

In summary then, based upon the number of zeros, the polynomial must be of degree at least 2, whereas based upon the number of turning points the polynomial must be of degree at least 6. Therefore, we conclude that the minimum possible degree of the polynomial is 6.

Keep in mind that although a sixth degree polynomial *may* have as many as six real zeros, it need not have that many. ■

Given a polynomial function such as $y = f(x) = x^3 + x^2 - 2x$, we could get a rough sketch of its graph by plotting a sufficient number of points (probably a large number of points would be needed to get a good idea of what the graph actually looks like). In fact, this is basically the way a graphics calculator plots the graph of a function. In order to get a detailed description of the graph of a polynomial and understand *why* it looks the way it does, it is necessary to employ the ideas of calculus. Nevertheless, for certain polynomials we can use some of the ideas about solving quadratic inequalities developed in Chapter 1. Let's illustrate with the following example.

EXAMPLE 3 Sketch the graph of $y = f(x) = x^3 + x^2 - 2x$. Label the intercepts.

Solution In Section 3.4 we discussed how we can use the graph of a function to determine the intervals where the function is positive and negative. In this example we are going to do the opposite. We find x-intercepts and the intervals where $f(x)$ is positive and negative to help us sketch the graph. We do this by performing a sign analysis on $f(x)$. We begin by finding the zeros of $f(x)$.

$$y = f(x) = x^3 + x^2 - 2x = 0$$

$$x(x^2 + x - 2) = 0$$

$$x(x + 2)(x - 1) = 0$$

Therefore, the cut points are $x = 0$, $x = -2$ and $x = 1$.

We draw a number line and complete the sign analysis by picking a test point on each interval. We have chosen -10, -1, $\frac{1}{2}$, and 5 as the test points.

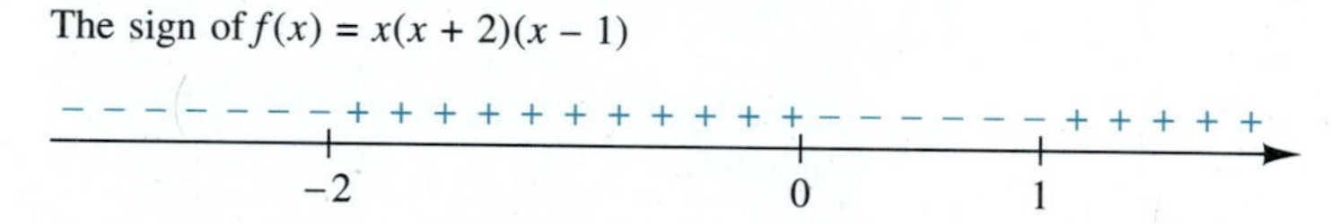

Wherever $f(x)$ is positive, its graph will be above the x-axis, and wherever $f(x)$ is negative, the graph will be below the x-axis. From this sign analysis we know the graph of $f(x)$ will fall below the x-axis for x less than -2 and between 0 and 1, whereas the graph of $f(x)$ will fall above the x-axis for x between -2 and 0 and for x greater than 1. Keep in mind that the cut points are the x-intercepts of the graph of $f(x)$. In other words, the graph will fall in the shaded regions in Figure 4.9.

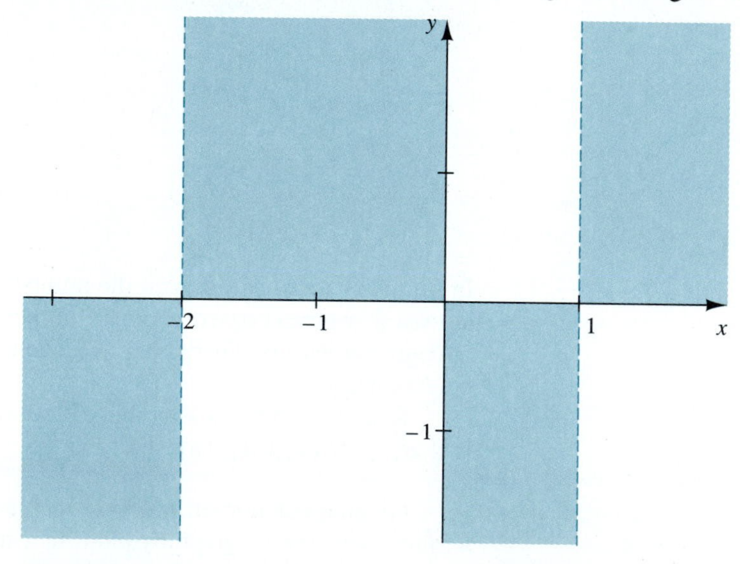

FIGURE 4.9 The graph of $f(x)$ must lie in the shaded regions.

Clearly, then, the graph must have a turning point for $-2 < x < 0$ and another turning point for $0 < x < 1$. Since $f(x)$ is a polynomial of degree 3, we know that it has at most 2 turning points, and so we have *all* the turning points. Since the precise x-values at which these turning points occur is a matter for calculus, we will content ourselves with finding several more values of $f(x)$ to help us sketch the graph. We include values to help us see the behavior of $f(x)$ to the extreme left and to the extreme right.

$$f(x) = x^3 + x^2 - 2x$$ *We choose several values for x. A calculator may be helpful here.*

$$f(-1) = (-1)^3 + (-1)^2 - 2(-1) = -1 + 1 + 2 = 2$$

$$f\left(\frac{1}{2}\right) = \left(\frac{1}{2}\right)^3 + \left(\frac{1}{2}\right)^2 - 2\left(\frac{1}{2}\right) = \frac{1}{8} + \frac{1}{4} - 1 = -\frac{5}{8}$$

$$f(-100) = (-100)^3 + (-100)^2 - 2(-100)$$
$$= -1{,}000{,}000 + 10{,}000 + 200 = -989{,}800$$

$$f(100) = (100)^3 + (100)^2 - 2(100)$$
$$= 1{,}000{,}000 + 10{,}000 - 200 = 1{,}009{,}800$$

These values agree with the polynomial property that $|f(x)|$ gets very large as $|x|$ gets large. We make a reasonable guess as to the turning points; the graph appears in Figure 4.10.

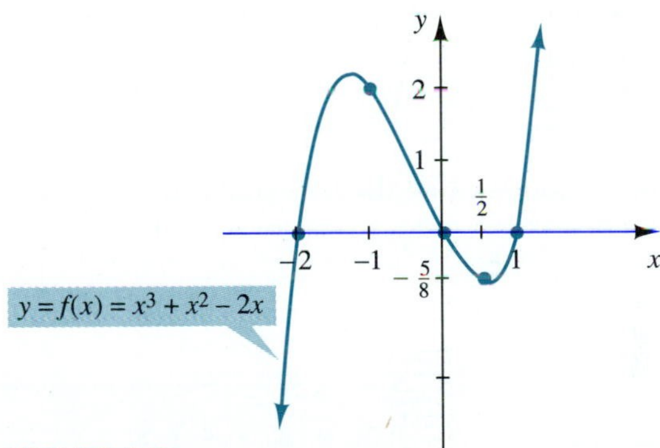

FIGURE 4.10

■

GRAFFIX

Use a graphics calculator or computer to sketch the graph of the function of Example 3, $f(x) = x^3 + x^2 - 2x$, and estimate the turning points of $f(x)$.

We have repeatedly emphasized that finding the zeros of a function corresponds to finding the x-intercepts of its graph, and vice versa. Finding the point(s) of intersection of two graphs can also be interpreted as finding the zeros of a function. For example, if we want to find the points of intersection of the graphs of $y = x^3$ and $y = -x^2 + 2x$, we want to solve the equation $x^3 = -x^2 + 2x$, which is equivalent to solving $x^3 + x^2 - 2x = 0$. Thus finding the points of intersection of $y = x^3$ and $y = -x^2 + 2x$ is equivalent to finding the zeros of $f(x) = x^3 + x^2 - 2x$, which we did in the last example.

DIFFERENT PERSPECTIVES: *Points of Intersection*

Finding the points of intersection of the graphs of two functions can be approached geometrically and algebraically.

GEOMETRIC DESCRIPTION

If we want to find the points of intersection of the graph of $y = g(x) = x^3$ and the parabola $y = h(x) = -x^2 + 2x$, we can sketch their graphs, which appear in the following figure.

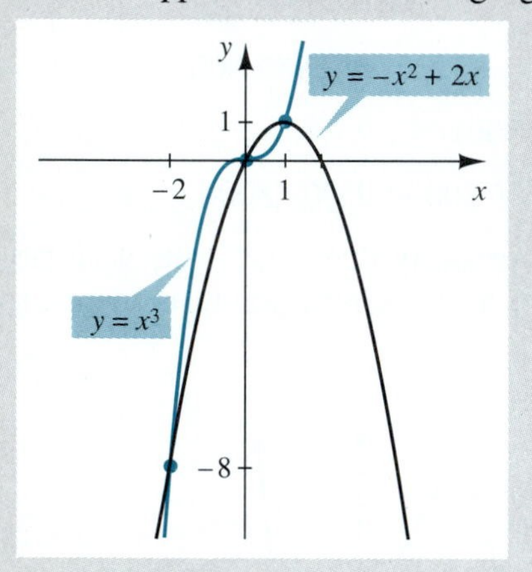

From the figure we can see that the two graphs intersect at three points: $(-2, -8)$, $(0, 0)$ and $(1, 1)$.

ALGEBRAIC DESCRIPTION

In order to find the points of intersection of $y = g(x) = x^3$ and $y = h(x) = -x^2 + 2x$ algebraically, we write $x^3 = -x^2 + 2x$ giving $x^3 + x^2 - 2x = 0$, which is exactly the equation we solved in Example 3. We found the solutions to be $x = -2$, 0 and, 1. Substituting these x-values we get the points of intersection to be $(-2, -8)$, $(0, 0)$ and $(1, 1)$.

GRAFFIX

Use a graphics calculator or computer to estimate the points of intersection of $y = f(x) = x^3 + 1$ and $y = g(x) = x^2 + 3x + 1$, first by graphing both $f(x)$ and $g(x)$ and noting their points of intersection and second by sketching the function $h(x) = f(x) - g(x)$ and noting its x-intercepts.

Concavity

Before we look at another example, let's take a moment to discuss the shape of a curve. Throughout our discussions we have always described a graph as we move from left to right. Figure 4.11 illustrates four graphs. The first two graphs are rising between points A and B and yet have different shapes, and second two graphs are falling between points A and B and also have different shapes.

The functions illustrated by the first two graphs are both increasing. How would you describe the difference between the two graphs?

As the first graph rises, it bends or curves upward; as the second graph rises, it bends or curves downward. Similarly, the third is falling and curved upward, whereas the fourth is falling and curved downward. A graph that curves upward as in Figure 4.11(a) and (c) is called *concave up;* a graph that curves downward as in Figure 4.11(b) and (d) is called *concave down*.

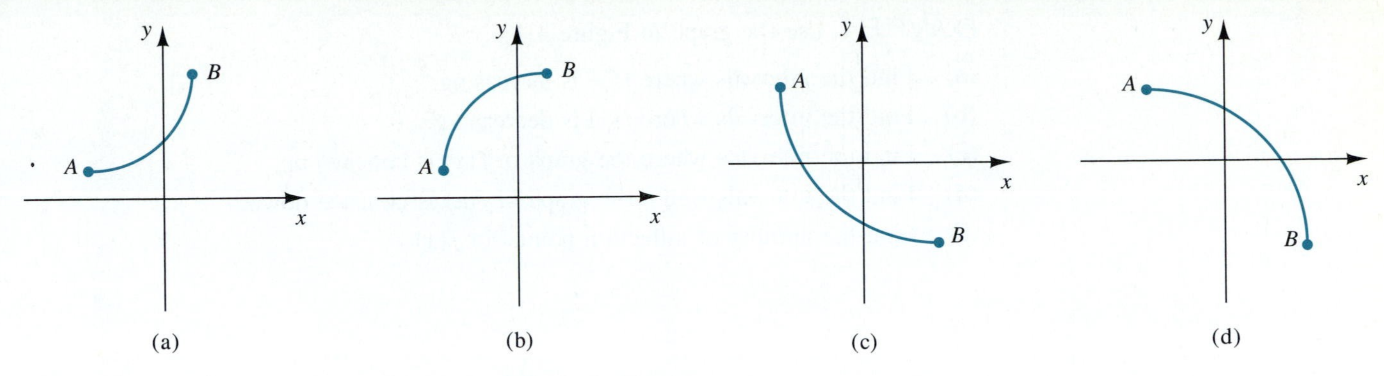

FIGURE 4.11 Rising and falling graphs with different shapes

Figure 4.12(a) illustrates two graphs that are concave up, and Figure 4.12(b) illustrates two graphs that are concave down.

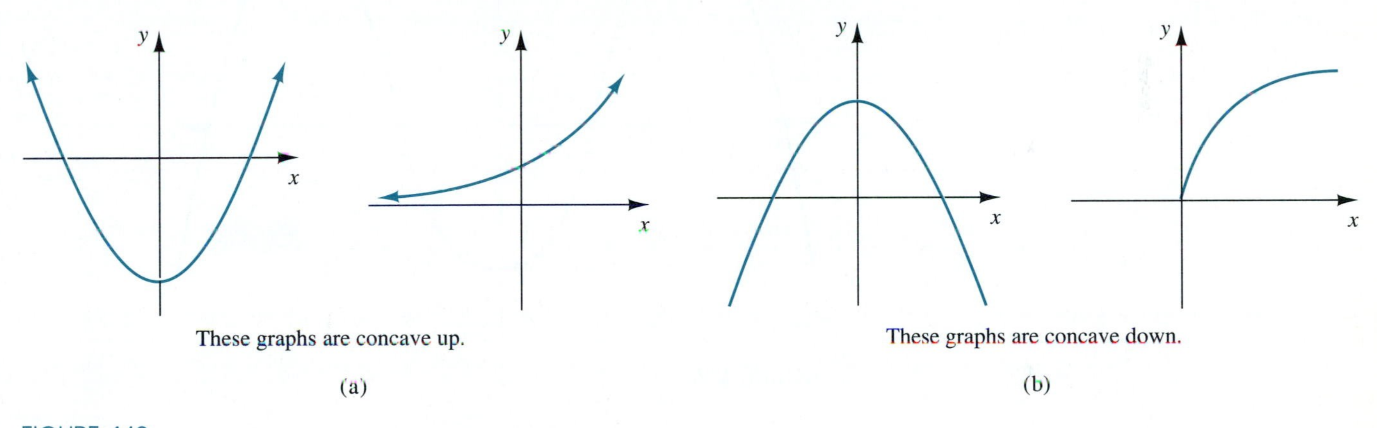

FIGURE 4.12

Figure 4.13 illustrates a graph that is rising from A to B and is concave up part of the way (up to point C) and then becomes concave down. This point at which the concavity changes is called a **point of inflection.**

Looking at the graph of $y = x^3 + x^2 - 2x$ of Example 3, which we repeat in Figure 4.14, we can see that the graph appears to have an inflection point a bit to the left of the origin.

How is a point of inflection related to the concavity of a graph?

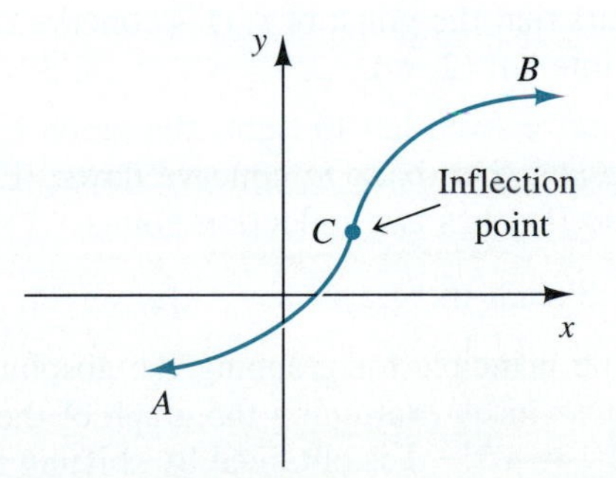

FIGURE 4.13 C is a point of inflection.

Inflection point

FIGURE 4.14 The graph of $y = f(x) = x^3 + x^2 - 2x$ has one infection point.

EXAMPLE 4 Use the graph in Figure 4.15.

(a) Find the intervals where $f(x)$ is increasing.

(b) Find the intervals where $f(x)$ is decreasing.

(c) Find the intervals where the graph of $f(x)$ is concave up.

(d) Find the intervals where the graph of $f(x)$ is concave down.

(e) Find the number of inflection points for $f(x)$.

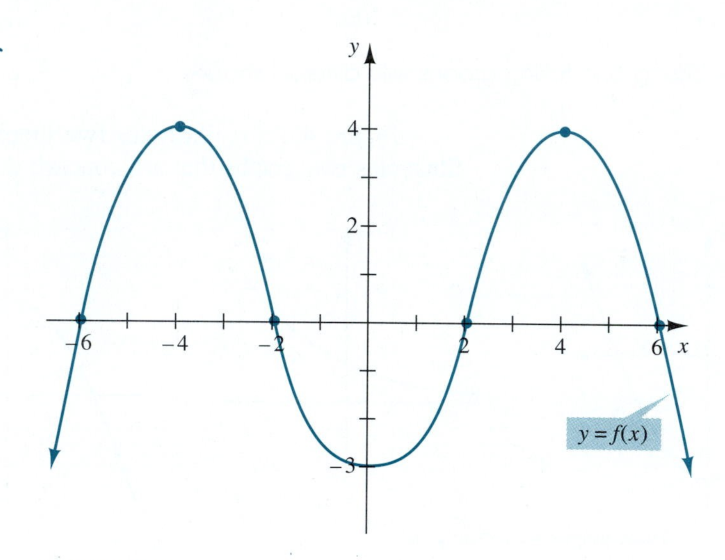

FIGURE 4.15

Solution

(a) The intervals where $f(x)$ is increasing are the x-intervals on which the graph is rising, which are $(-\infty, -4] \cup [0, 4]$.

(b) The intervals where $f(x)$ is decreasing are the x-intervals on which the graph is falling, which are $[-4, 0] \cup [4, \infty)$.

(c) It appears that the graph of $f(x)$ is concave up on the interval $(-2, 2)$.

(d) It appears that the graph of $f(x)$ is concave down on the interval $(-\infty, -2)$ and on the interval $(2, \infty)$.

(e) As we move from left to right, the graph changes from concave down to concave up and then back to concave down. Each change occurs at an inflection point, so $f(x)$ has two inflection points. ■

EXAMPLE 5 Sketch the graph of $y = |x^3 - 1|$.

Solution The principle for graphing the absolute value of a familiar function suggests that we begin by examining the graph of the underlying function, $y = x^3 - 1$. The graph of $y = x^3 - 1$ is obtained by shifting the graph of $y = x^3$ one unit down. See Figure 4.16(a). The absolute value takes that portion of the graph of $y = x^3 - 1$ which is below the x-axis and reflects it above the x-axis giving us the graph which appears in Figure 4.16(b). ■

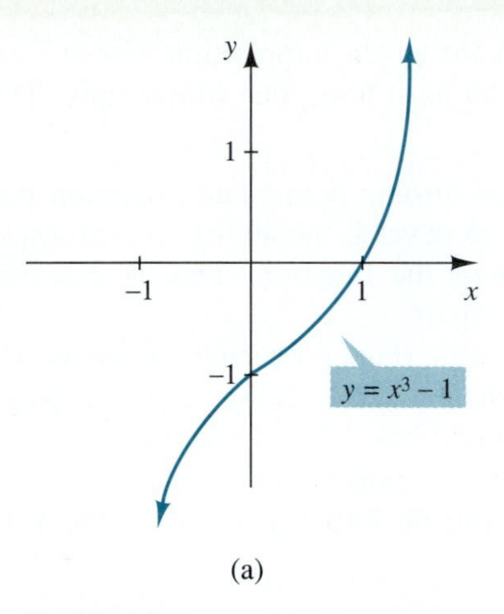

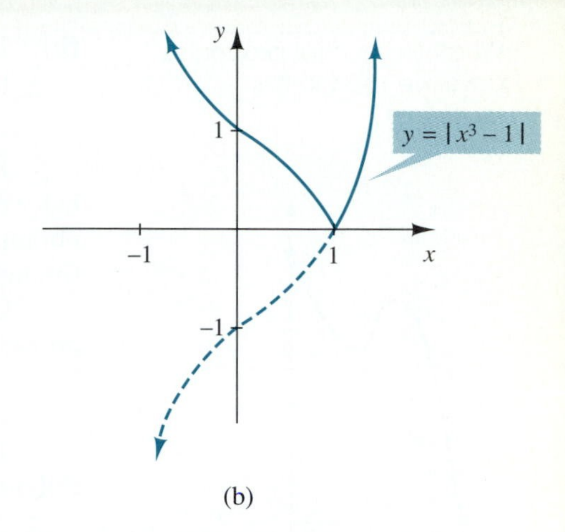

FIGURE 4.16

EXAMPLE 6 Suppose $p(x)$ is a polynomial function that satisfies the following conditions: $p(x)$ has exactly three turning points at $(-3, -4)$, $(0, 2)$, $(3, -4)$ and exactly two inflection points at $(-1, 0)$ and $(1, 0)$.

(a) Sketch a graph of $p(x)$ based upon this information.

(b) How many real zeros does $p(x)$ have?

Solution

(a) We begin by plotting the given points and labeling them as turning points or inflection points. See Figure 4.17(a). Let's analyze the given information logically.

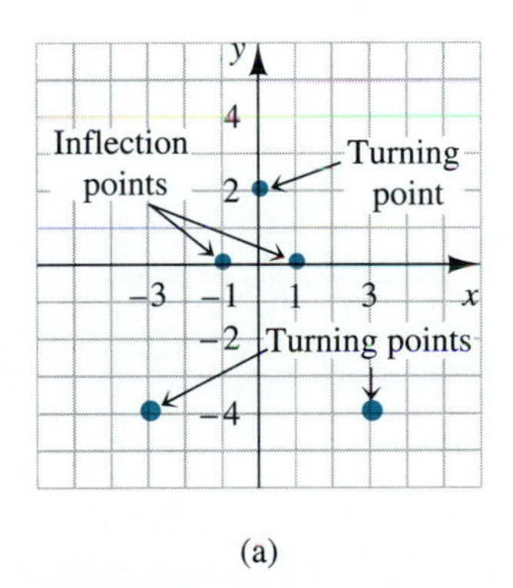

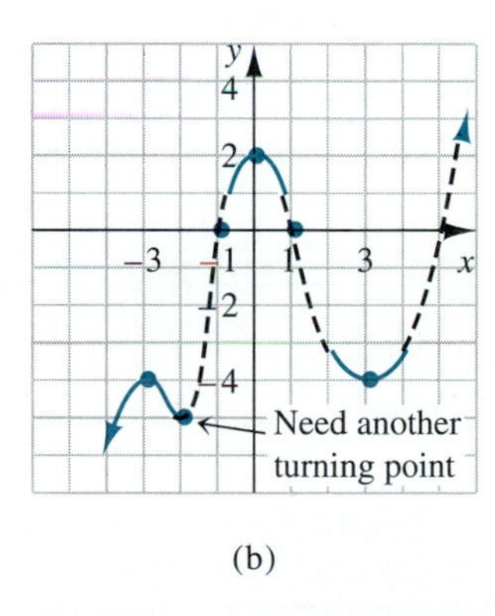

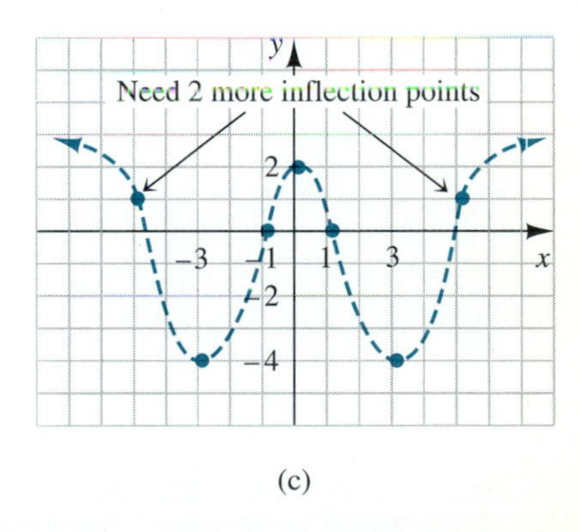

FIGURE 4.17

cally. Suppose we imagine the given turning points of $p(x)$ look like the colored graph in Figure 4.17(b). We would then connect the points with the dashed graph in the same figure. However, this would force the graph to have more than the three turning points we are given. Similarly, if we try to connect the points as in Figure 4.17(c), then the graph would have *four* inflection points instead of two.

Since we know that $p(x)$ is a polynomial, its graph will be a smooth curve with no breaks or corners, and since there are no turning points other than the ones we are given, the graph in Figure 4.18 is a fair representation of $p(x)$.

Note that the inflection points help us sketch the shape of the graph. Also note that the graph shows that the y-values on the graph are getting very large as $|x|$ gets large, as expected for a polynomial function.

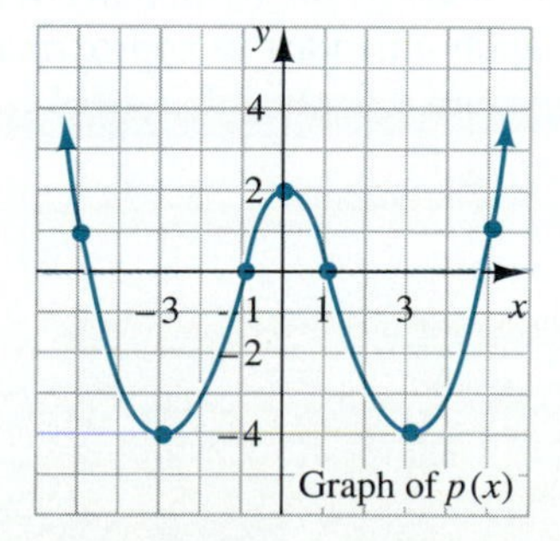

FIGURE 4.18

What polynomial properties have we used in this argument?

(b) Based upon the given information and our analysis of the graph, we can see that the graph must have four x-intercepts. Therefore, $p(x)$ has four real zeros. ■

Knowing the turning points and inflection points for a function is important, but we still need to develop the ability to synthesize all this information in order to obtain the graph of the function. That is exactly what we are attempting to do throughout this chapter.

One of the important properties of polynomials is that their graphs have no breaks. (The technical terminology is that a polynomial function is *continuous*.) Thus, if a polynomial has -3 and 4 as range values, then all the values in between -3 and 4 must also be range values. See Figure 4.19.

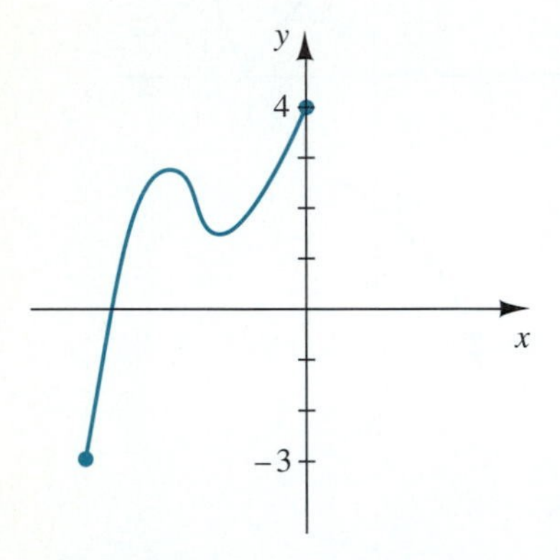

FIGURE 4.19 If -3 and 4 are range values of a polynomial, then all the values in between -3 and 4 must also be in the range.

This result can be formally stated as the **Intermediate Value Theorem for Polynomials.**

THEOREM 4.1 *Intermediate Value Theorem for Polynomials*
Consider a polynomial $p(x)$ and a closed interval $[a, b]$. For any value c between $p(a)$ and $p(b)$, there is at least one x-value in $[a, b]$ such that $p(x) = c$.

Figure 4.20 illustrates the content of the Intermediate Value Theorem.

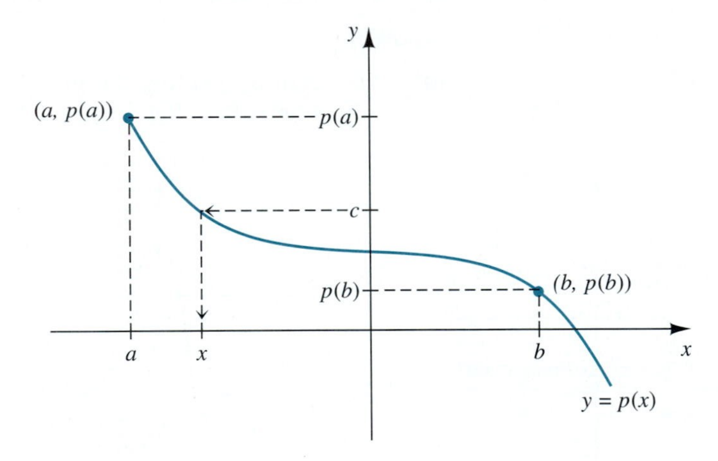

FIGURE 4.20 An illustration of the intermediate value theorem

In words, this theorem says that a polynomial function will take on all the values between any two range values. In particular, this theorem tells us that *if we know that $p(a)$ is negative and $p(b)$ is positive, then there must be an x-value between a and b for which $p(x) = 0$.* Let's see how we can we apply this idea to approximate the zero of a polynomial.

GRAFFIX

Use a graphics calculator to graph the following polynomials.

$$p(x) = x^4 - x^2 - 2 \qquad p(x) = x^5 - 2x^2 + 1$$

1. Use the graph to observe the location of the zeros between consecutive integers.
2. Use the **RANGE** key to change the viewing window and reset the scale on the x-axis to increments of 0.1. Now redraw the graph in this new viewing window and locate the zeros between consecutive tenths.

EXAMPLE 7 Approximate the real zero of $p(x) = x^3 - x^2 - 2$.

Solution Since we cannot factor $p(x)$, we begin by computing some values of $p(x)$ in the hope that we can find two values of x for which $p(x)$ changes from positive to negative. According to the Intermediate Value Theorem, we will know that $p(x)$ has a zero in between those two values.

x	-2	-1	0	1	2
$p(x)$	-14	-4	-2	-2	2

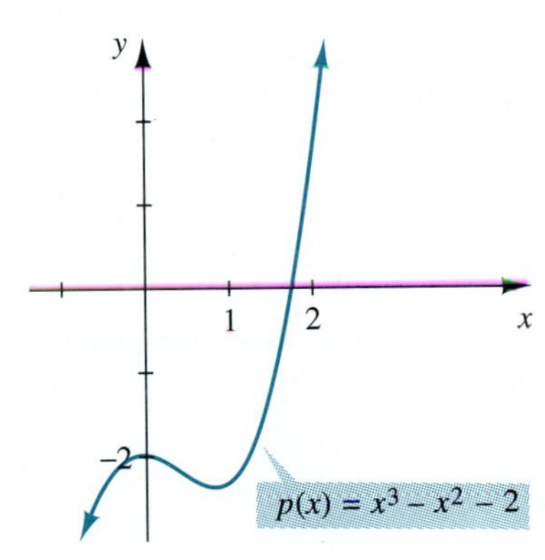

FIGURE 4.21 The graph of $p(x) = x^3 - x^2 - 2$ has a zero between 1.6 and 1.7.

Since $p(1)$ is negative and $p(2)$ is positive, we know that $p(x)$ has a zero between 1 and 2.

In order to get a better approximation to the zero, we subdivide the interval from 1 to 2 into tenths and evaluate $p(x)$ at 1.1, 1.2, 1.3, . . . , 1.9. Doing this we find

$$p(1.6) = -0.464 \quad \text{and} \quad p(1.7) = 0.023$$

and therefore we know that $p(x)$ has a zero between 1.6 and 1.7. Continuing in this manner we can approximate this zero of $p(x)$ to any desired degree of accuracy. We include a sketch of the graph of $p(x)$ for completeness. See Figure 4.21. ■

Much of the work in this section has been predicated upon finding the zeros of polynomial functions, which we have been able to do by factoring. In the next section we digress a bit to discuss polynomial division, which will allow us to develop some additional methods for finding the zeros of polynomial functions.

EXERCISES 4.1

In Exercises 1–8, if the given graph could be the graph of a polynomial, what is its minimum degree? If it cannot be the graph of a polynomial, explain why.

1.

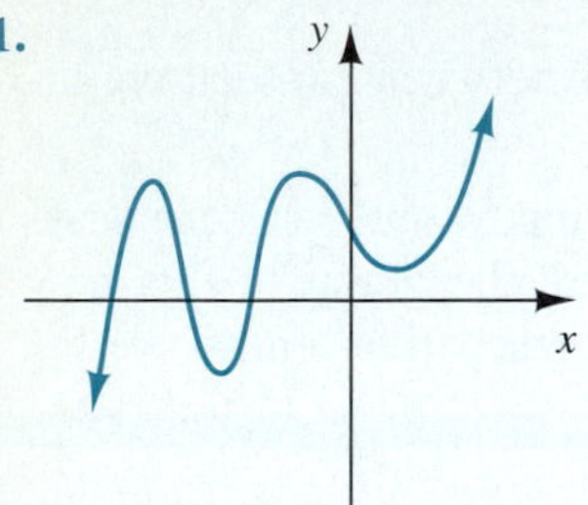

2.

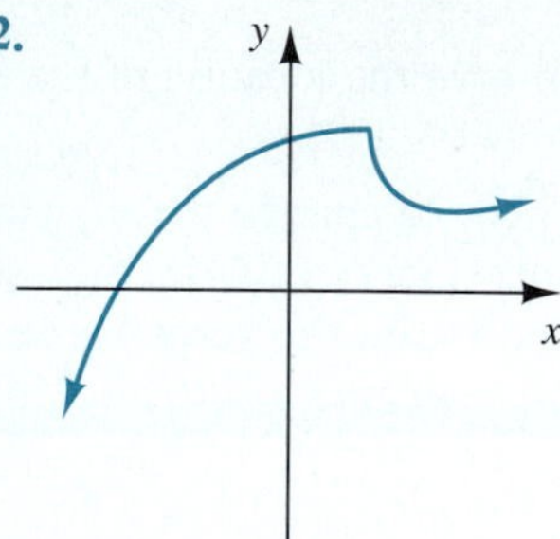

3.

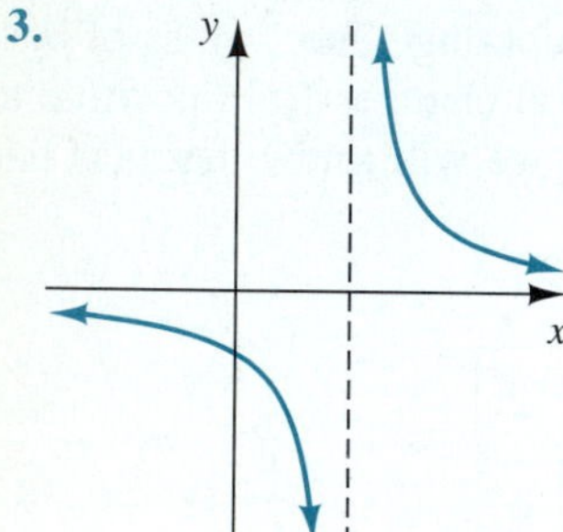

4.

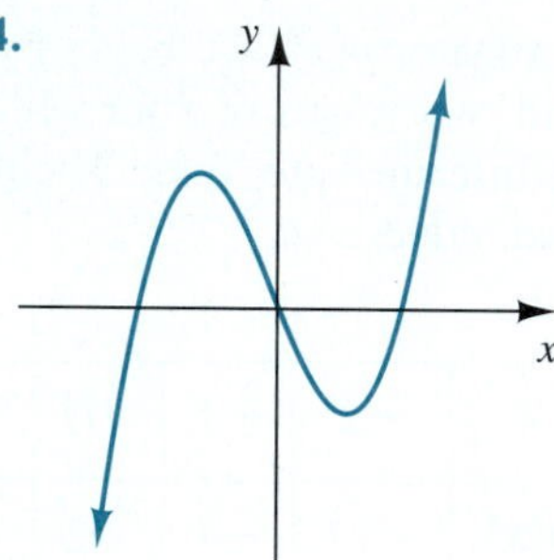

5.

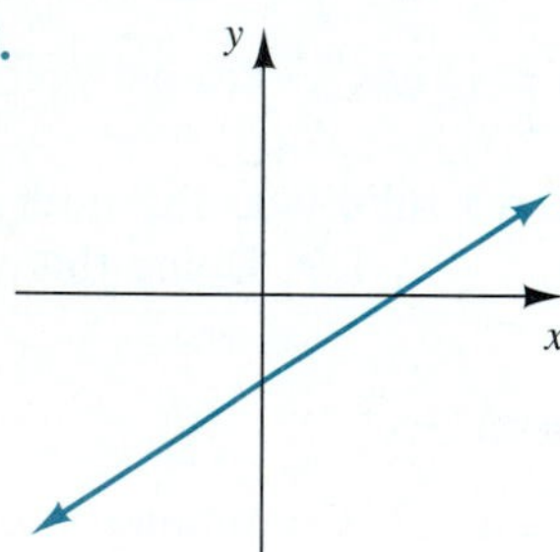

6.

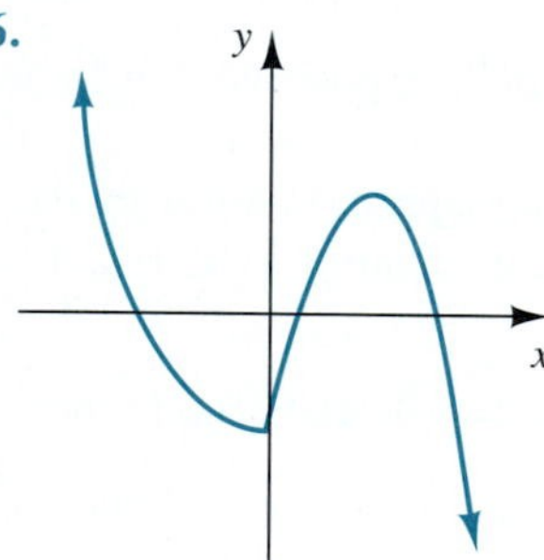

7.

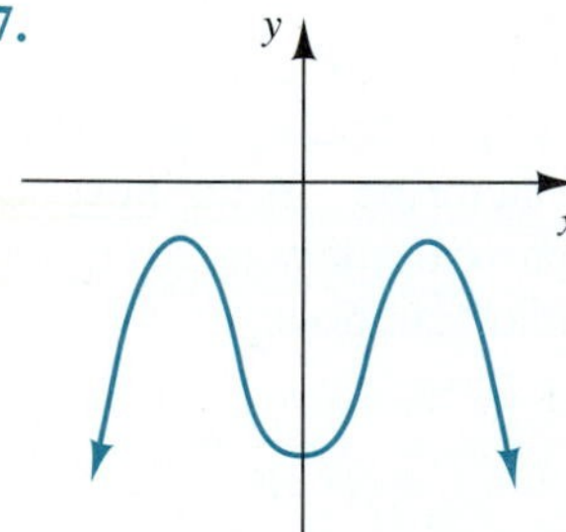

8.

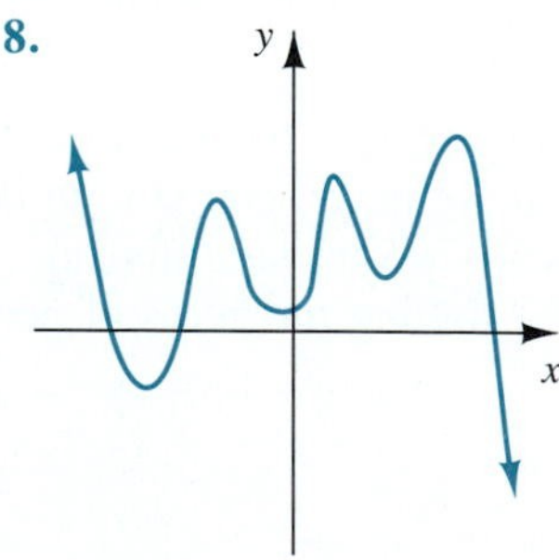

In Exercises 9–26, sketch the graph of the given function. Label the intercepts where appropriate.

9. $y = x^3 + 1$

10. $y = (x + 1)^3$

11. $y = (x - 1)^4$

12. $y = x^4 - 1$

13. $y = 8 - x^3$

14. $y = 9 - x^2$

15. $y = x^2 - 4$

16. $y = (x - 4)^2$

17. $y = -(x + 2)^5$

18. $y = (x - 2)^4 - 16$

19. $y = 2x^3 - 16$

20. $y = -4x^6 - 2$

21. $y = -(x - 1)^5 + 1$

22. $y = (x - 2)^3 - 1$

23. $y = |(x - 2)^3|$

24. $y = |x^3 - 8|$

25. $y = |1 - x^4|$

26. $y = |x^6 - 1|$

27. On the same set of coordinate axes, sketch and label the graphs of $y = x^3$, $y = 2x^3$, $y = 3x^3$, and $y = \frac{1}{2}x^3$.

28. On the same set of coordinate axes sketch and label the graphs of $y = -x^4$, $y = -2x^4$, $y = -3x^4$, and $y = -\frac{1}{2}x^4$.

29. On the same set of coordinate axes sketch and label the graphs of $y = x^2$, $y = x^4$, $y = x^6$, and $y = x^8$ for $x \geq 0$.

30. On the same set of coordinate axes sketch and label the graphs of $y = -x^3$, $y = -x^5$, $y = -x^7$, and $y = -x^9$ for $x \geq 0$.

In Exercises 31–44, use a sign analysis to sketch the graph of the given polynomial function.

31. $y = f(x) = x^3 - x^2 - 2x$

32. $y = f(x) = -x^3 + x^2 + 6x$

33. $y = g(x) = x^2 - 6x + 8$

34. $y = h(x) = 3x^2 + 2x - 8$

35. $y = p(x) = x^3 - 9x^2$

36. $y = h(x) = x^3 + 6x^2 + 8x$

37. $y = f(x) = x^4 + x^2$

38. $y = g(x) = 5x - 8$

39. $y = h(x) = x(x - 1)(x + 1)(x + 2)$

40. $y = r(x) = x^4 - 5x^2 + 4$

41. $y = f(x) = (x + 1)(x - 2)(x + 3)(x - 4)$

42. $y = g(x) = (x - 1)(x + 2)(3 - x)(x + 4)$

43. $y = f(x) = x^3 + x^2 - 2x - 2$

44. $y = g(x) = x^3 + 2x^2 - 4x - 8$

In Exercises 45–50, the given information pertains to a polynomial function $p(x)$. Sketch a graph of $p(x)$ consistent with this information.

45. $p(x)$ has exactly one turning point at $(2, 0)$ and is always concave down.

46. $p(x)$ has exactly three turning points at $(-3, 2)$, $(0, -3)$, and $(3, 2)$ and exactly two inflection points at $(-1, -1)$ and $(1, -1)$.

47. $p(x)$ has exactly one turning point at $(0, 0)$ and exactly two inflection points at $(1, 2)$ and $(3, 5)$.

48. $p(x)$ has exactly one turning point at $(2, -6)$, one inflection point at $(6, 0)$, and one additional x-intercept at $(-4, 0)$.

49. $p(x)$ has exactly two turning points at $(-1, 4)$ and $(2, -1)$ and exactly one inflection point at $(1, 0)$.

50. $p(x)$ has exactly three turning points at $(-1, 0)$, $(0, 3)$, and $(2, -10)$ and two inflection points at $(-0.5, 2)$ and $(1.5, -6)$.

In Exercises 51–56, place the real zeros of $p(x)$ within consecutive tenths of a unit; i.e., the zero is between 3.6 and 3.7. Each exercise indicates how many real zeros there are.

51. $p(x) = x^3 - 2$; 1 real zero

52. $p(x) = 4 - x^5$; 1 real zero

53. $p(x) = x^3 + x^2 - 8x + 8$; 1 real zero

54. $p(x) = \frac{1}{3}x^3 - x^2 - 3x - 6$; 1 real zero

55. $p(x) = x^4 + 2x^3 - 1$; 2 real zeros

56. $p(x) = x^4 - 3x^3 + 3x^2 - x - 1$; 2 real zeros

57. Find the points of intersection of the graph of $y = x^3$ with the graph of $y = 9x$.

58. Find the points of intersection of the graph of $y = 3x^2$ with the graph of $y = x^3 - 10x$.

59. Find the points of intersection of the graph of $y = x^5$ with the graph of $y = 2x^4 + 3x^3$.

60. Find the points of intersection of the graph of $y = x^2$ with the graph of $y = 49x^2 - 4x^3$.

61. Let $f(x) = x^3$; find and simplify the difference quotient $\dfrac{f(x) - f(2)}{x - 2}$.

62. Let $f(x) = x^4$; find and simplify the difference quotient $\dfrac{f(x) - f(3)}{x - 3}$.

63. Let $f(x) = x^4$; find and simplify the difference quotient $\dfrac{f(x + h) - f(x)}{h}$.

64. Let $f(x) = x^3 + 1$; find and simplify the difference quotient $\dfrac{f(x + h) - f(x)}{h}$.

65. Suppose n is a positive integer and $|x| < \frac{1}{2}$. What is the smallest value of n for which $|x^n| < 0.005$?

66. For what range of x-values will $|x^4| < 10^{-4}$?

67. For what range of x-values will $\left|\dfrac{x^3}{6}\right| < 0.001$?

68. Suppose n is a positive integer and $|x| < \frac{1}{3}$. What is the smallest value of n for which $\left|\dfrac{x^n}{n}\right| < 10^{-5}$?

QUESTIONS FOR THOUGHT

69. We have stated that a polynomial of degree n can have at most n zeros and $n - 1$ turning points. Show that if a polynomial has n distinct zeros, then it must have $n - 1$ distinct turning points. Also show that the converse of this statement is false.

70. Consider $p(x) = 2x^3 - 10x^2 - 3x - 100$ and compute $p(10)$, $p(100)$, $p(1000)$, $p(10{,}000)$, and $p(100{,}000)$. We may write

$$\begin{aligned} p(x) &= 2x^3 - 10x^2 - 3x - 100 \\ &= x^3\left(2 - \frac{10}{x} - \frac{3}{x^2} - \frac{100}{x^3}\right) \end{aligned}$$

Analyze this factored form of $p(x)$ to support the fact that polynomial functions have the property that $|p(x)|$ gets very large as $|x|$ gets large.

71. Consider the general third-degree polynomial

$$p(x) = ax^3 + bx^2 + cx + d$$

What determines whether $p(x) \to +\infty$ or $p(x) \to -\infty$ as $|x|$ gets large?

HINT: Use the factorization technique illustrated in Exercise 70.

72. Find a polynomial function that has exactly four real zeros at $x = -3, -1, 2, 4$. Are there any other such polynomials?

GRAFFIX

Use a graphics calculator or computer for the following exercises.

73. Sketch the graphs of $y = 2x^4$, $y = 2x^6$, and $y = 2x^8$ on the interval $[-2, 2]$.

74. Sketch the graphs of $y = 2x^3$, $y = 2x^5$, and $y = 2x^7$ on the interval $[-2, 2]$.

75. Sketch the graphs of $y = x^4$, $y = x^4 - 1$, and $y = x^4 + 2$ on the interval $[-2, 2]$.

76. Sketch the graph of $p(x) = x^4 - 2x^2 + 1$ and trace the function to approximate its real zeros to the nearest tenth.

77. Sketch the graph of $p(x) = x^5 - 5x^2 + 3$ and trace the function to approximate its real zeros to the nearest tenth.

78. **(a)** Sketch the graph of $y = 2x^3 - x^4$ using the method described in Example 3.
(b) Does your sketch of the graph indicate that there are any inflection points?
(c) Sketch the graph of this function using a graphics calculator.
(d) How does the calculator graph compare with the graph you obtained in part (a)?

In Exercises 79–81, sketch the graphs and comment on whether the graph exhibits the various properties of polynomial functions described in this section.

79. $y = x^5 - 5x^3 + 4x + 1$
80. $y = x^4 - 5x^2 + 6$
81. $y = x^6 - x^2 - 4$

82. Sketch the graph of $y = x^3 + 4x^2 - 3$
(a) Use this graph to locate the zeros between consecutive integers.
(b) Use the RANGE key to change the viewing window and reset the scale on the x-axis to increments of 0.1. Now redraw the graph in this new viewing window and locate the zeros between consecutive tenths.

83. Repeat the procedure of Exercise 82 for $y = x^4 - 5x^2 + 3$ to locate the zeros of this function between consecutive hundredths.

4.2 Division of Polynomials and Synthetic Division

In the last section we saw that the real zeros of a polynomial function provide valuable information that can be helpful in sketching its graph. We briefly discussed how to approximate the real zeros of a polynomial, but for the most part we focused on finding the zeros by factoring the polynomial. However, we have no general method for factoring polynomials of degree greater than two. In this section and the next we turn our attention to methods that will allow us to find zeros of higher-degree polynonials. In order to do this we first need to discuss the division process for polynomials.

The same four-step process of long division (divide, multiply, subtract, and bring down) used in dividing numbers can be applied to long division of polynomials.

Let's begin by illustrating the division of $\dfrac{2x^2 - 11x + 5}{x - 3}$. We set up the long division as we would for numbers and we carry out each of the four steps. Note that in setting up the long division process, we make sure that both dividend and divisor are in standard form (written from highest power to lowest).

1. First we divide $2x^2$ by x: $\dfrac{2x^2}{x} = 2x$. This becomes the first term in the quotient.

2. Multiply the divisor $x - 3$ by the term $2x$ obtained in step 1. This produces $2x^2 - 6x$, which is written below the dividend, as shown.

$$\begin{array}{r|l} & 2x \quad - \quad 5 \\ \hline x - 3 & 2x^2 - 11x + 5 \\ & -(2x^2 - 6x) \\ \hline & \qquad -5x + 5 \\ & \qquad -(-5x + 15) \\ \hline & \qquad\qquad -10 \end{array}$$

This is the long division **tableau.**

3. Next we subtract $2x^2 - 6x$ from $2x^2 - 11x$ in the dividend, obtaining $-5x$. Note that we indicate subtraction by enclosing the $2x^2 - 6x$ in parentheses with a subtraction sign in front of it.
4. Finally, we bring down the next term or terms in the dividend and continue by repeating steps 1 through 3.

Thus when $2x^2 - 11x + 5$ is divided by $x - 3$, the quotient is $2x - 5$ and the remainder is -10. We may write this result in two ways, either as

$$\frac{2x^2 - 11x + 5}{x - 3} = 2x - 5 + \frac{-10}{x - 3} \tag{1}$$

or, if we multiply both sides of the equation by $x - 3$,

$$\underbrace{2x^2 - 11x + 5}_{\text{Dividend}} = \underbrace{(x - 3)}_{\text{divisor}} \cdot \underbrace{(2x - 5)}_{\text{quotient}} + \underbrace{(-10)}_{\text{remainder}} \tag{2}$$

It is important to note that the division process ends when the degree of the remainder is less than the degree of the divisor, which is quite similar to division of natural numbers, where we stop when the remainder is less than the divisor. Also note that equation (1) is valid for all $x \neq 3$, whereas equation (2) is valid for all values of x.

We illustrate this process again in the next example.

EXAMPLE 1 Divide: $\dfrac{6x^3 + 19x^2 - 25}{3x + 5}$.

Solution As mentioned before, when we set up the long division, we want the dividend and divisor to be in standard form. In addition, if there are any "missing" terms in the *dividend,* we insert these missing terms with zero coefficients, as indicated here. These zero-coefficient terms serve as placeholders to make the "bookkeeping" a bit easier.

$$\begin{array}{r@{}l}
 & \;2x^2 + 3x - 5 \\
3x + 5 & \overline{)\;6x^3 + 19x^2 + 0x - 25} \\
 & \underline{-(6x^3 + 10x^2)} \\
 & \; 9x^2 + 0x \\
 & \underline{-(9x^2 + 15x)} \\
 & \; -15x - 25 \\
 & \underline{-(-15x - 25)} \\
 & \; 0
\end{array}$$

The fact that the remainder is 0 means that the division is exact. Therefore, we have

$$6x^3 + 19x^2 - 25 = (3x + 5)(2x^2 + 3x - 5)$$

and so $3x + 5$ is a factor of $6x^3 + 19x^2 - 25$. ■

EXAMPLE 2 Divide: $\dfrac{x^4 - 1}{x^2 + 2x}$.

Solution

$$
\begin{array}{r}
x^2 - 2x \;\; + 4 \qquad\qquad\qquad \\
x^2 + 2x \overline{) \; x^4 + 0x^3 + 0x^2 + 0x - 1} \\
\underline{-(x^4 + 2x^3)} \qquad\qquad\qquad\quad \\
-2x^3 + 0x^2 \qquad\qquad \\
\underline{-(-2x^3 - 4x^2)} \qquad\qquad \\
4x^2 + 0x \qquad \\
\underline{-(4x^2 + 8x)} \qquad \\
-8x - 1
\end{array}
$$

Again this long division means

$$
\underbrace{x^4 - 1}_{\text{Dividend}} = \underbrace{(x^2 + 2x)}_{\text{divisor}} \cdot \underbrace{(x^2 - 2x + 4)}_{\text{quotient}} + \underbrace{(-8x - 1)}_{\text{remainder}}
$$

It is left to the reader to verify this last equation. ■

An algorithm is a precise sequence of steps that can be followed to solve a specific problem.

The result of this division process for polynomials can be summarized in the following theorem, called the **Division Algorithm.**

THEOREM 4.2 *The Division Algorithm*
Let $p(x)$ and $d(x)$ be polynomials with $d(x) \neq 0$ and the degree of $d(x)$ be less than or equal to the degree of $p(x)$. Then there are polynomials $q(x)$ and $R(x)$ such that

$$
\underbrace{p(x)}_{\text{Dividend}} = \underbrace{d(x)}_{\text{divisor}} \cdot \underbrace{q(x)}_{\text{quotient}} + \underbrace{R(x)}_{\text{remainder}}
$$

where either $R(x) = 0$ or the degree of $R(x)$ is less than the degree of $d(x)$.

In the next section we will see that this theorem is instrumental in deriving important results about polynomials.

Synthetic Division

When we divide a polynomial by a divisor of the form $x - r$, there is quite a bit of repetitious writing. We can shorten this process by recognizing that all the essential information in the division process is carried by the various coefficients and their specific location in the tableau.

Let's twice rewrite the division problem with which we began this section, once as before and a second time with just the coefficients.

$$
\begin{array}{r}
2x - \quad 5 \quad \\
x - 3\overline{)2x^2 - 11x + 5} \\
\underline{2x^2 - 6x} \quad\;\; \\
-5x + 5 \\
\underline{-5x + 15} \\
-10
\end{array}
\qquad\qquad
\begin{array}{r}
2 - 5 \quad \\
-3\overline{)2 - 11 + 5} \\
\underline{\textcircled{2} - 6} \quad\;\; \\
\textcircled{-5} + \textcircled{5} \\
\underline{\textcircled{-5} + 15} \\
-10
\end{array}
$$

The circled coefficients are merely duplicates of those in the quotient or dividend, so we delete these and consolidate the tableau vertically as follows:

$$\begin{array}{r|l} & 2 - 5 \\ \hline -3 & 2 - 11 + 5 \\ & \quad - 6 \\ \hline & \qquad + 15 \\ \hline & \qquad -10 \end{array} \qquad \begin{array}{r|rrr} & 2 & -5 & \\ \hline -3 & 2 & -11 & 5 \\ & & -6 & 15 \\ \hline & & & -10 \end{array}$$

Next we move the coefficients of the quotient to the bottom row.

$$\begin{array}{r|rrr} -3 & 2 & -11 & 5 \\ & & -6 & 15 \\ \hline & 2 & -5 & -10 \end{array}$$

When we divide by $x - r$ we represent the divisor by r.

Finally, in order to avoid subtraction (which sometimes introduces careless errors), we change the sign of the divisor and the signs in the second row, which allows us to *add* in each column instead of subtract.

The final format which follows is called **synthetic division** for dividing $2x^2 - 11x + 5$ by $x - 3$. We include an explanation of how to construct this table.

$$\begin{array}{r|rrr} 3 & 2 & -11 & 5 \\ & & 6 & -15 \\ \hline & 2 & -5 & \boxed{-10} \end{array} \leftarrow \textit{Remainder}$$

1. Since the divisor is $x - 3$, we represent the divisor by 3, written at the left of the top row. We write the coefficients of the dividend in the first row of the table. Remember to hold a place for any missing terms by including a coefficient of 0.
2. Bring down the 2.
3. Multiply 3×2; enter 6 in second column and add to get -5.
4. Multiply $3 \times (-5)$; enter -15 in third column and add to get -10.
5. Recognizing the last number in the bottom row as the remainder, we read the coefficients of the quotient as 2 and -5. Therefore, the quotient is $2x - 5$ and the remainder is -10.

We can summarize the synthetic division procedure as follows:

SYNTHETIC DIVISION

To divide $a_4x^4 + a_3x^3 + a_2x^2 + a_1x + a_0$ by $x - r$ by the process of synthetic division, we set up the following table.

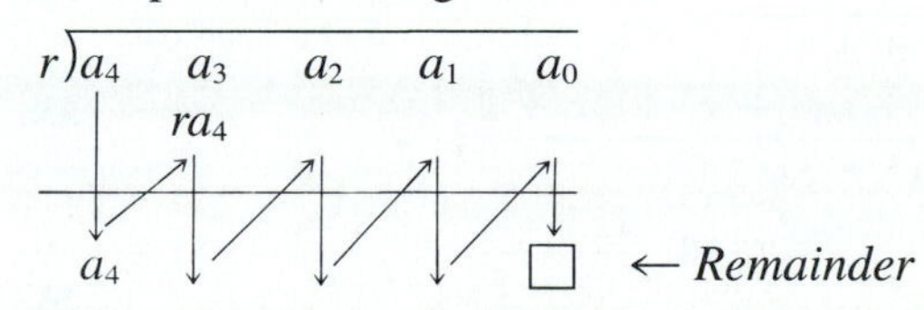

Note that the diagonal arrows indicate multiplication by r; the vertical arrows indicate addition in each column.

It is most important to remember that the synthetic division process works *only* when the divisor is of the form $x - r$.

EXAMPLE 3 Use synthetic division to find the quotient $\dfrac{x^5 + x^4 - 2x^2 + 24}{x + 2}$.

Solution Since the divisor $x + 2$ corresponds to $x - r$, we have $r = -2$ for the synthetic division process. (That is, $x - (-2) = x + 2$.) Also, we must remember to insert 0 for the coefficients of the missing terms. In other words, we view the dividend as $x^5 + x^4 + 0x^3 - 2x^2 + 0x + 24$.

$$\begin{array}{r|rrrrrr} -2 & 1 & 1 & 0 & -2 & 0 & 24 \\ & & -2 & 2 & -4 & 12 & -24 \\ \hline & 1 & -1 & 2 & -6 & 12 & \boxed{0} \end{array} \leftarrow \textit{Remainder}$$

Quotient: $\boxed{x^4 - x^3 + 2x^2 - 6x + 12}$ ■

In the next section we discuss some important properties of polynomials and polynomial equations.

EXERCISES 4.2

In Exercises 1–20, perform the required long division.

1. $\dfrac{x^2 - 5x - 14}{x - 4}$
2. $\dfrac{x^2 + 6x - 16}{x + 8}$
3. $\dfrac{x^2 + 7x - 18}{x + 9}$
4. $\dfrac{x^2 - 3x - 10}{x - 4}$
5. $\dfrac{6t^2 - 23t + 15}{3t - 4}$
6. $\dfrac{10a^2 + 7a - 10}{2a + 3}$
7. $\dfrac{y^2 + 2}{y^2 + 1}$
8. $\dfrac{x^3 - 4}{x^2 - 1}$
9. $\dfrac{2x^3 + x^2 - 5x + 2}{x + 2}$
10. $\dfrac{x^4 - 4x^3 - x^2 + x - 8}{x - 3}$
11. $\dfrac{3x^4 - x^3 + 6}{x - 1}$
12. $\dfrac{5x^3 + 3x^2 + 15}{x + 3}$
13. $\dfrac{8c^3 + 1}{2c + 1}$
14. $\dfrac{27z^3 - 1}{3z - 1}$
15. $\dfrac{2w^4 + w^3 - 3w^2 - 5w + 4}{w^2 + w + 1}$
16. $\dfrac{x^6 + 3x^5 - 2x^4 - 6x^3 + x^2 + 3x + 2}{x^2 + 3x}$
17. $\dfrac{x^4 - 16}{x - 2}$
18. $\dfrac{125 - x^3}{x - 5}$
19. $\dfrac{3x^2 + 6 - x + x^3}{x - 2}$
20. $\dfrac{8 - 2x^3 + 6x^2}{3 - x}$

In Exercises 21–40, use synthetic division to find the quotient and remainder.

21. $\dfrac{x^2 - 8x + 7}{x - 2}$
22. $\dfrac{5x^2 + 6x - 9}{x + 3}$
23. $\dfrac{2x^3 + x^2 - 3x + 15}{x + 2}$
24. $\dfrac{x^4 - 5x^3 + 3x^2 - 3x - 2}{x - 3}$
25. $\dfrac{2x^4 + x^3 - 6x^2 - 10x + 13}{x - 1}$
26. $\dfrac{6x^3 + 4x^2 - 3x - 5}{x + 1}$
27. $\dfrac{5x^2 - 23}{x + 4}$
28. $\dfrac{5x^3 - 4x - 12}{x - 2}$
29. $\dfrac{2x^3 + 4x^2 + 6}{x + 3}$
30. $\dfrac{x^4 - 7x^2 + 8x - 42}{x - 3}$
31. $\dfrac{x^4 - 64}{x - 4}$
32. $\dfrac{x^5 + 1}{x + 1}$
33. $\dfrac{6x^3 - 3x^2 - 8x + 4}{x - \frac{1}{2}}$
34. $\dfrac{12x^3 + 4x^2 - 15x - 5}{x + \frac{1}{3}}$

35. $\dfrac{2x^3 - 3x^2 + 4x - 6}{x - \dfrac{3}{2}}$

36. $\dfrac{3x^3 - 2x^2 + 12x - 8}{x - \dfrac{2}{3}}$

37. $\dfrac{x^3 + a^3}{x + a}$

38. $\dfrac{x^3 - a^3}{x - a}$

39. $\dfrac{x^4 - a^4}{x - a}$

40. $\dfrac{x^5 - a^5}{x - a}$

41. Given $p(x) = x^3 + x^2 + 4x - 5$ and $d(x) = x - 5$, use the Division Algorithm to find $q(x)$ and $R(x)$.

42. Given $p(x) = x^4 - 16$ and $d(x) = x - 2$, use the Division Algorithm to find $q(x)$ and $R(x)$.

43. When $x^3 - kx + 5$ is divided by $x - 2$, the remainder is 1. Find k.

44. When $2x^3 + kx^2 - 6$ is divided by $x + 1$, the remainder is -3. Find k.

45. When $p(x) = x^3 + kx + 4$ is divided by $x + 2$, the remainder is the same as when $q(x) = 2x^3 + kx^2 - 2$ is divided by $x - 2$. Find k.

46. When $p(x) = 3x^4 + kx^2 + 7$ is divided by $x - 1$, the remainder is the same as when $q(x) = x^4 + kx - 6$ is divided by $x - 2$. Find k.

QUESTIONS FOR THOUGHT

47. Let $p(x) = 2x^3 - 5x^2 + 6x + 4$. Use synthetic division to divide $p(x)$ by $x - 2$. What is the remainder? Evaluate $p(2)$. Now divide $p(x)$ by $x + 2$, find the remainder and also evaluate $p(-2)$. Do you notice anything special?

48. Analyze the Division Algorithm to determine what the remainder will be when we divide a polynomial $p(x)$ by $x - a$. HINT: See what happens when you compute $p(a)$.

4.3 Roots of Polynomial Equations: The Remainder Theorem and Factor Theorem

As we saw in Section 4.1, finding the zeros of a polynomial function can help in sketching its graph. With the aid of the division algorithm, we can derive two important theorems that will allow us to recognize the zeros of polynomials.

The division algorithm states that if we divide a polynomial $p(x)$ by a polynomial $d(x)$, where the degree of $d(x)$ is less than or equal to the degree of $p(x)$, then there are polynomials $q(x)$ and $R(x)$ such that

$$p(x) = d(x) \cdot q(x) + R(x)$$

where either $R(x) = 0$ or the degree of $R(x)$ is less than the degree of $d(x)$.

If we apply the division algorithm where the divisor, $d(x)$, is linear, that is, of the form $x - r$, we get

Why must R be a constant?

$$p(x) = (x - r)q(x) + R$$

Note that since the divisor is of the first degree, the remainder, R, must be a constant. If we now substitute $x = r$ into this equation, we get

$$p(r) = (r - r)q(r) + R = 0 \cdot q(r) + R \qquad \textit{Therefore,}$$

$$p(r) = R$$

The result we just proved is called the **Remainder Theorem.**

THEOREM 4.3 The Remainder Theorem
When a polynomial $p(x)$ of degree at least 1 is divided by $x - r$, the remainder is $p(r)$.

Use synthetic division to divide $p(x) = x^3 - x^2 + 3x - 1$ by $x - 2$ and verify the Remainder Theorem.

Solution We set up the synthetic division:

$$\begin{array}{r|rrrr} 2 & 1 & -1 & 3 & -1 \\ & & 2 & 2 & 10 \\ \hline & 1 & 1 & 5 & \boxed{9} \end{array} \leftarrow \textit{Remainder}$$

In this example $x - r = x - 2$ and so $r = 2$; therefore, according to the Remainder Theorem, the remainder should be equal to $p(2)$.

$$\begin{aligned} p(x) &= x^3 - x^2 + 3x - 1 && \textit{Substitute } x = 2. \\ p(2) &= 2^3 - 2^2 + 3(2) - 1 \\ &= 8 - 4 + 6 - 1 = 9 && \textit{Thus} \\ p(2) &= 9 && \textit{Which is the remainder, as required.} \end{aligned}$$

■

According to the remainder theorem, what does it mean that $p(2) = 9$?

As an immediate corollary to the remainder theorem, we recognize that if $x - r$ is a factor of $p(x)$, then the remainder must be 0; conversely, if the remainder is 0, then $x - r$ is a factor of $p(x)$. This is known as the **Factor Theorem.**

THEOREM 4.4 The Factor Theorem
$x - r$ is a factor of $p(x)$ if and only if $p(r) = 0$.

Recall that when we solve a quadratic equation $p(x) = x^2 - 3x - 10 = 0$, we get

$$\begin{aligned} p(x) = x^2 - 3x - 10 &= 0 \\ (x - 5)(x + 2) &= 0 && \textit{The factors of } p(x) \textit{ are } x - 5 \textit{ and } x + 2. \\ x = 5 \quad \text{or} \quad x &= -2 && \textit{The roots are } 5 \textit{ and } -2. \end{aligned}$$

Thus the Factor Theorem is merely a generalization of what we have already seen for quadratic polynomials.

EXAMPLE 2 Given that one root of the equation $px = x^3 - 23x + 10 = 0$ is $x = -5$, find the other roots.

If $x + 5$ is a factor of $p(x)$, what will the remainder be when $p(x)$ is divided by $x + 5$?

Solution Since $x = -5$ is given as a root of $p(x) = 0$, the Factor Theorem tells us that $x + 5$ is a factor. If we divide $p(x)$ by $x + 5$, we know the remainder will be 0 and the quotient will be of degree 2, to which we can apply the techniques for finding the roots of quadratic equations. We may use synthetic division to divide the polynomial $x^3 - 23x + 10$ by $x + 5$.

Note that if -5 is the root, then -5 appears as the divisor in the synthetic division tableau.

$$\begin{array}{r|rrrr} -5 & 1 & 0 & -23 & 10 \\ & & -5 & 25 & -10 \\ \hline & 1 & -5 & 2 & \boxed{0} \end{array}$$

The quotient polynomial is $q(x) = x^2 - 5x + 2$, and the original equation becomes

$$x^3 - 23x + 10 = (x + 5)(x^2 - 5x + 2) = 0$$

Thus any additional roots must be solutions to $x^2 - 5x + 2 = 0$. We can find the roots of $x^2 - 5x + 2 = 0$ by using the quadratic formula:

$$x = \frac{5 \pm \sqrt{17}}{2}$$

Recall that we stated in Section 4.1 that for a polynomial $p(x)$ of degree n, the equation $p(x) = 0$ has at most n solutions. Since we have three roots of the third-degree equation $x^3 - 23x + 10 = 0$, we have found *all* the roots of this equation. The roots are $x = -5$, $x = \dfrac{5 \pm \sqrt{17}}{2}$. ■

We discuss this fact in greater detail later in this section.

EXAMPLE 3 Factor $p(x) = 2x^4 + x^3 - 14x^2 - 19x - 6$ as completely as possible, given that $x = 3$ and $x = -\dfrac{1}{2}$ are roots of $p(x) = 0$.

Solution Since $x = 3$ is a root, we know that $x - 3$ is a factor and so we begin by dividing $p(x)$ by $x - 3$. $\left(\text{Note that we could also begin by dividing } p(x) \text{ by } x + \dfrac{1}{2}.\right)$ As before, we use synthetic division.

$$\begin{array}{r|rrrrr} 3 & 2 & 1 & -14 & -19 & -6 \\ & & 6 & 21 & 21 & 6 \\ \hline & 2 & 7 & 7 & 2 & \boxed{0} \end{array}$$

Thus we have

$$p(x) = 2x^4 + x^3 - 14x^2 - 19x - 6 = (x - 3)(2x^3 + 7x^2 + 7x + 2)$$

Since we are given that $x = -\dfrac{1}{2}$ is also a root of $p(x) = 0$, it follows that $x + \dfrac{1}{2}$ must be a factor of $2x^3 + 7x^2 + 7x + 2$.

We again use synthetic division to divide $2x^3 + 7x^2 + 7x + 2$ by $x + \dfrac{1}{2}$. Remember that the divisor for the synthetic division is $-\dfrac{1}{2}$.

$$\begin{array}{r|rrrr} -\frac{1}{2} & 2 & 7 & 7 & 2 \\ & & -1 & -3 & -2 \\ \hline & 2 & 6 & 4 & \boxed{0} \end{array}$$

Thus we have

$$\begin{aligned} p(x) = 2x^4 + x^3 - 14x^2 - 19x - 6 &= (x - 3)(2x^3 + 7x^2 + 7x + 2) \\ &= (x - 3)\left(x + \frac{1}{2}\right)(2x^2 + 6x + 4) \end{aligned}$$

We now factor $2x^2 + 6x + 4$ as $2(x^2 + 3x + 2) = 2(x + 2)(x + 1)$, and so the complete factorization of $p(x)$ is

$$p(x) = 2x^4 + x^3 - 14x^2 - 19x - 6 = 2(x + 2)(x + 1)(x - 3)\left(x + \frac{1}{2}\right)$$

We may incorporate the factor of 2 into $x + \frac{1}{2}$ and rewrite the factorization as

$$p(x) = 2x^4 + x^3 - 14x^2 - 19x - 6 = \boxed{(x + 2)(x + 1)(x - 3)(2x + 1)}$$

■

Zeros of Polynomials

The Factor and Remainder Theorems establish the intimate relationship between the factors of a polynomial $p(x)$ and the roots of the equation $p(x) = 0$. In Section 4.1 we mentioned several properties of polynomial functions, one of which was that a polynomial of degree n can have at most n zeros. We now elaborate on this idea.

In order to make the counting of the zeros of polynomials more uniform, we introduce the notion of *multiplicity*. For example, the following polynomial equation $x^2 - 2x + 1 = (x - 1)(x - 1) = 0$ has as its only root $x = 1$; however, since the factor $x - 1$ appears twice, we call 1 a **double root** or a repeated **root of multiplicity 2**. In general, if $(x - a)$ appears as a factor of a polynomial k times, we say that a is a repeated **root of multiplicity k**.

Does every polynomial equation (of degree at least one) have a root? Our answer depends upon the number system in which we are working. If we restrict ourselves to the real number system, then we are already familiar with the fact that an equation such as $x^2 + 1 = 0$ has no *real* solutions. However, this equation does have two roots in the complex number system. (The roots are i and $-i$.)

Carl Friedrich Gauss (1777–1855) is generally considered to be one of the greatest mathematicians that the world has known.* In 1799 Gauss, in his doctoral dissertation, proved the rather remarkable fact that within the complex number system every polynomial equation of degree greater than or equal to 1 has at least one root. This fact has such far-reaching significance that it is usually referred to as the **Fundamental Theorem of Algebra.**

THEOREM 4.5 The Fundamental Theorem of Algebra
If $p(x)$ is a polynomial of degree $n > 0$ whose coefficients are complex numbers, then the equation $p(x) = 0$ has at least one root in the complex number system.

Keep in mind that since all real numbers are also complex numbers, a polynomial with real coefficients also satisfies the Fundamental Theorem of Algebra.

As an immediate consequence of the Fundamental Theorem, we can prove the following.

*Although there have been many very great mathematicians, three—Archimedes, Newton, and Gauss—are generally considered to be in a class of their own.

THEOREM 4.6 The Linear Factorization Theorem

If $p(x) = a_n x^n + a_{n-1}x^{n-1} + \cdots + a_1 x + a_0$, where $n \geq 1$ and $a_n \neq 0$, then

$$p(x) = a_n(x - r_1)(x - r_2) \cdots (x - r_n)$$

where the r_i are complex numbers (possibly real and not necessarily distinct).

In words, this theorem says that any polynomial of degree at least 1 with complex number coefficients can be expressed as a product of linear factors.

Proof By virtue of the Fundamental Theorem we know that the equation $p(x) = 0$ has at least one complex root, call it r_1. By the Factor Theorem we know that $x - r_1$ is a factor of $p(x)$. Therefore, we can write

$$p(x) = (x - r_1)q_1(x)$$

where the degree of $q_1(x)$ is $n - 1$ and again has leading coefficient a_n. (This is true because we are dividing by $x - r_1$, which has a leading coefficient of 1. See Exercise 38.)

If the degree of $q_1(x)$ is zero, then we are done. (Why?) If not, then the Fundamental Theorem applied to $q_1(x)$ says that $q_1(x)$ has at least one zero, call it r_2. Applying the Factor Theorem to $q_1(x)$, we get

$$q_1(x) = (x - r_2)q_2(x)$$

where the degree of $q_2(x)$ is $n - 2$ and it again has leading coefficient a_n. Therefore, we have

$$p(x) = (x - r_1)q_1(x) = (x - r_1)(x - r_2)q_2(x)$$

We can continue this process until the last quotient $q_n(x)$ is a constant; as explained before, it must be a_n. Thus $q_n(x) = a_n$, which gives us

$$p(x) = (x - r_1)(x - r_2) \cdots (x - r_n)a_n$$

as required. ■

An immediate consequence of the Linear Factorization Theorem is the fact, quoted earlier, that a polynomial of degree n can have at most n distinct zeros. In fact, we can state even more.

THEOREM 4.7 Every polynomial equation of degree $n \geq 1$ has exactly n roots in the complex number system, where a root of multiplicity k is counted k times.

EXAMPLE 4 In each of the following equations express the polynomial in the form described by the Linear Factorization Theorem. List each root and its multiplicity.

(a) $p(x) = x^3 - 6x^2 - 16x = 0$ **(b)** $q(x) = 3x^2 - 10x + 8 = 0$

(c) $f(x) = 2x^4 + 8x^3 + 10x^2 = 0$

Solution

(a) We may factor $p(x)$ as follows:

$$p(x) = x^3 - 6x^2 - 16x = x(x^2 - 6x - 16)$$
$$= x(x - 8)(x + 2)$$

To fit the Linear Factorization Theorem, we write $x + 2 = x - (-2)$.

The factor $(x - 0)$ is usually just written as x.

$$= x(x - 8)(x - (-2))$$

The roots of $p(x) = 0$ are 0, 8, and -2, each of multiplicity one.

(b) We may factor $q(x)$ as follows:

$$q(x) = 3x^2 - 10x + 8 = (3x - 4)(x - 2)$$

To fit the Linear Factorization Theorem, we write $3x - 4$ as $3\left(x - \frac{4}{3}\right)$.

$$= 3\left(x - \frac{4}{3}\right)(x - 2)$$

The roots of $p(x) = 0$ are $\frac{4}{3}$ and 2, each of multiplicity one.

(c) We may factor $f(x)$ as follows:

$$f(x) = 2x^4 + 8x^3 + 10x^2 = 2x^2(x^2 + 4x + 5)$$

We can find the roots of $x^2 + 4x + 5$ by using the quadratic formula. The roots are $-2 \pm i$. Therefore, we have

$$= 2x^2[x - (-2 + i)][x - (-2 - i)]$$

The roots of $f(x) = 0$ are: 0 with multiplicity two and $-2 + i$ and $-2 - i$, each with multiplicity one.

■

EXAMPLE 5

(a) Find a polynomial equation $p(x) = 0$ with exactly the following roots and multiplicities:

Root	Multiplicity
-1	3
2	4
5	2

Are there any other polynomials which give these same roots and multiplicities?

(b) Find a polynomial $f(x)$ having the zeros described in part (a) such that $f(1) = 32$.

Solution

(a) Based upon the Factor Theorem we may write the polynomial

$$p(x) = (x - (-1))^3(x - 2)^4(x - 5)^2 = (x + 1)^3(x - 2)^4(x - 5)^2$$

which gives exactly the required roots and multiplicities.

Since these are the only roots and hence the only linear factors, any other polynomial must be a constant multiple of $p(x)$. Therefore, any polynomial of the form $kp(x)$, where k is a nonzero constant, will give the same roots and multiplicities.

(b) Based upon part (a) we know that $f(x) = k(x + 1)^3(x - 2)^4(x - 5)^2$. Since we want $f(1) = 32$, we have

$$f(1) = k(1 + 1)^3(1 - 2)^4(1 - 5)^2$$

$$32 = k(8)(1)(16) \Rightarrow 32 = 128k \Rightarrow \boxed{k = \frac{1}{4}}$$

Thus $\boxed{f(x) = \frac{1}{4}(x + 1)^3(x - 2)^4(x - 5)^2}$. ■

Our experience in using the quadratic formula on quadratic equations with real coefficients has shown us that complex roots always appear in conjugate pairs. For example, the roots of $x^2 - 2x + 5 = 0$ are $1 + 2i$ and $1 - 2i$. In fact, this property extends to all polynomial equations with real coefficients.

THEOREM 4.8 Conjugate Roots Theorem
Let $p(x)$ be a polynomial with real coefficients. If the complex number $a + bi$ (where a and b are real numbers) is a root of $p(x) = 0$, then so is its conjugate $a - bi$.

The proof of this theorem is a direct consequence of the properties of conjugates and is outlined in Exercise 39 at the end of this section.

EXAMPLE 6 Let $r(x) = x^4 + 2x^3 - 9x^2 + 26x - 20$. Given that $1 - i\sqrt{3}$ is a root, find the other roots of $r(x) = 0$.

Solution According to the Conjugate Roots Theorem, if $1 - i\sqrt{3}$ is a root, then its conjugate, $1 + i\sqrt{3}$, must also be a root. Therefore, $x - (1 - i\sqrt{3})$ and $x - (1 + i\sqrt{3})$ are both factors of $r(x)$, and so their product must be a factor of $r(x)$. The reader should verify that

$$[x - (1 - i\sqrt{3})][x - (1 + i\sqrt{3})] = x^2 - 2x + 4.$$

We divide $r(x)$ by $x^2 - 2x + 4$.

$$\begin{array}{r}
x^2 + 4x - 5 \\
x^2 - 2x + 4 \,\overline{\big)\, x^4 + 2x^3 - 9x^2 + 26x - 20} \\
-(x^4 - 2x^3 + 4x^2) \\
\hline
4x^3 - 13x^2 + 26x \\
-(4x^3 - 8x^2 + 16x) \\
\hline
-5x^2 + 10x - 20 \\
-(-5x^2 + 10x - 20) \\
\hline
0
\end{array}$$

Therefore, we have

$$r(x) = (x^2 - 2x + 4)(x^2 + 4x - 5) = (x^2 - 2x + 4)(x + 5)(x - 1)$$

and so the roots of $r(x) = 0$ are $1 - i\sqrt{3}$, $1 + i\sqrt{3}$, -5, and 1. ■

The theorems we have discussed in this section are called *existence theorems* because they ensure the existence of zeros and linear factors of polynomials. However, it is important to recognize that these theorems do not tell us *how* to find the zeros or the linear factors. We will discuss some techniques for actually finding the zeros of certain polynomials in the next section.

EXERCISES 4.3

In Exercises 1–8, perform the requested division. Find the quotient and remainder and verify the Remainder Theorem by computing $p(a)$.

1. Divide $p(x) = x^2 - 5x + 8$ by $x + 4$.
2. Divide $p(x) = 2x^2 - 7x + 3$ by $x - 3$.
3. Divide $p(x) = 2x^3 - 7x^2 + x + 4$ by $x - 4$.
4. Divide $p(x) = 4x^3 - x^2 + 1$ by $x + 2$.
5. Divide $p(x) = 1 - x^4$ by $x - 1$.
6. Divide $p(x) = x^3 + 27$ by $x + 3$.
7. Divide $p(x) = x^5 - 2x^2 - 3$ by $x + 1$.
8. Divide $p(x) = x^6 - 16x^3 + 64$ by $x - 2$.

In Exercises 9–18, use the Remainder Theorem to find $p(c)$.

9. $p(x) = x^2 - 5x + 3,\ c = 4$
10. $p(x) = 3x^2 + 7x - 2,\ c = -3$
11. $p(x) = x^3 - 2x^2 + 3x - 5,\ c = -2$
12. $p(x) = -2x^3 + 7x - 4,\ c = 3$
13. $p(x) = 3x^4 - 5x^2 - 6x - 16,\ c = 2$
14. $p(x) = x^5 + 2x^4 - 5x^3 + 6x^2 - 15x + 36,\ c = -4$
15. $p(x) = -x^6 + 6x^4 + 23,\ c = -3$
16. $p(x) = x^7 - 3x^5 - x^3 - x^2 - x + 18,\ c = -2$
17. $p(x) = 8x^4 + 4x^3 - 6x + 3,\ c = \frac{1}{2}$
18. $p(x) = 9x^3 + 3x^2 - 12x + 7,\ c = -\frac{1}{3}$
19. Given that $p(4) = 0$, factor

$$p(x) = 2x^3 - 11x^2 + 10x + 8$$

as completely as possible.
20. Given that $q(x) = 6x^3 - 5x^2 - 7x + 4$ and $q(-1) = 0$, find the remaining zeros of $q(x)$.
21. Given that $r(x) = 4x^3 - x^2 - 36x + 9$ and $r\left(\frac{1}{4}\right) = 0$, find the remaining zeros of $r(x)$.
22. Given that $s\left(-\frac{1}{5}\right) = 0$, factor

$$s(x) = 30x^3 - 19x^2 - 35x - 6$$

as completely as possible.
23. Given that 3 is a double root of

$$p(x) = x^4 - 3x^3 - 19x^2 + 87x - 90 = 0$$

find all the roots of $p(x) = 0$
24. Given that -2 is a double root of

$$q(x) = 3x^4 + 10x^3 - x^2 - 28x - 20 = 0$$

find all the roots of $q(x) = 0$.
25. Let $p(x) = x^3 - 3x^2 + x - 3$. Verify that $p(3) = 0$ and find the other roots of $p(x) = 0$.
26. Let $q(x) = 3x^3 + x^2 - 4x - 10$. Verify that $q\left(\frac{5}{3}\right) = 0$ and find the other roots of $q(x) = 0$.
27. Express $p(x) = x^3 - x^2 - 12x$ in the form described in the Linear Factorization Theorem. List each zero and its multiplicity.
28. Express $q(x) = 6x^5 - 33x^4 - 63x^3$ in the form described in the Linear Factorization Theorem. List each zero and its multiplicity.
29. **(a)** Write the general polynomial $p(x)$ whose only zeros are 1, 2, and 3, with multiplicities 3, 2, and 1, respectively. What is its degree?
(b) Find the $p(x)$ described in part (a) if $p(0) = 6$.
30. **(a)** Write the general polynomial $q(x)$ whose only zeros are -4 and -3 with multiplicities 4 and 6, respectively. What is its degree?
(b) Find the $q(x)$ described in part (a) if $q(1) = 48$.
31. Write a polynomial with zeros 0, -2, and 1 with multiplicities 2, 2, and 1, respectively. What is its degree?
32. Write a polynomial whose only zero is 6 with multiplicity 7. What is its degree?

In Exercises 33–36, use the given information to find the remaining roots.

33. $2 - 3i$ is a root of $2x^3 - 5x^2 + 14x + 39 = 0$
34. $1 + 2i$ is a root of $x^4 - x^3 - 9x^2 + 29x - 60 = 0$
35. $3 + 4i$ is a root of
$4x^4 - 28x^3 + 129x^2 - 130x + 125 = 0$

36. $i\sqrt{2}$ and $3i$ are roots of

$$x^6 - 2x^5 + 12x^4 - 22x^3 + 29x^2 - 36x + 18 = 0$$

QUESTIONS FOR THOUGHT

37. Let $p(x) = x^3 - 5x^2 + 3x - 2$. Compute $p\left(\frac{3}{4}\right)$ directly and by using synthetic division. Which method do you think is easier in this case?

38. Suppose a polynomial has a leading coefficient of $a_n \neq 0$, that is, suppose

$$p(x) = a_n x^n + a_{n-1}x^{n-1} + \cdots + a_1 x + a_0$$

Show that if $p(x)$ is divided by $x - a$, the leading coefficient of the quotient will also be a_n.

39. The conjugate of a complex number z is frequently denoted as $\overline{z}$. Thus if $z = a + bi$, then $\overline{z} = a - bi$. Use the following properties of conjugates to prove the Conjugate Roots Theorem.

(a) $\overline{r} = r$ for all real numbers r
(b) $\overline{u + v} = \overline{u} + \overline{v}$
(c) $\overline{uv} = \overline{u}\,\overline{v}$
(d) $\overline{u^m} = \overline{u}^m$

HINT: Show that for a polynomial, $p(\overline{z}) = \overline{p(z)}$ and so if $p(z) = 0$, then $p(\overline{z}) = 0$.

GRAFFIX

40. Sketch the graph of $y = p(x) = (x - 1)^2(x + 2)^4$. What do you notice about the graph at zeros of even multiplicity?

41. Sketch the graph of $y = p(x) = (x - 2)^3(x + 1)^5$. What do you notice about the graph at zeros of odd multiplicity?

42. Sketch the graph of $y = p(x) = (2x - 3)^2(3x + 1)^3$. What do you notice is the difference between how the polynomial behaves at a zero of even multiplicity as compared to how it behaves at a zero of odd multiplicity?

43. Sketch the graph of $y = p(x) = x^4 + x^2 + 1$. Does $p(x)$ have any zeros? Does this violate the Fundamental Theorem of Algebra? Explain.

4.4 More on the Roots of Polynomial Equations: The Rational Root Theorem and Descartes's Rule of Signs

The Linear Factorization Theorem we discussed in the last section guarantees that we can factor a polynomial of degree at least one into linear factors, but it doesn't tell us how!

We know from experience that if $p(x)$ happens to be a quadratic polynomial, then we may solve $p(x) = Ax^2 + Bx + C = 0$ by using the quadratic formula to obtain the roots

$$x = \frac{-B \pm \sqrt{B^2 - 4AC}}{2A}$$

A natural question one might then ask is whether there is some algebraic "formula" involving the coefficients, the four basic arithmetic operations, and various radicals with which to compute the roots of a polynomial equation, $p(x) = 0$, of degree greater than 2.

It turns out that this question interested mathematicians for many years. In the early sixteenth century significant progress was made, and formulas for the solutions to the general cubic and quartic (fourth-degree) equation were derived. For example, given the general cubic equation $p(x) = x^3 + a_2x^2 + a_1x + a_0 = 0$,* there is an explicit, albeit quite messy, formula for its three roots. It is known as **Cardan's formula,** and part of what it says is the following:

$$\text{Let } p = a_1 - \frac{(a_2)^2}{3} \quad \text{and} \quad q = \frac{2(a_2)^3}{27} - \frac{a_1 a_2}{3} + a_0$$

* Without loss of generality we may assume that $a_3 = 1$; otherwise we can divide both sides of the equation $a_3x^3 + a_2x^2 + a_1x + a_0 = 0$ by a_3.

and let

$$P = \sqrt[3]{-\frac{q}{2} + \sqrt{\frac{p^3}{27} + \frac{q^2}{4}}} \quad \text{and} \quad Q = \sqrt[3]{-\frac{q}{2} - \sqrt{\frac{p^3}{27} + \frac{q^2}{4}}}$$

Then one of the three roots of $p(x) = 0$ is given by $x = P + Q - \frac{a_2}{3}$. The other two roots are given by similar formulas.

If $p(x) = x^3 - 2x - 21$, we have $a_2 = 0$, $a_1 = -2$, and $a_0 = -21$ (the fact that $a_2 = 0$ simplifies the formula significantly) and yet it is still quite a chore to use the formula to obtain $x = 3$ as a root. (Try it!)

Mathematicians continued to search for a formula for the roots of the general fifth-degree (and higher) polynomial equation. Finally, in 1828, the Norwegian mathematician Niels Abel proved the remarkable fact that the search was futile. He proved that the problem in finding such a formula was not that mathematicians were not clever enough but rather that there can be no such formula for solving the general polynomial equation of degree 5 or higher. Keep in mind this does not mean that these equations don't have solutions (in fact, our previous theorems ensure that they do); rather, it just says that, *in general,* we cannot express the solutions in terms of the coefficients and radicals. Finally, in 1830 the French mathematician Evariste Galois (at the age of 18) settled the question by describing exactly which polynomials can be solved in terms of their coefficients and radicals.

The rest of this section is devoted to developing some special methods for finding the roots of polynomial equations.

As we have seen, even though we have no general techniques for factoring polynomials of degree greater than 2, if we happen to know a root, say r, we can use synthetic division to divide $p(x)$ by $x - r$ and obtain a quotient polynomial of *lower* degree. If we can get the quotient polynomial down to a quadratic, then we are able to determine all the roots. But how do we find a root to start this process along? The following theorem can be most helpful.

THEOREM 4.9 The Rational Root Theorem

Suppose that

$$f(x) = a_n x^n + a_{n-1}x^{n-1} + \cdots + a_2x^2 + a_1x + a_0 \qquad \text{where } n \geq 1,\ a_n \neq 0$$

is an nth-degree polynomial with *integer* coefficients. If $\frac{p}{q}$ is a rational root of $f(x) = 0$, where p and q have no common factors other than ± 1, then p is a factor of a_0 and q is a factor of a_n.

A general proof of this theorem is outlined in Exercise 53 at the end of this section; however, by looking at an example we can get a feeling as to why this theorem is true. Suppose $\frac{3}{2}$ is a root of the third-degree polynomial equation

$$a_3x^3 + a_2x^2 + a_1x + a_0 = 0$$

Then

$$a_3\left(\frac{3}{2}\right)^3 + a_2\left(\frac{3}{2}\right)^2 + a_1\left(\frac{3}{2}\right) + a_0 = 0$$

$$\frac{27a_3}{8} + \frac{9a_2}{4} + \frac{3a_1}{2} = -a_0 \quad \textit{Multiply both sides of the equation by 8.}$$

$$27a_3 + 18a_2 + 12a_1 = -8a_0 \tag{1}$$

Which can also be written as

$$27a_3 = -18a_2 - 12a_1 - 8a_0 \tag{2}$$

If we look carefully at equation (1) we can see that the left-hand side is divisible by 3, and therefore the right-hand side must also be divisible by 3. But if $-8a_0$ is divisible by 3, then a_0 must be divisible by 3 (because 8 is not divisible by 3). Similarly, by looking at the right-hand side of equation (2) we can see that it is divisible by 2, and since 27 is not divisible by 2, a_3 must be divisible by 2. This is exactly what the Rational Root Theorem asserts.

Why does the Rational Root Theorem tell us that if the leading coefficient of a polynomial is 1, then its rational roots must be a factor of the constant term?

It is very worthwhile to note that if a polynomial has a leading coefficient of 1, then the Rational Root Theorem tells us that any rational roots must be divisors of the constant term a_0.

EXAMPLE 1 Find all the roots of the equation $p(x) = 2x^3 + 3x^2 - 23x - 12 = 0$.

Solution We begin by searching for any possible rational roots to the given equation. According to the Rational Root Theorem, if $\frac{p}{q}$ is a rational root of the given equation, then p must be a factor of -12 and q must be a factor of 2. Thus we have

Possible values of p: $\pm 1, \pm 2, \pm 3, \pm 4, \pm 6, \pm 12$

Possible values of q: $\pm 1, \pm 2$

Possible rational roots $\frac{p}{q}$: $\pm 1, \pm\frac{1}{2}, \pm 2, \pm 3, \pm\frac{3}{2}, \pm 4, \pm 6, \pm 12$

Note that the possible quotients of $\frac{p}{q}$ repeat certain roots, which we only list once. For example $\frac{6}{2} = \frac{3}{1}$.

We may check these possible roots by substituting the values into $p(x)$; however, it is more efficient to check the values by using synthetic division, since in case we do find a root, we will already have the quotient available.

$$\begin{array}{r|rrrr} 1 & 2 & 3 & -23 & -12 \\ & & 2 & 5 & -18 \\ \hline & 2 & 5 & -18 & \boxed{-30} \end{array} \qquad \begin{array}{r|rrrr} -1 & 2 & 3 & -23 & -12 \\ & & -2 & -1 & 24 \\ \hline & 2 & 1 & -24 & \boxed{12} \end{array}$$

How do we know that $p(1) = -30$ and $p(-1) = 12$?

Therefore, we know that $p(1) = -30$ and $p(-1) = 12$. Since $p(1)$ is negative and $p(-1)$ is positive, the Intermediate Value Theorem guarantees that $p(x)$ has a zero between -1 and 1. Thus it seems reasonable next to check the rational roots on our list between -1 and 1.

$$\begin{array}{r|rrrr} \frac{1}{2} & 2 & 3 & -23 & -12 \\ & & 1 & 2 & -\frac{21}{2} \\ \hline & 2 & 4 & -21 & \boxed{-\frac{45}{2}} \end{array} \qquad \begin{array}{r|rrrr} -\frac{1}{2} & 2 & 3 & -23 & -12 \\ & & -1 & -1 & +12 \\ \hline & 2 & 2 & -24 & \boxed{0} \end{array}$$

Therefore, $-\dfrac{1}{2}$ is a root, and we can read the quotient from the bottom line of the synthetic division:

$$p(x) = 2x^3 + 3x^2 - 23x - 12 = \left(x + \frac{1}{2}\right)(2x^2 + 2x - 24)$$

Now we may try to find another root by using the Rational Root Theorem on $2x^2 + 2x - 24$. However, it is much easier to simply try to factor the quadratic directly.

$$2x^2 + 2x - 24 = 2(x^2 + x - 12) = 2(x + 4)(x - 3)$$

Therefore, the original equation becomes

$$\begin{aligned} p(x) = 2x^3 + 3x^2 - 23x - 12 &= \left(x + \frac{1}{2}\right)(2x^2 + 2x - 24) \\ &= 2\left(x + \frac{1}{2}\right)(x + 4)(x - 3) = 0 \end{aligned}$$

and all the roots are $-\dfrac{1}{2}$, -4, 3. ■

EXAMPLE 2 Find all the rational roots of

$$p(x) = 2x^4 - 12x^3 + 19x^2 - 6x + 9 = 0$$

Solution According to the Rational Root Theorem we consider

$$\frac{\text{possible values of } p}{\text{possible values of } q} \rightarrow \frac{\text{factors of } 9}{\text{factors of } 2} \rightarrow \frac{\pm 1, \pm 3, \pm 9}{\pm 1, \pm 2}$$

Therefore the possible rational roots of $p(x)$ are $\pm 1, \pm 3, \pm 9, \pm \dfrac{1}{2}, \pm \dfrac{3}{2}, \pm \dfrac{9}{2}$.

We leave it to the student to check that 1 and -1 are not roots of the equation. Next we check if $x = 3$ is a root by using synthetic division.

$$\begin{array}{r|rrrrr} 3 & 2 & -12 & 19 & -6 & 9 \\ & & 6 & -18 & 3 & -9 \\ \hline & 2 & -6 & 1 & -3 & \boxed{0} \end{array}$$

Therefore, 3 is a root and we have

$$p(x) = 2x^4 - 12x^3 + 19x^2 - 6x + 9 = (x - 3)(2x^3 - 6x^2 + x - 3)$$

and we now seek the rational roots of $q(x) = 2x^3 - 6x^2 + x - 3 = 0$. Rather than work with the list of possible rational roots of $p(x) = 0$, we can apply the Rational

Root Theorem to $q(x) = 0$ and, we hope, get a shorter list of possible rational roots. For $q(x)$ we consider roots of the form

$$\frac{\text{factors of } -3}{\text{factors of } 2} \quad \begin{matrix} \rightarrow \\ \rightarrow \end{matrix} \quad \frac{\pm 1, \pm 3}{\pm 1, \pm 2}$$

Therefore the possible rational roots of $q(x) = 0$ are $\pm 1, \pm 3, \pm \frac{1}{2}, \pm \frac{3}{2}$.

Why is it that if 1 is not a zero of $p(x)$, then it cannot be a zero of any factor of $p(x)$?

Since we have already seen that 1 and -1 are not roots of $p(x) = 0$, ± 1 cannot be roots of $q(x) = 0$. (Why not?) However, 3 could again be a root of $q(x) = 0$. We check this by again using synthetic division.

$$\begin{array}{r|rrrr} 3 & 2 & -6 & 1 & -3 \\ & & 6 & 0 & 3 \\ \hline & 2 & 0 & 1 & \boxed{0} \end{array}$$

and so 3 is a root of $q(x) = 0$. We have

$$\begin{aligned} p(x) = 2x^4 - 12x^3 + 19x^2 - 6x + 9 &= (x - 3)(2x^3 - 6x^2 + x - 3) \\ &= (x - 3)(x - 3)(2x^2 + 1) \end{aligned}$$

Since $2x^2 + 1 = 0$ has no real roots, $p(x) = 0$ has one rational root, 3, with multiplicity 2. ■

GRAFFIX

Use a graphics calculator or computer to graph $y = 6x^3 - 23x^2 - 29x + 12$. How can the information you obtain from this graph about the zeros of this function help you use the Rational Root Theorem more efficiently to find its exact zeros?

We have seen that an nth degree polynomial will have exactly n zeros (counting multiplicity) in the complex number system. However, if we are sketching the graph of a polynomial, it is only the *real* zeros in which we are interested. (Keep in mind that even if we use a graphics calculator or a computer to sketch the graph of a function, it will show us only the *real* zeros of the function.) The following two theorems are useful in determining the number of real zeros a polynomial has.

Since there is no difficulty determining whether or not 0 is a zero of a polynomial (we simply see if x is a factor of $p(x)$), for the purposes of this discussion, we may as well assume that $p(x) = a_n x^n + a_{n-1}x^{n-1} + \cdots + a_1 x + a_0$ where $a_0 \neq 0$. This assures us that 0 is not a zero of $p(x)$. Again, for the purposes of this discussion we will assume that all polynomials are written in standard form—that is, from highest power to lowest.

In order to state the first theorem, we need to introduce the following idea. By a *variation in sign* of the real coefficients of a polynomial, we mean that two consecutive coefficients have opposite signs. For example, the coefficients of the polynomial $p(x) = 5x^3 - 7x^2 - 4x + 6$ change from 5 to -7 (one sign variation) and then from -4 to 6 (a second variation in sign). So $p(x) = 5x^3 - 7x^2 - 4x + 6$ has two variations in sign.

We can now state **Descartes's Rule of Signs,** which deals with the number of positive and negative real zeros* a polynomial can have.

THEOREM 4.10 Descartes's Rule of Signs

Let $p(x)$ be a polynomial with real coefficients such that $p(0) \neq 0$. Then

1. The number of *positive real zeros* of $p(x)$ is either equal to the number of variations in sign of $p(x)$ or is less than that number by an even integer.
2. The number of *negative real zeros* of $p(x)$ is either equal to the number of variations in sign of $p(-x)$ or is less than that number by an even integer.

Keep in mind that this theorem tells us about the *real* zeros of $p(x)$, and in order to apply this theorem $p(x)$ must be in standard form and have a nonzero constant term. We will not prove this theorem.

When applying Descartes's Rule of Signs we count roots with their multiplicity. For example, in the equation $p(x) = x^2 - 6x + 9 = 0$, there are two variations in sign and so the equation has either two positive real roots or an even number fewer than 2, which would mean no roots. The factored form of this equation is $p(x) = (x - 3)^2 = 0$, so we see that 3 is a root of multiplicity 2, which agrees with the theorem.

EXAMPLE 3 Apply Descartes's Rule of Signs to find the real zeros of

$$p(x) = x^5 + 2x^4 + x^3 + 2x^2 + 3x + 6$$

Solution First we examine $p(x)$ and see that there are 0 variations in the signs of its coefficients. Therefore, according to Descartes's Rule of Signs, $p(x)$ must have 0 positive real zeros.

Next we examine

$$\begin{aligned} p(-x) &= (-x)^5 + 2(-x)^4 + (-x)^3 + 2(-x)^2 + 3(-x) + 6 \\ &= -x^5 + 2x^4 - x^3 + 2x^2 - 3x + 6 \end{aligned}$$

Why don't we need to test the possible positive rational roots?

and see that $p(-x)$ has 5 variations in the signs of its coefficients, so by the second part of Descartes's Rule of Signs, $p(x)$ must have 5, 3, or 1 negative real zeros.

Applying the Rational Root Theorem to $p(x)$, we see that the possible rational roots are $\pm 1, \pm 2, \pm 3, \pm 6$. Testing the possible negative rational roots by synthetic division we find that -2 is a root and that

$$p(x) = x^5 + 2x^4 + x^3 + 2x^2 + 3x + 6 = (x + 2)(x^4 + x^2 + 3)$$

Examining $q(x) = x^4 + x^2 + 3$, we find that both $q(x)$ and $q(-x)$ have 0 sign variations and so $q(x)$ has no real zeros. Therefore, the only real zero of $p(x)$ is -2.

*The terminology *positive real* number is actually redundant, since complex numbers cannot be positive or negative. Only real numbers can be positive or negative. Nevertheless, we feel the redundancy is useful in order to emphasize the fact that we are talking about the real zeros of $p(x)$.

As we have seen, the Rational Root Theorem can involve a fair number of possible rational roots. The following theorem can be helpful in reducing the number of possible rational roots that have to be considered.

We say that a real number U is an **upper bound** for the real zeros of a function f if all the real zeros of f are less than or equal to U. Similarly, a real number L is a lower bound for the real zeros of f if all the real zeros of f are greater than or equal to L. In other words, if a function has an upper bound U and a lower bound L for its real zeros, then all its real zeros must fall in the interval $[L, U]$ on the number line. See Figure 4.22.

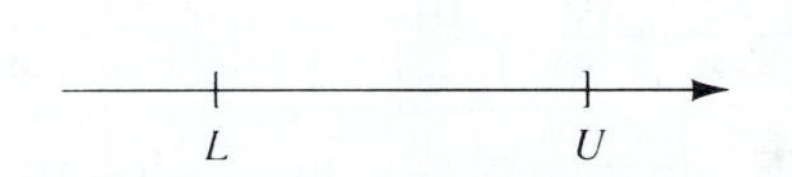

FIGURE 4.22 All the real zeros of f will fall in the closed interval $[L, U]$.

THEOREM 4.11 The Upper and Lower Bound Theorem
Let $p(x)$ be a polynomial with real coefficients whose leading coefficient is positive. Suppose we divide $p(x)$ by $x - r$ using synthetic division.

1. If $r > 0$ and all the numbers in the bottom row of the synthetic division table are positive or zero, then r is an upper bound for the real roots of $p(x) = 0$.
2. If $r < 0$ and the numbers in the bottom row of the synthetic division table are alternately positive and negative (zero entries count as either positive or negative), then r is a lower bound for the real roots of $p(x) = 0$.

A justification of part 2 of this theorem is outlined in Exercise 54.

EXAMPLE 4 Find the upper and lower bounds for the real roots of $p(x) = x^3 - 3x^2 - 2x + 10 = 0$.

Solution We test 1, 2, 3, 4, . . . as possible upper bounds by synthetically dividing $p(x)$ by $x - 1$, $x - 2$, etc.

$$\begin{array}{r|rrrr} 1 & 1 & -3 & -2 & 10 \\ & & 1 & -2 & -4 \\ \hline & 1 & -2 & -4 & \boxed{6} \end{array} \qquad \begin{array}{r|rrrr} 2 & 1 & -3 & -2 & 10 \\ & & 2 & -2 & -8 \\ \hline & 1 & -1 & -4 & \boxed{2} \end{array}$$

$$\begin{array}{r|rrrr} 3 & 1 & -3 & -2 & 10 \\ & & 3 & 0 & -6 \\ \hline & 1 & 0 & -2 & \boxed{4} \end{array} \qquad \begin{array}{r|rrrr} 4 & 1 & -3 & -2 & 10 \\ & & 4 & 4 & 8 \\ \hline & 1 & 1 & 2 & \boxed{18} \end{array}$$

Since the entries of the last row of the synthetic division of $p(x)$ by $x - 4$ are all positive, according to the Upper and Lower Bound Theorem, 4 is an upper bound on the real zeros of $p(x)$. (Let's take a moment to see what part 1 of this theorem is saying. Based upon this last division, we have

$$p(x) = x^3 - 3x^2 - 2x + 10 = (x - 4)(x^2 + x + 2) + 18$$

Looking at the right-hand side of this equation, we can see that if $x > 4$, then $x - 4$ is positive, as is $x^2 + x + 2$. Therefore, for $x > 4$, the right-hand side of the equation is positive. Hence $p(x)$ must be positive and so cannot be equal to zero for $x > 4$.)

In order to find a lower bound, we divide $p(x)$ synthetically by $x - (-1)$, $x - (-2)$, etc.

$$\begin{array}{r|rrrr} -1 & 1 & -3 & -2 & 10 \\ & & -1 & 4 & -2 \\ \hline & 1 & -4 & 2 & \boxed{8} \end{array} \qquad \begin{array}{r|rrrr} -2 & 1 & -3 & -2 & 10 \\ & & -2 & 10 & -16 \\ \hline & 1 & -5 & 8 & \boxed{-6} \end{array}$$

Using an analysis similar to the preceding one, which showed that 4 is an upper bound, show that -2 is a lower bound on the zeros of $p(x)$.

Since the entries of the last row of the synthetic division of $p(x)$ by $x - (-2)$ are alternately positive and negative, according to the Upper and Lower Bound Theorem -2 is a lower bound on the real zeros of $p(x)$. All the zeros of $p(x)$ lie in the interval $[-2, 4]$. ■

As we saw in Section 4.1, knowing the location of the zeros of a function can be very helpful in sketching its graph. We conclude this with an example in which we use the three theorems in this section dealing with finding the zeros of polynomials.

EXAMPLE 5 Find all the real zeros of $y = f(x) = x^4 - x^3 + x^2 - 3x - 6$ and sketch its graph.

Solution Starting with Descartes's Rule of Signs we see that $f(x)$ has 3 variations in sign, so that $f(x)$ has either 3 or 1 positive real zeros. Since

$$f(-x) = (-x)^4 - (-x)^3 + (-x)^2 - 3(-x) - 6 = x^4 + x^3 + x^2 + 3x - 6$$

has one variation in sign, $f(x)$ has exactly one negative real zero.

According to the Rational Root Theorem we consider zeros of the form

$$\frac{\text{factors of } 6}{\text{factors of } 1} \begin{array}{c}\rightarrow\\ \rightarrow\end{array} \frac{\pm 1, \pm 2, \pm 3, \pm 6}{\pm 1}$$

Therefore the possible rational zeros of $f(x)$ are $\pm 1, \pm 2, \pm 3, \pm 6$.

Trying $x = 1$ and $x = 2$ as roots by synthetic division, we get

If $x - 2$ is a factor of $p(x)$, what will the remainder be when $p(x)$ is divided by $x - 2$?

$$\begin{array}{r|rrrrr} 1 & 1 & -1 & 1 & -3 & -6 \\ & & 1 & 0 & 1 & -2 \\ \hline & 1 & 0 & 1 & -2 & \boxed{-8} \end{array} \qquad \begin{array}{r|rrrrr} 2 & 1 & -1 & 1 & -3 & -6 \\ & & 2 & 2 & 6 & 6 \\ \hline & 1 & 1 & 3 & 3 & \boxed{0} \end{array}$$

Not only do we see that 2 is a zero, but since all the numbers in the bottom row of the division by 2 are positive, the theorem on upper and lower bounds tells us that 2 is an upper bound on the positive zeros of $f(x)$. Therefore, we may ignore 3 and 6 as possible zeros, and we have $x = 2$ as the only positive rational zero.

As noted before, we know that $f(x)$ has exactly one negative zero, and so we now test the possible negative rational zeros in the quotient we obtained in the last synthetic division.

$$\begin{array}{r|rrr} -1 & 1 & 1 & 3 & 3 \\ & & -1 & 0 & -3 \\ \hline & 1 & 0 & 3 & \boxed{0} \end{array}$$

Therefore, $x = -1$ is also a root and we have 2 and -1 as the only real roots of $f(x)$.

Why doesn't the factor $(x^2 + 3)$ contribute a cut point?

Knowing that $x - 2$ and $x + 1$ are factors of $f(x)$, we can factor $f(x)$:

$$f(x) = (x - 2)(x + 1)(x^2 + 3)$$

Doing a a sign analysis on $f(x) = (x - 2)(x + 1)(x^2 + 3)$ and recognizing that $x^2 + 3$ is always positive, we get Figure 4.23.

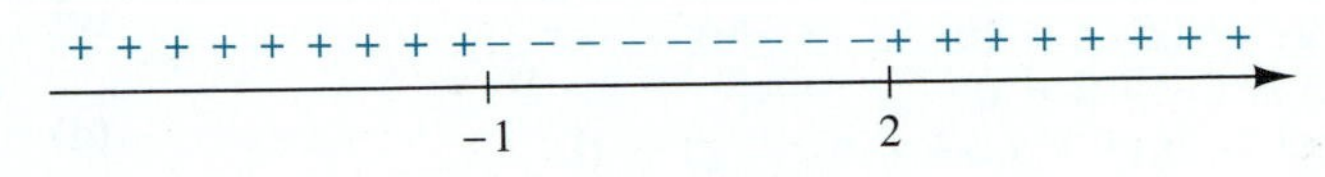

FIGURE 4.23 The sign analysis for $f(x) = (x - 2)(x + 1)(x^2 + 3)$

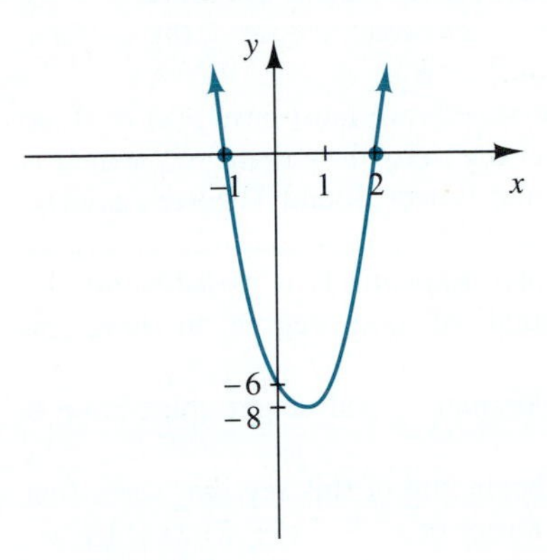

FIGURE 4.24 A rough sketch of the graph of $y = x^4 - x^3 + x^2 - 3x - 6$

We have $f(1) = -8$ (from the Remainder Theorem and the synthetic division) and the y-intercept is $f(0) = -6$. Checking some values, we see that y becomes positive as $|x|$ gets large. A rough sketch of the graph of $f(x)$ appears in Figure 4.24. ■

If after employing the Rational Root Theorem we determine that a polynomial has no rational zeros, a graphics calculator can be very helpful in estimating its zeros. Keep in mind that we can still use Descartes's Rule of Signs and the Upper and Lower Bound Theorem to obtain information which helps us pick the proper viewing rectangle and use the trace and zoom functions more effectively.

EXERCISES 4.4

In Exercises 1–18, determine the rational roots of the given equation.

1. $x^3 - 4x^2 - 7x + 10 = 0$
2. $x^4 + 7x^3 + 17x^2 + 17x + 6 = 0$
3. $2x^3 - 5x^2 + 15 = 28x$
4. $3x^3 + 13x^2 + 2x = 8$
5. $x^3 - 4x^2 + 5x - 2 = 0$
6. $2x^3 + 3x^2 = 1$
7. $4x^3 - 3x = 1$
8. $4y^3 + 3y^2 + 8y + 6 = 0$
9. $t^5 - 4t^3 + t^2 - 4 = 0$
10. $30z^3 - 31z^2 - 15z + 4 = 0$
11. $6u^3 + u^2 - 4u + 1 = 0$
12. $w^5 + 9w^3 - w^2 - 9 = 0$
13. $-x^4 + 8x^2 - 15 = 0$
14. $-x^6 + 4x^4 + x^2 - 4 = 0$
15. $8x^3 + 18x^2 + 45x + 27 = 0$
16. $9x^4 + 15x^3 - 20x^2 - 20x + 16 = 0$
17. $9x^6 - 18x^5 - 28x^4 + 38x^3 + 39x^2 - 4x - 4 = 0$
18. $4x^6 - 21x^4 + 11x^2 - 4 = 0$

In Exercises 19–26, use Descartes's Rule of Signs to determine the possible number of positive and negative real zeros of $p(x)$.

19. $p(x) = 8x^3 + 1$
20. $p(x) = x^3 - 5x^2 - 6x + 3$
21. $p(x) = 3x^4 + 5x^2 + 2$
22. $p(x) = x^4 - x^3 - 6$
23. $p(x) = x^4 - 3x^3 + 7x$
24. $p(x) = -x^4 + x^3 - 3x^2 - 2x + 1$
25. $p(x) = x^5 - x^4 + x^3 - x^2 + x - 8$
26. $p(x) = 4x^3 - 3x^2 + 7x - 4$

In Exercises 27–40, use the Upper and Lower Bound Theorem to find upper and lower bounds for the roots of the given equation.

27. $x^2 - 5x + 3 = 0$
28. $3x^2 + 9x - 2 = 0$
29. $x^3 - 3x^2 - 18x + 4 = 0$
30. $x^3 - 4x^2 - 5x + 8 = 0$
31. $x^4 - x^2 + 3x + 2 = 0$
32. $x^4 - 2x^3 + 4x - 3 = 0$
33. $2x^3 - 5x + 1 = 0$
34. $3x^3 - 6x^2 - 14 = 0$
35. $-x^4 - 6x^3 + 3x + 7 = 0$
36. $-2x^3 + 6x^2 - 4x + 3 = 0$
37. $x^3 + 3x^2 + 5 = 0$
38. $2x^3 - 3x^2 + 12x + 9 = 0$
39. $\frac{1}{3}x^3 - x^2 - 3x + 4 = 0$
40. $\frac{1}{2}x^3 - \frac{1}{3}x^2 - 3x + 4 = 0$

In Exercises 41–48, use the various theorems in this section to find all the roots of the given equation.

41. $3x^3 - 7x^2 + 5x - 1 = 0$
42. $2x^3 + 13x^2 - 2x - 4 = 0$
43. $2x^4 + 7x^3 - 8x + 3 = 0$
44. $x^4 + 4x^3 - x^2 - 20x - 20 = 0$
45. $6x^4 - 19x^3 + 21x^2 - 19x + 15 = 0$
46. $x^3 - 5x^2 + 6x - 8 = 0$
47. $3x^5 - 9x^4 - 28x^3 + 84x^2 + 9x - 27 = 0$
48. $2x^4 + 13x^3 + 4x^2 - 13x - 6 = 0$

In Exercises 49–52, use a sign analysis to determine the intervals on which $f(x)$ is positive and on which $f(x)$ is negative.

49. $f(x) = x^4 + 2x^3 - 13x^2 - 14x + 24$
50. $f(x) = x^4 - 2x^3 + 5x^2 - 8x + 4$
51. $f(x) = x^5 - 5x^3 - x^2 + 5$
52. $f(x) = x^5 + x^4 - 2x^3 + 8x^2 + 8x - 16$

QUESTIONS FOR THOUGHT

53. Look back at the discussion following the statement of the Rational Root Theorem. Then use the following outline to prove the theorem. Suppose

$$f(x) = a_nx^n + a_{n-1}x^{n-1} + \cdots + a_1x + a_0$$

and suppose that $\frac{p}{q}$ is a solution to $f(x) = 0$, where $\frac{p}{q}$ is reduced to lowest terms.

(a) Show that if $f\left(\frac{p}{q}\right) = 0$, then

$$a_np^n + a_{n-1}p^{n-1}q + \cdots + a_1pq^{n-1} + a_0q^n = 0$$

(b) Show that this implies that p must divide a_0 and that q must divide a_n.

54. Use the following outline to explain part 2 of the Upper and Lower Bound Theorem.

(a) Use synthetic division to divide

$$p(x) = 2x^3 + 5x^2 - x + 3 \text{ by } x + 3$$

(b) Examine the bottom line of the synthetic division tableau and see that it satisfies the condition of part 2 of the Upper and Lower Bound Theorem.

(c) Translate the division into the form $p(x) = (x + 3)q(x) + R$.

(d) Verify that for $x < -3$, we must have $p(x) < 0$, so $p(x)$ cannot have any zeros less than -3, which is what the Upper and Lower Bound Theorem asserts.

55. Is it possible for the graph of a polynomial of even degree to have no x-intercepts? Explain. Is it possible for the graph of a polynomial of odd degree to have no x-intercepts? Explain.

56. Prove that every polynomial of odd degree must have at least one real zero.

57. As mentioned at the beginning of this section, show that 3 can be obtained as a root of $x^3 - 2x - 21 = 0$ by using Cardan's formula. Now use the Rational Root Theorem to list the possible rational roots and then check that 3 is a root. Which approach do you think is easier?

GRAFFIX

58. **(a)** Sketch the graph of

$$y = p(x) = 6x^3 - 31x^2 + 25x + 12$$

and estimate its zeros.

(b) Use the Rational Root Theorem on $p(x)$ to find its zeros. Which approach was easier? More accurate?

59. Graph $y = p(x) = 12x^3 - 40x^2 + 13x + 30$. Use the graph to estimate the zeros of $p(x)$, and use this information to apply the Rational Root Theorem more efficiently and find the exact zeros of $p(x)$.

60. Repeat the process used in Exercise 59 for

$$y = p(x) = 48x^4 + 4x^3 - 128x^2 - 29x + 15$$

4.5 Rational Functions

A **rational function** is a function of the form

$$y = f(x) = \frac{p(x)}{q(x)} \quad \text{where } p(x) \text{ and } q(x) \text{ are polynomials and } q(x) \neq 0$$

Some examples of rational functions are

$$f(x) = \frac{3}{x + 5} \qquad g(x) = \frac{x - 1}{x^2 - 9} \qquad h(x) = \frac{1}{x^2 + 4}$$

Unlike a polynomial function, whose domain is always the set of all real numbers, a rational function may have a restricted domain, since we must exclude any values that make the denominator equal to zero. Consider the functions f, g, and h defined on the bottom of page 288. The domain of $f(x)$ excludes $x = -5$, and the domain of $g(x)$ excludes the values $x = \pm 3$. On the other hand, since $x^2 + 4$ cannot be equal to zero (for real values of x), the domain of $h(x)$ is the set of all real numbers.

EXAMPLE 1 Find the domain and zeros of the function $f(x) = \dfrac{3x - 5}{x^2 - x - 12}$.

Solution Those values of x for which $x^2 - x - 12 = 0$ are excluded from the domain of $f(x)$. Since $x^2 - x - 12 = (x - 4)(x + 3)$, we have $D_f = \{x \mid x \neq -3, 4\}$.

In order to find the zeros of $f(x)$, we recognize that the only way for a fraction to be equal to zero is if the numerator is zero. In general

$$\frac{p(x)}{q(x)} = 0 \qquad \text{if and only if} \qquad p(x) = 0 \quad \text{and} \quad q(x) \neq 0$$

Therefore, to find the zeros of $f(x)$ we solve $3x - 5 = 0$, giving $x = \dfrac{5}{3}$. Since $\dfrac{5}{3}$ does not make the denominator equal to zero, $f(x)$ has one zero, $x = \dfrac{5}{3}$. ■

Throughout the discussion of rational functions that follows, we assume that the numerator and denominator have no common factors. (See Example 5 in Section 3.1 to see what can happen if we do not insist that $p(x)$ and $q(x)$ have no common factors.)

In Section 4.1 we used a sign analysis on polynomial functions in order to sketch their graphs. Knowing where a polynomial is positive, negative, and zero, along with the behavior of the polynomial when $|x|$ gets large, allows us to get a fairly accurate sketch of its graph. In order to sketch the graph of rational functions we also need to investigate the behavior of the function near those values of x that make the denominator equal to zero.

Let us begin our discussion with the following example.

EXAMPLE 2 Sketch the graph of $y = f(x) = \dfrac{1}{x}$.

Solution As usual, rather than resorting to plotting points, we will try to analyze this function in order to obtain information that will help us sketch the graph.

The domain of $f(x)$ is all $x \neq 0$, and so the graph of $f(x)$ has no y-intercept. (Why?) Similarly, since there are no values of x for which $\dfrac{1}{x} = 0$ (remember that in order for a fraction to be zero, the numerator must be zero), the graph has no x-intercepts. $\left(\text{Try solving } \dfrac{1}{x} = 0.\right)$

Why does the graph of $y = \dfrac{1}{x}$ have neither x- nor y-intercepts?

Looking at the equation $y = \dfrac{1}{x}$, we can see that if x is positive, then y is positive and if x is negative, then y is negative. In terms of the graph, this means that the graph will appear only in quadrants I and III. However, since we have already estab-

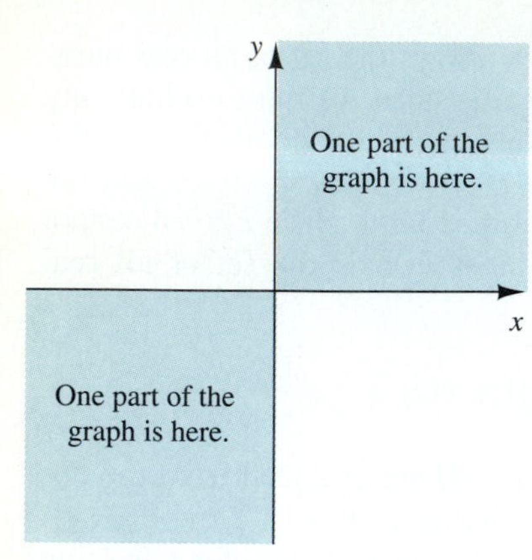

FIGURE 4.25 The graph of $y = \frac{1}{x}$ lies in quadrants I and III.

lished that there are no x- or y-intercepts, the graph must consist of two separate pieces, one in quadrant I and one in quadrant III. See Figure 4.25.

In other words, this graph has a break in it, unlike the graph of a polynomial, which cannot have any breaks.

We might also make note of the fact that for $f(x) = \frac{1}{x}$,

$$f(-x) = \frac{1}{-x} = -\frac{1}{x} = -f(x)$$

which means that the graph of this function exhibits origin symmetry. This agrees with our observation that the graph is in quadrants I and III.

Next let's examine what happens as $|x|$ gets very large. As we move along the x-axis farther and farther to the right, say for $x = 10$; 100; 1000; 10,000; etc., the values of $y = \frac{1}{x}$ are getting closer and closer to 0 $\left(\text{the } y\text{-values would be } \frac{1}{10}, \frac{1}{100}, \frac{1}{1000}, \frac{1}{10{,}000}, \text{etc.}\right)$. Similarly, as we move along the x-axis farther and farther to the left, say for $x = -10, -100, -1000, -10{,}000$, etc., the values of $y = \frac{1}{x}$ again become closer and closer to 0 $\left(\text{the } y\text{-values would be } -\frac{1}{10}, -\frac{1}{100}, -\frac{1}{1000}, -\frac{1}{10{,}000}, \text{etc.}\right)$.

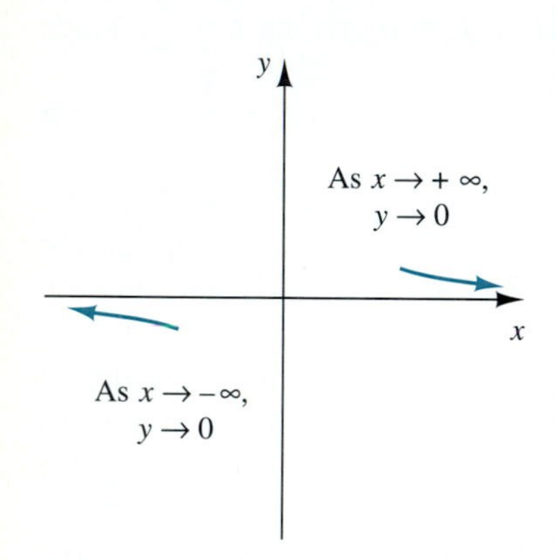

FIGURE 4.26

Thus "out at the ends," the graph gets closer and closer to the x-axis, but it never touches the x-axis, because—as we have already established—there are no x-intercepts. In such a situation the x-axis (which has equation $y = 0$) is called a **horizontal asymptote.** See Figure 4.26.

In general, we say that a line is an **asymptote** for a graph if the distance between the graph and the line approaches 0 as we move farther and farther out along the graph.*

It remains for us to determine the behavior of y for values of x near 0 where y is undefined. We want to determine what happens to y-values as the x-values get closer and closer to zero. Let's examine some values of y for values of x close to 0 (both positive and negative).

To indicate that x is approaching 0 *from the right*, we write $x \to 0^+$.

x	1	0.5	0.1	0.01	0.001	0.0001	$x \to 0^+$
y	1	2	10	100	1000	10,000	$y \to \infty$

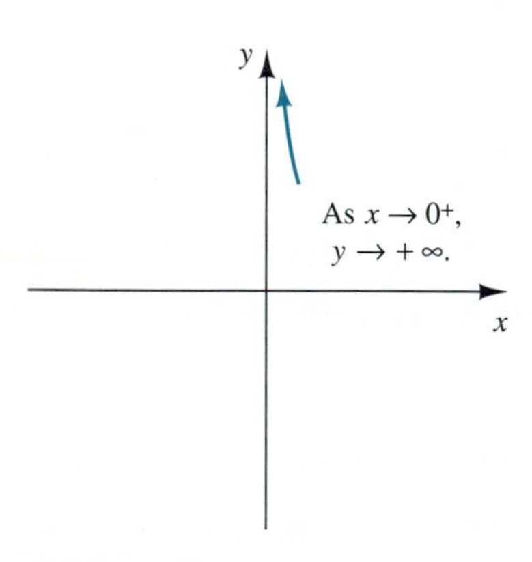

FIGURE 4.27

$y \to \infty$ means that y increases without bound. In other words, as x gets closer and closer to zero from the right side of the y-axis, the values of y are getting larger and larger. See Figure 4.27.

* Note that this description allows a graph to cross a horizontal asymptote *before* we get out to the ends.

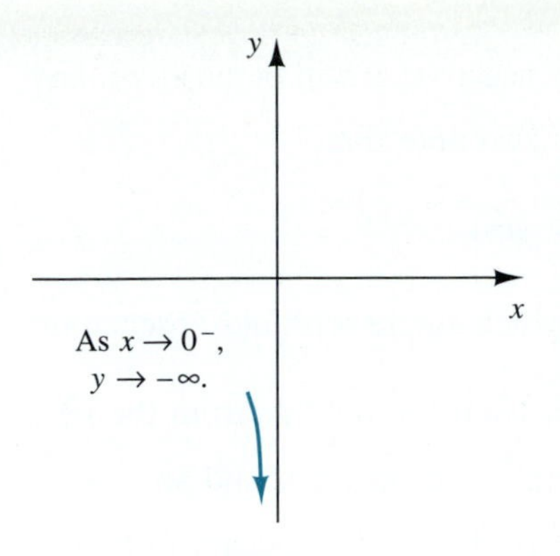

FIGURE 4.28

To indicate that x is approaching 0 *from the left*, we write $x \to 0^-$.

x	-1	-0.5	-0.1	-0.01	-0.001	-0.0001	$x \to 0^-$
y	-1	-2	-10	-100	-1000	$-10{,}000$	$y \to -\infty$

$y \to -\infty$ means that y decreases without bound; that is, as x gets closer and closer to zero from the left side of the y-axis, the values of y are negative but getting larger in absolute value. See Figure 4.28.

To summarize, as x gets close to zero from the right-hand side, the graph is getting higher and higher but never touches the y-axis (why not?), whereas as x gets closer to zero from the left-hand side, the graph is getting lower and lower but never touches the y-axis. Thus the y-axis (which has equation $x = 0$) is a **vertical asymptote.**

Putting together all the information we have gathered about the function and its graph, we obtain Figure 4.29.

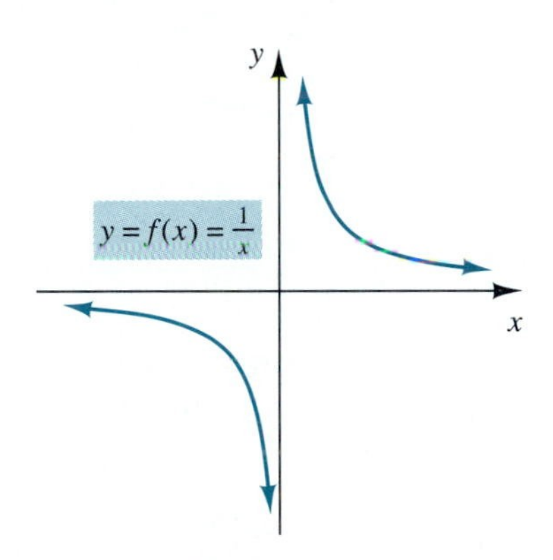

FIGURE 4.29 Notice that the graph approaches but never touches the x and y-axes.

Keep in mind that all our information does not tell us the exact shape of the curve. As mentioned previously, determining the concavity of a graph requires some techniques from calculus. It turns out that Figure 4.29 describes the shape of $y = \frac{1}{x}$ accurately. ■

Having gone through this detailed analysis for $y = \frac{1}{x}$, we can now go through a similar analysis much more quickly.

EXAMPLE 3 Sketch the graph of $y = f(x) = \frac{1}{x^2}$.

Solution Our analysis of this function will lead us to many of the same conclusions we came to for $y = \frac{1}{x}$. The graph will have no x- or y-intercepts and will have the x-axis as a horizontal asymptote and the y-axis as a vertical asymptote.

However, for $y = \frac{1}{x^2}$, whether x is positive or negative, y will be positive, and so the graph will be in quadrants I and II only. We also note that

$$f(-x) = \frac{1}{(-x)^2} = \frac{1}{x^2} = f(x)$$

and so the graph of $f(x)$ exhibits y-axis symmetry, which agrees with our observation that the graph is in quadrants I and II.

Notice that when x gets close to zero, whether from the right or from the left, x^2 will be a small positive number getting closer and closer to zero, and so $y = \frac{1}{x^2}$ will be increasing without bound. See Figure 4.30(a). The graph appears in Figure 4.30(b). ■

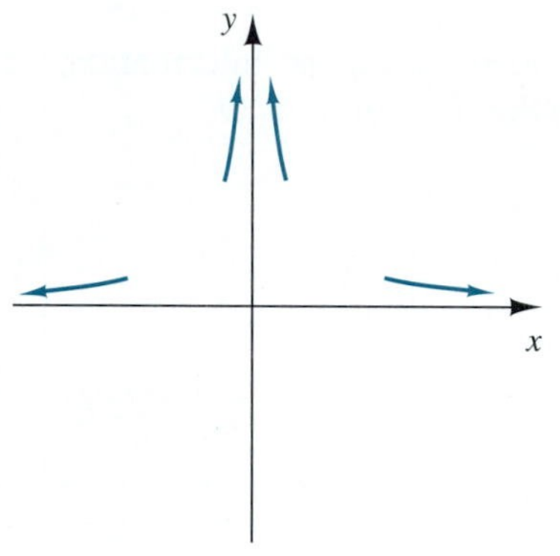

The behavior of $y = f(x) = \frac{1}{x^2}$

(a)

The graph of $y = f(x) = \frac{1}{x^2}$

(b)

FIGURE 4.30

GRAFFIX

Use a graphics calculator or computer.

1. Graph the following on the same set of coordinate axes:

$$y = \frac{1}{x}, \qquad y = \frac{1}{x^3}, \qquad y = \frac{1}{x^5}$$

For odd values of n, what can you conclude about how changing n affects the graph of $y = \frac{1}{x^n}$?

2. Graph the following on the same set of coordinate axes:

$$y = \frac{1}{x^2}, \qquad y = \frac{1}{x^4}, \qquad y = \frac{1}{x^6}$$

For even values of n, what can you conclude about how changing n affects the graph of $y = \frac{1}{x^n}$?

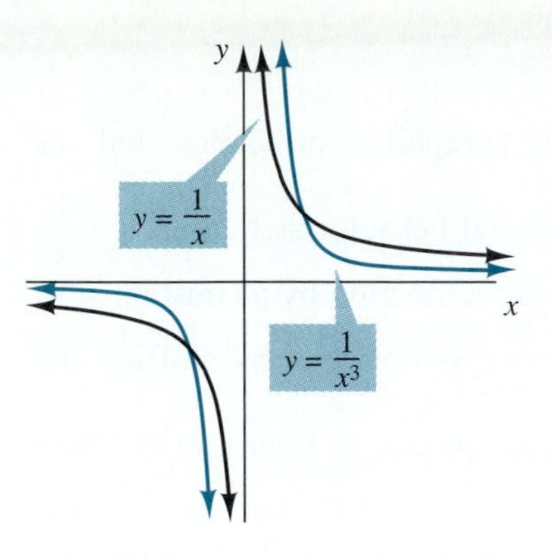

FIGURE 4.31

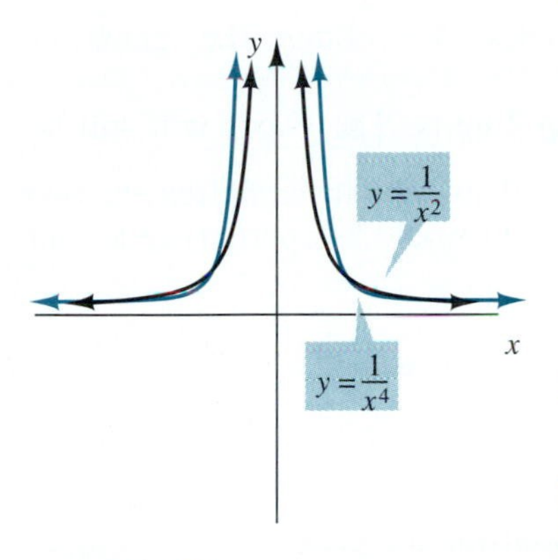

FIGURE 4.32

EXAMPLE 4 Sketch the graphs of **(a)** $y = f(x) = \frac{1}{x^3}$ **(b)** $y = g(x) = \frac{1}{x^4}$.

Solution

(a) It is left to the student to analyze the function $y = \frac{1}{x^3}$ and recognize that it has all the essential characteristics of $y = \frac{1}{x}$, including the same domain and range. This is directly due to the fact that the exponent 3 is odd, and hence when x is negative, y is negative, putting the graph in quadrants I and III. The graph appears in Figure 4.31. The graph of $y = \frac{1}{x}$ is also shown. Note that the higher exponent (3) pulls the graph closer to the x-axis.

(b) Again it is left to the student to analyze the function $y = \frac{1}{y^4}$ and recognize that it has all the essential characteristics of $y = \frac{1}{y^2}$, including the same domain and range. This is directly due to the fact that the exponent 4 is even, and hence when x is negative, y is positive, putting the graph in quadrants I and II. The graph appears in Figure 4.32. The graph of $y = \frac{1}{y^2}$ is also shown. Note that the higher exponent (4) pulls the graph closer to the x-axis. ■

Based upon these examples we can generalize as follows:

FOR POSITIVE INTEGERS $n \geq 1$

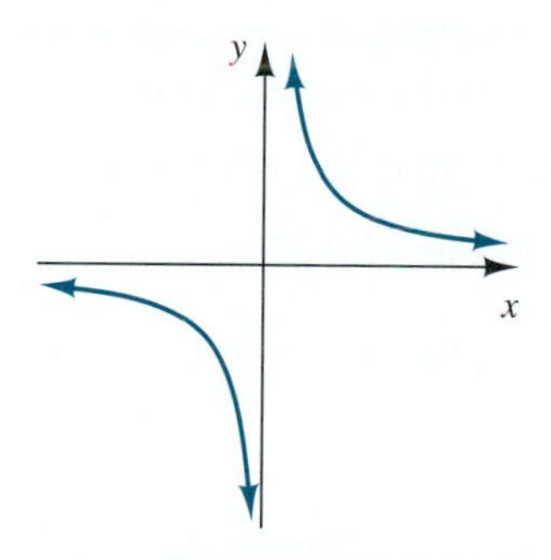

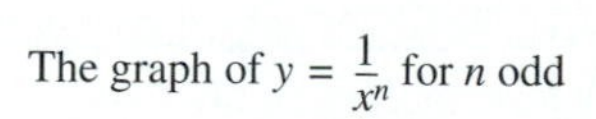
The graph of $y = \frac{1}{x^n}$ for n odd

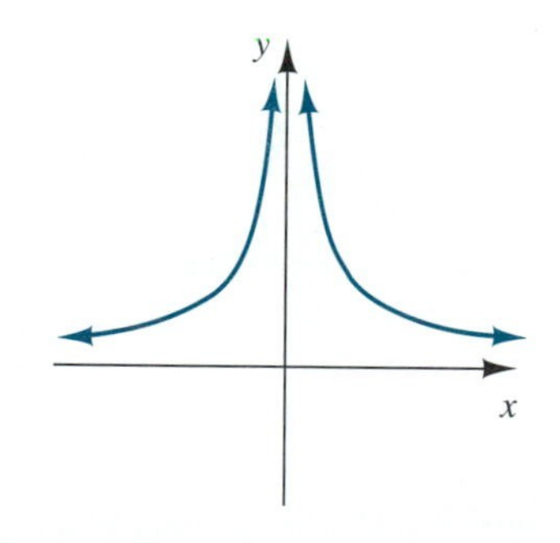

The graph of $y = \frac{1}{x^n}$ for n even

The vertical asymptote is $x = 0$ *and the horizontal asymptote is* $y = 0$.

These graphs should be added to our catalog of basic functions whose graphs we recognize immediately.

EXAMPLE 5 Sketch the graphs of

(a) $y = \frac{5}{x + 2}$ **(b)** $y = \frac{1}{x} + 2$ **(c)** $y = -\frac{1}{(x - 3)^2}$.

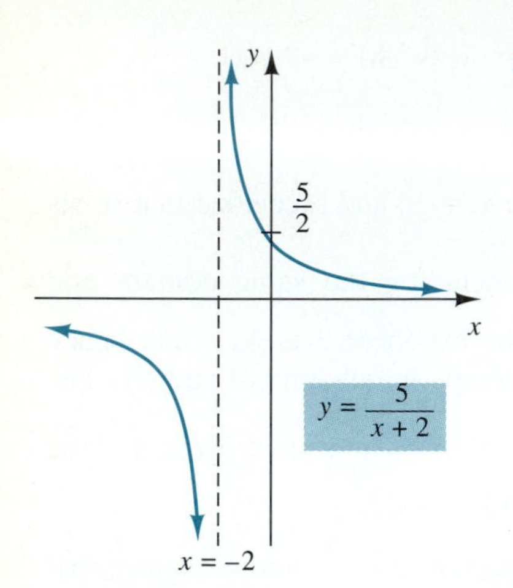

FIGURE 4.33

Why don't we bother to look for a y-intercept?

Solution

(a) Knowing the graph of $y = \frac{1}{x}$, our basic graphing principles tell us $y = \frac{5}{x} = 5\left(\frac{1}{x}\right)$ will have the same basic shape and behavior as that of $y = \frac{1}{x}$. As we saw with the power functions, multiplying a function by a constant simply stretches the graph. The graph of $y = \frac{5}{x+2}$ is obtained by shifting the graph of $y = \frac{5}{x}$ two units to the left. The graph appears in Figure 4.33. Note that the horizontal asymptote remains the x-axis ($y = 0$) but the vertical asymptote is shifted from the y-axis ($x = 0$) to the line $x = -2$. As usual, we find the y-intercept by setting $x = 0$.

(b) Again applying our basic graphing principles, we obtain the graph of $y = \frac{1}{x} + 2$ by shifting the graph of $y = \frac{1}{x}$ up 2 units. The y-axis will still be the vertical asymptote, but now the horizontal asymptote is shifted up two units from the x-axis ($y = 0$) to the line $y = 2$. It would be appropriate to find the x-intercept (we set $y = 0$).

$$0 = \frac{1}{x} + 2 \qquad \textit{Solving for x gives} \qquad x = -\frac{1}{2}$$

The graph appears in Figure 4.34.

(c) The graph of $y = -\frac{1}{(x-3)^2}$ is obtained by shifting the graph of $y = \frac{1}{x^2}$ three units to the right and then reflecting it about the x-axis. The horizontal asymptote remains the x-axis and the vertical asymptote is shifted to $x = 3$. It is appropriate to find the y-intercept. Setting $x = 0$, we get $y = -\frac{1}{9}$. The graph appears in Figure 4.35. ■

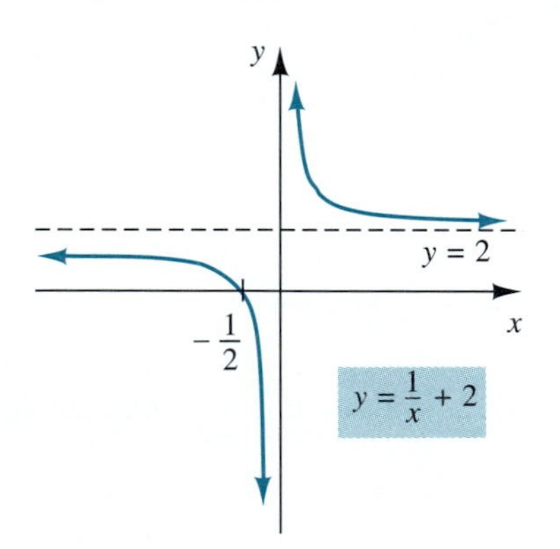

FIGURE 4.34

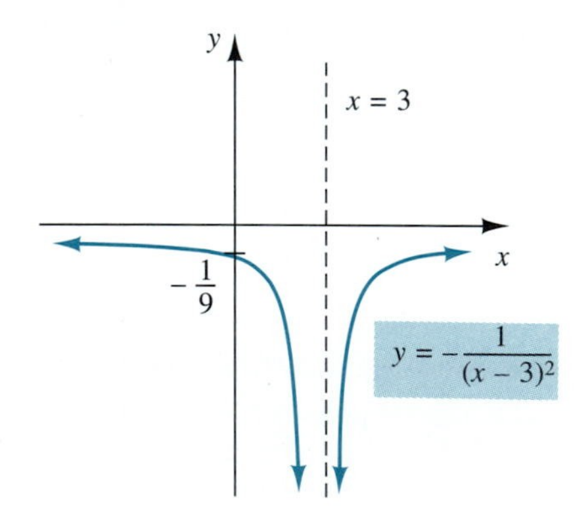

FIGURE 4.35

EXAMPLE 6 Sketch the graphs of **(a)** $y = \frac{1}{x^4} - 3$ **(b)** $y = \frac{1}{(x-2)^5} + 1$.

Solution

(a) We recognize that the underlying graph is of the form $y = \frac{1}{x^n}$ with n even, so the graph will have the $\frac{1}{x^2}$ look, and we will shift the underlying graph down 3 units. Thus the y-axis is the vertical asymptote and the horizontal asymptote is now $y = -3$. We find the x-intercepts.

$$0 = \frac{1}{x^4} - 3 \Rightarrow 3 = \frac{1}{x^4}$$

$$x = \pm\frac{1}{\sqrt[4]{3}} \approx \pm 0.76$$

The graph appears in Figure 4.36.

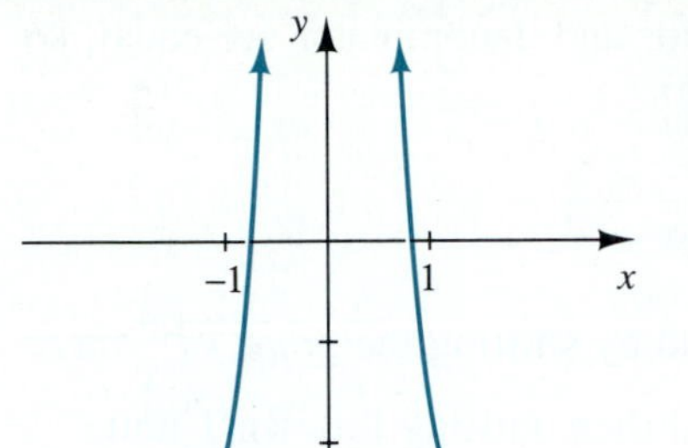

FIGURE 4.36

(b) We recognize that the underlying graph is of the form $y = \frac{1}{x^n}$ with n odd, so the graph will have the $\frac{1}{x}$ look, and we will shift the underlying graph 2 units to the right and 1 unit up. Thus the vertical asymptote is shifted 2 units to the right to $x = 2$ and the horizontal asymptote is shifted up 1 unit to $y = 1$. We find the x- and y-intercepts:

Set $y = 0$.

$$y = \frac{1}{(x-2)^5} + 1$$

$$0 = \frac{1}{(x-2)^5} + 1$$

$$-1 = \frac{1}{(x-2)^5}$$

$$-1 = (x-2)^5 \qquad \textit{Take fifth roots.}$$

$$-1 = x - 2$$

$1 = x$ The x-intercept is 1.

Set $x = 0$.

$$y = \frac{1}{(x-2)^5} + 1$$

$$y = \frac{1}{(0-2)^5} + 1$$

$$y = -\frac{1}{32} + 1 = \frac{31}{32}$$

The y-intercept is $\frac{31}{32}$.

The graph appears in Figure 4.37. ■

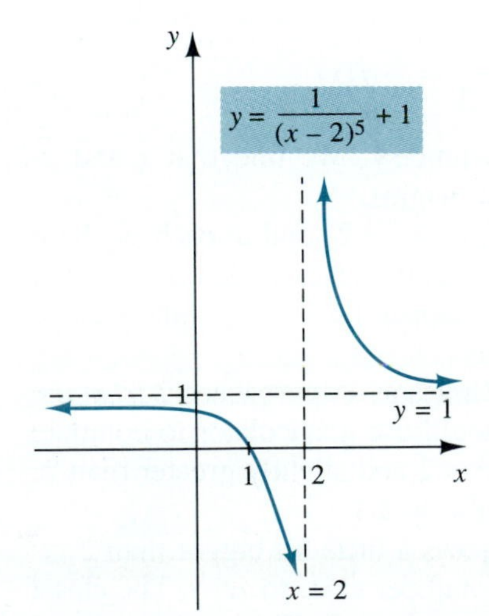

FIGURE 4.37

The next example illustrates that the same technique can sometimes be applied if we do some algebra first.

EXAMPLE 7 Sketch the graph of $y = \frac{1 - x}{x - 3}$.

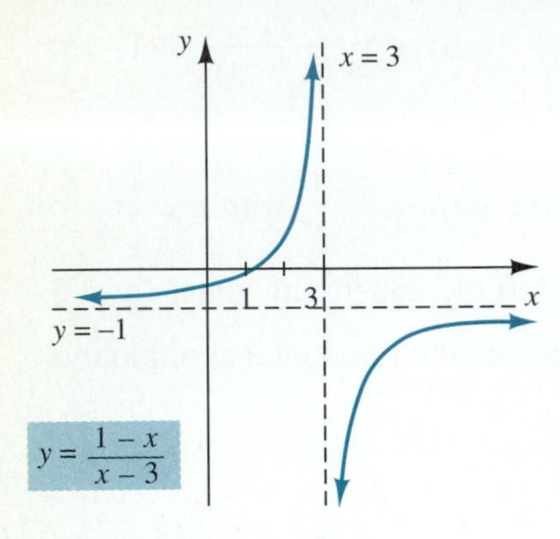

FIGURE 4.38

Solution We note that the degrees of the numerator and denominator are equal, so we can perform long division (or synthetic division):

$$\frac{1-x}{x-3} = -1 - \frac{2}{x-3} = -\frac{2}{x-3} - 1$$

We can now see that the graph can be obtained by shifting the graph of $\frac{2}{x}$ three units to the right, reflecting it about the x-axis, and then shifting it down 1 unit.

We leave it to the student to verify that the y-intercept is $-\frac{1}{3}$ and the x-intercept is $x = 1$. The graph appears in Figure 4.38. ■

This same type of approach used in Example 7 will work for any rational function of the form

$$f(x) = \frac{ax+b}{cx+d}$$

A rational function will have a vertical asymptote at every x-value that makes the denominator (but not the numerator) zero and only at those values, so a rational function need not have any vertical asymptotes. See Exercises 57 and 58 for more about asymptotes.

In order to determine the exact shape of the graph of a general rational function, we need to employ more sophisticated techniques, as alluded to in our discussion of polynomial functions. Nevertheless, the next example presents some ideas that allow us to get a reasonable sketch of the graph of a rational function.

EXAMPLE 8 Sketch the graph of $y = f(x) = \frac{x}{x^2 - 4}$.

Solution We begin by noting that

$$f(-x) = \frac{-x}{(-x)^2 - 4} = -\frac{x}{x^2-4} = -f(x)$$

which tells us that the graph will exhibit origin symmetry. We find that x and y-intercepts are both zero (the graph goes through the origin).

The domain consists of all real numbers except $x = \pm 2$, and at each of these values the graph will have a vertical asymptote. It remains for us to determine what type of vertical asymptote we have—that is, what is happening to the y-values as we approach the vertical asymptote from both directions.

Let's carefully analyze what happens to the y values as x approaches 2 *from the right*. (Recall that we write this as $x \to 2^+$. You should use a calculator to compute $f(x)$ for some values of x near 2, both slightly less than 2 and slightly greater than 2, to get some information as to how $f(x)$ behaves near $x = 2$.)

As x gets closer and closer to 2 from the right, x is a little bit bigger than 2, x^2 is a little bit bigger than 4, and $x^2 - 4$ is a positive number close to zero. The closer x gets to 2 from the right, the closer the denominator gets to 0. Thus

$$\text{As } x \to 2^+, \qquad y = \frac{\text{a number close to 2}}{\text{a positive number close to 0}}$$

and so as x gets closer to 2 from the right, y increases without bound. Symbolically,

$$\text{As } x \to 2^+, \qquad y \to +\infty$$

Let's also carefully analyze what happens to the y values as x approaches -2 *from the right*. (Recall that we write this $x \to -2^+$. Again use a calculator to see how $f(x)$ behaves for values near $x = -2$). As x gets closer and closer to -2 from the right, x is a little bit bigger than -2 (e.g., -1.99), x^2 is a little bit *smaller* than 4, and $x^2 - 4$ is a *negative* number close to zero. The closer x gets to -2 from the right, the closer the denominator gets to 0, but it is still negative. Therefore,

$$\text{As } x \to -2^+, \qquad y = \frac{\text{a number close to } -2}{\text{a negative number close to } 0}$$

and so as x gets closer to -2 from the right, y increases without bound. Symbolically,

$$\text{as } x \to -2^+, \qquad y \to +\infty$$

The reader is encouraged to carry out the same type of analysis as x approaches 2 and -2 from the left hand side, which we write as $x \to 2^-$ and $x \to -2^-$, respectively. The following table summarizes the behavior of $f(x)$ near the vertical asymptotes.

As $x \to 2^+$	$x^2 - 4$ is a positive number near 0	$y = \dfrac{x}{x^2 - 4} \to +\infty$
As $x \to 2^-$	$x^2 - 4$ is a negative number near 0	$y = \dfrac{x}{x^2 - 4} \to -\infty$
As $x \to -2^+$	$x^2 - 4$ is a negative number near 0	$y = \dfrac{x}{x^2 - 4} \to +\infty$
As $x \to -2^-$	$x^2 - 4$ is a positive number near 0	$y = \dfrac{x}{x^2 - 4} \to -\infty$

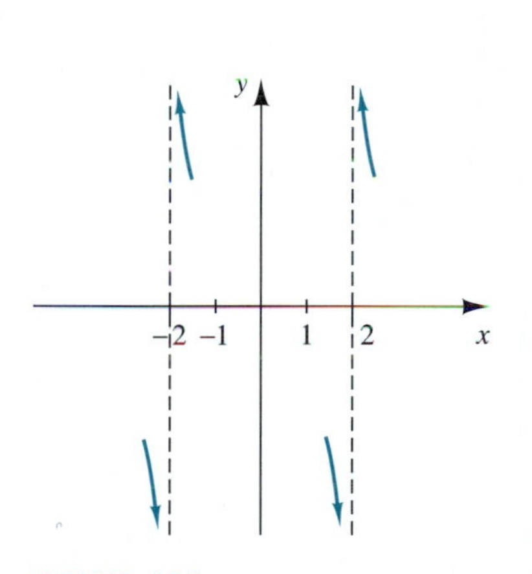

FIGURE 4.39

This information allows us to visualize those portions of the graph close to the vertical asymptotes, as shown in Figure 4.39.

Finally, we would like to examine the behavior of $f(x)$ "out at the ends"—that is, as $|x|$ gets larger and larger.

We use a calculator to compute some values of $f(x)$ to get a feeling for what is happening. (We record the display as it appeared on our calculator.)

Try computing $f(x)$ for some additional large values of x.

$$f(100) = \frac{100}{100^2 - 4} \approx 0.010004$$

$$f(1000) = \frac{1000}{1000^2 - 4} \approx 0.001$$

$$f(10{,}000) = \frac{10{,}000}{10{,}000^2 - 4} \approx 0.0001$$

It thus appears that as $x \to +\infty$, we have $y = f(x) \to 0$. We do not need to compute $f(x)$ as $x \to -\infty$, since we have already established that the graph will exhibit origin symmetry. It follows then that as $x \to -\infty$, we also have $f(x) \to 0$. Consequently, the x- axis is a horizontal asymptote.

By employing a bit of algebraic manipulation, we can make the behavior of a rational function as $|x| \to +\infty$ more apparent. We may divide the numerator and denominator of $f(x)$ by the highest power of x in the denominator. In the case of $y = \dfrac{x}{x^2 - 4}$, we divide each term in the numerator and denominator by x^2.

$$y = f(x) = \frac{x}{x^2 - 4} = \frac{\dfrac{x}{x^2}}{\dfrac{x^2}{x^2} - \dfrac{4}{x^2}} = \frac{\dfrac{1}{x}}{1 - \dfrac{4}{x^2}}$$

Now, when $|x|$ gets very large, $\dfrac{1}{x}$ and $\dfrac{4}{x^2}$ approach 0, and so $f(x)$ approaches $\dfrac{0}{1 - 0} = 0$. This agrees with the numerical evidence indicating that the x-axis is a horizontal asymptote.

Putting together all the information we have accumulated about $f(x)$ gives the graph in Figure 4.40.

Do a sign analysis for $\dfrac{x}{x^2 - 4}$. Do your results agree with the graph in Figure 4.40?

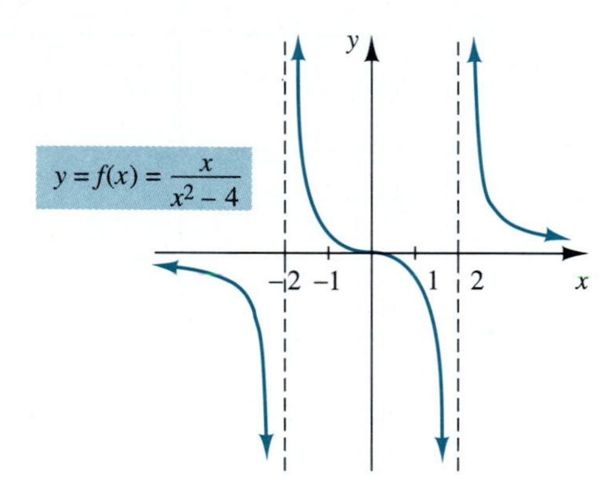

FIGURE 4.40

Many students find that they can obtain the various pieces of information about a function but have trouble sketching a graph consistent with all this information. It is important that you go back over this problem and make sure you see how all the information we gathered about $f(x)$—its intercepts and its behavior near the vertical asymptotes and the horizontal asymptote—has been integrated to produce the graph in Figure 4.40. Although we have not been able directly to address the question of concavity, the information we have about the horizontal and vertical asymptotes does suggest the concavity pictured in the graph. ■

GRAFFIX

Sketch the graph of $y = \dfrac{2x^2}{x^2 - 1}$.

Use the trace function to determine the y-values as $|x|$ gets very large.

What would you conclude about the horizontal asymptote of the graph?

We must point out that the techniques we have developed are somewhat limited in that they do not tell the entire story for many rational functions, as illustrated in Exercise 60. A number of the exercises deal with some additional ideas about asymptotes.

EXERCISES 4.5

In Exercises 1–6, find the domain and the real zeros of the given function.

1. $f(x) = \dfrac{3}{x^2 - 25}$

2. $g(x) = \dfrac{x - 3}{x^2 - 4x - 12}$

3. $h(x) = \dfrac{x^2 - 4x}{2x^2 + 3x - 20}$

4. $f(x) = \dfrac{x^2 - 16}{x^2 + 4}$

5. $g(x) = \dfrac{x^2 - 2x + 5}{x^2}$

6. $g(x) = \dfrac{(x - 3)^2}{x^3 - 3x^2 + 2x}$

7. Sketch the graphs.

(a) $y = \dfrac{2}{x}$ (b) $y = \dfrac{2}{x - 3}$ (c) $y = \dfrac{2}{x} - 3$

8. Sketch the graphs.

(a) $y = \dfrac{3}{x^2}$ (b) $y = \dfrac{3}{x^2} + 2$

(c) $y = \dfrac{3}{(x + 2)^2}$

9. Sketch the graphs.

(a) $y = -\dfrac{1}{x^4}$ (b) $y = -\dfrac{1}{(x - 1)^4}$

(c) $y = -\dfrac{1}{x^4} - 1$

10. Sketch the graphs.

(a) $y = \dfrac{2}{x^3}$ (b) $y = \dfrac{2}{x^3} - 4$

(c) $y = \dfrac{2}{(x + 4)^3}$

11. On the same set of coordinate axes, sketch the graphs of $y = \dfrac{1}{x}$, $y = \dfrac{6}{x}$, and $y = \dfrac{10}{x}$.

12. On the same set of coordinate axes, sketch the graphs of $y = -\dfrac{1}{x^2}$, $y = -\dfrac{4}{x^2}$, and $y = -\dfrac{9}{x^2}$.

In Exercises 13–44, sketch the graph of each function. Be sure to identify the intercepts and the horizontal and vertical asymptotes wherever appropriate.

13. $y = \dfrac{4}{x^6}$

14. $y = \dfrac{3}{x^5}$

15. $y = \dfrac{1}{x} - 5$

16. $y = \dfrac{1}{x - 5}$

17. $y = \dfrac{1}{(x - 2)^2}$

18. $y = \dfrac{1}{x^2} - 2$

19. $y = \dfrac{1}{x^3} - 8$

20. $y = \dfrac{1}{(x - 1)^4}$

21. $y = 9 - \dfrac{1}{x^2}$

22. $y = 4 - \dfrac{1}{x}$

23. $y = \dfrac{x}{x + 5}$

24. $y = \dfrac{-2x}{x - 2}$

25. $y = \dfrac{3x - 5}{x}$

26. $y = \dfrac{5x + 4}{2x}$

27. $y = \dfrac{x + 2}{x - 2}$

28. $y = \dfrac{x - 3}{x + 6}$

29. $y = \dfrac{5 - x}{x + 4}$

30. $y = \dfrac{6x - 3}{2x - 3}$

31. $y = \dfrac{2x}{x^2 - 1}$

32. $y = \dfrac{-x}{x^2 - 9}$

33. $y = \dfrac{x^2}{x^2 - 4}$

34. $y = \dfrac{3x^2}{1 - x^2}$

35. $y = \dfrac{x - 1}{x^2 - 2x - 3}$

36. $y = \dfrac{x + 1}{x^2 - x - 6}$

37. $y = \dfrac{2 - x}{2x^2 - x - 3}$

38. $y = \dfrac{x^2 - 4}{x^2 + 2x - 3}$

39. $y = \dfrac{x^2 - 4x + 3}{x^2 - 2x}$

40. $y = \dfrac{x^2 - 4x + 3}{x^2 - 4x}$

41. $y = \left|\dfrac{1}{x}\right| - 2$

42. $y = \left|\dfrac{1}{x} - 2\right|$

43. $y = \left|\dfrac{1}{x^2} - 4\right|$

44. $y = \left|\dfrac{1}{x^2}\right| - 4$

45. Let $f(x) = \dfrac{1}{x}$; compute and simplify the difference quotient $\dfrac{f(x) - f(5)}{x - 5}$.

46. Let $f(x) = \dfrac{1}{x + 1}$; compute and simplify the difference quotient $\dfrac{f(x) - f(2)}{x - 2}$.

47. Let $f(x) = \dfrac{1}{x^2}$; compute and simplify the difference quotient $\dfrac{f(x + h) - f(x)}{h}$.

48. Let $f(x) = \dfrac{1}{x^2 - 4}$; compute and simplify the difference quotient $\dfrac{f(x) - f(a)}{x - a}$.

49. Determine the behavior of $\frac{x^3 - 8x - 3}{x - 3}$ when x is near 3.

50. Determine the behavior of $\frac{2x^4 - 3x^2 + 1}{x + 1}$ when x is near -1.

51. A fence is to be set up to enclose a rectangular field and divide it down the middle, as indicated in the diagram. If the area of the field is 60,000 sq ft, express the length, L, of fence needed as a function of x.

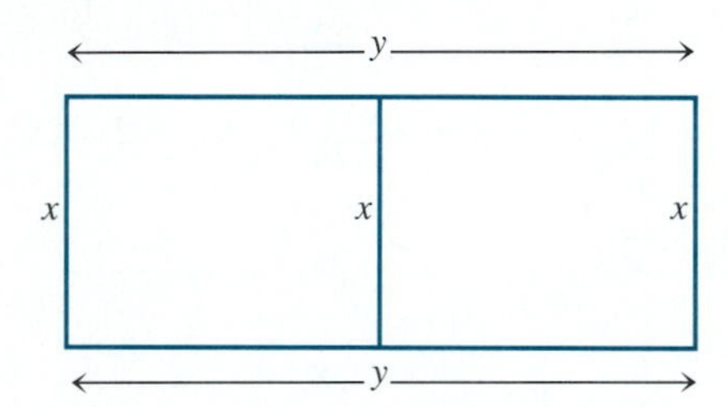

52. Use the fact that the volume of a closed right circular cylinder is 200 cu cm to express the surface area, S, of the cylinder as a function of r, the radius of the cylinder.

Certain procedures in calculus applied to rational functions give rise to the following types of problems.

53. Show that $\frac{(x^2 + 1)(2x) - (x^2 - 1)(2x)}{(x^2 + 1)^2} = \frac{4x}{(x^2 + 1)^2}$.

54. Show that
$$\frac{(x + 2)^3(2x) - (x^2 - 1)3(x + 2)^2}{(x + 2)^6} = \frac{3 + 4x - x^2}{(x + 2)^4}.$$

55. Show that
$$\frac{x^3(2x - 2) - (x^2 - 2x - 3)3x^2}{x^6} = \frac{-x^2 + 4x + 9}{x^4}.$$

56. Show that
$$\frac{(x + 3)^2 \cdot 3x^2 - x^3 \cdot 2(x + 3)}{(x + 3)^4} = \frac{x^2(x + 9)}{(x + 3)^3}.$$

QUESTIONS FOR THOUGHT

57. As we mentioned in this section, a rational function may have no horizontal or vertical asymptotes. Consider the function $y = f(x) = \frac{x^4 + 2}{x^2 + 1}$.

(a) Verify that $f(x)$ has no vertical asymptotes.

(b) Compute $f(x)$ for $x = 10$, 100, and 1000 and for $x = -10$, -100, and -1000. What do you think is happening to $f(x)$ as $x \to +\infty$? As $x \to -\infty$?

(c) Verify your conjecture in part (b). HINT: Divide the numerator and denominator of $f(x)$ by x^2 and then determine what happens to $f(x)$ as $|x| \to \infty$.

(d) Why does this imply that $f(x)$ does not have any horizontal asymptotes?

58. Although a rational function need not have any horizontal or vertical asymptotes, it may have what is known as an *oblique*, or *slant asymptote*. The accompanying graph $y = f(x) = \frac{x^2 + 1}{x}$.

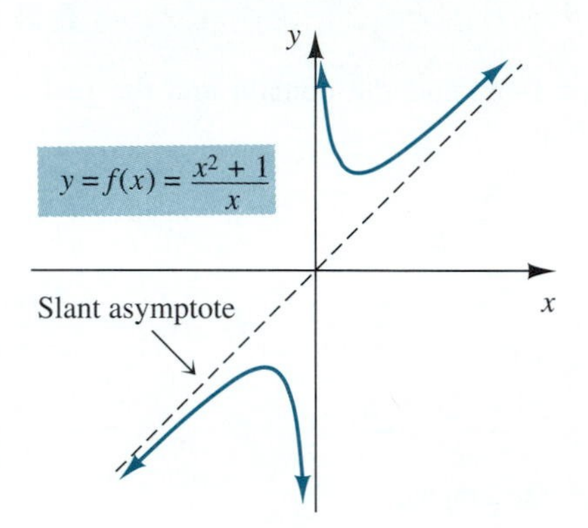

(a) Show that $y = f(x) = x + \frac{1}{x}$ and that as $|x| \to \infty$, we have y getting closer and closer to x. Thus the graph of $f(x)$ is getting closer and closer to the graph of $y = x$.

(b) Also show that as $x \to 0$, $y \approx \frac{1}{x}$, so that near the y-axis, the graph of $f(x)$ looks like the graph of $y = \frac{1}{x}$.

59. The graph of any rational function in which the degree of the numerator is exactly one more than the degree of the denominator will have a slant asymptote.

(a) Use long division to show that
$$y = f(x) = \frac{x^2 - x + 6}{x - 2} = x + 1 + \frac{8}{x - 2}$$

(b) Show that this means that the line $y = x + 1$ is a slant asymptote for the graph and sketch the graph of $y = f(x)$.

GRAFFIX

60. **(a)** Try to sketch the graph of $y = \frac{x^2 - 4}{x^2 + 1}$ using the techniques developed in this section.

(b) Does our approach develop enough information to get an accurate graph?

(c) Use a graphics calculator to sketch the graph of this function and see if you can justify the various features of the graph from its equation.

(d) Does the graph as exhibited on the calculator agree with your attempt at sketching the graph in part (a)?

61. **(a)** Try to sketch the graph of $y = \dfrac{x^2 - 1}{x + 2}$ using the techniques developed in this section.

(b) Does our approach develop enough information to get an accurate graph?

(c) Use a graphics calculator to sketch the graph of this function and see if you can justify the various features of the graph from its equation.

(d) Sketch the graph of $y = x - 2$ with a calculator on the same set of coordinate axes. What conclusion would you draw?

4.6 Radical Functions

A **radical function** is a function that contains roots of variables. The following are examples of radical functions.

$$f(x) = \sqrt{x} \qquad g(x) = \frac{\sqrt[3]{x + 1}}{x^2 + 2} \qquad h(x) = \frac{(x - 4)^{1/6}}{3}$$

Keep in mind that $h(x)$ is a radical function because $(x - 4)^{1/6} = \sqrt[6]{x - 4}$.

Given a radical function, the first consideration is, as usual, to determine its domain.

EXAMPLE 1 Find the domain of each function.

(a) $f(x) = \dfrac{1}{\sqrt{12 + 4x - x^2}}$ **(b)** $g(x) = \dfrac{\sqrt{x + 5}}{x - 3}$ **(c)** $h(x) = \sqrt[3]{x - 3}$

Solution

(a) The domain of $f(x)$ consists of those real numbers for which $\sqrt{12 + 4x - x^2}$ is a real number and not equal to zero. (Why?) Thus we need $12 + 4x - x^2 > 0$, which we can solve by doing a sign analysis.

$$12 + 4x - x^2 > 0 \qquad \textit{Multiply both sides by } -1.$$
$$x^2 - 4x - 12 < 0$$
$$(x + 2)(x - 6) < 0 \qquad \textit{The cut points are } -2 \textit{ and } 6.$$

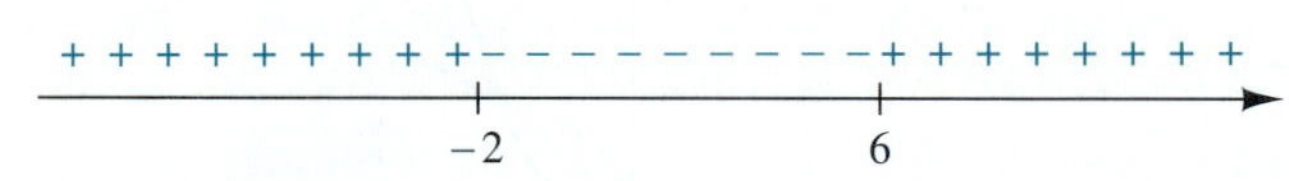

Thus $D_f = \{x \mid -2 < x < 6\}$.

(b) For x to be in the domain of $g(x)$, we require

$$x + 5 \geq 0 \quad \text{and} \quad x - 3 \neq 0$$

and so $D_g = \{x \mid x \geq -5 \text{ and } x \neq 3\}$.

(c) Since the cube root is defined for all real numbers, there are no restrictions on the domain of $h(x)$ and so D_h = all real numbers. ■

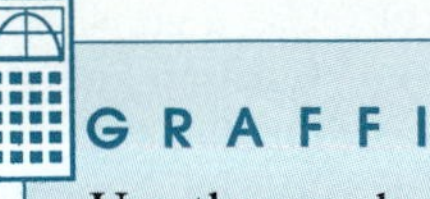

GRAFFIX

Use the graph of $y = \dfrac{\sqrt{x+3}}{x-1}$ to determine its domain and range.

Radical functions can exhibit rather complicated behavior, and it can be difficult to sketch their graphs even with the tools of calculus. However, using some of the ideas we have developed, we can describe the graphs of a fairly large class of radical functions.

Suppose we want to find the graph of the radical function $y = f(x) = \sqrt{x}$, whose domain is $\{x \mid x \geq 0\}$. At first glance, this function does not appear to be related to any of the functions we have studied thus far and so its graph is not immediately apparent. Consider the following strategy for obtaining the graph of $y = \sqrt{x}$. Looking at the equation $y = f(x) = \sqrt{x}$, we can see that $f(x)$ is a one-to-one function. (Each y value comes from a unique x-value). Therefore, we know that $f(x)$ has an inverse function. Perhaps the inverse function has a graph with which we *are* familiar and that we can reflect through the line $y = x$ to get the graph of $y = \sqrt{x}$.

So let's find the inverse of $y = \sqrt{x}$ keeping in mind that y is defined only for $x \geq 0$.

Are the equations $y^2 = x$ and $y = \sqrt{x}$ equivalent?

$$y = \sqrt{x} \quad \textit{for } x \geq 0 \qquad \textit{Interchange } x \textit{ and } y.$$

$$x = \sqrt{y} \quad \textit{for } x \geq 0 \qquad \textit{Now solve for } y.$$

$$x^2 = y \quad \textit{for } x \geq 0$$

Thus we see that the inverse function of $y = f(x) = \sqrt{x}$ for $x \geq 0$ is the function $y = f^{-1}(x) = x^2$ for $x \geq 0$. We can get the graph of $y = \sqrt{x}$ by reflecting the graph of $y = x^2$ for $x \geq 0$ about the line $y = x$. See Figure 4.41(a) and (b).

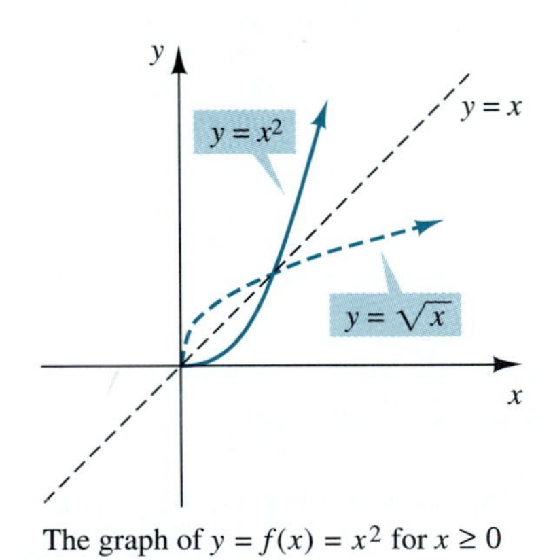

The graph of $y = f(x) = x^2$ for $x \geq 0$

(a)

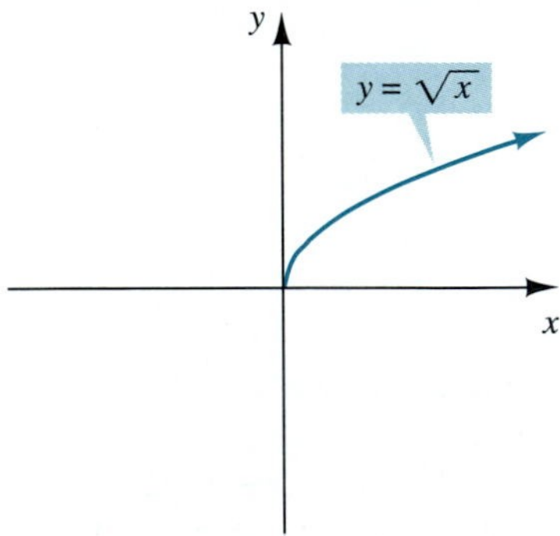

The graph of $y = f(x) = \sqrt{x}$

(b)

FIGURE 4.41

This is an important graph and should be added to your catalog of basic graphs.

EXAMPLE 2 Sketch the graph of $y = f(x) = \sqrt[3]{x}$.

Solution Using the same approach outlined above, we note that $y = f(x) = \sqrt[3]{x}$ is a one-to-one function. We will find the inverse function, sketch its graph, and use the reflection principle for inverse functions to obtain the graph we are interested in.

$$y = \sqrt[3]{x} \quad \textit{Interchange } x \textit{ and } y.$$

$$x = \sqrt[3]{y} \quad \textit{Now solve for } y.$$

$$x^3 = y$$

Thus the inverse function of $y = f(x) = \sqrt[3]{x}$ is the function $y = f^{-1}(x) = x^3$. We can get the graph of $y = \sqrt[3]{x}$ by reflecting the graph of $y = x^3$ about the line $y = x$. See Figure 4.42(a) and (b).

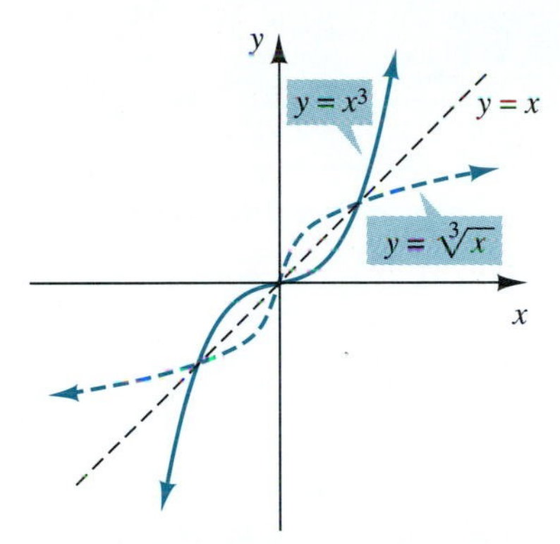

The graph of $y = x^3$

(a)

and its inverse

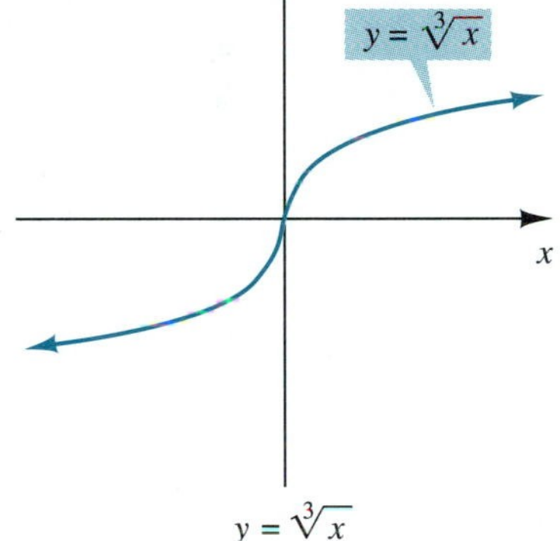

$y = \sqrt[3]{x}$

(b)

FIGURE 4.42

■

GRAFFIX

Use a graphics calculator or computer.

1. Graph the following on the same set of coordinate axes:

$$y = \sqrt{x} = x^{1/2}, \qquad y = \sqrt[4]{x} = x^{1/4}, \qquad y = \sqrt[6]{x} = x^{1/6}$$

For even values of n, what can you conclude about how changing n affects the graph of $y = x^{1/n}$?

2. Graph the following on the same set of coordinate axes:

$$y = \sqrt[3]{x} = x^{1/3}, \qquad y = \sqrt[5]{x} = x^{1/5}, \qquad y = \sqrt[7]{x} = x^{1/7}$$

For odd values of n, what can you conclude about how changing n affects the graph of $y = x^{1/n}$?

EXAMPLE 3 Sketch the graphs of **(a)** $y = \sqrt[4]{x}$ and **(b)** $y = \sqrt[5]{x}$.

Solution

(a) Proceeding as we did earlier for $y = \sqrt{x}$, we find that the inverse function of $y = \sqrt[4]{x}$ for $x \geq 0$ is $y = x^4$ for $x \geq 0$. We obtain the graph of $y = \sqrt[4]{x}$ by reflecting the graph of $y = x^4$ for $x \geq 0$ about the line $y = x$. See Figure 4.43(a) and (b). The graph of $y = \sqrt[4]{x}$ will be a bit flatter than the graph of $y = \sqrt{x}$.

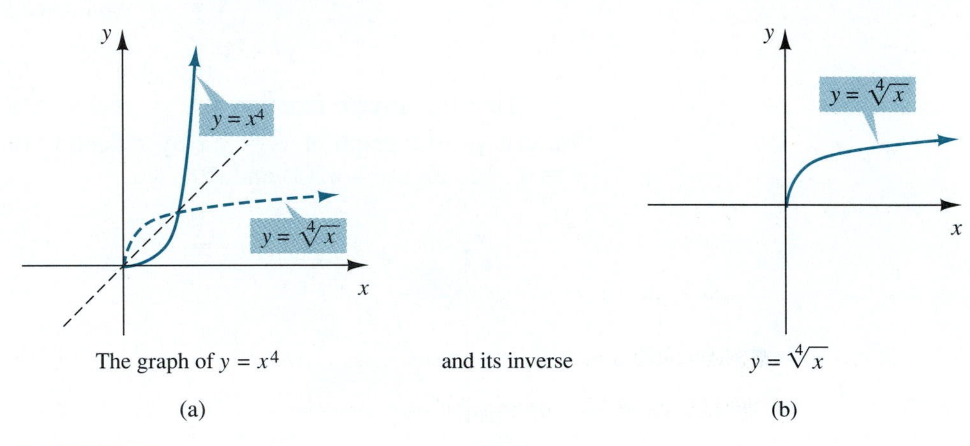

The graph of $y = x^4$ and its inverse $y = \sqrt[4]{x}$

(a) (b)

FIGURE 4.43

(b) Proceeding as we did for $y = \sqrt[3]{x}$, we find that the inverse function of $y = \sqrt[5]{x}$ is $y = x^5$. The graph of $y = \sqrt[5]{x}$ is obtained by reflecting the graph of $y = x^5$ about the line $y = x$. See Figure 4.44(a) and (b). The graph of $y = \sqrt[5]{x}$ will be a bit flatter than the graph of $y = \sqrt[3]{x}$.

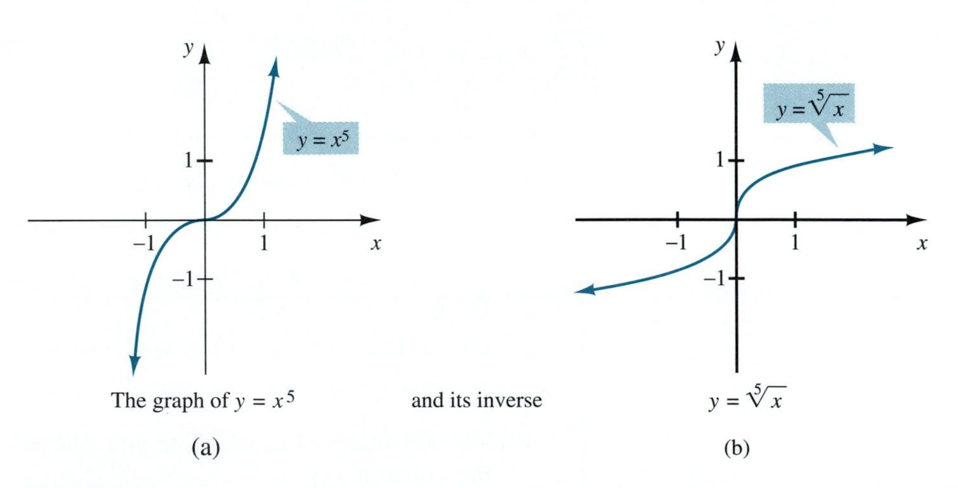

The graph of $y = x^5$ and its inverse $y = \sqrt[5]{x}$

(a) (b)

FIGURE 4.44

Based upon these examples we understand that the graph of $y = \sqrt[n]{x}$ will be similar to the graph of $y = \sqrt{x}$ when the index n is *even* and similar to the graph of $y = \sqrt[3]{x}$ when the index n is *odd*. These results are summarized below.

THE GRAPHS OF $y = f(x) = \sqrt[n]{x}$

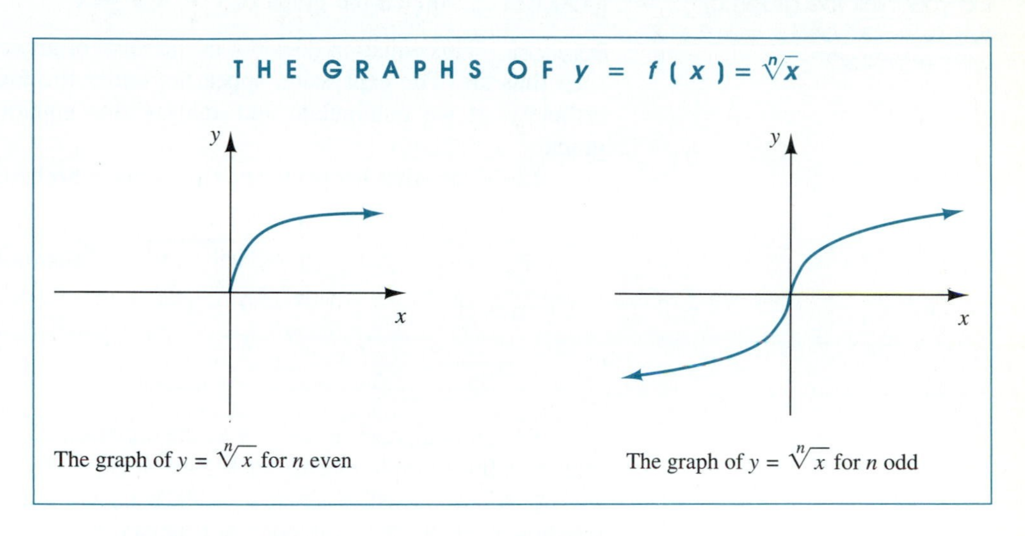

The graph of $y = \sqrt[n]{x}$ for n even

The graph of $y = \sqrt[n]{x}$ for n odd

These graphs should be added to your catalog of basic graphs.

EXAMPLE 4 Sketch the graphs of **(a)** $y = \sqrt{x - 4}$ and **(b)** $y = x^{1/3} - 2$.

Solution

(a) The graph of $y = \sqrt{x - 4}$ can be obtained by shifting the graph of $y = \sqrt{x}$ horizontally 4 units to the right. See Figure 4.45.

(b) We recognize that $y = x^{1/3} - 2$ is the same as $y = \sqrt[3]{x} - 2$. The graph of $y = \sqrt[3]{x} - 2$ can be obtained by shifting the graph of $y = \sqrt[3]{x}$ down 2 units. See Figure 4.46. We find the x-intercept of $y = \sqrt[3]{x} - 2$ by setting $y = 0$.

$$0 = \sqrt[3]{x} - 2$$

$$2 = \sqrt[3]{x} \quad \textit{Cube both sides of the equation.}$$

$$8 = x$$

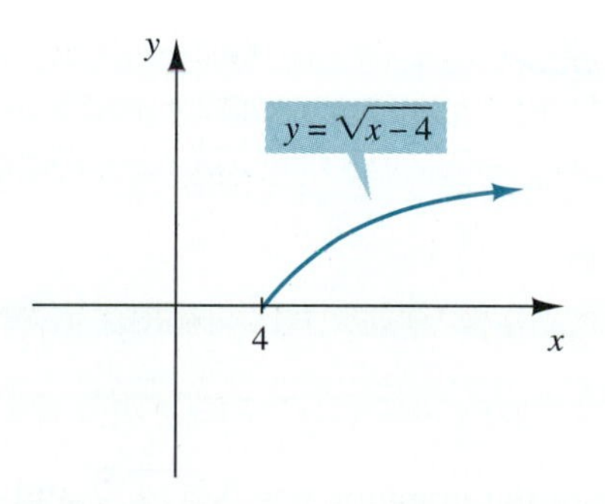

FIGURE 4.45

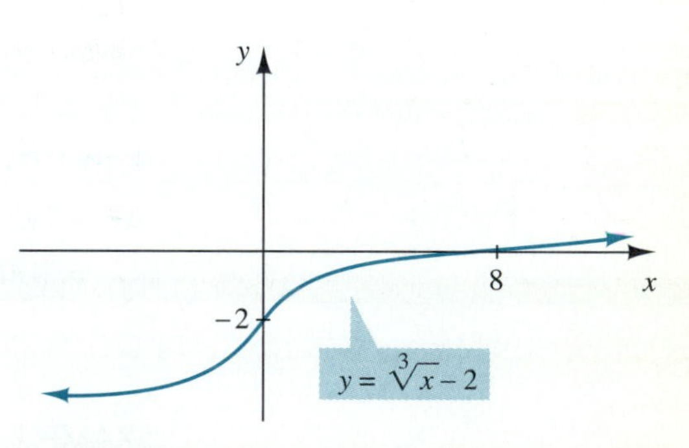

FIGURE 4.46

Often, we can use algebraic techniques to manipulate an equation into a more familiar form.

Do you think the graph of $y = \sqrt{9 - x^2}$ and $y = \sqrt{9 - x}$ will be similar?

EXAMPLE 5 Sketch the graph of $y = \sqrt{9 - x^2}$.

Solution This equation does not fit the form of any of the radical functions we have seen thus far. The expression appearing under the radical symbol is not linear. Nevertheless, if we manipulate and analyze this equation a bit we can determine its graph.

Let's see what happens when we square both sides of this equation.

$$y = \sqrt{9 - x^2} \qquad \textit{Square both sides.}$$
$$y^2 = 9 - x^2$$
$$x^2 + y^2 = 9$$

We recognize $x^2 + y^2 = 9$ as the equation of a circle with center (0, 0) and radius 3, whose graph appears in Figure 4.47(a). However, every point that satisfies $y = \sqrt{9 - x^2}$ satisfies $x^2 + y^2 = 9$ but *not* vice versa. For example, $(0, -3)$ satisfies $x^2 + y^2 = 9$ but does not satisfy $y = \sqrt{9 - x^2}$. In fact, in the original equation $y = \sqrt{9 - x^2}$, y must be nonnegative, since the square root is, by definition, nonnegative. Therefore, in order to get the graph of $y = \sqrt{9 - x^2}$, we delete that portion of the graph of $x^2 + y^2 = 9$ for which y is negative; that is, we delete that portion of the graph below the x-axis. The graph of $y = \sqrt{9 - x^2}$ appears in Figure 4.47(b).

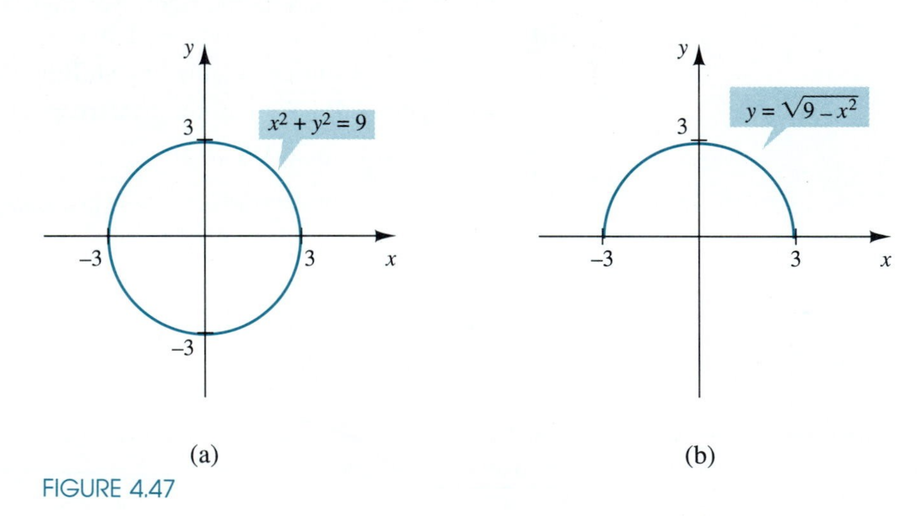

FIGURE 4.47

■

EXAMPLE 6 Sketch the graph of $y = \sqrt{x + 2}$ and $y = x$.

(a) Estimate the points of intersection of the two graphs.

(b) Find the points of intersection of the two graphs algebraically.

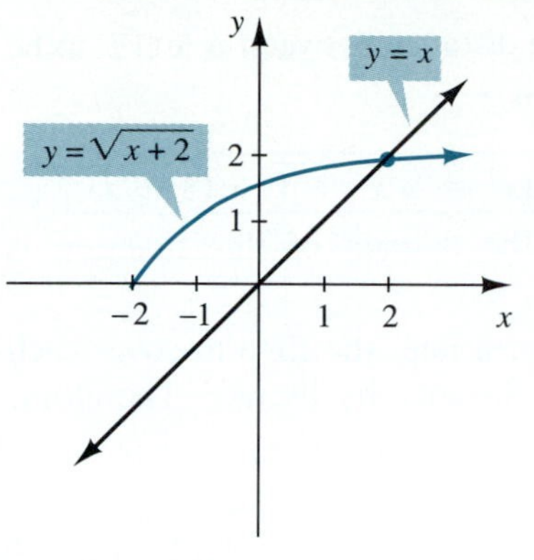

FIGURE 4.48

Solution We sketch the graphs of $y = \sqrt{x + 2}$ and $y = x$ on the same coordinate system. The graph of $y = \sqrt{x + 2}$ is obtained by shifting the graph of $y = \sqrt{x}$ to the left 2 units. The graph of $y = x$ is the familiar line of slope 1 passing through the origin. See Figure 4.48.

(a) By examining the graphs, we can see that the graphs intersect in one point. We estimate the point of intersection of the two graphs to be (2,2).

(b) We can find the points of intersection algebraically by recognizing that at any point of intersection, the y-values on the graphs must be equal. If we equate the y-values, we get

$$\sqrt{x + 2} = x \qquad \textit{Square both sides.}$$
$$x + 2 = x^2$$
$$x^2 - x - 2 = 0$$
$$(x - 2)(x + 1) = 0$$
$$x = 2 \quad \text{or} \quad x = -1$$

Substituting these x-values into the equation $y = x$, we get $x = 2$, $y = 2$ and $x = -1$, $y = -1$. However, if we check these points in the equation $y = \sqrt{x + 2}$ we see that (2, 2) satisfies it, but $(-1, -1)$ does not. (Remember that $\sqrt{x + 2}$ means the nonnegative square root.) This agrees with the result we found in part (a), where we saw that there is only one point of intersection. ■

We discuss finding points of intersection and solving systems of nonlinear equations in greater detail in Chapter 10.

Real-life situations can also give rise to radical functions.

EXAMPLE 7 Points A and B are on opposite sides of a straight river that is 1 mi wide. Point C is 3 mi down the river on the same side as B. A person swims from A to some point P between B and C and then runs from P to C.

(a) Express the total distance, D, covered as a function of one variable.

(b) If the person can swim at 1.5 mi/hr and run at 7 mi/hr, also express the total time, T, to get from A to C as a function of one variable.

Solution

(a) Figure 4.49 illustrates the given situation. Note that we have labeled the distance from B to P by x.

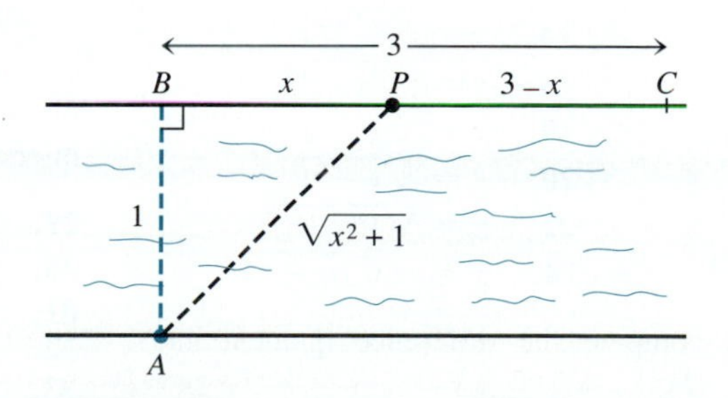

FIGURE 4.49

By using the Pythagorean theorem we find the distance between A and P to be $\sqrt{x^2+1}$, and the distance between P and C is $3-x$.

Therefore, we have the total distance $\boxed{D = D(x) = \sqrt{x^2+1} + (3-x)}$.

From the diagram we can clearly see that the domain of this function is $\{x \mid 0 \le x \le 3\}$.

(b) Since each distance is being covered at a uniform rate, the time to cover each part of the trip is computed by dividing each distance by its rate. Therefore, the total time needed to cover the distance D is

$$\boxed{T = T(x) = \frac{\sqrt{x^2+1}}{1.5} + \frac{(3-x)}{7} \quad \text{for} \quad 0 \le x \le 3}$$

Calculus techniques can be applied to this function to determine where the point P should be chosen to *minimize* the total time needed to get from A to C. ■

The types of functions we have discussed in this chapter, polynomial functions, rational functions, and radical functions, are part of a general category called *algebraic functions*.

DEFINITION An **algebraic function** is a function obtained by applying a finite number of additions, subtractions, multiplications, divisions and taking of roots of constants and/or variables.

In the next chapter we will begin to examine functions that are not algebraic.

EXERCISES 4.6

In Exercises 1–24, sketch the graph of the given equation. Label the intercepts where appropriate.

1. $y = \sqrt{x+3}$
2. $y = \sqrt{x} + 3$
3. $y = \sqrt{x} - 4$
4. $y = |\sqrt{x} - 4|$
5. $y = \sqrt{2x-6}$
6. $y = \sqrt{2x} - 6$
7. $y = 2\sqrt{x} - 6$
8. $y = -\sqrt{x}$
9. $y = \sqrt{-x}$
10. $y = \sqrt{|x|}$
11. $y = \sqrt[3]{x+2}$
12. $y = \sqrt[3]{x} + 2$
13. $y = \sqrt[3]{x} - 3$
14. $y = 2 - \sqrt[3]{x}$
15. $y = |\sqrt[3]{x} - 3|$
16. $y = |\sqrt[3]{x-3}|$
17. $y = \sqrt[5]{x} - 1$
18. $y = \sqrt[5]{x-1}$
19. $y = \sqrt[6]{x} - 1$
20. $y = \sqrt[6]{x-1}$
21. $y = \sqrt{16-x^2}$
22. $x = \sqrt{16-y^2}$
23. $y = -\sqrt{16-x^2}$
24. $x = -\sqrt{16-y^2}$

25. Let $f(x) = \sqrt{x}$; compute the difference quotient and show that $\dfrac{f(x) - f(4)}{x-4} = \dfrac{1}{\sqrt{x}+2}$.

26. Let $f(x) = \sqrt{x}$; compute the difference quotient and show that $\dfrac{f(9+h) - f(9)}{h} = \dfrac{1}{\sqrt{9+h}+3}$.

27. Let $f(x) = \dfrac{1}{\sqrt{x}}$; compute the difference quotient and show that $\dfrac{f(t) - f(3)}{t-3} = -\dfrac{1}{\sqrt{3t}(\sqrt{3}+\sqrt{t})}$.

28. Let $f(x) = 2\sqrt{x}$; compute the difference quotient and show that $\dfrac{f(u) - f(5)}{u-5} = \dfrac{2}{\sqrt{u}+\sqrt{5}}$.

In Exercises 29–34, algebraically find the points of intersection of the graphs of the two given functions.

29. $y = \sqrt{x}$, $y = x - 2$
30. $y = \sqrt{x-4}$, $y = 4 - x$
31. $y = \sqrt{x} - 2$, $y = -x$
32. $y = \sqrt{x+5}$, $y = x - 1$
33. $y = \sqrt{9-x}$, $y = x + 3$
34. $y = \sqrt{x} - 2$, $y = \frac{1}{6}(x-4)$

35. Exercises 25–28 illustrate how radical expressions, particularly those obtained from difference quotients, can be simplified. Why is it difficult to determine the behavior of the expression $\dfrac{\sqrt{x}-2}{x-4}$ when x is near 4? Rationalize the numerator of this expression. Is it now easier to analyze the behavior of this expression when x is near 4? What is happening to this expression when x is near 4?

36. Using the procedure outlined in Exercise 35, analyze the behavior of $\dfrac{\sqrt{x+3}-3}{x-6}$ when x is near 6.

37. **(a)** Write an equation that identifies those points equidistant from the points (2, 3) and (5, 1).
(b) Verify that the equation obtained in part (a) is the equation of the perpendicular bisector of the line segment joining the two points.

38. Write an equation that identifies the points equidistant from the point (0, 4) and the x-axis.

39. Points A and B are directly opposite each other along the banks of a straight river that is 3 miles wide. Point C is 8 miles down the river on the same side as B. An oil company wishes to lay a pipeline so that oil can be pumped from point A to point C. However, it costs twice as much to lay each mile of pipe underwater as it does on dry land. Therefore, the company considers laying the pipe from point A to some point P between B and C, say x miles from B.
(a) Express the total length of pipeline needed, L, as a function of x.
(b) If it costs D dollars per mile to lay pipe on land and $2D$ dollars per mile to lay pipe underwater, express the total cost of the pipeline, C, as a function of x.
(c) If the company chooses point P to be 5 mi down the shoreline from B, approximate the total cost of the pipeline in terms of D.
(The methods of calculus allow us actually to determine where the point P should be chosen to minimize the cost of the pipeline.)

40. Two poles, 20 feet and 30 feet high, are to be anchored to the ground with guy wires from the top of each to a point directly between the two poles, which are 50 feet apart. See the following figure. Express the total length of wire, L, needed as a function of x.

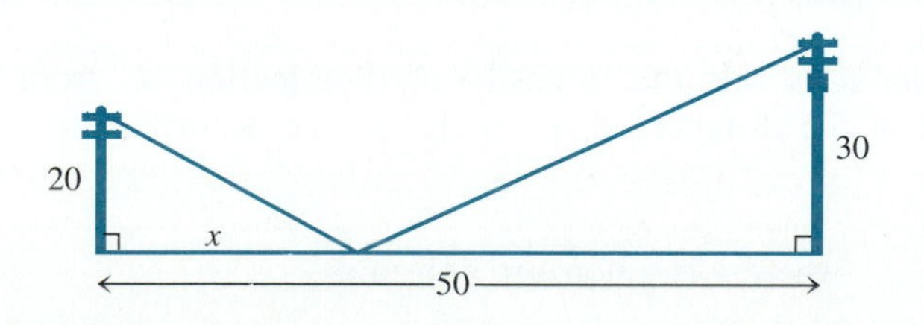

41. The slant height, s, of a right circular cone is as indicated in the following figure.

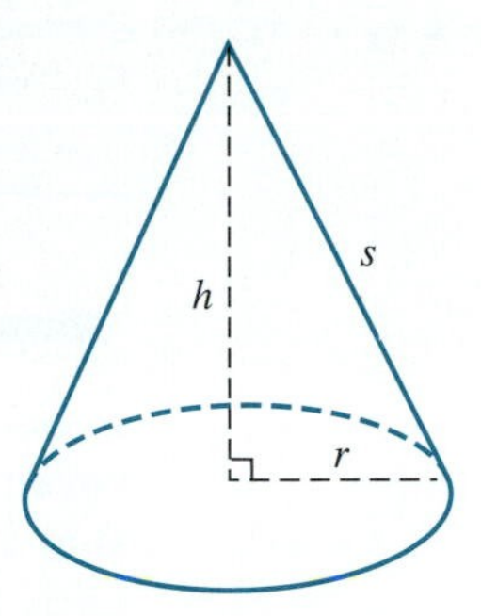

If the height of a cone is 8 cm, express the volume of the cone as a function of s.

42. An open right circular cylinder of height h is inscribed in a sphere of radius 12 cm. See the following figure.

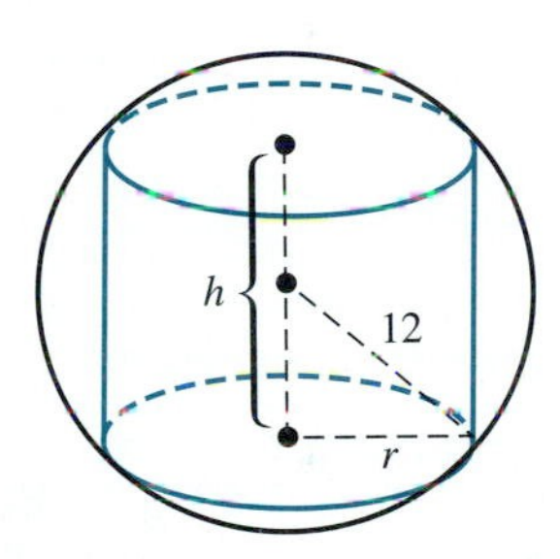

(a) Express the surface area of the cylinder as a function of r.
(b) Express the volume of the cylinder as a function of r.

43. A rectangle of base b and height h is inscribed in a circle of radius 18 meters. Express the area of the rectangle as a function of h.

44. Use the formula for the surface area, S, of a sphere to express the radius of the sphere as a function of S.

Certain procedures in calculus applied to radical functions give rise to the following types of problems.

45. Show that $x^{1/2} + (x+8)\frac{1}{2}x^{-1/2} = \dfrac{3x+8}{2\sqrt{x}}$.

46. Show that $\dfrac{x^{1/2} - (x-1)\frac{1}{2}x^{-1/2}}{x} = \dfrac{x+1}{2\sqrt{x^3}}$.

47. Show that

$$x\left(\frac{1}{2}\right)(x^2+4)^{-1/2}(2x) + (x^2+4)^{1/2} = \frac{2x^2+4}{\sqrt{x^2+4}}.$$

48. Show that
$$\frac{\sqrt{x^2+1} - x[\frac{1}{2}(x^2+1)^{-1/2}(2x)]}{(\sqrt{x^2+1})^2} = \frac{1}{(x^2+1)^{3/2}}$$

49. Show that $\sqrt[3]{x} + (x-3)x^{-2/3} = \dfrac{2x-3}{3\sqrt[3]{x^2}}$.

50. Show that
$$\frac{x^2\frac{1}{3}(x-1)^{-2/3} - (x-1)^{1/3}(2x)}{x^4} = \frac{6-5x}{3x^3\sqrt[3]{(x-1)^2}}.$$

Variation

In Section 3.3 we discussed how functions can be used to describe relationships between various physical quantities. We called the equation describing such a relationship a mathematical model of the real-life situation. In this section we extend these ideas to particular types of relationships using the polynomial, rational, and radical functions studied in this chapter.

We are familiar with the fact that if we travel at a constant rate of 50 mi/h for a period of t hours, then the distance traveled is $d = 50t$. Here d is a function of t, and we often say "The distance varies directly as the time."

Suppose someone drops a heavy metal ball from a high bridge and is able to determine the distance the object has fallen (in feet) after t seconds, the results of which are recorded in the following table.

t	1	2	3	4
d	16.3	64.7	144.5	256.8

Based upon these observed data, we might conjecture that a mathematical model for the distance fallen is $d = 16t^2$, which would give distances of 16, 64, 144, and 256 ft after 1, 2, 3, and 4 seconds, respectively. In fact, it turns out that if the ball were falling in a vacuum with no air resistance, the distance the ball falls is given exactly by $d = 16t^2$. This gives the distance fallen as a function of t.

In both of these cases, $d = 50t$ and $d = 16t^2$, the distance is a function of the time t. However, there is a special vocabulary used to describe these functions and indicate the way in which the distance is a function of time. In the case of $d = 50t$, we say that the distance varies *directly as the time,* whereas in the case $d = 16t^2$, we say that the distance varies *directly as the square of the time*.

A formula from physics called Boyle's law says that all other factors remaining the same, the volume, V, occupied by a specific amount of a gas is a function of the pressure, P, according to the formula

$$V = \frac{k}{P} \qquad \text{where } k \text{ is a nonzero constant}$$

Another formula from physics says that the intensity of illumination, I, from a source of light is a function of the distance, d, from the source according to the formula

$$I = \frac{k}{d^2} \qquad \text{where } k \text{ is a nonzero constant}$$

Again, although V is a function of P and I is a function of d, the language used to describe these functions is not the same. We often say "The volume of the gas varies inversely as its pressure" and "The intensity of illumination varies inversely as the square of the distance from the source of light."

The following makes this terminology precise.

Direct Variation

The following three statements are equivalent.

1. y **varies directly** as x.
2. y is **directly proportional** to x.
3. $y = kx$ for some constant k not equal to zero.

Inverse Variation

The following three statements are equivalent.

1. y **varies inversely** as x.
2. y is **inversely proportional** to x.
3. $y = \frac{k}{x}$ for some constant k not equal to zero.

k is called the **constant of variation** or the **constant of proportionality.**

For example, from the formula for the circumference of a circle, $C = 2\pi r$, we see that the circumference of a circle varies directly as its radius, and the constant of variation for this relationship is 2π. On the other hand, from the formula for the area of a circle, $A = \pi r^2$, we see that the area of a circle is directly proportional to the *square* of its radius, and the constant of proportionality is π.

If y varies directly as x, does it follow that x varies directly as y?

If y varies inversely as x, does it follow that x varies inversely as y?

EXAMPLE 1 Suppose that s varies directly as t and that $s = 12$ when $t = 5$. Find **(a)** s when $t = 9$ **(b)** t when $s = \frac{3}{4}$.

Solution

(a) Since s varies directly as t, we use the definition of direct variation to write

$$s = kt$$

Now we substitute $s = 12$ and $t = 5$ to find k.

$$12 = k(5) \Rightarrow k = \frac{12}{5}$$

Thus the variation equation becomes

$$s = \frac{12}{5}t$$

Now we substitute $t = 9$.

$$s = \frac{12}{5}(9) \quad \text{or} \quad \boxed{s = \frac{108}{5}}$$

An alternative approach is to recognize that if s varies directly as t, then we can write

$$s = kt \quad \text{or} \quad \frac{s}{t} = k$$

Thus the ratio $\frac{s}{t}$ is constant and we can set up the following proportion:

$$\frac{12}{5} = \frac{s}{9} \Rightarrow \boxed{s = \frac{108}{5}}$$

The first approach offers the advantage that we obtain the variation equation $s = \frac{12}{5}t$, which can be used to find additional values of s or t, as we illustrate in part (b).

(b) Using the variation equation $s = \frac{12}{5}t$ obtained in part (a), we can substitute $s = \frac{3}{4}$ to find t.

$$s = \frac{12}{5}t \qquad \textit{Substitute } s = \frac{3}{4}.$$

$$\frac{3}{4} = \frac{12}{5}t \Rightarrow \boxed{t = \frac{5}{16}}$$

■

EXAMPLE 2 As mentioned before, the intensity of illumination, I, of a light source is inversely proportional to the square of the distance from the source. If a light source has an intensity of 1000 lumens (lm) at a distance of 1.5 feet, find the intensity at a distance of 10 feet.

Solution The variation equation for this example is $I = \frac{k}{d^2}$. We use the given information to find the constant of proportionality.

$$I = \frac{k}{d^2} \qquad \textit{Substitute } I = 1000 \textit{ and } d = 1.5.$$

$$1000 = \frac{k}{1.5^2} \Rightarrow k = 2250 \qquad \textit{The variation equation becomes}$$

$$I = \frac{2250}{d^2} \qquad \textit{Now we substitute } d = 10.$$

$$I = \frac{2250}{10^2} \Rightarrow \boxed{I = 22.5 \text{ lumens}}$$

At a distance of 10 feet the light source has an intensity of 22.5 lumens. ■

Having the variation model for a particular situation allows us to analyze the relationship among the variables, as illustrated in the following example.

EXAMPLE 3 The volume of a sphere varies directly as the cube of its radius, according to the formula

$$V = \frac{4}{3}\pi r^3$$

See what happens to the volume of a sphere if the radius is doubled from 3 inches to 6 inches.

What do you get when you divide the volume of the sphere when the radius is $3r$ by the volume of the sphere when the radius is r?

What happens to the volume of a sphere if the radius is tripled?

Solution Tripling the radius means changing the radius from r to $3r$. When the radius is r, the volume is

$$V = \frac{4}{3}\pi r^3$$

When the radius is $3r$, the volume is

$$V = \frac{4}{3}\pi(3r)^3 = \frac{4}{3}\pi(27)r^3 = 27\left(\frac{4}{3}\pi r^3\right)$$

When the radius is $3r$, we can see that the volume is 27 times the volume when the radius is r. Therefore, we conclude that tripling the radius causes the volume to increase by a factor of 27. ■

There are many other types of variation. For example, if z varies directly as the product of x and y, then $z = kxy$ for some nonzero constant k and we say that z **varies jointly as x and y**. If z varies directly as x and inversely as y, we write

$$z = k\left(\frac{x}{y}\right) = \frac{kx}{y}$$

If z varies directly as the square of x and inversely as the square root of y, we write

$$z = \frac{kx^2}{\sqrt{y}}$$

EXAMPLE 4 The resistance, R, of a wire to an electrical current (measured in ohms) is directly proportional to the length, L, of the wire and inversely proportional to the square of the diameter, d, of the wire. If the resistance of a 50-foot-long wire of diameter 0.1 cm is 8 ohms, find the resistance of 100 meters of the same type of wire with a diameter of 0.05 cm.

Solution The given relationship can be translated into the following variation equation

$$R = \frac{kL}{d^2}$$ *We substitute $R = 8$, $L = 50$, and $d = 0.1$ and solve for k.*

$$8 = \frac{k(50)}{(0.1)^2} \Rightarrow k = 0.0016$$ *The variation equation becomes*

$$R = \frac{0.0016L}{d^2}$$ *Now we substitute $L = 100$ and $d = 0.05$.*

$$R = \frac{0.0016(100)}{(0.05)^2} = 64 \text{ ohms}$$

Note that doubling the length of the wire (from 50 meters to 100 meters) and halving the diameter (from 0.1 cm to 0.05 cm), increases the resistance by a factor of 8 (from 8 ohms to 64 ohms). ■

EXERCISES 4.7

In the following exercises, be sure to identify the constant of variation. Round answers to the nearest hundredth where necessary.

1. If y varies directly as x and $y = 15$ when $x = 8$, find y when $x = 25$.

2. If y varies inversely as x and $y = 15$ when $x = 8$, find y when $x = 25$.

3. If u varies inversely as the square of t and $u = 4$ when $t = 8$, find u when $t = 10$.

4. If a varies directly as the square root of b and $a = 5$ when $b = 12$, find b when $a = 6$.

5. If z varies jointly as m and p and $z = 20$ when $m = \frac{1}{2}$ and $p = 7$, find z when $m = -3$ and $p = \frac{2}{9}$.

6. If v varies directly as r and inversely as s and $v = -12$ when $r = 3$ and $s = 4$, find s when $v = 2$ and $r = -6$.

7. If z varies directly with the square of x and inversely with the cube of y and $z = 1$ when $x = 2$ and $y = 3$, find z when $x = 3$ and $y = 2$.

8. If z varies jointly with the cube of x and the fourth power of y and $z = 2$ when $x = 0.25$ and $y = 0.5$, find x when $z = 0.2$ and $y = 1.2$.

9. Suppose y varies inversely as x. What happens to y if x is multiplied by a factor of 4?

10. Suppose y varies directly as x. What happens to y if x is multiplied by a factor of 4?

11. Suppose y varies directly as the square root of x. What happens to y if x is multiplied by a factor of 9?

12. Suppose y varies inversely as the cube root of x. What happens to y if x is multiplied by a factor of 8?

13. Suppose s varies jointly as t and u. What happens to s if t is doubled and u is tripled?

14. Suppose r varies directly as c and inversely as d. What happens to r if c is tripled and d is halved?

15. Hooke's law for springs states that the force or weight required to stretch a spring x units beyond its natural length is directly proportional to x. If a weight of 5 pounds is necessary to stretch a spring from its natural length of 8 cm to 8.4 cm, find the weight necessary to stretch the spring to a length of 9.2 cm.

16. Hooke's law also applies to the force or weight necessary to compress a spring x units within its natural length. If a force of 25 pounds is needed to compress a spring 4 inches shorter than its natural length, find the force necessary to compress the spring 6 additional inches.

17. If a force of 30 pounds stretches a spring 6.5 inches, how far will a force of 42 pounds stretch the spring?

18. A force of 45 pounds stretches a spring 6 cm. If the spring has a natural length of 30 cm and can be stretched at most to twice its natural length, what is the maximum weight that the spring can support?

19. If a searchlight has an intensity of 50,000 lumens at a distance of 100 feet, what will the intensity be at a distance of 200 feet? (See Example 2.)

20. If a light source has an intensity of 1600 lumens at a distance of 10 feet, at what distance will the intensity be 2 lumens? (See Example 2.)

21. According to Boyle's law, if all other factors remain the same, the volume, V, occupied by a specific amount of a gas varies inversely as the pressure, P. If 100 cu in. of a gas exerts a pressure of 40 lb/sq in., what pressure will be exerted if the same amount of gas is compressed to 30 cu in.?

22. To what volume must the gas in Exercise 21 be allowed to expand if it is to exert a pressure of 15 lb/sq in.?

23. The electrical resistance of a wire varies directly with the length of the wire and inversely with the square of its diameter. If a wire 100 meters long with a diameter of 0.36 cm. has a resistance of 80 ohms, how much resistance will there be if only 40 meters of the wire is used?

24. If you want to use a wire made of the same material as that in Exercise 23 (which means it has the same variation constant) and you want 100 feet of this wire to have a resistance of 40 ohms, how thick must the wire be?

25. When analyzing an accident scene the police can sometimes estimate the speed of a vehicle (before the brakes were applied) by measuring the length of the skid marks and using the fact that the speed is directly proportional to the square root of the length of the skid marks. If skid marks of 50 feet are made at 35 mi/h, estimate the speed of a car that makes skid marks of 125 feet.

26. The driver of a car involved in an accident claims that he was driving at a legal speed limit of 55 mi/h. His car is tested and leaves a skid mark of 20 feet at 30 mi/h. If his vehicle left skid marks of 160 feet at the accident site is his claim plausible?

27. The destructive force, F, of a car in an automobile accident can be described approximately by saying that it varies jointly with the weight of the car, w, and the square of the speed of the car, v. How would F be affected if

(a) The speed of a car is doubled?
(b) The weight of a car is doubled?
(c) The speed and weight of a car are doubled?

28. The range of a projectile such as an artillery shell shot out of a cannon or a motorcycle stunt rider flying off a ramp is directly proportional to the square of the velocity at takeoff. If a motorcycle stunt rider has jumped a distance of 168 ft with a takeoff speed of 70 mi/h, how far can she expect to jump with a takeoff speed of 85 mi/h?

29. The load, L, that can be safely supported by a beam with a rectangular cross section varies directly with the product of the width, w, and the square of the depth, d, of the cross section and inversely with the length of the beam, ℓ. For the purposes of this discussion we assume that the width is the shorter dimension of the cross section. If a 3-inch by 5-inch beam that is 10 feet long can safely carry a load of 750 pounds, what is the maximum load that can be safely supported by a beam of the same material if it is 2.5 in. by 4 in. by 15 ft long?

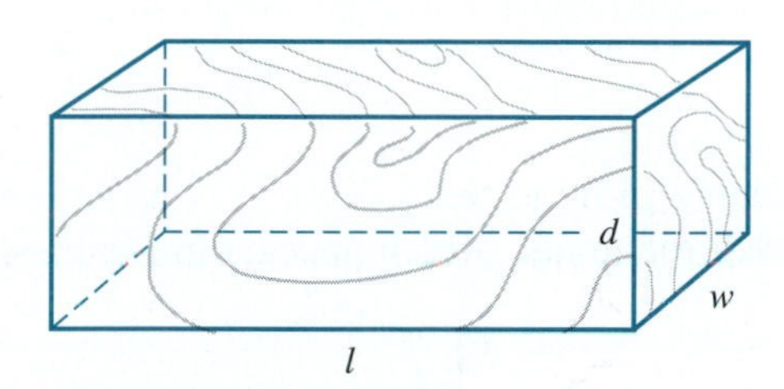

30. Using the information in Exercise 29, how is the safe load affected by doubling each of the dimensions of the beam individually? All together?

31. Newton's law of gravitation states that the force, F, of attraction between two objects of mass m_1 and m_2 is

$$F = G\frac{m_1 m_2}{d^2}$$

where G is a constant and d is the distance between the two masses. (In using this formula we perform the computations as if the entire mass were concentrated at the center of each object.)

(a) Use the given formula to describe how the force of attraction between two objects varies in relation to their masses and the distance between them.

(b) Suppose it is known that if two 1-kg masses are placed 1 m apart they will exert a force of attraction of 6.67×10^{-11} newtons (N). Find the variation constant G.

(c) The earth has a mass of approximately 5.98×10^{24} kg. Find the gravitational force exerted by the earth on a 1000 kg satellite orbiting the earth at an altitude of 400 km. (The radius of the earth is approximately 6400 km.)

32. The moon has a mass of 6.7×10^{22} kg and is a distance of approximately 385,000 km away from the earth. Use the information obtained in part (b) of the previous exercise to determine the gravitational attraction between the earth and the moon.

In Exercises 33–38, use the given equation to describe in words how the variable on the left-hand side of the equation varies with respect to the *variables* on the right hand side. In each case, k is a constant.

33. $S = 2\pi rh$

34. $V = \pi r^2 h$

35. $V = \frac{4}{3}\pi r^3$

36. $y = \frac{kx^4}{\sqrt{z}}$

37. $z = \frac{kx^2y^3}{4w}$

38. $s = \frac{\sqrt[3]{rt}}{kv^2}$

QUESTIONS FOR THOUGHT

39. In the formula $d = rt$, relating distance, rate, and time, describe which two of the three quantities vary directly and which vary inversely.

40. The electrical resistance of a wire varies directly with its length and inversely with its cross sectional area. Does this statement agree or disagree with the statement made in Example 4?

Chapter 4 SUMMARY

After completing this chapter you should:

1. Recognize the graphs of the basic power functions (Section 4.1)
For $n > 1$, the graph of $y = x^n$ looks like the graph of $y = x^2$ when n is even and like the graph of $y = x^3$ when n is odd.
For example:
Sketch the graphs of
(a) $y = (x - 1)^3$ **(b)** $y = (x + 2)^4 - 16$.
Solution:
(a) The graph of $y = (x - 1)^3$ is obtained by shifting the graph of $y = x^3$ one unit to the right.
(b) The graph of $y = (x + 2)^4 - 16$ is obtained by shifting the graph of $y = x^4$ two units to the left and 16 units down. The intercepts are found as usual. The graphs appear in Figure 4.50 (a) and (b).

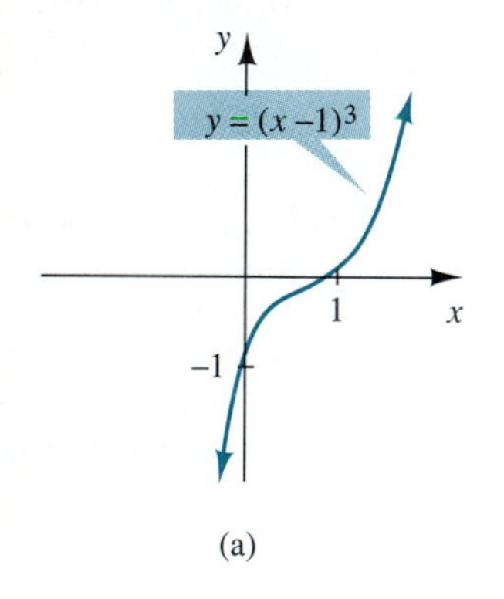

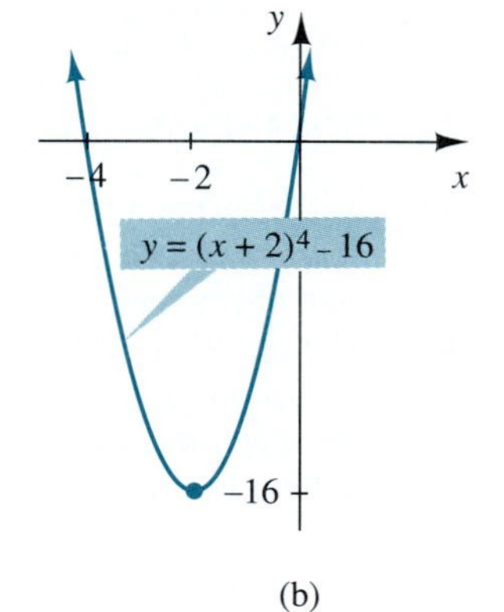

FIGURE 4.50

2. Be able to sketch the graph of a polynomial function when its zeros can be found. (Section 4.1)
We can use the properties of polynomial functions to sketch the graphs of more general polynomials. The zeros of f are the x-intercepts of the graph of $f(x)$. A sign analysis tells us where the graph is above the x-axis and where the graph is below the x-axis.
For example:
Sketch the graph of $y = f(x) = x^3 + 4x^2 + 3x$.
Solution:
We begin by finding the zeros of f and doing a sign analysis on $f(x)$.

$$f(x) = x^3 + 4x^2 + 3x = x(x^2 + 4x + 3) = 0$$
$$= x(x + 1)(x + 3) = 0$$

The cut points are $0, \ -1, -3$.

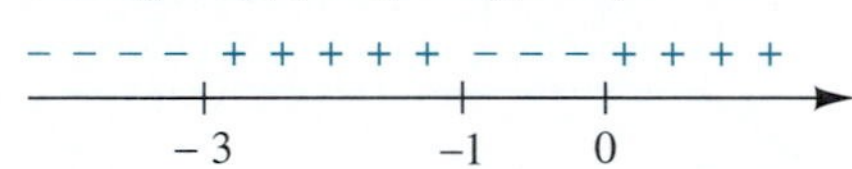

Keeping in mind that $|f(x)| \to \infty$ as $|x| \to \infty$, we sketch the graph, which appears in Figure 4.51.

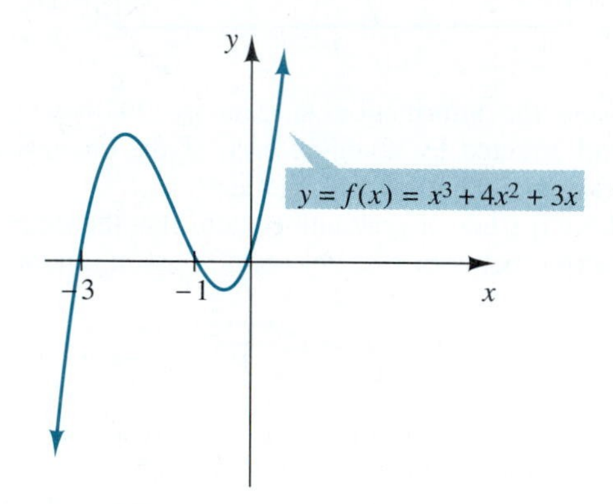

FIGURE 4.51

3. Be able to use long division to divide polynomials and synthetic division to divide a polynomial by $x - a$. (Section 4.2)

4. Understand the Factor and Remainder Theorems and use them to determine whether or not $x - a$ is a factor of $p(x)$. (Section 4.3)
According to the Factor Theorem $x - a$ is a factor of $p(x)$ if and only if $p(a) = 0$.
For example:
Determine whether or not $x + 2$ is a factor of $p(x) = 2x^4 + x^3 - 3x^2 - 12$.
Solution:
We look at $x + 2$ as $x - a$, which implies $a = -2$. Since

$$p(-2) = 2(-2)^4 + (-2)^3 - 3(-2)^2 - 12$$
$$= 32 - 8 - 12 - 12 = 0$$

we know that when $p(x)$ is divided by $x - (-2) = x + 2$, the remainder is 0, and so $x + 2$ is a factor of $p(x)$.

5. Understand the Rational Root Theorem, Descartes's Rule of Signs, and the Upper and Lower Bound Theorem, and use them to help factor a polynomial. (Section 4.4)
By determining the factors of a_0 and a_n in the polynomial equation

$$f(x) = a_n x^n + a_{n-1}x^{n-1} + \cdots + a_1 x + a_0 = 0$$

where $n \geq 1$, $a_n \neq 0$, we can determine the possible rational roots of $f(x) = 0$. We can then use synthetic division to check these possible roots.

For example:
Find all the real zeros of
$f(x) = x^3 - 3x^2 + 2x - 6$.
Solution:

According to the Rational Root Theorem, if $\frac{p}{q}$ is a rational root of $f(x) = 0$ then p must be a factor of -6 and q must be a factor of 1. Thus we have

possible values of p: $\pm 1, \pm 2, \pm 3, \pm 6$

possible values of q: ± 1

possible rational roots $\frac{p}{q}$: $\pm 1, \pm 2, \pm 3, \pm 6$

By using synthetic division, we can divide $f(x)$ by $x \pm 1$, $x \pm 2$, $x \pm 3$ and $x \pm 6$. We find that the remainder when $f(x)$ is divided by $x - 3$ is zero and so by the Factor Theorem, $x - 3$ is a factor and we have

$$f(x) = x^3 - 3x^2 + 2x - 6 = (x - 3)(x^2 + 2)$$

Since $x^2 + 2$ has no real zeros, 3 is the only real zero of $f(x)$.

6. Be familiar with the graphs of basic rational functions. (Section 4.5)
For $n \geq 1$, the graph of $y = \frac{1}{x^n}$ looks like the graph of $y = \frac{1}{x}$ when n is odd and like the graph of $y = \frac{1}{x^2}$ when n is even.

For example:
Sketch the graphs of
(a) $y = \frac{1}{x} + 1$ **(b)** $y = \frac{1}{(x - 3)^2}$

Solution:

(a) The graph of $y = \frac{1}{x} + 1$ can be obtained by shifting the graph of $y = \frac{1}{x}$ up 1 unit. The graph appears in Figure 4.52

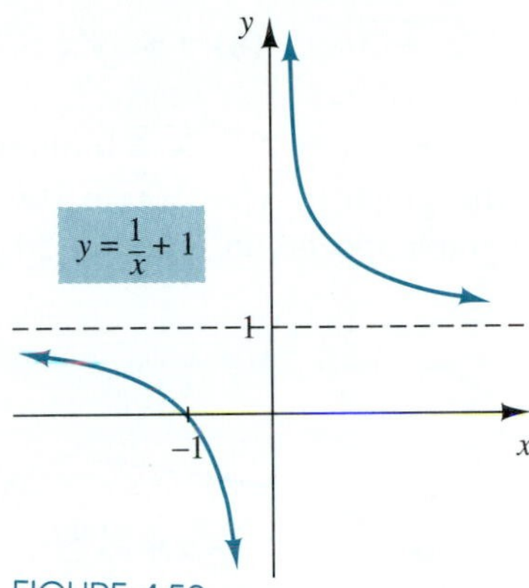

FIGURE 4.52

(b) The graph of $y = \frac{1}{(x - 3)^2}$ can be obtained by shifting the graph of $y = \frac{1}{x^2}$ to the right 3 units. The graph appears in Figure 4.53.

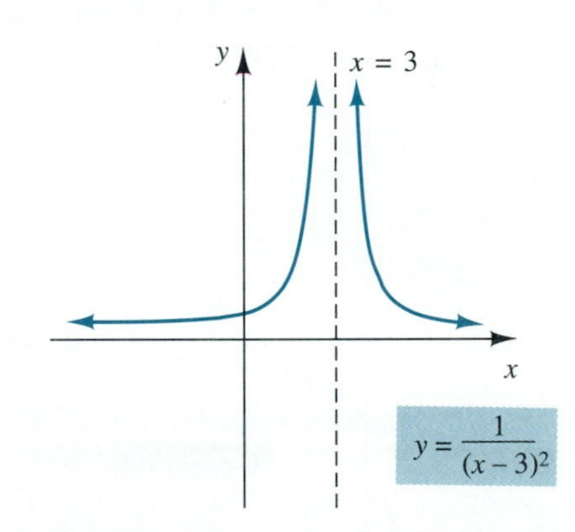

FIGURE 4.53

7. Be familiar with the graphs of basic radical functions. (Section 4.6)
The graph of $y = \sqrt[n]{x}$ will look like the graph of $y = \sqrt{x}$ when n is even and the graph of $y = \sqrt[3]{x}$ when n is odd.
For example:
Sketch the graphs of
(a) $y = \sqrt{x+3}$ **(b)** $y = \sqrt[5]{x} - 2$
Solution:
(a) The graph of $y = \sqrt{x+3}$ is obtained by shifting the graph of $y = \sqrt{x}$ to the left 3 units. The graph appears in Figure 4.54.

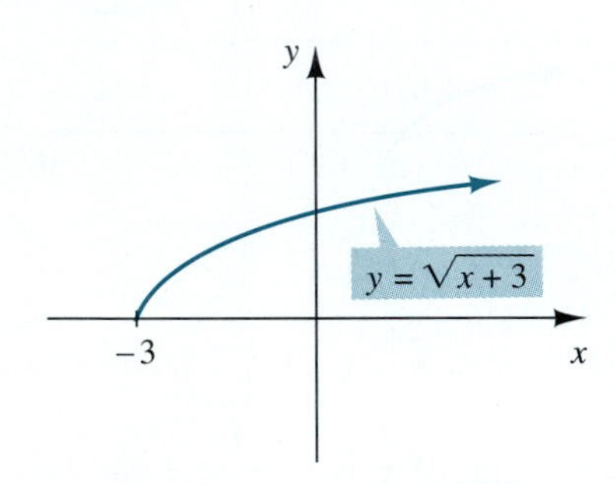

FIGURE 4.54

(b) The graph of $y = \sqrt[5]{x} - 2$ is obtained by shifting the graph of $y = \sqrt[5]{x}$ down 2 units. The graph appears in Figure 4.55.

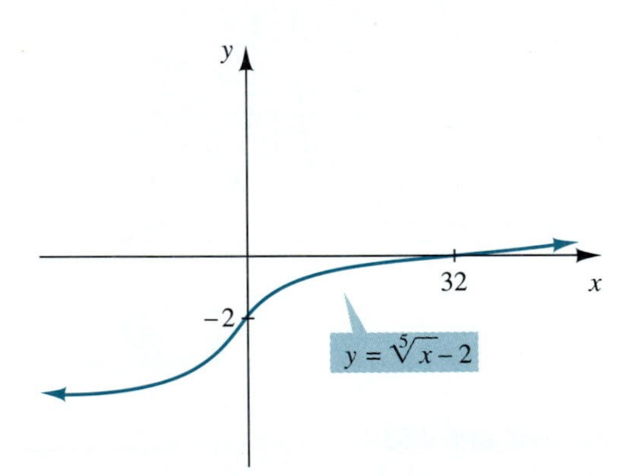

FIGURE 4.55

8. Be able to translate statements of variation and solve variation problems. (Section 4.7)
The relationship between some physical quantities can be expressed as a variation. The mathematical model may be a polynomial, rational or radical function which can be used to describe the real-life problem.
For example:
Suppose that z varies directly with the square root of x and inversely with the square of y. If $z = 20$ when $x = 4$ and $y = 3$, find z when $x = 10$ and $y = 2$.
Solution:
The given variation relationship is translated as $z = \dfrac{k\sqrt{x}}{y^2}$. We can find the constant of variation, k, by substituting the given values.

$$z = \frac{k\sqrt{x}}{y^2}$$ *Substitute $z = 20$, $x = 4$, and $y = 3$.*

$$20 = \frac{k\sqrt{4}}{3^2} \Rightarrow k = 90$$ *Therefore, we have*

$$z = \frac{90\sqrt{x}}{y^2}$$ *Substitute $x = 10$ and $y = 2$.*

$$z = \frac{90\sqrt{10}}{2^2}$$

$$= \boxed{\frac{45\sqrt{10}}{2}}$$

Chapter 4 REVIEW EXERCISES

In Exercises 1–22, sketch the graph of the given function. Be sure to indicate the intercepts and asymptotes where appropriate.

1. $y = x^3 + 8$

2. $y = (x - 2)^6$

3. $y = (x - 1)^4 - 16$

4. $y = (x + 2)^5 - 1$

5. $y = x^3 - x^2 - 6x$

6. $y = x^4 - 4x^2$

7. $y = x^2 - 6x + 9$

8. $y = 1 - x^4$

9. $y = \dfrac{1}{(x - 1)^2}$

10. $y = \dfrac{-1}{x} + 2$

11. $y = \dfrac{1}{x + 2} - 3$

12. $y = \dfrac{1}{(x - 3)^4} + 2$

13. $y = \dfrac{x + 3}{x + 2}$

14. $y = \dfrac{x}{x + 1}$

15. $y = \dfrac{x}{x^2 - 1}$

16. $y = \dfrac{x^2 + 2}{4 - x^2}$

17. $y = \sqrt{2x - 5}$

18. $y = \sqrt[3]{4 - x} + 1$

19. $y = \sqrt[5]{x} - 2$

20. $y = \sqrt[4]{3x + 6}$

21. $y = |x^3 - 8|$

22. $y = |(x - 2)^3|$

In Exercises 23–26, use long division to find the quotient and remainder.

23. $\dfrac{2x^3 - 3x^2 + 4x - 7}{2x - 3}$

24. $\dfrac{x^4 - x + 1}{x^2 + 1}$

25. $\dfrac{2x^4 + x^3 + 2x + 1}{2x + 1}$

26. $\dfrac{x^5 + 3x^4 - 4x^2 + 2}{x^2 + x - 1}$

In Exercises 27–30, use synthetic division to find the quotient and remainder.

27. $\dfrac{2x^3 - 3x^2 - 4x - 15}{x - 3}$

28. $\dfrac{x^4 - 5x + 3}{x + 2}$

29. $\dfrac{x^5 - 2x^4 + x^3 - 3x^2 + 5x - 4}{x - 2}$

30. $\dfrac{x^6 - x - 3}{x + 1}$

In Exercises 31–36, find all the zeros of the given polynomial.

31. $p(x) = x^3 - 13x - 12$

32. $p(x) = x^3 + 2x^2 - 3x - 6$

33. $p(x) = x^3 - 6x^2 + 7x - 10$

34. $p(x) = 3x^3 - 2x^2 - 27x + 18$

35. $p(x) = 2x^4 + 7x^3 - 2x^2 + 7x - 4$

36. $p(x) = x^4 - 2x^3 - 3x^2 + 4x + 4$

37. If -2 and $5 - 2i$ are two zeros of a third degree polynomial, how many more zeros does $p(x)$ have? What are they? What can $p(x)$ be?

38. If $2 + 3i$ and $1 - i$ are two zeros of a fourth degree polynomial $p(x)$, how many more zeros does $p(x)$ have? What are they? What can $p(x)$ be?

39. How many positive real zeros can the polynomial $p(x) = 2x^3 + 4x^2 + 5x + 3$ have?

40. How many negative real zeros can the polynomial $p(x) = x^4 - x^3 - 2x^2 + 3x + 4$ have?

41. How many positive real zeros can the polynomial $p(x) = 3x^5 - 4x^2 + 3x - 5$ have?

42. How many negative real zeros can the polynomial $p(x) = x^6 - x^5 + x^4 - x^3 + x^2 - x + 1$ have?

In Exercises 43–46, find an upper and lower bound for the real zeros of $p(x)$.

43. $p(x) = x^4 - 4x^3 + 15$

44. $p(x) = 2x^3 - 3x^2 - 12x + 9$

45. $p(x) = x^3 - \dfrac{2}{3}x^2 + \dfrac{1}{2}x - \dfrac{1}{3}$

46. $p(x) = x^4 - 4x^3 + 16x - 17$

47. The perimeter of the square $ABCD$ is 20 cm. Express the area of the triangle as a function of x. See the accompanying figure.

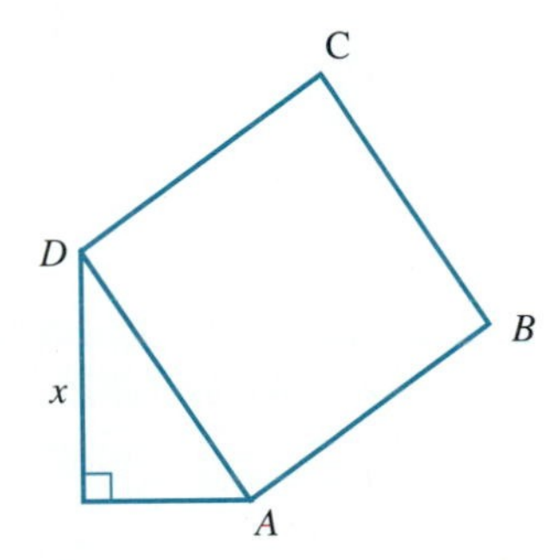

48. The base of a closed rectangular box is a square of side x. If the volume of the box is 100 cm³, express the surface area, S, of the box as a function of x.

In Exercises 49–52, find the points of intersection of the graphs of the two given functions algebraically.

49. $y = \sqrt{4 - x}$, $y = x + 2$

50. $y = \sqrt{x} - 3$, $y = x - 9$

51. $y = \sqrt{2x + 5} - 2$, $y = x - 7$

52. $y = 4 - \sqrt{x}$, $y = 2x + 1$

53. If x varies inversely as the cube root of y and $x = 27$ when $y = 27$, find x when $y = 8$.

54. If z varies jointly as x and the square of y and $z = 30$ when $x = 5$ and $y = 2$, find x when $z = 10$ and $y = 3$.

55. Suppose that on a certain planet the distance that an object falls is directly proportional to the length of time it falls raised to the $\frac{3}{2}$ power. If an object on this planet falls 350 feet in 4 seconds, how long will it take for an object to fall 500 feet?

56. The volume of a cone varies jointly as its height and the square of its radius. If a cone with a height of 3 cm and a radius of 2 cm has a volume of 4π cu cm, find the volume of a cone whose height is 2 cm and radius is 3 cm.

Chapter 4 PRACTICE TEST

1. Sketch the graph of the given function. Be sure to indicate the intercepts and asymptotes where appropriate.

(a) $y = \dfrac{1}{x - 2} + 3$ **(b)** $y = \dfrac{-3}{(x + 1)^2}$

(c) $y = 2 - \sqrt{x}$ **(d)** $y = \dfrac{x + 3}{x - 2}$

(e) $y = x^3 - x^2 - 12x$ **(f)** $y = \dfrac{2x}{9 - x^2}$

(g) $y = \sqrt[3]{x - 2} + 1$ **(h)** $y = (x + 1)^2(x - 2)^2$

2. Divide: $\dfrac{3x^4 - 10x^3 - 7x^2 + 17x + 3}{3x - 4}$.

3. Which of the following are factors of $p(x) = 2x^4 - x^3 - 11x^2 + 4x + 12$?

(a) $x + 3$ **(b)** $x - 2$ **(c)** $x - 1$ **(d)** $x + 2$

4. Find all the zeros of the following polynomials.

(a) $p(x) = 3x^3 - 20x^2 + 29x + 12$

(b) $p(x) = 2x^4 - 9x^3 + 19x^2 - 15x$

5. According to Poiseuille's law, the blood pressure, P, in a blood vessel varies directly as the length of the vessel and inversely as the fourth power of its radius. What happens to the blood pressure in a particular blood vessel if, due to a buildup in the blood vessel, the radius decreases from 3 mm to 2 mm?

CHAPTER 5

Exponential and Logarithmic Functions

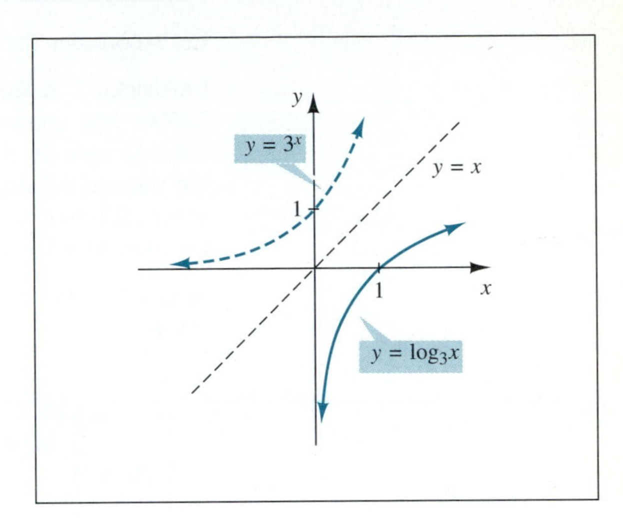

In the previous chapter we examined polynomial, rational, and radical functions, all of which are types of *algebraic functions,* as described in Chapter 2. In this chapter we introduce two new *nonalgebraic* functions, the exponential function and its inverse, the logarithmic function. As we shall see, these new functions have a wide variety of applications.

5.1 Exponential Functions

Let's consider the following example.

EXAMPLE 1 A student decides she wants to save up to buy a used car, which costs \$2600. She comes up with what she thinks is a very modest savings plan. She decides to save 2¢ the first day and double the amount she saves each day thereafter. On the second day she plans to save 4¢, on the third day, 8¢, and so on. Write an expression that represents the amount saved on day n and determine how long it will take her to save up enough money to buy the car. (The answer may surprise you.)

Solution We construct a table that records the amount of money earned on day n.

Day	A = Amount saved *on* that day	T = Total amount saved *by* that day
1	$2¢ = 2^1¢$	$2¢$
2	$4¢ = 2^2¢$	$2¢ + 4¢ = 6¢$
3	$8¢ = 2^3¢$	$2¢ + 4¢ + 8¢ = 14¢$
4	$16¢ = 2^4¢$	$2¢ + 4¢ + 8¢ + 16¢ = 30¢$
⋮	⋮	
n	$2^n¢$	⋮
⋮	⋮	
16	$65{,}536¢ = 2^{16}¢$	$131{,}070¢ = \$1310.70$
17	$131{,}072¢ = 2^{17}¢$	$262{,}142¢ = \$2621.42$

The expression $A = 2^n$ represents the amount saved on day n. As the table illustrates, the expression 2^n grows quite rapidly. If the student can manage to adhere to this rather aggressive savings plan, she would be able to purchase the car in just 17 days!

(In Chapter 11 we develop certain formulas that will allow us to compute directly the values of T, the total amount saved by day n, without having to actually add the amounts from the previous days.) ■

In Example 1 we should recognize that the amount, A, saved on day n is a *function* of n. This is our first exposure to a function where the variable appears in the exponent and our first example of a nonalgebraic function, usually called a **transcendental function.** The function $A = 2^n$ is a particular type of transcendental function called an **exponential function.**

In Section 4.1 we examined functions of the form $f(x) = x^n$, where n is constant. How is this different from $f(x) = n^x$?

DEFINITION A function of the form $y = f(x) = b^x$, where $b > 0$ and $b \neq 1$, is called an **exponential function.** b is called the **base** of the exponential function.

Exercise 75 discusses why we have restricted b to be nonnegative and not equal to 1.

As we have seen repeatedly, the first order of business when we encounter a new function is to determine its domain. Since rational exponents are well defined, we know that any rational number will be in the domain of an exponential function. For example, as x takes on the rational values 4, -2, $\frac{1}{2}$, and $\frac{4}{5}$,

$$\text{we have for} \quad f(x) = 3^x$$

$$f(4) = 3^4 = 3 \cdot 3 \cdot 3 \cdot 3 = 81$$

$$f(-2) = 3^{-2} = \frac{1}{3^2} = \frac{1}{9}$$

$$f\left(\frac{1}{2}\right) = 3^{1/2} = \sqrt{3}$$

$$f\left(\frac{4}{5}\right) = 3^{4/5} = \sqrt[5]{3^4} = \sqrt[5]{81}$$

Note that even though we do not know the *exact* value of $\sqrt{3}$ or $\sqrt[5]{81}$, we do know exactly what they mean.

However, what about $f(x)$ for irrational values of x?

$$f(\sqrt{2}) = 3^{\sqrt{2}} = ?$$

We have not defined the meaning of irrational exponents.

In fact, a precise formal definition of b^x where x is irrational requires the ideas of calculus. However, we can get an idea of what $3^{\sqrt{2}}$ should be by using successive *rational* approximations to $\sqrt{2}$. For example, we have

$$1.414 < \sqrt{2} < 1.415$$

Thus it would seem reasonable to expect that

$$3^{1.414} < 3^{\sqrt{2}} < 3^{1.415}$$

Since 1.414 and 1.415 are rational numbers, $3^{1.414}$ and $3^{1.415}$ are well defined, even though we cannot compute their values directly. With a calculator we get

$$4.727695 < 3^{\sqrt{2}} < 4.7328918$$

If we use better approximations to $\sqrt{2}$, we get

$$3^{1.4142} < 3^{\sqrt{2}} < 3^{1.4143}$$

Using a calculator again we get

$$4.7287339 < 3^{\sqrt{2}} < 4.7292535$$

Computing $3^{\sqrt{2}}$ directly on a calculator gives $3^{\sqrt{2}} \approx 4.7288044$.

This numerical evidence suggests that as x approaches $\sqrt{2}$, the values of 3^x approach a unique real number that we designate as $3^{\sqrt{2}}$, and so we will accept, without proof, the fact that the domain of the exponential function is the set of all real numbers.

In fact, we state even more.

> The exponential function $y = b^x$, where $b > 0$ and $b \neq 1$, is defined for all real values of x. In addition, all the rules for rational exponents hold for real number exponents as well.

Before we state some general facts about exponential functions, let's see if we can determine what the graph of an exponential function will look like.

GRAFFIX

Use a graphics calculator or computer.

1. Sketch the graphs of $y = 2^x$, $y = 4^x$, and $y = 6^x$ on the same set of axes. Discuss the similarities and differences among the 3 graphs. What would you conjecture about how the curve changes as the base b varies for $b > 1$?

2. Sketch the graphs of $y = \left(\frac{1}{2}\right)^x$, $y = \left(\frac{1}{3}\right)^x$, and $y = \left(\frac{1}{5}\right)^x$ on the same set of axes. Discuss the similarities and differences among the 3 graphs. What would you conjecture about how the curve changes as the base b varies for $0 < b < 1$?

EXAMPLE 2 Sketch the graph of the function $y = 2^x$ and identify its domain and range.

Solution To aid in our analysis, we set up a short table of values to give us a frame of reference (Table 5.1).

TABLE 5.1

x	$y = f(x) = 2^x$	y
-3	$y = 2^{-3}$	$\frac{1}{8}$
-2	$y = 2^{-2}$	$\frac{1}{4}$
-1	$y = 2^{-1}$	$\frac{1}{2}$
0	$y = 2^0$	1
1	$y = 2^1$	2
2	$y = 2^2$	4
3	$y = 2^3$	8

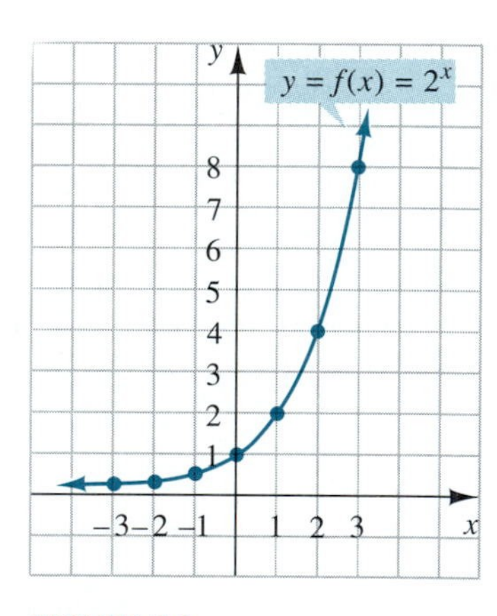

FIGURE 5.1

With these points in hand, we draw a smooth curve through the points, obtaining the graph appearing in Figure 5.1. It is important to recognize that by joining these points we are assuming that the exponential function is defined for *all real numbers*.

Recall that $x \to +\infty$ means that x gets extremely large (x takes on values to the extreme right), and $x \to -\infty$ means that x gets extremely small (x takes on values to the extreme left).

We note several important aspects of this graph. First as $x \to +\infty$, the y-values are increasing very rapidly, whereas as $x \to -\infty$ the y-values are getting closer and closer to 0. Thus the x-axis is a horizontal asymptote.

Second, there is no x-intercept. In fact, the graph of an exponential function $y = b^x$ will never have an x-intercept, since $b^x \neq 0$ for any value of x.

Are there any values of x for which $3^x = 0$? Are there any values of x for which $3^x < 0$?

Third, the y-intercept is 1. The graph of every exponential function of the form $y = f(x) = b^x$ will have a y-intercept of 1, since $b^0 = 1$ (remember $b \neq 0$).

Fourth, from the graph we can see that the range of $y = f(x) = 2^x$ is the *set of positive real numbers*. ■

EXAMPLE 3 Sketch the graph of $y = f(x) = \left(\frac{1}{2}\right)^x$.

Solution It would be instructive to compute a table of values as we did in Example 1 (you are urged to do so). However, we will take a different approach. We note that

$$y = f(x) = \left(\frac{1}{2}\right)^x = \frac{1}{2^x} = 2^{-x}$$

The graphing principle for $f(-x)$ is discussed in Section 3.2.

If $f(x) = 2^x$, then $f(-x) = 2^{-x}$. Thus by the graphing principle for $f(-x)$, we can obtain the graph of $y = 2^{-x}$ by reflecting the graph of $y = 2^x$, which we found in Example 2, about the y-axis. The graph appears in Figure 5.2.

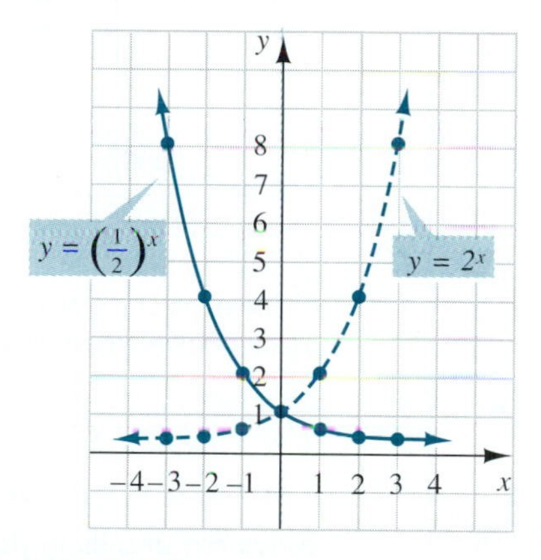

FIGURE 5.2

Here again the x-axis is a horizontal asymptote, there is no x-intercept, 1 is the y-intercept, and the range is the set of positive real numbers. However, the graph is now decreasing rather than increasing. ■

If we construct a table of values similar to Table 5.1 for $b = 3$ or 5 or $\frac{7}{2}$, in fact, for all $b > 1$, we can see that the graph of the exponential function $y = b^x$ will look very much like the graph in Example 2. This function is said to exhibit **exponential growth.** On the other hand, a similar table for $b = \frac{1}{3}$ or $\frac{4}{5}$, in fact, for all $0 < b < 1$ would reveal that the graph of the exponential function $y = b^x$ will look very much like the graph of $y = \left(\frac{1}{2}\right)^x$ in Example 3. This function is said to exhibit **exponential decay**. We also note that graphs in Examples 2 and 3 satisfy the horizontal line test and so the functions $y = 2^x$ and $y = \left(\frac{1}{2}\right)^x$ are one-to-one functions.

The following box summarizes the important facts about exponential functions and their graphs.

THE EXPONENTIAL FUNCTION $y = f(x) = b^x$

1. The domain of the exponential function is the set of all real numbers.
2. The range of the exponential function is the set of all *positive* real numbers.
3. The graph $y = b^x$ exhibits exponential growth if $b > 1$ or exponential decay if $0 < b < 1$.

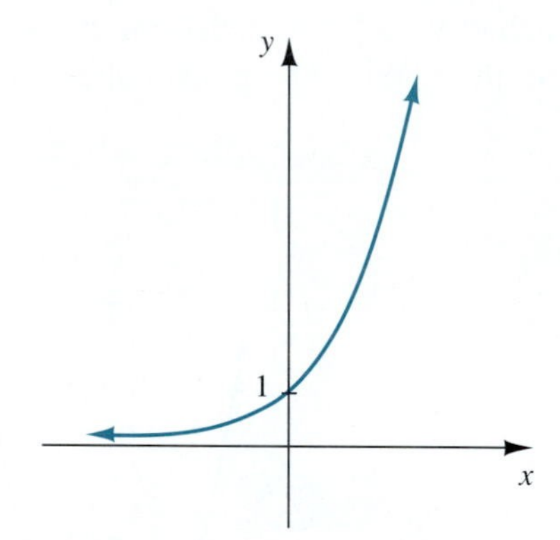

The graph of $y = b^x$ for $b > 1$
Exponential growth

The graph of $y = b^x$ for $0 < b < 1$
Exponential decay

4. The y-intercept is 1. There are no x-intercepts.
5. The x-axis is a horizontal asymptote.
6. Since the graphs in the preceding figure satisfy the horizontal line test, the exponential function is one-to-one. *Algebraically, this means if* $b^{x_1} = b^{x_2}$, *then* $x_1 = x_2$.

Example 5 on page 327 illustrates the algebraic meaning of the fact that exponential functions are one-to-one.

These graphs of exponential growth and decay should be added to our catalog of basic graphs.

EXAMPLE 4 Sketch the graph of each of the following. Find the domain, range, intercepts, and asymptotes.

(a) $y = f(x) = 3^x + 1$ **(b)** $y = g(x) = 3^{x+1}$ **(c)** $y = h(x) = \left(\frac{2}{3}\right)^x$

Solution

(a) In order to get the graph of $y = 3^x + 1$, we start with the graph of $y = 3^x$, which is the basic exponential growth graph, and shift it up 1 unit. The graph appears in Figure 5.3. From the graph we can see that the domain is the set of

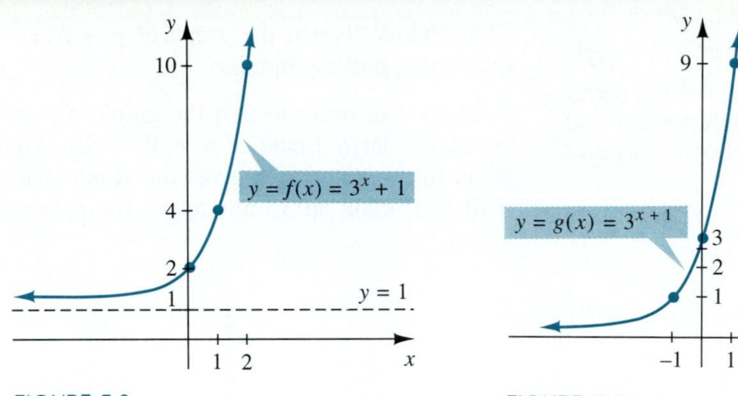

FIGURE 5.3

FIGURE 5.4

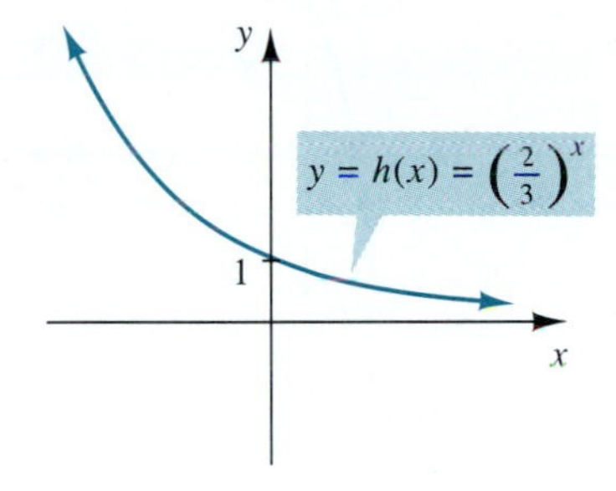

FIGURE 5.5

all real numbers; the range is the set of real numbers greater than 1; the y-intercept is 2; there are no x-intercepts; and the line $y = 1$ is a horizontal asymptote.

(b) In order to get the graph of $y = 3^{x+1}$, we start with the graph of $y = 3^x$, and shift it one unit to the left. The graph appears in Figure 5.4. From the graph we can see that the domain is the set of all real numbers; the range is the set of real numbers greater than 0; there is no x-intercept; by setting $x = 0$, we find that the y-intercept is 3; the x-axis is a horizontal asymptote.

(c) Since $\frac{2}{3}$ is less than 1, the graph of $y = \left(\frac{2}{3}\right)^x$ is the basic exponential decay curve with the same domain, range, intercepts, and asymptote as described in the preceding box. The graph appears in Figure 5.5. ■

The fact that all exponential functions are one-to-one allows us to solve equations in which the variable appears in the exponent, as illustrated in the next example. Such equations are called **exponential equations.**

EXAMPLE 5 Solve for x: $8^x = 16$.

Solution As we just noted, an exponential function is one-to-one. Recall that if a function is one-to-one, it means that we cannot have the same y-value for two different x-values. Symbolically, we may write that, for a one-to-one function, $f(x_1) = f(x_2)$ implies that $x_1 = x_2$. In the case of the exponential function, this means, for example, that $2^a = 2^b \Rightarrow a = b$. Thus if we can rewrite the given equation so that both sides are expressed in terms of the same base, we should be able to solve the equation. We proceed as follows.

$8^x = 16$ *We can rewrite both sides using base 2.*

$(2^3)^x = 2^4$

$2^{3x} = 2^4$ *Since the exponential function is one-to-one, this implies*

$3x = 4$ *and so* $\boxed{x = \frac{4}{3}}$. ■

We will discuss more general exponential equations in Section 5.4.

Keep in mind that $9^{-x} = \left(\frac{1}{9}\right)^x$.

Note that -9^x is not the same as $(-9)^x$, which is not defined because the base is negative.

EXAMPLE 6 Sketch the graph of $y = F(x) = -9^{-x} + 3$. Find the domain, range, intercepts, and asymptotes.

Solution In order to find the graph of $y = -9^{-x} + 3$, we start with the basic exponential decay graph of $y = 9^{-x}$. See Figure 5.6(a). We then reflect this graph about the x-axis, which gives the graph of $y = -9^{-x}$. See Figure 5.6(b). Finally we shift this graph up 3 units to get the required graph of $y = -9^{-x} + 3$. See Figure 5.6(c).

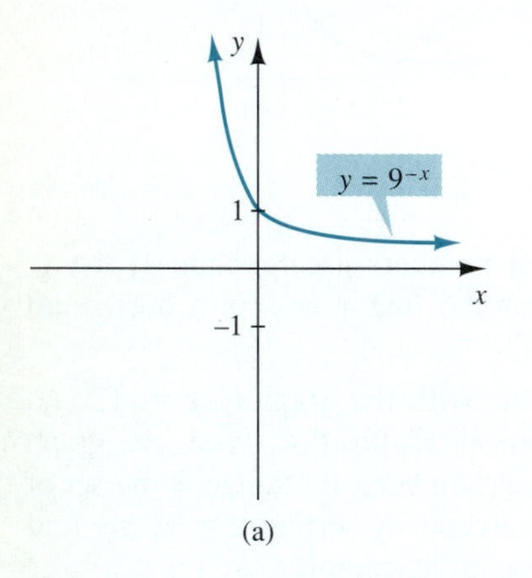

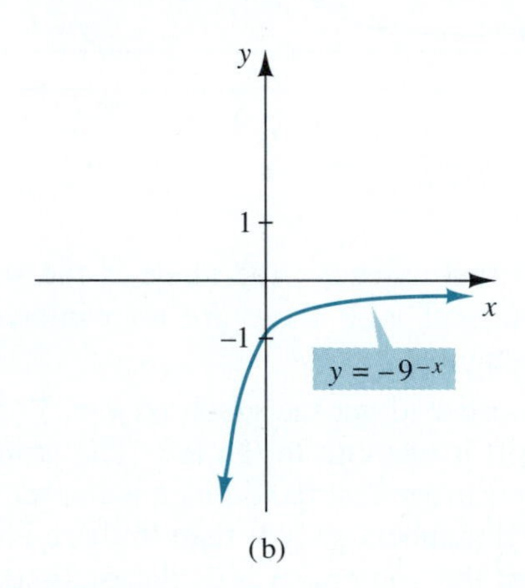

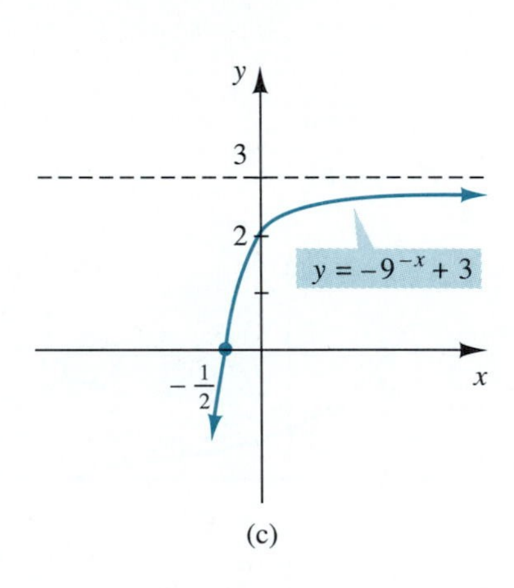

FIGURE 5.6 The steps leading to the graph of $y = -9^{-x} + 3$

From this graph we can see the following for $y = -9^{-x} + 3$:

1. The domain is the set of all real numbers.
2. The range is the set of real numbers less than 3.
3. The line $y = 3$ is a horizontal asymptote.
4. By setting $x = 0$, we find that the y-intercept is 2.

The x-intercept is not immediately apparent from the graph, so we substitute $y = 0$ and try to solve the resulting equation.

$$y = -9^{-x} + 3 \qquad \textit{Substitute } y = 0.$$

$$0 = -9^{-x} + 3$$

Again we solve this exponential equation by expressing both sides of the equation in terms of the *same* base.

$$9^{-x} = 3 \qquad \textit{We express both sides in terms of base 3.}$$

$$(3^2)^{-x} = 3$$

$$3^{-2x} = 3^1 \qquad \textit{Since the exponential function is one-to-one, } 3^a = 3^b \Rightarrow a = b\textit{; therefore}$$

$$-2x = 1$$

$$x = -\frac{1}{2} \qquad \textit{Thus the x-intercept is } -\frac{1}{2}.$$

■

We have already described how the base determines exponential growth if $b > 1$ and exponential decay if $0 < b < 1$. However, the size of b will also have a bearing on the relative shape of the graph of $y = b^x$. For $b > 1$, as b increases, the exponential growth graph increases more rapidly, whereas for $0 < b < 1$, as b gets closer to 0, the exponential decay graph decreases more rapidly. Figure 5.7 shows a number of graphs of exponential functions for a variety of values of b.

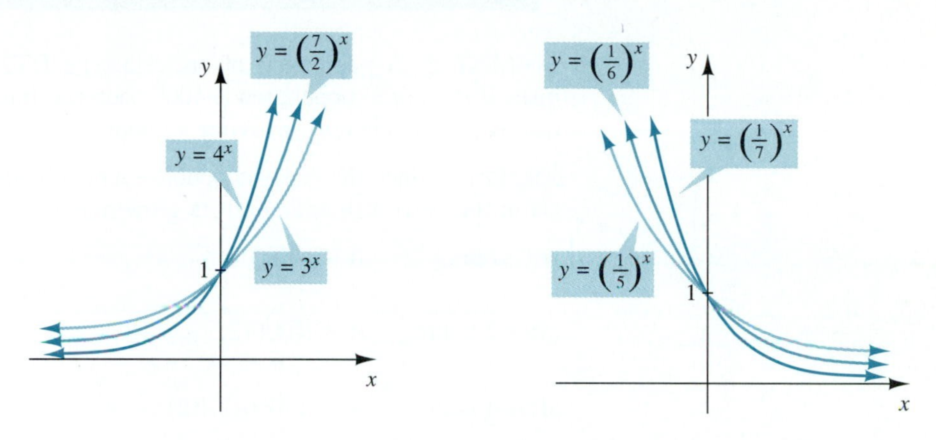

FIGURE 5.7 The graphs of $y = b^x$ for a variety of values of b

Knowing the behavior of exponential functions—in particular, the fact that the exponential function is never equal to 0—allows us to determine the zeros of related functions, as in the next example.

What does a zero of $f(x)$ mean?

EXAMPLE 7 Find the zeros of $f(x) = 2(x^3)5^x - (8x)5^x$.

Solution We solve $f(x) = 0$ by factoring $f(x)$.

$$2(x^3)5^x - (8x)5^x = 0 \qquad \textit{There is a common factor of } 2x5^x.$$

$$2x5^x(x^2 - 4) = 0$$

Is there a value of x for which $5^x = 0$?

We can ignore the factor of 5^x since it is never equal to 0. Alternatively, we may divide both sides of the equation by 5^x without losing any solutions. Therefore, we have

$$x = 0 \quad \text{or} \quad x^2 - 4 = 0$$

$$x = 0 \quad \text{or} \quad x = \pm 2$$

Therefore, the zeros of $f(x)$ are 0, 2, and -2. ■

We have seen that we can solve an equation such as $3^x = 9$ by expressing both sides of the equation as powers of the same base. However, an equation such as $3^x = 9x^2$ does not yield to this type of solution. In fact, we have no algebraic technique that will solve this equation. We noted in Chapter 4 that the solutions to the equation $3^x - 9x^2 = 0$ are the x-intercepts of the graph of $y = 3^x - 9x^2$. Although this equation is difficult to graph by hand, a graphics calculator can sketch such a graph and allows us to approximate the x-intercepts of the graph and hence the solutions to the given equation.

GRAFFIX

Use a graphics calculator or computer to sketch the graph of $y = 3^x - 9x^2$, estimate the x-intercepts, and in so doing obtain approximate solutions to the equation $3^x - 9x^2 = 0$.

EXAMPLE 8 A population of bacteria type DT3 doubles every 3 hours in a petri dish. If the initial population is 4000 bacteria, how many bacteria will be in the petri dish **(a)** after 6 hours? **(b)** After 12 hours? **(c)** After t hours? **(d)** After 17 hours?

Solution Since the bacteria double every 3 hours, we let A = the number of bacteria in the petri dish and chart its growth as follows:

at 0 hours, $A = 4000$ — *This is the initial number of bacteria.*

after 3 hours, $A = 4000(2) = 8{,}000$ — *The number of bacteria has doubled once.*

after 6 hours, $A = [4000(2)](2) = 4000(2)^2 = 16{,}000$ — *The number of bacteria has doubled twice.*

after 9 hours, $A = [4000(2)^2](2) = 4000(2)^3 = 32{,}000$ — *The number of bacteria has doubled three times.*

after 12 hours, $A = [4000(2)^3](2) = 4000(2)^4 = 64{,}000$ — *The number of bacteria has doubled four times.*

(a) By the chart we can see that the number of bacteria at the end of 6 hours is $\boxed{16{,}000}$.

(b) At the end of 12 hours it is $\boxed{64{,}000}$.

(c) Notice the pattern: When the 4000 bacteria have doubled 5 times, the number of bacteria present is $4000(2)^5$.

Since the number of bacteria doubles every 3 hours, in t hours the number will double $\frac{t}{3}$ times. Hence, if we continue the pattern, we see that the number of bacteria at the end of time t is:

$$\boxed{A = 4000(2)^{t/3}} \quad \text{where } t \text{ is the time in hours}$$

(d) To find the number of bacteria in the petri dish at the end of 17 hours, we need to use the formula found in part (c).

$A = 4000(2)^{t/3}$ — *Since $t = 17$, we get*

$= 4000(2)^{17/3}$ — *We first evaluate $2^{17/3}$ on a calculator to get*

$= 4000(50.796834) = 203{,}187.33$ — *Which we round to*

$= \boxed{203{,}187 \text{ bacteria}}$ ■

We noted that the exponential function is one-to-one, and we know that a one-to-one function has an inverse function. The inverse of the exponential function is the subject of the next section.

EXERCISES 5.1

In Exercises 1–10, solve the given exponential equation.

1. $2^{x-1} = 8$ **2.** $2^{x^2-1} = 8$

3. $3^{2x} = 243$ **4.** $25^{a+3} = 5$

5. $9^t = 27$ **6.** $4^x = \frac{1}{2}$

7. $\left(\frac{1}{2}\right)^{x+2} = 16$ **8.** $\left(\frac{1}{3}\right)^{s+1} = 9$

9. $8^x = \sqrt{2}$ **10.** $16^{3a-2} = \frac{1}{4}$

In Exercises 11–16, find the domain of the given function.

11. $f(x) = \frac{1}{6^x}$ **12.** $g(x) = \sqrt{3^x + 1}$

13. $h(x) = \sqrt{2^x - 8}$ **14.** $f(x) = \frac{5}{4^x - 1}$

15. $g(x) = \frac{1}{2^{3x} - 2}$ **16.** $h(x) = \frac{1}{2^{3x-2}}$

In Exercises 17–30, sketch the graph of the given function. Identify the domain, range, intercept(s), and asymptote.

17. $y = 5^{-x}$ **18.** $y = -5^x$

19. $y = 4^x - 1$ **20.** $y = 4^{x-1}$

21. $y = 9 - 3^x$ **22.** $y = 3^{-x} + 2$

23. $y = 2^{x-3} - 8$ **24.** $y = 3^{x+2} - 9$

25. $y = -3^{-x} + 1$ **26.** $y = \left(\frac{1}{2}\right)^x - 4$

27. $y = 2^{x^2}$ **28.** $y = 2^{|x|}$

29. $y = \frac{1}{2^{x-1}}$ **30.** $y = \frac{5}{3^{x+2}}$

In Exercises 31–36, let $f(x) = 5^x$, $g(x) = x^2 - 6x + 2$, $h(x) = \frac{1}{x+1}$, and $t(x) = \sqrt{x}$. Find each of the following.

31. $(f \circ g)(x)$ **32.** $(g \circ f)(x)$

33. $(f \circ t)(x)$ **34.** $(t \circ f)(x)$

35. $(h \circ f)(x)$ **36.** $(f \circ h)(x)$

In Exercises 37–42, find all the real zeros of the given function.

37. $f(x) = x4^x$ **38.** $g(x) = \frac{2^x - 8}{x}$

39. $h(x) = 9(2^x) - x^2 2^x$

40. $F(x) = 6(3^x) - 5x3^x + x^2 3^x$

41. $f(x) = \frac{1 - 3^x}{x^2}$ **42.** $g(x) = 5(2^x) - 10$

43. Solve the equation $3^x + 3^{-x} = 2$. HINT: Let $u = 3^x$, making this a quadratic equation in u.

44. Solve the equation $2^x - 4(2^{-x}) = 3$. HINT: Let $u = 2^x$, making this a quadratic equation in u.

45. Let $f(x) = 2^x$. Show that $f(x + 3) = 8f(x)$.

46. Let $g(x) = 5^x$. Show that $g(x - 2) = \frac{1}{25}g(x)$.

47. Let $f(x) = 3^x$. Show that $\frac{f(x + 2) - f(x)}{2} = 4(3^x)$.

48. Let $f(x) = 4^x$. Show that

$$\frac{f(x + h) - f(x)}{h} = 4^x\left(\frac{4^h - 1}{h}\right)$$

49. On the same coordinate system, sketch the graphs of $y = 2^x$, $y = \left(\frac{5}{2}\right)^x$, and $y = 3^x$.

50. On the same coordinate system, sketch the graphs of $y = \left(\frac{1}{2}\right)^x$, $y = \left(\frac{2}{5}\right)^x$, and $y = \left(\frac{1}{3}\right)^x$.

The following exercises deal with applications of exponential growth and decay. You may find a calculator helpful for many of the exercises. Round your answers to two decimal places.

51. The population of Rabbitville doubles every 2 months. If the initial population is 4, how many people will be in Rabbitville at the end of 6 months? At the end of t months? At the end of 2 years?

52. Since 1960, the population of Lewistown has doubled every 10 years. If this trend continues and if there were 800 people living in Lewistown in 1972, how many people would you expect to be living there in 1995?

53. The book value, B, of a used car t years after it was new is often computed using the formula

$$B = N\left(\frac{S}{N}\right)^{t/10} \quad \text{for } 0 \le t \le 10$$

where N is the value of the car when it was new and S is the salvage value of the car after 10 years. Suppose that a new car costs \$12,000 and will have a salvage value of \$600 after 10 years.

(a) Sketch the graph of B as a function of t.

(b) What will the book value of the car be when it is 3 years old?

54. Radioactive argon-39 has a half-life of 4 minutes. This means that every 4 minutes one-half of the amount of argon-39 present changes into another substance, due to radioactive decay.

(a) If we start with A_0 milligrams of argon-39, explain why the amount A remaining after t minutes is given by the formula

$$A = A_0(2^{-t/4})$$

(b) If $A_0 = 50$ mg, sketch the graph of

$$A = 50(2^{-t/4})$$

Note the values of A for $t = 0, 1, 2, 4, 8, 16$, and 20 minutes.

(c) If $A_0 = 50$ mg, how much argon-39 will be left after 1 hour?

55. The demand function for a certain product is given by the function

$$p = 600 - 0.4(3^{0.005x})$$

where p is the price in dollars that can be charged when x units are in demand. Find the price p that can be charged when **(a)** $x = 500$ and **(b)** $x = 900$.

56. A company estimates that the annual profit, P (in dollars), from the sales of a particular item x years after the product is first introduced can be computed as

$$P = P(x) = 120{,}000 - 80{,}000\left(\frac{1}{3}\right)^x$$

(a) What will the annual profit be after 3 years? After 6 years?

(b) Sketch the graph of $y = P(x)$ for $x \geq 0$.

(c) What is the maximum profit the company can expect from this item? Is this maximum profit ever actually achieved?

57. The atmospheric pressure, P, at a height x feet above sea level can be approximated by the function

$$P = P(x) = (14.69)3^{-0.000035x}$$

where the pressure is given in atmospheres. (One atmosphere (1 atm) is approximately equal to 14.69 pounds of pressure per square inch.)

(a) What is the atmospheric pressure at sea level?

(b) What is the atmospheric pressure at the top of Mount Everest, which is 29,028 feet above sea level?

(c) What is the atmospheric pressure at the Dead Sea, which is 1290 feet *below* sea level?

58. A certain book states that the atmospheric pressure x feet above sea level can be computed using the formula

$$P = P(x) = 2^{3.8777 - 0.0000555x}$$

Use this formula to answer the questions in Exercise 57. How do the answers compare with those obtained using the formula in Exercise 57?

59. The population, P, of a town is growing according to the formula

$$P = P(t) = P_0 2^{0.05t}$$

where P_0 is the population in a certain year and P is the population after t years have elapsed. Suppose the town's population in 1990 is 12,000.

(a) Find the population 4 years later and 5 years later.

(b) Find the population 19 years later and 20 years later.

(c) Find the ratio of the two populations found in part (a) and the two populations in part (b).

(d) What is the relationship between the population t years later and $t + 1$ years later?

60. The *E. coli* type of bacteria grows according to the formula

$$N = N_0 2^{t/25}$$

where N_0 is the number of bacteria present initially and N is the number of bacteria present after t minutes.

(a) If an *E. coli* colony begins with a population of 1000, how long will it take for the population to double to 2000?

(b) How long will it take the population to double from 2000 to 4000?

(c) Based upon your answers to parts (a) and (b), what might you conclude about the time it takes for the population of an *E. coli* bacterial colony to double?

(d) Show that the doubling time for *E. coli* bacteria is 25 minutes.

61. The fruit fly *Drosophila* is often used for genetic studies. A typical fruit fly population will double in 2.5 days.

(a) If we start with an initial population of N_0 fruit flies, explain why the number of fruit flies, N, after t days is given by the formula

$$N = N_0 2^{t/2.5}$$

(b) If we start with an initial population of 10 male and 10 female flies, how many flies will there be after 1 week?

62. The use of certain insecticides has been discontinued because of their long-lasting activity in the environment. Suppose that a certain insecticide has a half-life of 16 years; that is, it takes 16 years for one-half of a given amount of the insecticide to become inactive and harmless.

(a) If we start with an initial amount of insecticide, A_0, explain why the amount of insecticide, A, still active after t years is given by the formula

$$A = A_0 2^{-t/16}$$

(b) If a farmer uses 100 pounds of this insecticide on his crop, how much will still be active after 10 years? After 100 years?

63. Radioactive isotopes that have relatively short half-lives are often used in medical imaging procedures. Suppose that a certain radio isotope has a half-life of 4 hours.

(a) Write an equation that gives the amount, A, of this isotope still radioactive t hours after an initial amount A_0 is injected into the body.

(b) If 10 mg are injected into the body, how much would still be radioactive after 2 hours? After 24 hours?

64. Suppose that the charge remaining in a battery is decreasing exponentially according to the formula

$$C = C(T) = C_0(0.65)^T$$

where C is the charge in coulombs remaining T days after the battery receives an initial charge of C_0. If a battery has a charge of 3.5×10^{-5} coulombs remaining after 14 days, find the initial charge.

65. According to Newton's law of cooling, the rate at which an object cools is directly proportional to the difference in temperature between the object and the surrounding environment. A metal rod, whose initial temperature is 100°C, cools as it is placed in surrounding air, which is kept at a temperature of 15°C. Using calculus, we find that in such a situation the temperature, T, of the rod will decrease exponentially according to the equation

$$T = 15 + 85(3^{-m})$$

where m is the number of minutes the rod has been exposed to the cooler air. How long will it take for the rod to cool down to a temperature of 40°C?

66. The biological half-life of a medication is the amount of time it takes an initial amount of medication to lose half its effectiveness. Suppose that a certain medication has a half-life of 9 hours in the body of a nonsmoker but a half-life of only 5 hours in the body of a smoker. If equal doses of this drug are administered to a nonsmoker and a smoker, compare the amounts of drug remaining in each person after 24 hours.

67. The rate at which certain electrical items fail can often be described using exponential functions. For example, suppose that a large corporate office has 2400 light bulbs and that the maintenance department replaces all the old bulbs at the same time, with new bulbs that have an average life of 800 hours. In such a situation, the number of bulbs, N, that should be expected to burn out after t hours can be given by the equation

$$N = N(t) = 2400(1 - 0.998^t)$$

(a) Using appropriate units along the axes, sketch the graph of $y = N(t)$.

(b) How many light bulbs would be expected to have burned out after 400 hours? 600 hours? 800 hours?

(c) Interpret the results of part (b) in terms of the *percentage* of bulbs that have burned out after 400, 600, and 800 hours.

(d) Explain why the equation for N suggests that it might be more economical to replace all the light bulbs at once rather than as they burn out.

68. Suppose that 18 wolves (9 male and 9 female) are introduced into an uninhabited area. Further suppose that the wolf population increases by 20% per year for the next 15 years, at which time the wolf population becomes so large that it overwhelms the environment by destroying important elements of the food chain. As a result, the wolf population decreases by 15% per year for the next 25 years.

(a) Using a split function, express the wolf population, P, as a function of the time, t, in years over this 40-year period.

(b) Estimate the largest and smallest wolf populations and in which years they occur.

69. Medical technicians often give intravenous infusions of a particular nutrient or medication. The concentration, C, of the nutrient or medication in the blood t minutes after the infusion begins might be given by a formula such as

$$C = C(t) = F + (I - F)3^{-0.01t}$$

(a) Explain the significance of I. HINT: Try finding the concentration when $t = 0$.

(b) Explain the significance of F. HINT: Try finding the concentration when $t = 100$, 200, etc.

70. We have seen previously that if a heavy object (such as a rock) is dropped from a high place and we neglect air resistance, then its velocity, v, after t seconds is given by $v = -32t$ feet per second. On the other hand, if a sky diver jumps out of a plane with arms and legs spread out, then air resistance has a significant effect on the velocity. For example, after jumping from a plane at an altitude of 10,000 feet, the sky diver's velocity, v, in feet per second, would be given by an equation such as

$$v = v(t) = -220(1 - 0.9^t)$$

where t is the number of seconds after the jump but before the parachute is opened.

(a) Sketch the graph of $y = v(t)$.

(b) Find the sky diver's velocity after 4 seconds, and compare it to the velocity of a rock 4 seconds after it is dropped from an altitude of 10,000 feet.

(c) As t increases, what happens to the sky diver's velocity? Find v after 10 sec, 20 sec, 30 sec, etc. (This "limiting" velocity is often called the *terminal velocity*.)

(d) What is this terminal velocity? (You may be more impressed if you convert to miles per hour. 88 ft/s = 60 mi/h.

71. Oceanographers know that the amount of light that penetrates to x meters below the surface decays exponentially as a function of x. For example, at a certain location if the light intensity at the surface is 12 lumens (lm), then the intensity, I, of the light x meters below the surface can be given by

$$I = 12(0.45)^x$$

Find the intensity of the light at a depth of 5 meters.

72. Suppose that if 20 grams of sugar are added to a quantity of water, then the amount of sugar, S, that remains undissolved after t minutes is given by

$$S = S(t) = 20\left(\frac{5}{7}\right)^t$$

Find the amount of sugar that remains undissolved after 5 minutes.

QUESTIONS FOR THOUGHT

73. Earlier in this section we noted that the exponential function is one-to-one, and so we know that the exponential function has an inverse function. In fact, we have used this idea to solve some exponential equations.

(a) Solve for x: $2^x = 16$. Explain how the idea of the one-to-oneness of the exponential function is used in the solution.

(b) Is there a solution to $2^x = 15$? How would you go about finding it?

74. Try to find the inverse function of $y = 2^x$. Are there any difficulties in following our familiar outline for finding an inverse function: Interchange x and y and then solve for y?

75. In the definition of the exponential function, we restricted the base b to $b > 0$ and $b \neq 1$.

(a) If we allow $b = 1$, what happens to the function $y = b^x$?

(b) If we allow b to be negative, what happens to the domain of $y = b^x$? Think about what happens when we try to find the y-values corresponding to $x = \frac{1}{2}$, $\frac{3}{4}$, etc.

5.2 Logarithmic Functions

In the last section we noted that the exponential function $y = f(x) = b^x$ (where $b > 0$ and $b \neq 1$) is one-to-one. From our previous work with functions, we know this means that the exponential function has an inverse function.

Let's review the process for finding an inverse function by comparing the process for the polynomial function $y = x^3$ and the exponential function $y = 3^x$. Keep in mind that throughout this text we let x be the *independent* variable and y be the *dependent* variable, and so whenever possible we want a function solved explicitly for y.

To find the inverse of $y = x^3$:

$y = x^3$ *Interchange x and y.*

$x = y^3$ *Solve for y.*

$y = \sqrt[3]{x}$

To find the inverse of $y = 3^x$:

$y = 3^x$ *Interchange x and y.*

$x = 3^y$ *Solve for y.*

$y = ?$

There is no algebraic procedure we can use to solve $x = 3^y$ for y. This comparison brings into sharp focus the difference between an algebraic function, such as $y = x^3$, and a transcendental function, such as $y = 3^x$.

In fact, had we studied functions and their inverses before learning about radicals, we would have had to invent radical notation so that we could express the inverse of $y = x^3$ explicitly in the form $y = \sqrt[3]{x}$. In other words, $y^3 = x$ and $y = \sqrt[3]{x}$ both mean exactly the same thing: y is the number whose cube is x. The only difference is that $y = \sqrt[3]{x}$ is solved explicitly for y.

Similarly, if we want to express $x = 3^y$ explicitly as a function of x, we need to invent a special notation for this. The key idea is to take the equation $x = 3^y$ and express it verbally.

$x = 3^y$ means y is the exponent to which 3 must be raised to yield x

We introduce the following notation, which expresses this same idea in a much more compact form.

DEFINITION For $b > 0$ and $b \neq 1$, we write $y = \log_b x$ to mean y is the exponent to which b must be raised to yield x. In other words,

$$x = b^y \Leftrightarrow y = \log_b x$$

We read $y = \log_b x$ as "y equals log base b of x" or "y equals the logarithm of x to the base b."

> *REMEMBER:* $y = \log_b x$ is just an alternative way of writing $x = b^y$.

When an expression is written in the form $x = b^y$, it is said to be in **exponential form.** When an expression is written in the form $y = \log_b x$, it is said to be in **logarithmic form.** All our work with logarithms will be made easier if we keep in mind the fact that, according to the definition, *a logarithm is simply an exponent*.

Table 5.2 illustrates the equivalence of the exponential and logarithmic forms of a number of statements.

Be sure you understand the equivalence of the two forms on each line in Table 5.2.

TABLE 5.2

Exponential Form	Logarithmic Form
$4^2 = 16$	$\log_4 16 = 2$
$2^4 = 16$	$\log_2 16 = 4$
$5^{-3} = \frac{1}{125}$	$\log_5 \frac{1}{125} = -3$
$6^{1/2} = \sqrt{6}$	$\log_6 \sqrt{6} = \frac{1}{2}$
$7^0 = 1$	$\log_7 1 = 0$
$b^m = u$	$\log_b u = m$

EXAMPLE 1 Write each of the following in exponential form.

(a) $\log_3 \frac{1}{9} = -2$ **(b)** $\log_{16} 2 = \frac{1}{4}$

Solution We use the fact that $y = \log_b x$ is equivalent to $b^y = x$.

(a) $\log_3 \frac{1}{9} = -2$ means $\boxed{3^{-2} = \frac{1}{9}}$.

(b) $\log_{16} 2 = \frac{1}{4}$ means $\boxed{16^{1/4} = 2}$.

■

EXAMPLE 2 Write each of the following in logarithmic form.

(a) $10^{-3} = 0.001$ **(b)** $27^{2/3} = 9$

Solution We use the fact that $b^y = x$ is equivalent to $y = \log_b x$.

(a) $10^{-3} = 0.001$ means $\boxed{\log_{10} 0.001 = -3}$.

(b) $27^{2/3} = 9$ means $\boxed{\log_{27} 9 = \frac{2}{3}}$.

■

EXAMPLE 3 Evaluate each of the following.

(a) $\log_3 81$ **(b)** $\log_8 \frac{1}{64}$ **(c)** $\log_9 3$ **(d)** $\log_{16} 8$ **(e)** $\log_2(-2)$

Solution

(a) When we want to find $\sqrt[3]{243}$, we restate the question using the inverse operation: To find $\sqrt[3]{243}$ we ask "What number raised to the 3rd power yields 243?" In the same way, when we want to find $\log_3 81$ we may restate this as "What exponent of 3 will yield an answer of 81?" Once we have restated $\log_3 81$ in this way, we recognize that the answer is 4, because $3^4 = 81$. However, the answer is not always quite so obvious, so we offer a more formal approach, which hinges on an idea we discussed in the last section. We let $t = \log_3 81$ and then rewrite this equation in exponential form, obtaining $3^t = 81$. Now if we can express both sides in terms of the same base, we can solve the resulting exponential equation, as follows:

Let $t = \log_3 81$ *Translate into exponential form.*

$3^t = 81$ *Express both sides in terms of the same base.*

$3^t = 3^4$ *Since the exponential function is one-to-one*

$t = 4$

Therefore, $\boxed{\log_3 81 = 4}$.

(b) We apply the same procedure as in part (a).

$$\text{Let } t = \log_8 \frac{1}{64} \qquad \textit{Rewrite in exponential form.}$$

$$8^t = \frac{1}{64} \qquad \textit{Express both sides in terms of the same base.}$$

$$8^t = \frac{1}{8^2} = 8^{-2} \qquad \textit{Therefore,}$$

$$t = -2$$

Hence $\boxed{\log_8 \frac{1}{64} = -2}$.

(c) Let $t = \log_9 3$.

$$9^t = 3$$

$$(3^2)^t = 3^1$$

$$3^{2t} = 3^1$$

$$2t = 1 \quad \text{and so} \quad t = \frac{1}{2}$$

Therefore, $\boxed{\log_9 3 = \frac{1}{2}}$.

(d) Let $t = \log_{16} 8$.

$$16^t = 8 \qquad \textit{We express both sides in terms of the base 2.}$$

$$(2^4)^t = 2^3$$

$$2^{4t} = 2^3 \qquad \textit{Therefore,}$$

$$4t = 3 \quad \text{and so} \quad t = \frac{3}{4}$$

Therefore, $\boxed{\log_{16} 8 = \frac{3}{4}}$.

Remember that because an exponential function and its corresponding logarithmic function are inverses, the domain of the logarithmic function is the same as the range of the exponential function.

(e) Let $t = \log_2(-2)$, which means

$$2^t = -2$$

But no value of t can make 2^t negative; $2^t > 0$ for all values of t. Hence $\log_2(-2)$ is undefined. ■

The result of Example 3(e) illustrates that since we are restricting the base of the exponential and logarithmic functions to positive numbers, *the log of a nonpositive number is undefined.*

As was pointed out at the beginning of this section, logarithm notation was invented to express the inverse of the exponential function. Thus $\log_b x$ is a function of x. We usually write $f(x) = \log_b x$ rather than writing $f(x) = \log_b(x)$ and use parentheses when needed to clarify the input to the log function. In this context, the word *argument* is often used in place of the word *input*. For example, the function $f(x) = \log_5(4 - x)$, the parentheses indicate that $4 - x$ is the argument of the logarithmic function. Thus we may say that the argument of a logarithmic function must be positive.

If $f(x) = \log_5(4 - x)$, then $f(a) = \log_5(4 - a)$. a is the input to $f(x)$; $4 - a$ is the argument of the logarithmic function.

For example,

If $f(x) = \log_5(4 - x)$, then $f(-1) = \log_5[4 - (-1)] = \log_5 5 = 1$, whereas if $f(x) = 4 - \log_5 x$, then $f(-1) = 4 - \log_5(-1)$, which is undefined.

EXAMPLE 4 Find $\log_5 5^7$.

Solution Again, let

$$t = \log_5 5^7 \qquad \textit{We rewrite this in exponential form.}$$

$$5^t = 5^7 \Rightarrow t = 7$$

Therefore, $\boxed{\log_5 5^7 = 7}$. Alternatively, if we simply understand that $\log_5 5^7$ means "the exponent to which 5 must be raised to yield 5^7," then the answer of 7 should be immediately apparent. ■

Explain in words why $\log_b b^n = n$.

The following example illustrates several *logarithmic equations*. As in the last two examples, we will often find it helpful to rewrite a logarithmic equation in exponential form.

EXAMPLE 5 Solve each of the following equations for t.

(a) $\log_5 t = 4$ **(b)** $\log_8 \dfrac{1}{2} = t$ **(c)** $\log_t 216 = 3$

Solution

(a) We rewrite $\log_5 t = 4$ in exponential form to obtain $t = 5^4 = \boxed{625}$.

(b) Again we rewrite the given logarithmic equation in exponential form.

$$\log_8 \frac{1}{2} = t \Rightarrow 8^t = \frac{1}{2}$$

$$(2^3)^t = 2^{-1}$$

$$3t = -1 \Rightarrow \boxed{t = -\frac{1}{3}}$$

(c) $\log_t 216 = 3 \Rightarrow t^3 = 216 \Rightarrow t = \sqrt[3]{216} \Rightarrow \boxed{t = 6}$. ■

Since $b^1 = b$ and $b^0 = 1$, we make note of the following useful logarithmic statements (which are true for all $b > 0$).

$$\boxed{\log_b b = 1 \quad \text{and} \quad \log_b 1 = 0}$$

GRAFFIX

Use a graphics calculator or computer to graph the functions $y = 10^x$, $y = \log x$, and $y = x$ on the same set of coordinate axes. Keeping in mind that **log** on your calculator means $\log_{10}$, what do the three graphs suggest?

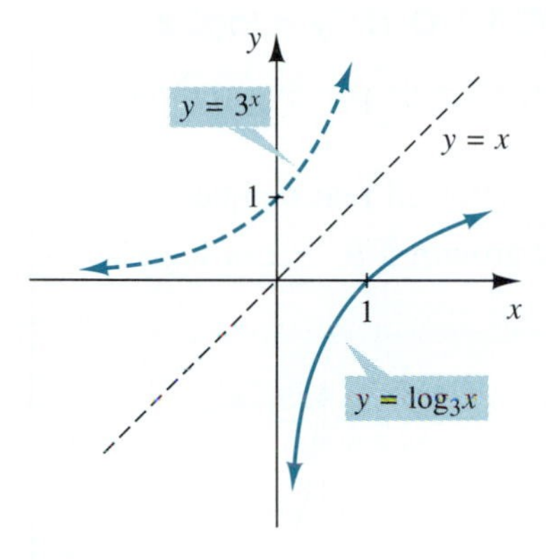

FIGURE 5.8

Acknowledging that the logarithmic and exponential functions are inverses, we can derive a great deal of information about the logarithmic function and its graph from the exponential function and its graph.

EXAMPLE 6 Sketch the graph of each of the following equations. Find the domain and range of each. **(a)** $y = \log_3 x$ **(b)** $y = \log_{1/2} x$

Solution

(a) Since $y = \log_3 x$ is the inverse of $y = 3^x$, we can obtain the graph of $y = \log_3 x$ by reflecting the graph of $y = 3^x$ about the line $y = x$. The graph appears in Figure 5.8.

(b) To get the graph of $y = \log_{1/2} x$, we reflect the graph of $y = \left(\frac{1}{2}\right)^x$ about the line $y = x$. The graph appears in Figure 5.9.

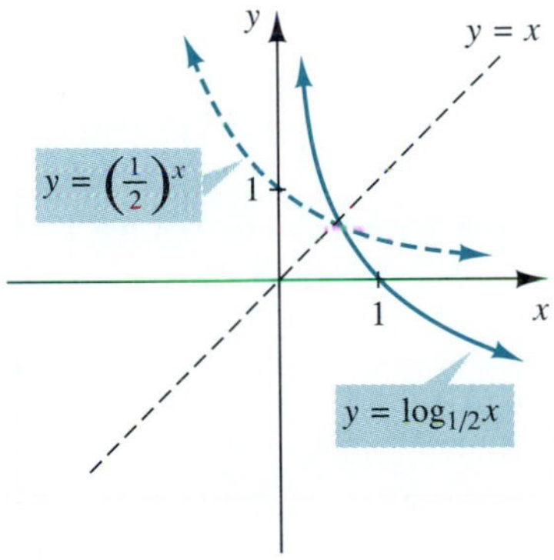

FIGURE 5.9

We take note of the features of these two graphs, which are the result of the logarithmic function being the inverse of the exponential function.

1. Since the graphs of $y = b^x$ all pass through the point $(0, 1)$, the graphs of $y = \log_b x$ all pass through $(1, 0)$.
2. When the graph of an exponential function has the negative x-axis as a horizontal asymptote, the graph of its inverse logarithmic function has the negative y-axis as a vertical asymptote. See Figure 5.8.

   When the graph of an exponential function has the positive x-axis as a horizontal asymptote, the graph of its inverse logarithmic function has the positive y-axis as a vertical asymptote. See Figure 5.9.
3. From its graph we can see that the domain of $y = \log_3 x$ is the set of positive real numbers and its range is the set of all real numbers (which are, respectively, the range and domain of the exponential function). A similar statement is true for $y = \log_{1/2} x$.

These statements agree with the fact that inverse functions interchange their domain and range. ■

We mentioned that the exponential function increases or decreases very rapidly; that is, a small change in x can cause a large change in y. For example, for the function $y = f(x) = 2^x$, $f(9) = 512$ and $f(10) = 1024$. Thus a 1-unit change in x causes a 512-unit change in y. If we consider the function $y = g(x) = \log_2 x$, we have $g(512) = 9$ and $g(1024) = 10$. For this logarithmic function, a 512-unit change in x causes a 1-unit change in y. It therefore follows that the logarithmic function increases or decreases very slowly. A large change in x causes a small change in y. For $b > 1$, the graph of $y = \log_b x$ is always rising, but very slowly. This is called *logarithmic growth*. If $0 < b < 1$, the graph exhibits *logarithmic decay*.

The following box summarizes some important information for the logarithmic function.

Notice that the domain of $f(x) = \log_b x$ is $(0, \infty)$, which agrees with our earlier comment that the logarithm of a nonpositive number is undefined.

THE LOGARITHMIC FUNCTION $y = \log_b x$

1. The domain of the logarithmic function is the set of positive real numbers.
2. The range of the logarithmic function is the set of all real numbers.
3. The graph of $y = \log_b x$ exhibits logarithmic growth if $b > 1$ or logarithmic decay if $0 < b < 1$.

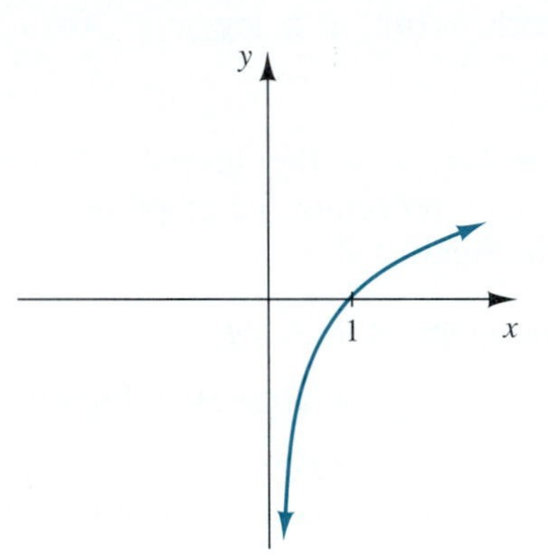

The graph of $y = \log_b x$ for $b > 1$
Logarithmic growth

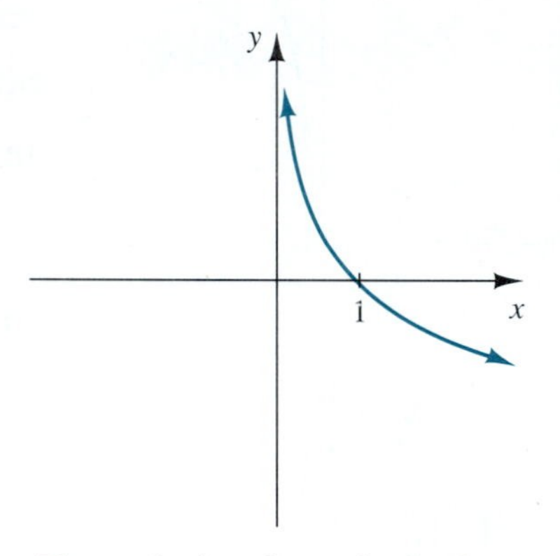

The graph of $y = \log_b x$ for $0 < b > 1$
Logarithmic decay

4. The x-intercept is 1. There is no y-intercept.
5. The y-axis is a vertical asymptote.

These logarithmic graphs should be added to our catalog of basic graphs.

EXAMPLE 7 Sketch the graphs of
(a) $y = \log_{10} x$ **(b)** $y = \log_{1/10} x$ **(c)** $y = -\log_{10} x$

Solution

(a) The graph of $y = \log_{10} x$ is the basic logarithmic graph for $b > 1$. The graph appears in Figure 5.10.

(b) The graph of $y = \log_{1/10} x$ is the basic logarithmic graph for $0 < b < 1$. The graph appears in Figure 5.11.

(c) We can obtain the graph of $y = -\log_{10} x$ by reflecting the graph of $y = \log_{10} x$ in Figure 5.10 about the x-axis. However, the graph we obtain from this

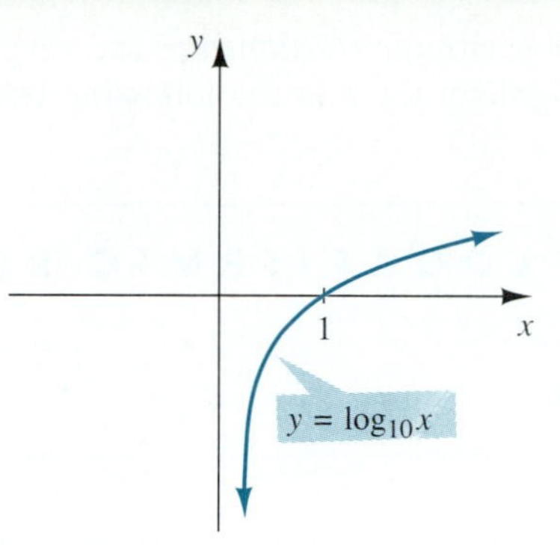

FIGURE 5.10

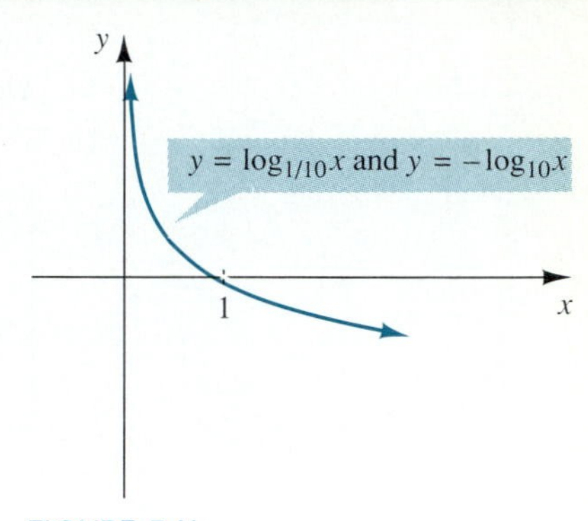

FIGURE 5.11

reflection is identical to the graph in Figure 5.11. As a result we know that $\log_{1/10} x = -\log_{10} x$. This idea is brought up again in the exercises. ■

EXAMPLE 8 Given $f(x) = \log_5 x$, find

(a) $f(25)$ **(b)** $f\left(\frac{1}{25}\right)$ **(c)** $f(0)$ **(d)** $f(-125)$

Solution Since $f(x) = \log_5 x$,

(a) $f(25) = \log_5 25 = \boxed{2}$ (since $5^2 = 25$)

(b) $f\left(\frac{1}{25}\right) = \log_5\left(\frac{1}{25}\right) = \boxed{-2}$ $\left(\text{since } 5^{-2} = \frac{1}{25}\right)$

(c) $f(0) = \log_5 0$ is not defined. (What power of 5 will yield 0?) We say that 0 is not in the domain of $f(x)$.

(d) $f(-125) = \log_5(-125)$ is not defined. (What power of 5 will yield -125?) We say that -125 is not in the domain of $f(x)$. ■

EXAMPLE 9 Find the domain of $f(x) = \log_6(2 - 5x)$.

Solution As we have stated previously, we cannot take the logarithm of a nonpositive number. Therefore, finding the domain of $\log_6(2 - 5x)$ can be reformulated as solving the inequality $2 - 5x > 0$, which gives us the domain $\left\{x \mid x < \frac{2}{5}\right\}$. In interval notation the domain is $\left(-\infty, \frac{2}{5}\right)$. ■

Why does the fact that the domain of $\log_6(2 - 5x)$ is $\left(-\infty, \frac{2}{5}\right)$ not contradict the earlier statement that the log of a negative number is undefined?

Once we clearly understand that a logarithm is simply an exponent and that the logarithmic function and exponential functions are inverses of each other, we can derive some very useful relationships.

Let $f(x) = \log_b b^x$ and $g(x) = b^x$. Then f and g are inverse functions, and by the definition of inverse functions we have $f[g(x)] = x$ and $g[f(x)] = x$. In other words, we have

$$f[g(x)] = f(b^x) = \log_b b^x = x$$

$$g[f(x)] = g(\log_b x) = b^{\log_b x} = x$$

These basic logarithmic relationships are very important for much of our future work and so we highlight them in the following box.

BASIC LOGARITHMIC RELATIONSHIPS

1. $\log_b b^x = x$
2. $b^{\log_b x} = x$

Although these two relationships may seem a bit obscure, they become more understandable if we analyze verbally what they are saying.

For the first, $\log_b b^x$ means the exponent to which b must be raised to yield b^x and so the answer is x. In other words, $\log_b b^x$ is asking $b^? = b^x$ and therefore, the answer is x.

For the second, $\log_b x$ means the exponent to which b must be raised to yield x. Therefore, if b is raised to this power the answer must be x. In other words, $b^{\log_b x} = x$.

EXAMPLE 10 Simplify **(a)** $3^{\log_3 7}$ **(b)** $\log_4(\log_2 16)$

Solution

Think about why $3^{\log_3 9} = 9$ and why $3^{\log_3 7} = 7$.

(a) Even though we do not know the value of $\log_3 7$, we can use the second basic logarithmic relationship to get $3^{\log_3 7} = \boxed{7}$

(b) $\log_4(\log_2 16) = \log_4 4$ *Because* $2^4 = 16$ *we have* $\log_2 16 = 4$
$= 1$ *Because* $4^1 = 4$

Thus $\log_4(\log_2 16) = \boxed{1}$. ■

In the next section we develop additional properties of the logarithmic function, which will allow us to apply logarithms to a variety of situations.

EXERCISES 5.2

In Exercises 1–20, translate the given logarithmic statement into exponential form and the given exponential statement into logarithmic form.

1. $\log_7 49 = 2$ **2.** $10^4 = 10{,}000$

3. $3^{-4} = \frac{1}{81}$ **4.** $\log_9 3 = \frac{1}{2}$

5. $\log_{1/4} 64 = -3$ **6.** $\log_8 4 = \frac{2}{3}$

7. $27^{-1/3} = \frac{1}{3}$ **8.** $\left(\frac{1}{5}\right)^{-2} = 25$

9. $\log_8 \frac{1}{2} = -\frac{1}{3}$ **10.** $7^0 = 1$

11. $5^{1/4} = \sqrt[4]{5}$ **12.** $\log_7 7 = 1$

13. $\log_2 \frac{1}{2} = -1$ **14.** $11^{1/2} = \sqrt{11}$

15. $\log_{2/3} \frac{27}{8} = -3$ **16.** $\log_{16} 8 = \frac{3}{4}$

17. $4^5 = 1024$ **18.** $\log_8 1 = 0$

19. $\left(\frac{1}{9}\right)^{-1/2} = 3$ **20.** $\log_4 \frac{1}{2} = -\frac{1}{2}$

In Exercises 21–46, evaluate the given logarithmic expression (where it is defined).

21. $\log_2 32$ **22.** $\log_{10} 1000$

23. $\log_6 \frac{1}{6}$ **24.** $\log_{1/3} 9$

25. $\log_8 16$ **26.** $\log_{16} 8$

27. $\log_{16} \frac{1}{8}$ **28.** $\log_8 \frac{1}{16}$

29. $\log_3(-9)$

30. $\log_9\left(\frac{1}{27}\right)$

31. $\log_4 \frac{1}{8}$

32. $\log_5(-25)$

33. $\log_{10} 0.0001$

34. $\log_{1/2} 8$

35. $\log_5 \sqrt[3]{25}$

36. $\log_4 \sqrt{128}$

37. $\log_5(\log_3 243)$

38. $\log_6(\log_7 7)$

39. $\log_2(\log_9 3)$

40. $\log_3(\log_8 2)$

41. $4^{\log_4 7}$

42. $\log_8 8^{0.3}$

43. $\log_6 \frac{1}{\sqrt{6}}$

44. $2^{\log_2 \sqrt{5}}$

45. $\log_b b^6$

46. $\log_b b^{2/3}$

In Exercises 47–52, solve the given equation for t.

47. $\log_3 t = 4$

48. $\log_{36} \frac{1}{6} = t$

49. $\log_t \frac{1}{8} = -3$

50. $\log_5 t = 0$

51. $\log_{32} 16 = t$

52. $\log_t 1 = 0$

53. Given $F(x) = \log_2(x^2 - 4)$, find $F(6)$ and the domain of $F(x)$.

54. Given $g(x) = \log_3(x^2 - 4x + 3)$, find $g(4)$ and the domain of $g(x)$.

In Exercises 55–61, sketch the graph of the given function and identify the domain, range, intercepts, and asymptotes.

55. $y = f(x) = \log_2(x - 3)$.

56. $y = f(x) = -3 + \log_2 x$.

57. $y = f(x) = \log_5(x + 4)$.

58. $y = f(x) = 4 + \log_5 x$.

59. $y = f(x) = -\log_3(-x)$.

60. $y = f(x) = 3\log_5 x$.

61. $y = f(x) = \log_3 |x|$.

QUESTIONS FOR THOUGHT

62. Use the meaning of $\log_b x$ to answer each question.
(a) Which is larger, $\log_2 35$ or $\log_2 41$? Explain.
(b) Which is larger, $\log_2 35$ or $\log_3 35$? Explain.
(c) Which is larger, $\log_4 60$ or $\log_3 40$? Explain.

63. Sketch the graphs of $y = f(x) = -\log_4 x$ and $y = g(x) = \log_{1/4} x$ on the same coordinate system. What do you notice?

64. Show that $\log_{1/6} x = -\log_6 x$. HINT: Write both sides in exponential form.

65. Show that $\log_{1/6} x = \log_6 \frac{1}{x}$. HINT: Write both sides in exponential form.

66. State and prove a generalization of the result of Exercise 63.

67. Suppose $\log_b u = m$ and $\log_b v = n$. Express
(a) $\log_b(uv)$ in terms of m and n.
(b) $\log_b \frac{u}{v}$ in terms of m and n.
(c) $\log_b(u^r)$ in terms of m and n, where r is a real number.

5.3 Properties of Logarithms; Logarithmic Equations

Historically, logarithms were first developed to simplify long and complex numerical computations. Inexpensive and powerful hand-held calculators have made the use of logarithms for computing virtually obsolete. Nevertheless, the logarithmic function, by virtue of the logarithmic properties we develop in this section, retains an important role in mathematics.

We have noted on many occasions that, in general, $f(s + t) \neq f(s) + f(t)$. This fact applies to the logarithmic function as well, $\log_b(s + t) \neq \log_b s + \log_b t$. If we encounter a function for which $f(s + t) = f(s) + f(t)$, we would consider this a "special property" exhibited by the function, which other functions, in general, do not have. In the last section we repeatedly emphasized that a logarithm is merely an exponent, and so it should come as no surprise that the logarithmic function does have some special properties that it "inherits" from the properties of exponents. The proofs of the properties listed here highlight the intimate connection between the properties of logarithms and the basic rules for exponents.

The following box contains the basic properties of logarithms. Properties 4 and 5 were derived in the last section as direct consequences of the definitions. We include them here for ease of reference.

PROPERTIES OF LOGARITHMS

The following properties assume b, u, and v are positive ($b \neq 1$).

1. $\log_b(uv) = \log_b u + \log_b v$
   In words this says the log of a product is equal to the sum of the logs of the factors.
2. $\log_b\left(\dfrac{u}{v}\right) = \log_b u - \log_b v$
   In words this says the log of a quotient is the log of the numerator minus the log of the denominator.
3. $\log_b(u^r) = r \log_b u$
   In words this says the log of a power is the exponent times the log.
4. $\log_b b^x = x$
5. $b^{\log_b x} = x$

How is property 3 of logarithms related to the property of exponents that states $(b^m)^n = b^{nm}$?

We prove these properties by using the fact that

$$x = b^y \Leftrightarrow \log_b x = y$$

Proof of Property 1 Let $u = b^m$ and $v = b^n$, and write these exponential statements in logarithmic form.

$$u = b^m \Leftrightarrow \log_b u = m \quad (1)$$

$$v = b^n \Leftrightarrow \log_b v = n \quad (2)$$

How is property 1 of logarithms related to the property of exponents that states $b^m b^n = b^{m+n}$?

How is property 2 of logarithms related to the property of exponents that states $\dfrac{b^m}{b^n} = b^{m-n}$?

Thus

$uv = b^m b^n = b^{m+n}$ — *Rewriting this in logarithmic form we get*

$uv = b^{m+n} \Leftrightarrow \log_b(uv) = m + n$ — *Using (1) and (2), we substitute for m and n to get*

$\log_b(uv) = \log_b u + \log_b v$ — *As required* ■

Property 1 for logarithms is, in fact, nothing more than a logarithmic statement of a property of exponents. The proof of property 2 is quite similar and is left as an exercise for the student.

Proof of Property 3 Again we let $u = b^m$ and write this exponential statement in logarithmic form.

$$u = b^m \Leftrightarrow \log_b u = m \quad (1)$$

Thus

$u^r = (b^m)^r = b^{rm}$ — *Rewriting this in logarithmic form, we get*

$u^r = b^{rm} \Leftrightarrow \log_b(u^r) = rm$ — *Using (1), we substitute for m to get*

$\log_b(u^r) = r \log_b u$ — *As required* ■

Again property 3 for logarithms is merely the logarithmic form a property of exponents.

As the following examples illustrate, these logarithmic properties allow us to rewrite fairly complex logarithmic expressions in terms of simpler logarithmic expressions. This process is especially useful in applications (Section 5.5) and in calculus.

EXAMPLE 1 Express in terms of simpler logarithms:

(a) $\log_b(x^3y)$ **(b)** $\log_b(x^3 + y)$

(a) $\log_b(x^3y)$ *Using log property 1, we get*

$$= \log_b x^3 + \log_b y \quad \textit{Using log property 3 we get}$$

$$= \boxed{3 \log_b x + \log_b y}$$

Note the difference between parts (a) and (b) of Example 1.

(b) Examining the three properties of logarithms, we see that they deal with the log of a product, quotient, and power. Thus $\log_b(x^3 + y)$, which is the log of a *sum*, *cannot* be simplified using the log properties. ■

EXAMPLE 2 Express in terms of simpler logarithms: $\log_b\left(\frac{\sqrt{xy}}{z^3}\right)$.

Solution Since a logarithm is an exponent, it is usually easier to simplify a logarithmic expression if we rewrite radical expressions in exponential form.

$$\log_b\left(\frac{\sqrt{xy}}{z^3}\right) = \log_b\left(\frac{(xy)^{1/2}}{z^3}\right) \quad \textit{Use property 2 on the quotient.}$$

$$= \log_b(xy)^{1/2} - \log_b z^3 \quad \textit{Use property 3 on the powers.}$$

$$= \frac{1}{2}\log_b(xy) - 3\log_b z \quad \textit{Use property 1 on the product.}$$

$$= \boxed{\frac{1}{2}(\log_b x + \log_b y) - 3\log_b z}$$

■

EXAMPLE 3 Show that $\log_b \frac{1}{2} = -\log_b 2$.

Solution We offer two approaches.

Approach 1:

$$\log_b \frac{1}{2} \quad \textit{Use log property 2 to get}$$

$$= \log_b 1 - \log_b 2 \quad \textit{But } \log_b 1 = 0$$

$$= \boxed{-\log_b 2}$$

Approach 2:

$$\log_b \frac{1}{2} = \log_b 2^{-1} \quad \textit{Now use log property 3.}$$

$$= \boxed{-\log_b 2}$$

■

The previous example can be generalized to the following useful result.

$$\log_b \frac{1}{x} = -\log_b x$$

It is also important to recognize what the log properties *do not* say.

Correct	Incorrect
$\log_b(uv) = \log_b u + \log_b v$	$\log_b(u + v) = \log_b u + \log_b v$
$\log_b\left(\frac{u}{v}\right) = \log_b u - \log_b v$	$\frac{\log_b u}{\log_b v} = \log_b u - \log_b v$
$\log_b(u^r) = r \log_b u$	$(\log_b u)^r = r \log_b u$

Thus

$$\log_b(x^2 + y^3) \neq 2 \log_b x + 3 \log_b y$$

and

$$\frac{\log_b x^3}{\log_b y^4} \neq 3 \log_b x - 4 \log_b y$$

■

The properties of logarithms can also be used to rewrite an expression involving several logarithmic expressions as a single logarithm. This process is useful in solving logarithmic equations.

EXAMPLE 4 Express as a single logarithm: $\log_b x + 3 \log_b y - \frac{1}{2}\log_b z$.

Solution We use the log properties in reverse.

$$\log_b x + \underbrace{3 \log_b y} - \frac{1}{2}\log_b z \qquad \textit{Use log property 3.}$$

$$= \underbrace{\log_b x + \log_b y^3} - \log_b z^{1/2} \qquad \textit{Use log property 1.}$$

$$= \log_b(xy^3) - \log_b z^{1/2} \qquad \textit{Use log property 2.}$$

$$= \boxed{\log_b\left(\frac{xy^3}{z^{1/2}}\right)}$$

■

The logarithm properties can be applied to solving logarithmic equations, as illustrated in the next three examples.

EXAMPLE 5 Solve for x: $\log_6 x - \log_6 3 = 2$.

Solution In the last section we solved simple logarithmic equations involving a single logarithm by translating it into exponential form. Thus our first step is to use the log properties to rewrite the left-hand side of the given equation as a single logarithm.

$$\log_6 x - \log_6 3 = 2 \quad \textit{Use log property 2.}$$

$$\log_6 \frac{x}{3} = 2 \quad \textit{Rewrite in exponential form.}$$

$$\frac{x}{3} = 6^2 = 36$$

$$\boxed{x = 108}$$

Check: $x = 108$

$$\log_6 108 - \log_6 3 = \log_6\left(\frac{108}{3}\right) = \log_6 36 \stackrel{\checkmark}{=} 2$$

Note that even though we do not know the exact values of $\log_6 108$ and $\log_6 3$, the log properties allow us to compute their difference exactly. ■

EXAMPLE 6 Solve for x: $\log_2 x + \log_2(x + 2) = 3$.

Solution Again we begin by rewriting the equation so that it involves a single logarithm.

$$\log_2 x + \log_2(x + 2) = 3 \quad \textit{Use log property 1.}$$

$$\log_2[x(x + 2)] = 3 \quad \textit{Rewrite in exponential form.}$$

$$x(x + 2) = 2^3 \Rightarrow x^2 + 2x - 8 = 0$$

$$(x + 4)(x - 2) = 0 \Rightarrow x = -4 \quad \text{or} \quad x = 2$$

Be careful! $x = -4$ does satisfy $\log_2[x(x + 2)] = 3$ (it makes the argument $-4(-2) = 8$), but it does not satisfy the original equation because $\log_2 x$ and $\log_2(x + 2)$ are undefined for $x = -4$. *Remember that the argument of a logarithmic function cannot be negative.* Thus the only solution to the given equation is $\boxed{x = 2}$. *Check:* $x = 2$

$$\log_2 x + \log_2(x + 2) = 3 \quad \textit{Substitute } x = 2.$$

$$\log_2 2 + \log_2 4 \stackrel{?}{=} 3$$

$$1 + 2 \stackrel{\checkmark}{=} 3$$

■

We have already made use of the fact that the exponential function is one-to-one. Of course, since the logarithmic function is the inverse of the exponential function, it too is one-to-one. We can also use this fact in solving logarithmic equations.

The log function being one-to-one means that $\log_b U = \log_b V \Rightarrow U = V$.

EXAMPLE 7 Solve for x: $\log_b(8 - x) - \log_b(2 - x) = \log_b 3$.

Solution

$$\log_b(8 - x) - \log_b(2 - x) = \log_b 3 \quad \textit{Use log property 2.}$$

$$\log_b\left(\frac{8 - x}{2 - x}\right) = \log_b 3 \quad \textit{Since the log function is one-to-one, we have}$$

$$\frac{8 - x}{2 - x} = 3$$

$$8 - x = 6 - 3x \Rightarrow 2x = -2 \Rightarrow \boxed{x = -1}$$

Note that even though $x = -1$ is a negative number, it is a perfectly valid solution to the original equation. When $x = -1$, both

$$\begin{aligned} \log_b(8 - x) &= \log_b (8 - (-1)) \\ &= \log_b 9 \\ \text{and } \log_b(2 - x) &= \log_b(2 - (-1)) \\ &= \log_b 3 \end{aligned}$$

are well defined. It is left to the student to complete the check. ■

EXAMPLE 8 Evaluate $2^{3 \log_2 4}$.

Solution We know that $b^{\log_b x} = x$. Unfortunately, the expression $2^{3 \log_2 4}$ is not quite in that form. However, we can use log property 3 to put the expression into the form $b^{\log_b x}$ as follows:

$2^{3 \log_2 4}$ *Use property 3 to rewrite $3 \log_2 4$ as $\log_2 4^3$.*

$= 2^{\log_2 4^3}$ *Since $b^{\log_b x} = x$*

$= 4^3 = \boxed{64}$ ■

As a final comment we want to point out that the log properties apply only when both sides make sense. That is $\log_b x^2$ is defined for all $x \neq 0$, whereas $2 \log_b x$ is defined for $x > 0$. Therefore, $\log_b x^2 = 2 \log_b x$ is true only for those values for which both sides are defined—i.e., for $x > 0$. Another way of saying the same thing is that the domain of $f(x) = \log_b x^2$ is all real numbers not equal to zero, whereas the domain of $g(x) = 2 \log_b x$ is all positive real numbers.

EXERCISES 5.3

In Exercises 1–14, use the properties of logarithms to write the given expression in terms of simpler logarithms. Express the answer so that it does not contain the logarithm of products, quotients, or powers. If the expression cannot be simplified, say so.

1. $\log_4(x^2yz^3)$

2. $\log_2\left(\frac{5a}{b^2}\right)$

3. $\log_b\left(\frac{x^3}{yz^4}\right)$

4. $\log_b b^5$

5. $\log_5 \sqrt{5x^3}$

6. $\log_t \sqrt[3]{\frac{4}{t}}$

7. $\log_b(x^3 + y^2 - z^5)$

8. $\log_2\left(\frac{4x}{\sqrt{y^3}}\right)$

9. $\log_b(x^2 - 4)$

10. $\log_b \sqrt{x^2 - x - 6}$

11. $\log_4 \sqrt[3]{\frac{x - 3}{2x^2}}$

12. $\log_9 \sqrt{\frac{x}{3}}$

13. $\log_b \sqrt{\frac{x^2 - 16}{x^2 - 2x - 8}}$

14. $\log_3 \sqrt{\frac{27r^5s}{t^3}}$

In Exercises 15–18, determine whether the given statement is true or false.

15. $\log_b(x^3 - y^4) = 3 \log_b x - 4 \log_b y$

16. $\frac{\log_b x^3}{\log_b y^4} = 3 \log_b x - 4 \log_b y$

17. $\log_b\left(\frac{x^3}{y^4}\right) = 3 \log_b x - 4 \log_b y$

18. $(\log_b x)^3 = 3 \log_b x$

In Exercises 19–30, write the given expression as a single logarithm with a coefficient of 1 and simplify as completely as possible.

19. $4 \log_b 2 + \log_b 3$

20. $2 \log_b x + \log_b y - 3 \log_b z$

21. $\frac{1}{2} \log_b s - \frac{3}{2} \log_b t$

22. $5 \log_b 3 - 2 \log_b 4$

23. $\log_{10} 50 - \log_{10} 5$

24. $\log_8 80 - \log_8 5$

25. $\log_6 9 + \log_6 24$

26. $\log_{10} 200 + \log_{10} 5$

27. $3 \log_b x - 4 \log_b y - 2 \log_b z$

28. $\frac{1}{3} \log_b(x + 1) - \frac{1}{3} \log_b(x + 2)$

29. $\frac{1}{4} \log_b x + \frac{1}{3} \log_b y - \frac{1}{2} \log_b z$

30. $\log_b \frac{x}{4} + \log_b \frac{y}{3} - \log_b \frac{z}{2}$

In Exercises 31–36, let $\log_{10} 2 = A$, $\log_{10} 3 = B$, and $\log_{10} 5 = C$. Verify each of the following statements.

31. $\log_{10} 16 = 4A$

32. $\log_{10} 18 = A + 2B$

33. $\log_{10} \sqrt{6} = \frac{1}{2}(A + B)$

34. $\log_{10} \frac{1}{20} = -A - 1$

35. $\log_{10} 300 = B + 2$

36. $\log_{10} \sqrt[3]{24} = A + \frac{1}{3}B$

In Exercises 37–40, use $\log_b 2 = 0.69$, $\log_b 3 = 1.09$, and $\log_b 5 = 1.61$ to evaluate the given logarithm.

37. $\log_b 12$

38. $\log_b 25$

39. $\log_b \frac{1}{18}$

40. $\log_b 60$

In Exercises 41–54, solve each logarithmic equation.

41. $\log_3 4 + \log_3 x = 2$

42. $\log_2 x = 4 + \log_2 3$

43. $2 \log_5 x = \log_5 49$

44. $\frac{1}{2} \log_4 x = \log_4 9$

45. $\log_2 t - \log_2(t - 2) = 3$

46. $\log_6 2 - \log_6 x = 1$

47. $\log_{10} 8 + \log_{10} x + \log_{10} x^2 = 3$

48. $\log_4 x + \log_4(x - 6) = 2$

49. $\log_9(2x + 7) - \log_9(x - 1) = \log_9(x - 7)$

50. $\log_2 6 + \log_2 x - \log_2(x + 2) = 2$

51. $\log_4 x - \log_4(x - 4) = \log_4(x - 6)$

52. $\log_3 7 - \log_3 x - \log_3(x - 2) = 2$

53. $\frac{1}{2} \log_3 x = \log_3(x - 6)$

54. $\frac{1}{2} \log_2(x + 1) = 2 + \frac{1}{2} \log_2 5$

55. Evaluate $3^{2 \log_3 5}$.

56. Evaluate $5^{3 \log_5 2}$.

QUESTION FOR THOUGHT

57. At the end of this section we explained why, despite the properties of logarithms, the functions $f(x) = \log_b x^2$ and $g(x) = 2 \log_b x$ are not the same. Sketch the graphs of these functions.

5.4 Common and Natural Logarithms; Exponential Equations and Change of Base

In Sections 5.1 and 5.2, we introduced the exponential and logarithmic functions without paying particular attention to the base chosen. However, there are two bases of special importance in mathematics.

Common Logarithms

We have already mentioned that, historically, logarithms were developed to simplify complex numerical computations. Since we do arithmetic in a base 10 number system, logarithms to the base 10, called **common logarithms,** play a significant role. In fact, $\log_{10} x$ is usually just written $\log x$. In other words, if a logarithmic function appears without a base, the base is assumed to be 10.

COMMON LOGARITHMS

$f(x) = \log_{10} x$ is called the **common logarithm function.** We write

$$\log_{10} x = \log x$$

EXAMPLE 1 Evaluate **(a)** log 1000 **(b)** log 0.01

Solution

(a) Let $a = \log 1000$. *Write this in exponential form. Remember the base is* 10.

$$10^a = 1000 = 10^3 \Rightarrow \boxed{a = 3}$$

(b) Let $t = \log 0.01$ *Write this in exponential form.*

$$10^t = 0.01 = 10^{-2} \Rightarrow \boxed{t = -2}$$

■

The Natural Exponential Function and the Natural Logarithmic Function

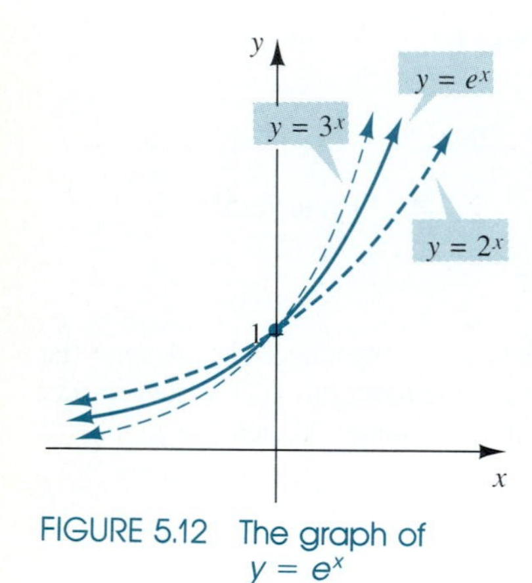

FIGURE 5.12 The graph of $y = e^x$

The second special base, and by far the most important for work in mathematics and the sciences, is the irrational number denoted by the letter e. We will have more to say about the number e in the next section, but for our immediate purpose it is sufficient to know that to seven decimal places, we have

$$e \approx 2.7182818$$

The function $y = f(x) = e^x$ is called the **natural exponential function.*** It may seem that using 2 or 3 or 10 as the base for the logarithmic function would be simpler and more natural than using the number e; however, once we are exposed to certain mathematical ideas in calculus, it will become quite clear what makes e the "natural" base.

Since e is a number between 2 and 3, the graph of $y = e^x$ exhibits exponential growth and falls between the graphs of $y = 2^x$ and $y = 3^x$. See Figure 5.12.

EXAMPLE 2 Sketch the graph of
(a) $y = f(x) = e^{x-1}$ **(b)** $y = g(x) = e^{-x} - 1$

Solution

(a) To find the graph of $y = e^{x-1}$, we shift the graph of $y = e^x$ horizontally 1 unit to the right. By setting $x = 0$, we find that the graph crosses the y-axis at $y = e^{-1} = \dfrac{1}{e} \approx 0.37$. The graph appears in Figure 5.13.

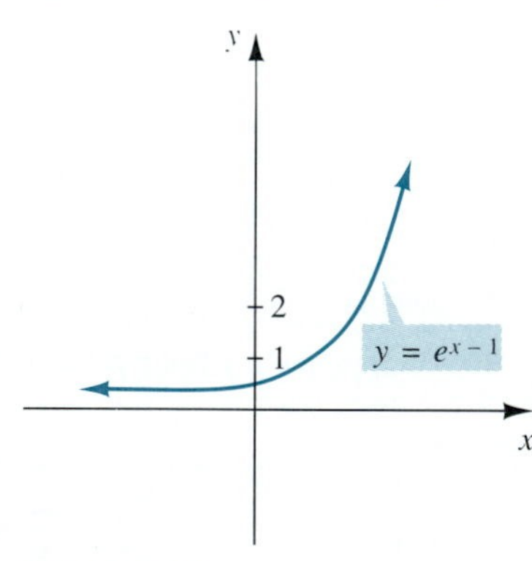

FIGURE 5.13

*The letter e is used to denote the base of the natural logarithmic function in honor of the brilliant Swiss mathematician Leonhard Euler (1707–1783), who is credited with first using it.

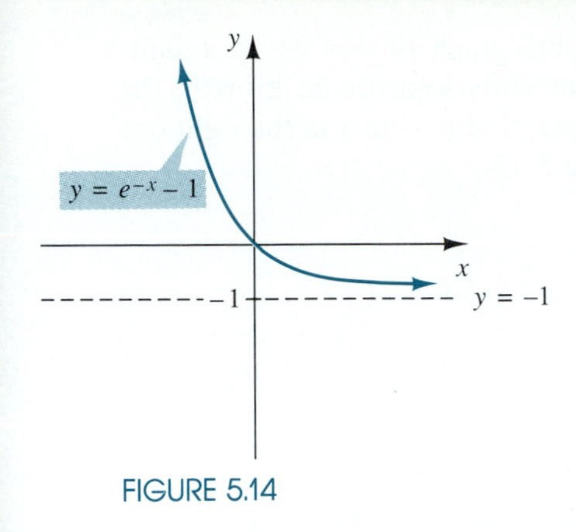

FIGURE 5.14

(b) Since e is greater than 1, the graph of $y = e^{-x}$ will exhibit exponential decay. The graph of $y = e^{-x} - 1$ is obtained by shifting the graph of $y = e^{-x}$ vertically down 1 unit. This makes the line $y = -1$ a horizontal asymptote. It is left to the reader to verify that this graph passes through the origin. The graph appears in Figure 5.14. ■

The inverse of the natural exponential function is called the **natural logarithmic function** and has its own special notation.

NATURAL LOGARITHMS

$f(x) = \log_e x$ is called the **natural logarithmic function.** We write

$$\log_e x = \ln x$$

EXAMPLE 3 Find **(a)** ln 1 **(b)** ln e.

Solution

(a) Since $\ln 1 = \log_e 1$, we are asking for the exponent of e that gives 1, so the answer is 0. Therefore, $\boxed{\ln 1 = 0}$.

(b) Since $\ln e = \log_e e$, we are asking for the exponent of e that gives e; clearly the answer is 1. Therefore, $\boxed{\ln e = 1}$. ■

GRAFFIX

Use a graphics calculator or computer to graph the functions $y = e^x$, $y = \ln x$, and $y = x$ on the same set of coordinate axes. Keeping in mind that **ln** on your calculator means $\log_e$, what do the three graphs suggest?

EXAMPLE 4 Find the inverse function of $y = f(x) = e^x + 1$ and sketch the graphs of $f(x)$ and $f^{-1}(x)$ on the same set of coordinate axes.

Solution In order to find $f^{-1}(x)$, we start with $y = e^x + 1$, interchange x and y, and then solve for y.

$$y = e^x + 1 \quad \textit{Interchange } x \textit{ and } y,$$

$$x = e^y + 1 \quad \textit{Solve for } y.$$

$$x - 1 = e^y \quad \textit{Rewrite in logarithmic form.}$$

$$\ln(x - 1) = y \quad \textit{This is } f^{-1}(x).$$

The graph of $y = e^x + 1$ is obtained by shifting the graph of $y = e^x$ up 1 unit. Since e is greater than 1, the graph of $y = \ln x$ exhibits logarithmic growth; the graph of $y = \ln(x - 1)$ is obtained by shifting the graph of $y = \ln x$ to the right one unit. The graphs of $f(x)$ and $f^{-1}(x)$ appear in Figure 5.15.

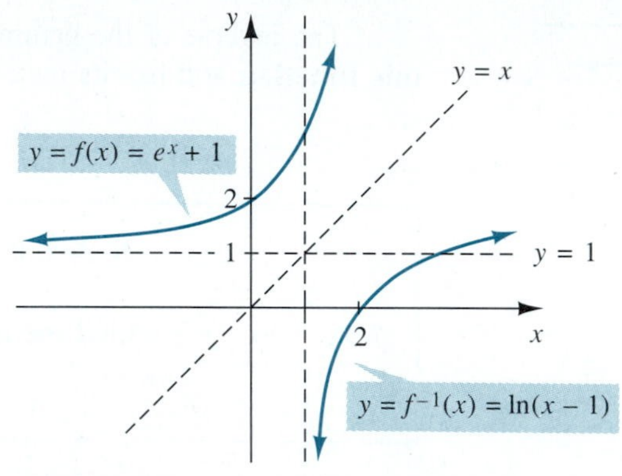

FIGURE 5.15

Since these are the graphs of inverse functions, they are symmetric about the line $y = x$. In fact, we could have gotten the graph of $y = \ln(x - 1)$ by reflecting the graph of $y = e^x + 1$ about the line $y = x$. ■

The next example deals with using a calculator to evaluate expressions involving e.

EXAMPLE 5 Use a calculator to approximate the value of e^5. Round to two decimal places.

Solution We offer several possible solutions. One possibility is to enter **2.7182818,**

the $\boxed{y^x}$ key to raise this quantity to the fifth power. The keystroke sequence would be to enter **2.7182818** and then press

$$\boxed{y^x}\ \boxed{5}\ \boxed{=}$$

The display would then read **148.41315**. Thus the answer would be $\boxed{148.41.}$ Some calculators have an $\boxed{e^x}$ key, which when pressed computes e to whatever power appears on the display. On such a calculator the keystroke sequence

$$\boxed{\mathbf{5}}\ \boxed{e^x}\ \boxed{=}$$

would cause the display to read **148.41316** and give us the same answer rounded to four decimal places.

Even if your calculator does not have an $\boxed{e^x}$ key, your calculator does have an $\boxed{\mathbf{inv}}$ key, which can be used in conjunction with the $\boxed{\mathbf{ln}}$ key to produce the same result. We know that the exponential and logarithmic functions are inverses of each other and that $\ln e = 1$. Therefore, if we use the $\boxed{\textbf{inv}}$ with the $\boxed{\textbf{ln}}$, it is equivalent to having an $\boxed{e^x}$ key. In other words, to get the value of e we can use the keystroke sequence

$$\boxed{1}\ \boxed{\textbf{inv}}\ \boxed{\textbf{ln}}$$

to obtain **2.7182818** on the display.

Similarly, to get e^5 we would key in

$$\boxed{5}\ \boxed{\textbf{inv}}\ \boxed{\textbf{ln}}$$

and get **148.41316,** as before. ■

Exponential Equations

We are able to solve certain exponential equations such as $5^x = \frac{1}{25}$ by expressing each side as a power of the same base ($5^x = 5^{-2}$), thus obtaining the solution $x = -2$. However, this approach does not work when solving an equation such as $5^x = 2$. How do we find, with some accuracy, the exponent x such that $5^x = 2$?

The definition of a function requires that each input x produces a unique output $f(x)$. Symbolically we say: If $a = b$, then $f(a) = f(b)$. Since the logarithm is a function, it follows that: If $U = V$, then $\log_b U = \log_b V$.

Thus if two expressions are equal, their logarithms are equal. We say that when we have an equation, we can "take the log of both sides." Once we have the equation in log form, we can often apply the log properties, which then allows us to solve the equation. In particular, log property 3 applied to an exponential equation allows us to get the variable out of the exponent and solve the equation more easily.

EXAMPLE 6 Solve $3^{x+1} = 7$ for x. Round to four decimal places.

Solution Since we cannot easily express both sides in terms of the same base, we take the logarithm of both sides of the equation. Although we may use any base we choose, most calculators come equipped with a log key and a ln key, and so we will generally use either base 10 logarithms (that is, use log) or base e logarithms (that is, use ln) in solving problems.

$$3^{x+1} = 7 \qquad \textit{Take } \ln \textit{ of both sides.}$$

$$\ln 3^{x+1} = \ln 7 \qquad \textit{Use log property } 3.$$

$$(x + 1)\ln 3 = \ln 7$$

$$x + 1 = \frac{\ln 7}{\ln 3}$$

$$x = \frac{\ln 7}{\ln 3} - 1 \qquad \textit{This is the exact answer.}$$

Using a calculator we can approximate this answer to be $\boxed{x = 0.7712}$. ■

EXAMPLE 7 Solve for x.

(a) $5^{x+2} = 3^{2x+1}$ **(b)** $5^{2x} = 25^{3x+1}$

Solution

(a) We proceed as we did before.

$$5^{x+2} = 3^{2x+1}$$ *Take the log of each side.*

$$\log(5^{x+2}) = \log(3^{2x+1})$$ *Apply log property 3.*

$$(x + 2)\log 5 = (2x + 1)\log 3$$ *Remember* log 5 *and* log 3 *are numbers. Now we have a linear equation similar to* $(x + 2)7 = (2x + 1)4$. *Multiply out both sides.*

$$x\log 5 + 2\log 5 = x(2\log 3) + \log 3$$ *Collect the x terms on one side and factor out x.*

$$x\log 5 - x(2\log 3) = \log 3 - 2\log 5$$

$$x(\log 5 - 2\log 3) = \log 3 - 2\log 5$$

$$\boxed{x = \frac{\log 3 - 2\log 5}{\log 5 - 2\log 3}}$$ *This is the exact answer.*

Using a calculator, this answer is approximately 3.6071991. Substituting this value into the original equation, the left-hand side becomes $5^{5.6071991}$, which a calculator computes as 8303.5533, whereas the right side becomes $3^{8.2143982}$, which the calculator also computes as 8303.5533.

(b) $5^{2x} = 25^{3x+1}$ *We could start by taking the log of each side of the equation, but in this equation we can get an exact rational answer by expressing each side in terms of the same base.*

$$5^{2x} = (5^2)^{3x+1}$$

$$5^{2x} = 5^{6x+2}$$ *Since the exponential function is one-to-one we have*

Parts (a) and (b) of Example 7 illustrate two approaches to solving exponential equations. How do you know which approach to take?

$$2x = 6x + 2 \Rightarrow \boxed{x = -\frac{1}{2}}$$ ■

A few things should be kept in mind when solving exponential equations. First, if we need or are asked to find the *exact* answer to an exponential equation, we may need to leave the answer in terms of logs rather than using the calculator to approximate the answer (just as we would give $\sqrt{2}$ as an exact answer rather than 1.414214 . . .).

Secondly, equations containing variables in their exponents may not lend themselves to algebraic methods of solution. For example, we have no *algebraic* method for solving the equation $2^x = 3x + 4$. However, we can always find an approximate solution by graphing. If we have access to a graphics calculator or graphics capability on a computer, we can approximate the solution to the equation $2^x = 3x + 4$ by graphing $y = 2^x$ and $y = 3x + 4$ and locating their point of intersection at $x = 4$ and $x = -1.17$ (see Figure 5.16(a). Alternatively we can find the solution by graphing $y = 2^x - (3x + 4)$ and locating the x-intercepts, as illustrated in Figure 5.16(b)).

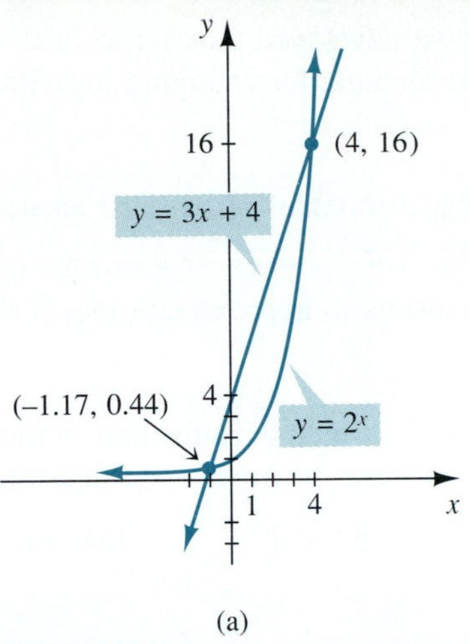

(a)

y
y = 2^x – (3x + 4)
–1.17
4
x

(b)

FIGURE 5.16

GRAFFIX

Use a graphics calculator or computer.

1. Approximate (to the nearest tenth) the solution(s) to the equation

$$3^x = 2x + 4$$

by graphing $y = 3^x$ and $y = 2x + 4$ on the same set of axes and using the [TRACE] key to find the x-values of their points of intersection.

2. Approximate (to the nearest tenth) the solution(s) to the equation

$$3^x = 2x + 4$$

by graphing $y = 3^x - 2x - 4$ and using the [TRACE] key to find the x-values to find the x-intercepts.

3. Approximate (to the nearest tenth) the solution(s) to the equation

$$5^x = 3x + 2$$

by graphing $y = 5^x$ and $y = 3x + 2$ on the same set of axes and using the [TRACE] key to find the x-values of their points of intersection.

4. Approximate (to the nearest tenth) the solution(s) to the equation

$$5^x = 3x + 2$$

by graphing $y = 5^x - 3x - 2$ and using the [TRACE] key to find the x-values to find the x-intercepts.

The method of taking the logarithm of both sides of an equation can be used to solve a broader range of exponential equations.

We are able to evaluate certain logarithmic expressions, such as $\log_5 25$, by recognizing that we are actually solving the exponential equation $5^x = 25$. However, if we want to evaluate $\log_5 2$, we run into difficulty. In the next example we utilize the technique we have developed thus far to deal with this problem, and in so doing we obtain a general formula for changing logarithms from one base to another.

Can you find $\log_5 2$ directly on your calculator?

EXAMPLE 8

(a) Express $\log_5 2$ in terms of base 10 logarithms.

(b) Express $\log_5 2$ in terms of natural logarithms.

(c) Use these results to approximate $\log_5 2$ rounded off to three decimal places.

Solution

(a) Let $x = \log_5 2$, which is equivalent to the exponential equation $5^x = 2$. We can now take the base 10 (or base e) logarithm of both sides of this equation.

$$5^x = 2 \quad \textit{Take the base 10 log of both sides.}$$

$$\log 5^x = \log 2 \quad \textit{Use log property 3.}$$

$$x \log 5 = \log 2$$

$$x = \frac{\log 2}{\log 5} \quad \textit{Thus}$$

$$\boxed{\log_5 2 = \frac{\log 2}{\log 5}}$$

(b) We now repeat this process with base e logarithms.

$$5^x = 2 \quad \textit{Take ln of both sides.}$$

$$\ln 5^x = \ln 2 \quad \textit{Use log property 3.}$$

$$x \ln 5 = \ln 2$$

$$x = \frac{\ln 2}{\ln 5} \quad \textit{Thus}$$

$$\boxed{\log_5 2 = \frac{\ln 2}{\ln 5}}$$

What we have done in parts (a) and (b) is convert an expression from base 5 logarithms to base 10 and base e logarithms.

(c) We first recognize that we cannot use a calculator to approximate $\log_5 2$ directly, because there is no $\log_5$ key on a calculator. In order to approximate $\log_5 2$, we use a calculator to compute the answers obtained in parts (a) and (b).

$$\log_5 2 = \frac{\log 2}{\log 5} = \frac{0.30103}{0.69897} = 0.4306766 = 0.431 \quad \textit{Rounded to 3 places}$$

$$\log_5 2 = \frac{\ln 2}{\ln 5} = \frac{0.6931472}{1.6094379} = 0.4306766 = 0.431 \quad \textit{Rounded to 3 places}$$

■

The method illustrated in the last example to convert from base 5 logarithms to base 10 logarithms or base e logarithms can easily be generalized to the following **change-of-base formula.** You are asked to prove this formula in Exercise 38.

CHANGE-OF-BASE FORMULA

$$\log_b c = \frac{\log_a c}{\log_a b}$$

EXAMPLE 9 Express $\log_4 x$ in terms of $\log_{16} x$.

Solution We can proceed as we did in the last example or we can use the change of base formula directly to get

$$\log_4 x = \frac{\log_{16} x}{\log_{16} 4} \qquad \textit{But } \log_{16} 4 = \tfrac{1}{2}.$$

$$= \frac{\log_{16} x}{\frac{1}{2}} = \boxed{2 \log_{16} x}$$

Think about how the base 4 and the base 16 are related, and explain why the answer to Example 9 makes sense.

In the next section we look at a variety of applications of exponential and logarithmic functions.

EXERCISES 5.4

In Exercises 1–6, simplify the given expression.

1. $\ln e^5$
2. $\log 0.001$
3. $\log \sqrt{10}$
4. $y = \ln \dfrac{1}{e}$
5. $e^{\ln(x+1)}$
6. $y = \log 10^{(2t+3)}$

In Exercises 7–14, sketch the graph of the given function. Indicate the domain, range, intercepts, and asymptotes, where appropriate.

7. $y = 10^{x+1}$
8. $y = \log(x + 1)$
9. $y = 2 - \ln x$
10. $y = e^{x-2}$
11. $y = \log_{10}(x - 3)$
12. $y = \log_{10}(2x + 5)$
13. $y = \log x^2$
14. $y = \ln(ex)$

15. Find the inverse function for $y = f(x) = e^{(3x-1)}$.
16. Find the inverse function for $y = f(x) = 10^{(5-x)}$.
17. Let $f(x) = e^{\sqrt{x}}$. Find a function $g(x)$ so that $(f \circ g)(x) = (g \circ f)(x) = x$.
18. Let $g(x) = 10^{(x-3)/2}$. Find a function $h(x)$ so that $(g \circ h)(x) = (h \circ g)(x) = x$.

In Exercises 19–24, evaluate the given logarithm using the change of base formula, once with base 10 logarithms and once with natural logarithms. Round the answer to two decimal places.

19. $\log_7 4$
20. $\log_4 7$
21. $\log_{1/2} 5$
22. $\log_8 30$
23. $\log_9 \dfrac{2}{3}$
24. $\log_6 14$

25. Express $\log_3 x$ in terms of $\log_9 x$.
26. Express $\log_8 x$ in terms of $\log_4 x$.
27. Express $\log_{10} x$ in terms of $\ln x$.
28. Express $\ln x$ in terms of $\log_{10} x$.

In Exercises 29–36, solve the given equation. You may leave your answer in terms of natural logarithms.

29. $e^{1-2x} = 5$
30. $3e^{x-3} = 10$
31. $\dfrac{e^{3x+2}}{4} = 5$
32. $6e^{x/5} = 2$
33. $3^{4-x} = 7$
34. $5^{2x+3} = 4^{x-2}$
35. $2^{2x-2} = 6^{-x}$
36. $3^{-x^2} = 4$

QUESTIONS FOR THOUGHT

37. Use a calculator to evaluate the expression $\left(1 + \frac{1}{n}\right)^n$ for $n = 10$; 20; 50; 100; 1000; and 10,000. Note that the values seem to be tending toward the value of e. In fact, this is another way of defining the number e.

38. Use the method outlined in Example 8 to prove the change of base formula $\log_b c = \dfrac{\log_a c}{\log_a b}$.

GRAFFIX

In Exercise 39–41, approximate the solution to the equation $f(x) = g(x)$ in 2 ways: first, by finding the points of intersection of the graphs of $y = f(x)$ and $y = f(x) - g(x)$; second, by finding the x-intercepts of the graph of $y = f(x) - g(x)$.

39. $f(x) = 3^x$ $\quad g(x) = 2 - 3x$

40. $f(x) = 2^x$ $\quad g(x) = 3x + 2$

41. $f(x) = x^2$ $\quad g(x) = 2^x$

5.5 Applications

There are many real-world quantities that increase or decrease with time in proportion to the amount of the quantity present. Some examples are human population, bacteria in a culture, radioactivity, the concentration of drugs in the bloodstream, and the value of certain types of investments.

Let's discuss some ideas from the world of finance, where it may come as a bit of a surprise that the number e makes a dramatic appearance. We begin by introducing some terminology. Suppose that \$1000 is invested in an account at an interest rate of 6% per year. The \$1000 is called the **principal** and is often designated as P. At the end of one year the amount in the account would be

$$\text{principal} + \text{interest} = \text{new principal}$$

$$1000 + 0.06(1000) = 1000(1 + 0.06) = 1.06(1000) \tag{1}$$

If the interest earned each year remains in the account so that in subsequent years the interest itself earns interest, the account is said to be earning **compound interest.**

Let's start with P dollars in an account paying an interest rate of r (expressed as a decimal) compounded annually (meaning that the interest is computed at the end of each year) and attempt to develop a formula for the amount, A, present in an account after t years.

Our computation will be simplified if we recognize that as in (1), *the amount in the account at the end of any year is* $(1 + r)$ *times the previous principal* (where, again, r is the interest rate expressed as a decimal). We therefore have the following:

at the end of 1 year: $A = P(1 + r)$ *This is now the new principal, as in* (1)

at the end of 2 years: $A = \underbrace{P(1 + r)}_{\text{Previous principal}}[1 + r] = P(1 + r)^2$

at the end of 3 years: $A = \underbrace{P(1 + r)^2}_{\text{Previous principal}}[1 + r] = P(1 + r)^3$

$\vdots$

at the end of t years: $A = P(1 + r)^t \quad (2)$

EXAMPLE 1 Suppose \$1000 is placed into a savings account paying 6% per year.

(a) How much money will be in the account after 10 years?

(b) How long will it take the initial principal of \$1000 to double?

(c) How long will it take an initial principal of \$1000 to double if the annual interest rate is 8%?

Solution

(a) Using the formula $A = P(1 + r)^t$, we substitute $P = 1000$, $r = 0.06$, and $t = 10$ to get

$$A = 1000(1 + 0.06)^{10} = 1000(1.06)^{10}$$ *Using a calculator we get*

$$= \$1790.85 \text{ in the account at the end of 10 years.}$$

(b) Since we are looking for the number of years it takes the money to double, we want the amount A to be \$2000. We substitute $A = 2000$, $P = 1000$, and $r = 0.06$ in the formula $A = P(1 + r)^t$ and solve for t.

$$2000 = 1000(1.06)^t$$ *To isolate t, first divide both sides of the equation by 1000.*

$$2 = 1.06^t$$ *Next, take ln (or $\log_{10}$) of both sides.*

$$\ln 2 = \ln(1.06^t)$$ *Use log property 3.*

$$\ln 2 = t \ln 1.06$$

$$t = \frac{\ln 2}{\ln 1.06} \approx \boxed{11.9 \text{ years}}$$

Since interest is paid at the end of each year, it will take 12 years for the money to double.

(c) If the annual interest rate is 8%, then $1 + r = 1 + 0.08 = 1.08$, and the same procedure used in part (b) gives

$$t = \frac{\ln 2}{\ln 1.08} \approx 9$$

Thus increasing the interest rate to 8% reduces the doubling time from almost 12 years to about 9 years. ■

Suppose we change the way the interest is compounded. Most banks do not compute the interest on an account once yearly; rather, they compound the interest a number of times per year. For example, if a bank compounds the interest quarterly, it computes and credits the interest 4 times per year (every 3 months) using one-quarter of the annual interest rate each time. Similarly, banks may compound monthly by computing and adding the interest to the account 12 times per year using one-twelfth the annual interest rate each time, or daily by computing $\frac{1}{365}$ of the interest rate each day.

Essentially what is happening is that we are making shorter and shorter interest periods with smaller and smaller interest rates. If we are compounding \$1000 quarterly at an annual interest rate of 6%, we are computing a new principal 4 times per year by using an interest rate of $\frac{0.06}{4} = 0.015$. (This situation will yield the same amount of money as if \$1000 were invested at 1.5% per year for 4 years.)

For illustrative purposes we again begin with $1000 invested at an annual interest rate of 6%. Let's see what happens if we compound this annual interest rate quarterly, monthly, and daily for one year.

$$\textit{compounding quarterly}: \quad A = 1000\left(1 + \frac{0.06}{4}\right)^4 = 1000(1.015)^4 = \$1061.36$$

$$\textit{compounding monthly}: \quad A = 1000\left(1 + \frac{0.06}{12}\right)^{12} = 1000(1.005)^{12} = \$1061.68$$

$$\textit{compounding daily}: \quad A = 1000\left(1 + \frac{0.06}{365}\right)^{365}$$

$$= 1000(1.0001644)^{365} = \$1061.83$$

Note that increasing the number of interest periods from 12 per year to 365 per year increases the annual interest by only 15¢.

Also note that if we were compounding the annual interest monthly for a period of 5 years, there would be $5(12) = 60$ interest periods and so

$$A = 1000\left(1 + \frac{0.06}{12}\right)^{5(12)} = 1000(1.005)^{60} = \$1348.85$$

As a result of this discussion, we state the following formula.

COMPOUND INTEREST FORMULA

If P dollars is invested at an annual interest rate r, compounded n times per year, then A, the amount of money present after t years, is given by

$$A = P\left(1 + \frac{r}{n}\right)^{nt}$$

EXAMPLE 2 In Example 1 we saw that it takes 11.9 years for $1000 invested at 6% per year to double. If the $1000 is invested at 6% compounded daily, how long will it take to double?

Solution Using the compound interest formula, substitute $A = 2000$, $P = 1000$, $r = 0.06$, and $n = 365$ and solve for t.

$$2000 = 1000\left(1 + \frac{0.06}{365}\right)^{365t}$$ *Divide both sides by 1000.*

$$2 = 1.0001644^{365t}$$ *Take ln of both sides (or $\log_{10}$ of both sides), and apply log property 3.*

$$\ln 2 = 365t(\ln 1.0001644)$$ *Isolate t.*

$$t = \frac{\ln 2}{365 \ln 1.0001644} \approx 11.55$$

Thus by compounding daily, the doubling time has been reduced from 11.9 years to a little more than $11\frac{1}{2}$ years. Compare this to the reduction in doubling time if the interest rate is increased from 6% to 8% (see Example 1(c)). ■

EXAMPLE 3 What simple interest rate would be equivalent to an interest rate of 6% compounded monthly? (Banks often call such a simple annual interest rate the *effective annual yield.*)

Solution Using the compound interest formula we have

$$A = P\left(1 + \frac{0.06}{12}\right)^{12} \approx P(1.0617)$$

which means that at the end of 1 year we have 1.0617 times the original principal. This is equivalent to a simple interest rate of 6.17% per year. Thus an account paying 6% per year compounded monthly has an effective annual yield of 6.17%. ■

A natural question one might ask is what happens if we start with P dollars and compound an interest rate of r every minute? Every second? Will the amount A keep growing larger? Is there any limit to how much the amount A will grow as we compound more and more frequently? As we saw before, increasing the number of interest periods does increase the amount A, but not by as much as we might have expected. It turns out there is a limit to how much A can grow regardless of how many times interest is compounded per year.

Let's suppose that \$1 is invested at an annual rate of 100% per year compounded n times per year. Using the formula for compound interest with $r = 1$, the amount A at the end of 1 year is given by

$$A = \left(1 + \frac{1}{n}\right)^n$$

if we compound it annually, then $n = 1$ $\quad A = \left(1 + \frac{1}{1}\right)^1 = 2$

semiannually, then $n = 2$ $\quad A = \left(1 + \frac{1}{2}\right)^2 = 2.25$

quarterly, then $n = 4$ $\quad A = \left(1 + \frac{1}{4}\right)^4 \approx 2.44$

monthly, then $n = 12$ $\quad A = \left(1 + \frac{1}{12}\right)^{12} \approx 2.61$

daily, then $n = 365$ $\quad A = \left(1 + \frac{1}{365}\right)^{365} \approx 2.7146$

each minute, then $n = 525{,}600$ $\quad A = \left(1 + \frac{1}{525{,}600}\right)^{525{,}600} \approx 2.71828$

each second, then $n = 31{,}536{,}000$ $\quad A = \left(1 + \frac{1}{31{,}536{,}000}\right)^{31{,}536{,}000} \approx 2.71828$

As soon as n gets larger than 365, the amount remains steady at \$2.72, assuming that the bank will round to the nearest cent. If we allow the number of interest periods to get larger and larger, we approach what we might describe as *continuous compounding*.

Note that as n increases, the value of $\left(1 + \frac{1}{n}\right)^n$ seems to be approaching e. In fact, this is an alternative way of defining e.

The following formula is derived in calculus.

COMPOUND INTEREST FORMULA (CONTINUOUS COMPOUNDING)

If P dollars is invested at an annual interest rate r compounded continuously, then A, the amount of money present after t years is given by

$$A = Pe^{rt}$$

EXAMPLE 4 At what annual rate, compounded continuously, must \$10,000 be invested to grow to \$25,000 in 8 years?

Solution We use the formula $A = Pe^{rt}$ with $A = 25{,}000$, $P = 10{,}000$, and $t = 8$ and solve for r.

$$25{,}000 = 10{,}000e^{8r} \qquad \textit{Divide both sides by } 10{,}000.$$

$$2.5 = e^{8r} \qquad \textit{Take } \ln \textit{ of both sides and use log property } 3.$$

$$\ln 2.5 = 8r \Rightarrow r = \frac{\ln 2.5}{8} \approx 0.1145$$

Thus the annual rate must be 11.45%. ■

When an equation describes a real-life situation (as the formula $A = Pe^{rt}$ describes the amount in an investment), we often say that the equation is a **model** of the situation. We can generalize this model to other situations where a quantity increases or decreases in proportion to the amount of the quantity present at time t.

If we start with a population or amount of a substance, A_0, which is growing at a continuous rate of r per unit of time (per year, per day, per minute, etc.), then after t units of time (t years, t days, t minutes, etc.) the number or amount present will have grown to A, where A is given by the exponential equation that appears in the following box. Populations or substances whose growth is described by such an equation are said to conform to an **exponential growth model.**

Keep in mind that if the growth rate r is given as a percent, it must be converted to decimal form to use the growth and decay models.

THE EXPONENTIAL GROWTH MODEL

Suppose a population or substance is growing at the continuous rate of r per unit of time. Let A_0 be the initial number or amount present. Then the amount, A, present after t units of time is given by

$$A = A_0 e^{rt} \qquad \text{where } r > 0$$

Similarly, if we start with a population or amount of a substance, A_0, that is decaying at a continuous rate of r per unit of time, then the number or amount still present after t units of time is said to conform to an **exponential decay model.**

THE EXPONENTIAL DECAY MODEL

Suppose a population or substance is decaying at the continuous rate of r per unit of time. Let A_0 be the initial number or amount present. Then the amount, A, still present after t units of time is given by

$$A = A_0 e^{-rt} \qquad \text{where } r > 0$$

Note that in both the exponential models the rate r is taken to be positive. The *sign* of the exponent determines whether the model is exponential growth or decay.

Note that in both the exponential models, the rate r is taken to be positive. In the growth model the exponent is positive, whereas in the decay model the exponent is negative.

EXAMPLE 5 According to a world almanac, the population of the world in 1986 was estimated to be 4.7 billion people. Assuming that the world's population is growing at the rate of 1.8% per year,

(a) Estimate the population of the world in the year 2000.

(b) In what year will the earth's population be 10 billion?

Solution We use the exponential growth model $A = A_0 e^{rt}$ with $A_0 = 4.7$ and $r = 0.018$ (the decimal equivalent of 1.8%). A represents the population (in billions) t years after 1986, and A_0 represents the population (in billions) of the earth in 1986 (this is the initial population). Thus we have

$$A = 4.7e^{0.018t}$$

(a) Since we want to estimate the earth's population in the year 2000, 14 years will have elapsed since the initial year of 1986. We substitute $t = 14$ into the growth equation.

$$A = 4.7e^{0.018t} \qquad \textit{Substitute } t = 14.$$

$$= 4.7e^{.018(14)} = 4.7e^{.252} \qquad \textit{Using a calculator we get}$$

$$A \approx 6.05$$

Thus the earth's population in the year 2000 will be approximately 6.05 billion people according to this model.

(b) In order to determine in what year the earth's population will be 10 billion, we again use the equation $A = 4.7e^{0.018t}$, but this time we want to know what value of t will make $A = 10$.

$$A = 4.7e^{0.018t}$$ *We let $A = 10$ and solve for t.*

$$10 = 4.7e^{0.018t}$$ *Divide both sides by 4.7.*

$$\frac{10}{4.7} = e^{0.018t}$$ *Take ln of both sides.*

$$\ln \frac{10}{4.7} = \ln e^{0.018t} = 0.018t$$

$$\frac{\ln \frac{10}{4.7}}{0.018} = t$$ *Using a calculator we get*

$$t \approx 41.95$$

It would take almost 42 years, or until the year 2028 (42 years after 1986), for the earth's population to reach 10 billi- ■

EXAMPLE 6 Using the exponential growth model, in terms of r how long will it take a population to double if it increases at a rate of r per year?

Solution We begin with the population model $A = A_0e^{rt}$ and note that we are looking for the time t it takes for the amount A to become $2A_0$, twice the initial amount.

$$A = A_0e^{rt}$$ *Substitute $2A_0$ for A.*

$$2A_0 = A_0e^{rt}$$ *Divide both sides of the equation by A_0.*

$$2 = e^{rt}$$ *Take the ln of both sides (or rewrite in ln form).*

$$\ln 2 = rt$$ *Solve for t.*

$$t = \frac{\ln 2}{r}$$

Hence, if a population were growing at a rate of 5% per year, then we have $r = 0.05$ and it would take $\frac{\ln 2}{0.05} \approx \frac{0.6931}{0.05} \approx 13.86$ years to double, whereas a population growing at a rate of 15% would take $\frac{\ln 2}{0.15} \approx 4.6$ years to double. ■

EXAMPLE 7 Suppose we have 100 grams of a radioactive substance that decays at the rate of 4% per hour.

(a) How long will it take until only 50 grams of radioactive substance remain?

(b) How long will it take until only 25 grams of radioactive substance remain?

(c) Find the time it takes for an initial amount A_0 of this radioactive substance to decay to half this amount. This time is called the *half-life* of the substance.

Solution The exponential decay equation for this substance is

$$A = 100e^{-0.04t}$$

(a) In order to determine how long it will take the 100 grams to decay to 50 grams, we set $A = 50$ and solve for t.

$$50 = 100e^{-0.04t} \qquad \textit{Divide both sides by 100.}$$

$$0.5 = e^{-0.04t} \qquad \textit{Take ln of both sides.}$$

$$\ln 0.5 = -0.04t$$

$$t = \frac{\ln 0.5}{-0.04} \approx 17.33$$

Thus it takes approximately 17.33 hours for the 100 grams to decay to 50 grams.

(b) We can determine the time it takes for 25 grams of radioactive substance to remain either by using exactly the same approach as in part (a) or by finding the time it takes for 50 grams to decay to 25 grams and then add this time to the answer obtained in part (a). We illustrate both approaches.

$$25 = 100e^{-0.04t}$$

$$0.25 = e^{-0.04t}$$

$$\ln 0.25 = -0.04t$$

$$t = \frac{\ln 0.25}{-0.04} \approx 34.66$$

Thus it takes 34.66 hours for 100 grams to decay to 25 grams.

$$25 = 50e^{-0.04t}$$

$$0.5 = e^{-0.04t}$$

$$\ln 0.5 = -0.04t$$

$$t = \frac{\ln 0.5}{-0.04} \approx 17.33$$

So it takes another 17.33 hours for 50 grams to decay to 25 grams. Thus it takes a total of 34.66 hours for 100 grams to decay to 25 grams.

(c) In order to determine the half-life of this substance, we want to know how long it takes an initial amount A_0 to decay to an amount $\frac{1}{2}A_0$.

$$\frac{1}{2}A_0 = A_0e^{-0.04t} \qquad \textit{Divide both sides by } A_0.$$

$$0.5 = e^{-0.04t} \Rightarrow \ln 0.5 = -0.04t$$

$$t = \frac{\ln 0.5}{-0.04} \approx 17.33$$

Thus the half-life of this substance is approximately 17.33 hours. Note that (as parts (a) and (b) suggested) this answer is independent of the initial amount. This fact is crucial for scientists' ability to use the radioactive decay of carbon 14 to determine the age of fossils and archeological artifacts. (See Exercise 47.) ■

EXAMPLE 8 Suppose a colony of bacteria grows from a population of approximately 600 to 4500 in 12 hours. Find an exponential growth model for these bacteria.

Solution We start with the exponential growth model $A = A_0e^{rt}$ and note that we are given that $A = 4500$, $A_0 = 600$, and $t = 12$, and we need to solve for r.

$$4500 = 600e^{12r} \qquad \textit{Divide both sides by 600.}$$

$$7.5 = e^{12r} \qquad \textit{Take ln of both sides.}$$

$$\ln 7.5 = 12r \Rightarrow r = \frac{\ln 7.5}{12} \approx 0.17$$

Thus an exponential growth model for these bacteria is $A = A_0e^{0.17t}$. ■

EXAMPLE 9 The half-life of a radioactive substance is 8 minutes. How long does it take 80 grams of this substance to decay to 7 grams?

Solution

WHAT DO WE NEED TO DO?	Find how long it takes for 80 grams of a substance to decay to 7 grams, given its half-life.
HOW DO WE START?	First write the model for exponential decay: $A = A_0e^{-rt}$. Then determine what variables are given and what needs to be found.
WE HAVE $A_0 = 7$ AND $A = 80$. WE NEED TO FIND t, BUT WE ARE NOT GIVEN r.	Is there information given in the problem that can help us to find r?
WE ARE GIVEN THE HALF-LIFE OF THE SUBSTANCE. HOW CAN WE USE THAT INFORMATION TO FIND r?	Remember that the half-life of a substance is the amount of time it takes for half the initial amount to decay. If we let $A = \frac{1}{2}A_0$ and $t = 8$ minutes, we can substitute these values into the exponential decay model to solve for r.

$$A = A_0e^{-rt}$$ *The half-life, the time it takes for half the substance to decay, is 8 minutes, which means it takes 8 minutes for A to become* $\frac{1}{2}A_0$, *so we substitute* $A = \frac{1}{2}A_0$ *and* $t = 8$ *to get*

$$\frac{1}{2}A_0 = A_0e^{-8r}$$ *Remember, we first need to solve for r. Divide both sides by* A_0.

$$\frac{1}{2} = e^{-8r}$$ *Now translate this into ln form.*

$$\ln 0.5 = -8r$$ *Solve for r.*

$$r = \frac{\ln 0.5}{-8} \approx 0.086643$$

NOW THAT WE HAVE r, WHAT DO WE NEED NEXT?

Given r, we now have the decay model for the substance: $A = A_0e^{-.086643t}$, where t is measured in minutes. Now we can substitute $A_0 = 80$ and $A = 7$ in the model to find t, the time it takes for 80 grams to decay to 7 grams:

$$A = A_0e^{-0.086643t} \quad \textit{Substitute } A_0 = 80 \textit{ and } A = 7.$$

$$7 = 80e^{-0.086643t} \quad \textit{Divide both sides by } 80.$$

$$0.0875 = e^{-0.086643t} \quad \textit{Rewrite in } \ln \textit{ form.}$$

$$\ln 0.0875 = -0.086643t \quad \textit{Solve for } t.$$

$$t = \frac{\ln 0.0875}{-0.086643} \approx 28.12 \text{ min}$$

See Exercise 59 for a slightly different approach to this example.

The exercises illustrate many areas in which exponential and logarithmic functions arise.

EXERCISES 5.5

1. If a store raises its prices by 20% and then reduces its prices by 20%, are the items back to their original prices? Explain.
2. Are successive price reductions of 25% and 15% equivalent to a price reduction of 40%? Explain.

In the following exercises, assume that the situations and populations described are governed by an exponential growth or decay model.

3. Plutonium-239 (^{239}Pu), one of the radioactive waste by-products resulting from the production of nuclear energy, has a half-life of approximately 25,000 years. Find the exponential decay model for ^{239}Pu.
4. The formula for the radioactive decay of radium is $A = A_0e^{-0.0004279t}$. Find the half-life of radium.
5. The half-life of carbon-14 (^{14}C) is approximately 5730 years. Of 100 grams of ^{14}C, how much will be left after 1000 years?
6. How much of 50 grams of ^{14}C will have decayed in 100 years?
7. Neptunium-239 (^{239}Np) has a short half-life of approximately 2.24 days. Find the decay model for neptunium and determine how much of an initial amount of 40 grams will be radioactive after 30 days.
8. Calculate how long it will take for 10 grams of neptunium to decay to less than 0.1 gram.
9. If it takes 100 years for 80 grams of a radioactive substance to decay to 60 grams, how long will it take for an amount of this substance to decay to one-fifth that amount?
10. Strontium-90, another waste by-product of nuclear fission reactors, has a half-life of approximately 28 years. How long will it take a sample of strontium 90 to be reduced by a factor of 100?
11. A population of a colony of bacteria increases according to the growth model $A = A_0 3^{t/20}$ (where t is measured in hours). How long will it take for the population to grow from 100 to 200? From 100 to 300? (You should be able to answer this second question without using a calculator.)
12. A certain type of bacteria grows according to the growth model $A = A_0e^{0.357t}$, where the time t is measured in hours. If a colony of bacteria starts with approximately 400 bacteria, how many will there be after 6 hours? After 2 days?
13. According to 1986 *World Almanac* census data, the population of the United States increased from approximately 203.3 million in 1970 to 226.5 million in 1980. Find the annual growth rate during this ten year period.
14. According to 1986 *World Almanac* census data, the population of the U.S. increased from approximately 39.8 million in 1870 to 50.2 million in 1880. Find the annual growth rate during this 10-year period. Compare this rate with the result of Exercise 11. Does the result surprise you? What do you think caused the significantly higher growth rate for the period between 1870 and 1880?
15. It is estimated that the world's population was 4.7 billion in 1986. Assuming a growth rate of 1.8% for the foreseeable future, what will the earth's population be in the year 2000?

16. It is estimated that in 1986, the population of the "third world" was 3.7 billion, or about 79% of the world's population. If the third world's population is growing at a rate of 2.1%, use the result of Exercise 15 to determine what part of the world's population the third world population will be in the year 2000.

17. According to United Nations' estimates, the population of the Republic of China was 686.4 *million* in 1960 and 1.032 *billion* in 1984. Find the growth rate of the Republic of China's population. Is this growth rate higher or lower than the world's current growth rate of 1.8%?

18. Bacteria B is known to satisfy the exponential growth model $A = A_0e^{0.37t}$, where t is the number of hours the initial number of bacteria are placed in a particular growth medium at a certain temperature. Suppose a lab technician places 800 bacteria of an unknown type in the same growth medium at the same temperature and observes that after 18 hours there are approximately 12,800 bacteria present. Would you conclude that these bacteria are of type B? Explain.

19. If $10,000 is deposited in a bank account paying an interest rate of 7.6% per year compounded quarterly, how long will it take for there to be $15,000 in the account?

20. If $8500 is invested at 6.5% per year compounded daily, how long will it take for the investment to double?

21. What is the effective annual yield if $3000 is invested at 6.7% per year compounded monthly? Does your answer depend upon the amount of money in the account?

22. How many years will it take $500 invested at 8.3% compounded weekly to grow to $1500?

23. What rate of continuous compounding gives an effective annual yield of 6.7%?

24. Find the doubling time for an investment at 6% compounded continuously. What must the interest rate be if the doubling time is to be cut in half?

25. How much money must be invested at 5.8% compounded continuously to yield $6,000 after 5 years?

26. How much money must be invested at 8% compounded continuously to yield A dollars after 10 years?

27. Find the number of years it takes $1000 invested at 6% and compounded continuously to double. Compare this to the doubling time if the interest rate is compounded daily, which was computed in Example 2. How does this reflect on the statement made in the text that increasing the number of compoundings does not significantly increase the amount of interest accumulated?

28. A common credit card company charges an annual interest rate of 18.5% compounded monthly. If you make a credit card purchase of $600 and don't make any payments for a year, how much do you owe at the end of the year?

In Exercises 29–32 use the information given in Exercise 29.

29. The repayment schedule for most long term loans, such as mortgages and car loans is calculated so that the amount borrowed plus the interest is paid back in equal monthly payments. The general formula for the monthly payment, M, required on an amount A borrowed at an annual rate r for t years is

$$M = \frac{Ar}{12\left[1 - \left(1 + \frac{r}{12}\right)^{-12t}\right]}$$

(a) What are the monthly payments required on a car loan of $8000 at a rate of 9% per year for a period of 4 years?

(b) Over this 4-year period, how much has been paid back to borrow the $8000?

30. If someone can afford a repayment schedule of $200 per month for 3 years, how large a car loan can they afford at a rate of 10%?

31. **(a)** What would the monthly payments be on a 30-year mortgage of $80,000 at 8.2%?

(b) Over the course of this 30-year mortgage, how much money has been paid back in total?

32. If a couple wishes to budget $700 per month for their mortgage payments, how large a 30-year mortgage can they afford at 7.8%?

THE RICHTER SCALE

33. Seismologists measure the magnitude of earthquakes using the Richter scale. This scale defines the magnitude, R, of an earthquake as

$$R = \log\frac{I}{I_0} \qquad \text{(remember that log means } \log_{10}\text{)}$$

where I is the intensity of the earthquake being measured and I_0 is the intensity of a *zero-level* earthquake.

(a) Find the Richter scale measure of an earthquake that is 1000 times as intense as a zero-level earthquake—that is, $I = 1000I_0$.

(b) The great 1906 earthquake in San Francisco had a Richter scale measure of approximately 8.3. Compare the intensity of this earthquake to a zero-level earthquake.

34. The Loma Prieta earthquake, which interrupted the 1989 World Series, registered 7.1 on the Richter scale. Compare the intensities of the Loma Prieta quake and the great 1906 quake, which registered 8.3 on the Richter scale.

35. Suppose that three earthquakes E_1, E_2, and E_3 register $R_1 = 2$, $R_2 = 6$, and $R_3 = 10$, respectively, on the

Richter scale. Note that the differences between R_1 and R_2, and R_2 and R_3 are both 4. Compare the intensities of the three earthquakes.

36. Show that if earthquake E_2 has a Richter scale magnitude of R_2 and twice the intensity of earthquake E_1, which has a Richter scale magnitude of R_1, then $R_2 - R_1 = \log 2$.

pH LEVELS

37. Chemists have defined the pH (which stands for *hydrogen potential*) of a solution to be

$$\text{pH} = -\log[H_3O^+]$$

where $[H_3O^+]$ stands for the concentration of the hydronium ion in the solution (measured in moles per liter). The pH is a measure of the acidity or alkalinity of the solution. Water, which is neutral, has a pH of 7. Solutions with a pH below 7 are acidic, whereas those with a pH above 7 are alkaline. What is the pH of a glass of orange juice if its hydronium ion concentration is 6.82×10^{-5}?

38. What is the hydronium ion concentration of a solution with a pH of 9.1?

39. Environmentalists are constantly monitoring the pH levels of rain and snow because of the destructive effects of "acid rain," which is caused primarily by sulfur dioxide emissions. Rain and snow have a natural concentration of $[H_3O^+] = 2.5 \times 10^{-6}$ due to dissolved carbon dioxide that is normally in the atmosphere. Find the natural pH of rain and snow.

40. If a rain sample shows that the concentration of $[H_3O^+]$ has increased by a factor of 100, what will its pH level be? Use the information in Exercise 39.

PSYCHOPHYSICS

41. The *Weber-Fechner* law in psychophysics relates the intensity S of a sensation (a psychological reaction) to the intensity P of the physical stimulus. This law was published in 1930 by Gustav Fechner and was based upon experiments performed by Ernst Weber a year earlier. It states that the change in the intensity of a sensation caused by a small change in the intensity of a stimulus is proportional not to the amount of change in P, as we might expect, but rather to the *percentage* change in P. Using calculus, this relationship gives the Weber-Fechner law, which says that

$$S = K \ln\left(\frac{P}{P_0}\right)$$

where P_0 is the minimum intensity that can be perceived, and K is some constant that determines the units in which S is measured.

The brightness of stars, as seen by the naked eye, is measured in units called *magnitudes*. The Greek astronomer Ptolemy set up six categories; the dimmest stars are of magnitude 6 and the brightest stars are of magnitude 1. If we let I be the brightness of a star of magnitude M and I_0 be the minimal brightness a star must have to be visible, then using the Weber-Fechner law we can derive the following formula for the magnitude M.

$$M = 6 - 2.5 \log\left(\frac{I}{I_0}\right)$$

(a) Calculate the ratio of the light intensity of a star of magnitude 2 to that of a star of magnitude 1.
(b) Calculate the ratio of the light intensity of a star of magnitude 5 to that of a star of magnitude 4.
(c) Based upon the results of (a) and (b), how much more intense is the light of a star of any magnitude than a star of magnitude one less?
(d) If a star's light intensity increases by a factor of 40 before it goes nova, how many magnitudes does it increase?

42. Suppose that the smallest electrical current you can feel is 1 milliampere and that the minimum *difference* in current you can detect is 25 milliamperes. Use the Weber-Fechner law to develop a scale of perceived electrical current. Call one unit on this scale a *jolt*, and write a formula for the number of jolts corresponding to a current of A milliamperes.

SOUND INTENSITY

43. The unit of measurement frequently used to measure sound levels is the *decibel* (dB). The number of decibels, N, of a sound with intensity I (usually measured in watts per square centimeter, or W/sq cm) is defined to be

$$N = 160 + 10 \log I$$

The faintest sound audible to the human ear is a sound with an intensity of approximately 10^{-16} W/sq cm. What is the decibel level of this sound?

44. A sound at the threshhold of pain to the human ear is a sound with an intensity of approximately 10^{-4} W/sq cm. What is the decibel level of this sound?

NATURAL RESOURCES

45. Many of the world's natural resources are being consumed at a rate that is increasing exponentially with respect to time. Assume an exponential growth model for the consumption of petroleum products.

(a) If the world consumed approximately 1.5 billion barrels of petroleum products in 1940 and 3.6 billion barrels in 1960, estimate the rate at which the consumption was increasing over this 20-year period.

(b) If the world consumed approximately 12.1 billion barrels of petroleum products in 1965 and 20.4 billion barrels in 1975, estimate the rate at which the consumption was increasing over this 10-year period.

46. It can be shown (using calculus) that if an amount A_0 is consumed in a certain year, and assuming an annual growth rate of r, then the amount A of oil consumed in the next T years is given by the formula

$$A = \frac{A_0}{r}(e^{rT} - 1)$$

(a) Show that solving this formula for T gives

$$T = \frac{\ln\left[\frac{rA}{A_0} + 1\right]}{r}.$$

(b) In 1990 it was estimated that confirmed available world oil reserves were 983.4 billion barrels of oil and that 21.3 billion barrels of oil were consumed that year. Assuming a rate of growth in oil consumption of 2.5%, use the formula derived in part (a) to determine how many years the 1990 oil reserves will last.

CARBON DATING

47. All living plant and animal tissue contains both carbon-12, which is not radioactive, and carbon-14, which is radioactive with a half-life of approximately 5730 years. While the organism is alive, the ratio of carbon-14 to carbon-12 remains constant. After the organism dies (we will call this $t = 0$), the carbon-14 begins to decay. Thus the smaller the ratio of carbon-14 to carbon-12, the older the remains.

(a) Verify that the decay model for carbon-14 is $A = A_0e^{-0.000121t}$.

(b) Given that the half-life of carbon-14 is 5730 years, explain why an alternative decay model is $A = A_0\left(\frac{1}{2}\right)^{t/5730}$. HINT: Examine what happens when $t = 5730$, $t = 11{,}460 = 2(5730)$, etc.

(c) Verify that both equations give the same result for the amount of carbon-14 left from an initial amount of 60 grams after 1000 years.

(d) Suppose that a fossil now contains 65% of the carbon-14 it contained originally. Estimate the age of the fossil.

48. Skeletal remains are found to contain 35% of their original carbon-14. Estimate the age of the skeleton.

49. In 1947 an Arab Bedouin herdsman climbed into a cave near Qumran on the shores of the Dead Sea in search of a stray goat. He found some earthenware jars containing what we now know as the *Dead Sea Scrolls*. The wrappings of the scrolls were analyzed and found to contain 76% of their original carbon-14. Estimate the age of the Dead Sea Scrolls.

50. If a fossil is estimated to be 1 million years old, what percentage of its original carbon-14 remains?

MISCELLANEOUS PROBLEMS

51. Suppose that water and chlorine are continuously being added to a swimming pool in such a way that the number of grams of chlorine in the pool at time t hours is given by

$$c(t) = 100 - 30e^{-t/10}$$

(a) How much chlorine is in the pool initially (at time $t = 0$)?

(b) How much chlorine is in the pool after 5 hours?

(c) How much chlorine is in the pool after 10 hours?

(d) How much chlorine is in the pool after 100 hours?

52. A basic cooling principle in physics (called Newton's law of cooling) states that if an object at temperature T_0 is placed into a surrounding environment, which is at a constant temperature C, then the temperature, T, of the object after t minutes is given by

$$T = C + (T_0 - C)e^{-kt}$$

where k is a constant that depends upon the particular object.

(a) Determine the constant k (to the nearest hundredth) for a bottle of orange juice that takes 10 minutes to cool from 70°F to 55°F after being placed in a refrigerator that is maintaining a constant temperature of 45°F.

(b) What will the temperature of the juice be $\frac{1}{2}$ hour after it is placed in the refrigerator?

(c) According to the given formula for T, is it possible for the temperature of the juice to reach 45°F? Explain.

53. If inflation is running at a steady rate of r per year over an extended period of time, then the value of A_0 dollars after t years is given by

$$A = A_0(1 - r)^t$$

Assume an inflation rate of 4.8%.

(a) How much will \$100 be worth after 5 years?

(b) How long will it take until \$100 is worth only \$50?

54. An altimeter is an instrument that measures altitude. The altimeters used in most aircraft measure the altitude by measuring the outside barometric pressure, P, and then displaying the altitude by means of a scale that is calibrated by using the following *barometric equation:*

$$a = (30T + 8000)\ln\left(\frac{P_0}{P}\right)$$

which relates the altitude a in meters above sea level, the air temperature T in degrees Celsius, the atmospheric pressure P_0 at sea level, and the atmospheric pressure P at altitude a. (Atmospheric pressure is measured in centimeters of mercury.) Suppose the atmospheric pressure at a certain altitude is 24.9 cm of mercury and the temperature is -3°C. If the atmospheric pressure at sea level is 76 cm of mercury, use the barometric equation to find the altitude to the nearest foot (1 meter is approximately 3.3 ft).

55. The number $n!$ (read *n factorial*) is defined as

$$n! = n(n - 1)(n - 2) \cdot \cdot \cdot 3 \cdot 2 \cdot 1$$

for all positive integers n. For example,

$$5! = 5 \cdot 4 \cdot 3 \cdot 2 \cdot 1 = 120.$$

Stirling's formula, which states

$$n! \approx \left(\frac{n}{e}\right)^n \sqrt{2\pi n}$$

can be used to approximate very large factorials. Use Stirling's formula to approximate 20! and compare it to the actual value.

QUESTIONS FOR THOUGHT

56. Develop a formula for the half-life of a substance that is decaying at the rate of r per year.

57. If we are exposed to two sounds with intensities I_1 and I_2 and decibel levels N_1 and N_2, respectively, then we are experiencing sound at an intensity level of $I_1 + I_2$. However, the decibel level of the combination of sounds is *not* $N_1 + N_2$. In other words, combining two sounds of 80 decibels and 90 decibels does not produce a sound of 170 decibels. Suppose that union regulations prohibit employers from constructing a work environment in which a worker is exposed to a sound level greater than 100 decibels. If an employee works at a machine that produces 90 decibels and then a new machine is moved nearby that produces 80 decibels, has the union regulation been violated?

58. Explain how the decibel scale discussed in Exercise 43 is an application of the Weber-Fechner law discussed in Exercise 41.

59. In Example 9 we used the exponential decay model to find how long it would take 80 grams of a substance with a half-life of 8 minutes to decay to 7 grams.

(a) Explain why an alternate exponential decay model for this substance is $A = A_0\left(\frac{1}{2}\right)^{t/8}$.

(b) Use this alternate decay model to answer the question of Example 9. Which model was easier to use?

Chapter 5 **SUMMARY**

After completing this chapter you should:

1. Recognize the graphs of the basic exponential and logarithmic functions. See Figure 5.17. (Sections 5.1 and 5.2)

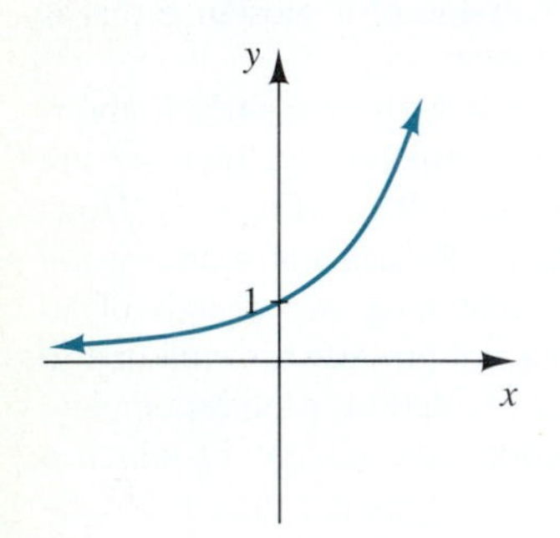

$y = b^x$ for $b > 1$

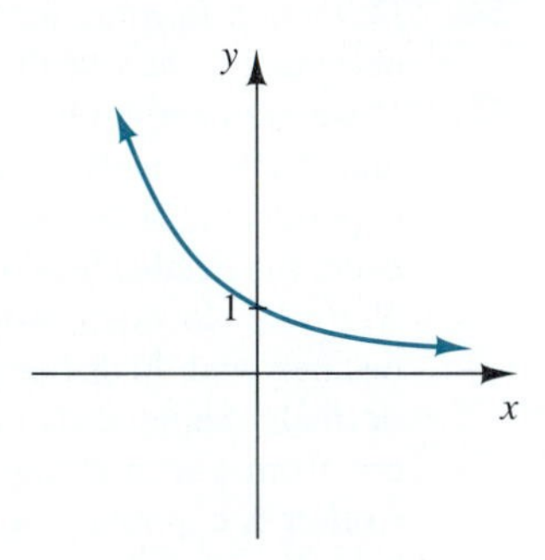

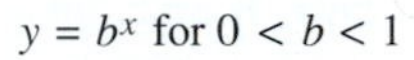

$y = b^x$ for $0 < b < 1$

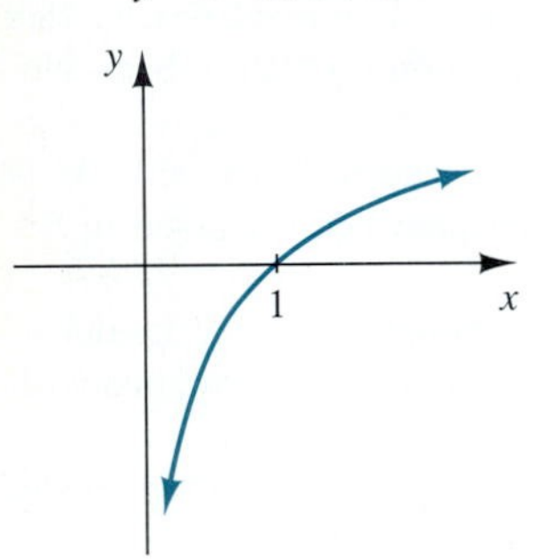

$y = \log_b x$ for $b > 1$

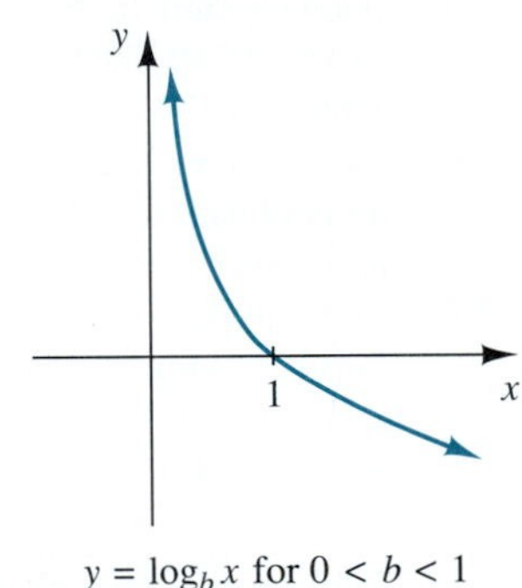

$y = \log_b x$ for $0 < b < 1$

FIGURE 5.17

Be able to translate statements from exponential form to logarithmic form, and vice versa. (Section 5.2)
A logarithm is simply an exponent

$$x = b^y \Leftrightarrow y = \log_b x$$

That is, y is the exponent to which b must be raised to yield x.
For example:

The exponential statement $4^{-3} = \frac{1}{64}$ is equivalent to the logarithmic statement $\log_4 \frac{1}{64} = -3$.

3. Be able to use the definition of logarithms to evaluate logarithmic expressions. (Section 5.2)

For example:
Find $\log_8 \frac{1}{4}$.

We let $\log_8 \frac{1}{4} = a$, translate this into exponential form, and solve the resulting exponential equation.

$$\log_8 \frac{1}{4} = a \quad \textit{is equivalent to}$$

$$8^a = \frac{1}{4} \quad \textit{Express both sides in terms of base 2.}$$

$$(2^3)^a = \frac{1}{2^2}$$

$$2^{3a} = 2^{-2} \Rightarrow 3a = -2 \Rightarrow a = -\frac{2}{3}$$

Thus $\log_8 \frac{1}{4} = -\frac{2}{3}$

4. Be able to use the properties of logarithms to rewrite logarithmic expressions. (Section 5.3)
The logarithm properties allow us to express the logarithm of products, quotients, and powers as the sum and difference of simpler logarithms.
For example:
Write in terms of simpler logarithmic expressions: $\log_b\left(\frac{\sqrt[3]{xy}}{z^4}\right)$.

$$\log_b\left(\frac{\sqrt[3]{xy}}{z^4}\right) = \log_b\left(\frac{(xy)^{1/3}}{z^4}\right) \quad \textit{Use property 2.}$$

$$= \log_b(xy)^{1/3} - \log_b z^4 \quad \textit{Use property 3.}$$

$$= \frac{1}{3}\log_b(xy) - 4\log_b z \quad \textit{Use property 1.}$$

$$= \frac{1}{3}(\log_b x + \log_b y) - 4\log_b z$$

5. Be able to use the logarithm properties to solve logarithmic equations. (Section 5.3)
For example:
Solve for x.

$$\log_2(3x - 1) + \log_2(x + 1) = 5$$

Write as a single logarithm.

$$\log_2[(3x - 1)(x + 1)] = 5$$

Write in exponential form.

$$(3x - 1)(x + 1) = 2^5 = 32$$

$$3x^2 + 2x - 33 = 0$$

$$(3x + 11)(x - 3) = 0$$

$$x = -\frac{11}{3} \quad \text{or} \quad x = 3$$

We reject $x = -\frac{11}{3}$ because it makes the argument negative; the solution to the equation is $\boxed{x = 3}$.

6. Be able to use logarithms to solve exponential equations. (Section 5.4)
If two expressions are equal, their logarithms are equal and we can take the log of both sides. We then apply the log properties to solve the equation.
For example:
Solve for t: $7^{2t-1} = 5$.
Since we cannot easily express both sides in terms of the same base, we can use logarithms to solve the equation.

$$7^{2t-1} = 5 \qquad \textit{Take the natural logarithm of both sides.}$$

$$\ln 7^{2t-1} = \ln 5 \qquad \textit{Use the log properties.}$$

$$(2t - 1)\ln 7 = \ln 5 \Rightarrow t = \frac{\ln 5}{2 \ln 7} + \frac{1}{2} \approx 0.914$$

7. Solve applied problems involving exponential or logarithmic functions. (Section 5.5)
Many real-life situations, such as continuous compounding, population growth, and radioactive decay, can be solved using the exponential growth model or the exponential decay model.
For example:
What sum of money must be invested at 7.3% compounded continuously to yield \$10,000 in 12 years?
Solution:
We use the formula $A = A_0e^{rt}$ with $A = 10{,}000$, $r = 0.073$, and $t = 12$ and solve for A_0.

$$10{,}000 = A_0e^{0.073(12)}$$

$$10{,}000 = A_0e^{0.876}$$

$$A_0 = \frac{10{,}000}{e^{0.876}} \approx \$4164 \quad \text{to the nearest dollar}$$

Chapter 5 REVIEW EXERCISES

In Exercises 1–6, sketch the graph of the given function. Be sure to indicate any asymptotes and label the intercepts.

1. $y = 2^{x-1}$
2. $f(x) = \left(\frac{1}{4}\right)^x - 1$
3. $f(x) = \log_{2/3}(x + 3)$
4. $y = 5 + \log_5 x$
5. $y = e^{x+2} - 1$
6. $f(x) = 2 + \ln x^2$

In Exercises 7–12, translate the logarithmic statements into exponential form and the exponential statements into logarithmic form.

7. $\log_6 \frac{1}{6} = -1$
8. $9^{1/2} = 3$
9. $8^{-2/3} = \frac{1}{4}$
10. $\log_3 81 = 4$
11. $\log_b b^6 = 6$
12. $b^{\log_b t} = t$

In Exercises 13–20, evaluate the given logarithm.

13. $\log_{10} 10{,}000$
14. $\log_{10} 0.000001$
15. $\log_3 \frac{1}{9}$
16. $\log_{1/2} 8$
17. $\log_{32} 16$
18. $\log_b \sqrt{b^3}$
19. $\log_b 1$
20. $\log_{1/5} 25$

In Exercises 21–26, express the given logarithm in terms of simpler logarithms where possible.

21. $\log_b(x^3y^4z^2)$
22. $\log_b\left(\frac{b^3}{\sqrt{xy}}\right)$
23. $\log_b \sqrt[3]{\frac{6x}{by^4}}$
24. $\log_b(x^2 + y^5)$
25. $\frac{\sqrt{\log_b x}}{\sqrt[3]{\log_b y}}$
26. $\log_b \frac{\sqrt{x}}{\sqrt[3]{y}}$

In Exercises 27–40, solve the given exponential or logarithmic equation.

27. $8^x = \frac{1}{64}$ **28.** $\left(\frac{1}{9}\right)^x = 29$

29. $\log_b(3x) + \log_b(x + 2) = \log_b 9$

30. $\log_2 x + \log_2(x + 1) = 1$

31. $7^{x-1} = 3$

32. $\log_2(t + 1) + \log_2(t - 1) = 3$

33. $\log_5(6x) - \log_5(x + 2) = 1$

34. $3^x = 5^{x+2}$ **35.** $3^x = 5(2^x)$

36. $\log_4 x - \log_4(x - 4) = \log_4(x - 6)$

37. $\frac{1}{2}\log_3 x = \log_3(x - 6)$

38. $2\log_b x = \log_b(6x - 5)$ **39.** $8^{3x-2} = 9^{x+2}$

40. $\log x = 2 + \log(x - 1)$ **41.** $\log_b 125 = 3$

42. $\log_8 128 = x$

43. Verify that $f(x) = e^{2x-3}$ and $g(x) = \frac{3 + \ln x}{2}$ are inverse functions.

44. Find the inverse function of $y = f(x) = 2^{x+3}$.

45. Find the inverse function of

$$y = f(x) = 5 + \log_3(x - 1)$$

46. Express $\log_6 x$ in terms of base 3 logarithms.

47. Estimate $\log_7 11$. Round your answer to two decimal places.

48. If $2000 is invested in an account paying 6.5% annual interest compounded daily, how much money will be in the account after 6 years?

49. How much money must be invested at 7.2% annual interest compounded continuously so that the investment will be worth $10,000 in 8 years?

50. What annual interest rate compounded continuously will make an investment of $1000 grow to $2000 in 6 years?

51. A colony of bacteria is growing according to the exponential growth model. Suppose 800 bacteria grow to 2000 in 16 hours.

(a) How many bacteria will be present after 10 hours?

(b) How long will it take for there to be 3000 bacteria?

52. Find the doubling time of a colony of bacteria with growth equation $A = A_0 2^{0.06t/4}$, where t is measured in hours.

53. A radioactive substance has a half-life of 53 days. Assuming an exponential decay model, find the decay equation for this model.

54. Certain manufactured radioactive substances have extremely short half-lives. Find the exponential decay model for a substance with a half-life of 5 minutes.

55. In the exponential decay model $A = A_0 e^{-rt}$, describe how the half-life is affected if r doubles from 0.1 to 0.2.

Chapter 5 PRACTICE TEST

1. Sketch the graphs of the following functions. Label the intercepts.

(a) $y = \left(\frac{1}{2}\right)^{x+1} - 4$ **(b)** $y = 1 + \log_3 x$

2. Given $y = f(x) = 2^{x+1}$, find its inverse function and sketch the graphs of both $f(x)$ and its inverse on the same coordinate system.

3. Evaluate each of the following. Round to the nearest hundredth where necessary

(a) $\log_9 \frac{1}{3}$ **(b)** $\log_7 4$

(c) $\log_8 16$ **(d)** $\log_{1/2} \frac{1}{4}$

4. Solve each of the following equations. Round your answer to the nearest hundredth where necessary.

(a) $3^{2x-1} = 9$

(b) $\log_3(x - 9) + \log_3(x + 1) = 2$

(c) $\frac{1}{2}\log_2 x - \log_2(x - 3) = 1$ **(d)** $5^{x+1} = 10$

5. A colony of bacteria is growing according to the exponential growth model. Suppose 600 bacteria grow to 1500 in 20 hours.

(a) How many bacteria will be present after 10 hours?

(b) How long will it take for there to be 6000 bacteria?

6. A radioactive substance has a half-life of 750 years. Assuming an exponential decay model, find the decay equation for this model, the half-life of this substance, and how much of 100 grams of this substance will be left in 100 years.

CHAPTER 6

Trigonometry

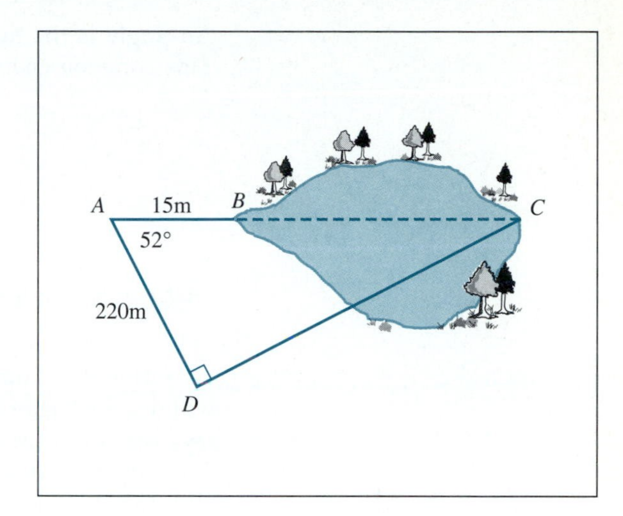

Trigonometry* is the branch of mathematics that focuses on the study of angles. Trigonometry was used by the ancient Greeks more than 2000 years ago to solve everyday problems involving angles in areas such as surveying, navigation, and engineering.

Our study of trigonometry in this chapter begins with a discussion of angle measurement. We then define the trigonometric functions of angles. Up to this point the domains of the functions with which we have dealt have been almost exclusively subsets of the real numbers. We will see that the trigonometric functions can be viewed as having domains consisting of either angles or sets of real numbers.

*The word trigonometry is derived from the Greek *trigōnon*, meaning triangle, and *metria*, meaning measurement.

6.1 Angle Measurement and Two Special Triangles

An **angle** is the figure formed by two half-lines or rays with a common endpoint. This common endpoint is called the **vertex** of the angle. (See Figure 6.1.)

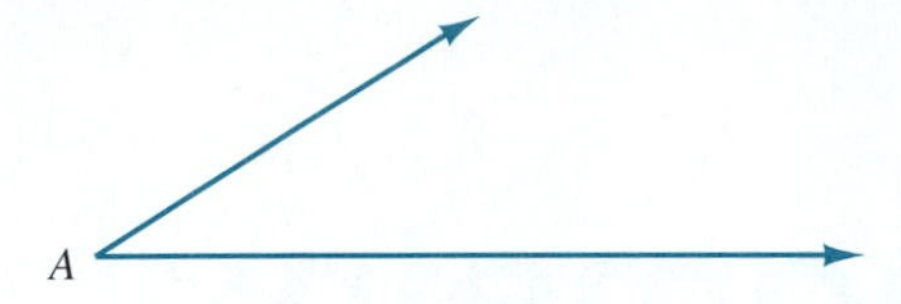

FIGURE 6.1 A is the vertex. This angle is referred to as angle A (often written $\angle A$).

It will suit our purposes to talk about angles from a dynamic point of view. We will discuss angles in terms of size and *orientation*—that is, we will think of the two rays that form the angle as being coincident (starting together). One side remains fixed and the other side rotates to form the angle. The fixed side is called the **initial side** and the side that rotates is called the **terminal side.**

If the terminal side rotates in a counterclockwise direction (as indicated by the arrow inside the angle in Figure 6.2 (a)), we call $\angle B$ a *positive* angle. If the terminal side rotates in a clockwise direction (as indicated by the arrow inside the angle in Figure 6.2 (b)), we call $\angle B$ a *negative* angle.

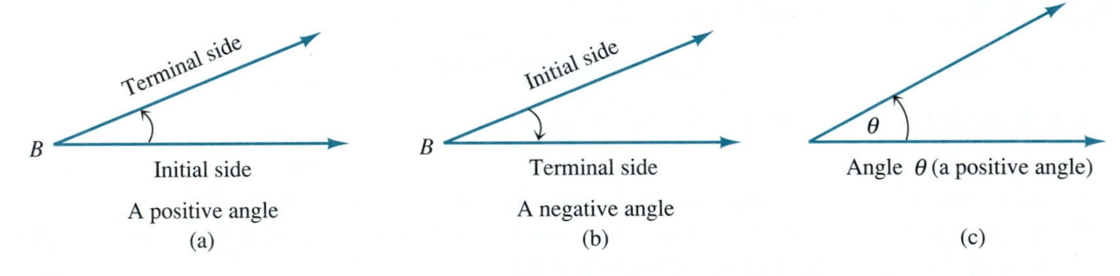

FIGURE 6.2

Besides naming an angle by its vertex, we often put a letter inside the angle and use that as its name. For example, Figure 6.2(c) illustrates angle θ (the Greek letter *theta,* a common name for angles). Some of the other Greek letters frequently used to name angles are α (alpha), β (beta), and γ (gamma).

Measuring Angles

What makes one angle bigger than another angle?

If we were to ask a small child to look at the two angles in Figure 6.3 and tell us which one is "bigger," the child would probably answer that $\angle A$ is bigger, but we know that $\angle B$ is bigger. What does it mean that $\angle B$ is bigger than $\angle A$? What attribute of an angle are we trying to measure when we measure the size of an angle? A moment's thought will lead us to the conclusion that when we measure an angle we are trying to answer the question: Through what part of a complete rotation has the terminal side rotated? *The more the terminal side has rotated, the bigger the angle.*

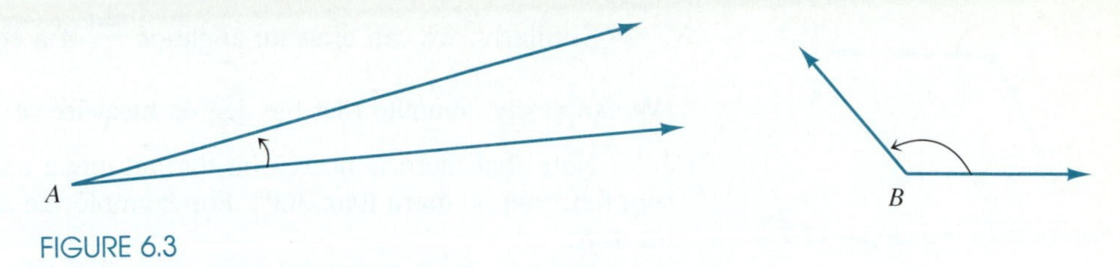

FIGURE 6.3

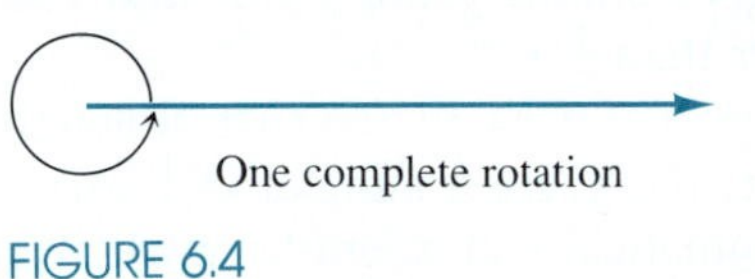

FIGURE 6.4

When we measure an angle we are trying to assign a number to the angle that indicates the size of the angle. The bigger the angle—that is, the more of a complete rotation we have—the bigger the number should be. If we think of the angle in Figure 6.4 as one complete rotation, we might assign it the number 1.

Then we might quite "naturally" be led to the angles and their numerical assignments shown in Figure 6.5.

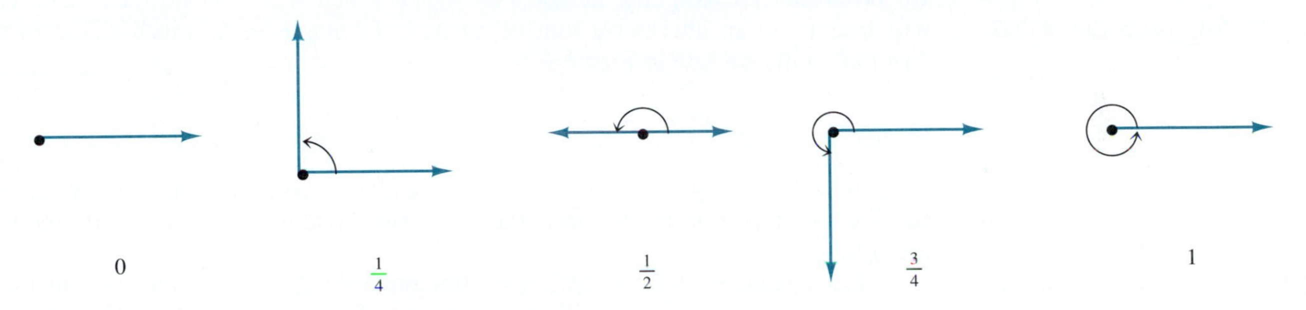

FIGURE 6.5 Possible numerical assignments for various angles

However, these are not the numbers most of us are accustomed to using when we measure angles. Most of us are familiar with using *degree* measure to describe the size of an angle.

Using degree measure simply means that instead of assigning the number 1 to one complete rotation, we assign the number 360; that is, we divide one complete rotation into 360 equal parts. Therefore, one degree (written 1°) is just $\frac{1}{360}$ of a complete rotation.

We can now redraw Figure 6.5 and include the degree measurements of the various angles. (See Figure 6.6.)

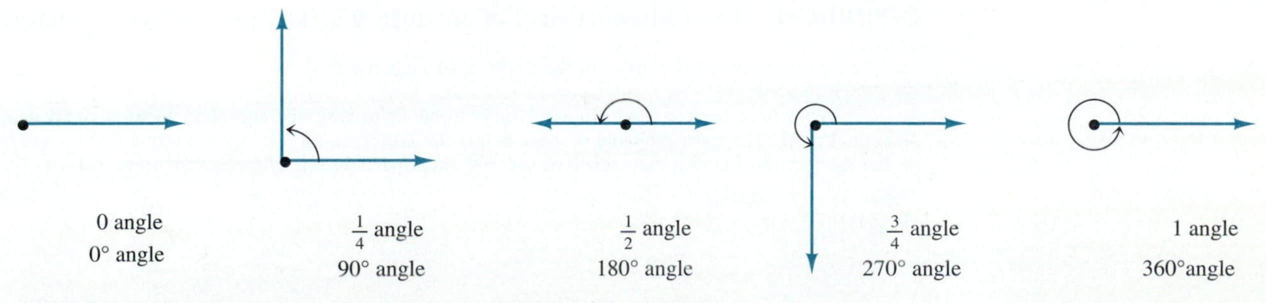

FIGURE 6.6 Possible numerical assignments for various angles

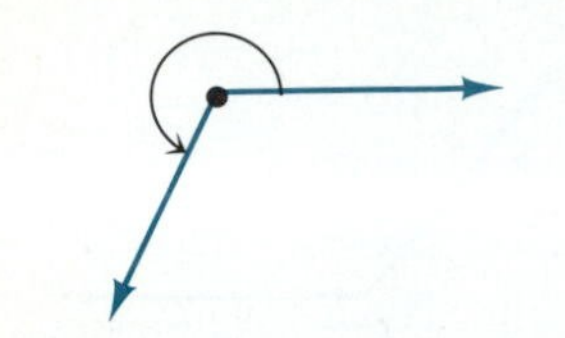

FIGURE 6.7 An angle of $\frac{2}{3}$ of a complete rotation, which is equivalent to 240°

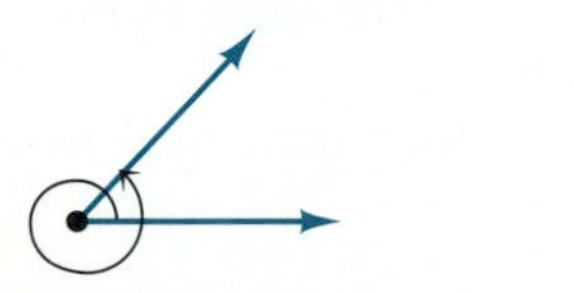

FIGURE 6.8 An angle of 400°

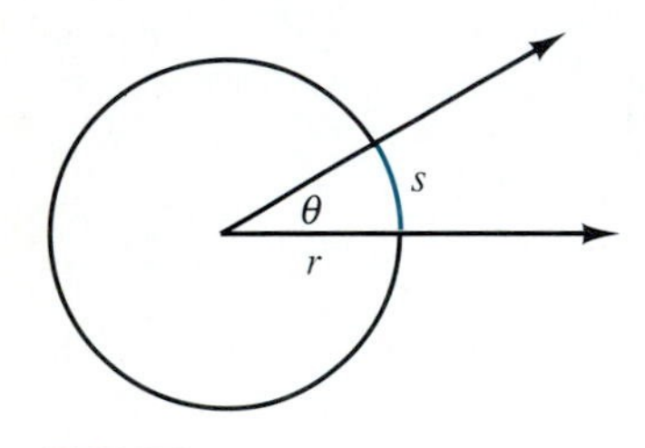

FIGURE 6.9

Similarly, we can draw an angle of $\frac{2}{3}$ of a complete rotation. See Figure 6.7. We can easily compute that the degree measure of this angle is $\frac{2}{3}(360°) = 240°$.

Note that there is no reason that an angle cannot be more than one complete rotation, that is, more than 360°. For example, an angle of 400° would look like Figure 6.8.

Keep in mind that dividing one complete rotation into 360 equal parts is completely arbitrary. In theory we could just as well have divided one complete rotation into 500 equal parts (in which case each part would be smaller than 1°) or into 100 equal parts (in which case each part would be larger than 1°). *

As we proceed with our discussion of trigonometry we will see that although degree measure has the advantage of being familiar, there are a number of reasons why degree measure is inappropriate for much mathematical and scientific work. We will have a bit more to say about this "inappropriateness" after we introduce a different procedure for assigning numbers to angles which will indicate their size. This will lead us to an alternative unit of measure for angles, one much closer to the "natural" units we saw in Figure 6.5.

Radian Measure

Let's consider an angle θ and draw a circle of radius r with the vertex of θ at its center. We let s represent the *length* of the arc of the circle intercepted by $\angle\theta$. See Figure 6.9.

Basic geometry tells us that the central angle θ will be the same fractional part of one complete rotation as s will be of the circumference of the circle. For example, if θ is $\frac{1}{4}$ of a complete rotation, then s will be $\frac{1}{4}$ of the circumference. (Recall that the formula for the circumference, C, of a circle is $C = 2\pi r$.)

In other words, we can set up the following proportion:

$$\frac{\theta}{\text{1 complete rotation}} = \frac{s}{\text{circumference of circle}} = \frac{s}{2\pi r}$$

If we use this ratio of $\frac{s}{2\pi r}$ to measure θ, we get exactly the numbers we saw in Figure 6.5. However, for reasons that will become clear as we proceed, we modify this ratio by multiplying it by 2π. Note that this does not alter the fact that the ratio still reflects the size of the angle. This ratio is then called the **radian** measure of an angle.

DEFINITION The **radian** measure of an angle θ is defined to be $\theta = \frac{s}{r}$, where θ, s, and r are as described in Figure 6.9.

* There have been several historical suggestions as to why one complete rotation is divided into 360 equal parts. The most probable of these is that the Babylonians, who used a numerical system based on the number 60, found it convenient to use 360 equal parts in one complete rotation.

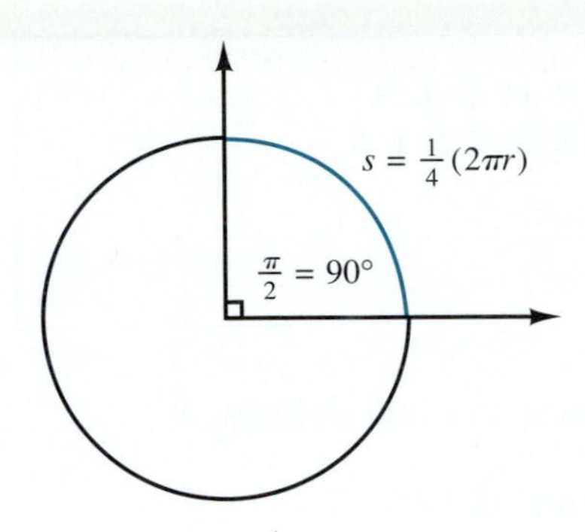

FIGURE 6.10 An angle of $90° = \frac{\pi}{2}$ radians

Thus an angle of 90°, which is $\frac{1}{4}$ of a complete rotation, subtends (cuts off) an arc s that is $\frac{1}{4}$ of the circumference of the circle. See Figure 6.10. Therefore, we have $s = \frac{1}{4}(2\pi r) = \frac{\pi r}{2}$. The radian measure of θ is thus

$$\theta = \frac{s}{r} = \frac{\frac{\pi r}{2}}{r} = \frac{\pi r}{2} \cdot \frac{1}{r} = \frac{\pi}{2}$$

Note that this result is independent of r.

Consequently, $90° = \frac{\pi}{2}$ radians; multiplying this result by 2 gives us

$$\boxed{180° = \pi \text{ radians}}$$

There are several important points to be made here. First, keep in mind that we are not saying that the numbers 180 and π are equal, any more than the fact that 36 inches is equal to 3 feet means that the numbers 3 and 36 are equal. As a number, π is approximately equal to 3.14 (remember that π is irrational). What we are saying is that an angle that measures π radians is the same size as an angle that measures 180°, in the same way that we might say that a table that is 2 meters long is the same length as a table that measures 6.56 feet long. Using degree measure, one complete rotation is 360°; using radian measure, one complete rotation is 2π radians.

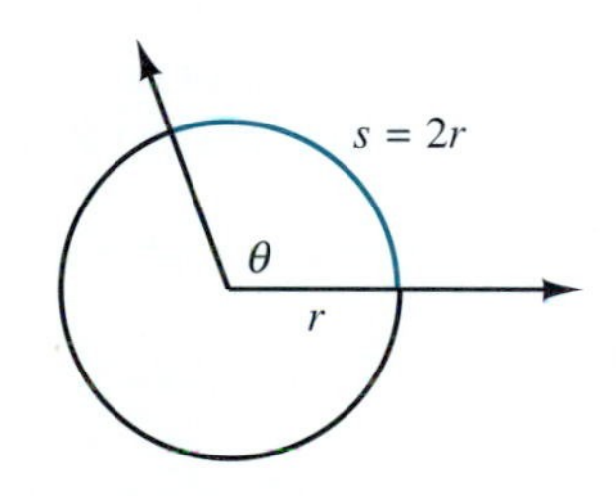

FIGURE 6.11 θ is an angle of 2 (radians).

Second, it is important to recognize that the radian measure of an angle is a *real number* with no units attached. In the definition $\theta = \frac{s}{r}$, both s and r are measured in units of length. For example, if $s = 6$ cm and $r = 3$ cm, then $\theta = \frac{6 \text{ cm}}{3 \text{ cm}} = 2$. The number 2 has no units attached. *Thus an angle of 2 (radians) means an angle that subtends an arc that is twice the length of the radius.* See Figure 6.11.

Third, if θ is an angle that subtends an arc whose length is the same as the radius—that is, $s = r$, then according to the definition,

$$\theta = \frac{s}{r} = \frac{r}{r} = 1$$

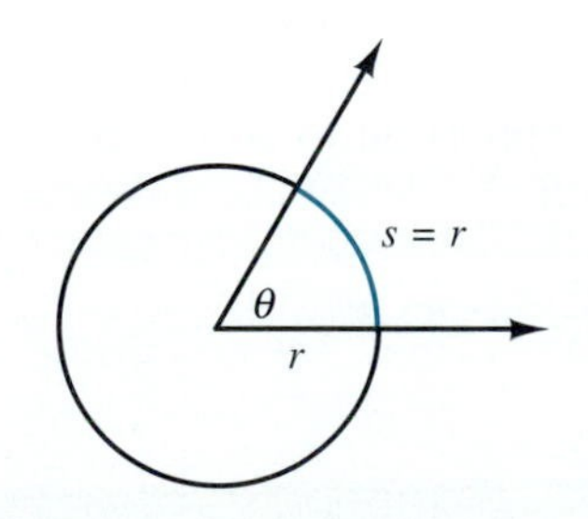

FIGURE 6.12 An angle of 1 radian

Thus an angle of 1 radian is a central angle that subtends an arc equal in length to the radius. See Figure 6.12. In fact, this is an alternative way to define the radian measure of an angle; that is, we can define an angle of 1 radian to be a central angle that subtends an arc equal to the length of the radius.

As with most problems on converting from one unit of measurement to another, in order to convert from radians to degrees or vice versa, we can use a proportion to carry out this conversion.

The proportion we will use is as follows.

CONVERSION FORMULA FOR RADIANS – DEGREES

$$\frac{\theta \text{ in degrees}}{180°} = \frac{\theta \text{ in radians}}{\pi}$$

EXAMPLE 1 Convert each of the following radian measures into degrees.

(a) $\frac{\pi}{6}$ **(b)** $\frac{\pi}{4}$ **(c)** $\frac{\pi}{3}$ **(d)** $-\frac{3\pi}{5}$ **(e)** 1

Solution We use the proportion just given in the box above to do the conversions. We let θ be the degree measure of each of the angles listed.

(a) $$\frac{\theta}{180°} = \frac{\frac{\pi}{6}}{\pi}$$

$$\frac{\theta}{180°} = \frac{\pi}{6} \cdot \frac{1}{\pi} = \frac{1}{6}$$ *Multiply both sides of the equation by 180.*

$$\theta = \frac{1}{6} \cdot 180° = \boxed{30°}$$

When converting the radian measure of an angle given in terms of π, rather than use the proportion as we just did, we can simply replace π by 180°. In other words,

$$\frac{\pi}{6} = \frac{180°}{6} = \boxed{30°}$$

(b) $$\frac{\pi}{4} = \frac{180°}{4} = \boxed{45°}$$

(c) $$\frac{\pi}{3} = \frac{180°}{3} = \boxed{60°}$$

(d) $$-\frac{3\pi}{5} = -\frac{3(180°)}{5} = \boxed{-108°}$$

(e) Since 1 radian is not in terms of π, we use the proportion to do the conversion. We will round our answer to the nearest tenth.

$$\frac{\theta}{180} = \frac{1}{\pi} \Rightarrow \theta = \frac{180}{\pi} \approx \frac{180}{3.14} \approx 57.3°$$

Thus 1 radian is approximately 57°. ■

EXAMPLE 2 Convert into radian measure. **(a)** 90° **(b)** 270°

Solution

(a) Let θ represent the radian measure of 90°. Using the conversion proportion, we obtain:

$$\frac{\theta}{\pi} = \frac{90^\circ}{180^\circ} = \frac{1}{2} \Rightarrow \theta = \boxed{\frac{\pi}{2}}$$

(b) Rather than using the conversion proportion, we notice that $270^\circ = 3(90^\circ)$. In part (a) we found that $90^\circ = \frac{\pi}{2}$, and so we have $270^\circ = \boxed{\frac{3\pi}{2}}$. ■

In fact, we will often be working with angles that are multiples of 30°, 45°, 60°, and 90°. If we know the radian measures of these angles, we can easily get the radian measure of their multiples.

For example, to find the radian measure of 135°, we recognize that $135^\circ = 3(45^\circ)$. Knowing that $45^\circ = \frac{\pi}{4}$, we get $135^\circ = \frac{3\pi}{4}$.

Having completed Examples 1 and 2, we note that the conversion proportion we have been using is equivalent to using the following conversion factors.

> To convert from radians to degrees, multiply the radian measure by $\frac{180^\circ}{\pi}$.
>
> To convert from degrees to radians, multiply the degree measure by $\frac{\pi}{180^\circ}$.

We mentioned before that radian measure is simply another unit of measure used for angles. Although we are basically free to choose any unit of measurement we wish, the unit of measurement we choose should be appropriate for what we are measuring. For example, if we are measuring the length of a table, we might tell someone that the table is 36 inches long or 3 feet long or 1 yard long. The table is still the same length; we are simply changing the unit of length that we are using to describe it. We could also tell someone that this same table is 0.0005682 miles long, and although this might be accurate, it is not particularly appropriate to measure the length of a table in miles; it doesn't give us any feeling for how big the table is.

When we get to the graphs of the trigonometric functions, we will see that it doesn't really make sense to measure angles in degrees and that radian measure is more appropriate.

Arc Length and Area

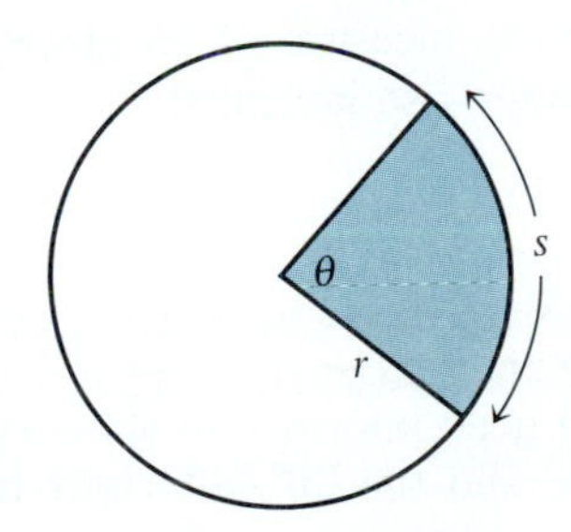

FIGURE 6.13 A sector of a circle

The shaded portion of Figure 6.13 is called a **sector.** From the definition of the radian measure of an angle θ, we have

$$\theta = \frac{s}{r} \qquad \textit{Solve for } s.$$

$$r\theta = s$$

Thus we have the following.

> **ARC LENGTH OF A SECTOR**
>
> $s = r\theta$ where θ is measured in radians

Using radian measure often simplifies formulas from geometry.

EXAMPLE 3 Derive formulas for the area of a sector with central angle θ and radius r.

(a) For the case where θ is measured in degrees

(b) For the case where θ is measured in radians

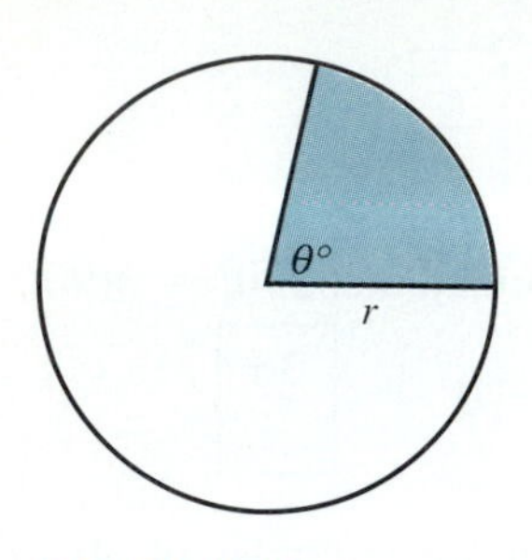

FIGURE 6.14

Solution We refer to Figure 6.13. We know that the area of the entire circle is πr^2. In order to find the area of the sector, we need to know what fraction of the entire circle we have. We then multiply the area of the entire circle by this fraction.

(a) We will write $\theta°$ to indicate that the central angle is being measured in degrees. See Figure 6.14. Since the entire central angle of a circle is 360°, the sector is $\frac{\theta°}{360°}$ of the entire circle. Therefore, the area of the sector is $A = \frac{\theta°}{360°}\pi r^2$.

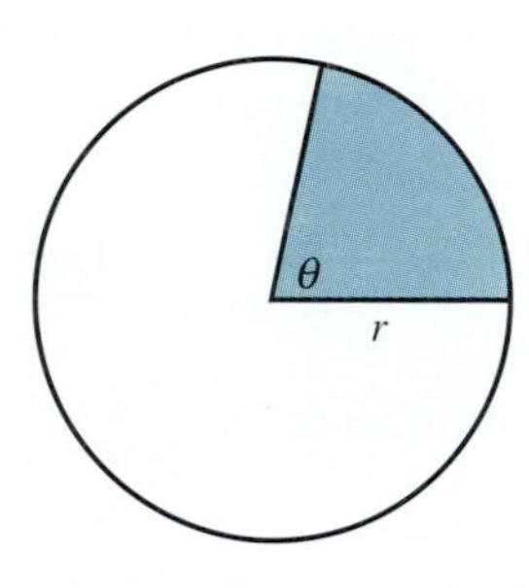

FIGURE 6.15

(b) In the case where θ is measured in radians, the entire central angle is 2π radians, and so the sector is $\frac{\theta}{2\pi}$ of the entire circle. See Figure 6.15. Therefore, the area of the sector is $A = \frac{\theta}{2\pi}\pi r^2 = \frac{1}{2}r^2\theta$. ■

We have thus derived the following formula.

AREA OF A SECTOR

$$A = \frac{1}{2}r^2\theta \qquad \text{where } \theta \text{ is measured in radians}$$

Note that the formula for the area of a sector is simpler when the central angle is expressed in radian measure.

Throughout the remainder of the text, whenever the measure of an angle is given, it will be assumed to be radian measure unless otherwise indicated.

Two Special Triangles

In the discussion that follows in the next section we will need information about two special triangles. They are the **isosceles right triangle** and the **30°-60° right triangle.**

The Isosceles Right Triangle Figure 6.16 shows an isosceles right triangle. Note that because the legs are equal, the base angles must be equal, and since the base angles must have a sum of 90°, they must be 45° each. We have labeled each leg s and the hypotenuse x. We solve for x by using the Pythagorean theorem.

$$x^2 = s^2 + s^2 \Rightarrow x^2 = 2s^2 \Rightarrow x = \pm s\sqrt{2}$$

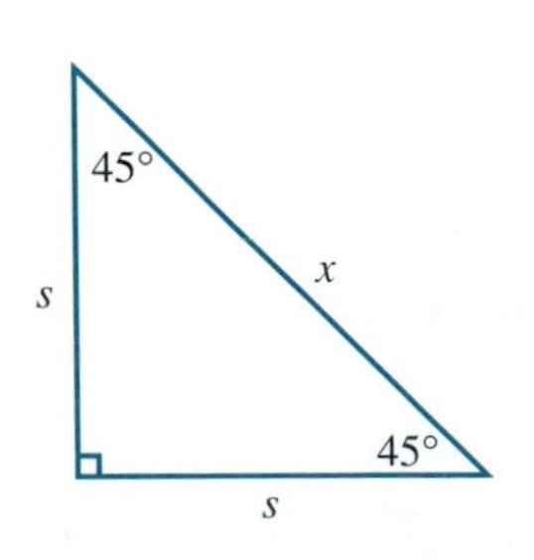

FIGURE 6.16 An isosceles right triangle

Since x is a length, we reject the negative solution, and so $x = s\sqrt{2}$. We have just derived the following.

THE ISOSCELES (45°) RIGHT TRIANGLE

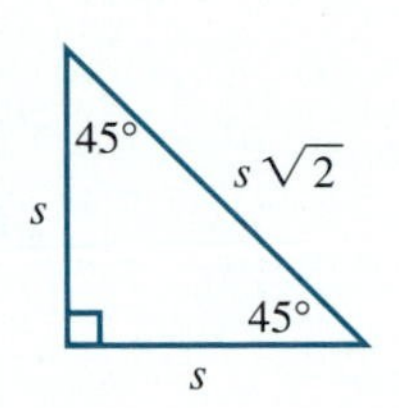

In words, the diagram says that in a 45° right triangle, the legs are equal and the hypotenuse is $\sqrt{2}$ times the length of the leg.

The 30°-60° Right Triangle Figure 6.17 (a) shows a 30°-60° right triangle. We label the hypotenuse h. If we duplicate the triangle as indicated by the dotted lines in Figure 6.17 (b), we can see that $\triangle ABD$ is equilateral (because each angle is 60°), so that $|\overline{AD}|$ is also h and $|\overline{AC}|$ must be $\frac{h}{2}$.

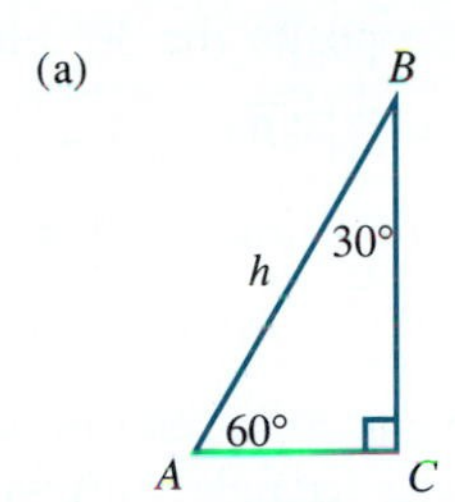

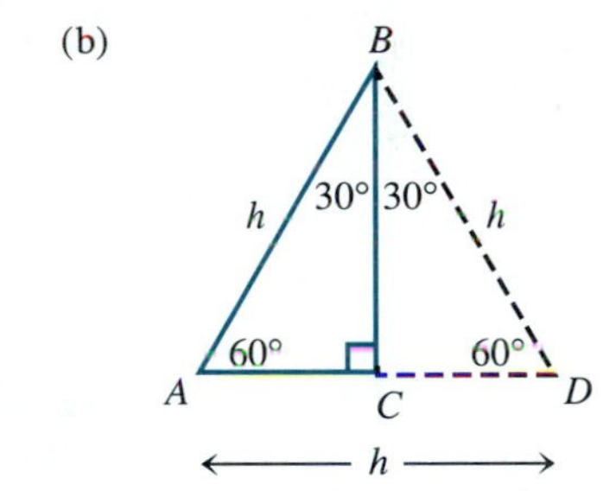

FIGURE 6.17

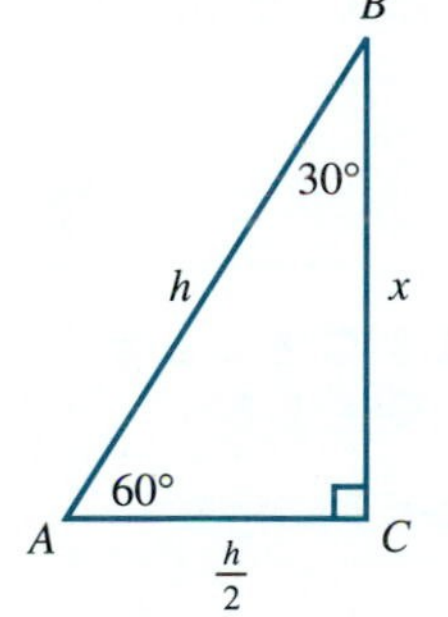

FIGURE 6.18

In Figure 6.18 we redraw Figure 6.17. We have labeled the hypotenuse h, $|\overline{AC}|$ as $\frac{h}{2}$, and the unknown side $|\overline{BC}|$ as x. We find x by again using the Pythagorean theorem. $x^2 + \left(\frac{h}{2}\right)^2 = h^2 \Rightarrow x^2 = \frac{3h^2}{4} \Rightarrow x = \pm\frac{h}{2}\sqrt{3}$. As before, we reject the negative solution, and so $x = \frac{h}{2}\sqrt{3}$. We have therefore derived the following.

Why is the name "30°-60° right triangle" actually redundant?

THE 30°–60° RIGHT TRIANGLE

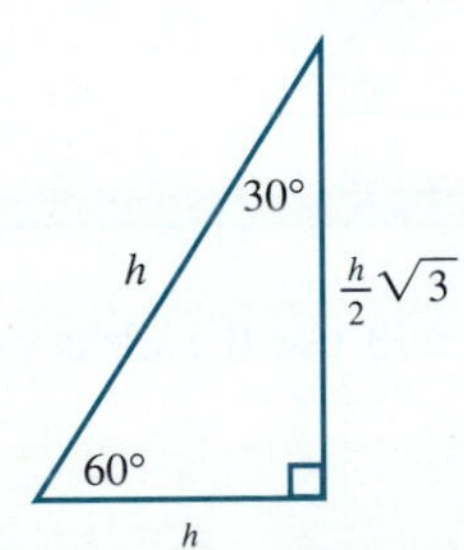

In words, the diagram says that in a 30°-60° right triangle, the side opposite the 30° angle is one-half the hypotenuse, and the side opposite the 60° angle is one-half the hypotenuse times $\sqrt{3}$.

EXAMPLE 4 Find the missing sides and angles in each of the following triangles.

(a)

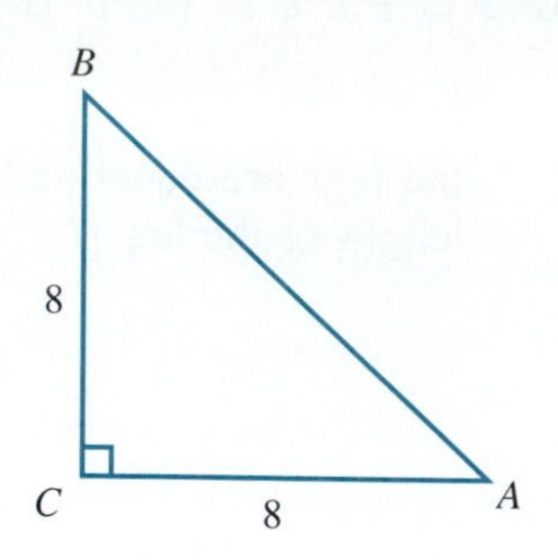

(b)

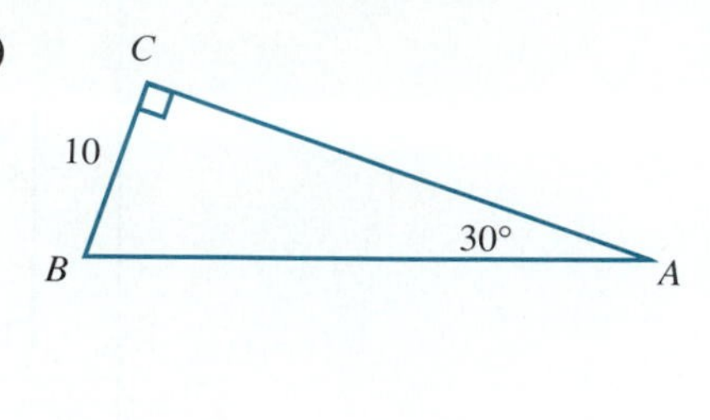

Solution

(a) Since the triangle is an isosceles right triangle, $\angle A = \angle B = 45°$.
Since the hypotenuse of a 45° right triangle is $\sqrt{2}$ times the leg, $|\overline{AB}| = 8\sqrt{2}$.

(b) From the diagram, we can see that $\angle B$ must be 60°. Since this is a 30°-60° right triangle, the side opposite the 30° angle is equal to half the hypotenuse. Therefore, $|\overline{BC}| = 10 = \frac{1}{2}|\overline{AB}|$, and so $|\overline{AB}| = 20$.
$\overline{AC}$ is the side opposite 60° and is, therefore, one-half the hypotenuse times $\sqrt{3}$, and so $|\overline{AC}| = 10\sqrt{3}$. ■

In many cases, when we need to refer to one of these special triangles, we will be able to choose the particular triangle with which we work. In such cases we usually choose one of the following two triangles as prototypes for the two special triangles.

In words, describe the relationships among the sides of an isosceles right triangle.

In words, describe the relationships among the sides of a 30°-60° right triangle.

PROTOTYPES FOR THE ISOSCELES AND 30° – 60° RIGHT TRIANGLES

Isosceles Right Triangle

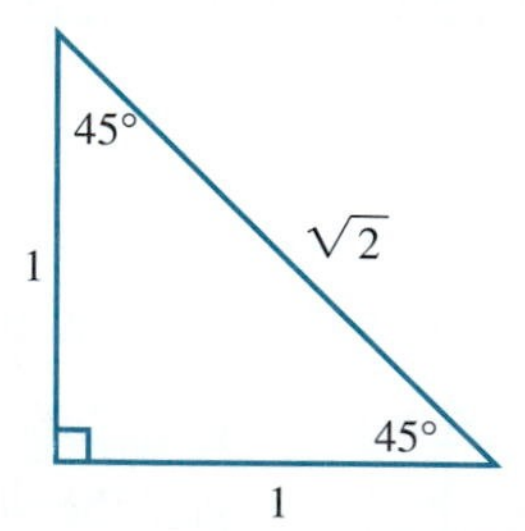

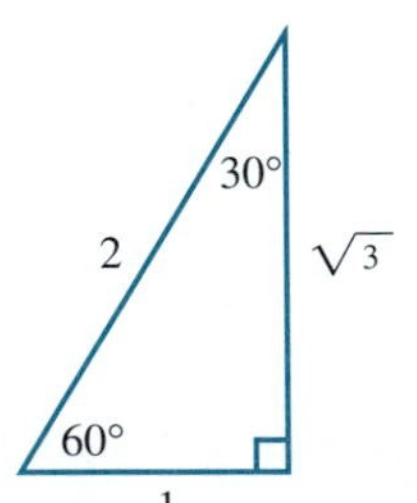

In the next section we will use the ideas presented here as we define the trigonometric functions.

EXERCISES 6.1

In Exercises 1–16, convert the given angle from radians to degrees.

1. $\frac{\pi}{6}$ **2.** $\frac{\pi}{4}$

3. $\frac{\pi}{3}$ **4.** $\frac{\pi}{2}$

5. $-\frac{5\pi}{2}$ **6.** $-\frac{4\pi}{3}$

7. $\frac{11\pi}{6}$ **8.** $\frac{7\pi}{4}$

9. $\frac{3\pi}{2}$ **10.** $-\frac{2\pi}{5}$

11. 3 **12.** 5

13. -2 **14.** -2π

15. $-\frac{\pi}{12}$ **16.** $\frac{\pi}{18}$

In Exercises 17–26, convert the given angle from degrees to radians.

17. 150° **18.** 315°

19. $-120°$ **20.** $-270°$

21. 18° **22.** 100°

23. $-40°$ **24.** 225°

25. 210° **26.** 330°

In Exercises 27–28, find the arc length s and the area A of the given sector.

27.

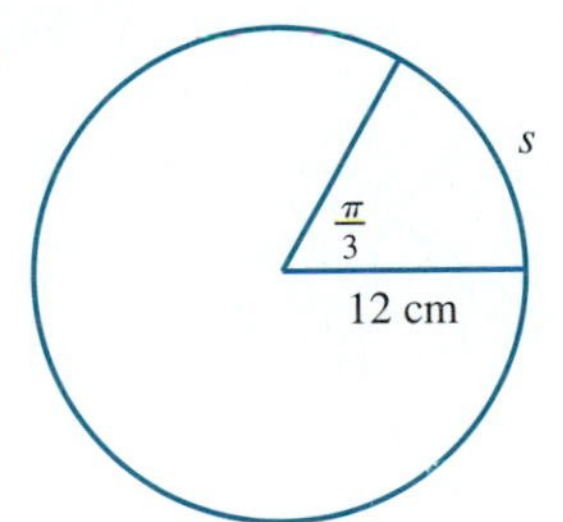

28.

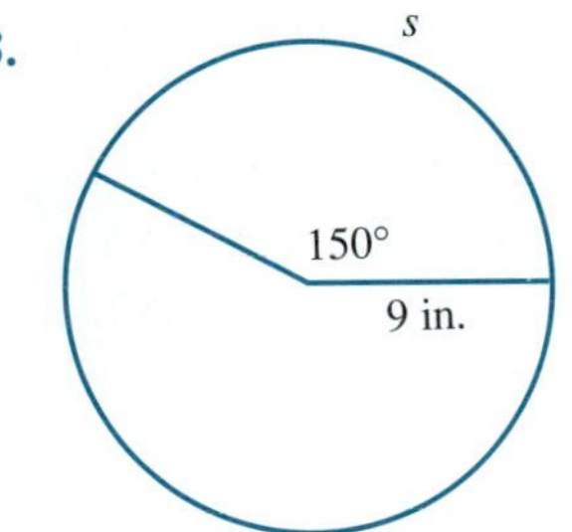

29. Find the radius of a sector with a central angle of $\frac{\pi}{6}$ and an arc length of π cm.

30. Find the central angle θ of a sector with a radius of 20 inches and an arc length of 4π inches.

31. Find the central angle θ of a sector with a radius of 8 cm and an area of 32π sq cm.

32. Find the radius of a sector with a central angle of 45° and an area of 25π sq ft.

33. Find the area A of a sector if its central angle is 120° and its arc length is 24π meters.

34. Find the arc length s of a sector if its central angle is $\frac{\pi}{9}$ and its area is 48π sq meters.

35. Find the central angle (rounded to the nearest tenth) of a sector with a radius of 4.3 inches and an arc length of 9.5 inches.

36. Find the radius (rounded to the nearest hundredth) of a sector with a central angle of $\frac{\pi}{7}$ and an area of 20 sq cm.

37. A wheel of a machine rotates 700 times per minute. Through how many radians does this wheel rotate in 1 minute? Through how many degrees does this wheel rotate in 1 minute?

38. The rate at which an object rotates is usually measured in *revolutions per minute (rpm)*. Convert each of the following rpm rates into radians per minute.
(a) 30 rpm **(b)** 10,000 rpm **(c)** 1 rpm

39. Convert each of the following rates of rotation into rpm.
(a) 180° per second **(b)** 180° per day

40. Convert each of the following rates of rotation into rpm.
(a) $\frac{\pi}{4}$ radians per second **(b)** 90π radians per day

41. A bicycle wheel has a radius of 1 foot. If the bicycle wheel is rotating at the rate of 150 rpm, approximately how far does the bicycle wheel travel in 1 hour?

42. Consider a point P on the rim of a wheel that is 20 inches in diameter. If the wheel makes 60 revolutions, through what distance has the point P traveled?

43. A 4-foot pendulum swings back and forth along a 2-foot arc. Find the number of degrees through which the pendulum passes in one swing. (See the accompanying figure.)

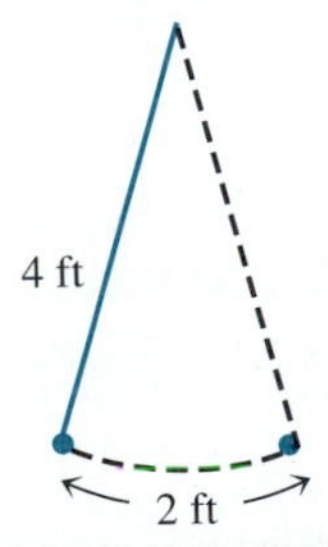

44. The wheels of a car turn at a rate of $\frac{100}{\pi}$ rev/sec when the car is traveling at 80 ft/sec. What is the diameter of the wheel?

45. A winch 10 inches in diameter, as illustrated in the accompanying figure, is used to pull up a weight. How far does the weight travel if the winch rotates $\frac{5\pi}{2}$ radians?

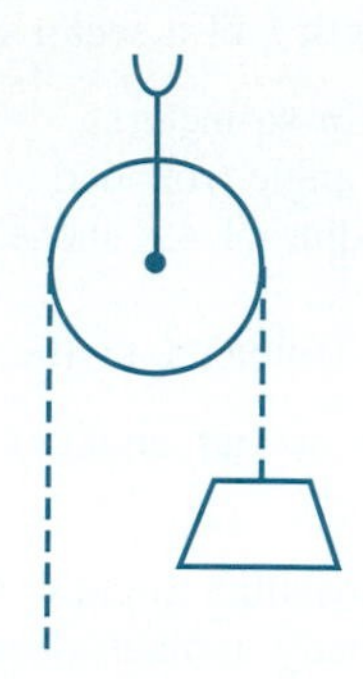

If the winch of Exercise 45 is turning at the rate of $\frac{\pi}{3}$ rad/sec, how fast is the weight traveling?

In Exercises 47–54, find all the sides and angles not given in the figure.

47.

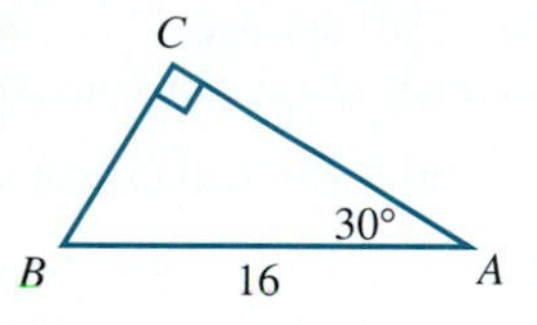

48.

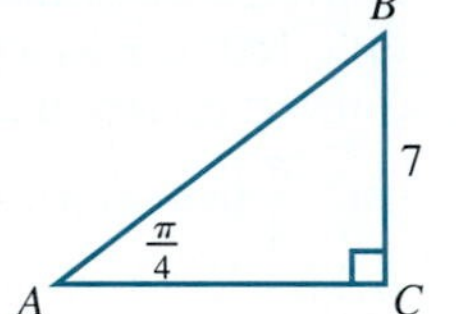

49.

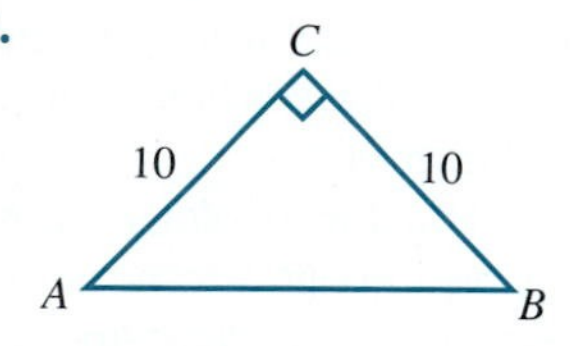

50.

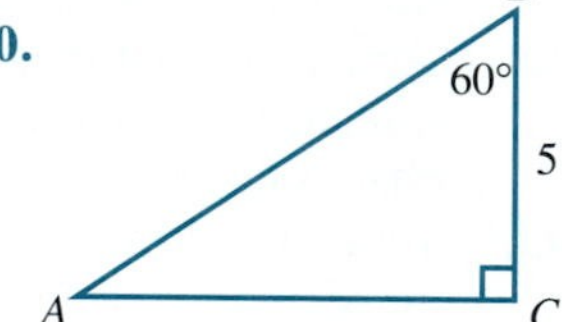

51.

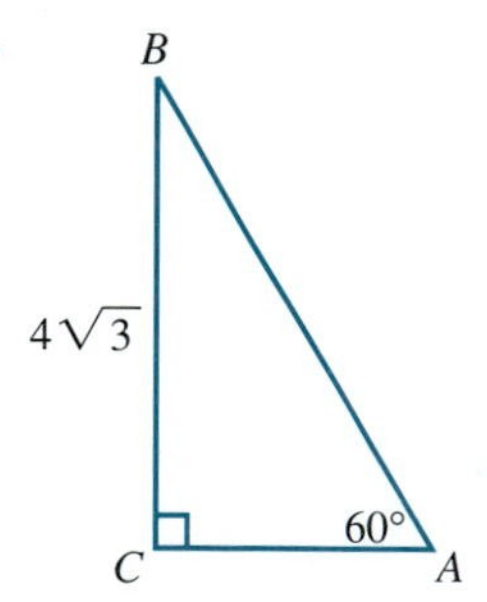

52.

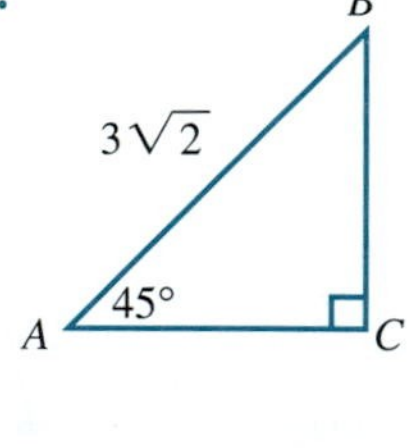

53.

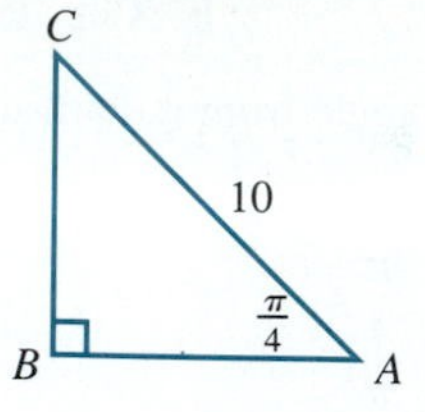

54.

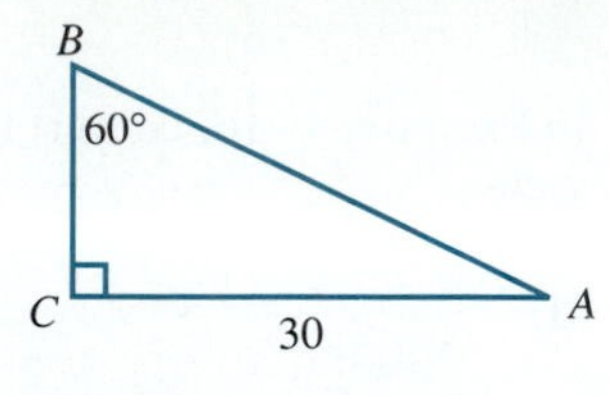

QUESTIONS FOR THOUGHT

55. Give two definitions for the radian measure of an angle.

56. Discuss the difference between degree measure and radian measure.

57. Convert the formula given for the arc length of a sector when θ is given in radians to a formula when θ is given in degrees.

58. Verify that if the angle, θ, of a sector is equal to 2π, then the arc length is the entire circumference of the circle.

59. Describe an angle of 5 radians.

60. If a point P is rotating around some center point, the rate at which the angle θ changes is called the *angular speed,* whereas the rate at which the distance d changes is called the *linear speed.* See the accompanying figure.

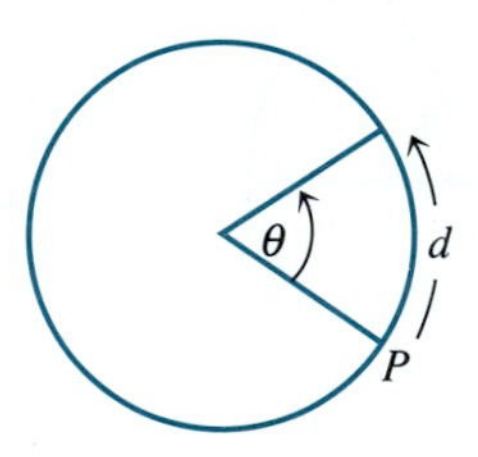

Suppose a record of radius 4 in. is revolving at 45 rpm and a fly is sitting on the record as it rotates. Find the angular and linear speed of a fly if it is sitting on the record:

(a) 1 inch from the center
(b) 3 inches from the center

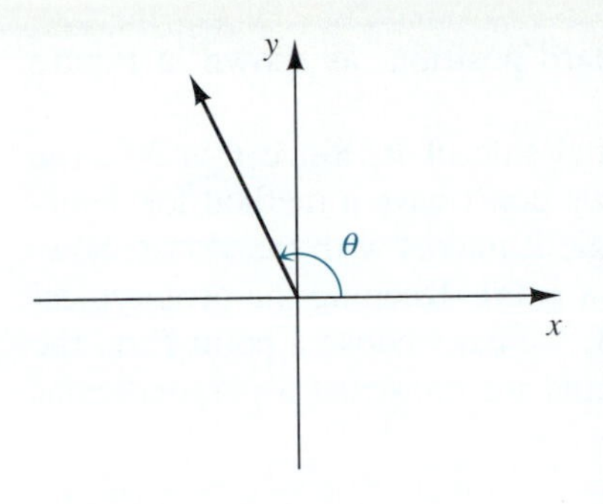

FIGURE 6.19 A typical positive angle in standard position

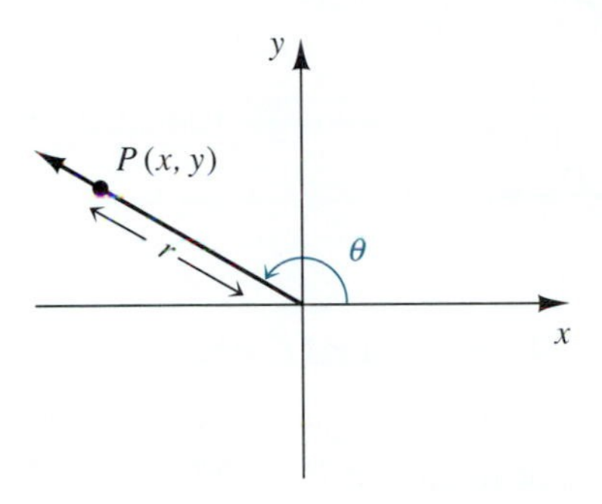

FIGURE 6.20 $P(x, y)$ is a point (other than the origin) on the terminal side and r is the distance from P to the origin.

6.2 The Trigonometric Functions of a General Angle

In the discussion that follows, we will be viewing all angles in the context of a Cartesian coordinate system; that is, given an angle θ, we begin by putting θ in **standard position,** meaning that the vertex of θ is placed at the origin and the initial side of θ is placed along the positive x-axis. The location of the terminal side of θ will, of course, depend upon the size of θ. Figure 6.19 illustrates a typical positive angle θ in standard position. The terminal side happens to be in the second quadrant.

Next we locate a point P (other than the origin) on the terminal side of θ and identify its coordinates (x, y) and its distance to the origin, which we denote by r. See Figure 6.20. Keep in mind that r is the distance from (x, y) to the origin, and so r must be positive.

With θ in standard position we now define the first three trigonometric functions of θ. (We define the other three functions later in this section.)

DEFINITION

Name of Function	Abbreviation	Definition
sine θ	$\sin\theta$	$\sin\theta = \frac{y}{r}$
cosine θ	$\cos\theta$	$\cos\theta = \frac{x}{r}$
tangent θ	$\tan\theta$	$\tan\theta = \frac{y}{x}$

As in the case of the logarithmic function, the trigonometric functions are often written without parentheses around the *argument* of the function. In other words,

$$\sin\theta \text{ means the same thing as } \sin(\theta)$$

Similar statements hold for all the trigonometric functions. Just as we have previously written functions such as $f(x) = x^2 - 3x + 5$, we may now write functions such as $f(\theta) = \sin\theta$ and $g(\theta) = \cos\theta$.

EXAMPLE 1 Find $\sin\frac{\pi}{6}$, $\cos\frac{\pi}{6}$, and $\tan\frac{\pi}{6}$.

Solution We are going to follow the procedure described at the beginning of this section in the specific case $\theta = \frac{\pi}{6}$. Since we are more familiar with degree measure, we can first convert $\frac{\pi}{6}$ into degree measure.

$$\frac{\pi}{6} = 30°$$

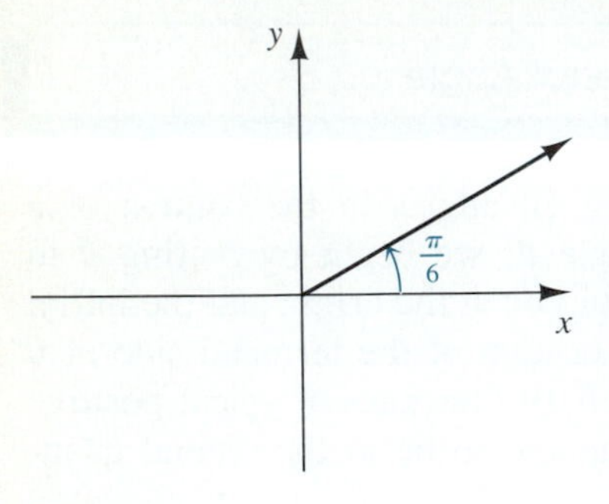

FIGURE 6.21 The angle $\frac{\pi}{6} = 30°$ in standard position

The first step is to put the 30° angle in standard position, as shown in Figure 6.21.

Next we need to locate a point on the terminal side of θ. Since θ is 30°, the terminal side falls in the first quadrant. Generally we don't have a method for determining a point on a line simply by knowing the angle it makes with the x-axis; however, because this is a 30° angle, we can find such a point. Keeping the prototypical 30°-60° right triangle from the last section in mind, we can choose a point P on the terminal side of θ that is 2 units from the origin, and we construct a perpendicular from this point to the x-axis. See Figure 6.22.

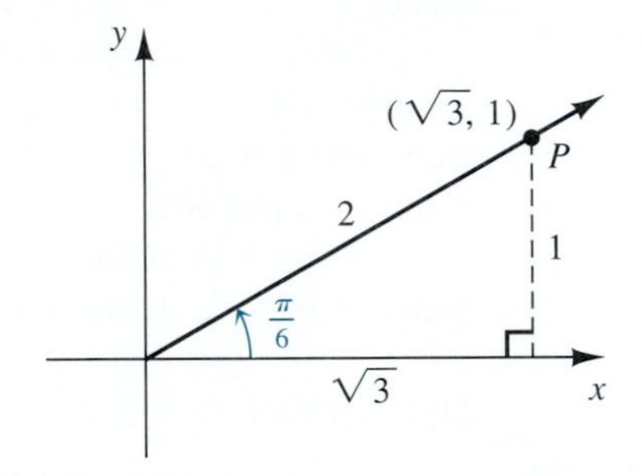

FIGURE 6.22 The diagram for a $\frac{\pi}{6} = 30°$ angle in standard position

From the triangle formed we can identify the coordinates of P as $(\sqrt{3}, 1)$ and $r = 2$. Therefore, in computing the sine, cosine, and tangent of $\frac{\pi}{6}$, we will use $x = \sqrt{3}$, $y = 1$, and $r = 2$. Consequently, we have

$$\sin\frac{\pi}{6} = \frac{y}{r} = \boxed{\frac{1}{2}}$$

Remember that $\sin\frac{\pi}{6}$ *means* $\sin\left(\frac{\pi}{6}\right)$*. This is function notation.*

$$\cos\frac{\pi}{6} = \frac{x}{r} = \boxed{\frac{\sqrt{3}}{2}}$$

$$\tan\frac{\pi}{6} = \frac{y}{x} = \boxed{\frac{1}{\sqrt{3}} = \frac{\sqrt{3}}{3}}$$

Both forms of the answer are acceptable. We will comment on rationalizing denominators later in this section.

Note that the triangle we formed was simply an aid in finding the coordinates of a point on the terminal side. ■

At this point you may be wondering a bit about the definitions of the sine, cosine, and tangent functions. We have supposedly defined the sine, cosine, and tangent "of an angle θ," yet our definition seems to depend on the point we choose on the terminal side of θ. If we have actually defined a function of θ, then the value of the function must be independent of the point we choose on the terminal side; otherwise the definition is ambiguous.

For instance, in Example 1 suppose we had chosen the point P to be 8 units from the origin instead of 2 units from the origin. Then our diagram would have

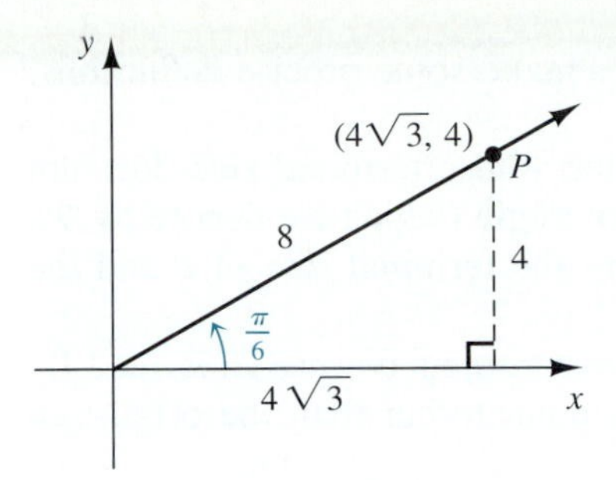

FIGURE 6.23 An alternative diagram for $\frac{\pi}{6}$

looked like Figure 6.23. We would get $\sin \frac{\pi}{6} = \frac{y}{r} = \frac{4}{8} = \frac{1}{2}$, which is the same answer we obtained before; similarly, for cosine and tangent.

In fact, a moment's thought should convince us that for any two points $P_1(x_1, y_1)$ and $P_2(x_2, y_2)$ chosen on the terminal side of θ, the triangles formed will be similar and thus their corresponding sides will be in proportion. Therefore, even though the values of x, y, and r may be different, the *ratios*—sine, cosine, and tangent—will remain the same.

EXAMPLE 2 Find $\sin \frac{4\pi}{3}$, $\cos \frac{4\pi}{3}$, and $\tan \frac{4\pi}{3}$.

Solution We follow the same procedure as in the previous example. Again, we may first convert $\frac{4\pi}{3}$ into degree measure.

$$\frac{4\pi}{3} = 4\left(\frac{\pi}{3}\right) = 4(60°) = 240°$$

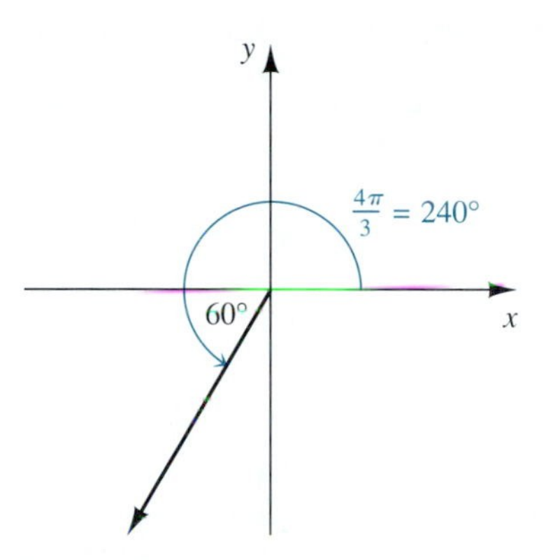

FIGURE 6.24 A 240° angle in standard position with 60° reference angle

We put a $\frac{4\pi}{3}$ angle in standard position, as shown in Figure 6.24. Next we need to locate a point on the terminal side of θ. We notice that since θ is 240°, the terminal side falls in the third quadrant and the acute angle formed by the terminal side and the negative x-axis is $240° - 180° = 60°$. This 60° angle is called the *reference angle*. See Figure 6.24. (We give a precise definition of the reference angle after completing this example.)

The fact that the reference angle is 60° allows us to use the information we have obtained about a 30°-60° right triangle, as follows. Keeping the prototypical 30°-60° right triangle in mind, we can choose a point P on the terminal side of θ that is 2 units from the origin, and we construct a perpendicular from this point to the x-axis. The triangle formed in this way is called a *reference triangle*. Note that this reference triangle is a 30°-60° right triangle, and so we know that the other two sides of the triangle are 1 and $\sqrt{3}$, as indicated in Figure 6.25. We can see that the coordinates of P are $(-1, -\sqrt{3})$. Therefore, in computing the sine, cosine, and tangent of $\frac{4\pi}{3}$, we use $x = -1$, $y = -\sqrt{3}$, and $r = 2$. Consequently, we have

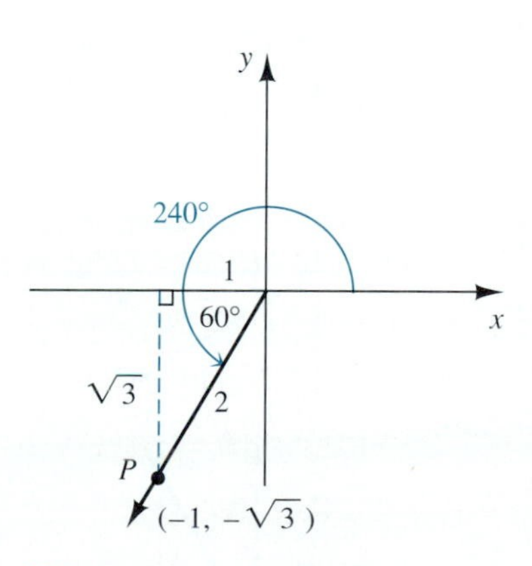

FIGURE 6.25 A reference triangle for a 240° angle

$$\sin \frac{4\pi}{3} = \frac{y}{r} = \boxed{\frac{-\sqrt{3}}{2}}$$

$$\cos \frac{4\pi}{3} = \frac{x}{r} = \boxed{\frac{-1}{2}}$$

$$\tan \frac{4\pi}{3} = \frac{y}{x} = \frac{-\sqrt{3}}{-1} = \boxed{\sqrt{3}}$$

Don't lose sight of the fact that we found the sine, cosine, and tangent of $\frac{4\pi}{3}$ (or 240°). The 60° reference angle was simply an aid in identifying x, y, and r. ■

Before we look at several more examples, let's make some precise definitions.

DEFINITION Let θ be an angle in standard position whose terminal side does not fall on the x or y-axis. The **reference angle** (which we denote by θ') is the *positive acute* angle formed by the terminal side of θ and the x-axis.

A **reference triangle** is a triangle containing θ' that is formed by constructing a perpendicular from a point (other than the origin) on the terminal side of θ to the *x-axis*.

EXAMPLE 3 Given each of the following angles, draw the angle in standard position and identify the reference angle in both radians and degrees.

(a) $\frac{3\pi}{4}$ **(b)** $\frac{7\pi}{6}$ **(c)** $-\frac{\pi}{6}$ **(d)** $\frac{\pi}{3}$

Solution

(a) $\frac{3\pi}{4} = 3\left(\frac{\pi}{4}\right) = 3(45°) = 135°$. We draw the angle in standard position. To find the reference angle, we compute $\pi - \frac{3\pi}{4} = 180° - 135°$; therefore, the reference angle is $\boxed{\frac{\pi}{4} = 45°}$.

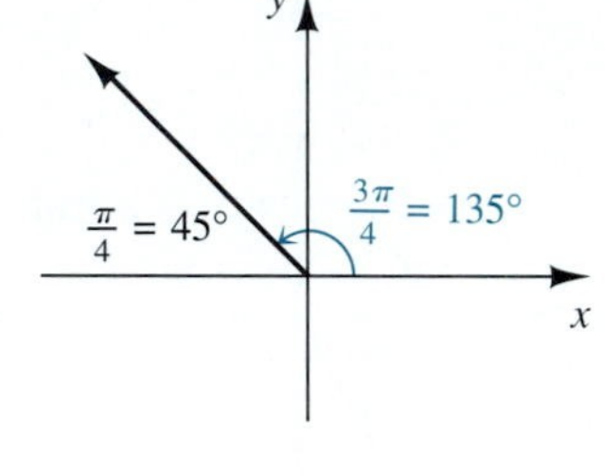

(b) $\frac{7\pi}{6} = 7\left(\frac{\pi}{6}\right) = 7(30°) = 210°$. We draw the angle in standard position. To find the reference angle, we compute $\frac{7\pi}{6} - \pi = 210° - 180°$; therefore, the reference angle is $\boxed{\frac{\pi}{6} = 30°}$.

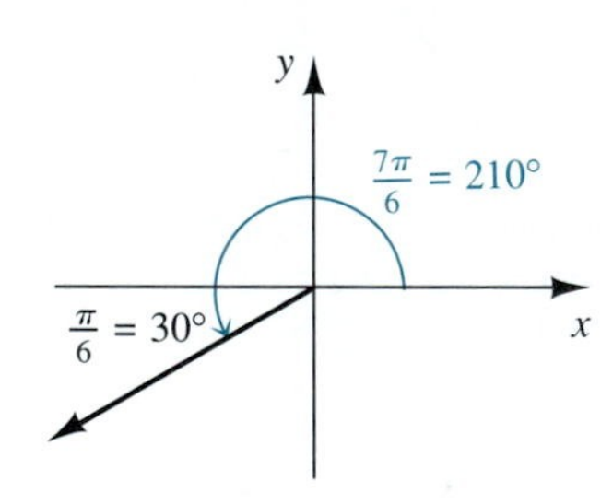

(c) $-\frac{\pi}{6} = -\frac{180°}{6} = -30°$. We draw the angle in standard position and we can see that the reference angle is $\boxed{\frac{\pi}{6} = 30°}$. Remember that according to the definition, the reference angle must be positive.

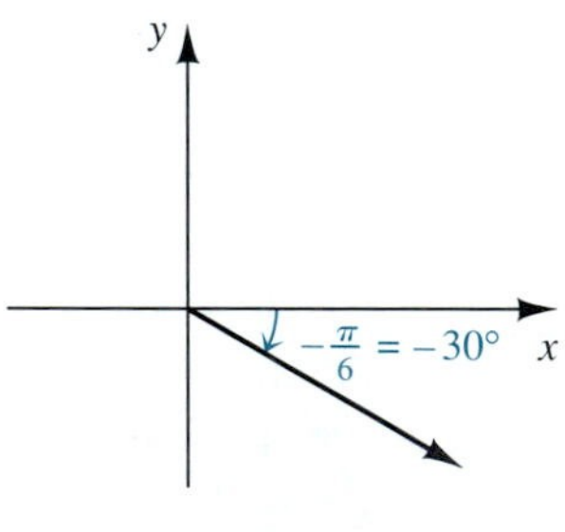

(d) $\frac{\pi}{3} = \frac{180°}{3} = 60°$. We draw the angle in standard position and we can see that the reference angle is the same as the angle itself, or $\boxed{\frac{\pi}{3} = 60°}$.

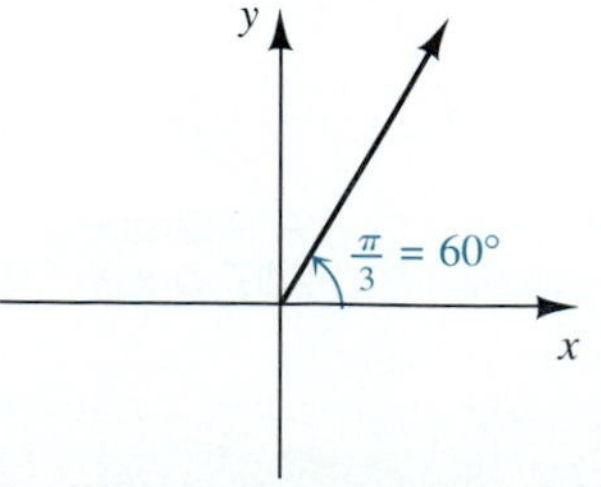

Note that for all first quadrant angles θ, the reference angle θ' is the same as θ. ■

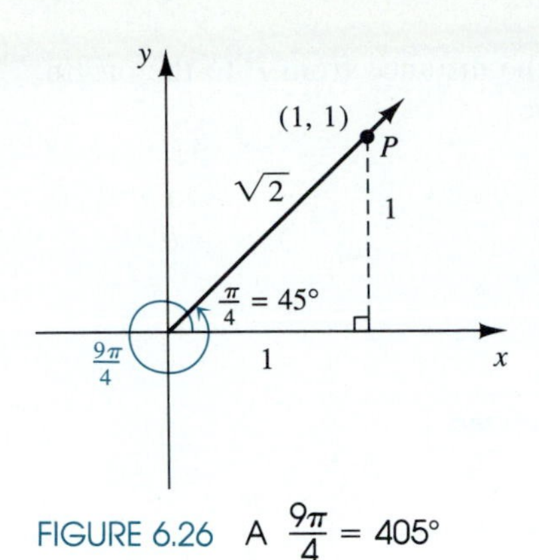

FIGURE 6.26 A $\frac{9\pi}{4} = 405°$ angle in standard position

EXAMPLE 4 Find the sine, cosine, and tangent of $\frac{9\pi}{4}$.

Solution Converting $\frac{9\pi}{4}$ to degrees, we get 405°. Putting a 405° angle in standard position gives us Figure 6.26. Note that, as indicated, the reference angle is $\frac{\pi}{4} = 45°$. We have chosen a point on the terminal side that is $\sqrt{2}$ units away from the origin and filled in the sides of the reference triangle according to the prototypical 45° right triangle. Accordingly, we have labeled the point P as (1, 1).

Therefore, we have $x = 1$, $y = 1$, and $r = \sqrt{2}$:

$$\sin \frac{9\pi}{4} = \frac{y}{r} = \boxed{\frac{1}{\sqrt{2}} = \frac{\sqrt{2}}{2}}$$

$$\cos \frac{9\pi}{4} = \frac{x}{r} = \boxed{\frac{1}{\sqrt{2}} = \frac{\sqrt{2}}{2}}$$

$$\tan \frac{9\pi}{4} = \frac{y}{x} = \frac{1}{1} = \boxed{1}$$

Note that the terminal side of $\frac{9\pi}{4}$ is in exactly the same location as the terminal side of $\frac{\pi}{4}$. Angles that have terminal sides in the same location are called **coterminal.** Since adding (or subtracting) 2π (or 360°) to (or from) any angle will bring you back around to the same location, adding any integer multiple of 2π to an angle will give a coterminal angle. It is also useful to note that since the values of the trigonometric functions are determined by the location of the terminal side of the angle (since that in turn determines the ratios), any two angles with the same terminal side will have the same trigonometric values. We will return to this idea repeatedly in our discussion of the trigonometric functions. ■

You will note that the definition of a reference angle excluded angles whose terminal side is on the x- or y-axis. The following example explains why.

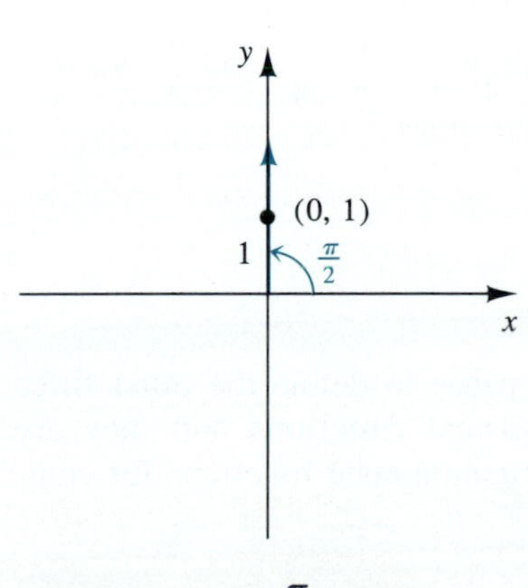

FIGURE 6.27 A $\frac{\pi}{2} = 90°$ angle in standard position

EXAMPLE 5 Find the sine, cosine, and tangent of $\frac{\pi}{2}$.

Solution As before, we begin by putting an angle of $\frac{\pi}{2}$ (or 90°) in standard position, as shown in Figure 6.27. Our next step is to locate a point on the terminal side of the angle. In previous examples we needed the reference triangle in order to identify x, y, and r. However, in this example, since the terminal side of $\frac{\pi}{2}$ lies along the positive y-axis, it is quite easy to pick a point P on the terminal side and identify x, y, and r. For example, we can see that the point (0, 1) is on the terminal side of $\frac{\pi}{2}$.

Therefore, we have $x = 0$ and $y = 1$; r, which is the distance from P to the origin, is also 1, as indicated in Figure 6.27. Thus we have

$$\sin\frac{\pi}{2} = \frac{y}{r} = \frac{1}{1} = \boxed{1}$$

$$\cos\frac{\pi}{2} = \frac{x}{r} = \frac{0}{1} = \boxed{0}$$

$$\tan\frac{\pi}{2} = \frac{y}{x} = \frac{1}{0} \rightarrow \boxed{\text{undefined}}$$

The fact that $\tan\frac{\pi}{2}$ is undefined is just another way of saying that $\frac{\pi}{2}$ is not in the domain of the tangent function. ■

Note that we didn't need a reference triangle to help us identify x, y, and r. In fact there was no reference angle in Example 5. If we follow the definition of a reference triangle, we construct a perpendicular from P on the terminal side of $\frac{\pi}{2}$ to the x-axis. Since this perpendicular coincides with the terminal side of $\frac{\pi}{2}$, there is no reference angle (since according to the definition, the reference angle must be acute); also, the perpendicular would *not* form a reference triangle.

Consequently, we see that a reference angle is defined only for an angle whose terminal side does not fall on the x- or y-axis. We will find the following definition useful.

DEFINITION A **quadrantal angle** is one whose terminal side falls on the x-axis or on the y-axis. Alternatively, we may say that a quadrantal angle is one that is a multiple of $\frac{\pi}{2}$ (or 90°).

Thus, in Example 5, $\frac{\pi}{2}$ is a quadrantal angle.

It is important that you become very familiar with the sine, cosine, and tangent of all the quadrantal angles as well as of the multiples of $\frac{\pi}{6}$, $\frac{\pi}{4}$, and $\frac{\pi}{3}$. See Exercises 73 and 74.

The Reciprocal Functions

Before we proceed to several more examples, let's pause to define the other three trigonometric functions. They are called the **reciprocal functions** and they are defined as follows. (We include the original three trigonometric functions for completeness.)

DEFINITION—THE TRIGONOMETRIC FUNCTIONS

Name of Function	Abbreviation	Definition	Reciprocal Functions: Name of Function	Reciprocal Functions: Abbreviation	Reciprocal Functions: Definition
sine θ	$\sin\theta$	$\sin\theta = \frac{y}{r}$	cosecant θ	$\csc\theta$	$\csc\theta = \frac{1}{\sin\theta} = \frac{r}{y}$
cosine θ	$\cos\theta$	$\cos\theta = \frac{x}{r}$	secant θ	$\sec\theta$	$\sec\theta = \frac{1}{\cos\theta} = \frac{r}{x}$
tangent θ	$\tan\theta$	$\tan\theta = \frac{y}{x}$	cotangent θ	$\cot\theta$	$\cot\theta = \frac{1}{\tan\theta} = \frac{x}{y}$

EXAMPLE 6 Find all six trigonometric functions of each angle.

(a) $\frac{7\pi}{4}$ **(b)** $-180°$ **(c)** $-\frac{5\pi}{6}$

Solution

(a) An angle of $\frac{7\pi}{4}$, which is equivalent to 315°, appears in Figure 6.28. We have drawn the reference triangle, which contains the reference angle of 45°. Thinking about the prototypical 45° right triangle, we may choose a point on the terminal side that is $\sqrt{2}$ units from the origin. This allows us to fill in the sides of the reference triangle as they appear in Figure 6.28. Thus for P we have $x = 1$ and $y = -1$; $r = \sqrt{2}$. Therefore,

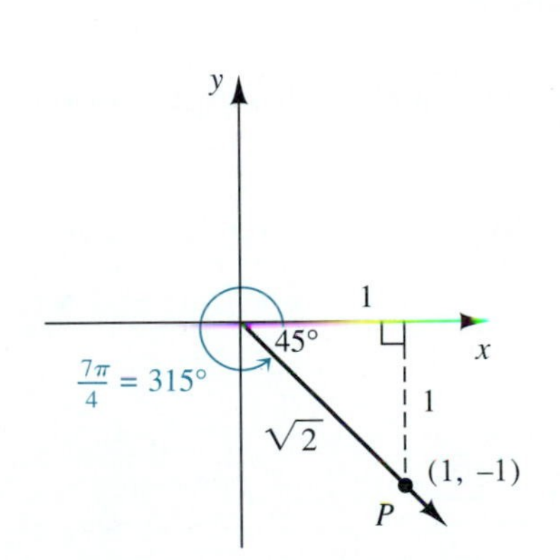

FIGURE 6.28 A $\frac{7\pi}{4}$ angle in standard position

$$\sin\frac{7\pi}{4} = \frac{y}{r} = \boxed{\frac{-1}{\sqrt{2}} = -\frac{\sqrt{2}}{2}}$$

$$\cos\frac{7\pi}{4} = \frac{x}{r} = \boxed{\frac{1}{\sqrt{2}} = \frac{\sqrt{2}}{2}}$$

$$\tan\frac{7\pi}{4} = \frac{y}{x} = \frac{-1}{1} = \boxed{-1}$$

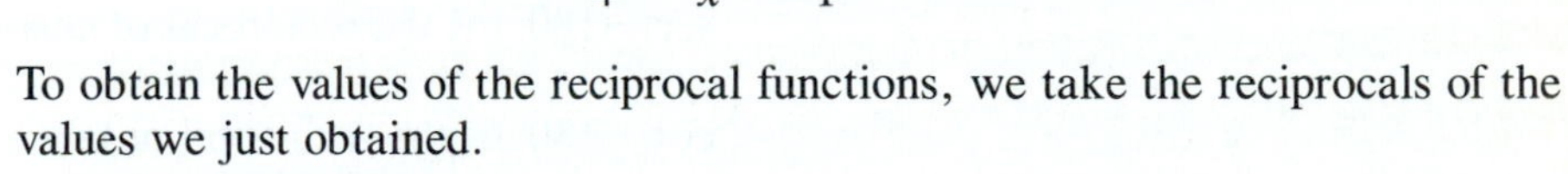

To obtain the values of the reciprocal functions, we take the reciprocals of the values we just obtained.

$$\csc\frac{7\pi}{4} \text{ is the reciprocal of } \sin\frac{7\pi}{4} \text{ and so } \boxed{\csc\frac{7\pi}{4} = -\sqrt{2}}$$

$$\sec\frac{7\pi}{4} \text{ is the reciprocal of } \cos\frac{7\pi}{4} \text{ and so } \boxed{\sec\frac{7\pi}{4} = \sqrt{2}}$$

$\cot \frac{7\pi}{4}$ is the reciprocal of $\tan \frac{7\pi}{4}$ and so $\boxed{\cot \frac{7\pi}{4} = -1}$ *

(b) Putting $-180°$ into standard position produces Figure 6.29. Note that since this is a negative angle, the rotation is clockwise. We can easily identify the point $P(-1, 0)$ as being on the terminal side; since this point is 1 unit from the origin, we have $x = -1$, $y = 0$, and $r = 1$.

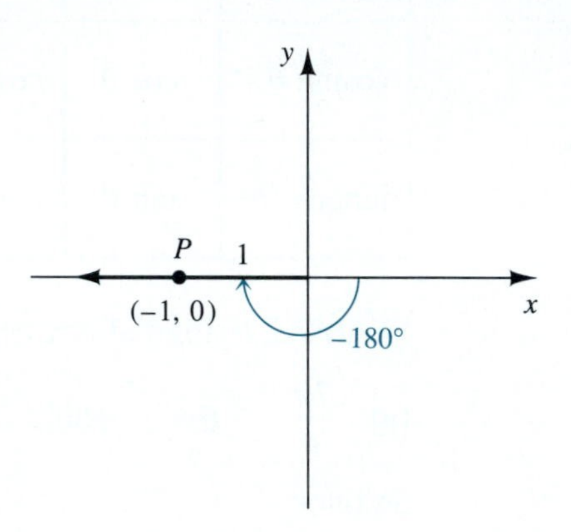

FIGURE 6.29 A −180° angle in standard position

Remember that r is the distance from P to the origin and is, therefore, *always* positive. Thus

$$\sin(-180°) = \frac{y}{r} = \frac{0}{1} = \boxed{0}$$

$$\cos(-180°) = \frac{x}{r} = \frac{-1}{1} = \boxed{-1}$$

$$\tan(-180°) = \frac{y}{x} = \frac{0}{-1} = \boxed{0}$$

$\csc(-180°)$ is the reciprocal of $\sin(-180°)$. Since $\sin(-180°) = 0$, $\csc(-180°)$ is $\boxed{\text{undefined}}$

$\sec(-180°)$ is the reciprocal of $\cos(-180°)$ and so $\boxed{\sec(-180°) = -1}$

$\cot(-180°)$is the reciprocal of $\tan(-180°)$. Since $\tan(-180°) = 0$, $\cot(-180°)$ is $\boxed{\text{undefined}}$

* When working with values of the trigonometric functions, it is not always helpful to rationalize denominators. In this example, if we use the rationalized value for $\sin \frac{7\pi}{4} = -\frac{1}{\sqrt{2}} = -\frac{\sqrt{2}}{2}$, then the reciprocal function cosecant will *not* have a rationalized denominator. Consequently, in this text we will not always insist that trigonometric values have rationalized denominators.

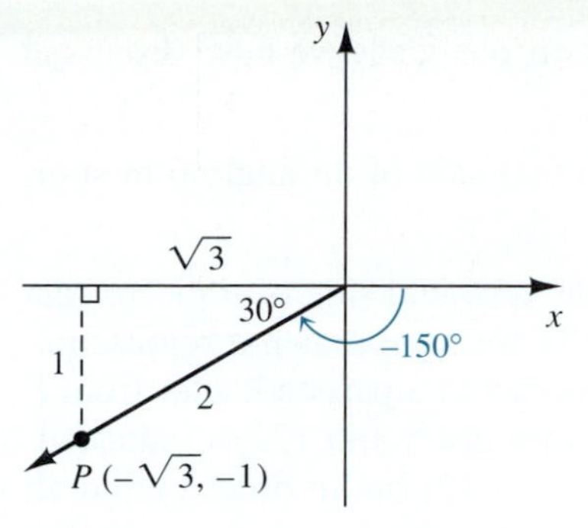

FIGURE 6.30 A $-\frac{5\pi}{6}$ angle in standard position

Remember that the reference angle is always a positive angle.

(c) An angle of $-\frac{5\pi}{6}$ is equivalent to $-150°$. Putting this angle in standard position gives Figure 6.30. Note that the reference angle (the acute angle between the terminal side and the x-axis) is $\frac{\pi}{6}$, or 30°. We have drawn the reference triangle and used the prototypical 30° right triangle to label the sides and thus for P we have $x = -\sqrt{3}$ and $y = -1$; $r = 2$. Therefore,

$$\sin\left(-\frac{5\pi}{6}\right) = \frac{y}{r} = \boxed{\frac{-1}{2}} \qquad \cos\left(-\frac{5\pi}{6}\right) = \frac{x}{r} = \boxed{\frac{-\sqrt{3}}{2}}$$

$$\tan\left(-\frac{5\pi}{6}\right) = \frac{y}{x} = \frac{-1}{-\sqrt{3}} = \boxed{\frac{1}{\sqrt{3}}}$$

To obtain the values of the reciprocal functions, we simply take the reciprocals of the values we just obtained.

$$\boxed{\csc\left(-\frac{5\pi}{6}\right) = -2} \qquad \boxed{\sec\left(-\frac{5\pi}{6}\right) = -\frac{2}{\sqrt{3}}} \qquad \boxed{\cot\left(-\frac{5\pi}{6}\right) = \sqrt{3}}$$ ■

Our work thus far has shown us that the *sign* of a trigonometric function depends upon the quadrant in which the terminal side falls. Keeping in mind that in the trigonometric ratios r is always positive, the sign of a trigonometric function depends upon the signs of x and y.

For θ:
In quadrant I, both x and y are positive, so *all* the trig functions are positive.
In quadrant II, x is negative and y is positive, so sine and its reciprocal cosecant are positive.
In quadrant III x is negative and y is negative, so tangent and its reciprocal cotangent are positive.
In quadrant IV x is positive and y is negative, so cosine and its reciprocal secant are positive.

See Figure 6.31.

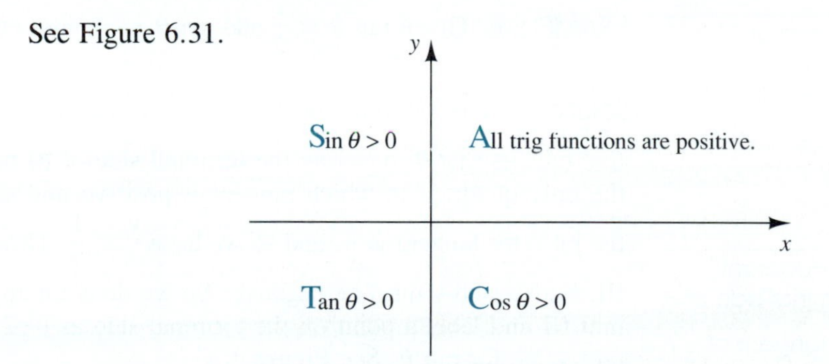

FIGURE 6.31 Determining the signs of the trig functions

Students often remember the signs by starting in quadrant I and using the mnemonic All Students Take Calculus.

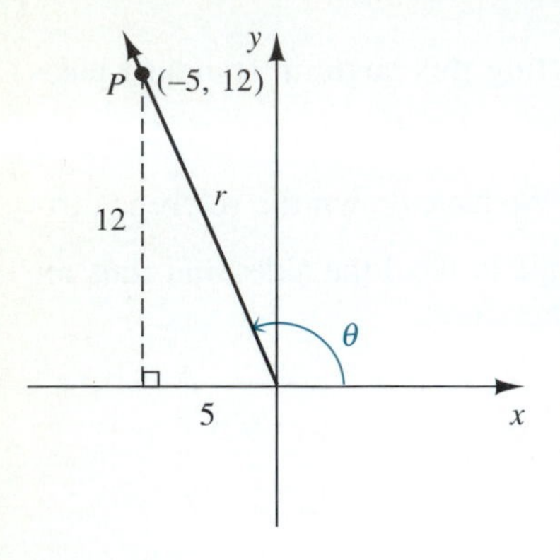

FIGURE 6.32

Let's look at a few examples that combine the various ideas we have discussed thus far.

EXAMPLE 7 If the point $P(-5, 12)$ lies on the terminal side of an angle θ in standard position, find all six trig functions of θ.

Solution The fact that the point $(-5, 12)$ lies on the terminal side of θ means that the terminal side of θ is in quadrant II. In Figure 6.32 we have drawn a representative diagram for θ in standard position. We have constructed a perpendicular from P to the x-axis, creating the reference triangle whose sides are 5 and 12, as indicated.

From Figure 6.32 we can see that $x = -5$ and $y = 12$, but in order to find all six trig functions of θ, we need to find r. We can find r by noting that r is the hypotenuse of the reference triangle.

$$r^2 = 5^2 + 12^2 = 169 \Rightarrow r = \pm 13$$ *Since r is the distance to the origin, it must be positive. Therefore,*

$$r = 13$$

Now that we know the values of x, y, and r, we have

$$\boxed{\sin\theta = \frac{12}{13}} \qquad \boxed{\csc\theta = \frac{13}{12}}$$

$$\boxed{\cos\theta = -\frac{5}{13}} \qquad \boxed{\sec\theta = -\frac{13}{5}}$$

$$\boxed{\tan\theta = -\frac{12}{5}} \qquad \boxed{\cot\theta = -\frac{5}{12}}$$

■

If we are not given that $\sin\theta < 0$, what can we conclude about the quadrant in which θ falls?

EXAMPLE 8 Given $\tan\theta = \frac{1}{2}$ and $\sin\theta < 0$, find $\cos\theta$.

Solution The fact that $\tan\theta = \frac{1}{2}$ and is, therefore, positive and that $\sin\theta$ is negative tells us that θ (meaning the terminal side of θ) must be in quadrant III. (This is the only quadrant in which tangent is positive and sine is negative.) We know that the ratio for tangent is $\frac{y}{x}$, and so we have $\frac{y}{x} = \frac{1}{2}$. However, since we are in quadrant III, both x and y must be negative. So we draw an angle with terminal side in quadrant III and label a point on the terminal side as $(-2, -1)$, which gives us the correct value for $\tan\theta$. See Figure 6.33.

What other third quadrant points on the terminal side of θ would give us a tangent of $\frac{1}{2}$?

In Figure 6.33 we label the hypotenuse of the reference triangle as r and we find r by using the Pythagorean theorem. (Be careful! Even though two of the sides of the triangle are 1 and 2, this is not a 30°-60° right triangle. Why not?)

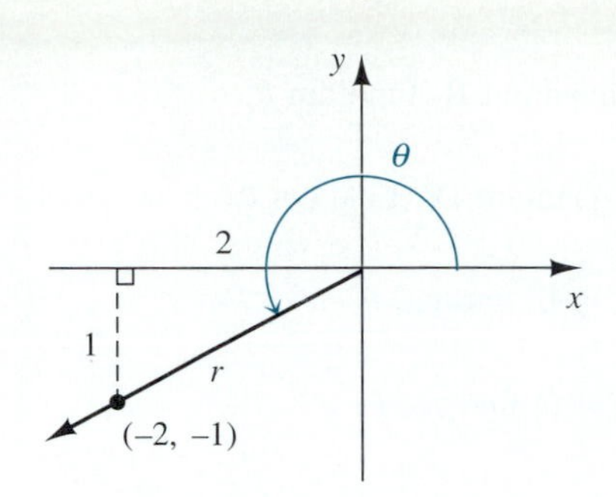

FIGURE 6.33 An angle in quadrant III with $\tan\theta = \frac{1}{2}$

$$r^2 = 1^2 + 2^2 = 5 \Rightarrow r = \pm\sqrt{5}$$ *As before, r must be positive.*

$$r = \sqrt{5}$$

Therefore, $\cos\theta = \frac{x}{r} = \boxed{\frac{-2}{\sqrt{5}} = \frac{-2\sqrt{5}}{5}}$. ■

EXERCISES 6.2

In Exercises 1–16, draw the given angle in standard position and find the reference angle. If the angle is quadrantal, indicate where the terminal side falls—e.g., positive y-axis or negative x-axis, etc.

1. $\frac{2\pi}{3}$ **2.** $\frac{5\pi}{4}$

3. $\frac{7\pi}{6}$ **4.** $\frac{4\pi}{3}$

5. $315°$ **6.** $150°$

7. $60°$ **8.** $45°$

9. $\frac{\pi}{2}$ **10.** π

11. $-\frac{5\pi}{6}$ **12.** $-\frac{3\pi}{4}$

13. 2π **14.** $\frac{3\pi}{2}$

15. $495°$ **16.** $\frac{13\pi}{4}$

In Exercises 17–36, find the indicated value.

17. $\sin\frac{5\pi}{6}$ **18.** $\cos\frac{3\pi}{4}$

19. $\tan\frac{4\pi}{3}$ **20.** $\sin\frac{11\pi}{6}$

21. $\sin\frac{3\pi}{2}$ **22.** $\cos\pi$

23. $\tan\frac{\pi}{2}$ **24.** $\cot 0$

25. $\sec 225°$ **26.** $\csc 300°$

27. $\cos 270°$ **28.** $\sin 180°$

29. $\cot\frac{3\pi}{4}$ **30.** $\sec\frac{2\pi}{3}$

31. $\sin 30°$ **32.** $\cos 60°$

33. $\sec(-225°)$ **34.** $\cot(-150°)$

35. $\sin\left(-\frac{3\pi}{2}\right)$ **36.** $\tan\left(-\frac{5\pi}{3}\right)$

In Exercises 37–54, use the given information to find the requested value.

37. $\cos\theta = \frac{2}{3}$; find $\sec\theta$.

38. $\sin\theta = -\frac{3}{7}$; find $\csc\theta$.

39. $\csc\theta = -3$; find $\sin\theta$.

40. $\cot\theta = -4$; find $\tan\theta$.

41. $\sin\theta = \frac{3}{5}$; θ is in quadrant II. Find $\cos\theta$.

42. $\cos\theta = -\frac{4}{5}$; θ is in quadrant III. Find $\sin\theta$.

43. $\tan\theta = -\frac{1}{3}$; θ is in quadrant IV. Find $\cos\theta$.

44. $\cot\theta = \frac{1}{5}$; θ is in quadrant I. Find $\sin\theta$.

45. $\csc\theta = \frac{2}{\sqrt{3}}$; θ is in quadrant II. Find $\tan\theta$.

46. $\sec\theta = \frac{3}{\sqrt{5}}$; θ is in quadrant IV. Find $\cot\theta$.

47. $\cos\theta = -\frac{5}{13}$; $\tan\theta < 0$; find $\tan\theta$.

48. $\sin\theta = -\frac{12}{13}$, $\cot\theta > 0$; find $\sec\theta$.

49. $\tan\theta = \frac{3}{4}$, $\cos\theta < 0$; find $\sin\theta$.

50. $\cot\theta = \frac{2}{5}$, $\sin\theta < 0$; find $\cos\theta$.

51. $\sec\theta = -\frac{5}{4}$, $\cot\theta < 0$; find $\tan\theta$.

52. $\csc\theta = -\frac{4}{3}$, $\tan\theta > 0$; find $\cos\theta$.

53. $\tan\theta = 3$, $\cos\theta < 0$; find $\cos\theta$.

54. $\cot\theta = 2$, $\sin\theta > 0$; find $\sin\theta$.

55. For what values of θ is $\sin\theta$ defined?

56. For what values of θ is $\cos\theta$ defined?

57. For what values of θ is $\tan\theta$ defined?

58. For what values of θ is $\cot\theta$ defined?

In Exercises 59–64, give two angles, one positive and one negative, that are coterminal with the given angle.

59. $\frac{2\pi}{3}$

60. $200°$

61. $110°$

62. $\frac{7\pi}{6}$

63. $\frac{3\pi}{2}$

64. $180°$

65. Find three angles θ for which $\sin\theta = \frac{1}{2}$.

66. Find three angles θ for which $\cos\theta = -\frac{\sqrt{2}}{2}$.

67. Find three angles θ for which $\tan\theta = -\sqrt{3}$.

68. Find three angles θ for which $\csc\theta = -1$.

69. In trigonometry we normally write $\sin^2\theta$ for $(\sin\theta)^2$; similar expressions are used for the other trigonometric functions. Find the value of $\sin^2\frac{\pi}{3} + \cos^2\frac{\pi}{3}$. a

70. Find the value of $\sin^2\frac{\pi}{6} + \cos^2\frac{\pi}{6}$.

71. Find the value of $\sin^2\frac{\pi}{2} + \cos^2\frac{\pi}{2}$.

72. Find the value of $\sec^2\frac{\pi}{4} - \tan^2\frac{\pi}{4}$.

In Exercises 73–74, complete the following tables.

73.

θ	$\sin\theta$	$\cos\theta$	$\tan\theta$
0			
$\frac{\pi}{6}$			
$\frac{\pi}{4}$			
$\frac{\pi}{3}$			
$\frac{\pi}{2}$			
$\frac{2\pi}{3}$			
$\frac{3\pi}{4}$			
$\frac{5\pi}{6}$			
π			

74.

θ	$\sin\theta$	$\cos\theta$	$\tan\theta$
π			
$\frac{7\pi}{6}$			
$\frac{5\pi}{4}$			
$\frac{4\pi}{3}$			
$\frac{3\pi}{2}$			
$\frac{5\pi}{3}$			
$\frac{7\pi}{4}$			
$\frac{11\pi}{6}$			
2π			

QUESTIONS FOR THOUGHT

75. Can $\sin \theta = 2$? Why or why not?

76. Can $\cos \theta = -4$? Why or why not?

77. In light of your answers to Exercises 75 and 76, what can you say about the possible values of $\csc \theta$ and $\sec \theta$?

78. Are there any limitations on the values of $\tan \theta$? Explain.

79. Compare the values of $\sin \frac{\pi}{4}$ and $\sin\left(-\frac{\pi}{4}\right)$, $\sin \frac{2\pi}{3}$ and $\sin\left(-\frac{2\pi}{3}\right)$, and $\sin \frac{11\pi}{6}$ and $\sin\left(-\frac{11\pi}{6}\right)$. Do you think this generalizes? Explain.

80. Compare the values of $\cos \frac{3\pi}{4}$ and $\cos\left(-\frac{3\pi}{4}\right)$, $\cos \frac{\pi}{3}$ and $\cos\left(-\frac{\pi}{3}\right)$, and $\cos \frac{7\pi}{6}$ and $\cos\left(-\frac{7\pi}{6}\right)$. Do you think this generalizes? Explain.

81. Based upon your generalizations in Exercises 79 and 80, what can you say regarding $\tan \theta$ and $\tan(-\theta)$?

6.3 Right Triangle Trigonometry and Applications

In the last section we discussed the trigonometry of a general angle θ when it is placed in standard position. This restriction makes it somewhat difficult to see how to apply trigonometry to real-world situations. In this section, we examine the trigonometry of the general right triangle (that is, right triangles other than the isosceles right triangle and the 30°-60° right triangle).

If θ happens to be a first-quadrant angle, then as we saw in the last section, θ is its own reference angle. See Figure 6.34.

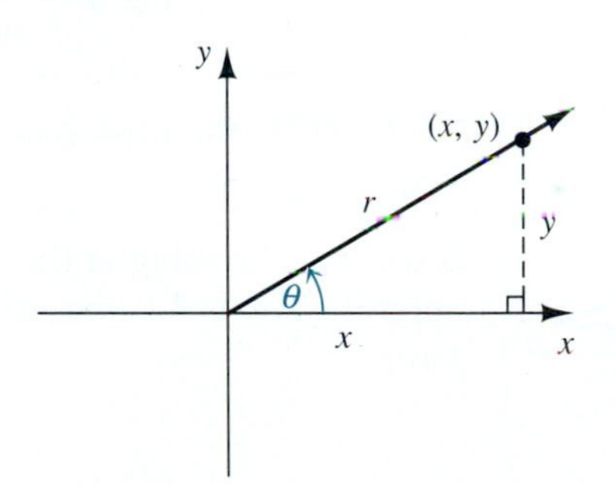

FIGURE 6.34 A first-quadrant angle

If we view the reference triangle from the perspective of angle θ, we see that we can rename the side labeled y as the side *opposite* θ, the side labeled x as the side *adjacent* to θ, and the side labeled r as the *hypotenuse*. See Figure 6.35. Note that this renaming works only for a first-quadrant (acute) angle θ.

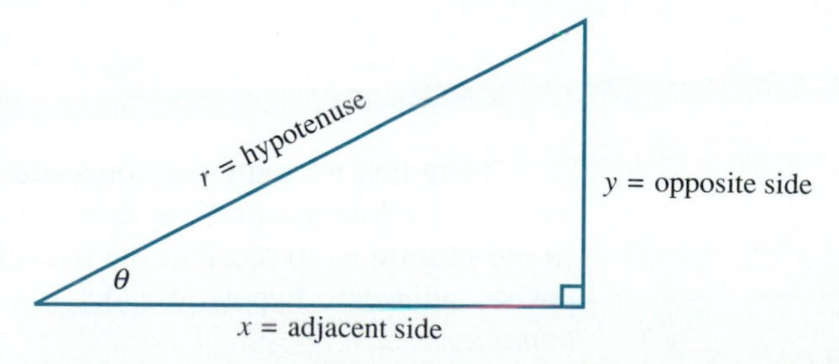

FIGURE 6.35 Naming the sides of the reference triangle

Thus for an acute angle θ in a right triangle, we can restate the definitions of the six trigonometric functions as follows:

RIGHT TRIANGLE DEFINITIONS OF THE TRIGONOMETRIC FUNCTIONS

For θ an acute angle in a right triangle, as in the accompanying figure,

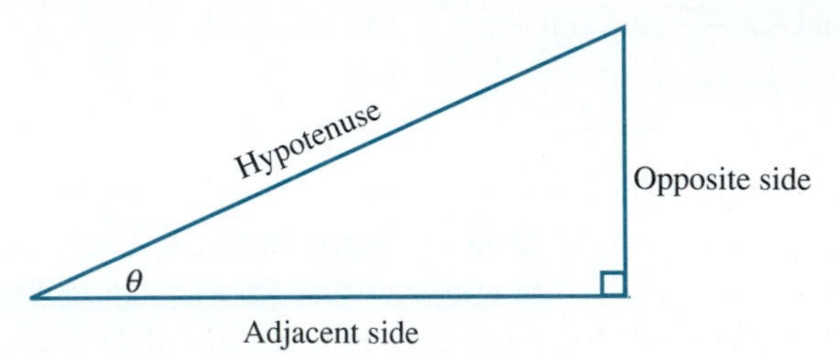

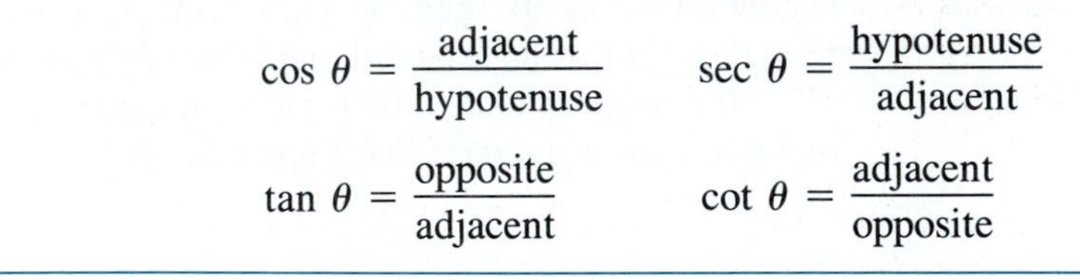

$$\sin\theta = \frac{\text{opposite}}{\text{hypotenuse}} \qquad \csc\theta = \frac{\text{hypotenuse}}{\text{opposite}}$$

$$\cos\theta = \frac{\text{adjacent}}{\text{hypotenuse}} \qquad \sec\theta = \frac{\text{hypotenuse}}{\text{adjacent}}$$

$$\tan\theta = \frac{\text{opposite}}{\text{adjacent}} \qquad \cot\theta = \frac{\text{adjacent}}{\text{opposite}}$$

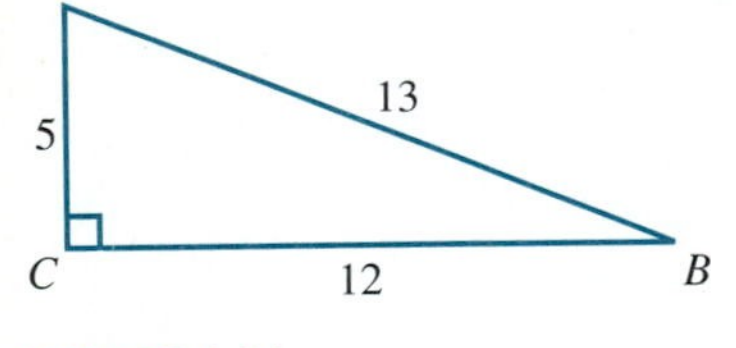

FIGURE 6.36

EXAMPLE 1 Find the sine, cosine, and tangent of angle A in the triangle in Figure 6.36.

Solution Looking at the sides of the triangle *from the perspective of* $\angle A$, the side opposite $\angle A$ is 12, the side adjacent to $\angle A$ is 5, and the hypotenuse is 13. Therefore,

$$\sin A = \frac{\text{opposite}}{\text{hypotenuse}} = \boxed{\frac{12}{13}}$$

$$\cos A = \frac{\text{adjacent}}{\text{hypotenuse}} = \boxed{\frac{5}{13}}$$

$$\tan A = \frac{\text{opposite}}{\text{adjacent}} = \boxed{\frac{12}{5}}$$

■

Note that by using the "opposite, adjacent, hypotenuse" definitions rather than the "x, y, r" definitions, we don't first have to put $\angle A$ in standard position, which would require us to reorient the triangle at the origin. But keep in mind that the opposite, adjacent, hypotenuse definitions *only apply to an acute angle of a right triangle*.

Using a Calculator to Find Trigonometric Values

Let's now return to our discussion of finding the trigonometric functions of various angles. Up to this point we have restricted our attention to the case where θ (or its reference angle θ') is a "special" angle—that is, where θ is a multiple of 30°, 45°, 60°, or θ is a quadrantal angle.

In case θ or θ' is not a special angle, we need a calculator or a table of values for the trigonometric functions. Given the widespread availability of inexpensive hand-held scientific calculators that can compute the values of trigonometric functions, where necessary, we will use a calculator to find required values of the trigonometric functions. However, if a calculator is not available, a table can be used to work out the examples in the text as well as the exercise sets. The appendix contains a table of trigonometric values along with an explanantion of how to use it.

To use a calculator, we typically first decide on the *mode* in which we want to enter the angle—i.e., degrees or radians. Then we enter the angle on the display and press the key labeled with the trigonometric function in which we are interested.

For example, to find sin 28° we first make sure the calculator is in degree mode and then press the following sequence of keys:

Be aware of the fact that some calculators require you to compute sin 28° by entering [sin] [2] [8].

[2] [8] [sin]

The display would read **0.4694716,** which when rounded to four decimal places is 0.4695 (the value you would obtain from the table).

To find cos 3 we must first make sure the calculator is in radian mode (remember, if there are no units, 3 means 3 radians) and then press the following sequence of keys:

[3] [cos]

The display will read **−0.9899925.**

Most calculators do not have a key for secant, cosecant, or cotangent. In order to evaluate these functions we must use the [1/x] key together with the appropriate reciprocal function. For example, to find sec 29° we enter 29° on the display, press the cosine key, and finally press [1/x] to take the reciprocal of cos 29° and get sec 29°. To summarize, to get sec 29° the keystroke sequence is (remember, the calculator must be in degree mode)

Does sin 10 mean the sine of 10 degrees or the sine of 10 radians? Find sin 10° and sin 10.

[2] [9] [cos] [1/x]

The display would read **1.1433541**.

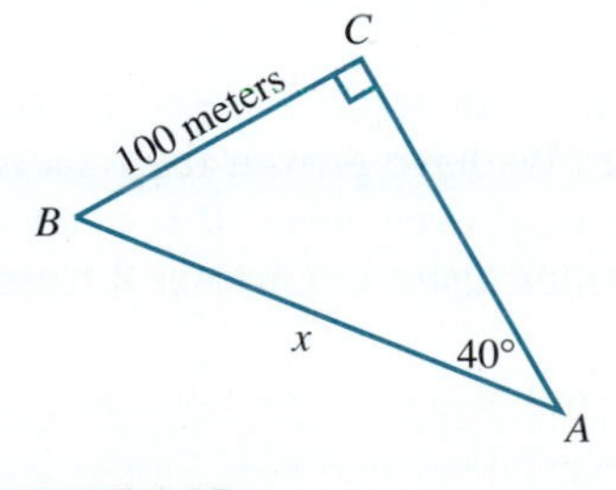

FIGURE 6.37

EXAMPLE 2 Find the length of side $\overline{AB}$ in the triangle in Figure 6.37. (Round the answer to the nearest tenth.)

Solution Looking at the triangle from the perspective of the 40° angle, we are given the length of side $\overline{BC}$, which is the side *opposite* the 40° angle, and we are looking for the length of side $\overline{AB}$ (labeled x), which is the *hypotenuse*. Since this is a right triangle, we can use the sine function to find x.

$$\sin 40° = \frac{\text{opposite}}{\text{hypotenuse}} = \frac{100}{x} \qquad \textit{We use a calculator to find } \sin 40°.$$

$$0.6427876 = \frac{100}{x} \qquad \textit{Now we solve for } x.$$

$$0.6427876x = 100 \Rightarrow x = \frac{100}{0.6427876} = 155.57238$$

Rounding to the nearest tenth we get $\boxed{x = 155.6 \text{ meters}}$. Thus the length of side $\overline{AB}$ is 155.6 meters. ■

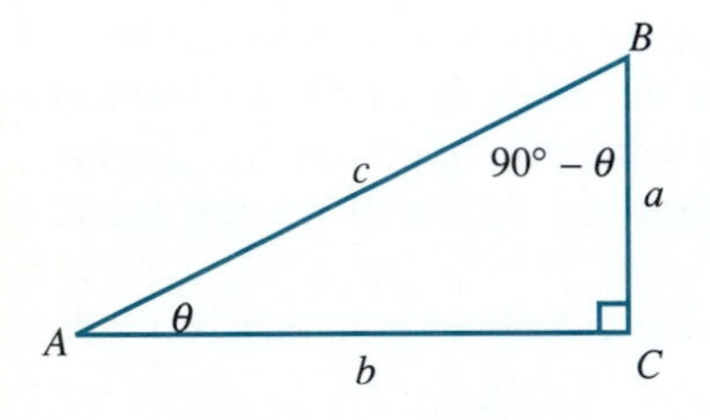

FIGURE 6.38 A typical right triangle

The Cofunctional Relationships

In Figure 6.38 we draw a typical right triangle. Since the two acute angles of a right triangle are complementary (add up to 90°), we have labeled the two acute angles as θ and $90° - \theta$. Thus we have

$$\sin \theta = \frac{a}{c} = \cos(90° - \theta)$$

$$\tan \theta = \frac{a}{b} = \cot(90° - \theta)$$

$$\sec \theta = \frac{c}{b} = \csc(90° - \theta)$$

These are called the **cofunctional** relationships. The prefix "co" that is used in the names of three of the functions comes from the word complementary.

We summarize the cofunctional relationships next.

THE COFUNCTIONAL RELATIONSHIPS

$$\sin \theta = \cos\left(\frac{\pi}{2} - \theta\right) \qquad \sin \theta = \cos(90° - \theta)$$

$$\tan \theta = \cot\left(\frac{\pi}{2} - \theta\right) \qquad \tan \theta = \cot(90° - \theta)$$

$$\sec \theta = \csc\left(\frac{\pi}{2} - \theta\right) \qquad \sec \theta = \csc(90° - \theta)$$

In words, this says that the trigonometric function of an angle is equal to the "co" trigonometric function of the complementary angle. We have proven these facts using the right triangle relationships, but, in fact, they hold even when θ is not an acute angle in a right triangle. We look at these relationships again in Chapter 8 from a different point of view.

EXAMPLE 3 Express the given trigonometric value as a trigonometric function of an acute angle less than 45° $\left(\text{or } \frac{\pi}{4}\right)$.

(a) sin 56° **(b)** cot 78° **(c)** $\sec \frac{2\pi}{5}$

Solution

(a) Using the cofunctional relationship for sine and cosine, we see that

$$\sin 56° = \cos(90 - 56)° = \boxed{\cos 34°}$$

(b) Using the cofunctional relationship for cotangent and tangent,

$$\cot 78° = \tan(90 - 78)° = \boxed{\tan 12°}$$

(c) Using the cofunctional relationship for secant and cosecant,

$$\sec \frac{2\pi}{5} = \csc\left(\frac{\pi}{2} - \frac{2\pi}{5}\right) = \boxed{\csc \frac{\pi}{10}}$$

Note that all the given angles were greater than 45° $\left(\text{or } \frac{\pi}{4}\right)$, and so their complements were all less than 45°, as required. ■

In Example 2 we used the trigonometric functions to find the missing side of a right triangle. The same idea can be used to find the missing angle of a right triangle as well.

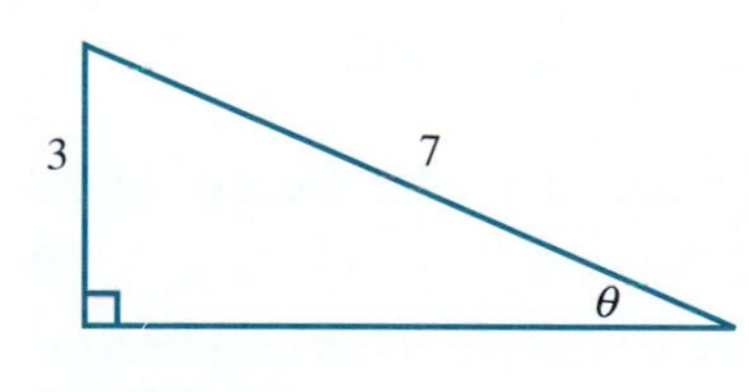

FIGURE 6.39

Some calculators have a [2nd] key or a [shift] key rather than an [inv] key.

EXAMPLE 4 Find the value of θ to the nearest degree. See Figure 6.39.

Solution From Figure 6.39 we see that $\sin \theta = \frac{3}{7}$. We have previously described how to use a calculator to find the trigonometric function of a given angle. A calculator also allows us to find the acute angle associated with a particular trigonometric value. (We discuss finding the angle associated with a negative trigonometric value in Section 7.5).

The following key sequence will give us θ:

[3] [÷] [7] [=] [inv] [sin] [=]

The display reads **25.376934** degrees. Rounding this answer to the nearest degree, we have $\boxed{\theta = 25°}$. ■

We discuss the inverse trigonometric functions in much greater detail in the next chapter.

As we indicated at the beginning of this chapter, trigonometry has many practical applications, as we begin to illustrate with the following examples.

EXAMPLE 5 In order to find the length of Lake Narrow, a surveyor wants to measure the distance between points A and C on opposite shores of the lake, as indicated in Figure 6.40. From point C, the surveyor measures a distance of 250 meters to point B so that $\overline{BC}$ is perpendicular to the sight line between A and C. Using an instrument called a *transit,* which measures angles, the surveyor finds the measure of $\angle B$ to be 82°. Find the distance between A and C (which we have labeled x) to the nearest meter.

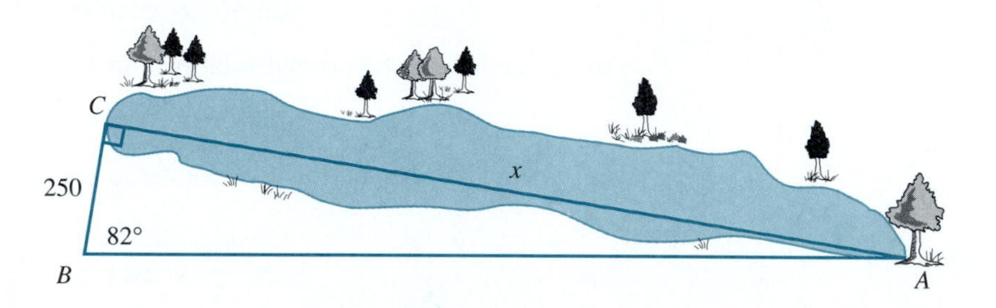

FIGURE 6.40 Diagram of Lake Narrow

Solution It is quite difficult to measure directly a fairly long distance across a body of water. Using trigonometry, however, it is quite easy to find such a distance. Looking at the diagram, we see that the side we are looking for is the side *opposite* the angle of 82°, and the side we know is the side *adjacent* to the angle of 82°. Therefore, we use the tangent function.

$$\tan 82^\circ = \frac{x}{250} \qquad \textit{We find } \tan 82^\circ.$$

$$7.1153697 = \frac{x}{250}. \Rightarrow x = 7.1153697(250) = 1778.8424$$

Thus the length of Lake Narrow is $\boxed{\text{1779 meters}}$ rounded to the nearest whole meter. ■

Let's look at several more situations in which we can apply the trigonometric ideas we have developed thus far. But first let's introduce some useful terminology.

In a situation in which two observers A and B are situated so that B is above the eye level of A we often refer to the **angle of elevation** or **angle of depression,** as indicated in Figure 6.41.

In a given situation what is the relationship between the angle of elevation and the angle of depression?

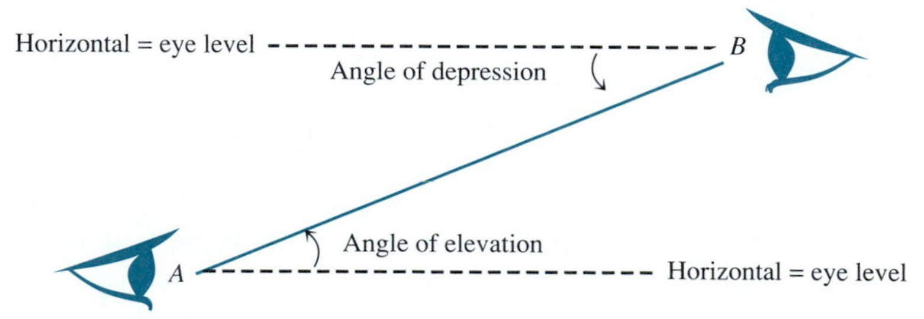

FIGURE 6.41 Angle of elevation and angle of depression

In other words, the angle of elevation is the angle through which an observer must elevate his or her eyes from eye level to focus on a particular point. Similarly, the angle of depression is the angle through which an observer must lower his or her eyes from eye level to focus on a particular point.

EXAMPLE 6 A pilot of a Navy jet is about to land on an aircraft carrier. At an altitude of 3000 feet, the pilot observes the carrier at an angle of depression of 15°. See Figure 6.42. To the nearest tenth of a mile, how far is the plane (in ground distance) from the carrier?

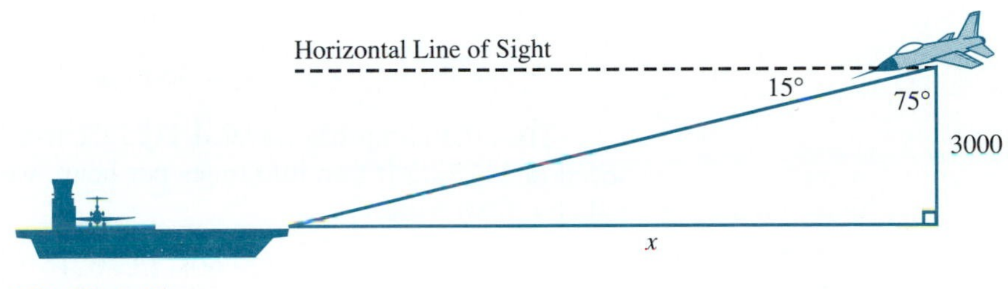

FIGURE 6.42

Solution We have labeled the ground distance from the plane to the carrier as x. Since the angle of depression is 15°, its complement is $90° - 15° = 75°$. Looking at the triangle in the figure, we see that we are looking for the side opposite the 75° angle and we know the side adjacent to the 75° angle. Therefore, we use the tangent function.

$$\tan 75° = \frac{x}{3000}$$

$$x = 3000 \tan 75°$$ *Using a calculator we get*

$$x = 11{,}196.152$$ *This answer is in feet, so we divide by 5280 to convert to miles.*

$$x = 2.1204834 \text{ miles}$$

Thus the ground distance from the jet to the carrier is $\boxed{2.1 \text{ miles}}$ to the nearest tenth of a mile. ■

EXAMPLE 7 The Goodyear blimp is flying at an altitude of 500 feet and passes directly over an observer on the ground. One minute later the angle of elevation to the blimp from the observer is 24°. Find the speed of the blimp to the nearest mile per hour.

Solution Figure 6.43 at the top of the next page summarizes the given information. A represents the point directly above the observer, and B is the position of the blimp 1 minute later. Once we find the distance the blimp travels in 1 minute (which we have labeled x), we will be able to compute the speed of the blimp.

How will finding the distance the blimp travels help us compute its speed?

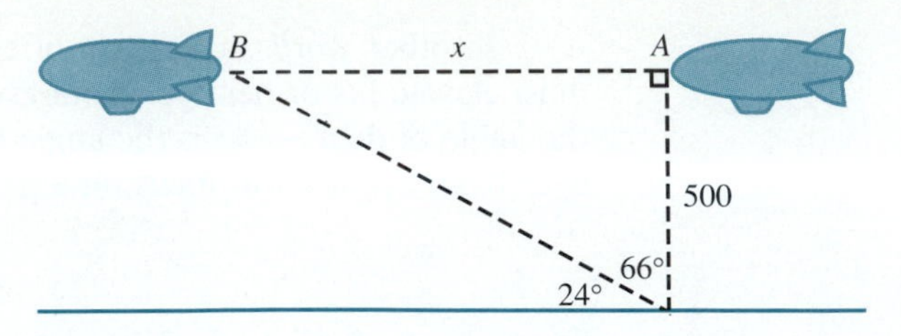

FIGURE 6.43

$$\tan 66^\circ = \frac{x}{500}$$

$$x = 500 \tan 66^\circ \approx 1123.02 \text{ feet}$$

Thus the blimp has traveled 1123.02 feet in 1 minute. In order to convert this speed of 1123.02 ft/min into miles per hour, we must multiply by 60 min/hr and divide by 5280 ft/mi.

$$\frac{60(1123.02)}{5280} \approx 12.76$$

Rounding to the nearest mile per hour, the speed of the blimp is 13 mi/hr. ■

EXAMPLE 8 A woman is standing at a window 80 feet above the ground. She observes a child walking directly toward her as the angle of depression to the child changes from 42° to 65°. How far has the child walked (to the nearest tenth of a foot)?

Solution We summarize the given information in Figure 6.44(a). The distance the child has walked is the distance between the points A and B, which we have labeled x. Since x is not the side of a right triangle, we cannot (at this point) find x directly. However, we observe *two* right triangles, $\triangle DCB$ and $\triangle DCA$. $\triangle DCB$ has side d and $\triangle DCA$ has side $d + x$. If we can find these sides, we can compute their difference: $(d + x) - d = x$.

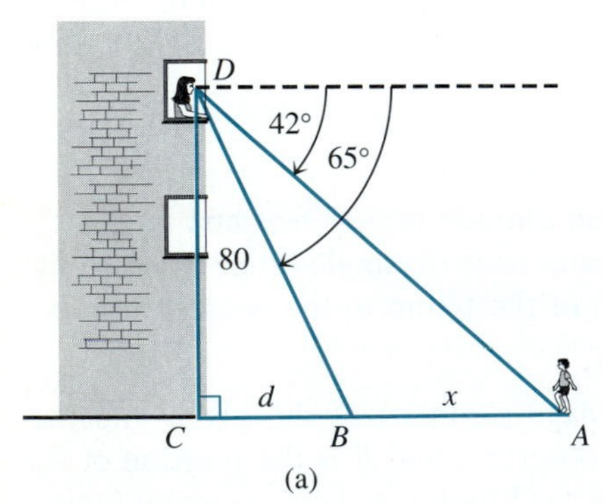

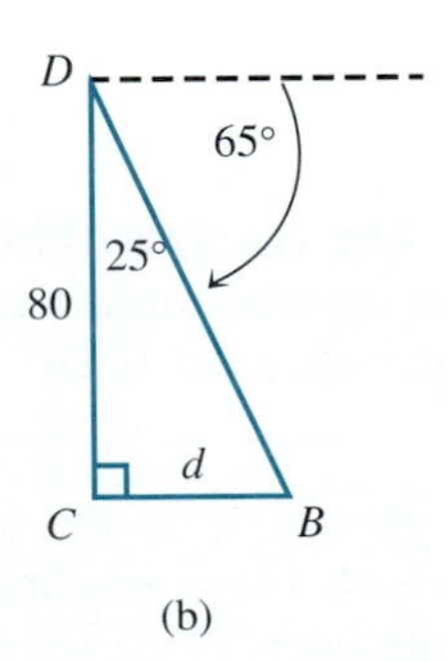

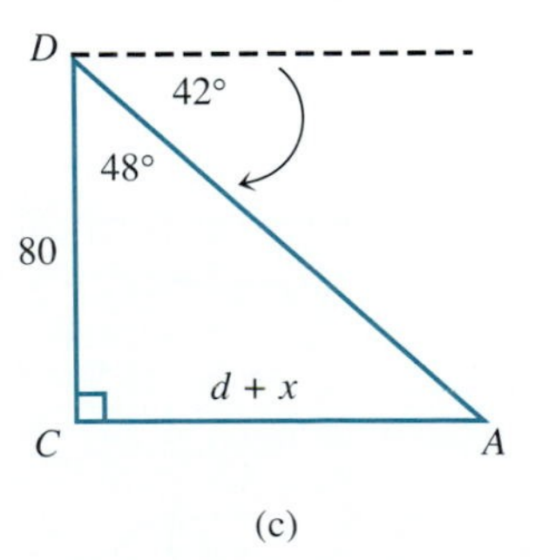

FIGURE 6.44

To find the sides of the two right triangles, we use the angles in the triangles that are complementary to the angles of depression. See Figures 6.44(b) and (c). We must first find $\overline{BC}$, which we have labeled d. Once we find d, we will be able to find x. In $\triangle DCB$ we have

$$\tan 25° = \frac{d}{80} \Rightarrow d = 80 \tan 25°$$

Now we can look at $\triangle DCA$. In $\triangle DCA$ we have

$$\tan 48° = \frac{x + d}{80} \qquad \textit{Substituting } d = 80 \tan 25° \textit{ we get}$$

$$\tan 48° = \frac{x + 80 \tan 25°}{80} \qquad \textit{Solving for } x \textit{ we get}$$

$$x = 80 \tan 48° - 80 \tan 25° \approx 51.54$$

Rounding to the nearest tenth of a foot, we see that the child has walked $\boxed{51.5 \text{ feet}}$. ■

EXAMPLE 9 An escalator makes an angle of 20° with the floor and carries people through a vertical distance of 38 feet between a subway platform and the street above. If it takes 30 seconds to carry a person from the bottom of the escalator to the top, how fast (to the nearest tenth) is the escalator moving?

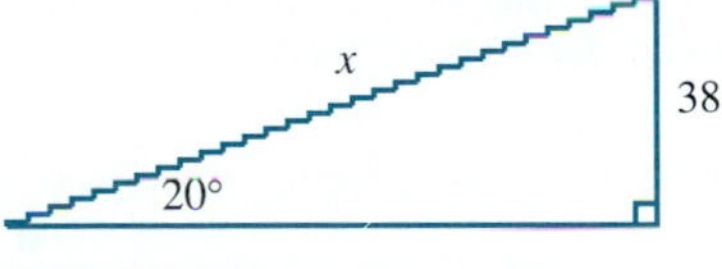

FIGURE 6.45 An escalator making an angle of 20° with the floor

Solution The given information gives rise to the diagram in Figure 6.45. In order to find the speed of the escalator, we need to find the distance (labeled x) that the escalator covers in 30 seconds.

$$\sin 20° = \frac{38}{x}$$

$$x = \frac{38}{\sin 20°} \approx 111.10457$$

Rounding to the nearest tenth, we see that the escalator travels 111.1 feet in 30 seconds, and so its rate, r, is

$$r = \frac{111.1 \text{ ft}}{30 \text{ sec}} = \boxed{3.7 \text{ ft/sec}}$$ ■

Let's take another look at an example of the type discussed in the last section in light of the new ideas we have introduced in this section.

EXAMPLE 10 Find cos 150°.

Solution We offer two solutions to this example. The first follows the approach outlined in the last section, while the second offers an alternative approach using the ideas developed in this section.

The first thing we must recognize is that we *cannot* apply the opposite, adjacent, hypotenuse definitions here, since 150° is **not** an acute angle in a right triangle.

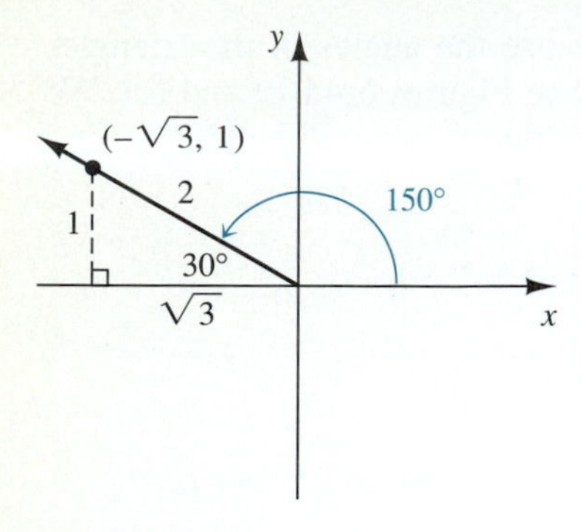

FIGURE 6.46

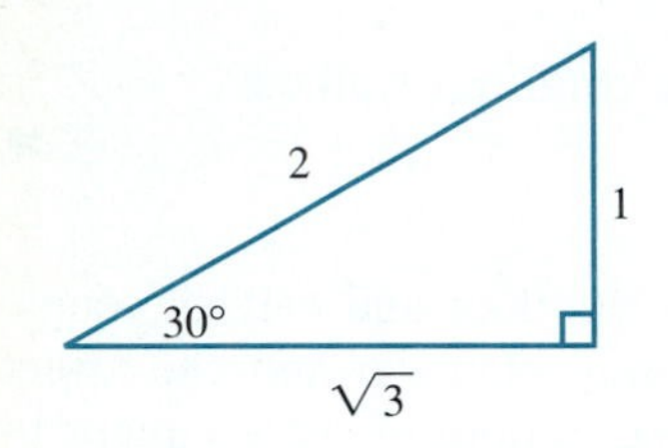

FIGURE 6.47 Reference triangle for a 30° reference angle

Following the approach of the last section, we draw an angle of 150° in standard position, find the reference angle of 30°, draw the reference triangle and fill in the sides, and finally label x, y, and r, as indicated in Figure 6.46. Thus we have

$$\boxed{\cos 150° = -\frac{\sqrt{3}}{2}}.$$

An alternative approach is to recognize that anytime an angle θ has a reference angle of 30°, regardless of which quadrant the angle is in, the lengths of the sides of the reference triangle will look like Figure 6.47. As the angle θ moves from quadrant to quadrant, all that will change will be the *signs* we use to fill in the coordinates (x, y). The reference triangle remains the same. Thus when the reference angle is 30°, the x-coordinate of the point will be either $\sqrt{3}$ or $-\sqrt{3}$ and the y-coordinate will be either 1 or -1.

In other words, any trigonometric function of an angle with a reference angle of 30° will involve the same numbers as the function of 30°. All that may differ is the sign of the answer.

Returning to the example, an angle of 150° is a second quadrant angle with a reference angle of 30°. Based upon our discussion in the last section, we know that the cosine function is negative in the second quadrant. Therefore, we have

$$\cos 150° = ?\cos 30°$$ *The value will be the same as* $\cos 30°$. *The ? indicates that the sign of the answer is still to be determined.*

$$= -\cos 30°$$ *The answer is negative, since 150° is a second quadrant angle. We find* $\cos 30°$ *by using the adjacent over hypotenuse ratio in a 30°-60° right triangle. See Figure 6.47.*

$$= \boxed{-\frac{\sqrt{3}}{2}}$$ ■

You may find this alternative approach more efficient, since it does not require completing the entire diagram of the angle in standard position.

Area of a Triangle

We conclude this section by deriving the trigonometric form of the formula for the area of a triangle. We know that the formula for the area of a triangle is

$$\text{Area} = \frac{1}{2}bh$$

Consider the triangle in Figure 6.48. Note that we have labeled the side opposite each angle with the lower case of the letter of the angle itself: a opposite $\angle A$, b opposite $\angle B$, and c opposite $\angle C$. This is the standard way to label a triangle.

We can express the height, h, in terms of the sine of $\angle C$ because in right triangle BCD we have

$$\sin C = \frac{h}{a}$$ *Solve for h.*

$$a \sin C = h$$

Therefore, we can substitute $a \sin C$ for h in the area formula $A = \frac{1}{2}bh$ to obtain $A = \frac{1}{2}ab \sin C$.

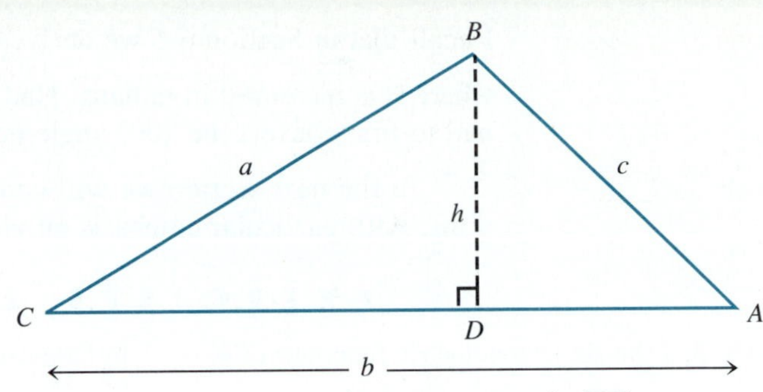

FIGURE 6.48 Triangle with height h to side $\overline{AC}$

If we want to find the area of a triangle, what advantage might the formula $A = \frac{1}{2}ab \sin C$ offer over the formula $A = \frac{1}{2}bh$?

THE TRIGONOMETRIC FORM OF THE FORMULA FOR THE AREA OF A TRIANGLE

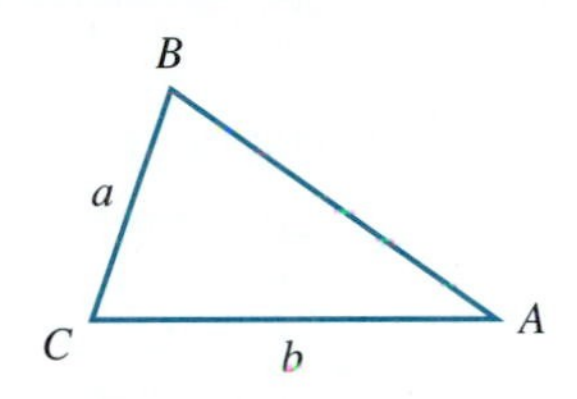

$$\text{Area} = \frac{1}{2}ab \sin C$$

In words, this formula says that the area of a triangle is equal to one-half the product of the lengths of any two sides of the triangle times the sine of the angle included between these two sides. (It is left to the student to verify this formula in the case where the height to side $\overline{AC}$ falls *outside* $\triangle ABC$.)

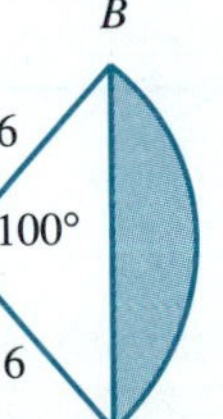

FIGURE 6.49

EXAMPLE 11 Figure 6.49 illustrates a 100° sector of a circle with radius 6 cm. A triangle is inscribed in the sector as indicated. Find the area of the shaded region to the nearest tenth.

Solution In sector AOB, $\overline{OA}$ and $\overline{OB}$ are both radii of a circle. Therefore, we have $|\overline{OA}| = |\overline{OB}| = 6$. The area of the shaded region is equal to the area of the sector minus the area of the triangle. Symbolically, we may write

$$A_{\text{shaded}} = A_{\text{sector}} - A_{\text{triangle}}$$

Note that the given information makes it easier to use the trigonometric form of the formula for the area of the triangle.

Try finding the area of the sector by using the formula $A = \frac{1}{2}r^2\theta$.

$$\begin{aligned} A_{\text{shaded}} &= \frac{\theta°}{360°} \cdot \pi r^2 - \frac{1}{2}|\overline{OA}| \cdot |\overline{OB}| \sin \theta \\ &= \frac{100}{360} \cdot \pi(6)^2 - \frac{1}{2}(6)(6) \sin 100° \\ &= 10\pi - 18 \sin 100° \approx 13.689387 \\ &= \boxed{13.7 \text{ sq cm}} \text{ rounded to the nearest tenth.} \end{aligned}$$

Recall that in Section 6.1 we derived the formula $A = \frac{1}{2}r^2\theta$ for the area of a sector where θ is measured in radians. Had we wanted to use this formula, we would have had to first convert the 100° angle into radian measure. ■

In the next section we will continue our discussion of the trigonometric functions, with particular emphasis on viewing them as functions of real numbers.

EXERCISES 6.3

In Exercises 1–10, find the six trigonometric functions of θ.

1.

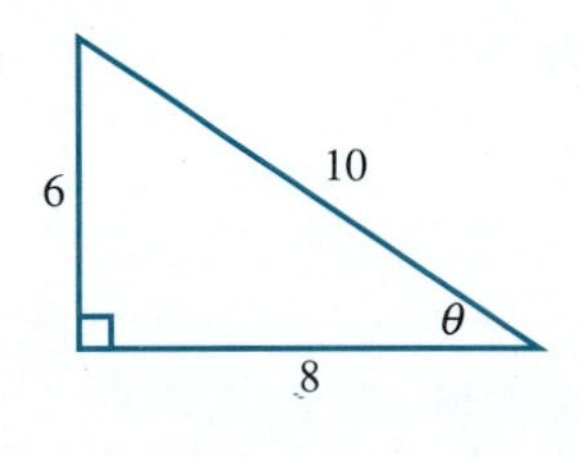

2.

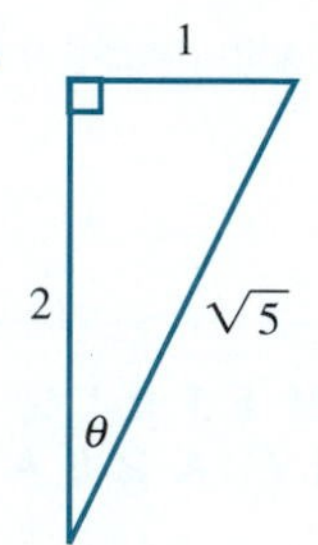

3.

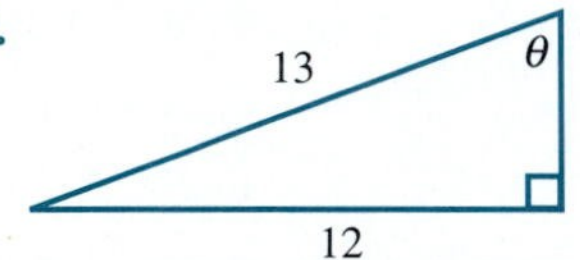

4.

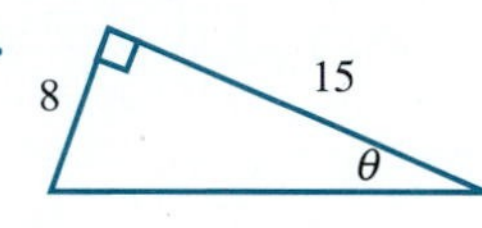

5.

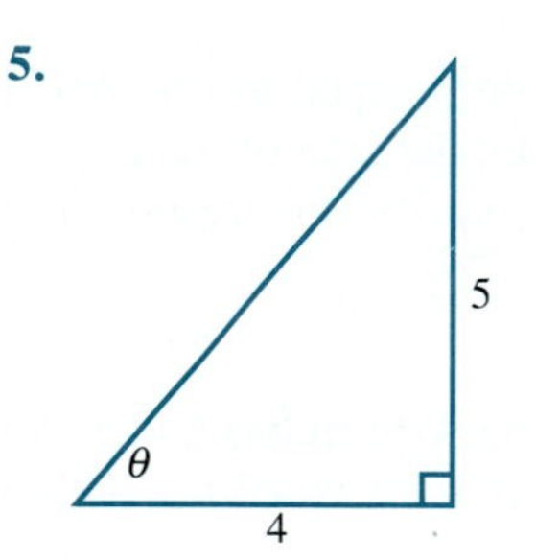

6.

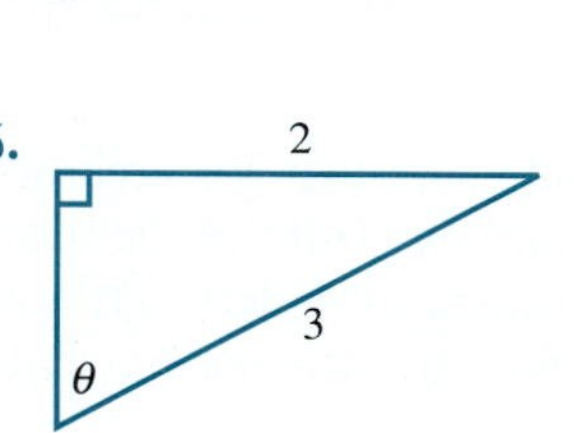

7.

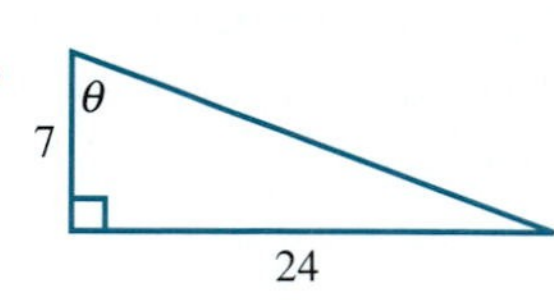

8.

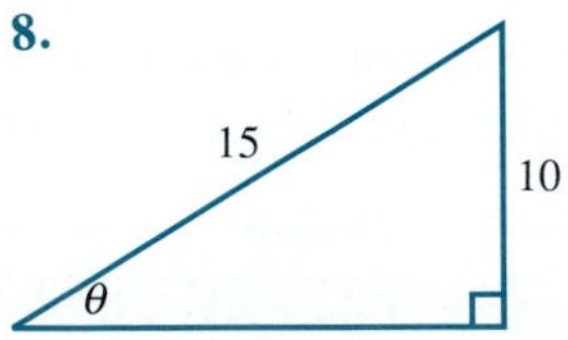

9.

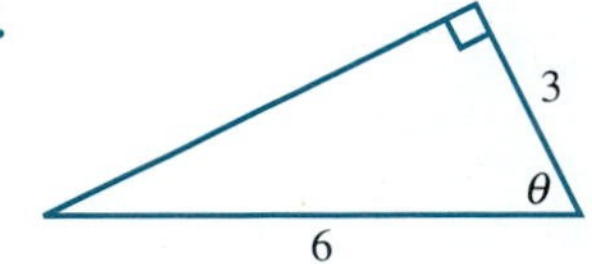

10.

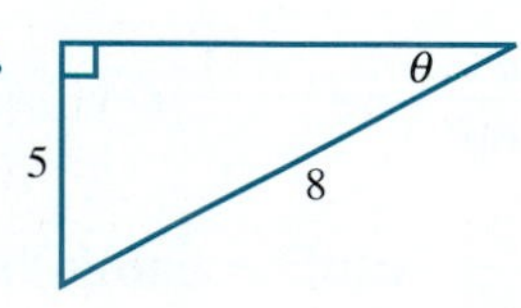

In Exercises 11–12, find $\sin A$ and $\cos B$.

11.

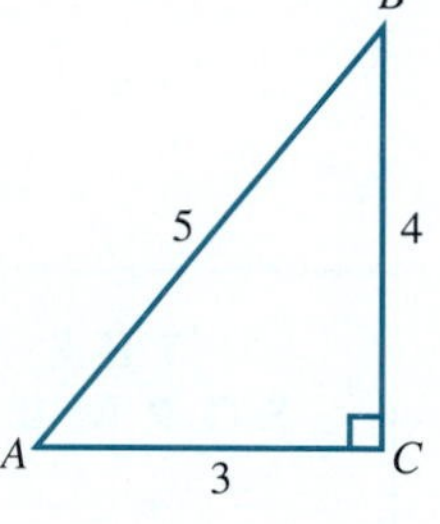
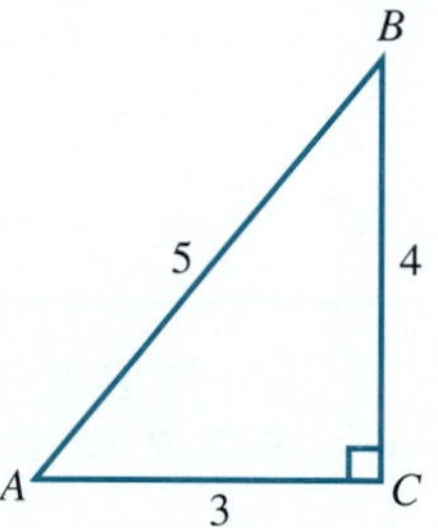

12.

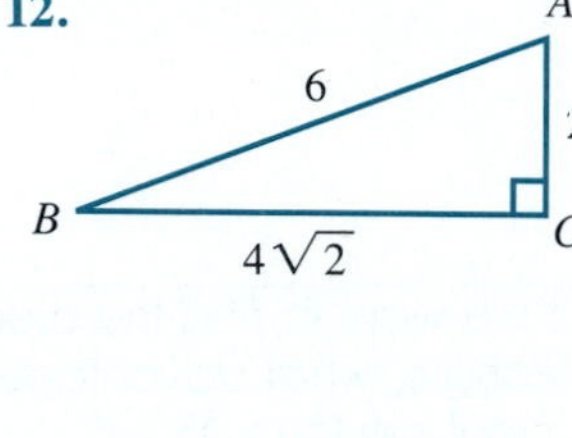
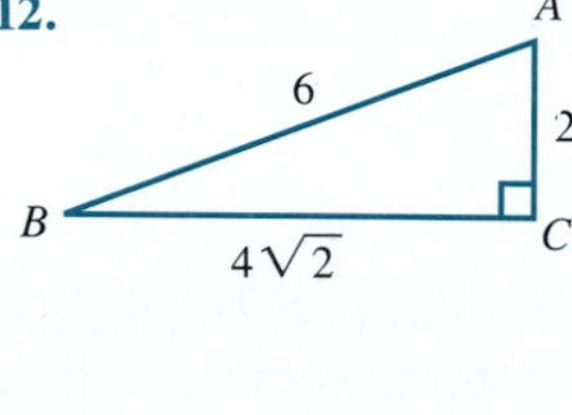

In Exercises 13–14, find $\tan \alpha$ and $\cot \beta$.

13.

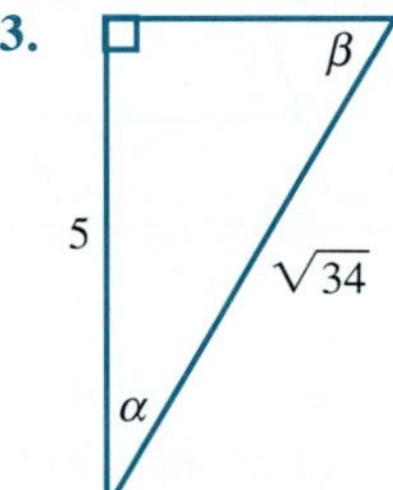

14.

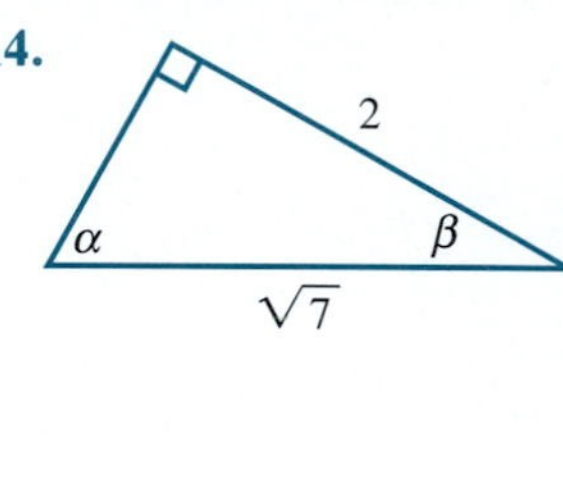

In Exercises 15–16, find $\sec P$ and $\csc Q$.

15.

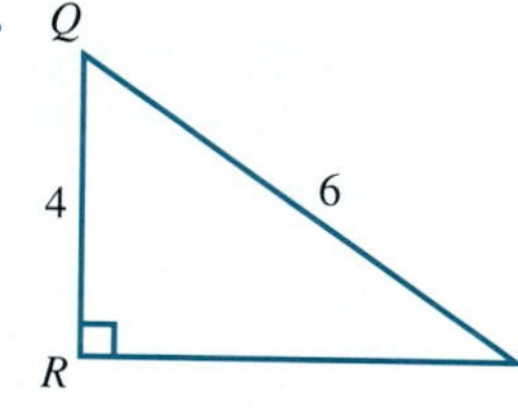

16.

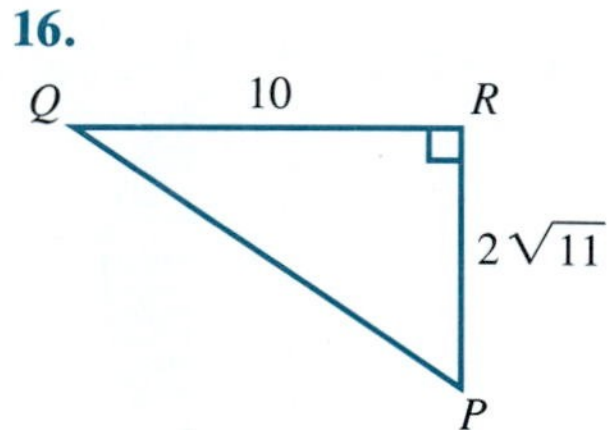

In Exercises 17–22, find x. Round your answer to the nearest tenth.

17.

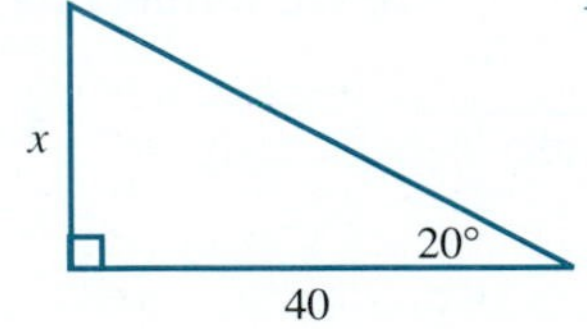

18.

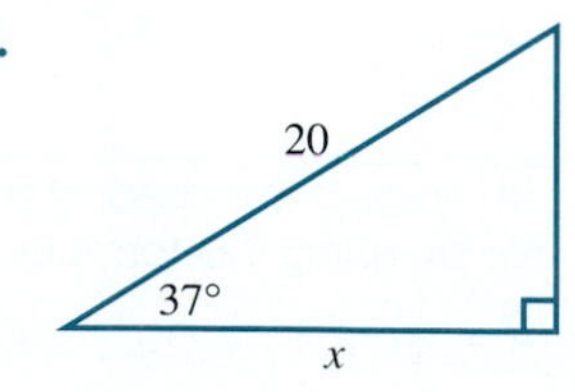

19. **20.**

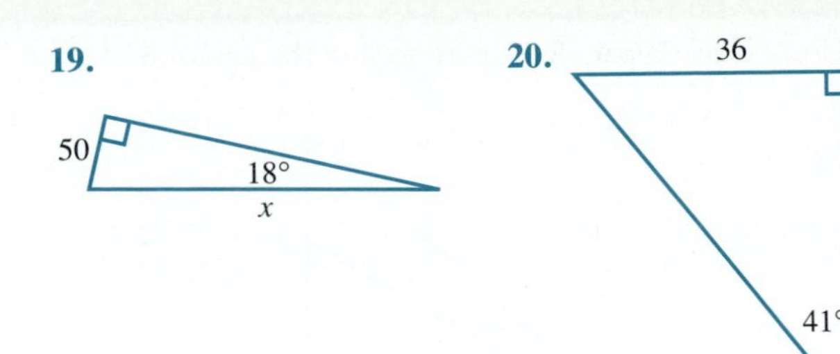

21. **22.**

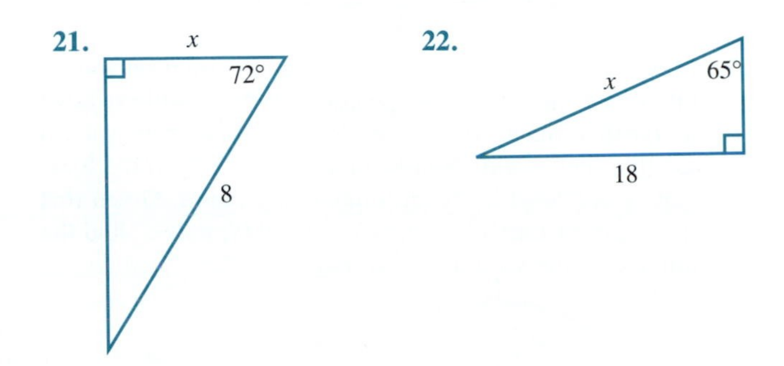

In Exercises 23–26, find θ to the nearest degree.

23. **24.** **25.** **26.**

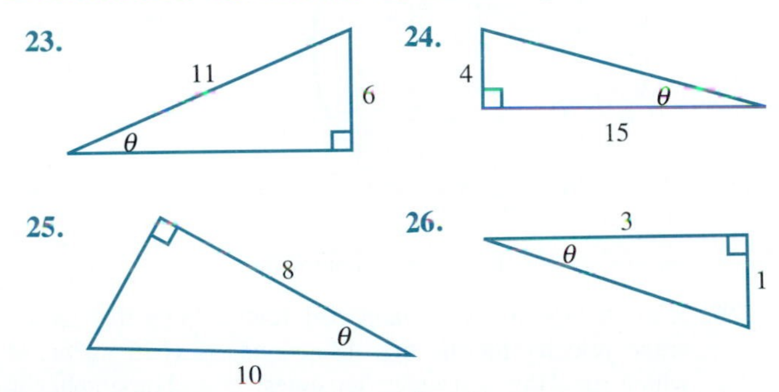

In Exercises 27–38, find the indicated value. Use a calculator only when necessary.

27. $\sin 123°$ **28.** $\cos 212°$
29. $\tan 351°$ **30.** $\cot 317°$
31. $\sec 225°$ **32.** $\csc 330°$
33. $\cos 158°$ **34.** $\sin 204°$
35. $\tan 111°$ **36.** $\cot 249°$
37. $\sec 150°$ **38.** $\csc 240°$

In Exercises 39–48, find θ or x to the nearest degree on the interval $[0, 90°]$.

39. $\sin \theta = 0.4384$ **40.** $\cos \theta = 0.7547$
41. $\tan x = 1.428$ **42.** $\cot x = 6.314$
43. $\sec \theta = 3.420$ **44.** $\csc x = 1.086$
45. $\sin x = 0.9630$ **46.** $\cos \theta = 0.7450$
47. $\tan \theta = 0.5275$ **48.** $\csc \theta = 1.320$

In Exercises 49–54, express the given value as a trigonometric function of an acute angle less than 45° or $\frac{\pi}{4}$ radians.

49. $\sin 76°$ **50.** $\cos 52°$
51. $\csc 88°$ **52.** $\tan 63°$
53. $\cos \frac{3\pi}{7}$ **54.** $\sin \frac{5\pi}{12}$

In Exercises 55–58, evaluate the given expression. Note that $\sin^2 \theta = (\sin \theta)^2$. Similarly for $\cos^2 \theta$ etc.

55. $\sin^2 30° + \cos^2 30°$ **56.** $\sec^2 60° - \tan^2 60°$
57. $\csc^2 45° - \cot^2 45°$ **58.** $\sin^2 60° + \cos^2 60°$

In Exercises 59–101, round your answers to the nearest tenth unless otherwise indicated.

59. A 25-foot ladder is leaning against a building. If the ladder makes an angle of 37° with the level ground, how high up on the building does the ladder reach?

60. A ladder is leaning up against a building. If the ladder makes an angle of 63° with the level ground and reaches 16 meters up on the building, how far away from the building is the base of the ladder?

61. An observer standing 50 feet away from the base of a flagpole finds that the angle of elevation to the top of the pole is 48°. How tall is the flagpole?

62. A forest ranger is stationed in a tower 40 meters above the ground. She spots a fire at an angle of depression of 6°. How far is the forest fire from the ranger tower?

63. A support wire (often called a *guy wire*) is to be attached to the top of a telephone pole 30 feet tall and anchored into the ground. How long a wire would be needed for it to make an angle of 50° with the level ground?

64. An escalator is to carry people through a vertical distance of 18 feet and make an angle of 20° with the floor. How long must the escalator be?

65. A regular pentagon (the word *regular* means that all the sides are the same length) is inscribed in a circle of radius 12 cm. Find the area of the pentagon.

66. A regular hexagon is inscribed in a circle of radius 10 inches. Find the area of the hexagon.

67. A hot air balloon is maintaining a constant altitude of 800 meters and passes directly over an observer. Two minutes later the observer sees the balloon at an angle of elevation of 70°. Find the speed of the balloon to the nearest kilometer per hour.

68. A motor boat is one-half mile directly offshore from point A and traveling parallel to the shore. Five minutes later the boat is observed at an angle of 34° away from the original line of sight. Find the speed of the boat to the nearest mile per hour.

69. If a man 6 feet tall casts a shadow 9 feet long along the level ground, approximate the angle of elevation of the sun to the nearest degree.

70. If the sun is at an angle of elevation of 62°, how long a shadow will be cast by a girl 5 feet tall?

71. A woman is driving directly toward Hoover Dam, which is 221 meters high. If she drives along a level road from a point at which the angle of elevation to the top of the dam is 20° to a point where the angle of elevation is 25°, how far has she driven?

72. The Empire State Building is 1250 feet high. If the corner of 33rd Street and Fifth Avenue is at an angle of depression of 79° from the top and the corner of 32nd Street and Fifth Avenue is at an angle of depression of 67° from the top, how far is it between the two corners?

73. In the accompanying figure a TV antenna is mounted at the edge of the roof of a house that is 30 feet tall. From a point 100 feet from the base of the house, the angle of elevation to the top of the antenna is 24°. How long is the antenna?

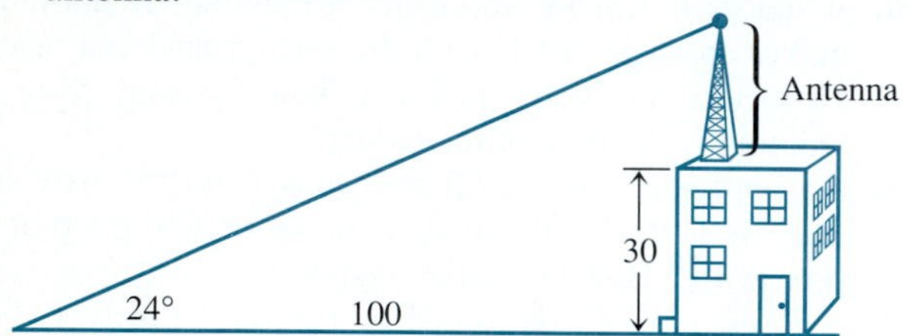

74. Use the information given in the accompanying figure to determine the distance from point B to C on opposite sides of Lake Small. Note that $\overline{AC}$ is a straight-line segment.

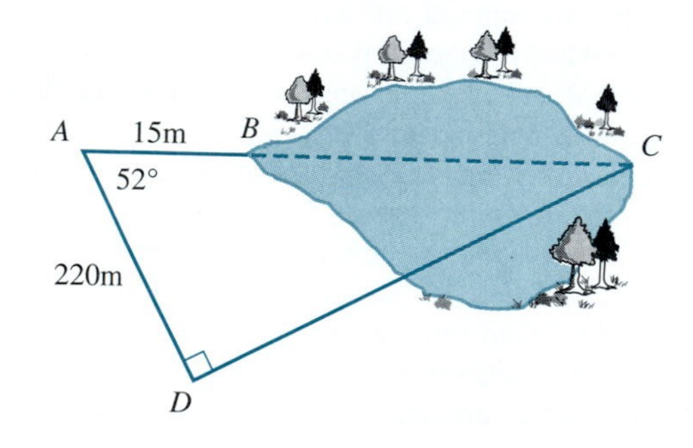

75. A Coast Guard observer is in a lighthouse 58 feet above the water level. He observes two ships on opposite sides of the lighthouse, along the same line of sight. One is at an angle of depression of 41° and heading directly toward the lighthouse, whereas the other is at an angle of depression of 28° and heading directly away from the lighthouse. How far apart are the two ships?

76. Use the accompanying figure to find the length of $\overline{AD}$.

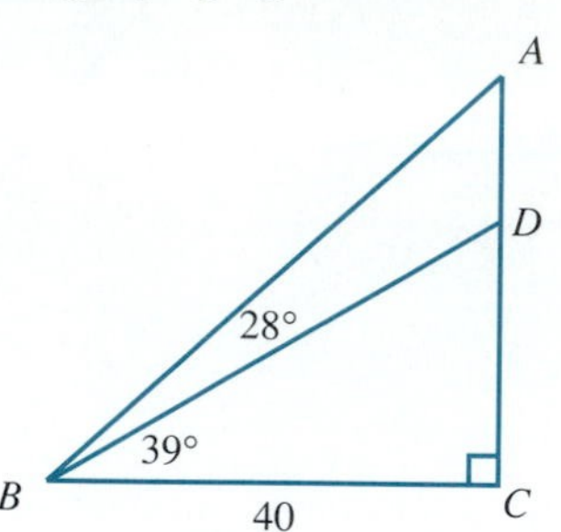

77. Many satellites are launched into a *geosynchronous* orbit, which means that the position of the satellite relative to Earth remains the same. Suppose that from such a satellite one would observe an angle of 41.4° to the horizon as indicated in the accompanying figure. Given that the radius of Earth is approximately 4000 miles, find the altitude of the satellite above Earth.

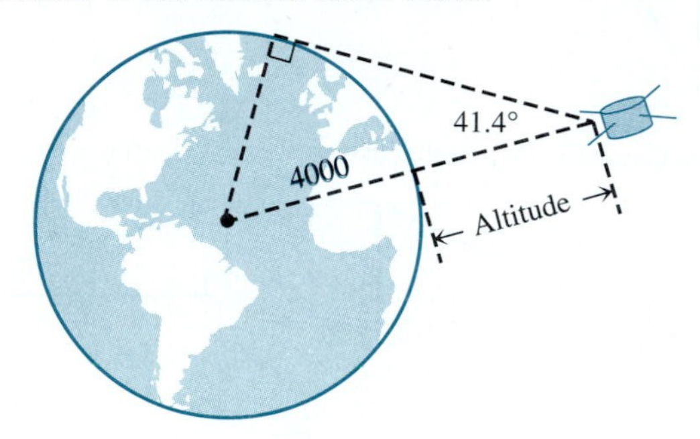

This figure is not to scale.

78. A multistage rocket is launched vertically so that its average velocity for the first 43 seconds is 1675 mi/hr, at which time the first stage separates. If a photographer is situated 1.3 mi from the launch site, at what angle of elevation should the photographer aim her camera so that she can photograph the separation?

79. Find the area of a regular pentagon with each side of length 20 mm. HINT: Draw line segments from the center of the pentagon to the vertices, creating five triangles.

80. A regular pentagon with the length of each side equal to 12 cm is inscribed in a circle. Find the area of the circle. HINT: Draw the altitude from the center of the pentagon to a side.

81. A regular hexagon of side 12 in. is inscribed in a circle. Find the area of the circle.

82. A circle is inscribed in a regular hexagon of side 12 in. Find the area of the circle.

83. The accompanying figure shows a regular pentagon of side 8 cm inscribed in a circle. Find the shaded area inside the circle and outside the pentagon.

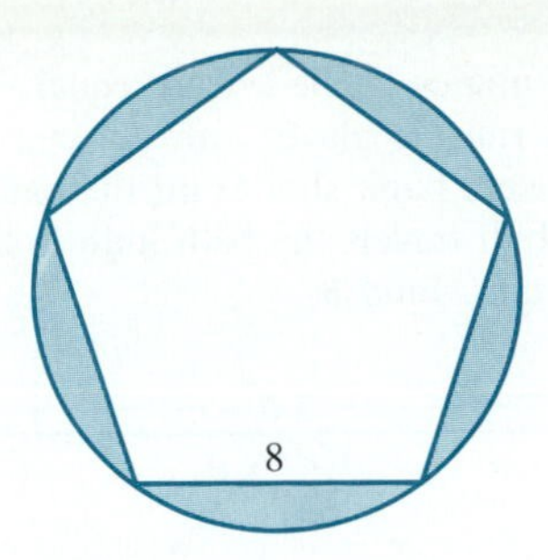

84. The accompanying figure shows a circle inscribed in a regular pentagon of side 8 cm. Find the shaded area inside the pentagon and outside the circle.

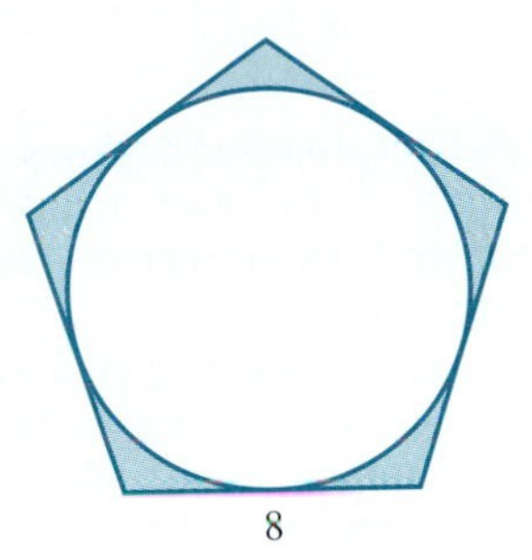

85. The accompanying figure consists of a right triangle and a semicircle. Express the area of the figure as a function of r, the radius of the semicircle.

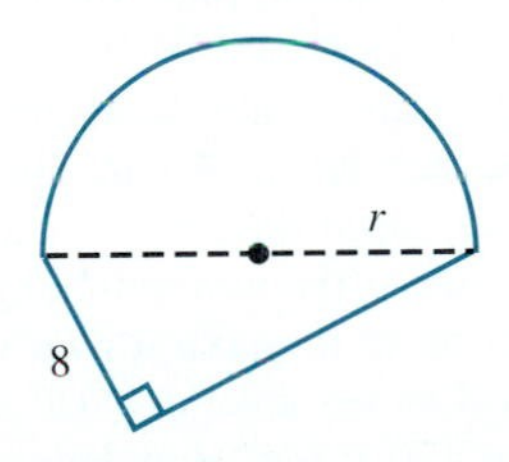

86. The accompanying figure shows a circle inscribed in a square of side s. Express the area of the shaded portion of the figure as a function of s.

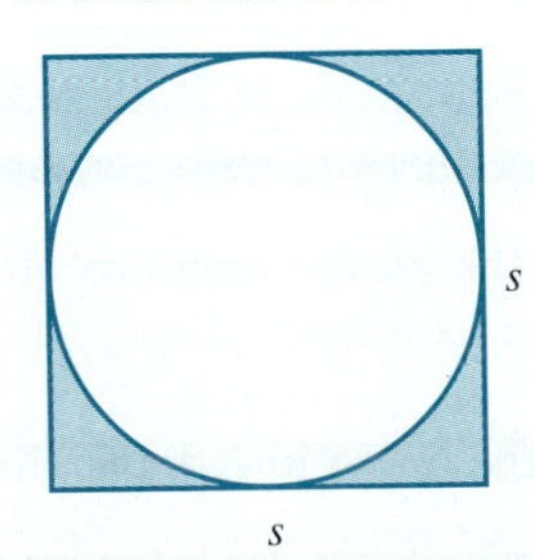

87. Given the accompanying figure, express the area of each of the following triangles in terms of θ.

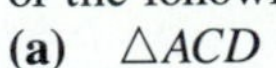

(a) $\triangle ACD$ **(b)** $\triangle ABC$ **(c)** $\triangle BCD$

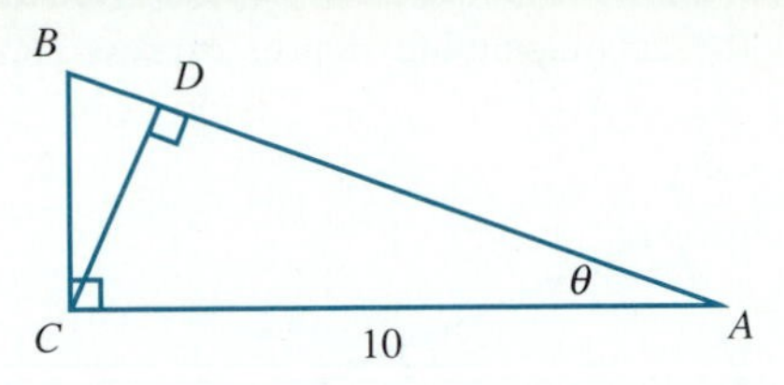

88. The formula for the volume, V, of a right circular cone is $V = \frac{1}{3}\pi r^2 h$. If $r = 6$, use the accompanying diagram to express the volume as a function of θ.

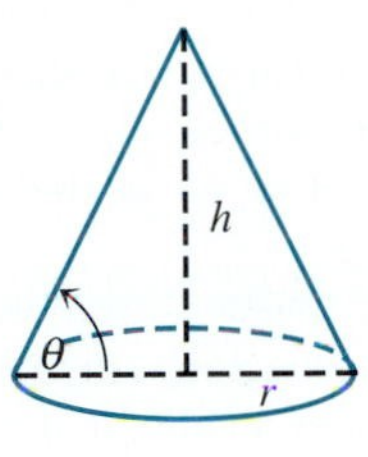

89. Given the accompanying figure, express $|\overline{AC}|$ in terms of θ.

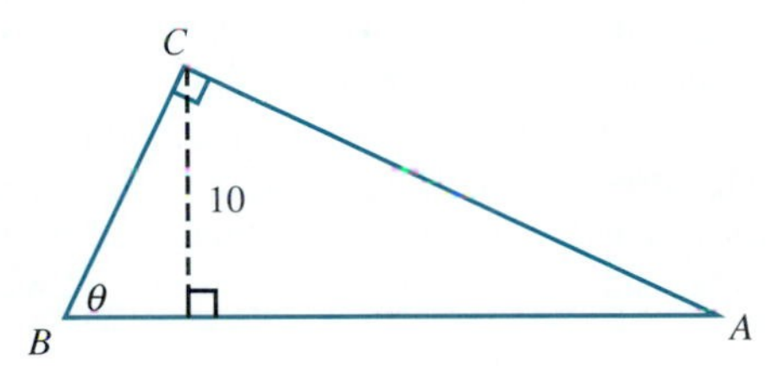

90. Given the accompanying triangle, express $\sin\theta$ and $\cos\theta$ in terms of x.

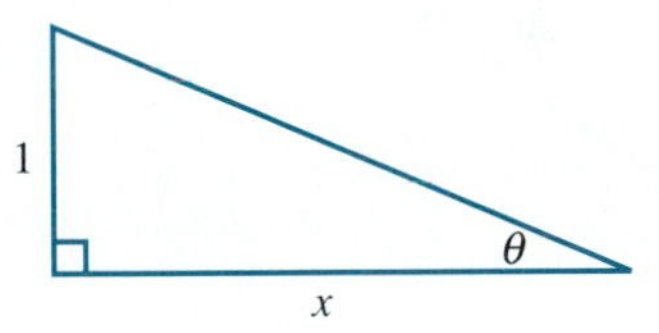

91. Given the accompanying figure, express $|\overline{PQ}|$ as a function of θ.

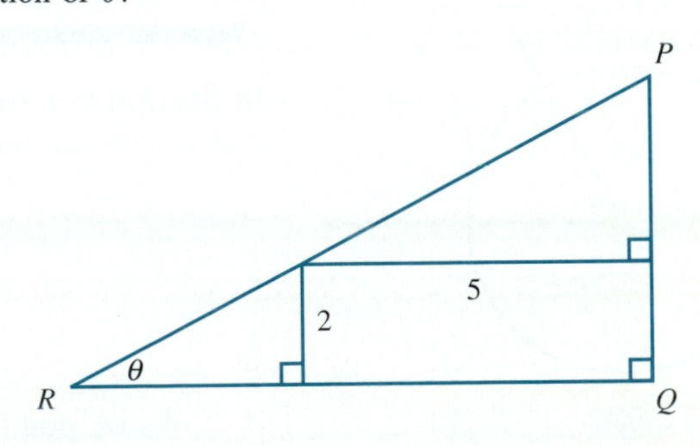

92. Given the accompanying figure, express $|\overline{RS}|$ in terms of θ.

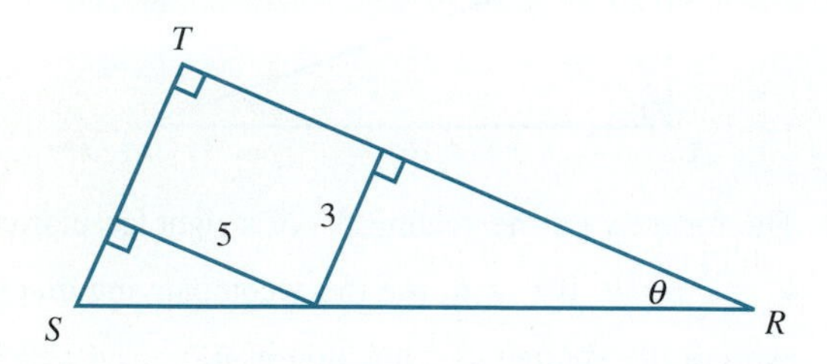

93. An airplane flying at a steady altitude of 1500 feet passes directly over a nuclear power plant, and the pilot notices that a fireworks factory is ahead of her at an angle of depression of 30°. How far is the nuclear plant from the fireworks factory?

94. Lew is relaxing under a tree when he notices that a weather balloon launching sight is 1200 feet away. He looks up and notices a balloon at a 20° angle of elevation. He begins to read a book and then, 10 minutes later, he looks up again and the angle of elevation of the same balloon is 70°. Assuming that the balloon goes straight up, how far did the balloon rise while he was reading the book and what was its rate of ascent?

95. A billiard table is set up with the cue ball (the ball that is initially struck) at position A and the ball that is to be hit by the cue ball at position B, as indicated in the accompanying diagram. A ball that strikes the rail will rebound so that the angles labeled α are equal. To which point P on the rail must a player aim the cue ball if he or she wishes to use a bank shot to hit the ball at B.

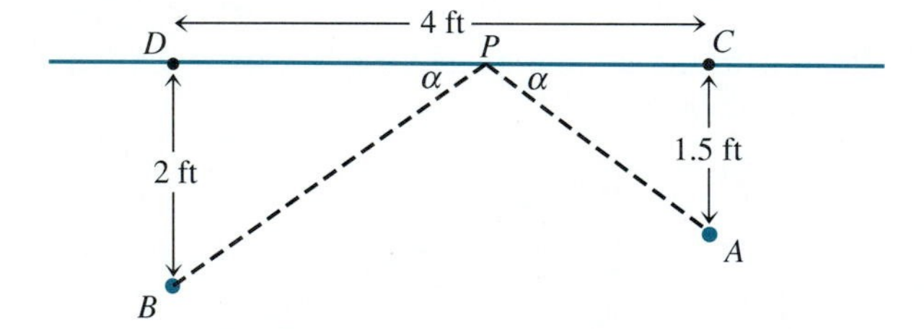

96. A billiard ball travels the path indicated in the accompanying diagram. Find θ.

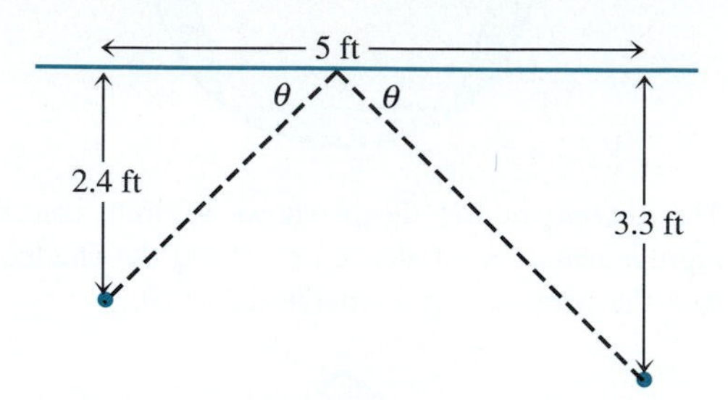

97. A sign on a straight stretch of highway warns drivers that the grade is 15° for the next 5 miles. Determine the change in elevation for a vehicle traveling this 5-mile stretch of graded highway.

98. At 2:00 A.M., a man is standing on a street 300 feet from a building. He suddenly notices a light flickering in a window of the building at an angle of elevation of 42°. Then he notices another flickering light directly above the window at an angle of elevation of 65°. How far apart are the flickering lights?

99. A 30-foot ramp must be built so that it rises 5 feet above ground level. What angle must the ramp make with the ground?

100. A pilot flying straight and level notes her altitude is 800 feet above sea level. About 3000 feet ahead is a mountain, which, according to her map, has an elevation of 2000 feet. What is the minimum angle she should nose the plane up in order to make it over the mountain?

101. A pilot takes off at sea level at a 15° angle traveling at a constant rate of 170 ft/sec. How long will it take the aircraft to reach an altitude of 5000 feet?

6.4 The Trigonometric Functions as Functions of Real Numbers

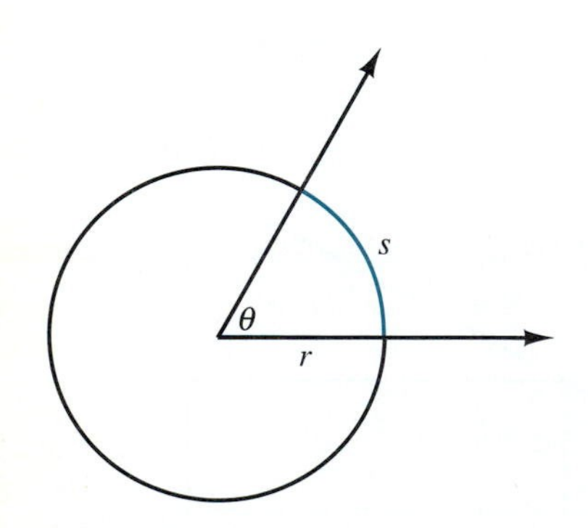

FIGURE 6.50

In Section 6.1 we introduced the notion of the radian measure of an angle. Recall that the radian measure of an angle is defined as follows (Figure 6.50):

$$\theta = \frac{s}{r}$$

where θ is the angle in radians

s is the length of the arc intercepted by θ

r is the length of the radius

As we pointed out at the time, since s and r are both lengths, the quotient $\frac{s}{r}$ is

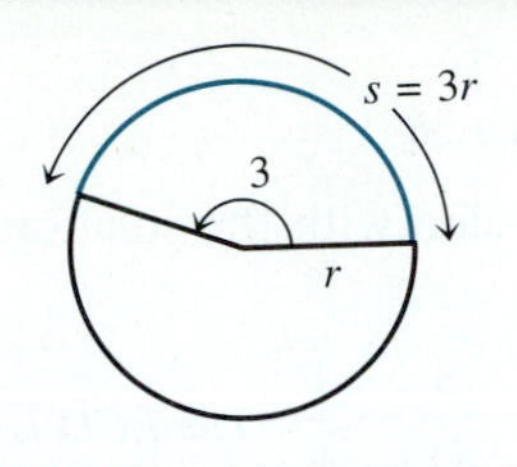

FIGURE 6.51 An angle of 3 radians, corresponding to the number 3

In Figure 6.51, keep in mind that an angle of π radians (approximately 3.14 radians) is $\frac{1}{2}$ of a complete rotation. Therefore, an angle of 3 radians is a bit less than $\frac{1}{2}$ of a complete rotation.

How would you visualize an angle of 2?

Find $\sin^2 3$ and $\sin 3^2$.

a pure number without any units attached. For example, if $s = 8$ cm and $r = 4$ cm, then

$$\theta = \frac{s}{r} = \frac{8 \text{ cm}}{4 \text{ cm}} = 2$$

Thus any angle can be interpreted as a real number. Conversely, any real number can be interpreted as an angle. For example, the real number 3 corresponds to an angle that intercepts an arc that is 3 times the length of the radius. We call such an angle an angle of 3 radians; it appears in Figure 6.51.

When we discuss $\sin \theta$ or any of the other trigonometric functions, θ may be any value in the domain of the particular function. As is the case with every function we have discussed in detail, we obviously need to determine the domain and range of each of the trigonometric functions. Before we do that in the next chapter, we will look at several examples that illustrate the algebraic manipulation of trigonometric expressions. In each example we will attempt to show how the trigonometric expression is handled along the same lines as a similar algebraic one.

EXAMPLE 1 Multiply and simplify: $(\sin \theta + 4)^2$.

Solution $(\sin \theta + 4)^2$ is handled in exactly the same way as $(A + 4)^2$.

$$(A + 4)^2 = (A + 4)(A + 4) = A^2 + 8A + 16$$
$$(\sin \theta + 4)^2 = (\sin \theta + 4)(\sin \theta + 4) = (\sin \theta)^2 + 8 \sin \theta + 16$$

Note that we wrote $(\sin \theta)^2$. It would be incorrect to write $\sin \theta^2$, which would be equivalent to $\sin(\theta^2)$ which means the sine of the square of θ rather than the square of the sine of θ.

In general, we use the notation $f^n(x) = [f(x)]^n$ except when $n = -1$ (which we reserve for inverse functions). Thus $f^2(x) = [f(x)]^2$ and, in the same way, we write

$$\sin^2 \theta = (\sin \theta)^2$$

which makes the extra set of parentheses unnecessary. Thus we may write the answer as

$$(\sin \theta + 4)^2 = \boxed{\sin^2 \theta + 8 \sin \theta + 16}$$ ■

EXAMPLE 2 Multiply and simplify.

(a) $(\sin 3\theta - 2)(\sin 3\theta + 1)$ **(b)** $\cos 4\theta(\cos 2\theta - 3)$

Solution

(a) We multiply as we would any two binomials.

$$(\sin 3\theta - 2)(\sin 3\theta + 1) = \underbrace{(\sin 3\theta)(\sin 3\theta)} + \sin 3\theta - 2 \sin 3\theta - 2$$
$$= \boxed{\sin^2 3\theta - \sin 3\theta - 2}$$

Note that when we multiply $(\sin 3\theta)(\sin 3\theta)$, we do not multiply 3θ times 3θ, just as we do not multiply t times t when we multiply $f(t)f(t) = [f(t)]^2$ (which we can also write as $f^2(t)$). We are multiplying the *functions, not* the *arguments* of the functions.

(b) $\cos 4\theta(\cos 2\theta - 3) = \boxed{\cos 4\theta \cos 2\theta - 3 \cos 4\theta}$ ■

EXAMPLE 3 Combine into a single fraction: $\dfrac{5}{4\sin\theta} - \dfrac{1}{\cos 3\theta}$.

Solution We offer the solution to this problem side by side with the solution to a parallel algebraic example involving x and y.

$$\frac{5}{4\sin\theta} - \frac{1}{\cos 3\theta} \quad \textit{The LCD is } 4\sin\theta\cos 3\theta.$$

$$= \frac{5(\cos 3\theta)}{4\sin\theta\cos 3\theta} - \frac{1(4\sin\theta)}{4\sin\theta\cos 3\theta}$$

$$= \boxed{\frac{5\cos 3\theta - 4\sin\theta}{4\sin\theta\cos 3\theta}}$$

$$\frac{5}{4x} - \frac{1}{y} \quad \textit{The LCD is } 4xy.$$

$$= \frac{5(y)}{4xy} - \frac{1(4x)}{4xy}$$

$$= \frac{5y - 4x}{4xy}$$

■

EXAMPLE 4 Factor $2\sin^2\theta - 3\sin\theta - 2$ as completely as possible.

Solution $2\sin^2\theta - 3\sin\theta - 2$ is of the form $2S^2 - 3S - 2$, which has no common factor. However, it can be factored as

$$2S^2 - 3S - 2 = (2S + 1)(S - 2)$$

Similarly,

$$2\sin^2\theta - 3\sin\theta - 2 = \boxed{(2\sin\theta + 1)(\sin\theta - 2)}$$

■

EXAMPLE 5 Reduce to lowest terms. $\dfrac{\cos^2\theta - \cos\theta}{\cos^2\theta - 2\cos\theta + 1}$

Solution This expression is similar to $\dfrac{C^2 - C}{C^2 - 2C + 1}$, which cannot be reduced in its present form because the numerator and denominator are made up of terms, not factors. We first factor both the numerator and the denominator, and then reduce the fraction.

$$\frac{C^2 - C}{C^2 - 2C + 1} = \frac{C(C - 1)}{(C - 1)(C - 1)} = \boxed{\frac{C}{C - 1}}$$

Similarly,

$$\frac{\cos^2\theta - \cos\theta}{\cos^2\theta - 2\cos\theta + 1} = \frac{\cos\theta(\cos\theta - 1)}{(\cos\theta - 1)(\cos\theta - 1)} = \boxed{\frac{\cos\theta}{\cos\theta - 1}}$$

■

In the following example we review an idea regarding functional notation in the context of the trigonometric functions.

EXAMPLE 6 Let $g(\theta) = \sin 2\theta$ and $h(\theta) = 2 \sin \theta$. Compare the values of $g\left(\frac{\pi}{6}\right)$ and $h\left(\frac{\pi}{6}\right)$.

Solution

$$g\left(\frac{\pi}{6}\right) = \sin 2\left(\frac{\pi}{6}\right) = \sin \frac{\pi}{3} = \boxed{\frac{\sqrt{3}}{2}}$$

$$h\left(\frac{\pi}{6}\right) = 2 \sin \frac{\pi}{6} = 2 \cdot \frac{1}{2} = \boxed{1}$$ ■

It is a very common error to believe that $\sin 2\theta$ is the same as $2 \sin \theta$. The last example clearly shows that, in general, $\sin 2\theta \neq 2 \sin \theta$.

This should come as no surprise, since in Chapter 2 we saw that in general $f(2x) \neq 2f(x)$. This is just a special case of the general fact that

$$f(nx) \neq nf(x)$$

This applies equally well to the trigonometric functions.

> *Remember:* In general, $\sin(n\theta) \neq n \sin \theta$.

Similar statements can be made for all the trigonometric functions.

EXAMPLE 7 Show that $\frac{\sin \theta}{\cos \theta} = \tan \theta$.

Solution We use the basic definitions of the sine and cosine functions.

$$\frac{\sin \theta}{\cos \theta} = \frac{\frac{y}{r}}{\frac{x}{r}} = \frac{y}{r} \cdot \frac{r}{x} = \frac{y}{x} = \tan \theta, \qquad \text{as required}$$

We use this result in the next chapter. ■

EXAMPLE 8 Use Figure 6.52 to express x as a function of θ.

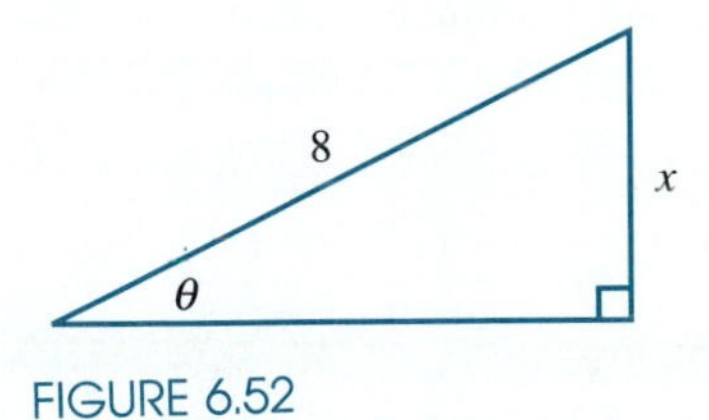

FIGURE 6.52

Solution Using the opposite over hypotenuse definition for the sine function, we get

$$\sin \theta = \frac{x}{8}$$ *Solving for x, we get*

$$\boxed{x = 8 \sin \theta}$$ *Which expresses x as a function of θ* ■

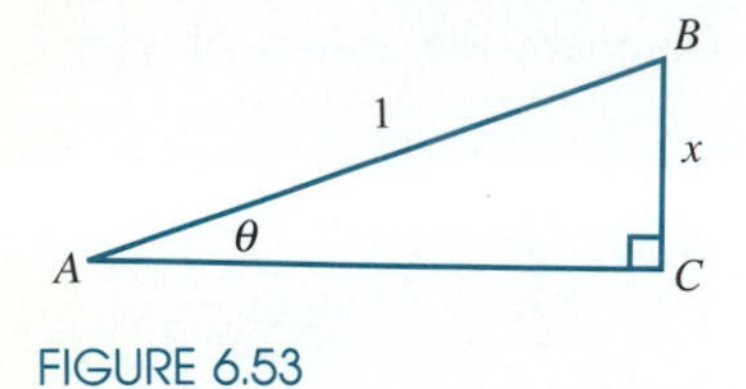

FIGURE 6.53

EXAMPLE 9 Use Figure 6.53 to express $\tan\theta$ as a function of x.

Solution Given this right triangle we would like to use the opposite over adjacent definition for the tangent function. In order to do so, we must find the missing side of the triangle by using the Pythagorean theorem.

$$|\overline{AC}|^2 + x^2 = 1^2$$
$$|\overline{AC}|^2 = 1 - x^2$$
$$|\overline{AC}| = \pm\sqrt{1 - x^2} \qquad \textit{We take the positive square root as the length.}$$
$$|\overline{AC}| = \sqrt{1 - x^2}$$

Therefore,

$$\tan\theta = \frac{\text{opposite}}{\text{adjacent}} = \frac{x}{|\overline{AC}|}$$

$$\boxed{\tan\theta = \frac{x}{\sqrt{1 - x^2}}} \qquad \textit{Which expresses } \tan\theta \textit{ as a function of } x$$ ■

What does it mean that $\tan\theta$ is a function of x?

The last two examples illustrate that certain algebraic expressions can be expressed in terms of trigonometric functions, and vice versa. This allows us to apply algebraic techniques to trigonometric expressions and techniques from trigonometry to algebraic expressions.

The algebraic techniques we have discussed in this section will be very useful in the next chapter as we explore more of the relationships among the various trigonometric functions.

EXERCISES 6.4

In Exercises 1–42, perform the indicated operations.

1. $2\cos\theta(\cos\theta - 1)$
2. $\tan\theta(3\tan\theta + 2)$
3. $\sin 5\theta(\sin 5\theta + 4)$
4. $\sec 2\theta(3 - \sec 2\theta)$
5. $(\csc\theta - 2)^2$
6. $(3 - \cot\theta)^2$
7. $(\cos 4\theta + 1)^2$
8. $(\sin 3\theta - 1)^2$
9. $(5\sin\theta - 2)(\sin\theta + 3)$
10. $(4\cos\theta + 3)(\cos\theta - 2)$
11. $(\cos 2\theta + 3)(\cos 2\theta - 3)$
12. $(\tan 4\theta - 1)(\tan 4\theta + 1)$
13. $(\sin 3\theta + 2)(\sin 5\theta - 4)$
14. $(\cos 6\theta - 3)(\cos 2\theta + 4)$
15. Factor: $\sin^2\theta - \sin\theta - 2$.
16. Factor: $\cos^2\theta - 2\cos\theta + 1$.
17. Factor: $\tan^2\theta + 2\tan\theta - 8$.
18. Factor: $\sec^2\theta + 5\sec\theta + 6$.
19. Factor: $2\csc^2\theta - 5\csc\theta - 3$.
20. Factor: $3\cot^2\theta + 7\cot\theta - 6$.

21. $\dfrac{2}{\sin\theta} + \dfrac{3}{\cos\theta}$
22. $\dfrac{1}{\sin^2\theta} - \dfrac{5}{\sin\theta}$
23. $\dfrac{\sin\theta}{\cos^2\theta} - \dfrac{1}{\sin\theta}$
24. $\dfrac{\cos\theta}{2\sin\theta} + \dfrac{2}{3\sin^2\theta}$
25. $\dfrac{1}{\cos^2\theta} + 1$
26. $\dfrac{1}{\tan^2\theta} - 1$
27. $\dfrac{1}{\tan\theta} + \sec\theta$
28. $\cot\theta - \dfrac{1}{\csc\theta}$
29. $1 - \dfrac{1}{\cot^2\theta}$
30. $\dfrac{1}{\tan^2\theta} + \dfrac{1}{\sec^2\theta}$
31. $\dfrac{3}{\sin 2\theta} + \dfrac{2}{\sin 3\theta}$
32. $\dfrac{5}{\cos 2\theta} - \dfrac{7}{2\cos\theta}$
33. $\dfrac{1}{\cos^2\theta} + \dfrac{\tan^2\theta}{2\cos\theta}$
34. $\dfrac{1}{\tan 3\theta} + \dfrac{1}{3\tan\theta}$

35. $\sin\theta + \dfrac{1}{\sin\theta}$

36. $\csc\theta + \dfrac{1}{\csc\theta}$

37. Simplify: $\dfrac{5\cos\theta}{3\cos^2\theta + 2\cos\theta}$

38. Simplify: $\dfrac{8\sin\theta}{4\sin\theta - 2\sin^2\theta}$

39. Simplify: $\dfrac{1 - \sin^2\theta}{(1 - \sin\theta)^2}$

40. Simplify: $\dfrac{4\cos^2\theta - 1}{2\cos^2\theta - 5\cos\theta - 3}$

41. Simplify: $\dfrac{\sin^2 4\theta}{\sin^2 4\theta - \sin 4\theta}$

42. Simplify: $\dfrac{9 - \cos^2 2\theta}{9 - 6\cos 2\theta + \cos^2 2\theta}$

In Exercises 43–54, use the given triangle.

43. Express:
$\sin\theta$ in terms of x.

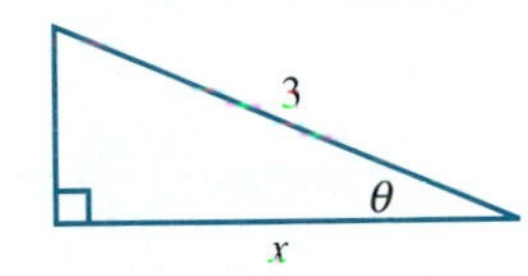

44. Express:
$\cos\theta$ in terms of a.

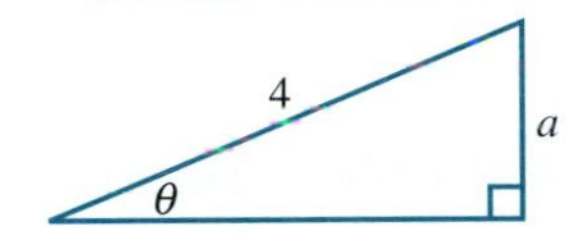

45. Express:
$\tan\theta$ in terms of y.

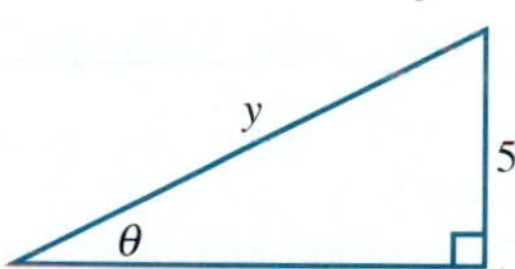

46. Express:
$\cot\theta$ in terms of x.

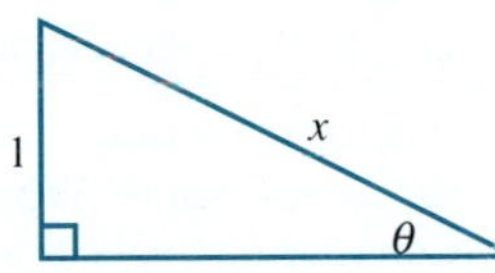

47. Express:
$\sec\theta$ in terms of a.

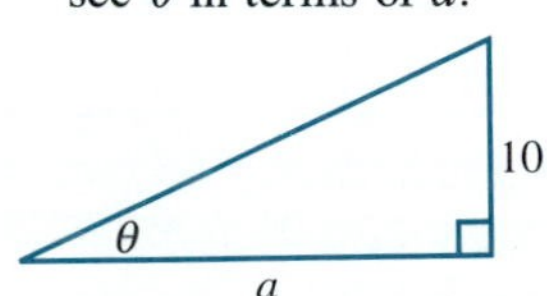

48. Express:
$\csc\theta$ in terms of x.

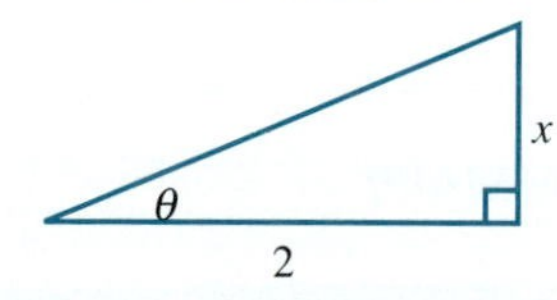

49. Express:
x in terms of $\sin\theta$.

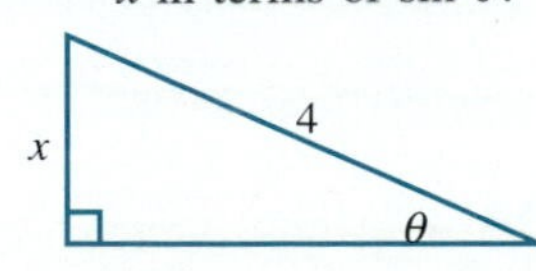

50. Express:
x in terms of $\sin\theta$.

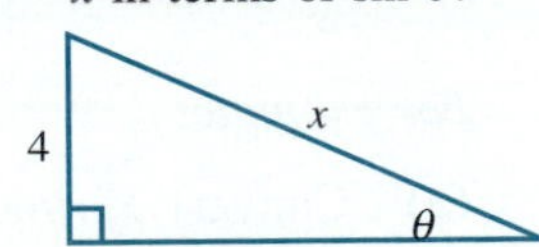

51. Express:
y in terms of $\tan\theta$.

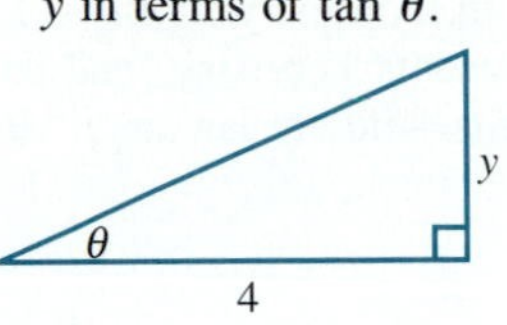

52. Express:
y in terms of $\tan\theta$.

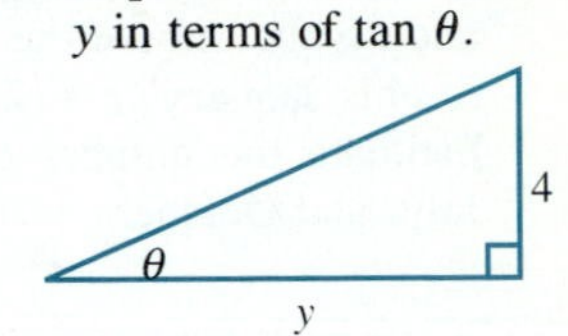

53. Express:
s in terms of $\sec\theta$.

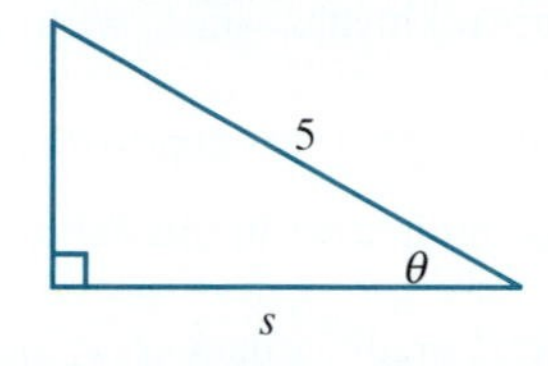

54. Express:
s in terms of $\sec\theta$

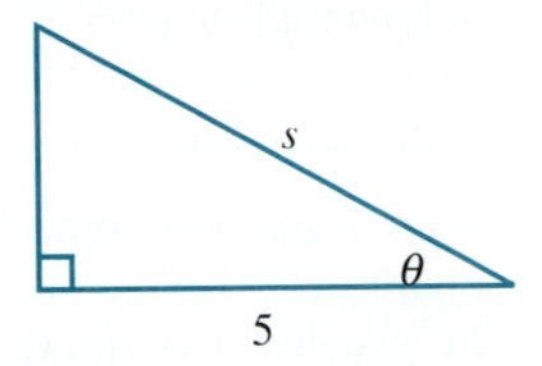

55. A 5-kg weight hanging from the end of a spring is pulled down *(displaced)* 6 inches below its position of equilibrium (see the accompanying figure). Once the weight is released it will oscillate up and down. Suppose that, ignoring air resistance and friction, the distance d of the weight from the position of equilibrium at time t in seconds is given by

$$d = 6\cos 4t$$

where d is positive if the weight is below the equilibrium position and negative if the weight is above the equilibrium position.

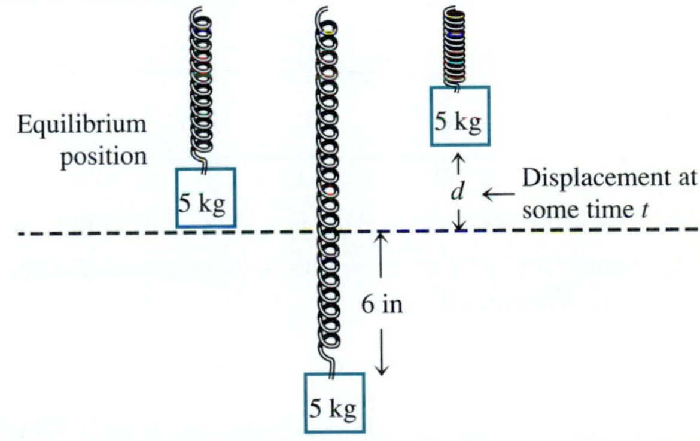

How far is the weight from its equilibrium position at each time?

(a) $t = 0$ seconds **(b)** $t = 2$ seconds
(c) $t = 5$ seconds

56. The Tru-form company manufactures a seasonal product. The monthly sales for their product is estimated by the equation

$$w = 27.5 + 6.4\sin\left(\frac{\pi t}{6}\right)$$

where w is the number of units sold (in the thousands) and t is the time of the year in months, where $t = 1$ represents January, $t = 2$ represents February, and so on. Estimate the number of units sold in January, March, July, and October.

QUESTIONS FOR THOUGHT

57. In calculus it is shown that for "small" values of x, the polynomial $p(x) = x - \frac{x^3}{6}$ approximates $\sin x$ with a maximum error of $\left|\frac{x^5}{120}\right|$. Use $p(x)$ to approximate $\sin 0.1$ and determine the maximum error in this approximation.

58. In calculus it is shown that for "small" values of x, the polynomial $p(x) = 1 - \frac{x^2}{2} + \frac{x^4}{4}$ approximates $\cos x$ with a maximum error of $\left|\frac{x^6}{720}\right|$. Use $p(x)$ to approximate $\cos(-0.2)$ and determine the maximum error in this approximation.

59. Use the following triangle to convert the algebraic expression $\frac{x^2}{\sqrt{9 - x^2}}$ into a trigonometric expression. HINT: Find $\tan \theta$.

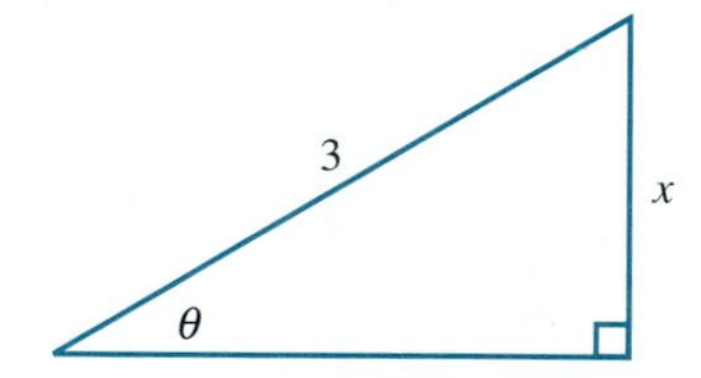

60. Use the following triangle to convert the algebraic expression $x^2\sqrt{x^2 + 9}$ into a trigonometric expression. HINT: Find $\tan \theta$.

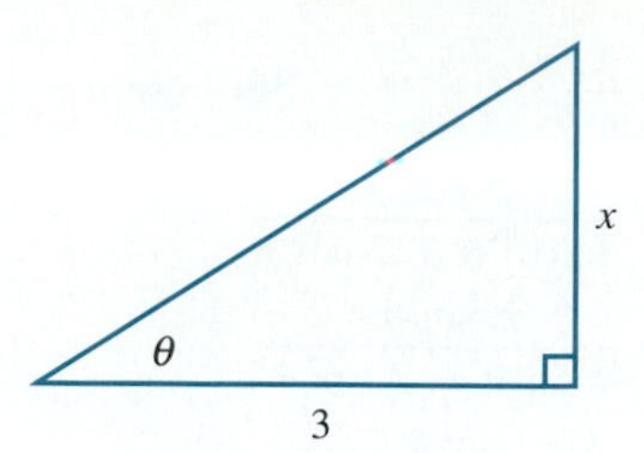

61. Use the following triangle to convert the algebraic expression $\frac{x\sqrt{x^2 - 9}}{9}$ into a trigonometric expression.

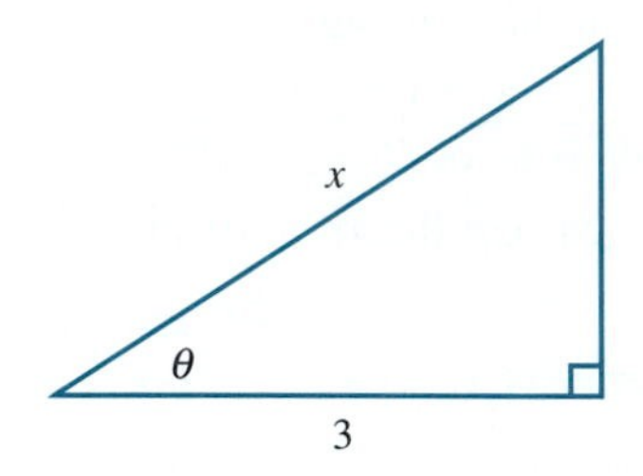

62. Let $f(\theta) = \cos 2\theta$ and $g(\theta) = 2 \cos \theta$. Compute $f\left(\frac{\pi}{4}\right)$ and $g\left(\frac{\pi}{4}\right)$. Are they equal? Should they be? Explain.

63. Let $f(\theta) = \sin 2\theta$ and $g(\theta) = 2 \sin \theta$. Compute $f\left(\frac{\pi}{3}\right)$ and $g\left(\frac{\pi}{3}\right)$. Are they equal? Does this result contradict the statement made on page 417 that in general $\sin 2\theta \neq 2 \sin \theta$? Why or why not?

64. Consider an angle θ in standard position and pick a point (x, y) on the terminal side of θ and on the unit circle. Show that in such a case, $\sin \theta = y$ and $\cos \theta = x$.

65. Use the result of Exercise 64 to prove for all values of θ that $\sin^2 \theta + \cos^2 \theta = 1$.

Chapter 6 **SUMMARY**

After completing this chapter you should:

1. Understand the meaning of the *radian* measure of an angle. (Section 6.1)

2. Be able to convert from radian measure to degree measure, and vice versa. (Section 6.1)
The following proportion gives the relationship between radian and degree measures, and can be used to convert from one to the other:

$$\frac{\theta \text{ in radians}}{\pi} = \frac{\theta \text{ in degrees}}{180^\circ}$$

For example:

(a) Convert $\frac{2\pi}{3}$ into degrees.

(b) Convert 20° into radians.

Solution:

(a) $\dfrac{\frac{2\pi}{3}}{\pi} = \dfrac{\theta}{180^\circ} \Rightarrow \theta = \dfrac{2\pi}{3} \cdot \dfrac{1}{\pi} \cdot 180^\circ = \boxed{120^\circ}$

(b) $\dfrac{\theta}{\pi} = \dfrac{20^\circ}{180^\circ} \Rightarrow \theta = \boxed{\dfrac{\pi}{9}}$

3. Be able to set up prototypes for the two basic triangles. (Section 6.1)

The 45° right triangle

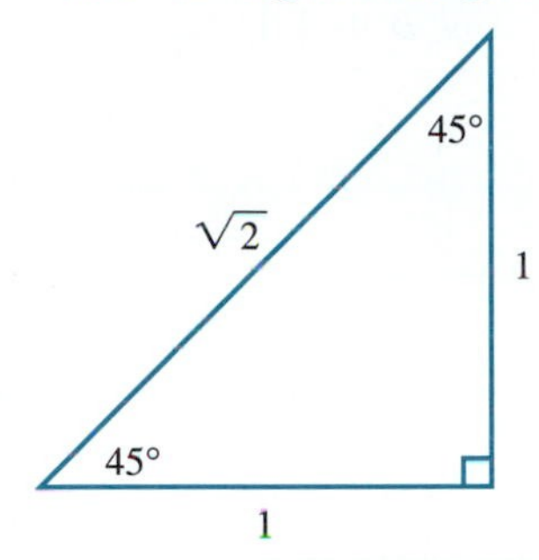

30°-60° right triangle

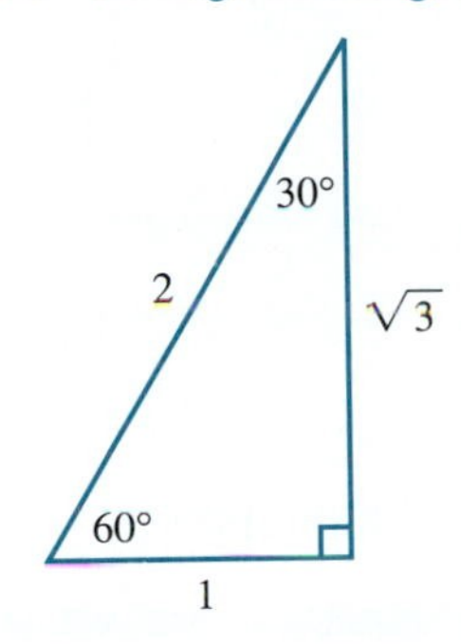

4. Know the definitions of the six trigonometric functions. (Section 6.2)
5. Be able to use the definitions to find the trigonometric functions of an angle in standard position. (Section 6.2)

For example:

(a) Find $\cos\dfrac{5\pi}{6}$. (b) Find $\sin\dfrac{3\pi}{2}$.

Solution:

(a) $\dfrac{5\pi}{6} = \dfrac{5(180^\circ)}{6} = 150^\circ$. As we see in Figure 6.54, the reference angle is 30°. We draw the reference triangle and fill in (x, y) and r.

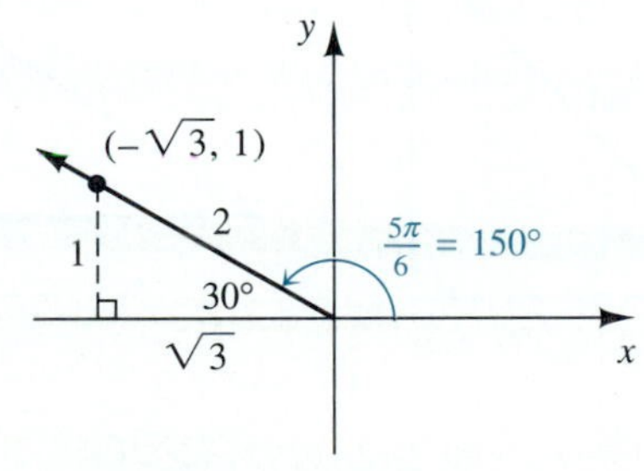

FIGURE 6.54

Therefore, we have $\cos\dfrac{5\pi}{6} = \dfrac{x}{r} = \boxed{-\dfrac{\sqrt{3}}{2}}$

(b) $\dfrac{3\pi}{2} = \dfrac{3(180^\circ)}{2} = 270^\circ$ (see Figure 6.55)

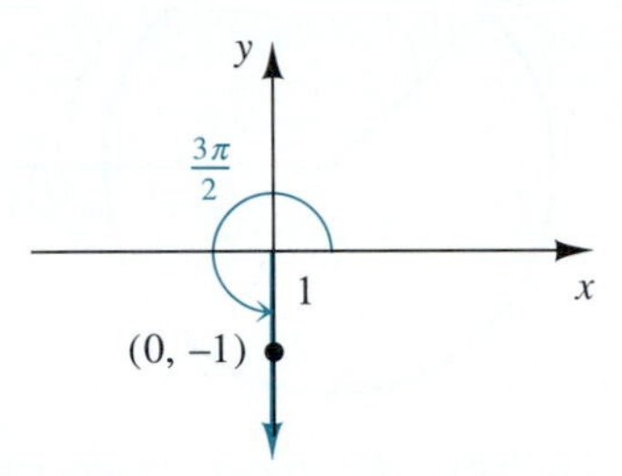

FIGURE 6.55

$\sin\dfrac{3\pi}{2} = \dfrac{y}{r} = \boxed{-1}$

6. Know how to use the trigonometric functions to find the missing parts of triangles. (Section 6.3) The right triangle definitions of the trigonometric functions, found on page 400, can be used to find the missing parts of triangles.

For example:

Find x (to the nearest tenth) in the triangle in Figure 6.56.

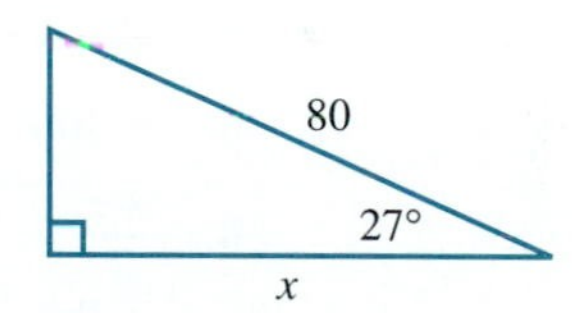

FIGURE 6.56

Solution:

$\cos 27^\circ = \dfrac{x}{80}$ *We use a calculator to find cos 27°*

$0.8910065 = \dfrac{x}{80}$

$x = 71.3$ *To the nearest tenth*

7. Understand how real numbers can be interpreted as angles. (Section 6.4)

For example:

An angle of 2 (radians) corresponds to a central angle that subtends an arc of a circle twice the length of its radius as illustrated in Figure 6.57. We can say that an angle of 2 radians corresponds to the real number 2.

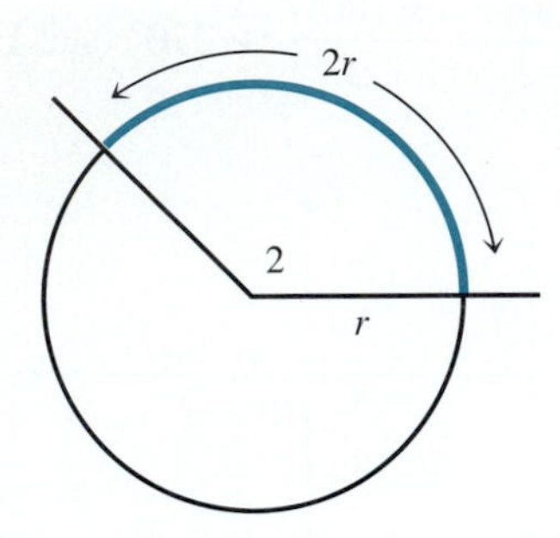

FIGURE 6.57 An angle corresponding to the real number 2

8. Be able to apply algebraic techniques to trigonometric expressions. (Section 6.4)
For example:

We can simplify $\dfrac{\cos^2\theta - \cos\theta}{\cos^2\theta - 1}$ as follows.

$$\frac{\cos^2\theta - \cos\theta}{\cos^2\theta - 1} = \frac{\cos\theta(\cos\theta - 1)}{(\cos\theta + 1)(\cos\theta - 1)}$$

$$= \boxed{\frac{\cos\theta}{\cos\theta + 1}}$$

Chapter 6 REVIEW EXERCISES

In Exercises 1–10, convert the given degree measure into radian measure or the given radian measure into degree measure.

1. $\dfrac{\pi}{12}$
2. $80°$
3. $200°$
4. $\dfrac{\pi}{9}$
5. $-\dfrac{3\pi}{5}$
6. $-400°$
7. $330°$
8. $\dfrac{3\pi}{2}$
9. $\dfrac{7\pi}{4}$
10. $-\dfrac{2\pi}{3}$

In Exercises 11–30, evaluate the given expression. Use a calculator only where necessary.

11. $\tan\dfrac{5\pi}{6}$
12. $\sec\dfrac{7\pi}{4}$
13. $\sin 240°$
14. $\cos 180°$
15. $\sin\dfrac{\pi}{2}$
16. $\tan 0$
17. $\csc 125°$
18. $\cos(-143°)$
19. $\cot(-300°)$
20. $\cos\dfrac{\pi}{3}$
21. $\sin\dfrac{\pi}{9}$
22. $\csc\dfrac{11\pi}{6}$
23. $\tan 3$
24. $\sin 2$
25. $\sec\dfrac{\pi}{2}$
26. $\tan\dfrac{3\pi}{2}$
27. $\cos\left(-\dfrac{\pi}{15}\right)$
28. $\sec 206°$
29. $\sin\dfrac{5\pi}{4}$
30. $\cos\dfrac{7\pi}{6}$

In Exercises 31–38, use the trigonometric functions to find the value of the indicated part of the triangle. Give exact values where possible; round to the nearest hundredth where necessary.

31. Find x.

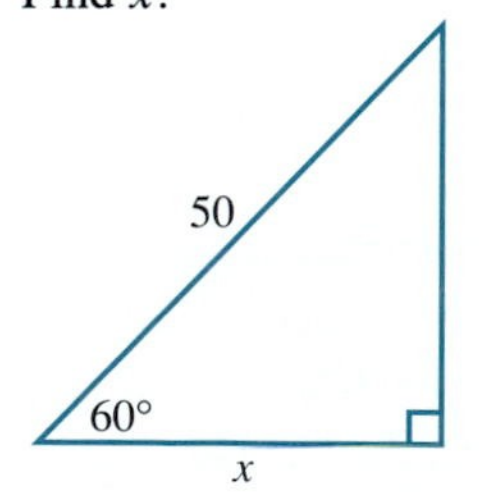

32. Find θ.

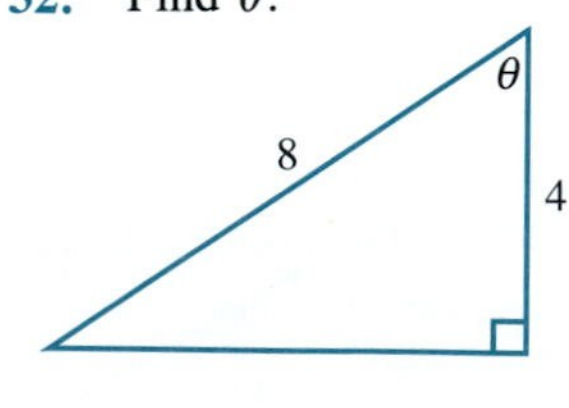

33. Find θ.

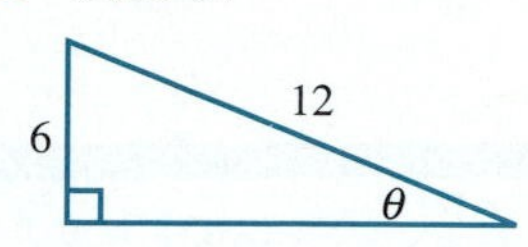

34. Find x.

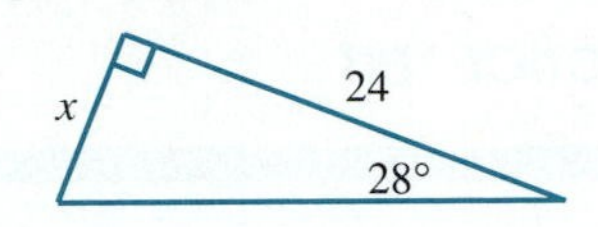

35. Find x.

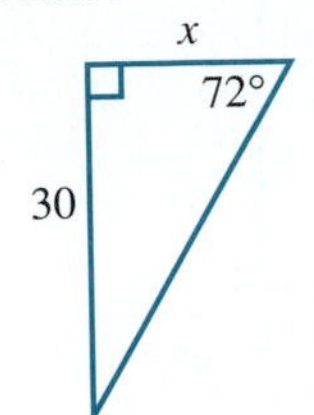

36. Find x.

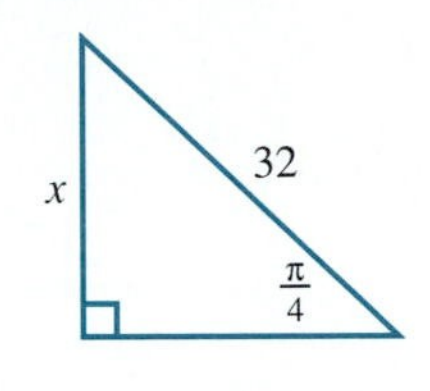

37. Find θ.

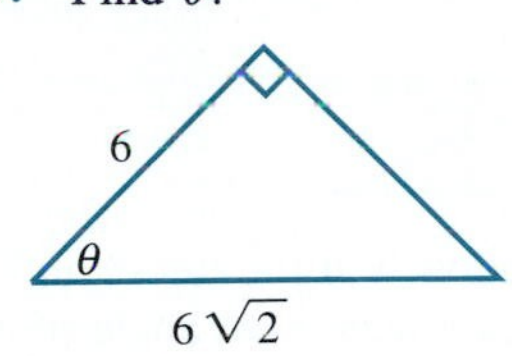

38. Find θ.

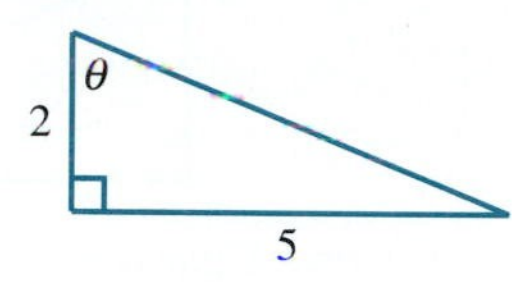

39. Find the area of $\triangle ABC$.

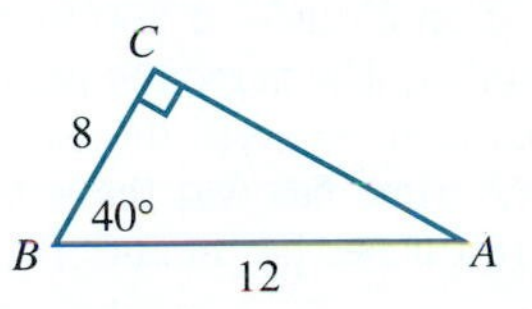

40. Find the area of a sector of a circle with a central angle of $\frac{\pi}{9}$ and a radius 18 cm.

41. Find the arc length of a sector with a central angle of 100° and a radius of 9 inches.

42. Find the radian measure of a central angle in a circle of radius 12 cm that cuts off an arc of length 3 cm.

43. A 22-foot ladder is leaning against a tall building. If the bottom of the ladder makes an angle of 65° with the level ground, how high up the building does the ladder reach?

44. A helicopter is hovering directly above point A. The pilot observes a point B 1000 yards due east of point A at an angle of depression of 27°. What is the altitude of the helicopter (to the nearest foot)?

45. A jet plane, flying at a constant altitude of 10,000 feet, passes directly over the head of an observer on the ground. Twenty seconds later the plane is seen by the observer at an angle of elevation of 15°. Find the speed of the jet, to the nearest mile per hour.

46. Jane is standing 30 meters from the base of a building. A TV antenna is mounted on the roof at the edge of the building. She observes the top of the building at an angle of elevation of 50° and the top of the antenna at an angle of elevation of 54°. Find the length of the antenna to the nearest tenth of a foot.

47. The following figure shows two telephone poles attached by guy wires to a point on the ground between them. Find the distance between the two poles.

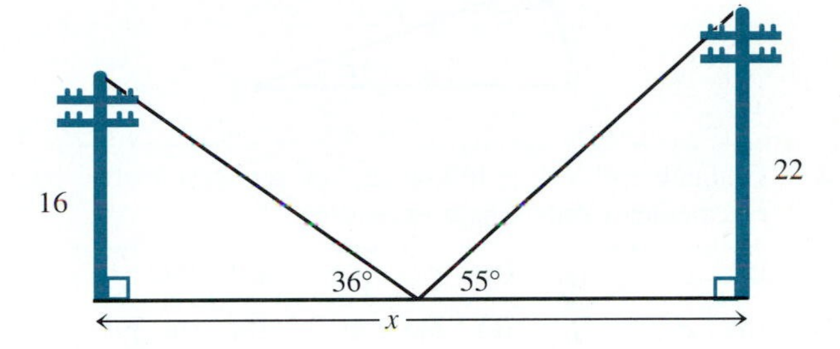

48. Find the area of the shaded portion of the following figure. Equilateral $\triangle ABC$ is inscribed in a circle of radius 4.

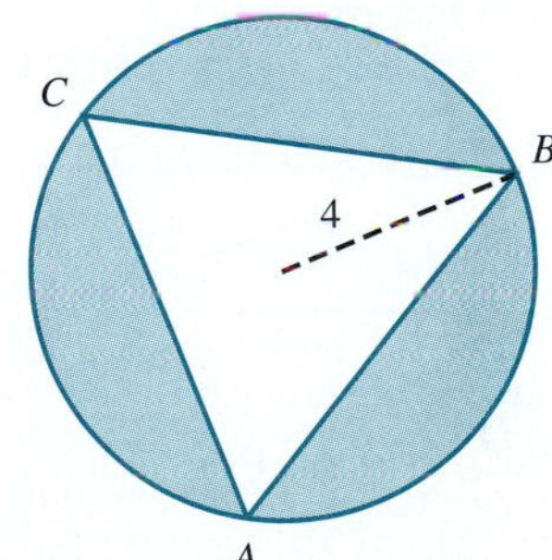

In Exercises 49–52, perform the indicated operations and simplify as completely as possible.

49. $\dfrac{1}{\sin\theta} + \dfrac{4}{\cos 2\theta}$

50. $(2\tan\theta - 3)(3\tan\theta + 4)$

51. $\dfrac{2\sec^2\theta - 9\sec\theta + 9}{\sec^2\theta - 9}$

52. $\dfrac{\cot\theta}{\cot\theta + 1} - \dfrac{1}{\cot^2\theta + \cot\theta}$

Chapter 6 PRACTICE TEST

1. Convert $\frac{3\pi}{5}$ radians into degree measure.
2. Write 80° in radian measure.
3. Find the length of $\overline{AB}$. Leave your answer in radical form.

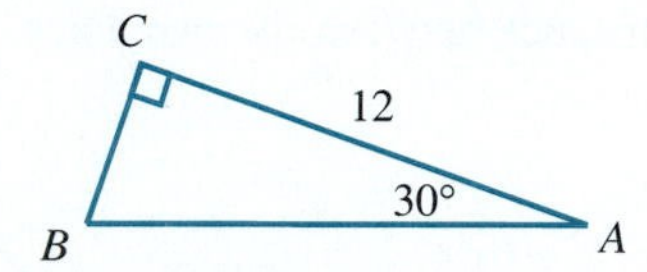

4. Evaluate each of the following. Use a trigonometric table or calculator only where necessary.
   (a) $\sin \frac{5\pi}{4}$ (b) $\cos \frac{2\pi}{3}$ (c) $\tan \frac{7\pi}{6}$
   (d) csc 180° (e) tan 270° (f) sec 39°
   (g) sin 84° (h) cot 134°
5. Find the arc length of a sector with a central angle of $\frac{\pi}{5}$ in a circle of radius 10 cm.
6. Find the area of $\triangle ABC$ to the nearest tenth.

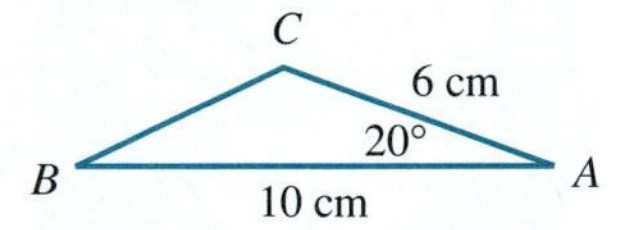

7. Find x to the nearest tenth.

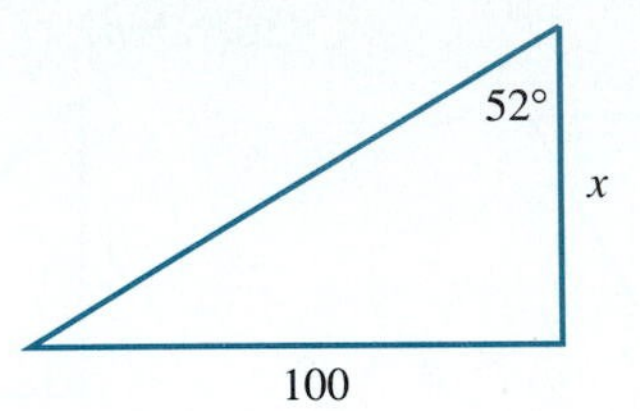

8. Find θ to the nearest degree.

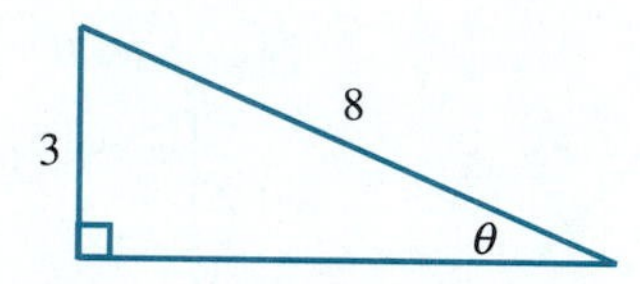

9. A Coast Guard observer is stationed at the top of an 80-foot lighthouse. If she observes a boat at an angle of depression of 18°, how far is the boat from the lighthouse? (Answer to the nearest foot.)
10. A hot air balloon is holding at an altitude of 500 meters directly above a person at point A. The person at point A begins running and 6 minutes later observes the balloon at an angle of elevation of 16°. How fast was the person running? (Answer to the nearest meter per minute.)
11. Simplify as completely as possible. $\frac{1}{\sin 2\theta} + \frac{1}{2 \sin \theta}$

CHAPTER 7

The Trigonometric Functions

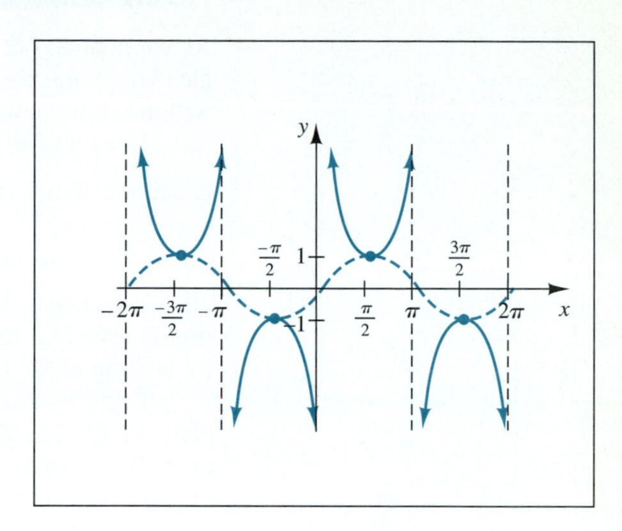

In this chapter we take a closer look at the trigonometric functions. We examine them from the perspective that we have developed and used to study functions in general. In the first two sections we analyze each of the six trigonometric functions in order to determine its domain, range, and graph. In Sections 7.3 and 7.4 we focus our attention on trigonometric identities and equations. In Section 7.5 we conclude the chapter with an introduction to the inverse trigonometric functions.

7.1 The Sine and Cosine Functions and Their Graphs

As we pointed out in the last chapter, any real number can be interpreted as an angle. Therefore, we can describe the "natural" domains of the trigonometric functions within the framework of the real number system.

If we let $f(\theta) = \sin\theta$, the domain consists of all real numbers θ for which $\sin\theta$ is defined. Since $\sin\theta = \dfrac{y}{r}$ and r is never equal to zero, the domain for $\sin\theta$ is the set of all real numbers. Similarly, the domain for $f(\theta) = \cos\theta = \dfrac{x}{r}$ is also the set of all real numbers. We have seen on many occasions that finding the range of a function is usually a much more subtle question, one that is often most easily answered by looking at the graph of the function.

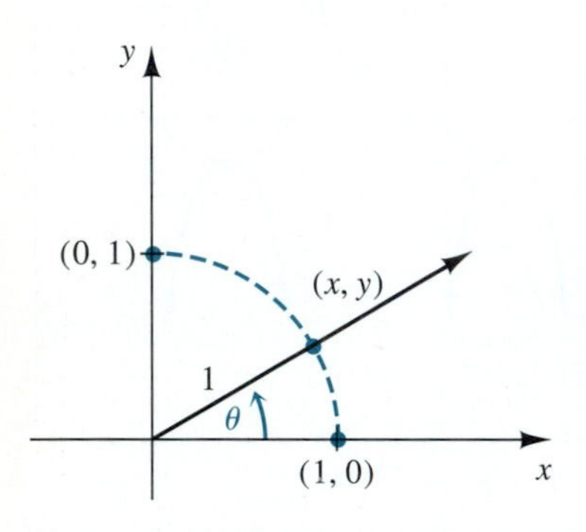

FIGURE 7.1

Begin by determining what the graph of $f(\theta) = \sin\theta$ looks like. We will complete a table of values for $f(\theta) = \sin\theta$ that will aid us in drawing the graph. However, let's analyze the sine function to see if we can determine the general behavior of its values and hence its graph. In the course of this analysis we will see one reason why radian measure is generally preferred to degree measure.

To analyze $f(\theta) = \sin\theta$, we keep in mind that once we choose a real number θ, we draw the angle, in standard position, that corresponds to θ. Recall that we label the point on the terminal side of the angle as (x, y) and the distance to the origin as r. To simplify our analysis, we choose this point (x, y) so that $r = 1$. That is, (x, y) is a point on the circle $x^2 + y^2 = 1$, which is called the **unit circle**. Thus, for a nonquadrantal angle θ, we have Figure 7.1. Note that $\sin\theta = \dfrac{y}{1} = y$.

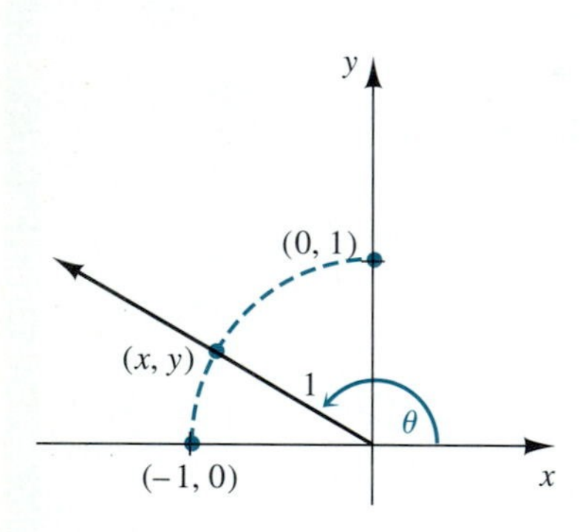

FIGURE 7.2

As the terminal side of θ moves through the first quadrant, y increases from 0 (when $\theta = 0$) to 1 $\left(\text{when } \theta = \dfrac{\pi}{2}\right)$. Thus as θ increases from 0 to $\dfrac{\pi}{2}$, $y = \sin\theta$ steadily increases from 0 to 1.

As θ increases through the second quadrant, that is, from $\dfrac{\pi}{2}$ to π, y decreases from 1 to 0. See Figure 7.2. Thus as θ increases from $\dfrac{\pi}{2}$ to π, $\sin\theta$ decreases from 1 to 0. A similar analysis (see Figures 7.3 and 7.4) reveals that as θ increases from π to $\dfrac{3\pi}{2}$, $\sin\theta$ decreases from 0 to -1; and as θ increases from $\dfrac{3\pi}{2}$ to 2π, $\sin\theta$ increases from -1 to 0.

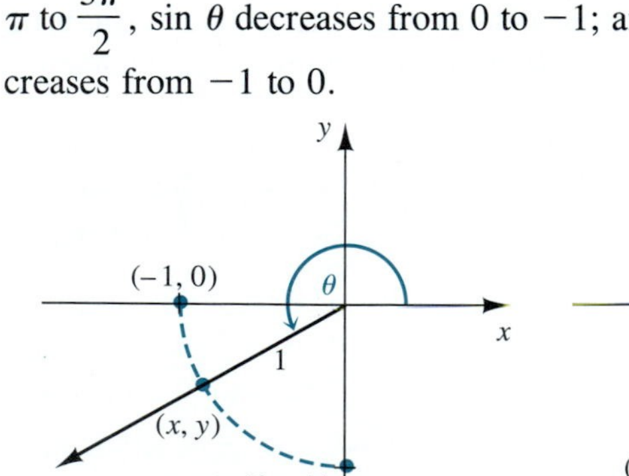

FIGURE 7.3

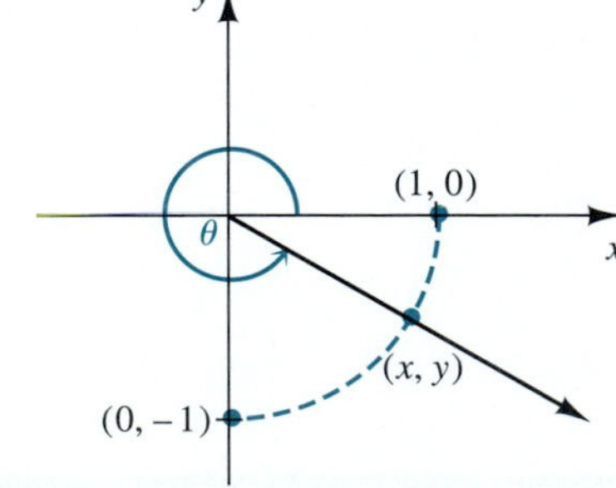

FIGURE 7.4

The following table fills in some additional values, which illustrate the same pattern we have just described. We use a calculator where necessary to approximate values.

θ	0	$\frac{\pi}{6}$	$\frac{\pi}{4}$	$\frac{\pi}{3}$	$\frac{\pi}{2}$	$\frac{2\pi}{3}$	$\frac{3\pi}{4}$	$\frac{5\pi}{6}$	π	$\frac{7\pi}{6}$	$\frac{5\pi}{4}$	$\frac{4\pi}{3}$	$\frac{3\pi}{2}$	$\frac{5\pi}{3}$	$\frac{7\pi}{4}$	$\frac{11\pi}{6}$	2π
sin θ	0	0.5	0.707	0.866	1	0.866	0.707	0.5	0	−0.5	−0.707	−0.866	−1	−0.866	−0.707	−0.5	0

Using these points as an aid, we have the graph for $f(\theta) = \sin \theta$ shown in Figure 7.5.

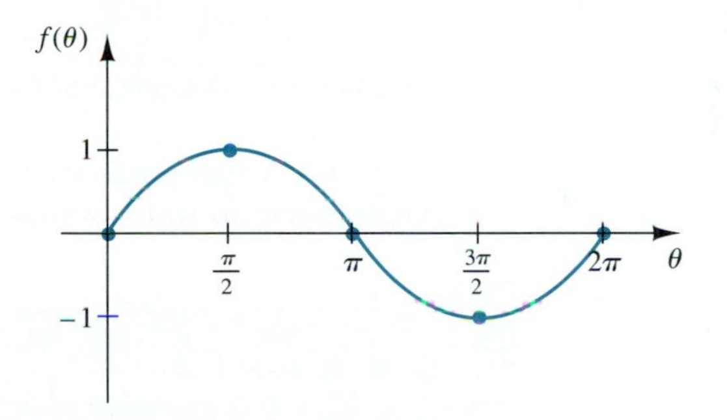

FIGURE 7.5 The graph of $f(\theta) = \sin \theta$ on the interval $[0, 2\pi]$

As we discussed in Section 6.4, the values of $f(\theta) = \sin \theta$ depend only on the position of the terminal side. Adding or subtracting multiples of 2π to θ will leave the value of $\sin \theta$ unchanged. Thus the values of $f(\theta) = \sin \theta$ will repeat every 2π units. The complete graph of $f(\theta) = \sin \theta$ appears below.

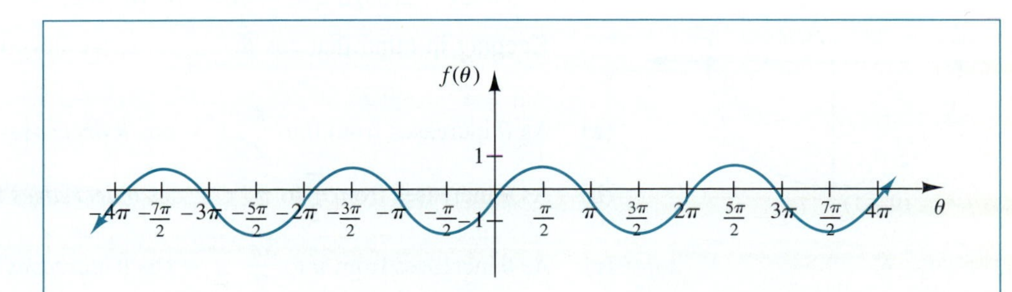

The graph of $f(\theta) = \sin \theta$, which is called the basic sine curve.

Note that the numerical value of $\frac{\pi}{2} \approx 1.57$, so that the graph of $f(\theta) = \sin\theta$ rises 1 unit as θ changes by about 1.57 units. This fact offers a perfect opportunity to pause a moment to discuss the point mentioned in Section 6.1—that it is often more appropriate to measure angles in radians rather than in degrees.

If we try to graph $f(\theta) = \sin\theta$ for θ measured in degrees, then $f(\theta)$ will increase from 0 to 1 as θ increases from 0° to 90°. In other words, it takes a 90-unit change in θ to cause a 1 unit change in $f(\theta)$. Thus the graph would look something like Figure 7.6.

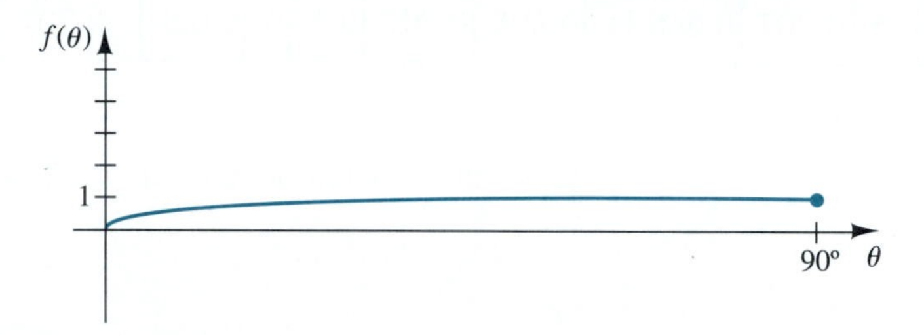

FIGURE 7.6 A portion of the graph of $f(\theta) = \sin\theta$ with θ measured in degrees

Although this graph might be accurate, it seems far less "readable" than our previous graph, in which we used radian measure.*

GRAFFIX

Use a graphics calculator or computer.

1. Graph $y = \sin x$ in degree mode as x varies from 0° to 90°.
2. Graph $y = \sin x$ in radian mode as x varies from 0 to $\frac{\pi}{2}$.

FIGURE 7.7

If we apply the same type of analysis to $f(\theta) = \cos\theta$, we will be able to get a good idea of what its graph looks like. Figure 7.7 shows the angle corresponding to θ as it increases through quadrants I, II, III, and IV.

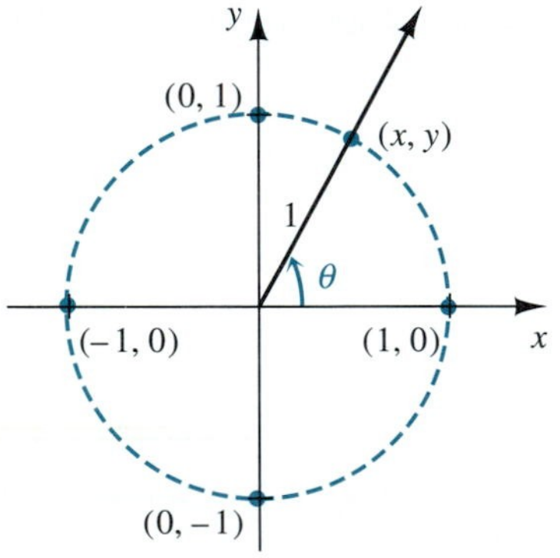

Keeping in mind that $\cos\theta = \frac{x}{1} = x$, we have the following:

(a) As θ increases from 0 to $\frac{\pi}{2}$, $x = \cos\theta$ *decreases* from 1 to 0. (See Figure 7.1)

(b) As θ increases from $\frac{\pi}{2}$ to π, $x = \cos\theta$ *decreases* from 0 to -1. (See Figure 7.2)

(c) As θ increases from π to $\frac{3\pi}{2}$, $x = \cos\theta$ *increases* from -1 to 0. (See Figure 7.3)

(d) As θ increases from $\frac{3\pi}{2}$ to 2π, $x = \cos\theta$ *increases* from 0 to 1. (See Figure 7.4)

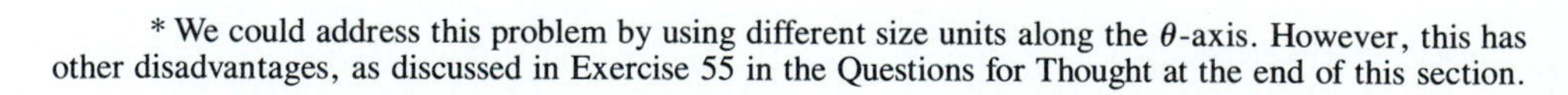

* We could address this problem by using different size units along the θ-axis. However, this has other disadvantages, as discussed in Exercise 55 in the Questions for Thought at the end of this section.

Based upon this analysis, we have the graph of $f(\theta) = \cos\theta$ in the interval $[0, 2\pi]$ as shown in Figure 7.8.

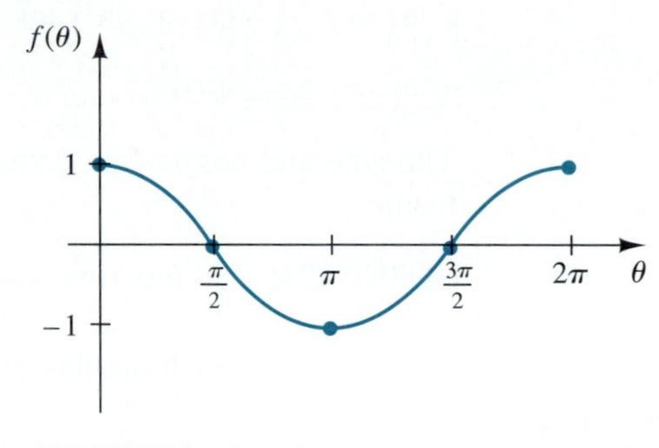

FIGURE 7.8

GRAFFIX

Use a graphics calculator or computer to graph the function $y \cos x$ on the interval $[-2\pi, 2\pi]$.

Again using the fact that the values of $f(\theta) = \cos\theta$, repeat every 2π units, we obtain the graph in the following box.

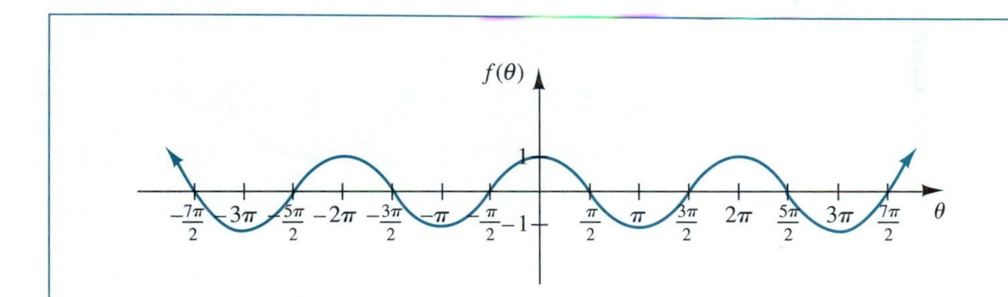

The graph of $f(\theta) = \cos\theta$, which is called the basic cosine curve.

The basic sine and cosine curves are very important and should be added to our catalog of basic graphs.

In our previous work with graphing functions we usually named the function $y = f(x)$. Thus far in our discussion of the graphs of the sine and cosine functions, we have named the functions $f(\theta) = \sin\theta$ and $f(\theta) = \cos\theta$. This was done to avoid confusion between the use of (x, y) as the point on the terminal side of the angle and the use of x as the angle and y as the value of the function. However, due to the fact that when we graph a function we normally label our axes as x and y rather than θ and $f(\theta)$, we will return to our usual convention and write the sine and cosine functions as $y = f(x) = \sin x$ and $y = f(x) = \cos x$.

Periodic Functions

The sine and cosine functions, and by extension their graphs, exhibit a very unique feature.

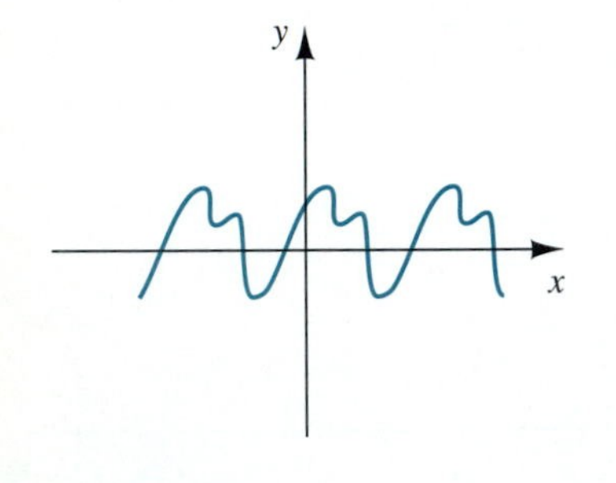

FIGURE 7.9
The graph of a periodic function

DEFINITION A function $y = f(x)$ is called **periodic** if there exists a number p such that $f(x + p) = f(x)$ for all x in the domain of f. The smallest such number p is called the **period** of the function.

A periodic function keeps repeating the same set of y-values over and over again. Thus the graph of a periodic function shows the same basic segment of its graph being repeated (see Figure 7.9).

In the case of the sine and cosine functions, the period is 2π. Let's call the portion of the graphs of $y = \sin x$ and $y = \cos x$ for $0 \leq x \leq 2\pi$ the **fundamental cycle** of the graph.

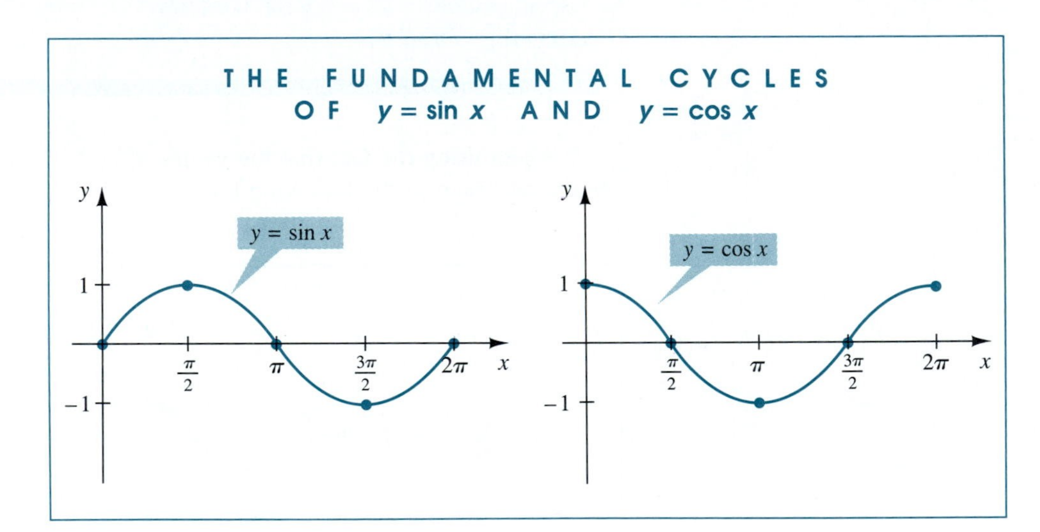

Many naturally occurring phenomena vary in a cyclic or periodic manner and can be described mathematically by using combinations of the various trigonometric functions. In more advanced mathematics courses these phenomena are often modeled by using the trigonometric functions and analyzed using the techniques of calculus.

The graphs of the fundamental cycles of $y = \sin x$ and $y = \cos x$ provide a very efficient means for remembering some important facts relating to the sine and cosine functions. For example, by looking at the graph of $y = \sin x$, we can very easily find the values of $\sin x$ for the quadrantal angles, such as $\sin 0 = 0$, $\sin \frac{\pi}{2} = 1$, and $\sin \pi = 0$.

As we mentioned earlier, once we have the graphs, we can easily see the domain and range. The domain of each is the set of all real numbers, and the range of each is $\{y \mid -1 \le y \le 1\}$.

Negative Angles

Up to this point we have done some computations involving negative angles, but we haven't described how to work with negative angles in general. Now that we have the graphs of $y = \sin x$ and $y = \cos x$, we can draw some general conclusions.

By looking at the graph of $y = \sin x$, we can see that the graph exhibits origin symmetry. Recall that such a function (which is called an *odd* function) is characterized as satisfying the functional relationship $F(-x) = -F(x)$. Thus we have

$$\sin(-x) = -\sin x$$

Similarly, by looking at the graph of $y = \cos x$, we can see that the graph exhibits y-axis symmetry. Recall that such a function (which is called an *even* function) is characterized as satisfying the functional relationship $F(-x) = F(x)$. Thus we have

$$\cos(-x) = \cos x$$

In Example 7 of Section 6.4 we showed that $\tan x = \dfrac{\sin x}{\cos x}$, and so

$$\tan(-x) = \frac{\sin(-x)}{\cos(-x)} = \frac{-\sin x}{\cos x} = -\tan x$$

The results for the cosecant, secant, and cotangent functions of negative angles can be derived from the reciprocal relationships.

$$\csc(-x) = \frac{1}{\sin(-x)} = \frac{1}{-\sin x} = -\csc x$$

$$\sec(-x) = \frac{1}{\cos(-x)} = \frac{1}{\cos x} = \sec x$$

$$\cot(-x) = \frac{1}{\tan(-x)} = \frac{1}{-\tan x} = -\cot x$$

We summarize these results in the following box.

TRIGONOMETRIC FUNCTIONS OF NEGATIVE ANGLES

$\sin(-x) = -\sin x$	$\csc(-x) = -\csc x$
$\cos(-x) = \cos x$	$\sec(-x) = \sec x$
$\tan(-x) = -\tan x$	$\cot(-x) = -\cot x$

The preceding results can be used to convert expressions involving negative angles to ones involving positive angles, as illustrated in the following example.

EXAMPLE 1 Express $\sin(-132°)$ as a function of a *positive* acute angle.

Solution

$$\sin(-132°) = -\sin 132° \qquad \textit{The reference angle is } 48°.$$

$$= \boxed{-\sin 48°}$$

■

Let us now turn our attention to applying some of our graphing techniques to the trigonometric functions.

EXAMPLE 2 Sketch the graph of **(a)** $y = \sin\left(x + \frac{\pi}{2}\right)$ **(b)** $y = 1 + \cos x$.

Solution

(a) Just as the graph of $y = (x + 2)^2$ is the graph of the basic parabola shifted 2 units to the *left*, the graph of $y = \sin\left(x + \frac{\pi}{2}\right)$ is the basic sine graph shifted $\frac{\pi}{2}$ units to the left. We say that the basic sine graph has undergone a *phase shift* of $\frac{\pi}{2}$ units to the left, or, equivalently, that the phase shift is $-\frac{\pi}{2}$ units. See Figure 7.10. Note that the graph of $y = \sin\left(x + \frac{\pi}{2}\right)$ is identical to the graph of $y = \cos x$. We often say that the graph of $y = \sin x$ and $y = \cos x$ are out of phase by $\frac{\pi}{2}$ units.

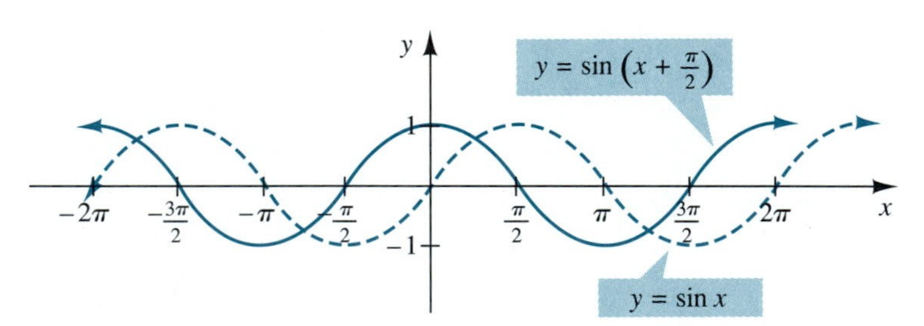

FIGURE 7.10

The fact that $\sin\left(x + \frac{\pi}{2}\right) = \cos x$ for *all* x makes it trigonometric identity. We will prove this identity, as well as many others like it, using other trigonometric relationships which are derived in the next chapter.

(b) The graph of $y = 1 + \cos x$ is obtained by shifting the basic cosine curve up 1 unit. See Figure 7.11. The graph of $y = \cos x$ appears as the dashed curve.

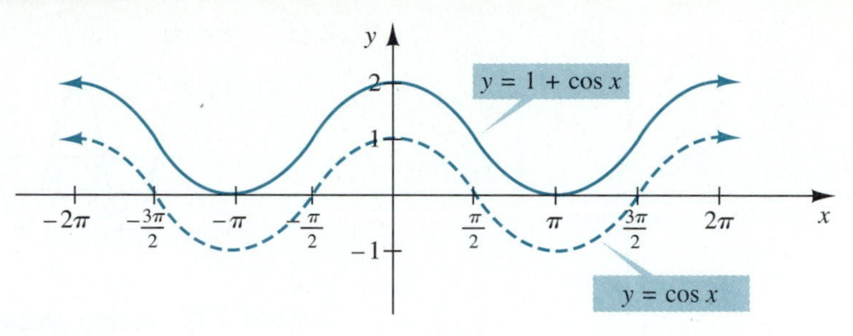

FIGURE 7.11

Let's introduce some terminology that is often used in conjunction with the trigonometric functions (due to their periodicity). Looking at the basic sine and cosine curves, we can see that the maximum value on either graph is 1 and the minimum value on either graph is -1. If we think of 0 as the value midway between the maximum and minimum values, the distance that the graph rises above this middle value is called the **amplitude** of the function. The amplitude is formally defined as follows.

DEFINITION The **amplitude** of a periodic function $f(x)$ is

$$A = \frac{1}{2}[\text{maximum value of } f(x) - \text{minimum value of } f(x)]$$

Thus the amplitude of the basic sine and cosine function is 1.

The portion of the graph of a sine or cosine functions over one period is called a **complete cycle** of the graph. In other words, the minimal portion of a sine or cosine graph that keeps repeating itself is called a complete cycle of the graph. In the case of a *basic* sine or cosine curve, a complete cycle corresponds to *any* portion of the graph on an interval of length 2π.

Recall that the portion of the basic sine and cosine curves on the interval $[0, 2\pi]$ is called the fundamental cycle of the graph.

One way of comparing sine and cosine curves is by comparing their amplitudes. Another way to compare other sine and cosine graphs to the basic curves is by comparing the number of cycles they make on an interval of length 2π.

DEFINITION The number of complete cycles a sine or cosine graph makes on an interval of length equal to 2π is called its **frequency**.

The frequency of the basic sine curve $y = \sin x$ and the basic cosine curve $y = \cos x$ is 1, because each graph makes 1 complete cycle in the interval $[0, 2\pi]$.

If a sine function has a period of $\frac{\pi}{2}$ (see Figure 7.12 on page 434), then the number of complete cycles its graph will make in an interval of length 2π is $\frac{2\pi}{\pi/2} = 4$. Thus if a sine function has a period of $\frac{\pi}{2}$, its frequency is 4 and its graph will make 4 complete cycles in an interval of length 2π.

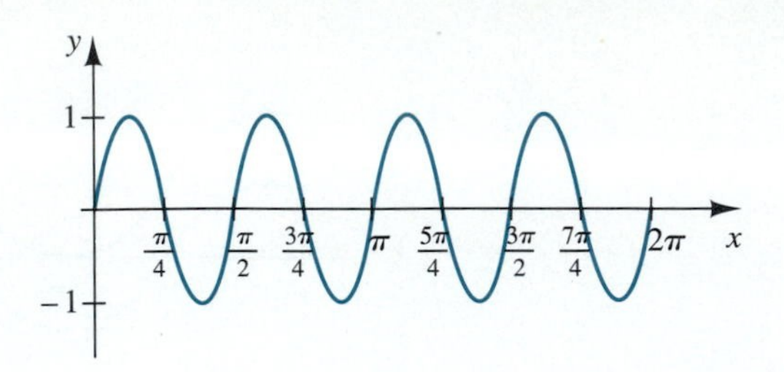

FIGURE 7.12 A sine graph of period $\frac{\pi}{2}$ and frequency 4

GRAFFIX

Use a graphics calculator or computer.

1. Graph the following on the same set of coordinate axes:

$$y = \sin x \quad y = 0.5 \sin x \quad y = 4 \sin x$$

What would you conclude about how the coefficient A affects the graph of $y = A \sin x$?

2. Graph the following on the same set of coordinate axes:

$$y = \sin x \quad y = \sin 2x \quad y = \sin\left(\frac{1}{2}\right)x$$

What would you conclude about how the coefficient B affects the graph of $y = \sin Bx$?

EXAMPLE 3 Sketch the graph of each of the following on the interval $[0, 2\pi]$. Find the amplitude, period, and frequency of each. **(a)** $y = \sin 2x$ **(b)** $y = 2 \cos x$

Solution We can obtain these graphs by applying our knowledge of the basic sine and cosine graphs.

(a) For the basic sine curve, we have

$$\sin 0 = 0 \qquad \sin\frac{\pi}{2} = 1 \qquad \sin \pi = 0 \qquad \sin\frac{3\pi}{2} = -1 \qquad \sin 2\pi = 0$$

These quadrantal values serve as guidepoints, which help us draw the graph. To obtain similar guidepoints for $y = \sin 2x$, we ask for what values of x is

$$2x = 0 \qquad 2x = \frac{\pi}{2} \qquad 2x = \pi \qquad 2x = \frac{3\pi}{2} \qquad 2x = 2\pi$$

and we get

$$x = 0 \qquad x = \frac{\pi}{4} \qquad x = \frac{\pi}{2} \qquad x = \frac{3\pi}{4} \qquad x = \pi$$

Thus $y = \sin 2x$ will have the values 0, 1, 0, -1, 0 at $x = 0, \frac{\pi}{4}, \frac{\pi}{2}, \frac{3\pi}{4}$, and π, respectively. These guidepoints appear in Figure 7.13(a). The graph of $y = \sin 2x$ will thus complete one cycle in the interval $[0, \pi]$, and it will repeat the same values in the interval $[\pi, 2\pi]$. $\left(\text{For example, at } x = \frac{5\pi}{4}, \sin 2\left(\frac{5\pi}{4}\right) = \sin \frac{5\pi}{2} = \sin\left(2\pi + \frac{\pi}{2}\right) = \sin \frac{\pi}{2} = 1\text{, which is the same value of } y = \sin 2x \text{ at } x = \frac{\pi}{4}.\right)$

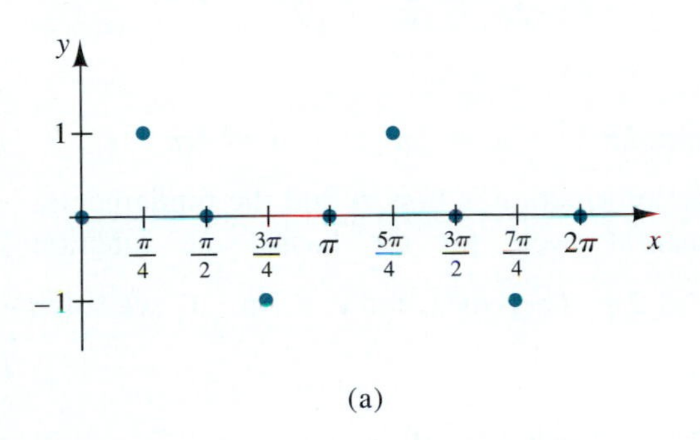

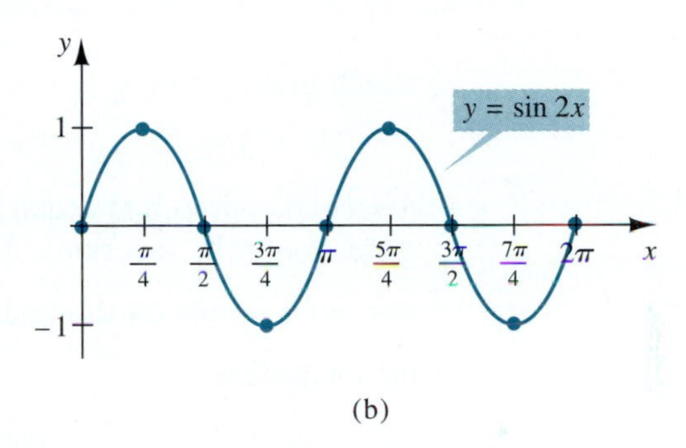

FIGURE 7.13 (a) The guidepoints for $y = \sin 2x$; (b) the graph of $y = \sin 2x$ on $[0, 2\pi]$

The graph of $y = \sin 2x$ appears in Figure 7.13(b). From this graph we can see that $y = \sin 2x$ has an amplitude of 1, a period of π (since it takes π units to make 1 complete cycle), and a frequency of 2 (since the graph makes 2 complete cycles in the interval $[0, 2\pi]$).

The graph of $y = 2 \cos x$ is obtained by stretching the graph of $y = \cos x$.

(b) We can easily graph $y = 2 \cos x$ by recognizing that the values of $y = 2 \cos x$ are twice the values of $y = \cos x$. Thus the values on the basic cosine curve are doubled. The graph appears in Figure 7.14. From this graph we can see that $y = 2 \cos x$ has an amplitude of 2, a period of 2π, and a frequency of 1.

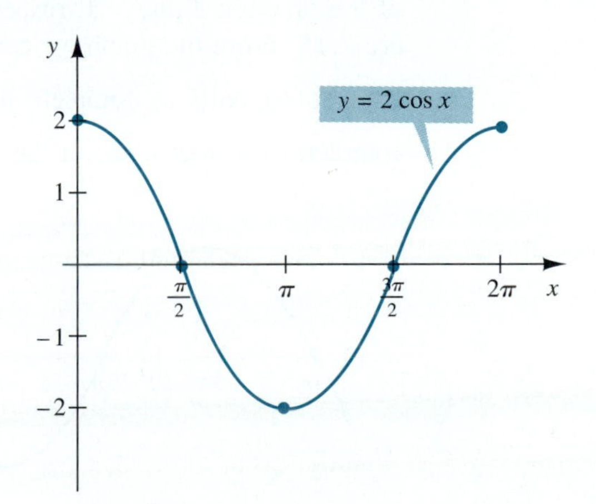

FIGURE 7.14 The graph of $y = 2 \cos x$ on $[0, 2\pi]$ ■

EXAMPLE 4 Sketch the graph of $y = 3 \sin \frac{1}{2}x$ and find its amplitude, period, and frequency.

Solution Our analysis of $y = 3 \sin \frac{1}{2}x$ combines the ideas we used in Example 3. We want to focus our attention on the key points that help us draw the graph, that is, the highest and lowest points on the graph as well as the x-intercepts. Thus we again seek those x-values for which

$$\frac{1}{2}x = 0 \qquad \frac{1}{2}x = \frac{\pi}{2} \qquad \frac{1}{2}x = \pi \qquad \frac{1}{2}x = \frac{3\pi}{2} \qquad \frac{1}{2}x = 2\pi$$

which gives

$$x = 0 \qquad x = \pi \qquad x = 2\pi \qquad x = 3\pi \qquad x = 4\pi$$

A second approach to acquiring the same information is first to find the fundamental cycle for this function. The fundamental cycle for the basic sine function $y = \sin x$ occurs on the interval $0 \le x \le 2\pi$. Therefore, for $y = \sin \frac{1}{2}x$, we solve the inequality

$$0 \le \frac{1}{2}x \le 2\pi \qquad \textit{which gives us}$$

$$0 \le x \le 4\pi$$

Thus the graph will make one complete cycle in the interval $[0, 4\pi]$. In order to get the guidepoints, we simply divide this interval into four equal parts. Since the interval $[0, 4\pi]$ has length 4π units, the guidepoints will be 0, π, 2π, 3π, and 4π, just as we obtained before. (We will find that this second approach is a bit more efficient and will use it again in the next example.)

The coefficient of 3 in $y = 3 \sin \frac{1}{2}x$ makes the maximum and minimum values of the function 3 and -3, respectively. Therefore, the graph looks as shown in Figure 7.15. From the graph we can see that the amplitude is 3, the period is 4π (since it takes 4π units to complete one cycle), and the frequency is $\frac{1}{2}$ (since the graph completes one-half cycle in the interval $[0, 2\pi]$).

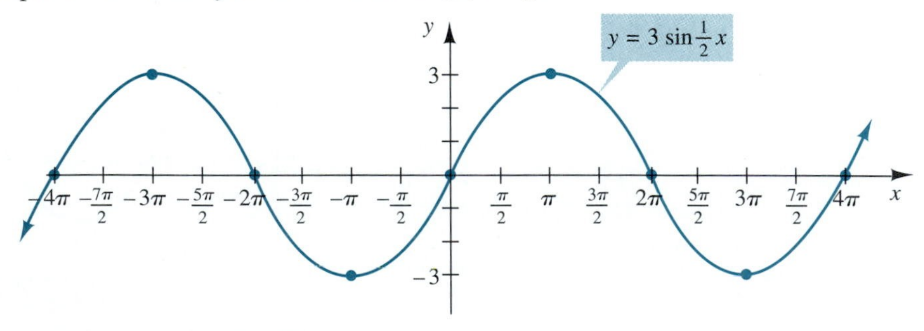

FIGURE 7.15 ■

In the previous examples we have seen basic sine and cosine functions undergo phase shifts as well as changes in amplitude and frequency. In the following example, the sine function undergoes all three changes.

GRAFFIX

Use a graphics calculator or computer.

1. Graph the following on the same set of coordinate axes: $y = \sin x$, $y = \sin\left(x + \frac{\pi}{2}\right)$, and $y = \sin\left(x - \frac{\pi}{4}\right)$. Describe how changing C affects the graph of $y = \sin(x + C)$.
2. Graph the following on the same set of coordinate axes: $y = \sin 2x$, $y = \sin(2x + \pi)$, and $y = \sin\left(2x - \frac{\pi}{2}\right)$. Describe how changing C affects the graph of $y = \sin(2x + C)$.

EXAMPLE 5 Sketch the graph of $y = 5 \sin(2x + \pi)$.

Solution Using the idea mentioned in the previous example, we find the fundamental cycle by solving the inequality

$$0 \le 2x + \pi \le 2\pi \qquad \textit{We solve this inequality for } x.$$

$$-\pi \le \quad 2x \quad \le \pi$$

$$-\frac{\pi}{2} \le \quad x \quad \le \frac{\pi}{2}$$

Thus the interval for the fundamental cycle is $\left[-\frac{\pi}{2}, \frac{\pi}{2}\right]$. We get the guidepoints by dividing this interval into four equal parts. (This can be done by applying the midpoint formula.) This gives us the x-values: $-\frac{\pi}{2}, -\frac{\pi}{4}, 0, \frac{\pi}{4}, \frac{\pi}{2}$.

The coefficient of 5 in $y = 5 \sin(2x + \pi)$ tells us that the maximum and minimum values of the function will be 5 and -5, respectively.

Knowing that the trigonometric functions are periodic, we can draw the fundamental cycle of the graph and complete the entire graph by repeating this fundamental cycle. See Figure 7.16.

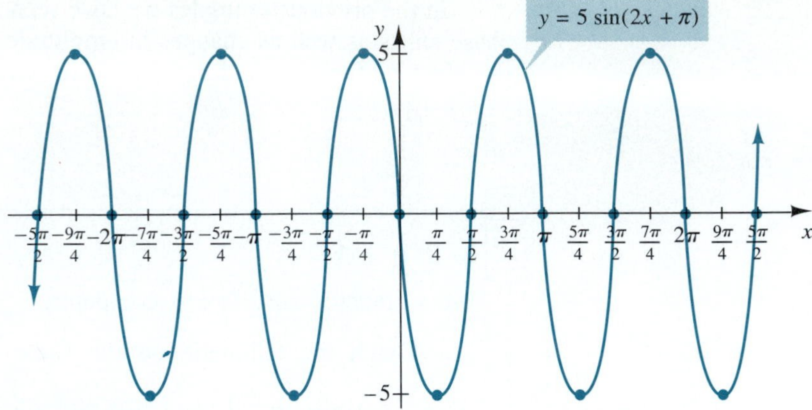

FIGURE 7.16 The fundamental cycle is on the interval $\left[-\frac{\pi}{2}, \frac{\pi}{2}\right]$ ■

Based upon the previous examples we can extract information from the equations of $y = A\sin(Bx + C)$ and $y = A\cos(Bx + C)$.

Describe the significance of the numbers 4 and 7 in the equation $y = 7\sin 4x$.

THE GRAPHS OF $y = A\sin(Bx + C)$ AND $y = A\cos(Bx + C)$

The graphs of $y = A\sin(Bx + C)$ and $y = A\cos(Bx + C)$ have

$$\text{Amplitude} = |A|$$

$$\text{Period} = \frac{2\pi}{|B|}$$

$$\text{Frequency} = |B|$$

$$\text{Phase shift} = -\frac{C}{B}$$

The interval for the fundamental cycle can be found by solving the inequality

$$0 \le Bx + C \le 2\pi$$

The guidepoints can be found by dividing this interval into four equal parts.

Note that we can recognize the phase shift as $-\frac{C}{B}$ most easily by writing $A\sin(Bx + C)$ in the equivalent form $A\sin\left[B\left(x + \frac{C}{B}\right)\right]$, which fits into our general description of the horizontal shift exhibited by $f(x + h)$ as compared to $f(x)$.

EXAMPLE 6 Sketch the graph of $y = \frac{1}{2}\cos\left(2x - \frac{\pi}{3}\right)$ over one period of the graph. Determine the amplitude, period, frequency, and phase shift.

Solution From the equation $y = \frac{1}{2}\cos\left(2x - \frac{\pi}{3}\right)$, we have $A = \frac{1}{2}$, $B = 2$, and $C = -\frac{\pi}{3}$. Thus the amplitude is $A = \frac{1}{2}$, the period is $\frac{2\pi}{B} = \frac{2\pi}{2} = \pi$, the frequency is $B = 2$, and the phase shift is $-\frac{C}{B} = -\frac{-\pi/3}{2} = \frac{\pi}{6}$ units to the right. We find the fundamental cycle by solving the inequality $0 \le 2x - \frac{\pi}{3} \le 2\pi$. The interval for the fundamental cycle is $\left[\frac{\pi}{6}, \frac{7\pi}{6}\right]$. We divide this interval into four equal parts to find the guidepoints. The graph appears in Figure 7.17. ■

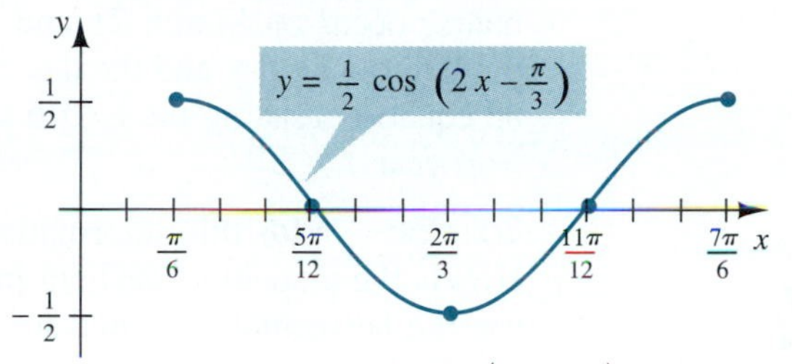

FIGURE 7.17 The graph of $y = \frac{1}{2}\cos\left(2x - \frac{\pi}{3}\right)$ over one period ■

EXAMPLE 7 Sketch the graph of $y = \sin\left(\frac{\pi}{2} - x\right)$ on the interval $[-\pi, \pi]$.

Solution We can use our knowledge of how the sine function behaves for negative angles to make this example a bit easier. We have

$$y = \sin\left(\frac{\pi}{2} - x\right) = \sin\left(-x + \frac{\pi}{2}\right) = \sin\left[-\left(x - \frac{\pi}{2}\right)\right] \quad \textit{Since } \sin(-t) = -\sin t, \textit{ we have}$$

$$= -\sin\left(x - \frac{\pi}{2}\right)$$

We now recognize the function $y = -\sin\left(x - \frac{\pi}{2}\right)$. We obtain its graph by taking the basic sine curve, shifting it $\frac{\pi}{2}$ units to the right, and then reflecting it about the x-axis. These steps are illustrated in Figure 7.18.

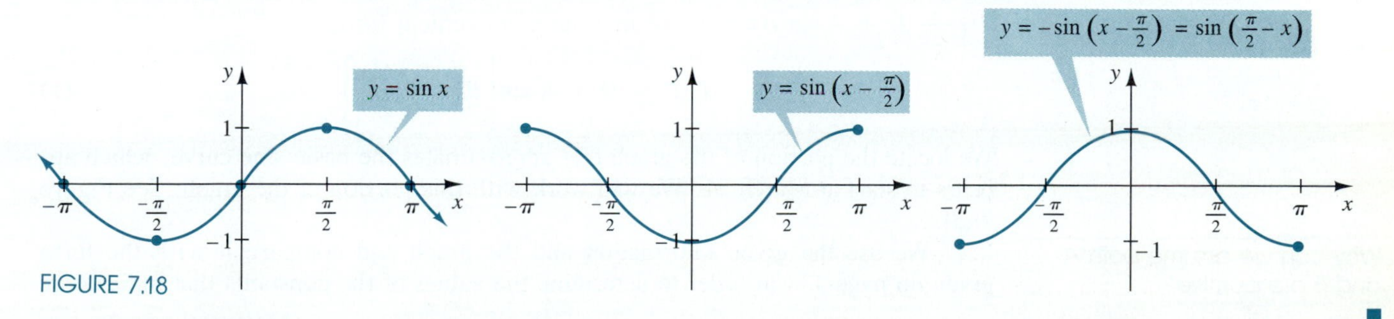

FIGURE 7.18

■

Some recurring or periodic phenomena can be described and approximated using the sine or cosine functions:

$$y = A\sin(Bx + C) \quad \text{or} \quad y = A\cos(Bx + C)$$

For example, the length of a day and the phases of the moon are periodic phenomena that can be approximated by use of a sine function. As demonstrated in the next example, we can interpret specific information in terms of the various characteristics of a sine function that we have discussed in this section.

EXAMPLE 8 In a particular region, the longest day of the year occurs on June 21, with 15 hours of daylight; the shortest day is December 21, with 9 hours of daylight. The equinoxes (the dates on which the length of day and night are both equal to 12 hours) occur on March 21 and September 21. Given that the relationship between the length of a day and the day of the year is approximated by a sine curve, construct an equation relating the length of a day to the day of the year. (Assume this is not a leap year.)

Solution With this information, we sketch a graph of the function $H(t)$, where $H(t)$ is the amount of daylight in hours (on the vertical axis), t is the day of the year (on the horizontal axis) and we count $t = 1$ as January 1. See Figure 7.19.

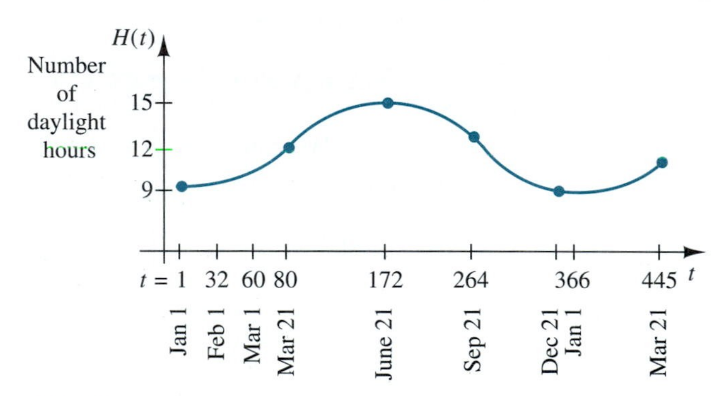

FIGURE 7.19

We start by noting that we are told that the function is (approximately) a sine function and can see that it is shifted vertically upward; therefore it has equation $H(t) = D + A\sin(Bx + C)$ or in a more convenient form,

$$H(t) = D + A\sin\left[B\left(x + \frac{C}{B}\right)\right] \tag{1}$$

We locate the portion of the graph that approximates the basic sine curve, which appears to start at March 21. We will work with this portion of the graph. See Figure 7.20.

Why can we assume both A and B are positive?

We use the given information and the graph and compare it with the form given on page 438 in order to determine the values of the constants that indicate its

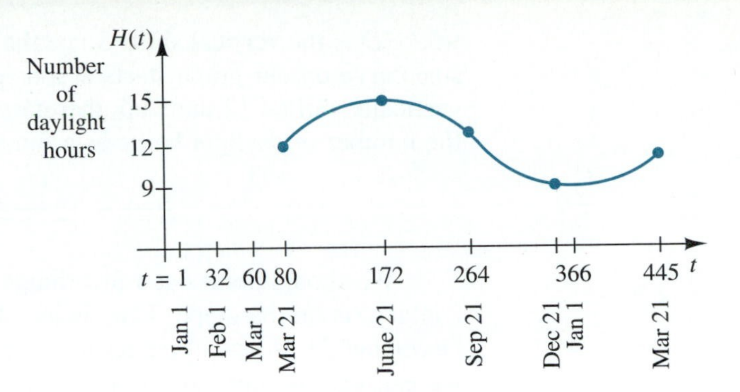

FIGURE 7.20

amplitude (A), period $\left(\frac{2\pi}{B}\right)$, phase shift $\left(-\frac{C}{B}\right)$, and vertical shift ($D$). (Since we are looking at the basic sine curve, both A and B are positive.) By definition, the amplitude of the sine function is $\frac{1}{2}$[maximum value of $H(t)$ − minimum value of $H(t)$], which we can see by the graph is

$$\frac{1}{2}[15 - 9] = 3$$

Hence the amplitude is 3; therefore, $A = 3$.

The period is given by $\frac{2\pi}{B}$. We know that the period (one complete cycle) for our problem is 365 days. Hence we have

$$\frac{2\pi}{B} = 365 \Rightarrow B = \frac{2\pi}{365}$$

Our next step is to determine the phase shift. The *basic* sine curve starts at the y-value halfway between the minimum and maximum y-values of the sine curve *where the curve is rising*. The sine curve for our graph has a median y-value of 12 hours (halfway between 9 and 15); it passes through this point as it is rising on March 21. We compare this curve with the basic sine curve and see that the sine curve is horizontally shifted to March 21, which is the 80th day of the year. Hence the sine curve is shifted 80 days to the right.

Since the phase shift of the sine curve is $-\frac{C}{B}$, we have

$$-\frac{C}{B} = 80 \Rightarrow \frac{C}{B} = -80$$

Substituting the values for A, B, and $\frac{C}{B}$ into equation (1) we have

$$H(t) = D + 3 \sin\left[\frac{2\pi}{365}(t - 80)\right]$$

where D is the vertical shift. Since the basic sine curve starts at a height of 0 and the sine curve on our graph starts at a height of 12, we can see that the sine curve was vertically shifted 12 units up; therefore, $D = 12$. Hence the equation of the graph of the number of daylight hours as a function of the day of the year is

$$H(t) = 12 + 3\sin\left[\frac{2\pi}{365}(t - 80)\right] \qquad \blacksquare$$

We should point out a few things about the way we originally labeled the horizontal axis on the graph. Our choice of 1 on the t-axis as January 1 (and, therefore, December 31 as $t = 0$) makes it convenient to read, but it is quite arbitrary. That is, we can choose any day to be the starting day, such as April 1, and although this would have no effect on the amplitude or the period, our choice of the 0 point *would* have an effect on the phase shift.

As a matter of fact, looking at the graph, and comparing it with the basic sine curve, we can see the most convenient starting point is March 21. If we choose this day as the 0 point on the t-axis, then there will be no phase shift (which means C would be 0).

EXERCISES 7.1

In Exercises 1–8, find the amplitude A, period P, and frequency f of the given function.

1. $y = 3\sin x$ **2.** $y = \cos 3x$
3. $f(x) = \cos 5x$ **4.** $y = 5\sin x$
5. $y = \frac{1}{4}\sin 2x$ **6.** $f(x) = 2\cos\frac{1}{4}x$
7. $y = -2\cos\frac{x}{7}$ **8.** $y = -\frac{1}{3}\sin\frac{2x}{5}$

In Exercises 9–12, find the amplitude A, period P, frequency f, and phase shift s of the given function.

9. $y = 3\cos\left(x - \frac{\pi}{3}\right)$ **10.** $y = -2\sin\left(x + \frac{\pi}{6}\right)$
11. $y = -4\sin\left(3x + \frac{\pi}{4}\right)$ **12.** $y = \frac{1}{3}\cos\left(2x - \frac{\pi}{2}\right)$

In Exercises 13–32, sketch the graph of the given function on the interval $[-2\pi, 2\pi]$.

13. $y = 2\sin x$ **14.** $y = 4\cos x$
15. $y = \sin 2x$ **16.** $y = \cos 4x$
17. $y = 1 + 3\cos\theta$ **18.** $y = -1 + 2\sin\theta$.
19. $y = 2 - \sin x$ **20.** $y = 3 - 2\cos x$
21. $y = \cos\frac{1}{3}x$ **22.** $y = \frac{1}{3}\sin x$
23. $y = -2\sin 2x$ **24.** $y = -\cos 2x$
25. $y = -2 + 4\cos 3x$ **26.** $y = 1 - 3\sin 2x$
27. $y = |\sin t|$ **28.** $y = |\cos t|$
29. $y = |1 + \sin x|$ **30.** $y = |\sin x| + 1$
31. $y = |-1 + 2\cos\theta|$ **32.** $y = |2\cos\theta| - 1$

In Exercises 33–44, sketch the graph of the given equation over one complete cycle.

33. $y = \sin\left(x + \frac{\pi}{4}\right)$ **34.** $y = \cos\left(x - \frac{\pi}{2}\right)$
35. $y = 2\cos\left(x - \frac{\pi}{8}\right)$ **36.** $y = 3\sin\left(x + \frac{\pi}{3}\right)$
37. $y = -\sin(\theta + \pi)$ **38.** $y = -\cos(\theta - \pi)$
39. $y = 3\cos(2x - \pi)$ **40.** $y = 4\sin(3x + \pi)$
41. $y = \sin(-x + \pi)$ **42.** $y = |\cos(x + \pi)|$
43. $y = \left|\cos\left(t - \frac{\pi}{2}\right)\right|$ **44.** $y = \sin\left(t - \frac{\pi}{2}\right)$

In Exercises 45–48, identify the equation of the given graph. Each is of the form $y = A\sin Bx$ or $y = A\cos Bx$.

45.

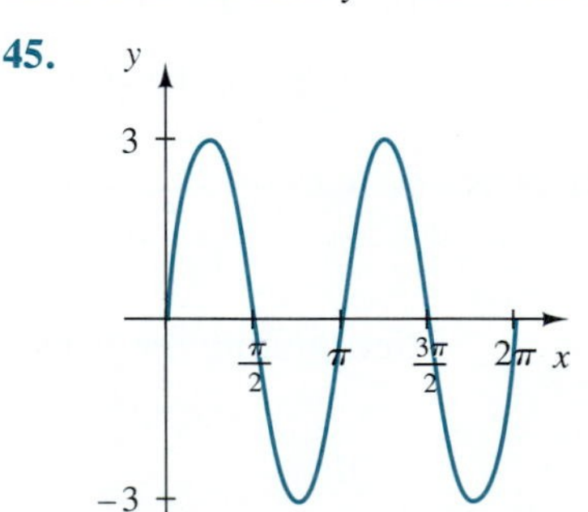

46.

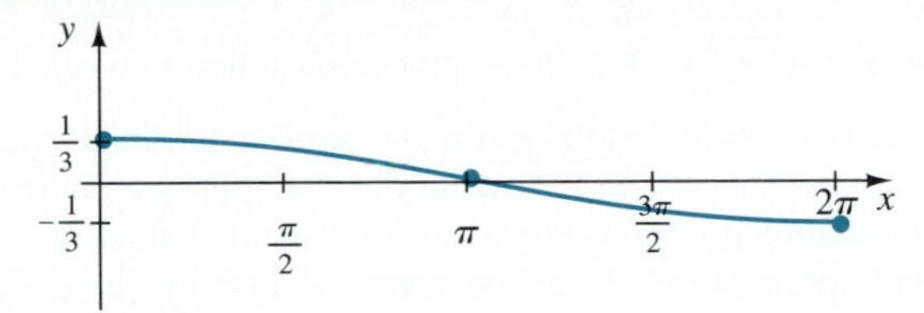

47.

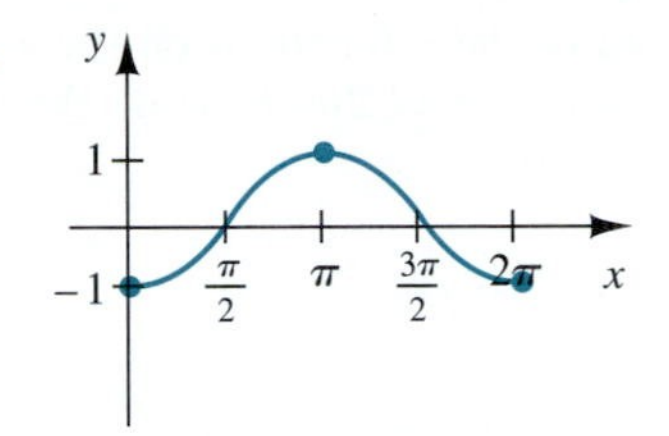

48.

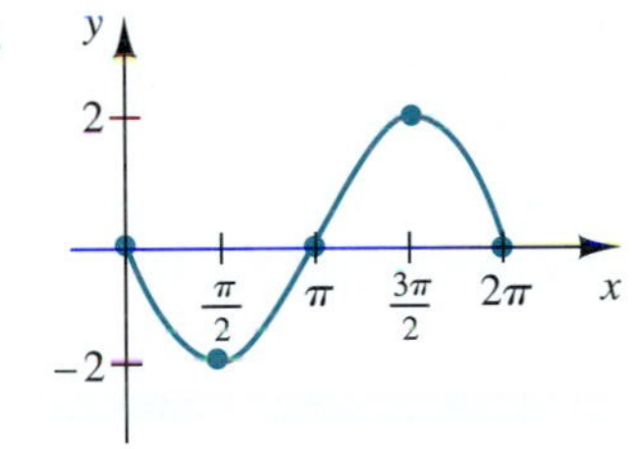

49. The number of hours of daylight for a particular area is related to the day of the year as follows

$$D = 12 + 2.5\sin\left[\frac{2\pi}{365}(t - 81)\right]$$

where D is the number of hours of daylight and t is the day of the year, with $t = 1$ being January 1. Sketch a graph of this function. Determine its period, amplitude, phase shift, and vertical shift.

50. The average daily temperature of a region is given by the equation

$$F = 68 + 18\cos\left[\frac{2\pi}{365}(t - 140)\right]$$

where F is the average daily temperature for that region (in degrees Farenheit) and t is the day of the year, with $t = 1$ being January 1. Sketch a graph of this function. Determine its period, amplitude, phase shift, and vertical shift.

51. A 5-kg weight hanging from the end of a spring is pulled down (*displaced*) 6 inches below its position of equilibrium (see the accompanying diagram). Suppose that, ignoring air resistance and friction, the distance d that the weight is from the position of equilibrium at time t in seconds is given by

$$d = 6\cos 4t$$

Sketch a graph of this function. Determine its period, amplitude, phase shift, and vertical shift.

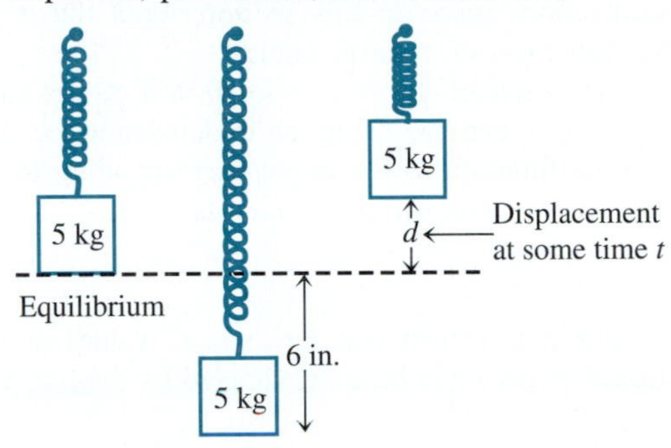

52. A *lunar month* consists of $29\frac{1}{2}$ days. We say that the moon has a period of 29.5 days meaning that it takes 29.5 days from one full moon to the next. Suppose that the portion of the moon which is seen in the sky (called the *phase* of the moon) can be expressed as a sinusoidal function of the day of the lunar month. Sketch a graph relating the phase of the moon to the day of the iunar month, and use the sine function to express the phase of the moon as a function of the day of the lunar month.

53. Suppose that the average daily temperature of a particular region is periodic. The lowest mean temperature of a region is 40°F, occurring on February 10, and the highest mean temperature is 96°F, occurring on August 10. Sketch a graph of the average daily temperature for that region, and express the average daily temperature as a function of the day of the year using the cosine function.

QUESTIONS FOR THOUGHT

54. On the same set of coordinate axes, sketch the graphs of $y = \sin x$ and $y = \cos\left(x - \frac{\pi}{2}\right)$. What does this imply about the graphs of $y = \sin x$ and $y = \cos x$?

55. In the course of our discussion in this section we mentioned that if we want to use degree measure, one way of making the graph of $y = \sin x$ more reasonable would be to use a different scale along the x- and y-axes. However, using units of different size along the x- and y-axes has other implications as well. For example, if we are not using units of the same size on both axes, what would a line with slope $\frac{2}{5}$ look like? Draw two lines with slope $\frac{2}{5}$, one assuming that x-units are larger than y-units and one assuming the reverse is true.

56. Suppose θ is a second quadrant angle. In which quadrant is $-\theta$? By consulting the reference angles for θ and $-\theta$, what can you say about $\sin\theta$ versus $\sin(-\theta)$? $\cos\theta$ versus $\cos(-\theta)$? $\tan\theta$ versus $\tan(-\theta)$? Does this agree with the generalizations made in this section about the trigonometric functions of negative angles?

57. Draw the graph of $y = \cos x$ for $0 \le x \le 2\pi$ and explain how you can use the graph to determine the sign of the cosine function as the angle corresponding to x assumes values in the various quadrants.

GRAFFIX

58. In calculus it is shown that for "small" values of x, the function $y = \sin x$ can be approximated by the polynomial $p(x) = x - \frac{x^3}{6}$. Use a graphics calculator to sketch the graphs of both functions on the same coordinate system for $-2\pi \le x \le 2\pi$. How do the two graphs compare?

59. In calculus it is also shown that for "small" values of x, the function $y = \cos x$ can be approximated by the polynomial $p(x) = 1 - \frac{x^2}{2} + \frac{x^4}{24}$. Use a graphics calculator to sketch the graphs of both of these functions on the same coordinate system for $-2\pi \le x \le 2\pi$. How do the two graphs compare?

7.2 The Tangent, Secant, Cosecant, and Cotangent Functions and Their Graphs

The Tangent Function

We have previously determined that there are some values of θ for which the function $f(\theta) = \tan\theta$ is undefined. Since $\tan\theta = \frac{y}{x}$ is undefined whenever $x = 0$, $\tan\theta$ is undefined whenever the terminal side of the angle corresponding to θ falls on the y-axis. This happens for $\theta = \frac{\pi}{2}$, to which we can add or subtract any multiple of π that will again bring the terminal side back to the y-axis. We usually write the domain of $\tan\theta$ as $\left\{\theta \mid \theta \ne \frac{\pi}{2} + n\pi\right\}$, where n is an integer.

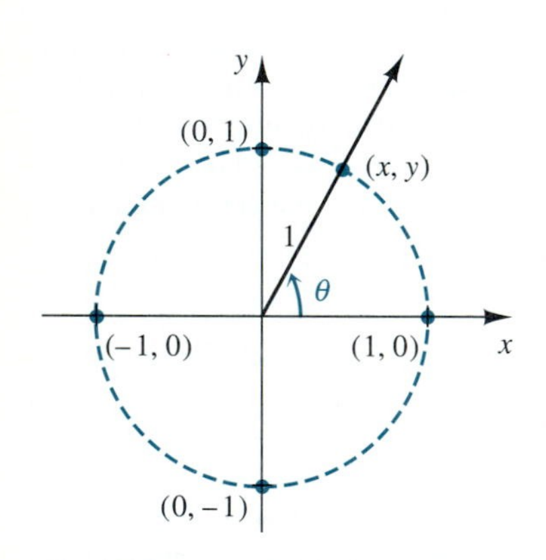

FIGURE 7.21

What happens to the value of the fraction $\frac{a}{b}$ as a increases from 0 to 1 and b decreases from 1 to 0?

In order to determine the graph of the tangent function, let's proceed as we did in the last section and examine what happens to the values of $f(\theta) = \tan\theta$ as θ increases from 0 to 2π. Again we let (x, y) be a point 1 unit from the origin that is on the terminal side of the angle corresponding to θ. See Figure 7.21. Since $\tan\theta = \frac{y}{x}$, as θ increases from 0 to $\frac{\pi}{2}$, y *increases* from 0 to 1 and x *decreases* from 1 to 0. Thus the fraction $\frac{y}{x}$ *increases* in the first quadrant for $0 \le \theta < \frac{\pi}{2}$.

Note that as θ gets closer and closer to $\frac{\pi}{2}$, y is very close to 1, whereas x is very close to 0, and so the ratio $\frac{y}{x}$ is growing larger and larger. In fact, for values of

θ very close to but less than $\frac{\pi}{2}$, tan θ is growing larger than any predetermined value. For example,

Check these values with a calculator. Make sure the calculator is in radian mode.

For $\theta = \frac{\pi}{2.01}$, $\tan\theta = 127.95798$

For $\theta = \frac{\pi}{2.001}$, $\tan\theta = 1273.8762$

Note that these values of θ are less than $\frac{\pi}{2}$.

For $\theta = \frac{\pi}{2.0001}$, $\tan\theta = 12733.032$, and so on

At $\theta = \frac{\pi}{2}$, $\tan\theta$ is undefined

Interestingly, for values of θ very close to but larger than $\frac{\pi}{2}$ (which correspond to angles in the second quadrant), y will remain very close to 1, and x will now be a *negative* number very close to 0. Therefore, the ratio $\frac{y}{x}$ will be negative but have a large absolute value. For example,

For $\theta = \frac{\pi}{1.9999}$, $\tan\theta = -12731.759$, and so on

For $\theta = \frac{\pi}{1.999}$, $\tan\theta = -1272.6024$

Note that these values are increasing.

For $\theta = \frac{\pi}{1.99}$, $\tan\theta = -126.68471$, and so on

As we continue through the second quadrant, the values will continue to increase. At $\theta = \pi$, $\tan\theta = 0$. Rather than continue our analysis in this way, let's summarize what is happening in a table.

θ	x	y	$\tan\theta = \frac{y}{x}$
As θ increases from 0 to $\frac{\pi}{2}$	x decreases from 1 to 0	y increases from 0 to 1	$\tan\theta$ increases from 0 to ∞
As θ increases from $\frac{\pi}{2}$ to π	x decreases from 0 to -1	y decreases from 1 to 0	$\tan\theta$ increases from $-\infty$ to 0
As θ increases from π to $\frac{3\pi}{2}$	x increases from -1 to 0	y decreases from 0 to -1	$\tan\theta$ increases from 0 to ∞
As θ increases from $\frac{3\pi}{2}$ to 2π	x increases from 0 to 1	y increases from -1 to 0	$\tan\theta$ increases from $-\infty$ to 0

Note that the graph will have vertical asymptotes at $\theta = \frac{\pi}{2}$ and $\theta = \frac{3\pi}{2}$ $\left(\text{in fact, at all odd multiples of } \frac{\pi}{2}\right)$.

You may want to add some more specific values to this analysis. In any case, we get the following as the graph of the tangent function. Note that just as we did with the graphs of the sine and cosine functions, having completed our analysis we label the coordinate axes x and y rather than θ and $f(\theta)$.

In words, describe the domain of the tangent function.

THE GRAPH OF $y = \tan x$

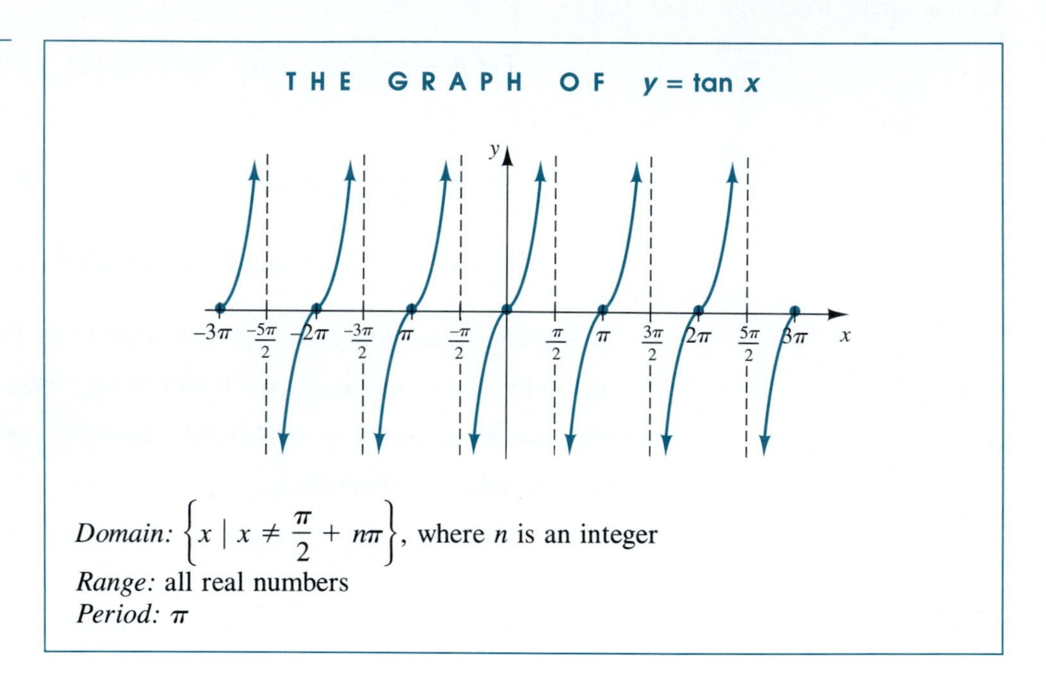

Domain: $\left\{x \mid x \neq \dfrac{\pi}{2} + n\pi\right\}$, where n is an integer

Range: all real numbers

Period: π

EXAMPLE 1 Sketch the graph of $y = 5 \tan 2x$ for $0 \leq x \leq 2\pi$.

Solution In the last section we saw that to get the period of $y = A \sin Bx$, we divided the period of the sine function by $|B|$. Similarly, since the tangent function has period π, $\tan 2x$ will have period $\dfrac{\pi}{2}$. The coefficient of 5 serves to stretch the basic tangent graph so that it rises more quickly. The graph appears in Figure 7.22.

Does it make any sense to talk about the *amplitude* of the tangent function?

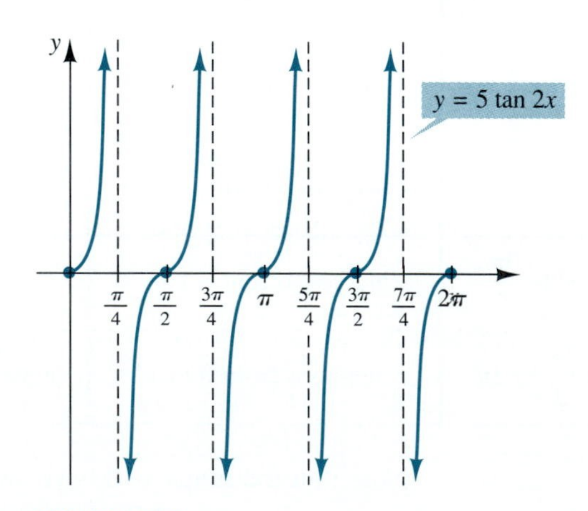

FIGURE 7.22

■

The Cotangent Function

$f(x) = \cot x = \dfrac{1}{\tan x}$ will be undefined whenever $\tan x = 0$. From the graph of $\tan x$, we know that $\tan x = 0$ whenever x is a multiple of π. Therefore, the domain of $\cot x$ is $\{x \mid x \neq n\pi\}$, where n is an integer.

In order to obtain the graph of $y = \cot x$, we could analyze the cotangent function in the same way we just analyzed the tangent function. However, we can obtain the same result by using the cofunctional relationship between the tangent and cotangent functions.

$$y = \cot x$$
$$= \tan\left(\frac{\pi}{2} - x\right) \qquad \textit{Because of the cofunctional relationship.}$$
$$= \tan\left(-x + \frac{\pi}{2}\right)$$

Recall that if we know the graph of $y = f(x)$, then the graph of $y = f(-x)$ is obtained by reflecting the graph of $f(x)$ about the y-axis. Thus, to obtain the graph of $y = \cot x = \tan\left(-x + \dfrac{\pi}{2}\right)$, shift the graph of $y = \tan x$ to the left $\dfrac{\pi}{2}$ units and then reflect it about the y-axis. Thus we have the following.

Analyze the behavior of the cotangent function by using the fact that $\cot \theta = \dfrac{x}{y}$, as we did for the tangent function. Does your analysis agree with the behavior exhibited by the graph of the cotangent function?

THE GRAPH OF $y = \cot x$

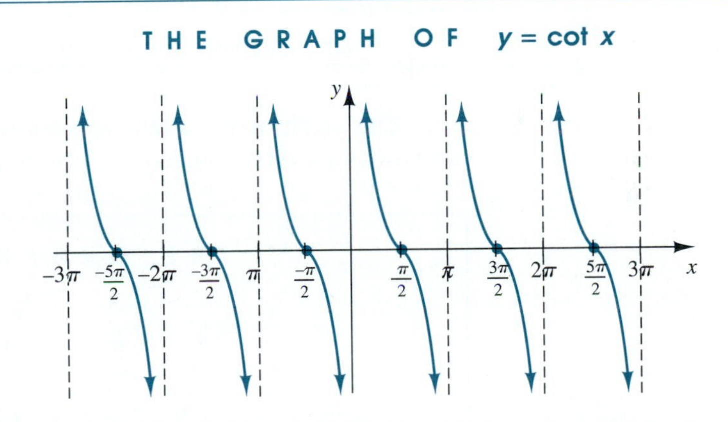

Domain: $\{x \mid x \neq n\pi\}$, when n is an integer.
Range: all real numbers
Period: π

GRAFFIX

Use a graphics calculator or computer to graph the functions $y = \tan x$ and $y = \cot x$.

The Cosecant Function

$f(x) = \csc x = \dfrac{1}{\sin x}$ will be undefined whenever $\sin x = 0$. From the basic sine graph, we know that $\sin x = 0$ whenever x is a multiple of π. We usually write the domain of $\csc x$ as $\{x \mid x \neq n\pi\}$, where n is an integer.

We can rewrite the fact that $-1 \leq \sin x \leq 1$ as $|\sin x| \leq 1$.

The values for $y = \csc x$ are the reciprocals of the values of $y = \sin x$. We know that

$$|\sin x| \leq 1$$

Since $|\sin x|$ is nonnegative, we can divide both sides of the inequality by $|\sin x|$.

$$1 \leq \frac{1}{|\sin x|} = |\csc x|$$

Therefore we have $|\csc x| \geq 1$.

x	$\sin x$	$\csc x$
As x increases from 0 to $\frac{\pi}{2}$	$\sin x$ increases from 0 to 1	$\csc x$ decreases from ∞ to 1
As x increases from $\frac{\pi}{2}$ to π	$\sin x$ decreases from 1 to 0	$\csc x$ increases from 1 to ∞
As x increases from π to $\frac{3\pi}{2}$	$\sin x$ decreases from 0 to -1	$\csc x$ increases from $-\infty$ to -1
As x increases from $\frac{3\pi}{2}$ to 2π	$\sin x$ increases from -1 to 0	$\csc x$ decreases from -1 to $-\infty$

The graph of $y = \csc x$ appears below. The graph of $y = \sin x$ is indicated with a dashed curve to make it easier to see how the two graphs fit together.

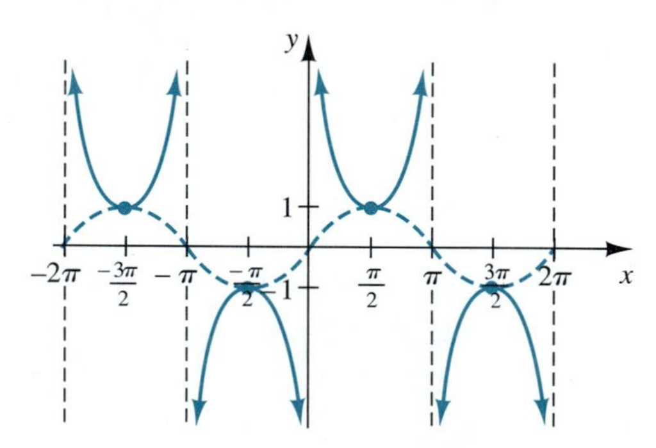

Domain: $\{x \mid x \neq n\pi\}$, when n is an integer.

Range: $\{y \mid y \leq -1 \text{ or } y \geq 1\}$ *Period:* 2π

The Secant Function

$f(x) = \sec x = \dfrac{1}{\cos x}$ is undefined whenever $\cos x = 0$. From the basic cosine graph, we know that $\cos x = 0$ whenever x is an odd multiple of $\frac{\pi}{2}$. Therefore, the domain of $\sec x$ is $\left\{x \mid x \neq \frac{\pi}{2} + n\pi\right\}$, where n is an integer.

We could use the same approach we just did for the cosecant function and obtain the graph of the secant function by analyzing it as the reciprocal of the cosine function. However, it is easier to obtain the graph of $y = \sec x$ from the graph of $y = \csc x$ by using the cofunctional relationship:

$$\sec x = \csc\left(\frac{\pi}{2} - x\right) = \csc\left(-x + \frac{\pi}{2}\right)$$

and so the graph of $y = \sec x$ is obtained by shifting the graph of $y = \csc x$ to the left $\frac{\pi}{2}$ units and then reflecting it about the y-axis. The graph appears in the following box. Note that the graph of $y = \cos x$ has been indicated with a dotted curve to make it easier to see how the two graphs fit together.

Analyze the behavior of the secant function by using the fact that $\sec x = \dfrac{1}{\cos x}$. Does your analysis agree with the behavior exhibited by the graph of the secant function?

THE GRAPH OF $y = \sec x$

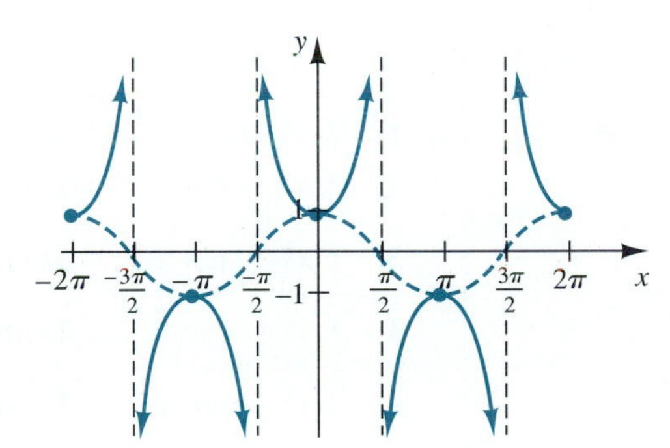

Domain: $\left\{x \mid x \neq \frac{\pi}{2} + n\pi\right\}$, where n is an integer

Range: $\{y \mid y \leq -1 \quad \text{or} \quad y \geq 1\}$

Period: 2π

GRAFFIX

Use a graphics calculator or computer.

1. Graph the functions $y = \sin x$ and $y = \csc x$ on the same set of coordinate axes. Keep in mind that on many calculators you will have to graph $y = \csc x$ as $y = \dfrac{1}{\sin x}$. Reset the viewing window so that y varies from -6 to 6.

2. Graph the functions $y = \cos x$ and $y = \sec x$ on the same set of coordinate axes. Keep in mind that on many calculators you will have to graph $y = \sec x$ as $y = \dfrac{1}{\cos x}$.

For convenience we summarize our discussion on the domains of the trigonometric functions in the following table.

DOMAINS OF THE TRIGONOMETRIC FUNCTIONS

n is an integer

	Function	Domain
1.	$f(x) = \sin x$	Domain = all real numbers
2.	$f(x) = \cos x$	Domain = all real numbers
3.	$f(x) = \tan x$	Domain = $\left\{x \mid x \neq \dfrac{\pi}{2} + n\pi\right\}$
4.	$f(x) = \csc x$	Domain = $\{x \mid x \neq n\pi\}$
5.	$f(x) = \sec x$	Domain = $\left\{x \mid x \neq \dfrac{\pi}{2} + n\pi\right\}$
6.	$f(x) = \cot x$	Domain = $\{x \mid x \neq n\pi\}$

EXAMPLE 2 Sketch the graph of $y = 3\sec(x - \pi)$ for $-2\pi \leq x \leq 2\pi$.

Solution The coefficient of 3 stretches the basic secant graph and causes the upper branch of the secant graph to have a minimum value of 3 (instead of 1) and the lower branch to have a maximum value of -3 (instead of -1). The fact that the argument of the cosecant function is $x - \pi$ shifts the basic secant graph π units to the right. The graph appears in Figure 7.23.

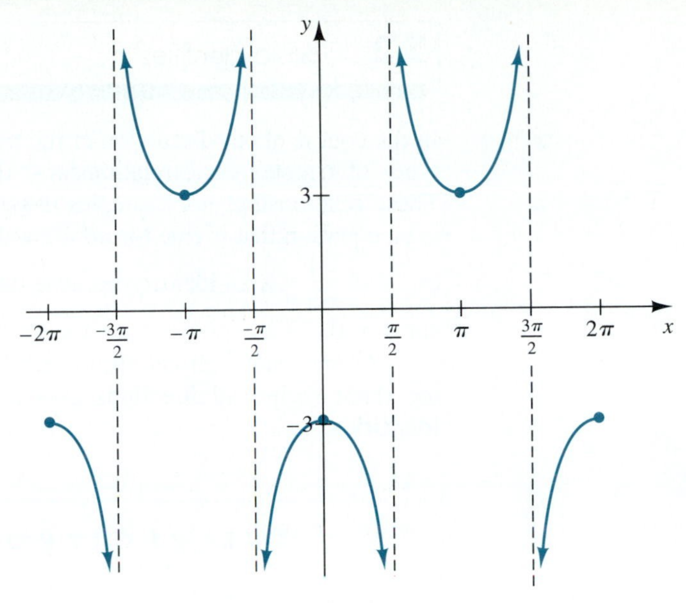

FIGURE 7.23 The graph of $y = 3\sec(x - \pi)$ for $-2\pi \le x \le 2\pi$

EXERCISES 7.2

In Exercises 1–22, sketch the graph of the given function on the interval $[-2\pi, 2\pi]$.

1. $y = \tan 2x$
2. $y = 2\tan x$
3. $y = 3\cot x$
4. $y = \cot 3x$
5. $y = -\tan x$
6. $y = \tan(-x)$
7. $y = \sec(-x)$
8. $y = -\sec x$
9. $y = \cos\left(x + \frac{\pi}{2}\right)$
10. $y = \sec\left(x + \frac{\pi}{2}\right)$
11. $y = \frac{1}{2}\sin 3x$
12. $y = 2\csc\frac{x}{3}$
13. $y = \cot(x - \pi)$
14. $y = \tan(\pi - x)$
15. $y = 2 + \csc x$
16. $y = 1 - \sec x$
17. $y = \sin\left(x + \frac{\pi}{2}\right)$
18. $y = \csc(x + \pi)$
19. $y = \tan\left(2x - \frac{\pi}{4}\right)$
20. $y = \cos(x - \pi)$
21. $y = |\tan x|$
22. $y = |\csc x|$

QUESTIONS FOR THOUGHT

23. Do the trigonometric functions have inverse functions? Explain.
24. Suppose $f(x)$ is a periodic function with period p. What will be the period of the function $f(px)$? Explain.

GRAFFIX

25. Sketch the graphs of $y = \tan x$ and $y = \cot\left(\frac{\pi}{2} - x\right)$ on the same set of coordinate axes to verify the cofunctional relationship for the tangent and cotangent functions.
26. Sketch the graphs of $y = \sec x$ and $y = \csc\left(\frac{\pi}{2} - x\right)$ on the same set of coordinate axes to verify the cofunctional relationship for the secant and cosecant functions.
27. Sketch the graph of $y = f(x) = \sin 2\pi x$. What is the period of $f(x)$?
28. Sketch the graph of $y = f(x) = \tan \pi x$. What is the period of $f(x)$?

7.3 Basic Identities

In the course of our discussion of the trigonometric functions, we have discussed two types of trigonometric relationships: the reciprocal and cofunctional relationships. These relationships are examples of *trigonometric identities*. Recall that an identity is an equation that is true for *all* allowable replacement values for the variable. Thus, $\csc\theta = \dfrac{1}{\sin\theta}$ is an identity because the equation is true for all values of θ such that $\sin\theta \neq 0$.

In fact, since all definitions are "if and only if" statements, the definitions of the three reciprocal functions give rise to three identities, called the **reciprocal identities.**

THE RECIPROCAL IDENTITIES

1. $\csc\theta = \dfrac{1}{\sin\theta}$ **2.** $\sec\theta = \dfrac{1}{\cos\theta}$ **3.** $\cot\theta = \dfrac{1}{\tan\theta}$

An identity stands in contrast to a conditional equation, which is true for (at most) some of the allowable replacement values for the variable. For example, $\cos\theta = \dfrac{1}{2}$ is a conditional trigonometric equation. As we have seen, one possible solution to this equation is $\theta = \dfrac{\pi}{3}$. However, this equation is not true for all possible replacement values of θ. (We discuss trigonometric equations in much greater detail in the next section.)

Although there are infinitely many trigonometric identities, there are a number of *basic* identities that will serve as our focus of attention both for their importance in calculus and their use in deriving other identities.

In Example 7 of Section 6.4 we derived the fact that

$$\frac{\sin\theta}{\cos\theta} = \tan\theta$$

This is an identity because it is true for all values of θ for which both sides are defined. The left-hand side is true for all values θ for which $\cos\theta \neq 0$, that is, for θ not equal to an odd multiple of $\dfrac{\pi}{2}$. These are exactly the same values of θ for which the right-hand side $\tan\theta$ is defined.

Using the reciprocal relationship for tangent, we immediately obtain another identity, namely,

$$\cot\theta = \frac{\cos\theta}{\sin\theta}$$

These last two identities are often called the **quotient identities**.

THE QUOTIENT IDENTITIES

4. $\tan\theta = \dfrac{\sin\theta}{\cos\theta}$ **5.** $\cot\theta = \dfrac{\cos\theta}{\sin\theta}$

Next we examine the expression $\sin^2\theta + \cos^2\theta$. Using the definition of $\sin\theta$ and $\cos\theta$, we have

$$\sin^2\theta + \cos^2\theta = \left(\frac{y}{r}\right)^2 + \left(\frac{x}{r}\right)^2 = \frac{y^2 + x^2}{r}$$

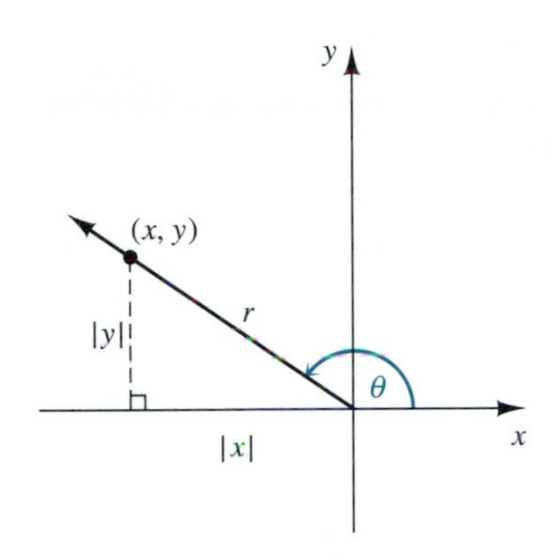

FIGURE 7.24 Typical reference triangle

In Figure 7.24 we have drawn a typical angle in standard position with the point (x, y) on the terminal side of θ and r units from the origin. Note that the sides of the reference triangle have been labeled $|x|$ and $|y|$, since, in general, we do not specify whether x and y are positive or negative.

In the reference triangle we have $|y|^2 + |x|^2 = r^2$, but $|y|^2 = y^2$ and $|x|^2 = x^2$; therefore, we have

$$\frac{y^2 + x^2}{r^2} = \frac{r^2}{r^2} = 1$$

Thus we have derived the fact that $\sin^2\theta + \cos^2\theta = 1$ for all values of θ.

This derivation is valid for θ corresponding to an angle in any of the four quadrants. We leave it as a Question for Thought for the student to verify this same result when θ is a quadrantal angle.

This identity is the first in a group of three identities called the **Pythagorean identities**. They are so called because each is fundamentally a restatement of the Pythagorean theorem in trigonometric form.

If we start with the identity $\sin^2\theta + \cos^2\theta = 1$ and divide each term by $\cos^2\theta$, we obtain the following.

$$\sin^2\theta + \cos^2\theta = 1$$ *Divide each term by $\cos^2\theta$.*

$$\frac{\sin^2\theta}{\cos^2\theta} + \frac{\cos^2\theta}{\cos^2\theta} = \frac{1}{\cos^2\theta}$$

$$\left(\frac{\sin\theta}{\cos\theta}\right)^2 + 1 = \left(\frac{1}{\cos\theta}\right)^2$$ *But $\dfrac{\sin\theta}{\cos\theta} = \tan\theta$ and $\dfrac{1}{\cos\theta} = \sec\theta$.*

$$\tan^2\theta + 1 = \sec^2\theta$$ *This is the second of the Pythagorean identities, which is true for all values of θ for which $\tan\theta$ and $\sec\theta$ are defined*

Derive the third Pythagorean identity $1 + \cot^2\theta = \csc^2\theta$.

If we repeat this process once again, this time dividing each term by $\sin^2\theta$, we obtain the third of the Pythagorean identities, which is

$1 + \cot^2\theta = \csc^2\theta$ for all values of θ for which $\cot\theta$ and $\csc\theta$ are defined.

If you know the identity $\sin^2\theta + \cos^2\theta = 1$, the other two Pythagorean identities can easily be derived from it.

THE PYTHAGOREAN IDENTITIES

6. $\sin^2\theta + \cos^2\theta = 1$ **7.** $\tan^2\theta + 1 = \sec^2\theta$ **8.** $1 + \cot^2\theta = \csc^2\theta$

It is important to be able to recognize these identities in different but equivalent forms. For example, the first Pythagorean identity often appears in the form $\sin^2\theta = 1 - \cos^2\theta$; the second in the form $\tan^2\theta = \sec^2\theta - 1$; and the third in the form $\csc^2\theta - 1 = \cot^2\theta$. It is also worthwhile noting that identity 6 is the most important, because the other two can easily be derived from it.

In the next few examples we will illustrate several approaches to verifying identities. These examples offer an excellent opportunity to practice the handling of trigonometric expressions algebraically, a skill you will find useful in calculus.

EXAMPLE 1 Verify the following identity: $\tan\theta + \cot\theta = \sec^2\theta\cot\theta$.

Solution 1 We offer two of a number of possible solutions. We will transform the left-hand side of the equation into the right-hand side.

$$\tan\theta + \cot\theta \stackrel{?}{=} \sec^2\theta\cot\theta$$

We use one of the quotient identities to substitute $\cot\theta = \dfrac{1}{\tan\theta}$.

$$\tan\theta + \frac{1}{\tan\theta} \stackrel{?}{=} \sec^2\theta\cot\theta$$

We combine the left-hand side into one fraction. The LCD is $\tan\theta$.

$$\frac{\tan^2\theta + 1}{\tan\theta} \stackrel{?}{=} \sec^2\theta\cot\theta$$

By one of the Pythagorean identities, $\tan^2\theta + 1 = \sec^2\theta$.

$$\frac{\sec^2\theta}{\tan\theta} \stackrel{?}{=} \sec^2\theta\cot\theta$$

$$\sec^2\theta\cdot\frac{1}{\tan\theta} \stackrel{\checkmark}{=} \sec^2\theta\cot\theta$$

Solution 2 This second solution illustrates a general strategy that many students find appealing. This strategy consists of using the reciprocal and quotient identities (wherever possible) to express both sides of the prospective identity in terms of $\sin\theta$ and $\cos\theta$ and then verifying the identity.

$$\tan\theta + \cot\theta \stackrel{?}{=} \sec^2\theta\cot\theta$$

We use the reciprocal and quotient identities to substitute for tangent, cotangent, and secant.

$$\frac{\sin\theta}{\cos\theta} + \frac{\cos\theta}{\sin\theta} \stackrel{?}{=} \frac{1}{\cos^2\theta}\cdot\frac{\cos\theta}{\sin\theta}$$

On the left-hand side we combine; the LCD is $\cos\theta\sin\theta$. *On the right-hand side we reduce using the common factor of* $\cos\theta$.

$$\frac{\sin^2\theta + \cos^2\theta}{\cos\theta\sin\theta} \stackrel{?}{=} \frac{1}{\cos\theta}\cdot\frac{1}{\sin\theta}$$

On the left-hand side, $\sin^2\theta + \cos^2\theta = 1$.

$$\frac{1}{\cos\theta\sin\theta} \stackrel{\checkmark}{=} \frac{1}{\cos\theta\sin\theta}$$

■

The second method, although it may sometimes be a bit longer, offers the advantage of being a somewhat more systematic method and also requires remembering only the first of the Pythagorean identities.

We might also mention that we could use the x, y, r definitions of the trigonometric functions and verify the identity in that way. However, most students seem to prefer either of the two approaches outlined in the last example.

Although there is no specific procedure which will work for all identities, we do offer the following suggestions.

OUTLINE FOR VERIFYING IDENTITIES

1. Know the basic identities listed in this section—that is, the reciprocal, quotient, and Pythagorean identities.
2. If one of the sides of the identity appears more complicated than the other, try to simplify the more complicated side and transform it, in a step-by-step fashion, until it looks exactly like the other side of the identity. You may find this step easier if you rewrite all the trigonometric expressions in terms of sine and cosine.
3. The simplification process will probably involve algebraic procedures such as combining fractions, factoring, and reducing to lowest terms.

Be Careful! The one method *not* to be used is to write down the identity to be verified as your first step and "do the same thing to both sides," as you would in solving an equation. For example, in trying to verify an identity, we may not multiply both sides of the equation by the same quantity. Working on the identity in this way can be done only when the identity is *assumed* to be true. When we are trying to *prove* that the given identity is true, we cannot assume that it is true. Thus when proving an identity, we must work on each side independently.

EXAMPLE 2 Verify the following identity:

$$(\sin\theta - \cos\theta)(\csc\theta + \sec\theta) = \tan\theta - \cot\theta.$$

Solution The outline suggests that since the left-hand side appears more complicated, we can begin by multiplying out the left-hand side.

$$(\sin\theta - \cos\theta)(\csc\theta + \sec\theta) \stackrel{?}{=} \tan\theta - \cot\theta$$

$$\sin\theta \cdot \csc\theta + \sin\theta\sec\theta - \cos\theta\csc\theta - \cos\theta\sec\theta \stackrel{?}{=} \tan\theta - \cot\theta$$

We use the reciprocal identities and simplify using the quotient idealities.

$$\underbrace{\sin\theta\cdot\frac{1}{\sin\theta}}_{\downarrow} + \underbrace{\sin\theta\cdot\frac{1}{\cos\theta}}_{\downarrow} - \underbrace{\cos\theta\cdot\frac{1}{\sin\theta}}_{\downarrow} - \underbrace{\cos\theta\cdot\frac{1}{\cos\theta}}_{\downarrow} \stackrel{?}{=} \tan\theta - \cot\theta$$

$$1 + \tan\theta - \cot\theta - 1 \stackrel{\checkmark}{=} \tan\theta - \cot\theta \quad \blacksquare$$

EXAMPLE 3 Verify the following identity: $\sec^2\theta - \csc^2\theta = \tan^2\theta - \cot^2\theta$.

Solution Since the right hand side involves $\tan\theta$ and $\cot\theta$, we can use Pythagorean identities to rewrite the left hand side in terms of $\tan\theta$ and $\cot\theta$ as well.

$$\sec^2\theta - \csc^2\theta \stackrel{?}{=} \tan^2\theta - \cot^2\theta$$

We use the identities $\tan^2\theta + 1 = \sec^2\theta$ and $1 + \cot^2\theta = \csc^2\theta$ on the left-hand side.

$$(\tan^2\theta + 1) - (1 + \cot^2\theta) \stackrel{?}{=} \tan^2\theta - \cot^2\theta$$

$$\tan^2\theta + 1 - 1 - \cot^2\theta \stackrel{?}{=} \tan^2\theta - \cot^2\theta$$

$$\tan^2\theta - \cot^2\theta \stackrel{\checkmark}{=} \tan^2\theta - \cot^2\theta$$

■

Try verifying this identity by first expressing everything in terms of $\sin\theta$ and $\cos\theta$.

GRAFFIX

Use a graphics calculator or computer to verify the identity

$$\cos x + \cos x \tan^2 x = \sec x$$

by using the calculator to sketch the graphs of

$$y = \cos x + \cos x \tan^2 x \qquad \text{and} \qquad y = \sec x$$

on the same coordinate system.

Sometimes verifying an identity requires a bit more algebraic manipulation.

EXAMPLE 4 Verify the following identity: $\dfrac{\sec\theta - \cos\theta}{\csc\theta - \sin\theta} = \tan^3\theta$.

Remember that we cannot prove the identity by multiplying each side by $\csc\theta - \sin\theta$ to clear the fractions.

Solution We begin by using the definitions of secant and cosecant (the reciprocal identities) to rewrite them in terms of sine and cosine.

$$\frac{\sec\theta - \cos\theta}{\csc\theta - \sin\theta} \stackrel{?}{=} \tan^3\theta$$

$$\frac{\dfrac{1}{\cos\theta} - \cos\theta}{\dfrac{1}{\sin\theta} - \sin\theta} \stackrel{?}{=} \tan^3\theta$$

The left-hand side is complex fraction, which we simplify.

$$\frac{\left(\dfrac{1}{\cos\theta} - \cos\theta\right)\cos\theta\sin\theta}{\left(\dfrac{1}{\sin\theta} - \sin\theta\right)\cos\theta\sin\theta} \stackrel{?}{=} \tan^3\theta$$

$$\frac{\sin\theta - \cos^2\theta\sin\theta}{\cos\theta - \sin^2\theta\cos\theta} \stackrel{?}{=} \tan^3\theta$$ *Factor both the numerator and and denominator.*

$$\frac{\sin\theta\,(1-\cos^2\theta)}{\cos\theta\,(1-\sin^2\theta)} \stackrel{?}{=} \tan^3\theta$$ *But* $1-\cos^2\theta = \sin^2\theta$ *and* $1-\sin^2\theta = \cos^2\theta$

$$\frac{\sin\theta\,\sin^2\theta}{\cos\theta\,\cos^2\theta} \stackrel{?}{=} \tan^3\theta$$

$$\frac{\sin^3\theta}{\cos^3\theta} = \left(\frac{\sin\theta}{\cos\theta}\right)^3 \stackrel{\checkmark}{=} \tan^3\theta$$ ■

As we develop more trigonometric relationships, we will be able to prove a greater variety of identities.

The trigonometric identities can also be used to compute values of the trigonometric functions. The next example appeared as Example 8 in Section 6.2. We repeat it here so that you can compare the solution offered here with the one given in Section 6.2.

EXAMPLE 5 Given $\tan\theta = \frac{1}{2}$ and $\sin\theta < 0$, find $\cos\theta$.

Solution As we noted in Section 6.2, since the example tells us that $\tan\theta$ is positive and $\sin\theta$ is negative, we know that θ must be a third quadrant angle. Using the identity $\tan^2\theta + 1 = \sec^2\theta$, we substitute the given value for $\tan\theta$ and solve for $\sec\theta$.

$$\sec^2\theta = \tan^2\theta + 1$$

$$\sec^2\theta = \left(\frac{1}{2}\right)^2 + 1 = \frac{5}{4}$$ *Take square roots.*

$$\sec\theta = \pm\sqrt{\frac{5}{4}} = \pm\frac{\sqrt{5}}{2}$$

We have already noted that θ must be in quadrant III, and so $\sec\theta$ must be negative. Therefore,

Compare this solution with the approach taken in Example 8 of Section 6.2.

$$\sec\theta = -\frac{\sqrt{5}}{2}$$ *To get* $\cos\theta$ *we take the reciprocal of* $\sec\theta$.

$$\boxed{\cos\theta = -\frac{2}{\sqrt{5}} = -\frac{2\sqrt{5}}{5}}$$ ■

As illustrated in the last example, we can use the Pythagorean identities to find specific values of the trigonometric functions. However, if we need to take square roots, the decision as to whether to take the positive or negative square root is determined by the quadrant in which θ falls.

We conclude this section by presenting a list of the fundamental trigonometric identities for ease of reference.

THE FUNDAMENTAL TRIGONOMETRIC IDENTITIES

The Reciprocal Identities

1. $\csc\theta = \dfrac{1}{\sin\theta}$
2. $\sec\theta = \dfrac{1}{\cos\theta}$
3. $\cot\theta = \dfrac{1}{\tan\theta}$

The Quotient Identities

4. $\tan\theta = \dfrac{\sin\theta}{\cos\theta}$
5. $\cot\theta = \dfrac{\cos\theta}{\sin\theta}$

The Pythagorean Identities

6. $\sin^2\theta + \cos^2\theta = 1$
7. $\tan^2\theta + 1 = \sec^2\theta$
8. $1 + \cot^2\theta = \csc^2\theta$

EXERCISES 7.3

In Exercises 1–10, simplify the given expression to one involving only constants, $\sin\theta$, and $\cos\theta$.

1. $\sin\theta\cot\theta$
2. $\cos\theta\tan\theta$
3. $\tan\theta\csc\theta$
4. $\cot\theta\sec\theta$
5. $\cos^2\theta(\tan^2\theta + 1)$
6. $\sin^2\theta(\cot^2\theta + 1)$
7. $\dfrac{\tan\theta}{\cot\theta}$
8. $\dfrac{\tan\theta}{\sec\theta}$
9. $\dfrac{1-\tan\theta}{\cos\theta}$
10. $\dfrac{1+\cot\theta}{\sin\theta}$

In Exercises 11–14, rewrite the given expression in terms of $\sin\theta$ only.

11. $\dfrac{\cos^2\theta}{\sin\theta}$
12. $\cos^2\theta - \sin^2\theta$
13. $\sin\theta - \csc\theta$
14. $\dfrac{\tan\theta + \sec\theta}{\cos\theta}$

In Exercises 15–18, rewrite the given expression in terms of $\cos\theta$ only.

$\cos\theta - \sec\theta$

16. $\dfrac{\sin^2\theta}{\cos\theta}$

$\cos^2\theta - \sin^2\theta$

18. $\dfrac{\cot\theta + \csc\theta}{\sin\theta}$

In Exercises 19–66, verify the given identity.

19. $\dfrac{\sin\theta}{\tan\theta} = \cos\theta$
20. $\sec\theta\cot\theta = \csc\theta$
21. $\sin\theta\cot\theta\sec\theta = 1$
22. $\cos\theta\tan\theta\csc\theta = 1$
23. $\sin\alpha + \cos\alpha\cot\alpha = \csc\alpha$
24. $\cos\beta + \cos\beta\tan^2\beta = \sec\beta$
25. $(1+\sin w)(1-\sin w) = \dfrac{1}{\sec^2 w}$
26. $1 - 2\sin^2\gamma = 2\cos^2\gamma - 1$
27. $\sin x(\csc x - \sin x) = \cos^2 x$
28. $\cot\theta + \tan\theta = \csc\theta\sec\theta$
29. $\sec\beta - \cos\beta = \tan\beta\sin\beta$
30. $(\cos t - \sin t)(\cos t + \sin t) = 1 - 2\sin^2 t$
31. $\sin^4\theta - \cos^4\theta = \sin^2\theta - \cos^2\theta$
32. $\sec^4\theta - \tan^4\theta = 1 + 2\tan^2\theta$
33. $\cot^2\theta - \cos^4\theta\csc^2\theta = \cos^2\theta$
34. $(\sec\alpha + \tan\alpha)(1 - \sin\alpha) = \cos\alpha$
35. $\dfrac{1}{\sec^2\theta} = 1 - \dfrac{1}{\csc^2\theta}$
36. $\tan u + \cot u = \dfrac{1}{\sin u\cos u}$
37. $\dfrac{\sin x + \cos x}{\cos x} = 1 + \tan x$

38. $\dfrac{\tan x + \cos x}{\sin x} = \sec x + \cot x$

39. $\dfrac{\sin t + \tan t}{\sin t} = 1 + \sec t$

40. $\dfrac{1 + \sec \theta}{\csc \theta} = \sin \theta + \tan \theta$

41. $\dfrac{\cos x}{1 + \sin x} + \dfrac{\cos x}{1 - \sin x} = 2 \sec x$

42. $\dfrac{\cos x + \tan x}{\sin x \cos x} = \csc x + \sec^2 x$

43. $\tan \theta(\tan \theta + \cot \theta) = \sec^2 \theta$

44. $\tan^2 \beta - \sin^2 \beta = \tan^2 \beta \sin^2 \beta$

45. $(\tan u + \cot u)(\cos u + \sin u) = \sec u + \csc u$

46. $(\cot \gamma + \csc \gamma)(\tan \gamma - \sin \gamma) = \sec \gamma - \cos \gamma$

47. $\csc x = \dfrac{\cot x + \tan x}{\sec x}$

48. $\dfrac{\sin t}{\csc t} + \dfrac{\cos t}{\sec t} = 1$

49. $\dfrac{1 + \cos u}{\sin u} + \dfrac{\sin u}{1 + \cos u} = 2 \csc u$

50. $\dfrac{\sec \theta - \cos \theta}{\tan \theta} = \dfrac{\tan \theta}{\sec \theta}$

51. $\dfrac{\cot A \cos A}{\csc^2 A - 1} = \sin A$

52. $\dfrac{\cot B - \tan B}{\sin B + \cos B} = \csc B - \sec B$

53. $(\sin^2 \theta + \cos^2 \theta)^6 = 1$

54. $(\sec^2 \theta - \tan^2 \theta)^{10} = 1$

55. $\dfrac{1}{\tan \alpha + \cot \alpha} = \sin \alpha \cos \alpha$

56. $1 - \sin C = \dfrac{\cot C - \cos C}{\cot C}$

57. $\dfrac{\cos \gamma}{1 - \sin \gamma} = \sec \gamma + \tan \gamma$

58. $\dfrac{1 + \sin \theta}{\cos \theta} = \dfrac{\cos \theta}{1 - \sin \theta}$

59. $\dfrac{\sin A + \tan A}{1 + \sec A} = \sin A$

60. $\dfrac{1 + \tan B}{1 - \tan B} = \dfrac{\cot B + 1}{\cot B - 1}$

61. $\dfrac{\tan^2 \beta}{\sec \beta + 1} = \dfrac{1 - \cos \beta}{\cos \beta}$

62. $\dfrac{\sin C}{1 - \cos C} = \csc C + \cot C$

63. $\dfrac{\sin A + \cos A}{\sin A - \cos A} = \dfrac{\sec A + \csc A}{\sec A - \csc A}$

64. $\dfrac{\cos B \cot B}{\cot B - \sin B} = \dfrac{\cot B + \cos B}{\cos B \cot B}$

65. $\ln |\sec x| = -\ln |\cos x|$

66. $\ln |\sin x| = -\ln |\csc x|$

In Exercises 67–72, find the required value.

67. If $\sin \theta = \dfrac{4}{5}$ and $\tan \theta$ is negative, find $\cos \theta$.

68. If $\cos A = -\dfrac{5}{13}$ and $\cot A$ is positive, find $\sin A$.

69. If $\tan B = -\dfrac{2}{3}$ and $\sin B > 0$, find $\sec B$.

70. If $\sec C = \dfrac{5}{3}$ and $\csc B < 0$, find $\tan B$.

71. If $\csc \theta = -\dfrac{7}{3}$ and $\cos \theta < 0$, find $\tan \theta$.

72. If $\cot \theta = \dfrac{3}{4}$ and $\cos \theta > 0$, find $\sec \theta$.

QUESTIONS FOR THOUGHT

73. Verify the identity $\sin^2 \theta + \cos^2 \theta = 1$ for the case in which θ is a quadrantal angle.

74. In the following figure, $\triangle PQR$ is a right triangle and $\overset{\frown}{RS}$ is an arc of a circle whose center is P.

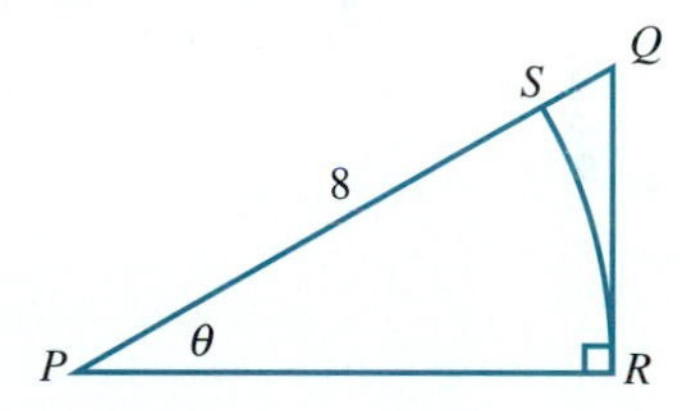

(a) Use the formula $A = \dfrac{1}{2}bh$ to express the area of the triangle as a function of θ.

(b) Use the formula $A = \dfrac{1}{2}ab \sin C$ to express the area of the triangle as a function of θ.

(c) Show that the functions obtained in **(a)** and **(b)** are actually the same.

75. A student was asked the following: Given $\tan \theta = \dfrac{4}{5}$ for a first quadrant angle θ, find $\sin \theta$ and $\cos \theta$. The student reasoned that since $\tan \theta = \dfrac{\sin \theta}{\cos \theta}$, it must follow that $\sin \theta = 4$ and $\cos \theta = 5$. Is the student's logic correct? Explain.

7.4 Trigonometric Equations

In previous sections we were given a specific angle (or number) and were asked to compute some trigonometric function of the given angle. We are now ready to reverse this process. Several examples will serve to illustrate this idea.

EXAMPLE 1 Solve the equation $2 \sin x - 1 = 0$, where

(a) x is in the interval $[0, 2\pi)$ **(b)** x is not restricted.

Solution

(a) In order to solve this equation we need to find value(s) of x for which $\sin x$ satisfies the given equation. We first solve the equation for $\sin x$.

$$2 \sin x - 1 = 0$$

$$2 \sin x = 1$$

$$\sin x = \frac{1}{2}$$

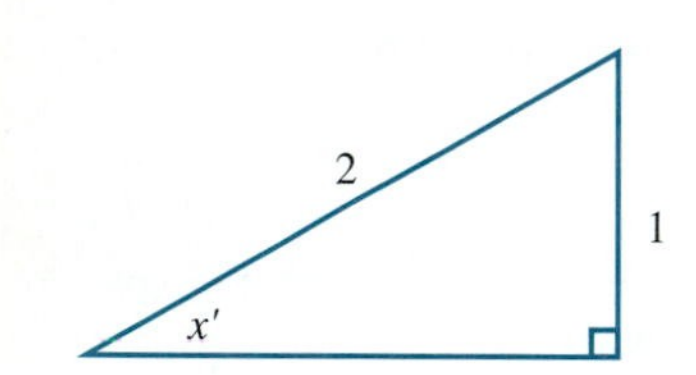

FIGURE 7.25 Reference angle x' with $\sin x' = \frac{1}{2}$

Now we seek those values of x for which $\sin x = \frac{1}{2}$. We can reason as follows: We draw a reference triangle with reference angle x' whose sine is $\frac{1}{2} = \frac{\text{opposite}}{\text{hypotenuse}}$ (see Figure 7.25). We recognize this as a 30°-60° right triangle and x' as $\frac{\pi}{6}$. Since the value of sine is positive, the reference angle x' may be in either quadrant I or II.

If $x' = \frac{\pi}{6}$ is in quadrant I, then $x = \frac{\pi}{6}$.

If $x' = \frac{\pi}{6}$ is in quadrant II, then $x = \frac{5\pi}{6}$.

Thus our answer is $\boxed{x = \frac{\pi}{6} \text{ or } x = \frac{5\pi}{6}}$.

(b) The solutions found in part (a) in the interval $[0, 2\pi)$ are called the *basic solutions,* or *fundamental solutions,* to the equation. As we already know, we may add multiples of 2π to these basic solutions and $\sin x$ will still have the same value. These are called the *general solutions* to the given equation.

Thus our general solutions are $\boxed{x = \frac{\pi}{6} + 2n\pi \text{ or } x = \frac{5\pi}{6} + 2n\pi}$, where n is an integer. This answer means, for example:

If $n = 3$, then $x = \frac{\pi}{6} + 2(3)\pi = \frac{\pi}{6} + 6\pi = \frac{13\pi}{6}$ is a solution.

If $n = -2$, then $x = \frac{5\pi}{6} - 2(2)\pi = \frac{5\pi}{6} - 4\pi = -\frac{19\pi}{6}$ is a solution. ■

DIFFERENT PERSPECTIVES: *Trigonometric Equations*

In order to determine the number of solutions to a trigonometric equation such as $2 \cos x = 1$ on the interval $[0, 2\pi]$, we can view the equation both geometrically and algebraically.

GEOMETRIC DESCRIPTION

We can draw the graphs of $y = 2 \cos x$ and $y = 1$ on the same coordinate system and count the number of points at which the graphs intersect.

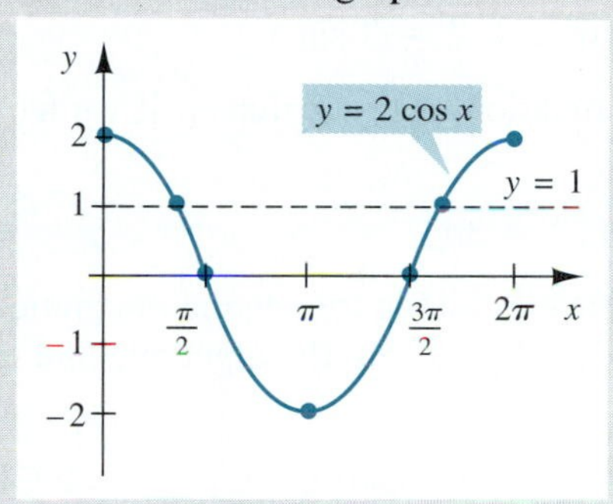

From the graphs, we can see that the equation $2 \cos x = 1$ has 2 solutions on the interval $[0, 2\pi]$. With a graphics calculator we can approximate these solutions fairly accurately.

ALGEBRAIC DESCRIPTION

In order to solve the equation $2 \cos x = 1$ we first need to find the value(s) of $\cos x$ which satisfy the equation.

$$2 \cos x = 1$$
$$\cos x = \frac{1}{2}$$

Now we find the values of x for which $\cos x = \frac{1}{2}$. We draw a reference triangle and determine that the reference angle is $\frac{\pi}{3}$. In order for the cosine to be positive, the reference angle must be in quadrant I or quadrant IV. Thus there are two solutions to the equation $2 \cos x = 1$ on the interval $[0, 2\pi]$. In fact, we have determined that these solutions are $\frac{\pi}{3}$ and $\frac{5\pi}{3}$.

EXAMPLE 2 Find the general solution to $\tan^2 \theta = 1$.

Solution We begin by taking square roots. Remember that $\tan^2\theta = (\tan \theta)^2$.

$$\tan^2 \theta = 1 \qquad \textit{Use the square root theorem.}$$
$$\tan \theta = \pm\sqrt{1} = \pm 1$$

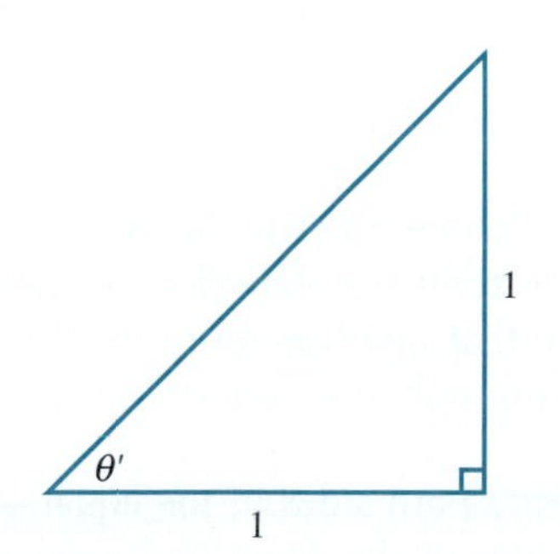

FIGURE 7.26 Reference angle θ' with $\tan \theta' = 1$

Now we seek those values of θ for which $\tan \theta = \pm 1$. We temporarily ignore the sign of the tangent, and we again draw a reference triangle with reference angle θ' whose tangent is 1. See Figure 7.26. We recognize this as a $\frac{\pi}{4}$ right triangle, and so $\theta' = \frac{\pi}{4}$. Since $\tan \theta = \pm 1$, θ' can be in any of the four quadrants.

Thus our basic solutions are $\theta = \frac{\pi}{4}, \frac{3\pi}{4}, \frac{5\pi}{4}, \frac{7\pi}{4}$, to which we can add or subtract multiples of π (because the tangent function has a period of π).

We write the general solutions as $\frac{\pi}{4} + n\pi, \frac{3\pi}{4} + n\pi, \frac{5\pi}{4} + n\pi, \frac{7\pi}{4} + n\pi$. However, if we notice that the four basic solutions can be obtained by successively adding $\frac{\pi}{2}$ to the first, we can write the general solution as $\boxed{\theta = \frac{\pi}{4} + \frac{n\pi}{2}}$, where n is any integer. ■

GRAFFIX

Use a graphics calculator or computer to graph the functions $y = \tan^2 x$ and $y = 1$ on the same coordinate system. Use these graphs to estimate the solutions to $\tan^2 x = 1$, and compare your answers to those obtained in the previous example.

Some of the algebraic procedures we have learned can be applied to trigonometric equations as well.

EXAMPLE 3 Solve for x in $[0, 2\pi)$: $\sin^2 x + 2 = 3 \sin x$.

Solution This equation is a quadratic equation in **sin *x***; that is, if we let $u = \sin x$, the equation becomes

$$u^2 + 2 = 3u$$

to which we can apply the methods we have available for solving quadratic equations. We illustrate the similarity in the procedure by solving the algebraic and trigonometric equations side by side.

$\sin^2 x + 2 = 3 \sin x$	← *is similar to* →	$u^2 + 2 = 3u$
$\sin^2 x - 3 \sin x + 2 = 0$	← *The left-hand side factors.* →	$u^2 - 3u + 2 = 0$
$(\sin x - 1)(\sin x - 2) = 0$		$(u - 1)(u - 2) = 0$
$\sin x = 1$ or $\sin x = 2$		$u = 1$ or $u = 2$

However, in the case of the trigonometric equation, we still need to solve for x, that is, find value(s) of x for which $\sin x = 1$ or $\sin x = 2$. Based upon the graph of $y = \sin x$, we know that $\sin \frac{\pi}{2} = 1$; therefore, $x = \frac{\pi}{2}$. From the graph we also know that there are no values for which $\sin x = 2$. Therefore, the only solution to the equation in the interval $[0, 2\pi)$ is $\boxed{x = \frac{\pi}{2}}$. ■

EXAMPLE 4 Solve for x to the nearest degree for $0° \le x < 360°$: $3 \cos x = 4 \sin x$.

Solution Trigonometric equations involving only one trigonometric function are generally easier to solve than equations involving more than one function. Consequently, one basic approach to a trigonometric equation that involves more than one function is to try to transform it into an equation involving only one function, if possible.

The easiest way to approach this equation is to divide both sides of the equation by $\sin x$ (or $\cos x$). However, in order to divide both sides of the equation by $\sin x$, we must be sure that $\sin x \neq 0$. If $\sin x = 0$ and x is to satisfy the equation, then $\cos x$ must also be equal to 0. But a brief glance at the graphs of $y = \sin x$ and $y = \cos x$ tells us that there is no value of x for which both the sine and cosine are equal to 0. Therefore, we may assume that $\sin x \neq 0$ and divide both sides of the equation by $\sin x$. The solution is as follows:

Considering the equation $3 \sin x = 4 \cos x$, why does $\sin x = 0$ imply that $\cos x = 0$ (and vice versa)?

$$3 \cos x = 4 \sin x \qquad \textit{Divide both sides by } \sin x.$$

$$3\left(\frac{\cos x}{\sin x}\right) = 4$$

$$3 \cot x = 4 \Rightarrow \cot x = \frac{4}{3} \text{ or equivalently } \tan x = \frac{3}{4}$$

Since this is not a familiar value, we use a calculator (set in degree mode) to find x. The keystroke sequence is

We have more to say about the [inv] key on the calculator when we discuss the inverse trigonometric functions in the next section.

[4] [÷] [3] [=] [1/x] [inv] [tan] = 36.869898 = 37° to the nearest degree

Since most calculators do not have a cotangent function key, we first use the reciprocal key on the value of the cotangent to get the tangent. Also note that using a calculator only gives the first-quadrant answer. A 37° reference angle in the third quadrant will also give the same result, and so x can also be 217°.

Thus the final answer is $\boxed{x = 37° \text{ or } x = 217°}$. ■

GRAFFIX

Use a graphics calculator or computer to approximate (to the nearest tenth) the solution(s) to the following equations:

1. $\sin x = \frac{1}{2}x$ by graphing $y = \sin x$ and $y = \frac{1}{2}x$ on the same set of axes and tracing the x-values of their points of intersection.
2. $2 \sin x = \cos x - 1$ by graphing $y = 2 \sin x - \cos x + 1$ and tracing the x-values to find the x-intercepts.
3. $\tan x = \cos x$ on the interval $\left[-\frac{\pi}{2}, \frac{\pi}{2}\right]$ by graphing $y = \tan x$ and $y = \cos x$ on the same set of axes and tracing the x values of their points of intersection.
4. $\tan x = \sin x - 1$ on the interval $\left[-\frac{\pi}{2}, \frac{\pi}{2}\right]$ by graphing $y = \tan x - \sin x + 1$ and tracing the x-values to find the x-intercepts.

Being able to solve trigonometric equations allows us to answer related questions.

EXAMPLE 5 Find the domain of $f(\theta) = \dfrac{5}{4 \sin^2 \theta - 3}$.

Solution We must exclude all values of θ for which $4 \sin^2 \theta - 3 = 0$. In other words, we seek the general solution to this equation.

$$4 \sin^2 \theta - 3 = 0 \Rightarrow \sin^2 \theta = \frac{3}{4} \qquad \textit{Take square roots.}$$

$$\sin \theta = \pm\sqrt{\frac{3}{4}} = \pm\frac{\sqrt{3}}{2}$$

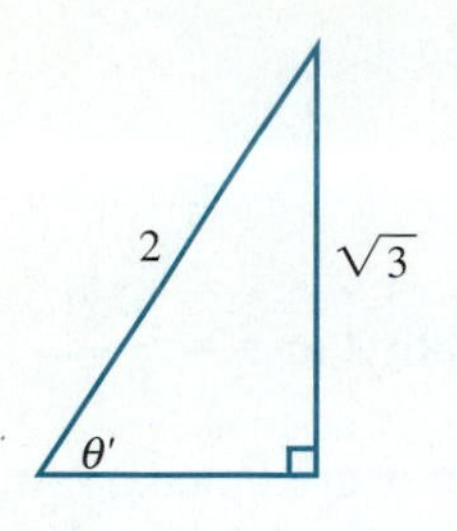

FIGURE 7.27 Reference triangle for reference angle whose sine is $\frac{\sqrt{3}}{2}$

As before, we temporarily ignore the sign of our answer, and we draw a reference triangle with reference angle θ' whose sine is $\frac{\sqrt{3}}{2}$. See Figure 7.27. We recognize that θ' is $\frac{\pi}{3}$. Since the $\sin\theta$ can be either positive or negative, the reference angle can be in any of the four quadrants. Thus we have $\theta = \frac{\pi}{3}, \frac{2\pi}{3}, \frac{4\pi}{3}, \frac{5\pi}{3}$. Since the first and third solutions differ by π, as do the second and fourth, we can write the domain as

$$\boxed{D_f = \left\{\theta \mid \theta \neq \frac{\pi}{3} + n\pi, \frac{2\pi}{3} + n\pi, \text{ where } n \text{ is an integer}\right\}}$$ ■

EXAMPLE 6 Find the domain of $g(\theta) = \sqrt{\csc\theta}$ for $-2\pi \le \theta \le 2\pi$.

Solution To begin with, in order for $g(\theta)$ to be defined, θ must be in the domain of the cosecant function. Thus θ cannot be an integral multiple of π. But in addition, in order for $\sqrt{\csc\theta}$ to be defined in the real number system, we must require that $\csc\theta \ge 0$. Looking at the graph of $y = \csc\theta$ on the given interval, we see that

$$\boxed{D_g = \{\theta \mid -2\pi < \theta < -\pi, 0 < \theta < \pi\}}$$ ■

EXAMPLE 7 Solve for x in the interval $[0, 2\pi)$: $\sin^2 x - \cos^2 x = 1$.

Solution Again we would prefer to work with an equation that involves only one of the trigonometric functions. We can do this by using one of our basic identities.

$$\begin{aligned}
\sin^2 x - \cos^2 x &= 1 && \textit{We substitute } \cos^2 x = 1 - \sin^2 x. \\
\sin^2 x - (1 - \sin^2 x) &= 1 \\
\sin^2 x - 1 + \sin^2 x &= 1 \\
2\sin^2 x - 1 &= 1 \\
2\sin^2 x &= 2 \\
\sin^2 x &= 1 \\
\sin x &= \pm 1 && \textit{From the sine graph we get}
\end{aligned}$$

$$\boxed{x = \frac{\pi}{2} \quad \text{or} \quad x = \frac{3\pi}{2}}$$ ■

EXAMPLE 8 Find the general solution to $2\cos 2\theta = -1$.

Solution The given equation $2\cos 2\theta = -1$ implies $\cos 2\theta = -\frac{1}{2}$. Thus 2θ must be a number whose cosine is $-\frac{1}{2}$. We recognize that the cosine function will be equal to $-\frac{1}{2}$ when the argument is $\frac{2\pi}{3}$ or $\frac{4\pi}{3}$, to which we know that we may add multiples of 2π. Hence the angle 2θ can be $\frac{2\pi}{3} + 2n\pi$ or $\frac{4\pi}{3} + 2n\pi$. Therefore, we have

$$2\theta = \frac{2\pi}{3} + 2n\pi \quad \text{or} \quad 2\theta = \frac{4\pi}{3} + 2n\pi$$

Dividing both sides of each equation by 2 gives

$$\boxed{\theta = \frac{\pi}{3} + n\pi} \qquad \boxed{\theta = \frac{2\pi}{3} + n\pi}$$

These are the general solutions. ■

We can obtain approximate solutions to trigonometric equations by using a calculator.

The equation $\sin^2 t - 4\sin t + 1 = 0$ is of the form $u^2 - 4u + 1 = 0$.

EXAMPLE 9 Solve for t to the nearest hundredth, where $0° \le t < 90°$: $\sin^2 t - 4\sin t + 1 = 0$.

Solution Since the left-hand side of this equation does not factor, we use the quadratic formula, with $A = 1$, $B = -4$, and $C = 1$.

$$\sin^2 t - 4\sin t + 1 = 0$$

$$\sin t = \frac{-(-4) \pm \sqrt{(-4)^2 - 4(1)(1)}}{2(1)} = \frac{4 \pm \sqrt{12}}{2}$$

Using a calculator we find

Why does $\sin t = 3.7320508$ have no solutions?

$$\sin t = 3.7320508 \quad \text{or} \quad \sin t = 0.2679492$$

The first equation has no solutions.

To find the value of t that satisfies $\sin t = 0.2679492$, we enter this value on the display and then use the following keystrokes (in degree mode):

inv **sin**

which gives **15.542268,** or $\boxed{15.54°}$ to the nearest hundredth. ■

In the next section our work with trigonometric equations will help us in analyzing the inverse trigonometric functions.

EXERCISES 7.4

In Exercises 1–44, find all solutions to the given equation in the interval $[0, 2\pi)$.

1. $\sin x = \frac{1}{2}$
2. $\cos x = -\frac{1}{2}$
3. $\cos x = -\frac{\sqrt{3}}{2}$
4. $\sin x = \frac{\sqrt{2}}{2}$
5. $\sin x = 1$
6. $\sec x = 1$
7. $\tan\theta + 1 = 0$
8. $1 - \cot\theta = 0$
9. $\csc x = 1$
10. $\csc x = -1$
11. $\tan\theta = -\sqrt{3}$
12. $\cot x = \sqrt{3}$
13. $3\cos x + 1 = 5$
14. $4\csc\theta + 3 = 6$
15. $\sqrt{3}\tan t = 1$
16. $4\sec t = 8$
17. $\cos x = 0$
18. $\cot x = 0$
19. $5\sec\theta = 2$
20. $\sec\theta = 0$
21. $\sin^2 x + 4 = 5$
22. $\tan^2 x = 3$
23. $4\cos^2 x - 3 = 0$
24. $5\sec^2 x - 4 = 6$
25. $6\cot^2 w + 1 = 3$
26. $3\csc^2 z - 4 = 0$
27. $\cos^2 x + 2 = 3\cos x$
28. $\sin^2 x - 2 + \sin x = 0$
29. $\tan^2\theta = \tan\theta$
30. $\sec^2\theta = 2\sec\theta$
31. $2\sin t = \sin t\cos t$
32. $\csc t = \csc t\sec t$
33. $\sin x = \cos x$
34. $\sin x = \tan x$
35. $\cos w - \sec w = 0$
36. $\tan u = \cot u$
37. $\sin x + 3 = -2\csc x$
38. $2\cos x - 3\sec x = 5$
39. $2\cos^2 x - \sin x - 1 = 0$
40. $2\sin^2\theta + 3\cos\theta = 0$
41. $3\tan^2 x - \sec^2 x = 5$

42. $\cot^2 t - \csc t = 1$

43. $\sin \theta + \cos \theta = 1$
HINT: Square both sides.

44. $\tan \theta + 1 = \sec \theta$
HINT: Square both sides.

In Exercises 45–50, solve the given equation to the nearest degree on $[0°, 360°)$.

45. $5 \sin x = 3$
46. $3 \cos x = -2$
47. $7 - 4 \tan x = 14$
48. $7 \sec x = 4$
49. $9 \csc x = -20$
50. $2 \cot x - 1 = 5$

In Exercises 51–54, use a calculator to solve the given equation on the indicated interval. Round your answer to the nearest hundredth.

51. $\cos^2 x - \cos x - 1 = 0$ on $[0, 2\pi)$
52. $3 \sin^2 \theta + 5 \sin \theta + 1 = 0$ on $[0°, 360°)$
53. $2 \tan^2 t - 3 \tan t - 4 = 0$ on $[0°, 360°)$
54. $\sec^2 \theta + 2 \sec \theta - 1 = 0$ on $[0, 2\pi)$

In Exercises 55–68, find the general solution to the given equation. Use radian measure.

55. $2 \sin x = 1$
56. $2 \cos x = -\sqrt{3}$
57. $\tan^2 x = 3$
58. $\sec^2 x = 4$
59. $3 \csc^2 x = 4$
60. $3 \cot^2 x = 1$
61. $5 \cos x = 0$
62. $-4 \sin x = 0$
63. $5 \sin x = -5$
64. $4 \cos x = -2$
65. $\sin 3x = 1$
66. $\tan 2x = \sqrt{3}$
67. $2 \cos 4x = -\sqrt{2}$
68. $4 \sin 2\theta + 5 = 1$

In Exercises 69–76, find the domain of the given function.

69. $f(\theta) = \dfrac{1}{\sin \theta - 1}$
70. $g(\theta) = -\dfrac{3}{2 \cos \theta + 1}$
71. $h(\theta) = 2 \sin \theta \cos \theta$
72. $F(\theta) = \dfrac{1}{\sec^2 \theta}$
73. $G(x) = \dfrac{6}{4 \cos^2 x - 3}$
74. $H(x) = \dfrac{7}{3 \tan^2 x - 1}$
75. $f(t) = \dfrac{5}{2 - \csc t}$
76. $g(u) = \dfrac{5}{2 - \sec^2 u}$

In Exercises 77–84, find the domain of the given function for θ restricted to $[0, 2\pi)$.

77. $f(\theta) = \sqrt{\sec \theta}$
78. $g(\theta) = \sqrt{\sin \theta}$
79. $g(\theta) = \dfrac{1}{\csc \theta}$
80. $f(\theta) = \dfrac{1}{\sqrt{\csc \theta}}$
81. $F(\theta) = \dfrac{1}{1 - \tan \theta}$
82. $G(\theta) = \dfrac{1}{2 - \sin \theta}$
83. $f(\theta) = \dfrac{1}{2 \cos^2 \theta - 1}$
84. $f(\theta) = \tan^2 \theta - 3$

QUESTION FOR THOUGHT

85. Discuss the differences between the equation $\sin \theta + \cos \theta = 1$ and the identity $\sin^2 \theta + \cos^2 \theta = 1$.

GRAFFIX

In Exercises 86–95, use a graphics calculator or computer to solve the given trigonometric equation. Estimate the solutions to the nearest tenth where necessary.

86. $3 \sin x = 1$
87. $5 \tan x + 4 = 3$
88. $2 \sec^2 x = 6$
89. $3 \cos^2 x - 1 = 0$
90. $\cot x = 4$
91. $4 \csc x - 1 = 6$
92. $\dfrac{1}{2} \sin x + \cos x = \dfrac{1}{2}$
93. $\sin x + \tan x = 1$
94. $\sin^2 x + \sin x = 1$
95. $\cos^2 x + 3 \cos x = 2$

7.5 The Inverse Trigonometric Functions

Whenever we encounter a new type of function, our study of the function has included the question of whether or not the new function has an inverse.

By virtue of the horizontal line test, we can look at the graphs of each of the six trigonometric functions, see that each one fails the horizontal line test, and therefore know that none of the trigonometric functions has an inverse function. For example, see Figure 7.28.

When we evaluate $y = f(x) = \sin x$ for $x = \dfrac{\pi}{6}$, we get $f\left(\dfrac{\pi}{6}\right) = \sin \dfrac{\pi}{6} = \dfrac{1}{2}$. However, when we try to reverse this process, as we did in Example 1 of the previous section, we obtain *two* solutions to the equation $\sin x = \dfrac{1}{2}$ in the interval $[0, 2\pi)$

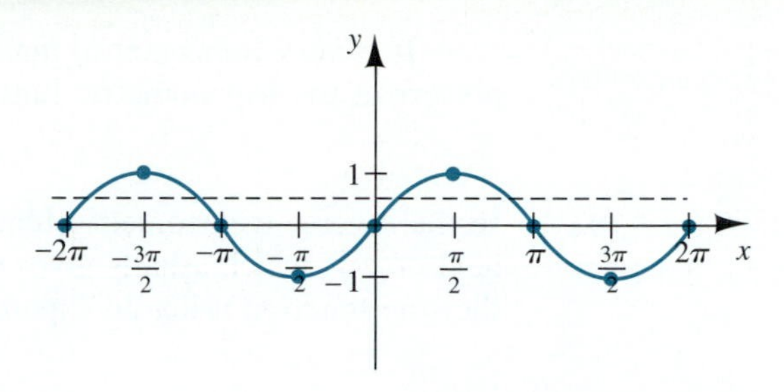

FIGURE 7.28 The graph of $y = \sin x$ fails the horizontal line test.

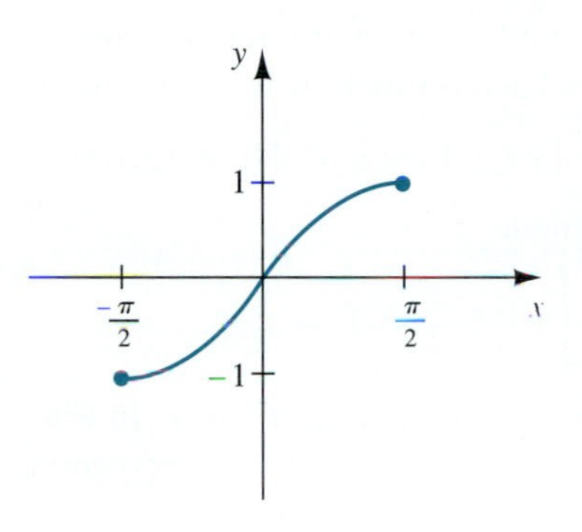

FIGURE 7.29 $y = \sin x$ restricted to the interval $\left[-\frac{\pi}{2}, \frac{\pi}{2}\right]$

alone. For $y = \sin x$ to have an inverse function, we need to be able to pick y and get x uniquely, which we have just seen we cannot do. In other words, when we interchange x and y in $y = \sin x$ to get $x = \sin y$ we do not get y as a function of x *unless we somehow restrict the set of possible y-values* so that the function becomes one-to-one.

We would encounter similar problems in attempting to find the inverses of any of the trigonometric functions. Let's focus our attention on the sine function and consider Figure 7.29.

Note that on this restricted domain, the sine function takes on *all* its range values. In other words, we have not lost any of the range values but we do now have a one-to-one function.

With this restricted domain, we know that the sine function is one-to-one and therefore has an inverse. Following our usual routine for finding the inverse of a function, we start with

$$y = \sin x \qquad \textit{We interchange } x \textit{ and } y.$$

$$x = \sin y \qquad \textit{Now we want to solve for } y.$$

Unfortunately, we have no algebraic method by which to solve for y. (Recall that we encountered exactly the same situation when we discussed the inverse of the exponential function in Section 5.2.)

Consequently, we will simply invent a name for the inverse function. Thus we will write

$$x = \sin y \iff y = \sin^{-1} x$$

In words, $y = \sin^{-1} x$ says that "y is the number (angle) in the interval $\left[-\frac{\pi}{2}, \frac{\pi}{2}\right]$ whose sine is x."

We should also mention that there is another commonly used notation for the inverse trigonometric functions.

$$y = \sin^{-1} x \qquad \text{is equivalent to} \qquad y = \arcsin x$$

For example, $\sin^{-1}(-1) = -\frac{\pi}{2}$ because $-\frac{\pi}{2}$ is the number in the interval $\left[-\frac{\pi}{2}, \frac{\pi}{2}\right]$ whose sine is -1. Similarly, $\arcsin \frac{1}{2} = 30°$, because $30°$ is the angle in the interval $[-90°, 90°]$ whose sine is $\frac{1}{2}$.

It is very important to make the following distinction in the way we represent powers of the trigonometric functions. Although $\sin^2 x = (\sin x)^2$,

$$\sin^{-1} x \neq (\sin x)^{-1}$$

In the inverse trigonometric function notation, the -1 *does not* represent an exponent of -1, which usually gives us a reciprocal. If we want to write the reciprocal of the sine function using an exponent of -1, we must write it as

$$(\sin x)^{-1} = \frac{1}{\sin x}$$

As we have seen, interchanging the roles of x and y causes the interchange of the domain and range as well. If we consider the restricted function $y = \sin x$ with domain $D = \left\{x \mid -\frac{\pi}{2} \leq x \leq \frac{\pi}{2}\right\}$ and range $R = \{y \mid -1 \leq y \leq 1\}$, it therefore follows that its inverse function $y = \sin^{-1} x$ has domain

$$D = \{x \mid -1 \leq x \leq 1\} \text{ and range } R = \left\{y \mid -\frac{\pi}{2} \leq y \leq \frac{\pi}{2}\right\}.$$

We can now approach the cosine function from the same point of view. In Figure 7.30 we see that the graph of the cosine function does not satisfy the horizontal line test.

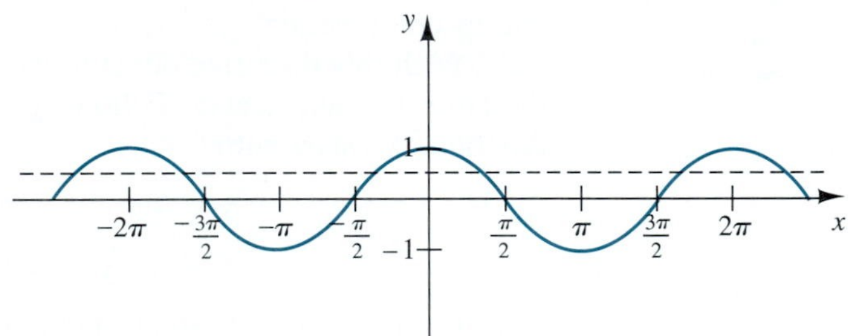

FIGURE 7.30 The graph of $y = \cos x$ fails the horizontal line test.

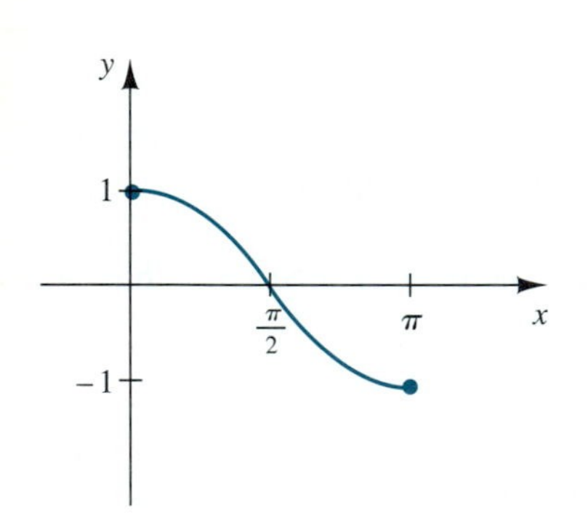

FIGURE 7.31 $y = \cos x$ restricted to the interval $[0, \pi]$

Why don't we restrict the cosine graph to the interval $\left[-\frac{\pi}{2}, \frac{\pi}{2}\right]$ as we did the sine graph?

In Figure 7.31 we have restricted the domain of the cosine function in such a way that the cosine function takes on all its range values yet satisfies the horizontal line test on this restricted domain and so has an inverse.

Using the same notation we introduced for the inverse sine function, we write

$$x = \cos y \iff y = \cos^{-1} x$$

In words, $y = \cos^{-1} x$ says that "y is the number (angle) in the interval $[0, \pi]$ whose cosine is x."

Also

$$y = \cos^{-1} x \qquad \text{is equivalent to} \qquad y = \arccos x$$

Similarly, the prefix "arc" is used to indicate the inverse of the other trigonometric functions.

Although we will use both notations so that you will be familiar with both, we will primarily use the $\sin^{-1}$ and $\cos^{-1}$ notation.

We can now formally define the inverse sine and inverse cosine functions.

DEFINITIONS OF THE INVERSE SINE AND INVERSE COSINE FUNCTIONS

$$y = \sin^{-1} x \iff x = \sin y \quad \text{Domain} = \{x \mid -1 \le x \le 1\}$$

$$\text{Range} = \left\{y \mid -\frac{\pi}{2} \le y \le \frac{\pi}{2}\right\}$$

$$y = \cos^{-1} x \iff x = \cos y \quad \text{Domain} = \{x \mid -1 \le x \le 1\}$$

$$\text{Range} = \{y \mid 0 \le y \le \pi\}$$

In words, $y = \sin^{-1} x$ means "y is the number whose sine is x." A similar statement holds for $y = \cos^{-1} x$.

EXAMPLE 1 Evaluate **(a)** $\sin^{-1}\left(\frac{\sqrt{2}}{2}\right)$ **(b)** $\cos^{-1}\left(-\frac{1}{2}\right)$.

Solution

(a) If we let $y = \sin^{-1}\left(\frac{\sqrt{2}}{2}\right)$, then it follows that $\sin y = \frac{\sqrt{2}}{2}$. In other words, y is the number (angle) in the interval $\left[-\frac{\pi}{2}, \frac{\pi}{2}\right]$ whose sine is $\frac{\sqrt{2}}{2}$. If we do not recognize the answer, we may draw a reference triangle with reference angle y' whose sine is $\frac{\sqrt{2}}{2}$. See Figure 7.32. We should then recognize that the reference angle is $\frac{\pi}{4}$. Thus $\boxed{\sin^{-1}\left(\frac{\sqrt{2}}{2}\right) = \frac{\pi}{4}}$. Note that there is only one solution in the interval $\left[-\frac{\pi}{2}, \frac{\pi}{2}\right]$.

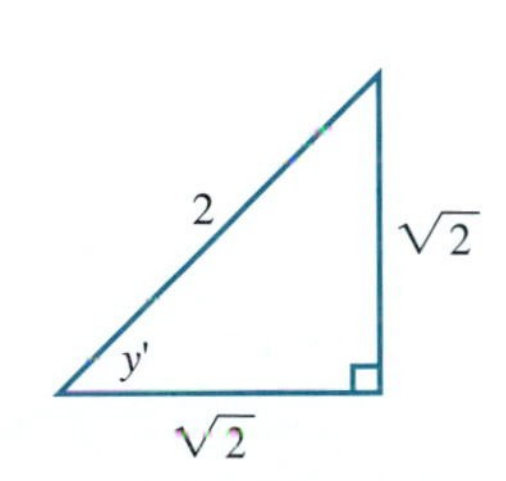

FIGURE 7.32 Reference triangle for an angle whose sine is $\frac{\sqrt{2}}{2}$

(b) If we let $y = \cos^{-1}\left(-\frac{1}{2}\right)$, then it follows that $\cos y = -\frac{1}{2}$. In other words, y is the number (angle) in the interval $[0, \pi]$ whose cosine is $-\frac{1}{2}$. Again, if we don't recognize the answer, we may draw a reference triangle with reference angle y' whose cosine is $\frac{1}{2}$. Looking at Figure 7.33, we should recognize that the reference angle is $\frac{\pi}{3}$. In order to have $\cos y = -\frac{1}{2}$ *and* have y be in the range of the inverse function, it must correspond to a second quadrant angle.

Therefore, $\cos^{-1}\left(-\frac{1}{2}\right) = \boxed{\frac{2\pi}{3}}$. ■

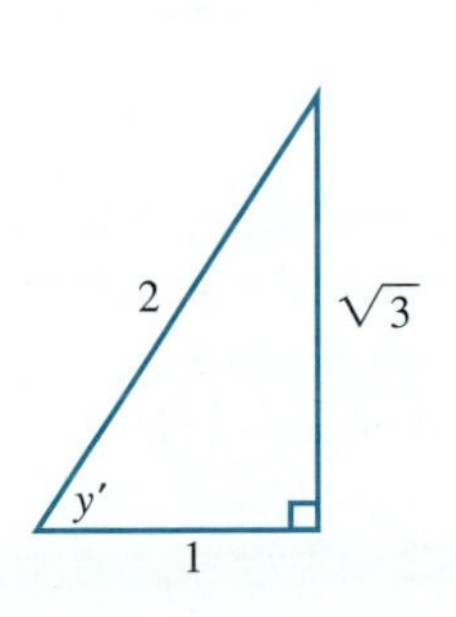

FIGURE 7.33 Reference triangle for an angle whose cosine is $-\frac{1}{2}$

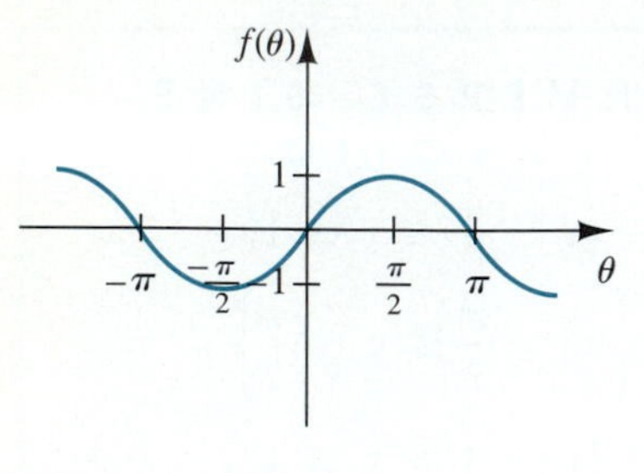

FIGURE 7.34

EXAMPLE 2 Evaluate $\arcsin(-1)$.

Solution Recall that $\arcsin(-1)$ means the same thing as $\sin^{-1}(-1)$; that is, the number (angle) in the interval $\left[-\frac{\pi}{2}, \frac{\pi}{2}\right]$ whose sine is -1. From the graph of the sine function (see Figure 7.34), we can see that the answer is $-\frac{\pi}{2}$.

Thus $\arcsin(-1) = \boxed{-\frac{\pi}{2}}$. ■

We now turn our attention to the other four trigonometric functions. The usual convention is to limit their domains as shown in Figure 7.35.

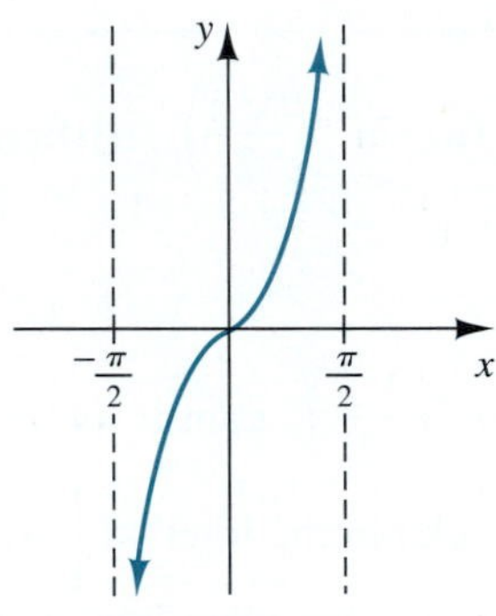

(a) $y = \tan x$
Domain $= \{x \mid -\frac{\pi}{2} < x < \frac{\pi}{2}\}$

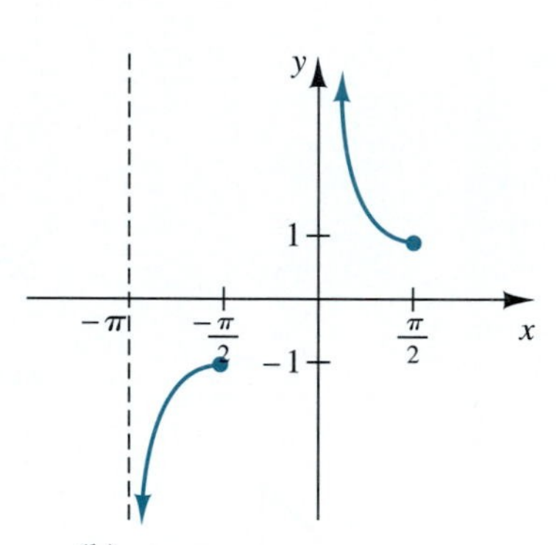

(b) $y = \csc x$
Domain $= \{x \mid -\pi < x \leq -\frac{\pi}{2}$ or $0 < x \leq \frac{\pi}{2}\}$

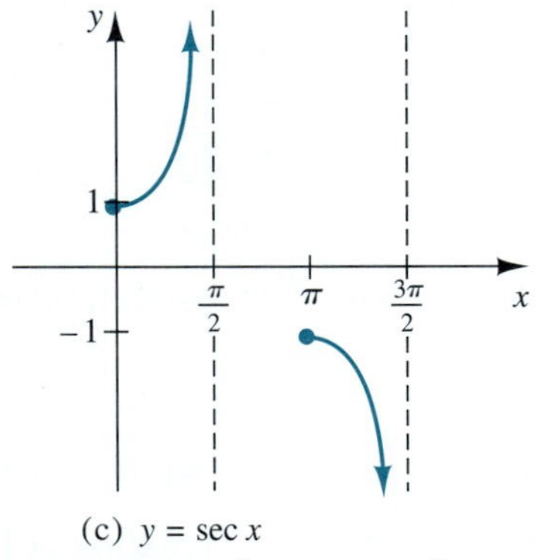

(c) $y = \sec x$
Domain $= \{x \mid 0 \leq x < \frac{\pi}{2}$ or $\pi \leq x < \frac{3\pi}{2}\}$

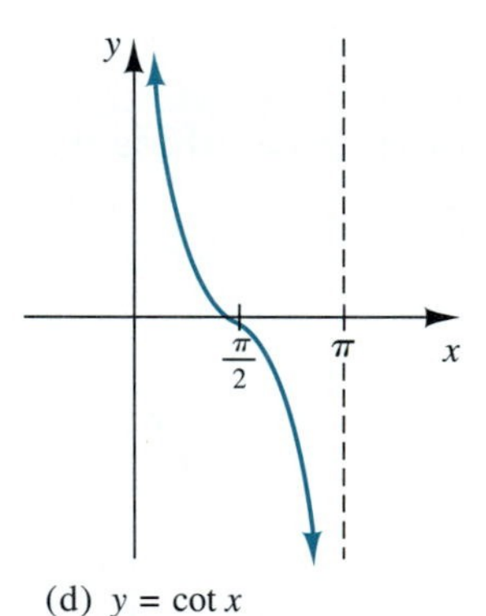

(d) $y = \cot x$
Domain $= \{x \mid 0 < x < \pi\}$

FIGURE 7.35

Based upon this discussion and Figures 7.29, 7.31, and 7.35, we can define the other four inverse trigonometric functions. For ease of reference, the following box contains the definitions of all six inverse trigonometric functions.

DEFINITIONS OF THE INVERSE TRIGONOMETRIC FUNCTIONS

$y = \sin^{-1} x \iff x = \sin y$ Domain $= \{x \mid -1 \le x \le 1\}$

Range $= \left\{y \mid -\frac{\pi}{2} \le y \le \frac{\pi}{2}\right\}$

$y = \cos^{-1} x \iff x = \cos y$ Domain $= \{x \mid -1 \le x \le 1\}$

Range $= \{y \mid 0 \le y \le \pi\}$

$y = \tan^{-1} x \iff x = \tan y$ Domain $=$ all real numbers

Range $= \left\{y \mid -\frac{\pi}{2} < y < \frac{\pi}{2}\right\}$

$y = \csc^{-1} x \iff x = \csc y$ Domain $= \{x \mid x \le -1 \text{ or } x \ge 1\}$

Range $= \left\{y \mid 0 < y \le \frac{\pi}{2} \text{ or } -\pi < y \le -\frac{\pi}{2}\right\}$

$y = \sec^{-1} x \iff x = \sec y$ Domain $= \{x \mid x \le -1 \text{ or } x \ge 1\}$

Range $= \left\{y \mid 0 \le y < \frac{\pi}{2} \text{ or } \pi \le y < \frac{3\pi}{2}\right\}$

$y = \cot^{-1} x \iff x = \cot y$ Domain $=$ all real numbers

Range $= \{y \mid 0 < y < \pi\}$

GRAFFIX

Using the definitions of the inverse trigonometric functions just given in the box and a graphics calculator or computer, sketch the graphs of the six inverse trigonometric functions.

EXAMPLE 3 Evaluate each of the following.

(a) $\sec^{-1} 2$ **(b)** $\arctan 1$ **(c)** $\csc^{-1}(-\sqrt{2})$ **(d)** $\cot^{-1} \sqrt{3}$

Solution

(a) According to the definition, $\sec^{-1} 2$ is the number in $\left[0, \frac{\pi}{2}\right) \cup \left[\pi, \frac{3\pi}{2}\right)$ whose secant is equal to 2 $\left(\text{or, equivalently, whose cosine is } \frac{1}{2}\right)$. Using a reference triangle, we find that the answer is $\boxed{\frac{\pi}{3}}$.

(b) arctan 1 is the number in the interval $\left(-\frac{\pi}{2}, \frac{\pi}{2}\right)$ whose tangent is 1. The answer is $\boxed{\frac{\pi}{4}}$.

(c) $\csc^{-1}(-\sqrt{2})$ is the number in $\left(-\pi, -\frac{\pi}{2}\right] \cup \left(0, \frac{\pi}{2}\right]$ whose cosecant is $-\sqrt{2}$ $\left(\text{or, equivalently, whose sine is } -\frac{1}{\sqrt{2}}\right)$. The answer is $\boxed{-\frac{3\pi}{4}}$.

(d) $\cot^{-1}\sqrt{3}$ is the number in the interval $(0, \pi)$ whose cotangent is $\sqrt{3}$ $\left(\text{or, equivalently, whose tangent is } \frac{1}{\sqrt{3}}\right)$. The answer is $\boxed{\frac{\pi}{6}}$. ■

EXAMPLE 4 Evaluate each of the following.

(a) $\sin^{-1}\left(\sin\frac{\pi}{6}\right)$ **(b)** $\cos(\tan^{-1} 0)$ **(c)** $\tan^{-1}\left(\sec\frac{\pi}{7}\right)$ **(d)** $\cos\left(\sin^{-1}\frac{2}{3}\right)$

Solution

(a) We begin by evaluating $\sin\frac{\pi}{6} = \frac{1}{2}$. Therefore, $\sin^{-1}\left(\sin\frac{\pi}{6}\right) = \sin^{-1}\left(\frac{1}{2}\right)$, which is the number in the interval $\left[-\frac{\pi}{2}, \frac{\pi}{2}\right]$ whose sine is $\frac{1}{2}$. The answer is $\boxed{\frac{\pi}{6}}$.

Alternatively, we can recognize that since $\frac{\pi}{6}$ is in the restricted domain of the sine function, it follows immediately that $\sin^{-1}\left(\sin\frac{\pi}{6}\right) = \frac{\pi}{6}$. This is just the fundamental relationship of inverse functions, i.e., $(f^{-1} \circ f)(x) = x$.

(b) We begin by finding that $\tan^{-1} 0 = 0$. Then $\cos(\tan^{-1} 0) = \cos(0) = \boxed{1}$.

(c) Since $\frac{\pi}{7}$ is not one of the basic angles with which we are familiar, we need a calculator (or a table) to evaluate the expression. Thus, to find the $\tan^{-1}\left(\sec\frac{\pi}{7}\right)$, the keystroke sequence is

What is the calculator doing in this keystroke sequence?

$$\boxed{\pi}\ \boxed{\div}\ \boxed{7}\ \boxed{=}\ \boxed{\textbf{cos}}\ \boxed{1/x}\ \boxed{\textbf{inv}}\ \boxed{\textbf{tan}}$$

and the final answer is $\boxed{0.8374}$ rounded to four decimal places.

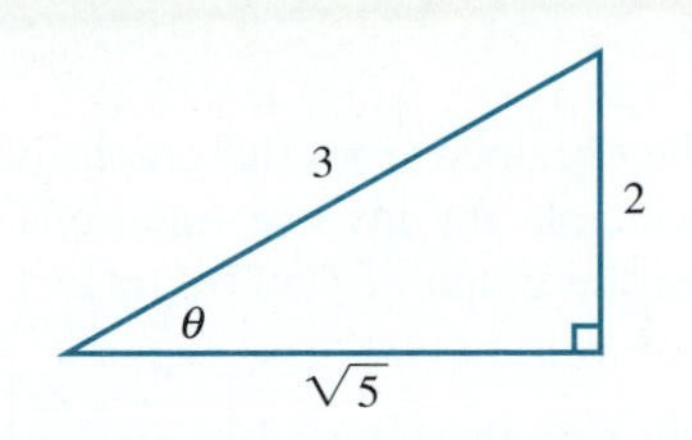

FIGURE 7.36 The reference triangle for $\sin\theta = \frac{2}{3}$

(d) We could begin by using a calculator to evaluate $\sin^{-1}\frac{2}{3}$, as we did in part (c). However, understanding the meaning of the inverse functions allows us to reason as follows: If $\theta = \sin^{-1}\frac{2}{3}$, this means that $\sin\theta = \frac{2}{3}$. Let's draw a reference triangle for θ and complete the triangle using the Pythagorean theorem. See Figure 7.36. Looking at the reference triangle we see that

$$\cos\left(\sin^{-1}\frac{2}{3}\right) = \cos\theta = \boxed{\frac{\sqrt{5}}{3}}$$

We note that since the sine is positive, the definition of the inverse sine function requires that θ be in quadrant I, and so the cosine will also be positive. ■

EXAMPLE 5 Evaluate each of the following. **(a)** $\sin^{-1}\left(\sin\frac{5\pi}{4}\right)$ **(b)** $\cos^{-1}\left(\cos\frac{5\pi}{7}\right)$

Solution

(a) We may begin by evaluating $\sin\frac{5\pi}{4} = -\frac{\sqrt{2}}{2}$. Therefore, we have $\sin^{-1}\left(\sin\frac{5\pi}{4}\right) = \sin^{-1}\left(-\frac{\sqrt{2}}{2}\right)$, which is, by definition, the number in the interval $\left[-\frac{\pi}{2}, \frac{\pi}{2}\right]$ whose sine is $-\frac{\sqrt{2}}{2}$. Thus the final answer is $\sin^{-1}\left(\sin\frac{5\pi}{4}\right) = \sin^{-1}\left(-\frac{\sqrt{2}}{2}\right) = \boxed{-\frac{\pi}{4}}$. Note that because $\frac{5\pi}{4}$ is *not* in the restricted domain of the sine function, we cannot use the inverse function relationship. In other words, $\sin^{-1}\left(\sin\frac{5\pi}{4}\right) \neq \frac{5\pi}{4}$.

Using a calculator, do you find that $\sin^{-1}[\sin(150°)] = 150°$? Should it be?

(b) Since $\frac{5\pi}{7}$ is in the interval $[0, \pi]$, the restricted domain of the cosine function, we may use the inverse function relationship to conclude that $\cos^{-1}\left(\cos\frac{5\pi}{7}\right) = \boxed{\frac{5\pi}{7}}$. Notice that by using the inverse function relationship, we do not need to find the value of $\cos\frac{5\pi}{7}$. ■

In our development of the inverse trigonometric functions, the choice of the restricted domain for each function was somewhat arbitrary. In order to get a complete set of range values for each of the trigonometric functions and still have an inverse, we could have chosen to restrict the domains in other ways. For example, for the tangent function we could have chosen the interval $\left(\frac{\pi}{2}, \frac{3\pi}{2}\right)$ as our domain. See Figure 7.37.

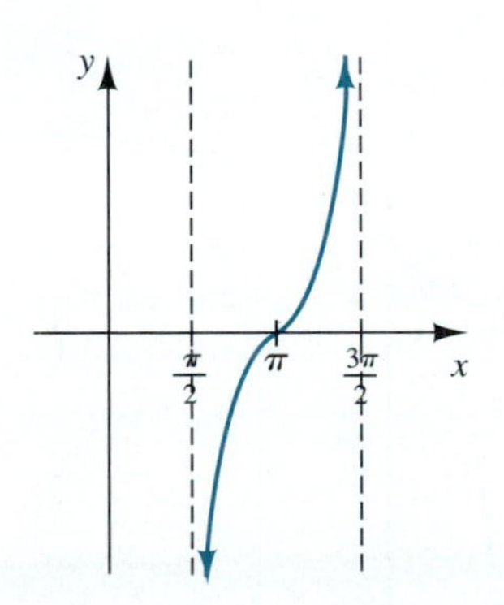

FIGURE 7.37 An alternative restriction on the domain of $y = \tan x$

The restricted domains we have been using are the commonly accepted ones.

The Graphs of the Inverse Trigonometric Functions

The graphs of the inverse trigonometric functions can be obtained from the graphs of the trigonometric functions by using the reflection principle for inverse functions. We know that the graph of $f^{-1}(x)$ can be obtained from the graph of $f(x)$ by reflecting it about the line $y = x$.

In Figure 7.38 we have drawn the graph of $y = \sin x$ on the interval $\left[-\frac{\pi}{2}, \frac{\pi}{2}\right]$ and reflected it about the line $y = x$ to obtain the graph of $y = \sin^{-1} x$.

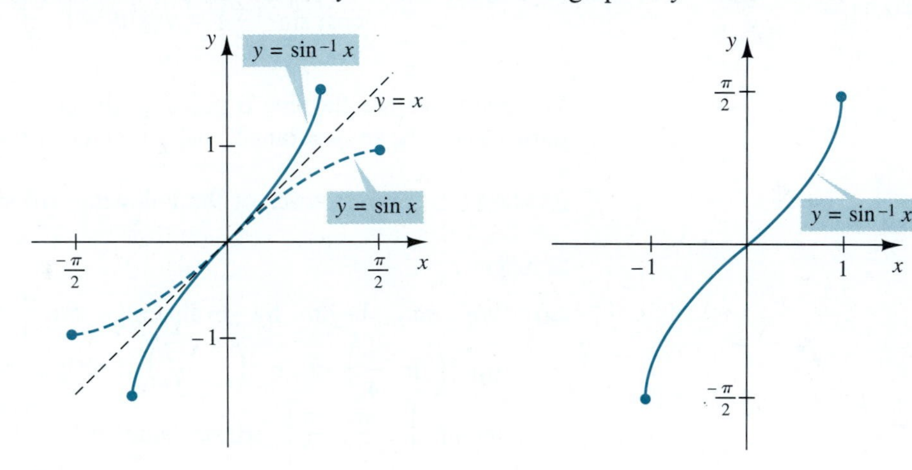

FIGURE 7.38

Figure 7.39(a) and (b) shows the graphs of $y = \cos^{-1} x$ and $y = \tan^{-1} x$ obtained by the reflection principle for inverse functions.

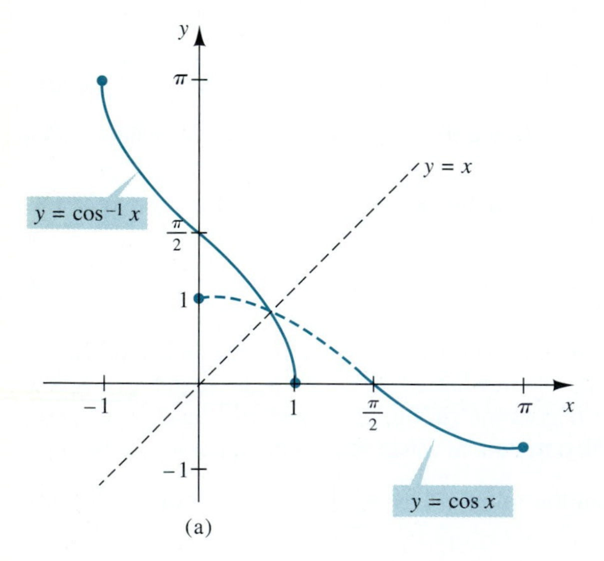

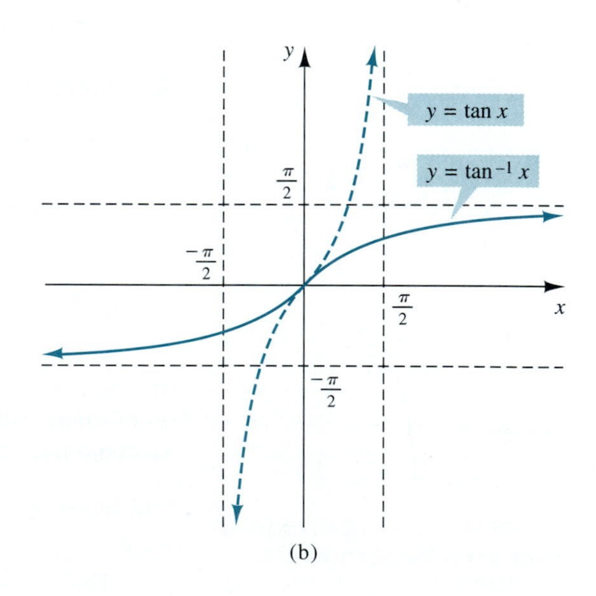

FIGURE 7.39

We leave it as an exercise to use the reflection principle to obtain the graphs of $y = \csc^{-1} x$, $y = \sec^{-1} x$, and $y = \cot^{-1} x$.

The following box summarizes the graphs of the inverse trigonometric functions.

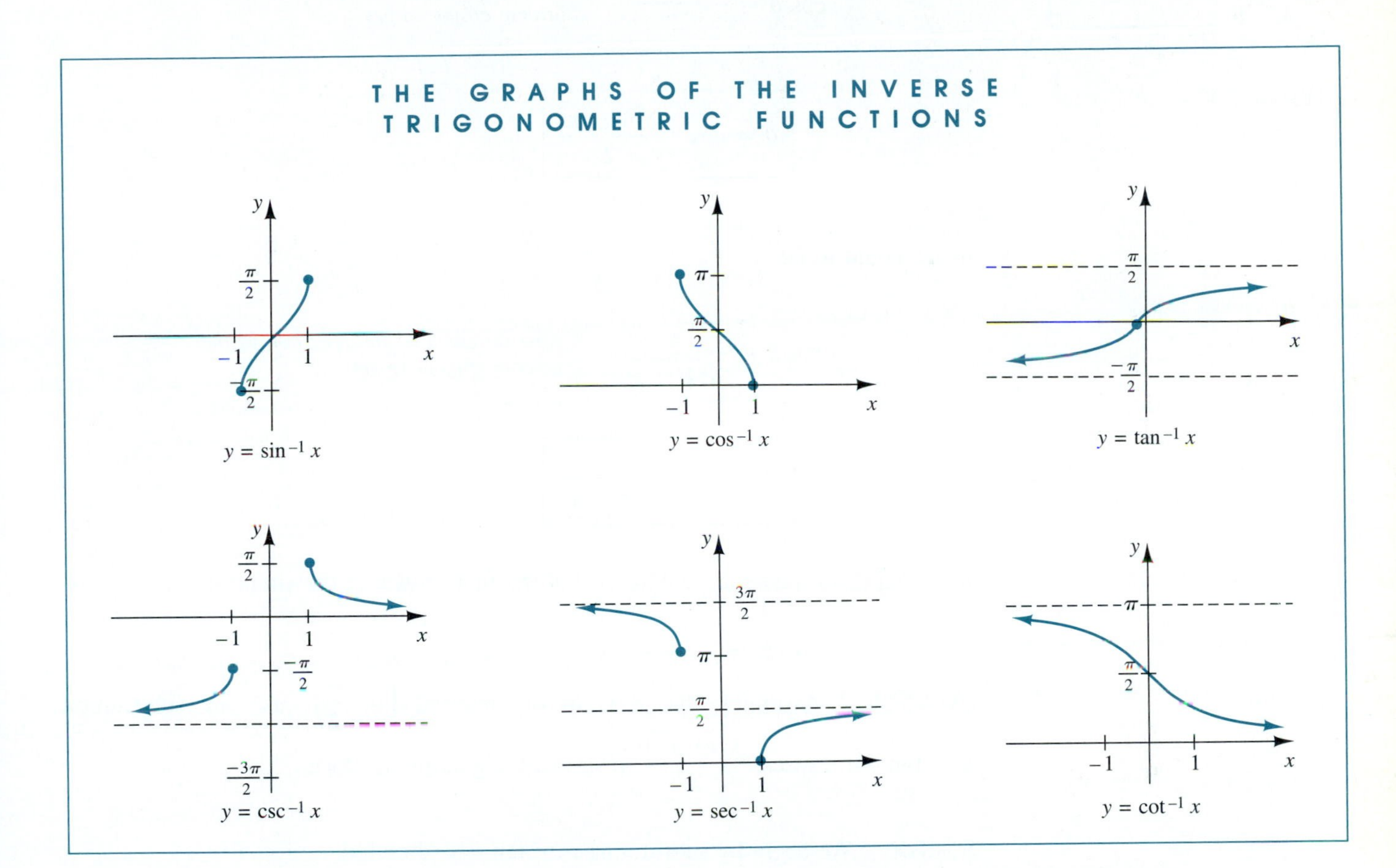

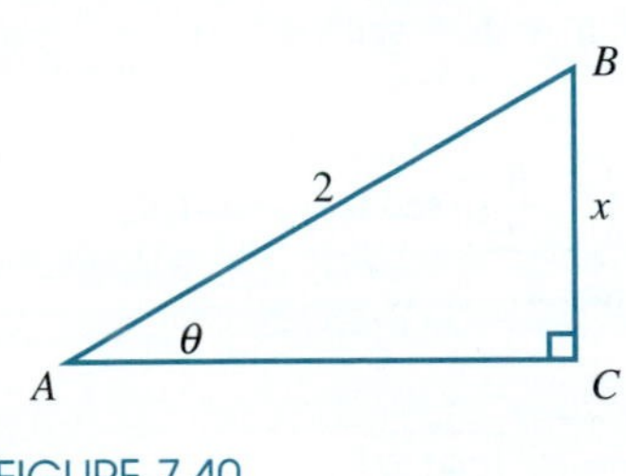

FIGURE 7.40

One of the most important uses of the inverse trigonometric functions is in allowing us to express certain algebraic expressions in trigonometric form, and vice versa, as the next two examples illustrate.

EXAMPLE 6 Use the diagram in Figure 7.40 to express θ as a function of x.

Solution From the diagram we see that

$$\sin \theta = \frac{x}{2}$$

Since θ is an acute angle, we can use the definition of the inverse sine function to get

$$\boxed{\theta = \sin^{-1} \frac{x}{2}}$$

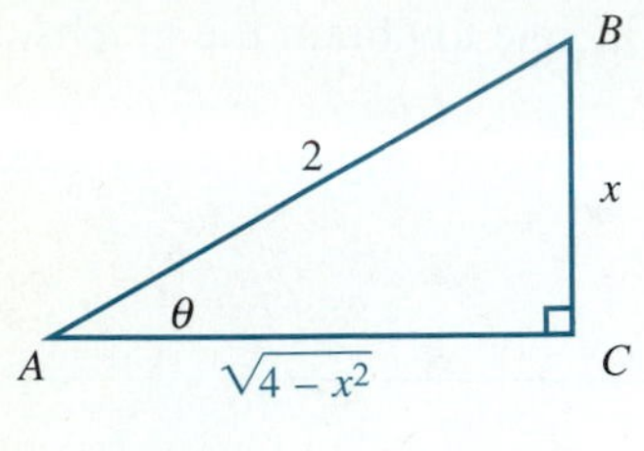

FIGURE 7.41

Note that we could have used the Pythagorean theoream to find $|\overline{AC}|$. The triangle would then look like Figure 7.41. Thus we have

$$\cos\theta = \frac{\sqrt{4-x^2}}{2}$$ *Now we can use the definition of the inverse cosine to get*

$$\boxed{\theta = \cos^{-1}\frac{\sqrt{4-x^2}}{2}}$$

or we could write

$$\tan\theta = \frac{x}{\sqrt{4-x^2}}$$ *Now we can use the definition of the inverse tangent to get*

$$\boxed{\theta = \tan^{-1}\frac{x}{\sqrt{4-x^2}}}$$

All three answers are correct, but the first answer is the simplest. ■

EXAMPLE 7 Assuming that θ is in the interval $\left(0, \frac{\pi}{2}\right)$, use the substitution $u = \tan\theta$ to express $\frac{\sqrt{u^2+1}}{u}$ in simplest trigonometric form.

Solution We begin by substituting $u = \tan\theta$ as directed.

$$\frac{\sqrt{u^2+1}}{u} = \frac{\sqrt{\tan^2\theta+1}}{\tan\theta}$$ *We use the identity* $\tan^2\theta + 1 = \sec^2\theta$

$$= \frac{\sqrt{\sec^2\theta}}{\tan\theta}$$ *Since* θ *is restricted to* $\left(0, \frac{\pi}{2}\right)$, $\sec\theta$ *is positive. Therefore,* $\sqrt{\sec^2\theta} = \sec\theta$.

$$= \frac{\sec\theta}{\tan\theta} = \frac{\frac{1}{\cos\theta}}{\frac{\sin\theta}{\cos\theta}} = \frac{1}{\cos\theta}\cdot\frac{\cos\theta}{\sin\theta} = \frac{1}{\sin\theta} = \boxed{\csc\theta}$$ ■

The ability to transform algebraic expressions into trigonometric ones, and vice versa, is very useful in order to carry out certain procedures in calculus.

EXERCISES 7.5

In Exercises 1–38, evaluate the given expression. Give exact values whenever possible; use a calculator only when necessary. When using a calculator, round your answer to four decimal places.

1. $\sin^{-1}\dfrac{\sqrt{3}}{2}$

2. $\cos^{-1}\left(-\dfrac{\sqrt{2}}{2}\right)$

3. $\arccos\left(-\dfrac{1}{2}\right)$

4. $\arccos\dfrac{1}{2}$

5. $\tan^{-1}(-\sqrt{3})$

6. $\cot^{-1} 1$

7. $\sec^{-1} 2$

8. $\csc^{-1}(-\sqrt{2})$

9. $\sin^{-1}\dfrac{2}{5}$

10. $\tan^{-1}\dfrac{1}{2}$

11. $\csc^{-1}(-1)$

12. $\sec^{-1} 1$

13. $\cot^{-1} 0$

14. $\tan^{-1}(-1)$

15. $\arcsin 1$

16. $\arccos(-1)$

17. $\sin\left(\cos^{-1}\dfrac{\sqrt{3}}{2}\right)$

18. $\cos\left(\sin^{-1}\left(-\dfrac{1}{\sqrt{2}}\right)\right)$

19. $\tan\left(\sin^{-1}\left(-\dfrac{1}{2}\right)\right)$

20. $\cot\left(\cos^{-1}\dfrac{1}{2}\right)$

21. $\cos(\cos^{-1} 1)$

22. $\sin\left(\arcsin\left(-\dfrac{\sqrt{3}}{2}\right)\right)$

23. $\cos\left(\tan^{-1}\dfrac{1}{2}\right)$

24. $\csc\left(\cos^{-1}\left(-\dfrac{3}{4}\right)\right)$

25. $\tan\left(\sin^{-1}\dfrac{2}{5}\right)$

26. $\sin\left(\cos^{-1}\dfrac{1}{4}\right)$

27. $\sec(\csc^{-1} 1)$

28. $\tan(\cos^{-1}(-1))$

29. $\sin^{-1}\left(\sin\dfrac{\pi}{3}\right)$

30. $\cos^{-1}\left(\cos\dfrac{7\pi}{4}\right)$

31. $\cos^{-1}\left(\cos\dfrac{5\pi}{3}\right)$

32. $\sin^{-1}\left(\sin\dfrac{7\pi}{4}\right)$

33. $\cos^{-1}\left(\cos\left(-\dfrac{\pi}{3}\right)\right)$

34. $\csc^{-1}\left(\csc\dfrac{\pi}{6}\right)$

35. $\tan(\csc^{-1} 1)$

36. $\sec(\sin^{-1} 1)$

37. $\sin^{-1}(\cos 82°)$

38. $\cot^{-1}(\tan 23°)$

39. Use the accompanying diagram to express θ as a function of x.

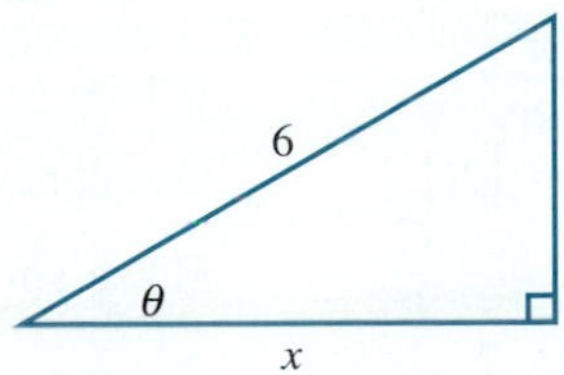

40. Use the figure accompanying Exercise 39 to express x as a function of θ.

41. Use the accompanying diagram to express θ as a function of x.

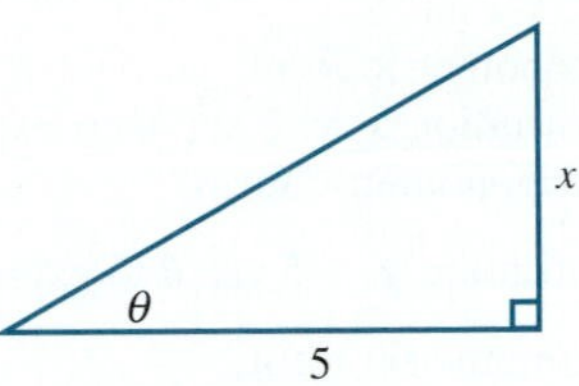

42. Use the figure accompanying Exercise 41 to express x as a function of θ.

43. Use the accompanying diagram to express $\sqrt{16 - x^2}$ as a function of θ.

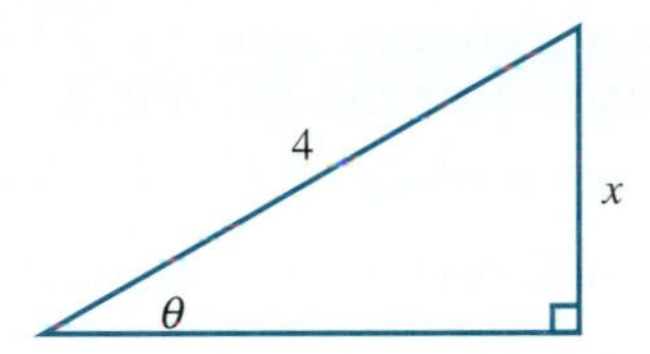

44. Use the accompanying diagram to express $\sqrt{x^2 - 16}$ as a function of θ.

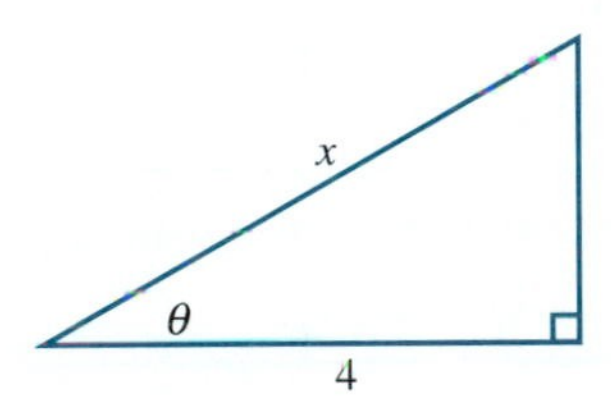

45. Use the accompanying diagram to express $\sqrt{x^2 + 16}$ as a function of θ.

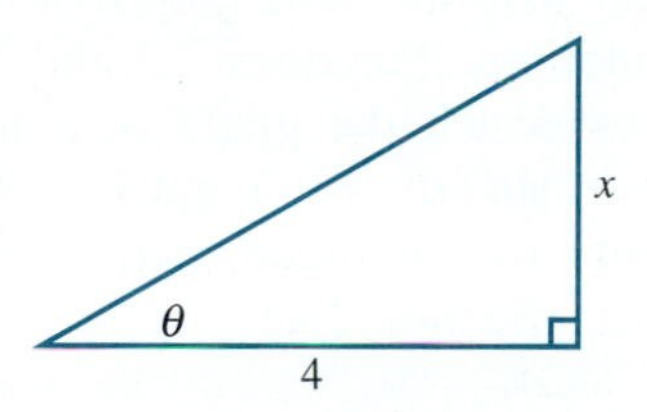

46. Use the accompanying diagram to express $\sqrt{36 + x^2}$ as a function of θ.

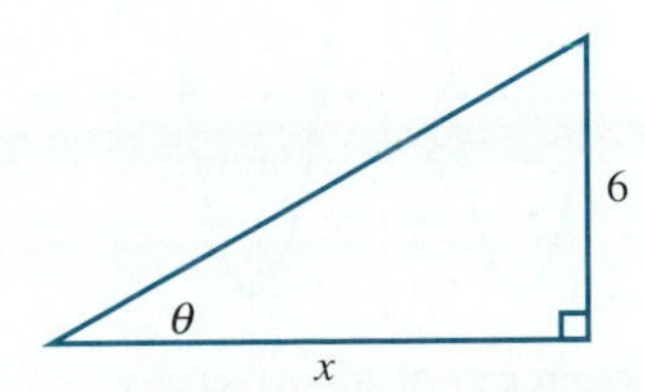

47. Use the substitution $u = 3 \sin \theta$ to express $u\sqrt{9 - u^2}$ in simplest trigonometric form.

48. Use the substitution $u = 8 \cos \theta$ to express $\dfrac{\sqrt{64 - u^2}}{u}$ in simplest trigonometric form.

49. Use the substitution $x = 2 \sec \theta$ to express $x^2\sqrt{x^2 - 4}$ in simplest trigonometric form.

50. Use the substitution $y = 5 \csc \theta$ to express $\dfrac{y}{\sqrt{y^2 - 25}}$ in simplest trigonometric form.

51. Use the substitution $u = 4 \tan \theta$ to express $\dfrac{u^2}{16 + u^2}$ in simplest trigonometric form.

52. Use the substitution $x = 9 \tan \theta$ to express $\dfrac{1}{x^2\sqrt{x^2 + 81}}$ in simplest trigonometric form.

53. Draw a suitable right triangle, with θ as one of its acute angles, so that $\theta = \sin^{-1} \dfrac{x}{5}$.

54. Draw a suitable right triangle, with θ as one of its acute angles, so that $\theta = \tan^{-1} \dfrac{x}{4}$.

55. Draw a suitable right triangle, with θ as one of its acute angles, so that $\theta = \sec^{-1} \dfrac{x}{a}$.

56. Draw a suitable right triangle, with θ as one of its acute angles, so that $\theta = \cos^{-1} \dfrac{x}{b}$.

57. Draw a suitable right triangle, with θ as one of its acute angles, so that two of the sides are x and $\sqrt{x^2 - 9}$; then express the ratio $\dfrac{\sqrt{x^2 - 9}}{x}$ as a function of θ.

QUESTIONS FOR THOUGHT

58. **(a)** Sketch the graph of $y = \sin 2x$.
(b) Find a suitable restricted domain so that $y = \sin 2x$ has an inverse function.
(c) Find the inverse of $y = \sin 2x$ on this restricted domain and sketch its graph.

59. Can you generalize the results of Exercise 58 to $y = \sin 3x$, $y = \sin 4x$, etc.—that is, to $y = \sin kx$?

60. In calculus it is shown that for "small" values of x, the polynomial $p(x) = x - \dfrac{x^3}{3} + \dfrac{x^5}{5}$ approximates $\tan^{-1} x$ with a maximum error of $\left|\dfrac{x^7}{7}\right|$. Use $p(x)$ to approximate $\tan^{-1} 0.5$ and determine the maximum error in this approximation.

Chapter 7 SUMMARY

After completing this chapter you should:

1. Be familiar with the basic graphs of the six trigonometric functions. (Sections 7.1 and 7.2)

2. Be able to sketch the graph of a function of the form $y = A \sin (Bx + C)$ and $y = A \cos(Bx + C)$ and identify its amplitude, period, frequency, and phase shift. (Section 7.1)

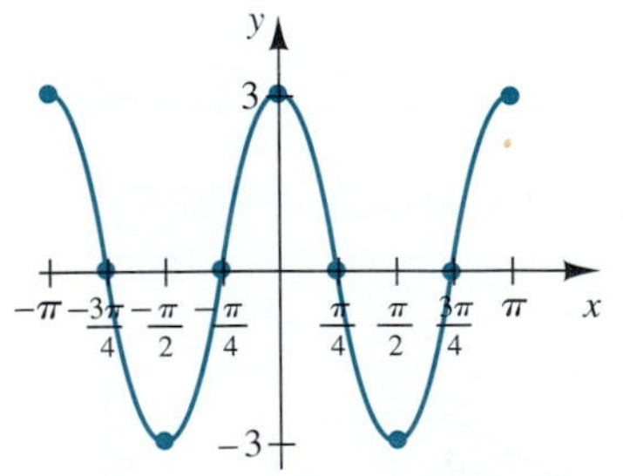

FIGURE 7.42 The graph of $y = 3 \sin(2x + \frac{\pi}{2})$

For example:

The graph of $y = 3 \sin\left(2x + \dfrac{\pi}{2}\right)$ on the interval $[-\pi, \pi]$ appears in Figure 7.42.

In general, for $y = A \sin(Bx + C)$, we have	***For $y = 3 \sin\left(2x + \frac{\pi}{2}\right)$, we have***
Amplitude $= \lvert A \rvert$	Amplitude $= 3$
Period $= \dfrac{2\pi}{\lvert B \rvert}$	Period $= \pi$
Frequency $= \lvert B \rvert$	Frequency $= 2$
Phase shift $= -\dfrac{C}{B}$	Phase shift $= -\dfrac{\pi}{4}$

3. Be able to express trigonometric functions of negative angles in terms of positive angles. (Section 7.2)

For example:
(a) $\sin(-28°) = -\sin 28°$
(b) $\cos(-145°) = \cos 145°$
(c) $\tan(-95°) = -\tan 95°$

4. Know the fundamental trigonometric identities. (Section 7.3)

5. Be able to use the fundamental trigonometric identities to verify other identities. (Section 7.3)
For example:
Verify the following identity:

$$\frac{\csc\theta - \sin\theta}{\sec\theta - \tan\theta} = \cot\theta + \cos\theta$$

We begin by expressing the left-hand side in terms of sine and cosine.

$$\frac{\frac{1}{\sin\theta} - \sin\theta}{\frac{1}{\cos\theta} - \frac{\sin\theta}{\cos\theta}} \stackrel{?}{=} \cot\theta + \cos\theta$$

We simplify the complex fraction.

$$\frac{\left(\frac{1}{\sin\theta} - \sin\theta\right)\sin\theta\cos\theta}{\left(\frac{1}{\cos\theta} - \frac{\sin\theta}{\cos\theta}\right)\sin\theta\cos\theta} \stackrel{?}{=} \cot\theta + \cos\theta$$

$$\frac{\cos\theta - \cos\theta\sin^2\theta}{\sin\theta - \sin^2\theta} \stackrel{?}{=} \cot\theta + \cos\theta$$

Factor and reduce.

$$\frac{\cos\theta(1 - \sin\theta)(1 + \sin\theta)}{\sin\theta(1 - \sin\theta)} \stackrel{?}{=} \cot\theta + \cos\theta$$

$$\frac{\cos\theta(1 + \sin\theta)}{\sin\theta} \stackrel{?}{=} \cot\theta + \cos\theta$$

$$\frac{\cos\theta}{\sin\theta} + \frac{\cos\theta\sin\theta}{\sin\theta} \stackrel{?}{=} \cot\theta + \cos\theta$$

$$\cot\theta + \cos\theta \stackrel{\checkmark}{=} \cot\theta + \cos\theta$$

6. Be able to solve trigonometric equations. (Section 7.4)
Some of the algebraic procedures we have learned can be applied to trigonometric equations. A reference triangle will help you to identify the reference angle.

For example:
(a) Find the general solution to $2\cos x = -\sqrt{3}$.

$$2\cos x = -\sqrt{3} \Rightarrow \cos x = -\frac{\sqrt{3}}{2}$$

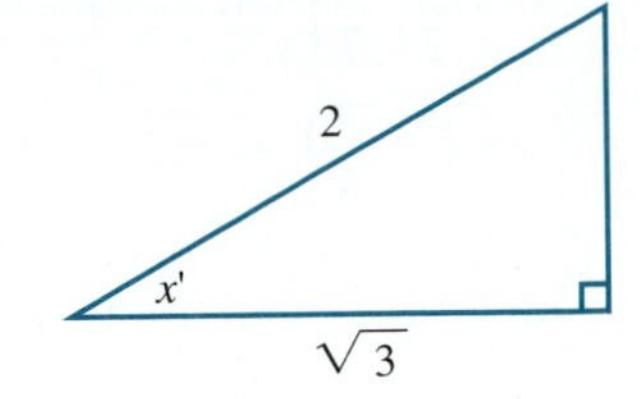

FIGURE 7.43

We draw a reference triangle, Figure 7.43, so that the cosine of the reference angle is $\frac{\sqrt{3}}{2}$ and we recognize that the reference angle is $\frac{\pi}{6}$. Since the cosine is negative, the reference angle must be in quadrant II or III. Therefore, $x = \frac{5\pi}{6}$ or $\frac{7\pi}{6}$.
Thus the general solution is

$$\boxed{x = \frac{5\pi}{6} \pm 2n\pi,\ \frac{7\pi}{6} \pm 2n\pi}$$

(b) Solve for x in the interval $[0, 2\pi)$: $2\sin^2 x + 3\sin x = 2$.

$$2\sin^2 x + 3\sin x - 2 = 0$$

$$(2\sin x - 1)(\sin x + 2) = 0$$

$$\sin x = \frac{1}{2} \quad \text{or} \quad \sin x = -2$$

$\sin x = -2$ has no solutions.

$$\sin x = \frac{1}{2} \Rightarrow \boxed{x = \frac{\pi}{6}, \frac{5\pi}{6}}$$

7. Know the domain and range of the inverse trigonometric functions. (Section 7.5)
By restricting the domain of each trigonometric function, we can define its inverse. $y = \sin^{-1} x$ means that y is the number (angle) whose sin is x.

8. Be able to evaluate expressions involving inverse trigonometric functions. (Section 7.5)

For example:

Evaluate **(a)** $\tan^{-1}(-1)$ **(b)** $\csc^{-1}\left(\csc \frac{\pi}{3}\right)$.

(a) $\tan^{-1}(-1)$ means the number in the interval $\left(-\frac{\pi}{2}, \frac{\pi}{2}\right)$ whose tangent is -1.

$\tan\left(-\frac{\pi}{4}\right) = -1$

therefore, $\tan^{-1}(-1) = \boxed{-\frac{\pi}{4}}$.

(b) We recognize that $\frac{\pi}{3}$ is in the restricted domain of the cosecant function, so that we have the composition of a function with its inverse.

Therefore, $\csc^{-1}\left(\csc \frac{\pi}{3}\right) = \boxed{\frac{\pi}{3}}$.

9. Be able to use the trigonometric functions and the inverse trigonometric functions to change the form of expressions. (Section 7.5)

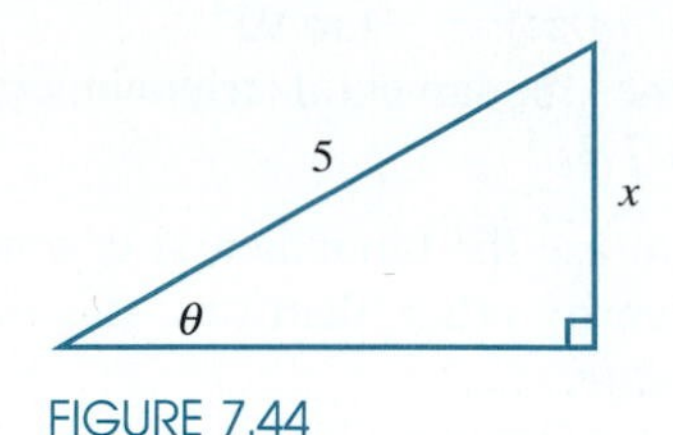

FIGURE 7.44

For example:
Use the right triangle in Figure 7.44 to express θ in terms of x.
From the triangle we see that $\tan \theta = \frac{x}{5}$ and so we have

$$\boxed{\theta = \tan^{-1} \frac{x}{5}}.$$

Chapter 7 **REVIEW EXERCISES**

In Exercises 1–12, sketch the graph of the given function on the interval $[0, 2\pi]$.

1. $y = \sin 2x$
2. $y = 2 \sin x$
3. $y = -\cos x$
4. $y = 2 \tan\left(\frac{1}{2}x\right)$
5. $y = \csc 3x$
6. $y = \sec\left(\frac{1}{3}x\right)$
7. $y = 3 \sin\left(x + \frac{\pi}{4}\right)$
8. $y = -2 \cos\left(x - \frac{\pi}{2}\right)$
9. $y = 4 \tan(2x - \pi)$
10. $y = -5 \sin\left(\frac{1}{2}x + \frac{\pi}{2}\right)$
11. $y = \sin \pi x$
12. $y = \cos \frac{\pi}{2} x$

In Exercises 13–18, express the given function of a negative angle as a function of a positive *acute* angle.

13. $\sin(-80°)$
14. $\cos\left(-\frac{5\pi}{7}\right)$
15. $\tan\left(-\frac{9\pi}{8}\right)$
16. $\cot(-329°)$
17. $\csc(-187°)$
18. $\sec(-256°)$

In Exercises 19–24, verify the given identity.

19. $\sin \theta + \sin \theta \cot^2 \theta = \frac{1}{\sin \theta}$

20. $\frac{\sec \beta}{\csc \beta} + \frac{\sin \beta}{\cos \beta} = 2 \tan \beta$

21. $\frac{1 + \sec x}{\sin x + \tan x} = \csc x$

22. $\frac{1}{\tan A + \cot A} = \sin A \cos A$

23. $\tan^2 \alpha - \sin^2 \alpha = \tan^2 \alpha \sin^2 \alpha$

24. $\sec \gamma - \frac{\cos \gamma}{1 + \sin \gamma} = \tan \gamma$

In Exercises 25–32, find the general solution to the given equation.

25. $2 \cos x = -1$
26. $\tan^2 x = 3$
27. $3 \csc^2 x = 6$
28. $\cot x = -1$
29. $4 \sin x = 2\sqrt{3}$
30. $5 \sec^2 x = 5$
31. $\sqrt{3} \sec x - 1 = 1$
32. $\sin^2 x + 1 = 2$

In Exercises 33–44, find the solution(s) to the given equation on the interval $[0, 2\pi)$. Use a calculator (or table) only when necessary, and round to four decimal places.

33. $2\sin^2 x = 1$

34. $4\cos^2 x = 3$

35. $\cos^2\theta \sin\theta = \cos\theta$

36. $\sin^2 x - \sin x - 2 = 0$

37. $1 - \sin^2 x - \cos x = 6$

38. $\tan^2\theta \cot\theta = \tan\theta$

39. $2\sin^2\theta + 1 = 3\sin\theta$

40. $\sec^2\theta + 2 = 3\sec\theta$

41. $2\csc^2\theta - 3\csc\theta = 2$

42. $12\sin^2\theta - 7\sin\theta + 1 = 0$

43. $5\sin^2\theta = 3$

44. $2\tan^2\theta = 14$

In Exercises 45–50, find the solution(s) to the given equation on the interval $[0°, 360°)$. Use a calculator (or table) only when necessary, and round to the nearest degree.

45. $4\cos x = 1$

46. $\cot^2 x = 10$

47. $\sec\theta \csc\theta = \sec\theta$

48. $2\cos^2\theta + 1 = 3\cos\theta$

49. $2\sin^2\theta + 1 = 3\sin\theta$

50. $2\csc^2 x + 3\csc x - 9 = 0$

In Exercises 51–64, evaluate the given expression.

51. $\sin^{-1}\left(-\frac{\sqrt{3}}{2}\right)$

52. $\tan^{-1}\left(\frac{1}{\sqrt{3}}\right)$

53. $\sec^{-1}(\sqrt{2})$

54. $\csc^{-1}(-2)$

55. $\cot^{-1}(0)$

56. $\cos^{-1}(-1)$

57. $\sin^{-1}\left(\cos\frac{2\pi}{3}\right)$

58. $\tan(\sin^{-1} 1)$

59. $\sec\left(\cot^{-1}\left(-\frac{\sqrt{3}}{3}\right)\right)$

60. $\csc^{-1}(\sin 45°)$

61. $\cos^{-1}\left(\cos\left(\frac{5\pi}{6}\right)\right)$

62. $\tan^{-1}\left(\tan\left(-\frac{\pi}{3}\right)\right)$

63. $\sin^{-1}\left(\sin\frac{5\pi}{3}\right)$

64. $\sec^{-1}\left(\sec\left(-\frac{\pi}{6}\right)\right)$

65. Use the following diagram to express θ as a function of x.

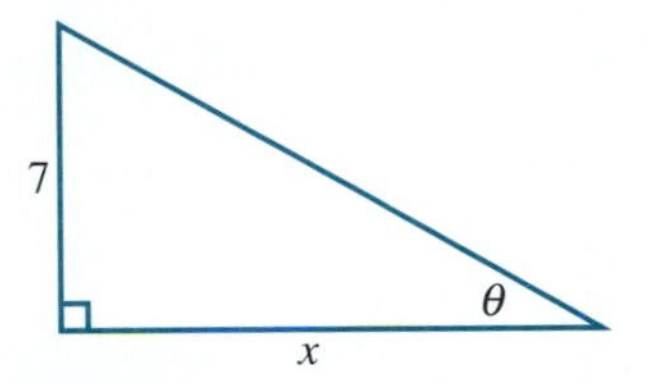

66. Use the substitution $x = 3\sin\theta$ to express $x\sqrt{9 - x^2}$ in simplest trigonometric form.

67. Use the following diagram to express $\sqrt{x^2 - 64}$ as a function of θ.

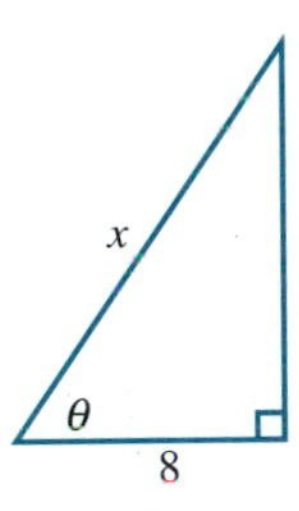

68. Draw a suitable right triangle so that $\theta = \sin^{-1}\frac{x}{7}$.

Chapter 7 PRACTICE TEST

1. What are the domain and range of the function $y = \tan 3x$?

2. Sketch the graph of $y = 3\cos 2x$ on the interval $[-2\pi, 2\pi]$.

3. Sketch the graph of $y = -\sin\left(x + \frac{\pi}{2}\right)$ on the interval $[-\pi, \pi]$.

4. Sketch the graph of $y = \tan\left(\frac{1}{2}x + \pi\right)$ on the interval $[-2\pi, 2\pi]$.

5. Sketch the graph of $y = \frac{1}{2}\csc 3x$ on the interval $(0, \pi)$.

6. Verify the following identities.

(a) $\frac{\sec\theta}{\tan\theta} = \frac{\tan\theta}{\sec\theta - \cos\theta}$

(b) $\frac{1 + \sec A}{\tan A + \sin A} = \csc A$

7. Find the general solutions to the following equations.

(a) $2\cos x + 1 = 0$

(b) $\tan^2\theta = 3$

8. Find all solutions in the interval $[0, 2\pi)$ to the following equations.

(a) $\sin^2 x = \sqrt{3}\sin x \cos x$

(b) $3\sin^2 x - \sin x = 4$

9. What is the domain of the function $f(x) = \dfrac{5}{\sec x + 1}$?

10. Express each of the following as a trigonometric function of a *positive acute* angle.

(a) $\sin(-127°)$ **(b)** $\cos\left(-\dfrac{7\pi}{8}\right)$

11. Evaluate each of the following.

(a) $\sin^{-1}\left(-\dfrac{\sqrt{3}}{2}\right)$ **(b)** $\arccos(-1)$

(c) $\tan^{-1} 0$ **(d)** $\sin(\sec^{-1}\sqrt{2})$

(e) $\csc^{-1}\left(\csc\left(-\dfrac{\pi}{3}\right)\right)$ **(f)** $\cos^{-1}\left(\cos\left(\dfrac{5\pi}{6}\right)\right)$

12. Use the following diagram to express θ as a function of x.

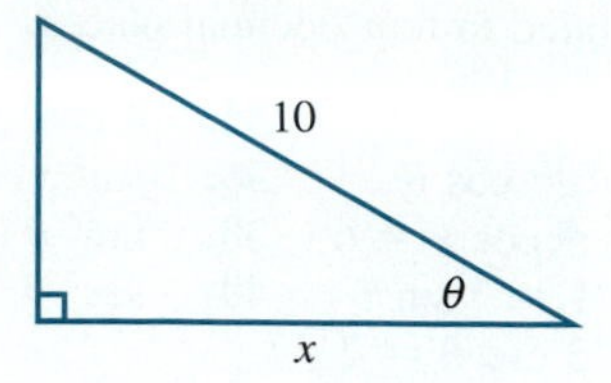

13. Use the substitution $x = 4\tan\theta$ to express $\dfrac{x^2}{x^2 + 16}$ in simplest trigonometric form.

CHAPTER 8

More Trigonometry and Its Applications

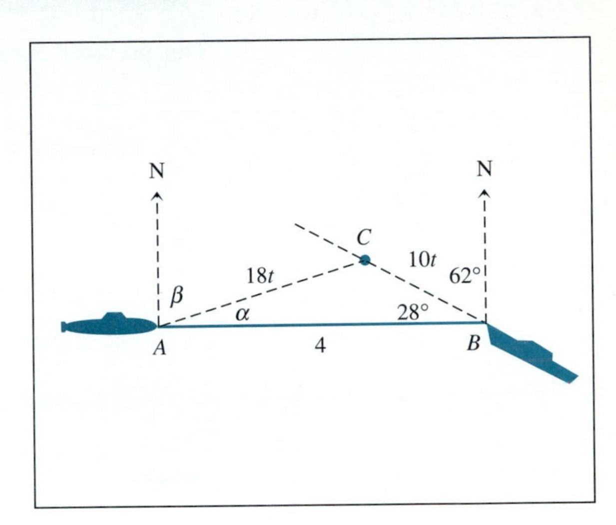

In the last two chapters we discovered and discussed many of the properties of the trigonometric functions. The first three sections of this chapter continue the development of a variety of trigonometric relationships. In the remaining three sections we apply our knowledge of trigonometry to discuss vectors in the plane; to develop the relationship between trigonometry and the complex numbers; and to introduce the polar coordinate system and some interesting graphs.

8.1 The Addition Formulas

Our previous experience with functions has shown us that, in general,

$$f(A + B) \neq f(A) + f(B)$$

For example, if $f(x) = x^2 + 7$, then

$$f(2 + 3) = f(5) = 5^2 + 7 = 32$$

but

$$f(2) + f(3) = (2^2 + 7) + (3^2 + 7) = 11 + 16 = 27$$

The trigonometric functions behave in a similar fashion. For example,

$$\sin\left(\frac{\pi}{3} + \frac{\pi}{6}\right) \neq \sin\left(\frac{\pi}{3}\right) + \sin\left(\frac{\pi}{6}\right)$$

because

$$\sin\left(\frac{\pi}{3} + \frac{\pi}{6}\right) = \sin\left(\frac{2\pi}{6} + \frac{\pi}{6}\right) = \sin\left(\frac{\pi}{2}\right) = 1$$

but

$$\sin\left(\frac{\pi}{3}\right) + \sin\left(\frac{\pi}{6}\right) = \frac{\sqrt{3}}{2} + \frac{1}{2} \neq 1$$

Thus we see that, in general, $\sin(A + B) \neq \sin A + \sin B$.

It turns out, however, that there is a way to express the trigonometric functions of $A \pm B$ in terms of trigonometric functions of A and B.

We begin by deriving a formula for $\cos(A - B)$, where $A \neq B$. Since A and B are unequal, let's assume that $0 < B < A < 2\pi$. In Figure 8.1(a) we have drawn a unit circle and labeled the angles corresponding to A and B (in standard position) and $A - B$. Figure 8.1(b) shows the angle corresponding to $A - B$ rotated so that it is in standard position. (R is rotated to P and S is rotated to Q.)

To find the x- and y-coordinates of P, we have $\cos(A - B) = \frac{x}{1} = x$ and $\sin(A - B) = \frac{y}{1} = y$.

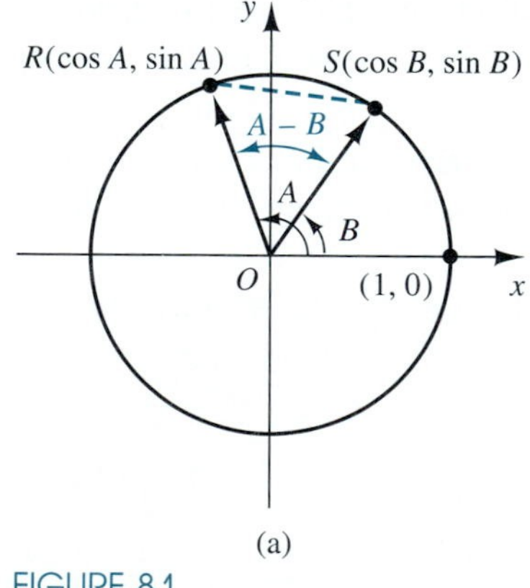

(a)

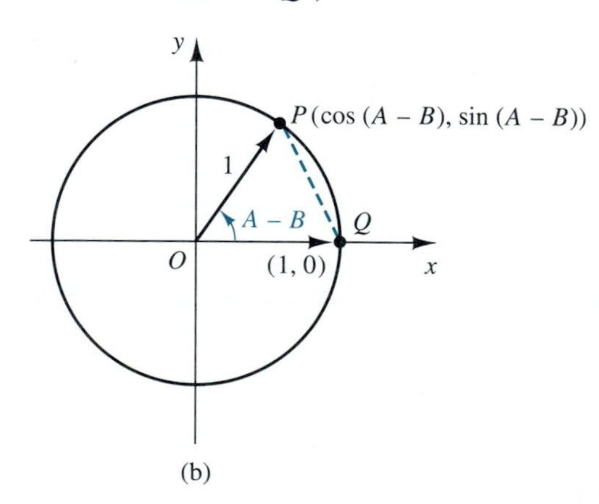

(b)

FIGURE 8.1

We have labeled the coordinates of P, Q, R, and S by using the definition of the sine and cosine functions. The fact that these points are on the unit circle means that $|\overline{OR}| = |\overline{OS}| = |\overline{OP}| = |\overline{OQ}| = 1$. We note that since the angle $A - B$ is the same in both figures, arcs PQ and RS have the same length. Therefore, line segments PQ and RS (called chords of the circle) are also of equal length. Therefore, applying the distance formula we get

$$\text{Length of chord } \overline{PQ} = \text{Length of chord } \overline{RS}$$

$$\sqrt{[\cos(A-B)-1]^2 + [\sin(A-B)-0]^2} = \sqrt{(\cos A - \cos B)^2 + (\sin A - \sin B)^2}$$ *Square both sides.*

$$[\cos(A-B)-1]^2 + [\sin(A-B)-0]^2 = (\cos A - \cos B)^2 + (\sin A - \sin B)^2$$

$$\cos^2(A-B) - 2\cos(A-B) + 1 + \sin^2(A-B) = \cos^2 A - 2\cos A\cos B + \cos^2 B + \sin^2 A - 2\sin A\sin B + \sin^2 B$$

But $\sin^2\theta + \cos^2\theta = 1$.

$$2 - 2\cos(A-B) = 2 - 2\cos A\cos B - 2\sin A\sin B$$ *Subtract 2 from both sides; then divide both sides by* -2.

What happens to $\cos(A-B)$ if $A = B$?

$$\boxed{\cos(A-B) = \cos A\cos B + \sin A\sin B} \qquad (1a)$$

This is one of the *addition formulas*.

We can use formula (1a) to derive other addition formulas.

To obtain a formula for $\cos(A + B)$, we can proceed as follows.

$$\cos(A+B) = \cos(A - [-B])$$ *Now we apply addition formula* (1a).

$$= \cos A\cos(-B) + \sin A\sin(-B)$$ *Recall that* $\cos(-B) = \cos B$ *and* $\sin(-B) = -\sin B$.

$$= \cos A\cos B + \sin A(-\sin B)$$

$$\boxed{\cos(A+B) = \cos A\cos B - \sin A\sin B} \qquad (1b)$$

Next we derive a formula for $\sin(A + B)$. We rewrite $\sin(A + B)$ using the cofunctional relationship and apply addition formula (1a). Recall that sine and cosine are cofunctions, which means that $\sin x = \cos\left(\frac{\pi}{2} - x\right)$.

$$\sin(A+B) = \cos\left(\frac{\pi}{2} - (A+B)\right)$$ *Regroup.*

$$= \cos\left(\left[\frac{\pi}{2} - A\right] - B\right)$$ *Now we use formula* (1a).

$$= \cos\left(\frac{\pi}{2} - A\right)\cos B + \sin\left(\frac{\pi}{2} - A\right)\sin B$$

Again we use the cofunctional relationship.

$$= \sin A\cos B + \cos A\sin B$$ *Hence,*

$$\boxed{\sin(A+B) = \sin A\cos B + \cos A\sin B} \qquad (2a)$$

What happens to $\sin(A - B)$ if $A = B$?

It is left to the student to rewrite $\sin(A - B)$ as $\sin(A + (-B))$ and use formula (2a) to derive the following formula for $\sin(A - B)$.

$$\boxed{\sin(A - B) = \sin A \cos B - \cos A \sin B} \qquad (2b)$$

Next we derive addition formulas for the tangent function. We begin by using the identity that expresses the tangent as the quotient of the sine and cosine so that we can use the addition formulas for the sine and cosine.

$$\begin{aligned}\tan(A + B) &= \frac{\sin(A + B)}{\cos(A + B)} \\ &= \frac{\sin A \cos B + \cos A \sin B}{\cos A \cos B - \sin A \sin B} && \textit{Divide each term by } \cos A \cos B. \\ &= \frac{\dfrac{\sin A \cos B}{\cos A \cos B} + \dfrac{\cos A \sin B}{\cos A \cos B}}{\dfrac{\cos A \cos B}{\cos A \cos B} - \dfrac{\sin A \sin B}{\cos A \cos B}} && \textit{Using the identity } \frac{\sin \theta}{\cos \theta} = \tan \theta,\end{aligned}$$

$$\boxed{\tan(A + B) = \frac{\tan A + \tan B}{1 - \tan A \tan B}} \qquad (3a)$$

We leave the derivation of the formula for $\tan(A - B)$ as an exercise.

We summarize the addition formulas, each of which is an identity, in the following box.

THE ADDITION FORMULAS

1. **(a)** $\sin(A + B) = \sin A \cos B + \cos A \sin B$

(b) $\sin(A - B) = \sin A \cos B - \cos A \sin B$

2. **(a)** $\cos(A + B) = \cos A \cos B - \sin A \sin B$

(b) $\cos(A - B) = \cos A \cos B + \sin A \sin B$

3. **(a)** $\tan(A + B) = \dfrac{\tan A + \tan B}{1 - \tan A \tan B}$

(b) $\tan(A - B) = \dfrac{\tan A - \tan B}{1 + \tan A \tan B}$

Instead of memorizing the formula for $\tan(A \pm B)$, you may use the addition formulas for sine and cosine together with the quotient identities.

We can use the addition formulas to evaluate certain trigonometric expressions and verify other identities.

EXAMPLE 1 Find the exact value of **(a)** $\sin 75°$ **(b)** $\tan \dfrac{\pi}{12}$.

Solution

Why do we rewrite 75° as 45° + 30° rather than say 40° + 35°?

(a) If we rewrite 75° as 45° + 30°, we can use the addition formula as follows.

$$\sin 75° = \sin(45° + 30°) = \sin 45° \cos 30° + \cos 45° \sin 30°$$
$$= \frac{1}{\sqrt{2}} \cdot \frac{\sqrt{3}}{2} + \frac{1}{\sqrt{2}} \cdot \frac{1}{2}$$
$$= \boxed{\frac{\sqrt{3}+1}{2\sqrt{2}}}$$

Use a calculator to compute the value of $\frac{\sqrt{3}+1}{2\sqrt{2}}$. Compare these values to the calculator value for sin 75°.

This answer is the exact value for sin 75°. Keep in mind that a calculator will give us only an approximate value (although a very accurate one) for sin 75°.

(b) Since $\frac{\pi}{12} = 15° = 60° - 45°$, we write $\frac{\pi}{12} = \frac{\pi}{3} - \frac{\pi}{4}$.

$$\tan \frac{\pi}{12} = \tan\left(\frac{\pi}{3} - \frac{\pi}{4}\right) \qquad \textit{Now use the addition formula for } \tan(A - B).$$

Can we express 15° using any other pairs of appropriate angles?

$$= \frac{\tan \frac{\pi}{3} - \tan \frac{\pi}{4}}{1 + \tan \frac{\pi}{3} \tan \frac{\pi}{4}} = \frac{\sqrt{3} - 1}{1 + (\sqrt{3})(1)} = \boxed{\frac{\sqrt{3}-1}{1+\sqrt{3}}}$$

■

EXAMPLE 2 Prove that $\cos(\pi - \theta) = -\cos \theta$.

Solution We may verify this identity by applying the addition formula for $\cos(A - B)$.

$$\cos(\pi - \theta) = \cos \pi \cos \theta + \sin \pi \sin \theta$$
$$= (-1)\cos \theta + 0 \cdot \sin \theta$$
$$= -\cos \theta \qquad \textit{As required}$$

If θ happens to be an acute angle, $\pi - \theta$ is a second quadrant angle with reference angle θ. The result of this example agrees with our analysis in Section 7.3. ■

EXAMPLE 3 Suppose that for α in quadrant II and β in quadrant III, we have $\sin \alpha = \frac{3}{4}$ and $\cos \beta = -\frac{2}{5}$. Find $\sin(\alpha - \beta)$.

Solution Using the addition formula for $\sin(\alpha - \beta)$, we get

$$\sin(\alpha - \beta) = \sin \alpha \cos \beta - \cos \alpha \sin \beta$$

We have the values of $\sin \alpha$ and $\cos \beta$; we need to find $\cos \alpha$ and $\sin \beta$. Figure 8.2 illustrates two reference triangles, one for α and one for β. We complete the reference triangles by using the Pythagorean theorem. The results appear in Figure 8.3. Since α is in quadrant II and β is in quadrant III, both $\cos \alpha$ and $\sin \beta$ are negative.

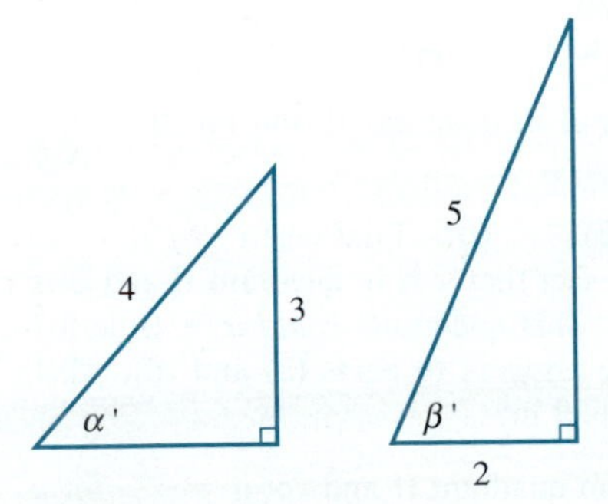

FIGURE 8.2 The reference triangles for α and β

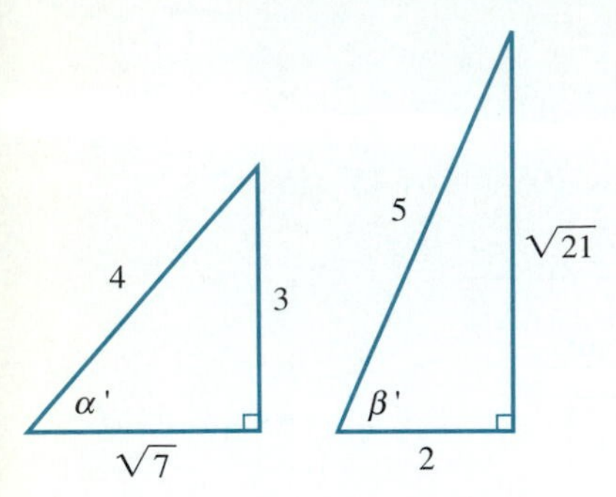

FIGURE 8.3 The completed reference triangles for α and β

We can now find the required value.

$$\sin(\alpha - \beta) = \sin\alpha\cos\beta - \cos\alpha\sin\beta$$

$$= \left(\frac{3}{4}\right)\left(-\frac{2}{5}\right) - \left(-\frac{\sqrt{7}}{4}\right)\left(-\frac{\sqrt{21}}{5}\right)$$

Keep in mind that $\cos\alpha$ is negative because α is in quadrant II. Similarly, $\sin\beta$ is negative because β is in quadrant III.

$$= -\frac{6}{20} - \frac{\sqrt{7}}{4}\cdot\frac{\sqrt{3}\sqrt{7}}{5} = \boxed{\frac{-6-7\sqrt{3}}{20}}$$

■

In the next section we will use the addition formulas to derive some useful special identities.

EXERCISES 8.1

In Exercises 1–8, use the addition formulas to find the exact value of the given expression.

1. $\cos 105°$ **2.** $\sin 15°$

3. $\tan\frac{5\pi}{12}$ **4.** $\cos\frac{11\pi}{12}$

5. $\sin 165°$ **6.** $\tan 195°$

7. $\cos\frac{17\pi}{12}$ **8.** $\sin\frac{19\pi}{12}$

In Exercises 9–14, use the addition formulas to simplify each expression.

9. $\sin 23°\cos 37° + \cos 23°\sin 37°$

10. $\cos 57°\cos 12° + \sin 57°\sin 12°$

11. $\cos\frac{4\pi}{5}\cos\frac{3\pi}{10} + \sin\frac{4\pi}{5}\sin\frac{3\pi}{10}$

12. $\sin\frac{3\pi}{7}\cos\frac{4\pi}{7} + \cos\frac{3\pi}{7}\sin\frac{4\pi}{7}$

13. $\dfrac{\tan 50° + \tan 10°}{1 - \tan 50°\tan 10°}$

14. $\dfrac{\tan\frac{2\pi}{5} - \tan\frac{3\pi}{20}}{1 + \tan\frac{2\pi}{5}\tan\frac{3\pi}{20}}$

In Exercises 15–28, verify the given identity.

15. $\sin(\pi - \theta) = \sin\theta$

16. $\tan(\pi - \theta) = -\tan\theta$

17. $\cos(\pi + \theta) = -\cos\theta$

18. $\sin(\pi + \theta) = -\sin\theta$

19. $\tan\left(x - \frac{\pi}{4}\right) = \dfrac{\tan x - 1}{\tan x + 1}$

20. $\sin\left(x + \frac{\pi}{2}\right) = \cos x$

21. $\cos\left(\alpha + \frac{\pi}{2}\right) = -\sin\alpha$

22. $\sin\left(\beta + \frac{\pi}{4}\right) = \dfrac{\sin\beta + \cos\beta}{\sqrt{2}}$

23. $\sin(2\pi - \theta) = -\sin\theta$

24. $\tan\left(w + \frac{\pi}{2}\right) = -\cot w$

25. $\cos\left(u + \frac{\pi}{4}\right) = \dfrac{\cos u - \sin u}{\sqrt{2}}$

26. $\sin\left(\gamma - \frac{3\pi}{2}\right) = \cos\gamma$

27. $\sin(A + B) - \sin(A - B) = 2\cos A\sin B$

28. $\cos(A + B) - \cos(A - B) = -2\sin A\sin B$

29. Given $\sin A = -\frac{3}{5}$ for A in quadrant III and $\cos B = \frac{5}{12}$ for B in quadrant IV:
(a) Find $\sin(A + B)$.
(b) Find $\cos(A + B)$.
(c) What quadrant is $A + B$ in?

30. Given $\sin\alpha = \frac{2}{3}$ for α in quadrant II and $\tan\beta = \frac{3}{\sqrt{5}}$ for β in quadrant III:
(a) Find $\tan(\alpha + \beta)$. **(b)** Find $\cos(\alpha + \beta)$.
(c) Based upon the fact that α is in quadrant II and β is in quadrant III, what quadrants *could* $\alpha + \beta$ be in?
(d) Based upon the answers to parts (a) and (b), what quadrant *is* $\alpha + \beta$ in?

31. If $\sec t = -\frac{5}{4}$ for t in quadrant II and $\cot u = \frac{2}{\sqrt{5}}$ for u in quadrant III, find
(a) $\sin(u - t)$ **(b)** $\tan(u - t)$

32. If $\sin u = -\frac{2}{\sqrt{7}}$ for u in quadrant III and $\tan v = -\frac{\sqrt{6}}{2}$ for v in quadrant IV, find
(a) $\sec(u + v)$ **(b)** $\csc(u - v)$

33. If $\sec t = -\frac{5}{4}$ with $\sin t > 0$ and $\cos r = \frac{4}{7}$ with $\tan r < 0$, find
(a) $\sin(r + t)$ **(b)** $\cos(t - r)$

34. If $\sin \alpha = -\frac{7}{25}$ with $\cos \alpha > 0$ and $\cos \beta = \frac{12}{13}$ with $\sin \beta < 0$, find
(a) $\tan(\alpha + \beta)$ **(b)** $\cot(\alpha - \beta)$

35. Show that for any triangle ABC,
(a) $\sin(A + B) = \sin C$
(b) $\cos(A + B) = -\cos C$
(c) $\tan(A + B) = -\tan C$

36. Prove that $\sin(2\theta) = 2 \sin \theta \cos \theta$. HINT: $2\theta = \theta + \theta$

37. Let $f(\theta) = \sin \theta$. Prove that

$$\frac{f(\theta + h) - f(\theta)}{h} = \sin \theta\left(\frac{\cos h - 1}{h}\right) + \cos \theta\left(\frac{\sin h}{h}\right)$$

38. Let $f(\theta) = \cos \theta$. Prove that

$$\frac{f(\theta + h) - f(\theta)}{h} = \cos \theta\left(\frac{\cos h - 1}{h}\right) - \sin \theta\left(\frac{\sin h}{h}\right)$$

In Exercises 39–42, verify the given identity.

39. $\sin(A + B) + \sin(A - B) = 2 \sin A \cos B$
40. $\cos(A + B) + \cos(A - B) = 2 \cos A \cos B$
41. $\sin(A + B) \sin(A - B) = \sin^2 A - \sin^2 B$
42. $\cos(A + B) \cos(A - B) = \cos^2 A - \sin^2 B$

QUESTIONS FOR THOUGHT

43. Derive a formula for $\sin(A + B + C)$ in terms of functions of A, B, and C. HINT: Write $A + B + C$ as $(A + B) + C$ and use the addition formula.

44. Use the formula for $\tan(A + B)$ to derive the formula for $\tan(A - B)$.

45. Use the accompanying figure to find $\tan \beta$.

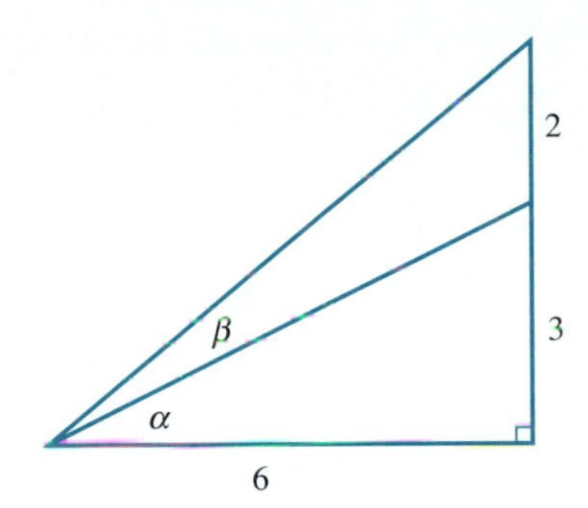

8.2 The Double-Angle and Half-Angle Formulas

We use the addition formulas derived in the last section to derive some additional identities that are particularly useful in calculus.

The Double-Angle Formulas

The **double-angle formulas,** which we derive in the first three examples, refer to formulas for $\sin 2\theta$, $\cos 2\theta$, and $\tan 2\theta$.

EXAMPLE 1 Derive a formula for $\sin 2\theta$.

Solution We rewrite 2θ as $\theta + \theta$ and use the addition formula for $\sin(A + B)$.

$$\sin 2\theta = \sin(\theta + \theta) \quad \textit{Apply the formula for } \sin(A + B).$$
$$= \sin \theta \cos \theta + \cos \theta \sin \theta$$
$$= \boxed{2 \sin \theta \cos \theta}$$

How would you use a similar approach to find $\sin 3\theta$?

This is called the *double-angle formula* for the sine function. ■

EXAMPLE 2 Derive a formula for $\cos 2\theta$.

Solution We proceed as we did in Example 1.

$$\begin{aligned}\cos 2\theta &= \cos(\theta + \theta) && \textit{Apply the formula for } \cos(A + B).\\ &= \cos\theta\cos\theta - \sin\theta\sin\theta\\ &= \boxed{\cos^2\theta - \sin^2\theta}\end{aligned}$$

This is one form of the double-angle formula for the cosine function. By using the fundamental Pythagorean identity $\sin^2\theta + \cos^2\theta = 1$, we can write two alternative forms of the double-angle formula for the cosine function.

$$\begin{aligned}\cos 2\theta &= \cos^2\theta - \sin^2\theta && \textit{Substitute } \cos^2\theta = 1 - \sin^2\theta.\\ &= 1 - \sin^2\theta - \sin^2\theta\\ &= \boxed{1 - 2\sin^2\theta}\end{aligned}$$

$$\begin{aligned}\cos 2\theta &= \cos^2\theta - \sin^2\theta && \textit{Substitute } \sin^2\theta = 1 - \cos^2\theta.\\ &= \cos^2\theta - (1 - \cos^2\theta)\\ &= \cos^2\theta - 1 + \cos^2\theta\\ &= \boxed{2\cos^2\theta - 1}\end{aligned}$$

Thus we have three equivalent versions of the double-angle formula for the cosine function. ■

EXAMPLE 3 Derive a formula for $\tan 2\theta$.

Solution

$$\begin{aligned}\tan 2\theta &= \tan(\theta + \theta) && \textit{We use the formula for } \tan(A + B).\\ &= \frac{\tan\theta + \tan\theta}{1 - \tan\theta\tan\theta}\\ &= \boxed{\frac{2\tan\theta}{1 - \tan^2\theta}}\end{aligned}$$

■

We summarize the double-angle formulas in the following box.

THE DOUBLE-ANGLE FORMULAS

1. $\sin 2\theta = 2 \sin \theta \cos \theta$
2. (a) $\cos 2\theta = \cos^2 \theta - \sin^2 \theta$
   (b) $\cos 2\theta = 1 - 2 \sin^2 \theta$
   (c) $\cos 2\theta = 2 \cos^2 \theta - 1$
3. $\tan 2\theta = \dfrac{2 \tan \theta}{1 - \tan^2 \theta}$

EXAMPLE 4 If θ is a second-quadrant angle for which $\cos \theta = -\dfrac{5}{13}$, find

(a) $\sin 2\theta$ **(b)** $\cos 2\theta$ **(c)** $\tan 2\theta$

Solution The sine, cosine, and tangent of 2θ all involve those same functions of θ. Therefore, we begin by completing the reference triangle for θ, which appears in Figure 8.4.

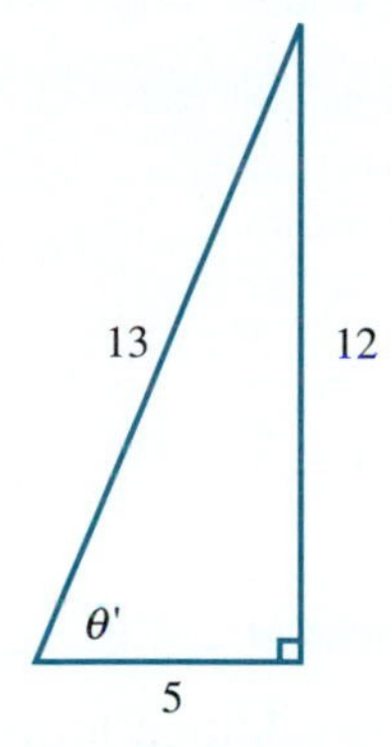

FIGURE 8.4 The reference triangle for θ

(a) 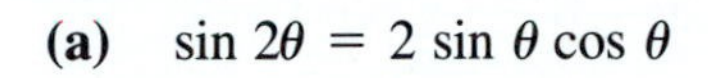

$\sin 2\theta = 2 \sin \theta \cos \theta$ *The fact that θ is in quadrant II tells us that* $\sin \theta$ *is positive, and* $\cos \theta$ *is negative. Using the reference triangle we get*

$$= 2\left(\frac{12}{13}\right)\left(-\frac{5}{13}\right)$$

$$= \boxed{-\frac{120}{169}}$$

(b) We may use any of the three formulas for $\cos 2\theta$. We use the third formula, since it involves only $\cos \theta$, which we are given.

$\cos 2\theta = 2 \cos^2 \theta - 1$ *We substitute the given value for* $\cos \theta$.

$$= 2\left(-\frac{5}{13}\right)^2 - 1$$

$$= 2\left(\frac{25}{169}\right) - 1$$

$$= \boxed{-\frac{119}{169}}$$

(c) $\tan 2\theta = \dfrac{2 \tan \theta}{1 - \tan^2 \theta}$ *Again we use the reference triangle and the fact that θ is in quadrant II to get*

$$= \frac{2\left(-\frac{12}{5}\right)}{1 - \left(-\frac{12}{5}\right)^2} = \frac{-\frac{24}{5}}{1 - \frac{144}{25}} = \frac{\left(-\frac{24}{5}\right)25}{\left(1 - \frac{144}{25}\right)25} = \boxed{\frac{120}{119}}$$

■

EXAMPLE 5 Suppose that θ is a first-quadrant angle and that $\sin \theta = \frac{a}{3}$. Express $\sin 2\theta$ as a function of a.

Solution The fact that $\sin \theta = \frac{a}{3}$ allows us to draw the reference triangle appearing in Figure 8.5 and find the missing side by using the Pythagorean theorem.

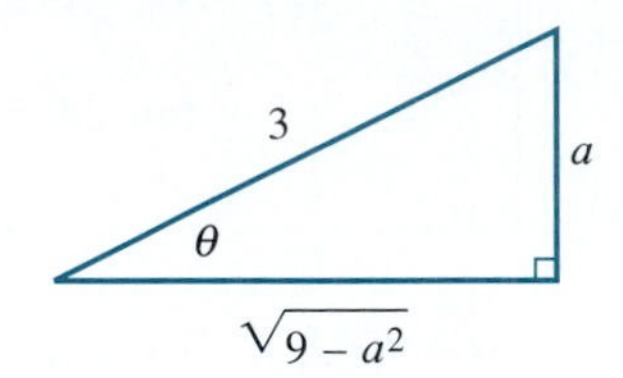

FIGURE 8.5

In order to express $\sin 2\theta$ as a function of a, we use the double-angle formula.

$$\sin 2\theta = 2 \sin \theta \cos \theta \qquad \textit{Using the reference triangle we get}$$

$$= 2\left(\frac{a}{3}\right)\left(\frac{\sqrt{9 - a^2}}{3}\right)$$

$$\boxed{\sin 2\theta = \frac{2a\sqrt{9 - a^2}}{9}}$$

■

EXAMPLE 6 Derive a "triple-angle formula" for $\cos 3\theta$ in terms of $\cos \theta$.

Solution We begin by writing $3\theta = 2\theta + \theta$ and applying the addition formula for the cosine function.

$$\cos 3\theta = \cos(2\theta + \theta)$$

$$= \cos 2\theta \cos \theta - \sin 2\theta \sin \theta$$

Use the double-angle formulas for sine and cosine. Since we want the answer in terms of $\cos \theta$*, we use the third version of the double-angle formula for cosine.*

$$= (2 \cos^2 \theta - 1) \cos \theta - (2 \sin \theta \cos \theta) \sin \theta$$

$$= 2 \cos^3 \theta - \cos \theta - 2 \sin^2 \theta \cos \theta$$

Substitute $1 - \cos^2 \theta$ *for* $\sin^2 \theta$.

$$= 2 \cos^3 \theta - \cos \theta - 2(1 - \cos^2 \theta) \cos \theta$$

$$= 2 \cos^3 \theta - \cos \theta - 2 \cos \theta + 2 \cos^3 \theta$$

$$= \boxed{4 \cos^3 \theta - 3 \cos \theta}$$

■

Keep in mind that these formulas are, in fact, trigonometric identities.

In calculus it is frequently advantageous to convert an expression involving higher powers of sine and cosine into one that is first degree in sine and cosine. If we start with the second and third double-angle formulas for cosine, which are $\cos 2\theta = 1 - 2 \sin^2 \theta$ and $\cos 2\theta = 2 \cos^2 \theta - 1$, and solve them for $\sin^2 \theta$ and $\cos^2 \theta$, respectively, we get the following.

ALTERNATIVE FORMS OF THE DOUBLE-ANGLE FORMULAS FOR SINE AND COSINE

2. (d) $\sin^2 \theta = \dfrac{1 - \cos 2\theta}{2}$

(e) $\cos^2 \theta = \dfrac{1 + \cos 2\theta}{2}$

EXAMPLE 7 Express $\sin^4 t$ in terms of first-degree trigonometric expressions.

Solution

$\sin^4 t = (\sin^2 t)(\sin^2 t)$ *Use the alternative form 2(d) of the double-angle formula for $\sin^2 \theta$ with $\theta = t$.*

$$= \left(\frac{1 - \cos 2t}{2}\right)\left(\frac{1 - \cos 2t}{2}\right)$$

$$= \frac{1 - 2\cos 2t + \cos^2 2t}{4} = \frac{1}{4} - \frac{1}{2}\cos 2t + \frac{1}{4}\cos^2 2t$$

Use formula 2(e) for $\cos^2 \theta$ with $\theta = 2t$.

$$= \frac{1}{4} - \frac{1}{2}\cos 2t + \frac{1}{4}\left(\frac{1 + \cos 4t}{2}\right) = \frac{1}{4} - \frac{1}{2}\cos 2t + \frac{1}{8} + \frac{1}{8}\cos 4t$$

$$= \boxed{\frac{3}{8} - \frac{1}{2}\cos 2t + \frac{1}{8}\cos 4t}$$

Note that the final answer is a first-degree expression in cosine. ■

We use these alternative forms of the double-angle formulas to derive **half-angle formulas.** Starting with Formula 2(d), we have

$$\sin^2 \theta = \frac{1 - \cos 2\theta}{2}$$ *Apply the square root theorem.*

$$\sin \theta = \pm\sqrt{\frac{1 - \cos 2\theta}{2}}$$ *We substitute $\dfrac{\theta}{2}$ for θ.*

$$\sin \frac{\theta}{2} = \pm\sqrt{\frac{1 - \cos 2\left(\frac{\theta}{2}\right)}{2}}$$

$$\sin \frac{\theta}{2} = \pm\sqrt{\frac{1 - \cos \theta}{2}}$$

The $\pm$ symbol indicates that we must choose the positive or negative value depending upon the quadrant in which the angle associated with $\frac{\theta}{2}$ terminates. The $\pm$ symbol is not being used in the same way as when we write $x^2 = 5 \Rightarrow x = \pm\sqrt{5}$, where the $\pm$ indicates that there are two possible answers.

The derivations of the other two half-angle formulas are left to the student as Exercises 53 and 54 at the end of this section. The half-angle formulas are recorded in the following box.recorded in the following box.

Remember that the $\pm$ here does not mean that we have a choice. The + or − sign must be chosen based upon the quadrant in which the terminal side of the angle will fall.

THE HALF-ANGLE FORMULAS

$$\sin\frac{\theta}{2} = \pm\sqrt{\frac{1-\cos\theta}{2}}$$

$$\cos\frac{\theta}{2} = \pm\sqrt{\frac{1+\cos\theta}{2}}$$

$$\tan\frac{\theta}{2} = \pm\sqrt{\frac{1-\cos\theta}{1+\cos\theta}}$$

EXAMPLE 8 Use a half-angle formula to evaluate $\sin 105°$.

Solution We use the half-angle formula for $\sin\frac{\theta}{2}$ with $\theta = 210°$.

$$\sin 105° = \sin\left(\frac{210°}{2}\right) = \pm\sqrt{\frac{1-\cos 210°}{2}}$$

The reference angle for 210° is 30°; 210° is in quadrant III, so the cosine is negative.

$$= \pm\sqrt{\frac{1-\left(-\frac{\sqrt{3}}{2}\right)}{2}} = \pm\sqrt{\frac{1+\left(\frac{\sqrt{3}}{2}\right)}{2}}$$

Simplify the complex fraction.

$$= \pm\sqrt{\frac{2+\sqrt{3}}{4}} = \frac{\pm\sqrt{2+\sqrt{3}}}{2}$$

As mentioned earlier, the choice of sign depends upon the quadrant. Since the terminal side of a 105° angle falls in the second quadrant, $\sin 105°$ is positive.

Thus our final answer is $\boxed{\sin 105° = \frac{\sqrt{2+\sqrt{3}}}{2}}$ ■

EXAMPLE 9 Solve for x: $\cos 2x = 3\cos x + 1$ for $0 \le x < 2\pi$.

Solution This trigonometric equation is a bit different than the ones we saw in Section 7.3. Looking back at our previous discussion of trigonometric equations, we can see that in all the equations there was only *one* argument throughout the equation. However, in this equation there are two arguments: x and $2x$. Generally, a trigonometric equation is easier to solve if it involves only one argument (and if possible one trigonometric function). By using the appropriate double-angle formula for $\cos 2x$, we can transform this equation into one involving only one argument and one function.

$$\cos 2x = 3\cos x + 1 \quad \textit{Substitute } \cos 2x = 2\cos^2 x - 1.$$

$$2\cos^2 x - 1 = 3\cos x + 1$$

$$2\cos^2 x - 3\cos x - 2 = 0$$

$$(2\cos x + 1)(\cos x - 2) = 0$$

$$\cos x = -\frac{1}{2} \quad \text{or} \quad \cos x = 2$$

Now we solve for x. For $\cos x = -\frac{1}{2}$, *we need x to correspond to a reference angle of* $\frac{\pi}{3}$ *in quadrants II or III (Figure 8.6).*

Since cos x *cannot be greater than* 1 *(or less than* -1), $\cos x = 2$ *has* no solutions. *Therefore, the solutions to the equation are*

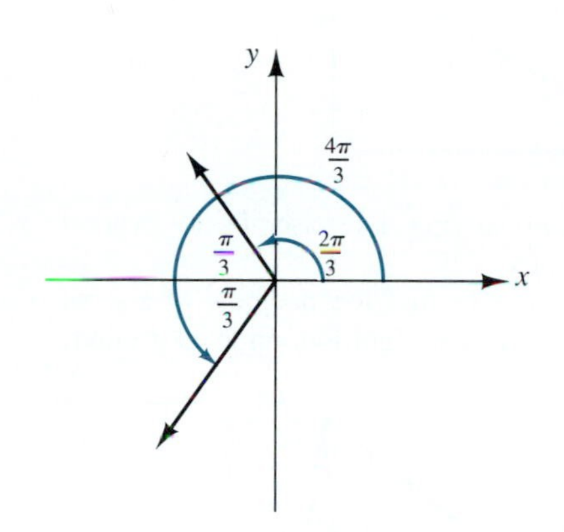

FIGURE 8.6

$$\boxed{x = \frac{2\pi}{3} \quad \text{or} \quad \frac{4\pi}{3}}$$

■

In the next section we develop more trigonometric relationships.

EXERCISES 8.2

1. Given that $\sin\theta = -\frac{24}{25}$ and θ is in quadrant IV, find $\sin 2\theta$.
2. Given that $\cos\theta = -\frac{12}{13}$ and θ is in quadrant III, find $\cos 2\theta$.
3. Given that $\tan\theta = \frac{3}{5}$, find $\tan 2\theta$.
4. Given that $\tan\theta = -\frac{1}{3}$ and $\frac{\pi}{2} < \theta < \pi$, find
   (a) $\sin 2\theta$ **(b)** $\cos 2\theta$
5. Given that $\sin\theta = -\frac{3}{5}$ and $\pi < \theta < \frac{3\pi}{2}$, find
   (a) $\sin\frac{\theta}{2}$ **(b)** $\cos\frac{\theta}{2}$ **(c)** $\tan\frac{\theta}{2}$
6. Given that $\cos\theta = \frac{3}{4}$ and θ is in quadrant IV, find
   (a) $\sin\frac{\theta}{2}$ **(b)** $\cos\frac{\theta}{2}$ **(c)** $\tan\frac{\theta}{2}$

In Exercises 7–22, verify the given identity.

7. $(\sin x + \cos x)^2 = 1 + \sin 2x$
8. $\cos 2x = \cos^4 x - \sin^4 x$
9. $\tan 2A = \dfrac{2}{\cot A - \tan A}$
10. $\tan x = \dfrac{\sin 2x}{1 + \cos 2x}$
11. $\sec 2\theta = \dfrac{\sec^2\theta}{2 - \sec^2\theta}$

12. $\cot 2u = \dfrac{\cot^2 u - 1}{2 \cot u}$
13. $2 \cos^2 x = \cot x \sin 2x$
14. $1 + \cos 2A = \cot A \sin 2A$
15. $\sin 3x = 3 \sin x - 4 \sin^3 x$
16. $\tan x = \csc 2x - \cot 2x$
17. $\cos 4x = 8 \cos^4 x - 8 \cos^2 x + 1$
18. $\cos^4 x = \dfrac{1}{8} \cos 4x + \dfrac{1}{2} \cos 2x + \dfrac{3}{8}$
19. $\sin^2 \dfrac{\theta}{2} = \dfrac{\sec \theta - 1}{2 \sec \theta}$
20. $\sin^2 \dfrac{\theta}{2} = \dfrac{\tan \theta - \sin \theta}{2 \tan \theta}$
21. $\tan \theta + \cot \theta = \dfrac{2}{\sin 2\theta}$
22. $\cos 2t = \dfrac{1 - \tan^2 t}{1 + \tan^2 t}$

In Exercises 23–36, find all solutions to the given equation on the interval $[0, 2\pi)$.

23. $\sin 2x = \sin x$
24. $\cos 2\theta + \cos \theta = 0$
25. $\tan 2x = \tan x$
26. $\tan 2t = 2 \cos t$
27. $\sin^3 2x = \sin 2x$
28. $3 \cos 2x + 2 \sin^2 x = 2$
29. $\cos 4x = \sin 2x$
30. $\sin 4x + \sin 2x = 0$
31. $\sin x + \cos x = 1$
    HINT: Square both sides.
32. $\sec x + \tan x = 1$
    HINT: Square both sides.
33. $2 \sin \dfrac{\theta}{2} = 1$
34. $2 \sin \theta \cos \theta = \sqrt{3}$
35. $\sin \dfrac{t}{2} = 1 - \cos t$
36. $4 \sin^2 \dfrac{x}{2} = 2 - \cos^2 x$
37. Find the exact value of $\cos \dfrac{\pi}{12}$.
38. Find the exact value of $\sin 22.5°$.
39. Find the exact value of $\tan 67.5°$.
40. Find the exact value of $\cos \dfrac{\pi}{8}$.
41. Express $\cos^4 t$ in terms of first-degree trigonometric expressions.
42. Express $\tan^4 t$ in terms of first-degree trigonometric expressions.
43. Express the volume of the following trough as a function of θ. The ends of the trough are isosceles triangles with sides of 2 feet, and the length of the trough is 8 feet.

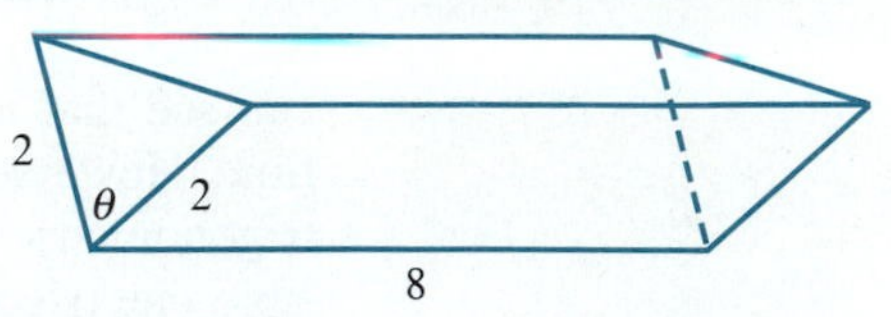

44. Use the following figure to express $\cos 2\theta$ as a function of x.

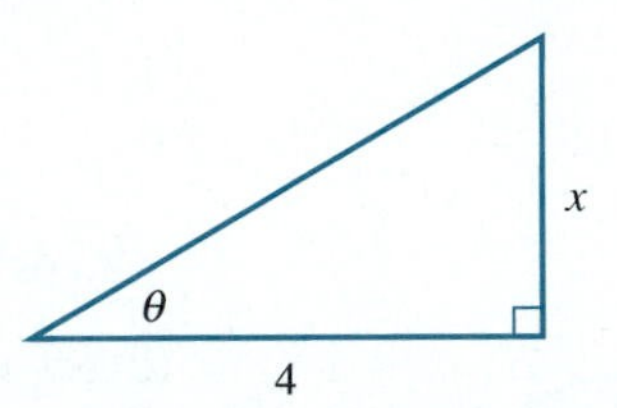

45. Use the figure accompanying Exercise 44 to express $\sin 2\theta$ as a function of x.
46. Use the following figure to find the height h of the oil well. Assume ABC forms a straight line on level ground.

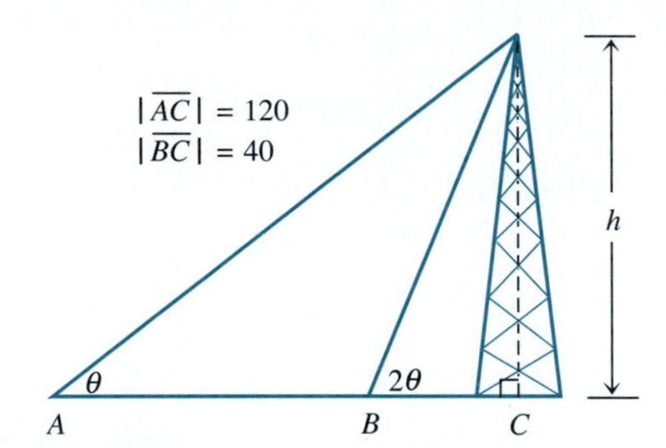

47. Use the following figure to find x.

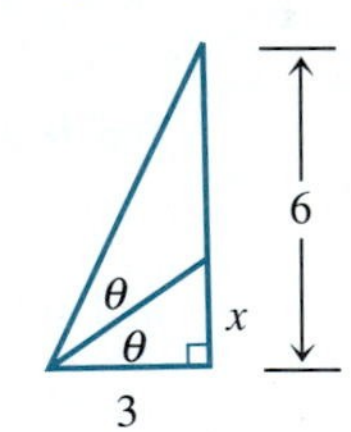

48. Use the following figure to find a.

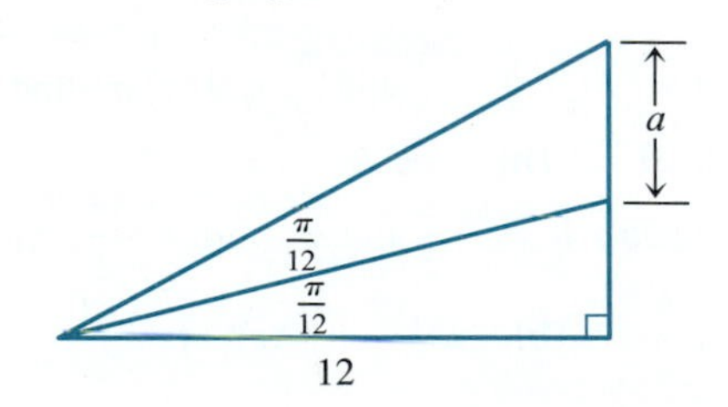

49. Given that $\sin \theta = \frac{a}{4}$, express $\cos 2\theta$ as a function of a.

50. Given that $\cos \theta = \frac{a}{4}$, express $\tan 2\theta$ as a function of a.

51. Derive a formula for $\sin 3\theta$ in terms of trigonometric functions of $\sin \theta$.

52. Derive a formula for $\tan 3\theta$ in terms of trigonometric functions of $\tan \theta$.

53. Derive the half-angle formula $\cos \frac{\theta}{2}$.

54. Derive the half-angle formula $\tan \frac{\theta}{2}$.

55. Derive the *product formula*

$$\sin A \sin B = \frac{1}{2}[\cos(A - B) - \cos(A + B)].$$

HINT: Subtract the formulas for $\cos(A - B)$ and $\cos(A + B)$.

56. Use a technique similar to that of Exercise 55 to derive the three other product formulas.

$$\cos A \cos B = \frac{1}{2}[\cos(A + B) + \cos(A - B)]$$

$$\sin A \cos B = \frac{1}{2}[\sin(A + B) + \sin(A - B)]$$

$$\cos A \sin B = \frac{1}{2}[\sin(A + B) - \sin(A - B)]$$

57. Use the results of Exercises 55 and 56 to express each of the following as a sum or difference.
(a) $\sin 5x \cos 3x$ **(b)** $\sin 5x \sin 3x$
(c) $\cos x \cos \frac{x}{2}$

58. Substitute $u = A + B$ and $v = A - B$ into the formulas of Exercises 55 and 56 to derive the following *sum formulas*.

$$\sin u + \sin v = 2 \sin \frac{u + v}{2} \cos \frac{u - v}{2}$$

$$\sin u - \sin v = 2 \cos \frac{u + v}{2} \sin \frac{u - v}{2}$$

$$\cos u + \cos v = 2 \cos \frac{u + v}{2} \cos \frac{u - v}{2}$$

$$\cos u - \cos v = -2 \sin \frac{u + v}{2} \sin \frac{u - v}{2}$$

59. Use the results of Exercise 58 to express each of the following as a product.
(a) $\sin 5x + \sin 3x$ **(b)** $\cos 7x - \cos 4x$
(c) $\sin 3x - \sin \frac{1}{2}x$

QUESTIONS FOR THOUGHT

60. Verify the identity $\tan \theta = \dfrac{\sin 2\theta}{1 + \cos 2\theta}$ and use this result to derive the following alternative form of a half-angle formula for the tangent function:

$$\tan \frac{\theta}{2} = \frac{\sin \theta}{1 + \cos \theta}$$

61. Find $\sin 105°$ using the addition formula for $\sin(A + B)$ as illustrated in Example 1 of Section 8.1. Look back to Example 8 of this section, where the $\sin 105°$ was found using the half-angle formula. How do the answers compare?

8.3 The Law of Sines and the Law of Cosines

In our previous discussion of the trigonometry of the right triangle, we saw that the trigonometric functions of acute angles can be defined as the ratios of the sides of a right triangle. A natural question that comes to mind is whether there are any trigonometric relationships among the sides of a general triangle. The answer is affirmative; the relationships are known as the *Law of Sines* and the *Law of Cosines*.

Let's consider the following right triangle situation, which we have encountered before.

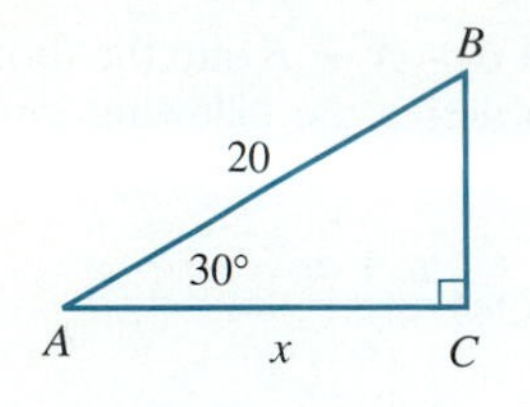

FIGURE 8.7

EXAMPLE 1 Find x (Figure 8.7).

Solution From Figure 8.7 we have

$$\cos 30° = \frac{x}{20} \Rightarrow \frac{\sqrt{3}}{2} = \frac{x}{20} \Rightarrow \boxed{x = 10\sqrt{3}}$$

Note that we could also have found the length of $\overline{AC}$ by using the sine function of angle B. ■

Every triangle consists of six parts: three sides and three angles. One way of viewing the previous example is that from certain given information about a right triangle, we can find any of the remaining parts, which quite naturally leads us to the following question: How much information about a *general* triangle is necessary for us to find any of the missing parts? Finding the missing sides or angles of a triangle from some given information is called **solving the triangle.**

In a course on plane geometry, a discussion of congruent triangles (triangles that are identical) gives rise to the fact that two triangles are congruent if they satisfy one of the following congruency conditions:

Notation	*Meaning*
SSS	Three sides of the triangle are given
SAS	Two sides and the angle included between them are given.
ASA	Two angles and the side included between them are given.
SAA	Two angles and one side (not the included one) are given.

Why is the SAA situation actually equivalent to the ASA situation?

From geometry we know that SSS, SAS, ASA, and SAA determine *unique* triangles. In trigonometry the Law of Sines and the Law of Cosines actually allow us to find the missing sides or angles.

The Law of Sines

We will continue to use the convention of labeling the angles of a triangle with A, B, and C and the sides opposite these angles as a, b, and c, respectively.

In Figure 8.8 we have drawn two triangles ABC (with $\angle A$ in standard position), one with $\angle A$ acute and the other with $\angle A$ obtuse.

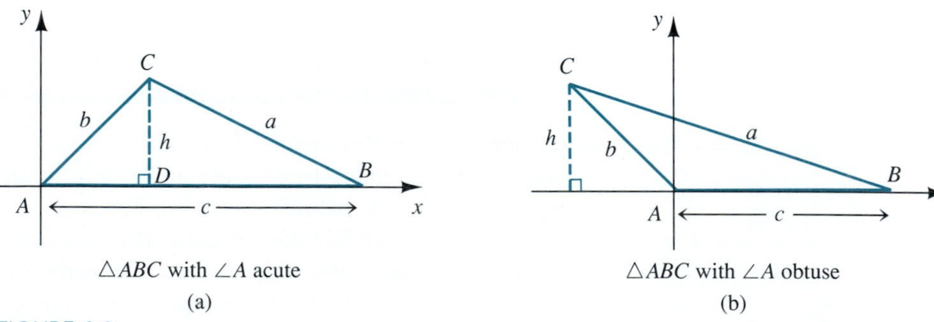

$\triangle ABC$ with $\angle A$ acute
(a)

$\triangle ABC$ with $\angle A$ obtuse
(b)

FIGURE 8.8

Looking at both parts (a) and (b) of Figure 8.8, we can see that

$$\sin A = \frac{h}{b} \quad \text{and} \quad \sin B = \frac{h}{a} \qquad \textit{And so we have}$$

$$b \sin A = h \quad \text{and} \quad a \sin B = h \qquad \textit{Which gives us}$$

$$b \sin A = a \sin B \qquad \textit{Dividing both sides by ab, we get}$$

$$\frac{\sin A}{a} = \frac{\sin B}{b}$$

If we reorient these triangles so that $\angle C$ is in standard position with a along the positive x-axis, we have

$$\frac{\sin B}{b} = \frac{\sin C}{c}$$

Combining these results, we have the **Law of Sines.**

THE LAW OF SINES

In $\triangle ABC$ with sides a, b, and c, we have

$$\frac{\sin A}{a} = \frac{\sin B}{b} = \frac{\sin C}{c}$$

Note that the first equality in the Law of Sines may be rewritten as

$$\frac{a}{b} = \frac{\sin A}{\sin B}$$

In words then, the Law of Sines says that in any triangle, *the ratio of any two sides is equal to the ratio of the sines of the angles opposite those sides.*

We also have an alternative form of the Law of Sines obtained by inverting each of the ratios. We will use whichever form seems to be the most appropriate.

From now on, unless otherwise specified, we will round all answers to the *nearest tenth.*

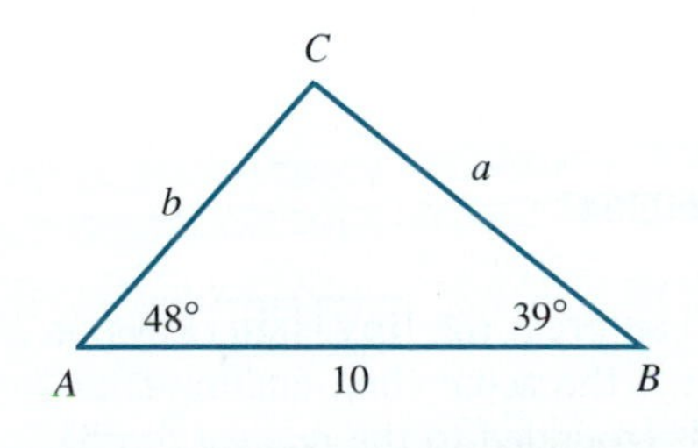

FIGURE 8.9

EXAMPLE 2 In Figure 8.9, solve $\triangle ABC$: $\angle A = 48°$, $\angle B = 39°$, $c = 10$.

Solution Recall that solving the triangle means finding all missing sides or angles. Since the angle sum of a triangle is 180°, we see that $\boxed{\angle C = 93°}$.

We find a by using the Law of Sines, which gives

$$\frac{a}{\sin 48°} = \frac{10}{\sin 93°}$$

$$a = \frac{10 \sin 48°}{\sin 93°} = 7.4416468$$

$$\boxed{a = 7.4} \qquad \textit{Rounded to the nearest tenth}$$

Similarly, to find b we have

$$\frac{b}{\sin 39^\circ} = \frac{10}{\sin 93^\circ}$$

$$b = \frac{10 \sin 39^\circ}{\sin 93^\circ} = 6.3018404$$

$$\boxed{b = 6.3} \quad \textit{Rounded to the nearest tenth}$$

The information given in this example corresponds to the ASA situation, and the Law of Sines allows us to find the missing sides of the triangle. ■

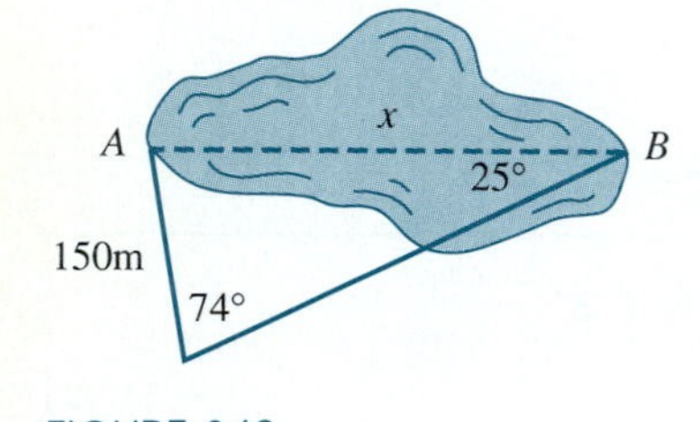

FIGURE 8.10

EXAMPLE 3 Find the distance between points A and B on opposite sides of a lake, as indicated in Figure 8.10. Round your answer to the nearest meter.

Solution The given information corresponds to the SAA situation.
Using the Law of Sines we have

$$\frac{x}{\sin 74^\circ} = \frac{150}{\sin 25^\circ}$$

$$x = \frac{150 \sin 74^\circ}{\sin 25^\circ} = 341.18084$$

$$\boxed{x = 341 \text{ meters}} \quad \textit{Rounded to the nearest meter}$$ ■

The Ambiguous Case

In the beginning of this section we listed the various congruence relations for triangles. You probably recall that there is no SSA relationship; that is, specifying two sides of a triangle and one of the nonincluded angles does not completely determine the triangle, as the following example illustrates.

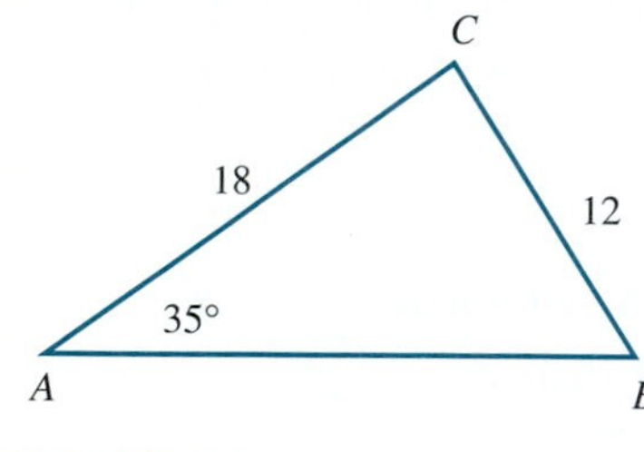

FIGURE 8.11

EXAMPLE 4 In $\triangle ABC$ find the measure of $\angle B$ (Figure 8.11).

Solution Applying the Law of Sines to find $\angle B$ we have

$$\frac{\sin B}{18} = \frac{\sin 35^\circ}{12}$$

$$\sin B = \frac{18 \sin 35^\circ}{12} = 0.8603647$$

With 0.8603647 on the calculator display, we can press the **inv** **sin** keys on a calculator to get angle B. However, we will only get the acute (first quadrant) angle whose sine is 0.8603647, which gives $B = 59.4^\circ$ (rounded to the nearest tenth). Since the sine function is also positive in quadrant II, a second quadrant angle with a reference angle of 59.4° will also work, and so B can also be 120.6°.

Therefore, the answer to this example is $\boxed{B = 59.4^\circ \text{ or } 120.6^\circ}$.

This answer means that we may draw two triangles with the given information. The two triangles appear in Figure 8.12(a) and (b).

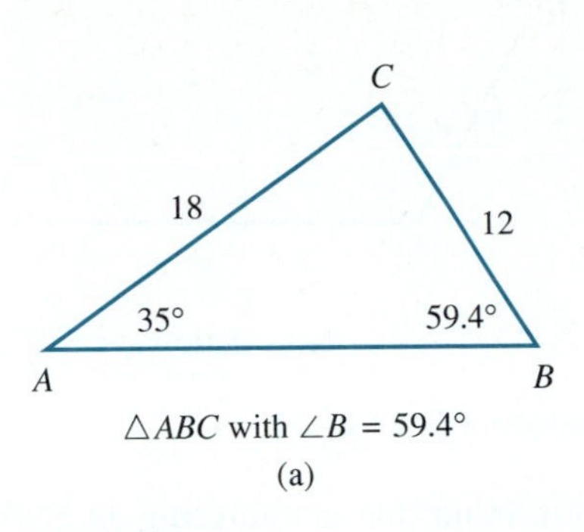

$\triangle ABC$ with $\angle B = 59.4°$

(a)

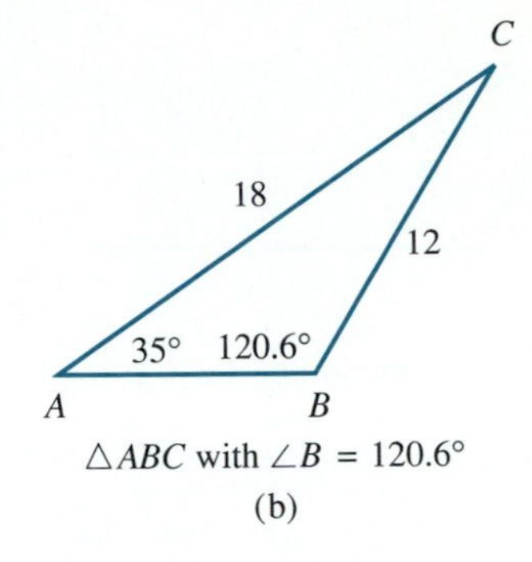

$\triangle ABC$ with $\angle B = 120.6°$

(b)

FIGURE 8.12

As this last example illustrates, there are certain cases of SSA where there may be two possible triangles. This situation is often referred to as the *ambiguous case*. We now illustrate the various possibilities that exist for the SSA situation in the case where angle A is acute. We label the given sides as a and b and the given angle as A.

Case (i) pertains to the situation depicted in Example 4.
Note that there are two possible triangles because a is greater than h but less than b.

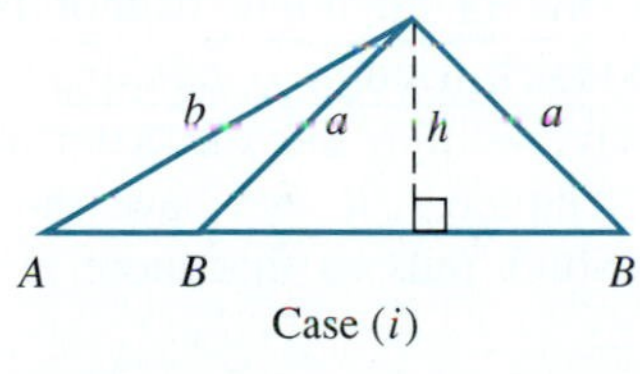

Case (i)

In case (ii) there is no possible triangle because a is less than h.

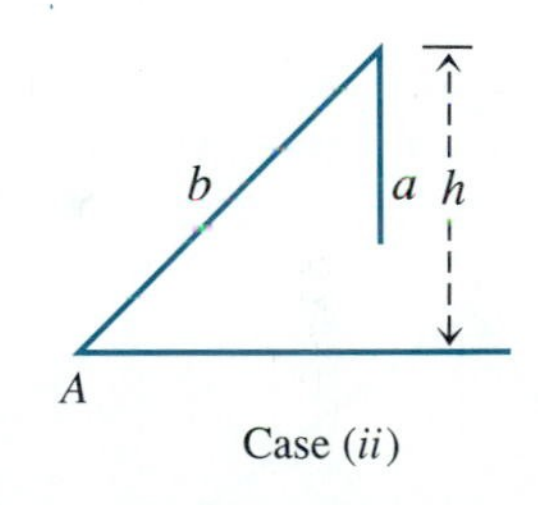

Case (ii)

In case (iii) there is only one possible triangle, because $a = h$.

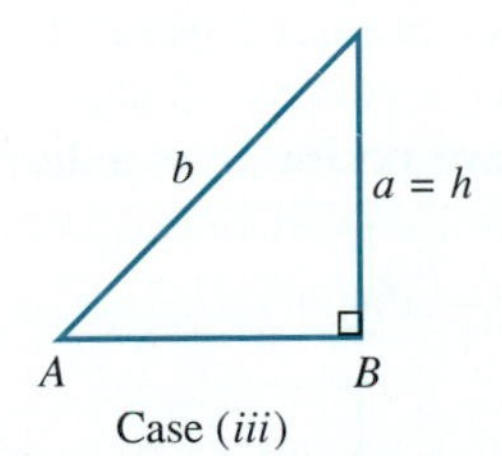

Case (iii)

In case (iv) there is also only one possible triangle, because a is greater than b.

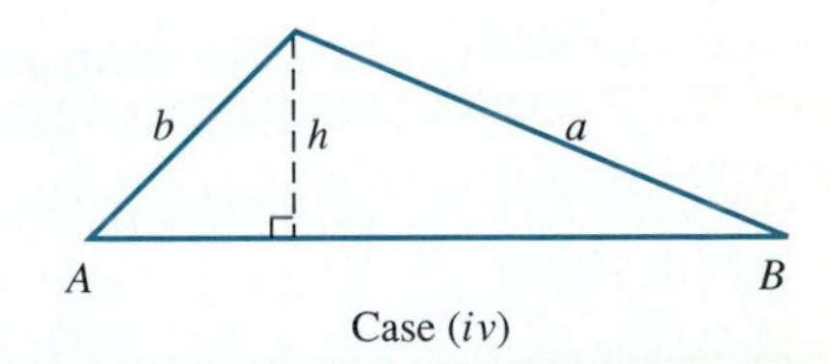

Case (iv)

In effect, we are comparing the relative sizes of a, b, and h. Since in each of these cases $\sin A = \dfrac{h}{b}$ and so $h = b \sin A$, we are actually comparing the size of a to the size of $b \sin A$.

If $a < b \sin A$ then $a < h$, which means that no such triangle is possible.

In Figure 8.13 we illustrate the three possibilities for the SSA case where $\angle A$ is obtuse.

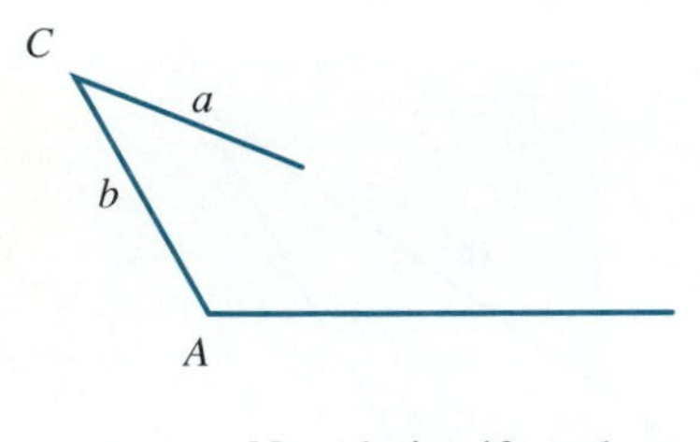

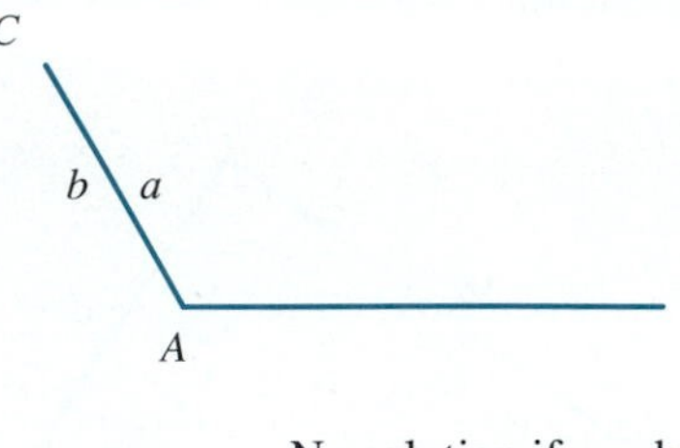

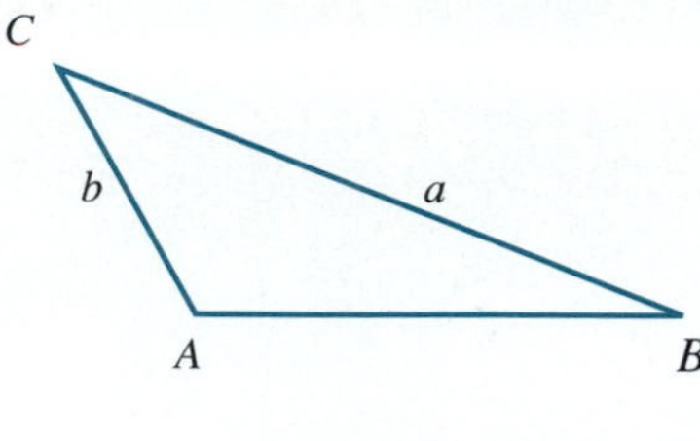

FIGURE 8.13 The three possibilities for *SSA* where $\angle A$ is obtuse

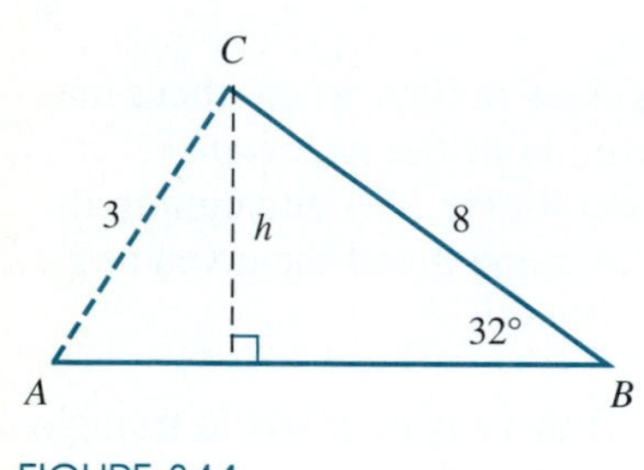

FIGURE 8.14

Keep in mind that in analyzing an SSA situation, it does not matter which side we designate as a and which as b, as long as the given angle is not included between the two given sides.

EXAMPLE 5 Solve $\triangle ABC$ with $a = 8$, $b = 3$, and $\angle B = 32°$.

Solution We draw a tentative sketch of $\triangle ABC$ (Figure 8.14) and use the Law of Sines to find $\angle A$.

$$\frac{\sin A}{8} = \frac{\sin 32°}{3}$$

$$\sin A = \frac{8 \sin 32°}{3} = 1.413118$$

Since the sine of an angle cannot be greater than 1, this equation has no solutions and no such triangle is possible.

Alternatively, we may use the criteria listed previously. From Figure 8.14, we can see that in triangle ABC we have the height $h = 8 \sin 32° \approx 4.24$, which is greater than 3, which tells us that there is no such triangle. (See case (*ii*) on page 501.) ■

Consider the following triangle

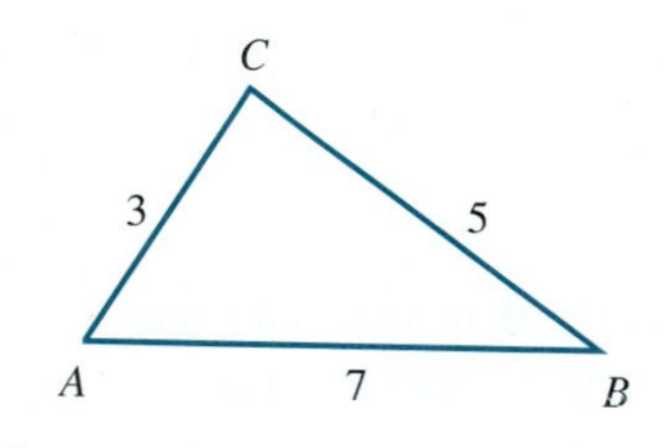

Can we use the Law of Sines to find $\angle A$?

The Law of Cosines

The Law of Cosines, which we shall derive in a moment, addresses the SAS and SSS situations, which are not covered by the Law of Sines.

Let us consider a general triangle ABC and place it in a coordinate system so that $\angle C$ is in standard position, as indicated in Figure 8.15.

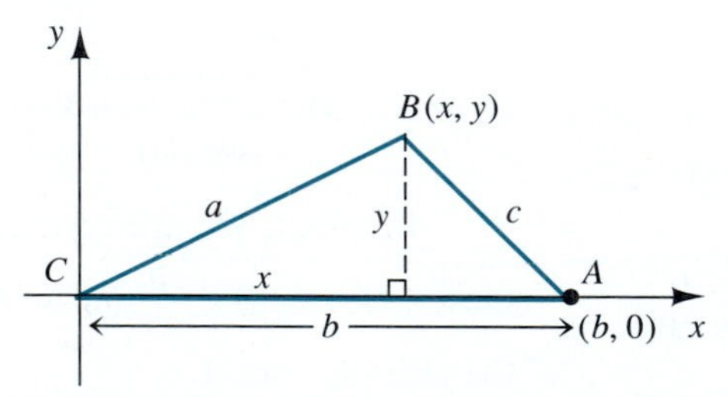

FIGURE 8.15 $\triangle ABC$ with $\angle C$ in standard position

Note that we have drawn $\triangle ABC$ with $\angle C$ acute. It is left to the reader to verify that the following argument is equally valid when $\angle C$ is obtuse (or right).

The altitude from vertex B to side $\overline{AC}$ has been labeled y. From Figure 8.15 we have

$$\cos C = \frac{x}{a} \qquad \sin C = \frac{y}{a}$$

$$x = a \cos C \qquad y = a \sin C$$

We now use the distance formula to compute the length of $\overline{AB}$, which we designate as c, using the coordinates we have just found.

$$c = \sqrt{(x - b)^2 + (y - 0)^2}$$

$$= \sqrt{(a \cos C - b)^2 + (a \sin C)^2}$$

$$= \sqrt{a^2(\cos C)^2 - 2ab \cos C + b^2 + a^2(\sin C)^2}$$

Group the first and last terms and factor out a^2.

$$= \sqrt{a^2(\sin^2 C + \cos^2 C) - 2ab \cos C + b^2}$$

Since $\sin^2 C + \cos^2 C = 1$ we have

$$c = \sqrt{a^2 - 2ab \cos C + b^2}$$

Square both sides of the equation.

$$c^2 = a^2 + b^2 - 2ab \cos C$$

This last equation is called the **Law of Cosines.**

Note that if $\angle C$ is a right angle, then $\cos C = 0$ and the Law of Cosines reduces to the Pythagorean theorem. In fact, we should view the Law of Cosines as a generalization of the Pythagorean theorem.

Keep in mind that we could have begun by putting either angle A or angle B in standard position, which would give us two more forms of the Law of Cosines.

THE LAW OF COSINES

In any triangle ABC, the square of any side is equal to the sum of the squares of the other two sides minus twice the product of the other two sides times the cosine of their included angle.

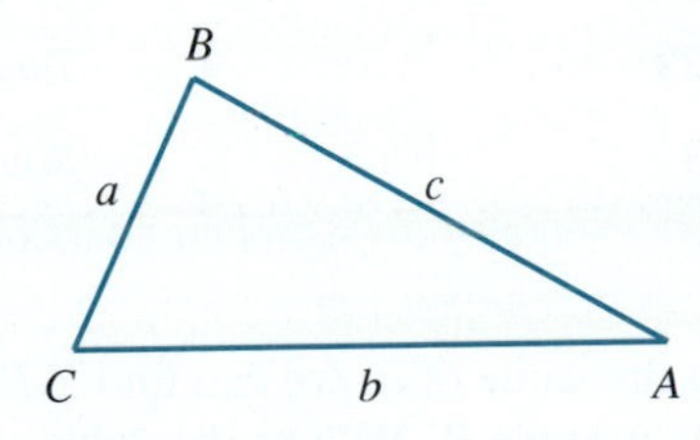

$$c^2 = a^2 + b^2 - 2ab \cos C$$

$$b^2 = a^2 + c^2 - 2ac \cos B$$

$$a^2 = b^2 + c^2 - 2bc \cos A$$

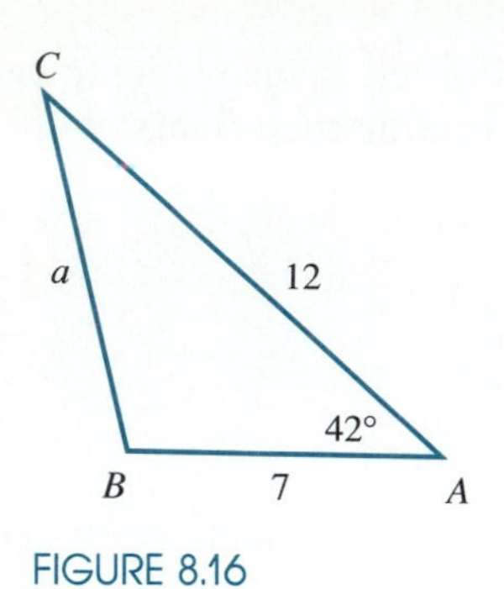

FIGURE 8.16

EXAMPLE 6 Use Figure 8.16 to find a to the nearest tenth.

Solution Using the Law of Cosines we have

$$a^2 = b^2 + c^2 - 2bc \cos A$$

Substituting the values for this triangle we get

$$a^2 = 12^2 + 7^2 - 2(12)(7)\cos 42^\circ$$

$$a^2 = 193 - 168 \cos 42^\circ = 68.151669$$

Taking square roots and rounding to the nearest tenth we get

$$\boxed{a = 8.3}$$

■

EXAMPLE 7 Use Figure 8.17 to find the value of a and $\angle B$.

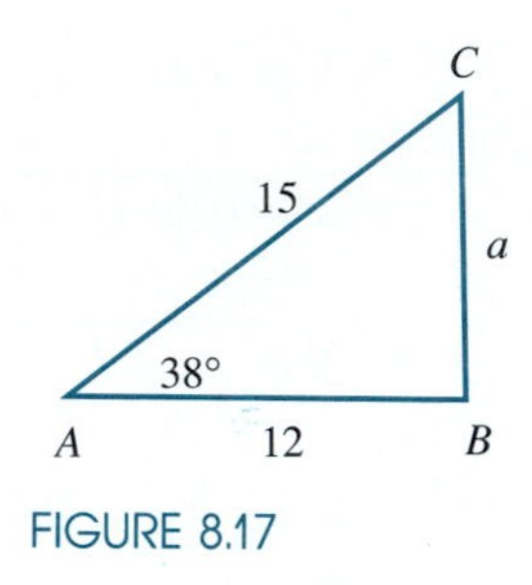

FIGURE 8.17

How do we know whether to use the Law of Sines or the Law of Cosines?

Solution Now that we have both the Law of Sines and the Law of Cosines available to us, we need to take a moment to determine which law to use in each particular situation. In order to solve a proportion, we must know three out of the four members of the proportion. Thus in order to use the Law of Sines to find a missing part of a triangle, we must know at least one angle and its corresponding side. This is precisely another way of saying that the Law of Sines applies to ASA, SAA, and SSA situations.

In order to employ the Law of Cosines to solve for a missing part of a triangle, we need to know two sides and the included angle or all three sides—that is, the Law of Cosines applies to SAS and SSS situations.

Since this example is an SAS situation, we begin by applying the Law of Cosines to find a.

$$a^2 = 12^2 + 15^2 - 2(12)(15) \cos 38^\circ$$

$$a^2 = 144 + 225 - 360(0.7880108)$$

$$a^2 = 85.316129$$

Take square roots.

$$a = 9.236673$$

Rounding to the nearest tenth we get

$$\boxed{a = 9.2}$$

Now that we have the value of a, we can find $\angle B$ by applying the form of the Law of Cosines that involves angle B. We use the value $a = 9.236673$ and round the answer at the end of the computation.

$$b^2 = a^2 + c^2 - 2ac\cos B$$

$$15^2 = (9.236673)^2 + 12^2 - 2(9.236673)(12)\cos B$$

$$225 = 85.316129 + 144 - 221.68015\cos B$$

$$225 = 229.31613 - 221.68015\cos B$$

$$-4.31613 = -221.68015\cos B$$

$$\cos B = \frac{-4.31613}{-221.68015} = 0.0194701$$

Using the **inv** **cos** *keys and rounding to the nearest tenth, we get*

$$\boxed{B = 88.9^\circ}$$

It is worthwhile noting that once we have found the value for a, we could have used the Law of Sines to find $\angle B$.

$$\frac{\sin B}{15} = \frac{\sin 38^\circ}{9.236673}$$

$$\sin B = \frac{15\sin 38^\circ}{9.236673}$$

Using a calculator we get

$$\sin B = 0.9998104$$

Again using a calculator and rounding to the nearest tenth, we have

Why can't $\measuredangle B = 91.1^\circ$?

$$\boxed{B = 88.9^\circ}$$

Finding angle B via the Law of Sines gives a simpler computation, but the choice of methods is up to you. ■

We can summarize our discussion on using the Law of Sines and the Law of Cosines as follows.

USING THE LAW OF SINES AND THE LAW OF COSINES

Use the **Law of Sines** when two sides and one of their corresponding angles or two angles and any side are known. (This says that the Law of Sines applies to the ASA, SAA, and SSA situations.)

Use the **Law of Cosines** when two sides and their included angle or all three sides are known. (This says that the Law of Cosines applies to the SAS and SSS situations.)

EXAMPLE 8 A forest ranger is located in an observation tower and notices two fires at distances of 4 and 7 miles, respectively, from the tower. If the angle between the lines of sight to the two fire points is 137°, how far apart are the fires?

Solution The sketch in Figure 8.18 represents the given information, with the observation tower at point C and the fires located at points A and B.

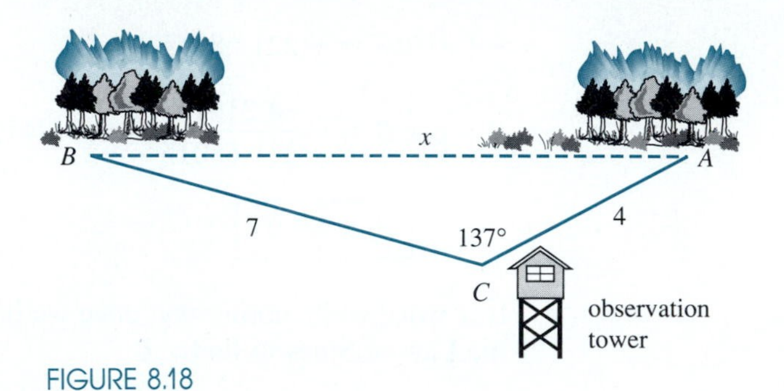

FIGURE 8.18

We may apply the Law of Cosines to find the length of $\overline{AB}$, which we have labeled as x.

$$x^2 = 4^2 + 7^2 - 2(4)(7)\cos 137°$$

$$x^2 = 16 + 49 - 56(\cos 137°)$$

Don't forget that 137° is a second-quadrant angle. Its cosine is negative.

$$x^2 = 105.95581$$

$$x = 10.293484$$

Therefore, the fires are approximately $\boxed{10.3 \text{ mi apart}}$. ■

Trigonometry can be very useful in solving navigation problems. The direction of travel of a ship or a plane, called its *course*, is usually indicated with reference to due north or due south. Thus a direction of N40°E means that the direction of travel is along a ray making an angle of 40° in the clockwise direction from due north. A direction of S35°E means that the direction of travel is along a ray making an angle of 35° in the counterclockwise direction from due south. Several such directions of travel are illustrated in Figure 8.19.

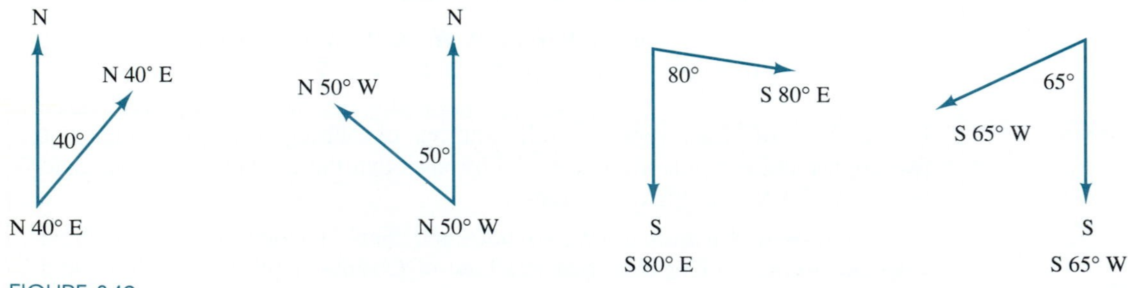

FIGURE 8.19

EXAMPLE 9 A submarine uses sonar to determine that a ship is 4 miles due east and traveling at 10 mi/hr in the direction N62°W.

(a) If the submarine travels at a speed of 18 mi/hr, approximate to the nearest tenth of a degree the direction the submarine should travel to intercept the ship.

(b) How long will it take the submarine to intercept the ship?

Solution

WHAT ARE WE TRYING TO FIND?

The direction the submarine should travel to intercept the ship and the time it will take to intercept the ship.

WHAT INFORMATION ARE WE GIVEN?

The distance from the submarine to the ship, the direction of travel of the ship, and the speed of both the ship and the submarine. We draw Figure 8.20. The submarine is at point A, the observed ship is at point B, and the point of intercept is at C. Note that $\angle CBA = 90° - 62° = 28°$.

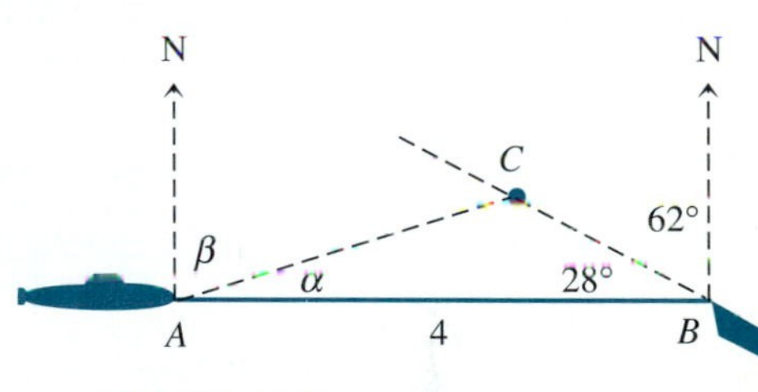

FIGURE 8.20

HOW CAN WE RESTATE THE PROBLEM?

In terms of the diagram, we are trying to find angle β and the time it takes to travel from A to C.

HOW DO WE USE THE SPEED OF THE SUB?

If we let t be the number of hours it takes the submarine to intercept the ship, then we can express $|\overline{BC}|$ and $|\overline{AC}|$ in terms of t. Since the ship is traveling at 10 mi/hr and distance equals rate times time, $|\overline{BC}| = 10t$. Similarly, since the submarine is traveling at 18 mi/hr, $|\overline{AC}| = 18t$. The diagram in Figure 8.21 on page 508 represents all the information we have thus far.

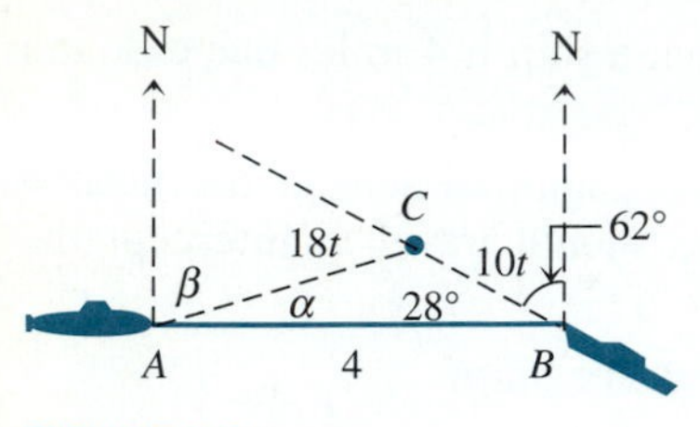

FIGURE 8.21

CAN WE NOW FIND ANGLE β?

(a) Not directly, but we can use the Law of Sines to find angle α, which we can then subtract from 90° to find angle β.

$$\frac{\sin \alpha}{|\overline{BC}|} = \frac{\sin 28^\circ}{|\overline{AC}|}$$

$$\frac{\sin \alpha}{10t} = \frac{\sin 28^\circ}{18t} \qquad \textit{Solve for } \sin \alpha.$$

$$\sin \alpha = \frac{10t \sin 28^\circ}{18t} = \frac{10 \sin 28^\circ}{18}$$

Note how the t's cancel out.

$$\sin \alpha = 0.2608175 \qquad \textit{And so}$$

$$\alpha = 15.1^\circ$$

Rounded to the nearest tenth.

Thus, $\beta = 90^\circ - 15.1^\circ = 74.9^\circ$, and so the submarine should travel in the approximate direction of $\boxed{\text{N}74.9^\circ\text{E}}$.

HOW DO WE FIND THE TIME IT TAKES THE SUB TO INTERCEPT THE SHIP?

If we can find $|\overline{AC}|$, then we can use the speed of the sub to compute the time needed to travel from A to C.

HOW DO WE FIND $|\overline{AC}|$?

(b) We first find that $\angle ACB = 180^\circ - 28^\circ - 15.1^\circ = 136.9^\circ$; then we again use the Law of Sines to find distance from A to C.

$$\frac{|\overline{AC}|}{\sin 28^\circ} = \frac{4}{\sin 136.9^\circ}$$

$$|\overline{AC}| = \frac{4 \sin 28^\circ}{\sin 136.9^\circ} \approx 2.7 \text{ miles}$$

Since $|\overline{AC}| = 18t$, we have $18t = 2.7$ and $t = 0.15$. Thus it takes approximately $\boxed{0.15 \text{ hours}}$ or about $\boxed{9 \text{ minutes}}$ for the submarine to intercept the ship. ■

EXERCISES 8.3

A scientific calculator will be helpful for this exercise set.

In Exercises 1–10, use the given information to find the missing parts of $\triangle ABC$. Approximate your answers to the nearest tenth.

1. $A = 80°$, $B = 35°$, $a = 12$
2. $A = 80°$, $B = 35°$, $c = 12$
3. $A = 72°$, $a = 24$, $b = 15$
4. $A = 72°$, $b = 24$, $c = 15$
5. $a = 6$, $b = 9$, $c = 10$
6. $B = 53°$, $b = 7$, $c = 10$
7. $B = 110°$, $C = 25°$, $c = 16$
8. $a = 15$, $b = 12$, $c = 5$
9. $A = 138°$, $b = 5$, $c = 11$
10. $A = 80°$, $C = 41°$, $b = 30$

In Exercises 11–14, determine how many triangles ABC can be constructed using the given information.

11. $A = 54°$, $a = 7$, $b = 10$
12. $A = 73°$, $a = 48$, $b = 50$
13. $C = 134°$, $a = 35$, $b = 40$
14. $B = 61°$, $a = 12$, $b = 15$
15. A surveyor wishes to measure the distance between points A and B on opposite sides of a river as indicated in the accompanying figure. The surveyor measures the distance between points A and C to be 200 meters, and uses an instrument called a transit to find that $\angle B = 63.8°$ and $\angle C = 84.2°$ Find the distance from A to B.

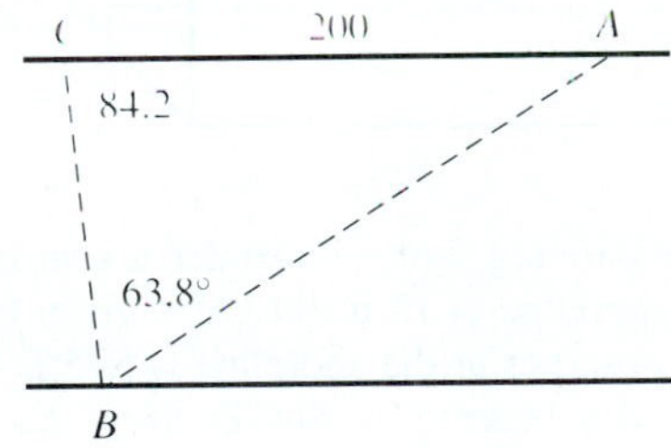

16. Two trains leave a train station at 10:00 A.M. traveling along straight tracks at 120 and 150 km/hr, respectively. If the angle between their directions of travel is 118°, how far apart are the trains at 10:40 A.M.?
17. A weather balloon is hovering directly above the line joining points A and B which are 4.6 km apart. If the angles of elevation to the balloon from points A and B are 28°50′ and 52°10′, respectively, find the altitude of the balloon.
18. Two observation towers A and B are located 15 miles apart in a national forest. Both observers sight a fire at point C so that $\angle CAB = 73°$ and $\angle CBA = 59°$. How far is the fire from tower B?
19. A straight tunnel with endpoints P and Q is to be constructed through a mountain. From point R the surveyor finds $\overline{PR} = 500$ meters, $\overline{RQ} = 600$ meters, and measures $\angle R = 76°40'$ as indicated in the following figure. Find the length of the tunnel.

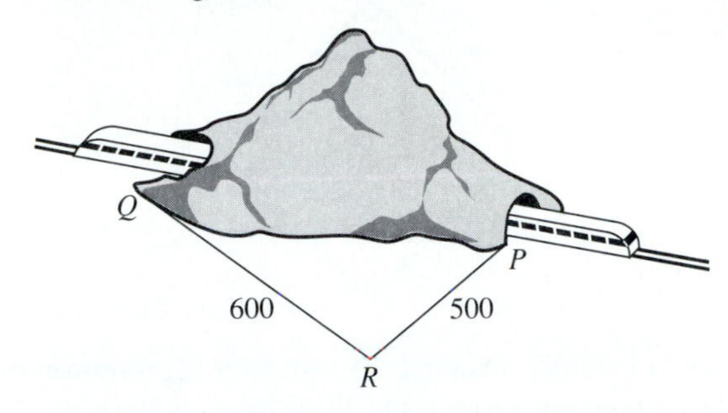

20. A telephone pole is being supported by two guy wires that are attached to the top of the pole and anchored into the ground on opposite sides of the pole at points A and B, which are 80 feet apart. If the angles of elevation at A and B are 70° and 58°, respectively, find the lengths of both wires and the height of the pole.
21. In the accompanying figure find the length of $\overline{CD}$. HINT: First find the length of $\overline{BD}$.

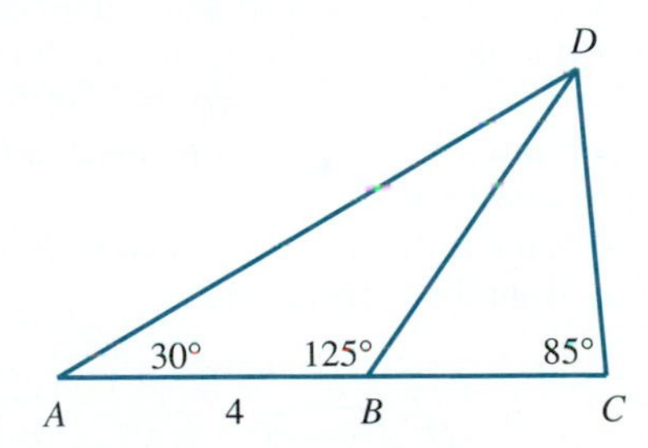

22. Suppose that, ideally, a solar panel should be tilted at a 45° angle toward the sun. An 8-foot long solar panel is being mounted on a roof that makes a 20° angle with the horizontal. Approximate the length, s, of a strut, perpendicular to the horizontal, needed to support the panel so that it makes a 45° angle with the horizontal. (See the accompanying figure.). HINT: The two dotted lines are horizontal.

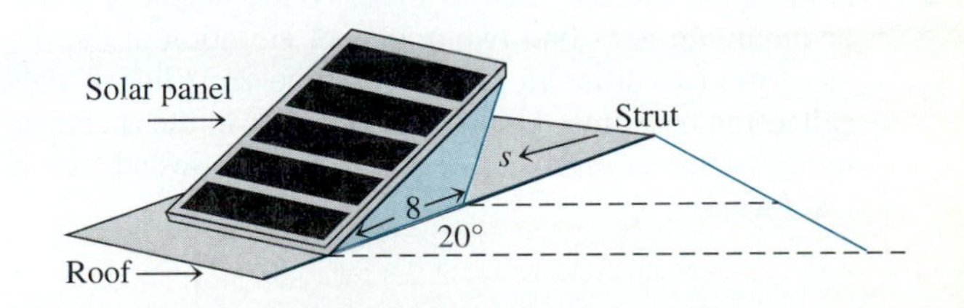

23. In the major leagues, a baseball "diamond" (it's actually a square) has its bases 90 feet apart and its pitcher's mound 60.5 feet from home plate. See the accompanying figure. Approximate the distance from the pitcher's mound to each of the other bases.

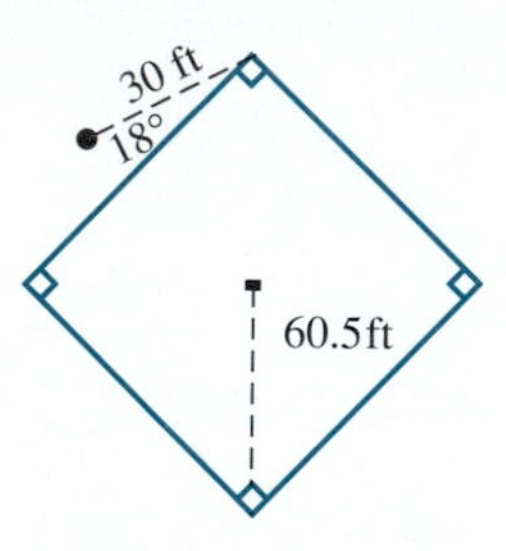

24. On a baseball diamond the shortstop is positioned in the area between second and third base. If the shortstop is "playing deep," that is, toward the outfield, 30 feet from second base and 18° off of the line joining second and third bases, find the distance between the shortstop and first base. See the figure accompanying Exercise 23.

25. A triangle has sides of length 42, 35, and 20. Approximate the size of the smallest angle of the triangle.

26. The leaning tower of Pisa is 179 feet long but due to an unstable foundation it tilts a bit more away from the perpendicular each year so that it leans at a certain angle θ as indicated in the accompanying figure. From a distance of 100 feet from the center of the base of the tower, the angle of elevation to the top of the tower is 64.7°.

(a) Approximate angle θ.

(b) Approximate the distance d that the tower is leaning away from the perpendicular.

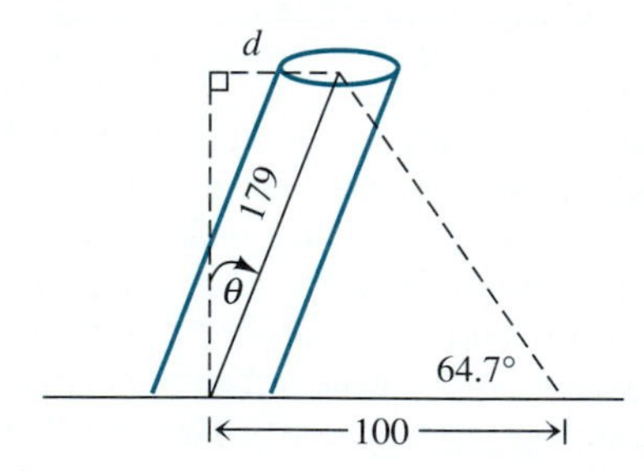

27. A common method used to measure the height of a tree or mountain is to find two angles of elevation of the object from two different points along the same line of sight called the base line. Use the information in the accompanying figure to estimate the height of a redwood tree in California.

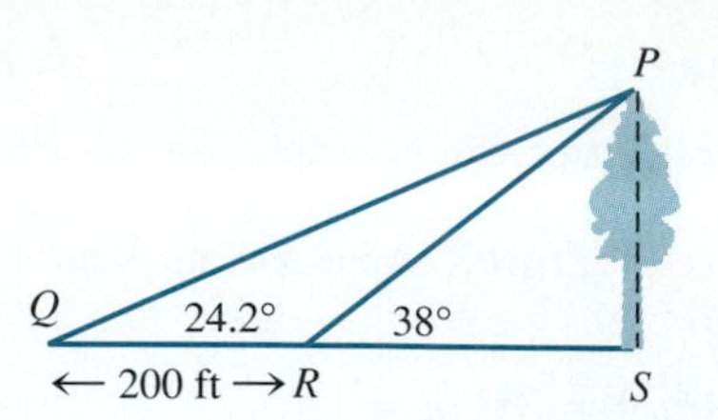

28. Using the technique described in Exercise 27, observers wanting to measure the height of a mountain above sea level set up a level baseline 1000 feet long at an altitude of 4000 feet. They measure the farther angle of elevation to be 36.4° and the closer angle of elevation to be 52.7°. Estimate how high the mountain rises above sea level.

29. One long range cannon is located at point A, and a second cannon at B is located 4 miles due east of A. From point A the direction to a target, T, is N54°E and the direction to T from B is N68°W. At what range should each of the cannons be set?

30. A rectangular box of sides 6, 8, and 10 is shown in the accompanying figure. Find the angle θ formed by the diagonal of the base with the diagonal of the 6×8 side.

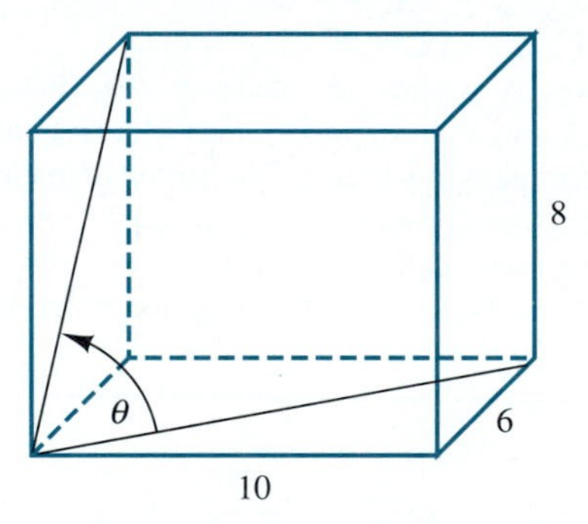

31. The accompanying figure illustrates a boat traveling parallel to a shoreline at 15 mi/hr. At a given time its bearing to an observer at the shoreline is S75°E and 10 minutes later the bearing is S69°E. How far is the boat offshore?

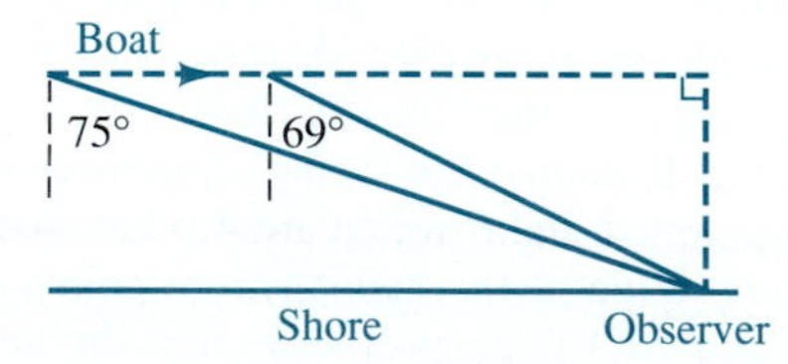

32. Three circles of radius 4, 8, and 12 are tangent to each other, as indicated in the following figure. Approximate angle α.

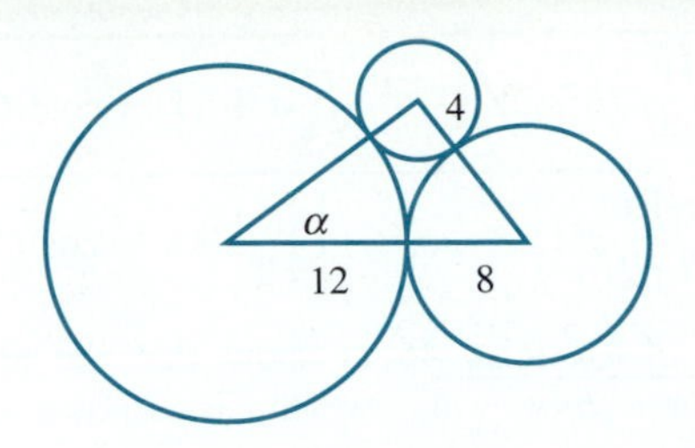

33. A ship travels from point A for 2 hours at a speed of 16 km/hr on a course of N65°E and then changes to a course of N15°E for 3 hours at the same speed. After this 5-hour period, how far is the ship from point A?
34. A plane is flying at a constant altitude. It flies from point P at a speed of 350 mi/hr on a course of S20°W for 4 hours and then changes to a course of S80°W for 5 hours at a speed of 400 mi/hr. After this 9-hour period, how far is the plane from point P?
35. In order to measure the height of a mountain, a surveyor stands at a marked spot and measures the angle of elevation to the top of the mountain to be 48.2°. She then moves 1000 feet further from the mountain and measures the angle of elevation to be 36.4°. How high is the mountain?
36. Two observers 1 mile apart notice an unidentified flying object hovering above a small town between them. Assuming they are standing at the same altitude, the angles of elevation from each observer to the UFO are 28° and 46°, respectively. At what altitude was the UFO hovering?
37. While driving in a car at 60 km/hr toward a mountain, a passenger notes that the angle of elevation to the top of the mountain is 12°. Five minutes later, he measures the angle of elevation at 18°. How tall is the mountain?
38. While driving in a car at 30 km/hr toward a radio tower, a passenger notes that the angle of elevation to the top of the tower is 7°. One minute later, she measures the angle of elevation at 10°. How far is she from the tower, and how tall is the tower?
39. To measure the distance between two markers on opposite sides of a house, Art stands at a position where he can see both markers. He knows that he is 180 feet from one marker and 220 feet from the other marker. If the angle between the two lines of sight is 73°, find the distance between the two markers.
40. A helicopter is hovering at an altitude of 1500 feet above mountain peak A of known altitude 4800 feet. A second, taller, nearby mountain peak B is viewed at an angle of depression of 50° from the helicopter and at an angle of elevation of 15° from A. See the accompanying figure. Find the distance between the two mountain peaks and the approximate altitude of B.

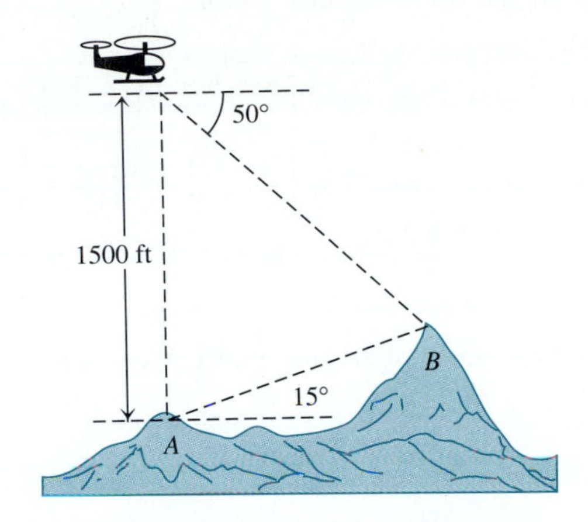

41. Find the area of the following triangle.

42. Find the area of the following triangle.

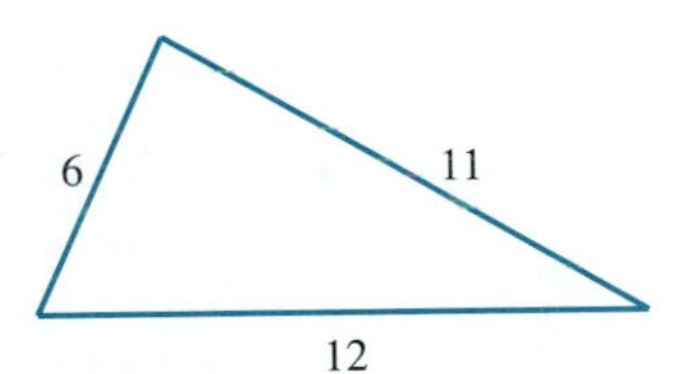

QUESTIONS FOR THOUGHT

43. In Section 6.3 we derived the fact that the area of a triangle is $A = \frac{1}{2}ab \sin C$. Show how this formula can be used to derive the Law of Sines. HINT: Use this area formula for each of the angles of the triangle.
44. The Law of Cosines says that for $\triangle ABC$, we have $c^2 = a^2 + b^2 - 2ab \cos C$. We have pointed out that if $\angle C$ is a right angle, then $\cos C = 0$ and we get $c^2 = a^2 + b^2$. Suppose a and b remain constant. Draw figures for $\triangle ABC$ where $\angle C$ is acute, right, and obtuse. What is happening to the size of c? How is this fact reflected by the Law of Cosines? HINT: What happens to the sign of $\cos C$?

The next few exercises derive a beautiful theorem called *Heron's Formula,* which expresses the area of a triangle in terms of the lengths of its sides.

45. For $\triangle ABC$ use the Law of Cosines to prove

(a) $\frac{1}{2}ab(1 + \cos C) = \left(\frac{a+b+c}{2}\right)\left(\frac{a+b-c}{2}\right)$

(b) $\frac{1}{2}ab(1 - \cos C) = \left(\frac{-a+b+c}{2}\right)\left(\frac{a-b+c}{2}\right)$

46. Let $s = \frac{a+b+c}{2}$; s is one-half of the perimeter of the triangle and is called the *semiperimeter*. Use the result of Exercise 45 to show that $\frac{1}{2}ab(1 + \cos C) = s(s - c)$ and $\frac{1}{2}ab(1 - \cos C) = (s - a)(s - b)$.

47. Starting with the formula for the area of $\triangle ABC$, $A = \frac{1}{2}ab \sin C$, we have the following:

$$\frac{1}{2}ab \sin C = \sqrt{\frac{1}{4}a^2b^2 \sin^2 C} = \sqrt{\frac{1}{4}a^2b^2(1 - \cos^2 C)}$$

$$= \sqrt{\left[\frac{1}{2}ab(1 + \cos C)\right]\left[\frac{1}{2}ab(1 - \cos C)\right]}$$

Use the results of Exercise 46 to show that the area $A = \sqrt{s(s-a)(s-b)(s-c)}$, which is known as *Heron's Formula*.

48. Use Heron's Formula to find the area of the following triangle.

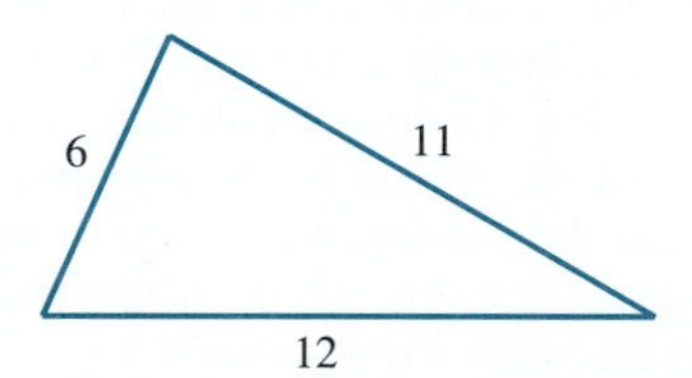

8.4 Vectors

Many physical quantities, such as length, area, volume, and temperature, can be completely specified by a single real number (provided that a system of units has been decided upon). Such a quantity is called a **scalar.** For example, in saying that a rectangle has an area of 26 sq meters or that the temperature is -3°C, the numbers 26 and -3 give a complete description of those quantities. Thus area and temperature are examples of scalar quantities. On the other hand when meteorologists talk about wind velocity, they are referring to *two* quantities: the wind's speed *and* its direction. A quantity such as velocity, for which we must specify a number, usually called its **magnitude,** and a direction, is called a **vector.**

Geometrically, we represent a vector by an arrow: The direction of the arrow indicates the direction in which the quantity is acting, whereas the length of the arrow represents the magnitude of the quantity. In most cases we refer to the arrow itself as the vector. Our discussion here is about two-dimensional vectors, or vectors in the plane.

In Figure 8.22 both arrows represent a wind of 10 mi/hr, and so they have the same magnitude. However, arrow **A** represents a 10-mi/hr wind blowing in a northeasterly direction, but arrow **B** represents a 10-mi/hr wind blowing in a northwesterly direction.

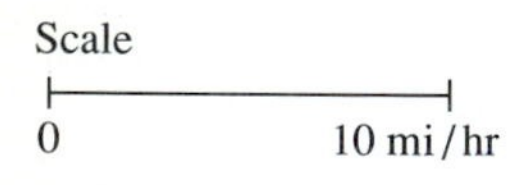

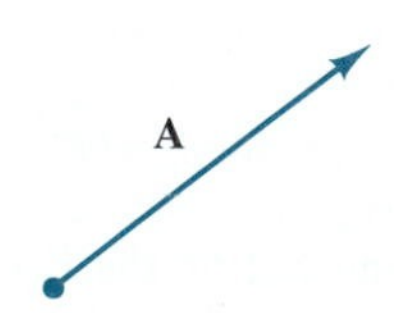

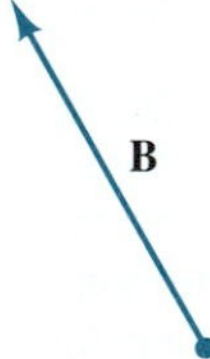

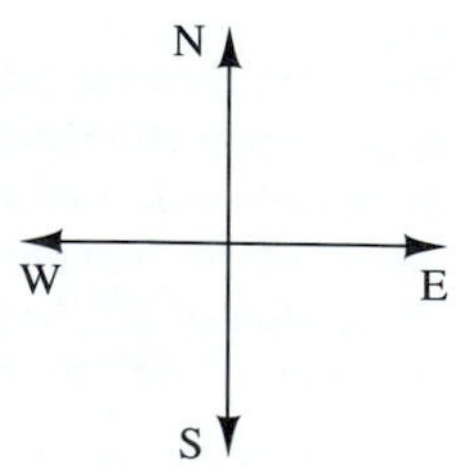

FIGURE 8.22 Two vectors representing a windspeed of 10 mi/hr. They have the same magnitude but different directions.

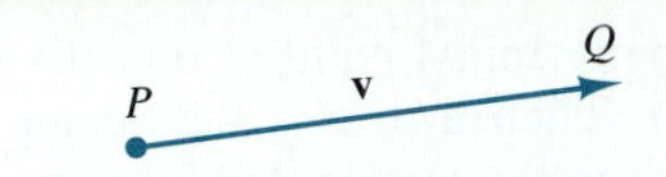

FIGURE 8.23 Vector $\overrightarrow{PQ}$, or **v**

If you use a single letter for a vector in your notes, it is advisable to write $\vec{v}$, since it is difficult to write in boldface.

Let's introduce some notation. A vector drawn from point P to point Q is denoted by $\overrightarrow{PQ}$. P is called the initial point and Q is called the terminal point. We may also denote a vector by using a boldface letter and write $\overrightarrow{PQ} = \mathbf{v}$. See Figure 8.23.

It is important to note that vectors $\overrightarrow{PQ}$ and $\overrightarrow{QP}$ are not the same. They have the same magnitude (length) but opposite directions. We denote the magnitude of a vector as $|\overrightarrow{PQ}|$ or $|\mathbf{v}|$. Thus we have $|\overrightarrow{PQ}| = |\overrightarrow{QP}|$.

Two vectors that have the same magnitude and the same direction are said to be **equal.** In Figure 8.24 vectors **A** and **B** are equal, even though they are not in the same location. We are free to move a vector from one location to another as long as we don't change its magnitude or its direction.

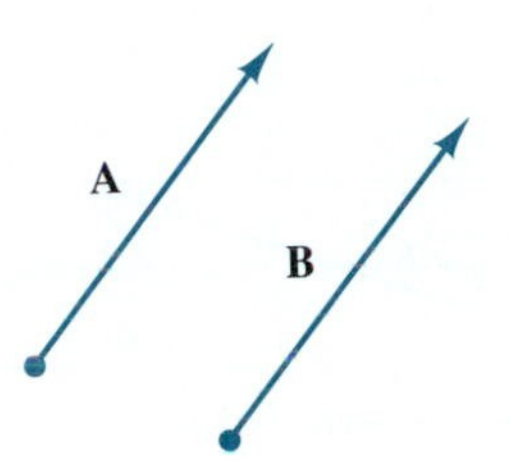

$\mathbf{A} = \mathbf{B}$

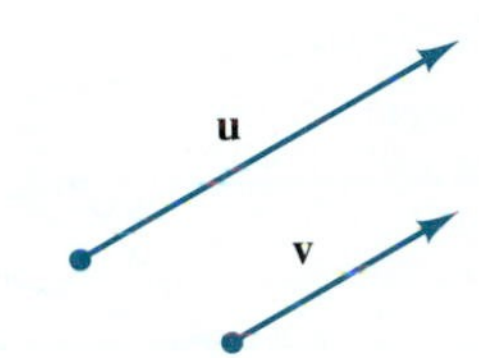

$\mathbf{u} \neq \mathbf{v}$. They have the same direction but not the same magnitude.

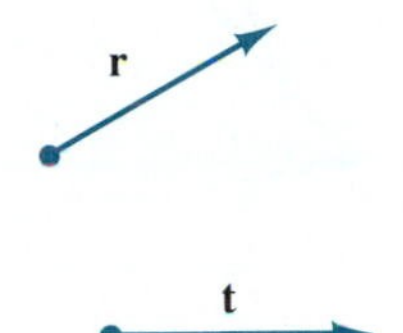

$\mathbf{r} \neq \mathbf{t}$. They have the same magnitude but not the same direction.

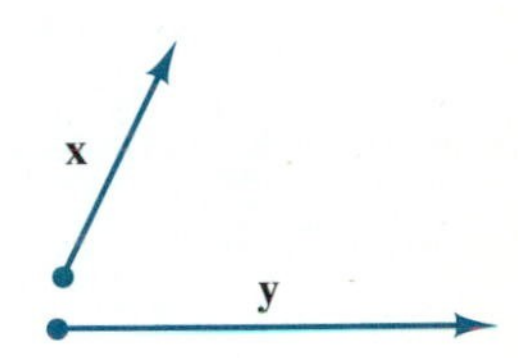

$\mathbf{x} \neq \mathbf{y}$. They have neither the same magnitude nor direction.

FIGURE 8.24 Some equal and unequal vectors

Remember that a vector must have both magnitude and direction.

Suppose we have two forces represented by vectors **u** and **v** acting on an object at point A, as represented in Figure 8.25(a). We can think of force **u** moving the object from A to the endpoint of **u** and then force **v** acting on the object; that is, we may move **v** to the endpoint of **u** and see that the object will react as if the single force represented by vector **w** is acting at point A. See Figure 8.25(b)

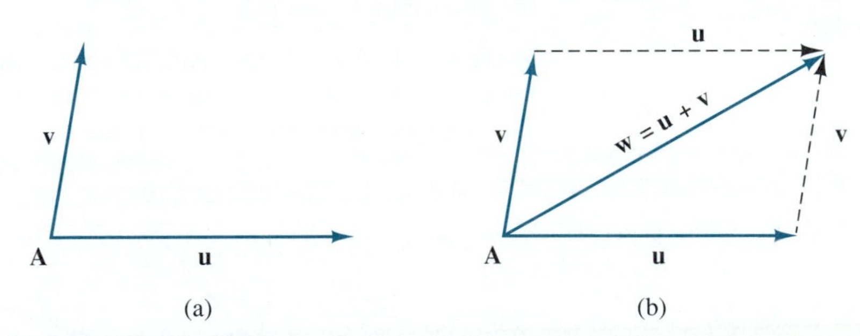

FIGURE 8.25 (a) Vectors **u** and **v** acting at point A; (b) the resultant of two vectors: **w** = **u** + **v**

The vector **w** is called the **resultant** vector of **u** and **v**. We call the resultant vector **w** the **sum** of vectors **u** and **v**.

DEFINITION Let **u** and **v** be two vectors. Move **v** to the terminal point of **u** (without changing its magnitude or direction). Then **u** + **v** is the vector that starts at the initial point of **u** and ends at the terminal point of **v**. The vector **u** + **v** is called the **sum**, or **resultant** of **u** and **v**.

If we consider two vectors **u** and **v**, Figure 8.26 illustrates how we would apply the above definition to find **u** + **v**.

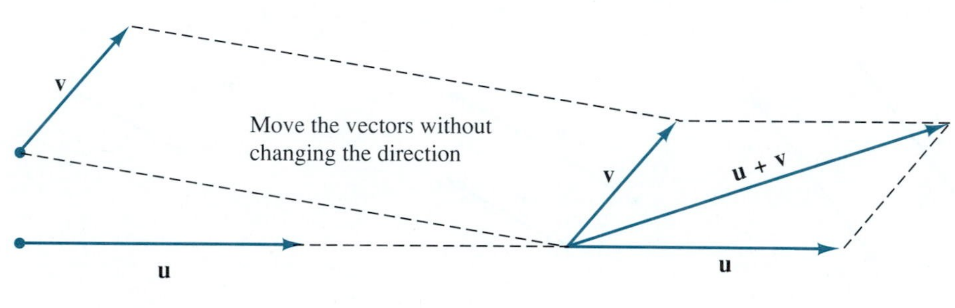

FIGURE 8.26

If we look at the lower triangle in Figure 8.25(b) we can see that the resultant vector is **u** + **v**, whereas the upper triangle shows that the resultant vector is **v** + **u**. In other words, to find **u** + **v** we can move **v** to the terminal point of **u** *or* we can move **u** to the terminal point of **v.** It therefore follows that addition of vectors is commutative: **u** + **v** = **v** + **u**. Figure 8.25(b) illustrates why the definition for the addition of two vectors is often called the **parallelogram law.** A number of additional properties of vectors are examined in the exercises at the end of this section.

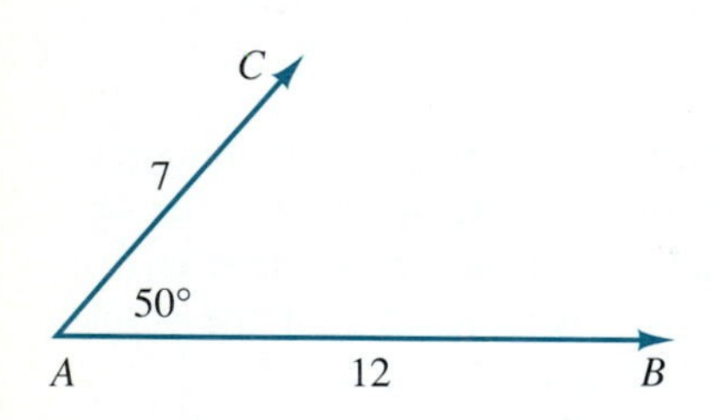

FIGURE 8.27 The force of 12 is horizontal

EXAMPLE 1 Determine the resultant of the two forces in Figure 8.27; that is find the magnitude (length) of the resultant and the angle that the resultant force makes with the horizontal vector $\overrightarrow{AB}$.

Solution Following the parallelogram law, we complete the parallelogram of forces, as indicated in Figure 8.28. Since the consecutive angles of a parallelogram are supplementary, we know that the angle at B is 130°.

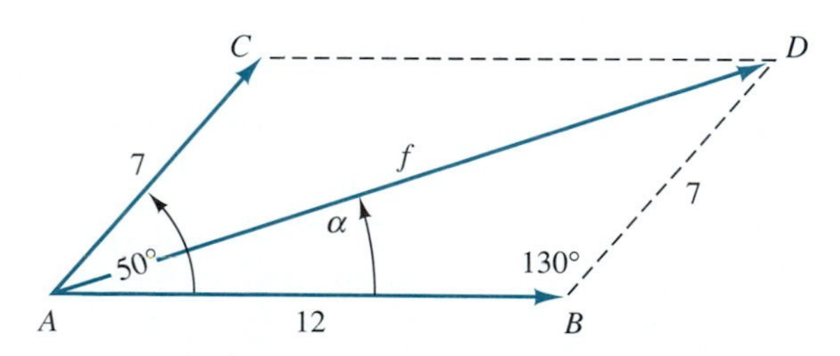

FIGURE 8.28 Since $\angle CAB = 50°$, $\angle ABD = 130°$

We can now use the Law of Cosines in $\triangle ABD$ to find the length, f, of the resultant.

$$f^2 = 12^2 + 7^2 - 2(12)(7)\cos 130^\circ$$ *Remember that* $\cos 130^\circ$ *is negative.*

$$f^2 = 144 + 49 - 168(\cos 130^\circ)$$

$$f = \sqrt{193 - 168\cos 130^\circ} = 17.34901 = \boxed{17.3}$$ *Rounded to the nearest tenth*

Using the approximate length of 17.3, we employ the Law of Sines to find α.

$$\frac{\sin\alpha}{7} = \frac{\sin 130^\circ}{17.3} \Rightarrow \sin\alpha = .3090847$$

$$\alpha = \sin^{-1}0.3090847$$ *Using a calculator we get*

$$\boxed{\alpha = 18.0^\circ}$$ *Rounded to the nearest tenth*

Thus the resultant force has a magnitude of approximately 17.3 and makes an approximate angle of 18.0° with the horizontal. ■

In the previous example, vectors $\overrightarrow{AB}$ and $\overrightarrow{AC}$ are called **component vectors,** which yield the resultant vector $\overrightarrow{AD}$. Although a vector may be the resultant of many component vectors, we will limit our attention to vectors with two components.

Figure 8.29 shows how a vector may have more than one set of component vectors. However, if we insist that the component vectors be horizontal and vertical vectors (as they are in the third illustration of Figure 8.29), then the component vectors are unique.

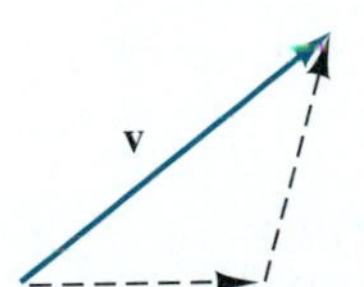

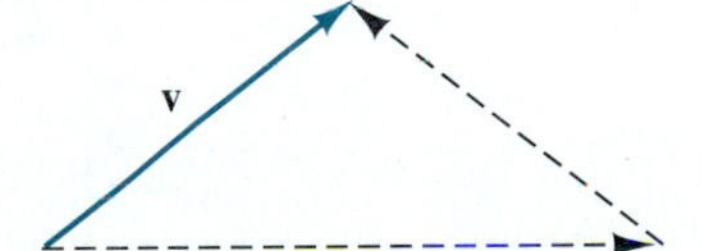

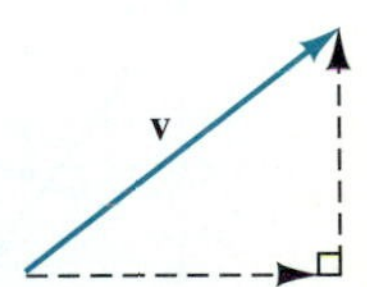

FIGURE 8.29 A vector **v** with several sets of component vectors

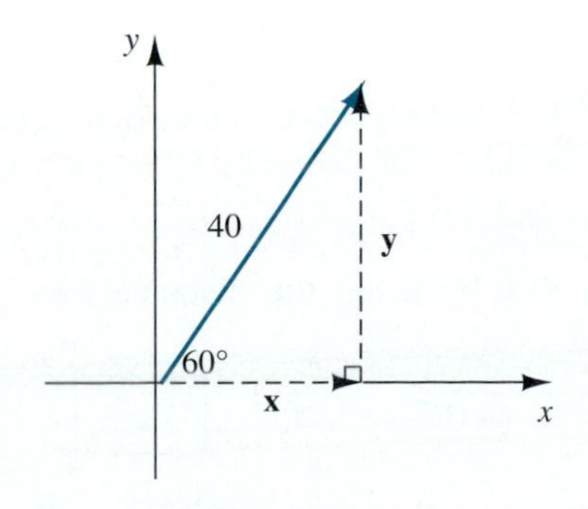

FIGURE 8.30

EXAMPLE 2 Determine the length of the horizontal and vertical component vectors of the vector in Figure 8.30.

Solution We label the horizontal component vector **x** and the vertical component vector **y** and find their lengths by using the right triangle definitions of sine and cosine. Remember $|\mathbf{v}|$ stands for the length of vector **v**.

$$\cos 60^\circ = \frac{|\mathbf{x}|}{40} \qquad \sin 60^\circ = \frac{|\mathbf{y}|}{40}$$

$$\frac{1}{2} = \frac{|\mathbf{x}|}{40} \qquad \frac{\sqrt{3}}{2} = \frac{|\mathbf{y}|}{40}$$

$$|\mathbf{x}| = 20 \qquad |\mathbf{y}| = 20\sqrt{3}$$

Thus the horizontal component vector has magnitude 20 and the vertical component vector has magnitude $20\sqrt{3}$.

This process is called **resolving** a vector into its horizontal and vertical components. ■

From now on when we talk about *the* components of a vector, we always mean its x- (horizontal) and y- (vertical) components.

Thus far our description of vectors has been geometric in nature. By placing vectors in a coordinate plane, we can also view vectors from an algebraic point of view.

EXAMPLE 3 Let vector $\mathbf{u} = \overrightarrow{PQ}$ and $\mathbf{v} = \overrightarrow{RS}$, where the points are $P(1, 6)$, $Q(4, 2)$, $R(3, 4)$, and $S(6, 5)$. **(a)** Find $\mathbf{u} + \mathbf{v}$. **(b)** Find $|\mathbf{u} + \mathbf{v}|$.

Solution

(a) In Figure 8.31(a) we have drawn vectors **u** and **v** in a rectangular coordinate system. Figure 8.31(b) shows that vector **v** has been moved so that its initial point coincides with the endpoint of **u**; the parallelogram law for the addition of vectors has then been followed. Looking at the horizontal and vertical components of **v**, we see that the vector $\mathbf{u} + \mathbf{v}$ has an initial point of (1, 6) and a terminal point of (7, 3). In Figure 8.31(c) we have found $\mathbf{u} + \mathbf{v}$ by moving the initial point of **u** to the terminal point of **v**. Note that the vector we obtain in this way is equal to the vector obtained in Figure 8.31(b): They both have the same magnitude and direction. The location (the initial point) of the vector does not matter.

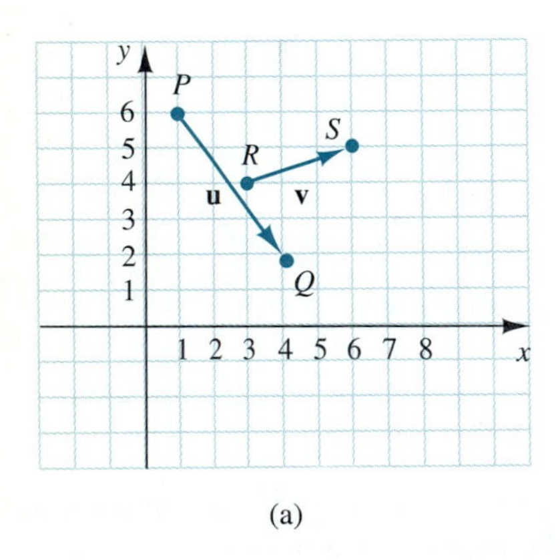

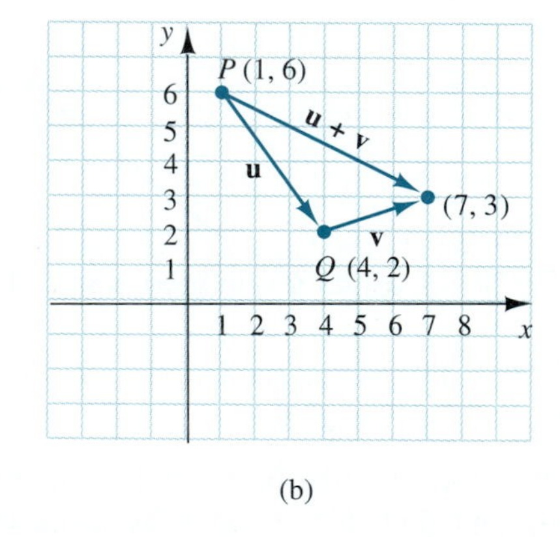

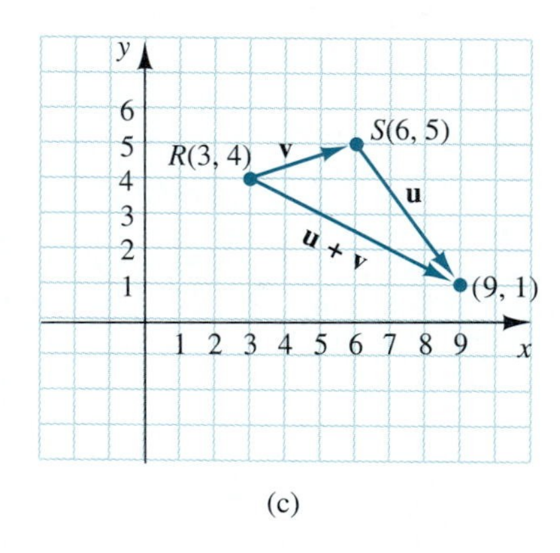

FIGURE 8.31

(b) We can compute the magnitude (length) of $\mathbf{u} + \mathbf{v}$ by using the distance formula in Figure 8.31(b).

$$|\mathbf{u} + \mathbf{v}| = \sqrt{(7-1)^2 + (3-6)^2} = \sqrt{(6)^2 + (-3)^2} = \sqrt{45} = \boxed{3\sqrt{5}}$$

It is left to the student to use Figure 8.31(c) to obtain the same result. ■

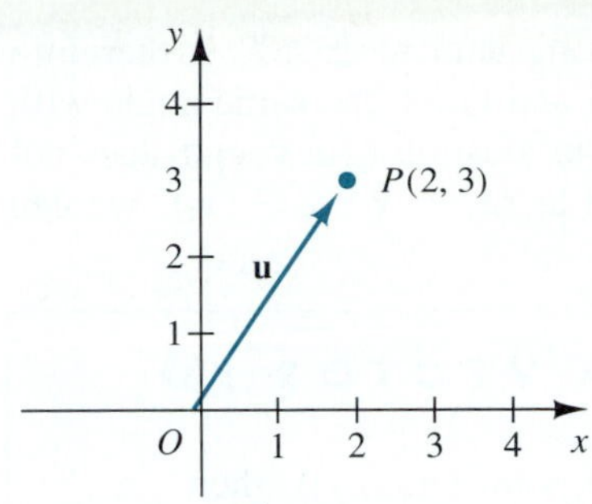

The vector $\mathbf{u} = \overrightarrow{OP}$ is denoted by <2, 3>.

FIGURE 8.32(a)

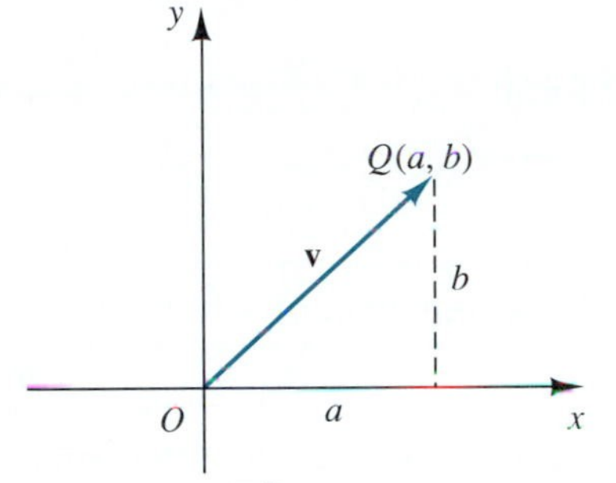

The vector $\mathbf{v} = \overrightarrow{OQ}$ is denoted by $<a, b>$.

FIGURE 8.32(b)

As we pointed out, it doesn't make any difference where a vector has its initial point; therefore, we may agree to specify vectors as having their initial points at the origin. Such vectors are said to be in **standard position.** By considering vectors in standard position, we may specify a vector simply by using the coordinates of its endpoint.

For example, the vector $\mathbf{u} = \overrightarrow{OP}$, which starts at the point $O(0, 0)$ and ends at the point $P(2, 3)$, as shown in Figure 8.32(a) will be denoted by the special vector notation $\langle 2, 3\rangle$. In general, in vector $\mathbf{v} = \langle a, b\rangle$, a is called the ***x*-**, or **horizontal component** and b is called the ***y*-**, or **vertical component,** so $\langle a, b\rangle$ is called the **component form** of the vector.

Looking at Figure 8.32(a), the Pythagorean theorem tells us that $|\mathbf{u}| = \sqrt{2^2 + 3^2} = \sqrt{13}$. In general, regardless of the quadrant in which the terminal point of $\mathbf{v} = \langle a, b\rangle$ falls, we have a simple formula for the length of a vector in standard position. See Figure 8.32(b).

THE LENGTH OF A VECTOR IN STANDARD POSITION

If $\mathbf{v} = \langle a, b\rangle$,

$$|\mathbf{v}| = |\langle a, b\rangle| = \sqrt{a^2 + b^2}$$

Another advantage to working with vectors in standard position is that it is easy to determine if two vectors are equal—that is, have the same length and the same direction.

$$\langle a, b\rangle = \langle c, d\rangle \quad \text{if and only if} \quad a = c \quad \text{and} \quad b = d$$

We define the **zero vector, 0**, to be $\langle 0, 0\rangle$, and so the zero vector has length 0.

Suppose we have the endpoints of a vector that is not in standard position. How do we find an equal vector in standard position? Let's look at the vectors $\overrightarrow{PQ}$ for $P(x_1, y_1)$ and $Q(x_2, y_2)$ and $\overrightarrow{OS}$ for $S(x_2 - x_1, y_2 - y_1)$, as indicated in Figure 8.33.

Why is $\triangle PQR$ congruent to $\triangle OST$?

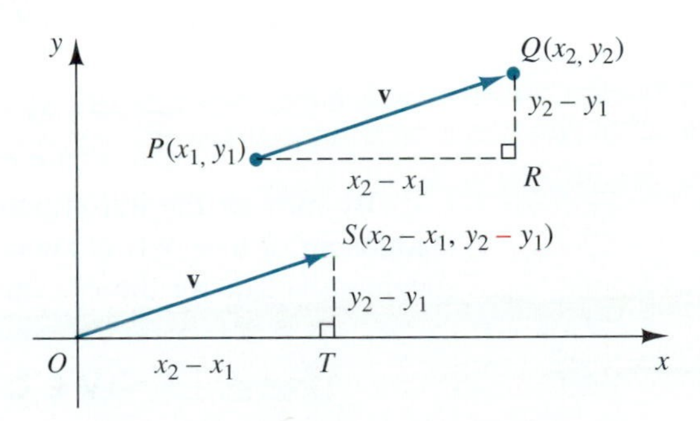

FIGURE 8.33 Vectors $\overrightarrow{PQ}$ and $\overrightarrow{OS}$ are equal.

From Figure 8.33 we can see that $\triangle PQR$ is congruent to $\triangle OST$. It therefore follows that vectors $\overrightarrow{PQ}$ and $\overrightarrow{OS}$ have the same length and make the same angle with the horizontal, so they are equal. (Remember that the location of a vector does not matter.) If we keep in mind that $\overrightarrow{OS}$ can be denoted as $\langle x_2 - x_1, y_2 - y_1 \rangle$, we can summarize this result as follows.

COMPONENT FORM OF A VECTOR $\overrightarrow{PQ}$

If vector **v** has initial point $P(x_1, y_1)$ and terminal point $Q(x_2, y_2)$, then

$$\mathbf{v} = \overrightarrow{PQ} = \langle x_2 - x_1, y_2 - y_1 \rangle$$

EXAMPLE 4 Let A and B be the points $A(-1, 4)$ and $B(6, -2)$. Draw $\mathbf{v} = \overrightarrow{AB}$ in standard position and find its components.

Solution We begin by putting $\overrightarrow{AB}$ into component form:

$$\overrightarrow{AB} = \langle 6 - (-1), -2 - 4 \rangle = \langle 7, -6 \rangle$$

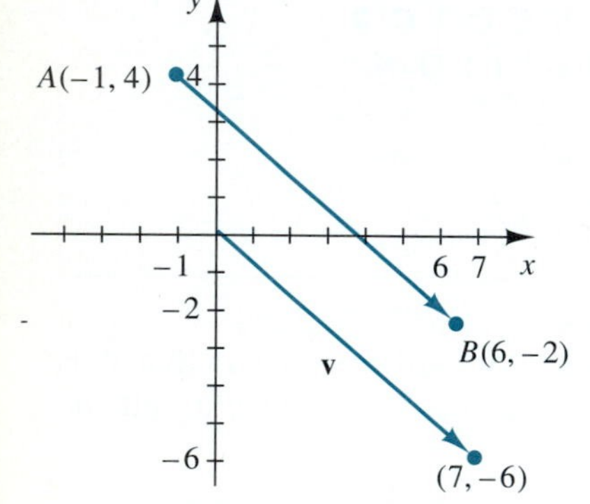

FIGURE 8.34 Vector $\mathbf{v} = \overrightarrow{AB}$ in standard position

Therefore, the x-component of $\overrightarrow{AB}$ is 7 and the y-component of $\overrightarrow{AB}$ is -6. Figure 8.34 shows vector $\mathbf{v} = \overrightarrow{AB}$ in standard position. ■

We have previously described how we can add two vectors by using the parallelogram law. However, if two vectors are in component form, the process of vector addition becomes quite simple.

Suppose vectors **u** and **v** are in standard position with $\mathbf{u} = \langle x_1, y_1 \rangle$ and $\mathbf{v} = \langle x_2, y_2 \rangle$. Let's look at $\mathbf{u} + \mathbf{v}$, as drawn in standard position in Figure 8.35.

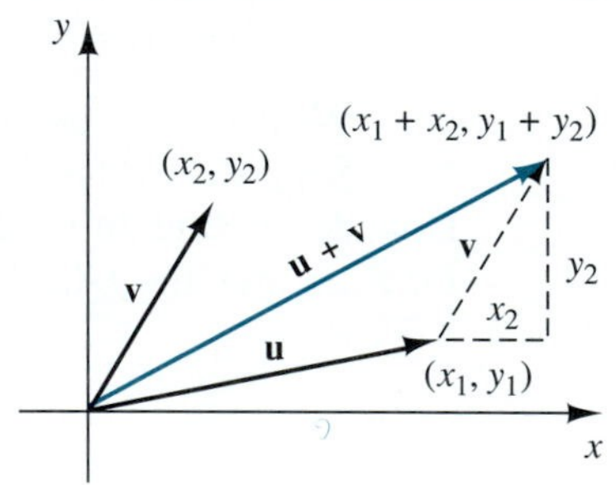

FIGURE 8.35 The components of **u** + **v** are the sums of the components of **u** and **v**.

By moving the initial point of **v** to the endpoint of **u**, we can see that the x-component of $\mathbf{u} + \mathbf{v}$ is obtained by adding the x-components of **u** and **v**. A similar statement is true for the y-component of $\mathbf{u} + \mathbf{v}$. See Figure 8.35.

VECTOR ADDITION

If $\mathbf{u} = \langle x_1, y_1 \rangle$ and $\mathbf{v} = \langle x_2, y_2 \rangle$, then $\mathbf{u} + \mathbf{v} = \langle x_1, + x_2, y_1 + y_2 \rangle$.

There is another operation with vectors, called **scalar multiplication,** defined as follows.

DEFINITION Given a vector $\mathbf{v} = \langle x_1, y_1 \rangle$ and a real number k, we define the vector $k\mathbf{v}$ as $k\mathbf{v} = k\langle x_1, y_1 \rangle = \langle kx_1, ky_1 \rangle$.

For example, if $\mathbf{v} = \langle 1, 5 \rangle$, then $2\mathbf{v} = \langle 2, 10 \rangle$ and $-4\mathbf{v} = \langle -4, -20 \rangle$. Note that the length of vector $k\mathbf{v}$ **is**

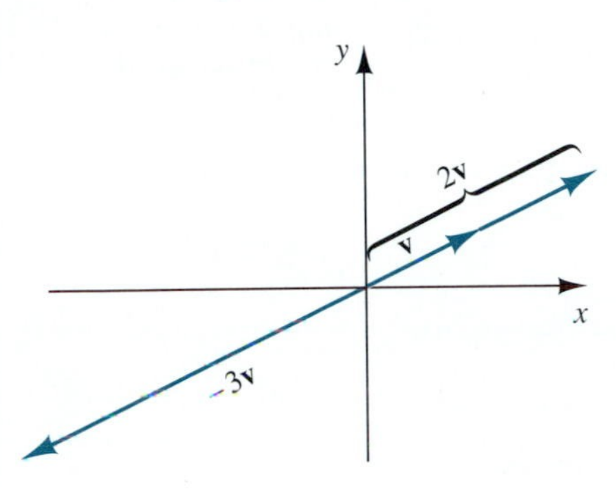

FIGURE 8.36 Scalar multiples of a vector

$$\begin{aligned} |k\mathbf{v}| = \langle kx_1, ky_1 \rangle &= \sqrt{(kx_1)^2 + (ky_1)^2} = \sqrt{k^2(x_1)^2 + k^2(y_1)^2} \\ &= \sqrt{k^2[(x_1)^2 + (y_1)^2]} = \sqrt{k^2}\sqrt{(x_1)^2 + (y_1)^2} \\ &= |k| \cdot |\mathbf{v}| \end{aligned}$$

which is $|k|$ times the length of vector $\mathbf{v}$. Thus scalar multiplication multiplies the length of the original vector by the absolute value of the scalar.

In Figure 8.36, we have drawn vector $\mathbf{v} = \langle 3, 1 \rangle$ and vectors $2\mathbf{v} = \langle 6, 2 \rangle$ and $-3\mathbf{v} = \langle -9, -3 \rangle$. Note that $2\mathbf{v}$ and $\mathbf{v}$ have the same direction, but the length of $2\mathbf{v}$ is twice the length of $\mathbf{v}$. On the other hand, the vectors $-3\mathbf{v}$ and $\mathbf{v}$ have *opposite* directions, with the length of $-3\mathbf{v}$ being three times the length of $\mathbf{v}$.

In general, the vectors $\mathbf{v}$ and $k\mathbf{v}$ have the same direction if $k > 0$ and opposite directions if $k < 0$. In particular, if $\mathbf{v} = \langle x_1, y_1 \rangle$, then $-\mathbf{v} = -1 \cdot \mathbf{v} = \langle -x_1, -y_1 \rangle$ is called the **negative** of $\mathbf{v}$. Thus the vectors $\mathbf{v}$ and $-\mathbf{v}$ have the same length but opposite directions. See Figure 8.37.

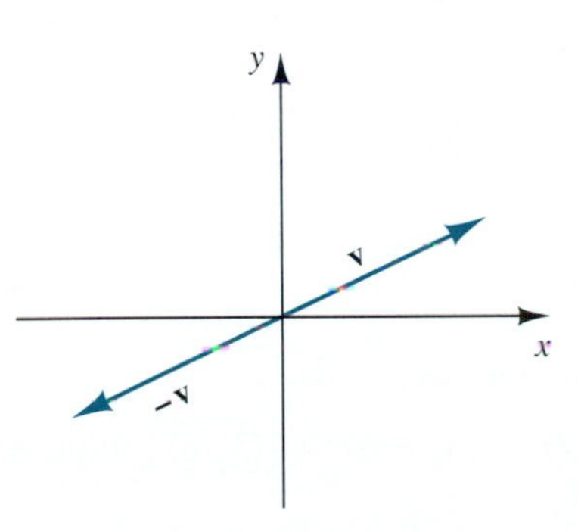

FIGURE 8.37 The vectors **v** and $-$**v**

We can now describe what we mean by subtracting vectors. Let $\mathbf{u} = \langle x_1, y_1 \rangle$ and $\mathbf{v} = \langle x_2, y_2 \rangle$; we have

$$\mathbf{u} - \mathbf{v} = \mathbf{u} + (-\mathbf{v}) = \langle x_1, y_1 \rangle + \langle -x_2, -y_2 \rangle = \langle x_1 - x_2, y_1 - y_2 \rangle \tag{1}$$

In other words, to subtract two vectors, we just subtract their corresponding components.

We can interpret vector subtraction geometrically by recalling the component form of a vector. The component form of a vector with initial point $Q(x_2, y_2)$ and endpoint $P(x_1, y_1)$ is $\langle x_1 - x_2, y_1 - y_2 \rangle$, which is exactly the same as the vector $\mathbf{u} - \mathbf{v}$ we have just described in (1). Therefore, $\mathbf{u} - \mathbf{v}$ is the vector drawn from the terminal point of $\mathbf{v}$ to the terminal point of $\mathbf{u}$. In other words, $\mathbf{u} - \mathbf{v}$ is the directed diagonal (not the resultant) of the parallelogram formed by vectors $\mathbf{u}$ and $\mathbf{v}$ indicated in Figure 8.38.

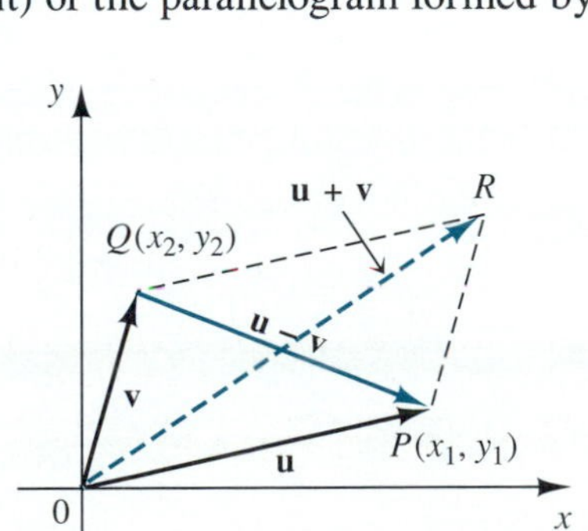

FIGURE 8.38 Parallelogram showing $\mathbf{u} + \mathbf{v} = \overrightarrow{OR}$ and $\mathbf{u} - \mathbf{v} = \overrightarrow{QP}$

EXAMPLE 5 Let $\mathbf{u} = \langle -2, 3\rangle$ and $\mathbf{v} = \langle 4, 1\rangle$. Find and sketch each of the following vectors. **(a)** $\mathbf{u} + \mathbf{v}$ **(b)** $\mathbf{u} - \mathbf{v}$ **(c)** $-\mathbf{u}$

Solution

(a) $\mathbf{u} + \mathbf{v} = \langle -2, 3\rangle + \langle 4, 1\rangle = \langle -2 + 4, 3 + 1\rangle = \boxed{\langle 2, 4\rangle}$

(b) $\mathbf{u} - \mathbf{v} = \langle -2, 3\rangle - \langle 4, 1\rangle = \langle -2 - 4, 3 - 1\rangle = \boxed{\langle -6, 2\rangle}$

(c) $-\mathbf{u} = -\langle -2, 3\rangle = \boxed{\langle 2, -3\rangle}$

Note that the arrow representing $\mathbf{u} - \mathbf{v}$ *is not in standard position.*

These vectors are sketched in Figure 8.39. Note that vector $\mathbf{u} - \mathbf{v}$ is equal to $\overrightarrow{QP}$.

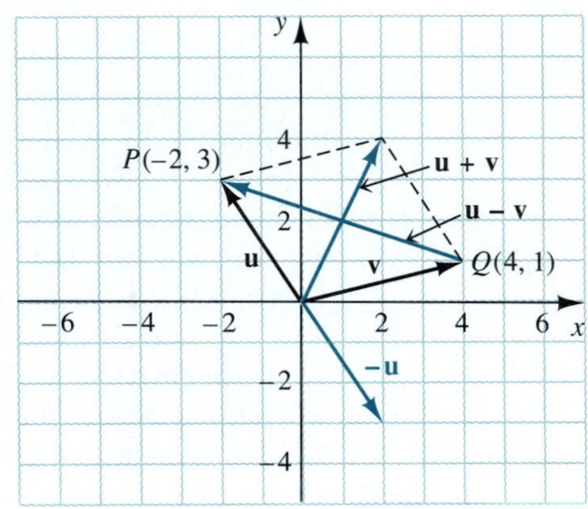

FIGURE 8.39

■

EXAMPLE 6 Let $\mathbf{A} = \langle 3, 1\rangle$ and $\mathbf{B} = \langle 1, -2\rangle$. Compute $2\mathbf{A} - 3\mathbf{B}$.

Solution $2\mathbf{A} - 3\mathbf{B} = 2\langle 3, 1\rangle - 3\langle 1, -2\rangle = \langle 6, 2\rangle - \langle 3, -6\rangle = \boxed{\langle 3, 8\rangle}$. Figure 8.40 illustrates $2\mathbf{A} - 3\mathbf{B}$.

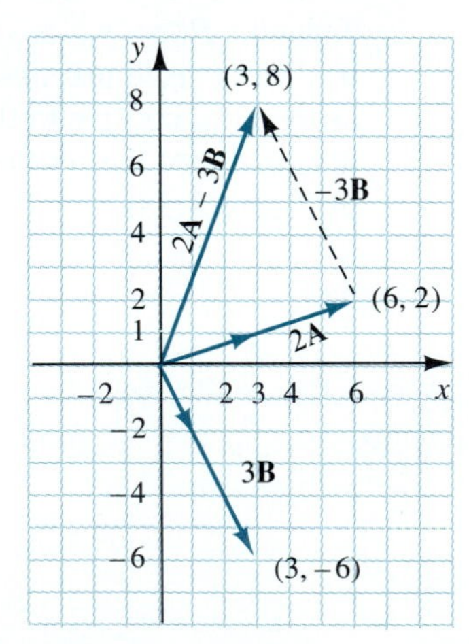

FIGURE 8.40

■

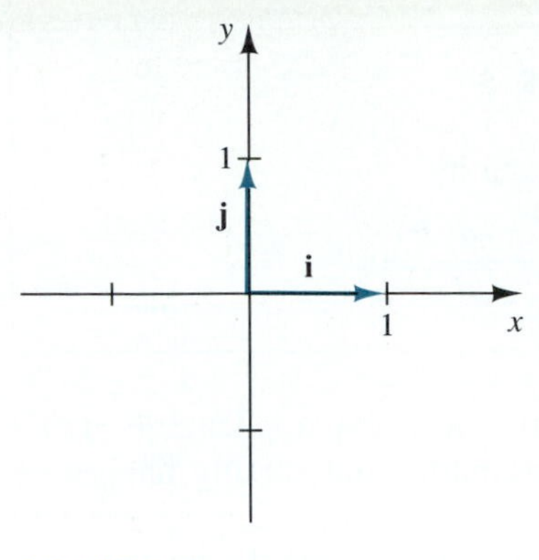

FIGURE 8.41 The unit vectors **i** and **j**

Any vector of length 1 is called a **unit vector**. Two particularly important unit vectors are

$$\mathbf{i} = \langle 1, 0 \rangle \quad \text{and} \quad \mathbf{j} = \langle 0, 1 \rangle$$

The vectors **i** and **j** are illustrated in Figure 8.41. These two vectors are important because they serve as the basic horizontal and vertical component vectors. *Any vector* $\langle a, b \rangle$ *can be expressed uniquely in terms of* **i** and **j**. We write

$$a\mathbf{i} + b\mathbf{j} = a\langle 1, 0 \rangle + b\langle 0, 1 \rangle = \langle a, 0 \rangle + \langle 0, b \rangle = \langle a, b \rangle$$

Thus we have, in general,

$$\boxed{\langle a, b \rangle = a\mathbf{i} + b\mathbf{j}}$$

This is called expressing a vector as a **linear combination** of the vectors **i** and **j**.

EXAMPLE 7

(a) Express the vector $\langle -5, 4 \rangle$ as a linear combination of **i** and **j**.

(b) Express the vector $\mathbf{u} = 6\mathbf{i} - 2\mathbf{j}$ in component form.

Solution

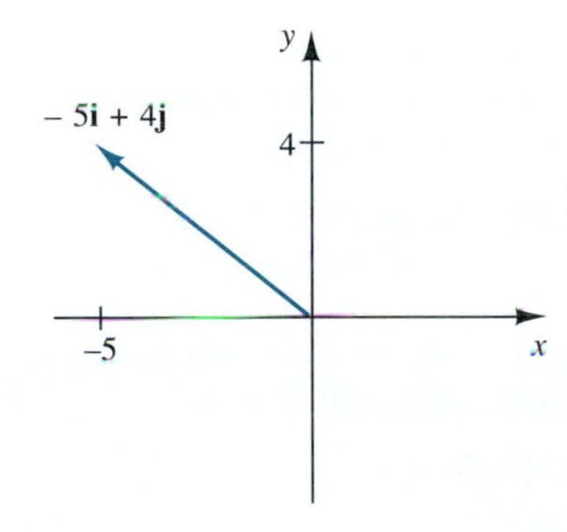

FIGURE 8.42 The vector $-5\mathbf{i} + 4\mathbf{j}$

(a) Using the result in the preceding box, we have $\langle -5, 4 \rangle = \boxed{-5\mathbf{i} + 4\mathbf{j}}$ This linear combination is illustrated in Figure 8.42.

(b) $\mathbf{u} = 6\mathbf{i} - 2\mathbf{j} = 6\langle 1, 0 \rangle - 2\langle 0, 1 \rangle = \langle 6, 0 \rangle + \langle 0, -2 \rangle = \boxed{\langle 6, -2 \rangle}$ ■

EXAMPLE 8 Find a unit vector in the same direction as $\mathbf{v} = \langle 4, 3 \rangle$.

Solution From our previous discussion, we know that for $k > 0$, any scalar multiple $k\mathbf{v}$ will have the same direction as **v** and have length $|k|$ times the length of **v**. Therefore, if we multiply **v** by the reciprocal of its length, we will have a vector of length 1 in the same direction as **v**.

We begin by finding the length of **v**:

$$|\mathbf{v}| = |\langle 4, 3 \rangle| = \sqrt{4^2 + 3^2} = \sqrt{25} = 5$$

We claim that $\frac{1}{5}\mathbf{v}$ is a unit vector in the direction of **v**. Since it is a positive scalar multiple of **v**, it has the same direction. Its length is

$$\left|\frac{1}{5}\mathbf{v}\right| = \left|\frac{1}{5}\langle 4, 3 \rangle\right| = \left|\left\langle \frac{4}{5}, \frac{3}{5} \right\rangle\right| = \sqrt{\left(\frac{4}{5}\right)^2 + \left(\frac{3}{5}\right)^2}$$

$$= \sqrt{\frac{16}{25} + \frac{9}{25}} = \sqrt{1} = 1 \qquad \textit{As required}$$ ■

The last example can be generalized to the following.

UNIT VECTORS

A unit vector in the direction of $\mathbf{v} = \langle a, b \rangle$ is given by

$$\frac{\mathbf{v}}{|\mathbf{v}|} \quad \text{or, alternatively,} \quad \frac{\langle a, b \rangle}{\sqrt{a^2 + b^2}}$$

Vectors and trigonometry play an important role in many applications. Let's introduce some terminology that is useful in understanding and solving navigation problems. In navigation we must understand the difference between *course* and *heading*. The *course* of a ship or plane is the direction of the actual path traveled, whereas the *heading* is the direction in which the ship or plane is pointed. The reason there is often a difference between course and heading is that often there is some other force, such as the wind or ocean currents, being applied to the ship or plane. When we speak of *speed,* we refer only to the magnitude of the vector. *Airspeed* refers to the speed of a plane in still air, and *ground speed* refers to the speed of a plane relative to the ground. The word *velocity* is used for a vector and so refers to both the speed *and* direction of the ship or plane.

EXAMPLE 9 A motorboat on one side of a river is traveling on a heading N20°E at a speed of 18 mi/hr. The river is flowing due east at a speed of 5 mi/hr. Find the velocity of the boat to the nearest tenth. (That is, find its course and speed.)

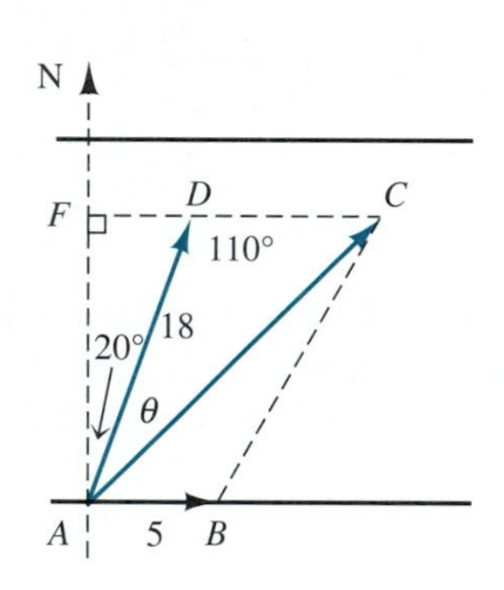

Solution We begin by drawing a diagram incorporating the given information. See Figure 8.43. Finding the speed and course of the boat means determining the length of $\overrightarrow{AC}$ and the size of $\angle FAC$.

We can easily determine that $\angle ADF = 70°$, so $\angle ADC = 110°$. We can find the length of $\overrightarrow{AC}$ by using the Law of Cosines in $\triangle ADC$. Note $|\overrightarrow{DC}| = 5$.

$$|\overrightarrow{AC}|^2 = 5^2 + 18^2 - 2(5)(18)\cos 110° = 410.56363$$

$$|\overrightarrow{AC}| = \sqrt{410.56363} = 20.3 \qquad \textit{Rounded to the nearest tenth}$$

We can now find angle θ by using the Law of Sines in $\triangle ADC$.

$$\frac{\sin\theta}{5} = \frac{\sin 110°}{20.262}$$

$$\sin\theta = \frac{5\sin 110°}{20.262} = 0.2318854$$

$$\theta = \sin^{-1} 0.2318854 \approx 13.4°$$

We have determined that the boat is traveling on an approximate course of N33.4°E (we add the 13.4° to the initial heading of N20°E) at an approximate speed of 20.3 mi/hr. ■

EXAMPLE 10 A barrel of oil weighing 300 pounds is lying on its side on a ramp that makes an angle of 9.5° with the horizontal. If we neglect friction, find the approximate force parallel to the ramp that is necessary to prevent the barrel from rolling down the ramp.

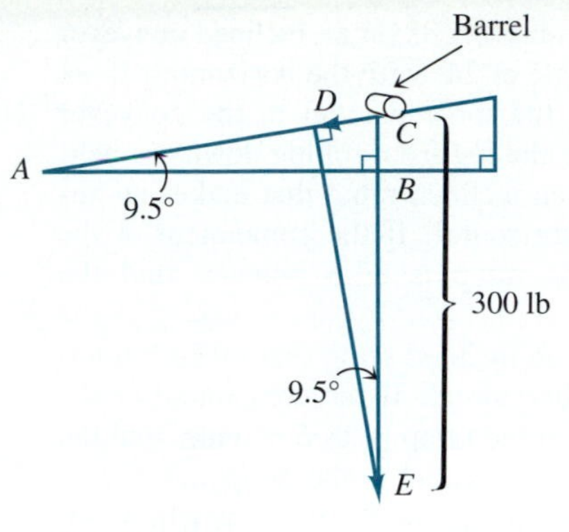

FIGURE 8.44

Solution Figure 8.44 illustrates the given information. The weight of the barrel is represented by $\overrightarrow{CE}$. We want to find the length of vector $\overrightarrow{CD}$, which represents the component of the weight parallel to the ramp that must be counteracted to prevent the barrel from rolling down the ramp.

We have $\triangle ABC$ is similar to $\triangle DCE$ (why?) and so $\angle E = 9.5°$. We may now solve for $|\overrightarrow{DC}|$ by using the right angle trigonometry of $\triangle CDE$.

$$\sin 9.5° = \frac{|\overrightarrow{DC}|}{300}$$

$$|\overrightarrow{DC}| = 300 \sin 9.5° = 49.5 \qquad \textit{Rounded to the nearest tenth}$$

Therefore, a force of approximately $\boxed{49.5 \text{ pounds}}$ is required to prevent the barrel from rolling down the ramp. ■

EXERCISES 8.4

In Exercises 1–6, use the vectors **u** and **v** as illustrated in the following figure to sketch the given vector.

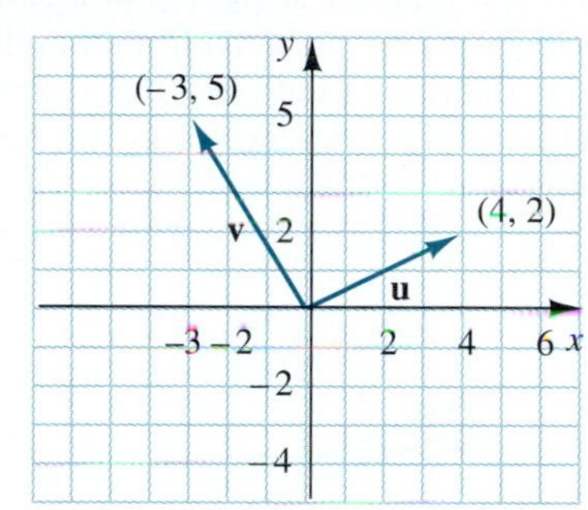

1. $2\mathbf{u}$
2. $-\mathbf{v}$
3. $\mathbf{u} + \mathbf{v}$
4. $-2\mathbf{u}$
5. $\mathbf{u} - \mathbf{v}$
6. $\frac{1}{2}\mathbf{v}$

In Exercises 7–14, let the points O, P, Q, R, and S be as follows:

$O(0, 0)$ $P(3, 4)$ $Q(-1, 2)$ $R(1, 5)$ $S(4, -1)$

Sketch the given vector and find its magnitude.

7. $\overrightarrow{OQ}$
8. $\overrightarrow{OR}$
9. $\overrightarrow{PQ}$
10. $\overrightarrow{RS}$
11. $\overrightarrow{OP} + \overrightarrow{OQ}$
12. $\overrightarrow{OS} + \overrightarrow{SQ}$
13. $\overrightarrow{OR} - \overrightarrow{OS}$
14. $\overrightarrow{PQ} - \overrightarrow{RS}$

In Exercises 15–20, let $\mathbf{u} = \langle -3, 2\rangle$ and $\mathbf{v} = \langle 2, 4\rangle$. Express each of the following in component form and as a linear combination of **i** and **j**.

15. $\mathbf{u} + \mathbf{v}$
16. $\mathbf{u} - \mathbf{v}$
17. $2\mathbf{u} - 3\mathbf{v}$
18. $5\mathbf{u} - 4\mathbf{v}$
19. $-6\mathbf{v} - \mathbf{u}$
20. $-(3\mathbf{u} + \mathbf{v})$

In Exercises 21–24, vectors **A** and **B** represent two forces acting at the origin. Find the magnitude of the resultant vector and the angle θ the resultant makes with **A**.

21. $\mathbf{A} = \langle 3, 0\rangle$, $\mathbf{B} = \langle 0, 2\rangle$
22. $\mathbf{A} = \langle -2, 0\rangle$, $\mathbf{B} = \langle 0, 5\rangle$
23. $\mathbf{A} = \langle 12, 0\rangle$, $\mathbf{B} = \langle 0, -1\rangle$
24. $\mathbf{A} = \langle 2.5, 0\rangle$, $\mathbf{B} = \langle 0, 4.7\rangle$

In Exercises 25–28, find the length of the resultant vector and the angle that the resultant makes with **v**.

25.
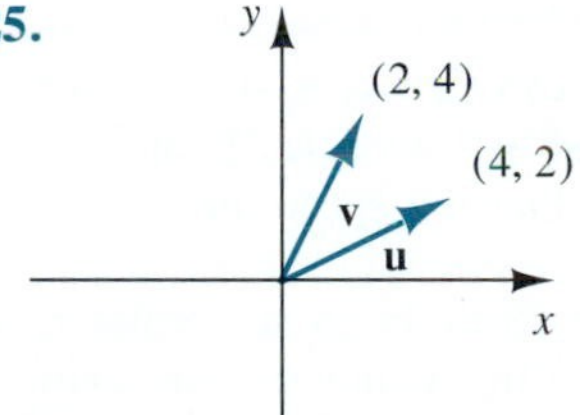

26.
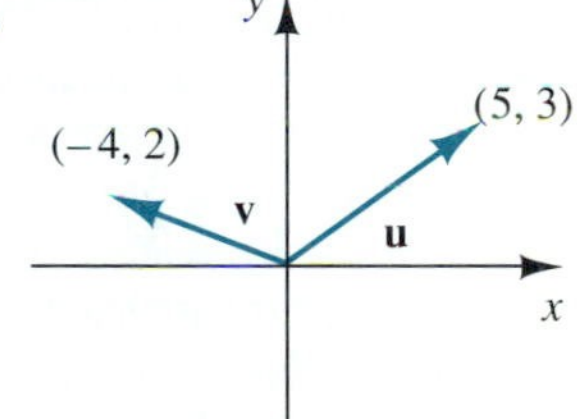

27.
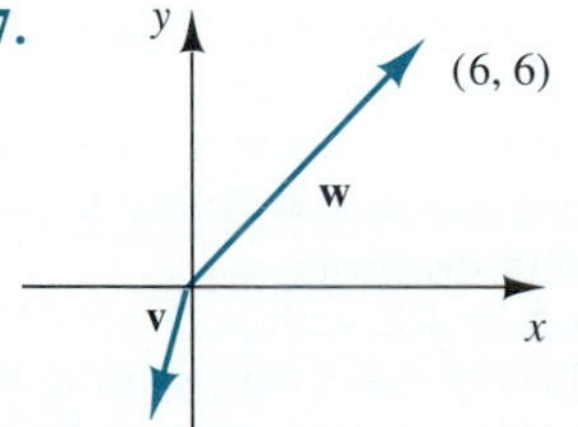

28.
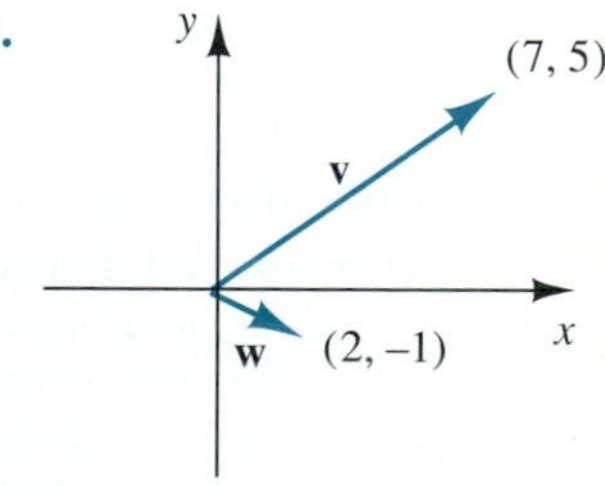

In Exercises 29–32, θ is the angle that the vector **v** makes with the positive x-axis. Given $|\mathbf{v}|$, resolve the given vector into horizontal and vertical components and express **v** in component form.

29. $|\mathbf{v}| = 7$, $\theta = 60°$
30. $|\mathbf{v}| = 12$, $\theta = 135°$
31. $|\mathbf{v}| = 100$, $\theta = 126°$
32. $|\mathbf{v}| = 6$, $\theta = 36°$

In Exercises 33–38, if the given vector is in **i, j** form, express it in component form; if the vector is given in component form, express it in **i, j** form.

33. $2\mathbf{i} + 7\mathbf{j}$
34. $\langle 4, -1\rangle + \langle 2, 9\rangle$
35. $5\langle 1, 0\rangle - 6\langle 0, 1\rangle$
36. $-3\mathbf{i} + 5\mathbf{j}$
37. $4\langle 2, -1\rangle + 7\langle -3, 2\rangle$
38. $\frac{2}{5}\mathbf{i} - \frac{3}{4}\mathbf{j}$

In Exercises 39–44, find a unit vector in the direction of **v**.

39. $\mathbf{v} = \langle 3, 5\rangle$
40. $\mathbf{v} = 2\mathbf{i} + 7\mathbf{j}$
41. $\mathbf{v} = \langle -1, -3\rangle$
42. $\mathbf{v} = 3\langle 2, 4\rangle + \langle 1, 3\rangle$
43. $\mathbf{v} = -2\langle 1, 0\rangle - 5\langle 0, 1\rangle$
44. $\mathbf{v} = \frac{2}{3}\mathbf{i} - \frac{1}{6}\mathbf{j}$

45. Two forces of 17 pounds and 30 pounds are acting at the same point in the plane. If the angle between the two vectors is 51°, find the magnitude of the resultant force.

46. Two forces of 6.3 pounds and 9.7 pounds are acting at the same point in the plane. If the angle between the two vectors is 38.2°, find the magnitude of the resultant force.

47. Two forces of 8 pounds and 11 pounds acting at the same point in the plane produce a resultant of magnitude 15. Find the angle between the forces.

48. Two forces acting at the same point in the plane produce a resultant of magnitude 22.3 pounds. If one of the forces is 18 pounds and its angle with the resultant is 64°, find the other force.

49. A rowboat is headed due east across a stream that is flowing due north at a speed of 1.8 mi/hr. If the boat is being rowed at 3.2 mi/hr, find the speed and course at which the boat is traveling.

50. A motorboat that travels at 16 mi/hr in still water is to travel due west across a river that is flowing due south at 5.8 mi/hr. On what heading should the boat be set?

51. A plane is on a heading of N28°W with an airspeed of 375 mi/hr. If the wind is blowing due west at 42 mi/hr, find the course and groundspeed of the plane.

52. A plane is on a heading of N63°E with an airspeed of 420 mi/hr. If the wind is blowing due east at 26 mi/hr, find the course and groundspeed of the plane.

53. A plane is on a heading of N28.4°E with an airspeed of 315 mi/hr. The wind is blowing due south, causing the plane to be on a course of N31.2°E with a groundspeed of 275 mi/hr. Find the speed of the wind.

54. A helicopter is on a heading of N68.6°W with an airspeed of 155 mi/hr. The wind is blowing due north, causing the plane to be on a course of N66.1°W with a groundspeed of 175 mi/hr. Find the speed of the wind.

55. A block weighing 20 pounds rests on an inclined ramp that makes an angle of 18° with the horizontal. Determine the components of the weight parallel and perpendicular to the ramp.

56. A log weighing 175 pounds rests on an inclined conveyor belt that makes an angle of 24° with the horizontal. If we neglect friction, find the force parallel to the conveyor belt necessary to keep the log from rolling down the belt.

57. A block is placed on an inclined ramp that makes an angle of 15° with the horizontal. If the component of the weight parallel to the ramp is 87.3 pounds, find the weight of the block.

58. A block is resting on an inclined ramp that makes an angle of 9.7° with the horizontal. If the component of the weight perpendicular to the ramp is 28.6 pounds, find the component of the weight parallel to the ramp.

59. A ship has a final destination of point C, which is 200 miles away at a heading of N52°W from its present position A. However, the ship must first stop at point B, which is 125 miles at a heading of N26°E from point A. Find the distance and heading of the ship from B to C.

60. Suppose that two weights are attached together by a rope passing through a pulley at the top of a two-sided ramp, as indicated in the accompanying figure. The 50-pound weight is on the side of the ramp that is inclined at a 20° angle, and the 20-pound weight is on the side of the ramp that is inclined at a 50° angle. If we neglect friction (and the weight of the rope), determine which way the weights will slide.

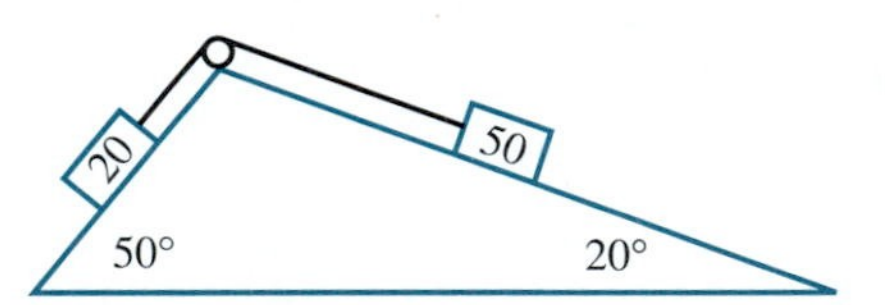

61. An object at rest is said to be in *static equilibrium*. In order for an object to be in static equilibrium at the origin, it is necessary that the sum of all the horizontal components of the forces acting at the origin be 0 and the sum of all the vertical components also be 0. Consider the following force diagram, which represents a 100-pound weight hanging from two wires at the indicated angles. If the tension in $\overrightarrow{OB}$ is 50 pounds, what must the tension in $\overrightarrow{OA}$ be to maintain equilibrium?

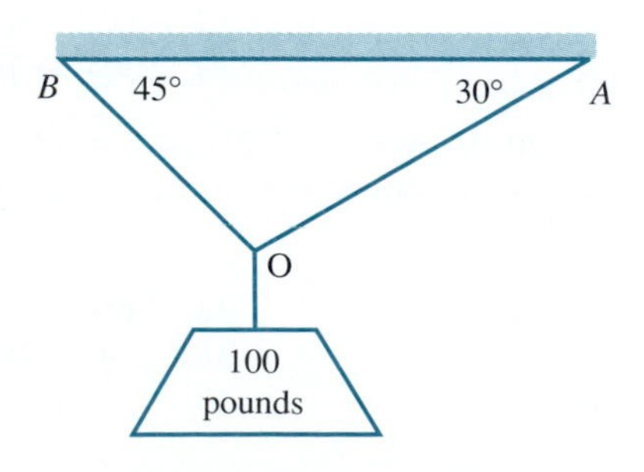

62. Repeat Exercise 61 if $\angle OAB = 20°$ and $\angle OBA = 55°$.

The next group of exercises refer to the following operation defined for vectors.

DEFINITION Given two vectors $\mathbf{u} = \langle x_1, y_1 \rangle = x_1\mathbf{i} + y_1\mathbf{j}$ and $\mathbf{v} = \langle x_2, y_2 \rangle = x_2\mathbf{i} + y_2\mathbf{j}$, the *dot product* of the vectors $\mathbf{u}$ and $\mathbf{v}$ is denoted by $\mathbf{u} \cdot \mathbf{v}$ and is defined to be

$$\mathbf{u} \cdot \mathbf{v} = x_1x_2 + y_1y_2$$

($\mathbf{u} \cdot \mathbf{v}$ is read "**u** dot **v**")

In Exercises 63–66, find $\mathbf{u} \cdot \mathbf{v}$.

63. $\mathbf{u} = \langle -1, 6 \rangle$, $\mathbf{v} = \langle 3, -4 \rangle$

64. $\mathbf{u} = 2\mathbf{i} - \mathbf{j}$, $\mathbf{v} = \frac{1}{2}\mathbf{i} + 3\mathbf{j}$

65. $\mathbf{u} = \left\langle -\frac{3}{4}, \frac{1}{10} \right\rangle$, $\mathbf{v} = \left\langle \frac{2}{9}, 5 \right\rangle$

66. $\mathbf{u} = 4\mathbf{i}$, $\mathbf{v} = -5\mathbf{i} + 4\mathbf{j}$

In Exercises 67–72, let $\mathbf{A} = \langle -3, 4 \rangle$, $\mathbf{B} = \langle 5, 2 \rangle$, and $\mathbf{C} = \langle 6, -1 \rangle$ and find

67. $\mathbf{A} \cdot (\mathbf{B} - \mathbf{C})$

68. $\mathbf{A} \cdot \mathbf{B} - \mathbf{A} \cdot \mathbf{C}$

69. $\mathbf{B} \cdot (\mathbf{A} + \mathbf{C})$

70. $2\mathbf{A} \cdot 3\mathbf{B}$

71. $(\mathbf{A} + \mathbf{B}) \cdot (\mathbf{B} - \mathbf{C})$

72. $(3\mathbf{A} + \mathbf{B}) \cdot (2\mathbf{C})$

QUESTIONS FOR THOUGHT

73. Three forces of 60 pounds, 90 pounds, and 100 pounds are acting at the origin at angles of 0°, 40°, and 65°, respectively, with the positive x-axis. Find the magnitude and direction of the resultant of these three forces.

74. Use the ideas introduced in this section to prove each of the following properties for all vectors $\mathbf{u}$, $\mathbf{v}$, and $\mathbf{w}$ and scalars a and b.

(a) $\mathbf{u} + \mathbf{v} = \mathbf{v} + \mathbf{u}$

(b) $(\mathbf{u} + \mathbf{v}) + \mathbf{w} = \mathbf{v} + (\mathbf{u} + \mathbf{w})$

(c) $\mathbf{u} + \mathbf{0} = \mathbf{u}$

(d) $\mathbf{v} + (-\mathbf{v}) = \mathbf{0}$

(e) $a(\mathbf{u} + \mathbf{v}) = a\mathbf{u} + a\mathbf{v}$

(f) $(a + b)\mathbf{v} = a\mathbf{v} + b\mathbf{v}$

(g) $(ab)\mathbf{v} = a(b\mathbf{v})$

(h) $\left| \frac{\mathbf{v}}{|\mathbf{v}|} \right| = 1$

75. Refer to the definition of the dot product preceding Exercise 63 and the following figure.

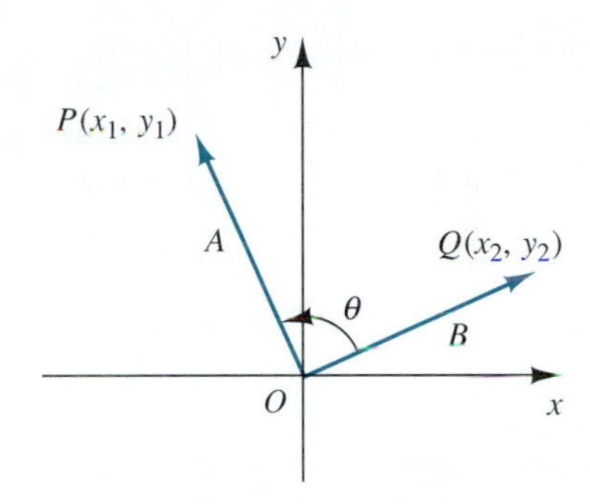

Let $\mathbf{A} = \langle x_1, y_1 \rangle$ and $\mathbf{B} = \langle x_2, y_2 \rangle$. Use the Law of Cosines on $\triangle POQ$ to show that $\mathbf{A} \cdot \mathbf{B} = |\mathbf{A}|\,|\mathbf{B}| \cos \theta$.

8.5 The Trigonometric Form of Complex Numbers and DeMoivre's Theorem

In this section we investigate and develop some of the interesting (and perhaps somewhat surprising) connections between trigonometry and the complex number system.

We obtain a geometric representation of the real number system by setting up a one-to-one correspondence between the set of real numbers and the points on a number line. We can obtain a geometric representation of the complex number system by using the points in a rectangular coordinate system. Specifically, we may associate each complex number $z = a + bi$ with the ordered pair (a, b), so we may label the point (a, b) as $a + bi$. See Figure 8.45 on page 526.

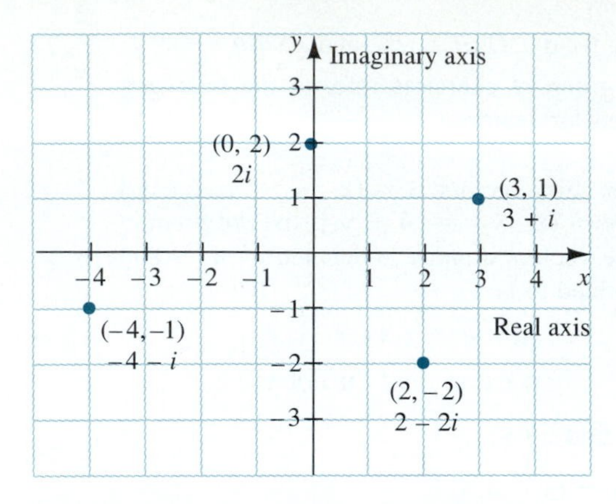

FIGURE 8.45 The complex plane

Note that the x-axis is referred to as the **real axis** and the y-axis is called the **imaginary axis.** When we associate a complex number with each point in the plane in this way, the xy-coordinate system is called the **complex plane.**

Just as $|x|$ represents the distance between the origin and the real number x, $|z| = |a + bi|$ represents the distance between the point (a, b), which corresponds to $a + bi$, and the origin of the complex plane. We often denote $|z| = |a + bi|$ as r, which is also called the **modulus** of the complex number $a + bi$.

Remember that in a different context, the word *argument* means the input to a function.

Figure 8.46 illustrates a typical complex number $z = a + bi$ (z need not be in the first quadrant). We call $a + bi$ the **rectangular form** of the complex number z. The angle θ pictured in Figure 8.46 may be *any* angle in standard position whose terminal side falls on the line passing through the origin and the point (a, b). The angle θ, which is called the **argument** of z, is usually chosen in the interval $[0, 2\pi)$.

From Figure 8.46 we have

$$\cos\theta = \frac{a}{r} \Rightarrow a = r\cos\theta \tag{1}$$

$$\sin\theta = \frac{b}{r} \Rightarrow b = r\sin\theta \tag{2}$$

$$|z| = r = \sqrt{a^2 + b^2} \tag{3}$$

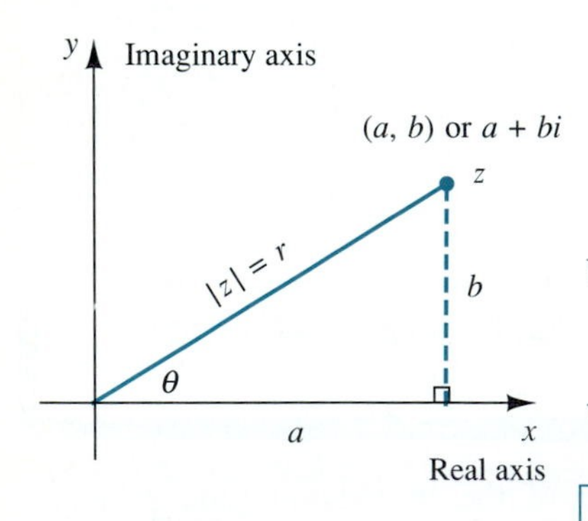

FIGURE 8.46

Using these results, we may rewrite any complex number $z = a + bi$ as

$$z = r\cos\theta + (r\sin\theta)i = r(\cos\theta + i\sin\theta)$$

which is called the **trigonometric form** of the complex number $a + bi$.

TRIGONOMETRIC FORM OF A COMPLEX NUMBER $a + bi$

Let $z = a + bi$, $r = \sqrt{a^2 + b^2}$, and θ be any argument of z. Then

$$z = r(\cos\theta + i\sin\theta)$$

EXAMPLE 1 Express the complex number $z = 3\left(\cos\frac{\pi}{6} + i\sin\frac{\pi}{6}\right)$ in rectangular form.

Solution Looking at the given complex number, we see that it is in the form $r(\cos\theta + i\sin\theta)$, so $r = 3$ and $\theta = \frac{\pi}{6}$. We can now use (1) and (2) to find that $a = r\cos\theta = 3\cos\frac{\pi}{6} = \frac{3\sqrt{3}}{2}$ and $b = r\sin\theta = 3\sin\frac{\pi}{6} = \frac{3}{2}$.

Therefore, in rectangular form we have $\boxed{z = \frac{3\sqrt{3}}{2} + \frac{3}{2}i}$. ■

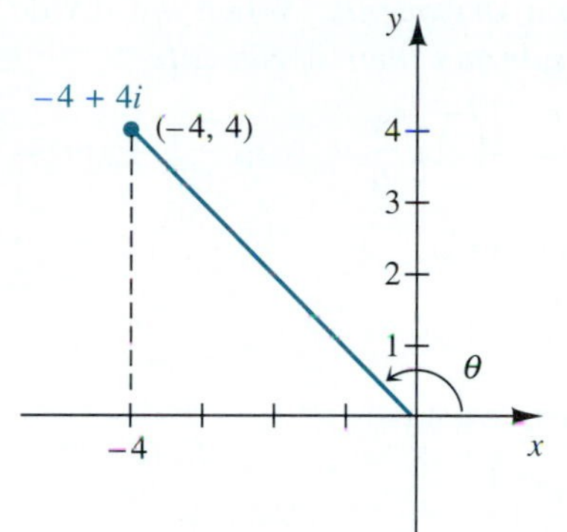

FIGURE 8.47 The complex number $-4 + 4i$

EXAMPLE 2 Express $-4 + 4i$ in trigonometric form.

Solution Since we are being asked to write the given number in the form $r(\cos\theta + i\sin\theta)$, we need to find the values of r and θ. If we draw the point corresponding to $-4 + 4i$ (see Figure 8.47), we see that $r = \sqrt{(-4)^2 + 4^2} = \sqrt{32} = 4\sqrt{2}$. From Figure 8.47 we can also see that

$$\sin\theta = \frac{4}{4\sqrt{2}} = \frac{1}{\sqrt{2}} \quad \text{and} \quad \cos\theta = \frac{-4}{4\sqrt{2}} = -\frac{1}{\sqrt{2}}$$

We recognize that $\theta = \frac{3\pi}{4}$. In fact, θ can be $\frac{3\pi}{4} + 2n\pi$ for any integer n; however, as mentioned before, we usually choose θ on the interval $[0, 2\pi)$. Therefore, we have $\boxed{4\sqrt{2}\left(\cos\frac{3\pi}{4} + i\sin\frac{3\pi}{4}\right)}$ as the requested trigonometric form. ■

When two complex numbers are written in trigonometric form, we can derive very simple formulas for their product and quotient. Suppose that the two complex numbers are

$$z_1 = r_1(\cos\theta_1 + i\sin\theta_1) \quad \text{and} \quad z_2 = r_2(\cos\theta_2 + i\sin\theta_2)$$

Then

$$z_1 \cdot z_2 = r_1(\cos\theta_1 + i\sin\theta_1) \cdot r_2(\cos\theta_2 + i\sin\theta_2)$$

We multiply out and separate the real and imaginary parts.

$$= r_1r_2[(\cos\theta_1\cos\theta_2 + i^2\sin\theta_1\sin\theta_2) + i(\sin\theta_1\cos\theta_2 + \cos\theta_1\sin\theta_2)]$$

$$= r_1r_2[(\cos\theta_1\cos\theta_2 - \sin\theta_1\sin\theta_2) + i(\sin\theta_1\cos\theta_2 + \cos\theta_1\sin\theta_2)]$$

Using the addition formulas for $\cos(\theta_1 + \theta_2)$ *and* $\sin(\theta_1 + \theta_2)$, *we get*

$$= r_1r_2[\cos(\theta_1 + \theta_2) + i\sin(\theta_1 + \theta_2)]$$

We have thus derived the first of the two formulas in the following box. The derivation of the second formula is left as an exercise for the student. (See Exercise 51.)

THE PRODUCT AND QUOTIENT OF TWO COMPLEX NUMBERS IN TRIGONOMETRIC FORM

Let $z_1 = r_1(\cos\theta_1 + i\sin\theta_1)$ and $z_2 = r_2(\cos\theta_2 + i\sin\theta_2)$, where $z_2 \neq 0$. Then

$$z_1 \cdot z_2 = r_1 r_2[\cos(\theta_1 + \theta_2) + i\sin(\theta_1 + \theta_2)]$$

$$\frac{z_1}{z_2} = \frac{r_1}{r_2}[\cos(\theta_1 - \theta_2) + i\sin(\theta_1 - \theta_2)]$$

In words, these formulas say that to multiply two complex numbers in trigonometric form, we *multiply their moduli and add their arguments*. When we divide two complex numbers, we *divide their moduli and subtract their arguments*.

EXAMPLE 3 Let $z = 2\left(\cos\frac{2\pi}{3} + i\sin\frac{2\pi}{3}\right)$ and $w = 5\left(\cos\frac{\pi}{6} + i\sin\frac{\pi}{6}\right)$. Express

(a) zw **(b)** $\frac{z}{w}$ in both trigonometric and rectangular form.

Solution

(a) Using the product formula we have just derived, we have

$$zw = (2)(5)\left[\cos\left(\frac{2\pi}{3} + \frac{\pi}{6}\right) + i\sin\left(\frac{2\pi}{3} + \frac{\pi}{6}\right)\right]$$

$$= \boxed{10\left(\cos\frac{5\pi}{6} + i\sin\frac{5\pi}{6}\right)}$$ *This is the answer in trigonometric form.*

$$= 10\left(\frac{-\sqrt{3}}{2} + i\left(\frac{1}{2}\right)\right)$$

$$= \boxed{-5\sqrt{3} + 5i}$$ *This is the answer in rectangular form.*

(b) Using the quotient formula,

$$\frac{z}{w} = \frac{2}{5}\left[\cos\left(\frac{2\pi}{3} - \frac{\pi}{6}\right) + i\sin\left(\frac{2\pi}{3} - \frac{\pi}{6}\right)\right]$$

$$= \boxed{\frac{2}{5}\left(\cos\frac{\pi}{2} + i\sin\frac{\pi}{2}\right)}$$ *This is the answer in trigonometric form.*

$$= \boxed{\frac{2}{5}i}$$ *This is the answer in rectangular form.* ■

Repeated use of the multiplication rule allows us easily to compute powers of a complex number. Remember that when we multiply two complex numbers, we multiply their moduli and add their arguments.

$$\text{Let}\quad z = r(\cos\theta + i\sin\theta) \qquad \textit{Then}$$

$$z^2 = z \cdot z = r^2(\cos 2\theta + i\sin 2\theta)$$

$$z^3 = z^2 \cdot z = r^3(\cos 3\theta + i\sin 3\theta) \qquad \textit{Similarly}$$

$$z^4 = r^4(\cos 4\theta + i\sin 4\theta)$$

$$\vdots$$

$$z^n = r^n(\cos n\theta + i\sin n\theta)$$

The following result, known as DeMoivre's theorem, generalizes this pattern. DeMoivre's theorem can be proven by the method of mathematical induction, which is discussed in Chapter 11.

DeMOIVRE'S THEOREM

Let $z = r(\cos\theta + i\sin\theta)$ and n be a natural number. Then

$$z^n = [r(\cos\theta + i\sin\theta)]^n = r^n(\cos n\theta + i\sin n\theta)$$

EXAMPLE 4 Find the following in rectangular form.

(a) $\left[3\left(\cos\frac{\pi}{3} + i\sin\frac{\pi}{3}\right)\right]^5$ **(b)** $(6 - 6i)^3$

Solution

(a) According to DeMoivre's theorem, we have

$$\left[3\left(\cos\frac{\pi}{3} + i\sin\frac{\pi}{3}\right)\right]^5 = 3^5\left(\cos\frac{5\pi}{3} + i\sin\frac{5\pi}{3}\right)$$

$$= 243\left[\frac{1}{2} + i\left(\frac{-\sqrt{3}}{2}\right)\right] = \boxed{\frac{243}{2} - \frac{243\sqrt{3}}{2}i}$$

What is $6 - 6i$ in trigonometric form?

(b) Rather than cube $(6 - 6i)$, we can use DeMoivre's theorem. However, in order to use DeMoivre's theorem, we must first convert $6 - 6i$ into trigonometric form.

$$6 - 6i = 6\sqrt{2}\left(\cos\frac{7\pi}{4} + i\sin\frac{7\pi}{4}\right) \quad \textit{Therefore,}$$

$$(6 - 6i)^3 = \left[6\sqrt{2}\left(\cos\frac{7\pi}{4} + i\sin\frac{7\pi}{4}\right)\right]^3 = (6\sqrt{2})^3\left(\cos\frac{21\pi}{4} + i\sin\frac{21\pi}{4}\right)$$

$$= 432\sqrt{2}\left(\cos\frac{5\pi}{4} + i\sin\frac{5\pi}{4}\right)$$

$$= 432\sqrt{2}\left(-\frac{\sqrt{2}}{2} - \frac{\sqrt{2}}{2}i\right)$$

$$= \boxed{-432 - 432i}$$

■

The following example illustrates how DeMoivre's theorem can be used to find roots of equations we were unable to find before. Suppose we want to find the solutions to $z^3 = 8$, or, in other words, to find the cube roots of 8. Although we know that there is only one real solution, which is $z = 2$, in Section 4.3 we saw that a third-degree equation will have exactly *three* roots in the *complex number system*. It turns out that finding the solutions to this equation in trigonometric form is quite straightforward, as the next example illustrates.

EXAMPLE 5 Find all the roots of the equation $z^3 = 8$.

Solution Suppose that $z = r(\cos\theta + i\sin\theta)$ is a solution to $z^3 = 8$. We may consider 8 a complex number and write its trigonometric form as

$$8 = 8 + 0i = 8(\cos 0 + i\sin 0)$$

By DeMoivre's theorem the original equation $z^3 = 8$ becomes

$$z^3 = r^3(\cos 3\theta + i\sin 3\theta) = 8(\cos 0 + i\sin 0) \tag{1}$$

From (1) it follows that $r^3 = 8$ and so $r = 2$. (Remember that r is a real number.) It also follows that 3θ can be 0 or any angle that is coterminal with 0. In other words, we have

$$3\theta = 0 + 2k\pi \quad \text{or} \quad \theta = \frac{2k\pi}{3} \text{ for } k = 0, 1, 2, \ldots$$

If you graph $f(x) = x^3 - 8$ on a graphics calculator, will you get the three zeros we found in Example 5? Why or why not?

If $k = 0$, we get $\theta = 0$, which gives $z = 2$, as we saw before.

If $k = 1$, we get $\theta = \frac{2\pi}{3}$, which gives $z = 2\left(\cos\frac{2\pi}{3} + i\sin\frac{2\pi}{3}\right)$ as a second cube root of 8.

If $k = 2$, we get $\theta = \frac{4\pi}{3}$, which gives $z = 2\left(\cos\frac{4\pi}{3} + i\sin\frac{4\pi}{3}\right)$ as a third cube root of 8.

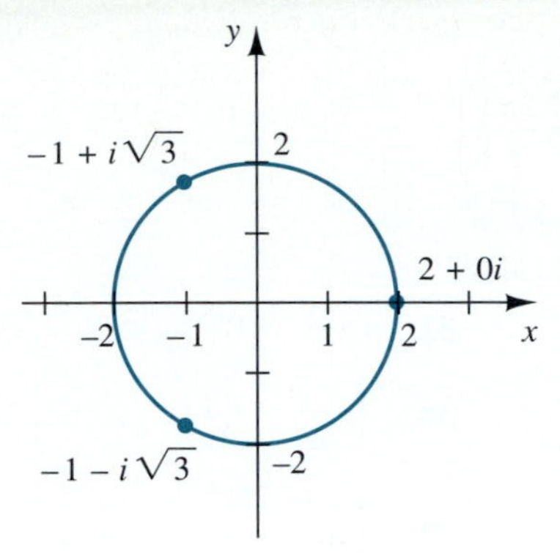

FIGURE 8.48 The three cube roots of 8

Since we know that a third-degree equation has exactly three complex roots, we have found the three solutions to $z^3 = 8$. It is left to the student to verify that if we let $k = 3, 4, 5, \ldots$, we will just repeat the solutions we have already obtained.

In rectangular form, the three cube roots of 8 are $z = 2$, $z = -1 + i\sqrt{3}$, and $z = -1 - i\sqrt{3}$. (Cube $-1 + i\sqrt{3}$ and verify that it is a cube root of 8.) Figure 8.48 illustrates the three cube roots of 8. Note that they are equally spaced on the circle of radius 2 about the origin. ■

The previous example illustrates the following theorem.

THE *n*th ROOTS OF A COMPLEX NUMBER

Let $z = r(\cos\theta + i\sin\theta)$ and let n be any positive integer. The n distinct nth roots of z are $w_0, w_1, w_2, \ldots, w_{n-1}$, where

$$w_k = \sqrt[n]{r}\left(\cos\frac{\theta + 2k\pi}{n} + i\sin\frac{\theta + 2k\pi}{n}\right)$$

for $k = 0, 1, 2, 3, \ldots, n - 1$.

We usually do not write $z^{1/n}$ because complex roots are not unique.

EXAMPLE 6 Find all the fifth roots of $z = 1 + i$.

Solution We begin by writing the complex number $z = 1 + i$ in trigonometric form as $z = \sqrt{2}\left(\cos\frac{\pi}{4} + i\sin\frac{\pi}{4}\right)$. Using the nth roots theorem with $n = 5$, we get

$$w_k = \sqrt[5]{\sqrt{2}}\left[\cos\left(\frac{\frac{\pi}{4} + 2k\pi}{5}\right) + i\sin\left(\frac{\frac{\pi}{4} + 2k\pi}{5}\right)\right] \qquad \text{for } k = 0, 1, 2, 3, 4$$

$\sqrt[5]{\sqrt{2}} = (2^{1/2})^{1/5} = 2^{1/10} = \sqrt[10]{2}$

So the five fifth roots are

$$w_0 = \sqrt[10]{2}\left[\cos\left(\frac{\pi}{20}\right) + i\sin\left(\frac{\pi}{20}\right)\right]$$

$$w_1 = \sqrt[10]{2}\left[\cos\left(\frac{9\pi}{20}\right) + i\sin\left(\frac{9\pi}{20}\right)\right]$$

$$w_2 = \sqrt[10]{2}\left[\cos\left(\frac{17\pi}{20}\right) + i\sin\left(\frac{17\pi}{20}\right)\right]$$

$$w_3 = \sqrt[10]{2}\left[\cos\left(\frac{25\pi}{20}\right) + i\sin\left(\frac{25\pi}{20}\right)\right] = \sqrt[10]{2}\left[\cos\left(\frac{5\pi}{4}\right) + i\sin\left(\frac{5\pi}{4}\right)\right]$$

$$w_4 = \sqrt[10]{2}\left[\cos\left(\frac{33\pi}{20}\right) + i\sin\left(\frac{33\pi}{20}\right)\right]$$

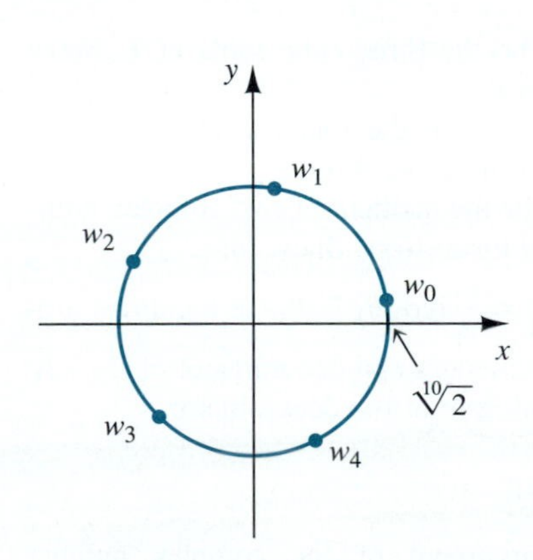

FIGURE 8.49 The five fifth roots of $1 + i$

These fifth roots of $1 + i$ are illustrated in Figure 8.49. Again notice that they are equally spaced on the circle of radius $\sqrt[10]{2}$ centered at the origin. ■

EXERCISES 8.5

In Exercises 1–16, if the given number is in rectangular form, convert it to trigonometric form; if it is in trigonometric form, convert it to rectangular form. Where necessary, round to two decimal places.

1. $\sqrt{3} - i$

2. $4\left(\cos \frac{5\pi}{6} + i \sin \frac{5\pi}{6}\right)$

3. $10\left(\cos \frac{7\pi}{4} + i \sin \frac{7\pi}{4}\right)$

4. $-5 + 5i$

5. $2 - 7i$

6. $9\left(\cos \frac{2\pi}{5} + i \sin \frac{2\pi}{5}\right)$

7. 6

8. $3i$

9. $2\left(\cos \frac{3\pi}{2} + i \sin \frac{3\pi}{2}\right)$

10. $8(\cos \pi + i \sin \pi)$

11. $\sqrt{5} + 4i$

12. $-\frac{\sqrt{3}}{4} - \frac{1}{4}i$

13. $9(\cos 300° + i \sin 300°)$

14. $\frac{1}{3}(\cos 88° + i \sin 88°)$

15. -4

16. $-2i$

In Exercises 17–20, use the product and quotient rules to find $z_1 \cdot z_2$ and $\frac{z_1}{z_2}$; simplify where possible.

17. $z_1 = 6\left(\cos \frac{\pi}{3} + i \sin \frac{\pi}{3}\right)$; $z_2 = 2\left(\cos \frac{\pi}{9} + i \sin \frac{\pi}{9}\right)$

18. $z_1 = 10(\cos 210° + i \sin 210°)$;
$z_2 = 4(\cos 100° + i \sin 100°)$

19. $z_1 = 5\left(\cos \frac{5\pi}{6} + i \sin \frac{5\pi}{6}\right)$; $z_2 = \left(\cos \frac{\pi}{8} + i \sin \frac{\pi}{8}\right)$

20. $z_1 = \sqrt{15}(\cos 210° + i \sin 210°)$;
$z_2 = \sqrt{5}(\cos 120° + i \sin 120°)$

In Exercises 21–24, perform the indicated operations first in rectangular form and then in trigonometric form. Verify that the answers obtained are equivalent. (Round to the nearest hundredth where necessary.)

21. $(3 + 3i)(1 - i)$

22. $6i(-2 - 2i)$

23. $\frac{i}{1 + i}$

24. $\frac{4}{-3 + 3i}$

In Exercises 25–38, compute the indicated power by using DeMoivre's theorem.

25. $\left[3\left(\cos \frac{\pi}{6} + i \sin \frac{\pi}{6}\right)\right]^5$

26. $\left[2\left(\cos \frac{\pi}{4} + i \sin \frac{\pi}{4}\right)\right]^6$

27. $\left[\sqrt{2}\left(\cos \frac{5\pi}{4} + i \sin \frac{5\pi}{4}\right)\right]^7$

28. $\left[\sqrt[3]{5}\left(\cos \frac{2\pi}{3} + i \sin \frac{2\pi}{3}\right)\right]^9$

29. $[\sqrt{6}(\cos 20° + i \sin 20°)]^6$

30. $[\sqrt[4]{3}(\cos 15° + i \sin 15°)]^{12}$

31. $(1 + i)^8$

32. $(\sqrt{3} - i)^4$

33. $(-3 + 3i)^5$

34. $\left(\frac{\sqrt{3}}{2} - \frac{1}{2}i\right)^6$

35. $\left(\frac{i}{1 - i}\right)^3$

36. $\left(\frac{2 + 2i}{3 - 3i}\right)^5$

37. $\left(\frac{1 - i\sqrt{3}}{-1 + i}\right)^4$

38. $[(1 - i)(1 + i)]^7$

In Exercises 39–44, use the nth roots theorem to find the requested roots of the given complex number.

39. Find the fourth roots of $3 + 3i$.

40. Find the cube roots of $1 - i$.

41. Find the sixth roots of 1.

42. Find the square roots of i.

43. Find the fifth roots of $-\sqrt{3} + i$.

44. Find the cube roots of 125.

45. Compute the cube roots of $1 + i\sqrt{3}$. Express your answers in both rectangular and trigonometric form. Round your answers to two decimal places.

46. Compute the fifth roots of i. Express your answers in both rectangular and trigonometric form. Round your answers to two decimal places.

47. Find the fourth roots of i in exact $a + bi$ form. You may find the half-angle formulas in Section 8.2 useful.

48. Find the fourth roots of $-6 - 6i\sqrt{3}$ in exact $a + bi$ form. You may find the half-angle formulas in Section 8.2 useful.

49. Let w_1, w_2, and w_3 be the three cube roots of 1. Show that $w_1 + w_2 + w_3 = 0$.

50. Let w_1, w_2, w_3, and w_4 be the four fourth roots of 1. Show that $w_1 + w_2 + w_3 + w_4 = 0$.

51. Derive the formula for the quotient of two complex numbers in trigonometric form. HINT: Show that

$$\frac{r_1(\cos \theta_1 + i \sin \theta_1)}{r_2(\cos \theta_2 + i \sin \theta_2)} = \frac{r_1}{r_2}[\cos(\theta_1 - \theta_2) + i \sin(\theta_1 - \theta_2)]$$

by multiplying the numerator and denominator of the left-hand side by the conjugate of the denominator.

QUESTIONS FOR THOUGHT

52. Explain why the argument of the complex number $0 = 0 + 0i$ is undefined.

53. Is it easier to find $|z|$ when z is written in rectangular form or trigonometric form? Explain.

8.6 Polar Coordinates

Up to this point we have used a rectangular coordinate system to label points in the plane, and of course the graph of each equation we have examined has its particular appearance because of the coordinate system we have chosen.

However, a rectangular coordinate system is not the only possible coordinate system that can be used. In this section we introduce another system called the **polar coordinate system** and discuss equations and their graphs in this new coordinate system.

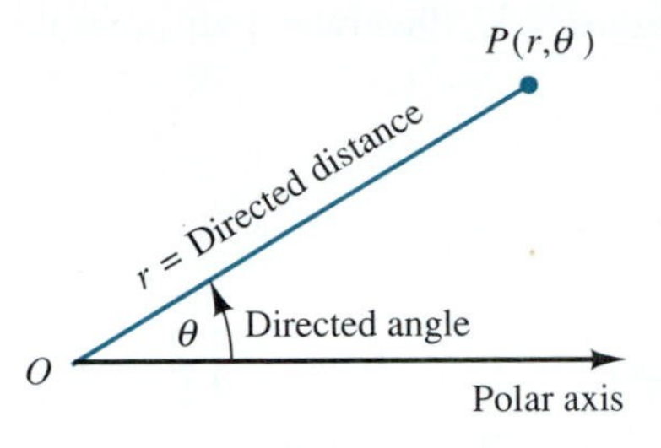

FIGURE 8.50 Polar coordinates

In order to construct the polar coordinate system we begin with a point O, called the **polar origin,** or **pole,** and a ray emanating from O called the **polar axis.** (The polar axis is usually drawn horizontally and extending to the right.) See Figure 8.50. We can then label each point P with a pair of coordinates of the form (r, θ), where

1. r is the *directed distance* from O to P. r can be any real number: positive, negative, or zero.
2. θ is the *directed angle* from the polar axis to the line segment $\overline{OP}$. We follow the conventions adopted in trigonometry that positive angles are measured in a counterclockwise fashion, whereas negative angles are measured in a clockwise fashion.

Remember that the polar origin is also called the pole.

The following example illustrates these ideas.

EXAMPLE 1 Plot each of the following points in a polar coordinate system.

(a) $\left(2, \frac{\pi}{3}\right)$ **(b)** $\left(3, \frac{-\pi}{4}\right)$ **(c)** $\left(-4, \frac{5\pi}{6}\right)$ **(d)** $\left(-2, \frac{-2\pi}{3}\right)$

Solution We can locate a given point (r, θ) in polar coordinates by finding the ray emanating from the pole that makes an angle of θ with the polar axis and then measure $|r|$ units along the terminal side of θ if r is positive. See Figure 8.51(a) and (b). If r is negative, then we measure $|r|$ units along the ray with endpoint O that has direction *opposite* to that of the terminal side of θ. See Figure 8.51(c) and (d).

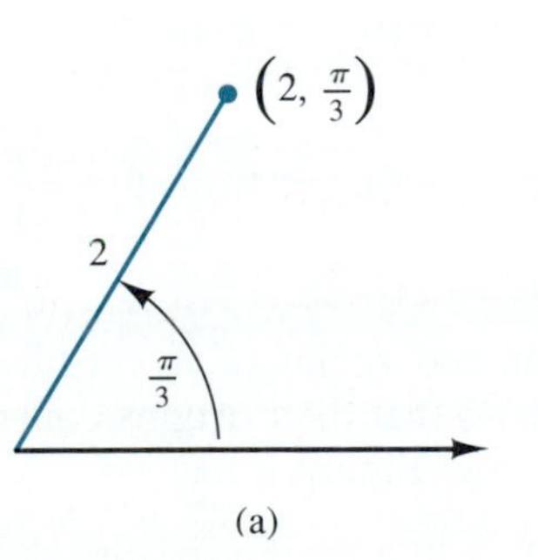

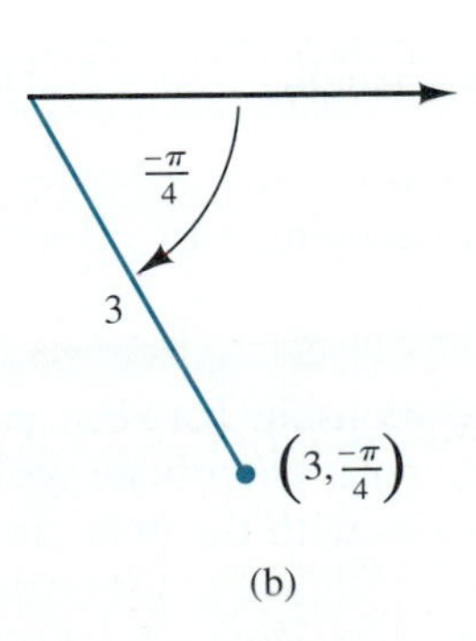

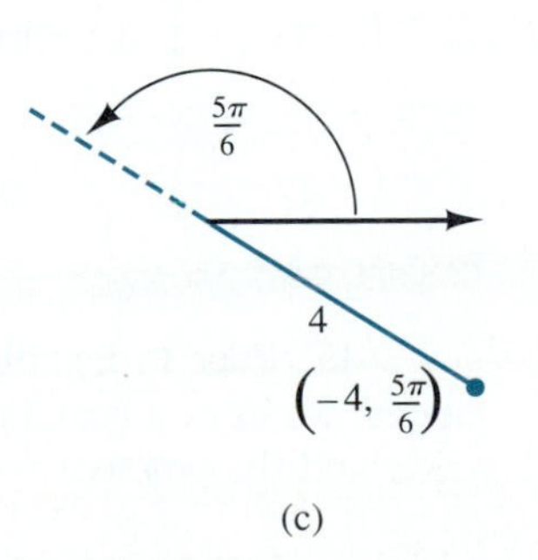

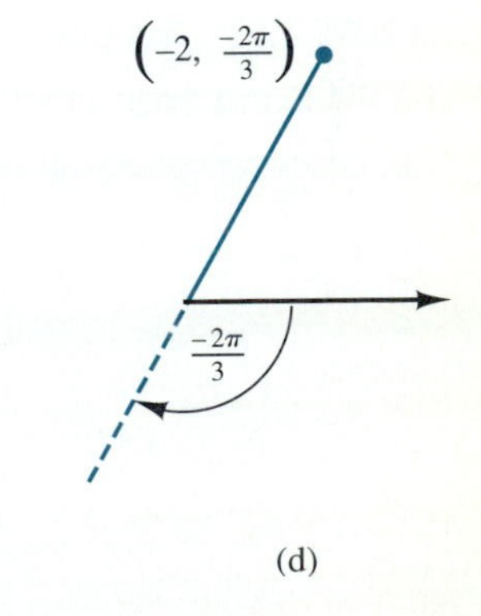

FIGURE 8.51

■

Several important points need to be made here. Unlike a rectangular coordinate system, in which the coordinates of a point are unique, in a polar coordinate system a point may have many pairs of coordinates. Since adding or subtracting multiples of 2π will bring us back to the same terminal side, (r, θ) and $(r, \theta + 2k\pi)$, where k is any integer, are coordinates of the same point. Similarly, (r, θ) and $(-r, \theta \pm \pi)$ are coordinates of the same point. Figure 8.52 illustrates four possible pairs of coordinates for the point $\left(2, \frac{\pi}{4}\right)$.

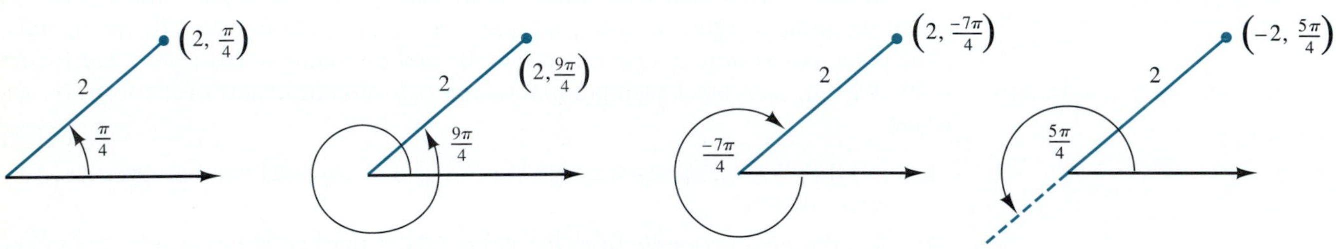

FIGURE 8.52 Other possible polar coordinates for the point $\left(2, \frac{\pi}{4}\right)$

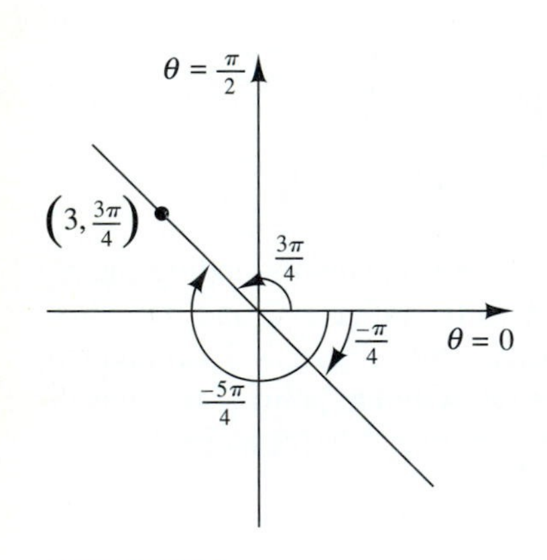

FIGURE 8.53 The point $\left(3, \frac{3\pi}{4}\right)$ and three other possible polar representations

In the case of the polar origin, we cannot really define an angle θ, since $(0, \theta)$ will represent the pole for *any* value of θ. Thus we agree that the coordinates of the pole are $(0, \theta)$ for any value of θ.

EXAMPLE 2 Plot the point $\left(3, \frac{3\pi}{4}\right)$ in a polar coordinate system and find three additional polar representations for this point, at least one of which has a negative r.

Solution The point $\left(3, \frac{3\pi}{4}\right)$ is plotted in Figure 8.53. Three other representations are

$\left(3, \frac{11\pi}{4}\right)$ Obtained by adding 2π to $\theta = \frac{3\pi}{4}$

$\left(-3, \frac{-\pi}{4}\right)$ Obtained by changing 3 to -3 and subtracting π from

$\left(-3, \frac{-\pi}{4}\right)$ Obtained by changing 3 to -3 and subtracting π from $\theta = \frac{3\pi}{4}$

■

In order to establish the relationship between polar and rectangular coordinates, we draw a rectangular and a polar coordinate system so that their origins coincide and the positive x-axis coincides with the polar axis. See Figure 8.54.

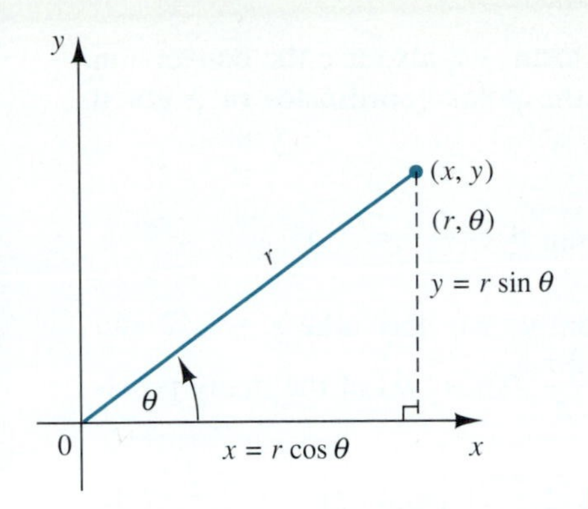

FIGURE 8.54 Rectangular and polar coordinates

If we now choose a point P whose rectangular coordinates are (x, y) and whose polar coordinates are (r, θ), as indicated in Figure 8.54, we obtain the following relationships.

$$x = r\cos\theta \qquad y = r\sin\theta \qquad x^2 + y^2 = r^2 \qquad \tan\theta = \frac{y}{x}$$

These relationships hold regardless of where the point P falls. We may use these relationships to convert from rectangular to polar coordinates, and vice versa. We summarize these conversion equations in the following box.

CONVERSION OF COORDINATES

To Convert from Polar to Rectangular Coordinates

Given a point (r, θ) in polar form, then

$$x = r\cos\theta \qquad y = r\sin\theta$$

To Convert from Rectangular to Polar Coordinates

Given a point (x, y) in rectangular form, then

$$r^2 = x^2 + y^2 \qquad \tan\theta = \frac{y}{x}$$

EXAMPLE 3

(a) Convert from polar form to rectangular form: **(i)** $\left(4, \frac{5\pi}{6}\right)$ **(ii)** $(3, 2)$.

(b) Convert from rectangular form to polar form: **(i)** $(1, -1)$ **(ii)** $(3, \pi)$.

Solution

(a) In order to convert from polar to rectangular form, we use the conversion equations in the box.

(i) Given the point $\left(4, \frac{5\pi}{6}\right)$ we get

$$x = 4\cos\frac{5\pi}{6} = 4\left(-\frac{\sqrt{3}}{2}\right) = -2\sqrt{3} \quad \text{and} \quad y = 4\sin\frac{5\pi}{6} = 4\left(\frac{1}{2}\right) = 2$$

Thus the rectangular coordinates are $\boxed{(-2\sqrt{3}, 2)}$.

(ii) Keep in mind that we are told that $(3, 2)$ is in *polar form*. Thus we have $r = 3$ and $\theta = 2$ (in radian measure).

$$x = 3\cos 2 \approx -1.25 \quad \text{and} \quad y = 3\sin 2 \approx 2.73$$

Thus the rectangular coordinates are approximately $\boxed{(-1.25, 2.73)}$.

(b) In order to convert from rectangular to polar form, we also use the conversion equations in the box, keeping in mind that the polar coordinates of a given point are not unique.

(i) For the point $(1, -1)$ we have

$$r^2 = 1^2 + (-1)^2 = 2 \quad \text{and} \quad \tan\theta = \frac{-1}{1} = -1$$

Since the point $(1, -1)$ is in the fourth quadrant, we may take $r = \sqrt{2}$ and $\theta = \frac{7\pi}{4}$ or we may take $r = -\sqrt{2}$ and $\theta = \frac{3\pi}{4}$. Thus two of the many possible answers for the polar coordinates are $\boxed{\left(\sqrt{2}, \frac{7\pi}{4}\right)}$ and $\boxed{\left(-\sqrt{2}, \frac{3\pi}{4}\right)}$.

(ii) Keeping in mind that we are given the point $(3, \pi)$ in rectangular coordinates, we have

$$r^2 = 3^2 + \pi^2 \quad \text{and} \quad \tan\theta = \frac{\pi}{3}$$

Since the point $(3, \pi)$ is in the first quadrant, we have $r = \sqrt{9 + \pi^2} \approx 4.34$ and $\theta \approx 0.81$ Thus one of the many possible answers for the polar coordinates is approximately $\boxed{(4.34, 0.81)}$. ■

By comparing parts (a) and (b) of Example 3 we can see that converting from polar form to rectangular form is quite straightforward, whereas converting from rectangular form to polar form is somewhat more involved. However, when converting equations, the opposite is true. If we have an equation in rectangular form, such as $y = 2x^2$, it is very easy to convert this into polar form by simply substituting $x = r\cos\theta$ and $y = r\sin\theta$. Thus

$$y = 2x^2 \qquad \textit{Becomes}$$

$$r\sin\theta = 2(r\cos\theta)^2$$

On the other hand, converting from polar form to rectangular form can sometimes require some ingenuity, but the rectangular form may make it easier to recognize the graph of the equation. By the graph of a polar equation we mean, of course, the set of all points (r, θ) that satisfy the equation.

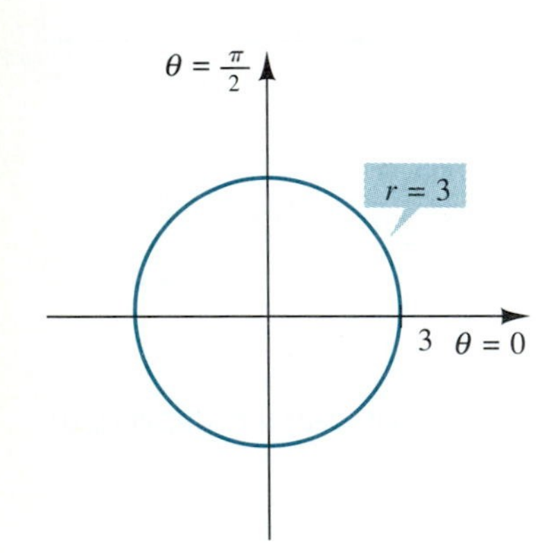

FIGURE 8.55

EXAMPLE 4 Graph the following polar equations and check the answer by finding the corresponding rectangular equation.

(a) $r = 3$ **(b)** $\theta = \frac{\pi}{4}$ **(c)** $r = 2\sec\theta$

Solution

(a) The equation $r = 3$ implies that a point will be on the graph if and only if it is of the form $(3, \theta)$, where θ can be any angle. This means that the graph consists of all points that are 3 units away from the pole—that is, a circle with center at the polar origin and a radius of 3. To convert the equation $r = 3$ into rectangular form, we want the equation to involve x and y, so we substitute $r = 3$ into the conversion equation $x^2 + y^2 = r^2$ to get $x^2 + y^2 = 3^2$, which we also recognize as a circle with center $(0, 0)$ and radius 3. The graph appears in Figure 8.55.

Which form of the equation is simpler, the polar form, $r = 3$, or the rectangular form, $x^2 + y^2 = 9$? Is it simpler to check a point of the form (r, θ) in the equation $r = 3$ or a point of the form (x, y) in the equation $x^2 + y^2 = 9$?

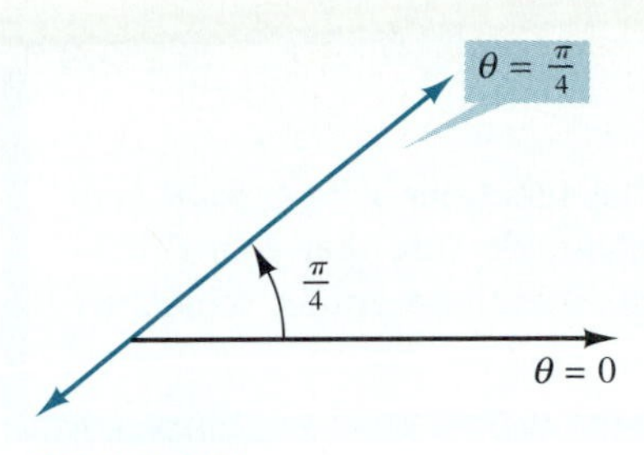

FIGURE 8.56

(b) The equation $\theta = \frac{\pi}{4}$ implies that a point will be on the graph if and only if it is of the form $\left(r, \frac{\pi}{4}\right)$, where r can be any real number. This means that the graph consists of all points on the line through the pole that makes an angle of $\frac{\pi}{4}$ with the polar axis. The graph appears in Figure 8.56.

In order to convert the equation $\theta = \frac{\pi}{4}$ into rectangular form, we can use the conversion equation $\tan\theta = \frac{y}{x}$ to get

$$\tan\frac{\pi}{4} = \frac{y}{x} \Rightarrow 1 = \frac{y}{x} \Rightarrow y = x$$

which we also recognize as the straight line appearing in Figure 8.56.

(c) The graph of the equation $r = 2\sec\theta$ is not apparent by analyzing the equation. However, if we transform it into rectangular form, we get

$$r = 2\sec\theta = \frac{2}{\cos\theta} \qquad \textit{Therefore}$$

$$r\cos\theta = 2 \Rightarrow x = 2$$

Why does $r\cos\theta = 2 \Rightarrow x = 2$?

which we recognize as the vertical line shown in Figure 8.57. ■

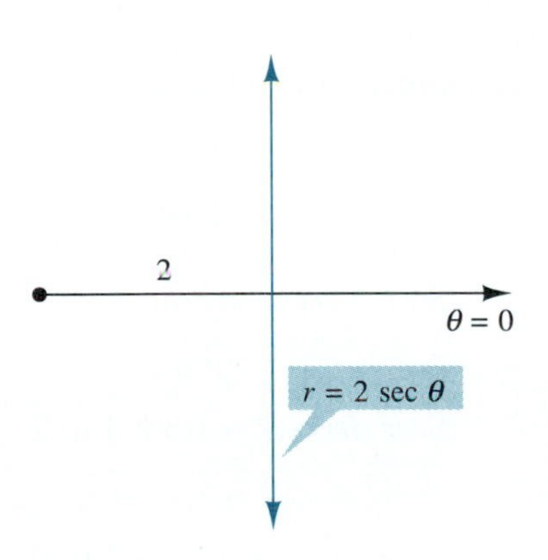

FIGURE 8.57

EXAMPLE 5 Identify and sketch the graph of $r = 2\sin\theta$.

Solution The graph is certainly not evident by simply analyzing the given polar equation. We might consider plotting a "sufficient" number of points to try to get an idea of what the graph looks like; however, if we can convert the given polar equation into rectangular form, we might be able to recognize its graph. Although there is no uniform method for doing this, it is important to keep the basic conversion equations in mind. We would like to have the equation involve $r\cos\theta$, $r\sin\theta$, and r^2 because they are equal to x, y, and $x^2 + y^2$, respectively.

We can multiply both sides of the given equation by r.

$$r = 2\sin\theta \qquad \textit{Multiply both sides of the equation by } r.$$

$$r^2 = 2r\sin\theta \qquad \textit{But } r^2 = x^2 + y^2 \textit{ and } y = r\sin\theta\textit{. So we get}$$

$$x^2 + y^2 = 2y \qquad \textit{Which we recognize as the equation of a circle. We may complete the square to get}$$

$$x^2 + (y - 1)^2 = 1$$

which is the equation of a circle with center (0, 1) and radius 1. The graph appears in Figure 8.58. ■

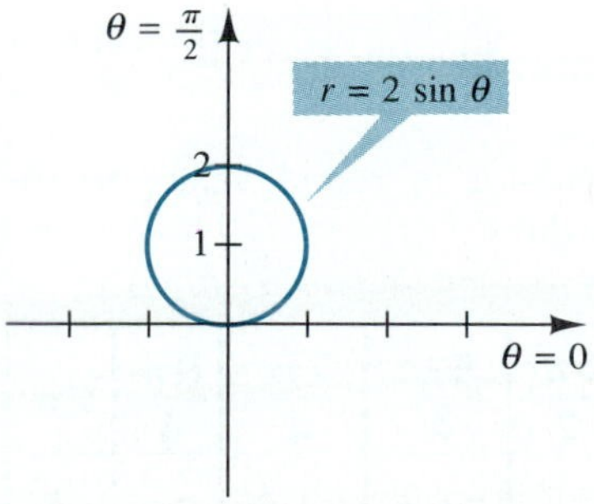

FIGURE 8.58

GRAFFIX

Use a graphics calculator or computer to graph the functions $r = 4 \cos\theta$ and $r = 4 + 4 \cos\theta$ in the same polar coordinate system. Be sure that your calculator has been set up to graph in polar coordinates. Are these graphs related to each other in the way that you expected?

Not only is it sometimes difficult to convert a polar equation into a familiar rectangular form, but many polar equations have graphs that are very different from the graphs with which we have become familiar.

EXAMPLE 6 Sketch the graph of the equation $r = 2 + 2 \sin \theta$.

Try converting the polar equation $r = 2 + 2 \sin \theta$ into rectangular form.

Solution In the last example we saw that the graph of $r = 2 \sin \theta$ is a circle, and so we might expect the graph of $r = 2 + 2 \sin \theta$ to simply be a 2-unit shift of the graph in the previous example. However (as we will shortly see), this is not the case. The graphing principles we developed are *specific* to a rectangular coordinate system and so do not apply to the polar coordinate system.

We can convert the given equation into rectangular form (in fact, into a variety of rectangular forms), but unfortunately none are of familiar graphs.

Since we are not very familiar with polar graphs, let's set up a table of values to plot some points. We note that since the sine function is periodic, it is sufficient for us to look at values on the interval $[0, 2\pi]$. If we recall how the values of $\sin \theta$ fluctuate (visualizing the graph of $y = \sin \theta$ would be helpful), we can analyze the graph of $r = 2 + 2 \sin \theta$ as follows:

θ	$\sin \theta$	$r = 2 + 2 \sin \theta$
As θ increases from 0 to $\frac{\pi}{2}$	$\sin \theta$ increases from 0 to 1	So r *increases* from 2 to 4
As θ increases from $\frac{\pi}{2}$ to π	$\sin \theta$ decreases from 1 to 0	So r *decreases* from 4 to 2
As θ increases from π to $\frac{3\pi}{2}$	$\sin \theta$ decreases from 0 to -1	So r *decreases* from 2 to 0
As θ increases from $\frac{3\pi}{2}$ to 2π	$\sin \theta$ increases from -1 to 0	So r *increases* from 0 to 2

Choosing some convenient values and using a calculator to round values to the nearest tenth when necessary, we get the following values.

θ	0	$\frac{\pi}{6}$	$\frac{\pi}{4}$	$\frac{\pi}{3}$	$\frac{\pi}{2}$	$\frac{2\pi}{3}$	$\frac{3\pi}{4}$	$\frac{5\pi}{6}$	π	$\frac{7\pi}{6}$	$\frac{5\pi}{4}$	$\frac{4\pi}{3}$	$\frac{3\pi}{2}$	$\frac{5\pi}{3}$	$\frac{7\pi}{4}$	$\frac{11\pi}{6}$	2π
$2 \sin \theta$	0	1	1.4	1.7	2	1.7	1.4	1	0	-1	-1.4	-1.7	-2	-1.7	-1.4	-1	0
$r = 2 + 2 \sin \theta$	2	3	3.4	3.7	4	3.7	3.4	3	2	1	0.6	0.3	0	0.3	0.6	1	2

Figure 8.59(a), (b), (c), (d) uses these tabulated values to sketch the portion of the curve in each quadrant.

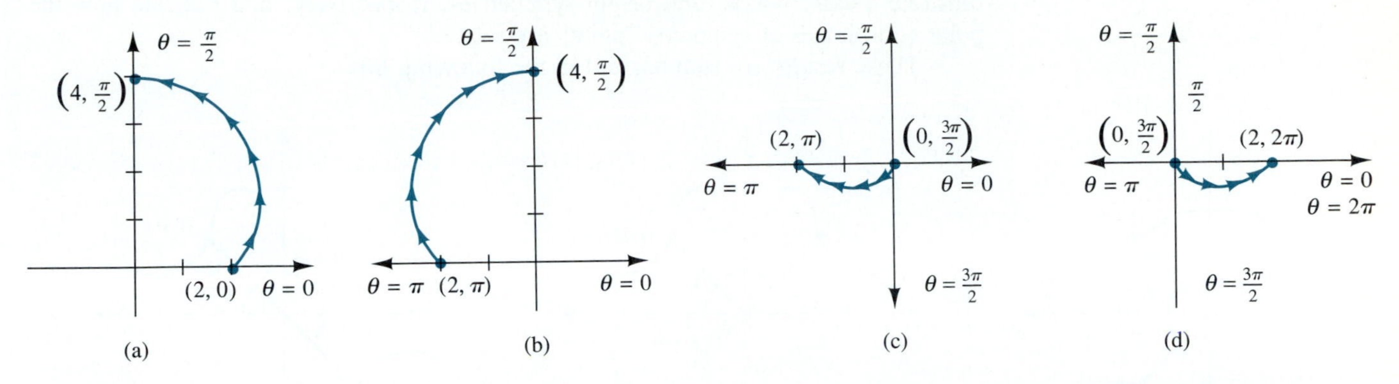

FIGURE 8.59

Putting these pieces together, we get the graph in Figure 8.60 which is called a *cardioid* because of its heartlike shape. Notice the arrows on the curve, which indicate how the curve is traced out starting at the point (2, 0).

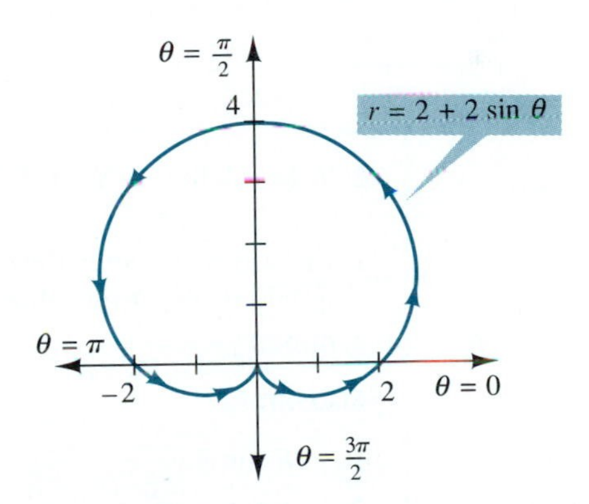

FIGURE 8.60

Several ideas to keep in mind: Remember that when r is negative for a given θ, the point will be along the ray determined by θ but on the opposite side of the pole. In graphing a polar curve, it is often helpful to determine the values of θ for which r is a maximum and for which $r = 0$. Every time $r = 0$, the graph revisits the pole.

■

Just as with graphs in a rectangular coordinate system, symmetry considerations can simplify the graphing of polar curves. In Figure 8.61(a), (b), and (c) we illustrate x-axis, y-axis, and origin symmetries, respectively, and indicate how the polar coordinates of symmetric points are related.

These results are summarized in the following box.

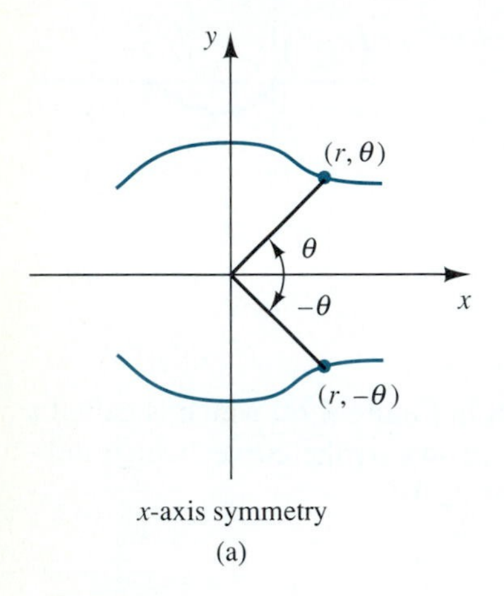

x-axis symmetry
(a)

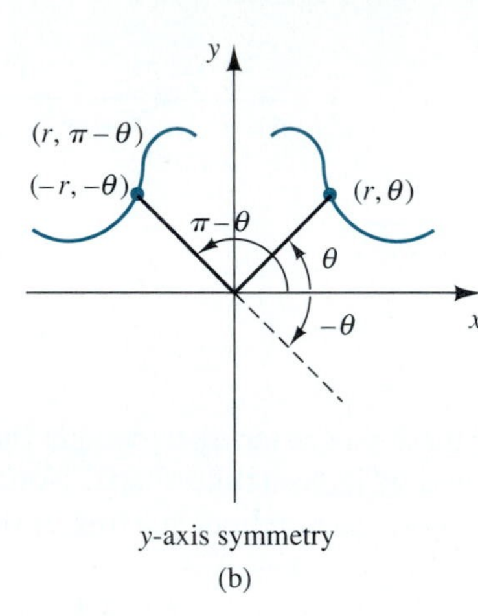

y-axis symmetry
(b)

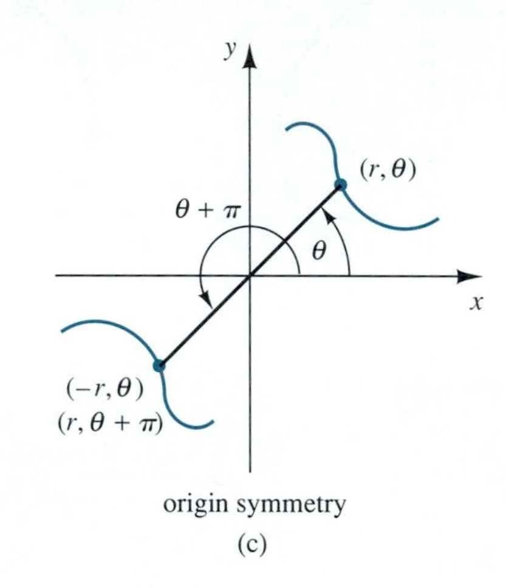

origin symmetry
(c)

FIGURE 8.61

SYMMETRY TESTS IN POLAR COORDINATES

If the following substitutions are made in a polar equation and they produce an equivalent equation, then the graph of the polar equation has the indicated symmetry.

Substitute	*Equivalence of Equation Indicates*
1. $-\theta$ for θ	Symmetry about the polar axis (x-axis symmetry)
2. $-r$ for r	Symmetry about the polar origin
3. $-r$ for r and $-\theta$ for θ	Symmetry about the line $\theta = \dfrac{\pi}{2}$ (y-axis symmetry)
4. $\pi - \theta$ for θ	Symmetry about the line $\theta = \dfrac{\pi}{2}$ (y-axis symmetry)

Reviewing Example 6, we note that if we replace θ by $\pi - \theta$ in the equation $r = 2 + 2 \sin \theta$, we get $r = 2 + 2 \sin(\pi - \theta)$. However, $\sin(\pi - \theta) = \sin \theta$ (*why*?), so the resulting equation is equivalent to the original. Thus the graph of this equation is symmetric about the line $\theta = \frac{\pi}{2}$ (it exhibits y-axis symmetry), which agrees with the graph we obtained.

EXAMPLE 7 Sketch the graph of $r^2 = 4 \cos 2\theta$.

Solution By analyzing the equation we can draw the following conclusions about its graph.

1. Replacing θ with $-\theta$ in the equation, we get

$$r^2 = 4 \cos 2(-\theta) = 4 \cos(-2\theta) = 4 \cos 2\theta$$

which implies that the graph will exhibit polar axis symmetry.

2. Replacing r with $-r$ produces the same equation, which implies that the graph is also symmetric with respect to the polar origin.
3. It is left to the student to use the remaining symmetry test(s) to verify that the graph of this equation also exhibits symmetry with respect to the line $\theta = \frac{\pi}{2}$.
4. Since r^2 cannot be negative, we can ignore any values of θ for which $\cos 2\theta$ is negative. Therefore, we do not need to consider values of θ in the interval $\frac{\pi}{4} < \theta < \frac{3\pi}{4}$, since these values of θ will place 2θ between $\frac{\pi}{2}$ and $\frac{3\pi}{2}$, where the cosine function is negative.
5. Therefore, we need to determine the graph only for $0 \le \theta \le \frac{\pi}{4}$, since the graph for $\frac{3\pi}{4} \le \theta \le \pi$ can be obtained by using the origin symmetry of the graph. As the following table shows, the portion of the graph for $\frac{3\pi}{4} \le \theta \le \pi$ actually comes from θ values between 0 and $\frac{\pi}{4}$, since r can be negative.
6. Since the cosine function varies from 1 to 0, $r^2 = 4 \cos 2\theta$ varies from 4 to 0 and so $|r|$ varies from 2 to 0.

The following table contains a few values to aid in sketching the graph.

θ	0	$\frac{\pi}{12}$	$\frac{\pi}{8}$	$\frac{\pi}{6}$	$\frac{\pi}{4}$
$r = \pm\sqrt{4 \cos 2\theta}$	± 2	± 1.9	± 1.7	± 1.4	0

The portion of the graph for $0 \leq \theta \leq \frac{\pi}{4}$ appears in Figure 8.62(a) and the complete graph obtained by symmetry appears in part (b) of the figure. This graph is called a *lemniscate*.

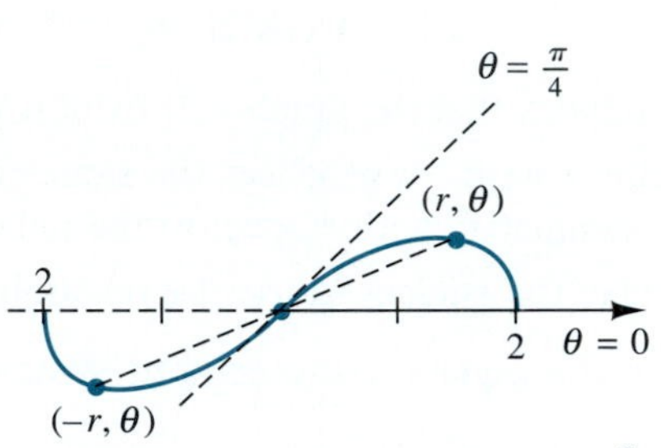

The graph of $r^2 = 4 \cos 2\theta$ for $0 \leq \theta \leq \frac{\pi}{4}$

(a)

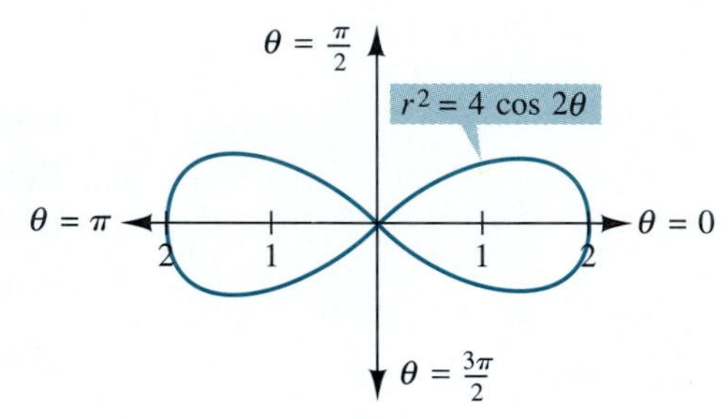

The graph of the lemniscate $r^2 = 4 \cos 2\theta$

(b)

FIGURE 8.62

■

EXAMPLE 8 Sketch the graph of $r = \theta$ for $\theta \geq 0$.

Solution As θ increases and we rotate around the pole, r also increases. (You may want to compute a table of values to help you visualize the graph.) The graph appears in Figure 8.63. This endless, ever-widening spiral is called the *spiral of Archimedes*.

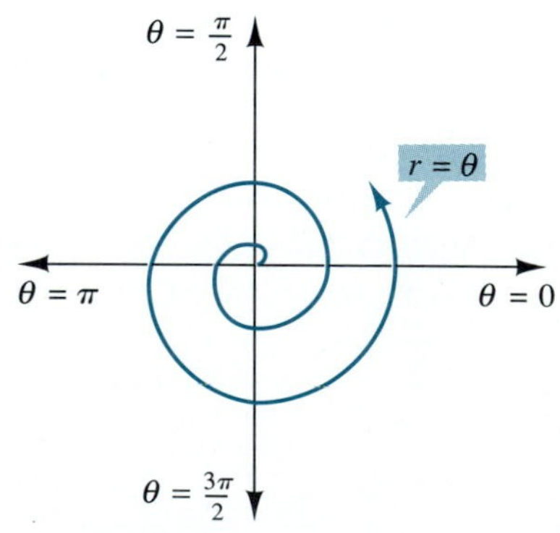

FIGURE 8.63

■

EXERCISES 8.6

In Exercises 1–6, plot the given point in a polar coordinate system.

1. $\left(3, \frac{\pi}{4}\right)$ **2.** $\left(2, -\frac{2\pi}{3}\right)$

3. $\left(-1, \frac{\pi}{2}\right)$ **4.** $\left(0, \frac{5\pi}{6}\right)$

5. $(4, -\pi)$ **6.** $\left(-2, -\frac{3\pi}{2}\right)$

In Exercises 7–10, find two additional polar representations of the given point, one with r negative and one with θ negative.

7. $\left(5, \frac{\pi}{3}\right)$ **8.** $\left(4, -\frac{5\pi}{6}\right)$

9. $\left(-2, \frac{\pi}{2}\right)$ **10.** $\left(-3, -\frac{3\pi}{4}\right)$

In Exercises 11–22, the given coordinates are listed as rectangular or polar. If they are rectangular, convert them to polar, and vice versa. Limit θ to the interval $[0, 2\pi)$, and round to 2 decimal places where necessary.

11. Polar: $\left(2, \frac{2\pi}{3}\right)$ **12.** Rectangular: $(3, -3)$

13. Rectangular: $(-4\sqrt{3}, -4)$

14. Polar: $\left(-5, -\frac{7\pi}{6}\right)$

15. Polar: $\left(3, \frac{\pi}{7}\right)$ **16.** Rectangular: $(5, -\pi)$

17. Rectangular: $(-6, 0)$ **18.** Polar: $\left(1, \frac{3\pi}{2}\right)$

19. Polar: $(2, 3)$ **20.** Rectangular: $(2, 3)$

21. Rectangular: $\left(-1, \frac{\pi}{4}\right)$ **22.** Polar: $\left(\frac{\pi}{4}, -1\right)$

In Exercises 23–32, convert the given equation from polar to rectangular form. If you are able to, identify the graph of the given equation.

23. $r = 5 \cos \theta$ **24.** $r = 5$

25. $\theta = \frac{\pi}{3}$ **26.** $r = 2 \csc \theta$

27. $r \sin \theta = -1$ **28.** $r \cos \theta = 5$

29. $r = \tan \theta$ **30.** $r = \cos 2\theta$

31. $r = \dfrac{2}{1 - \sin \theta}$ **32.** $r = \dfrac{4}{1 + \cos \theta}$

In Exercises 33–40, convert the given equation from rectangular to polar form.

33. $x^2 + y^2 = 16$ **34.** $4x - 5y = 7$

35. $2xy = 1$ **36.** $y = x^2$

37. $x^2 + 2x + y^2 + 2y = 0$ **38.** $y^2 = 4x$

39. $y = 4$ **40.** $x = -1$

In Exercises 41–59, sketch the graph of the given equation in a polar coordinate system. Some of these graphs have names (indicated in parentheses), which are suggestive of what the graph looks like.

41. $r \cos \theta = 4$ **42.** $r \sin \theta = -2$

43. $r = 2 \sin \theta$ **44.** $r = 3 \cos \theta$

45. $r = 1 - \sin \theta$ (cardioid)

46. $r = 1 + \cos \theta$ (cardioid)

47. $r = 2 \cos 3\theta$ (three-petal rose)

48. $r = 3 \sin 2\theta$ (four-petal rose)

49. $r^2 = \sin 2\theta$ (lemniscate)

50. $r^2 = 9 \cos 2\theta$ (lemniscate)

51. $r = |\sin 2\theta|$ **52.** $r = |2 \cos 3\theta|$

53. $r = 3 + 2 \cos \theta$ (limaçon; look up limaçine in the dictionary)

54. $r = 4 + 3 \sin \theta$ (limaçon)

55. $r = 1 + 2 \cos \theta$ (limaçon with an inner loop)

56. $r = 1 - 2 \sin \theta$ (limaçon with an inner loop)

57. $r = \frac{\theta}{\pi}$ for $\theta \geq 0$ (spiral)

58. $r = e^{\theta}$ for $\theta \geq 0$ (logarithmic spiral)

59. $r^2 = \theta$ (parabolic spiral)

QUESTIONS FOR THOUGHT

60. Show that the distance, d, between two points $P_1(r_1, \theta_1)$ and $P_2(r_2, \theta_2)$ in the polar plane is

$$d = \sqrt{(r_1)^2 + (r_2)^2 - 2r_1r_2\cos(\theta_2 - \theta_1)}$$

61. In a rectangular coordinate system, the graph of $y = 4 \sin x$ is a sine curve of amplitude 4 and period 2π. However, in a polar coordinate system, the graph of $r = 4 \sin \theta$ is a circle with center $(0, 2)$ and radius 2. Explain the difference.

62. Unlike in a rectangular coordinate system, the points of intersection of two polar curves cannot necessarily be found by solving the two equations simultaneously. For example, suppose we want to find the points of intersection of the polar curves $r = 2 \sin \theta$ and $r = 2 \cos \theta$.

(a) First use an *algebraic approach;* that is, solve the equations $r = 2 \sin \theta$ and $r = 2 \cos \theta$ simultaneously in order to find the point(s) of intersection of the two graphs of these equations.

(b) Now use a *geometric approach;* that is, sketch the graphs of $r = 2 \sin \theta$ and $r = 2 \cos \theta$ on the same polar coordinate system and determine the point(s) of intersection of the two graphs.

(c) Are the answers you obtained from parts (a) and (b) the same? Explain why you didn't get both solutions from the algebraic method. HINT: Think about what the coordinates of the pole are on the graph of $r = 2 \sin \theta$ and what the coordinates of the pole are on the graph of $r = 2 \cos \theta$.

63. Is there such a thing as a vertical line test for the graph of a function in polar coordinates?

64. We know that the function $f(\theta) = \sin 2\theta$ has a period of π. In order to sketch the graph of $r = |\sin 2\theta|$, is it sufficient to consider only $0 \le \theta \le \pi$?

Chapter 8 SUMMARY

After completing this chapter you should:

1. Understand the addition formulas and be able to apply them. (Section 8.1)
For example:

We can use the addition formula

$$\cos(A + B) = \cos A \cos B - \sin A \sin B$$

to verify that $\cos\left(\theta + \frac{3\pi}{2}\right) = \sin \theta$ as follows

$$\begin{aligned}\cos\left(\theta + \frac{3\pi}{2}\right) &= \cos \theta \cos \frac{3\pi}{2} - \sin \theta \sin \frac{3\pi}{2} \\ &= \cos \theta(0) - \sin \theta(-1) \\ &= \sin \theta \qquad \textit{as required}\end{aligned}$$

2. Understand the double-angle and half-angle formulas. (Section 8.2)
For example:

Given that θ is in quadrant II and $\sin \theta = \frac{1}{3}$, find $\sin 2\theta$.
We draw a reference triangle for θ and complete it by using the Pythagorean theorem (see Figure 8.64).

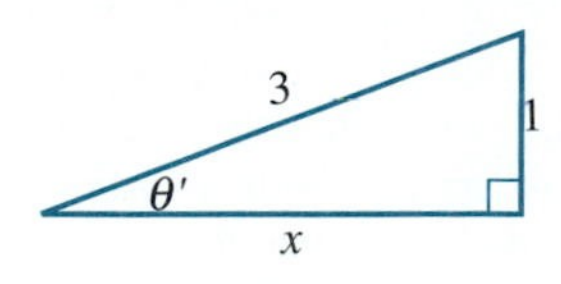

FIGURE 8.64

$$\begin{aligned}x^2 + 1^2 &= 3^2 \\ x^2 &= 8 \\ x &= \sqrt{8} = 2\sqrt{2} \\ \sin 2\theta &= 2 \sin \theta \cos \theta \qquad \textit{Cosine is negative in quadrant II.} \\ &= 2\left(\frac{1}{3}\right)\left(\frac{-2\sqrt{2}}{3}\right) \\ &= \boxed{\frac{-4\sqrt{2}}{9}}\end{aligned}$$

3. Understand and be able to use the Law of Sines or the Law of Cosines to complete a triangle. (Section 8.3)
For example:
(a) Use the triangle in Figure 8.65 to find x to the nearest hundredth.

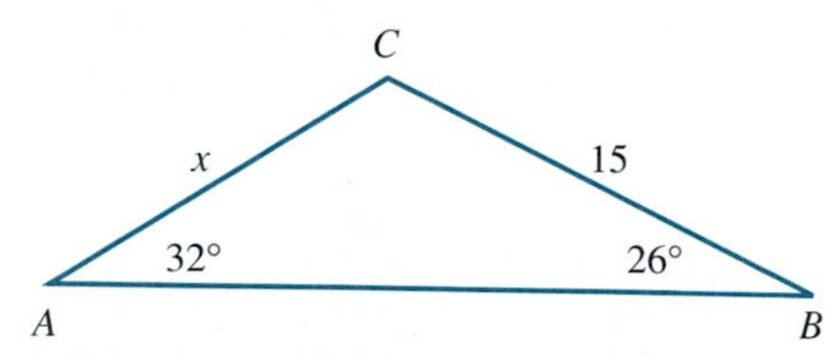

FIGURE 8.65 This triangle requires us to use the Law of Sines.

The given information corresponds to SAA and we must use the Law of Sines.

$$\frac{x}{\sin 26^\circ} = \frac{15}{\sin 32^\circ} \Rightarrow x = \frac{15 \sin 26^\circ}{\sin 32^\circ} \approx \boxed{12.41}$$

(b) Use the triangle in Figure 8.66 to find $\angle A$.

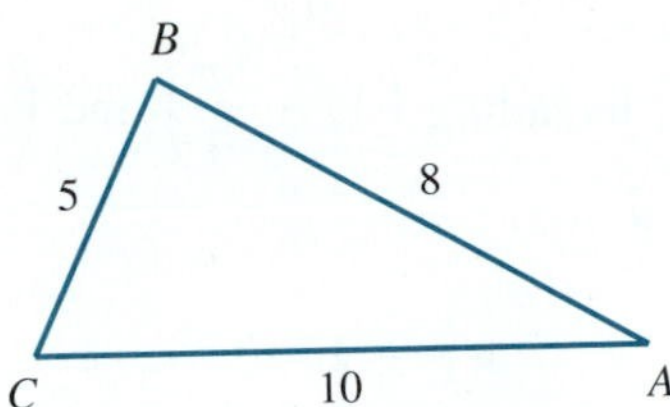

FIGURE 8.66 This triangle requires us to use the Law of Cosines.

The given information corresponds to SSS and we must use the Law of Cosines.

$$5^2 = 8^2 + 10^2 - 2(8)(10)\cos A$$

$$25 = 164 - 160\cos A$$

$$-139 = -160\cos A \Rightarrow \cos A = \frac{139}{160}$$

$$\text{Therefore } A = \cos^{-1}\frac{139}{160} \approx \boxed{29.7^\circ}$$

4. Understand vectors and vector notation. (Section 8.4).
A quantity which has magnitude and direction is called a vector. Vectors can be added by using the parallelogram law. Alternatively vectors can be broken up into their horizontal and vertical components and added by using their component form.
For example:
(a) Express $\overrightarrow{PQ}$ in component form and find $|\overrightarrow{PQ}|$ for $P(-1, 3)$ and $Q(2, 5)$.

$$\overrightarrow{PQ} = \langle 2-(-1), 5-3\rangle = \boxed{\langle 3, 2\rangle}$$

$$|\overrightarrow{PQ}| = |\langle 3, 2\rangle| = \sqrt{3^2+2^2} = \boxed{\sqrt{13}}$$

(b) Given $\mathbf{u} = \langle 1, 3\rangle$ and $\mathbf{v} = \langle 5, 2\rangle$, sketch $\mathbf{u} + \mathbf{v}$.
We may draw vectors **u** and **v** and then find **u** + **v** geometrically by using the parallelogram law, as shown in Figure 8.67, or we may find the component form of

$$\mathbf{u} + \mathbf{v} = \langle 1, 3\rangle + \langle 5, 2\rangle = \langle 6, 5\rangle$$

getting the same result.

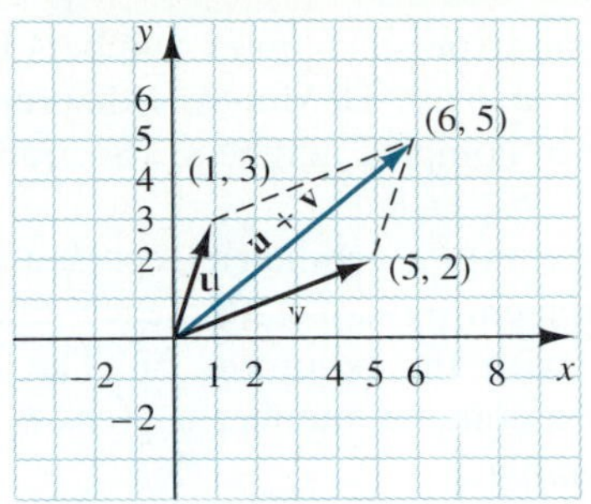

FIGURE 8.67 Using the Parallelogram Law or component form to find **u** + **v**

5. Use vectors in applied problems. (Section 8.4)
For example:
Two forces of 11 pounds and 6 pounds are acting at the same point in the plane. If the angle between the two vectors is 58°, find the magnitude of the resultant to the nearest hundredth.
Solution:
A diagram representing the given information appears in Figure 8.68.

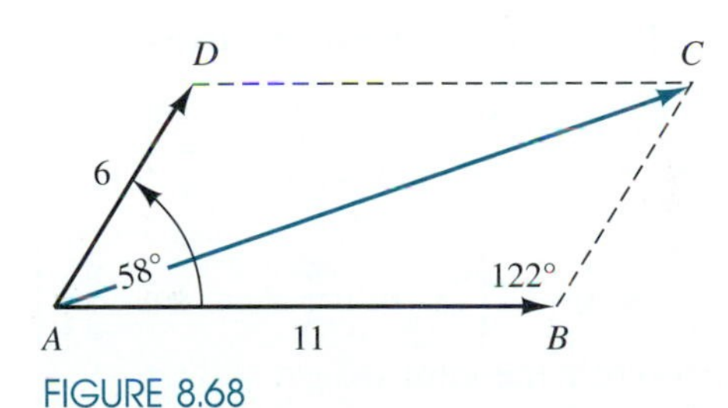

FIGURE 8.68

We have drawn the resultant by completing the parallelogram of forces and noted that $\angle ABC = 122^\circ$. The magnitude of the resultant may be found by using the Law of Cosines.

$$|\overrightarrow{AC}|^2 = 6^2 + 11^2 - 2(6)(11)\cos 122^\circ$$

$$|\overrightarrow{AC}|^2 = 36 + 121 - 132\cos 122^\circ$$

$$|\overrightarrow{AC}|^2 = 226.94934$$

$$|\overrightarrow{AC}| = \sqrt{226.94934} = \boxed{15.06 \text{ pounds}}$$

Rounded to the nearest hundredth

6. Be able to express complex numbers in trigonometric form and use DeMoivre's theorem to compute powers of complex numbers. (Section 8.5)

We can obtain a geometric representation of complex numbers by associating each complex number $a + bi$ with the point (a, b). The trigonometric form of a complex number is $r(\cos\theta + i\sin\theta)$.
For example:
In order to find $(1 + i\sqrt{3})^4$, we first convert $1 + i\sqrt{3}$ into trigonometric form.

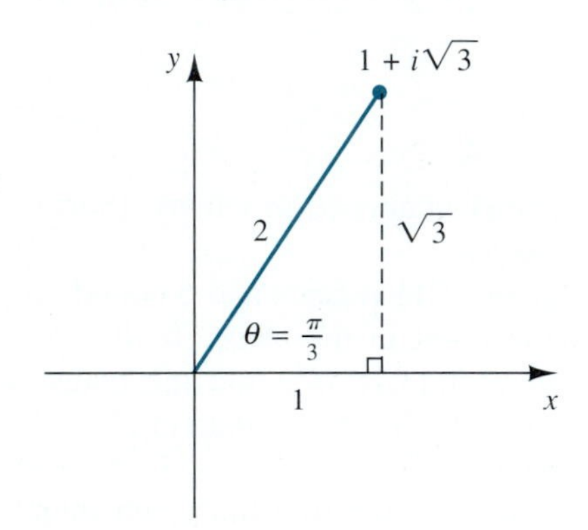

FIGURE 8.69

From Figure 8.69 we can see that $r = \sqrt{1^2 + (\sqrt{3})^2} = \sqrt{4} = 2$; since $\sin\theta = \frac{\sqrt{3}}{2}$, we can take $\theta = \frac{\pi}{3}$. Therefore, $(1 + i\sqrt{3})^4 = \left[2\left(\cos\frac{\pi}{3} + i\sin\frac{\pi}{3}\right)\right]^4$. Using DeMoivre's theorem we get

$$\left[2\left(\cos\frac{\pi}{3} + i\sin\frac{\pi}{3}\right)\right]^4$$
$$= 2^4\left[\cos 4\left(\frac{\pi}{3}\right) + i\sin 4\left(\frac{\pi}{3}\right)\right]$$
$$= 16\left(\cos\frac{4\pi}{3} + i\sin\frac{4\pi}{3}\right)$$

The answer may also be given in rectangular form $-8 - 8\sqrt{3}i$.

7. Be able to convert from rectangular coordinates to polar coordinates, and vice versa, and recognize the different ways a point can be represented using polar coordinates. (Section 8.6)

For example:

The polar point $\left(1, \frac{3\pi}{4}\right)$ has many other representations, including $\left(1, -\frac{5\pi}{4}\right)$ and $\left(-1, \frac{7\pi}{4}\right)$. See Figure 8.70.

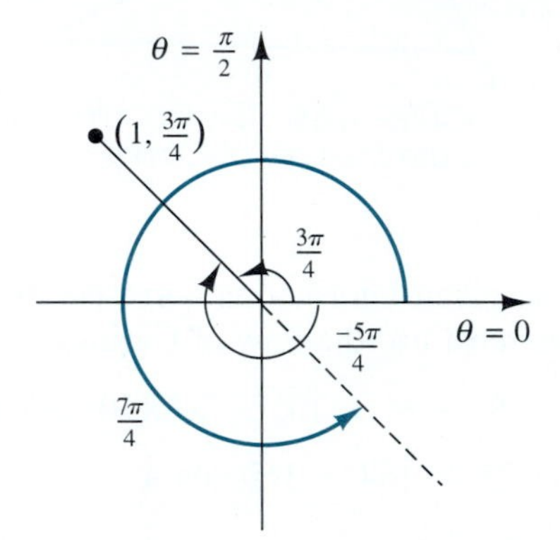

FIGURE 8.70 Various polar representations of $\left(1, \frac{3\pi}{4}\right)$

Applying the conversion equations $x = r\cos\theta$ and $y = r\sin\theta$ to the point $\left(1, \frac{3\pi}{4}\right)$, we get

$$x = 1\cos\frac{3\pi}{4} = -\frac{\sqrt{2}}{2} \quad \text{and } y = 1\sin\frac{3\pi}{4} = \frac{\sqrt{2}}{2}$$

and so the rectangular coordinates of $\left(1, \frac{3\pi}{4}\right)$ are $\left(-\frac{\sqrt{2}}{2}, \frac{\sqrt{2}}{2}\right)$.

8. Be able to sketch curves in polar form by plotting some points and using symmetry considerations wherever possible. (Section 8.6)
For example:
The graph of $r = 2\cos\theta$ will exhibit symmetry with respect to the polar axis because if we substitute $-\theta$ for θ, we get the same equation. Plotting a few points for $\theta = 0, \frac{\pi}{2}, \pi, \frac{3\pi}{2}$, we get the graph of the circle in Figure 8.71.

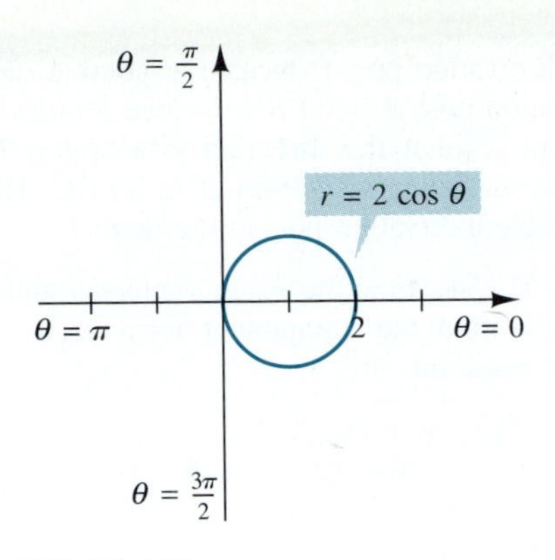

FIGURE 8.71

9. Recognize the graph of a polar equation by converting it into rectangular form. (Section 8.6)

For example:
The equation $r = 2\cos\theta$ can be converted into rectangular form as follows:

$r = 2\cos\theta$ *Multiply both sides of the equation by r.*

$r^2 = 2r\cos\theta$ *Since $r^2 = x^2 + y^2$ and $x = r\cos\theta$, we get*

$x^2 + y^2 = 2x$ *Which we recognize as the equation of a circle. We may complete the square to get*

$$(x - 1)^2 + y^2 = 1$$

We recognize this equation as the equation of a circle with center $(1, 0)$ and radius 1, which is exactly the graph we obtained in Figure 8.71.

Chapter 8 REVIEW EXERCISES

In Exercises 1–4, use the addition formulas to find the exact value of the given expression.

1. $\cos 15°$

2. $\sin\frac{5\pi}{12}$

3. $\tan\frac{7\pi}{12}$

4. $\cos 195°$

In Exercises 5–16, verify the given identity.

5. $\sin(3\pi - x) = \sin x$

6. $\cos(x + \pi) = -\cos x$

7. $\tan\left(\theta - \frac{\pi}{4}\right) = \frac{\tan\theta - 1}{\tan\theta + 1}$

8. $\sin(A + B)\sin(A - B) = \sin^2 A - \sin^2 B$

9. $\sin\left(\frac{\pi}{6} + x\right) = \frac{1}{2}(\cos x + \sqrt{3}\sin x)$

10. $\sin\left(\frac{\pi}{2} + \theta\right) + \cos(\pi - \theta) = 0$

11. $\sin(\alpha + \beta)\cos(\alpha - \beta) = \sin\alpha\cos\alpha + \sin\beta\cos\beta$

12. $\frac{1 + \cos 2x}{\sin^2 2x} = \frac{\csc^2 x}{2}$

13. $\frac{1 - \cos 2\theta}{\sin 2\theta} = \tan\theta$

14. $\frac{\sin 2\theta}{1 + \cos 2\theta} = \tan\theta$

15. $\tan 2A - \tan A = \tan A \sec 2A$

16. $\tan 2\beta = \frac{2}{\cot\beta - \tan\beta}$

In Exercises 17–26, solve each trigonometric equation on the interval $[0, 2\pi)$.

17. $\cos 2x = \sin x$

18. $2\cos^2 x + \cos 2x = 0$

19. $\cos 2x = 2\sin^2 x$

20. $\cos 2x + 2 = 3\cos x$

21. $\sin 4x = \cos 2x$

22. $\sin 2x = 2\cos x$

23. $\cos x = \cos\frac{x}{2}$

24. $\cos 4x - 7\cos 2x = 8$

25. $2 - \sin^2 x = 2\cos^2\frac{x}{2}$

26. $\sin 2x + \sqrt{2}\sin x = 0$

27. Given that $\sin\theta = -\frac{2}{5}$ and θ is in quadrant III, find $\sin 2\theta$.

28. Given that $\cos\theta = \frac{3}{7}$ and θ is in quadrant IV, find $\cos 2\theta$.

29. Given that $\tan\theta = \frac{4}{9}$, find $\tan 2\theta$.

30. Given that $\tan\theta = -\frac{1}{4}$ and $\frac{\pi}{2} < \theta < \pi$, find
(a) $\sin 2\theta$ **(b)** $\cos 2\theta$

31. Given that $\sin\theta = -\frac{3}{5}$ and $\frac{3\pi}{2} < \theta < 2\pi$, find

(a) $\sin\frac{\theta}{2}$ **(b)** $\cos\frac{\theta}{2}$ **(c)** $\tan\frac{\theta}{2}$

32. Use the following figure to express $\sin 2\theta$ as a function of x.

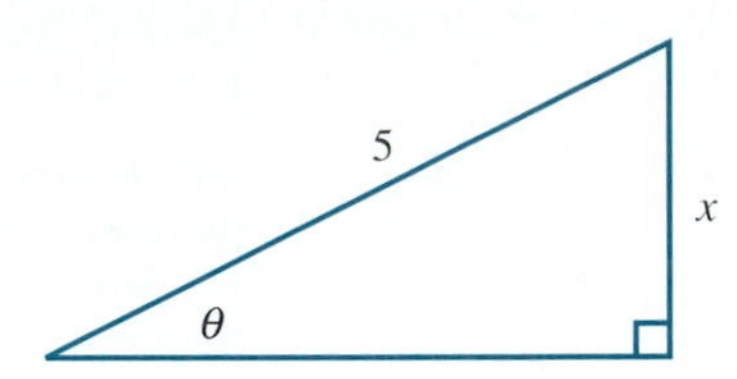

33. Use the figure for Exercise 32 to express $\cos 2\theta$ as a function of x.

34. Given that $\sin\theta = \frac{a}{7}$, express $\cos 2\theta$ as a function of a.

35. Given that $\cos\theta = \frac{a}{5}$, express $\tan 2\theta$ as a function of a.

36. Express $\cos 3\theta$ in terms of $\cos\theta$.

In Exercises 37–42, use the given information to find the missing parts of $\triangle ABC$. Approximate your answer to the nearest tenth.

37. $A = 70°, B = 32°, a = 14$
38. $A = 105°, b = 8, c = 15$
39. $B = 39°, a = 9, c = 16$
40. $B = 41°, C = 24°, c = 5$
41. $a = 3, b = 7, c = 9$
42. $C = 115°, b = 25, c = 40$

In Exercises 43–46, determine how many triangles ABC can be constructed using the given information.

43. $A = 52°, a = 8, b = 11$
44. $A = 71°, a = 30, b = 27$
45. $C = 130°, a = 65, b = 70$
46. $B = 47°, a = 12, b = 10$
47. Two trains leave a train station at 8:00 A.M. traveling along straight tracks at 90 and 110 km/hr, respectively. If the angle between their directions of travel is 106°, how far apart are the trains at 9:15 A.M.?
48. A helicopter is hovering directly above the line joining points A and B, which are 1.7 km apart. If the angles of elevation to the helicopter from points A and B are 46.33° and 38.83°, respectively, find the altitude of the helicopter.
49. The angle of elevation of a tree from point A is 53°, whereas the angle of elevation from a second point B along the same line of sight to the tree but 50 feet further away is 46°. Estimate the height of the tree.
50. One observation post is located at point A, and a second observation post at point B is located 3 miles due west of A. From point A the direction to a target, T, is N57°E and the direction to T from B is N71°E. How far is it from each observation post to the target?

In Exercises 51–54, draw the given vectors **u** and **v** and their resultant. Also find the component form of **u** + **v** and the length of the resultant.

51. $\mathbf{u} = \langle 2, 7\rangle, \quad \mathbf{v} = \langle 5, 3\rangle$
52. $\mathbf{u} = \langle -3, 2\rangle, \quad \mathbf{v} = \langle 4, -1\rangle$
53. $\mathbf{u} = 4\mathbf{i} - \mathbf{j}, \quad \mathbf{v} = \mathbf{i} + \mathbf{j}$
54. $\mathbf{u} = -3\mathbf{i} - 6\mathbf{j}, \quad \mathbf{v} = \mathbf{i} - \mathbf{j}$

In Exercises 55–58, let $\mathbf{u} = \langle 2, -5\rangle$ and $\mathbf{v} = \langle 4, 3\rangle$. Express each of the following vectors in component form and as a linear combination of **i** and **j**.

55. $\mathbf{u} + \mathbf{v}$
56. $\mathbf{u} - 5\mathbf{v}$
57. $2\mathbf{u} + 4\mathbf{v}$
58. $-3\mathbf{u} + 5\mathbf{v}$

In Exercises 59–62, vector **v** is in standard position and θ is the angle that **v** makes with the positive x-axis. Given $|\mathbf{v}|$, resolve the given vector into horizontal and vertical components and express **v** in component form.

59. $|\mathbf{v}| = 3, \theta = 45°$
60. $|\mathbf{v}| = 8, \theta = 30°$
61. $|\mathbf{v}| = 10, \theta = 118°$
62. $|\mathbf{v}| = 5, \theta = 25°$

In Exercises 63–66, find a unit vector in the direction of **v**.

63. $\mathbf{v} = \langle 5, 4\rangle$
64. $\mathbf{v} = 3\mathbf{i} + 2\mathbf{j}$
65. $\mathbf{v} = 4\langle 3, 5\rangle + \langle 2, 1\rangle$
66. $\mathbf{v} = \frac{1}{3}\mathbf{i} - \frac{3}{4}\mathbf{j}$

In Exercises 67–70, vectors **A** and **B** represent two forces acting at the origin whose magnitudes are given. Find the magnitude of the resultant vector and the angle θ the resultant makes with **A**.

67. $\mathbf{A} = \langle 4, 0\rangle, \mathbf{B} = \langle 0, 5\rangle$
68. $\mathbf{A} = \langle 1, 3\rangle, \mathbf{B} = \langle 2, 4\rangle$
69. $\mathbf{A} = \langle 6, 2\rangle, \mathbf{B} = \langle 1, 8\rangle$
70. $\mathbf{A} = \langle 10, 0\rangle, \mathbf{B} = \langle 0, 7\rangle$
71. Two forces of 19 pounds and 13 pounds are acting at the same point in the plane. If the angle between the two forces is 28°, find the magnitude of the resultant force.
72. Two forces of 28 pounds and 35 pounds acting at the same point in the plane produce a resultant of magnitude 46 pounds. Find the angle between the forces.
73. Two forces acting at the same point in the plane produce a resultant of magnitude 15.6 pounds. If one of the forces is 9 pounds and its angle with the resultant is 12°, find the other force.
74. A motorboat is headed due north across a stream that is flowing due east at a speed of 3.6 mi/hr. If the boat is

traveling at 25 mi/hr, find the speed and course at which the boat is traveling.

75. A plane is on a heading of N16°E with an airspeed of 475 mi/hr. If the wind is blowing due north at 35 mi/hr, find the course and groundspeed of the plane.

76. A plane is on a heading of N31.5°E with an airspeed of 230 mi/hr. The wind is blowing due south, causing the plane to be on a course of N32.8°E with a groundspeed of 210 mi/hr. Find the speed of the wind.

77. A block weighing 50 pounds rests on an inclined ramp that makes an angle of 12° with the horizontal. Determine the components of the weight parallel and perpendicular to the ramp.

78. A log weighing 215 pounds rests on an inclined conveyor belt that makes an angle of 14° with the horizontal. If we neglect friction, find the force parallel to the conveyor belt necessary to keep the log from rolling down the belt.

79. A block is placed on an inclined ramp that makes an angle of 9° with the horizontal. If the component of the weight parallel to the ramp is 46.7 pounds, find the weight of the block.

80. A block is resting on an inclined ramp that makes an angle of 11.2° with the horizontal. If the component of the weight perpendicular to the ramp is 13.5 pounds, find the component of the weight parallel to the ramp.

In Exercises 81–84, if the given number is in rectangular form, convert it to trigonometric form; if it is in trigonometric form, convert it to rectangular form. Where necessary, round to two decimal places.

81. $\sqrt{2} - i$

82. $6\left(\cos \frac{7\pi}{6} + i \sin \frac{7\pi}{6}\right)$

83. $8\left(\cos \frac{3\pi}{4} + i \sin \frac{3\pi}{4}\right)$

84. $3 + 3i$

In Exercises 85–86, use the product and quotient rules to find $z_1 \cdot z_2$ and $\frac{z_1}{z_2}$.

85. $z_1 = 4\left(\cos \frac{\pi}{3} + i \sin \frac{\pi}{3}\right)$; $z_2 = 3\left(\cos \frac{\pi}{5} + i \sin \frac{\pi}{5}\right)$

86. $z_1 = 9(\cos 120° + i \sin 120°)$;
$z_2 = 3(\cos 210° + i \sin 210°)$

In Exercises 87–92, compute the indicated power by using DeMoivre's theorem.

87. $\left[4\left(\cos \frac{\pi}{3} + i \sin \frac{\pi}{3}\right)\right]^4$

88. $\left[\sqrt[3]{4}\left(\cos \frac{5\pi}{6} + i \sin \frac{5\pi}{6}\right)\right]^6$

89. $(1 - i)^6$

90. $(\sqrt{3} + i)^8$

91. $(-2 - 2i)^5$

92. $\left(-\frac{1}{2} + \frac{\sqrt{3}}{2}i\right)^6$

In Exercises 93–96, use the nth roots theorem to find the requested roots of the given complex number.

93. Find the fourth roots of $2\sqrt{3} - 2i$.

94. Find the cube roots of $1 + i$.

95. Find the fifth roots of 1.

96. Find the square roots of $-i$.

In Exercises 97–100, find two additional polar representations of the given point, one with r positive and one with r negative.

97. $\left(3, \frac{\pi}{4}\right)$

98. $\left(5, -\frac{2\pi}{3}\right)$

99. $(-1, \pi)$

100. $\left(-2, -\frac{5\pi}{6}\right)$

In Exercises 101–104, the given coordinates are listed as rectangular or polar. If they are rectangular, convert them to polar, and vice versa. Limit θ to the interval $[0, 2\pi)$, and round to two decimal places where necessary.

101. Polar: $\left(3, \frac{3\pi}{4}\right)$

102. Rectangular: $(2, 2)$

103. Rectangular: $(-5\sqrt{3}, -5)$

104. Polar: $\left(-4, -\frac{5\pi}{6}\right)$

In Exercises 105–112, convert the given equation from polar to rectangular form. If you are able to, identify the graph of the given equation.

105. $r = 6 \sin \theta$

106. $r = 8$

107. $\theta = \frac{\pi}{4}$

108. $r = 4 \sec \theta$

109. $r \sin \theta = 4$

110. $r \cos \theta = -2$

111. $r = \frac{1}{1 - \cos \theta}$

112. $r = 2 \tan \theta$

In Exercises 113–124, sketch the graph of the given equation in a polar coordinate system.

113. $r \sin \theta = 2$

114. $r \cos \theta = -4$

115. $r = 5 \cos \theta$

116. $r = 4 \sin \theta$

117. $r = 1 + \sin \theta$ (cardioid)

118. $r = 1 - \cos \theta$ (cardioid)

119. $r = \sin 3\theta$ (three-petal rose)

120. $r = 4 \cos 2\theta$ (four-petal rose)
121. $r^2 = \cos 2\theta$ (lemniscate)
122. $r^2 = 9 \sin 2\theta$ (lemniscate)
123. $r = 2 + 4 \sin \theta$ (limaçon)
124. $r = 4 - 3 \cos \theta$ (limaçon)

Chapter 8 PRACTICE TEST

1. Use the addition formula to find the exact value of $\cos \frac{7\pi}{12}$.

2. Verify the following identities:
(a) $\sin(\theta + \pi) = -\sin \theta$
(b) $\tan^2 x = \frac{1 - \cos 2x}{1 + \cos 2x}$

3. Solve the following equation for θ in $[0, 2\pi)$.

$$3 - \cos^2\theta = 2 \sin^2 \frac{\theta}{2}$$

4. Given that $\tan \theta = \frac{1}{6}$ and $\pi < \theta < \frac{3\pi}{2}$, find
(a) $\sin 2\theta$ **(b)** $\cos 2\theta$

5. Given that $\cos \theta = \frac{2a}{3}$, express $\tan 2\theta$ as a function of a.

6. Given the following information about $\triangle ABC$, find the missing parts of the triangle.
(a) $A = 75°$, $B = 43°$, $a = 20$
(b) $B = 105°$, $a = 10$, $c = 20$

7. The angle of elevation to the top of a bridge from point A is 88.4°, and the angle of elevation from a second point B along the same line of sight to the bridge but 90 feet further away is 70.3°. Estimate the height of the bridge.

8. Draw the given vectors **u** and **v** and their resultant. Also find the component form of **u** + **v** and the length of the resultant.
(a) $\mathbf{u} = \langle 3, -1 \rangle$, $\mathbf{v} = \langle 2, 4 \rangle$
(b) $\mathbf{u} = 5\mathbf{i} - 2\mathbf{j}$, $\mathbf{v} = \mathbf{i} + 2\mathbf{j}$

9. Let $\mathbf{u} = \langle 3, -7 \rangle$ and $\mathbf{v} = \langle 2, 6 \rangle$.
(a) Express $5\mathbf{u} + 3\mathbf{v}$ in component form.
(b) Express $-4\mathbf{u} + 3\mathbf{v}$ in **i**, **j** form.

10. Suppose that vector **v** is in standard position, has length 8, and makes an angle of 56° with the positive x-axis. Resolve the given vector into horizontal and vertical components and express **v** in component form.

11. Find a unit vector in the direction of $\langle 4, 5 \rangle$.

12. Let the points O, P, and Q be $O(0, 0)$, $P(1, 4)$, and $Q(-5, 2)$. Find the resultant $\overrightarrow{OP} + \overrightarrow{OQ}$ and the angle that the resultant makes with $\overrightarrow{OP}$.

13. A plane has a heading of N35°E with an airspeed of 180 mi/hr. The wind is blowing due east, resulting in the plane having a course of N37°E and a groundspeed of 200 mi/hr. Find the speed of the wind.

14. A barrel of oil weighing 280 pounds is lying on its side on a ramp that makes an angle of 8° with the horizontal. If we neglect friction, find the force parallel to the ramp necessary to roll the barrel up the ramp.

15. Express the complex number $-1 - i\sqrt{3}$ in trigonometric form.

16. Use DeMoivre's theorem to compute $(1 + i)^8$.

17. Give two additional polar representations for the polar point $\left(5, \frac{7\pi}{6}\right)$, one with r negative and one with θ negative.

18. Sketch the graph of $r = 4 - 4 \cos \theta$ in a polar coordinate system.

19. Find the cube roots of $27i$.

CHAPTER 9

Systems of Linear Equations and Inequalities

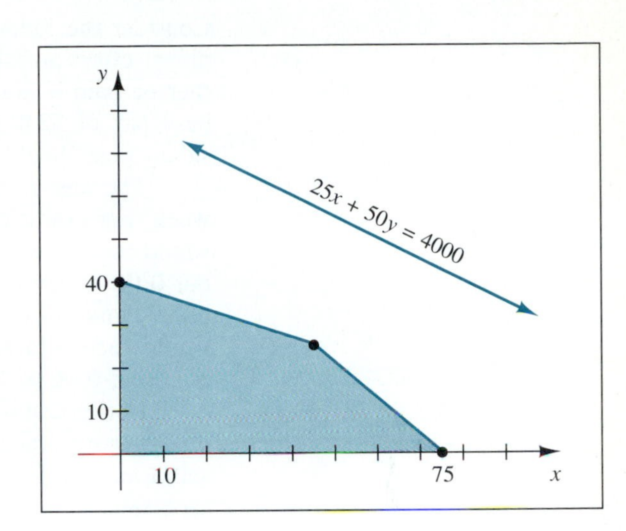

In this chapter, we will discuss systematic procedures for finding solutions to problems that involve several conditions or constraints. These conditions may be represented by systems of equations or systems of inequalities.

In the first section, we will discuss solving systems of two linear equations in two variables by manipulating the equations. While much of this material will be a review, it is important that you thoroughly understand these fundamental concepts as they will be generalized to larger systems in Section 9.2, and will form the basis for developing other systematic methods to be discussed in the remaining sections.

9.1 Elimination and Substitution: 2×2 Linear Systems

Consider the following situation: As an employee in a sales department, your employer offers you the option of receiving your salary two possible ways: You can either be paid a straight commission of 9% of your gross sales, or you can receive a base pay of \$240 per week plus a commission of 6% of your gross sales. Which salary plan should you choose?

The answer to this question depends upon how much you expect to sell each week. For example, if you expect to sell \$5000 worth of merchandise in a week, you would receive a salary of $0.09 \times 5000 = \$450$ taking the straight commission plan, but $0.06 \times 5000 + 240 = \$540$ taking the base pay + commission plan. On the other hand, if you expect to sell \$10,000 worth of merchandise in a week, you would receive a salary of $0.09 \times 10{,}000 = \$900$ taking the straight commission plan, but $0.06 \times 10{,}000 + 240 = \$840$ taking the base pay + commission plan.

Notice that if you expect to sell "a lot" of merchandise, the straight commission plan is more desirable than the base pay + commission plan. If you do not expect to sell a lot, the base pay + commission plan is better. We can write an equation for each salary plan expressing the salary s in terms of the gross sales g, as follows:

$$s = 0.09g \qquad \text{for straight 9\% commission}$$

$$s = 240 + 0.06g \qquad \text{for base pay + 6\% commission}$$

Each of these equations is a first-degree or linear equation in two variables and hence has a straight line as its graph. We let g represent the horizontal axis and s represent the vertical axis and graph each line on the same set of axes in Figure 9.1.

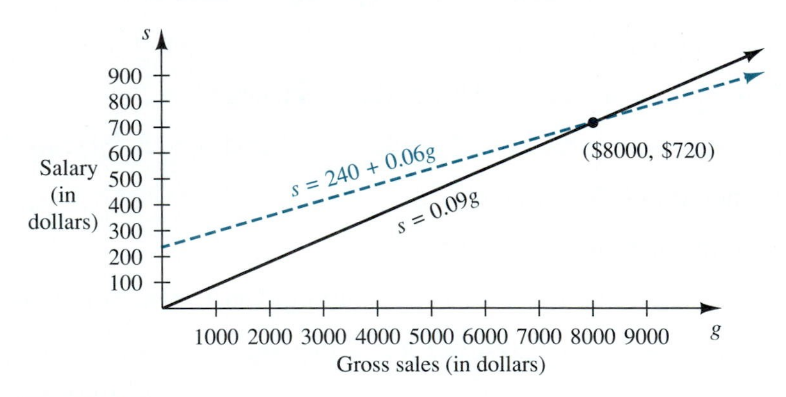

FIGURE 9.1

Throughout this text we have emphasized the importance of knowing the graph of an equation in order to better understand the relationship between the variables. Notice that the graph of each line tells us the salary, s, of each plan based on gross sales, g.

By determining the point of intersection of the lines, we can find the gross sales amount that yields the same income under both plans. For this example, gross sales of \$8000 will yield the same income, \$720, for both plans. The point of intersection of the two lines, (8000, 720), is important for our purposes because it tells us when one plan is more advantageous than the other. You can see from the graph in Figure 9.1 that if you sell less than \$8000 worth of merchandise per week, the base pay + commission plan yields a higher salary; if you sell more \$8000 worth of merchandise per week, the straight commission plan yields a higher salary. Your decision as to what plan to take reduces to whether or not you think you can average above or below \$8000 in gross sales per week.

For this problem, finding the point of intersection helped to clarify the information we are given in order to make a decision as to which plan of income to select. We are often interested in finding points of intersection, but determining such points from the graph can be imprecise. In this chapter we discuss more systematic procedures for finding solutions to problems like these, and then expand these methods to more general cases.

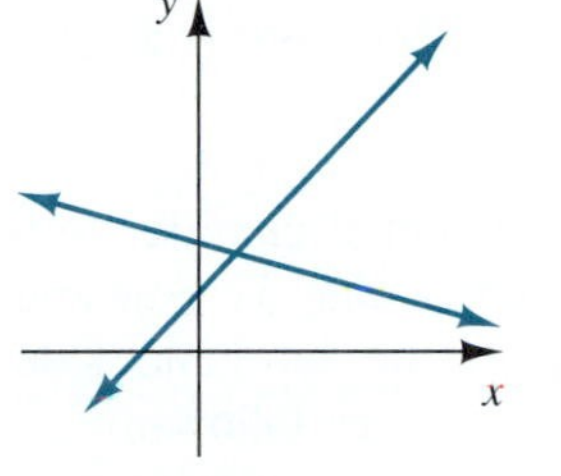

Lines intersect: consistent and independent;
exactly one solution.

(a)

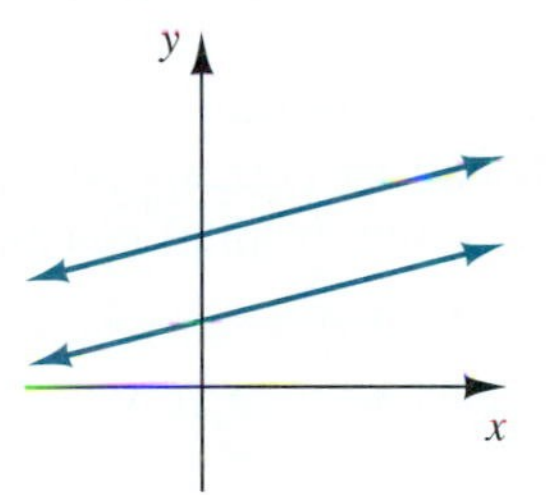

Lines are parallel: inconsistent;
no solution.

(b)

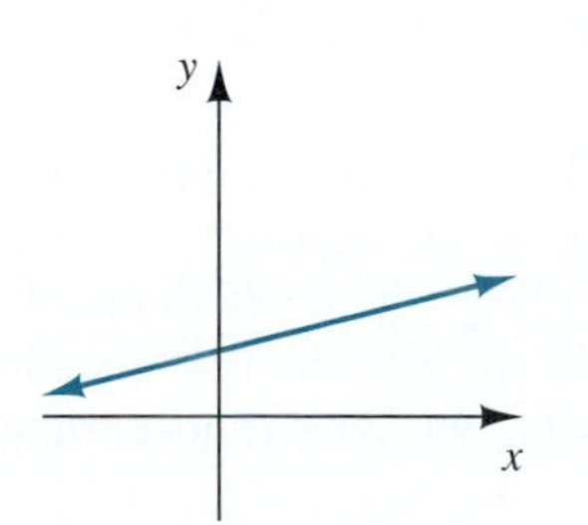

Lines coincide: dependent;
infinitely many solutions.

(c)

FIGURE 9.2

Linear Systems: Two Variables

First we review some terminology. Two or more equations considered together are called a **system of equations**. In particular, if the equations are of the first degree, it is called a **linear system**. The system

$$\begin{cases} x - 5y = 6 \\ 2x + 3y = 4 \end{cases}$$

is an example of a linear system in two variables called a **2 × 2 system**, since there are two equations and two variables.

Solving a system of equations means finding all ordered pairs that satisfy all the equations in the system.

Since the graph of an equation is a picture of its solutions and the graph of a first-degree equation in two variables is a straight line, we have the following three possibilities:

1. The lines intersect at exactly one point. The coordinates of this point are the solution to this system. In this case we say that the system is **consistent and independent**. See Figure 9.2(a).
2. The lines are parallel; that is, they never intersect. In this case we say that the system is **inconsistent** and that there is no solution. See Figure 9.2(b).
3. The lines coincide; that is, all solutions to one of the equations are also solutions to the other (the two equations are equivalent). There are an infinite number of solutions. In this case, we call the system **dependent**. See Figure 9.2(c).

These three conditions are illustrated in Figure 9.2.

In much of this chapter we will focus on algebraic techniques for solving these systems. In this section we will discuss the **elimination method** and the **substitution method** for solving systems of equations.

The idea behind the **elimination method**, as its name suggests, is to eliminate variables (and equations) until we have a single equation in one variable. We illustrate this method with some examples.

EXAMPLE 1 Solve the following systems of equations:

(a) $\begin{cases} 2x + y = 5 \\ 3x - y = 10 \end{cases}$ **(b)** $\begin{cases} 2x - 5y = -1.5 \\ 3x - 9y = -3 \end{cases}$ **(c)** $\begin{cases} 6x - 3y = 9 \\ 8x - 4y = 5 \end{cases}$

Solution

(a) The addition property of equality allows us to add the equal quantity to both sides of an equation. From the second equation we know that $3x - y$ and 10 are equal quantities. Hence we can add $3x - y$ to the left-hand side of $2x + y = 5$, and 10 to the right-hand side of $2x + y = 5$ as follows:

$$\begin{array}{rl} 2x + y = 5 & \\ \underline{3x - y = 10} & \textit{Add.} \\ 5x \quad = 15 & \end{array}$$

We *eliminated* one of the variables; we can now solve this first-degree equation in *one* variable, $5x = 15$, to get $x = 3$.

A solution to a system of equations in two variables consists of two numbers: an x-value and a y-value. We have found x. To find its corresponding y-value, we substitute the value found for x into either of the original equations and solve for y.

$$2x + y = 5 \qquad \textit{Substitute } x = 3 \textit{ in the first equation and solve for } y.$$

$$2(3) + y = 5 \qquad \textit{Which yields } y = -1$$

The solution is $\boxed{(3, -1)}$. We check this solution by substituting $x = 3$ and $y = -1$ in both of the original equations. We leave this check to the student.

(b) $\begin{cases} 2x - 5y = -1.5 \\ 3x - 9y = -3 \end{cases}$

Adding these two equations at this point will not eliminate either of the variables. However, we can transform one or both of the equations by using the multiplication property of equality in order to change the coefficients of either x or y so that they are exact opposites. Then we can add the two equations and eliminate the variable with opposite coefficients. We choose to eliminate the x variable in this problem.

$$\begin{array}{lcrl} 2x - 5y = -1.5 & \xrightarrow{\textit{Multiply by } 3} & 6x - 15y = -4.5 & \\ 3x - 9y = -3 & \xrightarrow{\textit{Multiply by } -2} & \underline{-6x + 18y = 6} & \textit{Add.} \\ & & 3y = 1.5 & \textit{Which yields } y = 0.5 \end{array}$$

Now we find x by substituting $y = 0.5$ into the first equation.

$$2x - 5y = -1.5 \qquad \textit{Substitute } y = 0.5 \textit{ to get}$$

$$2x - 5(0.5) = -1.5 \qquad \textit{Which yields } x = 0.5$$

The solution is $\boxed{(0.5, 0.5)}$. Again, the check is left to the student.

(c) $\begin{cases} 6x - 3y = 9 \\ 8x - 4y = 5 \end{cases}$

In this system we choose to eliminate y by multiplying the first equation by 4 and the second by -3.

$$\begin{array}{lcrl} 6x - 3y = 9 & \xrightarrow{\textit{Multiply by } 4} & 24x - 12y = 36 & \\ 8x - 4y = 5 & \xrightarrow{\textit{Multiply by } -3} & -24x + 12y = -15 & \textit{Add.} \\ \hline & & 0 = 21 & \textit{This is a contradiction.} \end{array}$$

For a system of two equations in two variables, what does it mean graphically when we arrive at no solution?

This means that it is impossible for both equations to have a common solution; therefore, there is $\boxed{\text{no solution}}$.

If you write the two equations in slope-intercept form, you will see that both equations have the same slope but different y-intercepts. This means that the two lines are parallel, and therefore the system has no solution. ■

We can also solve systems of equations by *substitution,* as we discussed in Section 1.9. Our goal remains the same, to obtain a single equation in one variable, but the approach is slightly different than the elimination method, as illustrated next.

EXAMPLE 2 Solve the following systems of equations:

(a) $\begin{cases} 5x + 3y = 5 \\ x = 2y - 10 \end{cases}$ **(b)** $\begin{cases} 6x - y = 9 \\ 8x - \frac{4}{3}y = 12 \end{cases}$

Solution

(a) The second equation, $x = 2y - 10$, is solved explicitly for x; hence we can substitute $2y - 10$ in place of x into the first equation as follows:

$$5x + 3y = 5 \qquad \textit{Replace } x \textit{ by } 2y - 10.$$

$$5(2y - 10) + 3y = 5 \qquad \textit{Now solve for } y.$$

$$10y - 50 + 3y = 5 \qquad \textit{To get } y = \frac{55}{13}$$

Now solve for x by substituting $\frac{55}{13}$ for y in the *second* equation, since it is already explicitly solved for x.

$$x = 2y - 10$$

$$x = 2\left(\frac{55}{13}\right) - 10 = -\frac{20}{13}$$

Try checking the solution in both equations using a calculator.

The solution is $\boxed{\left(-\frac{20}{13}, \frac{55}{13}\right)}$. The check is left to the student.

(b) $\begin{cases} 6x - y = 9 \\ 8x - \frac{4}{3}y = 12 \end{cases}$

We can easily solve the first equation explicitly for y:

$$6x - y = 9 \Rightarrow y = 6x - 9$$

Now substitute $6x - 9$ for y in the second equation:

$$8x - \frac{4}{3}y = 12 \qquad \textit{Replace } y \textit{ with } 6x - 9.$$

$$8x - \frac{4}{3}(6x - 9) = 12$$

$$8x - 8x + 12 = 12$$

$$12 = 12 \qquad \textit{This is an identity.}$$

If we examine the two equations closely, we find that they both have the same slope and y-intercept. If we put the second equation into slope-intercept form, we get the first equation: $y = 6x - 9$. Hence, both equations represent the same line. The equations are dependent. The solution set is all points lying on the line; that is, $\boxed{\{(x, y) \mid y = 6x - 9\}}$.

This notation means the solutions are all ordered pairs (x, y) such that $y = 6x - 9$.

There are infinitely many solutions to this system: For any value we choose for x, we can come up with a corresponding y-value, $6x - 9$, such that the ordered pair (x, y) satisfies the system. A more convenient way to express this type of solution is by letting x be a constant; for example, let $x = a$, where a is any real number. Then $y = 6x - 9 \Rightarrow y = 6a - 9$. This allows us to write the solution as $\boxed{(a, 6a - 9), \text{ where } a \text{ is any real number}}$. ■

For a system of two equations in two variables, what does a dependent solution mean graphically?

In the last example, we pointed out that a convenient way to express the dependent solution $\{(x, y) \mid y = 6x - 9\}$ is to write it as $(a, 6a - 9)$, where a is any real number. This notation stresses that there are infinitely many solutions: Pick any value for a, and we can generate a solution. For example, for $a = 1$, 2, and -2, the solutions are $(1, -3)$, $(2, 3)$, and $(-2, -21)$, respectively. What determines a solution to this system is the relationship between the two components. For an ordered pair of numbers to satisfy the system, the second component must be 9 less than 6 times the first component. Keep in mind, however, we can express a dependent solution in many ways: If we decide to let $y = b$, then $x = \frac{y + 9}{6} = \frac{b + 9}{6}$, and we

Notice that if you use the solution described by $(a, 6a - 9)$, then $(1, -3)$ is one solution, found by letting $a = 1$. On the other hand, if you use the solution $\left(\frac{b+9}{6}, b\right)$, then $(1, -3)$ is still a solution, found by letting $b = -3$.

can express the solution as $\left(\frac{b+9}{6}, b\right)$, where b is any real number. Notice that although the solutions $(a, 6a - 9)$ and $\left(\frac{b+9}{6}, b\right)$, where a and b are real numbers, have a different appearance, they represent the same solution set, since the second component is still 9 less than 6 times the first component. Keep this in mind when you check the answers in the back of the book.

GRAFFIX

Using a graphics calculator or computer, find the solution to the following systems by finding the intersection of the graphs of the two lines.

(a) $\begin{cases} 2x - 7y = 5 \\ x + y = 1 \end{cases}$ **(b)** $\begin{cases} x - 5y = 8 \\ x = 10y + 3 \end{cases}$

Solve the same problems algebraically and compare the solutions.

To summarize, we solve a 2×2 system by reducing it to a 1×1 system; that is, one equation in one variable. Now that we are able to solve systems of equations with two variables, we have more flexibility in solving verbal problems.

EXAMPLE 3 How much of each of a 30% alcohol solution and a 15% alcohol solution should be mixed together in order to get 50 liters of a 20% alcohol solution?

Solution We approach this problem using two variables. If we let x be the number of liters of the 30% alcohol solution and y be the number of liters of the 15% alcohol solution, then we need to write *two* equations expressing the relationships between x and y.

We know that $x + y = 50$, since we want the final mixture to be 50 liters. This is the first equation. The second equation is written in terms of the amounts of pure alcohol. The 30% solution contains $0.30x$ liters of pure alcohol, and the 15% solution contains $0.15y$ liters of pure alcohol. Since we want to end up with 50 liters of a 20% alcohol solution, this means we want to have $0.20(50) = 10$ liters of pure alcohol in the mixture when we are done. Hence our second equation is:

$$\underbrace{0.30x}_{\substack{\text{Amount pure alcohol} \\ \text{in 30\% solution}}} + \underbrace{0.15y}_{\substack{\text{Amount pure alcohol} \\ \text{in the 15\% solution}}} = \underbrace{(0.20)(50)}_{\substack{\text{Amount pure alcohol} \\ \text{in the final mixture}}}$$

So our system is

$$\begin{cases} x + y = 50 \\ 0.30x + 0.15y = 10 \end{cases}$$

We will use the substitution method and begin by solving for y in the first equation:

$$x + y = 50 \Rightarrow y = 50 - x$$

Now substitute $50 - x$ for y into the second equation.

$$0.30x + 0.15y = 10 \qquad \textit{Replace } y \textit{ with } 50 - x.$$

$$0.30x + 0.15(50 - x) = 10$$

This is the same equation we would arrive at if we used the single-variable approach. We clear the decimals by first multiplying each side by 100.

$$30x + 15(50 - x) = 1000$$

$$30x + 750 - 15x = 1000$$

$$x = 16.67$$

Given that factors such as air resistance and friction are negligible, this equation is a mathematical model of a real physical relationship.

Hence we have **16.67 liters of the 30% solution**. Since $y = 50 - x$, we have $y = 50 - 16.67 =$ **33.33 liters of the 15% solution**.

The student should check this solution. ■

EXAMPLE 4 An object is thrown up in the air. If v_0 is the upward velocity with which it is thrown and s_0 is the height of the object above the ground when it is first thrown (called its initial height), then the equation

$$s = -16t^2 + v_0t + s_0$$

gives the distance (s) in feet the object is above the ground t seconds after it is thrown. See Figure 9.3. If the object is 64 feet above the ground after 1 second, and 46 feet above the ground after 2 seconds, find its upward velocity (v_0) and initial height (s_0).

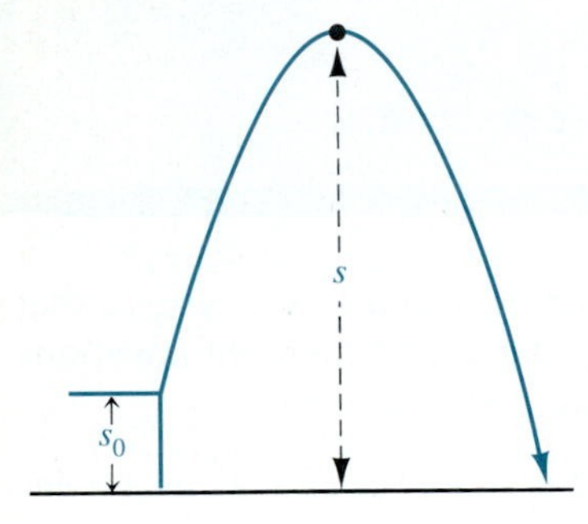

FIGURE 9.3

Solution In this example we are asked to find v_0 and s_0 for the equation $s = -16t^2 + v_0t + s_0$. It is important that you understand that this equation has a real meaning: It tells us how high above the ground, s, the object is (in feet) as t (the time in seconds) changes. We are given that the object is 64 feet above the ground after 1 second. This translates into $s = 64$ when $t = 1$. If we substitute these values into the equation, we get

$$s = -16t^2 + v_0t + s_0 \qquad \textit{Substitute } s = 64 \textit{ and } t = 1 \textit{ into the equation.}$$

$$64 = -16(1)^2 + v_0(1) + s_0$$

$$64 = -16 + v_0 + s_0 \qquad \textit{Which simplifies to } 80 = v_0 + s_0$$

Next we are given that after 2 seconds, the object is 46 feet above the ground. We translate this as $s = 46$ when $t = 2$. We substitute these values into the same equation to generate another equation in the variables v_0 and s_0.

$$s = -16t^2 + v_0t + s_0 \qquad \textit{Substitute } s = 46 \textit{ and } t = 2 \textit{ into the position equation.}$$

$$46 = -16(2)^2 + v_0(2) + s_0$$

$$46 = -64 + 2v_0 + s_0 \qquad \textit{Which simplifies to } 110 = 2v_0 + s_0$$

Hence, our two equations in two unknowns are:

$$\begin{cases} v_0 + s_0 = 80 \\ 2v_0 + s_0 = 110 \end{cases}$$

We solve this system as follows

$$\begin{array}{rcl} v_0 + s_0 = 80 & \xrightarrow{\text{Multiply by } -1} & -v_0 - s_0 = -80 \\ 2v_0 + s_0 = 110 & \xrightarrow{\text{As is}} & 2v_0 + s_0 = 110 \quad \textit{Add.} \\ & & \overline{v_0 \qquad = 30} \quad \textit{Now we find } s_0. \end{array}$$

$v_0 + s_0 = 80$ *Substitute* $v_0 = 30$ *into the first equation to get*

$30 + s_0 = 80$ *Which yields* $s_0 = 50$

Hence, the upward velocity of the object is $\boxed{v_0 = 30 \text{ ft/sec}}$ and the initial height of the object is $\boxed{s_0 = 50 \text{ feet.}}$ That is, the object starts out at 50 feet above the ground and is thrown with an upward velocity of 30 ft/sec. The equation for s is $s = -16t^2 + 30t + 50$. ■

Partial Fractions

In algebra, we are used to finding the sum of two simple fractions such as $\frac{3}{x-4} + \frac{1}{x+1}$ and arriving at the answer $\frac{4x-1}{(x-4)(x+1)}$.

In calculus, it is occasionally necessary to do the opposite—that is, to take a rational expression and express it as a sum of *simpler* rational expressions. For example, we may need to rewrite the expression $\frac{x-5}{(x-2)(x+1)}$ as the sum $\frac{A}{x-2} + \frac{B}{x+1}$, where A and B are constants. Rewriting a fraction as a sum of simpler fractions in this way is called **partial fraction decomposition**. It will be our goal to find the constants A and B as illustrated in the following example.

EXAMPLE 5 Find A and B such that $\frac{x-5}{(x-2)(x+1)} = \frac{A}{x-2} + \frac{B}{x+1}$.

Solution

$$\frac{x-5}{(x-2)(x+1)} = \frac{A}{x-2} + \frac{B}{x+1}$$

Rewrite the right-hand side as a single fraction.

$$\frac{x-5}{(x-2)(x+1)} = \frac{A(x+1) + B(x-2)}{(x-2)(x+1)}$$

Since the denominators are identical, the two fractions are equivalent if and only if the two numerators are equal, or

$$x - 5 = A(x+1) + B(x-2)$$

Multiply out the right-hand side.

$$x - 5 = Ax + A + Bx - 2B$$

Rewrite this equation so that the right-hand side is in polynomial form.

$$x - 5 = Ax + Bx + A - 2B$$

$$x - 5 = (A + B)x + (A - 2B)$$

This last equation must be true for *all* x. In effect, we are saying that the *polynomials* $x - 5$ and $(A + B)x + (A - 2B)$ are equal.

As we discussed in Chapter 1, two polynomials of the same degree are equal if and only if *the coefficients of each power of x are identical*. Hence, in order for these two polynomials to be equal, the coefficients of x for both polynomials must be equal and the numerical terms for both polynomials must be equal.

$$x - 5$$ *The coefficient of x is 1: the numerical term is -5.*

$$(A + B)x + (A - 2B)$$ *The coefficient of x is $A + B$: the numerical term is $A - 2B$.*

Hence we have $\begin{cases} 1 = A + B \\ -5 = A - 2B \end{cases}$ which is a 2×2 system solved as follows:

$$\begin{aligned} A + B &= 1 \xrightarrow{\text{As is}} & A + B &= 1 \\ A - 2B &= -5 \xrightarrow{\text{Multiply by } -1} & -A + 2B &= 5 \\ & & 3B &= 6 \end{aligned}$$

Add. *Which yields $B = 2$*

Now we find A by substituting $B = 2$ into the first equation.

$$A + B = 1 \Rightarrow A + (2) = 1 \qquad \text{which yields} \qquad A = -1.$$

Since $A = -1$ and $B = 2$, we have

$$\frac{x - 5}{(x - 2)(x + 1)} = \frac{A}{x - 2} + \frac{B}{x + 1} \Rightarrow \boxed{\frac{x - 5}{(x - 2)(x + 1)} = \frac{-1}{x - 2} + \frac{2}{x + 1}}$$

The approach we took in this partial fraction decomposition problem is to use methods of solving systems of linear equations. Another way to approach this problem is to realize that if the equation $x - 5 = A(x + 1) + B(x - 2)$ is true for all values of x, then, in particular, it must be true for $x = -1$. Therefore, if we substitute $x = -1$ into this equation we get

$$x - 5 = A(\ x + 1) + B(\ x - 2)$$ *Let $x = -1$. Then we have*

$$-1 - 5 = A(-1 + 1) + B(-1 - 2)$$ *Which becomes*

$$-6 = -3B$$ *Which yields $B = 2$*

If we substitute $x = 2$ into this equation we get

$$x - 5 = A(x + 1) + B(x - 2)$$ *Let $x = 2$. Then we have*

$$2 - 5 = A(2 + 1) + B(2 - 2)$$ *Which becomes*

$$-3 = 3A$$ *Which yields $A = -1$*

Note that each value we chose for x eliminated either A or B. ■

EXERCISES 9.1

In Exercises 1–16, solve the system of equations.

1. $\begin{cases} 2x + y = 12 \\ 3x - y = 8 \end{cases}$

2. $\begin{cases} -x + 2y = -4 \\ x - y = 3 \end{cases}$

3. $\begin{cases} 3x - 2y = 15 \\ 2x + y = 10 \end{cases}$

4. $\begin{cases} 5x + 2y = -8 \\ 3x + y = -4 \end{cases}$

5. $\begin{cases} 5x - 2y = 1 \\ x - 5y = -32 \end{cases}$

6. $\begin{cases} x = 3y - 1 \\ 2x - 5y = -3 \end{cases}$

7. $\begin{cases} y = 3x - 15 \\ 8x - 2y = 10 \end{cases}$

8. $\begin{cases} x + 7y = -1 \\ 3x + 21y = -3 \end{cases}$

9. $\begin{cases} \dfrac{2}{3}y + \dfrac{3}{2}x = 2 \\ \dfrac{3}{4}x + \dfrac{1}{12}y = 1 \end{cases}$

10. $\begin{cases} 5x + 2y = -8 \\ 3x + 5y = -20 \end{cases}$

11. $\begin{cases} \dfrac{a}{6} + \dfrac{b}{8} = \dfrac{3}{4} \\ \dfrac{a}{4} + \dfrac{b}{3} = \dfrac{17}{12} \end{cases}$

12. $\begin{cases} \dfrac{s}{5} + \dfrac{t}{3} = 1 \\ \dfrac{s}{4} + \dfrac{t}{3} = 2 \end{cases}$

13. $\begin{cases} \dfrac{x+3}{2} + \dfrac{y-4}{3} = \dfrac{10}{6} \\ \dfrac{x-2}{3} + \dfrac{y-2}{2} = 2 \end{cases}$

14. $\begin{cases} 0.01x + 0.003y = 6 \\ 0.2x = 0.05y - 2 \end{cases}$

15. $\begin{cases} \dfrac{x}{2} - 0.03y = 0.6 \\ 0.0004y + \dfrac{x}{2} = 0.2 \end{cases}$

16. $\begin{cases} 0.3x - 0.002y = 0.6 \\ 12x = 0.08y - 2 \end{cases}$

17. How much of each of a 30% alcohol solution and a 45% alcohol solution must be mixed together in order to get 30 liters of a 40% solution of alcohol?

18. How much of each of a 20% alcohol solution and a 45% alcohol solution must be mixed together in order to get 30 liters of a 30% solution of alcohol?

19. Carol wants to invest a total of $20,000 so that her yearly interest is $1690. If she invests part in a certificate yielding 8% and the other part in stock at 9.5%, how much should she invest at each rate?

20. Rob invests money at 10% and 8%, earning a yearly interest of $640. Had the amounts invested been reversed, he would have received $610. How much is invested altogether?

21. A car rental agency charges a flat fee plus a mileage rate for a 1-day rental. If the charge for a 1-day rental with 90 miles is $58.30 and the charge for a 1-day rental with 140 miles is $74.30, find the flat fee and the charge per mile.

22. A photocopy machine company charges a flat monthly fee plus a usage rate for their XS-8600 duplicating machine. The charge for the first month, when 3000 copies were made, was $220 and the charge for the second month, when 2600 copies were made, was $210. Find the flat monthly fee and charge per copy for the XS-8600.

23. A plane can cover a distance of 2520 miles in 4 hours with a tailwind (with the wind) and a distance of 2280 miles in the same time with a headwind (against the wind). Find the speed of the plane and the speed of the wind.

24. One train travels 30 km/hr faster than another. After 2 hours they have traveled a total of 430 km. Find the rate of each train.

25. A manufacturer produces two types of telephones. The more expensive model requires 1 hour to manufacture and 30 minutes to assemble. The less expensive model requires 45 minutes to manufacture and 15 minutes to assemble. If the company can allocate 150 hours for manufacturing and 60 hours for assembly, how many of each type can be produced?

26. The physics department hires graders and tutors. For January the department budgets $900 for 60 hours of grading and 35 hours of tutoring. The next month the department budgets $700 for 40 hours of grading and 30 hours of tutoring. How much does the department pay for each hour of tutoring and for each hour of grading?

27. The Goodman car rental agency charges a daily flat fee of $18, plus a mileage rate of $0.28 per mile. The Hirsch car rental agency charges a daily flat fee of $14 plus a mileage rate of $0.32 per mile for the same car. If you only need the car for a day, explain under what conditions one company may be better than the other.

28. The Goodman car rental agency charges a daily flat fee of $18 plus a mileage rate of $0.28 per mile. The Hirsch car rental agency charges a daily flat fee of $14 plus a mileage rate of $0.32 per mile for the same car. If you need the car for 3 days, explain under what conditions one company may be better than the other.

29. A car-telephone company has two plans: plan A and plan B. In plan A, the company charges air time of $0.90 per minute; for plan B, the company charges a flat fee of $20 a month, but the air charge is reduced to $0.70 per minute. Explain under what conditions one plan may be better than the other.

30. Ace Loan Company charges 18% simple annual interest on loans plus a $100 loan processing fee. Deuce Loan Company charges 20% simple annual interest but does not charge a processing fee. If you intend to borrow money and pay the entire principal and interest back at the end of 1 year, explain under what conditions one company might be giving a better deal than the other.

31. An object is thrown straight up in the air. If v_0 is its initial velocity and s_0 is its initial height, then the position equation

$$s = -16t^2 + v_0t + s_0$$

gives the distance (s) in feet the object is above the ground t seconds after it is thrown. If the object is 64 feet above the ground after 1 second and 96 feet above the ground after 2 seconds, find its initial velocity and initial height.

32. An object is thrown straight up in the air. If v_0 is its initial velocity and s_0 is its initial height, then the position equation

$$s = -16t^2 + v_0t + s_0$$

gives the distance (s) in feet the object is above the ground t seconds after it is thrown. If the object is 32 feet above the ground after $\frac{1}{2}$ second, and 36 feet above the ground after 1 second, find its initial velocity and initial height.

In Exercises 33–36, find A and B such that:

33. $\dfrac{10}{(x-2)(2x+1)} = \dfrac{A}{x-2} + \dfrac{B}{2x+1}$.

34. $\dfrac{7x+1}{(x+3)(x-1)} = \dfrac{A}{x+3} + \dfrac{B}{x-1}$.

35. $\dfrac{2x-1}{(x+1)^2} = \dfrac{A}{x+1} + \dfrac{B}{(x+1)^2}$.

36. $\dfrac{4x-15}{(x-3)^2} = \dfrac{A}{x-3} + \dfrac{B}{(x-3)^2}$

37. Find the point (x, y) so that the line passing through (x, y) and $(-1, 2)$ has a slope of 3 and intersects the line passing through (x, y) and $(2, -1)$ with a slope of 1.

38. Find the values of A and B so that the line whose equation is $Ax + By = 4$ passes through the points $(1, 3)$ and $(-2, 4)$.

39. Suppose we arrive at the following solution for a 2×2 system: $(a, 2a - 5)$ for all real a. How many solutions to this system are there? Give examples of some numerical solutions, and explain in words the conditions under which a pair of numbers can be a solution to the system.

40. Suppose we arrive at the following solution for a 2×2 system: $\left(\dfrac{a+5}{2}, a\right)$ for all real a. How does this dependent solution compare to the dependent solution in Exercise 39?

QUESTIONS FOR THOUGHT

41. Explain what it means geometrically when we find that there is no solution to a system, or that the system is inconsistent.

42. Explain what it means geometrically when we find that there are an infinite number of solutions to a 2×2 system—that is, if the solution is dependent.

43. Explain how you solve the following system of equations:

$$\begin{cases} 5x - 3y = 8 \\ \quad\;\; 2y = -10 \end{cases}$$

Do you need to eliminate a variable?

44. Explain how you solve the following system of equations:

$$\begin{cases} 2x - 3y + z = 1 \\ \quad\;\; 2y - z = 9 \\ \quad\quad\;\; 3z = -3 \end{cases}$$

Do you need to eliminate a variable?

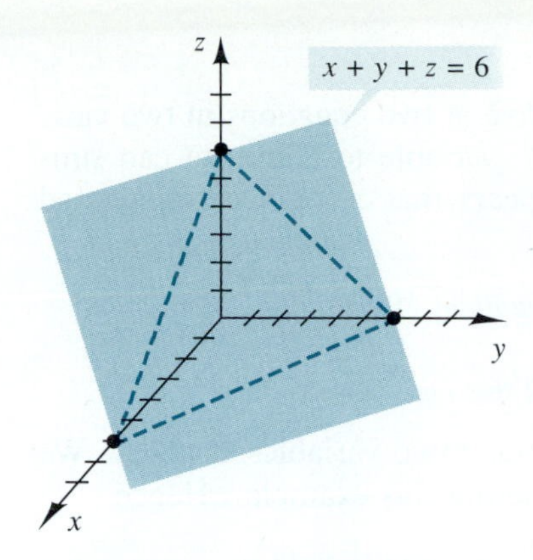

FIGURE 9.4 On a three-dimensional coordinate system, the graph of $x + y + z = 6$ is a plane.

9.2 Elimination and Gaussian Elimination: 3 × 3 Linear Systems

For a first-degree equation in three variables, a solution is an ordered triple (x, y, z). Thus, for example, a few solutions to the equation $3x - 2y + z = 11$ are $(3, -1, 0)$, $(2, 1, 7)$, and $(1, 1, 10)$.

Geometrically, we know that we can represent the solutions to a first-degree equation in two variables as a straight line in a two-dimensional coordinate system. A first-degree equation in three variables can be represented geometrically as a plane in a three-dimensional coordinate system, as shown in Figure 9.4.

A 3 × 3 system is a system of three equations in three variables. For example

$$\begin{cases} 2x + 3y - z = 13 \\ 5x - y + z = 0 \\ x - 3y - z = -6 \end{cases}$$

is a system satisfied by the ordered triple $(1, 3, -2)$. A natural question to ask is: Is this the only solution? As with a 2 × 2 system, a 3 × 3 system has three possible types of solutions, illustrated by the pictures of intersections of planes in Figure 9.5.

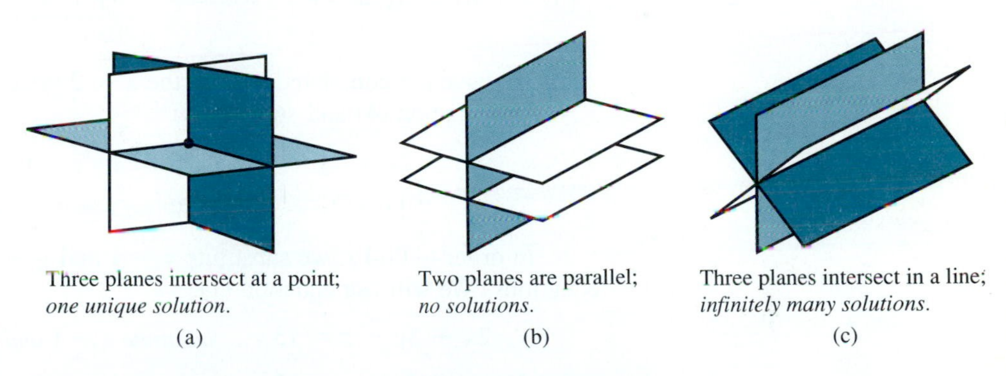

FIGURE 9.5 Three possible intersections of three planes

In general, we take the same approach solving larger systems of linear equations as we did in solving 2 × 2 systems; that is, we reduce a larger system to a smaller system that we already know how to solve. In this case, we can use elimination or substitution to reduce a 3 × 3 system (three equations in three variables) to a 2 × 2 system and solve the 2 × 2 system by the methods discussed in the previous section.

EXAMPLE 1 Solve the following system of equations:

$$\begin{cases} 2x + 3y - z = 13 & (1) \\ 5x - y + z = 0 & (2) \\ x - 3y - z = -6 & (3) \end{cases}$$

We number the equations for ease of reference.

Solution

Our goal is to eliminate one of the variables and arrive at two equations in two variables. As with 2×2 systems, choosing the "right" variable to eliminate can simplify the procedure. Looking at this system, it appears that z could be eliminated most easily.

$$\begin{array}{lrl} (1) & 2x + 3y - z = 13 & \textit{Add equations (1) and (2).} \\ (2) & \underline{5x - \ \ y + z = \ \ 0} & \\ & 7x + 2y \qquad = 13 & \textit{We call this equation (4).} \end{array}$$

Now we have to find another equation in the *same* two variables, x and y. We must use equation (3). (We explain why after completing this example.) Hence

$$\begin{array}{lrl} (2) & 5x - \ \ y + z = \ \ 0 & \textit{Add equations (2) and (3).} \\ (3) & \underline{x - 3y - z = -6} & \\ & 6x - 4y \qquad = -6 & \textit{We call this equation (5).} \end{array}$$

Equations (4) and (5) form a 2×2 system, which we can solve.

$$\begin{array}{llcrl} (4) & 7x + 2y = 13 & \xrightarrow{\textit{Multiply by 2}} & 14x + 4y = \ \ 26 & \\ (5) & 6x - 4y = -6 & \xrightarrow{\textit{As is}} & \underline{6x - 4y = -6} & \textit{Add.} \\ & & & 20x \qquad = \ \ 20 & \textit{Which yields } x = 1 \end{array}$$

We have not completed solving the 2×2 system until we find y. We substitute $x = 1$ in equation (4) and solve for y:

$$7x + 2y = 13 \qquad \textit{Substitute } x = 1 \textit{ and solve for } y.$$

$$7(1) + 2y = 13 \qquad \textit{We get } y = 3.$$

In order to find z, we substitute $x = 1$ and $y = 3$ into any of the three original equations. We will use equation (1):

$$2x + 3y - z = 13 \qquad \textit{Substitute } x = 1 \textit{ and } y = 3 \textit{ and solve for } z.$$

$$2(1) + 3(3) - z = 13 \qquad \textit{We get } z = -2.$$

Hence our solution is $\boxed{(1, 3, -2)}$. The student should check this solution in all three original equations. ■

We already noted that a *single* equation in *two* variables such as $x - 3y = 5$ does not have a unique solution. The same is true of a simultaneous system of two equations in three variables: If the system *has* a solution, the solution will not be unique. If we represent the two equations in three variables as planes in a three-dimensional coordinate system, we can see that if the planes intersect (if there is a solution), their intersection will be a line. (See Figure 9.6).

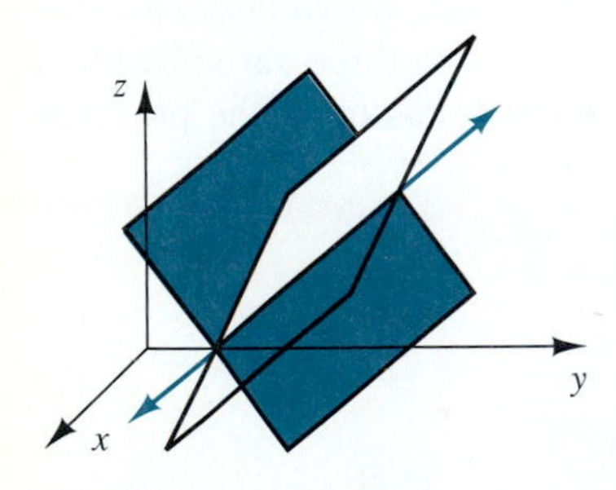

FIGURE 9.6 The intersection of two distinct planes is a line.

In general, if the number of variables is greater than the number of equations in a system, the system will not have a unique solution.

In Example 1, we started out by combining equations (1) and (2) to arrive at an equation in two variables. Then, in order to create a second equation in the same two variables, we noted that we *must* somehow use equation (3). If we did not use equation (3), we would be changing the problem, and rather than looking at the in-

What kind of solution would you expect to get if we had one equation in two variables?

tersection of the *three* planes described by the three equations, we would be looking only at the intersection of the two planes described by equations (1) and (2), which would produce a different solution.

In general, when solving 3×3 linear systems, we choose the most convenient variable to eliminate and produce a 2×2 system using all three equations. We use two equations to get one equation in two variables and then use a different pair of equations to get another equation in the same two variables. We solve the 2×2 system and then substitute these values into one of the original equations to find the third variable.

EXAMPLE 2 Solve the following system of equations:

$$\begin{cases} x - 3y + z = -1 & (1) \\ 2x - 6y + 2z = -2 & (2) \\ x - y - z = 1 & (3) \end{cases}$$

Solution We choose to eliminate x. We combine equations (1) and (2) as follows:

$$\begin{array}{ll} (1)\ x - 3y + z = -1 \xrightarrow{\text{Multiply by } -2} & -2x + 6y - 2z = 2 \\ (2)\ 2x - 6y + 2z = -2 \xrightarrow{\text{As is}} & \underline{2x - 6y + 2z = -2} \quad \textit{Add.} \\ & \qquad\qquad 0 = 0 \quad \textit{This is always true.} \end{array}$$

Equations (1) and (2) are equivalent. This means our system is really a system of two equations in three unknowns; therefore, if a solution to the system does exist, there must be an infinite number of them. Let's set aside equation (2) for a moment, since it is equivalent to equation (1), and examine the system consisting of equations (1) and (3):

$$\begin{cases} x - 3y + z = -1 & (1) \\ x - y - z = 1 & (3) \end{cases}$$

If, in the process of eliminating one variable from this system, we end up with an identity such as $0 = 0$, we would know that equations (1) and (3) are equivalent. Then all three equations in the 3×3 system would be equivalent, and we would choose one of the equations to represent the system and express the dependent solution as we did when we found two equations in two variables equivalent (see Example 2(b) in Section 9.1).

If all three equations were equivalent, how would we express the solution?

On the other hand, equations (1) and (3) may represent parallel planes, and there would be no solution to the system; if we tried to eliminate a variable, we would end up with a contradiction, such as in Example 1(c) in Section 9.1.

Let's see what happens when we try to eliminate z in this 2×3 system:

$$\begin{array}{lrl} (1) & x - 3y + z = -1 & \textit{Add equations (1) and (3).} \\ (3) & \underline{x - y - z = 1} & \\ & 2x - 4y \quad = 0 & \textit{Which we can simplify and rewrite as} \\ & x = 2y & \textit{We call this equation (4).} \end{array}$$

Since this equation is not a contradiction, there must be solutions (in fact, infinitely many solutions, since we have two equations in three unknowns).

How do we represent the solution? We know that whatever the solutions are, equation (4) tells us that x must be twice y. To find the restrictions on z and conveniently express the solutions, we proceed as follows: We let $y = a$, where a is a constant, and we substitute $y = a$ into equation (4) and solve for x, so that we may express x in terms of a.

$$x = 2y \qquad \textit{To express } x \textit{ in terms of } a, \textit{ let } y = a.$$

$$x = 2a$$

So far we have expressed x and y in terms of a. Next, we want to express z in terms of a. We can choose any one of the original equations, so we select equation (3); we substitute $y = a$ and $x = 2a$ into this equation and solve for z.

$$x - y - z = 1 \qquad \textit{Substitute } x = 2a \textit{ and } y = a.$$

$$2a - a - z = 1 \qquad \textit{Solve for } z \textit{ to get}$$

$$z = a - 1$$

Hence if $y = a$, then $x = 2a$ and $z = a - 1$, which can be written as the ordered triple $\boxed{(2a, a, a - 1)}$.

This means that for every real number a, we get a different solution for the system: any ordered triple where the x-coordinate is twice the y-coordinate and the z-coordinate is 1 less than the y-coordinate will be a solution to the system. For example, if we let $a = 3$, then (6, 3, 2) is a solution. As with 2×2 dependent solutions, there are an infinite number of solutions; what is being expressed in the preceding solution is how x, y, and z should be related if the ordered triple is to be one of the solutions. Had we decided to let $x = a$ and then solved for y and z in terms of a, the solution would have had a different appearance, but the relationships among the variables would have been the same: x would be twice y and z would be 1 less than y. ■

EXAMPLE 3 A VCR manufacturing company produces three models: model A, model B, and model C. The company knows how much time it takes for the production, assembly, and testing of each model. This information is found in the accompanying table. To minimize costs, the company decides that 385 hours should be allocated for production, 557 hours for assembly, and 128 hours for testing. How many of each model can the company produce if it wants to use up all the allotted time for each phase of the process?

Model	Production hours	Assembly hours	Testing hours
A	1.8	3.0	0.5
B	2.2	3.2	0.8
C	3.0	3.5	1.0

Solution If we let a, b, and c represent the number of VCRs of models A, B, and C, respectively, then we can create three equations in three variables; each equation represents *the number of hours allotted for each of the three phases*:

$$\begin{cases} 1.8a + 2.2b + 3c = 385 & (1) \quad \textit{No. hours for production} \\ 3a + 3.2b + 3.5c = 557 & (2) \quad \textit{No. hours for assembly} \\ 0.5a + 0.8b + c = 128 & (3) \quad \textit{No. hours for testing} \end{cases}$$

We choose to eliminate c. Combine equations (1) and (3) as follows:

$$\begin{array}{lll} (1)\ 1.8a + 2.2b + 3c = 385 & \xrightarrow{\text{As is}} & 1.8a + 2.2b + 3c = 385 \\ (3)\ 0.5a + 0.8b + c = 128 & \xrightarrow{\text{Multiply by } -3} & -1.5a - 2.4b - 3c = -384 \quad \textit{Add.} \\ \hline & & 0.3a - 0.2b = 1 \end{array}$$

Call this equation (4).

Combine equations (2) and (3) as follows:

$$\begin{array}{lll} (2)\ 3a + 3.2b + 3.5c = 557 & \xrightarrow{\text{As is}} & 3a + 3.2b + 3.5c = 557 \\ (3)\ 0.5a + 0.8b + c = 128 & \xrightarrow{\text{Multiply by } -3.5} & -1.75a - 2.8b - 3.5c = -448 \quad \textit{Add.} \\ \hline & & 1.25a + 0.4b = 109 \end{array}$$

Call this equation (5).

Now we solve the 2 × 2 system consisting of equations (4) and (5).

$$\begin{array}{lll} (4)\ \begin{cases} 0.3a - 0.2b = 1 \\ 1.25a + 0.4b = 109 \end{cases} & \begin{array}{l} \xrightarrow{\text{Multiply by } 2} \\ \xrightarrow{\text{As is}} \end{array} & \begin{array}{l} 0.6a - 0.4b = 2 \\ 1.25a + 0.4b = 109 \quad \textit{Add.} \\ \hline 1.85a = 111 \end{array} \end{array}$$

Which yields $a = 60$

Now find b by substituting $a = 60$ in equation (4):

$$0.3a - 0.2b = 1 \qquad \textit{Substitute } a = 60.$$

$$0.3(60) - 0.2b = 1 \qquad \textit{Which yields } b = 85$$

Finally, find c by substituting $a = 60$ and $b = 85$ in equation (3):

$$0.5a + 0.8b + c = 128 \qquad \textit{Substitute } a = 60 \textit{ and } b = 85.$$

$$0.5(60) + 0.8(85) + c = 128 \qquad \textit{Which yields } c = 30$$

Hence the company must produce

$$a = \boxed{\text{60 of model A}}, \quad b = \boxed{\text{85 of model B}}, \quad \text{and} \quad c = \boxed{\text{30 of model C}}.$$

The student should check these solutions in the original statement of the problem. ■

The procedures we have discussed in this section can be generalized to any size system of linear equations. If we had to solve a 5 × 5 system, we would start by eliminating a variable and an equation, reducing it to a 4 × 4 system. Then we would reduce the 4 × 4 system to a 3 × 3 system, and so on.

Gaussian Elimination

Gaussian elimination is a variation of the elimination method that uses the same algebraic procedures as elimination, but rather than reducing the number of equations and variables, as we did previously in reducing a 3×3 system to a 2×2 system, the goal is to reduce only the number of variables in each successive equation in the original system.

Consider the following two systems of equations:

$$\text{I.} \begin{cases} x + 3y - z = -4 \\ 2x + 8y + z = -4 \\ x + 5y + 6z = 8 \end{cases} \qquad \text{II.} \begin{cases} x + 3y - z = -4 \\ 2y + 3z = 4 \\ 4z = 8 \end{cases}$$

You may verify for yourself that these two systems are equivalent, since they each have the same unique solution: $(1, -1, 2)$. System II, however, is much easier to solve. Notice that last equation in system II is an equation in one variable, z, which can be solved easily:

$$4z = 8 \Rightarrow z = 2$$

Now that we have found z, we can find y by substituting the value for z in the second equation of system II:

$$2y + 3z = 4 \qquad \textit{Substitute } z = 2.$$

$$2y + 3(2) = 4 \qquad \textit{Which yields } y = -1$$

Finally, we can find x by substituting values for z and y in the first equation of system II:

$$x + 3y - z = -4 \qquad \textit{Substitute } z = 2 \textit{ and } y = -1.$$

$$x + 3(-1) - (2) = -4 \qquad \textit{Which yields } x = 1\textit{; the solution is } (1, -1, 2).$$

This process is called **back substitution**.

In order for us to solve a system by using substitution in this manner, we have to put the system into a form like that of system II, called (upper) **triangular form**: A form in which each successive equation in the system is an equation containing at least one less variable than the preceding equation. The process of putting a system into triangular form is called **Gaussian elimination**.

To put a system into triangular form, we use the same procedures as we did with the elimination method. Let's pause to summarize some of these procedures, since we will refer to them in subsequent sections.

We remind you that *two systems are equivalent if their solutions are identical.*

A system of equations can be changed into an *equivalent* system by the following transformations:

1. Interchanging any two equations.
2. Multiplying both sides of any equation by a nonzero constant.
3. Adding a multiple of one equation to another equation.

EXAMPLE 4 Solve the following system by Gaussian elimination:

$$\begin{cases} x - y + 3z = 11 & (1) \\ 3x - y + z = 5 & (2) \\ x + 3y - z = -9 & (3) \end{cases}$$

Solution Our approach is to put the system into triangular form and then use back substitution to find the solution. We start by transforming the system into an equivalent system in triangular form; that is, we first want to eliminate x from the last two equations and then eliminate y from the last equation. We start by eliminating x from equation (2) by multiplying each side of equation (1) by -3 and adding this equation to equation (2) to get equation (4):

$$\begin{array}{lll} (1)\quad x - y + 3z = 11 & \xrightarrow{\text{Multiply by } -3} & -3x + 3y - 9z = -33 \\ (2)\quad 3x - y + z = 5 & \xrightarrow{\text{As is}} & \underline{3x - y + z = 5} \quad \textit{Add.} \\ & & 2y - 8z = -28 \end{array}$$

This is equation (4).

Next, we eliminate x using equations (1) and (3): Multiply both sides of equation (1) by -1 and add the resultant equation to equation (3) to get equation (5):

$$\begin{array}{lll} (1)\quad x - y + 3z = 11 & \xrightarrow{\text{Multiply by } -1} & -x + y - 3z = -11 \\ (3)\quad x + 3y - z = -9 & \xrightarrow{\text{As is}} & \underline{x + 3y - z = -9} \quad \textit{Add.} \\ & & 4y - 4z = -20 \end{array}$$

This is equation (5).

So now we have the following system:

$$\begin{cases} x - y + 3z = 11 & (1) \\ 2y - 8z = -28 & (4) \\ 4y - 4z = -20 & (5) \end{cases}$$

This is almost in triangular form. What has to be done next?

Finally, to get the new system into triangular form, we still need to eliminate y from equation (5). We do this by multiplying both sides of equation (4) by -2 and adding the resultant equation to equation (5).

$$\begin{array}{lll} (4)\quad 2y - 8z = -28 & \xrightarrow{\text{Multiply by } -2} & -4y + 16z = 56 \\ (5)\quad 4y - 4z = -20 & \xrightarrow{\text{As is}} & \underline{4y - 4z = -20} \quad \textit{Add.} \\ & & 12z = 36 \quad \textit{This is equation (6).} \end{array}$$

We now have the following equivalent system:

$$\begin{cases} x - y + 3z = 11 & (1) \\ 2y - 8z = -28 & (4) \\ 12z = 36 & (6) \end{cases}$$

This system is now in triangular form. We can now solve the system using back substitution. Equation (6) in this system yields $z = 3$. We use equation (4) to find y.

$$2y - 8z = -28 \qquad \textit{Substitute } z = 3 \textit{ and solve for } y.$$

$$2y - 8(3) = -28 \qquad \textit{To get } y = -2.$$

We use equation (1) to find x:

$$x - y + 3z = 11 \qquad \textit{Substitute } z = 3 \textit{ and } y = -2 \textit{ and solve for } x.$$

$$x - (-2) + 3(3) = 11 \qquad \textit{To get } x = 0.$$

Hence the solution is $\boxed{x = 0,\ y = -2, \text{ and } z = 3}$, or $\boxed{(0, -2, 3)}$. ■

When a system is in triangular form and the coefficient of the leftmost variable in each equation is 1, we say that the system is in **echelon form**. For example, the following system is in echelon form.

What is the difference between a system in triangular form and one in echelon form?

$$\begin{cases} x - 5y + z = 0 \\ \quad y - 2z = -8 \\ \quad\quad z = 12 \end{cases}$$

The process of putting a system into echelon form lends itself well to programming. However, for all practical purposes, if you are transforming the system by hand, putting a system into triangular form usually requires less computation.

In the next sections we discuss systematic procedures for solving linear systems that are essentially variations on the elimination methods.

EXERCISES 9.2

In Exercises 1–10, solve the system of equations by elimination.

1. $\begin{cases} x + y + z = 6 \\ 3x + y - z = 6 \\ 2x + y - z = 4 \end{cases}$

2. $\begin{cases} x - 2y + z = -5 \\ 3x - y + z = 2 \\ x + 2y - z = 9 \end{cases}$

3. $\begin{cases} x + y - z = 1 \\ 2x + 2y + 2z = 0 \\ x - y + z = 3 \end{cases}$

4. $\begin{cases} 3x + y + 2z = 0 \\ x - 2y - z = -3 \\ x + y + z = 1 \end{cases}$

5. $\begin{cases} x + y + z = 6 \\ 3x + y - z = 4 \\ 2x + 2y + 2z = 4 \end{cases}$

6. $\begin{cases} x + y - z = 3 \\ 2x + 2y - 2z = 6 \\ x - y + z = 5 \end{cases}$

7. $\begin{cases} 0.01a + 0.2b + c = 0.45 \\ 0.3a - b + 0.2c = 1.06 \\ a + b + c = 5.8 \end{cases}$

8. $\begin{cases} x + y = 2x + z \\ x - 2y = z + 3 \\ 2x - y = 3z - 9 \end{cases}$

9. $\begin{cases} \dfrac{x}{4} + \dfrac{y}{6} - \dfrac{z}{3} = 1 \\ \dfrac{x}{2} + \dfrac{y}{3} + z = -3 \\ \dfrac{x}{8} + \dfrac{y}{4} - z = 5 \end{cases}$

10. $\begin{cases} x + y = 5 \\ x + 2z = 5 \\ y - z = 1 \end{cases}$

In Exercises 11–16, solve the system of equations by Gaussian elimination.

11. $\begin{cases} x + y - z = 7 \\ x + 2y + 2z = 4 \\ 2x - y + z = -1 \end{cases}$

12. $\begin{cases} x - y + 3z = 14 \\ x + 2y - z = -8 \\ 2x + y + z = 2 \end{cases}$

13. $\begin{cases} x + y - 2z = 3 \\ 2x + y + z = -4 \\ x + 2y - 3z = 5 \end{cases}$

14. $\begin{cases} x + z = 4 \\ x - y = 1 \\ y - z = 1 \end{cases}$

15. $\begin{cases} 2x + y + z = 4 \\ x - y - z = 5 \\ 2x - 2y - 2z = 1 \end{cases}$

16. $\begin{cases} x - y - z = 5 \\ 2x - 2y - 2z = 10 \\ 2x + y + z = 4 \end{cases}$

17. Elaine has a total of \$19,750 in three investments. She has a bank account paying 6%, a certificate of deposit paying 8%, and a stock yielding 10.5%. Her annual interest from the three investments is \$1680. If the interest from her stock is equal to the sum of the interest of the other two investments, how much was invested at each rate?

18. A theater group plans to sell 520 tickets for a play. They are charging \$12 for orchestra seats, \$10 for mezzanine seats, and \$6 for balcony seats, and they plan to collect \$5080. If there are 100 more mezzanine tickets than orchestra and balcony tickets combined, how many of each type are there?

19. A compact disc manufacturer produces three models of compact disc players, model A, model B, and model C.

The company knows how much production, assembly, and testing time is needed for each model. This information is found in the accompanying table. If the company has allocated 586 hours for production, 227 hours for assembly, and 147 hours for testing, how many of each model can the company produce if it wants to use up all the time allocated for each phase of the process?

Model	Production hours	Assembly hours	Testing hours
A	1.2	0.4	0.2
B	1.4	0.5	0.3
C	1.8	0.8	0.6

20. A nutritionist wants to create a food supplement out of three substances; X, Y, and Z. He wants the food supplement to have the following characteristics: 10 grams of the supplement should supply 3.5 grams of iron and cost 77¢. The iron content and cost of the substances X, Y, and Z are entered in the accompanying table. How many grams of each substance should be used to make such a food supplement?

Substance	Iron content per gram	Cost per gram
X	0.5	10¢
Y	0.2	6¢
Z	0.4	8¢

21. A chemist needs to mix three solutions containing 10%, 25%, and 40% adepic acid, respectively, in order to get 100 liters of a solution that is 30% acid. If she needs to use all three solutions and there are 20 liters more of the 25% solution than of the 10% solution in the mixture, how much of each solution will be needed?

22. A horticulturist wants to mix three types of fertilizers, which contain 20%, 30%, and 40% nitrogen, respectively. If the final mixture should be 1000 pounds of 32% nitrogen, all three types are used, and there is twice as much of the 40% type as of the 20% type, how much of each is in the final mixture?

23. A function of the form $f(x) = ax^2 + bx + c$, where a, b, and c are constants, is called a quadratic function. Find the quadratic function $f(x)$ such that $f(1) = 2$, $f(-1) = 6$, and $f(2) = 9$.

24. A function of the form $f(x) = ax^2 + bx + c$, where a, b, and c are constants, is called a quadratic function. Find the quadratic function $f(x)$ such that $f(1) = 0$, $f(-1) = 8$, and $f(2) = 2$.

In Exercises 25 and 26, find A, B, and C such that:

25. $\dfrac{2x + 5}{x(x - 1)^2} = \dfrac{A}{x} + \dfrac{B}{x - 1} + \dfrac{C}{(x - 1)^2}$.

26. $\dfrac{5x + 6}{x(x - 2)(x + 3)} = \dfrac{A}{x} + \dfrac{B}{x - 2} + \dfrac{C}{x + 3}$.

QUESTION FOR THOUGHT

27. Find A, B, C, and D such that

$$\frac{x^3 + 2x^2 + 3x + 5}{(x^2 + 1)^2} = \frac{Ax + B}{x^2 + 1} + \frac{Cx + D}{(x^2 + 1)^2}.$$

9.3 Solving Linear Systems Using Augmented Matrices

When we solve a system of linear equations using the elimination method, we first make sure that the equations are written in such a way that the same variables are aligned vertically. Once we have the system in this form, we concentrate on the coefficients of the variables, transforming equations so that the coefficients of the variables to be eliminated are "matched up." This process suggests that if we can develop some systematic procedure to keep track of the variables, we can ignore the variables and focus on the coefficients in order to solve a system of linear equations. The matrix methods we develop in this section provide such a procedure.

A **matrix** is a rectangular array of numbers. The numbers in the matrix are called the **elements**, or **entries**, of the matrix. The entries are arranged in rows and columns, as follows:

$$\begin{bmatrix} 3 & -1 & 4 & 0 \\ 5 & 2 & -7 & 5 \\ -6 & 3 & 8 & -1 \end{bmatrix}$$

When we refer to a matrix, we usually refer to its size by specifying the number of rows and the number of columns. The preceding matrix is called a 3 by 4 matrix (3 *rows* by 4 *columns*), which is usually referred to as a 3×4 matrix. A *square* matrix is a matrix with an equal number of rows and columns.

We rewrite the coefficients of a system of equations as an **augmented matrix** in the following way:

The system $\begin{cases} 3x - 2y + 3z = 9 \\ 2x + y - z = -4 \\ 5x - y + 2z = 0 \end{cases}$ is written as $\left[\begin{array}{ccc|c} 3 & -2 & 3 & 9 \\ 2 & 1 & -1 & -4 \\ 5 & -1 & 2 & 0 \end{array}\right]$

The matrix of the coefficients of the variables, $\begin{bmatrix} 3 & -2 & 3 \\ 2 & 1 & -1 \\ 5 & -1 & 2 \end{bmatrix}$, is called the **coefficient matrix** of the system; the first column of this matrix is the coefficients of the x-variables, the second column is the coefficients of the y-variables, and the third column is the coefficients of the z-variables. The augmented matrix is made up of the coefficient matrix, with the constants of the system adjoined as the last column on the right.

If necessary, we begin by rewriting the system so that the variables x, y, and z are vertically aligned and then rewrite the system as an augmented matrix of the coefficients and constants. Where there are missing variables, we fill in a 0 as the coefficient.

Recall that the goal of Gaussian elimination (discussed in Section 9.2) is to begin with any system and, using certain transformations, change it to a system in triangular form. Once we have the system in triangular form, we can solve it using back substitution. The following system is in triangular form. Beside this system, we write its associated augmented matrix:

Note the "triangle" of zeros below the main diagonal of the matrix.

$$\begin{cases} x + 2y + 2z = 12 \\ y - 3z = -5 \\ 2z = 6 \end{cases} \qquad \left[\begin{array}{ccc|c} 1 & 2 & 2 & 12 \\ 0 & 1 & -3 & -5 \\ 0 & 0 & 2 & 6 \end{array}\right]$$

We refer to this system's augmented matrix as being in (upper) *triangular* form. The elements 1, 1, and 2 outlined in the augmented matrix of the system are called the **main diagonal**. The main diagonal of an augmented matrix actually consists of the diagonal elements of the coefficient matrix. A matrix is said to be in (upper) **triangular form** if it has all zero entries below the main diagonal.

We saw in Section 9.2 that we can use certain operations on a system to transform it into a simpler equivalent system, one in triangular (or echelon) form. In this section, we will do exactly the same thing, but rather than working with the equations in the system, we will work instead with its augmented matrix of coefficients and constants. We begin with the following definition.

DEFINITION Two augmented matrices are **row equivalent** if and only if their associated systems of equations are equivalent.

In order to change a matrix into a row equivalent matrix, we are allowed to transform them in certain ways. These transformations are called the **elementary row operations** and are listed next.

ELEMENTARY ROW OPERATIONS

1. Multiply each entry in a given row by a nonzero constant.
2. Interchange any two rows.
3. Add a multiple of one row to another row.

The following are three illustrations of the elementary row operations.

$$\begin{bmatrix} 3 & 2 & 2 & -5 \\ 1 & 2 & -1 & 6 \end{bmatrix} \quad \xrightarrow{-3R_2 \to R_2} \quad \begin{bmatrix} 3 & 2 & 2 & -5 \\ -3 & -6 & 3 & -18 \end{bmatrix}$$

This notation means multiply each entry in row 2 by −3 to get the new row 2.

$$\begin{bmatrix} 3 & 2 & 2 \\ 2 & -6 & 0 \\ 1 & 2 & -1 \end{bmatrix} \quad \xrightarrow{R_1 \leftrightarrow R_3} \quad \begin{bmatrix} 1 & 2 & -1 \\ 2 & -6 & 0 \\ 3 & 2 & 2 \end{bmatrix}$$

This notation means interchange row 1 and row 3 (row 2 remains unchanged).

$$\begin{bmatrix} 3 & 2 & 2 \\ 2 & -6 & 0 \\ 1 & 2 & -1 \end{bmatrix} \quad \xrightarrow{2R_1 + R_3 \to R_3} \quad \begin{bmatrix} 3 & 2 & 2 \\ 2 & -6 & 0 \\ 7 & 6 & 3 \end{bmatrix}$$

This notation means multiply each entry in row 1 by 2 and add the resulting entries to each entry in row 3 to get a new row 3 (rows 1 and 2 remain unchanged).

We clarify this last illustration a bit.

$$\text{Since } R_1 = 3 \quad 2 \quad 2 \quad \text{then} \quad \begin{array}{rrrrl} 2 \times R_1 = & 6 & 4 & 4 & \\ R_3 = & 1 & 2 & -1 & \textit{Add to get} \\ \hline & 7 & 6 & 3 & \textit{This is the new } R_3. \end{array}$$

The elementary row operations are equivalent to the transformations we performed on systems of equations. For example,

$$\begin{array}{rcl} 5x + y = 7 & \xrightarrow{\text{Multiply by 3}} & 15x + 3y = 21 \\ x - 3y = -5 & \xrightarrow{\text{As is}} & x - 3y = -5 \end{array}$$

is the same as using elementary row operation 1:

$$\left[\begin{array}{rr|r} 5 & 1 & 7 \\ 1 & -3 & -5 \end{array}\right] \quad 3R_1 \to R_1 \quad \left[\begin{array}{rr|r} 15 & 3 & 21 \\ 1 & -3 & -5 \end{array}\right]$$

Hence elementary row operations produce matrices whose associated systems are equivalent to each other.

> Performing the elementary row operations on a matrix produces a row equivalent matrix.

Now we can use matrices to solve systems of equations by **Gaussian elimination**, a process in which we use the elementary row operations to convert the augmented matrix of a system into a row equivalent matrix in triangular form. Once we have the augmented matrix in triangular form, we write out its associated system and solve the system by back substitution. This process is illustrated in the next example.

EXAMPLE 1 Solve the following systems of equations:

(a) $\begin{cases} x - 5y = 4 \\ 2x - 3y = 7 \end{cases}$ **(b)** $\begin{cases} x + 2y + 3z = -5 \\ 2x + 6y + 7z = -10 \\ x + 8y + 2z = 3 \end{cases}$

Solution

(a) First we write the associated augmented matrix of the system:

$$\begin{cases} x - 5y = 4 \\ 2x - 3y = 7 \end{cases} \Rightarrow \left[\begin{array}{cc|c} 1 & -5 & 4 \\ 2 & -3 & 7 \end{array}\right]$$

Next we put it in triangular form.

Triangular form for a 2×2 matrix means the bottom left entry of the matrix (row 2, column 1) is 0.

$$\left[\begin{array}{cc|c} 1 & -5 & 4 \\ 2 & -3 & 7 \end{array}\right] \qquad -2R_1 + R_2 \rightarrow R_2 \qquad \left[\begin{array}{cc|c} 1 & -5 & 4 \\ 0 & 7 & -1 \end{array}\right]$$

Multiply the entries in row 1 by −2, add the result row 2, and replace row 2 with this result.

REMEMBER: $-2R_1 + R_2 \rightarrow R_2$ means

since $R_1 = 1 \quad -5 \quad 4$ then

$$\begin{array}{rrrr} -2R_1 = & -2 & 10 & -8 \\ R_2 = & 2 & -3 & 7 \\ \hline & 0 & 7 & -1 \end{array}$$

Add.

This is the new R_2.

This matrix is now in triangular form. If we write its associated system we get

$$\left[\begin{array}{cc|c} 1 & -5 & 4 \\ 0 & 7 & -1 \end{array}\right] \Rightarrow \begin{cases} x - 5y = 4 \\ \quad 7y = -1 \end{cases}$$

The last equation in this new (equivalent) system yields $y = -\frac{1}{7}$. We use the first equation in this new system to find x.

$$x - 5y = 4$$

Substitute $y = -\frac{1}{7}$ and solve for x.

$$x - 5\left(-\frac{1}{7}\right) = 4 \qquad \textit{To get } x = \frac{23}{7}$$

The student can verify that the solution is $\boxed{\left(\frac{23}{7}, -\frac{1}{7}\right)}$.

(b) First we write the associated augmented matrix of the system.

$$\begin{cases} x + 2y + 3z = -5 \\ 2x + 6y + 7z = -10 \\ x + 8y + 2z = 3 \end{cases} \Rightarrow \left[\begin{array}{ccc|c} 1 & 2 & 3 & -5 \\ 2 & 6 & 7 & -10 \\ 1 & 8 & 2 & 3 \end{array}\right]$$

Now use the elementary row operations to transform the matrix into an equivalent matrix in triangular form; that is, we want 0 to be the first entry in row 2 and the first two entries in row 3. First, we get 0 to be the first entry in row 2:

$$\left[\begin{array}{ccc|c} 1 & 2 & 3 & -5 \\ 2 & 6 & 7 & -10 \\ 1 & 8 & 2 & 3 \end{array}\right] \quad -2R_1 + R_2 \rightarrow R_2 \quad \left[\begin{array}{ccc|c} 1 & 2 & 3 & -5 \\ 0 & 2 & 1 & 0 \\ 1 & 8 & 2 & 3 \end{array}\right]$$

Multiply entries in row 1 by -2, add the result to row 2, and replace row 2 with this result.

Next, we get 0 as the first entry in row 3:

$$\left[\begin{array}{ccc|c} 1 & 2 & 3 & -5 \\ 0 & 2 & 1 & 0 \\ 1 & 8 & 2 & 3 \end{array}\right] \quad -R_1 + R_3 \rightarrow R_3 \quad \left[\begin{array}{ccc|c} 1 & 2 & 3 & -5 \\ 0 & 2 & 1 & 0 \\ 0 & 6 & -1 & 8 \end{array}\right]$$

Multiply row 1 by -1, add the result to row 3 to get the new row 3.

Finally, we get 0 as the second entry in row 3:

To get 0 as the second entry in row 8, why not multiply the entries in row 1 by -3, add the result to row 3, and replace row 3 with this result?

$$\left[\begin{array}{ccc|c} 1 & 2 & 3 & -5 \\ 0 & 2 & 1 & 0 \\ 0 & 6 & -1 & 8 \end{array}\right] \quad -3R_2 + R_3 \rightarrow R_3 \quad \left[\begin{array}{ccc|c} 1 & 2 & 3 & -5 \\ 0 & 2 & 1 & 0 \\ 0 & 0 & -4 & 8 \end{array}\right]$$

Multiply row 2 by -3, add the result to row 3 to get the new row 3.

This matrix is now in triangular form (zeros below the main diagonal). Now we write the associated system for the transformed matrix:

Remember that equivalent matrices produce equivalent associated systems.

$$\left[\begin{array}{ccc|c} 1 & 2 & 3 & -5 \\ 0 & 2 & 1 & 0 \\ 0 & 0 & -4 & 8 \end{array}\right] \Rightarrow \begin{cases} x + 2y + 3z = -5 \\ 2y + z = 0 \\ -4z = 8 \end{cases}$$

The third equation in this equivalent system yields $z = -2$. Now we find the rest of the variables by back substitution. We use the second equation to find y.

$$2y + z = 0 \qquad \textit{Substitute } z = -2 \textit{ and solve for } y.$$

$$2y + (-2) = 0 \qquad \textit{To get } y = 1.$$

We use the first equation to find x:

$$x + 2y + 3z = -5 \qquad \textit{Substitute } z = -2 \textit{ and } y = 1\textit{, and solve for } x.$$

$$x + 2(1) + 3(-2) = -5 \qquad \textit{To get } x = -1.$$

Hence the solution is $\boxed{x = -1, y = 1, \text{ and } z = -2}$, or $\boxed{(-1, 1, -2)}$. ■

EXAMPLE 2 Solve the following systems of equations:

(a) $\begin{cases} 2x - 6y = 8 \\ 3x - 9y = 12 \end{cases}$ **(b)** $\begin{cases} x + 2y + 3z = 2 \\ 2x + 4y + 6z = 8 \\ 2y - 5z = -7 \end{cases}$

Solution

(a) We first write its augmented matrix:

$$\begin{cases} 2x - 6y = 8 \\ 3x - 9y = 12 \end{cases} \Rightarrow \left[\begin{array}{cc|c} 2 & -6 & 8 \\ 3 & -9 & 12 \end{array}\right] \qquad \textit{and put it in triangular form.}$$

Getting 1 in the upper left-hand corner usually makes later computations easier.

$$\left[\begin{array}{cc|c} 2 & -6 & 8 \\ 3 & -9 & 12 \end{array}\right] \quad \tfrac{1}{2}R_1 \to R_1 \quad \left[\begin{array}{cc|c} 1 & -3 & 4 \\ 3 & -9 & 12 \end{array}\right]$$

Multiply row 1 by $\frac{1}{2}$ to get a new row 1.

Next we get 0 to be the first entry in row 2:

$$\left[\begin{array}{cc|c} 1 & -3 & 4 \\ 3 & -9 & 12 \end{array}\right] \quad -3R_1 + R_2 \to R_2 \quad \left[\begin{array}{cc|c} 1 & -3 & 4 \\ 0 & 0 & 0 \end{array}\right] \qquad \textit{Triangular form}$$

Multiply row 1 by -3, add the result to row 2 to get the new row 2.

We could have also arrived at triangular form in one step by applying the row operation $-\frac{3}{2}R_1 + R_2 \to R_2$.

This matrix is now in triangular form. If we write its associated system we get

$$\left[\begin{array}{cc|c} 1 & -3 & 4 \\ 0 & 0 & 0 \end{array}\right] \Rightarrow \begin{cases} x - 3y = 4 \\ 0 = 0 \end{cases}$$

The second equation in this system is always true, so our system consists of only one equation in two variables. Hence the system is dependent. Recall from Section 9.1, that we can put the dependent solution into ordered pair form as follows: If we let $y = a$ and solve for x, we get $x = 4 + 3a$. Hence, the solutions are all ordered pairs of the form $\boxed{(4 + 3a, a), \text{ where } a \text{ is any real number}}$.

Equivalently we could have let $x = a$, and the solution would have the form $\left(a, \dfrac{a-4}{3}\right)$

(b) First we write the system's augmented matrix:

$$\begin{cases} x + 2y + 3z = 2 \\ 2x + 4y + 6z = 8 \\ 2y - 5z = -7 \end{cases} \Rightarrow \left[\begin{array}{ccc|c} 1 & 2 & 3 & 2 \\ 2 & 4 & 6 & 8 \\ 0 & 2 & -5 & -7 \end{array}\right]$$

Now use the elementary row operations to transform the matrix into an equivalent matrix in triangular form. Get 0 as the first entry in row 2:

$$\left[\begin{array}{ccc|c} 1 & 2 & 3 & 2 \\ 2 & 4 & 6 & 8 \\ 0 & 2 & -5 & -7 \end{array}\right] \quad R_2 \leftrightarrow R3 \quad \left[\begin{array}{ccc|c} 1 & 2 & 3 & 2 \\ 0 & 2 & -5 & -7 \\ 2 & 4 & 6 & 8 \end{array}\right]$$

Switch rows 2 and 3

Next we need to get 0 as the first entry in row 3.

$$\left[\begin{array}{ccc|c} 1 & 2 & 3 & 2 \\ 0 & 2 & -5 & -7 \\ 2 & 4 & 6 & 8 \end{array}\right] \quad -2R_1 + R_3 \rightarrow R_3 \quad \left[\begin{array}{ccc|c} 1 & 2 & 3 & 2 \\ 0 & 2 & -5 & -7 \\ 0 & 0 & 0 & 4 \end{array}\right]$$

Multiply row 1 by −2, add the result to row 3 to get the new row 3.

We write the associated system of the matrix in triangular form.

$$\left[\begin{array}{ccc|c} 1 & 2 & 3 & 2 \\ 0 & 2 & -5 & -7 \\ 0 & 0 & 0 & 4 \end{array}\right] \Rightarrow \begin{cases} x + 2y + 3z = 2 \\ 2y - 5z = -7 \\ 0 = 4 \end{cases}$$

What does the row 0 0 0 4 in an augmented matrix mean for the associated system of equations?

The last equation in this system is a contradiction and therefore has no solution. Hence the *system* has no solution. ■

Another method of solving systems of linear equations using matrices is **Gauss-Jordan elimination**. As with Gaussian elimination, we use elementary row operations to transform a matrix into a row equivalent matrix, except rather than having triangular form as the final goal, Gauss-Jordan elimination requires that we put a system's associated matrix into **reduced echelon form**: A form where each entry on the main diagonal is 1 and all other entries in the coefficient matrix are 0.

$$\left[\begin{array}{ccc|c} 1 & 0 & 0 & a \\ 0 & 1 & 0 & b \\ 0 & 0 & 1 & c \end{array}\right] \quad \text{where } a,\ b, \text{ and } c \text{ are constants}$$

A matrix in this form has the associated system

$$\begin{cases} x \quad\quad = a \\ \quad y \quad = b \\ \quad\quad z = c \end{cases}$$

If we have a system's augmented matrix in this form, we can simply read the solution: $x = a$, $y = b$, and $z = c$. The following is a formal definition of reduced echelon form.

We mentioned echelon form for a system in the previous section. What would echelon form for a matrix look like, and how would it be different than reduced echelon form?

A matrix is in **reduced echelon form** if

1. The first nonzero entry in each row, called the pivot element, is 1.
2. The pivot element of each nonzero row occurs farther to the right than the pivot element in the preceding row.
3. In each column containing a pivot element, the remaining entries are 0.
4. Rows consisting of all zeros are at the bottom of the matrix.

The next example illustrates Gauss-Jordan elimination, or putting a system's augmented matrix into reduced echelon form.

EXAMPLE 3 Solve the following system using Gauss-Jordan elimination.

$$\begin{cases} 2x + 2y + 4z = 8 \\ 2x + 4y + 6z = 10 \\ x + 2y + 5z = 1 \end{cases}$$

Solution First we write the system's augmented matrix: $\left[\begin{array}{ccc|c} 2 & 2 & 4 & 8 \\ 2 & 4 & 6 & 10 \\ 1 & 2 & 5 & 1 \end{array}\right]$. Now we use the elementary row operations to transform the matrix into an equivalent matrix in reduced echelon form.

1. First we want to get a 1 in the upper left-hand corner. A quick way to do this is to interchange rows 1 and 3.

$$\left[\begin{array}{ccc|c} 2 & 2 & 4 & 8 \\ 2 & 4 & 6 & 10 \\ 1 & 2 & 5 & 1 \end{array}\right] \quad R_1 \leftrightarrow R_3 \quad \left[\begin{array}{ccc|c} 1 & 2 & 5 & 1 \\ 2 & 4 & 6 & 10 \\ 2 & 2 & 4 & 8 \end{array}\right]$$

2. Now we want 0's as the remaining entries of the first column. (We call this sweeping out the column.) We will take two steps at once.

$$\left[\begin{array}{ccc|c} 1 & 2 & 5 & 1 \\ 2 & 4 & 6 & 10 \\ 2 & 2 & 4 & 8 \end{array}\right] \quad \begin{array}{l} -2R_1 + R_2 \rightarrow R_2 \\ -2R_1 + R_3 \rightarrow R_3 \end{array} \quad \left[\begin{array}{ccc|c} 1 & 2 & 5 & 1 \\ 0 & 0 & -4 & 8 \\ 0 & -2 & -6 & 6 \end{array}\right]$$

3. Next we want to get 1 in the middle of the main diagonal (in row 2, column 2). First we interchange rows 2 and 3; then we divide row 2 by -2.

$$\left[\begin{array}{ccc|c} 1 & 2 & 5 & 1 \\ 0 & 0 & -4 & 8 \\ 0 & -2 & -6 & 6 \end{array}\right] \quad R_2 \leftrightarrow R_3 \quad \left[\begin{array}{ccc|c} 1 & 2 & 5 & 1 \\ 0 & -2 & -6 & 6 \\ 0 & 0 & -4 & 8 \end{array}\right]$$

$$\left[\begin{array}{ccc|c} 1 & 2 & 5 & 1 \\ 0 & -2 & -6 & 6 \\ 0 & 0 & -4 & 8 \end{array}\right] \quad -\tfrac{1}{2}R_2 \rightarrow R_2 \quad \left[\begin{array}{ccc|c} 1 & 2 & 5 & 1 \\ 0 & 1 & 3 & -3 \\ 0 & 0 & -4 & 8 \end{array}\right]$$

4. Now we want 0s in the remaining entries of the second column. (We sweep out the rest of the second column.)

$$\left[\begin{array}{ccc|c} 1 & 2 & 5 & 1 \\ 0 & 1 & 3 & -3 \\ 0 & 0 & -4 & 8 \end{array}\right] \quad -2R_2 + R_1 \rightarrow R_1 \quad \left[\begin{array}{ccc|c} 1 & 0 & -1 & 7 \\ 0 & 1 & 3 & -3 \\ 0 & 0 & -4 & 8 \end{array}\right]$$

5. Next, we want to get 1 in the lower right entry of the main diagonal.

$$\left[\begin{array}{ccc|c} 1 & 0 & -1 & 7 \\ 0 & 1 & 3 & -3 \\ 0 & 0 & -4 & 8 \end{array}\right] \quad -\tfrac{1}{4}R_3 \rightarrow R_3 \quad \left[\begin{array}{ccc|c} 1 & 0 & -1 & 7 \\ 0 & 1 & 3 & -3 \\ 0 & 0 & 1 & -2 \end{array}\right]$$

6. Finally, sweep out the third column.

$$\left[\begin{array}{ccc|c} 1 & 0 & -1 & 7 \\ 0 & 1 & 3 & -3 \\ 0 & 0 & 1 & -2 \end{array}\right] \quad \begin{array}{c} R_3 + R_1 \rightarrow R_1 \\ -3R_3 + R_2 \rightarrow R_2 \end{array} \quad \left[\begin{array}{ccc|c} 1 & 0 & 0 & 5 \\ 0 & 1 & 0 & 3 \\ 0 & 0 & 1 & -2 \end{array}\right]$$

This matrix is in reduced echelon form. We can simply read the solution (or write its associated system) to get $\boxed{x = 5,\ y = 3,\ z = -2.}$ ■

A matrix can have an infinite number of equivalent triangular forms, but its reduced echelon form is unique. This is one advantage of Gauss-Jordan elimination. Another advantage is that Gauss-Jordan elimination lends itself well to programming. However, putting a matrix into reduced echelon form can require tedious computations. If you are not required to use one particular approach, you may start out trying to put a matrix into reduced echelon form, but you may want to stop if you find that the matrix is in triangular form and you can use back substitution to find your solution. Actually, you can stop anytime you see that you can arrive at a solution fairly easily from the transformed matrix.

EXERCISES 9.3

In Exercises 1–4, set up the augmented matrix of the system of equations.

1. $\begin{cases} 7x + y = 11 \\ x - 3y = 4 \end{cases}$

2. $\begin{cases} 6x + 2y = -1 \\ 4x - y = 4 \end{cases}$

3. $\begin{cases} 3x - 2y + z = 15 \\ 2x + y - 3z = 10 \\ 5x - 3y + z = -2 \end{cases}$

4. $\begin{cases} 5x + 2y - z = -8 \\ 3x + y = -4 \\ 3y - 4z = 7 \end{cases}$

In Exercises 5–8, put the matrix into triangular form by following the given elementary row operations.

5. $\left[\begin{array}{cc|c} 1 & 3 & 2 \\ 5 & 4 & 0 \end{array}\right]$ Perform the row operation $-5R_1 + R_2 \rightarrow R_2$.

6. $\left[\begin{array}{cc|c} 5 & 4 & 0 \\ 1 & 3 & 2 \end{array}\right]$ First perform row operation $R_1 \leftrightarrow R_2$; then row operation $-5R_1 + R_2 \rightarrow R_2$ on the resultant matrix.

7. $\left[\begin{array}{ccc|c} 1 & 2 & 1 & 0 \\ 2 & 6 & 3 & 1 \\ 3 & 8 & 7 & -8 \end{array}\right]$ Perform the following row operations: $-2R_1 + R_2 \rightarrow R_2$ and $-3R_1 + R_3 \rightarrow R_3$; then perform row operation $-R_2 + R_3 \rightarrow R_3$ on the resultant matrix.

8. $\left[\begin{array}{ccc|c} 1 & 2 & 3 & 5 \\ 3 & 1 & 0 & 4 \\ -1 & 3 & 0 & 2 \end{array}\right]$ Perform the following row operations: First $-3R_1 + R_2 \rightarrow R_2$ and $R_1 + R_3 \rightarrow R_3$; then perform row operation $R_2 + R_3 \rightarrow R_3$ on the resultant matrix.

In Exercises 9–28, solve the system of equations by Gaussian elimination.

9. $\begin{cases} x - y = 4 \\ 2x + y = 11 \end{cases}$

10. $\begin{cases} x + 2y = -2 \\ x - y = 4 \end{cases}$

11. $\begin{cases} x + y = -1 \\ 3x - 2y = 12 \end{cases}$

12. $\begin{cases} 5x - 2y = 0 \\ x - 2y = 8 \end{cases}$

13. $\begin{cases} 5x - 2y = 1 \\ x - 5y = -32 \end{cases}$

14. $\begin{cases} x = 3y - 1 \\ 2x - 5y = -3 \end{cases}$

15. $\begin{cases} 8x - 2y = 10 \\ y = 3x - 15 \end{cases}$

16. $\begin{cases} x + 2y = -1 \\ 3x + 6y = 3 \end{cases}$

17. $\begin{cases} 2y + 3x = 2 \\ 3x + y = 1 \end{cases}$

18. $\begin{cases} 5x + 2y = -8 \\ 3x + 2y = -4 \end{cases}$

19. $\begin{cases} x + y + z = 6 \\ 2x + y - z = 4 \\ 3x + y - z = 6 \end{cases}$

20. $\begin{cases} x + y - z = 1 \\ 2x + 2y + 2z = 0 \\ x - y + z = 3 \end{cases}$

21. $\begin{cases} x + y - 2z = -5 \\ x - y + 2z = 9 \\ 3x + y - z = 2 \end{cases}$

22. $\begin{cases} x - 2y + z = 4 \\ x - y - 2z = -3 \\ 3x + 2y + z = 0 \end{cases}$

23. $\begin{cases} x + y + z = 6 \\ 3x + y - z = 4 \\ 2x + 2y + 2z = 4 \end{cases}$

24. $\begin{cases} x + y - z = 3 \\ 2x + 2y - 2z = 6 \\ x - y + z = 5 \end{cases}$

25. $\begin{cases} 2a - 2b + c = 0 \\ 4a - b + 2c = 6 \\ 2a + 3b - c = 4 \end{cases}$

26. $\begin{cases} x + 3y - z = 0 \\ 2x + y - z = 4 \\ 3x - y + 2z = 5 \end{cases}$

27. $\begin{cases} w + x + y - z = 4 \\ w + 2x + 2y + z = 3 \\ 2w + x + 3y - 2z = 7 \\ 3w + x - y + 2z = 5 \end{cases}$

28. $\begin{cases} w - x + y - z = 4 \\ w - 3y - z = 0 \\ w + 2x - 2z = 4 \\ -x + 2y - z = 3 \end{cases}$

In Exercises 29–30, put the given matrix into reduced echelon form by successively applying the given elementary row operations to the resultant matrix.

29. $\left[\begin{array}{cc|c} 3 & 6 & 0 \\ 3 & 1 & 5 \end{array}\right]$

(a) $\frac{1}{3}R_1 \to R_1$ **(b)** $-3R_1 + R_2 \to R_2$

(c) $-\frac{1}{5}R_2 \to R_2$ **(d)** $-2R_2 + R_1 \to R_1$

30. $\left[\begin{array}{ccc|c} 2 & 0 & 2 & -2 \\ 2 & 3 & 5 & 4 \\ 4 & 3 & 9 & 6 \end{array}\right]$

(a) $\frac{1}{2}R_1 \to R_1$

(b) $-2R_1 + R_2 \to R_2$ and $-4R_1 + R_3 \to R_3$

(c) $\frac{1}{3}R_2 \to R_2$

(d) $-3R_2 + R_3 \to R_3$

(e) $\frac{1}{2}R_3 \to R_3$

(f) $-R_3 + R_2 \to R_2$ and $-R_3 + R_1 \to R_1$

In Exercises 31–34, put the augmented matrix in reduced echelon form.

31. $\left[\begin{array}{cc|c} 1 & 4 & 0 \\ 2 & 3 & -5 \end{array}\right]$

32. $\left[\begin{array}{cc|c} 5 & 2 & 0 \\ 2 & 1 & 1 \end{array}\right]$

33. $\left[\begin{array}{ccc|c} 1 & 1 & 2 & 0 \\ 2 & 4 & 2 & 6 \\ 3 & 1 & 2 & 6 \end{array}\right]$

34. $\left[\begin{array}{ccc|c} 2 & 1 & 3 & 4 \\ 1 & 2 & 1 & 1 \\ 2 & 3 & 2 & 6 \end{array}\right]$

In Exercises 35–40, solve by the Gauss-Jordan method.

35. $\begin{cases} x - 2y = 4 \\ 3x + 5y = 1 \end{cases}$

36. $\begin{cases} x + 2y = 3 \\ 6x - y = 31 \end{cases}$

37. $\begin{cases} x + 2y + z = 8 \\ x + 4y - z = 12 \\ x - 2y + z = -4 \end{cases}$

38. $\begin{cases} x + z = 2 \\ 2x - 2y = -6 \\ 2y - z = 4 \end{cases}$

39. $\begin{cases} x + y + 2z = 1 \\ 2x + 4y + 2z = 6 \\ 3x + y + 2z = -4 \end{cases}$

40. $\begin{cases} w - x + y = -2 \\ x - z = -3 \\ -2x + 2y - z = -7 \\ w + 2y + z = -1 \end{cases}$

41. Find A, B, C, and D, such that

$$\frac{5x^3 + 3x^2 + 2}{(x^2 + 5)(x^2 + 4)} = \frac{Ax + B}{x^2 + 5} + \frac{Cx + D}{x^2 + 4}$$

for all values of x where the fraction is defined. HINT: See page 559.

42. Find A, B, C, and D, such that

$$\frac{3x^2 + 2x + 4}{(x^2 + 3)^2} = \frac{Ax + B}{x^2 + 3} + \frac{Cx + D}{(x^2 + 3)^2}$$

for all values of x where the fraction is defined. HINT: See page 559.

QUESTIONS FOR THOUGHT

43. In our approach using Gaussian elimination, our first goal was to put the system's augmented matrix into (upper) triangular form, a form where there are all zeros below the main diagonal. We could have just as easily solved the systems by putting the matrix into *lower* triangular form, a form where there are all zeros *above* the main diagonal, and using forward substitution. Solve Exercises 21 and 22 by putting the matrix into lower triangular form.

44. Discuss the differences among the following matrix forms: triangular form, echelon form, and reduced echelon form.

9.4 The Algebra of Matrices

In the previous section we defined and used matrices in developing methods for solving linear systems of equations. In this section we expand our discussion on matrices and discuss matrix operations.

Recall that a matrix is a rectangular array of numbers; the numbers in the matrix are called the *elements,* or *entries,* of the matrix. When we refer to a matrix, we usually refer to its size by specifying the number of rows by the number of columns. The following matrix is called an $m \times n$ matrix (m *rows* by n *columns*).

$$\begin{bmatrix} a_{11} & a_{12} & a_{13} & \cdots & a_{1n} \\ a_{21} & a_{22} & a_{23} & \cdots & a_{2n} \\ a_{31} & a_{32} & a_{33} & \cdots & a_{3n} \\ \vdots & \vdots & \vdots & & \vdots \\ a_{m1} & a_{m2} & a_{m3} & \cdots & a_{mn} \end{bmatrix}$$

Take a careful look at the subscripts of a, the entries of the above matrix. The subscripts indicate the position of the entry in the matrix. The first number indicates the row; the second indicates the column. For example a_{34} is the entry in the third row and fourth column. In general, a_{ij} is the entry in the ith row and jth column.

To designate matrices we will use either uppercase letters such as A, B, X, or Y or use the notation $[a_{ij}]$, whichever is more convenient. (*Remember:* Although a_{ij} refers to a single entry in a matrix, $[a_{ij}]$ refers to the whole matrix.)

When we introduced number systems, such as the real or complex numbers, we started out defining the elements of the system (what a real number or complex number looks like), we next defined operations on these elements, and then we examined the properties of the elements and their operations. We approach matrices in the same way. We have already defined what a matrix is; next we make sure these elements are well defined.

EQUIVALENCE OF MATRICES

If $A = [a_{ij}]$ and $B = [b_{ij}]$ are $m \times n$ matrices, then

$$A = B \qquad \text{if and only if} \qquad a_{ij} = b_{ij}$$

for i and j such that $1 \le i \le m$ and $1 \le j \le n$.

In words, two matrices are equal if and only if corresponding elements in each matrix are equal.

This requires, of course, that the two matrices have the same number of rows and columns. For example, $\begin{bmatrix} 2 & 5 \\ -3 & 8 \end{bmatrix}$ is equivalent to $\begin{bmatrix} 2 & \sqrt{25} \\ 2-5 & |-8| \end{bmatrix}$. Now we can define addition of matrices.

ADDITION OF MATRICES

If $A = [a_{ij}]$ and $B = [b_{ij}]$ are *both* $m \times n$ matrices, then

$$[a_{ij}] + [b_{ij}] = [a_{ij} + b_{ij}]$$

for i and j such that $1 \leq i \leq m$ and $1 \leq j \leq n$. This notation states that the sum of two matrices is found by adding their corresponding entries.

EXAMPLE 1 Find the following sum: $\begin{bmatrix} 3 & -2 & 0 & 1 \\ 2 & 8 & -1 & 4 \end{bmatrix} + \begin{bmatrix} 5 & 0 & -1 & 3 \\ -2 & 1 & 9 & 10 \end{bmatrix}$.

Solution First we observe that both matrices are the same size, 2×4; therefore, we can add them. (If they are not the same size, then addition is not defined.) To find their sum, all we do is add their corresponding elements.

$$\begin{bmatrix} 3 & -2 & 0 & 1 \\ 2 & 8 & -1 & 4 \end{bmatrix} + \begin{bmatrix} 5 & 0 & -1 & 3 \\ -2 & 1 & 9 & 10 \end{bmatrix}$$

$$= \begin{bmatrix} 3+5 & -2+0 & 0+(-1) & 1+3 \\ 2+(-2) & 8+1 & -1+9 & 4+10 \end{bmatrix}$$

$$= \begin{bmatrix} 8 & -2 & -1 & 4 \\ 0 & 9 & 8 & 14 \end{bmatrix}$$ ■

Using the definition of matrix addition, we can easily show that matrix addition, where it makes sense, is both commutative and associative.

As far as additive identities for matrices, keep in mind that matrices can be added only if they are the same size. So we must describe an additive identity for each class of $m \times n$ matrices. Let $\mathbf{O}$ be the $m \times n$ matrix consisting of all zero entries. Then the following can be proven.

THEOREM The $m \times n$ matrix $\mathbf{O}$, consisting of all zero entries, is the additive identity for all $m \times n$ matrices. That is, if A is any $m \times n$ matrix, then $A + \mathbf{O} = \mathbf{O} + A = A$.

For example, $\begin{bmatrix} 0 & 0 \\ 0 & 0 \end{bmatrix}$ is the $\mathbf{O}$ matrix for *all* 2×2 matrices. Hence

$$A + \mathbf{O} = \begin{bmatrix} -1 & 3 \\ 5 & -8 \end{bmatrix} + \begin{bmatrix} 0 & 0 \\ 0 & 0 \end{bmatrix} = \begin{bmatrix} -1+0 & 3+0 \\ 5+0 & -8+0 \end{bmatrix} = \begin{bmatrix} -1 & 3 \\ 5 & -8 \end{bmatrix} = A$$

The additive inverse of an $m \times n$ matrix A is the $m \times n$ matrix designated $-A$, consisting of entries that are opposite in sign to the entries of A. We summarize these properties as follows.

MATRIX ADDITION PROPERTIES

If A, B, and C are $m \times n$ matrices, then

1. $A + B = B + A$. **2.** $A + (B + C) = (A + B) + C$.

3. There exists a unique $m \times n$ matrix, called $\mathbf{O}$, such that $A + \mathbf{O} = \mathbf{O} + A = A$.

4. For each A, there exists a unique matrix, called $-A$, such that $A + (-A) = (-A) + A = \mathbf{O}$.

We define multiplying a matrix by a constant, which is called **scalar multiplication**, as follows.

SCALAR MULTIPLICATION

If $A = [a_{ij}]$ is an $m \times n$ matrix and c is a real number, then

$$cA = c[a_{ij}] = [ca_{ij}]$$

for i and j such that $1 \le i \le m$ and $1 \le j \le n$.

In words, multiplying a matrix by a constant yields a matrix in which *each* entry is multiplied by the constant.

For example,

Notice that the matrix $-A$ can be viewed as $(-1)A$, which is the matrix consisting of entries which are opposite in sign to the entries of A.

$$2\begin{bmatrix} 5 & 2 & 0 \\ -2 & 1 & -1 \\ 2 & -3 & 7 \end{bmatrix} = \begin{bmatrix} 2(5) & 2(2) & 2(0) \\ 2(-2) & 2(1) & 2(-1) \\ 2(2) & 2(-3) & 2(7) \end{bmatrix} = \begin{bmatrix} 10 & 4 & 0 \\ -4 & 2 & -2 \\ 4 & -6 & 14 \end{bmatrix}$$

We can similarly define Ac, where A is a matrix and c is a real number. We can easily show that scalar multiplication and addition have the following properties.

SCALAR MULTIPLICATION PROPERTIES

If A and B are $m \times n$ matrices and c and d are real numbers, then

1. $cA = Ac$ **2.** $(cd)A = c(dA)$

3. $(c + d)A = cA + dA$ **4.** $c(A + B) = cA + cB$

EXAMPLE 2 If $A = \begin{bmatrix} 3 & -2 \\ 0 & 1 \\ 2 & 8 \end{bmatrix}$ and $B = \begin{bmatrix} 7 & 0 \\ -1 & 3 \\ -2 & 1 \end{bmatrix}$, find $5A - B$.

Solution We make sure addition makes sense by noting that since A is a 3×2 matrix, $5A$ will also be a 3×2, which *can* be added to B, also a 3×2 matrix.

$$5A - B = 5A + (-1)B$$

$$= 5\begin{bmatrix} 3 & -2 \\ 0 & 1 \\ 2 & 8 \end{bmatrix} + (-1)\begin{bmatrix} 7 & 0 \\ -1 & 3 \\ -2 & 1 \end{bmatrix} \qquad \textit{Find } 5A \textit{ and } (-1)B.$$

$$= \begin{bmatrix} 15 & -10 \\ 0 & 5 \\ 10 & 40 \end{bmatrix} + \begin{bmatrix} -7 & 0 \\ 1 & -3 \\ 2 & -1 \end{bmatrix} = \begin{bmatrix} 8 & -10 \\ 1 & 2 \\ 12 & 39 \end{bmatrix}$$

■

EXAMPLE 3 If $A = \begin{bmatrix} -4 & 0 \\ 1 & 2 \end{bmatrix}$ and $B = \begin{bmatrix} 2 & -12 \\ 0 & 5 \end{bmatrix}$, find X such that $3X + A = B$.

Solution We begin by solving the equation $3X + A = B$ for X as we would any linear equation:

$$3X + A = B \qquad \textit{Solve for X to get}$$

$$X = \tfrac{1}{3}(B - A)$$

Since both A and B are 2×2 matrices, then their sum (and difference) is a 2×2 matrix and hence X is a 2×2 matrix.

$$X = \frac{1}{3}\left(\begin{bmatrix} 2 & -12 \\ 0 & 5 \end{bmatrix} - \begin{bmatrix} -4 & 0 \\ 1 & 2 \end{bmatrix}\right) = \frac{1}{3}\begin{bmatrix} 6 & -12 \\ -1 & 3 \end{bmatrix} \qquad \textit{Hence}$$

$$X = \begin{bmatrix} 2 & -4 \\ -\frac{1}{3} & 1 \end{bmatrix}$$

The student should check that the matrix X satisfies the equation $3X + A = B$. ■

Matrix Multiplication

The product of two matrices is not as easy to define as addition or scalar multiplication. In order to more easily describe the process of matrix multiplication, we first introduce a special operation called the *inner product*, defined for matrices with one row or matrices with one column. We call a matrix with one row a **row vector**, and a matrix with one column, a **column vector**.

The **inner product** of a $1 \times p$ row vector and a $p \times 1$ column vector is

$$[a_{11} \quad a_{12} \quad a_{13} \quad \cdots \quad a_{1p}] \bullet \begin{bmatrix} b_{11} \\ b_{21} \\ b_{31} \\ \vdots \\ b_{p1} \end{bmatrix} = a_{11}b_{11} + a_{12}b_{21} + a_{13}b_{31} + \cdots + a_{1p}b_{p1}$$

The inner product is sometimes called the dot product.

The inner product of a row vector and column vector is the sum of the products of the entries in the row vector with its corresponding entries in the column vector. For example, we can find the inner product

$$[3 \quad -2 \quad 0 \quad 4] \bullet \begin{bmatrix} 7 \\ -1 \\ 2 \\ 5 \end{bmatrix} = 3(7) + (-2)(-1) + 0(2) + 4(5) = 43.$$

It is important to note that (1) the inner product is a single number, and (2) we define inner products only for $1 \times p$ row vectors by $p \times 1$ column vectors, where the row vector is multiplied *left* of the column vector. Thus the *inner product* $\begin{bmatrix} 7 \\ -1 \\ 2 \\ 5 \end{bmatrix} \bullet [3 \quad -2 \quad 0 \quad 4]$ is not defined. Keep in mind that the symbol • used here is special notation used to indicate the inner product of a row vector by a column vector; it is not the usual dot symbol used to indicate numerical multiplications. Now we can define matrix multiplication.

MATRIX MULTIPLICATION

The product of A, an $m \times n$ matrix, and B, an $n \times p$ matrix, is the $m \times p$ matrix C, where the entry in the ith row and jth column is the inner product of the ith row vector of A with the jth column vector of B.

This definition requires that the number of columns of A, the matrix on the left, be equal to the number of rows of B, the matrix on the right. For example, if

$A = \begin{bmatrix} 6 & -2 & 8 \\ 1 & 4 & 5 \end{bmatrix}$, and $B = \begin{bmatrix} 1 & 3 \\ 5 & 0 \\ 2 & 7 \end{bmatrix}$, the product AB is found as follows:

$$\begin{bmatrix} 6 & -2 & 8 \\ 1 & 4 & 5 \end{bmatrix} \begin{bmatrix} 1 & 3 \\ 5 & 0 \\ 2 & 7 \end{bmatrix} \qquad 6(1) + (-2)5 + 8(2) = 12 \qquad \begin{bmatrix} 12 & \\ & \end{bmatrix}$$

The first entry (row 1, column 1) of the product AB is the inner product of row 1 of A and column 1 of B:

$$\begin{bmatrix} 6 & -2 & 8 \\ 1 & 4 & 5 \end{bmatrix} \begin{bmatrix} 1 & 3 \\ 5 & 0 \\ 2 & 7 \end{bmatrix} \qquad 6(3) + (-2)0 + 8(7) = 74 \qquad \begin{bmatrix} 12 & 74 \\ & \end{bmatrix}$$

The entry for row 1, column 2 of the product AB is the inner product of row 1 of A and column 2 of B:

$$\begin{bmatrix} 6 & -2 & 8 \\ 1 & 4 & 5 \end{bmatrix} \begin{bmatrix} 1 & 3 \\ 5 & 0 \\ 2 & 7 \end{bmatrix} \qquad \begin{bmatrix} 12 & 74 \\ 31 & \end{bmatrix}$$

The entry for row 2, column 1 of the product AB is the inner product of row 2 of A and column 1 of B:

$$1(1) + (4)5 + 5(2) = 31$$

$$\begin{bmatrix} 6 & -2 & 8 \\ 1 & 4 & 5 \end{bmatrix} \begin{bmatrix} 1 & 3 \\ 5 & 0 \\ 2 & 7 \end{bmatrix} \qquad \begin{bmatrix} 12 & 74 \\ 31 & 38 \end{bmatrix}$$

Finally, the entry in row 2, column 2 of the product AB is the inner product of row 2 of A and column 2 of B:

$$1(3) + 4(0) + 5(7) = 38$$

The product $AB = \begin{bmatrix} 6 & -2 & 8 \\ 1 & 4 & 5 \end{bmatrix} \begin{bmatrix} 1 & 3 \\ 5 & 0 \\ 2 & 7 \end{bmatrix} = \begin{bmatrix} 12 & 74 \\ 31 & 38 \end{bmatrix}$. Notice that we end up with a 2×2 matrix; the resulting matrix AB should have the same number of rows as A and the same number of columns as B. ■

On the other hand, we find a different result computing the matrix BA for the same A and B, as demonstrated in the next example.

EXAMPLE 4 If $A = \begin{bmatrix} 6 & -2 & 8 \\ 1 & 4 & 5 \end{bmatrix}$ and $B = \begin{bmatrix} 1 & 3 \\ 5 & 0 \\ 2 & 7 \end{bmatrix}$, find BA.

Solution

$$\begin{bmatrix} 1 & 3 \\ 5 & 0 \\ 2 & 7 \end{bmatrix} \begin{bmatrix} 6 & -2 & 8 \\ 1 & 4 & 5 \end{bmatrix} \qquad 1(6) + (3)1 = 9$$

The entry for row 1, column 1 of the product BA is the inner product of row 1 of B and column 1 of A:

$$\begin{bmatrix} 1 & 3 \\ 5 & 0 \\ 2 & 7 \end{bmatrix} \begin{bmatrix} 6 & -2 & 8 \\ 1 & 4 & 5 \end{bmatrix} \qquad \begin{bmatrix} 9 & & \\ & & 40 \\ & & \end{bmatrix}$$

The entry for row 2, column 3 of the product BA is the inner product of row 2 of B and column 3 of A:

$$5(8) + (0)5 = 40$$

The remaining entries are as follows:

$$BA = \begin{bmatrix} 1 & 3 \\ 5 & 0 \\ 2 & 7 \end{bmatrix} \begin{bmatrix} 6 & -2 & 8 \\ 1 & 4 & 5 \end{bmatrix} = \begin{bmatrix} 1(6) + 3(1) & 1(-2) + 3(4) & 1(8) + 3(5) \\ 5(6) + 0(1) & 5(-2) + 0(4) & 5(8) + 0(5) \\ 2(6) + 7(1) & 2(-2) + 7(4) & 2(8) + 7(5) \end{bmatrix}$$

$$= \begin{bmatrix} 9 & 10 & 23 \\ 30 & -10 & 40 \\ 19 & 24 & 51 \end{bmatrix}$$

Notice that BA is a 3×3 matrix (the number of rows of B by the number of columns of A). ■

As illustrated, if A is a 2×3 matrix and B is a 3×2 matrix, the product AB is a 2×2 matrix, but the product BA is a 3×3 matrix. In general we find that $AB \neq BA$; that is, matrix multiplication is not commutative. As a matter of fact, in many cases we may be able to compute a product of two matrices but not be able to compute the product of the same two matrices in reverse order. This is demonstrated in the next example.

EXAMPLE 5 Given $A = \begin{bmatrix} 1 & -2 \\ 3 & 1 \\ 0 & -1 \end{bmatrix}$ and $B = \begin{bmatrix} 2 & 5 & 3 & 0 \\ 4 & -1 & 2 & 1 \end{bmatrix}$.

(a) Find AB. **(b)** Find BA.

Solution

(a) AB means we multiply the 3×2 matrix A by the 2×4 matrix B, where A is left of B. We first check to see if the multiplication makes sense—that is, if the number of columns in the left matrix is equal to the number of rows in the right matrix (which is 2). Then we note that we should end up with a 3×4 matrix.

Matrix multiplication: Remember left-row vector by right-column vector.

$$\begin{bmatrix} 1 & -2 \\ 3 & 1 \\ 0 & -1 \end{bmatrix} \begin{bmatrix} 2 & 5 & 3 & 0 \\ 4 & -1 & 2 & 1 \end{bmatrix} \quad 1(3) + (-2)2 \quad = \begin{bmatrix} -6 & 7 & -1 & -2 \\ 10 & 14 & 11 & 1 \\ -4 & 1 & -2 & -1 \end{bmatrix}$$

Row 1 vector · column 3 vector = entry in row 1, column 3 of the product AB

$$\text{Hence, } AB = \begin{bmatrix} -6 & 7 & -1 & -2 \\ 10 & 14 & 11 & 1 \\ -4 & 1 & -2 & -1 \end{bmatrix}.$$

Matrix multiplication:

$$(A)(B) = C$$

$(m \times n)(n \times p)(m \times p)$

same

(b) BA means we multiply the 2×4 matrix B by the 3×2 matrix A, where B is left of A. We first check to see if the multiplication makes sense—that is, if the number of columns in the left matrix (which is 4) is equal to the number of rows in the right matrix (which is 3). Since they are not equal, the product BA is not defined. ■

Although matrix multiplication is not commutative, matrix multiplication is associative and does distribute over addition. The following box summarizes some useful properties:

MATRIX MULTIPLICATION PROPERTIES

Suppose A, B, and C are matrices (where multiplication makes sense) and k is a real number. Then

1. $A(BC) = (AB)C$
2. $A(B + C) = AB + AC$
3. $(B + C)A = BA + CA$
4. $(cA)B = c(AB)$

Matrices can be used in applications where we have to perform repetitive operations on groups or tables of numbers, as illustrated next.

EXAMPLE 6 Over a 6-month period, the Mathematics and English departments use office supplies as illustrated by the accompanying table.

This type of problem is ideal for a spreadsheet program.

Supplies	English Department	Mathematics Department
Cases of legal pads	8	6
Cases of duplicating paper	12	18
Gross of pencils	6	3

Tom's Office Supply Company charges \$230 for a case of legal pads, \$65 for a case of duplicating paper, and \$18 for a gross of pencils. Kuma's Office Supply Company charges \$250 for a case of legal pads, \$56 for a case of duplicating paper, and \$16 for a gross of pencils. If supplies must be ordered from one company, which office supply company is the least expensive for each department's 6-month order?

There is some ambiguity left in the problem. Must we order both math and English supplies from one supplier? Or can separate departments order from different suppliers? We will explore all possibilities.

Solution It is convenient to represent the department order and costs as matrices. The matrix representing the order from each department can be written as $\begin{bmatrix} 8 & 6 \\ 12 & 18 \\ 6 & 3 \end{bmatrix}$. We can represent the cost matrix for Tom's Office Supplies as $[\$230 \quad \$65 \quad \$18]$ and Kuma's Office Supplies as $[\$250 \quad \$56 \quad \$16]$.

To find the cost of the order for each department, we multiply the cost matrix by the order matrix:

The cost from Tom's Office Supplies is

$$\begin{array}{ccc} \text{Legal Pads} & \text{Paper} & \text{Pencils} \\ [\$230 & \$65 & \$18] \end{array} \begin{array}{c} \begin{array}{cc} \text{English} & \text{Math} \end{array} \\ \begin{bmatrix} 8 & 6 \\ 12 & 18 \\ 6 & 3 \end{bmatrix} \end{array} = \begin{array}{cc} \text{Total For English} & \text{Total For Math} \\ [\$2728 & \$2604] \end{array}$$

The cost from Kuma's Office Supplies is

$$[\$250 \quad \$56 \quad \$16] \begin{bmatrix} 8 & 6 \\ 12 & 18 \\ 6 & 3 \end{bmatrix} = [\$2768 \quad \$2556]$$

If the departments order separately, for the lowest cost the English department should order from Tom, but the mathematics department should order from Kuma. By ordering separately from their least expensive vendor, they would spend altogether:

$$\underbrace{\$2728}_{\text{English department from Tom}} + \underbrace{\$2556}_{\text{Math department from Kuma}} = \$5284$$

If both departments had to combine their orders, then the total cost from Tom's Office Supplies would be $2728 + $2604 = $5332; the total cost from Kuma's Office Supplies would be $2768 + $2556 = $5324. Hence, a combined order should be made from Kuma's Office Supplies. Notice that combining orders would cost more altogether than each department ordering separately from its "best"-priced supplier. ■

EXERCISES 9.4

In Exercises 1–8, perform the operations if possible.

1. $\begin{bmatrix} 1 & -3 \\ 5 & 2 \end{bmatrix} + \begin{bmatrix} 2 & 5 \\ 3 & 0 \end{bmatrix}$

2. $\begin{bmatrix} 2 & 9 \\ 15 & -23 \end{bmatrix} + \begin{bmatrix} -14 & 6 \\ 0 & -20 \end{bmatrix}$

3. $\begin{bmatrix} 1 & 3 \\ -5 & 0 \\ 2 & -7 \end{bmatrix} + \begin{bmatrix} -1 & -3 \\ 5 & 0 \\ -2 & 7 \end{bmatrix}$

4. $\begin{bmatrix} 2 & -3 & 8 \\ 5 & -1 & 0 \\ -2 & 7 & 18 \end{bmatrix} + \begin{bmatrix} -1 & -3 \\ 5 & 0 \\ -2 & 7 \end{bmatrix}$

5. $\begin{bmatrix} 0 & 3 & 6 \\ 80 & 0 & -19 \\ 2 & 12 & 4 \end{bmatrix} + \begin{bmatrix} 2 & 3 \\ 5 & 0 \end{bmatrix}$

6. $\begin{bmatrix} 2 & -3 & 8 \\ 5 & -1 & 0 \\ -2 & 7 & 18 \end{bmatrix} + \begin{bmatrix} 0 & 0 & 0 \\ 0 & 0 & 0 \\ 0 & 0 & 0 \end{bmatrix}$

7. $3\begin{bmatrix} 0 & 3 \\ 17 & 0 \end{bmatrix}$

8. $-4\begin{bmatrix} 2 & -3 & 8 \\ 5 & -1 & 0 \\ -2 & 7 & 18 \end{bmatrix}$

In Exercises 9–16, compute the matrix, if possible, given matrices A, B, C, and D.

$$A = \begin{bmatrix} 2 & -3 \\ -1 & 0 \end{bmatrix} \qquad B = \begin{bmatrix} 5 & 1 \\ -2 & 3 \end{bmatrix}$$

$$C = \begin{bmatrix} 1 & -1 & 3 \\ -2 & 3 & 0 \\ 5 & 8 & 2 \end{bmatrix} \qquad D = \begin{bmatrix} 3 & 1 & 0 \\ -2 & 4 & 6 \\ 3 & 8 & -10 \end{bmatrix}$$

9. $-A$

10. $2C$

11. $\frac{2}{3}B$

12. $-3D$

13. $A - B$

14. $B - A$

15. $2A - B$

16. $C - 3D$

In Exercises 17–20, solve for X, if possible, given the matrices A, B, C, and D defined above.

17. $2X = B$

18. $3X = B + C$

19. $2X - B = 2A$

20. $2(4D - X) = C$

In Exercises 21–36, compute the matrix products, if possible.

21. $\begin{bmatrix} 3 & -3 & 1 \end{bmatrix} \begin{bmatrix} 5 \\ 1 \\ -2 \end{bmatrix}$

22. $\begin{bmatrix} 5 & 0 & -1 \end{bmatrix} \begin{bmatrix} 3 \\ -9 \\ 0 \end{bmatrix}$

23. $\begin{bmatrix} 0 & -1 & 1 & 2 \end{bmatrix} \begin{bmatrix} 1 \\ -1 \\ -2 \\ 2 \end{bmatrix}$

24. $\begin{bmatrix} 1 \\ -1 \\ -2 \\ 2 \end{bmatrix} \begin{bmatrix} 3 & 2 & 5 & 1 \\ 0 & -1 & 1 & 2 \end{bmatrix}$

25. $\begin{bmatrix} 0 & -1 \\ 3 & 2 \end{bmatrix} \begin{bmatrix} -1 & 3 \\ 0 & -2 \end{bmatrix}$

26. $\begin{bmatrix} 1 & 0 \\ 0 & 1 \end{bmatrix} \begin{bmatrix} 3 & 5 \\ -4 & 6 \end{bmatrix}$

27. $\begin{bmatrix} 3 & -1 & 4 \\ 3 & -2 & 5 \end{bmatrix} \begin{bmatrix} -1 & 3 & 5 \\ 0 & -2 & 1 \end{bmatrix}$

28. $\begin{bmatrix} 1 & 0 & -3 \\ 2 & 1 & -4 \end{bmatrix} \begin{bmatrix} 3 & 5 \\ -4 & 6 \\ 3 & -1 \end{bmatrix}$

29. $\begin{bmatrix} 0 & 1 & -5 \\ -3 & 1 & 6 \end{bmatrix} \begin{bmatrix} -1 & 3 \\ 0 & -2 \\ 6 & 0 \end{bmatrix}$

30. $\begin{bmatrix} -1 & 3 \\ 0 & -2 \\ 6 & 0 \end{bmatrix} \begin{bmatrix} 0 & 1 & -5 \\ -3 & 1 & 6 \end{bmatrix}$

31. $\begin{bmatrix} 1 & 0 & -2 \\ 3 & 1 & -1 \\ 2 & 0 & 1 \end{bmatrix} \begin{bmatrix} 1 & -3 & 0 \\ 0 & 2 & -2 \\ -6 & 0 & 5 \end{bmatrix}$

32. $\begin{bmatrix} 1 & -3 & 0 \\ 0 & 2 & -2 \\ -6 & 0 & 5 \end{bmatrix} \begin{bmatrix} 1 & 0 & -2 \\ 3 & 1 & -1 \\ 2 & 0 & 1 \end{bmatrix}$

33. $\begin{bmatrix} 2 & 0 & -2 & 1 \\ -2 & 3 & -1 & 1 \end{bmatrix} \begin{bmatrix} 1 & 3 \\ 1 & 0 \\ -1 & 5 \\ 1 & -3 \end{bmatrix}$

34. $\begin{bmatrix} 1 & 3 \\ 1 & 0 \\ -1 & 5 \\ 1 & -3 \end{bmatrix} \begin{bmatrix} 2 & 0 & -2 & 1 \\ -2 & 3 & -1 & 1 \end{bmatrix}$

35. $\begin{bmatrix} 1 & 0 & 0 \\ 0 & 1 & 0 \\ 0 & 0 & 1 \end{bmatrix} \begin{bmatrix} 3 & 1 & 0 \\ -2 & 4 & 6 \\ 3 & 8 & -10 \end{bmatrix}$

36. $\begin{bmatrix} 3 & 1 & 0 \\ -2 & 4 & 6 \\ 3 & 8 & -10 \end{bmatrix} \begin{bmatrix} 1 & 0 & 0 \\ 0 & 1 & 0 \\ 0 & 0 & 1 \end{bmatrix}$

In Exercises 37–40, given matrices A, B, C, and D, demonstrate whether the given statement is true or false.

$$A = \begin{bmatrix} 2 & 1 & 0 \\ 3 & 1 & -2 \end{bmatrix} \qquad B = \begin{bmatrix} 3 & 2 \\ -5 & 0 \end{bmatrix}$$

$$C = \begin{bmatrix} 2 & -2 \\ 4 & 1 \end{bmatrix} \qquad D = \begin{bmatrix} 5 & -1 \\ 0 & -2 \end{bmatrix}$$

37. $CB = BC$

38. $A(B + C) = AB + AC$

39. $(B + C)A = BA + CA$

40. $D(BC) = (DB)C$

41. Show that

$$\left(\frac{1}{2}\begin{bmatrix} 3 & 5 \\ 2 & 4 \end{bmatrix}\right)\begin{bmatrix} 0 & 2 \\ 1 & -1 \end{bmatrix} = \frac{1}{2}\left(\begin{bmatrix} 3 & 5 \\ 2 & 4 \end{bmatrix}\begin{bmatrix} 0 & 2 \\ 1 & -1 \end{bmatrix}\right).$$

42. If A is a 3×4 matrix and B is a 4×3 matrix, what size is the matrix AB? What size is the matrix BA?

43. If A is a 5×4 matrix and B is a 3×5 matrix, what size is the matrix AB? What size is the matrix BA?

44. When does it make sense to discuss A^2 if A is a matrix?

45. Prove that if A and B are both 3×3 matrices, then $A + B = B + A$.

46. Prove that if A, B, and C are all 2×2 matrices, then $A + (B + C) = (A + B) + C$.

47. If an additive inverse exists, does each $n \times n$ matrix A have an additive inverse? (In other words, for each matrix A, does there exist a matrix B such that $A + B = B + A = \mathbf{O}$?) If so, explain what it should look like.

48. Prove that if A is an $m \times n$ matrix and c and d are real numbers, then $(c + d)A = cA + dA$.

49. Prove that if A and B are $m \times n$ matrices and c is a real number, then $c(A + B) = cA + cB$.

50. Over a 6-month period, the political science and the sociology departments use office supplies as illustrated by the accompanying table.

Supplies	Sociology Department	Political Science Department
Cases of legal pads	7	4
Cases of duplicating paper	6	10
Gross of pencils	2	9

Tom's Office Supply Company charges \$230 for a case of legal pads, \$65 for a case of duplicating paper, and \$18 for a gross of pencils. Kuma's Office Supply Company charges \$250 for a case of legal pads, \$56 for a case of duplicating paper, and \$16 for a gross of pencils. If supplies are ordered for each department, which office supply company is the least expensive for each 6-month order?

51. Three scientific supply companies each manufacture their own brand of pneumatic articulators: brand A, brand B, and brand C. Each brand contains a certain amount of plastic, rubber, and glass tubing, as illustrated by the accompanying table.

Type of tubing required	Brand A	Brand B	Brand C
Plastic tubing	12 inches	18 inches	14 inches
Rubber tubing	14 inches	6 inches	12 inches
Glass tubing	18 inches	12 inches	5 inches

Only two companies in the country manufacture tubing suitable for the pneumatic articulators: George's Tubing Company and Raju's Tubing Company. George's Tubing Company charges 1¢ per inch for plastic tubing, 1.5¢ per inch for rubber tubing, and 1.8¢ per inch for glass tubing. Raju's Tubing Company charges 1.2¢ per inch for plastic tubing, 1.4¢ per inch for rubber tubing, and 1.6¢ per inch for glass tubing. If tubing supplies are ordered for all three brands, which tubing company offers the least expensive tubing for each brand?

52. Evaluate the inner product $\begin{bmatrix} 2 & -3 & 1 \end{bmatrix} \cdot \begin{bmatrix} 1 \\ 0 \\ 4 \end{bmatrix}$ and the matrix product $\begin{bmatrix} 2 & -3 & 1 \end{bmatrix} \begin{bmatrix} 1 \\ 0 \\ 4 \end{bmatrix}$. What are the differences between the two?

53. Evaluate the inner product $\begin{bmatrix} 1 \\ 0 \\ 4 \end{bmatrix} \cdot \begin{bmatrix} 2 & -3 & 1 \end{bmatrix}$ and the matrix product $\begin{bmatrix} 1 \\ 0 \\ 4 \end{bmatrix} \begin{bmatrix} 2 & -3 & 1 \end{bmatrix}$. What are the differences between the two?

QUESTION FOR THOUGHT

54. If A is a matrix and c is a real number, how should Ac be defined such that $Ac = cA$?

9.5 Solving Linear Systems Using Matrix Inverses

In this section we examine another method for solving systems of linear equations. Using matrices, we approach and solve a linear system as though it were a simple linear equation. The process of solving simple linear equations requires the application of real number properties. In order to approach matrices in the same way, we need a bit more information regarding the algebra of matrices.

In Section 9.4 we mentioned that matrix addition, where it makes sense, is commutative and associative and that each matrix has an additive inverse; we pointed out that matrix multiplication, where it makes sense, is generally not commutative, but it is associative and does distribute over addition. Now we will discuss multiplicative identities and inverses.

Multiplicative Identities and Inverses

For all real numbers, the multiplicative identity is 1; that is, if we multiply any real number by 1, it remains unchanged. Is there a matrix that will act as an identity for matrix multiplication? That is, can we find the matrix I such that $AI = IA = A$ for *all* matrices A? First, we note that multiplication must make sense; that is, AI may not be defined if the number of columns of A is not equal to the number of rows of I. For simplicity, we will restrict our discussion to *square* matrices and define the following:

DEFINITION I_n is the $n \times n$ matrix with 1's in the main diagonal and 0's elsewhere.

For example, $I_4 = \begin{bmatrix} 1 & 0 & 0 & 0 \\ 0 & 1 & 0 & 0 \\ 0 & 0 & 1 & 0 \\ 0 & 0 & 0 & 1 \end{bmatrix}$. The following theorem can be proven:

THEOREM I_n is the identity matrix for all $n \times n$ matrices.
I_n is called the **multiplicative identity of order *n***. Hence, if A is an $n \times n$ matrix, then $AI_n = I_nA = A$.

In particular, I_4 is the identity matrix for all 4×4 matrices.

EXAMPLE 1 Prove that if A is a 2×2 matrix, then $AI_2 = I_2A = A$.

Solution The identity matrix of order 2 is $I_2 = \begin{bmatrix} 1 & 0 \\ 0 & 1 \end{bmatrix}$. If we let $A = \begin{bmatrix} a_{11} & a_{12} \\ a_{21} & a_{22} \end{bmatrix}$, then

$$AI_2 = \begin{bmatrix} a_{11} & a_{12} \\ a_{21} & a_{22} \end{bmatrix} \begin{bmatrix} 1 & 0 \\ 0 & 1 \end{bmatrix} = \begin{bmatrix} (a_{11})(1) + (a_{12})(0) & (a_{11})(0) + (a_{12})(1) \\ (a_{21})(1) + (a_{22})(0) & (a_{21})(0) + (a_{22})(1) \end{bmatrix}$$

$$= \begin{bmatrix} a_{11} & a_{12} \\ a_{21} & a_{22} \end{bmatrix} = A.$$

On the other hand,

$$I_2A = \begin{bmatrix} 1 & 0 \\ 0 & 1 \end{bmatrix} \begin{bmatrix} a_{11} & a_{12} \\ a_{21} & a_{22} \end{bmatrix} = \begin{bmatrix} (1)(a_{11}) + (0)(a_{21}) & (1)(a_{12}) + (0)(a_{22}) \\ (0)(a_{11}) + (1)(a_{21}) & (0)(a_{12}) + (1)(a_{22}) \end{bmatrix}$$

$$= \begin{bmatrix} a_{11} & a_{12} \\ a_{21} & a_{22} \end{bmatrix} = A.$$ ■

Now that we have defined and demonstrated the existence of multiplicative identities for square matrices, our next step is to determine what multiplicative inverses look like if they do exist. In other words, for each $n \times n$ matrix A, does there exist a matrix A^{-1} such that $AA^{-1} = A^{-1}A = I_n$? We call A^{-1} the inverse of the matrix A.

Throughout this section, whenever we refer to the inverse of a matrix A, we mean its *multiplicative* inverse A^{-1}.

It would be nice if we would find that A^{-1} is simply some sort of reciprocal of the elements of A; unfortunately, because of the complex way matrix multiplication is defined, finding the inverse of a matrix takes a bit more work. A discussion of the derivation of matrix inverses is beyond the scope of this text. We will, however, demonstrate how to find inverses of square matrices in the next few examples.

EXAMPLE 2 Find the inverse of the matrix $A = \begin{bmatrix} 1 & 1 & 2 \\ 1 & 2 & 4 \\ 3 & 4 & 10 \end{bmatrix}$.

Solution

1. We are looking for the inverse, A^{-1}, of this 3×3 matrix. We first form a new matrix in the following way:

We put the 3×3 matrix A together with the identity matrix of order 3, I_3, where the elements of I_3 are right of the elements of A to form a **grafted** 3×6 matrix $[A \mid I]$:

$$\left[\begin{array}{ccc|ccc} 1 & 1 & 2 & 1 & 0 & 0 \\ 1 & 2 & 4 & 0 & 1 & 0 \\ 3 & 4 & 10 & 0 & 0 & 1 \end{array}\right]$$

This is known as **grafting** I_3 *onto* a 3×3 *matrix.*

2. Our goal is to try to transform the left-hand square matrix (the A portion of the grafted matrix) into the identity matrix I_3. We use the elementary row operations discussed in Section 9.3 as we did when we transformed a matrix into reduced echelon form. *All transformations performed on the A portion are applied to the entire* 3×6 *grafted matrix.*

We already have 1 in the upper left-hand corner of the main diagonal of A, so our first step is to sweep out (get zeros in the remaining entries of) the first column.

$$\left[\begin{array}{ccc|ccc} 1 & 1 & 2 & 1 & 0 & 0 \\ 1 & 2 & 4 & 0 & 1 & 0 \\ 3 & 4 & 10 & 0 & 0 & 1 \end{array}\right] \begin{array}{c} \\ -R_1 + R_2 \rightarrow R_2 \\ -3R_1 + R_3 \rightarrow R_3 \end{array} \left[\begin{array}{ccc|ccc} 1 & 1 & 2 & 1 & 0 & 0 \\ 0 & 1 & 2 & -1 & 1 & 0 \\ 0 & 1 & 4 & -3 & 0 & 1 \end{array}\right]$$

We already have 1 in the second row, second column; now we sweep out the second column.

$$\left[\begin{array}{ccc|ccc} 1 & 1 & 2 & 1 & 0 & 0 \\ 0 & 1 & 2 & -1 & 1 & 0 \\ 0 & 1 & 4 & -3 & 0 & 1 \end{array}\right] \begin{array}{c} -R_2 + R_1 \rightarrow R_1 \\ \\ -R_2 + R_3 \rightarrow R_3 \end{array} \left[\begin{array}{ccc|ccc} 1 & 0 & 0 & 2 & -1 & 0 \\ 0 & 1 & 2 & -1 & 1 & 0 \\ 0 & 0 & 2 & -2 & -1 & 1 \end{array}\right]$$

Next get 1 in row 3, column 3.

$$\left[\begin{array}{ccc|ccc} 1 & 0 & 0 & 2 & -1 & 0 \\ 0 & 1 & 2 & -1 & 1 & 0 \\ 0 & 0 & 2 & -2 & -1 & 1 \end{array}\right] \quad \tfrac{1}{2}R_3 \to R_3 \quad \left[\begin{array}{ccc|ccc} 1 & 0 & 0 & 2 & -1 & 0 \\ 0 & 1 & 2 & -1 & 1 & 0 \\ 0 & 0 & 1 & -1 & -\frac{1}{2} & \frac{1}{2} \end{array}\right]$$

Finally, sweep out the third column.

$$\left[\begin{array}{ccc|ccc} 1 & 0 & 0 & 2 & -1 & 0 \\ 0 & 1 & 2 & -1 & 1 & 0 \\ 0 & 0 & 1 & -1 & -\frac{1}{2} & \frac{1}{2} \end{array}\right] \; -2R_3 + R_2 \to R_2 \; \left[\begin{array}{ccc|ccc} 1 & 0 & 0 & 2 & -1 & 0 \\ 0 & 1 & 0 & 1 & 2 & -1 \\ 0 & 0 & 1 & -1 & -\frac{1}{2} & \frac{1}{2} \end{array}\right]$$

3. The right-hand 3×3 matrix in the grafted matrix is A^{-1}.

$$A^{-1} = \begin{bmatrix} 2 & -1 & 0 \\ 1 & 2 & -1 \\ -1 & -\frac{1}{2} & \frac{1}{2} \end{bmatrix}$$

We will verify that $AA^{-1} = I_3$:

$$AA^{-1} = \begin{bmatrix} 1 & 1 & 2 \\ 1 & 2 & 4 \\ 3 & 4 & 10 \end{bmatrix} \begin{bmatrix} 2 & -1 & 0 \\ 1 & 2 & -1 \\ -1 & -\frac{1}{2} & \frac{1}{2} \end{bmatrix}$$

$$= \begin{bmatrix} (1)(2) + (1)(1) + (2)(-1) & (1)(-1) + (1)(2) + (2)(-\frac{1}{2}) & (1)(0) + (1)(-1) + (2)(\frac{1}{2}) \\ (1)(2) + (2)(1) + (4)(-1) & (1)(-1) + (2)(2) + (4)(-\frac{1}{2}) & (1)(0) + (2)(-1) + (4)(\frac{1}{2}) \\ (3)(2) + (4)(1) + (10)(-1) & (3)(-1) + (4)(2) + (10)(-\frac{1}{2}) & (3)(0) + (4)(-1) + (10)(\frac{1}{2}) \end{bmatrix}$$

$$= \begin{bmatrix} 1 & 0 & 0 \\ 0 & 1 & 0 \\ 0 & 0 & 1 \end{bmatrix} = I_3$$

We leave it to the student to verify $A^{-1}A$ is also equal to I_3. ■

In general, if A is an $n \times n$ matrix, to find A^{-1},

1. Graft I_n onto A to get $[A \mid I_n]$.
2. Using the elementary row operations on the entire matrix, transform the A portion of the grafted matrix into I_n.
3. When the left-hand portion of the grafted matrix is I_n, the right-hand $n \times n$ matrix is A^{-1}; that is; if $[A \mid I_n]$ is row equivalent to $[I_n \mid B]$, then B is the inverse of A.

EXAMPLE 3 If $A = \begin{bmatrix} 2 & 4 \\ 3 & 1 \end{bmatrix}$, find A^{-1}.

Solution

1. Graft I_2 onto A: $\left[\begin{array}{cc|cc} 2 & 4 & 1 & 0 \\ 3 & 1 & 0 & 1 \end{array}\right]$.

2. Try to transform the left-hand square matrix (the A portion of the grafted matrix) into the identity matrix, I_2, using the elementary row operations applied to the entire grafted matrix.

$$\left[\begin{array}{cc|cc} 2 & 4 & 1 & 0 \\ 3 & 1 & 0 & 1 \end{array}\right] \quad \tfrac{1}{2}R_1 \to R_1 \quad \left[\begin{array}{cc|cc} 1 & 2 & \frac{1}{2} & 0 \\ 3 & 1 & 0 & 1 \end{array}\right]$$

$$\left[\begin{array}{cc|cc} 1 & 2 & \frac{1}{2} & 0 \\ 3 & 1 & 0 & 1 \end{array}\right] \quad -3R_1 + R_2 \to R_2 \quad \left[\begin{array}{cc|cc} 1 & 2 & \frac{1}{2} & 0 \\ 0 & -5 & -\frac{3}{2} & 1 \end{array}\right]$$

$$\left[\begin{array}{cc|cc} 1 & 2 & \frac{1}{2} & 0 \\ 0 & -5 & -\frac{3}{2} & 1 \end{array}\right] \quad -\tfrac{1}{5}R_2 \to R_2 \quad \left[\begin{array}{cc|cc} 1 & 2 & \frac{1}{2} & 0 \\ 0 & 1 & \frac{3}{10} & -\frac{1}{5} \end{array}\right]$$

$$\left[\begin{array}{cc|cc} 1 & 2 & \frac{1}{2} & 0 \\ 0 & 1 & \frac{3}{10} & -\frac{1}{5} \end{array}\right] \quad -2R_2 + R_1 \to R_1 \quad \left[\begin{array}{cc|cc} 1 & 0 & -\frac{1}{10} & \frac{2}{5} \\ 0 & 1 & \frac{3}{10} & -\frac{1}{5} \end{array}\right]$$

3. Since the left-hand portion is now I_2, the right-hand portion is A^{-1}. Hence

$$A^{-1} = \begin{bmatrix} -\frac{1}{10} & \frac{2}{5} \\ \frac{3}{10} & -\frac{1}{5} \end{bmatrix} \quad \text{which can be written as} \quad \frac{1}{10}\begin{bmatrix} -1 & 4 \\ 3 & -2 \end{bmatrix}.$$

We factor the scalar $\frac{1}{10}$ out in order to simplify computations performed with this matrix. For example, we next demonstrate that $A^{-1}A = I_2$.

$$A^{-1}A = \left(\frac{1}{10}\begin{bmatrix} -1 & 4 \\ 3 & -2 \end{bmatrix}\right)\begin{bmatrix} 2 & 4 \\ 3 & 1 \end{bmatrix}$$

$(cA)B = c(AB)$ where c is a scalar.

$$= \frac{1}{10}\left(\begin{bmatrix} -1 & 4 \\ 3 & -2 \end{bmatrix}\begin{bmatrix} 2 & 4 \\ 3 & 1 \end{bmatrix}\right)$$

It is easier to multiply the matrices first.

$$= \frac{1}{10}\begin{bmatrix} (-1)(2) + (4)(3) & (-1)(4) + (4)(1) \\ (3)(2) + (-2)(3) & (3)(4) + (-2)(1) \end{bmatrix}$$

Then we multiply the scalar.

$$= \frac{1}{10}\begin{bmatrix} 10 & 0 \\ 0 & 10 \end{bmatrix} = \begin{bmatrix} 1 & 0 \\ 0 & 1 \end{bmatrix} = I_2$$

We leave it to the student to verify $AA^{-1} = I_2$. ■

We have yet to answer the question, Does every square matrix have an inverse? (In the system of real numbers we know that a^{-1} does not exist for $a = 0$.) Let's consider the square matrix $A = \begin{bmatrix} 1 & 3 \\ -2 & -6 \end{bmatrix}$. To try to find A^{-1} we proceed as follows:

1. Graft I_2 onto the matrix: $\left[\begin{array}{cc|cc} 1 & 3 & 1 & 0 \\ -2 & -6 & 0 & 1 \end{array}\right]$.

2. Then try to transform the left-hand square matrix into the identity matrix using the elementary row operations.

$$\left[\begin{array}{rr|rr} 1 & 3 & 1 & 0 \\ -2 & -6 & 0 & 1 \end{array}\right] \quad 2R_1 + R_2 \to R_2 \quad \left[\begin{array}{rr|rr} 1 & 3 & 1 & 0 \\ 0 & 0 & 2 & 1 \end{array}\right]$$

The entries of the bottom row of the left-hand A portion are all 0's. No matter what approach we take, we will find that we cannot transform A into I_2 using the elementary row operations. We assert the following:

THE EXISTENCE OF A^{-1}

If the $n \times n$ matrix A cannot be transformed into I_n using the elementary row operations, then A^{-1} does not exist.

If in the process of transforming A into I_n using elementary row operations, we arrive at either a row or column consisting entirely of 0's, then A cannot be transformed into I_n and therefore, A^{-1} does not exist.

Now we have the tools to approach linear systems in the same way we approach simple linear equations.

Solving Linear Systems Using Matrix Inverses

We return to the linear system $\begin{cases} a_{11}x + a_{12}y = k_1 \\ a_{21}x + a_{22}y = k_2 \end{cases}$

If we let $A = \begin{bmatrix} a_{11} & a_{12} \\ a_{21} & a_{22} \end{bmatrix}$, $X = \begin{bmatrix} x \\ y \end{bmatrix}$, and $K = \begin{bmatrix} k_1 \\ k_2 \end{bmatrix}$, then

$$AX = \begin{bmatrix} a_{11} & a_{12} \\ a_{21} & a_{22} \end{bmatrix} \begin{bmatrix} x \\ y \end{bmatrix}$$

$$= \begin{bmatrix} a_{11}x + a_{12}y \\ a_{21}x + a_{22}y \end{bmatrix}$$ *Since $a_{11}x + a_{12}y = k_1$ and $a_{21}x + a_{22}y = k_2$,*

$$= \begin{bmatrix} k_1 \\ k_2 \end{bmatrix} = K$$

Thus we can write this system in matrix terms as $AX = K$. We call A the **coefficient matrix**, X the **variable vector,** and K the **constant vector** of the system. Given $AX = K$, we can solve for X using the matrix properties as follows:

Notice the left-hand multiplication.

$AX = K$ *Multiply both sides of this equation by A^{-1} on the left.*

$A^{-1}(AX) = A^{-1}K$ *Use associativity of matrix multiplication.*

$(A^{-1}A)X = A^{-1}K$ *Since $A^{-1}A = I_2$,*

$I_2X = A^{-1}K$ *Since I_2 is the identity matrix for the matrix X, we have*

$X = A^{-1}K$

Hence if $AX = K$, then $X = A^{-1}K$ yields the matrix of solutions. In other words, the solution to $AX = K$ is the constant matrix *left-multiplied* by the inverse of the coefficient matrix. Although we approach a linear system as though we are solving a simple linear equation, unlike simple linear equations, matrix multiplication is not commutative; therefore, it is important that A^{-1} be multiplied *left* of K.

EXAMPLE 4 Using matrix inverses, find the solution to the following system:

$$\begin{cases} 2x - 3y = 7 \\ 3x + 5y = 20 \end{cases}$$

Solution We identify matrices A, X, and K:

$$A = \begin{bmatrix} 2 & -3 \\ 3 & 5 \end{bmatrix}, \qquad X = \begin{bmatrix} x \\ y \end{bmatrix}, \qquad K = \begin{bmatrix} 7 \\ 20 \end{bmatrix}$$

Hence $AX = K$, and our solution is found by $X = A^{-1}K$, which means we must find the inverse of the coefficient matrix and multiply it by the constant matrix to find the solution. First we find A^{-1}.

1. Graft I_2 onto A: $\left[\begin{array}{cc|cc} 2 & -3 & 1 & 0 \\ 3 & 5 & 0 & 1 \end{array}\right]$.
2. Transform the left 2×2 portion into I_2.

$$\left[\begin{array}{cc|cc} 2 & -3 & 1 & 0 \\ 3 & 5 & 0 & 1 \end{array}\right] \quad \tfrac{1}{2}R_1 \to R_1 \quad \left[\begin{array}{cc|cc} 1 & -\frac{3}{2} & \frac{1}{2} & 0 \\ 3 & 5 & 0 & 1 \end{array}\right]$$

$$\left[\begin{array}{cc|cc} 1 & -\frac{3}{2} & \frac{1}{2} & 0 \\ 3 & 5 & 0 & 1 \end{array}\right] \quad -3R_1 + R_2 \to R_2 \quad \left[\begin{array}{cc|cc} 1 & -\frac{3}{2} & \frac{1}{2} & 0 \\ 0 & \frac{19}{2} & -\frac{3}{2} & 1 \end{array}\right]$$

$$\left[\begin{array}{cc|cc} 1 & -\frac{3}{2} & \frac{1}{2} & 0 \\ 0 & \frac{19}{2} & -\frac{3}{2} & 1 \end{array}\right] \quad \tfrac{2}{19}R_2 \to R_2 \quad \left[\begin{array}{cc|cc} 1 & -\frac{3}{2} & \frac{1}{2} & 0 \\ 0 & 1 & -\frac{3}{19} & \frac{2}{19} \end{array}\right]$$

$$\left[\begin{array}{cc|cc} 1 & -\frac{3}{2} & \frac{1}{2} & 0 \\ 0 & 1 & -\frac{3}{19} & \frac{2}{19} \end{array}\right] \quad \tfrac{3}{2}R_2 + R_1 \to R_1 \quad \left[\begin{array}{cc|cc} 1 & 0 & \frac{5}{19} & \frac{3}{19} \\ 0 & 1 & -\frac{3}{19} & \frac{2}{19} \end{array}\right]$$

3. Hence, the right-hand portion is $A^{-1} = \begin{bmatrix} \frac{5}{19} & \frac{3}{19} \\ -\frac{3}{19} & \frac{2}{19} \end{bmatrix} = \frac{1}{19}\begin{bmatrix} 5 & 3 \\ -3 & 2 \end{bmatrix}$

Again, notice A^{-1} is multiplied on the left.

Now we can solve for X:

$$\begin{aligned} X &= A^{-1}K \\ &= \tfrac{1}{19}\begin{bmatrix} 5 & 3 \\ -3 & 2 \end{bmatrix}\begin{bmatrix} 7 \\ 20 \end{bmatrix} \\ &= \tfrac{1}{19}\begin{bmatrix} (5)(7) + (3)(20) \\ (-3)(7) + (2)(20) \end{bmatrix} = \tfrac{1}{19}\begin{bmatrix} 95 \\ 19 \end{bmatrix} && \textit{Perform matrix multiplication first.} \\ &= \begin{bmatrix} 5 \\ 1 \end{bmatrix} && \textit{Then do scalar multiplication.} \end{aligned}$$

Hence $X = \begin{bmatrix} x \\ y \end{bmatrix} = \begin{bmatrix} 5 \\ 1 \end{bmatrix}$, which means $\boxed{x = 5, y = 1}$. Check the solution. ■

If we were to use Gaussian elimination with back substitution on the last example, we would get to the solution much more quickly than by using matrix inverses. So a natural question to ask is, Why bother with this method? We answer this question by considering the following example.

EXAMPLE 5 Danny invests \$10,000, part at 6% per year in a conservative investment and part at 12% per year in a riskier investment. How much should he invest at each rate if he wants a return on his total investment of **(a)** 8%? **(b)** 9%? **(c)** 10.5%?

Solution If we let $x =$ the amount invested at 6% and $y =$ the amount invested at 12%, the first equation for all three parts of this problem is

$$x + y = 10{,}000 \qquad \textit{The total amount invested}$$

The second equation involves the interest received on each investment. In general, if Danny invests x dollars at 6% and y dollars at 12%, then his total interest for the year is $0.06x + 0.12y$. The three parts of this example differ on what this amount should be.

For part (a), interest is expressed in the equation $0.06x + 0.12y = 0.08(10{,}000) = 800$.

For part (b), interest is expressed in the equation $0.06x + 0.12y = 0.09(10{,}000) = 900$.

For part (c), interest is expressed in the equation $0.06x + 0.12y = 0.105(10{,}000) = 1050$.

Hence we are being asked to solve 3 systems of equations:

$$\textbf{(a)} \begin{cases} x + y = 10{,}000 \\ 0.06x + 0.12y = 800 \end{cases}$$

$$\textbf{(b)} \begin{cases} x + y = 10{,}000 \\ 0.06x + 0.12y = 900 \end{cases}$$

$$\textbf{(c)} \begin{cases} x + y = 10{,}000 \\ 0.06x + 0.12y = 1050 \end{cases}$$

Notice that all three parts of this example have the same matrix of coefficients. We could solve each system separately using Gaussian elimination; however, once we have the inverse of the coefficient matrix, we can solve all three quickly.

The coefficient matrix of all three systems above is the same:

$$A = \begin{bmatrix} 1 & 1 \\ 0.06 & 0.12 \end{bmatrix}.$$

The student should verify that the inverse of this matrix is $A^{-1} = \begin{bmatrix} 2 & -\frac{50}{3} \\ -1 & \frac{50}{3} \end{bmatrix}$.

Given that we have the inverse of the coefficient matrix, we can find the solution to all three systems easily:

(a) $\begin{cases} x + y = 10{,}000 \\ 0.06x + 0.12y = 800 \end{cases}$

The constant matrix for this problem is $\begin{bmatrix} 10{,}000 \\ 800 \end{bmatrix}$. We have

$$X = A^{-1}K$$

$$\begin{bmatrix} x \\ y \end{bmatrix} = \begin{bmatrix} 2 & -\frac{50}{3} \\ -1 & \frac{50}{3} \end{bmatrix}\begin{bmatrix} 10{,}000 \\ 800 \end{bmatrix} = \begin{bmatrix} 6666.67 \\ 3333.33 \end{bmatrix}$$

Hence $x = 6666.67$ and $y = 3333.33$; therefore, Danny should invest $\boxed{\$6666.67 \text{ at } 6\%}$ and $\boxed{\$3333.33 \text{ at } 12\%}$ in order to gain a return of 8% on his total investment.

(b) $\begin{cases} x + y = 10{,}000 \\ 0.06x + 0.12y = 900 \end{cases}$

The constant matrix for this problem is $\begin{bmatrix} 10{,}000 \\ 900 \end{bmatrix}$. We have

$$X = A^{-1}K$$

$$\begin{bmatrix} x \\ y \end{bmatrix} = \begin{bmatrix} 2 & -\frac{50}{3} \\ -1 & \frac{50}{3} \end{bmatrix}\begin{bmatrix} 10{,}000 \\ 900 \end{bmatrix} = \begin{bmatrix} 5000 \\ 5000 \end{bmatrix}$$

Hence $x = 5000$ and $y = 5000$; therefore, Danny should invest $\boxed{\$5000 \text{ at } 6\%}$ and $\boxed{\$5000 \text{ at } 12\%}$ in order to gain a return of 9% on his total investment.

(c) $\begin{cases} x + y = 10{,}000 \\ 0.06x + 0.12y = 1050 \end{cases}$

The constant matrix for this problem is $\begin{bmatrix} 10{,}000 \\ 1050 \end{bmatrix}$.

We have

$$X = A^{-1}K$$

$$\begin{bmatrix} x \\ y \end{bmatrix} = \begin{bmatrix} 2 & -\frac{50}{3} \\ -1 & \frac{50}{3} \end{bmatrix}\begin{bmatrix} 10{,}000 \\ 1050 \end{bmatrix} = \begin{bmatrix} 2500 \\ 7500 \end{bmatrix}$$

Hence $x = 2500$ and $y = 7500$; therefore, Danny should invest $\boxed{\$2500 \text{ at } 6\%}$ and $\boxed{\$7500 \text{ at } 12\%}$ in order to gain a return of 10.5% on his total investment. ∎

Some calculators have the capability of computing the inverse of a matrix. If you have access to this type of calculator or have computer software that performs this function, then you may find this method to be easiest of the methods discussed in this chapter.

In the course of solving the system $AX = K$, we may find that A^{-1} does not exist; that is, we may arrive at a column or row consisting entirely of zeros and hence A cannot be transformed into I_n using the elementary row operations. If this is the case, then there are no unique solutions: the system may either be dependent or inconsistent.

EXERCISES 9.5

1. Compute $\begin{bmatrix} 3 & -1 \\ 3 & -2 \end{bmatrix} \begin{bmatrix} 1 & 0 \\ 0 & 1 \end{bmatrix}$.

2. Compute $\begin{bmatrix} 1 & 0 & 0 \\ 0 & 1 & 0 \\ 0 & 0 & 1 \end{bmatrix} \begin{bmatrix} 2 & -2 & 1 \\ 3 & 0 & -4 \\ 7 & -1 & 5 \end{bmatrix}$.

3. Show that if A is a 3×3 square matrix, then $AI_3 = I_3A = A$.

4. If A is a 3×2 matrix, then does $AI_3 = I_3A = A$?

In Exercises 5–20, find the inverse of the matrix if it exists.

5. $\begin{bmatrix} 1 & 2 \\ 2 & 0 \end{bmatrix}$

6. $\begin{bmatrix} 3 & -2 \\ 1 & 1 \end{bmatrix}$

7. $\begin{bmatrix} -1 & 5 \\ 3 & 1 \end{bmatrix}$

8. $\begin{bmatrix} 2 & -1 \\ 0 & 1 \end{bmatrix}$

9. $\begin{bmatrix} 4 & 2 \\ 2 & 1 \end{bmatrix}$

10. $\begin{bmatrix} -1 & 5 \\ 2 & -10 \end{bmatrix}$

11. $\begin{bmatrix} -1 & 5 \\ -2 & 10 \end{bmatrix}$

12. $\begin{bmatrix} 3 & 0 \\ -5 & 1 \end{bmatrix}$

13. $\begin{bmatrix} 1 & 1 & 2 \\ 2 & 0 & 1 \\ 3 & -1 & 1 \end{bmatrix}$

14. $\begin{bmatrix} 1 & -1 & 0 \\ -2 & 2 & 1 \\ 4 & -1 & 0 \end{bmatrix}$

15. $\begin{bmatrix} 1 & 0 & 1 \\ -2 & 3 & 2 \\ 2 & -2 & 0 \end{bmatrix}$

16. $\begin{bmatrix} 1 & 1 & 1 \\ -2 & 2 & 3 \\ 5 & 0 & 1 \end{bmatrix}$

17. $\begin{bmatrix} 2 & 1 & 3 \\ 1 & 0 & 1 \\ 4 & -2 & 2 \end{bmatrix}$

18. $\begin{bmatrix} 1 & -1 & 2 \\ -2 & 2 & -3 \\ 1 & 1 & 0 \end{bmatrix}$

19. $\begin{bmatrix} 1 & -1 & 2 \\ 2 & -2 & 1 \\ 3 & -1 & 0 \end{bmatrix}$

20. $\begin{bmatrix} 1 & 0 & 0 \\ -2 & 0 & 0 \\ 4 & -3 & 1 \end{bmatrix}$

In Exercises 21–38, solve the given system using matrix inverses.

21. $\begin{cases} x + 2y = 3 \\ x - y = 6 \end{cases}$

22. $\begin{cases} x + y = 5 \\ 3x - y = 8 \end{cases}$

23. $\begin{cases} x - 2y = 4 \\ 2x + y = 13 \end{cases}$

24. $\begin{cases} 3x + 2y = 3 \\ 3x + y = 0 \end{cases}$

25. $\begin{cases} 5x - 2y = 11 \\ x - 5y = -7 \end{cases}$

26. $\begin{cases} x = 3y - 1 \\ 2x - 5y = -3 \end{cases}$

27. $\begin{cases} y = 3x - 15 \\ 8x - 2y = 10 \end{cases}$

28. $\begin{cases} x + 7y = -1 \\ 3x + 21y = -3 \end{cases}$

29. $\begin{cases} \frac{3}{2}x + \frac{1}{3}y = 1 \\ x + \frac{1}{12}y = \frac{1}{4} \end{cases}$

30. $\begin{cases} 5x + 2y = -8 \\ 3x + y = -4 \end{cases}$

31. $\begin{cases} x + y + z = 6 \\ 3x + y - z = 6 \\ 2x + y - z = 4 \end{cases}$

32. $\begin{cases} x + y - z = 1 \\ 2x + 2y + 2z = 0 \\ x - y + z = 3 \end{cases}$

33. $\begin{cases} x - 3y + z = 0 \\ 3x - y + z = 2 \\ x + 2y - z = 2 \end{cases}$

34. $\begin{cases} x + y + 2z = 7 \\ x - 2y - z = 1 \\ x + y + z = 4 \end{cases}$

35. $\begin{cases} x + y + z = 6 \\ 3x + y - z = 4 \\ 2x + 2y + 2z = 4 \end{cases}$

36. $\begin{cases} x + y - z = 3 \\ 2x + 2y - 2z = 6 \\ x - y + z = 5 \end{cases}$

37. $\begin{cases} x + 0.2y + z = 0.02 \\ 0.3x - y + 0.2z = 0 \\ x + y + z = 0.1 \end{cases}$

38. $\begin{cases} x + y = 2x + z \\ x - 2y = z - 1 \\ 4x - 2y = 6z - 5 \end{cases}$

39. Jenna has \$20,000 to split between a low-risk certificate of deposit yielding 6% per year and a high-risk oil stock yielding 18% per year. How much should she invest at each rate if she wants a return on her total investment of **(a)** 8%? **(b)** 10%? **(c)** 12%?

40. A nutritionist wants to create a food supplement out of three substances: X, Y, and Z. The iron content and cost of the substances X, Y, and Z are entered in the accompanying table.

Substance	Iron content per gram	Cost per gram
X	0.5	10¢
Y	0.2	6¢
Z	0.4	8¢

How many grams of each substance should be used to make a food supplement if she wants 10 grams of the food supplement to supply 3.5 grams of iron and cost **(a)** 77¢? **(b)** 78¢? **(c)** 79¢?

9.6 Determinants and Cramer's Rule: 2 × 2 and 3 × 3 Systems

In the first section we solved many systems of equations. You may have noticed that the same procedure was consistently repeated. This leads us to a natural question as to whether we can apply this procedure to a general case and devise a formula for solving systems. In this section we will discuss a "formula" for solving systems of equations called **Cramer's Rule**. We will begin by solving a general 2 × 2 system.

EXAMPLE 1 Solve the following system of equations for x and y.

$$\begin{cases} a_{11}x + a_{12}y = k_1 \\ a_{21}x + a_{22}y = k_2 \end{cases}$$

The advantages of this subscript notation will become apparent in the next few pages.

Solution The subscripts of a indicate their position in the system. The first number indicates the equation number; the second number indicates the variable. (Usually the leftmost variable, in this case x, is considered the "first.") For example, a_{12} appears in the first equation and is the coefficient of the second variable, whereas a_{21} appears in the second equation, and is the coefficient of the first variable.

We will solve this general 2 × 2 system by elimination. First we eliminate y:

$$\begin{array}{lcl} a_{11}x + a_{12}y = k_1 & \xrightarrow{\text{Multiply by } a_{22}} & a_{11}a_{22}x + a_{12}a_{22}y = k_1a_{22} \\ a_{21}x + a_{22}y = k_2 & \xrightarrow{\text{Multiply by } -a_{12}} & \underline{-a_{21}a_{12}x - a_{22}a_{12}y = -k_2a_{12}} \quad \textit{Add.} \\ & & a_{11}a_{22}x - a_{21}a_{12}x = k_1a_{22} - k_2a_{12} \end{array}$$

Now solve for x.

$$(a_{11}a_{22} - a_{21}a_{12})x = k_1a_{22} - k_2a_{12}$$

$$x = \frac{k_1a_{22} - k_2a_{12}}{a_{11}a_{22} - a_{21}a_{12}}$$

Note that in order to have a solution we must have $a_{11}a_{22} - a_{21}a_{12} \neq 0$. We can solve for y using the same method as above—i.e., eliminating x—to arrive at

$$y = \frac{a_{11}k_2 - a_{21}k_1}{a_{11}a_{22} - a_{21}a_{12}}$$

Again, this solution is valid only if $a_{11}a_{22} - a_{21}a_{12} \neq 0$. Hence the general solution for this 2 × 2 system is

$$\boxed{x = \frac{k_1a_{22} - k_2a_{12}}{a_{11}a_{22} - a_{21}a_{12}}, \qquad y = \frac{a_{11}k_2 - a_{21}k_1}{a_{11}a_{22} - a_{21}a_{12}}}.$$

■

This looks like a difficult formula to commit to memory. So we will digress a bit and define some notation that will help us to use this formula as well as provide us with a formula for larger systems of equations. We begin by defining a **determinant.**

DEFINITION The symbol $\begin{vmatrix} a & c \\ b & d \end{vmatrix}$ is called a **2 × 2 determinant**. It is a number whose value is defined to be

$$\begin{vmatrix} a & c \\ b & d \end{vmatrix} = ad - bc$$

EXAMPLE 2 Evaluate the following determinants.

(a) $\begin{vmatrix} 3 & 2 \\ 8 & 1 \end{vmatrix}$ **(b)** $\begin{vmatrix} 5 & -1 \\ -2 & 6 \end{vmatrix}$

Note that a determinant is a single number.

Solution

(a) $\begin{vmatrix} 3 & 2 \\ 8 & 1 \end{vmatrix} = 3(1) - 8(2) = \boxed{-13}$

(b) $\begin{vmatrix} 5 & -1 \\ -2 & 6 \end{vmatrix} = 5(6) - (-2)(-1) = \boxed{28}$ ■

Returning to Example 1, we found the general solution to the system

$$\begin{cases} a_{11}x + a_{12}y = k_1 \\ a_{21}x + a_{22}y = k_2 \end{cases} \text{ to be } \quad x = \frac{k_1a_{22} - k_2a_{12}}{a_{11}a_{22} - a_{21}a_{12}}, \qquad y = \frac{a_{11}k_2 - a_{21}k_1}{a_{11}a_{22} - a_{21}a_{12}}$$

which, using determinants, could be written as $x = \dfrac{\begin{vmatrix} k_1 & a_{12} \\ k_2 & a_{22} \end{vmatrix}}{\begin{vmatrix} a_{11} & a_{12} \\ a_{21} & a_{22} \end{vmatrix}}$, $y = \dfrac{\begin{vmatrix} a_{11} & k_1 \\ a_{21} & k_2 \end{vmatrix}}{\begin{vmatrix} a_{11} & a_{12} \\ a_{21} & a_{22} \end{vmatrix}}$.

Notice that each denominator of the general solution is the determinant of the coefficients of the variables. We usually denote this determinant by D:

$$D = \begin{vmatrix} a_{11} & a_{12} \\ a_{21} & a_{22} \end{vmatrix}$$

The numerators of the general solution are the determinants arrived at by taking D and replacing the coefficients of the variable for which we are solving by the constant terms. We denote these numerators by

$$D_x = \begin{vmatrix} k_1 & a_{12} \\ k_2 & a_{22} \end{vmatrix} \quad \text{and} \quad D_y = \begin{vmatrix} a_{11} & k_1 \\ a_{21} & k_2 \end{vmatrix}$$

Using this notation, we have derived the following, called **Cramer's Rule**:

CRAMER'S RULE FOR 2 × 2 SYSTEMS OF EQUATIONS

The solution to the general system

$$\begin{cases} a_{11}x + a_{12}y = k_1 \\ a_{21}x + a_{22}y = k_2 \end{cases} \quad \text{is} \quad x = \frac{D_x}{D} \quad \text{and} \quad y = \frac{D_y}{D}$$

where $D = \begin{vmatrix} a_{11} & a_{12} \\ a_{21} & a_{22} \end{vmatrix} \neq 0$, $\quad D_x = \begin{vmatrix} k_1 & a_{12} \\ k_2 & a_{22} \end{vmatrix}$, $\quad D_y = \begin{vmatrix} a_{11} & k_1 \\ a_{21} & k_2 \end{vmatrix}$.

EXAMPLE 3 Solve the following system using Cramer's Rule.

$$\begin{cases} 3x - 5y = 2 \\ 2x + 8y = 9 \end{cases}$$

Solution Using Cramer's Rule, we first find D, D_x, and D_y:

$$D = \begin{vmatrix} 3 & -5 \\ 2 & 8 \end{vmatrix} = 3(8) - 2(-5) = 34$$

$$D_x = \begin{vmatrix} 2 & -5 \\ 9 & 8 \end{vmatrix} = 2(8) - 9(-5) = 61$$

$$D_y = \begin{vmatrix} 3 & 2 \\ 2 & 9 \end{vmatrix} = 3(9) - 2(2) = 23$$

By Cramer's Rule we have

$$x = \frac{D_x}{D} = \frac{61}{34} \quad \text{and} \quad y = \frac{D_y}{D} = \frac{23}{34}$$

Hence the solution is $\boxed{\left(\frac{61}{34}, \frac{23}{34}\right)}$. The student should check this solution. ■

If $D = 0$, then there is no unique solution: The system is either inconsistent and has no solutions (if $D_x \neq 0$ or $D_y \neq 0$) or it is dependent and has infinitely many solutions (if $D_x = 0$ and $D_y = 0$).

Cramer's Rule can be extended to 3 × 3 linear systems as well. Before we do however, we must define a 3 × 3 determinant.

The subscripts of a, the entries of the 3 × 3 determinant, indicate their positions. The first number indicates the row; the second indicates the column. For example, a_{32} is the entry of the third row in the second column. In general, a_{ij} is the entry of the ith row and jth column.

DEFINITION A **3 × 3 determinant** is defined as

$$\begin{vmatrix} a_{11} & a_{12} & a_{13} \\ a_{21} & a_{22} & a_{23} \\ a_{31} & a_{32} & a_{33} \end{vmatrix} = a_{11}\begin{vmatrix} a_{22} & a_{23} \\ a_{32} & a_{33} \end{vmatrix} - a_{21}\begin{vmatrix} a_{12} & a_{13} \\ a_{32} & a_{33} \end{vmatrix} + a_{31}\begin{vmatrix} a_{12} & a_{13} \\ a_{22} & a_{23} \end{vmatrix}$$

The 2 × 2 determinants in this definition are called the **minors** of their coefficients.

A minor is found by choosing an element of a determinant and then crossing off the row and column that contain it. Hence the minor of a_{11} is found as follows:

$$\begin{vmatrix} a_{11} & a_{12} & a_{13} \\ a_{21} & a_{22} & a_{23} \\ a_{31} & a_{32} & a_{33} \end{vmatrix} \quad \text{which gives} \quad \begin{vmatrix} a_{22} & a_{23} \\ a_{32} & a_{33} \end{vmatrix}$$

EXAMPLE 4 Evaluate the determinant $\begin{vmatrix} 5 & 0 & -2 \\ 3 & 1 & -4 \\ 2 & -1 & 5 \end{vmatrix}$.

Solution By definition we get

$$\begin{vmatrix} 5 & 0 & -2 \\ 3 & 1 & -4 \\ 2 & -1 & 5 \end{vmatrix} = 5\begin{vmatrix} 5 & 0 & -2 \\ 3 & 1 & -4 \\ 2 & -1 & 5 \end{vmatrix} - 3\begin{vmatrix} 5 & 0 & -2 \\ 3 & 1 & -4 \\ 2 & -1 & 5 \end{vmatrix} + 2\begin{vmatrix} 5 & 0 & -2 \\ 3 & 1 & -4 \\ 2 & -1 & 5 \end{vmatrix}$$

$$= 5\begin{vmatrix} 1 & -4 \\ -1 & 5 \end{vmatrix} - 3\begin{vmatrix} 0 & -2 \\ -1 & 5 \end{vmatrix} + 2\begin{vmatrix} 0 & -2 \\ 1 & -4 \end{vmatrix}$$

$$= 5[(1)(5) - (-1)(-4)] - 3[(0)(5) - (-1)(-2)] + 2[(0)(-4) - (1)(-2)]$$

$$= 5(5 - 4) - 3(0 - 2) + 2(0 + 2) = \boxed{15}$$

■

Actually we could have defined a 3×3 determinant by using its minors in several ways. The procedure given in the definition above is called *expanding down the first column,* since we use the coefficients of the first column along with their minors. In general, a minor of any entry is the determinant formed by deleting the row and column containing that entry. Hence in the

$$\text{determinant } \begin{vmatrix} a_{11} & a_{12} & a_{13} \\ a_{21} & a_{22} & a_{23} \\ a_{31} & a_{32} & a_{33} \end{vmatrix}, \text{ the minor of } a_{32} \text{ is } \begin{vmatrix} a_{11} & a_{12} & a_{13} \\ a_{21} & a_{22} & a_{23} \\ a_{31} & a_{32} & a_{33} \end{vmatrix} = \begin{vmatrix} a_{11} & a_{13} \\ a_{21} & a_{23} \end{vmatrix}.$$

We can write a 3×3 determinant as an expansion of minors by expanding along any column or row provided we prefix each entry with the appropriate sign. The sign is determined by its position which is represented by the following **sign array**.

$$\begin{vmatrix} + & - & + \\ - & + & - \\ + & - & + \end{vmatrix}$$

Notice that starting with + at the top left, the signs alternate along each row and column.

For example, we can expand and evaluate the determinant of Example 4,

$$\begin{vmatrix} 5 & 0 & -2 \\ 3 & 1 & -4 \\ 2 & -1 & 5 \end{vmatrix}$$, *across the second row* as follows:

$$\begin{vmatrix} 5 & 0 & -2 \\ 3 & 1 & -4 \\ 2 & -1 & 5 \end{vmatrix} = -3\begin{vmatrix} 0 & -2 \\ -1 & 5 \end{vmatrix} + 1\begin{vmatrix} 5 & -2 \\ 2 & 5 \end{vmatrix} - (-4)\begin{vmatrix} 5 & 0 \\ 2 & -1 \end{vmatrix}$$

↑ ↑ ↑

Note we use the signs of the second row of the sign array.

$$= -3(0 - 2) + 1(25 + 4) + 4(-5 - 0) = 15$$

In general it is convenient to expand down columns or across rows that contain 0s or 1s.

EXAMPLE 5 Evaluate the determinant $\begin{vmatrix} 3 & 2 & 0 \\ 6 & -2 & 0 \\ 2 & 3 & 4 \end{vmatrix}$.

Solution Since we can expand the determinant along any row or column we choose, it would be convenient to expand this determinant along the third column, since there are two zeros in this column.

Why did we expand down the third column?

$$\begin{vmatrix} 3 & 2 & 0 \\ 6 & -2 & 0 \\ 2 & 3 & 4 \end{vmatrix} = +0\begin{vmatrix} 6 & -2 \\ 2 & 3 \end{vmatrix} - 0\begin{vmatrix} 3 & 2 \\ 2 & 3 \end{vmatrix} + 4\begin{vmatrix} 3 & 2 \\ 6 & -2 \end{vmatrix}$$

↑ ↑ ↑
Note we use the signs of the third column of the sign array.

$$= 0 - 0 + 4(-6 - 12) = \boxed{-72}$$

■

Now we can generalize Cramer's Rule for 3 × 3 systems of linear equations.

CRAMER'S RULE FOR 3×3 SYSTEMS OF EQUATIONS

The solution to the general system

$$\begin{cases} a_{11}x + a_{12}y + a_{13}z = k_1 \\ a_{21}x + a_{22}y + a_{23}z = k_2 \\ a_{31}x + a_{32}y + a_{33}z = k_3 \end{cases}$$

is

$$x = \frac{D_x}{D}, \qquad y = \frac{D_y}{D}, \qquad z = \frac{D_z}{D}$$

where

$$D = \begin{vmatrix} a_{11} & a_{12} & a_{13} \\ a_{21} & a_{22} & a_{23} \\ a_{31} & a_{32} & a_{33} \end{vmatrix} \neq 0, \qquad D_x = \begin{vmatrix} k_1 & a_{12} & a_{13} \\ k_2 & a_{22} & a_{23} \\ k_3 & a_{32} & a_{33} \end{vmatrix},$$

$$D_y = \begin{vmatrix} a_{11} & k_1 & a_{13} \\ a_{21} & k_2 & a_{23} \\ a_{31} & k_3 & a_{33} \end{vmatrix}, \qquad D_z = \begin{vmatrix} a_{11} & a_{12} & k_1 \\ a_{21} & a_{22} & k_2 \\ a_{31} & a_{32} & k_3 \end{vmatrix}.$$

EXAMPLE 6 Solve the following system of equations using Cramer's Rule.

$$\begin{cases} 2x + 2y + z = 7 \\ 4x - y + 3z =. -1 \\ 6x - y - z = 0 \end{cases}$$

Solution First we compute D, because if $D = 0$, then there is no unique solution. We compute D by expanding across the last row.

$$D = \begin{vmatrix} 2 & 2 & 1 \\ 4 & -1 & 3 \\ 6 & -1 & -1 \end{vmatrix} = 6\begin{vmatrix} 2 & 1 \\ -1 & 3 \end{vmatrix} - (-1)\begin{vmatrix} 2 & 1 \\ 4 & 3 \end{vmatrix} + (-1)\begin{vmatrix} 2 & 2 \\ 4 & -1 \end{vmatrix}$$
$$= 6(6 + 1) - (-1)(6 - 4) + (-1)(-2 - 8) = 54$$

Next we compute D_x by expanding across the last row.

$$D_x = \begin{vmatrix} 7 & 2 & 1 \\ -1 & -1 & 3 \\ 0 & -1 & -1 \end{vmatrix} = 0\begin{vmatrix} 2 & 1 \\ -1 & 3 \end{vmatrix} - (-1)\begin{vmatrix} 7 & 1 \\ -1 & 3 \end{vmatrix} + (-1)\begin{vmatrix} 7 & 2 \\ -1 & -1 \end{vmatrix}$$
$$= 0(6 + 1) - (-1)(21 + 1) + (-1)(-7 + 2) = 27$$

For variety, we compute D_y down the second column.

$$D_y = \begin{vmatrix} 2 & 7 & 1 \\ 4 & -1 & 3 \\ 6 & 0 & -1 \end{vmatrix} = -7\begin{vmatrix} 4 & 3 \\ 6 & -1 \end{vmatrix} + (-1)\begin{vmatrix} 2 & 1 \\ 6 & -1 \end{vmatrix} - 0\begin{vmatrix} 2 & 1 \\ 4 & 3 \end{vmatrix}$$
$$= -7(-4 - 18) + (-1)(-2 - 6) - 0(6 - 4) = 162$$

We compute D_z expanding down the third column.

$$D_z = \begin{vmatrix} 2 & 2 & 7 \\ 4 & -1 & -1 \\ 6 & -1 & 0 \end{vmatrix} = 7\begin{vmatrix} 4 & -1 \\ 6 & -1 \end{vmatrix} - (-1)\begin{vmatrix} 2 & 2 \\ 6 & -1 \end{vmatrix} + 0\begin{vmatrix} 2 & 2 \\ 4 & -1 \end{vmatrix}$$
$$= 7(-4 + 6) - (-1)(-2 - 12) + 0(-2 - 8) = 0$$

Hence $x = \dfrac{D_x}{D} = \dfrac{27}{54} = \dfrac{1}{2}$, $y = \dfrac{D_y}{D} = \dfrac{162}{54} = 3$, and $z = \dfrac{D_z}{D} = \dfrac{0}{54} = 0$.

The solution is $\boxed{\left(\frac{1}{2}, 3, 0\right)}$. The student should verify this solution. ■

EXERCISES 9.6

In Exercises 1–12, evaluate the determinant.

1. $\begin{vmatrix} 2 & 3 \\ 1 & 4 \end{vmatrix}$

2. $\begin{vmatrix} 1 & 0 \\ 5 & -2 \end{vmatrix}$

3. $\begin{vmatrix} 1 & -5 \\ 2 & 4 \end{vmatrix}$

4. $\begin{vmatrix} 3 & -1 \\ -1 & 3 \end{vmatrix}$

5. $\begin{vmatrix} 4 & -3 \\ -8 & 6 \end{vmatrix}$

6. $\begin{vmatrix} 1 & 0 \\ 0 & 1 \end{vmatrix}$

7. $\begin{vmatrix} 2 & 3 & -1 \\ 1 & 4 & 1 \\ 2 & 0 & 1 \end{vmatrix}$

8. $\begin{vmatrix} 1 & 0 & 2 \\ 5 & -2 & 1 \\ 3 & -1 & 0 \end{vmatrix}$

9. $\begin{vmatrix} 1 & 3 & -1 \\ 1 & 0 & 1 \\ 2 & 0 & 5 \end{vmatrix}$

10. $\begin{vmatrix} 1 & 0 & 0 \\ 3 & -3 & 1 \\ 2 & -1 & 5 \end{vmatrix}$

11. $\begin{vmatrix} 2 & 3 & 1 \\ 2 & 0 & 2 \\ 4 & 6 & 2 \end{vmatrix}$

12. $\begin{vmatrix} 3 & 1 & 0 \\ 3 & -3 & 1 \\ 2 & -1 & 0 \end{vmatrix}$

In Exercises 13–32, solve the system of equations using Cramer's Rule.

13. $\begin{cases} x - 2y = 4 \\ 3x + 5y = 1 \end{cases}$

14. $\begin{cases} x + 2y = -1 \\ 6x - y = 4 \end{cases}$

15. $\begin{cases} 3x - y = 7 \\ 3x + 5y = -1 \end{cases}$

16. $\begin{cases} 2x + 3y = 9 \\ 6x - y = -3 \end{cases}$

17. $\begin{cases} 3x + 4y = 4 \\ 5x - 5y = 1 \end{cases}$

18. $\begin{cases} -x + 2y = 1 \\ 6x - 2y = 4 \end{cases}$

19. $\begin{cases} 2x + 4y = 4 \\ x + 2y = 2 \end{cases}$

20. $\begin{cases} -x + 2y = 1 \\ 2x - 4y = 3 \end{cases}$

21. $\begin{cases} 5x - 10y = 4 \\ x - 2y = 1 \end{cases}$

22. $\begin{cases} x + y = 1 \\ 3x + 3y = 3 \end{cases}$

23. $\begin{cases} x + 2y + z = 8 \\ x + 4y - z = 12 \\ x - 2y + z = -4 \end{cases}$

24. $\begin{cases} x - y + z = 6 \\ 2x - 2y - z = 3 \\ 3x - 2y - z = 5 \end{cases}$

25. $\begin{cases} 3x + 2y + z = 1 \\ x - 2y - z = 3 \\ x - 2y + z = 3 \end{cases}$

26. $\begin{cases} 2x + y + z = -1 \\ 2x - y - z = -5 \\ 3x - 2y - z = -9 \end{cases}$

27. $\begin{cases} 2x - 2y + z = 8 \\ x - 2y - z = 6 \\ x - 2y + z = 4 \end{cases}$

28. $\begin{cases} x - y + z = 1 \\ x - 2y + z = 0 \\ x - 2y - z = 5 \end{cases}$

29. $\begin{cases} x + 2y + z = 1 \\ 2x + 4y + z = 2 \\ x - 2y + z = -4 \end{cases}$

30. $\begin{cases} x - y + z = 3 \\ x - 2y - z = 0 \\ 3x - 3y + 3z = 9 \end{cases}$

31. $\begin{cases} x + z = 1 \\ 2y - z = 2 \\ x - z = 4 \end{cases}$

32. $\begin{cases} x - y + z = 6 \\ x - z = 0 \\ 3y + z = 8 \end{cases}$

33. Brian invests \$10,000, part at 6% and part at 8.5%, earning a yearly interest equivalent to earning 7% on the entire amount. How much is invested at each rate?

34. A lawyer charges \$110 an hour for her time and \$50 an hour for her assistant's time. A client received a bill from the lawyer that included a \$5280 charge for their combined time. If the amount of time spent by the lawyer and her assistant had been reversed, the bill for their time would have been \$4320. How much time did the lawyer and her assistant spend on the client's case?

35. A manufacturer produces two types of computers: model A and model B. Model A requires 1.5 hours to manufacture and 1 hour to assemble: Model B requires 1 hour to manufacture and 30 minutes to assemble. If the company can allocate 100 hours for manufacture and 59 hours for assembly, how many of each type can be produced?

36. Raju splits up \$20,000 into three investments. He has a bank account paying 5%, a certificate of deposit paying 6.5%, and a stock paying 9%. His annual interest from the three investments is \$1365. If he invested \$2000 more in his certificate than in the other two investments combined, how much did he invest at each rate?

37. A radio manufacturer produces three models, model A, model B, and model C. The company knows how much time is needed for production and assembly and how much raw materials cost for each model. This information is found in the accompanying table. If the company has allocated 123 hours for production, 145 hours for assembly, and \$1280 for raw materials, how many of each model can the company produce if it wants to use up all the time allocated for each phase of the process and all the money for raw materials?

Model	Production hours	Assembly hours	Cost of raw materials
A	0.8	0.9	\$9.80
B	0.7	0.5	\$6.20
C	0.4	0.8	\$3.10

38. A quadratic function is a function of the form $f(x) = ax^2 + bx + c$, where a, b, and c are constants. Find the quadratic function $f(x)$ such that $f(1) = 5$, $f(-1) = 7$, and $f(2) = 19$.

QUESTION FOR THOUGHT

39. Discuss the differences between $\begin{vmatrix} 5 & 0 \\ -1 & 2 \end{vmatrix}$ and $\begin{bmatrix} 5 & 0 \\ -1 & 2 \end{bmatrix}$.

9.7 Properties of Determinants

In the last section, we defined 2×2 and 3×3 determinants and discussed Cramer's Rule applied to 2×2 and 3×3 systems. In this section we generalize our findings from the previous section to $n \times n$ determinants and discuss some of their properties. Then we generalize Cramer's Rule to $n \times n$ systems of linear equations. In order to do this, we need a more formal way to refer to determinants, minors, and the sign array. This requires that we introduce some new notation and incorporate some of the ideas and notation used previously with matrices.

When we refer to a matrix, we usually refer to its size by specifying the number of rows by the number of columns. In this section we will restrict our discussion to square matrices. We will refer to an $n \times n$ square matrix as a *matrix of order n*.

If A is a matrix of order n, then we can represent the determinant of the matrix A by $|A|$. Shown next is a matrix A of order n and its determinant, $|A|$.

A Matrix of Order n

$$A = \begin{bmatrix} a_{11} & a_{12} & a_{13} & \cdots & a_{1n} \\ a_{21} & a_{22} & a_{23} & \cdots & a_{2n} \\ a_{31} & a_{32} & a_{33} & \cdots & a_{3n} \\ \vdots & \vdots & \vdots & & \vdots \\ a_{n1} & a_{n2} & a_{n3} & \cdots & a_{nn} \end{bmatrix}$$

The Determinant of A

$$|A| = \begin{vmatrix} a_{11} & a_{12} & a_{13} & \cdots & a_{1n} \\ a_{21} & a_{22} & a_{23} & \cdots & a_{2n} \\ a_{31} & a_{32} & a_{33} & \cdots & a_{3n} \\ \vdots & \vdots & \vdots & & \vdots \\ a_{n1} & a_{n2} & a_{n3} & \cdots & a_{nn} \end{vmatrix}$$

Again, the subscripts of a indicate the position of the entry in the matrix; in general, a_{ij} is the entry in the ith row and jth column.

Although a matrix is a rectangular array of numbers, *the determinant of a matrix is a single real number* defined only when the matrix is square.

In the last section, we needed minors in order to define the determinant of a matrix of order 3 (a 3×3 matrix). Recall that the minor of an entry of a 3×3 determinant is a 2×2 determinant arrived at by deleting the row and column containing the entry. For example, for the determinant

$$|A| = \begin{vmatrix} 3 & 7 & -1 \\ 0 & 8 & 1 \\ 2 & 5 & 3 \end{vmatrix},$$ the minor of $2 = a_{31}$ is the 2×2 determinant $$\begin{vmatrix} 3 & 7 & -1 \\ 0 & 8 & 1 \\ 2 & 5 & 3 \end{vmatrix} = \begin{vmatrix} 7 & -1 \\ 8 & 1 \end{vmatrix}.$$

We use the symbol M_{ij} to designate the minor of a_{ij}. Hence, for A the minor of a_{31} is $M_{31} = \begin{vmatrix} 7 & -1 \\ 8 & 1 \end{vmatrix}$.

In general,

> If A is a matrix of order n (where $n \geq 2$), then M_{ij}, the minor of a_{ij}, is the determinant of the matrix of order $n - 1$ arrived at by deleting the row and column containing a_{ij}.

For the determinant $\begin{vmatrix} 6 & 3 & 7 & -1 \\ 0 & 0 & 8 & 1 \\ -3 & 2 & 5 & 3 \\ 9 & -1 & 4 & 1 \end{vmatrix}$, the minor of $8 = a_{23}$ is the determinant

$$M_{23} = \begin{vmatrix} 6 & 3 & 7 & -1 \\ 0 & 0 & 8 & 1 \\ -3 & 2 & 5 & 3 \\ 9 & -1 & 4 & 1 \end{vmatrix} = \begin{vmatrix} 6 & 3 & -1 \\ -3 & 2 & 3 \\ 9 & -1 & 1 \end{vmatrix}$$

The sign array tells us the sign prefixed to the minor when the determinant is expanded. As with 3×3 determinants, the sign array for an $n \times n$ determinant is constructed by starting with a + sign in the top left corner and alternating signs along the rows and columns.

$$\begin{vmatrix} + & - & + & - & + & \cdots \\ - & + & - & + & - & \cdots \\ + & - & + & - & + & \cdots \\ - & + & - & + & - & \cdots \\ + & - & + & - & + & \cdots \\ \vdots & \vdots & \vdots & \vdots & \vdots & \end{vmatrix}$$

Although the pattern may not be obvious, you can check that the following is true: In the sign array, the entry in the ith row and jth column is positive when the sum $i + j$ is an even number and negative when the sum $i + j$ is an odd number.

Recognizing that $(-1)^k = +1$ when k is even and $(-1)^k = -1$ when k is odd, we can express a minor prefixed with its sign from the sign array as $(-1)^{i+j}M_{ij}$; we call this term the **cofactor** of the element a_{ij} and designate it A_{ij}.

DEFINITION The **cofactor** A_{ij} of the element a_{ij} is defined by

$$A_{ij} = (-1)^{i+j}M_{ij}$$

What is the difference between a minor and a cofactor?

The cofactor A_{ij} is the minor M_{ij} prefixed with its sign. We can write the expansion of a 3×3 determinant down its first column as

$$|A| = \begin{vmatrix} a_{11} & a_{12} & a_{13} \\ a_{21} & a_{22} & a_{23} \\ a_{31} & a_{32} & a_{33} \end{vmatrix} = a_{11}A_{11} + a_{21}A_{21} + a_{31}A_{31} = a_{11}M_{11} - a_{21}M_{21} + a_{31}M_{31}$$

We can now define an $n \times n$ determinant by first-column expansion as we did with 3×3 determinants:

> If A is a matrix of order $n \geq 2$, then
>
> $$|A| = a_{11}A_{11} + a_{21}A_{21} + a_{31}A_{31} + \cdots + a_{n1}A_{n1}$$
> $$= a_{11}M_{11} - a_{21}M_{21} + a_{31}M_{31} + \cdots + a_{n1}(-1)^{n+1}M_{n1}$$

Again, this definition is based on a first-column expansion. We can define a determinant in general as follows: If A is a matrix of order n ($n \geq 2$), the determinant $|A|$ is found by multiplying elements in any row (or column) by their respective cofactors and adding the resulting products.

EXAMPLE 1 If $A = \begin{bmatrix} -2 & 3 & 0 & -1 \\ 9 & 3 & 1 & 2 \\ 2 & -1 & 0 & -4 \\ 0 & 1 & 4 & -7 \end{bmatrix}$, find $|A|$.

Solution Since the third column has two zeros, we can reduce the amount of computation by expanding down the third column to get

$$\begin{vmatrix} -2 & 3 & 0 & -1 \\ 9 & 3 & 1 & 2 \\ 2 & -1 & 0 & -4 \\ 0 & 1 & 4 & -7 \end{vmatrix} = 0A_{13} + 1A_{23} + 0A_{33} + 4A_{43}$$

$$= 0M_{13} - 1M_{23} + 0M_{33} - 4M_{43}$$

$$= -M_{23} - 4M_{43}$$

Now we have to expand M_{23} and M_{43} using 2×2 minors:

$$M_{23} = \begin{vmatrix} -2 & 3 & -1 \\ 2 & -1 & -4 \\ 0 & 1 & -7 \end{vmatrix} \quad \textit{Expand down the first column.}$$

$$= -2\begin{vmatrix} -1 & -4 \\ 1 & -7 \end{vmatrix} - 2\begin{vmatrix} 3 & -1 \\ 1 & -7 \end{vmatrix} + 0\begin{vmatrix} 3 & -1 \\ -1 & -4 \end{vmatrix} = -2(11) - 2(-20) = 18$$

and

$$M_{43} = \begin{vmatrix} -2 & 3 & -1 \\ 9 & 3 & 2 \\ 2 & -1 & -4 \end{vmatrix} \quad \textit{Expand across the last row.}$$

$$= 2\begin{vmatrix} 3 & -1 \\ 3 & 2 \end{vmatrix} - (-1)\begin{vmatrix} -2 & -1 \\ 9 & 2 \end{vmatrix} + (-4)\begin{vmatrix} -2 & 3 \\ 9 & 3 \end{vmatrix} = 2(9) + (5) - 4(-33) = 155$$

Hence

$$\begin{vmatrix} -2 & 3 & 0 & -1 \\ 9 & 3 & 1 & 2 \\ 2 & -1 & 0 & -4 \\ 0 & 1 & 4 & -7 \end{vmatrix} = -M_{23} - 4M_{43}$$

$$= -18 - 4(155) = \boxed{-638}$$

■

If we had to evaluate a 7×7 determinant, then we would have to expand it to seven 6×6 cofactors, then expand *each* of the seven 6×6 cofactor to six 5×5 cofactors, and so on. As you can see, evaluating determinants of higher-order matrices can be quite tedious.

As we have noted in examples before, when we expand determinants down columns or across rows that contain zeros, computations can be significantly reduced. The following example illustrates one condition under which the computation of a determinant is straightforward.

EXAMPLE 2 Prove that the determinant of a third-order matrix in triangular form is equal to the product of the entries in its main diagonal.

Solution We begin with a determinant of 3×3 matrix in triangular form:

$$\begin{vmatrix} a & b & c \\ 0 & d & e \\ 0 & 0 & f \end{vmatrix}$$ *If we expand down column 1, we get*

$$= a\begin{vmatrix} d & e \\ 0 & f \end{vmatrix} - 0\begin{vmatrix} b & c \\ 0 & f \end{vmatrix} + 0\begin{vmatrix} b & c \\ d & e \end{vmatrix}$$

$$= a(df - 0e) = adf$$

which is the product of the entries in its main diagonal. ■

We can generalize this result:

> The determinant of a matrix in triangular form is equal to the product of the entries in its main diagonal.

Hence, if we can somehow transform a determinant into one whose matrix is in triangular form, the computation of a determinant is greatly simplified.

As with matrices, we can define operations on rows that will help us to transform a determinant into one whose matrix is in triangular form. Unlike matrices, for determinants these operations apply to columns as well. Also unlike matrices, the row and column operations we define do not always produce equivalent forms of the determinant.

ELEMENTARY ROW AND COLUMN OPERATIONS FOR DETERMINANTS

1. Interchanging any two rows (or columns) in a determinant changes the sign of the determinant.

For example: $\begin{vmatrix} a & b \\ c & d \end{vmatrix} = -\begin{vmatrix} c & d \\ a & b \end{vmatrix}$ *Switching two rows.*

2. Multiplying a row (or column) by a nonzero constant k multiplies the value of the determinant by k. We can also view this operation as factoring a common factor from a row or column of a determinant.

For example: $\begin{vmatrix} a & b & kc \\ d & e & kf \\ g & h & ki \end{vmatrix} = k\begin{vmatrix} a & b & c \\ d & e & f \\ g & h & i \end{vmatrix}$ *Factoring $k \neq 0$ from column 3*

3. Adding a multiple of one row to another does not affect the value of the determinant; adding a multiple of one column to another does not affect the value of the determinant.

For example:

$\begin{vmatrix} a & b \\ c & d \end{vmatrix} = \begin{vmatrix} a & b \\ c + ka & d + kb \end{vmatrix}$ *Adding a multiple of row 1 to row 2.*

The following are some examples of the elementary row and column operations and how they affect the value of the determinant.

Column operation 1 for determinants: $C_2 \leftrightarrow C_3$

$$\text{If } \begin{vmatrix} 2 & -1 & 4 \\ 1 & -2 & 3 \\ 6 & 0 & 5 \end{vmatrix} = 15, \quad \text{then } \begin{vmatrix} 2 & 4 & -1 \\ 1 & 3 & -2 \\ 6 & 5 & 0 \end{vmatrix} = -15$$

Interchanging columns 2 and 3 changes the sign of the determinant.

The same is true interchanging *any* two columns.

Row operation 2 for determinants: $4R_2 \rightarrow R_2$

$$\text{If } \begin{vmatrix} 5 & 2 \\ -2 & 1 \end{vmatrix} = 9, \quad \text{then } \begin{vmatrix} 5 & 2 \\ 4(-2) & 4(1) \end{vmatrix} = 4\begin{vmatrix} 5 & 2 \\ -2 & 1 \end{vmatrix} = 4(9) = 36$$

Multiplying all entries in row 2 by 4 multiplies the determinant by 4.

The same is true if we multiply any row or column by a nonzero constant k: The determinant is multiplied by k.

EXAMPLE 3 Prove the third elementary row operation for 2×2 determinants.

Solution We are trying to show that for the constant $k \neq 0$,

$$\begin{vmatrix} a & b \\ c + ka & d + kb \end{vmatrix} = \begin{vmatrix} a & b \\ c & d \end{vmatrix} \quad \text{and} \quad \begin{vmatrix} a + kc & b + kd \\ c & d \end{vmatrix} = \begin{vmatrix} a & b \\ c & d \end{vmatrix}$$

We start with the general 2×2 determinant $\begin{vmatrix} a & b \\ c & d \end{vmatrix} = ad - bc$. If we multiply row 1 by a nonzero constant k and add it to row 2, we get

$$\begin{vmatrix} a & b \\ c + ka & d + kb \end{vmatrix} \quad \textit{which, when evaluated, is}$$

$$= a(d + kb) - b(c + ka) = ad + akb - bc - bka = ad - bc = \begin{vmatrix} a & b \\ c & d \end{vmatrix}$$

On the other hand, if we multiply row 2 of the determinant by a nonzero constant k and add it to row 1, we get

$$\begin{vmatrix} a + kc & b + kd \\ c & d \end{vmatrix} \quad \textit{which, when evaluated, is}$$

$$= (a + kc)d - c(b + kd) = ad + kcd - bc - ckd = ad - bc = \begin{vmatrix} a & b \\ c & d \end{vmatrix}$$ ■

Our goal is to use the elementary row or column operations to get the matrix of the determinant into triangular form; once we have it in this form, we can evaluate its determinant by simply computing the product of the diagonal entries.

EXAMPLE 4 Evaluate the determinant $\begin{vmatrix} 3 & -3 & 1 \\ 5 & 2 & 1 \\ 5 & 2 & 4 \end{vmatrix}$.

Solution We approach this problem in the same way we approached getting a matrix into triangular form. It would be convenient to have 1 (or -1) in the upper left-hand corner. We can interchange columns 3 and 1 provided we keep in mind that we must change the sign of the determinant:

$$\begin{vmatrix} 3 & -3 & 1 \\ 5 & 2 & 1 \\ 5 & 2 & 4 \end{vmatrix} = -\begin{vmatrix} 1 & -3 & 3 \\ 1 & 2 & 5 \\ 4 & 2 & 5 \end{vmatrix} \qquad C_1 \leftrightarrow C_3$$

Now that we have 1 in the upper left-hand corner, we can sweep out the rest of the first column. We use elementary row operation 3, which does not change the value of the determinant.

$$-\begin{vmatrix} 1 & -3 & 3 \\ 1 & 2 & 5 \\ 4 & 2 & 5 \end{vmatrix} = -\begin{vmatrix} 1 & -3 & 3 \\ 0 & 5 & 2 \\ 0 & 14 & -7 \end{vmatrix} \qquad \begin{matrix} -R_1 + R_2 \to R_2 \\ -4R_1 + R_3 \to R_3 \end{matrix}$$

We need to get the final zero in column 2 below diagonal entry 5. We could multiply entries in row 2 by $-\frac{14}{5}$ and add the result to row 3. Instead we will use a column operation: Multiply entries in column 3 by 2 and add the result to column 2:

$$-\begin{vmatrix} 1 & -3 & 3 \\ 0 & 5 & 2 \\ 0 & 14 & -7 \end{vmatrix} = -\begin{vmatrix} 1 & 3 & 3 \\ 0 & 9 & 2 \\ 0 & 0 & -7 \end{vmatrix} \qquad 2C_3 + C_2 \to C_2$$

The matrix of this determinant is in triangular form.

Since the matrix of this determinant is in triangular form, we can evaluate the determinant by multiplying the entries of the main diagonal together:

$$-\begin{vmatrix} 1 & 3 & 3 \\ 0 & 9 & 2 \\ 0 & 0 & -7 \end{vmatrix} = -(1)(9)(-7) = \boxed{63}$$

■

EXAMPLE 5 Evaluate $\begin{vmatrix} 3 & -3 & 6 & 15 \\ 2 & 1 & -3 & 1 \\ -2 & 2 & -4 & -10 \\ 6 & 1 & -4 & 2 \end{vmatrix}$.

Solution We could multiply row 1 by $-\frac{2}{3}$ and add it to row 2 to get a zero in row 2, column 1; however, we would like to avoid computations with fractions. We demonstrate elementary row operation 2 for determinants: *We factor* 3 *from row* 1 (and the determinant is therefore multiplied by 3).

$$\begin{vmatrix} 3 & -3 & 6 & 15 \\ 2 & 1 & -3 & 1 \\ -2 & 2 & -4 & -10 \\ 6 & 1 & -4 & 2 \end{vmatrix} = 3\begin{vmatrix} 1 & -1 & 2 & 5 \\ 2 & 1 & -3 & 1 \\ -2 & 2 & -4 & -10 \\ 6 & 1 & -4 & 2 \end{vmatrix}$$

$\frac{1}{3}R_1 \to R_1$

Factoring 3 from row 1 means we multiply row 1 by $\frac{1}{3}$.

Now we sweep out the rest of column 1 by applying elementary row operation 3.

$$3\begin{vmatrix} 1 & -1 & 2 & 5 \\ 2 & 1 & -3 & 1 \\ -2 & 2 & -4 & -10 \\ 6 & 1 & -4 & 2 \end{vmatrix} = 3\begin{vmatrix} 1 & -1 & 2 & 5 \\ 0 & 3 & -7 & -9 \\ 0 & 0 & 0 & 0 \\ 6 & 1 & -4 & 2 \end{vmatrix}$$

$-2R_1 + R_2 \to R_2$
$2R_1 + R_3 \to R_3$

At this point we notice that the third row consists entirely of zeros. If we were now to expand this determinant across the third row, we would get:

$$3\begin{vmatrix} 1 & -1 & 2 & 5 \\ 0 & 3 & -7 & -9 \\ 0 & 0 & 0 & 0 \\ 6 & 1 & -4 & 2 \end{vmatrix}$$

Expand across row 3.

$$= 3(0A_{31} + 0A_{32} + 0A_{33} + 0A_{34})$$

where A_{3i} are the cofactors for row 3 entries

$$= 0$$

■

We can see that if any row or column of a matrix consists entirely of zeros, then, when we expand the determinant along that row or column, all coefficients—and hence the determinant—will be zero. For Example 5, this occurred because row 3 is a multiple of row 1 ($R_3 = \frac{2}{3}R_1$). We can draw the following general conclusions.

> If A is a square matrix, then $|A| = 0$ if
>
> **1.** Any column or row consists entirely of zeros.
>
> **2.** Two rows are identical or two columns are identical.
>
> **3.** One row is a multiple of another or one column is a multiple of another.

Cramer's Rule for solving 3×3 systems of equations (page 605) generalizes to $n \times n$ systems as illustrated in the next example.

EXAMPLE 6 Solve the following system of equations using Cramer's Rule.

$$\begin{cases} x_1 + x_2 = 3 \\ x_2 + x_3 + x_4 = 3 \\ x_2 + x_4 = 2 \\ x_3 + x_4 = 0 \end{cases}$$

Solution (When a system involves several variables, we often use subscripted variables such as x_1, x_2, x_3, etc., rather than different letters of the alphabet.) This is a 4×4 system. First we compute D, the determinant of the matrix of coefficients, since if $D = 0$, there is no unique solution. We compute $D = \begin{vmatrix} 1 & 1 & 0 & 0 \\ 0 & 1 & 1 & 1 \\ 0 & 1 & 0 & 1 \\ 0 & 0 & 1 & 1 \end{vmatrix}$ by first putting its matrix into triangular form.

We note that the first column already has zeros below the first diagonal entry, so we will work on getting zeros below the second diagonal entry.

$$D = \begin{vmatrix} 1 & 1 & 0 & 0 \\ 0 & 1 & 1 & 1 \\ 0 & 1 & 0 & 1 \\ 0 & 0 & 1 & 1 \end{vmatrix} = \begin{vmatrix} 1 & 1 & 0 & 0 \\ 0 & 1 & 1 & 1 \\ 0 & 0 & -1 & 0 \\ 0 & 0 & 1 & 1 \end{vmatrix} \qquad -R_2 + R_3 \rightarrow R_3$$

Next we get zeros below the main diagonal in column 3.

$$\begin{vmatrix} 1 & 1 & 0 & 0 \\ 0 & 1 & 1 & 1 \\ 0 & 0 & -1 & 0 \\ 0 & 0 & 1 & 1 \end{vmatrix} = \begin{vmatrix} 1 & 1 & 0 & 0 \\ 0 & 1 & 1 & 1 \\ 0 & 0 & -1 & 0 \\ 0 & 0 & 0 & 1 \end{vmatrix} \qquad R_3 + R_4 \rightarrow R_4$$

The matrix of the determinant is in triangular form.

Hence $D = (1)(1)(-1)(1) = -1$.

We identify D_{x_1}, the determinant found by replacing the column of x_1 coefficients in D with the column of constants. Hence,

$$D_{x_1} = \begin{vmatrix} 3 & 1 & 0 & 0 \\ 3 & 1 & 1 & 1 \\ 2 & 1 & 0 & 1 \\ 0 & 0 & 1 & 1 \end{vmatrix}$$

To compute D_{x_1}, we start by interchanging columns 1 and 2 (and changing the sign of the determinant).

$$D_{x_1} = \begin{vmatrix} 3 & 1 & 0 & 0 \\ 3 & 1 & 1 & 1 \\ 2 & 1 & 0 & 1 \\ 0 & 0 & 1 & 1 \end{vmatrix} = -\begin{vmatrix} 1 & 3 & 0 & 0 \\ 1 & 3 & 1 & 1 \\ 1 & 2 & 0 & 1 \\ 0 & 0 & 1 & 1 \end{vmatrix} \qquad C_1 \leftrightarrow C_2$$

Then we get zeros below the first entry in column 1.

$$-\begin{vmatrix} 1 & 3 & 0 & 0 \\ 1 & 3 & 1 & 1 \\ 1 & 2 & 0 & 1 \\ 0 & 0 & 1 & 1 \end{vmatrix} = -\begin{vmatrix} 1 & 3 & 0 & 0 \\ 0 & 0 & 1 & 1 \\ 0 & -1 & 0 & 1 \\ 0 & 0 & 1 & 1 \end{vmatrix} \qquad \begin{matrix} -R_1 + R_2 \rightarrow R_2 \\ -R_1 + R_3 \rightarrow R_3 \end{matrix}$$

Can you see why $D_{x_1} = 0$?

At this point we note that rows 2 and 4 are identical. By the statements in the box on page 614, we can conclude that this determinant is 0. Hence $D_{x_1} = 0$.

Next we compute D_{x_2}, the determinant found by replacing the column of x_2 coefficients in D with the column of constants. Hence,

$$D_{x_2} = \begin{vmatrix} 1 & 3 & 0 & 0 \\ 0 & 3 & 1 & 1 \\ 0 & 2 & 0 & 1 \\ 0 & 0 & 1 & 1 \end{vmatrix}$$

Applying the elementary row and column operations for determinants, we arrive at $D_{x_2} = -3$. (Verify this value for D_{x_2}.)

Approaching D_{x_3} and D_{x_4} in the same manner, we get $D_{x_3} = -1$ and $D_{x_4} = 1$. (We leave it for the student to verify these values.)

The determinants are $D = -1$, $D_{x_1} = 0$, $D_{x_2} = -3$, $D_{x_3} = -1$, and $D_{x_4} = 1$. Now we can find x_1, x_2, x_3, and x_4 by Cramer's Rule:

$$x_1 = \frac{D_{x_1}}{D} = \frac{0}{-1} = 0, \qquad x_2 = \frac{D_{x_2}}{D} = \frac{-3}{-1} = 3,$$

$$x_3 = \frac{D_{x_3}}{D} = \frac{-1}{-1} = 1, \quad \text{and} \quad x_4 = \frac{D_{x_4}}{D} = \frac{1}{-1} = -1$$

Hence our solution is $\boxed{(0, 3, 1, -1)}$. ■

If $D = 0$ and all $D_{x_i} = 0$, the system has infinitely many solutions. If $D = 0$ and any $D_{x_i} \neq 0$, the system has no solutions.

As we mentioned in Section 9.6, if $D = 0$, then the system has no unique solutions: The system is dependent and has infinitely many solutions if *all* $D_{x_i} = 0$; or the system is inconsistent and has no solutions if *any* $D_{x_i} \neq 0$.

Let's step back for a moment and pull together some of our findings in the previous sections. In Section 9.1, we found that if at least one of the equations in an $n \times n$ system is a multiple of another, then we really have fewer equations than the number of unknowns: We will arrive at a dependent solution. If we tried to solve this dependent system by Cramer's Rule, we would find that *all* the determinants would contain at least one row that is a multiple of another and, therefore, all determinants would be 0. This is consistent with our statement regarding $D = 0$ and all $D_{x_i} = 0$.

In Section 9.5 we mentioned that a matrix does not have an inverse if we cannot transform the matrix into an identity matrix using the elementary row operations. This result occurs when, in the process of transforming a matrix into an identity, we obtain at least one row or column consisting entirely of zeros. Given that we now know that the value of a determinant with a row or column of zeros is 0, we can conclude the following.

> A square matrix A has an inverse if and only if $|A| \neq 0$.

We mentioned at the end of Section 9.4 that if the coefficient matrix of a system does not have an inverse, then there is no unique solution. Again, this is equivalent to $D = 0$ using Cramer's Rule.

EXERCISES 9.7

In Exercises 1–6, find the following given

$$A = \begin{vmatrix} 2 & 5 & 3 & -1 \\ 7 & 4 & 0 & -3 \\ -2 & 1 & 8 & 9 \\ -6 & -4 & 6 & -5 \end{vmatrix}.$$

1. The minor of 6
2. The cofactor of −6
3. The cofactor of 0
4. The minor of 1
5. A_{24}
6. M_{32}

In Exercises 7–14, evaluate the determinants.

7. $\begin{vmatrix} 1 & 2 & -1 \\ 0 & 3 & 4 \\ 5 & 1 & 2 \end{vmatrix}$

8. $\begin{vmatrix} 8 & -2 & 5 \\ 3 & 1 & 0 \\ 2 & -1 & 0 \end{vmatrix}$

9. $\begin{vmatrix} 1 & 1 & -3 & 4 \\ 3 & 1 & 0 & 2 \\ 3 & 1 & 0 & 4 \\ 2 & 2 & 0 & 5 \end{vmatrix}$

10. $\begin{vmatrix} 2 & 3 & 1 & 5 \\ 0 & 0 & 0 & 0 \\ 5 & 3 & 1 & 1 \\ 2 & 9 & 5 & 4 \end{vmatrix}$

11. $\begin{vmatrix} 2 & 1 & 3 & 1 \\ -1 & 2 & 4 & 0 \\ 3 & 1 & 1 & -2 \\ 2 & -1 & 0 & 5 \end{vmatrix}$

12. $\begin{vmatrix} 1 & 1 & 3 & -1 & 2 \\ 2 & 1 & 0 & 3 & -1 \\ 0 & 1 & 3 & 1 & 2 \\ 0 & 4 & 1 & 1 & -2 \\ 1 & 1 & 2 & 1 & 3 \end{vmatrix}$

13. $\begin{vmatrix} 1 & 4 & 0 & 3 & 8 \\ 2 & 3 & 0 & 4 & 1 \\ 5 & 6 & 0 & 2 & 3 \\ 4 & 2 & 0 & 3 & 1 \\ 5 & 1 & 0 & 3 & 6 \end{vmatrix}$

14. $\begin{vmatrix} 2 & 3 & 0 & 0 & 1 & 2 \\ 1 & 3 & 4 & 0 & -1 & 3 \\ 2 & 3 & 1 & 0 & -1 & 2 \\ 1 & 3 & 0 & 1 & 1 & 3 \\ 2 & 3 & 0 & 0 & 1 & 4 \\ 3 & 0 & 5 & -1 & 1 & 0 \end{vmatrix}$

In Exercises 15–20, solve the system of equations using Cramer's Rule.

15. $\begin{cases} w + x = -1 \\ w + y + z = 2 \\ x + y = 1 \\ y + z = 3 \end{cases}$

16. $\begin{cases} w + x = 1 \\ w + y = 1 \\ x + y = 0 \\ y + z = -1 \end{cases}$

17. $\begin{cases} w + x + 2y - z = 1 \\ 2x + 3y + z = 1 \\ w + y - z = 0 \\ w + x - y = 4 \end{cases}$

18. $\begin{cases} w + x + y + z = 2 \\ w + x + y = -1 \\ w + x + z = 1 \\ x - y + z = 1 \end{cases}$

19. $\begin{cases} w + x + y + z = 0 \\ w + x + y = 0 \\ 2w + 2x + 2y = 0 \\ 2x - y + z = 0 \end{cases}$

20. $\begin{cases} w + 2x + y + z = 5 \\ w - x - y = 0 \\ 3w - 3x - 3z = 1 \\ x - y + z = 0 \end{cases}$

QUESTION FOR THOUGHT

21. In a system of equations, two equations are equivalent. If you are trying to solve the system using Cramer's Rule, explain what kinds of results you could expect.

9.8 Systems of Linear Inequalities

In Section 2.1 we saw that solutions to linear inequalities in two variables can best be expressed geometrically—that is, by graphing the solution set. The graph of the solution set of the inequality $Ax + By \leq C$ is a half-plane bounded by the *line* $Ax + By = C$. In this section we solve **systems of inequalities** by graphing the solution set of each inequality in the system and then determining where the solution sets intersect.

Solve the following system of linear inequalities: $\begin{cases} x - 2y < 14 \\ x + y < 5 \end{cases}$

By the methods discussed in Section 2.1, we graph the first inequality, $x - 2y < 14$.

1. Graph the equation $x - 2y = 14$.
2. Draw a dashed line to indicate that the line is not included in the solution set.
3. Pick a test point not on the line and determine if it satisfies the inequality $x - 2y < 14$. The point (0, 0) is not on the line; since $0 - 2(0) < 14$ is true, (0, 0) satisfies the inequality.
4. Shade the half-plane on the side of the line containing (0, 0) to indicate that the half-plane is the solution set. The graph of $x - 2y < 14$ is shown in Figure 9.7(a).

Next we follow these same steps and graph the inequality $x + y < 5$, as shown in Figure 9.7(b). Notice that each individual inequality is a half-plane. The dashed lines indicate that the solutions do *not* include the lines.

Since we are looking for ordered pairs that satisfy both inequalities, this means we want the points of intersection of the two half-planes. This is shown as the crosshatched region in Figure 9.7(c).

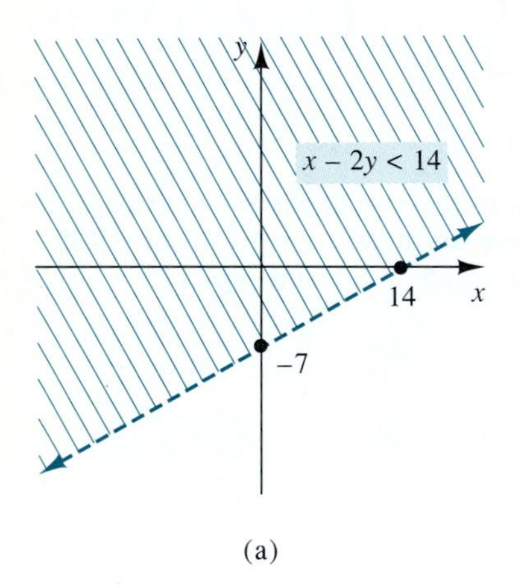

(a)

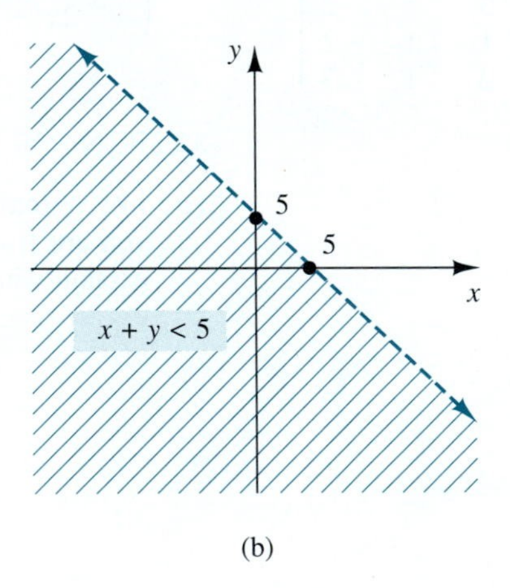

(b)

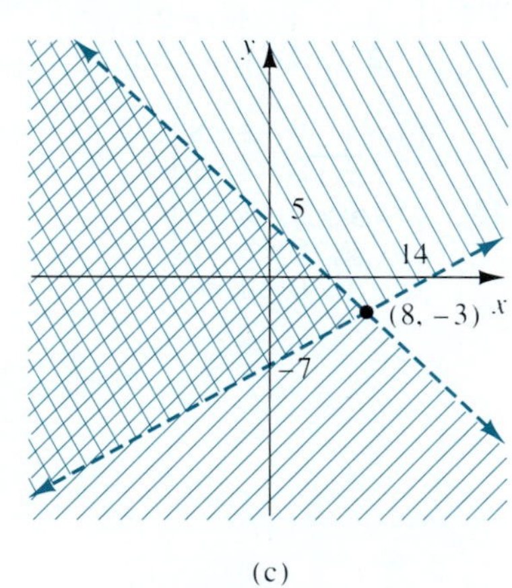

(c)

The crosshatched region is the solution to the system $\begin{cases} x - 2y < 14 \\ x + y < 5 \end{cases}$

FIGURE 9.7

The crosshatched region is where the solutions lie; that is, every ordered pair representing a point in this region satisfies *both* inequalities. For example, (−1, 3) and (2, 2) satisfy both inequalities and lie in this region. The points are solutions to the system of inequalities. On the other hand, points such as (4, 3), (5, −5), and (12, −2) do not satisfy *both* inequalities; they do not lie in this region, and therefore they are not solutions to the system.

To determine the points of intersection of the half-planes we need to solve the system of equations represented by the dashed lines. By using one of the methods discussed in this chapter, we find that the two lines intersect at the point (8, −3). ■

EXAMPLE 2 Solve the following system of linear inequalities: $\begin{cases} 3x - 2y \leq 12 \\ 5x + 2y > 4 \\ y \leq 0 \end{cases}$

Solution We graph each line and determine which half-plane to shade in by methods discussed in Section 2.1. See Figures 9.8(a), 9.8(b), and 9.8(c).

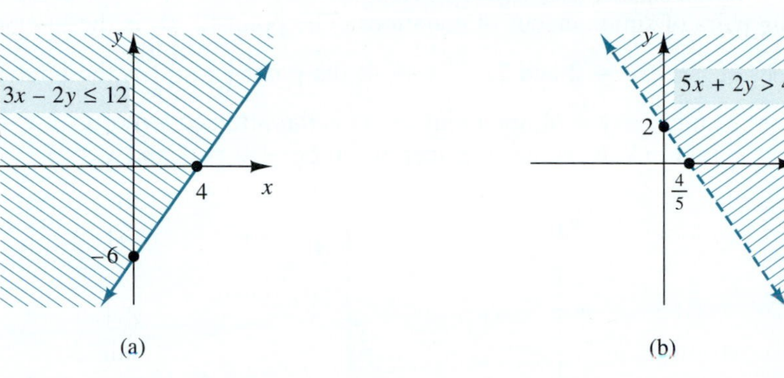

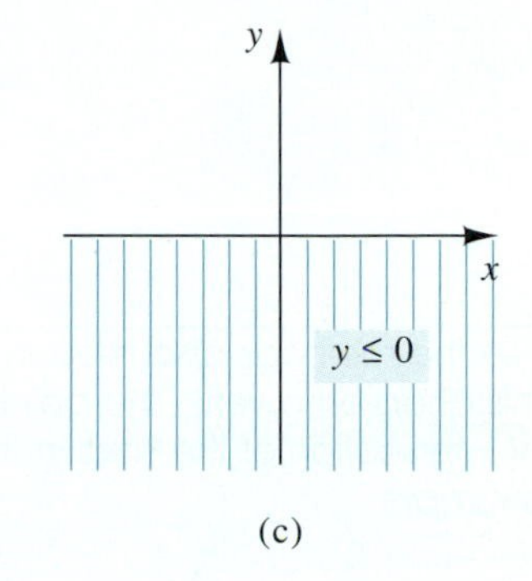

FIGURE 9.8

We use one of the methods discussed in this chapter to find the points of intersection of the lines: $(2, -3)$, $(4, 0)$, and $\left(\frac{4}{5}, 0\right)$. The solution is the region where all three half-planes intersect: the shaded region in Figure 9.9. ■

It's always a good idea to check the solution with a test point. Does $(2, -2)$ work?

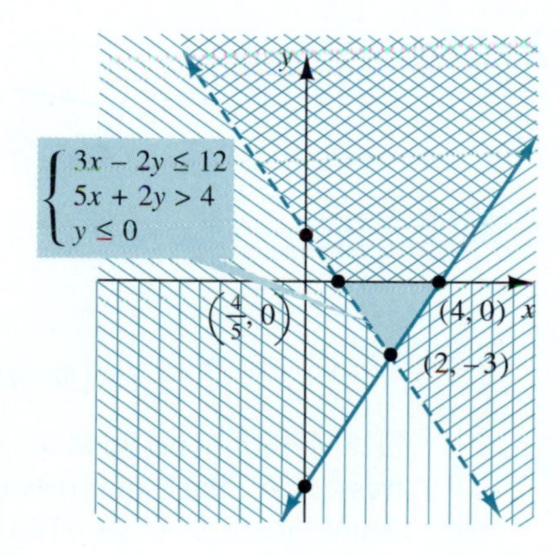

FIGURE 9.9

The points representing the corners of the shaded polygonal region (in this case a triangular region) are called the *vertices* of the solution set. They are the points of intersection of each pair of lines in Example 2.

EXAMPLE 3 Solve the following system of linear inequalities: $\begin{cases} x - 3y < 0 \\ 2x - 6y \geq 12 \end{cases}$

Solution We graph each inequality by graphing the lines and shading the appropriate regions. Since we are looking for ordered pairs that satisfy both inequalities, we are looking for the points of intersection of the two half-planes. When we attempt to find the intersection of the two lines, we find that they have no points in common. As Figure 9.10 shows, the shaded regions never intersect, so there is no solution.

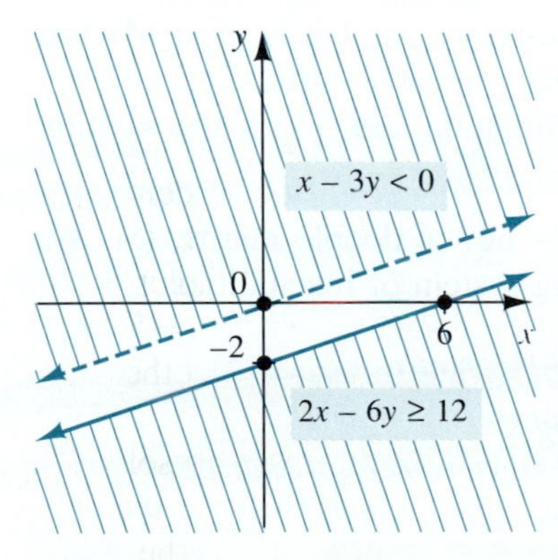

FIGURE 9.10

■

EXAMPLE 4 Solve the following system of linear inequalities: $\begin{cases} x - 5y \le 2 & (1) \\ 2x - y > 4 & (2) \\ x - y < 4 & (3) \\ x < 5 & (4) \end{cases}$

Solution The solution appears in Figure 9.11. Note the corners, or vertices, are found by solving pairs of simultaneous of equations: The point (2, 0) is the intersection of the graphs of $x - 5y = 2$ and $2x - y = 4$; the point $\left(\frac{9}{2}, \frac{1}{2}\right)$ is the intersection of $x - 5y = 2$ and $x - y = 4$; the point (5, 1) is the intersection of $x - y = 4$ and $x = 5$; and the point (5, 6) is the intersection of $2x - y = 4$ and $x = 5$.

Use the methods discussed in this chapter to verify the points of intersection of the lines in this example.

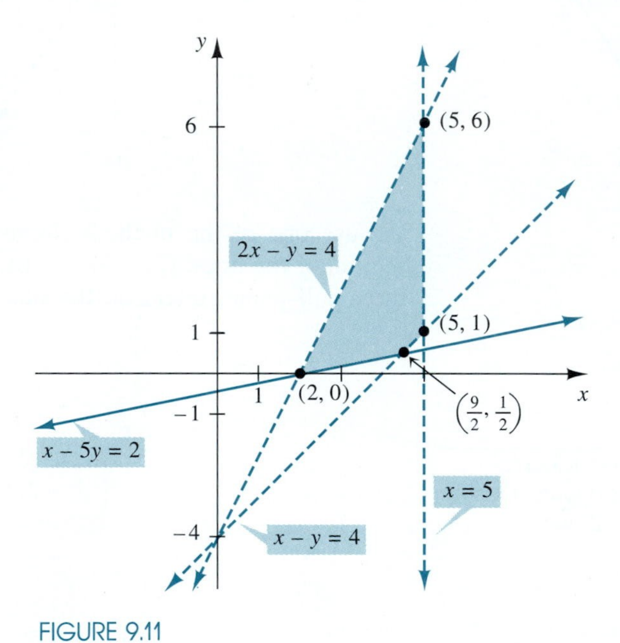

FIGURE 9.11 ■

EXAMPLE 5 The Bedding Store needs to order a shipment of single and double mattresses. A single mattress costs \$80 and takes up 16 cu ft of storage space, and a double mattress costs \$120 and takes up 36 cu ft of storage space. If the store manager wants to order no more than \$6000 worth of mattresses and has at most 1440 cu ft of storage space in the store's storage area, how many of each mattress can she order? Write a system of inequalities to describe this situation and sketch the solution set of the system.

Solution Let x = no. of single mattresses and y = no. of double mattresses. We can translate the information given into the following system of inequalities.

$80x + 120y \le 6000$ *Represents the total cost of the mattresses*

$16x + 36y \le 1440$ *Since the storage space is at most 1440 cu ft*

$x \ge 0$ *Since we cannot have a negative number of mattresses*

$y \ge 0$

We have chosen to let the horizontal axis represent the number of single mattresses and the vertical axis to represent the number of double mattresses. The sketch of the system appears in Figure 9.12.

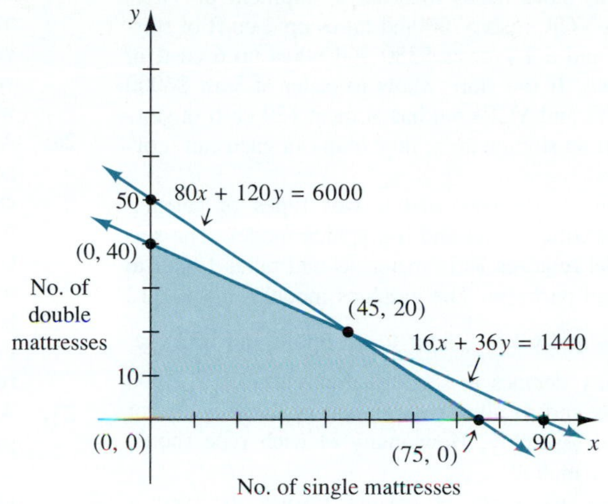

FIGURE 9.12

Noting the boundaries, we stay within the first quadrant: The solution set is the shaded region shown in Figure 9.12. The points of intersection are (0, 0), (0, 40), (45, 20), and (75, 0). Any point in the shaded region satisfies the restrictions. (Note that the point (75, 0) represents what would happen if the store manager ordered single mattresses only.) Keep in mind that since we cannot order a fraction of a mattress, only whole number coordinates make sense. For example, (12.5, 18.2) is a point in the region but is not a realistic option. ■

EXERCISES 9.8

In Exercises 1–20, sketch the solution set of the system of inequalities.

1. $\begin{cases} 2x + y \le 12 \\ 4x + y \ge 8 \end{cases}$

2. $\begin{cases} -x + 2y \le -4 \\ x - y < 3 \end{cases}$

3. $\begin{cases} 3x - 5y < 15 \\ 2x + y > 10 \end{cases}$

4. $\begin{cases} 3x - 5y < 15 \\ 3x - 5y > 10 \end{cases}$

5. $\begin{cases} 5x - 2y \le 10 \\ 6x - 5y > -30 \end{cases}$

6. $\begin{cases} x \le 3y - 12 \\ 2x + 5y > -10 \end{cases}$

7. $\begin{cases} 6x + 8y \le 24 \\ 9x + 12y \ge 18 \end{cases}$

8. $\begin{cases} 3x - 12y \ge 24 \\ 12x \le 8y + 24 \end{cases}$

9. $\begin{cases} 5y < 3x - 15 \\ 8x - 2y \le 10 \end{cases}$

10. $\begin{cases} x + 7y < -14 \\ 3x + 21y > -21 \end{cases}$

11. $\begin{cases} 2x + y \le 12 \\ 4x + y \ge 8 \\ y \ge 0 \end{cases}$

12. $\begin{cases} 5x + 2y \ge -10 \\ -2x + 5y \le -10 \\ x \le 1 \end{cases}$

13. $\begin{cases} x + y - 4 \le 2 \\ 2x - 6y > 12 \\ x \ge 0 \end{cases}$

14. $\begin{cases} 2x - 5y < -20 \\ 4x + 2y > -8 \\ y \le 8 \end{cases}$

15. $\begin{cases} x - y < 4 \\ 2x + y < 12 \\ x \ge 0 \\ y \ge 0 \end{cases}$

16. $\begin{cases} x + 2y > -1 \\ x - y < 4 \\ x \ge 0 \\ y \le 1 \end{cases}$

17. $\begin{cases} x + y \ge -1 \\ 3x - 2y < 12 \\ x \ge -1 \\ y \le 5 \end{cases}$

18. $\begin{cases} 5x - 2y > 0 \\ x - 2y \le 8 \\ x > -1 \\ y \le 0 \end{cases}$

19. $\begin{cases} 5x - 2y < 10 \\ 2x - 5y > -10 \\ x + y \ge -5 \\ x \ge -2 \end{cases}$

20. $\begin{cases} x < 3y - 1 \\ 2x - 5y \ge -10 \\ 2x - 2y < 1 \\ x \ge -1 \end{cases}$

For Exercises 21–28, write a system of inequalities to describe the conditions of the problem and sketch the solution set of the system.

21. An appliance store needs to order a shipment of VCRs and TVs. A VCR costs $200 and takes up 2 cu ft of storage space, and a TV costs $380 and takes up 6 cu ft of storage space. If the store wants to order at least $6000 worth of TVs and VCRs but has at most 120 cu ft of storage space in its storage area, how many of each can it order?

22. An electronics company makes two types of calculators—a scientific model and a graphics model. The scientific model requires $8 in materials and takes 1 hour to assemble and package. The graphics model requires $12 in materials and takes $\frac{1}{2}$ hour to assemble and package. The company decides to spend a maximum of $15,000 on materials and to allot a maximum of 1000 hours for packing and assembly. How many of each type should the company make?

23. Carol makes two kinds of jewelry boxes; one type is custom-made to specifications and the other type is a standard model made from precut parts. On the average, it takes her 3 hours to make a custom-made jewelry box and 1 hour to make the standard model. She can make a profit of $80 on the custom-made box and $20 on the standard model. If she can spend no more than 20 hours a week and wants to earn at least $200 per week from making jewelry boxes, how many of each type should she produce?

24. Joe wants to eat a more balanced breakfast. He reads the sides of two cereal boxes and finds out that 1 ounce of cereal X contains 8 grams of carbohydrates and 0.25 grams of sodium, whereas 1 ounce of cereal Y contains 12 grams of carbohydrates and 0.34 grams of sodium. He wants to create a mixture of cereals with at least 24 grams of carbohydrates and no more than 1 gram of sodium. How much of each cereal should he use?

25. Lana has no more than $60,000 available for investment. She can invest in two types of certificates of deposit: A 6-month certificate, which yields a 5% yearly return on investment, and a 1-year certificate, which yields a 6.5% yearly return. How much should she invest in each if she wants to earn at least $3300 interest in one year?

26. A judge is working on the weekly schedule of hearings in her courtroom, where she hears both minor criminal and civil complaints. She finds that the average civil complaint requires 45 minutes to settle; the average criminal complaint requires 30 minutes. Next week she wants to schedule at least 20 hours but no more than 35 hours of hearings. She prefers to adjudicate at least as many criminal complaints as civil complaints. How many of each type of case can she hear next week?

27. A lawyer handles both divorce cases and malpractice suits. The average divorce case takes 12 hours of his time and 20 hours of his assistants' time. The average malpractice suit takes 22 hours of his time and 18 hours of his assistants' time. The lawyer decided that he can put in no more than 60 hours per week and his assistants should put in at least 40 hours a week. Given these constraints, how many cases of each type should the lawyer handle weekly?

28. An orthodontist uses two types of appliances (A and B), both shown to be effective for overbites. Over a period of a year, each appliance A requires 6 hours of the doctor's labor and 12 hours of the dental assistants' labor; each appliance B requires 4 hours of the doctor's labor and 16 hours of the dental assistants' labor. Over the year, the doctor cannot put in more than 1200 hours of labor for appliances, and the dental assistants cannot put in more than 3200 hours. How many of each appliance should the orthodontist use?

9.9 An Introduction to Linear Programming: Geometric Solutions

Example 5 in Section 9.8 illustrates how we can take a set of restrictions, represent them mathematically, and then picture the restrictions so we may make the most appropriate decisions. Throughout this text we have shown how we can express relationships between two variables or functions of one variable in an effort to determine important points. For example, in Section 3.4, we expressed profit as a quadratic function of price and used what we knew about quadratic functions to determine the price that produces a maximum profit.

There are many problems in economics and the sciences where it is important to find the maximum or minimum value of a function of *several* variables, given a

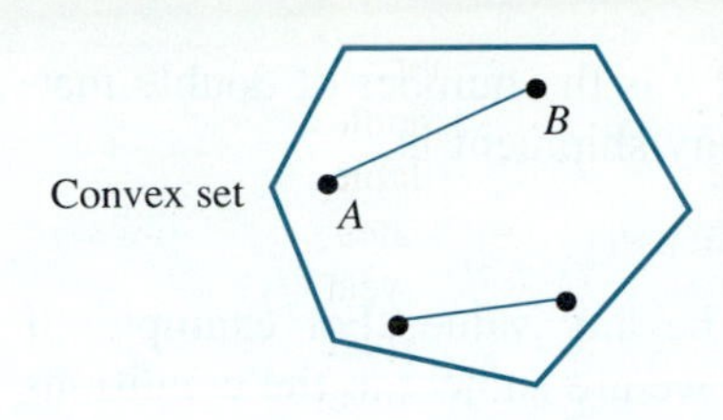

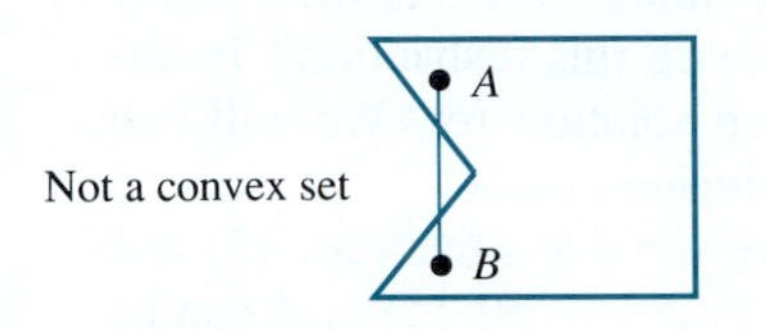

FIGURE 9.13

set of restrictions on these variables. In this section, we discuss how to find solutions to these problems when the function and the restrictions are linear; this is called **linear programming**. Although we can generalize most of our discussion to several variables, we will restrict our attention to linear functions of two variables. Before we continue, we will discuss some terminology.

The solution sets to some of the systems of linear inequalities we have seen so far have been polygonal regions, such as Figure 9.9, Figure 9.11, and Figure 9.12. These regions are called **convex sets**. A set is **convex** if for any two points in the set, the line segment joining them lies completely in the set. See Figure 9.13. In particular, a convex set is **bounded** if it can be enclosed within a circle.

With this terminology in hand, we can start by reexamining Example 5 in Section 9.8, which is restated here: The Bedding Store needs to order a shipment of single and double mattresses. A single mattress costs \$80 and takes up 16 cu ft of storage space, and a double mattress costs \$120 and takes up 36 cu ft of storage space. If the store manager wants to order no more than \$6000 worth of mattresses and has at most 1440 cu ft of storage space in its storage area, how many of each mattress can she order?

Recall that the solution set for the shipment of mattresses was represented by the system of inequalities

$$\begin{cases} 80x + 120y \leq 6000 \\ 16x + 36y \leq 1440 \\ x \geq 0 \\ y \geq 0 \end{cases}$$

The solution set appears in Figure 9.14; we note that this set is a bounded convex set.

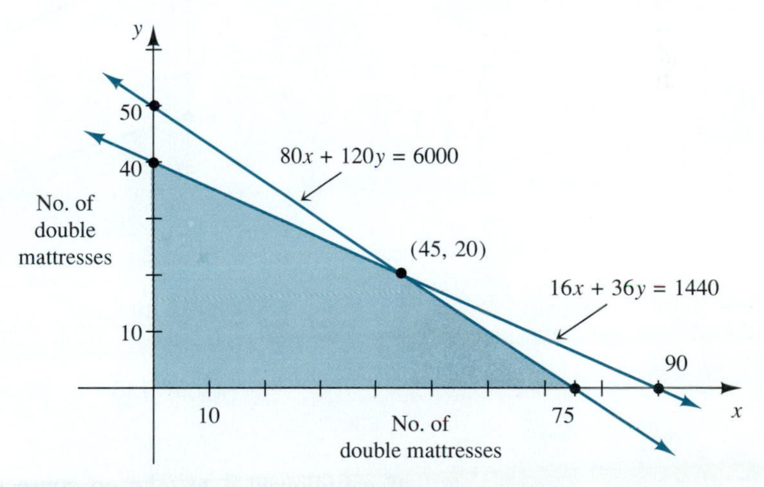

FIGURE 9.14

Suppose The Bedding Store makes a profit of \$25 on the sale of each single mattress and \$50 on the sale of each double mattress. Let's examine the amount of money the store can make on the sale of all mattresses in this shipment.

Since x is the number of single mattresses and y is the number of double mattresses, the total amount the store would make on any shipment is

$$A = 25x + 50y \quad \text{dollars}$$

If x and y were unrestricted, then A could be any value. For example, if $x = 50$ and $y = 100$, then $A = \$6250$. But suppose we are subject to the conditions of the example, where we must restrict the values of x and y so that (x, y) must lie in the solution set given before; how does that restrict the values of A? Is there some maximum or minimum value for profit $A = 25x + 50y$ with this restriction? To answer this question, we examine A restricted to the given solution set. We will call this solution set (the shaded area) the set of **feasible solutions** for A.

With (x, y) confined to the feasible solution set, we see that since (20, 15) and (50, 10) lie in the solution set, then A can be $\$1250 = 25(20) + 50(15)$ or A can be $\$1750 = 25(50) + 50(10)$. To determine if there is some maximum value for A, we start by examining the graph of equation $A = 25x + 50y$.

Suppose $A = 4000$; that is, suppose $25x + 50y = 4000$. The graph of $25x + 50y = 4000$ is a line, and if we graph $25x + 50y = 4000$ on the same set of coordinate axes as the set of feasible solutions, then we find that the line $25x + 50y = 4000$ does not intersect this set. This means that given the restrictions on (x, y) (lying in the shaded area), A can never be 4000. (The store manager cannot make \$4000 from this shipment.) See Figure 9.15.

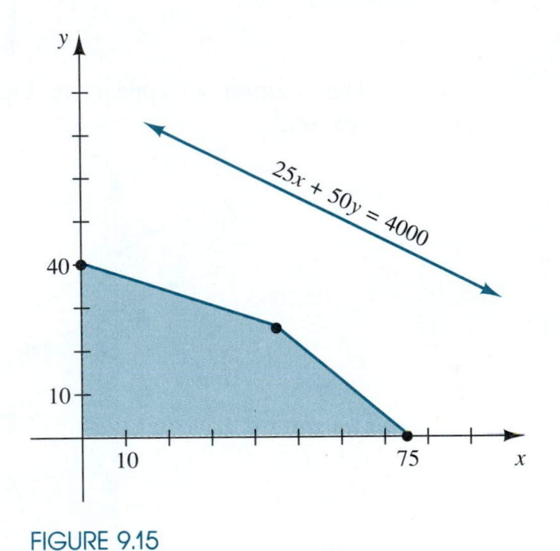

FIGURE 9.15

If we allowed A to take on consecutively lower values, such as 3000, 2000, 1500, 500, and -100, and graphed $A = 25x + 50y$ for each value of A, as illustrated in Figure 9.16, we would find that eventually the line would intersect the shaded region of feasible solutions. This means that A can take on those values where the line $A = 25x + 50y$ intersects the shaded region. For example, since $500 = 25x + 50y$ intersects the shaded region, A can be \$500.

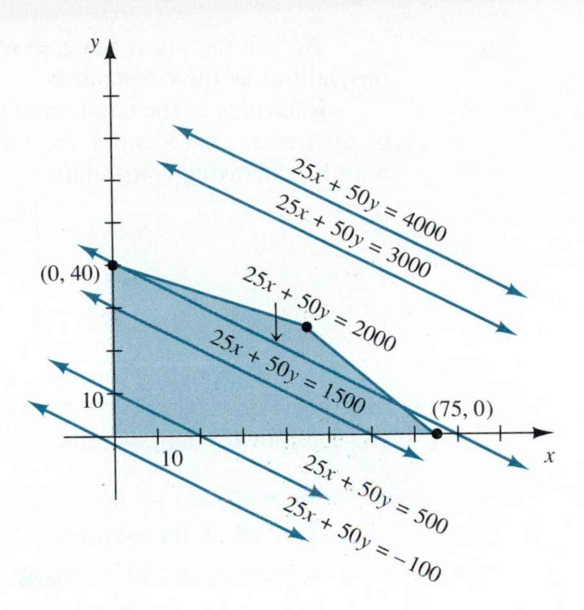

FIGURE 9.16

Looking at Figure 9.16, we see that as A changes, the lines $A = 25x + 50y$ are parallel (since they have the same slope), and, as the line $A = 25x + 50y$ moves toward and then through the shaded area, the values of A get progressively smaller. Moving in this direction, we can see that the first point where the line $A = 25x + 50y$ touches the shaded region will give the maximum value of A for that region, and the last point where the line touches before leaving the shaded region will give the minimum value of A for that region.

It seems intuitive that as A gets smaller, the *first point the line touches on the shaded region will be a vertex, and the last point will be a vertex*. Hence A has its maximum and minimum values in the region at one or more of the vertices. This will be true for any linear function $F = ax + by + c$ where the set of feasible solutions is a bounded convex set. We state the following without proof.

A LINEAR PROGRAMMING THEOREM

Given a linear function F,

$$F = ax + by + c$$

where a, b, and c are constants and x and y are subject to the constraints of a system of linear inequalities,

1. If the set of feasible solutions is a bounded convex set, then F will attain a maximum and a minimum value.
2. If F does have a maximum or a minimum value, it will occur at one of the vertices of the set of feasible solutions.

We call the linear function F the **objective function.** We refer to the system of inequalities as the **constraints**.

Returning to the question of the maximum the store can make on this shipment of mattresses, we identify the function $A = 25x + 50y$ as the objective function, with the following constraints:

$$\begin{cases} 80x + 120y \leq 6000 \\ 16x + 36y \leq 1440 \\ x \geq 0 \\ y \geq 0 \end{cases}$$

By the preceding theorem, the maximum the store can profit on the sale of the mattresses ordered will occur at a vertex. Therefore, we compute A at each vertex and look for the largest value of A. The vertices are (0, 0), (0, 40), (45, 20), and (75, 0).

If we check the vertex (0, 0) we get $A = 25x + 50y = 25(0) + 50(0) = \$0$.

The vertex (0, 40) yields $A = 25x + 50y = 25(0) + 50(40) = \$2000$.

The vertex (45, 20) yields $A = 25x + 50y = 25(45) + 50(20) = \$2125$. *Maximum value*

The vertex (75, 0) yields $A = 25x + 50y = 25(75) + 50(0) = \$1875$.

Hence the maximum the store could make from this shipment (given the constraints) is \$2125, which will occur if the store orders 45 single mattresses and 20 double mattresses.

EXAMPLE 1 A bank has \$90 million to invest for business loans and personal loans. A state regulation requires that the bank must invest at least twice as much money in business loans as in personal loans. If the interest rate for business loans is 10% yearly and the interest rate for personal loans is 11% yearly, what is the maximum profit the bank can make, given its constraints?

Solution Let x = the amount of dollars (in millions) lent by the bank for business loans and let y = the amount of dollars (in millions) lent by the bank for personal loans. The yearly profit made on loans can be expressed as a linear function of x and y, as follows:

$$P = 0.10x + 0.11y \qquad \textit{This is the objective function.}$$

This objective function, P, is what we want to maximize. The constraints are as follows:

$$\begin{cases} x + y \leq 90 & \textit{Represents the total amount the bank invests (in millions of dollars).} \\ 2y \leq x & \textit{Represents the state regulation (at least twice as much invested in business as in personal loans).} \\ x \geq 0 & \textit{Since the bank cannot invest a negative amount of money.} \\ y \geq 0 \end{cases}$$

Note that under the given constraints, the bank could invest \$65 million in business loans and \$10 million in personal loans (the ordered pair (65, 10) satisfies the constraints). The profit function would yield a profit of $P = 0.10x + 0.11y = 0.10(65) + 0.11(10) = 7.6$ million dollars.

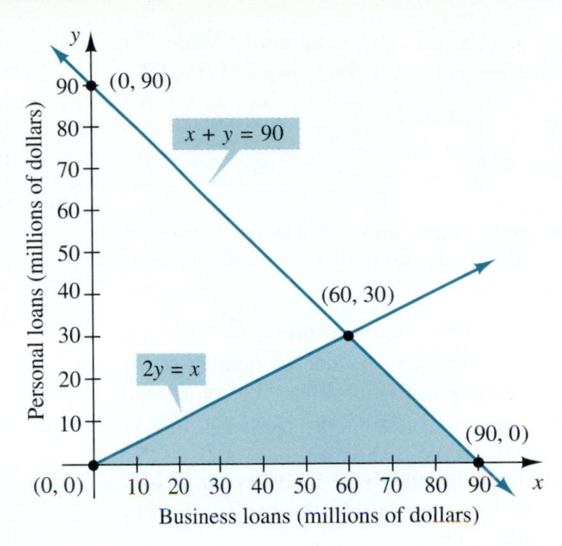

FIGURE 9.17

The graph of the constraints appears in Figure 9.17. Note the vertices are (0, 0), (60, 30), and (90, 0). We note that the shaded area is a bounded convex set; therefore, by the linear programming theorem, there is a maximum, and this maximum occurs at a vertex. So we only have to check the vertices to determine how much of each type of loan will maximize the bank's profit. If we check the vertex (0, 0), we get:

$$P = 0.10x + 0.11y = 0.10(0) + 0.11(0) = 0 \text{ dollars}$$

If we check the vertex (60, 30), we get:

$$P = 0.10x + 0.11y = 0.10(60) + 0.11(30)$$
$$= 9.3 \text{ million dollars} \qquad \textit{Maximum value}$$

If we check the vertex (90, 0), we get:

$$P = 0.10x + 0.11y = 0.10(90) + 0.11(0) = 9 \text{ million dollars}$$

Hence the maximum profit of \$9.3 million occurs if the bank can make \$60 million in business loans and \$30 million in personal loans. ■

EXERCISES 9.9

1. What are the maximum and minimum values of $f = 3x + 2y$ subject to the following constraints?

$$\begin{cases} x + 2y \le 6 \\ y \le x + 1 \\ x \ge 0 \\ y \ge 0 \end{cases}$$

2. What are the maximum and minimum values of $f = 3x + 9y$ subject to the same constraints as Exercise 1?

3. What are the maximum and minimum values of $f = 0.3x - y$ subject to the following constraints?

$$\begin{cases} x - 2y \le 4 \\ y + x \le 1 \\ x \ge 0 \end{cases}$$

4. What are the maximum and minimum values of $f = 3x - 0.5y$ subject to the constraints of Exercise 3?

5. What are the maximum and minimum values of $f = 2x + 5y - 1$ subject to the following constraints?

$$\begin{cases} 2x + 3y \geq 6 \\ x + y \leq 3 \\ x \geq 0 \end{cases}$$

6. What are the maximum and minimum values of $f = 8x - 2y + 4$ subject to the constraints of Exercise 5?

7. A bank has a total of $75 million to invest for high-risk and conservative investments. A federal regulation requires that the bank can put no more than 30% of its total investment funds into high-risk ventures. If the rate of return for conservative investments averages 10% yearly and the rate of return for high-risk investments is 14% yearly, what is the maximum profit the bank can make given its constraints?

8. A computer store needs to order two types of computers: Model A and model B. Model A costs $400 for the store to buy wholesale, and the wholesale price of model B is $200. Model A takes up 12 cu ft of space and model B takes up 16 cu ft of space. The store owner decides that he wants to buy at most $80,000 worth of the models wholesale, but he has only 6000 cu ft of storage space. If his profit is $300 on each model A he sells and $350 on each model B, how many of each model should he buy to maximize his profit?

9. A computer store needs to order two types of printers: Model X and model Y. Model X costs $400 for the store to buy wholesale, and the wholesale price of Model Y is $500. Model X takes up 16 cu ft of space and model Y takes up 12 cu ft of space. The store owner decides that she wants to buy at most $50,000 worth of the models wholesale, but she has only 1600 cu ft of storage space. Her profit is $90 on each model X she sells and $100 on each model Y she sells. How many of each model should she order to maximize her profit?

10. Farmer Jack has 100 acres available to plant crops. Crop A costs $8 an acre for seed and $80 an acre for labor. Crop B costs $10 an acre for seed and $60 an acre for labor. He does not want to spend more than $900 total for seed or more than $7000 for labor. If his profits are $120 per acre for crop A and $140 an acre for crop B, how many of each crop should he plant in order to maximize his profits?

11. An electronics company makes two types of calculators, a scientific model and a graphics model. The scientific model requires $8 in materials and takes 1 hour to assemble and package. The graphics model requires $15 in materials and takes $\frac{1}{2}$ hour to assemble and package. The company decides to spend a maximum of $15,000 on material and to allot a maximum of 1000 hours for packing and assembly. Under these constraints, if the electronics company profits $10 on each scientific model and $15 on each graphics model, how many of each should they produce if they want the maximum profit?

12. Under the constraints of Exercise 11, if the electronics company profits $10 on each scientific model and $30 on each graphics model, how many of each should they produce if they want the maximum profit?

13. Maria wants to eat a more balanced breakfast. She reads the sides of two cereal boxes and finds out that 1 ounce of cereal X contains 8 grams of carbohydrates and 0.25 grams of sodium, whereas 1 ounce of cereal Y contains 12 grams of carbohydrates and 0.34 grams of sodium. She wants to create a mixture of cereals that contains at least 24 grams of carbohydrates and no more than 1 gram of sodium. Under these constraints, if it costs Maria 25¢ an ounce for cereal X and 30¢ an ounce for cereal Y, what combination of cereals would be the least expensive?

14. Under the constraints of Exercise 13, if it costs Maria 25¢ an ounce for cereal X and 40¢ an ounce for cereal Y, what combination of cereals would be the least expensive?

15. A lawyer handles both divorce cases and malpractice suits. The average divorce case takes 12 hours of her time and 20 hours of her assistants' time. The average malpractice suit takes 22 hours of her time and 18 hours of her assistants' time. The lawyer decided that she can put in no more than 60 hours per week, and her assistants can put in no more than 80 hours a week. Under these constraints, if the lawyer averages $6000 on divorce cases and $8000 on malpractice suits, how many of each type of case should she take on weekly to maximize her income?

16. An orthodontist uses two types of appliances (A and B), both shown to be effective for overbites. Over a year, each appliance A requires 6 hours of the doctor's labor and 14 hours of the dental assistants' labor; each appliance B requires 4 hours of the doctor's labor and 16 hours of the dental assistants' labor. Over the year the doctor cannot put in more than 1200 hours labor for appliances, and the assistants cannot put in more than 3200 hours. Under these constraints, if the orthodontist makes a profit of $2000 on appliance A and $1500 on appliance B, how many of each appliance should he use in order to maximize profit?

17. A pharmaceutical company produces two types of drugs, Zomine X and Zomine Y, which help to reduce the effects of migraine headaches. But each drug contains substances which produce negative side-effects as well. Of each gram of Zomine X, 5 mg will directly relieve migraine pain; but 2 mg will produce a Type I negative side-effect and 4 mg will produce a Type II negative side-effect. On the other hand, of each gram of Zomine Y, 2 mg will provide pain relief; but 1 mg produces a Type I negative side-effect and 1 mg will produce a Type II negative side-effect. The body will not tolerate more than 6 mg daily of anything producing a Type I effect, nor will it tolerate more than 8 mg daily of anything producing a Type II side-effect. For the drug to have any affect, the user must take at least 1 gram of either or both Zomines. How many grams of each type of Zomine should be administered daily in order to maximize the pain relief?

Chapter 9 SUMMARY

After completing this chapter, you should be able to:

1. Solve linear systems by the elimination, Gaussian elimination, and substitution methods. (Sections 9.1 and 9.2)

The point of the elimination method is to combine equations in order to eliminate variables (and equations) until we have a single equation in one variable.

For example:
Solve the following system of equations.

$$\begin{cases} x - 2y + 3z = 13 & (1) \\ 2x + y - z = -5 & (2) \\ x - 2y + z = 7 & (3) \end{cases}$$

Solution:
The first step is to eliminate a variable and arrive at two equations in two variables. Look at equations (2) and (3). We will eliminate z.

$$\begin{array}{rl} 2x + y - z = -5 & \textit{Add equations (2) and (3).} \\ x - 2y + z = 7 & \\ \hline 3x - y \quad = 2 & (4) \end{array}$$

Now we have to find another equation in the same two variables, x and y. We will use equations (1) and (2).

$$\begin{array}{ll} (1)\ x - 2y + 3z = 13 & \textit{We leave equation (1) as is and} \\ (2)\ 2x + y - z = -5 & \textit{multiply equation (2) by 3 to get} \end{array}$$

$$\begin{array}{rl} x - 2y + 3z = 13 & \\ 6x + 3y - 3z = -15 & \textit{Add.} \\ \hline 7x + y \quad = -2 & (5) \end{array}$$

$$\begin{array}{rl} \begin{cases} 3x - y = 2 \\ 7x + y = -2 \end{cases} & \textit{Add equations (4) and (5).} \\ \hline 10x \quad = 0 & \textit{which yields } x = 0 \end{array}$$

Now substitute $x = 0$ back into equation (4) and solve for y.

$$3x - y = 2$$
$$3(0) - y = 2 \Rightarrow y = -2$$

Finally, substitute $x = 0$ and $y = -2$ back into equation (1) and solve for z.

$$x - 2y + 3z = 13$$
$$0 - 2(-2) + 3z = 13 \Rightarrow z = 3$$

Hence our solution is $(0, -2, 3)$. The student should check this solution.

2. Solve linear systems using augmented matrices, by either Gaussian elimination or the Gauss-Jordan method. (Section 9.3)

Using matrix methods, we need only focus on the coefficients of the variables in a system of equations. We use row operations to transform the augmented matrix to triangular or reduced echelon form, which yields either the solution or a simpler system to solve.

For example:
Solve the following system using Gauss-Jordan elimination.

Solution:
First we write the systems associated matrix:

$$\begin{cases} x + y + 4z = 3 \\ 4x + 2y + 4z = 8 \\ 2x + 3y + 12z = 8 \end{cases} \Rightarrow \left[\begin{array}{ccc|c} 1 & 1 & 4 & 3 \\ 4 & 2 & 4 & 8 \\ 2 & 3 & 12 & 8 \end{array}\right]$$

Now use the elementary row operations to transform the matrix into reduced echelon form.

a. Since 1 is in the upper left-hand corner, begin by sweeping out the rest of the first column.

$$\left[\begin{array}{ccc|c} 1 & 1 & 4 & 3 \\ 4 & 2 & 4 & 8 \\ 2 & 3 & 12 & 8 \end{array}\right]$$

$$\begin{array}{l} -4R_1 + R_2 \to R_2 \\ -2R_1 + R_3 \to R_3 \end{array} \quad \left[\begin{array}{ccc|c} 1 & 1 & 4 & 3 \\ 0 & -2 & -12 & -4 \\ 0 & 1 & 4 & 2 \end{array}\right]$$

b. Next we want to get 1 in the middle of the main diagonal (in row 2, column 2).

$$\left[\begin{array}{ccc|c} 1 & 1 & 4 & 3 \\ 0 & -2 & -12 & -4 \\ 0 & 1 & 4 & 2 \end{array}\right]$$

$$-\tfrac{1}{2}R_2 \to R_2 \quad \left[\begin{array}{ccc|c} 1 & 1 & 4 & 3 \\ 0 & 1 & 6 & 2 \\ 0 & 1 & 4 & 2 \end{array}\right]$$

c. Now we sweep out the rest of the second column.

$$\left[\begin{array}{ccc|c} 1 & 1 & 4 & 3 \\ 0 & 1 & 6 & 2 \\ 0 & 1 & 4 & 2 \end{array}\right]$$

$$\begin{array}{l} -R_2 + R_1 \to R_1 \\ -R_2 + R_3 \to R_3 \end{array} \quad \left[\begin{array}{ccc|c} 1 & 0 & -2 & 1 \\ 0 & 1 & 6 & 2 \\ 0 & 0 & -2 & 0 \end{array}\right]$$

d. Next, we want to get 1 in the lower right entry of the main diagonal.

$$\left[\begin{array}{ccc|c} 1 & 0 & -2 & 1 \\ 0 & 1 & 6 & 2 \\ 0 & 0 & -2 & 0 \end{array}\right]$$

$$-\tfrac{1}{2}R_3 \to R_3 \quad \left[\begin{array}{ccc|c} 1 & 0 & -2 & 1 \\ 0 & 1 & 6 & 2 \\ 0 & 0 & 1 & 0 \end{array}\right]$$

e. Finally sweep out the rest of the last column.

$$\left[\begin{array}{ccc|c} 1 & 0 & -2 & 1 \\ 0 & 1 & 6 & 2 \\ 0 & 0 & 1 & 0 \end{array}\right]$$

$$\begin{array}{l} 2R_3 + R_1 \to R_1 \\ -6R_3 + R_2 \to R_2 \end{array} \quad \left[\begin{array}{ccc|c} 1 & 0 & 0 & 1 \\ 0 & 1 & 0 & 2 \\ 0 & 0 & 1 & 0 \end{array}\right]$$

Read off the solution to get $x = 1$, $y = 2$, $z = 0$.

3. Add and multiply matrices, including scalar multiplication. (Section 9.4)

We can add two matrices of the same size by adding their corresponding elements. To multiply a matrix by a constant, called scalar multiplication, multiply each entry by the constant.

For example:

Given matrices

$$A = \begin{bmatrix} 1 & 4 & 2 \\ -1 & 0 & 3 \end{bmatrix}, \quad B = \begin{bmatrix} 2 & 1 \\ -1 & 3 \\ 0 & 5 \end{bmatrix}, \quad C = \begin{bmatrix} 1 & -1 \\ 4 & 0 \\ 3 & 1 \end{bmatrix},$$

to compute $2B - C$, notice that addition makes sense since B and C are the same size, 3×2.

$$2B - C = 2\begin{bmatrix} 2 & 1 \\ -1 & 3 \\ 0 & 5 \end{bmatrix} + (-1)\begin{bmatrix} 1 & -1 \\ 4 & 0 \\ 3 & 1 \end{bmatrix}.$$

Find $2B$ and $(-1)C$.

$$= \begin{bmatrix} 4 & 2 \\ -2 & 6 \\ 0 & 10 \end{bmatrix} + \begin{bmatrix} -1 & 1 \\ -4 & 0 \\ -3 & -1 \end{bmatrix} = \begin{bmatrix} 3 & 3 \\ -6 & 6 \\ -3 & 9 \end{bmatrix}$$

Two matrices can be multiplied if the number of rows of the matrix on the left is equal to the number of columns of the matrix on the right.

For example:
To compute the product AB using the matrices given above, we first note that since A is a 2×3 matrix and B is a 3×2 matrix, AB makes sense and is a 2×2 matrix.

$$AB = \begin{bmatrix} 1 & 4 & 2 \\ -1 & 0 & 3 \end{bmatrix}\begin{bmatrix} 2 & 1 \\ -1 & 3 \\ 0 & 5 \end{bmatrix}$$

$$= \begin{bmatrix} 1(2) + 4(-1) + 2(0) & 1(1) + 4(3) + 2(5) \\ -1(2) + 0(-1) + 3(0) & -1(1) + 0(3) + 3(5) \end{bmatrix}$$

$$= \begin{bmatrix} -2 & 23 \\ -2 & 14 \end{bmatrix}$$

4. Find the inverse of a matrix and use it to solve a system of linear equations. (Section 9.5)

By using the inverse of a matrix, we can approach

solving a system as though we were solving a simple linear equation. That is, if $AX = K$, then $X = A^{-1}K$.

For example:
Using a matrix inverse, find the solution to the following system of equations:

$$\begin{cases} x - 2y = 4 \\ x + 5y = -3 \end{cases}$$

Solution:
We identify matrices A, X, and K:

$$A = \begin{bmatrix} 1 & -2 \\ 1 & 5 \end{bmatrix} \quad X = \begin{bmatrix} x \\ y \end{bmatrix} \quad K = \begin{bmatrix} 4 \\ -3 \end{bmatrix}$$

Then $AX = K$. We find A^{-1} to be,

$$A^{-1} = \begin{bmatrix} \frac{5}{7} & \frac{2}{7} \\ -\frac{1}{7} & \frac{1}{7} \end{bmatrix} = \frac{1}{7}\begin{bmatrix} 5 & 2 \\ -1 & 1 \end{bmatrix}$$

Now we can solve for X

$$X = A^{-1}K$$

$$= \frac{1}{7}\begin{bmatrix} 5 & 2 \\ -1 & 1 \end{bmatrix}\begin{bmatrix} 4 \\ -3 \end{bmatrix} = \begin{bmatrix} 2 \\ -1 \end{bmatrix}$$

Hence $X = \begin{bmatrix} x \\ y \end{bmatrix} = \begin{bmatrix} 2 \\ -1 \end{bmatrix}$ which means $x = 2$ and $y = -1$.

5. Evaluate determinants. (Sections 9.6 and 9.7)

A 2×2 determinant is defined as $\begin{vmatrix} a & b \\ c & d \end{vmatrix} = ad - bc$.

A 3×3 determinant is evaluated by expansion of minors: We can expand down any column or across any row provided we prefix each entry with the appropriate sign from the sign array.

For example:
We evaluate determinants as follows:

(a) $\begin{vmatrix} 5 & 1 \\ 3 & -2 \end{vmatrix} = (5)(-2) - (3)(1) = -13$

(b) $\begin{vmatrix} 4 & 1 & -2 \\ -3 & 2 & 3 \\ 2 & 0 & -1 \end{vmatrix}$ *We expand across the third row to get*

$$= 2\begin{vmatrix} 1 & -2 \\ 2 & 3 \end{vmatrix} - 0\begin{vmatrix} 4 & -2 \\ -3 & 3 \end{vmatrix} + (-1)\begin{vmatrix} 4 & 1 \\ -3 & 2 \end{vmatrix}$$

$$= 2[3 - (-4)] - 0[12 - 6] - [8 - (-3)]$$

$$= 2(7) - 0(6) + (-1)(11) = 3$$

6. Solve systems of equations using Cramer's Rule. (Sections 9.6 and 9.7)

We can solve $n \times n$ linear systems of equations by using Cramer's Rule. Cramer's Rule utilizes determinants of the system's coefficients.

For example:
Solve the following system of equations using Cramer's Rule:

$$\begin{cases} 3x - y = -4 \\ 2x + y - 2z = 3 \\ y - 6z = 1 \end{cases}$$

Solution:
First we compute D, because if $D = 0$, then there is no unique solution. We compute D by expanding across the third row.

$$D = \begin{vmatrix} 3 & -1 & 0 \\ 2 & 1 & -2 \\ 0 & 1 & -6 \end{vmatrix}$$

$$= 0\begin{vmatrix} -1 & 0 \\ 1 & -2 \end{vmatrix} - 1\begin{vmatrix} 3 & 0 \\ 2 & -2 \end{vmatrix} + (-6)\begin{vmatrix} 3 & -1 \\ 2 & 1 \end{vmatrix}$$

$$= 0 - 1(-6) - 6(3 + 2) = -24$$

Next we compute D_x by expanding down the last column.

$$D_x = \begin{vmatrix} -4 & -1 & 0 \\ 3 & 1 & -2 \\ 1 & 1 & -6 \end{vmatrix}$$

$$= 0\begin{vmatrix} 3 & 1 \\ 1 & 1 \end{vmatrix} - (-2)\begin{vmatrix} -4 & -1 \\ 1 & 1 \end{vmatrix} + (-6)\begin{vmatrix} -4 & -1 \\ 3 & 1 \end{vmatrix}$$

$$= 0 + 2(-4 + 1) - 6(-4 + 3) = 0$$

We compute D_y by expanding down the first column.

$$D_y = \begin{vmatrix} 3 & -4 & 0 \\ 2 & 3 & -2 \\ 0 & 1 & -6 \end{vmatrix}$$

$$= 3\begin{vmatrix} 3 & -2 \\ 1 & -6 \end{vmatrix} - 2\begin{vmatrix} -4 & 0 \\ 1 & -6 \end{vmatrix} + 0\begin{vmatrix} -4 & 0 \\ 3 & -2 \end{vmatrix}$$

$$= 3(-18 + 2) - 2(24) + 0 = -96$$

We compute D_z by expanding across the third row.

$$D_z = \begin{vmatrix} 3 & -1 & -4 \\ 2 & 1 & 3 \\ 0 & 1 & 1 \end{vmatrix}$$

$$= 0\begin{vmatrix} -1 & -4 \\ 1 & 3 \end{vmatrix} - 1\begin{vmatrix} 3 & -4 \\ 2 & 3 \end{vmatrix} + 1\begin{vmatrix} 3 & -1 \\ 2 & 1 \end{vmatrix}$$

$$= 0 - 1(9 + 8) + 1(3 + 2) = -12$$

Hence $x = \frac{D_x}{D} = \frac{0}{-24} = 0$, $y = \frac{D_y}{D} = \frac{-96}{-24} = 4$, and $z = \frac{D_z}{D} = \frac{-12}{-24} = \frac{1}{2}$.

7. Solve a variety of verbal problems that give rise to systems of linear equations. (Sections 9.1–9.7)

8. Sketch the solution set of a system of linear inequalities. (Section 9.8)

On the same coordinate plane, we graph each linear inequality as a half-plane as described in Section 2.1. The intersection of the half-planes is the solution, which we represent as the darkest or cross-hatched region. You can determine the vertices of the region by solving pairs of simultaneous equations.

For example:
Find the solution set for the following system of linear inequalities.

$$\begin{cases} 3x - 2y \le 12 \\ 2x + y > -2 \\ y \le 2 \end{cases}$$

Solution:
Graph each inequality. That is, graph the line and determine which half-plane to shade. The solution is the region where all three half-planes intersect, which is represented by the darkest shaded region in Figure 9.18.

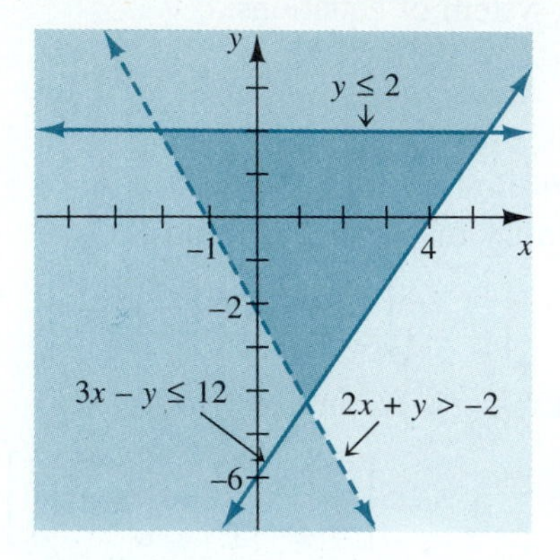

FIGURE 9.18

9. Use the linear programming theorem to solve verbal problems. (Section 9.9)

Chapter 9 REVIEW EXERCISES

In Exercises 1–12, solve the system of equations.

1. $\begin{cases} 2x + y = 12 \\ 3x - y = 8 \end{cases}$

2. $\begin{cases} -x + 2y = -4 \\ x - y = 3 \end{cases}$

3. $\begin{cases} 3x - 2y = 15 \\ 2x + y = 10 \end{cases}$

4. $\begin{cases} 5x + 2y = -8 \\ 3x + y = -4 \end{cases}$

5. $\begin{cases} 5x - 2y = 1 \\ x - 5y = -32 \end{cases}$

6. $\begin{cases} x = 3y - 1 \\ 2x - 5y = -3 \end{cases}$

7. $\begin{cases} x + y + z = 6 \\ 3x + y - z = 6 \\ 2x + y - z = 4 \end{cases}$

8. $\begin{cases} x - 2y + z = -5 \\ 3x - y + z = 2 \\ x + 2y - z = 9 \end{cases}$

9. $\begin{cases} x + y - z = 1 \\ 2x + 2y + 2z = 0 \\ x - y + z = 3 \end{cases}$

10. $\begin{cases} 3x + y + 2z = 0 \\ x - 2y - z = -3 \\ x + y + z = 1 \end{cases}$

11. $\begin{cases} 0.2a + 0.1b - c = 0.7 \\ a - 0.5b + 2c = 0.5 \\ a + 2b + c = 8 \end{cases}$

12. $\begin{cases} x - y = 2z + y \\ x - 3y = y - 1 \\ 2x - 3y = 3z \end{cases}$

In Exercises 13–16, put the augmented matrix into reduced echelon form.

13. $\left[\begin{array}{cc|c} -1 & 2 & 2 \\ 4 & 6 & 6 \end{array}\right]$

14. $\left[\begin{array}{cc|c} 3 & 0 & 6 \\ -6 & 1 & 3 \end{array}\right]$

15. $\left[\begin{array}{ccc|c} 1 & 1 & 2 & 2 \\ 2 & 0 & 2 & 0 \\ 3 & -1 & 1 & 1 \end{array}\right]$

16. $\left[\begin{array}{ccc|c} 1 & -1 & 0 & 5 \\ -2 & 2 & 1 & -1 \\ 4 & -1 & 0 & 2 \end{array}\right]$

In Exercises 17–20, solve the system of equations by Gaussian elimination.

17. $\begin{cases} x + y = 5 \\ 3x - y = 8 \end{cases}$

18. $\begin{cases} x + 2y = 9 \\ x - y = -6 \end{cases}$

19. $\begin{cases} x + y + z = -3 \\ 3x + y - z = 11 \\ 2x + y - z = 9 \end{cases}$

20. $\begin{cases} x + y - z = 1 \\ 2x + 2y + 2z = 1 \\ x - y + z = 3 \end{cases}$

In Exercises 21–24, solve the system of equations by Gauss-Jordan elimination.

21. $\begin{cases} x - 2y = 4 \\ 2x + y = 13 \end{cases}$

22. $\begin{cases} 3x + 2y = 3 \\ 3x + y = 0 \end{cases}$

23. $\begin{cases} x - 3y + z = 0 \\ x - y + z = 0 \\ x + 2y - z = 2 \end{cases}$

24. $\begin{cases} x + y + 2z = 7 \\ x - 2y - z = 1 \\ x + y + z = 4 \end{cases}$

In Exercises 25–34, perform the operations if possible.

25. $\begin{bmatrix} 3 & -2 \\ 5 & -8 \end{bmatrix} + \begin{bmatrix} 7 & -1 \\ 4 & 0 \end{bmatrix}$

26. $-2\begin{bmatrix} 2 & -3 & 8 \\ 5 & -1 & 0 \\ -2 & 7 & 18 \end{bmatrix}$

27. $2\begin{bmatrix} 1 & -3 \\ 5 & 2 \end{bmatrix} - \begin{bmatrix} 2 & 5 \\ 3 & 0 \end{bmatrix}$

28. $\begin{bmatrix} 5 & 3 & 6 \\ 3 & 9 & 0 \\ 2 & 12 & 6 \end{bmatrix} + \begin{bmatrix} 2 & -3 & 1 \\ 5 & 0 & 4 \end{bmatrix}$

29. $\begin{bmatrix} 1 & 0 & 2 & 1 \end{bmatrix}\begin{bmatrix} 1 \\ 0 \\ 2 \\ 1 \end{bmatrix}$

30. $\begin{bmatrix} 1 \\ 0 \\ 2 \\ 1 \end{bmatrix}\begin{bmatrix} 2 & 3 & 4 & 0 \\ 1 & 0 & 2 & 1 \end{bmatrix}$

31. $\begin{bmatrix} 1 & 0 \\ 0 & 1 \end{bmatrix}\begin{bmatrix} 5 & 8 \\ -4 & 1 \end{bmatrix}$

32. $\begin{bmatrix} -1 & 0 & 3 \\ 2 & 2 & 5 \end{bmatrix}\begin{bmatrix} 1 & 0 \\ -4 & 3 \\ 2 & -1 \end{bmatrix}$

33. $\begin{bmatrix} 1 & -1 & 0 \\ 2 & 3 & 1 \\ 5 & 1 & 3 \end{bmatrix}\begin{bmatrix} 1 & -2 & 1 \\ 2 & 0 & 1 \\ 3 & -1 & 4 \end{bmatrix}$

34. $\begin{bmatrix} 1 & -2 & 1 \\ 2 & 0 & 1 \\ 3 & -1 & 4 \end{bmatrix}\begin{bmatrix} 1 & -1 & 0 \\ 2 & 3 & 1 \\ 5 & 1 & 3 \end{bmatrix}$

In Exercises 35–36, find the inverse of the matrix if it exists.

35. $\begin{bmatrix} 1 & 3 \\ 2 & 2 \end{bmatrix}$

36. $\begin{bmatrix} 1 & -1 & 1 \\ 2 & -2 & 1 \\ 1 & 1 & 0 \end{bmatrix}$

In Exercises 37–40, solve the given system using matrix inverses.

37. $\begin{cases} x - 2y = -5 \\ 2x + y = 5 \end{cases}$

38. $\begin{cases} 3x + 2y = -4 \\ 3x + y = -2 \end{cases}$

39. $\begin{cases} x - y + z = 5 \\ x - 2y + z = 7 \\ x + 2y - z = -3 \end{cases}$

40. $\begin{cases} x - y + 2z = 1 \\ x - 2y + z = 2 \\ x - y + z = 2 \end{cases}$

In Exercises 41–44, evaluate the determinant.

41. $\begin{vmatrix} 1 & 0 \\ 5 & -2 \end{vmatrix}$

42. $\begin{vmatrix} 3 & 3 \\ 1 & 2 \end{vmatrix}$

43. $\begin{vmatrix} 1 & 0 & 2 \\ 5 & -2 & 1 \\ 3 & -1 & 0 \end{vmatrix}$

44. $\begin{vmatrix} 1 & 0 & 2 \\ 3 & -1 & 1 \\ 2 & -1 & 5 \end{vmatrix}$

In Exercises 45–48, solve the system of equations using Cramer's Rule.

45. $\begin{cases} x - 2y = -5 \\ 3x + 5y = 7 \end{cases}$

46. $\begin{cases} x + 2y = 5 \\ 6x - 2y = 23 \end{cases}$

47. $\begin{cases} x + 2y + z = 5 \\ x + 4y - z = -1 \\ x - 2y + z = 5 \end{cases}$

48. $\begin{cases} x - y + z = 2 \\ 2x - 2y - z = -11 \\ 3x - 2y + z = -2 \end{cases}$

49. A manufacturer produces two types of cameras. The more expensive model requires 2 hours to manufacture and 20 minutes to assemble. The less expensive model requires 1 hour to manufacture and 15 minutes to assemble. If the company can allocate 280 hours for manufacture and 60 hours for assembly, how many of each type can be produced?

50. The Physics Department hires graders and tutors. For February the department budgets \$855 for 20 hours of grading and 40 hours for tutoring. The next month the department budgets \$630 for 40 hours of grading and 20 hours of tutoring. How much does the department pay for each hour of tutoring and for each hour of grading?

51. Brian invests \$10,000, part at 6% and part at 8%, earning a yearly interest of \$750 on the entire amount. How much is invested at each rate?

52. A lawyer charges \$110 an hour for her time and \$50 an hour for her assistant's time. A client received a bill from the lawyer that included a \$2650 charge for their combined time. If the assistant worked 5 hours longer than the lawyer, how much time did the lawyer and her assistant spend on the client's case?

53. Josh has \$40,000 in three investments. He has a bank account paying 5%, a certificate of deposit paying 6.5%, and a stock yielding 9%. His annual interest from the three investments is \$2900. If he invested \$6000 more in his stock than in his certificate of deposit, how much did he invest at each rate?

54. A radio manufacturer produces three models, model A, model B, and model C. The company knows how much time is needed for production and assembly and how much raw materials cost for each model. This information is found in the accompanying table. If the company has allocated 90 hours for production, 87 hours for assembly, and \$895 for raw materials, how many of each model can the company produce if it wants to use up all the time allocated for each phase of the process and all the money allocated for raw materials?

Model	Production hours	Assembly hours	Cost of raw materials
A	0.8	0.9	\$9.80
B	0.7	0.5	\$6.20
C	0.4	0.8	\$4.20

In Exercises 55–58, sketch the solution set of the system of inequalities.

55. $\begin{cases} 2x + y \le 12 \\ 3x + y \ge 12 \end{cases}$

56. $\begin{cases} -x + 2y \le 8 \\ x - y < -3 \end{cases}$

57. $\begin{cases} y < 3x \\ 4x - y \le 5 \\ y \le 5 \end{cases}$

58. $\begin{cases} x + 3y > -1 \\ 3x - 6y > -3 \\ x \le 4 \end{cases}$

In Exercises 59–62, write a system of inequalities to describe the conditions of the problem and sketch the solution set of the system.

59. An appliance store needs to order a shipment of VCRs and TVs. A VCR costs \$200 and takes up 2 cu ft of storage space, and a TV costs \$250 and takes up 6 cu ft of storage space. If the store wants to order at least \$10,000 worth of TVs and VCRs but has at most 500 cu ft of storage space in its storage area, how many of each can it order?

60. Irena has at most \$20,000 available for investment. She can invest in two types of certificates of deposit: A 6-month certificate, which yields a 5% yearly return on investment, and a 1-year certificate, which yields a 6.5% yearly return. She decides that she wants to put at least twice as much in the 1-year certificate as in the 6-month certificate and she wants at least \$2000 but no more than \$4000 invested in the 6-month certificate. How much should she invest in each certificate?

61. A judge is working on the weekly schedule of hearings in her courtroom, where she hears both minor criminal and civil complaints. She finds that the average minor civil complaint requires 1 hour to settle; the average minor criminal complaint requires 45 minutes. Next week she wants to schedule at least 20 hours of hearings but no more than 35 hours. She prefers to adjudicate more criminal complaints than civil complaints. How many of each type of case should she hear next week?

62. An orthodontist uses two types of appliances (A and B), both shown to be effective for overbites. Over a period of a year, each appliance A requires 6 hours of the doctor's labor and 14 hours of the dental assistants' labor; each appliance B requires 4 hours of the doctor's labor and 16 hours of the dental assistants' labor. If the doctor wants to put in no more than 800 hours for appliances over the year and his assistants must put in at least 1000 hours labor on appliances, how many of each appliance should be used?

63. A bank has a total of \$80 million to invest for both high-risk and conservative investments. A federal regulation requires that the bank have no more than 25% of its total investment funds in high-risk ventures. If the rate of return for conservative investments averages 8% yearly and the rate of return for high-risk investments is 14% yearly, what is the maximum profit the bank can make, given its constraints?

64. The owner of a computer store needs to order two types of printers, model X and model Y. Model X costs \$300 wholesale, and the wholesale price of model Y is \$400.

Model X takes up 12 cu ft of space and model Y takes up 6 cu ft of space. The store owner decides that she wants to buy at least \$48,000 worth of the models but has only 1080 cu ft of storage space. If her profit is \$225 on each model X she sells and \$100 on each model Y she sells, how many of each model should she buy to maximize profit?

Chapter 9 PRACTICE TEST

1. Solve the system of equations.

(a) $\begin{cases} 2x + y = 12 \\ 3x - y = 8 \end{cases}$

(b) $\begin{cases} -x + 2y = -4 \\ 5x + 3y = 7 \end{cases}$

(c) $\begin{cases} x - y - z = 4 \\ x + 2y + 2z = -2 \\ x - y + z = 3 \end{cases}$

2. Solve the system of equations using augmented matrices.

$$\begin{cases} x + y - 2z = -7 \\ 2x + 2y + 2z = 10 \\ -x + y + z = -1 \end{cases}$$

3. Perform the operations.

(a) $3\begin{bmatrix} 1 & 0 \\ -2 & 5 \end{bmatrix} - \begin{bmatrix} 2 & 5 \\ 1 & 3 \end{bmatrix}$

(b) $\begin{bmatrix} 3 & 5 & 1 \\ -4 & 6 & 0 \\ 2 & 1 & -1 \end{bmatrix}\begin{bmatrix} 1 & 0 & 4 \\ 0 & 1 & -1 \\ 2 & 1 & -2 \end{bmatrix}$

4. Find the inverse of the matrix if it exists.

$$\begin{bmatrix} 2 & 8 \\ 1 & -4 \end{bmatrix}$$

5. Solve the given system using matrix inverses.

$$\begin{cases} x - 2y = 4 \\ 2x + y = 13 \end{cases}$$

6. Evaluate the determinants.

(a) $\begin{vmatrix} 2 & 5 \\ -8 & 1 \end{vmatrix}$ **(b)** $\begin{vmatrix} 3 & 0 & 2 \\ 2 & -1 & 4 \\ 1 & 2 & -3 \end{vmatrix}$

7. Solve the system of equations using Cramer's Rule.

$$\begin{cases} x - 2y + 3z = 6 \\ -x + y - 2z = -4 \\ 2x + y - z = -1 \end{cases}$$

8. A car rental agency charges a flat fee plus a mileage rate for a 1-day rental. If the charge for a 1-day rental with 120 miles is \$65.70 and the charge for a 1-day rental with 80 miles is \$51.30, find the flat fee and the charge per mile.

9. A horticulturist wants to mix three types of fertilizers, which contain 10%, 25%, and 30% nitrogen, respectively. If the final mixture should be 2000 lb of 21% nitrogen, all three types are used, and there is twice as much of the 30% type as of the 25% type, how much of each type is in the final mixture?

10. Sketch the solution set of the system of inequalities.

$$\begin{cases} 2y < 3x - 12 \\ 4x - y \le 16 \end{cases}$$

11. An appliance store needs to order two types of microwave ovens, model X and model Y. Model X costs \$100, wholesale, and the wholesale price of model Y is \$200. Model X takes up 4 cu ft of space and model Y takes up 3 cu ft of space. The store owner decides that he wants to buy at most \$10,000 worth of the models but has only 300 cu ft of storage space. Write a system of inequalities to describe this situation and sketch the solution set of the system.

12. The appliance store in Exercise 11 makes a profit of \$60 on each model X microwave sold and a profit of \$50 on each model Y microwave sold. How many of each type should be ordered to maximize the microwave profits within the given constraints? What is the maximum possible profit?

CHAPTER 10

Conic Sections and Nonlinear Systems

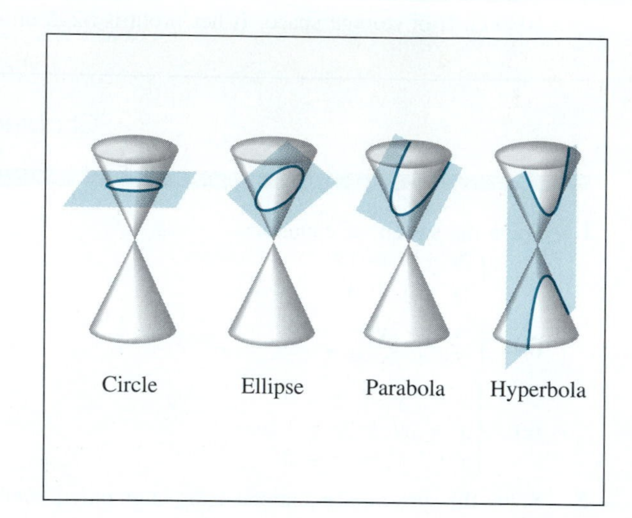

In Section 2.1, we discussed graphs of equations of the form $Ax + By = C$: first-degree equations in two variables. In this chapter we discuss the graphs of the *general second-degree equation in two variables,*

$$Ax^2 + Bxy + Cy^2 + Dx + Ey + F = 0 \tag{1}$$

If there are any ordered pairs satisfying such an equation, it can be shown that (with some exceptions) the graph will be one of the following four figures: a circle, a parabola, an ellipse, or a hyperbola. These figures are called **conic sections**, since they describe the intersection of a plane and a double-napped cone as illustrated above.

In this and in the next few sections, we concentrate on the cases where $B = 0$ in the general second-degree equation. In other words, we focus on equations of the form (1) that do not contain an xy term and where A and C are not both zero.

10.1 Conic Sections: Circles

In Section 2.1 we stated the geometric definition of a circle and derived the equation of a circle by applying the distance formula to this definition. We briefly review the discussion of the circle here, since we will take the same approach in deriving equations for the other conic sections.

Recall that the formula we use to find the distance between the points (x_1, y_1) and (x_2, y_2) in the Cartesian plane is $d = \sqrt{(x_2 - x_1)^2 + (y_2 - y_1)^2}$. The circle is defined to be the set of all points in a plane equidistant from a fixed point. We label the fixed point the center, $C(h, k)$, and the distance from the center to any point on the circle the radius, r. See Figure 10.1.

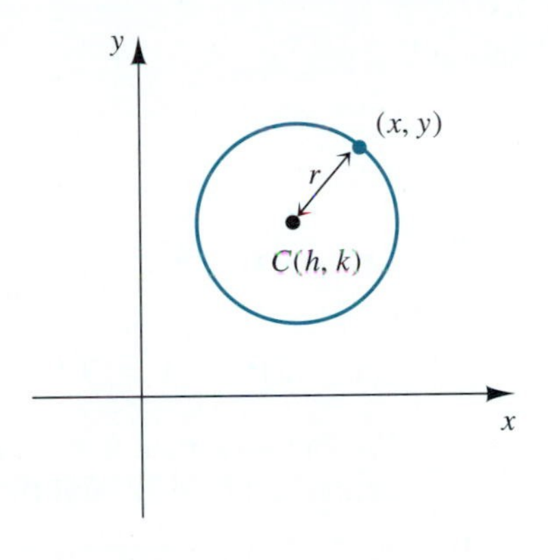

FIGURE 10.1

By the distance formula, any point (x, y) lies on the circle if and only if $\sqrt{(x - h)^2 + (y - k)^2} = r$, which, when we square each side, yields the standard form of a circle.

STANDARD FORM OF A CIRCLE WITH CENTER (h, k) AND RADIUS r

The equation of a circle with center (h, k) and radius r is

$$(x - h)^2 + (y - k)^2 = r^2$$

To find an equation of the circle with center $(-1, 5)$ and radius 4, we simply use the standard form of the circle with $h = -1$, $k = 5$, and $r = 4$ to get

$$[x - (-1)]^2 + (y - 5)^2 = 4^2 \quad \text{or} \quad (x + 1)^2 + (y - 5)^2 = 16$$

EXAMPLE 1 Find the center and radius of the circle

$$2x^2 + 2y^2 - 8x + 12y - 28 = 0.$$

Solution The equation $2x^2 + 2y^2 - 8x + 12y - 28 = 0$ is not in standard form, so we must complete the square to put the equation in standard form. Before we do, however, we divide each side of the equation by 2 to make the coefficients of the second-degree terms equal to 1.

$2x^2 + 2y^2 - 8x + 12y - 28 = 0$	*Divide both sides of the equation by 2.*
$x^2 + y^2 - 4x + 6y - 14 = 0$	*Add 14 to each side and group terms as shown.*
$(x^2 - 4x \quad) + (y^2 + 6y \quad) = 14$	*Complete the square for each quadratic expression:* $\left[\frac{1}{2}(-4)\right]^2 = 4, \quad \left[\frac{1}{2}(6)\right]^2 = 9$ *Add 4 and 9 to both sides of the equation.*
$(x^2 - 4x + 4) + (y^2 + 6y + 9) = 14 + 4 + 9$	*Rewrite the quadratic expressions in factored form.*
$(x - 2)^2 + (y + 3)^2 = 27$	*This is now in standard form.*

We can now read $h = 2$, $k = -3$, and $r^2 = 27 \Rightarrow r = \sqrt{27} = 3\sqrt{3}$. Hence, the center is $(2, -3)$, and the radius is $3\sqrt{3}$. See Figure 10.2.

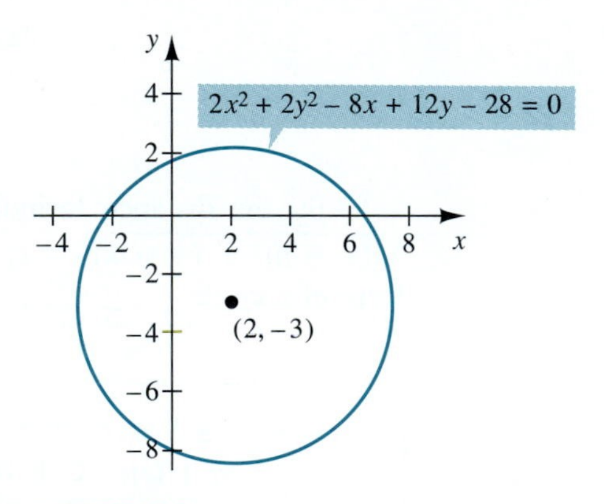

FIGURE 10.2

■

GRAFFIX

To use a graphics calculator to graph a relation that is not a function, such as $x^2 + y^2 = 9$, you must first solve for y to get: $y = \pm\sqrt{9 - x^2}$. Then graph *both* $y = \sqrt{9 - x^2}$ and $y = -\sqrt{9 - x^2}$ on the same set of coordinate axes.

1. Using this procedure, graph the circle $x^2 + y^2 = 12$ on your graphics calculator.
2. Using this procedure, graph the circle $(x + 5)^2 + (y - 3)^2 = 20$ on your graphics calculator.

Note: If, on the display, the horizontal and vertical units are not equal in length, the circle may not appear to be round.

Will the graph of the equation $x^2 + y^2 = 0$ or $x^2 + y^2 = -5$ be a circle?

At this point you may have recognized that if the coefficients of the squared terms in the general second-degree equation (1) are identical (that is, if $A = C$ in the equation $Ax^2 + Cy^2 + Dx + Ey + F = 0$), then, with the help of completing the square, we can always put the equation into the form $(x - h)^2 + (y - k)^2 = a$, where a is a constant. Although we expect this form to give us a circle, the graph will be a circle only if $a > 0$. We will have more to say about the exceptional cases in Section 10.5.

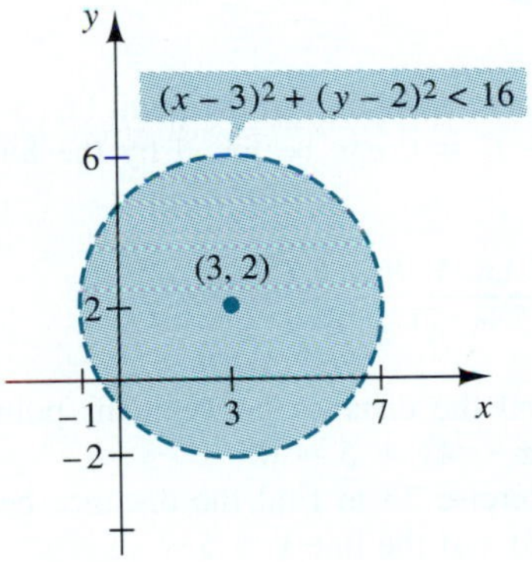

FIGURE 10.3

EXAMPLE 2 Graph the solution to the inequality $(x - 3)^2 + (y - 2)^2 < 16$.

Solution We graph quadratic inequalities using the same approach we took with linear inequalities. We first graph the equation $(x - 3)^2 + (y - 2)^2 = 16$, which is a circle centered at (3, 2) and with radius 4. This graph divides the plane into two regions: the region inside the circle and the region outside the circle. The solution is one of the regions.

For the circle, we can determine which region is the solution by realizing that by the definition of the circle, $(x - 3)^2 + (y - 2)^2 < 16$ means all points *less than* 4 units from the center, (3, 2). The solution is therefore the region inside the circle, and we shade in this region. We represent the circle by a dashed curve to indicate that the circle itself is not included in the solution. See Figure 10.3.

Alternatively, we can find the solution by using a test point; that is, by choosing a point in one of the regions and testing its coordinates to see if it satisfies the inequality $(x - 3)^2 + (y - 2)^2 < 16$. If the coordinates satisfy the inequality, then we shade the region containing the test point; otherwise, we shade the region that does not contain the test point. ■

How do we interpret $(x - 3)^2 + (y - 2)^2 \geq 16$?

EXERCISES 10.1

1. Find the equation of the circle with center $(3, -8)$ and radius 5.
2. Find the equation of the circle with center $\left(2, \frac{1}{3}\right)$ and radius 6.
3. Find the equation of the circle with center $(-3, 1)$ and radius 1. Put this equation into the general form of a second-degree equation.
4. Find the equation of the circle with center $\left(-\frac{1}{5}, \frac{1}{2}\right)$ and radius 3. Put this equation into the general form of a second-degree equation.

In Exercises 5–14, identify the center and radius of the circles described by the equation.

5. $(x - 2)^2 + (y + 3)^2 = 81$
6. $(x + 5)^2 + (y - 3)^2 = 16$
7. $\left(x - \frac{1}{2}\right)^2 + y^2 = 18$
8. $x^2 + \left(y - \frac{1}{3}\right)^2 = 12$
9. $x^2 + y^2 - 6x - 8y + 9 = 0$
10. $x^2 + y^2 + 10x - 6y + 10 = 0$
11. $x^2 + y^2 - 6x - 15 = 0$
12. $x^2 + y^2 - 14x + 31 = 0$
13. $4x^2 + 4y^2 - 4y - 47 = 0$
14. $9x^2 + 9y^2 - 6x - 6y - 79 = 0$
15. There is a theorem in geometry that any line tangent to a circle at a point, P, is perpendicular to the line passing through P and the center of the circle. (See the following figure.) Use this theorem to find the equation of the line tangent to the circle $x^2 + y^2 = 5$ at the point $(1, 2)$.

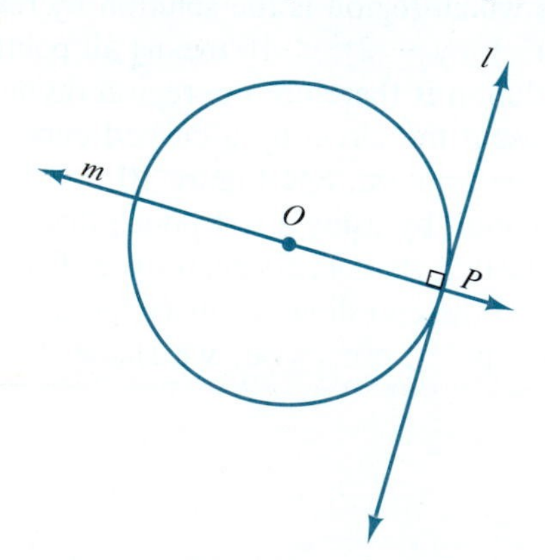

16. Use the theorem in Exercise 15 to find the equation of the line tangent to the circle $(x - 3)^2 + (y - 2)^2 = 40$ at the point $(5, -4)$.
17. Use the theorem in Exercise 15 to find the equation of the line tangent to the circle $x^2 + (y - 3)^2 = 8$ at the point $(2, 5)$.
18. Use the theorem in Exercise 15 to find the equation of the line tangent to the circle $(x - 1)^2 + y^2 = 32$ at the point $(5, -4)$.

In Exercises 19–22, sketch the solution for each inequality.

19. $x^2 + y^2 \leq 81$
20. $(x + 2)^2 + (y - 1)^2 \leq 25$
21. $x^2 + y^2 - 8x - 6y + 9 < 0$
22. $x^2 + y^2 - 2x + 10y > -10$
23. When we refer to the distance from a point to a line, we are referring to the shortest distance to the line, which is defined as the *perpendicular* distance: the length of the perpendicular line segment joining the point and the line. (See the following figure.)

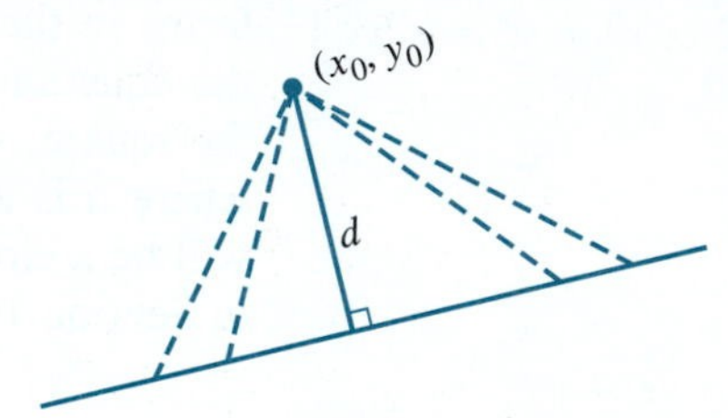

The (perpendicular) distance from the point (x_0, y_0) to the line $Ax + By + C = 0$ can be found by the formula

$$d = \frac{|Ax_0 + By_0 + C|}{\sqrt{A^2 + B^2}}$$

Use this formula to find the distance between the point $(3, -2)$ and the line $3x - 4y + 5 = 0$.
24. Use the formula in Exercise 23 to find the distance between the point $(0, -3)$ and the line $x = 5$.
25. Find the equation of the circle with center $(0, 0)$, if a tangent line to the circle has the equation $2x - 3y = 12$. HINT: Use the formula in Exercise 23.
26. Find the equation of the circle with center $(2, 6)$, if a tangent line to the circle has the equation $x - 3y = 9$. HINT: Use the formula in Exercise 23.
27. The (perpendicular) distance from the point (x_0, y_0) to the line $y = mx + b$ can be found by the formula

$$d = \frac{|mx_0 + b - y_0|}{\sqrt{1 + m^2}}$$

Use the formula to find the distance from the point $(2, 4)$ to the line $y = 3x - 4$.

28. Use the formula in Exercise 27 to find the distance from the point $(3, -1)$ to the line $y = 5x + 7$.

29. Use the formula in Exercise 27 to show that the (perpendicular) distance between two parallel lines $y = mx + b_1$ and $y = mx + b_2$ is found by the formula

$$d = \frac{|b_1 - b_2|}{\sqrt{1 + m^2}}$$

HINT: Choose any point on the line $y = mx + b_1$, and call it (x_0, y_0). Then use the formula for the distance from the point (x_0, y_0) to the line $y = mx + b_2$.

30. Use the formula in Exercise 29 to find the distance between the lines $y = 5x - 4$ and $y = 5x + 7$.

31. We will outline a way to demonstrate that the equation of the line tangent to the circle $x^2 + y^2 = r^2$ at the point (x_0, y_0) has the equation

$$xx_0 + yy_0 = r^2$$

(a) First draw a picture of the circle $x^2 + y^2 = r^2$, with tangent line passing through the point (x_0, y_0). Then draw a line passing through the center and through the point (x_0, y_0). What is the center of this circle? (See the accompanying figure.)

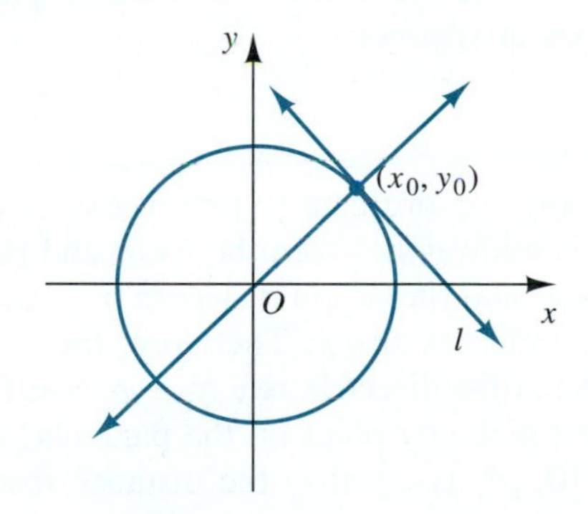

(b) In order to find the equation of a line, we need the point through which the line passes and the slope. We know that the tangent line passes through the point (x_0, y_0); we still need to find its slope.

(c) We know that the tangent line is perpendicular to the line segment joining the center and the point of tangency. Use this information to show that the tangent line has slope $-\dfrac{x_0}{y_0}$.

(d) Now we have the slope of the tangent line and a point through which the tangent line passes. Use the point-slope formula to show that the equation of the tangent line is $y - y_0 = -\dfrac{x_0}{y_0}(x - x_0)$.

(e) Multiplying both sides of the equation by y_0 and collecting the constant terms on the right-hand side of the equation, we get $xx_0 + yy_0 = x_0^2 + y_0^2$. How do we arrive at $xx_0 + yy_0 = r^2$?

32. Using the result of Exercise 31, find the equation of the line tangent to the circle $x^2 + y^2 = 29$ at the point $(2, 5)$.

GRAFFIX

In Exercises 33–36, use a graphics calculator or computer to graph the equation.

33. $x^2 + y^2 = 36$

34. $2x^2 + 2y^2 = 18$

35. $(x - 3)^2 + (y + 4)^2 = 18$

36. $(x + 2)^2 + (y + 1)^2 = 2$

10.2 The Parabola

In Section 3.4 we examined the general quadratic function $f(x) = ax^2 + bx + c$ in detail and mentioned that the graph of this function is called a **parabola**. In this section we start with the geometric definition of a parabola and, as we did with the circle, derive the equation of a parabola by applying the distance formula to its definition.

DEFINITION A **parabola** is the set of all points in a plane whose distance from a fixed point is equal to its distance from a fixed line. We call the fixed point the **focus** and the fixed line the **directrix**. See Figure 10.4(a).

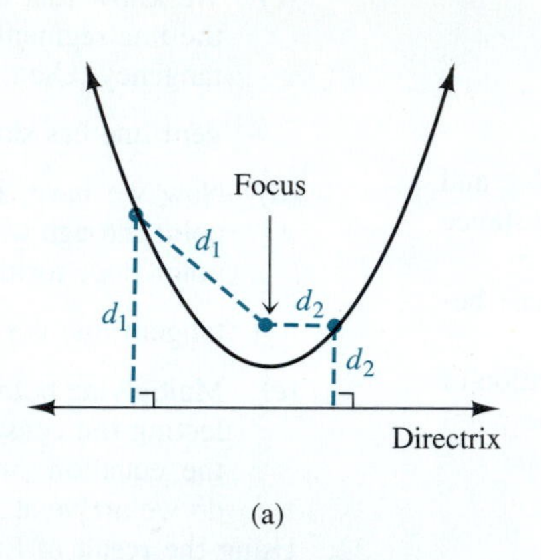

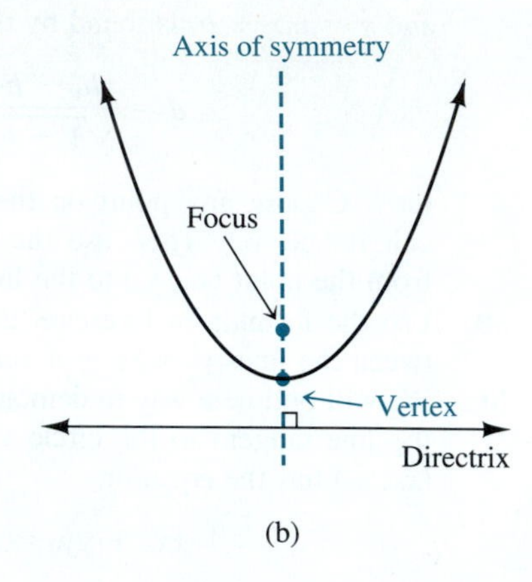

FIGURE 10.4 The parabola

The line passing through the focus and perpendicular to the directrix is called its **axis of symmetry**, and the point where the parabola intersects its axis of symmetry is called its **vertex**. See Figure 10.4(b). We will concentrate on parabolas that have either a horizontal or vertical axis of symmetry.

The Parabola with Vertex (0, 0)

Given the focus and directrix, we choose a coordinate system in such a way that the directrix is horizontal and the origin is midway between the focus and the directrix. Let's call the distance between the focus and the origin p (where $p > 0$), and so the distance between the origin and the directrix is also p. Therefore, the coordinates of the focus F are $(0, p)$, and the equation of the directrix is $y = -p$. See Figure 10.5.

By definition of a parabola, if we pick any point on the parabola, $P(x, y)$, the distance from $P(x, y)$ to its focus, $F(0, p)$, is equal to the distance from the point $P(x, y)$ to the point $L(x, -p)$. (Notice that $L(x, -p)$ is the point that yields the perpendicular distance to the line $y = -p$.)

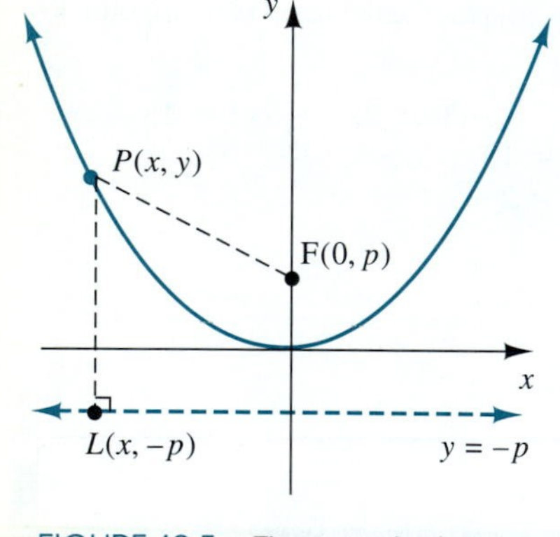

FIGURE 10.5 The parabola with focus $(0, p)$, $p > 0$, vertex $(0, 0)$, and directrix $y = -p$

$$|\overline{PF}| = |\overline{PL}|$$ *Using the distance formula we get*

$$\sqrt{(x-0)^2 + (y-p)^2} = \sqrt{(x-x)^2 + (y+p)^2}$$ *Squaring both sides we get*

$$(x-0)^2 + (y-p)^2 = (x-x)^2 + (y+p)^2$$ *Then simplify.*

$$x^2 + (y-p)^2 = (y+p)^2$$

$$x^2 + y^2 - 2py + p^2 = y^2 + 2py + p^2$$ *Which yields*

$$x^2 = 4py$$

We have just derived the following:

The "vertical" parabola with vertex at (0, 0)

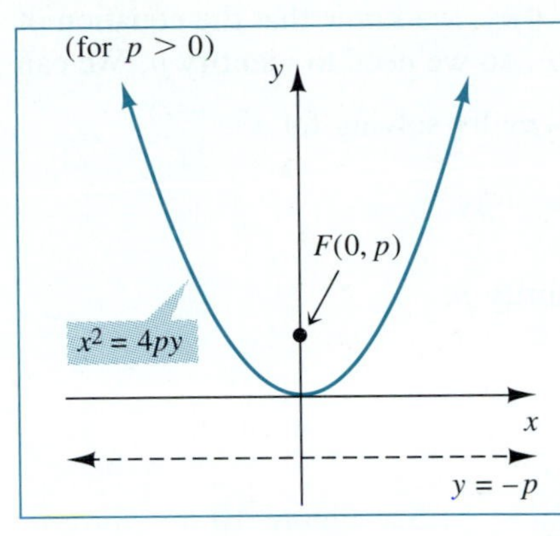

STANDARD FORM FOR THE EQUATION OF A PARABOLA WITH FOCUS (0, p) AND DIRECTRIX $y = -p$

The standard form for the equation of a parabola with focus $(0, p)$ and directrix $y = -p$ is

$$x^2 = 4py$$

This is a parabola with the vertex at the origin and having the y-axis as its axis of symmetry.

Figure 10.6 shows the relationship between the shape of the parabola and the distance between its focus and directrix. Notice that the further apart the focus and the directrix are, the wider the parabola.

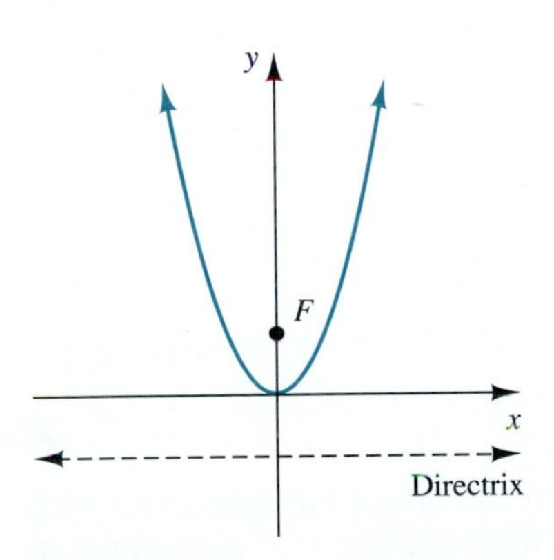

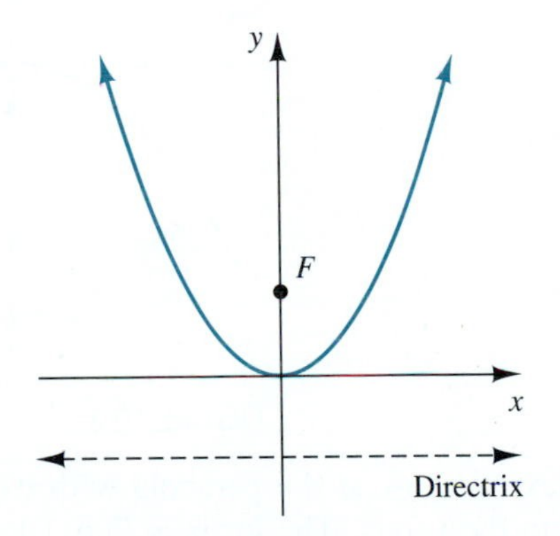

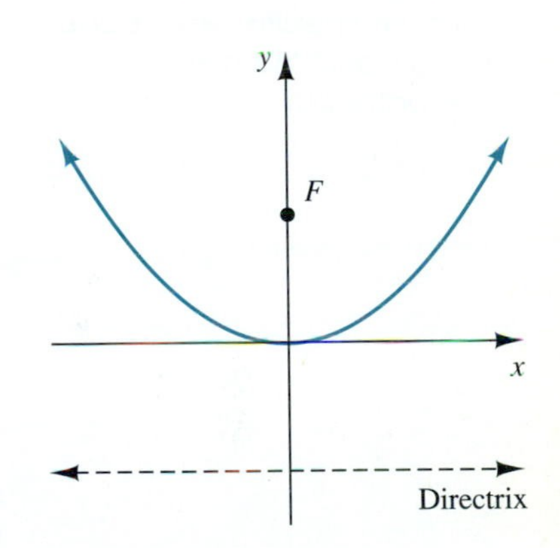

FIGURE 10.6

Keep in mind that if p is negative, then $-p$ is positive.

We arrive at the same result with $p < 0$. In this case, the focus of the parabola, $(0, p)$ lies below the x-axis, and the horizontal directrix, $y = -p$, lies above the x-axis. The parabola still has its vertex at the origin and the y-axis as its axis of symmetry, but it opens downward, as indicated in Figure 10.7.

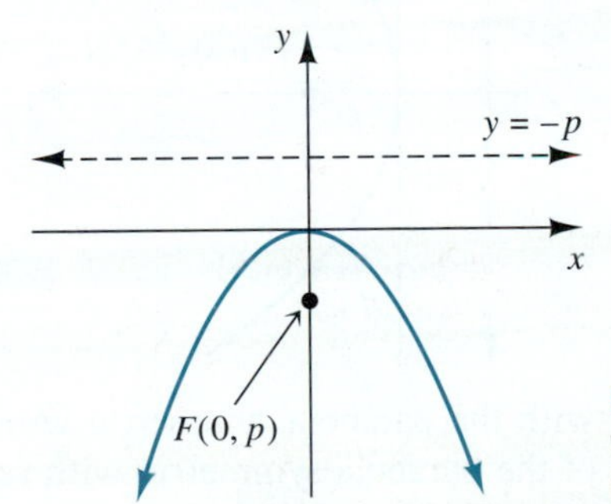

FIGURE 10.7 The parabola with focus $(0, p)$, $p < 0$, vertex $(0, 0)$, and directrix $y = -p$

EXAMPLE 1 Find the focus and directrix of the parabola $y = -\frac{1}{3}x^2$.

Solution For a parabola given in the form $x^2 = 4py$, we know that the equation of the directrix is $y = -p$ and that the focus is $(0, p)$, so we need to identify p. We can put the equation $y = -\frac{1}{3}x^2$ into the form $x^2 = 4py$ by solving for x^2:

$$y = -\frac{1}{3}x^2 \Rightarrow x^2 = -3y.$$

We compare this with standard form to identify p:

$$x^2 = 4py$$
$$x^2 = -3y \qquad \textit{Now we see } 4p = -3 \Rightarrow p = -\frac{3}{4}.$$

Hence the focus is $\left(0, -\frac{3}{4}\right)$ and the directrix is $y = \frac{3}{4}$. See Figure 10.8.

Why does this parabola open downward? If the focus is above the directrix, what can be said about the orientation of the parabola?

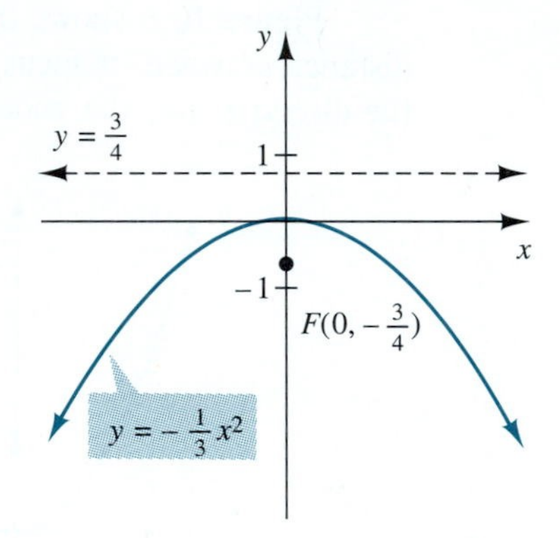

FIGURE 10.8 ■

Next we look at the parabola with its vertex at the origin but symmetric with respect to the x-axis. The focus is $F(p, 0)$, and the directrix is $x = -p$, pictured in Figure 10.9.

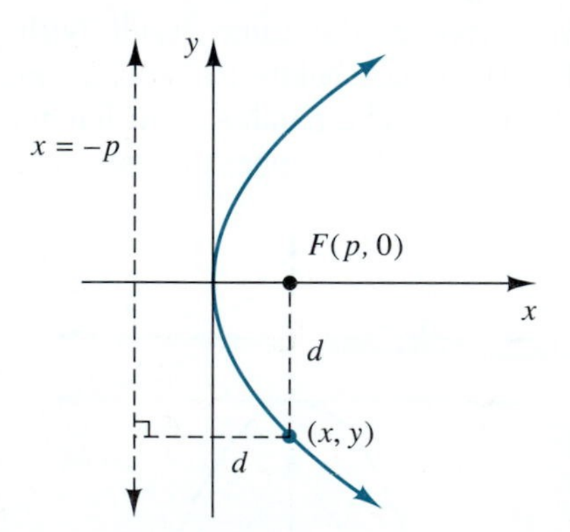

FIGURE 10.9 The parabola with focus $(p, 0)$, $p > 0$, vertex $(0, 0)$, and directrix $x = -p$

As with the parabola symmetric with respect to the y-axis, we can derive the equation of the parabola symmetric with respect to the x-axis using the distance formula.

We arrive at the following:

The "horizontal" parabola with vertex (0, 0)

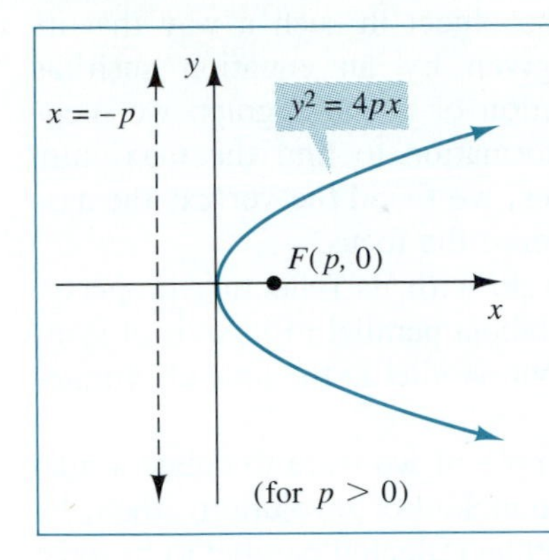

STANDARD FORM FOR THE EQUATION OF A PARABOLA WITH FOCUS $(p, 0)$ AND DIRECTRIX $x = -p$

The standard form for the equation of a parabola with focus $(p, 0)$ and directrix $x = -p$ is

$$y^2 = 4px$$

This is a parabola with the vertex at the origin and having the x-axis as its axis of symmetry.

Take a close look at the differences between the two standard forms: The key element for determining whether the parabola opens up (down) or right (left) is the second-degree (squared) term. There is only one second-degree term in the equation of the parabola; if there is an x^2 term, the parabola opens either up or down (y-axis symmetry), but if there is a y^2-term, the parabola opens either left or right (x-axis symmetry).

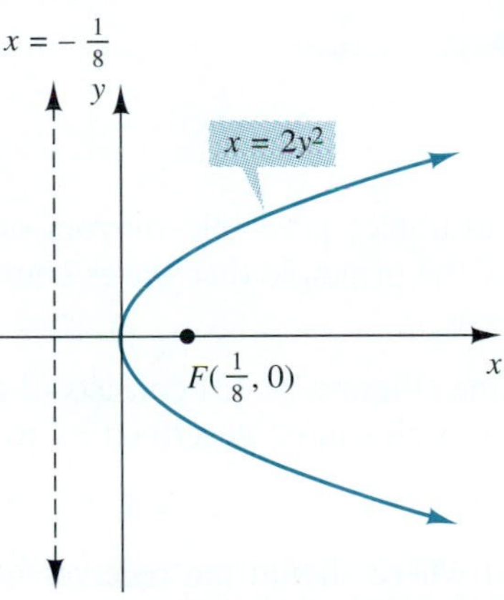

FIGURE 10.10

EXAMPLE 2 Find the focus and directrix of the parabola $x = 2y^2$.

Solution We first determine which standard form to use. We note that since there is a y^2 term (and no x^2-term), we have a parabola symmetric with respect to the x-axis and use the form $y^2 = 4px$.

We can put the equation $x = 2y^2$ into the form $y^2 = 4px$ by solving for y^2:

$$x = 2y^2 \Rightarrow y^2 = \frac{1}{2}x.$$

We compare this to the standard form for the parabola with the vertex at the origin and symmetric with respect to the x-axis to identify p:

$$y^2 = 4px$$

$$y^2 = \frac{1}{2}x \qquad \textit{Now we have } 4p = \frac{1}{2} \Rightarrow p = \frac{1}{8}.$$

The focus is $\left(\frac{1}{8}, 0\right)$ and the directrix is $x = -\frac{1}{8}$. See Figure 10.10. ■

Why does this parabola open right? If the focus is right of the directrix, what can be said about the "orientation" of the parabola?

EXAMPLE 3 Find the equation of the parabola with the vertex at the origin and symmetric with respect to the y-axis if its focus is $(0, -3)$.

Solution Given that the parabola has its vertex at the origin and is symmetric with respect to the y-axis, its equation will be of the form $x^2 = 4py$. Since we are given that the focus is $(0, -3)$, then $p = -3$. Hence the equation is $x^2 = 4(-3)y$, or $x^2 = -12y$. ■

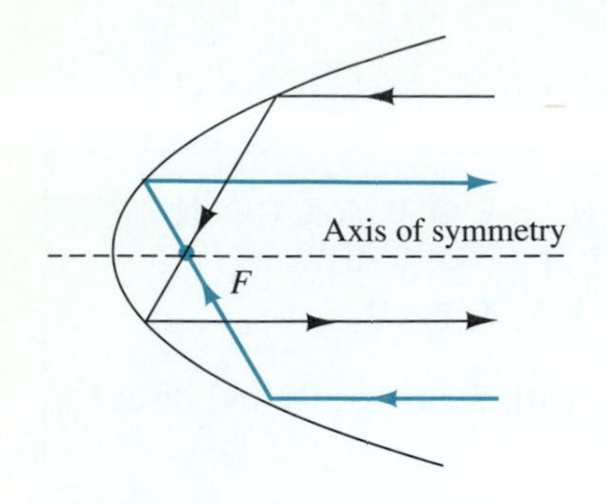

FIGURE 10.11 The reflecting property of a parabola

As we have seen in earlier chapters, there are many applications involving parabolas. If an object is fired or thrown in the air in a vertical direction and if other forces are negligible, gravity acts upon the object in such a way that its height s above the ground at time t can be given by an equation such as $s = -16t^2 + 32t + 10$, which is a quadratic function of t whose graph we mentioned is a parabola. We previously used this information to find the maximum height of an object and when it landed. In these cases, we found the vertex, the axis of symmetry, or intercepts to be important and ignored the focus.

Another important use of the parabola has to do with its reflecting property. Figure 10.11 illustrates that any ray entering the parabola parallel to its axis of symmetry will be reflected through the focus and back out parallel to the axis of symmetry. (See Exercise 51.)

This property has many applications. For example, if we were to create a mirror by rotating a parabola about its axis (called a paraboloid of revolution), then, by placing a source of light at the focus, the beam would be reflected parallel to its axis. This concentrates the beam of light in one direction. You have probably noticed that flashlights, car headlights, and spotlights contain paraboloid-shaped reflecting surfaces (see Figure 10.12). The light source or bulb is placed at its focus.

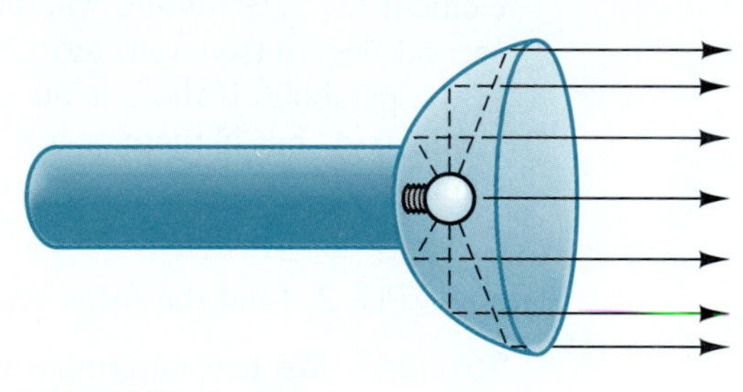

FIGURE 10.12

The same principle is used for receiving. For example, parabolic mirrors on telescopes, TV satellite dishes, and radiotelescopes use the principle that waves coming in parallel to the axis will be reflected into the focus.

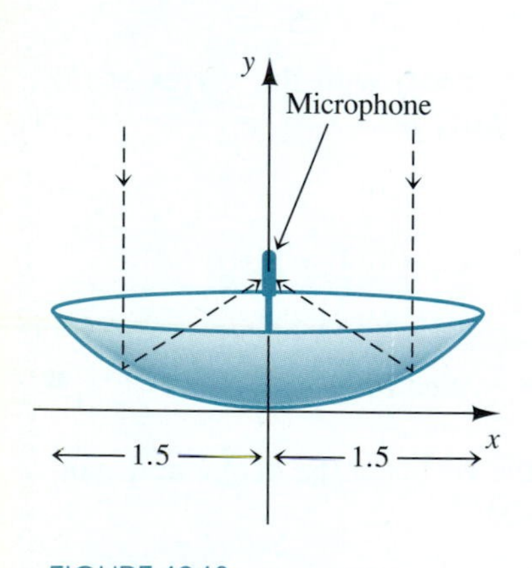

FIGURE 10.13

EXAMPLE 4 A field microphone used at a football game (Figure 10.13) consists of a parabolic dish with the receiver placed at its focus. The dish can be described by rotating the parabola $y = \frac{1}{9}x^2$ about its axis of symmetry, where $-1.5 \le x \le 1.5$, and x is measured in feet. How deep is the dish, and where should the receiver be placed in relation to the bottom (vertex) of the dish?

Solution By Figure 10.13, we see that the cross section of the dish is a parabola. We can place the parabola such that its vertex is at the origin and the axis of symmetry is the y-axis. Hence we can use the form $x^2 = 4py$, which has focus $(0, p)$ as shown in Figure 10.14.

Looking at the equation of the cross section of the dish, $y = \frac{1}{9}x^2$, we can identify p by solving for x^2 to get $9y = x^2$. Hence $4p = 9$, and so $p = \frac{9}{4} = 2.25$ feet. This means that the receiver must be placed 2.25 feet above the vertex.

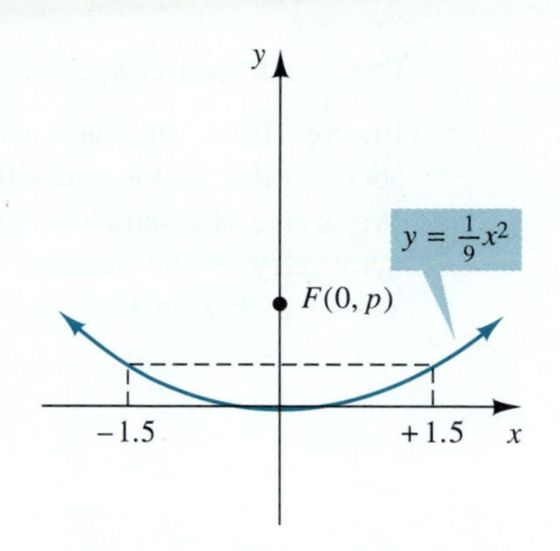

FIGURE 10.14

Notice that because $-1.5 \leq x \leq 1.5$, the dish has a diameter of 3 feet $[1.5 - (-1.5) = 3]$.

To find out how deep the dish is, we use the fact that $-1.5 \leq x \leq 1.5$. Looking at Figure 10.14, we see that if we substitute $x = 1.5$ (or $x = -1.5$) in the equation of the cross section, $y = \frac{1}{9}x^2$, we can find the "height," which will tell us how deep the dish is:

$$y = \frac{1}{9}x^2 \qquad \text{Let } x = 1.5 \text{ and find } y.$$

$$y = \frac{1}{9}(1.5)^2 = 0.25 \text{ feet, or 3 inches, deep}$$

■

GRAFFIX

Keep in mind that you must solve for y before sketching a graph with your graphics calculator.

1. Graph the following on the same set of coordinate axes:

$$x^2 = 12y, \qquad (x - 2)^2 = 12, \qquad (x + 2)^2 = 12y$$

What can you conclude about the effect of h on the graph of $(x - h)^2 = 4py$?

2. Graph the following on the same set of coordinate axes:

$$x^2 = 12y, \qquad x^2 = 12(y + 3), \qquad x^2 = 12(y - 3)$$

What can you conclude about the effect of k on the graph of $x^2 = 4p(y - k)$?

The Parabola Centered at (h, k)

Figure 10.15 illustrates what happens when we shift a parabola symmetric with respect to the y-axis and with vertex $(0, 0)$ *horizontally* h units and *vertically* k units. We arrive at a parabola with the following properties: Its vertex is (h, k); its axis of symmetry is the vertical line $x = h$; its focus is $F(h, k + p)$; and its directrix is $y = -p + k = k - p$.

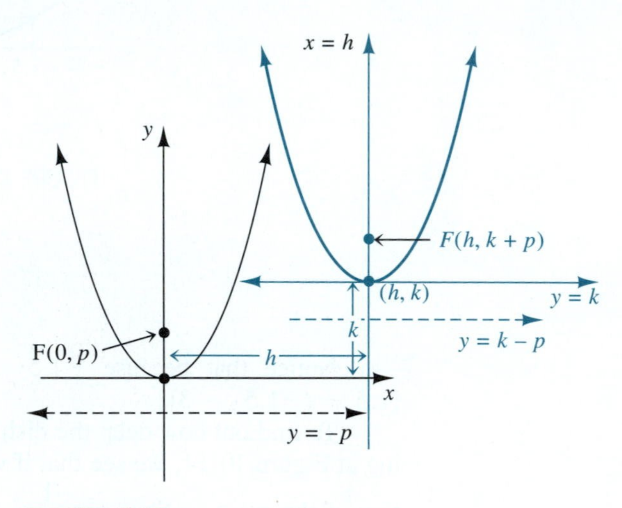

FIGURE 10.15

Horizontal and vertical shifting of the figure will produce a figure with axes of symmetry parallel to their original axes. We call this kind of shifting **translation of axes**. In the preceding case, the new axis of the parabola will be parallel to the y-axis.

GRAFFIX

On a graphics calculator, to graph a relation (that is not a function) such as $(y - 2)^2 = 24x$, you must graph both $y = 2 + \sqrt{24x}$ and $y = 2 - \sqrt{24x}$ on the same set of coordinate axes. (Why?) On your graphics calculator or computer, graph the following on the same set of coordinate axes:

$$y^2 = 20x, \qquad (y - 2)^2 = 20x, \qquad (y - 2)^2 = 20(x - 3).$$

What can you conclude about the effects of h and k on the graph of $(y - h)^2 = 4p(x - k)$?

If all three parabolas cannot be graphed on one set of axes, compare $y^2 = 20x$ with $(y - 2)^2 = 20x$ and then compare $(y - 2)^2 = 20x$ with $(y - 2)^2 = 20(x - 3)$.

In the same way we derived the equation for the parabola centered at the origin, we can derive the standard form of the equation for the parabola with vertex (h, k), focus $F(h, k + p)$, and axis of symmetry $x = h$. We arrive at the following standard form.

The "vertical" parabola with vertex (h, k)

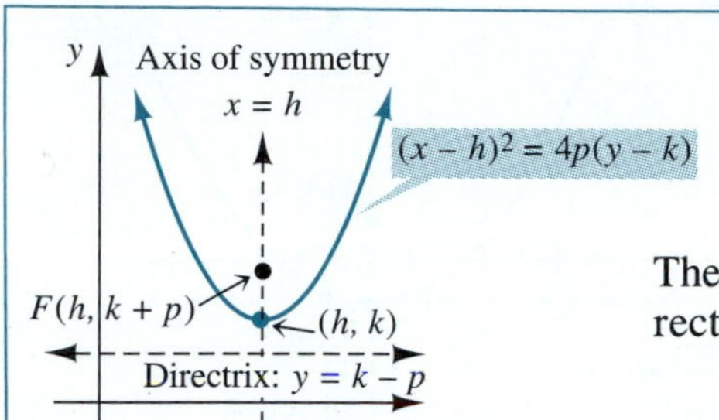

STANDARD FORM OF THE EQUATION FOR A PARABOLA WITH FOCUS $(h, k + p)$ AND DIRECTRIX $y = k - p$

The standard form for the equation of a parabola with focus $(h, k + p)$ and directrix $y = k - p$ is

$$(x - h)^2 = 4p(y - k).$$

This is a parabola with vertex (h, k) and axis of symmetry $x = h$.

Compare the form of the parabola with its vertex at $(0, 0)$ and symmetric with respect to the y-axis with the form of the parabola with its vertex at (h, k) and symmetric with respect to a line parallel to the y-axis.

To change the graph of the parabola $x^2 = 4py$ into $(x - h)^2 = 4p(y - k)$:

Shift the vertex from	$(0, 0)$ to	(h, k).
Shift the axis of symmetry from	$x = 0$ to	$x = h$.
Shift the focus from	$(0, p)$ to	$(h, k + p)$.

If we draw a new set of horizontal and vertical axes through the vertex, the new vertical axis is the axis of symmetry.

This suggests that if we have the parabola in the form $(x - h)^2 = 4p(y - k)$, then we can identify h, k (and p), *draw a new set of coordinate axes through the point (h, k), and graph the form $x^2 = 4py$ on the new set of axes.*

EXAMPLE 5 Sketch the graph of the parabola $(x - 3)^2 = 8(y + 1)$. Identify the vertex, focus, axis of symmetry, and directrix.

Solution We compare the graph of $(x - 3)^2 = 8(y + 1)$ with the standard form:

$$(x - h)^2 = 4p(y - k)$$

$$(x - 3)^2 = 8(y + 1) \qquad \textit{Now we can identify } h = 3,\ k = -1, \textit{ and } p = 2.$$

Hence the vertex is $(3, -1)$. We draw a new set of coordinate axes centered at $(3, -1)$ (and parallel to the x- and y-axes). See Figure 10.16(a) on page 650. The axis of symmetry is the new vertical axis; it is parallel to and 3 units right of the original y-axis and, therefore, has equation $x = 3$. Since $p = 2$, if we move 2 units up from the vertex along the new vertical axis, we locate the focus at $(3, 1)$.

If we move vertically down 2 units from the vertex $(3, -1)$ and draw a line through this point parallel to the x-axis, we have the directrix. Since the directrix is parallel to and 3 units below the x-axis, its equation is $y = -3$. See Figure 10.16(b).

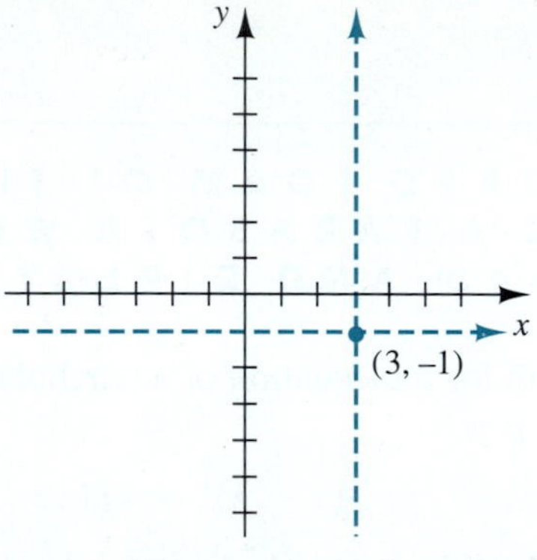

Draw a new set of horizontal and vertical axes through (3, – 1).
(a)

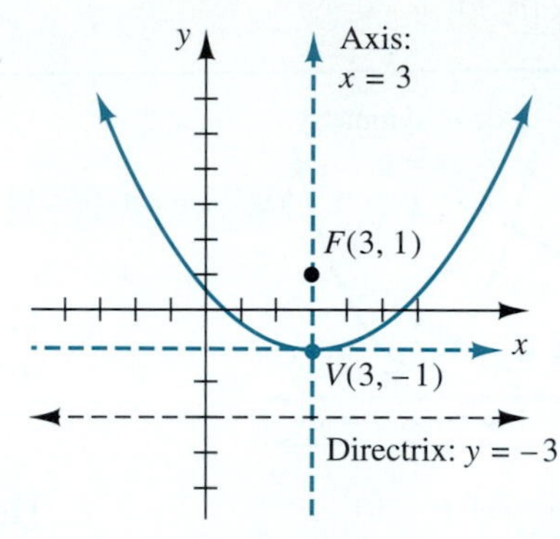

The graph of $(x-3)^2 = 8(y+1)$
(b)

FIGURE 10.16

The standard form for the equation of the parabola with vertex (h, k) and axis of symmetry parallel to the *x-axis* is summarized next.

The "horizontal" parabola with vertex (h, k)

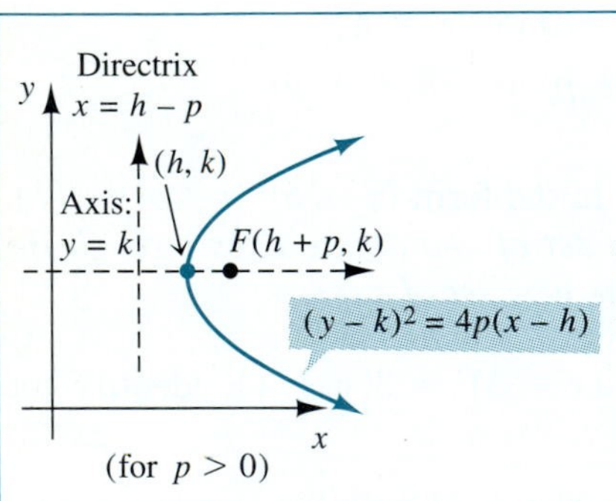

STANDARD FORM FOR THE EQUATION OF A PARABOLA WITH FOCUS $(h+p, k)$ AND DIRECTRIX $x = h - p$

The standard form for the equation of a parabola with focus $(h + p, k)$ and directrix $x = h - p$ is

$$(y - k)^2 = 4p(x - h).$$

This is a parabola with vertex (h, k) and axis of symmetry $y = k$.

EXAMPLE 6 Find the equation of the parabola with focus $(2, -1)$ and directrix $x = 6$.

Solution We plot the focus and sketch the directrix (Figure 10.17). Since the directrix has equation $x = 6$, which is a vertical line, the parabola must have a horizontal axis of symmetry, and therefore its general form is $(y - k)^2 = 4p(x - h)$, with focus $(h + p, k)$. Since $(2, -1)$ is the focus, we have $h + p = 2$ and $k = -1$. We found k, but we still need to find h and p.

Since the directrix is right of the focus, the parabola opens left. Why?

The directrix for the general form above is $x = h - p$. Since our directrix is $x = 6$, we have $h - p = 6$.

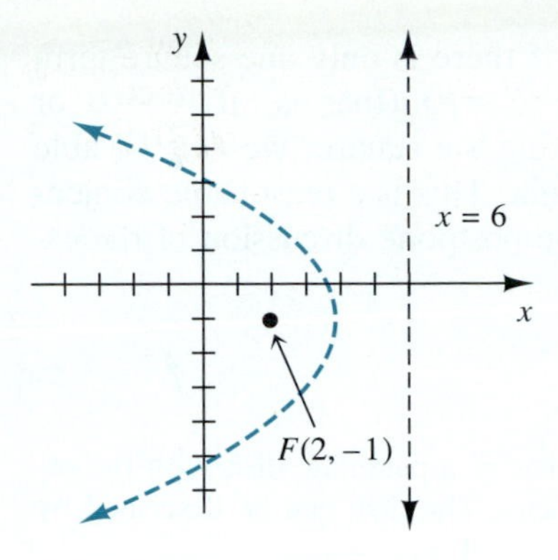

FIGURE 10.17 The directrix is $x = 6$ and the focus is $(2, -1)$.

Now we have two equations in two unknowns, $\begin{cases} h + p = 2 \\ h - p = 6 \end{cases}$ which we can solve:

$$\begin{array}{rl} h + p = 2 & \\ h - p = 6 & \textit{Add to get} \\ \hline 2h \quad = 8 & \textit{Which yields } h = 4 \end{array}$$

Substituting $h = 4$ into $h + p = 2$ yields $4 + p = 2 \Rightarrow p = -2$.

Hence $h = 4$, $k = -1$, and $p = -2$. We substitute these values into the general form:

$$(y - k)^2 = 4p(x - h) \qquad \textit{To get}$$

$$[y - (-1)]^2 = 4(-2)(x - 4) \quad \textit{or} \quad (y + 1)^2 = -8(x - 4)$$

■

FIGURE 10.18

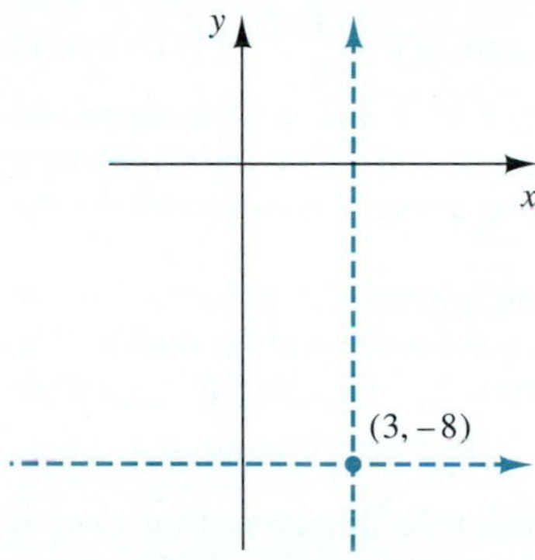

Draw a new set of horizontal and vertical axes through (3, −8).

(a)

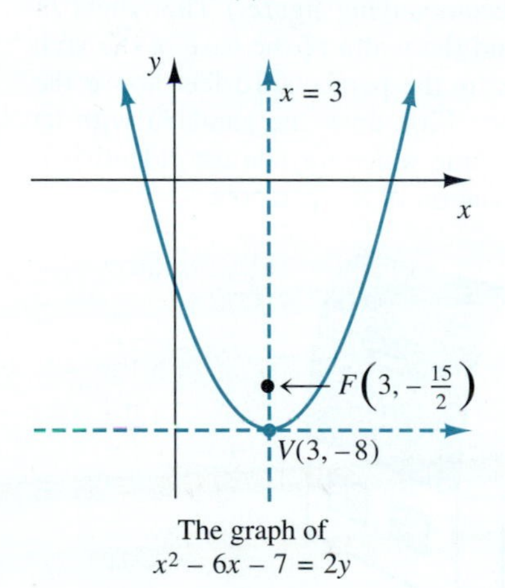

The graph of $x^2 - 6x - 7 = 2y$

(b)

EXAMPLE 7 Sketch the graph of the parabola $x^2 - 6x - 7 = 2y$, and identify its vertex, axis of symmetry, and focus.

Solution We recognize that there is an x^2-term, so the parabola has a vertical axis of symmetry. Thus we want the equation to be in the form $(x - h)^2 = 4p(y - k)$. We use completing the square as follows:

$$x^2 - 6x - 7 = 2y$$

Isolate the terms containing powers of x on one side of the equation.

$$x^2 - 6x \quad = 2y + 7$$

Complete the square for the left-hand side: Add $9 = \left[\frac{1}{2}(-6)\right]^2$ *to each side of the equation.*

$$x^2 - 6x + 9 = 2y + 7 + 9$$

Write the left-hand side in factored form.

$$(x - 3)^2 = 2y + 16$$

We want the right-hand side to look like $4p(y - k)$, *so we factor out 2, the coefficient of y, to get*

$$(x - 3)^2 = 2(y + 8)$$

Now we can see $4p = 2$, *so* $p = \frac{1}{2}$.

$$(x - 3)^2 = 4\left(\frac{1}{2}\right)[y - (-8)]$$

The equation is now in the form $(x - h)^2 = 4p(y - k)$ *and we can identify h, k, and p*

Thus $h = 3$, $k = -8$, and $p = \frac{1}{2}$. We draw a new set of coordinate axes through $(3, -8)$. The vertex is $(3, -8)$. See Figure 10.18(a). The axis of symmetry is the line $x = 3$. If we move up $\frac{1}{2}$ unit above the vertex along the parabola's axis, we locate the focus at $\left(3, -8 + \frac{1}{2}\right) = \left(3, -\frac{15}{2}\right)$. See Figure 10.18(b). ■

At this point it may have occurred to you that if there is only one square term in the general equation $Ax^2 + Cy^2 + Dx + Ey + F = 0$ (that is, if $A = 0$ or $C = 0$ but not both), then, with the help of completing the square, we *may* be able to put the equation into the standard form of a parabola. This is a reasonable conjecture, but as with the circle, there are exceptions. We postpone discussion of the exceptional cases until Section 10.5.

EXERCISES 10.2

In Exercises 1–6, identify the focus and the directrix of the parabolas.

1. $x^2 = 12y$ **2.** $y^2 = -16x$
3. $x = -8y^2$ **4.** $y = -15x^2$
5. $6x^2 - 3y = 0$ **6.** $2y^2 + 10x = 0$

In Exercises 7–10 graph the parabola and identify the focus, axis, and directrix.

7. $4x^2 - 12y = 0$ **8.** $6y^2 - 24x = 0$
9. $x + y^2 = 0$ **10.** $x^2 + 16y = 0$

In Exercises 11–16 write the equation of the parabola using the given information.

11. The parabola has focus (0, 4) and vertex at the origin.
12. The parabola has focus (−4, 0) and vertex at the origin.
13. The parabola has focus (0, −3) and directrix $y = 3$.
14. The parabola has focus (3, 0) and directrix $x = -3$.
15. The parabola has its vertex at the origin and directrix $x = -5$.
16. The parabola has its vertex at the origin and directrix $y = 8$.
17. A TV satellite antenna consists of a parabolic dish with the receiver placed at its focus. The dish can be described by rotating the parabola $y = \frac{1}{12}x^2$ about its axis of symmetry, where $-6 \le x \le 6$ and x is measured in feet. How deep is the dish, and where should the receiver be placed in relation to the bottom (vertex) of the dish? (See the accompanying figure.)

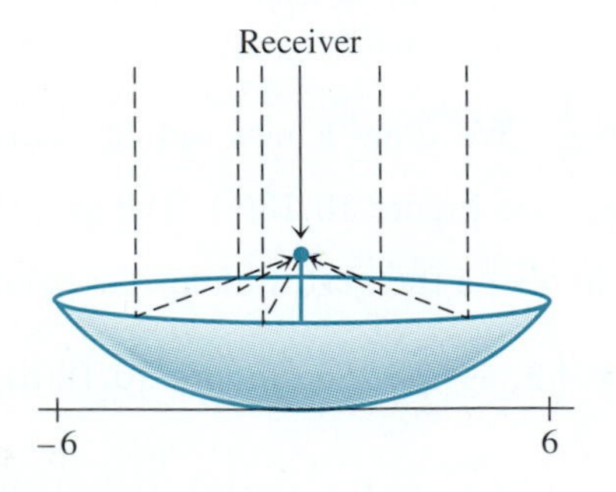

18. A radar antenna consists of a parabolic dish with the receiver placed at its focus. The dish can be described by rotating the parabola $y = \frac{1}{20}x^2$ about its axis of symmetry, where $-8 \le x \le 8$ and x is measured in feet. How deep is the dish, and where should the receiver be placed in relation to the bottom (vertex) of the dish?
19. A spotlight has a parabolic reflecting mirror with the light source placed at its focus. The mirror can be described by rotating the parabola $y = \frac{1}{6}x^2$ about its axis of symmetry, where $-2 \le x \le 2$ and x is measured in feet. How deep is the mirror, and where should the light source be placed in relation to the bottom (vertex) of the mirror?
20. A reflecting telescope has a parabolic reflecting mirror with the eyepiece focused on the focus of the mirror. The mirror can be described by rotating the parabola $y = \frac{1}{10}x^2$ about its axis of symmetry, where $-2 \le x \le 2$ and x is measured in inches. How deep is the mirror, and where should the eyepiece be focused in relation to the bottom (vertex) of the mirror?
21. An arch of a building is constructed in the shape of a parabola. (See the accompanying figure.) The height of the arch is 30 feet, and the width of the base of the arch is 20 feet. How wide is the parabola 10 feet above the base of the arch? HINT: First draw the parabola with its vertex at (0, 0), determine which form to use, identify p, and then write the equation of the parabola.

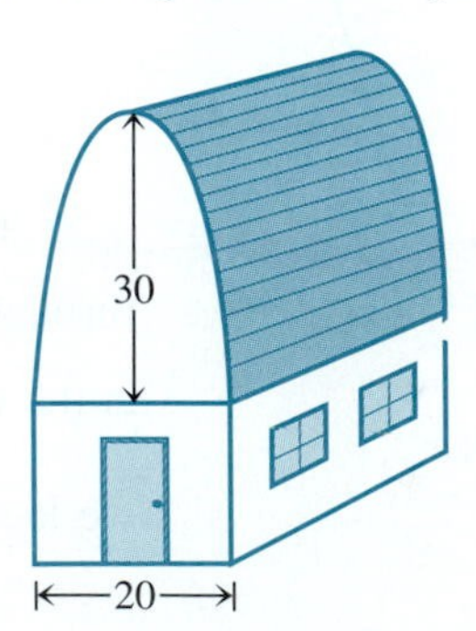

22. The cable of a suspension bridge is in the shape of a parabola. The cable extends over 200 feet of highway. The longest supporting wire, on either end of the cable, is 100 feet; the shortest, in the middle, is 30 feet. Find the length of the supporting wire that is 40 feet from the longest supporting wire. (See the figure below.)

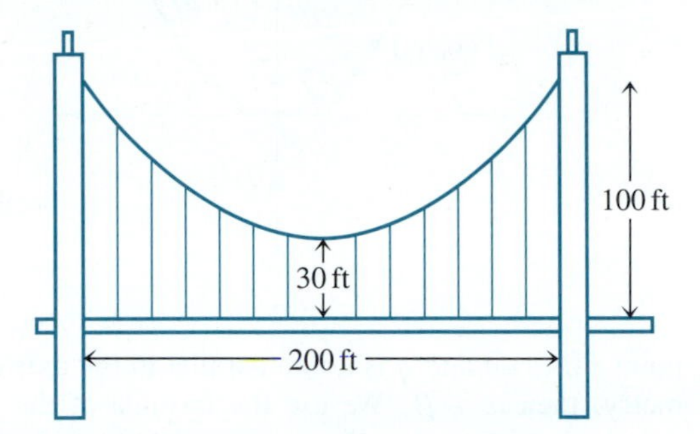

23. A flashlight is to be designed so that a cylindrical casing houses a parabolic reflecting mirror and the light source. (See the accompanying figure.) If the casing has a depth of 2 inches and a diameter of 4 inches, where should the light source be placed relative to the vertex of the mirror?

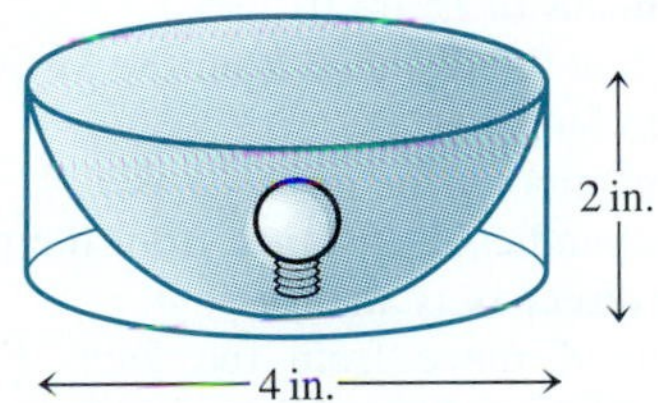

24. An engineer wants to design a parabolic microphone as described in Example 4. She wants the diameter of the dish to be 2 feet, but wants the receiver to be located below the lip of the dish. How deep can the dish be if the receiver is properly placed at the focus? HINT: Construct a parabola opening upward, with its vertex at the origin, and examine its equation in standard form. How would you translate the receiver being placed "below the lip of the dish" as an algebraic relationship between p and y?

In Exercises 25–30, identify the focus and the directrix of each parabola.

25. $(x - 2)^2 = 8y$ **26.** $(y + 5)^2 = -16x$
27. $8x = y^2 - 2y - 15$ **28.** $4y = x^2 - 6x + 5$
29. $-5y = x^2 - 4x + 19$ **30.** $2x = y^2 - 6y + 3$

In Exercises 31–34, graph the parabola and identify the focus, axis of symmetry, and directrix.

31. $8y = x^2 + 10x + 33$ **32.** $-12x = y^2 - 4y + 16$
33. $x = 3y^2 - 6y + 5$ **34.** $y = -2x^2 - 12x - 19$

In Exercises 35–40, write the equation of the parabola using the given information.

35. The parabola has focus $(2, 4)$ and vertex $(5, 4)$.
36. The parabola has focus $(-2, 3)$ and vertex $(-2, -1)$.
37. The parabola has focus $(2, -3)$ and directrix $y = 4$.
38. The parabola has focus $(3, 4)$ and directrix $x = 7$.
39. The parabola has its vertex at $(2, 5)$ and directrix $x = -3$.
40. The parabola has its vertex at $(3, -1)$, and directrix $y = 2$.

In Exercises 41–44 graph the solution to the inequality.

41. $y < 4x^2$ **42.** $-12x > y^2$
43. $x \geq y^2 - 6y + 5$ **44.** $y \geq -2x^2 - 12x - 8$

45. Use the distance formula to derive the equation for the parabola centered at the origin with focus $(p, 0)$ and directrix $x = -p$.

46. Use the distance formula to derive the equation for the parabola centered at (h, k) with focus $(h, k + p)$ and directrix $y = k - p$.

47. We define a line to be tangent to a parabola at point P if and only if it touches the parabola only at point P and the line is not parallel to its axis of symmetry. (See the accompanying figure.) The equation for the line tangent to the parabola $x^2 = 4py$ at the point (x_0, y_0) is given by

$$y = \frac{x_0}{2p}x - y_0$$

Use this formula to find the equation of the line tangent to the parabola $x^2 = 8y$ at the point $(-4, 2)$.

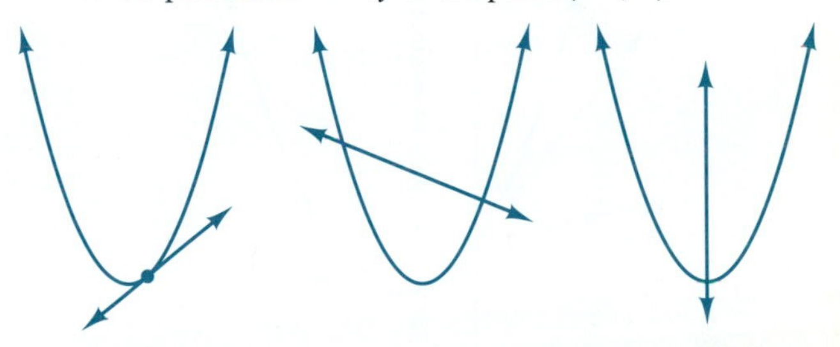

Tangent line Not a tangent line Not a tangent line

48. Use the formula for the equation of the line tangent to the parabola in Exercise 47 to find the equation of the line tangent to the parabola $x^2 = 8y$ at the point $(4, 2)$.

49. The *latus rectum* of a parabola is defined to be the line segment passing through the focus of the parabola which is perpendicular to its axis and has its endpoints on the parabola. (See the accompanying figure on page 654.) Show that the length of the latus rectum of the parabola $x^2 = 4py$ is $4p$.

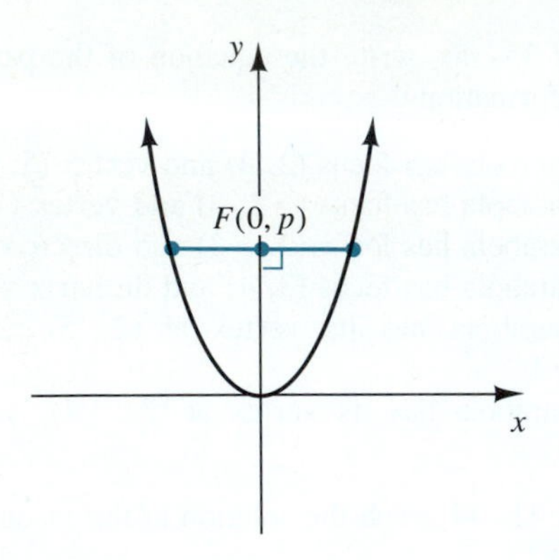

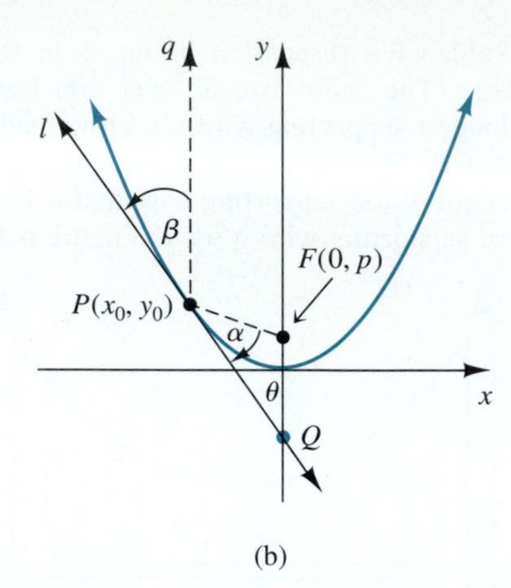

(b)

QUESTIONS FOR THOUGHT

50. The equations $x^2 + x - 6 = 0$ and $y^2 - 6y + 9 = 0$ are in the general form of a second-degree equation. If you graphed these equations on a rectangular coordinate system, what type of figure would you expect to have? Graph the equations on a set of coordinate axes. Is the graph what you expected?

51. We can state the reflecting property of the parabola in a more formal manner as follows: If l is a line tangent to a parabola at point P, then for any line q parallel to its axis of symmetry, the angle between l and q will equal the angle between l and the line joining P and the focus. (See the accompanying figure.)

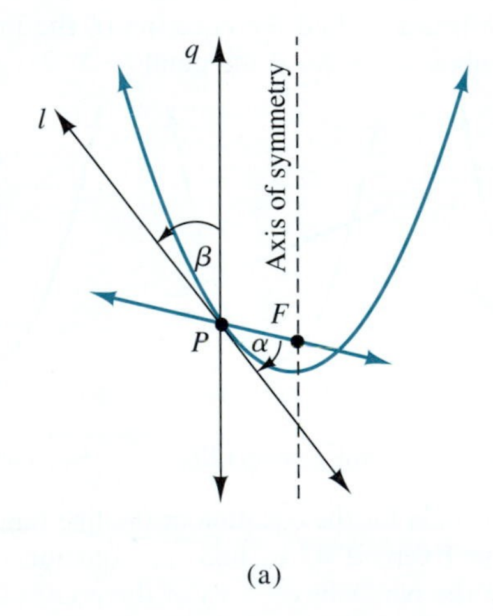

(a)

We outline a way to demonstrate the reflecting property of the parabola; that is, given the accompanying figure, where l is a line tangent to the parabola $x^2 = 4py$ at point $P(x_0, y_0)$ and q is a line parallel to the axis of symmetry, then $\alpha = \beta$. We use the formula of the tangent line given in Exercise 47.

(a) Exercise 47 states that the equation for the tangent line l is $y = \dfrac{x_0}{2p}x - y_0$. We designate the point where the tangent line crosses the y-axis as Q (forming the angle θ with the x-axis). Show that the coordinates of Q are $(0, -y_0)$.

(b) Show that the distance from the focus, $F(0, p)$ to Q is, therefore, $p + y_0$.

(c) The equation of the directrix is $y = -p$. Show that the perpendicular distance from the point $P(x_0, y_0)$ to the directrix is also $y_0 + p$.

(d) But the distance from the focus F to the point $P(x_0, y_0)$ must be equal to the distance from point $P(x_0, y_0)$ to the directrix. Explain why.

(e) Explain how we can now conclude that $|\overline{FQ}| = |\overline{FP}|$.

(f) But if $|\overline{FQ}| = |\overline{FP}|$, then $\alpha = \theta$. Why?

(g) Explain why $\theta = \beta$.

(h) Therefore, we conclude that $\alpha = \beta$.

GRAFFIX

In Exercises 52–55, use your graphics calculator or computer to sketch the graph.

52. $(x - 1)^2 = 18y$

53. $(y - 1)^2 = 24x$

54. $y^2 + 10y - x + 25 = 0$

55. $x^2 - 6x - 12y + 9 = 0$

10.3 The Ellipse

The **ellipse** is another conic section that is useful in providing a mathematical model for a variety of physical phenomena such as the orbits of planets.

DEFINITION An **ellipse** is the set of all points in a plane, the sum of whose distances from two fixed points is a constant (see Figure 10.19). The fixed points are called the **foci** of the ellipse.

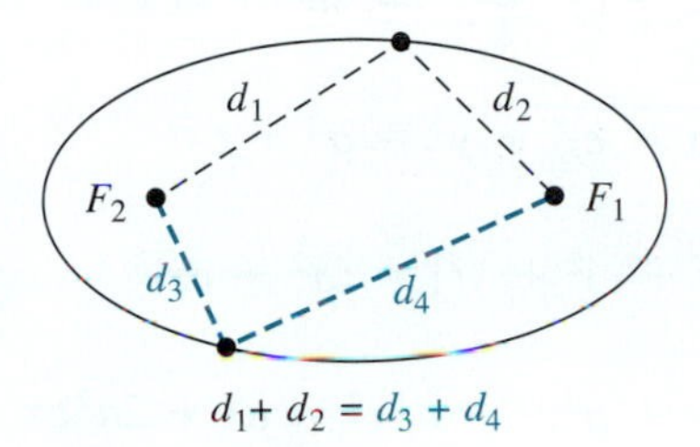

FIGURE 10.19 Constructing an ellipse

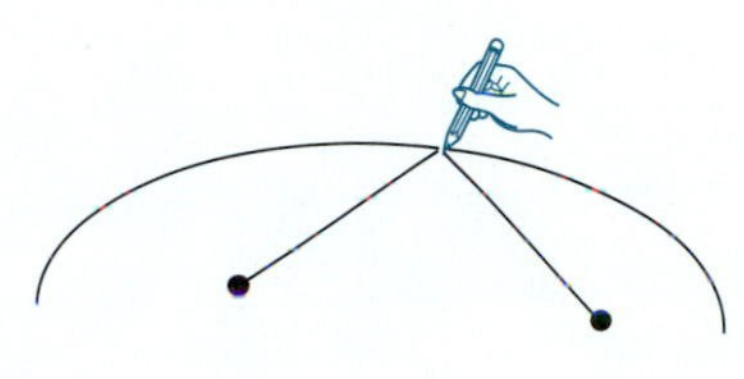

FIGURE 10.20

We can actually use this definition to draw an ellipse by fastening the ends of a string at two points (see Figure 10.20). Obviously, the string must be longer than the distance between the two points. If you move the pencil while holding it taut against the string, you will trace out an ellipse. The points where the string is fastened are the foci of the ellipse, and the length of the string is the sum of the distances from the foci, which is constant.

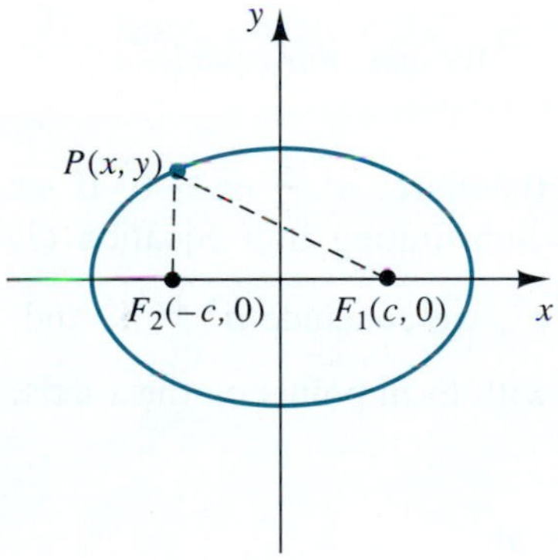

As with the parabola, we use the definition of the ellipse to derive its equation. We start by choosing a coordinate system in such a way that the foci lie on a horizontal line with the origin midway between the two foci. See Figure 10.21. Letting $c > 0$ be the distance from the origin to the right focus F_1, the distance to the left focus F_2, is also c. Hence the coordinates of F_1 are $(c, 0)$, and the coordinates of F_2 are $(-c, 0)$. The distance between the two foci, F_1 and F_2, is $2c$. We can derive an equation for this ellipse as follows.

FIGURE 10.21 Ellipse with foci $F_1(c, 0)$ and $F_2(-c, 0)$

By definition of an ellipse, if we pick any point on the ellipse, $P(x, y)$, the sum of the distance from $P(x, y)$ to $F_1(c, 0)$ and the distance from $P(x, y)$ to $F_2(-c, 0)$ is constant. To avoid working with fractions, we call this constant sum $2a$ and note that for any point on the ellipse, $2a > 2c$, or $a > c$. (This is equivalent to saying that the string is longer than the distance between the two foci.) Hence we have

$$|\overline{PF_1}| + |\overline{PF_2}| = 2a \qquad \textit{Using the distance formula we get}$$

$$\sqrt{(x-c)^2 + (y-0)^2} + \sqrt{(x+c)^2 + (y-0)^2} = 2a \qquad \textit{Or}$$

$$\sqrt{(x-c)^2 + y^2} + \sqrt{(x+c)^2 + y^2} = 2a \qquad \textit{Now isolate the first radical.}$$

$$\sqrt{(x-c)^2+y^2} = 2a - \sqrt{(x+c)^2+y^2}$$ *Square each side.*

$$(x-c)^2+y^2 = 4a^2 - 4a\sqrt{(x+c)^2+y^2} + (x+c)^2+y^2$$

$$x^2-2cx+c^2+y^2 = 4a^2 - 4a\sqrt{(x+c)^2+y^2} + x^2+2cx+c^2+y^2$$

Simplify and isolate the radical term again.

$$4a\sqrt{(x+c)^2+y^2} = 4a^2+4cx$$ *Dividing by 4 we get*

$$a\sqrt{(x+c)^2+y^2} = a^2+cx$$ *Squaring each side again yields*

$$a^2[(x+c)^2+y^2] = (a^2+cx)^2$$ *which yields*

$$a^2x^2+2a^2cx+a^2c^2+a^2y^2 = a^4+2a^2cx+c^2x^2$$ *which can be rewritten as*

$$a^2x^2-c^2x^2+a^2y^2 = a^4-a^2c^2$$ *or*

$$x^2(a^2-c^2)+a^2y^2 = a^2(a^2-c^2)$$ *We now divide both sides by $a^2(a^2-c^2)$ to get*

$$\frac{x^2}{a^2}+\frac{y^2}{a^2-c^2} = 1$$ *We label this equation (1)*

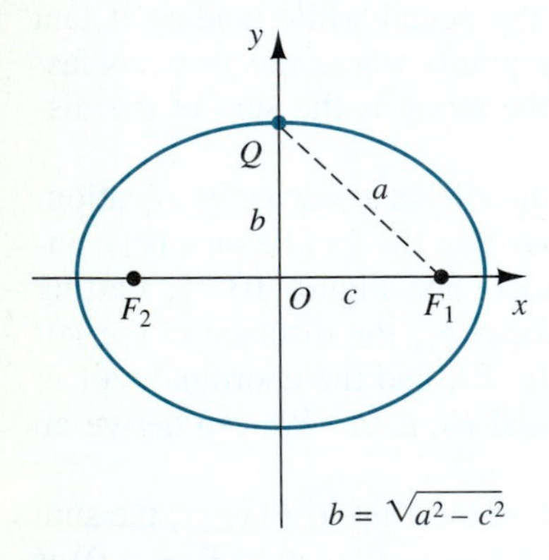

FIGURE 10.22

We noted that a and c are positive and $a > c$; therefore, $a^2 - c^2 > 0$. If we let $b = \sqrt{a^2-c^2}$, then $b > 0$ and $b^2 = a^2 - c^2$. Substituting into equation (1) gives us $\frac{x^2}{a^2}+\frac{y^2}{b^2} = 1$. Since $c > 0$ and $b^2 = a^2 - c^2$, we conclude $a^2 > b^2$ and, hence $a > b$. Thus the ellipse centered at the origin, with focal points on the x-axis, has the equation:

$$\frac{x^2}{a^2}+\frac{y^2}{b^2} = 1 \qquad (a > b)$$

To get the sense of how a, b, and c are related geometrically, consider the ellipse $\frac{x^2}{a^2}+\frac{y^2}{b^2} = 1$ in Figure 10.22.

Let Q be the point where the ellipse intersects the y-axis. Q lies on the ellipse and therefore, the sum of the distances from the foci F_1 and F_2 is $2a$. Since Q is equidistant from F_1 and F_2, then $|\overline{QF_1}| = a$. Since $|\overline{OF_1}| = c$, by the Pythagorean Theorem, $|\overline{OQ}| = \sqrt{a^2-c^2}$. Since we defined $\sqrt{a^2-c^2}$ as b, then $b = |\overline{OQ}|$.

The x-intercepts of the graph of the equation $\frac{x^2}{a^2}+\frac{y^2}{b^2} = 1$ are found by setting $y = 0$:

$$\frac{x^2}{a^2} + \frac{y^2}{b^2} = 1 \qquad \textit{Let } y = 0.$$

$$\frac{x^2}{a^2} + \frac{0^2}{b^2} = 1 \qquad \textit{Solve for } x.$$

$$\frac{x^2}{a^2} = 1 \Rightarrow x^2 = a^2 \quad \text{or} \quad x = \pm a. \qquad \textit{Hence the x-intercepts are } \pm a.$$

We call the points $V_1(a, 0)$, and $V_2(-a, 0)$ the **vertices** of this ellipse, and the line segment $\overline{V_1 V_2}$ is called the **major axis**. Similarly, if we let $x = 0$, then we find the y-intercepts to be $y = \pm b$. The line segment joining the points $(0, b)$ and $(0, -b)$ is called the **minor axis** of the ellipse. We summarize these facts as follows.

The ellipse with horizontal major axis, centered at (0, 0)

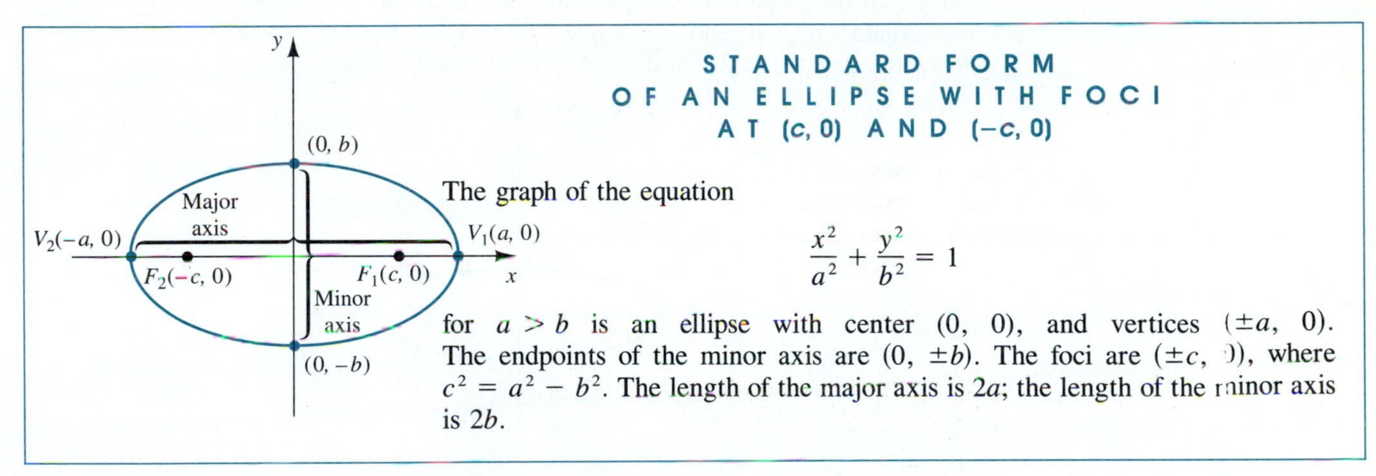

STANDARD FORM OF AN ELLIPSE WITH FOCI AT $(c, 0)$ AND $(-c, 0)$

The graph of the equation

$$\frac{x^2}{a^2} + \frac{y^2}{b^2} = 1$$

for $a > b$ is an ellipse with center (0, 0), and vertices $(\pm a, 0)$. The endpoints of the minor axis are $(0, \pm b)$. The foci are $(\pm c, 0)$, where $c^2 = a^2 - b^2$. The length of the major axis is $2a$; the length of the minor axis is $2b$.

How are the endpoints of the *major* axis related to the vertices?

EXAMPLE 1 Sketch the graph of the following ellipses. Identify the foci.

(a) $\frac{x^2}{16} + \frac{y^2}{9} = 1$ **(b)** $3x^2 + 15y^2 = 45$

Solution

(a) In order to sketch the graph of an ellipse centered at the origin, we compare our equation with the standard form of the ellipse $\frac{x^2}{a^2} + \frac{y^2}{b^2} = 1$ and we get

$$a^2 = 16 \Rightarrow a = 4 \quad \text{and} \quad b^2 = 9 \Rightarrow b = 3.$$ *(Remember a and b are positive.)*

We plot the vertices $(\pm 4, 0)$ and the endpoints of the minor axis, $(0, \pm 3)$, and sketch the graph of the ellipse, as shown in Figure 10.23.

Notice that $a > b$ and $c^2 = a^2 - b^2 = 16 - 9 = 7 \Rightarrow c = \sqrt{7}$. Hence the foci are $(\pm\sqrt{7}, 0)$.

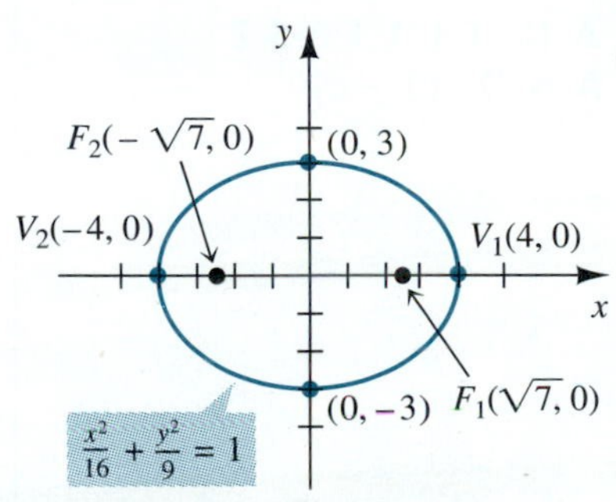

FIGURE 10.23

(b) First we put the equation into standard form. We observe that standard form requires that a 1 be on one side. In order for us to have 1 on the right-hand side, we must divide both sides of the equation by 45.

$$3x^2 + 15y^2 = 45 \qquad \textit{Divide both sides by 45.}$$

$$\frac{3x^2}{45} + \frac{15y^2}{45} = \frac{45}{45} \qquad \textit{Simplify.}$$

$$\frac{x^2}{15} + \frac{y^2}{3} = 1 \qquad \textit{The equation is now in standard form.}$$

Comparing our equation with the standard form of the ellipse, $\frac{x^2}{a^2} + \frac{y^2}{b^2} = 1$, we get

$$a^2 = 15 \Rightarrow a = \sqrt{15} \quad \text{and} \quad b^2 = 3 \Rightarrow b = \sqrt{3}$$

We plot the vertices $(\pm\sqrt{15}, 0)$ and the endpoints of the minor axis $(0, \pm\sqrt{3})$ and sketch the graph of the ellipse. See Figure 10.24.

Again, $a > b$ and $c^2 = a^2 - b^2 = 15 - 3 = 12 \Rightarrow c = \sqrt{12} = 2\sqrt{3}$. Hence the foci are $(\pm 2\sqrt{3}, 0)$.

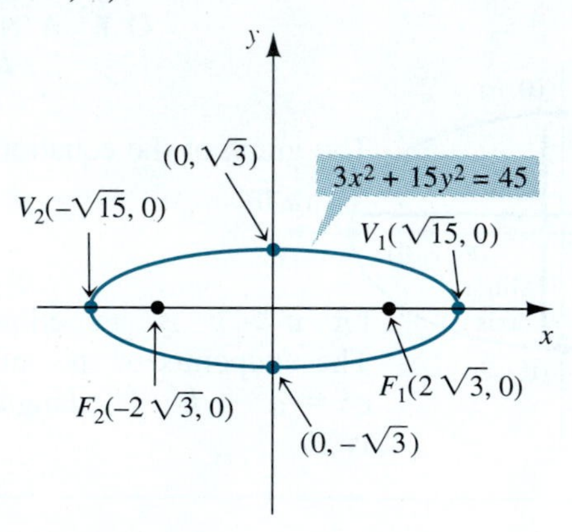

FIGURE 10.24 ■

The ellipse with vertical major axis, centered at (0, 0)

Using the same analysis (again with $a > b$), we use the distance formula to derive the equation of an ellipse with center (0, 0) and *foci on the y-axis*.

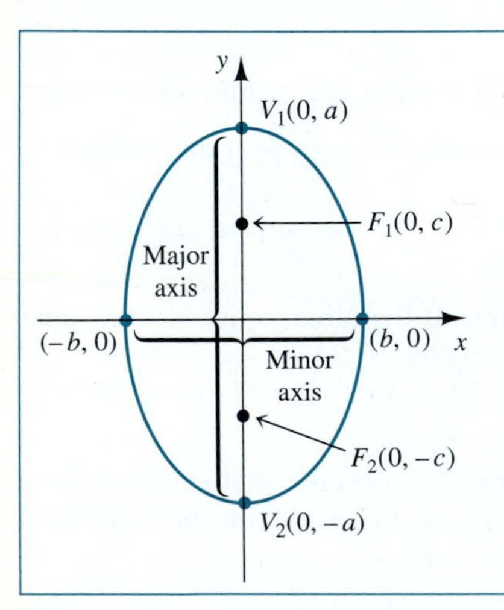

STANDARD FORM OF AN ELLIPSE WITH FOCI AT (0, c) AND (0, −c)

The graph of the equation

$$\frac{x^2}{b^2} + \frac{y^2}{a^2} = 1$$

for $a > b$ is an ellipse with center (0, 0), and vertices $(0, \pm a)$. The endpoints of the minor axis are $(\pm b, 0)$. The foci are $(0, \pm c)$, where $c^2 = a^2 - b^2$. The length of the major axis is $2a$; the length of the minor axis is $2b$.

Examine the two standard forms of the ellipse; observe their similarities and differences. The first has its foci on the x-axis; the second has its foci on the y-axis. Notice that the two forms switch a and b as denominators of x^2 and y^2. This ensures that a is larger than b, so the foci found by $c^2 = a^2 - b^2$ will exist. Also, as a consequence, the major axis (the axis containing the foci) is always the longer axis (of length $2a$), and the vertices, the endpoints of the longer axis, are always determined by a.

When the denominator of the y^2-term is larger than the denominator of the x^2-term, is the major axis along the x-axis or y-axis? What distinguishes the major axis from the minor axis?

The choice of which form to use is dependent upon which denominator is larger: If the denominator of x^2 is larger than the denominator of y^2, then the major axis lies on the x-axis, and we use the first form (since $a > b$). If the denominator of y^2 is greater than the denominator of x^2, then the major axis lies on the y-axis, and we use the second form (again, $a > b$). Thus a^2 is always the larger denominator.

EXAMPLE 2 Sketch the graph of $16x^2 + 4y^2 = 16$. Identify its foci.

Solution First we put the equation into standard form. In order for us to have 1 on the right-hand side, we must divide by 16.

$$16x^2 + 4y^2 = 16 \qquad \textit{Divide both sides by 16.}$$

$$\frac{16x^2}{16} + \frac{4y^2}{16} = \frac{16}{16} \qquad \textit{Simplify.}$$

$$\frac{x^2}{1} + \frac{y^2}{4} = 1 \qquad \textit{The equation is now in standard form.}$$

Comparing our equation with both standard forms of the ellipse,we note that the denominator of y^2 is greater than the denominator of x^2; therefore, since a must be greater than b, we use the second form, $\frac{x^2}{b^2} + \frac{y^2}{a^2} = 1$, an ellipse with foci on the y-axis. Hence

$$a^2 = 4 \Rightarrow a = 2 \quad \text{and} \quad b^2 = 1 \Rightarrow b = 1$$

We plot the vertices $(0, \pm 2)$ and the endpoints of the minor axis $(\pm 1, 0)$ and sketch the graph of the ellipse. See Figure 10.25.

Notice that $a > b$ and $c^2 = a^2 - b^2 = 4 - 1 = 3 \Rightarrow c = \sqrt{3}$. The foci are $(0, \pm\sqrt{3})$.

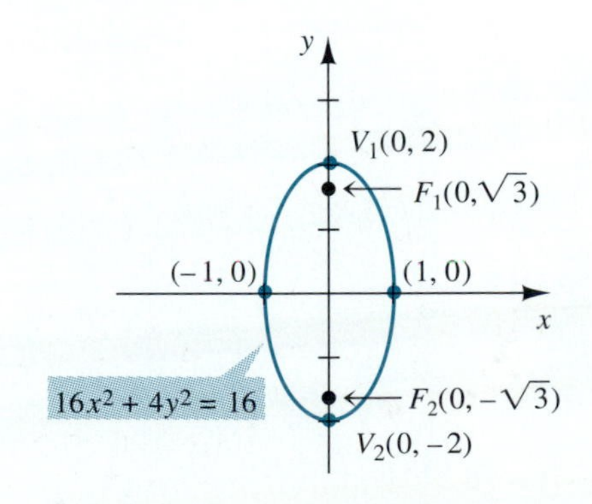

FIGURE 10.25

■

EXAMPLE 3 The arch of a bridge is semielliptical with a horizontal major axis. The base of the arch covers the entire 50-foot width of a two-lane road, and the highest part of the arch is 15 feet vertically above the centerline of the road. Can a truck 14 feet high pass through this bridge, staying right of the centerline, if the truck is 10 feet wide? See Figure 10.26.

FIGURE 10.26

Solution

WHAT DO WE NEED TO FIND?

Whether the bridge is high enough for the truck to pass through.

WHAT INFORMATION ARE WE GIVEN?

We note that the bridge opening is a semiellipse, and we can draw this ellipse on a rectangular coordinate system centered at the origin, with major axis lying on the x-axis. We sketch the graph and label the given information on the graph shown in Figure 10.27; the length of the major axis is 50 feet. Hence $a = 25$, and the vertices are $(\pm 25, 0)$. Since the highest part of the arch is 15 feet, we can label the top endpoint of the minor axis (0, 15).

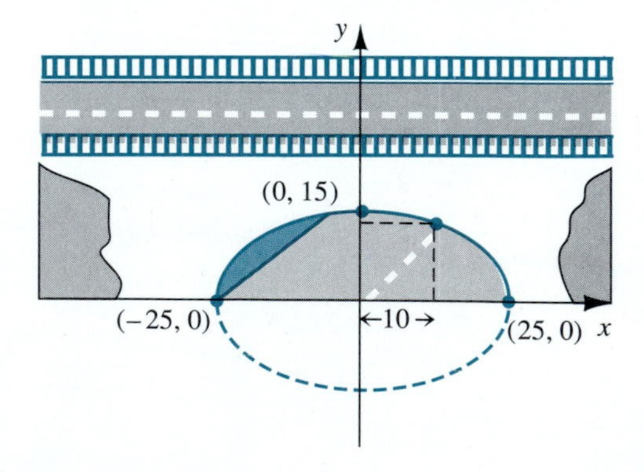

FIGURE 10.27

CAN WE RESTATE WHAT WE NEED TO FIND IN TERMS OF THE FIGURE?

The problem asks whether a truck that is 10 feet wide and 14 feet high can pass through the arch of the bridge staying right of the center line. This means that we need to determine the height of the arch 10 feet from the center. Hence, on the graph in Figure 10.27, this means we need to find y when $x = 10$. Algebraically, if we have the equation of the ellipse, then all we need to do is substitute $x = 10$ in the equation and find y. Hence our next step is to find the equation of the ellipse with the given information.

HOW DO WE FIND THE EQUATION OF THE ELLIPSE?

We use the form where the major axis is horizontal. Since $a = 25$, and $b = 15$, our equation is, therefore, $\dfrac{x^2}{25^2} + \dfrac{y^2}{15^2} = 1$.

(Note the foci lie on the horizontal axis, which means that the larger of a and b determines the denominator of x.) Now we just substitute $x = 10$ in the equation, and solve for y.

$$\frac{x^2}{25^2} + \frac{y^2}{15^2} = 1 \qquad \textit{Let } x = 10; \textit{ solve for } y.$$

$$\frac{10^2}{25^2} + \frac{y^2}{15^2} = 1$$

$$\frac{y^2}{15^2} = 1 - \frac{10^2}{25^2} = \frac{21}{25}$$

$$y^2 = 15^2\left(\frac{21}{25}\right) = 189 \qquad \textit{Hence } y = \pm\sqrt{189} = \pm 3\sqrt{21}.$$

Any suggestions as to how the truck can get through the tunnel without going into the left lane?

We ignore the negative value; the height of the arch, 10 feet from the center, is $3\sqrt{21} \approx 13.75$ feet. Hence the truck will not be able to make it through. ■

Eccentricity

Ellipses can range in shape from almost circular to elongated. Figure 10.28 depicts two ellipses, with the same vertices and centered at the origin, but with different foci.

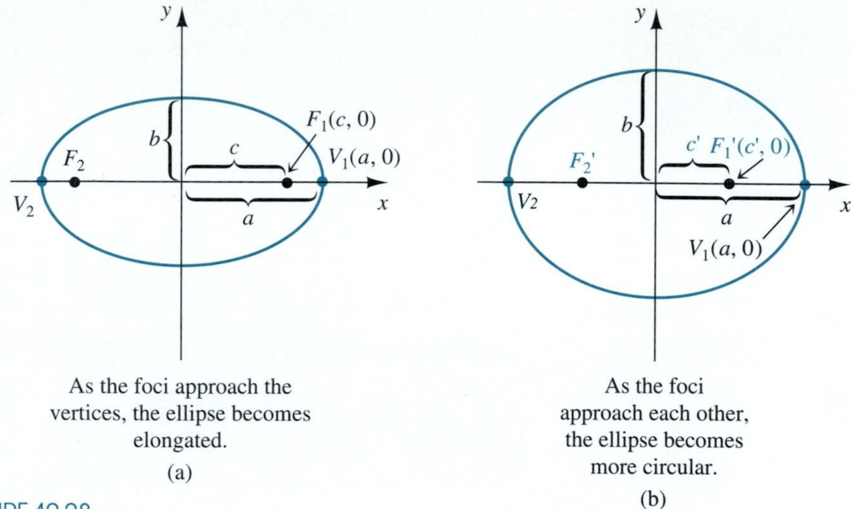

FIGURE 10.28

To understand how the foci affect the shape of the ellipse, let us reexamine the relationship among the foci, major, and minor axes algebraically defined by the equation $b^2 = a^2 - c^2$. See Figure 10.22 on page 656.

Looking at $b^2 = a^2 - c^2$, if we keep a constant and vary c, we find that as c gets closer to a, b becomes smaller. Geometrically, this means as the foci approach the vertices, the minor axis gets shorter. Note that the ellipse becomes elongated.

We use c and a to define the term "eccentricity," denoted by e, to numerically describe the "roundness" of an ellipse.

DEFINITION The **eccentricity**, e, of an ellipse is defined by

$$e = \frac{c}{a} = \frac{\sqrt{a^2 - b^2}}{a}$$

We point out here that the eccentricity, e, is not the constant $e \approx 2.71$ that is the base of natural logarithms, but a variable value that describes the shape of an ellipse.

Let's examine how the values of e are related to the shape of the ellipse.

Since $c > 0$ and $a > 0$, then $e = \frac{c}{a} > 0$. Also since $c < a$, then $e = \frac{c}{a} < 1$.

Putting these together, we have $0 < e < 1$.

Looking at the form of the eccentricity, $e = \frac{\sqrt{a^2 - b^2}}{a}$, keeping a constant, we see that as b gets smaller or closer to 0, algebraically e gets closer to 1. Hence, the flatter (or more elongated) the ellipse, the closer e gets to 1. (See the Different Perspectives box on page 663.)

On the other hand, as b gets closer to a (a must always be larger than b), algebraically $\sqrt{a^2 - b^2}$ gets closer to 0, and so e gets closer to 0. Hence, the ellipse becomes more circular as e approaches 0.

DIFFERENT PERSPECTIVES: *Eccentricity*

GEOMETRIC DESCRIPTION	ALGEBRAIC DESCRIPTION
The eccentricity, e:	Consider eccentricity e defined by $e = \dfrac{\sqrt{a^2 - b^2}}{a}$.
As e approaches 1, the ellipse becomes more elongated.	As the value of b approaches 0, e approaches 1.
As e approaches 0, the ellipse becomes more circular.	As the value of b approaches a, e approaches 0.

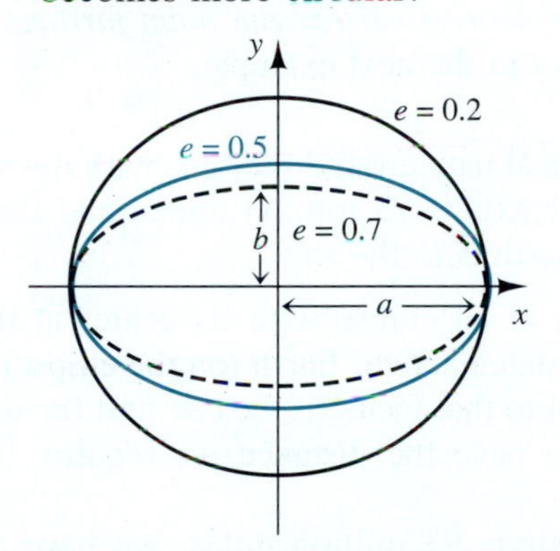

EXAMPLE 4 Find the eccentricity, e, for the ellipse $\dfrac{x^2}{25} + \dfrac{y^2}{8} = 1$.

Solution Since the eccentricity is $e = \dfrac{\sqrt{a^2 - b^2}}{a}$, we find e by first identifying a^2 and b^2. For the ellipse $\dfrac{x^2}{25} + \dfrac{y^2}{8} = 1$, we have $a^2 = 25$ (and, therefore, $a = 5$) and $b^2 = 8$. Hence

$$e = \frac{\sqrt{a^2 - b^2}}{a}$$

$$e = \frac{\sqrt{25 - 8}}{5} = \frac{\sqrt{17}}{5} \approx 0.825$$

■

The orbits of the planets are elliptical with the sun at one of the foci. The orbits of Earth and Mars are almost circular (for Earth, $e \approx 0.017$, for Mars, $e \approx 0.093$), whereas the orbits of other planets are less circular, such as Pluto ($e \approx 0.25$) and Mercury ($e \approx 0.21$). See Figure 10.29. Many comets have elliptical orbits with the sun as the focus; for example, Haley's comet has eccentricity $e \approx 0.967$.

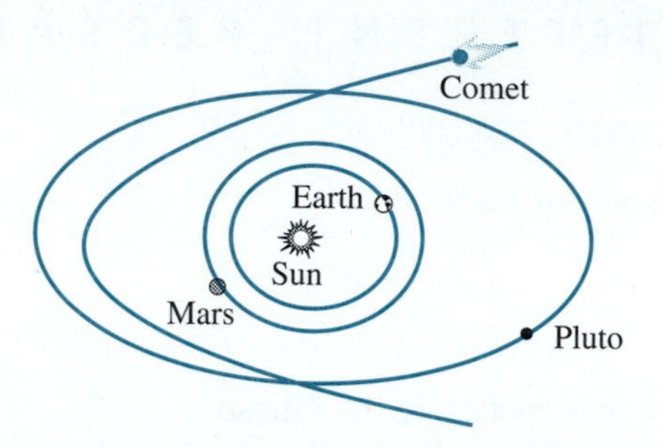

FIGURE 10.29 The solar system

Regardless of the eccentricity of the ellipse, it can be shown that *the point on the ellipse closest to a focus is the nearest vertex; the point farthest from a focus is the farthest vertex*. We use this fact in the next example.

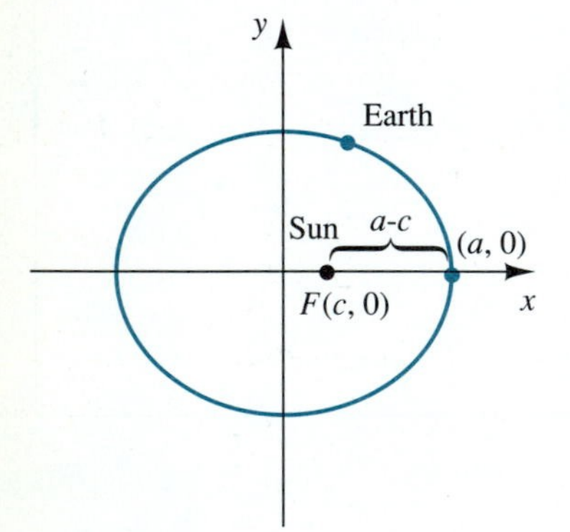

FIGURE 10.30 The earth traces an elliptical orbit with the sun at one focus.

EXAMPLE 5 Earth traces an elliptical path around the sun, with the sun at one of the foci. The length of half the major axis is 93 million miles, and the eccentricity is $e = 0.017$. Estimate the closest Earth is to the sun.

Solution If we sketch the graph of an ellipse with its center at the origin and a horizontal major axis, by the fact stated above, Earth (on the ellipse) is closest to the focus (the sun) at the vertex closest to that focus. If we can find the distance between the focus and its vertex, then we have the distance we require. This distance is $a - c$ in Figure 10.30.

Since one-half the major axis is 93 million miles, we have $a = 9.3 \times 10^7$. Now we need to find c. We are given $e = 0.017$, so we use the definition of e to find c, as follows

$$e = \frac{c}{a}$$ *Substitute $e = 0.017$ and $a = 9.3 \times 10^7$ to get*

$$0.017 = \frac{c}{9.3 \times 10^7}$$ *Solve for c.*

$$c = 0.017(9.3 \times 10^7)$$
$$= 1.581 \times 10^6 \text{ or approximately 1.6 million miles.}$$

Hence, the nearest distance that Earth is to the sun is
$a - c = 93$ million miles $- 1.6$ million miles $= 91.4$ million miles. ■

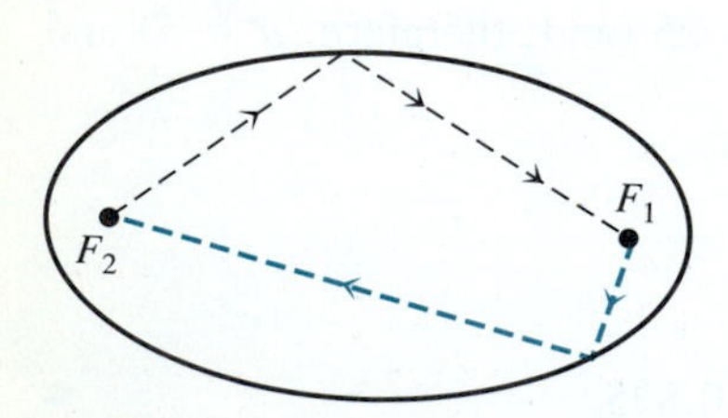

FIGURE 10.31 The reflecting properties of the ellipse

The ellipse also has a reflecting property similar to the parabola. In general, any ray emanating from one focus will be reflected off the ellipse to the other focus. This is illustrated in Figure 10.31.

This property is used in "whispering rooms," such as those found in St. Paul's Cathedral in London or the Capitol in Washington, D.C. In these rooms, the cross sections of the ceilings are made up of elliptical arcs. If you stand at one focus and whisper something that no one else can hear, sound is reflected off the ceiling so that whoever is standing at the other focus will be able to hear you.

GRAFFIX

On a graphics calculator, to graph the relation $\frac{x^2}{36} + \frac{y^2}{4} = 1$, you must graph both $y = \frac{1}{3}\sqrt{36 - x^2}$ and $y = -\frac{1}{3}\sqrt{36 - x^2}$ on the same set of coordinate axes.

On your graphics calculator or computer, graph the following on the same set of coordinate axes:

$$\frac{x^2}{36} + \frac{y^2}{4} = 1, \qquad \frac{(x-2)^2}{36} + \frac{y^2}{4} = 1, \qquad \frac{(x-2)^2}{36} + \frac{(y+5)^2}{4} = 1$$

What can you conclude about the effects of h and k on the graph of $\frac{(x-h)^2}{a^2} + \frac{(y-k)^2}{b^2} = 1$? If you cannot graph all three ellipses on one set of axes, compare the first with the second and then the second with the third.

The Ellipse Centered at (h, k)

As with the parabola, Figure 10.32 demonstrates that the ellipse centered at (0, 0) with foci on the x-axis can be shifted h units horizontally and k units vertically, to be centered at (h, k). This ellipse will have foci on the line $y = k$, the line parallel to the x-axis. Its vertices are $(h \pm a, k)$, minor axis endpoints, $(h, k \pm b)$, and foci, $(h \pm c, k)$, where $c^2 = a^2 - b^2$.

Using the distance formula, we can derive the general form for this ellipse.

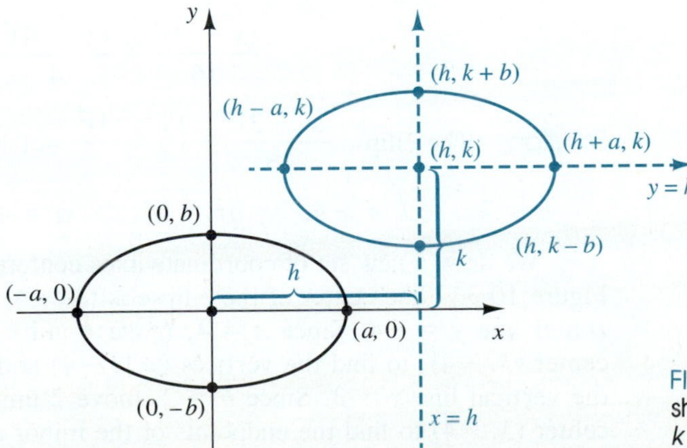

FIGURE 10.32 An ellipse shifted h units horizontally and k units vertically

The ellipse with horizontal major axis, centered at (h, k)

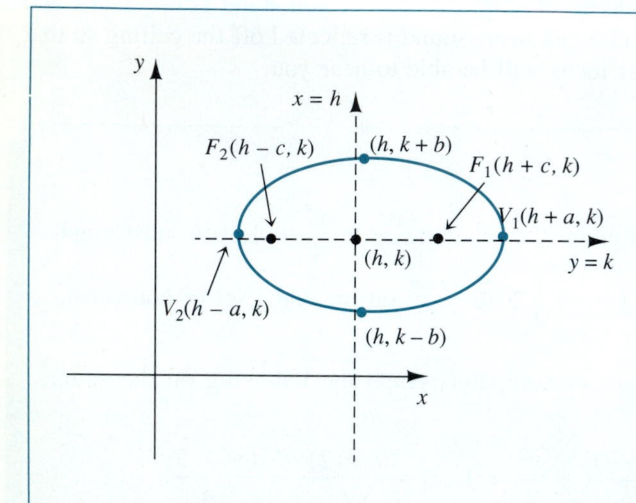

STANDARD FORM OF AN ELLIPSE CENTERED AT (h, k) WITH FOCI AT $(h + c, k)$ AND $(h - c, k)$

The graph of the equation

$$\frac{(x - h)^2}{a^2} + \frac{(y - k)^2}{b^2} = 1$$

for $a > b$ is an ellipse with center (h, k) and vertices $(h \pm a, k)$. The endpoints of the minor axis are $(h, k \pm b)$. The foci are $(h \pm c, k)$, where $c^2 = a^2 - b^2$. The length of the major axis is $2a$; the length of the minor axis is $2b$.

If we draw a new set of horizontal and vertical axes through the center, the new horizontal and vertical axes contain the axes of the ellipse.

Comparing $\frac{x^2}{a^2} + \frac{y^2}{b^2} = 1$, the standard form of the ellipse centered at $(0, 0)$, with $\frac{(x - h)^2}{a^2} + \frac{(y - k)^2}{b^2} = 1$, the standard form of the ellipse centered at (h, k), we see that both ellipses have the same shape, but the center is moved h units horizontally and k units vertically.

Rather than memorizing new vertices, axes, and foci, if we have the ellipse in the form $\frac{(x - h)^2}{a^2} + \frac{(y - k)^2}{b^2} = 1$, we can identify h and k, draw a new set of coordinate axes through (h, k) (parallel to the x- and y-axes), and graph the ellipse on the new set of axes as though it were in the form $\frac{x^2}{a^2} + \frac{y^2}{b^2} = 1$.

EXAMPLE 6 Sketch a graph of the following and identify its axes, center, vertices, and foci:

$$\frac{(x - 3)^2}{16} + \frac{(y + 4)^2}{4} = 1$$

Solution The ellipse $\frac{(x - 3)^2}{16} + \frac{(y + 4)^2}{4} = 1$ is in a standard form, so

$$h = 3, \quad k = -4, \quad a^2 = 16 \Rightarrow a = 4, \quad \text{and} \quad b^2 = 4 \Rightarrow b = 2$$

We draw a new set of coordinate axes centered at (h, k), which is $(3, -4)$. See Figure 10.33. The center of the ellipse is $(3, -4)$. The major axis lies on the horizontal line $y = -4$. Since $a = 4$, move 4 units horizontally left and right of the center $(3, -4)$, to find the vertices $(-1, -4)$ and $(7, -4)$. The minor axis lies on the vertical line $x = 3$. Since $b = 2$, move 2 units vertically above and below the center $(3, -4)$ to find the endpoints of the minor axis: $(3, -2)$ and $(3, -6)$.

Since $c^2 = a^2 - b^2 = 16 - 4 = 12 \Rightarrow c = \sqrt{12} = 2\sqrt{3}$, we can find the foci by moving $2\sqrt{3} \approx 3.46$ units horizontally left and right from the center $(3, -4)$ to get $F_2(3 - 2\sqrt{3}, -4)$ and $F_1(3 + 2\sqrt{3}, -4)$. The graph appears below.

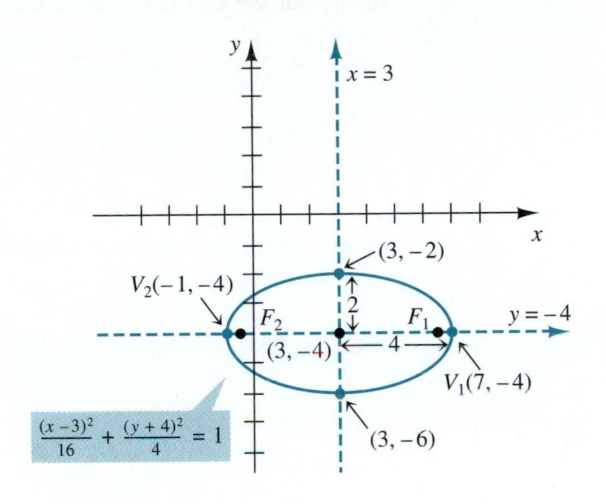

FIGURE 10.33

Similarly, we can shift an ellipse centered at the origin with foci on the y-axis horizontally and vertically so that it is centered at (h, k): The new foci will lie on the line $x = h$, the line parallel to the y-axis. We can derive the following.

The ellipse with vertical major axis, centered at (h, k)

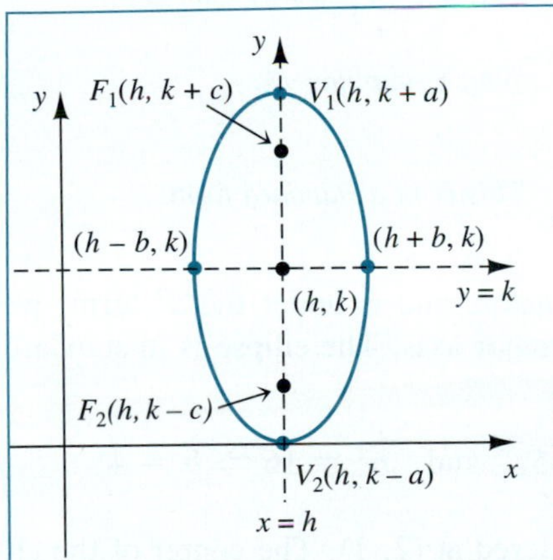

STANDARD FORM OF AN ELLIPSE CENTERED AT (h, k) WITH FOCI AT (h, k + c) AND (h, k − c)

The graph of the equation

$$\frac{(x-h)^2}{b^2} + \frac{(y-k)^2}{a^2} = 1$$

for $a > b$ is an ellipse with center (h, k), and vertices $(h, k \pm a)$. The endpoints of the minor axis are $(h \pm b, k)$. The foci are $(h, k \pm c)$, where $c^2 = a^2 - b^2$. The length of the major axis is $2a$; the length of the minor axis is $2b$.

The ellipse $\frac{(x-h)^2}{b^2} + \frac{(y-k)^2}{a^2} = 1$ has the same shape as the ellipse $\frac{x^2}{b^2} + \frac{y^2}{a^2} = 1$, but its center is shifted h units horizontally and k units vertically.

EXAMPLE 7 Sketch a graph of the ellipse $16x^2 + 25y^2 - 64x - 50y - 311 = 0$ and identify its axes, center, and vertices.

Solution The equation $16x^2 + 25y^2 - 64x - 50y - 311 = 0$ is not in standard form, but we can put it in standard form by completing the square.

$$16x^2 + 25y^2 - 64x - 50y - 311 = 0$$

Add 311 to each side, and group x-terms and y-terms together.

$$(16x^2 - 64x \quad) + (25y^2 - 50y \quad) = 311$$

The squared terms do not have coefficients equal to 1, and completing the square requires that the coefficients be 1. We can factor out the coefficient of each squared term, as follows.

$$(16x^2 - 64x \quad) + (25y^2 - 50y \quad) = 311$$

Factor 16 from the x-terms and 25 from the y-terms.

$$16(x^2 - 4x \quad) + 25(y^2 - 2y \quad) = 311$$

Complete the square for each expression within parentheses; add 4 and 1 within parentheses.

$$16(x^2 - 4x + 4) + 25(y^2 - 2y + 1) = 311 + 16 \cdot 4 + 25 \cdot 1$$

Note that we actually added $16 \cdot 4$ and $25 \cdot 1$ to both sides of the equation.

Now we write the left-hand expressions in factored form and simplify the numerical expression on the right-hand side:

$$16(x - 2)^2 + 25(y - 1)^2 = 400$$

Divide both sides by 400.

$$\frac{16(x - 2)^2}{400} + \frac{25(y - 1)^2}{400} = \frac{400}{400}$$

Which simplifies to

$$\frac{(x - 2)^2}{25} + \frac{(y - 1)^2}{16} = 1$$

This is in a standard form.

Since 25 is the larger of the two denominators and is under the x^2-term, we have $a^2 = 25$, and the ellipse has a horizontal major axis. The ellipse is in standard form, so

$$h = 2, \qquad k = 1, \qquad a^2 = 25 \Rightarrow a = 5, \quad \text{and} \quad b^2 = 16 \Rightarrow b = 4$$

Note that the major axis for this form is parallel to the x-axis.

We draw a new set of coordinate axes centered at $(2, 1)$. The center of the ellipse is $(2, 1)$. The major axis lies on the horizontal line $y = 1$. Since $a = 5$, move 5 units horizontally right and left of $(2, 1)$ to find the vertices $(7, 1)$, and $(-3, 1)$.

The minor axis lies on the vertical line $x = 2$. Since $b = 4$, move 4 units vertically above and below $(2, 1)$, to find the endpoints of the minor axis: $(2, 5)$ and $(2, -3)$. The graph is sketched in Figure 10.34. ■

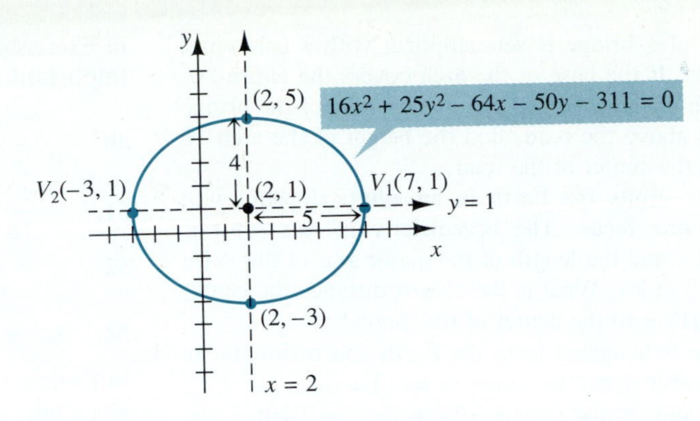

FIGURE 10.34

Graph the figure $\frac{x^2}{4} + \frac{y^2}{9} = 0$. Is it what you expected?

As with the circle and the parabola, we can manipulate the general equation $Ax^2 + Cy^2 + Dx + Ey + F = 0$ in such a way as to put it into one of the standard conic forms, depending upon the coefficients of the squared terms. If the coefficients of the squared terms are equal ($A = C$), we expect the graph to be a circle; if there is only one squared term (either $A = 0$ or $C = 0$ but not both), we expect a parabola. We can draw a similar conclusion for the ellipse: We expect an ellipse if $A \neq C$ and A and C have the same sign in the general equation

$$Ax^2 + Cy^2 + Dx + Ey + F = 0$$

We return to this idea in Section 10.5.

EXERCISES 10.3

In Exercises 1–10, identify the vertices and the foci of the ellipse.

1. $\frac{x^2}{16} + \frac{y^2}{9} = 1$
2. $\frac{x^2}{9} + \frac{y^2}{16} = 1$
3. $\frac{x^2}{12} + \frac{y^2}{18} = 1$
4. $\frac{x^2}{20} + \frac{y^2}{2} = 1$
5. $\frac{x^2}{24} + \frac{y^2}{9} = 1$
6. $\frac{y^2}{30} + \frac{x^2}{25} = 1$
7. $25x^2 + 9y^2 = 225$
8. $6x^2 + y^2 = 18$
9. $y^2 + 30x^2 = 30$
10. $x^2 + 2y^2 = 16$

In Exercises 11–16, graph the ellipse and identify the vertices and the foci.

11. $\frac{x^2}{49} + \frac{y^2}{9} = 1$
12. $\frac{y^2}{81} + \frac{x^2}{4} = 1$
13. $12x^2 + y^2 = 24$
14. $18y^2 + 6x^2 = 36$
15. $3x^2 + 8y^2 = 12$
16. $5x^2 + 10y^2 = 5$

In Exercises 17–20, identify the eccentricity of the ellipse.

17. $\frac{x^2}{25} + \frac{y^2}{36} = 1$
18. $\frac{y^2}{16} + \frac{x^2}{4} = 1$
19. $8x^2 + 3y^2 = 12$
20. $15x^2 + 10y^2 = 5$

In Exercises 21–24 use the test point method discussed in Example 2 on page 639 to sketch the solution set to the inequality.

21. $\frac{x^2}{36} + \frac{y^2}{24} > 1$
22. $\frac{y^2}{9} + \frac{x^2}{4} \leq 1$
23. $3x^2 + 4y^2 \leq 12$
24. $10x^2 + 15y^2 > 30$

In Exercises 25–28 write the equation of the ellipse using the given information.

25. The ellipse has foci (2, 0) and (−2, 0) and vertices (4, 0) and (−4, 0).
26. The ellipse has foci (0, 3) and (0, −3) and vertices (0, 5) and (0, −5).
27. The ellipse is centered at the origin; its major axis is horizontal, with length 8; the length of its minor axis is 4.
28. The ellipse is centered at the origin. The axes of the ellipse have lengths 5 and 9; the major axis is vertical.
29. The arch of a bridge is semielliptical with a horizontal major axis. If the base of the arch covers the entire 80-foot width of the road and the highest part of the bridge is 20 feet above the road, find the height of the arch 10 feet from the side of the road.

30. The arch of a bridge is semielliptical with a horizontal major axis. If the base of the arch covers the entire 80-foot width of the road and the highest part of the bridge is 20 feet above the road, find the height of the arch 10 feet from the center of the road.

31. The moon orbits the Earth in an elliptical path, with Earth at one focus. The eccentricity of this orbit is $e \approx 0.055$, and the length of the major axis of this orbit is 468,972 miles. What is the closest distance the center of the Earth is to the center of the moon?

32. A satellite is launched from the Earth and maintains an elliptical orbit with (the center of the) Earth as one of its foci. The longest and shortest distances from Earth's surface are 800 miles and 300 miles respectively. If the Earth's radius is 4000 miles, find the equation of the orbit.

In Exercises 33–40, identify the center, the vertices, and the foci of the ellipse.

33. $\dfrac{(x-2)^2}{16} + \dfrac{(y-3)^2}{4} = 1$

34. $\dfrac{(x+3)^2}{9} + \dfrac{(y-1)^2}{49} = 1$

35. $\dfrac{(y-3)^2}{18} + \dfrac{(x+1)^2}{12} = 1$

36. $\dfrac{(x+5)^2}{24} + \dfrac{y^2}{2} = 1$

37. $4x^2 + y^2 - 16x - 6y + 21 = 0$

38. $x^2 + 2y^2 + 4x + 12y + 6 = 0$

39. $6x^2 + 5y^2 - 60x - 10y + 65 = 0$

40. $x^2 + 2y^2 + 8x - 4y - 6 = 0$

In Exercises 41–48 graph the ellipse; identify the vertices and the foci.

41. $\dfrac{(x-2)^2}{9} + \dfrac{(y+3)^2}{8} = 1$

42. $\dfrac{(x+3)^2}{8} + \dfrac{y^2}{12} = 1$

43. $\dfrac{(x+2)^2}{16} + \dfrac{(y-1)^2}{24} = 1$

44. $\dfrac{x^2}{6} + \dfrac{(y-5)^2}{18} = 1$

45. $16x^2 + 9y^2 - 32x + 54y = 47$

46. $x^2 + 4y^2 + 4x + 40y + 100 = 0$

47. $x^2 + 4y^2 - 2x - 24y = -29$

48. $4x^2 + y^2 - 24x + 4y + 28 = 0$

In Exercises 49–54 identify and graph the figure labeling its important aspects.

49. $\dfrac{x^2}{9} + \dfrac{y^2}{4} = 1$

50. $\dfrac{x}{9} + \dfrac{y}{4} = 1$

51. $\dfrac{(x+2)^2}{16} + \dfrac{(y-1)^2}{16} = 1$

52. $y^2 - 6y - 2x + 1 = 0$

53. $x^2 - 9y = 0$

54. $4x^2 + 3y^2 - 8x - 6y - 5 = 0$

In Exercises 55–58, write the equation of the ellipse using the given information.

55. The ellipse has foci (1, 4) and (5, 4) and vertices (0, 4) and (6, 4).

56. The ellipse has foci $(-1, 2)$ and $(-1, 10)$ and vertices $(-1, 0)$ and $(-1, 12)$.

57. The major axis of the ellipse lies on the line $y = -5$ and has length 4; the minor axis lies on the line $x = 3$ and has length 2.

58. The axes of the ellipse have lengths 1 and 6, the major axis is vertical, and the center of the ellipse is $(-3, 5)$.

59. Use the distance formula to derive the equation for the ellipse centered at the origin with foci $(0, \pm c)$ and vertices $(0, \pm a)$.

60. Use the distance formula to derive the equation for the ellipse centered at (h, k), with foci $(h \pm c, k)$ and vertices $(h \pm a, k)$.

61. We define a line to be tangent to an ellipse at point P if and only if it touches the ellipse only at point P. The equation for the line tangent to the ellipse $\dfrac{x^2}{a^2} + \dfrac{y^2}{b^2} = 1$ at the point (x_0, y_0) is given by

$$\frac{xx_0}{a^2} + \frac{yy_0}{b^2} = 1.$$

(See the accompanying figure.) Use this formula to find the equation of the line tangent to the ellipse $\dfrac{x^2}{16} + \dfrac{y^2}{4} = 1$ at the point $(-2, \sqrt{3})$.

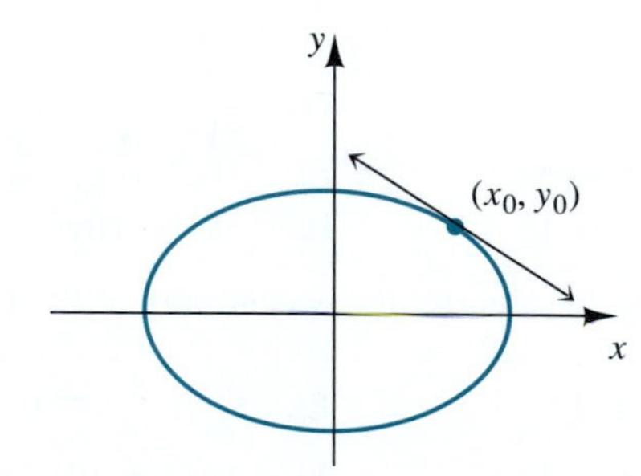

62. Use the formula for the equation of the line tangent to the ellipse in Exercise 61 to find the equation of the line tangent to the ellipse

$$\frac{x^2}{8} + \frac{y^2}{9} = 1$$

at the point (0, 3).

63. The line segment with end points on an ellipse, that passes through a focus and is perpendicular to the major axis is called *a latus rectum* of the ellipse. (See the accompanying figure.) Show that $\frac{2b^2}{a}$ is the length of each latus rectum of the ellipse $\frac{x^2}{a^2} + \frac{y^2}{b^2} = 1$.

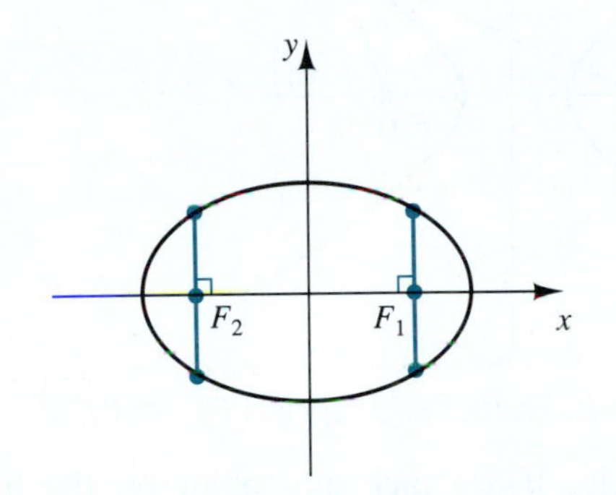

QUESTION FOR THOUGHT

64. The equations $x^2 + 2y^2 + 3 = 0$ and $y^2 + 3x^2 = 0$ are in the general form of a second-degree equation. If you graphed these figures, what type of figures would you expect to have? Graph each figure on a set of coordinate axes. Is it what you expected?

GRAFFIX

In Exercises 65–68, use your graphics calculator or computer to graph the equations.

65. $\frac{x^2}{12} + \frac{y^2}{4} = 1$

66. $8x^2 + 6y^2 = 24$

67. $x^2 + 2y^2 - 2x = 1$

68. $\frac{(x-3)^2}{12} + \frac{(y+2)^2}{4} = 1$

10.4 The Hyperbola

As with the other conic sections, we begin with the geometric definition of the **hyperbola**.

DEFINITION A **hyperbola** is the set of all points in a plane such that the absolute value of the difference of their distances from two fixed points is a positive constant. The fixed points are called the **foci** of the hyperbola. (See Figure 10.35.)

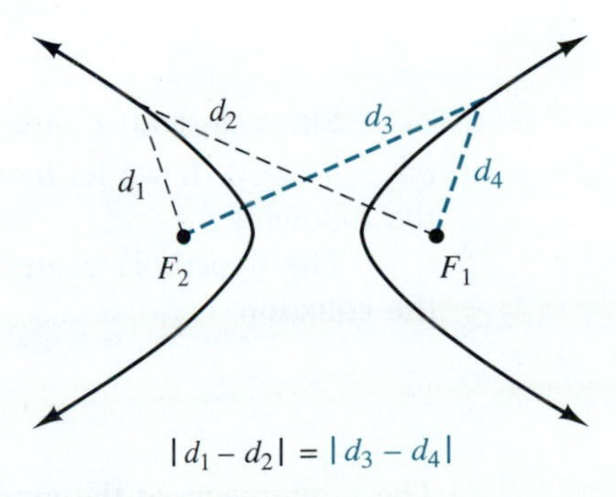

FIGURE 10.35

Whereas an ellipse is determined by the points whose sum of distances to the two foci is a constant, a hyperbola is determined by points whose *difference* of the distances to the two foci is constant.

As with the ellipse, given the definition of the hyperbola, we start by choosing a coordinate system in such a way that the foci lie on the horizontal axis with the origin midway between the two foci. Let $c > 0$ be the distance from the origin to each of the foci: F_1 and F_2. Hence the coordinates of F_1 are $(c, 0)$, and the coordinates of F_2 are $(-c, 0)$. The distance between the two foci, F_1 and F_2, is $2c$. (See Figure 10.36.) We will derive the equation for this hyperbola.

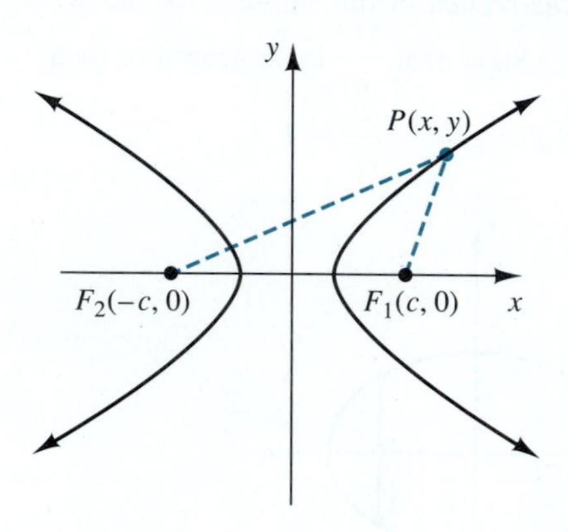

FIGURE 10.36 Hyperbola with foci $F_1(c, 0)$ and $F_2(-c, 0)$

To show $2c > 2a$, consider $\triangle PF_1F_2$ in Figure 10.36. We have $|\overline{PF_1}| + |\overline{F_1F_2}| > |\overline{PF_2}| \Rightarrow$ $|\overline{F_1F_2} > |\overline{PF_2}| - |\overline{PF_1}|$. But $|\overline{F_1F_2}| = 2c$. Hence $2c > 2a$. We can argue similarly for a point P on the left branch of the hyperbola.

By the definition of a hyperbola, if we pick any point on the hyperbola, $P(x, y)$, the absolute value of the difference of the distance from $P(x, y)$ to $F_1(c, 0)$ and the distance from $P(x, y)$ to $F_2(-c, 0)$ is constant. To avoid working with fractions, we call this constant distance $2a$ and note that $2c > 2a$, or $c > a$. Hence we have

$$|\overline{PF_1} - \overline{PF_2}| = 2a$$

Using the distance formula, we get

$$\left|\sqrt{(x-c)^2 + (y-0)^2} - \sqrt{(x+c)^2 + (y-0)^2}\right| = 2a$$

If we manipulate this equation in the same way we did for the ellipse, we get

$$\frac{x^2}{a^2} - \frac{y^2}{c^2 - a^2} = 1.$$

Since a and c are positive (they are distances), and $c > a$, we have $c^2 - a^2 > 0$. If we let $b = \sqrt{c^2 - a^2}$, then $b^2 = c^2 - a^2$; hence we have derived the following.

The hyperbola centered at the origin and with focal points on the x-axis has the equation

$$\frac{x^2}{a^2} - \frac{y^2}{b^2} = 1 \qquad (c > a).$$

The x-intercepts of the graph of this equation are found by setting $y = 0$:

$$\frac{x^2}{a^2} - \frac{y^2}{b^2} = 1$$ *Let $y = 0$; solve for x.*

$$\frac{x^2}{a^2} - \frac{0^2}{b^2} = 1$$

$$\frac{x^2}{a^2} = 1 \Rightarrow x^2 = a^2 \Rightarrow x = \pm a.$$ *The x-intercepts are $\pm a$.*

We call the points $V_1(a, 0)$, and $V_2(-a, 0)$ the **vertices** of this hyperbola, and we call the line segment $\overline{V_1V_2}$ the **transverse axis**.

Note that if we perform the same analysis to find the y-intercepts (let $x = 0$ and solve for y), we end up with $-\frac{y^2}{b^2} = 1 \Rightarrow y^2 = -b^2$, which is impossible (since $-b^2$ must be negative and y^2 must be nonnegative). Therefore, there are no y-intercepts. The line segment joining the points $P_1(0, b)$ and $P_2(0, -b)$ is called the **conjugate axis** of the hyperbola.

For example, Figure 10.37 illustrates the graph of the hyperbola $\frac{x^2}{16} - \frac{y^2}{9} = 1$. Since $a^2 = 16 \Rightarrow a = 4$, the vertices are $(\pm 4, 0)$. Also, $b^2 = 9 \Rightarrow b = 3$, and since $c^2 = a^2 + b^2$, we have $c^2 = 16 + 9 = 25 \Rightarrow c = 5$. Hence the foci are $(\pm 5, 0)$.

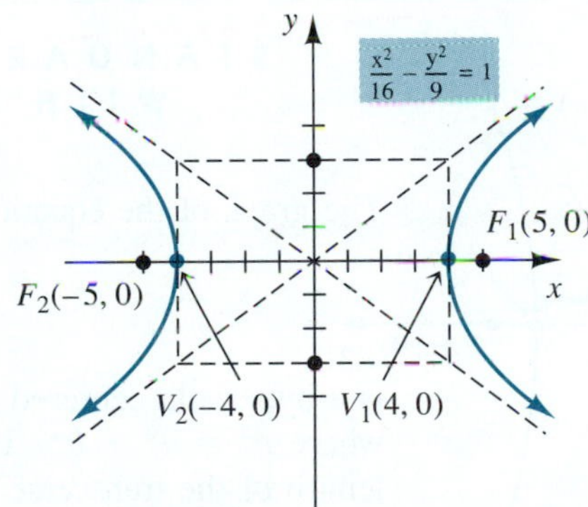

FIGURE 10.37

If we solve the equation $\frac{x^2}{16} - \frac{y^2}{9} = 1$ for y, we get

$$y = \pm\sqrt{\frac{9x^2 - 144}{16}}$$ *which can be rewritten as*

$$y = \pm\sqrt{\frac{9}{16}(x^2 - 16)} \quad \text{or} \quad y = \pm\frac{3}{4}\sqrt{x^2 - 16}$$

Why is $|x| \geq 4$?

Looking carefully at this form of the standard equation, we see that we must have $|x| \geq 4$. Notice that this is consistent with Figure 10.37; x never takes on values strictly between -4 and 4, and therefore there are no y-intercepts.

In our previous experience with graphs, we have worked with horizontal and vertical asymptotes. A distinguishing feature of the hyperbola is that it has asymptotes, although they are neither horizontal nor vertical. These asymptotes are shown in Figure 10.37. They determine the shape of the hyperbola and are helpful guides in sketching its graph. For the graph of the hyperbola $\frac{x^2}{16} - \frac{y^2}{9} = 1$ in Figure 10.37, the equations of the asymptotes are $y = \pm\frac{3}{4}x$.

In Exercise 51 we manipulate the standard form of the hyperbola to determine the asymptotes of the general hyperbola. We state this result here without proof.

The asymptotes of the hyperbola $\frac{x^2}{a^2} - \frac{y^2}{b^2} = 1$ are

$$y = \pm\frac{b}{a}x$$

We summarize this discussion as follows.

The "horizontal" hyperbola centered at (0, 0).

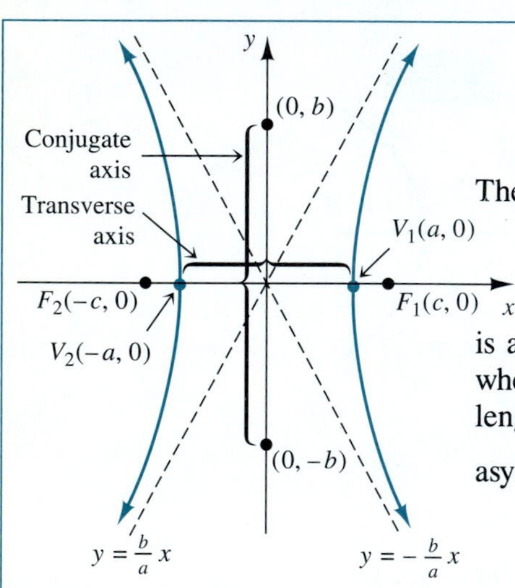

STANDARD FORM OF A HYPERBOLA WITH FOCI AT (c, 0) AND (−c, 0)

The graph of the equation

$$\frac{x^2}{a^2} - \frac{y^2}{b^2} = 1$$

is a hyperbola centered at (0, 0) with vertices $(\pm a, 0)$. The foci are $(\pm c, 0)$, where $c^2 = a^2 + b^2$. The endpoints of the conjugate axis are $(0, \pm b)$. The length of the transverse axis is $2a$; the length of the conjugate axis is $2b$. The asymptotes of the hyperbola are $y = \pm\frac{b}{a}x$.

How are the endpoints of the *transverse* axis related to the vertices?

We refer to the two "arcs" of the hyperbola as the *branches of the hyperbola*.

An efficient way to draw the hyperbola is to first plot the vertices, $(\pm a, 0)$, and the endpoints of the conjugate axis, $(0, \pm b)$. See Figure 10.38 on page 625. Draw a rectangle such that these points are the midpoints of the sides of the rectangle. Then draw the diagonals of the rectangle. *The lines containing the diagonals of the rectangle are the asymptotes of the hyperbola.* The vertices and the asymptotes are the only guides we need to draw the hyperbola.

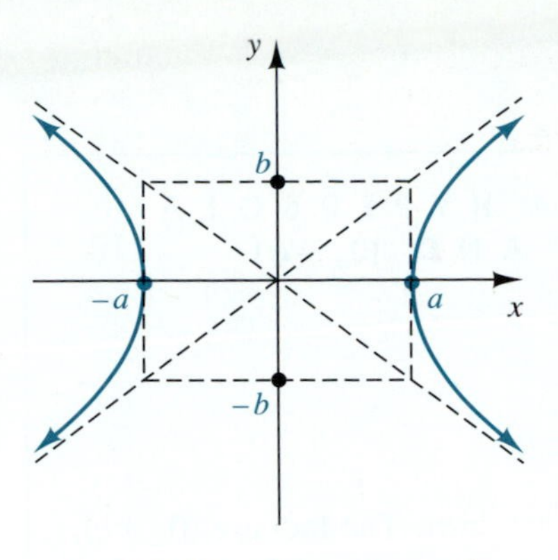

FIGURE 10.38 Draw the rectangle and the diagonals. The diagonals of the rectangle are contained in the asymptotes of the hyperbola.

Note that one diagonal of the rectangle passes through the origin and the point (a, b). Using the slope intercept form, you can verify that the equation of this diagonal is the asymptote $y = \frac{b}{a}x$.

EXAMPLE 1 Sketch the graph of the hyperbola $9x^2 - 25y^2 = 225$. Identify the foci and the asymptotes.

Solution

First we put the equation into standard form, which requires that a 1 be on one side of the equation. In order for us to have 1 on the right-hand side, we must divide by 225.

$$9x^2 - 25y^2 = 225 \qquad \textit{Divide both sides by } 225.$$

$$\frac{9x^2}{225} - \frac{25y^2}{225} = \frac{225}{225} \qquad \textit{Simplify.}$$

$$\frac{x^2}{25} - \frac{y^2}{9} = 1 \qquad \textit{This equation is in standard form.}$$

We compare our equation with the standard form of the hyperbola, and we get

$a^2 = 25 \Rightarrow a = 5$ and $b^2 = 9 \Rightarrow b = 3$ (a and b are both positive).

We plot the vertices $(\pm 5, 0)$ and the endpoints of the conjugate axis: $(0, \pm 3)$. We draw a rectangle such that the points we just plotted are the *midpoints of the sides of the rectangle*. The lines containing the diagonal of the rectangle are the asymptotes

$$y = \pm\frac{b}{a}x \Rightarrow y = \pm\frac{3}{5}x$$

Since $c^2 = a^2 + b^2 = 25 + 9 = 34$, $c = \sqrt{34}$. Hence the foci are $(\pm\sqrt{34}, 0)$. (The length of the transverse axis is $2(5) = 10$, and the length of the conjugate axis is $2(3) = 6$.) We sketch the graph as shown in Figure 10.39.

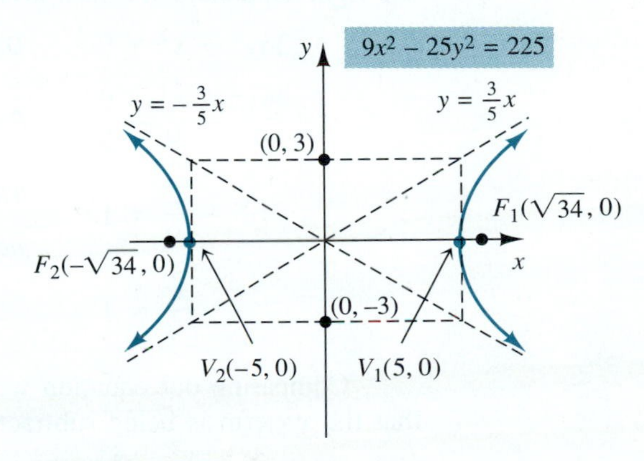

FIGURE 10.39

As with the hyperbola with foci on the x-axis, we can use the same analysis (again with $c > a$) to identify the standard form of the hyperbola with foci on the y-axis, which is summarized next.

The "vertical" hyperbola centered at (0, 0).

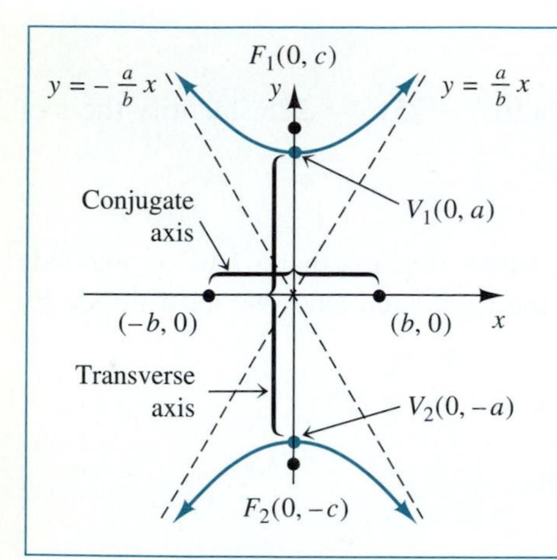

STANDARD FORM OF A HYPERBOLA WITH FOCI AT (0, *c*) AND (0, −*c*)

The graph of the equation

$$\frac{y^2}{a^2} - \frac{x^2}{b^2} = 1$$

is a hyperbola centered at (0, 0), and vertices $(0, \pm a)$. The foci are $(0, \pm c)$, where $c^2 = a^2 + b^2$. The endpoints of the conjugate axis are $(\pm b, 0)$. The length of the transverse axis is $2a$; the length of the conjugate axis is $2b$. The asymptotes of the hyperbola are $y = \pm\frac{a}{b}x$.

For the hyperbola centered at the origin with foci on the y-axis, the vertices $(0, \pm a)$ are the y-intercepts. There are no x-intercepts.

Compare the two forms of the hyperbola, noting the similarities and the differences: The first form has its foci on the x-axis and the second has its foci on the y-axis. Notice that the two forms switch a and b as denominators of x^2 and y^2. The choice of form to use depends upon which squared term is being subtracted: If the y^2-term is being subtracted, a^2 is the denominator of x^2, and the foci lie on the x-axis. On the other hand, if the x^2-term is being subtracted, a^2 is the denominator of y^2, and the foci lie on the y-axis.

EXAMPLE 2 Sketch the graph of the hyperbola $36y^2 - x^2 = 9$. Identify its foci.

Solution First we put the equation into standard form. In order for us to have 1 on the right-hand side, we must divide by 9.

$$36y^2 - x^2 = 9 \qquad \textit{Divide both sides by 9.}$$

$$\frac{36y^2}{9} - \frac{x^2}{9} = \frac{9}{9} \qquad \textit{Simplify.}$$

$$4y^2 - \frac{x^2}{9} = 1 \qquad \textit{The equation is not quite in standard form; however, we can rewrite } 4y^2 \textit{ as } \frac{y^2}{\frac{1}{4}}.$$

$$\frac{y^2}{\frac{1}{4}} - \frac{x^2}{9} = 1$$

Comparing our equation with both standard forms of the hyperbola, we note that the x^2 term is being subtracted; therefore a^2 is the denominator of y^2. We use the second form, $\frac{y^2}{a^2} - \frac{x^2}{b^2} = 1$, a hyperbola with foci on the y-axis. Hence

$$a^2 = \frac{1}{4} \Rightarrow a = \frac{1}{2} \quad \text{and} \quad b^2 = 9 \Rightarrow b = 3$$

We plot the vertices, $\left(0, \pm\frac{1}{2}\right)$, and the endpoints of the conjugate axis, $(\pm 3, 0)$ and draw the rectangle and the diagonals. The sketch of the graph appears in Figure 10.40. The asymptotes are $y = \pm\frac{a}{b}x \Rightarrow y = \pm\frac{1}{6}x$. Since $c^2 = a^2 + b^2 = 9 + \frac{1}{4} = \frac{37}{4} \Rightarrow c = \sqrt{\frac{37}{4}} = \frac{\sqrt{37}}{2}$, the foci are $\left(0, \pm\frac{\sqrt{37}}{2}\right)$.

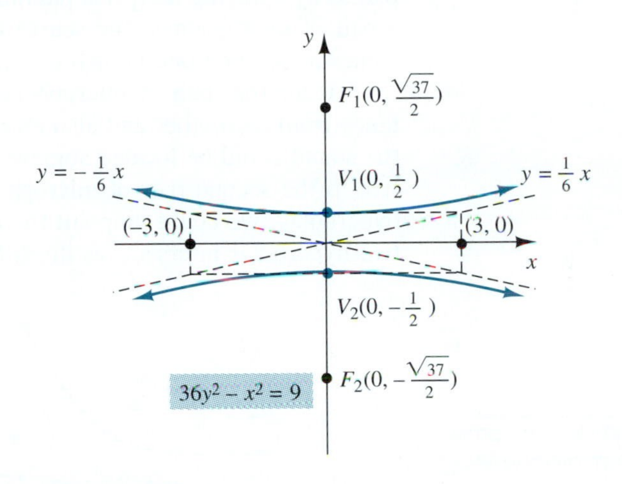

FIGURE 10.40

■

Hyperbolas are important in physics and engineering. The following example illustrates the hyperbola's use in navigation.

EXAMPLE 3 An explosion is recorded by two microphones, M_1 and M_2, which are 2 miles apart. Microphone M_1 receives the sound 4 seconds before microphone M_2. If the speed of sound is 1100 ft/sec, determine the possible locations of the explosion in relation to the location of the microphones.

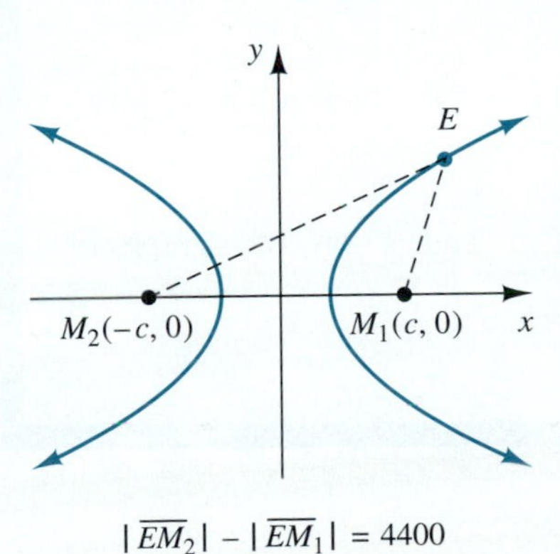

FIGURE 10.41

Solution Let's start by putting the microphones in a coordinate system, where M_1 and M_2 lie on a horizontal axis with the origin midway between. Although we are not sure how far away the explosion was from either microphone, we do know that M_2 received the sound 4 seconds after M_1. Since the speed of sound is 1100 ft/sec, this means that the explosion, E, took place 4400 feet further from M_2 than from M_1, or that *the difference between the distance from M_1 to E and the distance from M_2 to E is 4400 feet*. (See Figure 10.41.)

The possible points or locations of E that satisfy these conditions fit the definition of a hyperbola and therefore trace the path of a hyperbola, with the microphones at the foci.

To get the equation of the hyperbola, we note that the microphones are the foci, located 1 mile from the origin, and so $c = 5280$ feet. (We will express the equation in terms of feet.) The difference between the distances is 4400 feet; recalling the derivation of the hyperbola, this distance is $2a$. Hence $2a = 4400 \Rightarrow a = 2200 \Rightarrow a^2 = 4{,}840{,}000$.

Also,

$$b^2 = c^2 - a^2 = 5280^2 - 2200^2 = 23{,}038{,}400$$

How can we tell the explosion occurred on the branch with microphone M_1?

Hence the equation of the hyperbola (in feet) is $\dfrac{x^2}{4{,}840{,}000} - \dfrac{y^2}{23{,}038{,}400} = 1$. The explosion occurred somewhere on the right branch (the branch closest to M_1) of this hyperbola.

The long-range navigational system known as LORAN locates ships and planes by applying the principle illustrated by Example 3. We have shown that using a pair of microphones, the source of a sound, such as an explosion, can be located somewhere along one branch of a hyperbola.

If another pair of microphones were placed at another location at a fixed distance from each other and also recorded the sound, we would find that the source of the sound could be located somewhere along one branch of another hyperbola (relative to the second pair of microphones). Since the sound source would lie on both hyperbolas, we could pinpoint the exact location of the sound by finding where the two hyperbolas intersect, as illustrated in Figure 10.42.

Actually we can do the same thing with three microphones rather than four. How?

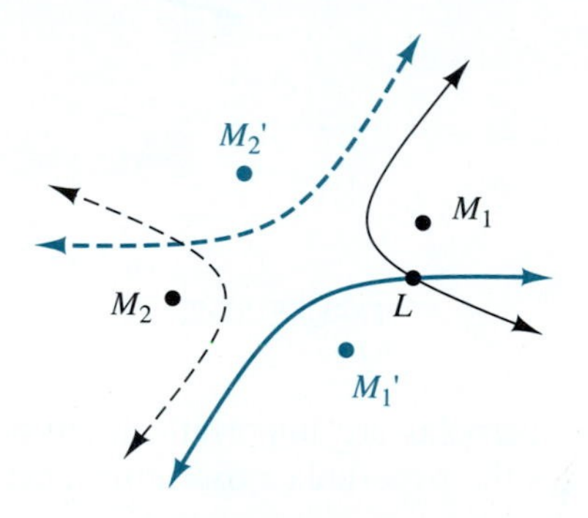

FIGURE 10.42 Pinpointing location L as the intersection of two hyperbolas

GRAFFIX

On your graphics calculator or computer, graph the following on the same set of coordinate axes:

$$\frac{x^2}{9} - \frac{y^2}{4} = 1, \qquad \frac{(x-2)^2}{9} - \frac{y^2}{4} = 1, \qquad \frac{(x-2)^2}{9} - \frac{(y+1)^2}{4} = 1$$

What can you conclude about the effects of h and k on the graph of $\dfrac{(x-h)^2}{a^2} - \dfrac{(y-k)^2}{b^2} = 1$?

The Hyperbola Centered at (h, k)

Figure 10.43 demonstrates that the hyperbola centered at (0, 0) with foci on the x-axis can be shifted h units horizontally and k units vertically to be centered at (h, k).

This hyperbola will have foci on the line $y = k$, the line parallel to the x-axis and shifted k units from the x-axis. Its vertices will be $(h \pm a, k)$, conjugate axis endpoints, $(h, k \pm b)$, and foci, $(h \pm c, k)$, where $c^2 = a^2 + b^2$.

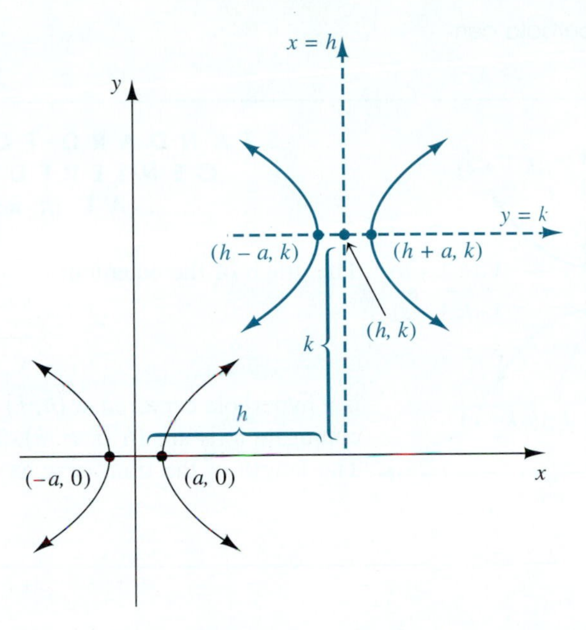

FIGURE 10.43 The hyperbola shifted h units horizontally and k units vertically

As with the other conics, we generalize the hyperbola centered at (h, k).

The "horizontal" hyperbola centered at (h, k)

STANDARD FORM OF A HYPERBOLA CENTERED AT (h, k) WITH FOCI AT $(h + c, k)$ AND $(h - c, k)$

The graph of the equation

$$\frac{(x-h)^2}{a^2} - \frac{(y-k)^2}{b^2} = 1$$

is a hyperbola centered at (h, k) with vertices $(h \pm a, k)$. The endpoints of the conjugate axis are $(h, k \pm b)$. The foci are $(h \pm c, k)$, where $c^2 = a^2 + b^2$. The length of the transverse axis is $2a$; the length of the conjugate axis is $2b$.

The asymptotes of the hyperbola $\dfrac{(x-h)^2}{a^2} - \dfrac{(y-k)^2}{b^2} = 1$ are given by $y - k = \pm \dfrac{b}{a}(x - h)$.

Similarly, we can shift a hyperbola with foci on the y-axis horizontally and vertically to be centered at (h, k), and the new foci will lie on the line $x = h$, the line parallel to the y-axis, h units from the y-axis, with the following result.

The "vertical" hyperbola centered at (h, k).

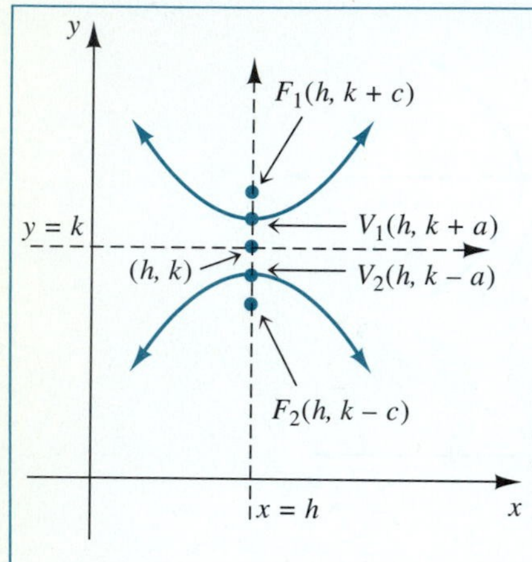

STANDARD FORM OF A HYPERBOLA CENTERED AT (h, k) WITH FOCI AT $(h, k + c)$ AND $(h, k - c)$

The graph of the equation

$$\frac{(y-k)^2}{a^2} - \frac{(x-h)^2}{b^2} = 1$$

is a hyperbola centered at (h, k) with vertices $(h, k \pm a)$. The endpoints of the conjugate axis are $(h \pm b, k)$. The foci are $(h, k \pm c)$, where $c^2 = a^2 + b^2$. The length of the transverse axis is $2a$; the length of the conjugate axis is $2b$.

The asymptotes of the hyperbola $\frac{(y-k)^2}{a^2} - \frac{(x-h)^2}{b^2} = 1$ are given by $y - k = \pm\frac{a}{b}(x - h)$. The hyperbola $\frac{(y-k)^2}{a^2} - \frac{(x-h)^2}{b^2} = 1$ has the same shape as the hyperbola $\frac{y^2}{a^2} - \frac{x^2}{b^2} = 1$, but its center is shifted h units horizontally, and k units vertically.

As with the ellipse, rather than memorizing the new set of vertices, foci, axes, and asymptotes of the hyperbola centered at (h, k), we use the fact that it has the same shape as the hyperbola centered at the origin and graph an equation of the form $\frac{(y-k)^2}{a^2} - \frac{(x-h)^2}{b^2} = 1$ by drawing a new set of coordinate axes through the point (h, k), and graphing the form of $\frac{y^2}{a^2} - \frac{x^2}{b^2} = 1$ on the new set of axes. This is demonstrated next.

EXAMPLE 4 Sketch a graph of the following and identify its center, axes, vertices, and asymptotes.

(a) $\frac{(x+2)^2}{9} - \frac{(y-4)^2}{36} = 1$ **(b)** $25y^2 - 4x^2 - 250y + 24x = -489$

Solution

(a) This equation is already in a standard form. Since the y^2-term is being subtracted from the x^2-term, we use the first standard form (the foci lie on a horizontal line) and the denominator of the x^2 term is a^2. Hence $a^2 = 9$. We can now read

$$h = -2, \qquad k = 4, \qquad a^2 = 9 \Rightarrow a = 3, \qquad b^2 = 36 \Rightarrow b = 6.$$

We draw a new set of coordinate axes centered at (h, k), which is $(-2, 4)$. The hyperbola is centered at $(-2, 4)$. The transverse axis lies on the horizontal line $y = 4$. Since $a = 3$, move 3 units horizontally left and right from its center $(-2, 4)$, to find the vertices $(-5, 4)$, and $(1, 4)$.

The conjugate axis lies on the vertical line $x = -2$. Since $b = 6$, move 6 units vertically up and down from its center $(-2, 4)$ to find the endpoints of the conjugate axis: $(-2, 10)$ and $(-2, -2)$. The rectangle is drawn such that the points plotted are the midpoints of the sides of the rectangle, and the asymptotes are the lines passing through the diagonals of the rectangle. The equations of the asymptotes are

$$y - k = \pm\frac{b}{a}(x - h) \Rightarrow y - 4 = \pm 2(x + 2) \qquad or$$

$$y = 2x + 8 \text{ and } y = -2x$$

The graph appears in Figure 10.44.

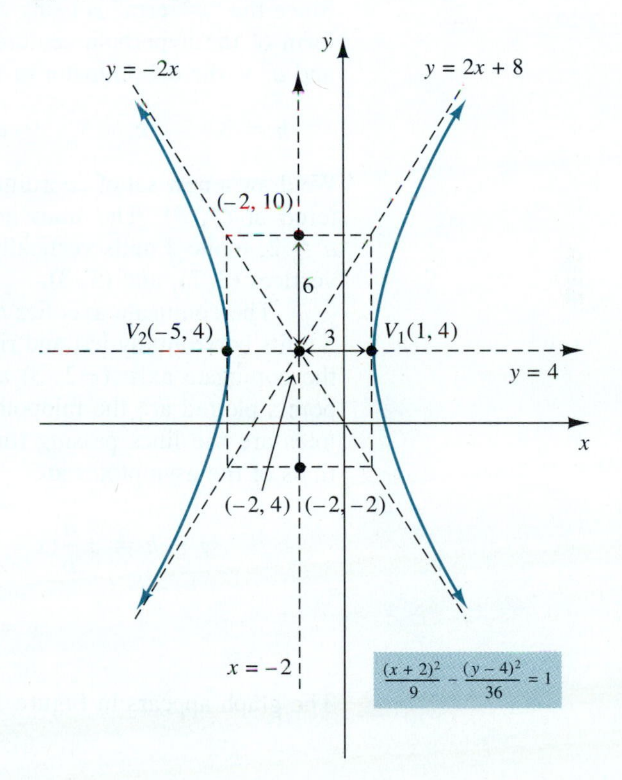

FIGURE 10.44

(b) This equation is not in standard form, but as with ellipses, we can put it in standard form by completing the square.

$$25y^2 - 4x^2 - 250y + 24x = -489$$ *Group terms containing powers of x; group terms containing powers of y.*

$$(25y^2 - 250y \quad) + (-4x^2 + 24x \quad) = -489$$ *Factor 25 from the first group and -4 from the second group.*

$$25(y^2 - 10y \quad) - 4(x^2 - 6x \quad) = -489$$ *We complete the square for each expression within parentheses, to get*

$$25(y^2 - 10y + 25) - 4(x^2 - 6x + 9) = -489 + 25 \cdot 25 - 4 \cdot 9$$ *Note we actually added $25 \cdot 25$ and $-4 \cdot 9$ to both sides of the equation.*

Now we write the expression in factored form and simplify the numerical expressions on the right-hand side to get $25(y - 5)^2 - 4(x - 3)^2 = 100$, which when written in standard form is

$$\frac{(y-5)^2}{4} - \frac{(x-3)^2}{25} = 1.$$

Since the "x^2-term" is being subtracted, we compare this form with the second form of the hyperbola centered at (h, k), where the foci lie on a vertical line and a^2 is the denominator of the y^2-term. We can now read

$$h = 3, \qquad k = 5, \qquad a^2 = 4 \Rightarrow a = 2, \qquad b^2 = 25 \Rightarrow b = 5.$$

We draw a new set of coordinate axes centered at (3, 5). The hyperbola is centered at (3, 5). The transverse axis lies on the vertical line $x = 3$. Since $a = 2$, move 2 units vertically up and down from its center (3, 5), to find the vertices (3, 7), and (3, 3).

The conjugate axis lies on the horizontal line $y = 5$. Since $b = 5$, move 5 units horizontally left and right from its center (3, 5) to find the endpoints of the conjugate axis: (−2, 5) and (8, 5). The rectangle is drawn such that the points plotted are the midpoints of the sides of the rectangle, and the asymptotes are the lines passing through the diagonals of the rectangle. The equations of the asymptotes are

$$y - k = \pm\frac{a}{b}(x - h) \Rightarrow y - 5 = \pm\frac{2}{5}(x - 3) \qquad or$$

$$y = \frac{2}{5}x + \frac{19}{5} \qquad \text{and} \qquad y = -\frac{2}{5}x + \frac{31}{5}.$$

The graph appears in Figure 10.45. ■

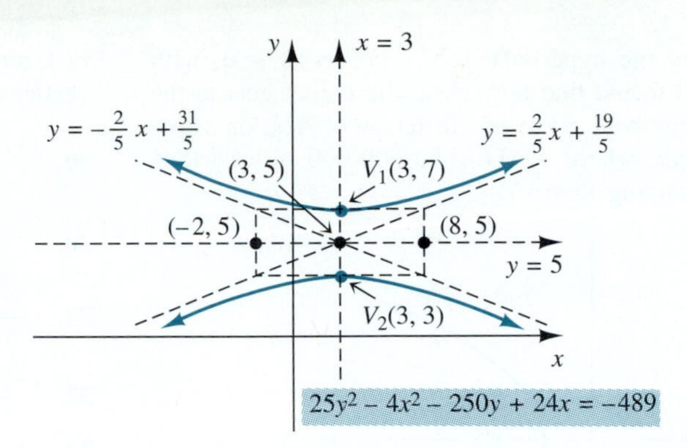

FIGURE 10.45

As with the other conic sections, we can manipulate the general equation $Ax^2 + Cy^2 + Dx + Ey + F = 0$ in such a way as to put it into one of the standard conic forms, depending upon the coefficients of the squared terms. For example, if the coefficients of the squared terms have the same sign (and are not equal to each other or 0), then we would expect the graph to be an ellipse. We can draw a similar conclusion for the hyperbola: If A and C have opposite signs in the equation $Ax^2 + Cy^2 + Dx + Ey + F = 0$, then we would expect the graph of this equation to be a hyperbola. We will discuss these conditions and their exceptions in the next section.

EXERCISES 10.4

In Exercises 1–10, identify the vertices, foci, and equations of the asymptotes of the hyperbola.

1. $\dfrac{x^2}{9} - \dfrac{y^2}{16} = 1$

2. $\dfrac{y^2}{16} - \dfrac{x^2}{16} = 1$

3. $\dfrac{y^2}{12} - \dfrac{x^2}{18} = 1$

4. $\dfrac{x^2}{2} - \dfrac{y^2}{20} = 1$

5. $\dfrac{x^2}{9} - \dfrac{y^2}{18} = 1$

6. $\dfrac{y^2}{36} - \dfrac{x^2}{15} = 1$

7. $25x^2 - 9y^2 = 225$

8. $6y^2 - 3x^2 = 18$

9. $30y^2 - x^2 = 30$

10. $x^2 - 8y^2 = 16$

In Exercises 11–16 graph the hyperbola and identify the vertices, foci, and equations of the asymptotes.

11. $\dfrac{x^2}{9} - \dfrac{y^2}{49} = 1$

12. $\dfrac{y^2}{81} - \dfrac{x^2}{4} = 1$

13. $8x^2 - y^2 = 24$

14. $20x^2 - 5y^2 = 40$

15. $3y^2 - 8x^2 = 24$

16. $3y^2 - 15x^2 = 30$

17. Coast Guard station A is located 100 miles east of Coast Guard station B. Radio signals are sent simultaneously from stations A and B, traveling at a rate of 980 ft/μsec (microsecond). A boat is sailing within range of both signals. If the boat receives the signal from station A 200 μsec after receiving the signal from station B, express the location of the boat as an equation in relation to the two stations. HINT: Create a coordinate system where the two stations are located on the x-axis, each station on opposite sides of and equidistant from the y-axis.

18. Sketch the graph of the equation found in Exercise 17. If this boat is sailing parallel to and 20 miles north of the two Coast Guard stations, use the graph to show its exact location relative to the two Coast Guard stations.

19. Some comets travel in a hyperbolic orbit with the sun as a focus (and we never see them again). It can be shown that the vertex of a branch of a hyperbola is the point on the hyperbola closest to the focus associated with that branch. Given this fact and that the path of the comet is described by the hyperbola $4x^2 - 3y^2 - 12 = 0$, with the sun as a focus, find how close the comet gets to the sun. (The numbers given are in terms of *AU*, or *astronomical units,* where 1 AU ≈ 93,000,000 miles.) (See the accompanying figure.)

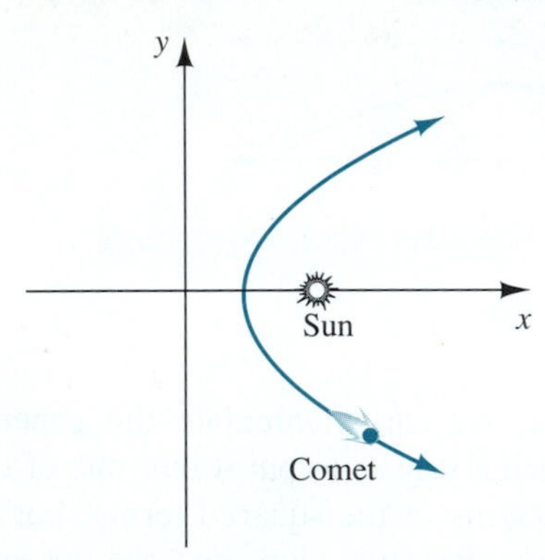

20. The impulse engines break down on the Starship Enterprise again, and the ship is traveling on momentum at a slow, constant rate of speed. Unfortunately, the ship is carrying medical supplies that a planet desperately needs within 24 hours; the shipboard computers calculate that at the current rate of speed, they would not arrive for another 4 days. The captain gets an idea: If they can approach a star (conveniently nearby) and maintain a hyperbolic orbit, the star's gravity will accelerate the ship and send it out at a rate of speed that will get them to the planet on time. Given their speed and the gravitational attraction of the star, the shipboard computer computes the angle of approach and comes up with the equation $5x^2 - 9y^2 - 45 = 0$ as the path they must travel using the star as a focus. Noting the comments in Exercise 19, how close does the ship get to the star?

In Exercises 21–28, identify the center, vertices, foci, and equations of the asymptotes of the hyperbola.

21. $\dfrac{(x-1)^2}{4} - \dfrac{(y-3)^2}{4} = 1$

22. $\dfrac{(y-3)^2}{9} - \dfrac{(x+2)^2}{16} = 1$

23. $\dfrac{(y+3)^2}{12} - \dfrac{(x+1)^2}{18} = 1$

24. $\dfrac{(x+5)^2}{2} - \dfrac{y^2}{24} = 1$

25. $x^2 - 4y^2 - 6x + 8y = 11$

26. $25y^2 - 9x^2 - 100y - 54x - 206 = 0$

27. $2y^2 - 3x^2 + 4y + 6x = 49$

28. $x^2 - 2y^2 - 2x - 12y = 35$

In Exercises 29–36 graph the hyperbola, identify the center, vertices, foci, and equations of the asymptotes.

29. $\dfrac{(x+4)^2}{9} - \dfrac{(y+3)^2}{16} = 1$

30. $\dfrac{(y-1)^2}{4} - \dfrac{x^2}{36} = 1$

31. $\dfrac{x^2}{5} - \dfrac{(y-3)^2}{15} = 1$

32. $\dfrac{(y-3)^2}{20} - \dfrac{(x+1)^2}{18} = 1$

33. $9x^2 - 16y^2 - 36x - 32y = 124$

34. $2y^2 - x^2 + 4y - 2x = 17$

35. $25y^2 - 36x^2 - 150y + 288x - 1251 = 0$

36. $18x^2 - y^2 - 72x + 6y - 45 = 0$

In Exercises 37–38, write the equation of the hyperbola using the given information.

37. The hyperbola has vertices $(2, -1)$ and $(10, -1)$ and foci $(0, -1)$ and $(12, -1)$.

38. The transverse axis of the hyperbola lies on the line $y = -2$ and has length 4; the conjugate axis lies on the line $x = 3$ and has length 6.

In Exercises 39–44, identify and graph the figure, labeling its important aspects.

39. $\dfrac{x^2}{12} + \dfrac{y^2}{4} = 1$

40. $\dfrac{x}{12} - \dfrac{y}{4} = 1$

41. $\dfrac{(x+2)^2}{9} - \dfrac{(y-1)^2}{25} = 1$

42. $3y^2 - 2x^2 - 12 = 0$

43. $x^2 - 2y - 12 = 0$

44. $4x^2 + 4y^2 - 8x - 16y + 4 = 0$

45. Use the distance formula to derive an equation for the hyperbola centered at the origin with foci $(\pm c, 0)$ and vertices $(\pm a, 0)$.

46. Use the distance formula to derive an equation for the hyperbola centered at (h, k) with foci $(h \pm c, k)$ and vertices $(h \pm a, k)$.

47. An equation for the line tangent to the hyperbola $\dfrac{x^2}{a^2} - \dfrac{y^2}{b^2} = 1$ at the point (x_0, y_o) is given by

$$\frac{xx_0}{a^2} - \frac{yy_0}{b^2} = 1$$

Use this formula to find an equation of the line tangent to the hyperbola $\dfrac{x^2}{16} - \dfrac{y^2}{4} = 1$ at the point $(4, 0)$.

48. Use the formula for the equation of the line tangent to the hyperbola in Exercise 47 to find an equation of the line tangent to the hyperbola

$$\frac{x^2}{6} - \frac{y^2}{9} = 1$$

at the point $(6, 3\sqrt{5})$.

49. The line segment AB passing through a focus of a hyperbola and perpendicular to the transverse axis has its endpoints on the hyperbola. See the accompanying figure.

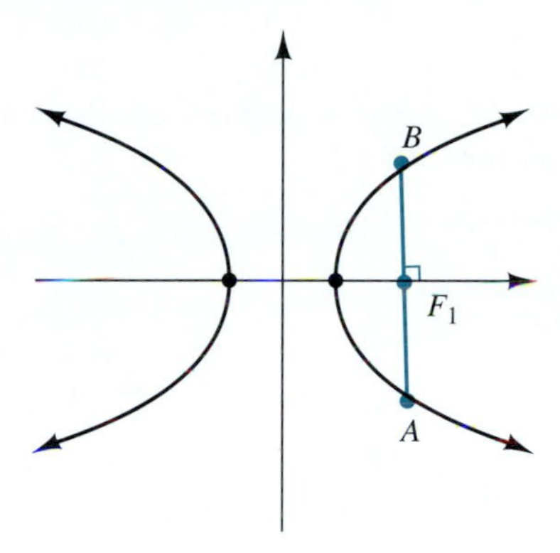

Show that, for the hyperbola $\frac{x^2}{a^2} - \frac{y^2}{b^2} = 1$, the length of the line segment AB is $\frac{2b^2}{a}$.

QUESTIONS FOR THOUGHT

50. **(a)** Show that if we start with the equation $\frac{x^2}{9} - \frac{y^2}{4} = 1$ and solve for y, we obtain

$$y = \pm\frac{2}{3}\sqrt{x^2 - 9}$$

(b) Compute y using the following four equations for these values of x: $x = 4, 10, 20, 100, 200, 1000, 2000$.

$$y = \pm\frac{2}{3}\sqrt{x^2 - 9} \qquad y = \pm\frac{2}{3}x$$

(c) What can you conclude about the relationship between the hyperbola and the lines

$$y = \pm\frac{2}{3}x.$$

51. **(a)** Show that if we start with the equation $\frac{x^2}{a^2} - \frac{y^2}{b^2} = 1$ and solve for y we obtain

$$y = \pm\frac{b}{a}\sqrt{x^2 - a^2}$$

(b) Show that we can factor x^2 in the radicand to get

$$y = \pm\frac{b}{a}\sqrt{x^2\left(1 - \frac{a^2}{x^2}\right)} \qquad \textit{which is}$$

$$y = \pm\frac{b}{a}x\sqrt{1 - \frac{a^2}{x^2}}$$

(c) What happens to the radical term in the last equation above as $x \to \infty$, or $x \to -\infty$?

(d) What can you conclude about y as $x \to \infty$, or $x \to -\infty$?

52. The eccentricity of the hyperbola $\frac{x^2}{a^2} - \frac{y^2}{b^2} = 1$ $(c > a)$ is defined to be

$$e = \frac{c}{a} = \frac{\sqrt{a^2 + b^2}}{a}$$

(See the accompanying figure.)

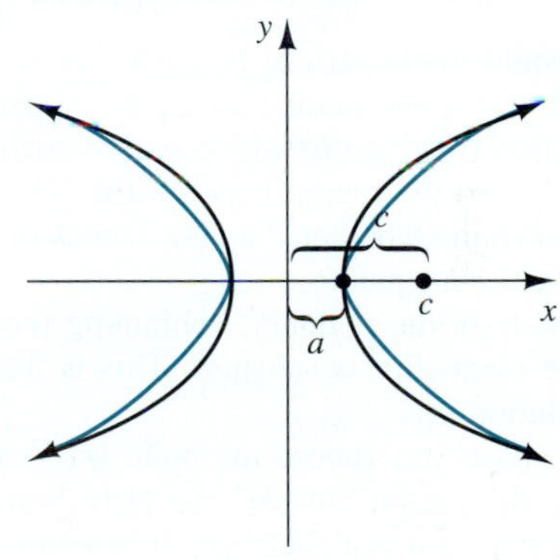

(a) What are the range of possible values for the eccentricity, e, of a hyperbola?

(b) What happens to the shape of the hyperbola as e gets larger or smaller?

53. The equation $x^2 - 4y^2 = 0$ is in the general form of a second-degree equation. If you graphed this figure, what type of figure would you expect to have? Graph the figure on a set of coordinate axes. Is it what you expected?

54. Finding the solutions of an inequality involving a hyperbola is not as obvious as for a circle or ellipse. The definition of the hyperbola is a bit more complicated, and the hyperbola splits the plane into three rather than two regions. The accompanying figure is an example of the graph of the second-degree inequality

$$\frac{(x-1)^2}{49} - \frac{(y-3)^2}{16} \le 1$$

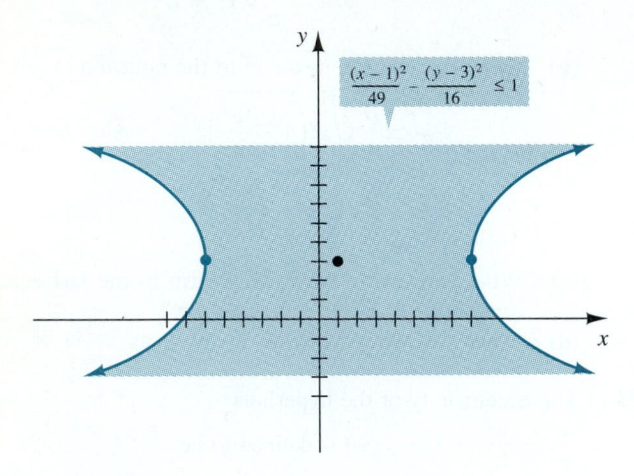

Sketch the graph of the inequality $\frac{x^2}{49} - \frac{y^2}{16} > 1$ using the following steps:

(a) Sketch the graph of the hyperbola $\frac{x^2}{49} - \frac{y^2}{16} = 1$. Should the hyperbola be dashed or solid?

(b) Choose a test point (not on the hyperbola) in each region (the regions within each branch of the hyperbola and the region between the two branches) and determine whether the coordinates of the point satisfy the inequality.

(c) Shade in the region(s) containing the points where the inequality is satisfied. This is the graph of the solution set.

(d) Suppose you choose the point (r, s) as a test point for the region "inside" the right branch of the hyperbola you just sketched. In which region does the point $(-r, s)$ lie?

(e) Looking at the inequality, if the test point (r, s) satisfies (or does not satisfy) the inequality, can you conclude anything about whether $(-r, s)$ satisfies the inequality?

(f) Putting together (d) and (e), if we find that solutions lie in the region inside one branch of the hyperbola, what can we conclude about the solutions with respect to the other branch?

(g) How many test points do we really need? Can we logically deduce the solution region(s) using one test point? Explain your answer.

In Exercises 55–58, use the results of Exercise 54 to graph the solutions of the inequality.

55. $\frac{x^2}{9} - \frac{y^2}{49} < 1$

56. $\frac{y^2}{81} - \frac{x^2}{4} \ge 1$

57. $8x^2 - y^2 \le 24$

58. $\frac{x^2}{5} + \frac{(y+3)^2}{16} > 1$

GRAFFIX

In Exercises 59–62, use your graphics calculator or computer to graph the equations.

59. $\frac{x^2}{8} - \frac{y^2}{4} = 1$

60. $9x^2 - 3y^2 = 24$

61. $x^2 - 3y^2 - 2x = 2$

62. $\frac{(x+3)^2}{8} - \frac{(y-2)^2}{2} = 1$

10.5 Identifying Conic Sections: Degenerate Forms

We summarize the equations of conic sections.

CONIC SECTIONS

1. The circle:

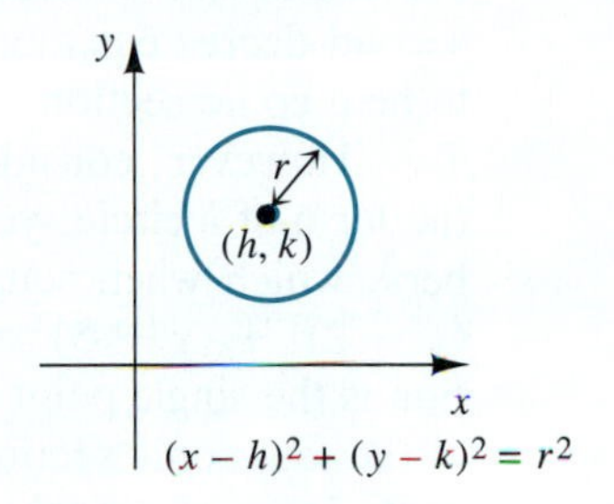

$$(x-h)^2 + (y-k)^2 = r^2$$

2. The parabola:

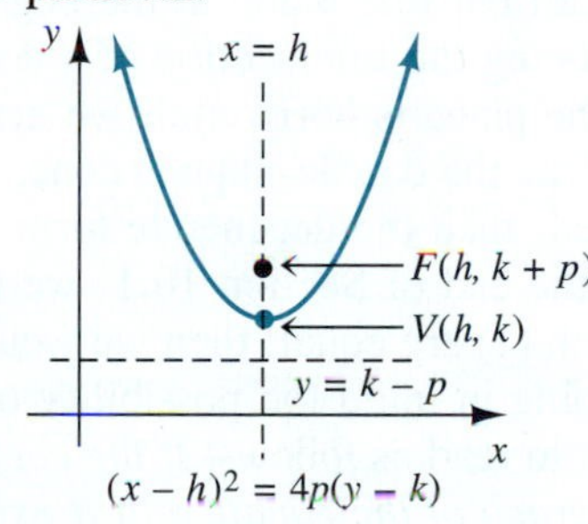

$$(x-h)^2 = 4p(y-k)$$

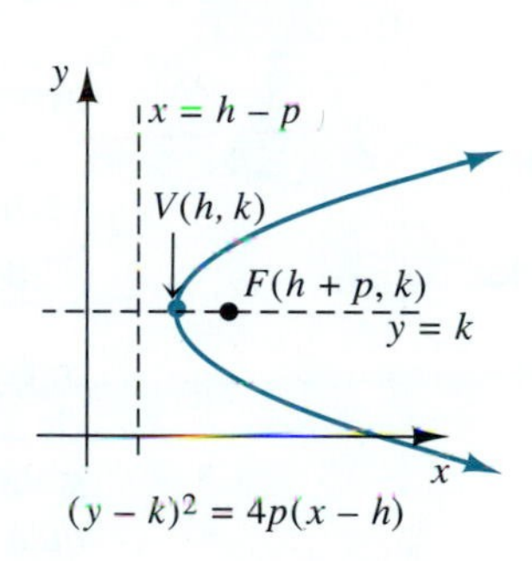

$$(y-k)^2 = 4p(x-h)$$

3. The ellipse:

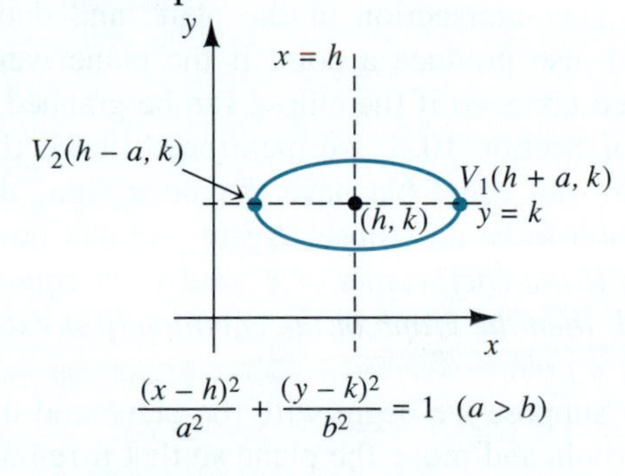

$$\frac{(x-h)^2}{a^2} + \frac{(y-k)^2}{b^2} = 1 \quad (a > b)$$

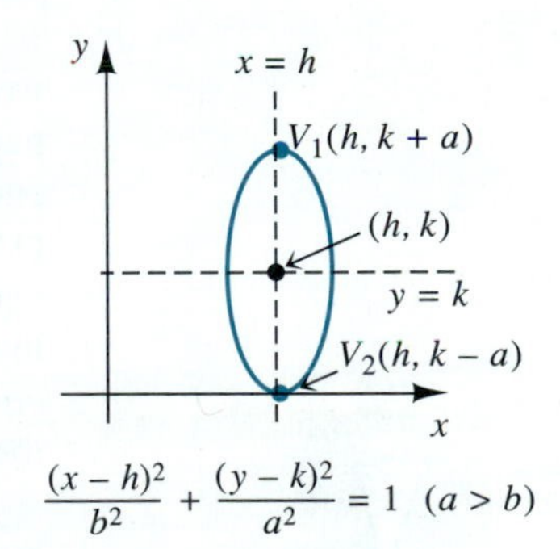

$$\frac{(x-h)^2}{b^2} + \frac{(y-k)^2}{a^2} = 1 \quad (a > b)$$

4. The hyperbola:

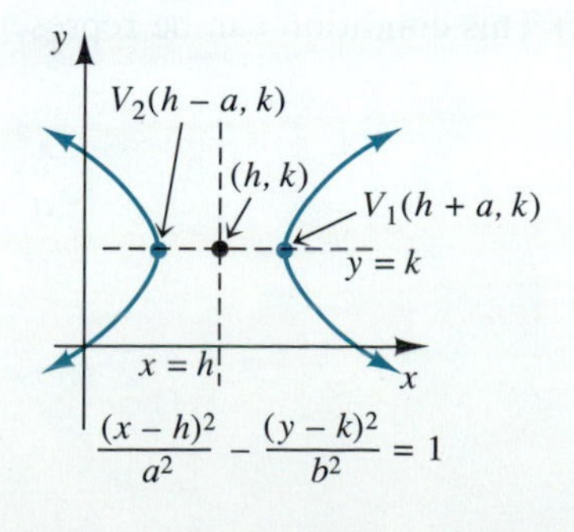

$$\frac{(x-h)^2}{a^2} - \frac{(y-k)^2}{b^2} = 1$$

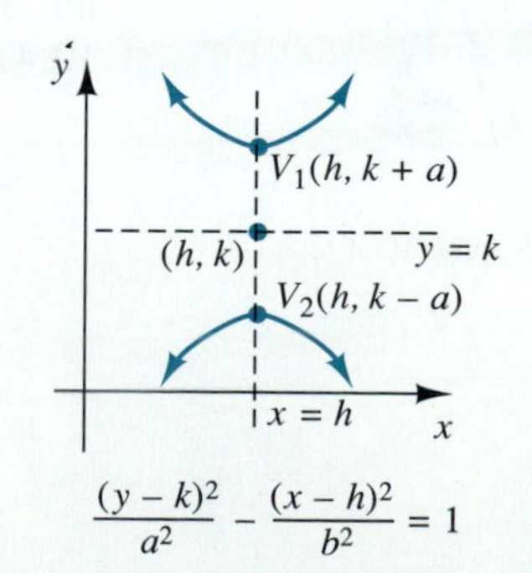

$$\frac{(y-k)^2}{a^2} - \frac{(x-h)^2}{b^2} = 1$$

As we moved through the sections in this chapter, we made mention of the general form

$$Ax^2 + Cy^2 + Dx + Ey + F = 0 \qquad (1)$$

where and A and C are not both zero, and its relationship to each standard form of the conic section being examined. It may have occurred to you that the general second-degree equation (with the help of completing the square) will always turn out to be a conic section.

However, consider the equation $x^2 + y^2 = -1$. This equation seems to be in the form of a circle, yet the equation cannot be graphed since there are no real numbers, which when squared and added, will yield a negative number. The equation $(x - 2)^2 + (y + 5)^2 = 0$ is also in the form of a circle, yet the graph of this equation is the single point $(2, -5)$. If the equation is of form (1), but its graph is not one of the four conic sections discussed in this chapter, we refer to the graph as a **degenerate form of a conic section**.

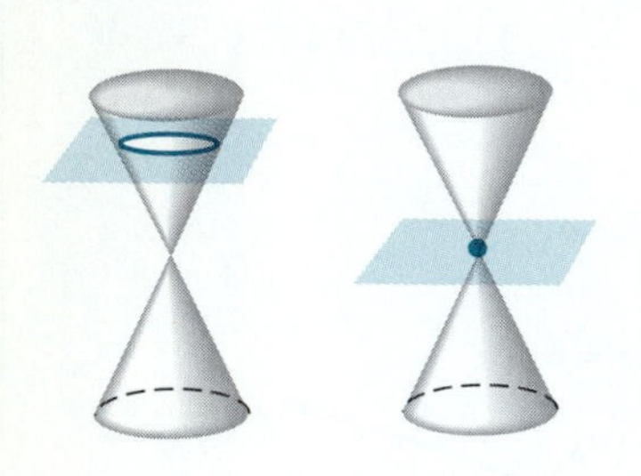

FIGURE 10.46

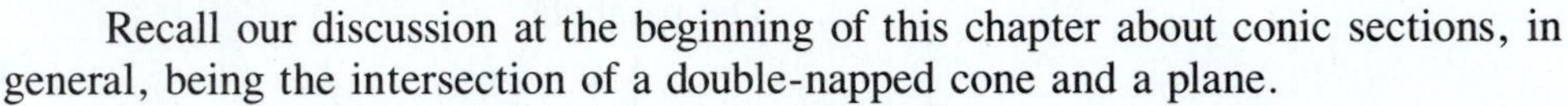

Recall our discussion at the beginning of this chapter about conic sections, in general, being the intersection of a double-napped cone and a plane.

If the plane is horizontal, we arrive at a circle, but if we move it down to the vertex of the the double-napped cone, we end up with a point. So if the equation can be graphed, then the degenerate form of a circle is a point. (See Figure 10.46.)

At the end of Section 10.1, we mentioned that if the coefficients of x^2 and y^2 in equation (1) are equal, then we would *expect* the graph of the equation to be a circle. Keeping in mind the possibility of a degenerate form, we can now modify this statement to read as follows: *If the coefficients of x^2 and y^2 in equation* (1) *are equal, then the graph of the equation (if it exists) is either a circle or its degenerate form, a point*.

The intersection of the plane and double-napped cone producing an ellipse would also produce a point if the plane were moved to the vertex of the double-napped cone, so if the ellipse can be graphed, the degenerate form is a point. At the end of Section 10.3, we mentioned that if the coefficients of x^2 and y^2 in equation (1) are not equal but have the same sign, then we would expect the graph of the equation to be an ellipse. Again, we can now modify this statement to read as follows: *If the coefficients of x^2 and y^2 in equation* (1) *have the same sign but are not equal, then the graph of the equation (if it exists) is either an ellipse, or it degenerate form, a point*.

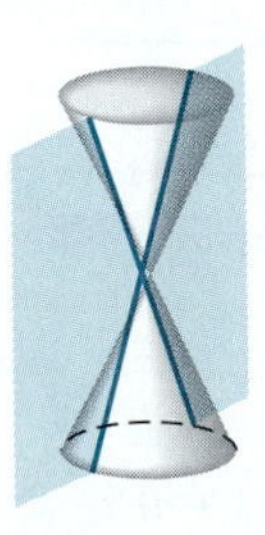

FIGURE 10.47

Suppose we begin with the plane and double-napped cone used to produce the hyperbola and move the plane so that it remains vertical and intersects the vertex of the double-napped cone. The figure it traces is two intersecting lines. (See Figure 10.47.) This condition can be represented by the form

$$\frac{x^2}{a^2} - \frac{y^2}{b^2} = 0$$

If we tried to graph this equation, we would get two intersecting lines. We can see this by solving for y in this equation to get

$$y = \pm\frac{b}{a}x$$

the equations of two intersecting lines. Hence the degenerate form of a hyperbola is two intersecting lines.

As with the circle and the ellipse, we can modify the closing statement made in Section 10.4 as follows: *If the coefficients of x^2 and y^2 in equation* (1) *have different signs, then the graph of the equation (if it exists) is either a hyperbola or its degenerate form, two intersecting lines.*

At the end of Section 10.2, we stated that if one of the coefficients of x^2 or y^2 in equation (1) is 0, then we expect the graph of the equation to be a parabola. The equation $x^2 - 16 = 0$ is a second-degree equation that satisfies this condition, but if we solve for x, we get $x = -4$, $x = 4$, which are the equations of two vertical lines. On the other hand, the equation $x^2 - 6x + 9 = 0$, which is $(x - 3)^2 = 0$, is the equation of the single vertical line $x = 3$. We refer to these forms as degenerate forms of the parabola and modify the statement made at the end of Section 10.2 to read as follows: *If one of the coefficients of x^2 or y^2 in equation* (1) *is* 0, *then the graph of the equation (if it exists) is either a parabola, or one of its degenerate forms, two parallel lines, or a single line.*

Now we can conclude the following.

The graph of the second-degree equation

$$Ax^2 + Cy^2 + Dx + Ey + F = 0$$

where A and C are not both zero, is a conic section or one of its degenerate forms.

We can summarize this discussion as follows.

Suppose the equation $Ax^2 + Cy^2 + Dx + Ey + F = 0$ (where A and C are not both zero) can be graphed:

1. If either $A = 0$, or $C = 0$, then the graph of the equation will be a parabola or one of its degenerate forms, a line or two parallel lines.
2. If the signs of A and C are the same, the graph of the equation will be a circle, if $A = C$; an ellipse, if $A \neq C$; or their degenerate form, a point.
3. If the signs of A and C are different, then the graph of the equation will be a hyperbola or its degenerate form, two intersecting lines.

EXAMPLE 1 Identify and graph each of the following.

(a) $-3x^2 - 3y^2 - 6x + 12y - 15 = 0$

(b) $x^2 - 9y^2 - 4x + 18y - 14 = 0$

(c) $9x^2 - 16y^2 = 0$

Solution

(a) $-3x^2 - 3y^2 - 6x + 12y - 15 = 0$

We look at the equation and note that the coefficients of the squared terms are identical, so we can conclude that if it can be graphed, it will be a circle or a point (its degenerate form). Using the methods discussed earlier (the student should verify this), we put it into standard form to get $(x + 1)^2 + (y - 2)^2 = 0$, which is the point $(-1, 2)$. See Figure 10.48.

(b) $x^2 - 9y^2 - 4x + 18y - 14 = 0$

We look at the equation and note that the coefficients of the squared terms are opposite in sign, so we can conclude that if it can be graphed, it will be a hyperbola or its degenerate form, two interesting lines. Using the methods discussed earlier (the student should verify this), we put it into standard form to get $\frac{(x - 2)^2}{9} - \frac{(y - 1)^2}{1} = 1$, which is the hyperbola graphed in Figure 10.49.

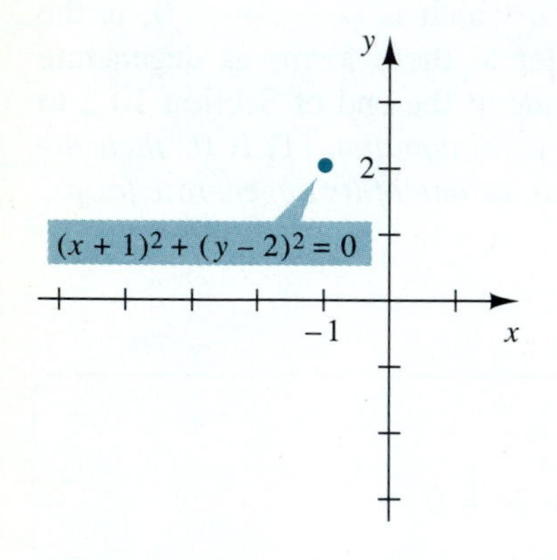

FIGURE 10.48

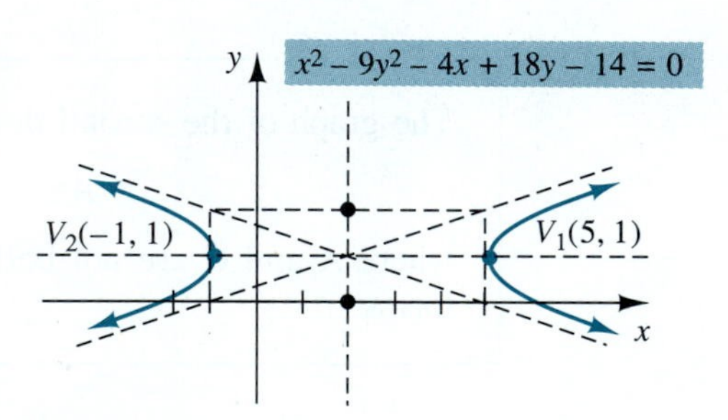

FIGURE 10.49

■

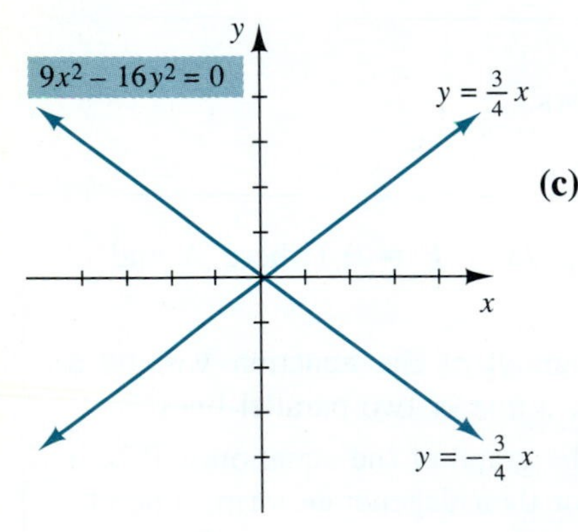

FIGURE 10.50

(c) $9x^2 - 16y^2 = 0$

Since the coefficients of the squared terms are opposite in sign, we conclude that if the equation can be graphed, it would describe a hyperbola or one of its degenerate forms. We cannot put this equation in standard form, but we note that we can solve for y to get two intersecting lines:

$$y = \pm\frac{3}{4}x$$

which is a degenerate form of the hyperbola, graphed in Figure 10.50.

EXERCISES 10.5

In Exercises 1–10 identify the type of conic section if it could be graphed.

1. $3x^2 + 2y^2 - 18x = 0$
2. $4x^2 + 3y - 8x - 37 = 0$
3. $5x^2 - 4y^2 - 10x - 8y + 25 = 0$
4. $3x^2 - 3y^2 - 18y = 0$
5. $6x^2 + 6y^2 - 8x - 2y - 7 = 0$
6. $2y^2 - 5x + 3y = 0$
7. $-x^2 + y^2 - 50x - 6y - 16 = 0$
8. $-2x^2 - 2y^2 - 12x - 6y + 16 = 0$
9. $25x^2 + 8y^2 - 225 = 0$
10. $x^2 + y - 18 = 0$

In Exercises 11–40 identify the type of conic section if it could be graphed, and graph it.

11. $y^2 + 16x = 0$
12. $2x^2 + 3y^2 - 8x - 6y - 37 = 0$
13. $x^2 - 10x + 25 = 0$
14. $9y^2 + 18 = 0$
15. $2x^2 + y^2 - 8x - 2y - 7 = 0$
16. $y^2 - 5x + 4y + 24 = 0$
17. $25x^2 + y^2 - 50x - 6y - 16 = 0$
18. $2x^2 + 3y^2 - 12x - 6y + 24 = 0$
19. $25x^2 + 9y^2 - 225 = 0$
20. $x^2 + y^2 - 18 = 0$
21. $x^2 - y^2 - 18 = 0$
22. $8x^2 + 2y^2 - 24 = 0$
23. $y^2 - 6y - 16 = 0$
24. $-3x^2 - 3y^2 - 30x + 12y - 91 = 0$
25. $-10x^2 + y = 0$
26. $-20x^2 + 9y^2 - 18y - 171 = 0$
27. $x^2 + y^2 + 2x - 6y - 5 = 0$
28. $y^2 - 4y + 8x = 0$
29. $x^2 + y^2 - 4x - 2y + 3 = 0$
30. $6x^2 - 8y^2 + 24 = 0$
31. $9x^2 - y^2 - 18x - 4y - 139 = 0$
32. $30x^2 + y^2 = 0$
33. $x^2 + y^2 + 36 = 0$
34. $3x^2 - 4y^2 - 8y - 52 = 0$
35. $9x^2 + 25y^2 + 18x + 9 = 0$
36. $x^2 + 14x + 49 = 0$
37. $16y^2 - x^2 = 0$
38. $2x^2 + 2y^2 + 12x - 4y + 20 = 0$
39. $x^2 - 2x + 12y - 47 = 0$
40. $9x^2 - 20y^2 - 225 = 0$

QUESTION FOR THOUGHT

41. Consider the general equation

$$Ax^2 + Cy^2 + Dx + Ey + F = 0$$

where A and C are not both 0.

(a) If $AC < 0$, what can you conclude about the graph of the general equation?

(b) If $AC > 0$, what can you conclude about the graph of the general equation?

(c) If $AC = 0$, what can you conclude about the graph of the general equation?

10.6 Translations and Rotations of Coordinate Axes

In previous sections of this chapter we discussed conics centered at the origin and conics centered at (h, k) arrived at by a translation, that is, horizontal and or vertical shifting of the figure. In this section we will discuss translations in a general and more formal manner and then move on to discuss rotations.

Translation of Axes

In Figure 10.51(a) we plot a point $P(h, k)$ on a coordinate axis system. We draw two perpendicular lines that intersect at P and are parallel to the x- and y-axes, and create a new coordinate system with the origin at the point P, which we label O': Figure 10.51(b). We label the new horizontal axis the x'-axis and the new vertical axis the y'-axis. Another way to view this new $x'y'$ coordinate system is that we are shifting the origin h units horizontally and k units vertically.

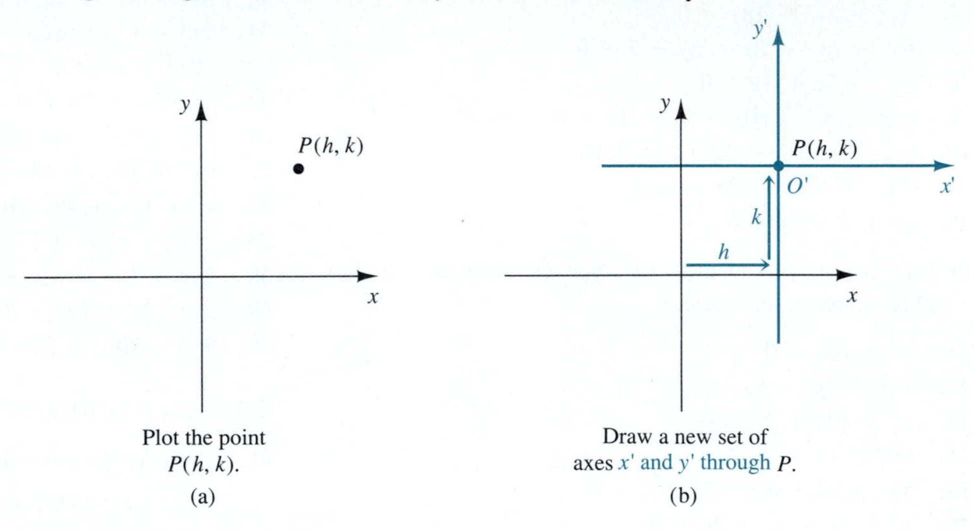

FIGURE 10.51

The new coordinate axes created by horizontal and vertical shifting are called the **translated axes**; the new axes are parallel to the original axes.

We develop a coordinate system in which a point is designated (x', y'), where the x'-coordinate gives us the (perpendicular) distance from the y'-axis and the y'-coordinate gives us the (perpendicular) distance from the x'-axis. Thus the new origin is designated $(0', 0')$. For example, $(3', -4')$ represents a point that is a distance of 3 units right of the y'-axis and 4 units below the x'-axis. (See Figure 10.52.)

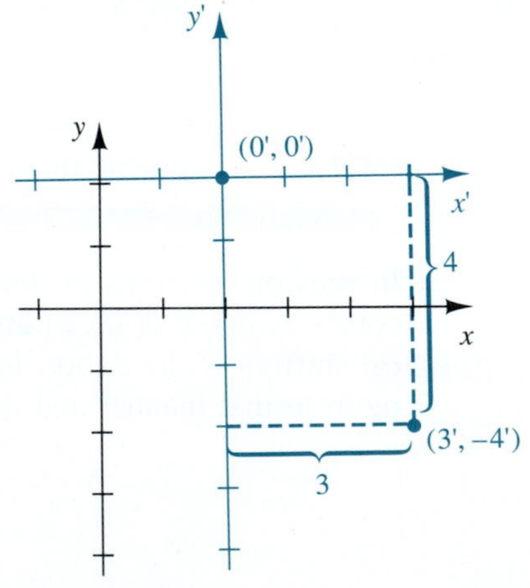

FIGURE 10.52

How is the new coordinate system related to the original coordinate system? For convenience, we refer to the original horizontal and vertical xy-axes as the **standard set of coordinate axes**.

Keeping in mind that the new coordinates are arrived at by horizontal and vertical shifting of the axes and looking at Figure 10.53, we can see how (x, y) and (x', y') are related.

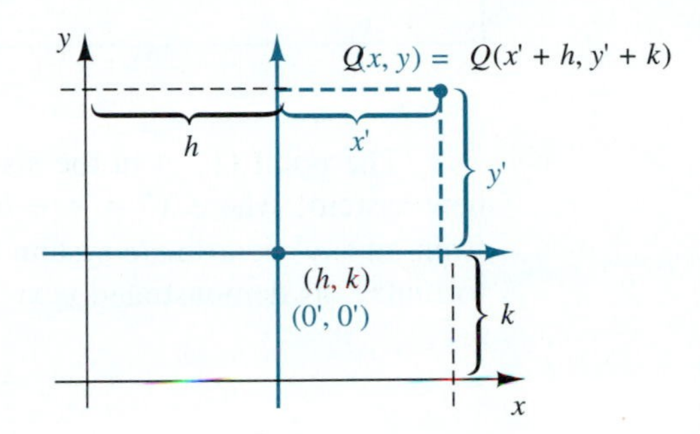

FIGURE 10.53
$Q(x, y)$ and $Q(x' + h, y' + k)$ represent the same point.
Note $x = x' + h$ and $y = y' + k$.

For example, if we shift the axes so that the translated set of coordinates are now centered on (3, 2), then $h = 3$ and $k = 2$. The point (5, 8) on the standard coordinate system has new coordinates found by $x' = 5 - 3 = 2$ and $y' = 8 - 2 = 6$; the new coordinates are $(2', 6')$. (See Figure 10.54.) Another example: The point $(8, -5)$ has new coordinates $x' = 8 - 3 = 5$ and $y' = -5 - 2 = -7$; the new coordinates are $(5', -7')$.

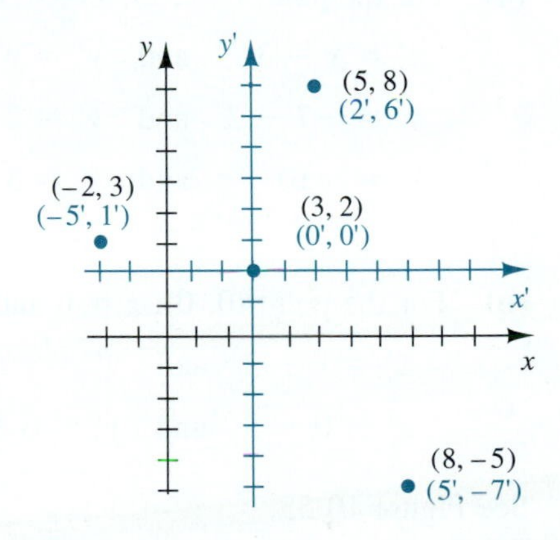

FIGURE 10.54 $x'y'$-axes centered at (3, 2).

By shifting all points h units horizontally, and k units vertically, we arrive at the following.

COORDINATE FORMULAS FOR TRANSLATION OF AXES

If a set of coordinate axes is moved h units horizontally and k units vertically, then the new coordinates (x', y') are related to the old coordinates (x, y) in the following way:

$$x' = x - h \quad \text{and} \quad y' = y - k$$

The point (x, y) in the standard coordinate system is the point (x', y') in the new system, where $x' = x - h$ and $y' = y - k$. We can convert back and forth from an $x'y'$ coordinate system to the standard xy system by using these coordinate formulas, as demonstrated next.

EXAMPLE 1 A set of axes is translated so that the new origin is $(3, -1)$ on the standard set of coordinate axes. Find the coordinates of the following points relative to the translated set of axes. **(a)** $(-7, 2)$ **(b)** $(0, 0)$

Solution Since the new origin of the translated axes is $(3, -1)$, then $h = 3$ and $k = -1$. We substitute these values for h and k in the coordinate formulas for translation of axes $x' = x - h$ and $y' = y - k$ to get

$$x' = x - 3 \quad \text{and} \quad y' = y + 1$$

We use these formulas to find the translated coordinates (x', y').

(a) For the point $(-7, 2)$, $x = -7$ and $y = 2$. The new coordinates are found by

$x' = x - 3$ and $y' = y + 1$ *Substitute $x = -7$ and $y = 2$ to get*

$x' = -7 - 3$ and $y' = 2 + 1$ *Hence*

$x' = -10$ and $y' = 3$ *The new coordinates relative to the $x'y'$-axes are $(-10', 3')$.*

(b) For the point $(0, 0)$, $x = 0$ and $y = 0$. The new coordinates are found by

$x' = x - 3$ and $y' = y + 1$ *Substitute $x = 0$ and $y = 0$ to get*

$x' = 0 - 3$ and $y' = 0 + 1$ *Hence the old origin has new coordinates $(-3', 1')$.*

See Figure 10.55. ■

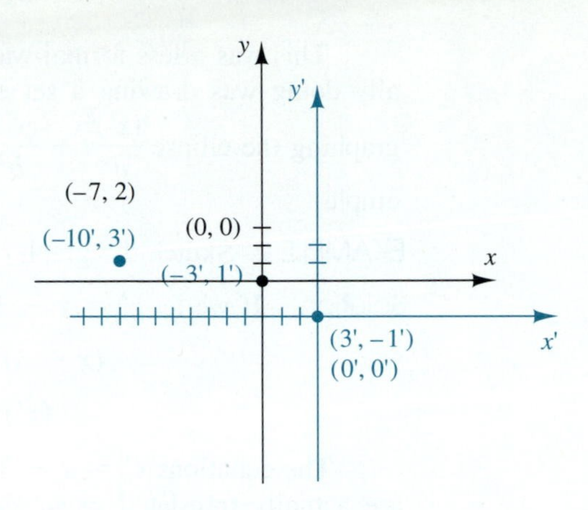

FIGURE 10.55

Since the y-axis has equation $x = 0$, the new y'-axis has equation $x = h$, which is what we would expect of a line parallel to the y-axis and passing through the point (h, k). By the same reasoning, the equation of the x'-axis is $y = k$. This is also consistent with the coordinate formulas. See Figure 10.56.

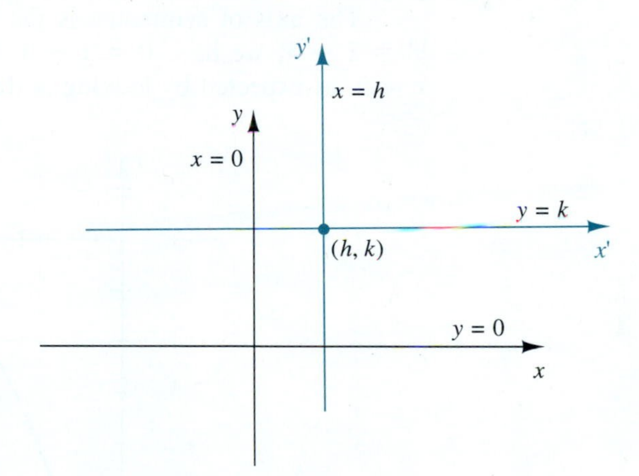

FIGURE 10.56

It is often convenient or necessary to translate a set of axes in order to identify or sketch certain graphs. We have already done this in the previous sections on conics. For example we learned to graph the ellipse $\frac{(x-h)^2}{a^2} + \frac{(y-k)^2}{b^2} = 1$ by *drawing a new set of axes through (h, k), ignoring the old axes, and graphing the form* $\frac{x^2}{a^2} + \frac{y^2}{b^2} = 1$ *on the new axes.*

This was a less formal way to approach translation of axes. What we were really doing was drawing a set of translated axes $x'y'$ through the point (h, k) and graphing the ellipse $\frac{(x')^2}{a^2} + \frac{(y')^2}{b^2} = 1$. We take this formal approach in the next example.

EXAMPLE 2 Sketch the graph of the parabola $(x - 3)^2 = -4(y + 2)$.

Solution If we let $x' = x - 3$ and $y' = y + 2$, then the equation

$$(x - 3)^2 = -4(y + 2) \qquad \textit{becomes}$$
$$(x')^2 = -4y'$$

The equations $x' = x - 3$ and $y' = y + 2$ tell us that the new x'- and y'-axes are actually translated axes, shifted horizontally and vertically to be centered on $(3, -2)$; hence $h = 3$ and $k = -2$.

The form $(x')^2 = -4y'$ tells us that we have a "vertical" parabola with focus on the y'-axis with $(0', 0')$ as its vertex, and since $p = -1$, the focus is $(0', -1')$. We graph it as shown in Figure 10.57. If you use the coordinate formulas $x' = x - 3$ and $y' = y + 2$, you can verify the vertex and focus labeled on the graph in both coordinate systems.

The axis of symmetry is the y'-axis, which has the equation $x' = 0$. Since $x' = x - 3$, we have $0 = x - 3$. Hence the equation of the axis of symmetry is $x = 3$, as expected by looking at the graph.

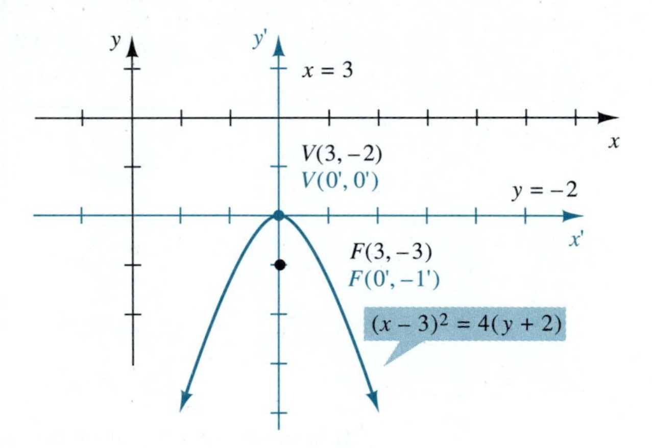

FIGURE 10.57 ■

Rotation of Axes

In Section 10.5 we pointed out that the graph of the general equation

$$Ax^2 + Bxy + Cy^2 + Dx + Ey + F = 0 \qquad \text{where } B = 0$$

will yield a conic section or one of its degenerate forms, where the axes of the conic are parallel to the x- and y-axes.

In this section we discuss the impact of the xy-term in the general form

$$Ax^2 + Bxy + Cy^2 + Dx + Ey + F = 0 \qquad \text{where } B \neq 0$$

We will find that the xy-term will produce a **rotation of the axes**. That is, the axes of the figure will *not* be parallel to the coordinate axes.

Let's begin by plotting any point $P(x, y)$ on the xy-coordinate system and then rotate the axes counterclockwise about the origin through an angle of ϕ (where $0° < \phi < 90°$), so that we have a new set of coordinate axes, which we designate as x' and y'. (See Figure 10.58.)

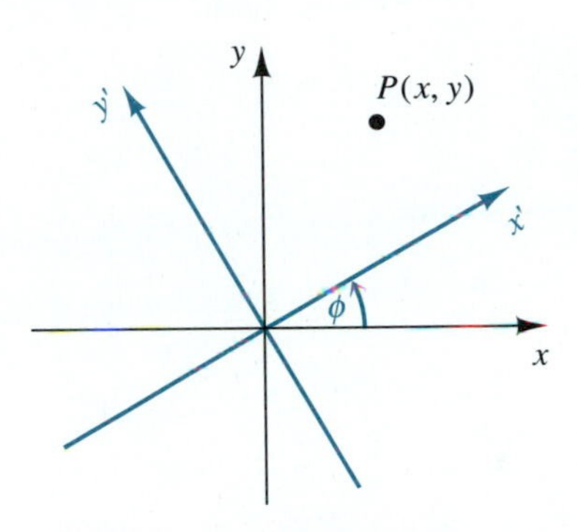

FIGURE 10.58 Rotation of axes ϕ degrees counterclockwise

Let's step back for a moment and recall that point S having coordinates (5, 4) means that the point S is a distance of 5 units from the y-axis and a distance of 4 units from the x-axis. See Figure 10.59(a). Similarly, if the point Q has coordinates $(3', 7')$, this means that the point Q is a distance of 3 units from the y'-axis and a distance of 7 units from the x'-axis. See Figure 10.59(b).

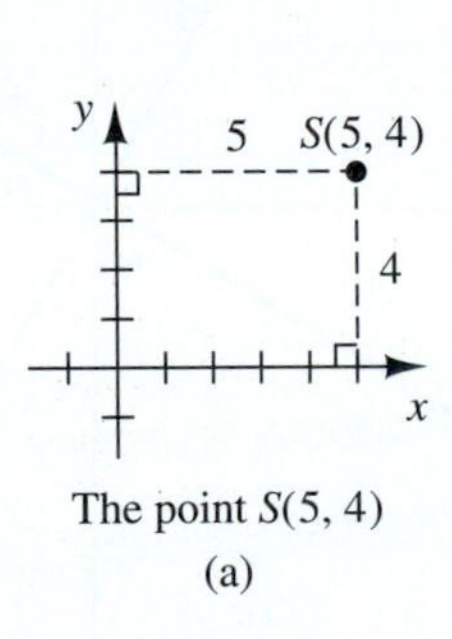

The point $S(5, 4)$

(a)

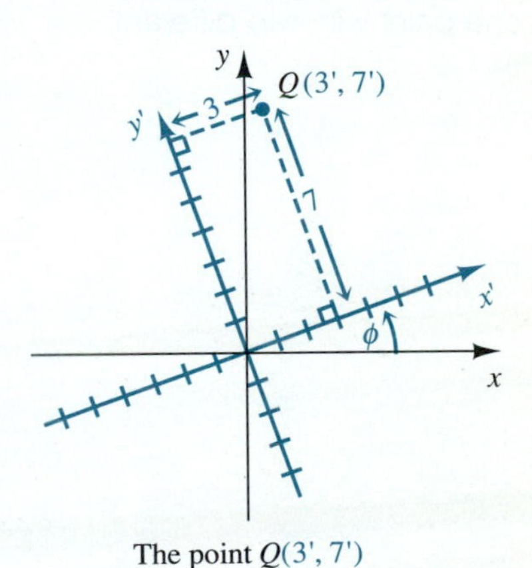

The point $Q(3', 7')$

(b)

FIGURE 10.59

Our goal is to find the new coordinates of P in relation to the new set of axes, $x'y'$. By Figure 10.60 we see that this means we are trying to find x', the distance P is from the y'-axis, and y', the distance P is from the x'-axis.

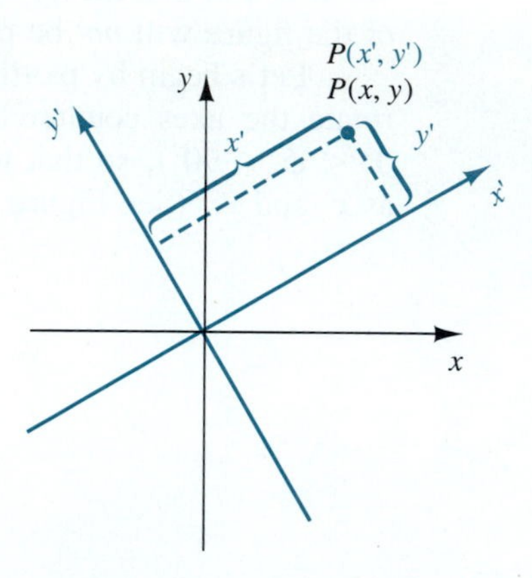

FIGURE 10.60 P has coordinates (x', y') relative to the $x'y'$-axes

Let r be the length of the line segment between $P(x, y)$ and the origin. If we rotate the axes ϕ degrees (ϕ is the angle of rotation) and let α be the angle between the x-axis and OP, we have Figure 10.61.

P is one point with two different labels.

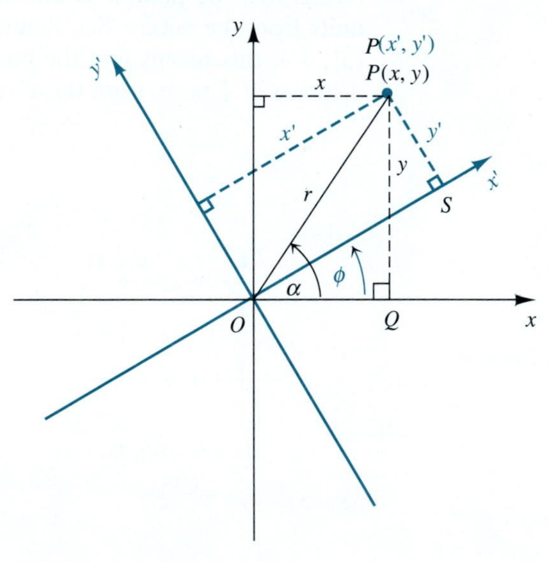

FIGURE 10.61

Using our knowledge of trigonometry, we look at ΔOPQ in Figure 10.61 and arrive at the following relationships for each set of coordinate axes.

(1) $x = r\cos\alpha \qquad y = r\sin\alpha$ *(x, y) are the coordinates of P relative to the xy-axes.*

We take the same approach expressing x' and y' in terms of r, α, and ϕ. We look at ΔOPS, and noting that the angle between the x'-axis and OP is $\alpha - \phi$, we have

(2) $x' = r\cos(\alpha - \phi) \qquad y' = r\sin(\alpha - \phi)$ *(x', y') are also the coordinates of P but relative to the $x'y'$-axes.*

Now we look for a relationship expressing x and y, the standard coordinates of P, in terms of x' and y', the rotated coordinates of P. We apply the formulas for $\sin(A - B)$ and $\cos(A - B)$ in Section 8.1 to the second set of equations and get the coordinates of $P(x', y')$. We begin with x':

$x' = r\cos(\alpha - \phi)$ *Apply the formula for $\cos(\alpha - \phi)$ to get*

$\quad = r\cos\alpha\cos\phi + r\sin\alpha\sin\phi$ *From (1) we substitute $x = r\cos\alpha$ and $y = r\sin\alpha$ to get*

(3) $x' = x\cos\phi + y\sin\phi$

Similarly for y':

$y' = r\sin(\alpha - \phi)$ *Apply the formula for $\sin(\alpha - \phi)$ to get*

$\quad = r\sin\alpha\cos\phi - r\cos\alpha\sin\phi$ *From (1) we substitute $x = r\cos\alpha$ and $y = r\sin\alpha$ to get*

(4) $y' = y\cos\phi - x\sin\phi$

Equations (3) and (4) tell us the new coordinates of the point P relative to the $x'y'$-axes in terms of x and y (the coordinates relative to the old axes) and ϕ, the angle of rotation of the new axes.

We can treat these two equations as a system and solve for x and y; this will give us the old coordinates, x and y, in terms of the new coordinates, x' and y', and the angle of rotation, ϕ. We leave this as an exercise (see Exercise 53). The equations are as follows:

$$x = x'\cos\phi - y'\sin\phi \qquad y = x'\sin\phi + y'\cos\phi$$

THE COORDINATE FORMULAS FOR ROTATION OF AXES

If a set of coordinate axes is rotated ϕ degrees, then the new coordinates (x', y') are related to the old coordinates (x, y) in the following way:

$$\begin{cases} x' = x\cos\phi + y\sin\phi \\ y' = -x\sin\phi + y\cos\phi \end{cases} \quad \text{and} \quad \begin{cases} x = x'\cos\phi - y'\sin\phi \\ y = x'\sin\phi + y'\cos\phi \end{cases}$$

EXAMPLE 3 On the standard set of coordinate axes, the coordinates of a point P are $(-2, 4)$. A new set of coordinate axes $x'y'$ is obtained by rotating the standard axes by 30°. (The x'-axis is rotated 30° counterclockwise from the x-axis.) Find the coordinates of the point P relative to the $x'y'$-axes.

Solution We are given the standard coordinates (x, y) of a point and need to find (x', y') when the angle of rotation is 30°. We use the left-hand formulas in the box.

$$x' = x\cos\phi + y\sin\phi \quad \text{and} \quad y' = -x\sin\phi + y\cos\phi$$

Substitute $x = -2$, $y = 4$, and $\phi = 30°$ to get

$$x' = -2\cos 30° + 4\sin 30° \quad \text{and} \quad y' = -(-2)\sin 30° + 4\cos 30° \qquad \textit{Hence,}$$

$$x' = -2\left(\frac{\sqrt{3}}{2}\right) + 4\left(\frac{1}{2}\right) \quad \text{and} \quad y' = 2\left(\frac{1}{2}\right) + 4\left(\frac{\sqrt{3}}{2}\right) \qquad \textit{or}$$

$$x' = -\sqrt{3} + 2 \quad \text{and} \quad y' = 1 + 2\sqrt{3}$$

Hence $(x', y') = \left((-\sqrt{3} + 2)', (1 + 2\sqrt{3})'\right)$. *These are the rotated coordinates of $P(-2, 4)$.*

See Figure 10.62. ■

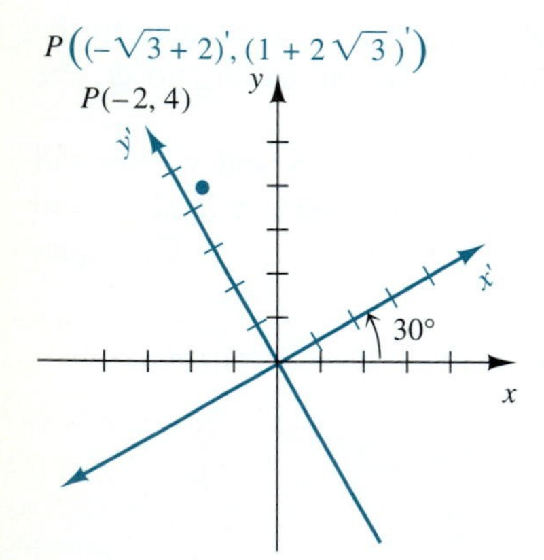

FIGURE 10.62

We examined the rational function $y = \dfrac{1}{x}$ in Section 4.5. In this next example, we will examine this function or its equivalent $xy = 1$ as a second-degree equation.

EXAMPLE 4 The angle of rotation for a set of coordinate axes is 45°. If the equation of a graph is $xy = 1$ on the standard set of coordinate axes, express this equation in terms of the rotated set of coordinates x' and y', and sketch the graph.

Solution Since we want to express the equation $xy = 1$ in terms of x' and y', we use the coordinate formulas for x and y.

$$x = x'\cos\phi - y'\sin\phi \quad \text{and} \quad y = x'\sin\phi + y'\cos\phi$$

Since $\phi = 45°$, we have

$$x = x'\cos 45° - y'\sin 45° \quad \text{and} \quad y = x'\sin 45° + y'\cos 45° \qquad \textit{Which yields}$$

$$x = x'\left(\frac{\sqrt{2}}{2}\right) - y'\left(\frac{\sqrt{2}}{2}\right) \quad \text{and} \quad y = x'\left(\frac{\sqrt{2}}{2}\right) + y'\left(\frac{\sqrt{2}}{2}\right) \qquad \textit{or}$$

$$x = \frac{\sqrt{2}}{2}(x' - y') \quad \text{and} \quad y = \frac{\sqrt{2}}{2}(x' + y')$$

We want to write the equation $xy = 1$ in terms of the coordinates x' and y', so we use the preceding equations and substitute for x and y in the equation $xy = 1$.

$$xy = 1 \qquad \textit{Substitute } x = \frac{\sqrt{2}}{2}(x' - y') \textit{ and } y = \frac{\sqrt{2}}{2}(x' + y') \textit{ to get}$$

$$\left[\frac{\sqrt{2}}{2}(x' - y')\right]\left[\frac{\sqrt{2}}{2}(x' + y')\right] = 1 \qquad \textit{Now simplify.}$$

$$\frac{1}{2}((x')^2 - (y')^2) = 1 \qquad \textit{which we can put into the form of a hyperbola to get}$$

$$\frac{(x')^2}{2} - \frac{(y')^2}{2} = 1$$

This is the equation of a hyperbola with $a = \sqrt{2}$, $b = \sqrt{2}$, and since $c^2 = a^2 + b^2 = 4$, we have $c = 2$. Before we graph the hyperbola, we first have to draw in the new 45° set of axes: the x'-axis is the line $y = x$ and the y'-axis is the line $y = -x$.

To graph the hyperbola, we start at the origin and find the vertices by moving $a = \sqrt{2}$ units in each direction along the x'-axis. The endpoints of the conjugate axis are found by starting at the origin and moving $b = \sqrt{2}$ units in each direction along the y'-axis. We draw the rectangle whose sides are bisected by these points, and the diagonals of this rectangle are the asymptotes: $y' = \pm\frac{b}{a}x' \Rightarrow y' = \pm x'$. We note that the vertices (on the new set of coodinate axes) are $(\pm\sqrt{2}', 0')$. Since $c = 2$, the foci are $(\pm 2', 0')$. (See Figure 10.63.)

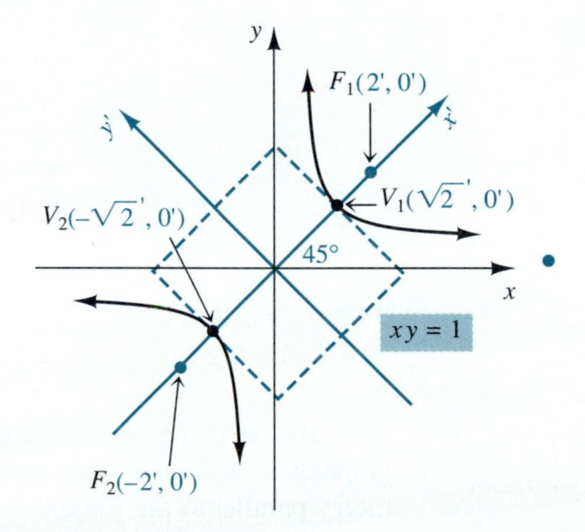

FIGURE 10.63 The graph of $xy = 1$, on the axes rotated 45°

Hence the rational function $y = \frac{1}{x}$ is a hyperbola rotated 45°. We refer to this equation as a *rectangular hyperbola*. ■

We used the new set of axes in order to help us to identify and draw the figure, as well as to locate its vertices, foci, and asymptotes. While these points and lines given in (x', y') form are helpful guides in allowing us to identify and sketch the graph, we may want them expressed in terms of the standard coordinates (x, y). To convert those points and equations back to the standard system, we would again use the coordinate formulas for rotations.

For example, we found in Example 4 that one vertex of $xy = 1$ is $(\sqrt{2}', 0')$. This point is written relative to the new rotated axes. To convert this point to a point given relative to the standard set of coordinate axes, we use the same coordinate formulas used to convert $xy = 1$ into an equation relative to the rotated coordinates: they are $x = \frac{\sqrt{2}}{2}(x' - y')$ and $y = \frac{\sqrt{2}}{2}(x' + y')$.

Hence, for $(\sqrt{2}', 0')$, we substitute $x' = \sqrt{2}$ and $y' = 0$ into the coordinate formulas as follows:

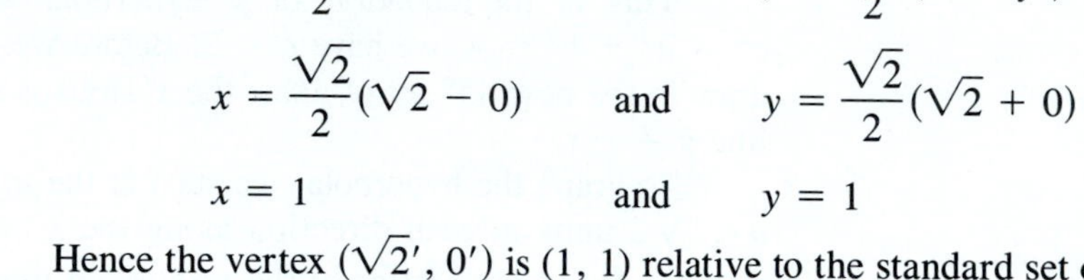

$$x = \frac{\sqrt{2}}{2}(x' - y') \quad \text{and} \quad y = \frac{\sqrt{2}}{2}(x' + y')$$

$$x = \frac{\sqrt{2}}{2}(\sqrt{2} - 0) \quad \text{and} \quad y = \frac{\sqrt{2}}{2}(\sqrt{2} + 0)$$

$$x = 1 \quad \text{and} \quad y = 1$$

In Chapter 4 we found that the asymptotes for $y = \frac{1}{x}$ are the x-axis and y-axis.

Hence the vertex $(\sqrt{2}', 0')$ is $(1, 1)$ relative to the standard set of axes. By doing the same type of transformation, we find that the other vertex is $(-1, -1)$; the foci are $(\sqrt{2}, \sqrt{2})$ and $(-\sqrt{2}, -\sqrt{2})$; and the equations for the asymptotes are $y = 0$, and $x = 0$, which are, respectively, the x- and y-axes. (See Figure 10.64.)

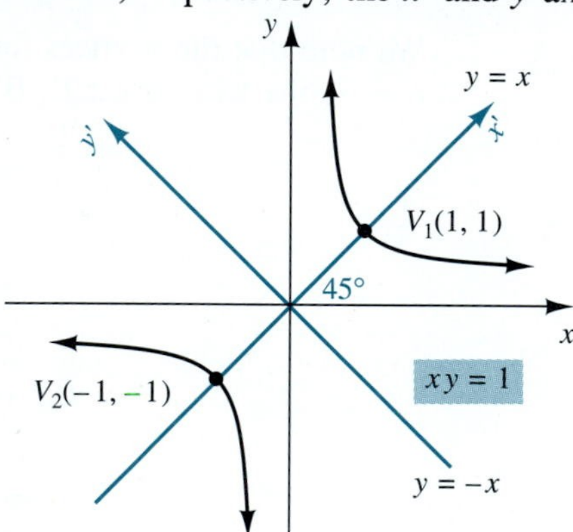

FIGURE 10.64 The graph of $xy = 1$ in standard coordinates

As we have seen, if the equation $Ax^2 + Cy^2 + Dx + Ey + F = 0$ can be graphed, it will be a conic section or one of its degenerate forms, with axes of symmetry parallel to the standard coordinate axes. The process of completing the square helped us to put the equation in one of the standard forms, so that we could quickly identify and graph the figure.

On the other hand, if the equation $Ax^2 + Bxy + Cy^2 + Dx + Ey + F = 0$ (where $B \neq 0$) can be graphed, it will also be a conic section or one of its degenerate forms. However, unlike the form where $B = 0$, the xy-term does not allow a convenient method or form for us to quickly identify and sketch the graph.

Example 2 demonstrates that if we find a suitable angle of rotation where the rotated axes are x' and y', then we can transform an equation of the form

$$Ax^2 + Bxy + Cy^2 + Dx + Ey + F = 0$$

into an equation of the form

$$A'(x')^2 + B'x'y' + C'(y')^2 + D'x' + E'y' + F' = 0 \quad \text{where } B' = 0.$$

that is, we can transform the original equation into an equation with new coefficients such that there is no $x'y'$-term. Then we can graph the transformed equation by using the standard forms developed in the previous sections of this chapter. This requires that we find the angle of rotation ϕ that will make the $x'y'$-term disappear. *Hence, we are looking for ϕ that will make $B' = 0$.*

We start with the equation $Ax^2 + Bxy + Cy^2 + Dx + Ey + F = 0$ and substitute the rotation formulas $x = x' \cos \phi - y' \sin \phi$ and $y = x' \sin \phi + y' \cos \phi$ for x and y in the equation to get

$$A(x' \cos \phi - y' \sin \phi)^2 + B(x' \cos \phi - y' \sin \phi)(x' \sin \phi + y' \cos \phi)$$
$$+ C(x' \sin \phi + y' \cos \phi)^2 + D(x' \cos \phi - y' \sin \phi)$$
$$+ E(x' \sin \phi + y' \cos \phi) + F = 0$$

We can simplify this expression above by performing the operations, collecting terms, and using the basic trigonometric identities to get the equation in the form

$$A'(x')^2 + B'x'y' + C'(y')^2 + D'x' + E'y' + F' = 0$$

where A', B', C', D', E', and F' are:

$$A' = A \cos^2 \phi + B \sin \phi \cos \phi + C \sin^2 \phi$$
$$B' = 2(C - A) \sin \phi \cos \phi + B(\cos^2 \phi - \sin^2 \phi)$$
$$C' = A \sin^2 \phi - B \sin \phi \cos \phi + C \cos^2 \phi$$
$$D' = D \cos \phi + E \sin \phi$$
$$E' = E \cos \phi - D \sin \phi$$
$$F' = F$$

Since we want $B' = 0$,

$$2(C - A) \sin \phi \cos \phi + B(\cos^2 \phi - \sin^2 \phi) = 0$$

Using the double-angle formulas for $\sin 2\phi$ and $\cos 2\phi$, we get

$$(C - A) \sin 2\phi + B \cos 2\phi = 0 \quad \textit{or}$$

$$B \cos 2\phi = (A - C) \sin 2\phi$$

Divide both sides by $B \sin 2\phi$ to get

Note $A - C$ is the negative of $C - A$.

$$\frac{B \cos 2\phi}{B \sin 2\phi} = \frac{(A - C) \sin 2\phi}{B \sin 2\phi}$$

Which gives us

$$\cot 2\phi = \frac{A - C}{B}$$

This is the condition we want.

We summarize this discussion in the following box.

> Given a conic section with equation
>
> $$Ax^2 + Bxy + Cy^2 + Dx + Ey + F = 0$$
>
> if the axes are rotated ϕ degrees where
>
> $$\cot 2\phi = \frac{A - C}{B},$$
>
> then the equation of the conic section in the rotated coordinate system will have no xy-term. (In other words, B' will be 0.)

It can be shown that there is always such a ϕ for $0° < \phi < 90°$. We demonstrate how we can use this in the next example.

If you have a graphics calculator, what do you need to do first in order to graph this equation?

EXAMPLE 5 Sketch the graph of the equation $7x^2 - 6\sqrt{3}xy + 13y^2 - 16 = 0$.

Solution Comparing this with the general equation, we have $A = 7$, $B = -6\sqrt{3}$, and $C = 13$. Since there is an xy-term, we want to find ϕ such that we can transform this equation into one in which there is no $x'y'$ term. As shown before, this happens when

$$\cot 2\phi = \frac{A - C}{B} \qquad \textit{Hence}$$

$$\cot 2\phi = \frac{7 - 13}{-6\sqrt{3}} = \frac{1}{\sqrt{3}} \qquad \textit{Which means that } 2\phi = 60°\text{, hence } \phi = 30°$$

We evaluate the rotation formulas for $\phi = 30°$ to get

$$x = x' \cos 30° - y' \sin 30° \quad \text{and} \quad y = x' \sin 30° + y' \cos 30° \qquad \textit{Or}$$

$$x = \frac{x'\sqrt{3} - y'}{2} \quad \text{and} \quad y = \frac{x' + y'\sqrt{3}}{2}$$

We substitute these formulas for x and y in the equation $7x^2 - 6\sqrt{3}xy + 13y^2 - 16 = 0$.

$$7x^2 - 6\sqrt{3}xy + 13y^2 - 16 = 0 \qquad \textit{Substitute } x = \frac{x'\sqrt{3} - y'}{2} \textit{ and } y = \frac{x' + y'\sqrt{3}}{2}.$$

$$7\left(\frac{x'\sqrt{3} - y'}{2}\right)^2 - 6\sqrt{3}\left(\frac{x'\sqrt{3} - y'}{2}\right)\left(\frac{x' + y'\sqrt{3}}{2}\right) + 13\left(\frac{x' + y'\sqrt{3}}{2}\right)^2 - 16 = 0$$

We simplify to get

$$4(x')^2 + 16(y')^2 - 16 = 0 \qquad \textit{Which can be put into the standard form of a hyperbola}$$

$$\frac{(x')^2}{4} + \frac{(y')^2}{1} = 1$$

Hence we have an ellipse with $a = 2'$ and $b = 1'$. The axes of symmetry are the $x'y'$-axes; the major axis is the x'-axis. The vertices are on the x'-axis and are $(\pm 2', 0')$. We plot the endpoints of the minor axis, which are $(0', \pm 1')$, and draw the ellipse as shown in Figure 10.65.

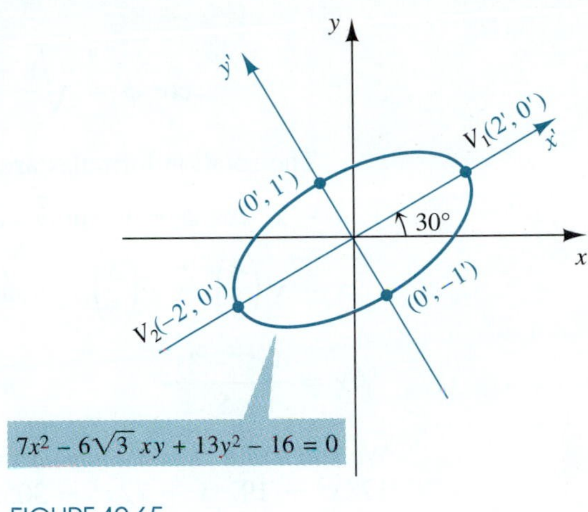

FIGURE 10.65

We can apply these same procedures to figures that have been both translated and rotated.

EXAMPLE 6 Identify and graph the following figure:

$$128x^2 + 192xy + 72y^2 - 305x - 260y = -150$$

Solution Comparing this with the general equation, we have $A = 128$, $B = 192$, and $C = 72$. Since there is an xy-term, we want to find ϕ such that we can transform this equation into the form $A'(x')^2 + C'(y')^2 + D'x + E'y + F' = 0$. For this to happen, we must have

$$\cot 2\phi = \frac{A - C}{B} \qquad \textit{Hence}$$

$$\cot 2\phi = \frac{128 - 72}{192} = \frac{7}{24}$$

The rotation formulas require that we find $\sin \phi$ and $\cos \phi$ (and we will need ϕ in order to graph the figure). In order to find these exact values, we note that we can find both $\sin \phi$ and $\cos \phi$ by using the half-angle formulas (Section 8.2) if we know the value of $\cos 2\phi$. To find the value of $\cos 2\phi$, we draw a reference triangle and find that if $\cot 2\phi = \frac{7}{24}$, then $\cos 2\phi = \frac{7}{25}$.

Using the half-angle formulas we get

$$\sin\phi = \sqrt{\frac{1-\cos 2\phi}{2}} = \sqrt{\frac{1-\frac{7}{25}}{2}} = \sqrt{\frac{9}{25}} = \frac{3}{5} \quad \textit{and}$$

$$\cos\phi = \sqrt{\frac{1+\cos 2\phi}{2}} = \sqrt{\frac{1+\frac{7}{25}}{2}} = \sqrt{\frac{16}{25}} = \frac{4}{5}$$

The rotation formulas are

$$x = x'\cos\phi - y'\sin\phi \quad \text{and} \quad y = x'\sin\phi + y'\cos\phi \qquad \textit{Which become}$$

$$x = x'\left(\frac{4}{5}\right) - y'\left(\frac{3}{5}\right) \quad \text{and} \quad y = x'\left(\frac{3}{5}\right) + y'\left(\frac{4}{5}\right) \qquad \textit{or}$$

$$x = \frac{4x' - 3y'}{5} \quad \text{and} \quad y = \frac{3x' + 4y'}{5}$$

We substitute the rotation formulas in the equation $128x^2 + 192xy + 72y^2 - 305x - 260y = -150$ and simplify to get

$$200(x')^2 - 400x' - 25y' + 150 = 0 \qquad \textit{Solving for } y' \textit{ we get}$$

$$8(x')^2 - 16x' + 6 = y' \qquad \textit{This is a translated parabola whose standard form is}$$

$$(x' - 1)^2 = \frac{1}{8}(y' + 2)$$

This is the parabola with the vertex at the $x'y'$-coordinates $(1', -2')$; the y'-axis is the axis of symmetry. Since $4p' = \frac{1}{8} \Rightarrow p' = \frac{1}{32}$, the focus is $\left(1', \left(-\frac{63}{32}\right)'\right)$.

The coordinates of the vertex relative to the standard axis system are

$$x = \frac{4x' - 3y'}{5} = \frac{4(1) - 3(-2)}{5} = 2 \qquad \textit{and}$$

$$y = \frac{3x' + 4y'}{5} = \frac{3(1) + 4(-2)}{5} = -1$$

The vertex is $(2, -1)$. By the same procedure, the standard coordinates of the focus are $\left(\frac{317}{160}, -\frac{39}{40}\right)$.

In order to draw the graph we must have the angle of rotation, ϕ. Since we know $\sin\phi = \frac{3}{5}$, we can use a calculator to find $\phi \approx 37°$. The sketch appears in Figure 10.66.

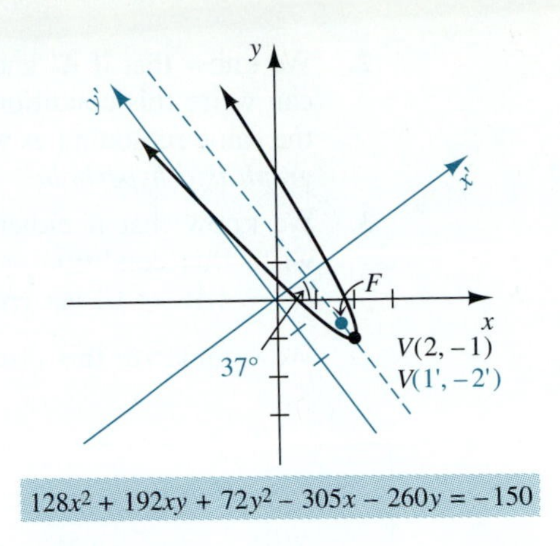

FIGURE 10.66

The Discriminant: Invariants Under Rotations

If we rotate the x- and y-axis through an angle ϕ, the general equation

$$Ax^2 + Bxy + Cy^2 + Dx + Ey + F = 0 \qquad \textit{becomes}$$
$$A'(x')^2 + B'x'y' + C'(y')^2 + D'x' + E'y' + F' = 0$$

where A, B, C, D, E, F and A', B', C', D', E', F, and ϕ are related as indicated on page 703. It can be shown (see Exercise 54) that no matter what the new rotated $x'y'$-axes are, for the new A', B' and C',

$$B^2 - 4AC = (B')^2 - 4A'C'.$$

We refer to quantities that do not change under rotations as **invariant under rotations**. $B^2 - 4AC$ is such an invariant under rotation.

Starting with the general equation $Ax^2 + Bxy + Cy^2 + Dx + Ey + F = 0$, where $B \neq 0$, we showed that we can rotate the axes in such a way that the xy-term is eliminated, or $B' = 0$, and arrive at the rotated equation $A'(x')^2 + C'(y')^2 + D'x' + E'y' + F' = 0$.

1. We know that if A' and C' have the same sign, the graph is an ellipse. We can write this condition as:

   $$A'C' > 0 \Rightarrow \text{ The graph is an ellipse (or its degenerate form).}$$

   But $\quad B^2 - 4AC = (B')^2 - 4A'C' \qquad$ *And since $B' = 0$, we have*

   $$B^2 - 4AC = -4A'C'$$

   Therefore, if $B^2 - 4AC < 0$, then $-4A'C' < 0$, which implies $A'C' > 0$; hence A' and C' have the same sign. Thus, *if $B^2 - 4AC < 0$, the graph is an ellipse*.

2. We know that if A' and C' have opposite signs, the graph is a hyperbola. We can write this condition as follows: If $A'C' < 0$, the graph is a hyperbola. By the same reasoning as with the ellipse, we conclude that *if* $B^2 - 4AC > 0$, *the graph is a hyperbola.*
3. We know that if either $A' = 0$ or $C' = 0$, the graph is a parabola. We can write this condition as $A'C' = 0$. Again, using the same logic we have: *If* $B^2 - 4AC = 0$, *the graph is a parabola.*

We summarize this discussion as follows.

The graph of the equation

$$Ax^2 + Bxy + Cy^2 + Dx + Ey + F = 0$$

is a conic or one of its degenerate forms.

1. If $B^2 - 4AC < 0$, the graph is an ellipse.
2. If $B^2 - 4AC > 0$, the graph is a hyperbola.
3. If $B^2 - 4AC = 0$, the graph is a parabola.

The quantity $B^2 - 4AC$ is referred to as the **discriminant** of the conic.

EXAMPLE 7 Determine what the following equation is if it could be graphed:

$$3x^2 - 5xy + 6y^2 - 2x + 8y - 5 = 0$$

Solution For the equation $3x^2 - 5xy + 6y^2 - 2x + 8y - 5 = 0$, we have $A = 3$, $B = -5$, and $C = 6$. Hence $B^2 - 4AC = (-5)^2 - 4(3)(6) = -47$. Since the discriminant is less than 0, if the equation could be graphed, it would be an ellipse or its degenerate form, a point. ■

EXERCISES 10.6

In Exercises 1–8, a set of axes is translated so that the new origin is $(2, -4)$ on the standard set of coordinate axes. Given the coordinates of a point relative to one set of axes, find the coordinates of the points relative to the other set of axes. (The translated set of axes is referred to as $x'y'$.)

1. $(x, y) = (3, 5)$; find (x', y').
2. $(x', y') = (3', 0')$; find (x, y).
3. $(x', y') = (2', 5')$; find (x, y).
4. $(x, y) = (2, 0)$; find (x', y').
5. $(x, y) = (-4, 4)$; find (x', y').
6. $(x, y) = (2, -8)$; find (x', y').
7. $(x', y') = (0', 0')$; find (x, y).
8. $(x', y') = (-2', 3')$; find (x, y).

In Exercises 9–16, find the translation $x' = x - h$ and $y' = y - k$ that will allow you to express the equation as a standard form of a conic section centered at the origin with the coordinate axes as the axes of symmetry.

9. $\frac{(x-3)^2}{9} - \frac{(y+4)^2}{18} = 1$
10. $\frac{(y+5)^2}{36} + \frac{(x-2)^2}{15} = 1$
11. $(y - 3)^2 = 18x$
12. $x^2 + (y - 2)^2 = 25$
13. $3x^2 + 4y^2 - 6x + 56y - 187 = 0$
14. $2x^2 + 2y^2 - 4x - 36y - 150 = 0$
15. $x^2 - 4y^2 - 10x + 16y + 1 = 0$
16. $y^2 - 4y - 5x - 26 = 0$

In Exercises 17–24, a standard set of axes is rotated an angle ϕ counterclockwise, producing a new set of axes, $x'y'$. Given ϕ and the point on one set of axes, find the coordinates of the point relative to the other set of axes.

17. $(x, y) = (2, 5)$, $\phi = 45°$; find (x', y').
18. $(x', y') = (3', 7')$, $\phi = 30°$; find (x, y).
19. $(x', y') = (2', -1')$, $\phi = 60°$; find (x, y).
20. $(x, y) = (-2, -3)$, $\phi = \frac{3\pi}{4}$; find (x', y').
21. $(x, y) = (2, -1)$, $\phi = 60°$; find (x', y').
22. $(x', y') = (-3', 0')$, $\phi = \frac{3\pi}{4}$; find (x, y).
23. $(x, y) = (-2, 5)$, $\phi = 45°$; find (x', y').
24. $(x', y') = (1', 3')$, $\phi = \frac{\pi}{6}$; find (x, y).

In Exercises 25–30, a standard set of axes is rotated an angle ϕ counterclockwise, producing a new set of axes, $x'y'$. Given ϕ and the equation relative to the standard set of axes, express the equation in terms of the rotated set of axes, $x'y'$.

25. $2x^2 = y$, $\phi = 60°$
26. $xy = 6$, $\phi = \frac{\pi}{4}$
27. $2x^2 - 4y^2 = 16$, $\phi = 30°$
28. $2xy - 3y - x = 8$, $\phi = 30°$
29. $x^2 + y^2 = 4$, $\phi = \frac{\pi}{4}$
30. $x^2 - 2xy - 3y^2 - 8 = 0$, $\phi = 30°$

In Exercises 31–34 a standard set of axes is rotated an angle ϕ counterclockwise, producing a new set of axes $x'y'$. Given ϕ and the equation relative to the rotated axes, $x'y'$, express the equation in terms of the standard set of axes, xy.

31. $3x'y' = 4$, $\phi = 45°$
32. $(x')^2 - (y')^2 = 8$, $\phi = 60°$
33. $(x')^2 - 2x'y' = 8$, $\phi = 30°$
34. $2(y')^2 = -16$, $\phi = 30°$

In Exercises 35–38, an equation is given in terms of x and y, relative to the standard set of coordinate axes. Find the angle ϕ that the axes should be rotated so that the equation can be written in terms of x' and y' (relative to the new set of axes) without an $x'y'$-term (that is, $B' = 0$).

35. $2x^2 - \sqrt{3}xy + 3y^2 = 0$
36. $5x^2 + xy + 5y^2 - 2x + 3y = 6$
37. $2x^2 + xy + y^2 = 5$
38. $2x^2 - xy + 8y^2 + x - y = 8$

In Exercises 39–48, sketch a graph of the equation.

39. $xy = 8$
40. $2x^2 + 4xy + 2y^2 + \sqrt{2}x - \sqrt{2}y = 0$
41. $x^2 - 10\sqrt{3}xy + 11y^2 - 16 = 0$
42. $9x^2 - 6xy + 17y^2 = 72$
43. $23x^2 + 18\sqrt{3}xy + 5y^2 - 256 = 0$
44. $3x^2 + 8xy - 3y^2 - 20 = 0$
45. $x^2 + 2xy + y^2 = 1$
46. $2xy + \sqrt{2}x - 3\sqrt{2}y = 12$
47. $31x^2 + 10\sqrt{3}xy + 21y^2 + 48x - 48\sqrt{3}y = 0$
48. $3x^2 - 6xy + 3y^2 - 16 = 0$

In Exercises 49–52, using the discriminant of the conic, identify the graph of the equation.

49. $5x^2 - 7xy + 3y^2 - 2x + 4y - 8 = 0$
50. $5x^2 + xy - 2y^2 + 5x + 3y = 6$
51. $2x^2 - 7xy = 5$
52. $2x^2 - 4xy + 2y^2 - x + y = 8$

QUESTIONS FOR THOUGHT

53. Treat the two equations $x' = x\cos\phi + y\sin\phi$ and $y' = -x\sin\phi + y\cos\phi$ as a system and express x and y in terms of x' and y'.

54. Prove that the quantity $B^2 - 4AC$ is invariant under rotations; that is, use the equations on page 703, and show that $B'^2 - 4A'C' = B^2 - 4AC$.

55. We can solve for y in the equation

$$Ax^2 + Bxy + Cy^2 + Dx + Ey + F = 0 \quad (C \neq 0)$$

if we look at it as a quadratic equation in y. We could rewrite it as follows:

$$Ax^2 + Bxy + Cy^2 + Dx + Ey + F = 0$$

Group the y^2-term, the y-terms, and the non-y terms.

$$Cy^2 + Bxy + Ey + Ax^2 + Dx + F = 0$$

Factor y from the y terms.

$$Cy^2 + (Bx + E)y + (Ax^2 + Dx + F) = 0$$

Looking at the last form as a quadratic equation in y with C as the coefficient of y^2, $Bx + E$ as the coefficient of y, and $Ax^2 + Dx + F$ as the numerical term, use the quadratic formula to show

$$y = \frac{-(Bx + E) \pm \sqrt{(Bx + E)^2 - 4(C)(Ax^2 + Dx + F)}}{2C}$$

GRAFFIX

In Exercises 56–57, use the result of Exercise 55 to sketch the graph on your graphics calculator or computer.

56. $3x^2 - 2xy + 5y^2 = 12$
57. $x^2 + 4xy - 3y^2 - 2y = 24$

10.7 Nonlinear Systems of Equations and Inequalities

We have devoted much time and attention to graphing a wide variety of functions and equations. In the course of our discussions we solved various types of equations, particularly in order to find the x- and y-intercepts of a graph. In fact, in Sections 4.1 and 5.1, we also saw how solutions to an equation can lead us to the points of intersection of two graphs.

In Chapter 9 we investigated linear systems of equations and inequalities. In this section we discuss nonlinear systems. In the course of this development we review many of the basic graphs in our catalog.

Up to this point we have been considering systems involving only first-degree equations (or inequalities). A **nonlinear system** is one in which at least one of the equations is not a linear equation. The following are two examples of nonlinear systems that we will consider in this section.

$$\begin{cases} x^2 + y^2 = 9 \\ y = x^2 - 3 \end{cases} \qquad \begin{cases} \sqrt{x} + \sqrt{y} = 3 \\ x + y = 5 \end{cases}$$

When the system involves one linear equation and one quadratic equation or two quadratic equations, it is called a **quadratic system**.

Before we discuss the algebraic techniques for solving such systems, let's take a moment to use our knowledge of the graphs of the conic sections to analyze what we might expect from a quadratic system. If we examine the first system, we recognize that the graph of the first equation, $x^2 + y^2 = 9$, is a circle and the graph of the second equation, $y = x^2 - 3$, is a parabola. Keeping in mind that each point of intersection of the graphs of the equations corresponds to a solution to the system of equations, it would be useful to know how many points of intersection a circle and a parabola can have.

The graphs in Figure 10.67 illustrate a variety of possibilities for a circle and parabola. As the figure shows, a circle and a parabola can intersect in 0, 1, 2, 3, or 4 points. Can there be more than 4 points of intersection? If you play around with the graphs of a circle and a parabola, you will quickly convince yourself that there can be no more than 4 points of intersection.

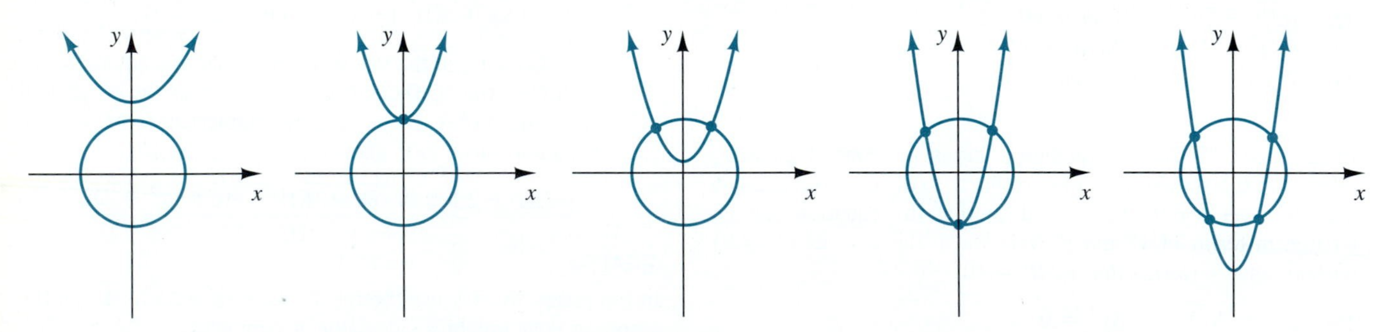

FIGURE 10.67 Possible intersections of a parabola and a circle

A similar analysis should convince us that any two distinct conic sections (straight line, parabola, circle, ellipse, and hyperbola) can intersect in *at most* 4 points. Exercise 55 suggests how we might prove this fact algebraically.

Knowing how many solutions a system of equations may have can be helpful in detecting errors or extraneous solutions.

GRAFFIX

Use a graphics calculator.

1. Graph the equations $x^2 + 3y^2 = 1$ and $x - y = 1$ on the same set of coordinate axes and use the trace function to find their points of intersection.
2. Graph the equations $x^2 - y = 3$ and $x + y = 3$ on the same set of coordinate axes and use the trace function to find their points of intersection.

Let us now discuss algebraic techniques for solving nonlinear systems. As with linear systems, there are two basic methods: elimination and substitution. As the following example illustrates, we don't always have an option as to which method to use.

EXAMPLE 1 Solve the following system of equations.

$$\begin{cases} 2x^2 - y^2 = 7 \\ x - y = 1 \end{cases}$$

Solution A moment's thought about using the elimination method should convince us that it will not be effective on this system. In order for the elimination method to work, we must eliminate *all* occurrences of one of the variables, which requires that we have like terms to eliminate. However, in this system the two equations have no like terms and so the elimination method will not work. Consequently, we will use the substitution method.

In order to make the substitution process as straightforward as possible, we try to use the "simpler" equation and explicitly solve for whichever variable appears to be easier to isolate.

In this example, the second equation is simpler, since it is linear in both variables, whereas the first is quadratic in both. Looking at the second equation, it should be apparent that it is easier to solve explicitly for y than for x.

We solve the second equation for y, obtaining $y = x - 1$. Let's substitute $y = x - 1$ into the first equation.

$$
\begin{aligned}
2x^2 - y^2 &= 7 && \textit{We substitute } y = 3x - 5 \textit{ and solve for } x. \\
2x^2 - (x-1)^2 &= 7 \\
2x^2 - (x^2 - 2x + 1) &= 7 \\
2x^2 - x^2 + 2x - 1 &= 7 \\
x^2 + 2x - 8 &= 0 && \textit{We can solve this equation by factoring.} \\
(x+4)(x-2) &= 0 \\
x = -4 \quad \text{or} \quad x &= 2
\end{aligned}
$$

In order to get the corresponding y-values, we take advantage of the equation solved explicitly for y.

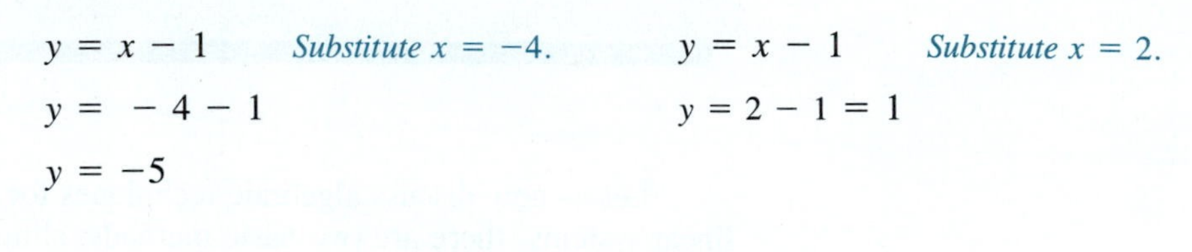

$$
\begin{aligned}
y &= x - 1 && \textit{Substitute } x = -4. \qquad & y &= x - 1 && \textit{Substitute } x = 2. \\
y &= -4 - 1 && & y &= 2 - 1 = 1 \\
y &= -5
\end{aligned}
$$

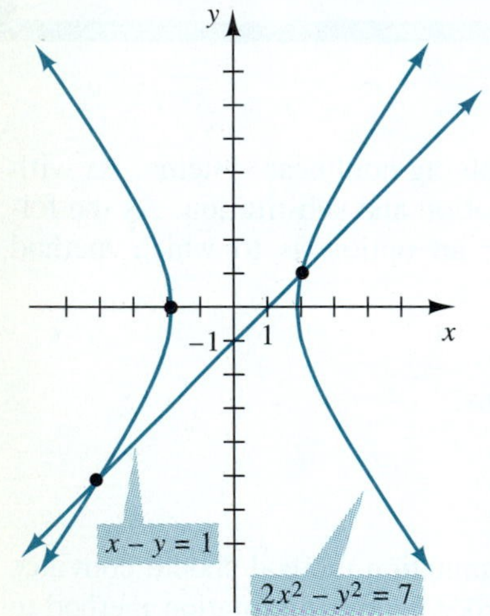

FIGURE 10.68

Therefore, the solutions to the system are $(-4, -5)$ and $(2, 1)$. It is left to the student to check that these ordered pairs satisfy *both* equations.

We note that the graph of $2x^2 - y^2 = 7$ is a hyperbola and the graph of $x - y = 1$ is a straight line, which is certainly consistent with having two solutions to the system. We conclude the example by making a rough sketch of the graphs and verifying that our picture agrees with our algebraic results in terms of the approximate location of the intersection of the two graphs (see Figure 10.68).

Knowing what the solutions are, we can see that it is unrealistic to think that, in general, we can find solutions to a system of equations by visually examining the graphs.

■

EXAMPLE 2 Solve the following system of equations.

$$
\begin{cases} x^2 + y^2 = 9 \\ y = x^2 - 3 \end{cases}
$$

We offer two possible approaches to solving this system.

Solution 1 Using the Substitution Method

The fact that the second equation is solved explicitly for y suggests that, perhaps, the substitution method will be the easiest. However, as we shall see, it raises some additional problems.

$$x^2 + y^2 = 9$$ *We use the second equation to substitute $y = x^2 - 3$.*

$$x^2 + (x^2 - 3)^2 = 9$$

$$x^2 + x^4 - 6x^2 + 9 = 9$$

$$x^4 - 5x^2 = 0$$

$$x^2(x^2 - 5) = 0$$

$$x = 0 \quad \text{or} \quad x = \pm\sqrt{5}$$ *We may solve for y by substituting into the second equation.*

If $x = 0$, *then*

$$y = x^2 - 3 = 0^2 - 3 = -3$$

If $x = \pm\sqrt{5}$, *then*

$$y = (\pm\sqrt{5})^2 - 3 = 5 - 3 = 2$$

Therefore we have three solutions: $(0, -3)$, $(\sqrt{5}, 2)$, $(-\sqrt{5}, 2)$. It is left to the student to verify that these ordered pairs satisfy *both* equations.

Before we look at the second approach, two important points need to be made.

1. When we substituted $y = x^2 - 3$ into the first equation, we obtained a *fourth-degree* equation. We know that when we encounter a fourth-degree equation, we may or may not be able to solve it, because we don't have a general method for solving fourth-degree equations. In this example, we were able to solve it because we could factor the fourth-degree polynomial. Thus it would seem to be to our advantage, as we manipulate the given system, to try to keep the degree of any resulting equations as small as possible.
2. In the past when we solved linear systems, once we obtained the value of one of the variables, it did not matter into which equation we substituted to obtain the other variable. However, in this example let's see what happens if we substitute the x-values we obtained into the first equation instead of the second equation. To solve for y, we substitute $x = 0$ and $x = \pm\sqrt{5}$ into the first equation.

$$x^2 + y^2 = 9$$ *Let $x = 0$.*

$$y^2 = 9$$

$$y = \pm 3$$

The "solutions" are $(0, 3)$, $(0, -3)$.

$$x^2 + y^2 = 9$$ *Let $x = \pm\sqrt{5}$.*

$$(\pm\sqrt{5})^2 + y^2 = 9$$

$$5 + y^2 = 9$$

$$y^2 = 4$$

$$y = \pm 2$$

This gives 4 *"solutions":* $(\sqrt{5}, 2)$, $(-\sqrt{5}, 2)$, $(\sqrt{5}, -2)$, $(-\sqrt{5}, -2)$.

Note that we have obtained a total of six apparent solutions, which we know, based upon our discussion above, is impossible for a quadratic system. In fact, if we substitute one of these ordered pairs that we did not obtain earlier, such as $(0, 3)$, into the second equation, we get

$$y = x^2 - 3$$ *Substitute* $(0, 3)$.

$$3 \stackrel{?}{=} (0)^2 - 3$$

$$3 \neq -3$$ *So* $(0, 3)$ *is not a solution to the system.*

The reason we obtained more apparent solutions is that by substituting into the first equation, which is quadratic in y, we generated more y-values than by substituting into the second equation, which is linear in y. Thus in order to avoid extraneous solutions, once we have found x-values and we want to find the associated y-values, it is a good idea to substitute into the equation of lowest degree in y.

Solution 2 Using the Elimination Method

We rewrite the second equation in the system to get

$$\begin{aligned} x^2 + y^2 &= 9 \\ -x^2 + y &= -3 \end{aligned} \qquad \textit{Adding the two equations eliminates } x^2.$$

$$\begin{aligned} y^2 + y &= 6 \\ y^2 + y - 6 &= 0 \\ (y + 3)(y - 2) &= 0 \\ y = -3 \quad &\text{or} \quad y = 2 \end{aligned}$$

We can now substitute these y-values into one of the original equations to obtain the corresponding x-values. Based upon the comments made in point 2, we note that both of the original equations are of degree 2 in x, so it makes no difference which equation we use. We substitute into the second equation.

$$\begin{aligned} y &= x^2 - 3 \qquad \textit{Substitute } y = -3. \\ -3 &= x^2 - 3 \\ x^2 &= 0 \\ x &= 0 \end{aligned} \qquad\qquad \begin{aligned} y &= x^2 - 3 \qquad \textit{Substitute } y = 2. \\ 2 &= x^2 - 3 \\ x^2 &= 5 \\ x &= \pm\sqrt{5} \end{aligned}$$

This result gives us the same three solutions obtained by the substitution method: $(0, -3)$, $(-\sqrt{5}, 2)$, $(\sqrt{5}, 2)$.

We include a sketch of the graph of the two equations, since the graph often gives us a fairly simple way to check if we have the correct number of solutions. See Figure 10.69.

Finally, we note that by using the elimination method, we avoided the potential problem of dealing with a fourth-degree equation, which we encountered when we used the substitution method. ■

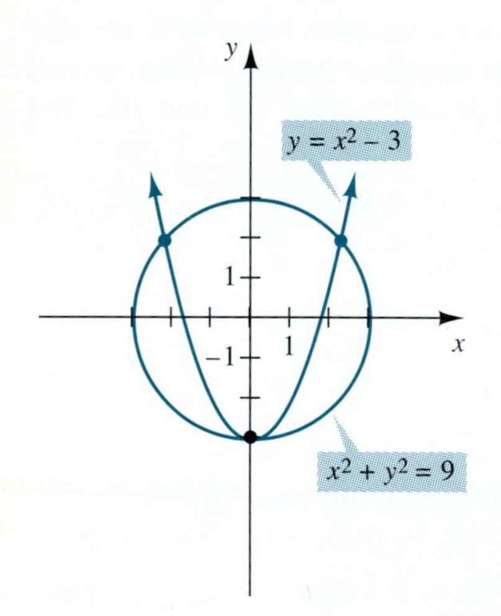

FIGURE 10.69

EXAMPLE 3 Solve and sketch a graph of the following system of equations.

$$\begin{cases} y = \dfrac{7}{x} \\ x = y + 3 \end{cases}$$

Round the answers to the nearest hundredth.

Solution Since each equation is solved explicitly for one of the variables, we may substitute the second equation into the first equation or vice versa. We will use the second equation to substitute into the first equation.

$$y = \frac{7}{x} \qquad \textit{Substitute } x = y + 3.$$

$$y = \frac{7}{y+3} \qquad \textit{Multiply both sides by } y + 3.$$

$$y^2 + 3y = 7$$

$$y^2 + 3y - 7 = 0 \qquad \textit{We use the quadratic formula to solve for } y.$$

$$y = \frac{-3 \pm \sqrt{37}}{2} \qquad \textit{Using a calculator we get}$$

$$y = 1.54, -4.54$$

We get the corresponding x-values by substituting into equation (2).

$$x = 1.54 + 3 = 4.54 \quad \text{or} \quad x = -4.54 + 3 = -1.54.$$

Thus the solutions are (4.54, 1.54) and (−1.54, −4.54). The graphs of the equations in this system appear in Figure 10.70.

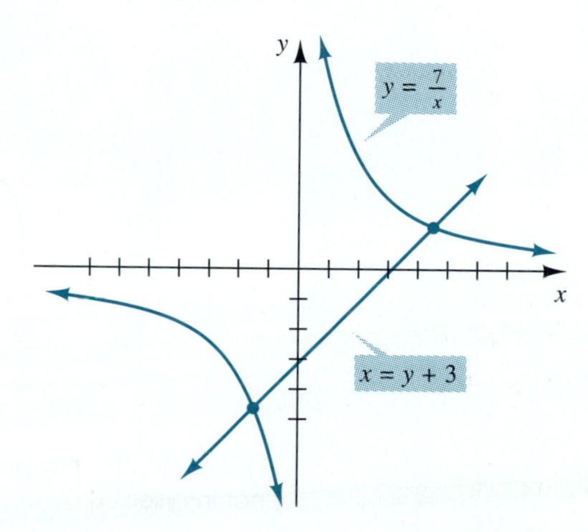

FIGURE 10.70

■

EXAMPLE 4 Find the points of intersection (if any) of the graphs of the equations $16x^2 + 25y^2 = 400$ and $x^2 + y^2 = 9$.

Solution The problem is asking us to solve the following system of equations.

$$\begin{cases} 16x^2 + 25y^2 = 400 \\ x^2 + y^2 = 9 \end{cases}$$

Since the system does have like terms, we can use the elimination method.

$$\begin{array}{rcr} 16x^2 + 25y^2 = 400 & \xrightarrow{\textit{As is}} & 16x^2 + 25y^2 = 400 \\ x^2 + y^2 = 9 & \xrightarrow{\textit{Multiply by } -25} & \underline{-25x^2 - 25y^2 = -225} \\ & & -9x^2 = 175 \end{array}$$

$$x^2 = -\frac{175}{9}$$

$$x = \pm\sqrt{-\frac{175}{9}}$$

Thus there are no real solutions to this system, which means that the graphs of the two equations do not intersect.

We could have come to the same conclusion had we first graphed the two equations. The graph of $16x^2 + 25y^2 = 400$ is an ellipse and the graph of $x^2 + y^2 = 9$ is a circle. Both have centers at the origin. Their graphs appear in Figure 10.71.

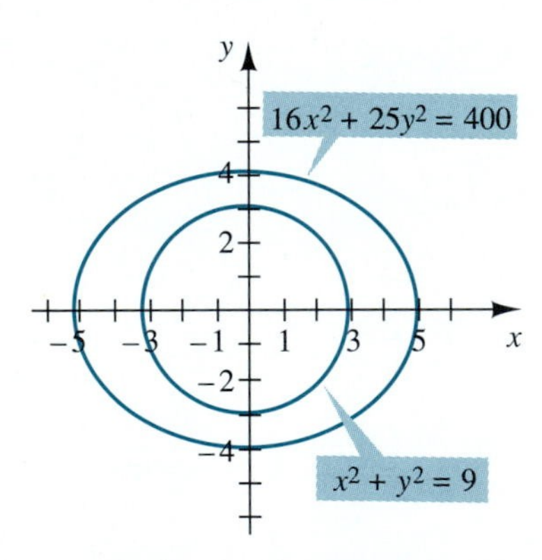

FIGURE 10.71 The graphs of $16x^2 + 25y^2 = 400$ and $x^2 + y^2 = 9$ do not intersect. ■

In Section 9.8 we saw that we could describe the solutions for a system of linear inequalities by graphing each of the inequalities separately. The same approach can be used for some nonlinear systems of inequalities.

EXAMPLE 5 Sketch the solutions to the following system of inequalities.

$$\begin{cases} y \le 4 - x^2 \\ x - y \le 2 \end{cases}$$

Solution As with a system of linear inequalities, we sketch the solutions of each inequality and see where those regions overlap.

To find the solutions to $y \le 4 - x^2$, we begin by sketching the graph of the boundary of the region, $y = 4 - x^2$. We see that the solution set will be the region on and below (inside) the parabola, since the inequality says that y is less then or equal to $4 - x^2$. Alternatively, we may choose a test point not on the graph of $y = 4 - x^2$ and obtain the same result. The solution set of the first inequality is indicated by the shaded region in Figure 10.72(a).

Proceeding in a similar manner, the solution to the second inequality $x - y \le 2$ is the shaded region on and above the line $x - y = 2$ in Figure 10.72(b). Thus the solutions of the system of inequalities consist of those points that are in *both* shaded regions. This region is indicated in Figure 10.72(b) by the crosshatching.

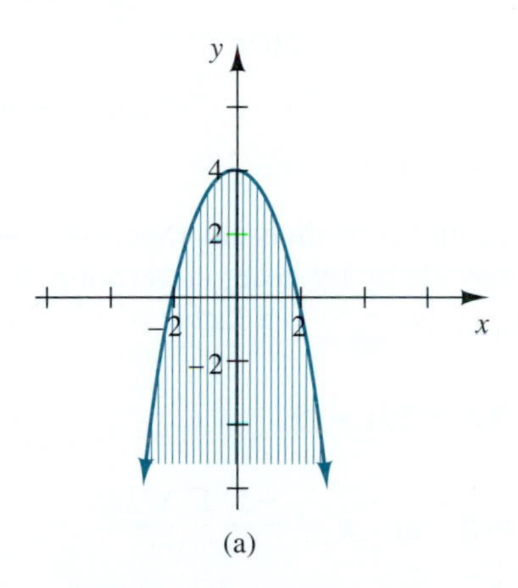

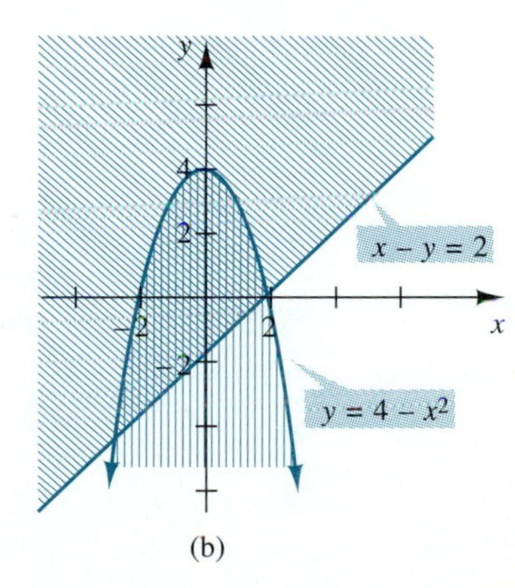

FIGURE 10.72

Real-life applications often give rise to nonlinear systems of equations.

EXAMPLE 6 Suppose that a manufacturer wants to produce rectangular boxes with square tops and bottoms, such that the volume of a box is 200 cu cm. The company has on hand a supply of metal sheets, each of which measures 210 sq cm in area. Assuming no waste of material, to the nearest hundredth square centimeter,what should the dimensions of the box be so that each metal sheet produces one rectangular box?

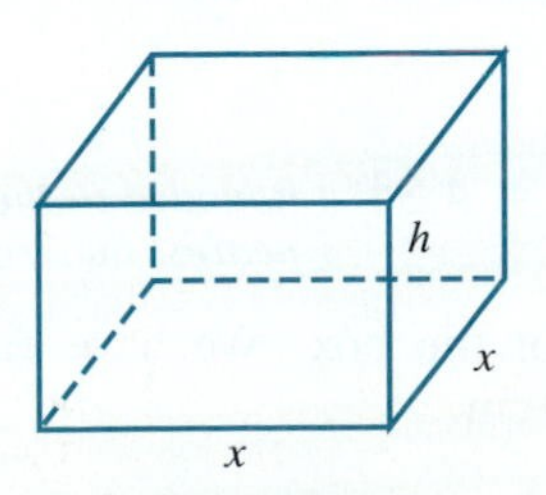

FIGURE 10.73

Solution We begin by drawing a diagram of the proposed box (see Figure 10.73). Since we are told that the top and bottom of the box are square, we label each edge of the top and bottom x. We also label the height of the box h.

The fact that we want the volume of each box to be 200 cu cm translates into the equation $x^2h = 200$, and the fact that each metal sheet has an area of 210 sq cm means that the surface area of the box must be 210 sq cm. The area of both the top and bottom is x^2, whereas the area of each of the four sides is xh, so the surface area of the box is $2x^2 + 4xh$. We thus have a second equation, $2x^2 + 4xh = 210$.

Therefore, finding the required dimensions of the box is equivalent to solving the following system of equations.

$$\begin{cases} x^2h = 200 \\ 2x^2 + 4xh = 210 \end{cases}$$

We may solve the first equation for h, obtaining $h = \frac{200}{x^2}$, and substitute it into the second equation.

$$2x^2 + 4xh = 210 \qquad \textit{We substitute } h = \frac{200}{x^2}.$$

$$2x^2 + 4x\left(\frac{200}{x^2}\right) = 210 \qquad \textit{Simplify.}$$

$$2x^2 + \frac{800}{x} = 210 \qquad \textit{Multiply both sides of the equation by } x.$$

$$2x^3 + 800 = 210x$$

$$2x^3 - 210x + 800 = 0 \qquad \textit{Divide both sides of the equation by 2.}$$

$$x^3 - 105x + 400 = 0$$

By the rational root theorem (Section 4.4), there are many possible rational roots that we may check by using either long division or synthetic division. We find that 5 is a root, so $(x - 5)$ is a factor of $x^3 - 105x + 400$ and we have

$$(x - 5)(x^2 + 5x - 80) = 0 \qquad \textit{We obtain the other roots by using the quadratic formula.}$$

$$x = 5 \quad \text{or} \quad x = \frac{-5 \pm \sqrt{345}}{2} \qquad \textit{Using a calculator we get}$$

$$x = 5 \quad \text{or} \quad x = 6.79 \quad \text{or} \quad x = -11.79 \qquad \textit{Rounded to the nearest hundredth.}$$

Since x represents the length of a side of the top and bottom of the box, it makes no sense for x to be negative. By substituting the two possible x-values into $h = \frac{200}{x^2}$ we get

$$h = \frac{200}{x^2} \qquad \textit{Substitute } x = 5. \qquad\qquad h = \frac{200}{x^2} \qquad \textit{Substitute } x = 6.79.$$

$$h = \frac{200}{25} = 8 \qquad\qquad h = \frac{200}{(6.79)^2} = 4.34 \qquad \textit{Rounded to the nearest hundredth.}$$

Therefore, we have two possible configurations for the box. We have either $x = 5$ cm and $h = 8$ cm or $x = 6.79$ cm and $h = 4.34$ cm. ■

EXERCISES 10.7

In Exercises 1–22, solve the system of equations.

1. $\begin{cases} x^2 + y^2 = 10 \\ x + y = 2 \end{cases}$

2. $\begin{cases} x^2 + y^2 = 8 \\ x^2 + y = 6 \end{cases}$

3. $\begin{cases} 4x^2 + y^2 = 36 \\ 2x - y = 6 \end{cases}$

4. $\begin{cases} x^2 + y^2 = 4 \\ x^2 - y = 9 \end{cases}$

5. $\begin{cases} x^2 - y^2 = 16 \\ 3x - y = 1 \end{cases}$

6. $\begin{cases} y = x^2 + 8x - 10 \\ y = 3x + 4 \end{cases}$

7. $\begin{cases} x^2 - y^2 = 8 \\ x^2 + 2y^2 = 11 \end{cases}$

8. $\begin{cases} y = 2 - x^2 \\ x^2 = y^2 - 4 \end{cases}$

9. $\begin{cases} x^2 + y^2 = 1 \\ \dfrac{x^2}{4} + y^2 = 1 \end{cases}$

10. $\begin{cases} 4x^2 + 5y^2 = 10 \\ 5x^2 + 4y^2 = 10 \end{cases}$

11. $\begin{cases} y = x^2 - 2x - 4 \\ x - y = 2 \end{cases}$

12. $\begin{cases} y = x^2 + x - 1 \\ y = -x^2 - x + 1 \end{cases}$

13. $\begin{cases} y = \sqrt{x} \\ 2x + y = 10 \end{cases}$

14. $\begin{cases} y = \sqrt{2x} - 1 \\ y = \sqrt{x} - 1 \end{cases}$

15. $\begin{cases} (x - 2)^2 + y^2 = 13 \\ y = \sqrt{3 - x} \end{cases}$

16. $\begin{cases} x = y^2 - 4 \\ y = -\sqrt{x^2 + 4} \end{cases}$

17. $\begin{cases} y = (x - 3)^2 \\ (x - 3)^2 + y^2 = 12 \end{cases}$

18. $\begin{cases} x^2 + (y + 2)^2 = 2 \\ x = (y + 2)^2 \end{cases}$

19. $\begin{cases} xy = 2 \\ x + 3y = 7 \end{cases}$

20. $\begin{cases} xy = 5 \\ 2y - 3x = 2 \end{cases}$

21. $\begin{cases} y = \log_2(x - 1) \\ y = \log_2(x + 3) - 1 \end{cases}$

22. $\begin{cases} y = \log_3(x - 3) \\ y = 2 - \log_3(x + 5) \end{cases}$

In Exercises 23–34, solve the system of equations and sketch the graphs.

23. $\begin{cases} x^2 + y^2 = 1 \\ x^2 + (y - 1)^2 = 1 \end{cases}$

24. $\begin{cases} 9x^2 + 4y^2 = 36 \\ x^2 - y^2 = 16 \end{cases}$

25. $\begin{cases} y = 2x^2 \\ y = x^2 - 1 \end{cases}$

26. $\begin{cases} y = 2x^2 - 1 \\ y = x^2 \end{cases}$

27. $\begin{cases} x^2 + y^2 = 9 \\ x - y = 3 \end{cases}$

28. $\begin{cases} y = x^2 - 4 \\ y = 4 - x^2 \end{cases}$

29. $\begin{cases} y = x^2 - 6x \\ 2x + 3y + 13 = 0 \end{cases}$

30. $\begin{cases} y = x^2 - 4x + 1 \\ 3x - 2y = 7 \end{cases}$

31. $\begin{cases} x^2 + y^2 = 9 \\ 2y = x^2 - 1 \end{cases}$

32. $\begin{cases} x^2 + 9y^2 = 9 \\ x^2 - y^2 = 4 \end{cases}$

33. $\begin{cases} x^2 + y^2 = 16 \\ x + y = 2 \end{cases}$

34. $\begin{cases} y = x^2 - 6x + 9 \\ 2x - y = 1 \end{cases}$

35. Solve the following system of equations: $\begin{cases} \dfrac{8}{x^2} - \dfrac{4}{y^2} = 3 \\ \dfrac{6}{x^2} + \dfrac{8}{y^2} = 5 \end{cases}$

HINT: Let $u = \dfrac{1}{x^2}$ and $v = \dfrac{1}{y^2}$ and rewrite the system in terms of u and v.

36. Solve the following system of equations: $\begin{cases} \dfrac{2}{\sqrt{x}} + \dfrac{5}{\sqrt{y}} = 6 \\ \dfrac{8}{\sqrt{x}} - \dfrac{3}{\sqrt{y}} = 1 \end{cases}$

HINT: Let $s = \dfrac{1}{\sqrt{x}}$ and $t = \dfrac{1}{\sqrt{y}}$ and rewrite the system in terms of s and t.

37. Solve the following system of equations: $\begin{cases} y = 5^x \\ y = 5^{2x} - 12 \end{cases}$

HINT: Note that $5^{2x} = (5^x)^2$. Estimate your answer to the nearest tenth and interpret your answer graphically.

38. Solve the following system of equations: $\begin{cases} y = 3^x \\ y = 9^x - 2 \end{cases}$

HINT: Note that $9^x = (3^2)^x = (3^x)^2$. Estimate your answer to the nearest tenth and interpret your answer graphically.

39. Find the dimensions of a rectangle whose area is 35 sq cm and whose perimeter is 27 cm.

40. If the diagonal of a rectangle is 10 in. and the perimeter of the rectangle is 28 in., find the dimensions of the rectangle.

41. If the hypotenuse of a right triangle is 25 cm and the area of the right triangle is 84 sq cm, find the lengths of the legs.

42. Find the dimensions of a piece of pipe (an open right circular cylinder) whose volume is 72π cu cm and surface area is 48π sq cm.

43. **(a)** If we want the system $\begin{cases} y = x^2 - 2x - 3 \\ y = K \end{cases}$ to have exactly one solution, what must the value of K be?
(b) If we want the system to have no real solutions, what can the value of K be?

44. Find the point(s) where the graph of $y = 9 - x^2$ intersects the graph of $2x + y = 6$.

45. Where do the circle $x^2 + y^2 = 16$ and the ellipse $4x^2 + y^2 = 64$ intersect?

46. What points are common to the graphs of $x^2 - 4y^2 = 12$ and $3y - x + 1 = 0$?

47. A farmer has 180 feet of fence with which to enclose a rectangular garden one side of which will be bounded by the side of a barn. (See the accompanying figure.) If no fence is used along the barn, is it possible to enclose an area of 4000 sq ft? If so, how?

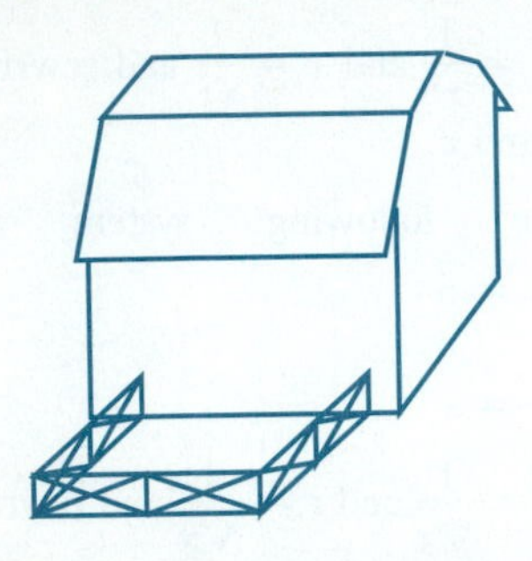

48. Is it possible for the farmer in Exercise 47 to enclose an area of 4200 sq ft with the 180 feet of fencing?

In Exercises 49–54, sketch the solution set to each system of inequalities.

49. $\begin{cases} x^2 + y^2 \le 4 \\ y \ge 1 - x^2 \end{cases}$

50. $\begin{cases} y \ge x^2 - 4 \\ x + y \le 2 \end{cases}$

51. $\begin{cases} 4x^2 + 9y^2 < 36 \\ x^2 + y^2 > 4 \end{cases}$

52. $\begin{cases} x \le \sqrt{9 - y^2} \\ y \le x \end{cases}$

53. $\begin{cases} 4x^2 - 25y^2 \le 100 \\ x^2 + y^2 < 100 \end{cases}$

54. $\begin{cases} y \ge x^2 - 4x + 1 \\ y \le -x^2 - 4x \end{cases}$

QUESTION FOR THOUGHT

55. Consider a quadratic system consisting of two equations in two variables. Suppose we use either the substitution or elimination methods to produce an equation involving only one variable. What is the maximum degree of such an equation? Why? What does this tell us about the maximum number of solutions to this system? Why?

GRAFFIX

In Exercises 56–59, use your graphics calculator or computer to find the solutions to each system of equations.

56. $\begin{cases} y = x^2 \\ y = 3x^2 - 2 \end{cases}$

57. $\begin{cases} y = 5x^2 - 4 \\ y = 4x^2 \end{cases}$

58. $\begin{cases} x^2 + y^2 = 16 \\ x - y = 3 \end{cases}$

59. $\begin{cases} y = x^2 - 3 \\ y = 3 - x^2 \end{cases}$

Chapter 10 SUMMARY

After completing this chapter you should be able to:

1. Graph a parabola, and identify its vertex, focus, axis of symmetry, and directrix. (Section 10.2)

For example:

Sketch the graph of the parabola

$$x^2 - 2x - 8y = 23$$

Solution:

We recognize that there is an x^2-term, so the parabola has a vertical axis of symmetry. Thus we want the equation to be in the form

$$(x - h)^2 = 4p(y - k)$$

To do this, we complete the square as follows:

$$x^2 - 2x - 8y = 23$$ *Isolate the terms containing powers of x on one side.*

$$x^2 - 2x = 8y + 23$$ *Complete the square for the left-hand side:*

$$x^2 - 2x + 1 = 8y + 23 + 1$$ *Write each side in factored form.*

$$(x - h)^2 = 4p(y - k)$$

$$(x - 1)^2 = 8(y + 3)$$ *Now we can identify h, k, and p.*

Thus $h = 1$, $k = -3$, and $p = 2$. We draw a new set of coordinate axes through $(1, -3)$. The vertex is $(1, -3)$. The axis of symmetry is the line $x = 1$. If we move up 2 units above the vertex along the parabola's axis, we locate the focus at $(1, -1)$. The directrix is the horizontal line located 2 units below the vertex and has equation $y = -5$. See Figure 10.74.

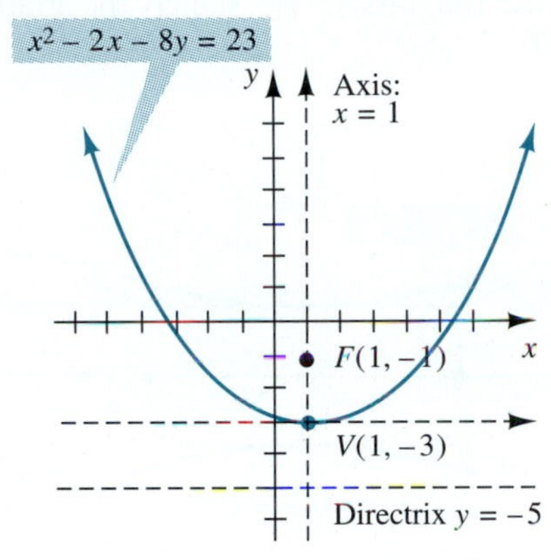

FIGURE 10.74

2. Write the equation of the parabola given certain conditions. (Section 10.2)
   For example:
   Find the equation of the parabola if its focus is (2, 0) and its vertex is (−4, 0).
   Solution:
   Given the parabola has vertex $(-4, 0)$ and focus $(2, 0)$, we note that the vertex and focus lie on the horizontal line $y = 0$, which is also its axis of symmetry. The form we use is $(y - k)^2 = 4p(x - h)$. Since its vertex is $(-4, 0)$, we have $h = -4$ and $k = 0$. The focus for this form is $(h + p, k)$. Since the focus is $(2, 0)$, we have

$$h + p = 2 \Rightarrow -4 + p = 2 \Rightarrow p = 6.$$

   (Alternatively we could have counted the number of units between the focus and vertex to get 6.) Hence the equation is

$$(y - 0)^2 = 4(6)[x - (-4)] \quad \text{or} \quad y^2 = 24(x + 4).$$

3. Graph an ellipse and identify its vertices, foci, and axes. (Section 10.3)
   For example:
   Sketch the graphs of the ellipses
   (a) $\dfrac{x^2}{12} + \dfrac{y^2}{4} = 1$ **(b)** $\dfrac{(x + 1)^2}{16} + \dfrac{(y - 2)^2}{25} = 1.$
   Solution:
   (a) The equation is in standard form with the x^2-term having the larger denominator, and therefore the major axis lies on the x-axis. Hence, we use the standard form of the ellipse $\dfrac{x^2}{a^2} + \dfrac{y^2}{b^2} = 1$. Comparing our equation with this general form we get

$$a^2 = 12 \Rightarrow a = 2\sqrt{3} \quad \text{and}$$

$$b^2 = 4 \Rightarrow b = 2$$

   We plot the vertices $(\pm 2\sqrt{3}, 0)$ and the endpoints of the minor axis $(0, \pm 2)$, and sketch the graph of the ellipse in Figure 10.75. Since $c^2 = a^2 - b^2 = 12 - 4 = 8$, we have $c = 2\sqrt{2}$. Hence the foci are $(\pm 2\sqrt{2}, 0)$.

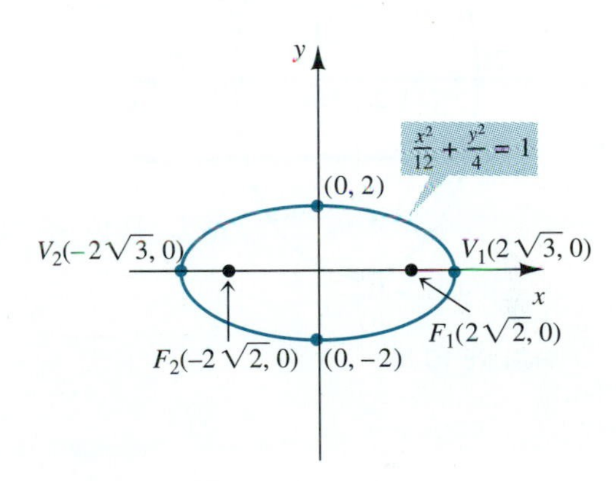

FIGURE 10.75

   (b) This ellipse is in a standard form with the "y^2-term" having the larger denominator; therefore, the major axis is a line parallel to the y-axis, and we use $\dfrac{(x - h)^2}{b^2} + \dfrac{(y - k)^2}{a^2} = 1.$ We compare our equation with this form and identify $h = -1$, $k = 2$, $a^2 = 25 \Rightarrow a = 5$, and $b^2 = 16 \Rightarrow b = 4$

We draw a new set of coordinate axes centered at (h, k), which is $(-1, 2)$, the center of the ellipse. The major axis lies on the vertical line $x = -1$. Since $a = 5$, move 5 units vertically above and below the center $(-1, 2)$, to find the vertices $(-1, 7)$, and $(-1, -3)$.

The minor axis lies on the horizontal line $y = 2$. Since $b = 4$, move 4 units left and right of the center $(-1, 2)$ to find the endpoints of the minor axis, $(-5, 2)$ and $(3, 2)$. Since

$$c^2 = a^2 - b^2 = 25 - 16 = 9 \Rightarrow c = 3,$$

we find the foci by moving 3 units vertically above and below the center $(-1, 2)$ to get $F_1(-1, 5)$ and $F_2(-1, -1)$. See Figure 10.76.

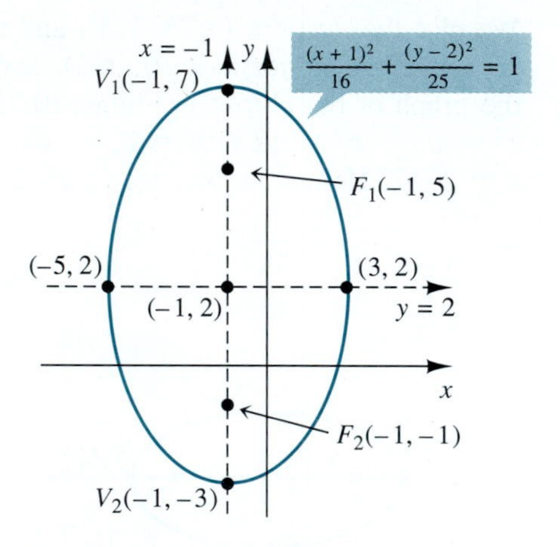

FIGURE 10.76

4. Graph a hyperbola and identify its vertices, foci, axes, and asymptotes. (Section 10.4)

For example:
Sketch the graph of the hyperbola $2y^2 - x^2 = 4$.
Solution:
First we put the equation into standard form, which requires that 1 be on one side. In order for us to have 1 on the right-hand side, we must divide by 4.

$2y^2 - x^2 = 4$ *Divide both sides by 4 to get*

$$\frac{y^2}{2} - \frac{x^2}{4} = 1$$

Comparing our equation with both standard forms of the hyperbola, we note that the x^2-term is being subtracted, and therefore a^2 is the denominator of y^2. We use the form $\frac{y^2}{a^2} - \frac{x^2}{b^2} = 1$, a hyperbola with foci on the y-axis. Hence

$$a^2 = 2 \Rightarrow a = \sqrt{2} \quad \text{and} \quad b^2 = 4 \Rightarrow b = 2$$

We plot the vertices $(0, \pm\sqrt{2})$ and the endpoints of the conjugate axis $(\pm 2, 0)$ and draw the rectangle and the diagonals. We sketch the graph in Figure 10.77.

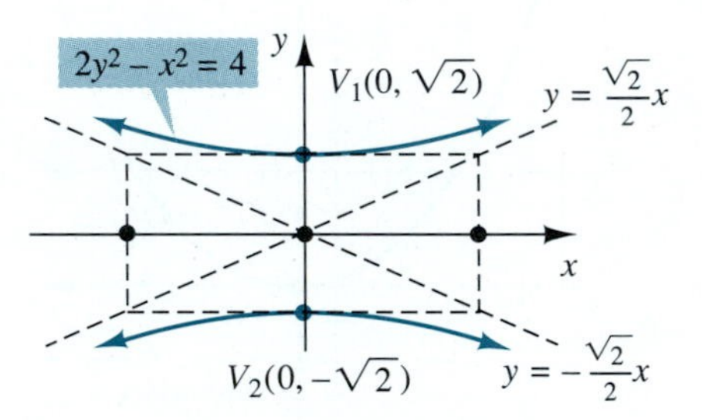

FIGURE 10.77

The asymptotes are $y = \pm\frac{a}{b}x \Rightarrow y = \pm\frac{\sqrt{2}}{2}x$. Since $c^2 = a^2 + b^2 = 2 + 4 = 6 \Rightarrow c = \sqrt{6}$, the foci are $(0, \pm\sqrt{6})$.

5. Identify conic sections (or their degenerate forms) by their equations in the general form $Ax^2 + Cy^2 + Dx + Ey + F = 0$. (Section 10.5.) The coefficients of the square terms of the equation can help you to identify the conic section determined by that equation.

For example:
Assuming the equation

$$3x^2 - 5y^2 - 6x + 8y - 9 = 0$$

can be graphed, to identify the figure that the equation will yield without putting the equation in standard form, we observe that the signs of the coefficients of the squared terms are opposite. Therefore, the equation will yield a hyperbola or its degenerate form.

6. Translate a set of coordinate axes and determine the coordinates of a given point relative to the original set of axes and the translated axes. (Section 10.6) The new coordinate axes are created by horizontal and vertical shifting. You can use the coordinate formulas for the translation of axes to determine the coordinates of a point relative to either set of axes.

7. Rotate a set of coordinate axes and, given a point relative to one set of coordinate axes, to write the point relative to the other set of coordinate axes. (Section 10.6)
In the general form of the equation

$$Ax^2 + Bxy + Cy^2 + Dx + Ey + F = 0$$

where $B \neq 0$, the xy term will produce a rotation of the axes. You can use the coordinate formulas for the rotation of axes to determine the coordinates of a point relative to either set of axes where ϕ is the angle of rotation.

For example:
A new set of coordinate axes $x'y'$ is formed by rotating the xy-axes 45° counterclockwise. Find (x', y') if $(x, y) = (1, 2)$.
Solution:
We are given the standard coordinates (x, y) of a point and need to find (x', y') when the angle of rotation is 45°. We use the coordinate formulas for rotation of axes

$$x' = x \cos \phi + y \sin \phi \quad \text{and}$$

$$y' = -x \sin \phi + y \cos \phi$$

Substitute $x = 1$, $y = 2$, and $\phi = 45°$ to get

$$x' = (1) \cos 45° + 2 \sin 45° \quad \text{and}$$

$$y' = (-1) \sin 45° + 2 \cos 45° \qquad \textit{Hence}$$

$$x' = \left(\frac{1}{\sqrt{2}}\right) + 2\left(\frac{1}{\sqrt{2}}\right) \quad \text{and}$$

$$y' = -\left(\frac{1}{\sqrt{2}}\right) + 2\left(\frac{1}{\sqrt{2}}\right) \qquad \textit{which gives}$$

$$x' = \frac{3}{\sqrt{2}} \quad \text{and} \quad y' = \frac{1}{\sqrt{2}}$$

The new coordinates (x', y') are $((\frac{3}{\sqrt{2}})', (\frac{1}{\sqrt{2}})')$.

8. Identify and sketch the graph of

$$Ax^2 + Bxy + Cy^2 + Dx + Ey + F = 0, B \neq 0$$

by using rotation of axes. (Section 10.6)
Evaluating the discriminant of the equation will help you to identify the conic. If you can then transform the original equation into an equation with new coefficients such that there is no $x'y'$-term, you can graph the transformed equation on the rotated axes using the techniques of graphing discussed in previous sections.
For example:
Identify and sketch the graph of $x^2 - \sqrt{3}xy + 2y^2 - 10 = 0$.
Solution:
Comparing this with the general equation, we have $A = 1$, $B = -\sqrt{3}$, and $C = 2$.
We examine the discriminant

$$B^2 - 4AC = (-\sqrt{3})^2 - 4(1)(2) = -5$$

Since the discriminant is negative, if the equation can be graphed it will be an ellipse or one of its degenerate forms. To find ϕ we have

$$\cot 2\phi = \frac{A - C}{B} = \frac{1}{\sqrt{3}} \Rightarrow \phi = 30°$$

We use the rotation formulas to transform the original equation into $\frac{(x')^2}{20} + \frac{(y')^2}{4} = 1$. The graph of the rotated ellipse appears in Figure 10.78.

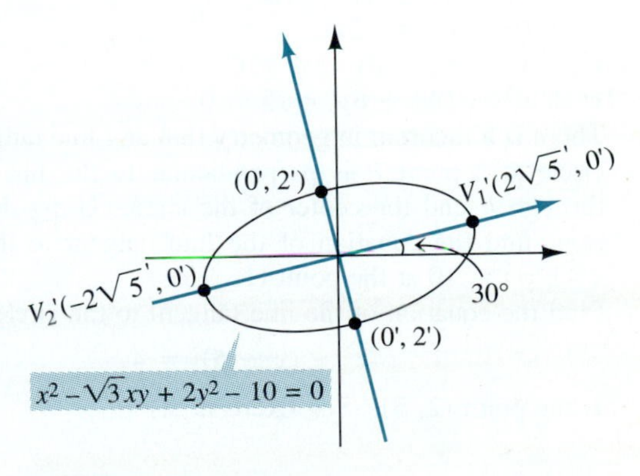

FIGURE 10.78

9. Solve a system of two nonlinear equations in two unknowns using the elimination and substitution methods. (Section 10.7)
For example:
Solve the following system of equations.

$$\begin{cases} x^2 + y^2 = 5 \\ y = 3x - 1 \end{cases}$$

The fact that the second equation is solved explicitly for y suggests that the substitution method will be the easiest.

$$x^2 + y^2 = 5$$

We use the second equation to substitute $y = 3x - 1$.

$$x^2 + (3x - 1)^2 = 5$$

$$x^2 + 9x^2 - 6x + 1 = 5$$

$$10x^2 - 6x - 4 = 0$$

$$5x^2 - 3x - 2 = 0$$

$$(5x + 2)(x - 1) = 0$$

$$x = -\frac{2}{5} \quad \text{or} \quad x = 1$$

We can solve for y by substituting into equation (2).
If $x = -\frac{2}{5}$, $y = 3x - 1 = 3\left(-\frac{2}{5}\right) - 1 = -\frac{11}{5}$.
If $x = 1$, $y = 3x - 1 = 3(1) - 1 = 2$. Therefore, we have two solutions: $\left(-\frac{2}{5}, -\frac{11}{5}\right)$ and $(1, 2)$. It is left to the student to verify that these ordered pairs satisfy both equations.

10. Sketch the solution set to a system of nonlinear inequalities. (Section 10.7)

Chapter 10 REVIEW EXERCISES

In Exercises 1–4, identify the center and radius of the circle described by the equation.

1. $(x - 3)^2 + (y + 4)^2 = 18$
2. $x^2 + (y - 1)^2 = 12$
3. $x^2 + y^2 - 8x - 8y - 4 = 0$
4. $x^2 + y^2 - 10x + 6y + 15 = 0$
5. There is a theorem in geometry that any line tangent to a circle at a point P is perpendicular to the line passing through P and the center of the circle. Using this theorem, find the equation of the line tangent to the circle $x^2 + y^2 = 10$ at the point $(1, -3)$.
6. Find the equation of the line tangent to the circle

$$x^2 + (y - 5)^2 = 4$$

at the point $(2, 5)$. (See Exercise 5.)

In Exercises 7–12, graph the parabola and identify the focus, axis and directrix.

7. $8x - y^2 = 0$
8. $y + 5x^2 = 0$
9. $(x - 4)^2 = 4y$
10. $(y - 2)^2 = -16x$
11. $4x = y^2 - 10y + 21$
12. $6y = x^2 - 4x - 2$

In Exercises 13–14, write the equation of the parabola using the given information.

13. The parabola has its vertex at the origin and directrix $x = 3$.
14. The parabola has its vertex at $(1, -1)$ and directrix $y = 2$.

15. A spotlight has a parabolic reflecting mirror with the light source placed at its focus. The mirror can be described by rotating the parabola $y = \frac{1}{8}x^2$ about its axis of symmetry, where $-3 \le x \le 3$ and x is measured in feet. How deep is the mirror, and where should the light source be placed in relation to the bottom (vertex) of the mirror?

16. An arch of a building is constructed in the shape of a parabola. (See the accompanying figure.) The height of the arch is 60 feet, and the width of the base parabola is 30 feet. How wide is the parabola 10 feet above the base?

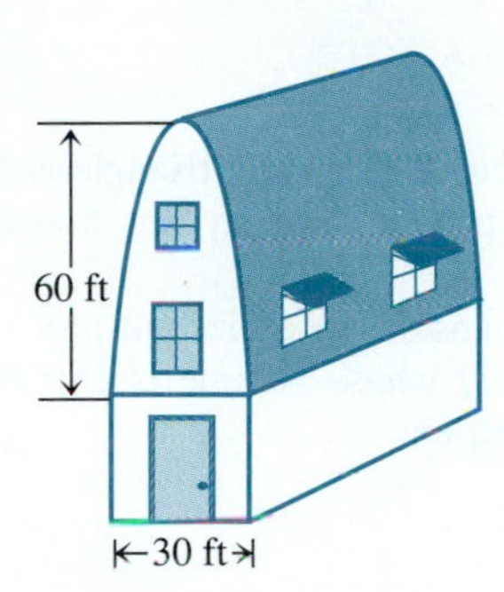

In Exercises 17–24, graph the ellipse and identify the vertices and foci.

17. $\dfrac{x^2}{9} + \dfrac{y^2}{36} = 1$

18. $\dfrac{x^2}{81} + \dfrac{y^2}{6} = 1$

19. $24x^2 + 2y^2 = 48$

20. $20y^2 + 10x^2 = 40$

21. $\dfrac{(x-2)^2}{6} + \dfrac{(y+4)^2}{4} = 1$

22. $\dfrac{x^2}{16} + \dfrac{(y+5)^2}{12} = 1$

23. $x^2 + 4y^2 - 6x + 8y = 3$

24. $6x^2 + 5y^2 + 12x - 20y - 4 = 0$

In Exercises 25–26, write the equation of the ellipse using the given information.

25. The ellipse is centered at the origin; the axes of the ellipse have lengths 6 and 8; the major axis is vertical.

26. The ellipse has foci (2, 5) and (8, 5) and vertices (0, 5) and (10, 5).

In Exercises 27–34 graph the hyperbola and identify the foci, transverse axis, vertices, and asymptotes.

27. $\dfrac{x^2}{49} - \dfrac{y^2}{4} = 1$

28. $\dfrac{y^2}{81} - \dfrac{x^2}{8} = 1$

29. $x^2 - 8y^2 = 24$

30. $10y^2 - 5x^2 = 40$

31. $\dfrac{(y-2)^2}{2} - \dfrac{(x-1)^2}{8} = 1$

32. $\dfrac{(x+3)^2}{12} - \dfrac{y^2}{24} = 1$

33. $x^2 - 4y^2 - 6x - 8y = 11$

34. $5y^2 - 6x^2 - 20y - 12x - 16 = 0$

In Exercises 35–38, identify the curve if it could be graphed.

35. $x^2 + 6y^2 - 8x - 2y - 7 = 0$

36. $2y^2 - 2x + 5y = 0$

37. $-x^2 + y^2 + 18x - 6y + 16 = 0$

38. $4x^2 + 4y^2 - 12x - 8y - 5 = 0$

In Exercises 39–52, identify the curve if it could be graphed and then graph it.

39. $\dfrac{x^2}{36} - \dfrac{y^2}{4} = 1$

40. $\dfrac{x^2}{49} + \dfrac{y^2}{4} = 1$

41. $\dfrac{x}{4} + \dfrac{y}{9} = 1$

42. $\dfrac{(x-1)^2}{4} + \dfrac{(y+1)^2}{8} = 1$

43. $\dfrac{(y-2)^2}{2} + \dfrac{(x-1)^2}{2} = 1$

44. $\dfrac{(y-2)^2}{2} - \dfrac{(x-1)^2}{2} = 1$

45. $x^2 + 64y = 0$

46. $x^2 + 4y^2 - 2x + 24y + 21 = 0$

47. $x^2 - 8x + 16 = 0$

48. $3y^2 - 3y - 18 = 0$

49. $x^2 + 2y^2 - 2x - 8y - 7 = 0$

50. $x^2 + 4x - 5y + 24 = 0$

51. $x^2 + 25y^2 - 6x - 50y + 34 = 0$

52. $3x^2 - y^2 - 6x - 6y - 6 = 0$

In Exercises 53–54, a set of axes is translated so that the new origin is (2, 5) on the standard set of coordinate axes. Given the coordinates of a point relative to one set of axes, find the coordinates of the point relative to the other set of axes. ((x, y) stands for the standard coordinates; (x', y') are the coordinates relative to the translated set of axes $x'y'$.)

53. $(x', y') = (2', 5')$; find (x, y).
54. $(x, y) = (3, -4)$; find (x', y').

In Exercises 55–56, a standard set of axes is rotated an angle ϕ counterclockwise, producing a new set of axes, $x'y'$. Given ϕ and the point on the standard set of axes, $P(x, y)$, find the coordinates of the point relative to the rotated set of axes, $P(x', y')$.

55. $(3, 5)$; $\phi = 30°$
56. $(-2, 3)$; $\phi = \frac{\pi}{4}$
57. A standard set of axes is rotated 30° counterclockwise, producing a new set of axes $x'y'$. Express the equation $x^2 - xy - 1 = 0$ in terms of the rotated set of axes, $x'y'$.
58. A standard set of axes is rotated 45° counterclockwise, producing a new set of axes, $x'y'$. Express the equation $3(y')^2 = -16x'$ in terms of the standard set of axes, xy.

In Exercises 59–62, sketch a graph of the equation.

59. $xy = 10$
60. $2x^2 - 2xy + 2y^2 = 9$
61. $11x^2 + 10\sqrt{3}xy + y^2 = 16$
62. $x^2 + xy + y^2 + \sqrt{2}x - \sqrt{2}y = 4$

In Exercises 63–68, solve the system of equations.

63. $\begin{cases} x^2 - y^2 = 8 \\ y = x - 2 \end{cases}$

64. $\begin{cases} y = 1 - x^2 \\ x^2 = y^2 - 5 \end{cases}$

65. $\begin{cases} x^2 + y^2 = 4 \\ \frac{y^2}{4} - \frac{x^2}{4} = 1 \end{cases}$

66. $\begin{cases} 2x^2 + 3y^2 = 8 \\ 5x^2 + 4y^2 = 7 \end{cases}$

67. $\begin{cases} x^2 + y^2 = 13 \\ y = x^2 - 1 \end{cases}$

68. $\begin{cases} x^2 + 2y^2 = 6 \\ x^2 - y^2 = 3 \end{cases}$

69. If the hypotenuse of a right triangle is 4 cm and the area of the right triangle is 4 sq cm, find the lengths of the legs.
70. Find the dimensions of a piece of pipe (an open right circular cylinder) whose volume is 18π cu cm and surface area is 21π sq cm.

In Exercises 71–72 sketch the solution set to each system of inequalities.

71. $\begin{cases} 25x^2 + 4y^2 \leq 100 \\ x^2 + y^2 > 9 \end{cases}$

72. $\begin{cases} y \geq x^2 + 1 \\ y \leq -x^2 + 4 \end{cases}$

Chapter 10 PRACTICE TEST

In Exercises 1–4, sketch the graph of the equation and label the important aspects of the graph.

1. $4x^2 + 4y = 0$
2. $4x^2 + 4y^2 = 36$
3. $\dfrac{x^2}{25} + \dfrac{y^2}{49} = 1$
4. $\dfrac{x^2}{25} - \dfrac{y^2}{49} = 1$

In Exercises 5–10, sketch the graph of the following equations and label the important aspects of the graph.

5. $(x - 3)^2 + (y + 5)^2 = 5$
6. $2(x - 3)^2 + 8y + 24 = 0$
7. $\dfrac{(x - 3)^2}{4} + \dfrac{(y - 2)^2}{12} = 1$
8. $\dfrac{(y + 2)^2}{25} - \dfrac{x^2}{18} = 1$
9. $4x^2 + 5y^2 - 24x + 30y + 61 = 0$
10. $3x^2 - 4y^2 - 6x = 45$
11. Find the equation of the parabola that has its vertex at the origin and directrix $x = -2$.
12. Find the equation of the ellipse with foci $(3, 1)$ and $(-1, 1)$ and vertices $(5, 1)$ and $(-3, 1)$.
13. A standard set of axes is rotated an angle 45° counterclockwise, producing a new set of axes, $x'y'$.
    (a) Given the point $(2, -1)$ on the standard set of coordinate axes, find (x', y'), the coordinates of the point relative to the rotated set of axes.
    (b) Express the equation $2y^2 - \sqrt{2}x = 1$ in terms of the rotated set of axes, $x'y'$.
14. Sketch a graph of the equation $x^2 + \sqrt{3}xy = 6$.
15. Solve the system

$$\begin{cases} x^2 + 2y^2 = 3 \\ 2x^2 - y^2 = 1 \end{cases}$$

16. Sketch the solution set of the system of inequalities.

$$\begin{cases} x^2 + y^2 \leq 4 \\ \quad\;\; y < x^2 - 2 \end{cases}$$

CHAPTER 11

Sequences, Series, and Related Topics

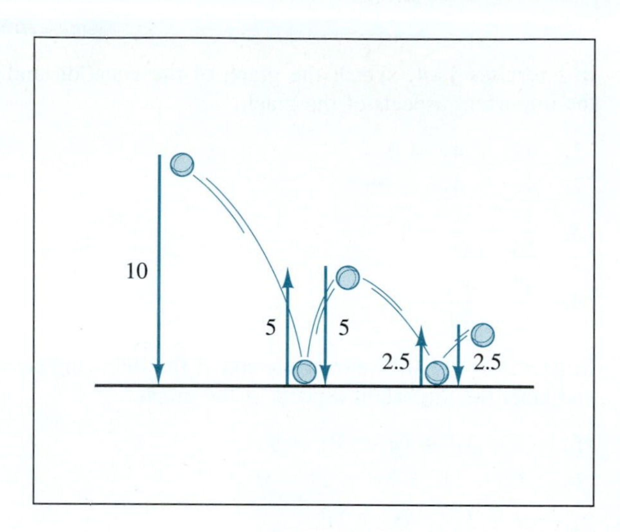

Recognizing and identifying patterns is one of the fundamental ideas that appear throughout mathematics and serves as a fruitful starting point for analyzing a wide variety of problems. In the first four sections we discuss sequences and series, two particularly important types of patterns in mathematics, that play a fundamental role in calculus and ultimately even in the way our scientific calculators compute values of certain functions. In Section 11.5 we introduce the method of mathematical induction, a powerful method of proof that can be used to show that certain patterns hold for all natural numbers. In Section 11.6 we discuss some basic counting principles, and in Section 11.7 we use a number of the ideas presented to prove the binomial theorem.

11.1 Sequences

The word sequence, used in everyday conversation, usually refers to a list of things occurring in a specific order. For example, if a person wants to build a house, he or she would probably have to arrange for the following sequence of events, which we have labeled with subscripts to indicate that they follow a particular order.

E_1: Get plans drawn up and obtain all required building permits.
E_2: Lay the foundation.
E_3: Erect frame and roof.
E_4: Install plumbing and wiring.
E_5: Put up interior walls.
E_6: Finish all interior surfaces.

Note that in order to construct the house, one is not only concerned with the events on the list, but also the order in which they occur: which is first, which is second, and so on.

In mathematics, a sequence is viewed in much the same way: an "ordered" set of things. For example, we could have the sequence of numbers 1, 3, 5, 7, where 1 is the first, 3 is the second, 5 is the third, and 7 is the fourth number. The following are examples of sequences:

1, 4, 9, 16, 25, 36, 49, 64, 81, 100
2, 4, 6, 8, 10
2, 4, 6, 8, 10, . . .
5, 25, 125, 625, . . .

The first two sequences are examples of **finite sequences** because they terminate. The third and fourth sequences are examples of **infinite sequences** because, as the three-dot notation indicates, they continue on without end in the same established pattern. For the most part, we focus our attention on infinite sequences.

Although we may describe a sequence intuitively as a list of things in a particular order, we still need to give a precise mathematical definition of a sequence. Since we are concerned with order, we can associate the first term in a sequence with the number 1, the second with the number 2, the third with the number 3, and so on. This approach lends itself quite naturally to using function notation to define a sequence:

DEFINITION A **sequence** is a function whose domain is a set of consecutive positive integers, usually beginning with 1.

The sequence 2, 4, 8, 16, 32, . . . can be viewed as a function, f, with domain {1, 2, 3, 4, 5, . . .} and range {2, 4, 8, 16, 32, . . .} where

$f(1) = 2$ *2 is the first number in the sequence.*
$f(2) = 4$ *4 is the second number in the sequence.*
$f(3) = 8$ *8 is the third number in the sequence.*
$f(4) = 16$ *16 is the fourth number in the sequence.*
$f(5) = 32$ *32 is the fifth number in the sequence.*
$\vdots$ *And so on*

Note that the subscripts are the domain elements.

In mathematics, rather than indicate the terms of the sequence as $f(1)$, $f(2)$, $f(3)$, and $f(4)$, we usually use subscript notation and write f_1, f_2, f_3, and f_4, respectively. Thus $f_1 = 2$, $f_2 = 4$, $f_3 = 8$, $f_4 = 16$, and $f_5 = 32$, etc. Thus the entire sequence can be denoted $\{f_i\}$.

In a sequence $\{a_i\}$, i is called the **index**. Unless otherwise indicated, we shall start our indices at 1. Thus we write $\{a_i\} = 1, \frac{1}{2}, \frac{1}{3}, \frac{1}{4}, \frac{1}{5}$ to mean $a_1 = 1$, $a_2 = \frac{1}{2}$, $a_3 = \frac{1}{3}$, $a_4 = \frac{1}{4}$, $a_5 = \frac{1}{5}$. The functional values a_1, a_2, a_3, a_4, a_5 are called the **terms** of the sequence.

Often, rather than writing out the terms of a sequence, we simply define a sequence by writing a formula for the general term of the sequence, $\{a_n\}$, in terms of n. For example, a general term for the sequence $\{a_n\} = 3, 6, 9, 12, 15, \ldots$ can be written as $a_n = 3n$, where it is understood that n can be any positive integer. *Unless otherwise indicated, we begin with $n = 1$ and find each term of the sequence by substituting successive positive integers for n.* Thus, for $a_n = 3n$,

$$a_1 = 3(1) = 3$$
$$a_2 = 3(2) = 6$$
$$a_3 = 3(3) = 9$$
$$a_4 = 3(4) = 12$$
$$\vdots$$

EXAMPLE 1 Write the first four terms and the seventh term of the sequence whose nth term is given by:

(a) $a_n = 3^n$ **(b)** $x_n = 2n - 3$ **(c)** $b_n = \dfrac{n-1}{n}$ **(d)** $y_n = \dfrac{(-1)^n}{n+2}$

Solution

(a) Since $a_n = 3^n$, the first term of the sequence, a_1, is found by substituting 1 for n. We get

$a_1 = 3^1 = 3$ *The second term, a_2, is found by substituting 2 for n in a_n. Hence,*

$a_2 = 3^2 = 9$ *The third term is*

$a_3 = 3^3 = 27$ *And the fourth term is*

$a_4 = 3^4 = 81$

So, the first four terms of the sequence are: $\boxed{3, 9, 27, \text{ and } 81}$. The seventh term is: $a_7 = 3^7 = \boxed{2187}$.

(b) $x_n = 2n - 3$ — *We substitute 1 for n in x_n to get the first term;*

$x_1 = 2(1) - 3 = -1$ — *For x_2, x_3, and x_4, we similarly substitute 2, 3, and 4, respectively, for n in x_n to get*

$$x_2 = 2(2) - 3 = 1$$

$$x_3 = 2(3) - 3 = 3$$

$$x_4 = 2(4) - 3 = 5$$

Hence, the first four terms of the sequence are: $\boxed{-1, 1, 3, 5}$. The seventh term is $x_7 = 2(7) - 3 = \boxed{11}$.

(c) $b_n = \dfrac{n-1}{n}$

$$b_1 = \frac{1-1}{1} = 0 \qquad b_2 = \frac{2-1}{2} = \frac{1}{2}$$

$$b_3 = \frac{3-1}{3} = \frac{2}{3} \qquad b_4 = \frac{4-1}{4} = \frac{3}{4}$$

Hence the first four terms are $\boxed{0, \frac{1}{2}, \frac{2}{3}, \frac{3}{4}}$, and the seventh term is given by $= \dfrac{7-1}{7} = \boxed{\dfrac{6}{7}}$.

(d) Since $y_n = \dfrac{(-1)^n}{n+2}$,

$$y_1 = \frac{(-1)^1}{1+2} = \frac{-1}{3} \qquad y_2 = \frac{(-1)^2}{2+2} = \frac{1}{4}$$

$$y_3 = \frac{(-1)^3}{3+2} = \frac{-1}{5} \qquad y_4 = \frac{(-1)^4}{4+2} = \frac{1}{6}$$

Thus the first four terms of the sequence are: $\boxed{-\frac{1}{3}, \frac{1}{4}, -\frac{1}{5}, \frac{1}{6}}$. The seventh term is: $y_7 = \dfrac{(-1)^7}{7+2} = \boxed{-\dfrac{1}{9}}$. ■

EXAMPLE 2 Suppose that a ball has the property that it always rebounds to a height equal to $\frac{1}{2}$ the distance from which it falls, and that such a ball is dropped from a height of 10 feet. Let d_n be the total vertical distance the ball has traveled when it hits the ground for the nth time. Find the first five terms of this sequence and graph them.

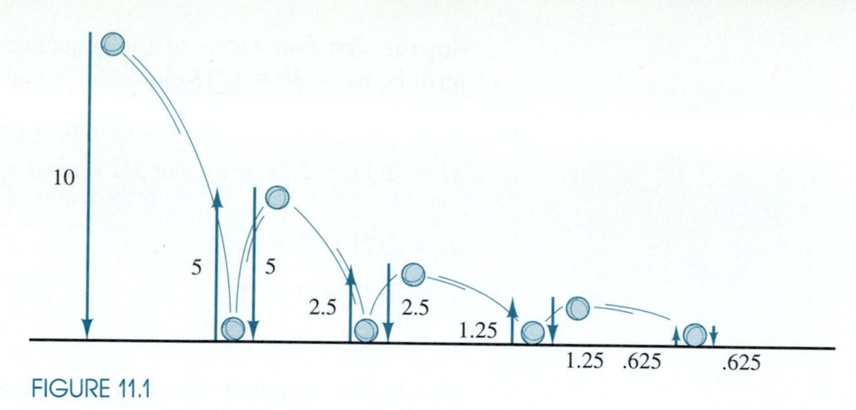

FIGURE 11.1

Solution Figure 11.1 illustrates the vertical distance traveled by the bouncing ball.

Using this figure and the fact that d_n is the total vertical distance the ball has traveled when it hits the ground for the nth time we have

$$d_1 = 10$$
$$d_2 = 10 + 2(5) = 20$$
$$d_3 = 10 + 2(5) + 2(2.5) = 25$$
$$d_4 = 10 + 2(5) + 2(2.5) + 2(1.25) = 27.5$$
$$d_5 = 10 + 2(5) + 2(2.5) + 2(1.25) + 2(.625) = 28.75$$

as the first five terms of the sequence.

As defined above, a sequence is a function whose domain is a subset of the natural numbers. Thus we are being asked to graph $d_n = f(n)$ for $n = 1, 2, 3, 4,$ and 5. As usual we use the horizotal axis for the inputs (n) and the vertical axis for the outputs ($d_n = f(n)$). The graph appears in Figure 11.2.

Do you think that the total distance the ball has traveled ever reaches 30 feet? This question is addressed in Example 7 of Section 11.4

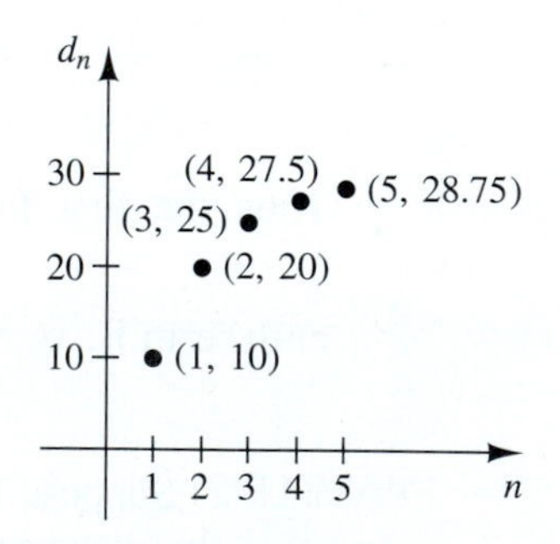

FIGURE 11.2 The graph of the sequence d_n for $n = 1, 2, 3, 4,$ and 5 ■

Examining a short sequence of numbers may allow us to identify a pattern and guess at the general term, but we must be careful: Two different sequences may start with the same terms. For example, if we examine the sequence $-1, 0, 1, \ldots$, we might guess that the general term of this sequence is $a_n = n - 2$. Then

$$a_1 = 1 - 2 = -1, \quad a_2 = 2 - 2 = 0, \quad a_3 = 3 - 2 = 1, \quad a_4 = 4 - 2 = 2,$$

so that the first three terms of $\{a_n\}$ agree with the first three terms in $-1, 0, 1, \ldots$. However, if we consider the sequence with general term $b_n = (n - 2)^3$, then

Can you find any other possible formulas for the general term?

$$b_1 = (1 - 2)^3 = -1 \quad b_2 = (2 - 2)^3 = 0 \quad b_3 = (3 - 2)^3 = 1$$

$$b_4 = (4 - 2)^3 = 8$$

Even though the first three terms of $\{a_n\}$ agree with the first three terms of $\{b_n\}$, they are not the same sequence. (Look at the fourth term of each sequence.)

Thus when we examine the pattern in the terms of a sequence to look for a formula for the general term we are only making a guess based upon the available data.

EXAMPLE 3 Find a possible general term for the following sequences:

(a) $6, 12, 18, 24, 30, \ldots$ **(b)** $10; 100; 1000; 10{,}000; 100{,}000; \ldots$

(c) $\frac{1}{3}, \frac{1}{5}, \frac{1}{7}, \frac{1}{9}, \ldots$

Solution We find the general term by examining the given terms and looking for a pattern.

(a) We can see that the first three terms are multiples of 6, so we can write $a_1 = 6 = 6 \cdot 1$, $a_2 = 12 = 6 \cdot 2$, $a_3 = 18 = 6 \cdot 3$, etc.; therefore, a general term is possibly $\boxed{a_n = 6n}$.

(b) We can see that the first three terms are powers of 10, so we can write $a_1 = 10 = 10^1$, $a_2 = 100 = 10^2$, $a_3 = 1000 = 10^3$, etc.; therefore, a general term is possibly $\boxed{a_n = 10^n}$.

(c) Given $a_1 = \frac{1}{3}$, $a_2 = \frac{1}{5}$, $a_3 = \frac{1}{7}$, $a_4 = \frac{1}{9}$, etc., we can see that the denominators are the odd numbers 3, 5, 7, 9. Thus a general term is possibly

$$\boxed{a_n = \frac{1}{2n + 1}}.$$

Describing certain real-life situations can lead us quite naturally to sequences.

EXAMPLE 4 Elizabeth deposits \$2000 in a bank that pays 5% simple annual interest. Write a sequence showing the amount of money Elizabeth has in the bank at the end of each year for 4 years.

Solution In Section 5.5, we developed the formula $A = P(1 + r)^t$ for the amount of money, A, in an account if P dollars is invested at an annual interest rate of r. (Remember that in the formula, r is the interest rate written as a decimal.) In this example, the interest rate is 5%, so $r = 0.05$, and $1 + r = 1.05$.

If we let A_k = the amount in Elizabeth's account at the end of year k, then using this formula gives

$$A_1 = 2000(1.05) = \$2100$$
$$A_2 = 2000(1.05)^2 = \$2205$$
$$A_3 = 2000(1.05)^3 = \$2315.25$$
$$A_4 = 2000(1.05)^4 = \$2431.01$$

Therefore, the sequence of amounts in the bank at the end of each year for 4 years is

$$\boxed{\$2100, \$2205, \$2315.25, \text{ and } \$2431.01}$$

■

All the sequences we have discussed thus far have had an explicit formula for the nth term—for example, $a_n = 3n + 2$. The next example illustrates a different way to specify the terms of a sequence. A **recursively defined sequence** is a sequence in which later terms are defined by referring to earlier terms.

EXAMPLE 5 Find the fourth term of each of the following recursively defined sequences:

(a) $a_1 = 5$ and $a_{k+1} = 3 + 2a_k$ **(b)** $a_1 = 4$, $a_2 = 3$ and $a_{k+2} = a_{k+1} - 2a_k$

Solution

(a) The given recursive relationship, $a_{k+1} = 3 + 2a_k$, means that each term after the first is 3 more than twice the previous term. We are given that $a_1 = 5$, so to find a_2 we substitute a_1 into the recursive formula.

a_{k+1} is the term following a_k and a_{k+2} is the term following a_{k+1}.

$$a_{k+1} = 3 + 2a_k \quad \textit{for } k = 1 \textit{ we have}$$
$$a_2 = 3 + 2a_1 = 3 + 2(5) = 13 \quad \textit{so } a_2 = 13$$
$$a_3 = 3 + 2a_2 = 3 + 2(13) = 29 \quad \textit{so } a_3 = 29$$
$$a_4 = 3 + 2a_3 = 3 + 2(29) = 61$$

Therefore, $\boxed{a_4 = 61}$.

(b) Here we are given the first two terms and that successive terms are obtained by taking the difference between the previous term and twice the term before that. Thus

$$a_1 = 4$$
$$a_2 = 3 \quad \textit{Given that } a_{k+2} = a_{k+1} - 2a_k\textit{, we have}$$
$$a_3 = a_2 - 2a_1 = 3 - 2(4) = -5 \quad \textit{So } a_3 = -5$$
$$a_4 = a_3 - 2a_2 = -5 - 2(3) = -11$$

Therefore, $\boxed{a_4 = -11}$.

■

Note that if we want to find the fifteenth term of a recursively defined sequence, we usually have to find the first fourteen terms of the sequence, whereas if we want to find the fifteenth term of a sequence for which we have a formula, we can just substitute $n = 15$. In the next few sections, we discuss some special recursively defined sequences.

Some very important sequences in mathematics involve terms that are particular products, indicated with the following special notation.

DEFINITION For n a positive integer, $\boldsymbol{n!}$ (read "n factorial") is defined as

$$n! = n(n-1)(n-2)\cdots 3\cdot 2\cdot 1$$

In words, $n!$ is the *product* of all the positive integers less than or equal to n. We also define $0! = 1$.

EXAMPLE 6 Compute each of the following:

(a) $5!$ **(b)** $\frac{8!}{4!}$ **(c)** $\frac{7! - 3!}{(7-3)!}$

Solution

(a) Using the definition of the factorial notation, we get

$$5! = 5\cdot 4\cdot 3\cdot 2\cdot 1 = \boxed{120}$$

(b) Again using the definition of $n!$, we get

$$\frac{8!}{4!} = \frac{8\cdot 7\cdot 6\cdot 5\cdot 4\cdot 3\cdot 2\cdot 1}{4\cdot 3\cdot 2\cdot 1}$$ *Reduce.*

$$= 8\cdot 7\cdot 6\cdot 5 = \boxed{1680}$$

(c) We must compute the factorials carefully.

$$\frac{7! - 3!}{(7-3)!} = \frac{7! - 3!}{4!}$$ *We could compute each factorial separately and then evaluate the fraction; however, we offer a more algebraic approach.*

$$= \frac{7\cdot 6\cdot 5\cdot 4\cdot 3\cdot 2\cdot 1 - 3\cdot 2\cdot 1}{4\cdot 3\cdot 2\cdot 1}$$ *Factor and reduce.*

$$= \frac{3\cdot 2\cdot 1(7\cdot 6\cdot 5\cdot 4 - 1)}{4\cdot 3\cdot 2\cdot 1} = \boxed{\frac{839}{4}}$$ ■

EXAMPLE 7 Find the fourth term of the sequence $a_n = \frac{2n!}{(2n)!}$.

Solution We must be careful to clearly distinguish $2n!$ from $(2n)!$ For $n = 4$ we have

$$a_4 = \frac{2 \cdot 4!}{(2 \cdot 4)!} = \frac{2 \cdot 4!}{8!} = \frac{2 \cdot 4 \cdot 3 \cdot 2 \cdot 1}{8 \cdot 7 \cdot 6 \cdot 5 \cdot 4 \cdot 3 \cdot 2 \cdot 1} = \frac{1}{840}$$ ■

We will have occasion to apply factorial notation in Sections 11.6 and 11.7.

EXERCISES 11.1

In Exercises 1–30, write the first four terms and the eighth term of the sequence whose general term is given.

1. $a_n = 5n - 3$
2. $a_n = 4n + 2$
3. $b_n = 3n + 2$
4. $b_n = 6n + 1$
5. $c_k = 3^k$
6. $c_k = 2^{k-1}$
7. $x_j = \dfrac{j+1}{j}$
8. $x_j = \dfrac{j-1}{j+1}$
9. $x_n = \dfrac{(-1)^{n+1}}{n+2}$
10. $y_n = \dfrac{(-1)^n}{2n}$
11. $a_n = \dfrac{(-1)^n n}{n+1}$
12. $a_n = \dfrac{(-1)^n n^2}{3n-2}$
13. $a_n = 2^n + |n - 3|$
14. $a_n = 3^n + |5 - n|$
15. $b_n = 1 + (-.1)^n$
16. $b_n = 2 - (.2)^n$
17. $x_n = \dfrac{2^n}{n^2}$
18. $y_n = \dfrac{n^3}{3^n}$
19. $a_n = \sin \dfrac{n\pi}{2}$
20. $b_n = \cos n\pi$
21. $c_n = \left(1 + \dfrac{1}{n}\right)^n$
22. $d_n = \dfrac{1}{n+1} - \dfrac{1}{n+2}$
23. $a_n = \sqrt{n^2 + 1}$
24. $b_n = 5$
25. $a_n = \dfrac{1}{n!}$
26. $b_n = \dfrac{(-1)^n}{(n+1)!}$
27. $c_n = \dfrac{n!}{n^n}$
28. $d_n = \dfrac{2^n}{(n-1)!}$
29. $t_n = \dfrac{(3n)!}{3^n n!}$
30. $s_n = \dfrac{(2n-1)!}{(2n+1)!}$

In Exercises 31–42, find the first six terms of the given recursive sequence.

31. $a_1 = 3;\ a_{n+1} = 2a_n - 1$
32. $a_1 = 1;\ a_n = \dfrac{1}{a_{n-1} + 1}$ for $n > 1$
33. $a_1 = -1;\ a_n = (1 - a_{n-1})^2$ for $n > 1$
34. $a_1 = 2;\ a_n = \sqrt{4 + (a_{n-1})^2}$ for $n > 1$
35. $a_1 = 1;\ a_{n+1} = 2^{a_n}$
36. $a_1 = 3;\ a_{n+1} = na_n$
37. $a_1 = 2;\ a_n = \dfrac{n}{a_{n-1}}$ for $n > 1$
38. $a_1 = 2;\ a_n = 3a_{n-1} - n$ for $n > 1$
39. $a_1 = 0;\ a_2 = 1;\ a_{n+2} = \dfrac{1}{2}(a_{n+1} + a_n)$
40. $a_1 = 1;\ a_2 = 2;\ a_n = \dfrac{a_{n-2}}{a_{n-1}}$ for $n \geq 3$
41. $a_1 = -1;\ a_2 = 1;\ a_n = (a_{n+2})(a_{n-1})$ for $n \geq 3$
42. $a_1 = 1;\ a_2 = 1;\ a_{n+2} = na_{n+1} + (n+1)a_n$

In Exercises 43–46, graph the first five terms of the sequence.

43. $y_n = 2^{n-1}$
44. $y_n = n^2 + 1$
45. $y_n = (-1)^n$
46. $y_n = \dfrac{(-1)^n}{n}$

47. A promotional firm begins a telemarketing campaign by making 100 phone calls on the first day. Each day the firm will make 20 more calls than were made the day before. Let c_n be the total number of calls made by the end of day n. Graph the first five terms of this sequence.

48. In a certain bacteria culture, the number of bacteria present triples every hour. Suppose that a culture begins with 10 bacteria. Let b_n be the number of bacteria present at the end of hour n. Graph the first 4 terms of this sequence.

In Exercises 49–58, find a possible formula for the general term of the given sequence.

49. 5, 10, 15, 20, . . .
50. 3, 8, 13, 18, . . .
51. 1, 3, 5, 7, 9, . . .
52. 5, 7, 9, 11, 13, . . .
53. $\dfrac{1}{2}, \dfrac{3}{4}, \dfrac{5}{6}, \dfrac{7}{8}, \ldots$
54. $\dfrac{2}{3}, \dfrac{4}{9}, \dfrac{8}{27}, \dfrac{16}{81}, \ldots$
55. $\dfrac{1}{2}, -\dfrac{1}{3}, \dfrac{1}{4}, -\dfrac{1}{5}, \ldots$
56. $-\dfrac{1}{2}, \dfrac{2}{3}, -\dfrac{3}{4}, \dfrac{4}{5}, \ldots$
57. −0.4, 0.04, −0.004, 0.0004, . . .
58. 0.07, −0.0007, 0.000007, −0.00000007, . . .

59. Ann gets a raise of \$650 at the end of each year she works for her company. If her starting salary is \$24,500, write a sequence to show her income at the end of each year for the first 3 years.

60. Sean gets a raise of \$600 every 6 months for the first 2 years he works for his firm. If his starting salary is \$22,750, write a sequence to show his income at the end of each 6-month period for the first 2 years.

61. The population of Centerville increases at a rate of 2% per year. If the population is currently 60,000 people, write a sequence to show the population at the end of each year for 3 years.

62. The population of Rivertown is decreasing at a rate of 1.5% per year. If the population is currently 130,000 people, write a sequence to show the population at the end of each year for 4 years.

63. Juan deposits \$1800 in a bank which is paying an effective annual yield of 3.5%. Write a sequence to show the amount of money Juan has in the bank at the end of each year for 5 years.

64. Elena deposits \$2750 in a bank that is paying an effective annual yield of 4.2%. Write a sequence to show how much Elena has in the bank at the end of each year for 3 years.

65. A country is depleting its oil reserves at the rate of 1.35% per year. If it starts out with reserves of 2.3 billion barrels of oil in 1992, write a sequence to show the amount of oil reserves left at the end of each year for 5 years.

66. A city finds that its electric power consumption on peak summer days is increasing at the rate of 4.2% per year. If the peak electric power consumption in the summer of 1993 is 8.3 megawatts, write a sequence to show the peak electric power consumption during the next six summers.

67. A ball rebounds to one-half the distance from which it falls. If it is dropped from a height of 60 feet, write a sequence to show how high it bounces for the first five bounces.

68. A ball rebounds to two-thirds the distance from which it falls. If it is dropped from a height of 60 feet, write a sequence to show how high it bounces for the first five bounces.

69. George borrows \$3000 and promises to pay back \$200 per month plus 1% of the remaining balance each month. Write a sequence showing how much George must pay each month for the first 4 months.

70. Lena borrows \$8000 and promises to pay back \$300 per month plus 2% of the remaining balance each month. Write a sequence showing how much Lena owes each month for the first 6 months.

QUESTIONS FOR THOUGHT

71. A very famous sequence is the Fibonacci sequence, F_n, where F_n is defined recursively as follows:

$$F_1 = 1; \quad F_2 = 1; \quad F_n = F_{n-2} + F_{n-1} \text{ for } n > 2$$

(a) Write the first eight terms of the Fibonacci sequence.

(b) Compute $F_1 + F_2 + F_3 + \cdots + F_n$ and $F_{n+2} - 1$ for $n = 5$, 7, and 10. What do you notice?

(c) Compute $(F_1)^2 + (F_2)^2 + (F_3)^2 + \cdots + (F_n)^2$ and $F_n \cdot F_{n+1}$ for $n = 4$, 6, and 9. What do you notice?

72. Suppose $\{a_n\}$ is a recursively defined sequence. Would you expect to be able to compute a_{16} directly? Explain.

73. Simplify each of the following factorial expressions.

(a) $\dfrac{(n+1)!}{n!}$ **(b)** $\dfrac{(2n+3)!}{(2n)!}$ **(c)** $\dfrac{(2n-1)!}{(2n+1)!}$

11.2 Series and Sigma Notation

In the last section we were interested in the individual terms of a sequence. In this section we describe the process of adding the terms of a sequence. In the next two sections, we discuss some specific applications of this process.

Given a sequence, $\{a_k\}$, we can associate a sum with it, obtained by adding the first n terms of the sequence. For example, for the sequence

$$5, 10, 15, 20, 25, 30, \ldots, 5k, \ldots$$

we can associate the sum S_n, where *S_n is the sum of the first n terms of the sequence* $\{a_k\}$. Thus

$S_1 = a_1 \quad = 5$ *S_1 is the first term of the sequence.*

$S_2 = a_1 + a_2 \quad = 5 + 10$ *S_2 is the sum of the first* two *terms of the sequence.*

$S_3 = a_1 + a_2 + a_3 \quad = 5 + 10 + 15$ *S_3 is the sum of the first* three *terms of the sequence.*

$S_4 = a_1 + a_2 + a_3 + a_4 \quad = 5 + 10 + 15 + 20$ *S_4 is the sum of the first* four *terms of the sequence.*

$\vdots$

$S_n = a_1 + a_2 + a_3 + a_4 + \cdots + a_n = 5 + 10 + 15 + 20 + \cdots + 5n$ *S_n is the sum of the first n terms of the sequence.*

We have the following definition:

DEFINITION The series S_n associated with the sequence $\{a_k\}$ is defined by

$$S_n = a_1 + a_2 + a_3 + \cdots + a_n$$

In words, this says that S_n is the sum of the first n terms of the sequence $\{a_k\}$. Note that we usually use uppercase letters to designate series and lowercase letters to designate sequences.

Verify that these are the first few terms of $\{a_n\}$.

In order to find a series associated with the sequence $\{a_n\}$, where $a_n = 7n - 8$, we first write out some of the terms of the sequence: $-1, 6, 13, 20, 27, \ldots$. We can then find the indicated sums of the given sequence:

$$S_1 = a_1 = -1$$

$$S_2 = a_1 + a_2 = -1 + 6 = 5$$

$$S_3 = a_1 + a_2 + a_3 = -1 + 6 + 13 = 18$$

$$S_4 = a_1 + a_2 + a_3 + a_4 = -1 + 6 + 13 + 20 = 38$$

EXAMPLE 1 Find S_5 for the sequences whose nth term is given by

(a) $a_n = 3n - 5$ **(b)** $b_n = 2^n - 1$ **(c)** $x_n = \dfrac{(-1)^n}{n}$

Solution

(a) S_5 is the sum of the first five terms of the sequence $\{a_n\}$, where $a_n = 3n - 5$. We find the first five terms by successively substituting $n = 1, 2, 3, 4, 5$ into the given formula for a_n. We find that $a_1 = -2$, $a_2 = 1$, $a_3 = 4$, $a_4 = 7$, and $a_5 = 10$. Thus

$S_5 = a_1 + a_2 + a_3 + a_4 + a_5 = -2 + 1 + 4 + 7 + 10 = \boxed{20}$.

(b) S_5 is the sum of the first five terms of the sequence $\{b_n\}$, where $b_n = 2^n - 1$. Again, we find the first five terms: $b_1 = 1$, $b_2 = 3$, $b_3 = 7$, $b_4 = 15$, and $b_5 = 31$. Hence,

$S_5 = b_1 + b_2 + b_3 + b_4 + b_5 = 1 + 3 + 7 + 15 + 31 = \boxed{57}$.

(c) We begin by finding the first five terms:

$$x_1 = \frac{(-1)^1}{1} = -1, \qquad x_2 = \frac{(-1)^2}{2} = \frac{1}{2}, \qquad x_3 = \frac{(-1)^3}{3} = \frac{-1}{3},$$

$$x_4 = \frac{(-1)^4}{4} = \frac{1}{4}, \qquad x_5 = \frac{(-1)^5}{5} = \frac{-1}{5}$$

Therefore, we have $S_5 = -1 + \frac{1}{2} + \frac{-1}{3} + \frac{1}{4} + \frac{-1}{5} = \boxed{-\frac{47}{60}}$. ■

Sigma Notation

When we have a formula for the general term of a sequence, we can express a series for the sequence in a more compact form by using a special notation for sums. The Greek (uppercase) letter sigma, Σ, often called the summation symbol, is used along with the general term of the sequence to indicate a series.

DEFINITION $\sum_{i=k}^{N} a_i = a_k + a_{k+1} + a_{k+2} + a_{k+3} + \cdots + a_N$ for $k \leq N$.

i is called the **index of summation**, or simply the **index**.

The Σ indicates that we are going to *add* the terms. The number k beneath the sigma is the number substituted for the index, i, in order to find the *first* term of the sum. Then, we increase the index by one and evaluate each subsequent term until the index reaches the number above the sigma, N, which is the last number used for i in the sum. k is called the lower limit of the summation and N is called the upper limit of the summation. In other words, we are adding the terms of the sequence $\{a_n\}$ starting with a_k up to and including a_N.

Remember that sigma notation is used to express a *sum*.

EXAMPLE 2 Evaluate each of the following.

(a) $\sum_{i=1}^{6} 4i$ **(b)** $\sum_{j=1}^{5} \frac{1}{j+1}$ **(c)** $\sum_{k=2}^{4} 10k^2$ **(d)** $\sum_{n=3}^{7} (4n - 3)$

Solution

(a) $\sum_{i=1}^{6} 4i$ *Each term in the sum is found by substituting consecutive integers for i in the general term 4i, starting with i = 1 and ending with i = 6. Remember: Σ is telling us to add the terms.*

$$= 4(1) + 4(2) + 4(3) + 4(4) + 4(5) + 4(6)$$

$$= \boxed{84}$$

(b) $\sum_{j=1}^{5} \frac{1}{j+1}$ *Find each term in the sum by substituting consecutive integers for j in the general term $\frac{1}{j+1}$, starting with j = 1 and ending with j = 5.*

$$\sum_{j=1}^{5} \frac{1}{j+1} = \frac{1}{1+1} + \frac{1}{2+1} + \frac{1}{3+1} + \frac{1}{4+1} + \frac{1}{5+1}$$

$$= \frac{1}{2} + \frac{1}{3} + \frac{1}{4} + \frac{1}{5} + \frac{1}{6}$$

$$= \frac{87}{60} = \boxed{\frac{29}{20}}$$

Remember that $\sum_{k=2}^{4}$ means that we start with $k = 2$ and continue adding terms up to and including $k = 4$.

(c) $\sum_{k=2}^{4} 10k^2$ *This time the first term is found by substituting $k = 2$ in $10k^2$. The next terms are found by substituting $k = 3$ and $k = 4$.*

$$= 10(2)^2 + 10(3)^2 + 10(4)^2$$

$$= 40 + 90 + 160$$

$$= \boxed{290}$$

(d) $\sum_{n=3}^{7} (4n - 3)$ *This time we are using n as the index of summation. The first term is found by evaluating the expression $4n - 3$ for $n = 3$. The next terms are found by evaluating the expression $4n - 3$ for $n = 4$, 5, 6, and 7, respectively.*

$$= [4(3) - 3] + [4(4) - 3] + [4(5) - 3] + [4(6) - 3] + [4(7) - 3]$$

$$= 9 + 13 + 17 + 21 + 25$$

$$= \boxed{85}$$

■

Note that the index is merely a "counter" and so can be replaced with any other letter. Hence

$$\sum_{i=k}^{N} a_i = \sum_{j=k}^{N} a_j = \sum_{h=k}^{N} a_h$$

Recall that we denote the sum of the first N terms of a sequence $\{a_i\}$ by S_N, so using sigma notation we may write

$$\boxed{S_N = a_1 + a_2 + a_3 + \cdots + a_N = \sum_{i=1}^{N} a_i}$$

EXAMPLE 3 Given a sequence for which $a_k = 2k^3$, evaluate S_4.

Solution

$$S_4 = \sum_{k=1}^{4} 2k^3 = 2(1^3) + 2(2^3) + 2(3^3) + 2(4^3) = 2 + 16 + 54 + 128 = \boxed{200}$$

■

EXAMPLE 4 Evaluate the following sum: $\sum_{n=1}^{5} \left(\frac{1}{n} - \frac{1}{n+1}\right)$.

Solution We replace n successively with the values 1, 2, 3, 4, and 5, in the expression $\frac{1}{n} - \frac{1}{n+1}$ and add the results.

$$\sum_{n=1}^{5}\left(\frac{1}{n} - \frac{1}{n+1}\right) = \left(\frac{1}{1} - \frac{1}{1+1}\right) + \left(\frac{1}{2} - \frac{1}{2+1}\right) + \left(\frac{1}{3} - \frac{1}{3+1}\right) + \left(\frac{1}{4} - \frac{1}{4+1}\right) + \left(\frac{1}{5} - \frac{1}{5+1}\right)$$

$$= \left(\frac{1}{1} - \frac{1}{2}\right) + \left(\frac{1}{2} - \frac{1}{3}\right) + \left(\frac{1}{3} - \frac{1}{4}\right) + \left(\frac{1}{4} - \frac{1}{5}\right) + \left(\frac{1}{5} - \frac{1}{6}\right)$$

$$= \frac{1}{1} - \frac{1}{2} + \frac{1}{2} - \frac{1}{3} + \frac{1}{3} - \frac{1}{4} + \frac{1}{4} - \frac{1}{5} + \frac{1}{5} - \frac{1}{6} = 1 - \frac{1}{6} = \boxed{\frac{5}{6}}$$

Notice that all but the first and last terms drop out.

Such a sum, where many of the terms drop out, is called a **telescoping sum**. ■

EXAMPLE 5 Given $f(x) = \sum_{n=1}^{4} nx^{n-1}$, find $f(2)$.

Solution We first write out $f(x)$ without sigma notation, and then substitute $x = 2$.

$$f(x) = \sum_{n=1}^{4} nx^{n-1}$$

$$= 1x^0 + 2x^1 + 3x^2 + 4x^3$$

So

$$f(x) = 1 + 2x + 3x^2 + 4x^3$$

Now that $f(x)$ is in more familiar form we compute $f(2)$.

$$f(2) = 1 + 2(2) + 3(2)^2 + 4(2)^3 = 1 + 4 + 12 + 32 = \boxed{49}$$ ■

Since sigma notation is merely a shorthand way of denoting a sum, we can restate some of the real number properties using sigma notation.

PROPERTIES OF Σ NOTATION

1. $\sum_{k=1}^{n} ca_k = c\left(\sum_{k=1}^{n} a_k\right)$, where c is any constant.

2. $\sum_{k=1}^{n} (a_k + b_k) = \sum_{k=1}^{n} a_k + \sum_{k=1}^{n} b_k$

3. $\sum_{k=1}^{n} (a_k - b_k) = \sum_{k=1}^{n} a_k - \sum_{k=1}^{n} b_k$

Each of these properties is a direct consequence of the commutative, associative, and distributive properties of the real numbers. See Exercise 68.

In the next section we investigate a special type of sequence and its corresponding series.

EXERCISES 11.2

In Exercises 1–26, use the given sequence (or general term) to find the indicated sum, S_n.

1. 3, 7, 11, 15, 19, 23, 27, . . . ; S_5
2. −8, −3, 2, 7, 12, 17, . . . ; S_4
3. −2, 0, 2, 4, 6, 8, . . . ; S_6
4. −1, 5, 11, 17, 23, . . . ; S_8
5. 10, 6, 2, −2, −6, −10, . . . ; S_7
6. 14, 11, 8, 5, 2, −1, . . . ; S_{10}
7. 1, 2, 4, 8, 16, . . . ; S_6
8. $1, \frac{1}{2}, \frac{1}{4}, \frac{1}{8}, \ldots$; S_5
9. $a_n = 3n + 1$; S_2
10. $a_n = 2n + 1$; S_3
11. $b_m = 5m - 1$; S_4
12. $b_m = 4m + 3$; S_5
13. $x_j = 2^j$; S_5
14. $x_j = 3^j$; S_6
15. $x_k = 2^k + 1$; S_6
16. $x_k = 2^{k+1}$; S_6
17. $x_k = 3^{k-2}$; S_5
18. $x_k = 3^k - 2$; S_5
19. $a_n = \frac{n+1}{n}$; S_4
20. $a_n = \frac{n}{n+2}$; S_4
21. $b_j = \frac{(-1)^{j+1}}{j+2}$; S_5
22. $b_n = \frac{(-1)^n n}{n+1}$; S_3
23. $r_m = \frac{1}{m!}$; S_4
24. $t_n = \frac{2^n}{(n+1)!}$; S_3
25. $a_n = \frac{1}{n} - \frac{1}{n+1}$; S_7
26. $b_n = \frac{n}{n+1} - \frac{n+1}{n+2}$; S_6

In Exercises 27–42, rewrite each sum without using sigma notation; then compute each sum.

27. $\sum_{n=1}^{5} n$
28. $\sum_{j=1}^{8} 3j$
29. $\sum_{i=1}^{6} 5(i-1)$
30. $\sum_{k=1}^{4} 4(i+2)$
31. $\sum_{k=2}^{6} k^2$
32. $\sum_{n=3}^{5} n^3$
33. $\sum_{m=3}^{7} (2m^2 - 5)$
34. $\sum_{k=2}^{6} 3k^2 - 4$
35. $\sum_{n=2}^{4} (n^2 - 3n + 1)$
36. $\sum_{j=3}^{5} (10j - j^2)$
37. $\sum_{j=1}^{5} \frac{1}{j+1}$
38. $\sum_{j=1}^{4} \left(\frac{1}{n} + 1\right)$
39. $\sum_{n=1}^{5} (\sqrt{n+1} - \sqrt{n})$
40. $\sum_{m=1}^{4} \left(\frac{2}{m} - \frac{2}{m+1}\right)$
41. $\sum_{n=1}^{5} 3$
42. $\sum_{k=3}^{8} 5$

In Exercises 43–46, use sigma notation to represent the sum of the first n terms of the given sequence.

43. 4, 8, 12, . . . , $4k$, . . . for $n = 5$
44. 2, 5, 8, . . . , $3k - 1$, . . . for $n = 8$
45. 2, 8, 18, . . . , $2k^2$, . . . for $n = 7$
46. 7, 9, 11, . . . , $(2k + 5)$, . . . for $n = 8$

In Exercises 47–52, express the given sum using sigma notation. (The answers to these exercises are not unique.)

47. 2 + 6 + 10 + 14 + 18
48. 5 + 25 + 125 + 625
49. 1 + 3 + 5 + 7 + · · · + 51
50. 4 + 7 + 10 + 13 + · · · + 52
51. $\frac{1}{2} + \frac{1}{2 \cdot 3} + \frac{1}{3 \cdot 4} + \cdots + \frac{1}{99 \cdot 100}$
52. $\frac{2}{5} + \frac{4}{9} + \frac{6}{13} + \cdots + \frac{20}{41}$

In Exercises 53–58, find S_6 for the given recursively defined sequence.

53. $a_1 = 3$; $a_n = 2a_{n-1} + 3$ for $n > 1$
54. $a_1 = 64$; $a_{k+1} = \frac{1}{2}a_k$
55. $a_1 = -2$; $a_{n+1} = (a_n + 1)^2$
56. $a_1 = 2$; $a_{k+1} = \frac{1}{a_k}$
57. $a_1 = 2$; $a_2 = 3$; $a_{n+2} = a_{n+1} \cdot a_n$
58. $a_1 = 5$; $a_k = ka_{k-1}$ for $k > 1$

59. Write out $\sum_{n=1}^{8} \log_3 \frac{n+1}{n}$ without using sigma notation and simplify the sum. HINT: Use the properties of logarithms.

60. Write out $\sum_{n=1}^{6} \log_8 2^n$ without using sigma notation and simplify the sum. HINT: Use the properties of logarithms.

61. Given that $f(x) = \sum_{k=1}^{5} x^k$, find $f(1)$, $f(-2)$, and $f\left(\frac{1}{2}\right)$.

62. Given that $g(x) = \sum_{k=1}^{4} \sin \frac{\pi x}{k}$, find $g(0)$, $g(1)$, and $g(2)$.

63. Given that $F(x) = \sum_{k=0}^{6} \frac{x^k}{k!}$.

(a) Find $F(1)$, $F(2)$, and $F(-1)$ correct to four decimal places. (Remember that, by definition, $0! = 1$.)

(b) Compare the values of $F(1)$, $F(2)$, and $F(-1)$ with the values of e, e^2, and e^{-1}, respectively (where e is the base for natural logarithms). What do you notice?

64. Given that $f(x) = \sum_{k=1}^{4} \frac{(-1)^{k+1}x^{2k-1}}{(2k-1)!}$.

(a) Find $f(1)$, $f(2)$, and $f(3)$ correct to 4 decimal places.

(b) Compare the values of $f(1)$, $f(2)$, and $f(3)$ with the values of sin 1, sin 2, and sin 3 respectively. What do you notice?

QUESTIONS FOR THOUGHT

65. Show that for c a constant, $\sum_{k=1}^{N} c = Nc$

66. Show that $\sum_{i=1}^{n} a_i = \sum_{i=2}^{n+1} a_{i-1}$.

67. Write the polynomial

$$p(x) = a_nx^n + a_{n-1}x^{n-1} + a_{n-2}x^{n-2} + \cdots + a_2x^2 + a_1x + a_0$$

using sigma notation.

68. Use the basic properties of the real numbers to prove the three properties of sigma notation listed at the end of this section.

11.3 Arithmetic Sequences and Series

Arithmetic Sequences

In the last section we saw that to compute the sum of a series, we actually had to add all the terms in the series. Sometimes, however, the special structure of a sequence allows us to derive a formula for its associated series.

Let's examine the following sequences of numbers:

$$\{a_k\} = 2, 5, 8, 11, 14, \ldots$$

$$\{b_k\} = 1, 6, 11, 16, 21, \ldots$$

$$\{c_k\} = 6, 4, 2, 0, -2, \ldots$$

Can you find the next three terms in each sequence?

In Section 11.1 we noted that different sequences may agree for the first few terms. Thus we cannot necessarily predict later terms of a sequence from the first few terms. Rather than trying to find a formula, we may find it fruitful to examine the relationship between successive terms of a sequence.

For example, in the sequence $\{a_k\} = 2, 5, 8, 11, 14, \ldots$ we observe that each term after the first is 3 more than the preceding term. Assuming that this same pattern continues for successive terms, the three terms after 14 would be 17, 20, 23. Symbolically, we can represent this relationship as

$$a_{k+1} = a_k + 3$$

Note that this means that any term, (say the $(k+1)$st term), is 3 more than the previous term (the kth term). For $k = 10$, $a_{11} = a_{10} + 3$. Equivalently, we could write $a_{k+1} = a_k + 3$ as a difference:

$$a_{k+1} - a_k = 3$$

This says that the *difference* between any two successive terms is 3.

$b_{k+1} - b_k = 5$ says that the difference between any two successive terms is 5.

In the second sequence, $\{b_k\} = 1, 6, 11, 16, 21, \ldots$, we observe that each term after the first is 5 more than the previous term. Assuming that this same pattern continues, we conclude that adding 5 will produce the next term in the sequence. Algebraically we write $b_{k+1} = b_k + 5$, which says that any term after the first is found by adding 5 to the preceeding term. Thus, the three terms following 21 would be 26, 31, 36. Again, we can write $b_{k+1} = b_k + 5$ as a difference:

$$b_{k+1} - b_k = 5$$

Notice that in $\{c_k\} = 6, 4, 2, 0, -2, \ldots$, the third sequence given, each term is 2 less than the previous term. Hence we could describe the relationship between successive terms in the sequence as

$$c_{k+1} = c_k - 2$$

So in $\{c_k\}$, the term -2 would be followed by $-4, -6, -8$, and so on. Rewriting this relationship as a difference we have

$$c_{k+1} - c_k = -2$$

The difference between two consecutive terms is -2 because $4 - 6 = -2$ and $-2 - (0) = -2$, etc.

Sequences in which the difference between successive terms is constant are called **arithmetic sequences**. This can be stated as follows.

DEFINITION An **arithmetic sequence** is one in which the difference between successive terms (called the **common difference**) is constant. This common difference is usually denoted by d.

An arithmetic sequence is sometimes called an *arithmetic progression*.

Hence, referring back to the sequences given at the beginning of this section, the common difference in the first sequence, $\{a_k\}$, is 3; the common difference in the second sequence, $\{b_k\}$, is 5; and the common difference in the third sequence, $\{c_k\}$, is -2.

Why is an arithmetic sequence a recursive sequence?

In other words, if $a_1, a_2, a_3, a_4, \ldots, a_{k-1}, a_k, \ldots$ is an arithmetic sequence, then $a_{k+1} - a_k = d$; or equivalently $a_{k+1} = a_k + d$ for all k.

Similarly, if an arithmetic sequence has a common difference of d and we start with a_1, then the next term in the sequence, a_2, is

$$a_2 = a_1 + d \quad \textit{The common difference added to the first term}$$

The next term, a_3, is

$$a_3 = a_2 + d = (a_1 + d) + d = a_1 + 2d$$

$$a_4 = a_3 + d = (a_1 + 2d) + d = a_1 + 3d$$

Note the pattern.

$$a_5 = a_4 + d = a_1 + 4d$$

$$a_6 = a_5 + d = a_1 + 5d$$

To get the fifth term, we add four *times the common difference to a_1. To get the sixth term, we add* five *times the common difference to a_1.*

$$\vdots$$

In general we have

$$a_n = a_{n-1} + d = a_1 + (n - 1)d$$

We have thus derived the following formula:

> If $\{a_n\}$ is an arithmetic sequence with first term a_1 and common difference d, then the nth term is given by
>
> $$a_n = a_1 + (n - 1)d \qquad \text{Formula 11.1}$$

Thus, for example, in any arithmetic sequence, $a_9 = a_1 + 8d$. In words, this says that to get the ninth term in an arithmetic sequence, we add eight times the common difference to the first term.

EXAMPLE 1 Given an arithmetic sequence with first term 5 and common difference 4, find the first five terms and the twentieth term.

Solution The first term of the arithmetic sequence is 5; hence

$a_1 = 5$ *In an arithmetic sequence with common difference d, each term is d more than the previous term. Since the common difference, d, is 4, a_2 is 4 more than a_1.*

$a_2 = 5 + 4 = 9$

$a_3 = 9 + 4 = 13$ *a_3 is 4 more than a_2.*

$a_4 = 13 + 4 = 17$

$a_5 = 17 + 4 = 21$

Thus the first five terms are $\boxed{5, 9, 13, 17, \text{ and } 21}$.

To find the twentieth term, we can use Formula 11.1:

$a_n = a_1 + (n - 1)d$ *Substituting $n = 20$, $a_1 = 5$, and $d = 4$, we get*

Notice that to get a_{20} we add 19 times the common difference to a_1.

$$a_{20} = 5 + (20 - 1)4 = 5 + 19(4) = \boxed{81}$$

EXAMPLE 2 Given an arithmetic sequence whose first two terms are -3 and 7, find the next three terms and the sixteenth term.

Solution Since the first two terms of the sequence are -3 and 7, we have $a_1 = -3$ and $a_2 = 7$; because the sequence is *arithmetic*, the difference between the first two terms is also the common difference between any two successive terms. Therefore, the common difference is $d = a_2 - a_1 = 7 - (-3) = 10$.

Since $a_2 = 7$ and $d = 10$, we have

$a_3 = 7 + 10 = 17$ *Because $a_3 = a_2 + d$*

$a_4 = 17 + 10 = 27$ *Because $a_4 = a_3 + d$*

$a_5 = 27 + 10 = 37$

Therefore, the three terms following -3 and 7 are $\boxed{17, 27, \text{ and } 37}$.

The sixteenth term is found by using Formula 11.1:

$a_n = a_1 + (n - 1)d$ *Where $n = 16$, $a_1 = -3$, and $d = 10$*

$a_{16} = -3 + (16 - 1)(10) = \boxed{147}$ ■

Given an arithmetic sequence a_j, let's examine the difference between the fourth term, a_4, and the tenth term, a_{10}. If we start with a_4, we have to add d to get a_5, another d to get a_6, another d to get a_7, and so on, until we get to a_{10}. Hence to get from a_4 to a_{10} we have to add d six times, or

$$a_{10} = a_4 + 6d$$

We can derive a formula for the relationship between *any* two terms in an arithmetic sequence as follows: We begin with Formula 11.1, giving the nth term of an arithmetic sequence with first term a_1 and common difference d:

$$a_n = a_1 + (n - 1)d$$

Now suppose we have two terms in an arithmetic sequence, a_j and a_k. Let's apply this formula to a_j and a_k by substituting $n = k$ and $n = j$; we get

$a_k = a_1 + (k - 1)d$ *And*

$a_j = a_1 + (j - 1)d$ *Subtracting the second equation from the first, we get*

$a_k - a_j = (k - 1)d - (j - 1)d$

$= kd - d - jd + d$

$= kd - jd$ *Factor out d.*

$= (k - j)d$ *Which gives us the following formula*

> The difference between any two terms, a_j and a_k, in an arithmetic sequence with common difference d is
>
> $$a_k - a_j = (k - j)d \qquad \text{Formula 11.2}$$

EXAMPLE 3 Given an arithmetic sequence with $a_3 = 12$ and $a_9 = 14$, find a_1 and a_{30}.

Solution The key feature of an arithmetic sequence is its common difference d. Formula 11.2 allows us to use any two terms in an arithmetic sequence to find d. We are given an arithmetic sequence with third term $a_3 = 12$ and ninth term $a_9 = 14$. Using Formula 11.2 with $j = 3$ and $k = 9$, the difference between these two terms is

$a_9 - a_3 = (9 - 3)d$ *Substitute for a_9 and a_3.*

$14 - 12 = (9 - 3)d$

$2 = 6d$ *So*

$\frac{1}{3} = d$

Once we know the value of d, we can find a_1 by substituting $d = \frac{1}{3}$ in Formula 11.1 for the nth term of an arithmetic sequence:

$a_n = a_1 + (n - 1)d$ *Because we know the value of a_3, we use $n = 3$ to obtain*

$a_3 = a_1 + (3 - 1)d$ *Substituting $a_3 = 12$, and $d = \frac{1}{3}$, we get*

$$12 = a_1 + 2\left(\frac{1}{3}\right)$$

$$12 = a_1 + \frac{2}{3}$$

$$\boxed{11\frac{1}{3} = a_1}$$

Knowing $a_1 = 11\frac{1}{3}$ and $d = \frac{1}{3}$, we can use Formula 11.1 to find a_{30}.

Knowing that $a_9 = 14$ and $d = \frac{1}{3}$, how can we find a_{30} without finding a_1?

$a_n = a_1 + (n - 1)d$ *Substituting $a_1 = 11\frac{1}{3}$, $d = \frac{1}{3}$, and $n = 30$, we get*

$a_{30} = 11\frac{1}{3} + (30 - 1)\frac{1}{3} = \frac{34}{3} + \frac{29}{3} = \frac{63}{3} = 21$. Therefore, $\boxed{a_{30} = 21}$ ■

EXAMPLE 4 Determine whether or not the sequence with the following general term is arithmetic. **(a)** $a_k = 4k - 7$ **(b)** $a_n = n^2 - n$

Solution

(a) In order to show that a sequence is arithmetic, we must show that the difference between *any* two consecutive terms is constant. Since $a_k = 4k - 7$, the first few terms of the sequence are $-3, 1, 5, 9, 13, \ldots$, which we can see are the terms of an arithmetic sequence with $a_1 = -3$ and $d = 4$.

More formally, we can prove that this is an arithmetic sequence by showing that the difference between *any* two consecutive terms is a constant. We have

Note that $a_{k-1} = 4(k - 1) - 7$.

$$a_k - a_{k-1} = 4k - 7 - [4(k - 1) - 7] = 4k - 7 - [4k - 4 - 7]$$
$$= 4k - 7 - 4k + 11 = 4$$

Thus the sequence has a common difference of 4 and is, therefore, arithmetic.

(b) If we list the first few terms of the sequence with general term $a_n = n^2 - n$, we get $0, 2, 6, 12, \ldots$. We can immediately see that the difference between a_1 and a_2 is $a_2 - a_1 = 2 - 0 = 2$, whereas the difference between a_2 and a_3 is $a_3 - a_2 = 6 - 2 = 4$. Since these differences are not the same, the sequence is *not* arithmetic. ■

Arithmetic Series

The particular structure of an arithmetic sequence allowed us to develop a formula for its nth term (Formula 11.1). This same structure allows us to develop formulas for S_n, the sum of the first n terms of an *arithmetic* sequence. The series obtained from an arithmetic sequence is called an **arithmetic series**.

We begin by examining a special arithmetic sequence and its associated series: Let $a_1 = 1$ and $d = 1$, then $a_2 = 2$, $a_3 = 3$, $a_4 = 4, \ldots, a_k = k$. In other words, this sequence is the sequence of positive integers. The series for this sequence is denoted by

$$A_n = 1 + 2 + 3 + \cdots + (n-2) + (n-1) + n$$

which is the sum of the first n positive integers. Let's see if we can find a way of computing $A_{100} = 1 + 2 + 3 + \cdots + 98 + 99 + 100$ without actually adding 100 numbers. We write A_{100}; below it we rewrite A_{100} in reverse order.

$$\begin{aligned} A_{100} &= 1 + 2 + 3 + \cdots + 98 + 99 + 100 \\ A_{100} &= 100 + 99 + 98 + \cdots + 3 + 2 + 1 \\ \hline 2A_{100} &= \underbrace{(101) + (101) + (101) + \cdots + (101) + (101) + (101)}_{100 \text{ times}} \end{aligned}$$

Note that there are 100 columns, and the sum in each column is 101. Add the two equations to get

Therefore, we have

$$2A_{100} = 100(101)$$

And so

$$A_{100} = \frac{100(101)}{2} = 5050$$

We can generalize this approach and derive a formula for the sum of the first k positive integers. We follow the same steps as we just did for A_{100}.

We write the series A_k; below it, we rewrite the same series in reverse order.

$$\begin{aligned} A_k &= 1 + 2 + 3 + \cdots + (k-2) + (k-1) + k \\ A_k &= k + (k-1) + (k-2) + \cdots + 3 + 2 + 1 \\ \hline 2A_k &= \underbrace{(k+1) + (k+1) + (k+1) + \cdots + (k+1) + (k+1) + (k+1)}_{k \text{ times}} \end{aligned}$$

Note that there are k columns and the sum in each column is k + 1.

Add the two equations to get

Therefore, we have

$$2A_k = k(k+1)$$

And so

$$A_k = \frac{k(k+1)}{2}$$

Thus we have derived Formula 11.3.

The sum of the first k positive integers is given by

$$A_k = 1 + 2 + 3 + \cdots + k = \frac{k(k+1)}{2} \qquad \text{Formula 11.3}$$

For example, using this formula for $k = 20$, we find that the sum of the first 20 positive integers is:

$$A_{20} = 1 + 2 + 3 + \cdots + 20 = \frac{20(20+1)}{2} = \frac{20(21)}{2} = 210$$

We can now derive a formula for the general arithmetic series S_n.

Recall that $S_n = a_1 + a_2 + a_3 + a_4 + \cdots + a_{n-2} + a_{n-1} + a_n$ is called the arithmetic series associated with the sequence $\{a_n\}$. We can use the formula $a_n = a_1 + (n-1)d$ to rewrite each term in the sequence in terms of a_1, n, and d. In other words, since $a_2 = a_1 + d$, $a_3 = a_1 + 2d$, . . . , $a_n = a_1 + (n-1)d$, we can write the entire sum S_n in terms of a_1, n, and d as follows:

$$S_n = a_1 + \quad a_2 \quad + \quad a_3 \quad + \quad a_4 \quad + \cdots + \quad a_{n-1} \quad + \quad a_n \qquad \textit{Becomes}$$

$$\downarrow \quad \downarrow \quad \downarrow \quad \downarrow \quad \cdots \quad \downarrow \quad \downarrow$$

$$S_n = a_1 + (a_1 + d) + (a_1 + 2d) + (a_1 + 3d) + \cdots + [a_1 + (n-2)d] + [a_1 + (n-1)d]$$

We collect all the a_1 terms (there are n of them).

$$S_n = na_1 + [d + 2d + 3d + \cdots + (n-2)d + (n-1)d]$$

Now factor out d from within the brackets.

$$= na_1 + d[1 + 2 + 3 + \cdots + (n-2) + (n-1)] \qquad (1)$$

Inside the brackets we have the sum of the first $n - 1$ positive integers. We can use Formula 11.3 to find the sum of the first $n - 1$ positive integers by substituting $n - 1$ for k. From Formula 11.3 we have

$$1 + 2 + 3 + \cdots + k = \frac{k(k+1)}{2} \qquad \textit{Substitute } n - 1 \textit{ for } k.$$

$$1 + 2 + 3 + \cdots + n - 1 = \frac{(n-1)[(n-1)+1]}{2} = \frac{(n-1)n}{2} \qquad (2)$$

Returning to equation (1) we have

$$S_n = na_1 + d[1 + 2 + 3 + \cdots + (n-2) + (n-1)] \qquad \textit{Substituting the result from (2) we get}$$

$$= na_1 + d\frac{(n-1)n}{2} \qquad \textit{Combine into a single fraction.}$$

$$= \frac{2na_1}{2} + \frac{n(n-1)d}{2} = \frac{2na_1 + n(n-1)d}{2} \qquad \textit{Factor out n in the numerator.}$$

$$= \frac{n[2a_1 + (n-1)d]}{2}$$

We finally rewrite this formula by separating out the factor of $\frac{n}{2}$.

SUM OF AN ARITHMETIC SERIES

The sum of the first n terms of an arithmetic sequence is given by

$$S_n = \sum_{i=1}^{n} a_i = \frac{n}{2}[2a_1 + (n-1)d] \qquad \text{Formula 11.4}$$

We can rewrite $2a_1$ as $a_1 + a_1$ and, since $a_n = a_1 + (n-1)d$, Formula 11.4 for S_n becomes

$$S_n = \frac{n}{2}[2a_1 + (n-1)d] = \frac{n}{2}[a_1 + \underbrace{a_1 + (n-1)d}_{a_n}] = \frac{n}{2}(a_1 + a_n) = n\left(\frac{a_1 + a_n}{2}\right)$$

Hence, we have derived an alternative (and simpler) version for Formula 11.4.

SUM OF AN ARITHMETIC SERIES

An alternative form for the sum of the first n terms of an arithmetic sequence is given by

$$S_n = \frac{n}{2}(a_1 + a_n) = n\left(\frac{a_1 + a_n}{2}\right) \qquad \text{Formula 11.5}$$

In words, Formula 11.5 says that the sum of the first n terms of an arithmetic sequence is equal to n times the average of the first term and the nth term.

We note that Formula 11.4 is most useful when we know n, a_1 (the first term), and d (the common difference), whereas Formula 11.5 is most useful when we know n, a_1, and a_n (the last term).

EXAMPLE 5 Given the arithmetic sequence: $\{a_k\} = 3, 7, 11, 15, \ldots$, find

(a) S_{12} **(b)** $\sum_{k=1}^{50} a_k$

Solution

How do you find the common difference of an arithmetic sequence?

(a) We could calculate S_{12} by writing out the first 12 terms in the sequence and adding them together, or we could use one of the formulas for S_n. The sequence $3, 7, 11, 15, \ldots$ has first term $a_1 = 3$ and common difference $d = 4$ (why?), and for S_{12} we have $n = 12$. Thus we can substitute these values into Formula 11.4 to get

You may want to check this answer by writing out the first 12 terms in the sequence and finding their sum.

$$S_n = \frac{n}{2}[2a_1 + (n-1)d] \qquad \textit{We substitute } n = 12,\ a_1 = 3, \textit{ and } d = 4.$$

$$S_{12} = \frac{12}{2}[2 \cdot 3 + (12-1)4] = 6[6 + 11(4)] = \boxed{300}$$

(b) To find the sum of the first 50 terms of the sequence, we are looking for the series S_{50}. Thus we use Formula 11.4 again, this time with $n = 50$.

$$S_{50} = \frac{50}{2}[2 \cdot 3 + (50-1)4] = \boxed{5050}$$

■

EXAMPLE 6 Find the sum of the first 25 terms of the sequence whose general term is $a_n = 7n$.

Solution This example is asking us to find the sum of the 25 terms 7, 14, 21, 28, 35, . . . , 175. We notice that this is an arithmetic sequence with $a_1 = 7(1)$ and $a_{25} = 7(25) = 175$. (What is the common difference for this sequence?) Since we can easily identify the first and last terms, we use the alternative formula for an arithmetic series (Formula 11.5).

Find the sum of the first 25 terms using Formula 11.4. Which formula is easier to use in this case?

$$S_n = \frac{n}{2}(a_1 + a_n)$$

$$S_{25} = 25\left(\frac{7 + 175}{2}\right) = 25(91) = \boxed{2275}$$

■

EXAMPLE 7 A job applicant finds that firm A offers a starting annual salary of \$22,500 with a guaranteed raise of \$1400 per year, whereas firm B offers a higher starting salary of \$26,000 but will guarantee a raise of only \$900 per year.

(a) What would the annual salary be in the tenth year at each firm?

(b) Over the first 10 years, how much would be earned at each firm?

Solution

(a) The annual salaries at firm A form the arithmetic sequence 22,500; 23,900; 25,300; . . . with first term \$22,500 and common difference \$1400. If we let a_n = the annual salary at firm A in year n, then the annual salary in year 10 is

$$a_{10} = a_1 + 9d = 22{,}500 + 9(1400) = \$35{,}100$$

Similarly, the annual salaries at Firm B form the arithmetic sequence 26,000; 26,900; 27,800; . . . with first term \$26,000 and common difference \$900. If we let b_n = the annual salary at firm B in year n, then the annual salary in year 10 is

$$b_{10} = b_1 + 9d = 26{,}000 + 9(900) = \$34{,}100$$

Therefore, in the tenth year the annual salary would be \$35,100 at firm A and \$34,100 at firm B.

(b) To determine the amount that would be earned at each firm over the first 10 years we need to add the first 10 annual salaries; $a_1 + a_2 + a_3 + \cdots + a_{10}$ represents the amount of money earned over the first 10 years at firm A and $b_1 + b_2 + b_3 + \cdots + b_{10}$ represents the amount of money earned over the first 10 years at firm B. In other words, we want to find the sum of the first 10 terms of each arithmetic sequence.

$$a_1 + a_2 + a_3 + \cdots + a_{10} = \frac{10}{2}(a_1 + a_{10}) \qquad \textit{We know } a_1 = 22{,}500 \textit{ and, from part (a), } a_{10} = 35{,}100.$$

$$= 5(22{,}500 + 35{,}100) = 5(57{,}600) = 288{,}000$$

$$b_1 + b_2 + b_3 + \cdots + b_{10} = \frac{10}{2}(b_1 + b_{10}) \qquad \textit{We know } b_1 = 26{,}000 \textit{ and, from part (a), } b_{10} = 34{,}100.$$

$$= 5(26{,}000 + 34{,}100) = 5(60{,}100) = 300{,}500$$

You might find it interesting to compute how much money would be earned at each firm during the next 10 years.

Therefore, over the first 10 years a total of \$288,000 would be earned at firm A, but a total of \$300,500 would be earned at firm B. ■

EXERCISES 11.3

In Exercises 1–6, determine whether or not the given sequence is arithmetic.

1. 4, 7, 10, 13, . . .

2. 1, 4, 9, 16, . . .

3. 2, 6, 10, 14, 20, 26, . . .

4. $\frac{1}{2}, \frac{5}{4}, 2, \frac{11}{4}, \ldots$

5. $a_n = 5 - 2n$

6. $a_k = \frac{1}{k}$

In Exercises 7–12, use the given information about an arithmetic sequence to find the common difference, d.

7. $a_1 = 3$ and $a_5 = 23$

8. $a_6 = -2$ and $a_{11} = 53$

9. $a_{20} - a_5 = 10$

10. $a_{12} - a_3 = -8$

11. $a_n = 4n + 1$

12. $a_k = \frac{k + 4}{5}$

In Exercises 13–24, use the given information about the arithmetic sequence to find the requested values.

13. $a_4 = 8$ and $a_8 = 10$. Find d and a_1.

14. $a_4 = 5$ and $d = 6$. Find a_1 and a_9.

15. $a_5 = 1$ and $d = \frac{2}{5}$. Find a_1 and a_{20}.

16. $a_7 - a_2 = 30$. Find d and $a_{12} - a_1$. Can you find a_1?

17. $\{a_n\} = 2, 5, 8, 11, \ldots$. Find d and a_{18}.

18. $\{a_n\} = -8, -1, 6, 13, \ldots$. Find a_8 and a_{100}.

19. $\{a_k\} = \frac{1}{2}, \frac{5}{6}, \frac{7}{6}, \ldots$. Find d and a_{10}.

20. $\{a_n\} = \frac{1}{4}, \frac{7}{12}, \frac{11}{12}, \ldots$. Find d and a_{21}.

21. $a_3 = -\frac{1}{3}$, $a_4 = -\frac{5}{6}$. Find a_1 and a_9.

22. $a_2 = -1$, $a_6 = -\frac{1}{2}$. Find a_4 and a_9.

23. $a_{10} = 11$ and $a_7 - a_4 = -\frac{5}{4}$. Find d and a_1.

24. $a_{61} = 102$ and $d = -\frac{5}{3}$. Find a_1.

25. Jim is scheduled to get a raise of \$250 every 6 months during his first 5 years on the job. If his starting salary is \$25,250 per year, what is his annual salary at the end of 3 years?

26. Rosa is scheduled to get a raise of \$150 every three months during her first 3 years on the job. If her starting salary is \$28,500, what is her yearly salary at the end of 2 years?

27. Olga is planning an exercise schedule in which she begins bench pressing 50 pounds and increases the weight by 2 pounds each week. How much will she be bench pressing at the end of the twentieth week?

28. Janice begins a savings program in which she will save \$1000 the first year, and each subsequent year will save \$200 more than she did the previous year. How much will she save during the eighth year?

29. Suppose that when an object is dropped from a hot air balloon, it falls 16 feet the first second, 48 feet the second second, 80 feet the third second, and so on. The distances it falls each second form an arithmetic sequence. How many feet will the object fall during the fifth second? During the sixth second?

30. How many feet will the object in Exercise 29 fall during the first 10 seconds?

In Exercises 31–42, use the given information about each arithmetic sequence to find the indicated arithmetic series, S_n.

31. $a_1 = 4$ and the common difference is 5; S_6

32. $a_1 = 9$ and the common difference is -1; S_8

33. $b_4 = 2$ and $b_7 = 17$; S_7

34. $c_9 = 12$ and $c_{15} = 30$; S_5

35. $a_3 = 2$ and the common difference is $\frac{1}{2}$; S_7

36. $a_5 = 6$ and the common difference is $-\frac{1}{3}$; S_{12}

37. $\{a_n\} = 3, -2, -7, \ldots; S_6$

38. $\{a_j\} = 4, 13, 22, \ldots; S_9$

39. $\{a_k\} = -4, -2, 0, 2, 4, \ldots; S_{10}$

40. $\{b_m\} = 6, 7, 8, \ldots; S_{17}$

41. $\{x_n\} = \frac{1}{4}, \frac{1}{3}, \frac{5}{12}, \ldots; S_8$

42. $\{t_n\} = -\frac{3}{2}, -1, -\frac{1}{2}, \ldots; S_7$

43. Find $\sum_{j=1}^{8} 4j$.

44. Find $\sum_{k=1}^{6} 3k$.

45. Find $\sum_{m=1}^{10} 10m$.

46. Find $\sum_{n=1}^{9} 8n$.

47. Find $\sum_{i=1}^{7} (3i + 4)$.

48. Find $\sum_{k=1}^{5} (4k - 9)$.

49. Find $\sum_{i=40}^{70} i$.

50. Find $\sum_{n=30}^{100} n$.

51. Find $\frac{1}{\pi} + \frac{3}{\pi} + \frac{5}{\pi} + \cdots + \frac{15}{\pi}$.

52. Find $\frac{e}{2} + e + \frac{3e}{2} + \cdots + 10e$.

53. A pile of logs is in the shape of a pyramid with 40 logs in the bottom level. Each level above the first has two fewer logs than the level below. If the pile of logs has 12 levels, how many logs are there in the pile?

54. A theater has 38 rows of seats. The first row has 17 seats, the second row has 20 seats, the third row has 23 seats, and so on. What is the seating capacity of the theater?

55. A contest offers a total of 18 prizes. The first prize is worth \$10,000, and each successive prize is worth \$500 less than the next higher prize. Find the value of the eighteenth prize and the total value of all the prizes.

56. A contest offers 10 prizes with a total value of \$13,250 in prizes. If the difference in value between successive prizes is \$250, what is the value of the first prize?

57. A grocer wants to empty 12 cases of breakfast cereal, with 48 boxes in each case. She wants to stack the boxes in a pyramid with each row containing two fewer boxes than the row below it and the top row containing one box.

(a) Can such a pyramid be constructed so that it uses up all the boxes in the 12 cases?

(b) How many boxes high must the pyramid be?

(c) With how many boxes must she start in the bottom row of the pyramid?

58. Find the sum of the first 100 positive even integers.

59. Find the sum of the first 100 positive odd integers.

60. Find the sum of the first 30 numbers in the arithmetic sequence 8, 16, 24, ...

61. Find the sum of the first 25 numbers in the arithmetic sequence $-3, 7, 17, 27, \ldots$.

62. Given an arithmetic sequence with $S_{10} = 165$ and $a_1 = 3$, find a_{10}.

63. Given an arithmetic sequence with $S_{20} = 910$ and $a_{20} = 95$, find a_1.

64. Given an arithmetic sequence with $S_{16} = 368$ and $a_1 = 1$, find a_8.

65. Given an arithmetic sequence with $S_{15} = 390$ and $a_{15} = 56$, find S_5.

66. Given an arithmetic sequence with $a_1 = 9$, $d = 6$, and $S_N = 969$, find N.

67. Given an arithmetic sequence with $a_1 = -5$ and $S_{22} = 2431$, find d, the common difference.

QUESTIONS FOR THOUGHT

68. Show that the terms $\frac{1}{1 + \sqrt{3}}, -\frac{1}{2}, \frac{1}{1 - \sqrt{3}}$ form an arithmetic sequence.

69. In Formula 11.2, does it matter whether $k > j$? Explain.

70. How does the fact that the sequence with $a_k = 4k - 7$ has a constant common difference of 4 relate to the fact that the line with equation $y = 4x - 7$ has a slope of 4?

11.4 Geometric Sequences and Series

Geometric Sequences

In the last section we saw that the structure of arithmetic sequences and series allows us to develop special formulas that pertain to them. In this section we analyze another type of special sequence and series.

Consider the following two sequences:

$$\{a_k\} = 4, 8, 16, 32, 64, \ldots \qquad \{b_j\} = 3, 15, 75, 375, 1875, \ldots$$

Neither of these sequences are arithmetic, because in each sequence the difference between consecutive terms is not constant. (For the sequence $\{a_k\}$, $8 - 4 = 4$, $16 - 8 = 8$, and $32 - 16 = 16$, and for the sequence $\{b_j\}$, $15 - 3 \neq 75 - 15$.

However, upon closer inspection of each sequence we find that there is a pattern: in $\{a_k\}$, each term after the first is 2 times the previous term,

$$8 = 2 \cdot 4, \qquad 16 = 2 \cdot 8, \qquad 32 = 2 \cdot 16, \ldots$$

and in $\{b_j\}$ each term after the first is 5 times the previous term:

$$15 = 5 \cdot 3, \qquad 75 = 5 \cdot 15, \qquad 375 = 5 \cdot 75, \ldots$$

Equivalently, we can describe these sequences by saying that the *quotient* of two consecutive terms (each term divided by its preceeding term) is constant. In the case of $\{a_k\}$, this quotient is $\dfrac{a_{k+1}}{a_k} = 2$. In the case of $\{b_j\}$, this quotient is $\dfrac{b_{j+1}}{b_j} = 5$.

A sequence for which this type of relationship holds for every pair of consecutive terms is called a **geometric sequence**.

DEFINITION A sequence $\{a_k\}$ in which the ratio of consecutive terms is constant is called a **geometric sequence**. This **common ratio** is usually denoted by r. In other words, a geometric sequence is defined by the fact that

$$\frac{a_{k+1}}{a_k} = r \qquad \text{or, equivalently,} \qquad a_{k+1} = ra_k$$

What are the similarities and differences between arithmetic and geometric sequences?

A geometric sequence is sometimes called a *geometric progression*.

In a geometric sequence, each successive term is found by *multiplying* the previous term by a constant. This makes a geometric sequence a recursively defined sequence.

In the sequence $\{a_k\}$, given at the beginning of this section, each successive term is found by multiplying the previous term by 2. Symbolically, we can represent this relationship as

$$a_{k+1} = 2a_k$$

If $k = 15$, we have $a_{15} = 2a_{14}$

Note the similarities and differences between an arithmetic sequence and a geometric sequence: in an arithmetic sequence, the *differences* between successive terms are constant, whereas in a geometric sequence, the *quotients* of successive terms are constant.

Suppose $a_1, a_2, a_3, a_4, \ldots, a_{k-1}, a_k, \ldots$ is a geometric sequence with common ratio r. Then if we start with a_1, we have

$a_2 = a_1 r$ — *To get the second term we multiplied the first term by the common ratio, r. The next term is*

$a_3 = ra_2 = r(a_1 r) = a_1 r^2$ — *Because $a_2 = a_1 r$; similarly*

$a_4 = ra_3 = r(a_1 r^2) = a_1 r^3$ — *Continuing the pattern, we have*

$a_5 = ra_4 = r(a_1 r^3) = a_1 r^4$

$a_6 = ra_5 = a_1 r^5$

$\vdots$ — *In general we have*

$a_n = ra_{n-1} = a_1 r^{n-1}$

This result gives us the following general formula.

THE nth TERM OF A GEOMETRIC SEQUENCE

The nth term of a geometric sequence with common ratio r is

$$a_n = a_1 r^{n-1} \qquad \text{for } n \geq 1 \qquad \text{Formula 11.6}$$

In other words, the terms of a geometric sequence may be written as

$$a_1, a_1 r, a_1 r^2, a_1 r^3, \ldots, a_1 r^n, \ldots$$

EXAMPLE 1 Given the geometric sequence 5, 15, 45, . . . , find the next two terms and the seventh term.

Solution Since we are given that the sequence is geometric, we first find the common ratio, r, which is $\frac{15}{5} = 3$. Note that we can use *any* two consecutive terms to find the common ratio.

$$\frac{15}{5} = \frac{45}{15} = 3$$

Therefore, the term following 45 in the sequence is found by multiplying 45 by the common ratio 3. The term following 45 is $45 \cdot 3 = \boxed{135}$ and the term following 135 is $135 \cdot 3 = \boxed{405}$. The seventh term is found by using Formula 11.6 for the nth term of a geometric sequence.

$$a_n = a_1 r^{n-1} \qquad \textit{Here we substitute } a_1 = 5,\ n = 7, \textit{ and } r = 3.$$

$$\begin{aligned} a_7 &= a_1 r^6 \\ &= 5 \cdot 3^6 \\ &= 5(729) \\ &= \boxed{3645} \end{aligned}$$

■

EXAMPLE 2 Find the eighth term of the geometric sequence whose first term is 5 and whose fourth term is $\frac{1}{25}$.

Solution First we need to find the common ratio, r. Given $a_1 = 5$ and $a_4 = \frac{1}{25}$, we use Formula 11.6 for the nth term of a geometric sequence as follows:

$$a_n = a_1 r^{n-1} \qquad \textit{Use the fact that we are given } a_1 \textit{ and } a_4 \textit{ and let } n = 4.$$

$$a_4 = a_1 r^3 \qquad \textit{Substitute } a_1 = 5 \textit{ and } a_4 = \frac{1}{25}.$$

$$\frac{1}{25} = 5r^3 \qquad \textit{Now solve for } r.$$

$$\frac{1}{125} = r^3 \Rightarrow \frac{1}{5} = r \qquad \textit{Now we use Formula 11.6 again to find } a_8.$$

$$a_8 = a_1 r^7 = 5\left(\frac{1}{5}\right)^7 = \boxed{\frac{1}{15{,}625}}$$

■

EXAMPLE 3 Ramona deposits \$3500 in a bank account paying an annual interest rate of 6%. Show that the amounts she has in the account at the end of each year form a geometric sequence.

Solution This example is very similar to Example 4 in Section 11.1. Recall that if P dollars is invested at an annual interest rate of r, then the amount of money A in the account after t years is given by the formula $A = P(1 + r)^t$. Therefore, if we let A_k be the amount of money in Ramona's account at the end of year k we have

$$A_k = 3500(1 + 0.06)^k = 3500(1.06)^k \quad \text{and so} \quad \frac{A_{k+1}}{A_k} = \frac{3500(1.06)^{k+1}}{3500(1.06)^k} = 1.06$$

Since the ratio of any two successive terms is a constant, this sequence is geometric.

■

EXAMPLE 4 Suppose a substance loses one-half its radioactive mass per year. If we start with 100 grams of radioactive substance, approximately how much is left after 10 years?

Solution If we record how much is left after each year and note that each term is $\frac{1}{2}$ the previous term, we obtain the following geometric sequence:

Let A_j be the amount left at the end of year j. Then

$$A_1 = \frac{1}{2}(100) \qquad \textit{is the amount left at the end of year 1}$$

$$A_2 = \frac{1}{2} \text{ of } A_1 = \frac{1}{2}\left[\frac{1}{2}(100)\right] = 100\left(\frac{1}{2}\right)^2 \qquad \textit{is the amount left at the end of year 2}$$

$$A_3 = \frac{1}{2} \text{ of } A_2 = \frac{1}{2}\left[100\left(\frac{1}{2}\right)^2\right] = 100\left(\frac{1}{2}\right)^3 \qquad \textit{is the amount left at the end of year 3}$$

In general we see that

$$A_j = 100\left(\frac{1}{2}\right)^j \qquad \textit{is the amount left at the end of year } j$$

Hence, at the end of 10 years there is

$$A_{10} = 100\left(\frac{1}{2}\right)^{10} = 100\left(\frac{1}{1024}\right) = \frac{100}{1024} \approx 0.09766$$

Therefore, there is approximately 0.1 gram of the radioactive substance remaining after 10 years. ■

Geometric Series

If $\{a_k\}$ is a geometric sequence, then its associated series, S_n, is

$$S_n = a_1 + a_2 + a_3 + \cdots + a_{n-1} + a_n$$

and S_n is called the **geometric series** associated with the geometric sequence. Throughout this discussion n is, as usual, a positive integer.

As with arithmetic sequences, we can find a formula to describe a geometric series associated with it. Before we begin to derive the formula, however, we first make note of the following product:

$$(1 - r)(1 + r + r^2 + r^3 + \cdots + r^{n-1}) = 1 - r^n \tag{1}$$

We leave the proof as an exercise (see Exercise 87), but we demonstrate this product below for $n = 6$.

$(1 - r)(1 + r + r^2 + r^3 + r^4 + r^5)$ *Use the distributive property to get*

$$= 1(1 + r + r^2 + r^3 + r^4 + r^5) - r(1 + r + r^2 + r^3 + r^4 + r^5)$$

$$= 1 + r + r^2 + r^3 + r^4 + r^5 - r - r^2 - r^3 - r^4 - r^5 - r^6$$

All but the first and last terms drop out.

$$= 1 - r^6$$

Hence,

$$(1 - r)(1 + r + r^2 + r^3 + r^4 + r^5) = 1 - r^6 \qquad \textit{As required}$$

If we divide both sides of equation (1) by $1 - r$ (assuming $r \neq 1$), we get

$$1 + r + r^2 + r^3 + \cdots + r^{n-2} + r^{n-1} = \frac{1 - r^n}{1 - r} \qquad (2)$$

We use this fact to find a formula for the sum of the first n terms of a geometric sequence as follows: Suppose $\{a_j\}$ is a geometric sequence. If we write out the terms of S_n we get

$$S_n = a_1 + a_2 + a_3 + a_4 + \cdots + a_{n-1} + a_n$$

and since $\{a_j\}$ is a geometric sequence, we can use Formula 11.6, which says that $a_j = a_1 r^{j-1}$, to rewrite each term using a_1 and r:

$$S_n = a_1 + a_1 r + a_1 r^2 + a_1 r^3 + \cdots + a_1 r^{n-2} + a_1 r^{n-1}$$ *Factoring out a_1, we get*

$$= a_1(1 + r + r^2 + r^3 + \cdots + r^{n-2} + r^{n-1})$$

Using the result in (2) on the expression inside the parentheses, we get

$$= a_1\left(\frac{1 - r^n}{1 - r}\right)$$

Thus we have derived Formula 11.7.

THE SUM OF A GEOMETRIC SERIES

The formula for the sum of the first n terms of a geometric sequence is

$$S_n = a_1 + a_1 r + a_1 r^2 + \cdots + a_1 r^{n-1} = \frac{a_1(1 - r^n)}{1 - r} \qquad \text{for } r \neq 1$$

Formula 11.7

Equivalently, we can write the formula for S_n using sigma notation: If $\{a_k\}$ is a geometric sequence with r $\neq$ 1, then

$$S_n = \sum_{k=1}^{n} a_k = \frac{a_1(1 - r^n)}{1 - r}$$

EXAMPLE 5 Given the geometric sequence $\{a_i\} = 6, 18, 54, \ldots$, find

(a) S_7 **(b)** $\sum_{i=1}^{12} a_i$

Solution

Find S_7 by finding the first seven terms of the sequence and adding them.

(a) To find S_7 we could simply find the next four terms of the sequence and add the first seven terms together. Instead, we use Formula 11.7, noting that the first term is $a_1 = 6$ and $r = 3$.

$$S_n = \frac{a_1(1 - r^n)}{1 - r}$$

$$S_7 = \frac{6(1 - 3^7)}{1 - 3}$$

$$= \frac{6(1 - 2187)}{-2} = \frac{6(-2186)}{-2} = \boxed{6558}$$

Does this answer agree with what you obtained by finding the first seven terms in the sequence and adding them together?

(b) Here we are looking for the sum of the first 12 terms in the sequence $\{a_i\}$. Again, $a_1 = 6$ and $r = 3$, but this time $n = 12$.

$$\sum_{i=1}^{12} a_i = S_{12} = \frac{6(1 - 3^{12})}{1 - 3}$$

$$= \frac{6(1 - 531,441)}{-2} = \frac{6(-531,440)}{-2} = \boxed{1,594,320}$$

In order to appreciate the usefulness of this formula, try to compute all 12 terms and find their sum. ■

As the last example shows, the terms in a geometric sequence can grow very rapidly.

Infinite Geometric Series

Thus far we have discussed finite geometric series—that is, we have computed the sum of a finite number of terms in a geometric sequence. We would now like to try to make sense out of the sum of an **infinite geometric sequence**. In other words, how are we to understand the sum

$$a_1 + a_1r + a_1r^2 + \cdots + a_1r^n + \cdots$$

where the intention is to add all the terms in the geometric sequence? Let's look at two examples of infinite geometric sequences:

$$\{a_k\} = 5, 10, 20, 40, \ldots \qquad \{b_k\} = 1, \frac{1}{2}, \frac{1}{4}, \ldots$$

For $\{a_k\}$, note that $a_1 = 5$ and $r = 2$. If we calculate S_5, S_{10}, and S_{20} using Formula 11.7, we get

$$S_5 = \frac{5(1 - 2^5)}{1 - 2} = 155$$

$$S_{10} = \frac{5(1 - 2^{10})}{1 - 2} = 5115$$

$$S_{20} = \frac{5(1 - 2^{20})}{1 - 2} = 5,242,875$$

The sums, S_k, are increasing very rapidly because the terms, a_k, are getting very large. (The expression 2^n gets very large as n increases.) Hence it makes no sense to try to find the sum of the *infinite* sequence $\{a_k\}$.

Looking at the sequence $\{b_k\} = 1, \frac{1}{2}, \frac{1}{4}, \ldots, \frac{1}{2^{k-1}}, \ldots$ we have $b_1 = 1$, and $r = \frac{1}{2}$. We find

$$S_n = \frac{1\left[1 - \left(\frac{1}{2}\right)^n\right]}{1 - \frac{1}{2}} = \frac{\left[1 - \left(\frac{1}{2}\right)^n\right]}{\frac{1}{2}} = 2\left[1 - \left(\frac{1}{2}\right)^n\right]$$

Use a calculator to compute the values of b_n and S_n for $n = 4, 5, 10$, and 20.

If we calculate b_n and S_n for a few values of n, a pattern does emerge. See Table 11.1.

TABLE 11.1

n	1	2	3	4	5	10	20
$b_n = \frac{1}{2^{n-1}}$	1	0.5	0.25	0.125	0.0625	0.0019531	0.00000191
$S_n = 1 + \frac{1}{2} + \cdots + \frac{1}{2^{n-1}}$	1	1.5	1.75	1.875	1.9375	1.9980469	1.99999810

We observe the following:

1. As n increases, it appears that b_n approaches, or gets close to, 0.
2. As n increases, it appears that S_n approaches, or gets close to, 2.

This suggests that as n gets larger and larger, S_n is getting closer and closer to 2. In fact, this is precisely what happens. (We leave the details to a course in calculus.)

Based upon this discussion, we say that the sum of the *infinite geometric series* $1 + \frac{1}{2} + \frac{1}{4} + \cdots$ is 2.

Indeed, we find that there are some cases where we can determine the sum of an infinite series and some cases where we cannot. We now show that this is dependent upon the size of r by generalizing the argument we made before for the sequence $\{b_k\}$.

If we start with a geometric sequence, $\{a_k\}$, we know that the associated series S_n can be found by Formula 11.7 (provided $r \neq 1$):

$$S_n = \frac{a_1(1 - r^n)}{1 - r}$$

Let's suppose $-1 < r < 1$ (usually written as $|r| < 1$) and take a closer look at what happens to the right-hand side of the last equation as n gets very large.

If $|r| < 1$, then r^n will get closer and closer to 0 as n gets larger and larger. $\left(\text{Table 11.1 illustrates this for } r = \frac{1}{2}.\right)$ If n gets "large enough," then r^n gets so close to 0 that it becomes "negligible"—that is, so small that we can ignore the term r^n being subtracted from 1 in the numerator. Thus we have the following.

SUM OF AN INFINITE SERIES

If $|r| < 1$, the sum S of an infinite geometric series $a_1 + a_1r + a_1r^2 + \cdots$ is given by

$$S = \frac{a_1}{1 - r} \qquad \text{Formula 11.8}$$

If $|r| \geq 1$, then the partial sums S_n increase (we say *increase without bound*) and thus we cannot find the sum S.

We can write S using sigma notation by writing the symbol ∞ above the Σ.

$$S = \sum_{n=1}^{\infty} a_1r^{n-1} = \frac{a_1}{1 - r} \qquad \text{for } |r| < 1$$

The symbol ∞ means that we keep increasing the index n without ever reaching an upper limit.

EXAMPLE 6 Find the sum of the given infinite geometric sequence, if possible

(a) $3, \frac{9}{4}, \frac{27}{16}, \frac{81}{64}, \ldots$ **(b)** $\frac{1}{25}, \frac{3}{25}, \frac{9}{25}, \ldots$

Solution

(a) $3, \frac{9}{4}, \frac{27}{16}, \frac{81}{64}, \ldots$ is an infinite geometric sequence with first term 3, and common ratio $r = \frac{\frac{9}{4}}{3} = \frac{3}{4}$. Since $|r| < 1$, the sum S can be found by using Formula 11.8.

$$S = \frac{3}{1 - \frac{3}{4}} = \frac{3}{\frac{1}{4}} = 12$$

Hence $S = 3 + \frac{9}{4} + \frac{27}{16} + \frac{81}{64} + \cdots = \boxed{12}$.

(b) $\frac{1}{25}, \frac{3}{25}, \frac{9}{25}, \ldots$ is an infinite sequence with first term $\frac{1}{25}$ and common ratio $r = \frac{\frac{3}{25}}{\frac{1}{25}} = 3$. Since $|r| > 1$, $\boxed{\text{the sum does not exist}}$.

You may find it instructive to compute S_{10}, S_{20}, S_{30} for the geometric sequences in this example to see that the sums in part (a) are getting closer and closer to 12, whereas the sums in part (b) are getting very large. ■

EXAMPLE 7 Suppose that a ball has the property that it always rebounds to a height equal to $\frac{1}{2}$ the distance from which it falls. Suppose you drop the ball from a height of 10 feet and let the ball bounce "forever." What is the total vertical distance the ball will travel?

Solution First the ball is dropped (or falls) 10 feet. Then the ball rebounds half the distance, 5 feet, and falls 5 feet to the ground; on the next rebound, the balls rises $\frac{5}{2}$ feet and falls $\frac{5}{2}$ feet. See Figure 11.3.

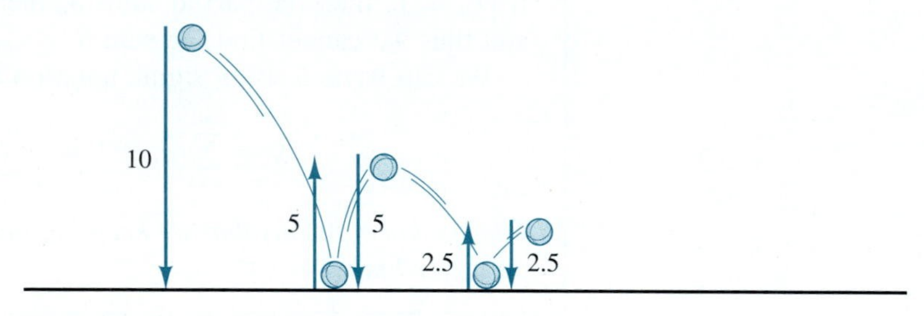

FIGURE 11.3

Continuing in this manner we find that the total vertical distance the ball has traveled is

$$10 + 2(5) + 2\left(\frac{5}{2}\right) + 2\left(\frac{5}{4}\right) + \cdots = 10 + 2\left(5 + \frac{5}{2} + \frac{5}{4} + \cdots\right) \tag{1}$$

Note that in each term after the first, we multiply by 2 because the ball is rising and falling the same distance. Also note that the terms inside the parentheses on the right-hand side form a geometric series with ratio $r = \frac{1}{2}$. Consequently, we can find the sum of the infinite geometric series appearing in equation (1) by using the formula $S = \frac{a_1}{1 - r}$ with $a_1 = 5$ and $r = \frac{1}{2}$.

$$5 + \frac{5}{2} + \frac{5}{4} + \cdots = \frac{5}{1 - \frac{1}{2}} = 10$$

And so equation (1) becomes

$$10 + 2\left(5 + \frac{5}{2} + \frac{5}{4} + \cdots\right) = 10 + 2(10) = 30$$

Thus the ball travels vertically a total of 30 feet before coming to rest. ■

EXAMPLE 8 Express the repeating decimal $0.3535\overline{35}$ as a fraction.

Solution In Example 1 of Section 1.1 we demonstrated one approach to solving this type of problem. Now we offer a different approach, using what we now know about infinite geometric series.

We can rewrite $0.3535\overline{35}$ as an infinite sum in the following way:

$$0.3535\overline{35} = 0.35 + 0.0035 + 0.000035 + \cdots$$

Note that the terms of this series, 0.35, 0.0035, 0.000035, . . . form an infinite geometric sequence; the first term is 0.35, and $r = 0.01$ (each term is 0.01 times the preceeding term). Since $|r| < 1$, we can find the sum using Formula 11.8:

$$0.3535\overline{35} = 0.35 + 0.0035 + 0.000035 + \cdots = \sum_{k=1}^{\infty} 0.35(0.01)^{k-1} = \frac{0.35}{1 - 0.01}$$

$$= \frac{0.35}{0.99} = \frac{35}{99}$$

Thus, $\boxed{0.3535\overline{35} = \frac{35}{99}}$. ■

EXERCISES 11.4

In Exercises 1–8, use the given pattern or general term to determine whether the given sequence is arithmetic, geometric, or neither.

1. 3, 6, 9, 12, . . .
2. 3, 9, 27, 81, . . .
3. $\frac{2}{5}, \frac{1}{10}, \frac{1}{40}, \ldots$
4. $\frac{4}{3}, 8, 48, \cdots$
5. $a_n = 2 + \frac{1}{n}$
6. $b_n = \frac{1}{n^2}$
7. $s_n = \frac{4^n}{7^{n+2}}$
8. $t_n = 7n - 2$

In Exercises 9–22, determine whether the given sequence $\{a_n\}$ is arithmetic or geometric (assume it is one or the other) and find the indicated a_k.

9. 1, 5, 25, 125, . . . ; find a_7.
10. −2, 6, −18, 54, . . . ; find a_6.
11. $\frac{1}{2}, -\frac{1}{3}, \frac{2}{9}, \ldots$; find a_5.
12. $\frac{5}{4}, 5, 20, \ldots$; find a_5.
13. 1, 8, 15, . . . ; find a_{10}.
14. 1, 8, 64, . . . ; find a_{10}.
15. 9, 3, 1, . . . ; find a_6.
16. 9, 3, −3, . . . ; find a_6.
17. $2, 1, \frac{1}{2}, \frac{1}{4}, \ldots$; find a_9.
18. $-5, \frac{5}{6}, -\frac{5}{36}, \ldots$; find a_5.
19. $8, 2, \frac{1}{2}, \frac{1}{8}, \ldots$; find a_9.
20. $9, \frac{9}{10}, \frac{9}{100}, \ldots$; find a_8.
21. $\sqrt{5}, 5, 5\sqrt{5}, 25, \ldots$; find a_7.
22. $1, -\frac{1}{\sqrt{2}}, \frac{1}{2}, \ldots$; find a_9.

In Exercises 23–28, use the given information about a geometric sequence to find the indicated a_n.

23. $a_1 = 10$ and $r = 2$; find a_4.
24. $a_1 = 4$ and $r = -3$; find a_6.
25. $a_3 = 1$ and $a_6 = 216$; find r and a_1.
26. $a_4 = -4$ and $a_9 = 128$; find a_7.
27. $a_2 = \frac{1}{\sqrt{3}}$ and $a_5 = -\frac{1}{9}$; find r and a_8.
28. $a_5 = 2$ and $a_9 = 4$; find a_7.

29. Find the fifth term of the geometric sequence whose first term is 1 and whose fourth term is 343.

30. Find the third term of the arithmetic sequence whose second term is 1 and whose fifth term is 64.

31. Find the eighth term of the arithmetic sequence whose first term is 5 and whose fourth term is -40.

32. Find the first term of the geometric sequence whose second term is 3 and whose seventh term is -96.

33. Find the seventh term of the geometric sequence whose first term is 1 and whose fifth term is $\frac{1}{81}$.

34. Find the sixth term of the geometric sequence whose first term is 2 and whose fourth term is $\frac{2}{27}$.

You may find the interest formula in Section 5.5 helpful in the next few exercises.

35. A certain luxury car loses one-tenth its value each year. If the car is worth \$28,000 today, how much will it be worth 4 years from now?

36. A boat is now worth \$34,000 and loses 12% of its value each year. What will it be worth in 5 years?

37. A machine depreciates in value by one-fifth each year. If the machine is now worth \$29,000, how much will it be worth 3 years from now?

38. An antique chair is now worth \$9500. If it appreciates in value at the rate of 8% per year, how much will it be worth 5 years from now?

39. The population of Porterville is increasing at a rate of 2.5% per year. If the population is currently 100,000, what will the population be 10 years from now?

40. The population of Hilltown is decreasing at a rate of 3% per year. If the population is currently 65,000, what will the population be 4 years from now?

41. Roberta invests \$10,000 in a bank that pays 5% simple annual interest. How much will Roberta have in the bank at the end of 5 years?

42. How much would Roberta (Exercise 41) have in the bank at the end of 5 years if the bank paid 5% annual interest compounded quarterly?

43. Teresa invests \$5000 in a bond that pays 6% interest compounded semiannually. How much is her bond worth at the end of 6 years?

44. How much will her bond be worth at the end of 10 years?

45. A ball bounces up one-half the distance from which it falls. How far does it bounce up on the fifth rebound if the ball is dropped from a height of 20 feet?

46. A ball bounces up one-third the distance from which it falls. How far does it bounce up on the fourth rebound if the ball is dropped from a height of 27 feet?

In Exercises 47–62, the given sequence is either arithmetic or geometric. Find the indicated S_n. Round to four decimal places where necessary.

47. $4, 12, 36, \ldots; \quad S_5$

48. $-3, 6, -12, \ldots; \quad S_8$

49. $\frac{2}{3}, -\frac{2}{5}, \frac{6}{25}, \ldots; \quad S_6$

50. $1, 4, 16, \ldots; \quad S_7$

51. $2, 6, 10, \ldots; \quad S_8$

52. $2, 6, 18, \ldots; \quad S_8$

53. $-\frac{3}{4}, \frac{1}{4}, -\frac{1}{12}, \ldots; \quad S_5$

54. $-\frac{3}{4}, \frac{1}{4}, \frac{5}{4}, \ldots; \quad S_5$

55. $\frac{1}{4}, \frac{1}{2}, 1, \ldots; \quad S_{10}$

56. $3, \frac{3}{10}, \frac{3}{100}, \ldots; \quad S_6$

57. $a_k = 2^k; \quad S_7$

58. $b_j = \left(\frac{2}{5}\right)^j; \quad S_9$

59. $a_k = (-1)^{k+1}\left(\frac{1}{3}\right)^k; \quad S_8$

60. $b_j = (-.4)^k; \quad S_{20}$

61. $a_n = 2n + 1; \quad S_7$

62. $a_n = 2^{n+1}; \quad S_7$

In Exercises 63–66, find each sum for the geometric sequence $\{a_i\} = 1, \frac{1}{10}, \frac{1}{100}, \ldots$.

63. $\sum_{i=1}^{6} a_i$

64. $\sum_{j=1}^{8} 5a_j$

65. $\sum_{k=3}^{10} a_k$

66. $\sum_{n=1}^{\infty} a_n$

67. A ball is dropped from a height of 60 feet and always rebounds one-third the distance from which it falls. Find the total distance the ball has traveled when it hits the ground for the fifth time.

68. Greg promises to pay Lamar a debt in the following way: Greg will pay 1¢ on the first day, 2¢ on the second day, 4¢ on the third day, and so on, where the payment on a particular day is double the previous day. At this rate, how much will Greg have paid Lamar altogether at the end of 30 days?

69. Suppose a chain letter is started by sending a letter to five people and the letter requires each person who receives the letter to send it on to five additional people. Assume that no person receives more than one letter.

(a) Show that the number of letters being sent at each stage form a geometric sequence.

(b) How many letters are sent at the fourth stage?

(c) How many letters are sent altogether if the chain reaches the eighth stage?

70. Repeat Exercise 69 if the letter requires you to send the letter on to six people.

In Exercises 71–80, find S, the sum of the infinite geometric series (if it exists).

71. $2 + 1 + \frac{1}{2} + \frac{1}{4} + \cdots$

72. $1 + \frac{2}{3} + \frac{4}{9} + \cdots$

73. $\sum_{k=1}^{\infty} \frac{9}{3^k}$

74. $\sum_{k=1}^{\infty} 4^{3-k}$

75. $\frac{1}{5} + \frac{1}{10} + \frac{1}{20} + \cdots$

76. $\frac{1}{5} + -\frac{1}{10} + \frac{1}{20} + \cdots$

77. $\sum_{n=1}^{\infty} 5\left(\frac{1}{3}\right)^{n-1}$

78. $\sum_{j=1}^{\infty} 2^{j-2}$

79. $\frac{1}{5} + \frac{4}{15} + \frac{16}{25} + \cdots$

80. $7 + \frac{7}{10} + \frac{7}{100} + \cdots$

81. Express $0.474\overline{747}$ as a fraction.

82. Express $3.139\overline{139}$ as a fraction.

83. Express $1.0681\overline{681}$ as a fraction.

84. Express $5.9871\overline{71}$ as a fraction.

85. Suppose a ball is dropped from a height of 15 ft and that it always rebounds to a height equal to two-thirds of the distance from which it falls. If you let the ball bounce "forever," what is the total distance it would travel until it comes to rest?

86. A pendulum sweeps out a 9-ft arc on its first pass; each succeeding time it passes, it travels seven-eighths of the distance of the previous arc. How far would it travel before coming to rest?

QUESTIONS FOR THOUGHT

87. Show that

$$(1 - r)(1 + r + r^2 + r^3 + \cdots + r^{n-1}) = 1 - r^n.$$

88. Show that the terms of the sum, $\sum_{i=1}^{N} cb^i$, where c is a constant, form a geometric sequence. What is the common ratio?

89. Given the result of Exercise 88, find each sum.

(a) $\sum_{i=1}^{10} 4 \cdot 5^i$ **(b)** $\sum_{i=1}^{10} 3 \cdot 2^i$

90. Explain what happens to the partial sums S_n for the series $\sum_{k=1}^{\infty} 2^k$ as n increases. Explain why the formula $S = \frac{1}{1-r}$ does not make any sense for this series.

11.5 Mathematical Induction

Throughout the text we have proven some of the facts (theorems) that we have stated and outlined the proofs of others. We have done algebraic proofs, geometric proofs, and trigonometric proofs. In this section we discuss a very different method for proving certain facts about natural numbers.

Let n be a natural number and consider the following collection of statements, which we call S_n.

$$S_n\text{: The sum of the first } n \text{ odd natural numbers is } n^2.$$

S_8 is the statement that the sum of the first 8 odd natural numbers is $8^2 = 64$; S_{12} is the statement that the sum of the first 12 odd natural numbers is $12^2 = 144$. For each natural number n, we obtain a different statement S_n. In other words, the S_n's form a sequence of statements.

We can easily verify that S_8 and S_{12} are true by adding the first 8 or the first 12 odd natural numbers. How might we prove or disprove S_n for all natural numbers n?

Next consider the following collection of statements, which we call T_n.

$$T_n\text{: } n^2 - n + 41 \text{ is a prime number.}$$

T_5 is the statement that $5^2 - 5 + 41 = 61$ is a prime number. T_9 is the statement that $9^2 - 9 + 41 = 113$ is a prime number. For each natural number n, we obtain a different statement T_n. In other words, the T_n's also form a sequence of statements.

We can check that both 61 and 113 are prime numbers, so that T_5 and T_9 are true statements. How might we prove or disprove T_n for all natural numbers n?

Before we prove a statement to be true or false, it is often called a *conjecture*. Let's begin by examining the truth of S_n and T_n for some values of n. The calculations appear in Table 11.2. (In effect, we are acting much like a scientist does by performing an experiment to help us decide whether or not to believe a particular statement.)

TABLE 11.2

	Statement			Statement	
n	$S_n: 1 + 3 + 5 + \cdots + (2n - 1) = n^2$		True?	$T_n: n^2 - n + 41$ is a prime number	True?
1	$S_1: 1$	$= 1^2$	True	$T_1: 1^2 - 1 + 41 = 41$ is prime.	True
2	$S_2: 1 + 3$	$= 2^2$	True	$T_2: 2^2 - 2 + 41 = 43$ is prime.	True
3	$S_3: 1 + 3 + 5$	$= 3^2$	True	$T_3: 3^2 - 3 + 41 = 47$ is prime.	True
4	$S_4: 1 + 3 + 5 + 7$	$= 4^2$	True	$T_4: 4^2 - 4 + 41 = 53$ is prime.	True
5	$S_5: 1 + 3 + 5 + 7 + 9$	$= 5^2$	True	$T_5: 5^2 - 5 + 41 = 61$ is prime.	True

Based upon this evidence we might easily conclude that both statements, S_n and T_n, are true. Keep in mind that for one of these conjectures to be true, it must be true for *all* natural numbers n, whereas for one of these conjectures to be false, it needs to be shown false for only one natural number n. It turns out that S_n is true (which we will prove a bit later in this section), but T_n is false.

In fact, T_n is true for all natural numbers up to and including $n = 40$. However, for $n = 41$ we get $41^2 - 41 + 41 = 41^2$, which is not a prime. Thus despite the fact that T_n is true for the first 40 natural numbers, it is not true in general. This "prime-generating" polynomial $n^2 - n + 41$ is generally credited to Leonhard Euler.

The behavior of T_n reinforces what we already know to be true: We certainly cannot check every possible natural number, and regardless of how many values of n for which we verify a conjecture, the conjecture still needs to be *proven* true. How then can we prove that a conjecture is true for *all natural numbers n?*

One of the most powerful methods for proving a conjecture true for all natural numbers is to use the *principle of mathematical induction,* which can be stated as follows.

THE PRINCIPLE OF MATHEMATICAL INDUCTION

Suppose P_n is a statement involving the natural number n for which the following two conditions hold.

1. P_1 is true, and
2. For each natural number k, if P_k is true, then P_{k+1} is true. (We often write this as $P_k \Rightarrow P_{k+1}$.)

Then P_n is true for all natural numbers n.

The idea underlying mathematical induction is a fairly simple one. Condition 1 says P_1 must be true. Condition 2 says that whenever P_k is true, it must follow that P_{k+1} is true. Therefore, by condition 2 if P_1 is true, then P_2 must be true. But again by condition 2, if P_2 is true, then P_3 must be true and if P_3 is true, then P_4 must also be true, and so on. In this way we know that P_n is true for all n. These two conditions are often called hypotheses, since we do not know whether or not they are true in a particular case.

We can think of the statements P_n as a sequence of dominoes, as indicated in Figure 11.4.

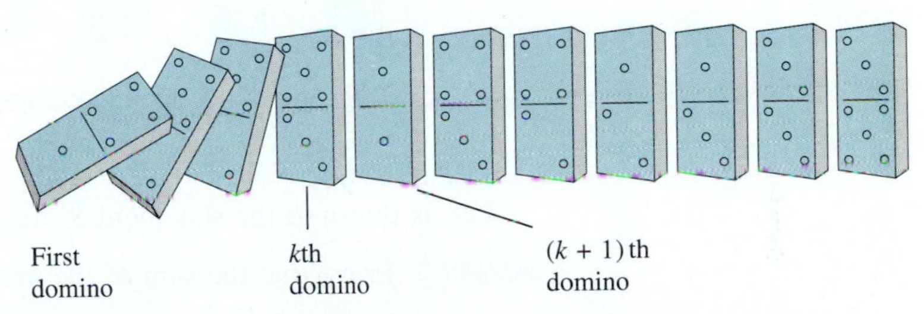

FIGURE 11.4 The domino analogy for mathematical induction

Here is how the steps in the principle of mathematical induction listed in the box correspond to the domino analogy.

Mathematical Induction

Hypotheses:
1. P_1 is true.
2. If P_k is true, then P_{k+1} is true for any k.

Conclusion: P_n is true for all n.

Domino Analogy

Hypotheses:
1. Knock over the first domino.
2. All the dominoes are set up in such a way that if any domino is knocked over, then the next domino is knocked over.

Conclusion: All the dominoes are knocked over

Note that in the domino analogy, in order to conclude that all the dominoes are knocked down, *both* hypotheses are necessary. If all we know is that the first domino is knocked over without knowing how closely the dominoes are set up, we cannot conclude that all the dominoes are knocked over. See Figure 11.5.

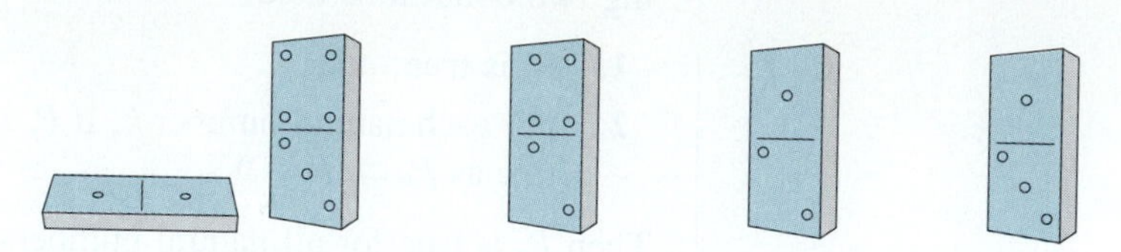

FIGURE 11.5 The domino analogy with hypothesis 2 not satisfied

Similarly, even if we know that the dominoes are set up closely enough so that if one is knocked over, the next one will fall also, unless we know that the first one is actually knocked over, we cannot conclude that they will all fall down. See Figure 11.6.

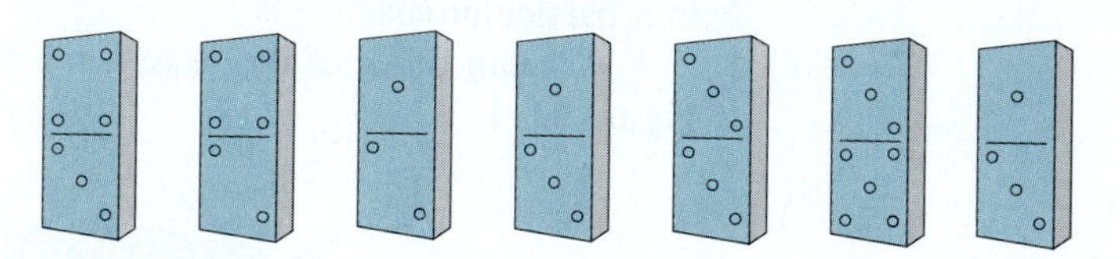

FIGURE 11.6 The domino analogy with hypothesis 1 not satisfied

Let us return to the statement S_n discussed earlier in this section.

EXAMPLE 1 Prove that the sum of the first n odd natural numbers is equal to n^2.

Solution We want to prove that

$$S_n: \quad 1 + 3 + 5 + \cdots + (2n - 1) = n^2$$

We use the principle of mathematical induction to prove the truth of S_n for all natural numbers n.

Step 1: We must verify that S_1 is true. However, S_1 is just the statement that $1 = 1^2$, which is true.

Step 2: *Assuming* that S_k is true, we must prove that S_{k+1} is true.

Let's examine S_k and S_{k+1}.

$$S_k: \quad 1 + 3 + 5 + \cdots + (2k - 1) = k^2 \quad \textit{This is called the induction hypothesis.} \quad (1)$$

S_{k+1} is obtained by substituting $k + 1$ for n in S_n.

S_{k+1}: $1 + 3 + 5 + \cdots + (2k - 1) + [2(k + 1) - 1] = (k + 1)^2$

Simplify $2(k + 1) - 1$.

$$S_{k+1}: \quad 1 + 3 + 5 + \cdots + (2k - 1) + (2k + 1) = (k + 1)^2 \tag{2}$$

Our goal in Step 2 is to show that S_{k+1} follows from S_k, or in other words, that we can derive equation (2) from equation (1). How do we derive equation (2) from equation (1)? We look at equations (1) and (2) and observe that the *left-hand sides* of equations (1) and (2) differ only by the quantity $(2k + 1)$, so we add $(2k + 1)$ to both sides of equation (1).

$1 + 3 + 5 + \cdots + (2k - 1) = k^2$ *This is equation (1). We add* $2k + 1$ *to both sides to obtain*

$1 + 3 + 5 + \cdots + (2k - 1) + (2k + 1) = k^2 + 2k + 1$

Factor the right-hand side to get

$1 + 3 + 5 + \cdots + (2k - 1) + (2k + 1) = (k + 1)^2$ *Which is equation (2)*

Since equation (2) follows from equation (1), we have demonstrated that $S_k \Rightarrow S_{k+1}$. Thus we have satisfied conditions (1) and (2) of the principle of mathematical induction and so we have proven that S_n is true for all n. ■

Students are sometimes a bit confused by the induction hypothesis because they think that we are assuming what we are supposed to prove, but this is not so. We are not proving that either P_k or P_{k+1} is true. Rather, in order to carry out step 2 in an induction proof, we must prove that *if* P_k *is true, it must necessarily follow that* P_{k+1} *is true*.

EXAMPLE 2 Prove that $1 + 4 + 9 + \cdots + n^2 = \dfrac{n(n + 1)(2n + 1)}{6}$ is true for all natural numbers n.

Solution Let P_n: $1 + 4 + 9 + \cdots + n^2 = \dfrac{n(n + 1)(2n + 1)}{6}$.

Step 1: We must verify that P_1 is true. P_1 is the statement that $1 = \dfrac{1(1 + 1)(2(1) + 1)}{6} = \dfrac{1(2)(3)}{6} = \dfrac{6}{6}$, which is true.

Step 2: *Assuming* that P_k is true, we must prove that P_{k+1} is true. Again, we begin by examining P_k and P_{k+1}.

$$P_k: \quad 1^2 + 2^2 + 3^2 + \cdots + k^2 = \frac{k(k + 1)(2k + 1)}{6} \quad \textit{This is the induction hypothesis.} \tag{1}$$

$$P_{k+1}: \quad 1^2 + 2^2 + 3^2 + \cdots + k^2 + (k + 1)^2 = \frac{(k + 1)[(k + 1) + 1][2(k + 1) + 1]}{6}$$

Simplifying, we get

$$P_{k+1}: \quad 1^2 + 2^2 + 3^2 + \cdots + k^2 + (k + 1)^2 = \frac{(k + 1)(k + 2)(2k + 3)}{6} \tag{2}$$

Step 2 requires us to derive equation (2) from equation (1). By examining equations (1) and (2), we notice that their left-hand sides differ by $(k + 1)^2$, so we add $(k + 1)^2$ to both sides of equation (1).

$$1^2 + 2^2 + 3^2 + \cdots + k^2 = \frac{k(k + 1)(2k + 1)}{6}$$

This is equation (1). Add $(k + 1)^2$ to both sides.

$$1^2 + 2^2 + 3^2 + \cdots + k^2 + (k + 1)^2 = \frac{k(k + 1)(2k + 1)}{6} + (k + 1)^2$$

$$= \frac{k(k + 1)(2k + 1) + 6(k + 1)^2}{6}$$

Factor out $(k + 1)$

$$= \frac{(k + 1)[k(2k + 1) + 6(k + 1)]}{6}$$

$$= \frac{(k + 1)[2k^2 + 7k + 6]}{6}$$

Factor again.

$$1^2 + 2^2 + 3^2 + \cdots + k^2 + (k + 1)^2 = \frac{(k + 1)(k + 2)(2k + 3)}{6}$$

This is equation (2).

Remember that in step 2 of an induction proof, we are not asserting that either P_k or P_{k+1} is true. We are proving only that $P_k \Rightarrow P_{k+1}$.

Thus we have derived equation (2) from equation (1), and we have verified both conditions of mathematical induction, so we may conclude that the given formula for the sum of the squares of the first n natural numbers is true for all n.

Formulas such as the one in this example and similar ones in the exercise set come up in calculus. ■

EXAMPLE 3 Prove that the statement S_n: $(x + y)^n = x^n + y^n$ is not true for *all* natural numbers n.

Recall that such a number n is called a counterexample.

Solution In order to prove that a statement is not true for all natural numbers n, it is sufficient to find *one* natural number for which the given statement is not true. For example S_2 says that $(x + y)^2 = x^2 + y^2$, but we know that

$$(x + y)^2 = x^2 + 2xy + y^2 \neq x^2 + y^2$$

Therefore, we have demonstrated that S_2 is false and so S_n is not true for all natural numbers n. ■

Sometimes a statement P_n is not true for some initial values of n but is then true for all larger values of n. For example, consider the statement

$$P_n: \quad 2^n > n^2$$

We can easily check that this statement is true for $n = 1$, false for $n = 2, 3, 4$, and true for $n = 5$. However, the logic of mathematical induction applies just as well if we start with $n = 1$, $n = 5$, or any other value of n. In the domino analogy, this corresponds to skipping the first few dominoes and knocking over the fifth domino, for instance, and then causing all dominoes from the fifth one on to fall down.

EXAMPLE 4 Prove that the statement P_n: $2^n > n^2$ is true for all $n \geq 5$.

Solution In order to prove P_n is true for $n > 5$, we will use the following fact, which you are asked to prove in Exercise 31.

$$n^2 > 2n + 1 \qquad \text{for } n \geq 3 \qquad (*)$$

Step 1: We must verify that P_5 is true. P_5 is the statement that $2^5 > 5^2$, which is true.

Step 2: *Assuming* that P_k is true, we must prove that P_{k+1} is true; that is, we assume $2^k > k^2$ and we must prove $2^{k+1} > (k+1)^2$, or, equivalently, that $2^{k+1} > k^2 + 2k + 1$.

We start with the induction hypothesis.

$$2^k > k^2 \qquad \textit{Multiply both sides of this inequality by 2.}$$

$$2(2^k) > 2k^2$$

$$2^{k+1} > 2k^2 = k^2 + k^2 > k^2 + 2k + 1 \qquad \textit{because by } (*)\ k^2 > 2k + 1. \textit{ Therefore } 2^{k+1} > k^2 + 2k + 1 \textit{ as required.}$$

We have thus satisfied both conditions of mathematical induction for $n \geq 5$ and so we have $2^n > n^2$ for $n \geq 5$. ■

We conclude this section with an example illustrating that mathematical induction can also be used to prove certain divisibility properties of natural numbers.

EXAMPLE 5 Prove that for all natural numbers n, $4^n - 1$ is divisible by 3.

Solution Before we proceed, let's keep in mind that in order for a natural number N to be divisible by 3, we must have $N = 3M$ for some natural number M. (For example, 21 is divisible by 3 because $21 = 3 \cdot 7$.)

Step 1: We must verify that P_1 is true. P_1 is the statement that $4^1 - 1 = 3$ is divisible by 3, which is true.

Step 2: *Assuming* that P_k is true, we must prove that P_{k+1} is true. That is, we assume that $4^k - 1$ is divisible by 3 and we must prove that $4^{k+1} - 1$ is also divisible by 3. Let's look at $4^{k+1} - 1$.

$$4^{k+1} - 1 = 4(4^k) - 1 \qquad \textit{We rewrite } -1 \textit{ as } -4 + 3.$$

$$= 4(4^k) - 4 + 3 \qquad \textit{Factor out 4 from the first two terms.}$$

$$= 4(4^k - 1) + 3 \qquad \textit{By the induction hypothesis, } 4^k - 1 \textit{ is divisible by 3. Therefore, } 4^k - 1 = 3N \textit{ for some natural number } N.$$

$$= 4(3N) + 3 \qquad \textit{By factoring out 3, we see that the right-hand side is a multiple of 3.}$$

$$4^{k+1} - 1 = 3(4N + 1) \qquad \textit{Which means that } 4^{k+1} - 1 \textit{ is divisible by 3 as required.}$$

Therefore, by the principle of mathematical induction, $4^n - 1$ is divisible by 3 for all natural numbers n. ■

Throughout mathematical history there have been many conjectures about natural numbers. Here are just a few of the more famous ones and their current status. rent status.

1. We are all familiar with natural numbers a, b, and c such that $a^2 + b^2 = c^2$. For example, $3^2 + 4^2 = 5^2$. We have encountered many such *Pythagorean triples* as the sides of a right triangle.

We may ask the following question: For $n > 2$, are there any natural numbers a, b, and c such that $a^n + b^n = c^n$? In other words, are there natural numbers a, b, and c such that $a^3 + b^3 = c^3$ or $a^8 + b^8 = c^8$, etc.? In 1637, the great mathematician Pierre de Fermat (1601–1665) wrote, in the margin of a book, that he had found a "marvelous" proof that there are no natural number solutions to $a^n + b^n = c^n$ for $n > 2$ but that "the margin was too narrow to contain it." This conjecture is known as "Fermat's Last Theorem."

There is much controversy about whether or not Fermat actually had a proof, but for more than 350 years mathematicians of the first rank have attempted unsuccessfully to prove Fermat's Last Theorem. In June of 1993, Dr. Andrew Wiles announced that he had proved certain results that in turn prove Fermat's Last Theorem. The mathematical world eagerly awaits the verification of this proof.

2. It was conjectured from the times of the early Greek mathematicians that every natural number can be expressed as the sum of the squares of four integers. For example, $39 = 1^2 + 2^2 + 3^2 + 5^2$ and $11 = 0^2 + 1^2 + 1^2 + 3^2$. In 1770, Joseph-Louis Lagrange, probably the greatest mathematician of the eighteenth century, finally gave the first complete proof of this fact.

3. In 1742 the mathematician Goldbach conjectured that every even natural number greater than 2 can be written as the sum of two prime numbers. For example, $20 = 3 + 17$ and $34 = 11 + 23$. Goldbach's conjecture still awaits a proof.

EXERCISES 11.5

In Exercises 1–20, use the principle of mathematical induction to prove the given statement is true for all natural numbers n.

1. P_n: $1 + 3 + 5 + \cdots + (2n - 1) = n^2$

2. S_n: $2 + 4 + 6 + \cdots + 2n = n^2 + n$

3. T_n: $3 + 7 + 11 + \cdots + (4n - 1) = n(2n + 1)$

4. R_n: $1 + 4 + 7 + \cdots + (3n - 2) = \dfrac{n(3n - 1)}{2}$

5. P_n: $2 + 5 + 8 + \cdots + (3n - 1) = \dfrac{n(3n + 1)}{2}$

6. S_n: $1^3 + 2^3 + 3^3 + \cdots + n^3 = \left(\dfrac{n(n + 1)}{2}\right)^2$

7. R_n: $2 + 9 + 16 + \cdots + (7n - 5) = \dfrac{n(7n - 3)}{2}$

8. T_n: $5 + 9 + 13 + \cdots + (4n + 1) = n(2n + 3)$

9. P_n: $2^2 + 4^2 + 6^2 + \cdots + (2n)^2 = \dfrac{2n(n + 1)(2n + 1)}{3}$

10. P_n: $1^2 + 3^2 + 5^2 + \cdots + (2n + 1)^2 = \dfrac{n(2n - 1)(2n + 1)}{3}$

11. P_n: $2 + 2^2 + 2^3 + \cdots + 2^n = 2^{n+1} - 2$

12. P_n: $3 + 3^2 + 3^3 + \cdots + 3^n = \dfrac{3^{n+1} - 3}{2}$

13. T_n: $\dfrac{1}{1 \cdot 2} + \dfrac{1}{2 \cdot 3} + \dfrac{1}{3 \cdot 4} + \cdots + \dfrac{1}{n(n + 1)} = \dfrac{n}{n + 1}$

14. T_n: $\dfrac{1}{1 \cdot 3} + \dfrac{1}{3 \cdot 5} + \dfrac{1}{5 \cdot 7} + \cdots + \dfrac{1}{(2n - 1)(2n + 1)} = \dfrac{n}{2n + 1}$

15. S_n: $1 \cdot 2 + 2 \cdot 3 + 3 \cdot 4 + \cdots + n(n + 1) = \dfrac{n(n + 1)(n + 2)}{3}$

16. S_n: $1 \cdot 2 + 3 \cdot 4 + 5 \cdot 6 + \cdots + (2n - 1)(2n) = \dfrac{n(n + 1)(4n - 1)}{3}$

17. T_n: $\left(1 + \dfrac{1}{1}\right)\left(1 + \dfrac{1}{2}\right)\left(1 + \dfrac{1}{3}\right) \cdots \left(1 + \dfrac{1}{n}\right) = n + 1$

18. T_n: $\left(1 - \dfrac{1}{2}\right)\left(1 - \dfrac{1}{3}\right)\left(1 - \dfrac{1}{4}\right) \cdots \left(1 - \dfrac{1}{n + 1}\right) = \dfrac{1}{n + 1}$

19. S_n: $a + (a + d) + (a + 2d) + \cdots + [a + (n - 1)d] = n\left(a + \dfrac{n - 1}{2}d\right)$

20. S_n: $a + ar + ar^2 + \cdots + ar^n = \dfrac{a(1 - r^n)}{1 - r}$ for $r \neq 1$

21. Prove that $n^3 - n + 6$ is divisible by 3 for all natural numbers n.

22. Prove that 5 is a factor of $6^n - 1$ for all natural numbers n.

23. Use mathematical induction to prove that $x - 1$ is a factor of $x^n - 1$ for all natural numbers n. Can you think of another way to prove this fact?

24. Use mathematical induction to prove that $n^2 + n$ is divisible by 2 for all natural numbers n. Can you think of another way to prove this fact?

25. Prove that $x - y$ is a factor of $x^n - y^n$ for all natural numbers n.

26. Use the result of Exercise 6 to show that
$$\sum_{k=1}^{n} k^3 = \left(\sum_{k=1}^{n} k\right)^2$$

27. Prove that $9^n - 1$ is divisible by 4 for all natural numbers n.

28. Prove that $n^3 + 3n^2 + 2n$ is divisible by 6 for all natural numbers n. Can you think of another way to prove this fact?

29. Use mathematical induction to prove DeMoivre's theorem, which says that $[r(\cos\theta + i\sin\theta)]^n = r^n(\cos n\theta + i\sin n\theta)$.

30. Use mathematical induction to prove that for all natural numbers n, $(a - bi)^n$ is the complex conjugate of $(a + bi)^n$.

In Exercises 31–40, use the principle of mathematical induction.

31. Prove that $n^2 > 2n + 1$ for $n \geq 3$.

32. Prove that $(n + 2)^2 < n^3$ for $n \geq 3$.

33. Prove that $\left(\dfrac{7}{5}\right)^n > n$ for $n \geq 5$.

34. Prove that $(1.1)^n > n$ for $n \geq 39$.

35. Prove that $n^2 > 10n$ for $n > 10$. Can you think of another way to prove this fact? HINT: Think about the graph of $y = x^2 - 10x$.

36. Prove that $n^2 > 3n + 10$ for $n > 5$. Can you think of another way to prove this fact? HINT: Think about the graph of $y = x^2 - 3x - 10$.

37. Prove that $n! > 2^n$ for $n \geq 4$.

38. Prove that $n^n > n!$ for $n > 1$.

39. Assuming the truth of the triangle inequality, which says that $|a + b| \leq |a| + |b|$, prove that $|a_1 + a_2 + \cdots + a_n| \leq |a_1| + |a_2| + \cdots + |a_n|$.

40. Prove each of the following inequalities is true for all natural numbers n.
   (a) $\dfrac{1}{\sqrt{1}} + \dfrac{1}{\sqrt{2}} + \dfrac{1}{\sqrt{3}} + \cdots + \dfrac{1}{\sqrt{n}} \geq \sqrt{n}$
   (b) $\dfrac{1}{\sqrt{1}} + \dfrac{1}{\sqrt{2}} + \dfrac{1}{\sqrt{3}} + \cdots + \dfrac{1}{\sqrt{n}} < 2\sqrt{n}$
   (c) Based upon the results of parts (a) and (b), what can you conclude about
   $\dfrac{1}{\sqrt{1}} + \dfrac{1}{\sqrt{2}} + \dfrac{1}{\sqrt{3}} + \cdots + \dfrac{1}{\sqrt{n}}$?

In Exercises 41–46, find the smallest natural number N for which the statement is true, and then prove the statement is true for all natural numbers greater than or equal to N.

41. $\left(\dfrac{4}{3}\right)^n > \dfrac{4}{3}n$

42. $n + 20 \leq n^2$

43. $n^3 > (n + 5)^2$

44. $n^2 > 2n$

45. $\log_2 n < \sqrt{n}$. HINT: $k + 1 < 2k$

46. $\dfrac{1}{n + 5} > \dfrac{1}{2n}$

In Exercises 47–58, determine what can be concluded from the given information about the sequence of statements P_n. For example, if P_6 is true and $P_k \Rightarrow P_{k+1}$ for all k, then we can conclude that P_n is true for all $n \geq 6$.

47. P_k is true for $1 \leq k < 50$.

48. P_1 is true but $P_k \not\Rightarrow P_{k+1}$.

49. P_{10} is true and $P_k \Rightarrow P_{k+1}$.

50. P_{10} is not true and $P_k \Rightarrow P_{k+1}$.

51. P_{10} is not true and $P_k \not\Rightarrow P_{k+1}$.

52. P_{10} is true and $P_k \not\Rightarrow P_{k+1}$.

53. P_1 is true and $P_k \Rightarrow P_{k+2}$.

54. P_1 and P_2 are true and P_k and P_{k+1} together imply P_{k+2}.

55. P_1 and P_2 are true and $P_k \Rightarrow P_{k+2}$.

56. P_{100} is true and $P_k \Rightarrow P_{k-1}$.

57. P_1 is true and $P_k \Rightarrow P_{3k}$.

58. P_1 is true and $P_k \Rightarrow P_{k+3}$.

QUESTIONS FOR THOUGHT

59. True or false: $5^n + 3$ is divisible by 4 for all natural numbers n. Justify your answer.

60. True or false: $n^2 + n + 11$ is a prime number for all natural numbers n. Justify your answer.

11.6 Permutations and Combinations

In this section we discuss some basic counting techniques. Let's consider the following situation.

EXAMPLE 1 Suppose Gerald wants to buy a computer, monitor, and printer. He is considering two choices for the computer, let's call them C_1 and C_2; two choices for the monitor, M_1 and M_2; and three choices for the printer, P_1, P_2, and P_3. Assuming that all the components are compatible, how many different computer systems of computer, monitor, and printer can Gerald put together?

Solution

In order to count the number of possible configurations we must consider the number of different ways we can choose a computer, monitor, and printer. One way to analyze the number of possibilities is by using a **tree diagram**, as illustrated in Figure 11.7. In the figure, $C_1M_2P_3$, for example, means that Gerald has chosen computer C_1, monitor M_2, and printer P_3.

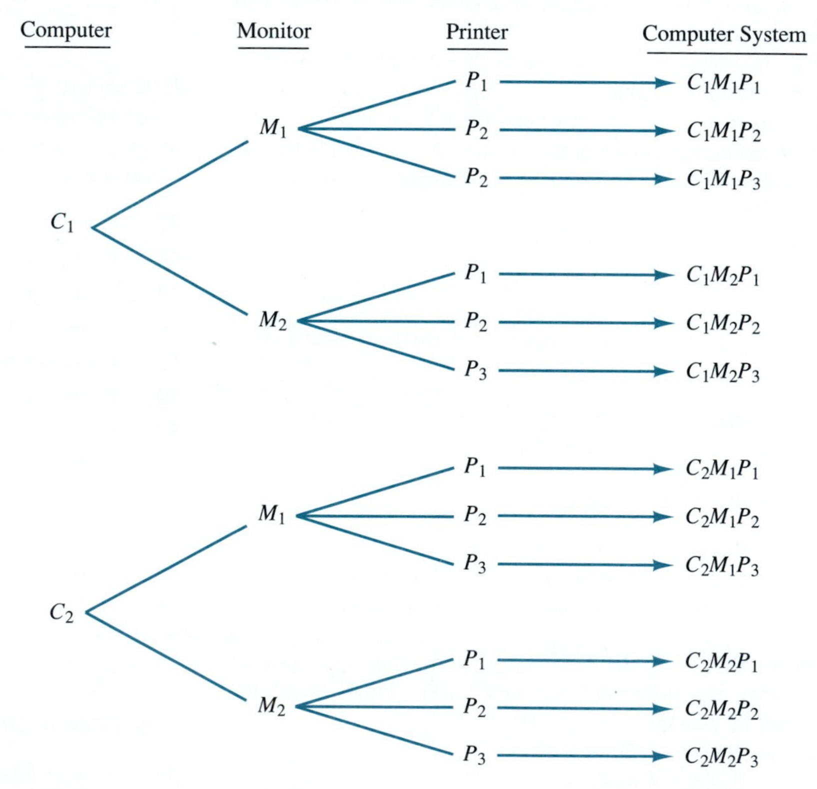

FIGURE 11.7

From this tree diagram we can see that there are $\boxed{12}$ possible configurations for the computer system. There are 2 choices for the computer, 2 choices for the monitor, and 3 choices for the printer, and we note that $12 = 2 \cdot 2 \cdot 3$. ■

Similarly, if Nina has 3 skirts and 5 blouses she can make $3 \cdot 5 = 15$ different outfits (assuming that she can wear any skirt with any blouse).

The situations we have just described are merely special cases of the *Fundamental Counting Principle*.

THE FUNDAMENTAL COUNTING PRINCIPLE

Let $E_1, E_2, E_3, \ldots, E_k$ be a sequence of k events such that E_1 can occur in n_1 ways, E_2 can occur in n_2 ways, E_3 can occur in n_3 ways, and so on. Then the total number of ways all the events can occur is

$$n_1 \cdot n_2 \cdot n_3 \cdots n_k$$

EXAMPLE 2 In a certain state a license plate number consists of 3 letters followed by 4 digits. How many different such license plates can there be?

Solution In Example 1 we were able to list all the possibilities with a tree diagram because there were only a relatively small number of possibilities. In this example, however, there are an enormous number of possibilities and it would be extremely impractical to try to list and count them all. Instead, we can view a license plate number as a series of seven blank spaces:

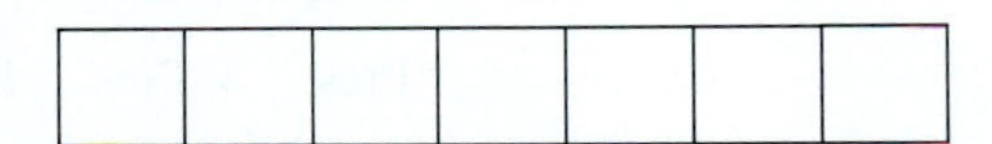

The first three spaces can be filled in with any of the 26 letters A–Z, and the last four spaces can be filled in with any of the ten digits 0–9. In the language of the Fundamental Counting Principle, E_1 is choosing the letter for the first space; E_2 is choosing the letter for the second space; and E_5 is choosing the digit for the fifth space. If we enter the number of ways each space can be filled in, we have

26	26	26	10	10	10	10

and so, by the Fundamental Counting Principle, there are

$$26 \cdot 26 \cdot 26 \cdot 10 \cdot 10 \cdot 10 \cdot 10 = \boxed{175{,}760{,}000}$$

different license plates. ■

EXAMPLE 3 If a fair coin is tossed ten times, how many possible different sequences of heads and tails are there?

Solution Each time we toss the coin (each toss is an **event**) there are two possible outcomes: heads or tails. Thus by the Fundamental Counting Principle, there are a total of

$$2 \cdot 2 \cdot 2 \cdot 2 \cdot 2 \cdot 2 \cdot 2 \cdot 2 \cdot 2 \cdot 2 = 2^{10} = \boxed{1024}$$

possible sequences of heads and tails.

Notice that in formulating this example, we are concerned with the order of the heads and tails. Thus the sequence HHTHTHTHHT is different from HTHTTH-HHTH, even though they both have 6 heads and 4 tails. ■

Permutations

One of the important applications of the Fundamental Counting Principle is in determining the number of ways that n distinct objects can be arranged, where the order of the objects matters.

EXAMPLE 4 Suppose an organization is having an election for president, vice-president, and secretary and the election is run so that the person receiving the most votes becomes president, second most becomes vice-president, and third most becomes secretary. If there are 6 candidates, John (J), Kita (K), Debra (D), Tom (T), Gina (G), and Lamar (L), competing for the three positions, how many outcomes to the election are there?

Solution Viewing the three positions to be filled as three vacant blanks, it is important we recognize that

Pres.	V.Pres.	Sec'y		Pres.	V.Pres.	Sec'y
K	**T**	**L**		**T**	**L**	**K**

are different outcomes. The first means that Kita is president, Tom is vice president, and Lamar is secretary, whereas the second means Tom is president, Lamar is vice president, and Kita is secretary. In other words, the *order* in which the top three finish is important. There are 6 possible presidents. Once the president is chosen, there are now 5 possible people left to be vice president, and then there are 4 people left to be secretary. Using the Fundamental Counting Principle and the box format, we would describe the possibilities as

6	5	4

and so there are

$$6 \cdot 5 \cdot 4 = \boxed{120}$$

different possible outcomes of the election. ■

Notice that in the license plate problem, we were allowed to use letters and digits more than once, but in the last example, once someone is elected to a position, he or she is no longer available for another position.

DEFINITION A **permutation** is an arrangement of distinct objects in a definite order.

When we speak of finding the number of permutations of n objects, we mean the number of ways in which the n objects can be rearranged.

EXAMPLE 5 Suppose that the Kentucky Derby has 9 horses entered in the race.

(a) In how many different orders can the horses finish?

(b) How many different 1-2-3 finishes can there be?

(c) How many different 7-8-9 finishes can there be?

(Assume that there are no ties.)

Solution

(a) We are being asked to find the number of permutations of the 9 distinct horses. First place can be any of the 9 horses; second place can then be any of the remaining 8 horses; third place can then be any of the remaining 7 horses, and so on. We can view the problem as

9	8	7	6	5	4	3	2	1

and so by the Fundamental Counting Principal we have

$$9 \cdot 8 \cdot 7 \cdot 6 \cdot 5 \cdot 4 \cdot 3 \cdot 2 \cdot 1 = \boxed{362{,}880}$$

The 9 horses can finish in any of 362,880 different orders.

(b) A similar analysis gives

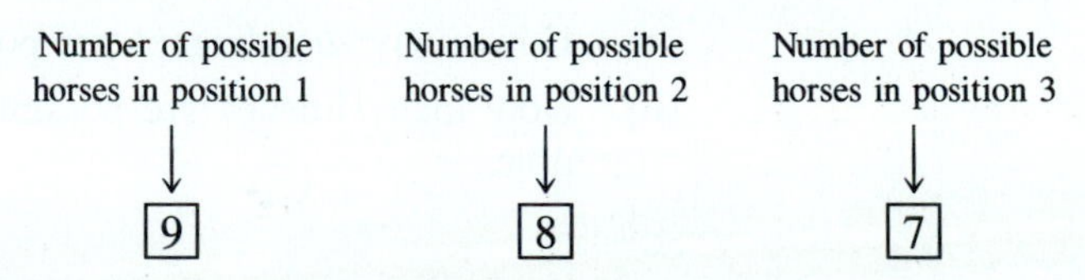

and so there are $9 \cdot 8 \cdot 7 = \boxed{504}$ possible 1-2-3 finishes.

(c) Interestingly, the result for positions 7-8-9 is the same as for positions 1-2-3. After all, just as any of the 9 horses can finish first, so, too, any of the 9 horses can finish seventh.

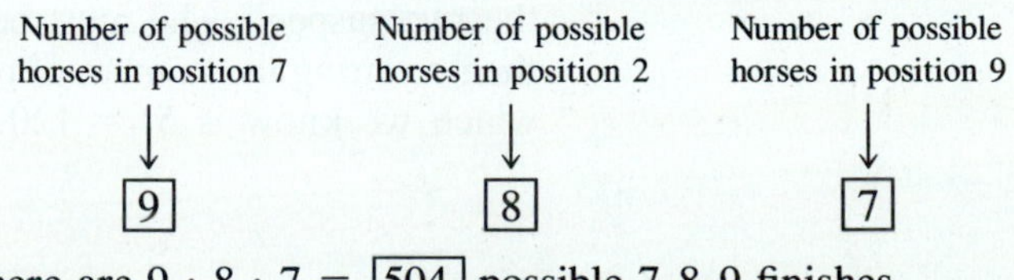

and so there are $9 \cdot 8 \cdot 7 = \boxed{504}$ possible 7-8-9 finishes. ■

Note that in analyzing the number of permutations of the 9 horses in Example 5(a), we encountered $9 \cdot 8 \cdot 7 \cdot 6 \cdot 5 \cdot 4 \cdot 3 \cdot 2 \cdot 1$. Recall from Section 11.1 that this product can be denoted as $9!$.

In Example 5, we were interested in the number of permutations of 3 distinct objects taken from a set of 9 distinct objects. In such a situation, we say that we are permuting 9 objects taken 3 at a time, and we denote this number as $P(9, 3)$. We will use the notation $P(n, r)$ to mean the number of permutations of n distinct objects taken r at a time. Using this notation, the results of Example 5 can be generalized to the following theorem.

THEOREM 11.1 **(a)** The number of permutations of n distinct objects is $P(n, n) = n!$.

(b) The number of permutations of n distinct objects taken r at a time is $P(n, r) = n(n-1)(n-2)\cdots(n-r+1)$.

Proof

(a) We want to fill n boxes with n choices for the first, $n-1$ for the second, $n-2$ for the third, and so on.

n	n − 1	n − 2	. . .	3	2	1

and so by the Fundamental Counting Principle the number of permutations is

$$n(n-1)(n-2)\cdots 3 \cdot 2 \cdot 1 = n! \qquad \textit{As required}$$

(b) In part (b) we are arranging only r out of the n objects. Thus we are filling r boxes. Note that in the second box we subtract 1, in the third box we subtract 2, in the fourth box, 3, and so on, until the rth box, where we subtract $r-1$. So the entry in the rth box is $n-(r-1) = n-r+1$. Therefore, the number of permutations is

$$P(n, r) = n(n-1)(n-2)\cdots(n-r+1) \qquad \textit{As required}$$

■

EXAMPLE 6 Suppose that the police are arranging a lineup of six suspects.

(a) How many such lineups are possible?

(b) How many lineups are possible if two specific suspects must be next to each other?

Solution

(a) The number of permutations of the six suspects is

$$P(6, 6) = 6! = 6 \cdot 5 \cdot 4 \cdot 3 \cdot 2 \cdot 1 = \boxed{720}.$$

(b) Let's refer to the six suspects as A, B, C, D, E, and F and let's suppose that A and B are the two suspects who must be next to each other. We can then view the two suspects who must be next to each other as one object (AB). We are then counting the number of permutations of the five objects (AB), C, D, E, F, which we know is $5! = 120$. However, we also get 120 arrangements with suspects A and B reversed. In other words, we also have 120 permutations of $(BA), C, D, E, F$.

Therefore, there are $120 + 120 = \boxed{240}$ lineups with two specific suspects next to each other. ■

EXAMPLE 7 How many different permutations are there of the letters of each word?

(a) DADDY **(b)** AARDVARK

Solution

(a) Up to this point we have been examining the permutations of distinct objects. The word DADDY has 5 letters, but they are not distinct. Let's temporarily distinguish the Ds in the word as $D_1AD_2D_3Y$, which Theorem 11.1 tells us has $P(5, 5) = 5! = 120$ permutations. Among these permutations we are counting

$$AY\,D_1D_2D_3 \qquad AY\,D_1D_3D_2 \qquad AY\,D_2D_1D_3$$
$$AY\,D_2D_3D_1 \qquad AY\,D_3D_1D_2 \qquad AY\,D_3D_2D_1$$

However, these six permutations are actually indistinguishable, since they all appear as $AYDDD$. Thus we see that for each permutation of $D_1AD_2D_3Y$, there are 6 permutations that contain the same letters in the same order. Note that $6 = 3!$, which is the number of permutations of the 3 D's that are repeated. In order to obtain the actual number of distinct permutations, we must divide the total number of permutations by 6.

Number of distinct permutations of DADDY

$$= \frac{\text{number of permutations}}{\textit{number of permutations of the } 3\ D\text{'s}} = \frac{P(5, 5)}{P(3, 3)} = \frac{5!}{3!} = \frac{120}{6} = \boxed{20}$$

(b) Applying the same type of analysis to the word AARDVARK, we can view it as $A_1A_2R_1DV\,A_3R_2K$, which has a total number of $8! = 40{,}320$ permutations. Due to the fact that there are $3A$'s, *each* permutation of AARDVARK is counted $6 = 3!$ times in the list of permutations of $A_1A_2R_1DVA_3R_2K$. But because of the fact that there are $2R$'s, *each* of these 6 repetitions is counted $2 = 2!$ times. Thus each permutation of AARDVARK appears $12 = 6 \times 2 = 3! \times 2!$ times in the list of permutations of $A_1A_2R_1DV\,A_3R_2K$. Therefore, we have

No. of distinct permutations of AARDVARK

$$= \frac{\text{total no. of permutations}}{\textit{no. of permutations of the } 3\ A\text{'s} \times \textit{no. of permutations of the } 2\ R\text{'s}}$$

$$= \frac{P(8, 8)}{P(3, 3) \cdot P(2, 2)} = \frac{8!}{3! \times 2!} = \frac{8 \cdot 7 \cdot 6 \cdot 5 \cdot 4 \cdot 3 \cdot 2 \cdot 1}{6 \cdot 2} = \boxed{3360}$$

■

The previous example generalizes to the following theorem.

THEOREM 11.2 Suppose a set of n objects has r_1 identical objects of one type, r_2 identical objects of a second type, r_3 identical objects of a third type, . . . , and r_k identical objects of a kth type. The number of *distinct* permutations of the n objects is

$$\frac{n!}{r_1! \cdot r_2! \cdot r_3! \cdots r_k!}$$

EXAMPLE 8 How many different 7-digit numbers can be made using all the digits 1, 1, 5, 5, 8, 8, and 8?

Solution Since we are using 7 digits (1, 1, 5, 5, 8, 8, 8), there are 7! permutations of these digits. Since there are two 1's, two 5's, and three 8's, Theorem 11.2 gives us

$$\frac{7!}{2! \cdot 2! \cdot 3!} = \frac{7 \cdot 6 \cdot 5 \cdot 4 \cdot 3 \cdot 2 \cdot 1}{2 \cdot 2 \cdot 6} = \boxed{210 \text{ distinct 7 digit numbers.}}$$

■

Combinations

In Example 4 we analyzed the number of ways in which a president, vice president, and secretary can be chosen from among a group of candidates. If there were 10 candidates for the 3 positions, then there would be $10 \cdot 9 \cdot 8 = 720$ different possible outcomes of the election. In this analysis we recognized that the *order* in which the candidates are chosen is important. The next example, which deals with a situation in which the order of the objects is *not* important, illustrates a particularly important application of Theorem 11.2.

EXAMPLE 9 A committee of 3 people is to be chosen from among 10 members of a club. How many such committees are there?

Solution This example is asking for the number of ways we can choose 3 distinct objects (people) out of 10 distinct objects *without regard to their order*. Suppose we think of the 10 people being numbered 1 through 10. If persons 3, 7, and 8 are chosen for the committee, it makes no difference if they were chosen in the order 3, 7, 8 or 8, 3, 7. The committee still consists of the same people. Therefore we take the total number of 3-person committees $P(10, 3)$ (that is the number of permutations of 10 objects taken 3 at a time, which we know is $10 \cdot 9 \cdot 8$) and divide that by $P(3, 3)$ (the number of permutations of the 3 members of the committee, which is 3!), and so the number of 3-person committees is

$$\frac{P(10, 3)}{P(3, 3)} = \frac{10 \cdot 9 \cdot 8}{3!} = \boxed{120}$$

■

Let's examine Example 9 from a slightly different perspective. We can think of the number of ways of choosing a 3-person committee as the number of permutations of a set of 10 people in which one subset consists of 3 "identical" people (the people who *are* chosen for the committee) and another subset of 7 identical people (the people who *are not* chosen for the committee). Therefore, according to Theorem 11.2, the number of such permutations is

$$\frac{10!}{3! \cdot 7!} = \frac{10 \cdot 9 \cdot 8 \cdot 7 \cdot 6 \cdot 5 \cdot 4 \cdot 3 \cdot 2 \cdot 1}{(7 \cdot 6 \cdot 5 \cdot 4 \cdot 3 \cdot 2 \cdot 1)(3 \cdot 2 \cdot 1)} = \frac{10 \cdot 9 \cdot 8}{3!} = \frac{P(10, 3)}{P(3, 3)} = 120$$

as we saw before.

The key feature of this example is that we were choosing a subset of 3 elements from a set of 10 elements *without regard to order*. A selection of a subset from a larger set where the order of the selection is not important is called a **combination**.

We will denote the number of ways in which r elements can be chosen from a set of n elements as

$$C(n, r) \quad \text{or} \quad \binom{n}{r}, \qquad \text{which are often read as "n choose r"}$$

Generalizing the discussion of the last example we have Theorem 11.3.

THEOREM 11.3 The number of combinations of n distinct objects taken r at a time is

$$C(n, r) = \frac{n!}{r! \cdot (n-r)!} = \frac{n(n-1)(n-2) \cdots (n-r+1)}{r!} = \frac{P(n, r)}{P(r, r)}$$

Recall that in Section 11.1, we defined $0! = 1$, so that based upon this theorem we have $C(n, 0) = \dfrac{n!}{0! \cdot (n-0)!} = \dfrac{n!}{n!} = 1.$

How would we interpret $C(n, 0)$? Based upon this discussion, we have n items and we are choosing 0 of them. There is only one way to choose 0 items: Do not choose any of them. Thus we would expect $C(n, 0) = 1$, which agrees with the result obtained using the theorem.

The following example puts to use a number of the ideas we have developed in this section.

EXAMPLE 10 A poker hand consists of 5 cards dealt from a standard deck of 52 playing cards

(a) How many different 5-card poker hands are there?

(b) How many different 5-card flushes are there? (A flush means 5 cards of the same suit.)

Solution

(a) Once the cards are dealt, the player can put them in any order that he or she wishes to, so the order in which the cards are dealt is irrelevant. The number of possible 5-card poker hands is the number of ways we can choose 5 distinct elements from a set of 52 distinct elements, which is precisely $C(52, 5)$. By Theorem 11.3 we have

$$C(52, 5) = \frac{52 \cdot 51 \cdot 50 \cdot 49 \cdot 48}{5 \cdot 4 \cdot 3 \cdot 2 \cdot 1} = \boxed{2{,}598{,}960}$$

(b) We can think of dealing a flush as a two-step process: First, choose one of the four suits (spades, hearts, clubs, or diamonds) and second, choose 5 out of the 13 cards of the suit. Therefore, by the Fundamental Counting Principle we have

No. of possible flushes

$$= \underbrace{(\textit{no. of ways to pick a suit})}_{C(4,\,1)} \cdot \underbrace{(\textit{no. of ways to pick 5 cards of that suit})}_{C(13,\,5)}$$

$$= C(4, 1) \cdot C(13, 5) = (4) \cdot \left(\frac{13 \cdot 12 \cdot 11 \cdot 10 \cdot 9}{5 \cdot 4 \cdot 3 \cdot 2 \cdot 1}\right) = \boxed{5{,}148}$$

■

We offer the following reminder to the reader.

> **GUIDELINES FOR SOLVING COUNTING PROBLEMS**
>
> When solving a counting problem, read the problem and ask yourself, "Is order important in the selection process?"
>
> 1. If the answer is yes, then the problem calls for permutations.
> 2. If the answer is no, then the problem calls for combinations.
> 3. Keep in mind that the Fundamental Counting Principle can often be used in conjunction with the various theorems in this section.

EXAMPLE 11 Show that $\binom{n}{k} = \binom{n}{n-k}$.

Solution According to Theorem 11.3 we have

$$\binom{n}{k} = \frac{n!}{k! \cdot (n-k)!}$$

And we also have

$$\binom{n}{n-k} = \frac{n!}{(n-k)! \cdot [n-(n-k)]!} = \frac{n!}{(n-k)! \cdot k!}$$

From these two equations we have

$$\binom{n}{k} = \binom{n}{n-k}$$

As required. ∎

We will find the result established in the last example useful in the next section, as we develop the Binomial Theorem.

EXERCISES 11.6

In Exercises 1–10, evaluate the given quantity and describe it verbally.

1. $P(8, 3)$

2. $C(10, 3)$

3. $\binom{12}{7}$

4. $\binom{12}{5}$

5. $P(6, 5)$

6. $P(6, 1)$

7. $\binom{18}{12}$

8. $\binom{18}{6}$

9. $\dfrac{P(8, 4)}{P(4, 4)}$

10. $\dfrac{12!}{3!} \cdot \dfrac{5!}{4!}$

11. If $C(n, 2) = 105$, find n.

12. If $P(n, 2) = 90$, find n.

13. A student has 5 pairs of slacks, 8 shirts, and 3 sweaters. Assuming that the student is a nonconformist and does not care about how the clothes match, how many different outfits consisting of slacks, shirt, and sweater are possible?

14. A car manufacturer offers a particular car model with 12 choices of paint color, 3 choices of body style, and 4 choices of interior trim. How many different configurations of color, body style, and interior trim are possible?

15. A mattress manufacturer offers 8 different choices in covering fabric (called the *ticking*), 5 different choices in firmness, and 4 different sizes. How many different types of mattresses are available?

16. A travel agent is offering a special discount on vacations to 12 different destinations with 9 different departure dates and 3 different categories of hotel accommodations. How many different possible discount vacations are available?

17. In how many ways can a student arrange 8 different books on a shelf?

18. In how many ways can a student arrange 9 different books consisting of 3 science books, 2 history books, and 4 mathematics books if
(a) The books can be arranged in any order?
(b) All books of the same type must be next to each other?

19. A CD player has a "shuffle" feature, which will play the selections in a random order. If a particular CD has 15 selections, in how many different orders can the CD be played?

20. A horse race has 10 entries. How many possible 1-2-3 finishes are there?

21. A jewelry manufacturer produces "initial" rings, which contain the first initials of the wearer's first and last names. If the manufacturer wants to produce every possible initial ring, how many rings must be made?

22. A company wants to buy initial rings for all its employees from the manufacturer of Exercise 21. If the company has 680 employees, explain why the company will have to order more than one copy of at least one ring.

23. Suppose a state's license plates consist of 4 letters followed by 3 digits, using only the digits 1 through 9.
(a) How many such license plates are possible?
(b) How many such license plates are possible if the first digit must be odd?
(c) How many such license plates are possible if the first digit must be even and the first letter must be a consonant? (Consider Y to be a consonant.)

24. A state lottery requires participants to choose 6 numbers from the numbers 1 through 40. If the order of the numbers is unimportant, how many ways are there to choose the 6 numbers?

25. A multiple-choice examination consists of 20 questions. If each question has 4 choices, how many possible ways are there to answer all the questions? What if there are 5 choices for each question?

26. A true-false examination has 15 questions. How many possible ways are there to answer the questions?

27. A student is required to answer 10 out of 12 questions on an examination. How many different sets of 10 questions can the student answer?

28. An audio component store has a special sale on 3 different amplifiers, 4 different tuners, 2 different cassette decks, 5 different CD players, and 8 different sets of speakers. How many different stereo systems can be made choosing one of each type of component?

29. A bag contains 9 red marbles and 7 blue marbles. In how many ways can you choose
(a) 5 blue marbles? **(b)** 4 red marbles?
(c) 9 marbles of any color?
(d) 3 blue and 6 red marbles?

30. A scrabble tournament has 16 participants. If the tournament committee wants each player to meet every other player exactly once, how many games must be scheduled?

31. A club has 25 members. A committee of 6 is to be chosen, and then this committee is to select two co-chairpersons. In how many ways can this be done?

32. Repeat Exercise 31 if the committee is to select one chairperson and a secretary.

33. Two teams are playing a best-of-seven series to determine the champion. How many different ways are there for the series to go the full seven games?

34. Four people are to be selected from a group of four couples. In how many ways can the four people be selected if
(a) There are no restrictions?
(b) One member from each couple is to be chosen?
(c) At least one couple is to be included?
(d) Exactly one couple is to be included?

35. An attache case has a three-digit-number combination lock. Each of the digits can be chosen from 1, 2, 3, 4, 5.
(a) How many such combinations are there?
(b) How many of the combinations are even numbers?

36. Using the digits 3, 4, 5, 6, 7, 8, 9 and assuming that digits can be repeated, determine how many four-digit numbers can be made if
(a) There are no restrictions.
(b) The number is greater than 5000.
(c) The number is less than 4000.
(d) The number is not divisible by 5.

37. Five cards are dealt from an ordinary deck of playing cards. In how many ways can the cards be dealt so that
(a) All the cards are spades?
(b) All the cards are the same suit?
(c) Exactly 3 are aces?
(d) The hand is a full house (3 of one kind and 2 of another)?

38. If a 5-card poker hand is dealt from a deck of 52 playing cards, how many hands will contain
(a) Exactly one pair? **(b)** Two pairs?
(c) A straight (five cards in sequence; e.g., 4 5 6 7 8)?

39. Suppose we have 10 points in the plane such that no 3 of them fall on the same line.
(a) How many different lines are determined by these 10 points?
(b) How many different triangles are determined by these 10 points?

40. Find the number of diagonals of **(a)** a pentagon; **(b)** an octagon.

41. How many different arrangements are there of the letters of each word?
(a) COMPUTER **(b)** ENCYCLOPEDIA
(c) EFFERVESCENT

42. In a family of 6 children, how many possible boy-girl orders of birth are there?

43. Suppose there are 3 routes connecting city A to city B, 5 routes connecting city B to city C, and 2 routes connecting city C to city D. How many routes are there from city A to city D?

44. A group of 20 basketball players wants to divide itself up into 4 teams of 5 players each. In how many ways can this be done?

In Exercises 45–46, prove the given identity.

45. $\binom{n}{r} + \binom{n}{r-1} = \binom{n+1}{r}$

46. Use mathematical induction to prove that

$$\binom{n}{0} + \binom{n}{1} + \binom{n}{2} + \cdots + \binom{n}{n} = 2^n$$

QUESTIONS FOR THOUGHT

47. How are the ideas of a permutation and a combination related?

48. Use the method of mathematical induction to prove that the number of permutations of n distinct objects is $n!$.

11.7 The Binomial Theorem

Recall that a sum of the form $A + B$ is called a binomial. In Chapter 1, we discussed special products of the form $(A + B)^2$ and $(A + B)^3$. In this section we develop a general formula called the **Binomial Theorem** for raising a binomial to a power.

We begin by actually multiplying out, or *expanding*, $(A + B)^n$ for $n = 0, 1, 2, 3, 4$, and 5.

	$(A + B)^n$ Expanded form
For $n = 0$:	$(A + B)^0 = 1$
For $n = 1$:	$(A + B)^1 = A + B$
For $n = 2$:	$(A + B)^2 = A^2 + 2AB + B^2$
For $n = 3$:	$(A + B)^3 = A^3 + 3A^2B + 3AB^2 + B^3$
For $n = 4$:	$(A + B)^4 = A^4 + 4A^3B + 6A^2B^2 + 4AB^3 + B^4$
For $n = 5$:	$(A + B)^5 = A^5 + 5A^4B + 10A^3B^2 + 10A^2B^3 + 5AB^4 + B^5$

Let's examine the bottom row of this table. We note that the terms (as we have arranged them) begin with A^5, which we can view as A^5B^0. In the subsequent terms, the exponent of A decreases by one each time, and the exponent of B increases by one each time. Also note that the sum of the exponents in each term is 5. We see this same pattern in the other rows of the table as well. It seems reasonable to believe that this pattern will hold in general; in fact, this is the case.

Let's now turn our attention to the coefficients and see if we can understand how they can be computed. Keep in mind that

$$(A + B)^5 = (A + B)(A + B)(A + B)(A + B)(A + B)$$

There are five binomials and the terms in the expanded form are the result of all possible products of the terms in the binomials. For example to get A^3B^2 in the expansion of $(A + B)^5$, we must multiply the A-term from three of the binomials with the B-term from the other two binomials. This can be done in exactly $\binom{5}{3}$ ways—the number of ways we can choose 3 of out the 5 binomials from which to get the factor

A. We could equally well reason that this can be done in exactly $\binom{5}{2}$ ways—the number of ways we can choose 2 out of the 5 binomials from which to get the factor B. We leave it to the reader to verify that $\binom{5}{3} = \binom{5}{2} = 10$.

Generalizing this argument to $(A + B)^n$, we conclude that the coefficient of the $A^{n-k}B^k$-term is $\binom{n}{k}$ or $\binom{n}{n-k}$. Recall that in the last example of Section 11.6 we proved algebraically that $\binom{n}{k} = \binom{n}{n-k}$.

Summarizing this discussion leads to the next theorem.

THEOREM 11.4 THE BINOMIAL THEOREM

$$(A + B)^n = A^n + \binom{n}{1}A^{n-1}B + \binom{n}{2}A^{n-2}B^2 + \cdots + \binom{n}{k}A^{n-k}B^k + \cdots + \binom{n}{n-1}AB^{n-1} + B^n$$

Note that we normally don't write the first and last coefficients in $\binom{n}{0}$ or $\binom{n}{n}$ form, but rather simply write them as 1. The proof of the Binomial Theorem is outlined in Exercise 53.

It is because of the Binomial Theorem that $\binom{n}{k}$ is often called a *binomial coefficient*.

EXAMPLE 1 Use the Binomial Theorem to expand $(x + y)^8$.

Solution By the Binomial Theorem we have

$$(x + y)^8 = x^8 + \binom{8}{1}x^7y + \binom{8}{2}x^6y^2 + \binom{8}{3}x^5y^3 + \binom{8}{4}x^4y^4 + \binom{8}{5}x^3y^5 + \binom{8}{6}x^2y^6 + \binom{8}{7}xy^7 + y^8$$

Computing the binomial coefficients we have

$$\binom{8}{1} = \frac{8}{1} = 8 \qquad \binom{8}{5} = \binom{8}{3} = 56$$

$$\binom{8}{2} = \frac{8 \cdot 7}{2 \cdot 1} = 28 \qquad \binom{8}{6} = \binom{8}{2} = 28$$

$$\binom{8}{3} = \frac{8 \cdot 7 \cdot 6}{3 \cdot 2 \cdot 1} = 56 \qquad \binom{8}{7} = \binom{8}{1} = 8$$

$$\binom{8}{4} = \frac{8 \cdot 7 \cdot 6 \cdot 5}{4 \cdot 3 \cdot 2 \cdot 1} = 70$$

Note that we have made use of the fact that $\binom{n}{k} = \binom{n}{n-k}$. Our final answer is

$$(x + y)^8 = x^8 + 8x^7y + 28x^6y^2 + 56x^5y^3 + 70x^4y^4 + 56x^3y^5 + 28x^2y^6 + 8xy^7 + y^8$$ ■

EXAMPLE 2 Expand $(2x - 3)^4$ using the Binomial Theorem.

Solution We view $(2x - 3)^4$ as $(2x + (-3))^4$, so according to the Binomial Theorem we have

$$(2x - 3)^4 = (2x)^4 + \binom{4}{1}(2x)^3(-3) + \binom{4}{2}(2x)^2(-3)^2 + \binom{4}{3}(2x)(-3)^3 + (-3)^4$$
$$= 16x^4 + 4(8)(-3)x^3 + 6(4)(9)x^2 + 4(2)(-27)x + 81$$
$$= 16x^4 - 96x^3 + 216x^2 - 216x + 81$$

Notice that because of the fact that the exponents alternate from even to odd, the signs in the expansion of $(A - B)^n$ will alternate from positive to negative. ■

There is another way to compute binomial coefficients. We write out the binomial coefficients of $(A + B)^n$ for $n = 0, 1, 2, 3, 4, 5, 6$, and tabulate them in the following triangular array.

$n = 0$: 1

$n = 1$: 1 1

$n = 2$: 1 2 1

$n = 3$: 1 3 3 1

$n = 4$: 1 4 6 4 1

$n = 5$: 1 5 10 10 5 1

$n = 6$: 1 6 15 20 15 6 1

If we examine the table carefully we can see that each row begins and ends with a 1 and that each of the other entries in a row can be obtained by adding the two numbers diagonally above it. Thus $10 = 6 + 4$ and $15 = 5 + 10$, as indicated by the triangles in the table.

The triangular array is called *Pascal's triangle,* named for the famous seventeenth-century French mathematician Blaise Pascal (1623–1662). We can use Pascal's triangle to generate the binomial coefficients for higher values of n by simply adding rows to the table, as described.

EXAMPLE 3 Expand $(x^2 + y^3)^7$.

Solution From the Binomial Theorem we know the pattern of the powers. We may compute the binomial coefficients by using the formula for $\binom{n}{k}$ or we may generate the line for $n = 7$ in Pascal's triangle. We choose the latter approach.
Using the row for $n = 6$ in the table, we have

$n = 6$: 1 6 15 20 15 6 1

$n = 7$: 1 7 21 35 35 21 7 1

Using these coefficients we get

$$(x^2 + y^3)^7 = (x^2)^7 + 7(x^2)^6y^3 + 21(x^2)^5(y^3)^2 + 35(x^2)^4(y^3)^3 + 35(x^2)^3(y^3)^4 + 21(x^2)^2(y^3)^5 + 7(x^2)(y^3)^6 + (y^3)^7$$
$$= x^{14} + 7x^{12}y^3 + 21x^{10}y^6 + 35x^8y^9 + 35x^6y^{12} + 21x^4y^{15} + 7x^2y^{18} + y^{21}$$ ■

For relatively small values of n, Pascal's triangle offers a convenient way to generate the binomial coefficients. However, if we want to compute the coefficient of a specific term in the binomial expansion for a fairly large value of n, it can be rather time consuming to generate all the rows of Pascal's triangle until we get to the row we need, as the following example illustrates.

EXAMPLE 4 Find the coefficient of x^8 in the expansion of $(\sqrt{x} - 3)^{20}$.

Solution Since we are dealing with a question about the exponent of x, it seems reasonable to expect that writing $\sqrt{x}$ as $x^{1/2}$ may be helpful. A typical term in the expansion of $(x^{1/2} - 3)^{20}$ will be

$$\binom{n}{k}(x^{1/2})^k(-3)^{n-k}$$

If we are looking for the term containing x^8, then we need

$$(x^{1/2})^k = x^8$$
$$x^{k/2} = x^8 \Rightarrow k = 16$$

Using the formula for the general term of the binomial expansion with $n = 20$ and $k = 16$ we get

$$\binom{20}{16}(x^{1/2})^{16}(-3)^4$$

To find the binomial coefficient $\binom{20}{16}$ or its equivalent, $\binom{20}{4}$, from Pascal's triangle would require us to complete the table all the way down to the row for $n = 20$. It seems much more efficient to directly compute the binomial coefficient we need.

$$\binom{20}{16} = \binom{20}{4} = \frac{20 \cdot 19 \cdot 18 \cdot 17}{4 \cdot 3 \cdot 2 \cdot 1} = 4845$$

Therefore, the term corresponding to $k = 16$ is

$$4845(x^{1/2})^{16}(-3)^4 = 4845(81)x^8 = 392{,}445x^8$$

and so the coefficient of x^8 is $\boxed{392{,}445}$. ■

EXERCISES 11.7

In Exercises 1–30, expand the given binomial and simplify.

1. $(x - 4)^3$
2. $(t + r)^5$
3. $(a + 2)^7$
4. $(y - z)^4$
5. $(2x + y)^6$
6. $(x + 2y)^6$
7. $(t - 3r)^5$
8. $(3t - r)^5$
9. $(w^2 + 1)^8$
10. $(1 - w^2)^7$
11. $(x^3 - y^3)^4$
12. $(x^2 + y^2)^3$
13. $(a^2 + 2b)^9$
14. $(c - 3d^2)^8$
15. $\left(x - \frac{1}{2}\right)^4$
16. $\left(y + \frac{1}{3}\right)^3$
17. $\left(\frac{x}{3} + 4\right)^3$
18. $\left(\frac{a}{4} - 3\right)^4$
19. $(x - \sqrt{x})^6$
20. $(2x - \sqrt{5})^4$
21. $(\sqrt{x} + \sqrt{y})^6$
22. $(\sqrt{a} - \sqrt{3})^5$
23. $\left(x - \frac{1}{x}\right)^5$
24. $\left(x - \frac{1}{\sqrt{x}}\right)^6$
25. $\left(\frac{x}{y} + \frac{y}{x}\right)^4$
26. $\left(\frac{2}{x} - \frac{x}{2}\right)^3$
27. $(x^{-1} - y^2)^5$
28. $(x + y^{-2})^4$
29. $(2r^{-1} + s^{-2})^6$
30. $(a^2 + 5b^{-2})^3$

31. What is the coefficient of x^6 in the expansion of $(x + 5)^{10}$?

32. What is the coefficient of x^6 in the expansion of $(x^2 + 5)^{10}$?

33. What is the coefficient of the term containing a^5 in the expansion of $(a - 3b)^7$?

34. What is the coefficient of the term containing s^4 in the expansion of $(2r - 3s)^{10}$?

35. What is the coefficient of z^5 in the expansion of $(2\sqrt{z} + 1)^{12}$?

36. What is the coefficient of z^5 in the expansion of $(\sqrt{2z} + 1)^{12}$?

37. What is the coefficient of the term containing c^3 in the expansion of $(\sqrt{c} - \sqrt{d})^8$?

38. What is the coefficient of x expansion of $\left(2x^2 - \dfrac{3}{\sqrt{x}}\right)^8$?

39. Use the Binomial Theorem to write out the expansion of $(a + b + c)^4$. HINT: Think of $(a + b + c)^4$ as $[(a + b) + c]^4$.

40. Use the Binomial Theorem to write out the expansion of $(x - 2 + y)^3$. HINT: Think of $(x - 2 + y)^3$ as $[(x - 2) + y]^3$.

In Exercises 41–44, simplify the given expression.

41. $\dbinom{9}{4}\dbinom{4}{2}$ **42.** $\dbinom{6}{3} + \dbinom{4}{3} - \dbinom{10}{3}$

43. $\dfrac{(n + 1)!}{n!}$ **44.** $\dfrac{(n + 1)!(n - 1)!}{(n!)^2}$

45. Is it true that $n! = n(n - 1)!$ for $n \geq 1$?

46. Is it true that $(n + m)! = n! + m!$?

In Exercises 47–50, expand the power of the given complex number. Simplify your answers by using the fact that $i^2 = -1$.

47. $(1 + i)^5$ **48.** $(2 - i)^4$

49. $(2 + \sqrt{-9})^3$ **50.** $(5 - \sqrt{-6})^6$

QUESTIONS FOR THOUGHT

51. In general, which do you think is greater, $(n!)^2$ or $(n^2)!$?

52. Show that $\dbinom{n}{0} + \dbinom{n}{1} + \dbinom{n}{2} + \cdots + \dbinom{n}{n} = 2^n$ by using the Binomial Theorem on $(1 + 1)^n$.

53. This exercise outlines a proof of the Binomial Theorem using the principle of mathematical induction.
Let P_n be the statement that

$$(A + B)^n = A^n + \binom{n}{1}A^{n-1}B + \binom{n}{2}A^{n-2}B^2 + \cdots + \binom{n}{n-1}AB^{n-1} + B^n$$

Step 1: First we verify that P_1 is true; that is, we need to verify that

$$(A + B)^1 = A^1 + \binom{1}{1}A^{1-1}B^1 = A^1 + 1A^0B^1 = A^1 + B^1, \quad \textit{which is true.}$$

Step 2: We want to prove that $P_K \Rightarrow P_{K+1}$. We assume that P_k is true—that is, we assume

$$(A + B)^k = \binom{k}{0}A^k + \binom{k}{1}A^{k-1}B + \binom{k}{2}A^{k-2}B^2 + \cdots + \binom{k}{k-1}AB^{k-1} + \binom{k}{k}B^k \quad \textit{This is the induction hypothesis.}$$

Multiply both sides of this equation by $(A + B)$ to obtain

$$(A + B)^{k+1} = (A + B)\left[\binom{k}{0}A^k + \binom{k}{1}A^{k-1}B + \binom{k}{2}A^{k-2}B^2 + \cdots + \binom{k}{k-1}AB^{k-1} + \binom{k}{k}B^k\right]$$

On the right-hand side, distribute the $(A + B)$, giving

$$(A + B)^{k+1} = A\left[\binom{k}{0}A^k + \binom{k}{1}A^{k-1}B + \binom{k}{2}A^{k-2}B^2 + \cdots + \binom{k}{k-1}AB^{k-1} + \binom{k}{k}B^k\right] +$$

$$B\left[\binom{k}{0}A^k + \binom{k}{1}A^{k-1}B + \binom{k}{2}A^{k-2}B^2 + \cdots + \binom{k}{k-1}AB^{k-1} + \binom{k}{k}B^k\right]$$

This equation becomes

$$(A + B)^{k+1} = \left[\binom{k}{0}A^{k+1} + \binom{k}{1}A^kB + \binom{k}{2}A^{k-1}B^2 + \cdots + \binom{k}{k-1}A^2B^{k-1} + \binom{k}{k}AB^k\right]$$
$$+ \left[\binom{k}{0}A^kB + \binom{k}{1}A^{k-1}B^2 + \binom{k}{2}A^{k-2}B^3 + \cdots + \binom{k}{k-1}AB^k + \binom{k}{k}B^{k+1}\right]$$

Regroup the like terms to get

$$(A + B)^{k+1} = \binom{k}{0}A^{k+1} + \left[\binom{k}{1} + \binom{k}{0}\right]A^kB + \left[\binom{k}{2} + \binom{k}{1}\right]A^{k-1}B^2 + \cdots + \left[\binom{k}{k} + \binom{k}{k-1}\right]AB^k + \binom{k}{k}B^{k+1}$$

Now use the fact that $\binom{k}{k} = \binom{k}{0} = 1$ and the result of Section 11.6, Exercise 45 that $\binom{n}{r-1} + \binom{n}{r} = \binom{n+1}{r}$ to rewrite the last equation as

$$(A + B)^{k+1} = A^{k+1} + \binom{k+1}{1}A^kB + \binom{k+1}{2}A^{k-1}B^2 + \cdots + \binom{k+1}{k}AB^k + B^{k+1}$$

which proves that P_{k+1} is true. Thus, by the principle of mathematical induction, P_n is true for all values of n, which proves the Binomial Theorem.

Chapter 11 SUMMARY

After having completed this chapter, you should:

1. Be able to find a specific term in a sequence. (Section 11.1)
   For example:
   Given a sequence with general term
   $$a_n = \frac{3n - 1}{n^2} \qquad \textit{then}$$
   $$a_7 = \frac{3(7) - 1}{7^2} = \frac{21 - 1}{49} = \frac{20}{49}$$

2. Understand and be able to use sigma (Σ) notation. (Section 11.2)
   For example:
   $$\sum_{k=1}^{5} (2k^2 - 3) = [2(1)^2 - 3] + [2(2)^2 - 3] + [2(3)^2 - 3] + [2(4)^2 - 3] + [2(5)^2 - 3]$$
   $$= -1 + 5 + 15 + 29 + 47$$
   $$= 95$$

3. Recognize arithmetic and geometric sequences. (Sections 11.3, 11.4)
   For example:
   (a) An *arithmetic* sequence is one in which the difference between successive terms is constant. Thus the numbers 4, 9, 14, 19, 24 are the first five terms of an arithmetic sequence with first term 4 and a common difference of 5.
   (b) A *geometric* sequence is one in which the quotient of successive terms is constant. Thus the numbers 4, 12, 36, 108, 324 are the first five terms of a geometric sequence with first term 4 and a common ratio of 3.

4. Be able to find a specific term of a given arithmetic or geometric sequence. (Sections 11.3, 11.4)

For example:

(a) To find the tenth term of the arithmetic sequence whose first term is 3 and whose common difference is 8, we use Formula 11.1 for the nth term of an arithmetic sequence, which is

$$a_n = a_1 + (n - 1)d$$ *We substitute $a_1 = 3$, $n = 10$, and $d = 8$.*

$$a_{10} = 3 + (10 - 1)8 = 75$$

(b) To find the seventh term of the geometric sequence whose first term is 2 and whose fourth term is -16, we can use Formula 11.6 for the nth term of a geometric sequence, which is

$$a_n = a_1 r^{n-1}$$ *We substitute $n - 4$, $a_1 = 2$ and $a_n = a_4 = -16$ to find r.*

$$a_4 = a_1 r^3$$

$$-16 = 2r^3 \Rightarrow r^3 = -8 \Rightarrow r = -2$$

Now we can use the formula again to find a_7.

$$a_7 = a_1 r^6 = 2(-2)^6 = 2(64) = 128$$

5. Be able to find the sum of an arithmetic or geometric series. (Sections 11.3, 11.4)

For example:

(a) To find the sum of the first eight terms of an *arithmetic* sequence whose first term is 5 and whose common difference is 4, we use Formula 11.4:

$$S_n = \frac{n}{2}[2a_1 + (n - 1)d]$$ *We substitute $a_1 = 5$, $n = 8$, and $d = 4$ to get*

$$S_8 = \frac{8}{2}[2(5) + (8 - 1)4] = \frac{8}{2}[10 + 28]$$
$$= 152$$

(b) To find the sum of the first eight terms of a *geometric* sequence whose first term is 5 and whose common ratio is 4, we use Formula 11.7:

$$S_n = \frac{a_1(1 - r^n)}{1 - r}$$ *We substitute $a_1 = 5$, 1 $n = 8$, and $r = 4$ to get*

$$S_8 = \frac{5(1 - 4^8)}{1 - 4} = 5\left(\frac{1 - 65{,}536}{1 - 4}\right)$$
$$= 5\left(\frac{-65{,}535}{-3}\right)$$
$$= 109{,}225$$

(c) To find the sum of an infinite geometric sequence $\{a_k\} = 2, \frac{2}{3}, \frac{2}{9}, \frac{2}{27}, \ldots$ (if it exists), we first identify a_1 and r. We are given $a_1 = 2$ and we find that $r = \frac{\frac{2}{3}}{2} = \frac{1}{3}$. Since $|r| < 1$, we know that the sum S exists and, by Formula 11.8, $S = \frac{a_1}{1 - r} = \frac{2}{1 - \frac{1}{3}} = 3$.

6. Understand the principle of mathematical induction and be able to apply it to prove results about sets of natural numbers. (Section 11.5)

For example:

Prove that the following statement P_n is true for all natural numbers n and $r \neq 1$.

$$P_n\colon 1 + r + r^2 + r^3 + \cdots + r^n = \frac{1 - r^{n+1}}{1 - r}$$

Step 1: We must verify that P_1 is true. P_1 is the statement

$$\frac{1 - r^2}{1 - r} = \frac{(1 + r)(1 - r)}{1 - r} = 1 + r$$

Thus P_1 is true.

Step 2: We must prove that $P_k \Rightarrow P_{k+1}$

$$P_k\colon 1 + r + r^2 + r^3 + \cdots + r^k = \frac{1 - r^{k+1}}{1 - r} \qquad (2)$$ *This is the induction hypothesis.*

$$P_{k+1}\text{: } 1 + r + r^2 + r^3 + \cdots + r^k + r^{k+1} = \frac{1 - r^{k+2}}{1 - r}$$

(2) To get P_{k+1} from P_k, we add r^{k+1} to both sides of (1).

$$1 + r + r^2 + r^3 + \cdots + r^k + r^{k+1} = \frac{1 - r^{k+1}}{1 - r} + r^{k+1} = \frac{1 - r^{k+1}}{1 - r} + \frac{r^{k+1}(1 - r)}{1 - r}$$

$$= \frac{1 - r^{k+1} + r^{k+1} - r^{k+2}}{1 - r} = \frac{1 - r^{k+2}}{1 - r}$$

which is Equation (2) as required. The proof by mathematical induction is complete.

7. Understand the fundamental counting principle and be able to apply it. (Section 11.6)
For example:
Marge is making up her class schedule, which is to contain 1 biology course, 1 math course, and 1 English course. If she chooses from 2 biology courses, 4 math courses, and 5 English courses, how many different possible schedules are there? (Assume none of the courses conflict.) According to the fundamental counting principle, she can make $2 \cdot 4 \cdot 5 = 40$ different schedules.

8. Understand the difference between permutations and combinations and be able to compute the number of permutations or combinations in a variety of situations. (Section 11.6)

When you are counting the number of ways that a selection can be made where order matters, you are dealing with permutations. When you are counting the number of ways that a selection can be made where order does not matter, you are dealing with combinations.
For example:
(a) How many 3-letter arrangements can be made using the letters of the word PENCIL?
Since the order of the letters clearly matters, we are looking for the number of *permutations* of 3 out of 6 distinct objects, which is $P(6, 3) = 6 \cdot 5 \cdot 4 = 120$. So there are 120 different 3 letter arrangements using the letters of PENCIL.
(b) A union local has 40 members, 5 of whom are to be chosen to attend a meeting of the national union. In how many ways can this 5-member group be chosen?
Since the order in which the 5-member group is chosen is not important, we want the number of *combinations* of 5 out of 40 distinct objects, which is

$$C(40, 5) = \frac{40 \cdot 39 \cdot 38 \cdot 37 \cdot 36}{5 \cdot 4 \cdot 3 \cdot 2 \cdot 1} = 658{,}008$$

Thus there are 658,008 different 5-member groups that can go to the meeting.

9. Understand and be able to apply the binomial theorem. (Section 11.7)
For example:
Use the binomial theorem to expand $(3x - 2)^5$.
According to the binomial theorem, we have

$$(3x - 2)^5 = (3x)^5 + \binom{5}{1}(3x)^4(-2) + \binom{5}{2}(3x)^3(-2)^2 + \binom{5}{3}(3x)^2(-2)^3 + \binom{5}{4}(3x)(-2)^4 + (-2)^5$$

We have $\binom{5}{1} = \binom{5}{4} = \frac{5}{1} = 5$ and $\binom{5}{2} = \binom{5}{3} = \frac{5 \cdot 4}{2 \cdot 1} = 10$

Thus

$$(3x - 2)^5 = 243x^5 + 5(81)x^4(-2) + 10(27)x^3(4) + 10(9)x^2(-8) + 5(3)x(16) + (-32)$$
$$= 243x^5 - 810x^4 + 1080x^3 - 720x^2 + 240x - 32$$

Chapter 11 REVIEW EXERCISES

In Exercises 1–8, write the first four terms and the twelfth term of the sequence whose nth term is given.

1. $a_n = 3n - 5$ **2.** $b_n = 7n + 1$
3. $x_n = 5n^2$ **4.** $y_n = 4n$
5. $a_n = \frac{(-1)^n}{n + 2}$ **6.** $b_n = \frac{(-1)^{n+2}}{n + 1}$
7. $x_n = 4 + (-1)^n$ **8.** $y_n = 2^n + |n - 5|$

In Exercises 9–12, find a possible general term for the given sequences.

9. $-2, 3, 8, 13, \ldots$ **10.** $3, 15, 75, 375, \ldots$
11. $\frac{1}{3}, -\frac{1}{6}, \frac{1}{9}, -\frac{1}{12}, \ldots$ **12.** $\frac{1}{2}, \frac{1}{4}, \frac{1}{8}, \frac{1}{16}, \ldots$

13. Byron gets a raise of \$8 in his weekly salary at the end of each month during his first two years on a job. Let p_n be his weekly salary at the beginning of month n. If his starting weekly salary is \$250 per week, write a formula for p_n during his first 2 years on the job.

14. The population of Sudyville increases at a rate of 3% per year. If the current population is 22,500, write a sequence to show how the population changes over the next 4 years.

15. Soojin invests \$5000 in a bond that increases in value by 5% each year. How much will her bond be worth in 5 years?

16. Jenna invests \$5000 in a bond that increases in value by 6% each year. How much is the bond worth in 5 years?

In Exercises 17–22, assume that the given sequence is either arithmetic or geometric and find the indicated sum, S_n.

17. $3, 6, 9, 12, \ldots; \quad S_7$ **18.** $3, 7, 11, 15, \ldots; \quad S_5$
19. $2, 4, 8, 16, \ldots; \quad S_6$
20. $3, 9, 81, 243, \ldots; \quad S_8$
21. $a_k = 5k - 2; \quad S_6$ **22.** $b_j = 3j + 4; \quad S_8$

In Exercises 23–28, rewrite each sum without using sigma notation; then compute each sum.

23. $\sum_{m=1}^{6} 2m$ **24.** $\sum_{j=4}^{9} 4j$
25. $\sum_{n=1}^{5} (n - 2)^2$ **26.** $\sum_{k=1}^{4} k^3$
27. $\sum_{i=1}^{7} 5i^2$ **28.** $\sum_{k=2}^{4} (k^2 - 6k + 1)$

In Exercises 29–34, use the given information to find the next two terms and the tenth term of the given *arithmetic* sequence.

29. $a_1 = 7, a_2 = 13$ **30.** $a_3 = 4, a_4 = -1$
31. $6, 11, 16, 21, \ldots$ **32.** $-4, -2, 0, 2, \ldots$
33. $2, 0, -2, -4, \ldots$ **34.** $\frac{1}{3}, 0, -\frac{1}{3}, \ldots$

35. Given the arithmetic sequence $\{x_i\}$ with $x_4 = 10$ and $x_8 = 22$, find x_1 and d.

36. Given the arithmetic sequence $\{y_i\}$ with $y_4 = -9$ and $y_6 = -13$, find y_2 and d.

37. Suppose that an object dropped from the top of a building falls 16 feet the first second, 48 feet the second second, and so on, the distances it falls each second forming an arithmetic sequence. How many feet will the object fall during the eighth second?

38. How many feet will the object described in Exercise 37 fall during the tenth second?

In Exercises 39–42, use a formula for S_n to find the indicated S_n for the given *arithmetic* sequence.

39. $\{a_k\}$, where the first term is 8 and the common difference is 6; find S_{12}.

40. $\{b_k\}$, where the first term is -9 and the common difference is -2; find S_9.

41. $4, 11, 18, \ldots; \quad S_7$ **42.** $-6, -2, 2, \ldots; \quad S_8$

43. Find $\sum_{i=1}^{40} 4i$. **44.** Find $\sum_{i=10}^{20} 2i$.

45. Find the sum of the first 25 positive odd numbers.

46. Find the sum of the first 25 positive even numbers.

47. Given an arithmetic sequence with $a_5 = 5$ and $a_9 = 8$, find S_{12}.

48. Given an arithmetic sequence with $a_2 = 2$ and $a_7 = 5$, find S_{10}.

In Exercises 49–54, find the required a_n for the given *geometric* sequence.

49. $2, -6, 18, \ldots; \quad a_5$ **50.** $1, 4, 16, \ldots; \quad a_6$
51. $1, \frac{1}{2}, \frac{1}{4}, \ldots; \quad a_7$ **52.** $4, -\frac{4}{3}, \frac{4}{9}, \ldots; \quad a_8$

53. $\{a_k\}$ with $a_1 = 4$ and $r = 3$; find a_5.

54. $\{a_j\}$ with $a_2 = 4$ and $r = \frac{1}{2}$; find a_6.

55. Jim promises to pay Cindy a debt in the following way: He will pay her \$3 the first month, \$6 the second month, \$12 the third month, and so on, where the payment in any month is double that of the previous month. How much would Cindy receive in the tenth month?

56. How much will Jim have paid Cindy of Exercise 55 altogether at the end of 1 year?

For Exercises 57–60, find the sum, S (if it exists), of the infinite geometric sequence.

57. $5, 1, \frac{1}{5}, \ldots$

58. $1, -\frac{1}{3}, \frac{1}{9}, \ldots$

59. $\frac{1}{4}, \frac{1}{2}, 1, \ldots$

60. $\frac{3}{7}, 1, \frac{7}{3}, \ldots$

61. Express $0.393\overline{939}$ as a fraction.

62. Express $2.1745745\overline{745}$ as a fraction.

63. A ball bounces back three-fourths the distance it falls on a rebound. If you could let the ball bounce forever, what is the total distance it would travel if it is dropped from a height of 80 feet?

64. Use mathematical induction to prove that

$$1 \cdot 2 + 2 \cdot 3 + 3 \cdot 4 + \cdots + (n-1)n = \frac{(n-1)n(n+1)}{3}$$

65. Use mathematical induction to prove that $2^n < n!$ for $n \geq 4$.

66. Use mathematical induction to prove that $n! > n^3$ for $n \geq 6$.

67. If there are 12 horses entered in a race, how many possible 1-2-3 finishes can there be?

68. The first race at a racetrack has 8 horses entered, and the second race has 10 horses entered. In how many ways can you pick the first 3 horses in the first two races?

69. How many different arrangements are there of the letters of the word HYPERBOLA?

70. How many different arrangements are there of the letters of the word PARABOLA?

71. How many different arrangements are there of the letters of the word BANANAS?

72. A briefcase has a 3-digit combination lock, where each digit can be any of 0, 1, 2, 3, 4, 5, 6, 7, 8.

(a) How many combinations are there?
(b) How many of these combinations are even numbers?
(c) How many of these combinations are greater than 500?

73. If a 5-card poker hand is dealt from a standard deck of 52 cards, in how many ways can you be dealt a

(a) Spade flush? **(b)** Straight?
(c) Straight flush?

74. How many boy-girl combinations are there in a family with 5 children?

75. How many boy-girl sequences are there in a family with 5 children?

76. Expand $(2x - y)^5$ and simplify.

77. Expand $\left(\frac{x}{3} + \frac{y}{4}\right)^3$ and simplify.

78. Expand $(x^{-2} + y^{-1})^6$ and simplify.

79. What is the coefficient of x^6 in the expansion of $(x + 5\sqrt{x})^8$?

80. What is the coefficient of a^7b^3 in the expansion of $(a - 3b)^{10}$?

Chapter 11 PRACTICE TEST

1. Identify the first three terms and the eighth term of the following sequences:
(a) $\{x_k\}$, where $x_k = k^2 - 3k$
(b) $\{y_j\}$, where $y_j = 5j^3 - 2$

2. Rewrite the following as a sum without using sigma notation; then compute the sum.

$$\sum_{k=4}^{7} (3k^2 - 5k + 1)$$

3. Find the 112th term of the arithmetic sequence with first term -4 and common difference 6.

4. Find the sixth term of a geometric sequence with first term 3 and common ratio -2.

5. Find S_{20} for the arithmetic sequence: 4, 8, 12, ...

6. Find S_5 for the geometric sequence: 3, 9, 27, ...

7. Use the fact that each of the following sequences is either arithmetic or geometric and the given information to find the requested values.

(a) Given the sequence 5, 9, 13, 17, . . . , find a_7.
(b) Given the sequence 1, 5, 25, 125, . . . , find a_7.
(c) Given $a_1 = \frac{1}{3}$ and $r = 6$, find a_5.
(d) Given $a_3 = 1$, $a_5 = 2$, and $a_7 = 3$, find S_{10}.

8. Find the sum of the first 20 positive multiples of 10.

9. A pendulum sweeps out an 8-foo t arc on its first pass, and each time it passes, it covers three-fourths the distance of the previous arc. How far would it travel before coming to rest?

10. Use the principle of mathematical induction to prove that $x - 2$ is a factor of $x^n - 2^n$.

11. Suppose a jar contains 10 yellow marbles, 8 blue marbles, and 5 green marbles.

(a) In how many ways can you choose 6 marbles from the jar?

(b) In how many ways can you select the 6 marbles choosing first 2 yellow, then 2 blue, and finally 2 green?

12. How many different arrangements are there of the letters in the word ABRACADABRA?

13. Use the binomial theorem to expand $(2x^2 - 3y)^8$.

TABLE 1: EXPONENTIAL FUNCTIONS

x	e^x	e^{-x}	x	e^x	e^{-x}
0.00	1.0000	1.0000	**1.5**	4.4817	0.2231
0.01	1.0101	0.9901	**1.6**	4.9530	0.2019
0.02	1.0202	0.9802	**1.7**	5.4739	0.1827
0.03	1.0305	0.9704	**1.8**	6.0496	0.1653
0.04	1.0408	0.9608	**1.9**	6.6859	0.1496
0.05	1.0513	0.9512	**2.0**	7.3891	0.1353
0.06	1.0618	0.9418	**2.1**	8.1662	0.1225
0.07	1.0725	0.9324	**2.2**	9.0250	0.1108
0.08	1.0833	0.9231	**2.3**	9.9742	0.1003
0.09	1.0942	0.9139	**2.4**	11.023	0.0907
0.10	1.1052	0.9048	**2.5**	12.182	0.0821
0.11	1.1163	0.8958	**2.6**	13.464	0.0743
0.12	1.1275	0.8869	**2.7**	14.880	0.0672
0.13	1.1388	0.8781	**2.8**	16.445	0.0608
0.14	1.1503	0.8694	**2.9**	18.174	0.0550
0.15	1.1618	0.8607	**3.0**	20.086	0.0498
0.16	1.1735	0.8521	**3.1**	22.198	0.0450
0.17	1.1853	0.8437	**3.2**	24.533	0.0408
0.18	1.1972	0.8353	**3.3**	27.113	0.0369
0.19	1.2092	0.8270	**3.4**	29.964	0.0334
0.20	1.2214	0.8187	**3.5**	33.115	0.0302
0.21	1.2337	0.8106	**3.6**	36.598	0.0273
0.22	1.2461	0.8025	**3.7**	40.447	0.0247
0.23	1.2586	0.7945	**3.8**	44.701	0.0224
0.24	1.2712	0.7866	**3.9**	49.402	0.0202
0.25	1.2840	0.7788	**4.0**	54.598	0.0183
0.30	1.3499	0.7408	**4.1**	60.340	0.0166
0.35	1.4191	0.7047	**4.2**	66.686	0.0150
0.40	1.4918	0.6703	**4.3**	73.700	0.0136
0.45	1.5683	0.6376	**4.4**	81.451	0.0123
0.50	1.6487	0.6065	**4.5**	90.017	0.0111
0.55	1.7333	0.5769	**4.6**	99.484	0.0101
0.60	1.8221	0.5488	**4.7**	109.95	0.0091
0.65	1.9155	0.5220	**4.8**	121.51	0.0082
0.70	2.0138	0.4966	**4.9**	134.29	0.0074
0.75	2.1170	0.4724	**5.0**	148.41	0.0067
0.80	2.2255	0.4493	**5.5**	244.69	0.0041
0.85	2.3396	0.4274	**6.0**	403.43	0.0025
0.90	2.4596	0.4066	**6.5**	665.14	0.0015
0.95	2.5857	0.3867	**7.0**	1096.6	0.0009
1.0	2.7183	0.3679	**7.5**	1808.0	0.0006
1.1	3.0042	0.3329	**8.0**	2981.0	0.0003
1.2	3.3201	0.3012	**8.5**	4914.8	0.0002
1.3	3.6693	0.2725	**9.0**	8103.1	0.0001
1.4	4.0552	0.2466	**10.0**	22026	0.00005

TABLE 2: COMMON LOGARITHMS

x	.00	.01	.02	.03	.04	.05	.06	.07	.08	.09
1.0	.0000	.0043	.0086	.0128	.0170	.0212	.0253	.0294	.0334	.0374
1.1	.0414	.0453	.0492	.0531	.0569	.0607	.0645	.0682	.0719	.0755
1.2	.0792	.0828	.0864	.0899	.0934	.0969	.1004	.1038	.1072	.1106
1.3	.1139	.1173	.1206	.1239	.1271	.1303	.1335	.1367	.1399	.1430
1.4	.1461	.1492	.1523	.1553	.1584	.1614	.1644	.1673	.1703	.1732
1.5	.1761	.1790	.1818	.1847	.1875	.1903	.1913	.1959	.1987	.2014
1.6	.2041	.2068	.2095	.2122	.2148	.2175	.2201	.2227	.2253	.2279
1.7	.2304	.2330	.2355	.2380	.2405	.2430	.2455	.2480	.2504	.2529
1.8	.2553	.2577	.2601	.2625	.2648	.2672	.2695	.2718	.2742	.2765
1.9	.2788	.2810	.2833	.2856	.2878	.2900	.2923	.2945	.2967	.2989
2.0	.3010	.3032	.3054	.3075	.3096	.3118	.3139	.3160	.3181	.3201
2.1	.3222	.3243	.3263	.3284	.3304	.3324	.3345	.3365	.3385	.3404
2.2	.3424	.3444	.3464	.3483	.3502	.3522	.3541	.3560	.3579	.3598
2.3	.3617	.3636	.3655	.3674	.3692	.3711	.3729	.3747	.3766	.3784
2.4	.3802	.3820	.3838	.3856	.3874	.3892	.3909	.3927	.3945	.3962
2.5	.3979	.3997	.4014	.4031	.4048	.4065	.4082	.4099	.4116	.4133
2.6	.4150	.4166	.4183	.4200	.4216	.4232	.4249	.4265	.4281	.4298
2.7	.4314	.4330	.4346	.4362	.4378	.4393	.4409	.4425	.4440	.4456
2.8	.4472	.4487	.4502	.4518	.4533	.4548	.4564	.4579	.4594	.4609
2.9	.4624	.4639	.4654	.4669	.4683	.4698	.4713	.4728	.4742	.4757
3.0	.4771	.4786	.4800	.4814	.4829	.4843	.4857	.4871	.4886	.4900
3.1	.4914	.4928	.4942	.4955	.4969	.4983	.4997	.5011	.5024	.5038
3.2	.5051	.5065	.5079	.5092	.5105	.5119	.5132	.5145	.5159	.5172
3.3	.5185	.5198	.5211	.5224	.5237	.5250	.5263	.5276	.5289	.5302
3.4	.5315	.5328	.5340	.5353	.5366	.5378	.5391	.5403	.5416	.5428
3.5	.5441	.5453	.5465	.5478	.5490	.5502	.5514	.5527	.5539	.5551
3.6	.5563	.5575	.5587	.5599	.5611	.5623	.5635	.5647	.5658	.5670
3.7	.5682	.5694	.5705	.5717	.5729	.5740	.5752	.5763	.5775	.5786
3.8	.5798	.5809	.5821	.5832	.5843	.5855	.5866	.5877	.5888	.5899
3.9	.5911	.5922	.5933	.5944	.5955	.5966	.5977	.5988	.5999	.6010
4.0	.6021	.6031	.6042	.6053	.6064	.6075	.6085	.6096	.6107	.6117
4.1	.6128	.6138	.6149	.6160	.6170	.6180	.6191	.6201	.6212	.6222
4.2	.6232	.6243	.6253	.6263	.6274	.6284	.6294	.6304	.6314	.6325
4.3	.6335	.6345	.6355	.6365	.6375	.6385	.6395	.6405	.6415	.6425
4.4	.6435	.6444	.6454	.6464	.6474	.6484	.6493	.6503	.6513	.6522
4.5	.6532	.6542	.6551	.6561	.6571	.6580	.6590	.6599	.6609	.6618
4.6	.6628	.6637	.6646	.6656	.6665	.6675	.6684	.6693	.6702	.6712
4.7	.6721	.6730	.6739	.6749	.6758	.6767	.6776	.6785	.6794	.6803
4.8	.6812	.6821	.6830	.6839	.6848	.6857	.6866	.6875	.6884	.6893
4.9	.6902	.6911	.6920	.6928	.6937	.6946	.6955	.6964	.6972	.6981
5.0	.6990	.6998	.7007	.7016	.7024	.7033	.7042	.7050	.7059	.7067
5.1	.7076	.7084	.7093	.7101	.7110	.7118	.7126	.7135	.7143	.7152
5.2	.7160	.7168	.7177	.7185	.7193	.7202	.7210	.7218	.7226	.7235
5.3	.7243	.7251	.7259	.7267	.7275	.7284	.7292	.7300	.7308	.7316
5.4	.7324	.7332	.7340	.7348	.7356	.7364	.7372	.7380	.7388	.7396

TABLE 2 *(continued)*

x	.00	.01	.02	.03	.04	.05	.06	.07	.08	.09
5.5	.7404	.7412	.7419	.7427	.7435	.7443	.7451	.7459	.7466	.7474
5.6	.7482	.7490	.7497	.7505	.7513	.7520	.7528	.7536	.7543	.7551
5.7	.7559	.7566	.7574	.7582	.7589	.7597	.7604	.7612	.7619	.7627
5.8	.7634	.7642	.7649	.7657	.7664	.7672	.7679	.7686	.7694	.7701
5.9	.7709	.7716	.7723	.7731	.7738	.7745	.7752	.7760	.7767	.7774
6.0	.7782	.7789	.7796	.7803	.7810	.7818	.7825	.7832	.7839	.7846
6.1	.7853	.7860	.7868	.7875	.7882	.7889	.7896	.7903	.7910	.7917
6.2	.7924	.7931	.7938	.7945	.7952	.7959	.7966	.7973	.7980	.7987
6.3	.7993	.8000	.8007	.8014	.8021	.8028	.8035	.8041	.8048	.8055
6.4	.8062	.8069	.8075	.8082	.8089	.8096	.8102	.8109	.8116	.8122
6.5	.8129	.8136	.8142	.8149	.8156	.8162	.8169	.8176	.8182	.8189
6.6	.8195	.8202	.8209	.8215	.8222	.8228	.8235	.8241	.8248	.8254
6.7	.8261	.8267	.8274	.8280	.8287	.8293	.8299	.8306	.8312	.8319
6.8	.8325	.8331	.8338	.8344	.8351	.8357	.8363	.8370	.8376	.8382
6.9	.8388	.8395	.8401	.8407	.8414	.8420	.8426	.8432	.8439	.8445
7.0	.8451	.8457	.8463	.8470	.8476	.8482	.8488	.8494	.8500	.8506
7.1	.8513	.8519	.8525	.8531	.8537	.8543	.8549	.8555	.8561	.8567
7.2	.8573	.8579	.8585	.8591	.8597	.8603	.8609	.8615	.8621	.8627
7.3	.8633	.8639	.8645	.8651	.8657	.8663	.8669	.8675	.8681	.8686
7.4	.8692	.8698	.8704	.8710	.8716	.8722	.8727	.8733	.8739	.8745
7.5	.8751	.8756	.8762	.8768	.8774	.8779	.8785	.8791	.8797	.8802
7.6	.8808	.8814	.8820	.8825	.8831	.8837	.8842	.8848	.8854	.8859
7.7	.8865	.8871	.8876	.8882	.8887	.8893	.8899	.8904	.8910	.8915
7.8	.8921	.8927	.8932	.8938	.8943	.8949	.8954	.8960	.8965	.8971
7.9	.8976	.8982	.8987	.8993	.8998	.9004	.9009	.9015	.9020	.9025
8.0	.9031	.9036	.9042	.9047	.9053	.9058	.9063	.9069	.9074	.9079
8.1	.9085	.9090	.9096	.9101	.9106	.9112	.9117	.9122	.9128	.9133
8.2	.9138	.9143	.9149	.9154	.9159	.9165	.9170	.9175	.9180	.9186
8.3	.9191	.9196	.9201	.9206	.9212	.9217	.9222	.9227	.9232	.9238
8.4	.9243	.9248	.9253	.9258	.9263	.9269	.9274	.9279	.9284	.9289
8.5	.9294	.9299	.9304	.9309	.9315	.9320	.9325	.9330	.9335	.9340
8.6	.9345	.9350	.9355	.9360	.9365	.9370	.9375	.9380	.9385	.9390
8.7	.9395	.9400	.9405	.9410	.9415	.9420	.9425	.9430	.9435	.9440
8.8	.9445	.9450	.9455	.9460	.9465	.9469	.9474	.9479	.9484	.9489
8.9	.9494	.9499	.9504	.9509	.9513	.9518	.9523	.9528	.9533	.9538
9.0	.9542	.9547	.9552	.9557	.9562	.9566	.9571	.9576	.9581	.9586
9.1	.9590	.9595	.9600	.9605	.9609	.9614	.9619	.9624	.9628	.9633
9.2	.9638	.9643	.9647	.9652	.9657	.9661	.9666	.9671	.9675	.9680
9.3	.9685	.9689	.9694	.9699	.9703	.9708	.9713	.9717	.9722	.9727
9.4	.9731	.9736	.9741	.9745	.9750	.9754	.9759	.9763	.9768	.9773
9.5	.9777	.9782	.9786	.9791	.9795	.9800	.9805	.9809	.9814	.9818
9.6	.9823	.9827	.9832	.9836	.9841	.9845	.9850	.9854	.9859	.9863
9.7	.9868	.9872	.9877	.9881	.9886	.9890	.9894	.9899	.9903	.9908
9.8	.9912	.9917	.9921	.9926	.9930	.9934	.9939	.9943	.9948	.9952
9.9	.9956	.9961	.9965	.9969	.9974	.9978	.9983	.9987	.9991	.9996

TABLE 3: NATURAL LOGARITHMS

$$\ln(a \times 10^n) = \ln a + n \ln 10, \ \ln 10 = 2.3026$$

x	.00	.01	.02	.03	.04	.05	.06	.07	.08	.09
1.0	0.0000	0.0100	0.0198	0.0296	0.0392	0.0488	0.0583	0.0677	0.0770	0.0862
1.1	0.0953	0.1044	0.1133	0.1222	0.1310	0.1398	0.1484	0.1570	0.1655	0.1740
1.2	0.1823	0.1906	0.1989	0.2070	0.2151	0.2231	0.2311	0.2390	0.2469	0.2546
1.3	0.2624	0.2700	0.2776	0.2852	0.2927	0.3001	0.3075	0.3148	0.3221	0.3293
1.4	0.3365	0.3436	0.3507	0.3577	0.3646	0.3716	0.3784	0.3853	0.3920	0.3988
1.5	0.4055	0.4121	0.4187	0.4253	0.4318	0.4383	0.4447	0.4511	0.4574	0.4637
1.6	0.4700	0.4762	0.4824	0.4886	0.4947	0.5008	0.5068	0.5128	0.5188	0.5247
1.7	0.5306	0.5365	0.5423	0.5481	0.5539	0.5596	0.5653	0.5710	0.5766	0.5822
1.8	0.5878	0.5933	0.5988	0.6043	0.6098	0.6152	0.6206	0.6259	0.6313	0.6366
1.9	0.6419	0.6471	0.6523	0.6575	0.6627	0.6678	0.6729	0.6780	0.6831	0.6881
2.0	0.6931	0.6981	0.7031	0.7080	0.7130	0.7178	0.7227	0.7275	0.7324	0.7372
2.1	0.7419	0.7467	0.7514	0.7561	0.7608	0.7655	0.7701	0.7747	0.7793	0.7839
2.2	0.7885	0.7930	0.7975	0.8020	0.8065	0.8109	0.8154	0.8198	0.8242	0.8286
2.3	0.8329	0.8372	0.8416	0.8459	0.8502	0.8544	0.8587	0.8629	0.8671	0.8713
2.4	0.8755	0.8796	0.8838	0.8879	0.8920	0.8961	0.9002	0.9042	0.9083	0.9123
2.5	0.9163	0.9203	0.9243	0.9282	0.9322	0.9361	0.9400	0.9439	0.9478	0.9517
2.6	0.9555	0.9594	0.9632	0.9670	0.9708	0.9746	0.9783	0.9821	0.9858	0.9895
2.7	0.9933	0.9969	1.0006	1.0043	1.0080	1.0116	1.0152	1.0188	1.0225	1.0260
2.8	1.0296	1.0332	1.0367	1.0403	1.0438	1.0473	1.0508	1.0543	1.0578	1.0613
2.9	1.0647	1.0682	1.0716	1.0750	1.0784	1.0818	1.0852	1.0886	1.0919	1.0953
3.0	1.0986	1.1019	1.1053	1.1086	1.1119	1.1151	1.1184	1.1217	1.1249	1.1282
3.1	1.1314	1.1346	1.1378	1.1410	1.1442	1.1474	1.1506	1.1537	1.1569	1.1600
3.2	1.1632	1.1663	1.1694	1.1725	1.1756	1.1787	1.1817	1.1848	1.1878	1.1909
3.3	1.1939	1.1970	1.2000	1.2030	1.2060	1.2090	1.2119	1.2149	1.2179	1.2208
3.4	1.2238	1.2267	1.2296	1.2326	1.2355	1.2384	1.2413	1.2442	1.2470	1.2499
3.5	1.2528	1.2556	1.2585	1.2613	1.2641	1.2669	1.2698	1.2726	1.2754	1.2782
3.6	1.2809	1.2837	1.2865	1.2892	1.2920	1.2947	1.2975	1.3002	1.3029	1.3056
3.7	1.3083	1.3110	1.3137	1.3164	1.3191	1.3218	1.3244	1.3271	1.3297	1.3324
3.8	1.3350	1.3376	1.3403	1.3429	1.3455	1.3481	1.3507	1.3533	1.3558	1.3584
3.9	1.3610	1.3635	1.3661	1.3686	1.3712	1.3737	1.3762	1.3788	1.3813	1.3838
4.0	1.3863	1.3888	1.3913	1.3938	1.3962	1.3987	1.4012	1.4036	1.4061	1.4085
4.1	1.4110	1.4134	1.4159	1.4183	1.4207	1.4231	1.4255	1.4279	1.4303	1.4327
4.2	1.4351	1.4375	1.4398	1.4422	1.4446	1.4469	1.4493	1.4516	1.4540	1.4563
4.3	1.4586	1.4609	1.4633	1.4656	1.4679	1.4702	1.4725	1.4748	1.4770	1.4793
4.4	1.4816	1.4839	1.4861	1.4884	1.4907	1.4929	1.4952	1.4974	1.4996	1.5019
4.5	1.5041	1.5063	1.5085	1.5107	1.5129	1.5151	1.5173	1.5195	1.5217	1.5239
4.6	1.5261	1.5282	1.5304	1.5326	1.5347	1.5369	1.5390	1.5412	1.5433	1.5454
4.7	1.5476	1.5497	1.5518	1.5539	1.5560	1.5581	1.5602	1.5623	1.5644	1.5665
4.8	1.5686	1.5707	1.5728	1.5748	1.5769	1.5790	1.5810	1.5831	1.5851	1.5872
4.9	1.5892	1.5913	1.5933	1.5953	1.5974	1.5994	1.6014	1.6034	1.6054	1.6074
5.0	1.6094	1.6114	1.6134	1.6154	1.6174	1.6194	1.6214	1.6233	1.6253	1.6273
5.1	1.6292	1.6312	1.6332	1.6351	1.6371	1.6390	1.6409	1.6429	1.6448	1.6467
5.2	1.6487	1.6506	1.6525	1.6544	1.6563	1.6582	1.6601	1.6620	1.6639	1.6658
5.3	1.6677	1.6696	1.6715	1.6734	1.6753	1.6771	1.6790	1.6808	1.6827	1.6845
5.4	1.6864	1.6882	1.6901	1.6919	1.6938	1.6956	1.6974	1.6993	1.7011	1.7029

TABLE 3 ***(continued)***

x	.00	.01	.02	.03	.04	.05	.06	.07	.08	.09
5.5	1.7047	1.7066	1.7084	1.7102	1.7120	1.7138	1.7156	1.7174	1.7192	1.7210
5.6	1.7228	1.7246	1.7263	1.7281	1.7299	1.7317	1.7334	1.7352	1.7370	1.7387
5.7	1.7405	1.7422	1.7440	1.7457	1.7475	1.7492	1.7509	1.7527	1.7544	1.7561
5.8	1.7579	1.7596	1.7613	1.7630	1.7647	1.7664	1.7682	1.7699	1.7716	1.7733
5.9	1.7750	1.7766	1.7783	1.7800	1.7817	1.7834	1.7851	1.7867	1.7884	1.7901
6.0	1.7918	1.7934	1.7951	1.7967	1.7984	1.8001	1.8017	1.8034	1.8050	1.8066
6.1	1.8083	1.8099	1.8116	1.8132	1.8148	1.8165	1.8181	1.8197	1.8213	1.8229
6.2	1.8245	1.8262	1.8278	1.8294	1.8310	1.8326	1.8342	1.8358	1.8374	1.8390
6.3	1.8406	1.8421	1.8437	1.8453	1.8469	1.8485	1.8500	1.8516	1.8532	1.8547
6.4	1.8563	1.8579	1.8594	1.8610	1.8625	1.8641	1.8656	1.8672	1.8687	1.8703
6.5	1.8718	1.8733	1.8749	1.8764	1.8779	1.8795	1.8810	1.8825	1.8840	1.8856
6.6	1.8871	1.8886	1.8901	1.8916	1.8931	1.8946	1.8961	1.8976	1.8991	1.9006
6.7	1.9021	1.9036	1.9051	1.9066	1.9081	1.9095	1.9110	1.9125	1.9140	1.9155
6.8	1.9169	1.9184	1.9199	1.9213	1.9228	1.9242	1.9257	1.9272	1.9286	1.9301
6.9	1.9315	1.9330	1.9344	1.9359	1.9373	1.9387	1.9402	1.9416	1.9430	1.9445
7.0	1.9459	1.9473	1.9488	1.9502	1.9516	1.9530	1.9544	1.9559	1.9573	1.9587
7.1	1.9601	1.9615	1.9629	1.9643	1.9657	1.9671	1.9685	1.9699	1.9713	1.9727
7.2	1.9741	1.9755	1.9769	1.9782	1.9796	1.9810	1.9824	1.9838	1.9851	1.9865
7.3	1.9879	1.9892	1.9906	1.9920	1.9933	1.9947	1.9961	1.9974	1.9988	2.0001
7.4	2.0015	2.0028	2.0042	2.0055	2.0069	2.0082	2.0096	2.0109	2.0122	2.0136
7.5	2.0149	2.0162	2.0176	2.0189	2.0202	2.0215	2.0229	2.0242	2.0255	2.0268
7.6	2.0282	2.0295	2.0308	2.0321	2.0334	2.0347	2.0360	2.0373	2.0386	2.0399
7.7	2.0412	2.0425	2.0438	2.0451	2.0464	2.0477	2.0490	2.0503	2.0516	2.0528
7.8	2.0541	2.0554	2.0567	2.0580	2.0592	2.0605	2.0618	2.0631	2.0643	2.0656
7.9	2.0669	2.0681	2.0694	2.0707	2.0719	2.0732	2.0744	2.0757	2.0769	2.0782
8.0	2.0794	2.0807	2.0819	2.0832	2.0844	2.0857	2.0869	2.0882	2.0894	2.0906
8.1	2.0919	2.0931	2.0943	2.0956	2.0968	2.0980	2.0992	2.1005	2.1017	2.1029
8.2	2.1041	2.1054	2.1066	2.1078	2.1090	2.1102	2.1114	2.1126	2.1138	2.1150
8.3	2.1163	2.1175	2.1187	2.1199	2.1211	2.1223	2.1235	2.1247	2.1258	2.1270
8.4	2.1282	2.1294	2.1306	2.1318	2.1330	2.1342	2.1353	2.1365	2.1377	2.1389
8.5	2.1401	2.1412	2.1424	2.1436	2.1448	2.1459	2.1471	2.1483	2.1494	2.1506
8.6	2.1518	2.1529	2.1541	2.1552	2.1564	2.1576	2.1587	2.1599	2.1610	2.1622
8.7	2.1633	2.1645	2.1656	2.1668	2.1679	2.1691	2.1702	2.1713	2.1725	2.1736
8.8	2.1748	2.1759	2.1770	2.1782	2.1793	2.1804	2.1815	2.1827	2.1838	2.1849
8.9	2.1861	2.1872	2.1883	2.1894	2.1905	2.1917	2.1928	2.1939	2.1950	2.1961
9.0	2.1972	2.1983	2.1994	2.2006	2.2017	2.2028	2.2039	2.2050	2.2061	2.2072
9.1	2.2083	2.2094	2.2105	2.2116	2.2127	2.2138	2.2148	2.2159	2.2170	2.2181
9.2	2.2192	2.2203	2.2214	2.2225	2.2235	2.2246	2.2257	2.2268	2.2279	2.2289
9.3	2.2300	2.2311	2.2322	2.2332	2.2343	2.2354	2.2364	2.2375	2.2386	2.2396
9.4	2.2407	2.2418	2.2428	2.2439	2.2450	2.2460	2.2471	2.2481	2.2492	2.2502
9.5	2.2513	2.2523	2.2534	2.2544	2.2555	2.2565	2.2576	2.2586	2.2597	2.2607
9.6	2.2618	2.2628	2.2638	2.2649	2.2659	2.2670	2.2680	2.2690	2.2701	2.2711
9.7	2.2721	2.2732	2.2742	2.2752	2.2762	2.2773	2.2783	2.2793	2.2803	2.2814
9.8	2.2824	2.2834	2.2844	2.2854	2.2865	2.2875	2.2885	2.2895	2.2905	2.2915
9.9	2.2925	2.2935	2.2946	2.2956	2.2966	2.2976	2.2986	2.2996	2.3006	2.3016

TABLE 4: TRIGONOMETRIC FUNCTIONS

For angles between 0° and 45°, read the angles on the left and the column headings across the top of the table. For angles between 45° and 90°, read the angles on the right and the column headings across the bottom of the table.

Degrees	Radians	sin	cos	tan	cot	sec	csc		
0°00′	.0000	.0000	1.0000	.0000	—	1.000	—	1.5708	**90°00′**
10	.0029	.0029	1.0000	.0029	343.8	1.000	343.8	1.5679	50
20	.0058	.0058	1.0000	.0058	171.9	1.000	171.9	1.5650	40
30	.0087	.0087	1.0000	.0087	114.6	1.000	114.6	1.5621	30
40	.0116	.0116	.9999	.0116	85.94	1.000	85.95	1.5592	20
50	.0145	.0145	.9999	.0145	68.75	1.000	68.76	1.5563	10
1°00′	.0175	.0175	.9998	.0175	57.29	1.000	57.30	1.5533	**89°00′**
10	.0204	.0204	.9998	.0204	49.10	1.000	49.11	1.5504	50
20	.0233	.0233	.9997	.0233	42.96	1.000	42.98	1.5475	40
30	.0262	.0262	.9997	.0262	38.19	1.000	38.20	1.5446	30
40	.0291	.0291	.9996	.0291	34.37	1.000	34.38	1.5417	20
50	.0320	.0320	.9995	.0320	31.24	1.001	31.26	1.5388	10
2°00′	.0349	.0349	.9994	.0349	28.64	1.001	28.65	1.5359	**88°00′**
10	.0378	.0378	.9993	.0378	26.43	1.001	26.45	1.5330	50
20	.0407	.0407	.9992	.0407	24.54	1.001	24.56	1.5301	40
30	.0436	.0436	.9990	.0437	22.90	1.001	22.93	1.5272	30
40	.0465	.0465	.9989	.0466	21.47	1.001	21.49	1.5243	20
50	.0495	.0494	.9988	.0495	20.21	1.001	20.23	1.5213	10
3°00′	.0524	.0523	.9986	.0524	19.08	1.001	19.11	1.5184	**87°00′**
10	.0553	.0552	.9985	.0553	18.07	1.002	18.10	1.5155	50
20	.0582	.0581	.9983	.0582	17.17	1.002	17.20	1.5126	40
30	.0611	.0610	.9981	.0612	16.35	1.002	16.38	1.5097	30
40	.0640	.0640	.9980	.0641	15.60	1.002	15.64	1.5068	20
50	.0669	.0669	.9978	.0670	14.92	1.002	14.96	1.5039	10
4°00′	.0698	.0698	.9976	.0699	14.30	1.002	14.34	1.5010	**86°00′**
10	.0727	.0727	.9974	.0729	13.73	1.003	13.76	1.5981	50
20	.0756	.0756	.9971	.0758	13.20	1.003	13.23	1.5952	40
30	.0785	.0785	.9969	.0787	12.71	1.003	12.75	1.4923	30
40	.0814	.0814	.9967	.0816	12.25	1.003	12.29	1.4893	20
50	.0844	.0843	.9964	.0846	11.83	1.004	11.87	1.4864	10
5°00′	.0873	.0872	.9962	.0875	11.43	1.004	11.47	1.4835	**85°00′**
10	.0902	.0901	.9959	.0904	11.06	1.004	11.10	1.4806	50
20	.0931	.0929	.9957	.0934	10.71	1.004	10.76	1.4777	40
30	.0960	.0958	.9954	.0963	10.39	1.005	10.43	1.4748	30
40	.0989	.0987	.9951	.0992	10.08	1.005	10.13	1.4719	20
50	.1018	.1016	.9948	.1022	9.788	1.005	9.839	1.4690	10
6°00′	.1047	.1045	.9945	.1051	9.514	1.006	9.567	1.4661	**84°00′**
10	.1076	.1074	.9942	.1080	9.255	1.006	9.309	1.4632	50
20	.1105	.1103	.9939	.1110	9.010	1.006	9.065	1.4603	40
30	.1134	.1132	.9936	.1139	8.777	1.006	8.834	1.4573	30
40	.1164	.1161	.9932	.1169	8.556	1.007	8.614	1.4544	20
50	.1193	.1190	.9929	.1198	8.345	1.007	8.405	1.4515	10
7°00′	.1222	.1219	.9925	.1228	8.144	1.008	8.206	1.4486	**83°00′**
10	.1251	.1248	.9922	.1257	7.953	1.008	8.016	1.4457	50
20	.1280	.1276	.9918	.1287	7.770	1.008	7.834	1.4428	40
30	.1309	.1305	.9914	.1317	7.596	1.009	7.661	1.4399	30
40	.1338	.1334	.9911	.1346	7.429	1.009	7.496	1.4370	20
50	.1367	.1363	.9907	.1376	7.269	1.009	7.337	1.4341	10
		cos	sin	cot	tan	csc	sec	Radians	Degrees

TABLE 4 *(continued)*

Degrees	Radians	sin	cos	tan	cot	sec	csc		
8°00′	.1396	.1392	.9903	.1405	7.115	1.010	7.185	1.4312	**82°00′**
10	.1425	.1421	.9899	.1435	6.968	1.010	7.040	1.4283	50
20	.1454	.1449	.9894	.1465	6.827	1.011	6.900	1.4254	40
30	.1484	.1478	.9890	.1495	6.691	1.011	6.765	1.4224	30
40	.1513	.1507	.9886	.1524	6.561	1.012	6.636	1.4195	20
50	.1542	.1536	.9881	.1554	6.435	1.012	6.512	1.4166	10
9°00′	.1571	.1564	.9877	.1584	6.314	1.012	6.392	1.4137	**81°00′**
10	.1600	.1593	.9872	.1614	6.197	1.013	6.277	1.4108	50
20	.1629	.1622	.9868	.1644	6.084	1.013	6.166	1.4079	40
30	.1658	.1650	.9863	.1673	5.976	1.014	6.059	1.4050	30
40	.1687	.1679	.9858	.1703	5.871	1.014	5.955	1.4021	20
50	.1716	.1708	.9853	.1733	5.769	1.015	5.855	1.4992	10
10°00′	.1745	.1736	.9848	.1763	5.671	1.015	5.759	1.3963	**80°00′**
10	.1774	.1765	.9843	.1793	5.576	1.016	5.665	1.3934	50
20	.1804	.1794	.9838	.1823	5.485	1.016	5.575	1.3904	40
30	.1833	.1822	.9833	.1853	5.396	1.017	5.487	1.3875	30
40	.1862	.1851	.9827	.1883	5.309	1.018	5.403	1.3846	20
50	.1891	.1880	.9822	.1914	5.226	1.018	5.320	1.3817	10
11°00′	.1920	.1908	.9816	.1944	5.145	1.019	5.241	1.3788	**79°00′**
10	.1949	.1937	.9811	.1974	5.066	1.019	5.164	1.3759	50
20	.1978	.1965	.9805	.2004	4.989	1.020	5.089	1.3730	40
30	.2007	.1994	.9799	.2035	4.915	1.020	5.016	1.3701	30
40	.2036	.2022	.9793	.2065	4.843	1.021	4.945	1.3672	20
50	.2065	.2051	.9787	.2095	4.773	1.022	4.876	1.3643	10
12°00′	.2094	.2079	.9781	.2126	4.705	1.022	4.810	1.3614	**78°00′**
10	.2123	.2108	.9775	.2156	4.638	1.023	4.745	1.3584	50
20	.2153	.2136	.9769	.2186	4.574	1.024	4.682	1.3555	40
30	.2182	.2164	.9763	.2217	4.511	1.024	4.620	1.3526	30
40	.2211	.2193	.9757	.2247	4.449	1.025	4.560	1.3497	20
50	.2240	.2221	.9750	.2278	4.390	1.026	4.502	1.3468	10
13°00′	.2269	.2250	.9744	.2309	4.331	1.026	4.445	1.3439	**77°00′**
10	.2298	.2278	.9737	.2339	4.275	1.027	4.390	1.3410	50
20	.2327	.2306	.9730	.2370	4.219	1.028	4.336	1.3381	40
30	.2356	.2334	.9724	.2401	4.165	1.028	4.284	1.3352	30
40	.2385	.2363	.9717	.2432	4.113	1.029	4.232	1.3323	20
50	.2414	.2391	.9710	.2462	4.061	1.030	4.182	1.3294	10
14°00′	.2443	.2419	.9703	.2493	4.011	1.031	4.134	1.3265	**76°00′**
10	.2473	.2447	.9696	.2524	3.962	1.031	4.086	1.3235	50
20	.2502	.2476	.9689	.2555	3.914	1.032	4.039	1.3206	40
30	.2531	.2504	.9681	.2586	3.867	1.033	3.994	1.3177	30
40	.2560	.2532	.9674	.2617	3.821	1.034	3.950	1.3148	20
50	.2589	.2560	.9667	.2648	3.776	1.034	3.906	1.3119	10
15°00′	.2618	.2588	.9659	.2679	3.732	1.035	3.864	1.3090	**75°00′**
10	.2647	.2616	.9652	.2711	3.689	1.036	3.822	1.3061	50
20	.2676	.2644	.9644	.2742	3.647	1.037	3.782	1.3032	40
30	.2705	.2672	.9636	.2773	3.606	1.038	3.742	1.3003	30
40	.2734	.2700	.9628	.2805	3.566	1.039	3.703	1.3974	20
50	.2763	.2728	.9621	.2836	3.526	1.039	3.665	1.3945	10
		cos	sin	cot	tan	csc	sec	Radians	Degrees

TABLE 4 ***(continued)***

Degrees	Radians	sin	cos	tan	cot	sec	csc		
16°00′	.2793	.2756	.9613	.2867	3.487	1.040	3.628	1.2915	**74°00′**
10	.2822	.2784	.9605	.2899	3.450	1.041	3.592	1.2886	50
20	.2851	.2812	.9596	.2931	3.412	1.042	3.556	1.2857	40
30	.2880	.2840	.9588	.2962	3.376	1.043	3.521	1.2828	30
40	.2909	.2868	.9580	.2994	3.340	1.044	3.487	1.2799	20
50	.2938	.2896	.9572	.3026	3.305	1.045	3.453	1.2770	10
17°00′	.2967	.2924	.9563	.3057	3.271	1.046	3.420	1.2741	**73°00′**
10	.2996	.2952	.9555	.3089	3.237	1.047	3.388	1.2712	50
20	.3025	.2979	.9546	.3121	3.204	1.048	3.356	1.2683	40
30	.3054	.3007	.9537	.3153	3.172	1.049	3.326	1.2654	30
40	.3083	.3035	.9528	.3185	3.140	1.049	3.295	1.2625	20
50	.3113	.3062	.9520	.3217	3.108	1.050	3.265	1.2595	10
18°00′	.3142	.3090	.9511	.3249	3.078	1.051	3.236	1.2566	**72°00′**
10	.3171	.3118	.9502	.3281	3.047	1.052	3.207	1.2537	50
20	.3200	.3145	.9492	.3314	3.018	1.053	3.179	1.2508	40
30	.3229	.3173	.9483	.3346	2.989	1.054	3.152	1.2479	30
40	.3258	.3201	.9474	.3378	2.960	1.056	3.124	1.2450	20
50	.3287	.3228	.9465	.3411	2.932	1.057	3.098	1.2421	10
19°00′	.3316	.3256	.9455	.3443	2.904	1.058	3.072	1.2392	**71°00′**
10	.3345	.3283	.9446	.3476	2.877	1.059	3.046	1.2363	50
20	.3374	.3311	.9436	.3508	2.850	1.060	3.021	1.2334	40
30	.3403	.3338	.9426	.3541	2.824	1.061	2.996	1.2305	30
40	.3432	.3365	.9417	.3574	2.798	1.062	2.971	1.2275	20
50	.3462	.3393	.9407	.3607	2.773	1.063	2.947	1.2246	10
20°00′	.3491	.3420	.9397	.3640	2.747	1.064	2.924	1.2217	**70°00′**
10	.3520	.3448	.9387	.3673	2.723	1.065	2.901	1.2188	50
20	.3549	.3475	.9377	.3706	2.699	1.066	2.878	1.2159	40
30	.3578	.3502	.9367	.3739	2.675	1.068	2.855	1.2130	30
40	.3607	.3529	.9356	.3772	2.651	1.069	2.833	1.2101	20
50	.3636	.3557	.9346	.3805	2.628	1.070	2.812	1.2072	10
21°00′	.3665	.3584	.9336	.3839	2.605	1.071	2.790	1.2043	**69°00′**
10	.3694	.3611	.9325	.3872	2.583	1.072	2.769	1.2014	50
20	.3723	.3638	.9315	.3906	2.560	1.074	2.749	1.2985	40
30	.3752	.3665	.9304	.3939	2.539	1.075	2.729	1.1956	30
40	.3782	.3692	.9293	.3973	2.517	1.076	2.709	1.1926	20
50	.3811	.3719	.9283	.4006	2.496	1.077	2.689	1.1897	10
22°00′	.3840	.3746	.9272	.4040	2.475	1.079	2.669	1.1868	**68°00′**
10	.3869	.3773	.9261	.4074	2.455	1.080	2.650	1.1839	50
20	.3898	.3800	.9250	.4108	2.434	1.081	2.632	1.1810	40
30	.3927	.3827	.9239	.4142	2.414	1.082	2.613	1.1781	30
40	.3956	.3854	.9228	.4176	2.394	1.084	2.595	1.1752	20
50	.3985	.3881	.9216	.4210	2.375	1.085	2.577	1.1723	10
23°00′	.4014	.3907	.9205	.4245	2.356	1.086	2.559	1.1694	**67°00′**
10	.4043	.3934	.9194	.4279	2.337	1.088	2.542	1.1665	50
20	.4072	.3961	.9182	.4314	2.318	1.089	2.525	1.1636	40
30	.4102	.3987	.9171	.4348	2.300	1.090	2.508	1.1606	30
40	.4131	.4014	.9159	.4383	2.282	1.092	2.491	1.1577	20
50	.4160	.4041	.9147	.4417	2.264	1.093	2.475	1.1548	10
		cos	sin	cot	tan	csc	sec	Radians	Degrees

TABLE 4 *(continued)*

Degrees	Radians	sin	cos	tan	cot	sec	csc		
24°00′	.4189	.4067	.9135	.4452	2.246	1.095	2.459	1.1519	**66°00′**
10	.4218	.4094	.9124	.4487	2.229	1.096	2.443	1.1490	50
20	.4247	.4120	.9112	.4522	2.211	1.097	2.427	1.1461	40
30	.4276	.4147	.9100	.4557	2.194	1.099	2.411	1.1432	30
40	.4305	.4173	.9088	.4592	2.177	1.100	2.396	1.1403	20
50	.4334	.4200	.9075	.4628	2.161	1.102	2.381	1.1374	10
25°00′	.4363	.4226	.9063	.4663	2.145	1.103	2.366	1.1345	**65°00′**
10	.4392	.4253	.9051	.4699	2.128	1.105	2.352	1.1316	50
20	.4422	.4279	.9038	.4734	2.112	1.106	2.337	1.1286	40
30	.4451	.4305	.9026	.4770	2.097	1.108	2.323	1.1257	30
40	.4480	.4331	.9013	.4806	2.081	1.109	2.309	1.1228	20
50	.4509	.4358	.9001	.4841	2.066	1.111	2.295	1.1199	10
26°00′	.4538	.4384	.8988	.4877	2.050	1.113	2.281	1.1170	**64°00′**
10	.4567	.4410	.8975	.4913	2.035	1.114	2.268	1.1141	50
20	.4596	.4436	.8962	.4950	2.020	1.116	2.254	1.1112	40
30	.4625	.4462	.8949	.4986	2.006	1.117	2.241	1.1083	30
40	.4654	.4488	.8936	.5022	1.991	1.119	2.228	1.1054	20
50	.4683	.4514	.8923	.5059	1.977	1.121	2.215	1.1025	10
27°00′	.4712	.4540	.8910	.5095	1.963	1.122	2.203	1.0996	**63°00′**
10	.4741	.4566	.8897	.5132	1.949	1.124	2.190	1.0966	50
20	.4771	.4592	.8884	.5169	1.935	1.126	2.178	1.0937	40
30	.4800	.4617	.8870	.5206	1.921	1.127	2.166	1.0908	30
40	.4829	.4643	.8857	.5243	1.907	1.129	2.154	1.0879	20
50	.4858	.4669	.8843	.5280	1.894	1.131	2.142	1.0850	10
28°00′	.4887	.4695	.8829	.5317	1.881	1.133	2.130	1.0821	**62°00′**
10	.4916	.4720	.8816	.5354	1.868	1.134	2.118	1.0792	50
20	.4945	.4746	.8802	.5392	1.855	1.136	2.107	1.0763	40
30	.4974	.4772	.8788	.5430	1.842	1.138	2.096	1.0734	30
40	.5003	.4797	.8774	.5467	1.829	1.140	2.085	1.0705	20
50	.5032	.4823	.8760	.5505	1.816	1.142	2.074	1.0676	10
29°00′	.5061	.4848	.8746	.5543	1.804	1.143	2.063	1.0647	**61°00′**
10	.5091	.4874	.8732	.5581	1.792	1.145	2.052	1.0617	50
20	.5120	.4899	.8718	.5619	1.780	1.147	2.041	1.0588	40
30	.5149	.4924	.8704	.5658	1.767	1.149	2.031	1.0559	30
40	.5178	.4950	.8689	.5696	1.756	1.151	2.020	1.0530	20
50	.5207	.4975	.8675	.5735	1.744	1.153	2.010	1.0501	10
30°00′	.5236	.5000	.8660	.5774	1.732	1.155	2.000	1.0472	**60°00′**
10	.5265	.5025	.8646	.5812	1.720	1.157	1.990	1.0443	50
20	.5294	.5050	.8631	.5851	1.709	1.159	1.980	1.0414	40
30	.5323	.5075	.8616	.5890	1.698	1.161	1.970	1.0385	30
40	.5352	.5100	.8601	.5930	1.686	1.163	1.961	1.0356	20
50	.5381	.5125	.8587	.5969	1.675	1.165	1.951	1.0327	10
31°00′	.5411	.5150	.8572	.6009	1.664	1.167	1.942	1.0297	**59°00′**
10	.5440	.5175	.8557	.6048	1.653	1.169	1.932	1.0268	50
20	.5469	.5200	.8542	.6088	1.643	1.171	1.923	1.0239	40
30	.5498	.5225	.8526	.6128	1.632	1.173	1.914	1.0210	30
40	.5527	.5250	.8511	.6168	1.621	1.175	1.905	1.0181	20
50	.5556	.5275	.8496	.6208	1.611	1.177	1.896	1.0152	10
		cos	sin	cot	tan	csc	sec	Radians	Degrees

TABLE 4 *(continued)*

Degrees	Radians	sin	cos	tan	cot	sec	csc		
32°00′	.5585	.5299	.8480	.6249	1.600	1.179	1.887	1.0123	**58°00′**
10	.5614	.5324	.8465	.6289	1.590	1.181	1.878	1.0094	50
20	.5643	.5348	.8450	.6330	1.580	1.184	1.870	1.0065	40
30	.5672	.5373	.8434	.6371	1.570	1.186	1.861	1.0036	30
40	.5701	.5398	.8418	.6412	1.560	1.188	1.853	1.0007	20
50	.5730	.5422	.8403	.6453	1.550	1.190	1.844	1.0977	10
33°00′	.5760	.5446	.8387	.6494	1.540	1.192	1.836	.9948	**57°00′**
10	.5789	.5471	.8371	.6536	1.530	1.195	1.828	.9919	50
20	.5818	.5495	.8355	.6577	1.520	1.197	1.820	.9890	40
30	.5847	.5519	.8339	.6619	1.511	1.199	1.812	.9861	30
40	.5876	.5544	.8323	.6661	1.501	1.202	1.804	.9832	20
50	.5905	.5568	.8307	.6703	1.492	1.204	1.796	.9803	10
34°00′	.5934	.5592	.8290	.6745	1.483	1.206	1.788	.9774	**56°00′**
10	.5963	.5616	.8274	.6787	1.473	1.209	1.781	.9745	50
20	.5992	.5640	.8258	.6830	1.464	1.211	1.773	.9716	40
30	.6021	.5664	.8241	.6873	1.455	1.213	1.766	.9687	30
40	.6050	.5688	.8225	.6916	1.446	1.216	1.758	.9657	20
50	.6080	.5712	.8208	.6959	1.437	1.218	1.751	.9628	10
35°00′	.6109	.5736	.8192	.7002	1.428	1.221	1.743	.9599	**55°00′**
10	.6138	.5760	.8175	.7046	1.419	1.223	1.736	.9570	50
20	.6167	.5783	.8158	.7089	1.411	1.226	1.729	.9541	40
30	.6196	.5807	.8141	.7133	1.402	1.228	1.722	.9512	30
40	.6225	.5831	.8124	.7177	1.393	1.231	1.715	.9483	20
50	.6254	.5854	.8107	.7221	1.385	1.233	1.708	.9454	10
36°00′	.6283	.5878	.8090	.7265	1.376	1.236	1.701	.9425	**54°00′**
10	.6312	.5901	.8073	.7310	1.368	1.239	1.695	.9396	50
20	.6341	.5925	.8056	.7355	1.360	1.241	1.688	.9367	40
30	.6370	.5948	.8039	.7400	1.351	1.244	1.681	.9338	30
40	.6400	.5972	.8021	.7445	1.343	1.247	1.675	.9308	20
50	.6429	.5995	.8004	.7490	1.335	1.249	1.668	.9279	10
37°00′	.6458	.6018	.7986	.7536	1.327	1.252	1.662	.9250	**53°00′**
10	.6487	.6041	.7969	.7581	1.319	1.255	1.655	.9221	50
20	.6516	.6065	.7951	.7627	1.311	1.258	1.649	.9192	40
30	.6545	.6088	.7934	.7673	1.303	1.260	1.643	.9163	30
40	.6574	.6111	.7916	.7720	1.295	1.263	1.636	.9134	20
50	.6603	.6134	.7898	.7766	1.288	1.266	1.630	.9105	10
38°00′	.6632	.6157	.7880	.7813	1.280	1.269	1.624	.9076	**52°00′**
10	.6661	.6180	.7862	.7860	1.272	1.272	1.618	.9047	50
20	.6690	.6202	.7844	.7907	1.265	1.275	1.612	.9018	40
30	.6720	.6225	.7826	.7954	1.257	1.278	1.606	.8988	30
40	.6749	.6248	.7808	.8002	1.250	1.281	1.601	.8959	20
50	.6778	.6271	.7790	.8050	1.242	1.284	1.595	.8930	10
39°00′	.6807	.6293	.7771	.8098	1.235	1.287	1.589	.8901	**51°00′**
10	.6836	.6316	.7753	.8146	1.228	1.290	1.583	.8872	50
20	.6865	.6338	.7735	.8195	1.220	1.293	1.578	.8843	40
30	.6894	.6361	.7716	.8243	1.213	1.296	1.572	.8814	30
40	.6923	.6383	.7698	.8292	1.206	1.299	1.567	.8785	20
50	.6952	.6406	.7679	.8342	1.199	1.302	1.561	.8756	10
		cos	sin	cot	tan	csc	sec	Radians	Degrees

TABLE 4 *(continued)*

Degrees	Radians	sin	cos	tan	cot	sec	csc		
40°00′	.6981	.6428	.7660	.8391	1.192	1.305	1.556	.8727	**50°00′**
10	.7010	.6450	.7642	.8441	1.185	1.309	1.550	.8698	50
20	.7039	.6472	.7623	.8491	1.178	1.312	1.545	.8668	40
30	.7069	.6494	.7604	.8541	1.171	1.315	1.540	.8639	30
40	.7098	.6517	.7585	.8591	1.164	1.318	1.535	.8610	20
50	.7127	.6539	.7566	.8642	1.157	1.322	1.529	.8581	10
41°00′	.7156	.6561	.7547	.8693	1.150	1.325	1.524	.8552	**49°00′**
10	.7185	.6583	.7528	.8744	1.144	1.328	1.519	.8523	50
20	.7214	.6604	.7509	.8796	1.137	1.332	1.514	.8494	40
30	.7243	.6626	.7490	.8847	1.130	1.335	1.509	.8465	30
40	.7272	.6648	.7470	.8899	1.124	1.339	1.504	.8436	20
50	.7301	.6670	.7451	.8952	1.117	1.342	1.499	.8407	10
42°00′	.7330	.6691	.7431	.9004	1.111	1.346	1.494	.8378	**48°00′**
10	.7359	.6713	.7412	.9057	1.104	1.349	1.490	.8348	50
20	.7389	.6734	.7392	.9110	1.098	1.353	1.485	.8319	40
30	.7418	.6756	.7373	.9163	1.091	1.356	1.480	.8290	30
40	.7447	.6777	.7353	.9217	1.085	1.360	1.476	.8261	20
50	.7476	.6799	.7333	.9271	1.079	1.364	1.471	.8232	10
43°00′	.7505	.6820	.7314	.9325	1.072	1.367	1.466	.8203	**47°00′**
10	.7534	.6841	.7294	.9380	1.066	1.371	1.462	.8174	50
20	.7563	.6862	.7274	.9435	1.060	1.375	1.457	.8145	40
30	.7592	.6884	.7254	.9490	1.054	1.379	1.453	.8116	30
40	.7621	.6905	.7234	.9545	1.048	1.382	1.448	.8087	20
50	.7650	.6926	.7214	.9601	1.042	1.386	1.444	.8058	10
44°00′	.7679	.6947	.7193	.9657	1.036	1.390	1.440	.8029	**46°00′**
10	.7709	.6967	.7173	.9713	1.030	1.394	1.435	.8999	50
20	.7738	.6988	.7153	.9770	1.024	1.398	1.431	.8970	40
30	.7767	.7009	.7133	.9827	1.018	1.402	1.427	.7941	30
40	.7796	.7030	.7112	.9884	1.012	1.406	1.423	.7912	20
50	.7825	.7050	.7092	.9942	1.006	1.410	1.418	.7883	10
45°00′	.7854	.7071	.7071	1.0000	1.000	1.414	1.414	.7854	**45°00′**
		cos	sin	cot	tan	csc	sec	Radians	Degrees

ANSWERS
TO SELECTED EXERCISES

CHAPTER 1

EXERCISES 1.1

1. $\frac{2}{9}$ **3.** $\frac{41}{9}$ **5.** $\frac{8230}{999}$

7. Associative property of addition **9.** False

11. Associative property of multiplication

13. Distributive property

15. Multiplicative inverse property

19. No, $\sqrt{2}$ and $-\sqrt{2}$ are examples of two irrational numbers whose sum (0) is not irrational.

21. 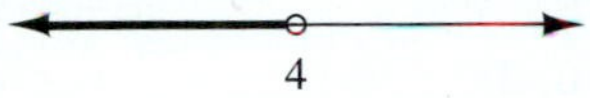 [number line: open circle at 4]

23. 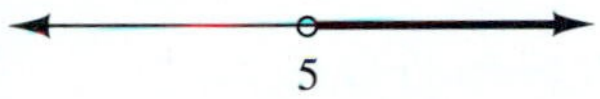 [number line: open circle at 5]

25. 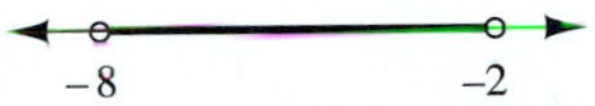 [number line: open circles at −8 and −2]

27. $(5, \infty)$ [number line: open circle at 5]

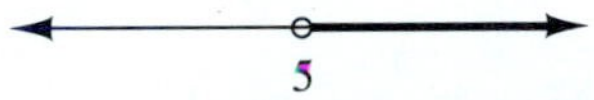

29. $[-5, \infty)$ [number line: closed circle at −5]

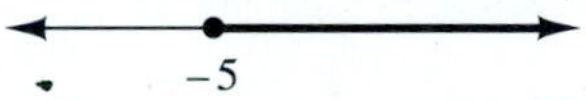

31. $[-8, -5)$ [number line: closed at −8, open at −5]

33. $(-\infty, -2]$ [number line: closed at −2]

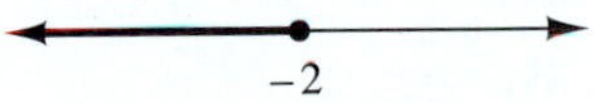

35. $(-9, -2]$ [number line: open at −9, closed at −2]

37. 2 **39.** $\sqrt{2} - 1$ **41.** $\begin{cases} x - 5 & \text{if } x \geq 5 \\ 5 - x & \text{if } x < 5 \end{cases}$

43. $x^2 + 1$ **45.** 1 unit **47.** 11 units

EXERCISES 1.2

1. -5 **3.** -36 **5.** -5.552 **7.** 3.29 **9.** -12

11. -7 **13.** 11 **15.** 10 **17.** $\frac{7}{2}$ **19.** $\frac{7}{12}$ **21.** $-\frac{59}{100}$

23. 0 **25.** $-\frac{82}{25}$ **27.** $\frac{48}{65}$ **29.** $\frac{20}{63}$ **31.** -14 **33.** 1

35. 7.0278 **37.** 1.15 **39.** 1.06 **41.** 0.087

EXERCISES 1.3

1. $-6x^3y^3$ **3.** $45x^3y^3$ **5.** $6x^2 - 5x - 6$

7. $6x^2 - 19x - 7$ **9.** $6x^3 - 13x^2 + 9x - 2$

11. $125x^3 + 27$ **13.** $x^3 - 4x^2 + x + 6$

15. $x^2 - 6xy + 9y^2$ **17.** $x^4 - 49$

19. $4x^2 - 12xy + 9y^2$ **21.** $x^2 - 2xy + y^2 - 49$

23. $x^2 - 2xy + y^2 + 4x - 4y + 4$ **25.** $-4x + 8$

27. $5y^3 - 51y^2 + 125y + 25$ **29.** $-12x + 18$

31. $4xh + 2h^2$ **33.** $\pi(x + 3)^2 - \pi(3)^2 = \pi x^2 + 6\pi x$

35. $6 - 4x^2$ sq feet **37.** $(x + 5)(x - 4)$

39. $2(3x - 2)(2x + 3)$ **41.** $(3a - 1)(b - 2)$

43. $5(3x - 2)(x - 1)^2$ **45.** $3(2x^2 + 1)(x^2 - 4x - 6)$

47. $(x - 2)(a + b)$ **49.** $(x - 3)(x^2 - 5)$

51. $(x - 3)(x + 3)$ **53.** $(x - 4)^2$

55. $(2x - 1)(4x^2 + 2x + 1)$ **57.** $3(3x - 2)(9x^2 + 6x + 4)$

59. $(x + 3)(x + 4)(x - 4)$ **61.** $(x^2 - 6)(x - 2)(x + 2)$

63. $(x - y - 4)(x - y + 4)$ **65.** $(a + 1)(a - 1)^2(a^2 + a + 1)$

67. $\frac{x + y}{2x(x - y)}$ **69.** 3 **71.** $\frac{12x^2 + 16}{(x^2 - 4)^3}$ **73.** $\frac{1}{x - 2}$

75. $\frac{1}{3x + 2}$ **77.** $\frac{2xy - x^2}{y(x - y)}$ **79.** $\frac{5a + 3}{a + 2}$

81. $\frac{2x^3 - 5x^2 + 9x - 9}{(x - 2)(x - 1)}$ **83.** $\frac{x^2 - 4x - 1}{(x - 2)^2}$

85. $\frac{x + 1}{x}$ **87.** $\frac{-1}{(x + 1)(x + 3)}$ **89.** $\frac{-5}{x(x + h)}$

91. $\frac{3x^2 + 7x + 2}{x^2 - 9x - 4}$

EXERCISES 1.4

1. x^6 **3.** $3x^5y^5$ **5.** $\frac{1 + 2x}{x^2 + 2x}$ **7.** $\frac{y + x}{x^2}$ **9.** 7.5×10^9

11. Approximately 8 minutes 20 seconds

13. $\frac{1}{8}$ **15.** $\frac{9}{4}$ **17.** $x^{5/6}y^{2/3}$ **19.** $(x^2 + 1)^{1/10}$

21. $x + 4x^{1/2} + 4$ **23.** $x - 4$ **25.** $\frac{3}{(x + 1)^2}$

27. $\frac{4x(2x^2 + 1)}{(x^2 + 1)^{1/2}}$ **29.** $\frac{5x^3 - 6}{2(x^3 - 3)^{1/2}}$

31. $\frac{3}{x(x^3 - 3)^{1/2}} = 3x^{-1}(x^3 - 3)^{-1/2}$

33. $2x^2y^3\sqrt{6}$ **35.** $\frac{4x^4}{3}$ **37.** $3x^3\sqrt[6]{3^5x}$ **39.** $\frac{5\sqrt{3x}}{3x^3}$

41. $\frac{\sqrt{x} + 4}{x - 16}$ **43.** $\frac{x - 2}{\sqrt{(x - 1)(x - 2)}}$ **45.** $\frac{1}{\sqrt{x} + 3}$

47. $\dfrac{1}{\sqrt{x+h}+\sqrt{x}}$ **49.** $18x^3y^3\sqrt{2xy}$ **51.** $\sqrt{x-1}$
53. $15\sqrt{3}-9\sqrt{2}$ **55.** $2-6\sqrt{3}$ **57.** $a-25$
59. $12-6\sqrt{x}$ **61.** $\dfrac{\sqrt{14}+\sqrt{10}}{2}$ **63.** 12
65. $\dfrac{2x\sqrt{x^2-2}-1}{\sqrt{x^2-2}}$ **67.** $\dfrac{3x^2-8}{\sqrt{x^2-2}}$
69. $\dfrac{\sqrt{2x}-\sqrt{2(x+h)}}{h\sqrt{x(x+h)}}$

EXERCISES 1.5

1. 1 **3.** 1 **5.** $-i$ **7.** $3+4i$ **9.** $5-\dfrac{\sqrt{3}}{6}i$
11. $9-6i$ **13.** $17+0i=17$ **15.** $13-13i$ **17.** 61
19. $-5-12i$ **21.** $1+4i$ **23.** $3-2i$ **25.** $\dfrac{6}{5}-\dfrac{3}{5}i$
27. $-\dfrac{1}{5}+\dfrac{2}{5}i$ **29.** $0-1i=-i$ **31.** $\dfrac{26}{29}-\dfrac{22}{29}i$ **33.** -5
35. $4-6i$ **39.** $-i$

EXERCISES 1.6

1. $x=\dfrac{16}{5}$ **3.** $(-\infty, -\frac{1}{3}]$ **5.** $(\frac{23}{9}, \infty)$ **7.** $x=-\dfrac{6}{29}$
9. $y=\dfrac{1}{2}$ **11.** $(-\infty, \frac{5}{9})$ **13.** No solution **15.** $x=-\dfrac{9}{5}$
17. $a=4$ **19.** $x=\dfrac{3}{2}$ **21.** No solution **23.** $y=\dfrac{2x+6}{5}$
25. $W=\dfrac{S-2LH}{2L+2H}$ **27.** $h=\dfrac{2A}{b_1+b_2}$ **29.** $y=\dfrac{1}{2}$ for $x\neq\dfrac{2}{3}$
31. $s>\dfrac{x-\mu}{1.96}$ **33.** $f=\dfrac{f_1f_2}{f_1+f_2}$ **35.** $x=\dfrac{2}{3-y}$
37.  Number line: open circle at $-\frac{2}{3}$, open circle at $\frac{7}{3}$, segment between
39. Number line: closed circle at $\frac{1}{2}$, open circle at $\frac{5}{2}$, segment between
41. 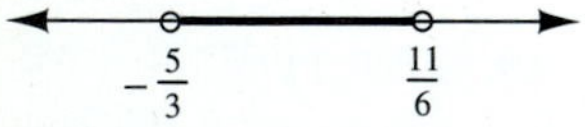 Number line: open circle at $-\frac{5}{3}$, open circle at $\frac{11}{6}$, segment between
43. 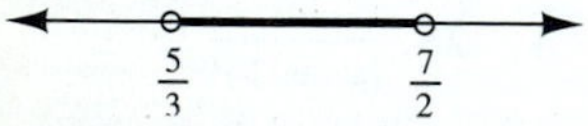 Number line: open circle at $\frac{5}{3}$, open circle at $\frac{7}{2}$, segment between
45. s is multiplied by 9 **47.** E is divided by 9
49. Let $x=$ no. of 20-lb boxes; then $50-x=$ no. of 25-lb boxes. $20x+25(50-x)=1175$; $x=15$; 35 25-lb boxes, 15 20-lb boxes
51. Let $x=$ no. of orchestra seats; then $56-x=$ no. of balcony seats. $48x+28(56-x)=2328$, $x=38$; 38 orchestra seats, 18 balcony seats
53. Let $t=$ no. of minutes older machine works; then $110-t=$ no. of minutes new machine works. $50t+70(110-t)=7200$, $t=25$; 1250 copies
55. Let $x=$ no. of ounces of 20% solution. $0.20x+0.50(5)=0.30(x+5)$, $x=10$; 10 oz.
57. Let $x=$ amount invested in the high-risk certificate; then $\$8000-x=$ amount invested in the lower-risk certificate. $0.14x+0.09(8000-x)\geq 890$, $x\geq \$3400$; at least \$3400.
59. Let $t=$ amount of time it takes for both pipes to fill the pool together. $\dfrac{t}{3}+\dfrac{t}{2}=1$, $t=\dfrac{6}{5}$; $1\frac{1}{5}$ days.
61. Let $t=$ amount of time it takes to fill the tub with the drain open. $\dfrac{t}{10}-\dfrac{t}{15}=1$, $t=30$; 30 minutes.

EXERCISES 1.7

1. $x=12, x=-12$ **3.** $(-9, 9)$ **5.** $[6, 10]$
7. $x=\dfrac{11}{2}, x=-\dfrac{3}{2}$ **9.** $(-4, -1)$
11. $(-\infty, 0)\cup(\frac{4}{3}, \infty)$ **13.** $(0, 3)$
15. $x=-\dfrac{3}{4}, x=-\dfrac{21}{4}$ **17.** $(-\infty, \frac{1}{3})\cup(\frac{7}{3}, \infty)$
19. $(-\infty, -3]\cup[3, \infty)$ **21.** $x=-\dfrac{5}{2}, x=\dfrac{11}{2}$
23. $x=\dfrac{5}{2}, x=\dfrac{15}{2}$ **25.** $x=\dfrac{7}{3}, x=-1$
27. $x=2, x=1$ **29.** $x=-\dfrac{1}{3}, x=\dfrac{5}{3}$ **31.** $(-\frac{7}{3}, \frac{17}{3})$
33. No solution **35.** x^2+8 **37.** $4x^2+9$
39. Not possible **41.** $\dfrac{5+x^6}{4}$

EXERCISES 1.8

1. $y=\pm\sqrt{65}$ **3.** $x=5, -\dfrac{3}{2}$ **5.** $x=\pm\sqrt{10}$
7. $x=-1\pm\sqrt{5}$ **9.** $a=\dfrac{4\pm\sqrt{43}}{2}$
11. $x=1\pm\sqrt{39}$ **13.** $x=3, -1$ **15.** $x=-\dfrac{4}{3}$
17. 1.60 **19.** 323 **21.** 30 **23.** $s=\sqrt{\dfrac{2gm}{K}}=\dfrac{\sqrt{2gmK}}{K}$
25. $a=\pm\sqrt{\dfrac{3b-2}{2x+3}}=\dfrac{\pm\sqrt{(3b-2)(2x+3)}}{2x+3}\left(x\neq-\dfrac{3}{2}\right)$
27. $\dfrac{s_e}{\sqrt{1-r_{xy}^2}}=\dfrac{s_e\sqrt{1-r_{xy}^2}}{1-r_{xy}^2}$
29. Let $x=$ the number. $x+\frac{1}{x}=\frac{13}{6}$; $x=\frac{2}{3}, \frac{3}{2}$
31. Let $w=$ width; then length $=2w+5$. $w(2w+5)=75$, $w=5$; 5 by 15

33. Let x = width of the path. $(21 + 2x)^2 - (21)^2 = 184$, $x = 2$; 2 feet
35. $x = 4$ **37.** $x = 9$ **39.** $x = \pm 1, \pm 4$
41. No solution **43.** $a = 81$ **45.** $x = \pm 3, -4$
47. $x = 3$

EXERCISES 1.9

1. $(-\infty, -6) \cup (4, \infty)$ **3.** $(-\frac{1}{2}, 2)$ **5.** $[-3, 8]$
7. $(-\infty, -1] \cup [\frac{5}{2}, \infty)$ **9.** $\left[-\frac{1}{5}, \frac{2}{3}\right]$
11. $(-\infty, -\frac{3}{2}) \cup (\frac{1}{2}, \infty)$ **13.** $(-\infty, \infty)$
15. $(-\infty, 1) \cup (1, \infty)$ **17.** $(-\infty, -1) \cup (1, \infty)$
19. $(-\infty, -4) \cup (-1, 1)$ **21.** $[2, \infty)$
23. $\left(\frac{3 - \sqrt{21}}{2}, \frac{3 + \sqrt{21}}{2}\right)$
25. $\left(-\infty, \frac{1 - \sqrt{17}}{4}\right] \cup \left[\frac{1 + \sqrt{17}}{4}, \infty\right)$
27. $(-\infty, -1)$ **29.** $(-\infty, -5) \cup [\frac{1}{2}, \infty)$ **31.** $(-1, 2)$
33. $(-\infty, \frac{1}{5}] \cup (2, \infty)$ **35.** $(-\infty, 0] \cup (1, \infty)$ **37.** $[-\frac{1}{2}, 1)$
39. $[-3, -1)$ **41.** $(-1, 2] \cup (3, \infty)$
43. $(-\infty, -1) \cup [0, 1)$ **45.** $(-\infty, -5) \cup (-3, 3)$
47. $[0, \infty)$ **49.** $(-\infty, -4) \cup (-1, 1) \cup (4, \infty)$
51. $\left(\frac{1}{3}, 1\right)$ **53.** $\left(\frac{1}{2}, \infty\right)$
55. Between (and including) 16°C and 26°C

EXERCISES 1.10

1. (a) Let s = length of a side. $P = 84 = 4s \Rightarrow s = 21$ in.; $A = s^2 = 441$ sq. in. **(b)** Let s = the length of a side. $P = x = 4s \Rightarrow s = \frac{x}{4}$; $A = s^2 = \frac{x^2}{16}$ sq. in. **3.** Let s = the length of a side. $s^2 + s^2 = 5^2 \Rightarrow s = \frac{5}{\sqrt{2}}$; $A = s^2 = \frac{25}{2} = 12\frac{1}{2}$ sq ft **5.** Let s = the length of a side and d = the length of a diagonal. $A = 84 = s^2 \Rightarrow s = \sqrt{84}$; $d^2 = s^2 + s^2 \Rightarrow d = \sqrt{168} \approx 12.96$ in. **7.** Let s = the length of a side and d = the length of a diagonal: $P = 4s \Rightarrow s = \frac{P}{4}$; $d^2 = s^2 + s^2 \Rightarrow d = s\sqrt{2} = \frac{P\sqrt{2}}{4}$
9. Let s = the length of a side and d = the length of a diagonal. $d^2 = s^2 + s^2 \Rightarrow s = \frac{d}{\sqrt{2}}$; $P = 4s = 2d\sqrt{2}$
11. Let d = the diameter and r = the radius. $r = \frac{d}{2} = \frac{8}{2} = 4$; $V = \frac{4}{3}\pi r^3 \Rightarrow V = \frac{256\pi}{3} \approx 268.08$ cu ft
13. Let d = the diameter and r = the radius. $r = \frac{d}{2}$; $S = 4\pi r^2 \Rightarrow S = \pi d^2$ **15.** $A = 100$ sq in.
17. $A = 4r^2$ **19. (a)** $A = 36\pi \approx 113.10$ sq in. **(b)** $A = \pi(3t)^2 = 9\pi t^2$ sq in. **21.** $4500\pi \approx 14{,}137.17$ cu in.
23. $V = \frac{4}{3}\pi(3t)^3 = 36\pi t^3$ cu in. **25.** Cost $= 8(500) = \$4000$
27. Let s = the length of a side. $200 = s^2 \Rightarrow s = 10\sqrt{2}$; cost $= 8(4)(10\sqrt{2}) = 320\sqrt{2} \approx \$452.55$
29. Let C = the circumference and r = the radius. $C = 2\pi r = 2\pi(5) = 10\pi$; cost $= 2(10\pi) = 20\pi \approx \$62.83$
31. Let A = the area, r = the radius, and C = the circumference. $A = 90 = \pi r^2 \Rightarrow r = \sqrt{\frac{90}{\pi}} = \frac{3\sqrt{10\pi}}{\pi}$; $C = 2\pi r = 6\sqrt{10\pi}$; cost $= 3(6\sqrt{10\pi}) = 18\sqrt{10\pi} \approx \$100.89$
33. 100 miles

CHAPTER 1 REVIEW EXERCISES

1. $\frac{817}{99}$ **3.** Commutative property of addition
5. $[-4, \infty)$ (number line: closed dot at -4, shaded to the right)
7. $(-\infty, 3)$ (number line: open dot at 3, shaded to the left)
9. $\sqrt{5} - 1$ **11.** 19
13. $\frac{38}{9}$ **15.** $-\frac{25}{6}$ **17.** $108x^5y^5$ **19.** $2x^2 + x - 21$
21. $25a^6 - b^2$ **23.** $27x^3 - 1$ **25.** $2x^3 - 9x^2 + 8x + 4$
27. $(x - 5)(x + 3)$ **29.** $(x - 2)(3xy - 6y + 5)$
31. $(5x - 3)^2$ **33.** $(a - 4)(a + 4)(a + 1)$
35. $(a - 1)(a^2 - a + 1)(a + 1)^2$
37. $\frac{1}{x + y}$ **39.** $\frac{3x^2 + x - 2}{x(x - 2)}$ **41.** $\frac{x + 2}{(x - 1)^2}$ **43.** $\frac{xy}{y - 6x}$
45. $\frac{y^2}{x^5}$ **47.** $\frac{1}{x^2}$ **49.** $\frac{3}{xy^3}$ **51.** $x^{49/60}$ **53.** $\frac{1}{16}$
55. $\frac{4x^5 + 8x^3 + x^2 + 4x}{2(x^2 + 1)^{3/2}}$ **57.** $3x^2(x^2 + 1)^{1/2}$
59. $2x\sqrt{x + 1} + \frac{x^2}{\sqrt{(x^2 + 1)^3}}$ **61.** $4x^6y^7\sqrt{3y}$
63. $\frac{5\sqrt{x}}{x^2}$ **65.** $2x - 4\sqrt{2x} + 4$ **67.** $8\sqrt{a} - 20$
69. $\frac{\sqrt{22} + \sqrt{6}}{4}$ **71.** $\frac{2\sqrt{x^2 - 4} - 2}{\sqrt{x^2 - 4}}$ **73.** $5 - i$
75. $24 + 10i$ **77.** $\frac{3}{2} - i$ **79.** $x = -\frac{2}{7}$ **81.** $\left[\frac{15}{38}, \infty\right)$
83. $x = \frac{38}{13}$ **85.** $\left[\frac{8}{9}, 1\right)$ **87.** $F = \frac{9C + 160}{5}$
89. Let x = the amount of pure water. $0x + 0.60(2) < 0.40(x + 2)$; $x > 1$; more than 1 liter.

91. $x = \dfrac{11}{7}, x = -1$ **93.** $(-\frac{1}{3}, \frac{11}{3})$

95. $(-\infty, -\frac{7}{2}] \cup [4, \infty)$ **97.** $x = -\dfrac{1}{3}, x = -1$

99. $7x^2 + 3$ **101.** $x = \dfrac{1 \pm \sqrt{33}}{2}$ **103.** $x = \pm\sqrt{5}$

105. $x = \dfrac{4}{3}, 5$ **107.** $x = -\dfrac{3}{2}, 5$ **109.** $x = 5$

111. $x = \dfrac{27}{8}, -1$

113. $x = \pm\sqrt{\dfrac{9 - 7y^2}{5}} = \dfrac{\pm\sqrt{5(9 - 7y^2)}}{5}$

115. $(-\infty, -\frac{1}{3}) \cup (5, \infty)$ **117.** No solution

119. $(-\infty, -1) \cup (\frac{2}{3}, \infty)$ **121.** $(0, 1]$ **123.** 10^4Å $= 1\ \mu$

125. Let x = width of the path.
$\pi(x + 10)^2 - \pi(10)^2 = 44\pi$, $x = 2$ feet.

127. Let C = the circumference. $C = 2\pi r$,
cost $= 3(2\pi r) = 6\pi r$ dollars

CHAPTER 1 PRACTICE TEST

1. multiplicative inverse property

2. (a) −2

(b) −6 −2

3. $\sqrt{5} - 1$ **4.** $\dfrac{26}{5}$ **5. (a)** $9a^6 - 4b^2$

(b) $3x^3 - 31x^2 + 75x + 25$ **6. (a)** $(5x - 2)(2x + 1)$

(b) $3(y + 3)(y^2 + 3y + 5)$ **(c)** $(x - 3)(x + 3)(x + 2)$

7. (a) $\dfrac{x + 3}{(x + 1)^2}$ **(b)** $\dfrac{ab}{a + 5b}$ **(c)** $\dfrac{(x^2 + y^2)(x + y)^2}{x^2y^2}$

(d) $\dfrac{3x^2}{x^3 - 2}$ **8. (a)** $\dfrac{4x^2\sqrt{xy}}{3y}$ **(b)** $\dfrac{5x\sqrt{2x}}{2}$

(c) $9x^2 - 6x\sqrt{2} + 2$ **(d)** $\sqrt{10} + \sqrt{7}$

9. (a) $-16 - 30i$; **(b)** $\dfrac{3}{10} - \dfrac{11}{10}i$ **10.** $x = \dfrac{5}{2}$ **11.** $(-17, \infty)$

12. $x = 0, -7$ **13.** $\left[-\dfrac{9}{2}, \dfrac{21}{2}\right]$ **14.** $(-\infty, \frac{3}{2}) \cup (\frac{11}{2}, \infty)$

15. $(-\infty, \infty)$ **16.** $x = 2, -1$

17. $x = \pm\sqrt{\dfrac{5}{3}} = \pm\dfrac{\sqrt{15}}{3}$ **18.** $x = \frac{7}{5}$ **19.** $x = 6, 5$

20. $[-1, \frac{3}{2}]$ **21.** $(-\infty, 1) \cup (4, \infty)$ **22.** $(-\infty, -\frac{1}{2}) \cup (2, \infty)$

23. $[-5, 0)$ **24.** Let t = time to process 200 forms.
$\dfrac{t}{3} + \dfrac{t}{2\frac{1}{3}} = 1$, $t = \dfrac{21}{16}$, $1\dfrac{5}{16}$ hours

25. Let $s = 0$; then $t = 4$: 4 seconds. **26.** Let A = the area, r = the radius, and C = the circumference.

$A = \pi r^2 \Rightarrow r = \sqrt{\dfrac{A}{\pi}} = \dfrac{\sqrt{A\pi}}{\pi}$, $C = 2\pi r = 2\sqrt{A\pi}$,
cost $= 6(2\sqrt{A\pi}) = 12\sqrt{A\pi}$ dollars.

CHAPTER 2

EXERCISES 2.1

1. x-int: -5; y-int: 4 **3.** x-int: 6; y-int: 2 **5.** x-int: $\dfrac{16}{3}$; y-int: -2 **7.** x-int: -3; y-int: $-\dfrac{2}{3}$ **9.** x-int: -3; no y-int

11. x-int: $-\dfrac{8}{5}$; y-int: $\dfrac{4}{3}$ **13.** x-int: 0; y-int: 0

15.

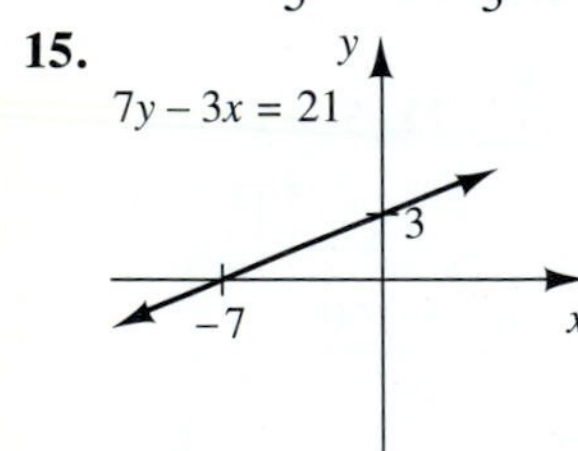

17.

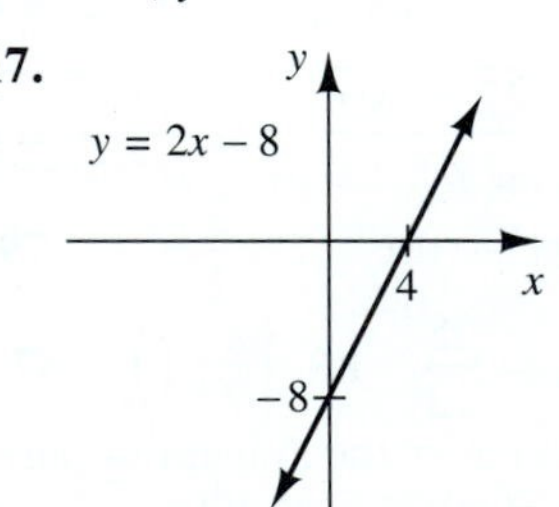

19.

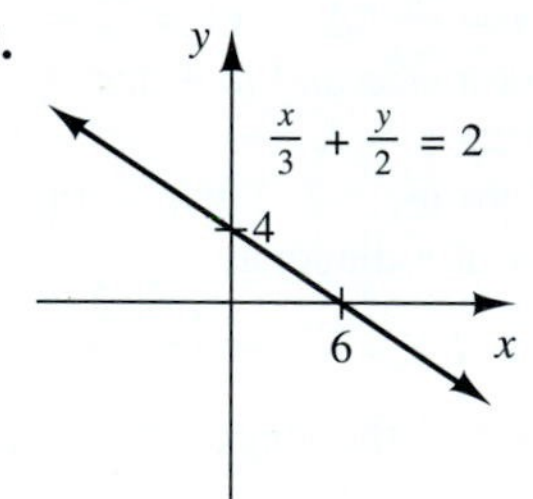

21.

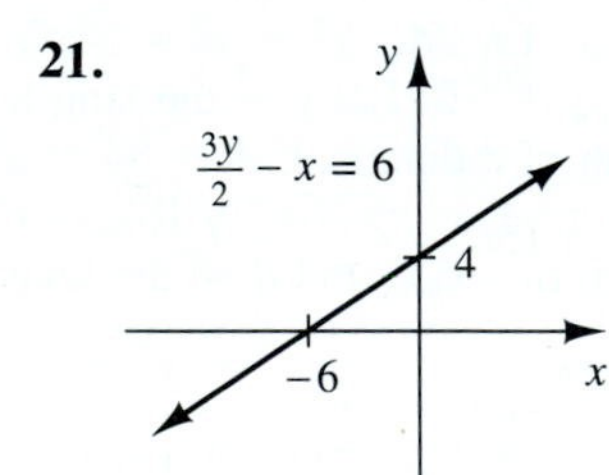

23.

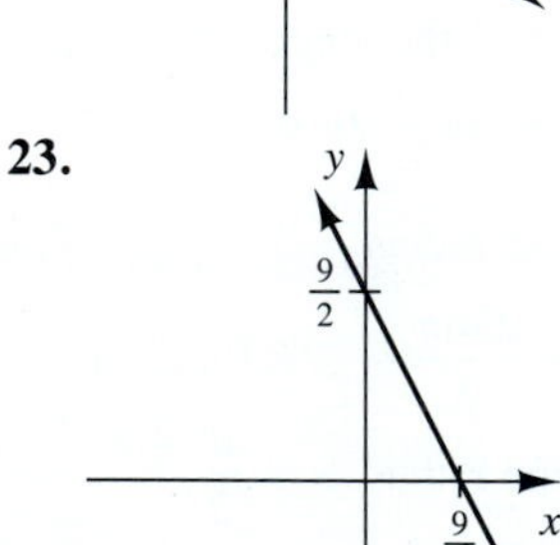

25.

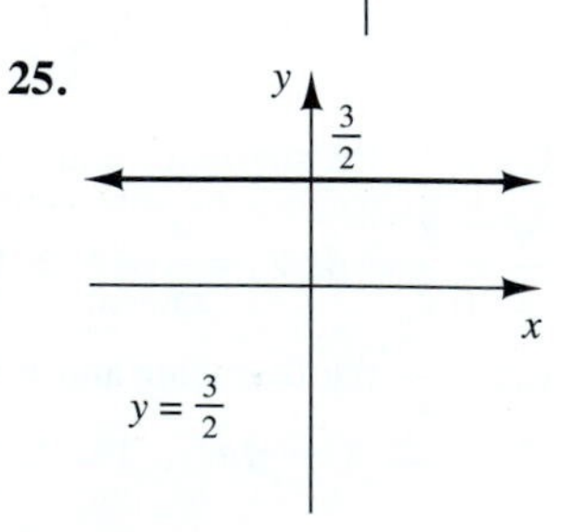

27.

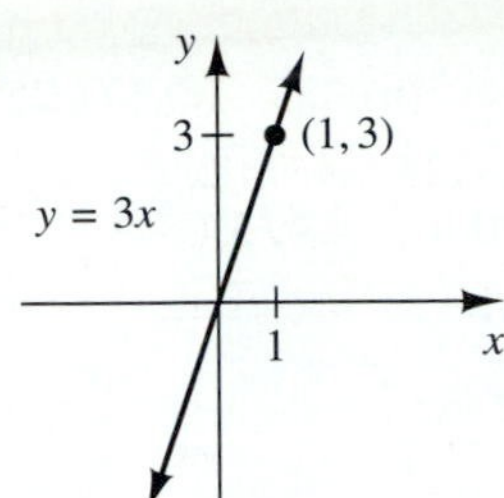

29.

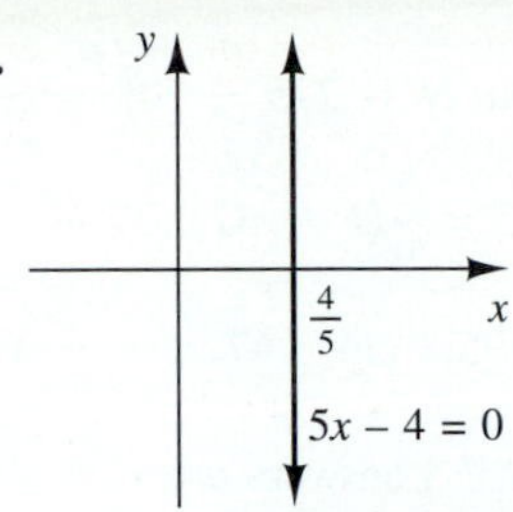

31.

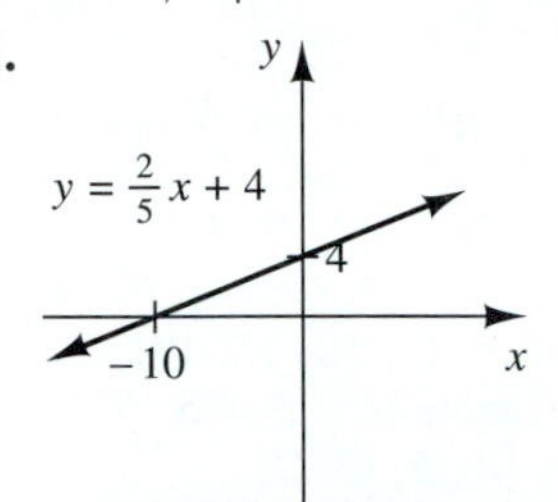

33.

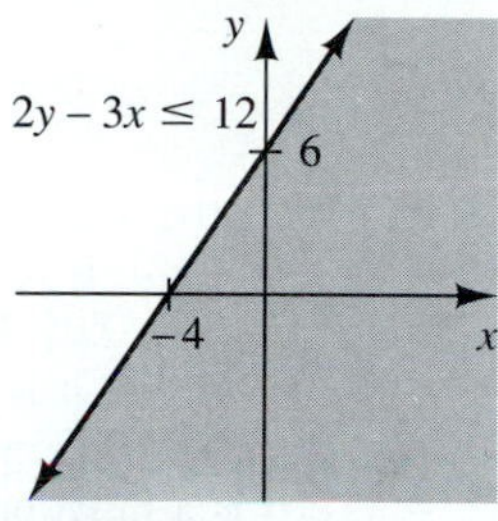

35.

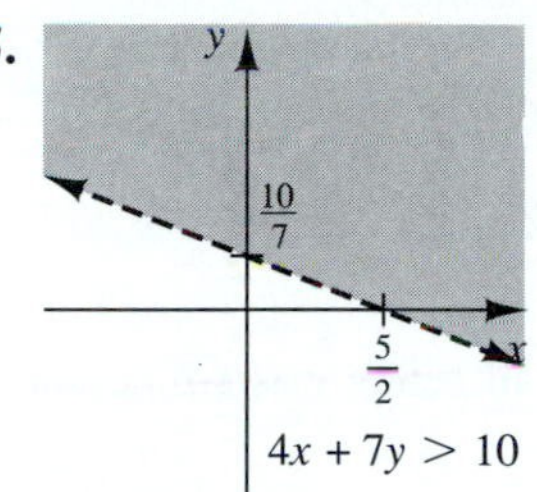

37.

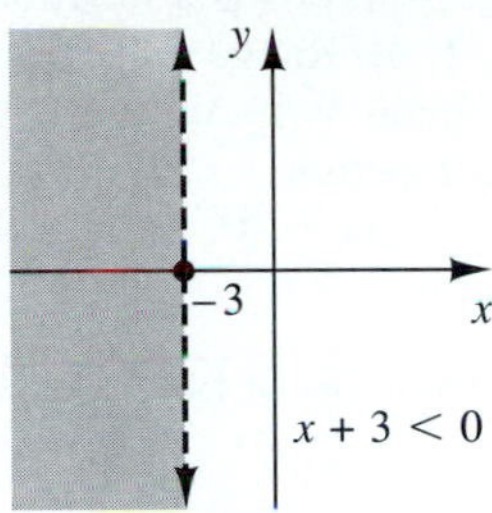

39.

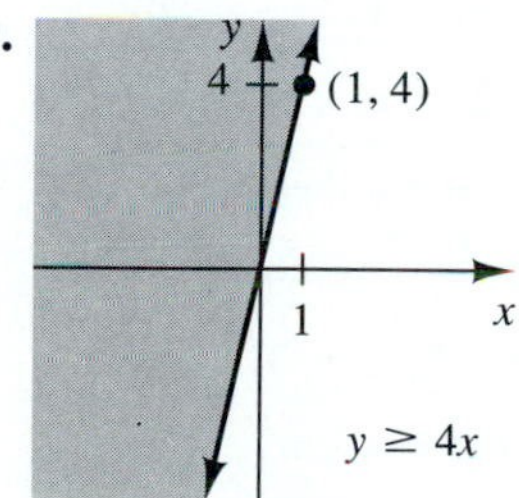

41.

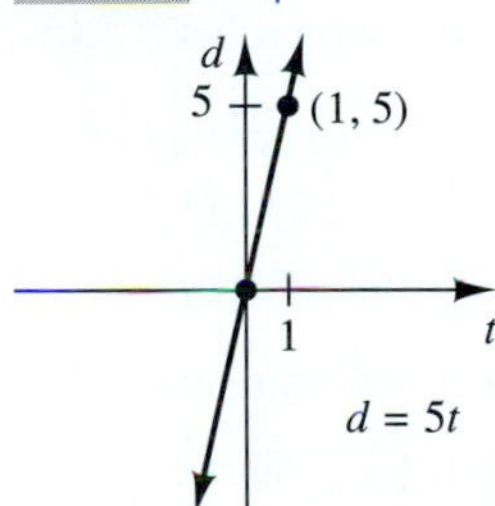

43.

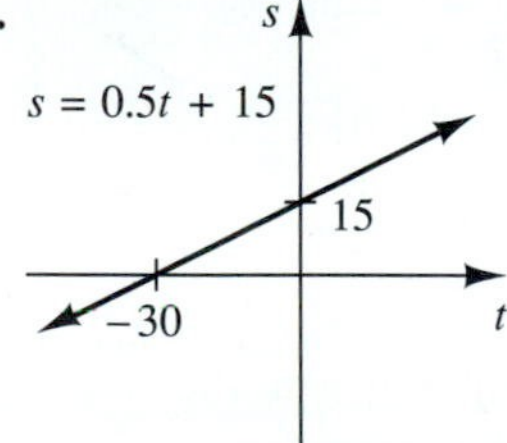

45. Length: $\sqrt{41}$; mp $= \left(-1, \frac{3}{2}\right)$ **47.** Length: $\sqrt{85}$; mp $= \left(0, \frac{1}{2}\right)$ **49.** Length: $|a|\sqrt{2}$; mp $= \left(\frac{a}{2}, \frac{a}{2}\right)$

51. $A = 15$ **53.** $A = \frac{13}{2}$ **55.** $w = 11$, $w = -5$ **57.** No

59. $(x - 2)^2 + (y - 3)^2 = 9$

61. $\left(x - \frac{1}{2}\right)^2 + (y - 4)^2 = 36$ **63.** Center: $(3, 2)$; $r = 4$

65. Center: $(0, 0)$; $r = 4$ **67.** Center: $(3, 5)$; $r = 5$

69. Center: $(0, -3)$; $r = 3$

71.

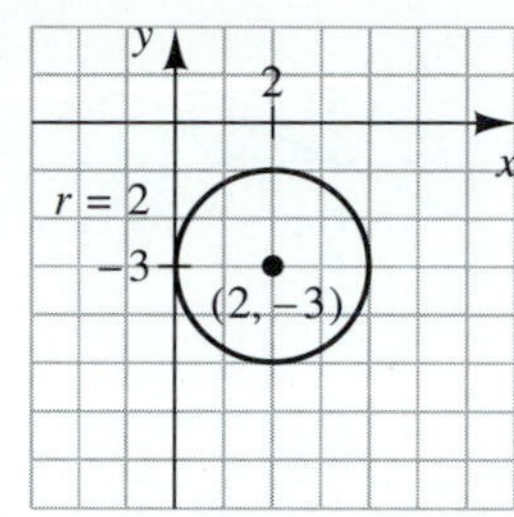

73. $(x - 1)^2 + \left(y - \frac{3}{2}\right)^2 = \frac{205}{4}$

75. $(x - 3)^2 + (y + 5)^2 = 122$ **77.** $C = 4\pi\sqrt{17}$

79. $(x - 3)^2 + (y + 2)^2 = 4$

81. $(x - 3)^2 + (y + 3)^2 = 9$

83. **(a)** $x^2 + y^2 = 1$ **(b)** $\left(\frac{3}{5}, -\frac{4}{5}\right)$ and $\left(-\frac{\sqrt{3}}{2}, \frac{1}{2}\right)$

EXERCISES 2.2

1. $m = \frac{5}{3}$ **3.** $m = -1$ **5.** $m = -\frac{2}{3}$ **7.** $m = \frac{3}{2}$

9. $m = \frac{2}{3}$ **11.** $m = 0$ **13.** $m = \frac{4}{15}$ **15.** $m = \sqrt{3}$

17. $m = a + b$ **19.** $m = 1$

21.

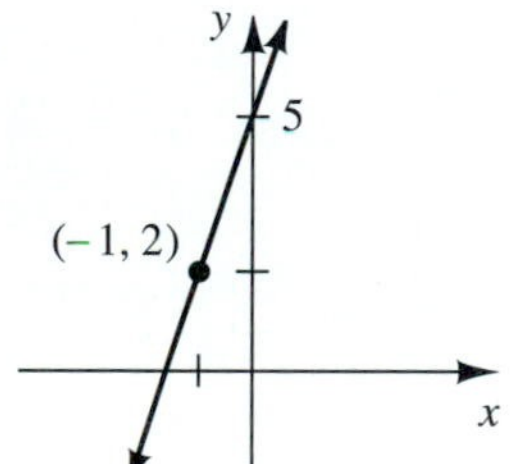

23.

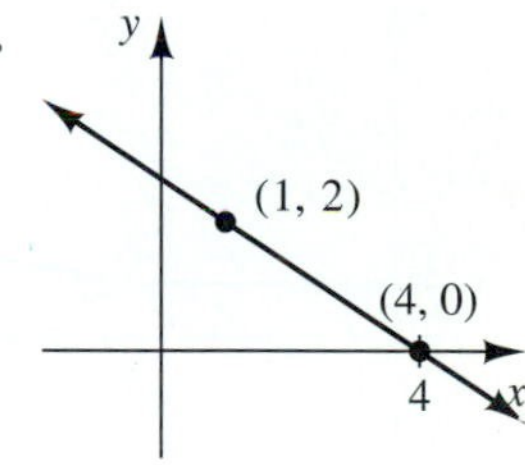

25.

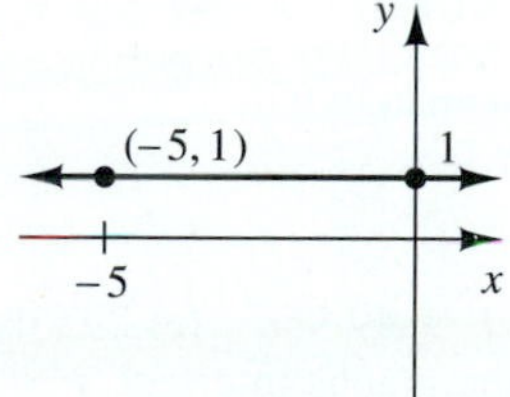

27. Parallel **29.** Neither **31.** Perpendicular **33.** $c = 8$

35. $t = \frac{1}{4}$ **37.** $h = -\frac{1}{2}$, $h = 5$ **45.** $m_4 < m_2 < m_3 < m_1$

EXERCISES 2.3

1. $y = 3x + 5$ **3.** $y = \frac{2}{5}x - \frac{11}{5}$

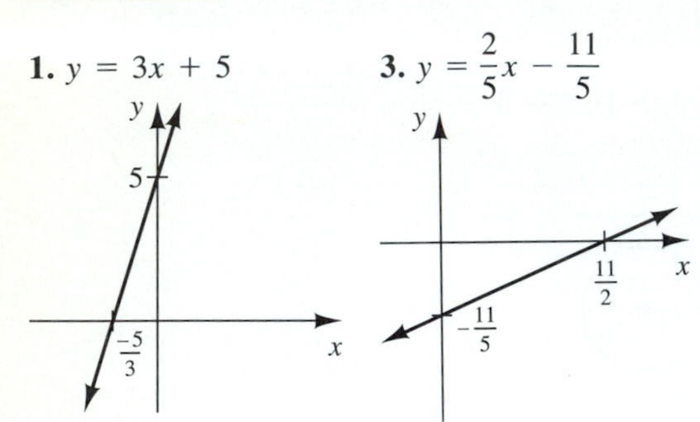

5. $y = 4x + 16$ **7.** $y = -\frac{1}{2}x + 7$ **9.** $y = 4x$

11. $y = -\frac{2}{3}x + 4$ **13.** $y = \frac{1}{5}x - 1$ **15.** $x = -3$

17. $y = -\frac{1}{3}x + 5$ **19.** $y = 5$ **21.** $y = -5$ **23.** $m = -2$

25. $m = \frac{2}{3}$ **27.** $m = \frac{1}{5}$ **29.** $m = -\frac{4}{3}$ **31.** $y = 4x - 9$

33. $y = \frac{4}{3}x + \frac{5}{3}$ **35.** $y = \frac{4}{5}x - 4$ **37.** $y = -\frac{7}{6}x + \frac{73}{6}$

39. $y = -x$ **41.** $y = 5$ **43.** $y = \frac{7}{2}x + 2$

45. $y = \frac{5}{3}x + 4$ **47.** $y = -\frac{5}{3}x - \frac{16}{45}$

49. $C = 0.30m + 29$

51. $A = \begin{cases} 7 \text{ if } 0 \le h \le 35 \\ 0.12(h - 35) + 7 \text{ if } h > 35 \end{cases}$

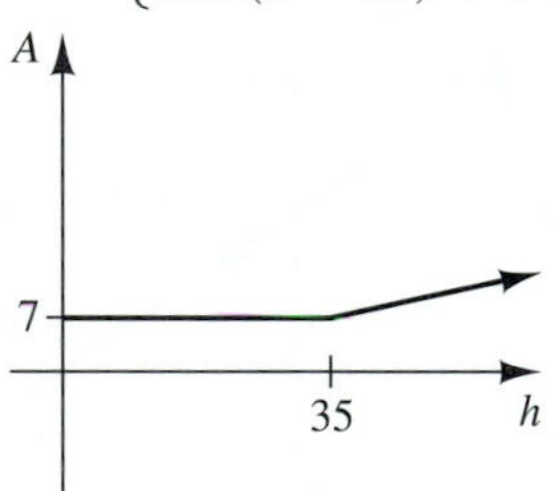

53. (a) $P = 10x + 100$ **(b)** $P = \$1000$

55. (a) $N = \frac{5}{3}s + 55$ **(b)** Approximately 97

57. $V = -\frac{2125}{3}t + 8500$

59. (a) $C = 8.25d + 12{,}375$ **(b)** $I = 49.95d$ **(c)** 297 days **(d)** The breakeven point is where the graphs intersect.

61. (a) $N = 275 - 10\left(\frac{P - 62.50}{2.50}\right)$ **(b)** 225 **(c)** \$71.25

63. $C = \frac{5}{9}(F - 32)$; 37°F **65. (a)** $h = 3.8f - 2.3$

(b) 195.3 cm **67.** $y = \frac{5}{12}x - \frac{169}{12}$

69. All 3 answers are $x < -\frac{3}{2}$.

EXERCISES 2.4

1. $\{(A, 10), (B, 20), (C, 30)\}$; Domain = $\{A, B, C\}$ Range = $\{10, 20, 30\}$; it is a function. **3.** $\{(6, A), (6, B), (10, C), (15, C), (19, D)\}$; Domain = $\{6, 10, 15, 19\}$ Range = $\{A, B, C, D\}$; it is not a function. **5.** $\{(A, 7), (B, 7), (C, 7), (D, 7)\}$; Domain = $\{A, B, C, D\}$ Range = $\{7\}$; it is a function. **7.** Domain = $\{-4, 2, 4\}$; Range = $\{-5, -3, 0, 6\}$; it is not a function.
9. Domain = $\{-\sqrt{3}, -1, 0, 1, \sqrt{3}\}$; Range = $\{0, 1, 3\}$; it is a function.
11. Domain = $\{0, 1, 2, 3, 4, 5\}$; Range = $\{5\}$; it is a function.

The equations in Exercises 13–42 all define y as a function of x.

13. All real numbers **15.** $\{x \mid x \neq 2\}$ **17.** $(-\infty, \infty)$

19. $[-2, \infty)$ **21.** $\left\{x \mid x \neq -\frac{4}{3}\right\}$ **23.** $\{x \mid x \neq 0\}$

25. $(-\infty, -3) \cup (3, \infty)$ **27.** All real numbers

29. All real numbers **31.** $\left\{x \mid x \le \frac{8}{5}\right\}$

33. All real numbers **35.** $\{x \mid x \ge -5, x \neq 3\}$

37. $\left(0, \frac{3}{2}\right)$ **39.** $\{x \mid x \neq 5\}$ **41.** $\{x \mid x \neq \pm 4\}$

43. Function **45.** Not a function **47.** Function
49. Domain = $(-\infty, \infty)$; Range = $[-3, \infty)$
51. Domain = $(-\infty, \infty)$; Range = $[-\infty, 6]$
53. Domain = $[-5, 5]$; Range = $[-3, 3]$
55. $C = 0.11m + 110$ **57.** $d = 60 - 2t$

EXERCISES 2.5

1. 28 **3.** $\sqrt{37}$ **5.** 161 **7.** $-\frac{1}{3}$ **9.** $5x + 8$

11. $3x^2 - 10x + 8$ **13.** $\sqrt{4x^2 - 3}$ **15.** $27x^2 - 12x + 1$
17. $20x + 33$ **19.** $3x^2 + 6xh + 3h^2 - 4x - 4h + 1$

21. $\frac{9}{10}$ **23.** $\frac{1}{\sqrt{14}} = \frac{\sqrt{14}}{14}$ **25.** $\frac{1}{3}$ **27.** $\frac{1}{\sqrt{t^2 - 1}}$

29. Undefined **31.** $\frac{x + 1}{x + 2}$ **33.** $\sqrt{5}$ **35.** $\frac{40}{41}$ **37.** $\frac{a}{a + 3}$

39. $\frac{x - 1}{x}$ **41.** $20x + 25$ **43.** $\frac{1}{x - a}$ **45.** $\frac{1}{x} - \frac{1}{a}$

47. $\frac{1}{x} - a$ **49.** 156 **51.** 45 **53.** $\frac{1}{x^2 - x}$ **55.** $k^2x^2 - kx$

57. $-\frac{7}{3}$ **59.** $\frac{1}{x^2} - \frac{1}{x} = \frac{1-x}{x^2}$ **61.** $x + 2$ **63.** -3
65. $2x + h - 3$ **67. (a)** -14 **(b)** 0 **(c)** 35 **69. (a)** 1
(b) Undefined **(c)** $\frac{3}{2}$ **(d)** Undefined **71. (a)** $h(1) = 34$; $h(2) = 36$; $h(3) = 6$ **(b)** 3.125 seconds

EXERCISES 2.6

1. (a) 2 **(b)** 2 **(c)** 1 **(d)** -1 **(e)** -1 **(f)** $x < 0$
(g) $x = 0$ **(h)** $x > 0$ **(i)** None **3. (a)** 5 **(b)** 1 **(c)** 3
(d) -1 **(e)** 2 **(f)** 3, 5 **(g)** $[-4, 3) \cup (5, 6]$ **(h)** $(3, 5)$
(i) $[-4, 6]$ **(j)** $[-1, 5]$ **5. (a)** Undefined **(b)** 4
(c) Undefined **(d)** -3 **(e)** 1 **(f)** $(-7, -3] \cup [1, 6)$
(g) $[-3, 2) \cup [3, 5)$ **7. (a)** -1.2 and 3.3 **(b)** -1.2 and 3.2 **9. (a)** -0.4 and 3.4 **(b)** -0.3 and 3.3
11. (a) -0.6, 0, and 1.6 **(b)** -0.6, 0, and 1.6
13. (a) $F(x) \to 1$ **(b)** $F(x) \to -4$ **(c)** $F(x) \to \infty$
15. (a) $[-\infty, -6] \cup [-2, \infty)$ **(b)** $[-6, -2]$
(c) $[-7, -3] \cup [-1, \infty)$ **(d)** $(-7, -3) \cup (-1, \infty)$
(e) $(-\infty, -7) \cup (-3, -1)$ **17. (a)** $[3, \infty)$
(b) $(-\infty, 3]$ **(c)** $(-\infty, \infty)$ **(d)** $(-\infty, 0) \cup (0, \infty)$
(e) Nowhere **19. (a)** $[-\infty, 0) \cup [0, 2]$ **(b)** $[5, \infty)$
(c) $[2, 5]$ **(d)** $(-\infty, 0) \cup (0, 7)$ **(e)** $(7, \infty)$ **21. (a)** 3
(b) 5 **(c)** 0 **(d)** $-5, -3, 5$ **(e)** 2 **23. (a)** 3 **(b)** 3
(c) None **25. (a)** $[-7, -6) \cup (-2, 4)$
(b) $(-6, -2) \cup (4, 6]$ **(c)** $[-7, -6] \cup [-2, 4]$
27. (a) 5 **(b)** $[-7, -6) \cup (-2, 4)$ **(c)** $[-6, -2] \cup [4, 6]$
29. (a) 3 **(b)** 4 **(c)** 5 **(d)** $-9, -2, 4$ **(e)** $-7, -3, 7$
31. (a) 3 **(b)** 3 **(c)** 1 **33. (a)** $(-\infty, -7) \cup (-3, 7)$
(b) $(-7, -3) \cup (7, \infty)$ **(c)** $(-\infty, -7] \cup [-3, 7]$
35. (a) -5 **(b)** $(-\infty, -7) \cup (-2, 7)$ **(c)** $[-7, -2] \cup [7, \infty)$
37. (a) The number of each type of paramecium increases and then levels off. **(b)** The number of aurelia increase, whereas the number of caudatum die out. **39. (a)** 13 **(b)** 700

EXERCISES 2.7

1. y-axis symmetry **3.** x-axis symmetry **5.** Origin symmetry **7.** None
9. (a)

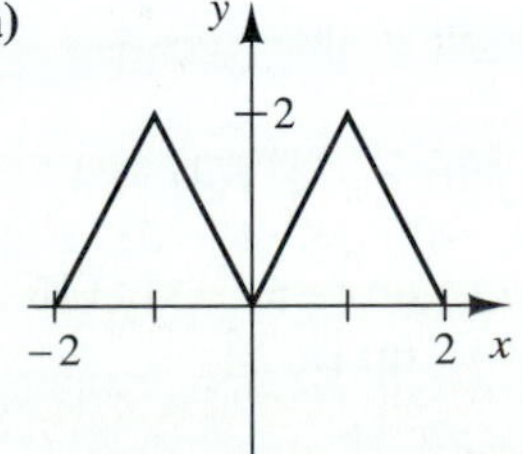

(b)

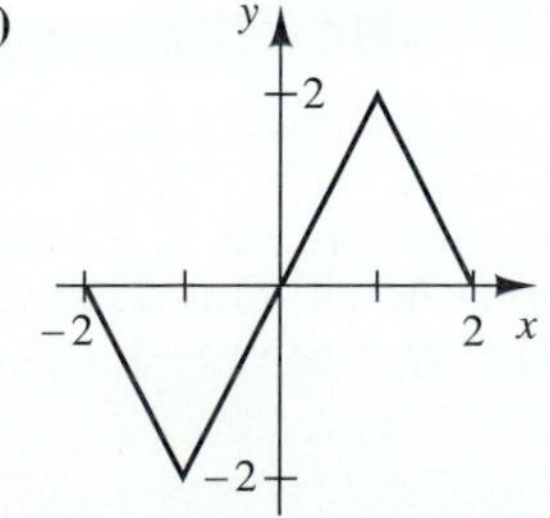

11. (a)

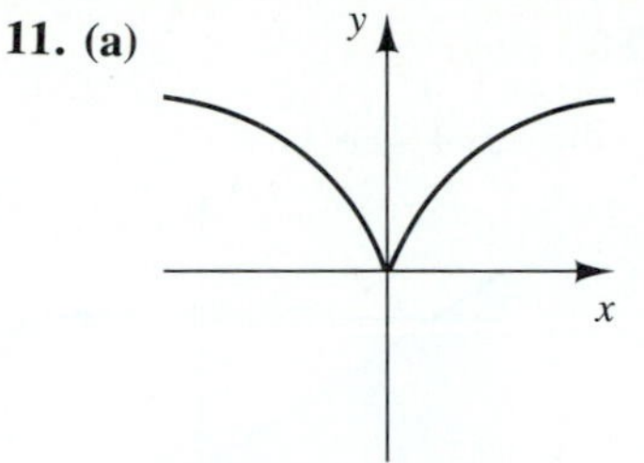

(b)

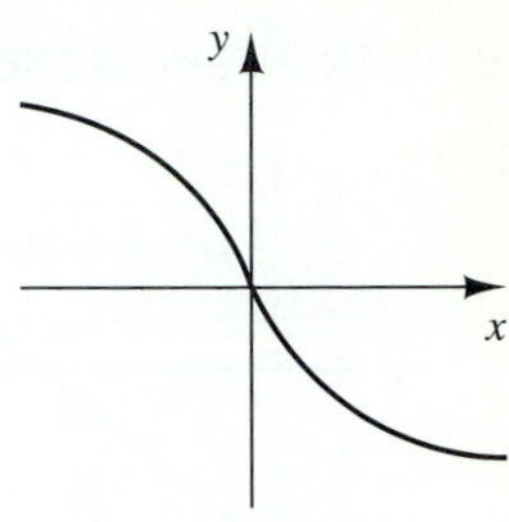

13. (a) y-axis symmetry

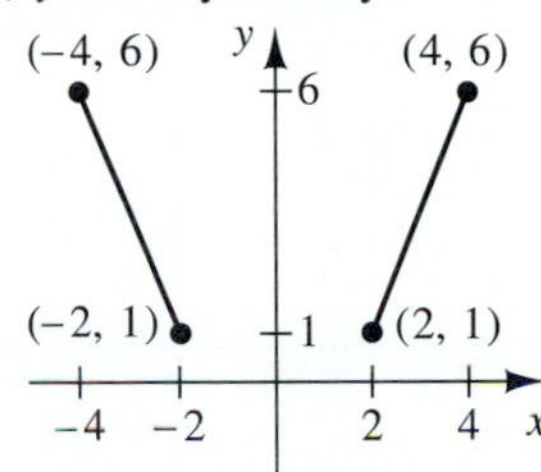

(b) x-axis symmetry

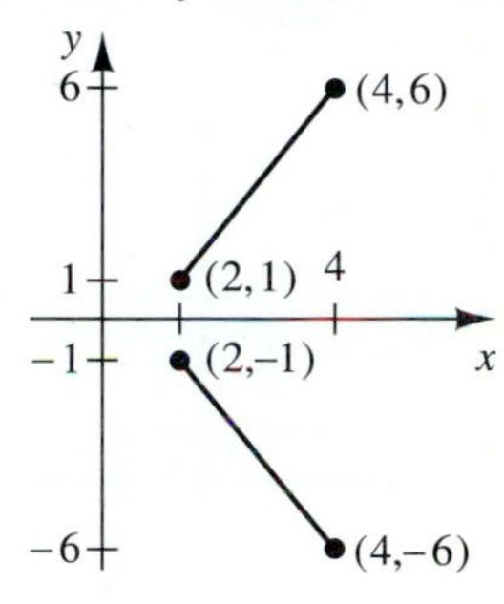

(c) Origin symmetry

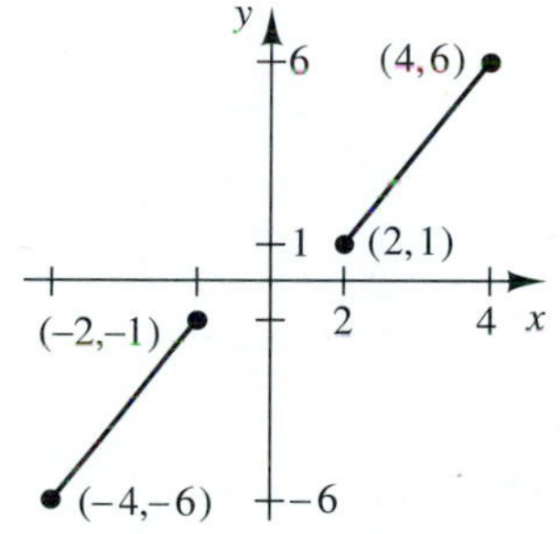

15. (a) y-axis symmetry

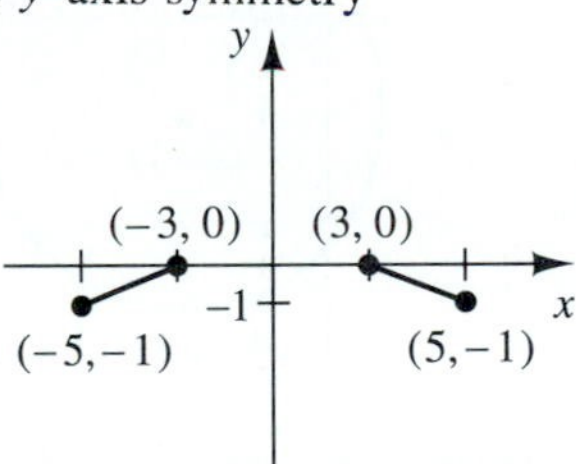

(b) x-axis symmetry

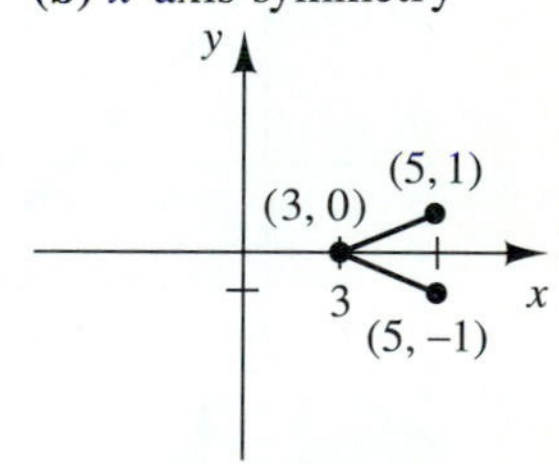

(c) Origin symmetry

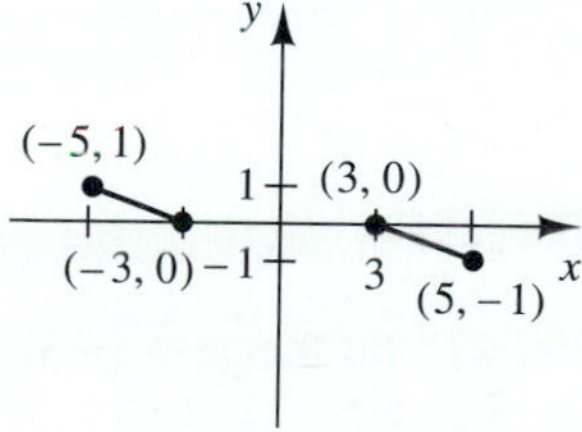

17. y-axis symmetry **19.** Origin symmetry **21.** Neither **23.** Origin symmetry **25.** y-axis symmetry **27.** Origin symmetry **29.** y-axis symmetry **31.** Origin symmetry **33.** Neither

CHAPTER 2 REVIEW EXERCISES

1.

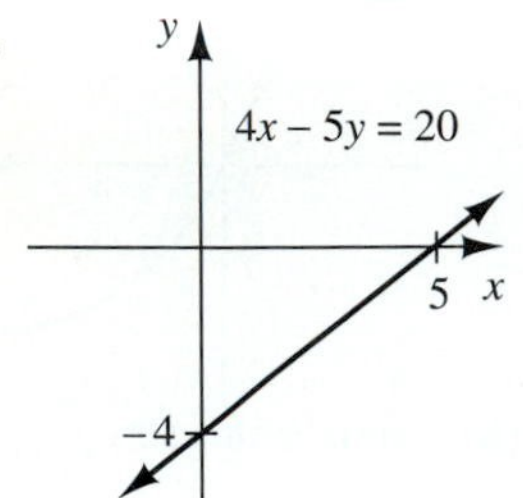

3.

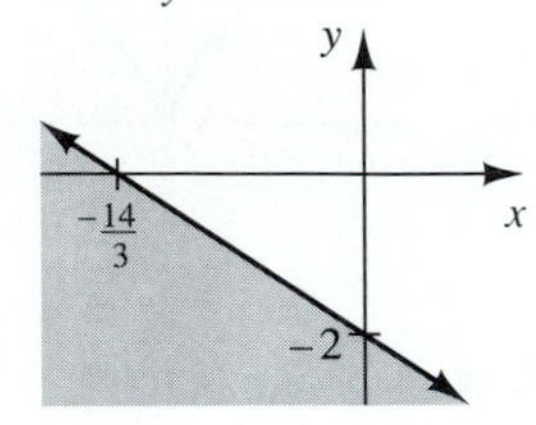

5.

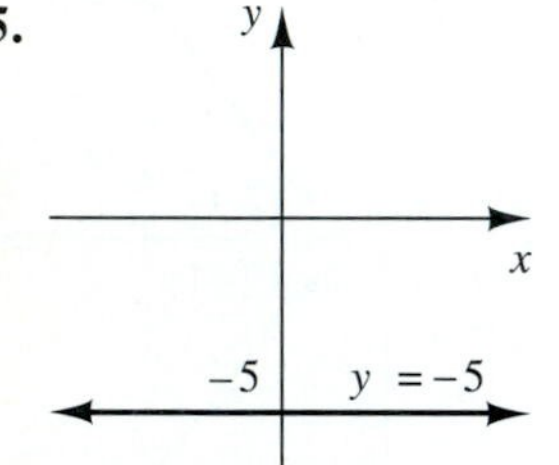

7.

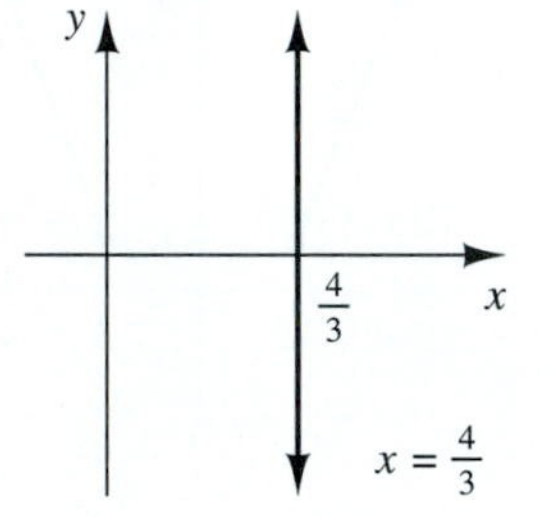

9.

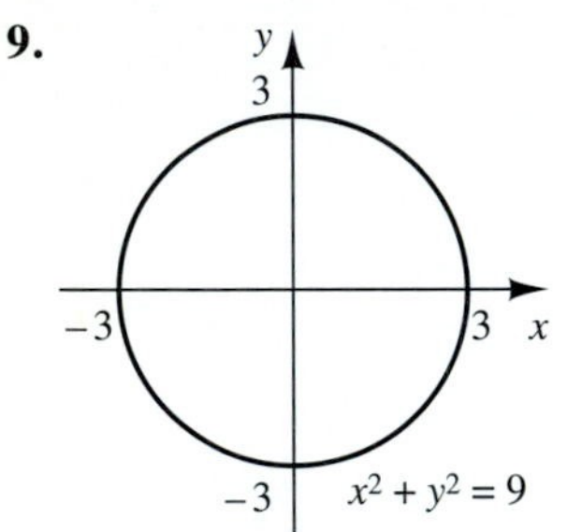

11.

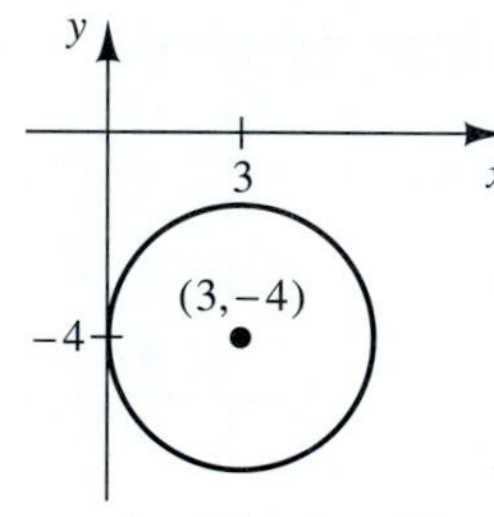

13.

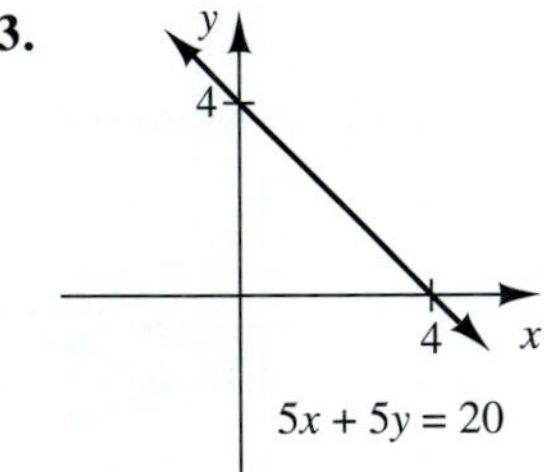

15.

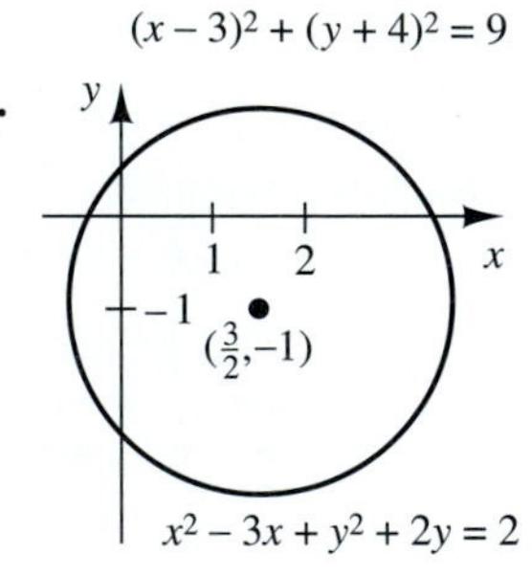

17.

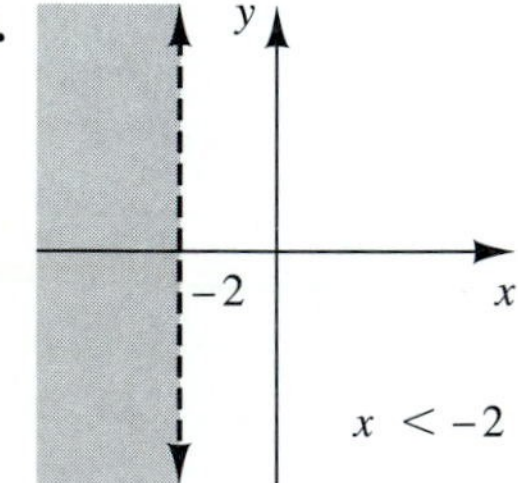

19. **(a)** Yes **(b)** Domain $= (-\infty, \infty)$ **(c)** Range $= [-5, \infty)$ **21.** **(a)** Yes **(b)** Domain $= \{x \mid -5 \le x \le -2, 2 < x < 5\}$ **(c)** Range $= \{y \mid -5 < y \le 4\}$ **23.** When $x = 2$, $y = 16$; when $x = -2$, $y = 16$; it is a function **25.** $y = -\frac{9}{5}x - \frac{7}{5}$

27. $t = \pm 9$ **29.** $(x + 4)^2 + (y - 1)^2 = 49$ **31.** Center: $(-4, 1)$; $r = 5$ **33.** $(x + 5)^2 + (y - 4)^2 = \frac{1}{4}$

35. $\left(x - \frac{1}{2}\right)^2 + \left(y - \frac{1}{2}\right)^2 = \frac{41}{2}$ **37.** $\left\{x \mid x \le \frac{5}{3}\right\}$

39. $\{x \mid x \ne -1, 4\}$ **41.** $[0, 6]$ **43.** $\{x \mid x > -4\}$ **45.** $f(-3) = -22$; $f(2x) = -4x^2 + 8x - 1$; $2f(x) = -2x^2 + 8x - 2$ **47.** $h(2) = h(0) = h(-3) = \sqrt{5}$

49. $g\left(\frac{1}{2}\right) = 0$; $g(-5)$ is undefined; $g(t + 1) = \frac{2t + 1}{t + 6}$

51. $f(0) = -1$; $f(-1) = 0$; $f(-5) = 17$ **53.** $-10x + 26$

55. $\frac{3}{t(t + h)}$ **57.** **(a)** $f(-4) = -2$; $f(-2) = 2$; $f(0) = 3$; $f(1) = 3$ **(b)** $-5, -3, 4$ **(c)** $(-\infty, -5) \cup (-3, 4)$ **(d)** $(-5, -3) \cup (4, \infty)$ **(e)** $[-4, -1]$ **(f)** $(-\infty, -4) \cup [2, \infty)$ **(g)** $[-1, 2]$ **59.** y-axis symmetry **61.** x-axis symmetry **63.** None **65.** Origin symmetry **67.** Origin symmetry **69.** 89.5 **71.** $A = (6 + 2.5t)^2$

CHAPTER 2 PRACTICE TEST

1. **(a)** -9 **(b)** $-2x^2 - 21x - 49$ **(c)** $-32x^2 - 20x + 3$ **(d)** $-2x^4 - 5x^2 + 3$ **(e)** $-4x - 2h - 5$ **2.** **(a)** $-\frac{7}{4}$ **(b)** $\frac{2 + t}{1 - 2t}$ **3.** $\frac{-3}{(x + 2)(x - 3)}$ **4.**

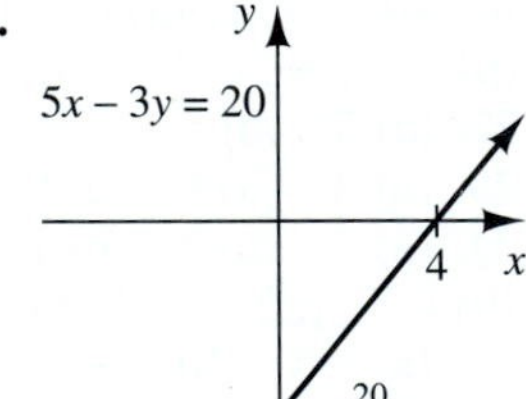

5. $y = \frac{3}{4}x - 3$ **6.** $y = -\frac{7}{3}x - \frac{29}{3}$

7. It is not a function. **8.** $(x + 1)^2 + \left(y - \frac{1}{2}\right)^2 = \frac{65}{4}$

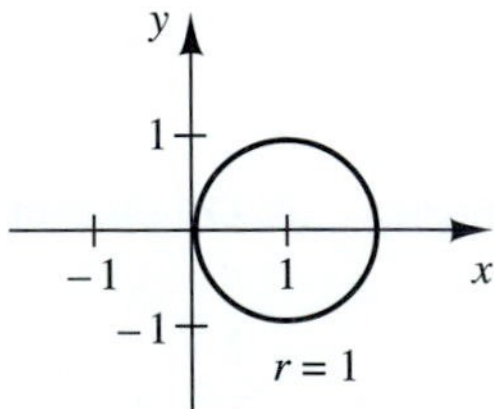

9. **(a)** $f(-5) = -2$; $f(-3) = 3$; $f(0) = -1$; $f(4) = 4$ **(b)** $-6, -4, -2, 2, 7$ **(c)** $(-\infty, -6) \cup (-4, -2) \cup (2, 7)$ **(d)** $(-6, -4) \cup (-2, 2) \cup (7, \infty)$ **(e)** $[-5, -3] \cup [0, 3]$ **(f)** $(-\infty, -5) \cup [-3, 0] \cup [5, \infty)$ **(g)** $[3, 5]$

10. (a)

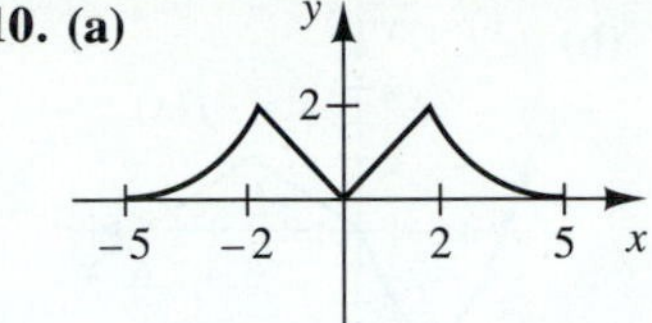

(b)

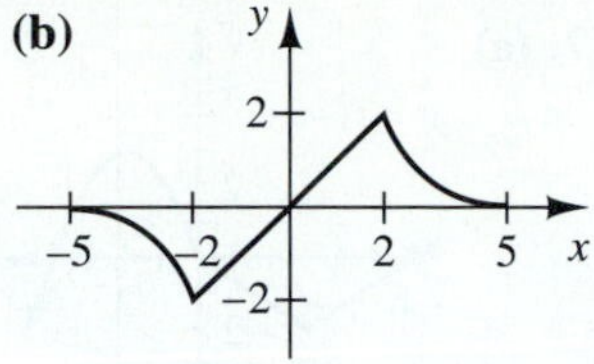

(c)

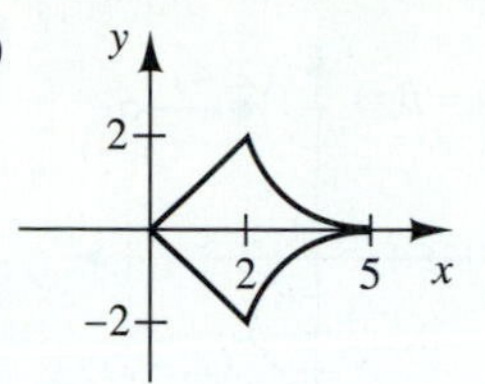

11. (a) Neither **(b)** y-axis symmetry **(c)** Origin symmetry

12. $C = 3.5[2w + 2(4w - 3)] = 3.5(10w - 6)$

CHAPTER 3

EXERCISES 3.1

1.

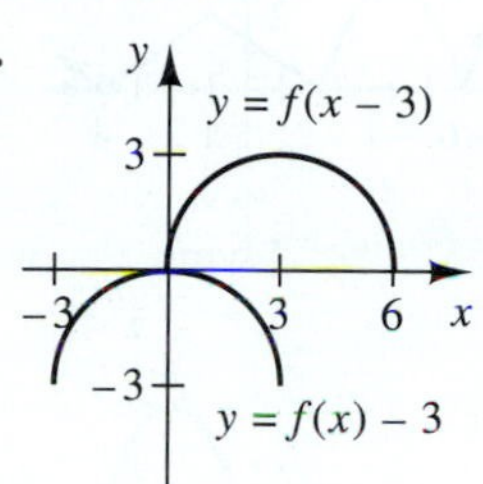

3.

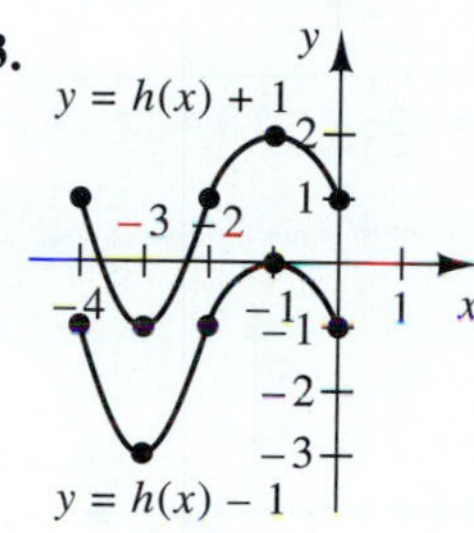

5.

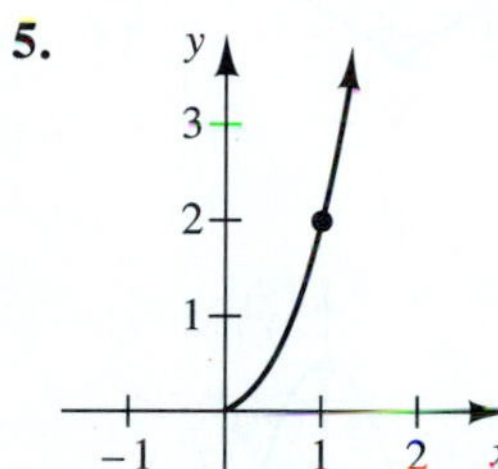

7.

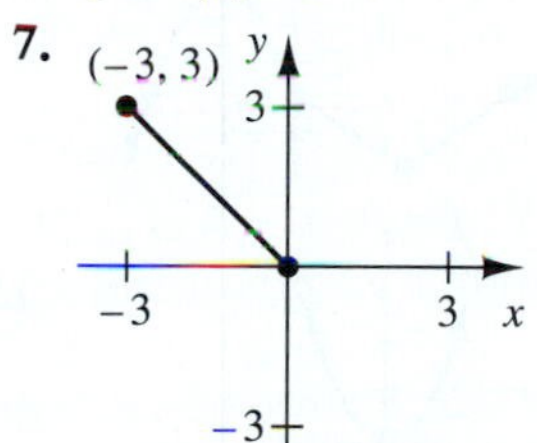

9.

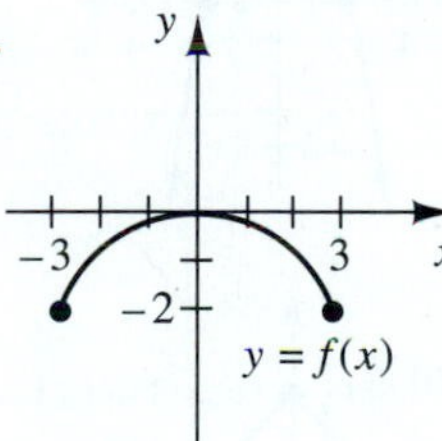

11.

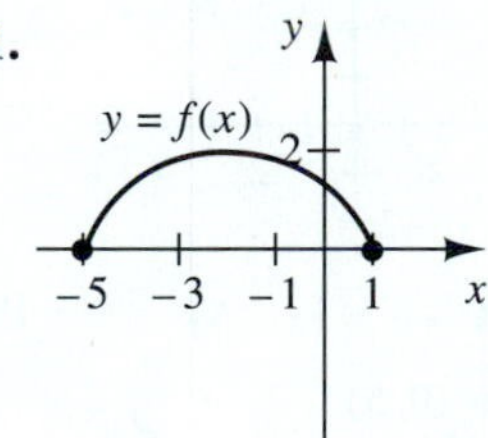

13.

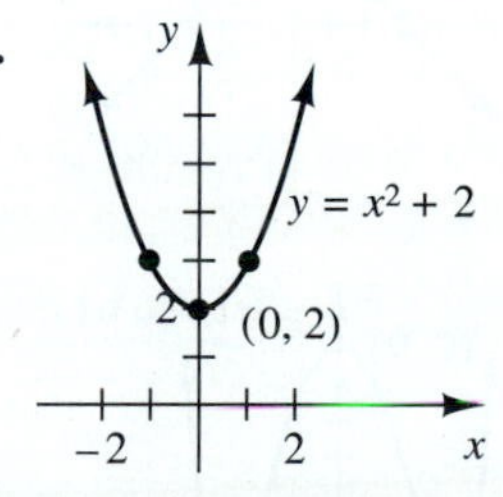

15.

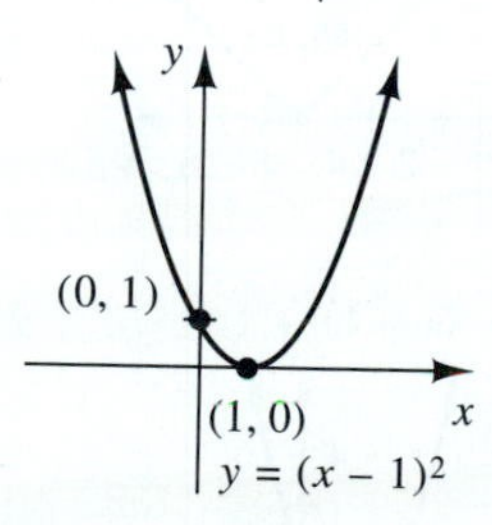

17.

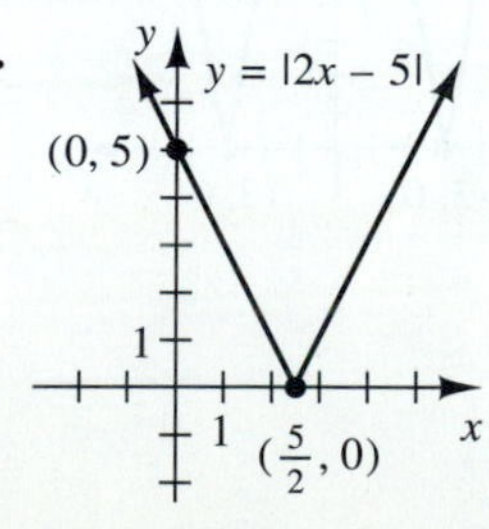

19.

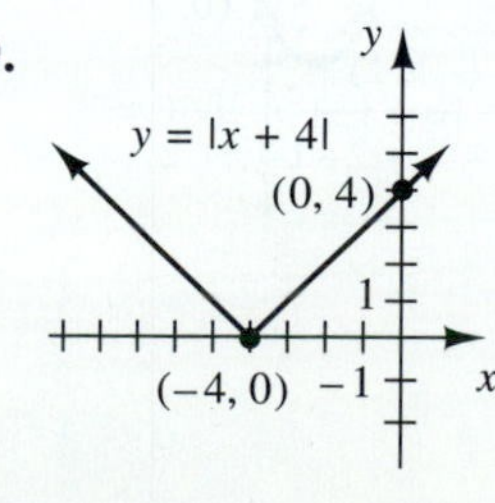

21.

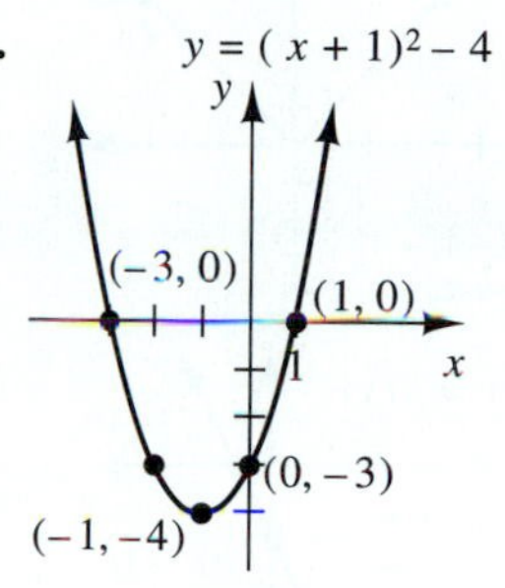

23.

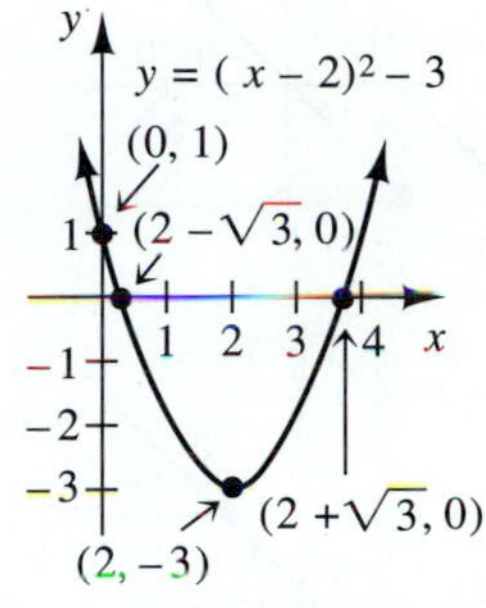

25.

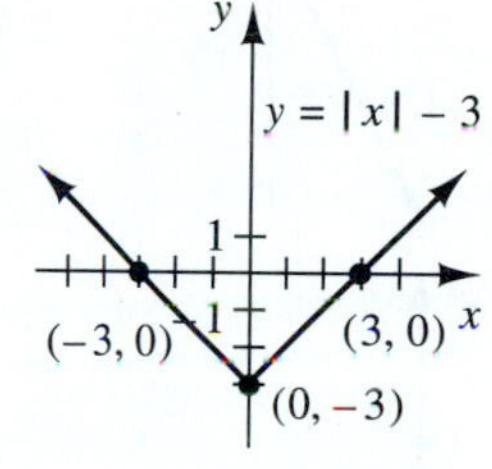

27.

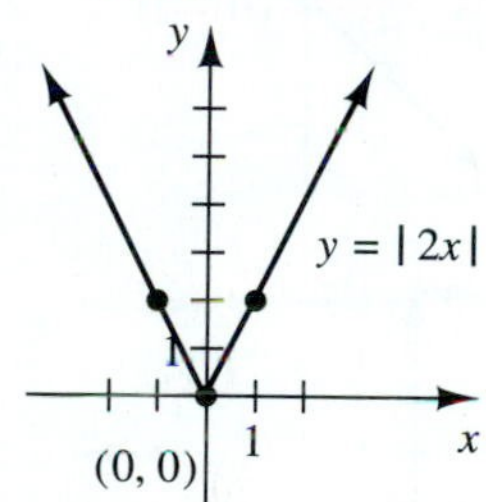

29.

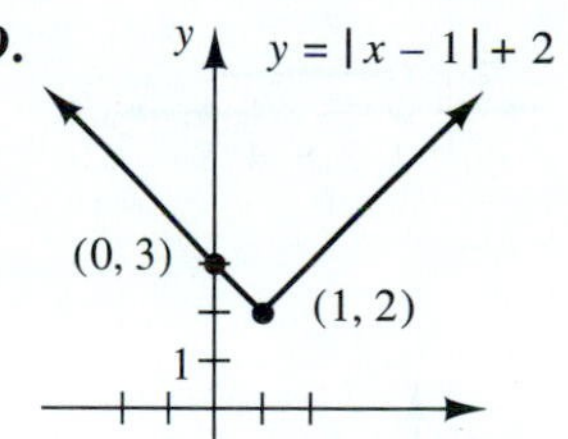

31.

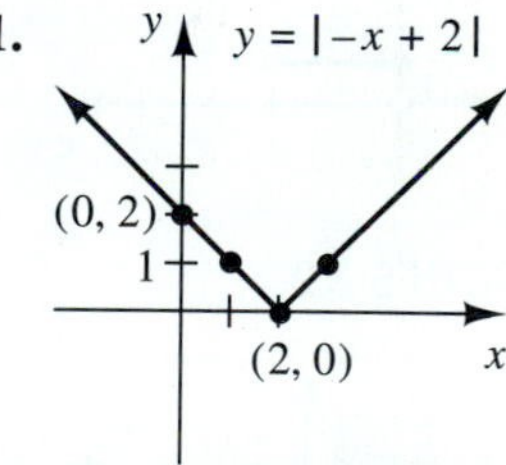

33.

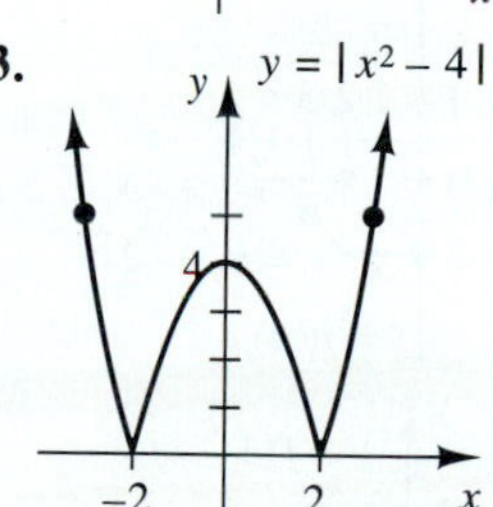

35.

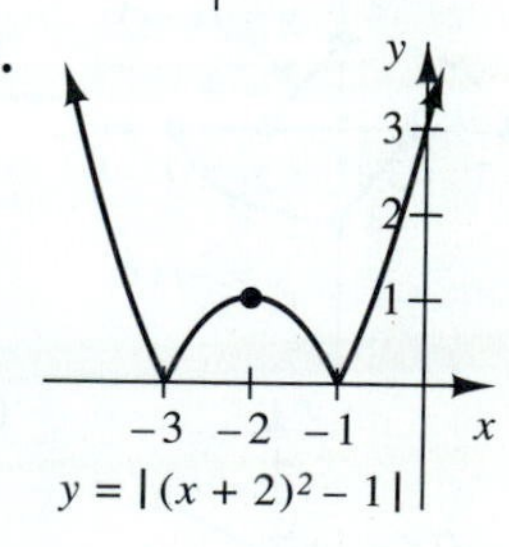

37.

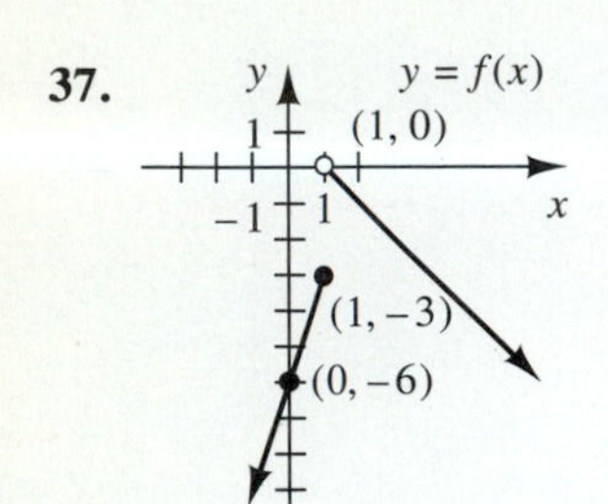

39.

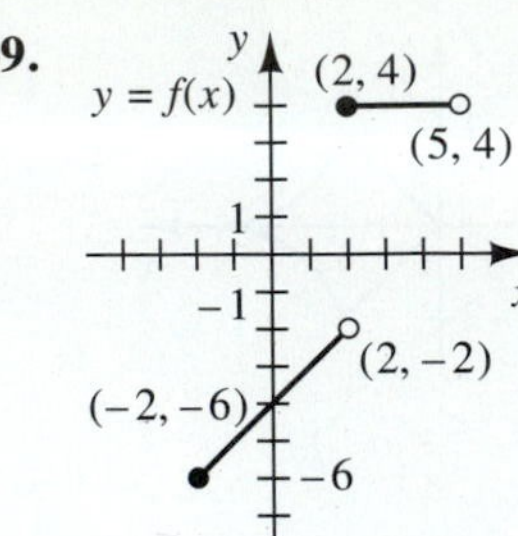

41.

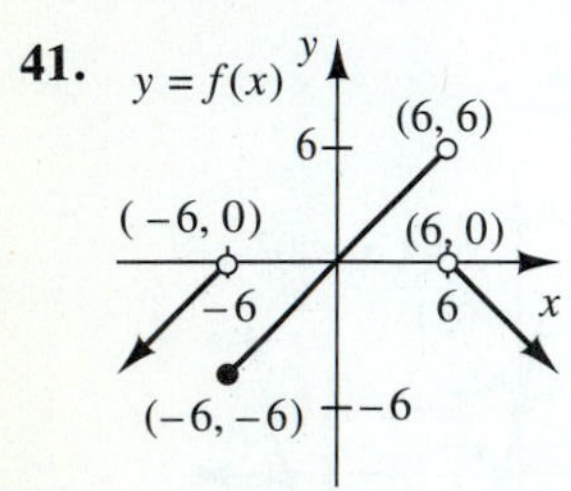

43.

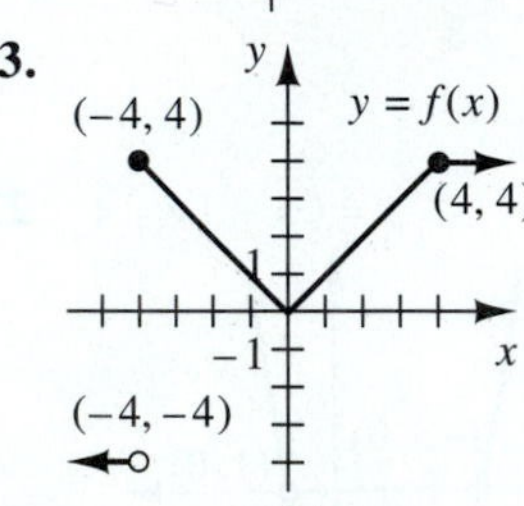

45.

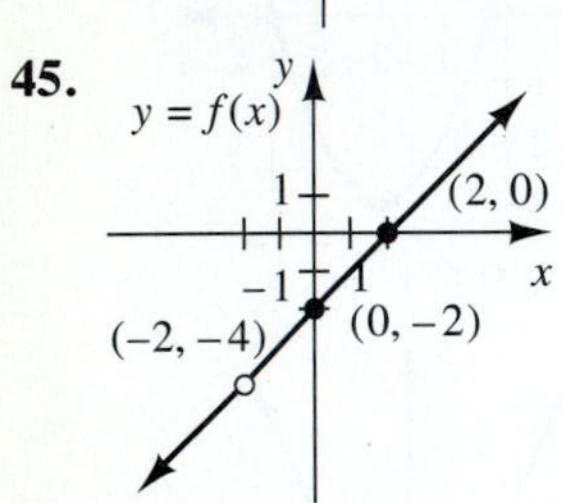

47.

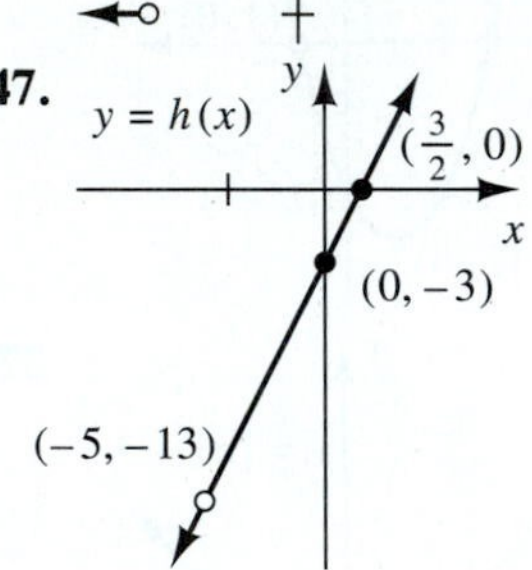

49.

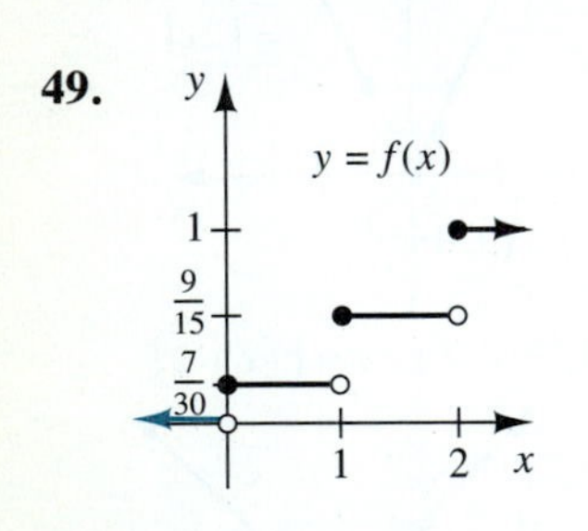

51.

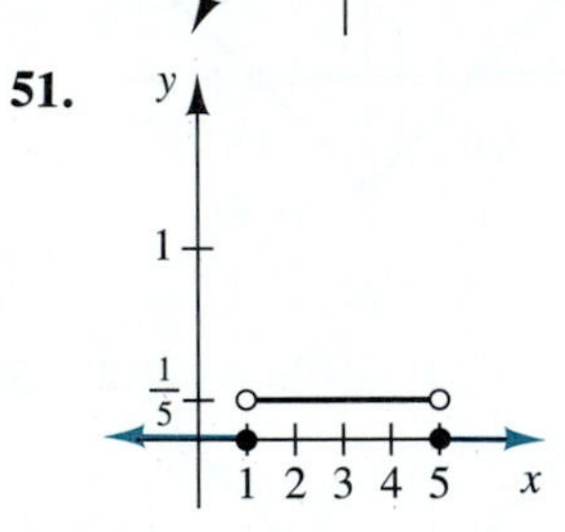

EXERCISES 3.2

1.

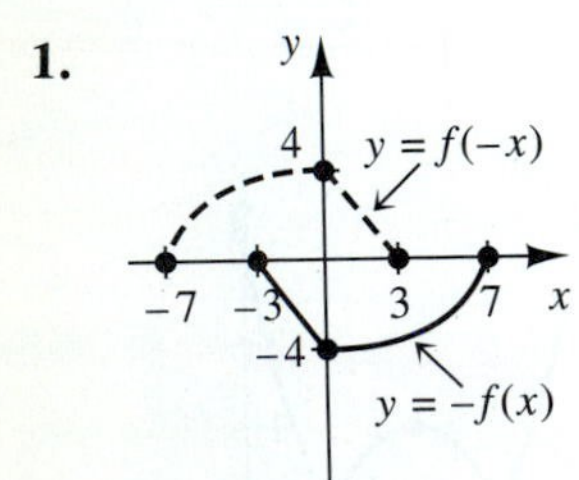

3.

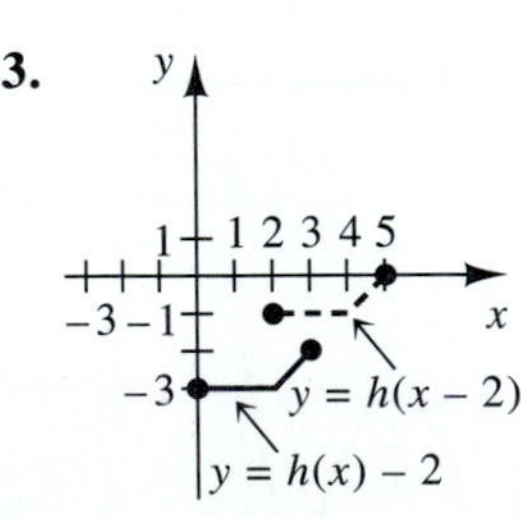

5. (a)

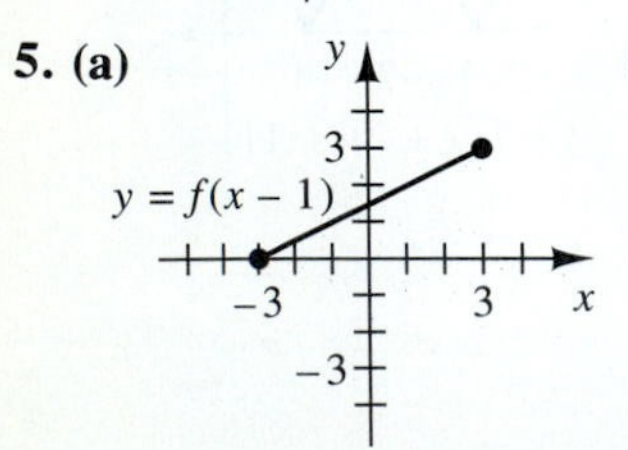

(b)

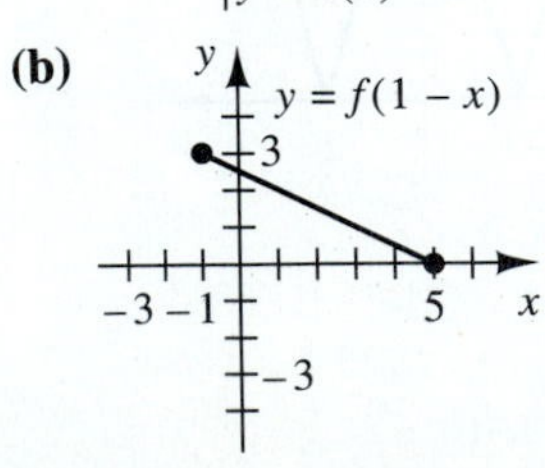

7. (a)

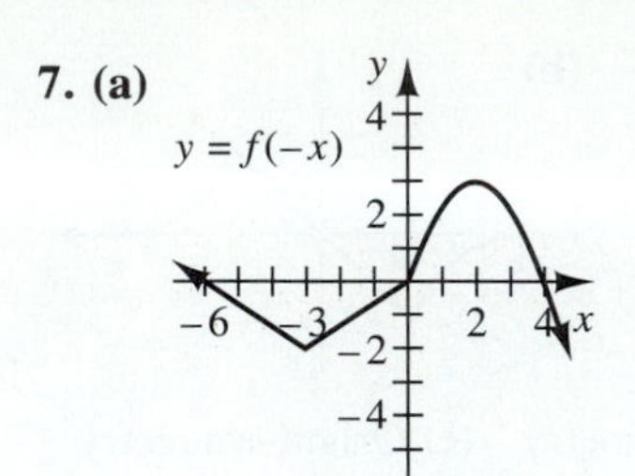

(b)

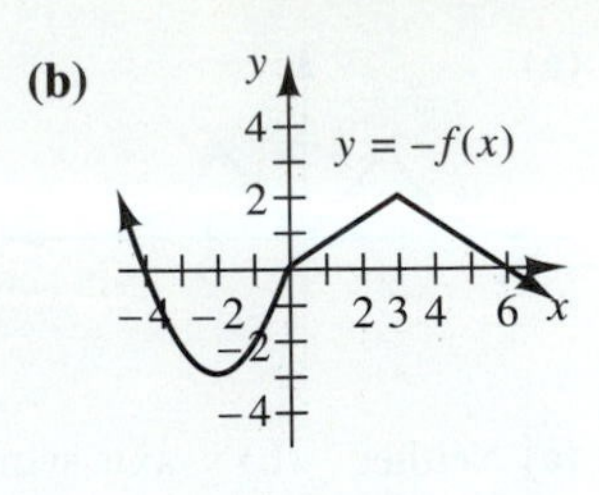

9. (a) $y = |f(x)| + 2$

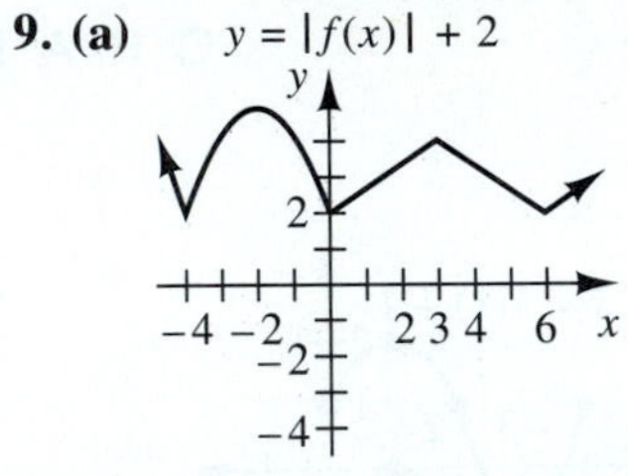

(b) $y = |f(x + 2)|$

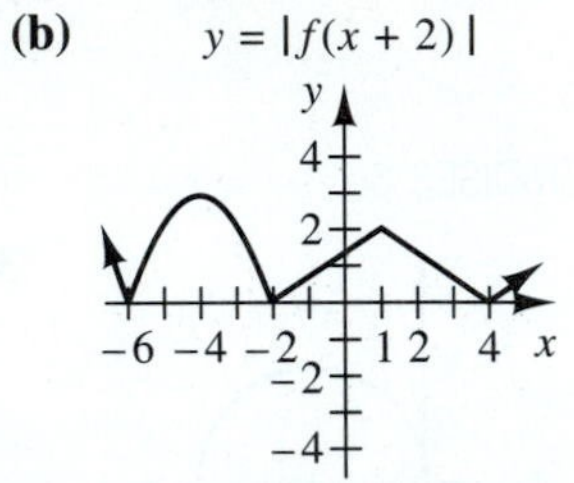

11. (a)

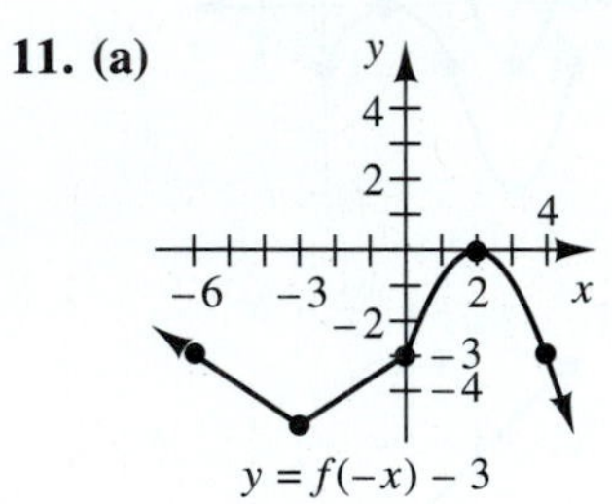

(b)

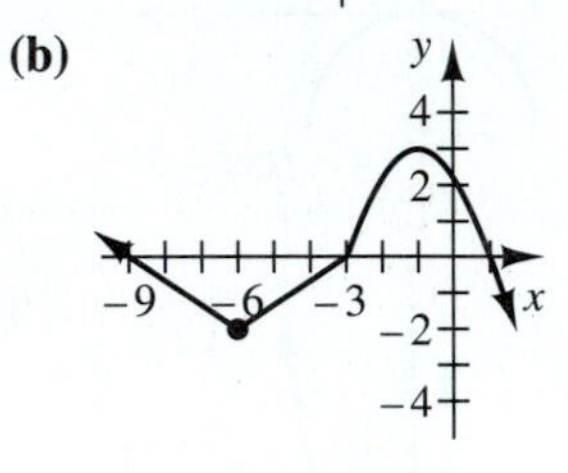

13.

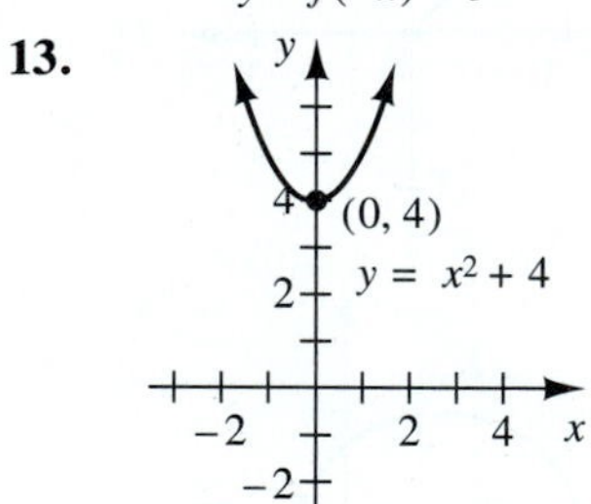

15.

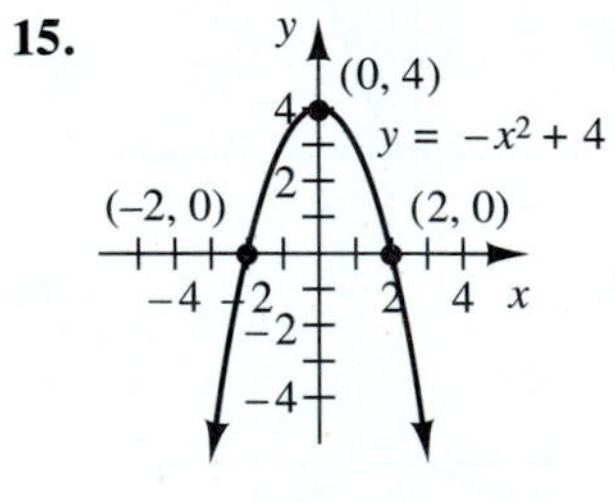

17.

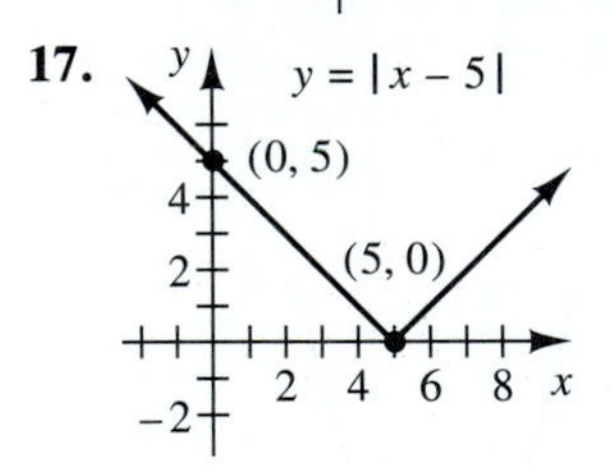

19.

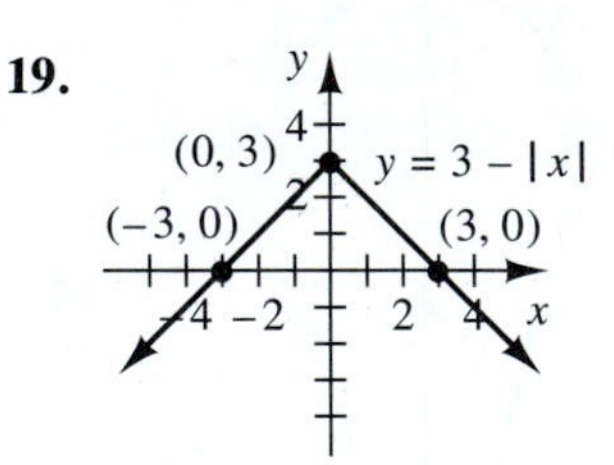

21. $y = (x + 1)^2 + 1$

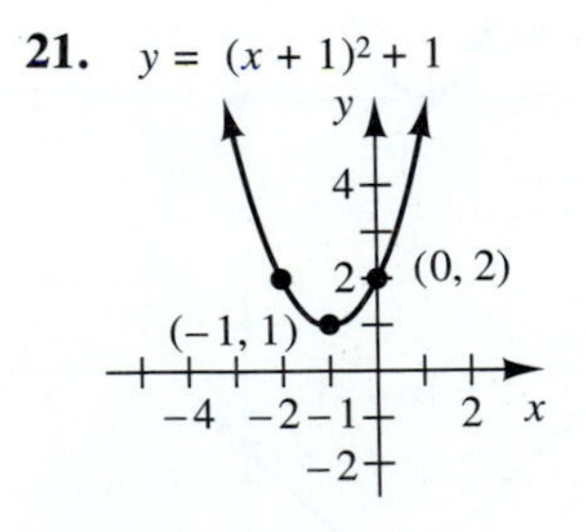

23.

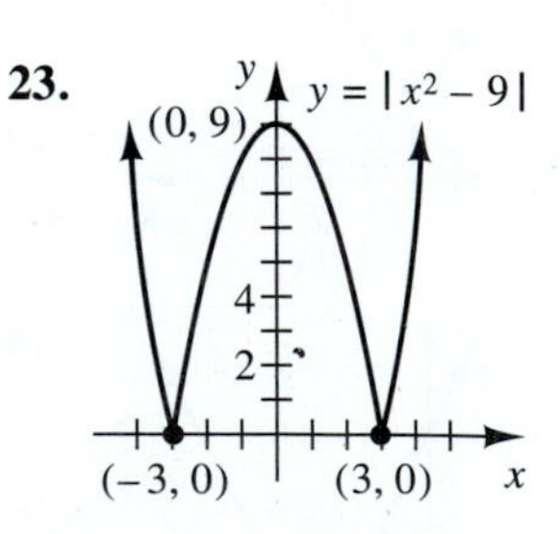

25.

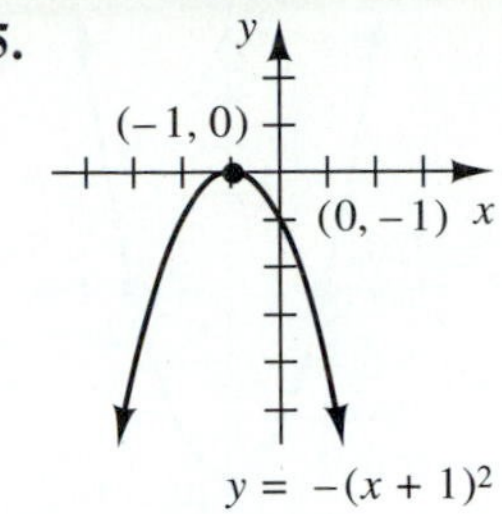

27.

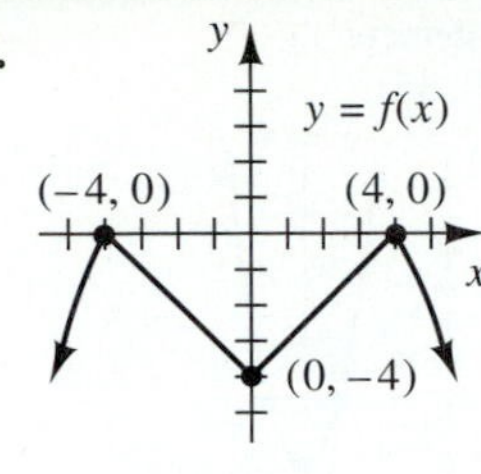

29.

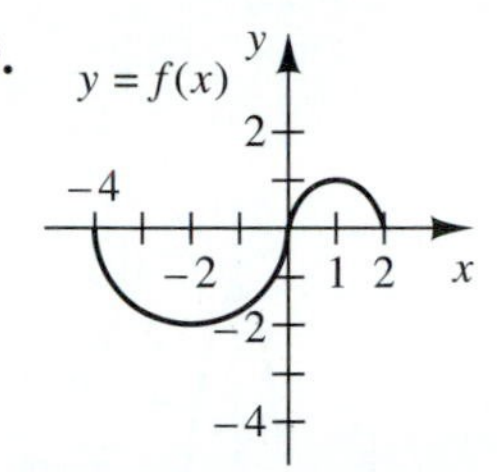

31.

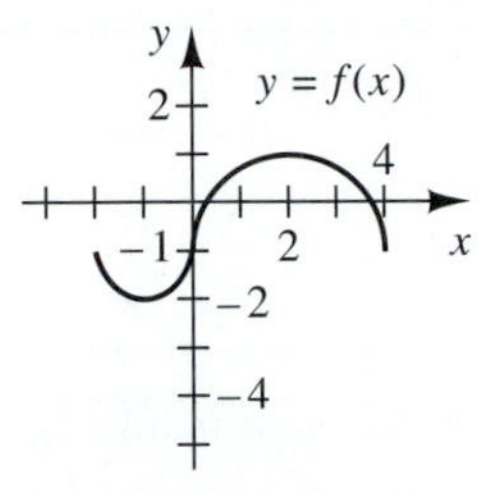

33. Shift $y = f(x)$ 4 units down. **35.** Reflect $y = f(x)$ about the y-axis. **37.** Graph $y = f(x)$; then take the portion of the graph below the x-axis and reflect it above the x-axis. **39.** Reflect $y = f(x)$ about the x-axis and shift it 3 units up. **41.** Shift $y = f(x)$ left 3 units and up 4 units. **43.** Reflect $y = f(x)$ about the y-axis and shift it 2 units left. **45.** 6 **47.** -3

49.

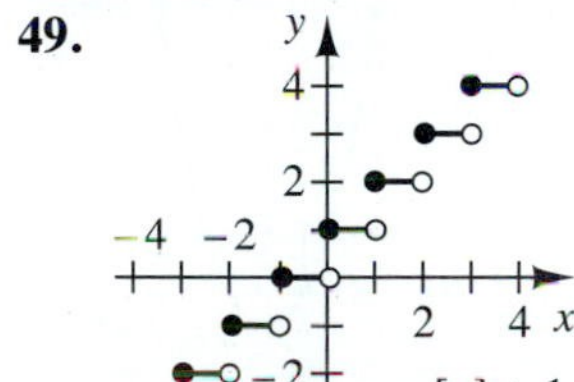

51.

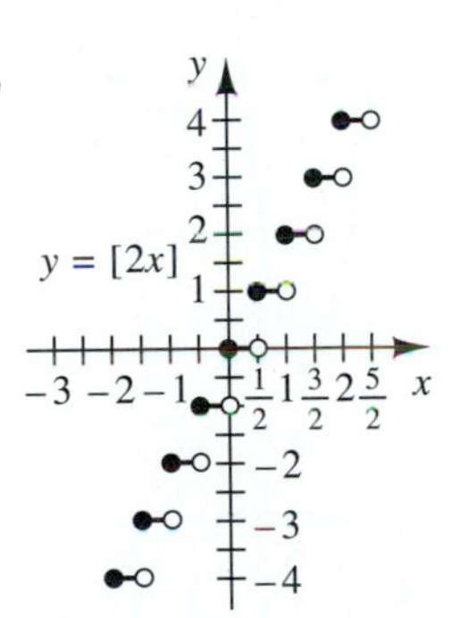

EXERCISES 3.3

1. (a) $y = \dfrac{6(9 + x)}{x}, x > 0$ **(b)** $A = \dfrac{3(9 + x)^2}{x}, x > 0$

3. $A = \dfrac{1}{2}\pi r^2 - \dfrac{1}{2}(5x) = \dfrac{1}{2}(\pi r^2 - 5\sqrt{4r^2 - 25}), r \geq \dfrac{5}{2}$

5. $V = 80 = x^2 h, \Rightarrow h = \dfrac{80}{x^2}$.

$$A = 2x^2 + 4xh = 2x^2 + 4x\left(\frac{80}{x^2}\right) = \frac{2x^3 + 320}{x}, x > 0$$

7. Since $C = 2\pi r$, $r = \dfrac{C}{2\pi}$. Hence

$$A = \pi r^2 = \pi\left(\frac{C}{2\pi}\right)^2 = \frac{C^2}{4\pi}, C > 0.$$

9. Since $\dfrac{20}{x + s} = \dfrac{6}{s}$, $s = \dfrac{3x}{7}$, $x > 0$.

11. Since $\dfrac{1 - 0}{4 - a} = \dfrac{1 - b}{4 - 0}$, $b = \dfrac{a}{a - 4}$. Hence

$$A = \frac{1}{2}ab = \frac{1}{2}a\left(\frac{a}{a - 4}\right) = \frac{a^2}{2(a - 4)}, a \neq 4.$$

13. $A =$ the area of the rectangle + the area of the two semicircles

$$= \frac{(200 - 2\pi r)(2r)}{2} + 2\left(\frac{1}{2}\right)\pi r^2$$

$$= 200r - \pi r^2, 0 < r < \frac{100}{\pi}.$$

15. If h is the height of the balloon, then $h = rt = 5t$.

$$d = \sqrt{120^2 + h^2} = \sqrt{120^2 + (5t)^2}$$

$$= \sqrt{14400 + 25t^2}, \qquad t \geq 0$$

17. (a) If $s =$ the length of the side of the square, then the length of the wire for the square is $4s$. Thus, the length of the wire for the circle is $30 - 4s$. Hence $2\pi r = 30 - 4s$ and so $r = \dfrac{30 - 4s}{2\pi}$. The area of the circle is $\pi\left(\dfrac{30 - 4s}{2\pi}\right)^2 = \dfrac{(15 - 2s)^2}{\pi}$. The area of the circle and the square is $s^2 + \dfrac{(15 - 2s)^2}{\pi}$, $0 \leq s \leq \dfrac{15}{2}$. **(b)** Using the same reasoning as in part (a), if $r =$ the radius of the circle, then the area of the square and the circle is $\pi r^2 + \dfrac{(15 - \pi r)^2}{4}$, $0 \leq r \leq \dfrac{15}{\pi}$.

19. $R(n) = (24 - .50n)(3500 + 80n), n \geq 0$

21. $R = R(n) = (9 + 0.5n)(8500 - 225n) = -112.5n^2 + 2225n + 76{,}500$, for $0 \leq n \leq 37$

23. Let $x =$ the side of the fence adjacent to the house; then the side opposite the house $= 80 - 2x$. $A = x(80 - 2x)$, $0 < x < 40$.

25. Let $x =$ the length of the vertical sides of the fence. If we let $a =$ the length of one entire horizontal side, then the cost, C, is $C = 3(2x) + 5(2x) + 5(2a)$, or $240 = 16x + 10a$. Hence $a = \dfrac{120 - 8x}{5}$. The area enclosed is

$$A = xa = x\left(\frac{120 - 8x}{5}\right) = -\frac{8}{5}x^2 + 24x, 0 < x < 15.$$

27. (a) $L = L(x) = 4x + \dfrac{1600}{3x}$, for $x > 0$

(b) $C = C(x) = 72x + \dfrac{9600}{x}$, for $x > 0$.

29. $L = \sqrt{x^2 + 1} + (6 - x)$, for $0 \leq x \leq 6$.

31. $I = \begin{cases} 0.0165b & \text{for } 0 \leq b \leq 2000 \\ 33 + 0.0125(b - 2000) & \text{for } b > 2000 \end{cases}$

33. $V = 18x(1 - 2x) = 18x - 36x^2$, for $0 < x < \frac{1}{2}$

35. Since $S = 4\pi r^2$, $r = \sqrt{\frac{S}{4\pi}}$. Hence

$$V = \frac{4}{3}\pi r^3 = \frac{4}{3}\pi\left(\sqrt{\frac{S}{4\pi}}\right)^3 = \frac{S\sqrt{\pi S}}{6\pi},\ S \geq 0.$$

37. $V = \pi r^2 h = \pi r^2(10) \Rightarrow r^2 = \frac{V}{10\pi}$. Therefore,

$$S = 2\pi rh + 2\pi r^2 = 20\pi\sqrt{\frac{V}{10\pi}} + 2\pi\left(\frac{V}{10\pi}\right)$$

$$= 2\sqrt{10\pi V} + \frac{V}{5},\ \text{for } V > 0.$$

39. $D = \sqrt{x^4 - 13x^2 + 2x + 50}$, for all real numbers x

41. $D = \sqrt{y^2 + 7y + 12}$, for $y \leq 3$

43. $D = \sqrt{x^2 + x}$, for $x \geq 0$.

EXERCISES 3.4

1.

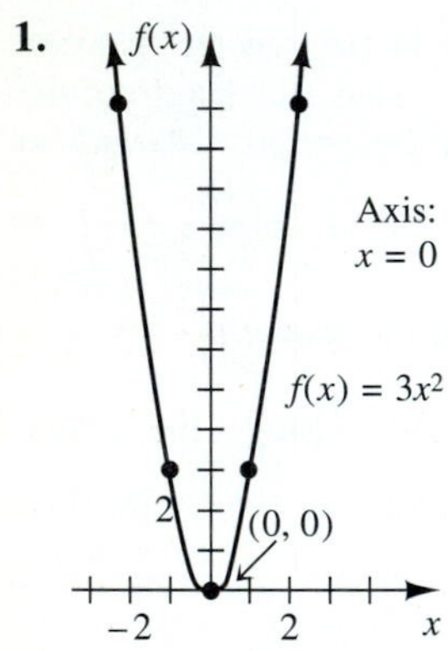

3.

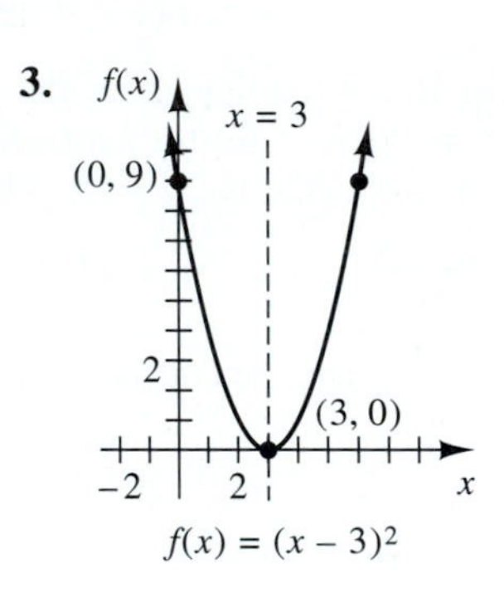

5.

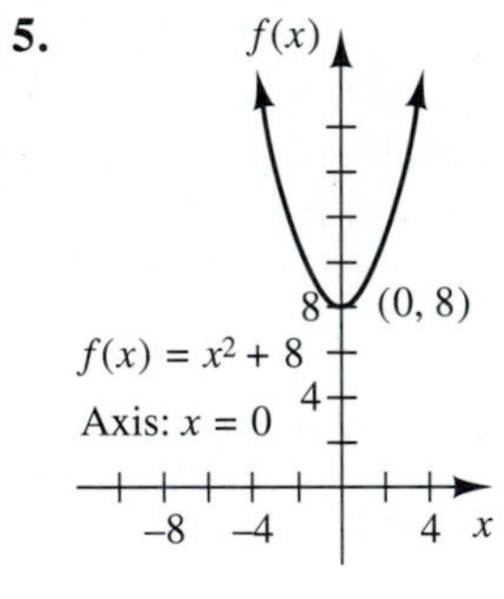

7.

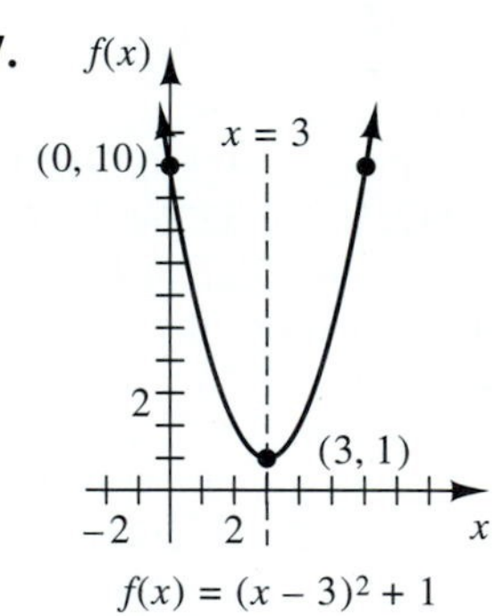

9.

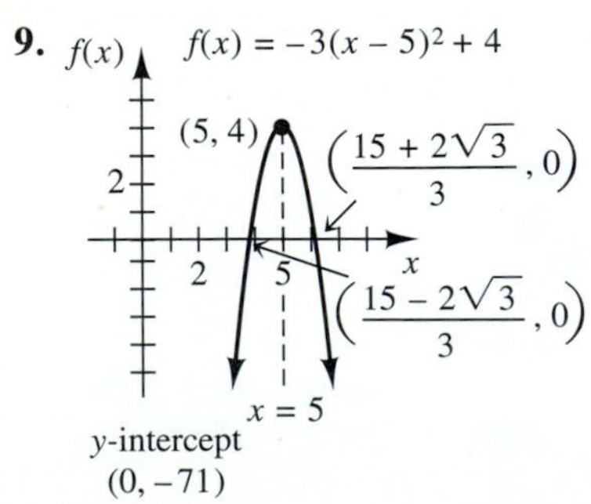

11.

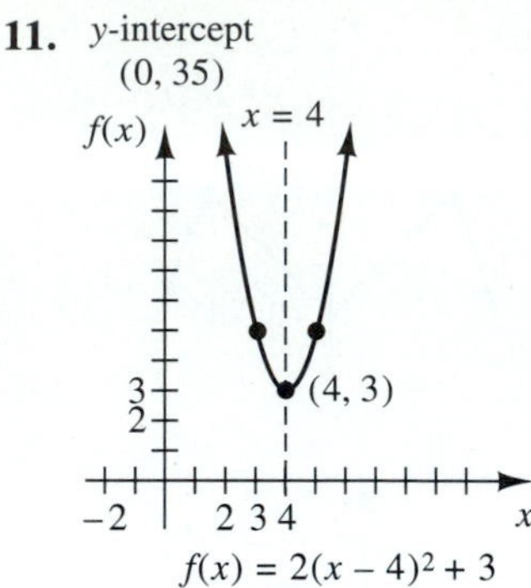

13.

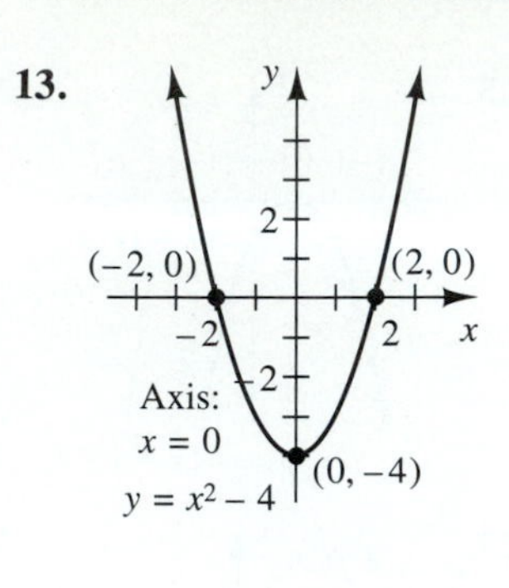

15.

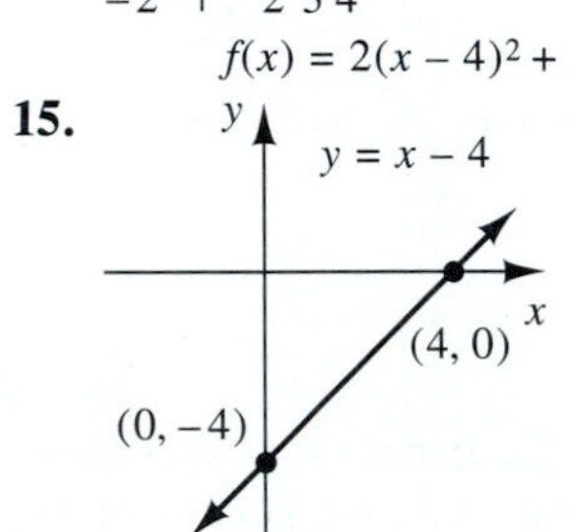

17.

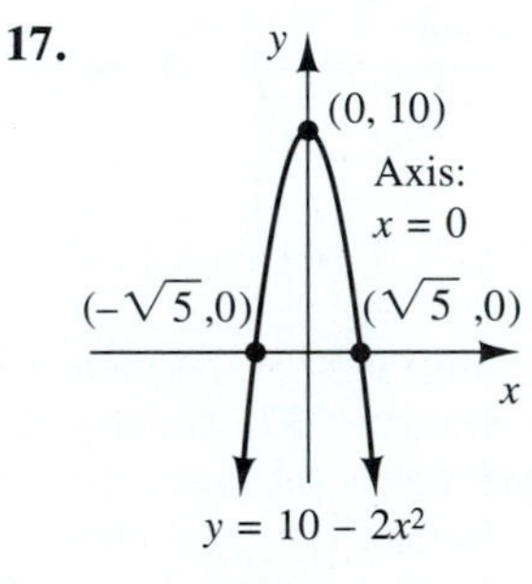

19.

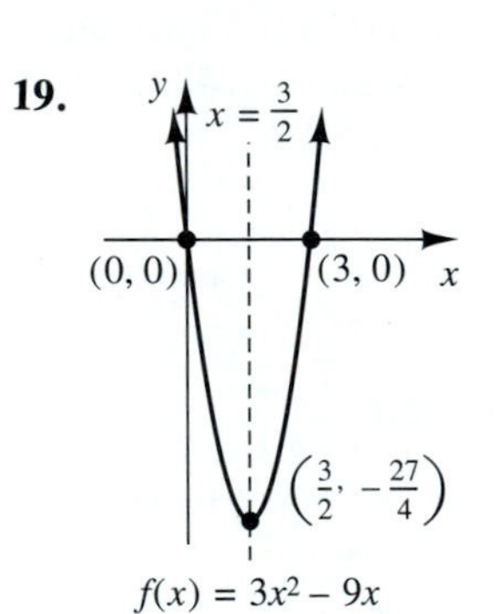

21.

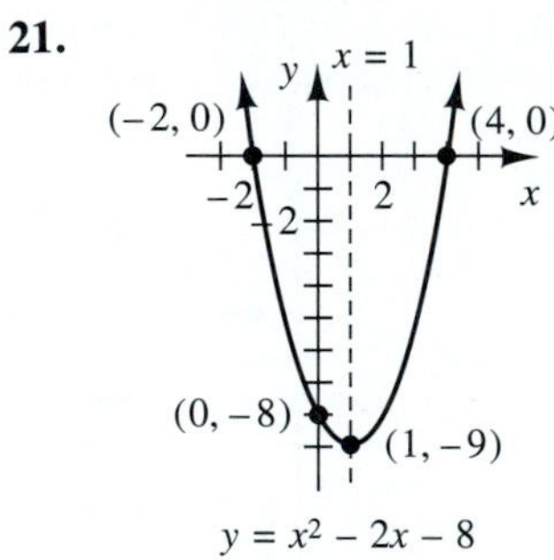

23.

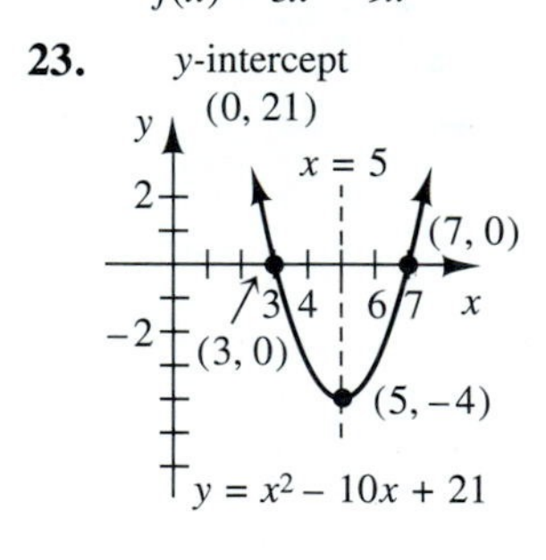

25.

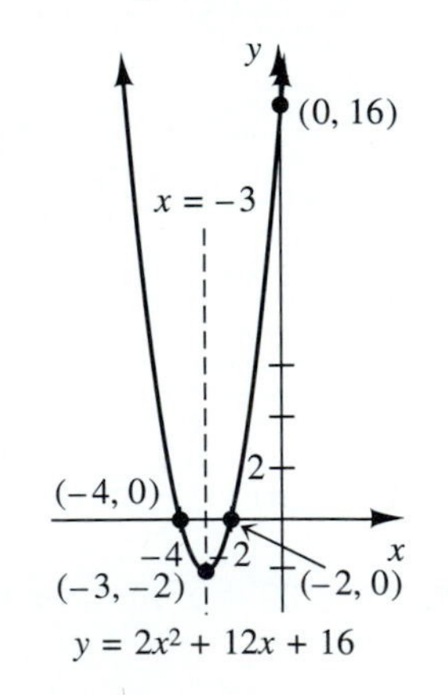

27.

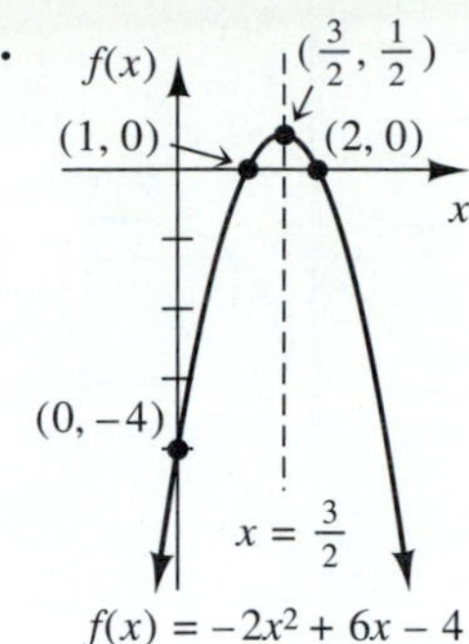

29.

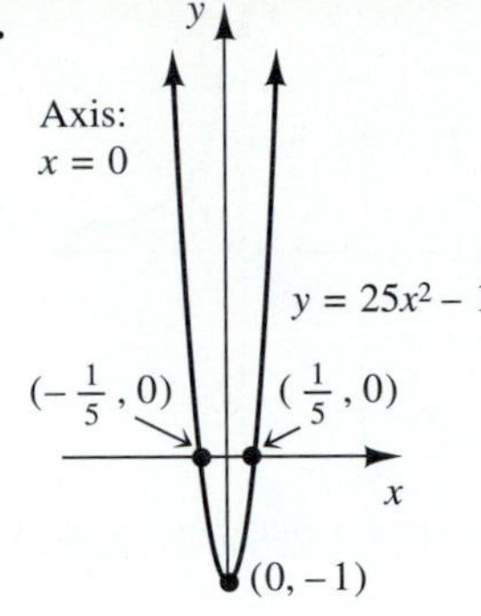

31.

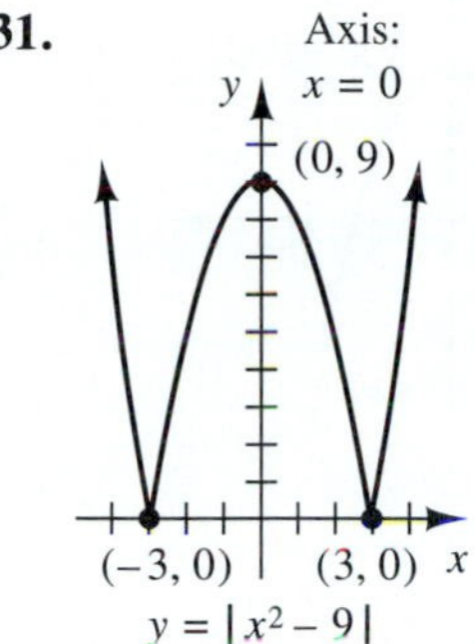

33.

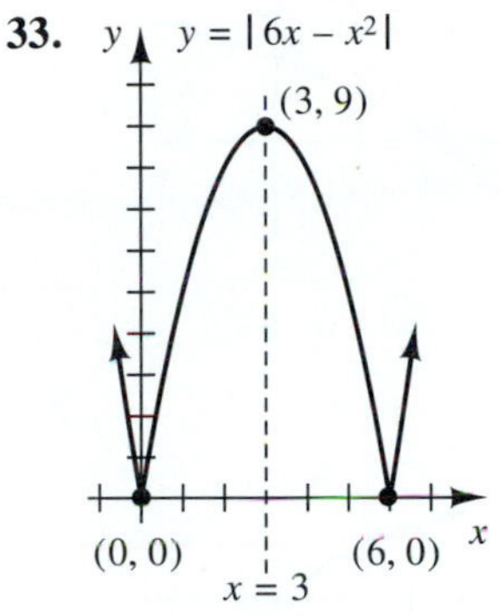

35. Maximum is 36 **37.** Maximum is $-\frac{27}{2}$

39. Minimum is -36 **41.** Minimum is $\frac{27}{2}$

43. **(a)**

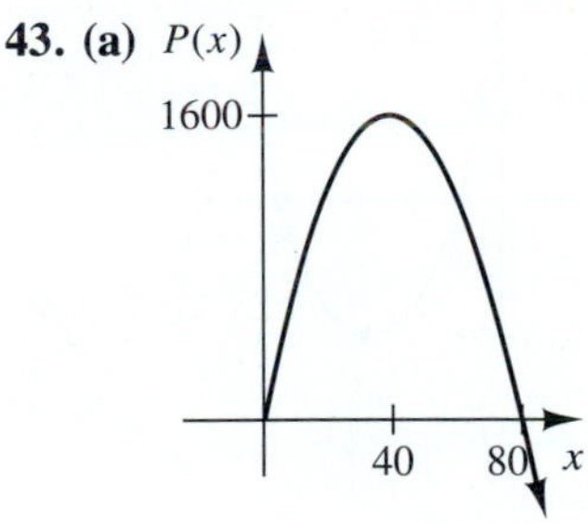

(b) $x = 40$ items
(c) Profit: $1600 daily

45. $x = 12$ tables: $425.92

47. $t = 27$ seconds; 11,664 feet

49. Let x = one side; then the adjacent side is $\frac{100 - 2x}{2} = 50 - x$. $A = x(50 - x) = -x^2 + 50x$. The maximum area is at $x = 25$: 25×25 feet.

51. Let x = one number; then the other number is $104 - x$. $P = x(104 - x) = -x^2 + 104x$. The maximum product is at $x = 52$: 52 and 52.

53. Since $800 = 3x + 2y$, then $y = \frac{800 - 3x}{2}$. Hence $A = xy = x\left(\frac{800 - 3x}{2}\right) = -\frac{3}{2}x^2 + 400x$. Maximum area at $x = \frac{400}{3}$ feet: $x = 133\frac{1}{3}$ feet, $y = 200$ feet.

55. **(a)** 15 **(b)** $5 **(c)**

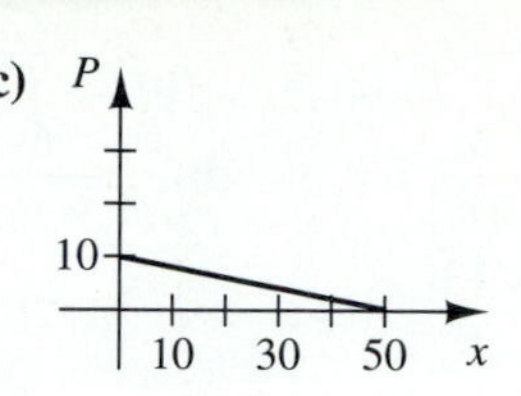

(d) $R(x) = 10x - \frac{x^2}{5}$ **(e)**

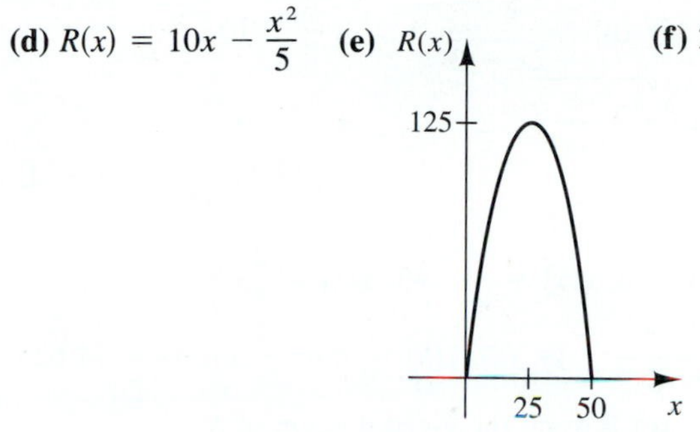

(f) 25 items; $5

57. $\frac{72}{\pi + 4}$ **59.** $1075 **61.** **(a)** $3.50 **(b)** $3.67

63. 10.5 lb

65. $f\left(\frac{B}{2A}\right) = \frac{4AC - B^2}{4A}$

67. **(a)**

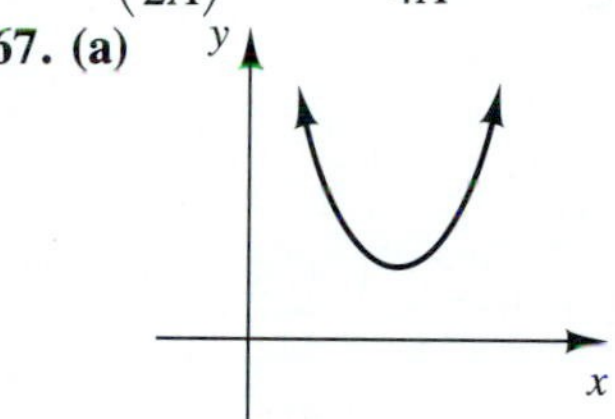

(b)

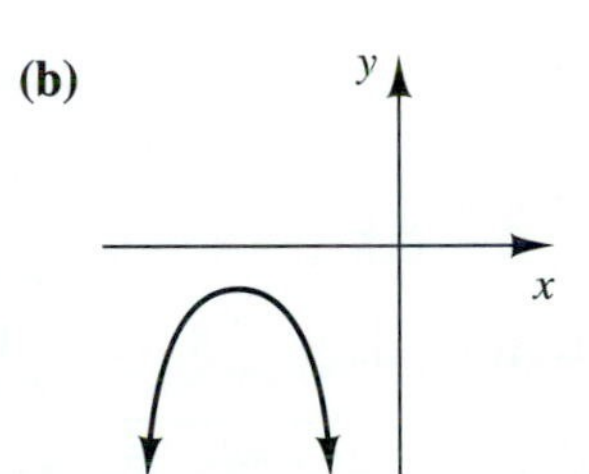

(c) No because $A > 0$ means the parabola opens up whereas $B^2 - 4AC < 0$ means that there are no x-intercepts.

69. **(a)** at $x = -1$ and $x = 4$ **(b)** $(-\infty, -1) \cup (4, \infty)$ **(c)** $(-1, 4)$

71. **(a)** 4 **(b)** 2 **73.** **(a)** 3 **(b)** $\frac{1}{3}$

EXERCISES 3.5

1. 11 **3.** $-\frac{5}{6}$ **5.** $\frac{2}{5}$ **7.** $-\frac{5}{2}$ **9.** 187 **11.** 5

13. $2x^2 + 2x - 5$ **15.** $2x^2 - 4x - 1$; Domain = all real numbers **17.** $\frac{1}{5x}$, $x \neq 0$ **19.** $x^3 - 2x^2 + x + 2$; Domain = all real numbers **21.** $\frac{4(3x - 2)}{x + 2}$, $x \neq -2$

23. $18x^2 - 27x + 7$; Domain = all real numbers

25. $\frac{x^3 - 1}{x^3}$, $x \neq 0$ **27.** 5, $x \neq -2$

29. $27x^3 - 54x^2 + 36x - 9$; Domain = all real numbers **31.** $\frac{x + 2}{4}$, $x \neq -2$

33. **(a)** $\sqrt{2x-6}$; $[3, \infty)$ **(b)** $2\sqrt{x+1}-7$; $[-1, \infty)$
35. **(a)** $\dfrac{18+6t}{(t-3)^2}$, $t \neq 3$ **(b)** $\dfrac{6}{t^2+t-3}$, $t \neq \dfrac{-1 \pm \sqrt{13}}{2}$ **(c)** $t^4+2t^3+2t^2+t$ **(d)** $\dfrac{2t-6}{5-t}$, $t \neq 5, 3$ **37.** **(a)** $\dfrac{1}{\sqrt{2x+3}}$, $x > -\dfrac{3}{2}$ **(b)** $\dfrac{3x+2\sqrt{x}}{x}$, $x > 0$ **(c)** $\dfrac{\sqrt{3x^2+2x}}{x}$, $x \leq -\dfrac{2}{3}$ or $x > 0$
39. $f(x) = x^3$; $g(x) = x+3$ **41.** $f(x) = x^2$; $g(x) = \dfrac{x+4}{x-1}$
43. $f(x) = x+6$; $g(x) = \dfrac{1}{x}$ **45.** $g(x) = \dfrac{2}{5}x + 2$
47. $g(x) = \dfrac{8-x}{5}$ **49.** **(a)** $N(h) = 75h^2 + 1700h + 14{,}625$ **(b)** 22,625 **(c)** Beyond the given domain of N
51. **(a)** 522.6 **(b)** 897 **(c)** $P = 41.6n + 481$

EXERCISES 3.6

1. Yes **3.** No **5.** Yes **17.** $f^{-1}(x) = \dfrac{x+9}{5}$
19. $F^{-1}(x) = \dfrac{\sqrt[3]{4x-4}}{2}$ **21.** $h^{-1}(x) = \dfrac{1-4x}{x}$
23. $f^{-1}(x) = \dfrac{(x+7)^2}{4}$, $x \geq 7$ **25.** $h^{-1}(x) = \sqrt{x+1}$
27. $g^{-1}(x) = \dfrac{2x+5}{3x-2}$ **29.** $f^{-1}(x) = x^{5/3}$
31. $f^{-1}(x) = (x-1)^{-7/5}$, $x \neq 1$ **33.** $f^{-1}(x) = \dfrac{x+6}{7}$
35. $f^{-1}(x) = \dfrac{3}{1-x}$ **37.** $f^{-1}(x) = \dfrac{x-3}{2x+5}$
39. $f^{-1}(x) = \sqrt{x}$; $D_f = D_{f^{-1}} = \{x \mid x \geq 0\}$; $R_f = R_{f^{-1}} = \{y \mid y \geq 0\}$

41.

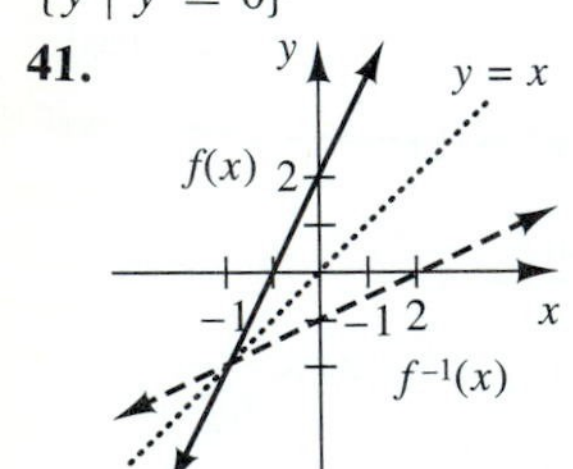

43.

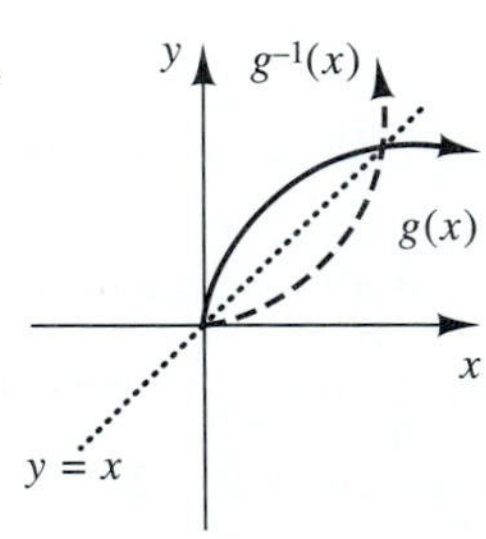

45.

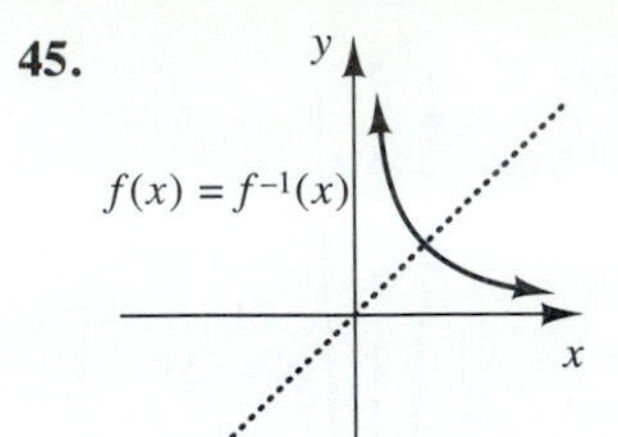

47.

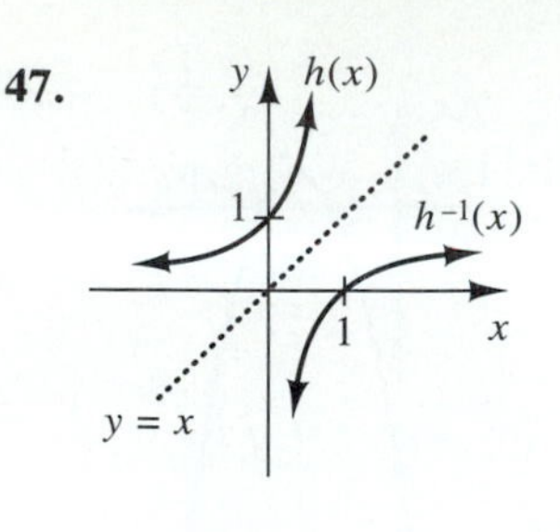

CHAPTER 3 REVIEW EXERCISES

1.

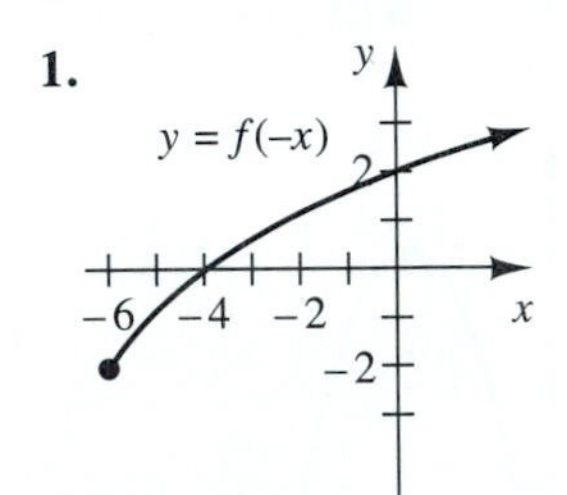

3.

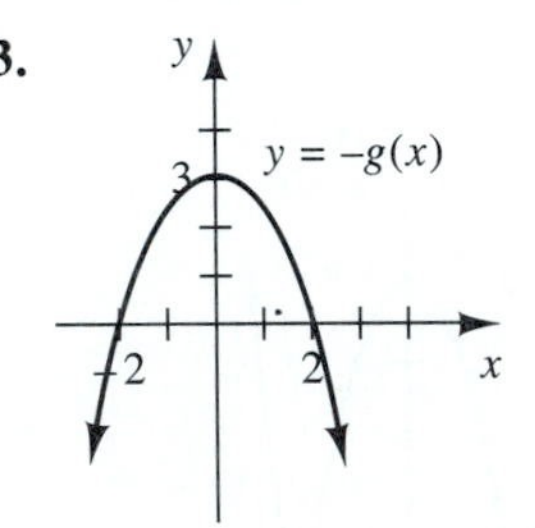

5.

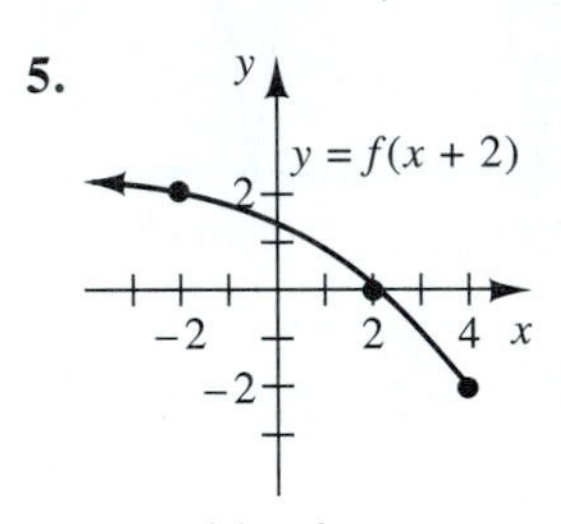

7.

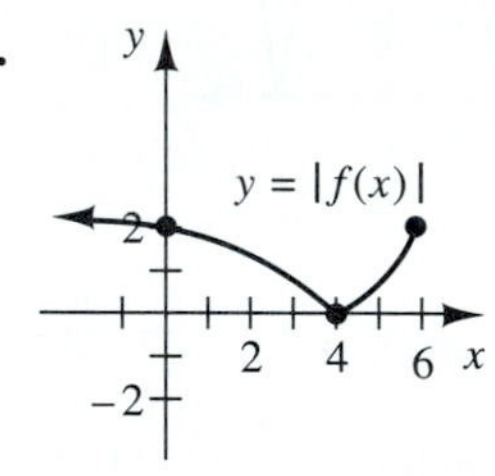

9.

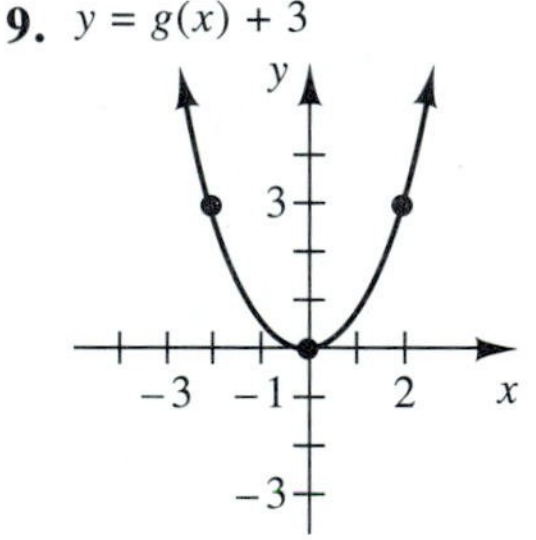

11.

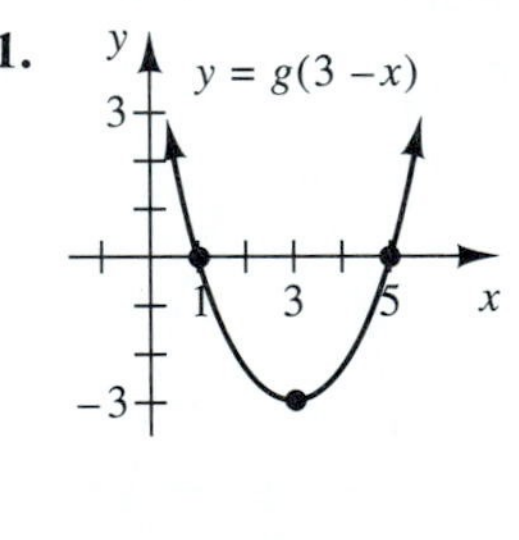

13.

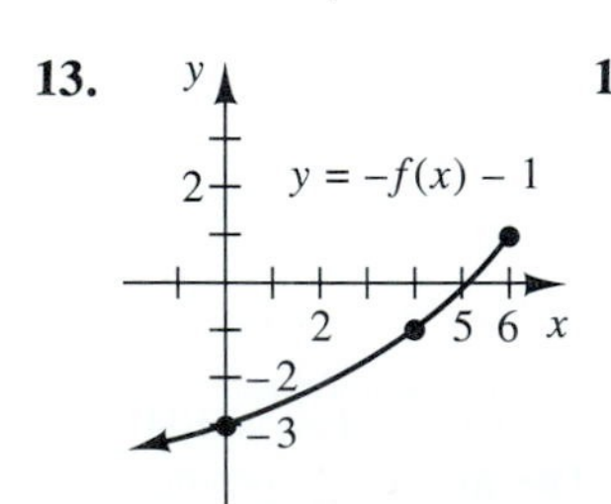

15.

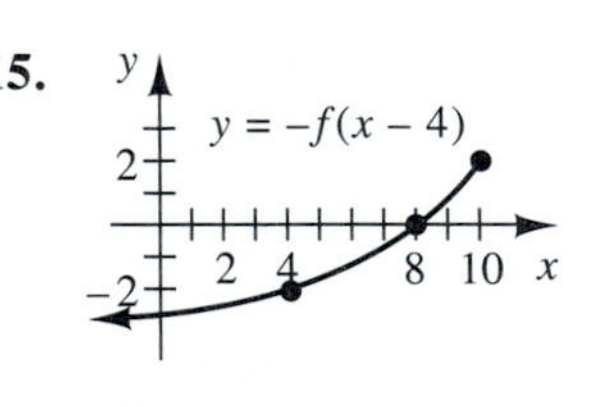

17. $y = f(-x)$ **19.** $y = f(x - 2)$

21. Vertex: (4,6); axis: $x = 4$ **23.** $\frac{81}{8}$

25.

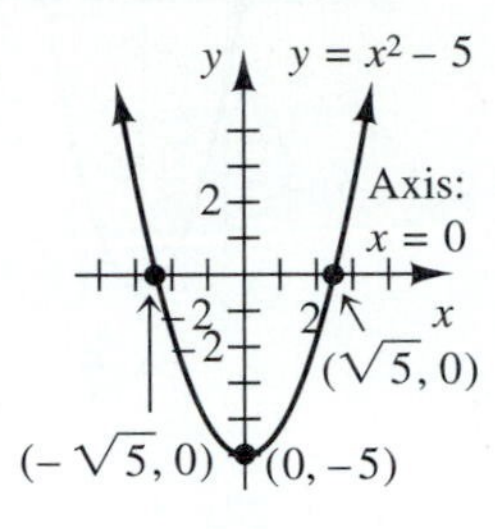

27.

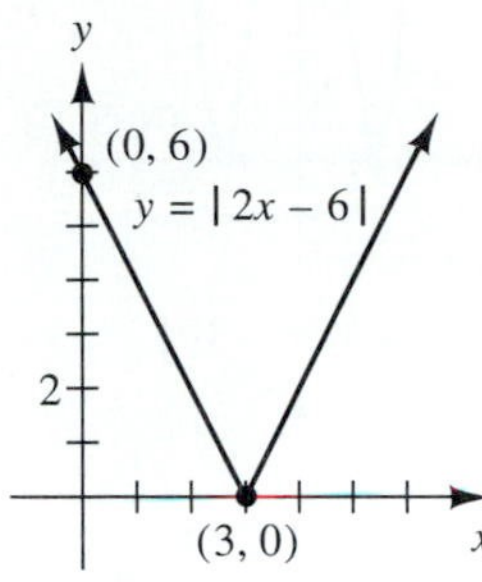

29.

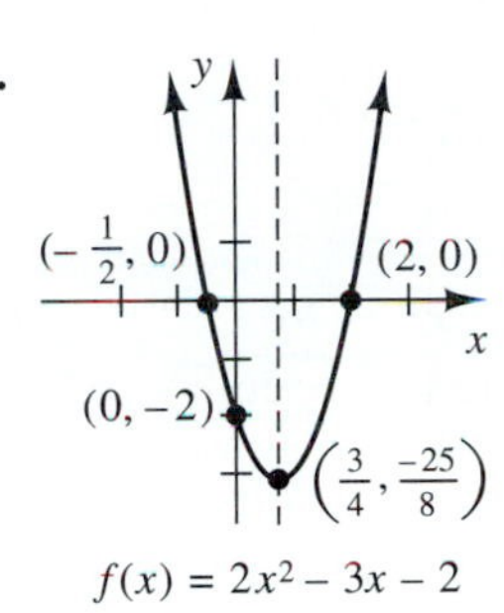

31.

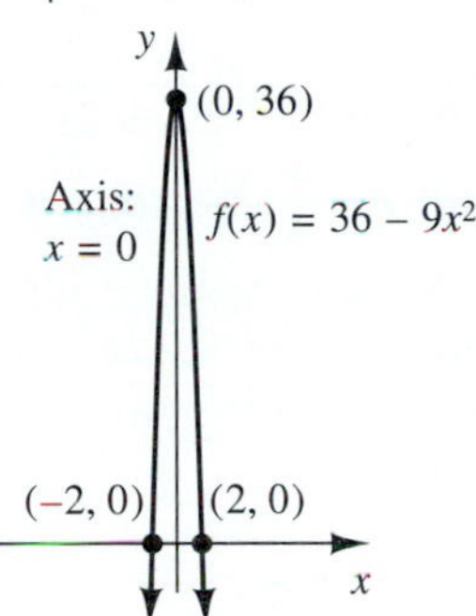

33.

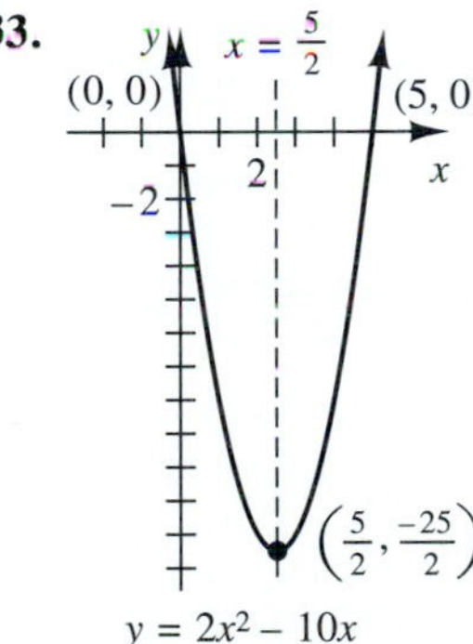

35.

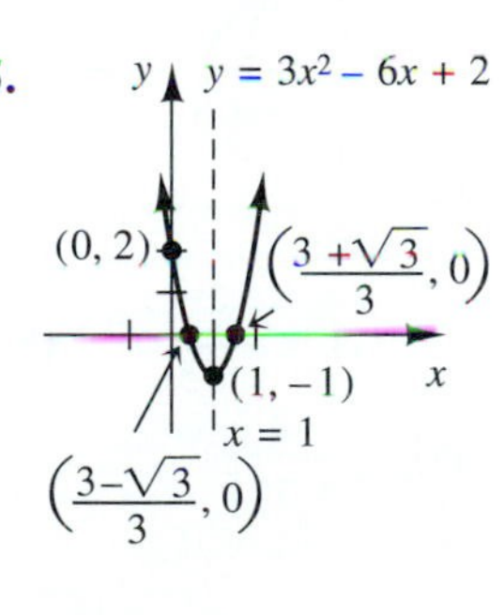

37. -2 **39.** $\frac{133}{3}$ **41.** $-\frac{1}{11}$ **43.** $-\frac{7}{3}$ **45.** 40

47. -3 **49.** $-8a^3 + 60a^2 - 150a + 133$

51. $3x^2 - 2x - 4$; Domain = all real numbers

53. $\frac{2}{3x}, x \neq 0$ **55.** $x^3 - 3x^2 + 4x + 7$; Domain = all real numbers **57.** $\frac{15 - 6x}{x - 4}, x \neq 4$ **59.** $-6x^2 + 8x + 3$; Domain = all real numbers **61.** $-\frac{2}{x^3 + 8}, x \neq -2$

63. $-\frac{3}{7}$; Domain = all real numbers **65.** $4x - 5$; Domain = all real numbers

67. $\frac{3x}{-2 - 4x}, x \neq -\frac{1}{2}, 0$ **69.** $\frac{8 - 2x}{5x - 26}, x \neq \frac{26}{5}, 4$

71. $f(x) = \frac{1}{\sqrt{x}}$, $g(x) = x + 2$ **73.** $f(x) = x^{-3}$; $g(x) = 5x - 7$ **75. (a)** $N(h) = 16h^2 + 580h + 2350$ **(b)** 4926 **(c)** Value is out of the domain of $N(h)$.

77. $A = 8x$

79. Since $S = 180 = 4x^2 + 6xh$, $h = \frac{90 - 2x^2}{3x}$.

Thus $V = 2x^2\left(\frac{90 - 2x^2}{3x}\right) = \frac{2x}{3}(90 - 2x^2)$. **81.** No

83. $D_f = R_f$ = all real numbers; $D_{f^{-1}} = R_{f^{-1}}$ = all real numbers

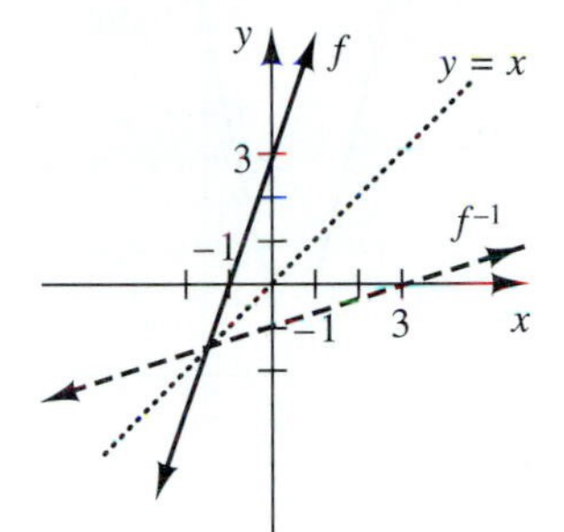

85. $D_f = R_f$ = all real numbers; $D_{f^{-1}} = R_{f^{-1}}$ = all real numbers

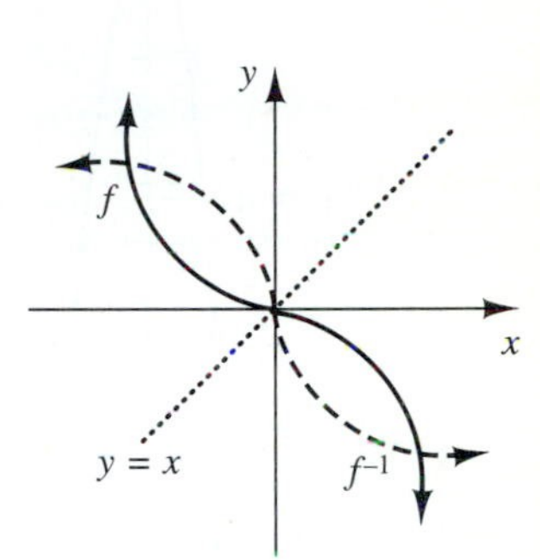

87. $f^{-1}(x) = 4x - 20$ **89.** $f^{-1}(x) = \frac{3x + 4}{x - 1}$

91. $f^{-1}(x) = \frac{3 - 6x}{x}$

CHAPTER 3 PRACTICE TEST

1. (a)

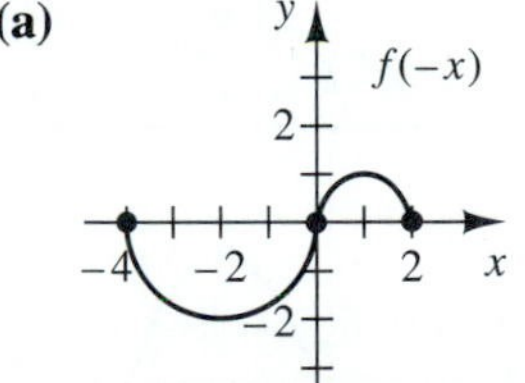

(b)

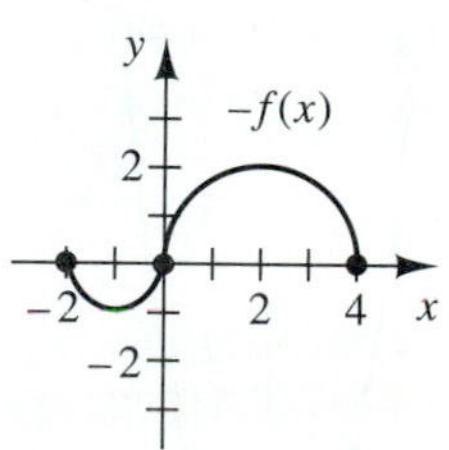

(c) $|f(x)|$

(d)

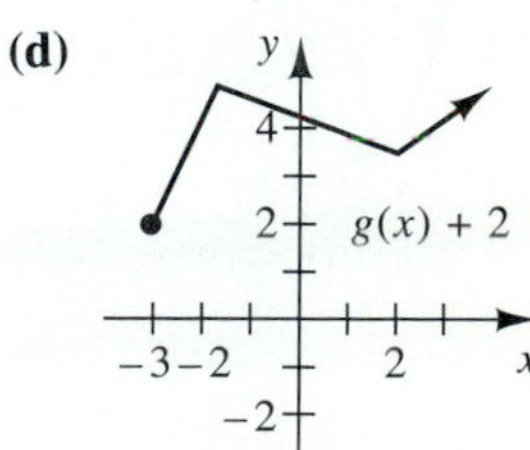

(e)

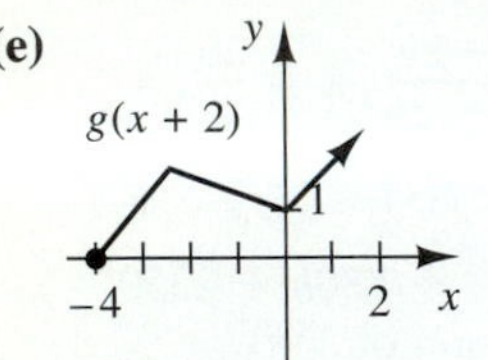

(f)

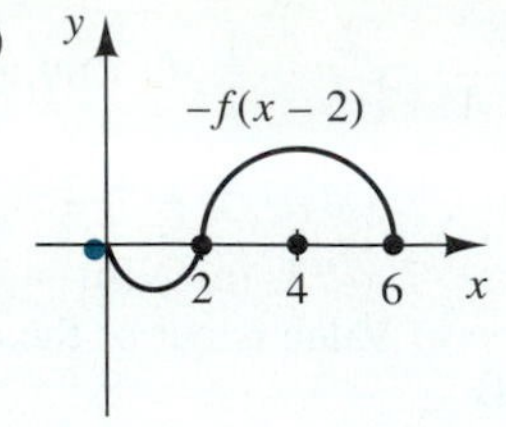

(g) **(h)**

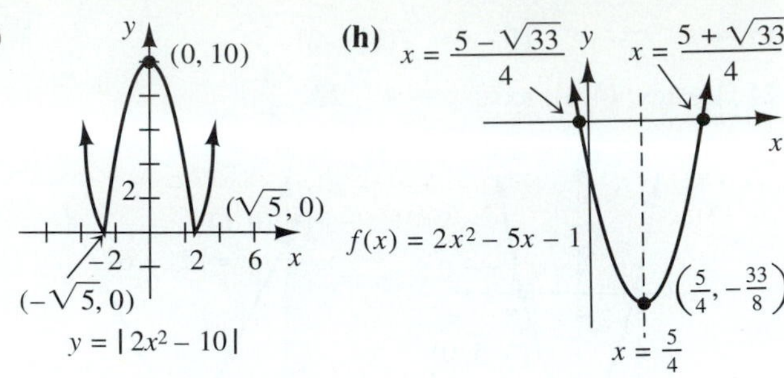

2. **(a)** Vertex: $(-5, -3)$; axis: $x = -5$

(b) Vertex: $\left(-\frac{5}{4}, \frac{33}{8}\right)$; axis: $x = -\frac{5}{4}$

3. **(a)**

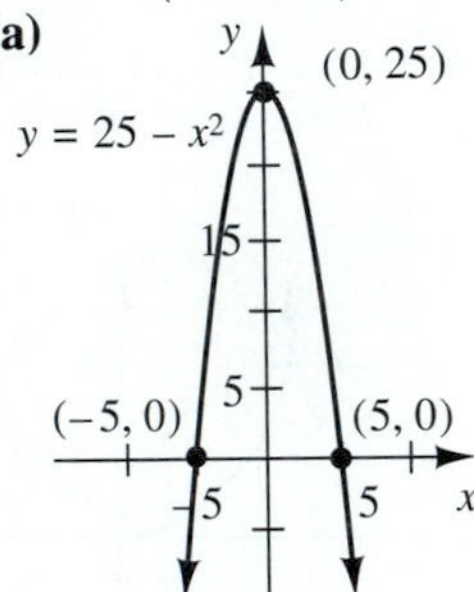

(b) $f(x) = |4x + 8|$

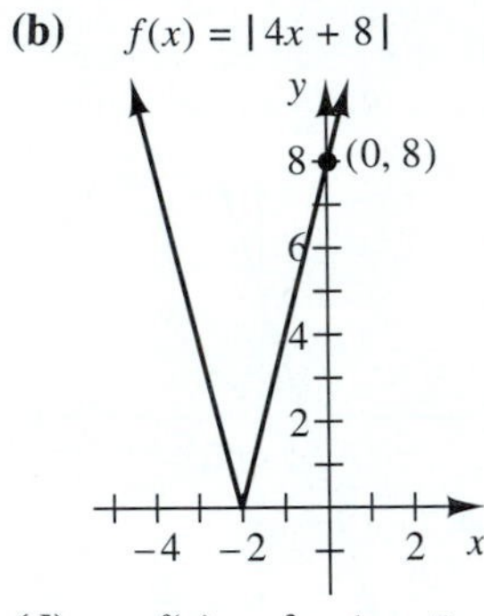

(c) $y = -2(x + 3)^2 + 8$

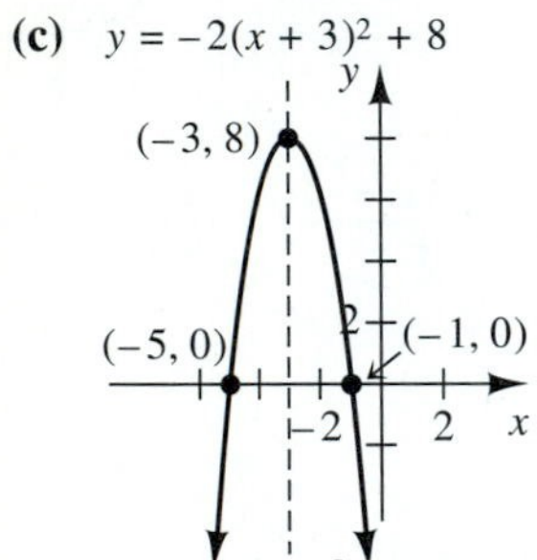

y-intercept $(0, -10)$

(d) $f(x) = x^2 + 4x - 5$

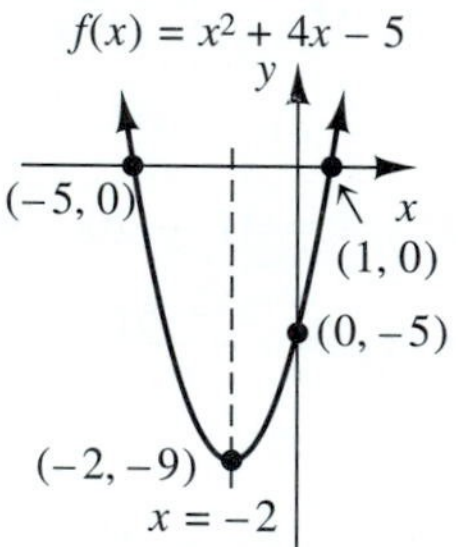

(e) $y = 3x^2 - 48x$

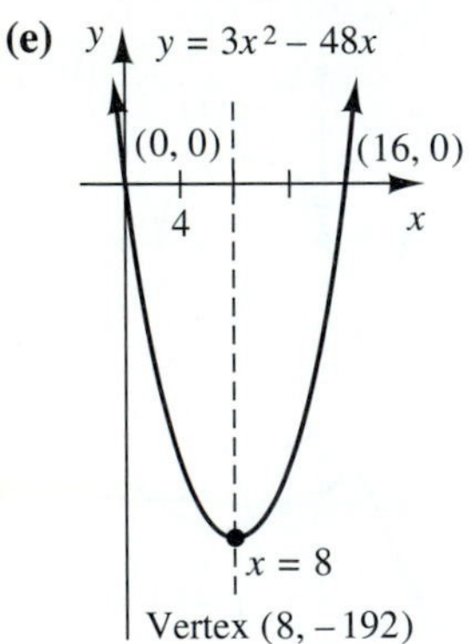

Vertex $(8, -192)$

(f)

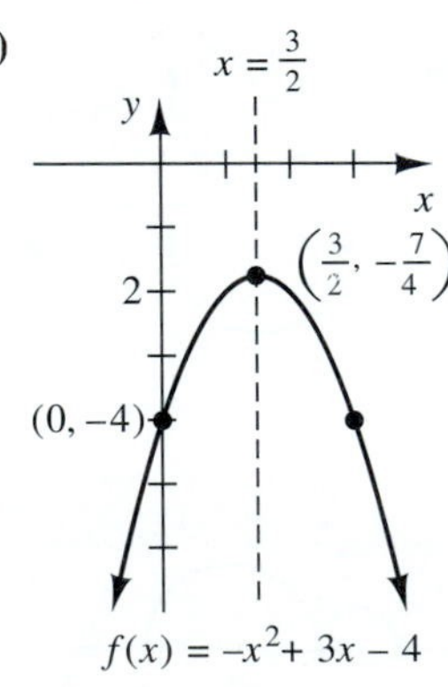

$f(x) = -x^2 + 3x - 4$

4. **(a)** $-2\sqrt{2}$ **(b)** $\frac{13}{4}$ **(c)** $\frac{2}{3}$ **(d)** $\frac{7x^2 + 28x + 19}{(x + 2)^2}$

(e) $\frac{3}{9 - x^2}$ **(f)** $\frac{3x + 6}{7 + 2x}$ **(g)** $\frac{3}{10 - x}$, $x \geq 1$

5. **(a)** $x \neq \frac{1}{2}, 1$ **(b)** $x \neq \pm 1$

6. Since $x^2 + 12^2 = (2r)^2$, $x = 2\sqrt{r^2 - 36}$. Thus $A = \pi r^2 - 12x = \pi r^2 - 12(2\sqrt{r^2 - 36}) = \pi r^2 - 24\sqrt{r^2 - 36}$.

7.

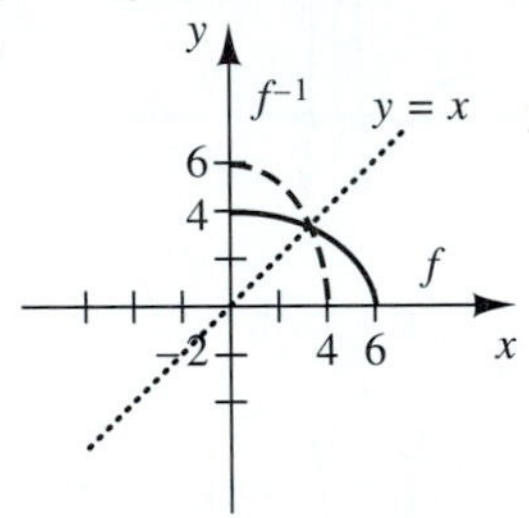

8. **(a)** $f^{-1}(x) = \frac{3x + 4}{5}$ **(b)** $f^{-1}(x) = \frac{x^3 - 9}{2}$

(c) $f^{-1}(x) = \frac{2}{x + 5}$

CHAPTER 4

EXERCISES 4.1

1. Degree is at least 5 **3.** Not a polynomial; has a break **5.** Degree is 1 **7.** Degree is at least 4

9. $y = x^3 + 1$, 1, −1, x, y

11.

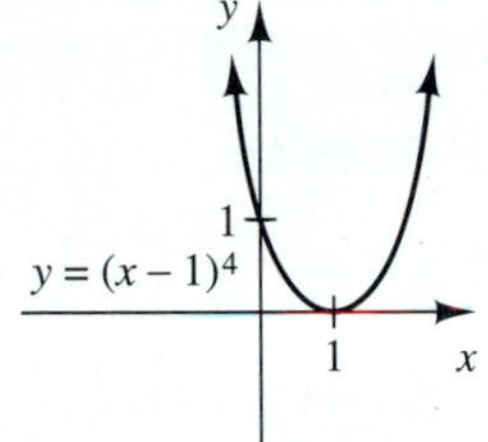

13.

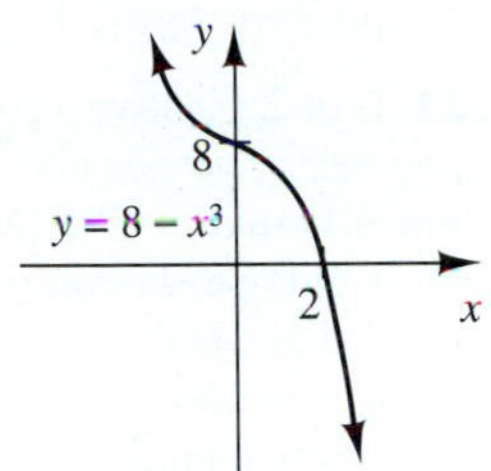

15.

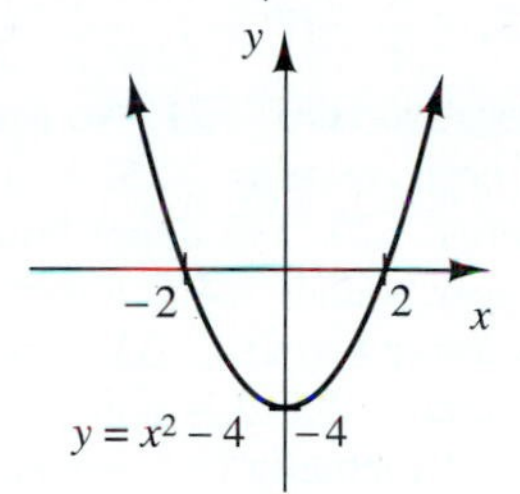

17.

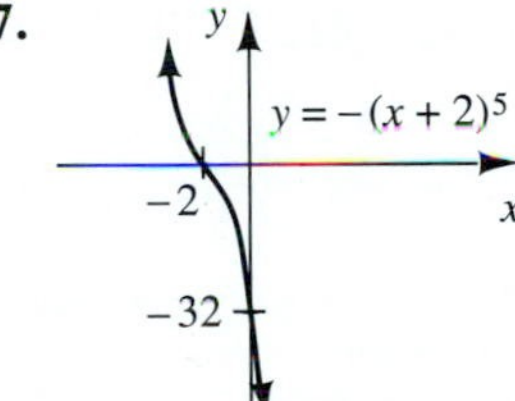

19.

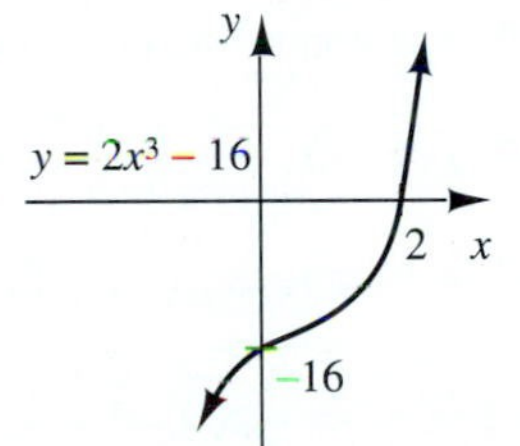

21.

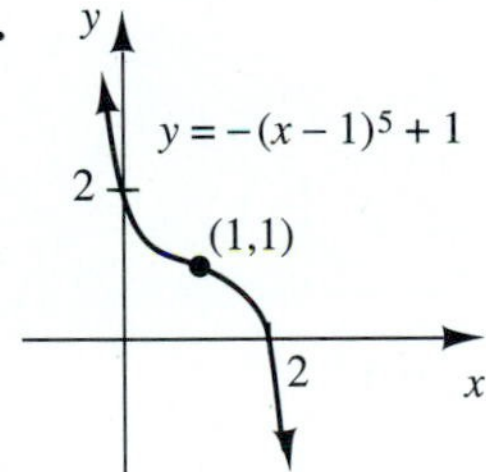

23.

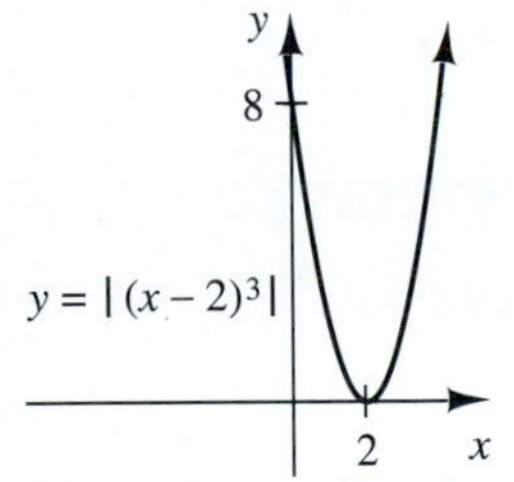

25. $y = |1 - x^4|$

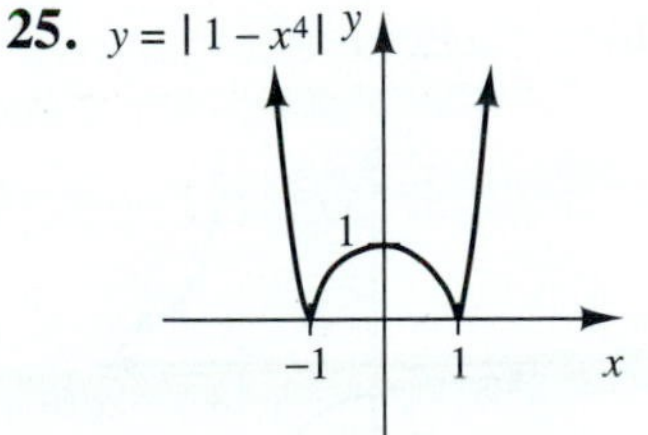

27.

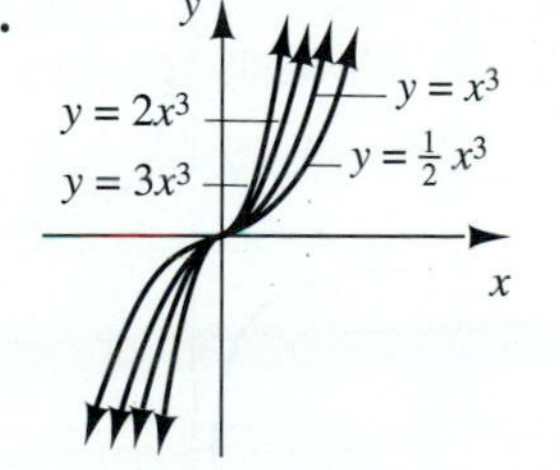

29.

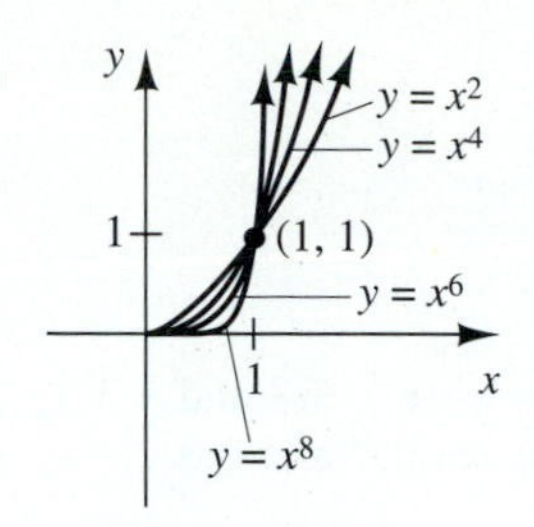

31. $y = x^3 - x^2 - 2x$

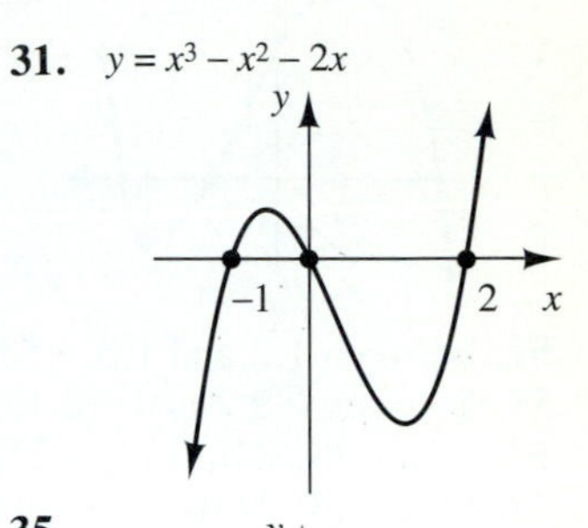

33. $y = x^2 - 6x + 8$

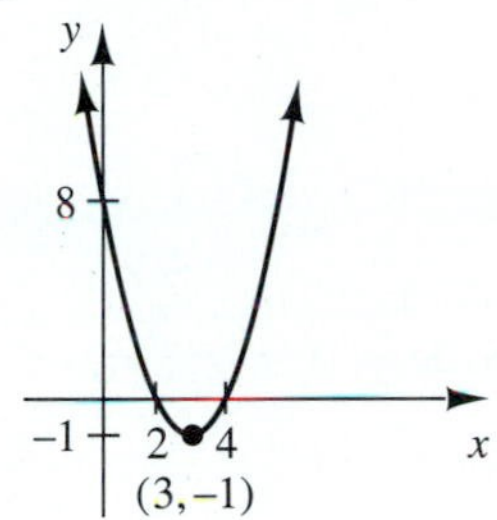

35.

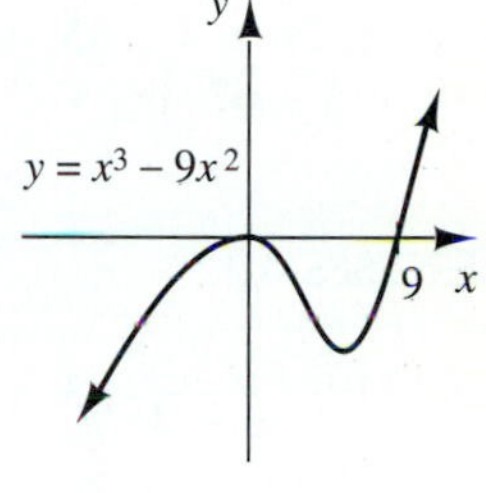

37.

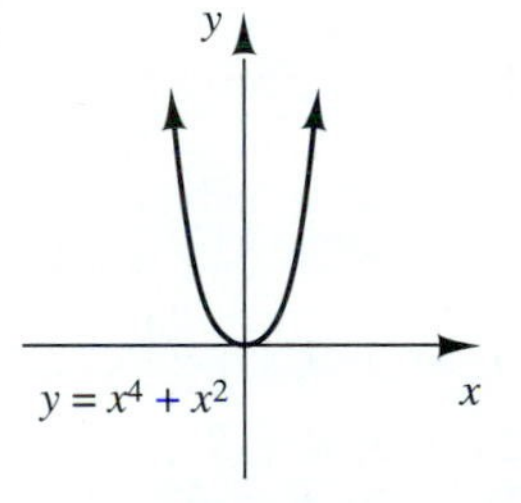

39. $y = x(x-1)(x+1)(x+2)$

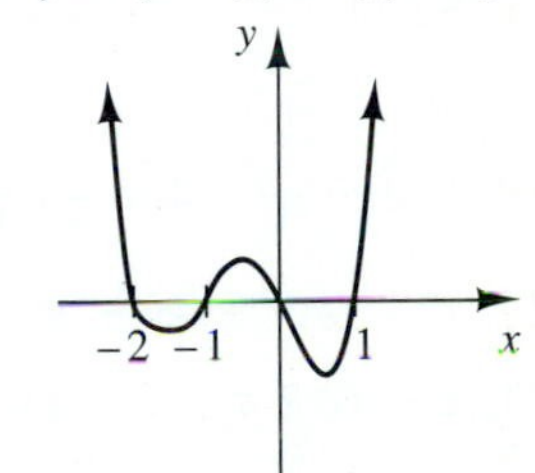

41. $y = (x+1)(x-2)(x+3)(x-4)$

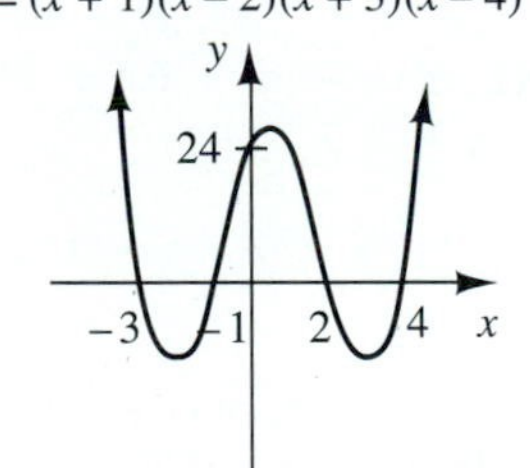

43.

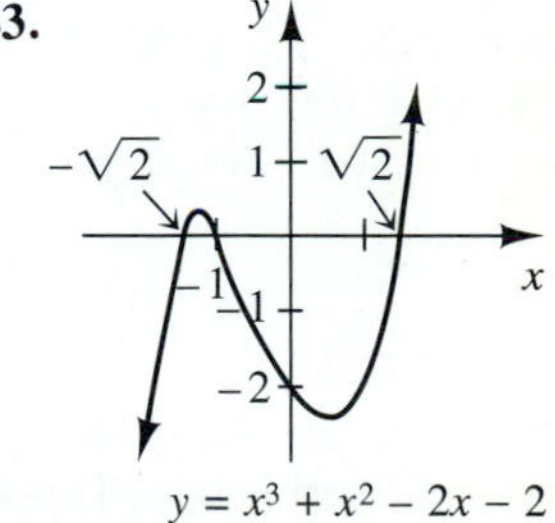

45.

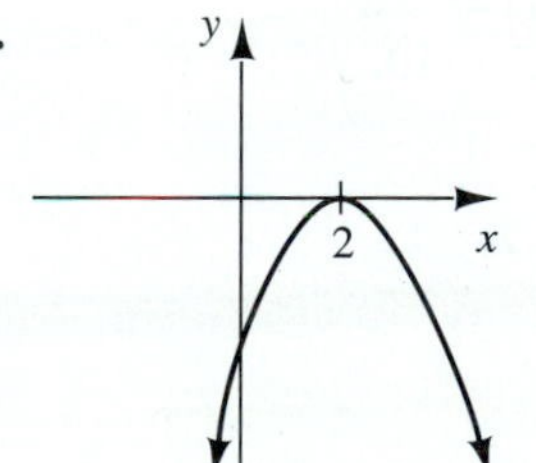

47.

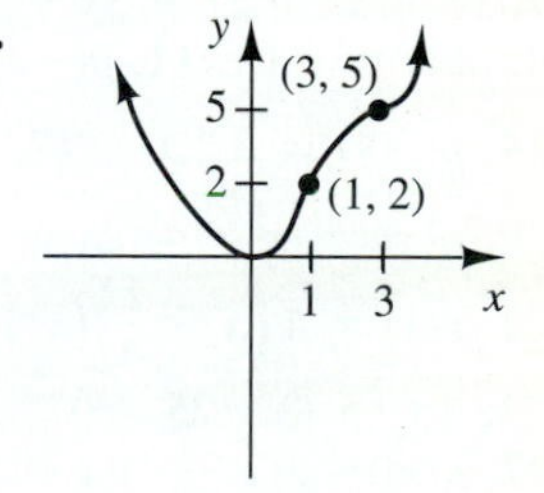

49.

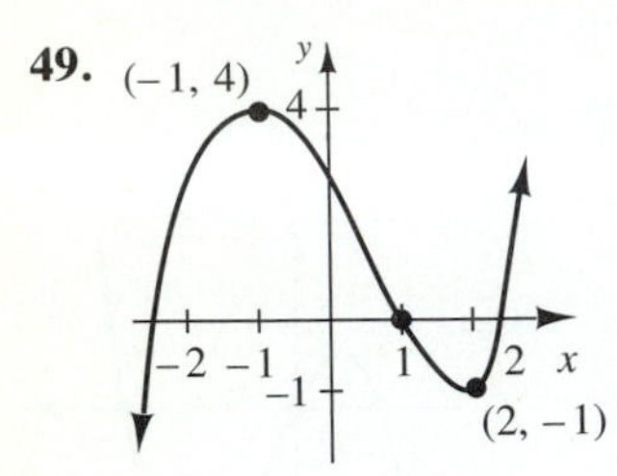

51. Between 1.2 and 1.3 **53.** Between −3.8 and −3.7
55. Between −2.2 and −2.1; between 0.7 and 0.8
57. (0, 0), (3, 27), (−3, −27)
59. (0, 0), (3, 243), (−1, −1)
61. $x^2 + 2x + 4$ **63.** $4x^3 + 6x^2h + 4xh^2 + h^3$
65. $n = 8$ **67.** $|x| < 0.182$

EXERCISES 4.2

1. $x - 1 - \dfrac{18}{x-4}$ **3.** $x - 2$ **5.** $2t - 5 - \dfrac{5}{3t-4}$
7. $1 + \dfrac{1}{y^2+1}$ **9.** $2x^2 - 3x + 1$
11. $3x^3 + 2x^2 + 2x + 2 + \dfrac{8}{x-1}$
13. $4c^2 - 2c + 1$
15. $2w^2 - w - 4 + \dfrac{8}{w^2+w+1}$
17. $x^3 + 2x^2 + 4x + 8$ **19.** $x^2 + 5x + 9 + \dfrac{24}{x-2}$
21. $x - 6$; $R = -5$ **23.** $2x^2 - 3x + 3$; $R = 9$
25. $2x^3 + 3x^2 - 3x - 13$ **27.** $5x - 20$; $R = 57$
29. $2x^2 - 2x + 6$; $R = -12$ **31.** $x^3 + 4x^2 + 16x + 64$; $R = 192$ **33.** $6x^2 - 8$ **35.** $2x^2 + 4$ **37.** $x^2 - ax + a^2$
39. $x^3 + ax^2 + a^2x + a^3$ **41.** $q(x) = x^2 + 6x + 34$; $R(x) = 165$ **43.** $k = 6$ **45.** $k = -3$

EXERCISES 4.3

1. $x - 9$; $p(-4) = 44$ **3.** $2x^2 + x + 5$; $p(4) = 24$
5. $-x^3 - x^2 - x - 1$; $p(1) = 0$
7. $x^4 - x^3 + x^2 - 3x + 3$; $p(-1) = -6$
9. $p(4) = -1$ **11.** $p(-2) = -27$ **13.** $p(2) = 0$
15. $p(-3) = -220$ **17.** $p\left(\dfrac{1}{2}\right) = 1$
19. $p(x) = (x - 4)(2x + 1)(x - 2)$
21. $r(x) = 4\left(x - \dfrac{1}{4}\right)(x + 3)(x - 3)$
23. 3, −5, 2 **25.** $\pm i$
27. $p(x) = x(x - 4)(x - (-3))$; the roots are 0, 4, −3, all with multiplicity 1.

In Exercises 29–32, a constant multiple $k[p(x)]$ $(k \neq 0)$ will also be a solution.

29. $p(x) = (x - 1)^3(x - 2)^2(x - 3)$; deg. = 6
31. $p(x) = x^2(x + 2)^2(x - 1)$; deg. = 5
33. $2 + 3i, -\dfrac{3}{2}$ **35.** $3 - 4i, \dfrac{1}{2} \pm i$

EXERCISES 4.4

1. −2, 1, 5 **3.** $-3, \dfrac{1}{2}, 5$ **5.** 1, 2 **7.** $-\dfrac{1}{2}, 1$
9. −2, −1, 2 **11.** $\dfrac{1}{3}, \dfrac{1}{2}, -1$ **13.** No rational roots
15. $-\dfrac{3}{4}$ **17.** $-1, -\dfrac{1}{3}, \dfrac{1}{3}, 2$ **19.** No positive zeros; 1 negative zero **21.** No real zeros **23.** 0 or 2 positive zeros; 1 negative zero **25.** 1, 3, or 5 positive zeros; no negative zeros **27.** 5 is upper bound; −1 is lower bound. **29.** 6 is upper bound; −4 is lower bound. **31.** 1 is upper bound; −2 is lower bound. **33.** 2 is upper bound; −2 is lower bound. **35.** 2 is upper bound; −6 is lower bound.
37. 1 (actually 0) is upper bound; −4 is lower bound.
39. 5 is upper bound; −3 is lower bound **41.** $x = 1, \dfrac{1}{3}$
43. $x = -3, \dfrac{-1 \pm \sqrt{5}}{2}, \dfrac{1}{2}$ **45.** $x = \dfrac{3}{2}, \dfrac{5}{3}, \pm i$
47. $x = \pm 3, \pm\dfrac{\sqrt{3}}{3}$ **49.** Positive for: $x < -4$, $-2 < x < 1$, $x > 3$; Negative for: $-4 < x < -2$, $1 < x < 3$ **51.** Positive for: $-\sqrt{5} < x < 1$, $x > \sqrt{5}$; negative for: $x < -\sqrt{5}$, $1 < x < \sqrt{5}$

EXERCISES 4.5

1. $D_f = \{x \mid x \neq \pm 5\}$; no zeros
3. $D_h = \left\{x \mid x \neq -4, \dfrac{5}{2}\right\}$; zeros: $x = 0, 4$
5. $D_g = \{x \mid x \neq 0\}$; no real zeros
7. **(a)** **(b)**

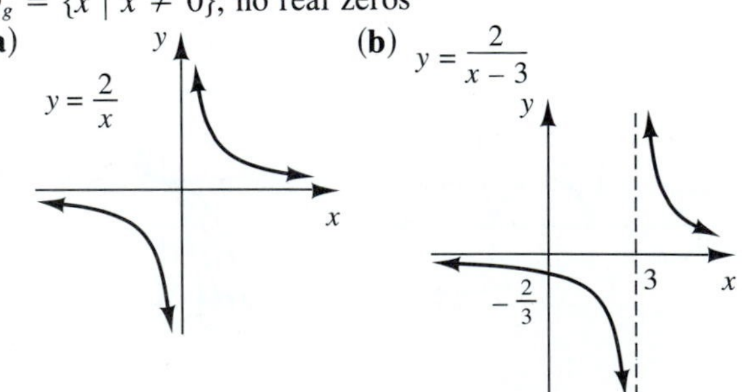

(c) $y = \frac{2}{x} - 3$

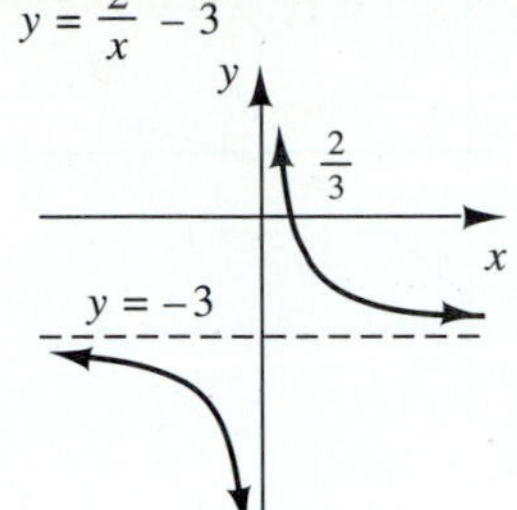

9. (a) $y = -\frac{1}{x^4}$

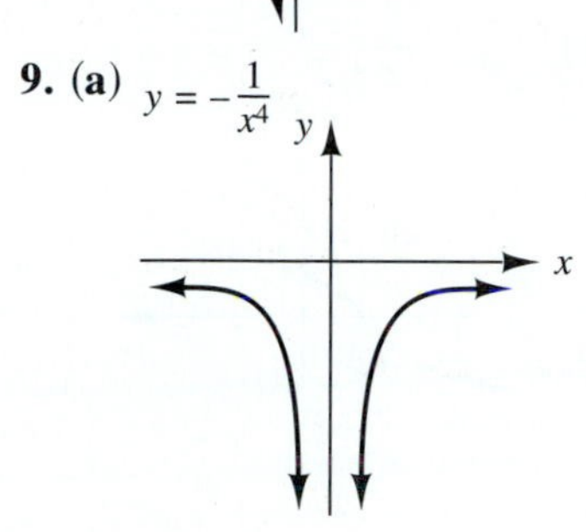

(b) $y = \frac{1}{(x-1)^4}$

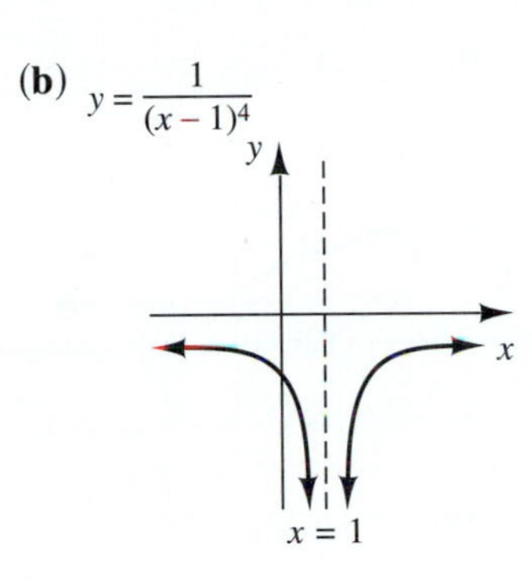

(c) $y = -\frac{1}{x^4} - 1$

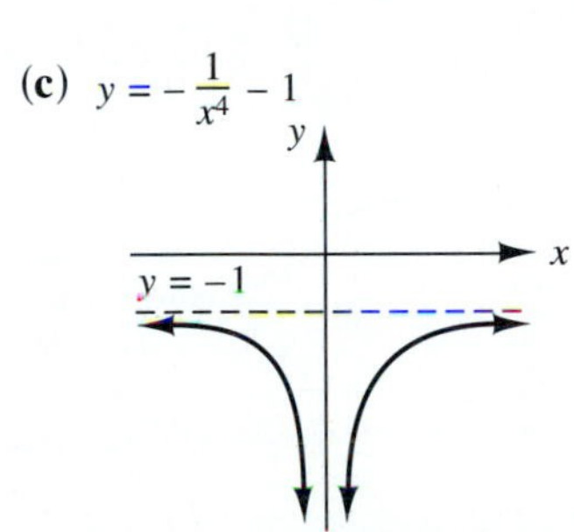

11.

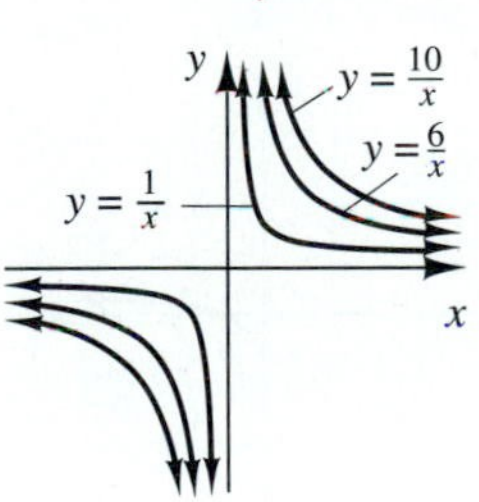

13. $y = \frac{4}{x^6}$

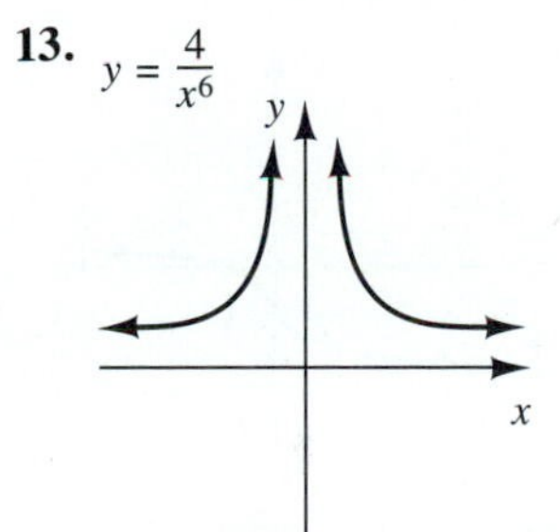

15. $y = \frac{1}{x} - 5$

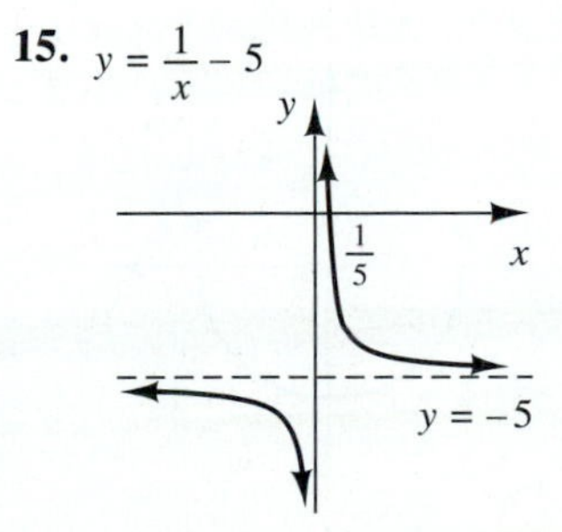

17. $y = \frac{1}{(x-2)^2}$

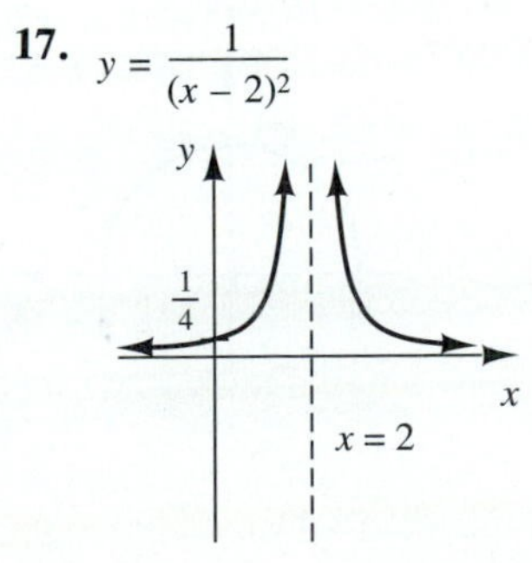

19. $y = \frac{1}{x^3} - 8$

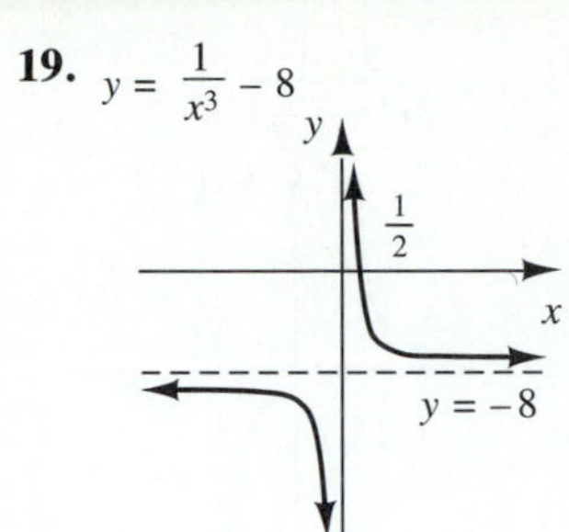

21. $y = 9 - \frac{1}{x^2}$

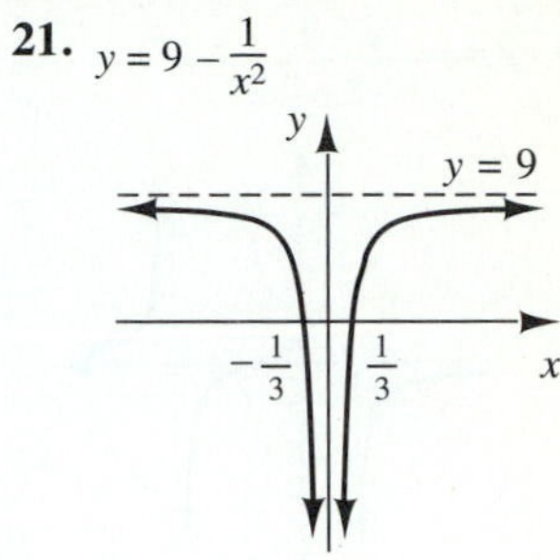

23. $y = \frac{x}{x+5}$

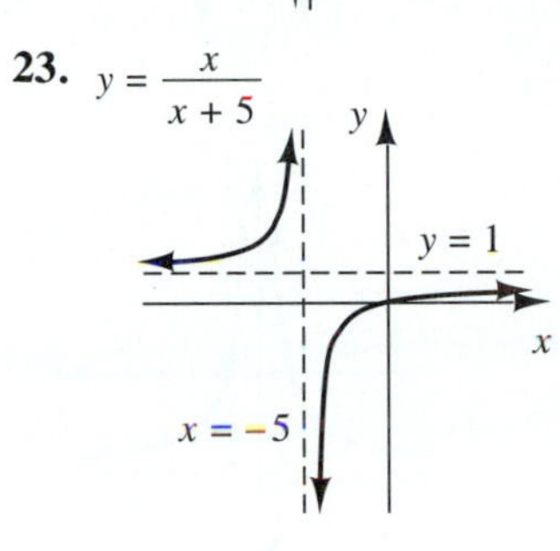

25. $y = \frac{3x-5}{x}$

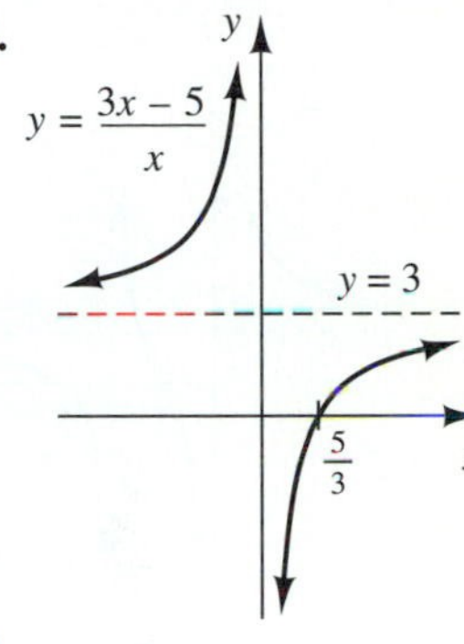

27. $y = \frac{x+2}{x-2}$

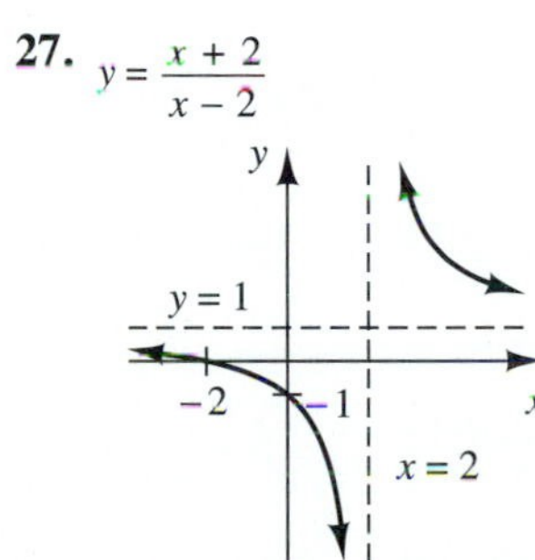

29. $y = \frac{5-x}{x+4}$

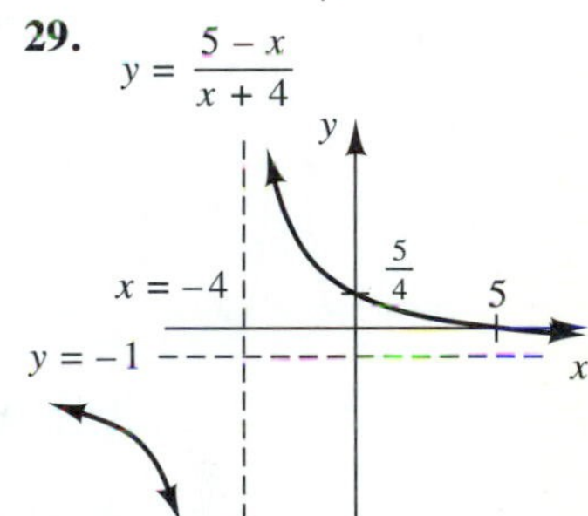

31. $y = \frac{2x}{x^2-1}$

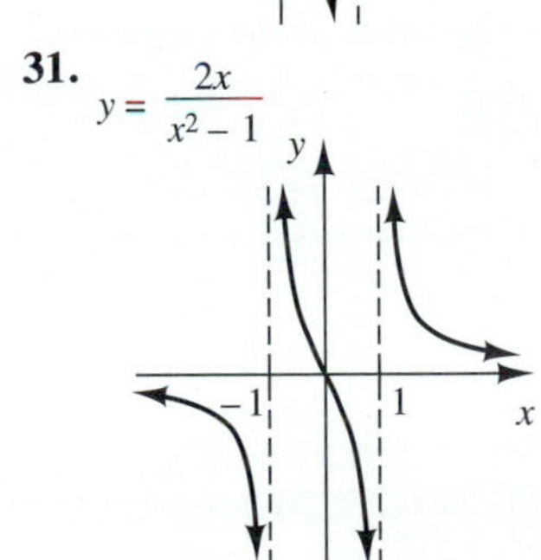

33. $y = \frac{x^2}{x^2-4}$

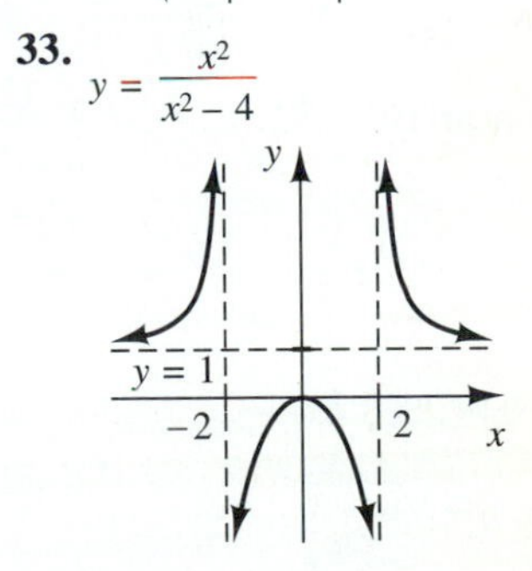

35. $y = \dfrac{x-1}{x^2-2x-3}$

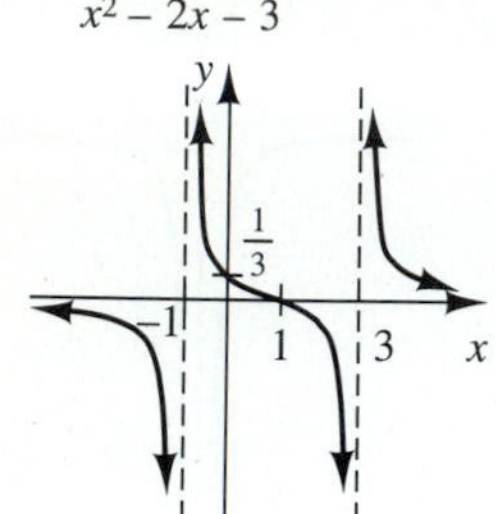

37. $y = \dfrac{2-x}{2x^2-x-3}$

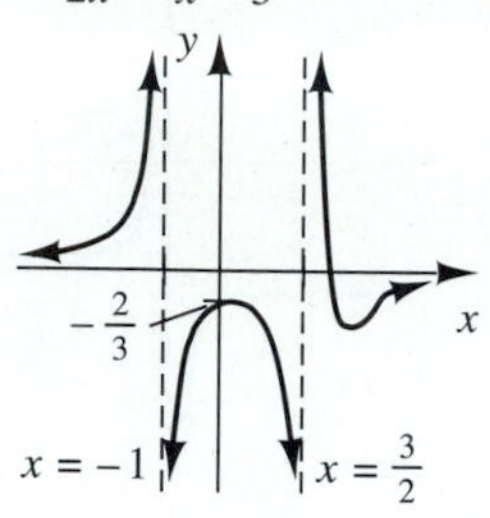

39. $y = \dfrac{x^2-4x+3}{x^2-2x}$

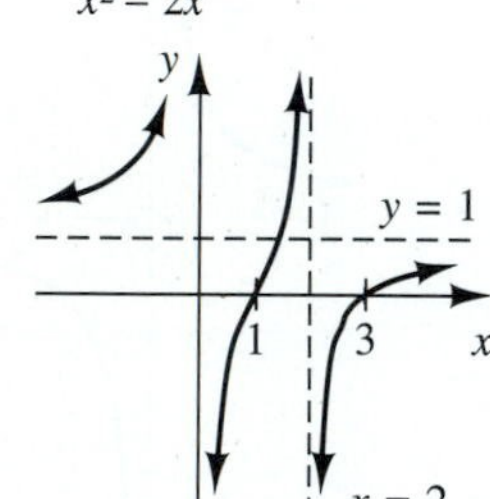

41. $y = \left|\dfrac{1}{x}\right| - 2$

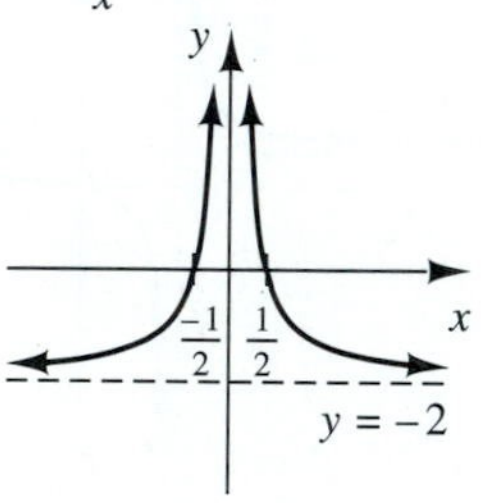

43. $y = \left|\dfrac{1}{x^2} - 4\right|$

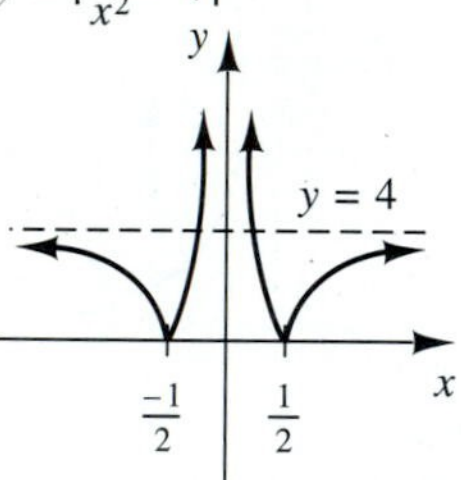

45. $-\dfrac{1}{5x}$ **47.** $\dfrac{-2x-h}{x^2(x+h)^2}$ **49.** The value of the expression is near 19. **51.** $L = 3x + \dfrac{120{,}000}{x}$

EXERCISES 4.6

1. $y = \sqrt{x+3}$

3. $y = \sqrt{x} - 4$

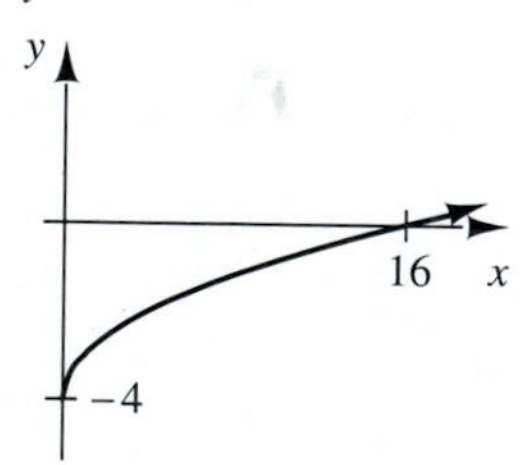

5. $y = \sqrt{2x-6}$

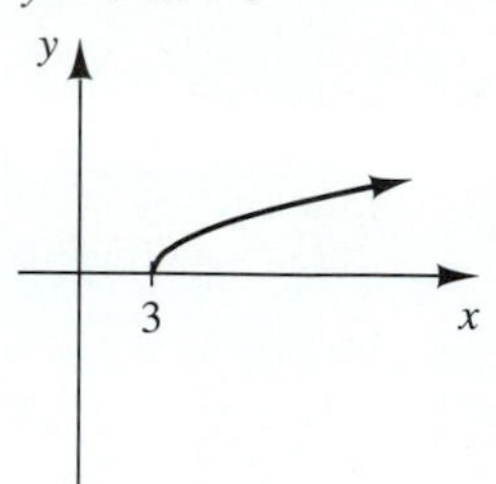

7. $y = 2\sqrt{x} - 6$

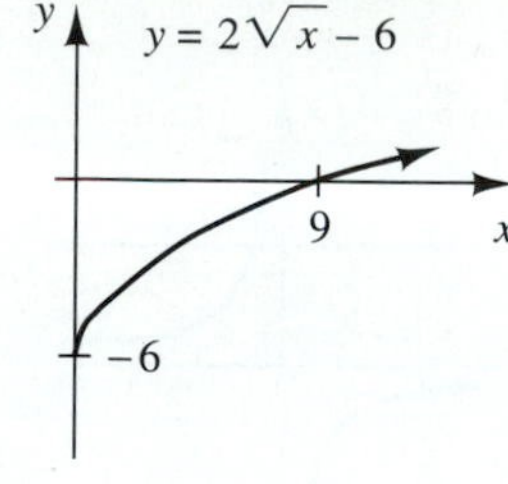

9. $y = \sqrt{-x}$

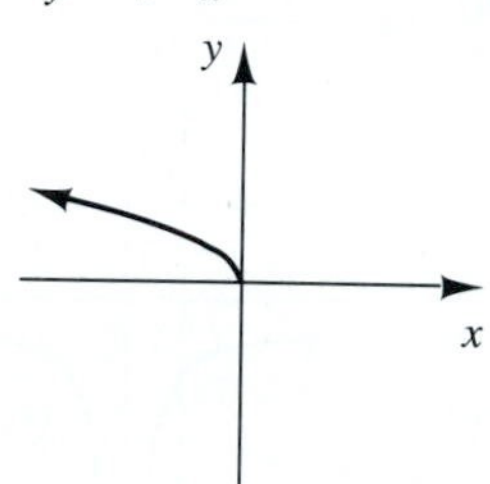

11. $y = \sqrt[3]{x+2}$

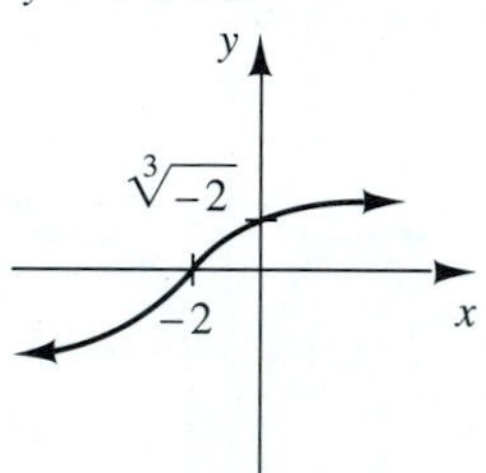

13. $y = \sqrt[3]{x} - 3$

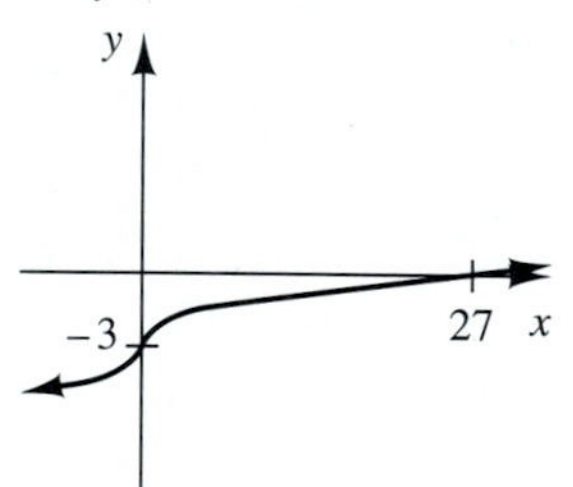

15. $y = |\sqrt[3]{x} - 3|$

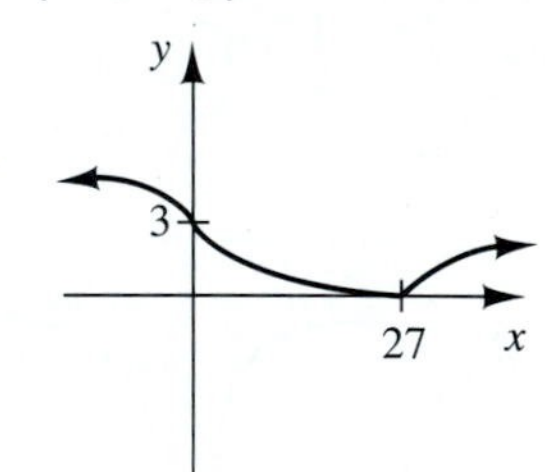

17. $y = \sqrt[5]{x-1}$

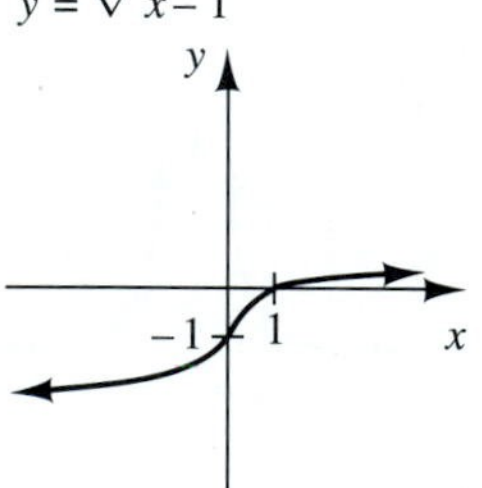

19. $y = \sqrt[6]{x} - 1$

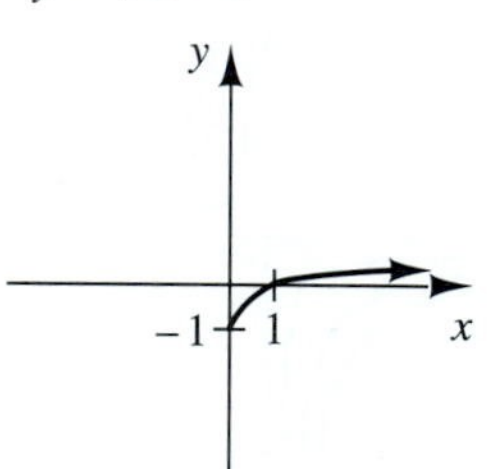

21. $y = \sqrt{16-x^2}$

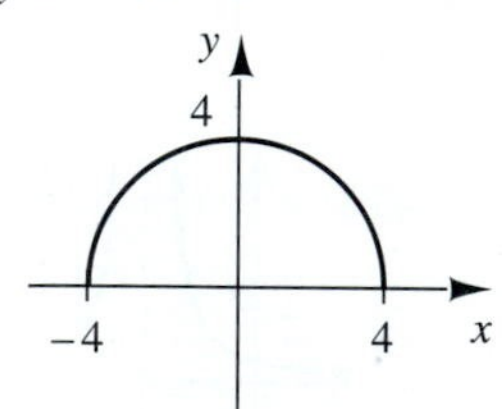

23. $y = -\sqrt{16-x^2}$

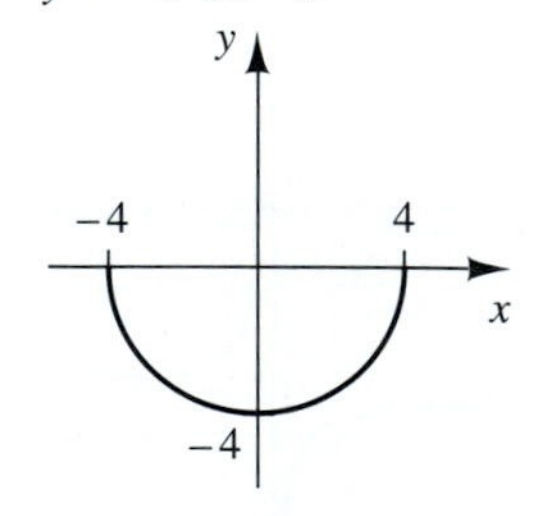

29. (4, 2) **31.** (1, −1) **33.** (0, 3)

35. $\dfrac{\sqrt{x}-2}{x-4} = \dfrac{1}{\sqrt{x}+2}$, which is near $\dfrac{1}{4}$ when x is near 4

37. (a) $\sqrt{(x-2)^2+(y-3)^2} = \sqrt{(x-5)^2+(y-1)^2}$
(b) Both equations become $6x - 4y = 13$.

39. (a) $L = \sqrt{x^2+9} + 8 - x$
(b) $C = 2D\sqrt{x^2+9} + D(8-x)$ **(c)** Approximately 14.7D dollars

41. (b) $V = \dfrac{8}{3}\pi(s^2 - 64)$ **43.** $A = h\sqrt{1296 - h^2}$

EXERCISES 4.7

1. $y = \dfrac{375}{8}$ **3.** $u = \dfrac{64}{25}$ **5.** $z = -\dfrac{80}{21}$ **7.** $z = \dfrac{243}{32}$

9. y is multiplied by a factor of $\dfrac{1}{4}$. **11.** y is multiplied by a factor of 3. **13.** s is multiplied by a factor of 6. **15.** 15 pounds **17.** 9.1 inches **19.** 12,500 lumens **21.** 133.33 lb/sq in. **23.** 32 ohms **25.** 55 mi/hr **27. (a)** Force is quadrupled. **(b)** Force is doubled. **(c)** Force is multiplied by a factor of 8. **29.** 266.67 pounds **31. (a)** The force varies jointly with the two masses and inversely with the square of the distance. **(b)** 6.67×10^{-11} **(c)** 8.63×10^{9} newtons **33.** S varies jointly as r and h. **35.** V varies directly as the cube of r. **37.** z varies jointly as the square of x and the cube of y and inversely as w.

CHAPTER 4 REVIEW EXERCISES

1. $y = x^3 + 8$

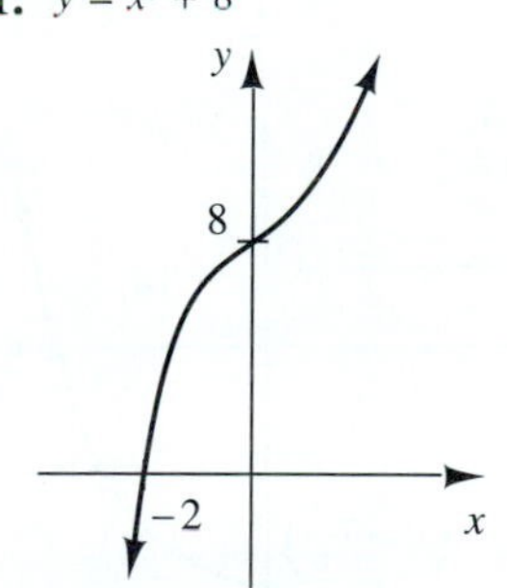

3. $y = (x-1)^4 - 16$

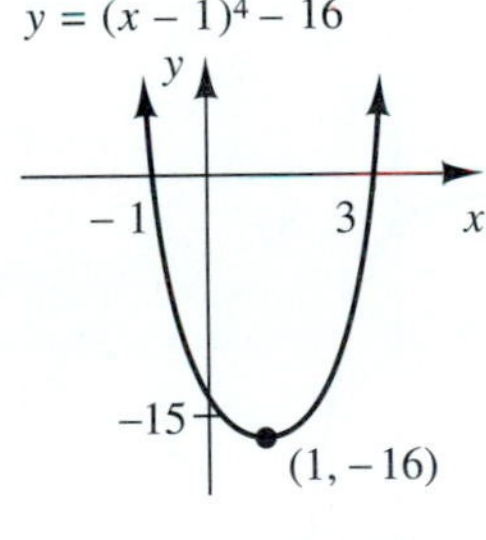

5. $y = x^3 - x^2 - 6x$

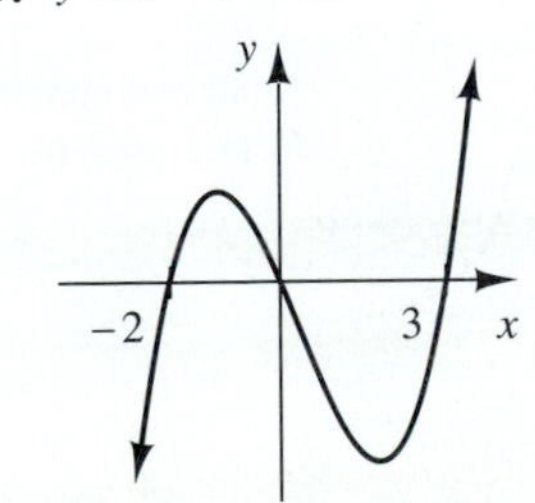

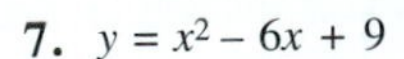

7. $y = x^2 - 6x + 9$

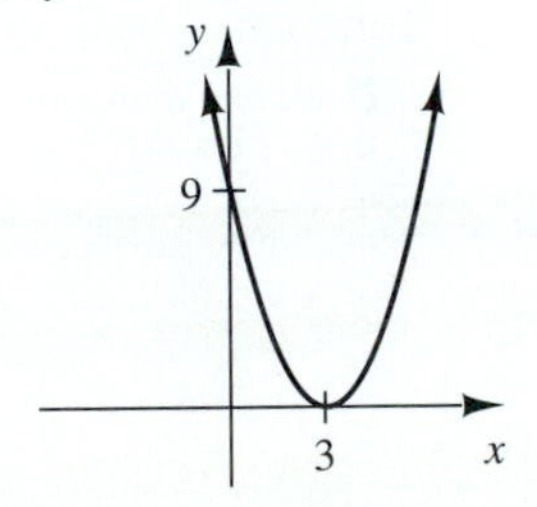

9. $y = \dfrac{1}{(x-1)^2}$

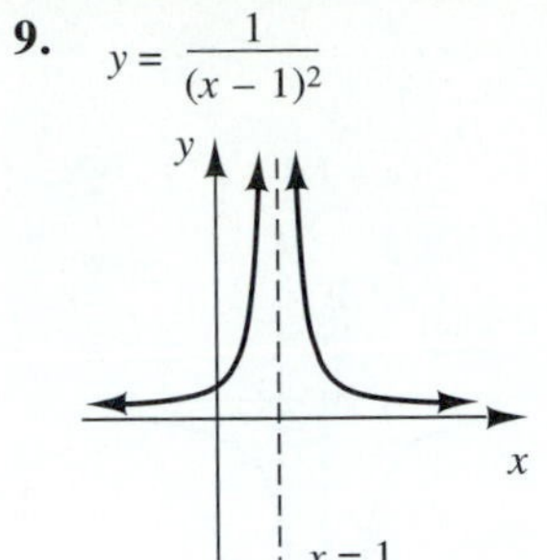

11. $y = \dfrac{1}{x+2} - 3$

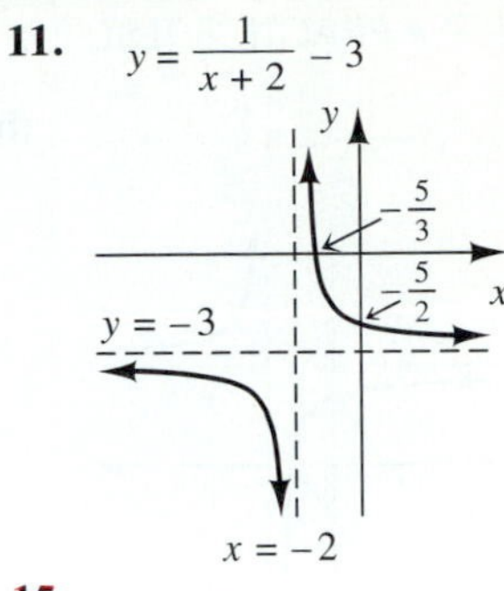

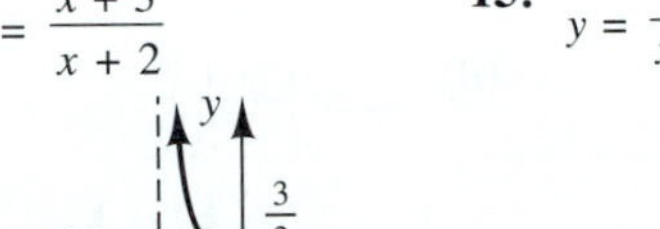

13. $y = \dfrac{x+3}{x+2}$

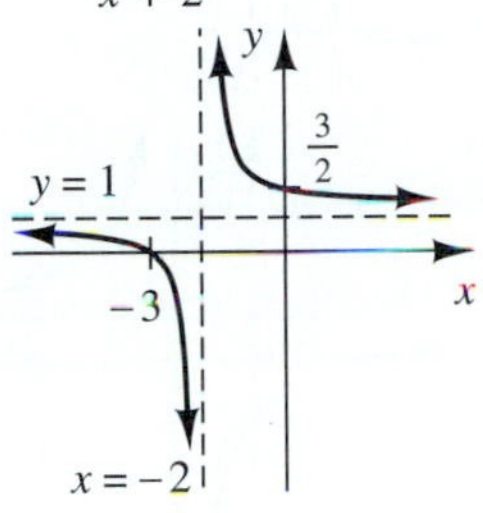

15. $y = \dfrac{x}{x^2-1}$

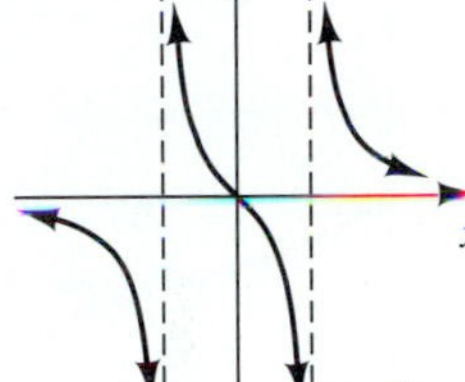

17. $y = \sqrt{2x-5}$

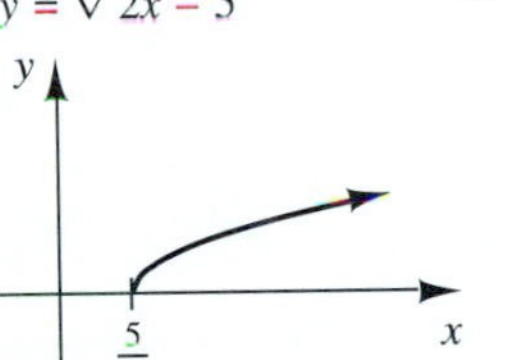

19. $y = \sqrt[5]{x} - 2$

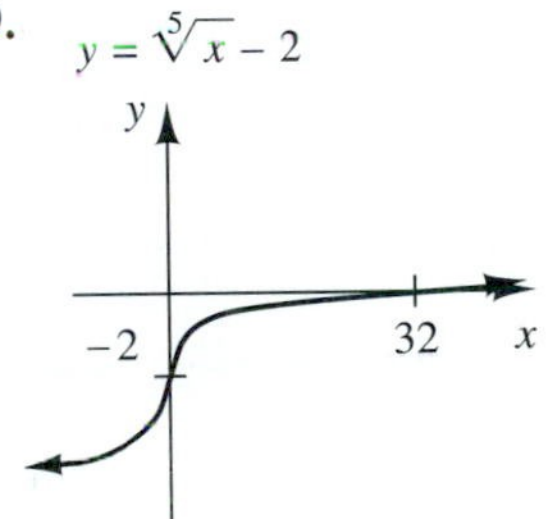

21. $y = |x^3 - 8|$

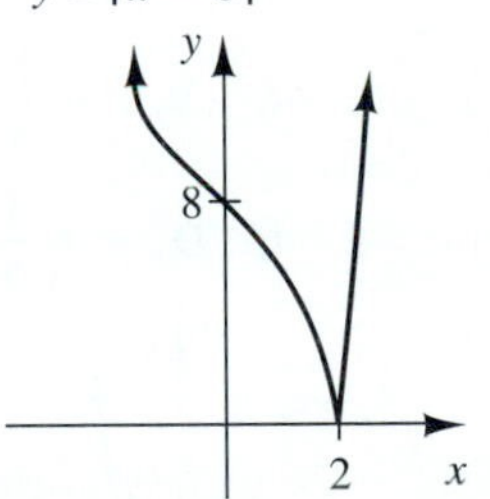

23. $x^2 + 2$; $R = -1$ **25.** $x^3 + 1$ **27.** $2x^2 + 3x + 5$ **29.** $x^4 + x^2 - x + 3$; $R = 2$ **31.** −3, −1, 4 **33.** 5, $\dfrac{1 \pm i\sqrt{7}}{2}$ **35.** $-4, \dfrac{1}{2}, \pm i$ **37.** 1 more; $5 + 2i$; $p(x) = x^3 - 8x^2 + 9x + 58$ **39.** 0 **41.** 3 or 1 **43.** Upper bound: 4; lower bound: −1 **45.** Upper bound: 1; lower bound: −1 **47.** $A = \dfrac{1}{2}x\sqrt{25 - x^2}$ **49.** (0, 2) **51.** (10, 3) **53.** $x = \dfrac{81}{2}$ **55.** Approximately 5.1 seconds

CHAPTER 4 PRACTICE TEST

1. (a)

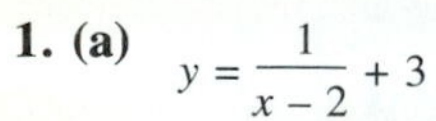

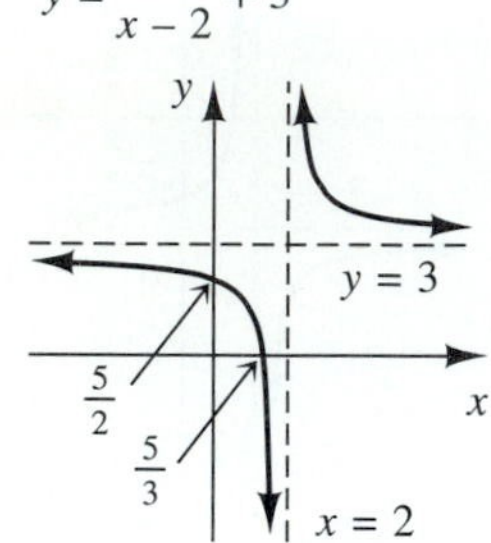

(b) $y = \frac{-3}{(x+1)^2}$

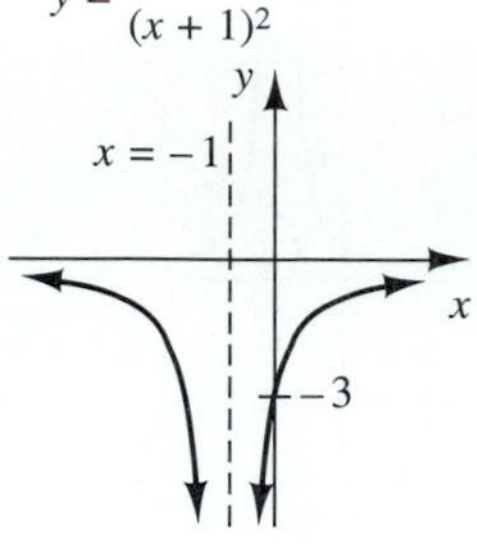

(c) $y = 2 - \sqrt{x}$

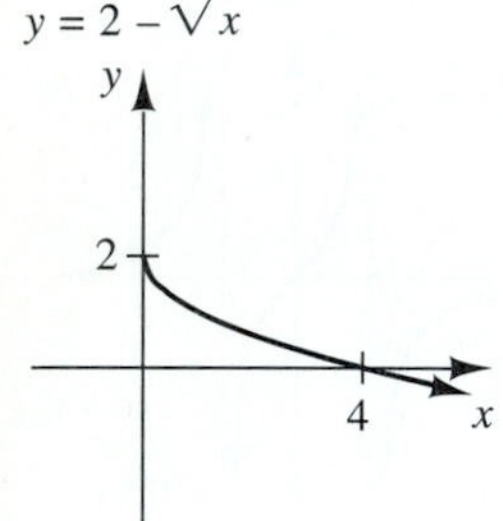

(d) $y = \frac{x+3}{x-2}$

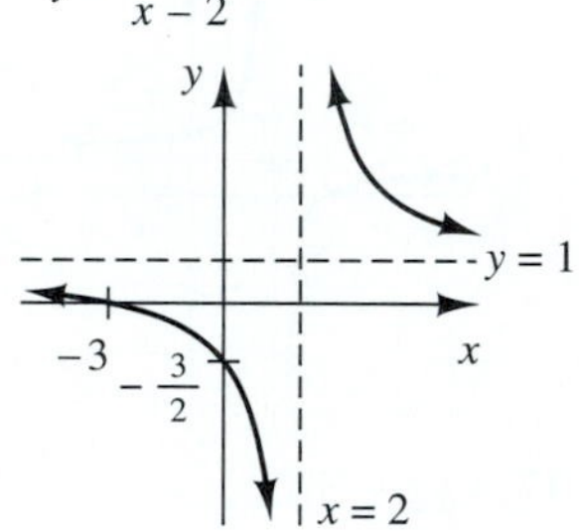

(e) $y = x^3 - x^2 - 12x$

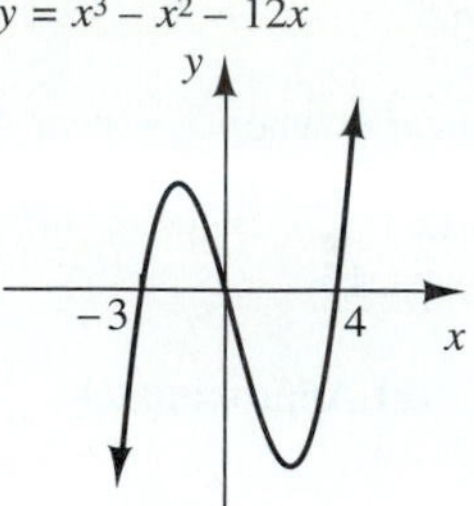

(f) $y = \frac{2x}{9 - x^2}$

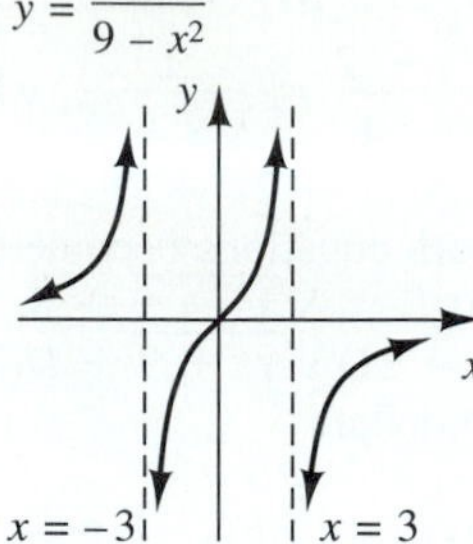

(g) $y = \sqrt[3]{x-2} + 1$

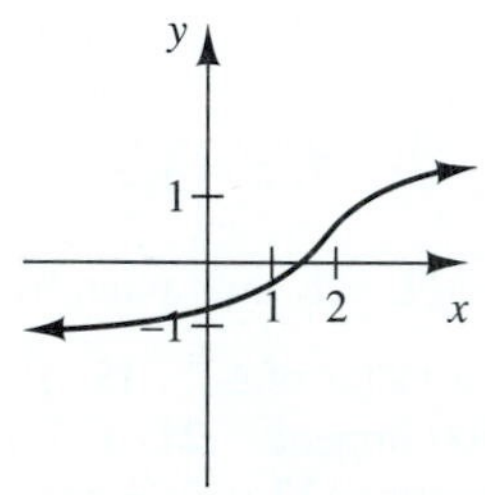

2. $x^3 - 2x^2 - 5x - 1 - \frac{1}{3x - 4}$

3. $x - 2$ and $x + 2$

4. (a) $x = 3, 4, -\frac{1}{3}$ **(b)** $x = 0, \frac{3}{2}, \frac{3 \pm i\sqrt{11}}{2}$

5. The blood pressure would increase by a factor of approximately 5.

CHAPTER 5

EXERCISES 5.1

1. $x = 4$ **3.** $x = \frac{5}{2}$ **5.** $t = \frac{3}{2}$ **7.** $x = -6$ **9.** $x = \frac{1}{6}$

11. D_f: all real numbers **13.** D_h: $[3, \infty)$ **15.** D_g: $x \neq \frac{1}{3}$

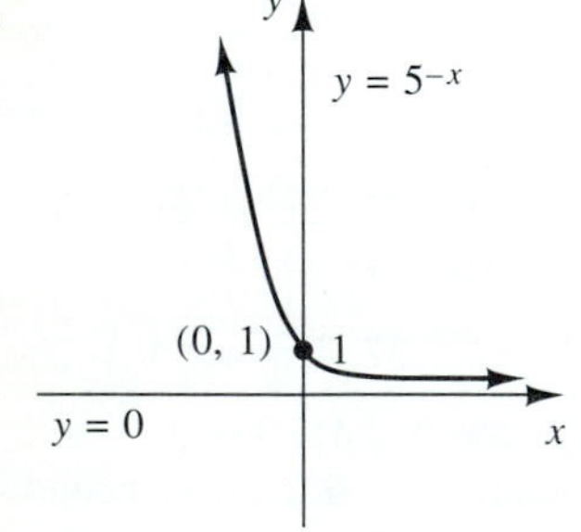

D: all real numbers
R: $\{y \mid y > 0\}$

19.

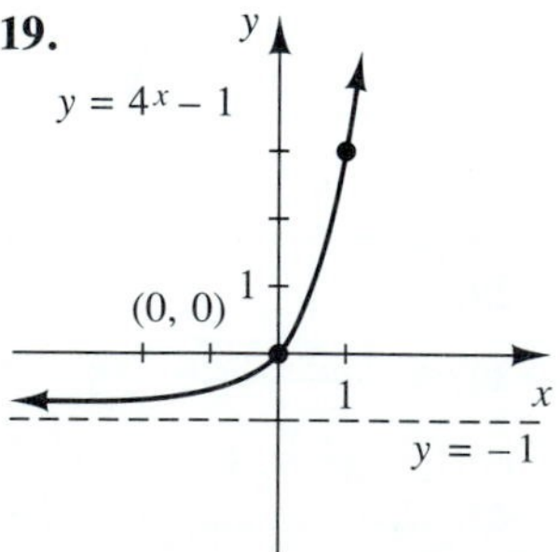

D: all real numbers
R: $\{y \mid y > -1\}$

21.

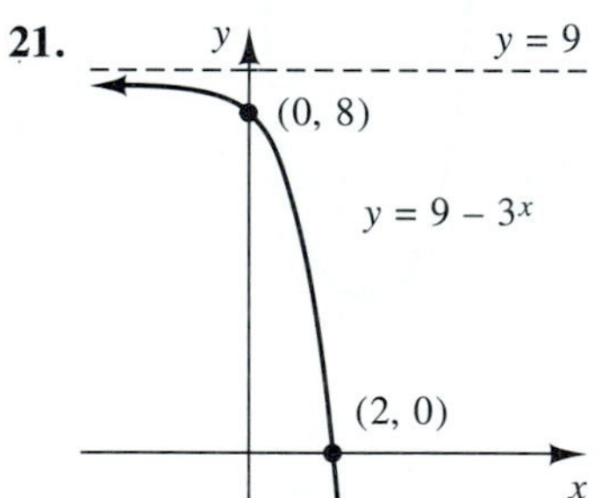

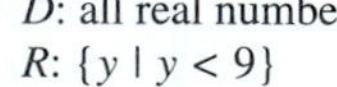

D: all real numbers
R: $\{y \mid y < 9\}$

23.

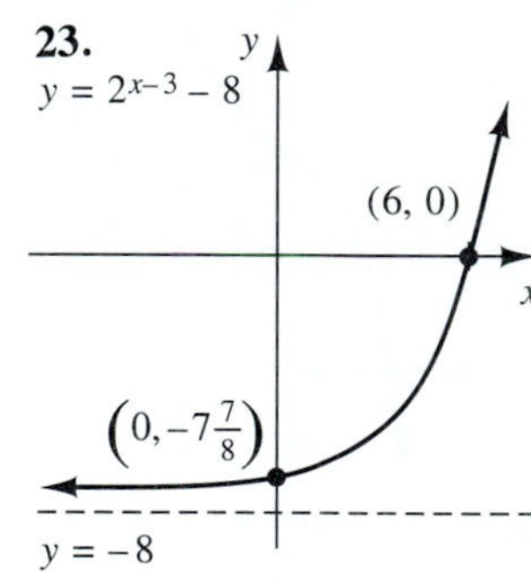

D: all real numbers
R: $\{y \mid y > -8\}$

25.

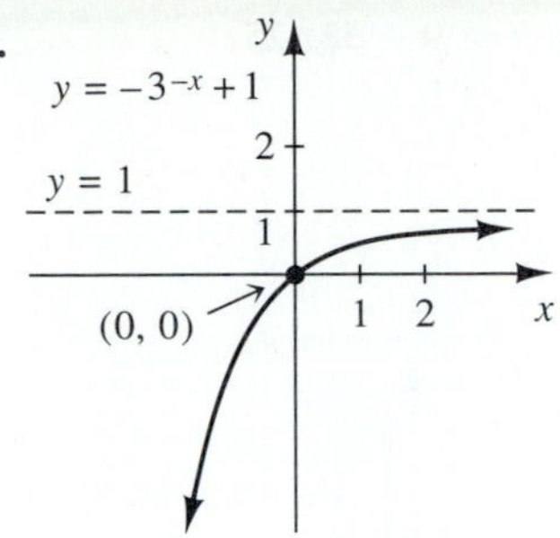

D: all real numbers
R: $\{y \mid y < 1\}$

27.

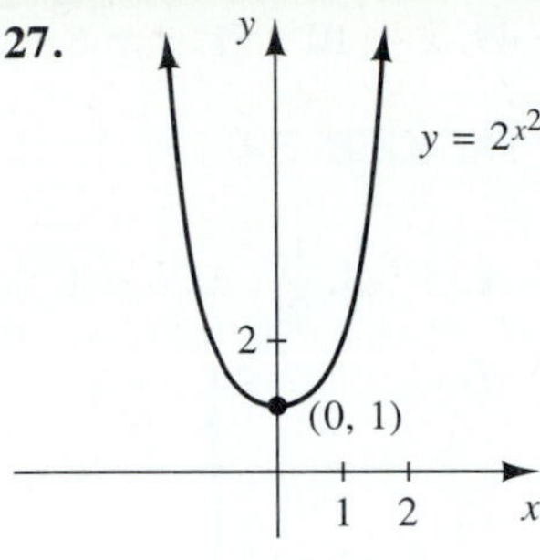

D: all real numbers
R: $\{y \mid y \geq 1\}$

29.

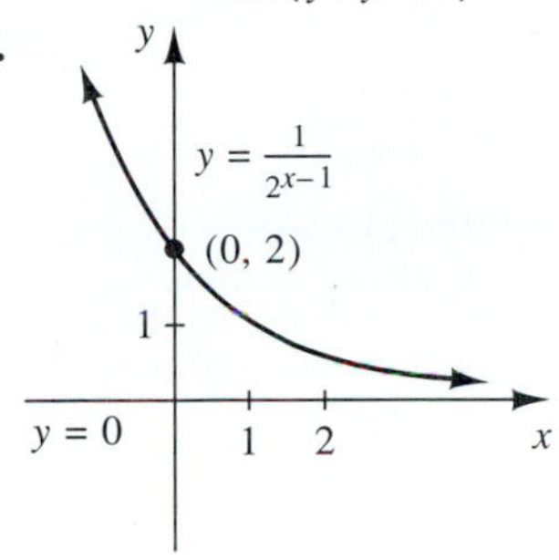

D: all real numbers
R: $\{y \mid y > 0\}$

31. 5^{x^2-6x+2} **33.** $5^{\sqrt{x}}$ for $x \geq 0$ **35.** $\dfrac{1}{5^x+1}$ **37.** $x = 0$

39. $x = \pm 3$ **41.** None **43.** $x = 0$

49.

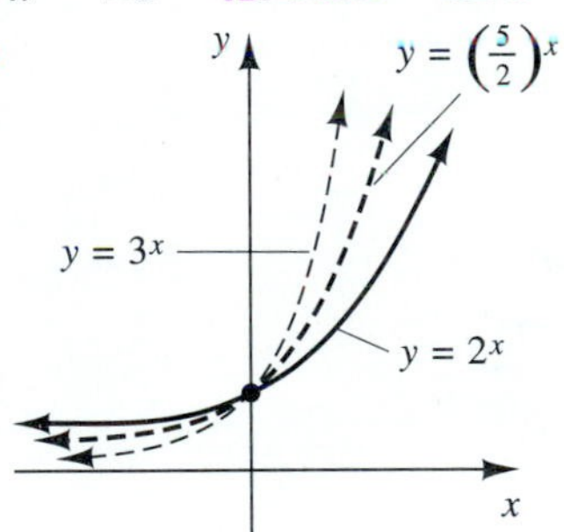

51. 32; $4(2^t)$; 16,384

53. (a)

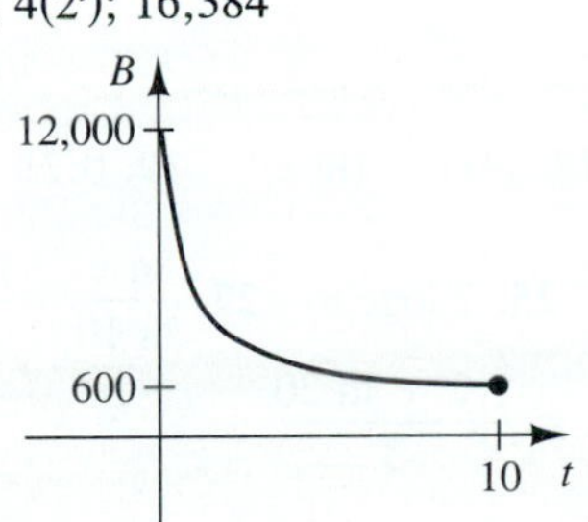

(b) Approx. $4885

55. (a) $593.76 **(b)** $543.88

57. (a) 1 atmosphere; 14.69 lb/sq in. **(b)** 0.33 atmosphere; 4.81 lb/sq in. **(c)** 1.05 atmosphere; 15.44 lb/sq in.

59. (a) 13,784; 14,270 **(b)** 23,182; 24,000 **(c)** 1.04; 1.04 **(d)** The population in year $t + 1$ is $2^{0.05}$ times the population in year t. **61. (b)** Approx. 139 **63. (a)** $A = A_0 2^{-t/4}$; **(b)** 7.07 mg; 0.16 mg **65.** Approx. 1.11 min

67. (a)

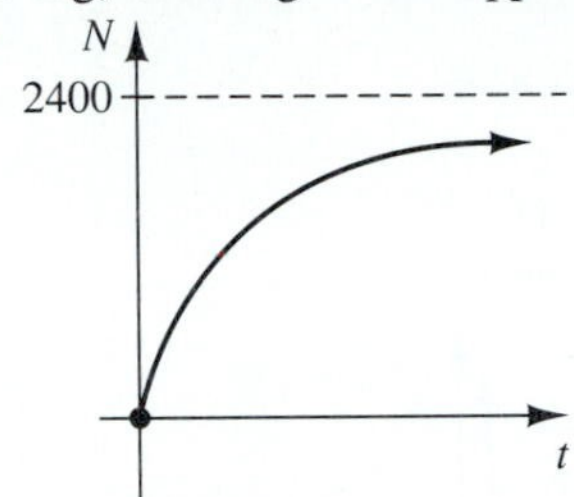

(b) 1322; 1678; 1916 **(c)** 55%, 70%, 80%

69. (a) I is the initial concentration. **(b)** F is the concentration level that is approached as more and more time has elapsed. **71.** 0.22 lumens

EXERCISES 5.2

1. $7^2 = 49$ **3.** $\log_3 \dfrac{1}{81} = -4$ **5.** $\left(\dfrac{1}{4}\right)^{-3} = 64$

7. $\log_{27} \dfrac{1}{3} = -\dfrac{1}{3}$ **9.** $8^{-1/3} = \dfrac{1}{2}$ **11.** $\log_5 \sqrt[4]{5} = \dfrac{1}{4}$

13. $2^{-1} = \dfrac{1}{2}$ **15.** $\left(\dfrac{2}{3}\right)^{-3} = \dfrac{27}{8}$ **17.** $\log_4 1024 = 5$

19. $\log_{1/9} 3 = -\dfrac{1}{2}$ **21.** 5 **23.** −1 **25.** $\dfrac{4}{3}$ **27.** $-\dfrac{3}{4}$

29. Not defined **31.** $-\dfrac{3}{2}$ **33.** −4 **35.** $\dfrac{2}{3}$ **37.** 1

39. −1 **41.** 7 **43.** $-\dfrac{1}{2}$ **45.** 6 **47.** $t = 81$

49. $t = 2$ **51.** $t = \dfrac{4}{5}$ **53.** $(-\infty, -2) \cup (2, \infty)$; $F(6) = 5$

55.

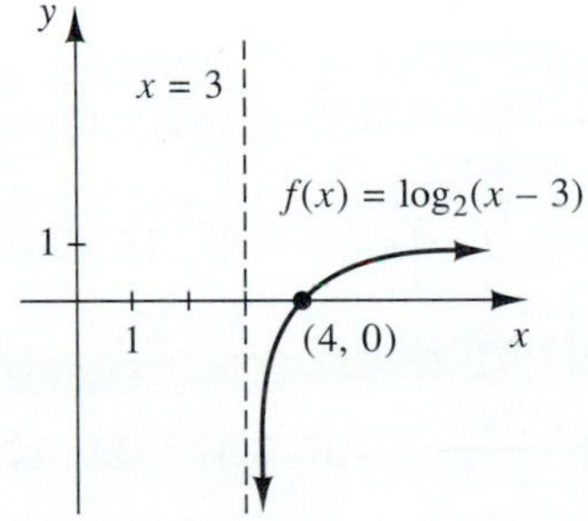

D: $\{x \mid x > 3\}$
R: all real numbers

57.

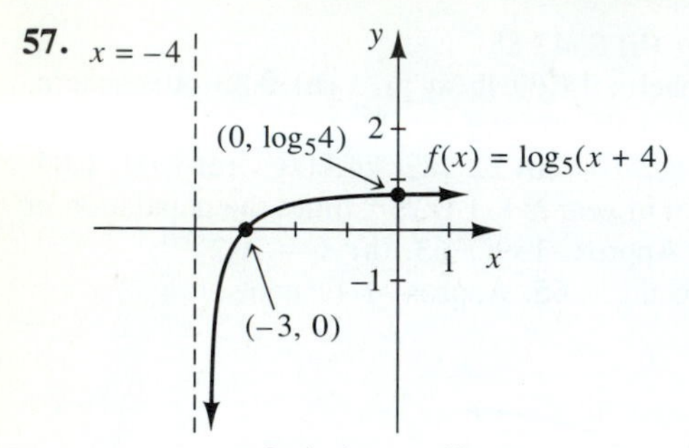

D: $\{x \mid x > -4\}$
R: all real numbers

59.

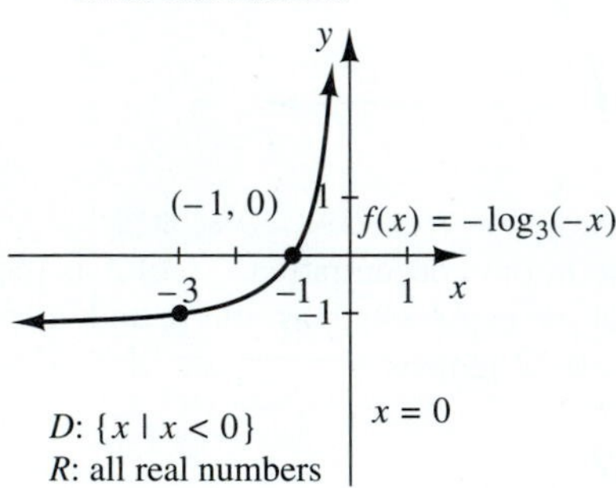

D: $\{x \mid x < 0\}$
R: all real numbers

61.

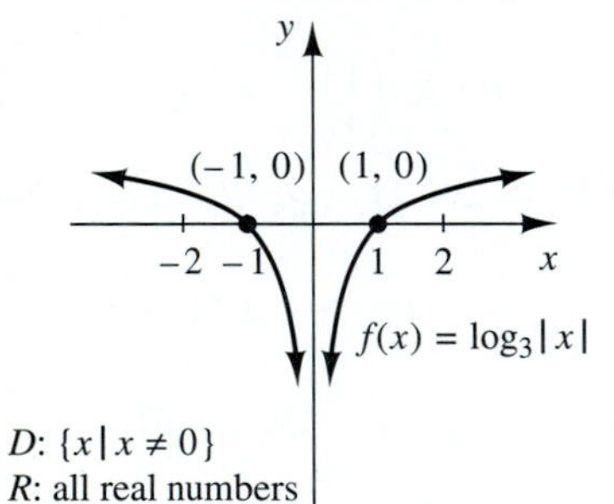

D: $\{x \mid x \neq 0\}$
R: all real numbers

EXERCISES 5.3

1. $2 \log_4 x + \log_4 y + 3 \log_4 z$
3. $3 \log_b x - \log_b y - 4 \log_b z$
5. $\frac{1}{2} + \frac{3}{2} \log_5 x$ **7.** Cannot be simplified
9. $\log_b(x - 2) + \log_b(x + 2)$
11. $\frac{1}{3}\log_4(x - 3) - \frac{1}{6} - \frac{2}{3} \log_4 x$
13. $\frac{1}{2}\log_b(x + 4) - \frac{1}{2}\log_b(x + 2)$ where $x \neq 4$ **15.** False
17. True **19.** $\log_b 48$ **21.** $\log_b \frac{s^{1/2}}{t^{3/2}}$ **23.** 1 **25.** 3
27. $\log_b\left(\frac{x^3}{y^4 z^2}\right)$ **29.** $\log_b\left(\frac{x^{1/4} y^{1/3}}{z^{1/2}}\right)$ **37.** 2.47 **39.** −2.87
41. $x = \frac{9}{4}$ **43.** $x = 7$ **45.** $t = \frac{16}{7}$ **47.** $x = 5$
49. $x = 10$ **51.** $x = 8$ **53.** $x = 9$ **55.** 25

EXERCISES 5.4

1. 5 **3.** $\frac{1}{2}$ **5.** $x + 1$

7.

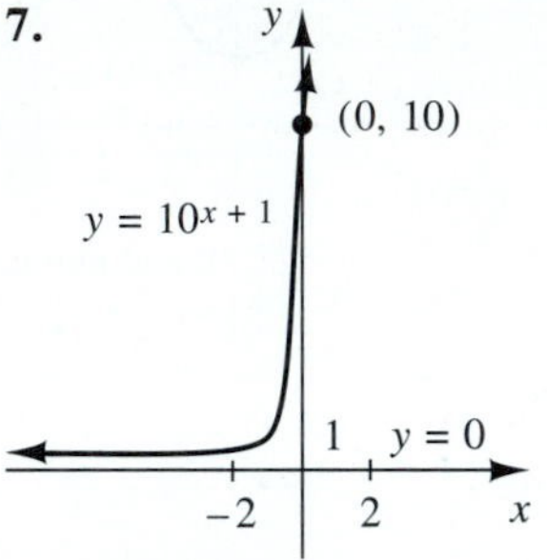

D: all real numbers
R: $\{y \mid y > 0\}$

9.

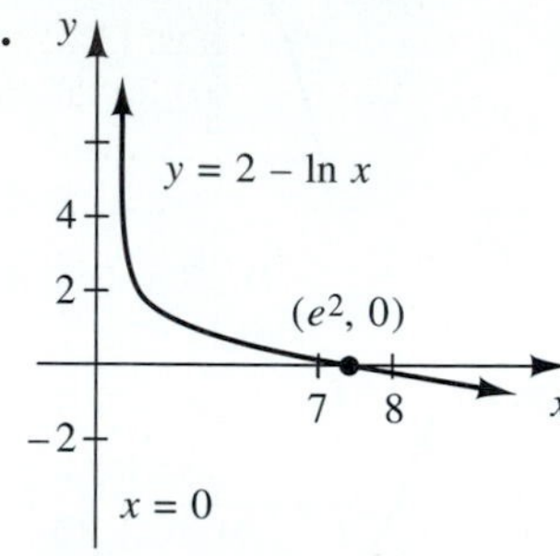

D: $\{x \mid x > 0\}$
R: all real numbers

11.

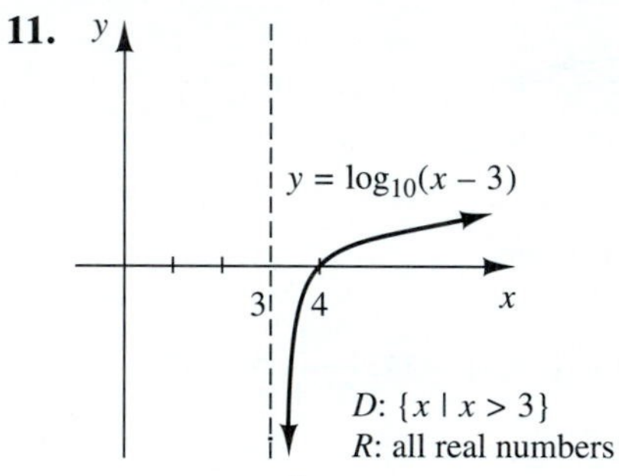

D: $\{x \mid x > 3\}$
R: all real numbers

$x = 3$

13.

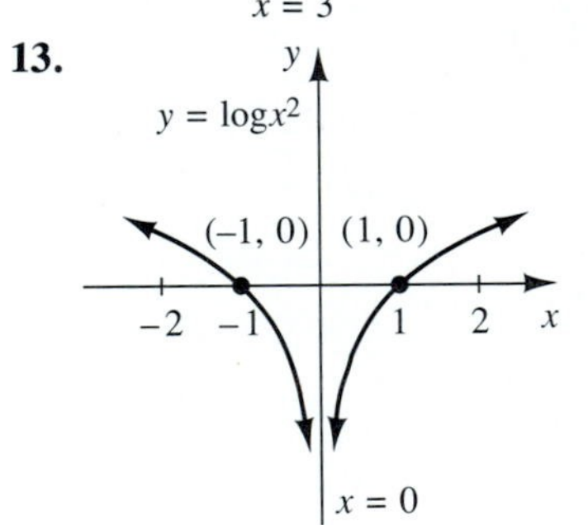

D: $\{x \mid x \neq 0\}$
R: all real numbers

15. $f^{-1}(x) = \frac{1 + \ln x}{3}$ **17.** $g(x) = (\ln x)^2$ **19.** 0.71
21. −2.32 **23.** −0.18 **25.** $2 \log_9 x$ **27.** $\frac{\ln x}{\ln 10} = \frac{\ln x}{2.303}$
29. $x = \frac{1 - \ln 5}{2}$ **31.** $x = \frac{-2 + \ln 20}{3}$
33. $x = \frac{4 \ln 3 - \ln 7}{\ln 3}$ **35.** $x = \frac{2 \ln 2}{2 \ln 2 + \ln 6}$

EXERCISES 5.5

1. No **3.** $A = A_0e^{-0.00002773t}$ **5.** 88.6 grams **7.** $A = A_0e^{-0.30944t}$; .00372 grams **9.** 559 years **11.** 12.6 hours; 20 hours **13.** 1.08% **15.** 6.05 billion **17.** 1.7%; lower **19.** 5.39 years **21.** 6.91%; no **23.** 6.49% **25.** \$4489.59 **27.** 11.552 years compounded continuously; 11.553 years compounded daily **29. (a)** \$199.08 **(b)** \$9555.86 **31. (a)** \$598.20 **(b)** \$215,353.28 **33. (a)** 3 **(b)** $10^{8.3} = 199{,}526{,}231.5$ times as intense. **35.** E_3 is 10,000 times as intense as E_2, which is 10,000 times as intense as E_1. **37.** 4.2 **39.** 5.6 **41. (a)** 2.51 **(b)** 2.51 **(c)** 2.5 **(d)** 4 **43.** $N = 0$ decibels **45. (a)** 4.38% **(b)** 5.22% **47. (d)** Approx. 3560 years old **49.** Approx. 2268 years old **51. (a)** 70 grams **(b)** 81.8 grams **(c)** 89.0 grams **(d)** 100.0 grams **53. (a)** \$ 78.20 **(b)** 14.09 years **55.** Stirling's formula: $2.422786844 \times 10^{18}$; actual value: $2.432902008 \times 10^{18}$

CHAPTER 5 REVIEW EXERCISES

1.

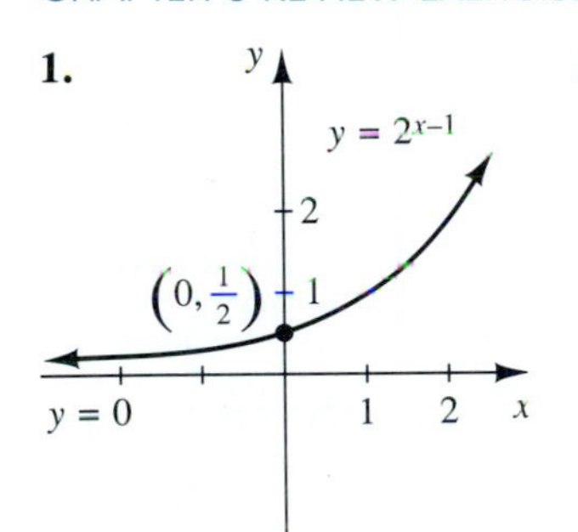

3.

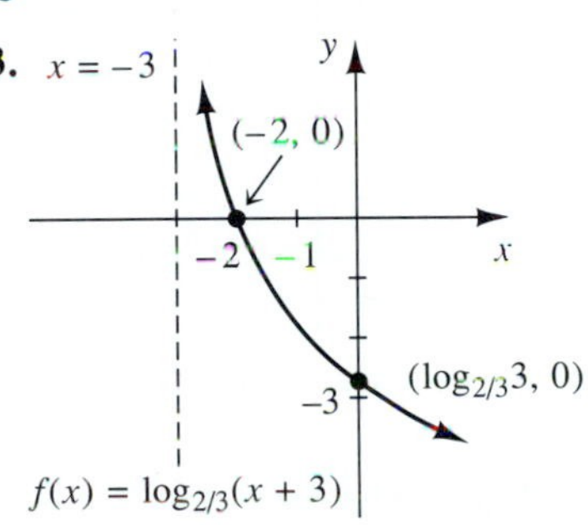

5.

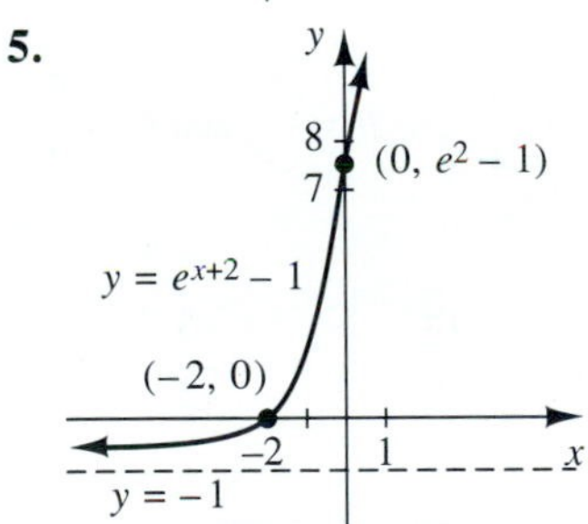

7. $6^{-1} = \frac{1}{6}$ **9.** $\log_8\left(\frac{1}{4}\right) = -\frac{2}{3}$ **11.** $b^6 = b^6$ **13.** 4 **15.** −2 **17.** $\frac{4}{5}$ **19.** 0 **21.** $3 \log_b x + 4 \log_b y + 2 \log_b z$ **23.** $\frac{1}{3}\log_b 6 + \frac{1}{3}\log_b x - \frac{1}{3} - \frac{4}{3}\log_b y$ **25.** Cannot be simplified **27.** $x = -2$ **29.** $x = 1$ **31.** $x = \frac{\ln 3 + \ln 7}{\ln 7} \approx 1.565$ **33.** $x = 10$ **35.** $x = \frac{\ln 5}{\ln 3 - \ln 2} \approx 3.969$ **37.** $x = 9$ **39.** $x = \frac{2 \ln 8 + 2 \ln 9}{3 \ln 8 - \ln 9} \approx 2.117$ **41.** $b = 5$ **45.** $f^{-1}(x) = 1 + 3^{x-5}$ **47.** 1.23 **49.** \$5621.42 **51. (a)** 1418 bacteria **(b)** 23 hours **53.** $A = A_0e^{-0.013078t}$ (t in days) **55.** The half-life is halved from 6.9 to 3.45 years.

CHAPTER 5 PRACTICE TEST

1. (a)

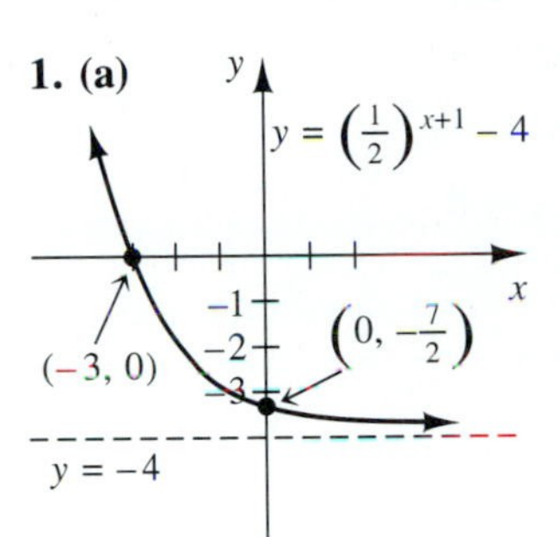

(b)

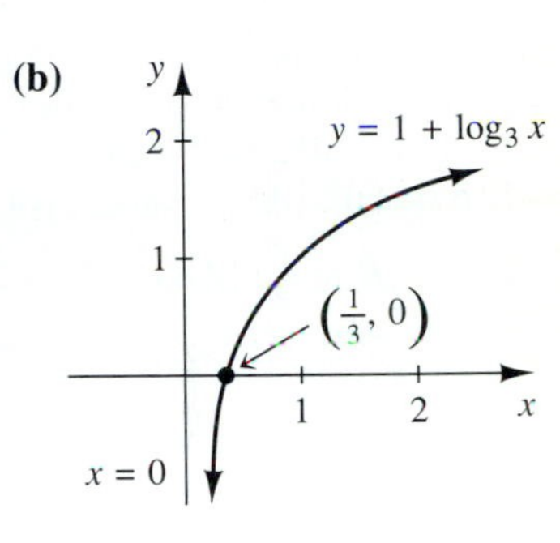

2.

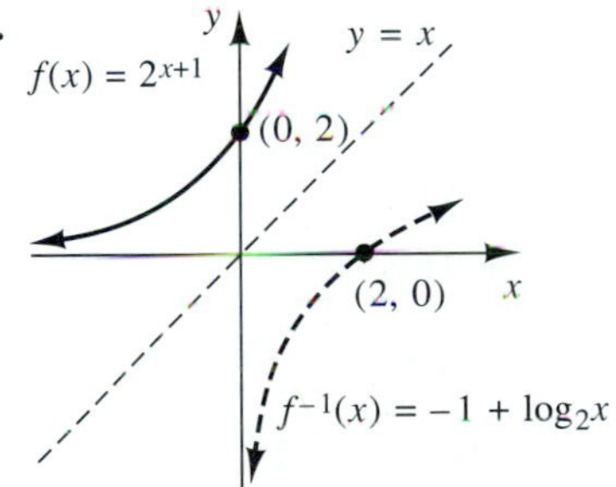

3. (a) $-\frac{1}{2}$ **(b)** 0.71 **(c)** $\frac{4}{3}$ **(d)** 2

4. (a) $x = \frac{3}{2}$ **(b)** No solution **(c)** $x = 4$

(d) $x = \frac{\ln 10 - \ln 5}{\ln 5} \approx 0.43$

5. (a) 949 **(b)** 50.26 hours

6. $A = A_0e^{-0.000924196t}$ for t in years; 91.17 grams

CHAPTER 6

EXERCISES 6.1

1. $30°$ **3.** $60°$ **5.** $-450°$ **7.** $330°$ **9.** $270°$

11. $171.9°$ **13.** $-114.6°$ **15.** $-15°$ **17.** $\frac{5\pi}{6}$

19. $-\frac{2\pi}{3}$ **21.** $\frac{\pi}{10}$ **23.** $-\frac{2\pi}{9}$ **25.** $\frac{7\pi}{6}$

27. $s = 4\pi$ cm; $A = 24\pi$ cm^2 **29.** $r = 6$ cm

31. $\theta = \pi$ **33.** $A = 432\pi$ square meters

35. 2.2 radians **37.** 1400π; $252{,}000°$

39. **(a)** 30 rpm **(b)** 0.00035 rpm

41. $18{,}000\pi$ feet ≈ 10.7 miles

43. $28.6°$ **45.** 39.3 inches

47. $B = 60°$; $|\overline{AC}| = 8\sqrt{3}$; $|\overline{BC}| = 8$

49. $A = B = \frac{\pi}{4}$; $|\overline{AB}| = 10\sqrt{2}$

51. $B = 30°$; $|\overline{AB}| = 8$; $|\overline{AC}| = 4$

53. $C = \frac{\pi}{4}$; $|\overline{AB}| = |\overline{BC}| = 5\sqrt{2}$

EXERCISES 6.2

1. Ref. angle is $\frac{\pi}{3}$. **3.** Ref. angle is $\frac{\pi}{6}$.

5. Ref. angle is $45°$. **7.** Ref. angle is $60°$.

9. Pos. y-axis **11.** Ref. angle is $\frac{\pi}{6}$. **13.** Pos. x-axis

15. Ref. angle is $45°$. **17.** $\frac{1}{2}$ **19.** $\sqrt{3}$ **21.** -1

23. Undefined **25.** $-\sqrt{2}$ **27.** 0 **29.** -1 **31.** $\frac{1}{2}$

33. $-\sqrt{2}$ **35.** 1 **37.** $\frac{3}{2}$ **39.** $-\frac{1}{3}$ **41.** $-\frac{4}{5}$

43. $\frac{3}{\sqrt{10}}$ **45.** $-\sqrt{3}$ **47.** $-\frac{12}{5}$ **49.** $-\frac{3}{5}$ **51.** $-\frac{3}{4}$

53. $-\frac{1}{\sqrt{10}}$ **55.** All real numbers

57. $\theta \neq$ an odd multiple of $\frac{\pi}{2}$ **59.** $\frac{8\pi}{3}, -\frac{4\pi}{3}$

61. $470°, -250°$ **63.** $\frac{7\pi}{2}, -\frac{\pi}{2}$

In 65 and 67, these are three among many possible answers.

65. $\theta = \frac{\pi}{6}, \frac{5\pi}{6}, \frac{13\pi}{6}$ **67.** $\theta = \frac{2\pi}{3}, \frac{5\pi}{3}, -\frac{\pi}{3}$

69. 1 **71.** 1

73.

θ	$\sin\theta$	$\cos\theta$	$\tan\theta$
0	0	1	0
$\frac{\pi}{6}$	$\frac{1}{2}$	$\frac{\sqrt{3}}{2}$	$\frac{1}{\sqrt{3}} = \frac{\sqrt{3}}{3}$
$\frac{\pi}{4}$	$\frac{1}{\sqrt{2}} = \frac{\sqrt{2}}{2}$	$\frac{1}{\sqrt{2}} = \frac{\sqrt{2}}{2}$	1
$\frac{\pi}{3}$	$\frac{\sqrt{3}}{2}$	$\frac{1}{2}$	$\sqrt{3}$
$\frac{\pi}{2}$	1	0	Undef.
$\frac{2\pi}{3}$	$\frac{\sqrt{3}}{2}$	$-\frac{1}{2}$	$-\sqrt{3}$
$\frac{3\pi}{4}$	$\frac{1}{\sqrt{2}}$	$-\frac{1}{\sqrt{2}}$	-1
$\frac{5\pi}{6}$	$\frac{1}{2}$	$-\frac{\sqrt{3}}{2}$	$-\frac{1}{\sqrt{3}}$
π	0	-1	0

EXERCISES 6.3

1. $\sin\theta = \frac{3}{5}$; $\cos\theta = \frac{4}{5}$; $\tan\theta = \frac{3}{4}$; $\csc\theta = \frac{5}{3}$; $\sec\theta = \frac{5}{4}$; $\cot\theta = \frac{4}{3}$

3. $\sin\theta = \frac{12}{13}$; $\cos\theta = \frac{5}{13}$; $\tan\theta = \frac{12}{5}$; $\csc\theta = \frac{13}{12}$; $\sec\theta = \frac{13}{5}$; $\cot\theta = \frac{5}{12}$

5. $\sin\theta = \frac{5}{\sqrt{41}}$; $\cos\theta = \frac{4}{\sqrt{41}}$; $\tan\theta = \frac{5}{4}$; $\csc\theta = \frac{\sqrt{41}}{5}$; $\sec\theta = \frac{\sqrt{41}}{4}$; $\cot\theta = \frac{4}{5}$

7. $\sin\theta = \frac{24}{25}$; $\cos\theta = \frac{7}{25}$; $\tan\theta = \frac{24}{7}$; $\csc\theta = \frac{25}{24}$; $\sec\theta = \frac{25}{7}$; $\cot\theta = \frac{7}{24}$

9. $\sin\theta = \frac{\sqrt{3}}{2}$; $\cos\theta = \frac{1}{2}$; $\tan\theta = \sqrt{3}$; $\csc\theta = \frac{2}{\sqrt{3}}$; $\sec\theta = 2$; $\cot\theta = \frac{1}{\sqrt{3}}$

11. $\sin A = \cos B = \frac{4}{5}$ **13.** $\tan\alpha = \cot\beta = \frac{3}{5}$

15. $\sec P = \csc Q = \frac{3}{\sqrt{5}}$ **17.** $x = 14.6$ **19.** $x = 161.8$

21. $x = 2.5$ **23.** 33° **25.** 37° **27.** 0.8387
29. −0.1584 **31.** $-\sqrt{2}$ **33.** −0.9272 **35.** −2.6051

37. $-\frac{2}{\sqrt{3}}$ **39.** $\theta = 26°$ **41.** $x = 55°$ **43.** $\theta = 73°$

45. $x = 74°$ **47.** $\theta = 28°$ **49.** cos 14° **51.** sec 2°

53. $\sin\frac{\pi}{14}$ **55.** 1 **57.** 1 **59.** 15.0 feet

61. 55.5 feet **63.** 39.2 feet **65.** $A = 342.4$ sq cm
67. 9 km/hr **69.** 34° **71.** 133.3 meters
73. 14.5 feet **75.** 175.8 feet **77.** 537.1 miles
79. 688.2 sq mm **81.** 452.4 sq in.

83. 35.4 sq cm **85.** $A = \frac{1}{2}\pi r^2 + 8\sqrt{r^2 - 16}$

87. (a) $A = 50\sin\theta\cos\theta$ (b) $A = 50\tan\theta$
(c) $A = 50(\tan\theta - \sin\theta\cos\theta)$
89. $|\overline{AC}| = 10\sec\theta$ **91.** $|\overline{PQ}| = 2 + 5\tan\theta$
93. 2598.1 feet **95.** 1.7 feet from point C
97. 1.3 miles **99.** 9.5° **101.** 1 minute 53.7 seconds

EXERCISES 6.4

1. $2\cos^2\theta - 2\cos\theta$ **3.** $\sin^2 5\theta + 4\sin 5\theta$
5. $\csc^2\theta - 4\csc\theta + 4$ **7.** $\cos^2 4\theta + 2\cos 4\theta + 1$
9. $5\sin^2\theta + 13\sin\theta - 6$ **11.** $\cos^2 2\theta - 9$
13. $\sin 3\theta\sin 5\theta - 4\sin 3\theta + 2\sin 5\theta - 8$
15. $(\sin\theta - 2)(\sin\theta + 1)$
17. $(\tan\theta + 4)(\tan\theta - 2)$ **19.** $(2\csc\theta + 1)(\csc\theta - 3)$

21. $\frac{2\cos\theta + 3\sin\theta}{\sin\theta\cos\theta}$ **23.** $\frac{\sin^2\theta - \cos^2\theta}{\sin\theta\cos^2\theta}$

25. $\frac{1 + \cos^2\theta}{\cos^2\theta}$ **27.** $\frac{1 + \tan\theta\sec\theta}{\tan\theta}$ **29.** $\frac{\cot^2\theta - 1}{\cot^2\theta}$

31. $\frac{3\sin 3\theta + 2\sin 2\theta}{\sin 2\theta\sin 3\theta}$ **33.** $\frac{2 + \cos\theta\tan^2\theta}{2\cos^2\theta}$

35. $\frac{\sin^2\theta + 1}{\sin\theta}$ **37.** $\frac{5}{3\cos\theta + 2}$ **39.** $\frac{1 + \sin\theta}{1 - \sin\theta}$

41. $\frac{\sin 4\theta}{\sin 4\theta - 1}$ **43.** $\sin\theta = \frac{\sqrt{9 - x^2}}{3}$

45. $\tan\theta = \frac{5}{\sqrt{y^2 - 25}}$ **47.** $\sec\theta = \frac{\sqrt{a^2 + 100}}{a}$

49. $x = 4\sin\theta$ **51.** $x = 4\tan\theta$ **53.** $s = \frac{5}{\sec\theta}$

55. (a) $d = 6$ inches (b) $d = -0.87$ inches
(c) $d = 2.45$ inches

CHAPTER 6 REVIEW EXERCISES

1. 15° **3.** $\frac{10\pi}{9}$ **5.** −108° **7.** $\frac{11\pi}{6}$ **9.** 315°

11. $-\frac{1}{\sqrt{3}}$ **13.** $-\frac{\sqrt{3}}{2}$ **15.** 1 **17.** 1.2208 **19.** $\frac{1}{\sqrt{3}}$

21. 0.3420 **23.** −0.1425 **25.** Undefined **27.** 0.9781

29. $-\frac{1}{\sqrt{2}}$ **31.** $x = 25$ **33.** $\theta = \frac{\pi}{6}$ **35.** $x = 9.7$

37. $\theta = \frac{\pi}{4}$ **39.** 30.9 **41.** 5π inches **43.** 19.9 feet

45. 1272 mi/hr **47.** 37.4 feet **49.** $\frac{\cos 2\theta - 4\sin\theta}{\sin\theta\cos 2\theta}$

51. $\frac{2\sec\theta - 3}{\sec\theta + 3}$

CHAPTER 6 PRACTICE TEST

1. 108° **2.** $\frac{8\pi}{9}$ **3.** $|\overline{AB}| = 8\sqrt{3}$ **4.** (a) $-\frac{1}{\sqrt{2}}$

(b) $-\frac{1}{2}$ (c) $\frac{1}{\sqrt{3}}$ (d) Undefined (e) Undefined (f) 1.2868
(g) 0.9945 (h) 0.9657 **5.** 2π cm **6.** 10.3 sq cm
7. $x = 78.1$ **8.** $\theta = 22°$ **9.** 246 feet

10. 291 meters/min **11.** $\frac{2\sin\theta + \sin 2\theta}{2\sin\theta\sin 2\theta}$

CHAPTER 7

EXERCISES 7.1

1. $A = 3$; $P = 2\pi$; $f = 1$ **3.** $A = 1$; $P = \frac{2\pi}{5}$; $f = 5$

5. $A = \frac{1}{4}$; $P = \pi$; $f = 2$ **7.** $A = 2$; $P = 14\pi$; $f = \frac{1}{7}$

9. $A = 3$; $P = 2\pi$; $f = 1$; $s = \frac{\pi}{3}$

11. $A = 4$; $P = \frac{2\pi}{3}$; $f = 3$; $s = -\frac{\pi}{12}$

13.

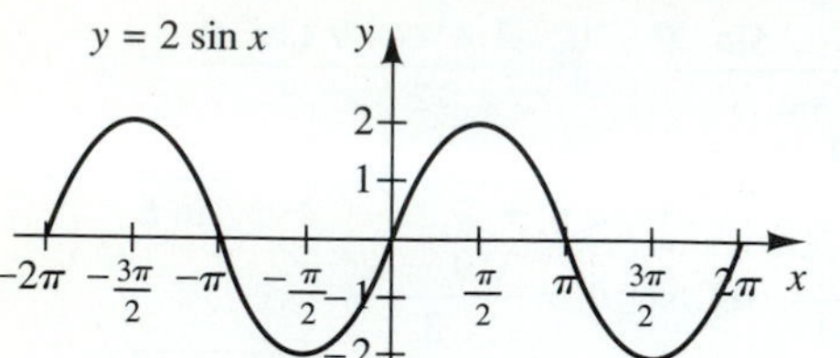

15.

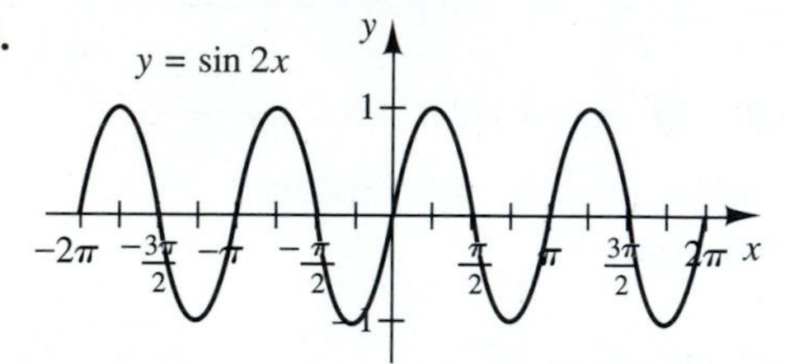

17.

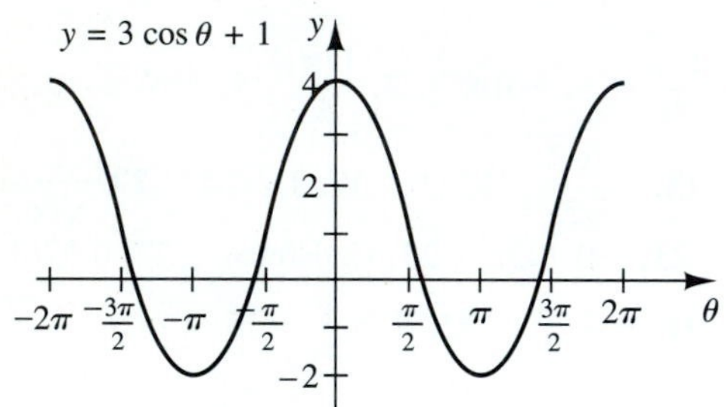

19.

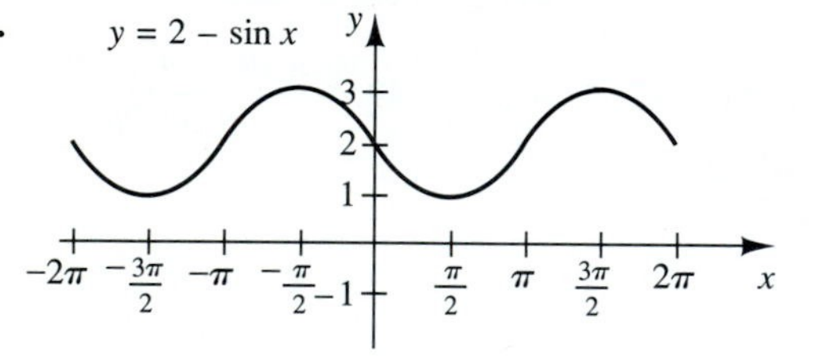

21.

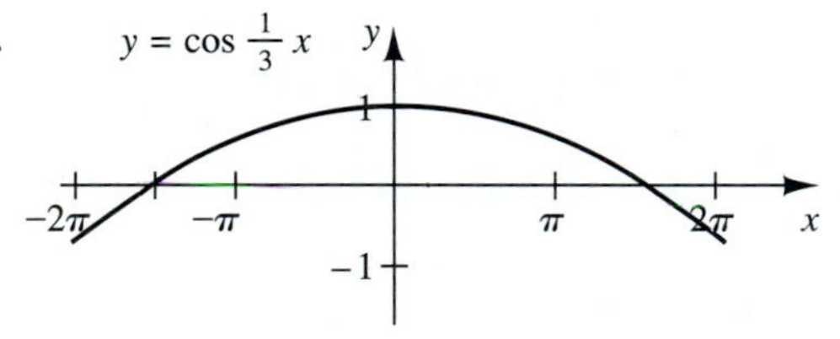

23.

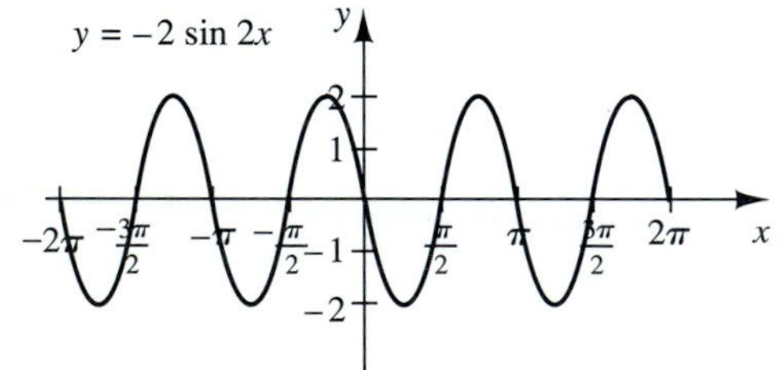

25.

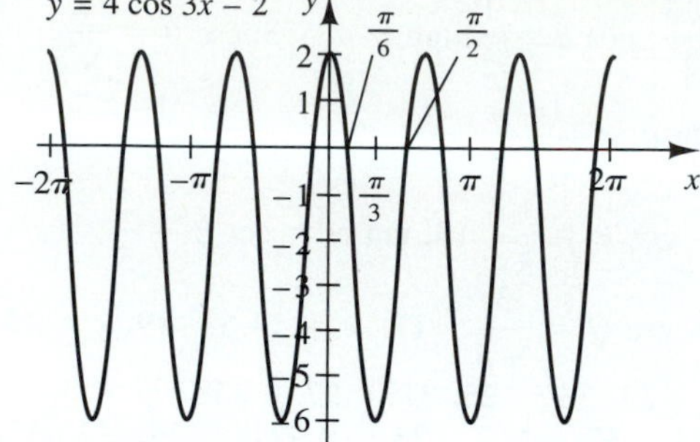

27.

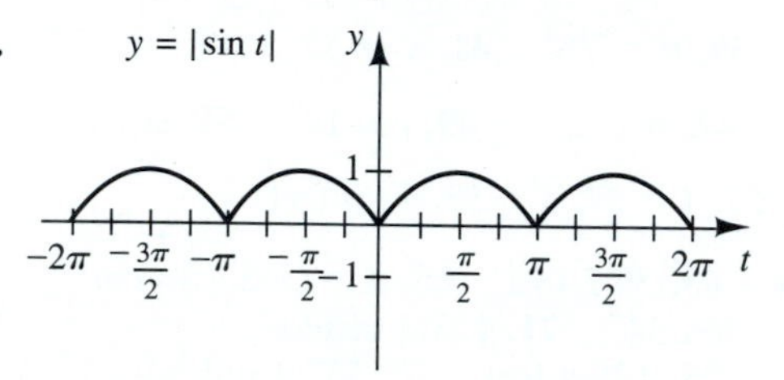

29.

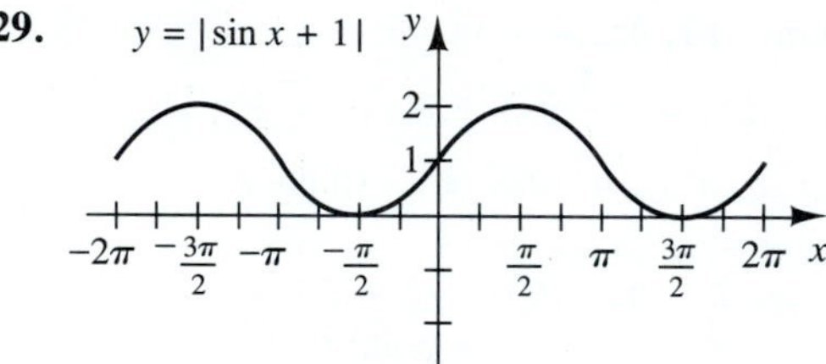

31.

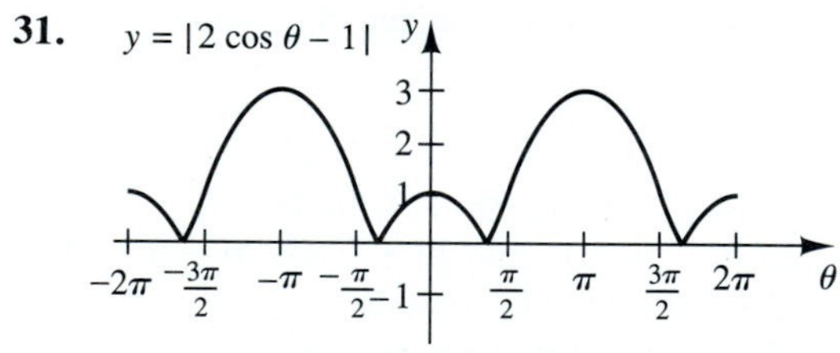

33.

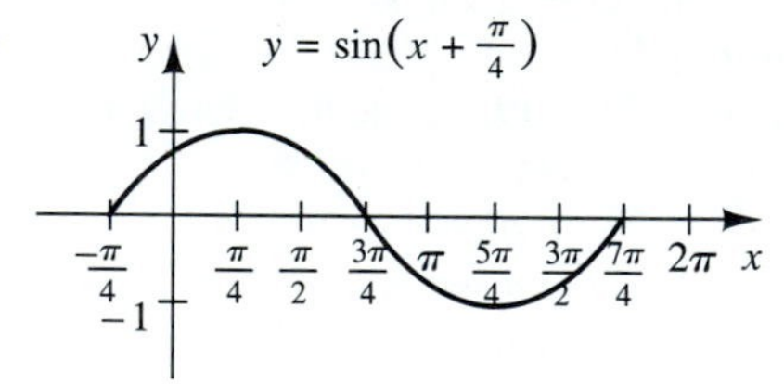

35.

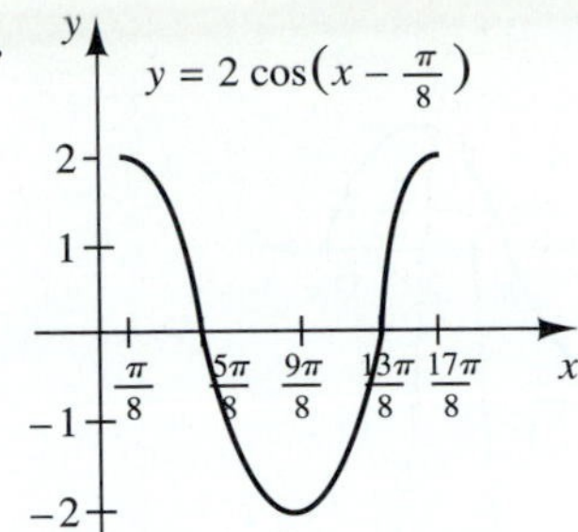

37.

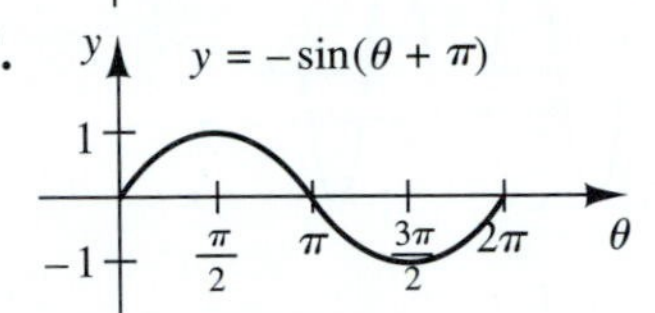

39. $y = 3\cos(2x - \pi)$

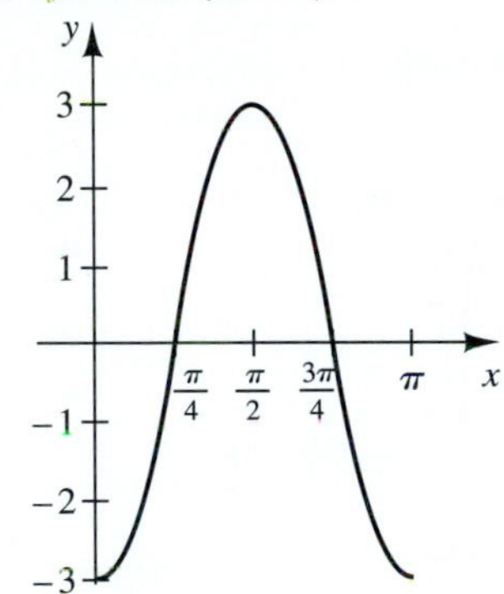

41. $y = \sin(-x + \pi)$

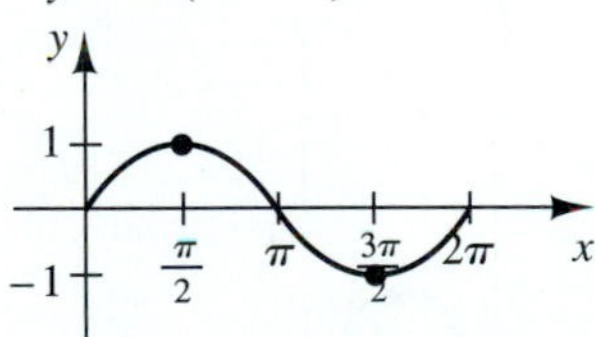

43. $y = \left|\cos\left(t - \frac{\pi}{2}\right)\right|$

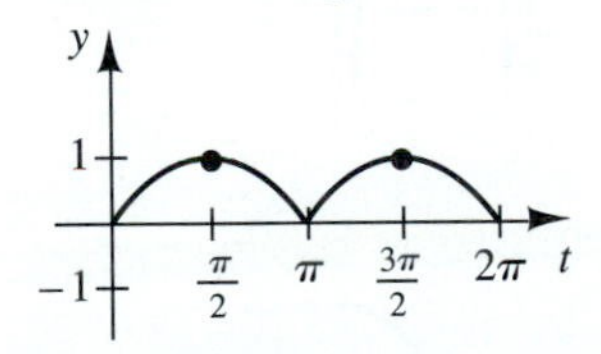

45. $y = 3 \sin 2x$ **47.** $y = -\cos x$

49. Amplitude = 2.5
Period = 365
Phase shift = 81
Vertical shift = 12

$$D = 12 + 2.5 \sin\left(\frac{2\pi}{365}(t - 81)\right)$$

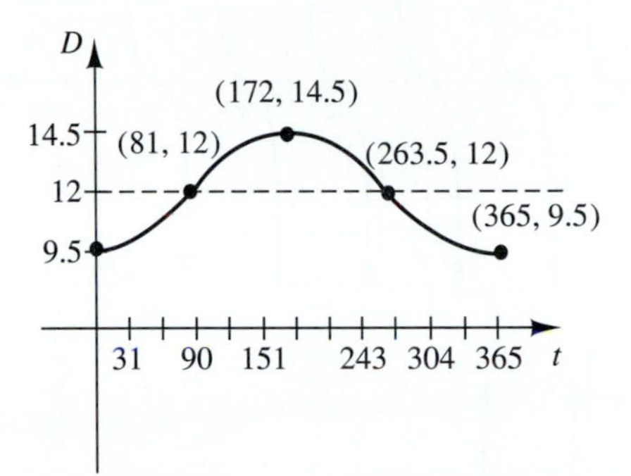

51. Amplitude = 6
Period = $\frac{\pi}{2}$
Phase shift = 0
Vertical shift = 0

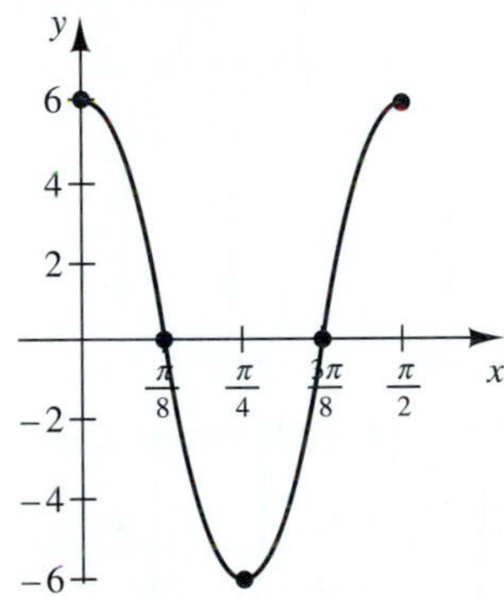

53. Let T = average daily temperature; d = day of the year

$$T = 68 + 28 \cos\left(\frac{2\pi}{362}(t - 222)\right)$$

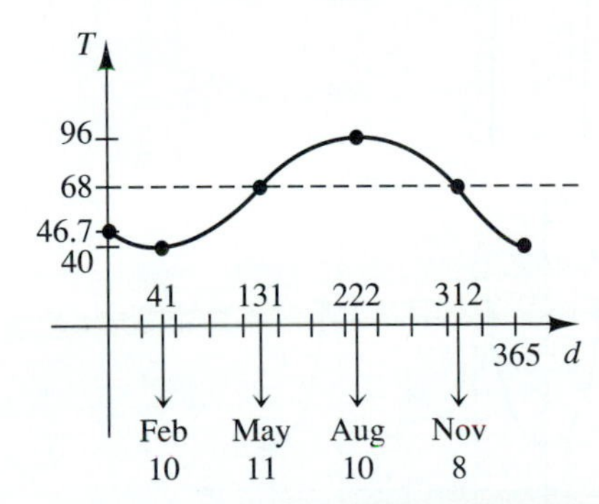

EXERCISES 7.2

1. $y = \tan 2x$

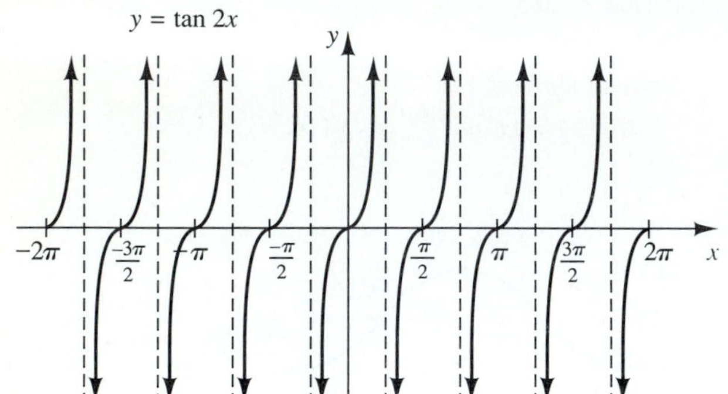

3. $y = 3 \cot x$

5. $y = -\tan x$

7. $y = \sec(-x)$

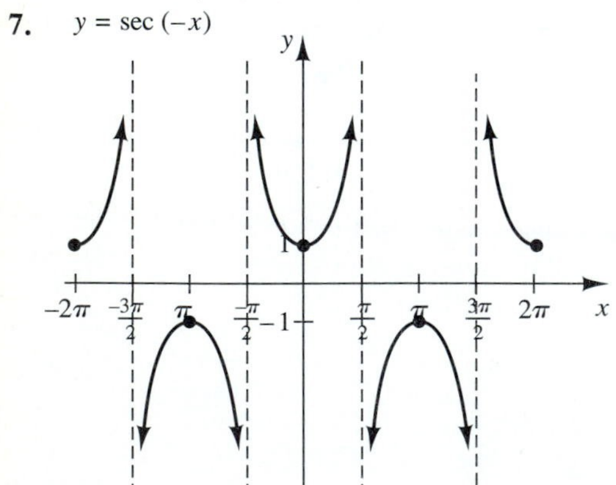

9. $y = \cos\left(x + \frac{\pi}{2}\right)$

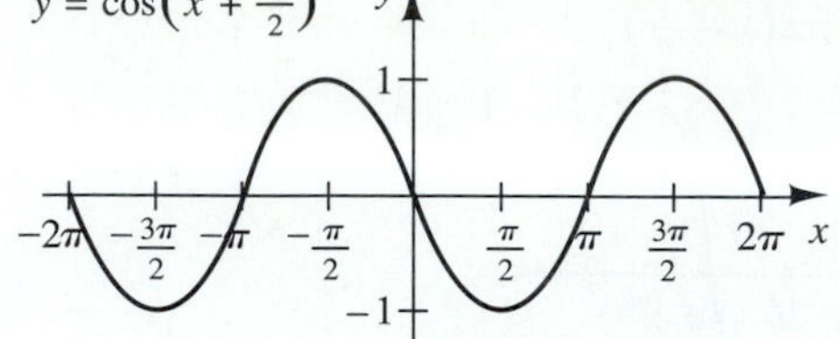

11. $y = \frac{1}{2} \sin 3x$

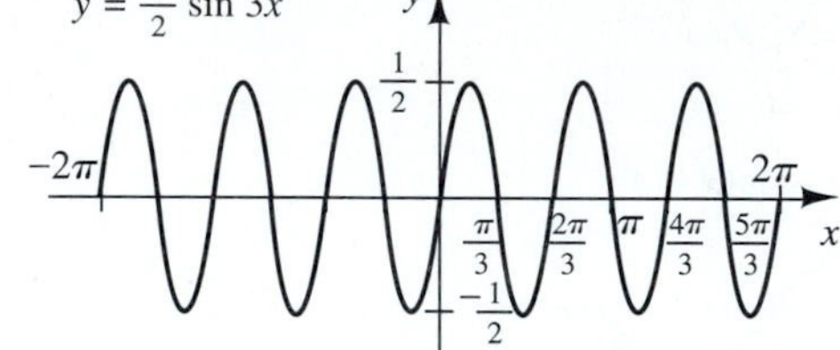

13.

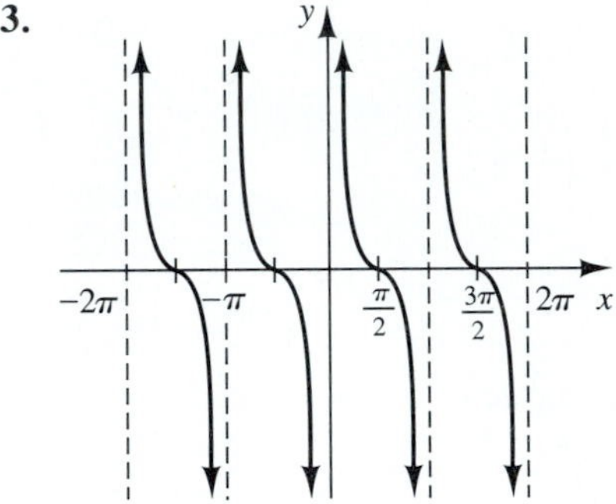

$y = \cot(x - \pi)$

15. $y = 2 + \csc x$

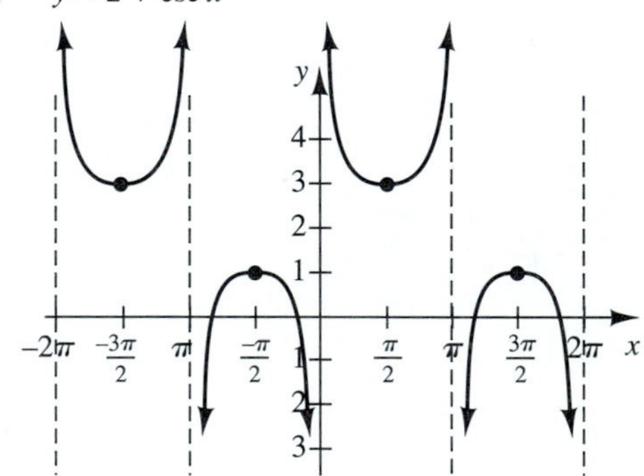

17. $y = \sin\left(x + \frac{\pi}{2}\right)$

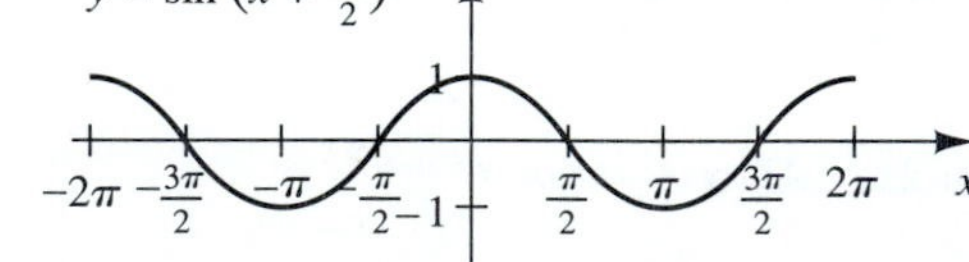

19.

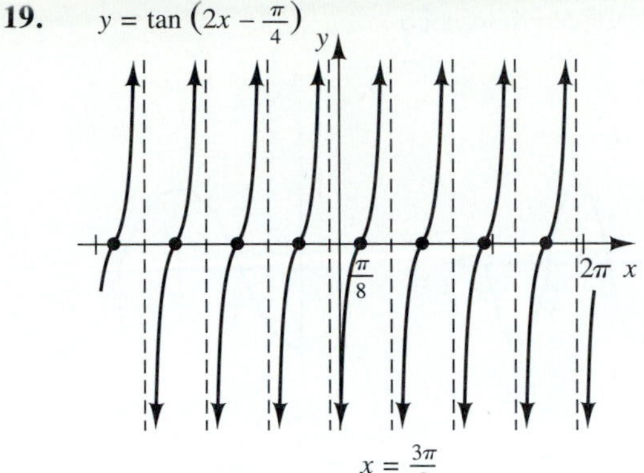

21.

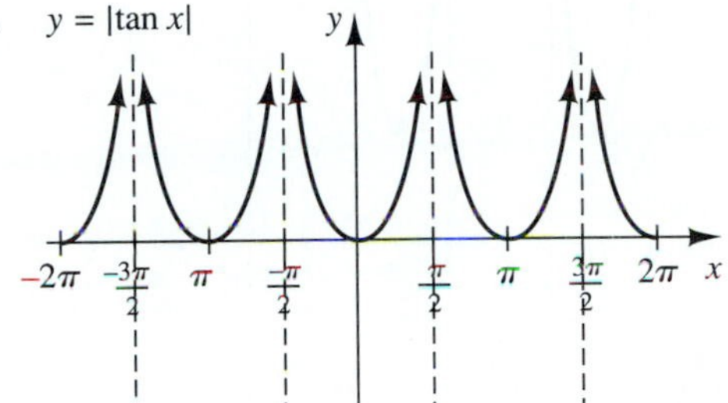

25.

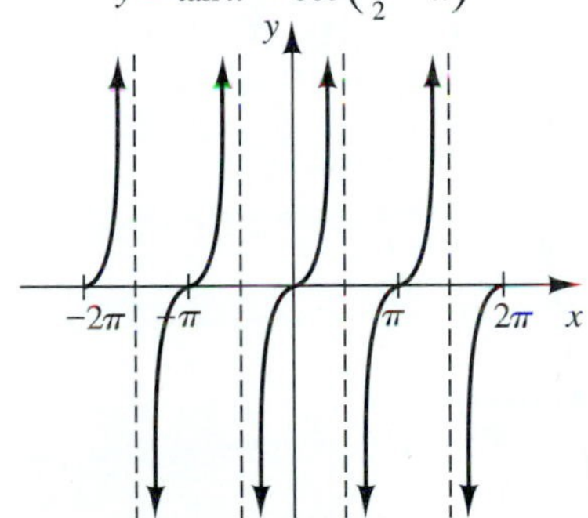

27.

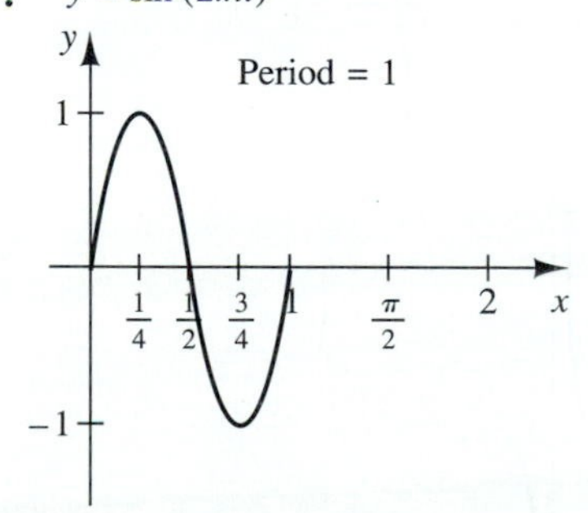

EXERCISES 7.3

1. $\cos\theta$ **3.** $\dfrac{1}{\cos\theta}$ **5.** 1 **7.** $\dfrac{\sin^2\theta}{\cos^2\theta}$ **9.** $\dfrac{\cos\theta - \sin\theta}{\cos^2\theta}$

11. $\dfrac{1-\sin^2\theta}{\sin\theta}$ **13.** $\sin\theta - \dfrac{1}{\sin\theta} = \dfrac{\sin^2\theta - 1}{\sin^2\theta}$

15. $\dfrac{\cos^2\theta - 1}{\cos\theta}$ **17.** $2\cos^2\theta - 1$ **67.** $\cos\theta = -\dfrac{3}{5}$

69. $\sec B = -\dfrac{\sqrt{13}}{3}$ **71.** $\tan\theta = \dfrac{3\sqrt{10}}{20}$

EXERCISES 7.4

1. $x = \dfrac{\pi}{6}, \dfrac{5\pi}{6}$ **3.** $x = \dfrac{5\pi}{6}, \dfrac{7\pi}{6}$ **5.** $x = \dfrac{\pi}{2}$

7. $\theta = \dfrac{3\pi}{4}, \dfrac{7\pi}{4}$ **9.** $x = \dfrac{\pi}{2}$ **11.** $\theta = \dfrac{2\pi}{3}, \dfrac{5\pi}{3}$

13. No solution **15.** $t = \dfrac{\pi}{6}, \dfrac{7\pi}{6}$ **17.** $x = \dfrac{\pi}{2}, \dfrac{3\pi}{2}$

19. No solution **21.** $x = \dfrac{\pi}{2}, \dfrac{3\pi}{2}$

23. $x = \dfrac{\pi}{6}, \dfrac{5\pi}{6}, \dfrac{7\pi}{6}, \dfrac{11\pi}{6}$ **25.** $w = \dfrac{\pi}{3}, \dfrac{2\pi}{3}, \dfrac{4\pi}{3}, \dfrac{5\pi}{3}$

27. $x = 0$ **29.** $\theta = 0, \dfrac{\pi}{4}, \pi, \dfrac{5\pi}{4}$ **31.** $t = 0, \pi$

33. $x = \dfrac{\pi}{4}, \dfrac{5\pi}{4}$ **35.** $w = 0, \pi$ **37.** $x = \dfrac{3\pi}{2}$

39. $x = \dfrac{\pi}{6}, \dfrac{5\pi}{6}, \dfrac{3\pi}{2}$ **41.** $\theta = \dfrac{\pi}{3}, \dfrac{2\pi}{3}, \dfrac{4\pi}{3}, \dfrac{5\pi}{3}$

43. $\theta = 0, \dfrac{\pi}{2}, \pi, \dfrac{3\pi}{2}$ **45.** $x = 37°, 143°$

47. $x = 120°, 300°$ **49.** $x = 207°, 333°$

51. $x = 2.24, 4.05$

53. $t = 66.96°, 139.61°, 246.96°, 319.61°$

55. $x = \dfrac{\pi}{6} + 2n\pi, \dfrac{5\pi}{6} + 2n\pi$ **57.** $x = \dfrac{\pi}{3} + n\pi, \dfrac{2\pi}{3} + n\pi$

59. $x = \dfrac{\pi}{3} + n\pi, \dfrac{2\pi}{3} + n\pi$ **61.** $x = \dfrac{\pi}{2} + n\pi,$

63. $x = \dfrac{3\pi}{2} + 2n\pi$ **65.** $x = \dfrac{\pi}{6} + \dfrac{2n\pi}{3}$

67. $x = \dfrac{3\pi}{16} + \dfrac{n\pi}{2}, \dfrac{5\pi}{16} + \dfrac{n\pi}{2}$

69. $D_f = \left\{\theta \mid \theta \neq \dfrac{\pi}{2} + 2n\pi\right\}$ **71.** D_h = all real numbers

73. $D_G = \left\{\theta \mid \theta \neq \dfrac{\pi}{6} + n\pi, \dfrac{5\pi}{6} + n\pi\right\}$

75. $D_f = \left\{\theta \mid \theta \neq \frac{\pi}{6} + n\pi, \frac{5\pi}{6} + n\pi\right\}$

77. $D_f = \left[0, \frac{\pi}{2},\right) \cup \left(\frac{3\pi}{2}, 2\pi\right)$ **79.** $D_g = \{\theta \mid \theta \neq 0, \pi\}$

81. $D_F = \left\{\theta \mid \theta \neq \frac{\pi}{4}, \frac{5\pi}{4}\right\}$

83. $D_f = \left\{\theta \mid \theta \neq \frac{\pi}{4}, \frac{3\pi}{4}, \frac{5\pi}{4}, \frac{7\pi}{4}\right\}$

EXERCISES 7.5

1. $\frac{\pi}{3}$ **3.** $\frac{2\pi}{3}$ **5.** $-\frac{\pi}{3}$ **7.** $\frac{\pi}{3}$ **9.** 0.4115 **11.** $-\frac{\pi}{2}$

13. $\frac{\pi}{2}$ **15.** $\frac{\pi}{2}$ **17.** $\frac{1}{2}$ **19.** $-\frac{\sqrt{3}}{3}$ **21.** 1 **23.** $\frac{2}{\sqrt{5}}$

25. $\frac{2}{\sqrt{21}}$ **27.** Not defined **29.** $\frac{\pi}{3}$ **31.** $\frac{\pi}{3}$ **33.** $\frac{\pi}{3}$

35. Not defined **37.** 8° **39.** $\theta = \cos^{-1}\left(\frac{x}{6}\right)$

41. $\theta = \tan^{-1}\left(\frac{x}{5}\right)$ **43.** $\sqrt{16 - x^2} = 4 \cos \theta$

45. $\sqrt{x^2 + 16} = 4 \sec \theta$ **47.** $9 \sin \theta \cos \theta$

49. $8 \tan \theta \sec^2 \theta$ **51.** $\frac{\tan^2 \theta}{\sec^2 \theta} = \sin^2 \theta$

53.

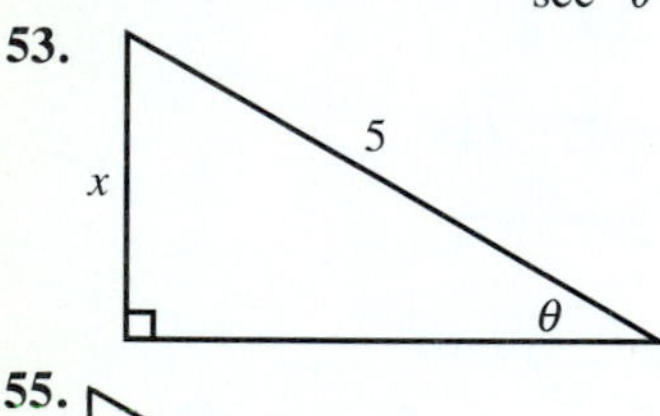

55.

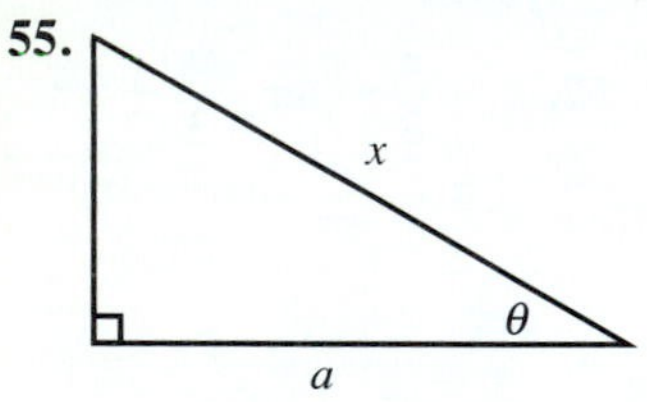

57. $\frac{\sqrt{x^2 - 9}}{x} = \sin \theta$

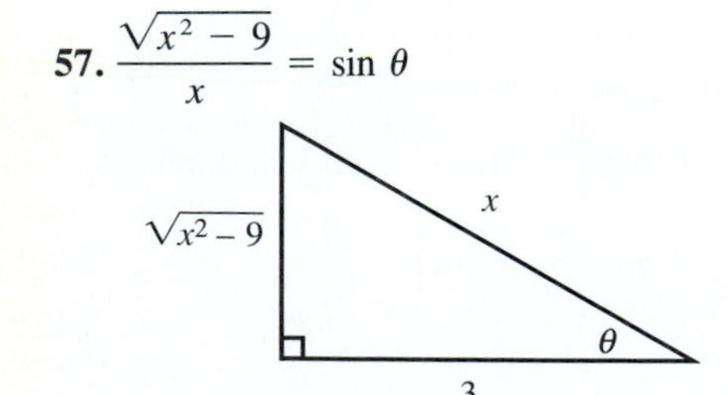

CHAPTER 7 REVIEW EXERCISES

1.

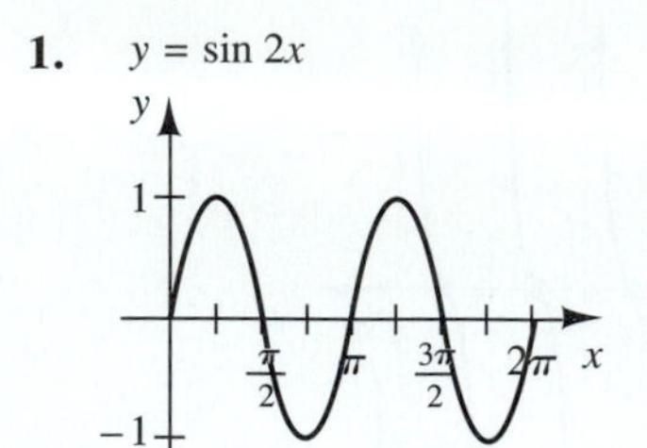

3.

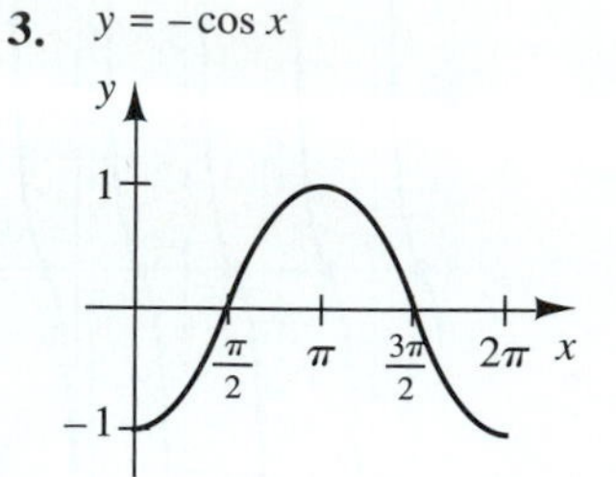

5.

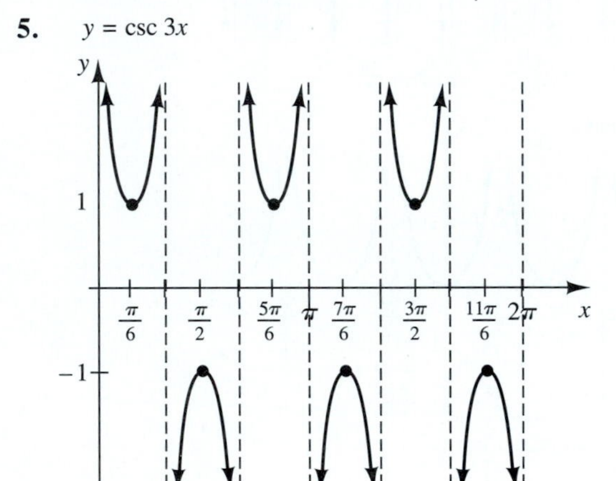

7.

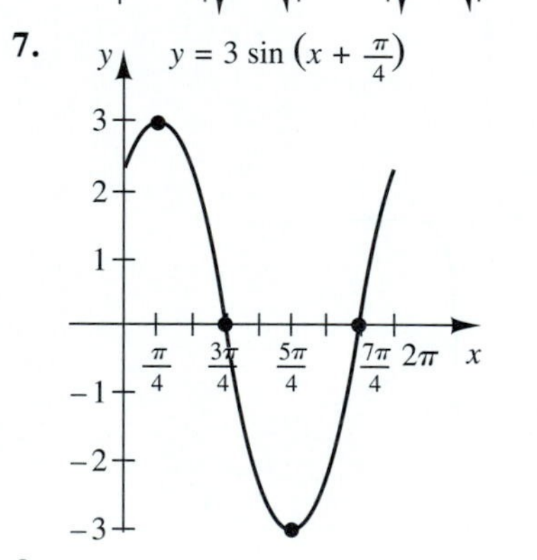

9.

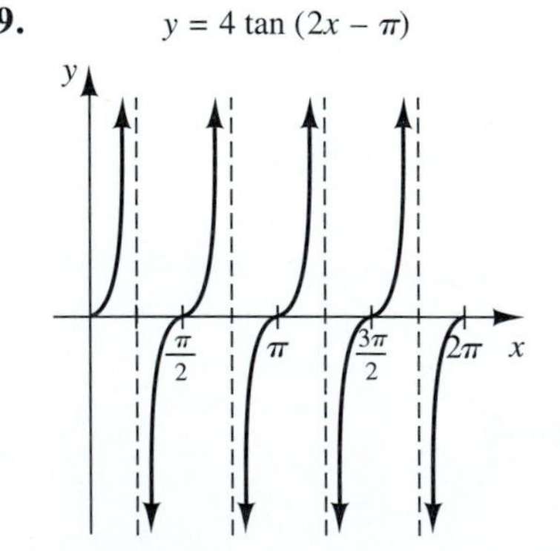

11.

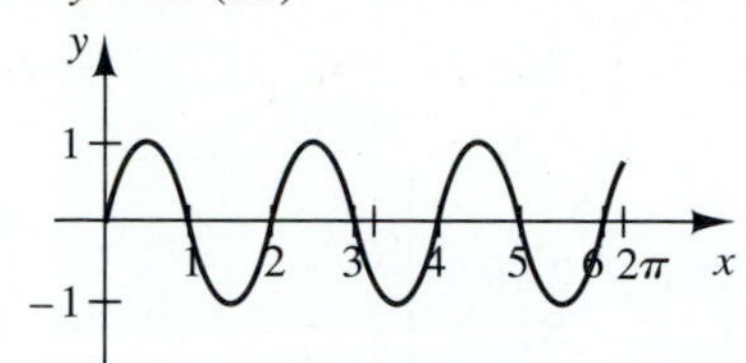

13. $-\sin 80°$ **15.** $-\tan\frac{\pi}{8}$ **17.** $\csc 7°$

25. $x = \frac{2\pi}{3} + 2n\pi, \frac{4\pi}{3} + 2n\pi$ **27.** $x = \frac{\pi}{4} + n\pi, \frac{3\pi}{4} + n\pi$

29. $x = \frac{\pi}{3} + 2n\pi, \frac{2\pi}{3} + 2n\pi$

31. $x = \frac{\pi}{6} + 2n\pi, \frac{11\pi}{6} + 2n\pi$ **33.** $x = \frac{\pi}{4}, \frac{3\pi}{4}, \frac{5\pi}{4}, \frac{7\pi}{4}$

35. $\theta = \frac{\pi}{2}, \frac{3\pi}{2}$ **37.** No solution **39.** $\theta = \frac{\pi}{6}, \frac{\pi}{2}, \frac{5\pi}{6}$

41. $\theta = \frac{\pi}{6}, \frac{5\pi}{6}$ **43.** $\theta = 0.8861, 2.2555, 4.0277, 5.3971$

45. $x = 76°, 284°$ **47.** $\theta = 90°$ **49.** $\theta = 30°, 90°, 150°$

51. $-\frac{\pi}{3}$ **53.** $\frac{\pi}{4}$ **55.** $\frac{\pi}{2}$ **57.** $-\frac{\pi}{6}$ **59.** 2

61. $\frac{5\pi}{6}$ **63.** $-\frac{\pi}{3}$ **65.** $\theta = \cot^{-1}\left(\frac{x}{7}\right)$

67. $\sqrt{x^2 - 64} = 8\tan\theta$

CHAPTER 7 PRACTICE TEST

1. Domain $= \left\{x \mid x \neq \frac{\pi}{6} + \frac{n\pi}{3}\right\}$ Range = all real numbers

2.

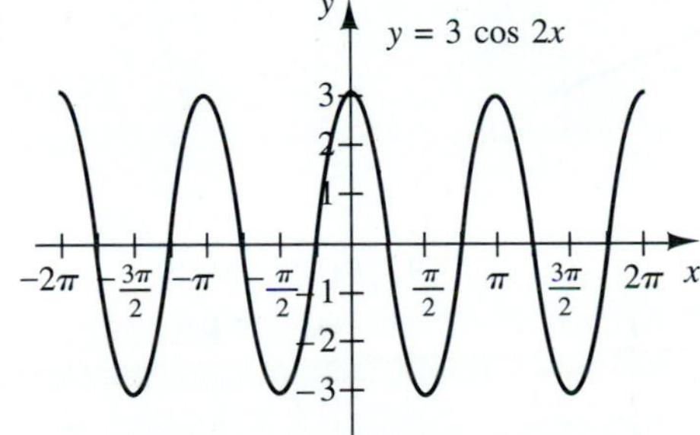

3.

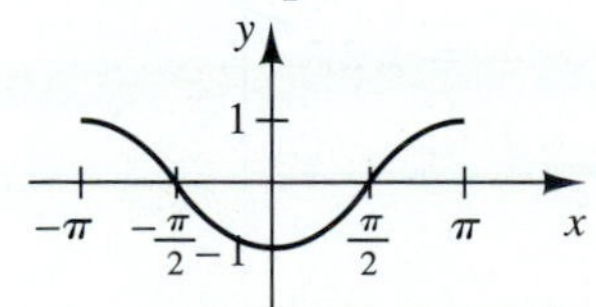

4.

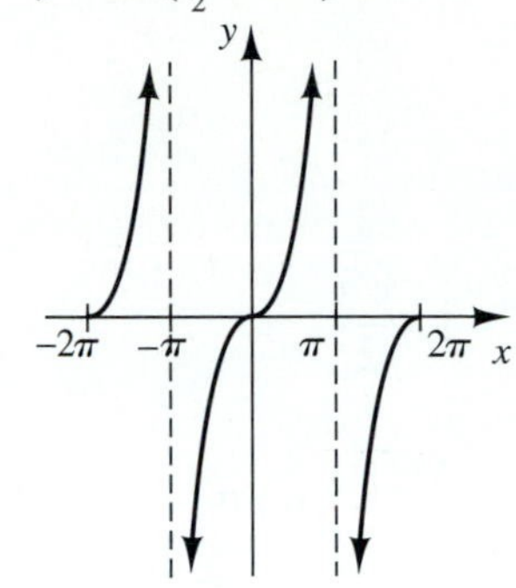

5.

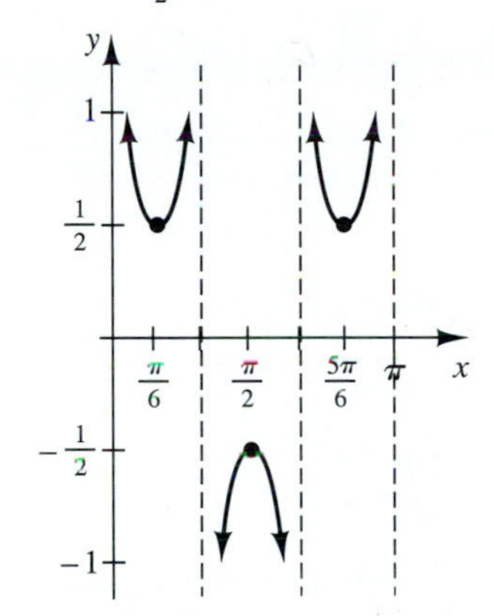

7. (a) $x = \frac{2\pi}{3} + 2n\pi, \frac{4\pi}{3} + 2n\pi$

(b) $\theta = \frac{\pi}{3} + n\pi, \frac{2\pi}{3} + n\pi$

8. (a) $x = 0, \frac{\pi}{3}, \pi, \frac{4\pi}{3}$ **(b)** $x = \frac{3\pi}{2}$

9. $D_f = \{x \mid x \neq \pi + 2n\pi, x \in R\}$

10. (a) $-\sin 53°$ **(b)** $-\cos\frac{\pi}{8}$

11. (a) $-\frac{\pi}{3}$ **(b)** π **(c)** **0**

(d) $\frac{\sqrt{2}}{2}$ **(e)** $-\frac{\pi}{3}$ **(f)** $\frac{5\pi}{6}$

12. $\theta = \cos^{-1}\left(\frac{x}{10}\right)$ **13.** $\sin^2\theta$

EXERCISES 8.1

1. $\dfrac{\sqrt{2}-\sqrt{6}}{4}$ **3.** $\dfrac{3+\sqrt{3}}{3-\sqrt{3}}$
5. $\dfrac{\sqrt{6}-\sqrt{2}}{4}$ **7.** $\dfrac{\sqrt{2}-\sqrt{6}}{4}$ **9.** $\dfrac{\sqrt{3}}{2}$
11. 0 **13.** $\sqrt{3}$
29. (a) $\dfrac{4\sqrt{119}-15}{60}$ **(b)** $\dfrac{-20-3\sqrt{119}}{60}$ **(c)** II
31. (a) $\dfrac{4\sqrt{5}+6}{15}$ **(b)** $\dfrac{4\sqrt{5}+6}{8-3\sqrt{5}}$
33. (a) $\dfrac{4\sqrt{33}+12}{35}$ **(b)** $\dfrac{-16-3\sqrt{33}}{35}$

EXERCISES 8.2

1. $-\dfrac{336}{625}$ **3.** $\dfrac{15}{8}$ **5. (a)** $\dfrac{3}{\sqrt{10}}$ **(b)** $-\dfrac{1}{\sqrt{10}}$ **(c)** -3
23. $x = 0, \dfrac{\pi}{3}, \pi, \dfrac{5\pi}{3}$ **25.** $x = 0, \pi$
27. $x = 0, \dfrac{\pi}{2}, \pi, \dfrac{3\pi}{2}, \dfrac{\pi}{4}, \dfrac{3\pi}{4}, \dfrac{5\pi}{4}, \dfrac{7\pi}{4}$
29. $x = \dfrac{\pi}{12}, \dfrac{5\pi}{12}, \dfrac{13\pi}{12}, \dfrac{17\pi}{12}, \dfrac{3\pi}{4}, \dfrac{7\pi}{4}$
31. $x = 0, \dfrac{\pi}{2}$ **33.** $\theta = \dfrac{\pi}{3}, \dfrac{5\pi}{3}$ **35.** $t = 0, \dfrac{\pi}{3}, \dfrac{5\pi}{3}$
37. $\dfrac{\sqrt{2+\sqrt{3}}}{2}$ **39.** $\sqrt{\dfrac{2+\sqrt{2}}{2-\sqrt{2}}}$
41. $\dfrac{1}{8}\cos 4t + \dfrac{1}{2}\cos 2t + \dfrac{3}{8}$ **43.** $V = 16 \sin \theta$
45. $\sin 2\theta = \dfrac{8x}{x^2+16}$ **47.** $x = \dfrac{-3+3\sqrt{5}}{2} \approx 1.9$
49. $\cos 2\theta = 1 - \dfrac{a^2}{8}$
51. $\sin 3\theta = 2 \sin \theta \cos^2 \theta + \sin \theta - 2 \sin^3 \theta$
57. (a) $\dfrac{1}{2}(\sin 8x + \sin 2x)$ **(b)** $\dfrac{1}{2}(\cos 2x - \cos 8x)$
(c) $\dfrac{1}{2}\left(\cos \dfrac{3x}{2} + \cos \dfrac{x}{2}\right)$
59. (a) $2 \sin 4x \cos x$ **(b)** $-2 \sin \dfrac{11x}{2} \sin \dfrac{3x}{2}$
(c) $2 \cos \dfrac{7x}{4} \sin \dfrac{5x}{4}$

EXERCISES 8.3

1. $C = 65°$; $b = 7.0$; $c = 11.0$
3. $B = 36.5°$; $C = 71.5°$; $c = 23.9$
5. $A = 36.3°$; $B = 62.6°$; $C = 81.1°$
7. $A = 45°$; $a = 26.8$; $b = 35.6$
9. $B = 12.8°$; $C = 29.2°$; $a = 15.1$ **11.** None
13. 1 triangle **15.** 221.8 meters **17.** 154.6 km
19. 1.8 km **21.** 3.9 **23.** 63.7 feet to first and third base; 66.8 feet to second base **25.** 28.3° **27.** 211.6 feet
29. Range for cannon A = 1.77 miles; range for cannon B = 2.77 miles **31.** 2.2 miles **33.** 72.8 miles
35. 2163.3 feet **37.** 3.1 km **39.** 240 feet
41. $A = 68.2$

EXERCISES 8.4

1. **3.**

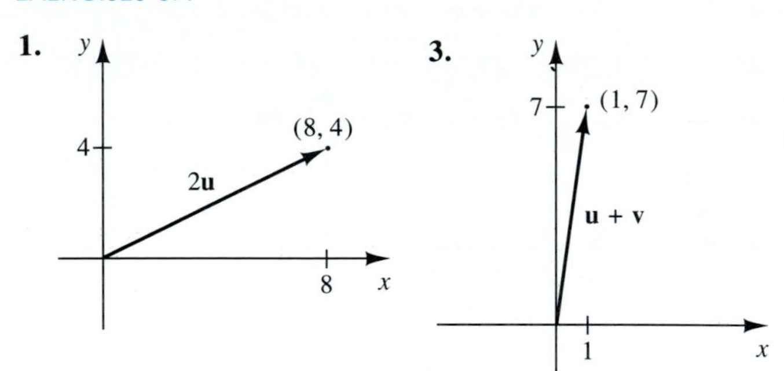

5.

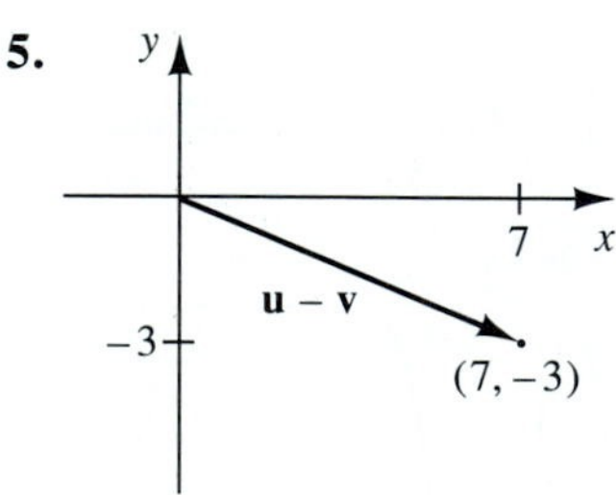

7. $|\overrightarrow{OQ}| = \sqrt{5}$ **9.** $|\overrightarrow{PQ}| = 2\sqrt{5}$

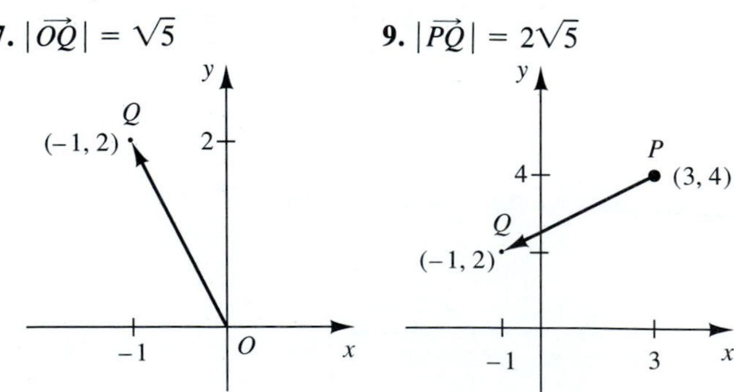

11. $|\overrightarrow{OP} + \overrightarrow{OQ}| = 2\sqrt{10}$

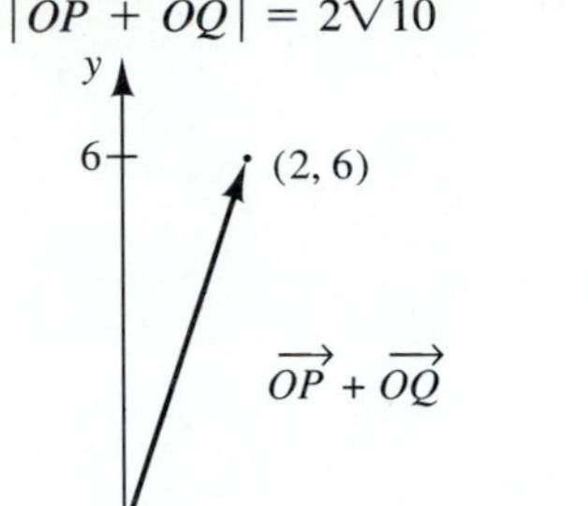

13. $|\overrightarrow{OR} - \overrightarrow{OS}| = 3\sqrt{5}$

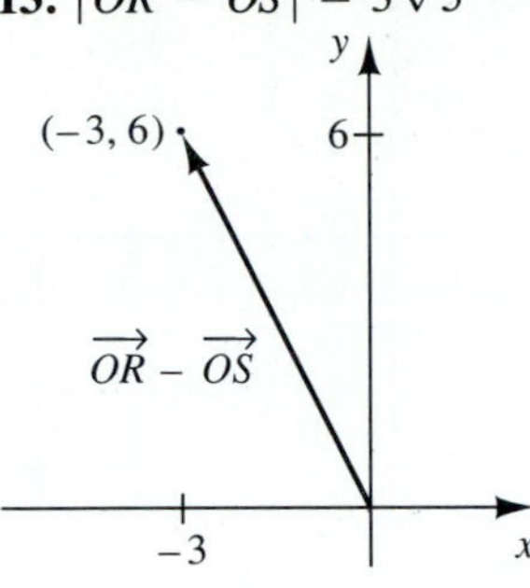

15. $\langle -1, 6\rangle$; $-\mathbf{i} + 6\mathbf{j}$ **17.** $\langle -12, -8\rangle$; $-12\mathbf{i} - 8\mathbf{j}$
19. $\langle -9, -26\rangle$; $-9\mathbf{i} - 26\mathbf{j}$
21. $|\mathbf{A} + \mathbf{B}| = \sqrt{13}$; $\theta = 33.7°$
23. $|\mathbf{A} + \mathbf{B}| = \sqrt{145}$; $\theta = 4.8°$
25. $|\mathbf{u} + \mathbf{v}| = 6\sqrt{2}$; $\theta = 18.4°$
27. $|\mathbf{v} + \mathbf{w}| = \sqrt{34}$; $\theta = 139.4°$ **29.** $\left\langle \frac{7}{2}, \frac{7\sqrt{3}}{2}\right\rangle$
31. $\langle -58.8, 80.9\rangle$ **33.** $\langle 2, 7\rangle$ **35.** $5\mathbf{i} - 6\mathbf{j}$
37. $-13\mathbf{i} + 10\mathbf{j}$ **39.** $\frac{1}{\sqrt{34}}\langle 3, 5\rangle$ **41.** $\frac{1}{\sqrt{10}}\langle -1, -3\rangle$
43. $\frac{1}{\sqrt{29}}\langle -2, -5\rangle$ **45.** 42.8 pounds **47.** 76.9°
49. Speed = 3.7 mi/hr; course = N60.6°E **51.** Ground-speed: 396.5 mi/hr; course: N33.4°W
53. 42.4 mi/hr **55.** Parallel component: 6.2 pounds; perpendicular component: 19 pounds **57.** 337.3 pounds
59. 212.7 miles at a heading of N87°W
61. $200 - 50\sqrt{2} \approx 129.3$ pounds **63.** -27 **65.** $\frac{1}{3}$
67. 15 **69.** 21 **71.** 16

EXERCISES 8.5

1. $2\left(\cos\frac{11\pi}{6} + i\sin\frac{11\pi}{6}\right)$ **3.** $5\sqrt{2} - i5\sqrt{2}$
5. $\sqrt{53}(\cos 285.95° + i\sin 285.95°)$
7. $6(\cos 0 + i\sin 0)$ **9.** $-2i$
11. $\sqrt{21}(\cos 60.79° + i\sin 60.79°)$ **13.** $\frac{9}{2} - \frac{9\sqrt{3}}{2}i$
15. $4(\cos\pi + i\sin\pi)$
17. $z_1 \cdot z_2 = 12\left(\cos\frac{4\pi}{9} + i\sin\frac{4\pi}{9}\right)$;
$\frac{z_1}{z_2} = 3\left(\cos\frac{2\pi}{9} + i\sin\frac{2\pi}{9}\right)$
19. $z_1 \cdot z_2 = 5\left(\cos\frac{23\pi}{24} + i\sin\frac{23\pi}{24}\right)$;
$\frac{z_1}{z_2} = 5\left(\cos\frac{17\pi}{24} + i\sin\frac{17\pi}{24}\right)$
21. 6; $6(\cos 360° + i\sin 360°)$
23. $\frac{1}{2} + \frac{1}{2}i$; $\frac{1}{\sqrt{2}}\left(\cos\frac{\pi}{4} + i\sin\frac{\pi}{4}\right)$
25. $243\left(\cos\frac{5\pi}{6} + i\sin\frac{5\pi}{6}\right)$ **27.** $8\sqrt{2}\left(\cos\frac{3\pi}{4} + i\sin\frac{3\pi}{4}\right)$
29. $216(\cos 120° + i\sin 120°)$
31. $16(\cos 0 + i\sin 0)$
33. $972\sqrt{2}\left(\cos\frac{7\pi}{4} + i\sin\frac{7\pi}{4}\right)$
35. $\frac{\sqrt{2}}{4}\left(\cos\frac{\pi}{4} + i\sin\frac{\pi}{4}\right)$
37. $4\left(\cos\frac{2\pi}{3} + i\sin\frac{2\pi}{3}\right)$
39. $w_0 = \sqrt[4]{6}\left(\cos\frac{\pi}{16} + i\sin\frac{\pi}{16}\right)$;
$w_1 = \sqrt[4]{6}\left(\cos\frac{9\pi}{16} + i\sin\frac{9\pi}{16}\right)$;
$w_2 = \sqrt[4]{6}\left(\cos\frac{17\pi}{16} + i\sin\frac{17\pi}{16}\right)$;
$w_3 = \sqrt[4]{6}\left(\cos\frac{25\pi}{16} + i\sin\frac{25\pi}{16}\right)$
41. $w_0 = \cos 0 + i\sin 0$; $w_1 = \cos\frac{\pi}{3} + i\sin\frac{\pi}{3}$;
$w_2 = \cos\frac{2\pi}{3} + i\sin\frac{2\pi}{3}$; $w_3 = \cos\pi + i\sin\pi$;
$w_4 = \cos\frac{4\pi}{3} + i\sin\frac{4\pi}{3}$; $w_5 = \cos\frac{5\pi}{3} + i\sin\frac{5\pi}{3}$
43. $w_0 = \sqrt[5]{2}\left(\cos\frac{\pi}{6} + i\sin\frac{\pi}{6}\right)$;
$w_1 = \sqrt[5]{2}\left(\cos\frac{17\pi}{30} + i\sin\frac{17\pi}{30}\right)$;
$w_2 = \sqrt[5]{2}\left(\cos\frac{29\pi}{30} + i\sin\frac{29\pi}{30}\right)$;
$w_3 = \sqrt[5]{2}\left(\cos\frac{41\pi}{30} + i\sin\frac{41\pi}{30}\right)$;
$w_4 = \sqrt[5]{2}\left(\cos\frac{53\pi}{30} + i\sin\frac{53\pi}{30}\right)$

45. $w_0 = \sqrt[3]{2}\left(\cos\frac{\pi}{9} + i\sin\frac{\pi}{9}\right) = 1.18 + 0.43i;$

$w_1 = \sqrt[3]{2}\left(\cos\frac{7\pi}{9} + i\sin\frac{7\pi}{9}\right) = -0.97 + 0.81i;$

$w_2 = \sqrt[3]{2}\left(\cos\frac{13\pi}{9} + i\sin\frac{13\pi}{9}\right) = -0.22 - 1.24i$

47. $w_0 = \frac{\sqrt{2+\sqrt{2}}}{2} + i\frac{\sqrt{2-\sqrt{2}}}{2};$

$w_1 = -\frac{\sqrt{2-\sqrt{2}}}{2} + i\frac{\sqrt{2+\sqrt{2}}}{2};$

$w_2 = -\frac{\sqrt{2+\sqrt{2}}}{2} - i\frac{\sqrt{2-\sqrt{2}}}{2};$

$w_3 = \frac{\sqrt{2-\sqrt{2}}}{2} - i\frac{\sqrt{2+\sqrt{2}}}{2}$

EXERCISES 8.6

1.–5.

$\theta = \frac{\pi}{2}$, $\theta = \frac{\pi}{3}$, $\theta = \frac{\pi}{4}$, $(3, \frac{\pi}{4})$, $\theta = \pi$, $(0, \frac{5\pi}{6})$, 3, 4, $(-1, \frac{\pi}{2})$, $\theta = 0$, $(4, -\pi)$, $(2, -\frac{2\pi}{3})$, 2, $(-2, \frac{-3\pi}{2})$, $\theta = \frac{3\pi}{2}$

7. $\left(-5, -\frac{2\pi}{3}\right); \left(5, -\frac{5\pi}{3}\right)$ **9.** $\left(-2, \frac{5\pi}{2}\right); \left(2, -\frac{\pi}{2}\right)$

11. $(-1, \sqrt{3})$ **13.** $\left(8, \frac{7\pi}{6}\right)$ **15.** $(2.70, 1.30)$

17. $(6, \pi)$ **19.** $(-1.98, 0.28)$

21. $\left(\frac{\sqrt{\pi^2 + 16}}{4}, 2.48\right)$ **23.** $x^2 + y^2 = 5x$; circle

25. $y = \sqrt{3}\,x$; straight line **27.** $y = -1$; straight line

29. $y^2 = \frac{x^4}{1 - x^2}$

31. $y = \frac{1}{4}x^2 - 1$; parabola **33.** $r = 4$

35. $r^2 = \csc 2\theta$ **37.** $r = -2\cos\theta - 2\sin\theta$

39. $r = \frac{4}{\sin\theta}$

41.

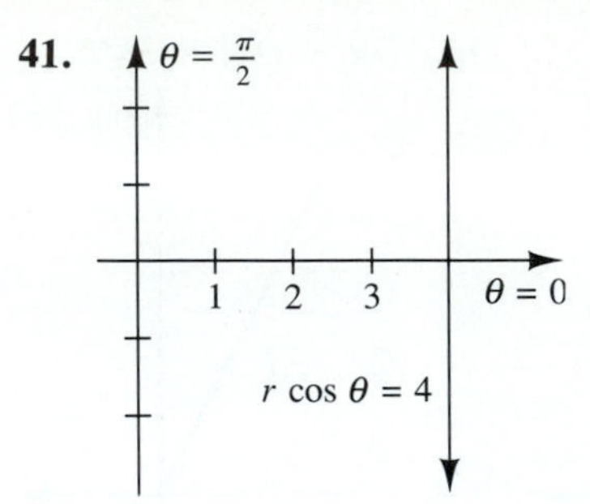

43.

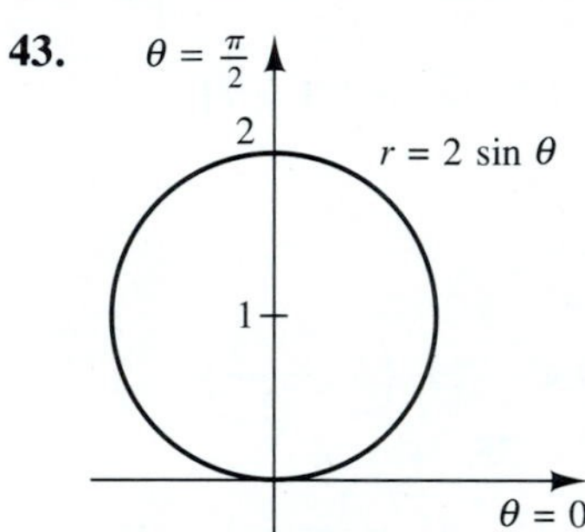

45.

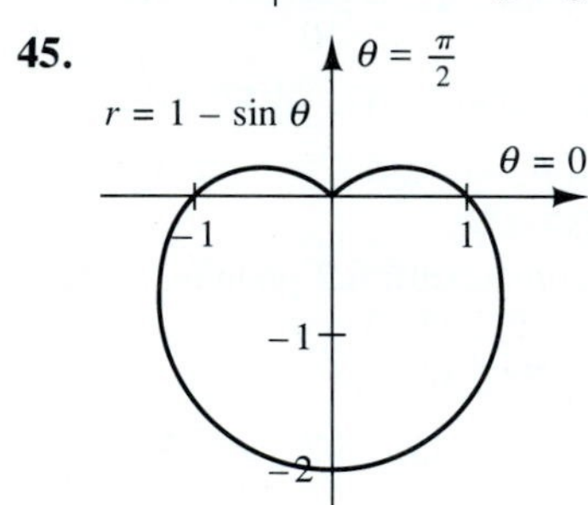

47.

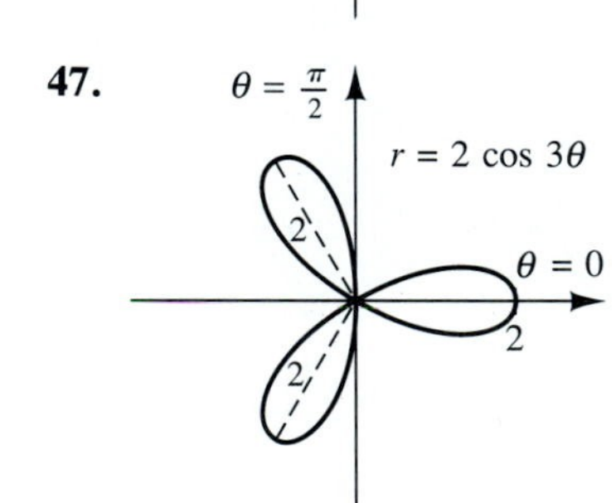

49.

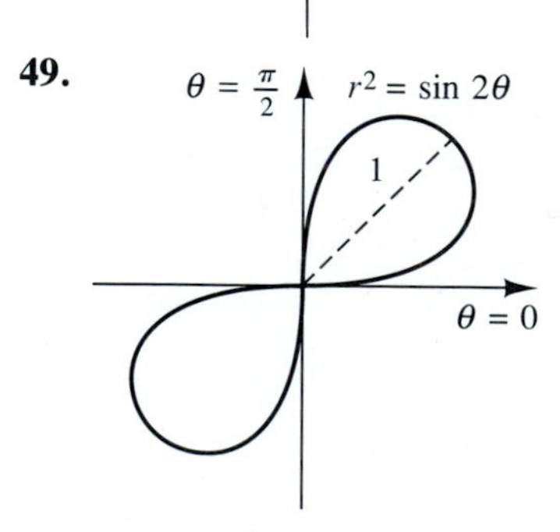

51.

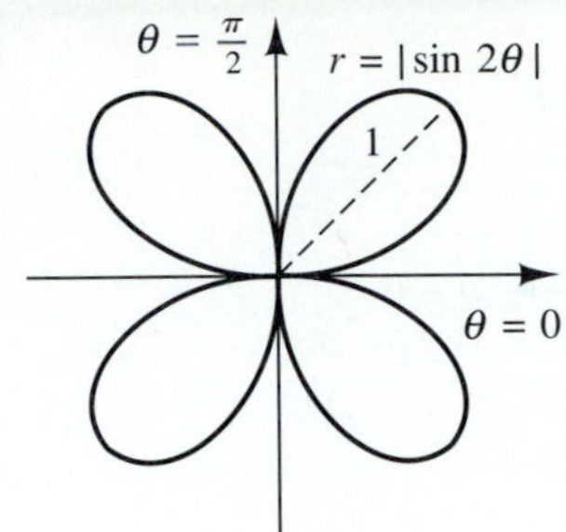

53.

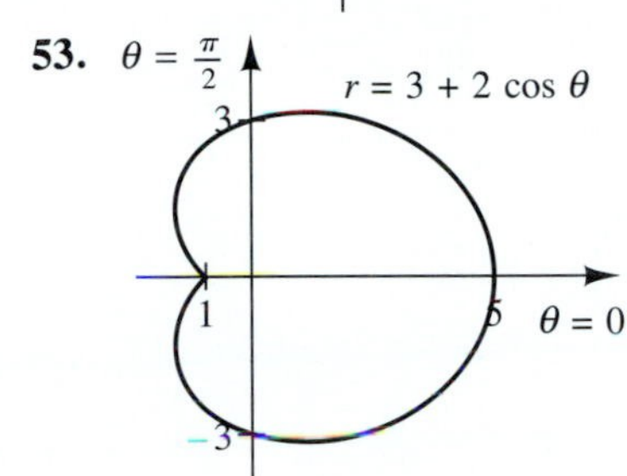

55.

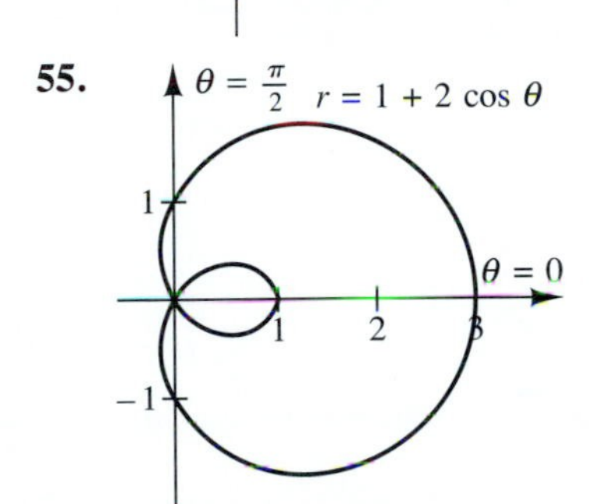

57.

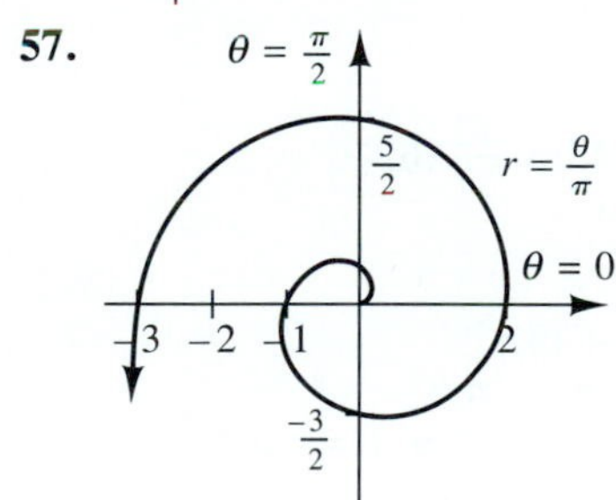

59.

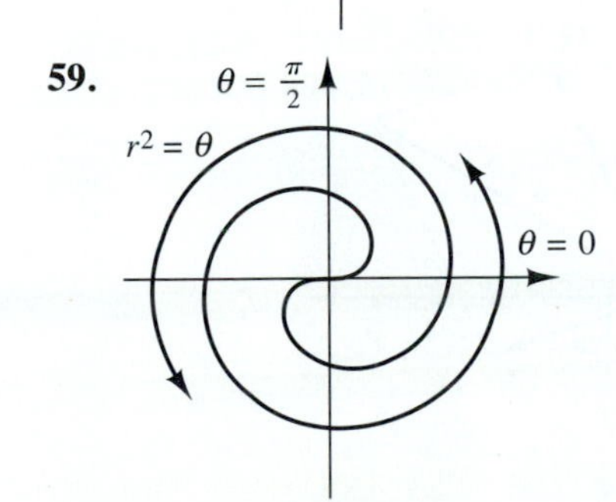

CHAPTER 8 REVIEW EXERCISES

1. $\frac{\sqrt{6}+\sqrt{2}}{4}$ **3.** $\frac{1+\sqrt{3}}{1-\sqrt{3}}$ **17.** $x = \frac{\pi}{6}; \frac{5\pi}{6}; \frac{3\pi}{2}$
19. $x = \frac{\pi}{6}, \frac{5\pi}{6}, \frac{7\pi}{6}, \frac{11\pi}{6}$
21. $x = \frac{\pi}{4}, \frac{3\pi}{4}, \frac{5\pi}{4}, \frac{7\pi}{4}, \frac{\pi}{12}, \frac{5\pi}{12}, \frac{13\pi}{12}, \frac{17\pi}{12}$
23. $x = 0, \frac{4\pi}{3}$ **25.** $x = 0, \frac{\pi}{2}, \frac{3\pi}{2}$
27. $\frac{4\sqrt{21}}{25}$ **29.** $\frac{72}{65}$ **31. (a)** $\frac{3}{\sqrt{10}}$
(b) $-\frac{1}{\sqrt{10}}$ **(c)** -3 **33.** $\cos 2\theta = 1 - \frac{2x^2}{25}$
35. $\tan 2\theta = \frac{2a\sqrt{25-a^2}}{2a^2-25}$ **37.** $C = 78°$; $b = 7.9$; $c = 14.6$ **39.** $A = 32.3°$; $C = 108.7°$; $b = 10.6$
41. $A = 16.2°$; $B = 40.6°$; $C = 123.2$ **43.** 0 **45.** 1
47. 200.2 miles **49.** 235.6 feet **51.** $\mathbf{u} + \mathbf{v} = \langle 7, 10\rangle$; $|\mathbf{u} + \mathbf{v}| = \sqrt{149}$ **53.** $\mathbf{u} + \mathbf{v} = \langle 5, 0\rangle$; $|\mathbf{u} + \mathbf{v}| = 5$
55. $\langle 6, -2\rangle$; $6\mathbf{i} - 2\mathbf{j}$ **57.** $\langle 20, 2\rangle$; $20\mathbf{i} + 2\mathbf{j}$
59. Horizontal component $= \frac{3\sqrt{2}}{2}$; vertical component $= \frac{3\sqrt{2}}{2}$; $\left\langle \frac{3\sqrt{2}}{2}, \frac{3\sqrt{2}}{2} \right\rangle$
61. Horizontal component $= -4.7$; vertical component $= 8.8$; $\langle -4.7, 8.8\rangle$
63. $\frac{1}{\sqrt{41}}\langle 5, 4\rangle$ **65.** $\frac{1}{\sqrt{13}}\langle 2, 3\rangle$ **67.** $|\mathbf{A} + \mathbf{B}| = \sqrt{41}$; $\theta = 51.3°$ **69.** $|\mathbf{A} + \mathbf{B}| = \sqrt{149}$; $\theta = 36.6°$
71. 31.1 pounds **73.** 7.0 pounds **75.** Groundspeed: 508.7 mi/hr; course: N14.9°E **77.** Parallel component = 10.4 pounds; perpendicular component = 48.9 pounds **79.** 298.5 pounds **81.** $\sqrt{3}(\cos 324.74° + i \sin 324.74°)$
83. $-4\sqrt{2} + 4i\sqrt{2}$ **85.** $z_1 \cdot z_2 = 12\left(\cos\frac{8\pi}{15} + i\sin\frac{8\pi}{15}\right)$; $\frac{z_1}{z_2} = \frac{4}{3}\left(\cos\frac{2\pi}{15} + i\sin\frac{2\pi}{15}\right)$ **87.** $256\left(\cos\frac{4\pi}{3} + i\sin\frac{4\pi}{3}\right)$
89. $8\left(\cos\frac{\pi}{2} + i\sin\frac{\pi}{2}\right)$ **91.** $128\sqrt{2}\left(\cos\frac{\pi}{4} + i\sin\frac{\pi}{4}\right)$
93. $w_0 = \sqrt{2}\left(\cos\frac{11\pi}{24} + i\sin\frac{11\pi}{24}\right)$;
$w_1 = \sqrt{2}\left(\cos\frac{23\pi}{24} + i\sin\frac{23\pi}{24}\right)$;
$w_2 = \sqrt{2}\left(\cos\frac{35\pi}{24} + i\sin\frac{35\pi}{24}\right)$;
$w_3 = \sqrt{2}\left(\cos\frac{47\pi}{24} + i\sin\frac{47\pi}{24}\right)$

95. $w_0 = \cos 0 + i \sin 0$;
$w_1 = \cos \frac{2\pi}{5} + i \sin \frac{2\pi}{5}$; $w_2 = \cos \frac{4\pi}{5} + i \sin \frac{4\pi}{5}$;
$w_3 = \cos \frac{6\pi}{5} + i \sin \frac{6\pi}{5}$; $w_4 = \cos \frac{8\pi}{5} + i \sin \frac{8\pi}{5}$

97. $\left(3, -\frac{7\pi}{4}\right)$; $\left(-3, \frac{5\pi}{4}\right)$; **99.** $(1, 0)$; $(-1, -\pi)$

101. $\left(-\frac{3\sqrt{2}}{2}, \frac{3\sqrt{2}}{2}\right)$ **103.** $\left(10, \frac{7\pi}{6}\right)$

105. $x^2 + y^2 = 6y$; circle **107.** $y = x$; straight line

109. $y = 4$; straight line **111.** $x = \frac{1}{2}y^2 - \frac{1}{2}$

113.

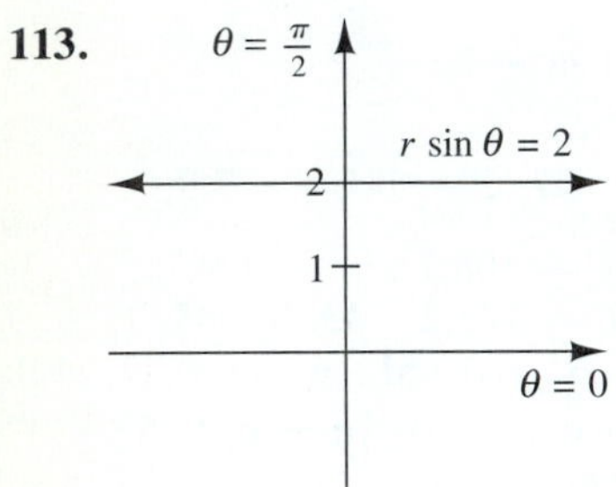

115.

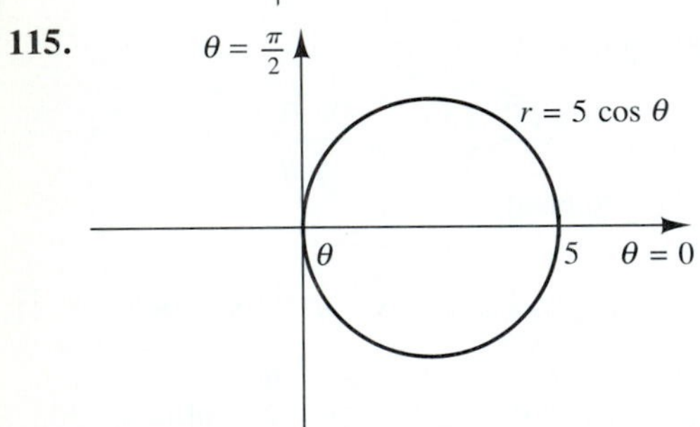

117.

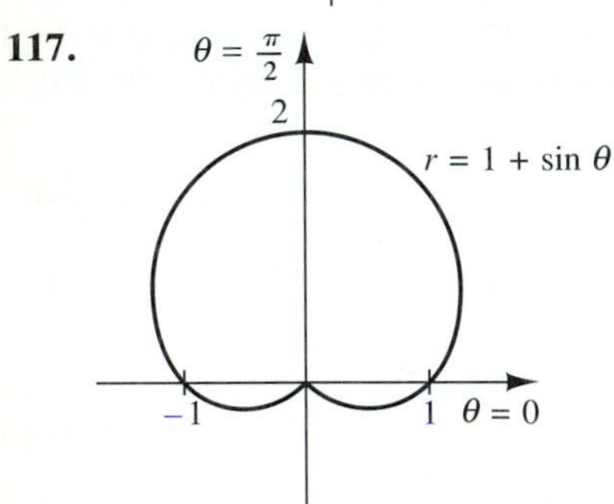

119.

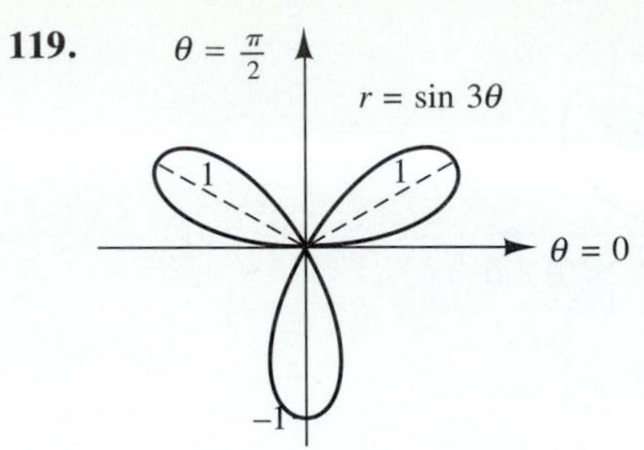

121.

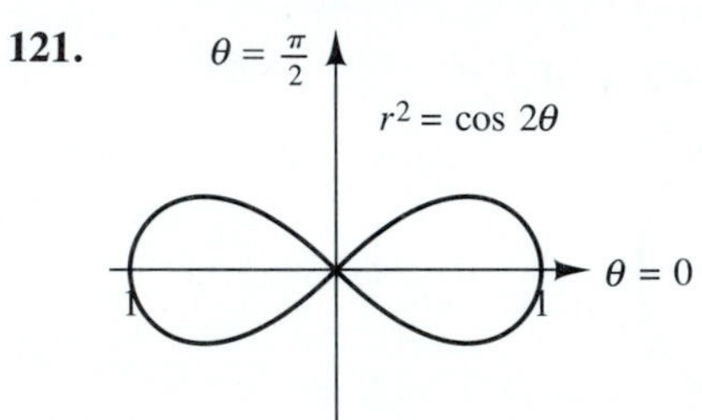

123.

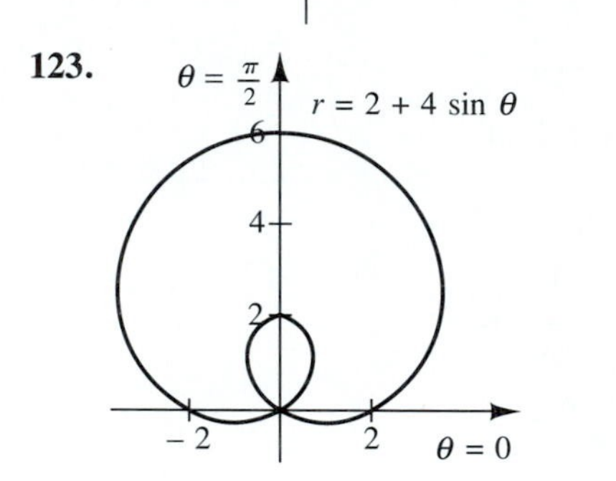

CHAPTER 8 PRACTICE TEST

1. $\frac{\sqrt{2} - \sqrt{6}}{4}$ **3.** $\theta = \pi$ **4. (a)** $\frac{12}{37}$ **(b)** $\frac{35}{37}$

5. $\tan 2\theta = \frac{4a\sqrt{9 - 4a^2}}{8a^2 - 9}$

6. (a) $C = 62°$; $b = 14.1$; $c = 18.3°$
(b) $A = 23.1°$; $C = 51.9°$; $b = 24.6$

7. 272.6 ft **8. (a)** $u + v = \langle 5, 3\rangle$; $|\mathbf{u} + \mathbf{v}| = \sqrt{34}$

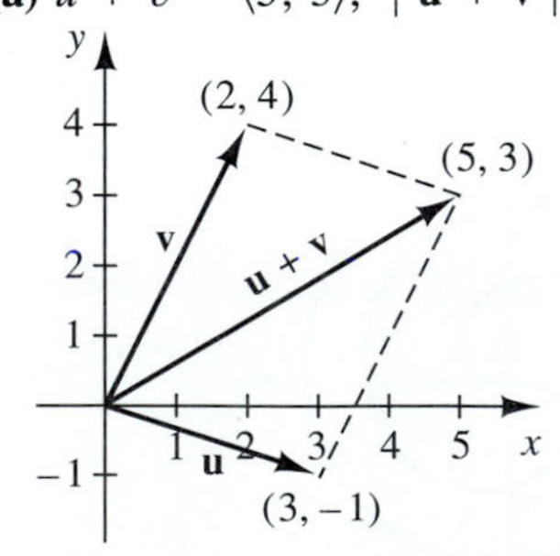

(b) $\mathbf{u} + \mathbf{v} = \langle 6,0 \rangle$: $|\mathbf{u} + \mathbf{v}| = 6$

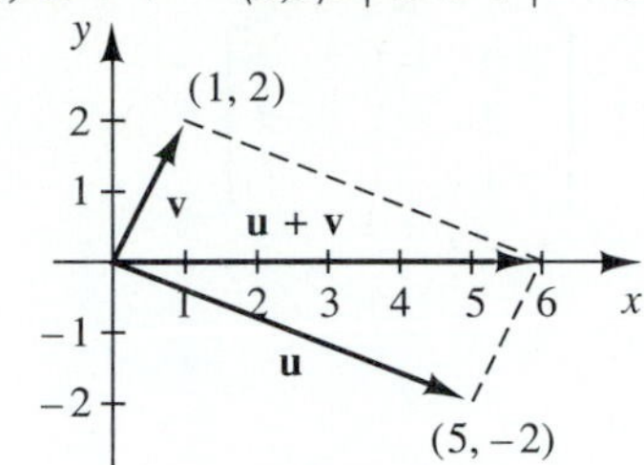

9. (a) $\langle 21, -17 \rangle$ **(b)** $-6\mathbf{i} + 46\mathbf{j}$
10. Horizontal component = 4.5; vertical component = 6.6
11. $\frac{1}{\sqrt{41}}\langle 4, 5 \rangle$ **12.** Resultant = $\langle -4,6 \rangle$; angle = 47.7°
13. 21.1 mi/hr **14.** A force greater than 38.9 pounds
15. $2\left(\cos \frac{4\pi}{3} + i \sin \frac{4\pi}{3}\right)$ **16.** $16(\cos 0 + i \sin 0) = 16$
17. $\left(5, -\frac{5\pi}{6}\right); \left(-5, \frac{\pi}{6}\right)$ **18.**

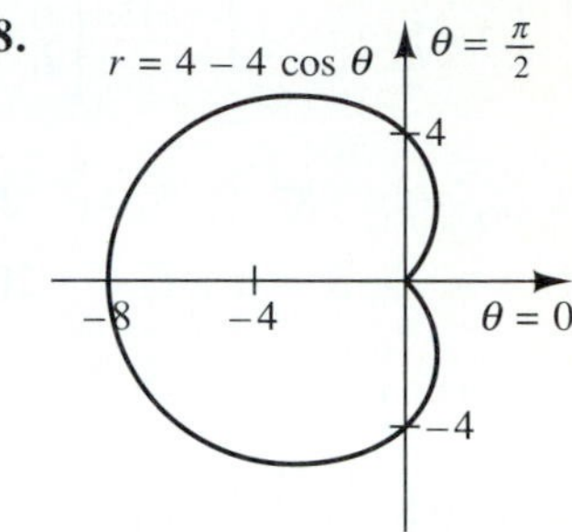

CHAPTER 9

EXERCISES 9.1

1. (4, 4) **3.** (5, 0) **5.** (3, 7) **7.** (−10, −45)
9. $\left(\frac{4}{3}, 0\right)$ **11.** $a = 3, b = 2$ **13.** $\left(\frac{-17}{5}, \frac{48}{5}\right)$
15. $\left(\frac{39}{95}, \frac{-250}{19}\right)$
17. 10 liters of the 30% alcohol solution: 20 liters of the 45% alcohol solution
19. \$6000 at 9.5%: \$14,000 at 8%
21. Flat fee = \$29.50; mileage fee = 32¢ per mile
23. Speed of the plane = 600 mi/hr; speed of wind = 30 mi/hr
25. 60 of the more expensive model; 120 of the less expensive model
27. Let m = no. of miles and C = the daily cost. For the Goodman Agency, $C = 18 + 0.28m$; for the Hirsch Agency, $C = 14 + 0.32m$. For mileage below 100 miles, the Hirsch Agency is less expensive; for mileage above 100 miles, the Goodman Agency is less expensive.
29. For air time of more than 100 minutes per month, Plan B is less expensive; for air time of less than 100 minutes, Plan A is less expensive.
31. At $t = 1$, $64 = -16 + v_0 + s_0$; at $t = 2$, $96 = -64 + 2v_0 + s_0$; $v_0 = 80$ ft/sec; $s_0 = 0$ feet
33. $A = 2, B = -4$ **35.** $A=2, B = -3$
37. (−4, −7)

EXERCISES 9.2

1. (2, 2, 2) **3.** $\left(2, -\frac{3}{2}, -\frac{1}{2}\right)$ **5.** Inconsistent
7. (5, 0.5, 0.3) **9.** (−8, 12, −3) **11.** (2, 3, −2)
13. (−1, 0, −2) **15.** Inconsistent
17. Let b = the amount in the bank accout, c = the amount invested in the certificate of deposit, and s = the amount invested in the stock. The three equations are $b + c + s = 19{,}750$, $0.06b + 0.08c + 0.105s = 1680$, and $0.06b + 0.08c = 0.105s$: \$5000 in the bank account, \$6750 in the certificate, and \$8000 in the stock.
19. 150 of model A, 110 of model B, and 140 of model C
21. $15\frac{5}{9} \approx 15.56$ liters of the 10% solution, $35\frac{5}{9} \approx 35.56$ liters of the 25% solution, and $48\frac{8}{9} \approx 48.89$ liters of the 40% solution
23. $f(x) = 3x^2 - 2x + 1$ **25.** $A = 5, B = -5, C = 7$

EXERCISES 9.3

1. $\left[\begin{array}{cc|c} 7 & 1 & 11 \\ 1 & -3 & 4 \end{array}\right]$
3. $\left[\begin{array}{ccc|c} 3 & -2 & 1 & 15 \\ 2 & 1 & -3 & 10 \\ 5 & -3 & 1 & -2 \end{array}\right]$
5. $\left[\begin{array}{cc|c} 1 & 3 & 2 \\ 0 & -11 & -10 \end{array}\right]$
7. $\left[\begin{array}{ccc|c} 1 & 2 & 1 & 0 \\ 0 & 2 & 1 & 1 \\ 0 & 0 & 3 & -9 \end{array}\right]$
9. (5, 1) **11.** (2, −3) **13.** (3, 7) **15.** (10, −45)
17. (0, 1) **19.** (2, 2, 2) **21.** (2, −1, 3)
23. Inconsistent **25.** $a = \frac{1}{2}, b = 2, c = 3$

27. (2, 1, 0, −1) **29.** $\left[\begin{array}{cc|c} 1 & 0 & 2 \\ 0 & 1 & -1 \end{array}\right]$

31. $\left[\begin{array}{cc|c} 1 & 0 & -4 \\ 0 & 1 & 1 \end{array}\right]$ **33.** $\left[\begin{array}{ccc|c} 1 & 0 & 0 & 3 \\ 0 & 1 & 0 & 1 \\ 0 & 0 & 1 & -2 \end{array}\right]$

35. (2, −1) **37.** (1, 3, 1) **39.** $\left(-\frac{5}{2}, \frac{5}{2}, \frac{1}{2}\right)$

41. $A = 25, B = 13, C = -20, D = -10$

EXERCISES 9.4

1. $\begin{bmatrix} 3 & 2 \\ 8 & 2 \end{bmatrix}$ **3.** $\begin{bmatrix} 0 & 0 \\ 0 & 0 \\ 0 & 0 \end{bmatrix}$

5. Not defined

7. $\begin{bmatrix} 0 & 9 \\ 51 & 0 \end{bmatrix}$ **9.** $\begin{bmatrix} -2 & 3 \\ 1 & 0 \end{bmatrix}$ **11.** $\begin{bmatrix} \frac{10}{3} & \frac{2}{3} \\ -\frac{4}{3} & 2 \end{bmatrix}$

13. $\begin{bmatrix} -3 & -4 \\ 1 & -3 \end{bmatrix}$ **15.** $\begin{bmatrix} -1 & -7 \\ 0 & -3 \end{bmatrix}$

17. $\begin{bmatrix} \frac{5}{2} & \frac{1}{2} \\ -1 & \frac{3}{2} \end{bmatrix}$ **19.** $\begin{bmatrix} \frac{9}{2} & \frac{5}{2} \\ -2 & \frac{3}{2} \end{bmatrix}$ **21.** [10] **23.** [3]

25. $\begin{bmatrix} 0 & 2 \\ -3 & 5 \end{bmatrix}$ **27.** Not defined **29.** $\begin{bmatrix} -30 & -2 \\ 39 & -11 \end{bmatrix}$

31. $\begin{bmatrix} 13 & -3 & -10 \\ 9 & -7 & -7 \\ -4 & -6 & 5 \end{bmatrix}$ **33.** $\begin{bmatrix} 5 & -7 \\ 3 & -14 \end{bmatrix}$

35. $\begin{bmatrix} 3 & 1 & 0 \\ -2 & 4 & 6 \\ 3 & 8 & -10 \end{bmatrix}$ **37.** False **39.** True

43. AB is not defined; BA is a 3×4 matrix.

47. Yes. The additive inverse for an $n \times n$ matrix A is the $n \times n$ matrix $-A = -1(A)$.

51. Brand A is less expensive when ordered from Raju (62.8¢); brands B and C are less expensive when ordered from George (48.6¢ and 41¢, respectively). If all three brands must be ordered from one company, then the least expensive company is Raju ($1.536).

53. The inner product is not defined. The matrix product is

$$\begin{bmatrix} 2 & -3 & 1 \\ 0 & 0 & 0 \\ 8 & -12 & 4 \end{bmatrix}$$

EXERCISES 9.5

1. $\begin{bmatrix} 3 & -1 \\ 3 & -2 \end{bmatrix}$ **5.** $\begin{bmatrix} 0 & \frac{1}{2} \\ \frac{1}{2} & -\frac{1}{4} \end{bmatrix}$ **7.** $\begin{bmatrix} -\frac{1}{16} & \frac{5}{16} \\ \frac{3}{16} & \frac{1}{16} \end{bmatrix} = \frac{1}{16}\begin{bmatrix} -1 & 5 \\ 3 & 1 \end{bmatrix}$

9. No inverse **11.** No inverse

13. $\frac{1}{2}\begin{bmatrix} -1 & 3 & -1 \\ -1 & 5 & -3 \\ 2 & -4 & 2 \end{bmatrix}$ **15.** $\begin{bmatrix} 2 & -1 & -\frac{3}{2} \\ 2 & -1 & -2 \\ -1 & 1 & \frac{3}{2} \end{bmatrix}$

17. No inverse

19. $\frac{1}{6}\begin{bmatrix} 1 & -2 & 3 \\ 3 & -6 & 3 \\ 4 & -2 & 0 \end{bmatrix}$

21. (5, −1) **23.** (6, 1) **25.** (3, 2) **27.** (−10, −45)
29. (0, 3) **31.** (2, 2, 2) **33.** (1, 0, −1)
35. Inconsistent **37.** (1, 0.1, −1)

39. **(a)** $16,666.67 invested in the certificate; $3333.33 invested in the stock. **(b)** $13,333.33 invested in the certificate; $6666.67 invested in the stock. **(c)** $10,000 invested in the certificate, $10,000 invested in the stock.

EXERCISES 9.6

1. 5 **3.** 14 **5.** 0 **7.** 19 **9.** −9 **11.** 0

13. (2, −1) **15.** $\left(\frac{17}{9}, -\frac{4}{3}\right)$ **17.** $\left(\frac{24}{35}, \frac{17}{35}\right)$

19. Dependent: $\left(a, \frac{2-a}{2}\right)$ **21.** Inconsistent

23. (1, 3, 1) **25.** (1, −1, 0)

27. $\left(4, -\frac{1}{2}, -1\right)$ **29.** $\left(-\frac{3}{2}, \frac{5}{4}, 0\right)$

31. $\left(\frac{5}{2}, \frac{1}{4}, -\frac{3}{2}\right)$

33. $6000 invested at 6%; $4000 invested at 8.5%

35. 36 of model A; 46 of model B

37. 80 of model A; 50 of model B; 60 of model C

EXERCISES 9.7

1. $\begin{vmatrix} 2 & 5 & -1 \\ 7 & 4 & -3 \\ -2 & 1 & 9 \end{vmatrix}$ **3.** $-\begin{vmatrix} 2 & 5 & -1 \\ -2 & 1 & 9 \\ -6 & -4 & -5 \end{vmatrix}$

5. $\begin{vmatrix} 2 & 5 & 3 \\ -2 & 1 & 8 \\ -6 & -4 & 6 \end{vmatrix}$

7. 57 **9.** 24 **11.** −29 **13.** 0

15. (−1, 0, 1, 2) **17.** (1, 2, −1, 0)

19. Dependent: $\left(a, -\frac{a}{3}, -\frac{2a}{3}, 0\right)$

EXERCISES 9.8

1.

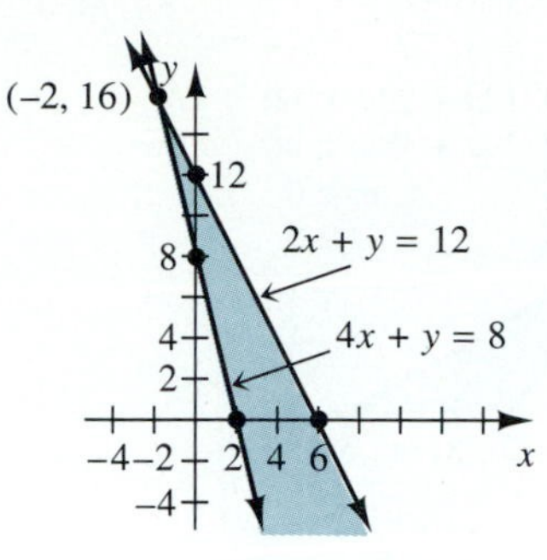

3.

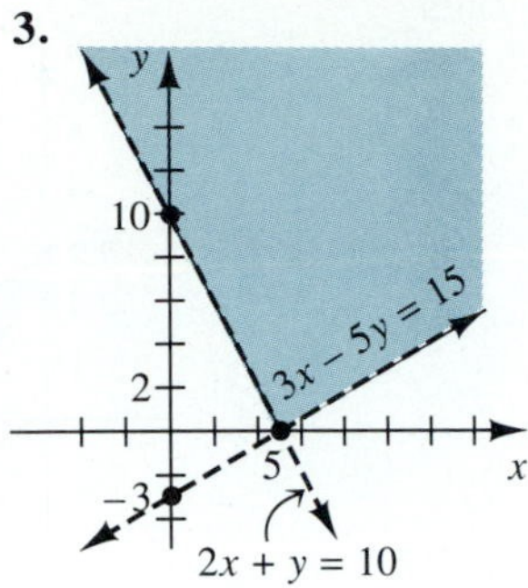

5.

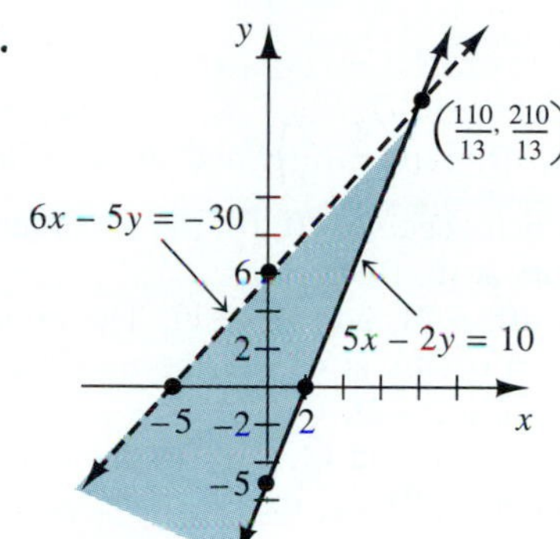

7.

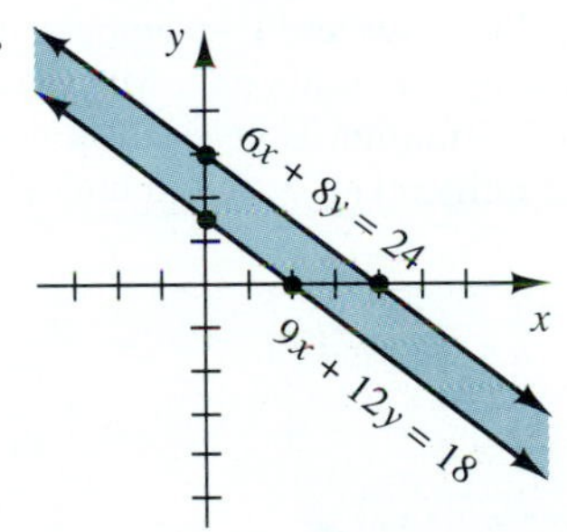

9.

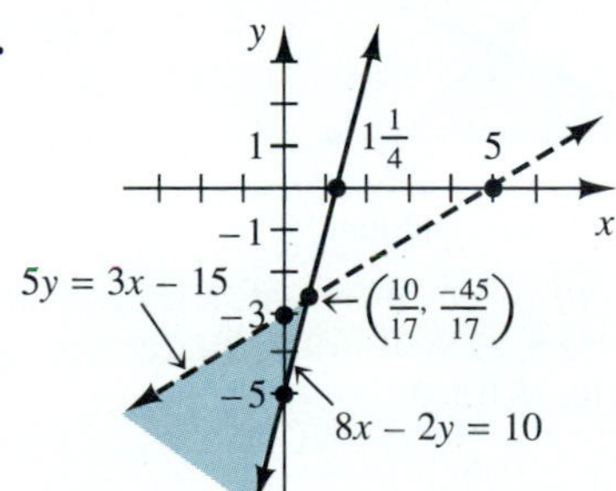

11.

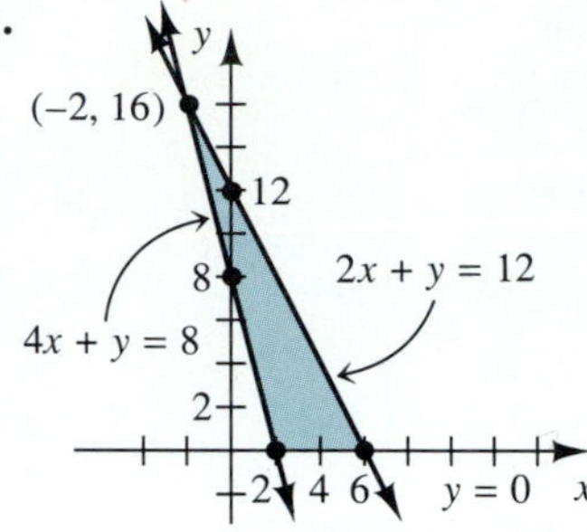

13.

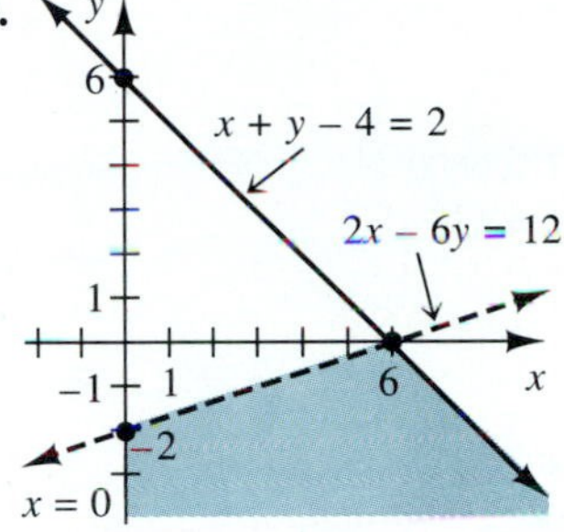

15.

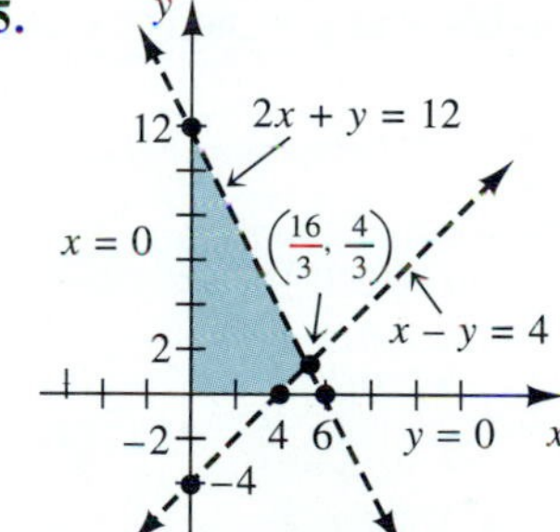

17.

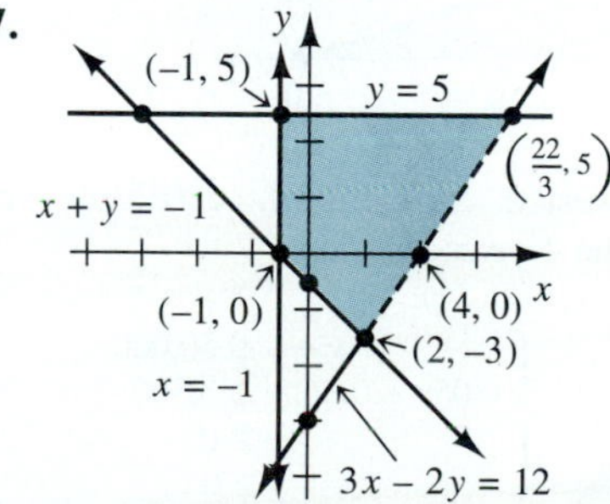

19.

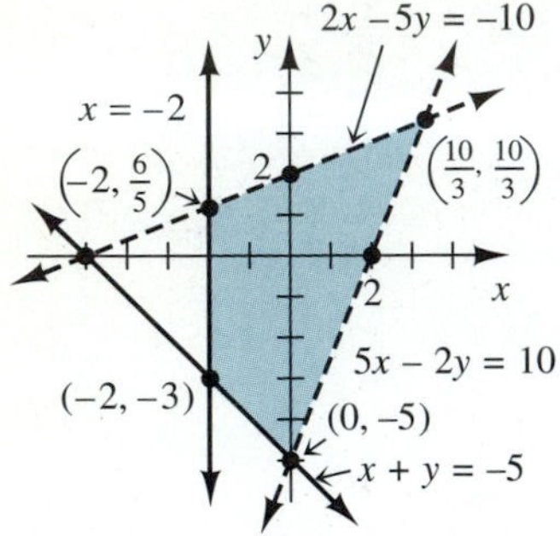

21. Let V = no. of VCRs and T = no. of TVs.

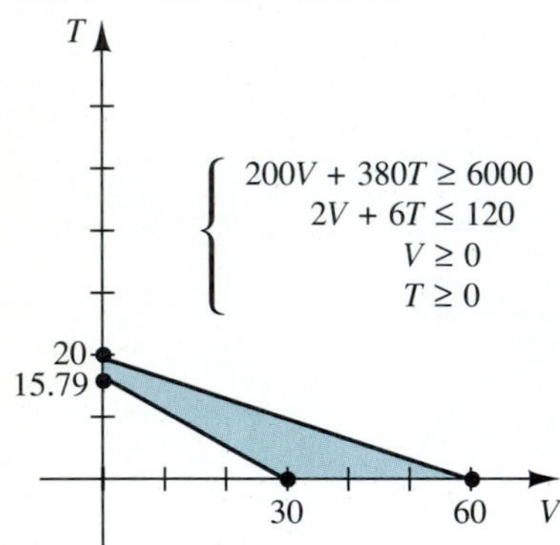

23. Let s = no. of standard models and c = no. of custom models.

$$\begin{cases} 3c + s \le 20 \\ 80c + 20s \ge 200 \\ c \ge 0 \\ s \ge 0 \end{cases}$$

c
10
$6\frac{2}{3}$
$2\frac{1}{2}$
1 10 20 s

25. Let s = amount invested in 6-month certificate and y = amount invested in the 1-year certificate.

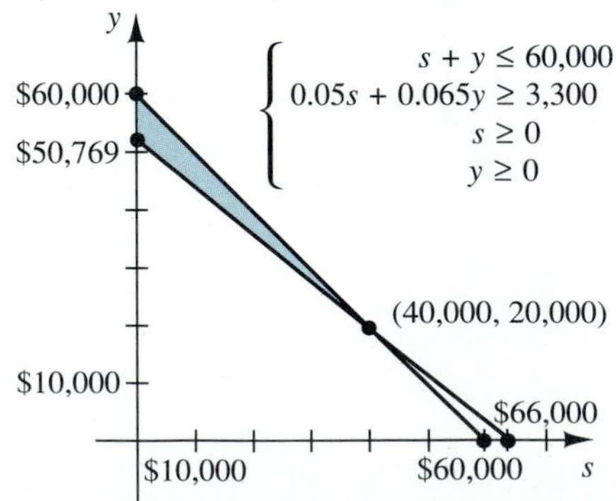

27. Let d = no. of divorce cases and m = no. of malpractice suits.

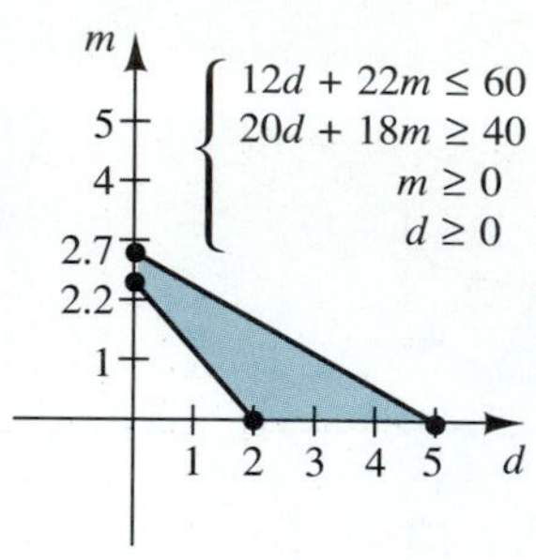

EXERCISES 9.9

1. The vertices are (0, 0), (0, 1), $\left(\frac{4}{3}, \frac{7}{3}\right)$, and (6, 0). The minimum value of f is 0, which occurs at (0, 0); the maximum value of f is 18, which occurs at (6, 0).
3. The vertices are (0, 1), (0, −2), and (2, −1). The minimum value of f is −1, which occurs at (0, 1); the maximum value of f is 2, which occurs at (0, −2).
5. The vertices are (0, 2), (0, 3), and (3, 0). The minimum value of f is 5, which occurs at (3, 0); the maximum value of f is 14, which occurs at (0, 3).
7. Let h = amount invested in high-risk and c = amount invested in conservative investments (in millions). Maximum profit is \$8.4 million, where \$52.5 million is invested in conservative investments and \$22.5 million is invested in high-risk investments.

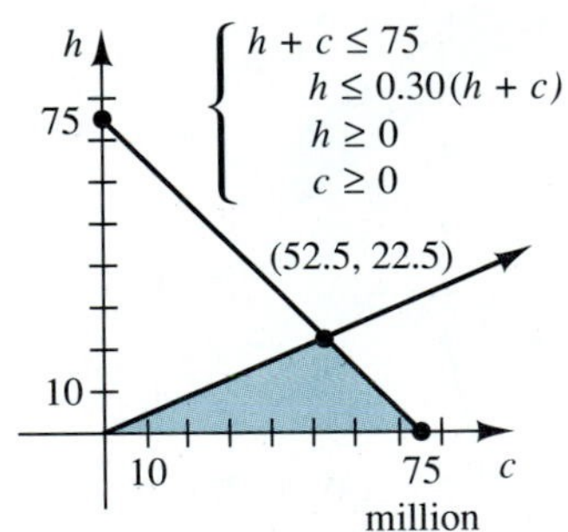

Vertices : (0, 0), (75, 0), and (52.5, 22.5)
$P = 0.10c + 0.14h$

9. Let x = no. of model X and y = no. of model Y. Maximum value of P is at (62.5, 50), where P = 10,625. The closest integral value within constraints is (63, 49). Maximum profit is \$10,570 for 63 of model X and 49 of model Y.

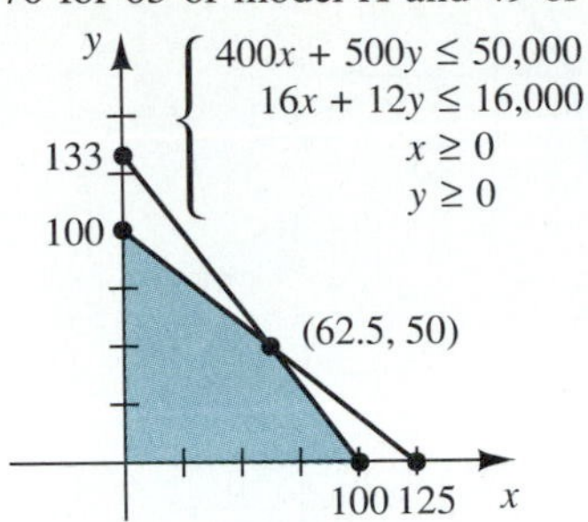

Vertices : (0, 0), (0, 100), (62.5, 50), and (100, 0)
$P = 90x + 100y$

11. The maximum profit would be \$16,363.60 for g (graphic calculators) = 636.36 and s (scientific calculators) = 681.82; however, constrained to integral values, the maximum profit is \$16,360 for 636 graphics calculators and 682 scientific calculators.

13. The least expensive combination of cereals is 60¢ for 0 oz of cereal X and 2 oz of Cereal Y.

15. The maximum profit would be \$26,800 for 3.04 divorce cases and 1.07 criminal cases; however, constrained to integral values, the maximum profit is \$26,000 for 3 divorce cases and 1 criminal case.

17. Let x = no. of grams of Zomine X and y = no. of grams of Zomine Y. Vertices are (0, 1), (0, 6), (1, 4), (2, 0), and (1, 0). R = $0.005x + 0.002y$. The *best* dosage is 1 gram of Zomine X and 4 grams of Zomine Y yielding 0.013 grams of pain relief, given the constraints.

CHAPTER 9 REVIEW EXERCISES

1. (4, 4) **3.** (5, 0) **5.** (3, 7) **7.** (2, 2, 2)

9. $\left(2, -\frac{3}{2}, -\frac{1}{2}\right)$ **11.** (2, 3, 0) **13.** $\left[\begin{array}{cc|c} 1 & 0 & 0 \\ 0 & 1 & 1 \end{array}\right]$

15. $\left[\begin{array}{ccc|c} 1 & 0 & 0 & 3 \\ 0 & 1 & 0 & 5 \\ 0 & 0 & 1 & -3 \end{array}\right]$ **17.** $\left(\frac{13}{4}, \frac{7}{4}\right)$ **19.** (2, 0, −5)

21. (6, 1) **23.** (1, 0, −1)

25. $\begin{bmatrix} 10 & -3 \\ 9 & -8 \end{bmatrix}$ **27.** $\begin{bmatrix} 0 & -11 \\ 7 & 4 \end{bmatrix}$ **29.** 6 **31.** $\begin{bmatrix} 5 & 8 \\ -4 & 1 \end{bmatrix}$

33. $\begin{bmatrix} -1 & -2 & 0 \\ 11 & -5 & 9 \\ 16 & -13 & 18 \end{bmatrix}$ **35.** $\frac{1}{4}\begin{bmatrix} -2 & 3 \\ 2 & -1 \end{bmatrix}$ **37.** (1, 3)

39. (2, −2, 1) **41.** −2 **43.** 3 **45.** (−1, 2) **47.** (2, 0, 3)

49. 60 of the more expensive cameras; 160 of the less expensive cameras.

51. \$2500 invested at 6%; \$7500 invested at 8%.

53. \$10,000 in the bank account; \$12,000 in the certificate of deposit; \$18,000 in the stock

55.

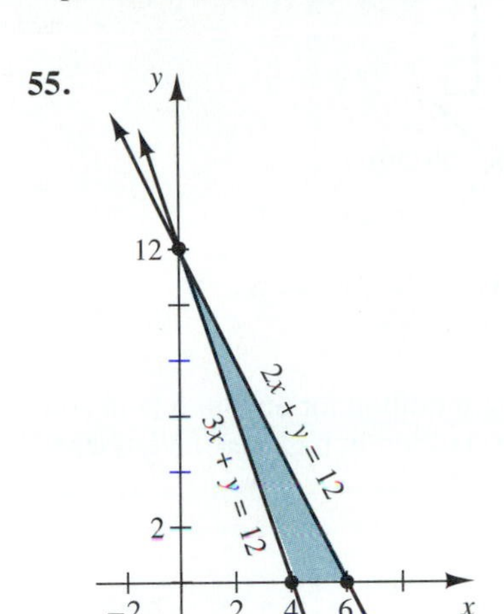

57.

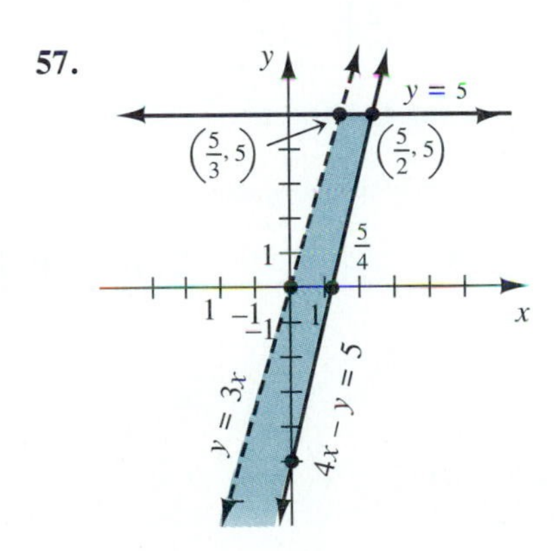

59. Let V = no. of VCRs and T = no. of TVs.

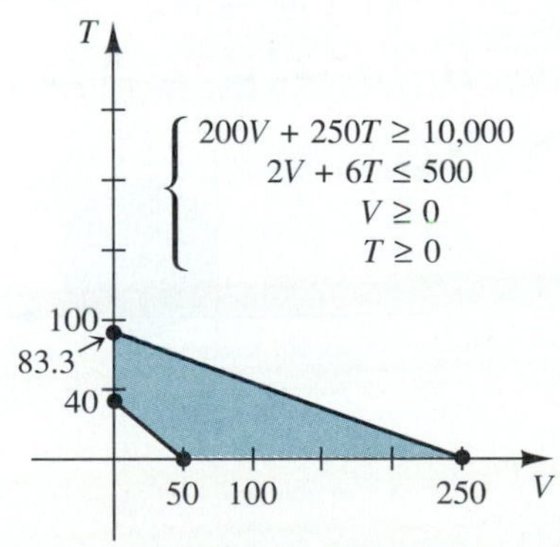

61. Let x = no. of civil complaints and y = no. of criminal complaints.

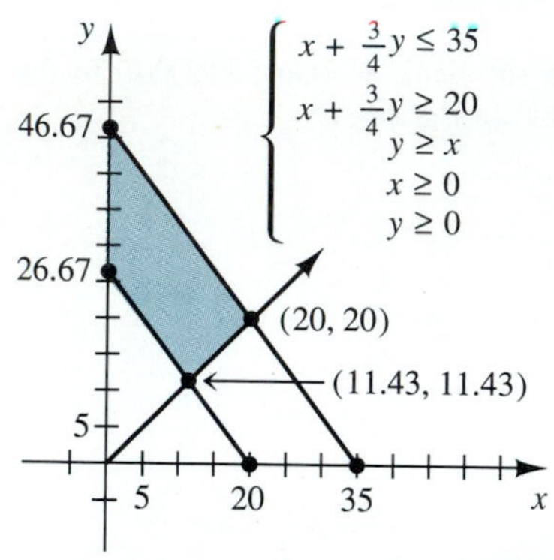

63. The maximum profit is \$7.6 million for \$60 million in conservative investments and \$20 million in high-risk investments.

CHAPTER 9 PRACTICE TEST

1. (a) (4, 4) **(b)** (2, −1) **(c)** $\left(2, -\frac{3}{2}, -\frac{1}{2}\right)$ **2.** (3, −2, 4)

3. (a) $\begin{bmatrix} 1 & -5 \\ -7 & 12 \end{bmatrix}$ **(b)** $\begin{bmatrix} 5 & 6 & 5 \\ -4 & 6 & -22 \\ 0 & 0 & 9 \end{bmatrix}$

4. $\frac{1}{16}\begin{bmatrix} 4 & 8 \\ 1 & -2 \end{bmatrix}$ **5.** (6, 1) **6. (a)** 42 **(b)** −5

7. $\left(\frac{1}{2}, -\frac{1}{2}, \frac{3}{2}\right)$

8. The flat fee is \$22.50 per day; mileage fee is 36¢ per mile.
9. 800 lb of the 10% mixture, 400 lb of the 25% mixture, and 800 lb of the 30% mixture.

10.

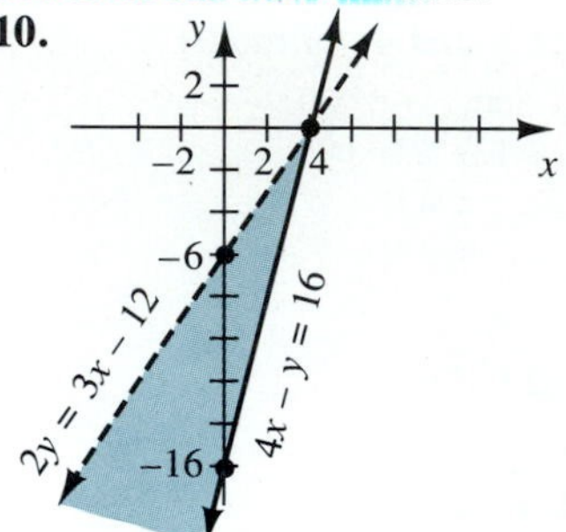

11. Let x = no. of model X and y = no. of model Y.

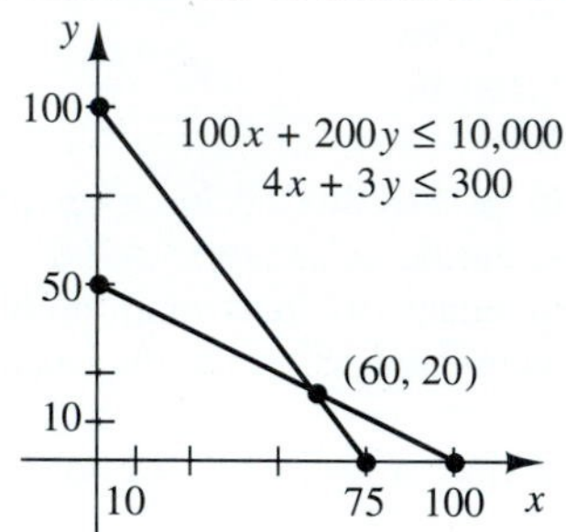

12. The maximum profit is \$4600 for 60 model X and 20 model Y.

CHAPTER 10

EXERCISES 10.1

1. $(x - 3)^2 + (y + 8)^2 = 25$
3. $x^2 + y^2 - 6x - 2y + 9 = 0$
5. $C(2, -3)$; $r = 9$ **7.** $C\left(\frac{1}{2}, 0\right)$; $r = 3\sqrt{2}$

9. $C(3, 4)$; $r = 4$ **11.** $C(3, 0)$; $r = 2\sqrt{6}$ **13.** $C\left(0, \frac{1}{2}\right)$; $r = 2\sqrt{3}$ **15.** $y = -\frac{1}{2}x + \frac{5}{2}$ **17.** $y = -x + 7$

19.

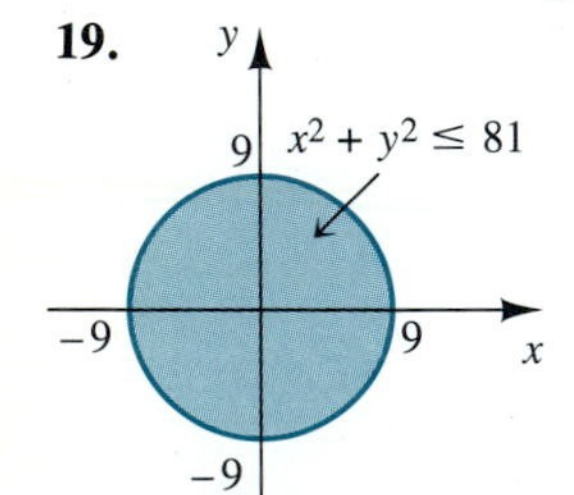

21. $x^2 + y^2 - 8x - 6y + 9 < 0$

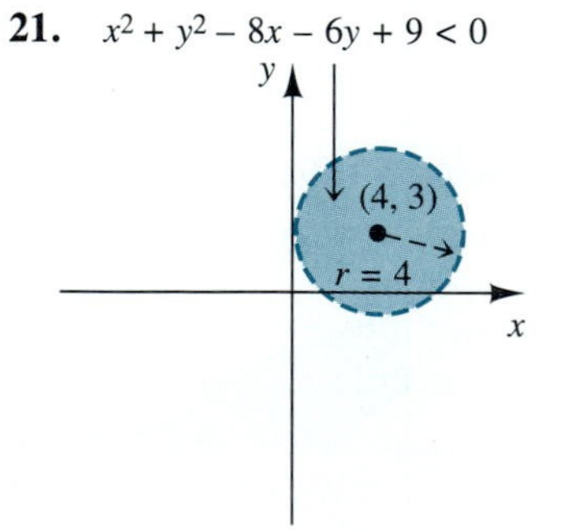

23. $\frac{22}{5}$ **25.** $x^2 + y^2 = \frac{144}{13}$ **27.** $\frac{\sqrt{10}}{5}$

EXERCISES 10.2

1. $F(0, 3)$; directrix: $y = -3$ **3.** $F\left(-\frac{1}{32}, 0\right)$; directrix: $x = \frac{1}{32}$ **5.** $F\left(0, \frac{1}{8}\right)$; directrix: $y = -\frac{1}{8}$
7. $4x^2 - 12y = 0$

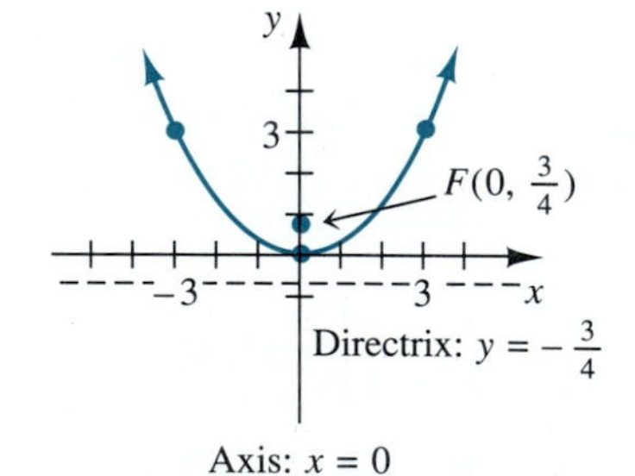

9. $x + y^2 = 0$

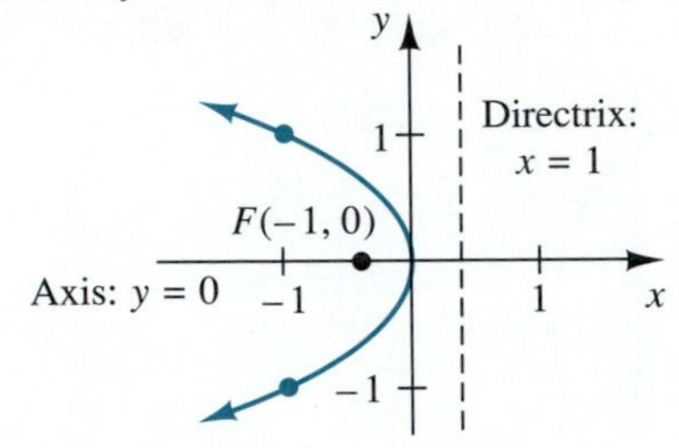

11. $x^2 = 16y$

13. $x^2 = -12y$ **15.** $y^2 = 20x$

17. The dish is 3 feet deep. The receiver should be placed 3 feet directly above the vertex.

19. The mirror is 8 inches deep. The light source should be placed $\frac{1}{2}$ inch directly above the vertex.

21. $\frac{20\sqrt{6}}{3} \approx 16.33$ feet **23.** $\frac{1}{2}$ inch above the vertex

25. $F(2, 2)$; directrix: $y = -2$ **27.** $F(0, 1)$; directrix: $x = -4$ **29.** $F\left(2, -\frac{17}{4}\right)$; directrix: $y = -\frac{7}{4}$

31. $8y = x^2 + 10x + 33$

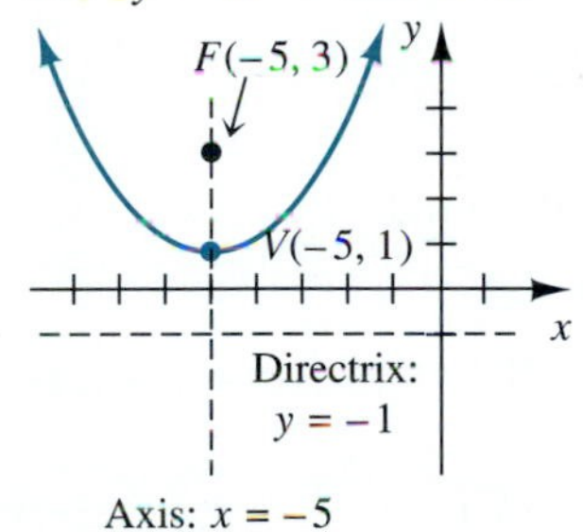

33. $x = 3y^2 - 6y + 5$

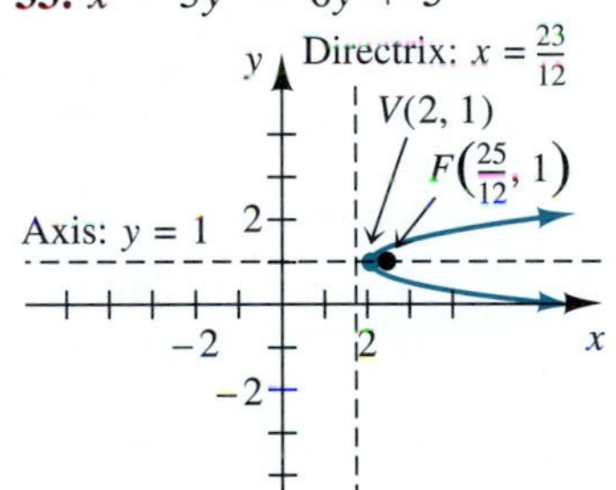

35. $(y - 4)^2 = -12(x - 5)$ **37.** $(x - 2)^2 = -14\left(y - \frac{1}{2}\right)$

39. $(y - 5)^2 = 20(x - 2)$

41. $y < 4x^2$

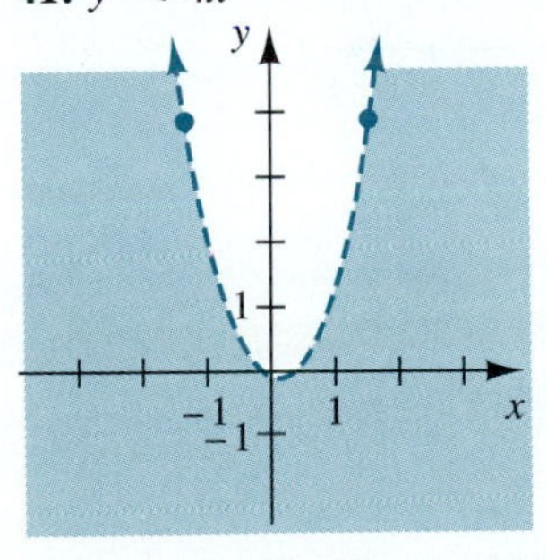

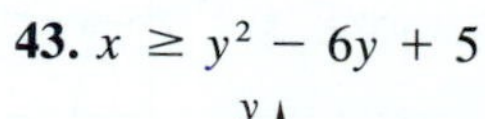

43. $x \geq y^2 - 6y + 5$

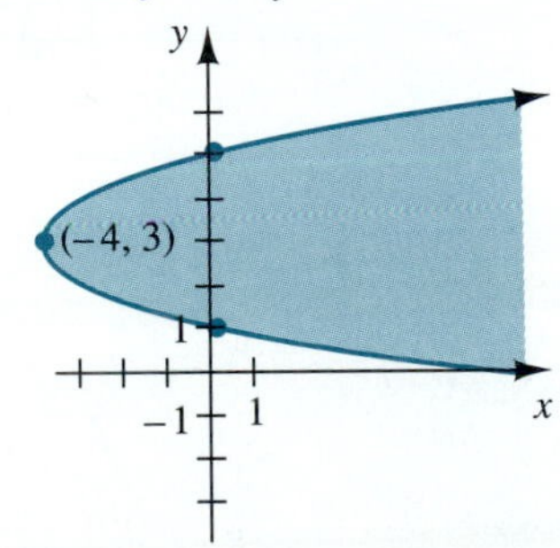

47. $y = -x - 2$

EXERCISES 10.3

1. $V_1(4, 0)$, $V_2(-4, 0)$, $F_1(\sqrt{7}, 0)$, $F_2(-\sqrt{7}, 0)$

3. $V_1(0, 3\sqrt{2})$, $V_2(0, -3\sqrt{2})$, $F_1(0, \sqrt{6})$, $F_2(0, -\sqrt{6})$

5. $V_1(2\sqrt{6}, 0)$, $V_2(-2\sqrt{6}, 0)$, $F_1(\sqrt{15}, 0)$, $F_2(-\sqrt{15}, 0)$

7. $V_1(0, 5)$, $V_2(0, -5)$, $F_1(0, 4)$, $F_2(0, -4)$

9. $V_1(0, \sqrt{30})$, $V_2(0, -\sqrt{30})$, $F_1(0, \sqrt{29})$, $F_2(0, -\sqrt{29})$

11. $\frac{x^2}{49} + \frac{y^2}{9} = 1$

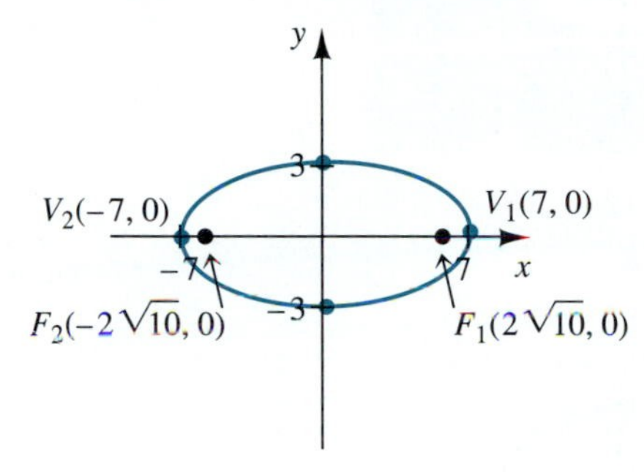

13. $12x^2 + y^2 = 24$

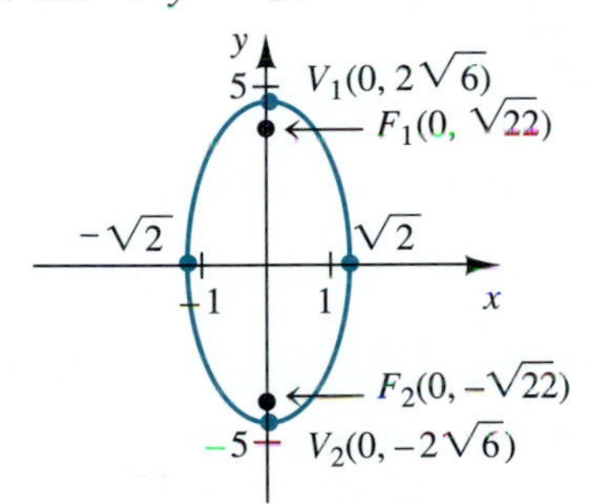

15. $3x^2 + 8y^2 = 12$

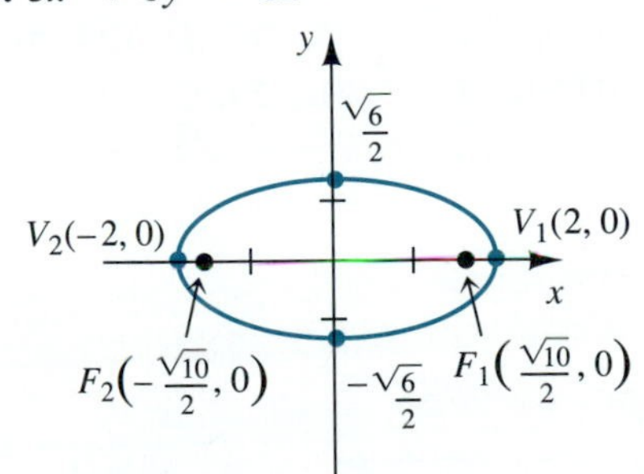

17. $e = \frac{\sqrt{11}}{6}$ **19.** $e = \frac{\sqrt{10}}{4}$

21. $\frac{x^2}{36} + \frac{y^2}{24} > 1$

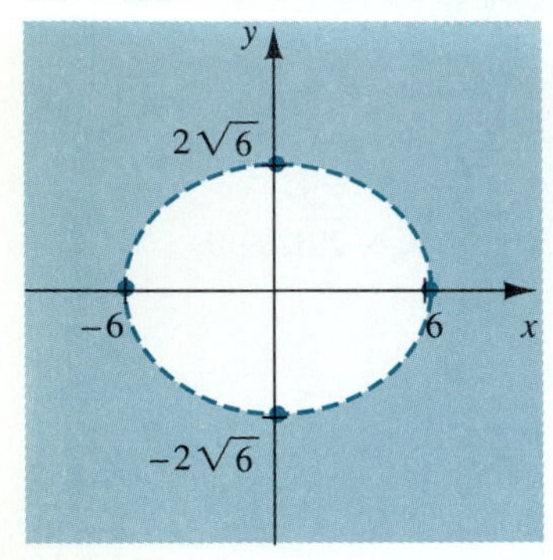

23. $3x^2 + 4y^2 \le 12$

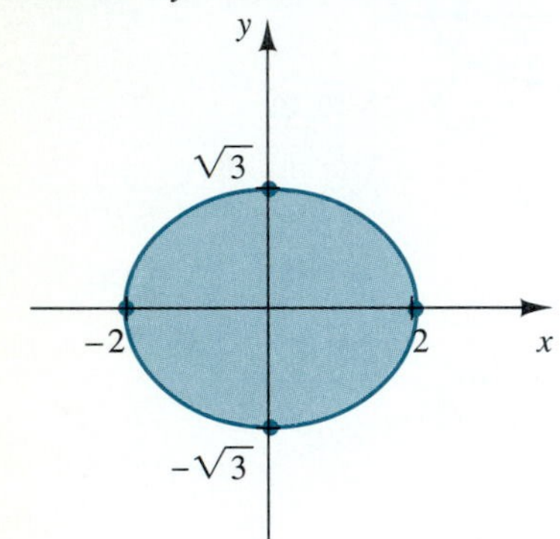

25. $\frac{x^2}{16} + \frac{y^2}{12} = 1$ **27.** $\frac{x^2}{16} + \frac{y^2}{4} = 1$

29. $5\sqrt{7} \approx 13.23$ feet **31.** 221,589 miles **33.** $C(2, 3)$, $V_1(6, 3)$, $V_2(-2, 3)$, $F_1(2 + 2\sqrt{3}, 3)$, $F_2(2 - 2\sqrt{3}, 3)$ **35.** $C(-1, 3)$, $V_1(-1, 3 + 3\sqrt{2})$, $V_2(-1, 3 - 3\sqrt{2})$, $F_1(-1, 3 + \sqrt{6})$, $F_2(-1, 3 - \sqrt{6})$ **37.** $C(2, 3)$, $V_1(2, 5)$, $V_2(2, 1)$, $F_1(2, 3 + \sqrt{3})$, $F_2(2, 3 - \sqrt{3})$ **39.** $C(5, 1)$, $V_1(5, 1 + 3\sqrt{2})$, $V_2(5, 1 - 3\sqrt{2})$, $F_1(5, 1 + \sqrt{3})$, $F_2(5, 1 - \sqrt{3})$

41. $\frac{(x - 2)^2}{9} + \frac{(y + 3)^2}{8} = 1$

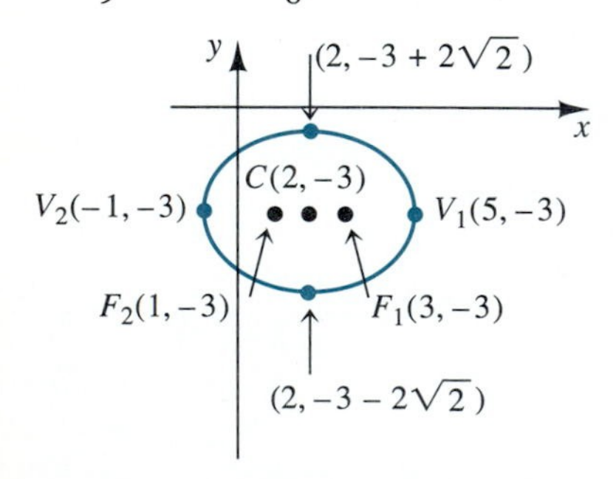

43. $\frac{(x + 2)^2}{16} + \frac{(y - 1)^2}{24} = 1$

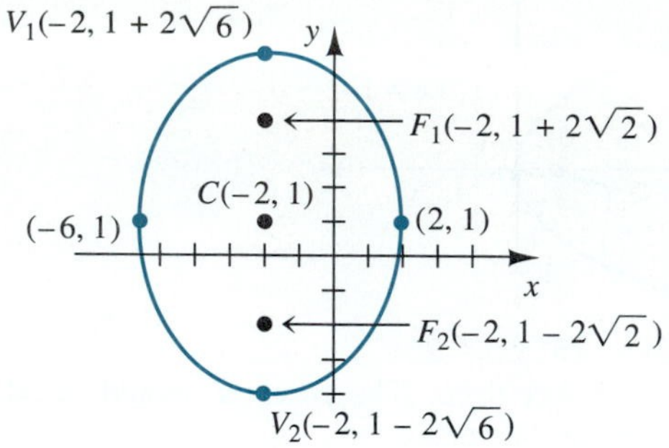

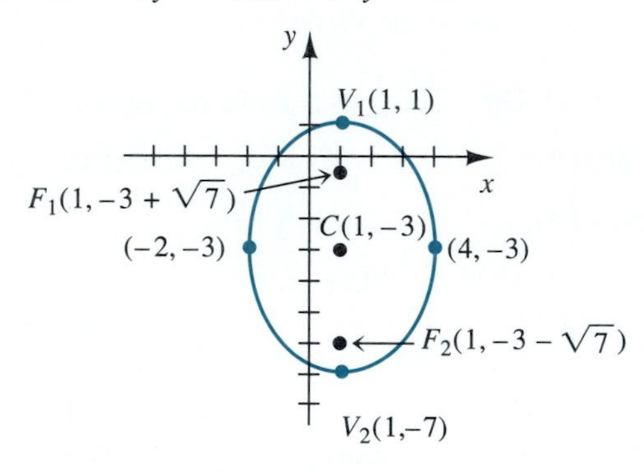

45. $16x^2 + 9y^2 - 32x + 54y = 47$

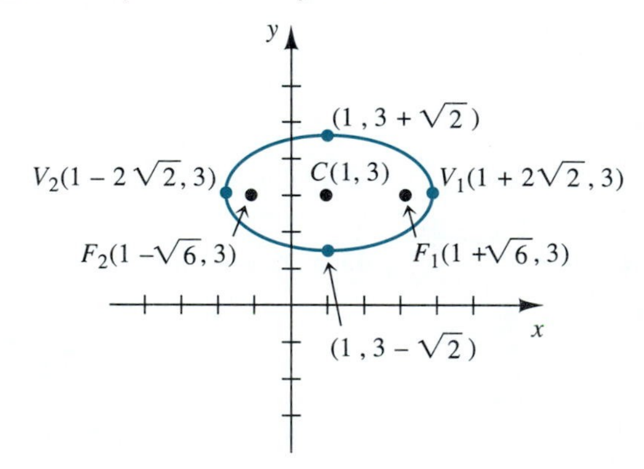

47. $x^2 + 4y^2 - 2x - 24y = -29$

49. $\dfrac{x^2}{9} + \dfrac{y^2}{4} = 1$; Ellipse

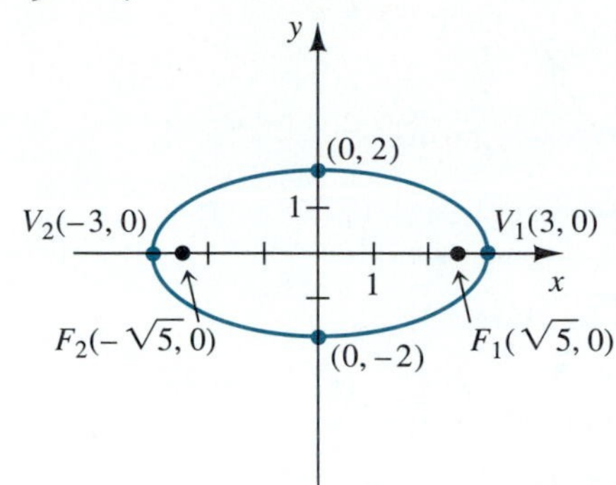

51. $\dfrac{(x + 2)^2}{16} + \dfrac{(y - 1)^2}{16} = 1$; Circle

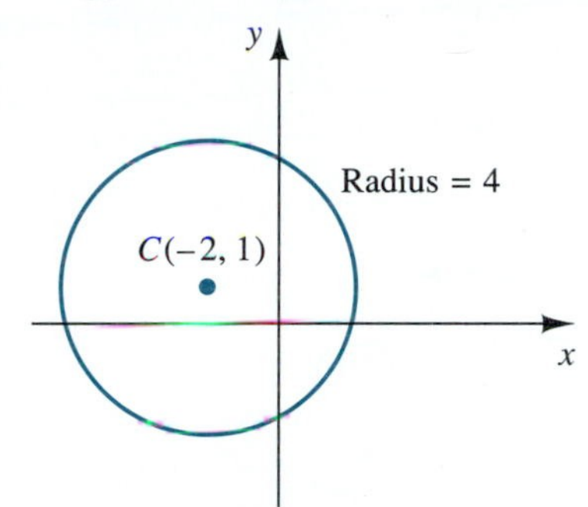

53. $x^2 - 9y = 0$; Parabola

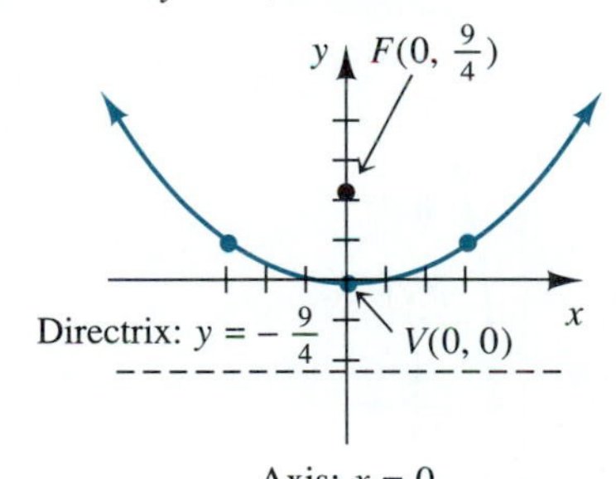

55. $\dfrac{(x - 3)^2}{9} + \dfrac{(y - 4)^2}{5} = 1$ **57.** $\dfrac{(x - 3)^2}{4} + (y + 5)^2 = 1$

61. $-x + 2\sqrt{3}\,y = 8$

EXERCISES 10.4

1. $V_1(3, 0)$, $V_2(-3, 0)$, $F_1(5, 0)$, $F_2(-5, 0)$; asymptotes: $y = \pm\frac{4}{3}x$ **3.** $V_1(0, 2\sqrt{3})$, $V_2(0, -2\sqrt{3})$, $F_1(0, \sqrt{30})$, $F_2(0, -\sqrt{30})$; asymptotes: $y = \pm\frac{\sqrt{6}}{3}x$ **5.** $V_1(3, 0)$, $V_2(-3, 0)$, $F_1(3\sqrt{3}, 0)$, $F_2(-3\sqrt{3}, 0)$; asymptotes: $y = \pm\sqrt{2}x$ **7.** $V_1(3, 0)$, $V_2(-3, 0)$, $F_1(\sqrt{34}, 0)$, $F_2(-\sqrt{34}, 0)$; asymptotes: $y = \pm\frac{5}{3}x$ **9.** $V_1(0, 1)$, $V_2(0, -1)$, $F_1(0, \sqrt{31})$, $F_2(0, -\sqrt{31})$; asymptotes: $y = \pm\frac{\sqrt{30}}{30}x$

11. $\dfrac{x^2}{9} - \dfrac{y^2}{49} = 1$

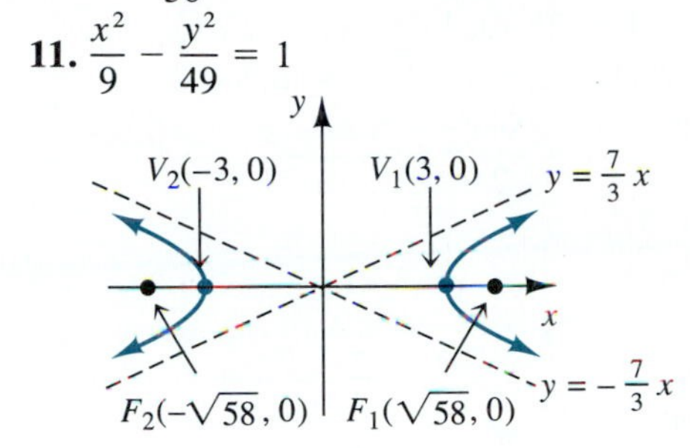

13. $8x^2 - y^2 = 24$

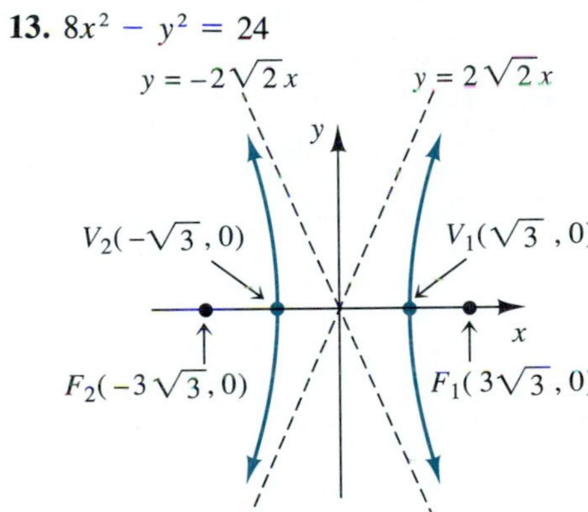

15. $3y^2 - 8x^2 = 24$

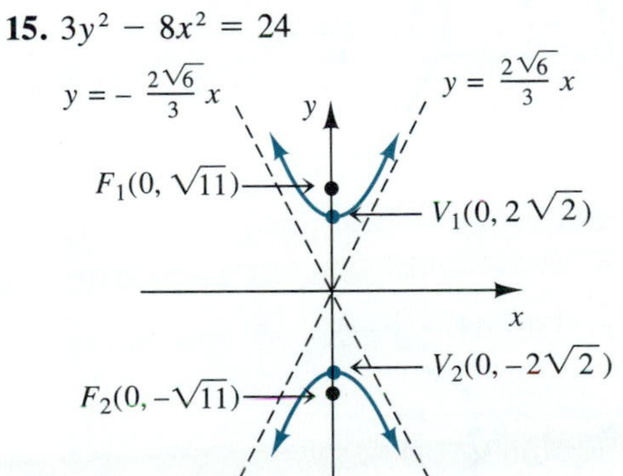

17. In miles (to one decimal place): $\dfrac{x^2}{344.5} - \dfrac{y^2}{2155.5} = 1$

19. $\sqrt{7} - \sqrt{3} \approx 0.9137$ AU

21. $C(1, 3)$, $V_1(3, 3)$, $V_2(-1, 3)$, $F_1(1 + 2\sqrt{2}, 3)$, $F_2(1 - 2\sqrt{2}, 3)$; asymptotes: $y = x + 2$, $y = -x + 4$

23. $C(-1, -3)$, $V_1(-1, -3 + 2\sqrt{3})$, $V_2(-1, -3 - 2\sqrt{3})$, $F_1(-1, -3 + \sqrt{30})$; $F_2(-1, -3 - \sqrt{30})$; asymptotes: $y = \dfrac{\sqrt{6}}{3}x + \dfrac{\sqrt{6} - 9}{3}$, $y = -\dfrac{\sqrt{6}}{3}x - \dfrac{\sqrt{6} + 9}{3}$

25. $C(3, 1)$, $V_1(7, 1)$, $V_2(-1, 1)$, $F_1(3 + 2\sqrt{5}, 1)$, $F_2(3 - 2\sqrt{5}, 1)$; asymptotes: $y = \dfrac{1}{2}x - \dfrac{1}{2}$, $y = -\dfrac{1}{2}x + \dfrac{5}{2}$

27. $C(1, -1)$, $V_1(1, -1 + 2\sqrt{6})$, $V_2(1, -1 - 2\sqrt{6})$, $F_1(1, -1 + 2\sqrt{10})$, $F_2(1, -1 - 2\sqrt{10})$; asymptotes: $y = \dfrac{\sqrt{6}}{2}x - \dfrac{\sqrt{6}+2}{2}$, $y = -\dfrac{\sqrt{6}}{2}x + \dfrac{\sqrt{6} - 2}{2}$

29. $\dfrac{(x + 4)^2}{9} - \dfrac{(y + 3)^2}{16} = 1$

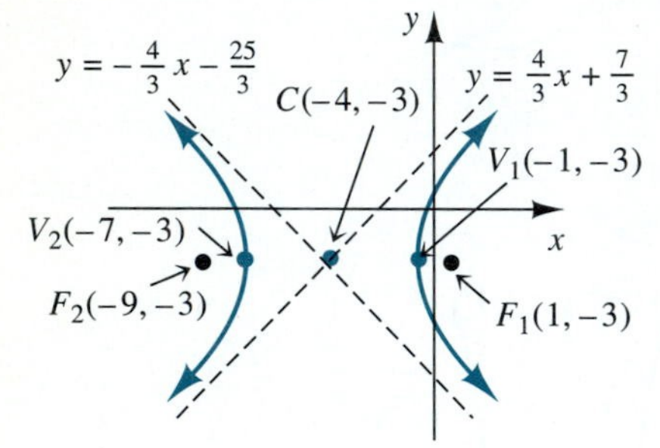

31. $\dfrac{x^2}{5} - \dfrac{(y - 3)^2}{15} = 1$

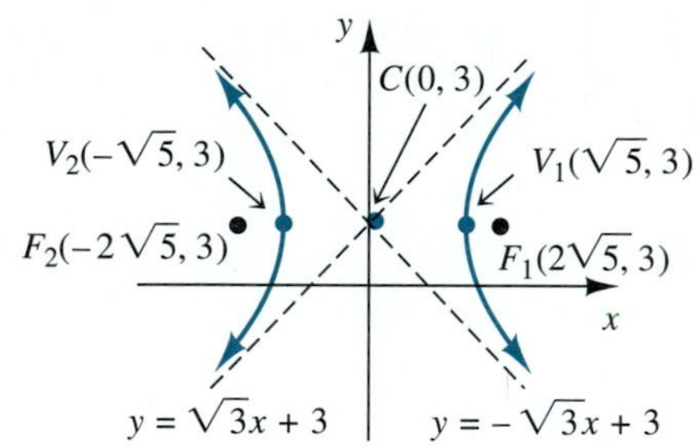

33. $9x^2 - 16y^2 - 36x - 32y = 124$

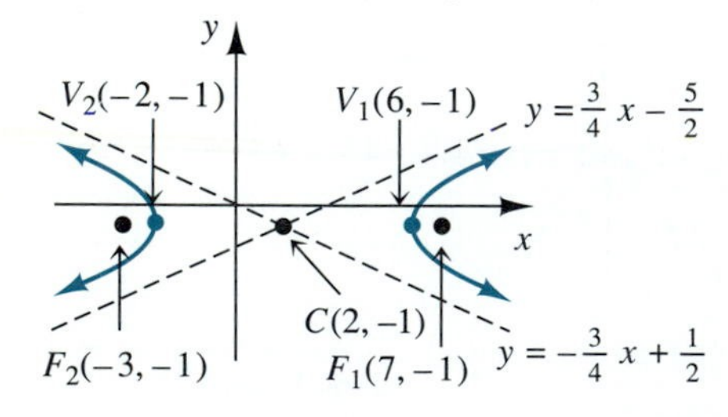

35. $25y^2 - 36x^2 - 150y + 288x - 1251 = 0$

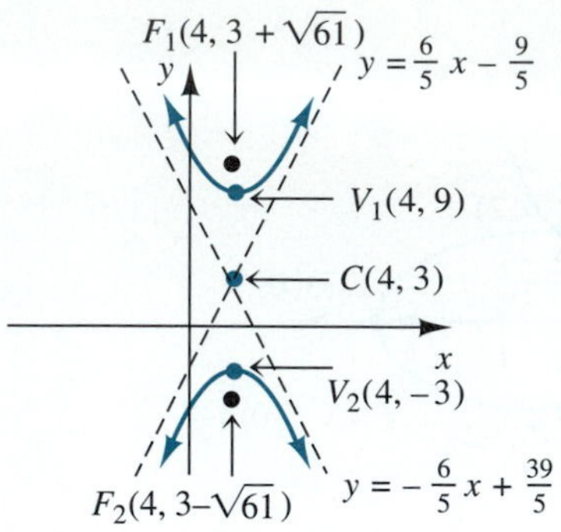

37. $\dfrac{(x - 6)^2}{16} - \dfrac{(y + 1)^2}{20} = 1$

39. $\dfrac{x^2}{12} + \dfrac{y^2}{4} = 1$; Ellipse

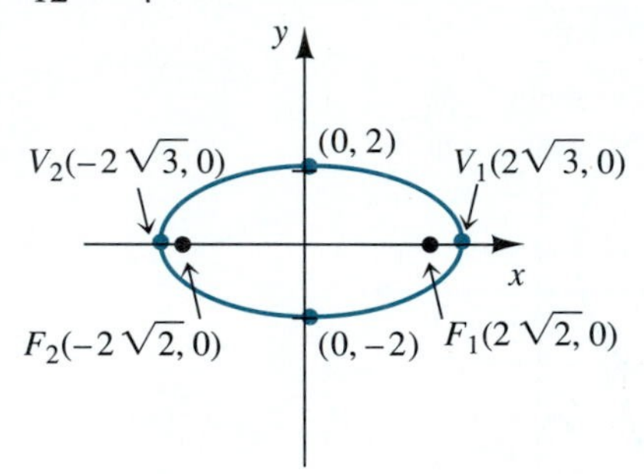

41. $\dfrac{(x + 2)^2}{9} - \dfrac{(y - 1)^2}{25} = 1$; Hyperbola

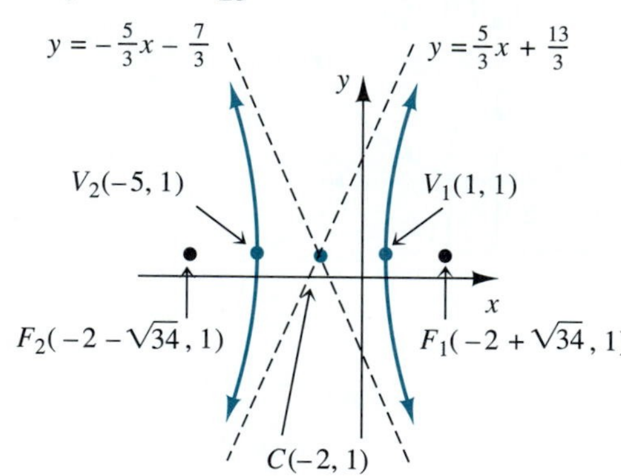

43. $x^2 - 2y - 12 = 0$ Parabola

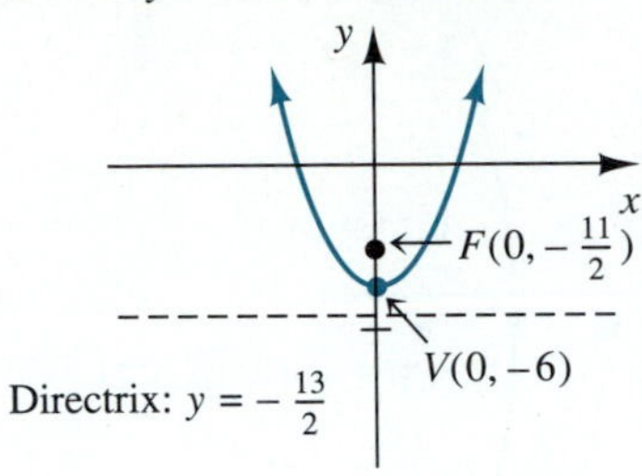

47. $x = 4$

EXERCISES 10.5

1. An ellipse or its degenerate form **3.** A hyperbola or its degenerate form **5.** A circle or its degenerate form **7.** A hyperbola or its degenerate form **9.** An ellipse or its degenerate form
11. $y^2 + 16x = 0$; Parabola

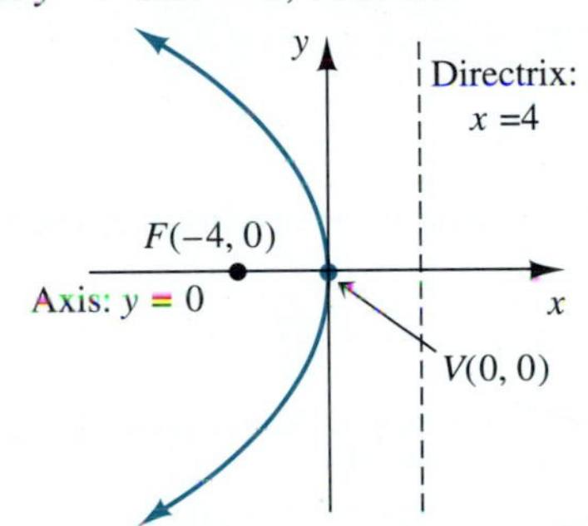

13. $x^2 - 10x + 25 = 0$ Degenerate parabola: a line $x = 5$

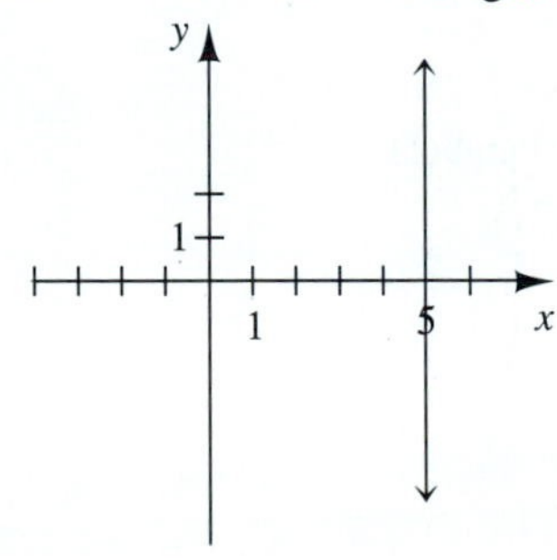

15. $2x^2 + y^2 - 8x - 2y - 7 = 0$; Ellipse

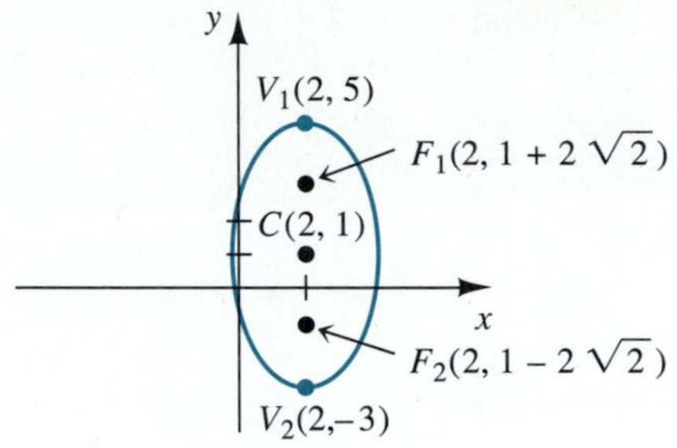

17. $25x^2 + y^2 - 50x - 6y - 16 = 0$; Ellipse

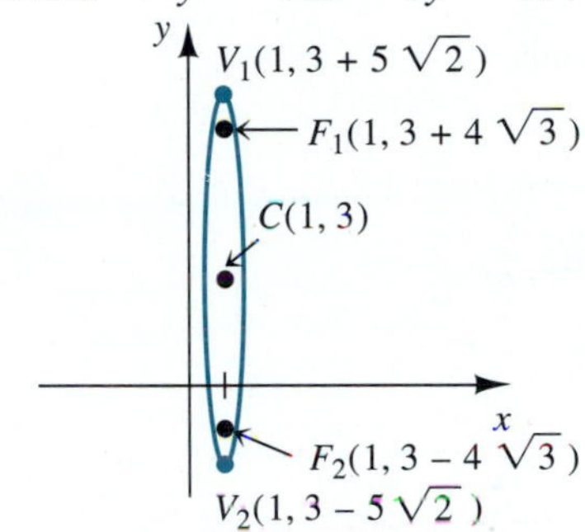

19. $25x^2 + 9y^2 - 225 = 0$; Ellipse

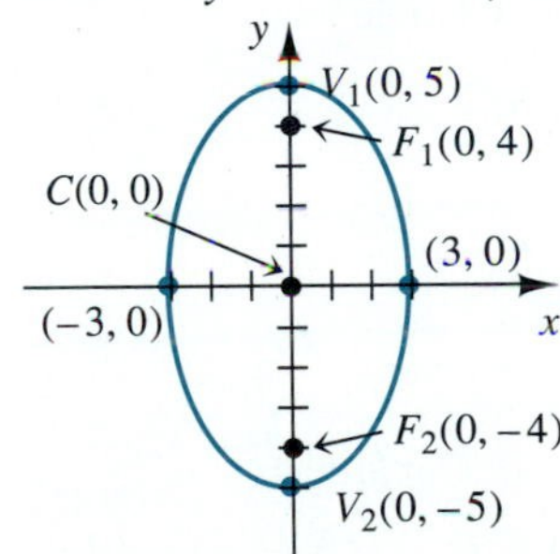

21. $x^2 - y^2 - 18 = 0$; Hyperbola

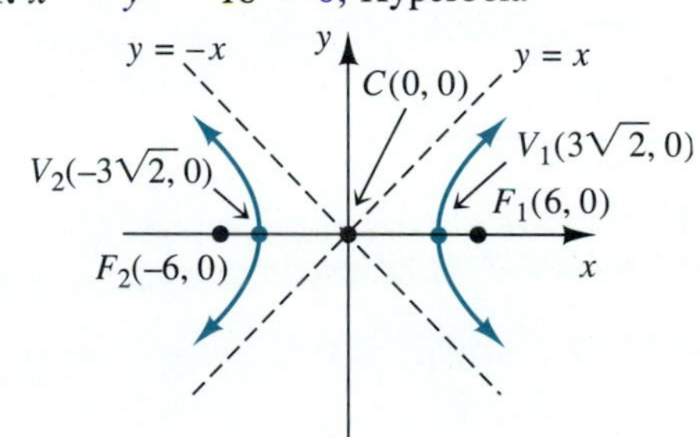

23. $y^2 - 6y - 16 = 0$; Degenerate Parabola: 2 parallel lines. $y = 8$ and $y = -2$

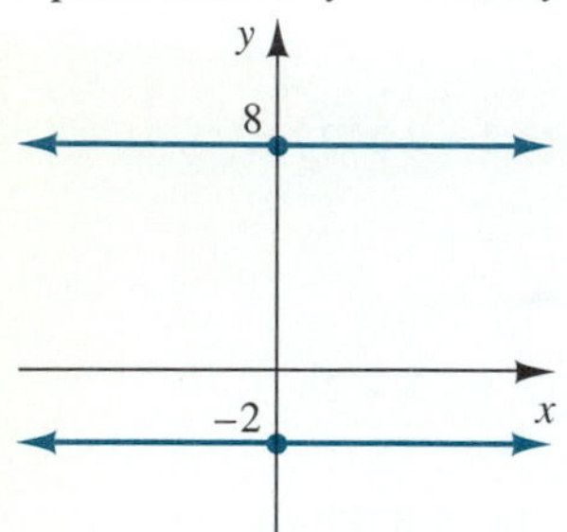

25. $-10x^2 + y = 0$; Parabola

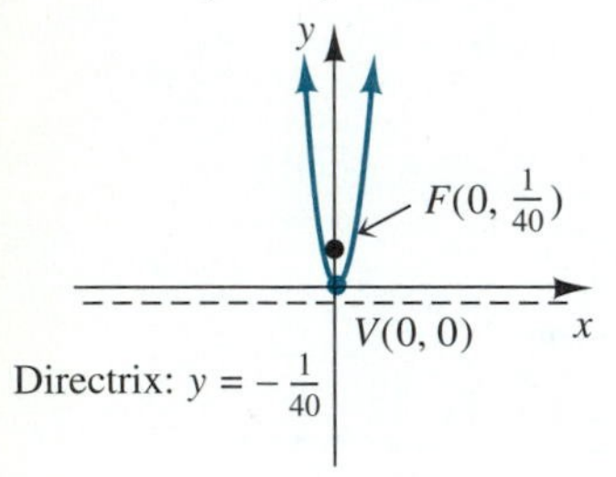

27. $x^2 + y^2 + 2x - 6y - 5 = 0$; Circle

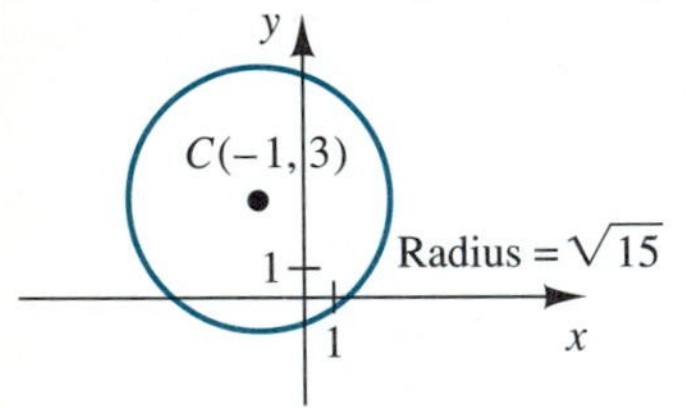

29. $x^2 + y^2 - 4x - 2y + 3 = 0$; Circle

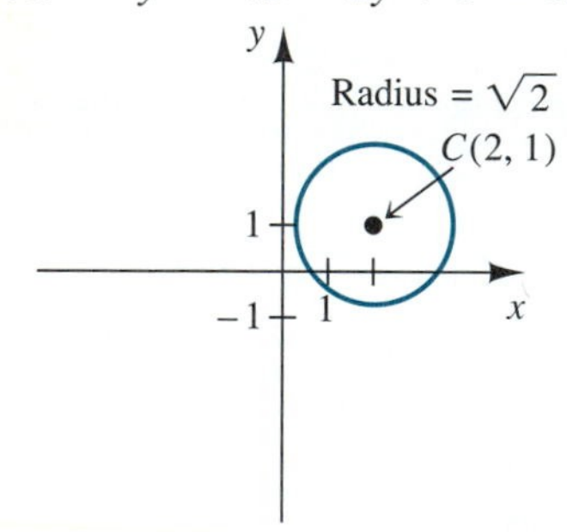

31. $9x^2 - y^2 - 18x - 4y - 139 = 0$ Hyperbola

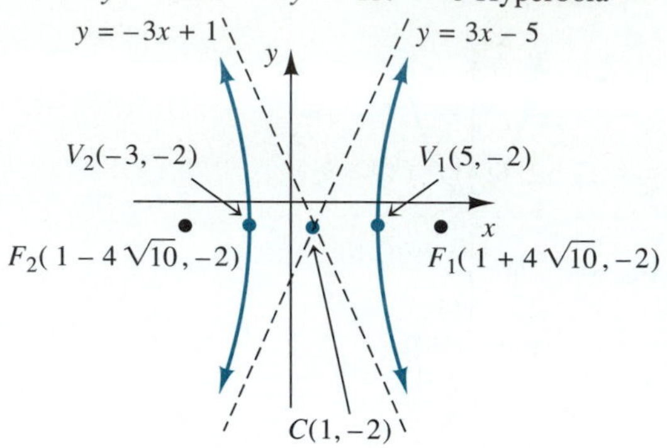

33. $x^2 + y^2 + 36 = 0$; degenerate circle (not graphable)

35. $9x^2 + 25y^2 + 18x + 9 = 0$ Degenerate ellipse: the point $(-1, 0)$

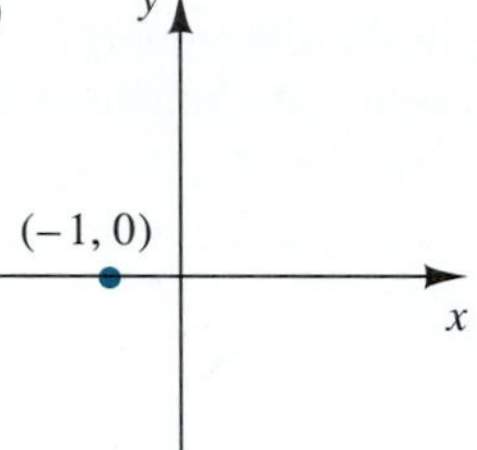

37. $16y^2 - x^2 = 0$ Degenerate hyperbola: two intersecting lines

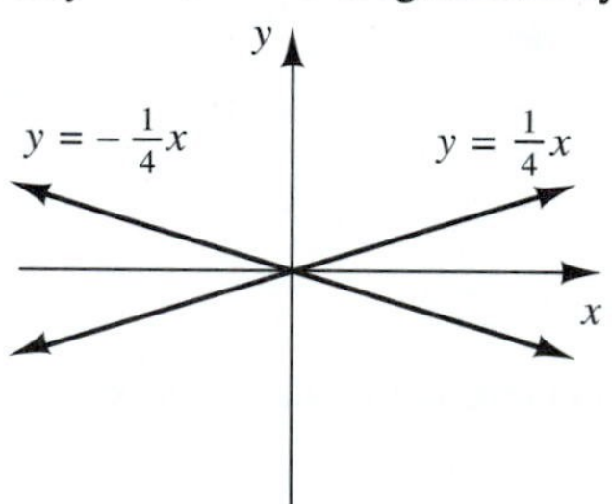

39. $x^2 - 2x + 12y - 47 = 0$ Parabola

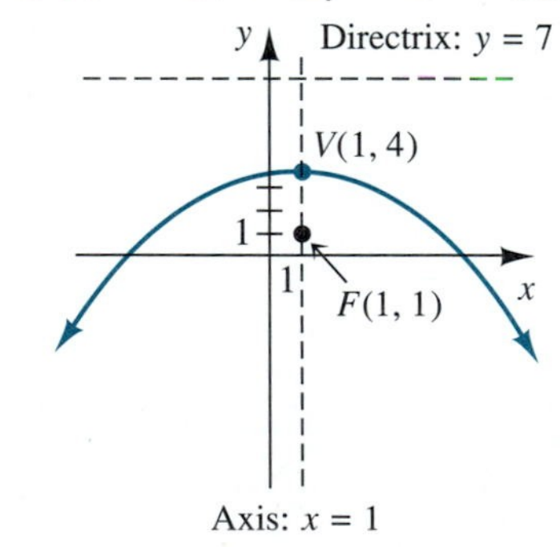

EXERCISES 10.6

1. $(1', 9')$ **3.** $(4, 1)$ **5.** $(-6', 8')$ **7.** $(2, -4)$
9. $x' = x - 3, y' = y + 4$ **11.** $x' = x, y' = y - 3$
13. $x' = x - 1, y' = y + 7$ **15.** $x' = x - 5, y' = y - 2$
17. $\left(\frac{7}{\sqrt{2}}, \frac{3}{\sqrt{2}}\right)$ **19.** $\left(\frac{2+\sqrt{3}}{2}, \frac{2\sqrt{3}-1}{2}\right)$
21. $\left(\frac{2-\sqrt{3}}{2}, \frac{-1-2\sqrt{3}}{2}\right)$ **23.** $\left(\frac{3}{\sqrt{2}}, \frac{7}{\sqrt{2}}\right)$
25. $(x')^2 + 3(y')^2 - \sqrt{3}x' - (2\sqrt{3}+1)y' = 0$
27. $(x')^2 - 6\sqrt{3}x'y' - 5(y')^2 - 32 = 0$
29. $(x')^2 + (y')^2 - 4 = 0$ **31.** $-3x^2 + 3y^2 - 8 = 0$
33. $(3 + 2\sqrt{3})x^2 + (2\sqrt{3} - 4)xy + (1 - 2\sqrt{3})y^2 - 32 = 0$
35. $\phi = 30°$ **37.** $\phi = 22.5°$
39. $xy = 8$
$$\frac{(x')^2}{16} - \frac{(y')^2}{16} = 1$$

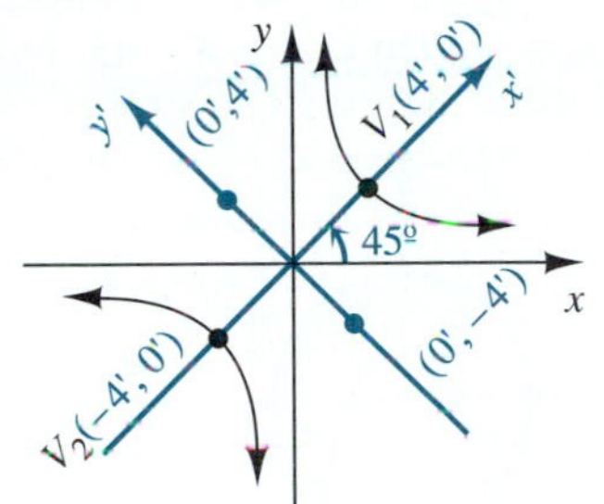

41. $x^2 - 10\sqrt{3}xy + 11y^2 - 16 = 0$
$$(y')^2 - \frac{(x')^2}{4} = 1$$

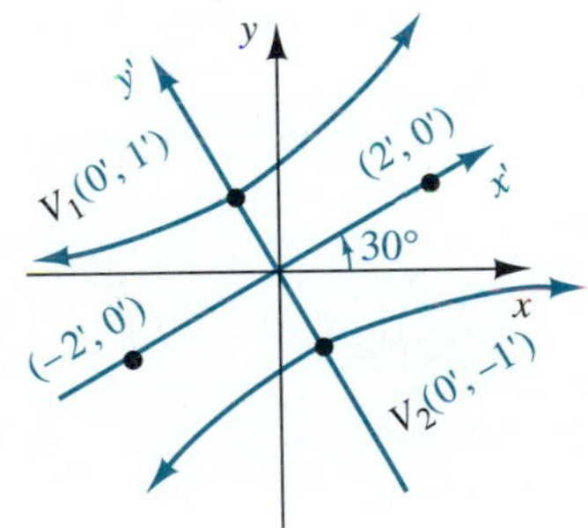

43. $23x^2 + 18\sqrt{3}xy + 5y^2 - 256 = 0$
$$\frac{(x')^2}{8} - \frac{(y')^2}{64} = 1$$

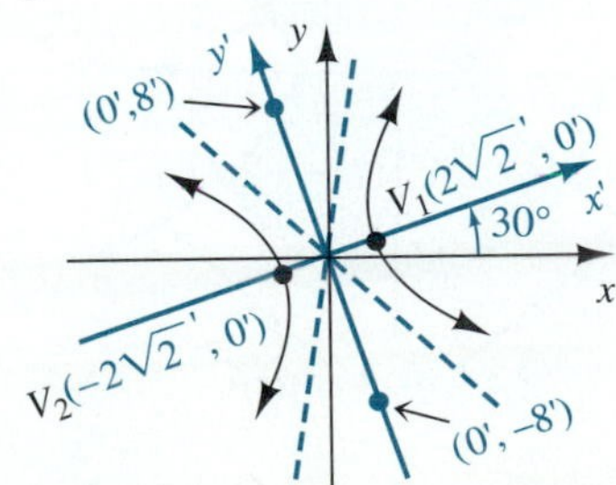

45. $x^2 + 2xy + y^2 = 1$

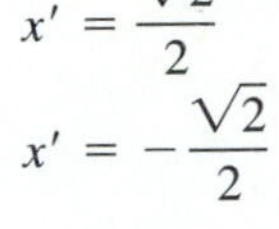
$$x' = \frac{\sqrt{2}}{2}$$
$$x' = -\frac{\sqrt{2}}{2}$$

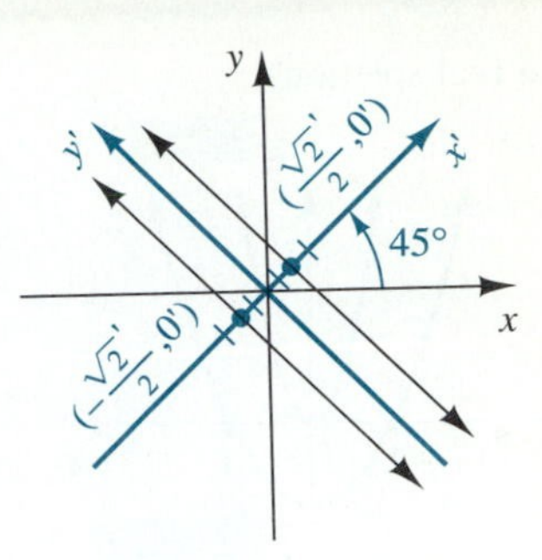

47. $31x^2 + 10\sqrt{3}xy + 21y^2 + 48x - 48\sqrt{3}y = 0$
$$\frac{(x')^2}{4} + \frac{(y' - 3)^2}{9} = 1$$

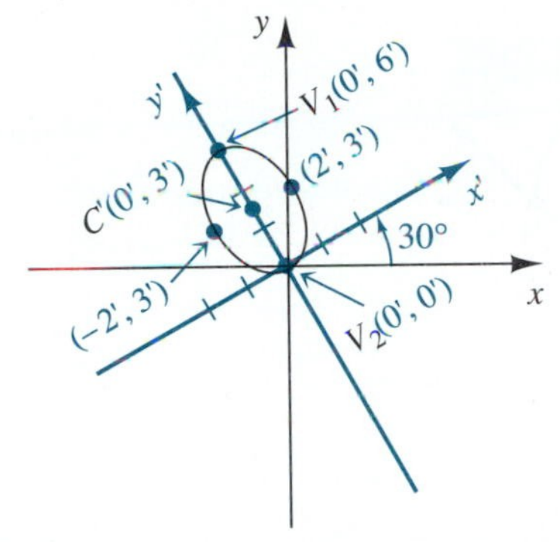

49. An ellipse or its degenerate form
51. A hyperbola or its degenerate form

EXERCISES 10.7

1. $(3, 1), (-1, 3)$ **3.** $(0, -6), (3, 0)$ **5.** No real solutions **7.** $(3, 1), (3, -1), (-3, 1), (-3, -1)$
9. $(0, 1), (0, -1)$
11. $\left(\frac{3+\sqrt{17}}{2}, \frac{-1+\sqrt{17}}{2}\right), \left(\frac{3-\sqrt{17}}{2}, \frac{-1-\sqrt{17}}{2}\right)$
13. $(4, 2)$ **15.** $(-1, 2)$ **17.** $(3 + \sqrt{3}, 3), (3 - \sqrt{3}, 3)$ **19.** $\left(6, \frac{1}{3}\right), (1, 2)$ **21.** $(5, 2)$
23. $\left(\frac{\sqrt{3}}{2}, \frac{1}{2}\right), \left(-\frac{\sqrt{3}}{2}, \frac{1}{2}\right)$

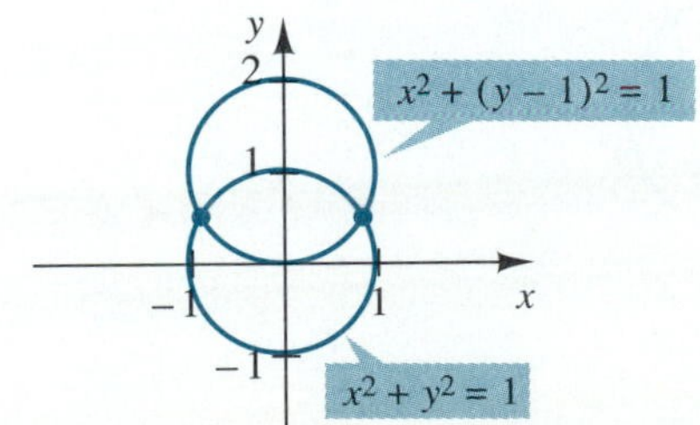

25. No real solutions

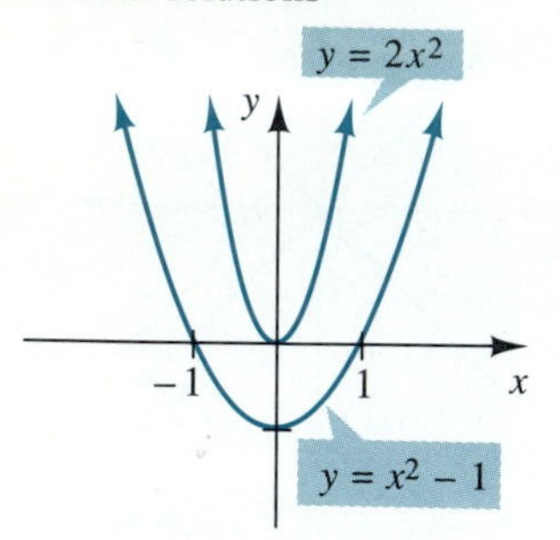

27. $(0, -3)$, $(3, 0)$

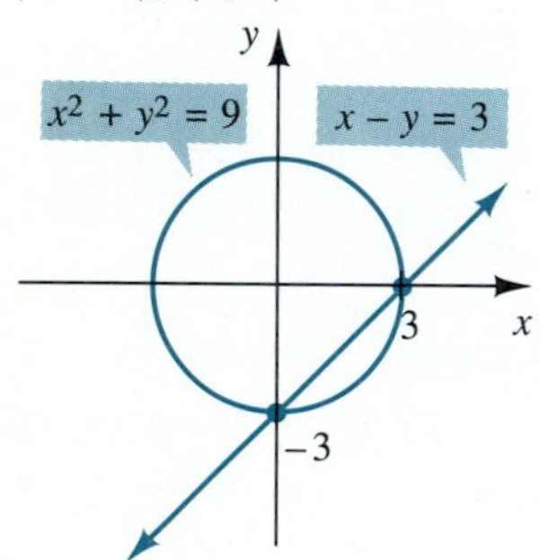

29. $\left(\frac{13}{3}, -\frac{65}{9}\right)$, $(1, -5)$

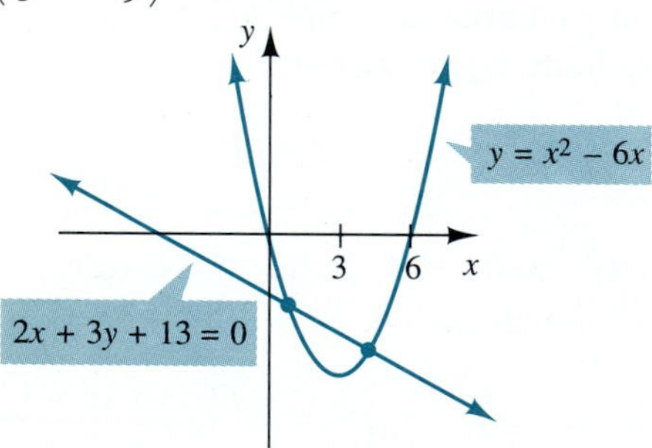

31. $(\sqrt{5}, 2)$, $(-\sqrt{5}, 2)$

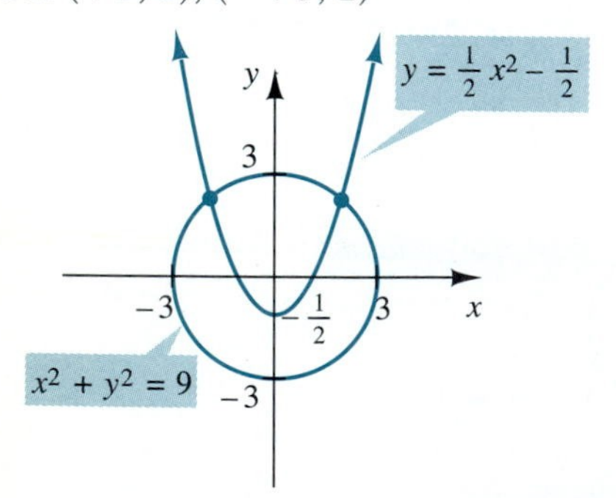

33. $(1 + \sqrt{7}, 1 - \sqrt{7})$, $(1 - \sqrt{7}, 1 + \sqrt{7})$

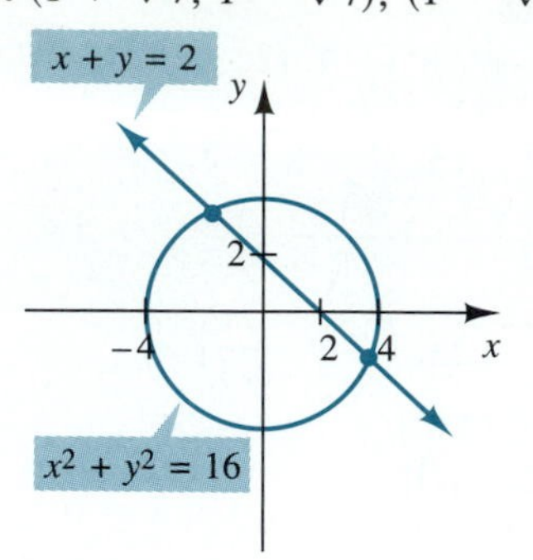

35. $(\sqrt{2}, 2)$, $(\sqrt{2}, -2)$, $(-\sqrt{2}, 2)$, $(-\sqrt{2}, -2)$
37. $(\log_5 4, 4) \approx (0.9, 4)$
39. 3.5 cm by 10 cm **41.** 7 cm by 24 cm
43. **(a)** $k = -4$ **(b)** $k < -4$ **45.** $(4, 0)$, $(-4, 0)$
47. Yes; 40×100 or 50×80 ft
49.

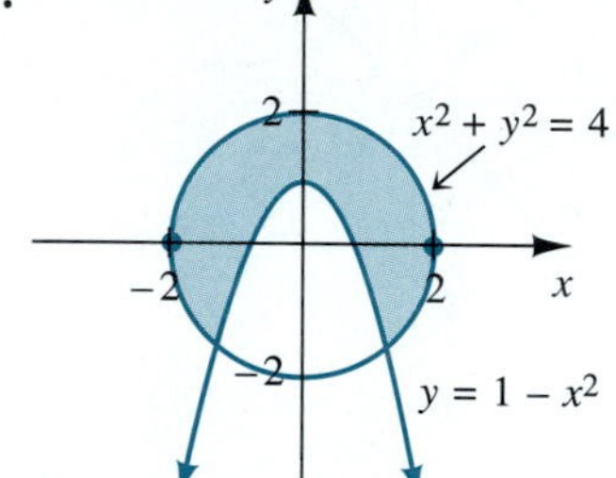

51.

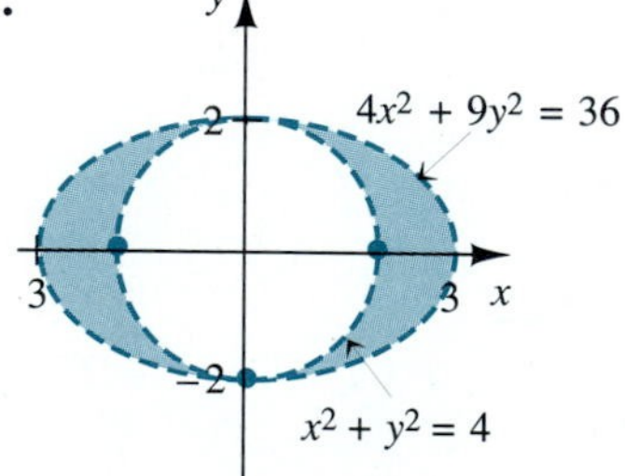

53.

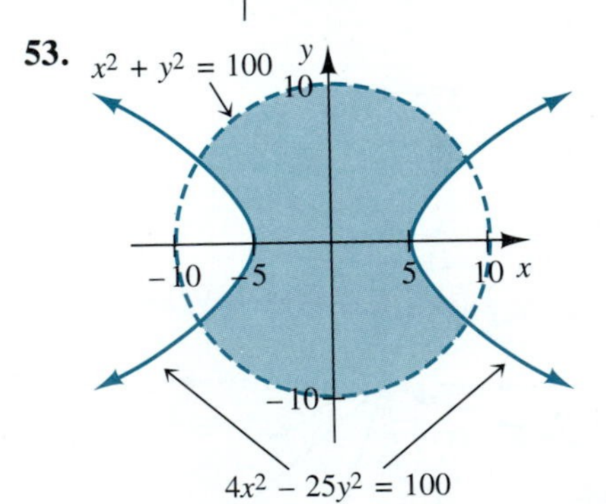

CHAPTER 10 REVIEW EXERCISES

1. $C(3, -4)$; $r = 3\sqrt{2}$ **3.** $C(4, 4)$; $r = 6$

5. $y = \frac{1}{3}x - \frac{10}{3}$

7. $8x - y^2 = 0$

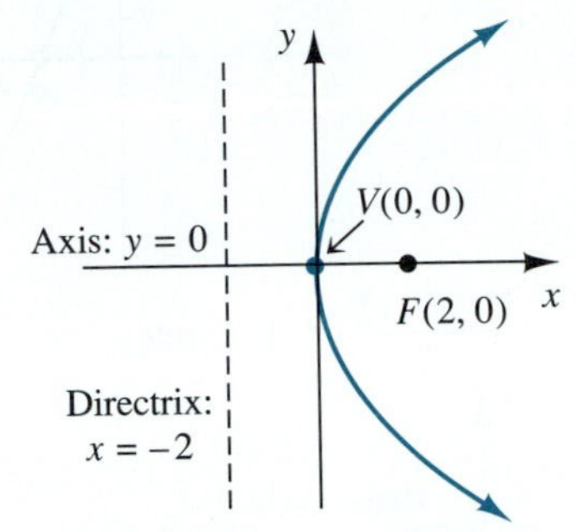

9. $(x - 4)^2 = 4y$

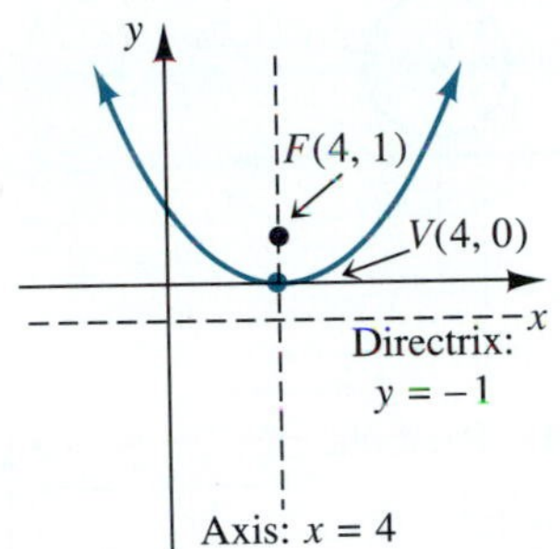

11. $4x = y^2 - 10y + 21$

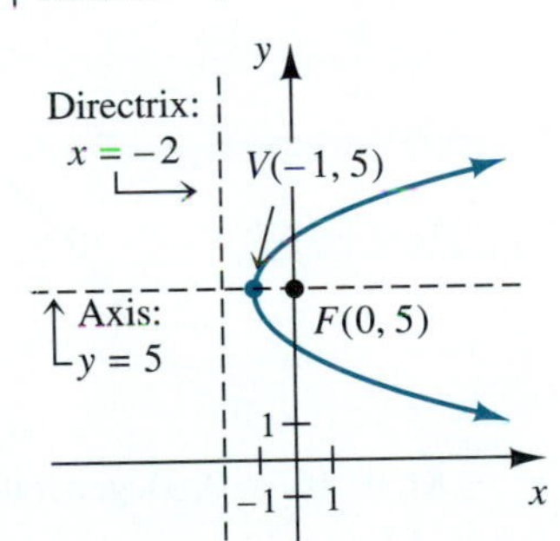

13. $y^2 = -12x$

15. The mirror is $1\frac{1}{8}$ ft deep. The light source should be placed 2 ft above the vertex.

17. $\frac{x^2}{9} + \frac{y^2}{36} = 1$

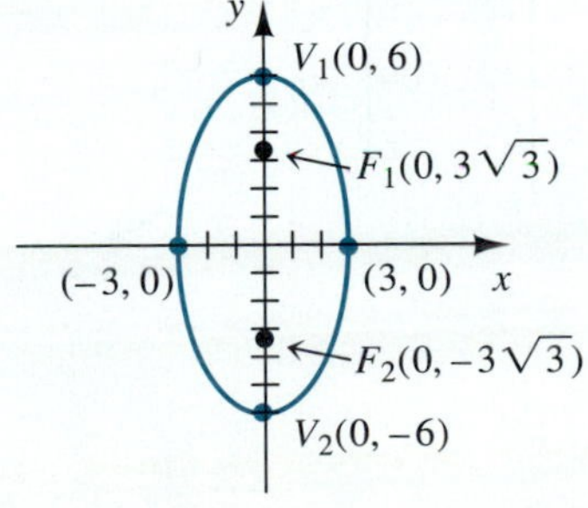

19. $24x^2 + 2y^2 = 48$

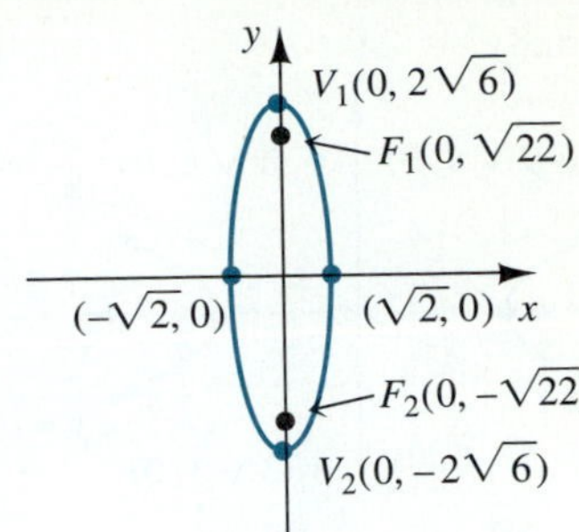

21. $\frac{(x - 2)^2}{6} + \frac{(y + 4)^2}{4} = 1$

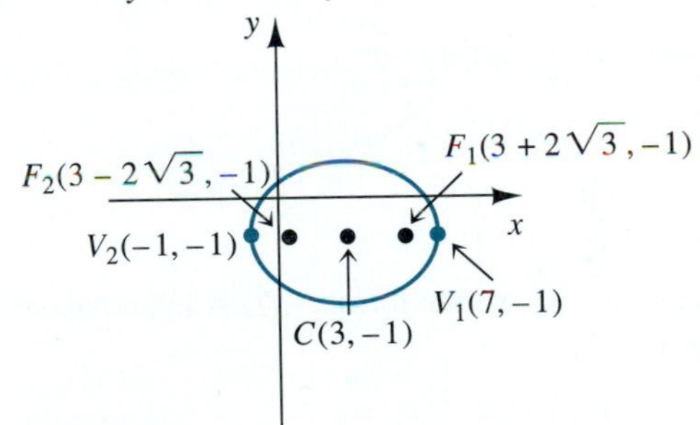

23. $x^2 + 4y^2 - 6x + 8y = 3$

25. $\frac{x^2}{9} + \frac{y^2}{16} = 1$

27. $\frac{x^2}{49} - \frac{y^2}{4} = 1$

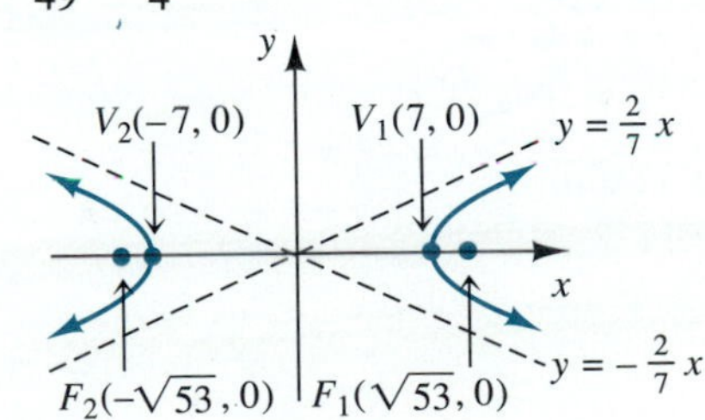

29. $x^2 - 8y^2 = 24$

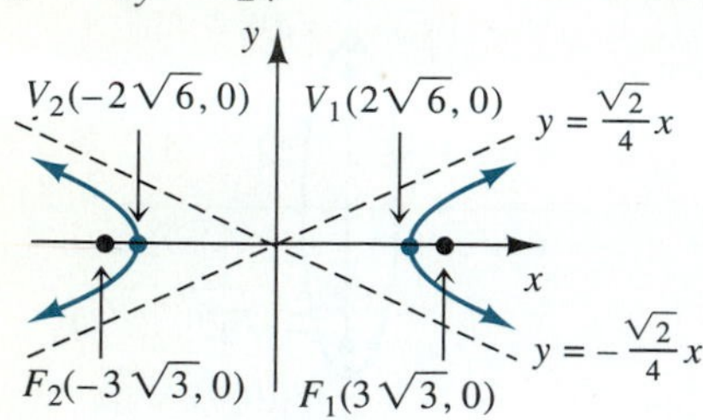

31. $\frac{(y-2)^2}{2} - \frac{(x-1)^2}{8} = 1$

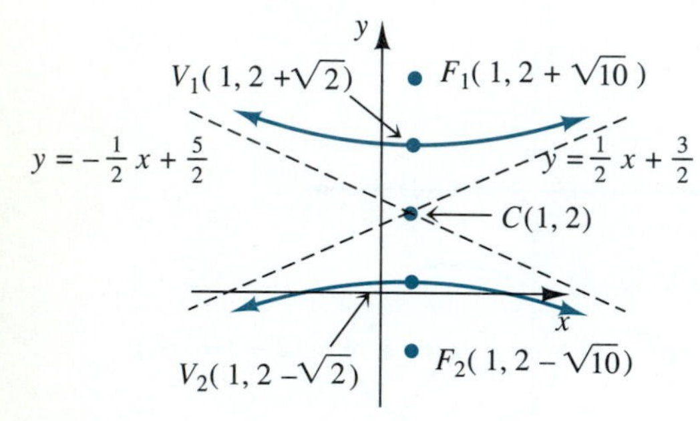

33. $x^2 - 4y^2 - 6x - 8y = 11$

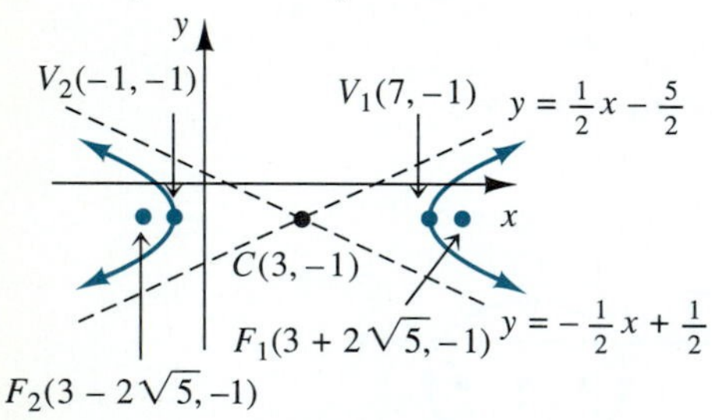

35. An ellipse or its degenerate form **37.** A hyperbola or its degenerate form

39. $\frac{x^2}{36} - \frac{y^2}{4} = 1$; Hyperbola

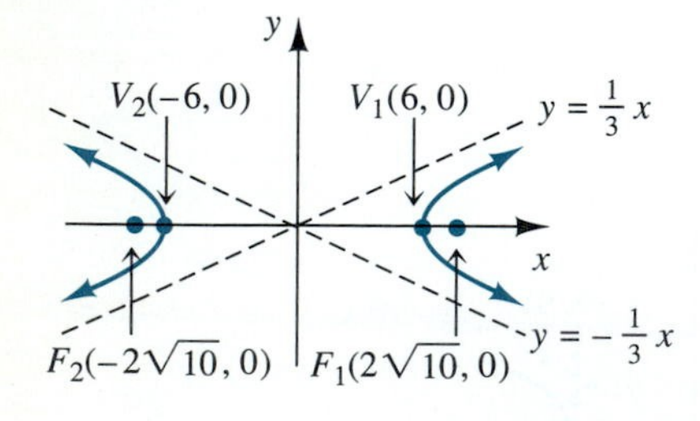

41. $\frac{x}{4} + \frac{y}{9} = 1$; Line

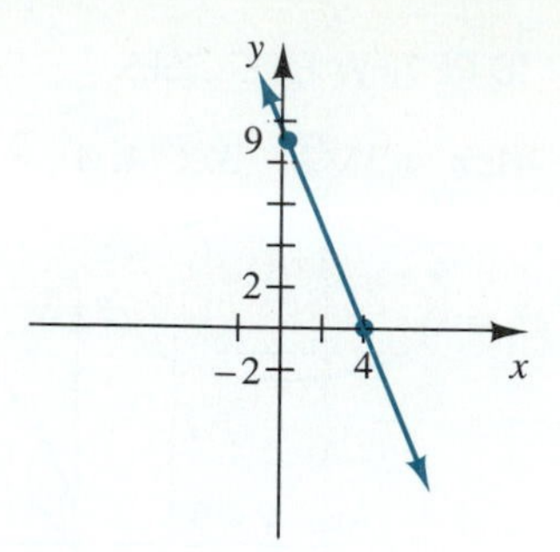

43. $\frac{(y-2)^2}{2} + \frac{(x-1)^2}{2} = 1$; Circle

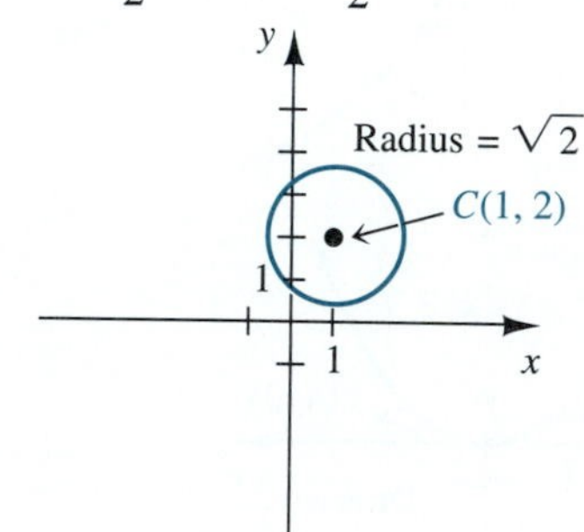

45. $x^2 = -64y$; Parabola

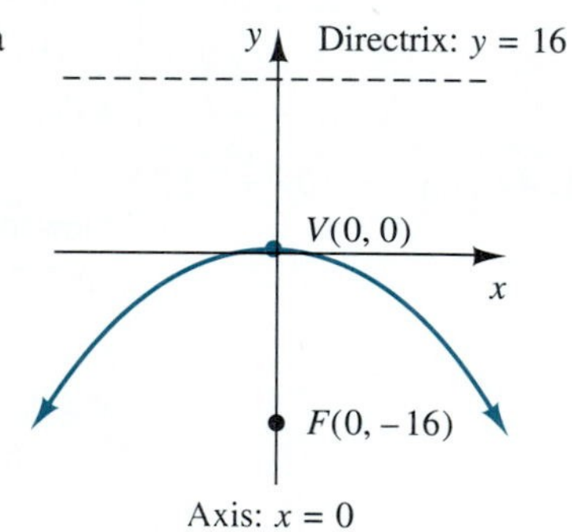

47. $x^2 - 8x + 16 = 0$; Degenerate parabola: a line

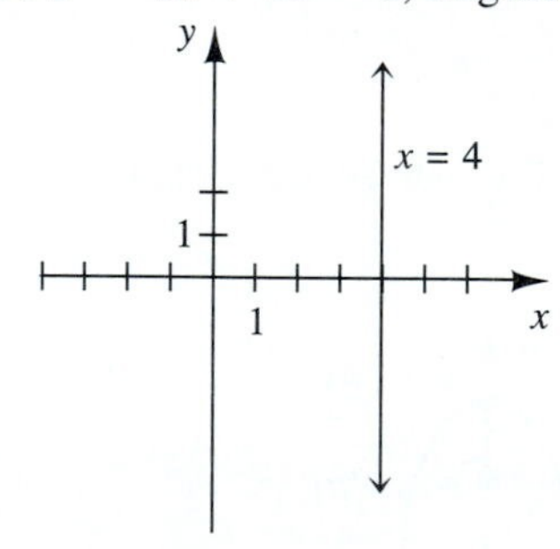

49. $x^2 + 2y^2 - 2x - 8y - 7 = 0$; Ellipse

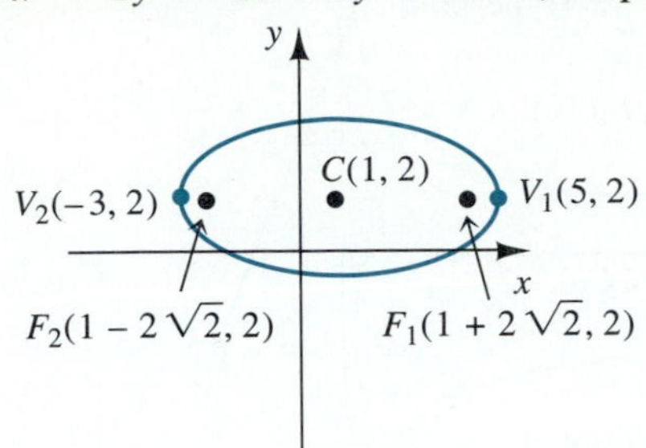

51. $x^2 + 25y^2 - 6x - 50y + 34 = 0$; Degenerate ellipse: a point

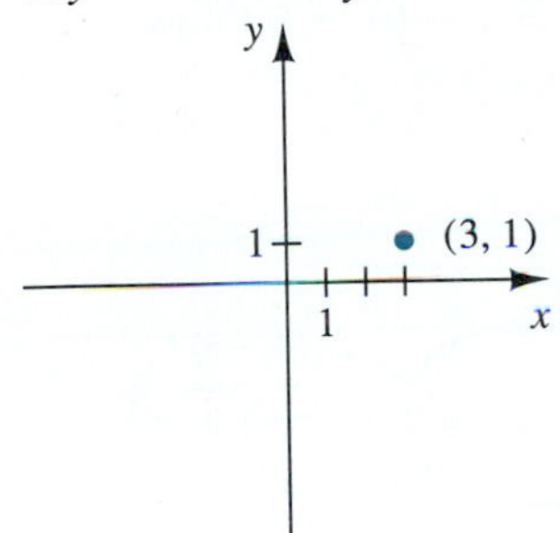

53. $(4, 10)$ **55.** $\left(\frac{3\sqrt{3}+5}{2}, \frac{5\sqrt{3}-3}{2}\right)$

57. $(3 - \sqrt{3})(x')^2 + (2 - 2\sqrt{3})x'y' + (1 + \sqrt{3})(y')^2 - 4 = 0$

59. $xy = 10$

$\frac{(x')^2}{20} - \frac{(y')^2}{20} = 1$

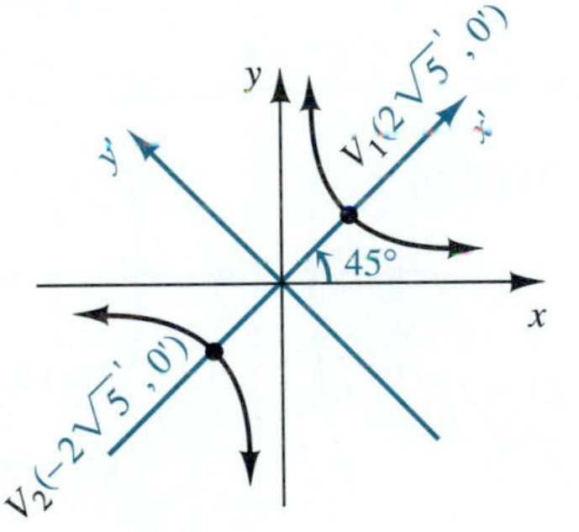

61. $11x^2 + y^2 + 10\sqrt{3}xy = 16$

$(x')^2 - \frac{(y')^2}{4} = 1$

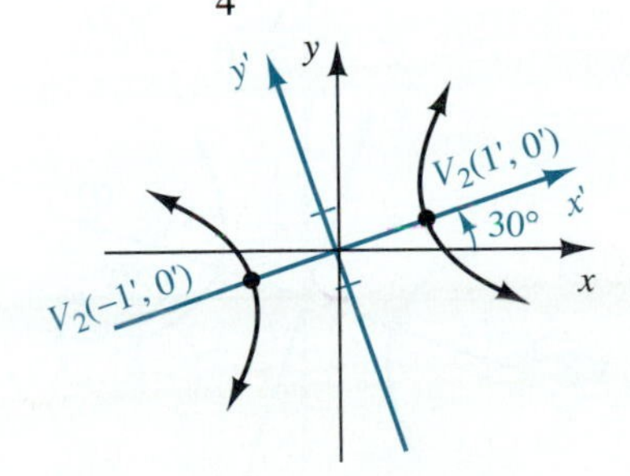

63. $(3, 1)$ **65.** $(0, 2)$, $(0, -2)$ **67.** $(2, 3)$, $(-2, 3)$

69. Both legs are $2\sqrt{2} \approx 2.83$ cm

71. $\begin{cases} 25x^2 + 4y^2 \le 100 \\ x^2 + y^2 > 9 \end{cases}$

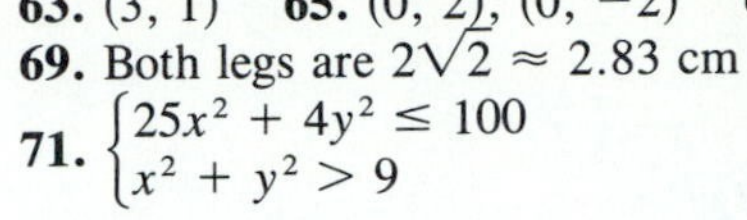

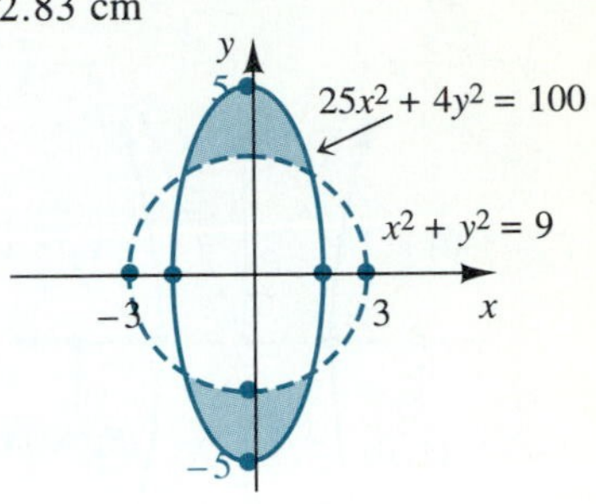

CHAPTER 10 PRACTICE TEST

1. $4x^2 + 4y = 0$ Parabola

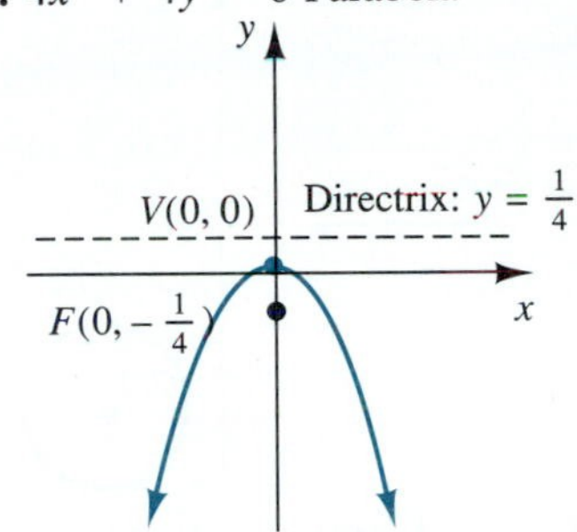

2. $4x^2 + 4y^2 = 36$ Circle

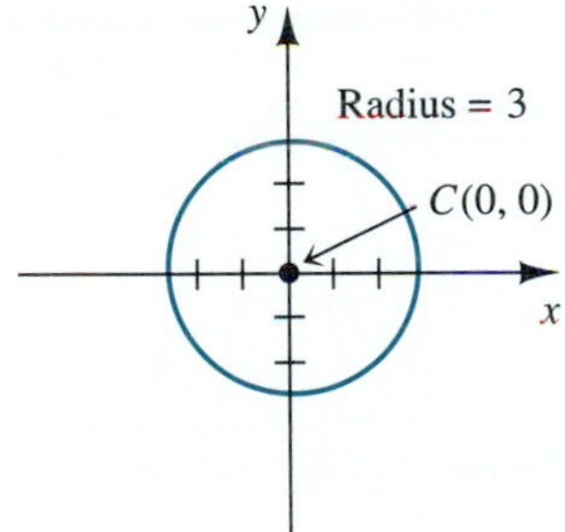

3. $\frac{x^2}{25} + \frac{y^2}{49} = 1$ Ellipse

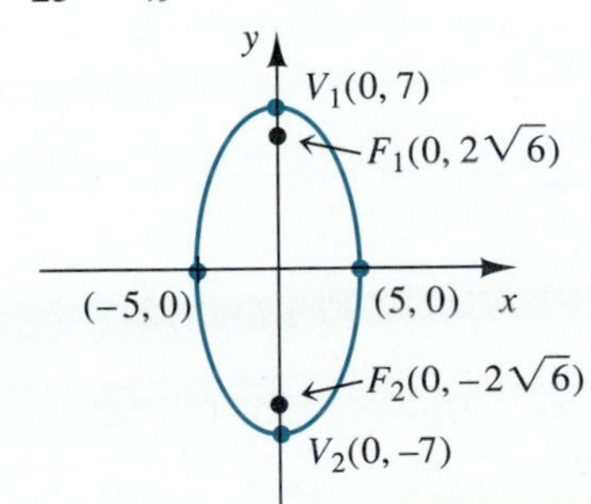

4. $\frac{x^2}{25} - \frac{y^2}{49} = 1$; Hyperbola

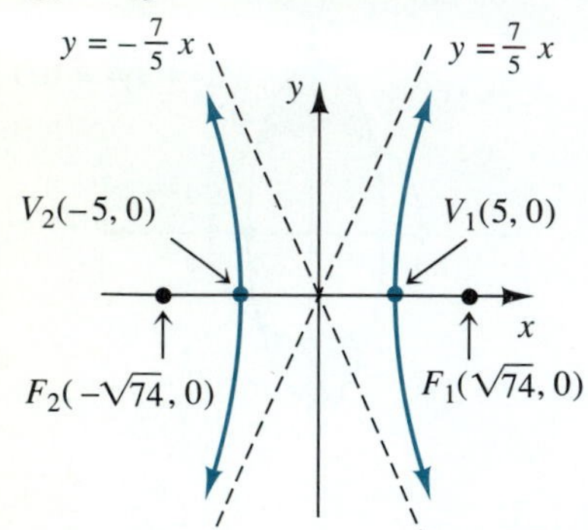

5. $(x - 3)^2 + (y + 5)^2 = 5$; Circle

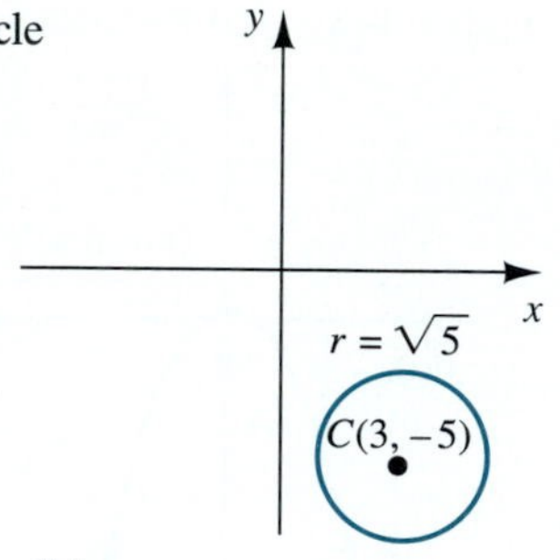

6. $2(x - 3)^2 + 8y + 24 = 0$; Parabola

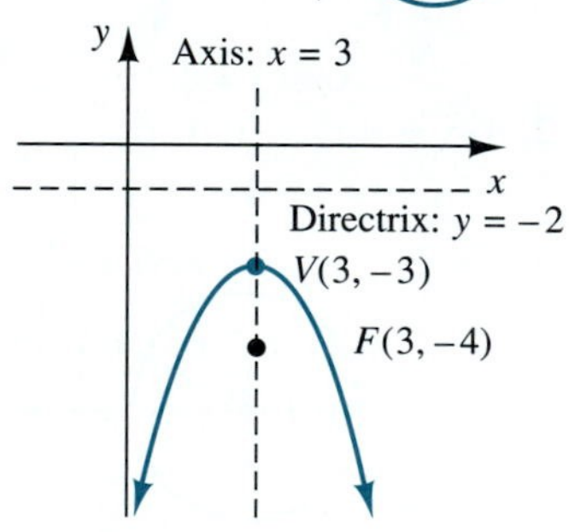

7. $\frac{(x - 3)^2}{4} + \frac{(y - 2)^2}{12} = 1$; Ellipse

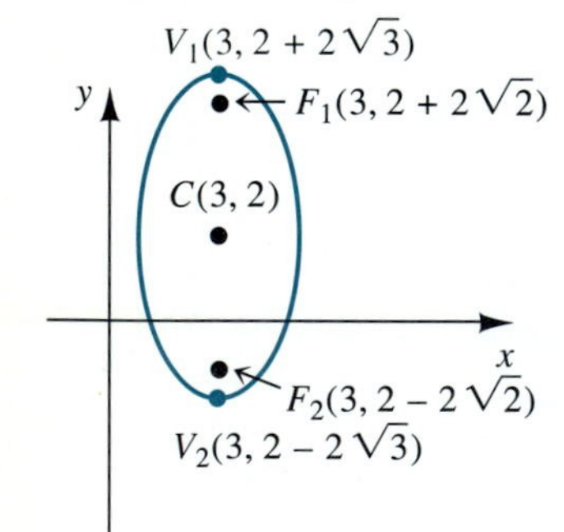

8. $\frac{(y + 2)^2}{25} - \frac{x^2}{18} = 1$; Hyperbola

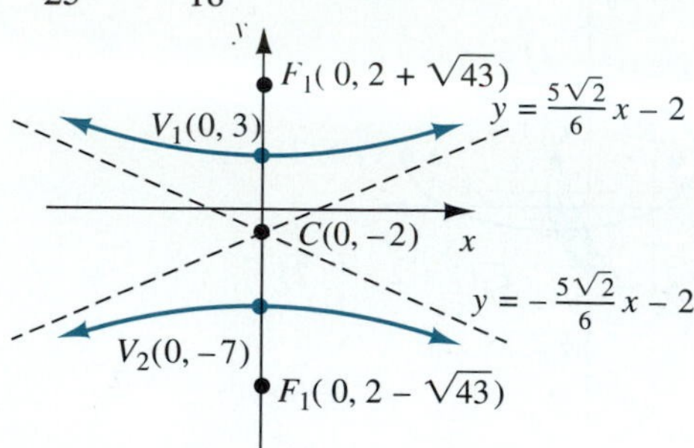

9. $4x^2 + 5y^2 - 24x + 30y + 61 = 0$ Ellipse

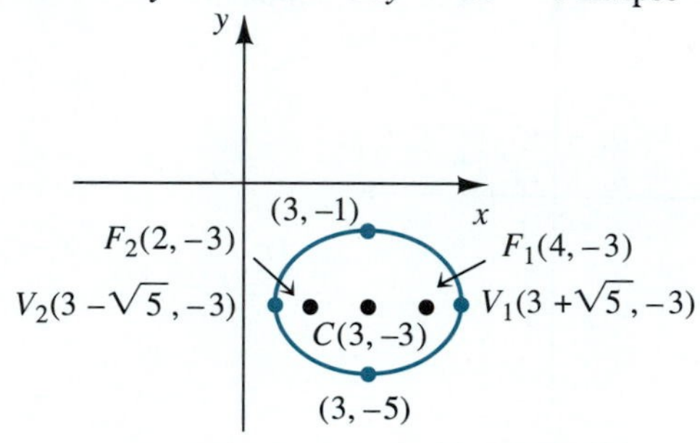

10. $3x^2 - 4y^2 - 6x = 45$; Hyperbola

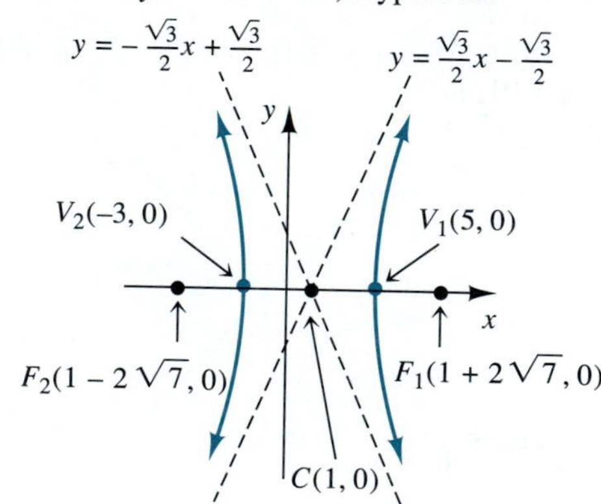

11. $y^2 = 8x$

12. $\frac{(x - 1)^2}{16} + \frac{(y - 1)^2}{12} = 1$ **13. (a)** $\left(\frac{1}{\sqrt{2}}, -\frac{3}{\sqrt{2}}\right)$

(b) $(x')^2 + 2x'y' + (y')^2 - x' + y' - 1 = 0$

14. $x^2 + \sqrt{3}\,xy = 6$

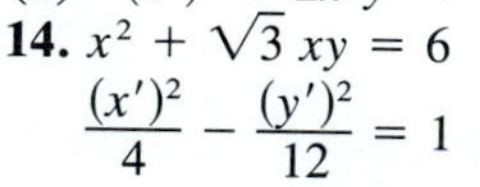

$\frac{(x')^2}{4} - \frac{(y')^2}{12} = 1$

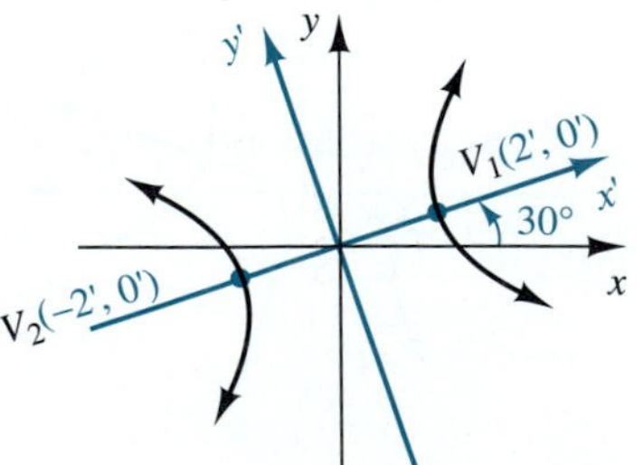

15. (1, 1), (1, −1), (−1, 1), (−1, −1)

16.

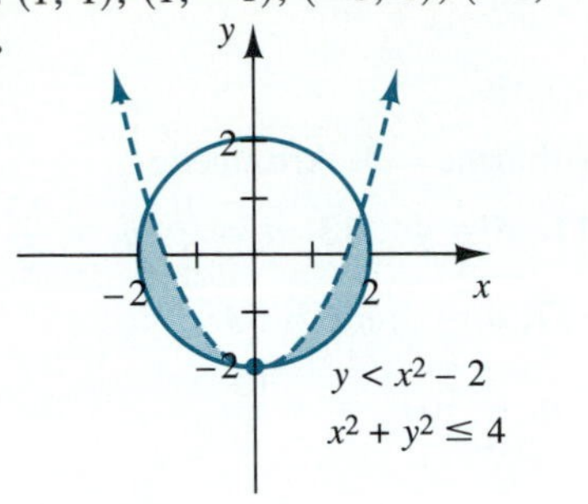

CHAPTER 11

EXERCISES 11.1

1. $a_1 = 2$; $a_2 = 7$; $a_3 = 12$; $a_4 = 17$; $a_8 = 37$

3. $b_1 = 5$; $b_2 = 8$; $b_3 = 11$; $b_4 = 14$; $b_8 = 26$

5. $c_1 = 3$; $c_2 = 9$; $c_3 = 27$; $c_4 = 81$; $c_8 = 6561$

7. $x_1 = 2$; $x_2 = \frac{3}{2}$; $x_3 = \frac{4}{3}$; $x_4 = \frac{5}{4}$; $x_8 = \frac{9}{8}$

9. $x_1 = \frac{1}{3}$; $x_2 = -\frac{1}{4}$; $x_3 = \frac{1}{5}$; $x_4 = -\frac{1}{6}$; $x_8 = -\frac{1}{10}$

11. $a_1 = -\frac{1}{2}$; $a_2 = \frac{2}{3}$; $a_3 = -\frac{3}{4}$; $a_4 = \frac{4}{5}$; $a_8 = \frac{8}{9}$

13. $a_1 = 4$; $a_2 = 5$; $a_3 = 8$; $a_4 = 17$; $a_8 = 261$

15. $b_1 = 0.9$; $b_2 = 1.01$; $b_3 = 0.999$; $b_4 = 1.0001$; $b_8 = 1.00000001$

17. $x_1 = 2$; $x_2 = 1$; $x_3 = \frac{8}{9}$; $x_4 = 1$; $x_8 = 4$

19. $a_1 = 1$; $a_2 = 0$; $a_3 = -1$; $a_4 = 0$; $a_8 = 0$

21. $c_1 = 2$; $c_2 = \frac{9}{4}$; $c_3 = \frac{64}{27}$; $c_4 = \frac{625}{256}$; $c_8 = \frac{43{,}046{,}721}{16{,}777{,}216}$

23. $a_1 = \sqrt{2}$; $a_2 = \sqrt{5}$; $a_3 = \sqrt{10}$; $a_4 = \sqrt{17}$; $a_8 = \sqrt{65}$

25. $a_1 = 1$; $a_2 = \frac{1}{2}$; $a_3 = \frac{1}{6}$; $a_4 = \frac{1}{24}$; $a_8 = \frac{1}{40{,}320}$

27. $c_1 = 1$; $c_2 = \frac{1}{2}$; $c_3 = \frac{2}{9}$; $c_4 = \frac{3}{32}$; $c_8 = \frac{315}{131{,}072}$

29. $t_1 = 2$; $t_2 = 40$; $t_3 = 2240$; $t_4 = 246{,}400$; $t_8 = \frac{7.0362 \times 10^{15}}{3}$

31. 3, 5, 9, 17, 33, 65

33. −1; 4; 9; 64; 3969; 15,745,024

35. 1; 2; 4; 16; 65,536; $2^{65{,}536}$

37. 2, 1, 3, $\frac{4}{3}$, $\frac{15}{4}$, $\frac{8}{5}$

39. 0, 1, $\frac{1}{2}$, $\frac{3}{4}$, $\frac{5}{8}$, $\frac{11}{16}$

41. −1, 1, −1, −1, 1, −1

43.

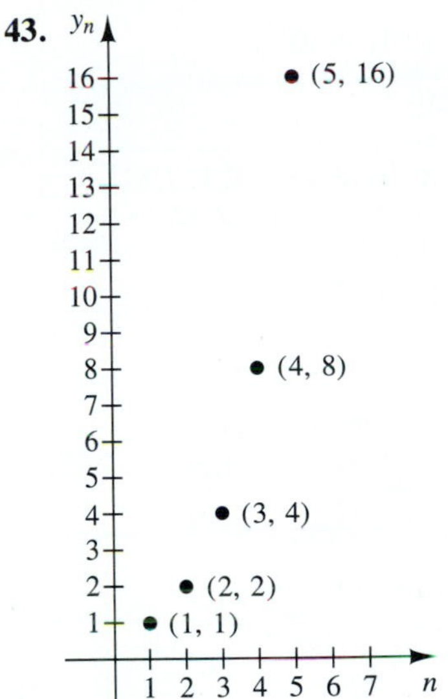

45.

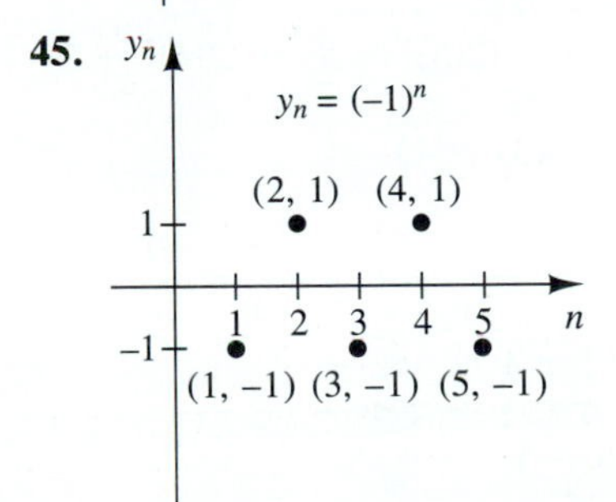

47.

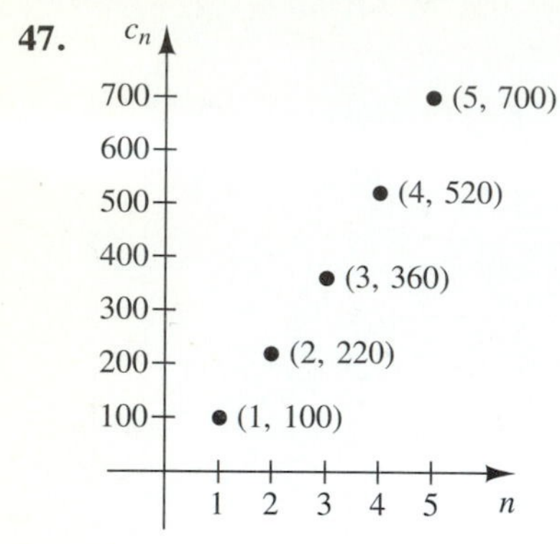

49. $a_n = 5n$ **51.** $a_n = 2n - 1$ **53.** $a_n = \frac{2n-1}{2n}$
55. $a_n = \frac{(-1)^{n+1}}{n+1}$ **57.** $a_n = (-1)^n 4(10^{-n})$
59. \$25,150; \$25,800; \$26,450
61. 61,200; 62,424; 63,672
63. \$1863; \$1928.21; \$1995.70; \$2065.55; \$2137.84
65. In billions: 2.27; 2.24; 2.21; 2.18; 2.15; 2.12
67. In feet: 30; 15; 7.5; 3.75; 1.875
69. \$230; \$227.70; \$225.42; \$223.17

EXERCISES 11.2

1. 55 **3.** 18 **5.** −14 **7.** 63 **9.** 11 **11.** 46
13. 62 **15.** 132 **17.** $\frac{121}{3}$ **19.** $\frac{73}{12}$ **21.** $\frac{109}{420}$
23. $\frac{41}{24}$ **25.** $\frac{7}{8}$ **27.** $1 + 2 + 3 + 4 + 5 = 15$
29. $5(1-1) + 5(2-1) + 5(3-1) + 5(4-1) + 5(5-1) = 50$
31. $2^2 + 3^2 + 4^2 + 5^2 + 6^2 = 90$
33. $(2(3)^2 - 5) + (2(4)^2 - 5) + (2(5)^2 - 5) + (2(6)^2 - 5) + (2(7)^2 - 5) = 245$
35. $(2^2 - 3(2) + 1) + (3^2 - 3(3) + 1) + (4^2 - 3(4) + 1) = 5$
37. $\frac{1}{1+1} + \frac{1}{2+1} + \frac{1}{3+1} + \frac{1}{4+1} + \frac{1}{5+1} = \frac{29}{20}$
39. $(\sqrt{1+1} - \sqrt{1}) + (\sqrt{2+1} - \sqrt{2}) + (\sqrt{3+1} - \sqrt{3}) + (\sqrt{4+1} - \sqrt{4}) + (\sqrt{5+1} - \sqrt{5}) = \sqrt{6} - 1$
41. $3 + 3 + 3 + 3 + 3 = 15$
43. $\sum_{k=1}^{5} 4k$ **45.** $\sum_{k=1}^{7} 2k^2$ **47.** $\sum_{n=1}^{5} (4n - 2)$
49. $\sum_{k=1}^{26} (2k - 1)$ **51.** $\sum_{n=1}^{99} \frac{1}{n(n+1)}$ **53.** 360
55. 459,033 **57.** 2081 **59.** 2
61. $f(1) = 5; f(-2) = -22; f\left(\frac{1}{2}\right) = \frac{31}{32}$
63. (a) $F(1) = 2.7181; F(2) = 7.3356; F(-1) = .3679$

EXERCISES 11.3

1. Arithmetic **3.** Not arithmetic **5.** Arithmetic
7. $d = 5$ **9.** $d = \frac{2}{3}$ **11.** $d = 4$ **13.** $d = \frac{1}{2}; a_1 = \frac{13}{2}$
15. $a_1 = -\frac{3}{5}; a_{20} = 7$ **17.** $d = 3; a_{18} = 53$
19. $d = \frac{1}{3}; a_{10} = \frac{7}{2}$ **21.** $a_1 = \frac{2}{3}; a_9 = -\frac{10}{3}$
23. $d = -\frac{5}{12}; a_1 = \frac{59}{4}$
25. \$26,750 **27.** 88 lb **29.** 144 ft.; 180 ft.
31. 99 **33.** 14 **35.** $\frac{35}{2}$ **37.** −57
39. 50 **41.** $\frac{13}{3}$ **43.** 144 **45.** 550 **47.** 112
49. 1705 **51.** $\frac{64}{\pi}$ **53.** 348 **55.** \$1500; \$103,500
57. (a) Yes **(b)** 24 **(c)** 47 **59.** 10,000 **61.** 2925
63. $a_1 = -4$ **65.** $S_5 = \frac{160}{7}$ **67.** $d = 11$

EXERCISES 11.4

1. Arithmetic **3.** Geometric **5.** Neither
7. Geometric **9.** 15,625 **11.** $\frac{8}{81}$ **13.** 64
15. $\frac{1}{27}$ **17.** $\frac{1}{128}$ **19.** $\frac{1}{8192}$ **21.** $125\sqrt{5}$ **23.** 80
25. $r = 6; a_1 = \frac{1}{36}$ **27.** $r = -\frac{1}{\sqrt{3}}; a_8 = \frac{1}{27\sqrt{3}}$
29. 2401 **31.** −100 **33.** $\frac{1}{729}$
35. \$18,370.80 **37.** \$14,848 **39.** 128,008
41. \$12,762.82 **43.** \$7128.80 **45.** 0.625 ft **47.** 484
49. $\frac{3724}{9375} \approx 0.397$ **51.** 128 **53.** $-\frac{244}{243} \approx -1.0041$
55. $\frac{1023}{4} = 255.75$ **57.** 254 **59.** $\frac{1640}{6561} \approx 0.2500$
61. 63 **63.** 1.111111 **65.** 0.011111111 **67.** 119.26 feet
69. (a) $5, 5^2, 5^3, 5^4, \ldots$ **(b)** 625 **(c)** 488,280
71. 4 **73.** $\frac{9}{2}$ **75.** $\frac{2}{5}$ **77.** $\frac{15}{2}$
79. S does not exist **81.** $\frac{47}{99}$ **83.** $\frac{10,671}{9990}$
85. 75 feet

EXERCISES 11.5

41. $N = 9$ **43.** $N = 5$ **45.** $N = 1$
47. $P_1, P_2, P_3, \ldots, P_{50}$ are true.

49. P_n is true for all $n \geq 10$.
51. P_{10} is not true. **53.** P_n is true for all odd n.
55. P_n is true for all n. **57.** P_n is true for all multiples of 3.

EXERCISES 11.6

1. 336 **3.** 792 **5.** 720 **7.** 18,564 **9.** 70
11. 15 **13.** 120 **15.** 160 **17.** 40,320
19. Approximately $1.3 \times 10^{12} = 1.3$ trillion **21.** 676
23. **(a)** 333,135,504 **(b)** 185,075,280 **(c)** 119,587,104
25. Approximately $1.1 \times 10^{12} = 1.1$ trillion; approximately $9.5 \times 10^{13} = 95$ trillion
27. 66 **29.** **(a)** 21 **(b)** 126 **(c)** 11,440 **(d)** 2940
31. 2,656,500 **33.** 35 **35.** **(a)** 125 **(b)** 50
37. **(a)** 1287 **(b)** 5148 **(c)** 4512 **(d)** 3744
39. **(a)** 45 **(b)** 120
41. **(a)** 40,320 **(b)** 119,750,400 **(c)** 9,979,200
43. 30

EXERCISES 11.7

1. $x^3 - 12x^2 + 48x - 64$
3. $a^7 + 14a^6 + 84a^5 + 280a^4 + 560a^3 + 672a^2 + 448a + 128$
5. $64x^6 + 192x^5y + 240x^4y^2 + 160x^3y^3 + 60x^2y^4 + 12xy^5 + y^6$
7. $t^5 - 15t^4r + 90t^3r^2 - 270t^2r^3 + 405tr^4 - 243r^5$
9. $w^{16} + 8w^{14} + 28w^{12} + 56w^{10} + 70w^8 + 56w^6 + 28w^4 + 8w^2 + 1$
11. $x^{12} - 4x^9y^3 + 6x^6y^6 - 4x^3y^9 + y^{12}$
13. $a^{18} + 18a^{16}b + 144a^{14}b^2 + 672a^{12}b^3 + 2016a^{10}b^4 + 4032a^8b^5 + 5376a^6b^6 + 4608a^4b^7 + 2304a^2b^8 + 512b^9$
15. $x^4 - 2x^3 + \frac{3}{2}x^2 - \frac{1}{2}x + \frac{1}{16}$
17. $\frac{1}{27}x^3 + \frac{4}{3}x^2 + 16x + 64$
19. $x^6 - 6x^5\sqrt{x} + 15x^5 - 20x^4\sqrt{x} + 15x^4 - 6x^3\sqrt{x} + x^3$
21. $x^3 + 6x^2\sqrt{xy} + 15x^2y + 20xy\sqrt{xy} + 15xy^2 + 6y^2\sqrt{xy} + y^3$
23. $x^5 - 5x^3 + 10x - \frac{10}{x} + \frac{5}{x^3} - \frac{1}{x^5}$
25. $\frac{x^4}{y^4} + \frac{4x^2}{y^2} + 6 + \frac{4y^2}{x^2} + \frac{y^4}{x^4}$
27. $x^{-5} - 5x^{-4}y^2 + 10x^{-3}y^4 - 10x^{-2}y^6 + 5x^{-1}y^8 - y^{10}$
29. $64r^{-6} + 192r^{-5}s^{-2} + 240r^{-4}s^{-4} + 160r^{-3}s^{-6} + 60r^{-2}s^{-8} + 12r^{-1}s^{-10} + s^{-12}$
31. 210 **33.** 21 **35.** 67,584 **37.** 28
39. $a^4 + 4a^3b + 6a^2b^2 + 4ab^3 + b^4 + 4a^3c + 12a^2bc + 12ab^2c + 4b^3c + 6a^2c^2 + 12abc^2 + 6b^2c^2 + 4ac^3 + 4bc^3 + c^4$

41. 756 **43.** $n + 1$
45. Yes **47.** $-4 - 4i$ **49.** $-46 + 9i$

CHAPTER 11 REVIEW EXERCISES

1. $a_1 = -2$; $a_2 = 1$; $a_3 = 4$; $a_4 = 7$; $a_{12} = 31$
3. $x_1 = 5$; $x_2 = 20$; $x_3 = 45$; $x_4 = 80$; $x_{12} = 720$
5. $a_1 = \frac{-1}{3}$; $a_2 = \frac{1}{4}$; $a_3 = \frac{-1}{5}$; $a_4 = \frac{1}{6}$; $a_{12} = \frac{1}{14}$
7. $x_1 = 3$; $x_2 = 5$; $x_3 = 3$; $x_4 = 5$; $x_{12} = 5$
9. $a_n = 5n - 7$ **11.** $a_n = \frac{(-1)^{n+1}}{3n}$
13. $p_n = 250 + 8(n - 1)$ **15.** \$6381.41 **17.** 84
19. 126 **21.** 93
23. $2(1) + 2(2) + 2(3) + 2(4) + 2(5) + 2(6) = 42$
25. $(1 - 2)^2 + (2 - 2)^2 + (3 - 2)^2 + (4 - 2)^2 + (5 - 2)^2 = 15$
27. $5(1)^2 + 5(2)^2 + 5(3)^2 + 5(4)^2 + 5(5)^2 + 5(6)^2 + 5(7)^2 = 700$
29. $a_3 = 19$; $a_4 = 25$; $a_{10} = 61$
31. $a_5 = 26$; $a_6 = 31$; $a_{10} = 51$
33. $a_5 = -6$; $a_6 = -8$; $a_{10} = -16$
35. $x_1 = 1$; $d = 3$ **37.** 240 ft **39.** 492 **41.** 175
43. 3280 **45.** 625 **47.** $\frac{147}{2}$ **49.** 162 **51.** $\frac{1}{64}$
53. 324 **55.** \$1536 **57.** $\frac{25}{4}$ **59.** S does not exist
61. $\frac{39}{99}$ **63.** 320 ft **67.** 504 **69.** 362,880
71. 420 **73.** **(a)** 1287 **(b)** 10,240 **(c)** 40 **75.** 32
77. $\frac{x^3}{27} + \frac{x^2y}{12} + \frac{xy^2}{16} + \frac{y^3}{64}$
79. 43,750

CHAPTER 11 PRACTICE TEST

1. **(a)** $x_1 = -2$; $x_2 = -2$; $x_3 = 0$; $x_8 = 40$
(b) $y_1 = 3$; $y_2 = 38$; $y_3 = 133$; $y_8 = 2558$
2. $(3(4)^2 - 5(4) + 1) + (3(5)^2 - 5(5) + 1) + (3(6)^2 - 5(6) + 1) + (3(7)^2 - 5(7) + 1) = 272$
3. 662 **4.** -96 **5.** 840 **6.** 363
7. **(a)** 29 **(b)** 15,625 **(c)** 432 **(d)** $\frac{45}{2}$ **8.** 2100
9. 32 ft **11.** **(a)** 100,947 **(b)** 12,600 **12.** 83,160
13. $256x^{16} - 3072x^{14}y + 16{,}128x^{12}y^2 - 48{,}384x^{10}y^3 + 90{,}720x^8y^4 - 108{,}864x^6y^5 + 81{,}648x^4y^6 - 34{,}492x^2y^7 + 6561y^8$

INDEX

A

B

C

D

E

F

G

H

I

J

L

S

U

V

W

X

Y

Z

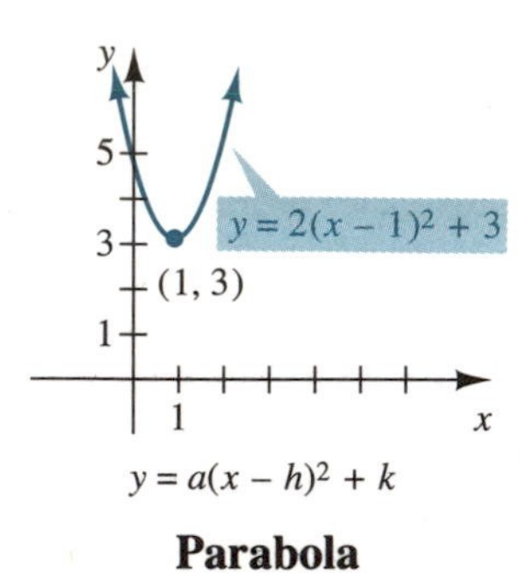

Parabola

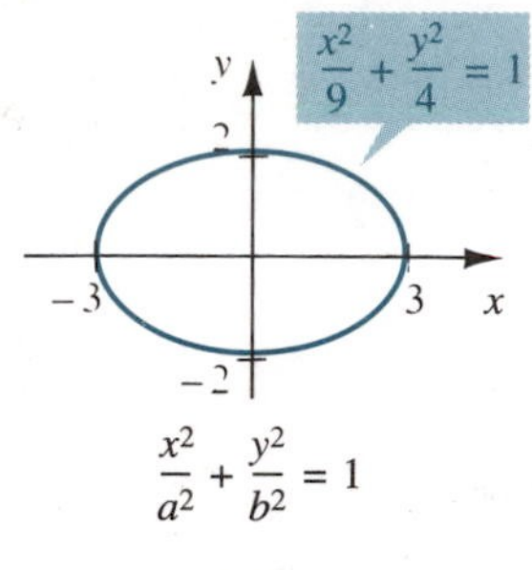

Ellipse

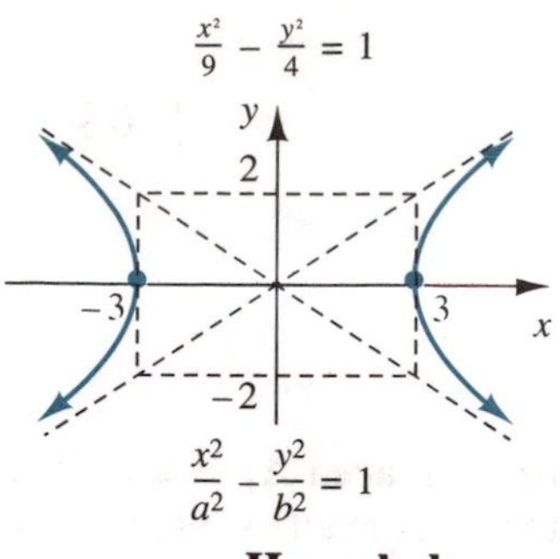

Hyperbola

GRAPHS OF THE TRIGONOMETRIC FUNCTIONS ON THE INTERVAL $[0, 2\pi]$

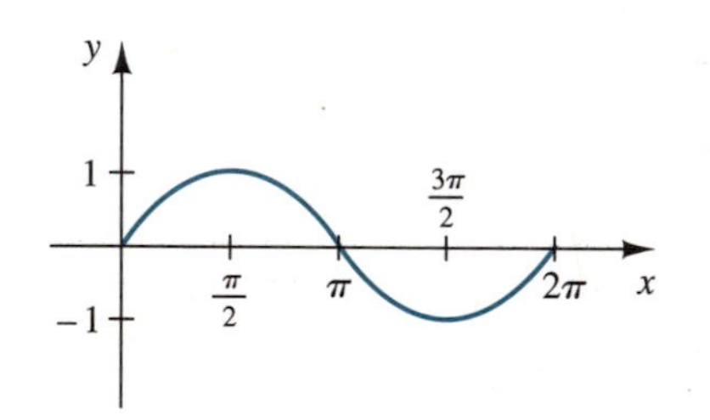

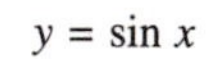
$y = \sin x$

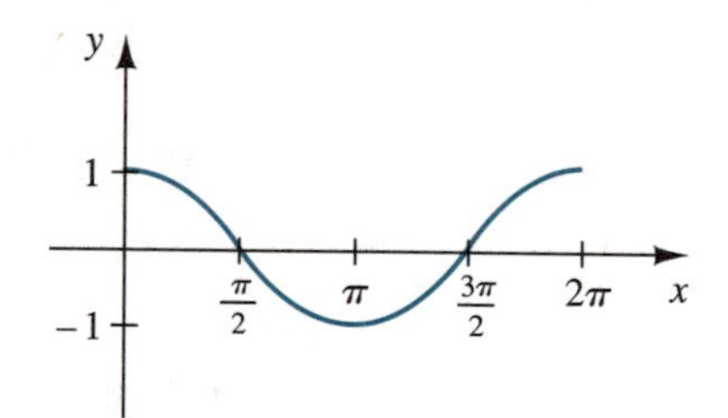

$y = \cos x$

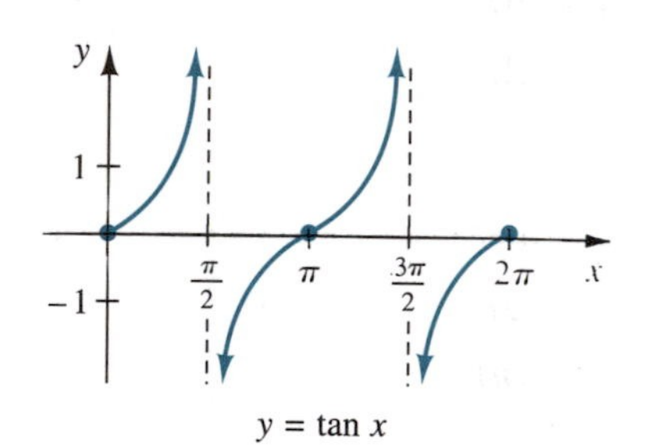

$y = \tan x$

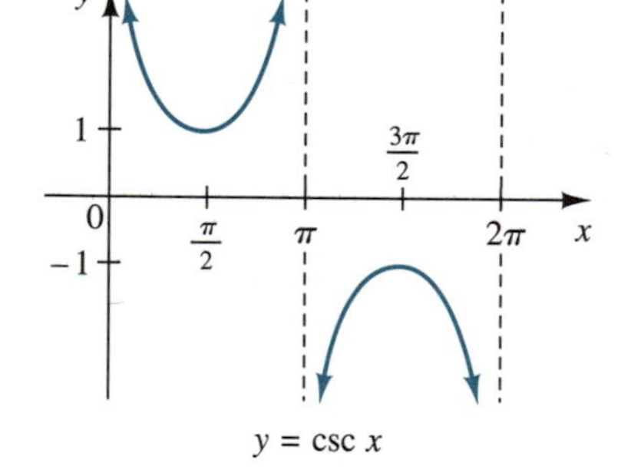

$y = \csc x$

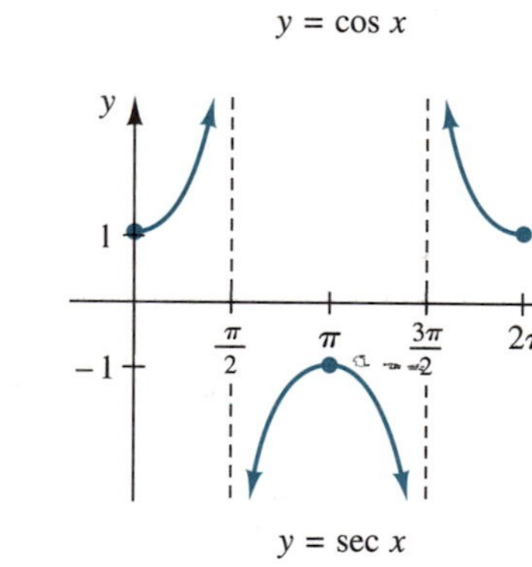

$y = \sec x$

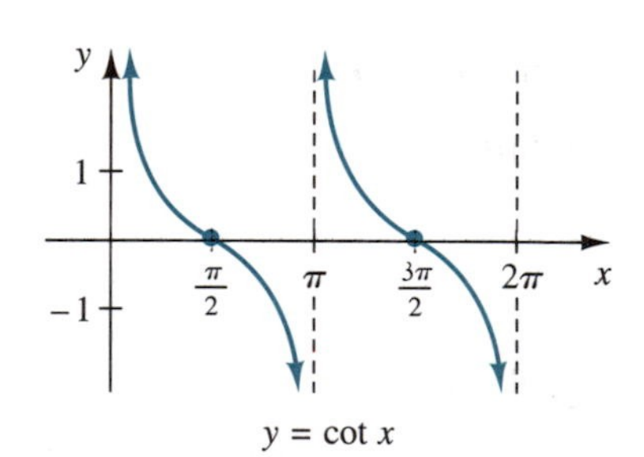

$y = \cot x$

GRAPHS OF THE INVERSE TRIGONOMETRIC FUNCTIONS

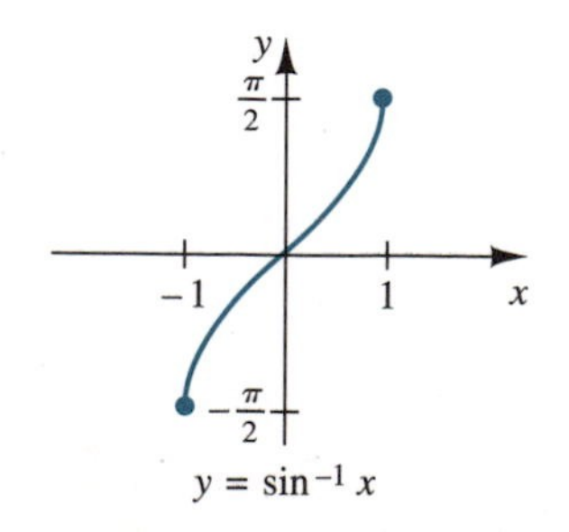

$y = \sin^{-1} x$

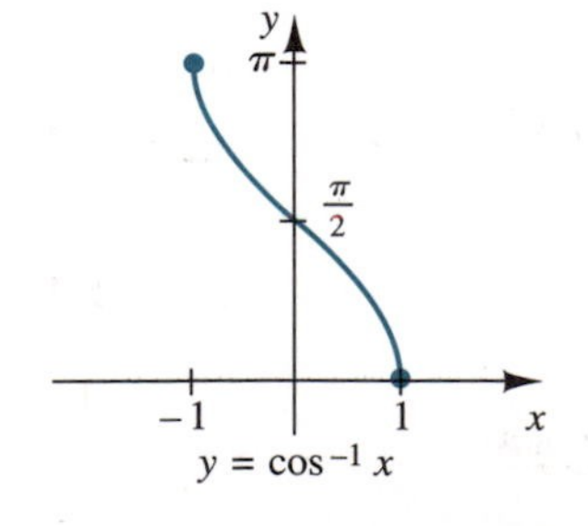

$y = \cos^{-1} x$

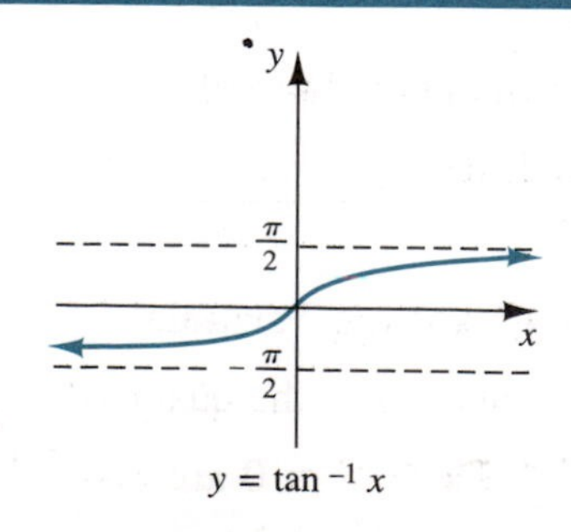

$y = \tan^{-1} x$